COMPLETE SOLUTIONS MANUAL

for

Stewart's

ESSENTIAL CALCULUS

Australia · Brazil · Japan · Korea · Mexico · Singapore · Spain · United Kingdom · United States

For product information and technology assistance, contact us at **Cengage Learning Customer & Sales Support, 1-800-354-9706**

For permission to use material from this text or product, submit all requests online at **www.cengage.com/permissions.** Further permissions questions can be e-mailed to **permissionrequest@cengage.com.**

ISBN-13: 978-0-495-01445-4
ISBN-10: 0-495-01445-1

Brooks/Cole
10 Davis Drive
Belmont, CA 94002-3098
USA

Cengage Learning is a leading provider of customized learning solutions with office locations around the globe, including Singapore, the United Kingdom, Australia, Mexico, Brazil, and Japan. Locate your local office at **www.cengage.com/global.**

Cengage Learning products are represented in Canada by Nelson Education, Ltd.

To learn more about Brooks/Cole, visit **www.cengage.com/brookscole**

Purchase any of our products at your local college store or at our preferred online store **www.ichapters.com.**

Printed in the United States of America
3 4 5 6 7 8 9 13 12 11 10 09

ABBREVIATIONS AND SYMBOLS

CD	concave downward
CU	concave upward
D	the domain of f
FDT	First Derivative Test
HA	horizontal asymptote(s)
I	interval of convergence
IP	inflection point(s)
R	radius of convergence
VA	vertical asymptote(s)
$\overset{\text{CAS}}{=}$	indicates the use of a computer algebra system.
$\overset{\text{H}}{=}$	indicates the use of l'Hospital's Rule.
$\overset{j}{=}$	indicates the use of Formula j in the Table of Integrals in the back endpapers.
$\overset{\text{s}}{=}$	indicates the use of the substitution $\{u = \sin x, du = \cos x\,dx\}$.
$\overset{\text{c}}{=}$	indicates the use of the substitution $\{u = \cos x, du = -\sin x\,dx\}$.

CONTENTS

1 FUNCTIONS AND LIMITS 1

2 DERIVATIVES 61

3 APPLICATIONS OF DIFFERENTIATION 127

4 INTEGRALS 213

5 INVERSE FUNCTIONS: Exponential, Logarithmic, and Inverse Trigonometric Functions 249

6 TECHNIQUES OF INTEGRATION 307

7 APPLICATIONS OF INTEGRATION 371

8 SERIES 435

9 PARAMETRIC EQUATIONS AND POLAR COORDINATES 507

10 VECTORS AND THE GEOMETRY OF SPACE 561

11 PARTIAL DERIVATIVES 639

1 ☐ FUNCTIONS AND LIMITS

1.1 Functions and Their Representations

In exercises requiring estimations or approximations, your answers may vary slightly from the answers given here.

1. (a) The point $(-1, -2)$ is on the graph of f, so $f(-1) = -2$.

(b) When $x = 2$, y is about 2.8, so $f(2) \approx 2.8$.

(c) $f(x) = 2$ is equivalent to $y = 2$. When $y = 2$, we have $x = -3$ and $x = 1$.

(d) Reasonable estimates for x when $y = 0$ are $x = -2.5$ and $x = 0.3$.

(e) The domain of f consists of all x-values on the graph of f. For this function, the domain is $-3 \le x \le 3$, or $[-3, 3]$. The range of f consists of all y-values on the graph of f. For this function, the range is $-2 \le y \le 3$, or $[-2, 3]$.

(f) As x increases from -1 to 3, y increases from -2 to 3. Thus, f is increasing on the interval $[-1, 3]$.

2. (a) The point $(-4, -2)$ is on the graph of f, so $f(-4) = -2$. The point $(3, 4)$ is on the graph of g, so $g(3) = 4$.

(b) We are looking for the values of x for which the y-values are equal. The y-values for f and g are equal at the points $(-2, 1)$ and $(2, 2)$, so the desired values of x are -2 and 2.

(c) $f(x) = -1$ is equivalent to $y = -1$. When $y = -1$, we have $x = -3$ and $x = 4$.

(d) As x increases from 0 to 4, y decreases from 3 to -1. Thus, f is decreasing on the interval $[0, 4]$.

(e) The domain of f consists of all x-values on the graph of f. For this function, the domain is $-4 \le x \le 4$, or $[-4, 4]$. The range of f consists of all y-values on the graph of f. For this function, the range is $-2 \le y \le 3$, or $[-2, 3]$.

(f) The domain of g is $[-4, 3]$ and the range is $[0.5, 4]$.

3. No, the curve is not the graph of a function because a vertical line intersects the curve more than once. Hence, the curve fails the Vertical Line Test.

4. Yes, the curve is the graph of a function because it passes the Vertical Line Test. The domain is $[-2, 2]$ and the range is $[-1, 2]$.

5. Yes, the curve is the graph of a function because it passes the Vertical Line Test. The domain is $[-3, 2]$ and the range is $[-3, -2) \cup [-1, 3]$.

6. No, the curve is not the graph of a function since for $x = 0$, ± 1, and ± 2, there are infinitely many points on the curve.

7. The person's weight increased to about 160 pounds at age 20 and stayed fairly steady for 10 years. The person's weight dropped to about 120 pounds for the next 5 years, then increased rapidly to about 170 pounds. The next 30 years saw a gradual increase to 190 pounds. Possible reasons for the drop in weight at 30 years of age: diet, exercise, health problems.

8. The salesman travels away from home from 8 to 9 AM and is then stationary until 10:00. The salesman travels farther away from 10 until noon. There is no change in his distance from home until 1:00, at which time the distance from home decreases until 3:00. Then the distance starts increasing again, reaching the maximum distance away from home at 5:00. There is no change from 5 until 6, and then the distance decreases rapidly until 7:00 PM, at which time the salesman reaches home.

9. The water will cool down almost to freezing as the ice melts. Then, when the ice has melted, the water will slowly warm up to room temperature.

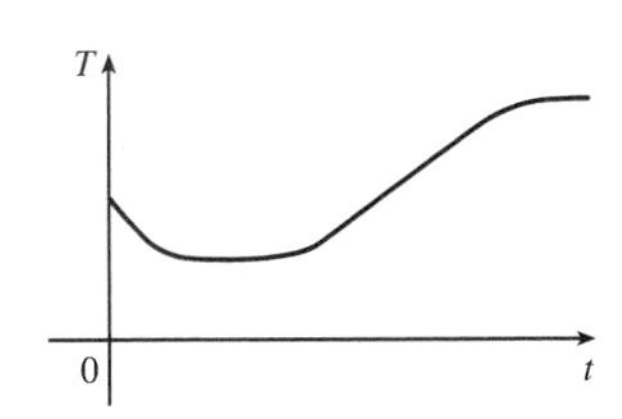

10. The summer solstice (the longest day of the year) is around June 21, and the winter solstice (the shortest day) is around December 22.

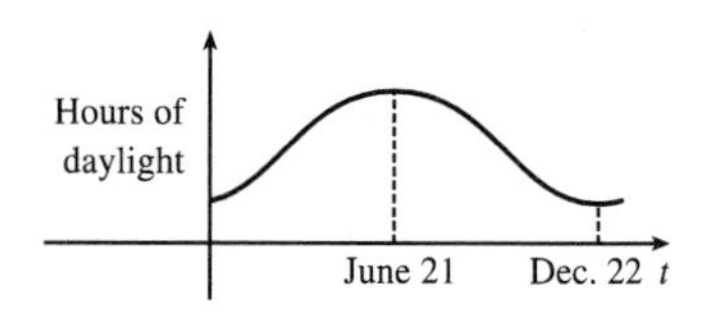

11. Of course, this graph depends strongly on the geographical location!

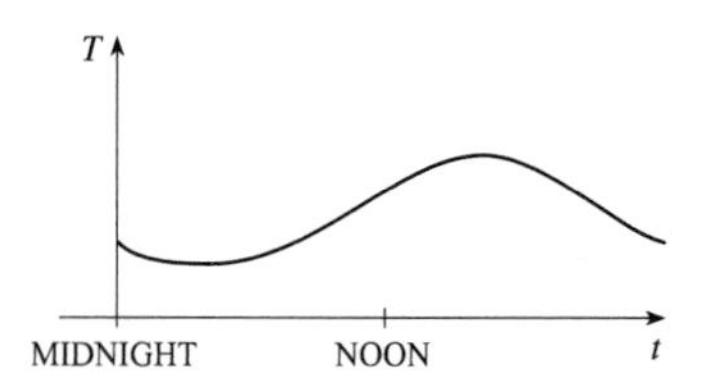

12. The value of the car decreases fairly rapidly initially, then somewhat less rapidly.

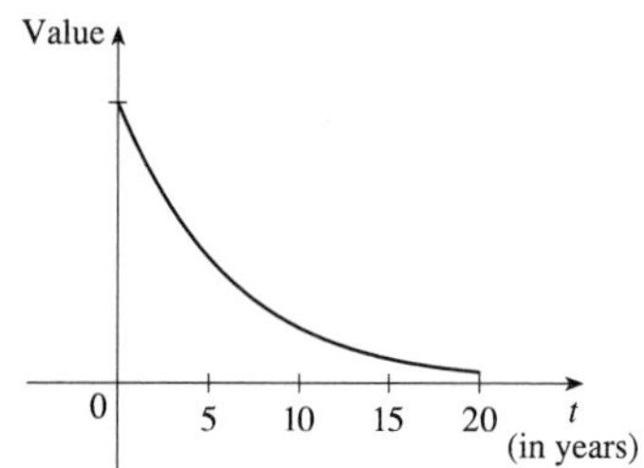

13. As the price increases, the amount sold decreases.

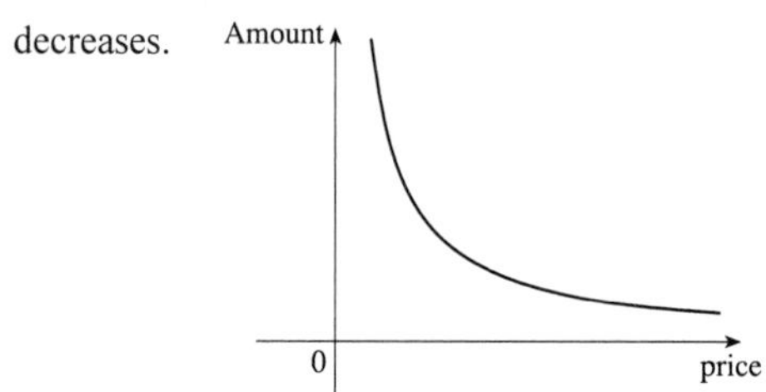

14. The temperature of the pie would increase rapidly, level off to oven temperature, decrease rapidly, and then level off to room temperature.

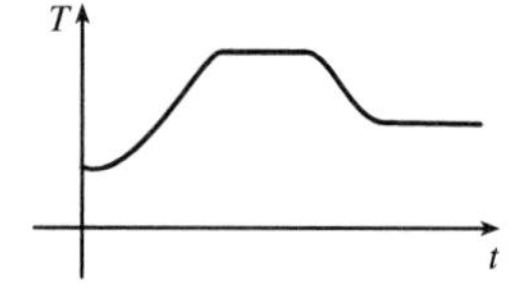

15. Height of grass

Wed. Wed. Wed. Wed. Wed. t

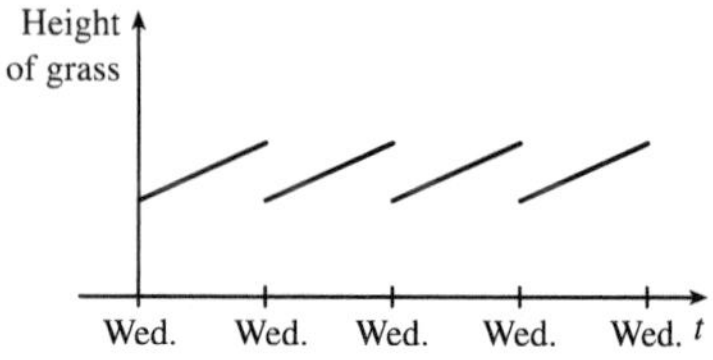

16. (a)

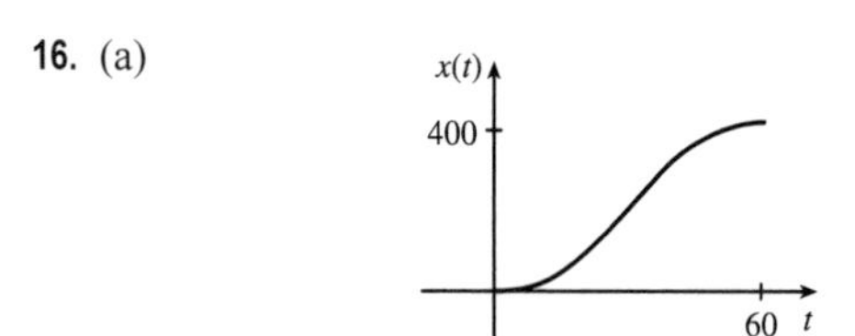

(b)

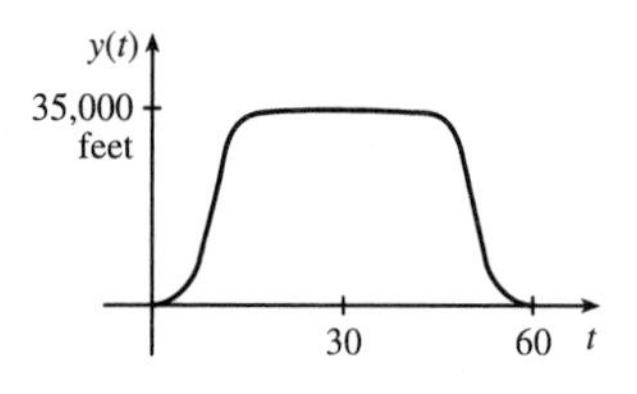

(c)

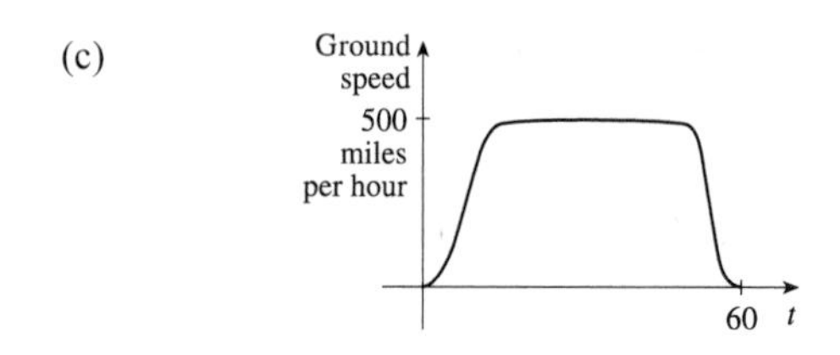

(d)

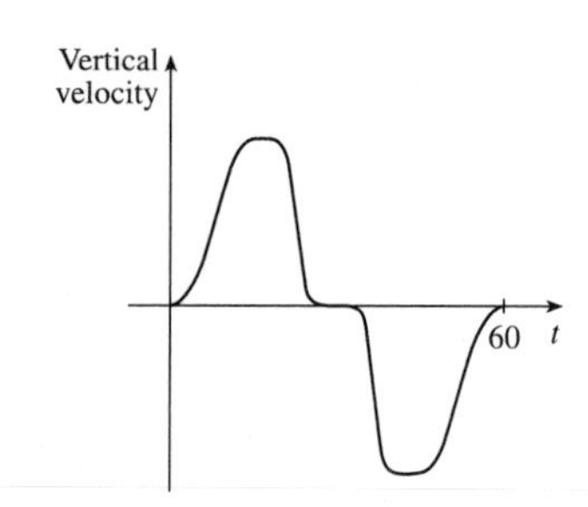

17. $f(x) = 3x^2 - x + 2.$

$f(2) = 3(2)^2 - 2 + 2 = 12 - 2 + 2 = 12.$

$f(-2) = 3(-2)^2 - (-2) + 2 = 12 + 2 + 2 = 16.$

$f(a) = 3a^2 - a + 2.$

$f(-a) = 3(-a)^2 - (-a) + 2 = 3a^2 + a + 2.$

$f(a+1) = 3(a+1)^2 - (a+1) + 2 = 3(a^2 + 2a + 1) - a - 1 + 2 = 3a^2 + 6a + 3 - a + 1 = 3a^2 + 5a + 4.$

$2f(a) = 2 \cdot f(a) = 2(3a^2 - a + 2) = 6a^2 - 2a + 4.$

$f(2a) = 3(2a)^2 - (2a) + 2 = 3(4a^2) - 2a + 2 = 12a^2 - 2a + 2.$

$f(a^2) = 3(a^2)^2 - (a^2) + 2 = 3(a^4) - a^2 + 2 = 3a^4 - a^2 + 2.$

$$\begin{aligned}[f(a)]^2 &= \left[3a^2 - a + 2\right]^2 = \left(3a^2 - a + 2\right)\left(3a^2 - a + 2\right) \\ &= 9a^4 - 3a^3 + 6a^2 - 3a^3 + a^2 - 2a + 6a^2 - 2a + 4 = 9a^4 - 6a^3 + 13a^2 - 4a + 4.\end{aligned}$$

$f(a+h) = 3(a+h)^2 - (a+h) + 2 = 3(a^2 + 2ah + h^2) - a - h + 2 = 3a^2 + 6ah + 3h^2 - a - h + 2.$

18. A spherical balloon with radius $r + 1$ has volume $V(r+1) = \frac{4}{3}\pi(r+1)^3 = \frac{4}{3}\pi(r^3 + 3r^2 + 3r + 1)$. We wish to find the amount of air needed to inflate the balloon from a radius of r to $r + 1$. Hence, we need to find the difference $V(r+1) - V(r) = \frac{4}{3}\pi(r^3 + 3r^2 + 3r + 1) - \frac{4}{3}\pi r^3 = \frac{4}{3}\pi(3r^2 + 3r + 1)$.

19. $f(x) = 4 + 3x - x^2$, so $f(3+h) = 4 + 3(3+h) - (3+h)^2 = 4 + 9 + 3h - (9 + 6h + h^2) = 4 - 3h - h^2$,

and $\dfrac{f(3+h) - f(3)}{h} = \dfrac{(4 - 3h - h^2) - 4}{h} = \dfrac{h(-3-h)}{h} = -3 - h.$

20. $f(x) = x^3$, so $f(a+h) = (a+h)^3 = a^3 + 3a^2h + 3ah^2 + h^3$,

and $\dfrac{f(a+h) - f(a)}{h} = \dfrac{(a^3 + 3a^2h + 3ah^2 + h^3) - a^3}{h} = \dfrac{h(3a^2 + 3ah + h^2)}{h} = 3a^2 + 3ah + h^2.$

21. $$\frac{f(x) - f(a)}{x - a} = \frac{\frac{1}{x} - \frac{1}{a}}{x - a} = \frac{\frac{a - x}{xa}}{x - a} = \frac{a - x}{xa(x - a)} = \frac{-1(x - a)}{xa(x - a)} = -\frac{1}{ax}$$

22. $$\begin{aligned}\frac{f(x) - f(1)}{x - 1} &= \frac{\frac{x+3}{x+1} - 2}{x - 1} = \frac{\frac{x + 3 - 2(x+1)}{x+1}}{x - 1} = \frac{x + 3 - 2x - 2}{(x+1)(x-1)} \\ &= \frac{-x + 1}{(x+1)(x-1)} = \frac{-(x-1)}{(x+1)(x-1)} = -\frac{1}{x+1}\end{aligned}$$

23. $f(x) = x/(3x - 1)$ is defined for all x except when $0 = 3x - 1 \Leftrightarrow x = \frac{1}{3}$, so the domain is $\left\{x \in \mathbb{R} \mid x \neq \frac{1}{3}\right\} = \left(-\infty, \frac{1}{3}\right) \cup \left(\frac{1}{3}, \infty\right)$.

24. $f(x) = (5x + 4)/(x^2 + 3x + 2)$ is defined for all x except when $0 = x^2 + 3x + 2 \Leftrightarrow 0 = (x+2)(x+1) \Leftrightarrow x = -2$ or -1, so the domain is $\{x \in \mathbb{R} \mid x \neq -2, -1\} = (-\infty, -2) \cup (-2, -1) \cup (-1, \infty)$.

25. $f(t) = \sqrt{t} + \sqrt[3]{t}$ is defined when $t \geq 0$. These values of t give real number results for $\sqrt{t}$, whereas any value of t gives a real number result for $\sqrt[3]{t}$. The domain is $[0, \infty)$.

26. $g(u) = \sqrt{u} + \sqrt{4 - u}$ is defined when $u \geq 0$ and $4 - u \geq 0 \Leftrightarrow u \leq 4$. Thus, the domain is $0 \leq u \leq 4 = [0, 4]$.

27. $h(x) = 1 / \sqrt[4]{x^2 - 5x}$ is defined when $x^2 - 5x > 0 \quad \Leftrightarrow \quad x(x-5) > 0$. Note that $x^2 - 5x \neq 0$ since that would result in division by zero. The expression $x(x-5)$ is positive if $x < 0$ or $x > 5$. (See *Review of Algebra* at www.stewartcalculus.com for methods for solving inequalities.) Thus, the domain is $(-\infty, 0) \cup (5, \infty)$.

28. $h(x) = \sqrt{4 - x^2}$. Now $y = \sqrt{4 - x^2} \quad \Rightarrow \quad y^2 = 4 - x^2 \quad \Leftrightarrow \quad x^2 + y^2 = 4$, so the graph is the top half of a circle of radius 2 with center at the origin. The domain is $\{x \mid 4 - x^2 \geq 0\} = \{x \mid 4 \geq x^2\} = \{x \mid 2 \geq |x|\} = [-2, 2]$. From the graph, the range is $0 \leq y \leq 2$, or $[0, 2]$.

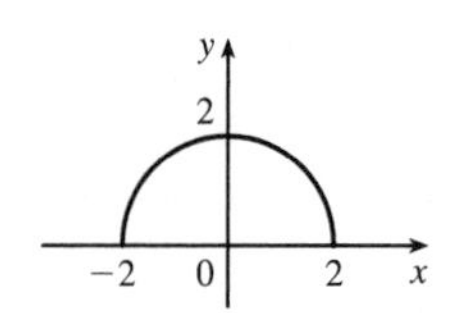

29. $f(x) = 5$ is defined for all real numbers, so the domain is $\mathbb{R}$, or $(-\infty, \infty)$. The graph of f is a horizontal line with y-intercept 5.

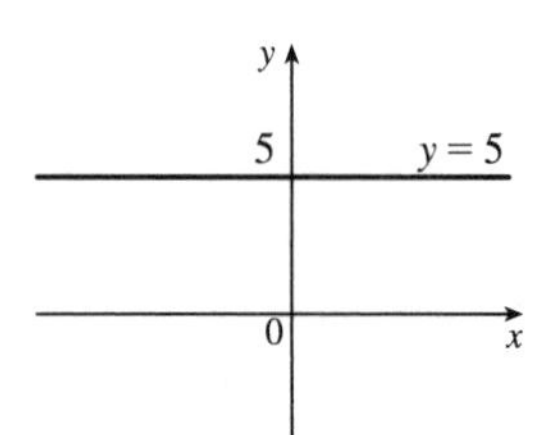

30. $F(x) = \frac{1}{2}(x + 3)$ is defined for all real numbers, so the domain is $\mathbb{R}$, or $(-\infty, \infty)$. The graph of F is a line with x-intercept -3 and y-intercept $\frac{3}{2}$.

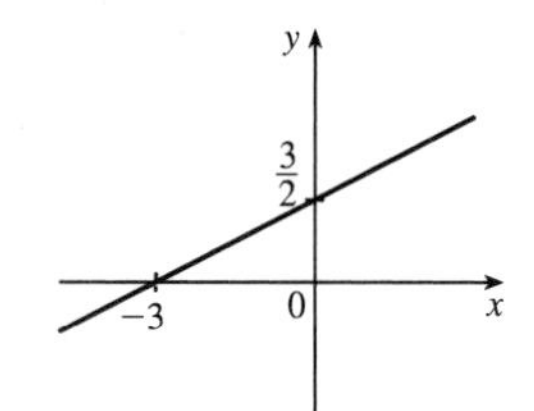

31. $f(t) = t^2 - 6t$ is defined for all real numbers, so the domain is $\mathbb{R}$, or $(-\infty, \infty)$. The graph of f is a parabola opening upward since the coefficient of t^2 is positive. To find the t-intercepts, let $y = 0$ and solve for t. $0 = t^2 - 6t = t(t-6) \quad \Rightarrow \quad t = 0$ and $t = 6$. The t-coordinate of the vertex is halfway between the t-intercepts, that is, at $t = 3$. Since $f(3) = 3^2 - 6 \cdot 3 = -9$, the vertex is $(3, -9)$.

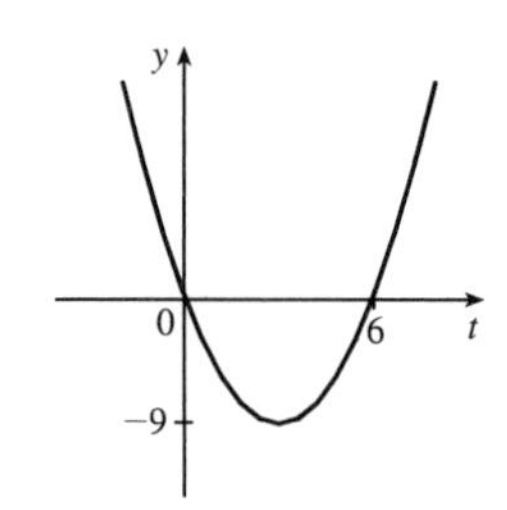

32. $H(t) = \dfrac{4 - t^2}{2 - t} = \dfrac{(2+t)(2-t)}{2-t}$, so for $t \neq 2$, $H(t) = 2 + t$. The domain is $\{t \mid t \neq 2\}$. So the graph of H is the same as the graph of the function $f(t) = t + 2$ (a line) except for the hole at $(2, 4)$.

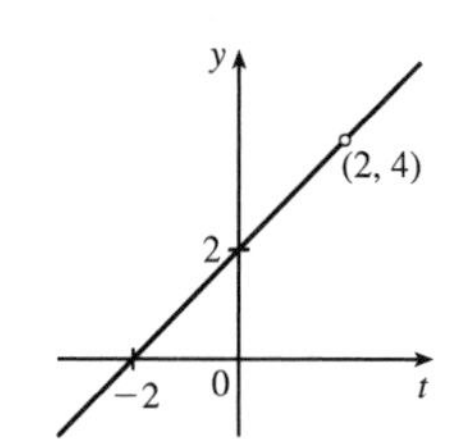

33. $g(x) = \sqrt{x - 5}$ is defined when $x - 5 \geq 0$ or $x \geq 5$, so the domain is $[5, \infty)$. Since $y = \sqrt{x - 5} \quad \Rightarrow \quad y^2 = x - 5 \quad \Rightarrow \quad x = y^2 + 5$, we see that g is the top half of a parabola.

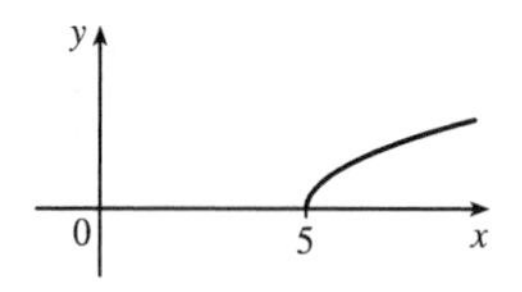

34. $F(x) = |2x + 1| = \begin{cases} 2x + 1 & \text{if } 2x + 1 \geq 0 \\ -(2x + 1) & \text{if } 2x + 1 < 1 \end{cases}$

$= \begin{cases} 2x + 1 & \text{if } x \geq -\frac{1}{2} \\ -2x - 1 & \text{if } x < -\frac{1}{2} \end{cases}$

The domain is $\mathbb{R}$, or $(-\infty, \infty)$.

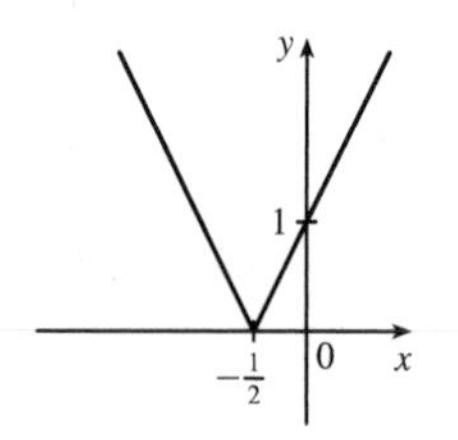

35. $G(x) = \dfrac{3x + |x|}{x}$. Since $|x| = \begin{cases} x & \text{if } x \geq 0 \\ -x & \text{if } x < 0 \end{cases}$, we have

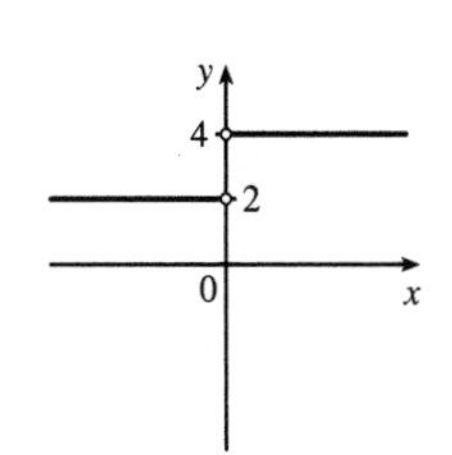

$$G(x) = \begin{cases} \dfrac{3x + x}{x} & \text{if } x > 0 \\ \dfrac{3x - x}{x} & \text{if } x < 0 \end{cases} = \begin{cases} \dfrac{4x}{x} & \text{if } x > 0 \\ \dfrac{2x}{x} & \text{if } x < 0 \end{cases} = \begin{cases} 4 & \text{if } x > 0 \\ 2 & \text{if } x < 0 \end{cases}$$

Note that G is not defined for $x = 0$. The domain is $(-\infty, 0) \cup (0, \infty)$.

36. $g(x) = \dfrac{|x|}{x^2}$. Since $|x| = \begin{cases} x & \text{if } x \geq 0 \\ -x & \text{if } x < 0 \end{cases}$, we have

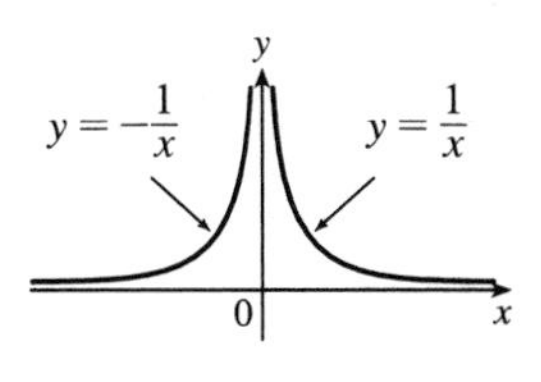

$$g(x) = \begin{cases} \dfrac{x}{x^2} & \text{if } x > 0 \\ \dfrac{-x}{x^2} & \text{if } x < 0 \end{cases} = \begin{cases} \dfrac{1}{x} & \text{if } x > 0 \\ \dfrac{1}{-x} & \text{if } x < 0 \end{cases}$$

Note that g is not defined for $x = 0$. The domain is $(-\infty, 0) \cup (0, \infty)$.

37. $f(x) = \begin{cases} x + 2 & \text{if } x < 0 \\ 1 - x & \text{if } x \geq 0 \end{cases}$

The domain is $\mathbb{R}$.

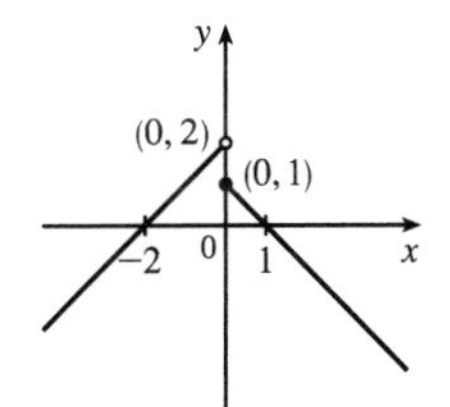

38. $f(x) = \begin{cases} 3 - \frac{1}{2}x & \text{if } x \leq 2 \\ 2x - 5 & \text{if } x > 2 \end{cases}$

The domain is $\mathbb{R}$.

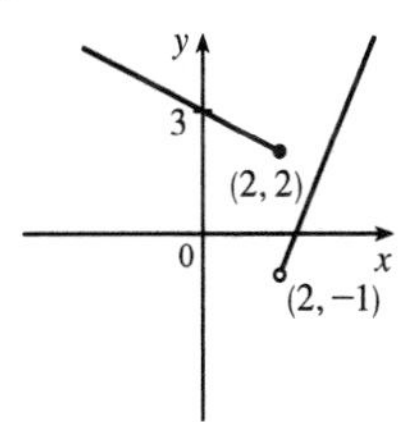

39. $f(x) = \begin{cases} x + 2 & \text{if } x \leq -1 \\ x^2 & \text{if } x > -1 \end{cases}$

Note that for $x = -1$, both $x + 2$ and x^2 are equal to 1.

The domain is $\mathbb{R}$.

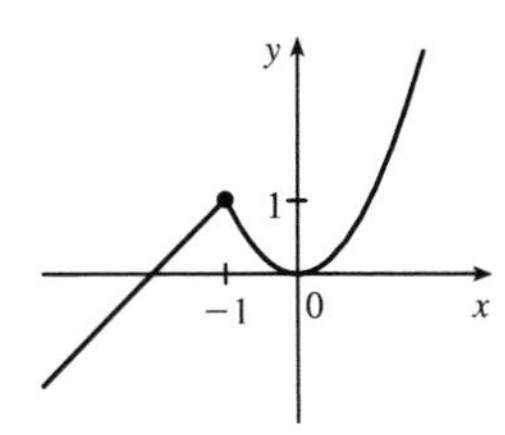

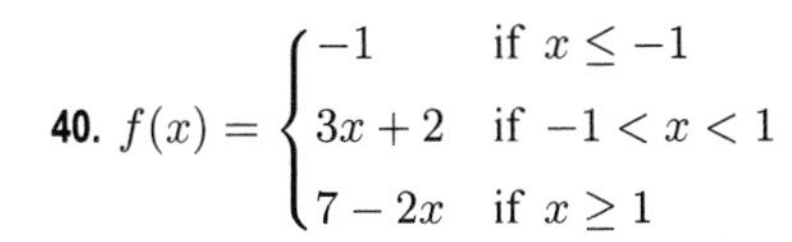

40. $f(x) = \begin{cases} -1 & \text{if } x \leq -1 \\ 3x + 2 & \text{if } -1 < x < 1 \\ 7 - 2x & \text{if } x \geq 1 \end{cases}$

The domain is $\mathbb{R}$.

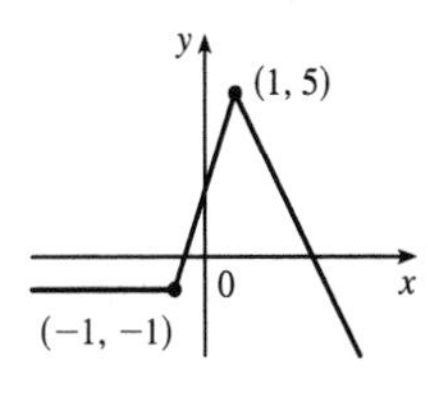

41. Recall that the slope m of a line between the two points (x_1, y_1) and (x_2, y_2) is $m = \dfrac{y_2 - y_1}{x_2 - x_1}$ and an equation of the line connecting those two points is $y - y_1 = m(x - x_1)$. The slope of this line segment is $\dfrac{-6 - 1}{4 - (-2)} = -\dfrac{7}{6}$, so an equation is $y - 1 = -\frac{7}{6}(x + 2)$. The function is $f(x) = -\frac{7}{6}x - \frac{4}{3}$, $-2 \leq x \leq 4$.

42. The slope of this line segment is $\dfrac{3-(-2)}{6-(-3)} = \dfrac{5}{9}$, so an equation is $y + 2 = \frac{5}{9}(x+3)$.

The function is $f(x) = \frac{5}{9}x - \frac{1}{3}$, $-3 \le x \le 6$.

43. We need to solve the given equation for y. $x + (y-1)^2 = 0 \;\Leftrightarrow\; (y-1)^2 = -x \;\Leftrightarrow\; y - 1 = \pm\sqrt{-x} \;\Leftrightarrow$ $y = 1 \pm \sqrt{-x}$. The expression with the positive radical represents the top half of the parabola, and the one with the negative radical represents the bottom half. Hence, we want $f(x) = 1 - \sqrt{-x}$. Note that the domain is $x \le 0$.

44. $(x-1)^2 + y^2 = 1 \;\Leftrightarrow\; y = \pm\sqrt{1-(x-1)^2} = \pm\sqrt{2x - x^2}$.

The top half is given by the function $f(x) = \sqrt{2x - x^2}$, $0 \le x \le 2$.

45. Let the length and width of the rectangle be L and W. Then the perimeter is $2L + 2W = 20$ and the area is $A = LW$. Solving the first equation for W in terms of L gives $W = \dfrac{20-2L}{2} = 10 - L$. Thus, $A(L) = L(10-L) = 10L - L^2$. Since lengths are positive, the domain of A is $0 < L < 10$. If we further restrict L to be larger than W, then $5 < L < 10$ would be the domain.

46. Let the length and width of the rectangle be L and W. Then the area is $LW = 16$, so that $W = 16/L$. The perimeter is $P = 2L + 2W$, so $P(L) = 2L + 2(16/L) = 2L + 32/L$, and the domain of P is $L > 0$, since lengths must be positive quantities. If we further restrict L to be larger than W, then $L > 4$ would be the domain.

47. Let the length of a side of the equilateral triangle be x. Then by the Pythagorean Theorem, the height y of the triangle satisfies $y^2 + \left(\frac{1}{2}x\right)^2 = x^2$, so that $y^2 = x^2 - \frac{1}{4}x^2 = \frac{3}{4}x^2$ and $y = \frac{\sqrt{3}}{2}x$. Using the formula for the area A of a triangle, $A = \frac{1}{2}(\text{base})(\text{height})$, we obtain $A(x) = \frac{1}{2}(x)\left(\frac{\sqrt{3}}{2}x\right) = \frac{\sqrt{3}}{4}x^2$, with domain $x > 0$.

48. Let the volume of the cube be V and the length of an edge be L. Then $V = L^3$ so $L = \sqrt[3]{V}$, and the surface area is $S(V) = 6\left(\sqrt[3]{V}\right)^2 = 6V^{2/3}$, with domain $V > 0$.

49. Let each side of the base of the box have length x, and let the height of the box be h. Since the volume is 2, we know that $2 = hx^2$, so that $h = 2/x^2$, and the surface area is $S = x^2 + 4xh$. Thus, $S(x) = x^2 + 4x(2/x^2) = x^2 + (8/x)$, with domain $x > 0$.

50.

$$C(x) = \begin{cases} \$2.00 & \text{if } 0.0 < x \le 1.0 \\ 2.20 & \text{if } 1.0 < x \le 1.1 \\ 2.40 & \text{if } 1.1 < x \le 1.2 \\ 2.60 & \text{if } 1.2 < x \le 1.3 \\ 2.80 & \text{if } 1.3 < x \le 1.4 \\ 3.00 & \text{if } 1.4 < x \le 1.5 \\ 3.20 & \text{if } 1.5 < x \le 1.6 \\ 3.40 & \text{if } 1.6 < x \le 1.7 \\ 3.60 & \text{if } 1.7 < x \le 1.8 \\ 3.80 & \text{if } 1.8 < x \le 1.9 \\ 4.00 & \text{if } 1.9 < x < 2.0 \end{cases}$$

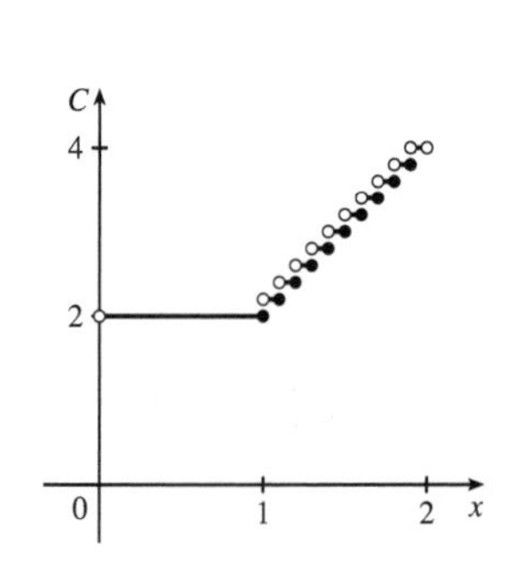

51. (a)

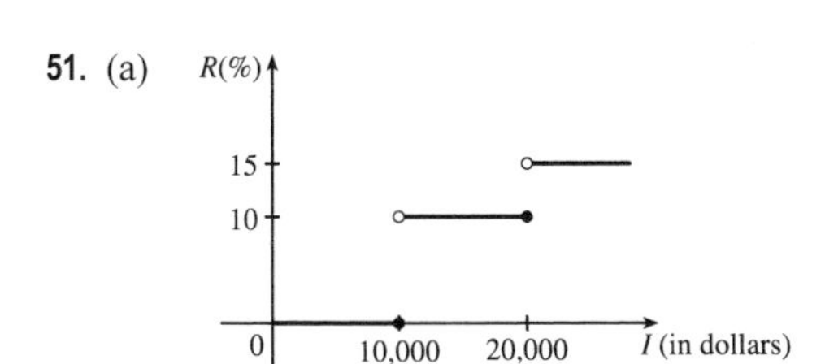

(b) On \$14,000, tax is assessed on \$4000, and $10\%(\$4000) = \$400$.

On \$26,000, tax is assessed on \$16,000, and

$10\%(\$10{,}000) + 15\%(\$6000) = \$1000 + \$900 = \$1900$.

(c) As in part (b), there is \$1000 tax assessed on \$20,000 of income, so the graph of T is a line segment from $(10{,}000, 0)$ to $(20{,}000, 1000)$. The tax on \$30,000 is \$2500, so the graph of T for $x > 20{,}000$ is the ray with initial point $(20{,}000, 1000)$ that passes through $(30{,}000, 2500)$.

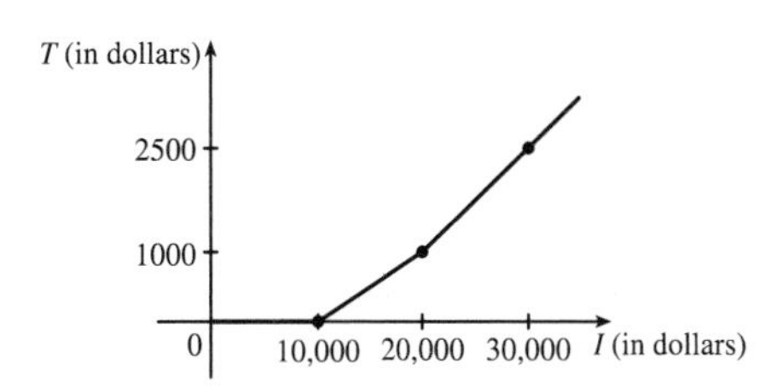

52. One example is the amount paid for cable or telephone system repair in the home, usually measured to the nearest quarter hour. Another example is the amount paid by a student in tuition fees, if the fees vary according to the number of credits for which the student has registered.

53. f is an odd function because its graph is symmetric about the origin. g is an even function because its graph is symmetric with respect to the y-axis.

54. f is not an even function since it is not symmetric with respect to the y-axis. f is not an odd function since it is not symmetric about the origin. Hence, f is *neither* even nor odd. g is an even function because its graph is symmetric with respect to the y-axis.

55. (a) Because an even function is symmetric with respect to the y-axis, and the point $(5, 3)$ is on the graph of this even function, the point $(-5, 3)$ must also be on its graph.

(b) Because an odd function is symmetric with respect to the origin, and the point $(5, 3)$ is on the graph of this odd function, the point $(-5, -3)$ must also be on its graph.

56. (a) If f is even, we get the rest of the graph by reflecting about the y-axis.

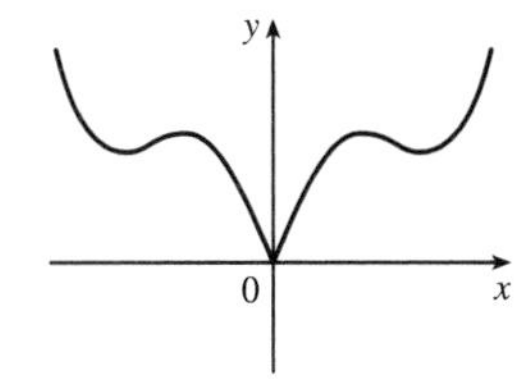

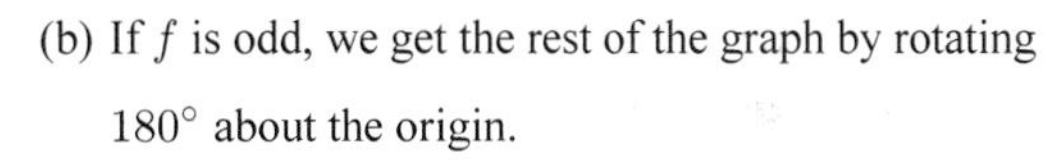

(b) If f is odd, we get the rest of the graph by rotating 180° about the origin.

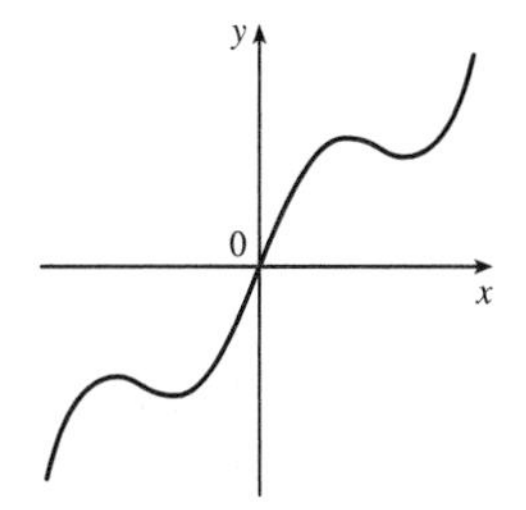

57. $f(x) = \dfrac{x}{x^2 + 1}$.

$$f(-x) = \frac{-x}{(-x)^2 + 1} = \frac{-x}{x^2 + 1} = -\frac{x}{x^2 + 1} = -f(x).$$

So f is an odd function.

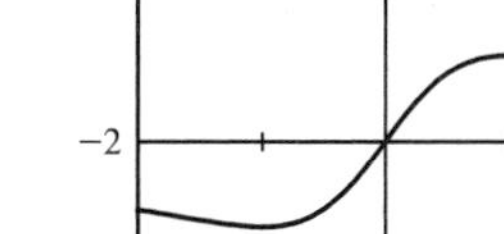

58. $f(x) = \dfrac{x^2}{x^4+1}$.

$$f(-x) = \frac{(-x)^2}{(-x)^4+1} = \frac{x^2}{x^4+1} = f(x).$$

So f is an even function.

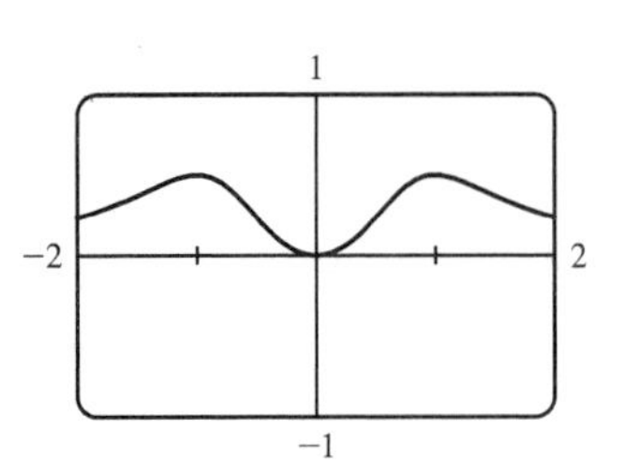

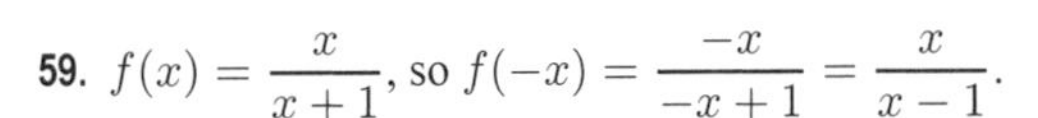

59. $f(x) = \dfrac{x}{x+1}$, so $f(-x) = \dfrac{-x}{-x+1} = \dfrac{x}{x-1}$.

Since this is neither $f(x)$ nor $-f(x)$, the function f is neither even nor odd.

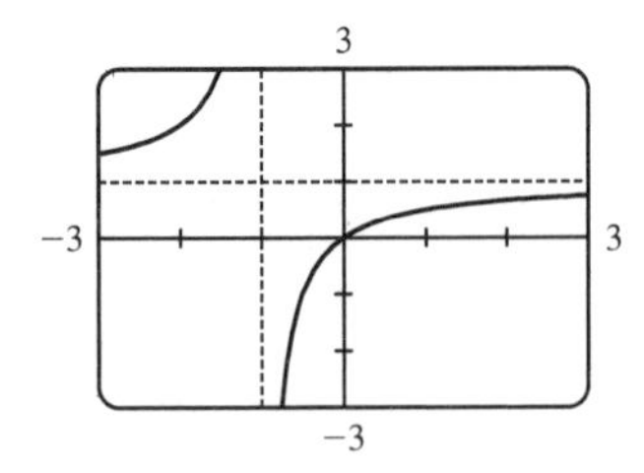

60. $f(x) = x\,|x|$.

$$f(-x) = (-x)\,|-x| = (-x)\,|x| = -(x\,|x|)$$
$$= -f(x)$$

So f is an odd function.

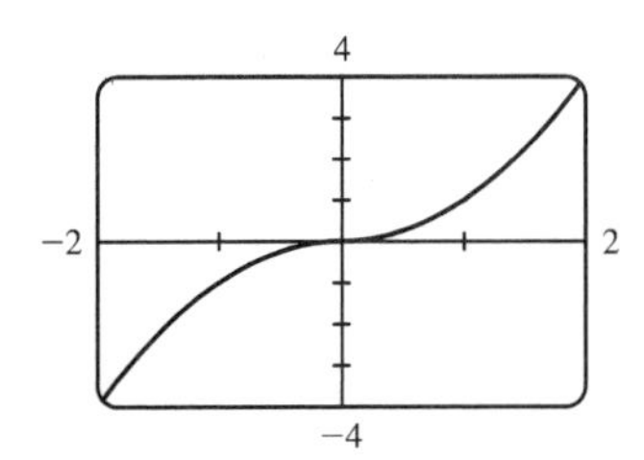

61. $f(x) = 1 + 3x^2 - x^4$.

$$f(-x) = 1 + 3(-x)^2 - (-x)^4 = 1 + 3x^2 - x^4 = f(x).$$

So f is an even function.

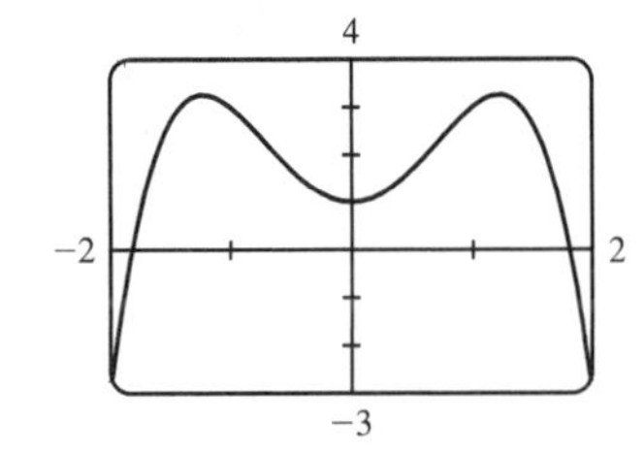

62. $f(x) = 1 + 3x^3 - x^5$, so

$$f(-x) = 1 + 3(-x)^3 - (-x)^5 = 1 + 3(-x^3) - (-x^5)$$
$$= 1 - 3x^3 + x^5$$

Since this is neither $f(x)$ nor $-f(x)$, the function f is neither even nor odd.

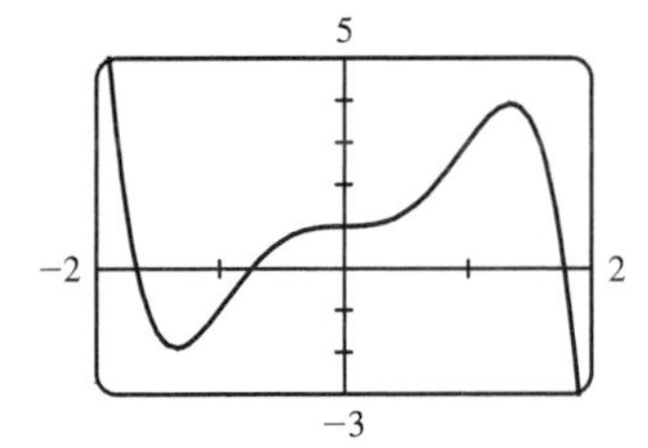

1.2 A Catalog of Essential Functions

1. (a) An equation for the family of linear functions with slope 2 is $y = f(x) = 2x + b$, where b is the y-intercept.

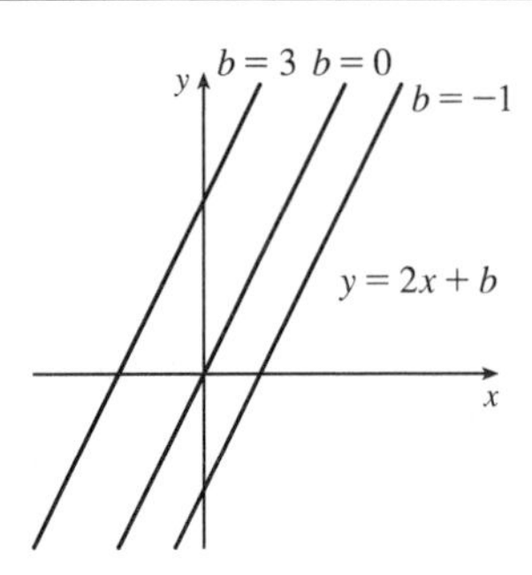

(b) $f(2) = 1$ means that the point $(2, 1)$ is on the graph of f. We can use the point-slope form of a line to obtain an equation for the family of linear functions through the point $(2, 1)$. $y - 1 = m(x - 2)$, which is equivalent to $y = mx + (1 - 2m)$ in slope-intercept form.

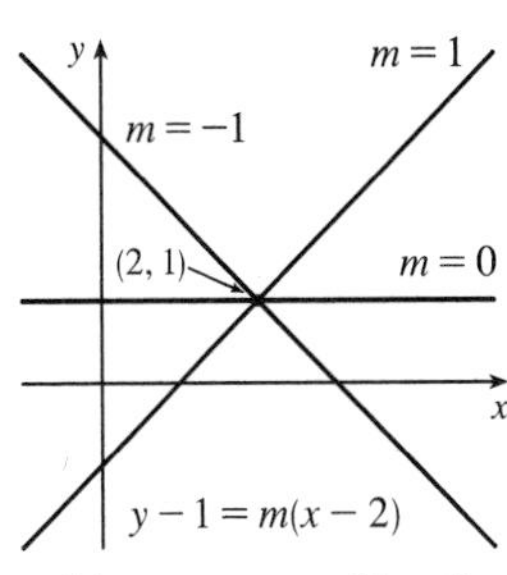

(c) To belong to both families, an equation must have slope $m = 2$, so the equation in part (b), $y = mx + (1 - 2m)$, becomes $y = 2x - 3$. It is the *only* function that belongs to both families.

2. All members of the family of linear functions $f(x) = 1 + m(x + 3)$ have graphs that are lines passing through the point $(-3, 1)$.

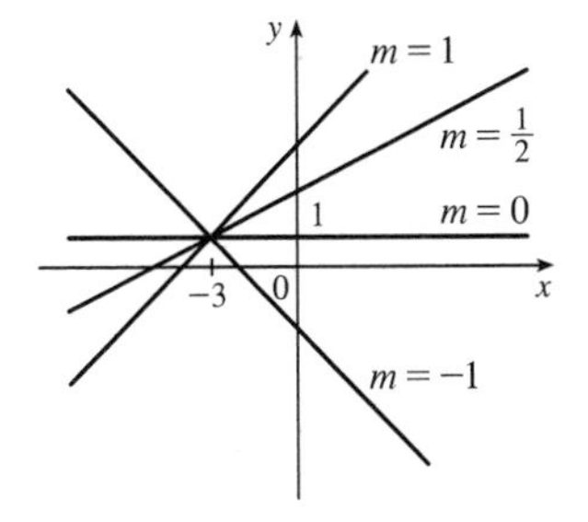

3. All members of the family of linear functions $f(x) = c - x$ have graphs that are lines with slope -1. The y-intercept is c.

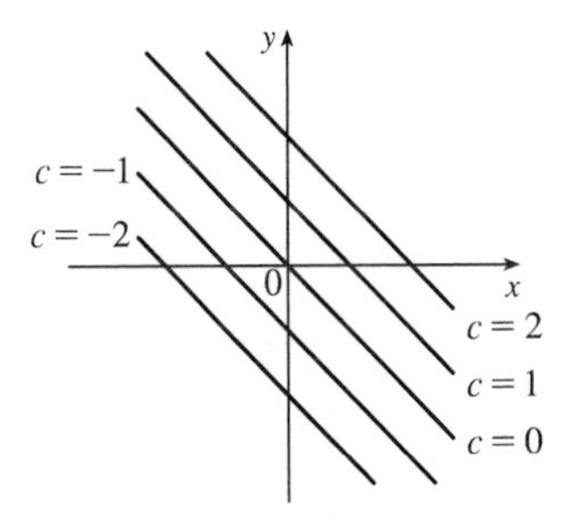

4. The vertex of the parabola on the left is $(3, 0)$, so an equation is $y = a(x - 3)^2 + 0$. Since the point $(4, 2)$ is on the parabola, we'll substitute 4 for x and 2 for y to find a. $2 = a(4 - 3)^2 \Rightarrow a = 2$, so an equation is $f(x) = 2(x - 3)^2$.

The y-intercept of the parabola on the right is $(0, 1)$, so an equation is $y = ax^2 + bx + 1$. Since the points $(-2, 2)$ and $(1, -2.5)$ are on the parabola, we'll substitute -2 for x and 2 for y as well as 1 for x and -2.5 for y to obtain two equations with the unknowns a and b.

$$(-2, 2): \quad 2 = 4a - 2b + 1 \Rightarrow 4a - 2b = 1 \quad \textbf{(1)}$$

$$(1, -2.5): \quad -2.5 = a + b + 1 \Rightarrow a + b = -3.5 \quad \textbf{(2)}$$

$2 \cdot$ **(2)** + **(1)** gives us $6a = -6 \Rightarrow a = -1$. From **(2)**, $-1 + b = -3.5 \Rightarrow b = -2.5$, so an equation is $g(x) = -x^2 - 2.5x + 1$.

5. Since $f(-1) = f(0) = f(2) = 0$, f has zeros of -1, 0, and 2, so an equation for f is $f(x) = a[x - (-1)](x - 0)(x - 2)$, or $f(x) = ax(x + 1)(x - 2)$. Because $f(1) = 6$, we'll substitute 1 for x and 6 for $f(x)$.
$6 = a(1)(2)(-1) \Rightarrow -2a = 6 \Rightarrow a = -3$, so an equation for f is $f(x) = -3x(x + 1)(x - 2)$.

6. (a) For $T = 0.02t + 8.50$, the slope is 0.02, which means that the average surface temperature of the world is increasing at a rate of $0.02°\text{C}$ per year. The T-intercept is 8.50, which represents the average surface temperature in $°\text{C}$ in the year 1900.

(b) $t = 2100 - 1900 = 200 \quad\Rightarrow\quad T = 0.02(200) + 8.50 = 12.50°\text{C}$

7. (a) $D = 200$, so $c = 0.0417D(a+1) = 0.0417(200)(a+1) = 8.34a + 8.34$. The slope is 8.34, which represents the change in mg of the dosage for a child for each change of 1 year in age.

(b) For a newborn, $a = 0$, so $c = 8.34$ mg.

8. (a)

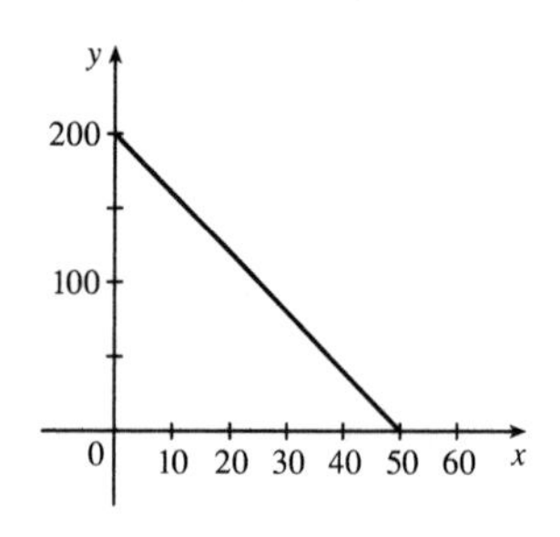

(b) The slope of -4 means that for each increase of 1 dollar for a rental space, the number of spaces rented *decreases* by 4. The y-intercept of 200 is the number of spaces that would be occupied if there were no charge for each space. The x-intercept of 50 is the smallest rental fee that results in no spaces rented.

9. (a)

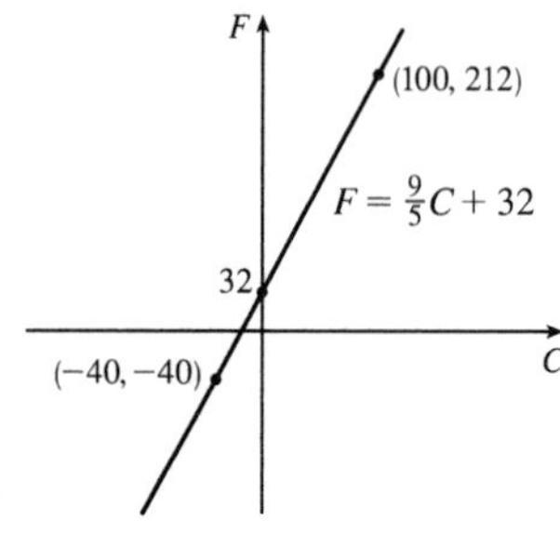

(b) The slope of $\frac{9}{5}$ means that F increases $\frac{9}{5}$ degrees for each increase of $1°\text{C}$. (Equivalently, F increases by 9 when C increases by 5 and F decreases by 9 when C decreases by 5.) The F-intercept of 32 is the Fahrenheit temperature corresponding to a Celsius temperature of 0.

10. (a) Let d = distance traveled (in miles) and t = time elapsed (in hours). At $t = 0$, $d = 0$ and at $t = 50$ minutes $= 50 \cdot \frac{1}{60} = \frac{5}{6}$ h, $d = 40$. Thus we have two points: $(0, 0)$ and $\left(\frac{5}{6}, 40\right)$, so $m = \dfrac{40 - 0}{\frac{5}{6} - 0} = 48$ and so $d = 48t$.

(b)

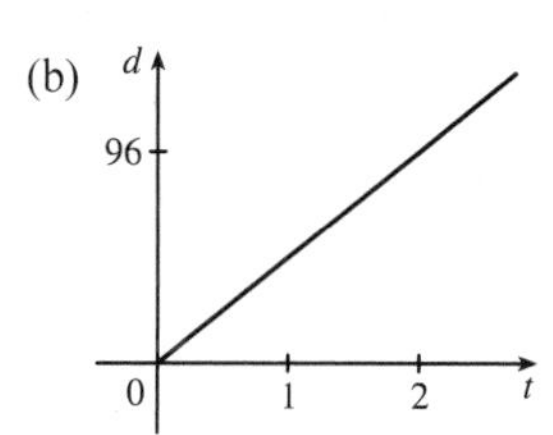

(c) The slope is 48 and represents the car's speed in mi/h.

11. (a) Using N in place of x and T in place of y, we find the slope to be $\dfrac{T_2 - T_1}{N_2 - N_1} = \dfrac{80 - 70}{173 - 113} = \dfrac{10}{60} = \dfrac{1}{6}$. So a linear equation is $T - 80 = \frac{1}{6}(N - 173) \;\Leftrightarrow\; T - 80 = \frac{1}{6}N - \frac{173}{6} \;\Leftrightarrow\; T = \frac{1}{6}N + \frac{307}{6}$ $\left[\frac{307}{6} = 51.1\overline{6}\right]$.

(b) The slope of $\frac{1}{6}$ means that the temperature in Fahrenheit degrees increases one-sixth as rapidly as the number of cricket chirps per minute. Said differently, each increase of 6 cricket chirps per minute corresponds to an increase of $1°\text{F}$.

(c) When $N = 150$, the temperature is given approximately by $T = \frac{1}{6}(150) + \frac{307}{6} = 76.1\overline{6}°\text{F} \approx 76°\text{F}$.

12. (a) Let x denote the number of chairs produced in one day and y the associated cost. Using the points $(100, 2200)$ and $(300, 4800)$, we get the slope $\frac{4800-2200}{300-100} = \frac{2600}{200} = 13$. So $y - 2200 = 13(x - 100) \;\Leftrightarrow$ $y = 13x + 900$.

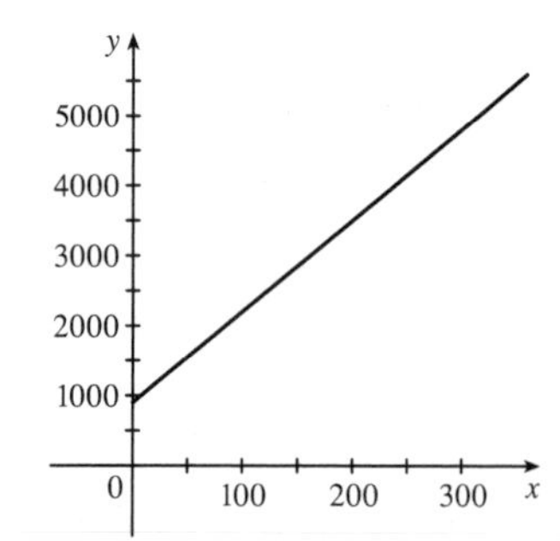

(b) The slope of the line in part (a) is 13 and it represents the cost (in dollars) of producing each additional chair.

(c) The y-intercept is 900 and it represents the fixed daily costs of operating the factory.

13. (a) We are given $\dfrac{\text{change in pressure}}{10 \text{ feet change in depth}} = \dfrac{4.34}{10} = 0.434$. Using P for pressure and d for depth with the point $(d, P) = (0, 15)$, we have the slope-intercept form of the line, $P = 0.434d + 15$.

(b) When $P = 100$, then $100 = 0.434d + 15 \Leftrightarrow 0.434d = 85 \Leftrightarrow d = \frac{85}{0.434} \approx 195.85$ feet. Thus, the pressure is 100 lb/in^2 at a depth of approximately 196 feet.

14. (a) Using d in place of x and C in place of y, we find the slope to be $\dfrac{C_2 - C_1}{d_2 - d_1} = \dfrac{460 - 380}{800 - 480} = \dfrac{80}{320} = \dfrac{1}{4}$.

So a linear equation is $C - 460 = \frac{1}{4}(d - 800) \Leftrightarrow C - 460 = \frac{1}{4}d - 200 \Leftrightarrow C = \frac{1}{4}d + 260$.

(b) Letting $d = 1500$ we get $C = \frac{1}{4}(1500) + 260 = 635$.

The cost of driving 1500 miles is \$635.

(c)

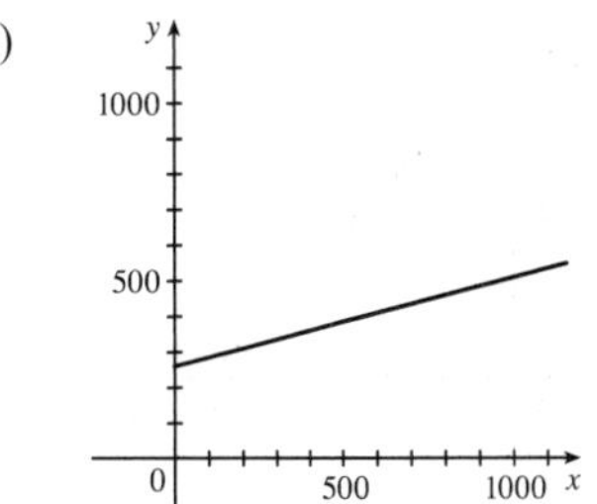

The slope of the line represents the cost per mile, \$0.25.

(d) The y-intercept represents the fixed cost, \$260.

(e) A linear function gives a suitable model in this situation because you have fixed monthly costs such as insurance and car payments, as well as costs that increase as you drive, such as gasoline, oil, and tires, and the cost of these for each additional mile driven is a constant.

15. (a) If the graph of f is shifted 3 units upward, its equation becomes $y = f(x) + 3$.

(b) If the graph of f is shifted 3 units downward, its equation becomes $y = f(x) - 3$.

(c) If the graph of f is shifted 3 units to the right, its equation becomes $y = f(x - 3)$.

(d) If the graph of f is shifted 3 units to the left, its equation becomes $y = f(x + 3)$.

(e) If the graph of f is reflected about the x-axis, its equation becomes $y = -f(x)$.

(f) If the graph of f is reflected about the y-axis, its equation becomes $y = f(-x)$.

(g) If the graph of f is stretched vertically by a factor of 3, its equation becomes $y = 3f(x)$.

(h) If the graph of f is shrunk vertically by a factor of 3, its equation becomes $y = \frac{1}{3}f(x)$.

16. (a) To obtain the graph of $y = 5f(x)$ from the graph of $y = f(x)$, stretch the graph vertically by a factor of 5.

(b) To obtain the graph of $y = f(x - 5)$ from the graph of $y = f(x)$, shift the graph 5 units to the right.

(c) To obtain the graph of $y = -f(x)$ from the graph of $y = f(x)$, reflect the graph about the x-axis.

(d) To obtain the graph of $y = -5f(x)$ from the graph of $y = f(x)$, stretch the graph vertically by a factor of 5 and reflect it about the x-axis.

(e) To obtain the graph of $y = f(5x)$ from the graph of $y = f(x)$, shrink the graph horizontally by a factor of 5.

(f) To obtain the graph of $y = 5f(x) - 3$ from the graph of $y = f(x)$, stretch the graph vertically by a factor of 5 and shift it 3 units downward.

17. (a) (graph 3) The graph of f is shifted 4 units to the right and has equation $y = f(x - 4)$.

(b) (graph 1) The graph of f is shifted 3 units upward and has equation $y = f(x) + 3$.

(c) (graph 4) The graph of f is shrunk vertically by a factor of 3 and has equation $y = \frac{1}{3}f(x)$.

(d) (graph 5) The graph of f is shifted 4 units to the left and reflected about the x-axis. Its equation is $y = -f(x + 4)$.

(e) (graph 2) The graph of f is shifted 6 units to the left and stretched vertically by a factor of 2. Its equation is $y = 2f(x + 6)$.

18. (a) To graph $y = f(x+4)$ we shift the graph of f, 4 units to the left.

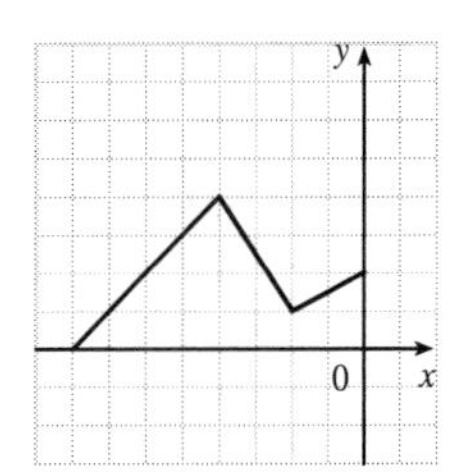

The point $(2,1)$ on the graph of f corresponds to the point $(2-4,1) = (-2,1)$.

(b) To graph $y = f(x) + 4$ we shift the graph of f, 4 units upward.

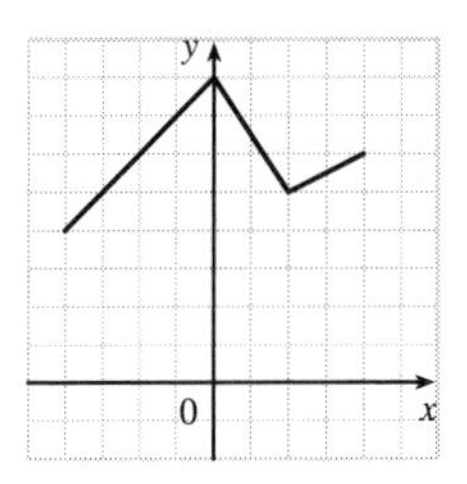

The point $(2,1)$ on the graph of f corresponds to the point $(2, 1+4) = (2,5)$.

(c) To graph $y = 2f(x)$ we stretch the graph of f vertically by a factor of 2.

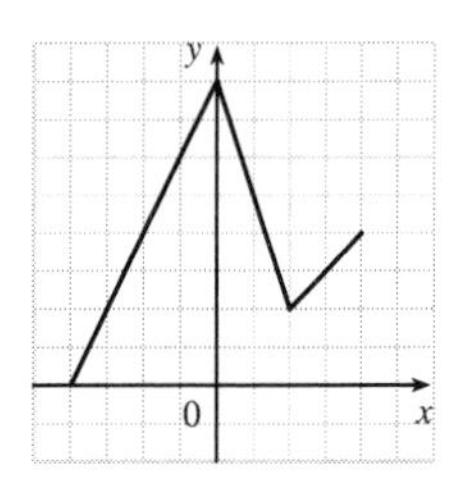

The point $(2,1)$ on the graph of f corresponds to the point $(2, 2\cdot 1) = (2,2)$.

(d) To graph $y = -\frac{1}{2}f(x) + 3$, we shrink the graph of f vertically by a factor of 2, then reflect the resulting graph about the x-axis, then shift the resulting graph 3 units upward.

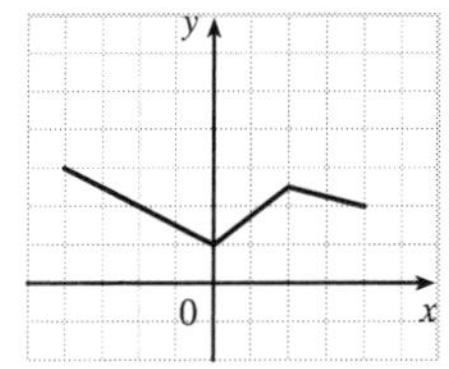

The point $(2,1)$ on the graph of f corresponds to the point $\left(2, -\frac{1}{2}\cdot 1 + 3\right) = (2, 2.5)$.

19. (a) To graph $y = f(2x)$ we shrink the graph of f horizontally by a factor of 2.

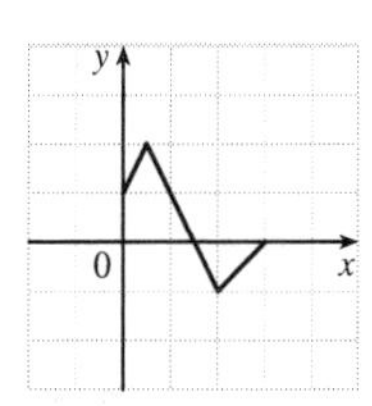

The point $(4,-1)$ on the graph of f corresponds to the point $\left(\frac{1}{2}\cdot 4, -1\right) = (2,-1)$.

(b) To graph $y = f\left(\frac{1}{2}x\right)$ we stretch the graph of f horizontally by a factor of 2.

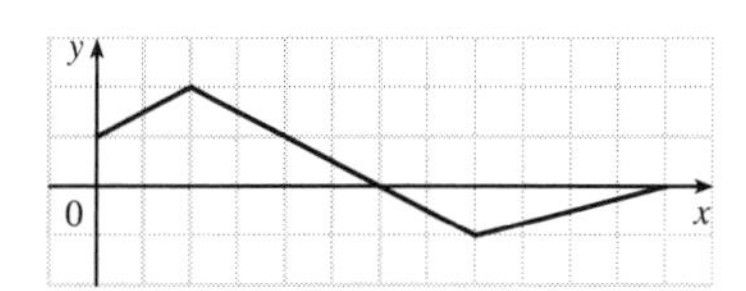

The point $(4,-1)$ on the graph of f corresponds to the point $(2\cdot 4, -1) = (8,-1)$.

(c) To graph $y = f(-x)$ we reflect the graph of f about the y-axis.

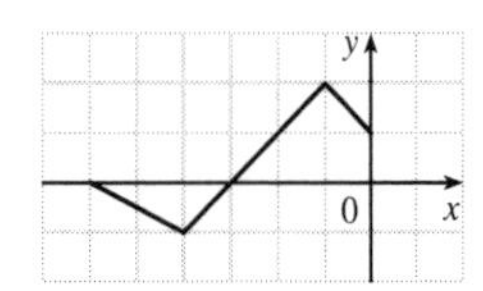

The point $(4,-1)$ on the graph of f corresponds to the point $(-1\cdot 4, -1) = (-4,-1)$.

(d) To graph $y = -f(-x)$ we reflect the graph of f about the y-axis, then about the x-axis.

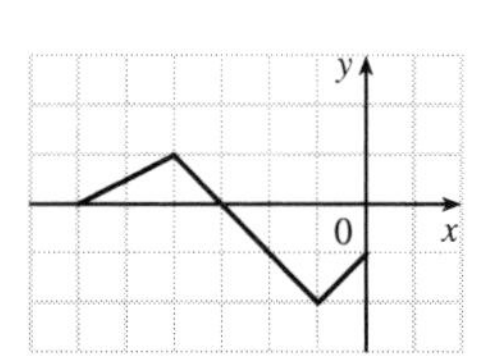

The point $(4,-1)$ on the graph of f corresponds to the point $(-1\cdot 4, -1\cdot -1) = (-4,1)$.

20. (a) The graph of $y = 2\sin x$ can be obtained from the graph of $y = \sin x$ by stretching it vertically by a factor of 2.

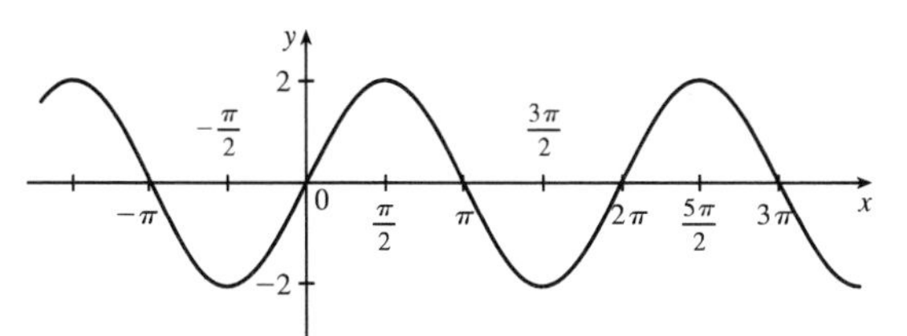

(b) The graph of $y = 1 + \sqrt{x}$ can be obtained from the graph of $y = \sqrt{x}$ by shifting it upward 1 unit.

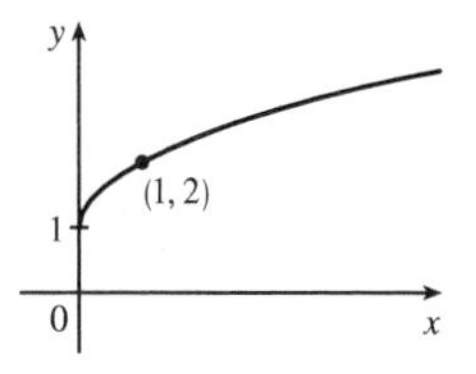

21. $y = -x^3$: Start with the graph of $y = x^3$ and reflect about the x-axis. Note: Reflecting about the y-axis gives the same result since substituting $-x$ for x gives us $y = (-x)^3 = -x^3$.

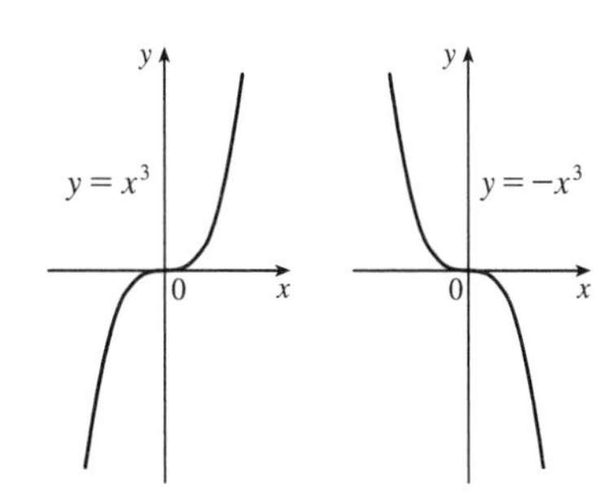

22. $y = 1 - x^2 = -x^2 + 1$: Start with the graph of $y = x^2$, reflect about the x-axis, and then shift 1 unit upward.

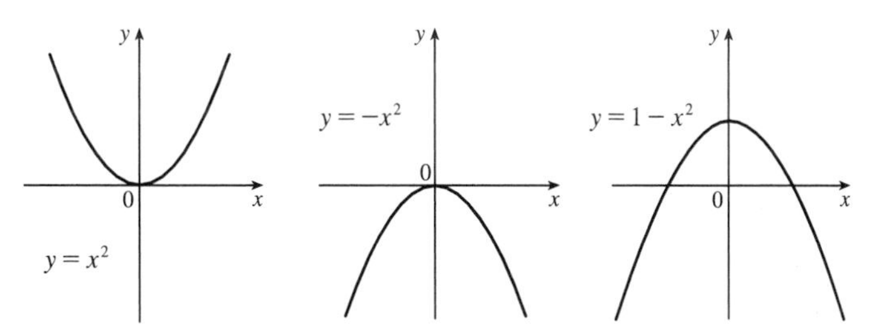

23. $y = (x + 1)^2$: Start with the graph of $y = x^2$ and shift 1 unit to the left.

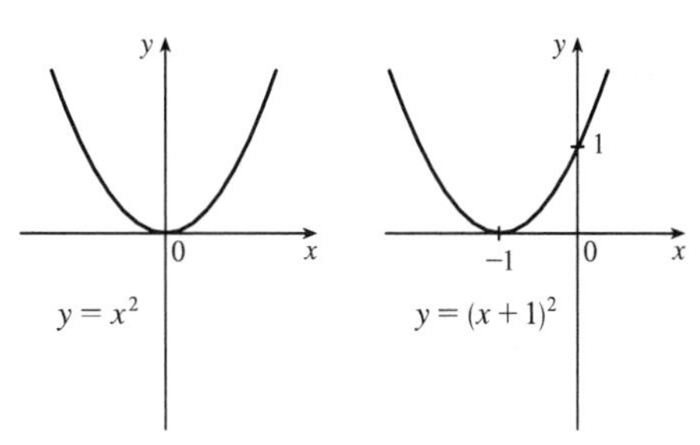

24. $y = x^2 - 4x + 3 = (x^2 - 4x + 4) - 1 = (x - 2)^2 - 1$: Start with the graph of $y = x^2$, shift 2 units to the right, and then shift 1 unit downward.

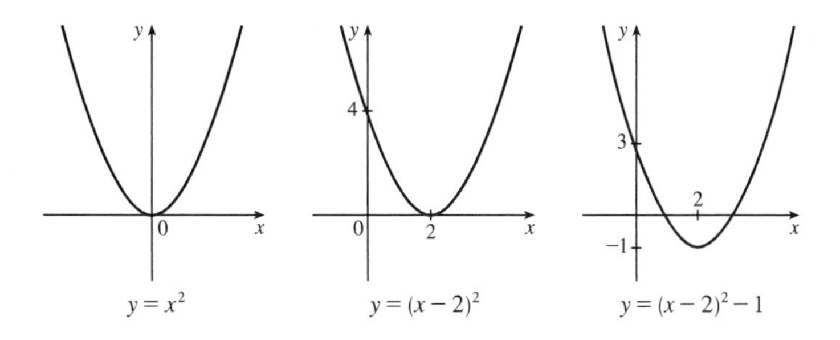

25. $y = 1 + 2\cos x$: Start with the graph of $y = \cos x$, stretch vertically by a factor of 2, and then shift 1 unit upward.

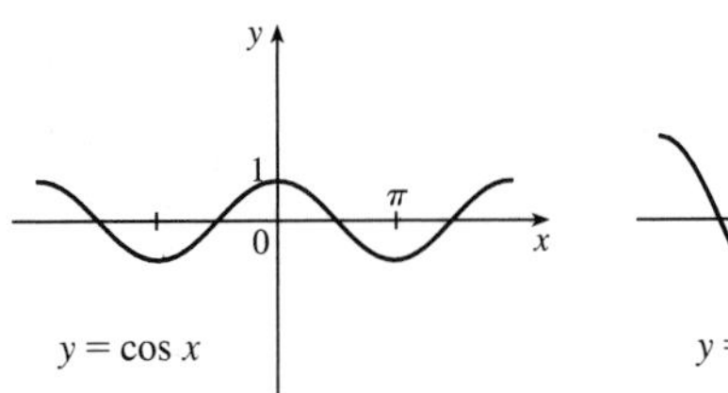

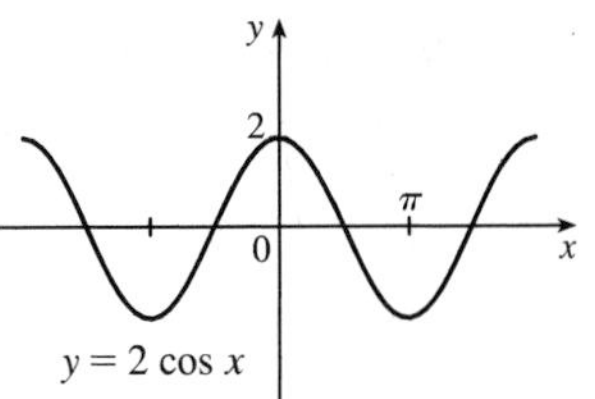

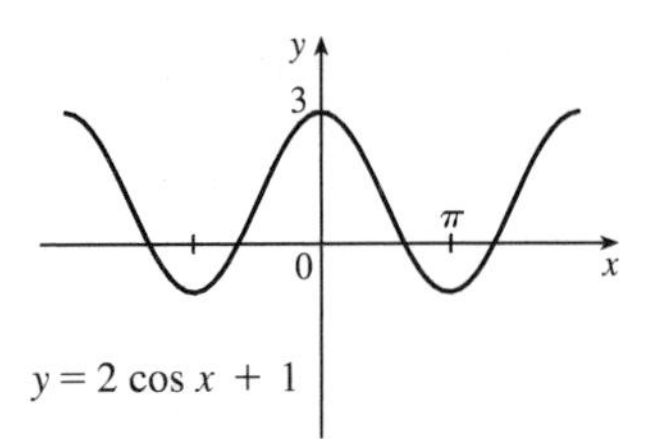

26. $y = 4\sin 3x$: Start with the graph of $y = \sin x$, compress horizontally by a factor of 3, and then stretch vertically by a factor of 4.

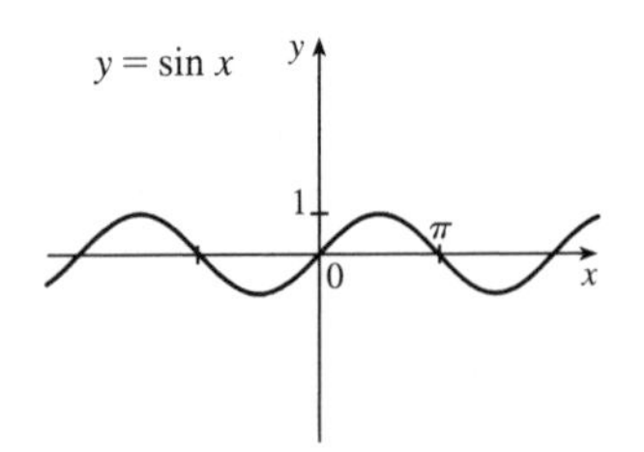

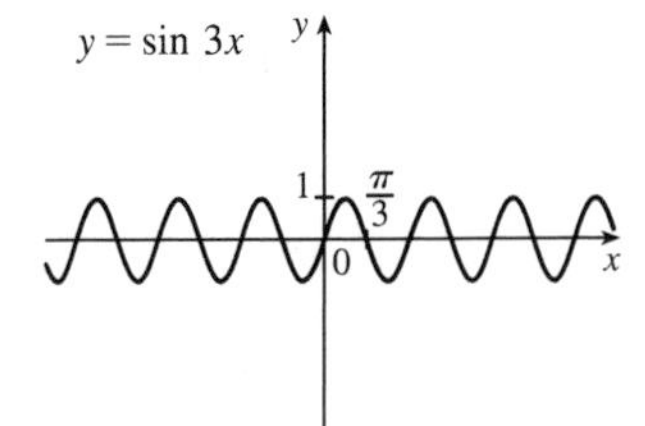

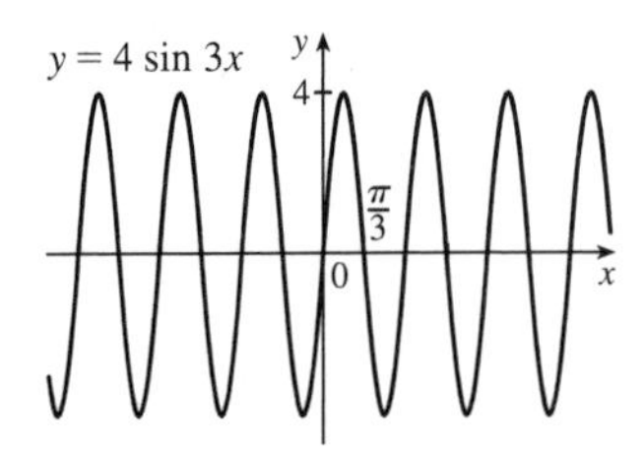

27. $y = \sin(x/2)$: Start with the graph of $y = \sin x$ and stretch horizontally by a factor of 2.

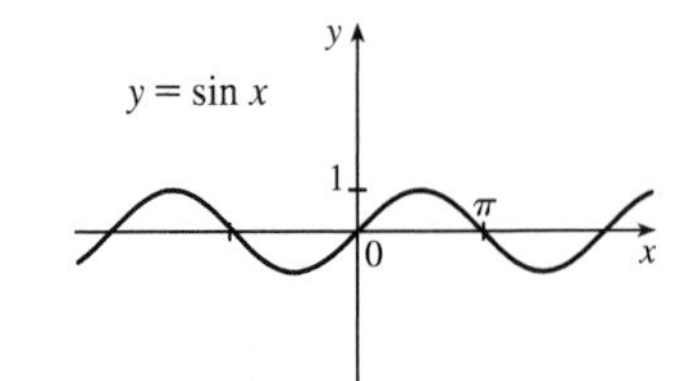

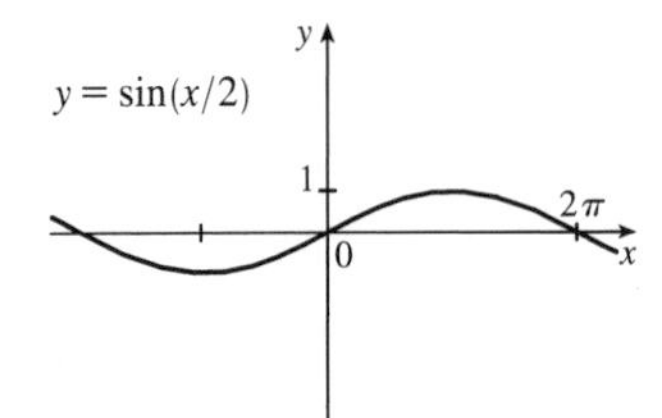

28. $y = \dfrac{1}{x-4}$: Start with the graph of $y = 1/x$ and shift 4 units to the right.

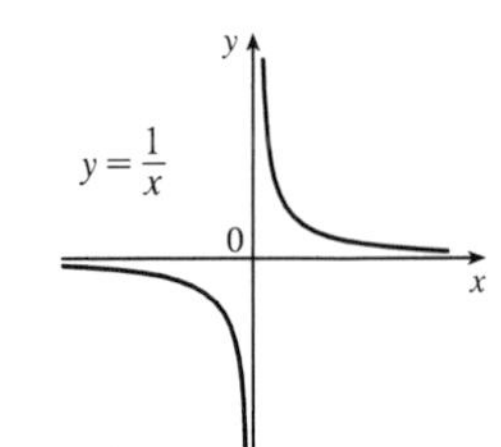

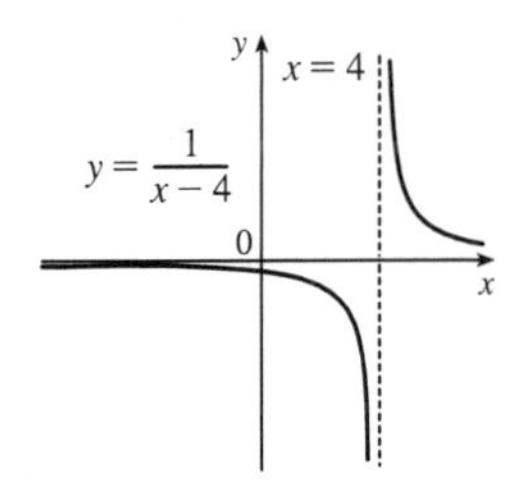

29. $y = \sqrt{x+3}$: Start with the graph of $y = \sqrt{x}$ and shift 3 units to the left.

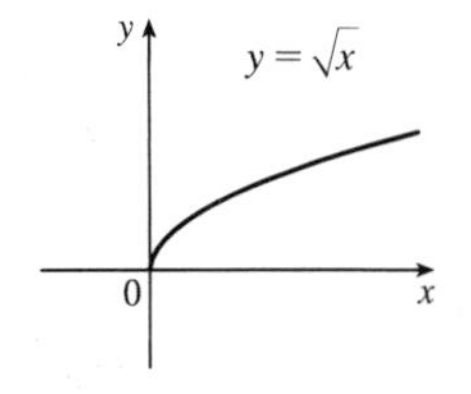

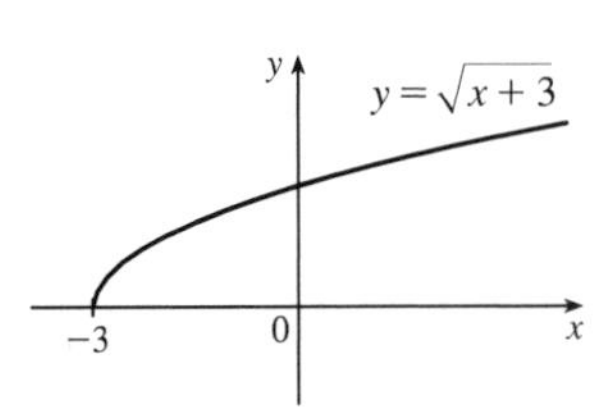

30. $y = (x+2)^4 + 3$: Start with the graph of $y = x^4$, shift 2 units to the left, and then shift 3 units upward.

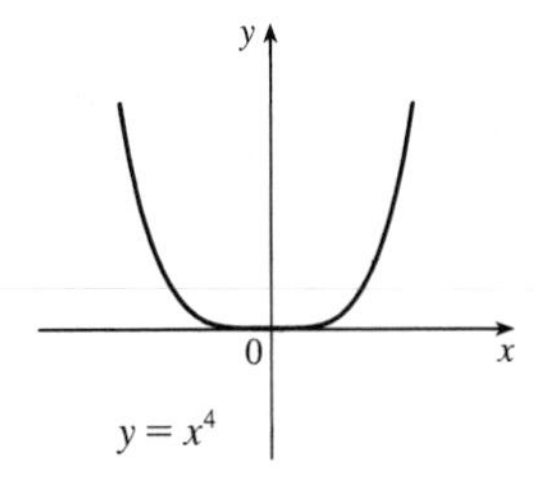

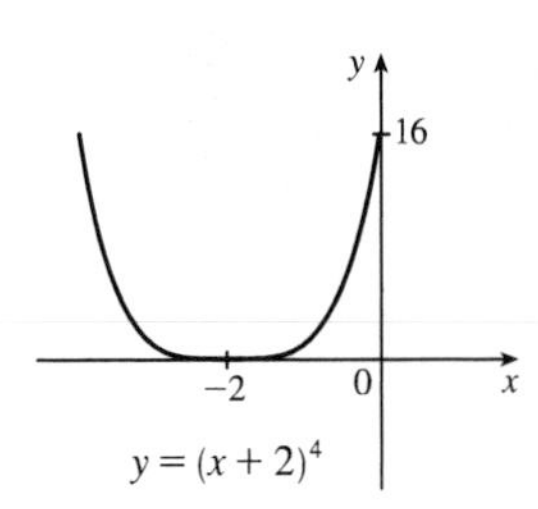

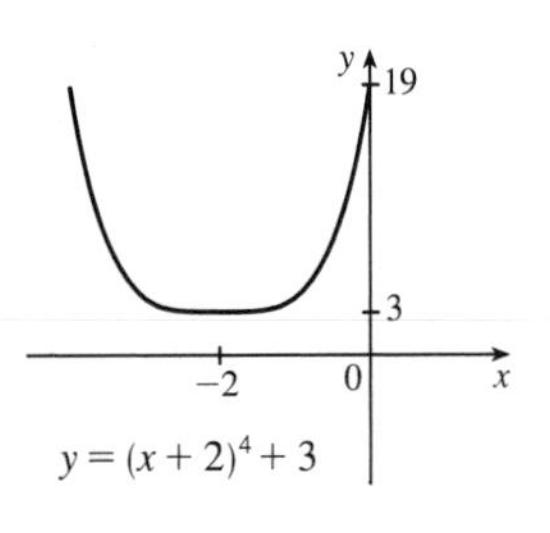

31. $y = \frac{1}{2}(x^2 + 8x) = \frac{1}{2}(x^2 + 8x + 16) - 8 = \frac{1}{2}(x + 4)^2 - 8$: Start with the graph of $y = x^2$, compress vertically by a factor of 2, shift 4 units to the left, and then shift 8 units downward.

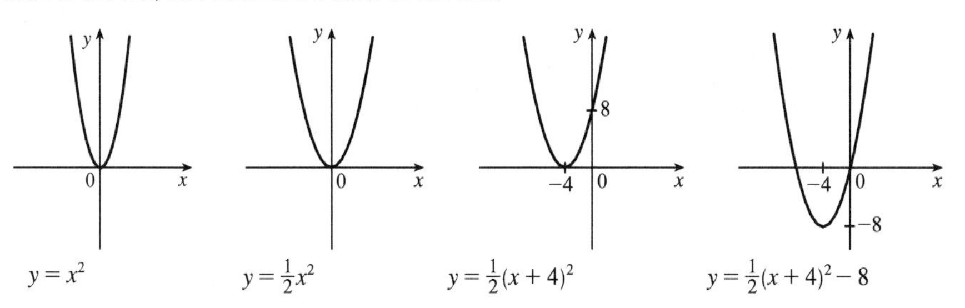

32. $y = 1 + \sqrt[3]{x - 1}$: Start with the graph of $y = \sqrt[3]{x}$, shift 1 unit to the right, and then shift 1 unit upward.

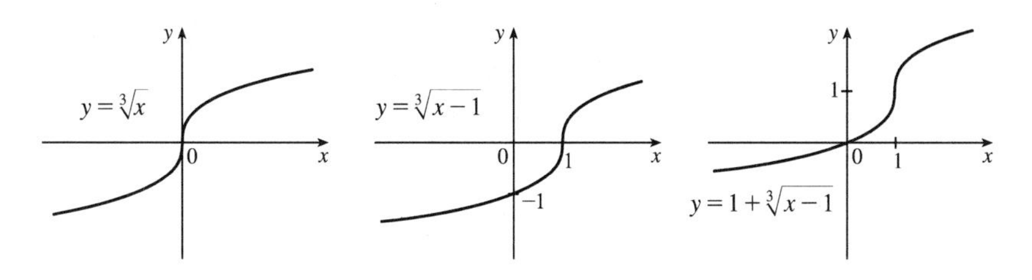

33. $y = 2/(x + 1)$: Start with the graph of $y = 1/x$, shift 1 unit to the left, and then stretch vertically by a factor of 2.

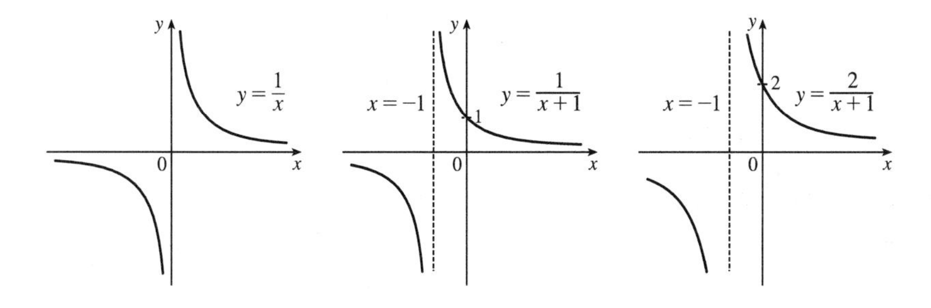

34. $y = \frac{1}{4}\tan\left(x - \frac{\pi}{4}\right)$: Start with the graph of $y = \tan x$, shift $\frac{\pi}{4}$ units to the right, and then compress vertically by a factor of 4.

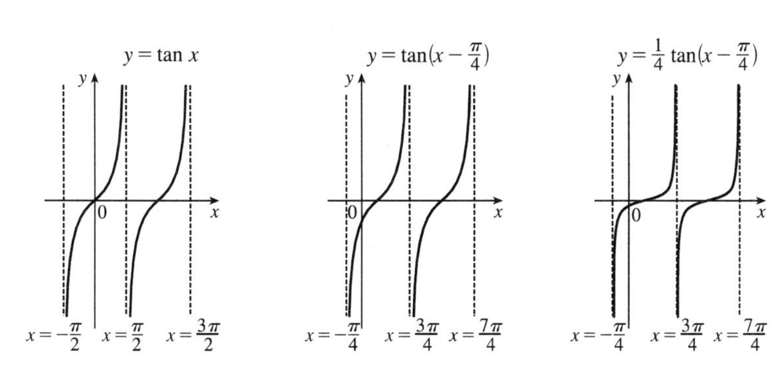

35. $f(x) = x^3 + 2x^2$; $g(x) = 3x^2 - 1$. $D = \mathbb{R}$ for both f and g.

$(f + g)(x) = (x^3 + 2x^2) + (3x^2 - 1) = x^3 + 5x^2 - 1$, $D = \mathbb{R}$.

$(f - g)(x) = (x^3 + 2x^2) - (3x^2 - 1) = x^3 - x^2 + 1$, $D = \mathbb{R}$.

$(fg)(x) = (x^3 + 2x^2)(3x^2 - 1) = 3x^5 + 6x^4 - x^3 - 2x^2$, $D = \mathbb{R}$.

$\left(\dfrac{f}{g}\right)(x) = \dfrac{x^3 + 2x^2}{3x^2 - 1}$, $D = \left\{x \mid x \neq \pm\dfrac{1}{\sqrt{3}}\right\}$ since $3x^2 - 1 \neq 0$.

36. $f(x) = \sqrt{1 + x}$, $D = [-1, \infty)$; $g(x) = \sqrt{1 - x}$, $D = (-\infty, 1]$.

$(f + g)(x) = \sqrt{1 + x} + \sqrt{1 - x}$, $D = (-\infty, 1] \cap [-1, \infty) = [-1, 1]$.

$(f - g)(x) = \sqrt{1 + x} - \sqrt{1 - x}$, $D = [-1, 1]$.

$(fg)(x) = \sqrt{1 + x} \cdot \sqrt{1 - x} = \sqrt{1 - x^2}$, $D = [-1, 1]$.

$\left(\dfrac{f}{g}\right)(x) = \dfrac{\sqrt{1 + x}}{\sqrt{1 - x}}$, $D = [-1, 1)$. We must exclude $x = 1$ since it would make $\dfrac{f}{g}$ undefined.

37. $f(x) = x^2 - 1$, $D = \mathbb{R}$; $g(x) = 2x + 1$, $D = \mathbb{R}$.

(a) $(f \circ g)(x) = f(g(x)) = f(2x + 1) = (2x + 1)^2 - 1 = (4x^2 + 4x + 1) - 1 = 4x^2 + 4x$, $D = \mathbb{R}$.

(b) $(g \circ f)(x) = g(f(x)) = g(x^2 - 1) = 2(x^2 - 1) + 1 = (2x^2 - 2) + 1 = 2x^2 - 1$, $D = \mathbb{R}$.

(c) $(f \circ f)(x) = f(f(x)) = f(x^2 - 1) = (x^2 - 1)^2 - 1 = (x^4 - 2x^2 + 1) - 1 = x^4 - 2x^2$, $D = \mathbb{R}$.

(d) $(g \circ g)(x) = g(g(x)) = g(2x + 1) = 2(2x + 1) + 1 = (4x + 2) + 1 = 4x + 3$, $D = \mathbb{R}$.

38. $f(x) = 1 - x^3$, $D = \mathbb{R}$; $g(x) = 1/x$, $D = \{x \mid x \neq 0\}$.

(a) $(f \circ g)(x) = f(g(x)) = f(1/x) = 1 - (1/x)^3 = 1 - 1/x^3$, $D = \{x \mid x \neq 0\}$.

(b) $(g \circ f)(x) = g(f(x)) = g(1 - x^3) = 1/(1 - x^3)$, $D = \{x \mid 1 - x^3 \neq 0\} = \{x \mid x \neq 1\}$.

(c) $(f \circ f)(x) = f(f(x)) = f(1 - x^3) = 1 - (1 - x^3)^3$ $[= x^9 - 3x^6 + 3x^3]$, $D = \mathbb{R}$.

(d) $(g \circ g)(x) = g(g(x)) = g(1/x) = 1/(1/x) = x$, $D = \{x \mid x \neq 0\}$ because 0 is not in the domain of g.

39. $f(x) = \sin x$, $D = \mathbb{R}$; $g(x) = 1 - \sqrt{x}$, $D = [0, \infty)$.

(a) $(f \circ g)(x) = f(g(x)) = f(1 - \sqrt{x}) = \sin(1 - \sqrt{x})$, $D = [0, \infty)$.

(b) $(g \circ f)(x) = g(f(x)) = g(\sin x) = 1 - \sqrt{\sin x}$.
For $\sqrt{\sin x}$ to be defined, we must have $\sin x \geq 0 \;\Leftrightarrow$
$x \in [0, \pi] \cup [2\pi, 3\pi] \cup [-2\pi, -\pi] \cup [4\pi, 5\pi] \cup [-4\pi, -3\pi] \cup \ldots$, so $D = \{x \mid x \in [2n\pi, \pi + 2n\pi]$, where n is an integer$\}$.

(c) $(f \circ f)(x) = f(f(x)) = f(\sin x) = \sin(\sin x)$, $D = \mathbb{R}$.

(d) $(g \circ g)(x) = g(g(x)) = g(1 - \sqrt{x}) = 1 - \sqrt{1 - \sqrt{x}}$,
$D = \{x \geq 0 \mid 1 - \sqrt{x} \geq 0\} = \{x \geq 0 \mid 1 \geq \sqrt{x}\} = \{x \geq 0 \mid \sqrt{x} \leq 1\} = [0, 1]$.

40. $f(x) = 1 - 3x$, $D = \mathbb{R}$; $g(x) = 5x^2 + 3x + 2$, $D = \mathbb{R}$.

(a) $(f \circ g)(x) = f(g(x)) = f(5x^2 + 3x + 2) = 1 - 3(5x^2 + 3x + 2) = 1 - 15x^2 - 9x - 6 = -15x^2 - 9x - 5$,
$D = \mathbb{R}$.

(b) $(g \circ f)(x) = g(f(x)) = g(1 - 3x) = 5(1 - 3x)^2 + 3(1 - 3x) + 2 = 5(1 - 6x + 9x^2) + 3 - 9x + 2$
$= 5 - 30x + 45x^2 - 9x + 5 = 45x^2 - 39x + 10$, $D = \mathbb{R}$.

(c) $(f \circ f)(x) = f(f(x)) = f(1-3x) = 1 - 3(1-3x) = 1 - 3 + 9x = 9x - 2, \quad D = \mathbb{R}.$

(d) $(g \circ g)(x) = g(g(x)) = g(5x^2 + 3x + 2) = 5(5x^2 + 3x + 2)^2 + 3(5x^2 + 3x + 2) + 2$

$$= 5(25x^4 + 30x^3 + 29x^2 + 12x + 4) + 15x^2 + 9x + 6 + 2$$

$$= 125x^4 + 150x^3 + 145x^2 + 60x + 20 + 15x^2 + 9x + 8$$

$$= 125x^4 + 150x^3 + 160x^2 + 69x + 28, \quad D = \mathbb{R}.$$

41. $f(x) = x + \dfrac{1}{x}, \quad D = \{x \mid x \neq 0\}; \quad g(x) = \dfrac{x+1}{x+2}, \quad D = \{x \mid x \neq -2\}.$

(a) $(f \circ g)(x) = f(g(x)) = f\left(\dfrac{x+1}{x+2}\right) = \dfrac{x+1}{x+2} + \dfrac{1}{\dfrac{x+1}{x+2}} = \dfrac{x+1}{x+2} + \dfrac{x+2}{x+1}$

$$= \frac{(x+1)(x+1) + (x+2)(x+2)}{(x+2)(x+1)} = \frac{(x^2 + 2x + 1) + (x^2 + 4x + 4)}{(x+2)(x+1)} = \frac{2x^2 + 6x + 5}{(x+2)(x+1)}$$

Since $g(x)$ is not defined for $x = -2$ and $f(g(x))$ is not defined for $x = -2$ and $x = -1$, the domain of $(f \circ g)(x)$ is $D = \{x \mid x \neq -2, -1\}$.

(b) $(g \circ f)(x) = g(f(x)) = g\left(x + \dfrac{1}{x}\right) = \dfrac{\left(x + \dfrac{1}{x}\right) + 1}{\left(x + \dfrac{1}{x}\right) + 2} = \dfrac{\dfrac{x^2 + 1 + x}{x}}{\dfrac{x^2 + 1 + 2x}{x}} = \dfrac{x^2 + x + 1}{x^2 + 2x + 1} = \dfrac{x^2 + x + 1}{(x+1)^2}$

Since $f(x)$ is not defined for $x = 0$ and $g(f(x))$ is not defined for $x = -1$, the domain of $(g \circ f)(x)$ is $D = \{x \mid x \neq -1, 0\}$.

(c) $(f \circ f)(x) = f(f(x)) = f\left(x + \dfrac{1}{x}\right) = \left(x + \dfrac{1}{x}\right) + \dfrac{1}{x + \frac{1}{x}} = x + \dfrac{1}{x} + \dfrac{1}{\frac{x^2+1}{x}} = x + \dfrac{1}{x} + \dfrac{x}{x^2 + 1}$

$$= \frac{x(x)(x^2+1) + 1\,(x^2+1) + x(x)}{x(x^2+1)} = \frac{x^4 + x^2 + x^2 + 1 + x^2}{x(x^2+1)}$$

$$= \frac{x^4 + 3x^2 + 1}{x(x^2+1)}, \quad D = \{x \mid x \neq 0\}.$$

(d) $(g \circ g)(x) = g(g(x)) = g\left(\dfrac{x+1}{x+2}\right) = \dfrac{\dfrac{x+1}{x+2} + 1}{\dfrac{x+1}{x+2} + 2} = \dfrac{\dfrac{x + 1 + 1(x+2)}{x+2}}{\dfrac{x + 1 + 2(x+2)}{x+2}} = \dfrac{x + 1 + x + 2}{x + 1 + 2x + 4} = \dfrac{2x+3}{3x+5}$

Since $g(x)$ is not defined for $x = -2$ and $g(g(x))$ is not defined for $x = -\frac{5}{3}$, the domain of $(g \circ g)(x)$ is $D = \left\{x \mid x \neq -2, -\frac{5}{3}\right\}$.

42. $f(x) = \sqrt{2x+3}, \quad D = \left\{x \mid x \geq -\frac{3}{2}\right\}; \quad g(x) = x^2 + 1, \quad D = \mathbb{R}.$

(a) $(f \circ g)(x) = f(x^2 + 1) = \sqrt{2(x^2+1) + 3} = \sqrt{2x^2 + 5}, \quad D = \mathbb{R}.$

(b) $(g \circ f)(x) = g\left(\sqrt{2x+3}\right) = \left(\sqrt{2x+3}\right)^2 + 1 = (2x+3) + 1 = 2x + 4, \quad D = \left\{x \mid x \geq -\frac{3}{2}\right\}.$

(c) $(f \circ f)(x) = f\left(\sqrt{2x+3}\right) = \sqrt{2\left(\sqrt{2x+3}\right) + 3} = \sqrt{2\sqrt{2x+3} + 3}, \quad D = \left\{x \mid x \geq -\frac{3}{2}\right\}.$

(d) $(g \circ g)(x) = g(x^2 + 1) = (x^2+1)^2 + 1 = (x^4 + 2x^2 + 1) + 1 = x^4 + 2x^2 + 2, \quad D = \mathbb{R}.$

43. $(f \circ g \circ h)(x) = f(g(h(x))) = f(g(x+3)) = f((x+3)^2 + 2) = f(x^2 + 6x + 11) = \sqrt{(x^2 + 6x + 11) - 1} = \sqrt{x^2 + 6x + 10}$

44. $(f \circ g \circ h)(x) = f(g(h(x))) = f\left(g\left(\sqrt{x+3}\right)\right) = f\left(\cos\sqrt{x+3}\right) = \dfrac{2}{\cos\sqrt{x+3} + 1}$

45. Let $g(x) = x^2 + 1$ and $f(x) = x^{10}$. Then $(f \circ g)(x) = f(g(x)) = f(x^2 + 1) = (x^2 + 1)^{10} = F(x)$.

46. Let $g(x) = \sqrt{x}$ and $f(x) = \sin x$. Then $(f \circ g)(x) = f(g(x)) = f(\sqrt{x}) = \sin(\sqrt{x}) = F(x)$.

47. Let $g(t) = \cos t$ and $f(t) = \sqrt{t}$. Then $(f \circ g)(t) = f(g(t)) = f(\cos t) = \sqrt{\cos t} = u(t)$.

48. Let $g(t) = \tan t$ and $f(t) = \dfrac{t}{1+t}$. Then $(f \circ g)(t) = f(g(t)) = f(\tan t) = \dfrac{\tan t}{1 + \tan t} = u(t)$.

49. Let $h(x) = x^2$, $g(x) = 3^x$, and $f(x) = 1 - x$. Then

$(f \circ g \circ h)(x) = f(g(h(x))) = f(g(x^2)) = f\left(3^{x^2}\right) = 1 - 3^{x^2} = H(x)$.

50. Let $h(x) = |x|$, $g(x) = 2 + x$, and $f(x) = \sqrt[8]{x}$. Then

$(f \circ g \circ h)(x) = f(g(h(x))) = f(g(|x|)) = f(2 + |x|) = \sqrt[8]{2 + |x|} = H(x)$.

51. Let $h(x) = \sqrt{x}$, $g(x) = \sec x$, and $f(x) = x^4$. Then

$(f \circ g \circ h)(x) = f(g(h(x))) = f(g(\sqrt{x})) = f(\sec\sqrt{x}) = (\sec\sqrt{x})^4 = \sec^4(\sqrt{x}) = H(x)$.

52. (a) $f(g(1)) = f(6) = 5$

(b) $g(f(1)) = g(3) = 2$

(c) $f(f(1)) = f(3) = 4$

(d) $g(g(1)) = g(6) = 3$

(e) $(g \circ f)(3) = g(f(3)) = g(4) = 1$

(f) $(f \circ g)(6) = f(g(6)) = f(3) = 4$

53. (a) $g(2) = 5$, because the point $(2, 5)$ is on the graph of g. Thus, $f(g(2)) = f(5) = 4$, because the point $(5, 4)$ is on the graph of f.

(b) $g(f(0)) = g(0) = 3$

(c) $(f \circ g)(0) = f(g(0)) = f(3) = 0$

(d) $(g \circ f)(6) = g(f(6)) = g(6)$. This value is not defined, because there is no point on the graph of g that has x-coordinate 6.

(e) $(g \circ g)(-2) = g(g(-2)) = g(1) = 4$

(f) $(f \circ f)(4) = f(f(4)) = f(2) = -2$

54. (a) The radius r of the balloon is increasing at a rate of 2 cm/s, so $r(t) = (2\text{ cm/s})(t\text{ s}) = 2t$ (in cm).

(b) Using $V = \frac{4}{3}\pi r^3$, we get $(V \circ r)(t) = V(r(t)) = V(2t) = \frac{4}{3}\pi(2t)^3 = \frac{32}{3}\pi t^3$.

The result, $V = \frac{32}{3}\pi t^3$, gives the volume of the balloon (in cm^3) as a function of time (in s).

55. (a) Using the relationship *distance* = *rate* · *time* with the radius r as the distance, we have $r(t) = 60t$.

(b) $A = \pi r^2 \;\Rightarrow\; (A \circ r)(t) = A(r(t)) = \pi(60t)^2 = 3600\pi t^2$. This formula gives us the extent of the rippled area (in cm^2) at any time t.

56. (a) $d = rt \;\Rightarrow\; d(t) = 350t$

(b) There is a Pythagorean relationship involving the legs with lengths d and 1 and the hypotenuse with length s: $d^2 + 1^2 = s^2$. Thus, $s(d) = \sqrt{d^2 + 1}$.

(c) $(s \circ d)(t) = s(d(t)) = s(350t) = \sqrt{(350t)^2 + 1}$

57. (a)

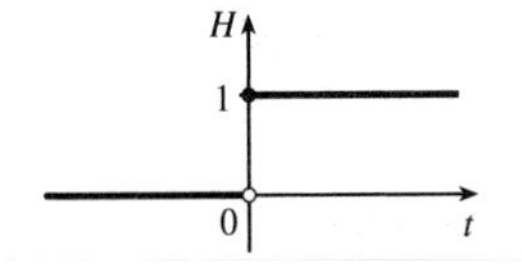

$$H(t) = \begin{cases} 0 & \text{if } t < 0 \\ 1 & \text{if } t \geq 0 \end{cases}$$

(b)

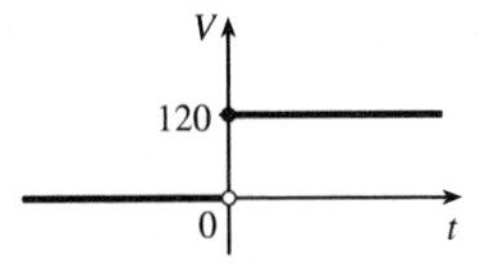

$$V(t) = \begin{cases} 0 & \text{if } t < 0 \\ 120 & \text{if } t \geq 0 \end{cases}$$ so $V(t) = 120H(t)$.

(c) 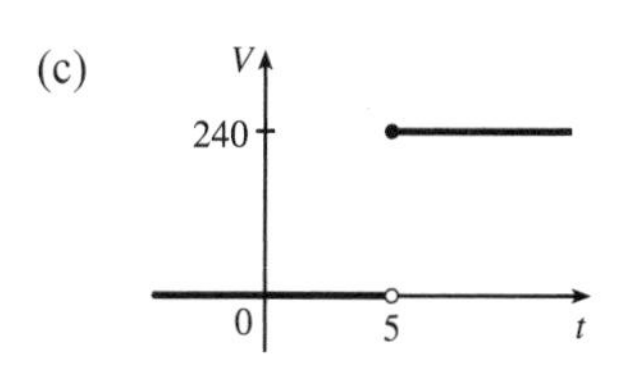

Starting with the formula in part (b), we replace 120 with 240 to reflect the different voltage. Also, because we are starting 5 units to the right of $t = 0$, we replace t with $t - 5$. Thus, the formula is $V(t) = 240H(t - 5)$.

58. (a) $R(t) = tH(t)$

$$= \begin{cases} 0 & \text{if } t < 0 \\ t & \text{if } t \geq 0 \end{cases}$$

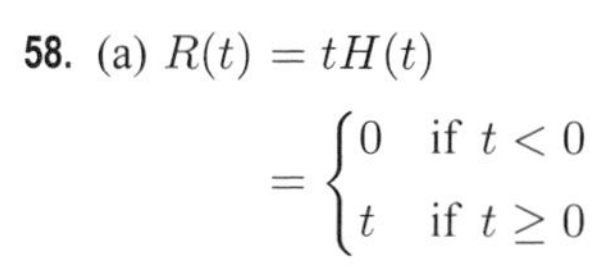

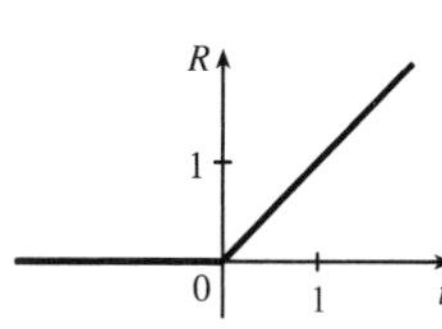

(b) $V(t) = \begin{cases} 0 & \text{if } t < 0 \\ 2t & \text{if } 0 \leq t \leq 60 \end{cases}$

so $V(t) = 2tH(t),\ t \leq 60$.

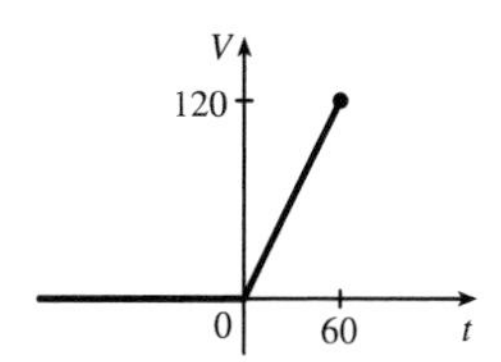

(c) $V(t) = \begin{cases} 0 & \text{if } t < 7 \\ 4\,(t-7) & \text{if } 7 \leq t \leq 32 \end{cases}$

so $V(t) = 4(t - 7)H(t - 7),\ t \leq 32$.

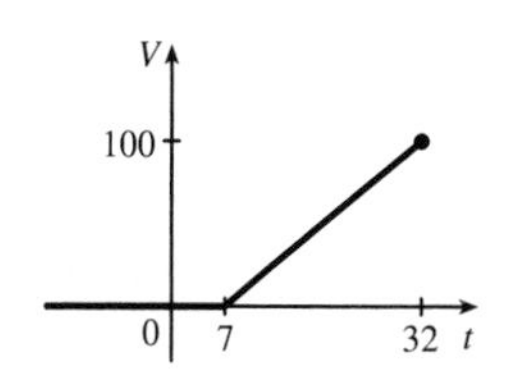

59. If $f(x) = m_1x + b_1$ and $g(x) = m_2x + b_2$, then

$(f \circ g)(x) = f(g(x)) = f(m_2x + b_2) = m_1(m_2x + b_2) + b_1 = m_1m_2x + m_1b_2 + b_1$.

So $f \circ g$ is a linear function with slope m_1m_2.

60. If $A(x) = 1.04x$, then $(A \circ A)(x) = A(A(x)) = A(1.04x) = 1.04(1.04x) = (1.04)^2x$,

$(A \circ A \circ A)(x) = A((A \circ A)(x)) = A((1.04)^2x) = 1.04(1.04)^2x = (1.04)^3x$, and

$(A \circ A \circ A \circ A)(x) = A((A \circ A \circ A)(x)) = A((1.04)^3x) = 1.04(1.04)^3x, = (1.04)^4x$.

These compositions represent the amount of the investment after 2, 3, and 4 years.

Based on this pattern, when we compose n copies of A, we get the formula $\underbrace{(A \circ A \circ \cdots \circ A)}_{n\ A's}(x) = (1.04)^nx$.

61. (a) By examining the variable terms in g and h, we deduce that we must square g to get the terms $4x^2$ and $4x$ in h. If we let $f(x) = x^2 + c$, then $(f \circ g)(x) = f(g(x)) = f(2x + 1) = (2x + 1)^2 + c = 4x^2 + 4x + (1 + c)$. Since $h(x) = 4x^2 + 4x + 7$, we must have $1 + c = 7$. So $c = 6$ and $f(x) = x^2 + 6$.

(b) We need a function g so that $f(g(x)) = 3(g(x)) + 5 = h(x)$. But $h(x) = 3x^2 + 3x + 2 = 3(x^2 + x) + 2 = 3(x^2 + x - 1) + 5$, so we see that $g(x) = x^2 + x - 1$.

62. We need a function g so that $g(f(x)) = g(x + 4) = h(x) = 4x - 1 = 4(x + 4) - 17$. So we see that the function g must be $g(x) = 4x - 17$.

63. (a) If f and g are even functions, then $f(-x) = f(x)$ and $g(-x) = g(x)$.

(i) $(f + g)(-x) = f(-x) + g(-x) = f(x) + g(x) = (f + g)(x)$, so $f + g$ is an *even* function.

(ii) $(fg)(-x) = f(-x) \cdot g(-x) = f(x) \cdot g(x) = (fg)(x)$, so fg is an *even* function.

(b) If f and g are odd functions, then $f(-x) = -f(x)$ and $g(-x) = -g(x)$.

(i) $(f + g)(-x) = f(-x) + g(-x) = -f(x) + [-g(x)] = -[f(x) + g(x)] = -(f + g)(x)$, so $f + g$ is an *odd* function.

(ii) $(fg)(-x) = f(-x) \cdot g(-x) = -f(x) \cdot [-g(x)] = f(x) \cdot g(x) = (fg)(x)$, so fg is an *even* function.

64. If f is even and g is odd, then $f(-x) = f(x)$ and $g(-x) = -g(x)$. Now

$(fg)(-x) = f(-x) \cdot g(-x) = f(x) \cdot [-g(x)] = -[f(x) \cdot g(x)] = -(fg)(x)$, so fg is an *odd* function.

65. We need to examine $h(-x)$: $h(-x) = (f \circ g)(-x) = f(g(-x)) = f(g(x))$ [because g is even] $= h(x)$

Because $h(-x) = h(x)$, h is an even function.

66. $h(-x) = f(g(-x)) = f(-g(x))$. At this point, we can't simplify the expression, so we might try to find a counterexample to show that h is not an odd function. Let $g(x) = x$, an odd function, and $f(x) = x^2 + x$. Then $h(x) = x^2 + x$, which is neither even nor odd.

Now suppose f is an odd function. Then $f(-g(x)) = -f(g(x)) = -h(x)$. Hence, $h(-x) = -h(x)$, and so h is odd if both f and g are odd.

Now suppose f is an even function. Then $f(-g(x)) = f(g(x)) = h(x)$. Hence, $h(-x) = h(x)$, and so h is even if g is odd and f is even.

1.3 The Limit of a Function

1. (a) $y = y(t) = 40t - 16t^2$. At $t = 2$, $y = 40(2) - 16(2)^2 = 16$. The average velocity between times 2 and $2 + h$ is

$$v_{\text{ave}} = \frac{y(2+h) - y(2)}{(2+h) - 2} = \frac{\left[40(2+h) - 16(2+h)^2\right] - 16}{h} = \frac{-24h - 16h^2}{h} = -24 - 16h, \text{ if } h \neq 0.$$

(i) $[2, 2.5]$: $h = 0.5$, $v_{\text{ave}} = -32$ ft/s

(ii) $[2, 2.1]$: $h = 0.1$, $v_{\text{ave}} = -25.6$ ft/s

(iii) $[2, 2.05]$: $h = 0.05$, $v_{\text{ave}} = -24.8$ ft/s

(iv) $[2, 2.01]$: $h = 0.01$, $v_{\text{ave}} = -24.16$ ft/s

(b) The instantaneous velocity when $t = 2$ (h approaches 0) is -24 ft/s.

2. The average velocity between t and $t + h$ seconds is

$$\frac{58(t+h) - 0.83(t+h)^2 - \left(58t - 0.83t^2\right)}{h} = \frac{58h - 1.66th - 0.83h^2}{h} = 58 - 1.66t - 0.83h \text{ if } h \neq 0.$$

(a) Here $t = 1$, so the average velocity is $58 - 1.66 - 0.83h = 56.34 - 0.83h$.

(i) $[1, 2]$: $h = 1$, 55.51 m/s

(ii) $[1, 1.5]$: $h = 0.5$, 55.925 m/s

(iii) $[1, 1.1]$: $h = 0.1$, 56.257 m/s

(iv) $[1, 1.01]$: $h = 0.01$, 56.3317 m/s

(v) $[1, 1.001]$: $h = 0.001$, 56.33917 m/s

(b) The instantaneous velocity after 1 second is 56.34 m/s.

3. (a) $f(x)$ approaches 2 as x approaches 1 from the left, so $\lim_{x \to 1^-} f(x) = 2$.

(b) $f(x)$ approaches 3 as x approaches 1 from the right, so $\lim_{x \to 1^+} f(x) = 3$.

(c) $\lim_{x \to 1} f(x)$ does not exist because the limits in part (a) and part (b) are not equal.

(d) $f(x)$ approaches 4 as x approaches 5 from the left and from the right, so $\lim_{x \to 5} f(x) = 4$.

(e) $f(5)$ is not defined, so it doesn't exist.

4. (a) $\lim_{x \to 0} f(x) = 3$

(b) $\lim_{x \to 3^-} f(x) = 4$

(c) $\lim_{x \to 3^+} f(x) = 2$

(d) $\lim_{x \to 3} f(x)$ does not exist because the limits in part (b) and part (c) are not equal.

(e) $f(3) = 3$

5. (a) $\lim_{t \to 0^-} g(t) = -1$

(b) $\lim_{t \to 0^+} g(t) = -2$

(c) $\lim_{t \to 0} g(t)$ does not exist because the limits in part (a) and part (b) are not equal.

(d) $\lim_{t \to 2^-} g(t) = 2$

(e) $\lim_{t \to 2^+} g(t) = 0$

(f) $\lim_{t \to 2} g(t)$ does not exist because the limits in part (d) and part (e) are not equal.

(g) $g(2) = 1$

(h) $\lim_{t \to 4} g(t) = 3$

6. $\lim_{x \to a} f(x)$ exists for all a except $a = \pm 1$.

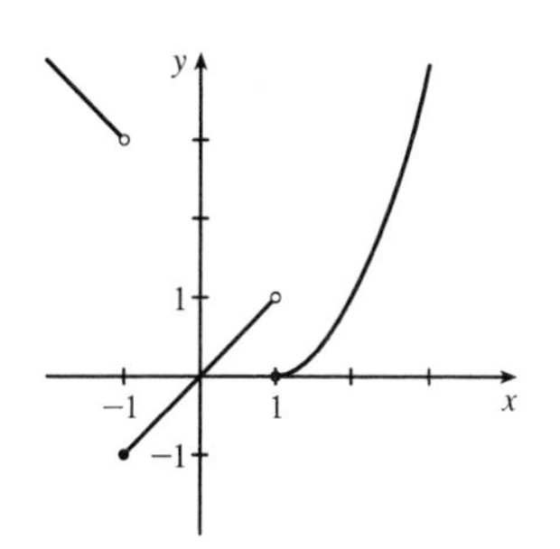

7. $\lim_{x \to 1^-} f(x) = 2, \quad \lim_{x \to 1^+} f(x) = -2, \quad f(1) = 2$

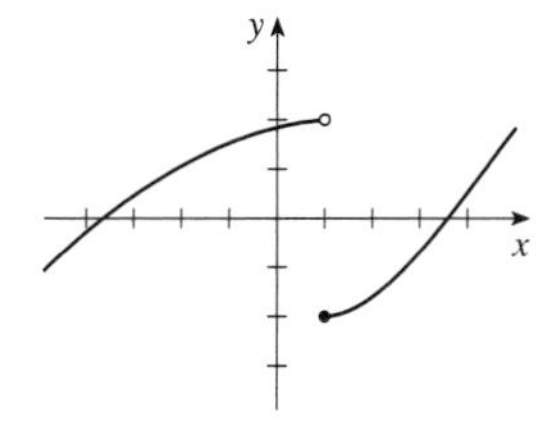

8. $\lim_{x \to 0^-} f(x) = 1, \quad \lim_{x \to 0^+} f(x) = -1, \quad \lim_{x \to 2^-} f(x) = 0,$

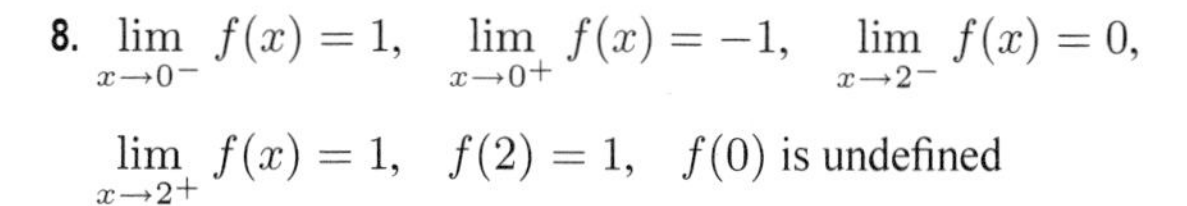

$\lim_{x \to 2^+} f(x) = 1, \quad f(2) = 1, \quad f(0)$ is undefined

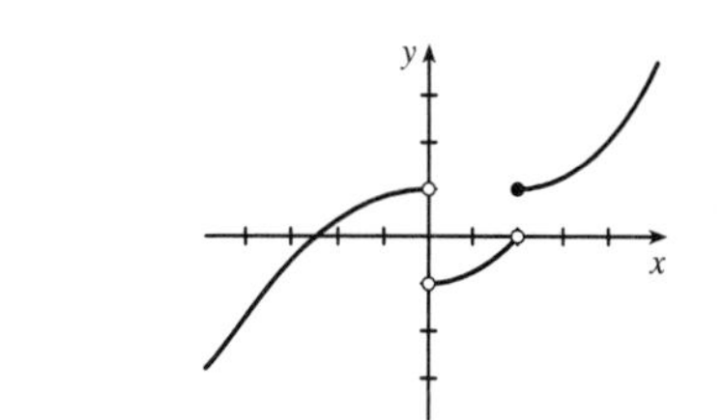

9. $\lim_{x \to 3^+} f(x) = 4, \quad \lim_{x \to 3^-} f(x) = 2, \quad \lim_{x \to -2} f(x) = 2,$

$f(3) = 3, \quad f(-2) = 1$

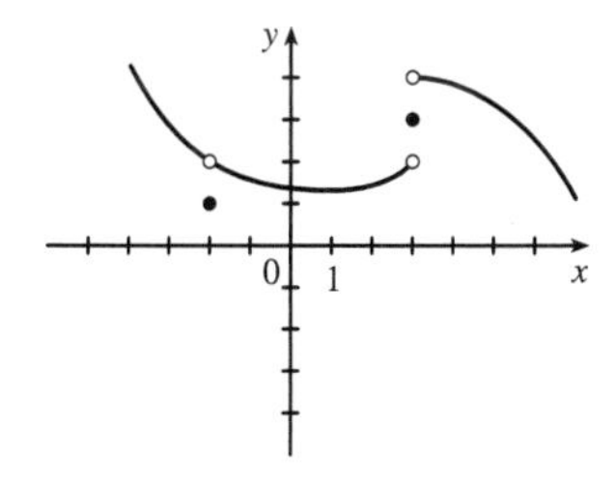

10. $\lim_{x \to 1} f(x) = 3, \quad \lim_{x \to 4^-} f(x) = 3, \quad \lim_{x \to 4^+} f(x) = -3,$

$f(1) = 1, \quad f(4) = -1$

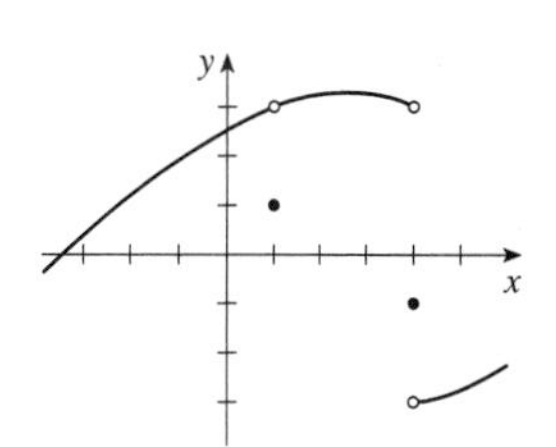

11. For $f(x) = \dfrac{x^2 - 2x}{x^2 - x - 2}$:

x	$f(x)$
2.5	0.714286
2.1	0.677419
2.05	0.672131
2.01	0.667774
2.005	0.667221
2.001	0.666778

x	$f(x)$
1.9	0.655172
1.95	0.661017
1.99	0.665552
1.995	0.666110
1.999	0.666556

It appears that $\lim_{x \to 2} \dfrac{x^2 - 2x}{x^2 - x - 2} = 0.\bar{6} = \frac{2}{3}$.

12. For $f(x) = \dfrac{x^2 - 2x}{x^2 - x - 2}$:

x	$f(x)$
0	0
−0.5	−1
−0.9	−9
−0.95	−19
−0.99	−99
−0.999	−999

x	$f(x)$
−2	2
−1.5	3
−1.1	11
−1.01	101
−1.001	1001

It appears that $\lim_{x \to -1} \dfrac{x^2 - 2x}{x^2 - x - 2}$ *does not exist* since

$f(x) \to \infty$ as $x \to -1^-$ and $f(x) \to -\infty$ as $x \to -1^+$.

13. For $f(x) = \dfrac{\sin x}{x + \tan x}$:

x	$f(x)$
± 1	0.329033
± 0.5	0.458209
± 0.2	0.493331
± 0.1	0.498333
± 0.05	0.499583
± 0.01	0.499983

It appears that $\lim\limits_{x \to 0} \dfrac{\sin x}{x + \tan x} = 0.5 = \dfrac{1}{2}$.

14. For $f(x) = \dfrac{\sqrt{x} - 4}{x - 16}$:

x	$f(x)$
17	0.123106
16.5	0.124038
16.1	0.124805
16.05	0.124902
16.01	0.124980

x	$f(x)$
15	0.127017
15.5	0.125992
15.9	0.125196
15.95	0.125098
15.99	0.125020

It appears that $\lim\limits_{x \to 16} \dfrac{\sqrt{x} - 4}{x - 16} = 0.125 = \dfrac{1}{8}$.

15. For $f(x) = \dfrac{\sqrt{x + 4} - 2}{x}$:

x	$f(x)$
1	0.236068
0.5	0.242641
0.1	0.248457
0.05	0.249224
0.01	0.249844

x	$f(x)$
-1	0.267949
-0.5	0.258343
-0.1	0.251582
-0.05	0.250786
-0.01	0.250156

It appears that $\lim\limits_{x \to 0} \dfrac{\sqrt{x + 4} - 2}{x} = 0.25 = \frac{1}{4}$.

16. For $f(x) = \dfrac{\tan 3x}{\tan 5x}$:

x	$f(x)$
± 0.2	0.439279
± 0.1	0.566236
± 0.05	0.591893
± 0.01	0.599680
± 0.001	0.599997

It appears that $\lim\limits_{x \to 0} \dfrac{\tan 3x}{\tan 5x} = 0.6 = \frac{3}{5}$.

17. For $f(x) = \dfrac{x^6 - 1}{x^{10} - 1}$:

x	$f(x)$
0.5	0.985337
0.9	0.719397
0.95	0.660186
0.99	0.612018
0.999	0.601200

x	$f(x)$
1.5	0.183369
1.1	0.484119
1.05	0.540783
1.01	0.588022
1.001	0.598800

It appears that $\lim\limits_{x \to 1} \dfrac{x^6 - 1}{x^{10} - 1} = 0.6 = \frac{3}{5}$.

18. For $f(x) = \dfrac{9^x - 5^x}{x}$:

x	$f(x)$
0.5	1.527864
0.1	0.711120
0.05	0.646496
0.01	0.599082
0.001	0.588906

x	$f(x)$
-0.5	0.227761
-0.1	0.485984
-0.05	0.534447
-0.01	0.576706
-0.001	0.586669

It appears that $\lim\limits_{x \to 0} \dfrac{9^x - 5^x}{x} = 0.59$. Later we will be able to show that the exact value is $\ln(9/5)$.

19. (a) From the graphs, it seems that $\lim_{x \to 0} \frac{\tan 4x}{x} = 4$.

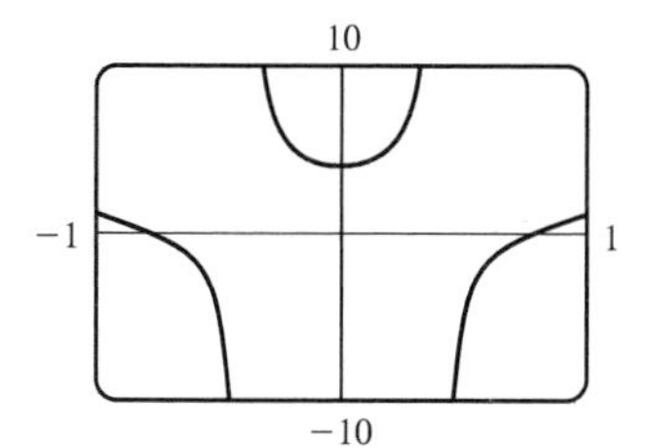

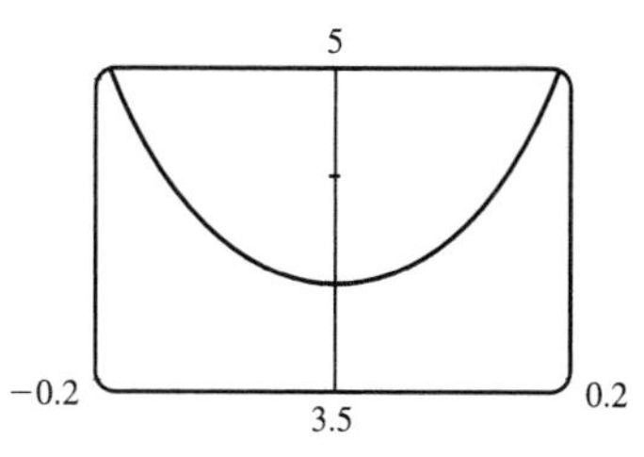

(b)

x	$f(x)$
± 0.1	4.227932
± 0.01	4.002135
± 0.001	4.000021
± 0.0001	4.000000

20. (a) From the following graphs, it seems that $\lim_{x \to 0} \frac{6^x - 2^x}{x} \approx 1.10$.

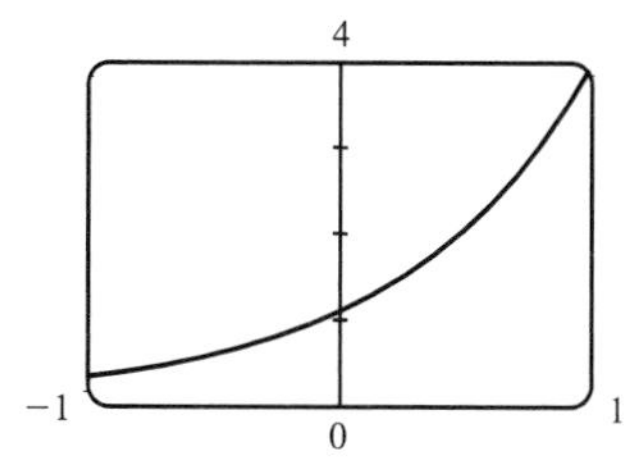

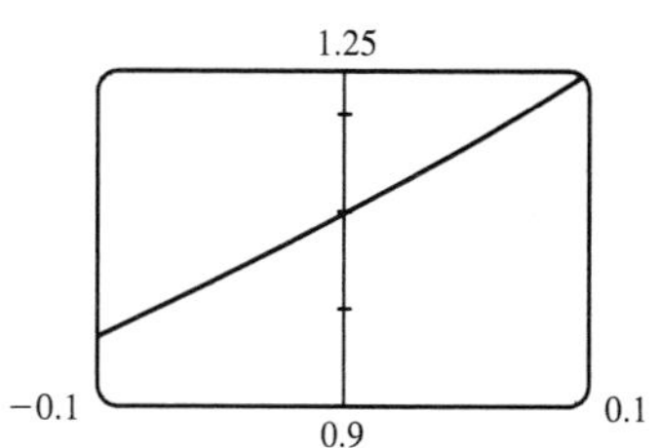

(b)

x	$f(x)$
-0.01	1.085052
-0.001	1.097248
-0.0001	1.098476
0.0001	1.098749
0.001	1.099978
0.01	1.112353

21. For $f(x) = x^2 - (2^x/1000)$:

(a)

x	$f(x)$
1	0.998000
0.8	0.638259
0.6	0.358484
0.4	0.158680
0.2	0.038851
0.1	0.008928
0.05	0.001465

It appears that $\lim_{x \to 0} f(x) = 0$.

(b)

x	$f(x)$
0.04	0.000572
0.02	-0.000614
0.01	-0.000907
0.005	-0.000978
0.003	-0.000993
0.001	-0.001000

It appears that $\lim_{x \to 0} f(x) = -0.001$.

22. For $h(x) = \frac{\tan x - x}{x^3}$:

(a)

x	$h(x)$
1.0	0.55740773
0.5	0.37041992
0.1	0.33467209
0.05	0.33366700
0.01	0.33334667
0.005	0.33333667

(b) It seems that $\lim_{x \to 0} h(x) = \frac{1}{3}$.

(c)

x	$h(x)$
0.001	0.33333350
0.0005	0.33333344
0.0001	0.33333000
0.00005	0.33333600
0.00001	0.33300000
0.000001	0.00000000

Here the values will vary from one calculator to another. Every calculator will eventually give *false values*.

(d) As in part (c), when we take a small enough viewing rectangle we get incorrect output.

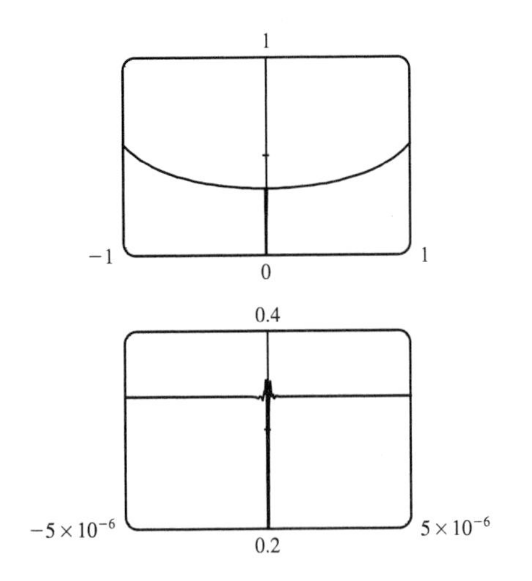

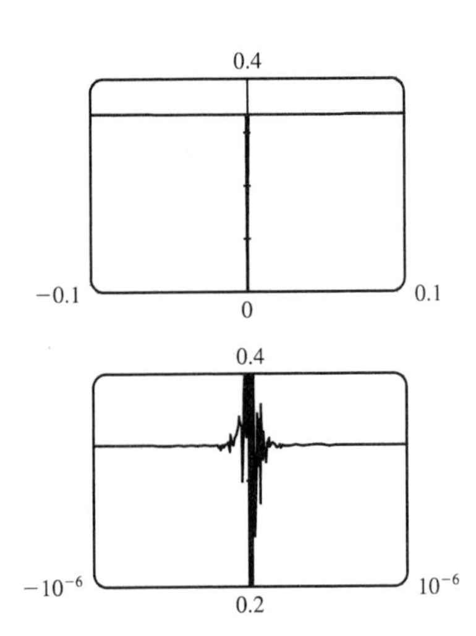

23. On the left side of $x = 2$, we need $|x - 2| < \left|\frac{10}{7} - 2\right| = \frac{4}{7}$. On the right side, we need $|x - 2| < \left|\frac{10}{3} - 2\right| = \frac{4}{3}$. For both of these conditions to be satisfied at once, we need the more restrictive of the two to hold, that is, $|x - 2| < \frac{4}{7}$. So we can choose $\delta = \frac{4}{7}$, or any smaller positive number.

24. The left-hand question mark is the positive solution of $x^2 = \frac{1}{2}$, that is, $x = \frac{1}{\sqrt{2}}$, and the right-hand question mark is the positive solution of $x^2 = \frac{3}{2}$, that is, $x = \sqrt{\frac{3}{2}}$. On the left side, we need $|x - 1| < \left|\frac{1}{\sqrt{2}} - 1\right| \approx 0.292$ (rounding down to be safe). On the right side, we need $|x - 1| < \left|\sqrt{\frac{3}{2}} - 1\right| \approx 0.224$. The more restrictive of these two conditions must apply, so we choose $\delta = 0.224$ (or any smaller positive number).

25. $\left|\sqrt{4x+1} - 3\right| < 0.5 \quad \Leftrightarrow \quad 2.5 < \sqrt{4x+1} < 3.5$. We plot the three parts of this inequality on the same screen and identify the x-coordinates of the points of intersection using the cursor. It appears that the inequality holds for $1.3125 \le x \le 2.8125$. Since $|2 - 1.3125| = 0.6875$ and $|2 - 2.8125| = 0.8125$, we choose $0 < \delta < \min\{0.6875, 0.8125\} = 0.6875$.

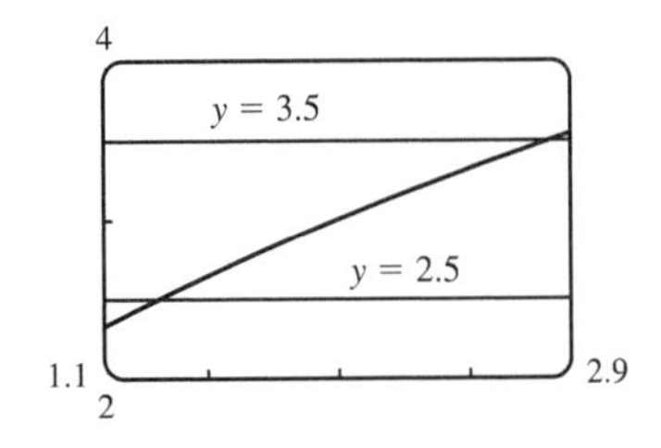

26. $\left|\sin x - \frac{1}{2}\right| < 0.1 \quad \Leftrightarrow \quad 0.4 < \sin x < 0.6$. From the graph, we see that for this inequality to hold, we need $a \le x \le b$, where $a \approx 0.4115$ and $b \approx 0.6435$. So since $|\pi/6 - a| \approx 0.11$ and $|\pi/6 - b| \approx 0.12$, we choose $0 < \delta \le \min\{0.11, 0.12\} = 0.11$.

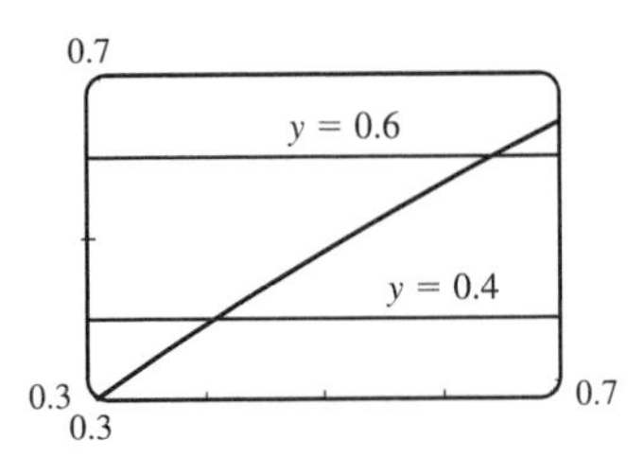

27. (a) $A = \pi r^2$ and $A = 1000\text{ cm}^2 \quad\Rightarrow\quad \pi r^2 = 1000 \quad\Rightarrow\quad r^2 = \frac{1000}{\pi} \quad\Rightarrow\quad r = \sqrt{\frac{1000}{\pi}} \quad [r > 0] \quad \approx 17.8412$ cm.

(b) $|A - 1000| \leq 5 \quad \Rightarrow \quad -5 \leq \pi r^2 - 1000 \leq 5 \quad \Rightarrow \quad 1000 - 5 \leq \pi r^2 \leq 1000 + 5 \quad \Rightarrow$

$\sqrt{\frac{995}{\pi}} \leq r \leq \sqrt{\frac{1005}{\pi}} \quad \Rightarrow \quad 17.7966 \leq r \leq 17.8858.$ $\sqrt{\frac{1000}{\pi}} - \sqrt{\frac{995}{\pi}} \approx 0.04466$ and $\sqrt{\frac{1005}{\pi}} - \sqrt{\frac{1000}{\pi}} \approx 0.04455.$

So if the machinist gets the radius within 0.0445 cm of 17.8412, the area will be within 5 cm^2 of 1000.

(c) x is the radius, $f(x)$ is the area, a is the target radius given in part (a), L is the target area (1000), ε is the tolerance in the area (5), and δ is the tolerance in the radius given in part (b).

28. (a) $T = 0.1w^2 + 2.155w + 20$ and $T = 200 \quad \Rightarrow$

$0.1w^2 + 2.155w + 20 = 200 \quad \Rightarrow$ [by the quadratic formula or from the graph] $\;w \approx 33.0$ watts $(w > 0)$

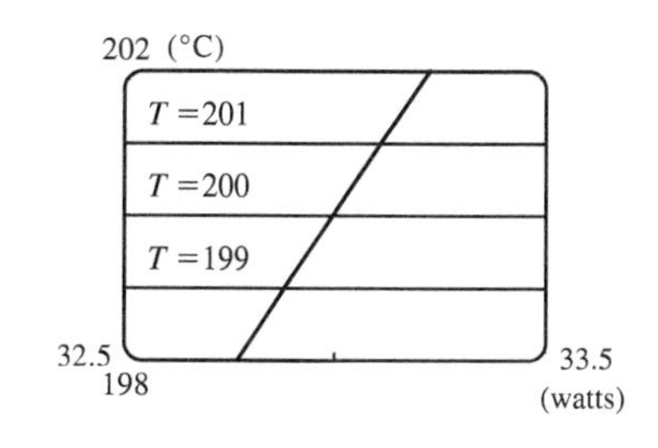

(b) From the graph, $199 \leq T \leq 201 \quad \Rightarrow \quad 32.89 < w < 33.11.$

(c) x is the input power, $f(x)$ is the temperature, a is the target input power given in part (a), L is the target temperature (200), ε is the tolerance in the temperature (1), and δ is the tolerance in the power input in watts indicated in part (b) (0.11 watts).

29. Given $\varepsilon > 0$, we need $\delta > 0$ such that if $0 < |x - 1| < \delta$, then $|(2x + 3) - 5| < \varepsilon$. But $|(2x + 3) - 5| < \varepsilon \quad \Leftrightarrow \quad |2x - 2| < \varepsilon$ $\Leftrightarrow \quad 2\,|x - 1| < \varepsilon \quad \Leftrightarrow \quad |x - 1| < \varepsilon/2$. So if we choose $\delta = \varepsilon/2$, then $0 < |x - 1| < \delta \quad \Rightarrow \quad |(2x + 3) - 5| < \varepsilon$. Thus, $\lim\limits_{x \to 1} (2x + 3) = 5$ by the definition of a limit.

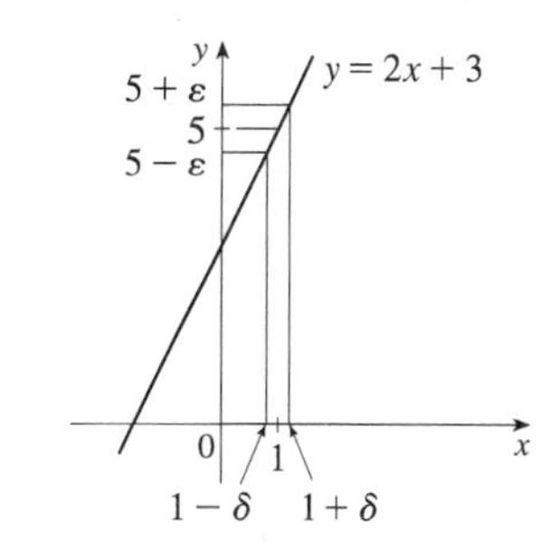

30. Given $\varepsilon > 0$, we need $\delta > 0$ such that if $0 < |x - (-2)| < \delta$, then $\left|\left(\frac{1}{2}x + 3\right) - 2\right| < \varepsilon$. But $\left|\left(\frac{1}{2}x + 3\right) - 2\right| < \varepsilon \quad \Leftrightarrow$ $\left|\frac{1}{2}x + 1\right| < \varepsilon \quad \Leftrightarrow \quad \frac{1}{2}\,|x + 2| < \varepsilon \quad \Leftrightarrow \quad |x - (-2)| < 2\varepsilon$. So if we choose $\delta = 2\varepsilon$, then $0 < |x - (-2)| < \delta \quad \Rightarrow \quad \left|\left(\frac{1}{2}x + 3\right) - 2\right| < \varepsilon$. Thus, $\lim\limits_{x \to -2} \left(\frac{1}{2}x + 3\right) = 2$ by the definition of a limit.

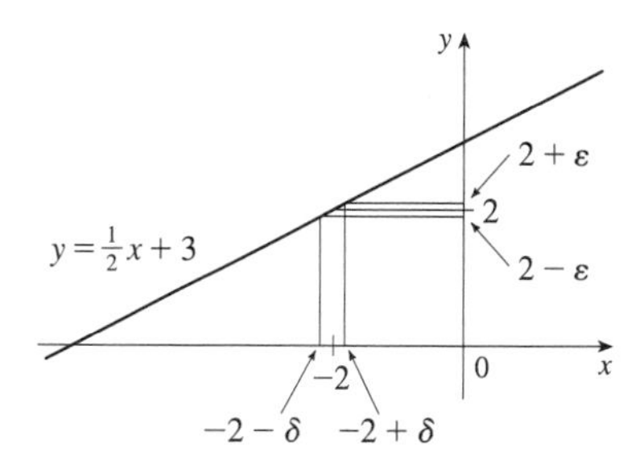

31. Given $\varepsilon > 0$, we need $\delta > 0$ such that if $0 < |x - (-3)| < \delta$, then $|(1 - 4x) - 13| < \varepsilon$. But $|(1 - 4x) - 13| < \varepsilon \quad \Leftrightarrow$ $|-4x - 12| < \varepsilon \quad \Leftrightarrow \quad |-4|\,|x + 3| < \varepsilon \quad \Leftrightarrow \quad |x - (-3)| < \varepsilon/4$. So if we choose $\delta = \varepsilon/4$, then $0 < |x - (-3)| < \delta \quad \Rightarrow$ $|(1 - 4x) - 13| < \varepsilon$. Thus, $\lim\limits_{x \to -3} (1 - 4x) = 13$ by the definition of a limit.

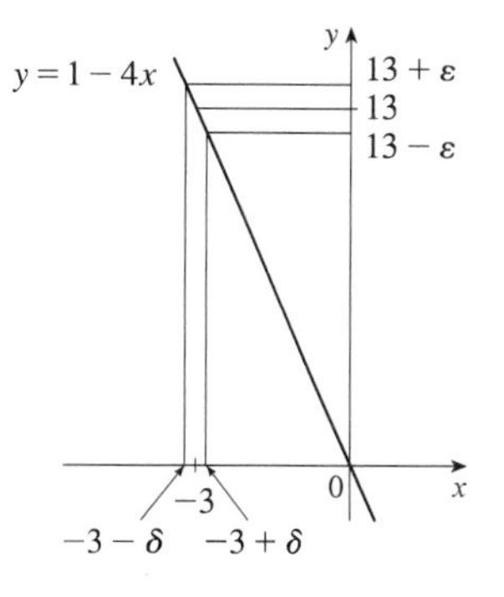

32. Given $\varepsilon > 0$, we need $\delta > 0$ such that if $0 < |x - 4| < \delta$, then $|(7 - 3x) - (-5)| < \varepsilon$. But $|(7 - 3x) - (-5)| < \varepsilon \quad \Leftrightarrow \quad |-3x + 12| < \varepsilon \quad \Leftrightarrow \quad |-3|\,|x - 4| < \varepsilon \quad \Leftrightarrow \quad |x - 4| < \varepsilon/3$. So if we choose $\delta = \varepsilon/3$, then $0 < |x - 4| < \delta \quad \Rightarrow \quad |(7 - 3x) - (-5)| < \varepsilon$. Thus, $\lim\limits_{x \to 4} (7 - 3x) = -5$ by the definition of a limit.

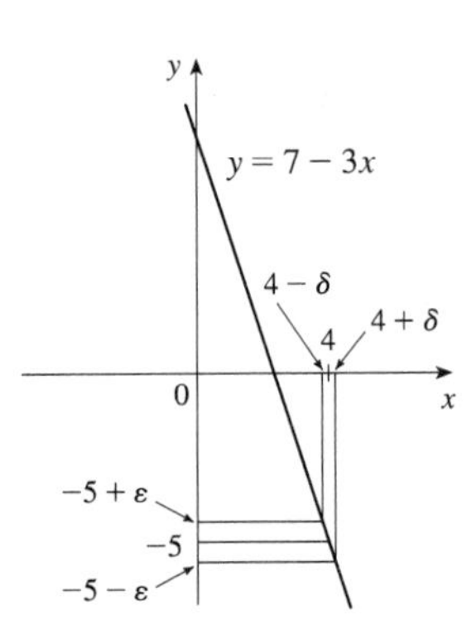

33. Given $\varepsilon > 0$, we need $\delta > 0$ such that if $0 < |x - 3| < \delta$, then $\left|\dfrac{x}{5} - \dfrac{3}{5}\right| < \varepsilon \quad \Leftrightarrow \quad \frac{1}{5}|x - 3| < \varepsilon \quad \Leftrightarrow \quad |x - 3| < 5\varepsilon$. So choose $\delta = 5\varepsilon$. Then $0 < |x - 3| < \delta \quad \Rightarrow \quad |x - 3| < 5\varepsilon \quad \Rightarrow \quad \dfrac{|x - 3|}{5} < \varepsilon \quad \Rightarrow \quad \left|\dfrac{x}{5} - \dfrac{3}{5}\right| < \varepsilon$. By the definition of a limit, $\lim\limits_{x \to 3} \dfrac{x}{5} = \dfrac{3}{5}$.

34. Given $\varepsilon > 0$, we need $\delta > 0$ such that if $0 < |x - 6| < \delta$, then $\left|\left(\frac{x}{4} + 3\right) - \frac{9}{2}\right| < \varepsilon \quad \Leftrightarrow \quad \left|\frac{x}{4} - \frac{3}{2}\right| < \varepsilon \quad \Leftrightarrow \quad \frac{1}{4}|x - 6| < \varepsilon \quad \Leftrightarrow \quad |x - 6| < 4\varepsilon$. So choose $\delta = 4\varepsilon$. Then $0 < |x - 6| < \delta \quad \Rightarrow \quad |x - 6| < 4\varepsilon \quad \Rightarrow \quad \dfrac{|x - 6|}{4} < \varepsilon \quad \Rightarrow \quad \left|\frac{x}{4} - \frac{6}{4}\right| < \varepsilon \quad \Rightarrow \quad \left|\left(\frac{x}{4} + 3\right) - \frac{9}{2}\right| < \varepsilon$. By the definition of a limit, $\lim\limits_{x \to 6} \left(\frac{x}{4} + 3\right) = \frac{9}{2}$.

35. Given $\varepsilon > 0$, we need $\delta > 0$ such that if $0 < |x - (-5)| < \delta$, then $\left|\left(4 - \frac{3}{5}x\right) - 7\right| < \varepsilon \quad \Leftrightarrow \quad \left|-\frac{3}{5}x - 3\right| < \varepsilon \quad \Leftrightarrow \quad \frac{3}{5}|x + 5| < \varepsilon \quad \Leftrightarrow \quad |x - (-5)| < \frac{5}{3}\varepsilon$. So choose $\delta = \frac{5}{3}\varepsilon$. Then $|x - (-5)| < \delta \quad \Rightarrow \quad \left|\left(4 - \frac{3}{5}x\right) - 7\right| < \varepsilon$. Thus, $\lim\limits_{x \to -5} \left(4 - \frac{3}{5}x\right) = 7$ by the definition of a limit.

36. Given $\varepsilon > 0$, we need $\delta > 0$ such that if $0 < |x - 3| < \delta$, then $\left|\dfrac{x^2 + x - 12}{x - 3} - 7\right| < \varepsilon$. Notice that if $0 < |x - 3|$, then $x \neq 3$, so $\dfrac{x^2 + x - 12}{x - 3} = \dfrac{(x + 4)(x - 3)}{x - 3} = x + 4$. Thus, when $0 < |x - 3|$, we have $\left|\dfrac{x^2 + x - 12}{x - 3} - 7\right| < \varepsilon \quad \Leftrightarrow \quad |(x + 4) - 7| < \varepsilon \quad \Leftrightarrow \quad |x - 3| < \varepsilon$. We take $\delta = \varepsilon$ and see that $0 < |x - 3| < \delta \quad \Rightarrow \quad \left|\dfrac{x^2 + x - 12}{x - 3} - 7\right| < \varepsilon$. By the definition of a limit, $\lim\limits_{x \to 3} \dfrac{x^2 + x - 12}{x - 3} = 7$.

37. Given $\varepsilon > 0$, we need $\delta > 0$ such that if $0 < |x - a| < \delta$, then $|x - a| < \varepsilon$. So $\delta = \varepsilon$ will work.

38. Given $\varepsilon > 0$, we need $\delta > 0$ such that if $0 < |x - a| < \delta$, then $|c - c| < \varepsilon$. But $|c - c| = 0$, so this will be true no matter what δ we pick.

39. Given $\varepsilon > 0$, we need $\delta > 0$ such that if $0 < |x - 0| < \delta$, then $\left|x^2 - 0\right| < \varepsilon \quad \Leftrightarrow \quad x^2 < \varepsilon \quad \Leftrightarrow \quad |x| < \sqrt{\varepsilon}$. Take $\delta = \sqrt{\varepsilon}$. Then $0 < |x - 0| < \delta \quad \Rightarrow \quad \left|x^2 - 0\right| < \varepsilon$. Thus, $\lim\limits_{x \to 0} x^2 = 0$ by the definition of a limit.

40. Given $\varepsilon > 0$, we need $\delta > 0$ such that if $0 < |x - 0| < \delta$, then $|x^3 - 0| < \varepsilon \Leftrightarrow |x|^3 < \varepsilon \Leftrightarrow |x| < \sqrt[3]{\varepsilon}$. Take $\delta = \sqrt[3]{\varepsilon}$. Then $0 < |x - 0| < \delta \Rightarrow |x^3 - 0| < \delta^3 = \varepsilon$. Thus, $\lim\limits_{x \to 0} x^3 = 0$ by the definition of a limit.

41. Given $\varepsilon > 0$, we need $\delta > 0$ such that if $0 < |x - 0| < \delta$, then $\big||x| - 0\big| < \varepsilon$. But $\big||x|\big| = |x|$. So this is true if we pick $\delta = \varepsilon$. Thus, $\lim\limits_{x \to 0} |x| = 0$ by the definition of a limit.

42. Given $\varepsilon > 0$, we need $\delta > 0$ such that if $9 - \delta < x < 9$, then $\left|\sqrt[4]{9 - x} - 0\right| < \varepsilon \Leftrightarrow \sqrt[4]{9 - x} < \varepsilon \Leftrightarrow 9 - x < \varepsilon^4 \Leftrightarrow 9 - \varepsilon^4 < x < 9$. So take $\delta = \varepsilon^4$. Then $9 - \delta < x < 9 \Rightarrow \left|\sqrt[4]{9 - x} - 0\right| < \varepsilon$. Thus, $\lim\limits_{x \to 9^-} \sqrt[4]{9 - x} = 0$ by the definition of a limit.

43. Given $\varepsilon > 0$, we need $\delta > 0$ such that if $0 < |x - 3| < \delta$, then $|x^2 - 9| < \varepsilon \Leftrightarrow |(x - 3)(x + 3)| < \varepsilon$. Notice that if $|x - 3| < 1$, then $-1 < x - 3 < 1 \Rightarrow 5 < x + 3 < 7 \Rightarrow |x + 3| < 7$. So take $\delta = \min\{1, \varepsilon/7\}$. Then $0 < |x - 3| < \delta \Leftrightarrow |(x - 3)(x + 3)| < |7(x - 3)| = 7 \cdot |x - 3| < 7\delta \leq \varepsilon$. Thus, $\lim\limits_{x \to 3} x^2 = 9$ by the definition of a limit.

44. Given $\varepsilon > 0$, we need $\delta > 0$ such that if $0 < |x - 3| < \delta$, then $\left|(x^2 + x - 4) - 8\right| < \varepsilon \Leftrightarrow |x^2 + x - 12| < \varepsilon \Leftrightarrow |(x - 3)(x + 4)| < \varepsilon$. Notice that if $|x - 3| < 1$, then $-1 < x - 3 < 1 \Rightarrow 6 < x + 4 < 8 \Rightarrow |x + 4| < 8$. So take $\delta = \min\{1, \varepsilon/8\}$. Then $0 < |x - 3| < \delta \Leftrightarrow |(x - 3)(x + 4)| \leq |8(x - 3)| = 8 \cdot |x - 3| < 8\delta \leq \varepsilon$. Thus, $\lim\limits_{x \to 3} (x^2 + x - 4) = 8$ by the definition of a limit.

45. (a) The points of intersection in the graph are $(x_1, 2.6)$ and $(x_2, 3.4)$ with $x_1 \approx 0.891$ and $x_2 \approx 1.093$. Thus, we can take δ to be the smaller of $1 - x_1$ and $x_2 - 1$. So $\delta = x_2 - 1 \approx 0.093$.

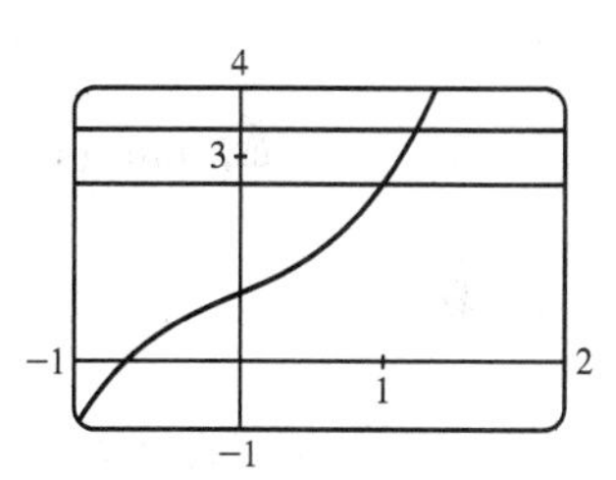

(b) Solving $x^3 + x + 1 = 3 + \varepsilon$ gives us two nonreal complex roots and one real root, which is

$$x(\varepsilon) = \frac{\left(216 + 108\varepsilon + 12\sqrt{336 + 324\varepsilon + 81\varepsilon^2}\right)^{2/3} - 12}{6\left(216 + 108\varepsilon + 12\sqrt{336 + 324\varepsilon + 81\varepsilon^2}\right)^{1/3}}. \text{ Thus, } \delta = x(\varepsilon) - 1.$$

(c) If $\varepsilon = 0.4$, then $x(\varepsilon) \approx 1.093\,272\,342$ and $\delta = x(\varepsilon) - 1 \approx 0.093$, which agrees with our answer in part (a).

46. Suppose that $\lim\limits_{t \to 0} H(t) = L$. Given $\varepsilon = \frac{1}{2}$, there exists $\delta > 0$ such that $0 < |t| < \delta \Rightarrow |H(t) - L| < \frac{1}{2} \Leftrightarrow L - \frac{1}{2} < H(t) < L + \frac{1}{2}$. For $0 < t < \delta$, $H(t) = 1$, so $1 < L + \frac{1}{2} \Rightarrow L > \frac{1}{2}$. For $-\delta < t < 0$, $H(t) = 0$, so $L - \frac{1}{2} < 0 \Rightarrow L < \frac{1}{2}$. This contradicts $L > \frac{1}{2}$. Therefore, $\lim\limits_{t \to 0} H(t)$ does not exist.

1.4 Calculating Limits

1. (a) $\lim_{x\to a}[f(x)+h(x)] = \lim_{x\to a} f(x) + \lim_{x\to a} h(x) = -3+8=5$

(b) $\lim_{x\to a}[f(x)]^2 = \left[\lim_{x\to a} f(x)\right]^2 = (-3)^2 = 9$

(c) $\lim_{x\to a}\sqrt[3]{h(x)} = \sqrt[3]{\lim_{x\to a} h(x)} = \sqrt[3]{8} = 2$

(d) $\lim_{x\to a}\frac{1}{f(x)} = \frac{1}{\lim_{x\to a} f(x)} = \frac{1}{-3} = -\frac{1}{3}$

(e) $\lim_{x\to a}\frac{f(x)}{h(x)} = \frac{\lim_{x\to a} f(x)}{\lim_{x\to a} h(x)} = \frac{-3}{8} = -\frac{3}{8}$

(f) $\lim_{x\to a}\frac{g(x)}{f(x)} = \frac{\lim_{x\to a} g(x)}{\lim_{x\to a} f(x)} = \frac{0}{-3} = 0$

(g) The limit does not exist, since $\lim_{x\to a} g(x) = 0$ but $\lim_{x\to a} f(x) \neq 0$.

(h) $\lim_{x\to a}\frac{2f(x)}{h(x)-f(x)} = \frac{2\lim_{x\to a} f(x)}{\lim_{x\to a} h(x) - \lim_{x\to a} f(x)} = \frac{2(-3)}{8-(-3)} = -\frac{6}{11}$

2. (a) $\lim_{x\to 2}[f(x)+g(x)] = \lim_{x\to 2} f(x) + \lim_{x\to 2} g(x) = 2+0=2$

(b) $\lim_{x\to 1} g(x)$ does not exist since its left- and right-hand limits are not equal, so the given limit does not exist.

(c) $\lim_{x\to 0}[f(x)g(x)] = \lim_{x\to 0} f(x)\cdot\lim_{x\to 0} g(x) = 0\cdot 1.3 = 0$

(d) Since $\lim_{x\to -1} g(x) = 0$ and g is in the denominator, but $\lim_{x\to -1} f(x) = -1 \neq 0$, the given limit does not exist.

(e) $\lim_{x\to 2} x^3 f(x) = \left[\lim_{x\to 2} x^3\right]\left[\lim_{x\to 2} f(x)\right] = 2^3\cdot 2 = 16$

(f) $\lim_{x\to 1}\sqrt{3+f(x)} = \sqrt{3+\lim_{x\to 1} f(x)} = \sqrt{3+1} = 2$

3.
$$\begin{aligned}\lim_{x\to -2}(3x^4+2x^2-x+1) &= \lim_{x\to -2}3x^4 + \lim_{x\to -2}2x^2 - \lim_{x\to -2}x + \lim_{x\to -2}1 && \text{[Limit Laws 1 and 2]}\\ &= 3\lim_{x\to -2}x^4 + 2\lim_{x\to -2}x^2 - \lim_{x\to -2}x + \lim_{x\to -2}1 && [3]\\ &= 3(-2)^4 + 2(-2)^2 - (-2) + (1) && \text{[9, 8, and 7]}\\ &= 48+8+2+1 = 59\end{aligned}$$

4.
$$\begin{aligned}\lim_{t\to -1}(t^2+1)^3(t+3)^5 &= \lim_{t\to -1}(t^2+1)^3 \cdot \lim_{t\to -1}(t+3)^5 && \text{[Limit Law 4]}\\ &= \left[\lim_{t\to -1}(t^2+1)\right]^3 \cdot \left[\lim_{t\to -1}(t+3)\right]^5 && [6]\\ &= \left[\lim_{t\to -1}t^2 + \lim_{t\to -1}1\right]^3 \cdot \left[\lim_{t\to -1}t + \lim_{t\to -1}3\right]^5 && [1]\\ &= \left[(-1)^2+1\right]^3 \cdot [-1+3]^5 = 8\cdot 32 = 256 && \text{[9, 7, and 8]}\end{aligned}$$

5.
$$\begin{aligned}\lim_{x\to 8}(1+\sqrt[3]{x})(2-6x^2+x^3) &= \lim_{x\to 8}(1+\sqrt[3]{x})\cdot\lim_{x\to 8}(2-6x^2+x^3) && \text{[Limit Law 4]}\\ &= \left(\lim_{x\to 8}1 + \lim_{x\to 8}\sqrt[3]{x}\right)\cdot\left(\lim_{x\to 8}2 - 6\lim_{x\to 8}x^2 + \lim_{x\to 8}x^3\right) && \text{[1, 2, and 3]}\\ &= (1+\sqrt[3]{8})\cdot(2-6\cdot 8^2+8^3) && \text{[7, 10, 9]}\\ &= (3)(130) = 390\end{aligned}$$

6.
$$\begin{aligned}\lim_{u\to -2}\sqrt{u^4+3u+6} &= \sqrt{\lim_{u\to -2}(u^4+3u+6)} && [11]\\ &= \sqrt{\lim_{u\to -2}u^4 + 3\lim_{u\to -2}u + \lim_{u\to -2}6} && \text{[1, 2, and 3]}\\ &= \sqrt{(-2)^4+3(-2)+6} && \text{[9, 8, and 7]}\\ &= \sqrt{16-6+6} = \sqrt{16} = 4\end{aligned}$$

7. $$\lim_{x\to 1}\left(\frac{1+3x}{1+4x^2+3x^4}\right)^3 = \left(\lim_{x\to 1}\frac{1+3x}{1+4x^2+3x^4}\right)^3 \qquad [6]$$

$$= \left[\frac{\lim_{x\to 1}(1+3x)}{\lim_{x\to 1}(1+4x^2+3x^4)}\right]^3 \qquad [5]$$

$$= \left[\frac{\lim_{x\to 1}1+3\lim_{x\to 1}x}{\lim_{x\to 1}1+4\lim_{x\to 1}x^2+3\lim_{x\to 1}x^4}\right]^3 \qquad [2, 1, \text{and } 3]$$

$$= \left[\frac{1+3(1)}{1+4(1)^2+3(1)^4}\right]^3 = \left[\frac{4}{8}\right]^3 = \left(\frac{1}{2}\right)^3 = \frac{1}{8} \qquad [7, 8, \text{and } 9]$$

8. $$\lim_{x\to 0}\frac{\cos^4 x}{5+2x^3} = \frac{\lim_{x\to 0}\cos^4 x}{\lim_{x\to 0}(5+2x^3)} \qquad [5]$$

$$= \frac{\left(\lim_{x\to 0}\cos x\right)^4}{\lim_{x\to 0}5+2\lim_{x\to 0}x^3} \qquad [6, 1, \text{and } 3]$$

$$= \frac{1^4}{5+2(0)^3} = \frac{1}{5} \qquad [7, 9, \text{and Equation } 1]$$

9. $$\lim_{\theta\to\pi/2}\theta\sin\theta = \left(\lim_{\theta\to\pi/2}\theta\right)\left(\lim_{\theta\to\pi/2}\sin\theta\right) \qquad [4]$$

$$= \frac{\pi}{2}\cdot\sin\frac{\pi}{2} \qquad [8 \text{ and Direct Substitution Property}]$$

$$= \frac{\pi}{2}$$

10. (a) The left-hand side of the equation is not defined for $x = 2$, but the right-hand side is.

(b) Since the equation holds for all $x \neq 2$, it follows that both sides of the equation approach the same limit as $x \to 2$, just as in Example 3. Remember that in finding $\lim_{x\to a} f(x)$, we never consider $x = a$.

11. $$\lim_{x\to 2}\frac{x^2+x-6}{x-2} = \lim_{x\to 2}\frac{(x+3)(x-2)}{x-2} = \lim_{x\to 2}(x+3) = 2+3 = 5$$

12. $$\lim_{x\to -4}\frac{x^2+5x+4}{x^2+3x-4} = \lim_{x\to -4}\frac{(x+4)(x+1)}{(x+4)(x-1)} = \lim_{x\to -4}\frac{x+1}{x-1} = \frac{-4+1}{-4-1} = \frac{-3}{-5} = \frac{3}{5}$$

13. $\lim_{x\to 2}\dfrac{x^2-x+6}{x-2}$ does not exist since $x-2\to 0$ but $x^2-x+6\to 8$ as $x\to 2$.

14. $$\lim_{x\to 4}\frac{x^2-4x}{x^2-3x-4} = \lim_{x\to 4}\frac{x(x-4)}{(x-4)(x+1)} = \lim_{x\to 4}\frac{x}{x+1} = \frac{4}{4+1} = \frac{4}{5}$$

15. $$\lim_{t\to -3}\frac{t^2-9}{2t^2+7t+3} = \lim_{t\to -3}\frac{(t+3)(t-3)}{(2t+1)(t+3)} = \lim_{t\to -3}\frac{t-3}{2t+1} = \frac{-3-3}{2(-3)+1} = \frac{-6}{-5} = \frac{6}{5}$$

16. $\lim_{x\to -1}\dfrac{x^2-4x}{x^2-3x-4}$ does not exist since $x^2-3x-4\to 0$ but $x^2-4x\to 5$ as $x\to -1$.

17. $$\lim_{h\to 0}\frac{(4+h)^2-16}{h} = \lim_{h\to 0}\frac{(16+8h+h^2)-16}{h} = \lim_{h\to 0}\frac{8h+h^2}{h} = \lim_{h\to 0}\frac{h(8+h)}{h} = \lim_{h\to 0}(8+h) = 8+0 = 8$$

18. $$\lim_{h\to 0}\frac{\sqrt{1+h}-1}{h} = \lim_{h\to 0}\frac{\sqrt{1+h}-1}{h}\cdot\frac{\sqrt{1+h}+1}{\sqrt{1+h}+1} = \lim_{h\to 0}\frac{(1+h)-1}{h\left(\sqrt{1+h}+1\right)} = \lim_{h\to 0}\frac{h}{h\left(\sqrt{1+h}+1\right)}$$

$$= \lim_{h\to 0}\frac{1}{\sqrt{1+h}+1} = \frac{1}{\sqrt{1}+1} = \frac{1}{2}$$

19. By the formula for the sum of cubes, we have

$$\lim_{x\to-2}\frac{x+2}{x^3+8}=\lim_{x\to-2}\frac{x+2}{(x+2)(x^2-2x+4)}=\lim_{x\to-2}\frac{1}{x^2-2x+4}=\frac{1}{4+4+4}=\frac{1}{12}.$$

20. $$\lim_{x\to-1}\frac{x^2+2x+1}{x^4-1}=\lim_{x\to-1}\frac{(x+1)^2}{(x^2+1)(x^2-1)}=\lim_{x\to-1}\frac{(x+1)^2}{(x^2+1)(x+1)(x-1)}=\lim_{x\to-1}\frac{x+1}{(x^2+1)(x-1)}=\frac{0}{2(-2)}=0$$

21. $$\lim_{x\to7}\frac{\sqrt{x+2}-3}{x-7}=\lim_{x\to7}\frac{\sqrt{x+2}-3}{x-7}\cdot\frac{\sqrt{x+2}+3}{\sqrt{x+2}+3}=\lim_{x\to7}\frac{(x+2)-9}{(x-7)\left(\sqrt{x+2}+3\right)}$$
$$=\lim_{x\to7}\frac{x-7}{(x-7)\left(\sqrt{x+2}+3\right)}=\lim_{x\to7}\frac{1}{\sqrt{x+2}+3}=\frac{1}{\sqrt{9}+3}=\frac{1}{6}$$

22. $$\lim_{h\to0}\frac{(3+h)^{-1}-3^{-1}}{h}=\lim_{h\to0}\frac{\frac{1}{3+h}-\frac{1}{3}}{h}=\lim_{h\to0}\frac{3-(3+h)}{h(3+h)3}=\lim_{h\to0}\frac{-h}{h(3+h)3}$$
$$=\lim_{h\to0}\left[-\frac{1}{3(3+h)}\right]=-\frac{1}{\lim\limits_{h\to0}[3(3+h)]}=-\frac{1}{3(3+0)}=-\frac{1}{9}$$

23. $$\lim_{x\to-4}\frac{\frac{1}{4}+\frac{1}{x}}{4+x}=\lim_{x\to-4}\frac{\frac{x+4}{4x}}{4+x}=\lim_{x\to-4}\frac{x+4}{4x(4+x)}=\lim_{x\to-4}\frac{1}{4x}=\frac{1}{4(-4)}=-\frac{1}{16}$$

24. $$\lim_{t\to0}\left(\frac{1}{t}-\frac{1}{t^2+t}\right)=\lim_{t\to0}\frac{(t^2+t)-t}{t(t^2+t)}=\lim_{t\to0}\frac{t^2}{t\cdot t(t+1)}=\lim_{t\to0}\frac{1}{t+1}=\frac{1}{0+1}=1$$

25. (a)

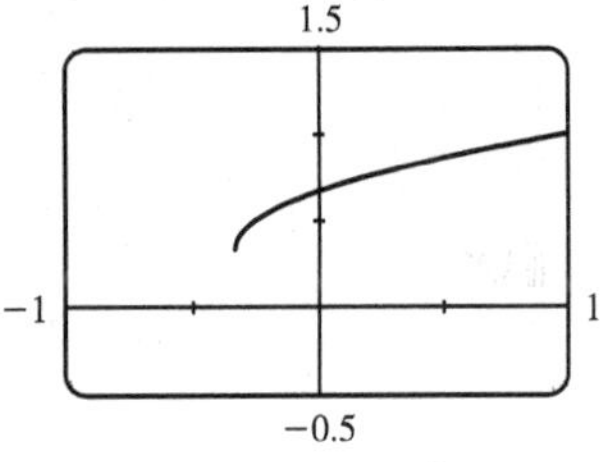

$$\lim_{x\to0}\frac{x}{\sqrt{1+3x}-1}\approx\frac{2}{3}$$

(b)

x	$f(x)$
−0.001	0.6661663
−0.0001	0.6666167
−0.00001	0.6666617
−0.000001	0.6666662
0.000001	0.6666672
0.00001	0.6666717
0.0001	0.6667167
0.001	0.6671663

The limit appears to be $\frac{2}{3}$.

(c) $$\lim_{x\to0}\left(\frac{x}{\sqrt{1+3x}-1}\cdot\frac{\sqrt{1+3x}+1}{\sqrt{1+3x}+1}\right)=\lim_{x\to0}\frac{x\left(\sqrt{1+3x}+1\right)}{(1+3x)-1}=\lim_{x\to0}\frac{x\left(\sqrt{1+3x}+1\right)}{3x}$$

$$=\frac{1}{3}\lim_{x\to0}\left(\sqrt{1+3x}+1\right)$$ [Limit Law 3]

$$=\frac{1}{3}\left[\sqrt{\lim_{x\to0}(1+3x)}+\lim_{x\to0}1\right]$$ [1 and 11]

$$=\frac{1}{3}\left(\sqrt{\lim_{x\to0}1+3\lim_{x\to0}x}+1\right)$$ [1, 3, and 7]

$$=\frac{1}{3}\left(\sqrt{1+3\cdot0}+1\right)$$ [7 and 8]

$$=\frac{1}{3}(1+1)=\frac{2}{3}$$

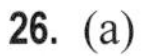

26. (a)

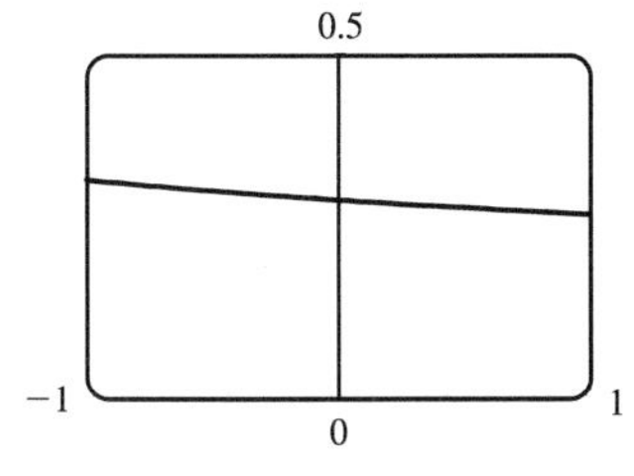

$$\lim_{x \to 0} \frac{\sqrt{3+x}-\sqrt{3}}{x} \approx 0.29$$

(b)

x	$f(x)$
-0.001	0.2886992
-0.0001	0.2886775
-0.00001	0.2886754
-0.000001	0.2886752
0.000001	0.2886751
0.00001	0.2886749
0.0001	0.2886727
0.001	0.2886511

The limit appears to be approximately 0.2887.

(c) $$\lim_{x \to 0} \left(\frac{\sqrt{3+x}-\sqrt{3}}{x} \cdot \frac{\sqrt{3+x}+\sqrt{3}}{\sqrt{3+x}+\sqrt{3}} \right) = \lim_{x \to 0} \frac{(3+x)-3}{x\left(\sqrt{3+x}+\sqrt{3}\right)} = \lim_{x \to 0} \frac{1}{\sqrt{3+x}+\sqrt{3}}$$

$$= \frac{\lim_{x \to 0} 1}{\lim_{x \to 0} \sqrt{3+x} + \lim_{x \to 0} \sqrt{3}} \qquad \text{[Limit Laws 5 and 1]}$$

$$= \frac{1}{\sqrt{\lim_{x \to 0} (3+x)} + \sqrt{3}} \qquad \text{[7 and 11]}$$

$$= \frac{1}{\sqrt{3+0}+\sqrt{3}} \qquad \text{[1, 7, and 8]}$$

$$= \frac{1}{2\sqrt{3}}$$

27. Let $f(x) = -x^2$, $g(x) = x^2 \cos 20\pi x$, and $h(x) = x^2$.
Then $-1 \le \cos 20\pi x \le 1 \;\Rightarrow\; -x^2 \le x^2 \cos 20\pi x \le x^2 \;\Rightarrow$
$f(x) \le g(x) \le h(x)$. So since $\lim_{x \to 0} f(x) = \lim_{x \to 0} h(x) = 0$, by the
Squeeze Theorem we have $\lim_{x \to 0} g(x) = 0$.

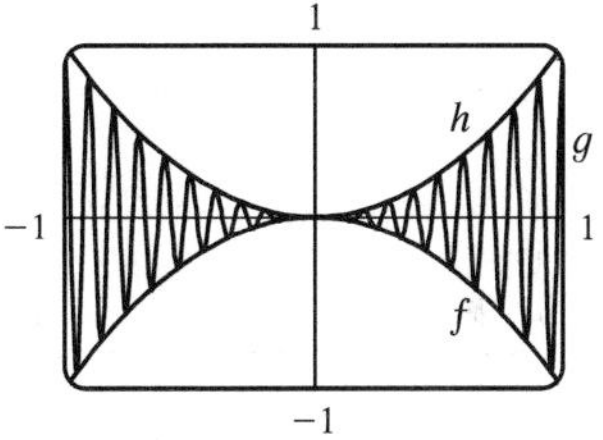

28. Let $f(x) = -\sqrt{x^3+x^2}$, $g(x) = \sqrt{x^3+x^2}\,\sin(\pi/x)$, and
$h(x) = \sqrt{x^3+x^2}$. Then $-1 \le \sin(\pi/x) \le 1 \;\Rightarrow$
$-\sqrt{x^3+x^2} \le \sqrt{x^3+x^2}\,\sin(\pi/x) \le \sqrt{x^3+x^2} \;\Rightarrow$
$f(x) \le g(x) \le h(x)$. So since $\lim_{x \to 0} f(x) = \lim_{x \to 0} h(x) = 0$, by the
Squeeze Theorem we have $\lim_{x \to 0} g(x) = 0$.

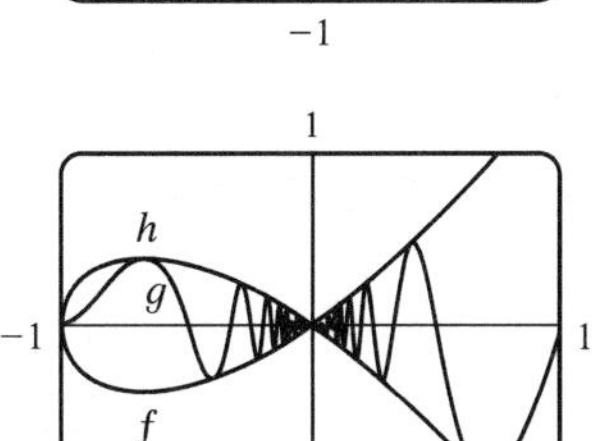

29. We have $\lim_{x \to 4} (4x-9) = 4(4)-9 = 7$ and $\lim_{x \to 4} \left(x^2-4x+7\right) = 4^2-4(4)+7 = 7$. Since $4x-9 \le f(x) \le x^2-4x+7$ for $x \ge 0$, $\lim_{x \to 4} f(x) = 7$ by the Squeeze Theorem.

30. We have $\lim_{x \to 1} (2x) = 2(1) = 2$ and $\lim_{x \to 1} (x^4-x^2+2) = 1^4-1^2+2 = 2$. Since $2x \le g(x) \le x^4-x^2+2$ for all x, $\lim_{x \to 1} g(x) = 2$ by the Squeeze Theorem.

31. $-1 \le \cos(2/x) \le 1 \;\Rightarrow\; -x^4 \le x^4 \cos(2/x) \le x^4$. Since $\lim_{x \to 0} \left(-x^4\right) = 0$ and $\lim_{x \to 0} x^4 = 0$, we have $\lim_{x \to 0} \left[x^4 \cos(2/x)\right] = 0$ by the Squeeze Theorem.

32. $-1 \le \sin(2\pi/x) \le 1 \;\Rightarrow\; 0 \le \sin^2(2\pi/x) \le 1 \;\Rightarrow\; 1 \le 1 + \sin^2(2\pi/x) \le 2 \;\Rightarrow$

$\sqrt{x} \le \sqrt{x}\left[1 + \sin^2(2\pi/x)\right] \le 2\sqrt{x}$. Since $\lim\limits_{x \to 0^+} \sqrt{x} = 0$ and $\lim\limits_{x \to 0^+} 2\sqrt{x} = 0$, we have

$\lim\limits_{x \to 0^+} \left[\sqrt{x}\left(1 + \sin^2(2\pi/x)\right)\right] = 0$ by the Squeeze Theorem.

33. $$|x-3| = \begin{cases} x-3 & \text{if } x-3 \ge 0 \\ -(x-3) & \text{if } x-3 < 0 \end{cases} = \begin{cases} x-3 & \text{if } x \ge 3 \\ 3-x & \text{if } x < 3 \end{cases}$$

Thus, $\lim\limits_{x \to 3^+} (2x + |x-3|) = \lim\limits_{x \to 3^+} (2x + x - 3) = \lim\limits_{x \to 3^+} (3x - 3) = 3(3) - 3 = 6$ and

$\lim\limits_{x \to 3^-} (2x + |x-3|) = \lim\limits_{x \to 3^-} (2x + 3 - x) = \lim\limits_{x \to 3^-} (x + 3) = 3 + 3 = 6$. Since the left and right limits are equal,

$\lim\limits_{x \to 3} (2x + |x-3|) = 6$.

34. $$|x+6| = \begin{cases} x+6 & \text{if } x+6 \ge 0 \\ -(x+6) & \text{if } x+6 < 0 \end{cases} = \begin{cases} x+6 & \text{if } x \ge -6 \\ -(x+6) & \text{if } x < -6 \end{cases}$$

We'll look at the one-sided limits.

$$\lim_{x \to -6^+} \frac{2x+12}{|x+6|} = \lim_{x \to -6^+} \frac{2(x+6)}{x+6} = 2 \quad \text{and} \quad \lim_{x \to -6^-} \frac{2x+12}{|x+6|} = \lim_{x \to -6^-} \frac{2(x+6)}{-(x+6)} = -2$$

The left and right limits are different, so $\lim\limits_{x \to -6} \dfrac{2x+12}{|x+6|}$ does not exist.

35. Since $|x| = -x$ for $x < 0$, we have $\lim\limits_{x \to 0^-} \left(\dfrac{1}{x} - \dfrac{1}{|x|}\right) = \lim\limits_{x \to 0^-} \left(\dfrac{1}{x} - \dfrac{1}{-x}\right) = \lim\limits_{x \to 0^-} \dfrac{2}{x}$, which does not exist since the denominator approaches 0 and the numerator does not.

36. Since $|x| = x$ for $x > 0$, we have $\lim\limits_{x \to 0^+} \left(\dfrac{1}{x} - \dfrac{1}{|x|}\right) = \lim\limits_{x \to 0^+} \left(\dfrac{1}{x} - \dfrac{1}{x}\right) = \lim\limits_{x \to 0^+} 0 = 0$.

37. (a) (i) If $x \to 1^+$, then $x > 1$ and $g(x) = x - 1$. Thus, $\lim\limits_{x \to 1^+} g(x) = \lim\limits_{x \to 1^+} (x - 1) = 1 - 1 = 0$.

(ii) If $x \to 1^-$, then $x < 1$ and $g(x) = 1 - x^2$. Thus, $\lim\limits_{x \to 1^-} g(x) = \lim\limits_{x \to 1^-} (1 - x^2) = 1 - 1^2 = 0$.

Since the left- and right-hand limits of g at 1 are equal, $\lim\limits_{x \to 1} g(x) = 0$.

(iii) If $x \to 0$, then $-1 < x < 1$ and $g(x) = 1 - x^2$. Thus, $\lim\limits_{x \to 0} g(x) = \lim\limits_{x \to 0} (1 - x^2) = 1 - 0^2 = 1$.

(iv) If $x \to -1^-$, then $x < -1$ and $g(x) = -x$. Thus, $\lim\limits_{x \to -1^-} g(x) = \lim\limits_{x \to -1^-} (-x) = -(-1) = 1$.

(v) If $x \to -1^+$, then $-1 < x < 1$ and $g(x) = 1 - x^2$. Thus,

$$\lim_{x \to -1^+} g(x) = \lim_{x \to -1^+} (1 - x^2) = 1 - (-1)^2 = 1 - 1 = 0$$

(vi) $\lim\limits_{x \to -1} g(x)$ does not exist because the limits in part (iv) and part (v) are not equal.

(b)

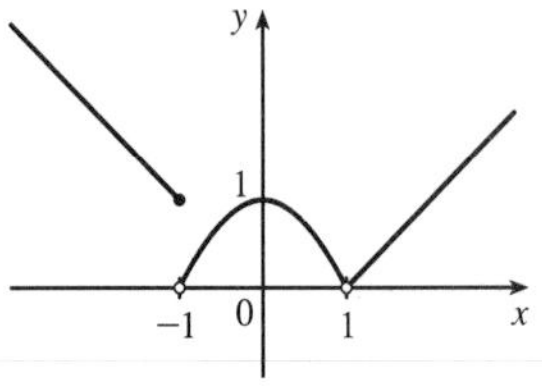

38. (a) (i) $\lim\limits_{x\to 1^+} \dfrac{x^2-1}{|x-1|} = \lim\limits_{x\to 1^+} \dfrac{x^2-1}{x-1} = \lim\limits_{x\to 1^+} (x+1) = 2$

(ii) $\lim\limits_{x\to 1^-} \dfrac{x^2-1}{|x-1|} = \lim\limits_{x\to 1^-} \dfrac{x^2-1}{-(x-1)} = \lim\limits_{x\to 1^-} -(x+1) = -2$

(b) No, $\lim\limits_{x\to 1} F(x)$ does not exist since $\lim\limits_{x\to 1^+} F(x) \neq \lim\limits_{x\to 1^-} F(x)$.

(c)

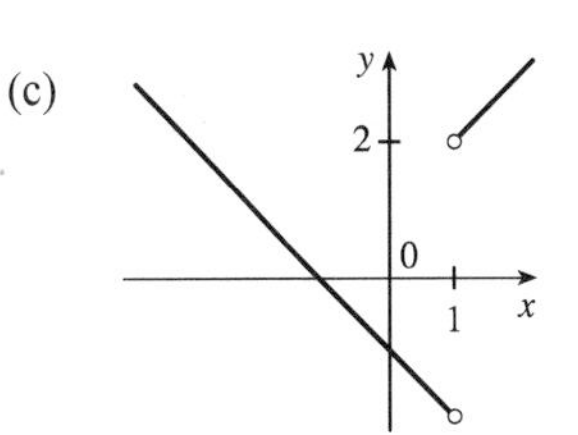

39. (a) (i) $[\![x]\!] = -2$ for $-2 \le x < -1$, so $\lim\limits_{x\to -2^+} [\![x]\!] = \lim\limits_{x\to -2^+} (-2) = -2$

(ii) $[\![x]\!] = -3$ for $-3 \le x < -2$, so $\lim\limits_{x\to -2^-} [\![x]\!] = \lim\limits_{x\to -2^-} (-3) = -3$.

The right and left limits are different, so $\lim\limits_{x\to -2} [\![x]\!]$ does not exist.

(iii) $[\![x]\!] = -3$ for $-3 \le x < -2$, so $\lim\limits_{x\to -2.4} [\![x]\!] = \lim\limits_{x\to -2.4} (-3) = -3$.

(b) (i) $[\![x]\!] = n-1$ for $n-1 \le x < n$, so $\lim\limits_{x\to n^-} [\![x]\!] = \lim\limits_{x\to n^-} (n-1) = n-1$.

(ii) $[\![x]\!] = n$ for $n \le x < n+1$, so $\lim\limits_{x\to n^+} [\![x]\!] = \lim\limits_{x\to n^+} n = n$.

(c) $\lim\limits_{x\to a} [\![x]\!]$ exists $\Leftrightarrow$ a is not an integer.

40. (a)

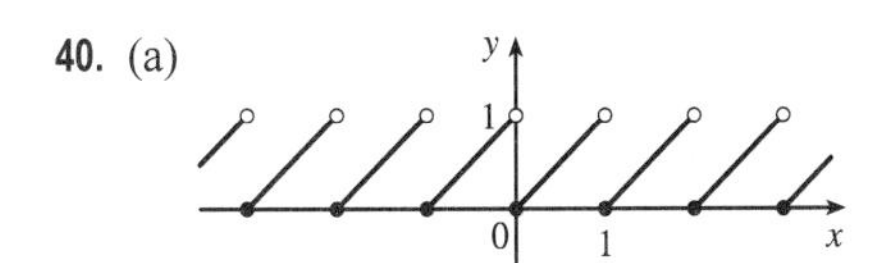

(b) (i) $\lim\limits_{x\to n^-} f(x) = \lim\limits_{x\to n^-} (x - [\![x]\!]) = \lim\limits_{x\to n^-} [x-(n-1)]$
$= n-(n-1) = 1$

(ii) $\lim\limits_{x\to n^+} f(x) = \lim\limits_{x\to n^+} (x - [\![x]\!]) = \lim\limits_{x\to n^+} (x-n) = n-n = 0$

(c) $\lim\limits_{x\to a} f(x)$ exists $\Leftrightarrow$ a is not an integer.

41. The graph of $f(x) = [\![x]\!] + [\![-x]\!]$ is the same as the graph of $g(x) = -1$ with holes at each integer, since $f(a) = 0$ for any integer a. Thus, $\lim\limits_{x\to 2^-} f(x) = -1$ and $\lim\limits_{x\to 2^+} f(x) = -1$, so $\lim\limits_{x\to 2} f(x) = -1$. However,

$f(2) = [\![2]\!] + [\![-2]\!] = 2 + (-2) = 0$, so $\lim\limits_{x\to 2} f(x) \neq f(2)$.

42. $\lim\limits_{v\to c^-} \left(L_0\sqrt{1-\dfrac{v^2}{c^2}}\right) = L_0\sqrt{1-1} = 0$. As the velocity approaches the speed of light, the length approaches 0.

A left-hand limit is necessary since L is not defined for $v > c$.

43.
$$\begin{aligned}
\lim_{x\to 0} \frac{\sin 3x}{x} &= \lim_{x\to 0} \frac{3\sin 3x}{3x} && \text{[multiply numerator and denominator by 3]}\\
&= 3\lim_{3x\to 0} \frac{\sin 3x}{3x} && [\text{as } x\to 0,\ 3x\to 0]\\
&= 3\lim_{\theta\to 0} \frac{\sin\theta}{\theta} && [\text{let } \theta = 3x]\\
&= 3(1) && \text{[Equation 2]}\\
&= 3
\end{aligned}$$

44.
$$\begin{aligned}
\lim_{x\to 0} \frac{\sin 4x}{\sin 6x} &= \lim_{x\to 0}\left(\frac{\sin 4x}{x}\cdot\frac{x}{\sin 6x}\right) = \lim_{x\to 0}\frac{4\sin 4x}{4x}\cdot\lim_{x\to 0}\frac{6x}{6\sin 6x}\\
&= 4\lim_{x\to 0}\frac{\sin 4x}{4x}\cdot\frac{1}{6}\lim_{x\to 0}\frac{6x}{\sin 6x} = 4(1)\cdot\frac{1}{6}(1) = \frac{2}{3}
\end{aligned}$$

45. $\displaystyle \lim_{t\to 0}\frac{\tan 6t}{\sin 2t} = \lim_{t\to 0}\left(\frac{\sin 6t}{t}\cdot\frac{1}{\cos 6t}\cdot\frac{t}{\sin 2t}\right) = \lim_{t\to 0}\frac{6\sin 6t}{6t}\cdot\lim_{t\to 0}\frac{1}{\cos 6t}\cdot\lim_{t\to 0}\frac{2t}{2\sin 2t}$

$$= 6\lim_{t\to 0}\frac{\sin 6t}{6t}\cdot\lim_{t\to 0}\frac{1}{\cos 6t}\cdot\frac{1}{2}\lim_{t\to 0}\frac{2t}{\sin 2t} = 6(1)\cdot\frac{1}{1}\cdot\frac{1}{2}(1) = 3$$

46. $\displaystyle \lim_{t\to 0}\frac{\sin^2 3t}{t^2} = \lim_{t\to 0}\left(\frac{\sin 3t}{t}\cdot\frac{\sin 3t}{t}\right) = \lim_{t\to 0}\frac{\sin 3t}{t}\cdot\lim_{t\to 0}\frac{\sin 3t}{t}$

$$= \left(\lim_{t\to 0}\frac{\sin 3t}{t}\right)^2 = \left(3\lim_{t\to 0}\frac{\sin 3t}{3t}\right)^2 = (3\cdot 1)^2 = 9$$

47. Divide numerator and denominator by θ. ($\sin\theta$ also works.)

$$\lim_{\theta\to 0}\frac{\sin\theta}{\theta+\tan\theta} = \lim_{\theta\to 0}\frac{\dfrac{\sin\theta}{\theta}}{1+\dfrac{\sin\theta}{\theta}\cdot\dfrac{1}{\cos\theta}} = \frac{\displaystyle\lim_{\theta\to 0}\frac{\sin\theta}{\theta}}{1+\displaystyle\lim_{\theta\to 0}\frac{\sin\theta}{\theta}\lim_{\theta\to 0}\frac{1}{\cos\theta}} = \frac{1}{1+1\cdot 1} = \frac{1}{2}$$

48. $\displaystyle \lim_{x\to 0}x\cot x = \lim_{x\to 0}x\cdot\frac{\cos x}{\sin x} = \lim_{x\to 0}\frac{x\cos x}{\sin x} = \lim_{x\to 0}\frac{\dfrac{x\cos x}{x}}{\dfrac{\sin x}{x}} = \lim_{x\to 0}\frac{\cos x}{\dfrac{\sin x}{x}} = \frac{\displaystyle\lim_{x\to 0}\cos x}{\displaystyle\lim_{x\to 0}\frac{\sin x}{x}} = \frac{1}{1} = 1$

49. Since $p(x)$ is a polynomial, $p(x) = a_0 + a_1x + a_2x^2 + \cdots + a_nx^n$. Thus, by the Limit Laws,

$$\lim_{x\to a}p(x) = \lim_{x\to a}\left(a_0 + a_1x + a_2x^2 + \cdots + a_nx^n\right) = a_0 + a_1\lim_{x\to a}x + a_2\lim_{x\to a}x^2 + \cdots + a_n\lim_{x\to a}x^n$$
$$= a_0 + a_1a + a_2a^2 + \cdots + a_na^n = p(a)$$

Thus, for any polynomial p, $\lim_{x\to a}p(x) = p(a)$.

50. Let $r(x) = \dfrac{p(x)}{q(x)}$ where $p(x)$ and $q(x)$ are any polynomials, and suppose that $q(a)\neq 0$. Thus,

$$\lim_{x\to a}r(x) = \lim_{x\to a}\frac{p(x)}{q(x)} = \frac{\displaystyle\lim_{x\to a}p(x)}{\displaystyle\lim_{x\to a}q(x)} \quad \text{[Limit Law 5]} \quad = \frac{p(a)}{q(a)} \quad \text{[Exercise 49]} \quad = r(a).$$

51. $\displaystyle \lim_{h\to 0}\sin(a+h) = \lim_{h\to 0}(\sin a\cos h + \cos a\sin h) = \lim_{h\to 0}(\sin a\cos h) + \lim_{h\to 0}(\cos a\sin h)$

$$= \left(\lim_{h\to 0}\sin a\right)\left(\lim_{h\to 0}\cos h\right) + \left(\lim_{h\to 0}\cos a\right)\left(\lim_{h\to 0}\sin h\right) = (\sin a)(1) + (\cos a)(0) = \sin a$$

52. As in the previous exercise, we must show that $\lim_{h\to 0}\cos(a+h) = \cos a$ to prove that the cosine function has the Direct Substitution Property.

$$\lim_{h\to 0}\cos(a+h) = \lim_{h\to 0}(\cos a\cos h - \sin a\sin h) = \lim_{h\to 0}(\cos a\cos h) - \lim_{h\to 0}(\sin a\sin h)$$
$$= \left(\lim_{h\to 0}\cos a\right)\left(\lim_{h\to 0}\cos h\right) - \left(\lim_{h\to 0}\sin a\right)\left(\lim_{h\to 0}\sin h\right) = (\cos a)(1) - (\sin a)(0) = \cos a$$

53. Let $f(x) = [\![x]\!]$ and $g(x) = -[\![x]\!]$. Then $\lim_{x\to 3}f(x)$ and $\lim_{x\to 3}g(x)$ do not exist [Example 8]

but $\lim_{x\to 3}[f(x)+g(x)] = \lim_{x\to 3}([\![x]\!] - [\![x]\!]) = \lim_{x\to 3}0 = 0$.

54. Let $f(x) = H(x)$ and $g(x) = 1 - H(x)$, where H is the Heaviside function defined in Example 6 in Section 1.3. Thus, either f or g is 0 for any value of x. Then $\lim_{x\to 0}f(x)$ and $\lim_{x\to 0}g(x)$ do not exist, but $\lim_{x\to 0}[f(x)g(x)] = \lim_{x\to 0}0 = 0$.

55. Since the denominator approaches 0 as $x \to -2$, the limit will exist only if the numerator also approaches 0 as $x \to -2$. In order for this to happen, we need $\lim\limits_{x \to -2} (3x^2 + ax + a + 3) = 0 \Leftrightarrow$ $3(-2)^2 + a(-2) + a + 3 = 0 \Leftrightarrow 12 - 2a + a + 3 = 0 \Leftrightarrow a = 15$. With $a = 15$, the limit becomes

$$\lim_{x \to -2} \frac{3x^2 + 15x + 18}{x^2 + x - 2} = \lim_{x \to -2} \frac{3(x+2)(x+3)}{(x-1)(x+2)} = \lim_{x \to -2} \frac{3(x+3)}{x-1} = \frac{3(-2+3)}{-2-1} = \frac{3}{-3} = -1.$$

56. *Solution 1:* First, we find the coordinates of P and Q as functions of r. Then we can find the equation of the line determined by these two points, and thus find the x-intercept (the point R), and take the limit as $r \to 0$. The coordinates of P are $(0, r)$. The point Q is the point of intersection of the two circles $x^2 + y^2 = r^2$ and $(x-1)^2 + y^2 = 1$. Eliminating y from these equations, we get $r^2 - x^2 = 1 - (x-1)^2 \Leftrightarrow r^2 = 1 + 2x - 1 \Leftrightarrow x = \frac{1}{2}r^2$. Substituting back into the equation of the shrinking circle to find the y-coordinate, we get $\left(\frac{1}{2}r^2\right)^2 + y^2 = r^2 \Leftrightarrow y^2 = r^2\left(1 - \frac{1}{4}r^2\right) \Leftrightarrow y = r\sqrt{1 - \frac{1}{4}r^2}$ (the positive y-value). So the coordinates of Q are $\left(\frac{1}{2}r^2, r\sqrt{1 - \frac{1}{4}r^2}\right)$. The equation of the line joining P and Q is thus

$$y - r = \frac{r\sqrt{1 - \frac{1}{4}r^2} - r}{\frac{1}{2}r^2 - 0}(x - 0).$$ We set $y = 0$ in order to find the x-intercept, and get

$$x = -r\frac{\frac{1}{2}r^2}{r\left(\sqrt{1 - \frac{1}{4}r^2} - 1\right)} = \frac{-\frac{1}{2}r^2\left(\sqrt{1 - \frac{1}{4}r^2} + 1\right)}{1 - \frac{1}{4}r^2 - 1} = 2\left(\sqrt{1 - \frac{1}{4}r^2} + 1\right).$$

Now we take the limit as $r \to 0^+$: $\lim\limits_{r \to 0^+} x = \lim\limits_{r \to 0^+} 2\left(\sqrt{1 - \frac{1}{4}r^2} + 1\right) = \lim\limits_{r \to 0^+} 2\left(\sqrt{1} + 1\right) = 4$.

So the limiting position of R is the point $(4, 0)$.

Solution 2: We add a few lines to the diagram, as shown. Note that $\angle PQS = 90°$ (subtended by diameter PS). So $\angle SQR = 90° = \angle OQT$ (subtended by diameter OT). It follows that $\angle OQS = \angle TQR$. Also $\angle PSQ = 90° - \angle SPQ = \angle ORP$. Since $\triangle QOS$ is isosceles, so is $\triangle QTR$, implying that $QT = TR$. As the circle C_2 shrinks, the point Q plainly approaches the origin, so the point R must approach a point twice as far from the origin as T, that is, the point $(4, 0)$, as above.

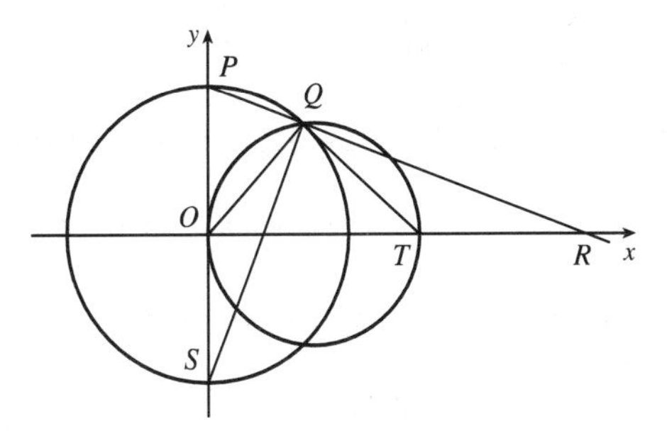

1.5 Continuity

1. From Definition 1, $\lim\limits_{x \to 4} f(x) = f(4)$.

2. The graph of f has no hole, jump, or vertical asymptote.

3. (a) The following are the numbers at which f is discontinuous and the type of discontinuity at that number: -4 (removable), -2 (jump), 2 (jump), 4 (infinite).

   (b) f is continuous from the left at -2 since $\lim\limits_{x \to -2^-} f(x) = f(-2)$. f is continuous from the right at 2 and 4 since $\lim\limits_{x \to 2^+} f(x) = f(2)$ and $\lim\limits_{x \to 4^+} f(x) = f(4)$. It is continuous from neither side at -4 since $f(-4)$ is undefined.

4. g is continuous on $[-4,-2)$, $(-2,2)$, $[2,4)$, $(4,6)$, and $(6,8)$.

5. The graph of $y = f(x)$ must have a discontinuity at $x = 3$ and must show that $\lim\limits_{x\to 3^-} f(x) = f(3)$.

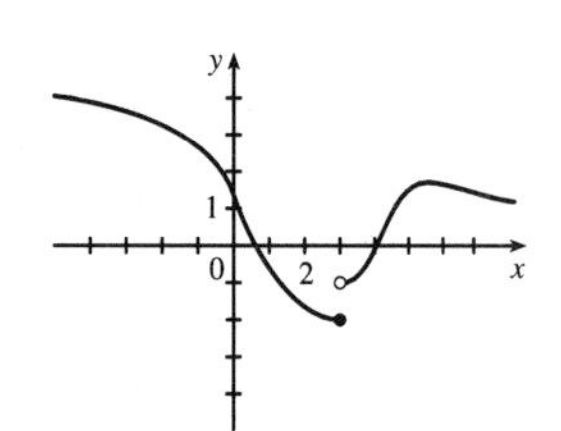

6.

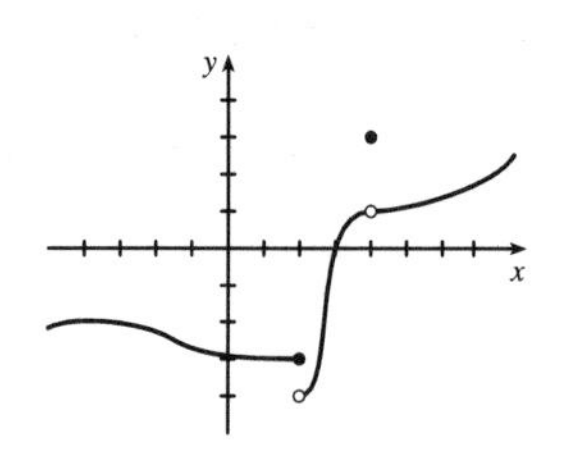

7. (a)

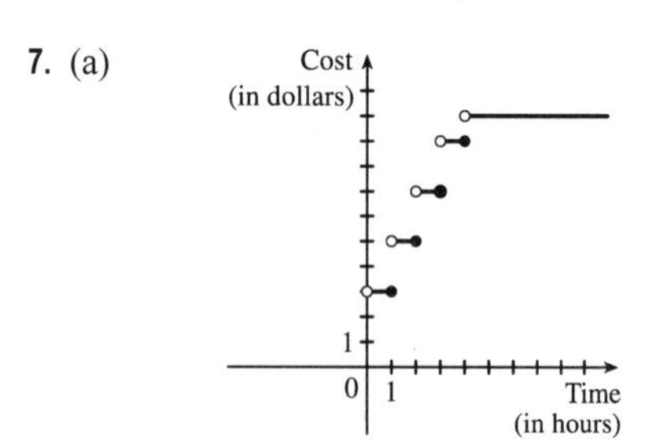

(b) There are discontinuities at times $t = 1$, 2, 3, and 4. A person parking in the lot would want to keep in mind that the charge will jump at the beginning of each hour.

8. (a) Continuous; at the location in question, the temperature changes smoothly as time passes, without any instantaneous jumps from one temperature to another.

(b) Continuous; the temperature at a specific time changes smoothly as the distance due west from New York City increases, without any instantaneous jumps.

(c) Discontinuous; as the distance due west from New York City increases, the altitude above sea level may jump from one height to another without going through all of the intermediate values—at a cliff, for example.

(d) Discontinuous; as the distance traveled increases, the cost of the ride jumps in small increments.

(e) Discontinuous; when the lights are switched on (or off), the current suddenly changes between 0 and some nonzero value, without passing through all of the intermediate values. This is debatable, though, depending on your definition of current.

9. Since f and g are continuous functions,

$$\begin{aligned}\lim_{x\to 3}\,[2f(x)-g(x)] &= 2\lim_{x\to 3} f(x) - \lim_{x\to 3} g(x) && \text{[by Limit Laws 2 and 3]}\\ &= 2f(3)-g(3) && \text{[by continuity of } f \text{ and } g \text{ at } x=3]\\ &= 2\cdot 5 - g(3) = 10 - g(3)\end{aligned}$$

Since it is given that $\lim\limits_{x\to 3}\,[2f(x)-g(x)] = 4$, we have $10 - g(3) = 4$, so $g(3) = 6$.

10. $\lim\limits_{x\to 4} f(x) = \lim\limits_{x\to 4}\left(x^2+\sqrt{7-x}\right) = \lim\limits_{x\to 4} x^2 + \sqrt{\lim\limits_{x\to 4} 7 - \lim\limits_{x\to 4} x} = 4^2+\sqrt{7-4} = 16+\sqrt{3} = f(4)$.

By the definition of continuity, f is continuous at $a = 4$.

11. $\lim\limits_{x\to -1} f(x) = \lim\limits_{x\to -1}(x+2x^3)^4 = \left(\lim\limits_{x\to -1} x + 2\lim\limits_{x\to -1} x^3\right)^4 = [-1+2(-1)^3]^4 = (-3)^4 = 81 = f(-1)$.

By the definition of continuity, f is continuous at $a = -1$.

12. For $-4 < a < 4$ we have $\lim\limits_{x \to a} f(x) = \lim\limits_{x \to a} x\sqrt{16 - x^2} = \lim\limits_{x \to a} x \sqrt{\lim\limits_{x \to a} 16 - \lim\limits_{x \to a} x^2} = a\sqrt{16 - a^2} = f(a)$,

so f is continuous on $(-4, 4)$. Similarly, we get $\lim\limits_{x \to 4^-} f(x) = 0 = f(4)$ and $\lim\limits_{x \to -4^+} f(x) = 0 = f(-4)$,

so f is continuous from the left at 4 and from the right at -4. Thus, f is continuous on $[-4, 4]$.

13. $f(x) = -\dfrac{1}{(x-1)^2}$ is discontinuous at 1 since $f(1)$ is not defined.

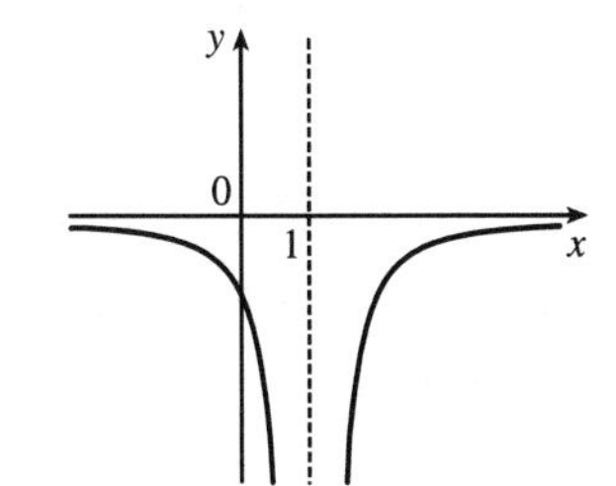

14. $f(x) = \begin{cases} 1/(x-1) & \text{if } x \neq 1 \\ 2 & \text{if } x = 1 \end{cases}$ is discontinuous at 1 because $\lim\limits_{x \to 1} f(x)$ does not exist.

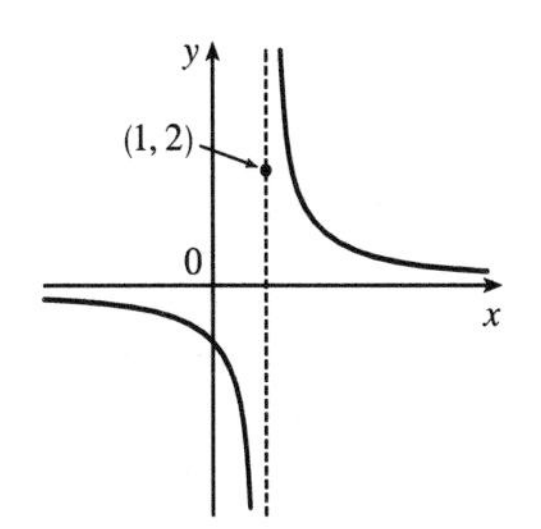

15. $f(x) = \begin{cases} 1 - x^2 & \text{if } x < 1 \\ 1/x & \text{if } x \geq 1 \end{cases}$

The left-hand limit of f at $a = 1$ is

$\lim\limits_{x \to 1^-} f(x) = \lim\limits_{x \to 1^-} (1 - x^2) = 0$. The right-hand limit of f at $a = 1$ is

$\lim\limits_{x \to 1^+} f(x) = \lim\limits_{x \to 1^+} (1/x) = 1$. Since these limits are not equal, $\lim\limits_{x \to 1} f(x)$

does not exist and f is discontinuous at 1.

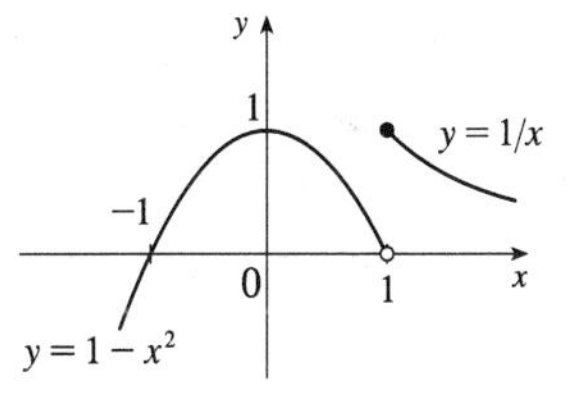

16. $f(x) = \begin{cases} \dfrac{x^2 - x}{x^2 - 1} & \text{if } x \neq 1 \\ 1 & \text{if } x = 1 \end{cases}$

$\lim\limits_{x \to 1} f(x) = \lim\limits_{x \to 1} \dfrac{x^2 - x}{x^2 - 1} = \lim\limits_{x \to 1} \dfrac{x(x-1)}{(x+1)(x-1)} = \lim\limits_{x \to 1} \dfrac{x}{x+1} = \dfrac{1}{2}$,

but $f(1) = 1$, so f is discontinous at 1.

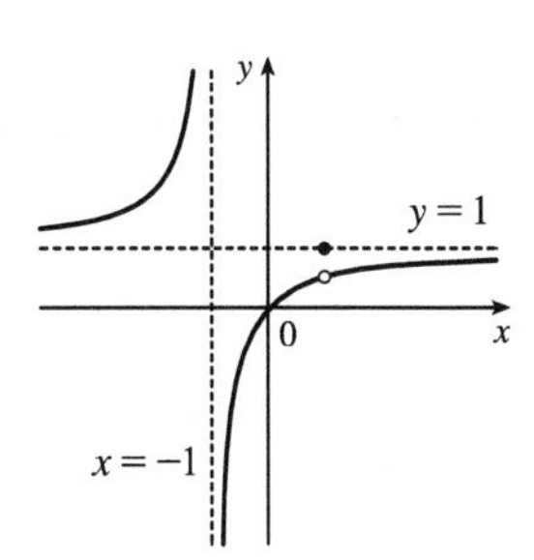

17. $F(x) = \dfrac{x}{x^2 + 5x + 6}$ is a rational function. So by Theorem 5 (or Theorem 6), F is continuous at every number in its domain,

$\{x \mid x^2 + 5x + 6 \neq 0\} = \{x \mid (x+3)(x+2) \neq 0\} = \{x \mid x \neq -3, \ -2\}$ or $(-\infty, -3) \cup (-3, -2) \cup (-2, \infty)$.

18. By Theorem 6, the root function $\sqrt[3]{x}$ and the polynomial function $1 + x^3$ are continuous on $\mathbb{R}$. By part 4 of Theorem 4, the product $G(x) = \sqrt[3]{x}\,(1 + x^3)$ is continuous on its domain, $\mathbb{R}$.

19. By Theorem 5, the polynomials x^2 and $2x - 1$ are continuous on $(-\infty, \infty)$. By Theorem 6, the root function $\sqrt{x}$ is continuous on $[0, \infty)$. By Theorem 8, the composite function $\sqrt{2x - 1}$ is continuous on its domain, $[\frac{1}{2}, \infty)$. By part 1 of Theorem 4, the sum $R(x) = x^2 + \sqrt{2x - 1}$ is continuous on $[\frac{1}{2}, \infty)$.

20. By Theorem 6, the trigonometric function $\sin x$ and the polynomial function $x + 1$ are continuous on $\mathbb{R}$. By part 5 of Theorem 4, $h(x) = \dfrac{\sin x}{x + 1}$ is continuous on its domain, $\{x \mid x \neq -1\}$.

21. By Theorem 6, the root function $\sqrt{x}$ and the trigonometric function $\sin x$ are continuous on their domains, $[0, \infty)$ and $(-\infty, \infty)$, respectively. Thus, the product $F(x) = \sqrt{x}\sin x$ is continuous on the intersection of those domains, $[0, \infty)$, by part 4 of Theorem 4.

22. The sine and cosine functions are continuous everywhere by Theorem 6, so $F(x) = \sin(\cos(\sin x))$, which is the composite of sine, cosine, and (once again) sine, is continuous everywhere by Theorem 8.

23.

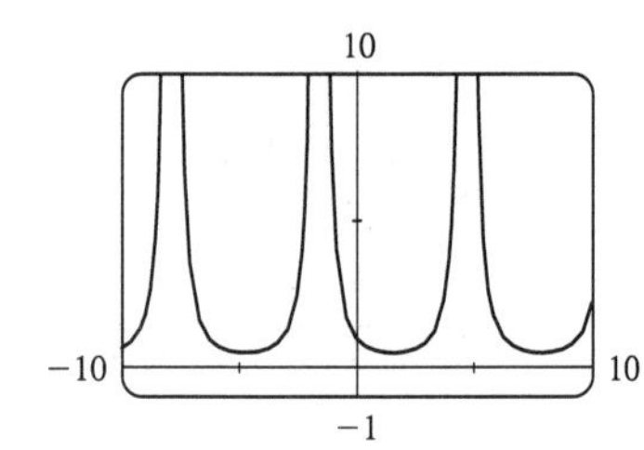

$y = \dfrac{1}{1 + \sin x}$ is undefined and hence discontinuous when $1 + \sin x = 0 \;\Leftrightarrow\; \sin x = -1 \;\Leftrightarrow\; x = -\frac{\pi}{2} + 2\pi n$, n an integer. The figure shows discontinuities for $n = -1$, 0, and 1; that is, $-\dfrac{5\pi}{2} \approx -7.85$, $-\dfrac{\pi}{2} \approx -1.57$, and $\dfrac{3\pi}{2} \approx 4.71$.

24.

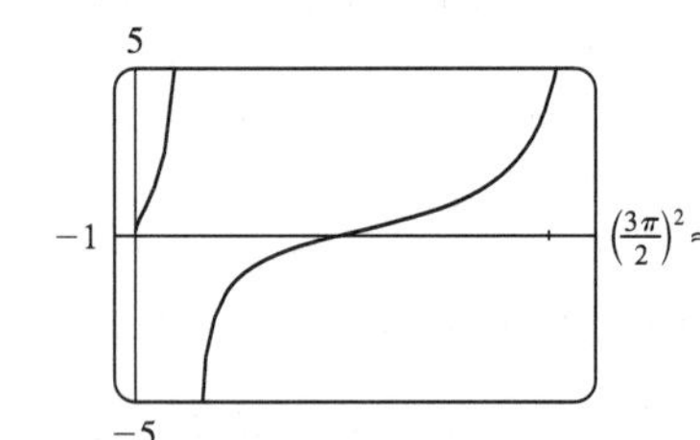

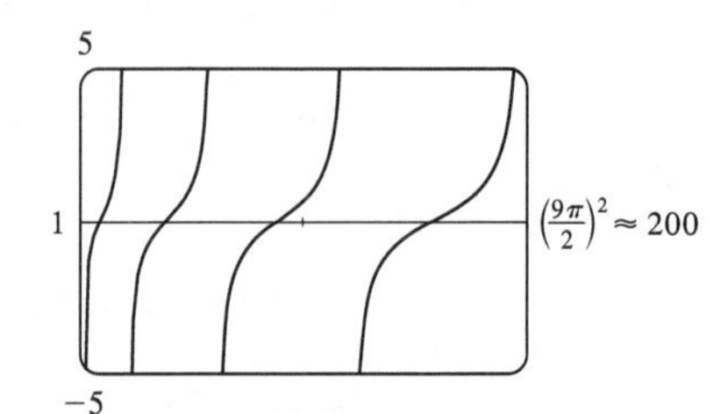

The function $y = f(x) = \tan\sqrt{x}$ is continuous throughout its domain because it is the composite of a trigonometric function and a root function. The square root function has domain $[0, \infty)$ and the tangent function has domain $\left\{x \mid x \neq \frac{\pi}{2} + \pi n\right\}$. So f is discontinuous when $x < 0$ and when $\sqrt{x} = \frac{\pi}{2} + \pi n \;\Rightarrow\; x = \left(\frac{\pi}{2} + \pi n\right)^2$, where n is a nonnegative integer. Note that as x increases, the distance between discontinuities increases.

25. Because we are dealing with root functions, $5 + \sqrt{x}$ is continuous on $[0, \infty)$, $\sqrt{x + 5}$ is continuous on $[-5, \infty)$, so the quotient $f(x) = \dfrac{5 + \sqrt{x}}{\sqrt{5 + x}}$ is continuous on $[0, \infty)$. Since f is continuous at $x = 4$, $\lim\limits_{x \to 4} f(x) = f(4) = \frac{7}{3}$.

26. Because x is continuous on $\mathbb{R}$, $\sin x$ is continuous on $\mathbb{R}$, and $x + \sin x$ is continuous on $\mathbb{R}$, the composite function $f(x) = \sin(x + \sin x)$ is continuous on $\mathbb{R}$, so $\lim\limits_{x \to \pi} f(x) = f(\pi) = \sin(\pi + \sin \pi) = \sin \pi = 0$.

27. $f(x) = \begin{cases} x^2 & \text{if } x < 1 \\ \sqrt{x} & \text{if } x \geq 1 \end{cases}$

By Theorem 5, since $f(x)$ equals the polynomial x^2 on $(-\infty, 1)$, f is continuous on $(-\infty, 1)$. By Theorem 6, since $f(x)$ equals the root function $\sqrt{x}$ on $(1, \infty)$, f is continuous on $(1, \infty)$. At $x = 1$, $\lim\limits_{x \to 1^-} f(x) = \lim\limits_{x \to 1^-} x^2 = 1$ and $\lim\limits_{x \to 1^+} f(x) = \lim\limits_{x \to 1^+} \sqrt{x} = 1$. Thus, $\lim\limits_{x \to 1} f(x)$ exists and equals 1. Also, $f(1) = \sqrt{1} = 1$. Thus, f is continuous at $x = 1$. We conclude that f is continuous on $(-\infty, \infty)$.

28. $f(x) = \begin{cases} \sin x & \text{if } x < \pi/4 \\ \cos x & \text{if } x \geq \pi/4 \end{cases}$

By Theorem 6, the trigonometric functions are continuous. Since $f(x) = \sin x$ on $(-\infty, \pi/4)$ and $f(x) = \cos x$ on $(\pi/4, \infty)$, f is continuous on $(-\infty, \pi/4) \cup (\pi/4, \infty)$. $\lim\limits_{x \to (\pi/4)^-} f(x) = \lim\limits_{x \to (\pi/4)^-} \sin x = \sin \frac{\pi}{4} = 1/\sqrt{2}$ since the sine function is continuous at $\pi/4$. Similarly, $\lim\limits_{x \to (\pi/4)^+} f(x) = \lim\limits_{x \to (\pi/4)^+} \cos x = 1/\sqrt{2}$ by continuity of the cosine function at $\pi/4$. Thus, $\lim\limits_{x \to (\pi/4)} f(x)$ exists and equals $1/\sqrt{2}$, which agrees with the value $f(\pi/4)$. Therefore, f is continuous at $\pi/4$, so f is continuous on $(-\infty, \infty)$.

29. 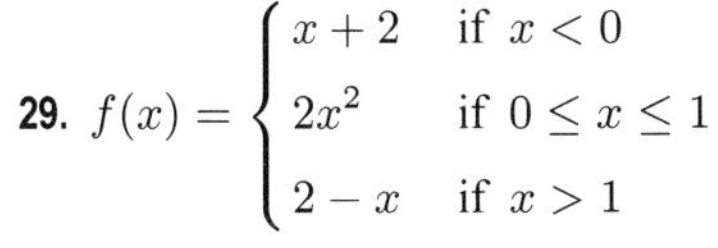

$f(x) = \begin{cases} x + 2 & \text{if } x < 0 \\ 2x^2 & \text{if } 0 \leq x \leq 1 \\ 2 - x & \text{if } x > 1 \end{cases}$

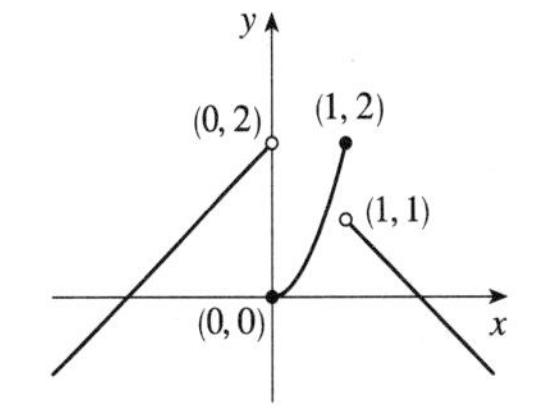

f is continuous on $(-\infty, 0)$, $(0, 1)$, and $(1, \infty)$ since on each of these intervals it is a polynomial. Now $\lim\limits_{x \to 0^-} f(x) = \lim\limits_{x \to 0^-} (x + 2) = 2$ and $\lim\limits_{x \to 0^+} f(x) = \lim\limits_{x \to 0^+} 2x^2 = 0$, so f is discontinuous at 0. Since $f(0) = 0$, f is continuous from the right at 0. Also $\lim\limits_{x \to 1^-} f(x) = \lim\limits_{x \to 1^-} 2x^2 = 2$ and $\lim\limits_{x \to 1^+} f(x) = \lim\limits_{x \to 1^+} (2 - x) = 1$, so f is discontinuous at 1. Since $f(1) = 2$, f is continuous from the left at 1.

30. By Theorem 5, each piece of F is continuous on its domain. We need to check for continuity at $r = R$. $\lim\limits_{r \to R^-} F(r) = \lim\limits_{r \to R^-} \dfrac{GMr}{R^3} = \dfrac{GM}{R^2}$ and $\lim\limits_{r \to R^+} F(r) = \lim\limits_{r \to R^+} \dfrac{GM}{r^2} = \dfrac{GM}{R^2}$, so $\lim\limits_{r \to R} F(r) = \dfrac{GM}{R^2}$. Since $F(R) = \dfrac{GM}{R^2}$, F is continuous at R. Therefore, F is a continuous function of r.

31. $f(x) = \begin{cases} cx^2 + 2x & \text{if } x < 2 \\ x^3 - cx & \text{if } x \geq 2 \end{cases}$

f is continuous on $(-\infty, 2)$ and $(2, \infty)$. Now $\lim\limits_{x \to 2^-} f(x) = \lim\limits_{x \to 2^-} \left(cx^2 + 2x\right) = 4c + 4$ and $\lim\limits_{x \to 2^+} f(x) = \lim\limits_{x \to 2^+} \left(x^3 - cx\right) = 8 - 2c$. So f is continuous $\Leftrightarrow$ $4c + 4 = 8 - 2c$ $\Leftrightarrow$ $6c = 4$ $\Leftrightarrow$ $c = \frac{2}{3}$. Thus, for f to be continuous on $(-\infty, \infty)$, $c = \frac{2}{3}$.

32. The functions $x^2 - c^2$ and $cx + 20$, considered on the intervals $(-\infty, 4)$ and $[4, \infty)$ respectively, are continuous for any value of c. So the only possible discontinuity is at $x = 4$. For the function to be continuous at $x = 4$, the left-hand and right-hand limits must be the same. Now $\lim\limits_{x \to 4^-} g(x) = \lim\limits_{x \to 4^-} \left(x^2 - c^2\right) = 16 - c^2$ and $\lim\limits_{x \to 4^+} g(x) = \lim\limits_{x \to 4^+} (cx + 20) = 4c + 20 = g(4)$. Thus, $16 - c^2 = 4c + 20$ $\Leftrightarrow$ $c^2 + 4c + 4 = 0$ $\Leftrightarrow$ $c = -2$.

33. (a) $f(x) = \dfrac{x^2 - 2x - 8}{x + 2} = \dfrac{(x - 4)(x + 2)}{x + 2}$ has a removable discontinuity at -2 because $g(x) = x - 4$ is continuous on $\mathbb{R}$ and $f(x) = g(x)$ for $x \neq -2$. [The discontinuity is removed by defining $f(-2) = -6$.]

(b) $f(x) = \dfrac{x - 7}{|x - 7|}$ $\Rightarrow$ $\lim\limits_{x \to 7^-} f(x) = -1$ and $\lim\limits_{x \to 7^+} f(x) = 1$. Thus, $\lim\limits_{x \to 7} f(x)$ does not exist, so the discontinuity is not removable. (It is a jump discontinuity.)

(c) $f(x) = \dfrac{x^3 + 64}{x + 4} = \dfrac{(x + 4)(x^2 - 4x + 16)}{x + 4}$ has a removable discontinuity at -4 because $g(x) = x^2 - 4x + 16$ is continuous on $\mathbb{R}$ and $f(x) = g(x)$ for $x \neq -4$. [The discontinuity is removed by defining $f(-4) = 48$.]

(d) $f(x) = \dfrac{3 - \sqrt{x}}{9 - x} = \dfrac{3 - \sqrt{x}}{(3 - \sqrt{x}\,)(3 + \sqrt{x}\,)}$ has a removable discontinuity at 9 because $g(x) = \dfrac{1}{3 + \sqrt{x}}$ is continuous on $[0, \infty)$ and $f(x) = g(x)$ for $x \neq 9$. [The discontinuity is removed by defining $f(9) = \frac{1}{6}$.]

34. 

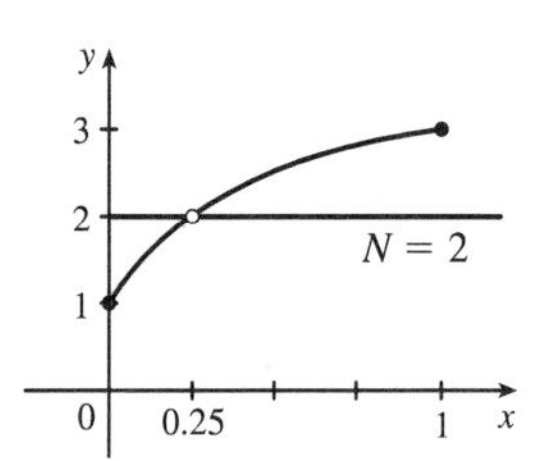

f does not satisfy the conclusion of the Intermediate Value Theorem.

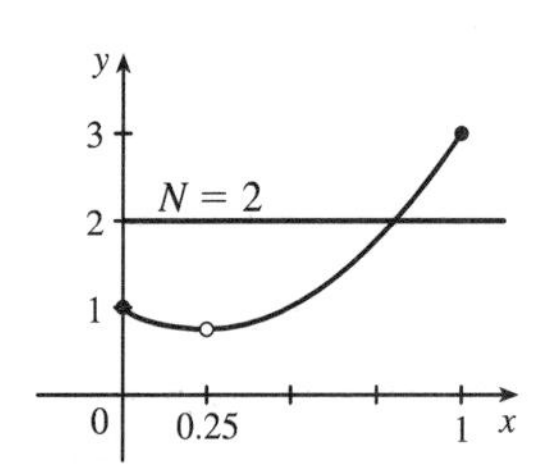

f does satisfy the conclusion of the Intermediate Value Theorem.

35. $f(x) = x^2 + 10 \sin x$ is continuous on the interval $[31, 32]$, $f(31) \approx 957$, and $f(32) \approx 1030$. Since $957 < 1000 < 1030$, there is a number c in $(31, 32)$ such that $f(c) = 1000$ by the Intermediate Value Theorem.

36. $f(x) = x^2$ is continuous on the interval $[1, 2]$, $f(1) = 1$, and $f(2) = 4$. Since $1 < 2 < 4$, there is a number c in $(1, 2)$ such that $f(c) = c^2 = 2$ by the Intermediate Value Theorem.

37. $f(x) = x^4 + x - 3$ is continuous on the interval $[1, 2]$, $f(1) = -1$, and $f(2) = 15$. Since $-1 < 0 < 15$, there is a number c in $(1, 2)$ such that $f(c) = 0$ by the Intermediate Value Theorem. Thus, there is a root of the equation $x^4 + x - 3 = 0$ in the interval $(1, 2)$.

38. $f(x) = \sqrt[3]{x} + x - 1$ is continuous on the interval $[0, 1]$, $f(0) = -1$, and $f(1) = 1$. Since $-1 < 0 < 1$, there is a number c in $(0, 1)$ such that $f(c) = 0$ by the Intermediate Value Theorem. Thus, there is a root of the equation $\sqrt[3]{x} + x - 1 = 0$, or $\sqrt[3]{x} = 1 - x$, in the interval $(0, 1)$.

39. $f(x) = \cos x - x$ is continuous on the interval $[0, 1]$, $f(0) = 1$, and $f(1) = \cos 1 - 1 \approx -0.46$. Since $-0.46 < 0 < 1$, there is a number c in $(0, 1)$ such that $f(c) = 0$ by the Intermediate Value Theorem. Thus, there is a root of the equation $\cos x - x = 0$, or $\cos x = x$, in the interval $(0, 1)$.

40. $f(x) = \tan x - 2x$ is continuous on the interval $[0, 1.4]$, $f(1) = \tan 1 - 2 \approx -0.44$, and $f(1.4) = \tan 1.4 - 2.8 \approx 3.00$. Since $-0.44 < 0 < 3.00$, there is a number c in $(0, 1.4)$ such that $f(c) = 0$ by the Intermediate Value Theorem. Thus, there is a root of the equation $\tan x - 2x = 0$, or $\tan x = 2x$, in the interval $(0, 1.4)$.

41. (a) $f(x) = \cos x - x^3$ is continuous on the interval $[0, 1]$, $f(0) = 1 > 0$, and $f(1) = \cos 1 - 1 \approx -0.46 < 0$. Since $1 > 0 > -0.46$, there is a number c in $(0, 1)$ such that $f(c) = 0$ by the Intermediate Value Theorem. Thus, there is a root of the equation $\cos x - x^3 = 0$, or $\cos x = x^3$, in the interval $(0, 1)$.

(b) $f(0.86) \approx 0.016 > 0$ and $f(0.87) \approx -0.014 < 0$, so there is a root between 0.86 and 0.87, that is, in the interval $(0.86, 0.87)$.

42. (a) $f(x) = x^5 - x^2 + 2x + 3$ is continuous on $[-1, 0]$, $f(-1) = -1 < 0$, and $f(0) = 3 > 0$. Since $-1 < 0 < 3$, there is a number c in $(-1, 0)$ such that $f(c) = 0$ by the Intermediate Value Theorem. Thus, there is a root of the equation $x^5 - x^2 + 2x + 3 = 0$ in the interval $(-1, 0)$.

(b) $f(-0.88) \approx -0.062 < 0$ and $f(-0.87) \approx 0.0047 > 0$, so there is a root between -0.88 and -0.87.

43. (a) Let $f(x) = x^5 - x^2 - 4$. Then $f(1) = 1^5 - 1^2 - 4 = -4 < 0$ and $f(2) = 2^5 - 2^2 - 4 = 24 > 0$. So by the Intermediate Value Theorem, there is a number c in $(1, 2)$ such that $f(c) = c^5 - c^2 - 4 = 0$.

(b) We can see from the graphs that, correct to three decimal places, the root is $x \approx 1.434$.

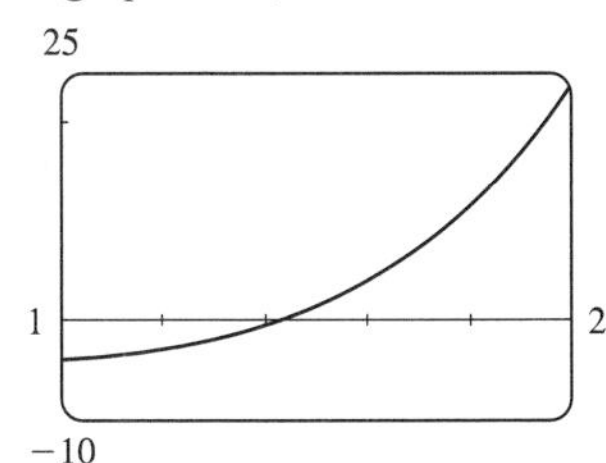

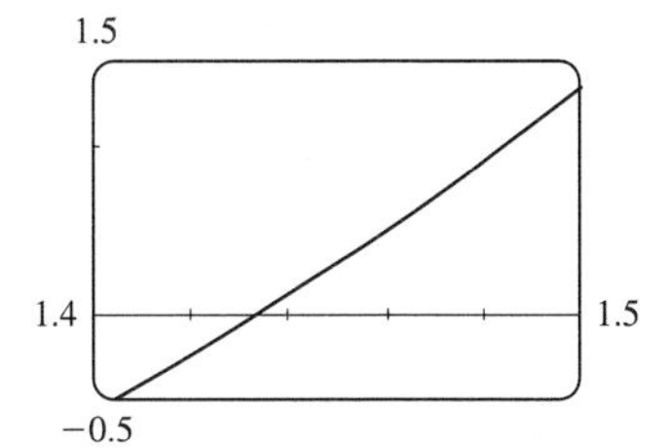

44. (a) Let $f(x) = \sqrt{x-5} - \dfrac{1}{x+3}$. Then $f(5) = -\frac{1}{8} < 0$ and $f(6) = \frac{8}{9} > 0$, and f is continuous on $[5, \infty)$. So by the Intermediate Value Theorem, there is a number c in $(5, 6)$ such that $f(c) = 0$. This implies that $\dfrac{1}{c+3} = \sqrt{c-5}$.

(b) Using the intersect feature of the graphing device, we find that the root of the equation is $x = 5.016$, correct to three decimal places.

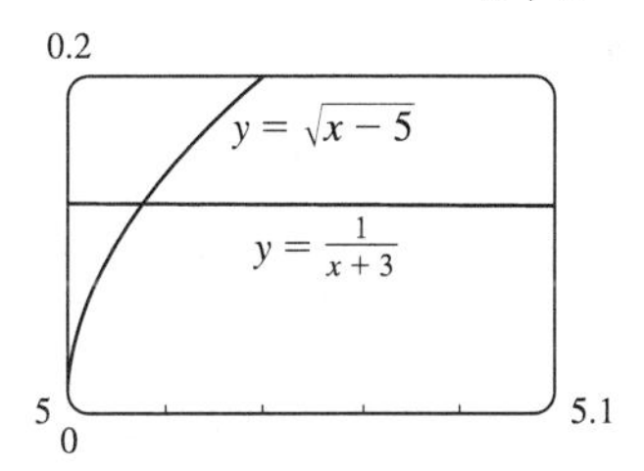

45. If there is such a number, it satisfies the equation $x^3 + 1 = x \iff x^3 - x + 1 = 0$. Let the left-hand side of this equation be called $f(x)$. Now $f(-2) = -5 < 0$, and $f(-1) = 1 > 0$. Note also that $f(x)$ is a polynomial, and thus continuous. So by the Intermediate Value Theorem, there is a number c between -2 and -1 such that $f(c) = 0$, so that $c = c^3 + 1$.

46. (a) $\lim\limits_{x\to 0^+} F(x) = 0$ and $\lim\limits_{x\to 0^-} F(x) = 0$, so $\lim\limits_{x\to 0} F(x) = 0$, which is $F(0)$, and hence F is continuous at $x = a$ if $a = 0$. For $a > 0$, $\lim\limits_{x\to a} F(x) = \lim\limits_{x\to a} x = a = F(a)$. For $a < 0$, $\lim\limits_{x\to a} F(x) = \lim\limits_{x\to a}(-x) = -a = F(a)$. Thus, F is continuous at $x = a$; that is, continuous everywhere.

(b) Assume that f is continuous on the interval I. Then for $a \in I$, $\lim\limits_{x\to a} |f(x)| = \left|\lim\limits_{x\to a} f(x)\right| = |f(a)|$ by Theorem 8. (If a is an endpoint of I, use the appropriate one-sided limit.) So $|f|$ is continuous on I.

(c) No, the converse is false. For example, the function $f(x) = \begin{cases} 1 & \text{if } x \geq 0 \\ -1 & \text{if } x < 0 \end{cases}$ is not continuous at $x = 0$, but $|f(x)| = 1$ is continuous on $\mathbb{R}$.

47. Define $u(t)$ to be the monk's distance from the monastery, as a function of time, on the first day, and define $d(t)$ to be his distance from the monastery, as a function of time, on the second day. Let D be the distance from the monastery to the top of the mountain. From the given information we know that $u(0) = 0$, $u(12) = D$, $d(0) = D$ and $d(12) = 0$. Now consider the function $u - d$, which is clearly continuous. We calculate that $(u - d)(0) = -D$ and $(u - d)(12) = D$. So by the Intermediate Value Theorem, there must be some time t_0 between 0 and 12 such that $(u - d)(t_0) = 0 \iff u(t_0) = d(t_0)$. So at time t_0 after 7:00 AM, the monk will be at the same place on both days.

1.6 Limits Involving Infinity

1. (a) $\lim_{x\to 2} f(x) = \infty$ (b) $\lim_{x\to -1^-} f(x) = \infty$ (c) $\lim_{x\to -1^+} f(x) = -\infty$

(d) $\lim_{x\to \infty} f(x) = 1$ (e) $\lim_{x\to -\infty} f(x) = 2$ (f) Vertical: $x = -1$, $x = 2$; Horizontal: $y = 1$, $y = 2$

2. (a) $\lim_{x\to \infty} g(x) = 2$ (b) $\lim_{x\to -\infty} g(x) = -2$ (c) $\lim_{x\to 3} g(x) = \infty$

(d) $\lim_{x\to 0} g(x) = -\infty$ (e) $\lim_{x\to -2^+} g(x) = -\infty$ (f) Vertical: $x = -2$, $x = 0$, $x = 3$; Horizontal: $y = -2$, $y = 2$

3. $f(0) = 0$, $f(1) = 1$,

$\lim_{x\to \infty} f(x) = 0$,

f is odd

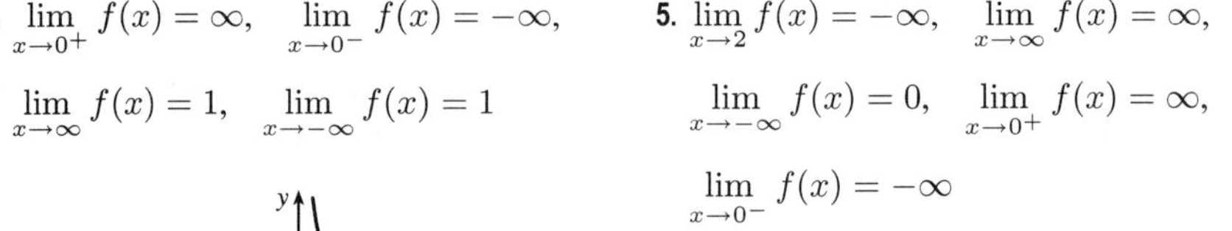

4. $\lim_{x\to 0^+} f(x) = \infty$, $\lim_{x\to 0^-} f(x) = -\infty$,

$\lim_{x\to \infty} f(x) = 1$, $\lim_{x\to -\infty} f(x) = 1$

5. $\lim_{x\to 2} f(x) = -\infty$, $\lim_{x\to \infty} f(x) = \infty$,

$\lim_{x\to -\infty} f(x) = 0$, $\lim_{x\to 0^+} f(x) = \infty$,

$\lim_{x\to 0^-} f(x) = -\infty$

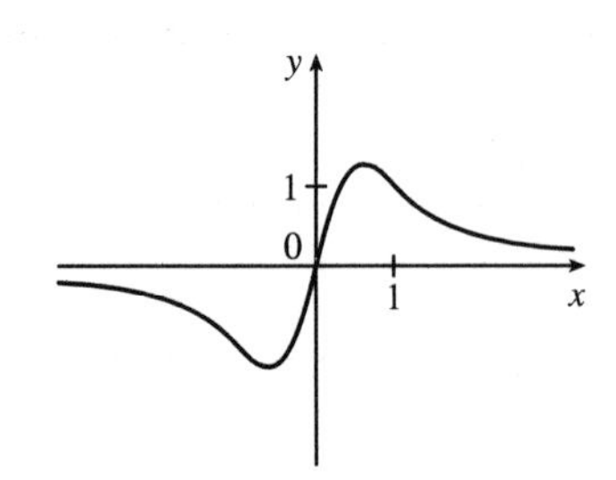

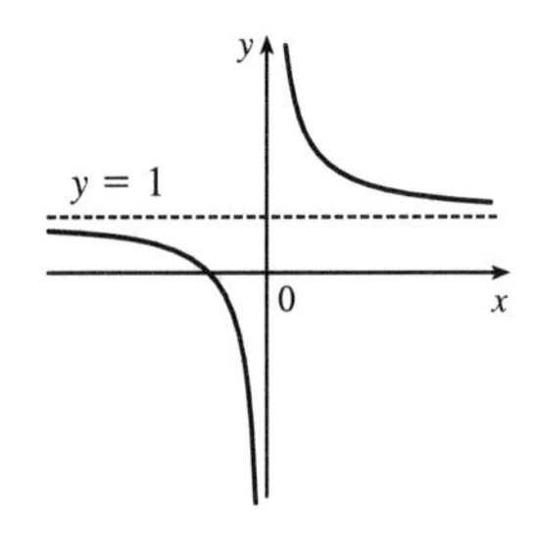

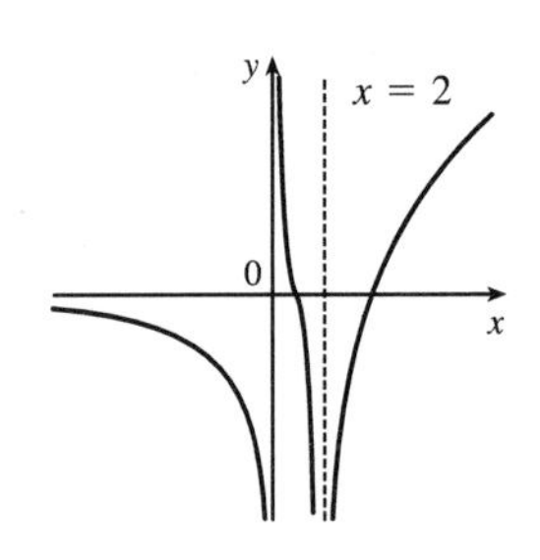

6. $\lim_{x\to -2} f(x) = \infty$,

$\lim_{x\to -\infty} f(x) = 3$,

$\lim_{x\to \infty} f(x) = -3$

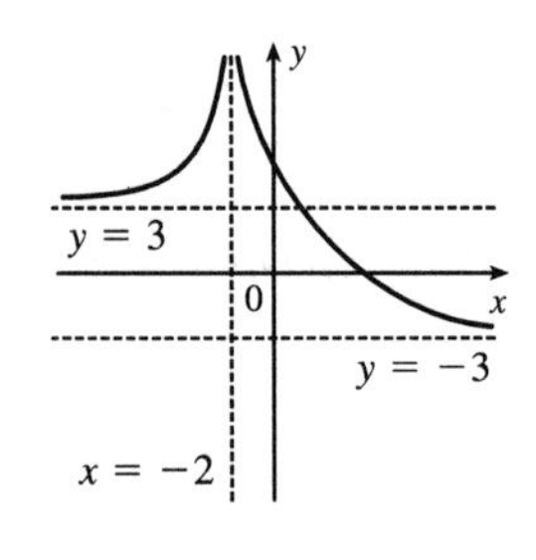

7. $f(0) = 3$, $\lim_{x\to 0^-} f(x) = 4$,

$\lim_{x\to 0^+} f(x) = 2$,

$\lim_{x\to -\infty} f(x) = -\infty$, $\lim_{x\to 4^-} f(x) = -\infty$,

$\lim_{x\to 4^+} f(x) = \infty$, $\lim_{x\to \infty} f(x) = 3$

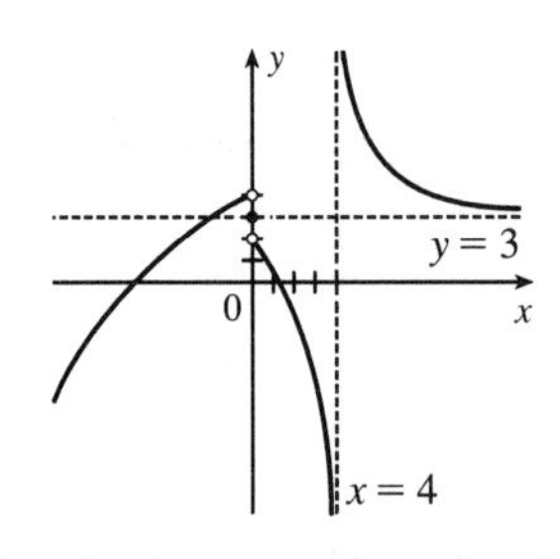

8. $\lim_{x\to 3} f(x) = -\infty$, $\lim_{x\to \infty} f(x) = 2$,

$f(0) = 0$, f is even

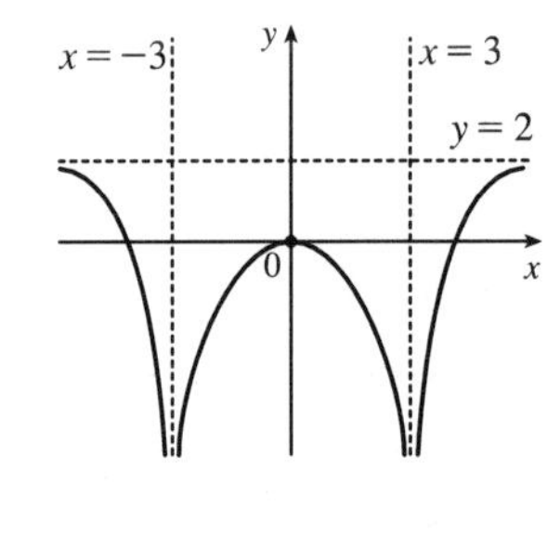

9. If $f(x) = x^2/2^x$, then a calculator gives $f(0) = 0$, $f(1) = 0.5$, $f(2) = 1$, $f(3) = 1.125$, $f(4) = 1$, $f(5) = 0.78125$, $f(6) = 0.5625$, $f(7) = 0.3828125$, $f(8) = 0.25$, $f(9) = 0.158203125$, $f(10) = 0.09765625$, $f(20) \approx 0.00038147$, $f(50) \approx 2.2204 \times 10^{-12}$, $f(100) \approx 7.8886 \times 10^{-27}$. It appears that $\lim_{x\to \infty} (x^2/2^x) = 0$.

10. (a) $f(x) = \dfrac{1}{x^3 - 1}$

x	$f(x)$
0.5	−1.14
0.9	−3.69
0.99	−33.7
0.999	−333.7
0.9999	−3333.7
0.99999	−33,333.7

x	$f(x)$
1.5	0.42
1.1	3.02
1.01	33.0
1.001	333.0
1.0001	3333.0
1.00001	33,333.3

From these calculations, it seems that $\lim\limits_{x \to 1^-} f(x) = -\infty$ and $\lim\limits_{x \to 1^+} f(x) = \infty$.

(b) If x is slightly smaller than 1, then $x^3 - 1$ will be a negative number close to 0, and the reciprocal of $x^3 - 1$, that is, $f(x)$, will be a negative number with large absolute value. So $\lim\limits_{x \to 1^-} f(x) = -\infty$.

If x is slightly larger than 1, then $x^3 - 1$ will be a small positive number, and its reciprocal, $f(x)$, will be a large positive number. So $\lim\limits_{x \to 1^+} f(x) = \infty$.

(c) It appears from the graph of f that $\lim\limits_{x \to 1^-} f(x) = -\infty$ and $\lim\limits_{x \to 1^+} f(x) = \infty$.

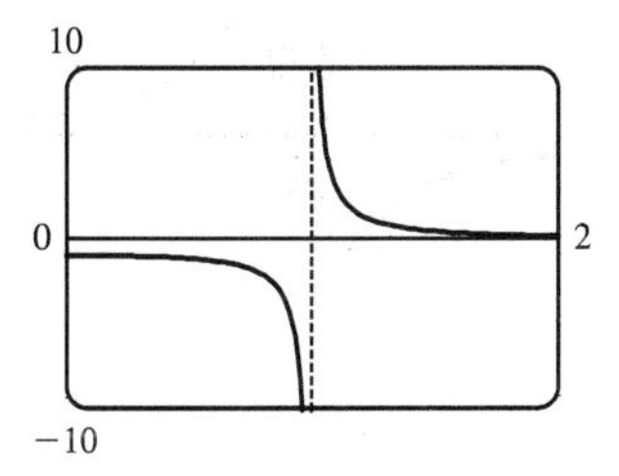

11. Vertical: $x \approx -1.62$, $x \approx 0.62$, $x = 1$;
Horizontal: $y = 1$

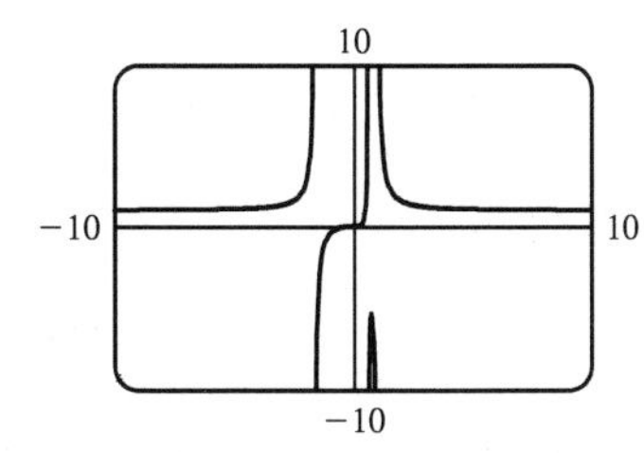

12. (a) From a graph of $f(x) = (1 - 2/x)^x$ in a window of $[0, 10{,}000]$ by $[0, 0.2]$, we estimate that $\lim\limits_{x \to \infty} f(x) = 0.14$ (to two decimal places.)

(b)

x	$f(x)$
10,000	0.135308
100,000	0.135333
1,000,000	0.135335

From the table, we estimate that $\lim\limits_{x \to \infty} f(x) = 0.1353$ (to four decimal places.)

13. $\lim\limits_{x \to -3^+} \dfrac{x + 2}{x + 3} = -\infty$ since the numerator is negative and the denominator approaches 0 from the positive side as $x \to -3^+$.

14. $\lim\limits_{x \to 5^-} \dfrac{6}{x - 5} = -\infty$ since $(x - 5) \to 0$ as $x \to 5^-$ and $\dfrac{6}{x - 5} < 0$ for $x < 5$.

15. $\lim\limits_{x \to 1} \dfrac{2 - x}{(x - 1)^2} = \infty$ since the numerator is positive and the denominator approaches 0 through positive values as $x \to 1$.

16. $\lim\limits_{x \to \pi^-} \cot x = \lim\limits_{x \to \pi^-} \dfrac{\cos x}{\sin x} = -\infty$ since the numerator is negative and the denominator approaches 0 through positive values as $x \to \pi^-$.

17. $\lim\limits_{x\to(-\pi/2)^-} \sec x = \lim\limits_{x\to(-\pi/2)^-} (1/\cos x) = -\infty$ since $\cos x \to 0$ as $x \to (-\pi/2)^-$ and $\cos x < 0$ for $-\pi < x < -\pi/2$.

18. $\lim\limits_{x\to\infty} \dfrac{3x+5}{x-4} = \lim\limits_{x\to\infty} \dfrac{(3x+5)/x}{(x-4)/x} = \lim\limits_{x\to\infty} \dfrac{3+5/x}{1-4/x} = \dfrac{\lim\limits_{x\to\infty} 3 + 5 \lim\limits_{x\to\infty} \frac{1}{x}}{\lim\limits_{x\to\infty} 1 - 4 \lim\limits_{x\to\infty} \frac{1}{x}} = \dfrac{3+5(0)}{1-4(0)} = 3$

19. Divide both the numerator and denominator by x^3 (the highest power of x that occurs in the denominator).

$$\lim_{x\to\infty} \frac{x^3+5x}{2x^3-x^2+4} = \lim_{x\to\infty} \frac{\frac{x^3+5x}{x^3}}{\frac{2x^3-x^2+4}{x^3}} = \lim_{x\to\infty} \frac{1+\frac{5}{x^2}}{2-\frac{1}{x}+\frac{4}{x^3}} = \frac{\lim\limits_{x\to\infty}\left(1+\frac{5}{x^2}\right)}{\lim\limits_{x\to\infty}\left(2-\frac{1}{x}+\frac{4}{x^3}\right)}$$

$$= \frac{\lim\limits_{x\to\infty} 1 + 5\lim\limits_{x\to\infty} \frac{1}{x^2}}{\lim\limits_{x\to\infty} 2 - \lim\limits_{x\to\infty} \frac{1}{x} + 4 \lim\limits_{x\to\infty} \frac{1}{x^3}} = \frac{1+5(0)}{2-0+4(0)} = \frac{1}{2}$$

20. $\lim\limits_{t\to-\infty} \dfrac{t^2+2}{t^3+t^2-1} = \lim\limits_{t\to-\infty} \dfrac{(t^2+2)/t^3}{(t^3+t^2-1)/t^3} = \lim\limits_{t\to-\infty} \dfrac{1/t+2/t^3}{1+1/t-1/t^3} = \dfrac{0+0}{1+0-0} = 0$

21. First, multiply the factors in the denominator. Then divide both the numerator and denominator by u^4.

$$\lim_{u\to\infty} \frac{4u^4+5}{(u^2-2)(2u^2-1)} = \lim_{u\to\infty} \frac{4u^4+5}{2u^4-5u^2+2} = \lim_{u\to\infty} \frac{\frac{4u^4+5}{u^4}}{\frac{2u^4-5u^2+2}{u^4}} = \lim_{u\to\infty} \frac{4+\frac{5}{u^4}}{2-\frac{5}{u^2}+\frac{2}{u^4}}$$

$$= \frac{\lim\limits_{u\to\infty}\left(4+\frac{5}{u^4}\right)}{\lim\limits_{u\to\infty}\left(2-\frac{5}{u^2}+\frac{2}{u^4}\right)} = \frac{\lim\limits_{u\to\infty} 4 + 5\lim\limits_{u\to\infty} \frac{1}{u^4}}{\lim\limits_{u\to\infty} 2 - 5\lim\limits_{u\to\infty} \frac{1}{u^2} + 2\lim\limits_{u\to\infty} \frac{1}{u^4}} = \frac{4+5(0)}{2-5(0)+2(0)} = \frac{4}{2} = 2$$

22. $\lim\limits_{x\to\infty} \dfrac{x+2}{\sqrt{9x^2+1}} = \lim\limits_{x\to\infty} \dfrac{(x+2)/x}{\sqrt{9x^2+1}/\sqrt{x^2}} = \lim\limits_{x\to\infty} \dfrac{1+2/x}{\sqrt{9+1/x^2}} = \dfrac{1+0}{\sqrt{9+0}} = \dfrac{1}{3}$

23. $\lim\limits_{x\to\infty} \left(\sqrt{9x^2+x}-3x\right) = \lim\limits_{x\to\infty} \dfrac{\left(\sqrt{9x^2+x}-3x\right)\left(\sqrt{9x^2+x}+3x\right)}{\sqrt{9x^2+x}+3x} = \lim\limits_{x\to\infty} \dfrac{\left(\sqrt{9x^2+x}\right)^2-(3x)^2}{\sqrt{9x^2+x}+3x}$

$$= \lim_{x\to\infty} \frac{(9x^2+x)-9x^2}{\sqrt{9x^2+x}+3x} = \lim_{x\to\infty} \frac{x}{\sqrt{9x^2+x}+3x} \cdot \frac{1/x}{1/x}$$

$$= \lim_{x\to\infty} \frac{x/x}{\sqrt{9x^2/x^2+x/x^2}+3x/x} = \lim_{x\to\infty} \frac{1}{\sqrt{9+1/x}+3} = \frac{1}{\sqrt{9}+3} = \frac{1}{3+3} = \frac{1}{6}$$

24. $\lim\limits_{x\to\infty} \left(\sqrt{x^2+ax}-\sqrt{x^2+bx}\right) = \lim\limits_{x\to\infty} \dfrac{\left(\sqrt{x^2+ax}-\sqrt{x^2+bx}\right)\left(\sqrt{x^2+ax}+\sqrt{x^2+bx}\right)}{\sqrt{x^2+ax}+\sqrt{x^2+bx}}$

$$= \lim_{x\to\infty} \frac{(x^2+ax)-(x^2+bx)}{\sqrt{x^2+ax}+\sqrt{x^2+bx}} = \lim_{x\to\infty} \frac{[(a-b)x]/x}{\left(\sqrt{x^2+ax}+\sqrt{x^2+bx}\right)/\sqrt{x^2}}$$

$$= \lim_{x\to\infty} \frac{a-b}{\sqrt{1+a/x}+\sqrt{1+b/x}} = \frac{a-b}{\sqrt{1+0}+\sqrt{1+0}} = \frac{a-b}{2}$$

25. $\lim\limits_{x\to\infty} \cos x$ does not exist because as x increases $\cos x$ does not approach any one value, but oscillates between 1 and -1.

26. Since $0 \le \sin^2 x \le 1$, we have $0 \le \dfrac{\sin^2 x}{x^2} \le \dfrac{1}{x^2}$. Now $\lim\limits_{x\to\infty} 0 = 0$ and $\lim\limits_{x\to\infty} \dfrac{1}{x^2} = 0$, so by the Squeeze Theorem,

$\lim\limits_{x\to\infty} \dfrac{\sin^2 x}{x^2} = 0$.

27. $\lim\limits_{x\to\infty} (x - \sqrt{x}\,) = \lim\limits_{x\to\infty} \sqrt{x}\,(\sqrt{x} - 1) = \infty$ since $\sqrt{x} \to \infty$ and $\sqrt{x} - 1 \to \infty$ as $x \to \infty$.

28. $\lim\limits_{x\to\infty} \dfrac{x^3 - 2x + 3}{5 - 2x^2} = \lim\limits_{x\to\infty} \dfrac{(x^3 - 2x + 3)/x^2}{(5 - 2x^2)/x^2}$ [divide by the highest power of x in the denominator]

$= \lim\limits_{x\to\infty} \dfrac{x - 2/x + 3/x^2}{5/x^2 - 2} = -\infty$ because $x - 2/x + 3/x^2 \to \infty$ and $5/x^2 - 2 \to -2$ as $x \to \infty$.

29. $\lim\limits_{x\to-\infty} (x^4 + x^5) = \lim\limits_{x\to-\infty} x^5(\frac{1}{x} + 1)$ [factor out the largest power of x] $= -\infty$ because $x^5 \to -\infty$ and $1/x + 1 \to 1$ as $x \to -\infty$.

30. $\lim\limits_{x\to\infty} (x^2 - x^4) = \lim\limits_{x\to\infty} x^2(1 - x^2) = -\infty$ since $x^2 \to \infty$ and $1 - x^2 \to -\infty$.

31. $\lim\limits_{x\to\infty} \dfrac{x + x^3 + x^5}{1 - x^2 + x^4} = \lim\limits_{x\to\infty} \dfrac{(x + x^3 + x^5)/x^4}{(1 - x^2 + x^4)/x^4}$ [divide by the highest power of x in the denominator]

$= \lim\limits_{x\to\infty} \dfrac{1/x^3 + 1/x + x}{1/x^4 - 1/x^2 + 1} = \infty$

because $(1/x^3 + 1/x + x) \to \infty$ and $(1/x^4 - 1/x^2 + 1) \to 1$ as $x \to \infty$.

32. (a)

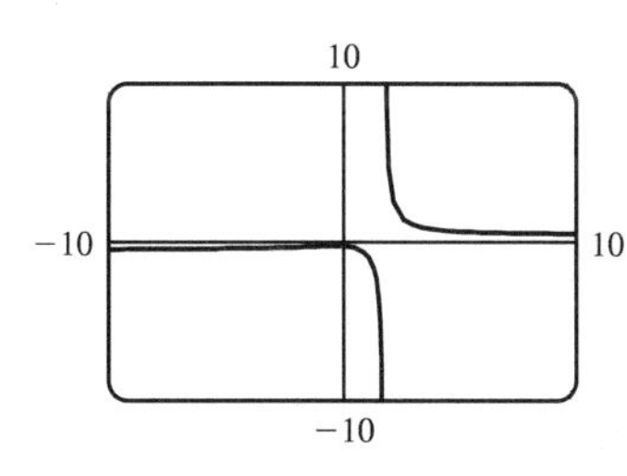

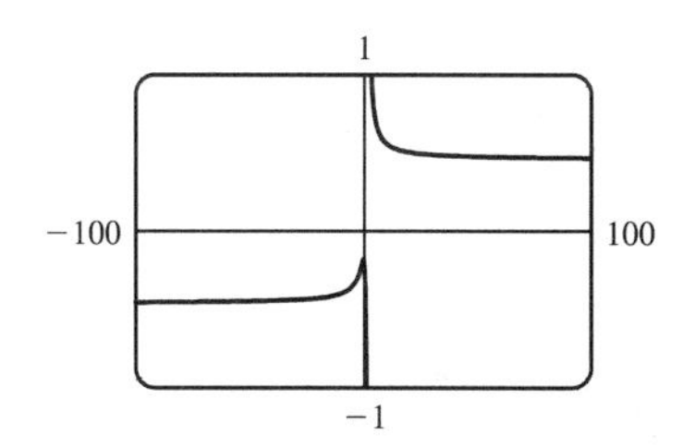

From the graph, it appears at first that there is only one horizontal asymptote, at $y \approx 0$, and a vertical asymptote at $x \approx 1.7$. However, if we graph the function with a wider viewing rectangle, we see that in fact there seem to be two horizontal asymptotes: one at $y \approx 0.5$ and one at $y \approx -0.5$. So we estimate that

$$\lim_{x\to\infty} \frac{\sqrt{2x^2+1}}{3x-5} \approx 0.5 \qquad \text{and} \qquad \lim_{x\to-\infty} \frac{\sqrt{2x^2+1}}{3x-5} \approx -0.5$$

(b) $f(1000) \approx 0.4722$ and $f(10{,}000) \approx 0.4715$, so we estimate that $\lim\limits_{x\to\infty} \dfrac{\sqrt{2x^2+1}}{3x-5} \approx 0.47$.

$f(-1000) \approx -0.4706$ and $f(-10{,}000) \approx -0.4713$, so we estimate that $\lim\limits_{x\to-\infty} \dfrac{\sqrt{2x^2+1}}{3x-5} \approx -0.47$.

(c) $\lim\limits_{x\to\infty} \dfrac{\sqrt{2x^2+1}}{3x-5} = \lim\limits_{x\to\infty} \dfrac{\sqrt{2+1/x^2}}{3-5/x}$ [since $\sqrt{x^2} = x$ for $x > 0$] $= \dfrac{\sqrt{2}}{3} \approx 0.471404$.

For $x < 0$, we have $\sqrt{x^2} = |x| = -x$, so when we divide the numerator by x, with $x < 0$, we get $\dfrac{1}{x}\sqrt{2x^2+1} = -\dfrac{1}{\sqrt{x^2}}\sqrt{2x^2+1} = -\sqrt{2+1/x^2}$. Therefore,

$$\lim_{x\to-\infty} \frac{\sqrt{2x^2+1}}{3x-5} = \lim_{x\to-\infty} \frac{-\sqrt{2+1/x^2}}{3-5/x} = -\frac{\sqrt{2}}{3} \approx -0.471405.$$

33. $\displaystyle \lim_{x\to\infty} \frac{2x^2+x-1}{x^2+x-2} = \lim_{x\to\infty} \frac{\dfrac{2x^2+x-1}{x^2}}{\dfrac{x^2+x-2}{x^2}} = \lim_{x\to\infty} \frac{2+\dfrac{1}{x}-\dfrac{1}{x^2}}{1+\dfrac{1}{x}-\dfrac{2}{x^2}} = \frac{\lim\limits_{x\to\infty}\left(2+\dfrac{1}{x}-\dfrac{1}{x^2}\right)}{\lim\limits_{x\to\infty}\left(1+\dfrac{1}{x}-\dfrac{2}{x^2}\right)}$

$\displaystyle = \frac{\lim\limits_{x\to\infty} 2 + \lim\limits_{x\to\infty}\dfrac{1}{x} - \lim\limits_{x\to\infty}\dfrac{1}{x^2}}{\lim\limits_{x\to\infty} 1 + \lim\limits_{x\to\infty}\dfrac{1}{x} - 2\lim\limits_{x\to\infty}\dfrac{1}{x^2}} = \frac{2+0-0}{1+0-2(0)} = 2$, so $y = 2$ is a horizontal asymptote.

$\displaystyle y = f(x) = \frac{2x^2+x-1}{x^2+x-2} = \frac{(2x-1)(x+1)}{(x+2)(x-1)}$, so

$\displaystyle \lim_{x\to-2^-} f(x) = \infty$, $\displaystyle \lim_{x\to-2^+} f(x) = -\infty$, $\displaystyle \lim_{x\to1^-} f(x) = -\infty$, and

$\displaystyle \lim_{x\to1^+} f(x) = \infty$. Thus, $x = -2$ and $x = 1$ are vertical asymptotes.

The graph confirms our work.

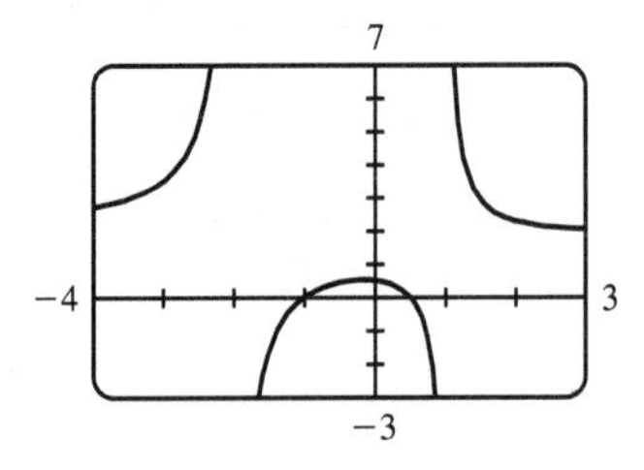

34. $\displaystyle \lim_{x\to\infty} \frac{x-9}{\sqrt{4x^2+3x+2}} = \lim_{x\to\infty} \frac{1-9/x}{\sqrt{4+(3/x)+(2/x^2)}} = \frac{1-0}{\sqrt{4+0+0}} = \frac{1}{2}.$

Using the fact that $\sqrt{x^2} = |x| = -x$ for $x < 0$, we divide the numerator by $-x$ and the denominator by $\sqrt{x^2}$.

Thus, $\displaystyle \lim_{x\to-\infty} \frac{x-9}{\sqrt{4x^2+3x+2}} = \lim_{x\to-\infty} \frac{-1+9/x}{\sqrt{4+(3/x)+(2/x^2)}} = \frac{-1+0}{\sqrt{4+0+0}} = -\frac{1}{2}.$

The horizontal asymptotes are $y = \pm\frac{1}{2}$. The polynomial $4x^2+3x+2$ is positive for all x, so the denominator never approaches zero, and thus there is no vertical asymptote.

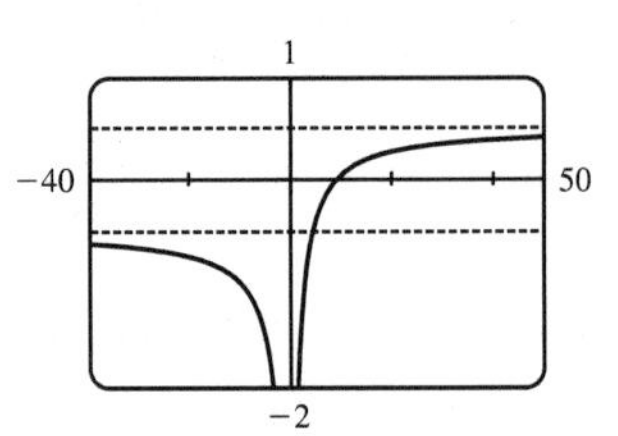

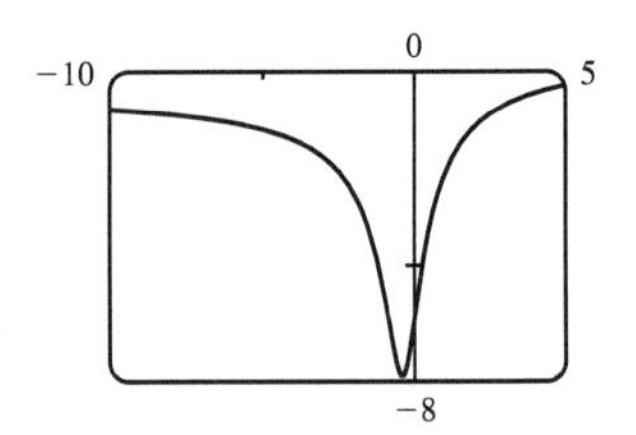

35. (a)

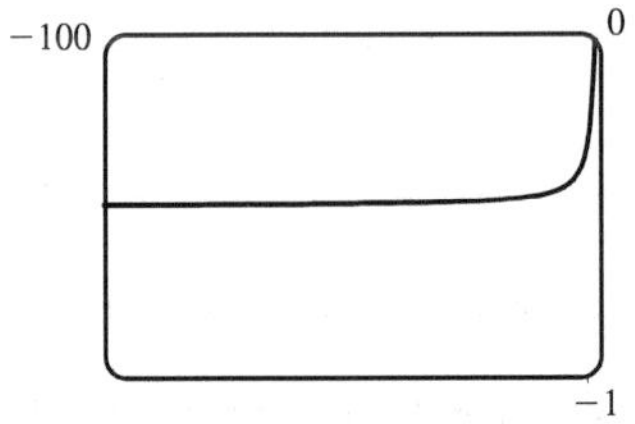

From the graph of $f(x) = \sqrt{x^2+x+1}+x$, we estimate the value of $\displaystyle \lim_{x\to-\infty} f(x)$ to be -0.5.

(b)

x	$f(x)$
$-10{,}000$	-0.4999625
$-100{,}000$	-0.4999962
$-1{,}000{,}000$	-0.4999996

From the table, we estimate the limit to be -0.5.

(c) $\displaystyle \lim_{x\to-\infty}\left(\sqrt{x^2+x+1}+x\right) = \lim_{x\to-\infty}\left(\sqrt{x^2+x+1}+x\right)\left[\frac{\sqrt{x^2+x+1}-x}{\sqrt{x^2+x+1}-x}\right] = \lim_{x\to-\infty}\frac{(x^2+x+1)-x^2}{\sqrt{x^2+x+1}-x}$

$\displaystyle = \lim_{x\to-\infty}\frac{(x+1)(1/x)}{\left(\sqrt{x^2+x+1}-x\right)(1/x)} = \lim_{x\to-\infty}\frac{1+(1/x)}{-\sqrt{1+(1/x)+(1/x^2)}-1}$

$\displaystyle = \frac{1+0}{-\sqrt{1+0+0}-1} = -\frac{1}{2}$

Note that for $x < 0$, we have $\sqrt{x^2} = |x| = -x$, so when we divide the radical by x, with $x < 0$, we get $\displaystyle \frac{1}{x}\sqrt{x^2+x+1} = -\frac{1}{\sqrt{x^2}}\sqrt{x^2+x+1} = -\sqrt{1+(1/x)+(1/x^2)}.$

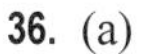

36. (a)

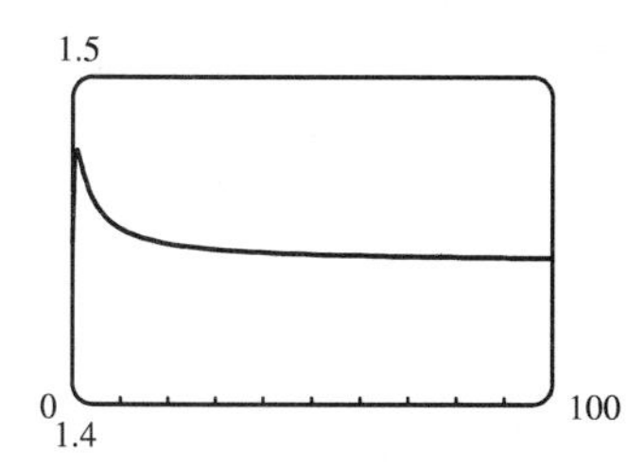

From the graph of $f(x) = \sqrt{3x^2 + 8x + 6} - \sqrt{3x^2 + 3x + 1}$, we estimate (to one decimal place) the value of $\lim_{x \to \infty} f(x)$ to be 1.4.

(b)

x	$f(x)$
10,000	1.44339
100,000	1.44338
1,000,000	1.44338

From the table, we estimate (to four decimal places) the limit to be 1.4434.

(c)
$$\lim_{x \to \infty} f(x) = \lim_{x \to \infty} \frac{\left(\sqrt{3x^2 + 8x + 6} - \sqrt{3x^2 + 3x + 1}\right)\left(\sqrt{3x^2 + 8x + 6} + \sqrt{3x^2 + 3x + 1}\right)}{\sqrt{3x^2 + 8x + 6} + \sqrt{3x^2 + 3x + 1}}$$

$$= \lim_{x \to \infty} \frac{\left(3x^2 + 8x + 6\right) - \left(3x^2 + 3x + 1\right)}{\sqrt{3x^2 + 8x + 6} + \sqrt{3x^2 + 3x + 1}} = \lim_{x \to \infty} \frac{(5x + 5)(1/x)}{\left(\sqrt{3x^2 + 8x + 6} + \sqrt{3x^2 + 3x + 1}\right)(1/x)}$$

$$= \lim_{x \to \infty} \frac{5 + 5/x}{\sqrt{3 + 8/x + 6/x^2} + \sqrt{3 + 3/x + 1/x^2}} = \frac{5}{\sqrt{3} + \sqrt{3}} = \frac{5}{2\sqrt{3}} = \frac{5\sqrt{3}}{6} \approx 1.443376$$

37. From the graph, it appears $y = 1$ is a horizontal asymptote.

$$\lim_{x \to \infty} \frac{3x^3 + 500x^2}{x^3 + 500x^2 + 100x + 2000} = \lim_{x \to \infty} \frac{\dfrac{3x^3 + 500x^2}{x^3}}{\dfrac{x^3 + 500x^2 + 100x + 2000}{x^3}}$$

$$= \lim_{x \to \infty} \frac{3 + (500/x)}{1 + (500/x) + (100/x^2) + (2000/x^3)} = \frac{3 + 0}{1 + 0 + 0 + 0} = 3,$$

so $y = 3$ is a horizontal asymptote. The discrepancy can be explained by the choice in the viewing window. Try $[-100{,}000, 100{,}000]$ by $[-1, 4]$ to get a graph that lends credibility to our calculation that $y = 3$ is a horizontal asymptote.

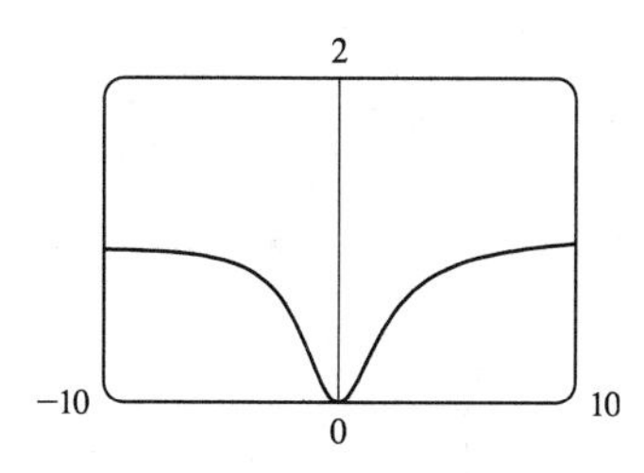

38. Since the function has vertical asymptotes $x = 1$ and $x = 3$, the denominator of the rational function we are looking for must have factors $(x - 1)$ and $(x - 3)$. Because the horizontal asymptote is $y = 1$, the degree of the numerator must equal the degree of the denominator, and the ratio of the leading coefficients must be 1. One possibility is $f(x) = \dfrac{x^2}{(x - 1)(x - 3)}$.

39. Let's look for a rational function.

(1) $\lim_{x \to \pm\infty} f(x) = 0 \quad \Rightarrow \quad$ degree of numerator < degree of denominator

(2) $\lim_{x \to 0} f(x) = -\infty \quad \Rightarrow \quad$ there is a factor of x^2 in the denominator (not just x, since that would produce a sign change at $x = 0$), and the function is negative near $x = 0$.

(3) $\lim_{x \to 3^-} f(x) = \infty$ and $\lim_{x \to 3^+} f(x) = -\infty \quad \Rightarrow \quad$ vertical asymptote at $x = 3$; there is a factor of $(x - 3)$ in the denominator.

(4) $f(2) = 0 \Rightarrow$ 2 is an x-intercept; there is at least one factor of $(x - 2)$ in the numerator.

Combining all of this information and putting in a negative sign to give us the desired left- and right-hand limits gives us $f(x) = \dfrac{2 - x}{x^2(x - 3)}$ as one possibility.

40. (a) In both viewing rectangles,

$$\lim_{x\to\infty} P(x) = \lim_{x\to\infty} Q(x) = \infty \text{ and}$$

$$\lim_{x\to-\infty} P(x) = \lim_{x\to-\infty} Q(x) = -\infty.$$

In the larger viewing rectangle, P and Q become less distinguishable.

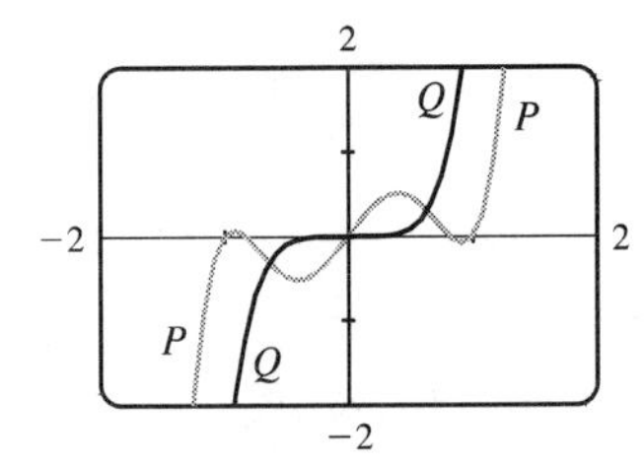

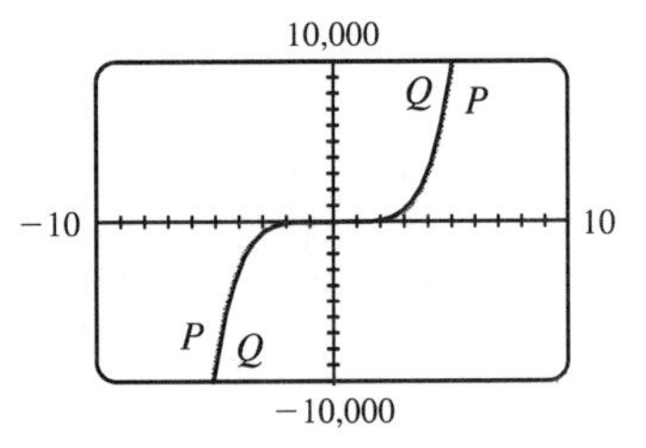

(b) $\displaystyle\lim_{x\to\infty} \frac{P(x)}{Q(x)} = \lim_{x\to\infty} \frac{3x^5 - 5x^3 + 2x}{3x^5} = \lim_{x\to\infty} \left(1 - \frac{5}{3} \cdot \frac{1}{x^2} + \frac{2}{3} \cdot \frac{1}{x^4}\right) = 1 - \tfrac{5}{3}(0) + \tfrac{2}{3}(0) = 1 \Rightarrow$ P and Q have the same end behavior.

41. Divide the numerator and the denominator by the highest power of x in $Q(x)$.

(a) If $\deg P < \deg Q$, then the numerator $\to 0$ but the denominator doesn't. So $\lim_{x\to\infty} [P(x)/Q(x)] = 0$.

(b) If $\deg P > \deg Q$, then the numerator $\to \pm\infty$ but the denominator doesn't, so $\lim_{x\to\infty} [P(x)/Q(x)] = \pm\infty$ (depending on the ratio of the leading coefficients of P and Q).

42.

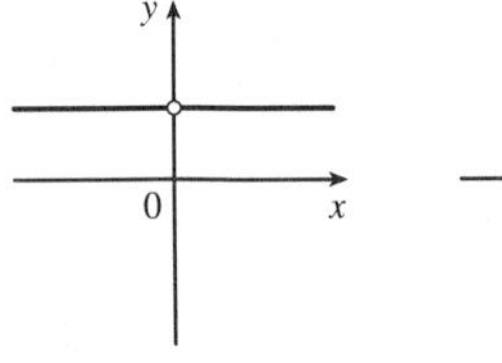

(i) $n = 0$

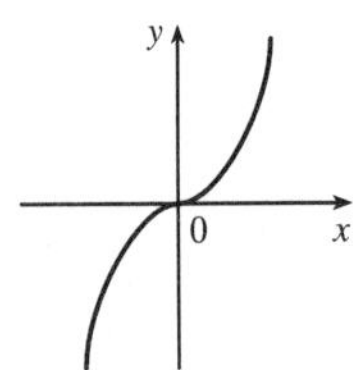

(ii) $n > 0$ (n odd)

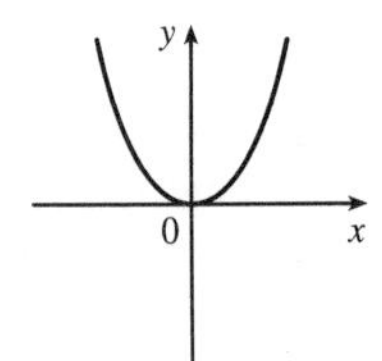

(iii) $n > 0$ (n even)

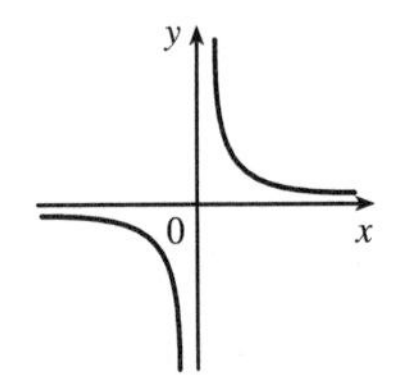

(iv) $n < 0$ (n odd)

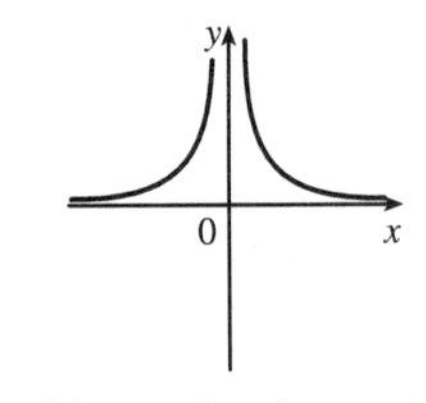

(v) $n < 0$ (n even)

From these sketches we see that

(a) $\displaystyle\lim_{x\to0^+} x^n = \begin{cases} 1 & \text{if } n = 0 \\ 0 & \text{if } n > 0 \\ \infty & \text{if } n < 0 \end{cases}$

(b) $\displaystyle\lim_{x\to0^-} x^n = \begin{cases} 1 & \text{if } n = 0 \\ 0 & \text{if } n > 0 \\ -\infty & \text{if } n < 0,\ n \text{ odd} \\ \infty & \text{if } n < 0,\ n \text{ even} \end{cases}$

(c) $\displaystyle\lim_{x\to\infty} x^n = \begin{cases} 1 & \text{if } n = 0 \\ \infty & \text{if } n > 0 \\ 0 & \text{if } n < 0 \end{cases}$

(d) $\displaystyle\lim_{x\to-\infty} x^n = \begin{cases} 1 & \text{if } n = 0 \\ -\infty & \text{if } n > 0,\ n \text{ odd} \\ \infty & \text{if } n > 0,\ n \text{ even} \\ 0 & \text{if } n < 0 \end{cases}$

43. $\displaystyle\lim_{x\to\infty} \frac{4x - 1}{x} = \lim_{x\to\infty} \left(4 - \frac{1}{x}\right) = 4$ and $\displaystyle\lim_{x\to\infty} \frac{4x^2 + 3x}{x^2} = \lim_{x\to\infty} \left(4 + \frac{3}{x}\right) = 4$. Therefore, by the Squeeze Theorem, $\displaystyle\lim_{x\to\infty} f(x) = 4$.

44. $\displaystyle\lim_{v\to c^-} m = \lim_{v\to c^-} \frac{m_0}{\sqrt{1 - v^2/c^2}}$. As $v \to c^-$, $\sqrt{1 - v^2/c^2} \to 0^+$, and $m \to \infty$.

45. (a) After t minutes, $25t$ liters of brine with 30 g of salt per liter has been pumped into the tank, so it contains $(5000 + 25t)$ liters of water and $25t \cdot 30 = 750t$ grams of salt. Therefore, the salt concentration at time t will be

$$C(t) = \frac{750t}{5000 + 25t} = \frac{30t}{200 + t}\ \frac{\text{g}}{\text{L}}.$$

(b) $\lim\limits_{t\to\infty} C(t) = \lim\limits_{t\to\infty} \dfrac{30t}{200 + t} = \lim\limits_{t\to\infty} \dfrac{30t/t}{200/t + t/t} = \dfrac{30}{0 + 1} = 30$. So the salt concentration approaches that of the brine being pumped into the tank.

46. (a) $\lim\limits_{x\to\infty} f(x) = \lim\limits_{x\to\infty} \dfrac{4x^2 - 5x}{2x^2 + 1} = \lim\limits_{x\to\infty} \dfrac{4 - 5/x}{2 + 1/x^2} = \dfrac{4}{2} = 2$

(b) $f(x) = 1.9 \;\Rightarrow\; x \approx 25.3744$, so $f(x) > 1.9$ when $x > N = 25.4$.

$f(x) = 1.99 \;\Rightarrow\; x \approx 250.3974$, so $f(x) > 1.99$ when $x > N = 250.4$.

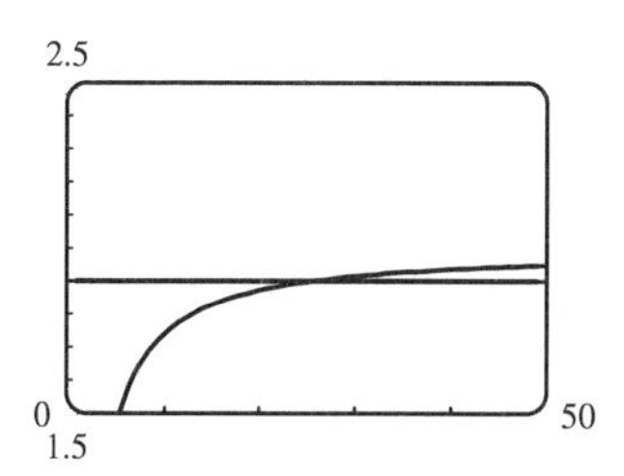

47. $\dfrac{1}{(x+3)^4} > 10{,}000 \;\Leftrightarrow\; (x+3)^4 < \dfrac{1}{10{,}000} \;\Leftrightarrow\; |x+3| < \dfrac{1}{\sqrt[4]{10{,}000}} \;\Leftrightarrow\; |x - (-3)| < \dfrac{1}{10}$

48. Given $M > 0$, we need $\delta > 0$ such that $0 < |x+3| < \delta \;\Rightarrow\; 1/(x+3)^4 > M$. Now $\dfrac{1}{(x+3)^4} > M \;\Leftrightarrow$ $(x+3)^4 < \dfrac{1}{M} \;\Leftrightarrow\; |x+3| < \dfrac{1}{\sqrt[4]{M}}$. So take $\delta = \dfrac{1}{\sqrt[4]{M}}$. Then $0 < |x+3| < \delta = \dfrac{1}{\sqrt[4]{M}} \;\Rightarrow\; \dfrac{1}{(x+3)^4} > M$, so $\lim\limits_{x\to-3} \dfrac{1}{(x+3)^4} = \infty$.

49. Let $N < 0$ be given. Then, for $x < -1$, we have $\dfrac{5}{(x+1)^3} < N \;\Leftrightarrow\; \dfrac{5}{N} < (x+1)^3 \;\Leftrightarrow\; \sqrt[3]{\dfrac{5}{N}} < x + 1$. Let $\delta = -\sqrt[3]{\dfrac{5}{N}}$. Then $-1 - \delta < x < -1 \;\Rightarrow\; \sqrt[3]{\dfrac{5}{N}} < x + 1 < 0 \;\Rightarrow\; \dfrac{5}{(x+1)^3} < N$, so $\lim\limits_{x\to-1^-} \dfrac{5}{(x+1)^3} = -\infty$.

50. For $\varepsilon = 0.5$, we must find N such that whenever $x \geq N$, we have $\left| \dfrac{\sqrt{4x^2+1}}{x+1} - 2 \right| < 0.5 \;\Leftrightarrow\; 1.5 < \dfrac{\sqrt{4x^2+1}}{x+1} < 2.5$. We graph the three parts of this inequality on the same screen, and find that it holds whenever $x \geq 3$. So we choose $N = 3$ (or any larger number). For $\varepsilon = 0.1$, we must have $1.9 < \dfrac{\sqrt{4x^2+1}}{x+1} < 2.1$, and the graphs show that this holds whenever $x \geq 19$. So we choose $N = 19$ (or any larger number).

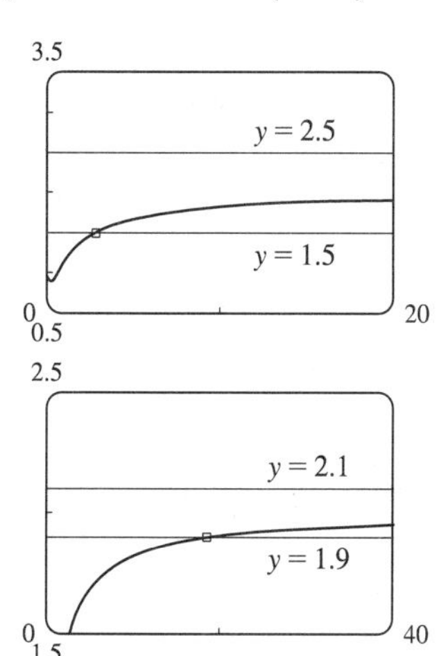

51. $\left|\dfrac{6x^2+5x-3}{2x^2-1}-3\right|<0.2 \quad\Leftrightarrow\quad 2.8<\dfrac{6x^2+5x-3}{2x^2-1}<3.2$. So we graph the three parts of this inequality on the same screen, and find that the curve $y=\dfrac{6x^2+5x-3}{2x^2-1}$ seems to lie between the lines $y=2.8$ and $y=3.2$ whenever $x>12.8$. So we can choose $N=13$ (or any larger number) so that the inequality holds whenever $x\geq N$.

52. We need N such that $\dfrac{2x+1}{\sqrt{x+1}}>100$ whenever $x\geq N$. From the graph, we see that this inequality holds for $x\geq 2500$. So we choose $N=2500$ (or any larger number).

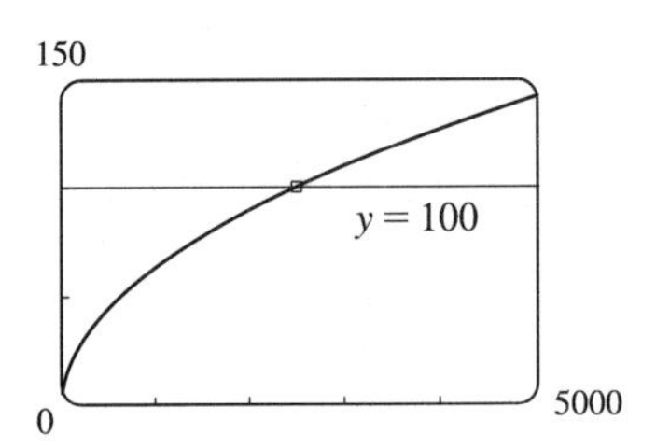

53. (a) $1/x^2<0.0001 \quad\Leftrightarrow\quad x^2>1/0.0001=10{,}000 \quad\Leftrightarrow\quad x>100 \quad (x>0)$

(b) If $\varepsilon>0$ is given, then $1/x^2<\varepsilon \quad\Leftrightarrow\quad x^2>1/\varepsilon \quad\Leftrightarrow\quad x>1/\sqrt{\varepsilon}$. Let $N=1/\sqrt{\varepsilon}$.

Then $x>N \quad\Rightarrow\quad x>\dfrac{1}{\sqrt{\varepsilon}} \quad\Rightarrow\quad \left|\dfrac{1}{x^2}-0\right|=\dfrac{1}{x^2}<\varepsilon$, so $\displaystyle\lim_{x\to\infty}\frac{1}{x^2}=0$.

54. Given $M>0$, we need $N>0$ such that $x>N \quad\Rightarrow\quad x^3>M$. Now $x^3>M \quad\Leftrightarrow\quad x>\sqrt[3]{M}$, so take $N=\sqrt[3]{M}$. Then $x>N=\sqrt[3]{M} \quad\Rightarrow\quad x^3>M$, so $\displaystyle\lim_{x\to\infty}x^3=\infty$.

55. Suppose that $\displaystyle\lim_{x\to\infty}f(x)=L$. Then for every $\varepsilon>0$ there is a corresponding positive number N such that $|f(x)-L|<\varepsilon$ whenever $x>N$. If $t=1/x$, then $x>N \quad\Leftrightarrow\quad 0<1/x<1/N \quad\Leftrightarrow\quad 0<t<1/N$. Thus, for every $\varepsilon>0$ there is a corresponding $\delta>0$ (namely $1/N$) such that $|f(1/t)-L|<\varepsilon$ whenever $0<t<\delta$. This proves that $\displaystyle\lim_{t\to0^+}f(1/t)=L=\lim_{x\to\infty}f(x)$.

Now suppose that $\displaystyle\lim_{x\to-\infty}f(x)=L$. Then for every $\varepsilon>0$ there is a corresponding negative number N such that $|f(x)-L|<\varepsilon$ whenever $x<N$. If $t=1/x$, then $x<N \quad\Leftrightarrow\quad 1/N<1/x<0 \quad\Leftrightarrow\quad 1/N<t<0$. Thus, for every $\varepsilon>0$ there is a corresponding $\delta>0$ (namely $-1/N$) such that $|f(1/t)-L|<\varepsilon$ whenever $-\delta<t<0$. This proves that $\displaystyle\lim_{t\to0^-}f(1/t)=L=\lim_{x\to-\infty}f(x)$.

1 Review

CONCEPT CHECK

1. (a) A **function** f is a rule that assigns to each element x in a set A exactly one element, called $f(x)$, in a set B. The set A is called the **domain** of the function. The **range** of f is the set of all possible values of $f(x)$ as x varies throughout the domain.

(b) If f is a function with domain A, then its **graph** is the set of ordered pairs $\{(x,f(x))\mid x\in A\}$.

(c) Use the Vertical Line Test on page 4.

2. The four ways to represent a function are: verbally, numerically, visually, and algebraically. An example of each is given below.

Verbally: An assignment of students to chairs in a classroom (a description in words)

Numerically: A tax table that assigns an amount of tax to an income (a table of values)

Visually: A graphical history of the Dow Jones average (a graph)

Algebraically: A relationship between distance, rate, and time: $d = rt$ (an explicit formula)

3. (a) An **even function** f satisfies $f(-x) = f(x)$ for every number x in its domain. It is symmetric with respect to the y-axis.

(b) An **odd function** g satisfies $g(-x) = -g(x)$ for every number x in its domain. It is symmetric with respect to the origin.

4. A **mathematical model** is a mathematical description (often by means of a function or an equation) of a real-world phenomenon.

5. (a) Linear function: $f(x) = 2x + 1$, $f(x) = ax + b$

(b) Power function: $f(x) = x^2$, $f(x) = x^a$

(c) Exponential function: $f(x) = 2^x$, $f(x) = a^x$

(d) Quadratic function: $f(x) = x^2 + x + 1$, $f(x) = ax^2 + bx + c$

(e) Polynomial of degree 5: $f(x) = x^5 + 2$

(f) Rational function: $f(x) = \dfrac{x}{x+2}$, $f(x) = \dfrac{P(x)}{Q(x)}$ where $P(x)$ and $Q(x)$ are polynomials

6.

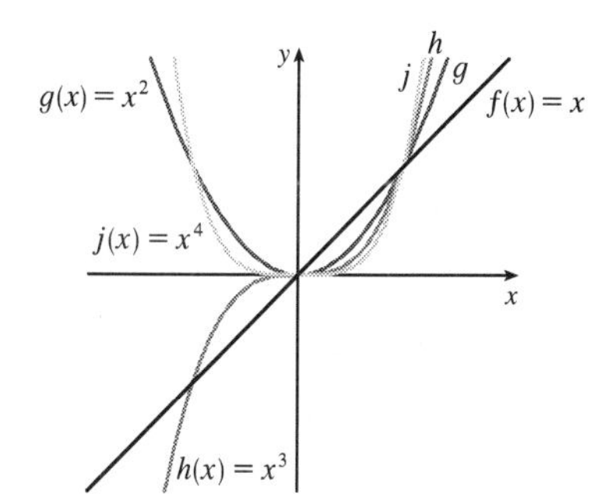

7. (a)

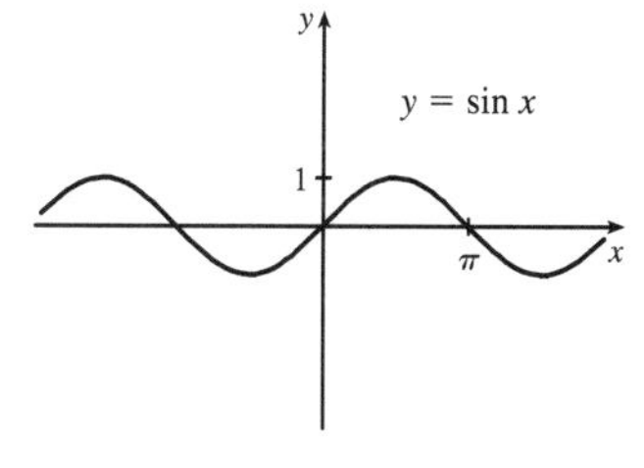

(b)

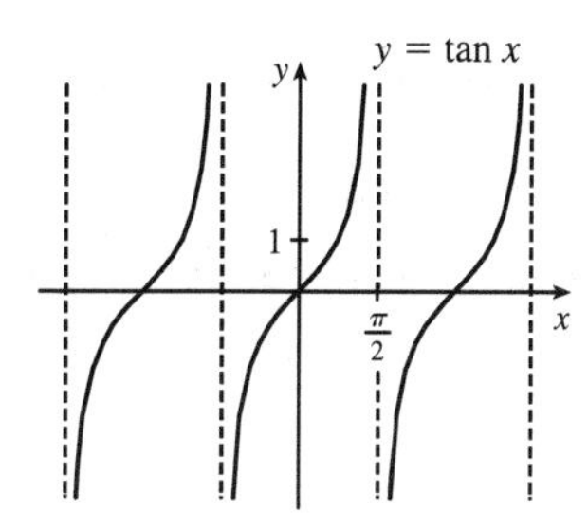

(c)

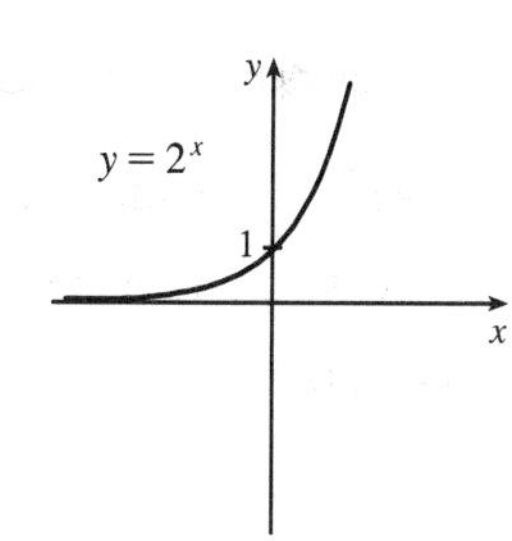

(d)

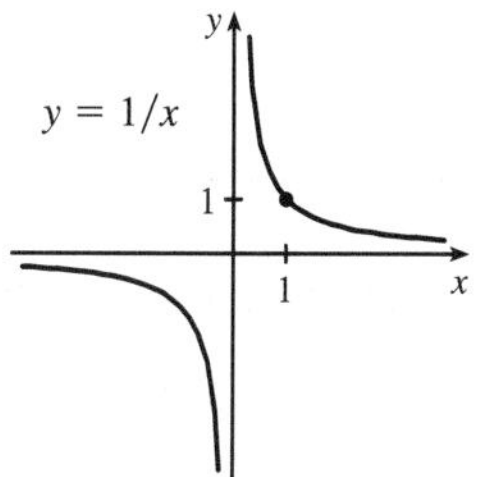

(e)

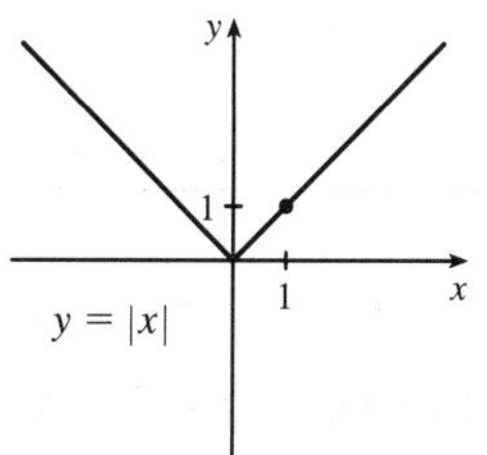

(f)

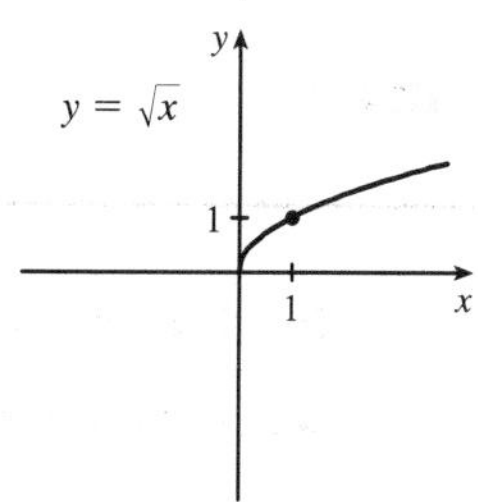

8. (a) The domain of $f + g$ is the intersection of the domain of f and the domain of g; that is, $A \cap B$.

(b) The domain of fg is also $A \cap B$.

(c) The domain of f/g must exclude values of x that make g equal to 0; that is, $\{x \in A \cap B \mid g(x) \neq 0\}$.

9. Given two functions f and g, the **composite function** $f \circ g$ is defined by $(f \circ g)(x) = f(g(x))$. The domain of $f \circ g$ is the set of all x in the domain of g such that $g(x)$ is in the domain of f.

10. (a) If the graph of f is shifted 2 units upward, its equation becomes $y = f(x) + 2$.

(b) If the graph of f is shifted 2 units downward, its equation becomes $y = f(x) - 2$.

(c) If the graph of f is shifted 2 units to the right, its equation becomes $y = f(x - 2)$.

(d) If the graph of f is shifted 2 units to the left, its equation becomes $y = f(x + 2)$.

(e) If the graph of f is reflected about the x-axis, its equation becomes $y = -f(x)$.

(f) If the graph of f is reflected about the y-axis, its equation becomes $y = f(-x)$.

(g) If the graph of f is stretched vertically by a factor of 2, its equation becomes $y = 2f(x)$.

(h) If the graph of f is shrunk vertically by a factor of 2, its equation becomes $y = \frac{1}{2}f(x)$.

(i) If the graph of f is stretched horizontally by a factor of 2, its equation becomes $y = f(\frac{1}{2})x$.

(j) If the graph of f is shrunk horizontally by a factor of 2, its equation becomes $y = f(2x)$.

11. (a) $\lim_{x \to a} f(x) = L$: See Definition 1.3.1 and Figures 1 and 2 in Section 1.3.

(b) $\lim_{x \to a^+} f(x) = L$: See the paragraph after Definition 1.3.2 and Figure 9(b) in Section 1.3.

(c) $\lim_{x \to a^-} f(x) = L$: See Definition 1.3.2 and Figure 9(a) in Section 1.3.

(d) $\lim_{x \to a} f(x) = \infty$: See Definition 1.6.1 and Figure 2 in Section 1.6.

(e) $\lim_{x \to \infty} f(x) = L$: See Definition 1.6.3 and Figure 8 in Section 1.6.

12. In general, the limit of a function fails to exist when the function does not approach a fixed number. For each of the following functions, the limit fails to exist at $x = 2$.

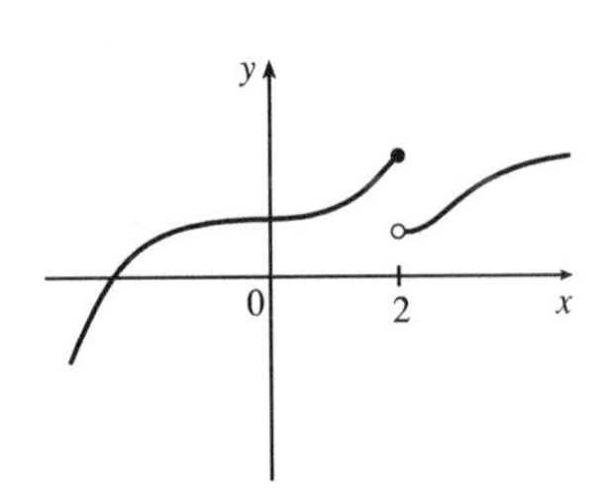

The left- and right-hand limits are not equal.

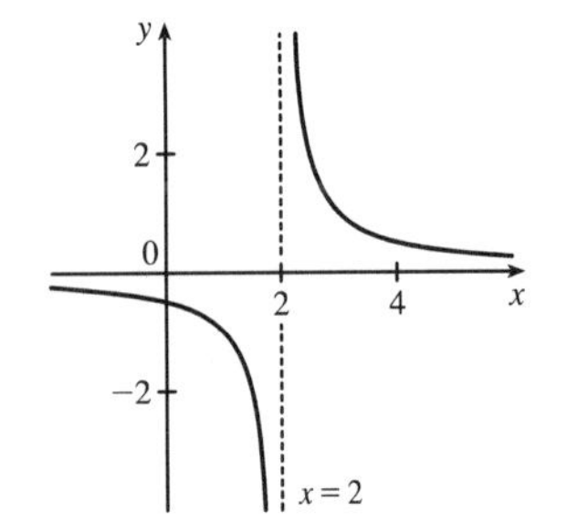

There is an infinite discontinuity.

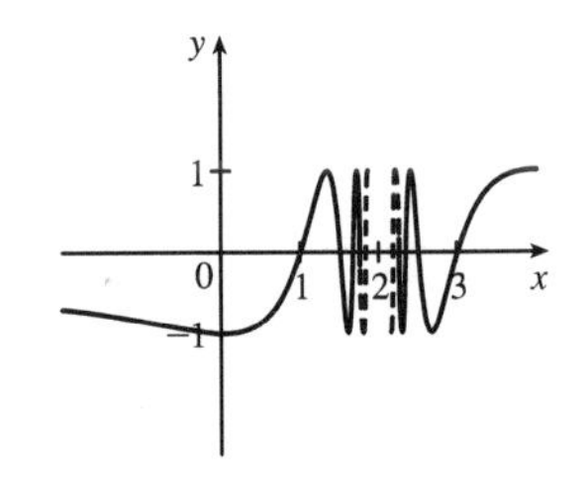

There are an infinite number of oscillations.

13. (a)–(g) See the statements of Limit Laws 1–6 and 11 in Section 1.4.

14. See Theorem 1.4.4.

15. (a) A function f is continuous at a number a if $f(x)$ approaches $f(a)$ as x approaches a; that is, $\lim_{x \to a} f(x) = f(a)$.

(b) A function f is continuous on the interval $(-\infty, \infty)$ if f is continuous at every real number a. The graph of such a function has no break and every vertical line crosses it.

16. See Theorem 1.5.9.

17. (a) See Definition 1.6.2 and Figures 2–4 in Section 1.6.

(b) See Definition 1.6.4 and Figures 8 and 9 in Section 1.6.

TRUE-FALSE QUIZ

1. False. Let $f(x) = x^2$, $s = -1$, and $t = 1$. Then $f(s+t) = (-1+1)^2 = 0^2 = 0$, but $f(s) + f(t) = (-1)^2 + 1^2 = 2 \neq 0 = f(s+t)$.

2. False. Let $f(x) = x^2$. Then $f(-2) = 4 = f(2)$, but $-2 \neq 2$.

3. True. See the Vertical Line Test.

4. False. Let $f(x) = x^2$ and $g(x) = 2x$. Then $(f \circ g)(x) = f(g(x)) = f(2x) = (2x)^2 = 4x^2$ and $(g \circ f)(x) = g(f(x)) = g(x^2) = 2x^2$. So $f \circ g \neq g \circ f$.

5. False. Limit Law 2 applies only if the individual limits exist (these don't).

6. False. Limit Law 5 cannot be applied if the limit of the denominator is 0 (it is).

7. True. Limit Law 5 applies.

8. True. The limit doesn't exist since $f(x)/g(x)$ doesn't approach any real number as x approaches 5. (The denominator approaches 0 and the numerator doesn't.)

9. False. Consider $\lim\limits_{x \to 5} \dfrac{x(x-5)}{x-5}$ or $\lim\limits_{x \to 5} \dfrac{\sin(x-5)}{x-5}$. The first limit exists and is equal to 5. By Equation 1.4.5, we know that the latter limit exists (and it is equal to 1).

10. False. Consider $\lim\limits_{x \to 6} [f(x)g(x)] = \lim\limits_{x \to 6} \left[(x-6)\dfrac{1}{x-6}\right]$. It exists (its value is 1) but $f(6) = 0$ and $g(6)$ does not exist, so $f(6)g(6) \neq 1$.

11. True. A polynomial is continuous everywhere, so $\lim\limits_{x \to b} p(x)$ exists and is equal to $p(b)$.

12. False. Consider $\lim\limits_{x \to 0} [f(x) - g(x)] = \lim\limits_{x \to 0} \left(\dfrac{1}{x^2} - \dfrac{1}{x^4}\right)$. This limit is $-\infty$ (not 0), but each of the individual functions approaches ∞.

13. True. See Figure 10 in Section 1.6.

14. False. Consider $f(x) = \sin x$ for $x \geq 0$. $\lim\limits_{x \to \infty} f(x) \neq \pm\infty$ and f has no horizontal asymptote.

15. False. Consider $f(x) = \begin{cases} 1/(x-1) & \text{if } x \neq 1 \\ 2 & \text{if } x = 1 \end{cases}$

16. False. The function f must be *continuous* in order to use the Intermediate Value Theorem. For example, let $f(x) = \begin{cases} 1 & \text{if } 0 \leq x < 3 \\ -1 & \text{if } x = 3 \end{cases}$ There is no number $c \in [0, 3]$ with $f(c) = 0$.

17. True. Use Theorem 1.5.7 with $a = 2$, $b = 5$, and $g(x) = 4x^2 - 11$. Note that $f(4) = 3$ is not needed.

18. True. Use the Intermediate Value Theorem with $a = -1$, $b = 1$, and $N = \pi$, since $3 < \pi < 4$.

19. True, by the definition of a limit with $\varepsilon = 1$.

20. False. For example, let $f(x) = \begin{cases} x^2 + 1 & \text{if } x \neq 0 \\ 2 & \text{if } x = 0 \end{cases}$

Then $f(x) > 1$ for all x, but $\lim\limits_{x \to 0} f(x) = \lim\limits_{x \to 0} (x^2 + 1) = 1$.

EXERCISES

1. (a) When $x = 2$, $y \approx 2.7$. Thus, $f(2) \approx 2.7$.

   (b) $f(x) = 3 \quad \Rightarrow \quad x \approx 2.3, 5.6$

   (c) The domain of f is $-6 \leq x \leq 6$, or $[-6, 6]$.

   (d) The range of f is $-4 \leq y \leq 4$, or $[-4, 4]$.

   (e) f is increasing on $[-4, 4]$, that is, on $-4 \leq x \leq 4$.

   (f) f is odd since its graph is symmetric about the origin.

2. (a) This curve *is not* the graph of a function of x since it *fails* the Vertical Line Test.

   (b) This curve *is* the graph of a function of x since it *passes* the Vertical Line Test. The domain is $[-3, 3]$ and the range is $[-2, 3]$.

3. $f(x) = \sqrt{4 - 3x^2}$. Domain: $4 - 3x^2 \geq 0 \quad \Rightarrow \quad 3x^2 \leq 4 \quad \Rightarrow \quad x^2 \leq \frac{4}{3} \quad \Rightarrow \quad |x| \leq \frac{2}{\sqrt{3}}$.

   Range: $y \geq 0$ and $y \leq \sqrt{4} \quad \Rightarrow \quad 0 \leq y \leq 2$.

4. $g(x) = \dfrac{1}{x+1}$. Domain: $x + 1 \neq 0 \quad \Rightarrow \quad x \neq -1$. Range: all reals except 0 ($y = 0$ is the horizontal asymptote for g.)

5. $y = 1 + \sin x$. Domain: $\mathbb{R}$. Range: $-1 \leq \sin x \leq 1 \quad \Rightarrow \quad 0 \leq 1 + \sin x \leq 2 \quad \Rightarrow \quad 0 \leq y \leq 2$.

6. $y = \tan 2x$. Domain: $2x \neq \frac{\pi}{2} + \pi n \quad \Rightarrow \quad x \neq \frac{\pi}{4} + \frac{\pi}{2} n$. Range: the tangent function takes on all real values, so the range is $\mathbb{R}$.

7. (a) To obtain the graph of $y = f(x) + 8$, we shift the graph of $y = f(x)$ up 8 units.

   (b) To obtain the graph of $y = f(x + 8)$, we shift the graph of $y = f(x)$ left 8 units.

   (c) To obtain the graph of $y = 1 + 2f(x)$, we stretch the graph of $y = f(x)$ vertically by a factor of 2, and then shift the resulting graph 1 unit upward.

   (d) To obtain the graph of $y = f(x - 2) - 2$, we shift the graph of $y = f(x)$ right 2 units (for the "-2" inside the parentheses), and then shift the resulting graph 2 units downward.

   (e) To obtain the graph of $y = -f(x)$, we reflect the graph of $y = f(x)$ about the x-axis.

   (f) To obtain the graph of $y = 3 - f(x)$, we reflect the graph of $y = f(x)$ about the x-axis, and then shift the resulting graph 3 units upward.

8. (a) To obtain the graph of $y = f(x - 8)$, we shift the graph of $y = f(x)$ right 8 units.

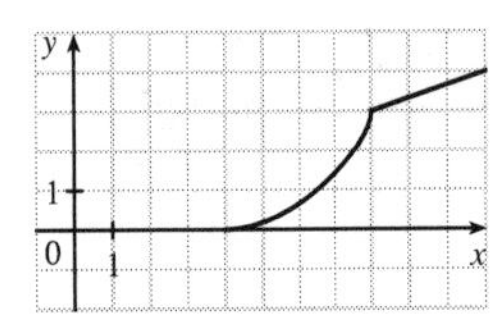

   (b) To obtain the graph of $y = -f(x)$, we reflect the graph of $y = f(x)$ about the x-axis.

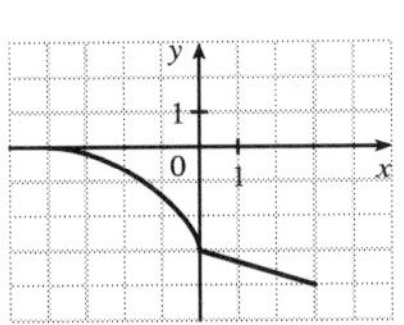

(c) To obtain the graph of $y = 2 - f(x)$, we reflect the graph of $y = f(x)$ about the x-axis, and then shift the resulting graph 2 units upward.

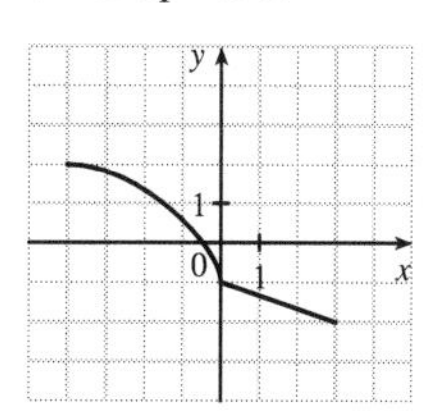

(d) To obtain the graph of $y = \frac{1}{2}f(x) - 1$, we shrink the graph of $y = f(x)$ by a factor of 2, and then shift the resulting graph 1 unit downward.

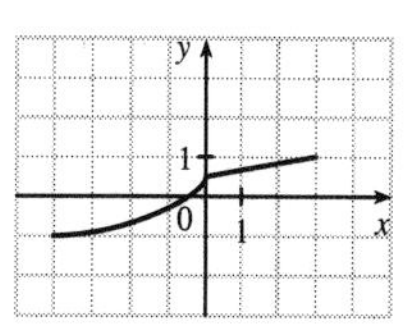

9. $y = -\sin 2x$: Start with the graph of $y = \sin x$, compress horizontally by a factor of 2, and reflect about the x-axis.

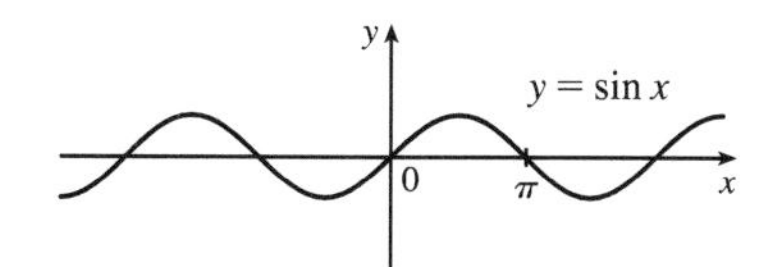

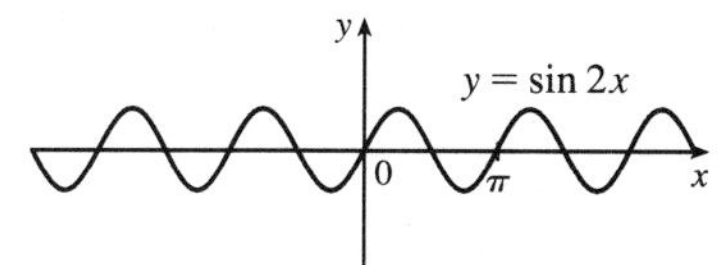

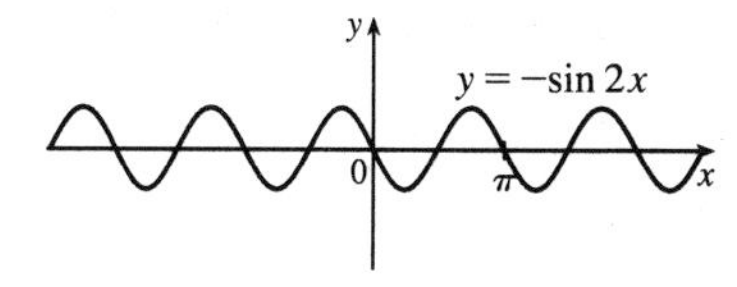

10. $y = (x - 2)^2$: Start with the graph of $y = x^2$ and shift 2 units to the right.

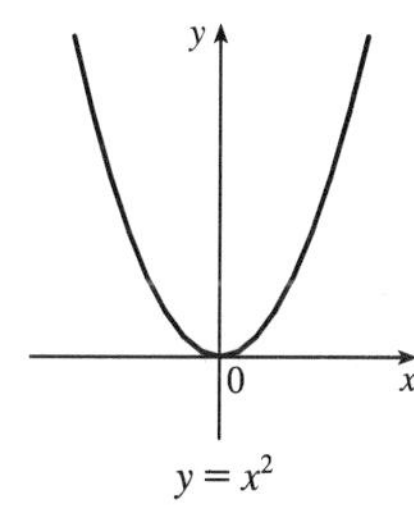

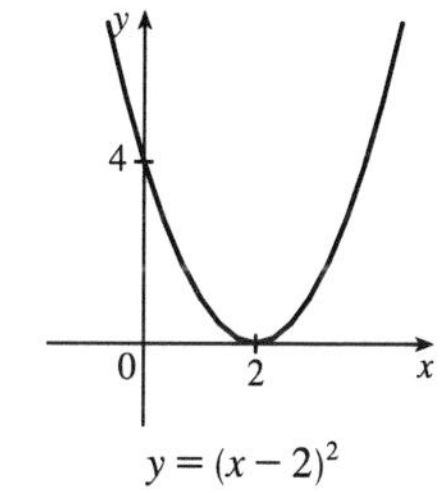

11. $y = 1 + \frac{1}{2}x^3$: Start with the graph of $y = x^3$, compress vertically by a factor of 2, and shift 1 unit upward.

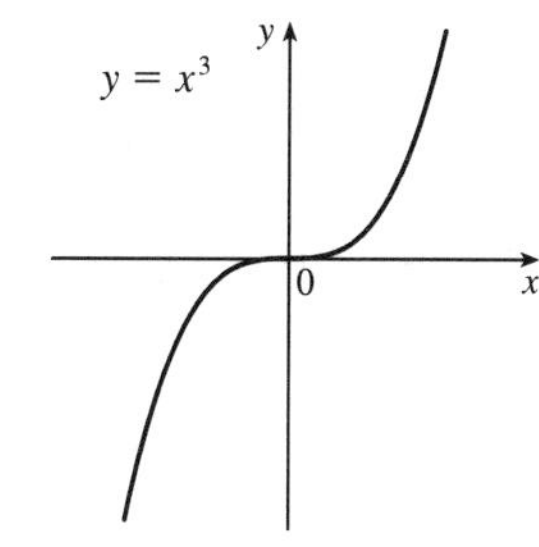

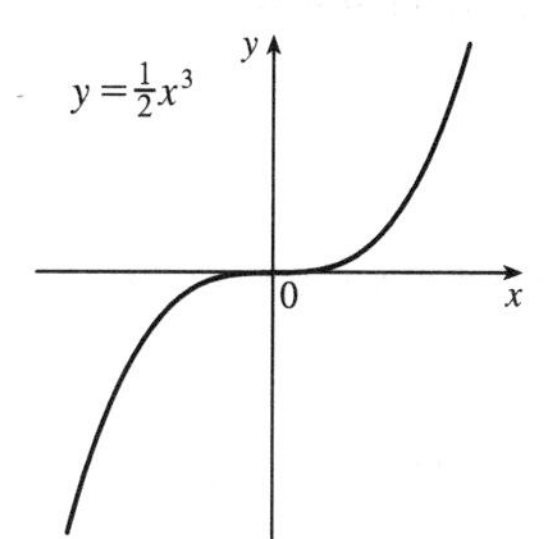

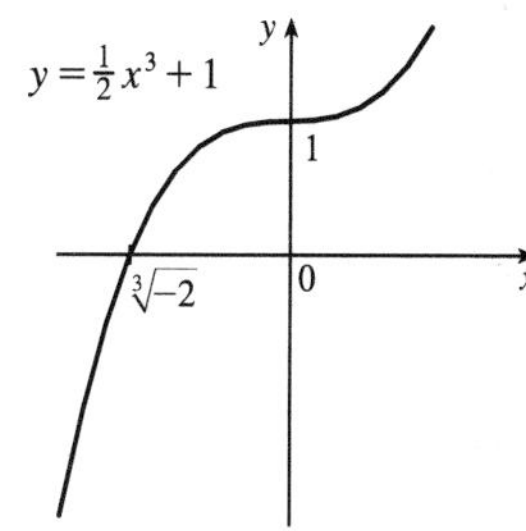

12. $y = 2 - \sqrt{x}$: Start with the graph of $y = \sqrt{x}$, reflect about the x-axis, and shift 2 units upward.

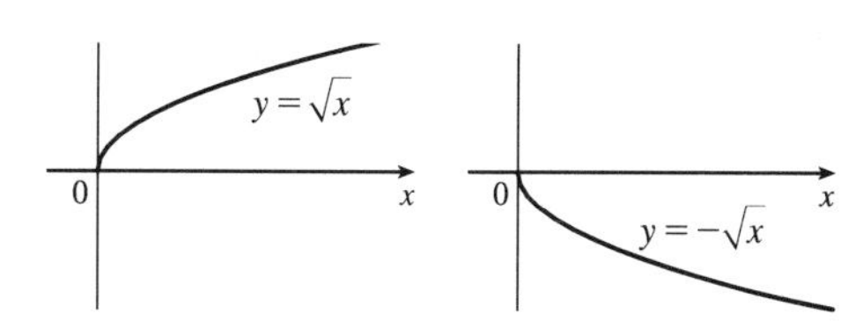

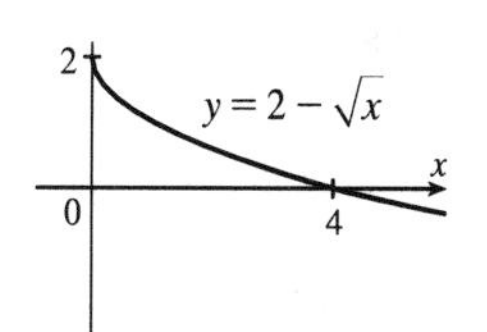

13. $f(x) = \dfrac{1}{x+2}$:

Start with the graph of $f(x) = 1/x$ and shift 2 units to the left.

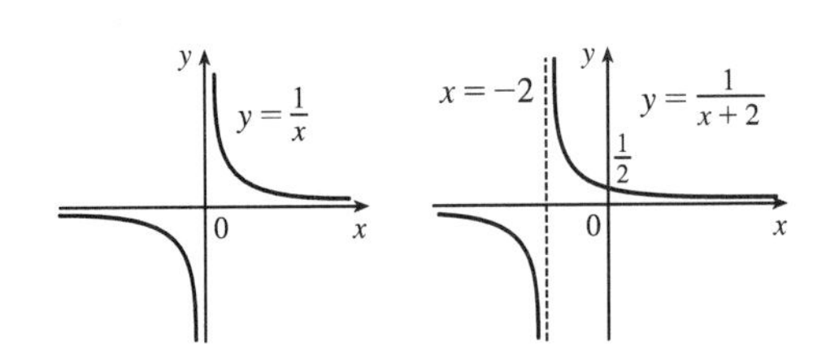

14. $f(x) = \begin{cases} 1 + x & \text{if } x < 0 \\ 1 + x^2 & \text{if } x \geq 0 \end{cases}$

On $(-\infty, 0)$, graph $y = 1 + x$ (the line with slope 1 and y-intercept 1) with open endpoint $(0, 1)$.

On $[0, \infty)$, graph $y = 1 + x^2$ (the rightmost half of the parabola $y = x^2$ shifted 1 unit upward) with closed endpoint $(0, 1)$.

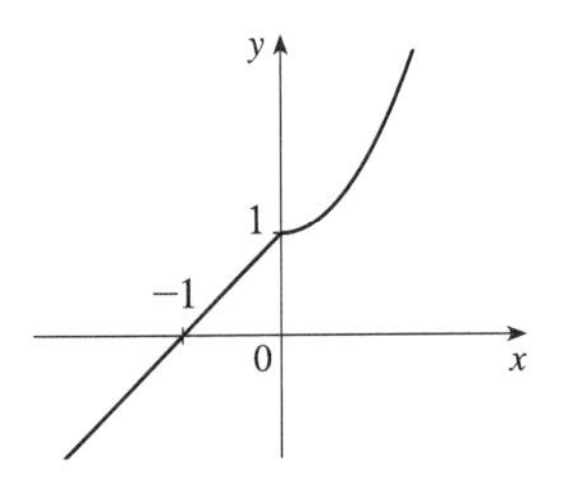

15. (a) The terms of f are a mixture of odd and even powers of x, so f is neither even nor odd.

(b) The terms of f are all odd powers of x, so f is odd.

(c) $f(-x) = \cos\left((-x)^2\right) = \cos\left(x^2\right) = f(x)$, so f is even.

(d) $f(-x) = 1 + \sin(-x) = 1 - \sin x$. Now $f(-x) \neq f(x)$ and $f(-x) \neq -f(x)$, so f is neither even nor odd.

16. For the line segment from $(-2, 2)$ to $(-1, 0)$, the slope is $\dfrac{0-2}{-1+2} = -2$, and an equation is $y - 0 = -2\,(x+1)$ or, equivalently, $y = -2x - 2$. The circle has equation $x^2 + y^2 = 1$; the top half has equation $y = \sqrt{1 - x^2}$ (we have solved for positive y.) Thus, $f(x) = \begin{cases} -2x - 2 & \text{if } -2 \leq x \leq -1 \\ \sqrt{1 - x^2} & \text{if } -1 < x \leq 1 \end{cases}$

17. $f(x) = \sqrt{x}$, $D = [0, \infty)$; $g(x) = \sin x$, $D = \mathbb{R}$.

(a) $(f \circ g)(x) = f(g(x)) = f(\sin x) = \sqrt{\sin x}$. For $\sqrt{\sin x}$ to be defined, we must have $\sin x \geq 0 \quad \Leftrightarrow \quad x \in [0, \pi]$, $[2\pi, 3\pi]$, $[-2\pi, -\pi]$, $[4\pi, 5\pi]$, $[-4\pi, -3\pi]$, $\ldots$, so $D = \{x \mid x \in [2n\pi, \pi + 2n\pi]$, where n is an integer$\}$.

(b) $(g \circ f)(x) = g(f(x)) = g(\sqrt{x}\,) = \sin\sqrt{x}$. x must be greater than or equal to 0 for $\sqrt{x}$ to be defined, so $D = [0, \infty)$.

(c) $(f \circ f)(x) = f(f(x)) = f(\sqrt{x}\,) = \sqrt{\sqrt{x}} = \sqrt[4]{x}$. $D = [0, \infty)$.

(d) $(g \circ g)(x) = g(g(x)) = g(\sin x) = \sin(\sin x)$. $D = \mathbb{R}$.

18. Let $h(x) = x + \sqrt{x}$, $g(x) = \sqrt{x}$, and $f(x) = 1/x$. Then $(f \circ g \circ h)(x) = \dfrac{1}{\sqrt{x + \sqrt{x}}} = F(x)$.

19. The graphs of $f(x) = \sin^n x$, where n is a positive integer, all have domain $\mathbb{R}$. For odd n, the range is $[-1, 1]$ and for even n, the range is $[0, 1]$. For odd n, the functions are odd and symmetric with respect to the origin. For even n, the functions are even and symmetric with respect to the y-axis. As n becomes large, the graphs become less rounded and more "spiky."

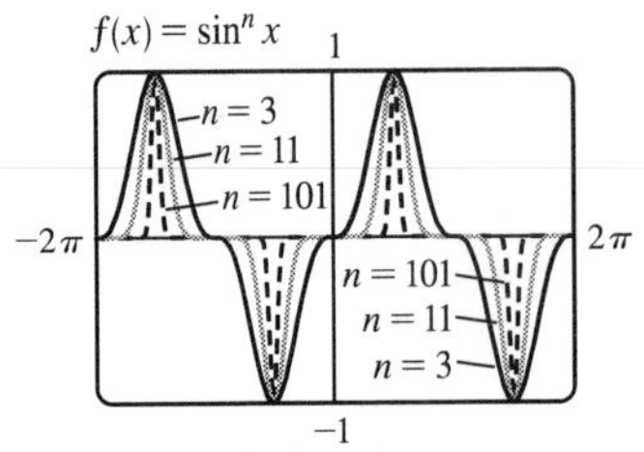

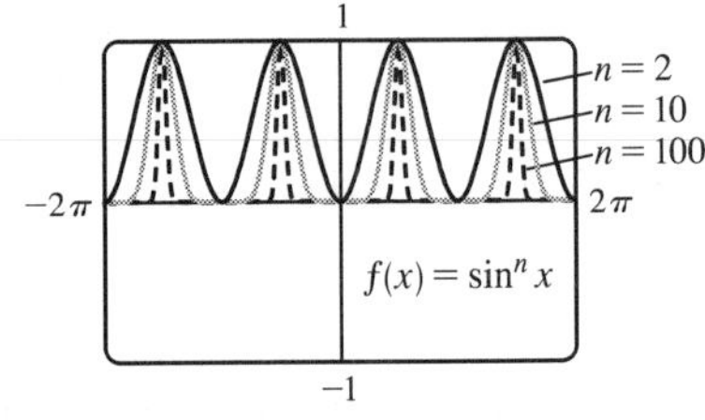

20. (a) Let x denote the number of toaster ovens produced in one week and y the associated cost. Using the points $(1000, 9000)$ and $(1500, 12{,}000)$, we get an equation of a line:

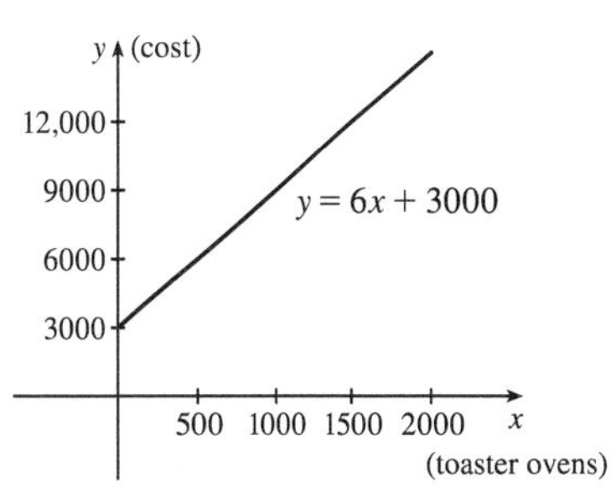

$$y - 9000 = \frac{12{,}000 - 9000}{1500 - 1000}(x - 1000) \quad \Rightarrow$$

$$y = 6(x - 1000) + 9000 \quad \Rightarrow \quad y = 6x + 3000.$$

(b) The slope of 6 means that each additional toaster oven produced adds \$6 to the weekly production cost.

(c) The y-intercept of 3000 represents the overhead cost—the cost incurred without producing anything.

21. (a) (i) $\lim\limits_{x\to 2^+} f(x) = 3$ (ii) $\lim\limits_{x\to -3^+} f(x) = 0$

(iii) $\lim\limits_{x\to -3} f(x)$ does not exist since the left and right limits are not equal. (The left limit is -2.)

(iv) $\lim\limits_{x\to 4} f(x) = 2$

(v) $\lim\limits_{x\to 0} f(x) = \infty$ (vi) $\lim\limits_{x\to 2^-} f(x) = -\infty$

(vii) $\lim\limits_{x\to \infty} f(x) = 4$ (viii) $\lim\limits_{x\to -\infty} f(x) = -1$

(b) The equations of the horizontal asymptotes are $y = -1$ and $y = 4$.

(c) The equations of the vertical asymptotes are $x = 0$ and $x = 2$.

(d) f is discontinuous at $x = -3$, 0, 2, and 4. The discontinuities are jump, infinite, infinite, and removable, respectively.

22. $\lim\limits_{x\to -\infty} f(x) = -2$, $\lim\limits_{x\to \infty} f(x) = 0$, $\lim\limits_{x\to -3} f(x) = \infty$,

$\lim\limits_{x\to 3^-} f(x) = -\infty$, $\lim\limits_{x\to 3^+} f(x) = 2$,

f is continuous from the right at 3

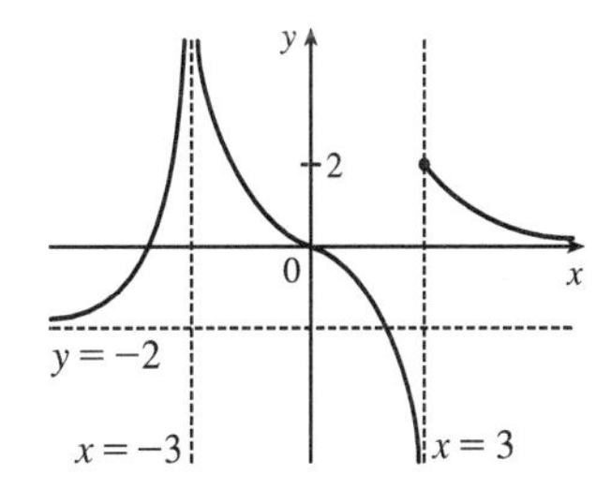

23. $\lim\limits_{x\to 0} \cos(x + \sin x) = \cos\left[\lim\limits_{x\to 0}(x + \sin x)\right]$ [by Theorem 1.5.7] $= \cos 0 = 1$

24. Since rational functions are continuous, $\lim\limits_{x\to 3} \dfrac{x^2 - 9}{x^2 + 2x - 3} = \dfrac{3^2 - 9}{3^2 + 2(3) - 3} = \dfrac{0}{12} = 0.$

25. $$\lim_{x\to -3} \frac{x^2 - 9}{x^2 + 2x - 3} = \lim_{x\to -3} \frac{(x + 3)(x - 3)}{(x + 3)(x - 1)} = \lim_{x\to -3} \frac{x - 3}{x - 1} = \frac{-3 - 3}{-3 - 1} = \frac{-6}{-4} = \frac{3}{2}$$

26. $\lim\limits_{x\to 1^+} \dfrac{x^2 - 9}{x^2 + 2x - 3} = -\infty$ since $x^2 + 2x - 3 \to 0$ as $x \to 1^+$ and $\dfrac{x^2 - 9}{x^2 + 2x - 3} < 0$ for $1 < x < 3$.

27. $$\lim_{h\to 0} \frac{(h - 1)^3 + 1}{h} = \lim_{h\to 0} \frac{(h^3 - 3h^2 + 3h - 1) + 1}{h} = \lim_{h\to 0} \frac{h^3 - 3h^2 + 3h}{h} = \lim_{h\to 0}(h^2 - 3h + 3) = 3$$

Another solution: Factor the numerator as a sum of two cubes and then simplify.

$$\begin{aligned}\lim_{h\to 0} \frac{(h - 1)^3 + 1}{h} &= \lim_{h\to 0} \frac{(h - 1)^3 + 1^3}{h} = \lim_{h\to 0} \frac{[(h - 1) + 1]\left[(h - 1)^2 - 1(h - 1) + 1^2\right]}{h} \\ &= \lim_{h\to 0}\left[(h - 1)^2 - h + 2\right] = 1 - 0 + 2 = 3\end{aligned}$$

28. $\lim\limits_{t\to 2}\dfrac{t^2-4}{t^3-8}=\lim\limits_{t\to 2}\dfrac{(t+2)(t-2)}{(t-2)(t^2+2t+4)}=\lim\limits_{t\to 2}\dfrac{t+2}{t^2+2t+4}=\dfrac{2+2}{4+4+4}=\dfrac{4}{12}=\dfrac{1}{3}$

29. $\lim\limits_{r\to 9}\dfrac{\sqrt{r}}{(r-9)^4}=\infty$ since $(r-9)^4\to 0$ as $r\to 9$ and $\dfrac{\sqrt{r}}{(r-9)^4}>0$ for $r\neq 9$.

30. $\lim\limits_{v\to 4^+}\dfrac{4-v}{|4-v|}=\lim\limits_{v\to 4^+}\dfrac{4-v}{-(4-v)}=\lim\limits_{v\to 4^+}\dfrac{1}{-1}=-1$

31. $\lim\limits_{s\to 16}\dfrac{4-\sqrt{s}}{s-16}=\lim\limits_{s\to 16}\dfrac{4-\sqrt{s}}{(\sqrt{s}+4)(\sqrt{s}-4)}=\lim\limits_{s\to 16}\dfrac{-1}{\sqrt{s}+4}=\dfrac{-1}{\sqrt{16}+4}=-\dfrac{1}{8}$

32. $\lim\limits_{v\to 2}\dfrac{v^2+2v-8}{v^4-16}=\lim\limits_{v\to 2}\dfrac{(v+4)(v-2)}{(v+2)(v-2)(v^2+4)}=\lim\limits_{v\to 2}\dfrac{v+4}{(v+2)(v^2+4)}=\dfrac{2+4}{(2+2)(2^2+4)}=\dfrac{3}{16}$

33. $\lim\limits_{x\to\infty}\dfrac{1+2x-x^2}{1-x+2x^2}=\lim\limits_{x\to\infty}\dfrac{(1+2x-x^2)/x^2}{(1-x+2x^2)/x^2}=\lim\limits_{x\to\infty}\dfrac{1/x^2+2/x-1}{1/x^2-1/x+2}=\dfrac{0+0-1}{0-0+2}=-\dfrac{1}{2}$

34. $\lim\limits_{x\to-\infty}\dfrac{1-2x^2-x^4}{5+x-3x^4}=\lim\limits_{x\to-\infty}\dfrac{(1-2x^2-x^4)/x^4}{(5+x-3x^4)/x^4}=\lim\limits_{x\to-\infty}\dfrac{1/x^4-2/x^2-1}{5/x^4+1/x^3-3}=\dfrac{0-0-1}{0+0-3}=\dfrac{-1}{-3}=\dfrac{1}{3}$

35. $\lim\limits_{x\to\infty}\left(\sqrt{x^2+4x+1}-x\right)=\lim\limits_{x\to\infty}\left[\dfrac{\sqrt{x^2+4x+1}-x}{1}\cdot\dfrac{\sqrt{x^2+4x+1}+x}{\sqrt{x^2+4x+1}+x}\right]=\lim\limits_{x\to\infty}\dfrac{(x^2+4x+1)-x^2}{\sqrt{x^2+4x+1}+x}$

$=\lim\limits_{x\to\infty}\dfrac{(4x+1)/x}{\left(\sqrt{x^2+4x+1}+x\right)/x}$ $\quad$ [divide by $x=\sqrt{x^2}$ for $x>0$]

$=\lim\limits_{x\to\infty}\dfrac{4+1/x}{\sqrt{1+4/x+1/x^2}+1}=\dfrac{4+0}{\sqrt{1+0+0}+1}=\dfrac{4}{2}=2$

36. $\lim\limits_{x\to 1}\left(\dfrac{1}{x-1}+\dfrac{1}{x^2-3x+2}\right)=\lim\limits_{x\to 1}\left[\dfrac{1}{x-1}+\dfrac{1}{(x-1)(x-2)}\right]=\lim\limits_{x\to 1}\left[\dfrac{x-2}{(x-1)(x-2)}+\dfrac{1}{(x-1)(x-2)}\right]$

$=\lim\limits_{x\to 1}\left[\dfrac{x-1}{(x-1)(x-2)}\right]=\lim\limits_{x\to 1}\dfrac{1}{x-2}=\dfrac{1}{1-2}=-1$

37. $\lim\limits_{x\to 0}\dfrac{\cot 2x}{\csc x}=\lim\limits_{x\to 0}\dfrac{\cos 2x\sin x}{\sin 2x}=\lim\limits_{x\to 0}\cos 2x\left[\dfrac{(\sin x)/x}{(\sin 2x)/x}\right]=\lim\limits_{x\to 0}\cos 2x\left[\dfrac{\lim\limits_{x\to 0}[(\sin x)/x]}{2\lim\limits_{x\to 0}[(\sin 2x)/2x]}\right]=1\cdot\dfrac{1}{2\cdot 1}=\dfrac{1}{2}$

38. $\lim\limits_{t\to 0}\dfrac{t^3}{\tan^3 2t}=\lim\limits_{t\to 0}\dfrac{t^3\cos^3 2t}{\sin^3 2t}=\lim\limits_{t\to 0}\cos^3 2t\cdot\dfrac{1}{8\dfrac{\sin^3 2t}{(2t)^3}}=\lim\limits_{t\to 0}\dfrac{\cos^3 2t}{8\left(\lim\limits_{t\to 0}\dfrac{\sin 2t}{2t}\right)^3}=\dfrac{1}{8\cdot 1^3}=\dfrac{1}{8}$

39. From the graph of $y=(\cos^2 x)/x^2$, it appears that $y=0$ is the horizontal asymptote and $x=0$ is the vertical asymptote. Now $0\le(\cos x)^2\le 1 \Rightarrow$ $\dfrac{0}{x^2}\le\dfrac{\cos^2 x}{x^2}\le\dfrac{1}{x^2} \Rightarrow 0\le\dfrac{\cos^2 x}{x^2}\le\dfrac{1}{x^2}$. But $\lim\limits_{x\to\pm\infty}0=0$ and $\lim\limits_{x\to\pm\infty}\dfrac{1}{x^2}=0$, so by the Squeeze Theorem, $\lim\limits_{x\to\pm\infty}\dfrac{\cos^2 x}{x^2}=0$.

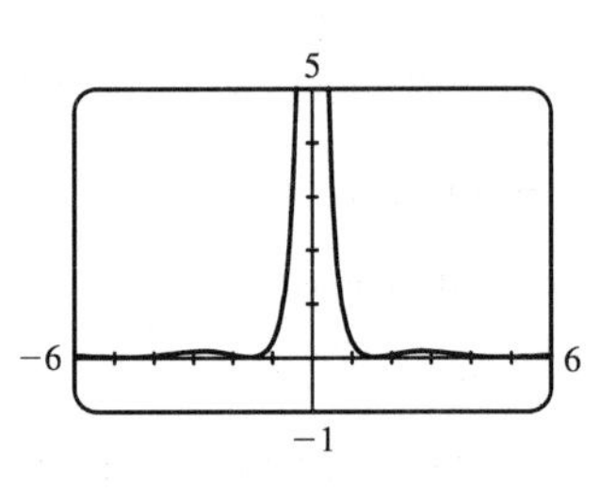

Thus, $y=0$ is the horizontal asymptote. $\lim\limits_{x\to 0}\dfrac{\cos^2 x}{x^2}=\infty$ because $\cos^2 x\to 1$ and $x^2\to 0$ as $x\to 0$, so $x=0$ is the vertical asymptote.

40. From the graph of $y = f(x) = \sqrt{x^2+x+1} - \sqrt{x^2-x}$, it appears that there are 2 horizontal asymptotes and possibly 2 vertical asymptotes. To obtain a different form for f, let's multiply and divide it by its conjugate.

$$f_1(x) = \left(\sqrt{x^2+x+1} - \sqrt{x^2-x}\right)\frac{\sqrt{x^2+x+1}+\sqrt{x^2-x}}{\sqrt{x^2+x+1}+\sqrt{x^2-x}} = \frac{(x^2+x+1)-(x^2-x)}{\sqrt{x^2+x+1}+\sqrt{x^2-x}} = \frac{2x+1}{\sqrt{x^2+x+1}+\sqrt{x^2-x}}$$

Now
$$\begin{aligned}\lim_{x\to\infty} f_1(x) &= \lim_{x\to\infty}\frac{2x+1}{\sqrt{x^2+x+1}+\sqrt{x^2-x}}\\ &= \lim_{x\to\infty}\frac{2+(1/x)}{\sqrt{1+(1/x)+(1/x^2)}+\sqrt{1-(1/x)}} \qquad [\text{since } \sqrt{x^2} = x \text{ for } x > 0]\\ &= \frac{2}{1+1} = 1,\end{aligned}$$

so $y = 1$ is a horizontal asymptote. For $x < 0$, we have $\sqrt{x^2} = |x| = -x$, so when we divide the denominator by x, with $x < 0$, we get

$$\frac{\sqrt{x^2+x+1}+\sqrt{x^2-x}}{x} = -\frac{\sqrt{x^2+x+1}+\sqrt{x^2-x}}{\sqrt{x^2}} = -\left[\sqrt{1+\frac{1}{x}+\frac{1}{x^2}}+\sqrt{1-\frac{1}{x}}\right]$$

Therefore,

$$\lim_{x\to-\infty} f_1(x) = \lim_{x\to-\infty}\frac{2x+1}{\sqrt{x^2+x+1}+\sqrt{x^2-x}} = \lim_{x\to\infty}\frac{2+(1/x)}{-\left[\sqrt{1+(1/x)+(1/x^2)}+\sqrt{1-(1/x)}\right]} = \frac{2}{-(1+1)} = -1,$$

so $y = -1$ is a horizontal asymptote. As $x \to 0^-$, $f(x) \to 1$, so $x = 0$ is *not* a vertical asymptote. As $x \to 1^+$, $f(x) \to \sqrt{3}$, so $x = 1$ is *not* a vertical asymptote and hence there are no vertical asymptotes.

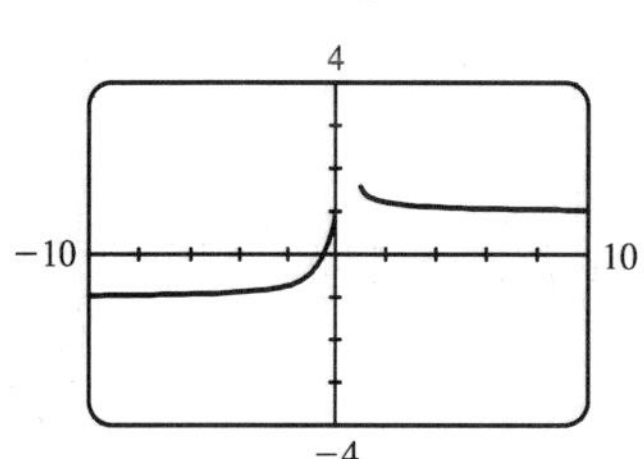

41. Since $2x - 1 \le f(x) \le x^2$ for $0 < x < 3$ and $\lim_{x\to1}(2x-1) = 1 = \lim_{x\to1} x^2$, we have $\lim_{x\to1} f(x) = 1$ by the Squeeze Theorem.

42. Let $f(x) = -x^2$, $g(x) = x^2\cos(1/x^2)$ and $h(x) = x^2$. Then since $\left|\cos(1/x^2)\right| \le 1$ for $x \ne 0$, we have $f(x) \le g(x) \le h(x)$ for $x \ne 0$, and so $\lim_{x\to0} f(x) = \lim_{x\to0} h(x) = 0 \;\Rightarrow\; \lim_{x\to0} g(x) = 0$ by the Squeeze Theorem.

43. Given $\varepsilon > 0$, we need $\delta > 0$ so that if $0 < |x-5| < \delta$, then $|(7x-27)-8| < \varepsilon \;\Leftrightarrow\; |7x-35| < \varepsilon \;\Leftrightarrow\; |x-5| < \varepsilon/7$. So take $\delta = \varepsilon/7$. Then $0 < |x-5| < \delta \;\Rightarrow\; |(7x-27)-8| < \varepsilon$. Thus, $\lim_{x\to5}(7x-27) = 8$ by the definition of a limit.

44. Given $\varepsilon > 0$ we must find $\delta > 0$ so that if $0 < |x-0| < \delta$, then $|\sqrt[3]{x}-0| < \varepsilon$. Now $|\sqrt[3]{x}-0| = |\sqrt[3]{x}| < \varepsilon \;\Rightarrow\; |x| = |\sqrt[3]{x}|^3 < \varepsilon^3$. So take $\delta = \varepsilon^3$. Then $0 < |x-0| = |x| < \varepsilon^3 \;\Rightarrow\; |\sqrt[3]{x}-0| = |\sqrt[3]{x}| = \sqrt[3]{|x|} < \sqrt[3]{\varepsilon^3} = \varepsilon$. Therefore, by the definition of a limit, $\lim_{x\to0}\sqrt[3]{x} = 0$.

45. If $\varepsilon > 0$ is given, then $1/x^4 < \varepsilon \;\Leftrightarrow\; x^4 > 1/\varepsilon \;\Leftrightarrow\; x > 1/\sqrt[4]{\varepsilon}$. Let $N = 1/\sqrt[4]{\varepsilon}$.

Then $x > N \;\Rightarrow\; x > \dfrac{1}{\sqrt[4]{\varepsilon}} \;\Rightarrow\; \left|\dfrac{1}{x^4} - 0\right| = \dfrac{1}{x^4} < \varepsilon$, so $\lim_{x\to\infty}\dfrac{1}{x^4} = 0$.

46. Given $M > 0$, we need $\delta > 0$ such that if $0 < x - 4 < \delta$, then $2/\sqrt{x-4} > M$. This is true $\Leftrightarrow$ $\sqrt{x-4} < 2/M$ $\Leftrightarrow$ $x - 4 < 4/M^2$. So if we choose $\delta = 4/M^2$, then $0 < x - 4 < \delta$ $\Rightarrow$ $2/\sqrt{x-4} > M$. So by the definition of a limit, $\lim_{x \to 4^+} \left(2/\sqrt{x-4}\right) = \infty$.

47. (a) $f(x) = \sqrt{-x}$ if $x < 0$, $f(x) = 3 - x$ if $0 \le x < 3$, $f(x) = (x-3)^2$ if $x > 3$.

(i) $\lim_{x \to 0^+} f(x) = \lim_{x \to 0^+} (3 - x) = 3$

(ii) $\lim_{x \to 0^-} f(x) = \lim_{x \to 0^-} \sqrt{-x} = 0$

(iii) Because of (i) and (ii), $\lim_{x \to 0} f(x)$ does not exist.

(iv) $\lim_{x \to 3^-} f(x) = \lim_{x \to 3^-} (3 - x) = 0$

(v) $\lim_{x \to 3^+} f(x) = \lim_{x \to 3^+} (x-3)^2 = 0$

(vi) Because of (iv) and (v), $\lim_{x \to 3} f(x) = 0$.

(b) f is discontinuous at 0 since $\lim_{x \to 0} f(x)$ does not exist.

f is discontinuous at 3 since $f(3)$ does not exist.

(c)

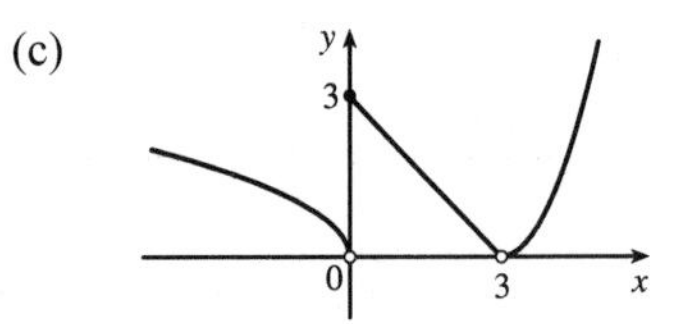

48. (a) $x^2 - 9$ is continuous on $\mathbb{R}$ since it is a polynomial and $\sqrt{x}$ is continuous on $[0, \infty)$, so the composition $\sqrt{x^2 - 9}$ is continuous on $\{x \mid x^2 - 9 \ge 0\} = (-\infty, -3] \cup [3, \infty)$. Note that $x^2 - 2 \ne 0$ on this set and so the quotient function

$$g(x) = \frac{\sqrt{x^2 - 9}}{x^2 - 2}$$ is continuous on its domain, $(-\infty, -3] \cup [3, \infty)$.

(b) x^3 is continuous on $\mathbb{R}$ since it is a polynomial and $\cos x$ is also continuous on $\mathbb{R}$, so the product $x^3 \cos x$ is continuous on $\mathbb{R}$. The root function $\sqrt[4]{x}$ is continuous on its domain, $[0, \infty)$, and so the sum, $h(x) = \sqrt[4]{x} + x^3 \cos x$, is continuous on its domain, $[0, \infty)$.

49. $f(x) = 2x^3 + x^2 + 2$ is a polynomial, so it is continuous on $[-2, -1]$ and $f(-2) = -10 < 0 < 1 = f(-1)$. So by the Intermediate Value Theorem there is a number c in $(-2, -1)$ such that $f(c) = 0$, that is, the equation $2x^3 + x^2 + 2 = 0$ has a root in $(-2, -1)$.

50. Let $f(x) = 2 \sin x - 3 + 2x$. Now f is continuous on $[0, 1]$ and $f(0) = -3 < 0$ and $f(1) = 2 \sin 1 - 1 \approx 0.68 > 0$. So by the Intermediate Value Theorem there is a number c in $(0, 1)$ such that $f(c) = 0$, that is, the equation $2 \sin x = 3 - 2x$ has a root in $(0, 1)$.

2 □ DERIVATIVES

2.1 Derivatives and Rates of Change

1. (a) (i) Using Definition 1,

$$m = \lim_{x \to a} \frac{f(x) - f(a)}{x - a} \lim_{x \to -3} \frac{f(x) - f(-3)}{x - (-3)} = \lim_{x \to -3} \frac{(x^2 + 2x) - (3)}{x - (-3)} = \lim_{x \to -3} \frac{(x+3)(x-1)}{x+3}$$

$$= \lim_{x \to -3} (x - 1) = -4$$

(ii) Using Equation 2,

$$m = \lim_{h \to 0} \frac{f(a+h) - f(a)}{h} = \lim_{h \to 0} \frac{f(-3+h) - f(-3)}{h} = \lim_{h \to 0} \frac{[(-3+h)^2 + 2(-3+h)] - (3)}{h}$$

$$= \lim_{h \to 0} \frac{9 - 6h + h^2 - 6 + 2h - 3}{h} = \lim_{h \to 0} \frac{h(h-4)}{h} = \lim_{h \to 0} (h - 4) = -4$$

(b) Using the point-slope form of the equation of a line, an equation of the tangent line is $y - 3 = -4(x + 3)$. Solving for y gives us $y = -4x - 9$, which is the slope-intercept form of the equation of the tangent line.

(c)

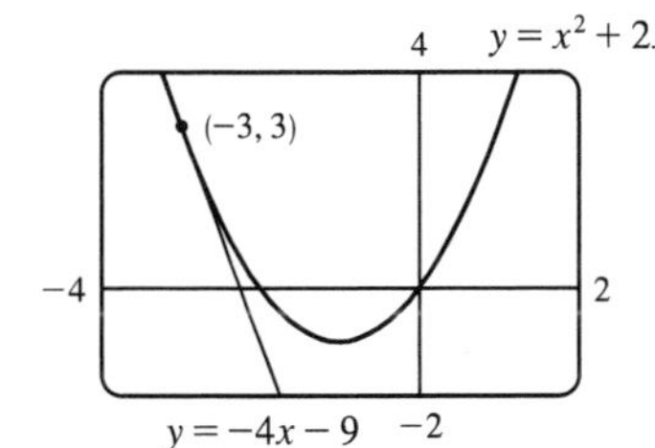

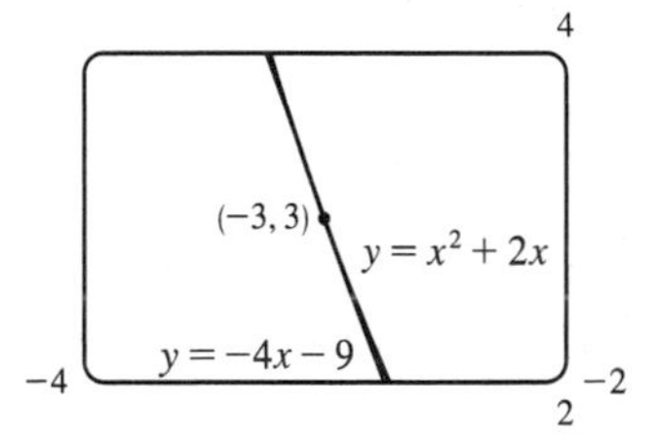

2. (a) (i) $m = \displaystyle\lim_{x \to -1} \frac{f(x) - f(-1)}{x - (-1)} = \lim_{x \to -1} \frac{x^3 - (-1)}{x + 1} = \lim_{x \to -1} \frac{(x+1)(x^2 - x + 1)}{x + 1} = \lim_{x \to -1} (x^2 - x + 1) = 3$

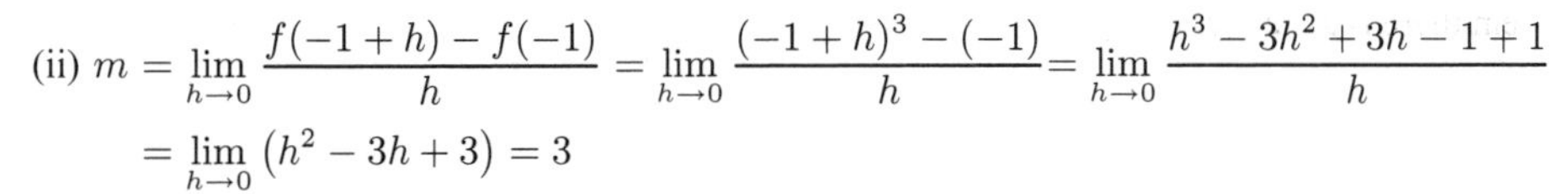

(ii) $m = \displaystyle\lim_{h \to 0} \frac{f(-1+h) - f(-1)}{h} = \lim_{h \to 0} \frac{(-1+h)^3 - (-1)}{h} = \lim_{h \to 0} \frac{h^3 - 3h^2 + 3h - 1 + 1}{h}$

$$= \lim_{h \to 0} (h^2 - 3h + 3) = 3$$

(b) $y - (-1) = 3[x - (-1)] \;\Leftrightarrow\; y + 1 = 3x + 3 \;\Leftrightarrow\; y = 3x + 2$

(c)

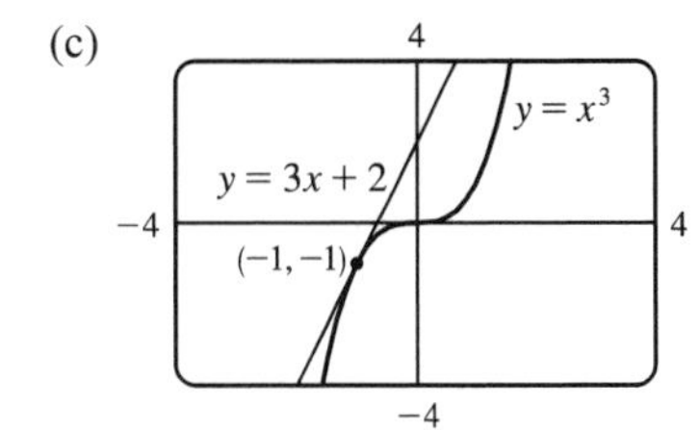

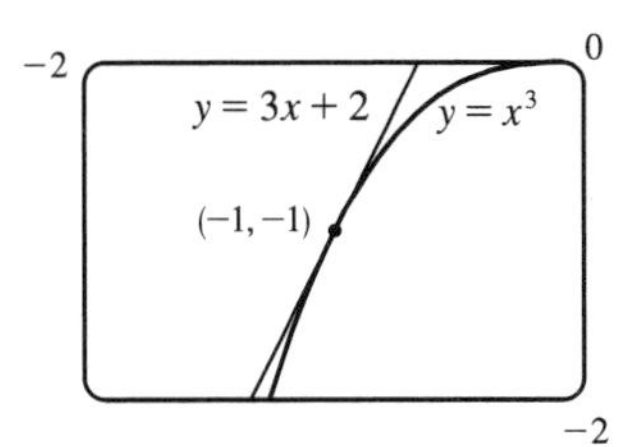

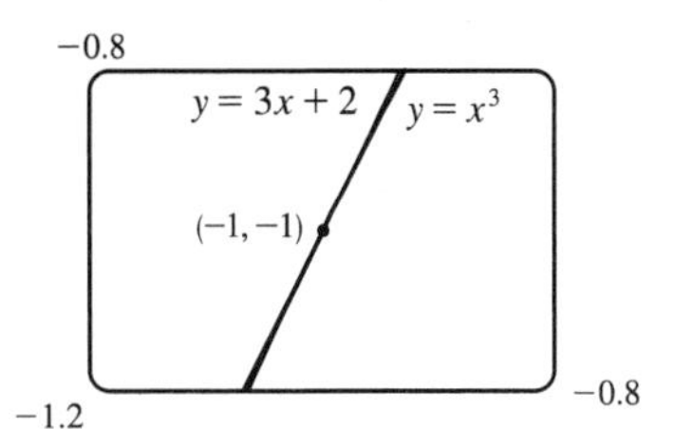

3. Using (1) with $f(x) = \dfrac{x-1}{x-2}$ and $P(3, 2)$,

$$m = \lim_{x \to a} \frac{f(x) - f(a)}{x - a} = \lim_{x \to 3} \frac{\frac{x-1}{x-2} - 2}{x - 3} = \lim_{x \to 3} \frac{\frac{x - 1 - 2(x-2)}{x-2}}{x - 3} = \lim_{x \to 3} \frac{3 - x}{(x-2)(x-3)} = \lim_{x \to 3} \frac{-1}{x - 2} = \frac{-1}{1} = -1.$$

Tangent line: $y - 2 = -1(x - 3) \;\Leftrightarrow\; y - 2 = -x + 3 \;\Leftrightarrow\; y = -x + 5$

4. Using (1),

$$m=\lim_{x\to-1}\frac{(2x^3-5x)-3}{x-(-1)}=\lim_{x\to-1}\frac{2x^3-5x-3}{x+1}=\lim_{x\to-1}\frac{(2x^2-2x-3)(x+1)}{x+1}=\lim_{x\to-1}(2x^2-2x-3)=1.$$

Tangent line: $y-3=1\,[x-(-1)] \quad\Leftrightarrow\quad y=x+4$

5. Using (1), $m=\lim\limits_{x\to1}\dfrac{\sqrt{x}-\sqrt{1}}{x-1}=\lim\limits_{x\to1}\dfrac{(\sqrt{x}-1)(\sqrt{x}+1)}{(x-1)(\sqrt{x}+1)}=\lim\limits_{x\to1}\dfrac{x-1}{(x-1)(\sqrt{x}+1)}=\lim\limits_{x\to1}\dfrac{1}{\sqrt{x}+1}=\dfrac{1}{2}.$

Tangent line: $y-1=\frac{1}{2}(x-1) \quad\Leftrightarrow\quad y=\frac{1}{2}x+\frac{1}{2}$

6. Using (1), $m=\lim\limits_{x\to0}\dfrac{\dfrac{2x}{(x+1)^2}-0}{x-0}=\lim\limits_{x\to0}\dfrac{2x}{x(x+1)^2}=\lim\limits_{x\to0}\dfrac{2}{(x+1)^2}=\dfrac{2}{1^2}=2.$

Tangent line: $y-0=2(x-0) \quad\Leftrightarrow\quad y=2x$

7. (a) Using (2) with $y=f(x)=3+4x^2-2x^3$,

$$\begin{aligned}m&=\lim_{h\to0}\frac{f(a+h)-f(a)}{h}=\lim_{h\to0}\frac{3+4(a+h)^2-2(a+h)^3-(3+4a^2-2a^3)}{h}\\&=\lim_{h\to0}\frac{3+4(a^2+2ah+h^2)-2(a^3+3a^2h+3ah^2+h^3)-3-4a^2+2a^3}{h}\\&=\lim_{h\to0}\frac{3+4a^2+8ah+4h^2-2a^3-6a^2h-6ah^2-2h^3-3-4a^2+2a^3}{h}\\&=\lim_{h\to0}\frac{8ah+4h^2-6a^2h-6ah^2-2h^3}{h}=\lim_{h\to0}\frac{h(8a+4h-6a^2-6ah-2h^2)}{h}\\&=\lim_{h\to0}(8a+4h-6a^2-6ah-2h^2)=8a-6a^2\end{aligned}$$

(b) At $(1,5)$: $m=8(1)-6(1)^2=2$, so an equation of the tangent line is $y-5=2(x-1) \quad\Leftrightarrow\quad y=2x+3.$

At $(2,3)$: $m=8(2)-6(2)^2=-8$, so an equation of the tangent line is $y-3=-8(x-2) \quad\Leftrightarrow\quad y=-8x+19.$

(c)

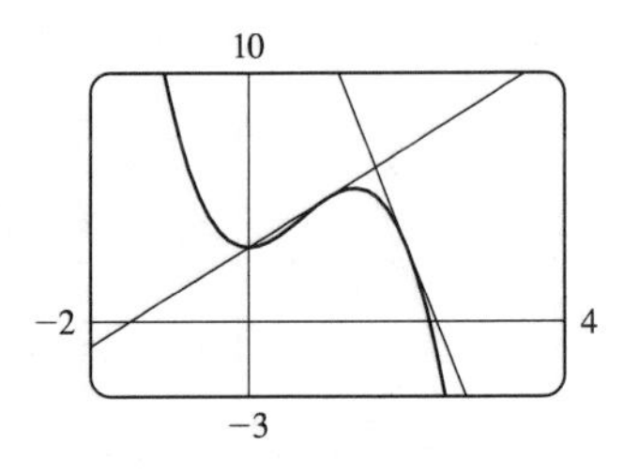

8. (a) Using (1),

$$\begin{aligned}m&=\lim_{x\to a}\frac{\dfrac{1}{\sqrt{x}}-\dfrac{1}{\sqrt{a}}}{x-a}=\lim_{x\to a}\frac{\dfrac{\sqrt{a}-\sqrt{x}}{\sqrt{ax}}}{x-a}=\lim_{x\to a}\frac{(\sqrt{a}-\sqrt{x})(\sqrt{a}+\sqrt{x})}{\sqrt{ax}\,(x-a)(\sqrt{a}+\sqrt{x})}\\&=\lim_{x\to a}\frac{a-x}{\sqrt{ax}\,(x-a)(\sqrt{a}+\sqrt{x})}=\lim_{x\to a}\frac{-1}{\sqrt{ax}\,(\sqrt{a}+\sqrt{x})}=\frac{-1}{\sqrt{a^2}\,(2\sqrt{a})}=-\frac{1}{2a^{3/2}}\text{ or }-\frac{1}{2}a^{-3/2}\end{aligned}$$

(b) At $(1,1)$: $m=-\frac{1}{2}$, so an equation of the tangent line is

$y-1=-\frac{1}{2}(x-1) \quad\Leftrightarrow\quad y=-\frac{1}{2}x+\frac{3}{2}.$

At $\left(4,\frac{1}{2}\right)$: $m=-\frac{1}{16}$, so an equation of the tangent line is

$y-\frac{1}{2}=-\frac{1}{16}(x-4) \quad\Leftrightarrow\quad y=-\frac{1}{16}x+\frac{3}{4}.$

(c)

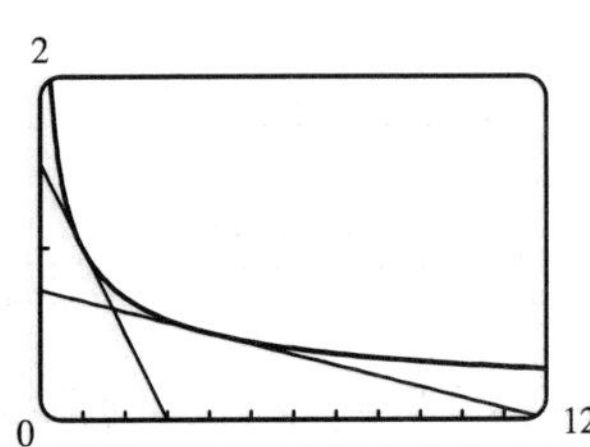

9. (a) Since the slope of the tangent at $t = 0$ is 0, the car's initial velocity was 0.

(b) The slope of the tangent is greater at C than at B, so the car was going faster at C.

(c) Near A, the tangent lines are becoming steeper as x increases, so the velocity was increasing, so the car was speeding up. Near B, the tangent lines are becoming less steep, so the car was slowing down. The steepest tangent near C is the one at C, so at C the car had just finished speeding up, and was about to start slowing down.

(d) Between D and E, the slope of the tangent is 0, so the car did not move during that time.

10. (a) Runner A runs the entire 100-meter race at the same velocity since the slope of the position function is constant. Runner B starts the race at a slower velocity than runner A, but finishes the race at a faster velocity.

(b) The distance between the runners is the greatest at the time when the largest vertical line segment fits between the two graphs—this appears to be somewhere between 9 and 10 seconds.

(c) The runners had the same velocity when the slopes of their respective position functions are equal—this also appears to be at about 9.5 s. Note that the answers for parts (b) and (c) must be the same for these graphs because as soon as the velocity for runner B overtakes the velocity for runner A, the distance between the runners starts to decrease.

11. Let $s(t) = 40t - 16t^2$.

$$v(2) = \lim_{t\to 2} \frac{s(t) - s(2)}{t - 2} = \lim_{t\to 2} \frac{(40t - 16t^2) - 16}{t - 2} = \lim_{t\to 2} \frac{-16t^2 + 40t - 16}{t - 2} = \lim_{t\to 2} \frac{-8(2t^2 - 5t + 2)}{t - 2}$$

$$= \lim_{t\to 2} \frac{-8(t-2)(2t-1)}{t-2} = -8 \lim_{t\to 2}(2t - 1) = -8(3) = -24$$

Thus, the instantaneous velocity when $t = 2$ is -24 ft/s.

12. (a) $v(1) = \lim\limits_{h\to 0} \dfrac{H(1+h) - H(1)}{h} = \lim\limits_{h\to 0} \dfrac{(58 + 58h - 0.83 - 1.66h - 0.83h^2) - 57.17}{h}$

$= \lim\limits_{h\to 0} (56.34 - 0.83h) = 56.34 \text{ m/s}$

(b) $v(a) = \lim\limits_{h\to 0} \dfrac{H(a+h) - H(a)}{h} = \lim\limits_{h\to 0} \dfrac{(58a + 58h - 0.83a^2 - 1.66ah - 0.83h^2) - (58a - 0.83a^2)}{h}$

$= \lim\limits_{h\to 0} (58 - 1.66a - 0.83h) = 58 - 1.66a \text{ m/s}$

(c) The arrow strikes the moon when the height is 0, that is, $58t - 0.83t^2 = 0 \Leftrightarrow t(58 - 0.83t) = 0 \Leftrightarrow$ $t = \dfrac{58}{0.83} \approx 69.9$ s (since t can't be 0).

(d) Using the time from part (c), $v\left(\dfrac{58}{0.83}\right) = 58 - 1.66\left(\dfrac{58}{0.83}\right) = -58$ m/s. Thus, the arrow will have a velocity of -58 m/s.

13. $$v(a) = \lim_{h\to 0} \frac{s(a+h) - s(a)}{h} = \lim_{h\to 0} \frac{\dfrac{1}{(a+h)^2} - \dfrac{1}{a^2}}{h} = \lim_{h\to 0} \frac{\dfrac{a^2 - (a+h)^2}{a^2(a+h)^2}}{h} = \lim_{h\to 0} \frac{a^2 - (a^2 + 2ah + h^2)}{ha^2(a+h)^2}$$

$$= \lim_{h\to 0} \frac{-(2ah + h^2)}{ha^2(a+h)^2} = \lim_{h\to 0} \frac{-h(2a+h)}{ha^2(a+h)^2} = \lim_{h\to 0} \frac{-(2a+h)}{a^2(a+h)^2} = \frac{-2a}{a^2 \cdot a^2} = \frac{-2}{a^3} \text{ m/s}$$

So $v(1) = \dfrac{-2}{1^3} = -2$ m/s, $v(2) = \dfrac{-2}{2^3} = -\dfrac{1}{4}$ m/s, and $v(3) = \dfrac{-2}{3^3} = -\dfrac{2}{27}$ m/s.

14. (a) The average velocity between times t and $t+h$ is

$$\frac{s(t+h)-s(t)}{(t+h)-t} = \frac{(t+h)^2 - 8(t+h) + 18 - (t^2 - 8t + 18)}{h} = \frac{t^2 + 2th + h^2 - 8t - 8h + 18 - t^2 + 8t - 18}{h}$$

$$= \frac{2th + h^2 - 8h}{h} = (2t + h - 8) \ \text{m/s}.$$

(i) $[3, 4]$: $t = 3$, $h = 4 - 3 = 1$, so the average velocity is $2(3) + 1 - 8 = -1$ m/s.

(ii) $[3.5, 4]$: $t = 3.5$, $h = 0.5$, so the average velocity is $2(3.5) + 0.5 - 8 = -0.5$ m/s.

(iii) $[4, 5]$: $t = 4$, $h = 1$, so the average velocity is $2(4) + 1 - 8 = 1$ m/s.

(iv) $[4, 4.5]$: $t = 4$, $h = 0.5$, so the average velocity is $2(4) + 0.5 - 8 = 0.5$ m/s.

(b) $v(t) = \lim\limits_{h\to 0} \dfrac{s(t+h)-s(t)}{h} = \lim\limits_{h\to 0}(2t + h - 8) = 2t - 8$, so $v(4) = 0$.

(c)

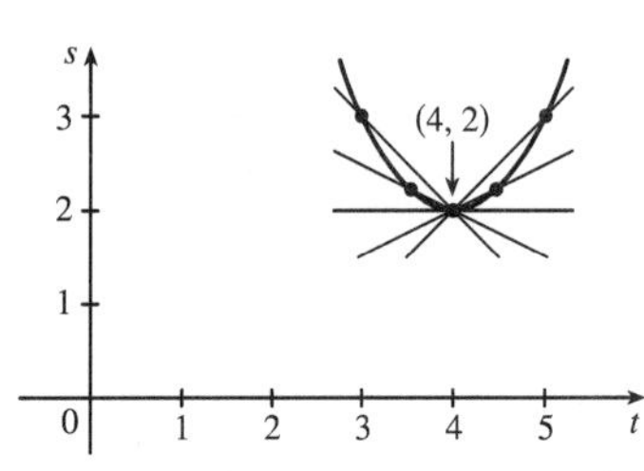

15. $g'(0)$ is the only negative value. The slope at $x = 4$ is smaller than the slope at $x = 2$ and both are smaller than the slope at $x = -2$. Thus, $g'(0) < 0 < g'(4) < g'(2) < g'(-2)$.

16. (a) Since $g(5) = -3$, the point $(5, -3)$ is on the graph of g. Since $g'(5) = 4$, the slope of the tangent line at $x = 5$ is 4. Using the point-slope form of a line gives us $y - (-3) = 4(x - 5)$, or $y = 4x - 23$.

(b) Since $(4, 3)$ is on $y = f(x)$, $f(4) = 3$. The slope of the tangent line between $(0, 2)$ and $(4, 3)$ is $\frac{1}{4}$, so $f'(4) = \frac{1}{4}$.

17. We begin by drawing a curve through the origin with a slope of 3 to satisfy $f(0) = 0$ and $f'(0) = 3$. Since $f'(1) = 0$, we will round off our figure so that there is a horizontal tangent directly over $x = 1$. Last, we make sure that the curve has a slope of -1 as we pass over $x = 2$. Two of the many possibilities are shown.

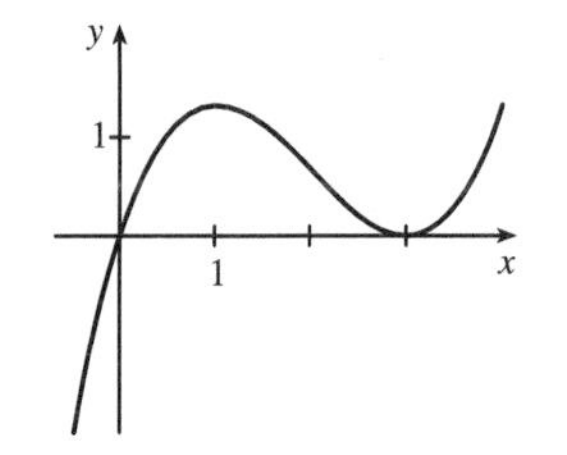

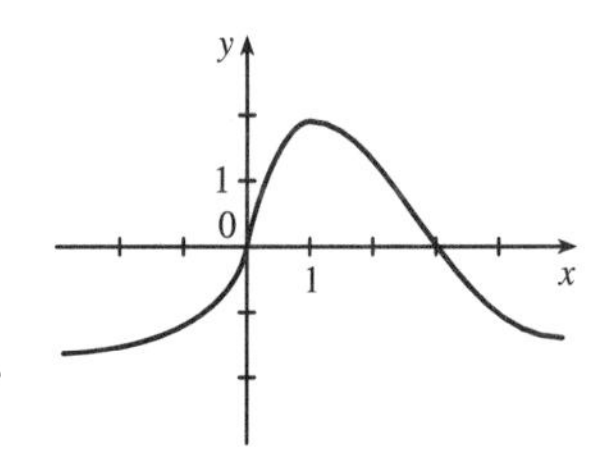

18. We begin by drawing a curve through the origin with a slope of 0 to satisfy $g(0) = 0$ and $g'(0) = 0$. The curve should have a slope of -1, 3, and 1 as we pass over $x = -1$, 1, and 2, respectively.

Note: In the figure, $y' = 0$ when $x \approx -1.27$ or 2.13.

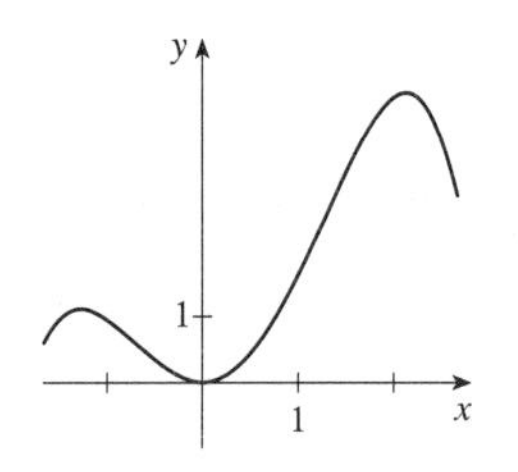

19. Using Definition 4 with $f(x) = 3x^2 - 5x$ and the point $(2, 2)$, we have

$$f'(2) = \lim_{h\to 0} \frac{f(2+h) - f(2)}{h} = \lim_{h\to 0} \frac{\left[3(2+h)^2 - 5(2+h)\right] - 2}{h}$$

$$= \lim_{h\to 0} \frac{(12 + 12h + 3h^2 - 10 - 5h) - 2}{h} = \lim_{h\to 0} \frac{3h^2 + 7h}{h} = \lim_{h\to 0}(3h + 7) = 7$$

So an equation of the tangent line at $(2, 2)$ is $y - 2 = 7(x - 2)$ or $y = 7x - 12$.

20. Using Definition 4 with $g(x) = 1 - x^3$ and the point $(0, 1)$, we have

$$g'(0) = \lim_{h \to 0} \frac{g(0+h) - g(0)}{h} = \lim_{h \to 0} \frac{[1 - (0+h)^3] - 1}{h} = \lim_{h \to 0} \frac{(1 - h^3) - 1}{h} = \lim_{h \to 0} (-h^2) = 0$$

So an equation of the tangent line is $y - 1 = 0(x - 0)$ or $y = 1$.

21. (a) Using Definition 4 with $F(x) = 5x/(1 + x^2)$ and the point $(2, 2)$, we have

$$\begin{aligned} F'(2) &= \lim_{h \to 0} \frac{F(2+h) - F(2)}{h} = \lim_{h \to 0} \frac{\dfrac{5(2+h)}{1 + (2+h)^2} - 2}{h} \\ &= \lim_{h \to 0} \frac{\dfrac{5h + 10}{h^2 + 4h + 5} - 2}{h} = \lim_{h \to 0} \frac{\dfrac{5h + 10 - 2(h^2 + 4h + 5)}{h^2 + 4h + 5}}{h} \\ &= \lim_{h \to 0} \frac{-2h^2 - 3h}{h(h^2 + 4h + 5)} = \lim_{h \to 0} \frac{h(-2h - 3)}{h(h^2 + 4h + 5)} = \lim_{h \to 0} \frac{-2h - 3}{h^2 + 4h + 5} = \frac{-3}{5} \end{aligned}$$

So an equation of the tangent line at $(2, 2)$ is $y - 2 = -\frac{3}{5}(x - 2)$ or $y = -\frac{3}{5}x + \frac{16}{5}$.

(b)

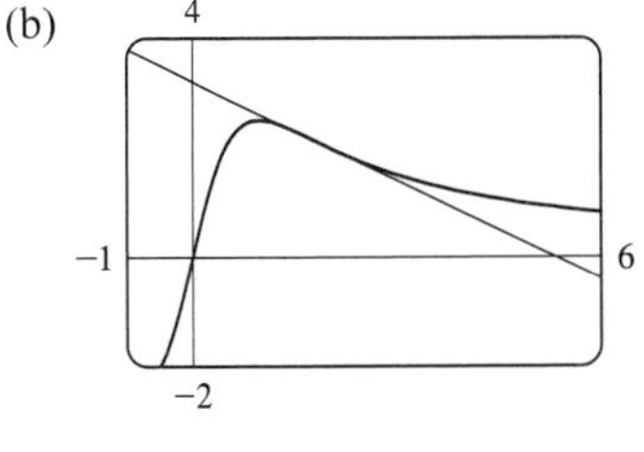

22. (a) Using Definition 4 with $G(x) = 4x^2 - x^3$, we have

$$\begin{aligned} G'(a) &= \lim_{h \to 0} \frac{G(a+h) - G(a)}{h} = \lim_{h \to 0} \frac{[4(a+h)^2 - (a+h)^3] - (4a^2 - a^3)}{h} \\ &= \lim_{h \to 0} \frac{4a^2 + 8ah + 4h^2 - (a^3 + 3a^2h + 3ah^2 + h^3) - 4a^2 + a^3}{h} \\ &= \lim_{h \to 0} \frac{8ah + 4h^2 - 3a^2h - 3ah^2 - h^3}{h} \\ &= \lim_{h \to 0} \frac{h(8a + 4h - 3a^2 - 3ah - h^{2)}}{h} = \lim_{h \to 0} (8a + 4h - 3a^2 - 3ah - h^2) = 8a - 3a^2 \end{aligned}$$

At the point $(2, 8)$, $G'(2) = 16 - 12 = 4$, and an equation of the tangent line is $y - 8 = 4(x - 2)$, or $y = 4x$.

At the point $(3, 9)$, $G'(3) = 24 - 27 = -3$, and an equation of the tangent line is $y - 9 = -3(x - 3)$, or $y = -3x + 18$.

(b)

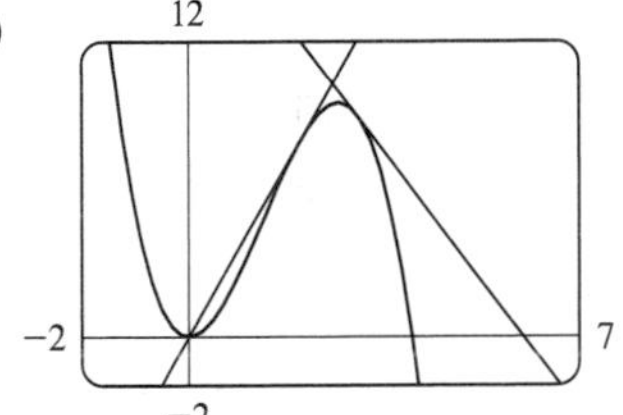

23. Use Definition 4 with $f(x) = 3 - 2x + 4x^2$.

$$\begin{aligned} f'(a) &= \lim_{h \to 0} \frac{f(a+h) - f(a)}{h} = \lim_{h \to 0} \frac{[3 - 2(a+h) + 4(a+h)^2] - (3 - 2a + 4a^2)}{h} \\ &= \lim_{h \to 0} \frac{(3 - 2a - 2h + 4a^2 + 8ah + 4h^2) - (3 - 2a + 4a^2)}{h} \\ &= \lim_{h \to 0} \frac{-2h + 8ah + 4h^2}{h} = \lim_{h \to 0} \frac{h(-2 + 8a + 4h)}{h} = \lim_{h \to 0} (-2 + 8a + 4h) = -2 + 8a \end{aligned}$$

24.
$$\begin{aligned} f'(a) &= \lim_{h \to 0} \frac{f(a+h) - f(a)}{h} = \lim_{h \to 0} \frac{[(a+h)^4 - 5(a+h)] - (a^4 - 5a)}{h} \\ &= \lim_{h \to 0} \frac{(a^4 + 4a^3h + 6a^2h^2 + 4ah^3 + h^4 - 5a - 5h) - (a^4 - 5a)}{h} = \lim_{h \to 0} \frac{4a^3h + 6a^2h^2 + 4ah^3 + h^4 - 5h}{h} \\ &= \lim_{h \to 0} \frac{h(4a^3 + 6a^2h + 4ah^2 + h^3 - 5)}{h} = \lim_{h \to 0} (4a^3 + 6a^2h + 4ah^2 + h^3 - 5) = 4a^3 - 5 \end{aligned}$$

25. Use Definition 4 with $f(t) = (2t+1)/(t+3)$.

$$f'(a) = \lim_{h\to 0} \frac{f(a+h)-f(a)}{h} = \lim_{h\to 0} \frac{\dfrac{2(a+h)+1}{(a+h)+3} - \dfrac{2a+1}{a+3}}{h} = \lim_{h\to 0} \frac{(2a+2h+1)(a+3)-(2a+1)(a+h+3)}{h(a+h+3)(a+3)}$$

$$= \lim_{h\to 0} \frac{(2a^2+6a+2ah+6h+a+3)-(2a^2+2ah+6a+a+h+3)}{h(a+h+3)(a+3)}$$

$$= \lim_{h\to 0} \frac{5h}{h(a+h+3)(a+3)} = \lim_{h\to 0} \frac{5}{(a+h+3)(a+3)} = \frac{5}{(a+3)^2}$$

26. $$f'(a) = \lim_{h\to 0} \frac{f(a+h)-f(a)}{h} = \lim_{h\to 0} \frac{\dfrac{(a+h)^2+1}{(a+h)-2} - \dfrac{a^2+1}{a-2}}{h}$$

$$= \lim_{h\to 0} \frac{(a^2+2ah+h^2+1)(a-2)-(a^2+1)(a+h-2)}{h(a+h-2)(a-2)}$$

$$= \lim_{h\to 0} \frac{(a^3-2a^2+2a^2h-4ah+ah^2-2h^2+a-2)-(a^3+a^2h-2a^2+a+h-2)}{h(a+h-2)(a-2)}$$

$$= \lim_{h\to 0} \frac{a^2h-4ah+ah^2-2h^2-h}{h(a+h-2)(a-2)} = \lim_{h\to 0} \frac{h(a^2-4a+ah-2h-1)}{h(a+h-2)(a-2)} = \lim_{h\to 0} \frac{a^2-4a+ah-2h-1}{(a+h-2)(a-2)}$$

$$= \frac{a^2-4a-1}{(a-2)^2}$$

27. Use Definition 4 with $f(x) = 1/\sqrt{x+2}$.

$$f'(a) = \lim_{h\to 0} \frac{f(a+h)-f(a)}{h} = \lim_{h\to 0} \frac{\dfrac{1}{\sqrt{(a+h)+2}} - \dfrac{1}{\sqrt{a+2}}}{h} = \lim_{h\to 0} \frac{\dfrac{\sqrt{a+2}-\sqrt{a+h+2}}{\sqrt{a+h+2}\,\sqrt{a+2}}}{h}$$

$$= \lim_{h\to 0} \left[\frac{\sqrt{a+2}-\sqrt{a+h+2}}{h\,\sqrt{a+h+2}\,\sqrt{a+2}} \cdot \frac{\sqrt{a+2}+\sqrt{a+h+2}}{\sqrt{a+2}+\sqrt{a+h+2}}\right] = \lim_{h\to 0} \frac{(a+2)-(a+h+2)}{h\sqrt{a+h+2}\,\sqrt{a+2}\left(\sqrt{a+2}+\sqrt{a+h+2}\right)}$$

$$= \lim_{h\to 0} \frac{-h}{h\sqrt{a+h+2}\,\sqrt{a+2}\left(\sqrt{a+2}+\sqrt{a+h+2}\right)} = \lim_{h\to 0} \frac{-1}{\sqrt{a+h+2}\,\sqrt{a+2}\left(\sqrt{a+2}+\sqrt{a+h+2}\right)}$$

$$= \frac{-1}{\left(\sqrt{a+2}\right)^2\left(2\sqrt{a+2}\right)} = -\frac{1}{2(a+2)^{3/2}}$$

28. $$f'(a) = \lim_{h\to 0} \frac{f(a+h)-f(a)}{h} = \lim_{h\to 0} \frac{\sqrt{3(a+h)+1}-\sqrt{3a+1}}{h}$$

$$= \lim_{h\to 0} \frac{\left(\sqrt{3a+3h+1}-\sqrt{3a+1}\right)\left(\sqrt{3a+3h+1}+\sqrt{3a+1}\right)}{h\left(\sqrt{3a+3h+1}+\sqrt{3a+1}\right)} = \lim_{h\to 0} \frac{(3a+3h+1)-(3a+1)}{h\left(\sqrt{3a+3h+1}+\sqrt{3a+1}\right)}$$

$$= \lim_{h\to 0} \frac{3h}{h\left(\sqrt{3a+3h+1}+\sqrt{3a+1}\right)} = \lim_{h\to 0} \frac{3}{\sqrt{3a+3h+1}+\sqrt{3a+1}} = \frac{3}{2\sqrt{3a+1}}$$

Note that the answers to Exercises 29–34 are not unique.

29. By Definition 4, $\displaystyle\lim_{h\to 0} \frac{(1+h)^{10}-1}{h} = f'(1)$, where $f(x) = x^{10}$ and $a = 1$.

Or: By Definition 4, $\displaystyle\lim_{h\to 0} \frac{(1+h)^{10}-1}{h} = f'(0)$, where $f(x) = (1+x)^{10}$ and $a = 0$.

30. By Definition 4, $\displaystyle\lim_{h\to 0} \frac{\sqrt[4]{16+h}-2}{h} = f'(16)$, where $f(x) = \sqrt[4]{x}$ and $a = 16$.

Or: By Definition 4, $\displaystyle\lim_{h\to 0} \frac{\sqrt[4]{16+h}-2}{h} = f'(0)$, where $f(x) = \sqrt[4]{16+x}$ and $a = 0$.

31. By Equation 5, $\lim\limits_{x\to 5} \dfrac{2^x - 32}{x - 5} = f'(5)$, where $f(x) = 2^x$ and $a = 5$.

32. By Equation 5, $\lim\limits_{x\to \pi/4} \dfrac{\tan x - 1}{x - \pi/4} = f'(\pi/4)$, where $f(x) = \tan x$ and $a = \pi/4$.

33. By Definition 4, $\lim\limits_{h\to 0} \dfrac{\cos(\pi + h) + 1}{h} = f'(\pi)$, where $f(x) = \cos x$ and $a = \pi$.

Or: By Definition 4, $\lim\limits_{h\to 0} \dfrac{\cos(\pi + h) + 1}{h} = f'(0)$, where $f(x) = \cos(\pi + x)$ and $a = 0$.

34. By Equation 5, $\lim\limits_{t\to 1} \dfrac{t^4 + t - 2}{t - 1} = f'(1)$, where $f(t) = t^4 + t$ and $a = 1$.

35. The sketch shows the graph for a room temperature of $72°$ and a refrigerator temperature of $38°$. The initial rate of change is greater in magnitude than the rate of change after an hour.

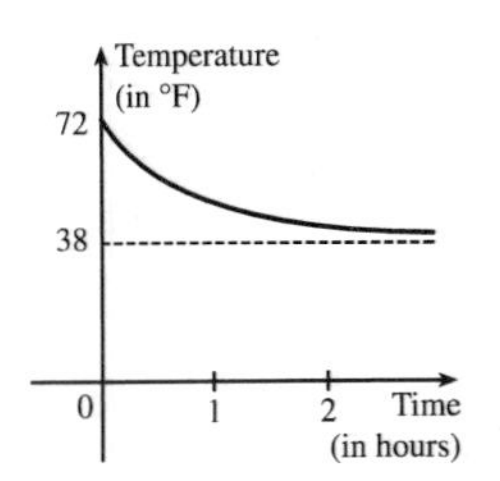

36. The slope of the tangent (that is, the rate of change of temperature with respect to time) at $t = 1$ h seems to be about $\dfrac{75 - 168}{132 - 0} \approx -0.7\,°\text{F/min}$.

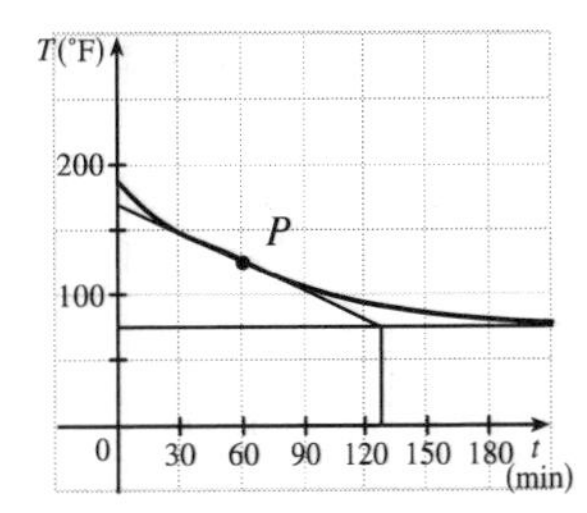

37. (a) (i) $[2000, 2002]$: $\dfrac{P(2002) - P(2000)}{2002 - 2000} = \dfrac{77 - 55}{2} = \dfrac{22}{2} = 11$ percent/year

(ii) $[2000, 2001]$: $\dfrac{P(2001) - P(2000)}{2001 - 2000} = \dfrac{68 - 55}{1} = 13$ percent/year

(iii) $[1999, 2000]$: $\dfrac{P(2000) - P(1999)}{2000 - 1999} = \dfrac{55 - 39}{1} = 16$ percent/year

(b) Using the values from (ii) and (iii), we have $\dfrac{13 + 16}{2} = 14.5$ percent/year.

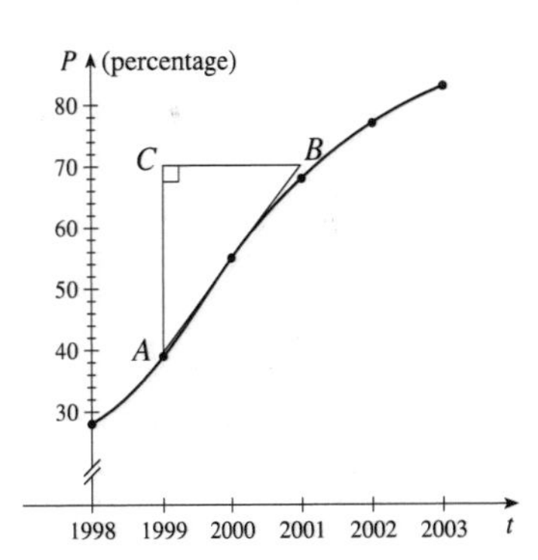

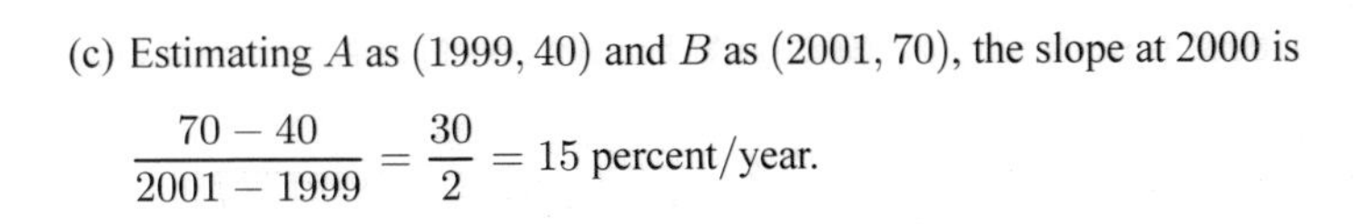
(c) Estimating A as $(1999, 40)$ and B as $(2001, 70)$, the slope at 2000 is

$$\frac{70 - 40}{2001 - 1999} = \frac{30}{2} = 15 \text{ percent/year.}$$

38. (a) (i) $[2000, 2002]$: $\dfrac{P(2002) - P(2000)}{2002 - 2000} = \dfrac{5886 - 3501}{2} = \dfrac{2385}{2} = 1192.5$ locations/year

(ii) $[2000, 2001]$: $\dfrac{P(2001) - P(2000)}{2001 - 2000} = \dfrac{4709 - 3501}{1} = 1208$ locations/year

(iii) $[1999, 2000]$: $\dfrac{P(2000) - P(1999)}{2000 - 1999} = \dfrac{3501 - 2135}{1} = 1366$ locations/year

(b) Using the values from (ii) and (iii), we have $\dfrac{1208 + 1366}{2} = 1287$ locations/year.

(c) Estimating A as $(1999, 2035)$ and B as $(2001, 4960)$, the slope at 2000 is

$$\frac{4960 - 2035}{2001 - 1999} = \frac{2925}{2} = 1462.5 \text{ locations/year.}$$

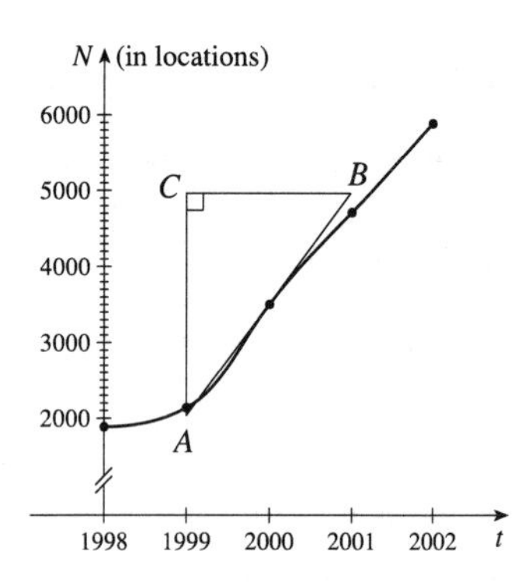

39. (a) (i) $\dfrac{\Delta C}{\Delta x} = \dfrac{C(105) - C(100)}{105 - 100} = \dfrac{6601.25 - 6500}{5} = \$20.25/\text{unit}.$

(ii) $\dfrac{\Delta C}{\Delta x} = \dfrac{C(101) - C(100)}{101 - 100} = \dfrac{6520.05 - 6500}{1} = \$20.05/\text{unit}.$

(b) $\dfrac{C(100 + h) - C(100)}{h} = \dfrac{\left[5000 + 10(100 + h) + 0.05(100 + h)^2\right] - 6500}{h} = \dfrac{20h + 0.05h^2}{h}$

$= 20 + 0.05h,\ h \neq 0$

So the instantaneous rate of change is $\displaystyle\lim_{h \to 0} \frac{C(100 + h) - C(100)}{h} = \lim_{h \to 0} (20 + 0.05h) = \$20/\text{unit}.$

40. $\Delta V = V(t + h) - V(t) = 100{,}000\left(1 - \dfrac{t + h}{60}\right)^2 - 100{,}000\left(1 - \dfrac{t}{60}\right)^2$

$= 100{,}000\left[\left(1 - \dfrac{t + h}{30} + \dfrac{(t + h)^2}{3600}\right) - \left(1 - \dfrac{t}{30} + \dfrac{t^2}{3600}\right)\right] = 100{,}000\left(-\dfrac{h}{30} + \dfrac{2th}{3600} + \dfrac{h^2}{3600}\right)$

$= \dfrac{100{,}000}{3600} h\,(-120 + 2t + h) = \dfrac{250}{9} h\,(-120 + 2t + h)$

Dividing ΔV by h and then letting $h \to 0$, we see that the instantaneous rate of change is $\frac{500}{9}(t - 60)$ gal/min.

t	Flow rate (gal/min)	Water remaining $V(t)$ (gal)
0	$-3333.\overline{3}$	$100,000$
10	$-2777.\overline{7}$	$69,444.\overline{4}$
20	$-2222.\overline{2}$	$44,444.\overline{4}$
30	$-1666.\overline{6}$	$25,000$
40	$-1111.\overline{1}$	$11,111.\overline{1}$
50	$-555.\overline{5}$	$2,777.\overline{7}$
60	0	0

The magnitude of the flow rate is greatest at the beginning and gradually decreases to 0.

41. (a) $f'(x)$ is the rate of change of the production cost with respect to the number of ounces of gold produced. Its units are dollars per ounce.

(b) After 800 ounces of gold have been produced, the rate at which the production cost is increasing is \$17/ounce. So the cost of producing the 800th (or 801st) ounce is about \$17.

(c) In the short term, the values of $f'(x)$ will decrease because more efficient use is made of start-up costs as x increases. But eventually $f'(x)$ might increase due to large-scale operations.

42. (a) $f'(5)$ is the rate of growth of the bacteria population when $t = 5$ hours. Its units are bacteria per hour.

(b) With unlimited space and nutrients, f' should increase as t increases; so $f'(5) < f'(10)$. If the supply of nutrients is limited, the growth rate slows down at some point in time, and the opposite may be true.

43. $T'(10)$ is the rate at which the temperature is changing at 10:00 AM. To estimate the value of $T'(10)$, we will average the difference quotients obtained using the times $t = 8$ and $t = 12$. Let $A = \dfrac{T(8) - T(10)}{8 - 10} = \dfrac{72 - 81}{-2} = 4.5$ and $B = \dfrac{T(12) - T(10)}{12 - 10} = \dfrac{88 - 81}{2} = 3.5$. Then $T'(10) = \lim\limits_{t \to 10} \dfrac{T(t) - T(10)}{t - 10} \approx \dfrac{A + B}{2} = \dfrac{4.5 + 3.5}{2} = 4°\text{F/h}$.

44. (a) $f'(8)$ is the rate of change of the quantity of coffee sold with respect to the price per pound when the price is \$8 per pound. The units for $f'(8)$ are pounds/(dollars/pound).

(b) $f'(8)$ is negative since the quantity of coffee sold will decrease as the price charged for it increases. People are generally less willing to buy a product when its price increases.

45. (a) $S'(T)$ is the rate at which the oxygen solubility changes with respect to the water temperature. Its units are $(\text{mg/L})/°\text{C}$.

(b) For $T = 16°\text{C}$, it appears that the tangent line to the curve goes through the points $(0, 14)$ and $(32, 6)$. So $S'(16) \approx \dfrac{6 - 14}{32 - 0} = -\dfrac{8}{32} = -0.25\ (\text{mg/L})/°\text{C}$. This means that as the temperature increases past $16°\text{C}$, the oxygen solubility is decreasing at a rate of $0.25\ (\text{mg/L})/°\text{C}$.

46. (a) $S'(T)$ is the rate of change of the maximum sustainable speed of Coho salmon with respect to the temperature. Its units are $(\text{cm/s})/°\text{C}$.

(b) For $T = 15°\text{C}$, it appears the tangent line to the curve goes through the points $(10, 25)$ and $(20, 32)$. So $S'(15) \approx \dfrac{32 - 25}{20 - 10} = 0.7\ (\text{cm/s})/°\text{C}$. This tells us that at $T = 15°\text{C}$, the maximum sustainable speed of Coho salmon is changing at a rate of $0.7\ (\text{cm/s})/°\text{C}$. In a similar fashion for $T = 25°\text{C}$, we can use the points $(20, 35)$ and $(25, 25)$ to obtain $S'(25) \approx \dfrac{25 - 35}{25 - 20} = -2\ (\text{cm/s})/°\text{C}$. As it gets warmer than $20°\text{C}$, the maximum sustainable speed decreases rapidly.

47. Since $f(x) = x\sin(1/x)$ when $x \neq 0$ and $f(0) = 0$, we have

$f'(0) = \lim\limits_{h \to 0} \dfrac{f(0 + h) - f(0)}{h} = \lim\limits_{h \to 0} \dfrac{h\sin(1/h) - 0}{h} = \lim\limits_{h \to 0} \sin(1/h)$. This limit does not exist since $\sin(1/h)$ takes the values -1 and 1 on any interval containing 0. (Compare with Example 5 in Section 1.3.)

48. Since $f(x) = x^2\sin(1/x)$ when $x \neq 0$ and $f(0) = 0$, we have

$f'(0) = \lim\limits_{h \to 0} \dfrac{f(0 + h) - f(0)}{h} = \lim\limits_{h \to 0} \dfrac{h^2\sin(1/h) - 0}{h} = \lim\limits_{h \to 0} h\sin(1/h)$. Since $-1 \leq \sin\dfrac{1}{h} \leq 1$, we have

$-|h| \leq |h|\sin\dfrac{1}{h} \leq |h| \quad \Rightarrow \quad -|h| \leq h\sin\dfrac{1}{h} \leq |h|$. Because $\lim\limits_{h \to 0}(-|h|) = 0$ and $\lim\limits_{h \to 0}|h| = 0$, we know that

$\lim\limits_{h \to 0}\left(h\sin\dfrac{1}{h}\right) = 0$ by the Squeeze Theorem. Thus, $f'(0) = 0$.

2.2 The Derivative as a Function

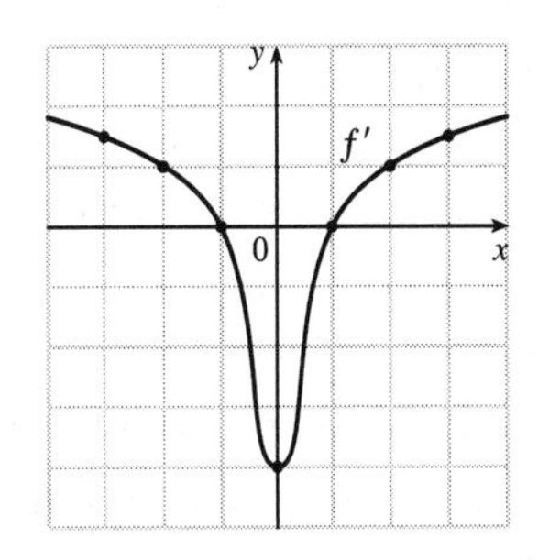

1. It appears that f is an odd function, so f' will be an even function—that is, $f'(-a) = f'(a)$.

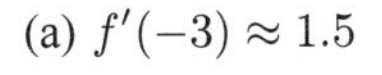

(a) $f'(-3) \approx 1.5$ (b) $f'(-2) \approx 1$

(c) $f'(-1) \approx 0$

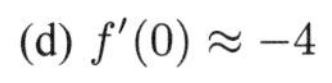

(d) $f'(0) \approx -4$

(e) $f'(1) \approx 0$

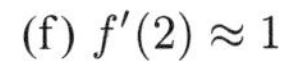

(f) $f'(2) \approx 1$

(g) $f'(3) \approx 1.5$

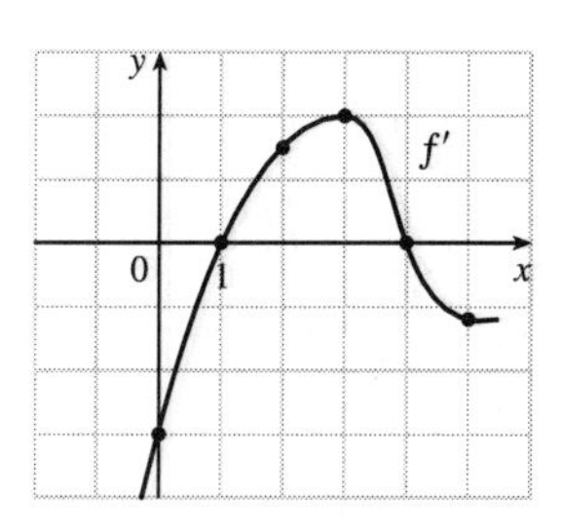

2. *Note:* Your answers may vary depending on your estimates. By estimating the slopes of tangent lines on the graph of f, it appears that

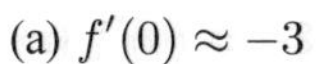

(a) $f'(0) \approx -3$ (b) $f'(1) \approx 0$

(c) $f'(2) \approx 1.5$ (d) $f'(3) \approx 2$

(e) $f'(4) \approx 0$ (f) $f'(5) \approx -1.2$

3. (a)$'$ = II, since from left to right, the slopes of the tangents to graph (a) start out negative, become 0, then positive, then 0, then negative again. The actual function values in graph II follow the same pattern.

(b)$'$ = IV, since from left to right, the slopes of the tangents to graph (b) start out at a fixed positive quantity, then suddenly become negative, then positive again. The discontinuities in graph IV indicate sudden changes in the slopes of the tangents.

(c)$'$ = I, since the slopes of the tangents to graph (c) are negative for $x < 0$ and positive for $x > 0$, as are the function values of graph I.

(d)$'$ = III, since from left to right, the slopes of the tangents to graph (d) are positive, then 0, then negative, then 0, then positive, then 0, then negative again, and the function values in graph III follow the same pattern.

Hints for Exercises 4–11: First plot x-intercepts on the graph of f' for any horizontal tangents on the graph of f. Look for any corners on the graph of f—there will be a discontinuity on the graph of f'. On any interval where f has a tangent with positive (or negative) slope, the graph of f' will be positive (or negative). If the graph of the function is linear, the graph of f' will be a horizontal line.

4.

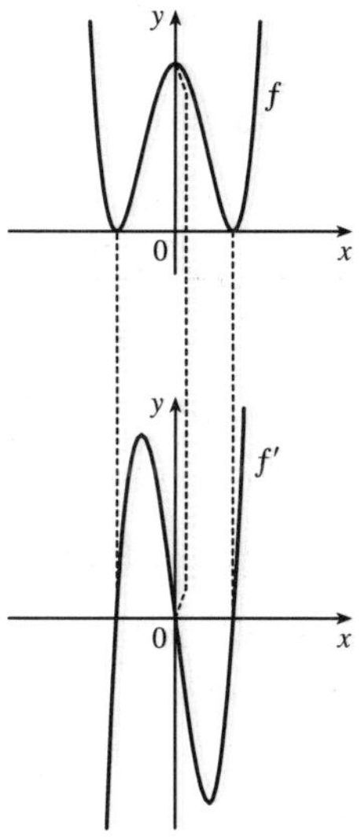

5.

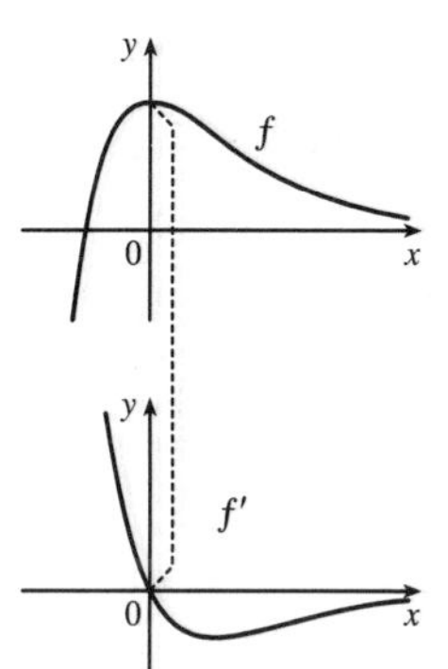

6.

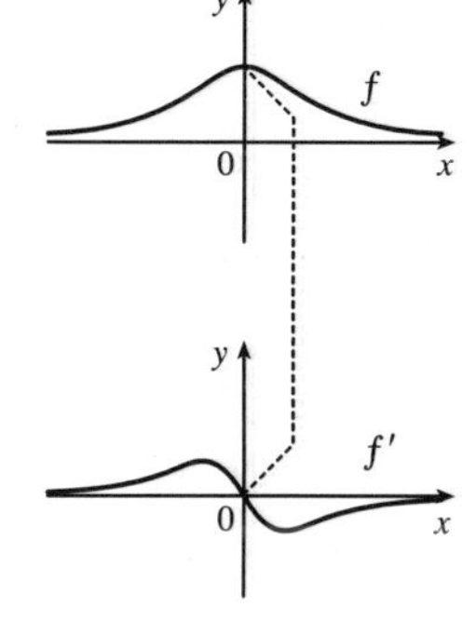

7.

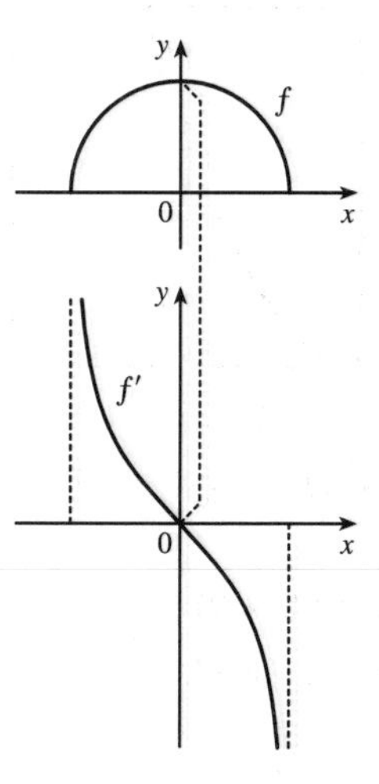

8.

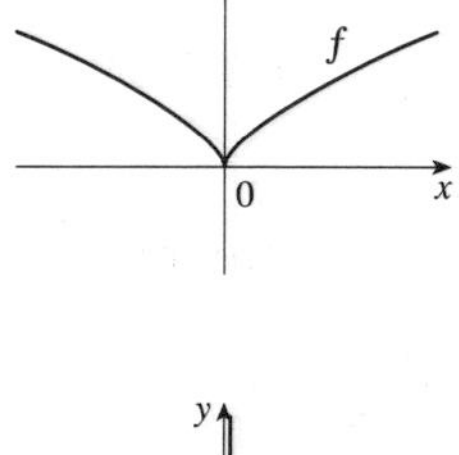

9.

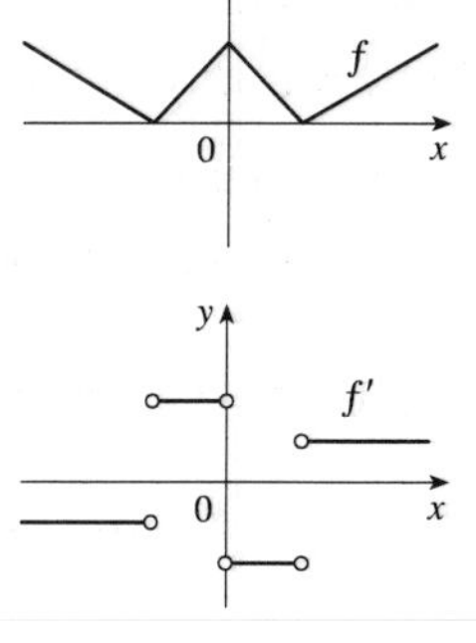

10.

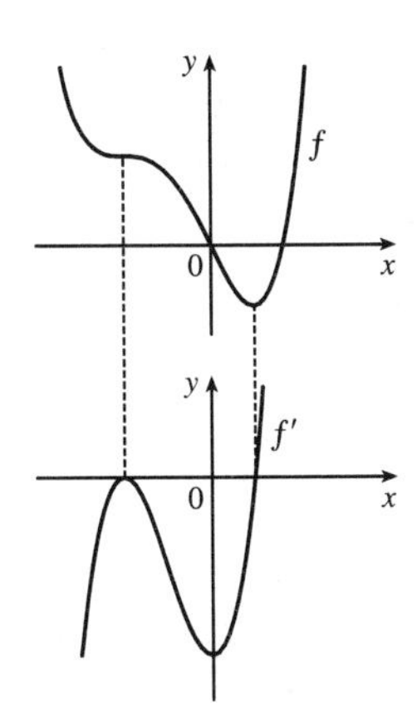

11.

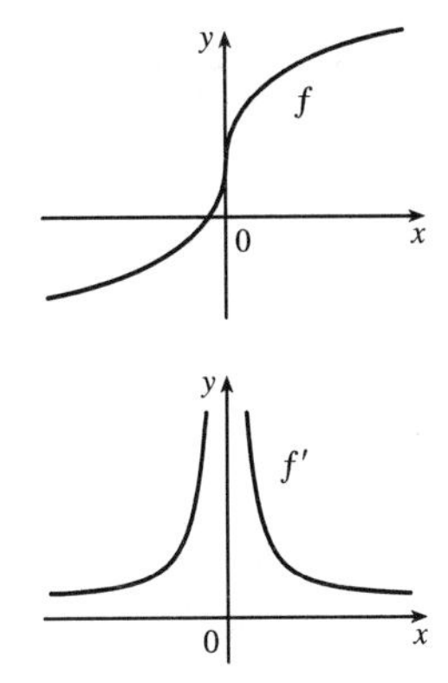

12. The slopes of the tangent lines on the graph of $y = P(t)$ are always positive, so the y-values of $y = P'(t)$ are always positive. These values start out relatively small and keep increasing, reaching a maximum at about $t = 6$. Then the y-values of $y = P'(t)$ decrease and get close to zero. The graph of P' tells us that the yeast culture grows most rapidly after 6 hours and then the growth rate declines.

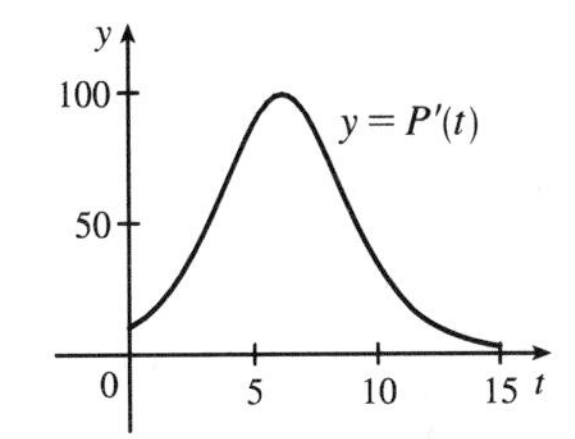

13. It appears that there are horizontal tangents on the graph of M for $t = 1963$ and $t = 1971$. Thus, there are zeros for those values of t on the graph of M'. The derivative is negative for the years 1963 to 1971.

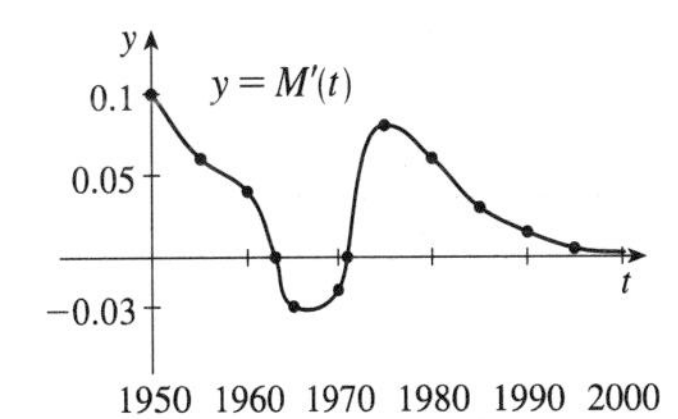

14.

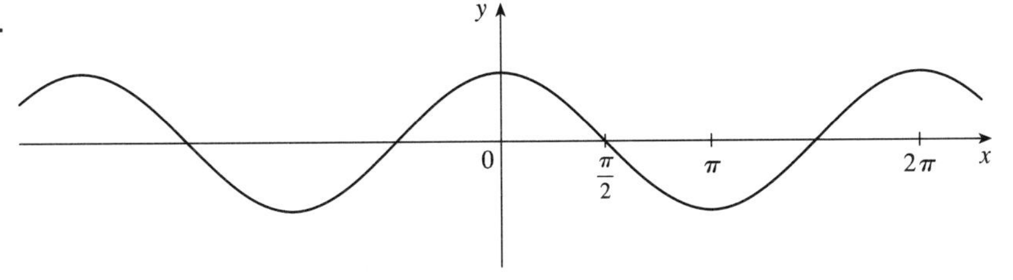

The graph of the derivative looks like the graph of the cosine function.

15. (a) By zooming in, we estimate that $f'(0) = 0$, $f'\left(\frac{1}{2}\right) = 1$, $f'(1) = 2$, and $f'(2) = 4$.

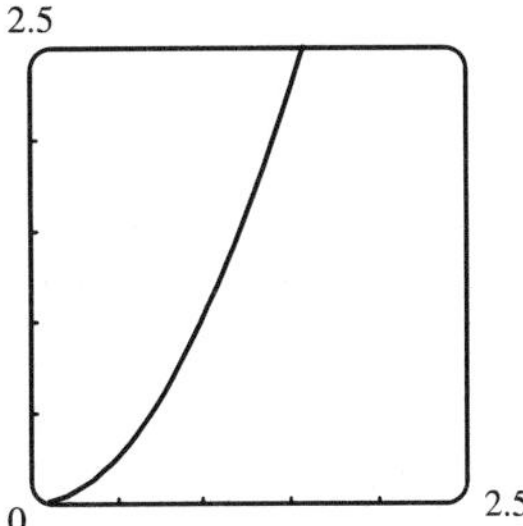

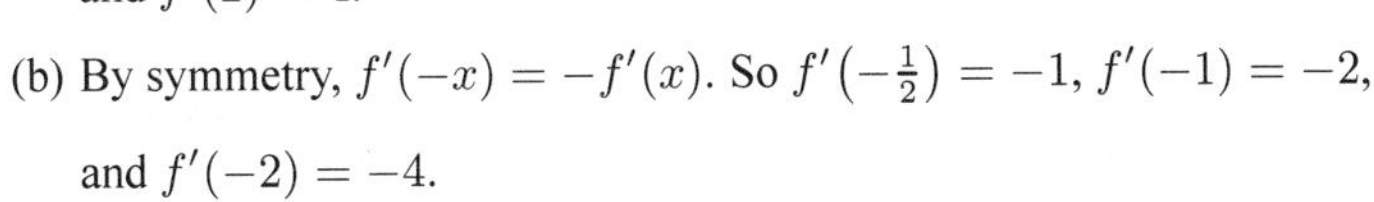

(b) By symmetry, $f'(-x) = -f'(x)$. So $f'\left(-\frac{1}{2}\right) = -1$, $f'(-1) = -2$, and $f'(-2) = -4$.

(c) It appears that $f'(x)$ is twice the value of x, so we guess that $f'(x) = 2x$.

(d) $f'(x) = \lim\limits_{h\to 0} \dfrac{f(x+h) - f(x)}{h} = \lim\limits_{h\to 0} \dfrac{(x+h)^2 - x^2}{h}$

$$= \lim_{h\to 0} \frac{(x^2 + 2hx + h^2) - x^2}{h} = \lim_{h\to 0} \frac{2hx + h^2}{h} = \lim_{h\to 0} \frac{h(2x+h)}{h} = \lim_{h\to 0}(2x+h) = 2x$$

16. (a) By zooming in, we estimate that $f'(0) = 0$, $f'\left(\frac{1}{2}\right) \approx 0.75$, $f'(1) \approx 3$, $f'(2) \approx 12$, and $f'(3) \approx 27$.

(b) By symmetry, $f'(-x) = f'(x)$. So $f'\left(-\frac{1}{2}\right) \approx 0.75$, $f'(-1) \approx 3$, $f'(-2) \approx 12$, and $f'(-3) \approx 27$.

(c)

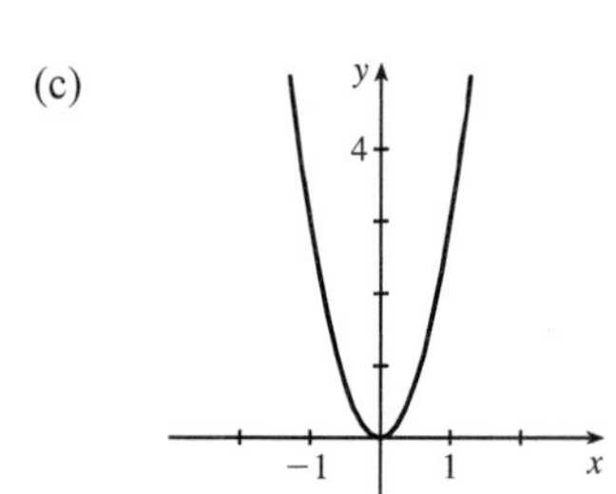

(d) Since $f'(0) = 0$, it appears that f' may have the form $f'(x) = ax^2$. Using $f'(1) = 3$, we have $a = 3$, so $f'(x) = 3x^2$.

(e) $f'(x) = \lim\limits_{h \to 0} \dfrac{f(x+h) - f(x)}{h} = \lim\limits_{h \to 0} \dfrac{(x+h)^3 - x^3}{h} = \lim\limits_{h \to 0} \dfrac{(x^3 + 3x^2h + 3xh^2 + h^3) - x^3}{h}$

$= \lim\limits_{h \to 0} \dfrac{3x^2h + 3xh^2 + h^3}{h} = \lim\limits_{h \to 0} \dfrac{h(3x^2 + 3xh + h^2)}{h} = \lim\limits_{h \to 0}(3x^2 + 3xh + h^2) = 3x^2$

17. $f'(x) = \lim\limits_{h \to 0} \dfrac{f(x+h) - f(x)}{h} = \lim\limits_{h \to 0} \dfrac{\left[\frac{1}{2}(x+h) - \frac{1}{3}\right] - \left(\frac{1}{2}x - \frac{1}{3}\right)}{h} = \lim\limits_{h \to 0} \dfrac{\frac{1}{2}x + \frac{1}{2}h - \frac{1}{3} - \frac{1}{2}x + \frac{1}{3}}{h}$

$= \lim\limits_{h \to 0} \dfrac{\frac{1}{2}h}{h} = \lim\limits_{h \to 0} \frac{1}{2} = \frac{1}{2}$

Domain of f = domain of $f' = \mathbb{R}$.

18. $f'(x) = \lim\limits_{h \to 0} \dfrac{f(x+h) - f(x)}{h} = \lim\limits_{h \to 0} \dfrac{\left[1.5(x+h)^2 - (x+h) + 3.7\right] - \left(1.5x^2 - x + 3.7\right)}{h}$

$= \lim\limits_{h \to 0} \dfrac{1.5x^2 + 3xh + 1.5h^2 - x - h + 3.7 - 1.5x^2 + x - 3.7}{h} = \lim\limits_{h \to 0} \dfrac{3xh + 1.5h^2 - h}{h}$

$= \lim\limits_{h \to 0}(3x + 1.5h - 1) = 3x - 1$

Domain of f = domain of $f' = \mathbb{R}$.

19. $f'(x) = \lim\limits_{h \to 0} \dfrac{f(x+h) - f(x)}{h} = \lim\limits_{h \to 0} \dfrac{\left[(x+h)^3 - 3(x+h) + 5\right] - (x^3 - 3x + 5)}{h}$

$= \lim\limits_{h \to 0} \dfrac{(x^3 + 3x^2h + 3xh^2 + h^3 - 3x - 3h + 5) - (x^3 - 3x + 5)}{h} = \lim\limits_{h \to 0} \dfrac{3x^2h + 3xh^2 + h^3 - 3h}{h}$

$= \lim\limits_{h \to 0} \dfrac{h(3x^2 + 3xh + h^2 - 3)}{h} = \lim\limits_{h \to 0}(3x^2 + 3xh + h^2 - 3) = 3x^2 - 3$

Domain of f = domain of $f' = \mathbb{R}$.

20. $f'(x) = \lim\limits_{h \to 0} \dfrac{f(x+h) - f(x)}{h} = \lim\limits_{h \to 0} \dfrac{\left(x + h + \sqrt{x+h}\right) - \left(x + \sqrt{x}\right)}{h}$

$= \lim\limits_{h \to 0} \left(\dfrac{h}{h} + \dfrac{\sqrt{x+h} - \sqrt{x}}{h} \cdot \dfrac{\sqrt{x+h} + \sqrt{x}}{\sqrt{x+h} + \sqrt{x}}\right) = \lim\limits_{h \to 0} \left[1 + \dfrac{(x+h) - x}{h\left(\sqrt{x+h} + \sqrt{x}\right)}\right]$

$= \lim\limits_{h \to 0} \left(1 + \dfrac{1}{\sqrt{x+h} + \sqrt{x}}\right) = 1 + \dfrac{1}{\sqrt{x} + \sqrt{x}} = 1 + \dfrac{1}{2\sqrt{x}}$

Domain of $f = [0, \infty)$, domain of $f' = (0, \infty)$.

21. $g'(x) = \lim\limits_{h \to 0} \dfrac{g(x+h) - g(x)}{h} = \lim\limits_{h \to 0} \dfrac{\sqrt{1 + 2(x+h)} - \sqrt{1 + 2x}}{h} \left[\dfrac{\sqrt{1 + 2(x+h)} + \sqrt{1 + 2x}}{\sqrt{1 + 2(x+h)} + \sqrt{1 + 2x}}\right]$

$= \lim\limits_{h \to 0} \dfrac{(1 + 2x + 2h) - (1 + 2x)}{h\left[\sqrt{1 + 2(x+h)} + \sqrt{1 + 2x}\right]} = \lim\limits_{h \to 0} \dfrac{2}{\sqrt{1 + 2x + 2h} + \sqrt{1 + 2x}} = \dfrac{2}{2\sqrt{1 + 2x}} = \dfrac{1}{\sqrt{1 + 2x}}$

Domain of $g = \left[-\frac{1}{2}, \infty\right)$, domain of $g' = \left(-\frac{1}{2}, \infty\right)$.

22. $f'(x) = \lim\limits_{h \to 0} \dfrac{f(x+h) - f(x)}{h} = \lim\limits_{h \to 0} \dfrac{\dfrac{3+(x+h)}{1-3(x+h)} - \dfrac{3+x}{1-3x}}{h} = \lim\limits_{h \to 0} \dfrac{(3+x+h)(1-3x) - (3+x)(1-3x-3h)}{h(1-3x-3h)(1-3x)}$

$= \lim\limits_{h \to 0} \dfrac{(3 - 9x + x - 3x^2 + h - 3hx) - (3 - 9x - 9h + x - 3x^2 - 3hx)}{h(1-3x-3h)(1-3x)}$

$= \lim\limits_{h \to 0} \dfrac{10h}{h(1-3x-3h)(1-3x)} = \lim\limits_{h \to 0} \dfrac{10}{(1-3x-3h)(1-3x)} = \dfrac{10}{(1-3x)^2}$

Domain of f = domain of $f' = \left(-\infty, \frac{1}{3}\right) \cup \left(\frac{1}{3}, \infty\right)$.

23. $G'(t) = \lim\limits_{h \to 0} \dfrac{G(t+h) - G(t)}{h} = \lim\limits_{h \to 0} \dfrac{\dfrac{4(t+h)}{(t+h)+1} - \dfrac{4t}{t+1}}{h} = \lim\limits_{h \to 0} \dfrac{\dfrac{4(t+h)(t+1) - 4t(t+h+1)}{(t+h+1)(t+1)}}{h}$

$= \lim\limits_{h \to 0} \dfrac{(4t^2 + 4ht + 4t + 4h) - (4t^2 + 4ht + 4t)}{h(t+h+1)(t+1)} = \lim\limits_{h \to 0} \dfrac{4h}{h(t+h+1)(t+1)}$

$= \lim\limits_{h \to 0} \dfrac{4}{(t+h+1)(t+1)} = \dfrac{4}{(t+1)^2}$

Domain of G = domain of $G' = (-\infty, -1) \cup (-1, \infty)$.

24. (a)

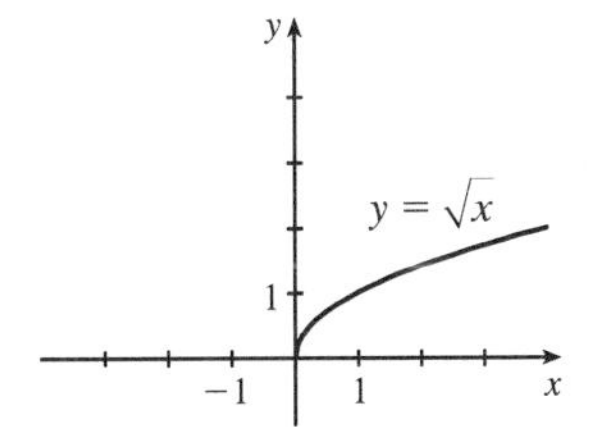

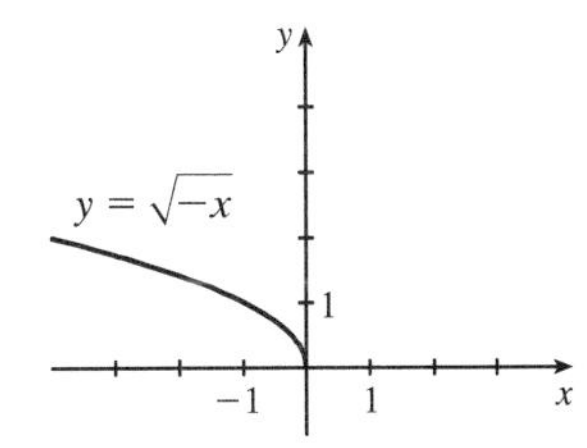

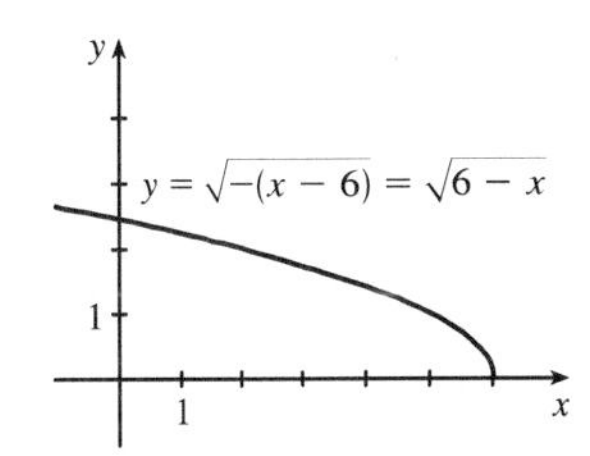

(b) Note that the third graph in part (a) has small negative values for its slope, f'; but as $x \to 6^-$, $f' \to -\infty$.

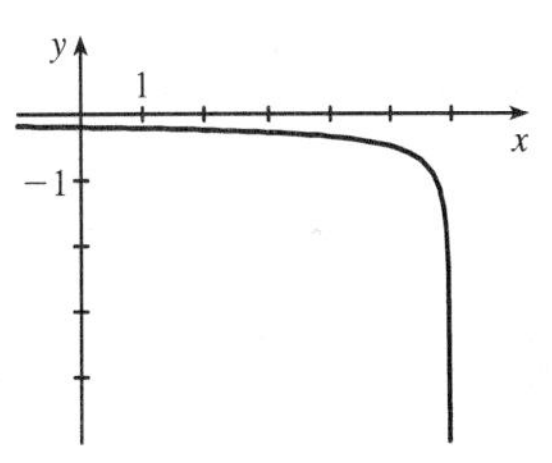

(c) $f'(x) = \lim\limits_{h \to 0} \dfrac{f(x+h) - f(x)}{h} = \lim\limits_{h \to 0} \dfrac{\sqrt{6-(x+h)} - \sqrt{6-x}}{h} \left[\dfrac{\sqrt{6-(x+h)} + \sqrt{6-x}}{\sqrt{6-(x+h)} + \sqrt{6-x}}\right]$

$= \lim\limits_{h \to 0} \dfrac{[6-(x+h)] - (6-x)}{h\left[\sqrt{6-(x+h)} + \sqrt{6-x}\right]} = \lim\limits_{h \to 0} \dfrac{-h}{h\left(\sqrt{6-x-h} + \sqrt{6-x}\right)}$

$= \lim\limits_{h \to 0} \dfrac{-1}{\sqrt{6-x-h} + \sqrt{6-x}} = \dfrac{-1}{2\sqrt{6-x}}$

Domain of $f = (-\infty, 6]$, domain of $f' = (-\infty, 6)$.

(d) The graph produced by a graphing device looks similar to the one in part (b).

25. (a) $f'(x) = \lim\limits_{h \to 0} \dfrac{f(x+h) - f(x)}{h} = \lim\limits_{h \to 0} \dfrac{[(x+h)^4 + 2(x+h)] - (x^4 + 2x)}{h}$

$= \lim\limits_{h \to 0} \dfrac{x^4 + 4x^3h + 6x^2h^2 + 4xh^3 + h^4 + 2x + 2h - x^4 - 2x}{h}$

$= \lim\limits_{h \to 0} \dfrac{4x^3h + 6x^2h^2 + 4xh^3 + h^4 + 2h}{h} = \lim\limits_{h \to 0} \dfrac{h(4x^3 + 6x^2h + 4xh^2 + h^3 + 2)}{h}$

$= \lim\limits_{h \to 0} (4x^3 + 6x^2h + 4xh^2 + h^3 + 2) = 4x^3 + 2$

(b) Notice that $f'(x) = 0$ when f has a horizontal tangent, $f'(x)$ is positive when the tangents have positive slope, and $f'(x)$ is negative when the tangents have negative slope

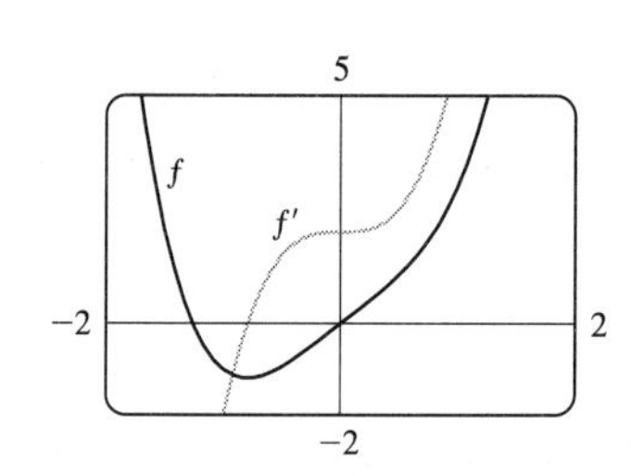

26. (a) $f'(t) = \lim\limits_{h\to 0} \dfrac{f(t+h) - f(t)}{h} = \lim\limits_{h\to 0} \dfrac{\left[(t+h)^2 - \sqrt{t+h}\,\right] - (t^2 - \sqrt{t}\,)}{h}$

$$= \lim_{h\to 0} \frac{t^2 + 2ht + h^2 - \sqrt{t+h} - t^2 + \sqrt{t}}{h} = \lim_{h\to 0}\left(\frac{2ht + h^2}{h} + \frac{\sqrt{t} - \sqrt{t+h}}{h}\right)$$

$$= \lim_{h\to 0}\left(\frac{h(2t+h)}{h} + \frac{\sqrt{t} - \sqrt{t+h}}{h} \cdot \frac{\sqrt{t} + \sqrt{t+h}}{\sqrt{t} + \sqrt{t+h}}\right)$$

$$= \lim_{h\to 0}\left(2t + h + \frac{t - (t+h)}{h\left(\sqrt{t} + \sqrt{t+h}\,\right)}\right) = \lim_{h\to 0}\left(2t + h + \frac{-h}{h\left(\sqrt{t} + \sqrt{t+h}\,\right)}\right)$$

$$= \lim_{h\to 0}\left(2t + h + \frac{-1}{\sqrt{t} + \sqrt{t+h}}\right) = 2t - \frac{1}{2\sqrt{t}}$$

(b) Notice that $f'(t) = 0$ when f has a horizontal tangent, $f'(t)$ is positive when the tangents have positive slope, and $f'(t)$ is negative when the tangents have negative slope.

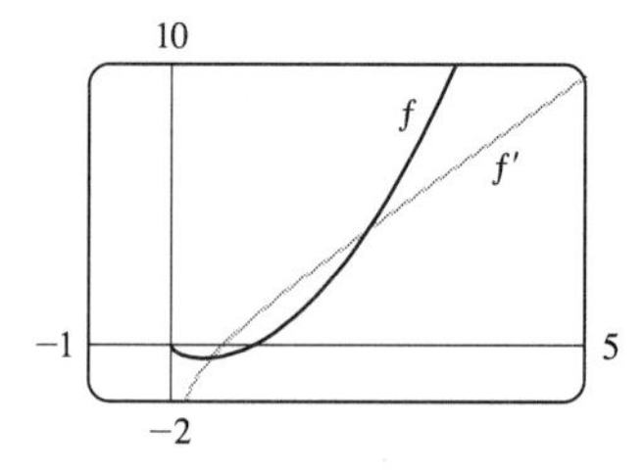

27. f is not differentiable at $x = -4$, because the graph has a corner there, and at $x = 0$, because there is a discontinuity there.

28. f is not differentiable at $x = 0$, because there is a discontinuity there, and at $x = 3$, because the graph has a vertical tangent there.

29. f is not differentiable at $x = -1$, because the graph has a vertical tangent there, and at $x = 4$, because the graph has a corner there.

30. f is not differentiable at $x = -1$, because there is a discontinuity there, and at $x = 2$, because the graph has a corner there.

31. As we zoom in toward $(-1, 0)$, the curve appears more and more like a straight line, so $f(x) = x + \sqrt{|x|}$ is differentiable at $x = -1$. But no matter how much we zoom in toward the origin, the curve doesn't straighten out—we can't eliminate the sharp point (a cusp). So f is not differentiable at $x = 0$.

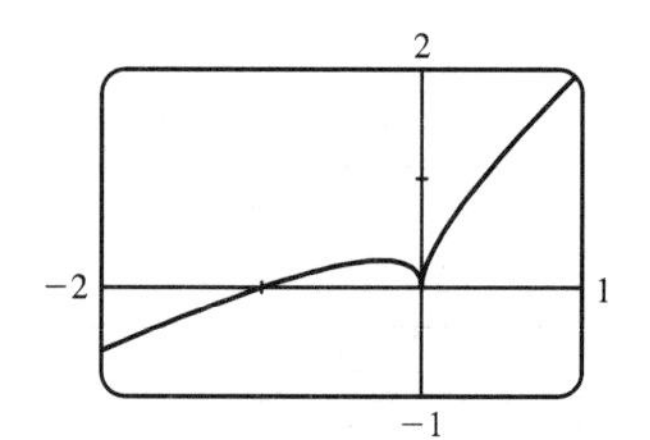

32. As we zoom in toward $(0, 1)$, the curve appears more and more like a straight line, so f is differentiable at $x = 0$. But no matter how much we zoom in toward $(1, 0)$ or $(-1, 0)$, the curve doesn't straighten out—we can't eliminate the sharp point (a cusp). So f is not differentiable at $x = \pm 1$.

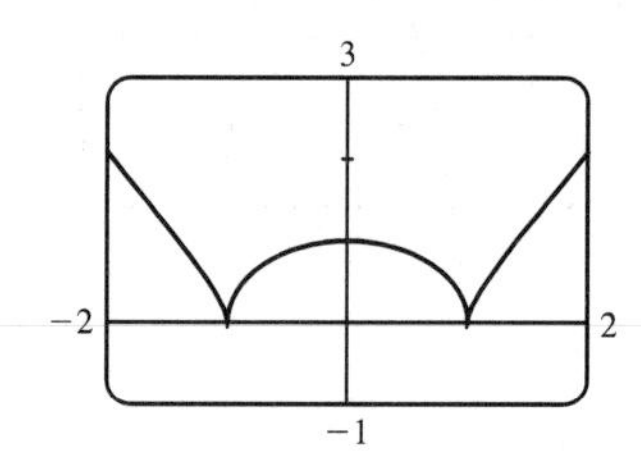

33. $a = f, b = f', c = f''$. We can see this because where a has a horizontal tangent, $b = 0$, and where b has a horizontal tangent, $c = 0$. We can immediately see that c can be neither f nor f', since at the points where c has a horizontal tangent, neither a nor b is equal to 0.

34. Where d has horizontal tangents, only c is 0, so $d' = c$. Curve c has negative tangents for $x < 0$ and b is the only graph that is negative for $x < 0$, so $c' = b$. Curve b has positive tangents on $\mathbb{R}$ (except at $x = 0$), and the only graph that is positive on the same domain is a, so $b' = a$. We conclude that $d = f, c = f', b = f''$, and $a = f'''$.

35. We can immediately see that a is the graph of the acceleration function, since at the points where a has a horizontal tangent, neither c nor b is equal to 0. Next, we note that $a = 0$ at the point where b has a horizontal tangent, so b must be the graph of the velocity function, and hence, $b' = a$. We conclude that c is the graph of the position function.

36. $$f'(x) = \lim_{h \to 0} \frac{f(x+h) - f(x)}{h} = \lim_{h \to 0} \frac{\dfrac{1}{x+h} - \dfrac{1}{x}}{h} = \lim_{h \to 0} \frac{x - (x+h)}{hx(x+h)} = \lim_{h \to 0} \frac{-h}{hx(x+h)} = \lim_{h \to 0} \frac{-1}{x(x+h)} = -\frac{1}{x^2}$$

$$f''(x) = \lim_{h \to 0} \frac{f'(x+h) - f'(x)}{h} = \lim_{h \to 0} \frac{-\dfrac{1}{(x+h)^2} - \left(-\dfrac{1}{x^2}\right)}{h} = \lim_{h \to 0} \frac{-x^2 + (x+h)^2}{hx^2(x+h)^2} = \lim_{h \to 0} \frac{2hx + h^2}{hx^2(x+h)^2}$$
$$= \lim_{h \to 0} \frac{2x + h}{x^2(x+h)^2} = \frac{2x}{x^4} = \frac{2}{x^3}$$

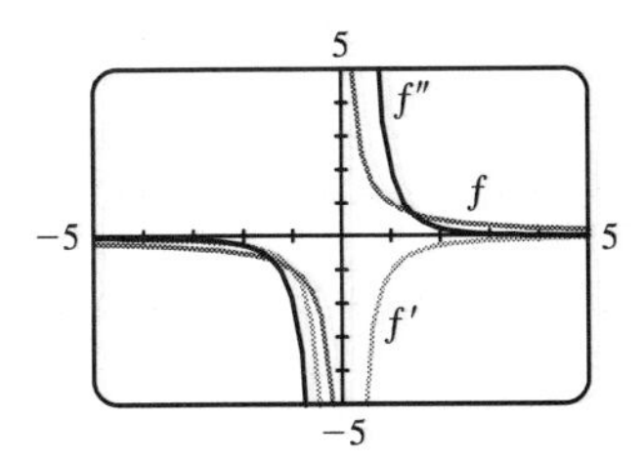

We see from the graph that our answers are reasonable because the graph of f' is that of an even function and is negative for all $x \neq 0$, and the graph of f'' is that of an odd function (negative for $x < 0$ and positive for $x > 0$).

37. $$f'(x) = \lim_{h \to 0} \frac{f(x+h) - f(x)}{h} = \lim_{h \to 0} \frac{\left[1 + 4(x+h) - (x+h)^2\right] - (1 + 4x - x^2)}{h}$$
$$= \lim_{h \to 0} \frac{(1 + 4x + 4h - x^2 - 2xh - h^2) - (1 + 4x - x^2)}{h} = \lim_{h \to 0} \frac{4h - 2xh - h^2}{h} = \lim_{h \to 0} (4 - 2x - h) = 4 - 2x$$

$$f''(x) = \lim_{h \to 0} \frac{f'(x+h) - f'(x)}{h} = \lim_{h \to 0} \frac{[4 - 2(x+h)] - (4 - 2x)}{h} = \lim_{h \to 0} \frac{-2h}{h} = \lim_{h \to 0} (-2) = -2$$

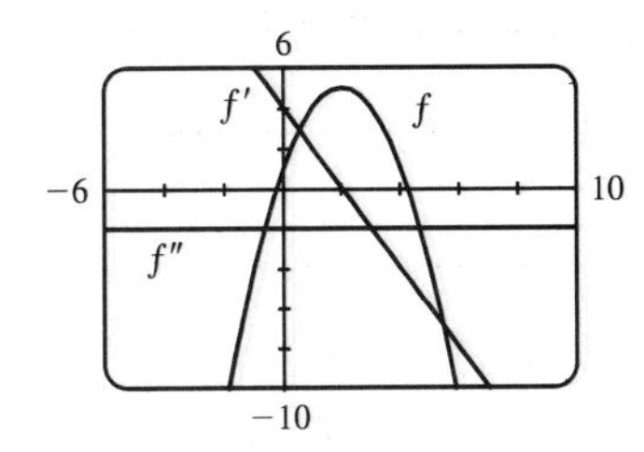

We see from the graph that our answers are reasonable because the graph of f' is that of a linear function and the graph of f'' is that of a constant function.

38. $f'(x) = \lim_{h\to 0} \dfrac{f(x+h) - f(x)}{h} = \lim_{h\to 0} \dfrac{\left[2(x+h)^2 - (x+h)^3\right] - (2x^2 - x^3)}{h}$

$= \lim_{h\to 0} \dfrac{h(4x + 2h - 3x^2 - 3xh - h^2)}{h} = \lim_{h\to 0}(4x + 2h - 3x^2 - 3xh - h^2) = 4x - 3x^2$

$$f''(x) = \lim_{h\to 0} \frac{f'(x+h) - f'(x)}{h} = \lim_{h\to 0} \frac{\left[4(x+h) - 3(x+h)^2\right] - (4x - 3x^2)}{h} = \lim_{h\to 0} \frac{h(4 - 6x - 3h)}{h}$$
$$= \lim_{h\to 0}(4 - 6x - 3h) = 4 - 6x$$

$$f'''(x) = \lim_{h\to 0} \frac{f''(x+h) - f''(x)}{h} = \lim_{h\to 0} \frac{\left[4 - 6(x+h)\right] - (4 - 6x)}{h} = \lim_{h\to 0} \frac{-6h}{h} = \lim_{h\to 0}(-6) = -6$$

$$f^{(4)}(x) = \lim_{h\to 0} \frac{f'''(x+h) - f'''(x)}{h} = \lim_{h\to 0} \frac{-6 - (-6)}{h} = \lim_{h\to 0} \frac{0}{h} = \lim_{h\to 0}(0) = 0$$

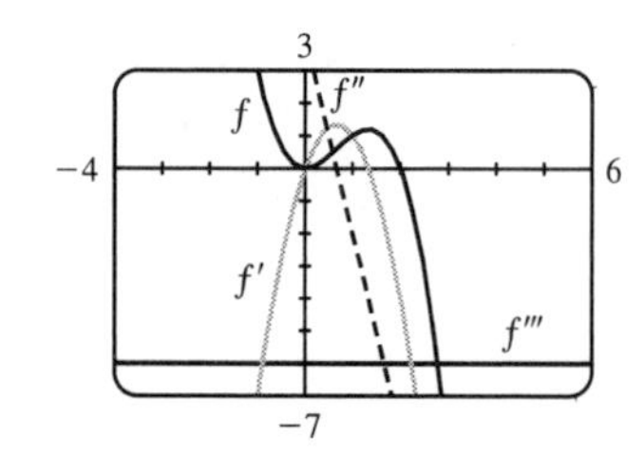

The graphs are consistent with the geometric interpretations of the derivatives because f' has zeros where f has a local minimum and a local maximum, f'' has a zero where f' has a local maximum, and f''' is a constant function equal to the slope of f''.

39. (a) Note that we have factored $x - a$ as the difference of two cubes in the third step.

$$f'(a) = \lim_{x\to a} \frac{f(x) - f(a)}{x - a} = \lim_{x\to a} \frac{x^{1/3} - a^{1/3}}{x - a} = \lim_{x\to a} \frac{x^{1/3} - a^{1/3}}{(x^{1/3} - a^{1/3})(x^{2/3} + x^{1/3}a^{1/3} + a^{2/3})}$$
$$= \lim_{x\to a} \frac{1}{x^{2/3} + x^{1/3}a^{1/3} + a^{2/3}} = \frac{1}{3a^{2/3}} \text{ or } \tfrac{1}{3}a^{-2/3}$$

(b) $f'(0) = \lim_{h\to 0} \dfrac{f(0+h) - f(0)}{h} = \lim_{h\to 0} \dfrac{\sqrt[3]{h} - 0}{h} = \lim_{h\to 0} \dfrac{1}{h^{2/3}}$. This function increases without bound, so the limit does not exist, and therefore $f'(0)$ does not exist.

(c) $\lim_{x\to 0} |f'(x)| = \lim_{x\to 0} \dfrac{1}{3x^{2/3}} = \infty$ and f is continuous at $x = 0$ (root function), so f has a vertical tangent at $x = 0$.

40. (a) $g'(0) = \lim_{x\to 0} \dfrac{g(x) - g(0)}{x - 0} = \lim_{x\to 0} \dfrac{x^{2/3} - 0}{x} = \lim_{x\to 0} \dfrac{1}{x^{1/3}}$, which does not exist.

(b) $g'(a) = \lim_{x\to a} \dfrac{g(x) - g(a)}{x - a} = \lim_{x\to a} \dfrac{x^{2/3} - a^{2/3}}{x - a} = \lim_{x\to a} \dfrac{(x^{1/3} - a^{1/3})(x^{1/3} + a^{1/3})}{(x^{1/3} - a^{1/3})(x^{2/3} + x^{1/3}a^{1/3} + a^{2/3})}$

$$= \lim_{x\to a} \frac{x^{1/3} + a^{1/3}}{x^{2/3} + x^{1/3}a^{1/3} + a^{2/3}} = \frac{2a^{1/3}}{3a^{2/3}} = \frac{2}{3a^{1/3}} \text{ or } \tfrac{2}{3}a^{-1/3}$$

(c) $g(x) = x^{2/3}$ is continuous at $x = 0$ and

$\lim_{x\to 0} |g'(x)| = \lim_{x\to 0} \dfrac{2}{3\,|x|^{1/3}} = \infty$. This shows that

g has a vertical tangent line at $x = 0$.

(d)

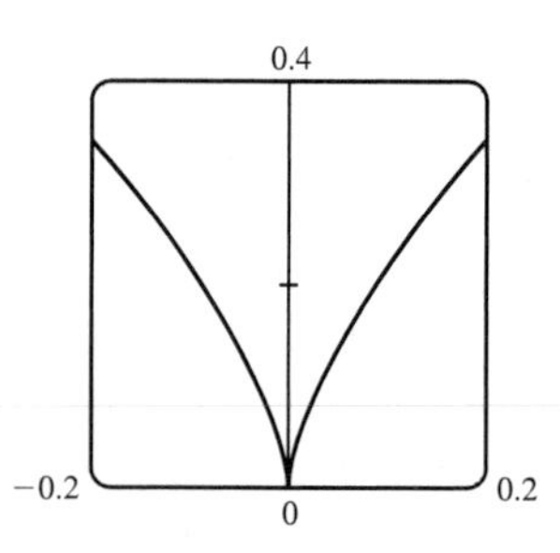

41. $f(x) = |x-6| = \begin{cases} x-6 & \text{if } x-6 \geq 6 \\ -(x-6) & \text{if } x-6 < 0 \end{cases} = \begin{cases} x-6 & \text{if } x \geq 6 \\ 6-x & \text{if } x < 6 \end{cases}$

So the right-hand limit is $\lim\limits_{x\to 6^+} \dfrac{f(x)-f(6)}{x-6} = \lim\limits_{x\to 6^+} \dfrac{|x-6|-0}{x-6} = \lim\limits_{x\to 6^+} \dfrac{x-6}{x-6} = \lim\limits_{x\to 6^+} 1 = 1$, and the left-hand limit is $\lim\limits_{x\to 6^-} \dfrac{f(x)-f(6)}{x-6} = \lim\limits_{x\to 6^-} \dfrac{|x-6|-0}{x-6} = \lim\limits_{x\to 6^-} \dfrac{6-x}{x-6} = \lim\limits_{x\to 6^-} (-1) = -1$. Since these limits are not equal,

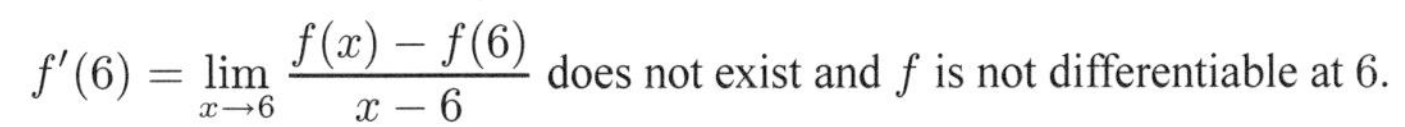

$f'(6) = \lim\limits_{x\to 6} \dfrac{f(x)-f(6)}{x-6}$ does not exist and f is not differentiable at 6.

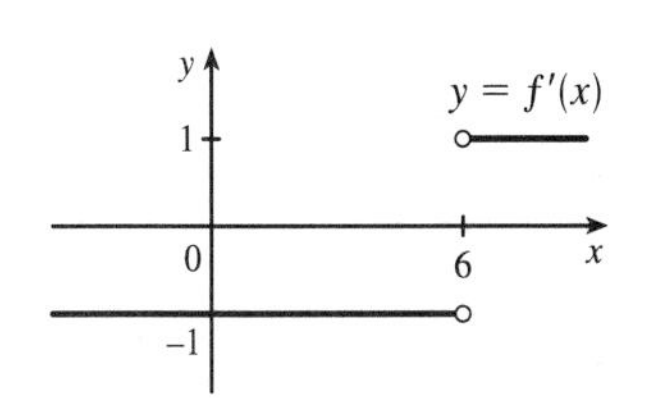

However, a formula for f' is $f'(x) = \begin{cases} 1 & \text{if } x > 6 \\ -1 & \text{if } x < 6 \end{cases}$

Another way of writing the formula is $f'(x) = \dfrac{x-6}{|x-6|}$.

42. $f(x) = [\![x]\!]$ is not continuous at any integer n, so f is not differentiable at n by the contrapositive of Theorem 4. If a is not an integer, then f is constant on an open interval containing a, so $f'(a) = 0$. Thus, $f'(x) = 0$, x not an integer.

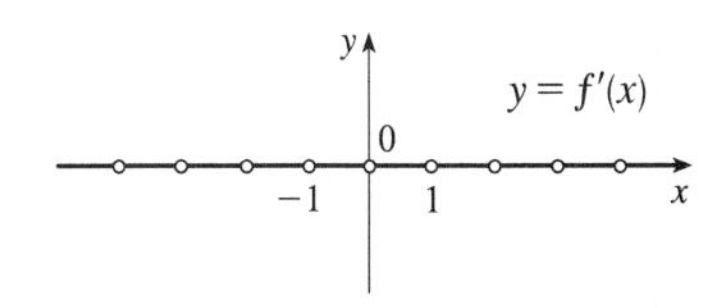

43. (a) If f is even, then

$$\begin{aligned} f'(-x) &= \lim_{h\to 0} \frac{f(-x+h)-f(-x)}{h} = \lim_{h\to 0} \frac{f[-(x-h)]-f(-x)}{h} \\ &= \lim_{h\to 0} \frac{f(x-h)-f(x)}{h} = -\lim_{h\to 0} \frac{f(x-h)-f(x)}{-h} \quad [\text{let } \Delta x = -h] \\ &= -\lim_{\Delta x\to 0} \frac{f(x+\Delta x)-f(x)}{\Delta x} = -f'(x) \end{aligned}$$

Therefore, f' is odd.

(b) If f is odd, then

$$\begin{aligned} f'(-x) &= \lim_{h\to 0} \frac{f(-x+h)-f(-x)}{h} = \lim_{h\to 0} \frac{f[-(x-h)]-f(-x)}{h} \\ &= \lim_{h\to 0} \frac{-f(x-h)+f(x)}{h} = \lim_{h\to 0} \frac{f(x-h)-f(x)}{-h} \quad [\text{let } \Delta x = -h] \\ &= \lim_{\Delta x\to 0} \frac{f(x+\Delta x)-f(x)}{\Delta x} = f'(x) \end{aligned}$$

Therefore, f' is even.

44. (a)

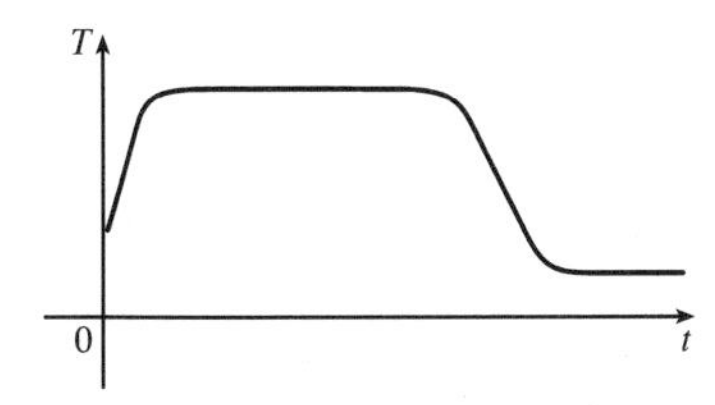

(b) The initial temperature of the water is close to room temperature because of the water that was in the pipes. When the water from the hot water tank starts coming out, dT/dt is large and positive as T increases to the temperature of the water in the tank. In the next phase, $dT/dt = 0$ as the water comes out at a constant, high temperature. After some time, dT/dt becomes small and negative as the contents of the hot water tank are exhausted. Finally, when the hot water has run out, dT/dt is once again 0 as the water maintains its (cold) temperature.

(c) 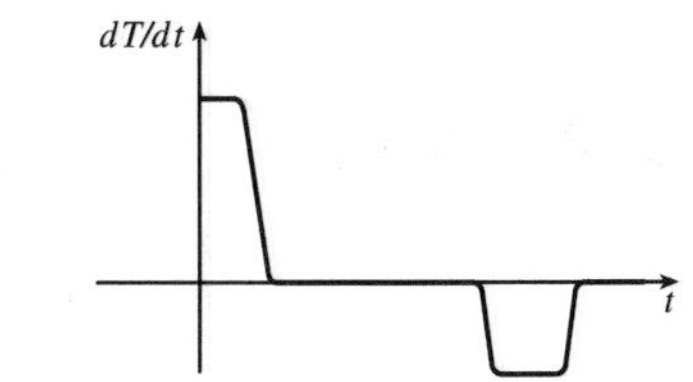

45. 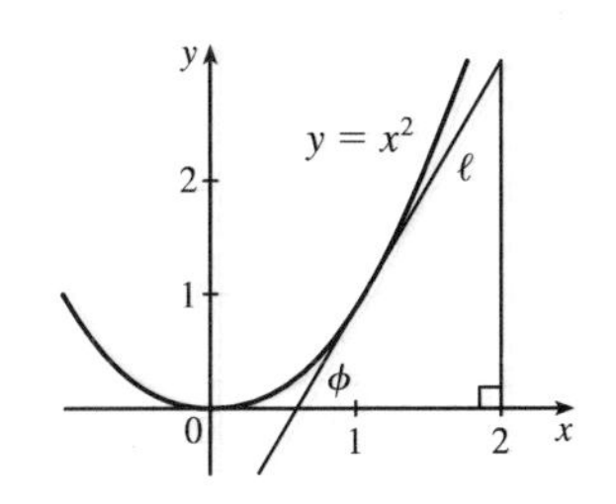

In the right triangle in the diagram, let Δy be the side opposite angle ϕ and Δx the side adjacent angle ϕ. Then the slope of the tangent line ℓ is $m = \Delta y/\Delta x = \tan\phi$. Note that $0 < \phi < \frac{\pi}{2}$. We know (see Exercise 15) that the derivative of $f(x) = x^2$ is $f'(x) = 2x$. So the slope of the tangent to the curve at the point $(1, 1)$ is 2. Thus, ϕ is the angle between 0 and $\frac{\pi}{2}$ whose tangent is 2; that is, $\phi = \tan^{-1} 2 \approx 63^\circ$.

2.3 Basic Differentiation Formulas

1. $f(x) = 186.5$ is a constant function, so its derivative is 0, that is, $f'(x) = 0$.

2. $f(x) = \sqrt{30}$ is a constant function, so its derivative is 0, that is, $f'(x) = 0$.

3. $f(x) = 5x - 1 \quad\Rightarrow\quad f'(x) = 5 - 0 = 5$

4. $F(x) = -4x^{10} \quad\Rightarrow\quad F'(x) = -4(10x^{10-1}) = -40x^9$

5. $f(x) = x^3 - 4x + 6 \quad\Rightarrow\quad f'(x) = 3x^2 - 4(1) + 0 = 3x^2 - 4$

6. $f(t) = \frac{1}{2}t^6 - 3t^4 + t \quad\Rightarrow\quad f'(t) = \frac{1}{2}(6t^5) - 3(4t^3) + 1 = 3t^5 - 12t^3 + 1$

7. $f(x) = x - 3\sin x \quad\Rightarrow\quad f'(x) = 1 - 3\cos x$

8. $y = \sin t + \pi\cos t \quad\Rightarrow\quad y' = \cos t + \pi(-\sin t) = \cos t - \pi\sin t$

9. $f(t) = \frac{1}{4}(t^4 + 8) \quad\Rightarrow\quad f'(t) = \frac{1}{4}(t^4 + 8)' = \frac{1}{4}(4t^{4-1} + 0) = t^3$

10. $h(x) = (x - 2)(2x + 3) = 2x^2 - x - 6 \quad\Rightarrow\quad h'(x) = 2(2x) - 1 - 0 = 4x - 1$

11. $y = x^{-2/5} \quad\Rightarrow\quad y' = -\frac{2}{5}x^{(-2/5)-1} = -\frac{2}{5}x^{-7/5} = -\dfrac{2}{5x^{7/5}}$

12. $R(t) = 5t^{-3/5} \quad\Rightarrow\quad R'(t) = 5\left[-\frac{3}{5}t^{(-3/5)-1}\right] = -3t^{-8/5}$

13. $V(r) = \frac{4}{3}\pi r^3 \quad\Rightarrow\quad V'(r) = \frac{4}{3}\pi(3r^2) = 4\pi r^2$

14. $R(x) = \dfrac{\sqrt{10}}{x^7} = \sqrt{10}\,x^{-7} \quad\Rightarrow\quad R'(x) = -7\sqrt{10}\,x^{-8} = -\dfrac{7\sqrt{10}}{x^8}$

15. $F(x) = (\frac{1}{2}x)^5 = (\frac{1}{2})^5 x^5 = \frac{1}{32}x^5 \quad\Rightarrow\quad F'(x) = \frac{1}{32}(5x^4) = \frac{5}{32}x^4$

16. $y = \sqrt{x}\,(x - 1) = x^{3/2} - x^{1/2} \quad\Rightarrow\quad y' = \frac{3}{2}x^{1/2} - \frac{1}{2}x^{-1/2} = \frac{1}{2}x^{-1/2}(3x - 1)$ [factor out $\frac{1}{2}x^{-1/2}$]

or $y' = \dfrac{3x - 1}{2\sqrt{x}}$.

17. $y = 4\pi^2 \Rightarrow y' = 0$ since $4\pi^2$ is a constant.

18. $g(u) = \sqrt{2}\,u + \sqrt{3u} = \sqrt{2}\,u + \sqrt{3}\sqrt{u} \Rightarrow g'(u) = \sqrt{2}\,(1) + \sqrt{3}\left(\frac{1}{2}u^{-1/2}\right) = \sqrt{2} + \sqrt{3}/(2\sqrt{u})$

19. $y = \dfrac{x^2 + 4x + 3}{\sqrt{x}} = x^{3/2} + 4x^{1/2} + 3x^{-1/2} \Rightarrow$

$y' = \frac{3}{2}x^{1/2} + 4\left(\frac{1}{2}\right)x^{-1/2} + 3\left(-\frac{1}{2}\right)x^{-3/2} = \frac{3}{2}\sqrt{x} + \dfrac{2}{\sqrt{x}} - \dfrac{3}{2x\sqrt{x}}$

$\left[\text{note that } x^{3/2} = x^{2/2} \cdot x^{1/2} = x\sqrt{x}\right]$

20. $y = \dfrac{x^2 - 2\sqrt{x}}{x} = x - 2x^{-1/2} \Rightarrow y' = 1 - 2\left(-\frac{1}{2}\right)x^{-3/2} = 1 + 1/(x\sqrt{x})$

21. $v = t^2 - \dfrac{1}{\sqrt[4]{t^3}} = t^2 - t^{-3/4} \Rightarrow v' = 2t - \left(-\frac{3}{4}\right)t^{-7/4} = 2t + \dfrac{3}{4t^{7/4}} = 2t + \dfrac{3}{4t\sqrt[4]{t^3}}$

22. $y = \dfrac{\sin\theta}{2} + \dfrac{c}{\theta} = \frac{1}{2}\sin\theta + c\theta^{-1} \Rightarrow y' = \frac{1}{2}\cos\theta + c(-1)\theta^{-2} = \dfrac{\cos\theta}{2} - \dfrac{c}{\theta^2}$

23. $z = \dfrac{A}{y^{10}} + B\cos y = Ay^{-10} + B\cos y \Rightarrow \dfrac{dz}{dy} = A(-10)y^{-11} + B(-\sin y) = -\dfrac{10A}{y^{11}} - B\sin y$

24. $u = \sqrt[3]{t^2} + 2\sqrt{t^3} = t^{2/3} + 2t^{3/2} \Rightarrow u' = \frac{2}{3}t^{-1/3} + 2\left(\frac{3}{2}\right)t^{1/2} = \dfrac{2}{3\sqrt[3]{t}} + 3\sqrt{t}$

25. $y = 6\cos x \Rightarrow y' = -6\sin x$. At $(\pi/3, 3)$, $y' = -6\sin(\pi/3) = -6\left(\sqrt{3}/2\right) = -3\sqrt{3}$ and an equation of the tangent line is $y - 3 = -3\sqrt{3}\,(x - \pi/3)$ or $y = -3\sqrt{3}x + 3 + \pi\sqrt{3}$. The slope of the normal line is $1/\left(3\sqrt{3}\right)$ (the negative reciprocal of $-3\sqrt{3}$) and an equation of the normal line is $y - 3 = \dfrac{1}{3\sqrt{3}}\left(x - \dfrac{\pi}{3}\right)$ or $y = \dfrac{1}{3\sqrt{3}}x + 3 - \dfrac{\pi}{9\sqrt{3}}$.

26. $y = (1 + 2x)^2 = 1 + 4x + 4x^2 \Rightarrow y' = 4 + 8x$. At $(1, 9)$, $y' = 12$ and an equation of the tangent line is $y - 9 = 12(x - 1)$ or $y = 12x - 3$. The slope of the normal line is $-\frac{1}{12}$ (the negative reciprocal of 12) and an equation of the normal line is $y - 9 = -\frac{1}{12}(x - 1)$ or $y = -\frac{1}{12}x + \frac{109}{12}$.

27. $y = f(x) = x + \sqrt{x} \Rightarrow f'(x) = 1 + \frac{1}{2}x^{-1/2}$.

So the slope of the tangent line at $(1, 2)$ is $f'(1) = 1 + \frac{1}{2}(1) = \frac{3}{2}$ and its equation is $y - 2 = \frac{3}{2}(x - 1)$ or $y = \frac{3}{2}x + \frac{1}{2}$.

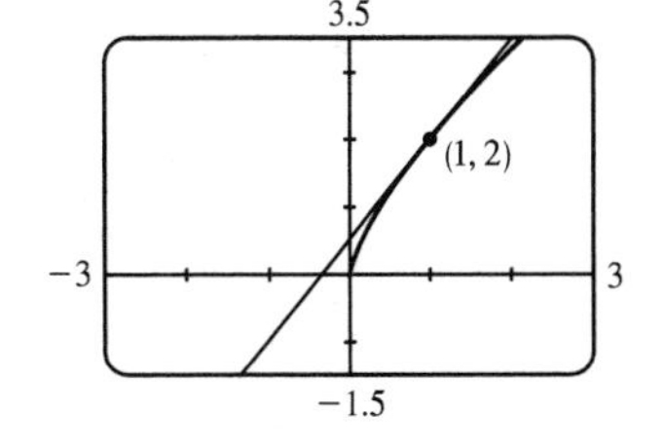

28. $y = 3x^2 - x^3 \Rightarrow y' = 6x - 3x^2$. At $(1, 2)$, $y' = 6 - 3 = 3$, so an equation of the tangent line is $y - 2 = 3(x - 1)$, or $y = 3x - 1$.

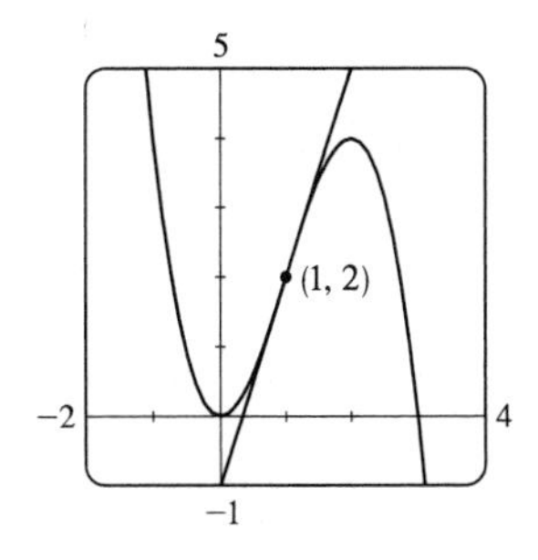

29. $f(x) = x^4 - 3x^3 + 16x \Rightarrow f'(x) = 4x^3 - 9x^2 + 16 \Rightarrow f''(x) = 12x^2 - 18x$

30. $G(r) = \sqrt{r} + \sqrt[3]{r} \Rightarrow G'(r) = \frac{1}{2}r^{-1/2} + \frac{1}{3}r^{-2/3} \Rightarrow G''(r) = -\frac{1}{4}r^{-3/2} - \frac{2}{9}r^{-5/3}$

31. $g(t) = 2\cos t - 3\sin t \Rightarrow g'(t) = -2\sin t - 3\cos t \Rightarrow g''(t) = -2\cos t + 3\sin t$

32. $h(t) = \sqrt{t} + 5\sin t \Rightarrow h'(t) = \frac{1}{2}t^{-1/2} + 5\cos t \Rightarrow h''(t) = -\frac{1}{4}t^{-3/2} - 5\sin t$

33. $\dfrac{d}{dx}(\sin x) = \cos x \Rightarrow \dfrac{d^2}{dx^2}(\sin x) = -\sin x \Rightarrow \dfrac{d^3}{dx^3}(\sin x) = -\cos x \Rightarrow \dfrac{d^4}{dx^4}(\sin x) = \sin x.$

The derivatives of $\sin x$ occur in a cycle of four. Since $99 = 4(24) + 3$, we have $\dfrac{d^{99}}{dx^{99}}(\sin x) = \dfrac{d^3}{dx^3}(\sin x) = -\cos x.$

34. (a) $f(x) = x^n \Rightarrow f'(x) = nx^{n-1} \Rightarrow f''(x) = n(n-1)x^{n-2} \Rightarrow \cdots \Rightarrow$
$f^{(n)}(x) = n(n-1)(n-2)\cdots 2 \cdot 1x^{n-n} = n!$

(b) $f(x) = x^{-1} \Rightarrow f'(x) = (-1)x^{-2} \Rightarrow f''(x) = (-1)(-2)x^{-3} \Rightarrow \cdots \Rightarrow$
$f^{(n)}(x) = (-1)(-2)(-3)\cdots(-n)x^{-(n+1)} = (-1)^n n! x^{-(n+1)}$ or $\dfrac{(-1)^n\, n!}{x^{n+1}}$

35. $f(x) = x + 2\sin x$ has a horizontal tangent when $f'(x) = 0 \Leftrightarrow 1 + 2\cos x = 0 \Leftrightarrow \cos x = -\frac{1}{2} \Leftrightarrow$
$x = \frac{2\pi}{3} + 2\pi n$ or $\frac{4\pi}{3} + 2\pi n$, where n is an integer. Note that $\frac{4\pi}{3}$ and $\frac{2\pi}{3}$ are $\pm\frac{\pi}{3}$ units from π. This allows us to write the solutions in the more compact equivalent form $(2n+1)\pi \pm \frac{\pi}{3}$, n an integer.

36. $f(x) = x^3 + 3x^2 + x + 3$ has a horizontal tangent when $f'(x) = 3x^2 + 6x + 1 = 0 \Leftrightarrow$
$$x = \frac{-6 \pm \sqrt{36 - 12}}{6} = -1 \pm \tfrac{1}{3}\sqrt{6}.$$

37. $y = 6x^3 + 5x - 3 \Rightarrow m = y' = 18x^2 + 5$, but $x^2 \geq 0$ for all x, so $m \geq 5$ for all x.

38. $y = x\sqrt{x} = x^{3/2} \Rightarrow y' = \frac{3}{2}x^{1/2}$. The slope of the line $y = 1 + 3x$ is 3, so the slope of any line parallel to it is also 3. Thus, $y' = 3 \Rightarrow \frac{3}{2}x^{1/2} = 3 \Rightarrow \sqrt{x} = 2 \Rightarrow x = 4$, which is the x-coordinate of the point on the curve at which the slope is 3. The y-coordinate is $y = 4\sqrt{4} = 8$, so an equation of the tangent line is $y - 8 = 3(x - 4)$ or $y = 3x - 4$.

39. The slope of $y = x^2 - 5x + 4$ is given by $m = y' = 2x - 5$. The slope of $x - 3y = 5 \Leftrightarrow y = \frac{1}{3}x - \frac{5}{3}$ is $\frac{1}{3}$, so the desired normal line must have slope $\frac{1}{3}$, and hence, the tangent line to the parabola must have slope -3. This occurs if $2x - 5 = -3 \Rightarrow 2x = 2 \Rightarrow x = 1$. When $x = 1$, $y = 1^2 - 5(1) + 4 = 0$, and an equation of the normal line is $y - 0 = \frac{1}{3}(x - 1)$ or $y = \frac{1}{3}x - \frac{1}{3}$.

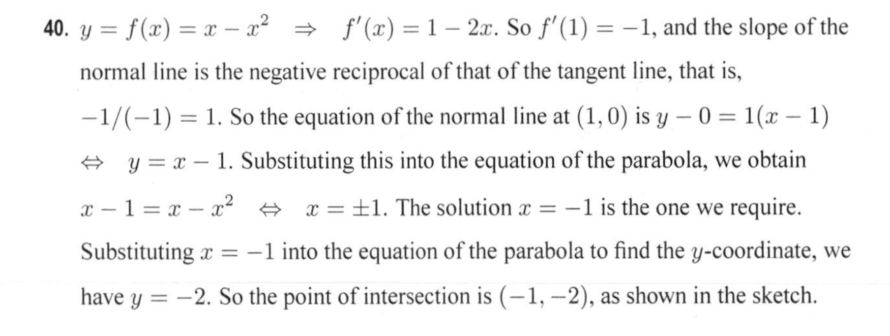

40. $y = f(x) = x - x^2 \Rightarrow f'(x) = 1 - 2x$. So $f'(1) = -1$, and the slope of the normal line is the negative reciprocal of that of the tangent line, that is, $-1/(-1) = 1$. So the equation of the normal line at $(1, 0)$ is $y - 0 = 1(x - 1)$ $\Leftrightarrow y = x - 1$. Substituting this into the equation of the parabola, we obtain $x - 1 = x - x^2 \Leftrightarrow x = \pm 1$. The solution $x = -1$ is the one we require. Substituting $x = -1$ into the equation of the parabola to find the y-coordinate, we have $y = -2$. So the point of intersection is $(-1, -2)$, as shown in the sketch.

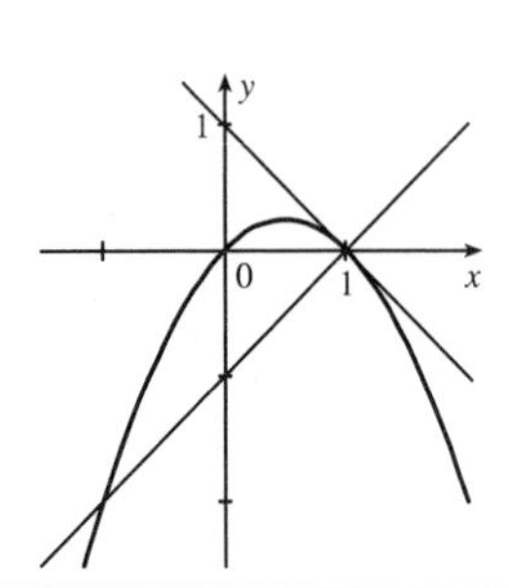

41. (a) $s = t^3 - 3t \quad \Rightarrow \quad v(t) = s'(t) = 3t^2 - 3 \quad \Rightarrow \quad a(t) = v'(t) = 6t$

(b) $a(2) = 6(2) = 12 \text{ m/s}^2$

(c) $v(t) = 3t^2 - 3 = 0$ when $t^2 = 1$, that is, $t = 1$ and $a(1) = 6 \text{ m/s}^2$.

42. (a) $s = 2t^3 - 7t^2 + 4t + 1 \quad \Rightarrow \quad v(t) = s'(t) = 6t^2 - 14t + 4 \quad \Rightarrow \quad a(t) = v'(t) = 12t - 14$

(b) $a(1) = 12 - 14 = -2 \text{ m/s}^2$

(c)

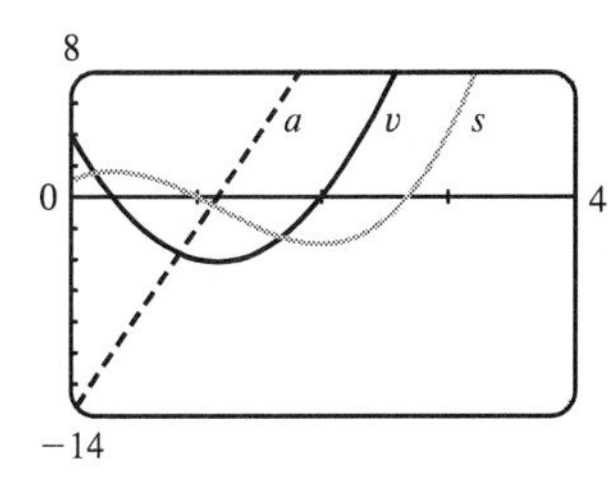

43. (a) $s = f(t) = t^3 - 12t^2 + 36t \quad \Rightarrow \quad v(t) = f'(t) = 3t^2 - 24t + 36$

(b) $v(3) = 27 - 72 + 36 = -9$ ft/s

(c) The particle is at rest when $v(t) = 0$. $3t^2 - 24t + 36 = 0 \quad \Rightarrow \quad 3(t-2)(t-6) = 0 \quad \Rightarrow \quad t = 2$ s or 6 s.

(d) The particle is moving in the positive direction when $v(t) > 0$. $3(t-2)(t-6) > 0 \quad \Leftrightarrow \quad 0 \le t < 2$ or $t > 6$.

(e) Since the particle is moving in the positive direction and in the negative direction, we need to calculate the distance traveled in the intervals $[0, 2]$, $[2, 6]$, and $[6, 8]$ separately.

$|f(2) - f(0)| = |32 - 0| = 32.$

$|f(6) - f(2)| = |0 - 32| = 32.$

$|f(8) - f(6)| = |32 - 0| = 32.$

The total distance is $32 + 32 + 32 = 96$ ft.

(f)

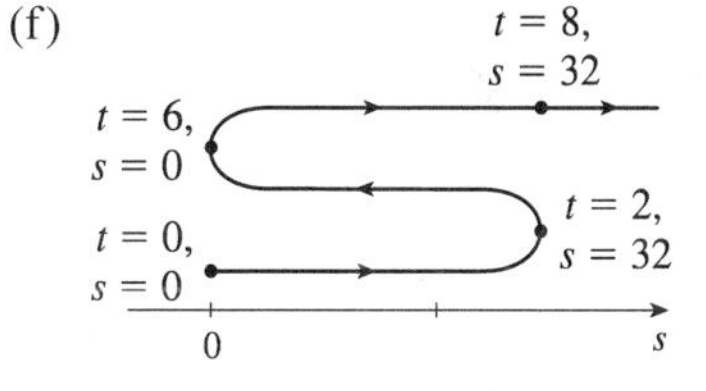

44. (a) $s = f(t) = t^3 - 9t^2 + 15t + 10 \quad \Rightarrow \quad v(t) = f'(t) = 3t^2 - 18t + 15 = 3(t-1)(t-5)$

(b) $v(3) = 3(2)(-2) = -12$ ft/s

(c) $v(t) = 0 \quad \Leftrightarrow \quad t = 1$ s or 5 s

(d) $v(t) > 0 \quad \Leftrightarrow \quad 0 \le t < 1$ or $t > 5$

(e) $|f(1) - f(0)| = |17 - 10| = 7,$

$|f(5) - f(1)| = |-15 - 17| = 32,$ and

$|f(8) - f(5)| = |66 - (-15)| = 81.$

Total distance $= 7 + 32 + 81 = 120$ ft.

(f)

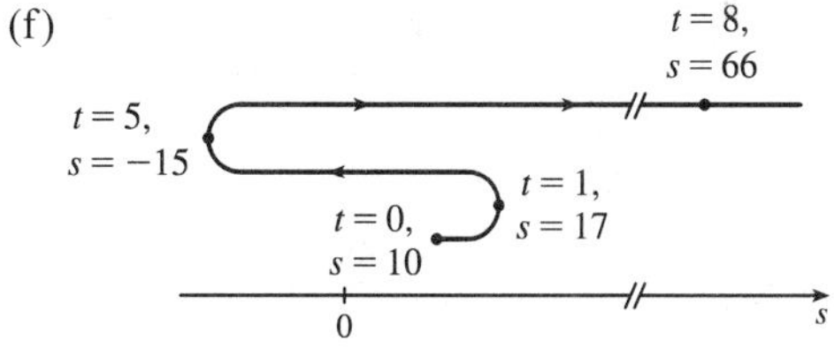

45. (a) $s(t) = t^3 - 4.5t^2 - 7t \quad \Rightarrow \quad v(t) = s'(t) = 3t^2 - 9t - 7 = 5 \quad \Leftrightarrow \quad 3t^2 - 9t - 12 = 0 \quad \Leftrightarrow$

$3(t-4)(t+1) = 0 \quad \Leftrightarrow \quad t = 4$ or -1. Since $t \ge 0$, the particle reaches a velocity of 5 m/s at $t = 4$ s.

(b) $a(t) = v'(t) = 6t - 9 = 0 \quad \Leftrightarrow \quad t = 1.5$. The acceleration changes from negative to positive, so the velocity changes from decreasing to increasing. Thus, at $t = 1.5$ s, the velocity has its minimum value.

46. (a) $s = 5t + 3t^2 \quad \Rightarrow \quad v(t) = \dfrac{ds}{dt} = 5 + 6t$, so $v(2) = 5 + 6(2) = 17 \text{ m/s}$.

(b) $v(t) = 35 \quad \Rightarrow \quad 5 + 6t = 35 \quad \Rightarrow \quad 6t = 30 \quad \Rightarrow \quad t = 5$ s.

47. (a) $h = 10t - 0.83t^2 \quad \Rightarrow \quad v(t) = \dfrac{dh}{dt} = 10 - 1.66t$, so $v(3) = 10 - 1.66(3) = 5.02 \text{ m/s}$.

(b) $h = 25 \quad\Rightarrow\quad 10t - 0.83t^2 = 25 \quad\Rightarrow\quad 0.83t^2 - 10t + 25 = 0 \quad\Rightarrow\quad t = \dfrac{10 \pm \sqrt{17}}{1.66} \approx 3.54$ or 8.51.

The value $t_1 = \left(10 - \sqrt{17}\right)/1.66$ corresponds to the time it takes for the stone to rise 25 m and $t_2 = \left(10 + \sqrt{17}\right)/1.66$ corresponds to the time when the stone is 25 m high on the way down. Thus, $v(t_1) = 10 - 1.66\left[\left(10 - \sqrt{17}\right)/1.66\right] = \sqrt{17} \approx 4.12$ m/s.

48. (a) At maximum height the velocity of the ball is 0 ft/s. $v(t) = s'(t) = 80 - 32t = 0 \quad\Leftrightarrow\quad 32t = 80 \quad\Leftrightarrow\quad t = \frac{5}{2}$.

So the maximum height is $s\left(\frac{5}{2}\right) = 80\left(\frac{5}{2}\right) - 16\left(\frac{5}{2}\right)^2 = 200 - 100 = 100$ ft.

(b) $s(t) = 80t - 16t^2 = 96 \quad\Leftrightarrow\quad 16t^2 - 80t + 96 = 0 \quad\Leftrightarrow\quad 16(t^2 - 5t + 6) = 0 \quad\Leftrightarrow\quad 16(t-3)(t-2) = 0$.

So the ball has a height of 96 ft on the way up at $t = 2$ and on the way down at $t = 3$. At these times the velocities are $v(2) = 80 - 32(2) = 16$ ft/s and $v(3) = 80 - 32(3) = -16$ ft/s, respectively.

49. (a) $C(x) = 2000 + 3x + 0.01x^2 + 0.0002x^3 \quad\Rightarrow\quad C'(x) = 3 + 0.02x + 0.0006x^2$

(b) $C'(100) = 3 + 0.02(100) + 0.0006(10{,}000) = 3 + 2 + 6 = \$11$/pair. $C'(100)$ is the rate at which the cost is increasing as the 100th pair of jeans is produced. It predicts the cost of the 101st pair.

(c) The cost of manufacturing the 101st pair of jeans is

$C(101) - C(100) = (2000 + 303 + 102.01 + 206.0602) - (2000 + 300 + 100 + 200) = 11.0702 \approx \$11.07$.

50. (a) $C(x) = 84 + 0.16x - 0.0006x^2 + 0.000003x^3 \quad\Rightarrow\quad C'(x) = 0.16 - 0.0012x + 0.000009x^2 \quad\Rightarrow\quad C'(100) = 0.13$.

This is the rate at which the cost is increasing as the 100th item is produced.

(b) $C(101) - C(100) = 97.13030299 - 97 \approx \$0.13$.

51. $S(r) = 4\pi r^2 \quad\Rightarrow\quad S'(r) = 8\pi r \quad\Rightarrow$

(a) $S'(1) = 8\pi$ ft²/ft (b) $S'(2) = 16\pi$ ft²/ft (c) $S'(3) = 24\pi$ ft²/ft

As the radius increases, the surface area grows at an increasing rate. In fact, the rate of change is linear with respect to the radius.

52. $V(t) = 5000\left(1 - \frac{1}{40}t\right)^2 = 5000\left(1 - \frac{1}{20}t + \frac{1}{1600}t^2\right) \quad\Rightarrow\quad V'(t) = 5000\left(-\frac{1}{20} + \frac{1}{800}t\right) = -250\left(1 - \frac{1}{40}t\right)$

(a) $V'(5) = -250\left(1 - \frac{5}{40}\right) = -218.75$ gal/min (b) $V'(10) = -250\left(1 - \frac{10}{40}\right) = -187.5$ gal/min

(c) $V'(20) = -250\left(1 - \frac{20}{40}\right) = -125$ gal/min (d) $V'(40) = -250\left(1 - \frac{40}{40}\right) = 0$ gal/min

The water is flowing out the fastest at the beginning—when $t = 0$, $V'(t) = -250$ gal/min. The water is flowing out the slowest at the end—when $t = 40$, $V'(t) = 0$. As the tank empties, the water flows out more slowly.

53. (a) To find the rate of change of volume with respect to pressure, we first solve for V in terms of P.

$$PV = C \quad\Rightarrow\quad V = \frac{C}{P} \quad\Rightarrow\quad \frac{dV}{dP} = -\frac{C}{P^2}.$$

(b) From the formula for dV/dP in part (a), we see that as P increases, the absolute value of dV/dP decreases. Thus, the volume is decreasing more rapidly at the beginning.

54. (a) $F = \dfrac{GmM}{r^2} = (GmM)r^{-2} \quad\Rightarrow\quad \dfrac{dF}{dr} = -2(GmM)r^{-3} = -\dfrac{2GmM}{r^3}$, which is the rate of change of the force with respect to the distance between the bodies. The minus sign indicates that as the distance r between the bodies increases, the magnitude of the force F exerted by the body of mass m on the body of mass M is decreasing.

(b) Given $F'(20{,}000) = -2$, find $F'(10{,}000)$. $-2 = -\dfrac{2GmM}{20{,}000^3} \Rightarrow GmM = 20{,}000^3$.

$$F'(10,000) = -\frac{2(20{,}000^3)}{10{,}000^3} = -2 \cdot 2^3 = -16 \text{ N/km}$$

55. $f'(x) = \lim\limits_{h\to 0} \dfrac{f(x+h)-f(x)}{h} = \lim\limits_{h\to 0} \dfrac{\dfrac{1}{x+h} - \dfrac{1}{x}}{h} = \lim\limits_{h\to 0} \dfrac{x-(x+h)}{hx(x+h)} = \lim\limits_{h\to 0} \dfrac{-h}{hx(x+h)} = \lim\limits_{h\to 0} \dfrac{-1}{x(x+h)} = -\dfrac{1}{x^2}$

56. $f(x) = \cos x \Rightarrow$

$$f'(x) = \lim_{h\to 0} \frac{f(x+h)-f(x)}{h} = \lim_{h\to 0} \frac{\cos(x+h)-\cos x}{h} = \lim_{h\to 0} \frac{\cos x\cos h - \sin x \sin h - \cos x}{h}$$

$$= \lim_{h\to 0}\left(\cos x\,\frac{\cos h - 1}{h} - \sin x\,\frac{\sin h}{h}\right) = \cos x \lim_{h\to 0} \frac{\cos h - 1}{h} - \sin x \lim_{h\to 0} \frac{\sin h}{h}$$

$$= (\cos x)(0) - (\sin x)(1) = -\sin x$$

57. $y = A\sin x + B\cos x \Rightarrow y' = A\cos x - B\sin x \Rightarrow y'' = -A\sin x - B\cos x$. Substituting these expressions for y, y', and y'' into the given differential equation $y'' + y' - 2y = \sin x$ gives us $(-A\sin x - B\cos x) + (A\cos x - B\sin x) - 2(A\sin x + B\cos x) = \sin x \Leftrightarrow$ $-3A\sin x - B\sin x + A\cos x - 3B\cos x = \sin x \Leftrightarrow (-3A - B)\sin x + (A - 3B)\cos x = 1\sin x$, so we must have $-3A - B = 1$ and $A - 3B = 0$ (since 0 is the coefficient of $\cos x$ on the right side). Solving for A and B, we add the first equation to three times the second to get $B = -\frac{1}{10}$ and $A = -\frac{3}{10}$.

58. $y = Ax^2 + Bx + C \Rightarrow y' = 2Ax + B \Rightarrow y'' = 2A$. We substitute these expressions into the equation $y'' + y' - 2y = x^2$ to get

$$(2A) + (2Ax + B) - 2(Ax^2 + Bx + C) = x^2$$
$$2A + 2Ax + B - 2Ax^2 - 2Bx - 2C = x^2$$
$$(-2A)x^2 + (2A - 2B)x + (2A + B - 2C) = (1)x^2 + (0)x + (0)$$

The coefficients of x^2 on each side must be equal, so $-2A = 1 \Rightarrow A = -\frac{1}{2}$. Similarly, $2A - 2B = 0 \Rightarrow$ $A = B = -\frac{1}{2}$ and $2A + B - 2C = 0 \Rightarrow -1 - \frac{1}{2} - 2C = 0 \Rightarrow C = -\frac{3}{4}$.

59.

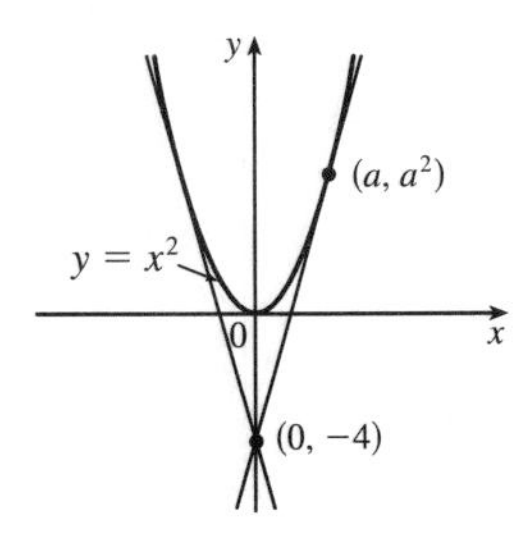

Let (a, a^2) be a point on the parabola at which the tangent line passes through the point $(0, -4)$. The tangent line has slope $2a$ and equation $y - (-4) = 2a(x - 0)$ $\Leftrightarrow y = 2ax - 4$. Since (a, a^2) also lies on the line, $a^2 = 2a(a) - 4$, or $a^2 = 4$. So $a = \pm 2$ and the points are $(2, 4)$ and $(-2, 4)$.

60. (a) If $y = x^2 + x$, then $y' = 2x + 1$. If the point at which a tangent meets the parabola is $(a, a^2 + a)$, then the slope of the tangent is $2a + 1$. But since it passes through $(2, -3)$, the slope must also be $\dfrac{\Delta y}{\Delta x} = \dfrac{a^2 + a + 3}{a - 2}$.

Therefore, $2a + 1 = \dfrac{a^2 + a + 3}{a - 2}$. Solving this equation for a we get $a^2 + a + 3 = 2a^2 - 3a - 2 \Leftrightarrow$ $a^2 - 4a - 5 = (a - 5)(a + 1) = 0 \Leftrightarrow a = 5$ or -1. If $a = -1$, the point is $(-1, 0)$ and the slope is -1, so the equation is $y - 0 = (-1)(x + 1)$ or $y = -x - 1$. If $a = 5$, the point is $(5, 30)$ and the slope is 11, so the equation is $y - 30 = 11(x - 5)$ or $y = 11x - 25$.

(b) As in part (a), but using the point $(2,7)$, we get the equation $2a+1=\dfrac{a^2+a-7}{a-2}$ $\Rightarrow$ $2a^2-3a-2=a^2+a-7$ $\Leftrightarrow$ $a^2-4a+5=0$. The last equation has no real solution (discriminant $=-16<0$), so there is no line through the point $(2,7)$ that is tangent to the parabola. The diagram shows that the point $(2,7)$ is "inside" the parabola, but tangent lines to the parabola do not pass through points inside the parabola.

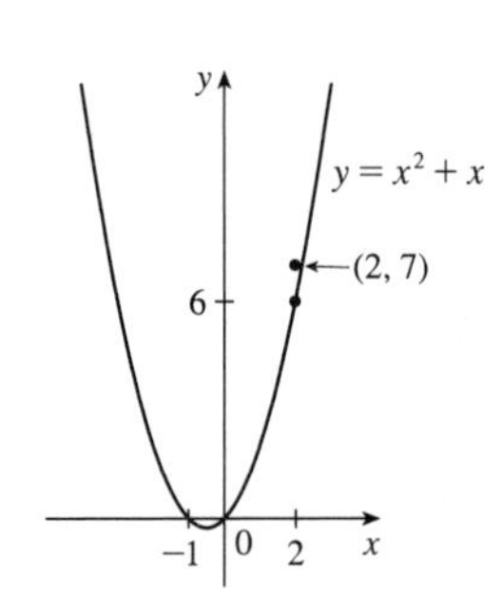

61. $y=f(x)=ax^2$ $\Rightarrow$ $f'(x)=2ax$. So the slope of the tangent to the parabola at $x=2$ is $m=2a(2)=4a$. The slope of the given line, $2x+y=b$ $\Leftrightarrow$ $y=-2x+b$, is seen to be -2, so we must have $4a=-2$ $\Leftrightarrow$ $a=-\frac{1}{2}$. So when $x=2$, the point in question has y-coordinate $-\frac{1}{2}\cdot 2^2=-2$. Now we simply require that the given line, whose equation is $2x+y=b$, pass through the point $(2,-2)$: $2(2)+(-2)=b$ $\Leftrightarrow$ $b=2$. So we must have $a=-\frac{1}{2}$ and $b=2$.

62. $y=ax^2+bx+c$ $\Rightarrow$ $y'(x)=2ax+b$. The parabola has slope 4 at $x=1$ and slope -8 at $x=-1$, so $y'(1)=4$ $\Rightarrow$ $2a+b=4$ **(1)** and $y'(-1)=-8$ $\Rightarrow$ $-2a+b=-8$ **(2)**. Adding **(1)** and **(2)** gives us $2b=-4$ $\Leftrightarrow$ $b=-2$. From **(1)**, $2a-2=4$ $\Leftrightarrow$ $a=3$. Thus, the equation of the parabola is $y=3x^2-2x+c$. Since it passes through the point $(2,15)$, we have $15=3(2)^2-2(2)+c$ $\Rightarrow$ $c=7$, so the equation is $y=3x^2-2x+7$.

63. $y=f(x)=ax^3+bx^2+cx+d$ $\Rightarrow$ $f'(x)=3ax^2+2bx+c$. The point $(-2,6)$ is on f, so $f(-2)=6$ $\Rightarrow$ $-8a+4b-2c+d=6$ **(1)**. The point $(2,0)$ is on f, so $f(2)=0$ $\Rightarrow$ $8a+4b+2c+d=0$ **(2)**. Since there are horizontal tangents at $(-2,6)$ and $(2,0)$, $f'(\pm 2)=0$. $f'(-2)=0$ $\Rightarrow$ $12a-4b+c=0$ **(3)** and $f'(2)=0$ $\Rightarrow$ $12a+4b+c=0$ **(4)**. Subtracting equation **(3)** from **(4)** gives $8b=0$ $\Rightarrow$ $b=0$. Adding **(1)** and **(2)** gives $8b+2d=6$, so $d=3$ since $b=0$. From **(3)** we have $c=-12a$, so **(2)** becomes $8a+4(0)+2(-12a)+3=0$ $\Rightarrow$ $3=16a$ $\Rightarrow$ $a=\frac{3}{16}$. Now $c=-12a=-12\left(\frac{3}{16}\right)=-\frac{9}{4}$ and the desired cubic function is $y=\frac{3}{16}x^3-\frac{9}{4}x+3$.

64. (a) $xy=c$ $\Rightarrow$ $y=\dfrac{c}{x}$. Let $P=\left(a,\dfrac{c}{a}\right)$. The slope of the tangent line at $x=a$ is $y'(a)=-\dfrac{c}{a^2}$. Its equation is $y-\dfrac{c}{a}=-\dfrac{c}{a^2}(x-a)$ or $y=-\dfrac{c}{a^2}x+\dfrac{2c}{a}$, so its y-intercept is $\dfrac{2c}{a}$. Setting $y=0$ gives $x=2a$, so the x-intercept is $2a$. The midpoint of the line segment joining $\left(0,\dfrac{2c}{a}\right)$ and $(2a,0)$ is $\left(a,\dfrac{c}{a}\right)=P$.

(b) We know the x- and y-intercepts of the tangent line from part (a), so the area of the triangle bounded by the axes and the tangent is $\frac{1}{2}(\text{base})(\text{height})=\frac{1}{2}xy=\frac{1}{2}(2a)(2c/a)=2c$, a constant.

65. *Solution 1:* Let $f(x)=x^{1000}$. Then, by the definition of a derivative, $f'(1)=\lim\limits_{x\to 1}\dfrac{f(x)-f(1)}{x-1}=\lim\limits_{x\to 1}\dfrac{x^{1000}-1}{x-1}$. But this is just the limit we want to find, and we know (from the Power Rule) that $f'(x)=1000x^{999}$, so $f'(1)=1000(1)^{999}=1000$. So $\lim\limits_{x\to 1}\dfrac{x^{1000}-1}{x-1}=1000$.

Solution 2: Note that $(x^{1000}-1)=(x-1)(x^{999}+x^{998}+x^{997}+\cdots+x^2+x+1)$. So

$$\lim_{x\to 1}\frac{x^{1000}-1}{x-1}=\lim_{x\to 1}\frac{(x-1)(x^{999}+x^{998}+x^{997}+\cdots+x^2+x+1)}{x-1}=\lim_{x\to 1}(x^{999}+x^{998}+x^{997}+\cdots+x^2+x+1)$$

$$=\underbrace{1+1+1+\cdots+1+1+1}_{1000\text{ ones}}=1000\text{, as above.}$$

66. 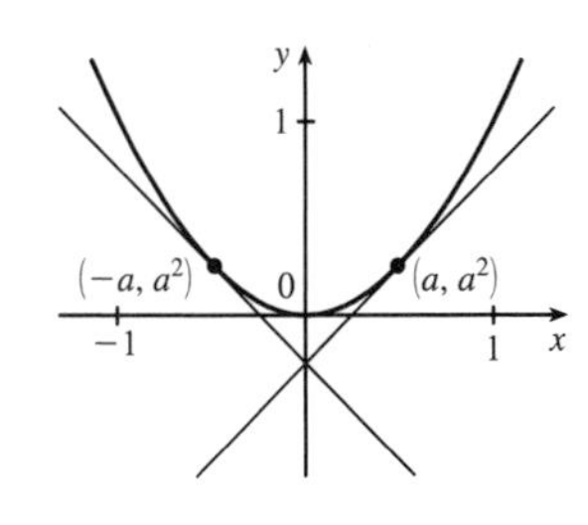

In order for the two tangents to intersect on the y-axis, the points of tangency must be at equal distances from the y-axis, since the parabola $y = x^2$ is symmetric about the y-axis. Say the points of tangency are (a, a^2) and $(-a, a^2)$, for some $a > 0$. Then since the derivative of $y = x^2$ is $dy/dx = 2x$, the left-hand tangent has slope $-2a$ and equation $y - a^2 = -2a(x + a)$, or $y = -2ax - a^2$, and similarly the right-hand tangent line has equation $y - a^2 = 2a(x - a)$, or $y = 2ax - a^2$. So the two lines intersect at $(0, -a^2)$. Now if the lines are perpendicular, then the product of their slopes is -1, so $(-2a)(2a) = -1 \Leftrightarrow a^2 = \frac{1}{4} \Leftrightarrow a = \frac{1}{2}$. So the lines intersect at $(0, -\frac{1}{4})$.

67. $y = x^2 \Rightarrow y' = 2x$, so the slope of a tangent line at the point (a, a^2) is $y' = 2a$ and the slope of a normal line is $-1/(2a)$, for $a \neq 0$. The slope of the normal line through the points (a, a^2) and $(0, c)$ is $\dfrac{a^2 - c}{a - 0}$, so $\dfrac{a^2 - c}{a} = -\dfrac{1}{2a} \Rightarrow$ $a^2 - c = -\frac{1}{2} \Rightarrow a^2 = c - \frac{1}{2}$. The last equation has two solutions if $c > \frac{1}{2}$, one solution if $c = \frac{1}{2}$, and no solution if $c < \frac{1}{2}$. Since the y-axis is normal to $y = x^2$ regardless of the value of c (this is the case for $a = 0$), we have three normal lines if $c > \frac{1}{2}$ and one normal line if $c \leq \frac{1}{2}$.

68. 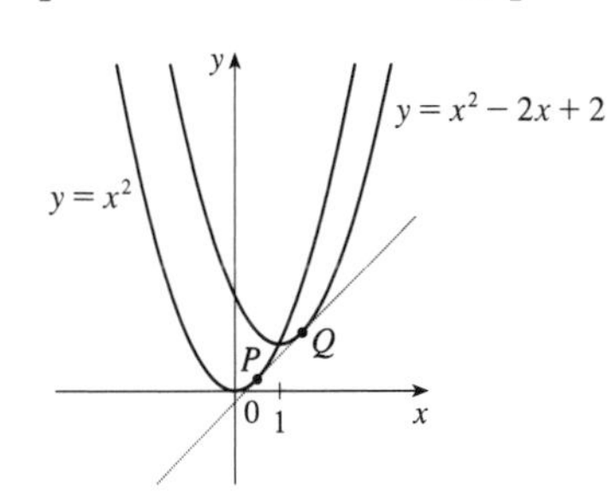

From the sketch, it appears that there may be a line that is tangent to both curves. The slope of the line through the points $P(a, a^2)$ and $Q(b, b^2 - 2b + 2)$ is $\dfrac{b^2 - 2b + 2 - a^2}{b - a}$. The slope of the tangent line at P is $2a$ $[y' = 2x]$ and at Q is $2b - 2$ $[y' = 2x - 2]$. All three slopes are equal, so $2a = 2b - 2 \Leftrightarrow a = b - 1$.

Also, $2b - 2 = \dfrac{b^2 - 2b + 2 - a^2}{b - a} \Rightarrow 2b - 2 = \dfrac{b^2 - 2b + 2 - (b-1)^2}{b - (b-1)} \Rightarrow 2b - 2 = b^2 - 2b + 2 - b^2 + 2b - 1 \Rightarrow$ $2b = 3 \Rightarrow b = \frac{3}{2}$ and $a = \frac{3}{2} - 1 = \frac{1}{2}$. Thus, an equation of the tangent line at P is $y - \left(\frac{1}{2}\right)^2 = 2\left(\frac{1}{2}\right)\left(x - \frac{1}{2}\right)$ or $y = x - \frac{1}{4}$.

2.4 The Product and Quotient Rules

1. Product Rule: $y = (x^2 + 1)(x^3 + 1) \Rightarrow$

$$y' = (x^2 + 1)(3x^2) + (x^3 + 1)(2x) = 3x^4 + 3x^2 + 2x^4 + 2x = 5x^4 + 3x^2 + 2x.$$

Multiplying first: $y = (x^2 + 1)(x^3 + 1) = x^5 + x^3 + x^2 + 1 \Rightarrow y' = 5x^4 + 3x^2 + 2x$ (equivalent).

2. Quotient Rule: $F(x) = \dfrac{x - 3x\sqrt{x}}{\sqrt{x}} = \dfrac{x - 3x^{3/2}}{x^{1/2}} \Rightarrow$

$$F'(x) = \frac{x^{1/2}\left(1 - \frac{9}{2}x^{1/2}\right) - \left(x - 3x^{3/2}\right)\left(\frac{1}{2}x^{-1/2}\right)}{\left(x^{1/2}\right)^2} = \frac{x^{1/2} - \frac{9}{2}x - \frac{1}{2}x^{1/2} + \frac{3}{2}x}{x} = \frac{\frac{1}{2}x^{1/2} - 3x}{x} = \frac{1}{2}x^{-1/2} - 3$$

Simplifying first: $F(x) = \dfrac{x - 3x\sqrt{x}}{\sqrt{x}} = \sqrt{x} - 3x = x^{1/2} - 3x \Rightarrow F'(x) = \frac{1}{2}x^{-1/2} - 3$ (equivalent).

For this problem, simplifying first seems to be the better method.

3. $g(t) = t^3 \cos t \quad \Rightarrow \quad g'(t) = t^3(-\sin t) + (\cos t) \cdot 3t^2 = 3t^2 \cos t - t^3 \sin t$ or $t^2(3\cos t - t \sin t)$

4. $f(x) = \sqrt{x} \sin x \quad \Rightarrow \quad f'(x) = \sqrt{x} \cos x + \sin x \left(\frac{1}{2}x^{-1/2}\right) = \sqrt{x} \cos x + \dfrac{\sin x}{2\sqrt{x}}$

5. $F(y) = \left(\dfrac{1}{y^2} - \dfrac{3}{y^4}\right)(y + 5y^3) = (y^{-2} - 3y^{-4})(y + 5y^3) \overset{\text{PR}}{\Rightarrow}$

$$\begin{aligned} F'(y) &= (y^{-2} - 3y^{-4})(1 + 15y^2) + (y + 5y^3)(-2y^{-3} + 12y^{-5}) \\ &= (y^{-2} + 15 - 3y^{-4} - 45y^{-2}) + (-2y^{-2} + 12y^{-4} - 10 + 60y^{-2}) \\ &= 5 + 14y^{-2} + 9y^{-4} \text{ or } 5 + 14/y^2 + 9/y^4 \end{aligned}$$

6. $Y(u) = (u^{-2} + u^{-3})(u^5 - 2u^2) \overset{\text{PR}}{\Rightarrow}$

$$\begin{aligned} Y'(u) &= (u^{-2} + u^{-3})(5u^4 - 4u) + (u^5 - 2u^2)(-2u^{-3} - 3u^{-4}) \\ &= (5u^2 - 4u^{-1} + 5u - 4u^{-2}) + (-2u^2 - 3u + 4u^{-1} + 6u^{-2}) = 3u^2 + 2u + 2u^{-2} \end{aligned}$$

7. $f(x) = \sin x + \frac{1}{2}\cot x \quad \Rightarrow \quad f'(x) = \cos x - \frac{1}{2}\csc^2 x$

8. $y = 2\csc x + 5\cos x \quad \Rightarrow \quad y' = -2\csc x \cot x - 5\sin x$

9. $h(\theta) = \theta \csc\theta - \cot\theta \quad \Rightarrow \quad h'(\theta) = \theta(-\csc\theta \cot\theta) + (\csc\theta) \cdot 1 - \left(-\csc^2\theta\right) = \csc\theta - \theta\csc\theta \cot\theta + \csc^2\theta$

10. $y = u(a\cos u + b\cot u) \quad \Rightarrow$

$y' = u(-a\sin u - b\csc^2 u) + (a\cos u + b\cot u) \cdot 1 = a\cos u + b\cot u - au\sin u - bu\csc^2 u$

11. $g(x) = \dfrac{3x - 1}{2x + 1} \overset{\text{QR}}{\Rightarrow} g'(x) = \dfrac{(2x+1)(3) - (3x-1)(2)}{(2x+1)^2} = \dfrac{6x + 3 - 6x + 2}{(2x+1)^2} = \dfrac{5}{(2x+1)^2}$

12. $f(t) = \dfrac{2t}{4 + t^2} \overset{\text{QR}}{\Rightarrow} f'(t) = \dfrac{(4 + t^2)(2) - (2t)(2t)}{(4 + t^2)^2} = \dfrac{8 + 2t^2 - 4t^2}{(4 + t^2)^2} = \dfrac{8 - 2t^2}{(4 + t^2)^2}$

13. $y = \dfrac{t^2}{3t^2 - 2t + 1} \overset{\text{QR}}{\Rightarrow}$

$$y' = \frac{(3t^2 - 2t + 1)(2t) - t^2(6t - 2)}{(3t^2 - 2t + 1)^2} = \frac{2t[3t^2 - 2t + 1 - t(3t - 1)]}{(3t^2 - 2t + 1)^2} = \frac{2t(3t^2 - 2t + 1 - 3t^2 + t)}{(3t^2 - 2t + 1)^2} = \frac{2t(1 - t)}{(3t^2 - 2t + 1)^2}$$

14. $y = \dfrac{t^3 + t}{t^4 - 2} \overset{\text{QR}}{\Rightarrow}$

$$\begin{aligned} y' &= \frac{(t^4 - 2)(3t^2 + 1) - (t^3 + t)(4t^3)}{(t^4 - 2)^2} = \frac{(3t^6 + t^4 - 6t^2 - 2) - (4t^6 + 4t^4)}{(t^4 - 2)^2} \\ &= \frac{-t^6 - 3t^4 - 6t^2 - 2}{(t^4 - 2)^2} = -\frac{t^6 + 3t^4 + 6t^2 + 2}{(t^4 - 2)^2} \end{aligned}$$

15. $y = \dfrac{v^3 - 2v\sqrt{v}}{v} = v^2 - 2\sqrt{v} = v^2 - 2v^{1/2} \quad \Rightarrow \quad y' = 2v - 2\left(\frac{1}{2}\right)v^{-1/2} = 2v - v^{-1/2}$.

We can change the form of the answer as follows: $2v - v^{-1/2} = 2v - \dfrac{1}{\sqrt{v}} = \dfrac{2v\sqrt{v} - 1}{\sqrt{v}} = \dfrac{2v^{3/2} - 1}{\sqrt{v}}$

16. $y = \dfrac{\sqrt{x} - 1}{\sqrt{x} + 1} \quad \Rightarrow \quad y' = \dfrac{(\sqrt{x} + 1)\left(\dfrac{1}{2\sqrt{x}}\right) - (\sqrt{x} - 1)\left(\dfrac{1}{2\sqrt{x}}\right)}{\left(\sqrt{x} + 1\right)^2} = \dfrac{\dfrac{1}{2} + \dfrac{1}{2\sqrt{x}} - \dfrac{1}{2} + \dfrac{1}{2\sqrt{x}}}{\left(\sqrt{x} + 1\right)^2} = \dfrac{1}{\sqrt{x}\left(\sqrt{x} + 1\right)^2}$

17. $y = \dfrac{r^2}{1+\sqrt{r}} \Rightarrow$

$$y' = \frac{(1+\sqrt{r})(2r) - r^2\left(\frac{1}{2}r^{-1/2}\right)}{(1+\sqrt{r})^2} = \frac{2r + 2r^{3/2} - \frac{1}{2}r^{3/2}}{(1+\sqrt{r})^2} = \frac{2r + \frac{3}{2}r^{3/2}}{(1+\sqrt{r})^2} = \frac{\frac{1}{2}r(4+3r^{1/2})}{(1+\sqrt{r})^2} = \frac{r(4+3\sqrt{r})}{2(1+\sqrt{r})^2}$$

18. $y = \dfrac{cx}{1+cx} \Rightarrow y' = \dfrac{(1+cx)(c)-(cx)(c)}{(1+cx)^2} = \dfrac{c + c^2x - c^2x}{(1+cx)^2} = \dfrac{c}{(1+cx)^2}$

19. $y = \dfrac{x}{\cos x} \Rightarrow y' = \dfrac{(\cos x)(1) - (x)(-\sin x)}{(\cos x)^2} = \dfrac{\cos x + x\sin x}{\cos^2 x}$

20. $y = \dfrac{1+\sin x}{x+\cos x} \Rightarrow$

$$y' = \frac{(x+\cos x)(\cos x) - (1+\sin x)(1-\sin x)}{(x+\cos x)^2} = \frac{x\cos x + \cos^2 x - (1-\sin^2 x)}{(x+\cos x)^2}$$

$$= \frac{x\cos x + \cos^2 x - (\cos^2 x)}{(x+\cos x)^2} = \frac{x\cos x}{(x+\cos x)^2}$$

21. $f(\theta) = \dfrac{\sec\theta}{1+\sec\theta} \Rightarrow$

$$f'(\theta) = \frac{(1+\sec\theta)(\sec\theta\,\tan\theta) - (\sec\theta)(\sec\theta\,\tan\theta)}{(1+\sec\theta)^2} = \frac{(\sec\theta\,\tan\theta)[(1+\sec\theta)-\sec\theta]}{(1+\sec\theta)^2} = \frac{\sec\theta\,\tan\theta}{(1+\sec\theta)^2}$$

22. $y = \dfrac{1-\sec x}{\tan x} \Rightarrow$

$$y' = \frac{\tan x\,(-\sec x\,\tan x) - (1-\sec x)(\sec^2 x)}{(\tan x)^2} = \frac{\sec x\left(-\tan^2 x - \sec x + \sec^2 x\right)}{\tan^2 x} = \frac{\sec x\,(1-\sec x)}{\tan^2 x}$$

23. $y = \dfrac{\sin x}{x^2} \Rightarrow y' = \dfrac{x^2\cos x - (\sin x)(2x)}{(x^2)^2} = \dfrac{x(x\cos x - 2\sin x)}{x^4} = \dfrac{x\cos x - 2\sin x}{x^3}$

24. $y = \dfrac{u^6 - 2u^3 + 5}{u^2} = u^4 - 2u + 5u^{-2} \Rightarrow y' = 4u^3 - 2 - 10u^{-3} = 2u^{-3}(2u^6 - u^3 - 5) = 2(2u^6 - u^3 - 5)/u^3$

25. $f(x) = \dfrac{x}{x + c/x} \Rightarrow f'(x) = \dfrac{(x+c/x)(1) - x(1-c/x^2)}{\left(x + \frac{c}{x}\right)^2} = \dfrac{x + c/x - x + c/x}{\left(\frac{x^2+c}{x}\right)^2} = \dfrac{2c/x}{\frac{(x^2+c)^2}{x^2}} \cdot \dfrac{x^2}{x^2} = \dfrac{2cx}{(x^2+c)^2}$

26. $f(x) = \dfrac{ax+b}{cx+d} \Rightarrow f'(x) = \dfrac{(cx+d)(a) - (ax+b)(c)}{(cx+d)^2} = \dfrac{acx + ad - acx - bc}{(cx+d)^2} = \dfrac{ad-bc}{(cx+d)^2}$

27. $y = \dfrac{2x}{x+1} \Rightarrow y' = \dfrac{(x+1)(2) - (2x)(1)}{(x+1)^2} = \dfrac{2}{(x+1)^2}$. At $(1,1)$, $y' = \frac{1}{2}$, and an equation of the tangent line is $y - 1 = \frac{1}{2}(x-1)$, or $y = \frac{1}{2}x + \frac{1}{2}$.

28. $y = \dfrac{\sqrt{x}}{x+1} \Rightarrow y' = \dfrac{(x+1)\left(\dfrac{1}{2\sqrt{x}}\right) - \sqrt{x}\,(1)}{(x+1)^2} = \dfrac{(x+1)-(2x)}{2\sqrt{x}\,(x+1)^2} = \dfrac{1-x}{2\sqrt{x}\,(x+1)^2}$. At $(4, 0.4)$, $y' = \frac{-3}{100} = -0.03$, and an equation of the tangent line is $y - 0.4 = -0.03(x-4)$, or $y = -0.03x + 0.52$.

29. $y = \tan x \quad \Rightarrow \quad y' = \sec^2 x \quad \Rightarrow$ the slope of the tangent line at $\left(\frac{\pi}{4}, 1\right)$ is $\sec^2 \frac{\pi}{4} = \left(\sqrt{2}\right)^2 = 2$ and an equation of the tangent line is $y - 1 = 2\left(x - \frac{\pi}{4}\right)$ or $y = 2x + 1 - \frac{\pi}{2}$.

30. $y = (1 + x)\cos x \quad \Rightarrow \quad y' = (1 + x)(-\sin x) + \cos x \cdot 1$. At $(0, 1)$, $y' = 1$, and an equation of the tangent line is $y - 1 = 1(x - 0)$, or $y = x + 1$.

31. (a) $y = f(x) = \dfrac{1}{1 + x^2} \quad \Rightarrow$

$$f'(x) = \frac{(1 + x^2)(0) - 1(2x)}{(1 + x^2)^2} = \frac{-2x}{(1 + x^2)^2}.$$ So the slope of the tangent line at the point $\left(-1, \frac{1}{2}\right)$ is $f'(-1) = \dfrac{2}{2^2} = \frac{1}{2}$ and its equation is $y - \frac{1}{2} = \frac{1}{2}(x + 1)$ or $y = \frac{1}{2}x + 1$.

(b)

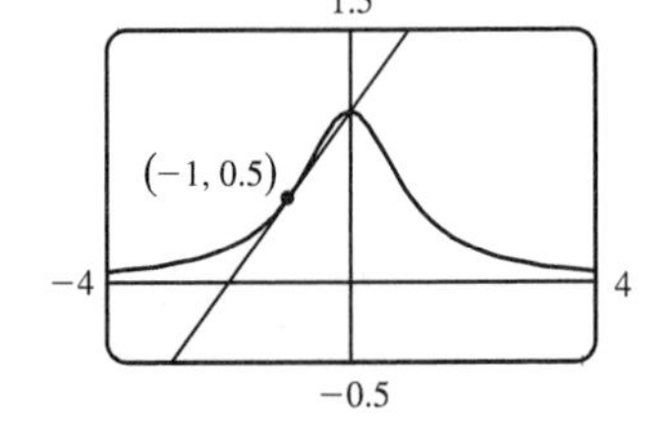

32. (a) $y = f(x) = \dfrac{x}{1 + x^2} \quad \Rightarrow$

$$f'(x) = \frac{(1 + x^2)1 - x(2x)}{(1 + x^2)^2} = \frac{1 - x^2}{(1 + x^2)^2}.$$ So the slope of the tangent line at the point $(3, 0.3)$ is $f'(3) = \frac{-8}{100}$ and its equation is $y - 0.3 = -0.08(x - 3)$ or $y = -0.08x + 0.54$.

(b)

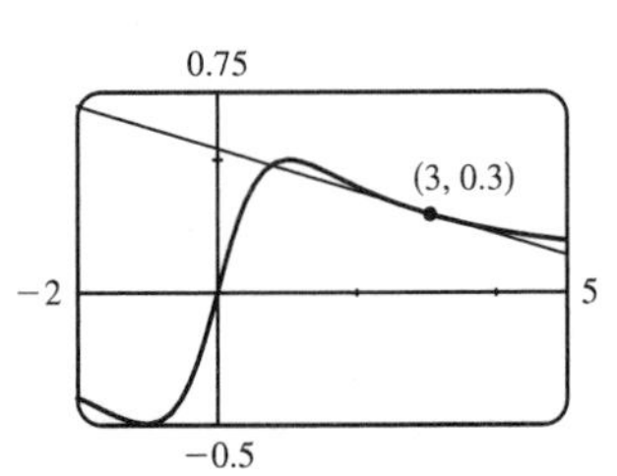

33. $$f(x) = \frac{x^2}{1 + x} \quad \Rightarrow \quad f'(x) = \frac{(1 + x)(2x) - x^2(1)}{(1 + x)^2} = \frac{2x + 2x^2 - x^2}{(1 + x)^2} = \frac{x^2 + 2x}{x^2 + 2x + 1} \quad \Rightarrow$$

$$f''(x) = \frac{(x^2 + 2x + 1)(2x + 2) - (x^2 + 2x)(2x + 2)}{(x^2 + 2x + 1)^2} = \frac{(2x + 2)(x^2 + 2x + 1 - x^2 - 2x)}{[(x + 1)^2]^2}$$

$$= \frac{2(x + 1)(1)}{(x + 1)^4} = \frac{2}{(x + 1)^3},$$

so $f''(1) = \dfrac{2}{(1 + 1)^3} = \dfrac{2}{8} = \dfrac{1}{4}$.

34. $f(x) = \sec x \quad \Rightarrow \quad f'(x) = \sec x \, \tan x \quad \Rightarrow \quad f''(x) = \sec x\,(\sec^2 x) + \tan x\,(\sec x \, \tan x) = \sec x\,(\sec^2 x + \tan^2 x)$.

$f''\left(\frac{\pi}{4}\right) = \sqrt{2}\left[\left(\sqrt{2}\right)^2 + 1^2\right] = \sqrt{2}\,(2 + 1) = 3\sqrt{2}$

35. $H(\theta) = \theta \sin\theta \quad \Rightarrow \quad H'(\theta) = \theta\,(\cos\theta) + (\sin\theta) \cdot 1 = \theta\cos\theta + \sin\theta \quad \Rightarrow$

$H''(\theta) = \theta\,(-\sin\theta) + (\cos\theta) \cdot 1 + \cos\theta = -\theta\sin\theta + 2\cos\theta$

36. Let $f(x) = x \sin x$ and $h(x) = \sin x$, so $f(x) = xh(x)$. Then $f'(x) = h(x) + xh'(x)$,

$f''(x) = h'(x) + h'(x) + xh''(x) = 2h'(x) + xh''(x)$,

$f'''(x) = 2h''(x) + h''(x) + xh'''(x) = 3h''(x) + xh'''(x), \cdots, f^{(n)}(x) = nh^{(n-1)}(x) + xh^{(n)}(x)$.

Since $34 = 4(8) + 2$, we have $h^{(34)}(x) = h^{(2)}(x) = \dfrac{d^2}{dx^2}(\sin x) = -\sin x$ and $h^{(35)}(x) = -\cos x$.

Thus, $\dfrac{d^{35}}{dx^{35}}(x \sin x) = 35h^{(34)}(x) + xh^{(35)}(x) = -35\sin x - x\cos x$.

37. $$\frac{d}{dx}(\csc x) = \frac{d}{dx}\left(\frac{1}{\sin x}\right) = \frac{(\sin x)(0) - 1(\cos x)}{\sin^2 x} = \frac{-\cos x}{\sin^2 x} = -\frac{1}{\sin x} \cdot \frac{\cos x}{\sin x} = -\csc x \, \cot x$$

38. $\frac{d}{dx}(\sec x) = \frac{d}{dx}\left(\frac{1}{\cos x}\right) = \frac{(\cos x)(0) - 1(-\sin x)}{\cos^2 x} = \frac{\sin x}{\cos^2 x} = \frac{1}{\cos x} \cdot \frac{\sin x}{\cos x} = \sec x \tan x$

39. $\frac{d}{dx}(\cot x) = \frac{d}{dx}\left(\frac{\cos x}{\sin x}\right) = \frac{(\sin x)(-\sin x) - (\cos x)(\cos x)}{\sin^2 x} = -\frac{\sin^2 x + \cos^2 x}{\sin^2 x} = -\frac{1}{\sin^2 x} = -\csc^2 x$

40. (a) $g(x) = f(x)\sin x \quad \Rightarrow \quad g'(x) = f(x)\cos x + \sin x \cdot f'(x)$, so

$g'(\frac{\pi}{3}) = f(\frac{\pi}{3})\cos\frac{\pi}{3} + \sin\frac{\pi}{3} \cdot f'(\frac{\pi}{3}) = 4 \cdot \frac{1}{2} + \frac{\sqrt{3}}{2} \cdot (-2) = 2 - \sqrt{3}$

(b) $h(x) = \frac{\cos x}{f(x)} \quad \Rightarrow \quad h'(x) = \frac{f(x) \cdot (-\sin x) - \cos x \cdot f'(x)}{[f(x)]^2}$, so

$h'(\frac{\pi}{3}) = \frac{f(\frac{\pi}{3}) \cdot (-\sin\frac{\pi}{3}) - \cos\frac{\pi}{3} \cdot f'(\frac{\pi}{3})}{[f(\frac{\pi}{3})]^2} = \frac{4\left(-\frac{\sqrt{3}}{2}\right) - \left(\frac{1}{2}\right)(-2)}{4^2} = \frac{-2\sqrt{3} + 1}{16} = \frac{1 - 2\sqrt{3}}{16}$

41. We are given that $f(5) = 1$, $f'(5) = 6$, $g(5) = -3$, and $g'(5) = 2$.

(a) $(fg)'(5) = f(5)g'(5) + g(5)f'(5) = (1)(2) + (-3)(6) = 2 - 18 = -16$

(b) $\left(\frac{f}{g}\right)'(5) = \frac{g(5)f'(5) - f(5)g'(5)}{[g(5)]^2} = \frac{(-3)(6) - (1)(2)}{(-3)^2} = -\frac{20}{9}$

(c) $\left(\frac{g}{f}\right)'(5) = \frac{f(5)g'(5) - g(5)f'(5)}{[f(5)]^2} = \frac{(1)(2) - (-3)(6)}{(1)^2} = 20$

42. We are given that $f(3) = 4$, $g(3) = 2$, $f'(3) = -6$, and $g'(3) = 5$.

(a) $(f + g)'(3) = f'(3) + g'(3) = -6 + 5 = -1$

(b) $(fg)'(3) = f(3)g'(3) + g(3)f'(3) = (4)(5) + (2)(-6) = 20 - 12 = 8$

(c) $\left(\frac{f}{g}\right)'(3) = \frac{g(3)f'(3) - f(3)g'(3)}{[g(3)]^2} = \frac{(2)(-6) - (4)(5)}{(2)^2} = \frac{-32}{4} = -8$

43. (a) From the graphs of f and g, we obtain the following values: $f(1) = 2$ since the point $(1, 2)$ is on the graph of f; $g(1) = 1$ since the point $(1, 1)$ is on the graph of g; $f'(1) = 2$ since the slope of the line segment between $(0, 0)$ and $(2, 4)$ is $\frac{4 - 0}{2 - 0} = 2$; $g'(1) = -1$ since the slope of the line segment between $(-2, 4)$ and $(2, 0)$ is $\frac{0 - 4}{2 - (-2)} = -1$.

Now $u(x) = f(x)g(x)$, so $u'(1) = f(1)g'(1) + g(1)\,f'(1) = 2 \cdot (-1) + 1 \cdot 2 = 0$.

(b) $v(x) = f(x)/g(x)$, so $v'(5) = \frac{g(5)f'(5) - f(5)g'(5)}{[g(5)]^2} = \frac{2\left(-\frac{1}{3}\right) - 3 \cdot \frac{2}{3}}{2^2} = \frac{-\frac{8}{3}}{4} = -\frac{2}{3}$

44. (a) $P(x) = F(x)\,G(x)$, so $P'(2) = F(2)\,G'(2) + G(2)\,F'(2) = 3 \cdot \frac{2}{4} + 2 \cdot 0 = \frac{3}{2}$.

(b) $Q(x) = F(x)/G(x)$, so $Q'(7) = \frac{G(7)\,F'(7) - F(7)\,G'(7)}{[G(7)]^2} = \frac{1 \cdot \frac{1}{4} - 5 \cdot \left(-\frac{2}{3}\right)}{1^2} = \frac{1}{4} + \frac{10}{3} = \frac{43}{12}$

45. (a) $y = xg(x) \quad \Rightarrow \quad y' = xg'(x) + g(x) \cdot 1 = xg'(x) + g(x)$

(b) $y = \frac{x}{g(x)} \quad \Rightarrow \quad y' = \frac{g(x) \cdot 1 - xg'(x)}{[g(x)]^2} = \frac{g(x) - xg'(x)}{[g(x)]^2}$

(c) $y = \frac{g(x)}{x} \quad \Rightarrow \quad y' = \frac{xg'(x) - g(x) \cdot 1}{(x)^2} = \frac{xg'(x) - g(x)}{x^2}$

46. (a) $y = x^2 f(x) \Rightarrow y' = x^2 f'(x) + f(x)(2x)$

(b) $y = \dfrac{f(x)}{x^2} \Rightarrow y' = \dfrac{x^2 f'(x) - f(x)(2x)}{(x^2)^2} = \dfrac{xf'(x) - 2f(x)}{x^3}$

(c) $y = \dfrac{x^2}{f(x)} \Rightarrow y' = \dfrac{f(x)(2x) - x^2 f'(x)}{[f(x)]^2}$

(d) $y = \dfrac{1 + xf(x)}{\sqrt{x}} \Rightarrow$

$$y' = \frac{\sqrt{x}\,[xf'(x) + f(x)] - [1 + xf(x)]\dfrac{1}{2\sqrt{x}}}{(\sqrt{x}\,)^2}$$

$$= \frac{x^{3/2} f'(x) + x^{1/2} f(x) - \frac{1}{2}x^{-1/2} - \frac{1}{2}x^{1/2} f(x)}{x} \cdot \frac{2x^{1/2}}{2x^{1/2}} = \frac{xf(x) + 2x^2 f'(x) - 1}{2x^{3/2}}$$

47. (a) $x(t) = 8\sin t \Rightarrow v(t) = x'(t) = 8\cos t \Rightarrow a(t) = x''(t) = -8\sin t$

(b) The mass at time $t = \frac{2\pi}{3}$ has position $x\left(\frac{2\pi}{3}\right) = 8\sin\frac{2\pi}{3} = 8\left(\frac{\sqrt{3}}{2}\right) = 4\sqrt{3}$, velocity $v\left(\frac{2\pi}{3}\right) = 8\cos\frac{2\pi}{3} = 8\left(-\frac{1}{2}\right) = -4$, and acceleration $a\left(\frac{2\pi}{3}\right) = -8\sin\frac{2\pi}{3} = -8\left(\frac{\sqrt{3}}{2}\right) = -4\sqrt{3}$. Since $v\left(\frac{2\pi}{3}\right) < 0$, the particle is moving to the left. Because v and a have the same sign, the particle is speeding up.

48. (a) $F = \dfrac{\mu W}{\mu\sin\theta + \cos\theta} \Rightarrow \dfrac{dF}{d\theta} = \dfrac{(\mu\sin\theta + \cos\theta)(0) - \mu W(\mu\cos\theta - \sin\theta)}{(\mu\sin\theta + \cos\theta)^2} = \dfrac{\mu W(\sin\theta - \mu\cos\theta)}{(\mu\sin\theta + \cos\theta)^2}$

(b) $\dfrac{dF}{d\theta} = 0 \Rightarrow \mu W(\sin\theta - \mu\cos\theta) = 0 \Rightarrow \sin\theta = \mu\cos\theta \Rightarrow \tan\theta = \mu \Rightarrow \theta = \tan^{-1}\mu$

(c)

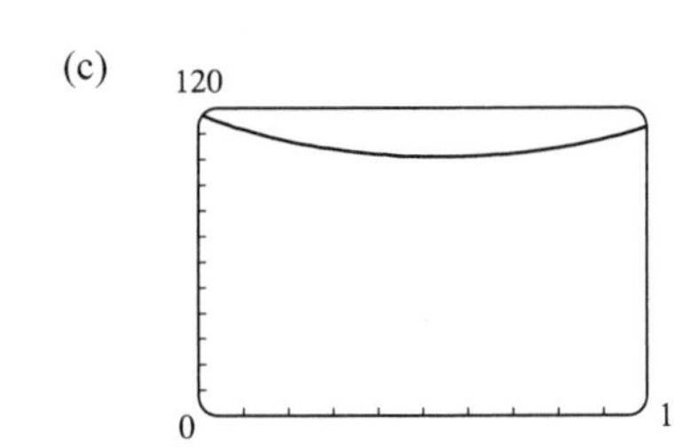

From the graph of $F = \dfrac{0.6(50)}{0.6\sin\theta + \cos\theta}$ for $0 \le \theta \le 1$, we see that $\dfrac{dF}{d\theta} = 0 \Rightarrow \theta \approx 0.54$. Checking this with part (b) and $\mu = 0.6$, we calculate $\theta = \tan^{-1} 0.6 \approx 0.54$. So the value from the graph is consistent with the value in part (b).

49. $PV = nRT \Rightarrow T = \dfrac{PV}{nR} = \dfrac{PV}{(10)(0.0821)} = \dfrac{1}{0.821}(PV)$. Using the Product Rule, we have

$$\frac{dT}{dt} = \frac{1}{0.821}\left[P(t)V'(t) + V(t)P'(t)\right] = \frac{1}{0.821}\left[(8)(-0.15) + (10)(0.10)\right] \approx -0.2436 \text{ K/min}.$$

50. (a) $S = \dfrac{dR}{dx} = \dfrac{(1 + 4x^{0.4})(9.6x^{-0.6}) - (40 + 24x^{0.4})(1.6x^{-0.6})}{(1 + 4x^{0.4})^2}$

$$= \frac{9.6x^{-0.6} + 38.4x^{-0.2} - 64x^{-0.6} - 38.4x^{-0.2}}{(1 + 4x^{0.4})^2} = -\frac{54.4x^{-0.6}}{(1 + 4x^{0.4})^2}$$

(b)

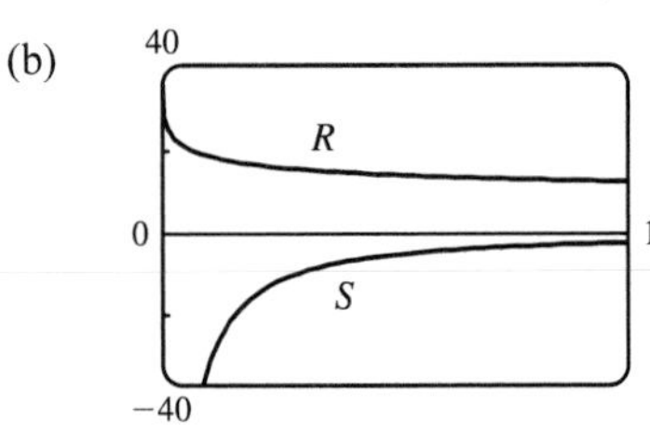

At low levels of brightness, R is quite large $[R(0) = 40]$ and is quickly decreasing, that is, S is negative with large absolute value. This is to be expected: at low levels of brightness, the eye is more sensitive to slight changes than it is at higher levels of brightness.

51. If $y = f(x) = \dfrac{x}{x+1}$, then $f'(x) = \dfrac{(x+1)(1) - x(1)}{(x+1)^2} = \dfrac{1}{(x+1)^2}$. When $x = a$, the equation of the tangent line is

$y - \dfrac{a}{a+1} = \dfrac{1}{(a+1)^2}(x-a)$. This line passes through $(1,2)$ when $2 - \dfrac{a}{a+1} = \dfrac{1}{(a+1)^2}(1-a) \Leftrightarrow$

$2(a+1)^2 - a(a+1) = 1 - a \quad \Leftrightarrow \quad 2a^2 + 4a + 2 - a^2 - a - 1 + a = 0 \quad \Leftrightarrow \quad a^2 + 4a + 1 = 0.$

The quadratic formula gives the roots of this equation as $a = \dfrac{-4 \pm \sqrt{4^2 - 4(1)(1)}}{2(1)} = \dfrac{-4 \pm \sqrt{12}}{2} = -2 \pm \sqrt{3}$,

so there are two such tangent lines. Since

$$\begin{aligned} f\left(-2 \pm \sqrt{3}\right) &= \frac{-2 \pm \sqrt{3}}{-2 \pm \sqrt{3} + 1} = \frac{-2 \pm \sqrt{3}}{-1 \pm \sqrt{3}} \cdot \frac{-1 \mp \sqrt{3}}{-1 \mp \sqrt{3}} \\ &= \frac{2 \pm 2\sqrt{3} \mp \sqrt{3} - 3}{1 - 3} = \frac{-1 \pm \sqrt{3}}{-2} = \frac{1 \mp \sqrt{3}}{2}, \end{aligned}$$

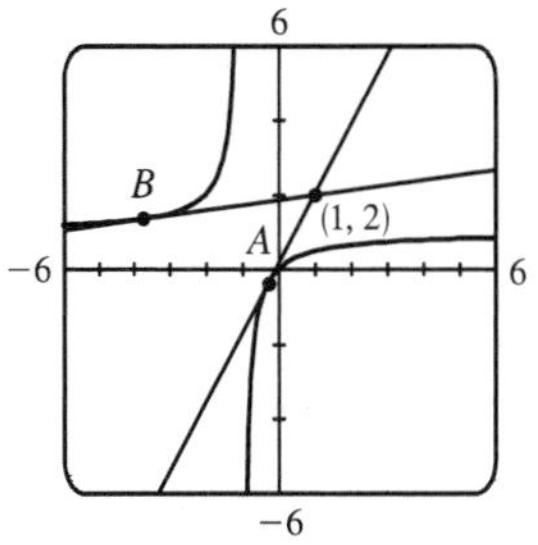

the lines touch the curve at $A\left(-2 + \sqrt{3}, \frac{1 - \sqrt{3}}{2}\right) \approx (-0.27, -0.37)$

and $B\left(-2 - \sqrt{3}, \frac{1 + \sqrt{3}}{2}\right) \approx (-3.73, 1.37)$.

6
B
A
(1, 2)
−6
6
−6

52. $y = \dfrac{\cos x}{2 + \sin x} \quad \Rightarrow \quad y' = \dfrac{(2 + \sin x)(-\sin x) - \cos x \cos x}{(2 + \sin x)^2} = \dfrac{-2\sin x - \sin^2 x - \cos^2 x}{(2 + \sin x)^2} = \dfrac{-2\sin x - 1}{(2 + \sin x)^2} = 0$ when

$-2\sin x - 1 = 0 \quad \Leftrightarrow \quad \sin x = -\frac{1}{2} \quad \Leftrightarrow \quad x = \frac{11\pi}{6} + 2\pi n$ or $x = \frac{7\pi}{6} + 2\pi n$, n an integer. So $y = \frac{1}{\sqrt{3}}$ or $y = -\frac{1}{\sqrt{3}}$ and the

points on the curve with horizontal tangents are: $\left(\frac{11\pi}{6} + 2\pi n, \frac{1}{\sqrt{3}}\right)$, $\left(\frac{7\pi}{6} + 2\pi n, -\frac{1}{\sqrt{3}}\right)$, n an integer.

53. (a) $(fgh)' = [(fg)h]' = (fg)'h + (fg)h' = (f'g + fg')h + (fg)h' = f'gh + fg'h + fgh'$

(b) $y = x\sin x \cos x \quad \Rightarrow \quad \dfrac{dy}{dx} = \sin x \cos x + x\cos x \cos x + x\sin x(-\sin x) = \sin x \cos x + x\cos^2 x - x\sin^2 x$

54. (a) We use the Product Rule repeatedly: $F = fg \quad \Rightarrow \quad F' = f'g + fg' \quad \Rightarrow$

$F'' = (f''g + f'g') + (f'g' + fg'') = f''g + 2f'g' + fg''$.

(b) $F''' = f'''g + f''g' + 2(f''g' + f'g'') + f'g'' + fg''' = f'''g + 3f''g' + 3f'g'' + fg''' \quad \Rightarrow$

$$\begin{aligned} F^{(4)} &= f^{(4)}g + f'''g' + 3(f'''g' + f''g'') + 3(f''g'' + f'g''') + f'g''' + fg^{(4)} \\ &= f^{(4)}g + 4f'''g' + 6f''g'' + 4f'g''' + fg^{(4)} \end{aligned}$$

(c) By analogy with the Binomial Theorem, we make the guess:

$$F^{(n)} = f^{(n)}g + nf^{(n-1)}g' + \binom{n}{2}f^{(n-2)}g'' + \cdots + \binom{n}{k}f^{(n-k)}g^{(k)} + \cdots + nf'g^{(n-1)} + fg^{(n)},$$

where $\dbinom{n}{k} = \dfrac{n!}{k!\,(n-k)!} = \dfrac{n(n-1)(n-2)\cdots(n-k+1)}{k!}$.

55. (a) $\dfrac{d}{dx}\left(\dfrac{1}{g(x)}\right) = \dfrac{g(x) \cdot \frac{d}{dx}(1) - 1 \cdot \frac{d}{dx}[g(x)]}{[g(x)]^2}$ [Quotient Rule] $= \dfrac{g(x) \cdot 0 - 1 \cdot g'(x)}{[g(x)]^2} = \dfrac{0 - g'(x)}{[g(x)]^2} = -\dfrac{g'(x)}{[g(x)]^2}$

(b) $y = \dfrac{1}{x^4 + x^2 + 1} \quad \Rightarrow \quad y' = -\dfrac{2x(2x^2 + 1)}{(x^4 + x^2 + 1)^2}$

(c) $\dfrac{d}{dx}(x^{-n}) = \dfrac{d}{dx}\left(\dfrac{1}{x^n}\right) = -\dfrac{(x^n)'}{(x^n)^2}$ [by the Reciprocal Rule] $= -\dfrac{nx^{n-1}}{x^{2n}} = -nx^{n-1-2n} = -nx^{-n-1}$

2.5 The Chain Rule

1. Let $u = g(x) = 4x$ and $y = f(u) = \sin u$. Then $\dfrac{dy}{dx} = \dfrac{dy}{du}\dfrac{du}{dx} = (\cos u)(4) = 4\cos 4x$.

2. Let $u = g(x) = 4 + 3x$ and $y = f(u) = \sqrt{u} = u^{1/2}$. Then $\dfrac{dy}{dx} = \dfrac{dy}{du}\dfrac{du}{dx} = \frac{1}{2}u^{-1/2}(3) = \dfrac{3}{2\sqrt{u}} = \dfrac{3}{2\sqrt{4+3x}}$.

3. Let $u = g(x) = 1 - x^2$ and $y = f(u) = u^{10}$. Then $\dfrac{dy}{dx} = \dfrac{dy}{du}\dfrac{du}{dx} = (10u^9)(-2x) = -20x(1-x^2)^9$.

4. Let $u = g(x) = \sin x$ and $y = f(u) = \tan u$. Then $\dfrac{dy}{dx} = \dfrac{dy}{du}\dfrac{du}{dx} = (\sec^2 u)(\cos x) = \sec^2(\sin x)\cdot\cos x$,

or equivalently, $[\sec(\sin x)]^2 \cos x$.

5. Let $u = g(x) = \sin x$ and $y = f(u) = \sqrt{u}$. Then $\dfrac{dy}{dx} = \dfrac{dy}{du}\dfrac{du}{dx} = \frac{1}{2}u^{-1/2}\cos x = \dfrac{\cos x}{2\sqrt{u}} = \dfrac{\cos x}{2\sqrt{\sin x}}$.

6. Let $u = g(x) = \sqrt{x}$ and $y = f(u) = \sin u$. Then $\dfrac{dy}{dx} = \dfrac{dy}{du}\dfrac{du}{dx} = (\cos u)\left(\frac{1}{2}x^{-1/2}\right) = \dfrac{\cos u}{2\sqrt{x}} = \dfrac{\cos\sqrt{x}}{2\sqrt{x}}$.

7. $F(x) = \sqrt[4]{1+2x+x^3} = (1+2x+x^3)^{1/4} \quad\Rightarrow$

$$F'(x) = \tfrac{1}{4}(1+2x+x^3)^{-3/4}\cdot\frac{d}{dx}\left(1+2x+x^3\right) = \frac{1}{4(1+2x+x^3)^{3/4}}\cdot(2+3x^2)$$
$$= \frac{2+3x^2}{4(1+2x+x^3)^{3/4}} = \frac{2+3x^2}{4\sqrt[4]{(1+2x+x^3)^3}}$$

8. $F(x) = (x^2-x+1)^3 \quad\Rightarrow\quad F'(x) = 3(x^2-x+1)^2(2x-1)$

9. $g(t) = \dfrac{1}{(t^4+1)^3} = (t^4+1)^{-3} \quad\Rightarrow\quad g'(t) = -3(t^4+1)^{-4}(4t^3) = -12t^3(t^4+1)^{-4} = \dfrac{-12t^3}{(t^4+1)^4}$

10. $f(t) = \sqrt[3]{1+\tan t} = (1+\tan t)^{1/3} \quad\Rightarrow\quad f'(t) = \frac{1}{3}(1+\tan t)^{-2/3}\sec^2 t = \dfrac{\sec^2 t}{3\sqrt[3]{(1+\tan t)^2}}$

11. $y = \cos(a^3+x^3) \quad\Rightarrow\quad y' = -\sin(a^3+x^3)\cdot 3x^2$ [a^3 is just a constant] $= -3x^2\sin(a^3+x^3)$

12. $y = a^3 + \cos^3 x \quad\Rightarrow\quad y' = 3(\cos x)^2(-\sin x)$ [a^3 is just a constant] $= -3\sin x\,\cos^2 x$

13. $y = \cot(x/2) \quad\Rightarrow\quad y' = -\csc^2(x/2)\cdot\frac{1}{2} = -\frac{1}{2}\csc^2(x/2)$

14. $y = 4\sec 5x \quad\Rightarrow\quad y' = 4\sec 5x\,\tan 5x(5) = 20\sec 5x\,\tan 5x$

15. $g(x) = (1+4x)^5(3+x-x^2)^8 \quad\Rightarrow$

$$g'(x) = (1+4x)^5\cdot 8(3+x-x^2)^7(1-2x) + (3+x-x^2)^8\cdot 5(1+4x)^4\cdot 4$$
$$= 4(1+4x)^4(3+x-x^2)^7[2(1+4x)(1-2x) + 5(3+x-x^2)]$$
$$= 4(1+4x)^4(3+x-x^2)^7[(2+4x-16x^2)+(15+5x-5x^2)] = 4(1+4x)^4(3+x-x^2)^7(17+9x-21x^2)$$

16. $h(t) = (t^4-1)^3(t^3+1)^4 \quad\Rightarrow$

$$h'(t) = (t^4-1)^3\cdot 4(t^3+1)^3(3t^2) + (t^3+1)^4\cdot 3(t^4-1)^2(4t^3)$$
$$= 12t^2(t^4-1)^2(t^3+1)^3[(t^4-1)+t(t^3+1)] = 12t^2(t^4-1)^2(t^3+1)^3(2t^4+t-1)$$

17. $y = (2x-5)^4(8x^2-5)^{-3} \Rightarrow$

$$y' = 4(2x-5)^3(2)(8x^2-5)^{-3} + (2x-5)^4(-3)(8x^2-5)^{-4}(16x)$$
$$= 8(2x-5)^3(8x^2-5)^{-3} - 48x(2x-5)^4(8x^2-5)^{-4}$$

[This simplifies to $8(2x-5)^3(8x^2-5)^{-4}(-4x^2+30x-5)$.]

18. $y = (x^2+1)(x^2+2)^{1/3} \Rightarrow$

$$y' = 2x(x^2+2)^{1/3} + (x^2+1)\left(\tfrac{1}{3}\right)(x^2+2)^{-2/3}(2x) = 2x(x^2+2)^{1/3}\left[1 + \frac{x^2+1}{3(x^2+2)}\right]$$

19. $y = x^3 \cos nx \Rightarrow y' = x^3(-\sin nx)(n) + \cos nx\,(3x^2) = x^2(3\cos nx - nx\sin nx)$

20. $y = x\sin\sqrt{x} \Rightarrow y' = x\cos\sqrt{x}\cdot\tfrac{1}{2}x^{-1/2} + \sin\sqrt{x}\cdot 1 = \tfrac{1}{2}\sqrt{x}\cos\sqrt{x} + \sin\sqrt{x}$

21. $y = \sin(x\cos x) \Rightarrow y' = \cos(x\cos x)\cdot[x(-\sin x) + \cos x\cdot 1] = (\cos x - x\sin x)\cos(x\cos x)$

22. $f(x) = \dfrac{x}{\sqrt{7-3x}} \Rightarrow$

$$f'(x) = \frac{\sqrt{7-3x}(1) - x\cdot\frac{1}{2}(7-3x)^{-1/2}\cdot(-3)}{(\sqrt{7-3x}\,)^2} = \frac{\sqrt{7-3x} + \dfrac{3x}{2\sqrt{7-3x}}}{(7-3x)^1}$$
$$= \frac{2(7-3x)+3x}{2(7-3x)^{3/2}} = \frac{14-3x}{2(7-3x)^{3/2}}$$

23. $F(z) = \sqrt{\dfrac{z-1}{z+1}} = \left(\dfrac{z-1}{z+1}\right)^{1/2} \Rightarrow$

$$F'(z) = \frac{1}{2}\left(\frac{z-1}{z+1}\right)^{-1/2}\cdot\frac{d}{dz}\left(\frac{z-1}{z+1}\right) = \frac{1}{2}\left(\frac{z+1}{z-1}\right)^{1/2}\cdot\frac{(z+1)(1)-(z-1)(1)}{(z+1)^2}$$
$$= \frac{1}{2}\frac{(z+1)^{1/2}}{(z-1)^{1/2}}\cdot\frac{z+1-z+1}{(z+1)^2} = \frac{1}{2}\frac{(z+1)^{1/2}}{(z-1)^{1/2}}\cdot\frac{2}{(z+1)^2} = \frac{1}{(z-1)^{1/2}(z+1)^{3/2}}$$

24. $G(y) = \left(\dfrac{y^2}{y+1}\right)^5 \Rightarrow G'(y) = 5\left(\dfrac{y^2}{y+1}\right)^4\cdot\dfrac{(y+1)(2y)-y^2(1)}{(y+1)^2} = 5\cdot\dfrac{y^8}{(y+1)^4}\cdot\dfrac{y(2y+2-y)}{(y+1)^2} = \dfrac{5y^9(y+2)}{(y+1)^6}$

25. $y = \dfrac{r}{\sqrt{r^2+1}} \Rightarrow$

$$y' = \frac{\sqrt{r^2+1}\,(1) - r\cdot\frac{1}{2}(r^2+1)^{-1/2}(2r)}{\left(\sqrt{r^2+1}\right)^2} = \frac{\sqrt{r^2+1} - \dfrac{r^2}{\sqrt{r^2+1}}}{\left(\sqrt{r^2+1}\right)^2} = \frac{\dfrac{\sqrt{r^2+1}\sqrt{r^2+1}-r^2}{\sqrt{r^2+1}}}{\left(\sqrt{r^2+1}\right)^2}$$
$$= \frac{(r^2+1)-r^2}{\left(\sqrt{r^2+1}\right)^3} = \frac{1}{(r^2+1)^{3/2}} \text{ or } (r^2+1)^{-3/2}$$

Another solution: Write y as a product and make use of the Product Rule. $y = r(r^2+1)^{-1/2} \Rightarrow$

$y' = r\cdot -\tfrac{1}{2}(r^2+1)^{-3/2}(2r) + (r^2+1)^{-1/2}\cdot 1 = (r^2+1)^{-3/2}[-r^2 + (r^2+1)^1] = (r^2+1)^{-3/2}(1) = (r^2+1)^{-3/2}$.

The step that students usually have trouble with is factoring out $(r^2+1)^{-3/2}$. But this is no different than factoring out x^2 from $x^2 + x^5$; that is, we are just factoring out a factor with the *smallest* exponent that appears on it. In this case, $-\frac{3}{2}$ is smaller than $-\frac{1}{2}$.

26. $y = \dfrac{\sin^2 x}{\cos x} \quad \Rightarrow$

$$y' = \frac{\cos x\,(2\sin x\,\cos x) - \sin^2 x\,(-\sin x)}{\cos^2 x} = \frac{\sin x\,(2\cos^2 x + \sin^2 x)}{\cos^2 x} = \frac{\sin x\,(1+\cos^2 x)}{\cos^2 x}$$
$$= \sin x\,(1+\sec^2 x)$$

Another method: $y = \tan x\,\sin x \quad \Rightarrow \quad y' = \sec^2 x\,\sin x + \tan x\,\cos x = \sec^2 x\,\sin x + \sin x$

27. $y = \tan(\cos x) \quad \Rightarrow \quad y' = \sec^2(\cos x)\cdot(-\sin x) = -\sin x \sec^2(\cos x)$

28. $y = \tan^2(3\theta) = (\tan 3\theta)^2 \quad \Rightarrow \quad y' = 2(\tan 3\theta)\cdot \dfrac{d}{d\theta}(\tan 3\theta) = 2\tan 3\theta \cdot \sec^2 3\theta \cdot 3 = 6\tan 3\theta\,\sec^2 3\theta$

29. $y = \sin\sqrt{1+x^2} \quad \Rightarrow \quad y' = \cos\sqrt{1+x^2}\cdot \frac{1}{2}(1+x^2)^{-1/2}\cdot 2x = \left(x\,\cos\sqrt{1+x^2}\,\right)/\sqrt{1+x^2}$

30. $y = x\sin\dfrac{1}{x} \quad \Rightarrow \quad y' = \sin\dfrac{1}{x} + x\cos\dfrac{1}{x}\left(-\dfrac{1}{x^2}\right) = \sin\dfrac{1}{x} - \dfrac{1}{x}\cos\dfrac{1}{x}$

31. $y = (1+\cos^2 x)^6 \quad \Rightarrow \quad y' = 6(1+\cos^2 x)^5\cdot 2\cos x\,(-\sin x) = -12\,\cos x\,\sin x\,(1+\cos^2 x)^5$

32. $y = \cot(x^2) + \cot^2 x = \cot(x^2) + (\cot x)^2 \quad \Rightarrow$

$y' = -\csc^2(x^2)\cdot 2x + 2(\cot x)^1(-\csc^2 x) = -2x\csc^2(x^2) - 2\cot x\csc^2 x$

33. $y = \sec^2 x + \tan^2 x = (\sec x)^2 + (\tan x)^2 \quad \Rightarrow$

$y' = 2(\sec x)(\sec x\,\tan x) + 2(\tan x)(\sec^2 x) = 2\sec^2 x\,\tan x + 2\sec^2 x\,\tan x = 4\sec^2 x\,\tan x$

34. $y = \sin(\sin(\sin x)) \quad \Rightarrow \quad y' = \cos(\sin(\sin x))\,\dfrac{d}{dx}(\sin(\sin x)) = \cos(\sin(\sin x))\,\cos(\sin x)\,\cos x$

35. $y = \cot^2(\sin\theta) = [\cot(\sin\theta)]^2 \quad \Rightarrow$

$y' = 2[\cot(\sin\theta)]\cdot\dfrac{d}{d\theta}[\cot(\sin\theta)] = 2\cot(\sin\theta)\cdot[-\csc^2(\sin\theta)\cdot\cos\theta] = -2\cos\theta\,\cot(\sin\theta)\,\csc^2(\sin\theta)$

36. $y = \sqrt{x+\sqrt{x+\sqrt{x}}} \quad \Rightarrow \quad y' = \frac{1}{2}\left(x+\sqrt{x+\sqrt{x}}\right)^{-1/2}\left[1+\frac{1}{2}(x+\sqrt{x})^{-1/2}\left(1+\frac{1}{2}x^{-1/2}\right)\right]$

37. $y = \sin\left(\tan\sqrt{\sin x}\right) \quad \Rightarrow$

$$y' = \cos\left(\tan\sqrt{\sin x}\right)\cdot\frac{d}{dx}\left(\tan\sqrt{\sin x}\right) = \cos\left(\tan\sqrt{\sin x}\right)\sec^2\sqrt{\sin x}\cdot\frac{d}{dx}(\sin x)^{1/2}$$
$$= \cos\left(\tan\sqrt{\sin x}\right)\sec^2\sqrt{\sin x}\cdot\tfrac{1}{2}(\sin x)^{-1/2}\cdot\cos x = \cos\left(\tan\sqrt{\sin x}\right)\left(\sec^2\sqrt{\sin x}\right)\left(\frac{1}{2\sqrt{\sin x}}\right)(\cos x)$$

38. $y = \sqrt{\cos(\sin^2 x)} \quad \Rightarrow \quad y' = \frac{1}{2}\left(\cos(\sin^2 x)\right)^{-1/2}\left[-\sin(\sin^2 x)\right](2\sin x\,\cos x) = -\dfrac{\sin(\sin^2 x)\,\sin x\,\cos x}{\sqrt{\cos(\sin^2 x)}}$

39. $y = (1+2x)^{10} \quad \Rightarrow \quad y' = 10(1+2x)^9\cdot 2 = 20(1+2x)^9$. At $(0,1)$, $y' = 20(1+0)^9 = 20$, and an equation of the tangent line is $y-1 = 20(x-0)$, or $y = 20x+1$.

40. $y = \sin x + \sin^2 x \quad \Rightarrow \quad y' = \cos x + 2\sin x\cos x$. At $(0,0)$, $y' = 1$, and an equation of the tangent line is $y - 0 = 1(x-0)$, or $y = x$.

41. (a) $y = f(x) = \tan\left(\frac{\pi}{4}x^2\right) \Rightarrow f'(x) = \sec^2\left(\frac{\pi}{4}x^2\right)\left(2 \cdot \frac{\pi}{4}x\right)$.

The slope of the tangent at $(1, 1)$ is thus

$f'(1) = \sec^2 \frac{\pi}{4}\left(\frac{\pi}{2}\right) = 2 \cdot \frac{\pi}{2} = \pi$, and its equation

is $y - 1 = \pi(x - 1)$ or $y = \pi x - \pi + 1$.

(b)

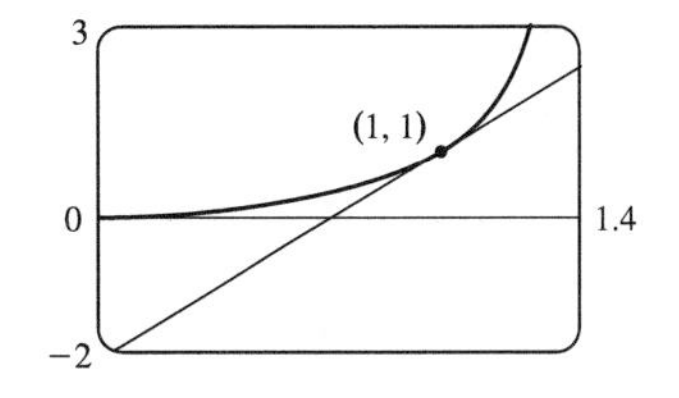

42. (a) For $x > 0$, $|x| = x$, and $y = f(x) = \dfrac{x}{\sqrt{2 - x^2}} \Rightarrow$

$$f'(x) = \frac{\sqrt{2 - x^2}\,(1) - x\left(\frac{1}{2}\right)(2 - x^2)^{-1/2}(-2x)}{\left(\sqrt{2 - x^2}\right)^2} \cdot \frac{(2 - x^2)^{1/2}}{(2 - x^2)^{1/2}}$$

$$= \frac{(2 - x^2) + x^2}{(2 - x^2)^{3/2}} = \frac{2}{(2 - x^2)^{3/2}}$$

So at $(1, 1)$, the slope of the tangent line is $f'(1) = 2$ and its equation is

$y - 1 = 2(x - 1)$ or $y = 2x - 1$.

(b)

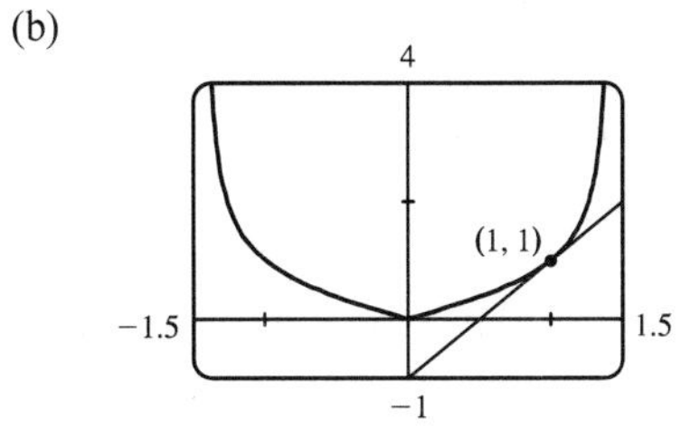

43. $F(t) = (1 - 7t)^6 \Rightarrow F'(t) = 6(1 - 7t)^5(-7) = -42(1 - 7t)^5 \Rightarrow F''(t) = -42 \cdot 5(1 - 7t)^4(-7) = 1470(1 - 7t)^4$

44. $h(x) = \sqrt{x^2 + 1} \Rightarrow h'(x) = \dfrac{1}{2}(x^2 + 1)^{-1/2}(2x) = \dfrac{x}{\sqrt{x^2 + 1}} \Rightarrow$

$$h''(x) = \frac{\sqrt{x^2 + 1} \cdot 1 - x\left[\frac{1}{2}(x^2 + 1)^{-1/2}(2x)\right]}{\left(\sqrt{x^2 + 1}\right)^2} = \frac{(x^2 + 1)^{-1/2}\left[(x^2 + 1) - x^2\right]}{(x^2 + 1)^1} = \frac{1}{(x^2 + 1)^{3/2}}$$

45. $y = (x^3 + 1)^{2/3} \Rightarrow y' = \frac{2}{3}(x^3 + 1)^{-1/3}(3x^2) = 2x^2(x^3 + 1)^{-1/3} \Rightarrow$

$y'' = 2x^2\left(-\frac{1}{3}\right)(x^3 + 1)^{-4/3}(3x^2) + (x^3 + 1)^{-1/3}(4x) = 4x(x^3 + 1)^{-1/3} - 2x^4(x^3 + 1)^{-4/3}$

46. $H(t) = \tan 3t \Rightarrow H'(t) = 3\sec^2 3t \Rightarrow$

$H''(t) = 2 \cdot 3\sec 3t \dfrac{d}{dt}(\sec 3t) = 6\sec 3t\,(3\sec 3t\,\tan 3t) = 18\sec^2 3t\,\tan 3t$

47. $F(x) = f(g(x)) \Rightarrow F'(x) = f'(g(x)) \cdot g'(x)$, so $F'(5) = f'(g(5)) \cdot g'(5) = f'(-2) \cdot 6 = 4 \cdot 6 = 24$

48. $h(x) = \sqrt{4 + 3f(x)} \Rightarrow h'(x) = \frac{1}{2}(4 + 3f(x))^{-1/2} \cdot 3f'(x)$, so

$h'(1) = \frac{1}{2}(4 + 3f(1))^{-1/2} \cdot 3f'(1) = \frac{1}{2}(4 + 3 \cdot 7)^{-1/2} \cdot 3 \cdot 4 = \frac{6}{\sqrt{25}} = \frac{6}{5}$

49. (a) $h(x) = f(g(x)) \Rightarrow h'(x) = f'(g(x)) \cdot g'(x)$, so $h'(1) = f'(g(1)) \cdot g'(1) = f'(2) \cdot 6 = 5 \cdot 6 = 30$.

(b) $H(x) = g(f(x)) \Rightarrow H'(x) = g'(f(x)) \cdot f'(x)$, so $H'(1) = g'(f(1)) \cdot f'(1) = g'(3) \cdot 4 = 9 \cdot 4 = 36$.

50. (a) $F(x) = f(f(x)) \Rightarrow F'(x) = f'(f(x)) \cdot f'(x)$, so

$F'(2) = f'(f(2)) \cdot f'(2) = f'(1) \cdot 5 = 4 \cdot 5 = 20$.

(b) $G(x) = g(g(x)) \Rightarrow G'(x) = g'(g(x)) \cdot g'(x)$, so $G'(3) = g'(g(3)) \cdot g'(3) = g'(2) \cdot 9 = 7 \cdot 9 = 63$.

51. (a) $u(x) = f(g(x)) \Rightarrow u'(x) = f'(g(x))g'(x)$. So $u'(1) = f'(g(1))g'(1) = f'(3)g'(1)$. To find $f'(3)$, note that f is linear from $(2, 4)$ to $(6, 3)$, so its slope is $\dfrac{3 - 4}{6 - 2} = -\dfrac{1}{4}$. To find $g'(1)$, note that g is linear from $(0, 6)$ to $(2, 0)$, so its slope is $\dfrac{0 - 6}{2 - 0} = -3$. Thus, $f'(3)g'(1) = \left(-\frac{1}{4}\right)(-3) = \frac{3}{4}$.

(b) $v(x) = g(f(x)) \Rightarrow v'(x) = g'(f(x))f'(x)$. So $v'(1) = g'(f(1))f'(1) = g'(2)f'(1)$, which does not exist since $g'(2)$ does not exist.

(c) $w(x) = g(g(x)) \Rightarrow w'(x) = g'(g(x))g'(x)$. So $w'(1) = g'(g(1))g'(1) = g'(3)g'(1)$. To find $g'(3)$, note that g is linear from $(2, 0)$ to $(5, 2)$, so its slope is $\frac{2-0}{5-2} = \frac{2}{3}$. Thus, $g'(3)g'(1) = \left(\frac{2}{3}\right)(-3) = -2$.

52. (a) $h(x) = f(f(x)) \Rightarrow h'(x) = f'(f(x))f'(x)$. So $h'(2) = f'(f(2))f'(2) = f'(1)f'(2) \approx (-1)(-1) = 1$.

(b) $g(x) = f(x^2) \Rightarrow g'(x) = f'(x^2) \cdot \frac{d}{dx}(x^2) = f'(x^2)(2x)$. So $g'(2) = f'(2^2)(2 \cdot 2) = 4f'(4) \approx 4(1.5) = 6$.

53. (a) $F(x) = f(\cos x) \Rightarrow F'(x) = f'(\cos x)\frac{d}{dx}(\cos x) = -\sin x\, f'(\cos x)$

(b) $G(x) = \cos(f(x)) \Rightarrow G'(x) = -\sin(f(x))\, f'(x)$

54. (a) $F(x) = f(x^\alpha) \Rightarrow F'(x) = f'(x^\alpha)\frac{d}{dx}(x^\alpha) = f'(x^\alpha)\alpha x^{\alpha-1}$

(b) $G(x) = [f(x)]^\alpha \Rightarrow G'(x) = \alpha\,[f(x)]^{\alpha-1} f'(x)$

55. $r(x) = f(g(h(x))) \Rightarrow r'(x) = f'(g(h(x))) \cdot g'(h(x)) \cdot h'(x)$, so
$r'(1) = f'(g(h(1))) \cdot g'(h(1)) \cdot h'(1) = f'(g(2)) \cdot g'(2) \cdot 4 = f'(3) \cdot 5 \cdot 4 = 6 \cdot 5 \cdot 4 = 120$

56. $f(x) = xg(x^2) \Rightarrow f'(x) = xg'(x^2)\,2x + g(x^2) \cdot 1 = 2x^2g'(x^2) + g(x^2) \Rightarrow$
$f''(x) = 2x^2g''(x^2)\,2x + g'(x^2)\,4x + g'(x^2)\,2x = 4x^3g''(x^2) + 4xg'(x^2) + 2xg'(x^2) = 6xg'(x^2) + 4x^3g''(x^2)$

57. For the tangent line to be horizontal, $f'(x) = 0$. $f(x) = 2\sin x + \sin^2 x \Rightarrow f'(x) = 2\cos x + 2\sin x\, \cos x = 0 \Leftrightarrow$ $2\cos x\,(1 + \sin x) = 0 \Leftrightarrow \cos x = 0$ or $\sin x = -1$, so $x = \frac{\pi}{2} + 2n\pi$ or $\frac{3\pi}{2} + 2n\pi$, where n is any integer. Now $f\left(\frac{\pi}{2}\right) = 3$ and $f\left(\frac{3\pi}{2}\right) = -1$, so the points on the curve with a horizontal tangent are $\left(\frac{\pi}{2} + 2n\pi, 3\right)$ and $\left(\frac{3\pi}{2} + 2n\pi, -1\right)$, where n is any integer.

58. The use of D, D^2, …, D^n is just a derivative notation (see text page 86). In general, $Df(2x) = 2f'(2x)$, $D^2f(2x) = 4f''(2x)$, …, $D^nf(2x) = 2^nf^{(n)}(2x)$. Since $f(x) = \cos x$ and $50 = 4(12) + 2$, we have $f^{(50)}(x) = f^{(2)}(x) = -\cos x$, so $D^{50}\cos 2x = -2^{50}\cos 2x$.

59. $s(t) = 10 + \frac{1}{4}\sin(10\pi t) \Rightarrow$ the velocity after t seconds is $v(t) = s'(t) = \frac{1}{4}\cos(10\pi t)(10\pi) = \frac{5\pi}{2}\cos(10\pi t)$ cm/s.

60. (a) $s = A\cos(\omega t + \delta) \Rightarrow$ velocity $= s' = -\omega A\sin(\omega t + \delta)$.

(b) If $A \neq 0$ and $\omega \neq 0$, then $s' = 0 \Leftrightarrow \sin(\omega t + \delta) = 0 \Leftrightarrow \omega t + \delta = n\pi \Leftrightarrow t = \frac{n\pi - \delta}{\omega}$, n an integer.

61. (a) $B(t) = 4.0 + 0.35\sin\frac{2\pi t}{5.4} \Rightarrow \frac{dB}{dt} = \left(0.35\cos\frac{2\pi t}{5.4}\right)\left(\frac{2\pi}{5.4}\right) = \frac{0.7\pi}{5.4}\cos\frac{2\pi t}{5.4} = \frac{7\pi}{54}\cos\frac{2\pi t}{5.4}$

(b) At $t = 1$, $\frac{dB}{dt} = \frac{7\pi}{54}\cos\frac{2\pi}{5.4} \approx 0.16$.

62. $L(t) = 12 + 2.8\sin\left(\frac{2\pi}{365}(t - 80)\right) \Rightarrow L'(t) = 2.8\cos\left(\frac{2\pi}{365}(t - 80)\right)\left(\frac{2\pi}{365}\right)$.
On March 21, $t = 80$, and $L'(80) \approx 0.0482$ hours per day. On May 21, $t = 141$, and $L'(141) \approx 0.02398$, which is approximately one-half of $L'(80)$.

63. By the Chain Rule, $a(t) = \dfrac{dv}{dt} = \dfrac{dv}{ds}\dfrac{ds}{dt} = \dfrac{dv}{ds}v(t) = v(t)\dfrac{dv}{ds}$. The derivative dv/dt is the rate of change of the velocity with respect to time (in other words, the acceleration) whereas the derivative dv/ds is the rate of change of the velocity with respect to the displacement.

64. (a) The derivative dV/dr represents the rate of change of the volume with respect to the radius and the derivative dV/dt represents the rate of change of the volume with respect to time.

(b) Since $V = \frac{4}{3}\pi r^3$, $\dfrac{dV}{dt} = \dfrac{dV}{dr}\dfrac{dr}{dt} = 4\pi r^2 \dfrac{dr}{dt}$.

65. (a)

$$\begin{aligned}
\frac{d}{dx}(\sin^n x \cos nx) &= n\sin^{n-1} x \cos x \cos nx + \sin^n x\,(-n\sin nx) && \text{[Product Rule]}\\
&= n\sin^{n-1} x\,(\cos nx \cos x - \sin nx \sin x) && [\text{factor out } n\sin^{n-1} x]\\
&= n\sin^{n-1} x \cos(nx + x) && \text{[Addition Formula for cosine]}\\
&= n\sin^{n-1} x \cos[(n+1)x] && [\text{factor out } x]
\end{aligned}$$

(b)

$$\begin{aligned}
\frac{d}{dx}(\cos^n x \cos nx) &= n\cos^{n-1} x\,(-\sin x)\cos nx + \cos^n x\,(-n\sin nx) && \text{[Product Rule]}\\
&= -n\cos^{n-1} x\,(\cos nx \sin x + \sin nx \cos x) && [\text{factor out } -n\cos^{n-1} x]\\
&= -n\cos^{n-1} x \sin(nx + x) && \text{[Addition Formula for sine]}\\
&= -n\cos^{n-1} x \sin[(n+1)x] && [\text{factor out } x]
\end{aligned}$$

66. (a) If f is even, then $f(x) = f(-x)$. Using the Chain Rule to differentiate this equation, we get $f'(x) = f'(-x)\dfrac{d}{dx}(-x) = -f'(-x)$. Thus, $f'(-x) = -f'(x)$, so f' is odd.

(b) If f is odd, then $f(x) = -f(-x)$. Differentiating this equation, we get $f'(x) = -f'(-x)(-1) = f'(-x)$, so f' is even.

67. Since $\theta^\circ = \left(\frac{\pi}{180}\right)\theta$ rad, we have $\dfrac{d}{d\theta}(\sin\theta^\circ) = \dfrac{d}{d\theta}\left(\sin\frac{\pi}{180}\theta\right) = \frac{\pi}{180}\cos\frac{\pi}{180}\theta = \frac{\pi}{180}\cos\theta^\circ$.

68. "The rate of change of y^5 with respect to x is eighty times the rate of change of y with respect to x" $\Leftrightarrow$

$\dfrac{d}{dx}y^5 = 80\dfrac{dy}{dx}$ $\Leftrightarrow$ $5y^4\dfrac{dy}{dx} = 80\dfrac{dy}{dx}$ $\Leftrightarrow$ $5y^4 = 80$ (Note that $dy/dx \neq 0$ since the curve never has a horizontal tangent) $\Leftrightarrow$ $y^4 = 16$ $\Leftrightarrow$ $y = 2$ (since $y > 0$ for all x)

69.

$$\begin{aligned}
\frac{d^2y}{dx^2} &= \frac{d}{dx}\left(\frac{dy}{dx}\right) && \text{[Leibniz notation for the second derivative]}\\
&= \frac{d}{dx}\left(\frac{dy}{du}\frac{du}{dx}\right) && \text{[Chain Rule]}\\
&= \frac{dy}{du}\cdot\frac{d}{dx}\left(\frac{du}{dx}\right) + \frac{du}{dx}\cdot\frac{d}{dx}\left(\frac{dy}{du}\right) && \text{[Product Rule]}\\
&= \frac{dy}{du}\cdot\frac{d^2u}{dx^2} + \frac{du}{dx}\cdot\frac{d}{du}\left(\frac{dy}{du}\right)\cdot\frac{du}{dx} && [dy/du \text{ is a function of } u]\\
&= \frac{dy}{du}\frac{d^2u}{dx^2} + \frac{d^2y}{du^2}\left(\frac{du}{dx}\right)^2
\end{aligned}$$

Or: Using function notation for $y = f(u)$ and $u = g(x)$, we have $y = f(g(x))$, so

$y' = f'(g(x))\cdot g'(x)$ [by the Chain Rule] $\Rightarrow$

$(y')' = [f'(g(x))\cdot g'(x)]' = f'(g(x))\cdot g''(x) + g'(x)\cdot f''(g(x))\cdot g'(x) = f'(g(x))\cdot g''(x) + f''(g(x))\cdot [g'(x)]^2$.

70. (a) $f(x) = |x| = \sqrt{x^2} = (x^2)^{1/2} \Rightarrow f'(x) = \frac{1}{2}(x^2)^{-1/2}(2x) = x/\sqrt{x^2} = x/|x|$ for $x \neq 0$.

f is not differentiable at $x = 0$.

(b) $f(x) = |\sin x| = \sqrt{\sin^2 x} \Rightarrow$

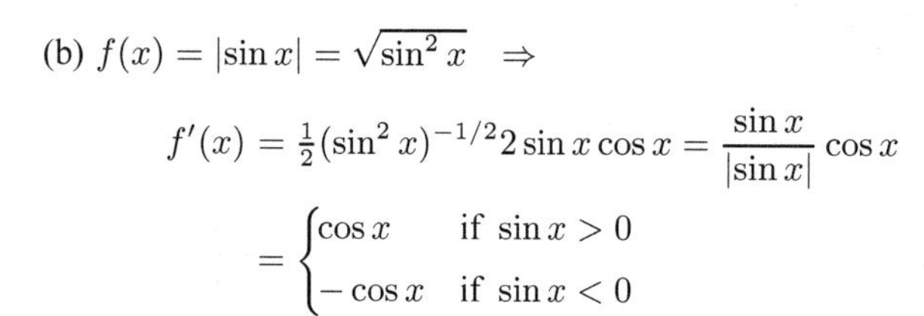

$$f'(x) = \tfrac{1}{2}(\sin^2 x)^{-1/2} 2 \sin x \cos x = \frac{\sin x}{|\sin x|}\cos x$$

$$= \begin{cases} \cos x & \text{if } \sin x > 0 \\ -\cos x & \text{if } \sin x < 0 \end{cases}$$

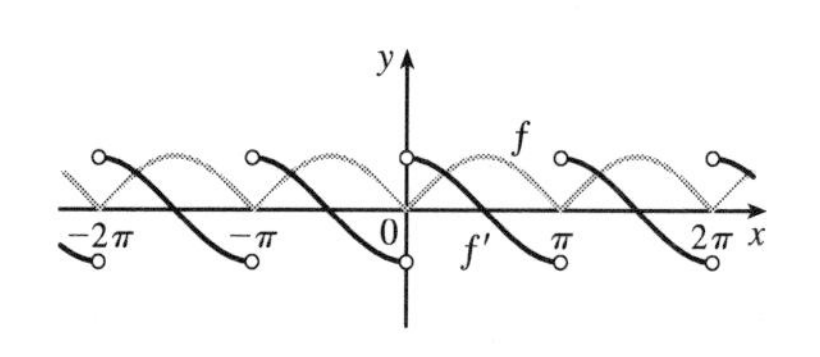

f is not differentiable when $x = n\pi$, n an integer.

(c) $g(x) = \sin|x| = \sin\sqrt{x^2} \Rightarrow$

$$g'(x) = \cos|x| \cdot \frac{x}{|x|} = \frac{x}{|x|}\cos x = \begin{cases} \cos x & \text{if } x > 0 \\ -\cos x & \text{if } x < 0 \end{cases}$$

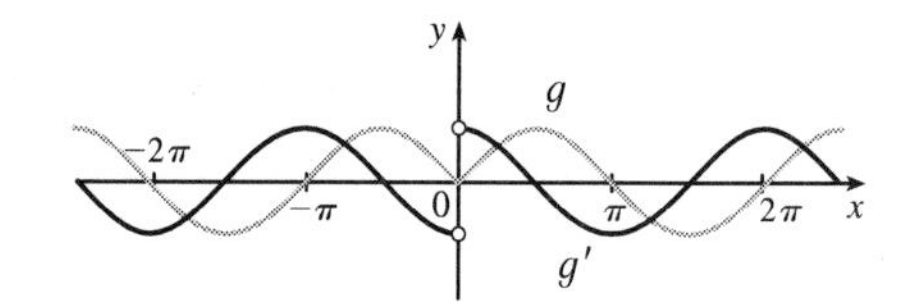

g is not differentiable at 0.

2.6 Implicit Differentiation

1. (a) $\frac{d}{dx}(xy + 2x + 3x^2) = \frac{d}{dx}(4) \Rightarrow (x \cdot y' + y \cdot 1) + 2 + 6x = 0 \Rightarrow xy' = -y - 2 - 6x \Rightarrow$

$y' = \frac{-y - 2 - 6x}{x}$ or $y' = -6 - \frac{y+2}{x}$.

(b) $xy + 2x + 3x^2 = 4 \Rightarrow xy = 4 - 2x - 3x^2 \Rightarrow y = \frac{4 - 2x - 3x^2}{x} = \frac{4}{x} - 2 - 3x$, so $y' = -\frac{4}{x^2} - 3$.

(c) From part (a), $y' = \frac{-y - 2 - 6x}{x} = \frac{-(4/x - 2 - 3x) - 2 - 6x}{x} = \frac{-4/x - 3x}{x} = -\frac{4}{x^2} - 3$.

2. (a) $\frac{d}{dx}(4x^2 + 9y^2) = \frac{d}{dx}(36) \Rightarrow 8x + 18y \cdot y' = 0 \Rightarrow y' = -\frac{8x}{18y} = -\frac{4x}{9y}$

(b) $4x^2 + 9y^2 = 36 \Rightarrow 9y^2 = 36 - 4x^2 \Rightarrow y^2 = \frac{4}{9}(9 - x^2) \Rightarrow y = \pm\frac{2}{3}\sqrt{9 - x^2}$, so

$y' = \pm\frac{2}{3} \cdot \frac{1}{2}(9 - x^2)^{-1/2}(-2x) = \mp\frac{2x}{3\sqrt{9 - x^2}}$

(c) From part (a), $y' = -\frac{4x}{9y} = -\frac{4x}{9\left(\pm\frac{2}{3}\sqrt{9 - x^2}\right)} = \mp\frac{2x}{3\sqrt{9 - x^2}}$.

3. $\frac{d}{dx}(x^3 + x^2y + 4y^2) = \frac{d}{dx}(6) \Rightarrow 3x^2 + (x^2y' + y \cdot 2x) + 8yy' = 0 \Rightarrow x^2y' + 8yy' = -3x^2 - 2xy \Rightarrow$

$(x^2 + 8y)y' = -3x^2 - 2xy \Rightarrow y' = -\frac{3x^2 + 2xy}{x^2 + 8y} = -\frac{x(3x + 2y)}{x^2 + 8y}$

4. $\frac{d}{dx}(x^2 - 2xy + y^3) = \frac{d}{dx}(c) \Rightarrow 2x - 2(xy' + y \cdot 1) + 3y^2y' = 0 \Rightarrow 2x - 2y = 2xy' - 3y^2y' \Rightarrow$

$2x - 2y = y'(2x - 3y^2) \Rightarrow y' = \frac{2x - 2y}{2x - 3y^2}$

5. $\frac{d}{dx}(x^2y + xy^2) = \frac{d}{dx}(3x) \Rightarrow (x^2y' + y \cdot 2x) + (x \cdot 2yy' + y^2 \cdot 1) = 3 \Rightarrow x^2y' + 2xyy' = 3 - 2xy - y^2 \Rightarrow$

$y'(x^2 + 2xy) = 3 - 2xy - y^2 \Rightarrow y' = \frac{3 - 2xy - y^2}{x^2 + 2xy}$

6. $\frac{d}{dx}\left(y^5 + x^2y^3\right) = \frac{d}{dx}\left(1 + x^4y\right) \Rightarrow 5y^4y' + x^2 \cdot 3y^2y' + y^3 \cdot 2x = 0 + x^4y' + y \cdot 4x^3 \Rightarrow$

$y'\left(5y^4 + 3x^2y^2 - x^4\right) = 4x^3y - 2xy^3 \Rightarrow y' = \frac{4x^3y - 2xy^3}{5y^4 + 3x^2y^2 - x^4}$

7. $\frac{d}{dx}\left(x^2y^2 + x\sin y\right) = \frac{d}{dx}(4) \Rightarrow x^2 \cdot 2yy' + y^2 \cdot 2x + x\cos y \cdot y' + \sin y \cdot 1 = 0 \Rightarrow$

$2x^2yy' + x\cos y \cdot y' = -2xy^2 - \sin y \Rightarrow (2x^2y + x\cos y)y' = -2xy^2 - \sin y \Rightarrow y' = \frac{-2xy^2 - \sin y}{2x^2y + x\cos y}$

8. $\frac{d}{dx}(1 + x) = \frac{d}{dx}\left[\sin(xy^2)\right] \Rightarrow 1 = [\cos(xy^2)](x \cdot 2yy' + y^2 \cdot 1) \Rightarrow 1 = 2xy\cos(xy^2)y' + y^2\cos(xy^2) \Rightarrow$

$1 - y^2\cos(xy^2) = 2xy\cos(xy^2)y' \Rightarrow y' = \frac{1 - y^2\cos(xy^2)}{2xy\cos(xy^2)}$

9. $\frac{d}{dx}(4\cos x\, \sin y) = \frac{d}{dx}(1) \Rightarrow 4\left[\cos x \cdot \cos y \cdot y' + \sin y \cdot (-\sin x)\right] = 0 \Rightarrow$

$y'(4\cos x\, \cos y) = 4\sin x\, \sin y \Rightarrow y' = \frac{4\sin x\, \sin y}{4\cos x\, \cos y} = \tan x\, \tan y$

10. $\frac{d}{dx}\left[y\sin(x^2)\right] = \frac{d}{dx}\left[x\sin(y^2)\right] \Rightarrow y\cos(x^2) \cdot 2x + \sin(x^2) \cdot y' = x\cos(y^2) \cdot 2yy' + \sin(y^2) \cdot 1 \Rightarrow$

$y'\left[\sin(x^2) - 2xy\cos(y^2)\right] = \sin(y^2) - 2xy\cos(x^2) \Rightarrow y' = \frac{\sin(y^2) - 2xy\cos(x^2)}{\sin(x^2) - 2xy\cos(y^2)}$

11. $\frac{d}{dx}\left[\tan(x/y)\right] = \frac{d}{dx}(x + y) \Rightarrow \sec^2(x/y) \cdot \frac{y \cdot 1 - x \cdot y'}{y^2} = 1 + y' \Rightarrow$

$y\sec^2(x/y) - x\sec^2(x/y) \cdot y' = y^2 + y^2y' \Rightarrow y\sec^2(x/y) - y^2 = y^2y' + x\sec^2(x/y) \Rightarrow$

$y\sec^2(x/y) - y^2 = \left[y^2 + x\sec^2(x/y)\right] \cdot y' \Rightarrow y' = \frac{y\sec^2(x/y) - y^2}{y^2 + x\sec^2(x/y)}$

12. $\frac{d}{dx}\left(\sqrt{x + y}\right) = \frac{d}{dx}\left(1 + x^2y^2\right) \Rightarrow \frac{1}{2}(x + y)^{-1/2}(1 + y') = x^2 \cdot 2yy' + y^2 \cdot 2x \Rightarrow$

$\frac{1}{2\sqrt{x + y}} + \frac{y'}{2\sqrt{x + y}} = 2x^2yy' + 2xy^2 \Rightarrow 1 + y' = 4x^2y\sqrt{x + y}\, y' + 4xy^2\sqrt{x + y} \Rightarrow$

$y' - 4x^2y\sqrt{x + y}\, y' = 4xy^2\sqrt{x + y} - 1 \Rightarrow y'(1 - 4x^2y\sqrt{x + y}) = 4xy^2\sqrt{x + y} - 1 \Rightarrow y' = \frac{4xy^2\sqrt{x + y} - 1}{1 - 4x^2y\sqrt{x + y}}$

13. $\sqrt{xy} = 1 + x^2y \Rightarrow \frac{1}{2}(xy)^{-1/2}(xy' + y \cdot 1) = 0 + x^2y' + y \cdot 2x \Rightarrow \frac{x}{2\sqrt{xy}}y' + \frac{y}{2\sqrt{xy}} = x^2y' + 2xy \Rightarrow$

$y'\left(\frac{x}{2\sqrt{xy}} - x^2\right) = 2xy - \frac{y}{2\sqrt{xy}} \Rightarrow y'\left(\frac{x - 2x^2\sqrt{xy}}{2\sqrt{xy}}\right) = \frac{4xy\sqrt{xy} - y}{2\sqrt{xy}} \Rightarrow y' = \frac{4xy\sqrt{xy} - y}{x - 2x^2\sqrt{xy}}$

14. $\sin x + \cos y = \sin x\, \cos y \Rightarrow \cos x - \sin y \cdot y' = \sin x\,(-\sin y \cdot y') + \cos y\, \cos x \Rightarrow$

$(\sin x\, \sin y - \sin y)\, y' = \cos x\, \cos y - \cos x \Rightarrow y' = \frac{\cos x\,(\cos y - 1)}{\sin y\,(\sin x - 1)}$

15. $\frac{d}{dx}\left\{f(x) + x^2[f(x)]^3\right\} = \frac{d}{dx}(10) \;\Rightarrow\; f'(x) + x^2 \cdot 3[f(x)]^2 \cdot f'(x) + [f(x)]^3 \cdot 2x = 0$. If $x = 1$, we have

$f'(1) + 1^2 \cdot 3[f(1)]^2 \cdot f'(1) + [f(1)]^3 \cdot 2(1) = 0 \;\Rightarrow\; f'(1) + 1 \cdot 3 \cdot 2^2 \cdot f'(1) + 2^3 \cdot 2 = 0 \;\Rightarrow$

$f'(1) + 12f'(1) = -16 \;\Rightarrow\; 13f'(1) = -16 \;\Rightarrow\; f'(1) = -\frac{16}{13}$.

16. $\frac{d}{dx}[g(x) + x\sin g(x)] = \frac{d}{dx}(x^2) \;\Rightarrow\; g'(x) + x\cos g(x) \cdot g'(x) + \sin g(x) \cdot 1 = 2x$. If $x = 0$, we have

$g'(0) + 0 + \sin g(0) = 2(0) \;\Rightarrow\; g'(0) + \sin 0 = 0 \;\Rightarrow\; g'(0) + 0 = 0 \;\Rightarrow\; g'(0) = 0$.

17. $x^2 + xy + y^2 = 3 \;\Rightarrow\; 2x + xy' + y \cdot 1 + 2yy' = 0 \;\Rightarrow\; xy' + 2yy' = -2x - y \;\Rightarrow\; y'(x + 2y) = -2x - y \;\Rightarrow$

$y' = \frac{-2x - y}{x + 2y}$. When $x = 1$ and $y = 1$, we have $y' = \frac{-2 - 1}{1 + 2 \cdot 1} = \frac{-3}{3} = -1$, so an equation of the tangent line is

$y - 1 = -1(x - 1)$ or $y = -x + 2$.

18. $x^2 + 2xy - y^2 + x = 2 \;\Rightarrow\; 2x + 2(xy' + y \cdot 1) - 2yy' + 1 = 0 \;\Rightarrow\; 2xy' - 2yy' = -2x - 2y - 1 \;\Rightarrow$

$y'(2x - 2y) = -2x - 2y - 1 \;\Rightarrow\; y' = \frac{-2x - 2y - 1}{2x - 2y}$. When $x = 1$ and $y = 2$, we have $y' = \frac{-2 - 4 - 1}{2 - 4} = \frac{-7}{-2} = \frac{7}{2}$,

so an equation of the tangent line is $y - 2 = \frac{7}{2}(x - 1)$ or $y = \frac{7}{2}x - \frac{3}{2}$.

19. $x^2 + y^2 = (2x^2 + 2y^2 - x)^2 \;\Rightarrow\; 2x + 2yy' = 2(2x^2 + 2y^2 - x)(4x + 4yy' - 1)$. When $x = 0$ and $y = \frac{1}{2}$, we have

$0 + y' = 2(\frac{1}{2})(2y' - 1) \;\Rightarrow\; y' = 2y' - 1 \;\Rightarrow\; y' = 1$, so an equation of the tangent line is $y - \frac{1}{2} = 1(x - 0)$

or $y = x + \frac{1}{2}$.

20. $x^{2/3} + y^{2/3} = 4 \;\Rightarrow\; \frac{2}{3}x^{-1/3} + \frac{2}{3}y^{-1/3}y' = 0 \;\Rightarrow\; \frac{1}{\sqrt[3]{x}} + \frac{y'}{\sqrt[3]{y}} = 0 \;\Rightarrow\; y' = -\frac{\sqrt[3]{y}}{\sqrt[3]{x}}$. When $x = -3\sqrt{3}$

and $y = 1$, we have $y' = -\frac{1}{(-3\sqrt{3})^{1/3}} = -\frac{(-3\sqrt{3})^{2/3}}{-3\sqrt{3}} = \frac{3}{3\sqrt{3}} = \frac{1}{\sqrt{3}}$, so an equation of the tangent line is

$y - 1 = \frac{1}{\sqrt{3}}(x + 3\sqrt{3})$ or $y = \frac{1}{\sqrt{3}}x + 4$.

21. $2(x^2 + y^2)^2 = 25(x^2 - y^2) \;\Rightarrow\; 4(x^2 + y^2)(2x + 2yy') = 25(2x - 2yy') \;\Rightarrow$

$4(x + yy')(x^2 + y^2) = 25(x - yy') \;\Rightarrow\; 4yy'(x^2 + y^2) + 25yy' = 25x - 4x(x^2 + y^2) \;\Rightarrow$

$y' = \frac{25x - 4x(x^2 + y^2)}{25y + 4y(x^2 + y^2)}$. When $x = 3$ and $y = 1$, we have $y' = \frac{75 - 120}{25 + 40} = -\frac{45}{65} = -\frac{9}{13}$, so an equation of the tangent line

is $y - 1 = -\frac{9}{13}(x - 3)$ or $y = -\frac{9}{13}x + \frac{40}{13}$.

22. $y^2(y^2 - 4) = x^2(x^2 - 5) \;\Rightarrow\; y^4 - 4y^2 = x^4 - 5x^2 \;\Rightarrow\; 4y^3y' - 8yy' = 4x^3 - 10x$. When $x = 0$ and $y = -2$, we

have $-32y' + 16y' = 0 \;\Rightarrow\; -16y' = 0 \;\Rightarrow\; y' = 0$, so an equation of the tangent line is $y + 2 = 0(x - 0)$ or $y = -2$.

23. $9x^2 + y^2 = 9 \;\Rightarrow\; 18x + 2yy' = 0 \;\Rightarrow\; 2yy' = -18x \;\Rightarrow\; y' = -9x/y \;\Rightarrow$

$y'' = -9\left(\frac{y \cdot 1 - x \cdot y'}{y^2}\right) = -9\left(\frac{y - x(-9x/y)}{y^2}\right) = -9 \cdot \frac{y^2 + 9x^2}{y^3} = -9 \cdot \frac{9}{y^3}$ [since x and y must satisfy the original

equation, $9x^2 + y^2 = 9$]. Thus, $y'' = -81/y^3$.

24. $\sqrt{x}+\sqrt{y}=1 \;\Rightarrow\; \dfrac{1}{2\sqrt{x}}+\dfrac{y'}{2\sqrt{y}}=0 \;\Rightarrow\; y'=-\sqrt{y}/\sqrt{x} \;\Rightarrow$

$$y''=-\frac{\sqrt{x}\,[1/(2\sqrt{y})]\,y'-\sqrt{y}\,[1/(2\sqrt{x})]}{x}=-\frac{\sqrt{x}\,(1/\sqrt{y})\,(-\sqrt{y}/\sqrt{x})-\sqrt{y}\,(1/\sqrt{x})}{2x}=\frac{1+\sqrt{y}/\sqrt{x}}{2x}$$

$=\dfrac{\sqrt{x}+\sqrt{y}}{2x\sqrt{x}}=\dfrac{1}{2x\sqrt{x}}$ since x and y must satisfy the original equation, $\sqrt{x}+\sqrt{y}=1$.

25. $x^3+y^3=1 \;\Rightarrow\; 3x^2+3y^2y'=0 \;\Rightarrow\; y'=-\dfrac{x^2}{y^2} \;\Rightarrow$

$$y''=-\frac{y^2(2x)-x^2\cdot 2yy'}{(y^2)^2}=-\frac{2xy^2-2x^2y(-x^2/y^2)}{y^4}=-\frac{2xy^4+2x^4y}{y^6}=-\frac{2xy(y^3+x^3)}{y^6}=-\frac{2x}{y^5},$$

since x and y must satisfy the original equation, $x^3+y^3=1$.

26. $x^4+y^4=a^4 \;\Rightarrow\; 4x^3+4y^3y'=0 \;\Rightarrow\; 4y^3y'=-4x^3 \;\Rightarrow\; y'=-x^3/y^3 \;\Rightarrow$

$$y''=-\left(\frac{y^3\cdot 3x^2-x^3\cdot 3y^2y'}{(y^3)^2}\right)=-3x^2y^2\cdot\frac{y-x(-x^3/y^3)}{y^6}=-3x^2\cdot\frac{y^4+x^4}{y^4y^3}=-3x^2\cdot\frac{a^4}{y^7}=\frac{-3a^4x^2}{y^7}$$

27. (a) $y^2=5x^4-x^2 \;\Rightarrow\; 2yy'=5(4x^3)-2x \;\Rightarrow\; y'=\dfrac{10x^3-x}{y}$.

So at the point $(1,2)$ we have $y'=\dfrac{10(1)^3-1}{2}=\dfrac{9}{2}$, and an equation of the tangent line is $y-2=\frac{9}{2}(x-1)$ or $y=\frac{9}{2}x-\frac{5}{2}$.

(b)

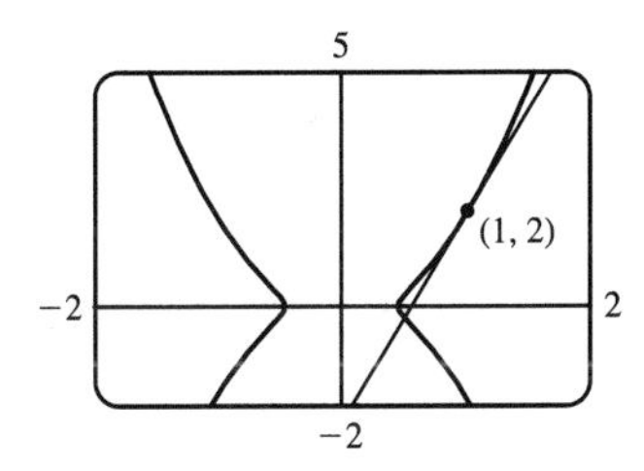

28. (a) $y^2=x^3+3x^2 \;\Rightarrow\; 2yy'=3x^2+3(2x) \;\Rightarrow\; y'=\dfrac{3x^2+6x}{2y}$. So at the point $(1,-2)$ we have

$y'=\dfrac{3(1)^2+6(1)}{2(-2)}=-\dfrac{9}{4}$, and an equation of the tangent line is $y+2=-\frac{9}{4}(x-1)$ or $y=-\frac{9}{4}x+\frac{1}{4}$.

(b) The curve has a horizontal tangent where $y'=0 \;\Leftrightarrow$

$3x^2+6x=0 \;\Leftrightarrow\; 3x(x+2)=0 \;\Leftrightarrow\; x=0$ or $x=-2$.

But note that at $x=0$, $y=0$ also, so the derivative does not exist.

At $x=-2$, $y^2=(-2)^3+3(-2)^2=-8+12=4$, so $y=\pm 2$.

So the two points at which the curve has a horizontal tangent are $(-2,-2)$ and $(-2,2)$.

(c)

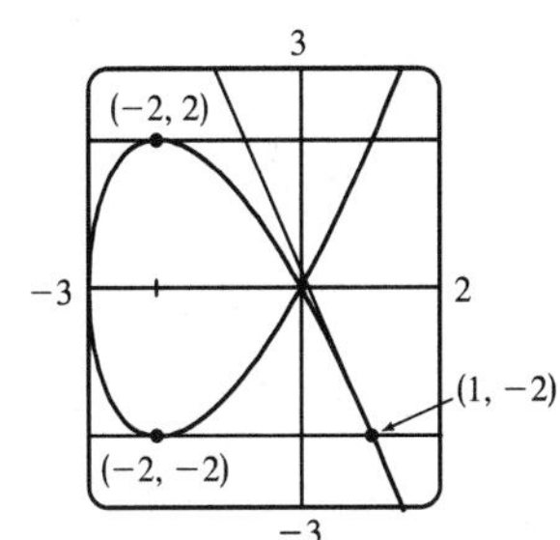

29. (a) There are eight points with horizontal tangents: four at $x\approx 1.57735$ and four at $x\approx 0.42265$.

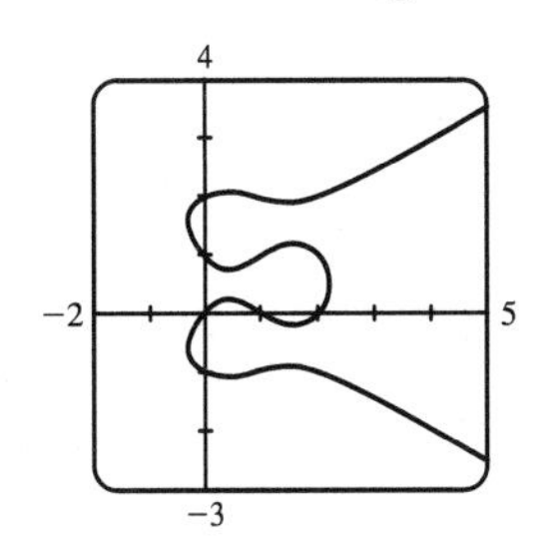

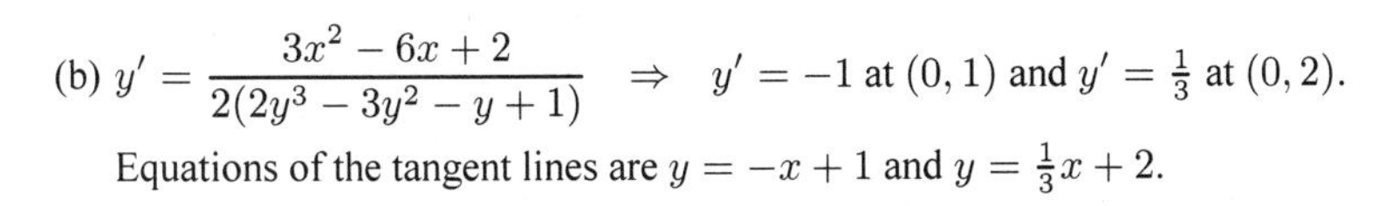

(b) $y'=\dfrac{3x^2-6x+2}{2(2y^3-3y^2-y+1)} \;\Rightarrow\; y'=-1$ at $(0,1)$ and $y'=\frac{1}{3}$ at $(0,2)$.

Equations of the tangent lines are $y=-x+1$ and $y=\frac{1}{3}x+2$.

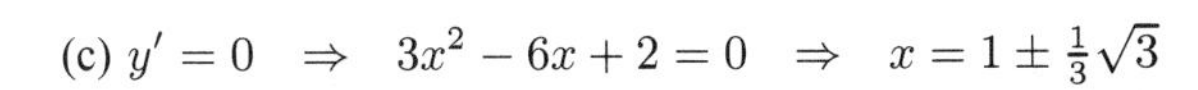

(c) $y'=0 \;\Rightarrow\; 3x^2-6x+2=0 \;\Rightarrow\; x=1\pm\frac{1}{3}\sqrt{3}$

(d) By multiplying the right side of the equation by $x - 3$, we obtain the first graph.By modifying the equation in other ways, we can generate the other graphs.

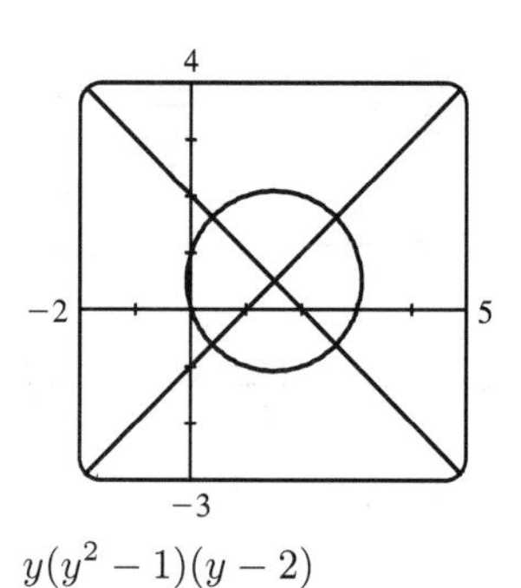

$y(y^2 - 1)(y - 2)$
$= x(x - 1)(x - 2)(x - 3)$

$y(y^2 - 4)(y - 2)$
$= x(x - 1)(x - 2)$

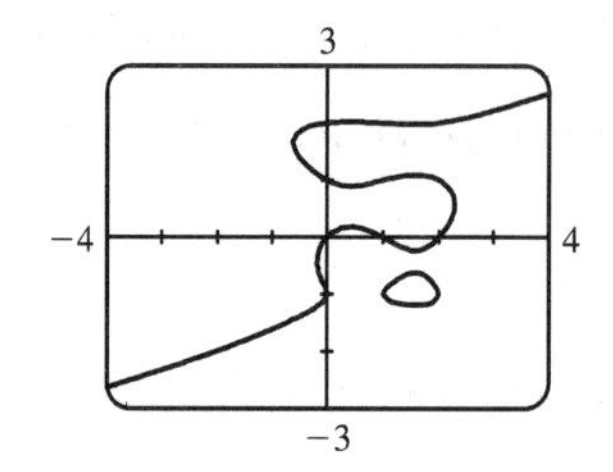

$y(y + 1)(y^2 - 1)(y - 2)$
$= x(x - 1)(x - 2)$

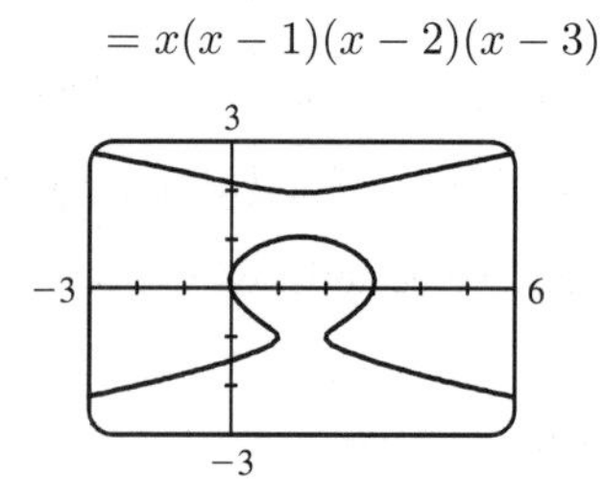

$(y + 1)(y^2 - 1)(y - 2)$
$= (x - 1)(x - 2)$

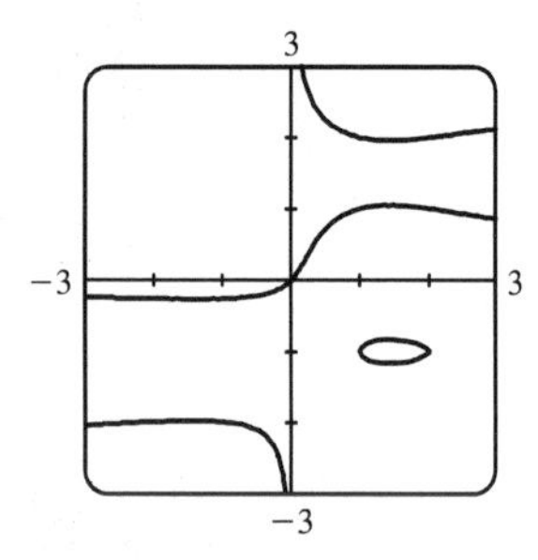

$x(y + 1)(y^2 - 1)(y - 2)$
$= y(x - 1)(x - 2)$

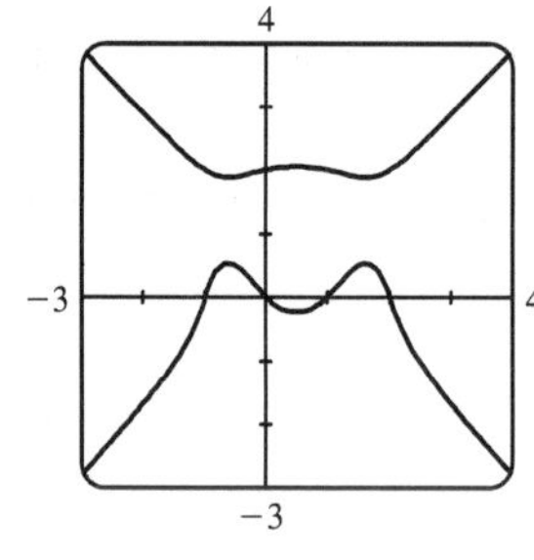

$y(y^2 + 1)(y - 2)$
$= x(x^2 - 1)(x - 2)$

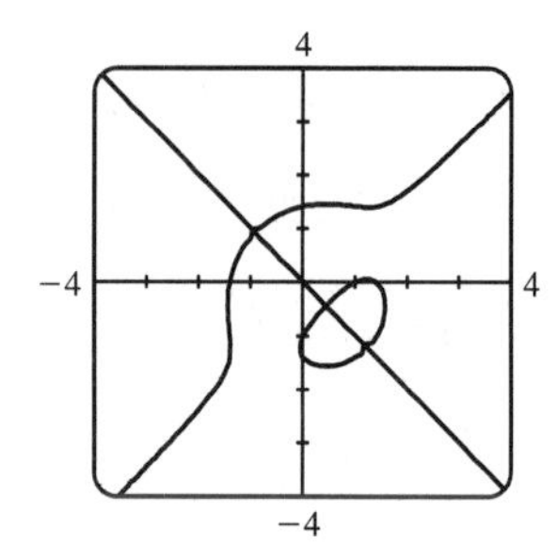

$y(y + 1)(y^2 - 2)$
$= x(x - 1)(x^2 - 2)$

30. (a)

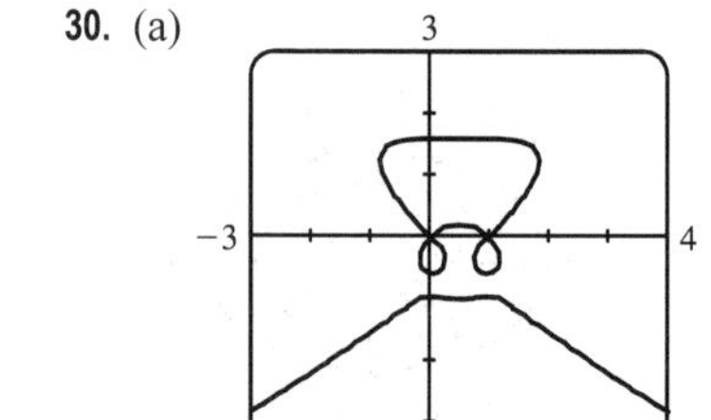

(b) There are 9 points with horizontal tangents: 3 at $x = 0$, 3 at $x = \frac{1}{2}$, and 3 at $x = 1$. The three horizontal tangents along the top of the wagon are hard to find, but by limiting the y-range of the graph (to $[1.6, 1.7]$, for example) they are distinguishable.

31. From Exercise 21, a tangent to the lemniscate will be horizontal if $y' = 0 \;\Rightarrow\; 25x - 4x(x^2 + y^2) = 0 \;\Rightarrow\; x[25 - 4(x^2 + y^2)] = 0 \;\Rightarrow\; x^2 + y^2 = \frac{25}{4}$ (**1**). (Note that when x is 0, y is also 0, and there is no horizontal tangent at the origin.) Substituting $\frac{25}{4}$ for $x^2 + y^2$ in the equation of the lemniscate, $2(x^2 + y^2)^2 = 25(x^2 - y^2)$, we get $x^2 - y^2 = \frac{25}{8}$ (**2**). Solving (**1**) and (**2**), we have $x^2 = \frac{75}{16}$ and $y^2 = \frac{25}{16}$, so the four points are $\left(\pm\frac{5\sqrt{3}}{4}, \pm\frac{5}{4}\right)$.

32. $\dfrac{x^2}{a^2} + \dfrac{y^2}{b^2} = 1 \;\Rightarrow\; \dfrac{2x}{a^2} + \dfrac{2yy'}{b^2} = 0 \;\Rightarrow\; y' = -\dfrac{b^2x}{a^2y} \;\Rightarrow$ an equation of the tangent line at (x_0, y_0) is

$y - y_0 = \dfrac{-b^2x_0}{a^2y_0}(x - x_0)$. Multiplying both sides by $\dfrac{y_0}{b^2}$ gives $\dfrac{y_0y}{b^2} - \dfrac{y_0^2}{b^2} = -\dfrac{x_0x}{a^2} + \dfrac{x_0^2}{a^2}$. Since (x_0, y_0) lies on the ellipse,

we have $\dfrac{x_0x}{a^2} + \dfrac{y_0y}{b^2} = \dfrac{x_0^2}{a^2} + \dfrac{y_0^2}{b^2} = 1$.

33. $x^2 + y^2 = r^2$ is a circle with center O and $ax + by = 0$ is a line through O.

$x^2 + y^2 = r^2 \;\Rightarrow\; 2x + 2yy' = 0 \;\Rightarrow\; y' = -x/y$, so the slope of the tangent line at $P_0\,(x_0, y_0)$ is $-x_0/y_0$. The slope of the line OP_0 is y_0/x_0, which is the negative reciprocal of $-x_0/y_0$. Hence, the curves are orthogonal, and the families of curves are orthogonal trajectories of each other.

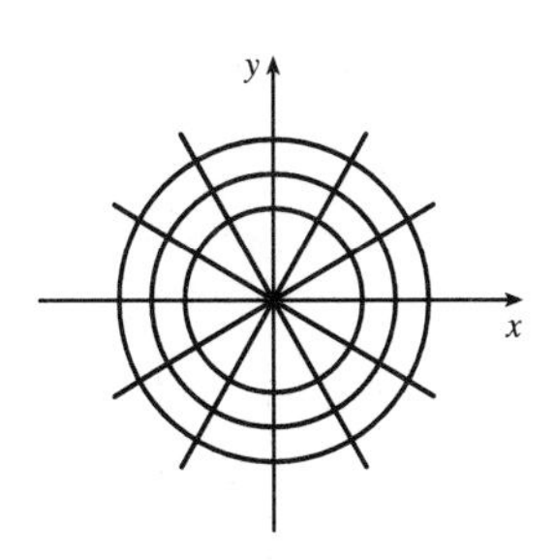

34. The circles $x^2 + y^2 = ax$ and $x^2 + y^2 = by$ intersect at the origin where the tangents are vertical and horizontal. If (x_0, y_0) is the other point of intersection, then $x_0^2 + y_0^2 = ax_0$ **(1)** and $x_0^2 + y_0^2 = by_0$ **(2)**.

Now $x^2 + y^2 = ax \;\Rightarrow\; 2x + 2yy' = a \;\Rightarrow\; y' = \dfrac{a - 2x}{2y}$ and $x^2 + y^2 = by \;\Rightarrow$

$2x + 2yy' = by' \;\Rightarrow\; y' = \dfrac{2x}{b - 2y}$. Thus, the curves are orthogonal at $(x_0, y_0) \;\Leftrightarrow$

$\dfrac{a - 2x_0}{2y_0} = -\dfrac{b - 2y_0}{2x_0} \;\Leftrightarrow\; 2ax_0 - 4x_0^2 = 4y_0^2 - 2by_0 \;\Leftrightarrow\; ax_0 + by_0 = 2(x_0^2 + y_0^2)$,

which is true by **(1)** and **(2)**.

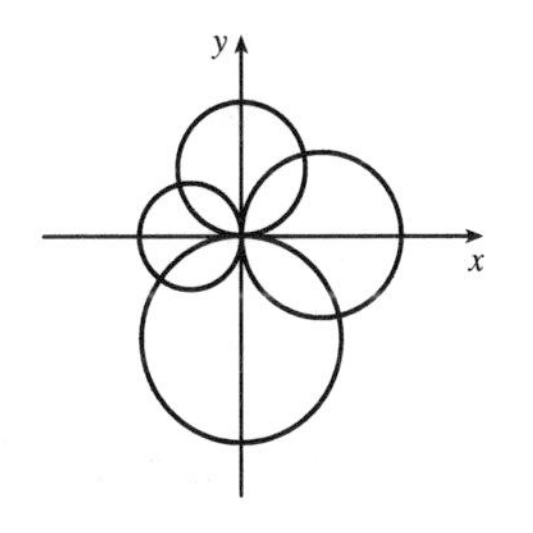

35. $y = cx^2 \;\Rightarrow\; y' = 2cx$ and $x^2 + 2y^2 = k \;\Rightarrow\; 2x + 4yy' = 0 \;\Rightarrow$

$2yy' = -x \;\Rightarrow\; y' = -\dfrac{x}{2(y)} = -\dfrac{x}{2(cx^2)} = -\dfrac{1}{2cx}$, so the curves are orthogonal.

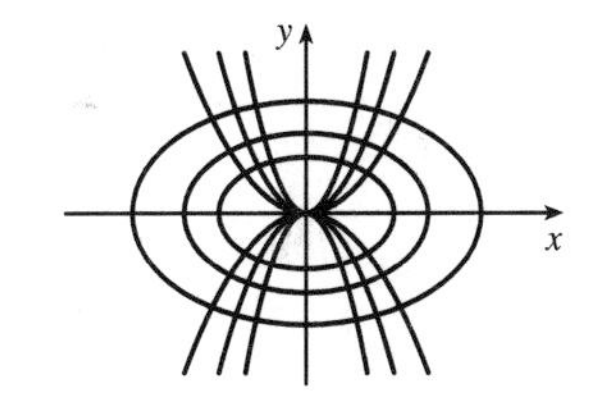

36. $y = ax^3 \;\Rightarrow\; y' = 3ax^2$ and $x^2 + 3y^2 = b \;\Rightarrow\; 2x + 6yy' = 0 \;\Rightarrow$

$3yy' = -x \;\Rightarrow\; y' = -\dfrac{x}{3(y)} = -\dfrac{x}{3(ax^3)} = -\dfrac{1}{3ax^2}$, so the curves are orthogonal.

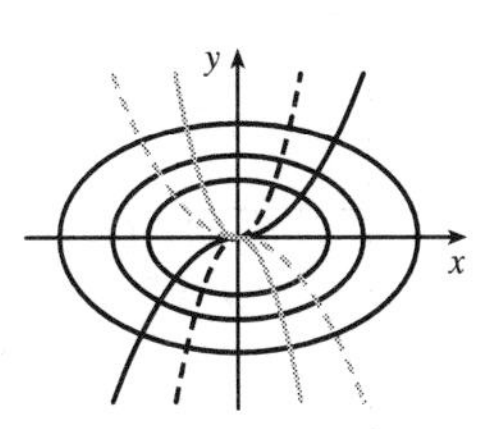

37. If the circle has radius r, its equation is $x^2 + y^2 = r^2 \;\Rightarrow\; 2x + 2yy' = 0 \;\Rightarrow\; y' = -\dfrac{x}{y}$, so the slope of the tangent line at $P(x_0, y_0)$ is $-\dfrac{x_0}{y_0}$. The negative reciprocal of that slope is $\dfrac{-1}{-x_0/y_0} = \dfrac{y_0}{x_0}$, which is the slope of OP, so the tangent line at P is perpendicular to the radius OP.

38. $\sqrt{x}+\sqrt{y}=\sqrt{c} \;\Rightarrow\; \dfrac{1}{2\sqrt{x}}+\dfrac{y'}{2\sqrt{y}}=0 \;\Rightarrow\; y'=-\dfrac{\sqrt{y}}{\sqrt{x}} \;\Rightarrow$ an equation of the tangent line at (x_0,y_0)

is $y-y_0=-\dfrac{\sqrt{y_0}}{\sqrt{x_0}}(x-x_0)$. Now $x=0 \;\Rightarrow\; y=y_0-\dfrac{\sqrt{y_0}}{\sqrt{x_0}}(-x_0)=y_0+\sqrt{x_0}\sqrt{y_0}$, so the y-intercept is

$y_0+\sqrt{x_0}\sqrt{y_0}$. And $y=0 \;\Rightarrow\; -y_0=-\dfrac{\sqrt{y_0}}{\sqrt{x_0}}(x-x_0) \;\Rightarrow\; x-x_0=\dfrac{y_0\sqrt{x_0}}{\sqrt{y_0}} \;\Rightarrow$

$x=x_0+\sqrt{x_0}\sqrt{y_0}$, so the x-intercept is $x_0+\sqrt{x_0}\sqrt{y_0}$. The sum of the intercepts is

$\left(y_0+\sqrt{x_0}\sqrt{y_0}\right)+\left(x_0+\sqrt{x_0}\sqrt{y_0}\right)=x_0+2\sqrt{x_0}\sqrt{y_0}+y_0=\left(\sqrt{x_0}+\sqrt{y_0}\right)^2=\left(\sqrt{c}\right)^2=c.$

39. To find the points at which the ellipse $x^2-xy+y^2=3$ crosses the x-axis, let $y=0$ and solve for x.

$y=0 \;\Rightarrow\; x^2-x(0)+0^2=3 \;\Leftrightarrow\; x=\pm\sqrt{3}$. So the graph of the ellipse crosses the x-axis at the points $\left(\pm\sqrt{3},0\right)$.

Using implicit differentiation to find y', we get $2x-xy'-y+2yy'=0 \;\Rightarrow\; y'(2y-x)=y-2x \;\Leftrightarrow\; y'=\dfrac{y-2x}{2y-x}$.

So y' at $\left(\sqrt{3},0\right)$ is $\dfrac{0-2\sqrt{3}}{2(0)-\sqrt{3}}=2$ and y' at $\left(-\sqrt{3},0\right)$ is $\dfrac{0+2\sqrt{3}}{2(0)+\sqrt{3}}=2$. Thus, the tangent lines at these points are parallel.

40. (a) We use implicit differentiation to find $y'=\dfrac{y-2x}{2y-x}$ as in Exercise 39. The slope of the tangent line at $(-1,1)$ is $m=\dfrac{1-2(-1)}{2(1)-(-1)}=\dfrac{3}{3}=1$, so the slope of the normal line is $-\dfrac{1}{m}=-1$, and its equation is $y-1=-1(x+1) \;\Leftrightarrow$ $y=-x$. Substituting $-x$ for y in the equation of the ellipse, we get $x^2-x(-x)+(-x)^2=3 \;\Rightarrow\; 3x^2=3 \;\Leftrightarrow\; x=\pm1$. So the normal line must intersect the ellipse again at $x=1$, and since the equation of the line is $y=-x$, the other point of intersection must be $(1,-1)$.

(b)

41. $x^2y^2+xy=2 \;\Rightarrow\; x^2\cdot 2yy'+y^2\cdot 2x+x\cdot y'+y\cdot 1=0 \;\Leftrightarrow\; y'(2x^2y+x)=-2xy^2-y \;\Leftrightarrow$

$y'=-\dfrac{2xy^2+y}{2x^2y+x}$. So $-\dfrac{2xy^2+y}{2x^2y+x}=-1 \;\Leftrightarrow\; 2xy^2+y=2x^2y+x \;\Leftrightarrow\; y(2xy+1)=x(2xy+1) \;\Leftrightarrow$

$y(2xy+1)-x(2xy+1)=0 \;\Leftrightarrow\; (2xy+1)(y-x)=0 \;\Leftrightarrow\; xy=-\frac{1}{2}$ or $y=x$. But $xy=-\frac{1}{2} \;\Rightarrow$

$x^2y^2+xy=\frac{1}{4}-\frac{1}{2}\neq 2$, so we must have $x=y$. Then $x^2y^2+xy=2 \;\Rightarrow\; x^4+x^2=2 \;\Leftrightarrow\; x^4+x^2-2=0 \;\Leftrightarrow$

$(x^2+2)(x^2-1)=0$. So $x^2=-2$, which is impossible, or $x^2=1 \;\Leftrightarrow\; x=\pm1$. Since $x=y$, the points on the curve where the tangent line has a slope of -1 are $(-1,-1)$ and $(1,1)$.

42. $x^2+4y^2=36 \;\Rightarrow\; 2x+8yy'=0 \;\Rightarrow\; y'=-\dfrac{x}{4y}$. Let (a,b) be a point on $x^2+4y^2=36$ whose tangent line passes

through $(12,3)$. The tangent line is then $y-3=-\dfrac{a}{4b}(x-12)$, so $b-3=-\dfrac{a}{4b}(a-12)$. Multiplying both sides by $4b$

gives $4b^2-12b=-a^2+12a$, so $4b^2+a^2=12(a+b)$. But $4b^2+a^2=36$, so $36=12(a+b) \;\Rightarrow\; a+b=3 \;\Rightarrow$

$b=3-a$. Substituting $3-a$ for b into $a^2+4b^2=36$ gives $a^2+4(3-a)^2=36 \;\Leftrightarrow\; a^2+36-24a+4a^2=36 \;\Leftrightarrow$

$5a^2 - 24a = 0 \Leftrightarrow a(5a - 24) = 0$, so $a = 0$ or $a = \frac{24}{5}$. If $a = 0$, $b = 3 - 0 = 3$, and if $a = \frac{24}{5}$, $b = 3 - \frac{24}{5} = -\frac{9}{5}$.

So the two points on the ellipse are $(0, 3)$ and $\left(\frac{24}{5}, -\frac{9}{5}\right)$. Using

$y - 3 = -\dfrac{a}{4b}(x - 12)$ with $(a, b) = (0, 3)$ gives us the tangent line

$y - 3 = 0$ or $y = 3$. With $(a, b) = \left(\frac{24}{5}, -\frac{9}{5}\right)$, we have

$y - 3 = -\frac{24/5}{4(-9/5)}(x - 12) \Leftrightarrow y - 3 = \frac{2}{3}(x - 12) \Leftrightarrow y = \frac{2}{3}x - 5.$

A graph of the ellipse and the tangent lines confirms our results.

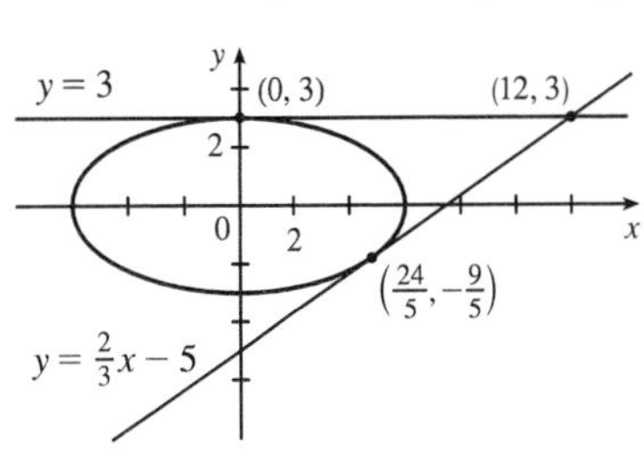

43. (a) $y = J(x)$ and $xy'' + y' + xy = 0 \Rightarrow xJ''(x) + J'(x) + xJ(x) = 0$. If $x = 0$, we have $0 + J'(0) + 0 = 0$, so $J'(0) = 0$.

(b) Differentiating $xy'' + y' + xy = 0$ implicitly, we get $xy''' + y'' \cdot 1 + y'' + xy' + y \cdot 1 = 0 \Rightarrow$ $xy''' + 2y'' + xy' + y = 0$, so $xJ'''(x) + 2J''(x) + xJ'(x) + J(x) = 0$. If $x = 0$, we have $0 + 2J''(0) + 0 + 1$ $[J(0) = 1$ is given$]$ $= 0 \Rightarrow 2J''(0) = -1 \Rightarrow J''(0) = -\frac{1}{2}$.

44. $x^2 + 4y^2 = 5 \Rightarrow 2x + 4(2yy') = 0 \Rightarrow y' = -\dfrac{x}{4y}$. Now let h be the height of the lamp, and let (a, b) be the point of tangency of the line passing through the points $(3, h)$ and $(-5, 0)$. This line has slope $(h - 0)/[3 - (-5)] = \frac{1}{8}h$. But the slope of the tangent line through the point (a, b) can be expressed as $y' = -\dfrac{a}{4b}$, or as $\dfrac{b - 0}{a - (-5)} = \dfrac{b}{a + 5}$ [since the line passes through $(-5, 0)$ and (a, b)], so $-\dfrac{a}{4b} = \dfrac{b}{a + 5} \Leftrightarrow 4b^2 = -a^2 - 5a \Leftrightarrow a^2 + 4b^2 = -5a$. But $a^2 + 4b^2 = 5$ [since (a, b) is on the ellipse], so $5 = -5a \Leftrightarrow a = -1$. Then $4b^2 = -a^2 - 5a = -1 - 5(-1) = 4 \Rightarrow b = 1$, since the point is on the top half of the ellipse. So $\dfrac{h}{8} = \dfrac{b}{a + 5} = \dfrac{1}{-1 + 5} = \dfrac{1}{4} \Rightarrow h = 2$. So the lamp is located 2 units above the x-axis.

2.7 Related Rates

1. $V = x^3 \Rightarrow \dfrac{dV}{dt} = \dfrac{dV}{dx}\dfrac{dx}{dt} = 3x^2\dfrac{dx}{dt}$

2. (a) $A = \pi r^2 \Rightarrow \dfrac{dA}{dt} = \dfrac{dA}{dr}\dfrac{dr}{dt} = 2\pi r\dfrac{dr}{dt}$

(b) $\dfrac{dA}{dt} = 2\pi r\dfrac{dr}{dt} = 2\pi(30\text{ m})(1\text{ m/s}) = 60\pi\text{ m}^2/\text{s}$

3. Let s denote the side of a square. The square's area A is given by $A = s^2$. Differentiating with respect to t gives us $\dfrac{dA}{dt} = 2s\dfrac{ds}{dt}$. When $A = 16$, $s = 4$. Substitution 4 for s and 6 for $\dfrac{ds}{dt}$ gives us $\dfrac{dA}{dt} = 2(4)(6) = 48\text{ cm}^2/\text{s}$.

4. $A = \ell w \Rightarrow \dfrac{dA}{dt} = \ell \cdot \dfrac{dw}{dt} + w \cdot \dfrac{d\ell}{dt} = 20(3) + 10(8) = 140\text{ cm}^2/\text{s}$.

5. $y = x^3 + 2x \Rightarrow \dfrac{dy}{dt} = \dfrac{dy}{dx}\dfrac{dx}{dt} = (3x^2 + 2)(5) = 5(3x^2 + 2)$. When $x = 2$, $\dfrac{dy}{dt} = 5(14) = 70$.

6. $x^2 + y^2 = 25 \Rightarrow 2x\,\frac{dx}{dt} + 2y\,\frac{dy}{dt} = 0 \Rightarrow x\,\frac{dx}{dt} = -y\,\frac{dy}{dt} \Rightarrow \frac{dx}{dt} = -\frac{y}{x}\,\frac{dy}{dt}$.

When $y = 4$, $x^2 + 4^2 = 25 \Rightarrow x = \pm 3$. For $\frac{dy}{dt} = 6$, $\frac{dx}{dt} = -\frac{4}{\pm 3}(6) = \mp 8$.

7. $z^2 = x^2 + y^2 \Rightarrow 2z\,\frac{dz}{dt} = 2x\,\frac{dx}{dt} + 2y\,\frac{dy}{dt} \Rightarrow \frac{dz}{dt} = \frac{1}{z}\left(x\,\frac{dx}{dt} + y\,\frac{dy}{dt}\right)$. When $x = 5$ and $y = 12$,

$z^2 = 5^2 + 12^2 \Rightarrow z^2 = 169 \Rightarrow z = \pm 13$. For $\frac{dx}{dt} = 2$ and $\frac{dy}{dt} = 3$, $\frac{dz}{dt} = \frac{1}{\pm 13}(5 \cdot 2 + 12 \cdot 3) = \pm\frac{46}{13}$.

8. $y = \sqrt{1 + x^3} \Rightarrow \frac{dy}{dt} = \frac{dy}{dx}\,\frac{dx}{dt} = \frac{1}{2}(1 + x^3)^{-1/2}(3x^2)\,\frac{dx}{dt} = \frac{3x^2}{2\sqrt{1 + x^3}}\,\frac{dx}{dt}$. With $\frac{dy}{dt} = 4$ when $x = 2$ and $y = 3$,

we have $4 = \frac{3(4)}{2(3)}\,\frac{dx}{dt} \Rightarrow \frac{dx}{dt} = 2$ cm/s.

9. (a) Given: the rate of decrease of the surface area is 1 cm²/min. If we let t be time (in minutes) and S be the surface area (in cm²), then we are given that $dS/dt = -1$ cm²/s.

(b) Unknown: the rate of decrease of the diameter when the diameter is 10 cm. If we let x be the diameter, then we want to find dx/dt when $x = 10$ cm.

(c)

(d) If the radius is r and the diameter $x = 2r$, then $r = \frac{1}{2}x$ and

$S = 4\pi r^2 = 4\pi\left(\frac{1}{2}x\right)^2 = \pi x^2 \Rightarrow \frac{dS}{dt} = \frac{dS}{dx}\,\frac{dx}{dt} = 2\pi x\,\frac{dx}{dt}$.

(e) $-1 = \frac{dS}{dt} = 2\pi x\,\frac{dx}{dt} \Rightarrow \frac{dx}{dt} = -\frac{1}{2\pi x}$. When $x = 10$, $\frac{dx}{dt} = -\frac{1}{20\pi}$. So the rate of decrease is $\frac{1}{20\pi}$ cm/min.

10. 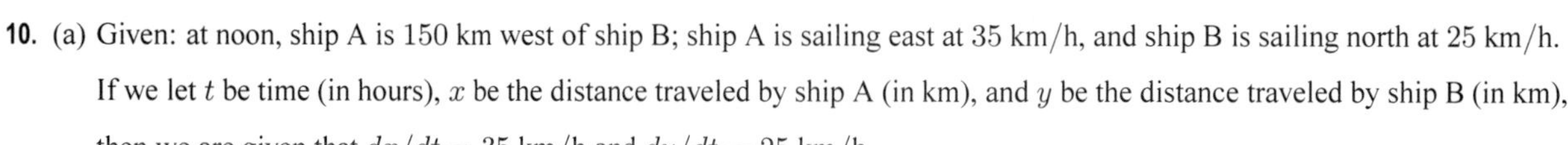
(a) Given: at noon, ship A is 150 km west of ship B; ship A is sailing east at 35 km/h, and ship B is sailing north at 25 km/h. If we let t be time (in hours), x be the distance traveled by ship A (in km), and y be the distance traveled by ship B (in km), then we are given that $dx/dt = 35$ km/h and $dy/dt = 25$ km/h.

(b) Unknown: the rate at which the distance between the ships is changing at 4:00 PM. If we let z be the distance between the ships, then we want to find dz/dt when $t = 4$ h.

(c)

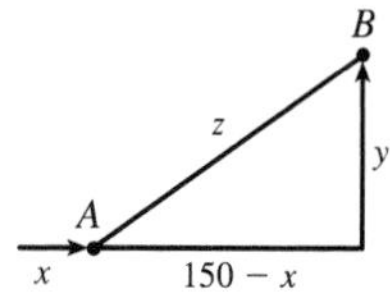

(d) $z^2 = (150 - x)^2 + y^2 \Rightarrow 2z\,\frac{dz}{dt} = 2(150 - x)\left(-\frac{dx}{dt}\right) + 2y\,\frac{dy}{dt}$

(e) At 4:00 PM, $x = 4(35) = 140$ and $y = 4(25) = 100 \Rightarrow z = \sqrt{(150 - 140)^2 + 100^2} = \sqrt{10{,}100}$.

So $\frac{dz}{dt} = \frac{1}{z}\left[(x - 150)\,\frac{dx}{dt} + y\,\frac{dy}{dt}\right] = \frac{-10(35) + 100(25)}{\sqrt{10{,}100}} = \frac{215}{\sqrt{101}} \approx 21.4$ km/h.

11. 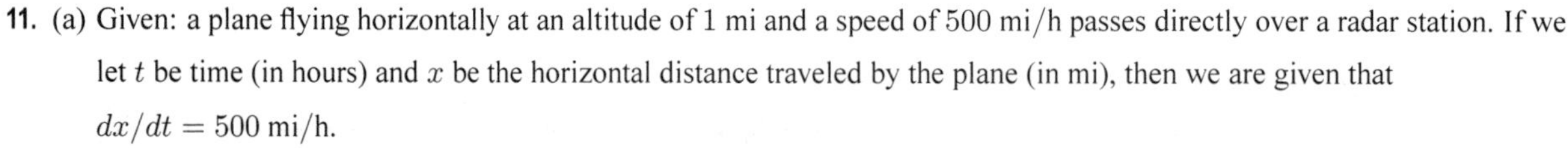
(a) Given: a plane flying horizontally at an altitude of 1 mi and a speed of 500 mi/h passes directly over a radar station. If we let t be time (in hours) and x be the horizontal distance traveled by the plane (in mi), then we are given that $dx/dt = 500$ mi/h.

(b) Unknown: the rate at which the distance from the plane to the station is increasing when it is 2 mi from the station. If we let y be the distance from the plane to the station, then we want to find dy/dt when $y = 2$ mi.

(c)

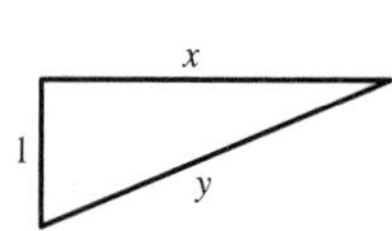

(d) By the Pythagorean Theorem, $y^2 = x^2 + 1 \Rightarrow 2y\,(dy/dt) = 2x\,(dx/dt)$.

(e) $\frac{dy}{dt} = \frac{x}{y}\,\frac{dx}{dt} = \frac{x}{y}(500)$. Since $y^2 = x^2 + 1$, when $y = 2$, $x = \sqrt{3}$, so $\frac{dy}{dt} = \frac{\sqrt{3}}{2}(500) = 250\sqrt{3} \approx 433$ mi/h.

12. (a) Given: a man 6 ft tall walks away from a street light mounted on a 15-ft-tall pole at a rate of 5 ft/s. If we let t be time (in s) and x be the distance from the pole to the man (in ft), then we are given that $dx/dt = 5$ ft/s.

(b) Unknown: the rate at which the tip of his shadow is moving when he is 40 ft from the pole. If we let y be the distance from the man to the tip of his shadow (in ft), then we want to find $\frac{d}{dt}(x+y)$ when $x = 40$ ft.

(c)

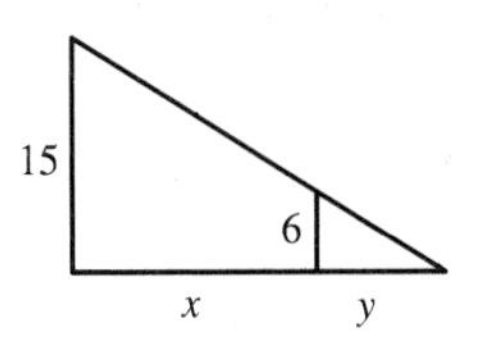

(d) By similar triangles, $\frac{15}{6} = \frac{x+y}{y} \Rightarrow 15y = 6x + 6y \Rightarrow 9y = 6x \Rightarrow y = \frac{2}{3}x$.

(e) The tip of the shadow moves at a rate of $\frac{d}{dt}(x+y) = \frac{d}{dt}\left(x + \frac{2}{3}x\right) = \frac{5}{3}\frac{dx}{dt} = \frac{5}{3}(5) = \frac{25}{3}$ ft/s.

13.

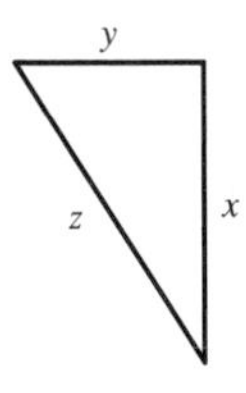

We are given that $\frac{dx}{dt} = 60$ mi/h and $\frac{dy}{dt} = 25$ mi/h. $z^2 = x^2 + y^2 \Rightarrow$

$$2z\frac{dz}{dt} = 2x\frac{dx}{dt} + 2y\frac{dy}{dt} \Rightarrow z\frac{dz}{dt} = x\frac{dx}{dt} + y\frac{dy}{dt} \Rightarrow \frac{dz}{dt} = \frac{1}{z}\left(x\frac{dx}{dt} + y\frac{dy}{dt}\right).$$

After 2 hours, $x = 2\,(60) = 120$ and $y = 2\,(25) = 50 \Rightarrow z = \sqrt{120^2 + 50^2} = 130$,

so $\frac{dz}{dt} = \frac{1}{z}\left(x\frac{dx}{dt} + y\frac{dy}{dt}\right) = \frac{120(60) + 50(25)}{130} = 65$ mi/h.

14.

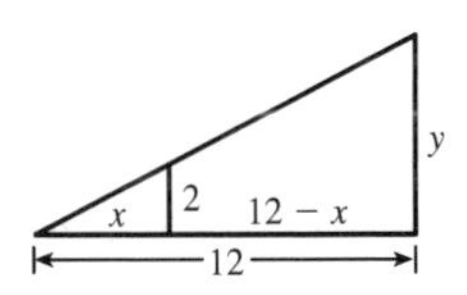

We are given that $\frac{dx}{dt} = 1.6$ m/s. By similar triangles, $\frac{y}{12} = \frac{2}{x} \Rightarrow y = \frac{24}{x} \Rightarrow$

$\frac{dy}{dt} = -\frac{24}{x^2}\frac{dx}{dt} = -\frac{24}{x^2}(1.6)$. When $x = 8$, $\frac{dy}{dt} = -\frac{24(1.6)}{64} = -0.6$ m/s, so the shadow is decreasing at a rate of 0.6 m/s

15.

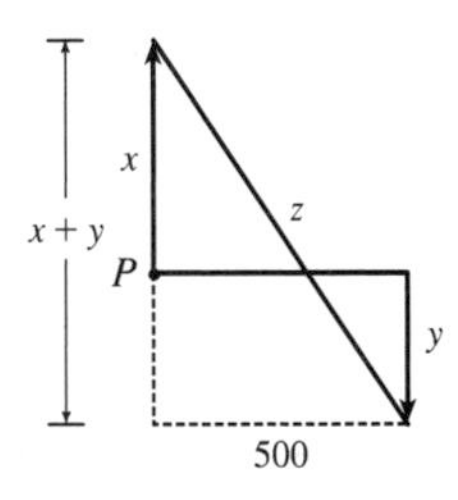

We are given that $\frac{dx}{dt} = 4$ ft/s and $\frac{dy}{dt} = 5$ ft/s. $z^2 = (x+y)^2 + 500^2 \Rightarrow$

$2z\frac{dz}{dt} = 2(x+y)\left(\frac{dx}{dt} + \frac{dy}{dt}\right)$. 15 minutes after the woman starts, we have

$x = (4 \text{ ft/s})(20 \text{ min})(60 \text{ s/min}) = 4800$ ft and $y = 5 \cdot 15 \cdot 60 = 4500 \Rightarrow$

$z = \sqrt{(4800 + 4500)^2 + 500^2} = \sqrt{86{,}740{,}000}$, so

$$\frac{dz}{dt} = \frac{x+y}{z}\left(\frac{dx}{dt} + \frac{dy}{dt}\right) = \frac{4800 + 4500}{\sqrt{86{,}740{,}000}}(4+5) = \frac{837}{\sqrt{8674}} \approx 8.99 \text{ ft/s}.$$

16. We are given that $\frac{dx}{dt} = 24$ ft/s.

(a)

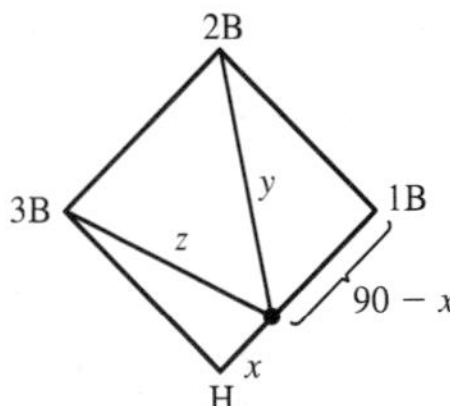

$y^2 = (90 - x)^2 + 90^2 \Rightarrow 2y\frac{dy}{dt} = 2(90 - x)\left(-\frac{dx}{dt}\right)$. When $x = 45$,

$y = \sqrt{45^2 + 90^2} = 45\sqrt{5}$, so $\frac{dy}{dt} = \frac{90 - x}{y}\left(-\frac{dx}{dt}\right) = \frac{45}{45\sqrt{5}}(-24) = -\frac{24}{\sqrt{5}}$,

so the distance from second base is decreasing at a rate of $\frac{24}{\sqrt{5}} \approx 10.7$ ft/s.

(b) Due to the symmetric nature of the problem in part (a), we expect to get the same answer—and we do.

$z^2 = x^2 + 90^2 \Rightarrow 2z\frac{dz}{dt} = 2x\frac{dx}{dt}$. When $x = 45$, $z = 45\sqrt{5}$, so $\frac{dz}{dt} = \frac{45}{45\sqrt{5}}(24) = \frac{24}{\sqrt{5}} \approx 10.7$ ft/s.

17. $A = \frac{1}{2}bh$, where b is the base and h is the altitude. We are given that $\frac{dh}{dt} = 1$ cm/min and $\frac{dA}{dt} = 2$ cm^2/min. Using the Product Rule, we have $\frac{dA}{dt} = \frac{1}{2}\left(b\,\frac{dh}{dt} + h\,\frac{db}{dt}\right)$. When $h = 10$ and $A = 100$, we have $100 = \frac{1}{2}b(10) \Rightarrow \frac{1}{2}b = 10 \Rightarrow$ $b = 20$, so $2 = \frac{1}{2}\left(20 \cdot 1 + 10\,\frac{db}{dt}\right) \Rightarrow 4 = 20 + 10\,\frac{db}{dt} \Rightarrow \frac{db}{dt} = \frac{4 - 20}{10} = -1.6$ cm/min.

18.

Given $\frac{dy}{dt} = -1$ m/s, find $\frac{dx}{dt}$ when $x = 8$ m. $y^2 = x^2 + 1 \Rightarrow 2y\,\frac{dy}{dt} = 2x\,\frac{dx}{dt} \Rightarrow$ $\frac{dx}{dt} = \frac{y}{x}\frac{dy}{dt} = -\frac{y}{x}$. When $x = 8$, $y = \sqrt{65}$, so $\frac{dx}{dt} = -\frac{\sqrt{65}}{8}$. Thus, the boat approaches the dock at $\frac{\sqrt{65}}{8} \approx 1.01$ m/s.

19.

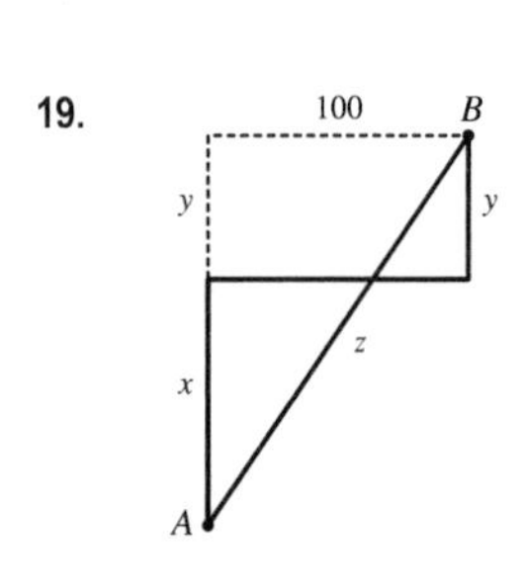

We are given that $\frac{dx}{dt} = 35$ km/h and $\frac{dy}{dt} = 25$ km/h. $z^2 = (x + y)^2 + 100^2 \Rightarrow$

$$2z\,\frac{dz}{dt} = 2(x + y)\left(\frac{dx}{dt} + \frac{dy}{dt}\right).$$ At 4:00 PM, $x = 4(35) = 140$ and $y = 4(25) = 100 \Rightarrow z = \sqrt{(140 + 100)^2 + 100^2} = \sqrt{67{,}600} = 260$, so

$$\frac{dz}{dt} = \frac{x + y}{z}\left(\frac{dx}{dt} + \frac{dy}{dt}\right) = \frac{140 + 100}{260}(35 + 25) = \frac{720}{13} \approx 55.4 \text{ km/h}.$$

20. Let D denote the distance from the origin $(0, 0)$ to the point on the curve $y = \sqrt{x}$.

$$D = \sqrt{(x - 0)^2 + (y - 0)^2} = \sqrt{x^2 + (\sqrt{x}\,)^2} = \sqrt{x^2 + x} \Rightarrow \frac{dD}{dt} = \tfrac{1}{2}(x^2 + x)^{-1/2}(2x + 1)\,\frac{dx}{dt} = \frac{2x + 1}{2\sqrt{x^2 + x}}\,\frac{dx}{dt}.$$

With $\frac{dx}{dt} = 3$ when $x = 4$, $\frac{dD}{dt} = \frac{9}{2\sqrt{20}}(3) = \frac{27}{4\sqrt{5}} \approx 3.02$ cm/s.

21. Using Q for the origin, we are given $\frac{dx}{dt} = -2$ ft/s and need to find $\frac{dy}{dt}$ when $x = -5$.

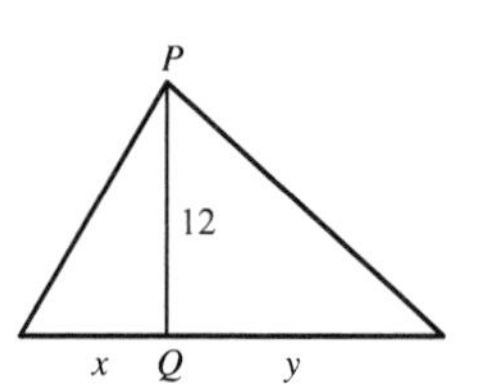

Using the Pythagorean Theorem twice, we have $\sqrt{x^2 + 12^2} + \sqrt{y^2 + 12^2} = 39$, the total length of the rope. Differentiating with respect to t, we get

$$\frac{x}{\sqrt{x^2 + 12^2}}\,\frac{dx}{dt} + \frac{y}{\sqrt{y^2 + 12^2}}\,\frac{dy}{dt} = 0, \text{ so } \frac{dy}{dt} = -\frac{x\sqrt{y^2 + 12^2}}{y\sqrt{x^2 + 12^2}}\,\frac{dx}{dt}.$$

Now when $x = -5$, $39 = \sqrt{(-5)^2 + 12^2} + \sqrt{y^2 + 12^2} = 13 + \sqrt{y^2 + 12^2} \Leftrightarrow \sqrt{y^2 + 12^2} = 26$, and $y = \sqrt{26^2 - 12^2} = \sqrt{532}$. So when $x = -5$, $\frac{dy}{dt} = -\frac{(-5)(26)}{\sqrt{532}\,(13)}(-2) = -\frac{10}{\sqrt{133}} \approx -0.87$ ft/s. So cart B is moving towards Q at about 0.87 ft/s.

22. If $C =$ the rate at which water is pumped in, then $\frac{dV}{dt} = C - 10{,}000$, where

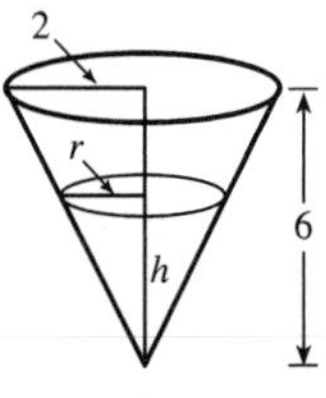

$V = \frac{1}{3}\pi r^2 h$ is the volume at time t. By similar triangles, $\frac{r}{2} = \frac{h}{6} \Rightarrow r = \frac{1}{3}h \Rightarrow$

$V = \frac{1}{3}\pi\left(\frac{1}{3}h\right)^2 h = \frac{\pi}{27}h^3 \Rightarrow \frac{dV}{dt} = \frac{\pi}{9}h^2\,\frac{dh}{dt}$. When $h = 200$ cm, $\frac{dh}{dt} = 20$ cm/min, so $C - 10{,}000 = \frac{\pi}{9}(200)^2(20) \Rightarrow C = 10{,}000 + \frac{800{,}000}{9}\pi \approx 289{,}253$ cm^3/min.

23. By similar triangles, $\dfrac{3}{1} = \dfrac{b}{h}$, so $b = 3h$. The trough has volume

$V = \frac{1}{2}bh(10) = 5(3h)h = 15h^2 \quad \Rightarrow \quad 12 = \dfrac{dV}{dt} = 30h\dfrac{dh}{dt} \quad \Rightarrow \quad \dfrac{dh}{dt} = \dfrac{2}{5h}$.

When $h = \frac{1}{2}$, $\dfrac{dh}{dt} = \dfrac{2}{5 \cdot \frac{1}{2}} = \dfrac{4}{5}$ ft/min.

24. The figure is drawn without the top 3 feet.

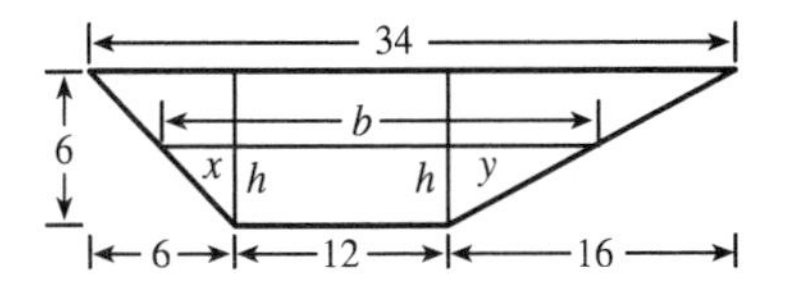

$V = \frac{1}{2}(b + 12)h(20) = 10(b + 12)h$ and, from similar triangles,

$\dfrac{x}{h} = \dfrac{6}{6}$ and $\dfrac{y}{h} = \dfrac{16}{6} = \dfrac{8}{3}$, so $b = x + 12 + y = h + 12 + \dfrac{8h}{3} = 12 + \dfrac{11h}{3}$.

Thus, $V = 10\left(24 + \dfrac{11h}{3}\right)h = 240h + \dfrac{110h^2}{3}$ and so $0.8 = \dfrac{dV}{dt} = \left(240 + \dfrac{220}{3}h\right)\dfrac{dh}{dt}$. When $h = 5$,

$\dfrac{dh}{dt} = \dfrac{0.8}{240 + 5(220/3)} = \dfrac{3}{2275} \approx 0.00132$ ft/min.

25. We are given that $\dfrac{dV}{dt} = 30$ ft^3/min. $V = \frac{1}{3}\pi r^2 h = \frac{1}{3}\pi\left(\dfrac{h}{2}\right)^2 h = \dfrac{\pi h^3}{12} \quad \Rightarrow$

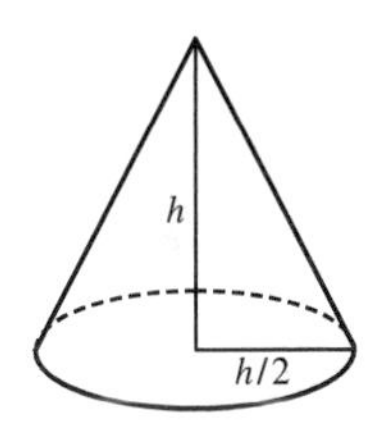

$\dfrac{dV}{dt} = \dfrac{dV}{dh}\dfrac{dh}{dt} \quad \Rightarrow \quad 30 = \dfrac{\pi h^2}{4}\dfrac{dh}{dt} \quad \Rightarrow \quad \dfrac{dh}{dt} = \dfrac{120}{\pi h^2}$. When $h = 10$ ft,

$\dfrac{dh}{dt} = \dfrac{120}{10^2\pi} = \dfrac{6}{5\pi} \approx 0.38$ ft/min.

26. We are given $dx/dt = 8$ ft/s. $\cot\theta = \dfrac{x}{100} \quad \Rightarrow \quad x = 100\cot\theta \quad \Rightarrow$

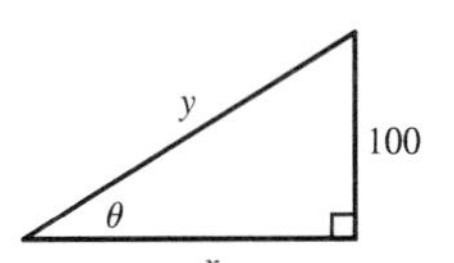

$\dfrac{dx}{dt} = -100\csc^2\theta\,\dfrac{d\theta}{dt} \quad \Rightarrow \quad \dfrac{d\theta}{dt} = -\dfrac{\sin^2\theta}{100} \cdot 8$. When $y = 200$, $\sin\theta = \dfrac{100}{200} = \dfrac{1}{2} \quad \Rightarrow$

$\dfrac{d\theta}{dt} = -\dfrac{(1/2)^2}{100} \cdot 8 = -\dfrac{1}{50}$ rad/s. The angle is decreasing at a rate of $\frac{1}{50}$ rad/s.

27. $A = \frac{1}{2}bh$, but $b = 5$ m and $\sin\theta = \dfrac{h}{4} \quad \Rightarrow \quad h = 4\sin\theta$, so $A = \frac{1}{2}(5)(4\sin\theta) = 10\sin\theta$.

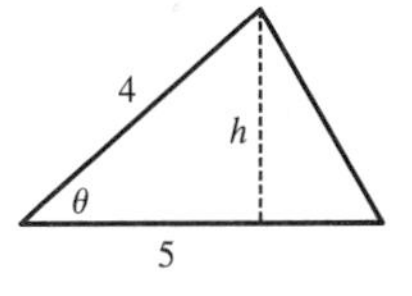

We are given $\dfrac{d\theta}{dt} = 0.06$ rad/s, so $\dfrac{dA}{dt} = \dfrac{dA}{d\theta}\dfrac{d\theta}{dt} = (10\cos\theta)(0.06) = 0.6\cos\theta$.

When $\theta = \frac{\pi}{3}$, $\dfrac{dA}{dt} = 0.6\left(\cos\frac{\pi}{3}\right) = (0.6)\left(\frac{1}{2}\right) = 0.3$ m^2/s.

28. We are given $d\theta/dt = 2°/\text{min} = \frac{\pi}{90}$ rad/min. By the Law of Cosines,

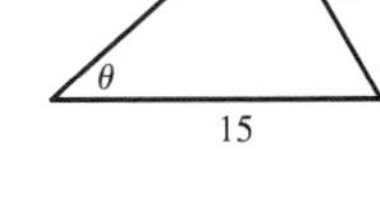

$x^2 = 12^2 + 15^2 - 2(12)(15)\cos\theta = 369 - 360\cos\theta \quad \Rightarrow$

$2x\dfrac{dx}{dt} = 360\sin\theta\,\dfrac{d\theta}{dt} \quad \Rightarrow \quad \dfrac{dx}{dt} = \dfrac{180\sin\theta}{x}\dfrac{d\theta}{dt}$. When $\theta = 60°$,

$x = \sqrt{369 - 360\cos 60°} = \sqrt{189} = 3\sqrt{21}$, so $\dfrac{dx}{dt} = \dfrac{180\sin 60°}{3\sqrt{21}}\dfrac{\pi}{90} = \dfrac{\pi\sqrt{3}}{3\sqrt{21}} = \dfrac{\sqrt{7}\,\pi}{21} \approx 0.396$ m/min.

29. Differentiating both sides of $PV = C$ with respect to t and using the Product Rule gives us $P\dfrac{dV}{dt} + V\dfrac{dP}{dt} = 0 \quad \Rightarrow$

$\dfrac{dV}{dt} = -\dfrac{V}{P}\dfrac{dP}{dt}$. When $V = 600$, $P = 150$ and $\dfrac{dP}{dt} = 20$, so we have $\dfrac{dV}{dt} = -\dfrac{600}{150}(20) = -80$. Thus, the volume is decreasing at a rate of 80 cm^3/min.

30. $PV^{1.4} = C \quad \Rightarrow \quad P \cdot 1.4V^{0.4}\,\dfrac{dV}{dt} + V^{1.4}\,\dfrac{dP}{dt} = 0 \quad \Rightarrow \quad \dfrac{dV}{dt} = -\dfrac{V^{1.4}}{P \cdot 1.4V^{0.4}}\,\dfrac{dP}{dt} = -\dfrac{V}{1.4P}\,\dfrac{dP}{dt}$. When $V = 400$, $P = 80$ and $\dfrac{dP}{dt} = -10$, so we have $\dfrac{dV}{dt} = -\dfrac{400}{1.4(80)}(-10) = \dfrac{250}{7}$. Thus, the volume is increasing at a rate of $\frac{250}{7} \approx 36\ \text{cm}^3/\text{min}$.

31. With $R_1 = 80$ and $R_2 = 100$, $\dfrac{1}{R} = \dfrac{1}{R_1} + \dfrac{1}{R_2} = \dfrac{1}{80} + \dfrac{1}{100} = \dfrac{180}{8000} = \dfrac{9}{400}$, so $R = \dfrac{400}{9}$. Differentiating $\dfrac{1}{R} = \dfrac{1}{R_1} + \dfrac{1}{R_2}$ with respect to t, we have $-\dfrac{1}{R^2}\dfrac{dR}{dt} = -\dfrac{1}{R_1^2}\dfrac{dR_1}{dt} - \dfrac{1}{R_2^2}\dfrac{dR_2}{dt} \quad \Rightarrow \quad \dfrac{dR}{dt} = R^2\left(\dfrac{1}{R_1^2}\dfrac{dR_1}{dt} + \dfrac{1}{R_2^2}\dfrac{dR_2}{dt}\right)$. When $R_1 = 80$ and $R_2 = 100$, $\dfrac{dR}{dt} = \dfrac{400^2}{9^2}\left[\dfrac{1}{80^2}(0.3) + \dfrac{1}{100^2}(0.2)\right] = \dfrac{107}{810} \approx 0.132\ \Omega/\text{s}$.

32. We want to find $\dfrac{dB}{dt}$ when $L = 18$ using $B = 0.007W^{2/3}$ and $W = 0.12L^{2.53}$.

$$\frac{dB}{dt} = \frac{dB}{dW}\frac{dW}{dL}\frac{dL}{dt} = \left(0.007 \cdot \tfrac{2}{3}W^{-1/3}\right)\left(0.12 \cdot 2.53 \cdot L^{1.53}\right)\left(\frac{20-15}{10{,}000{,}000}\right)$$

$$= \left[0.007 \cdot \tfrac{2}{3}\left(0.12 \cdot 18^{2.53}\right)^{-1/3}\right]\left(0.12 \cdot 2.53 \cdot 18^{1.53}\right)\left(\frac{5}{10^7}\right) \approx 1.045 \times 10^{-8}\ \text{g/yr}$$

33. (a) By the Pythagorean Theorem, $4000^2 + y^2 = \ell^2$. Differentiating with respect to t, we obtain $2y\,\dfrac{dy}{dt} = 2\ell\,\dfrac{d\ell}{dt}$. We know that $\dfrac{dy}{dt} = 600$ ft/s, so when $y = 3000$ ft,

$\ell = \sqrt{4000^2 + 3000^2} = \sqrt{25{,}000{,}000} = 5000$ ft

and $\dfrac{d\ell}{dt} = \dfrac{y}{\ell}\dfrac{dy}{dt} = \dfrac{3000}{5000}(600) = \dfrac{1800}{5} = 360$ ft/s.

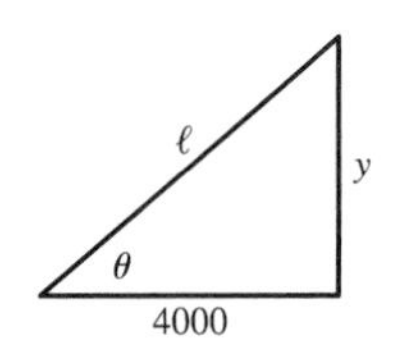

(b) Here $\tan\theta = \dfrac{y}{4000} \quad \Rightarrow \quad \dfrac{d}{dt}(\tan\theta) = \dfrac{d}{dt}\left(\dfrac{y}{4000}\right) \quad \Rightarrow \quad \sec^2\theta\,\dfrac{d\theta}{dt} = \dfrac{1}{4000}\dfrac{dy}{dt} \quad \Rightarrow \quad \dfrac{d\theta}{dt} = \dfrac{\cos^2\theta}{4000}\dfrac{dy}{dt}$. When $y = 3000$ ft, $\dfrac{dy}{dt} = 600$ ft/s, $\ell = 5000$ and $\cos\theta = \dfrac{4000}{\ell} = \dfrac{4000}{5000} = \dfrac{4}{5}$, so $\dfrac{d\theta}{dt} = \dfrac{(4/5)^2}{4000}(600) = 0.096$ rad/s.

34. We are given that $\dfrac{d\theta}{dt} = 4(2\pi) = 8\pi$ rad/min. $x = 3\tan\theta \quad \Rightarrow$

$\dfrac{dx}{dt} = 3\sec^2\theta\,\dfrac{d\theta}{dt}$. When $x = 1$, $\tan\theta = \frac{1}{3}$, so $\sec^2\theta = 1 + \left(\frac{1}{3}\right)^2 = \frac{10}{9}$

and $\dfrac{dx}{dt} = 3\left(\frac{10}{9}\right)(8\pi) = \frac{80\pi}{3} \approx 83.8$ km/min.

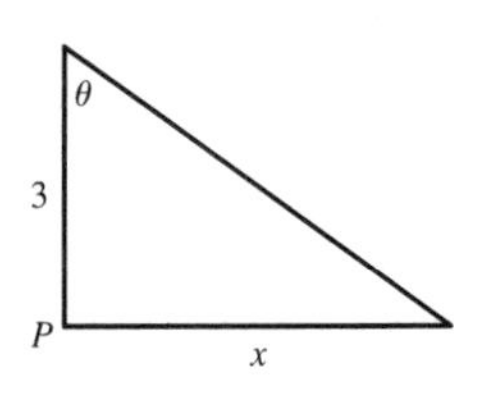

35. We are given that $\dfrac{dx}{dt} = 300$ km/h. By the Law of Cosines,

$y^2 = x^2 + 1^2 - 2(1)(x)\cos 120^\circ = x^2 + 1 - 2x\left(-\frac{1}{2}\right) = x^2 + x + 1$, so

$2y\,\dfrac{dy}{dt} = 2x\,\dfrac{dx}{dt} + \dfrac{dx}{dt} \quad \Rightarrow \quad \dfrac{dy}{dt} = \dfrac{2x+1}{2y}\dfrac{dx}{dt}$. After 1 minute, $x = \frac{300}{60} = 5$ km $\Rightarrow$

$y = \sqrt{5^2 + 5 + 1} = \sqrt{31}$ km $\quad \Rightarrow \quad \dfrac{dy}{dt} = \dfrac{2(5)+1}{2\sqrt{31}}(300) = \dfrac{1650}{\sqrt{31}} \approx 296$ km/h.

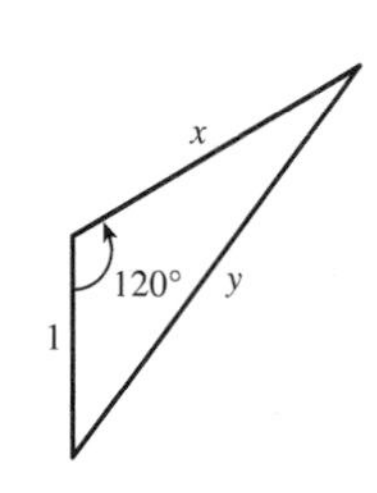

36. We are given that $\dfrac{dx}{dt} = 3$ mi/h and $\dfrac{dy}{dt} = 2$ mi/h. By the Law of Cosines,

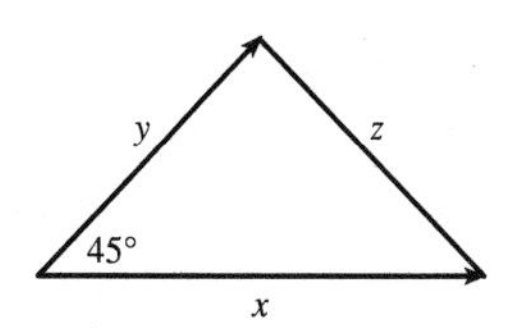

$z^2 = x^2 + y^2 - 2xy\cos 45° = x^2 + y^2 - \sqrt{2}\,xy \quad\Rightarrow$

$2z\dfrac{dz}{dt} = 2x\dfrac{dx}{dt} + 2y\dfrac{dy}{dt} - \sqrt{2}\,x\dfrac{dy}{dt} - \sqrt{2}\,y\dfrac{dx}{dt}$. After 15 minutes $\left[= \frac{1}{4}\text{ h}\right]$,

we have $x = \frac{3}{4}$ and $y = \frac{2}{4} = \frac{1}{2} \quad\Rightarrow\quad z^2 = \left(\frac{3}{4}\right)^2 + \left(\frac{2}{4}\right)^2 - \sqrt{2}\left(\frac{3}{4}\right)\left(\frac{2}{4}\right) \quad\Rightarrow\quad z = \dfrac{\sqrt{13 - 6\sqrt{2}}}{4}$ and

$$\begin{aligned}\frac{dz}{dt} &= \frac{2}{\sqrt{13 - 6\sqrt{2}}}\left[2\left(\tfrac{3}{4}\right)3 + 2\left(\tfrac{1}{2}\right)2 - \sqrt{2}\left(\tfrac{3}{4}\right)2 - \sqrt{2}\left(\tfrac{1}{2}\right)3\right] \\ &= \frac{2}{\sqrt{13 - 6\sqrt{2}}}\,\frac{13 - 6\sqrt{2}}{2} = \sqrt{13 - 6\sqrt{2}} \approx 2.125 \text{ mi/h}.\end{aligned}$$

37. Let the distance between the runner and the friend be ℓ. Then by the Law of Cosines,

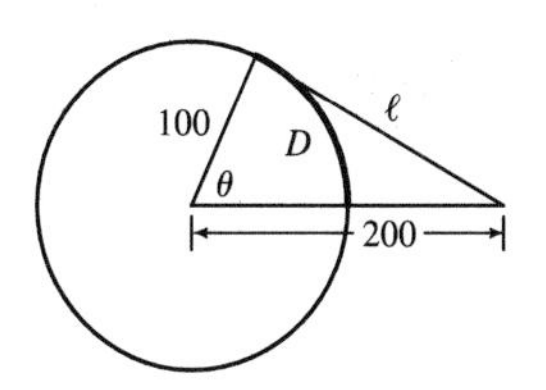

$\ell^2 = 200^2 + 100^2 - 2\cdot 200\cdot 100\cdot\cos\theta = 50{,}000 - 40{,}000\cos\theta$ $(\star)$. Differentiating implicitly with respect to t, we obtain $2\ell\dfrac{d\ell}{dt} = -40{,}000(-\sin\theta)\dfrac{d\theta}{dt}$. Now if D is the distance run when the angle is θ radians, then by the formula for the length of an arc on a circle, $s = r\theta$, we have $D = 100\theta$, so $\theta = \dfrac{1}{100}D \quad\Rightarrow\quad \dfrac{d\theta}{dt} = \dfrac{1}{100}\dfrac{dD}{dt} = \dfrac{7}{100}$. To substitute into the expression for $\dfrac{d\ell}{dt}$, we must know $\sin\theta$ at the time when $\ell = 200$, which we find from $(\star)$: $200^2 = 50{,}000 - 40{,}000\cos\theta \quad\Leftrightarrow$

$\cos\theta = \dfrac{1}{4} \quad\Rightarrow\quad \sin\theta = \sqrt{1 - \left(\frac{1}{4}\right)^2} = \dfrac{\sqrt{15}}{4}$. Substituting, we get $2(200)\dfrac{d\ell}{dt} = 40{,}000\dfrac{\sqrt{15}}{4}\left(\dfrac{7}{100}\right) \quad\Rightarrow$

$d\ell/dt = \frac{7\sqrt{15}}{4} \approx 6.78$ m/s. Whether the distance between them is increasing or decreasing depends on the direction in which the runner is running.

38. The hour hand of a clock goes around once every 12 hours or, in radians per hour, $\frac{2\pi}{12} = \frac{\pi}{6}$ rad/h. The minute hand goes around once an hour, or at the rate of 2π rad/h. So the angle θ between them (measuring clockwise from the minute hand to the hour hand) is changing at the rate of $d\theta/dt = \frac{\pi}{6} - 2\pi = -\frac{11\pi}{6}$ rad/h. Now, to relate θ to ℓ, we use the Law of Cosines: $\ell^2 = 4^2 + 8^2 - 2\cdot 4\cdot 8\cdot\cos\theta = 80 - 64\cos\theta$ $(\star)$.

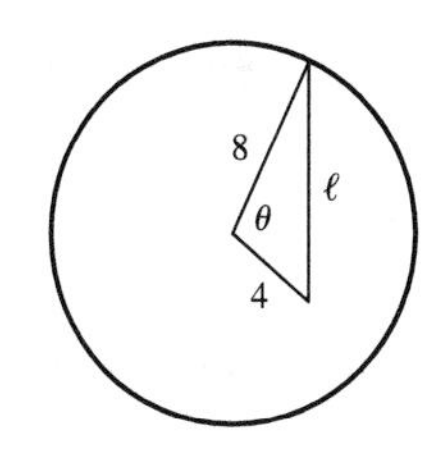

Differentiating implicitly with respect to t, we get $2\ell\dfrac{d\ell}{dt} = -64(-\sin\theta)\dfrac{d\theta}{dt}$. At 1:00, the angle between the two hands is one-twelfth of the circle, that is, $\frac{2\pi}{12} = \frac{\pi}{6}$ radians. We use $(\star)$ to find ℓ at 1:00: $\ell = \sqrt{80 - 64\cos\frac{\pi}{6}} = \sqrt{80 - 32\sqrt{3}}$.

Substituting, we get $2\ell\dfrac{d\ell}{dt} = 64\sin\frac{\pi}{6}\left(-\frac{11\pi}{6}\right) \quad\Rightarrow\quad \dfrac{d\ell}{dt} = \dfrac{64\left(\frac{1}{2}\right)\left(-\frac{11\pi}{6}\right)}{2\sqrt{80 - 32\sqrt{3}}} = -\dfrac{88\pi}{3\sqrt{80 - 32\sqrt{3}}} \approx -18.6.$

So at 1:00, the distance between the tips of the hands is decreasing at a rate of 18.6 mm/h ≈ 0.005 mm/s.

2.8 Linear Approximations and Differentials

1. $f(x) = x^4 + 3x^2 \Rightarrow f'(x) = 4x^3 + 6x$, so $f(-1) = 4$ and $f'(-1) = -10$.

   Thus, $L(x) = f(-1) + f'(-1)(x - (-1)) = 4 + (-10)(x + 1) = -10x - 6$.

2. $f(x) = 1/\sqrt{2+x} = (2+x)^{-1/2} \Rightarrow f'(x) = -\frac{1}{2}(2+x)^{-3/2}$ so $f(0) = \frac{1}{\sqrt{2}}$ and $f'(0) = -1/(4\sqrt{2})$. So

   $L(x) = f(0) + f'(0)(x - 0) = \frac{1}{\sqrt{2}} - \frac{1}{4\sqrt{2}}(x - 0) = \frac{1}{\sqrt{2}}\left(1 - \frac{1}{4}x\right)$.

3. $f(x) = \cos x \Rightarrow f'(x) = -\sin x$, so $f\left(\frac{\pi}{2}\right) = 0$ and $f'\left(\frac{\pi}{2}\right) = -1$.

   Thus, $L(x) = f\left(\frac{\pi}{2}\right) + f'\left(\frac{\pi}{2}\right)\left(x - \frac{\pi}{2}\right) = 0 - 1\left(x - \frac{\pi}{2}\right) = -x + \frac{\pi}{2}$.

4. $f(x) = x^{3/4} \Rightarrow f'(x) = \frac{3}{4}x^{-1/4}$, so $f(16) = 8$ and $f'(16) = \frac{3}{8}$.

   Thus, $L(x) = f(16) + f'(16)(x - 16) = 8 + \frac{3}{8}(x - 16) = \frac{3}{8}x + 2$.

5. $f(x) = \sqrt{1-x} \Rightarrow f'(x) = \dfrac{-1}{2\sqrt{1-x}}$, so $f(0) = 1$ and $f'(0) = -\frac{1}{2}$.

   Therefore,

   $\sqrt{1-x} = f(x) \approx f(0) + f'(0)(x - 0) = 1 + \left(-\frac{1}{2}\right)(x - 0) = 1 - \frac{1}{2}x$.

   So $\sqrt{0.9} = \sqrt{1 - 0.1} \approx 1 - \frac{1}{2}(0.1) = 0.95$

   and $\sqrt{0.99} = \sqrt{1 - 0.01} \approx 1 - \frac{1}{2}(0.01) = 0.995$.

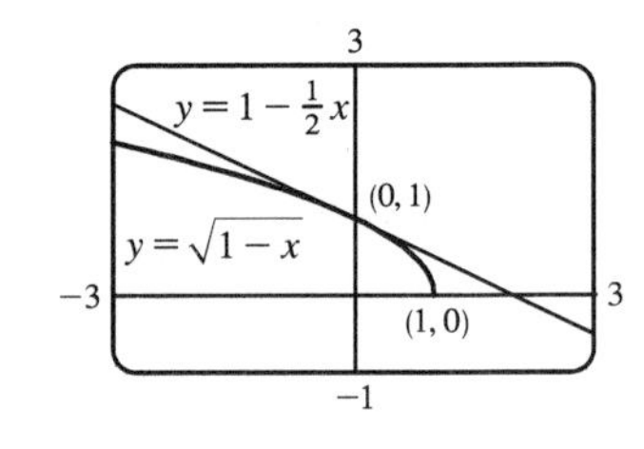

6. $g(x) = \sqrt[3]{1+x} = (1+x)^{1/3} \Rightarrow g'(x) = \frac{1}{3}(1+x)^{-2/3}$, so $g(0) = 1$ and

   $g'(0) = \frac{1}{3}$. Therefore, $\sqrt[3]{1+x} = g(x) \approx g(0) + g'(0)(x - 0) = 1 + \frac{1}{3}x$.

   So $\sqrt[3]{0.95} = \sqrt[3]{1 + (-0.05)} \approx 1 + \frac{1}{3}(-0.05) = 0.98\overline{3}$,

   and $\sqrt[3]{1.1} = \sqrt[3]{1 + 0.1} \approx 1 + \frac{1}{3}(0.1) = 1.0\overline{3}$.

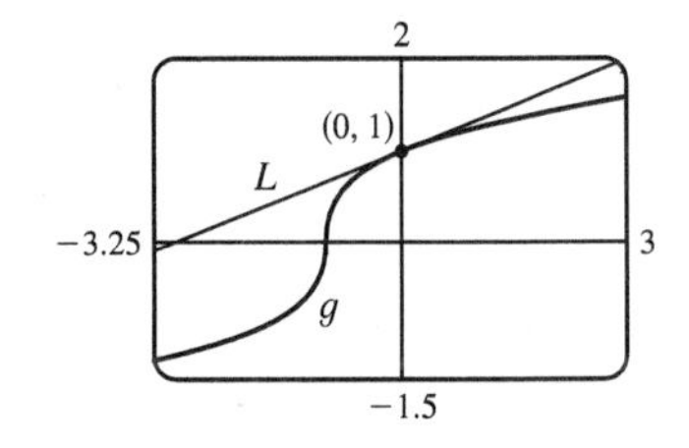

7. $f(x) = \sqrt[3]{1-x} = (1-x)^{1/3} \Rightarrow f'(x) = -\frac{1}{3}(1-x)^{-2/3}$, so $f(0) = 1$

   and $f'(0) = -\frac{1}{3}$. Thus, $f(x) \approx f(0) + f'(0)(x - 0) = 1 - \frac{1}{3}x$. We need

   $\sqrt[3]{1-x} - 0.1 < 1 - \frac{1}{3}x < \sqrt[3]{1-x} + 0.1$, which is true when

   $-1.204 < x < 0.706$.

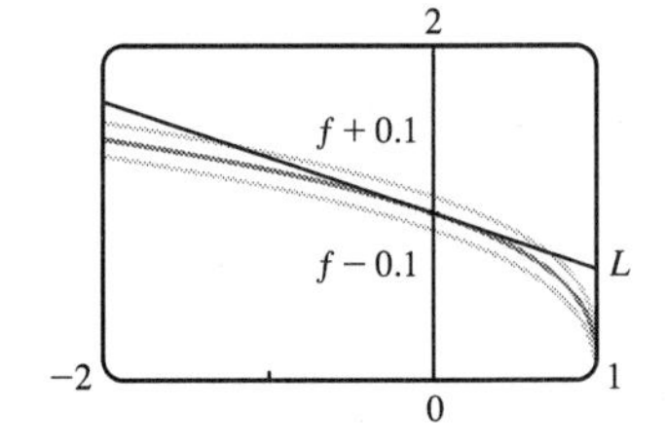

8. $f(x) = \tan x \Rightarrow f'(x) = \sec^2 x$, so $f(0) = 0$ and $f'(0) = 1$.

   Thus, $f(x) \approx f(0) + f'(0)(x - 0) = 0 + 1(x - 0) = x$.

   We need $\tan x - 0.1 < x < \tan x + 0.1$, which is true when

   $-0.63 < x < 0.63$.

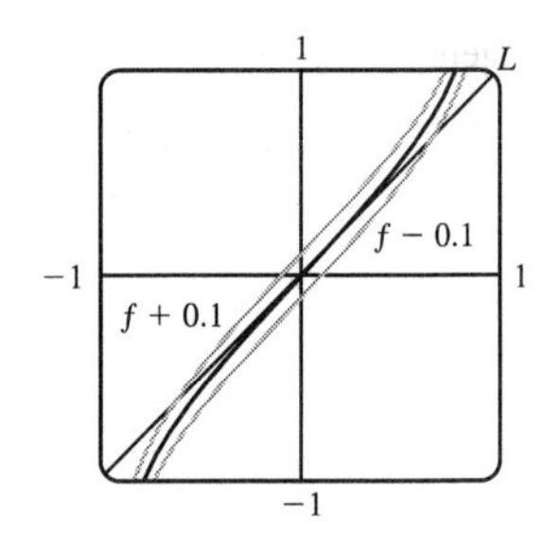

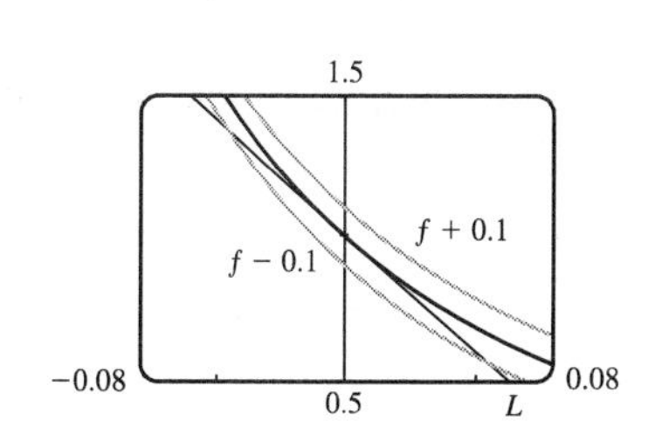

9. $f(x) = \dfrac{1}{(1+2x)^4} = (1+2x)^{-4} \Rightarrow$

$f'(x) = -4(1+2x)^{-5}(2) = \dfrac{-8}{(1+2x)^5}$, so $f(0) = 1$ and $f'(0) = -8$.

Thus, $f(x) \approx f(0) + f'(0)(x-0) = 1 + (-8)(x-0) = 1 - 8x$.

We need $1/(1+2x)^4 - 0.1 < 1 - 8x < 1/(1+2x)^4 + 0.1$, which is true when $-0.045 < x < 0.055$.

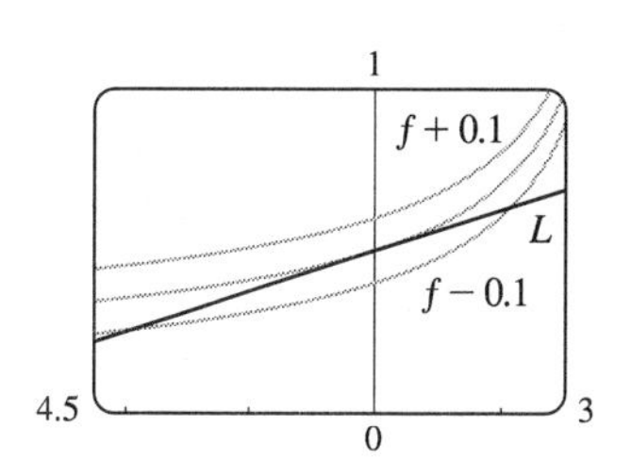

10. $f(x) = \dfrac{1}{\sqrt{4-x}} \Rightarrow f'(x) = \dfrac{1}{2(4-x)^{3/2}}$ so $f(0) = \frac{1}{2}$ and $f'(0) = \frac{1}{16}$. So $f(x) \approx \frac{1}{2} + \frac{1}{16}(x-0) = \frac{1}{2} + \frac{1}{16}x$. We need $\dfrac{1}{\sqrt{4-x}} - 0.1 < \frac{1}{2} + \frac{1}{16}x < \dfrac{1}{\sqrt{4-x}} + 0.1$, which is true when $-3.91 < x < 2.14$.

11. To estimate $(2.001)^5$, we'll find the linearization of $f(x) = x^5$ at $a = 2$. Since $f'(x) = 5x^4$, $f(2) = 32$, and $f'(2) = 80$, we have $L(x) = 32 + 80(x-2) = 80x - 128$. Thus, $x^5 \approx 80x - 128$ when x is near 2 , so $(2.001)^5 \approx 80(2.001) - 128 = 160.08 - 128 = 32.08$.

12. $y = f(x) = \sqrt{x} \Rightarrow dy = \dfrac{1}{2\sqrt{x}}\,dx$. When $x = 100$ and $dx = -0.2$, $dy = \dfrac{1}{2\sqrt{100}}(-0.2) = -0.01$, so $\sqrt{99.8} = f(99.8) \approx f(100) + dy = 10 - 0.01 = 9.99$.

13. To estimate $(8.06)^{2/3}$, we'll find the linearization of $f(x) = x^{2/3}$ at $a = 8$. Since $f'(x) = \frac{2}{3}x^{-1/3} = 2/(3\sqrt[3]{x})$, $f(8) = 4$, and $f'(8) = \frac{1}{3}$, we have $L(x) = 4 + \frac{1}{3}(x-8) = \frac{1}{3}x + \frac{4}{3}$. Thus, $x^{2/3} \approx \frac{1}{3}x + \frac{4}{3}$ when x is near 8, so $(8.06)^{2/3} \approx \frac{1}{3}(8.06) + \frac{4}{3} = \frac{12.06}{3} = 4.02$.

14. To estimate $1/1002$, we'll find the linearization of $f(x) = 1/x$ at $a = 1000$. Since $f'(x) = -1/x^2$, $f(1000) = 0.001$, and $f'(1000) = -0.000001$, we have $L(x) = 0.001 - 0.000001(x - 1000) = -0.000001x + 0.002$. Thus, $1/x \approx -0.000001x + 0.002$ when x is near 1000, so $1/1002 \approx -0.000001(1002) + 0.002 = 0.000998$.

15. $y = f(x) = \sec x \Rightarrow f'(x) = \sec x \tan x$, so $f(0) = 1$ and $f'(0) = 1 \cdot 0 = 0$. The linear approximation of f at 0 is $f(0) + f'(0)(x-0) = 1 + 0(x) = 1$. Since 0.08 is close to 0, approximating $\sec 0.08$ with 1 is reasonable.

16. If $y = x^6$, $y' = 6x^5$ and the tangent line approximation at $(1,1)$ has slope 6. If the change in x is 0.01, the change in y on the tangent line is 0.06, and approximating $(1.01)^6$ with 1.06 is reasonable.

17. (a) The differential dy is defined in terms of dx by the equation $dy = f'(x)\,dx$. For $y = f(x) = x^2 \sin 2x$, $f'(x) = x^2 \cos 2x \cdot 2 + \sin 2x \cdot 2x = 2x(x\cos 2x + \sin 2x)$, so $dy = 2x(x\cos 2x + \sin 2x)\,dx$.

(b) $y = \sqrt{4+5x} \Rightarrow dy = \frac{1}{2}(4+5x)^{-1/2} \cdot 5\,dx = \dfrac{5}{2\sqrt{4+5x}}\,dx$

18. (a) For $y = f(s) = \dfrac{s}{1+2s}$, $f'(s) = \dfrac{(1+2s)(1) - s(2)}{(1+2s)^2} = \dfrac{1}{(1+2s)^2}$, so $dy = \dfrac{1}{(1+2s)^2}\,ds$.

(b) $y = 1/(x+1) \Rightarrow dy = -\dfrac{1}{(x+1)^2}dx$

19. (a) $y = \tan x \quad \Rightarrow \quad dy = \sec^2 x\, dx$

(b) When $x = \pi/4$ and $dx = -0.1$, $dy = [\sec(\pi/4)]^2(-0.1) = \left(\sqrt{2}\right)^2(-0.1) = -0.2$.

$\Delta y = f(x + \Delta x) - f(x) = \tan\left(\frac{\pi}{4} - 0.1\right) - \tan\left(\frac{\pi}{4}\right) \approx -0.18237$.

20. (a) $y = \sqrt{x} \quad \Rightarrow \quad dy = \frac{1}{2}x^{-1/2}dx = \dfrac{1}{2\sqrt{x}}\,dx$

(b) $x = 1$ and $dx = 1 \quad \Rightarrow \quad dy = \frac{1}{2(1)}(1) = \frac{1}{2}$. $\Delta y = f(x + \Delta x) - f(x) = \sqrt{1+1} - \sqrt{1} = \sqrt{2} - 1 \approx 0.414$.

(c)

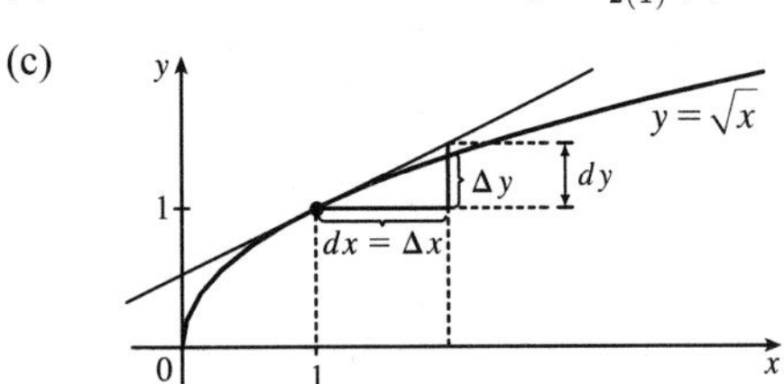

Remember, Δy represents the amount that the curve $y = f(x)$ rises or falls when x changes by an amount dx, whereas dy represents the amount that the tangent line rises or falls (the change in the linearization).

21. (a) If x is the edge length, then $V = x^3 \quad \Rightarrow \quad dV = 3x^2\,dx$. When $x = 30$ and $dx = 0.1$, $dV = 3(30)^2(0.1) = 270$, so the maximum possible error in computing the volume of the cube is about 270 cm^3. The relative error is calculated by dividing the change in V, ΔV, by V. We approximate ΔV with dV.

Relative error $= \dfrac{\Delta V}{V} \approx \dfrac{dV}{V} = \dfrac{3x^2\,dx}{x^3} = 3\,\dfrac{dx}{x} = 3\left(\dfrac{0.1}{30}\right) = 0.01$.

Percentage error = relative error $\times\, 100\% = 0.01 \times 100\% = 1\%$.

(b) $S = 6x^2 \quad \Rightarrow \quad dS = 12x\,dx$. When $x = 30$ and $dx = 0.1$, $dS = 12(30)(0.1) = 36$, so the maximum possible error in computing the surface area of the cube is about 36 cm^2.

Relative error $= \dfrac{\Delta S}{S} \approx \dfrac{dS}{S} = \dfrac{12x\,dx}{6x^2} = 2\,\dfrac{dx}{x} = 2\left(\dfrac{0.1}{30}\right) = 0.00\overline{6}$.

Percentage error = relative error $\times 100\% = 0.00\overline{6} \times 100\% = 0.\overline{6}\%$.

22. (a) $A = \pi r^2 \quad \Rightarrow \quad dA = 2\pi r\,dr$. When $r = 24$ and $dr = 0.2$, $dA = 2\pi(24)(0.2) = 9.6\pi$, so the maximum possible error in the calculated area of the disk is about $9.6\pi \approx 30\text{ cm}^2$.

(b) Relative error $= \dfrac{\Delta A}{A} \approx \dfrac{dA}{A} = \dfrac{2\pi r\,dr}{\pi r^2} = \dfrac{2\,dr}{r} = \dfrac{2(0.2)}{24} = \dfrac{0.2}{12} = \dfrac{1}{60} = 0.01\overline{6}$.

Percentage error = relative error $\times 100\% = 0.01\overline{6} \times 100\% = 1.\overline{6}\%$.

23. (a) For a sphere of radius r, the circumference is $C = 2\pi r$ and the surface area is $S = 4\pi r^2$, so

$r = \dfrac{C}{2\pi} \quad \Rightarrow \quad S = 4\pi\left(\dfrac{C}{2\pi}\right)^2 = \dfrac{C^2}{\pi} \quad \Rightarrow \quad dS = \dfrac{2}{\pi}C\,dC$. When $C = 84$ and $dC = 0.5$, $dS = \dfrac{2}{\pi}(84)(0.5) = \dfrac{84}{\pi}$,

so the maximum error is about $\dfrac{84}{\pi} \approx 27\text{ cm}^2$. Relative error $\approx \dfrac{dS}{S} = \dfrac{84/\pi}{84^2/\pi} = \dfrac{1}{84} \approx 0.012$

(b) $V = \dfrac{4}{3}\pi r^3 = \dfrac{4}{3}\pi\left(\dfrac{C}{2\pi}\right)^3 = \dfrac{C^3}{6\pi^2} \quad \Rightarrow \quad dV = \dfrac{1}{2\pi^2}C^2\,dC$. When $C = 84$ and $dC = 0.5$,

$dV = \dfrac{1}{2\pi^2}(84)^2(0.5) = \dfrac{1764}{\pi^2}$, so the maximum error is about $\dfrac{1764}{\pi^2} \approx 179\text{ cm}^3$.

The relative error is approximately $\dfrac{dV}{V} = \dfrac{1764/\pi^2}{(84)^3/(6\pi^2)} = \dfrac{1}{56} \approx 0.018$.

24. For a hemispherical dome, $V = \frac{2}{3}\pi r^3 \quad \Rightarrow \quad dV = 2\pi r^2\,dr$. When $r = \frac{1}{2}(50) = 25$ m and $dr = 0.05$ cm $= 0.0005$ m, $dV = 2\pi(25)^2(0.0005) = \frac{5\pi}{8}$, so the amount of paint needed is about $\frac{5\pi}{8} \approx 2\text{ m}^3$.

25. $F = kR^4 \;\Rightarrow\; dF = 4kR^3\,dR \;\Rightarrow\; \dfrac{dF}{F} = \dfrac{4kR^3\,dR}{kR^4} = 4\left(\dfrac{dR}{R}\right)$. Thus, the relative change in F is about 4 times the relative change in R. So a 5% increase in the radius corresponds to a 20% increase in blood flow.

26. (a) $f(x) = \sin x \;\Rightarrow\; f'(x) = \cos x$, so $f(0) = 0$ and $f'(0) = 1$. Thus, $f(x) \approx f(0) + f'(0)(x - 0) = 0 + 1(x - 0) = x$.

(b)

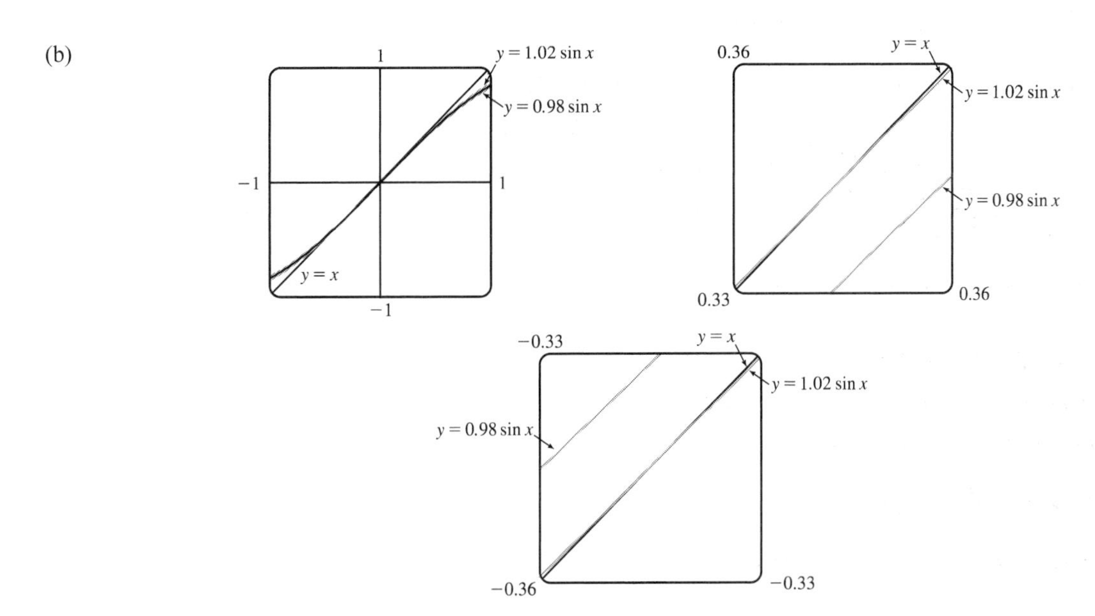

We want to know the values of x for which $y = x$ approximates $y = \sin x$ with less than a 2% difference; that is, the values of x for which

$$\left|\frac{x - \sin x}{\sin x}\right| < 0.02 \quad\Leftrightarrow\quad -0.02 < \frac{x - \sin x}{\sin x} < 0.02 \quad\Leftrightarrow$$

$$\begin{cases} -0.02\sin x < x - \sin x < 0.02\sin x & \text{if } \sin x > 0 \\ -0.02\sin x > x - \sin x > 0.02\sin x & \text{if } \sin x < 0 \end{cases} \quad\Leftrightarrow\quad \begin{cases} 0.98\sin x < x < 1.02\sin x & \text{if } \sin x > 0 \\ 1.02\sin x < x < 0.98\sin x & \text{if } \sin x < 0 \end{cases}$$

In the first figure, we see that the graphs are very close to each other near $x = 0$. Changing the viewing rectangle and using an intersect feature (see the second figure) we find that $y = x$ intersects $y = 1.02\sin x$ at $x \approx 0.344$. By symmetry, they also intersect at $x \approx -0.344$ (see the third figure). Converting 0.344 radians to degrees, we get $0.344\left(\frac{180^\circ}{\pi}\right) \approx 19.7^\circ \approx 20^\circ$, which verifies the statement.

27. (a) The graph shows that $f'(1) = 2$, so $L(x) = f(1) + f'(1)(x - 1) = 5 + 2(x - 1) = 2x + 3$.
$f(0.9) \approx L(0.9) = 4.8$ and $f(1.1) \approx L(1.1) = 5.2$.

(b) From the graph, we see that $f'(x)$ is positive and decreasing. This means that the slopes of the tangent lines are positive, but the tangents are becoming less steep. So the tangent lines lie *above* the curve. Thus, the estimates in part (a) are too large.

28. (a) $g'(x) = \sqrt{x^2 + 5} \;\Rightarrow\; g'(2) = \sqrt{9} = 3$. $g(1.95) \approx g(2) + g'(2)(1.95 - 2) = -4 + 3(-0.05) = -4.15$.
$g(2.05) \approx g(2) + g'(2)(2.05 - 2) = -4 + 3(0.05) = -3.85$.

(b) The formula $g'(x) = \sqrt{x^2 + 5}$ shows that $g'(x)$ is positive and increasing. This means that the slopes of the tangent lines are positive and the tangents are getting steeper. So the tangent lines lie *below* the graph of g. Hence, the estimates in part (a) are too small.

2 Review

CONCEPT CHECK

1. See Definition 2.1.1.

2. See the paragraph containing Formula 3 in Section 2.1.

3. See Definition 2.1.4. The pages following the definition discuss interpretations of $f'(a)$ as the slope of the tangent line to the graph of f at $x = a$ and as the instantaneous rate of change of $f(x)$ with respect to x when $x = a$.

4. (a) The average rate of change of y with respect to x over the interval $[x_1, x_2]$ is $\dfrac{f(x_2) - f(x_1)}{x_2 - x_1}$.

   (b) The instantaneous rate of change of y with respect to x at $x = x_1$ is $\lim\limits_{x_2 \to x_1} \dfrac{f(x_2) - f(x_1)}{x_2 - x_1}$.

5. See the paragraphs before and after Example 7 in Section 2.2.

6. (a) A function f is differentiable at a number a if its derivative f' exists at $x = a$; that is, if $f'(a)$ exists.

   (b) See Theorem 2.2.4. This theorem also tells us that if f is *not* continuous at a, then f is *not* differentiable at a.

   (c)

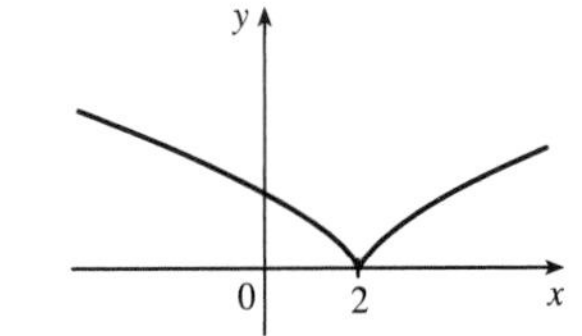

7. See the discussion and Figure 7 on page 89.

8. (a) The Power Rule: If n is any real number, then $\dfrac{d}{dx}(x^n) = nx^{n-1}$. The derivative of a variable base raised to a constant power is the power times the base raised to the power minus one.

   (b) The Constant Multiple Rule: If c is a constant and f is a differentiable function, then $\dfrac{d}{dx}[cf(x)] = c\dfrac{d}{dx}f(x)$. The derivative of a constant times a function is the constant times the derivative of the function.

   (c) The Sum Rule: If f and g are both differentiable, then $\dfrac{d}{dx}[f(x) + g(x)] = \dfrac{d}{dx}f(x) + \dfrac{d}{dx}g(x)$. The derivative of a sum of functions is the sum of the derivatives.

   (d) The Difference Rule: If f and g are both differentiable, then $\dfrac{d}{dx}[f(x) - g(x)] = \dfrac{d}{dx}f(x) - \dfrac{d}{dx}g(x)$. The derivative of a difference of functions is the difference of the derivatives.

   (e) The Product Rule: If f and g are both differentiable, then $\dfrac{d}{dx}[f(x)g(x)] = f(x)\dfrac{d}{dx}g(x) + g(x)\dfrac{d}{dx}f(x)$. The derivative of a product of two functions is the first function times the derivative of the second function plus the second function times the derivative of the first function.

   (f) The Quotient Rule: If f and g are both differentiable, then $\dfrac{d}{dx}\left[\dfrac{f(x)}{g(x)}\right] = \dfrac{g(x)\dfrac{d}{dx}f(x) - f(x)\dfrac{d}{dx}g(x)}{[g(x)]^2}$.

   The derivative of a quotient of functions is the denominator times the derivative of the numerator minus the numerator times the derivative of the denominator, all divided by the square of the denominator.

   (g) The Chain Rule: If f and g are both differentiable and $F = f \circ g$ is the composite function defined by $F(x) = f(g(x))$, then F is differentiable and F' is given by the product $F'(x) = f'(g(x))g'(x)$. The derivative of a composite function is the derivative of the outer function evaluated at the inner function times the derivative of the inner function.

9. (a) $y = x^n \Rightarrow y' = nx^{n-1}$ (b) $y = \sin x \Rightarrow y' = \cos x$

(c) $y = \cos x \Rightarrow y' = -\sin x$ (d) $y = \tan x \Rightarrow y' = \sec^2 x$

(e) $y = \csc x \Rightarrow y' = -\csc x \cot x$ (f) $y = \sec x \Rightarrow y' = \sec x \tan x$

(g) $y = \cot x \Rightarrow y' = -\csc^2 x$

10. Implicit differentiation consists of differentiating both sides of an equation involving x and y with respect to x, and then solving the resulting equation for y'.

11. (a) The linearization L of f at $x = a$ is $L(x) = f(a) + f'(a)(x - a)$.

(b) If $y = f(x)$, then the differential dy is given by $dy = f'(x)\,dx$.

(c) See Figure 5 in Section 2.8.

TRUE-FALSE QUIZ

1. False. See the note on page 88.

2. True. This is the Sum Rule.

3. False. See the warning before the Product Rule on page 106.

4. True. This is the Chain Rule.

5. True by the Chain Rule.

6. False. $\dfrac{d}{dx} f(\sqrt{x}) = \dfrac{f'(\sqrt{x})}{2\sqrt{x}}$ by the Chain Rule.

7. False. $f(x) = |x^2 + x| = x^2 + x$ for $x \geq 0$ or $x \leq -1$ and $|x^2 + x| = -(x^2 + x)$ for $-1 < x < 0$.
So $f'(x) = 2x + 1$ for $x > 0$ or $x < -1$ and $f'(x) = -(2x + 1)$ for $-1 < x < 0$. But $|2x + 1| = 2x + 1$ for $x \geq -\frac{1}{2}$ and $|2x + 1| = -2x - 1$ for $x < -\frac{1}{2}$.

8. True. $f'(r)$ exists $\Rightarrow$ f is differentiable at r $\Rightarrow$ f is continuous at r $\Rightarrow$ $\lim\limits_{x \to r} f(x) = f(r)$.

9. True. $g(x) = x^5 \Rightarrow g'(x) = 5x^4 \Rightarrow g'(2) = 5(2)^4 = 80$, and by the definition of the derivative,

$$\lim_{x \to 2} \frac{g(x) - g(2)}{x - 2} = g'(2) = 80.$$

10. False. $\dfrac{d^2y}{dx^2}$ is the second derivative while $\left(\dfrac{dy}{dx}\right)^2$ is the first derivative squared. For example, if $y = x$, then $\dfrac{d^2y}{dx^2} = 0$, but $\left(\dfrac{dy}{dx}\right)^2 = 1$.

11. False. A tangent line to the parabola $y = x^2$ has slope $dy/dx = 2x$, so at $(-2, 4)$ the slope of the tangent is $2(-2) = -4$ and an equation of the tangent line is $y - 4 = -4(x + 2)$. [The given equation, $y - 4 = 2x(x + 2)$, is not even linear!]

12. True. $\dfrac{d}{dx}(\tan^2 x) = 2\tan x \sec^2 x$, and $\dfrac{d}{dx}(\sec^2 x) = 2\sec x\,(\sec x \tan x) = 2\tan x \sec^2 x$.

EXERCISES

1. Estimating the slopes of the tangent lines at $x = 2$, 3, and 5, we obtain approximate values 0.4, 2, and 0.1. Since the graph is concave downward at $x = 5$, $f''(5)$ is negative. Arranging the numbers in increasing order, we have: $f''(5) < 0 < f'(5) < f'(2) < 1 < f'(3)$.

2. $2^6 = 64$, so $f(x) = x^6$ and $a = 2$.

3. (a) $f'(r)$ is the rate at which the total cost changes with respect to the interest rate. Its units are dollars/(percent per year).

   (b) The total cost of paying off the loan is increasing by \$1200/(percent per year) as the interest rate reaches 10%. So if the interest rate goes up from 10% to 11%, the cost goes up approximately \$1200.

   (c) As r increases, C increases. So $f'(r)$ will always be positive.

4. 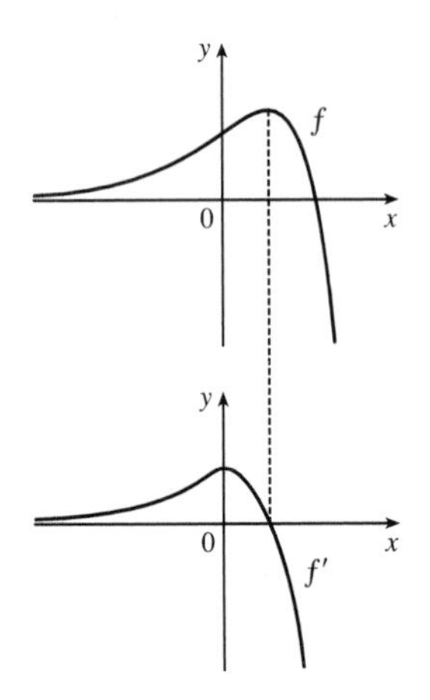

5. 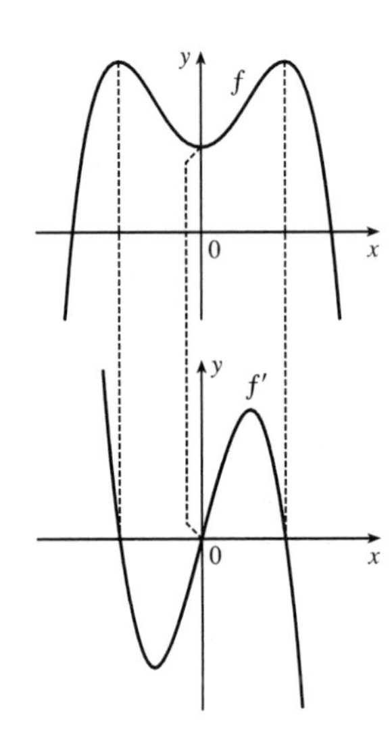

6. 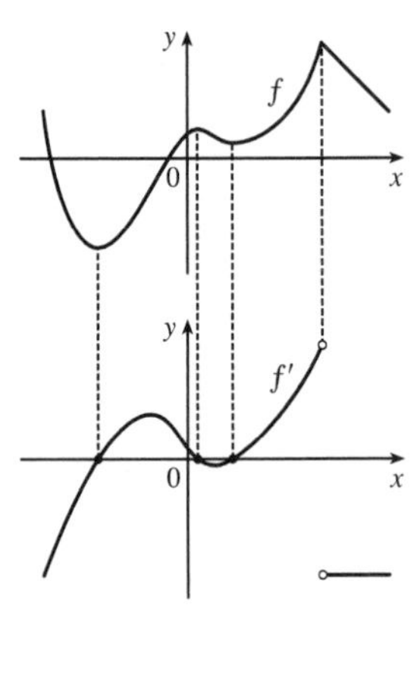

7. The graph of a has tangent lines with positive slope for $x < 0$ and negative slope for $x > 0$, and the values of c fit this pattern, so c must be the graph of the derivative of the function for a. The graph of c has horizontal tangent lines to the left and right of the x-axis and b has zeros at these points. Hence, b is the graph of the derivative of the function for c. Therefore, a is the graph of f, c is the graph of f', and b is the graph of f''.

8. (a) Drawing slope triangles, we obtain the following estimates: $F'(1950) \approx \frac{1.1}{10} = 0.11$, $F'(1965) \approx \frac{-1.6}{10} = -0.16$, and $F'(1987) \approx \frac{0.2}{10} = 0.02$.

   (b) The rate of change of the average number of children born to each woman was increasing by 0.11 in 1950, decreasing by 0.16 in 1965, and increasing by 0.02 in 1987.

   (c) There are many possible reasons:
   - In the baby-boom era (post-WWII), there was optimism about the economy and family size was rising.
   - In the baby-bust era, there was less economic optimism, and it was considered less socially responsible to have a large family.
   - In the baby-boomlet era, there was increased economic optimism and a return to more conservative attitudes.

9. $C'(1990)$ is the rate at which the total value of US currency in circulation is changing in billions of dollars per year. To estimate the value of $C'(1990)$, we will average the difference quotients obtained using the times $t = 1985$ and $t = 1995$.

Let $A = \dfrac{C(1985) - C(1990)}{1985 - 1990} = \dfrac{187.3 - 271.9}{-5} = \dfrac{-84.6}{-5} = 16.92$ and

$$B = \frac{C(1995) - C(1990)}{1995 - 1990} = \frac{409.3 - 271.9}{5} = \frac{137.4}{5} = 27.48. \text{ Then}$$

$$C'(1990) = \lim_{t \to 1990} \frac{C(t) - C(1990)}{t - 1990} \approx \frac{A + B}{2} = \frac{16.92 + 27.48}{2} = \frac{44.4}{2} = 22.2 \text{ billion dollars/year.}$$

10. $f(x) = \dfrac{4 - x}{3 + x} \quad \Rightarrow$

$$f'(x) = \lim_{h \to 0} \frac{f(x+h) - f(x)}{h} = \lim_{h \to 0} \frac{\dfrac{4 - (x+h)}{3 + (x+h)} - \dfrac{4 - x}{3 + x}}{h}$$

$$= \lim_{h \to 0} \frac{(4 - x - h)(3 + x) - (4 - x)(3 + x + h)}{h(3 + x + h)(3 + x)} = \lim_{h \to 0} \frac{-7h}{h(3 + x + h)(3 + x)}$$

$$= \lim_{h \to 0} \frac{-7}{(3 + x + h)(3 + x)} = -\frac{7}{(3 + x)^2}$$

11. $f(x) = x^3 + 5x + 4 \quad \Rightarrow$

$$f'(x) = \lim_{h \to 0} \frac{f(x+h) - f(x)}{h} = \lim_{h \to 0} \frac{(x+h)^3 + 5(x+h) + 4 - (x^3 + 5x + 4)}{h}$$

$$= \lim_{h \to 0} \frac{3x^2h + 3xh^2 + h^3 + 5h}{h} = \lim_{h \to 0} (3x^2 + 3xh + h^2 + 5) = 3x^2 + 5$$

12. (a) $f'(x) = \lim\limits_{h \to 0} \dfrac{f(x+h) - f(x)}{h} = \lim\limits_{h \to 0} \dfrac{\sqrt{3 - 5(x+h)} - \sqrt{3 - 5x}}{h} \dfrac{\sqrt{3 - 5(x+h)} + \sqrt{3 - 5x}}{\sqrt{3 - 5(x+h)} + \sqrt{3 - 5x}}$

$$= \lim_{h \to 0} \frac{[3 - 5(x+h)] - (3 - 5x)}{h\left(\sqrt{3 - 5(x+h)} + \sqrt{3 - 5x}\right)} = \lim_{h \to 0} \frac{-5}{\sqrt{3 - 5(x+h)} + \sqrt{3 - 5x}} = \frac{-5}{2\sqrt{3 - 5x}}$$

(b) Domain of f: (the radicand must be nonnegative) $3 - 5x \geq 0 \quad \Rightarrow$

$5x \leq 3 \quad \Rightarrow \quad x \in \left(-\infty, \frac{3}{5}\right]$

Domain of f': exclude $\frac{3}{5}$ because it makes the denominator zero;

$x \in \left(-\infty, \frac{3}{5}\right)$

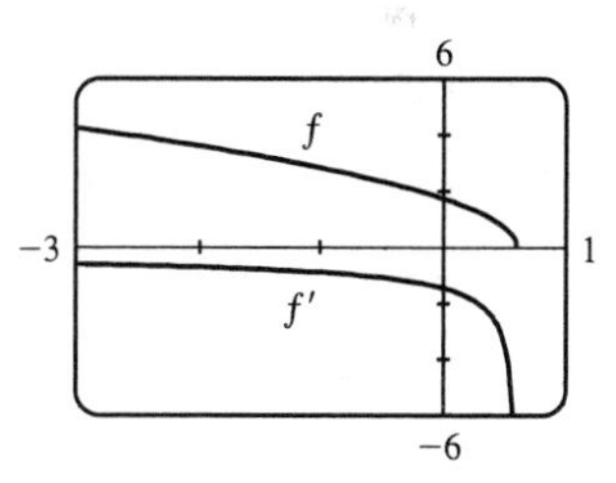

(c) Our answer to part (a) is reasonable because $f'(x)$ is always negative and f is always decreasing.

13. $y = (x^4 - 3x^2 + 5)^3 \quad \Rightarrow$

$$y' = 3(x^4 - 3x^2 + 5)^2 \frac{d}{dx}(x^4 - 3x^2 + 5) = 3(x^4 - 3x^2 + 5)^2(4x^3 - 6x) = 6x(x^4 - 3x^2 + 5)^2(2x^2 - 3)$$

14. $y = \cos(\tan x) \quad \Rightarrow \quad y' = -\sin(\tan x) \dfrac{d}{dx}(\tan x) = -\sin(\tan x)(\sec^2 x)$

15. $y = \sqrt{x} + \dfrac{1}{\sqrt[3]{x^4}} = x^{1/2} + x^{-4/3} \quad \Rightarrow \quad y' = \frac{1}{2}x^{-1/2} - \frac{4}{3}x^{-7/3} = \dfrac{1}{2\sqrt{x}} - \dfrac{4}{3\sqrt[3]{x^7}}$

16. $y = \dfrac{3x-2}{\sqrt{2x+1}} \quad \Rightarrow$

$$y' = \frac{\sqrt{2x+1}\,(3) - (3x-2)\frac{1}{2}(2x+1)^{-1/2}(2)}{\left(\sqrt{2x+1}\right)^2} \cdot \frac{(2x+1)^{1/2}}{(2x+1)^{1/2}} = \frac{3(2x+1)-(3x-2)}{(2x+1)^{3/2}} = \frac{3x+5}{(2x+1)^{3/2}}$$

17. $y = 2x\sqrt{x^2+1} \quad \Rightarrow$

$$y' = 2x \cdot \tfrac{1}{2}(x^2+1)^{-1/2}(2x) + \sqrt{x^2+1}\,(2) = \frac{2x^2}{\sqrt{x^2+1}} + 2\sqrt{x^2+1} = \frac{2x^2+2(x^2+1)}{\sqrt{x^2+1}} = \frac{2(2x^2+1)}{\sqrt{x^2+1}}$$

18. $y = \left(x + 1/x^2\right)^{\sqrt{7}} \quad \Rightarrow \quad y' = \sqrt{7}\left(x + 1/x^2\right)^{\sqrt{7}-1}\left(1 - 2/x^3\right)$

19. $y = \dfrac{t}{1-t^2} \quad \Rightarrow \quad y' = \dfrac{(1-t^2)(1) - t(-2t)}{(1-t^2)^2} = \dfrac{1-t^2+2t^2}{(1-t^2)^2} = \dfrac{t^2+1}{(1-t^2)^2}$

20. $y = \sin(\cos x) \quad \Rightarrow \quad y' = \cos(\cos x)(-\sin x) = -\sin x\,\cos(\cos x)$

21. $y = \tan\sqrt{1-x} \quad \Rightarrow \quad y' = \left(\sec^2\sqrt{1-x}\right)\left(\dfrac{1}{2\sqrt{1-x}}\right)(-1) = -\dfrac{\sec^2\sqrt{1-x}}{2\sqrt{1-x}}$

22. Using the Reciprocal Rule, $g(x) = \dfrac{1}{f(x)} \quad \Rightarrow \quad g'(x) = -\dfrac{f'(x)}{[f(x)]^2}$, we have $y = \dfrac{1}{\sin(x - \sin x)} \quad \Rightarrow$

$$y' = -\frac{\cos(x-\sin x)(1-\cos x)}{\sin^2(x-\sin x)}.$$

23. $\dfrac{d}{dx}\left(xy^4 + x^2y\right) = \dfrac{d}{dx}(x+3y) \quad \Rightarrow \quad x \cdot 4y^3y' + y^4 \cdot 1 + x^2 \cdot y' + y \cdot 2x = 1 + 3y' \quad \Rightarrow$

$$y'(4xy^3 + x^2 - 3) = 1 - y^4 - 2xy \quad \Rightarrow \quad y' = \frac{1-y^4-2xy}{4xy^3+x^2-3}$$

24. $y = \sec(1+x^2) \quad \Rightarrow \quad y' = 2x\sec(1+x^2)\,\tan(1+x^2)$

25. $y = \dfrac{\sec 2\theta}{1+\tan 2\theta} \quad \Rightarrow$

$$y' = \frac{(1+\tan 2\theta)(\sec 2\theta\,\tan 2\theta \cdot 2) - (\sec 2\theta)(\sec^2 2\theta \cdot 2)}{(1+\tan 2\theta)^2} = \frac{2\sec 2\theta\,[(1+\tan 2\theta)\tan 2\theta - \sec^2 2\theta]}{(1+\tan 2\theta)^2}$$

$$= \frac{2\sec 2\theta\,(\tan 2\theta + \tan^2 2\theta - \sec^2 2\theta)}{(1+\tan 2\theta)^2} = \frac{2\sec 2\theta\,(\tan 2\theta - 1)}{(1+\tan 2\theta)^2} \qquad \left[1 + \tan^2 x = \sec^2 x\right]$$

26. $\dfrac{d}{dx}\left(x^2\cos y + \sin 2y\right) = \dfrac{d}{dx}(xy) \quad \Rightarrow \quad x^2(-\sin y \cdot y') + (\cos y)(2x) + \cos 2y \cdot 2y' = x \cdot y' + y \cdot 1 \quad \Rightarrow$

$$y'(-x^2\sin y + 2\cos 2y - x) = y - 2x\cos y \quad \Rightarrow \quad y' = \frac{y - 2x\cos y}{2\cos 2y - x^2\sin y - x}$$

27. $y = \left(1 - x^{-1}\right)^{-1} \quad \Rightarrow$

$$y' = -1(1-x^{-1})^{-2}[-(-1x^{-2})] = -(1-1/x)^{-2}x^{-2} = -((x-1)/x)^{-2}x^{-2} = -(x-1)^{-2}$$

28. $y = \left(x + \sqrt{x}\right)^{-1/3} \quad \Rightarrow \quad y' = -\frac{1}{3}\left(x + \sqrt{x}\right)^{-4/3}\left(1 + \dfrac{1}{2\sqrt{x}}\right)$

29. $\sin(xy) = x^2 - y \quad \Rightarrow \quad \cos(xy)(xy' + y \cdot 1) = 2x - y' \quad \Rightarrow \quad x\cos(xy)y' + y' = 2x - y\cos(xy) \quad \Rightarrow$

$$y'[x\cos(xy) + 1] = 2x - y\cos(xy) \quad \Rightarrow \quad y' = \frac{2x - y\cos(xy)}{x\cos(xy)+1}$$

30. $y = \sqrt{\sin\sqrt{x}} \Rightarrow y' = \frac{1}{2}(\sin\sqrt{x})^{-1/2}(\cos\sqrt{x})\left(\frac{1}{2\sqrt{x}}\right) = \frac{\cos\sqrt{x}}{4\sqrt{x\sin\sqrt{x}}}$

31. $y = \cot(3x^2+5) \Rightarrow y' = -\csc^2(3x^2+5)(6x) = -6x\csc^2(3x^2+5)$

32. $y = \frac{(x+\lambda)^4}{x^4+\lambda^4} \Rightarrow y' = \frac{(x^4+\lambda^4)(4)(x+\lambda)^3 - (x+\lambda)^4(4x^3)}{(x^4+\lambda^4)^2} = \frac{4(x+\lambda)^3(\lambda^4-\lambda x^3)}{(x^4+\lambda^4)^2}$

33. $y = \sin\left(\tan\sqrt{1+x^3}\right) \Rightarrow y' = \cos\left(\tan\sqrt{1+x^3}\right)\left(\sec^2\sqrt{1+x^3}\right)\left[3x^2/\left(2\sqrt{1+x^3}\right)\right]$

34. $y = (\sin mx)/x \Rightarrow y' = (mx\cos mx - \sin mx)/x^2$

35. $y = \tan^2(\sin\theta) = [\tan(\sin\theta)]^2 \Rightarrow y' = 2[\tan(\sin\theta)]\cdot\sec^2(\sin\theta)\cdot\cos\theta = 2\cos\theta\tan(\sin\theta)\sec^2(\sin\theta)$

36. $x\tan y = y - 1 \Rightarrow \tan y + (x\sec^2 y)\,y' = y' \Rightarrow y' = \frac{\tan y}{1 - x\sec^2 y}$

37. $y = (x\tan x)^{1/5} \Rightarrow y' = \frac{1}{5}(x\tan x)^{-4/5}(\tan x + x\sec^2 x)$

38. $y = \frac{(x-1)(x-4)}{(x-2)(x-3)} = \frac{x^2-5x+4}{x^2-5x+6} \Rightarrow y' = \frac{(x^2-5x+6)(2x-5)-(x^2-5x+4)(2x-5)}{(x^2-5x+6)^2} = \frac{2(2x-5)}{(x-2)^2(x-3)^2}$

39. $f(t) = \sqrt{4t+1} \Rightarrow f'(t) = \frac{1}{2}(4t+1)^{-1/2}\cdot 4 = 2(4t+1)^{-1/2} \Rightarrow$

$f''(t) = 2(-\frac{1}{2})(4t+1)^{-3/2}\cdot 4 = -4/(4t+1)^{3/2}$, so $f''(2) = -4/9^{3/2} = -\frac{4}{27}$.

40. $g(\theta) = \theta\sin\theta \Rightarrow g'(\theta) = \theta\cos\theta + \sin\theta\cdot 1 \Rightarrow g''(\theta) = \theta(-\sin\theta) + \cos\theta\cdot 1 + \cos\theta = 2\cos\theta - \theta\sin\theta$,

so $g''(\pi/6) = 2\cos(\pi/6) - (\pi/6)\sin(\pi/6) = 2\left(\sqrt{3}/2\right) - (\pi/6)(1/2) = \sqrt{3} - \pi/12$.

41. $x^6 + y^6 = 1 \Rightarrow 6x^5 + 6y^5y' = 0 \Rightarrow y' = -x^5/y^5 \Rightarrow$

$$y'' = -\frac{y^5(5x^4) - x^5(5y^4y')}{(y^5)^2} = -\frac{5x^4y^4[y - x(-x^5/y^5)]}{y^{10}} = -\frac{5x^4[(y^6+x^6)/y^5]}{y^6} = -\frac{5x^4}{y^{11}}$$

42. $f(x) = (2-x)^{-1} \Rightarrow f'(x) = (2-x)^{-2} \Rightarrow f''(x) = 2(2-x)^{-3} \Rightarrow f'''(x) = 2\cdot 3(2-x)^{-4} \Rightarrow$

$f^{(4)}(x) = 2\cdot 3\cdot 4(2-x)^{-5}$. In general, $f^{(n)}(x) = 2\cdot 3\cdot 4\cdot\cdots\cdot n(2-x)^{-(n+1)} = \frac{n!}{(2-x)^{(n+1)}}$.

43. $y = 4\sin^2 x \Rightarrow y' = 4\cdot 2\sin x\cos x$. At $\left(\frac{\pi}{6}, 1\right)$, $y' = 8\cdot\frac{1}{2}\cdot\frac{\sqrt{3}}{2} = 2\sqrt{3}$, so an equation of the tangent line is $y - 1 = 2\sqrt{3}\left(x - \frac{\pi}{6}\right)$, or $y = 2\sqrt{3}\,x + 1 - \pi\sqrt{3}/3$.

44. $y = \frac{x^2-1}{x^2+1} \Rightarrow y' = \frac{(x^2+1)(2x)-(x^2-1)(2x)}{(x^2+1)^2} = \frac{4x}{(x^2+1)^2}$.

At $(0,-1)$, $y' = 0$, so an equation of the tangent line is $y + 1 = 0(x-0)$, or $y = -1$.

45. $y = \sqrt{1+4\sin x} \Rightarrow y' = \frac{1}{2}(1+4\sin x)^{-1/2}\cdot 4\cos x = \frac{2\cos x}{\sqrt{1+4\sin x}}$. At $(0,1)$, $y' = \frac{2}{\sqrt{1}} = 2$, so an equation of the tangent line is $y - 1 = 2(x-0)$, or $y = 2x+1$.

46. $x^2 + 4xy + y^2 = 13 \Rightarrow 2x + 4(xy' + y\cdot 1) + 2yy' = 0 \Rightarrow x + 2xy' + 2y + yy' = 0 \Rightarrow 2xy' + yy' = -x - 2y$

$\Rightarrow y'(2x+y) = -x-2y \Rightarrow y' = \frac{-x-2y}{2x+y}$. At $(2,1)$, $y' = \frac{-2-2}{4+1} = -\frac{4}{5}$, so an equation of the tangent line is

$y - 1 = -\frac{4}{5}(x-2)$, or $y = -\frac{4}{5}x + \frac{13}{5}$.

47. $y = \sin x + \cos x \;\Rightarrow\; y' = \cos x - \sin x = 0 \;\Leftrightarrow\; \cos x = \sin x$ and $0 \le x \le 2\pi \;\Leftrightarrow\; x = \frac{\pi}{4}$ or $\frac{5\pi}{4}$, so the points are $\left(\frac{\pi}{4}, \sqrt{2}\right)$ and $\left(\frac{5\pi}{4}, -\sqrt{2}\right)$.

48. $x^2 + 2y^2 = 1 \;\Rightarrow\; 2x + 4yy' = 0 \;\Rightarrow\; y' = -x/(2y) = 1 \;\Leftrightarrow\; x = -2y$. Since the points lie on the ellipse, we have $(-2y)^2 + 2y^2 = 1 \;\Rightarrow\; 6y^2 = 1 \;\Rightarrow\; y = \pm\frac{1}{\sqrt{6}}$. The points are $\left(-\frac{2}{\sqrt{6}}, \frac{1}{\sqrt{6}}\right)$ and $\left(\frac{2}{\sqrt{6}}, -\frac{1}{\sqrt{6}}\right)$.

49. (a) $h(x) = f(x)g(x) \;\Rightarrow\; h'(x) = f(x)g'(x) + g(x)f'(x) \;\Rightarrow$

$h'(2) = f(2)g'(2) + g(2)f'(2) = (3)(4) + (5)(-2) = 12 - 10 = 2$

(b) $F(x) = f(g(x)) \;\Rightarrow\; F'(x) = f'(g(x))g'(x) \;\Rightarrow\; F'(2) = f'(g(2))g'(2) = f'(5)(4) = 11 \cdot 4 = 44$

50. (a) $P(x) = f(x)g(x) \;\Rightarrow\; P'(x) = f(x)g'(x) + g(x)f'(x) \;\Rightarrow$

$P'(2) = f(2)g'(2) + g(2)f'(2) = (1)\left(\frac{6-0}{3-0}\right) + (4)\left(\frac{0-3}{3-0}\right) = (1)(2) + (4)(-1) = 2 - 4 = -2$

(b) $Q(x) = \dfrac{f(x)}{g(x)} \;\Rightarrow\; Q'(x) = \dfrac{g(x)f'(x) - f(x)g'(x)}{[g(x)]^2} \;\Rightarrow$

$$Q'(2) = \frac{g(2)f'(2) - f(2)g'(2)}{[g(2)]^2} = \frac{(4)(-1) - (1)(2)}{4^2} = \frac{-6}{16} = -\frac{3}{8}$$

(c) $C(x) = f(g(x)) \;\Rightarrow\; C'(x) = f'(g(x))g'(x) \;\Rightarrow$

$C'(2) = f'(g(2))g'(2) = f'(4)g'(2) = \left(\frac{6-0}{5-3}\right)(2) = (3)(2) = 6$

51. $f(x) = x^2 g(x) \;\Rightarrow\; f'(x) = x^2 g'(x) + g(x)(2x) = x[xg'(x) + 2g(x)]$

52. $f(x) = g(x^2) \;\Rightarrow\; f'(x) = g'(x^2)(2x) = 2xg'(x^2)$

53. $f(x) = [g(x)]^2 \;\Rightarrow\; f'(x) = 2[g(x)]^1 \cdot g'(x) = 2g(x)g'(x)$

54. $f(x) = x^a g(x^b) \;\Rightarrow\; f'(x) = ax^{a-1}g(x^b) + x^a g'(x^b)(bx^{b-1}) = ax^{a-1}g(x^b) + bx^{a+b-1}g'(x^b)$

55. $f(x) = g(g(x)) \;\Rightarrow\; f'(x) = g'(g(x))g'(x)$

56. $f(x) = \sin(g(x)) \;\Rightarrow\; f'(x) = \cos(g(x)) \cdot g'(x)$

57. $f(x) = g(\sin x) \;\Rightarrow\; f'(x) = g'(\sin x) \cdot \cos x$

58. $f(x) = g(\tan\sqrt{x}) \;\Rightarrow\; f'(x) = g'(\tan\sqrt{x}) \cdot \dfrac{d}{dx}(\tan\sqrt{x}) = g'(\tan\sqrt{x}) \cdot \sec^2\sqrt{x} \cdot \dfrac{d}{dx}(\sqrt{x}) = \dfrac{g'(\tan\sqrt{x})\sec^2\sqrt{x}}{2\sqrt{x}}$

59. $h(x) = \dfrac{f(x)g(x)}{f(x) + g(x)} \;\Rightarrow$

$$\begin{aligned} h'(x) &= \frac{[f(x) + g(x)]\,[f(x)g'(x) + g(x)f'(x)] - f(x)g(x)\,[f'(x) + g'(x)]}{[f(x) + g(x)]^2} \\ &= \frac{[f(x)]^2 g'(x) + f(x)g(x)f'(x) + f(x)g(x)g'(x) + [g(x)]^2 f'(x) - f(x)g(x)f'(x) - f(x)g(x)g'(x)}{[f(x) + g(x)]^2} \\ &= \frac{f'(x)[g(x)]^2 + g'(x)[f(x)]^2}{[f(x) + g(x)]^2} \end{aligned}$$

60. Using the Chain Rule repeatedly, $h(x) = f(g(\sin 4x)) \Rightarrow$

$$h'(x) = f'(g(\sin 4x)) \cdot \frac{d}{dx}(g(\sin 4x)) = f'(g(\sin 4x)) \cdot g'(\sin 4x) \cdot \frac{d}{dx}(\sin 4x) = f'(g(\sin 4x))g'(\sin 4x)(\cos 4x)(4).$$

61. f is not differentiable: at $x = -4$ because f is not continuous, at $x = -1$ because f has a corner, at $x = 2$ because f is not continuous, and at $x = 5$ because f has a vertical tangent.

62. (a) $V = \frac{1}{3}\pi r^2 h \Rightarrow dV/dh = \frac{1}{3}\pi r^2$ [r constant]

(b) $V = \frac{1}{3}\pi r^2 h \Rightarrow dV/dr = \frac{2}{3}\pi rh$ [h constant]

63. (a) $y = t^3 - 12t + 3 \Rightarrow v(t) = y' = 3t^2 - 12 \Rightarrow a(t) = v'(t) = 6t$

(b) $v(t) = 3(t^2 - 4) > 0$ when $t > 2$, so it moves upward when $t > 2$ and downward when $0 \le t < 2$.

(c) Distance upward $= y(3) - y(2) = -6 - (-13) = 7$, distance downward $= y(0) - y(2) = 3 - (-13) = 16$.
Total distance $= 7 + 16 = 23$.

64. (a) $C(x) = 920 + 2x - 0.02x^2 + 0.00007x^3 \Rightarrow C'(x) = 2 - 0.04x + 0.00021x^2$

(b) $C'(100) = 2 - 4 + 2.1 = \$0.10/\text{unit}$. This value represents the rate at which costs are increasing as the hundredth unit is produced, and is the approximate cost of producing the 101st unit.

(c) The cost of producing the 101st item is $C(101) - C(100) = 990.10107 - 990 = \$0.10107$, slightly larger than $C'(100)$.

65. If x = edge length, then $V = x^3 \Rightarrow dV/dt = 3x^2\,dx/dt = 10 \Rightarrow dx/dt = 10/(3x^2)$ and $S = 6x^2 \Rightarrow$
$dS/dt = (12x)\,dx/dt = 12x[10/(3x^2)] = 40/x$. When $x = 30$, $dS/dt = \frac{40}{30} = \frac{4}{3}\text{ cm}^2/\text{min}$.

66. Given $dV/dt = 2$, find dh/dt when $h = 5$. $V = \frac{1}{3}\pi r^2 h$ and, from similar triangles, $\frac{r}{h} = \frac{3}{10} \Rightarrow V = \frac{\pi}{3}\left(\frac{3h}{10}\right)^2 h = \frac{3\pi}{100}h^3$, so

$$2 = \frac{dV}{dt} = \frac{9\pi}{100}h^2\frac{dh}{dt} \Rightarrow \frac{dh}{dt} = \frac{200}{9\pi h^2} = \frac{200}{9\pi(5)^2} = \frac{8}{9\pi}\text{ cm/s}$$

when $h = 5$.

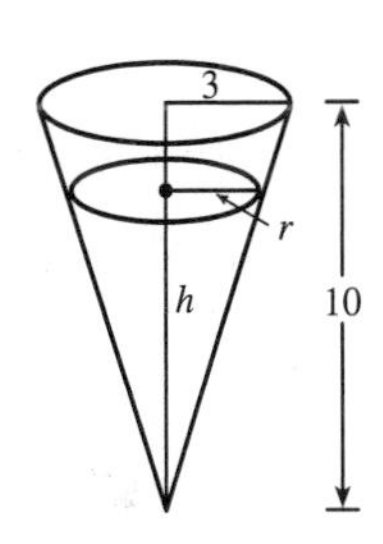

67. Given $dh/dt = 5$ and $dx/dt = 15$, find dz/dt. ${}^2 = x^2 + h^2 \Rightarrow$

$$2z\frac{dz}{dt} = 2x\frac{dx}{dt} + 2h\frac{dh}{dt} \Rightarrow \frac{dz}{dt} = \frac{1}{z}(15x + 5h).$$ When $t = 3$,

$h = 45 + 3(5) = 60$ and $x = 15(3) = 45 \Rightarrow z = \sqrt{45^2 + 60^2} = 75$,

so $\frac{dz}{dt} = \frac{1}{75}[15(45) + 5(60)] = 13\text{ ft/s}$.

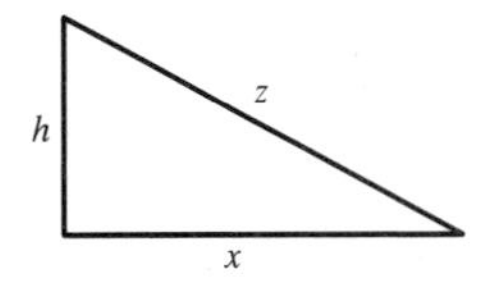

68. We are given $dz/dt = 30$ ft/s. By similar triangles, $\frac{y}{z} = \frac{4}{\sqrt{241}} \Rightarrow$

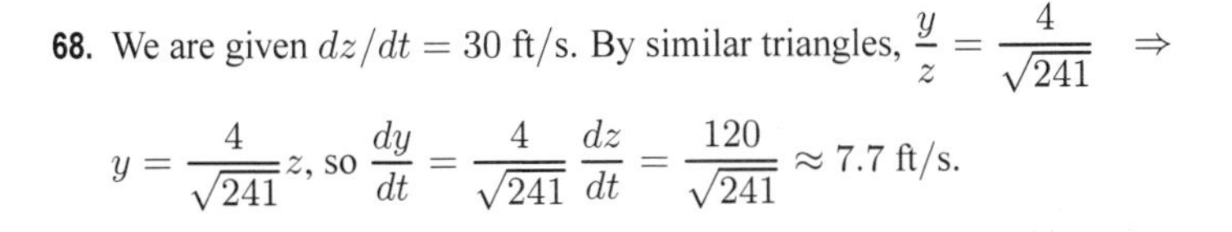

$$y = \frac{4}{\sqrt{241}}z, \text{ so } \frac{dy}{dt} = \frac{4}{\sqrt{241}}\frac{dz}{dt} = \frac{120}{\sqrt{241}} \approx 7.7\text{ ft/s}.$$

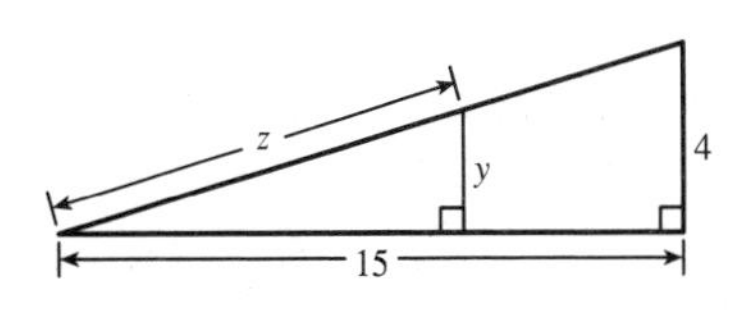

69. We are given $d\theta/dt = -0.25$ rad/h. $\tan\theta = 400/x \Rightarrow$

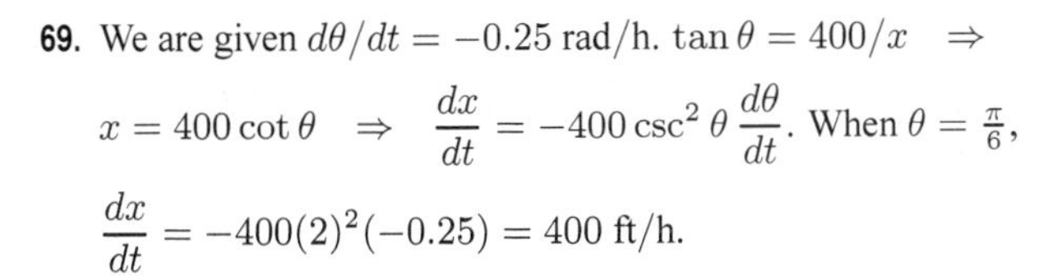

$$x = 400\cot\theta \Rightarrow \frac{dx}{dt} = -400\csc^2\theta\,\frac{d\theta}{dt}.$$ When $\theta = \frac{\pi}{6}$,

$$\frac{dx}{dt} = -400(2)^2(-0.25) = 400\text{ ft/h}.$$

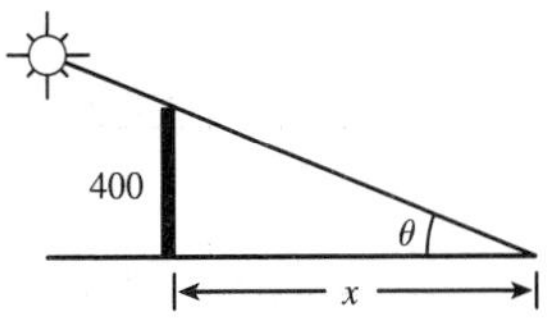

70. (a) $f(x) = \sqrt{25 - x^2} \;\Rightarrow$

$$f'(x) = \frac{-2x}{2\sqrt{25 - x^2}} = -x(25 - x^2)^{-1/2}.$$

So the linear approximation to $f(x)$ near 3 is

$f(x) \approx f(3) + f'(3)(x - 3) = 4 - \frac{3}{4}(x - 3)$.

(b)

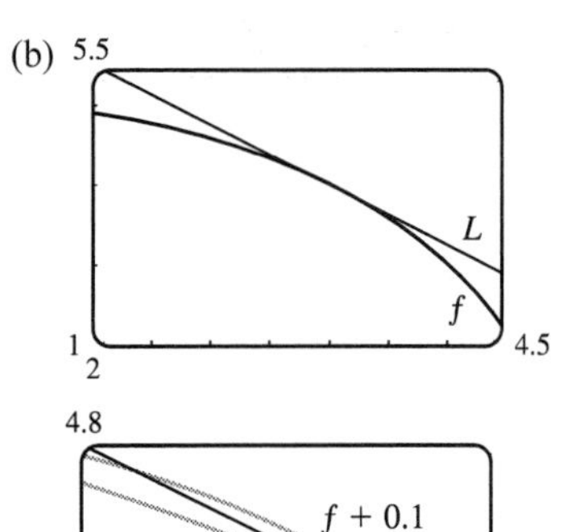

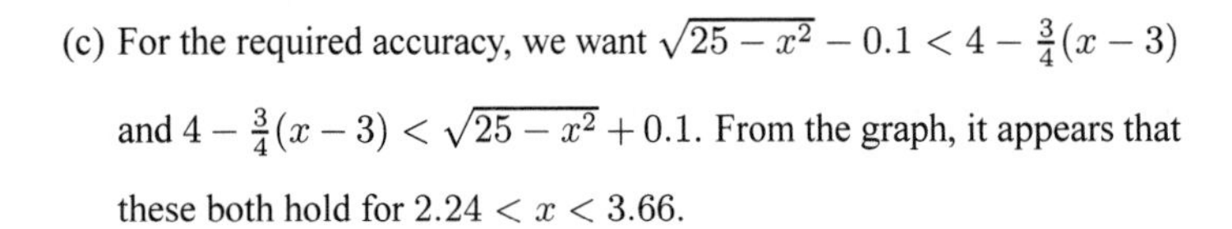

(c) For the required accuracy, we want $\sqrt{25 - x^2} - 0.1 < 4 - \frac{3}{4}(x - 3)$ and $4 - \frac{3}{4}(x - 3) < \sqrt{25 - x^2} + 0.1$. From the graph, it appears that these both hold for $2.24 < x < 3.66$.

71. (a) $f(x) = \sqrt[3]{1 + 3x} = (1 + 3x)^{1/3} \;\Rightarrow\; f'(x) = (1 + 3x)^{-2/3}$, so the linearization of f at $a = 0$ is $L(x) = f(0) + f'(0)(x - 0) = 1^{1/3} + 1^{-2/3}x = 1 + x$. Thus, $\sqrt[3]{1 + 3x} \approx 1 + x \;\Rightarrow$ $\sqrt[3]{1.03} = \sqrt[3]{1 + 3(0.01)} \approx 1 + (0.01) = 1.01$.

(b) The linear approximation is $\sqrt[3]{1 + 3x} \approx 1 + x$, so for the required accuracy we want $\sqrt[3]{1 + 3x} - 0.1 < 1 + x < \sqrt[3]{1 + 3x} + 0.1$. From the graph, it appears that this is true when $-0.23 < x < 0.40$.

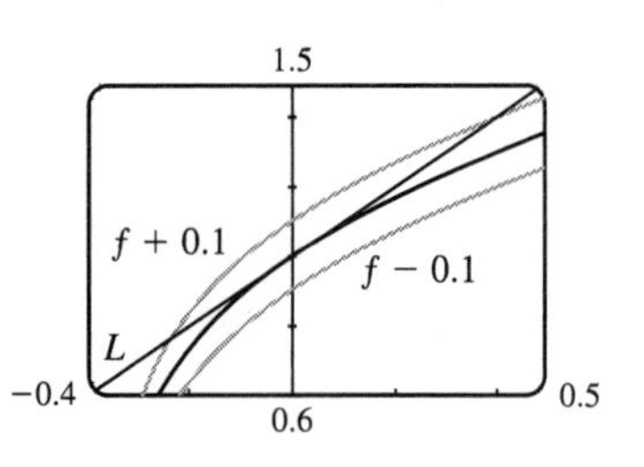

72. $y = x^3 - 2x^2 + 1 \;\Rightarrow\; dy = (3x^2 - 4x)\,dx$. When $x = 2$ and $dx = 0.2$, $dy = \left[3(2)^2 - 4(2)\right](0.2) = 0.8$.

73. $A = x^2 + \frac{1}{2}\pi\left(\frac{1}{2}x\right)^2 = \left(1 + \frac{\pi}{8}\right)x^2 \;\Rightarrow\; dA = \left(2 + \frac{\pi}{4}\right)x\,dx$. When $x = 60$ and $dx = 0.1$, $dA = \left(2 + \frac{\pi}{4}\right)60(0.1) = 12 + \frac{3\pi}{2}$, so the maximum error is approximately $12 + \frac{3\pi}{2} \approx 16.7\text{ cm}^2$.

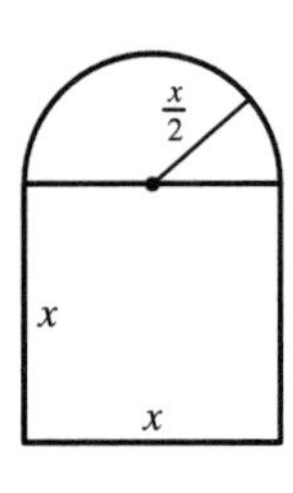

74. $\displaystyle\lim_{x\to 1} \frac{x^{17} - 1}{x - 1} = \left[\frac{d}{dx}x^{17}\right]_{x=1} = 17(1)^{16} = 17$

75. $\displaystyle\lim_{h\to 0} \frac{\sqrt[4]{16 + h} - 2}{h} = \left[\frac{d}{dx}\sqrt[4]{x}\right]_{x=16} = \frac{1}{4}x^{-3/4}\Big|_{x=16} = \frac{1}{4\left(\sqrt[4]{16}\right)^3} = \frac{1}{32}$

76. $\displaystyle\lim_{\theta\to\pi/3} \frac{\cos\theta - 0.5}{\theta - \pi/3} = \left[\frac{d}{d\theta}\cos\theta\right]_{\theta=\pi/3} = -\sin\frac{\pi}{3} = -\frac{\sqrt{3}}{2}$

77. $\displaystyle\lim_{x\to 0}\frac{\sqrt{1+\tan x}-\sqrt{1+\sin x}}{x^3}=\lim_{x\to 0}\frac{\left(\sqrt{1+\tan x}-\sqrt{1+\sin x}\right)\left(\sqrt{1+\tan x}+\sqrt{1+\sin x}\right)}{x^3\left(\sqrt{1+\tan x}+\sqrt{1+\sin x}\right)}$

$$=\lim_{x\to 0}\frac{(1+\tan x)-(1+\sin x)}{x^3\left(\sqrt{1+\tan x}+\sqrt{1+\sin x}\right)}=\lim_{x\to 0}\frac{\sin x\,(1/\cos x-1)}{x^3\left(\sqrt{1+\tan x}+\sqrt{1+\sin x}\right)}\cdot\frac{\cos x}{\cos x}$$

$$=\lim_{x\to 0}\frac{\sin x\,(1-\cos x)}{x^3\left(\sqrt{1+\tan x}+\sqrt{1+\sin x}\right)\cos x}\cdot\frac{1+\cos x}{1+\cos x}$$

$$=\lim_{x\to 0}\frac{\sin x\cdot\sin^2 x}{x^3\left(\sqrt{1+\tan x}+\sqrt{1+\sin x}\right)\cos x\,(1+\cos x)}$$

$$=\left(\lim_{x\to 0}\frac{\sin x}{x}\right)^3\lim_{x\to 0}\frac{1}{\left(\sqrt{1+\tan x}+\sqrt{1+\sin x}\right)\cos x\,(1+\cos x)}$$

$$=1^3\cdot\frac{1}{\left(\sqrt{1}+\sqrt{1}\right)\cdot 1\cdot(1+1)}=\frac{1}{4}$$

78. Let (b,c) be on the curve, that is, $b^{2/3}+c^{2/3}=a^{2/3}$. Now $x^{2/3}+y^{2/3}=a^{2/3}\;\Rightarrow\;\frac{2}{3}x^{-1/3}+\frac{2}{3}y^{-1/3}\dfrac{dy}{dx}=0$, so

$\dfrac{dy}{dx}=-\dfrac{y^{1/3}}{x^{1/3}}=-\left(\dfrac{y}{x}\right)^{1/3}$, so at (b,c) the slope of the tangent line is $-(c/b)^{1/3}$ and an equation of the tangent line is

$y-c=-(c/b)^{1/3}(x-b)$ or $y=-(c/b)^{1/3}x+(c+b^{2/3}c^{1/3})$. Setting $y=0$, we find that the x-intercept is

$b^{1/3}c^{2/3}+b=b^{1/3}(c^{2/3}+b^{2/3})$ and setting $x=0$ we find that the y-intercept is $c+b^{2/3}c^{1/3}=c^{1/3}(c^{2/3}+b^{2/3})$.

So the length of the tangent line between these two points is

$$\sqrt{[b^{1/3}(c^{2/3}+b^{2/3})]^2+[c^{1/3}(c^{2/3}+b^{2/3})]^2}=\sqrt{b^{2/3}(a^{2/3})^2+c^{2/3}(a^{2/3})^2}=\sqrt{(b^{2/3}+c^{2/3})a^{4/3}}$$

$$=\sqrt{a^{2/3}a^{4/3}}=\sqrt{a^2}=a=\text{constant}$$

3 ☐ APPLICATIONS OF DIFFERENTIATION

3.1 Maximum and Minimum Values

1. A function f has an **absolute minimum** at $x = c$ if $f(c)$ is the smallest function value on the entire domain of f, whereas f has a **local minimum** at c if $f(c)$ is the smallest function value when x is near c.

2. (a) The Extreme Value Theorem

(b) See the Closed Interval Method.

3. Absolute maximum at b; absolute minimum at d; local maxima at b and e; local minima at d and s; neither a maximum nor a minimum at a, c, r, and t.

4. Absolute maximum at e; absolute minimum at t; local maxima at c, e, and s; local minima at b, c, d, and r; neither a maximum nor a minimum at a.

5. Absolute maximum value is $f(4) = 4$; absolute minimum value is $f(7) = 0$; local maximum values are $f(4) = 4$ and $f(6) = 3$; local minimum values are $f(2) = 1$ and $f(5) = 2$.

6. Absolute maximum value is $f(7) = 5$; absolute minimum value is $f(1) = 0$; local maximum values are $f(0) = 2$, $f(3) = 4$, and $f(5) = 3$; local minimum values are $f(1) = 0$, $f(4) = 2$, and $f(6) = 1$.

7. Absolute minimum at 2, absolute maximum at 3, local minimum at 4

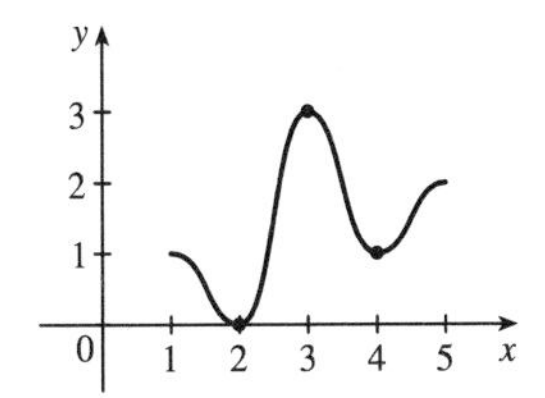

8. Absolute minimum at 1, absolute maximum at 5, local maximum at 2, local minimum at 4

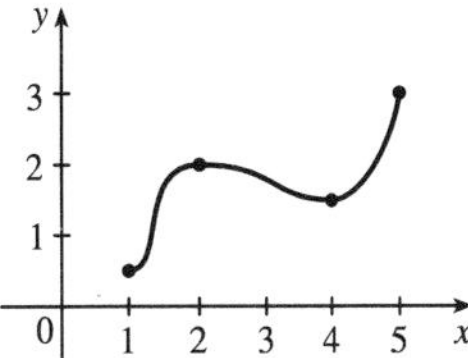

9. Absolute maximum at 5, absolute minimum at 2, local maximum at 3, local minima at 2 and 4

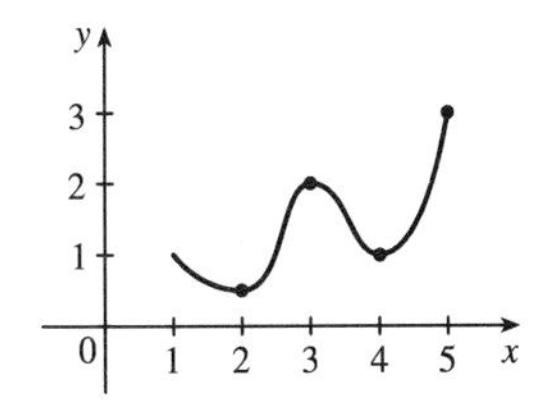

10. f has no local maximum or minimum, but 2 and 4 are critical numbers

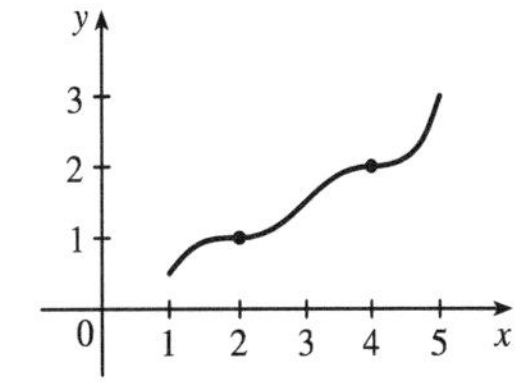

11. (a)

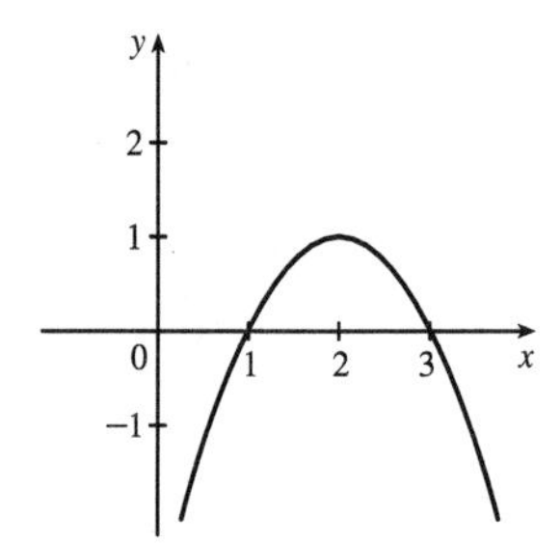

(b)

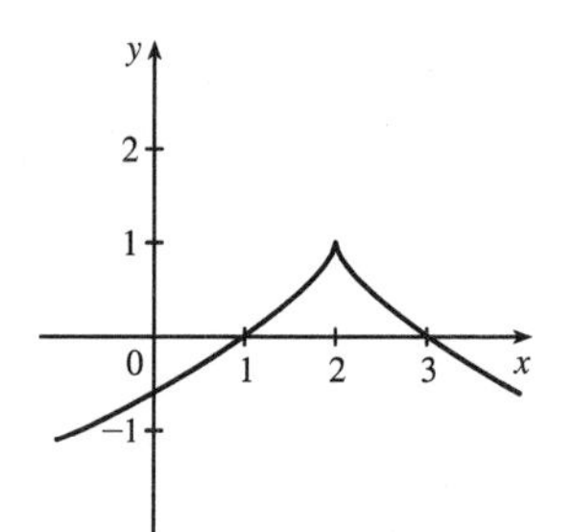

(c)

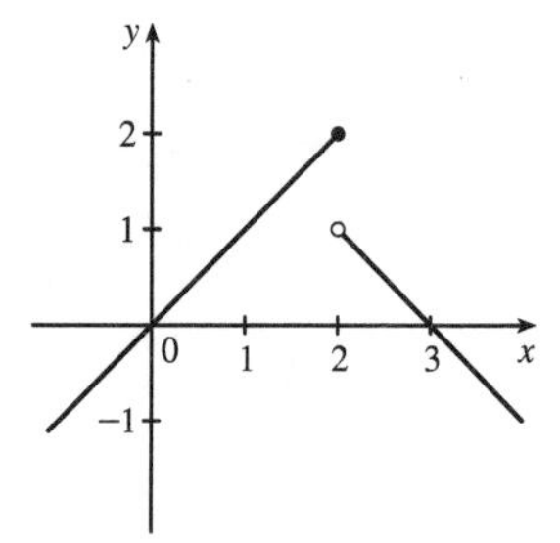

12. (a) Note that a local maximum cannot occur at an endpoint.

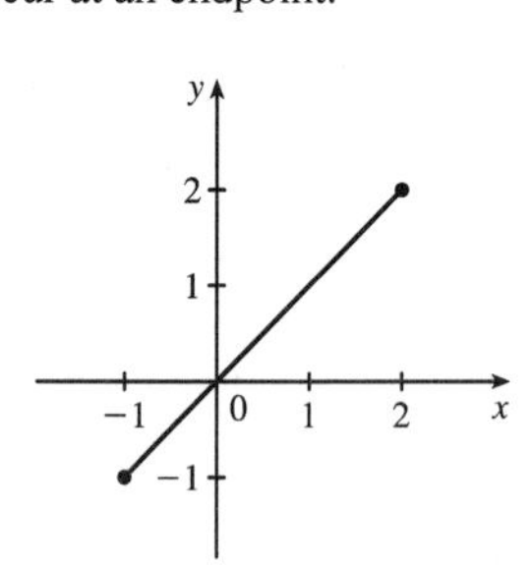

(b)

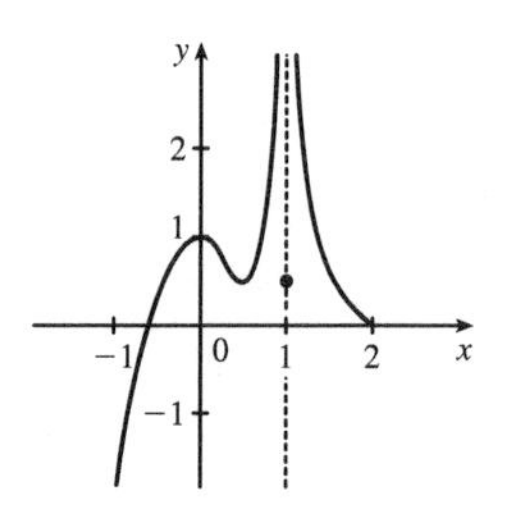

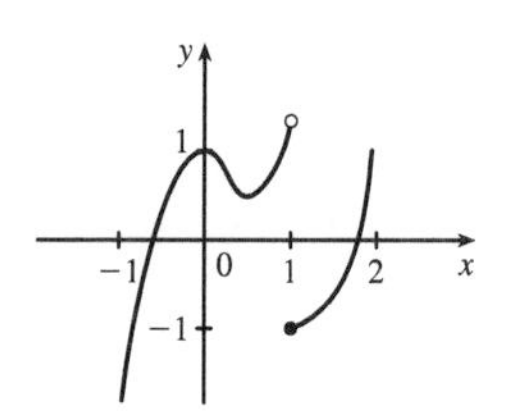

Note: By the Extreme Value Theorem, f must *not* be continuous.

13. (a) *Note:* By the Extreme Value Theorem, f must *not* be continuous; because if it were, it would attain an absolute minimum.

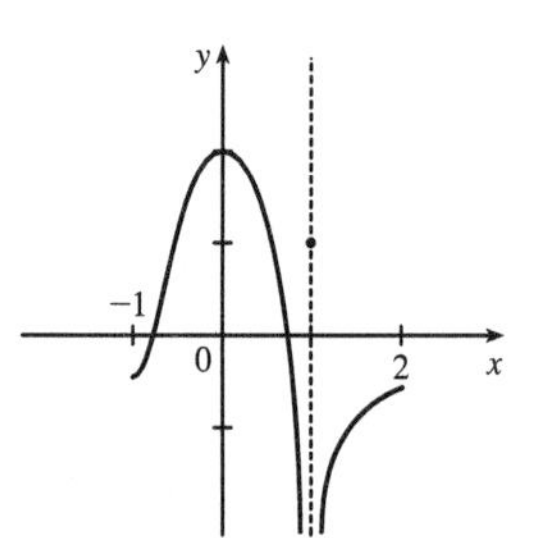

(b)

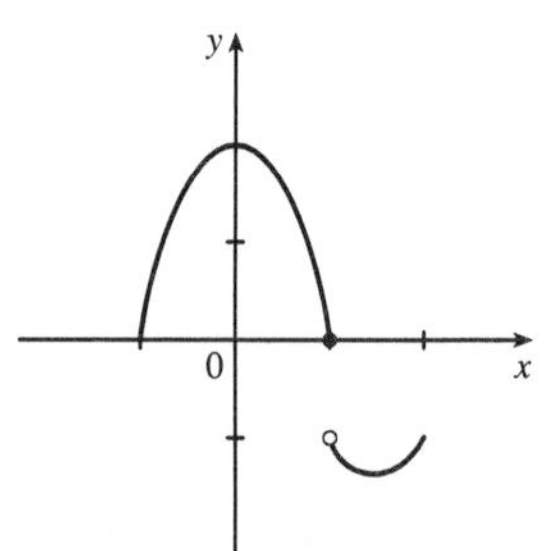

14. (a)

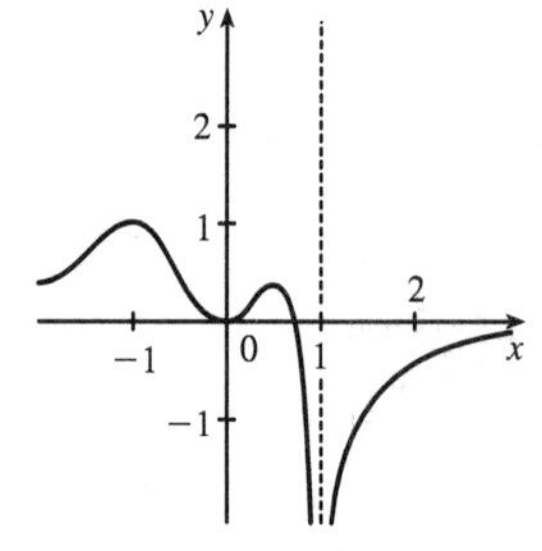

(b)

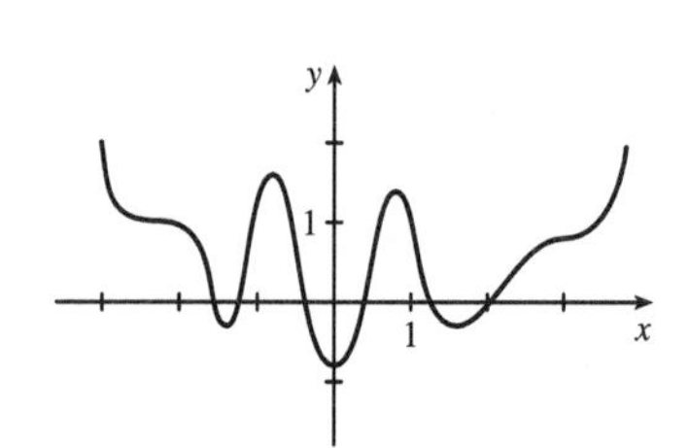

15. $f(x) = 8 - 3x$, $x \geq 1$. Absolute maximum $f(1) = 5$; no local maximum. No absolute or local minimum.

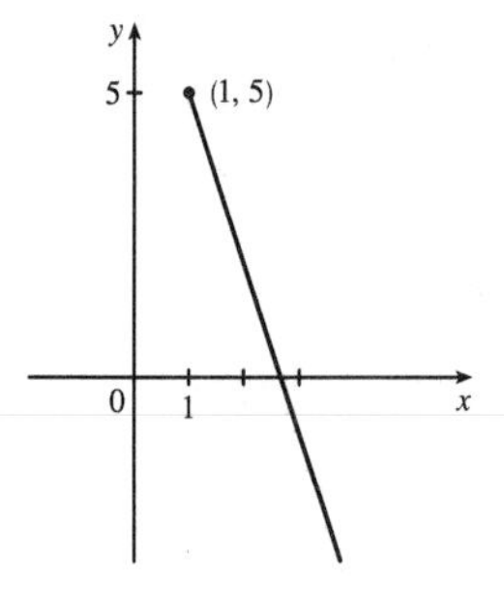

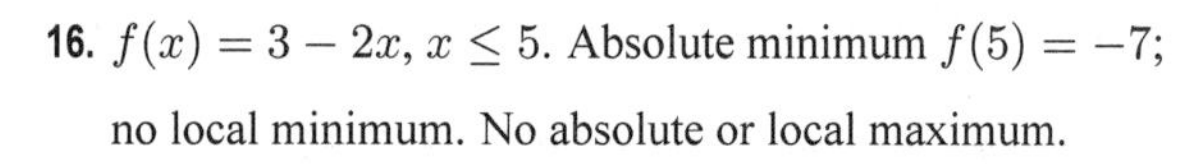

16. $f(x) = 3 - 2x$, $x \leq 5$. Absolute minimum $f(5) = -7$; no local minimum. No absolute or local maximum.

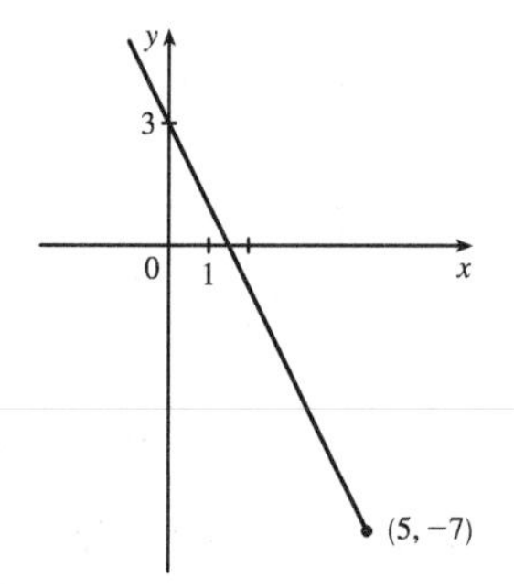

17. $f(x) = x^2$, $0 < x < 2$. No absolute or local maximum or minimum value.

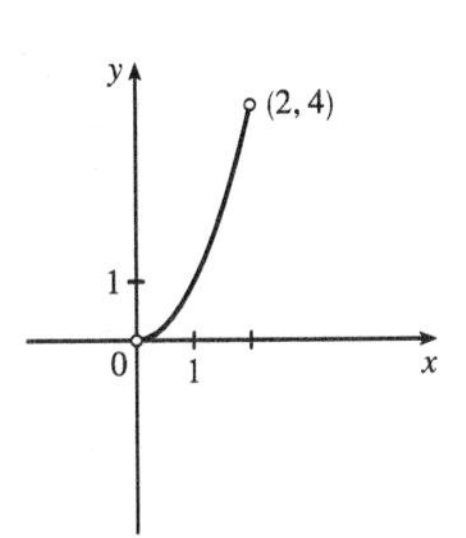

18. $f(t) = 1/t$, $0 < t \le 1$. Absolute minimum $f(1) = 1$; no local minimum. No local or absolute maximum.

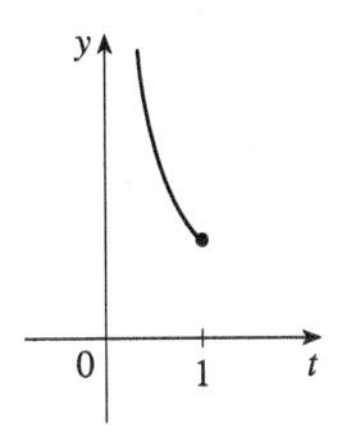

19. $f(\theta) = \sin\theta$, $-2\pi \le \theta \le 2\pi$. Absolute and local maxima $f\left(-\frac{3\pi}{2}\right) = f\left(\frac{\pi}{2}\right) = 1$. Absolute and local minima $f\left(-\frac{\pi}{2}\right) = f\left(\frac{3\pi}{2}\right) = -1$.

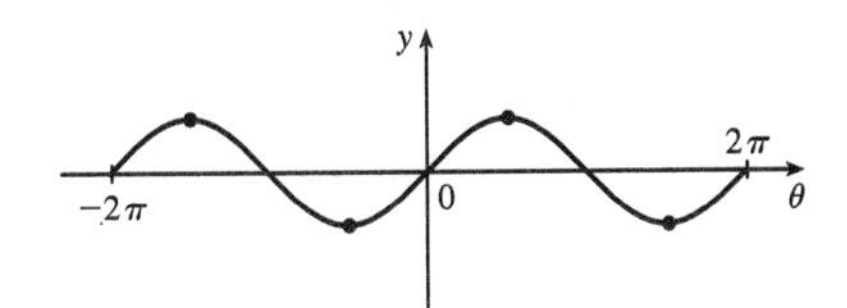

20. $f(\theta) = \tan\theta$, $-\frac{\pi}{4} \le \theta < \frac{\pi}{2}$. Absolute minimum $f\left(-\frac{\pi}{4}\right) = -1$; no local minimum. No absolute or local maximum.

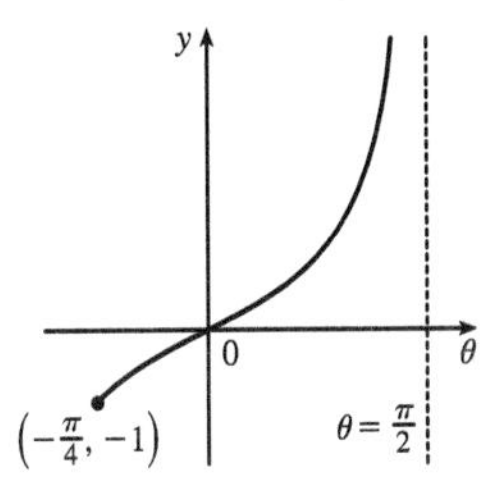

21. $f(x) = 1 - \sqrt{x}$. Absolute maximum $f(0) = 1$; no local maximum. No absolute or local minimum.

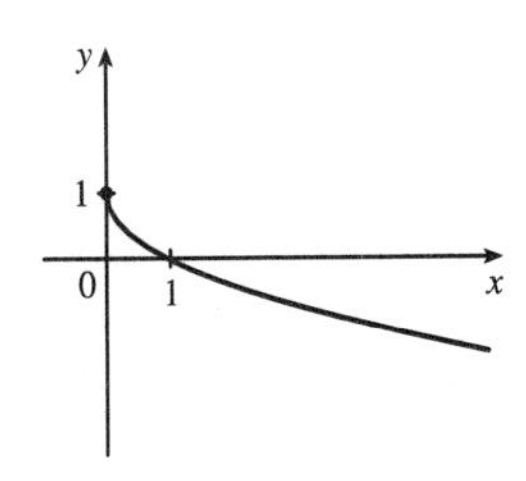

22. $f(x) = \begin{cases} 4 - x^2 & \text{if } -2 \le x < 0 \\ 2x - 1 & \text{if } 0 \le x \le 2 \end{cases}$

Abolute minimum $f(0) = -1$; no local minimim. No absolute or local maximum.

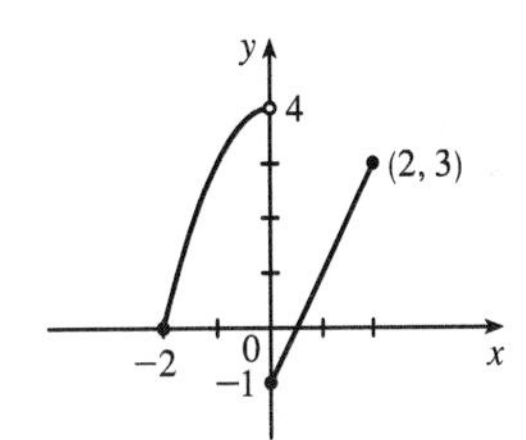

23. $f(x) = 5x^2 + 4x \;\Rightarrow\; f'(x) = 10x + 4$. $f'(x) = 0 \;\Rightarrow\; x = -\frac{2}{5}$, so $-\frac{2}{5}$ is the only critical number.

24. $f(x) = x^3 + x^2 - x \;\Rightarrow\; f'(x) = 3x^2 + 2x - 1$.

$f'(x) = 0 \;\Rightarrow\; (x + 1)(3x - 1) = 0 \;\Rightarrow\; x = -1, \frac{1}{3}$. These are the only critical numbers.

25. $f(x) = x^3 + 3x^2 - 24x \;\Rightarrow\; f'(x) = 3x^2 + 6x - 24 = 3(x^2 + 2x - 8)$.

$f'(x) = 0 \;\Rightarrow\; 3(x + 4)(x - 2) = 0 \;\Rightarrow\; x = -4, 2$. These are the only critical numbers.

26. $f(x) = x^3 + x^2 + x \;\Rightarrow\; f'(x) = 3x^2 + 2x + 1$. $f'(x) = 0 \;\Rightarrow\; 3x^2 + 2x + 1 = 0 \;\Rightarrow\; x = \dfrac{-2 \pm \sqrt{4 - 12}}{6}$.

Neither of these is a real number. Thus, there are no critical numbers.

27. $s(t) = 3t^4 + 4t^3 - 6t^2 \Rightarrow s'(t) = 12t^3 + 12t^2 - 12t.$ $s'(t) = 0 \Rightarrow 12t(t^2 + t - 1) \Rightarrow$

$t = 0$ or $t^2 + t - 1 = 0$. Using the quadratic formula to solve the latter equation gives us

$t = \dfrac{-1 \pm \sqrt{1^2 - 4(1)(-1)}}{2(1)} = \dfrac{-1 \pm \sqrt{5}}{2} \approx 0.618, -1.618$. The three critical numbers are 0, $\dfrac{-1 \pm \sqrt{5}}{2}$.

28. $g(t) = |3t - 4| = \begin{cases} 3t - 4 & \text{if } 3t - 4 \geq 0 \\ -(3t - 4) & \text{if } 3t - 4 < 0 \end{cases} = \begin{cases} 3t - 4 & \text{if } t \geq \frac{4}{3} \\ 4 - 3t & \text{if } t < \frac{4}{3} \end{cases}$

$g'(t) = \begin{cases} 3 & \text{if } t > \frac{4}{3} \\ -3 & \text{if } t < \frac{4}{3} \end{cases}$ and $g'(t)$ does not exist at $t = \frac{4}{3}$, so $t = \frac{4}{3}$ is a critical number.

29. $g(y) = \dfrac{y - 1}{y^2 - y + 1} \Rightarrow$

$$g'(y) = \frac{(y^2 - y + 1)(1) - (y - 1)(2y - 1)}{(y^2 - y + 1)^2} = \frac{y^2 - y + 1 - (2y^2 - 3y + 1)}{(y^2 - y + 1)^2} = \frac{-y^2 + 2y}{(y^2 - y + 1)^2} = \frac{y(2 - y)}{(y^2 - y + 1)^2}.$$

$g'(y) = 0 \Rightarrow y = 0, 2$. The expression $y^2 - y + 1$ is never equal to 0, so $g'(y)$ exists for all real numbers. The critical numbers are 0 and 2.

30. $h(p) = \dfrac{p - 1}{p^2 + 4} \Rightarrow h'(p) = \dfrac{(p^2 + 4)(1) - (p - 1)(2p)}{(p^2 + 4)^2} = \dfrac{p^2 + 4 - 2p^2 + 2p}{(p^2 + 4)^2} = \dfrac{-p^2 + 2p + 4}{(p^2 + 4)^2}.$

$h'(p) = 0 \Rightarrow p = \dfrac{-2 \pm \sqrt{4 + 16}}{-2} = 1 \pm \sqrt{5}$. The critical numbers are $1 \pm \sqrt{5}$. [$h'(p)$ exists for all real numbers.]

31. $F(x) = x^{4/5}(x - 4)^2 \Rightarrow$

$$\begin{aligned} F'(x) &= x^{4/5} \cdot 2(x - 4) + (x - 4)^2 \cdot \tfrac{4}{5}x^{-1/5} = \tfrac{1}{5}x^{-1/5}(x - 4)\left[5 \cdot x \cdot 2 + (x - 4) \cdot 4\right] \\ &= \frac{(x - 4)(14x - 16)}{5x^{1/5}} = \frac{2(x - 4)(7x - 8)}{5x^{1/5}} \end{aligned}$$

$F'(x) = 0 \Rightarrow x = 4, \frac{8}{7}$. $F'(0)$ does not exist. Thus, the three critical numbers are 0, $\frac{8}{7}$, and 4.

32. $G(x) = \sqrt[3]{x^2 - x} \Rightarrow G'(x) = \frac{1}{3}(x^2 - x)^{-2/3}(2x - 1)$. $G'(x)$ does not exist when $x^2 - x = 0$, that is, when $x = 0$ or 1. $G'(x) = 0 \Leftrightarrow 2x - 1 = 0 \Leftrightarrow x = \frac{1}{2}$. So the critical numbers are $x = 0, \frac{1}{2}, 1$.

33. $f(\theta) = 2\cos\theta + \sin^2\theta \Rightarrow f'(\theta) = -2\sin\theta + 2\sin\theta\cos\theta$. $f'(\theta) = 0 \Rightarrow 2\sin\theta(\cos\theta - 1) = 0 \Rightarrow \sin\theta = 0$ or $\cos\theta = 1 \Rightarrow \theta = n\pi$ (n an integer) or $\theta = 2n\pi$. The solutions $\theta = n\pi$ include the solutions $\theta = 2n\pi$, so the critical numbers are $\theta = n\pi$.

34. $g(\theta) = 4\theta - \tan\theta \Rightarrow g'(\theta) = 4 - \sec^2\theta$. $g'(\theta) = 0 \Rightarrow \sec^2\theta = 4 \Rightarrow \sec\theta = \pm 2 \Rightarrow \cos\theta = \pm\frac{1}{2} \Rightarrow$ $\theta = \frac{\pi}{3} + 2n\pi, \frac{5\pi}{3} + 2n\pi, \frac{2\pi}{3} + 2n\pi$, and $\frac{4\pi}{3} + 2n\pi$ are critical numbers.

Note: The values of θ that make $g'(\theta)$ undefined are not in the domain of g.

35. $f(x) = 3x^2 - 12x + 5$, $[0, 3]$. $f'(x) = 6x - 12 = 0 \Leftrightarrow x = 2$. Applying the Closed Interval Method, we find that $f(0) = 5$, $f(2) = -7$, and $f(3) = -4$. So $f(0) = 5$ is the absolute maximum value and $f(2) = -7$ is the absolute minimum value.

36. $f(x) = x^3 - 3x + 1$, $[0, 3]$. $f'(x) = 3x^2 - 3 = 0 \Leftrightarrow x = \pm 1$, but -1 is not in $[0, 3]$. $f(0) = 1$, $f(1) = -1$, and $f(3) = 19$. So $f(3) = 19$ is the absolute maximum value and $f(1) = -1$ is the absolute minimum value.

37. $f(x) = 2x^3 - 3x^2 - 12x + 1$, $[-2, 3]$. $f'(x) = 6x^2 - 6x - 12 = 6(x^2 - x - 2) = 6(x - 2)(x + 1) = 0 \Leftrightarrow x = 2, -1$. $f(-2) = -3$, $f(-1) = 8$, $f(2) = -19$, and $f(3) = -8$. So $f(-1) = 8$ is the absolute maximum value and $f(2) = -19$ is the absolute minimum value.

38. $f(x) = x^3 - 6x^2 + 9x + 2$, $[-1, 4]$. $f'(x) = 3x^2 - 12x + 9 = 3(x^2 - 4x + 3) = 3(x - 1)(x - 3) = 0 \Leftrightarrow x = 1, 3$. $f(-1) = -14$, $f(1) = 6$, $f(3) = 2$, and $f(4) = 6$. So $f(1) = f(4) = 6$ is the absolute maximum value and $f(-1) = -14$ is the absolute minimum value.

39. $f(x) = x^4 - 2x^2 + 3$, $[-2, 3]$. $f'(x) = 4x^3 - 4x = 4x(x^2 - 1) = 4x(x + 1)(x - 1) = 0 \Leftrightarrow x = -1, 0, 1$. $f(-2) = 11$, $f(-1) = 2$, $f(0) = 3$, $f(1) = 2$, $f(3) = 66$. So $f(3) = 66$ is the absolute maximum value and $f(\pm 1) = 2$ is the absolute minimum value.

40. $f(x) = (x^2 - 1)^3$, $[-1, 2]$. $f'(x) = 3(x^2 - 1)^2(2x) = 6x(x + 1)^2(x - 1)^2 = 0 \Leftrightarrow x = -1, 0, 1$. $f(\pm 1) = 0$, $f(0) = -1$, and $f(2) = 27$. So $f(2) = 27$ is the absolute maximum value and $f(0) = -1$ is the absolute minimum value.

41. $f(t) = t\sqrt{4 - t^2}$, $[-1, 2]$.

$$f'(t) = t \cdot \tfrac{1}{2}\left(4 - t^2\right)^{-1/2}(-2t) + \left(4 - t^2\right)^{1/2} \cdot 1 = \frac{-t^2}{\sqrt{4 - t^2}} + \sqrt{4 - t^2} = \frac{-t^2 + \left(4 - t^2\right)}{\sqrt{4 - t^2}} = \frac{4 - 2t^2}{\sqrt{4 - t^2}}.$$

$f'(t) = 0 \Rightarrow 4 - 2t^2 = 0 \Rightarrow t^2 = 2 \Rightarrow t = \pm\sqrt{2}$, but $t = -\sqrt{2}$ is not in the given interval, $[-1, 2]$. $f'(t)$ does not exist if $4 - t^2 = 0 \Rightarrow t = \pm 2$, but -2 is not in the given interval. $f(-1) = -\sqrt{3}$, $f(\sqrt{2}) = 2$, and $f(2) = 0$. So $f(\sqrt{2}) = 2$ is the absolute maximum value and $f(-1) = -\sqrt{3}$ is the absolute minimum value.

42. $f(x) = \dfrac{x}{x^2 + 4}$, $[0, 3]$. $f'(x) = \dfrac{(x^2 + 4)1 - x(2x)}{(x^2 + 4)^2} = \dfrac{4 - x^2}{(x^2 + 4)^2} = 0 \Leftrightarrow x = \pm 2$, but -2 is not in the interval $[0, 3]$. $f(0) = 0$, $f(2) = \frac{1}{4} = 0.25$, $f(3) = \frac{3}{13} \approx 0.23$. So $f(2) = \frac{1}{4}$ is the absolute maximum and $f(0) = 0$ is the absolute minimum.

43. $f(x) = \sin x + \cos x$, $\left[0, \frac{\pi}{3}\right]$. $f'(x) = \cos x - \sin x = 0 \Leftrightarrow \sin x = \cos x \Rightarrow \dfrac{\sin x}{\cos x} = 1 \Rightarrow \tan x = 1 \Rightarrow x = \frac{\pi}{4}$. $f(0) = 1$, $f\left(\frac{\pi}{4}\right) = \sqrt{2} \approx 1.41$, $f\left(\frac{\pi}{3}\right) = \frac{\sqrt{3} + 1}{2} \approx 1.37$. So $f\left(\frac{\pi}{4}\right) = \sqrt{2}$ is the absolute maximum value and $f(0) = 1$ is the absolute minimum value.

44. $f(x) = x - 2\cos x$, $[-\pi, \pi]$. $f'(x) = 1 + 2\sin x = 0 \iff \sin x = -\frac{1}{2} \iff x = -\frac{5\pi}{6}, -\frac{\pi}{6}$.

$f(-\pi) = 2 - \pi \approx -1.14$, $f\left(-\frac{5\pi}{6}\right) = \sqrt{3} - \frac{5\pi}{6} \approx -0.886$, $f\left(-\frac{\pi}{6}\right) = -\frac{\pi}{6} - \sqrt{3} \approx -2.26$, $f(\pi) = \pi + 2 \approx 5.14$.

So $f(\pi) = \pi + 2$ is the absolute maximum value and $f\left(-\frac{\pi}{6}\right) = -\frac{\pi}{6} - \sqrt{3}$ is the absolute minimum value.

45. $f(x) = x^a(1-x)^b$, $0 \le x \le 1$, $a > 0$, $b > 0$.

$$\begin{aligned} f'(x) &= x^a \cdot b(1-x)^{b-1}(-1) + (1-x)^b \cdot ax^{a-1} = x^{a-1}(1-x)^{b-1}\left[x \cdot b(-1) + (1-x) \cdot a\right] \\ &= x^{a-1}(1-x)^{b-1}(a - ax - bx) \end{aligned}$$

At the endpoints, we have $f(0) = f(1) = 0$ [the minimum value of f]. In the interval $(0,1)$, $f'(x) = 0 \iff x = \dfrac{a}{a+b}$.

$$f\left(\frac{a}{a+b}\right) = \left(\frac{a}{a+b}\right)^a \left(1 - \frac{a}{a+b}\right)^b = \frac{a^a}{(a+b)^a}\left(\frac{a+b-a}{a+b}\right)^b = \frac{a^a}{(a+b)^a} \cdot \frac{b^b}{(a+b)^b} = \frac{a^a b^b}{(a+b)^{a+b}}.$$

So $f\left(\dfrac{a}{a+b}\right) = \dfrac{a^a b^b}{(a+b)^{a+b}}$ is the absolute maximum value.

46.

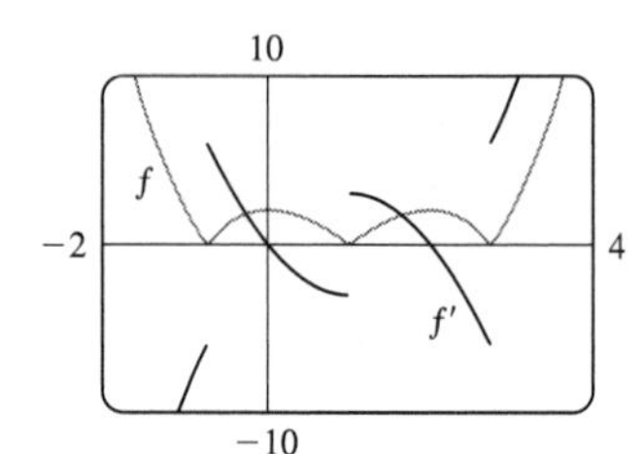

We see that $f'(x) = 0$ at about $x = 0.0$ and 2.0, and that $f'(x)$ does not exist at about $x = -0.7$, 1.0, and 2.7, so the critical numbers of f are about -0.7, 0.0, 1.0, 2.0, and 2.7.

47. (a)

From the graph, it appears that the absolute maximum value is about $f(-0.77) = 2.19$, and the absolute minimum value is about $f(0.77) = 1.81$.

(b) $f(x) = x^5 - x^3 + 2 \Rightarrow f'(x) = 5x^4 - 3x^2 = x^2(5x^2 - 3)$. So $f'(x) = 0 \Rightarrow x = 0, \pm\sqrt{\frac{3}{5}}$.

$$f\left(-\sqrt{\tfrac{3}{5}}\right) = \left(-\sqrt{\tfrac{3}{5}}\right)^5 - \left(-\sqrt{\tfrac{3}{5}}\right)^3 + 2 = -\left(\tfrac{3}{5}\right)^2\sqrt{\tfrac{3}{5}} + \tfrac{3}{5}\sqrt{\tfrac{3}{5}} + 2 = \left(\tfrac{3}{5} - \tfrac{9}{25}\right)\sqrt{\tfrac{3}{5}} + 2 = \tfrac{6}{25}\sqrt{\tfrac{3}{5}} + 2 \text{ (maximum)}$$

and similarly, $f\left(\sqrt{\frac{3}{5}}\right) = -\frac{6}{25}\sqrt{\frac{3}{5}} + 2$ (minimum).

48. (a)

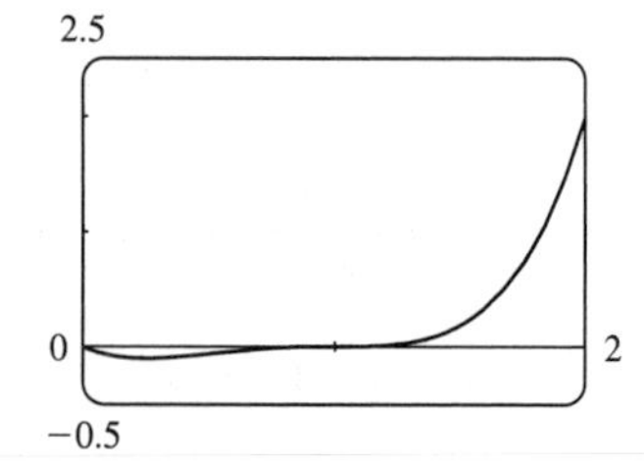

From the graph, it appears that the absolute maximum value is $f(2) = 2$, and that the absolute minimum value is about $f(0.25) = -0.11$.

(b) $f(x) = x^4 - 3x^3 + 3x^2 - x \quad \Rightarrow \quad f'(x) = 4x^3 - 9x^2 + 6x - 1 = (4x-1)(x-1)^2$.

So $f'(x) = 0 \quad \Rightarrow \quad x = \frac{1}{4}$ or $x = 1$. Now $f(1) = 1^4 - 3 \cdot 1^3 + 3 \cdot 1^2 - 1 = 0$ (not an extremum)

and $f\left(\frac{1}{4}\right) = \left(\frac{1}{4}\right)^4 - 3\left(\frac{1}{4}\right)^3 + 3\left(\frac{1}{4}\right)^2 - \frac{1}{4} = -\frac{27}{256}$ (minimum). At the right endpoint we have

$f(2) = 2^4 - 3 \cdot 2^3 + 3 \cdot 2^2 - 2 = 2$ (maximum).

49. (a)

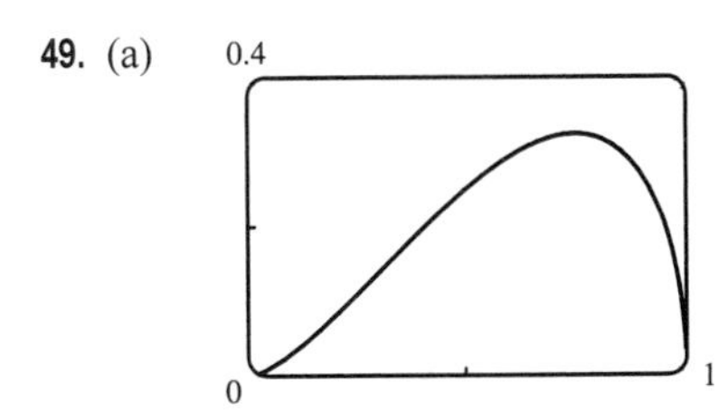

From the graph, it appears that the absolute maximum value is about $f(0.75) = 0.32$, and the absolute minimum value is $f(0) = f(1) = 0$; that is, at both endpoints.

(b) $f(x) = x\sqrt{x - x^2} \quad \Rightarrow \quad f'(x) = x \cdot \dfrac{1 - 2x}{2\sqrt{x - x^2}} + \sqrt{x - x^2} = \dfrac{(x - 2x^2) + (2x - 2x^2)}{2\sqrt{x - x^2}} = \dfrac{3x - 4x^2}{2\sqrt{x - x^2}}$.

So $f'(x) = 0 \quad \Rightarrow \quad 3x - 4x^2 = 0 \quad \Rightarrow \quad x(3 - 4x) = 0 \quad \Rightarrow \quad x = 0$ or $\frac{3}{4}$.

$f(0) = f(1) = 0$ (minimum), and $f\left(\frac{3}{4}\right) = \frac{3}{4}\sqrt{\frac{3}{4} - \left(\frac{3}{4}\right)^2} = \frac{3}{4}\sqrt{\frac{3}{16}} = \frac{3\sqrt{3}}{16}$ (maximum).

50. (a)

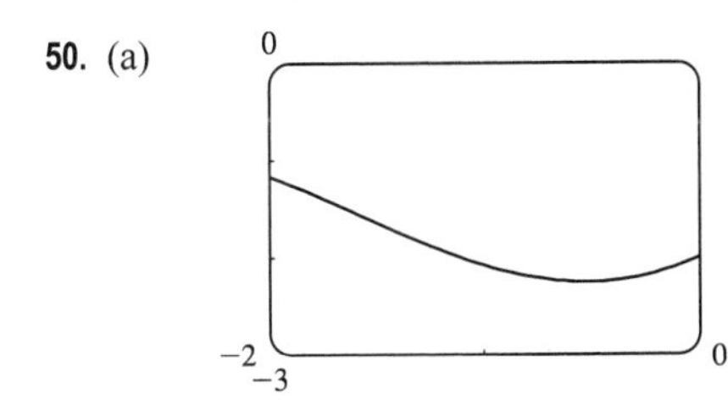

From the graph , it appears that the absolute maximum value is about $f(-2) = -1.17$, and the absolute minimum value is about $f(-0.52) = -2.26$.

(b) $f(x) = x - 2\cos x \quad \Rightarrow \quad f'(x) = 1 + 2\sin x$. So $f'(x) = 0 \quad \Rightarrow \quad \sin x = -\frac{1}{2} \quad \Rightarrow \quad x = -\frac{\pi}{6}$ on $[-2, 0]$.

$f(-2) = -2 - 2\cos(-2)$ (maximum) and $f\left(-\frac{\pi}{6}\right) = -\frac{\pi}{6} - 2\cos\left(-\frac{\pi}{6}\right) = -\frac{\pi}{6} - 2\left(\frac{\sqrt{3}}{2}\right) = -\frac{\pi}{6} - \sqrt{3}$ (minimum).

51. The density is defined as $\rho = \dfrac{\text{mass}}{\text{volume}} = \dfrac{1000}{V(T)}$ (in g/cm^3). But a critical point of ρ will also be a critical point of V

$\left[\text{since } \dfrac{d\rho}{dT} = -1000V^{-2}\dfrac{dV}{dT} \text{ and } V \text{ is never } 0\right]$, and V is easier to differentiate than ρ.

$V(T) = 999.87 - 0.06426T + 0.0085043T^2 - 0.0000679T^3 \quad \Rightarrow \quad V'(T) = -0.06426 + 0.0170086T - 0.0002037T^2$.

Setting this equal to 0 and using the quadratic formula to find T, we get

$T = \dfrac{-0.0170086 \pm \sqrt{0.0170086^2 - 4 \cdot 0.0002037 \cdot 0.06426}}{2(-0.0002037)} \approx 3.9665°\text{C}$ or $79.5318°\text{C}$. Since we are only interested in

the region $0°\text{C} \leq T \leq 30°\text{C}$, we check the density ρ at the endpoints and at 3.9665°C: $\rho(0) \approx \dfrac{1000}{999.87} \approx 1.00013$;

$\rho(30) \approx \dfrac{1000}{1003.7628} \approx 0.99625$; $\rho(3.9665) \approx \dfrac{1000}{999.7447} \approx 1.000255$. So water has its maximum density at

about 3.9665°C.

52. $F = \dfrac{\mu W}{\mu \sin\theta + \cos\theta} \quad\Rightarrow\quad \dfrac{dF}{d\theta} = \dfrac{(\mu\sin\theta + \cos\theta)(0) - \mu W(\mu\cos\theta - \sin\theta)}{(\mu\sin\theta + \cos\theta)^2} = \dfrac{-\mu W(\mu\cos\theta - \sin\theta)}{(\mu\sin\theta+\cos\theta)^2}$.

So $\dfrac{dF}{d\theta} = 0 \;\Rightarrow\; \mu\cos\theta - \sin\theta = 0 \;\Rightarrow\; \mu = \dfrac{\sin\theta}{\cos\theta} = \tan\theta$. Substituting $\tan\theta$ for μ in F gives us

$$F = \frac{(\tan\theta)W}{(\tan\theta)\sin\theta + \cos\theta} = \frac{W\tan\theta}{\dfrac{\sin^2\theta}{\cos\theta} + \cos\theta} = \frac{W\tan\theta\cos\theta}{\sin^2\theta + \cos^2\theta} = \frac{W\sin\theta}{1} = W\sin\theta.$$

If $\tan\theta = \mu$, then $\sin\theta = \dfrac{\mu}{\sqrt{\mu^2+1}}$ (see the figure), so $F = \dfrac{\mu}{\sqrt{\mu^2+1}}W$.

We compare this with the value of F at the endpoints: $F(0) = \mu W$ and $F\left(\frac{\pi}{2}\right) = W$.

Now because $\dfrac{\mu}{\sqrt{\mu^2+1}} \le 1$ and $\dfrac{\mu}{\sqrt{\mu^2+1}} \le \mu$, we have that $\dfrac{\mu}{\sqrt{\mu^2+1}}W$ is less than or equal to each of $F(0)$ and $F\left(\frac{\pi}{2}\right)$.

Hence, $\dfrac{\mu}{\sqrt{\mu^2+1}}W$ is the absolute minimum value of $F(\theta)$, and it occurs when $\tan\theta = \mu$.

53. We apply the Closed Interval Method to the continuous function

$I(t) = 0.00009045t^5 + 0.001438t^4 - 0.06561t^3 + 0.4598t^2 - 0.6270t + 99.33$ on $[0, 10]$. Its derivative is

$I'(t) = 0.00045225t^4 + 0.005752t^3 - 0.19683t^2 + 0.9196t - 0.6270$. Since I' exists for all t, the only critical numbers of I occur when $I'(t) = 0$. We use a rootfinder on a computer algebra system (or a graphing device) to find that $I'(t) = 0$ when $t \approx -29.7186, 0.8231, 5.1309$, or 11.0459, but only the second and third roots lie in the interval $[0, 10]$. The values of I at these critical numbers are $I(0.8231) \approx 99.09$ and $I(5.1309) \approx 100.67$. The values of I at the endpoints of the interval are $I(0) = 99.33$ and $I(10) \approx 96.86$. Comparing these four numbers, we see that food was most expensive at $t \approx 5.1309$ (corresponding roughly to August, 1989) and cheapest at $t = 10$ (midyear 1994).

54. $v(t) = 0.001302t^3 - 0.09029t^2 + 23.61t - 3.083 \;\Rightarrow\; a(t) = v'(t) = 0.003906t^2 - 0.18058t + 23.61 \;\Rightarrow$

$a'(t) = 0.007812t - 0.18058$. $a'(t) = 0 \;\Rightarrow\; t_1 = \frac{0.18058}{0.007812} \approx 23.12$. Evaluating $a(t)$ at the critical number and the endpoints give us $a(0) = 23.61$, $a(t_1) \approx 21.52$, and $a(126) \approx 62.87$. The absolute maximum is about $62.87\ \text{ft/s}^2$ and the absolute minimum is about $21.52\ \text{ft/s}^2$.

55. (a) $v(r) = k(r_0 - r)r^2 = kr_0r^2 - kr^3 \;\Rightarrow\; v'(r) = 2kr_0r - 3kr^2$. $v'(r) = 0 \;\Rightarrow\; kr(2r_0 - 3r) = 0 \;\Rightarrow$

$r = 0$ or $\frac{2}{3}r_0$ (but 0 is not in the interval). Evaluating v at $\frac{1}{2}r_0$, $\frac{2}{3}r_0$, and r_0, we get $v\left(\frac{1}{2}r_0\right) = \frac{1}{8}kr_0^3$, $v\left(\frac{2}{3}r_0\right) = \frac{4}{27}kr_0^3$, and $v(r_0) = 0$. Since $\frac{4}{27} > \frac{1}{8}$, v attains its maximum value at $r = \frac{2}{3}r_0$. This supports the statement in the text.

(b) From part (a), the maximum value of v is $\frac{4}{27}kr_0^3$.

(c)

56. $g(x) = 2 + (x - 5)^3 \quad \Rightarrow \quad g'(x) = 3(x - 5)^2 \quad \Rightarrow \quad g'(5) = 0$, so 5 is a critical number. But $g(5) = 2$ and g takes on values > 2 and values < 2 in any open interval containing 5, so g does not have a local maximum or minimum at 5.

57. $f(x) = x^{101} + x^{51} + x + 1 \quad \Rightarrow \quad f'(x) = 101x^{100} + 51x^{50} + 1 \geq 1$ for all x, so $f'(x) = 0$ has no solution. Thus, $f(x)$ has no critical number, so $f(x)$ can have no local maximum or minimum.

58. Suppose that f has a minimum value at c, so $f(x) \geq f(c)$ for all x near c. Then $g(x) = -f(x) \leq -f(c) = g(c)$ for all x near c, so $g(x)$ has a maximum value at c.

59. If f has a local minimum at c, then $g(x) = -f(x)$ has a local maximum at c, so $g'(c) = 0$ by the case of Fermat's Theorem proved in the text. Thus, $f'(c) = -g'(c) = 0$.

60. (a) $f(x) = ax^3 + bx^2 + cx + d$, $a \neq 0$. So $f'(x) = 3ax^2 + 2bx + c$ is a quadratic and hence has either 2, 1, or 0 real roots, so $f(x)$ has either 2, 1 or 0 critical number(s).

Case (i) [2 critical numbers]:
$f(x) = x^3 - 3x \quad \Rightarrow$
$f'(x) = 3x^2 - 3$, so $x = -1, 1$
are critical numbers.

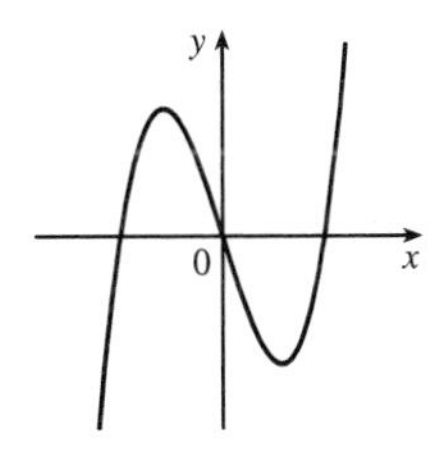

Case (ii) [1 critical number]:
$f(x) = x^3 \quad \Rightarrow$
$f'(x) = 3x^2$, so $x = 0$
is the only critical number.

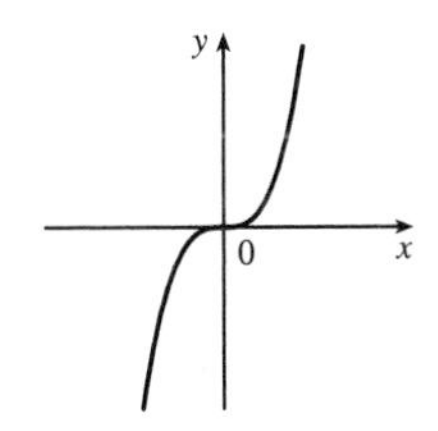

Case (iii) [no critical number]:
$f(x) = x^3 + 3x \quad \Rightarrow$
$f'(x) = 3x^2 + 3$,
so there is no critical number.

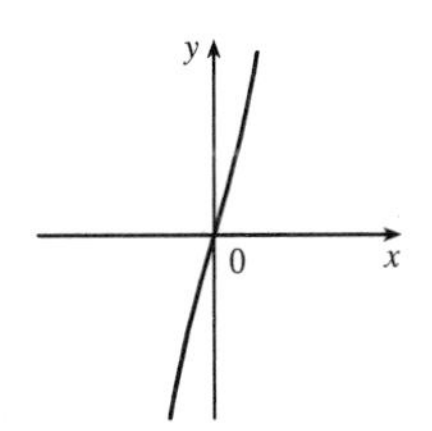

(b) Since there are at most two critical numbers, it can have at most two local extreme values and by (i) this can occur. By (iii) it can have no local extreme value. However, if there is only one critical number, then there is no local extreme value.

3.2 The Mean Value Theorem

1. $f(x) = x^2 - 4x + 1, \quad [0, 4]$. Since f is a polynomial, it is continuous and differentiable on $\mathbb{R}$, so it is continuous on $[0, 4]$ and differentiable on $(0, 4)$. Also, $f(0) = 1 = f(4)$. $f'(c) = 0 \quad \Leftrightarrow \quad 2c - 4 = 0 \quad \Leftrightarrow \quad c = 2$, which is in the open interval $(0, 4)$, so $c = 2$ satisfies the conclusion of Rolle's Theorem.

2. $f(x) = x^3 - 3x^2 + 2x + 5, \quad [0, 2]$. f is continuous on $[0, 2]$ and differentiable on $(0, 2)$. Also, $f(0) = 5 = f(2)$.

$f'(c) = 0 \quad \Leftrightarrow \quad 3c^2 - 6c + 2 = 0 \quad \Leftrightarrow \quad c = \dfrac{6 \pm \sqrt{36 - 24}}{6} = 1 \pm \frac{1}{3}\sqrt{3}$, both in $(0, 2)$.

3. $f(x) = \sin 2\pi x, \quad [-1, 1]$. f, being the composite of the sine function and the polynomial $2\pi x$, is continuous and differentiable on $\mathbb{R}$, so it is continuous on $[-1, 1]$ and differentiable on $(-1, 1)$. Also, $f(-1) = 0 = f(1)$.

$f'(c) = 0 \quad \Leftrightarrow \quad 2\pi \cos 2\pi c = 0 \quad \Leftrightarrow \quad \cos 2\pi c = 0 \quad \Leftrightarrow \quad 2\pi c = \pm\frac{\pi}{2} + 2\pi n \quad \Leftrightarrow \quad c = \pm\frac{1}{4} + n$. If $n = 0$ or ± 1, then $c = \pm\frac{1}{4}, \pm\frac{3}{4}$ is in $(-1, 1)$.

4. $f(x) = x\sqrt{x+6}$, $[-6, 0]$. f is continuous on its domain, $[-6, \infty)$, and differentiable on $(-6, \infty)$, so it is continuous on $[-6, 0]$ and differentiable on $(-6, 0)$. Also, $f(-6) = 0 = f(0)$. $f'(c) = 0 \;\Leftrightarrow\; \dfrac{3c+12}{2\sqrt{c+6}} = 0 \;\Leftrightarrow\; c = -4$, which is in $(-6, 0)$.

5. $f(x) = 1 - x^{2/3}$. $f(-1) = 1 - (-1)^{2/3} = 1 - 1 = 0 = f(1)$. $f'(x) = -\frac{2}{3}x^{-1/3}$, so $f'(c) = 0$ has no solution. This does not contradict Rolle's Theorem, since $f'(0)$ does not exist, and so f is not differentiable on $(-1, 1)$.

6. $f(x) = (x-1)^{-2}$. $f(0) = (0-1)^{-2} = 1 = (2-1)^{-2} = f(2)$. $f'(x) = -2(x-1)^{-3} \;\Rightarrow\; f'(x)$ is never 0. This does not contradict Rolle's Theorem since $f'(1)$ does not exist.

7. $\dfrac{f(8) - f(0)}{8 - 0} = \dfrac{6-4}{8} = \dfrac{1}{4}$. The values of c which satisfy $f'(c) = \frac{1}{4}$ seem to be about $c = 0.8, 3.2, 4.4$, and 6.1.

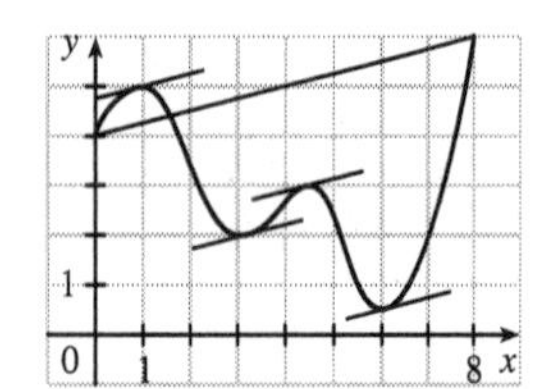

8. $\dfrac{f(7) - f(1)}{7 - 1} = \dfrac{2-5}{6} = -\dfrac{1}{2}$. The values of c which satisfy $f'(c) = -\frac{1}{2}$ seem to be about $c = 1.1, 2.8, 4.6$, and 5.8.

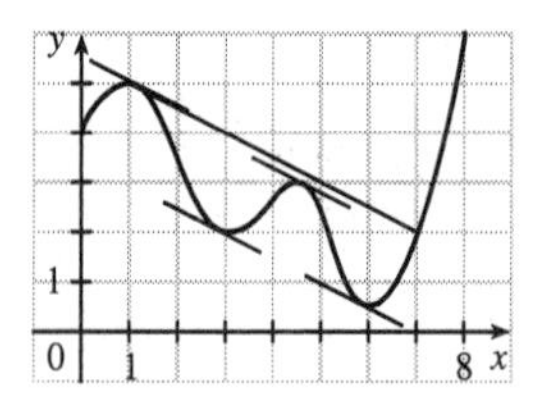

9. (a), (b) The equation of the secant line is

$$y - 5 = \frac{8.5 - 5}{8 - 1}(x - 1) \;\Leftrightarrow\; y = \tfrac{1}{2}x + \tfrac{9}{2}.$$

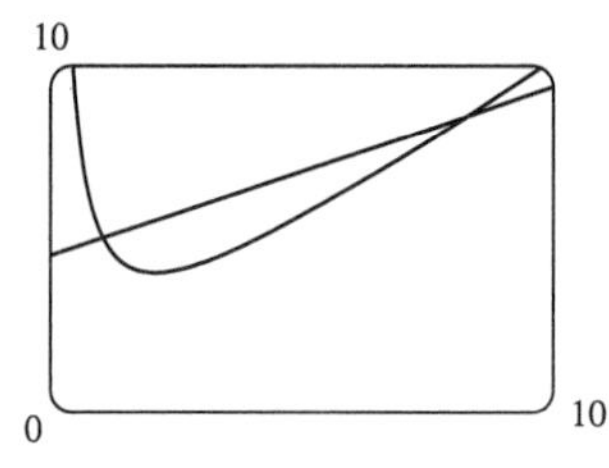

(c) $f(x) = x + 4/x \;\Rightarrow\; f'(x) = 1 - 4/x^2$.
So $f'(c) = \frac{1}{2} \;\Rightarrow\; c^2 = 8 \;\Rightarrow\; c = 2\sqrt{2}$, and $f(c) = 2\sqrt{2} + \frac{4}{2\sqrt{2}} = 3\sqrt{2}$. Thus, an equation of the tangent line is $y - 3\sqrt{2} = \frac{1}{2}(x - 2\sqrt{2}) \;\Leftrightarrow\; y = \frac{1}{2}x + 2\sqrt{2}$.

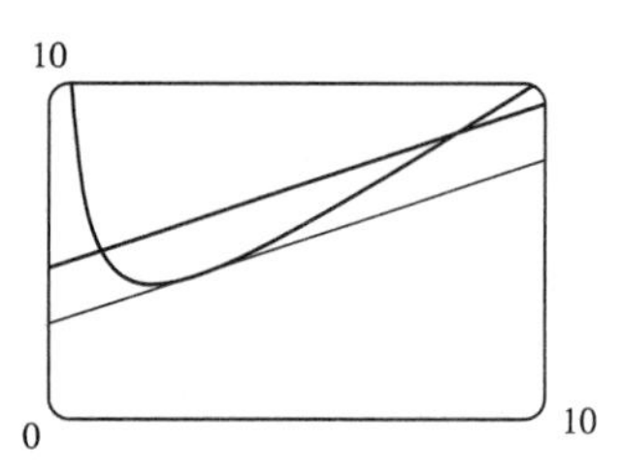

10. (a)

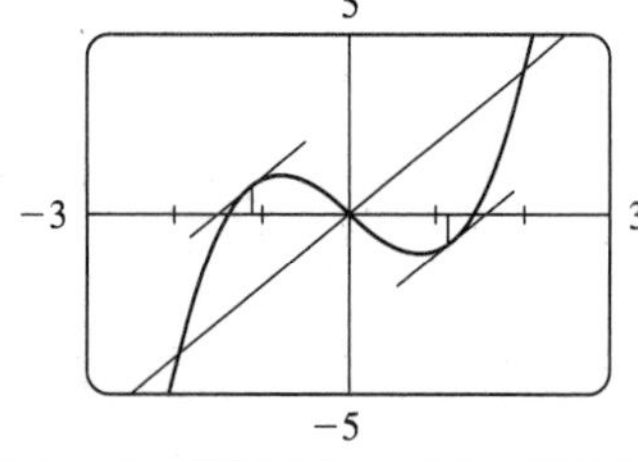

It seems that the tangent lines are parallel to the secant at $x \approx \pm 1.2$.

(b) The slope of the secant line is 2, and its equation is $y = 2x$.
$f(x) = x^3 - 2x \;\Rightarrow\; f'(x) = 3x^2 - 2$, so we solve $f'(c) = 2$ $\Rightarrow\; 3c^2 = 4 \;\Rightarrow\; c = \pm\frac{2\sqrt{3}}{3} \approx 1.155$. Our estimates were off by about 0.045 in each case.

11. $f(x) = 3x^2 + 2x + 5$, $[-1, 1]$. f is continuous on $[-1, 1]$ and differentiable on $(-1, 1)$ since polynomials are continuous and differentiable on $\mathbb{R}$. $f'(c) = \dfrac{f(b) - f(a)}{b - a} \Leftrightarrow 6c + 2 = \dfrac{f(1) - f(-1)}{1 - (-1)} = \dfrac{10 - 6}{2} = 2 \Leftrightarrow 6c = 0 \Leftrightarrow c = 0$, which is in $(-1, 1)$.

12. $f(x) = x^3 + x - 1$, $[0, 2]$. f is continuous on $[0, 2]$ and differentiable on $(0, 2)$. $f'(c) = \dfrac{f(2) - f(0)}{2 - 0} \Leftrightarrow$ $3c^2 + 1 = \dfrac{9 - (-1)}{2} \Leftrightarrow 3c^2 = 5 - 1 \Leftrightarrow c^2 = \frac{4}{3} \Leftrightarrow c = \pm\frac{2}{\sqrt{3}}$, but only $\frac{2}{\sqrt{3}}$ is in $(0, 2)$.

13. $f(x) = \sqrt[3]{x}$, $[0, 1]$. f is continuous on $\mathbb{R}$ and differentiable on $(-\infty, 0) \cup (0, \infty)$, so f is continuous on $[0, 1]$ and differentiable on $(0, 1)$. $f'(c) = \dfrac{f(b) - f(a)}{b - a} \Leftrightarrow \dfrac{1}{3c^{2/3}} = \dfrac{f(1) - f(0)}{1 - 0} \Leftrightarrow \dfrac{1}{3c^{2/3}} = \dfrac{1 - 0}{1} \Leftrightarrow 3c^{2/3} = 1$ $\Leftrightarrow c^{2/3} = \frac{1}{3} \Leftrightarrow c^2 = \left(\frac{1}{3}\right)^3 = \frac{1}{27} \Leftrightarrow c = \pm\sqrt{\frac{1}{27}} = \pm\frac{\sqrt{3}}{9}$, but only $\frac{\sqrt{3}}{9}$ is in $(0, 1)$.

14. $f(x) = \dfrac{x}{x + 2}$, $[1, 4]$. f is continuous on $[1, 4]$ and differentiable on $(1, 4)$. $f'(c) = \dfrac{f(b) - f(a)}{b - a} \Leftrightarrow$ $\dfrac{2}{(c + 2)^2} = \dfrac{\frac{2}{3} - \frac{1}{3}}{4 - 1} \Leftrightarrow (c + 2)^2 = 18 \Leftrightarrow c = -2 \pm 3\sqrt{2}$. $-2 + 3\sqrt{2} \approx 2.24$ is in $(1, 4)$.

15. $f(x) = |x - 1|$. $f(3) - f(0) = |3 - 1| - |0 - 1| = 1$. Since $f'(c) = -1$ if $c < 1$ and $f'(c) = 1$ if $c > 1$, $f'(c)(3 - 0) = \pm 3$ and so is never equal to 1. This does not contradict the Mean Value Theorem since $f'(1)$ does not exist.

16. $f(x) = \dfrac{x + 1}{x - 1}$. $f(2) - f(0) = 3 - (-1) = 4$. $f'(x) = \dfrac{1(x - 1) - 1(x + 1)}{(x - 1)^2} = \dfrac{-2}{(x - 1)^2}$. Since $f'(x) < 0$ for all x (except $x = 1$), $f'(c)(2 - 0)$ is always < 0 and hence cannot equal 4. This does not contradict the Mean Value Theorem since f is not continuous at $x = 1$.

17. Let $f(x) = 1 + 2x + x^3 + 4x^5$. Then $f(-1) = -6 < 0$ and $f(0) = 1 > 0$. Since f is a polynomial, it is continuous, so the Intermediate Value Theorem says that there is a number c between -1 and 0 such that $f(c) = 0$. Thus, the given equation has a real root. Suppose the equation has distinct real roots a and b with $a < b$. Then $f(a) = f(b) = 0$. Since f is a polynomial, it is differentiable on (a, b) and continuous on $[a, b]$. By Rolle's Theorem, there is a number r in (a, b) such that $f'(r) = 0$. But $f'(x) = 2 + 3x^2 + 20x^4 \geq 2$ for all x, so $f'(x)$ can never be 0. This contradiction shows that the equation can't have two distinct real roots. Hence, it has exactly one real root.

18. Let $f(x) = 2x - 1 - \sin x$. Then $f(0) = -1 < 0$ and $f(\pi/2) = \pi - 2 > 0$. f is the sum of the polynomial $2x - 1$ and the scalar multiple $(-1) \cdot \sin x$ of the trigonometric function $\sin x$, so f is continuous (and differentiable) for all x. By the Intermediate Value Theorem, there is a number c in $(0, \pi/2)$ such that $f(c) = 0$. Thus, the given equation has at least one real root. If the equation has distinct real roots a and b with $a < b$, then $f(a) = f(b) = 0$. Since f is continuous on $[a, b]$ and differentiable on (a, b), Rolle's Theorem implies that there is a number r in (a, b) such that $f'(r) = 0$. But $f'(r) = 2 - \cos r > 0$ since $\cos r \leq 1$. This contradiction shows that the given equation can't have two distinct real roots, so it has exactly one real root.

19. Let $f(x) = x^3 - 15x + c$ for x in $[-2, 2]$. If f has two real roots a and b in $[-2, 2]$, with $a < b$, then $f(a) = f(b) = 0$. Since the polynomial f is continuous on $[a, b]$ and differentiable on (a, b), Rolle's Theorem implies that there is a number r in (a, b) such that $f'(r) = 0$. Now $f'(r) = 3r^2 - 15$. Since r is in (a, b), which is contained in $[-2, 2]$, we have $|r| < 2$, so $r^2 < 4$. It follows that $3r^2 - 15 < 3 \cdot 4 - 15 = -3 < 0$. This contradicts $f'(r) = 0$, so the given equation can't have two real roots in $[-2, 2]$. Hence, it has at most one real root in $[-2, 2]$.

20. $f(x) = x^4 + 4x + c$. Suppose that $f(x) = 0$ has three distinct real roots a, b, d where $a < b < d$. Then $f(a) = f(b) = f(d) = 0$. By Rolle's Theorem there are numbers c_1 and c_2 with $a < c_1 < b$ and $b < c_2 < d$ and $0 = f'(c_1) = f'(c_2)$, so $f'(x) = 0$ must have at least two real solutions. However $0 = f'(x) = 4x^3 + 4 = 4(x^3 + 1) = 4(x + 1)(x^2 - x + 1)$ has as its only real solution $x = -1$. Thus, $f(x)$ can have at most two real roots.

21. (a) Suppose that a cubic polynomial $P(x)$ has roots $a_1 < a_2 < a_3 < a_4$, so $P(a_1) = P(a_2) = P(a_3) = P(a_4)$. By Rolle's Theorem there are numbers c_1, c_2, c_3 with $a_1 < c_1 < a_2$, $a_2 < c_2 < a_3$ and $a_3 < c_3 < a_4$ and $P'(c_1) = P'(c_2) = P'(c_3) = 0$. Thus, the second-degree polynomial $P'(x)$ has three distinct real roots, which is impossible.

(b) We prove by induction that a polynomial of degree n has at most n real roots. This is certainly true for $n = 1$. Suppose that the result is true for all polynomials of degree n and let $P(x)$ be a polynomial of degree $n + 1$. Suppose that $P(x)$ has more than $n + 1$ real roots, say $a_1 < a_2 < a_3 < \cdots < a_{n+1} < a_{n+2}$. Then $P(a_1) = P(a_2) = \cdots = P(a_{n+2}) = 0$. By Rolle's Theorem there are real numbers $c_1, \ldots, c_{n+1}$ with $a_1 < c_1 < a_2, \ldots, a_{n+1} < c_{n+1} < a_{n+2}$ and $P'(c_1) = \cdots = P'(c_{n+1}) = 0$. Thus, the nth degree polynomial $P'(x)$ has at least $n + 1$ roots. This contradiction shows that $P(x)$ has at most $n + 1$ real roots.

22. (a) Suppose that $f(a) = f(b) = 0$ where $a < b$. By Rolle's Theorem applied to f on $[a, b]$ there is a number c such that $a < c < b$ and $f'(c) = 0$.

(b) Suppose that $f(a) = f(b) = f(c) = 0$ where $a < b < c$. By Rolle's Theorem applied to $f(x)$ on $[a, b]$ and $[b, c]$ there are numbers $a < d < b$ and $b < e < c$ with $f'(d) = 0$ and $f'(e) = 0$. By Rolle's Theorem applied to $f'(x)$ on $[d, e]$ there is a number g with $d < g < e$ such that $f''(g) = 0$.

(c) Suppose that f is n times differentiable on $\mathbb{R}$ and has $n + 1$ distinct real roots. Then $f^{(n)}$ has at least one real root.

23. By the Mean Value Theorem, $f(4) - f(1) = f'(c)(4 - 1)$ for some $c \in (1, 4)$. But for every $c \in (1, 4)$ we have $f'(c) \geq 2$. Putting $f'(c) \geq 2$ into the above equation and substituting $f(1) = 10$, we get $f(4) = f(1) + f'(c)(4 - 1) = 10 + 3f'(c) \geq 10 + 3 \cdot 2 = 16$. So the smallest possible value of $f(4)$ is 16.

24. If $3 \leq f'(x) \leq 5$ for all x, then by the Mean Value Theorem, $f(8) - f(2) = f'(c) \cdot (8 - 2)$ for some c in $[2, 8]$. (f is differentiable for all x, so, in particular, f is differentiable on $(2, 8)$ and continuous on $[2, 8]$. Thus, the hypotheses of the Mean Value Theorem are satisfied.) Since $f(8) - f(2) = 6f'(c)$ and $3 \leq f'(c) \leq 5$, it follows that $6 \cdot 3 \leq 6f'(c) \leq 6 \cdot 5 \;\Rightarrow\; 18 \leq f(8) - f(2) \leq 30$.

25. Suppose that such a function f exists. By the Mean Value Theorem there is a number $0 < c < 2$ with

$f'(c) = \dfrac{f(2) - f(0)}{2 - 0} = \dfrac{5}{2}$. But this is impossible since $f'(x) \leq 2 < \frac{5}{2}$ for all x, so no such function can exist.

26. Let $h = f - g$. Then since f and g are continuous on $[a, b]$ and differentiable on (a, b), so is h, and thus h satisfies the assumptions of the Mean Value Theorem. Therefore, there is a number c with $a < c < b$ such that

$h(b) = h(b) - h(a) = h'(c)(b - a)$. Since $h'(c) < 0$, $h'(c)(b - a) < 0$, so $f(b) - g(b) = h(b) < 0$ and hence $f(b) < g(b)$.

27. We use Exercise 26 with $f(x) = \sqrt{1 + x}$, $g(x) = 1 + \frac{1}{2}x$, and $a = 0$. Notice that $f(0) = 1 = g(0)$ and

$f'(x) = \dfrac{1}{2\sqrt{1 + x}} < \dfrac{1}{2} = g'(x)$ for $x > 0$. So by Exercise 26, $f(b) < g(b) \Rightarrow \sqrt{1 + b} < 1 + \frac{1}{2}b$ for $b > 0$.

Another method: Apply the Mean Value Theorem directly to either $f(x) = 1 + \frac{1}{2}x - \sqrt{1 + x}$ or $g(x) = \sqrt{1 + x}$ on $[0, b]$.

28. f satisfies the conditions for the Mean Value Theorem, so we use this theorem on the interval $[-b, b]$: $\dfrac{f(b) - f(-b)}{b - (-b)} = f'(c)$

for some $c \in (-b, b)$. But since f is odd, $f(-b) = -f(b)$. Substituting this into the above equation, we get

$\dfrac{f(b) + f(b)}{2b} = f'(c) \Rightarrow \dfrac{f(b)}{b} = f'(c)$.

29. Let $f(x) = \sin x$ and let $b < a$. Then $f(x)$ is continuous on $[b, a]$ and differentiable on (b, a). By the Mean Value Theorem, there is a number $c \in (b, a)$ with $\sin a - \sin b = f(a) - f(b) = f'(c)(a - b) = (\cos c)(a - b)$. Thus,

$|\sin a - \sin b| \leq |\cos c|\,|b - a| \leq |a - b|$. If $a < b$, then $|\sin a - \sin b| = |\sin b - \sin a| \leq |b - a| = |a - b|$.

If $a = b$, both sides of the inequality are 0.

30. Suppose that $f'(x) = c$. Let $g(x) = cx$, so $g'(x) = c$. Then, by Corollary 7, $f(x) = g(x) + d$, where d is a constant, so $f(x) = cx + d$.

31. For $x > 0$, $f(x) = g(x)$, so $f'(x) = g'(x)$. For $x < 0$, $f'(x) = (1/x)' = -1/x^2$ and $g'(x) = (1 + 1/x)' = -1/x^2$, so again $f'(x) = g'(x)$. However, the domain of $g(x)$ is not an interval [it is $(-\infty, 0) \cup (0, \infty)$] so we cannot conclude that $f - g$ is constant (in fact it is not).

32. Let $v(t)$ be the velocity of the car t hours after 2:00 PM. Then $\dfrac{v(1/6) - v(0)}{1/6 - 0} = \dfrac{50 - 30}{1/6} = 120$. By the Mean Value Theorem, there is a number c such that $0 < c < \frac{1}{6}$ with $v'(c) = 120$. Since $v'(t)$ is the acceleration at time t, the acceleration c hours after 2:00 PM is exactly $120\ \text{mi/h}^2$.

33. Let $g(t)$ and $h(t)$ be the position functions of the two runners and let $f(t) = g(t) - h(t)$. By hypothesis, $f(0) = g(0) - h(0) = 0$ and $f(b) = g(b) - h(b) = 0$, where b is the finishing time. Then by the Mean Value Theorem, there is a time c, with $0 < c < b$, such that $f'(c) = \dfrac{f(b) - f(0)}{b - 0}$. But $f(b) = f(0) = 0$, so $f'(c) = 0$. Since $f'(c) = g'(c) - h'(c) = 0$, we have $g'(c) = h'(c)$. So at time c, both runners have the same speed $g'(c) = h'(c)$.

34. Assume that f is differentiable (and hence continuous) on $\mathbb{R}$ and that $f'(x) \neq 1$ for all x. Suppose f has more than one fixed point. Then there are numbers a and b such that $a < b$, $f(a) = a$, and $f(b) = b$. Applying the Mean Value Theorem to the function f on $[a, b]$, we find that there is a number c in (a, b) such that $f'(c) = \dfrac{f(b) - f(a)}{b - a}$. But then $f'(c) = \dfrac{b - a}{b - a} = 1$, contradicting our assumption that $f'(x) \neq 1$ for every real number x. This shows that our supposition was wrong, that is, that f cannot have more than one fixed point.

3.3 Derivatives and the Shapes of Graphs

Abbreviations: CU = Concave upward, CD = Concave downward, IP = Inflection point, HA = Horizontal asymptote, VA = Vertical asymptote.

1. (a) $f(x) = x^3 - 12x + 1 \quad \Rightarrow \quad f'(x) = 3x^2 - 12 = 3(x + 2)(x - 2)$.

We don't need to include "3" in the chart to determine the sign of $f'(x)$.

Interval	$x + 2$	$x - 2$	$f'(x)$	f
$x < -2$	$-$	$-$	$+$	increasing on $(-\infty, -2)$
$-2 < x < 2$	$+$	$-$	$-$	decreasing on $(-2, 2)$
$x > 2$	$+$	$+$	$+$	increasing on $(2, \infty)$

So f is increasing on $(-\infty, -2)$ and $(2, \infty)$ and f is decreasing on $(-2, 2)$.

(b) f changes from increasing to decreasing at $x = -2$ and from decreasing to increasing at $x = 2$. Thus, $f(-2) = 17$ is a local maximum value and $f(2) = -15$ is a local minimum value.

(c) $f''(x) = 6x$. $f''(x) > 0 \quad \Leftrightarrow \quad x > 0$ and $f''(x) < 0 \quad \Leftrightarrow \quad x < 0$. Thus, f is CU on $(0, \infty)$ and CD on $(-\infty, 0)$. There is an IP where the concavity changes, at $(0, f(0)) = (0, 1)$.

2. (a) $f(x) = x^4 - 4x - 1 \quad \Rightarrow \quad f'(x) = 4x^3 - 4 = 4(x^3 - 1) = 4(x - 1)(x^2 + x + 1)$. So $f'(x) > 0 \quad \Leftrightarrow$ $x - 1 > 0 \quad [4(x^2 + x + 1) > 0] \quad \Leftrightarrow \quad x > 1$. Thus, f is increasing on $(1, \infty)$ and decreasing on $(-\infty, 1)$.

(b) f changes from decreasing to increasing at its only critical number, $x = 1$. Thus, $f(1) = -4$ is a local minimum value.

(c) $f'(x) = 4x^3 - 4 \quad \Rightarrow \quad f''(x) = 12x^2$. $f''(x) > 0$ for all x except $x = 0$. Thus, f is CU on $(-\infty, 0)$ and $(0, \infty)$. Moreover, since f' is increasing on $(-\infty, \infty)$, f is CU on $(-\infty, \infty)$. There are no IPs.

3. (a) $f(x) = x^4 - 2x^2 + 3 \quad \Rightarrow \quad f'(x) = 4x^3 - 4x = 4x(x^2 - 1) = 4x(x + 1)(x - 1)$.

Interval	$x + 1$	x	$x - 1$	$f'(x)$	f
$x < -1$	$-$	$-$	$-$	$-$	decreasing on $(-\infty, -1)$
$-1 < x < 0$	$+$	$-$	$-$	$+$	increasing on $(-1, 0)$
$0 < x < 1$	$+$	$+$	$-$	$-$	decreasing on $(0, 1)$
$x > 1$	$+$	$+$	$+$	$+$	increasing on $(1, \infty)$

So f is increasing on $(-1, 0)$ and $(1, \infty)$ and f is decreasing on $(-\infty, -1)$ and $(0, 1)$.

(b) f changes from increasing to decreasing at $x = 0$ and from decreasing to increasing at $x = -1$ and $x = 1$. Thus, $f(0) = 3$ is a local maximum value and $f(\pm 1) = 2$ are local minimum values.

(c) $f''(x) = 12x^2 - 4 = 12\left(x^2 - \frac{1}{3}\right) = 12\left(x + 1/\sqrt{3}\right)\left(x - 1/\sqrt{3}\right)$. $f''(x) > 0 \quad \Leftrightarrow \quad x < -1/\sqrt{3}$ or $x > 1/\sqrt{3}$ and $f''(x) < 0 \quad \Leftrightarrow \quad -1/\sqrt{3} < x < 1/\sqrt{3}$. Thus, f is concave upward on $\left(-\infty, -\sqrt{3}/3\right)$ and $\left(\sqrt{3}/3, \infty\right)$ and concave downward on $\left(-\sqrt{3}/3, \sqrt{3}/3\right)$. There are inflection points at $\left(\pm\sqrt{3}/3, \frac{22}{9}\right)$.

4. (a) $f(x) = \dfrac{x^2}{x^2 + 3} \quad \Rightarrow \quad f'(x) = \dfrac{(x^2 + 3)(2x) - x^2(2x)}{(x^2 + 3)^2} = \dfrac{6x}{(x^2 + 3)^2}$. The denominator is positive so the sign of $f'(x)$ is determined by the sign of x. Thus, $f'(x) > 0 \quad \Leftrightarrow \quad x > 0$ and $f'(x) < 0 \quad \Leftrightarrow \quad x < 0$. So f is increasing on $(0, \infty)$ and f is decreasing on $(-\infty, 0)$.

(b) f changes from decreasing to increasing at $x = 0$. Thus, $f(0) = 0$ is a local minimum value.

(c) $f''(x) = \dfrac{(x^2 + 3)^2(6) - 6x \cdot 2(x^2 + 3)(2x)}{[(x^2 + 3)^2]^2} = \dfrac{6(x^2 + 3)\left[x^2 + 3 - 4x^2\right]}{(x^2 + 3)^4} = \dfrac{6(3 - 3x^2)}{(x^2 + 3)^3} = \dfrac{-18(x + 1)(x - 1)}{(x^2 + 3)^3}$.

$f''(x) > 0 \quad \Leftrightarrow \quad -1 < x < 1$ and $f''(x) < 0 \quad \Leftrightarrow \quad x < -1$ or $x > 1$. Thus, f is CU on $(-1, 1)$ and CD on $(-\infty, -1)$ and $(1, \infty)$. There are IPs at $\left(\pm 1, \frac{1}{4}\right)$.

5. (a) $f(x) = x - 2\sin x$ on $(0, 3\pi) \quad \Rightarrow \quad f'(x) = 1 - 2\cos x$. $f'(x) > 0 \quad \Leftrightarrow \quad 1 - 2\cos x > 0 \quad \Leftrightarrow \quad \cos x < \frac{1}{2} \quad \Leftrightarrow$ $\frac{\pi}{3} < x < \frac{5\pi}{3}$ or $\frac{7\pi}{3} < x < 3\pi$. $f'(x) < 0 \quad \Leftrightarrow \quad \cos x > \frac{1}{2} \quad \Leftrightarrow \quad 0 < x < \frac{\pi}{3}$ or $\frac{5\pi}{3} < x < \frac{7\pi}{3}$. So f is increasing on $\left(\frac{\pi}{3}, \frac{5\pi}{3}\right)$ and $\left(\frac{7\pi}{3}, 3\pi\right)$, and f is decreasing on $\left(0, \frac{\pi}{3}\right)$ and $\left(\frac{5\pi}{3}, \frac{7\pi}{3}\right)$.

(b) f changes from increasing to decreasing at $x = \frac{5\pi}{3}$, and from decreasing to increasing at $x = \frac{\pi}{3}$ and at $x = \frac{7\pi}{3}$. Thus, $f\left(\frac{5\pi}{3}\right) = \frac{5\pi}{3} + \sqrt{3} \approx 6.97$ is a local maximum value and $f\left(\frac{\pi}{3}\right) = \frac{\pi}{3} - \sqrt{3} \approx -0.68$ and $f\left(\frac{7\pi}{3}\right) = \frac{7\pi}{3} - \sqrt{3} \approx 5.60$ are local minimum values.

(c) $f''(x) = 2\sin x > 0 \quad \Leftrightarrow \quad 0 < x < \pi$ and $2\pi < x < 3\pi$, $f''(x) < 0 \quad \Leftrightarrow \quad \pi < x < 2\pi$. Thus, f is CU on $(0, \pi)$ and $(2\pi, 3\pi)$, and f is CD on $(\pi, 2\pi)$. There are IPs at (π, π) and $(2\pi, 2\pi)$.

6. (a) $f(x) = \cos^2 x - 2\sin x$, $0 \leq x \leq 2\pi$. $f'(x) = -2\cos x\sin x - 2\cos x = -2\cos x\,(1 + \sin x)$. Note that $1 + \sin x \geq 0$ [since $\sin x \geq -1$], with equality $\Leftrightarrow \quad \sin x = -1 \quad \Leftrightarrow \quad x = 3\pi/2$ [since $0 \leq x \leq 2\pi$] $\Rightarrow$ $\cos x = 0$. Thus, $f'(x) > 0 \quad \Leftrightarrow \quad \cos x < 0 \quad \Leftrightarrow \quad \pi/2 < x < 3\pi/2$ and $f'(x) < 0 \quad \Leftrightarrow \quad \cos x > 0 \quad \Leftrightarrow$ $0 < x < \pi/2$ or $3\pi/2 < x < 2\pi$. Thus, f is increasing on $(\pi/2, 3\pi/2)$ and f is decreasing on $(0, \pi/2)$ and $(3\pi/2, 2\pi)$.

(b) f changes from decreasing to increasing at $x = \pi/2$ and from increasing to decreasing at $x = 3\pi/2$. Thus, $f(\pi/2) = -2$ is a local minimum value and $f(3\pi/2) = 2$ is a local maximum value.

(c)
$$\begin{aligned} f''(x) &= 2\sin x\,(1 + \sin x) - 2\cos^2 x = 2\sin x + 2\sin^2 x - 2\left(1 - \sin^2 x\right) \\ &= 4\sin^2 x + 2\sin x - 2 = 2(2\sin x - 1)(\sin x + 1) \end{aligned}$$

so $f''(x) > 0 \quad \Leftrightarrow \quad \sin x > \frac{1}{2} \quad \Leftrightarrow \quad \frac{\pi}{6} < x < \frac{5\pi}{6}$, and $f''(x) < 0 \quad \Leftrightarrow \quad \sin x < \frac{1}{2}$ and $\sin x \neq -1 \quad \Leftrightarrow \quad 0 < x < \frac{\pi}{6}$ or $\frac{5\pi}{6} < x < \frac{3\pi}{2}$ or $\frac{3\pi}{2} < x < 2\pi$. Thus, f is concave upward on $\left(\frac{\pi}{6}, \frac{5\pi}{6}\right)$ and concave downward on $\left(0, \frac{\pi}{6}\right)$, $\left(\frac{5\pi}{6}, \frac{3\pi}{2}\right)$, and $\left(\frac{3\pi}{2}, 2\pi\right)$. There are inflection points at $\left(\frac{\pi}{6}, -\frac{1}{4}\right)$ and $\left(\frac{5\pi}{6}, -\frac{1}{4}\right)$.

7. $f(x) = x + \sqrt{1-x} \Rightarrow f'(x) = 1 + \frac{1}{2}(1-x)^{-1/2}(-1) = 1 - \dfrac{1}{2\sqrt{1-x}}$. Note that f is defined for $1 - x \geq 0$; that is, for $x \leq 1$. $f'(x) = 0 \Rightarrow 2\sqrt{1-x} = 1 \Rightarrow \sqrt{1-x} = \frac{1}{2} \Rightarrow 1 - x = \frac{1}{4} \Rightarrow x = \frac{3}{4}$. f' does not exist at $x = 1$, but we can't have a local maximum or minimum at an endpoint.

First Derivative Test: $f'(x) > 0 \Rightarrow x < \frac{3}{4}$ and $f'(x) < 0 \Rightarrow \frac{3}{4} < x < 1$. Since f' changes from positive to negative at $x = \frac{3}{4}$, $f\left(\frac{3}{4}\right) = \frac{5}{4}$ is a local maximum value.

Second Derivative Test: $f''(x) = -\frac{1}{2}\left(-\frac{1}{2}\right)(1-x)^{-3/2}(-1) = -\dfrac{1}{4\left(\sqrt{1-x}\right)^3}$.

$f''\left(\frac{3}{4}\right) = -2 < 0 \Rightarrow f\left(\frac{3}{4}\right) = \frac{5}{4}$ is a local maximum value.

Preference: The First Derivative Test may be slightly easier to apply in this case.

8. $f(x) = \dfrac{x}{x^2+4} \Rightarrow f'(x) = \dfrac{(x^2+4)\cdot 1 - x(2x)}{(x^2+4)^2} = \dfrac{4-x^2}{(x^2+4)^2} = \dfrac{(2+x)(2-x)}{(x^2+4)^2}$.

First Derivative Test: $f'(x) > 0 \Rightarrow -2 < x < 2$ and $f'(x) < 0 \Rightarrow x > 2$ or $x < -2$. Since f' changes from positive to negative at $x = 2$, $f(2) = \frac{1}{4}$ is a local maximum value; and since f' changes from negative to positive at $x = -2$, $f(-2) = -\frac{1}{4}$ is a local minimum value.

Second Derivative Test:

$$f''(x) = \frac{(x^2+4)^2(-2x) - (4-x^2)\cdot 2(x^2+4)(2x)}{[(x^2+4)^2]^2} = \frac{-2x(x^2+4)\left[(x^2+4) + 2(4-x^2)\right]}{(x^2+4)^4} = \frac{-2x(12-x^2)}{(x^2+4)^3}$$

$f'(x) = 0 \Leftrightarrow x = \pm 2$. $f''(-2) = \frac{1}{16} > 0 \Rightarrow f(-2) = -\frac{1}{4}$ is a local minimum value.

$f''(2) = -\frac{1}{16} < 0 \Rightarrow f(2) = \frac{1}{4}$ is a local maximum value.

Preference: Since calculating the second derivative is fairly difficult, the First Derivative Test is easier to use for this function.

9. (a) By the Second Derivative Test, if $f'(2) = 0$ and $f''(2) = -5 < 0$, f has a local maximum at $x = 2$.

(b) If $f'(6) = 0$, we know that f has a horizontal tangent at $x = 6$. Knowing that $f''(6) = 0$ does not provide any additional information since the Second Derivative Test fails. For example, the first and second derivatives of $y = (x-6)^4$, $y = -(x-6)^4$, and $y = (x-6)^3$ all equal zero for $x = 6$, but the first has a local minimum at $x = 6$, the second has a local maximum at $x = 6$, and the third has an inflection point at $x = 6$.

10. (a) $f(x) = x^4(x-1)^3 \Rightarrow f'(x) = x^4 \cdot 3(x-1)^2 + (x-1)^3 \cdot 4x^3 = x^3(x-1)^2\,[3x + 4(x-1)] = x^3(x-1)^2(7x-4)$

The critical numbers are 0, 1, and $\frac{4}{7}$.

(b)
$$\begin{aligned} f''(x) &= 3x^2(x-1)^2(7x-4) + x^3 \cdot 2(x-1)(7x-4) + x^3(x-1)^2 \cdot 7 \\ &= x^2(x-1)\,[3(x-1)(7x-4) + 2x(7x-4) + 7x(x-1)] \end{aligned}$$

Now $f''(0) = f''(1) = 0$, so the Second Derivative Test gives no information for $x = 0$ or $x = 1$.

$f''\left(\frac{4}{7}\right) = \left(\frac{4}{7}\right)^2\left(\frac{4}{7} - 1\right)\left[0 + 0 + 7\left(\frac{4}{7}\right)\left(\frac{4}{7} - 1\right)\right] = \left(\frac{4}{7}\right)^2\left(-\frac{3}{7}\right)(4)\left(-\frac{3}{7}\right) > 0$, so there is a local minimum at $x = \frac{4}{7}$.

(c) f' is positive on $(-\infty, 0)$, negative on $\left(0, \frac{4}{7}\right)$, positive on $\left(\frac{4}{7}, 1\right)$, and positive on $(1, \infty)$. So f has a local maximum at $x = 0$, a local minimum at $x = \frac{4}{7}$, and no local maximum or minimum at $x = 1$.

11. (a) There is an IP at $x = 3$ because the graph of f changes from CD to CU there. There is an IP at $x = 5$ because the graph of f changes from CU to CD there.

(b) There is an IP at $x = 2$ and at $x = 6$ because $f'(x)$ has a maximum value there, and so $f''(x)$ changes from positive to negative there. There is an IP at $x = 4$ because $f'(x)$ has a minimum value there and so $f''(x)$ changes from negative to positive there.

(c) There is an inflection point at $x = 1$ because $f''(x)$ changes from negative to positive there, and so the graph of f changes from CD to CU. There is an IP at $x = 7$ because $f''(x)$ changes from positive to negative there, and so the graph of f changes from CU to CD.

12. (a) f is increasing on the intervals where $f'(x) > 0$, namely, $(2, 4)$ and $(6, 9)$.

(b) f has a local maximum where it changes from increasing to decreasing, that is, where f' changes from positive to negative (at $x = 4$). Similarly, where f' changes from negative to positive, f has a local minimum (at $x = 2$ and at $x = 6$).

(c) When f' is increasing, its derivative f'' is positive and hence, f is CU. This happens on $(1, 3)$, $(5, 7)$, and $(8, 9)$. Similarly, f is CD when f' is decreasing—that is, on $(0, 1)$, $(3, 5)$, and $(7, 8)$.

(d) f has IPs at $x = 1, 3, 5, 7$, and 8, since the direction of concavity changes at each of these values.

13. The function must be always decreasing (since the first derivative is always negative) and concave downward (since the second derivative is always negative).

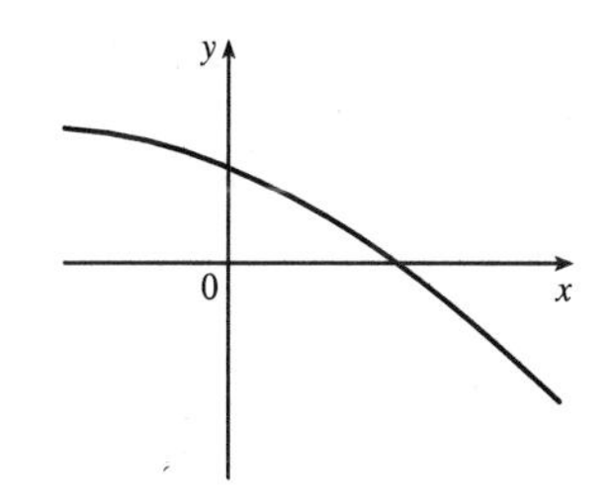

14. $f'(x) > 0$ for all $x \neq 1$ with vertical asymptote $x = 1$, so f is increasing on $(-\infty, 1)$ and $(1, \infty)$. $f''(x) > 0$ if $x < 1$ or $x > 3$, and $f''(x) < 0$ if $1 < x < 3$, so f is CU on $(-\infty, 1)$ and $(3, \infty)$, and CD on $(1, 3)$. There is an IP when $x = 3$.

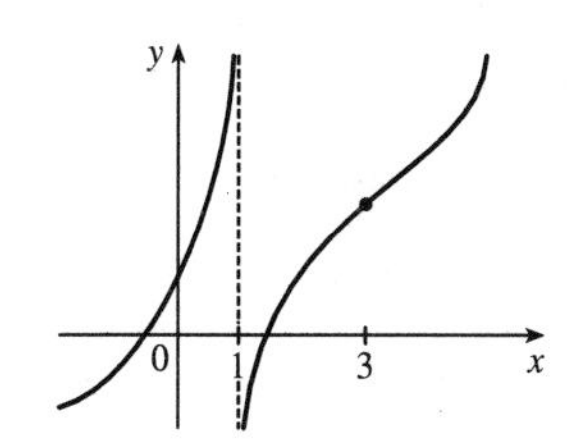

15. $f'(0) = f'(2) = f'(4) = 0 \;\Rightarrow\;$ horizontal tangents at $x = 0, 2, 4$.
$f'(x) > 0$ if $x < 0$ or $2 < x < 4 \;\Rightarrow\; f$ is increasing on $(-\infty, 0)$ and $(2, 4)$.
$f'(x) < 0$ if $0 < x < 2$ or $x > 4 \;\Rightarrow\; f$ is decreasing on $(0, 2)$ and $(4, \infty)$.
$f''(x) > 0$ if $1 < x < 3 \;\Rightarrow\; f$ is CU on $(1, 3)$. $f''(x) < 0$ if $x < 1$ or $x > 3 \;\Rightarrow\; f$ is CD on $(-\infty, 1)$ and $(3, \infty)$. There are IPs when $x = 1$ and 3.

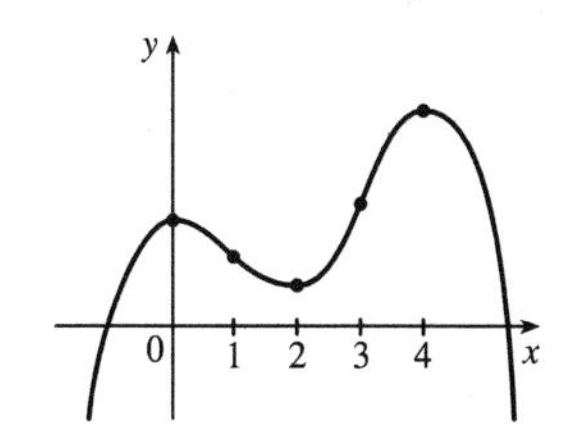

16. $f'(1) = f'(-1) = 0 \;\Rightarrow\;$ horizontal tangents at $x = \pm 1$. $f'(x) < 0$ if $|x| < 1 \;\Rightarrow\; f$ is decreasing on $(-1, 1)$. $f'(x) > 0$ if $1 < |x| < 2 \;\Rightarrow\; f$ is increasing on $(-2, -1)$ and $(1, 2)$. $f'(x) = -1$ if $|x| > 2 \;\Rightarrow\;$ the graph of f has constant slope -1 on $(-\infty, -2)$ and $(2, \infty)$.
$f''(x) < 0$ if $-2 < x < 0 \;\Rightarrow\; f$ is CD on $(-2, 0)$. The point $(0, 1)$ is an IP.

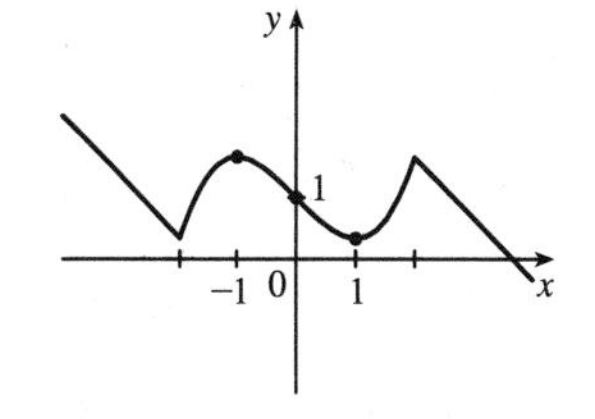

17. $f'(x) > 0$ if $|x| < 2 \;\Rightarrow\; f$ is increasing on $(-2, 2)$. $f'(x) < 0$ if $|x| > 2 \;\Rightarrow\; f$ is decreasing on $(-\infty, -2)$ and $(2, \infty)$. $f'(-2) = 0 \;\Rightarrow\;$ horizontal tangent at $x = -2$. $\lim\limits_{x \to 2} |f'(x)| = \infty \;\Rightarrow\;$ there is a vertical asymptote or vertical tangent (cusp) at $x = 2$. $f''(x) > 0$ if $x \neq 2 \;\Rightarrow\; f$ is CU on $(-\infty, 2)$ and $(2, \infty)$.

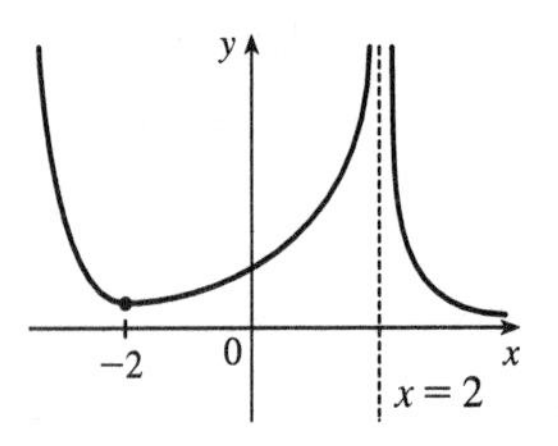

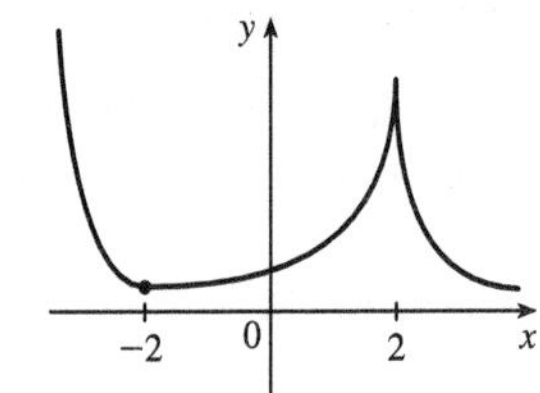

18. $f'(x) > 0$ if $|x| < 2 \;\Rightarrow\; f$ is increasing on $(-2, 2)$. $f'(x) < 0$ if $|x| > 2 \;\Rightarrow\; f$ is decreasing on $(-\infty, -2)$ and $(2, \infty)$. $f'(2) = 0$, so f has a horizontal tangent (and local maximum) at $x = 2$. $\lim\limits_{x \to \infty} f(x) = 1 \;\Rightarrow\; y = 1$ is a horizontal asymptote. $f(-x) = -f(x) \;\Rightarrow\; f$ is an odd function (its graph is symmetric about the origin). Finally, $f''(x) < 0$ if $0 < x < 3$ and $f''(x) > 0$ if $x > 3$, so f is CD on $(0, 3)$ and CU on $(3, \infty)$.

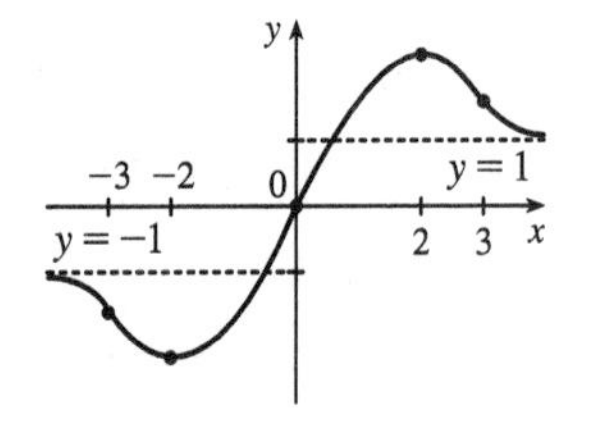

19. (a) f is increasing where f' is positive, that is, on $(0, 2)$, $(4, 6)$, and $(8, \infty)$; and decreasing where f' is negative, that is, on $(2, 4)$ and $(6, 8)$.

(b) f has local maxima where f' changes from positive to negative, at $x = 2$ and at $x = 6$, and local minima where f' changes from negative to positive, at $x = 4$ and at $x = 8$.

(c) f is CU where f' is increasing, that is, on $(3, 6)$ and $(6, \infty)$, and CD where f' is decreasing, that is, on $(0, 3)$.

(d) There is an IP where f changes from being CD to being CU, that is, at $x = 3$.

(e)

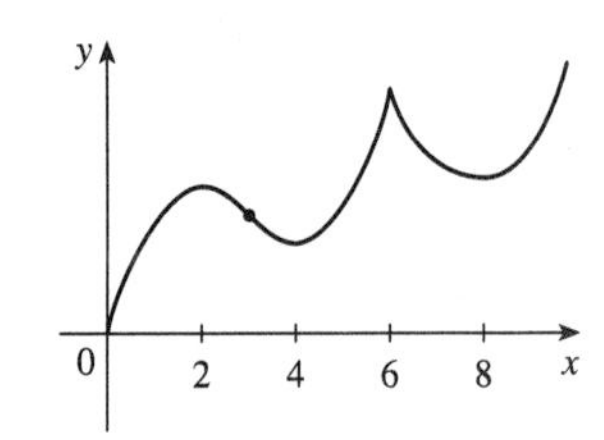

20. (a) f is increasing where f' is positive, on $(1, 6)$ and $(8, \infty)$, and decreasing where f' is negative, on $(0, 1)$ and $(6, 8)$.

(b) f has a local maximum where f' changes from positive to negative, at $x = 6$, and local minima where f' changes from negative to positive, at $x = 1$ and at $x = 8$.

(c) f is CU where f' is increasing, that is, on $(0, 2)$, $(3, 5)$, and $(7, \infty)$, and CD where f' is decreasing, that is, on $(2, 3)$ and $(5, 7)$.

(d) There are IPs where f changes its direction of concavity, at $x = 2$, $x = 3$, $x = 5$ and $x = 7$.

(e)

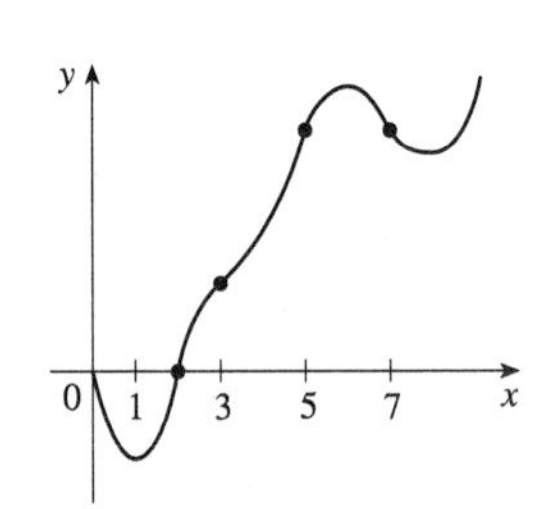

21. (a) $f(x) = 2x^3 - 3x^2 - 12x \;\Rightarrow\; f'(x) = 6x^2 - 6x - 12 = 6(x^2 - x - 2) = 6(x - 2)(x + 1)$.

$f'(x) > 0 \;\Leftrightarrow\; x < -1$ or $x > 2$ and $f'(x) < 0 \;\Leftrightarrow\; -1 < x < 2$.

So f is increasing on $(-\infty, -1)$ and $(2, \infty)$, and f is decreasing on $(-1, 2)$.

(b) Since f changes from increasing to decreasing at $x = -1$, $f(-1) = 7$ is a local maximum value. Since f changes from decreasing to increasing at $x = 2$, $f(2) = -20$ is a local minimum value.

(c) $f''(x) = 6(2x - 1) \Rightarrow f''(x) > 0$ on $\left(\frac{1}{2}, \infty\right)$ and $f''(x) < 0$ on $\left(-\infty, \frac{1}{2}\right)$. So f is CU on $\left(\frac{1}{2}, \infty\right)$ and CD on $\left(-\infty, \frac{1}{2}\right)$. There is a change in concavity at $x = \frac{1}{2}$, and we have an IP at $\left(\frac{1}{2}, -\frac{13}{2}\right)$.

(d)

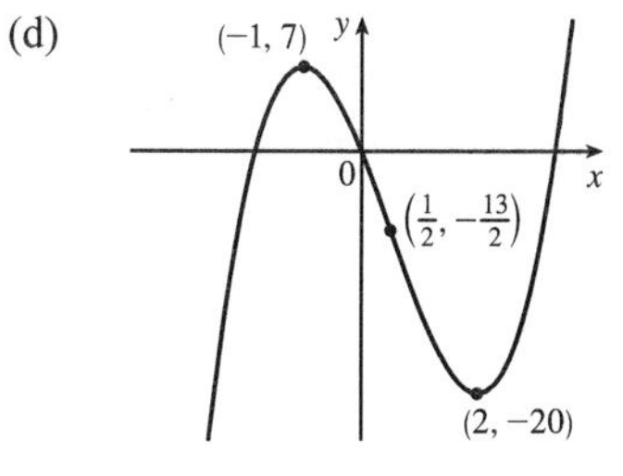

22. (a) $f(x) = 2 + 3x - x^3 \Rightarrow$

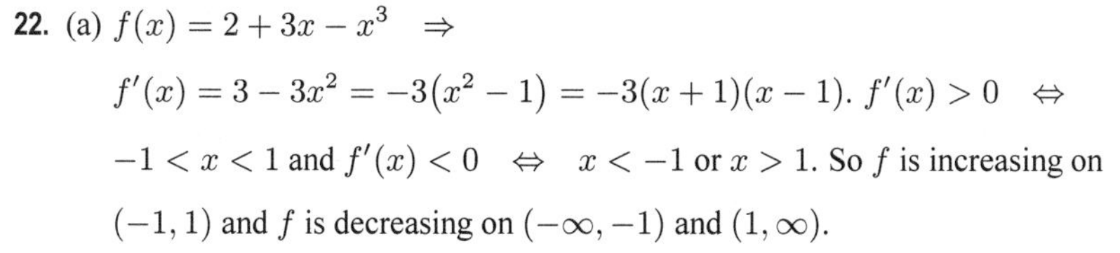

$f'(x) = 3 - 3x^2 = -3(x^2 - 1) = -3(x + 1)(x - 1)$. $f'(x) > 0 \Leftrightarrow$ $-1 < x < 1$ and $f'(x) < 0 \Leftrightarrow x < -1$ or $x > 1$. So f is increasing on $(-1, 1)$ and f is decreasing on $(-\infty, -1)$ and $(1, \infty)$.

(b) $f(-1) = 0$ is a local minimum value and $f(1) = 4$ is a local maximum value.

(c) $f''(x) = -6x \Rightarrow f''(x) > 0$ on $(-\infty, 0)$ and $f''(x) < 0$ on $(0, \infty)$. So f is CU on $(-\infty, 0)$ and CD on $(0, \infty)$. There is an IP at $(0, 2)$.

(d)

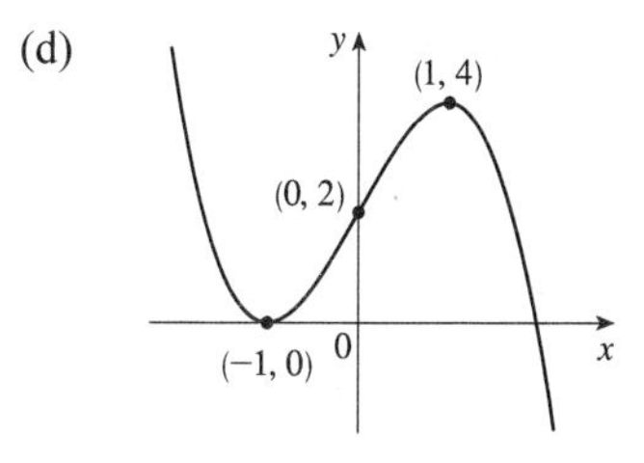

23. (a) $f(x) = x^4 - 6x^2 \Rightarrow f'(x) = 4x^3 - 12x = 4x(x^2 - 3) = 0$ when $x = 0, \pm\sqrt{3}$.

Interval	$4x$	$x^2 - 3$	$f'(x)$	f
$x < -\sqrt{3}$	$-$	$+$	$-$	decreasing on $(-\infty, -\sqrt{3})$
$-\sqrt{3} < x < 0$	$-$	$-$	$+$	increasing on $(-\sqrt{3}, 0)$
$0 < x < \sqrt{3}$	$+$	$-$	$-$	decreasing on $(0, \sqrt{3})$
$x > \sqrt{3}$	$+$	$+$	$+$	increasing on $(\sqrt{3}, \infty)$

(b) Local minimum values $f(\pm\sqrt{3}) = -9$, local maximum value $f(0) = 0$

(c) $f''(x) = 12x^2 - 12 = 12(x^2 - 1) > 0 \Leftrightarrow x^2 > 1 \Leftrightarrow |x| > 1 \Leftrightarrow$ $x > 1$ or $x < -1$, so f is CU on $(-\infty, -1)$, $(1, \infty)$ and CD on $(-1, 1)$. There are IPs at $(\pm 1, -5)$.

(d)

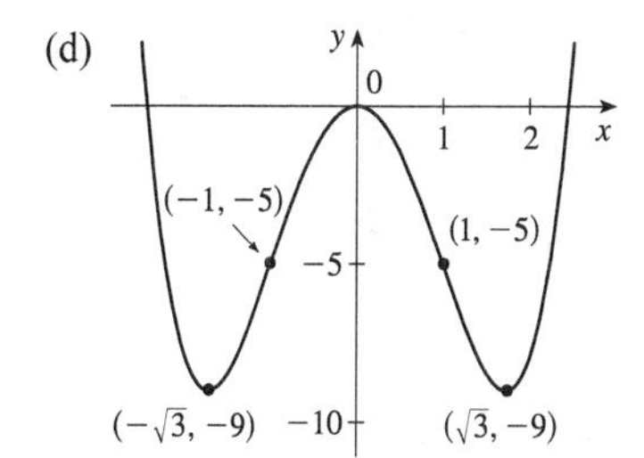

24. (a) $g(x) = 200 + 8x^3 + x^4 \Rightarrow g'(x) = 24x^2 + 4x^3 = 4x^2(6 + x) = 0$ when $x = -6$ and when $x = 0$. $g'(x) > 0 \Leftrightarrow$ $x > -6$ $(x \neq 0)$ and $g'(x) < 0 \Leftrightarrow x < -6$, so g is decreasing on $(-\infty, -6)$ and g is increasing on $(-6, \infty)$, with a horizontal tangent at $x = 0$.

(b) $g(-6) = -232$ is a local minimum value. There is no local maximum value.

(c) $g''(x) = 48x + 12x^2 = 12x(4 + x) = 0$ when $x = -4$ and when $x = 0$. $g''(x) > 0 \Leftrightarrow x < -4$ or $x > 0$ and $g''(x) < 0 \Leftrightarrow -4 < x < 0$, so g is CU on $(-\infty, -4)$ and $(0, \infty)$, and g is CD on $(-4, 0)$. There are IPs at $(-4, -56)$ and $(0, 200)$.

(d)

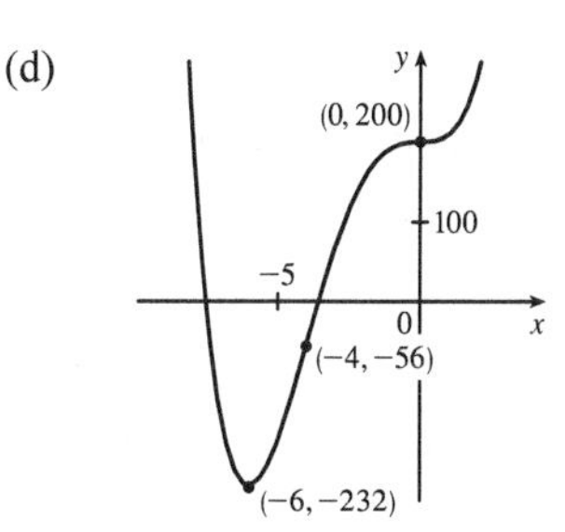

25. (a) $h(x) = 3x^5 - 5x^3 + 3 \Rightarrow h'(x) = 15x^4 - 15x^2 = 15x^2(x^2 - 1) = 0$ when $x = 0, \pm 1$. Since $15x^2$ is nonnegative, $h'(x) > 0 \Leftrightarrow x^2 > 1 \Leftrightarrow |x| > 1 \Leftrightarrow x > 1$ or $x < -1$, so h is increasing on $(-\infty, -1)$ and $(1, \infty)$ and decreasing on $(-1, 1)$, with a horizontal tangent at $x = 0$.

(b) Local maximum value $h(-1) = 5$, local minimum value $h(1) = 1$

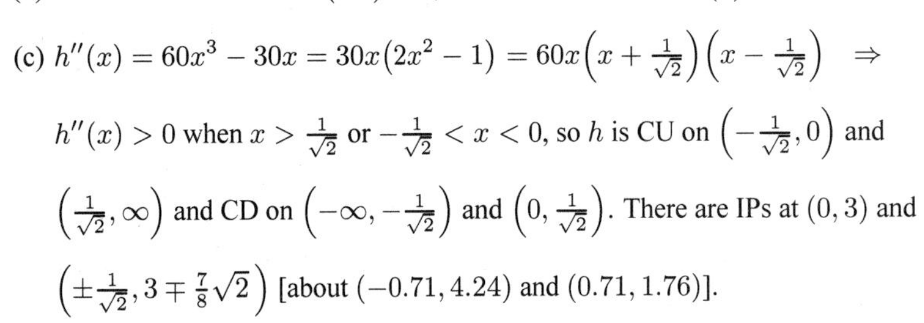

(c) $h''(x) = 60x^3 - 30x = 30x(2x^2 - 1) = 60x\left(x + \frac{1}{\sqrt{2}}\right)\left(x - \frac{1}{\sqrt{2}}\right) \Rightarrow$

$h''(x) > 0$ when $x > \frac{1}{\sqrt{2}}$ or $-\frac{1}{\sqrt{2}} < x < 0$, so h is CU on $\left(-\frac{1}{\sqrt{2}}, 0\right)$ and $\left(\frac{1}{\sqrt{2}}, \infty\right)$ and CD on $\left(-\infty, -\frac{1}{\sqrt{2}}\right)$ and $\left(0, \frac{1}{\sqrt{2}}\right)$. There are IPs at $(0, 3)$ and $\left(\pm\frac{1}{\sqrt{2}}, 3 \mp \frac{7}{8}\sqrt{2}\right)$ [about $(-0.71, 4.24)$ and $(0.71, 1.76)$].

(d)

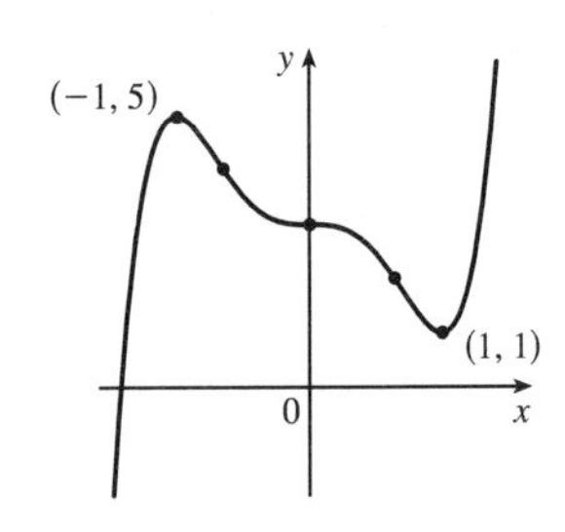

26. (a) $h(x) = (x^2 - 1)^3 \Rightarrow h'(x) = 6x(x^2 - 1)^2 \geq 0 \Leftrightarrow x > 0$ $(x \neq 1)$, so h is increasing on $(0, \infty)$ and decreasing on $(-\infty, 0)$.

(b) $h(0) = -1$ is a local minimum value.

(c) $h''(x) = 6(x^2 - 1)^2 + 24x^2(x^2 - 1) = 6(x^2 - 1)(5x^2 - 1)$. The roots ± 1 and $\pm\frac{1}{\sqrt{5}}$ divide $\mathbb{R}$ into five intervals.

Interval	$x^2 - 1$	$5x^2 - 1$	$h''(x)$	Concavity
$x < -1$	$+$	$+$	$+$	upward
$-1 < x < -\frac{1}{\sqrt{5}}$	$-$	$+$	$-$	downward
$-\frac{1}{\sqrt{5}} < x < \frac{1}{\sqrt{5}}$	$-$	$-$	$+$	upward
$\frac{1}{\sqrt{5}} < x < 1$	$-$	$+$	$-$	downward
$x > 1$	$+$	$+$	$+$	upward

From the table, we see that h is CU on $(-\infty, -1)$, $\left(-\frac{1}{\sqrt{5}}, \frac{1}{\sqrt{5}}\right)$ and $(1, \infty)$, and CD on $\left(-1, -\frac{1}{\sqrt{5}}\right)$ and $\left(\frac{1}{\sqrt{5}}, 1\right)$. There are IPs at $(\pm 1, 0)$ and $\left(\pm\frac{1}{\sqrt{5}}, -\frac{64}{125}\right)$.

(d)

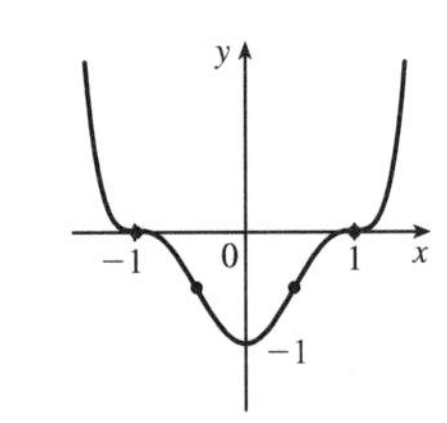

27. (a) $A(x) = x\sqrt{x+3} \Rightarrow A'(x) = x \cdot \frac{1}{2}(x+3)^{-1/2} + \sqrt{x+3} \cdot 1 = \dfrac{x}{2\sqrt{x+3}} + \sqrt{x+3} = \dfrac{x + 2(x+3)}{2\sqrt{x+3}} = \dfrac{3x+6}{2\sqrt{x+3}}$.

The domain of A is $[-3, \infty)$. $A'(x) > 0$ for $x > -2$ and $A'(x) < 0$ for $-3 < x < -2$, so A is increasing on $(-2, \infty)$ and decreasing on $(-3, -2)$.

(b) $A(-2) = -2$ is a local minimum value.

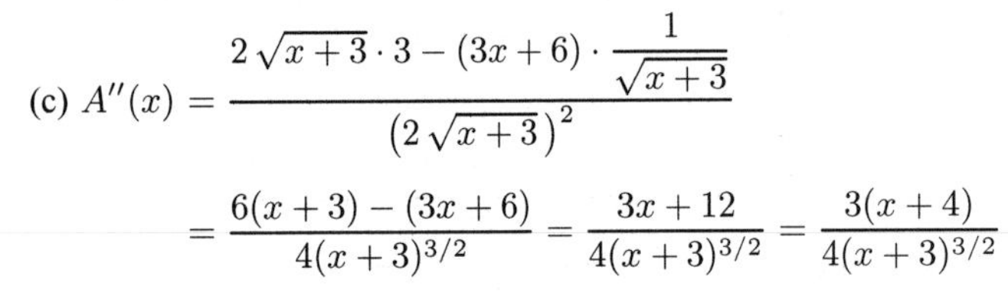

(c) $$A''(x) = \frac{2\sqrt{x+3} \cdot 3 - (3x+6) \cdot \dfrac{1}{\sqrt{x+3}}}{\left(2\sqrt{x+3}\right)^2} = \frac{6(x+3) - (3x+6)}{4(x+3)^{3/2}} = \frac{3x+12}{4(x+3)^{3/2}} = \frac{3(x+4)}{4(x+3)^{3/2}}$$

$A''(x) > 0$ for all $x > -3$, so A is CU on $(-3, \infty)$. No IP

(d)

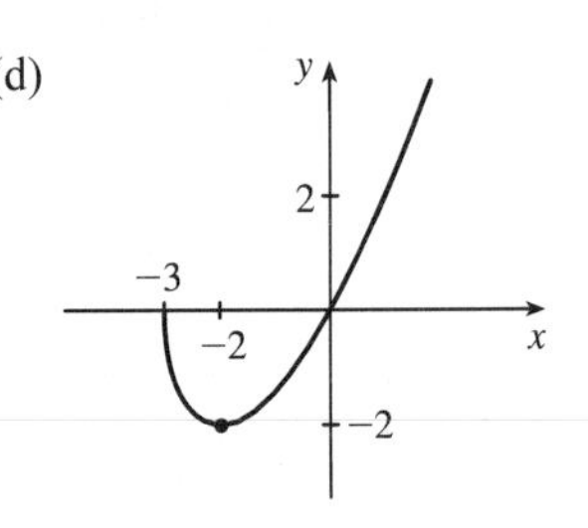

28. (a) $G(x) = x - 4\sqrt{x} \Rightarrow G'(x) = 1 - \dfrac{2}{\sqrt{x}} = \dfrac{1}{\sqrt{x}}(\sqrt{x} - 2)$. $G'(x) > 0 \Leftrightarrow x > 4$ and $G'(x) < 0 \Leftrightarrow$ $0 < x < 4$, so G is decreasing on $(0, 4)$ and increasing on $(4, \infty)$.

(b) Local minimum $G(4) = -4$. No local maximum.

(c) $G'(x) = 1 - 2x^{-1/2} \Rightarrow G''(x) = x^{-3/2} = 1/\sqrt{x^3}$, so $G''(x) > 0$ for $x > 0$. Thus, G is CU on $(0, \infty)$.

G has no IPs.

(d)

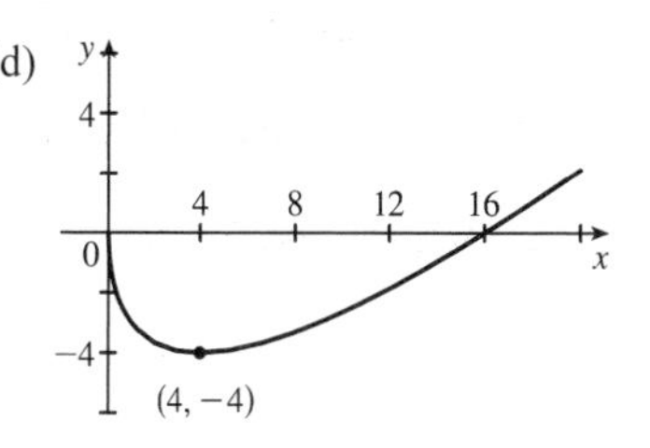

29. (a) $C(x) = x^{1/3}(x+4) = x^{4/3} + 4x^{1/3} \Rightarrow C'(x) = \frac{4}{3}x^{1/3} + \frac{4}{3}x^{-2/3} = \frac{4}{3}x^{-2/3}(x+1) = \dfrac{4(x+1)}{3\sqrt[3]{x^2}}$. $C'(x) > 0$ if $-1 < x < 0$ or $x > 0$ and $C'(x) < 0$ for $x < -1$, so C is increasing on $(-1, \infty)$ and C is decreasing on $(-\infty, -1)$.

(b) $C(-1) = -3$ is a local minimum value.

(c) $C''(x) = \frac{4}{9}x^{-2/3} - \frac{8}{9}x^{-5/3} = \frac{4}{9}x^{-5/3}(x-2) = \dfrac{4(x-2)}{9\sqrt[3]{x^5}}$.

$C''(x) < 0$ for $0 < x < 2$ and $C''(x) > 0$ for $x < 0$ and $x > 2$, so C is CD on $(0, 2)$ and CU on $(-\infty, 0)$ and $(2, \infty)$.

There are IPs at $(0, 0)$ and $(2, 6\sqrt[3]{2}) \approx (2, 7.56)$.

(d)

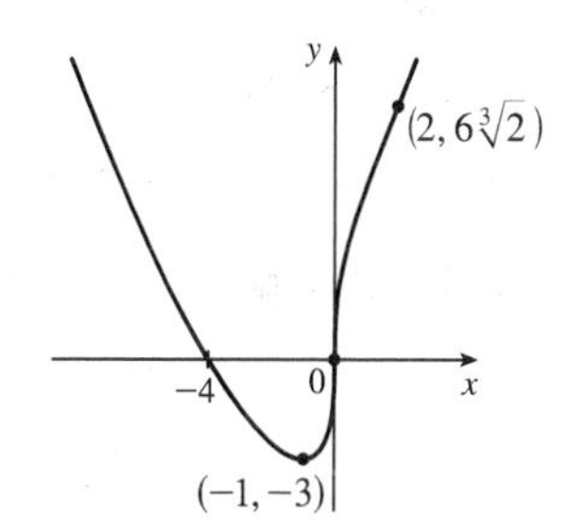

30. (a) $B(x) = 3x^{2/3} - x \Rightarrow B'(x) = 2x^{-1/3} - 1 = \dfrac{2}{\sqrt[3]{x}} - 1 = \dfrac{2 - \sqrt[3]{x}}{\sqrt[3]{x}}$. $B'(x) > 0$ if $0 < x < 8$ and $B'(x) < 0$ if $x < 0$ or $x > 8$, so B is decreasing on $(-\infty, 0)$ and $(8, \infty)$, and B is increasing on $(0, 8)$.

(b) $B(0) = 0$ is a local minimum value. $B(8) = 4$ is a local maximum value.

(c) $B''(x) = -\frac{2}{3}x^{-4/3} = \dfrac{-2}{3x^{4/3}}$, so $B''(x) < 0$ for all $x \neq 0$.

B is CD on $(-\infty, 0)$ and $(0, \infty)$. No IP

(d)

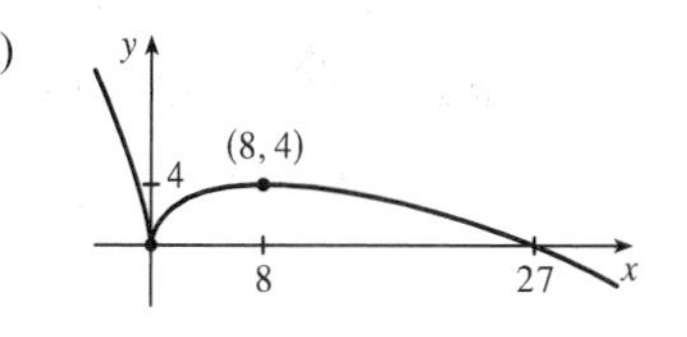

31. (a) $f(\theta) = 2\cos\theta - \cos 2\theta$, $0 \leq \theta \leq 2\pi$.

$f'(\theta) = -2\sin\theta + 2\sin 2\theta = -2\sin\theta + 2(2\sin\theta\cos\theta) = 2\sin\theta\,(2\cos\theta - 1)$.

Interval	$\sin\theta$	$2\cos\theta - 1$	$f'(\theta)$	f
$0 < \theta < \frac{\pi}{3}$	+	+	+	increasing on $(0, \frac{\pi}{3})$
$\frac{\pi}{3} < \theta < \pi$	+	−	−	decreasing on $(\frac{\pi}{3}, \pi)$
$\pi < \theta < \frac{5\pi}{3}$	−	−	+	increasing on $(\pi, \frac{5\pi}{3})$
$\frac{5\pi}{3} < \theta < 2\pi$	−	+	−	decreasing on $(\frac{5\pi}{3}, 2\pi)$

(b) $f(\frac{\pi}{3}) = \frac{3}{2}$ and $f(\frac{5\pi}{3}) = \frac{3}{2}$ are local maximum values and $f(\pi) = -3$ is a local minimum value.

(c) $f'(\theta) = -2\sin\theta + 2\sin 2\theta \;\Rightarrow$

$$f''(\theta) = -2\cos\theta + 4\cos 2\theta = -2\cos\theta + 4(2\cos^2\theta - 1) = 2(4\cos^2\theta - \cos\theta - 2)$$

$$f''(\theta) = 0 \;\Leftrightarrow\; \cos\theta = \frac{1 \pm \sqrt{33}}{8} \;\Leftrightarrow\; \theta = \cos^{-1}\left(\frac{1 \pm \sqrt{33}}{8}\right) \;\Leftrightarrow$$

$\theta = \cos^{-1}\left(\frac{1+\sqrt{33}}{8}\right) \approx 0.5678$, $2\pi - \cos^{-1}\left(\frac{1+\sqrt{33}}{8}\right) \approx 5.7154$,

$\cos^{-1}\left(\frac{1-\sqrt{33}}{8}\right) \approx 2.2057$, or $2\pi - \cos^{-1}\left(\frac{1-\sqrt{33}}{8}\right) \approx 4.0775$. Denote these four values of θ by $\theta_1, \theta_4, \theta_2$, and θ_3, respectively. Then f is CU on $(0, \theta_1)$, CD on (θ_1, θ_2), CU on (θ_2, θ_3), CD on (θ_3, θ_4), and CU on $(\theta_4, 2\pi)$. To find the *exact* y-coordinate for $\theta = \theta_1$, we have

$$f(\theta_1) = 2\cos\theta_1 - \cos 2\theta_1 = 2\cos\theta_1 - (2\cos^2\theta_1 - 1) = 2\left(\frac{1+\sqrt{33}}{8}\right) - 2\left(\frac{1+\sqrt{33}}{8}\right)^2 + 1$$

$$= \tfrac{1}{4} + \tfrac{1}{4}\sqrt{33} - \tfrac{1}{32} - \tfrac{1}{16}\sqrt{33} - \tfrac{33}{32} + 1 = \tfrac{3}{16} + \tfrac{3}{16}\sqrt{33} = \tfrac{3}{16}(1 + \sqrt{33}) = y_1 \approx 1.26.$$

Similarly, $f(\theta_2) = \frac{3}{16}(1 - \sqrt{33}) = y_2 \approx -0.89$. So f has IPs at (θ_1, y_1), (θ_2, y_2), (θ_3, y_2), and (θ_4, y_1).

(d)

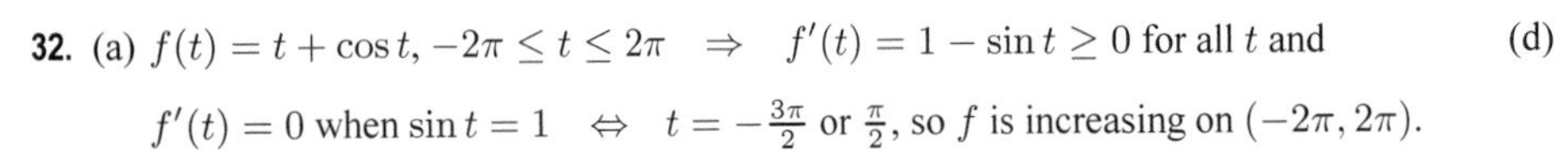

32. (a) $f(t) = t + \cos t$, $-2\pi \le t \le 2\pi \;\Rightarrow\; f'(t) = 1 - \sin t \ge 0$ for all t and $f'(t) = 0$ when $\sin t = 1 \;\Leftrightarrow\; t = -\frac{3\pi}{2}$ or $\frac{\pi}{2}$, so f is increasing on $(-2\pi, 2\pi)$.

(b) No maximum or minimum

(c) $f''(t) = -\cos t > 0 \;\Leftrightarrow\; t \in \left(-\frac{3\pi}{2}, -\frac{\pi}{2}\right) \cup \left(\frac{\pi}{2}, \frac{3\pi}{2}\right)$, so f is CU on these intervals and CD on $\left(-2\pi, -\frac{3\pi}{2}\right)$, $\left(-\frac{\pi}{2}, \frac{\pi}{2}\right)$, and $\left(\frac{3\pi}{2}, 2\pi\right)$.

There are IPs at $\left(\pm\frac{3\pi}{2}, \pm\frac{3\pi}{2}\right)$ and $\left(\pm\frac{\pi}{2}, \pm\frac{\pi}{2}\right)$.

(d)

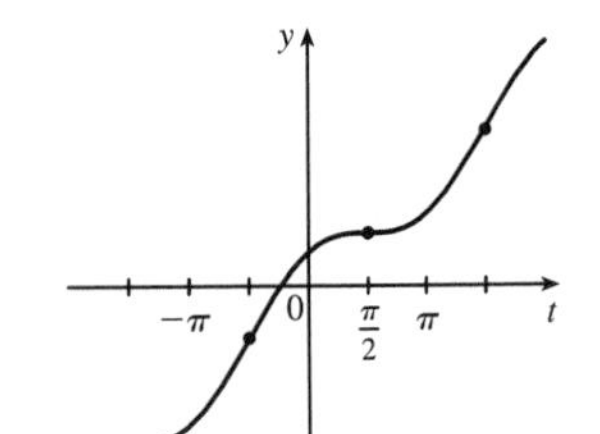

33. $f(x) = \dfrac{x^2}{x^2 - 1} = \dfrac{x^2}{(x+1)(x-1)}$ has domain $(-\infty, -1) \cup (-1, 1) \cup (1, \infty)$.

(a) $\displaystyle\lim_{x\to\pm\infty} f(x) = \lim_{x\to\pm\infty} \frac{x^2/x^2}{(x^2-1)/x^2} = \lim_{x\to\pm\infty} \frac{1}{1 - 1/x^2} = \frac{1}{1-0} = 1$, so $y = 1$ is a HA.

$\displaystyle\lim_{x\to -1^-} \frac{x^2}{x^2-1} = \infty$ since $x^2 \to 1$ and $(x^2 - 1) \to 0^+$ as $x \to -1^-$, so $x = -1$ is a VA.

$\displaystyle\lim_{x\to 1^+} \frac{x^2}{x^2-1} = \infty$ since $x^2 \to 1$ and $(x^2 - 1) \to 0^+$ as $x \to 1^+$, so $x = 1$ is a VA.

(b) $f(x) = \dfrac{x^2}{x^2-1} \;\Rightarrow\; f'(x) = \dfrac{(x^2-1)(2x) - x^2(2x)}{(x^2-1)^2} = \dfrac{2x[(x^2-1) - x^2]}{(x^2-1)^2} = \dfrac{-2x}{(x^2-1)^2}$. Since $(x^2-1)^2$ is positive for all x in the domain of f, the sign of the derivative is determined by the sign of $-2x$. Thus, $f'(x) > 0$ if $x < 0$ $(x \ne -1)$ and $f'(x) < 0$ if $x > 0$ $(x \ne 1)$. So f is increasing on $(-\infty, -1)$ and $(-1, 0)$, and f is decreasing on $(0, 1)$ and $(1, \infty)$.

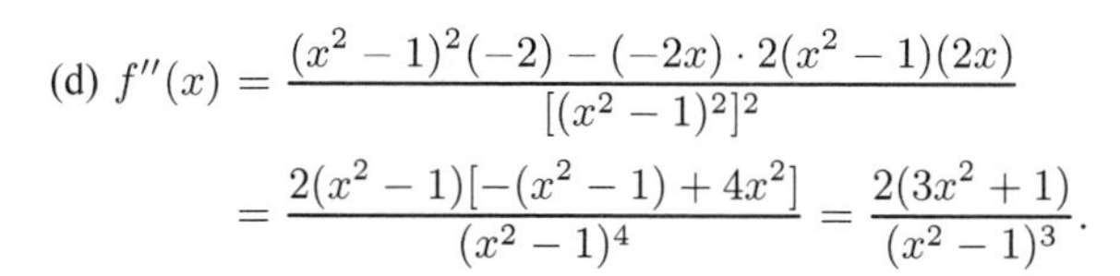

(c) $f'(x) = 0 \;\Rightarrow\; x = 0$ and $f(0) = 0$ is a local maximum value.

(d) $$f''(x) = \frac{(x^2-1)^2(-2) - (-2x)\cdot 2(x^2-1)(2x)}{[(x^2-1)^2]^2} = \frac{2(x^2-1)[-(x^2-1)+4x^2]}{(x^2-1)^4} = \frac{2(3x^2+1)}{(x^2-1)^3}.$$

The sign of $f''(x)$ is determined by the denominator; that is, $f''(x) > 0$ if $|x| > 1$ and $f''(x) < 0$ if $|x| < 1$. Thus, f is CU on $(-\infty, -1)$and $(1, \infty)$, and f is CD on $(-1, 1)$. No IP

(e)

34. $f(x) = \dfrac{x^2}{(x-2)^2}$ has domain $(-\infty, 2) \cup (2, \infty)$.

(a) $$\lim_{x\to\pm\infty} \frac{x^2}{x^2-4x+4} = \lim_{x\to\pm\infty} \frac{x^2/x^2}{(x^2-4x+4)/x^2} = \lim_{x\to\pm\infty} \frac{1}{1-4/x+4/x^2} = \frac{1}{1-0+0} = 1,$$

so $y = 1$ is a HA. $\displaystyle\lim_{x\to 2^+} \frac{x^2}{(x-2)^2} = \infty$ since $x^2 \to 4$ and $(x-2)^2 \to 0^+$ as $x \to 2^+$, so $x = 2$ is a VA.

(b) $$f(x) = \frac{x^2}{(x-2)^2} \;\Rightarrow\; f'(x) = \frac{(x-2)^2(2x) - x^2\cdot 2(x-2)}{[(x-2)^2]^2} = \frac{2x(x-2)[(x-2)-x]}{(x-2)^4} = \frac{-4x}{(x-2)^3}.$$ $f'(x) > 0$ if $0 < x < 2$ and $f'(x) < 0$ if $x < 0$ or $x > 2$, so f is increasing on $(0, 2)$ and f is decreasing on $(-\infty, 0)$ and $(2, \infty)$.

(c) $f(0) = 0$ is a local minimum value.

(d) $$f''(x) = \frac{(x-2)^3(-4) - (-4x)\cdot 3(x-2)^2}{[(x-2)^3]^2} = \frac{4(x-2)^2[-(x-2)+3x]}{(x-2)^6} = \frac{8(x+1)}{(x-2)^4}$$

$f''(x) > 0$ if $x > -1$ $(x \neq 2)$ and $f''(x) < 0$ if $x < -1$. Thus, f is CU on $(-1, 2)$ and $(2, \infty)$, and f is CD on $(-\infty, -1)$. There is an IP at $\left(-1, \frac{1}{9}\right)$.

(e)

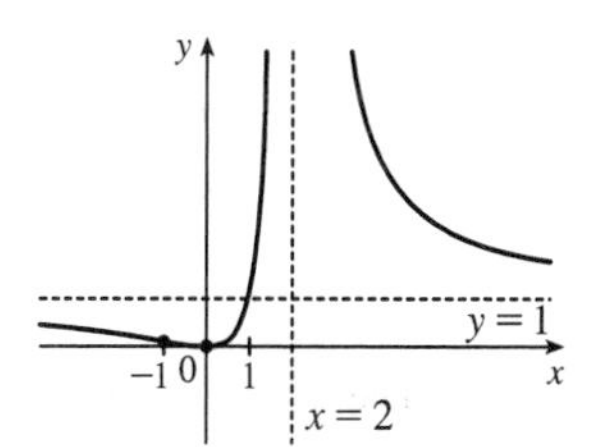

35. (a) $\displaystyle\lim_{x\to-\infty} \left(\sqrt{x^2+1} - x\right) = \infty$ and

$$\lim_{x\to\infty} \left(\sqrt{x^2+1} - x\right) = \lim_{x\to\infty} \left(\sqrt{x^2+1} - x\right) \frac{\sqrt{x^2+1}+x}{\sqrt{x^2+1}+x} = \lim_{x\to\infty} \frac{1}{\sqrt{x^2+1}+x} = 0,$$ so $y = 0$ is a HA.

(b) $f(x) = \sqrt{x^2+1} - x \;\Rightarrow\; f'(x) = \dfrac{x}{\sqrt{x^2+1}} - 1$. Since $\dfrac{x}{\sqrt{x^2+1}} < 1$ for all x, $f'(x) < 0$, so f is decreasing on $\mathbb{R}$.

(c) No minimum or maximum

(d) $$f''(x) = \frac{(x^2+1)^{1/2}(1) - x\cdot\frac{1}{2}(x^2+1)^{-1/2}(2x)}{\left(\sqrt{x^2+1}\right)^2} = \frac{(x^2+1)^{1/2} - \dfrac{x^2}{(x^2+1)^{1/2}}}{x^2+1} = \frac{(x^2+1)-x^2}{(x^2+1)^{3/2}} = \frac{1}{(x^2+1)^{3/2}} > 0,$$

so f is CU on $\mathbb{R}$. No IP

(e)

36. (a) $\lim\limits_{x \to \pi/2^-} x \tan x = \infty$ and $\lim\limits_{x \to -\pi/2^+} x \tan x = \infty$, so $x = \frac{\pi}{2}$ and $x = -\frac{\pi}{2}$ are VAs.

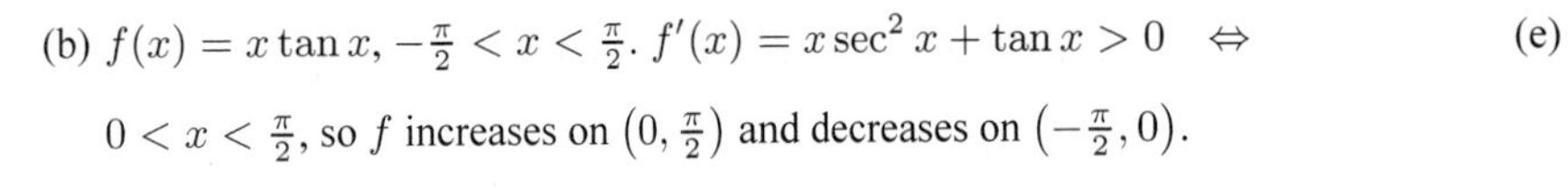

(b) $f(x) = x \tan x$, $-\frac{\pi}{2} < x < \frac{\pi}{2}$. $f'(x) = x \sec^2 x + \tan x > 0 \quad \Leftrightarrow$

$0 < x < \frac{\pi}{2}$, so f increases on $\left(0, \frac{\pi}{2}\right)$ and decreases on $\left(-\frac{\pi}{2}, 0\right)$.

(e)

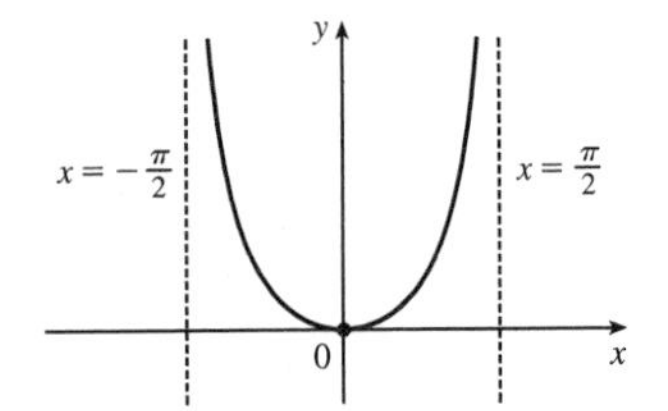

(c) $f(0) = 0$ is a local minimum value.

(d) $f''(x) = 2 \sec^2 x + 2x \tan x \sec^2 x > 0$ for $-\frac{\pi}{2} < x < \frac{\pi}{2}$,

so f is CU on $\left(-\frac{\pi}{2}, \frac{\pi}{2}\right)$. No IP

37. The nonnegative factors $(x+1)^2$ and $(x-6)^4$ do not affect the sign of $f'(x) = (x+1)^2(x-3)^5(x-6)^4$.

So $f'(x) > 0 \quad \Rightarrow \quad (x-3)^5 > 0 \quad \Rightarrow \quad x - 3 > 0 \quad \Rightarrow \quad x > 3$. Thus, f is increasing on the interval $(3, \infty)$.

38. $y = f(x) = x^3 - 3a^2x + 2a^3$, $a > 0$. The y-intercept is $f(0) = 2a^3$. $y' = 3x^2 - 3a^2 = 3(x^2 - a^2) = 3(x+a)(x-a)$.

The critical numbers are $-a$ and a. $f' < 0$ on $(-a, a)$, so f is decreasing on $(-a, a)$ and f is increasing on $(-\infty, -a)$ and (a, ∞). $f(-a) = 4a^3$ is a local maximum value and $f(a) = 0$ is a local minimum value. Since $f(a) = 0$, a is an x-intercept, and $x - a$ is a factor of f. Synthetically dividing $y = x^3 - 3a^2x + 2a^3$ by $x - a$ gives us the following result:

$$y = x^3 - 3a^2x + 2a^3 = (x-a)(x^2 + ax - 2a^2) = (x-a)(x-a)(x+2a) = (x-a)^2(x+2a),$$

which tells us that the only x-intercepts are $-2a$ and a. $y' = 3x^2 - 3a^2 \quad \Rightarrow$ $y'' = 6x$, so $y'' > 0$ on $(0, \infty)$ and $y'' < 0$ on $(-\infty, 0)$. This tells us that f is CU on $(0, \infty)$ and CD on $(-\infty, 0)$. There is an IP at $(0, 2a^3)$. The graph illustrates these features.

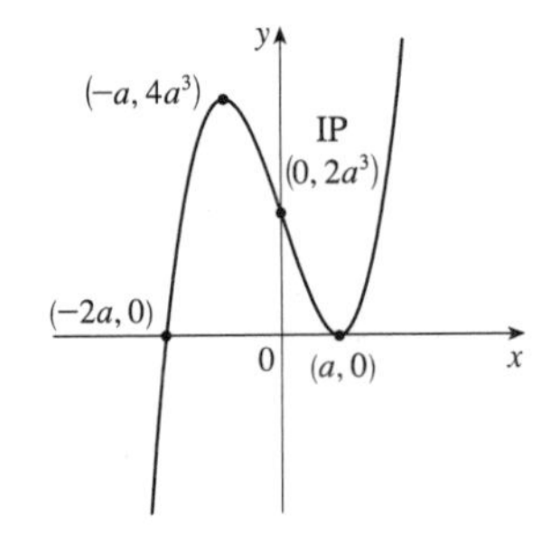

What the curves in the family have in common is that they are all CD on $(-\infty, 0)$, CU on $(0, \infty)$, and have the same basic shape. But as a increases, the four key points shown in the figure move further away from the origin.

39. (a)

From the graph, we get an estimate of $f(1) \approx 1.41$ as a local maximum value, and no local minimum value. $f(x) = \dfrac{x+1}{\sqrt{x^2+1}} \quad \Rightarrow \quad f'(x) = \dfrac{1-x}{(x^2+1)^{3/2}}$.

$f'(x) = 0 \quad \Leftrightarrow \quad x = 1$. $f(1) = \frac{2}{\sqrt{2}} = \sqrt{2}$ is the exact value.

(b) From the graph in part (a), f increases most rapidly somewhere between $x = -\frac{1}{2}$ and $x = -\frac{1}{4}$. To find the exact value, we need to find the maximum value of f', which we can do by finding the critical numbers of f'.

$f''(x) = \dfrac{2x^2 - 3x - 1}{(x^2+1)^{5/2}} = 0 \quad \Leftrightarrow \quad x = \dfrac{3 \pm \sqrt{17}}{4}$. $x = \dfrac{3 + \sqrt{17}}{4}$ corresponds to the *minimum* value of f'.

The maximum value of f' occurs at $\left(x = \frac{3 - \sqrt{17}}{4}\right) \approx -0.28$.

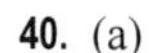

40. (a)

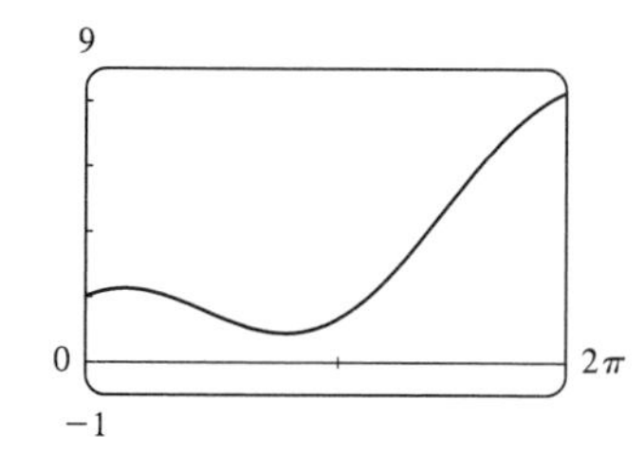

From the graph, we get estimates of $f(2.61) \approx 0.89$ as a local and absolute minimum, $f(0.53) \approx 2.26$ as a local maximum, and $f(2\pi) \approx 8.28$ as an absolute maximum. $f(x) = x + 2\cos x$ $(0 \le x \le 2\pi)$ $\Rightarrow$ $f'(x) = 1 - 2\sin x$. $f'(x) = 0 \Leftrightarrow \sin x = \frac{1}{2} \Leftrightarrow x = \frac{\pi}{6}, \frac{5\pi}{6}$. $f\left(\frac{\pi}{6}\right) = \frac{\pi}{6} + \sqrt{3}$ is the exact value of the local maximum, $f\left(\frac{5\pi}{6}\right) = \frac{5\pi}{6} - \sqrt{3}$ is the exact value of the local and absolute minimum, and $f(2\pi) = 2\pi + 2$ is the exact value of the absolute maximum.

(b) From the graph in part (a), f increases most rapidly somewhere between $x = 4.5$ and $x = 5$. Now f increases most rapidly when $f'(x) = 1 - 2\sin x$ has its maximum value. $f''(x) = -2\cos x = 0 \Leftrightarrow x = \frac{\pi}{2}, \frac{3\pi}{2}$. $f'(0) = f'(2\pi) = 1$, $f'\left(\frac{\pi}{2}\right) = -1$, and $f'\left(\frac{3\pi}{2}\right) = 3$. The maximum value of f' occurs at $\left(\frac{3\pi}{2}, \frac{3\pi}{2}\right)$.

41. $f(x) = ax^3 + bx^2 + cx + d \Rightarrow f'(x) = 3ax^2 + 2bx + c$. We are given that $f(1) = 0$ and $f(-2) = 3$, so $f(1) = a + b + c + d = 0$ and $f(-2) = -8a + 4b - 2c + d = 3$. Also $f'(1) = 3a + 2b + c = 0$ and $f'(-2) = 12a - 4b + c = 0$ by Fermat's Theorem. Solving these four equations, we get $a = \frac{2}{9}$, $b = \frac{1}{3}$, $c = -\frac{4}{3}$, $d = \frac{7}{9}$, so the function is $f(x) = \frac{1}{9}\left(2x^3 + 3x^2 - 12x + 7\right)$.

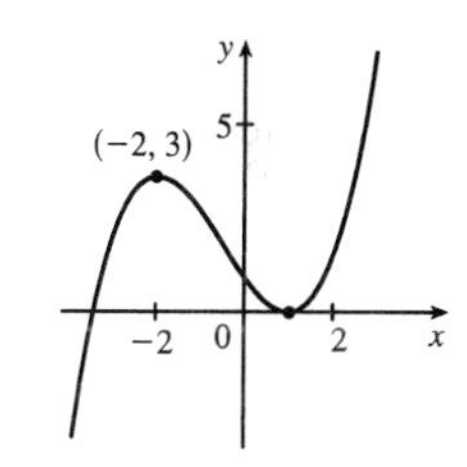

42. $f(x) = \tan x - x \Rightarrow f'(x) = \sec^2 x - 1 > 0$ for $0 < x < \frac{\pi}{2}$ since $\sec^2 x > 1$ for $0 < x < \frac{\pi}{2}$. So f is increasing on $\left(0, \frac{\pi}{2}\right)$. Thus, $f(x) > f(0) = 0$ for $0 < x < \frac{\pi}{2} \Rightarrow \tan x - x > 0 \Rightarrow \tan x > x$ for $0 < x < \frac{\pi}{2}$.

43. Let the cubic function be $f(x) = ax^3 + bx^2 + cx + d \Rightarrow f'(x) = 3ax^2 + 2bx + c \Rightarrow f''(x) = 6ax + 2b$. So f is CU when $6ax + 2b > 0 \Leftrightarrow x > -b/(3a)$, CD when $x < -b/(3a)$, and so the only IP occurs when $x = -b/(3a)$. If the graph has three x-intercepts x_1, x_2 and x_3, then the expression for $f(x)$ must factor as $f(x) = a(x - x_1)(x - x_2)(x - x_3)$. Multiplying these factors together gives us $f(x) = a\left[x^3 - (x_1 + x_2 + x_3)x^2 + (x_1x_2 + x_1x_3 + x_2x_3)x - x_1x_2x_3\right]$. Equating the coefficients of the x^2-terms for the two forms of f gives us $b = -a(x_1 + x_2 + x_3)$. Hence, the x-coordinate of the point of inflection is $-\dfrac{b}{3a} = -\dfrac{-a(x_1 + x_2 + x_3)}{3a} = \dfrac{x_1 + x_2 + x_3}{3}$.

44. $P(x) = x^4 + cx^3 + x^2 \Rightarrow P'(x) = 4x^3 + 3cx^2 + 2x \Rightarrow P''(x) = 12x^2 + 6cx + 2$. The graph of $P''(x)$ is a parabola. If $P''(x)$ has two roots, then it changes sign twice and so has two inflection points. This happens when the discriminant of $P''(x)$ is positive, that is, $(6c)^2 - 4 \cdot 12 \cdot 2 > 0 \quad \Leftrightarrow \quad 36c^2 - 96 > 0 \quad \Leftrightarrow \quad |c| > \frac{2\sqrt{6}}{3} \approx 1.63$.

If $36c^2 - 96 = 0 \quad \Leftrightarrow \quad c = \pm\frac{2\sqrt{6}}{3}$, $P''(x)$ is 0 at one point, but there is still no inflection point since $P''(x)$ never changes sign, and if $36c^2 - 96 < 0 \quad \Leftrightarrow \quad |c| < \frac{2\sqrt{6}}{3}$, then $P''(x)$ never changes sign, and so there is no IP.

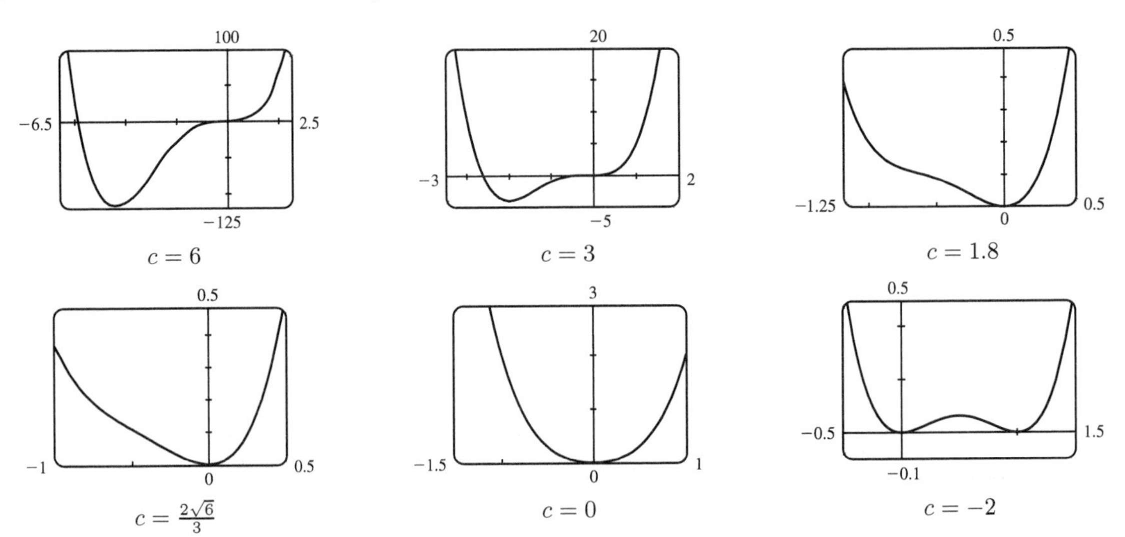

For large positive c, the graph of f has two inflection points and a large dip to the left of the y-axis. As c decreases, the graph of f becomes flatter for $x < 0$, and eventually the dip rises above the x-axis, and then disappears entirely, along with the inflection points. As c continues to decrease, the dip and the IPs reappear, to the right of the origin.

45. By hypothesis $g = f'$ is differentiable on an open interval containing c. Since $(c, f(c))$ is an IP, the concavity changes at $x = c$, so $f''(x)$ changes signs at $x = c$. Hence, by the First Derivative Test, f' has a local extremum at $x = c$. Thus, by Fermat's Theorem $f''(c) = 0$.

46. $f(x) = x^4 \quad \Rightarrow \quad f'(x) = 4x^3 \quad \Rightarrow \quad f''(x) = 12x^2 \quad \Rightarrow \quad f''(0) = 0$. For $x < 0$, $f''(x) > 0$, so f is CU on $(-\infty, 0)$; for $x > 0$, $f''(x) > 0$, so f is also CU on $(0, \infty)$. Since f does not change concavity at 0, $(0, 0)$ is not an IP.

47. Using the fact that $|x| = \sqrt{x^2}$, we have that $g(x) = x\sqrt{x^2} \quad \Rightarrow \quad g'(x) = \sqrt{x^2} + \sqrt{x^2} = 2\sqrt{x^2} = 2|x| \quad \Rightarrow$

$g''(x) = 2x\left(x^2\right)^{-1/2} = \dfrac{2x}{|x|} < 0$ for $x < 0$ and $g''(x) > 0$ for $x > 0$, so $(0, 0)$ is an IP. But $g''(0)$ does not exist.

48. There must exist some interval containing c on which f''' is positive, since $f'''(c)$ is positive and f''' is continuous. On this interval, f'' is increasing (since f''' is positive), so $f'' = (f')'$ changes from negative to positive at c. So by the First Derivative Test, f' has a local minimum at $x = c$ and thus cannot change sign there, so f has no maximum or minimum at c. But since f'' changes from negative to positive at c, f has an IP at c (it changes from concave down to concave up).

3.4 Curve Sketching

1. $y = f(x) = x^3 + x = x(x^2 + 1)$ **A.** f is a polynomial, so $D = \mathbb{R}$. **B.** x-intercept $= 0$, y-intercept $= f(0) = 0$ **C.** $f(-x) = -f(x)$, so f is odd; the curve is symmetric about the origin. **D.** f is a polynomial, so there is no asymptote. **E.** $f'(x) = 3x^2 + 1 > 0$, so f is increasing on $(-\infty, \infty)$. **F.** There is no critical number and hence, no local maximum or minimum value. **G.** $f''(x) = 6x > 0$ on $(0, \infty)$ and $f''(x) < 0$ on $(-\infty, 0)$, so f is CU on $(0, \infty)$ and CD on $(-\infty, 0)$. Since the concavity changes at $x = 0$, there is an IP at $(0, 0)$.

H.

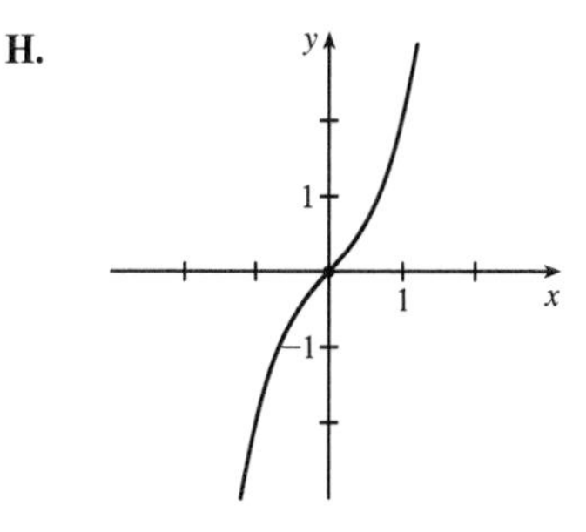

2. $y = f(x) = x^3 + 6x^2 + 9x = x(x + 3)^2$ **A.** $D = \mathbb{R}$ **B.** x-intercepts are -3 and 0, y-intercept $= 0$ **C.** No symmetry **D.** No asymptote **E.** $f'(x) = 3x^2 + 12x + 9 = 3(x + 1)(x + 3) < 0 \quad \Leftrightarrow \quad -3 < x < -1$, so f is decreasing on $(-3, -1)$ and increasing on $(-\infty, -3)$ and $(-1, \infty)$. **F.** Local maximum value $f(-3) = 0$, local minimum value $f(-1) = -4$ **G.** $f''(x) = 6x + 12 = 6(x + 2) > 0 \quad \Leftrightarrow \quad x > -2$, so f is CU on $(-2, \infty)$ and CD on $(-\infty, -2)$. IP at $(-2, -2)$

H.

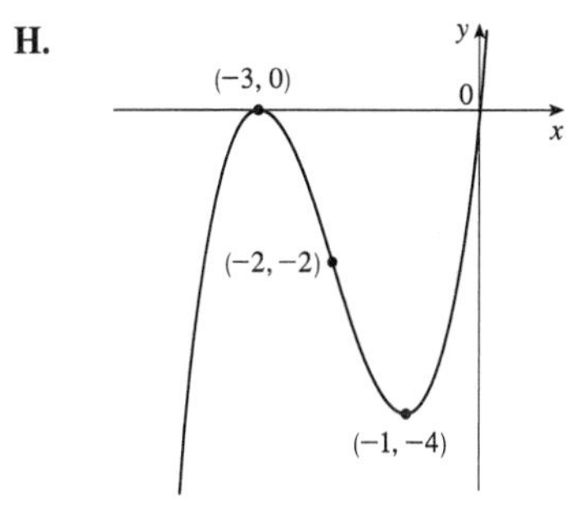

3. $y = f(x) = 2 - 15x + 9x^2 - x^3 = -(x - 2)(x^2 - 7x + 1)$ **A.** $D = \mathbb{R}$ **B.** y-intercept: $f(0) = 2$; x-intercepts: $f(x) = 0 \quad \Rightarrow \quad x = 2$ or [by the quadratic formula] $x = \frac{7 \pm \sqrt{45}}{2} \approx 0.15, 6.85$ **C.** No symmetry **D.** No asymptote **E.** $f'(x) = -15 + 18x - 3x^2 = -3(x^2 - 6x + 5)$ $= -3(x - 1)(x - 5) > 0 \quad \Leftrightarrow \quad 1 < x < 5$, so f is increasing on $(1, 5)$ and decreasing on $(-\infty, 1)$ and $(5, \infty)$. **F.** Local maximum value $f(5) = 27$, local minimum value $f(1) = -5$ **G.** $f''(x) = 18 - 6x = -6(x - 3) > 0 \quad \Leftrightarrow \quad x < 3$, so f is CU on $(-\infty, 3)$ and CD on $(3, \infty)$. IP at $(3, 11)$

H.

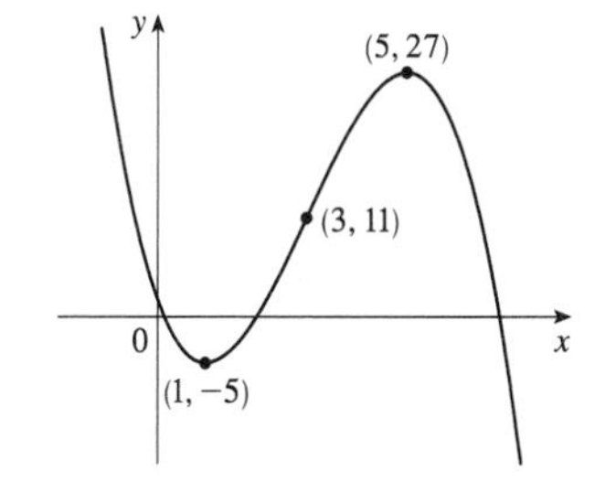

4. $y = f(x) = 8x^2 - x^4 = x^2(8 - x^2)$ **A.** $D = \mathbb{R}$ **B.** y-intercept: $f(0) = 0$; x-intercepts: $f(x) = 0 \quad \Rightarrow \quad x = 0$, $\pm 2\sqrt{2}$ ($\approx \pm 2.83$) **C.** $f(-x) = f(x)$, so f is even and symmetric about the y-axis. **D.** No asymptote **E.** $f'(x) = 16x - 4x^3 = 4x(4 - x^2) = 4x(2 + x)(2 - x) > 0 \quad \Leftrightarrow \quad x < -2$ or $0 < x < 2$, so f is increasing on $(-\infty, -2)$ and $(0, 2)$ and decreasing on $(-2, 0)$ and $(2, \infty)$. **F.** Local maximum value $f(\pm 2) = 16$, local minimum value $f(0) = 0$ **G.** $f''(x) = 16 - 12x^2 = 4(4 - 3x^2) = 0 \quad \Leftrightarrow \quad x = \pm\frac{2}{\sqrt{3}}$. $f''(x) > 0 \quad \Leftrightarrow \quad -\frac{2}{\sqrt{3}} < x < \frac{2}{\sqrt{3}}$, so f is CU on $\left(-\frac{2}{\sqrt{3}}, \frac{2}{\sqrt{3}}\right)$ and CD on $\left(-\infty, -\frac{2}{\sqrt{3}}\right)$ and $\left(\frac{2}{\sqrt{3}}, \infty\right)$. IP at $\left(\pm\frac{2}{\sqrt{3}}, \frac{80}{9}\right)$

H.

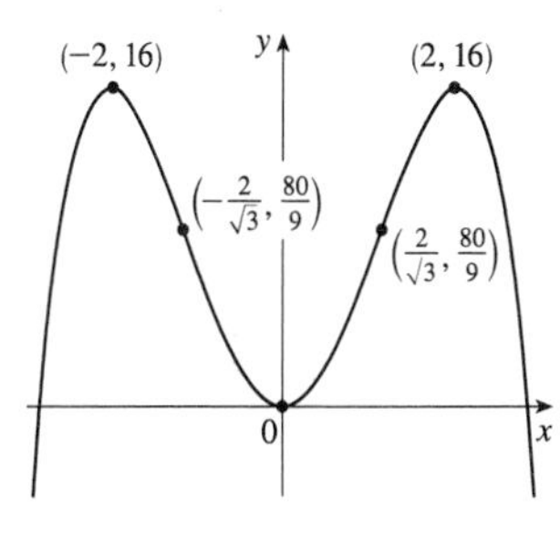

5. $y = f(x) = x^4 + 4x^3 = x^3(x+4)$ **A.** $D = \mathbb{R}$ **B.** y-intercept: $f(0) = 0$; x-intercepts: $f(x) = 0 \Leftrightarrow x = -4, 0$ **C.** No symmetry **D.** No asymptote **E.** $f'(x) = 4x^3 + 12x^2 = 4x^2(x+3) > 0 \Leftrightarrow$ $x > -3$, so f is increasing on $(-3, \infty)$ and decreasing on $(-\infty, -3)$. **F.** Local minimum value $f(-3) = -27$, no local maximum **G.** $f''(x) = 12x^2 + 24x = 12x(x+2) < 0 \Leftrightarrow -2 < x < 0$, so f is CD on $(-2, 0)$ and CU on $(-\infty, -2)$ and $(0, \infty)$. IP at $(0, 0)$ and $(-2, -16)$

H.

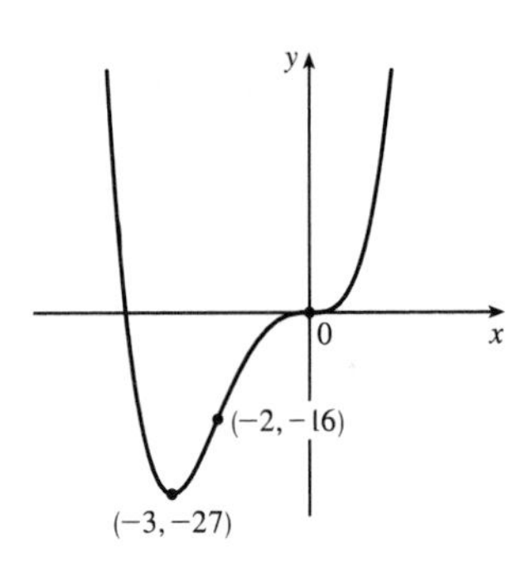

6. $y = f(x) = x(x+2)^3$ **A.** $D = \mathbb{R}$ **B.** y-intercept: $f(0) = 0$; x-intercepts: $f(x) = 0 \Leftrightarrow x = -2, 0$ **C.** No symmetry **D.** No asymptote **E.** $f'(x) = 3x(x+2)^2 + (x+2)^3 = (x+2)^2\,[3x + (x+2)] = (x+2)^2(4x+2)$. $f'(x) > 0 \Leftrightarrow x > -\frac{1}{2}$, and $f'(x) < 0 \Leftrightarrow x < -2$ or $-2 < x < -\frac{1}{2}$, so f is increasing on $\left(-\frac{1}{2}, \infty\right)$ and decreasing on $(-\infty, -2)$ and $\left(-2, -\frac{1}{2}\right)$. [Hence f is decreasing on $\left(-\infty, -\frac{1}{2}\right)$ by the analogue of Exercise 4.3.53 for decreasing functions.] **F.** Local minimum value $f\left(-\frac{1}{2}\right) = -\frac{27}{16}$, no local maximum

G.
$$\begin{aligned} f''(x) &= (x+2)^2(4) + (4x+2)(2)(x+2) \\ &= 2(x+2)[(x+2)(2) + 4x + 2] \\ &= 2(x+2)(6x+6) = 12(x+1)(x+2) \end{aligned}$$

$f''(x) < 0 \Leftrightarrow -2 < x < -1$, so f is CD on $(-2, -1)$ and CU on $(-\infty, -2)$ and $(-1, \infty)$. IP at $(-2, 0)$ and $(-1, -1)$

H.

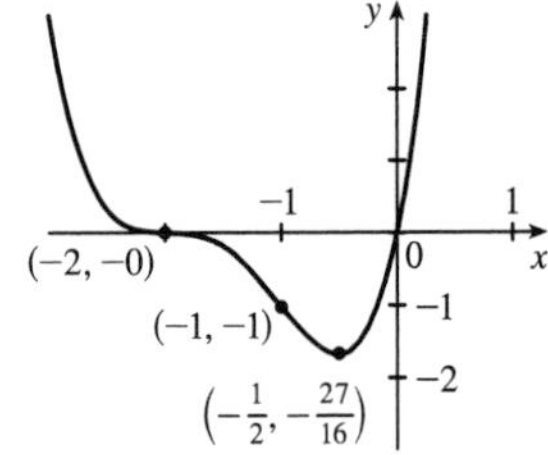

7. $y = f(x) = 2x^5 - 5x^2 + 1$ **A.** $D = \mathbb{R}$ **B.** y-intercept: $f(0) = 1$ **C.** No symmetry **D.** No asymptote **E.** $f'(x) = 10x^4 - 10x = 10x(x^3 - 1) = 10x(x-1)(x^2+x+1)$, so $f'(x) < 0 \Leftrightarrow 0 < x < 1$ and $f'(x) > 0 \Leftrightarrow$ $x < 0$ or $x > 1$. Thus, f is increasing on $(-\infty, 0)$ and $(1, \infty)$ and decreasing on $(0, 1)$. **F.** Local maximum value $f(0) = 1$, local minimum value $f(1) = -2$

G. $f''(x) = 40x^3 - 10 = 10(4x^3 - 1)$ so $f''(x) = 0 \Leftrightarrow x = 1/\sqrt[3]{4}$. $f''(x) > 0 \Leftrightarrow x > 1/\sqrt[3]{4}$ and $f''(x) < 0 \Leftrightarrow x < 1/\sqrt[3]{4}$, so f is CD on $\left(-\infty, 1/\sqrt[3]{4}\right)$ and CU on $\left(1/\sqrt[3]{4}, \infty\right)$.

IP at $\left(\dfrac{1}{\sqrt[3]{4}}, 1 - \dfrac{9}{2\left(\sqrt[3]{4}\right)^2}\right) \approx (0.630, -0.786)$.

H.

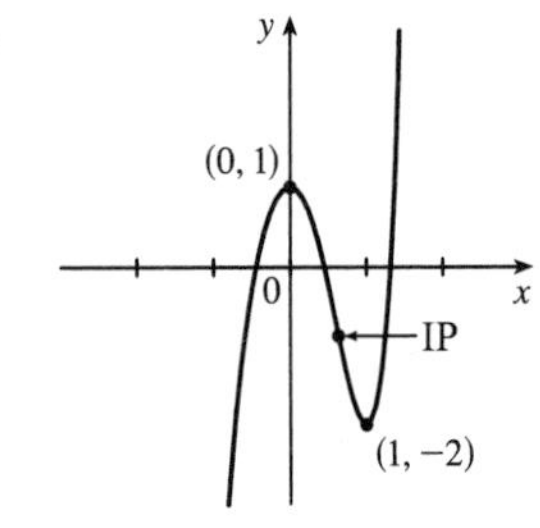

8. $y = f(x) = 20x^3 - 3x^5$ **A.** $D = \mathbb{R}$ **B.** y-intercept: $f(0) = 0$; x-intercepts: $f(x) = 0 \Leftrightarrow -3x^3\left(x^2 - \frac{20}{3}\right) = 0 \Leftrightarrow$ $x = 0$ or $\pm\sqrt{20/3} \approx \pm 2.582$ **C.** $f(-x) = -f(x)$, so f is odd; the curve is symmetric about the origin. **D.** No asymptote **E.** $f'(x) = 60x^2 - 15x^4 = -15x^2(x^2 - 4) = -15x^2(x+2)(x-2)$, so $f'(x) > 0 \Leftrightarrow$ $-2 < x < 0$ or $0 < x < 2$ and $f'(x) < 0 \Leftrightarrow x < -2$ or $x > 2$. Thus, f is increasing on $(-2, 0)$ and $(0, 2)$ [hence on $(-2, 2)$ by Exercise 3.4.43] and f is decreasing on $(-\infty, -2)$ and $(2, \infty)$. **F.** Local minimum value $f(-2) = -64$, local maximum value $f(2) = 64$

G. $f''(x) = 120x - 60x^3 = -60x(x^2 - 2)$. $f''(x) > 0 \;\Leftrightarrow\; x < -\sqrt{2}$ or $0 < x < \sqrt{2}$; $f''(x) < 0 \;\Leftrightarrow\; -\sqrt{2} < x < 0$ or $x > \sqrt{2}$. Thus, f is CU on $(-\infty, -\sqrt{2})$ and $(0, \sqrt{2})$, and f is CD on $(-\sqrt{2}, 0)$ and $(\sqrt{2}, \infty)$.

IP at $(-\sqrt{2}, -28\sqrt{2}) \approx (-1.414, -39.598)$, $(0, 0)$, and $(\sqrt{2}, 28\sqrt{2})$

H.

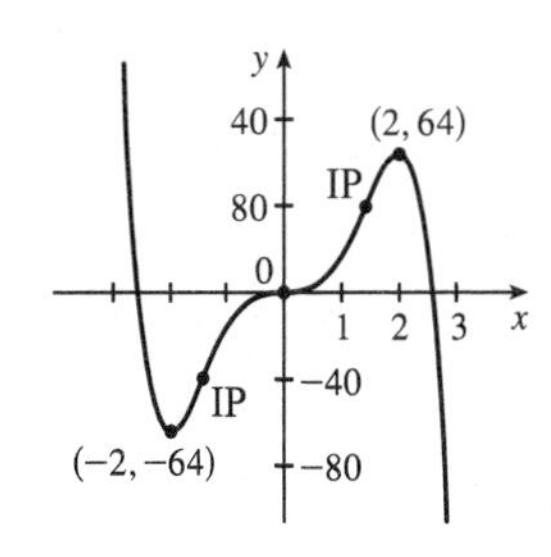

9. $y = f(x) = x/(x-1)$ **A.** $D = \{x \mid x \neq 1\} = (-\infty, 1) \cup (1, \infty)$ **B.** x-intercept $= 0$, y-intercept $= f(0) = 0$

C. No symmetry **D.** $\lim\limits_{x \to \pm\infty} \dfrac{x}{x-1} = 1$, so $y = 1$ is a HA. $\lim\limits_{x \to 1^-} \dfrac{x}{x-1} = -\infty$, $\lim\limits_{x \to 1^+} \dfrac{x}{x-1} = \infty$, so $x = 1$ is a VA.

E. $f'(x) = \dfrac{(x-1) - x}{(x-1)^2} = \dfrac{-1}{(x-1)^2} < 0$ for $x \neq 1$, so f is decreasing on $(-\infty, 1)$ and $(1, \infty)$. **F.** No extreme values **G.** $f''(x) = \dfrac{2}{(x-1)^3} > 0$ $\Leftrightarrow$ $x > 1$, so f is CU on $(1, \infty)$ and CD on $(-\infty, 1)$. No IP

H.

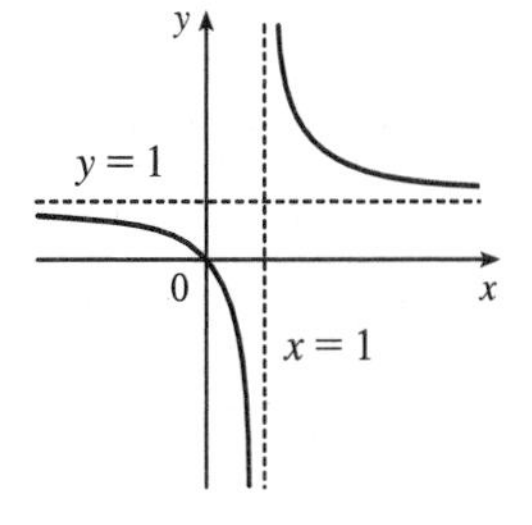

10. $y = x/(x-1)^2$ **A.** $D = \{x \mid x \neq 1\} = (-\infty, 1) \cup (1, \infty)$ **B.** x-intercept $= 0$, y-intercept $= f(0) = 0$

C. No symmetry **D.** $\lim\limits_{x \to \pm\infty} \dfrac{x}{(x-1)^2} = 0$, so $y = 0$ is a HA. $\lim\limits_{x \to 1} \dfrac{x}{(x-1)^2} = \infty$, so $x = 1$ is a VA.

E. $f'(x) = \dfrac{(x-1)^2(1) - x(2)(x-1)}{(x-1)^4} = \dfrac{-x-1}{(x-1)^3}$. This is negative on $(-\infty, -1)$ and $(1, \infty)$ and positive on $(-1, 1)$, so $f(x)$ is decreasing on $(-\infty, -1)$ and $(1, \infty)$ and increasing on $(-1, 1)$.

F. Local minimum value $f(-1) = -\frac{1}{4}$, no local maximum.

G. $f''(x) = \dfrac{(x-1)^3(-1) + (x+1)(3)(x-1)^2}{(x-1)^6} = \dfrac{2(x+2)}{(x-1)^4}$. This is negative on $(-\infty, -2)$, and positive on $(-2, 1)$ and $(1, \infty)$. So f is CD on $(-\infty, -2)$ and CU on $(-2, 1)$ and $(1, \infty)$. IP at $\left(-2, -\frac{2}{9}\right)$

H.

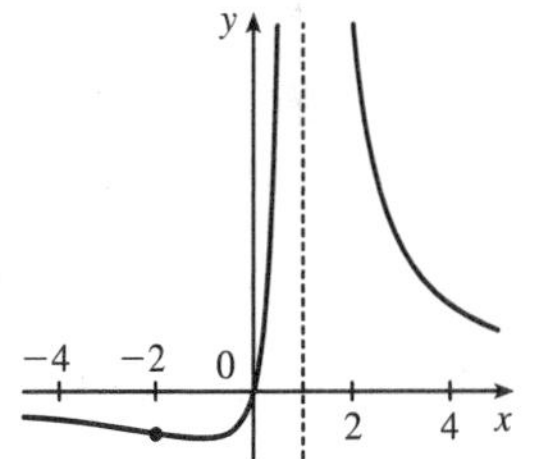

11. $y = f(x) = 1/(x^2 - 9)$ **A.** $D = \{x \mid x \neq \pm 3\} = (-\infty, -3) \cup (-3, 3) \cup (3, \infty)$ **B.** y-intercept $= f(0) = -\frac{1}{9}$, no x-intercept **C.** $f(-x) = f(x) \;\Rightarrow\; f$ is even; the curve is symmetric about the y-axis. **D.** $\lim\limits_{x \to \pm\infty} \dfrac{1}{x^2 - 9} = 0$, so $y = 0$ is a HA. $\lim\limits_{x \to 3^-} \dfrac{1}{x^2 - 9} = -\infty$, $\lim\limits_{x \to 3^+} \dfrac{1}{x^2 - 9} = \infty$, $\lim\limits_{x \to -3^-} \dfrac{1}{x^2 - 9} = \infty$, $\lim\limits_{x \to -3^+} \dfrac{1}{x^2 - 9} = -\infty$, so $x = 3$ and $x = -3$ are VAs. **E.** $f'(x) = -\dfrac{2x}{(x^2 - 9)^2} > 0 \;\Leftrightarrow\; x < 0$ $(x \neq -3)$ so f is increasing on

$(-\infty,-3)$ and $(-3,0)$ and decreasing on $(0,3)$ and $(3,\infty)$.

F. Local maximum value $f(0)=-\frac{1}{9}$.

G. $y''=\dfrac{-2(x^2-9)^2+(2x)2(x^2-9)(2x)}{(x^2-9)^4}=\dfrac{6(x^2+3)}{(x^2-9)^3}>0 \;\Leftrightarrow$

$x^2>9 \;\Leftrightarrow\; x>3$ or $x<-3$, so f is CU on $(-\infty,-3)$ and $(3,\infty)$ and

CD on $(-3,3)$. No IP

H.

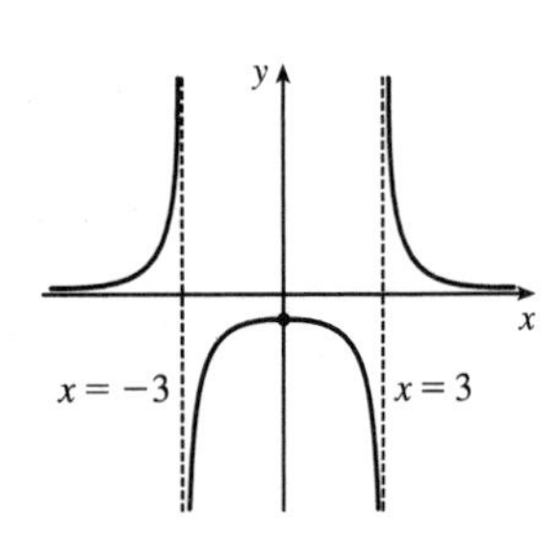

12. $y=f(x)=x/(x^2-9)$ **A.** $D=\{x \mid x\neq\pm 3\}=(-\infty,-3)\cup(-3,3)\cup(3,\infty)$ **B.** x-intercept $=0$, y-intercept $=f(0)=0$. **C.** $f(-x)=-f(x)$, so f is odd; the curve is symmetric about the origin.

D. $\displaystyle\lim_{x\to\pm\infty}\frac{x}{x^2-9}=0$, so $y=0$ is a HA. $\displaystyle\lim_{x\to 3^+}\frac{x}{x^2-9}=\infty$, $\displaystyle\lim_{x\to 3^-}\frac{x}{x^2-9}=-\infty$, $\displaystyle\lim_{x\to -3^+}\frac{x}{x^2-9}=\infty$,

$\displaystyle\lim_{x\to -3^-}\frac{x}{x^2-9}=-\infty$, so $x=3$ and $x=-3$ are VAs. **E.** $f'(x)=\dfrac{(x^2-9)-x(2x)}{(x^2-9)^2}=-\dfrac{x^2+9}{(x^2-9)^2}<0\;(x\neq\pm 3)$

so f is decreasing on $(-\infty,-3)$, $(-3,3)$, and $(3,\infty)$.

F. No extreme values

G. $f''(x)=-\dfrac{2x(x^2-9)^2-(x^2+9)\cdot 2(x^2-9)(2x)}{(x^2-9)^4}=\dfrac{2x(x^2+27)}{(x^2-9)^3}>0$

when $-3<x<0$ or $x>3$, so f is CU on $(-3,0)$ and $(3,\infty)$;

CD on $(-\infty,-3)$ and $(0,3)$. IP at $(0,0)$

H.

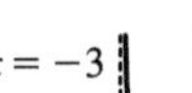

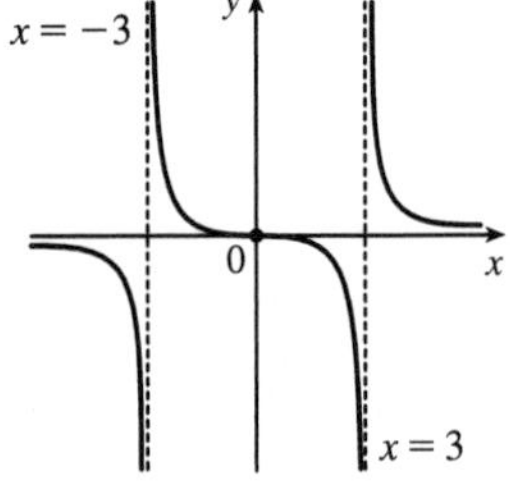

13. $y=f(x)=x/(x^2+9)$ **A.** $D=\mathbb{R}$ **B.** y-intercept: $f(0)=0$; x-intercept: $f(x)=0 \;\Leftrightarrow\; x=0$

C. $f(-x)=-f(x)$, so f is odd and the curve is symmetric about the origin. **D.** $\displaystyle\lim_{x\to\pm\infty}\frac{x}{x^2+9}=0$, so $y=0$ is a HA;

no VA **E.** $f'(x)=\dfrac{(x^2+9)(1)-x(2x)}{(x^2+9)^2}=\dfrac{9-x^2}{(x^2+9)^2}=\dfrac{(3+x)(3-x)}{(x^2+9)^2}>0 \;\Leftrightarrow\; -3<x<3$,

so f is increasing on $(-3,3)$ and decreasing on $(-\infty,-3)$ and $(3,\infty)$. **F.** Local minimum value $f(-3)=-\frac{1}{6}$, local maximum value $f(3)=\frac{1}{6}$

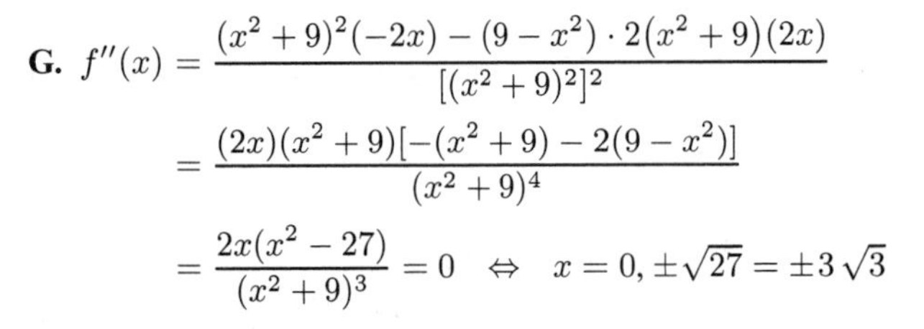

$$\textbf{G. } f''(x)=\frac{(x^2+9)^2(-2x)-(9-x^2)\cdot 2(x^2+9)(2x)}{[(x^2+9)^2]^2}$$
$$=\frac{(2x)(x^2+9)[-(x^2+9)-2(9-x^2)]}{(x^2+9)^4}$$
$$=\frac{2x(x^2-27)}{(x^2+9)^3}=0 \;\Leftrightarrow\; x=0,\pm\sqrt{27}=\pm 3\sqrt{3}$$

H.

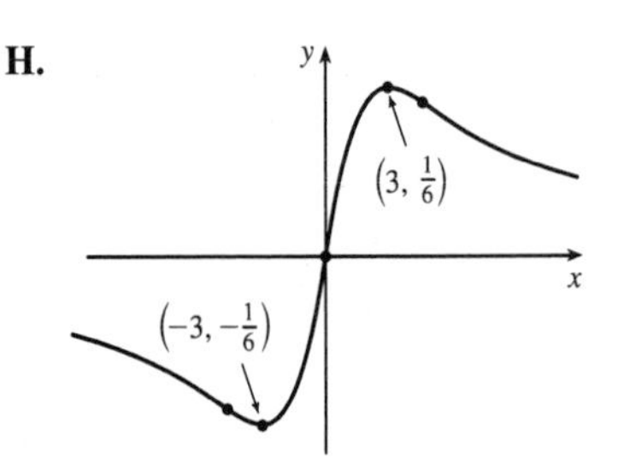

$f''(x)>0 \;\Leftrightarrow\; -3\sqrt{3}<x<0$ or $x>3\sqrt{3}$, so f is CU on $(-3\sqrt{3},0)$ and $(3\sqrt{3},\infty)$, and CD on $(-\infty,-3\sqrt{3})$ and $(0,3\sqrt{3})$. There are three IPs: $(0,0)$ and $(\pm 3\sqrt{3},\pm\frac{1}{12}\sqrt{3})$.

14. $y = f(x) = x^2/(x^2+9)$ **A.** $D = \mathbb{R}$ **B.** y-intercept: $f(0) = 0$; x-intercept: $f(x) = 0 \Leftrightarrow x = 0$

C. $f(-x) = f(x)$, so f is even and symmetric about the y-axis. **D.** $\displaystyle\lim_{x\to\pm\infty} \frac{x^2}{x^2+9} = 1$, so $y = 1$ is a HA;

no VA **E.** $f'(x) = \dfrac{(x^2+9)(2x) - x^2(2x)}{(x^2+9)^2} = \dfrac{18x}{(x^2+9)^2} > 0 \Leftrightarrow x > 0$, so f is increasing on $(0,\infty)$

and decreasing on $(-\infty, 0)$. **F.** Local minimum value $f(0) = 0$; no local maximum

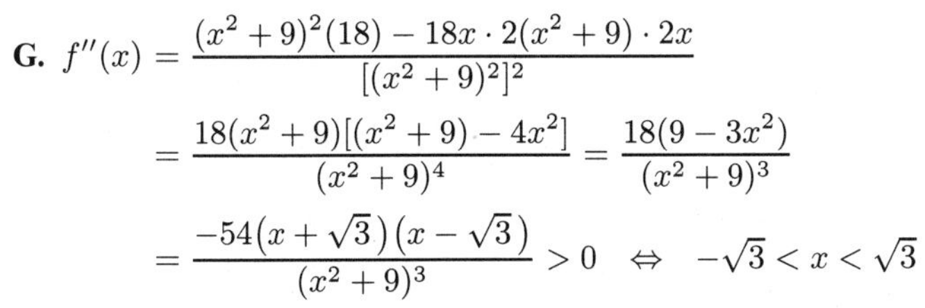

G.
$$\begin{aligned} f''(x) &= \frac{(x^2+9)^2(18) - 18x\cdot 2(x^2+9)\cdot 2x}{[(x^2+9)^2]^2} \\ &= \frac{18(x^2+9)[(x^2+9) - 4x^2]}{(x^2+9)^4} = \frac{18(9-3x^2)}{(x^2+9)^3} \\ &= \frac{-54(x+\sqrt{3})(x-\sqrt{3})}{(x^2+9)^3} > 0 \Leftrightarrow -\sqrt{3} < x < \sqrt{3} \end{aligned}$$

H.

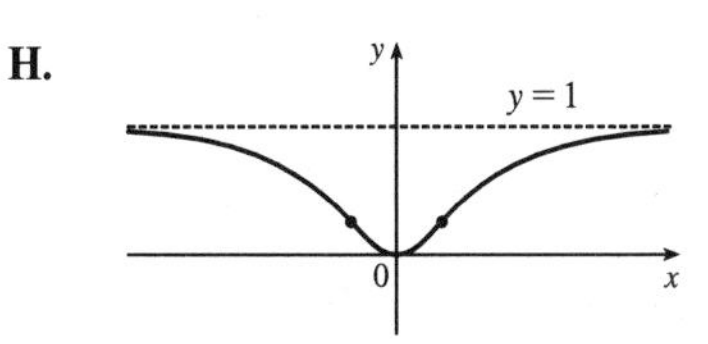

so f is CU on $(-\sqrt{3}, \sqrt{3})$ and CD on $(-\infty, -\sqrt{3})$ and $(\sqrt{3}, \infty)$. IPs at $\left(\pm\sqrt{3}, \frac{1}{4}\right)$

15. $y = f(x) = (x-1)/x^2$ **A.** $D = \{x \mid x \neq 0\} = (-\infty, 0) \cup (0, \infty)$ **B.** No y-intercept; x-intercept: $f(x) = 0 \Leftrightarrow$

$x = 1$ **C.** No symmetry **D.** $\displaystyle\lim_{x\to\pm\infty} \frac{x-1}{x^2} = 0$, so $y = 0$ is a HA. $\displaystyle\lim_{x\to 0} \frac{x-1}{x^2} = -\infty$, so $x = 0$ is a VA.

E. $f'(x) = \dfrac{x^2\cdot 1 - (x-1)\cdot 2x}{(x^2)^2} = \dfrac{-x^2+2x}{x^4} = \dfrac{-(x-2)}{x^3}$, so $f'(x) > 0 \Leftrightarrow 0 < x < 2$ and $f'(x) < 0 \Leftrightarrow$

$x < 0$ or $x > 2$. Thus, f is increasing on $(0, 2)$ and decreasing on $(-\infty, 0)$

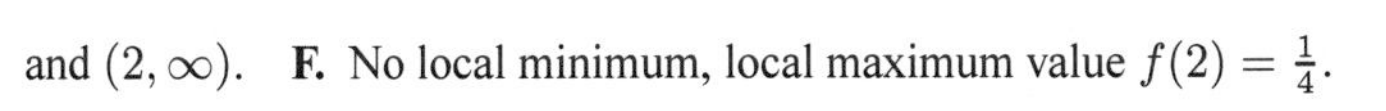

and $(2, \infty)$. **F.** No local minimum, local maximum value $f(2) = \frac{1}{4}$.

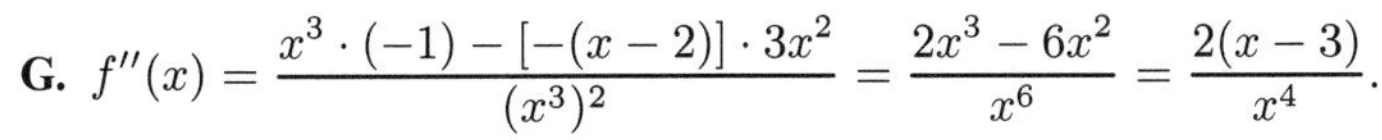

G. $f''(x) = \dfrac{x^3\cdot(-1) - [-(x-2)]\cdot 3x^2}{(x^3)^2} = \dfrac{2x^3 - 6x^2}{x^6} = \dfrac{2(x-3)}{x^4}$.

H.

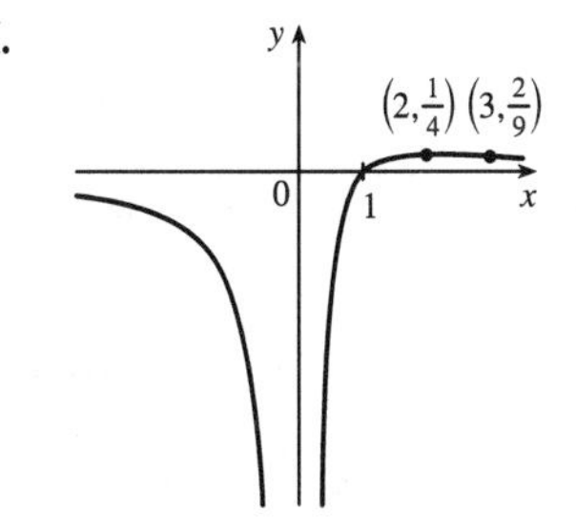

$f''(x)$ is negative on $(-\infty, 0)$ and $(0, 3)$ and positive on $(3, \infty)$, so f is

CD on $(-\infty, 0)$ and $(0, 3)$ and CU on $(3, \infty)$. IP at $\left(3, \frac{2}{9}\right)$

16. $y = f(x) = (x^3-1)/(x^3+1)$ **A.** $D = \{x \mid x \neq -1\} = (-\infty, -1) \cup (-1, \infty)$ **B.** x-intercept $= 1$,

y-intercept $= f(0) = -1$ **C.** No symmetry **D.** $\displaystyle\lim_{x\to\pm\infty} \frac{x^3-1}{x^3+1} = \lim_{x\to\pm\infty} \frac{1 - 1/x^3}{1 + 1/x^3} = 1$, so $y = 1$ is a HA.

$\displaystyle\lim_{x\to -1^-} \frac{x^3-1}{x^3+1} = \infty$ and $\displaystyle\lim_{x\to -1^+} \frac{x^3-1}{x^3+1} = -\infty$, so $x = -1$ is a VA.

E. $f'(x) = \dfrac{(x^3+1)(3x^2) - (x^3-1)(3x^2)}{(x^3+1)^2} = \dfrac{6x^2}{(x^3+1)^2} > 0$ $(x \neq -1)$ so f is increasing on $(-\infty, -1)$ and $(-1, \infty)$.

F. No extreme values

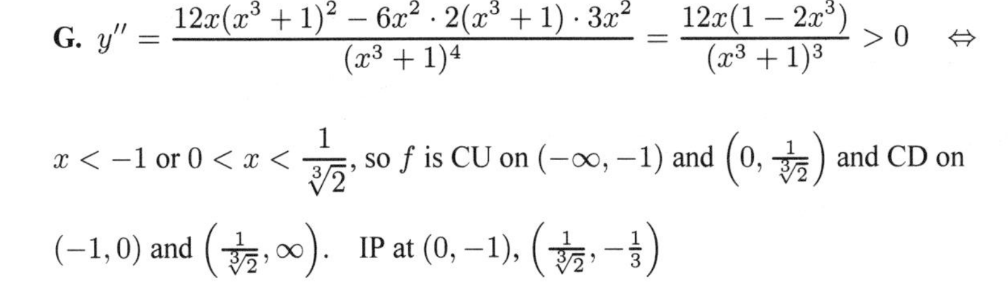

G. $y'' = \dfrac{12x(x^3+1)^2 - 6x^2\cdot 2(x^3+1)\cdot 3x^2}{(x^3+1)^4} = \dfrac{12x(1-2x^3)}{(x^3+1)^3} > 0 \Leftrightarrow$

$x < -1$ or $0 < x < \dfrac{1}{\sqrt[3]{2}}$, so f is CU on $(-\infty, -1)$ and $\left(0, \frac{1}{\sqrt[3]{2}}\right)$ and CD on

$(-1, 0)$ and $\left(\frac{1}{\sqrt[3]{2}}, \infty\right)$. IP at $(0, -1)$, $\left(\frac{1}{\sqrt[3]{2}}, -\frac{1}{3}\right)$

H.

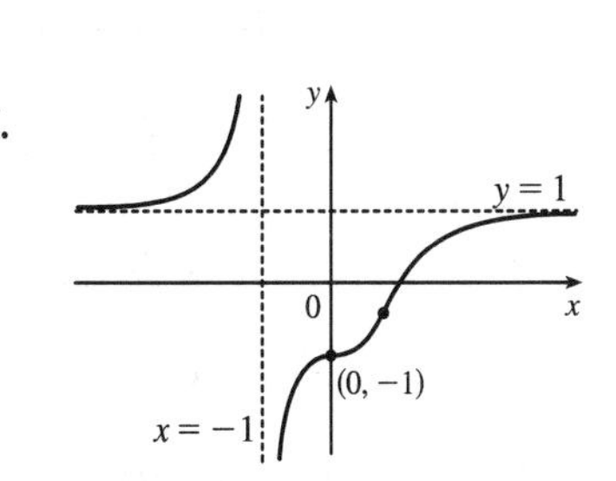

17. $y = f(x) = x\sqrt{5-x}$ **A.** The domain is $\{x \mid 5 - x \geq 0\} = (-\infty, 5]$ **B.** y-intercept: $f(0) = 0$;
x-intercepts: $f(x) = 0 \Leftrightarrow x = 0, 5$ **C.** No symmetry **D.** No asymptote

E. $f'(x) = x \cdot \frac{1}{2}(5-x)^{-1/2}(-1) + (5-x)^{1/2} \cdot 1 = \frac{1}{2}(5-x)^{-1/2}[-x + 2(5-x)] = \dfrac{10-3x}{2\sqrt{5-x}} > 0 \Leftrightarrow x < \frac{10}{3}$,

so f is increasing on $\left(-\infty, \frac{10}{3}\right)$ and decreasing on $\left(\frac{10}{3}, 5\right)$.

F. Local maximum value $f\left(\frac{10}{3}\right) = \frac{10}{9}\sqrt{15} \approx 4.3$; no local minimum

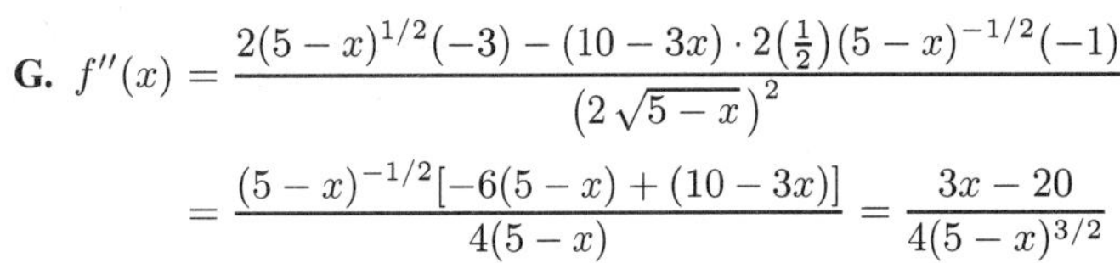

G. $$f''(x) = \frac{2(5-x)^{1/2}(-3) - (10-3x) \cdot 2\left(\frac{1}{2}\right)(5-x)^{-1/2}(-1)}{\left(2\sqrt{5-x}\right)^2}$$
$$= \frac{(5-x)^{-1/2}[-6(5-x) + (10-3x)]}{4(5-x)} = \frac{3x-20}{4(5-x)^{3/2}}$$

$f''(x) < 0$ for $x < 5$, so f is CD on $(-\infty, 5)$. No IP

H.

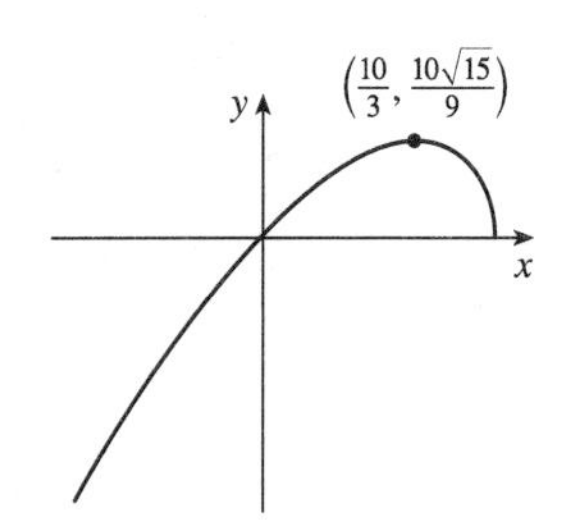

18. $y = f(x) = 2\sqrt{x} - x$ **A.** $D = [0, \infty)$ **B.** y-intercept: $f(0) = 0$; x-intercepts: $f(x) = 0 \Rightarrow 2\sqrt{x} = x \Rightarrow$
$4x = x^2 \Rightarrow 4x - x^2 = 0 \Rightarrow x(4-x) = 0 \Rightarrow x = 0, 4$ **C.** No symmetry **D.** No asymptote

E. $f'(x) = \dfrac{1}{\sqrt{x}} - 1 = \dfrac{1}{\sqrt{x}}(1 - \sqrt{x})$. This is positive for $x < 1$ and negative for $x > 1$, so f is increasing on $(0, 1)$
and decreasing on $(1, \infty)$. **F.** Local maximum value $f(1) = 1$, no local
minimum. **G.** $f''(x) = (x^{-1/2} - 1)' = -\frac{1}{2}x^{-3/2} = \dfrac{-1}{2x^{3/2}} < 0$ for $x > 0$,
so f is CD on $(0, \infty)$. No IP

H.

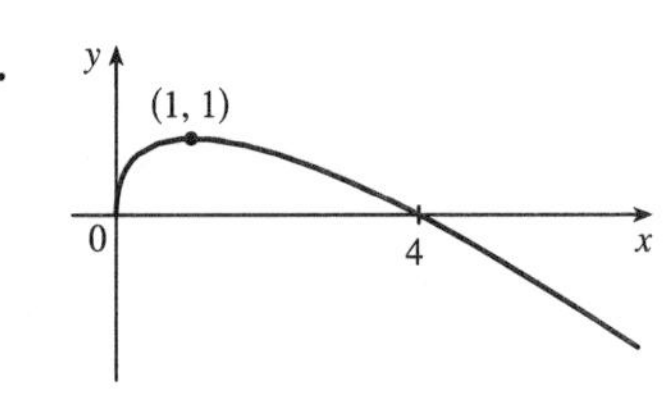

19. $y = f(x) = x/\sqrt{x^2+1}$ **A.** $D = \mathbb{R}$ **B.** y-intercept: $f(0) = 0$; x-intercept: $f(x) = 0 \Rightarrow x = 0$
C. $f(-x) = -f(x)$, so f is odd; the graph is symmetric about the origin.

D. $$\lim_{x\to\infty} f(x) = \lim_{x\to\infty} \frac{x}{\sqrt{x^2+1}} = \lim_{x\to\infty} \frac{x/x}{\sqrt{x^2+1}/x} = \lim_{x\to\infty} \frac{x/x}{\sqrt{x^2+1}/\sqrt{x^2}} = \lim_{x\to\infty} \frac{1}{\sqrt{1+1/x^2}} = \frac{1}{\sqrt{1+0}} = 1 \text{ and}$$

$$\lim_{x\to-\infty} f(x) = \lim_{x\to-\infty} \frac{x}{\sqrt{x^2+1}} = \lim_{x\to-\infty} \frac{x/x}{\sqrt{x^2+1}/x} = \lim_{x\to-\infty} \frac{x/x}{\sqrt{x^2+1}\Big/\left(-\sqrt{x^2}\right)}$$
$$= \lim_{x\to-\infty} \frac{1}{-\sqrt{1+1/x^2}} = \frac{1}{-\sqrt{1+0}} = -1$$

so $y = \pm 1$ are HAs. No VA. **E.** $f'(x) = \dfrac{\sqrt{x^2+1} - x \cdot \dfrac{2x}{2\sqrt{x^2+1}}}{[(x^2+1)^{1/2}]^2} = \dfrac{x^2+1-x^2}{(x^2+1)^{3/2}} = \dfrac{1}{(x^2+1)^{3/2}} > 0$ for all x,

so f is increasing on $\mathbb{R}$. **F.** No extreme values

G. $f''(x) = -\frac{3}{2}(x^2+1)^{-5/2} \cdot 2x = \dfrac{-3x}{(x^2+1)^{5/2}}$, so $f''(x) > 0$ for $x < 0$
and $f''(x) < 0$ for $x > 0$. Thus, f is CU on $(-\infty, 0)$ and CD on $(0, \infty)$.
IP at $(0, 0)$

H.

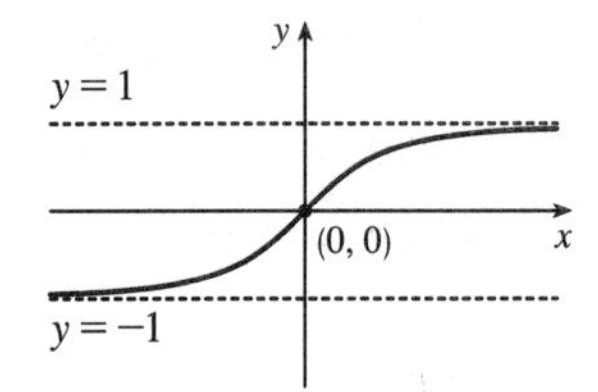

20. $y = f(x) = \sqrt{x/(x-5)}$ **A.** $D = \{x \mid x/(x-5) \geq 0\} = (-\infty, 0] \cup (5, \infty)$. **B.** Intercepts are 0.

C. No symmetry **D.** $\lim\limits_{x \to \pm\infty} \sqrt{\dfrac{x}{x-5}} = \lim\limits_{x \to \pm\infty} \sqrt{\dfrac{1}{1-5/x}} = 1$, so $y = 1$ is a HA. $\lim\limits_{x \to 5^+} \sqrt{\dfrac{x}{x-5}} = \infty$, so $x = 5$

is a VA. **E.** $f'(x) = \dfrac{1}{2}\left(\dfrac{x}{x-5}\right)^{-1/2} \dfrac{(-5)}{(x-5)^2} = -\frac{5}{2}\left[x(x-5)^3\right]^{-1/2} < 0$, so f is decreasing on $(-\infty, 0)$ and $(5, \infty)$.

F. No extreme values **G.** $f''(x) = \frac{5}{4}[x(x-5)^3]^{-3/2}(x-5)^2(4x-5) > 0$ for $x > 5$, and $f''(x) < 0$ for $x < 0$, so f is CU on $(5, \infty)$ and CD on $(-\infty, 0)$. No IP

H.

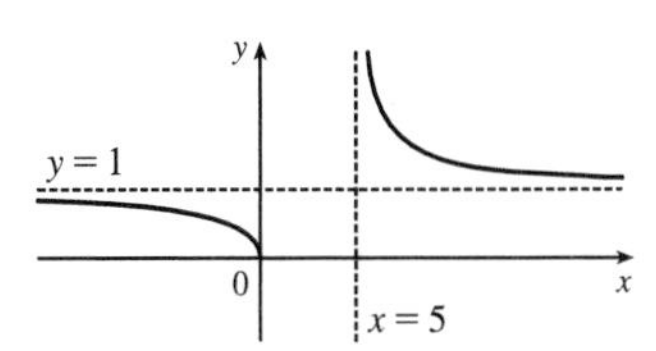

21. $y = f(x) = \sqrt{1-x^2}/x$ **A.** $D = \{x \mid |x| \leq 1, x \neq 0\} = [-1, 0) \cup (0, 1]$ **B.** x-intercepts ± 1, no y-intercept

C. $f(-x) = -f(x)$, so the curve is symmetric about $(0, 0)$. **D.** $\lim\limits_{x \to 0^+} \dfrac{\sqrt{1-x^2}}{x} = \infty$, $\lim\limits_{x \to 0^-} \dfrac{\sqrt{1-x^2}}{x} = -\infty$,

so $x = 0$ is a VA. **E.** $f'(x) = \dfrac{\left(-x^2/\sqrt{1-x^2}\right) - \sqrt{1-x^2}}{x^2} = -\dfrac{1}{x^2\sqrt{1-x^2}} < 0$, so f is decreasing on $(-1, 0)$

and $(0, 1)$. **F.** No extreme values **G.** $f''(x) = \dfrac{2-3x^2}{x^3(1-x^2)^{3/2}} > 0 \Leftrightarrow$

$-1 < x < -\sqrt{\frac{2}{3}}$ or $0 < x < \sqrt{\frac{2}{3}}$, so f is CU on $\left(-1, -\sqrt{\frac{2}{3}}\right)$ and $\left(0, \sqrt{\frac{2}{3}}\right)$

and CD on $\left(-\sqrt{\frac{2}{3}}, 0\right)$ and $\left(\sqrt{\frac{2}{3}}, 1\right)$. IP at $\left(\pm\sqrt{\frac{2}{3}}, \pm\frac{1}{\sqrt{2}}\right)$

H.

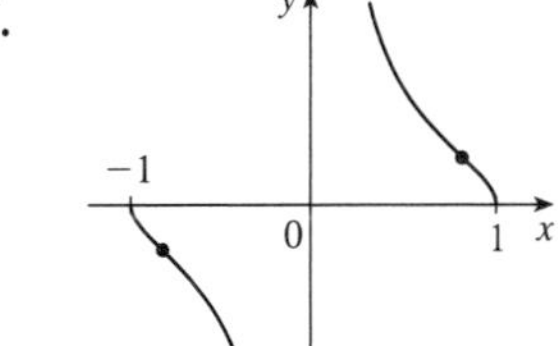

22. $y = f(x) = x\sqrt{2-x^2}$ **A.** $D = \left[-\sqrt{2}, \sqrt{2}\right]$ **B.** y-intercept: $f(0) = 0$; x-intercepts: $f(x) = 0 \Rightarrow$ $x = 0, \pm\sqrt{2}$. **C.** $f(-x) = -f(x)$, so f is odd; the graph is symmetric about the origin. **D.** No asymptote

E. $f'(x) = x \cdot \dfrac{-x}{\sqrt{2-x^2}} + \sqrt{2-x^2} = \dfrac{-x^2 + 2 - x^2}{\sqrt{2-x^2}} = \dfrac{2(1+x)(1-x)}{\sqrt{2-x^2}}$. $f'(x)$ is negative for $-\sqrt{2} < x < -1$ and $1 < x < \sqrt{2}$, and positive for $-1 < x < 1$, so f is decreasing on $\left(-\sqrt{2}, -1\right)$ and $\left(1, \sqrt{2}\right)$ and increasing on $(-1, 1)$.

F. Local minimum value $f(-1) = -1$, local maximum value $f(1) = 1$.

G. $f''(x) = \dfrac{\sqrt{2-x^2}(-4x) - (2-2x^2)\dfrac{-x}{\sqrt{2-x^2}}}{[(2-x^2)^{1/2}]^2}$

$= \dfrac{(2-x^2)(-4x) + (2-2x^2)x}{(2-x^2)^{3/2}} = \dfrac{(2-x^2)(-4x) + (2-2x^2)x}{(2-x^2)^{3/2}}$

Since $x^2 - 3 < 0$ for x in $\left[-\sqrt{2}, \sqrt{2}\right]$, $f''(x) > 0$ for $-\sqrt{2} < x < 0$ and $f''(x) < 0$ for $0 < x < \sqrt{2}$. Thus, f is CU on $\left(-\sqrt{2}, 0\right)$ and CD on $\left(0, \sqrt{2}\right)$. The only IP is $(0, 0)$.

H.

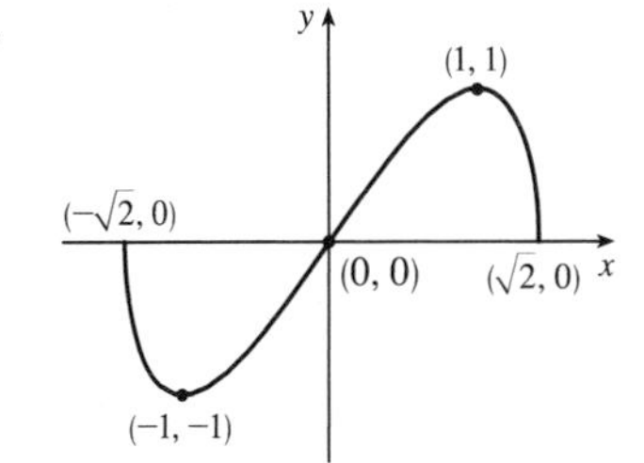

23. $y = f(x) = x - 3x^{1/3}$ **A.** $D = \mathbb{R}$ **B.** y-intercept: $f(0) = 0$; x-intercepts: $f(x) = 0 \Rightarrow x = 3x^{1/3} \Rightarrow$ $x^3 = 27x \Rightarrow x^3 - 27x = 0 \Rightarrow x(x^2 - 27) = 0 \Rightarrow x = 0, \pm 3\sqrt{3}$ **C.** $f(-x) = -f(x)$, so f is odd;

the graph is symmetric about the origin. **D.** No asymptote **E.** $f'(x) = 1 - x^{-2/3} = 1 - \dfrac{1}{x^{2/3}} = \dfrac{x^{2/3} - 1}{x^{2/3}}$.

$f'(x) > 0$ when $|x| > 1$ and $f'(x) < 0$ when $0 < |x| < 1$, so f is increasing on $(-\infty, -1)$ and $(1, \infty)$, and decreasing on $(-1, 0)$ and $(0, 1)$ [hence decreasing on $(-1, 1)$ since f is continuous on $(-1, 1)$]. **F.** Local maximum value $f(-1) = 2$, local minimum value $f(1) = -2$ **G.** $f''(x) = \frac{2}{3}x^{-5/3} < 0$ when $x < 0$ and $f''(x) > 0$ when $x > 0$, so f is CD on $(-\infty, 0)$ and CU on $(0, \infty)$. IP at $(0, 0)$

H.

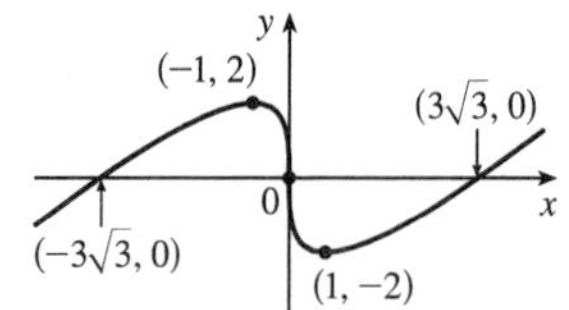

24. $y = f(x) = x^{5/3} - 5x^{2/3} = x^{2/3}(x - 5)$ **A.** $D = \mathbb{R}$ **B.** x-intercepts 0, 5; y-intercept 0 **C.** No symmetry **D.** $\lim\limits_{x\to\pm\infty} x^{2/3}(x - 5) = \pm\infty$, so there is no asymptote **E.** $f'(x) = \frac{5}{3}x^{2/3} - \frac{10}{3}x^{-1/3} = \frac{5}{3}x^{-1/3}(x - 2) > 0 \Leftrightarrow$ $x < 0$ or $x > 2$, so f is increasing on $(-\infty, 0)$, $(2, \infty)$ and decreasing on $(0, 2)$. **F.** Local maximum value $f(0) = 0$, local minimum value $f(2) = -3\sqrt[3]{4}$ **G.** $f''(x) = \frac{10}{9}x^{-1/3} + \frac{10}{9}x^{-4/3} = \frac{10}{9}x^{-4/3}(x + 1) > 0$ $\Leftrightarrow$ $x > -1$, so f is CU on $(-1, 0)$ and $(0, \infty)$, CD on $(-\infty, -1)$. IP at $(-1, -6)$

H.

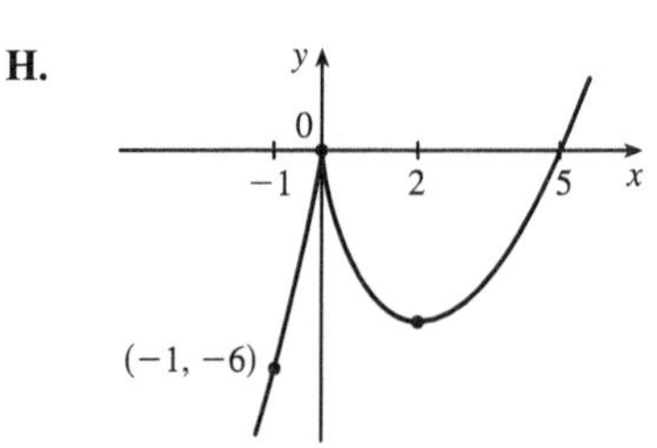

25. $y = f(x) = x + \sqrt{|x|}$ **A.** $D = \mathbb{R}$ **B.** x-intercepts 0, -1; y-intercept 0 **C.** No symmetry **D.** $\lim\limits_{x\to\infty}\left(x + \sqrt{|x|}\right) = \infty$, $\lim\limits_{x\to-\infty}\left(x + \sqrt{|x|}\right) = -\infty$. No asymptote **E.** For $x > 0$, $f(x) = x + \sqrt{x} \Rightarrow f'(x) = 1 + \dfrac{1}{2\sqrt{x}} > 0$, so f increases on $(0, \infty)$. For $x < 0$, $f(x) = x + \sqrt{-x} \Rightarrow f'(x) = 1 - \dfrac{1}{2\sqrt{-x}} > 0 \Leftrightarrow 2\sqrt{-x} > 1 \Leftrightarrow -x > \frac{1}{4} \Leftrightarrow$ $x < -\frac{1}{4}$, so f increases on $\left(-\infty, -\frac{1}{4}\right)$ and decreases on $\left(-\frac{1}{4}, 0\right)$.

F. Local maximum value $f\left(-\frac{1}{4}\right) = \frac{1}{4}$, local minimum value $f(0) = 0$

G. For $x > 0$, $f''(x) = -\frac{1}{4}x^{-3/2} \Rightarrow f''(x) < 0$, so f is CD on $(0, \infty)$.

For $x < 0$, $f''(x) = -\frac{1}{4}(-x)^{-3/2} \Rightarrow f''(x) < 0$, so f is CD on $(-\infty, 0)$.

No IP

H.

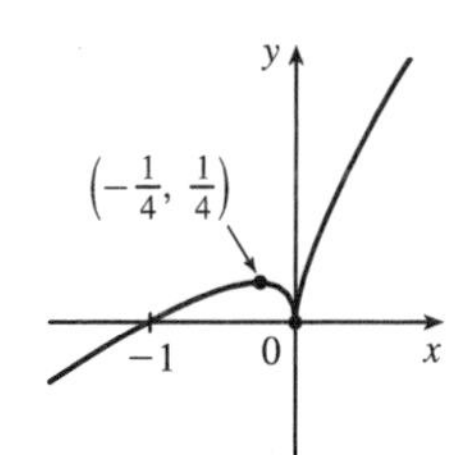

26. $y = f(x) = \sqrt[3]{(x^2 - 1)^2} = (x^2 - 1)^{2/3}$ **A.** $D = \mathbb{R}$ **B.** x-intercepts ± 1, y-intercept 1 **C.** $f(-x) = f(x)$, so the curve is symmetric about the y-axis. **D.** $\lim\limits_{x\to\pm\infty}(x^2 - 1)^{2/3} = \infty$, no asymptote **E.** $f'(x) = \frac{4}{3}x(x^2 - 1)^{-1/3} \Rightarrow$ $f'(x) > 0 \Leftrightarrow x > 1$ or $-1 < x < 0$, $f'(x) < 0 \Leftrightarrow x < -1$ or $0 < x < 1$. So f is increasing on $(-1, 0)$, $(1, \infty)$ and decreasing on $(-\infty, -1)$, $(0, 1)$. **F.** Local minimum values $f(-1) = f(1) = 0$, local maximum value $f(0) = 1$

G. $f''(x) = \frac{4}{3}(x^2 - 1)^{-1/3} + \frac{4}{3}x\left(-\frac{1}{3}\right)3(x^2 - 1)^{-4/3}(2x)$
$= \frac{4}{9}(x^2 - 3)(x^2 - 1)^{-4/3} > 0 \Leftrightarrow |x| > \sqrt{3}$

so f is CU on $\left(-\infty, -\sqrt{3}\right)$, $\left(\sqrt{3}, \infty\right)$ and CD on $\left(-\sqrt{3}, -1\right)$, $(-1, 1), (1, \sqrt{3})$. IPs at $\left(\pm\sqrt{3}, \sqrt[3]{4}\right)$

H.

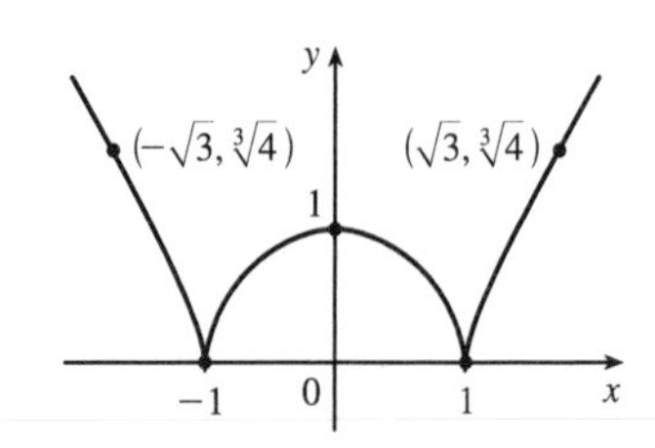

27. $y = f(x) = 3\sin x - \sin^3 x$ **A.** $D = \mathbb{R}$ **B.** y-intercept: $f(0) = 0$; x-intercepts: $f(x) = 0 \Rightarrow$ $\sin x\,(3 - \sin^2 x) = 0 \Rightarrow \sin x = 0$ [since $\sin^2 x \le 1 < 3$] $\Rightarrow x = n\pi$, n an integer. **C.** $f(-x) = -f(x)$, so f is odd; the graph (shown for $-2\pi \le x \le 2\pi$) is symmetric about the origin and periodic with period 2π. **D.** No asymptote **E.** $f'(x) = 3\cos x - 3\sin^2 x\cos x = 3\cos x\,(1 - \sin^2 x) = 3\cos^3 x$. $f'(x) > 0 \Leftrightarrow \cos x > 0 \Leftrightarrow$ $x \in \left(2n\pi - \frac{\pi}{2}, 2n\pi + \frac{\pi}{2}\right)$ for each integer n, and $f'(x) < 0 \Leftrightarrow \cos x < 0 \Leftrightarrow x \in \left(2n\pi + \frac{\pi}{2}, 2n\pi + \frac{3\pi}{2}\right)$ for each integer n. Thus, f is increasing on $\left(2n\pi - \frac{\pi}{2}, 2n\pi + \frac{\pi}{2}\right)$ for each integer n, and f is decreasing on $\left(2n\pi + \frac{\pi}{2}, 2n\pi + \frac{3\pi}{2}\right)$ for each integer n. **F.** f has local maximum values $f(2n\pi + \frac{\pi}{2}) = 2$ and local minimum values $f(2n\pi + \frac{3\pi}{2}) = -2$. **G.** $f''(x) = -9\sin x\cos^2 x = -9\sin x\,(1 - \sin^2 x) = -9\sin x\,(1 - \sin x)(1 + \sin x)$.
$f''(x) < 0 \Leftrightarrow \sin x > 0$ and $\sin x \ne \pm 1 \Leftrightarrow x \in \left(2n\pi, 2n\pi + \frac{\pi}{2}\right) \cup \left(2n\pi + \frac{\pi}{2}, 2n\pi + \pi\right)$ for some integer n.
$f''(x) > 0 \Leftrightarrow \sin x < 0$ and $\sin x \ne \pm 1 \Leftrightarrow x \in \left((2n-1)\pi, (2n-1)\pi + \frac{\pi}{2}\right) \cup \left((2n-1)\pi + \frac{\pi}{2}, 2n\pi\right)$ for some integer n. Thus, f is CD on the intervals $\left(2n\pi, \left(2n + \frac{1}{2}\right)\pi\right)$ and $\left(\left(2n + \frac{1}{2}\right)\pi, (2n+1)\,\pi\right)$ [hence CD on the intervals $(2n\pi, (2n+1)\,\pi)$] for each integer n, and f is CU on the intervals $\left((2n-1)\pi, \left(2n - \frac{1}{2}\right)\pi\right)$ and $\left(\left(2n - \frac{1}{2}\right)\pi, 2n\pi\right)$ [hence CU on the intervals $((2n-1)\pi, 2n\pi)$] for each integer n. f has IPs at $(n\pi, 0)$ for each integer n.

H.

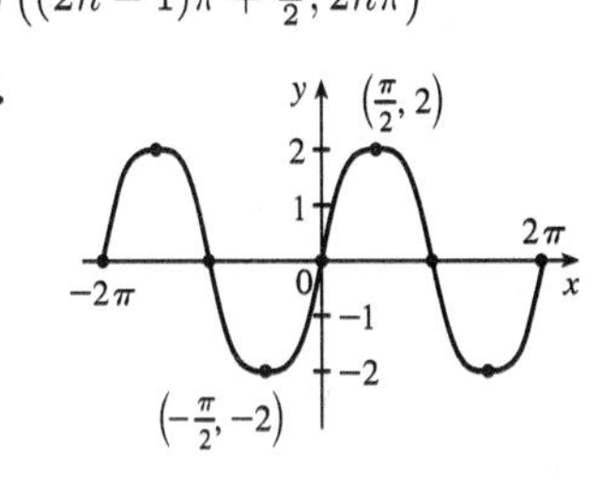

28. $y = f(x) = \sin x - \tan x$ **A.** $D = \left\{x \mid x \ne (2n+1)\frac{\pi}{2}\right\}$ **B.** $y = 0 \Leftrightarrow \sin x = \tan x = \dfrac{\sin x}{\cos x} \Leftrightarrow \sin x = 0$ or $\cos x = 1 \Leftrightarrow x = n\pi$ (x-intercepts), y-intercept $= f(0) = 0$ **C.** $f(-x) = -f(x)$, so the curve is symmetric about $(0, 0)$. Also periodic with period 2π **D.** $\lim\limits_{x \to (\pi/2)^-} (\sin x - \tan x) = -\infty$ and $\lim\limits_{x \to (\pi/2)^+} (\sin x - \tan x) = \infty$, so $x = n\pi + \frac{\pi}{2}$ are VAs. **E.** $f'(x) = \cos x - \sec^2 x \le 0$, so f decreases on each interval in its domain, that is, on $\left((2n-1)\frac{\pi}{2}, (2n+1)\frac{\pi}{2}\right)$. **F.** No extreme values **G.** $f''(x) = -\sin x - 2\sec^2 x\tan x = -\sin x\,(1 + 2\sec^3 x)$. Note that $1 + 2\sec^3 x \ne 0$ since $\sec^3 x \ne -\frac{1}{2}$. $f''(x) > 0$ for $-\frac{\pi}{2} < x < 0$ and $\frac{3\pi}{2} < x < 2\pi$, so f is CU on $\left(\left(n - \frac{1}{2}\right)\pi, n\pi\right)$ and CD on $\left(n\pi, \left(n + \frac{1}{2}\right)\pi\right)$. f has IPs at $(n\pi, 0)$. Note also that $f'(0) = 0$, but $f'(\pi) = -2$.

H.

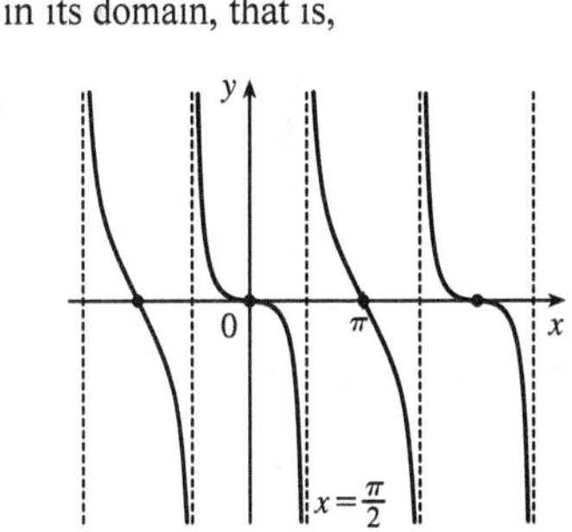

29. $y = f(x) = x\tan x$, $-\frac{\pi}{2} < x < \frac{\pi}{2}$ **A.** $D = \left(-\frac{\pi}{2}, \frac{\pi}{2}\right)$ **B.** Intercepts are 0 **C.** $f(-x) = f(x)$, so the curve is symmetric about the y-axis. **D.** $\lim\limits_{x \to (\pi/2)^-} x\tan x = \infty$ and $\lim\limits_{x \to -(\pi/2)^+} x\tan x = \infty$, so $x = \frac{\pi}{2}$ and $x = -\frac{\pi}{2}$ are VAs. **E.** $f'(x) = \tan x + x\sec^2 x > 0 \Leftrightarrow 0 < x < \frac{\pi}{2}$, so f increases on $\left(0, \frac{\pi}{2}\right)$ and decreases on $\left(-\frac{\pi}{2}, 0\right)$. **F.** Absolute and local minimum value $f(0) = 0$. **G.** $y'' = 2\sec^2 x + 2x\tan x\sec^2 x > 0$ for $-\frac{\pi}{2} < x < \frac{\pi}{2}$, so f is CU on $\left(-\frac{\pi}{2}, \frac{\pi}{2}\right)$. No IP

H.

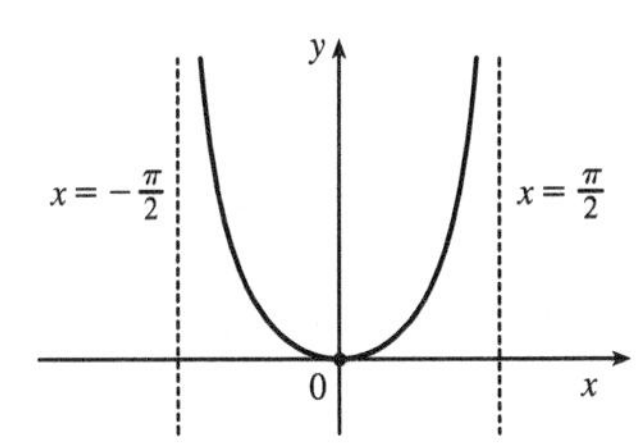

30. $y = f(x) = 2x - \tan x$, $-\frac{\pi}{2} < x < \frac{\pi}{2}$ **A.** $D = \left(-\frac{\pi}{2}, \frac{\pi}{2}\right)$ **B.** y-intercept: $f(0) = 0$; x-intercepts: $f(0) = 0 \Leftrightarrow 2x = \tan x \Leftrightarrow x = 0$ or $x \approx \pm 1.17$ **C.** $f(-x) = -f(x)$, so f is odd; the graph is symmetric about the origin.

D. $\lim\limits_{x \to (-\pi/2)^+} (2x - \tan x) = \infty$ and $\lim\limits_{x \to (\pi/2)^-} (2x - \tan x) = -\infty$, so $x = \pm\frac{\pi}{2}$ are VAs. No HA.

E. $f'(x) = 2 - \sec^2 x < 0 \Leftrightarrow |\sec x| > \sqrt{2}$ and $f'(x) > 0 \Leftrightarrow |\sec x| < \sqrt{2}$, so f is decreasing on $\left(-\frac{\pi}{2}, -\frac{\pi}{4}\right)$, increasing on $\left(-\frac{\pi}{4}, \frac{\pi}{4}\right)$, and decreasing again on $\left(\frac{\pi}{4}, \frac{\pi}{2}\right)$ **F.** Local maximum value $f\left(\frac{\pi}{4}\right) = \frac{\pi}{2} - 1$, local minimum value $f\left(-\frac{\pi}{4}\right) = -\frac{\pi}{2} + 1$

G. $f''(x) = -2\sec x \cdot \sec x \tan x = -2\tan x \sec^2 x = -2\tan x(\tan^2 x + 1)$

so $f''(x) > 0 \Leftrightarrow \tan x < 0 \Leftrightarrow -\frac{\pi}{2} < x < 0$, and $f''(x) < 0 \Leftrightarrow \tan x > 0 \Leftrightarrow 0 < x < \frac{\pi}{2}$. Thus, f is CU on $\left(-\frac{\pi}{2}, 0\right)$ and CD on $\left(0, \frac{\pi}{2}\right)$.

IP at $(0, 0)$

H.

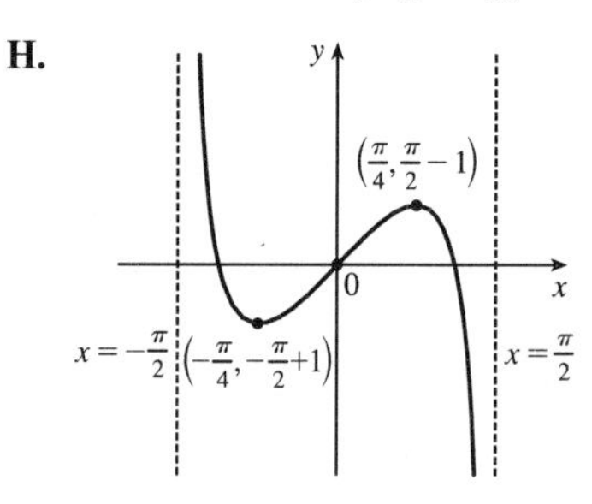

31. $y = f(x) = \frac{1}{2}x - \sin x$, $0 < x < 3\pi$ **A.** $D = (0, 3\pi)$ **B.** No y-intercept. The x-intercept, approximately 1.9, can be found using Newton's method. **C.** No symmetry **D.** No asymptote **E.** $f'(x) = \frac{1}{2} - \cos x > 0 \Leftrightarrow \cos x < \frac{1}{2} \Leftrightarrow \frac{\pi}{3} < x < \frac{5\pi}{3}$ or $\frac{7\pi}{3} < x < 3\pi$, so f is increasing on $\left(\frac{\pi}{3}, \frac{5\pi}{3}\right)$ and $\left(\frac{7\pi}{3}, 3\pi\right)$ and decreasing on $\left(0, \frac{\pi}{3}\right)$ and $\left(\frac{5\pi}{3}, \frac{7\pi}{3}\right)$.

F. Local minimum value $f\left(\frac{\pi}{3}\right) = \frac{\pi}{6} - \frac{\sqrt{3}}{2}$, local maximum value $f\left(\frac{5\pi}{3}\right) = \frac{5\pi}{6} + \frac{\sqrt{3}}{2}$, local minimum value $f\left(\frac{7\pi}{3}\right) = \frac{7\pi}{6} - \frac{\sqrt{3}}{2}$

G. $f''(x) = \sin x > 0 \Leftrightarrow 0 < x < \pi$ or $2\pi < x < 3\pi$, so f is CU on $(0, \pi)$ and $(2\pi, 3\pi)$ and CD on $(\pi, 2\pi)$. IPs at $\left(\pi, \frac{\pi}{2}\right)$ and $(2\pi, \pi)$

H.

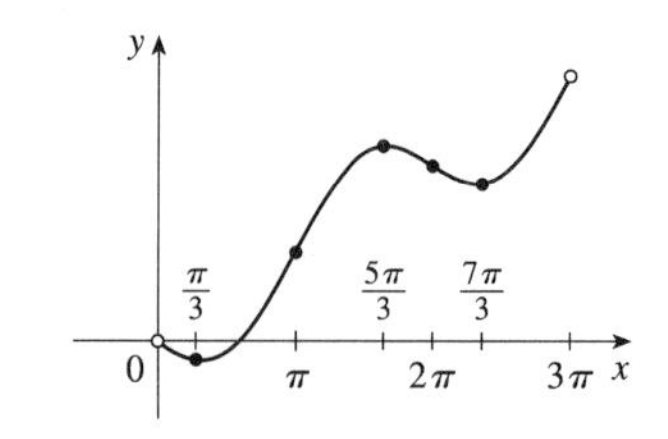

32. $y = f(x) = \cos^2 x - 2\sin x$ **A.** $D = \mathbb{R}$ **B.** y-intercept: $f(0) = 1$ **C.** No symmetry, but f has period 2π.

D. No asymptote **E.** $y' = 2\cos x\,(-\sin x) - 2\cos x = -2\cos x\,(\sin x + 1)$. $y' = 0 \Leftrightarrow \cos x = 0$ or $\sin x = -1 \Leftrightarrow x = (2n+1)\frac{\pi}{2}$. $y' > 0$ when $\cos x < 0$ since $\sin x + 1 \geq 0$ for all x. So $y' > 0$ and f is increasing on $\left((4n+1)\frac{\pi}{2}, (4n+3)\frac{\pi}{2}\right)$; $y' < 0$ and f is decreasing on $\left((4n-1)\frac{\pi}{2}, (4n+1)\frac{\pi}{2}\right)$. **F.** Local maximum values $f\left((4n+3)\frac{\pi}{2}\right) = 2$, local minimum values $f\left((4n+1)\frac{\pi}{2}\right) = -2$ **G.** $y' = -2\cos x\,(\sin x + 1) = -\sin 2x - 2\cos x \Rightarrow$

$y'' = -2\cos 2x + 2\sin x = -2(1 - 2\sin^2 x) + 2\sin x = 4\sin^2 x + 2\sin x - 2 = 2(2\sin x - 1)(\sin x + 1)$

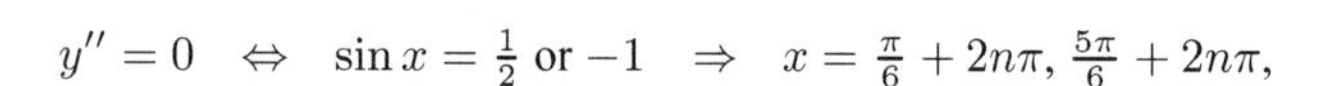

$y'' = 0 \Leftrightarrow \sin x = \frac{1}{2}$ or $-1 \Rightarrow x = \frac{\pi}{6} + 2n\pi, \frac{5\pi}{6} + 2n\pi$, or $\frac{3\pi}{2} + 2n\pi$. $y'' > 0$ and f is CU on $\left(\frac{\pi}{6} + 2n\pi, \frac{5\pi}{6} + 2n\pi\right)$; $y'' \leq 0$ and f is CD on $\left(\frac{5\pi}{6} + 2n\pi, \frac{\pi}{6} + 2(n+1)\pi\right)$.

IPs at $\left(\frac{\pi}{6} + 2n\pi, -\frac{1}{4}\right)$ and $\left(\frac{5\pi}{6} + 2n\pi, -\frac{1}{4}\right)$.

H.

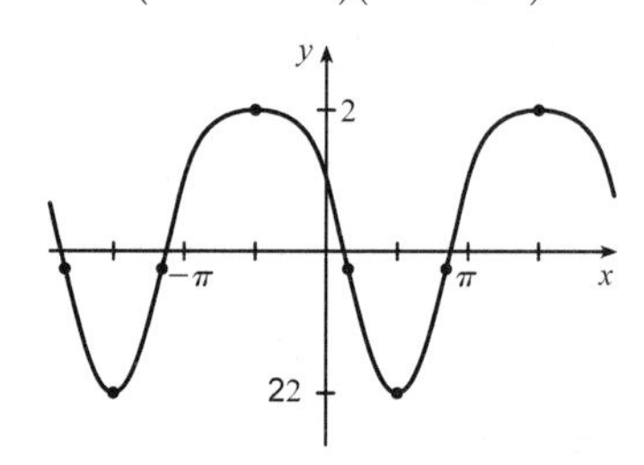

33. $y = f(x) = \dfrac{\sin x}{1 + \cos x}$ $\left[\overset{\text{when } \cos x \neq 1}{=} \dfrac{\sin x}{1 + \cos x} \cdot \dfrac{1 - \cos x}{1 - \cos x} = \dfrac{\sin x\,(1 - \cos x)}{\sin^2 x} = \dfrac{1 - \cos x}{\sin x} = \csc x - \cot x \right]$

A. The domain of f is the set of all real numbers except odd integer multiples of π. **B.** y-intercept: $f(0) = 0$; x-intercepts: $x = n\pi$, n an even integer. **C.** $f(-x) = -f(x)$, so f is an odd function; the graph is symmetric about the origin and has period 2π. **D.** When n is an odd integer, $\lim_{x\to(n\pi)^-} f(x) = \infty$ and $\lim_{x\to(n\pi)^+} f(x) = -\infty$, so $x = n\pi$ is a VA for each odd integer n. No HA. **E.** $f'(x) = \dfrac{(1+\cos x)\cdot\cos x - \sin x(-\sin x)}{(1+\cos x)^2} = \dfrac{1+\cos x}{(1+\cos x)^2} = \dfrac{1}{1+\cos x}$.

$f'(x) > 0$ for all x except odd multiples of π, so f is increasing on $((2k-1)\pi, (2k+1)\pi)$ for each integer k.

F. No extreme values **G.** $f''(x) = \dfrac{\sin x}{(1+\cos x)^2} > 0 \;\Rightarrow\; \sin x > 0 \;\Rightarrow\; x \in (2k\pi, (2k+1)\pi)$ and $f''(x) < 0$ on $((2k-1)\pi, 2k\pi)$ for each integer k. f is CU on $(2k\pi, (2k+1)\pi)$ and CD on $((2k-1)\pi, 2k\pi)$ for each integer k. f has IPs at $(2k\pi, 0)$ for each integer k.

H.

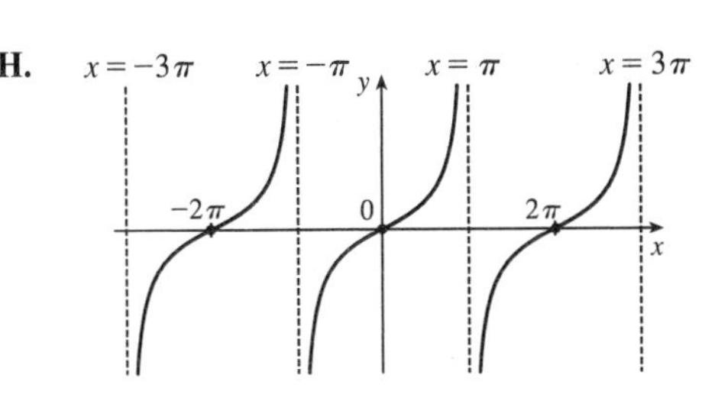

34. $f(x) = \sin x - x$ **A.** $D = \mathbb{R}$ **B.** x-intercept $= 0 = y$-intercept

C. $f(-x) = \sin(-x) - (-x) = -(\sin x - x) = -f(x)$, so f is odd.

D. No asymptote **E.** $f'(x) = \cos x - 1 \le 0$ for all x, so f is decreasing on $(-\infty, \infty)$. **F.** No extreme values **G.** $f''(x) = -\sin x \;\Rightarrow$

$f''(x) > 0 \;\Leftrightarrow\; \sin x < 0 \;\Leftrightarrow\; (2n-1)\pi < x < 2n\pi$, so f is CU on $((2n-1)\pi, 2n\pi)$ and CD on $(2n\pi, (2n+1)\pi)$, n an integer. IPs occur when $x = n\pi$.

H.

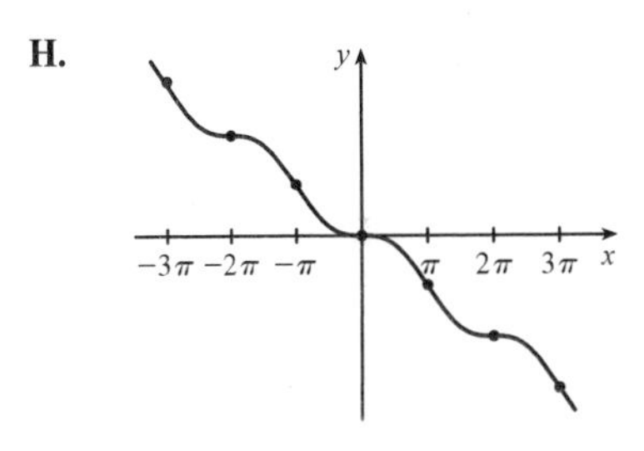

35. $y = -\dfrac{W}{24EI}x^4 + \dfrac{WL}{12EI}x^3 - \dfrac{WL^2}{24EI}x^2 = -\dfrac{W}{24EI}x^2\left(x^2 - 2Lx + L^2\right) = \dfrac{-W}{24EI}x^2(x-L)^2 = cx^2(x-L)^2$

where $c = -\dfrac{W}{24EI}$ is a negative constant and $0 \le x \le L$. We sketch $f(x) = cx^2(x-L)^2$ for $c = -1$.

$f(0) = f(L) = 0$

$f'(x) = cx^2[2(x-L)] + (x-L)^2(2cx) = 2cx(x-L)[x + (x-L)]$

$\quad = 2cx(x-L)(2x-L)$

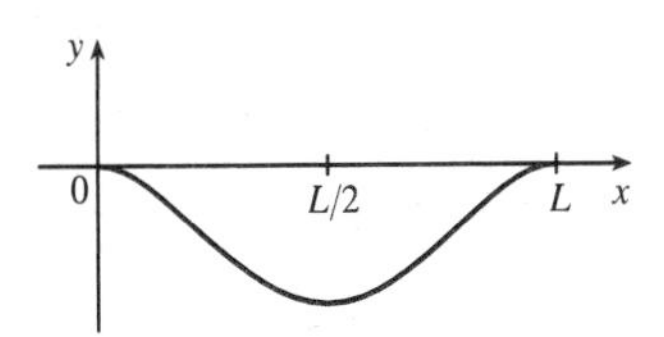

So for $0 < x < L$, $f'(x) > 0 \;\Leftrightarrow\; x(x-L)(2x-L) < 0$ [since $c < 0$] $\;\Leftrightarrow\; L/2 < x < L$ and $f'(x) < 0 \;\Leftrightarrow\; 0 < x < L/2$. Thus, f is increasing on $(L/2, L)$ and decreasing on $(0, L/2)$, and there is a local and absolute minimum at the point $(L/2, f(L/2)) = \left(L/2, cL^4/16\right)$.

$f'(x) = 2c\,[x(x-L)(2x-L)] \;\Rightarrow$

$f''(x) = 2c\,[1(x-L)(2x-L) + x(1)(2x-L) + x(x-L)(2)] = 2c\left(6x^2 - 6Lx + L^2\right) = 0 \;\Leftrightarrow$

$x = \dfrac{6L \pm \sqrt{12L^2}}{12} = \frac{1}{2}L \pm \frac{\sqrt{3}}{6}L$, and these are the x-coordinates of the two inflection points.

36. $F(x) = -\dfrac{k}{x^2} + \dfrac{k}{(x-2)^2}$, where $k > 0$ and $0 < x < 2$. For $0 < x < 2$, $x - 2 < 0$, so

$F'(x) = \dfrac{2k}{x^3} - \dfrac{2k}{(x-2)^3} > 0$ and F is increasing. $\lim\limits_{x \to 0^+} F(x) = -\infty$ and

$\lim\limits_{x \to 2^-} F(x) = \infty$, so $x = 0$ and $x = 2$ are vertical asymptotes. Notice that when the middle particle is at $x = 1$, the net force acting on it is 0. When $x > 1$, the net force is positive, meaning that it acts to the right. And if the particle approaches $x = 2$, the force on it rapidly becomes very large. When $x < 1$, the net force is negative, so it acts to the left. If the particle approaches 0, the force becomes very large to the left.

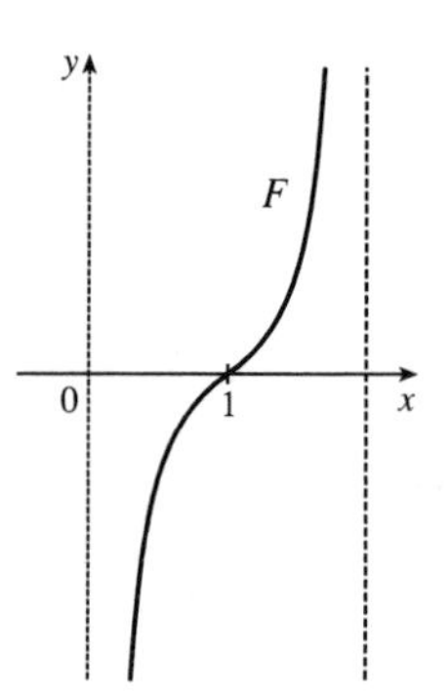

37. $y = f(x) = \dfrac{-2x^2 + 5x - 1}{2x - 1} = -x + 2 + \dfrac{1}{2x - 1}$ **A.** $D = \{x \in \mathbb{R} \mid x \neq \frac{1}{2}\} = \left(-\infty, \frac{1}{2}\right) \cup \left(\frac{1}{2}, \infty\right)$

B. y-intercept: $f(0) = 1$; x-intercepts: $f(x) = 0 \;\Rightarrow\; -2x^2 + 5x - 1 = 0 \;\Rightarrow\; x = \dfrac{-5 \pm \sqrt{17}}{-4} \;\Rightarrow\; x \approx 0.22, 2.28.$

C. No symmetry **D.** $\lim\limits_{x \to (1/2)^-} f(x) = -\infty$ and $\lim\limits_{x \to (1/2)^+} f(x) = \infty$, so $x = \frac{1}{2}$ is a VA.

$\lim\limits_{x \to \pm\infty} [f(x) - (-x + 2)] = \lim\limits_{x \to \pm\infty} \dfrac{1}{2x - 1} = 0$, so the line $y = -x + 2$ is a SA.

E. $f'(x) = -1 - \dfrac{2}{(2x - 1)^2} < 0$ for $x \neq \frac{1}{2}$, so f is decreasing on $\left(-\infty, \frac{1}{2}\right)$ and $\left(\frac{1}{2}, \infty\right)$. **F.** No extreme values **G.** $f'(x) = -1 - 2(2x - 1)^{-2} \;\Rightarrow$

$f''(x) = -2(-2)(2x - 1)^{-3}(2) = \dfrac{8}{(2x - 1)^3}$, so $f''(x) > 0$ when $x > \frac{1}{2}$ and $f''(x) < 0$ when $x < \frac{1}{2}$. Thus, f is CU on $\left(\frac{1}{2}, \infty\right)$ and CD on $\left(-\infty, \frac{1}{2}\right)$.
No IP

H.

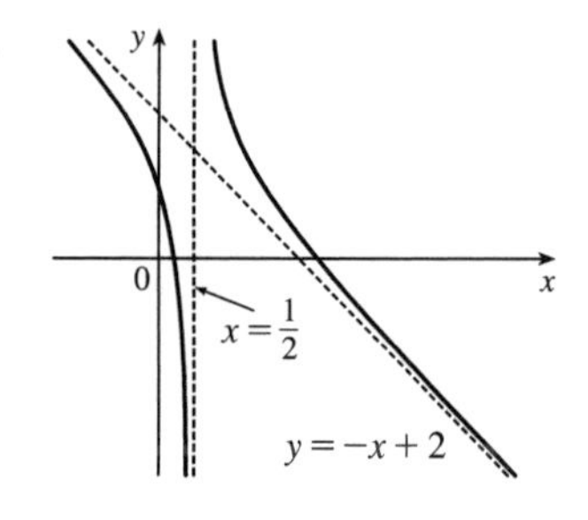

38. $y = f(x) = \dfrac{x^2 + 12}{x - 2} = x + 2 + \dfrac{16}{x - 2}$ **A.** $D = \{x \in \mathbb{R} \mid x \neq 2\} = (-\infty, 2) \cup (2, \infty)$ **B.** y-intercept: $f(0) = -6$; no x-intercepts. **C.** No symmetry **D.** $\lim\limits_{x \to 2^-} f(x) = -\infty$ and $\lim\limits_{x \to 2^+} f(x) = \infty$, so $x = 2$ is a VA.

$\lim\limits_{x \to \pm\infty} [f(x) - (x + 2)] = \lim\limits_{x \to \pm\infty} \dfrac{16}{x - 2} = 0$, so the line $y = x + 2$ is a SA.

E. $f'(x) = 1 - \dfrac{16}{(x - 2)^2} = \dfrac{x^2 - 4x - 12}{(x - 2)^2} = \dfrac{(x - 6)(x + 2)}{(x - 2)^2}$, so $f'(x) > 0$ when $x < -2$ or $x > 6$ and $f'(x) < 0$ when $-2 < x < 2$ or $2 < x < 6$. Thus, f is increasing on $(-\infty, -2)$ and $(6, \infty)$ and decreasing on $(-2, 2)$ and $(2, 6)$.

F. Local maximum value $f(-2) = -4$,

local minimum value $f(6) = 12$

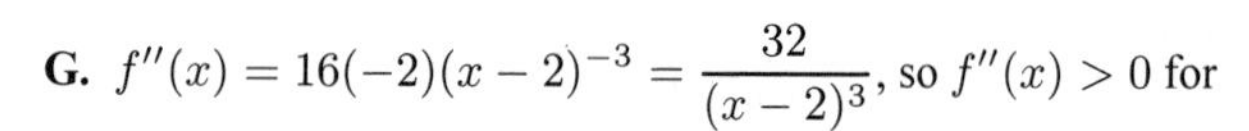

G. $f''(x) = 16(-2)(x - 2)^{-3} = \dfrac{32}{(x - 2)^3}$, so $f''(x) > 0$ for $x > 2$ and $f''(x) < 0$ for $x < 2$. f is CU on $(2, \infty)$ and CD on $(-\infty, 2)$. No IP

H.

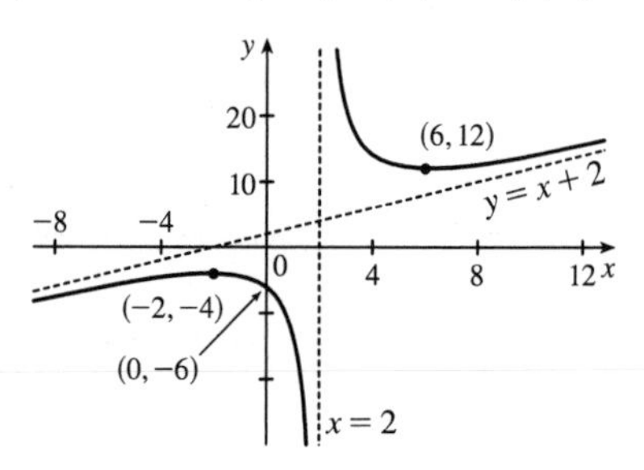

39. $y = f(x) = (x^2 + 4)/x = x + 4/x$ **A.** $D = \{x \mid x \neq 0\} = (-\infty, 0) \cup (0, \infty)$ **B.** No intercept

C. $f(-x) = -f(x) \Rightarrow$ symmetry about the origin **D.** $\lim\limits_{x \to \infty} (x + 4/x) = \infty$ but $f(x) - x = 4/x \to 0$ as $x \to \pm\infty$, so $y = x$ is a SA. $\lim\limits_{x \to 0^+} (x + 4/x) = \infty$ and $\lim\limits_{x \to 0^-} (x + 4/x) = -\infty$, so $x = 0$ is a VA.

E. $f'(x) = 1 - 4/x^2 > 0 \Leftrightarrow x^2 > 4 \Leftrightarrow x > 2$ or $x < -2$, so f is increasing on $(-\infty, -2)$ and $(2, \infty)$ and decreasing on $(-2, 0)$ and $(0, 2)$.

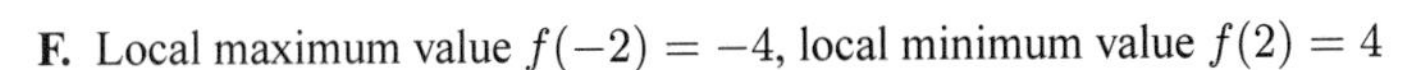
F. Local maximum value $f(-2) = -4$, local minimum value $f(2) = 4$

G. $f''(x) = 8/x^3 > 0 \Leftrightarrow x > 0$ so f is CU on $(0, \infty)$ and CD on $(-\infty, 0)$. No IP

H.

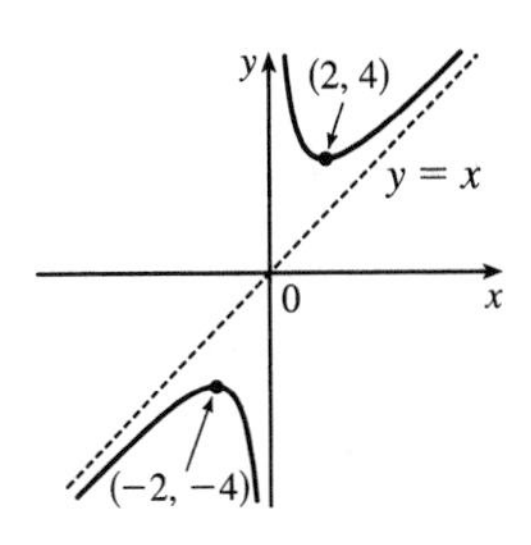

40. $y = f(x) = \dfrac{(x+1)^3}{(x-1)^2} = \dfrac{x^3 + 3x^2 + 3x + 1}{x^2 - 2x + 1} = x + 5 + \dfrac{12x - 4}{(x-1)^2}$

A. $D = \{x \in \mathbb{R} \mid x \neq 1\} = (-\infty, 1) \cup (1, \infty)$ **B.** y-intercept: $f(0) = 1$; x-intercept: $f(x) = 0 \Rightarrow x = -1$ **C.** No symmetry **D.** $\lim\limits_{x \to 1} f(x) = \infty$, so $x = 1$ is a VA.

$$\lim_{x \to \pm\infty} [f(x) - (x + 5)] = \lim_{x \to \pm\infty} \frac{12x - 4}{x^2 - 2x + 1} = \lim_{x \to \pm\infty} \frac{\dfrac{12}{x} - \dfrac{4}{x^2}}{1 - \dfrac{2}{x} + \dfrac{1}{x^2}} = 0, \text{ so the line } y = x + 5 \text{ is a SA.}$$

E.
$$f'(x) = \frac{(x-1)^2 \cdot 3(x+1)^2 - (x+1)^3 \cdot 2(x-1)}{[(x-1)^2]^2} = \frac{(x-1)(x+1)^2[3(x-1) - 2(x+1)]}{(x-1)^4} = \frac{(x+1)^2(x-5)}{(x-1)^3}$$

so $f'(x) > 0$ when $x < -1$, $-1 < x < 1$, or $x > 5$, and $f'(x) < 0$ when $1 < x < 5$. f is increasing on $(-\infty, 1)$ and $(5, \infty)$ and decreasing on $(1, 5)$.

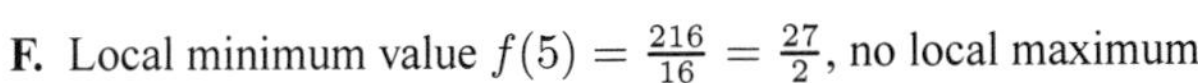
F. Local minimum value $f(5) = \frac{216}{16} = \frac{27}{2}$, no local maximum

H.

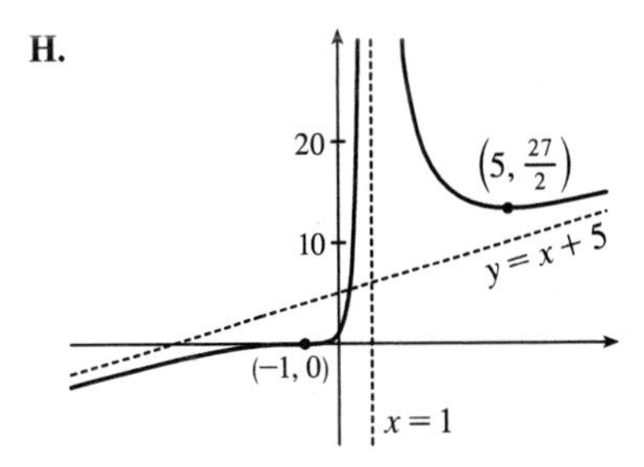

G.
$$f''(x) = \frac{(x-1)^3[(x-1)^2 + (x-5) \cdot 2(x+1)] - (x+1)^2(x-5) \cdot 3(x-1)^2}{[(x-1)^3]^2}$$
$$= \frac{(x-1)^2(x+1)\{(x-1)[(x+1) + 2(x-5)] - 3(x+1)(x-5)\}}{(x-1)^6}$$
$$= \frac{(x+1)\{(x-1)[3x - 9] - 3(x^2 - 4x - 5)\}}{(x-1)^4} = \frac{(x+1)(24)}{(x-1)^4}$$

so $f''(x) > 0$ if $-1 < x < 1$ or $x > 1$, and $f''(x) < 0$ if $x < -1$. Thus, f is CU on $(-1, 1)$ and $(1, \infty)$ and CD on $(-\infty, -1)$. IP at $(-1, 0)$

41. $y = f(x) = \sqrt{4x^2 + 9} \Rightarrow f'(x) = \dfrac{4x}{\sqrt{4x^2 + 9}} \Rightarrow$

$$f''(x) = \frac{\sqrt{4x^2 + 9} \cdot 4 - 4x \cdot 4x/\sqrt{4x^2 + 9}}{4x^2 + 9} = \frac{4(4x^2 + 9) - 16x^2}{(4x^2 + 9)^{3/2}} = \frac{36}{(4x^2 + 9)^{3/2}}.$$ f is defined on $(-\infty, \infty)$.

$f(-x) = f(x)$, so f is even, which means its graph is symmetric about the y-axis. The y-intercept is $f(0) = 3$. There are no

x-intercepts since $f(x) > 0$ for all x.

$$\lim_{x\to\infty}\left(\sqrt{4x^2+9}-2x\right) = \lim_{x\to\infty}\frac{\left(\sqrt{4x^2+9}-2x\right)\left(\sqrt{4x^2+9}+2x\right)}{\sqrt{4x^2+9}+2x}$$

$$= \lim_{x\to\infty}\frac{(4x^2+9)-4x^2}{\sqrt{4x^2+9}+2x} = \lim_{x\to\infty}\frac{9}{\sqrt{4x^2+9}+2x} = 0$$

and, similarly, $\lim\limits_{x\to-\infty}\left(\sqrt{4x^2+9}+2x\right) = \lim\limits_{x\to-\infty}\dfrac{9}{\sqrt{4x^2+9}-2x} = 0,$

so $y = \pm 2x$ are slant asymptotes. f is decreasing on $(-\infty, 0)$ and increasing on $(0, \infty)$ with local minimum $f(0) = 3$. $f''(x) > 0$ for all x, so f is CU on $\mathbb{R}$.

42. $y = f(x) = \sqrt{x^2+4x} = \sqrt{x(x+4)}$. $x(x+4) \geq 0 \quad\Leftrightarrow\quad x \leq -4$ or $x \geq 0$, so $D = (-\infty, -4] \cup [0, \infty)$.

y-intercept: $f(0) = 0$; x-intercepts: $f(x) = 0 \quad\Rightarrow\quad x = -4, 0$.

$$\sqrt{x^2+4x} \mp (x+2) = \frac{\sqrt{x^2+4x} \mp (x+2)}{1}\cdot\frac{\sqrt{x^2+4x} \pm (x+2)}{\sqrt{x^2+4x} \pm (x+2)} = \frac{(x^2+4x)-(x^2+4x+4)}{\sqrt{x^2+4x} \pm (x+2)}$$

$$= \frac{-4}{\sqrt{x^2+4x} \pm (x+2)}$$

so $\lim\limits_{x\to\pm\infty}[f(x) \mp (x+2)] = 0$. Thus, the graph of f approaches the slant asymptote $y = x + 2$ as $x \to \infty$ and it approaches the slant asymptote $y = -(x+2)$ as $x \to -\infty$. $f'(x) = \dfrac{x+2}{\sqrt{x^2+4x}}$, so $f'(x) < 0$ for $x < -4$ and $f'(x) > 0$ for $x > 0$; that is, f is decreasing on $(-\infty, -4)$ and increasing on $(0, \infty)$. There are no local extreme values.

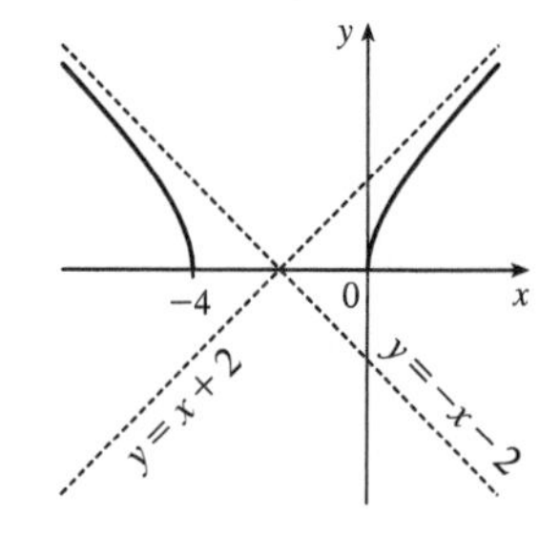

$$f'(x) = (x+2)(x^2+4x)^{-1/2} \quad\Rightarrow$$

$$f''(x) = (x+2)\cdot\left(-\tfrac{1}{2}\right)(x^2+4x)^{-3/2}\cdot(2x+4) + (x^2+4x)^{-1/2}$$

$$= (x^2+4x)^{-3/2}\left[-(x+2)^2 + (x^2+4x)\right]$$

$$= -4(x^2+4x)^{-3/2} < 0 \text{ on } D$$

so f is CD on $(-\infty, -4)$ and $(0, \infty)$. No IP

43. $f(x) = 4x^4 - 32x^3 + 89x^2 - 95x + 29 \quad\Rightarrow\quad f'(x) = 16x^3 - 96x^2 + 178x - 95 \quad\Rightarrow\quad f''(x) = 48x^2 - 192x + 178$.

$f(x) = 0 \quad\Leftrightarrow\quad x \approx 0.5,\ 1.60$; $f'(x) = 0 \quad\Leftrightarrow\quad x \approx 0.92,\ 2.5,\ 2.58$ and $f''(x) = 0 \quad\Leftrightarrow\quad x \approx 1.46,\ 2.54$.

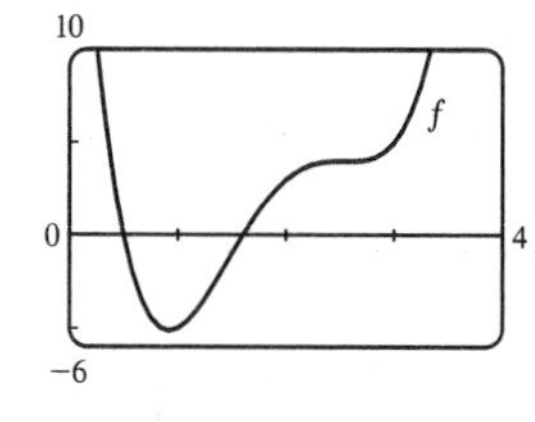

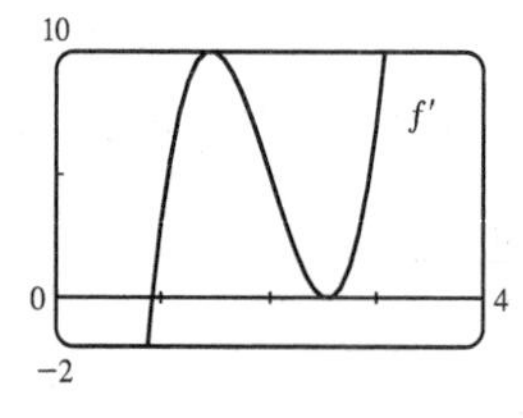

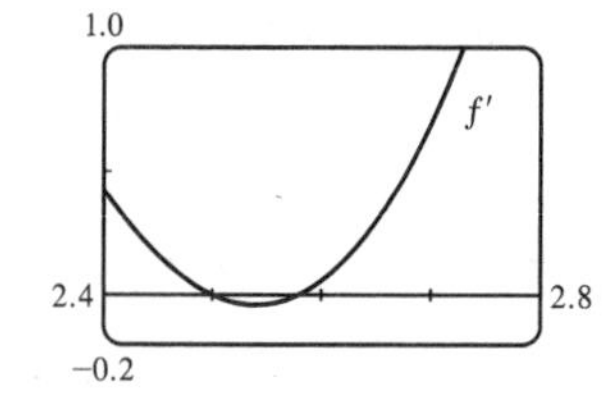

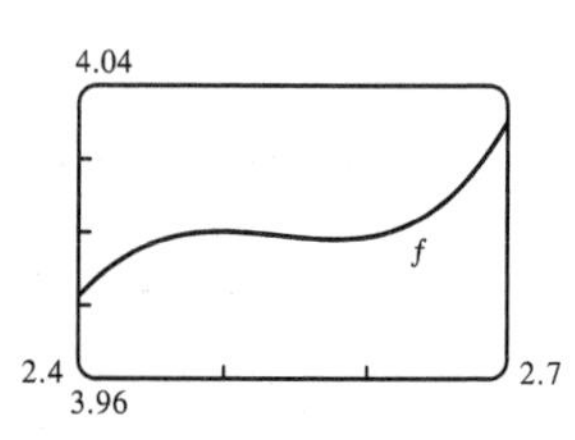

From the graphs of f', we estimate that $f' < 0$ and that f is decreasing on $(-\infty, 0.92)$ and $(2.5, 2.58)$, and that $f' > 0$ and f is increasing on $(0.92, 2.5)$ and $(2.58, \infty)$ with local minimum values $f(0.92) \approx -5.12$ and $f(2.58) \approx 3.998$ and local maximum value $f(2.5) \approx 4$. The graphs of f' make it clear that f has a maximum and a minimum near $x = 2.5$, shown more clearly in the fourth graph.

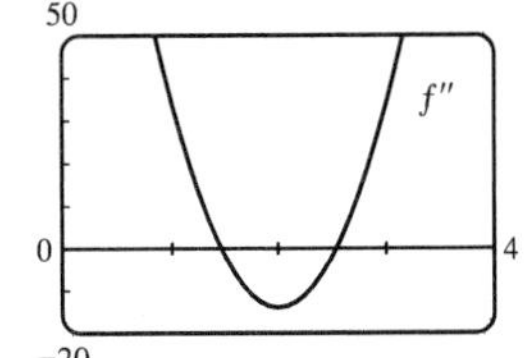

From the graph of f'', we estimate that $f'' > 0$ and that f is CU on $(-\infty, 1.46)$ and $(2.54, \infty)$, and that $f'' < 0$ and f is CD on $(1.46, 2.54)$. There are IPs at about $(1.46, -1.40)$ and $(2.54, 3.999)$.

44. $f(x) = x^6 - 15x^5 + 75x^4 - 125x^3 - x \quad \Rightarrow \quad f'(x) = 6x^5 - 75x^4 + 300x^3 - 375x^2 - 1 \quad \Rightarrow$

$f''(x) = 30x^4 - 300x^3 + 900x^2 - 750x$.

$f(x) = 0 \quad \Leftrightarrow \quad x = 0$ or $x \approx 5.33$; $\quad f'(x) = 0 \quad \Leftrightarrow \quad x \approx 2.50, 4.95$, or 5.05;

$f''(x) = 0 \quad \Leftrightarrow \quad x = 0, 5$ or $x \approx 1.38, 3.62$.

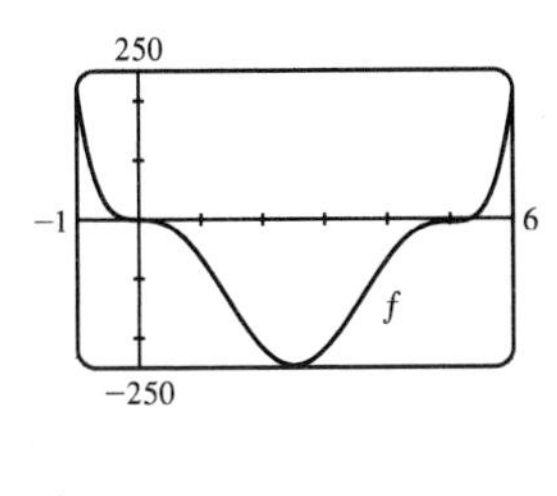

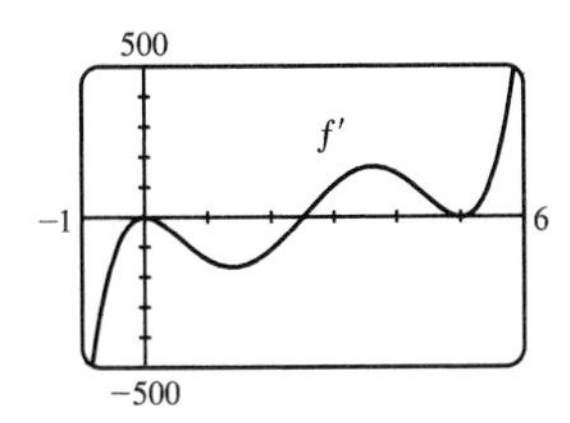

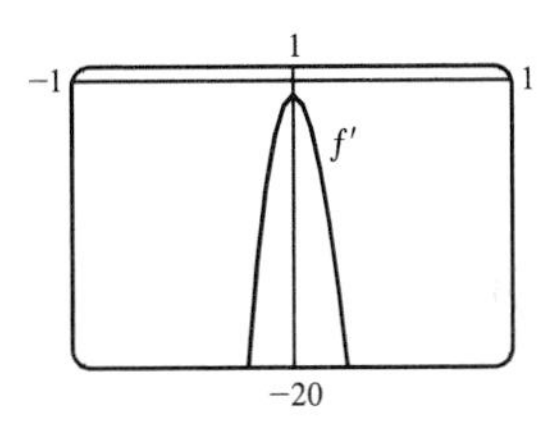

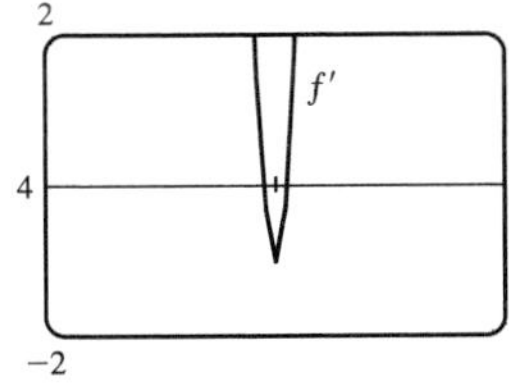

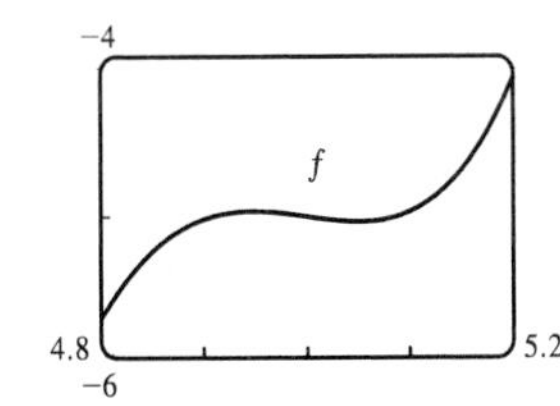

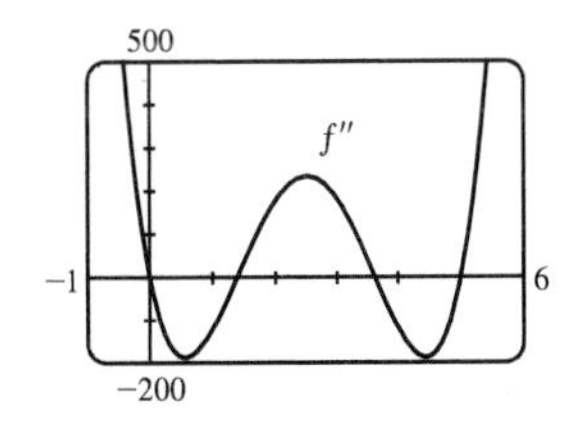

From the graphs of f', we estimate that f is decreasing on $(-\infty, 2.50)$, increasing on $(2.50, 4.95)$, decreasing on $(4.95, 5.05)$, and increasing on $(5.05, \infty)$, with local minimum values $f(2.50) \approx -246.6$ and $f(5.05) \approx -5.03$, and local maximum value $f(4.95) \approx -4.965$ (notice the second graph of f). From the graph of f'', we estimate that f is CU on $(-\infty, 0)$, CD on $(0, 1.38)$, CU on $(1.38, 3.62)$, CD on $(3.62, 5)$, and CU on $(5, \infty)$. There are IPs at $(0, 0)$ and $(5, -5)$, and at about $(1.38, -126.38)$ and $(3.62, -128.62)$.

45. $f(x) = x^2 - 4x + 7\cos x$, $\ -4 \le x \le 4$. $\ f'(x) = 2x - 4 - 7\sin x \quad \Rightarrow \quad f''(x) = 2 - 7\cos x$.

$f(x) = 0 \quad \Leftrightarrow \quad x \approx 1.10$; $f'(x) = 0 \quad \Leftrightarrow \quad x \approx -1.49, -1.07$, or 2.89; $f''(x) = 0 \quad \Leftrightarrow \quad x = \pm\cos^{-1}\left(\frac{2}{7}\right) \approx \pm 1.28$.

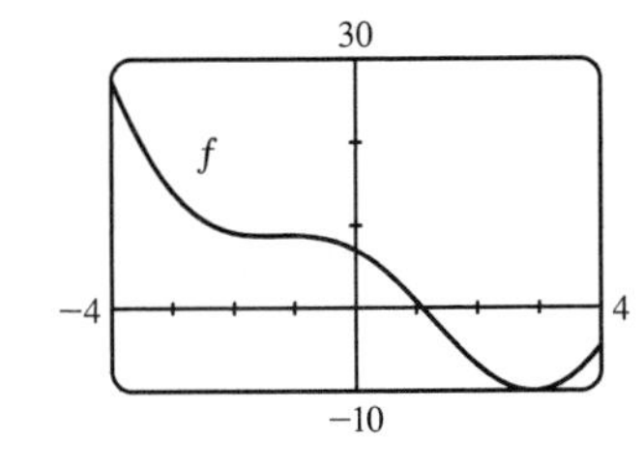

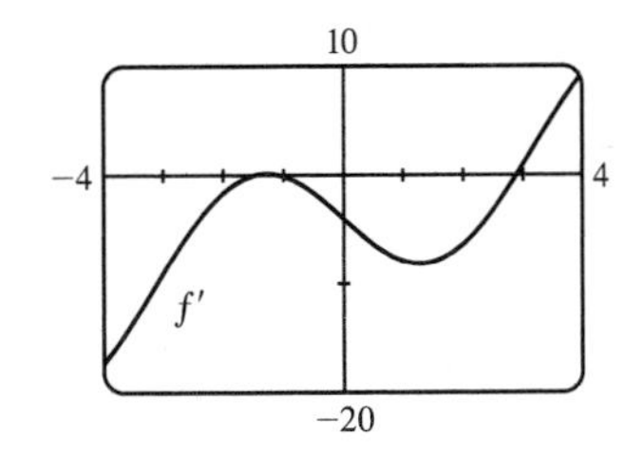

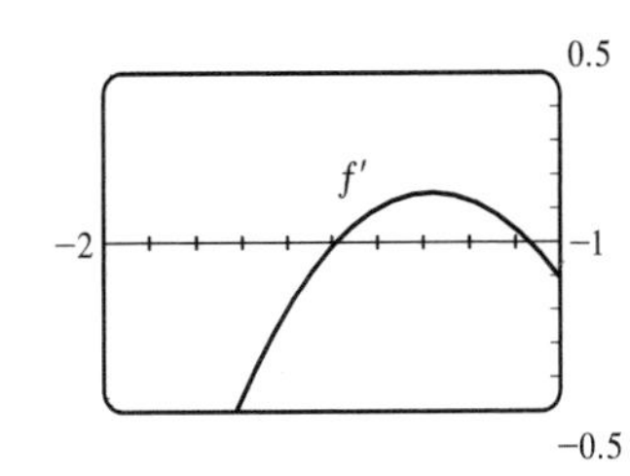

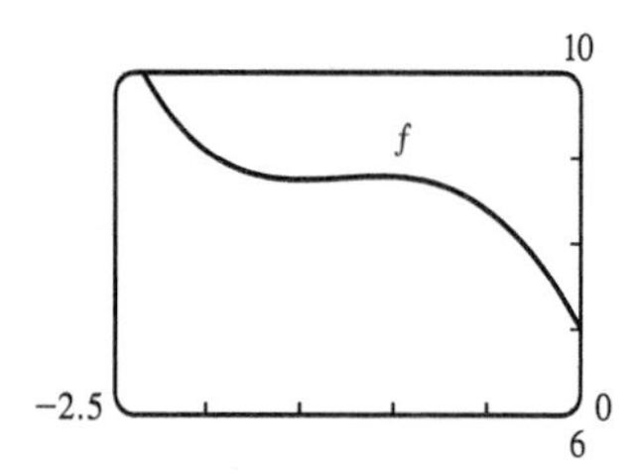

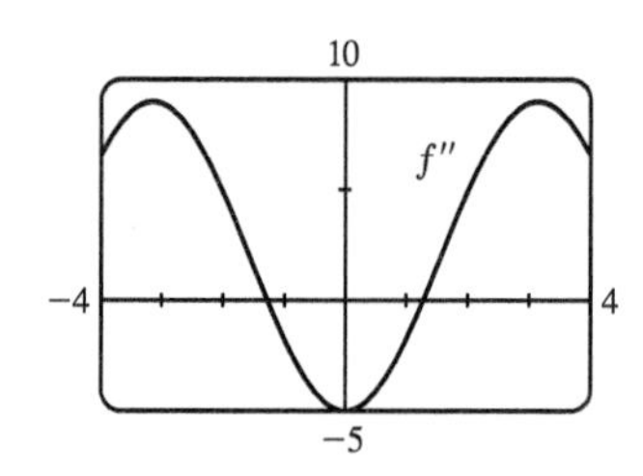

From the graphs of f', we estimate that f is decreasing ($f' < 0$) on $(-4, -1.49)$, increasing on $(-1.49, -1.07)$, decreasing on $(-1.07, 2.89)$, and increasing on $(2.89, 4)$, with local minimum values $f(-1.49) \approx 8.75$ and $f(2.89) \approx -9.99$ and local maximum value $f(-1.07) \approx 8.79$ (notice the second graph of f). From the graph of f'', we estimate that f is CU ($f'' > 0$) on $(-4, -1.28)$, CD on $(-1.28, 1.28)$, and CU on $(1.28, 4)$. There are IPs at about $(-1.28, 8.77)$ and $(1.28, -1.48)$.

46. $f(x) = \tan x + 5\cos x \quad \Rightarrow \quad f'(x) = \sec^2 x - 5\sin x \quad \Rightarrow \quad f''(x) = 2\sec^2 x\, \tan x - 5\, \cos x$. Since f is periodic with period 2π, and defined for all x except odd multiples of $\frac{\pi}{2}$, we graph f and its derivatives on $\left[-\frac{\pi}{2}, \frac{3\pi}{2}\right]$.

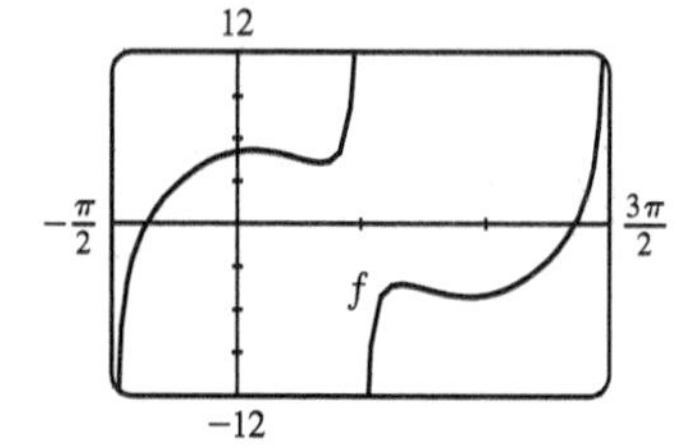

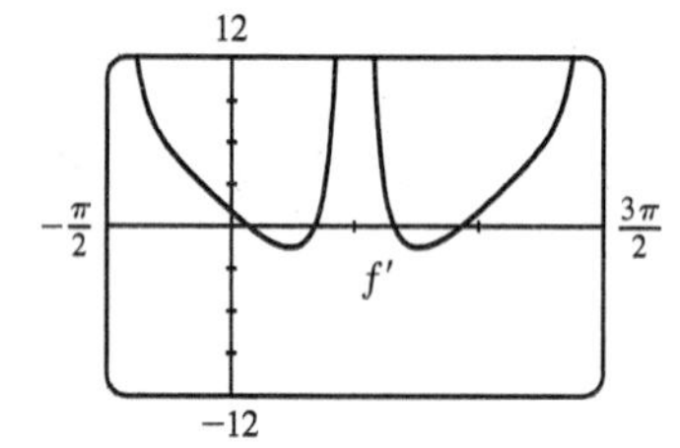

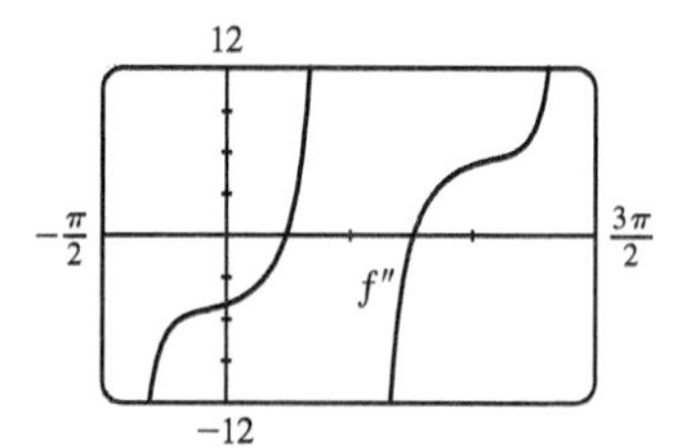

We estimate from the graph of f' that f is increasing on $\left(-\frac{\pi}{2}, 0.21\right)$, $\left(1.07, \frac{\pi}{2}\right)$, $\left(\frac{\pi}{2}, 2.07\right)$, and $\left(2.93, \frac{3\pi}{2}\right)$, and decreasing on $(0.21, 1.07)$ and $(2.07, 2.93)$. Local minimum values: $f(1.07) \approx 4.23$, $f\,(2.93) \approx -5.10$. Local maximum values: $f(0.21) \approx 5.10$, $f(2.07) \approx -4.23$. From the graph of f'', we estimate that f is CU on $\left(0.76, \frac{\pi}{2}\right)$ and $\left(2.38, \frac{3\pi}{2}\right)$, and CD on $\left(-\frac{\pi}{2}, 0.76\right)$ and $\left(\frac{\pi}{2}, 2.38\right)$. f has an IP at $(0.76, 4.57)$ and $(2.38, -4.57)$.

47. $f(x) = 1 + \dfrac{1}{x} + \dfrac{8}{x^2} + \dfrac{1}{x^3} \quad \Rightarrow \quad f'(x) = -\dfrac{1}{x^2} - \dfrac{16}{x^3} - \dfrac{3}{x^4} = -\dfrac{1}{x^4}(x^2 + 16x + 3) \quad \Rightarrow$

$f''(x) = \dfrac{2}{x^3} + \dfrac{48}{x^4} + \dfrac{12}{x^5} = \dfrac{2}{x^5}(x^2 + 24x + 6)$.

From the graphs, it appears that f increases on $(-15.8, -0.2)$ and decreases on $(-\infty, -15.8)$, $(-0.2, 0)$, and $(0, \infty)$; that f has a local minimum value of $f(-15.8) \approx 0.97$ and a local maximum value of $f(-0.2) \approx 72$; that f is CD on $(-\infty, -24)$ and $(-0.25, 0)$ and is CU on $(-24, -0.25)$ and $(0, \infty)$; and that f has IPs at $(-24, 0.97)$ and $(-0.25, 60)$.

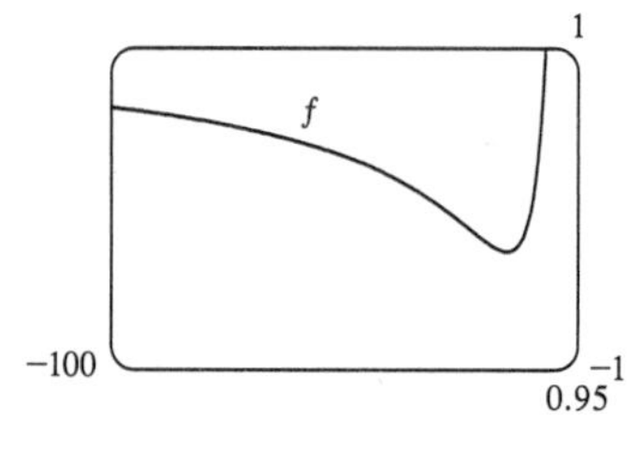

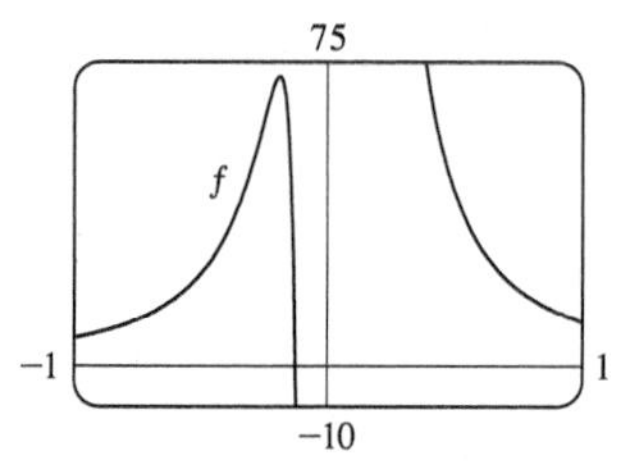

To find the exact values, note that $f' = 0 \quad \Rightarrow \quad x = \dfrac{-16 \pm \sqrt{256 - 12}}{2} = -8 \pm \sqrt{61} \quad [\approx -0.19 \text{ and } -15.81]$.

f' is positive (f is increasing) on $\left(-8 - \sqrt{61}, -8 + \sqrt{61}\right)$ and f' is negative (f is decreasing) on $\left(-\infty, -8 - \sqrt{61}\right)$,

$(-8+\sqrt{61}, 0)$, and $(0, \infty)$. $f''=0 \Rightarrow x = \dfrac{-24 \pm \sqrt{576-24}}{2} = -12 \pm \sqrt{138}$ [≈ -0.25 and -23.75].

f'' is positive (f is CU) on $(-12-\sqrt{138}, -12+\sqrt{138})$ and $(0, \infty)$ and f'' is negative (f is CD) on $(-\infty, -12-\sqrt{138})$ and $(-12+\sqrt{138}, 0)$.

48. $f(x) = \dfrac{1}{x^8} - \dfrac{c}{x^4}$ [$c = 2 \times 10^8$] $\Rightarrow$ $f'(x) = -\dfrac{8}{x^9} + \dfrac{4c}{x^5} = -\dfrac{4}{x^9}(2 - cx^4)$ $\Rightarrow$

$f''(x) = \dfrac{72}{x^{10}} - \dfrac{20c}{x^6} = \dfrac{4}{x^{10}}(18 - 5cx^4)$

From the graph, it appears that f increases on $(-0.01, 0)$ and $(0.01, \infty)$ and decreases on $(-\infty, -0.01)$ and $(0, 0.01)$; that f has a local minimum value of $f(\pm 0.01) = -10^{16}$; and that f is CU on $(-0.012, 0)$ and $(0, 0.012)$ and f is CD on $(-\infty, -0.012)$ and $(0.012, \infty)$.

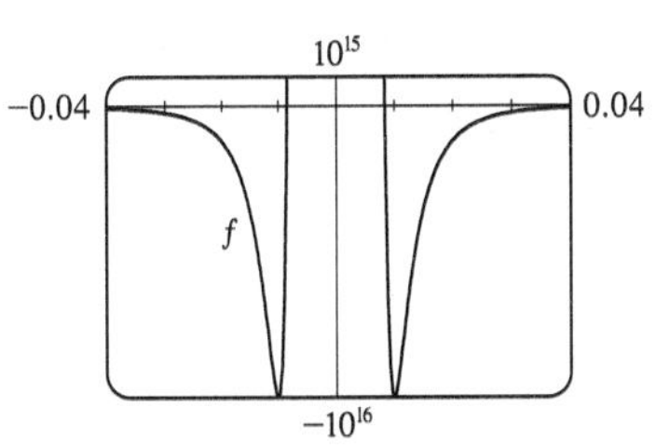

To find the exact values, note that $f' = 0 \Rightarrow x^4 = \dfrac{2}{c} \Rightarrow$

$x \pm \sqrt[4]{\dfrac{2}{c}} = \pm\dfrac{1}{100}$ [$c = 2 \times 10^8$]. f' is positive (f is increasing) on $(-0.01, 0)$ and $(0.01, \infty)$ and f' is negative (f is decreasing) on $(-\infty, -0.01)$ and $(0, 0.01)$. $f''=0 \Rightarrow x^4 = \dfrac{18}{5c} \Rightarrow x = \pm\sqrt[4]{\dfrac{18}{5c}} = \pm\dfrac{1}{100}\sqrt[4]{1.8}$ [$\approx \pm 0.0116$].

f'' is positive (f is CU) on $\left(-\frac{1}{100}\sqrt[4]{1.8}, 0\right)$ and $\left(0, \frac{1}{100}\sqrt[4]{1.8}\right)$ and f'' is negative (f is CD) on $\left(-\infty, -\frac{1}{100}\sqrt[4]{1.8}\right)$ and $\left(\frac{1}{100}\sqrt[4]{1.8}, \infty\right)$.

49. $f(x) = x^4 + cx^2 = x^2(x^2 + c)$. Note that f is an even function. For $c \geq 0$, the only x-intercept is the point $(0, 0)$. We calculate $f'(x) = 4x^3 + 2cx = 4x\left(x^2 + \frac{1}{2}c\right) \Rightarrow f''(x) = 12x^2 + 2c$. If $c \geq 0$, $x = 0$ is the only critical point and there is no inflection point. As we can see from the examples, there is no change in the basic shape of the graph for $c \geq 0$; it merely becomes steeper as c increases. For $c = 0$, the graph is the simple curve $y = x^4$. For $c < 0$, there are x-intercepts at 0 and at $\pm\sqrt{-c}$. Also, there is a maximum at $(0, 0)$, and there are minima at $\left(\pm\sqrt{-\frac{1}{2}c}, -\frac{1}{4}c^2\right)$. As $c \to -\infty$, the x-coordinates of these minima get larger in absolute value, and the minimum points move downward. There are IPs at $\left(\pm\sqrt{-\frac{1}{6}c}, -\frac{5}{36}c^2\right)$, which also move away from the origin as $c \to -\infty$.

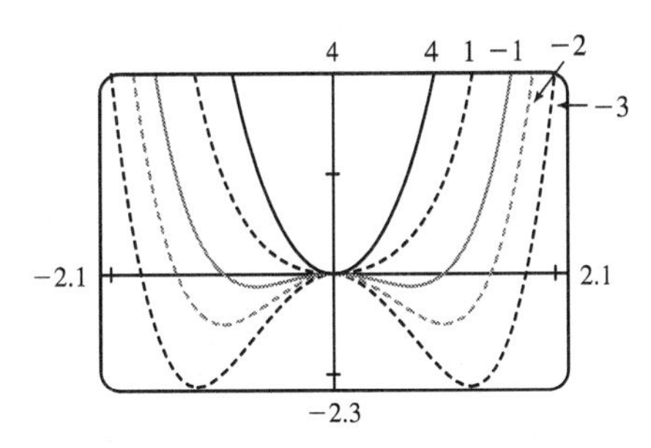

50. $f(x) = x^3 + cx = x(x^2 + c) \Rightarrow f'(x) = 3x^2 + c \Rightarrow f''(x) = 6x$

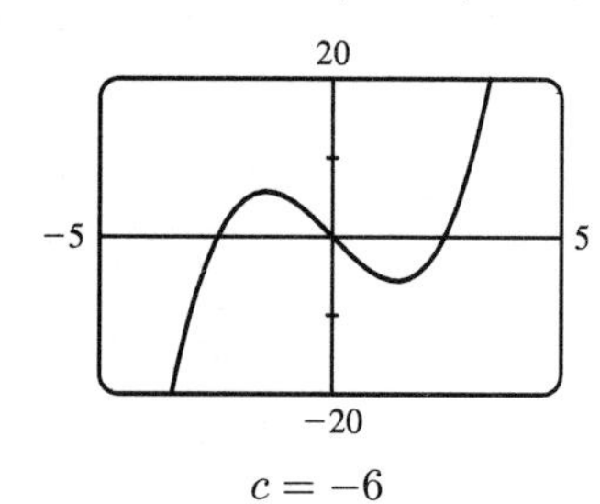

$c = -6$

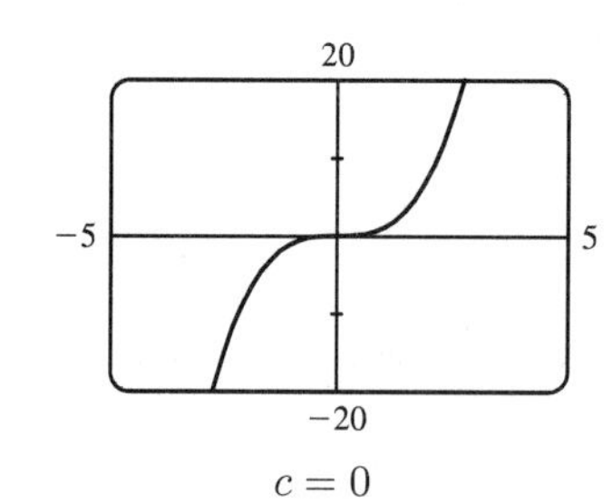

$c = 0$

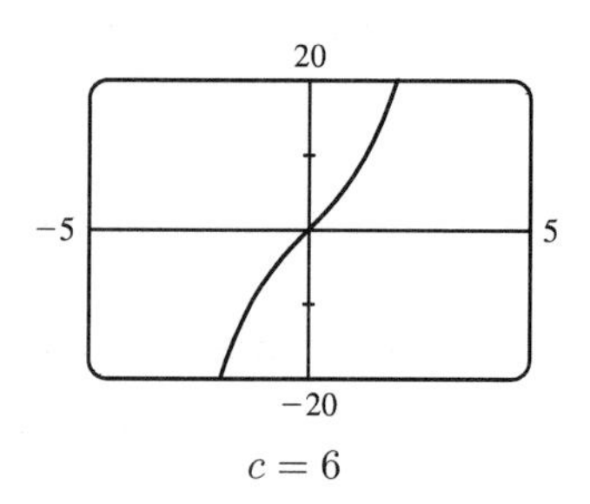

$c = 6$

x-intercepts: When $c \geq 0$, 0 is the only x-intercept. When $c < 0$, the x-intercepts are 0 and $\pm\sqrt{-c}$.

y-intercept $= f(0) = 0$. f is odd, so the graph is symmetric with respect to the origin. $f''(x) < 0$ for $x < 0$ and $f''(x) > 0$ for $x > 0$, so f is CD on $(-\infty, 0)$ and CU on $(0, \infty)$. The origin is the only inflection point.

If $c > 0$, then $f'(x) > 0$ for all x, so f is increasing and has no local maximum or minimum.

If $c = 0$, then $f'(x) \geq 0$ with equality at $x = 0$, so again f is increasing and has no local maximum or minimum.

If $c < 0$, then $f'(x) = 3[x^2 - (-c/3)] = 3\left(x + \sqrt{-c/3}\right)\left(x - \sqrt{-c/3}\right)$, so $f'(x) > 0$ on $\left(-\infty, -\sqrt{-c/3}\right)$ and $\left(\sqrt{-c/3}, \infty\right)$; $f'(x) < 0$ on $\left(-\sqrt{-c/3}, \sqrt{-c/3}\right)$. It follows that $f\left(-\sqrt{-c/3}\right) = -\frac{2}{3}c\sqrt{-c/3}$ is a local maximum value and $f\left(\sqrt{-c/3}\right) = \frac{2}{3}c\sqrt{-c/3}$ is a local minimum value. As c decreases (toward more negative values), the local maximum and minimum move further apart.

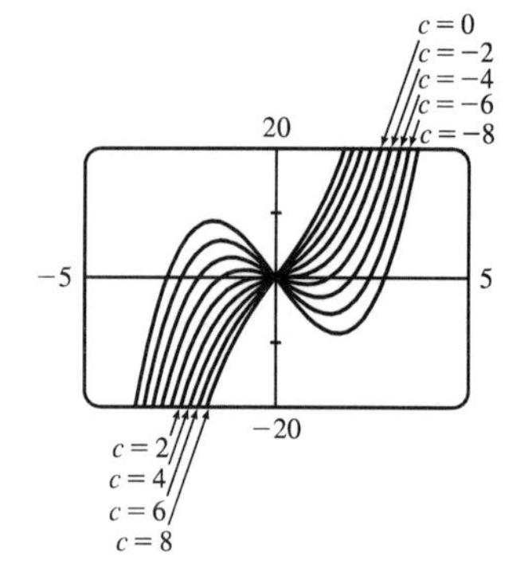

There is no absolute maximum or minimum value. The only transitional value of c corresponding to a change in character of the graph is $c = 0$.

51. Note that $c = 0$ is a transitional value at which the graph consists of the x-axis. Also, we can see that if we substitute $-c$ for c, the function $f(x) = \dfrac{cx}{1 + c^2x^2}$ will be reflected in the x-axis, so we investigate only positive values of c (except $c = -1$, as a demonstration of this reflective property). Also, f is an odd function. $\lim\limits_{x \to \pm\infty} f(x) = 0$, so $y = 0$ is a horizontal asymptote for all c. We calculate $f'(x) = \dfrac{(1 + c^2x^2)c - cx(2c^2x)}{(1 + c^2x^2)^2} = -\dfrac{c(c^2x^2 - 1)}{(1 + c^2x^2)^2}$. $f'(x) = 0 \Leftrightarrow c^2x^2 - 1 = 0 \Leftrightarrow x = \pm 1/c$.

So there is an absolute maximum value of $f(1/c) = \frac{1}{2}$ and an absolute minimum value of $f(-1/c) = -\frac{1}{2}$. These extrema have the same value regardless of c, but the maximum points move closer to the y-axis as c increases.

$$f''(x) = \frac{(-2c^3x)(1 + c^2x^2)^2 - (-c^3x^2 + c)[2(1 + c^2x^2)(2c^2x)]}{(1 + c^2x^2)^4}$$

$$= \frac{(-2c^3x)(1 + c^2x^2) + (c^3x^2 - c)(4c^2x)}{(1 + c^2x^2)^3} = \frac{2c^3x(c^2x^2 - 3)}{(1 + c^2x^2)^3}$$

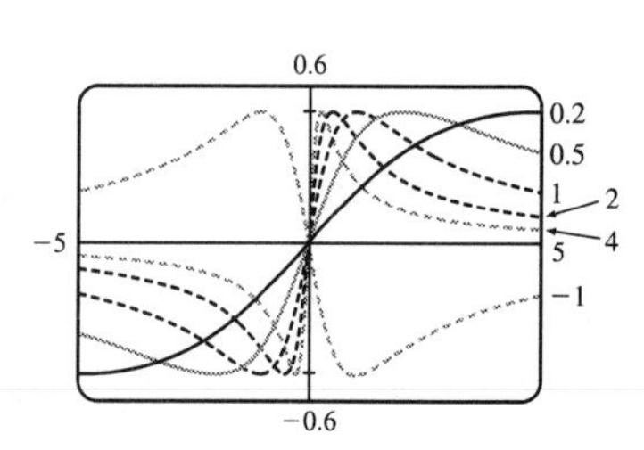

$f''(x) = 0 \Leftrightarrow x = 0$ or $\pm\sqrt{3}/c$, so there are inflection points at $(0, 0)$ and at $(\pm\sqrt{3}/c, \pm\sqrt{3}/4)$.

Again, the y-coordinate of the inflection points does not depend on c, but as c increases, both inflection points approach the y-axis.

52. We need only consider the function $f(x) = x^2\sqrt{c^2 - x^2}$ for $c \geq 0$, because if c is replaced by $-c$, the function is unchanged. For $c = 0$, the graph consists of the single point $(0, 0)$. The domain of f is $[-c, c]$, and the graph of f is symmetric about the y-axis.

$$f'(x) = 2x\sqrt{c^2 - x^2} + x^2\frac{-2x}{2\sqrt{c^2 - x^2}} = 2x\sqrt{c^2 - x^2} - \frac{x^3}{\sqrt{c^2 - x^2}} = \frac{2x(c^2 - x^2) - x^3}{\sqrt{c^2 - x^2}} = -\frac{3x\left(x^2 - \frac{2}{3}c^2\right)}{\sqrt{c^2 - x^2}}$$

So we see that all members of the family of curves have horizontal tangents at $x = 0$, since $f'(0) = 0$ for all $c > 0$. Also, the tangents to all the curves become very steep as $x \to \pm c$, since $\lim\limits_{x \to -c^+} f'(x) = \infty$ and $\lim\limits_{x \to c^-} f'(x) = -\infty$.

We set $f'(x) = 0 \quad \Leftrightarrow \quad x = 0$ or $x^2 - \frac{2}{3}c^2 = 0$, so the absolute maximum values are $f\left(\pm\sqrt{\frac{2}{3}}c\right) = \frac{2}{3\sqrt{3}}c^3$.

$$f''(x) = \frac{\left(-9x^2 + 2c^2\right)\sqrt{c^2 - x^2} - \left(-3x^3 + 2c^2x\right)\left(-x/\sqrt{c^2 - x^2}\right)}{c^2 - x^2} = \frac{6x^4 - 9c^2x^2 + 2c^4}{\left(c^2 - x^2\right)^{3/2}}$$

Using the quadratic formula, we find that $f''(x) = 0 \quad \Leftrightarrow \quad x^2 = \dfrac{9c^2 \pm c^2\sqrt{33}}{12}$. Since $-c < x < c$, we take $x^2 = \frac{9 - \sqrt{33}}{12}c^2$, so the inflection points are $\left(\pm\sqrt{\frac{9 - \sqrt{33}}{12}}c, \frac{(9 - \sqrt{33})(\sqrt{33} - 3)}{144}c^3\right)$.

From these calculations we can see that the maxima and the points of inflection get both horizontally and vertically further from the origin as c increases. Since all of the functions have two maxima and two inflection points, we see that the basic shape of the curve does not change as c changes.

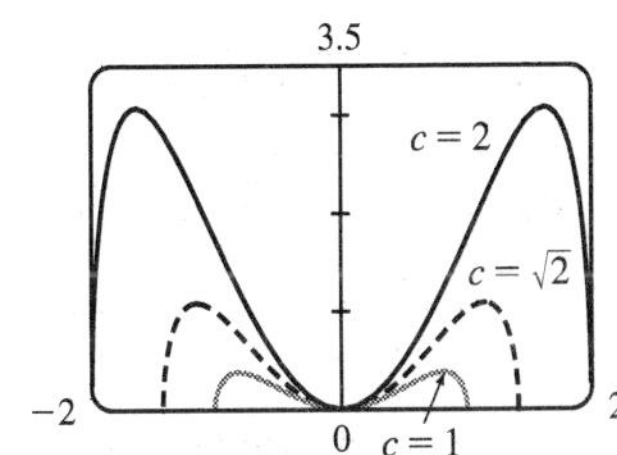

53. $f(x) = cx + \sin x \quad \Rightarrow \quad f'(x) = c + \cos x \quad \Rightarrow \quad f''(x) = -\sin x$

$f(-x) = -f(x)$, so f is an odd function and its graph is symmetric with respect to the origin.

$f(x) = 0 \quad \Leftrightarrow \quad \sin x = -cx$, so 0 is always an x-intercept.

$f'(x) = 0 \quad \Leftrightarrow \quad \cos x = -c$, so there is no critical number when $|c| > 1$. If $|c| \leq 1$, then there are infinitely many critical numbers. If x_1 is the unique solution of $\cos x = -c$ in the interval $[0, \pi]$, then the critical numbers are $2n\pi \pm x_1$, where n ranges over the integers. (Special cases: When $c = 1$, $x_1 = 0$; when $c = 0$, $x = \frac{\pi}{2}$; and when $c = -1$, $x_1 = \pi$.)

$f''(x) < 0 \quad \Leftrightarrow \quad \sin x > 0$, so f is CD on intervals of the form $(2n\pi, (2n + 1)\pi)$. f is CU on intervals of the form $((2n - 1)\pi, 2n\pi)$. The IPs of f are the points $(2n\pi, 2n\pi c)$, where n is an integer.

If $c \geq 1$, then $f'(x) \geq 0$ for all x, so f is increasing and has no extremum. If $c \leq -1$, then $f'(x) \leq 0$ for all x, so f is decreasing and has no extremum. If $|c| < 1$, then $f'(x) > 0 \quad \Leftrightarrow \quad \cos x > -c \quad \Leftrightarrow \quad x$ is in an interval of the form $(2n\pi - x_1, 2n\pi + x_1)$ for some integer n. These are the intervals on which f is increasing. Similarly, we find that f is decreasing on the intervals of the form $(2n\pi + x_1, 2(n + 1)\pi - x_1)$. Thus, f has local maxima at the points $2n\pi + x_1$, where f has the values $c(2n\pi + x_1) + \sin x_1 = c(2n\pi + x_1) + \sqrt{1 - c^2}$, and f has local minima at the points $2n\pi - x_1$, where we have $f(2n\pi - x_1) = c(2n\pi - x_1) - \sin x_1 = c(2n\pi - x_1) - \sqrt{1 - c^2}$.

The transitional values of c are -1 and 1. The IPs move vertically, but not horizontally, when c changes. When $|c| \geq 1$, there is no extremum. For $|c| < 1$, the maxima are spaced 2π apart horizontally, as are the minima. The horizontal spacing between maxima and adjacent minima is regular (and equals π) when $c = 0$, but the horizontal space between a local maximum and the nearest local minimum shrinks as $|c|$ approaches 1.

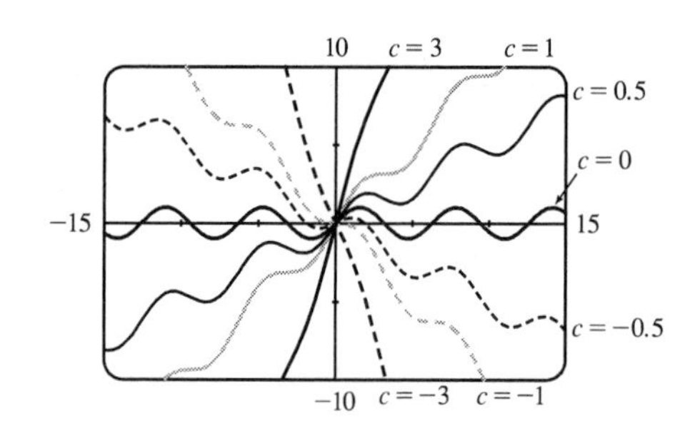

54. For $c = 0$, there is no IP; the curve is CU everywhere. If c increases, the curve simply becomes steeper, and there are still no IPs. If c starts at 0 and decreases, a slight upward bulge appears near $x = 0$, so that there are two IPs for any $c < 0$. This can be seen algebraically by calculating the second derivative: $f(x) = x^4 + cx^2 + x \Rightarrow f'(x) = 4x^3 + 2cx + 1 \Rightarrow$ $f''(x) = 12x^2 + 2c$. Thus, $f''(x) > 0$ when $c > 0$. For $c < 0$, there are IPs when $x = \pm\sqrt{-\frac{1}{6}c}$. For $c = 0$, the graph has one critical number, at the absolute minimum somewhere around $x = -0.6$. As c increases, the number of critical points does not change. If c instead decreases from 0, we see that the graph eventually sprouts another local minimum, to the right of the origin, somewhere between $x = 1$ and $x = 2$. Consequently, there is also a maximum near $x = 0$.

After a bit of experimentation, we find that at $c = -1.5$, there appear to be two critical numbers: the absolute minimum at about $x = -1$, and a horizontal tangent with no extremum at about $x = 0.5$. For any c smaller than this there will be 3 critical points, as shown in the graphs with $c = -3$ and with $c = -5$. To prove this algebraically, we calculate $f'(x) = 4x^3 + 2cx + 1$. Now if we substitute our value of $c = -1.5$, the formula for $f'(x)$ becomes $4x^3 - 3x + 1 = (x + 1)(2x - 1)^2$. This has a double root at $x = \frac{1}{2}$, indicating that the function has two critical points: $x = -1$ and $x = \frac{1}{2}$, just as we had guessed from the graph.

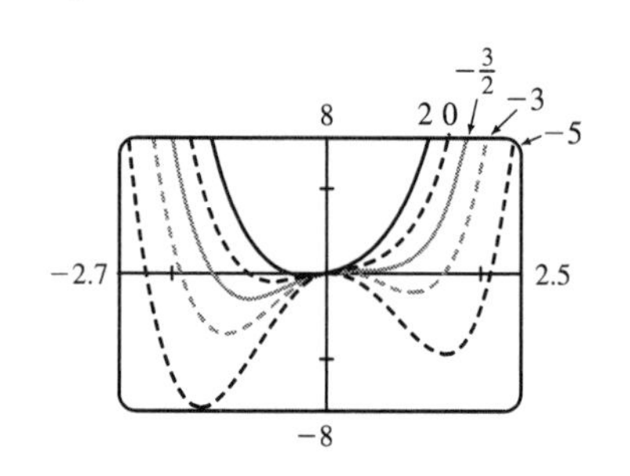

3.5 Optimization Problems

1. (a)

First Number	Second Number	Product
1	22	22
2	21	42
3	20	60
4	19	76
5	18	90
6	17	102
7	16	112
8	15	120
9	14	126
10	13	130
11	12	132

We needn't consider pairs where the first number is larger than the second, since we can just interchange the numbers in such cases. The answer appears to be 11 and 12, but we have considered only integers in the table.

(b) Call the two numbers x and y. Then $x + y = 23$, so $y = 23 - x$. Call the product P. Then $P = xy = x(23 - x) = 23x - x^2$, so we wish to maximize the function $P(x) = 23x - x^2$. Since $P'(x) = 23 - 2x$, we see that $P'(x) = 0 \Leftrightarrow x = \frac{23}{2} = 11.5$. Thus, the maximum value of P is $P(11.5) = (11.5)^2 = 132.25$ and it occurs when $x = y = 11.5$.

Or: Note that $P''(x) = -2 < 0$ for all x, so P is everywhere concave downward and the local maximum at $x = 11.5$ must be an absolute maximum.

2. The two numbers are $x + 100$ and x. Minimize $f(x) = (x + 100)x = x^2 + 100x$. $f'(x) = 2x + 100 = 0 \Rightarrow x = -50$. Since $f''(x) = 2 > 0$, there is an absolute minimum at $x = -50$. The two numbers are 50 and -50.

3. The two numbers are x and $\dfrac{100}{x}$, where $x > 0$. Minimize $f(x) = x + \dfrac{100}{x}$. $f'(x) = 1 - \dfrac{100}{x^2} = \dfrac{x^2 - 100}{x^2}$. The critical number is $x = 10$. Since $f'(x) < 0$ for $0 < x < 10$ and $f'(x) > 0$ for $x > 10$, there is an absolute minimum at $x = 10$. The numbers are 10 and 10.

4. Let $x > 0$ and let $f(x) = x + 1/x$. We wish to minimize $f(x)$. Now $f'(x) = 1 - \dfrac{1}{x^2} = \dfrac{1}{x^2}(x^2 - 1) = \dfrac{1}{x^2}(x + 1)(x - 1)$, so the only critical number in $(0, \infty)$ is 1. $f'(x) < 0$ for $0 < x < 1$ and $f'(x) > 0$ for $x > 1$, so f has an absolute minimum at $x = 1$, and $f(1) = 2$.

Or: $f''(x) = 2/x^3 > 0$ for all $x > 0$, so f is concave upward everywhere and the critical point $(1, 2)$ must correspond to a local minimum for f.

5. If the rectangle has dimensions x and y, then its perimeter is $2x + 2y = 100$ m, so $y = 50 - x$. Thus, the area is $A = xy = x(50 - x)$. We wish to maximize the function $A(x) = x(50 - x) = 50x - x^2$, where $0 < x < 50$. Since $A'(x) = 50 - 2x = -2(x - 25)$, $A'(x) > 0$ for $0 < x < 25$ and $A'(x) < 0$ for $25 < x < 50$. Thus, A has an absolute maximum at $x = 25$, and $A(25) = 25^2 = 625\text{ m}^2$. The dimensions of the rectangle that maximize its area are $x = y = 25$ m. (The rectangle is a square.)

6. If the rectangle has dimensions x and y, then its area is $xy = 1000\text{ m}^2$, so $y = 1000/x$. The perimeter $P = 2x + 2y = 2x + 2000/x$. We wish to minimize the function $P(x) = 2x + 2000/x$ for $x > 0$. $P'(x) = 2 - 2000/x^2 = (2/x^2)(x^2 - 1000)$, so the only critical number in the domain of P is $x = \sqrt{1000}$. $P''(x) = 4000/x^3 > 0$, so P is concave upward throughout its domain and $P(\sqrt{1000}) = 4\sqrt{1000}$ is an absolute minimum value. The dimensions of the rectangle with minimal perimeter are $x = y = \sqrt{1000} = 10\sqrt{10}$ m. (The rectangle is a square.)

7. (a)

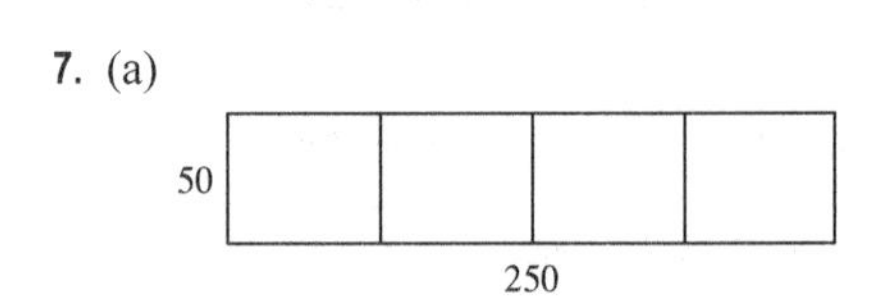

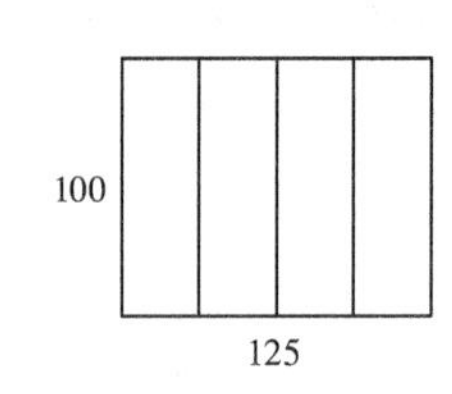

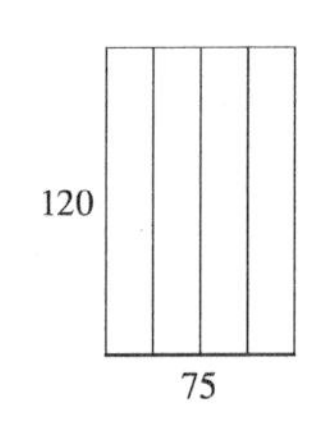

The areas of the three figures are 12,500, 12,500, and 9000 ft^2. There appears to be a maximum area of at least 12,500 ft^2.

(b) Let x denote the length of each of two sides and three dividers. Let y denote the length of the other two sides.

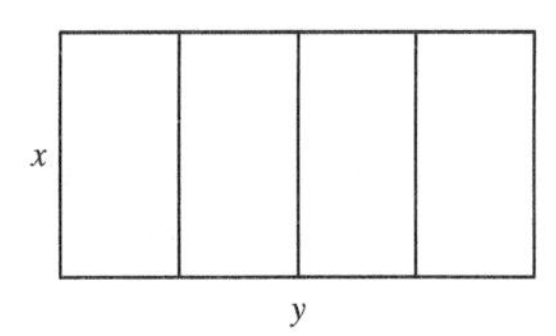

(c) Area $A = \text{length} \times \text{width} = y \cdot x$

(d) Length of fencing $= 750 \quad\Rightarrow\quad 5x + 2y = 750$

(e) $5x + 2y = 750 \quad\Rightarrow\quad y = 375 - \frac{5}{2}x \quad\Rightarrow\quad A(x) = \left(375 - \frac{5}{2}x\right)x = 375x - \frac{5}{2}x^2$

(f) $A'(x) = 375 - 5x = 0 \quad\Rightarrow\quad x = 75$. Since $A''(x) = -5 < 0$ there is an absolute maximum when $x = 75$. Then $y = \frac{375}{2} = 187.5$. The largest area is $75\left(\frac{375}{2}\right) = 14{,}062.5$ ft^2. These values of x and y are between the values in the first and second figures in part (a). Our original estimate was low.

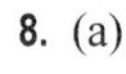

8. (a)

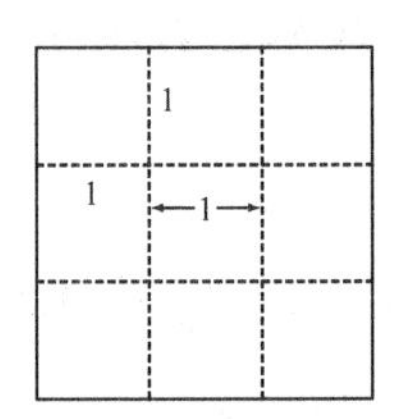

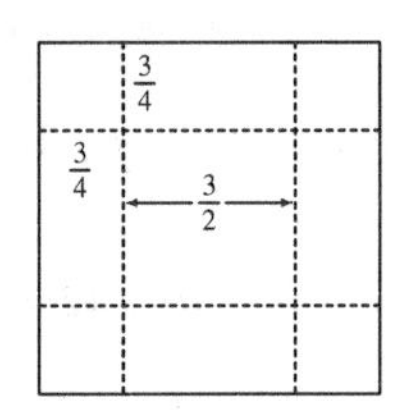

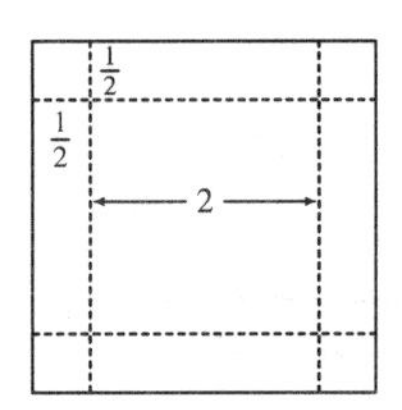

The volumes of the resulting boxes are 1, 1.6875, and 2 ft^3. There appears to be a maximum volume of at least 2 ft^3.

(b) Let x denote the length of the side of the square being cut out. Let y denote the length of the base.

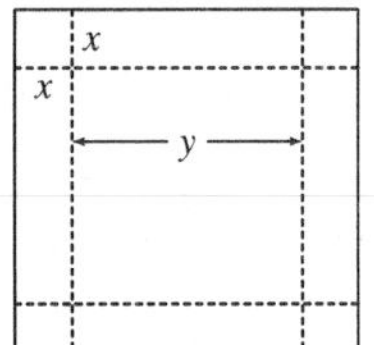

(c) Volume $V = \text{length} \times \text{width} \times \text{height} \quad\Rightarrow\quad V = y \cdot y \cdot x = xy^2$

(d) Length of cardboard $= 3 \quad\Rightarrow\quad x + y + x = 3 \quad\Rightarrow\quad y + 2x = 3$

(e) $y + 2x = 3 \;\Rightarrow\; y = 3 - 2x \;\Rightarrow\; V(x) = x(3 - 2x)^2$

(f) $V(x) = x(3 - 2x)^2 \;\Rightarrow$

$V'(x) = x \cdot 2(3 - 2x)(-2) + (3 - 2x)^2 \cdot 1 = (3 - 2x)[-4x + (3 - 2x)] = (3 - 2x)(-6x + 3)$,

so the critical numbers are $x = \frac{3}{2}$ and $x = \frac{1}{2}$. Now $0 \le x \le \frac{3}{2}$ and $V(0) = V\left(\frac{3}{2}\right) = 0$, so the maximum is

$V\left(\frac{1}{2}\right) = \left(\frac{1}{2}\right)(2)^2 = 2 \text{ ft}^3$, which is the value found from our third figure in part (a).

9. Let b be the length of the base of the box and h the height. The surface area is $1200 = b^2 + 4hb \;\Rightarrow\; h = (1200 - b^2)/(4b)$.

The volume is $V = b^2h = b^2(1200 - b^2)/4b = 300b - b^3/4 \;\Rightarrow\; V'(b) = 300 - \frac{3}{4}b^2$.

$V'(b) = 0 \;\Rightarrow\; 300 = \frac{3}{4}b^2 \;\Rightarrow\; b^2 = 400 \;\Rightarrow\; b = \sqrt{400} = 20$. Since $V'(b) > 0$ for $0 < b < 20$ and $V'(b) < 0$ for $b > 20$, there is an absolute maximum when $b = 20$ by the First Derivative Test for Absolute Extreme Values (see page 229).

If $b = 20$, then $h = (1200 - 20^2)/(4 \cdot 20) = 10$, so the largest possible volume is $b^2h = (20)^2(10) = 4000 \text{ cm}^3$.

10. Let b be the length of the base of the box and h the height. The volume is $32{,}000 = b^2h \;\Rightarrow\; h = 32{,}000/b^2$.

The surface area of the open box is $S = b^2 + 4hb = b^2 + 4(32{,}000/b^2)b = b^2 + 4(32{,}000)/b$. So

$S'(b) = 2b - 4(32{,}000)/b^2 = 2(b^3 - 64{,}000)/b^2 = 0 \;\Leftrightarrow\; b = \sqrt[3]{64{,}000} = 40$. This gives an absolute minimum since

$S'(b) < 0$ if $0 < b < 40$ and $S'(b) > 0$ if $b > 40$. The box should be $40 \times 40 \times 20$.

11. (a) Let the rectangle have sides x and y and area A, so $A = xy$ or $y = A/x$. The problem is to minimize the

perimeter $= 2x + 2y = 2x + 2A/x = P(x)$. Now $P'(x) = 2 - 2A/x^2 = 2(x^2 - A)/x^2$. So the critical number is

$x = \sqrt{A}$. Since $P'(x) < 0$ for $0 < x < \sqrt{A}$ and $P'(x) > 0$ for $x > \sqrt{A}$, there is an absolute minimum at $x = \sqrt{A}$.

The sides of the rectangle are $\sqrt{A}$ and $A/\sqrt{A} = \sqrt{A}$, so the rectangle is a square.

(b) Let p be the perimeter and x and y the lengths of the sides, so $p = 2x + 2y \;\Rightarrow\; 2y = p - 2x \;\Rightarrow\; y = \frac{1}{2}p - x$.

The area is $A(x) = x\left(\frac{1}{2}p - x\right) = \frac{1}{2}px - x^2$. Now $A'(x) = 0 \;\Rightarrow\; \frac{1}{2}p - 2x = 0 \;\Rightarrow\; 2x = \frac{1}{2}p \;\Rightarrow\; x = \frac{1}{4}p$.

Since $A''(x) = -2 < 0$, there is an absolute maximum for A when $x = \frac{1}{4}p$ by the Second Derivative Test. The sides of the rectangle are $\frac{1}{4}p$ and $\frac{1}{2}p - \frac{1}{4}p = \frac{1}{4}p$, so the rectangle is a square.

12. $V = lwh \;\Rightarrow\; 10 = (2w)(w)h = 2w^2h$, so $h = 5/w^2$. The cost is

$10(2w^2) + 6[2(2wh) + 2(hw)] = 20w^2 + 36wh$, so

$C(w) = 20w^2 + 36w(5/w^2) = 20w^2 + 180/w$.

$C'(w) = 40w - 180/w^2 = 40\left(w^3 - \frac{9}{2}\right)/w^2 \;\Rightarrow\; w = \sqrt[3]{\frac{9}{2}}$ is the critical number. There is an absolute minimum

for C when $w = \sqrt[3]{\frac{9}{2}}$ since $C'(w) < 0$ for $0 < w < \sqrt[3]{\frac{9}{2}}$ and $C'(w) > 0$ for $w > \sqrt[3]{\frac{9}{2}}$.

$C\left(\sqrt[3]{\frac{9}{2}}\right) = 20\left(\sqrt[3]{\frac{9}{2}}\right)^2 + \frac{180}{\sqrt[3]{9/2}} \approx \$163.54$.

13.

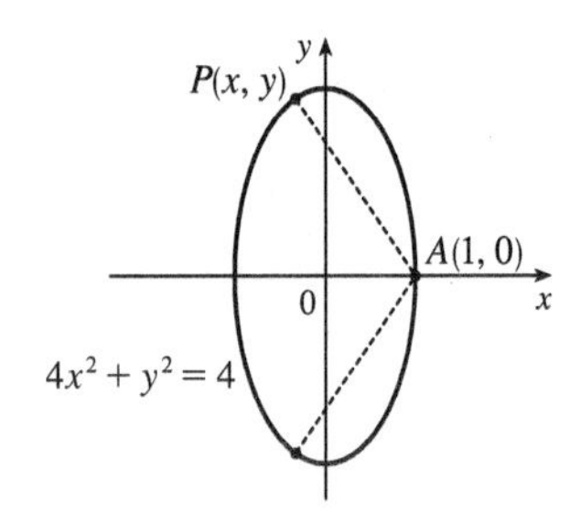

From the figure, we see that there are two points that are farthest away from $A(1,0)$. The distance d from A to an arbitrary point $P(x,y)$ on the ellipse is $d = \sqrt{(x-1)^2 + (y-0)^2}$ and the square of the distance is $S = d^2 = x^2 - 2x + 1 + y^2 = x^2 - 2x + 1 + (4 - 4x^2) = -3x^2 - 2x + 5$. $S' = -6x - 2$ and $S' = 0 \Rightarrow x = -\frac{1}{3}$. Now $S'' = -6 < 0$, so we know that S has a maximum at $x = -\frac{1}{3}$. Since $-1 \leq x \leq 1$, $S(-1) = 4$, $S\left(-\frac{1}{3}\right) = \frac{16}{3}$, and $S(1) = 0$, we see that the maximum distance is $\sqrt{\frac{16}{3}}$. The corresponding y-values are $y = \pm\sqrt{4 - 4\left(-\frac{1}{3}\right)^2} = \pm\sqrt{\frac{32}{9}} = \pm\frac{4}{3}\sqrt{2} \approx \pm 1.89$. The points are $\left(-\frac{1}{3}, \pm\frac{4}{3}\sqrt{2}\right)$.

14.

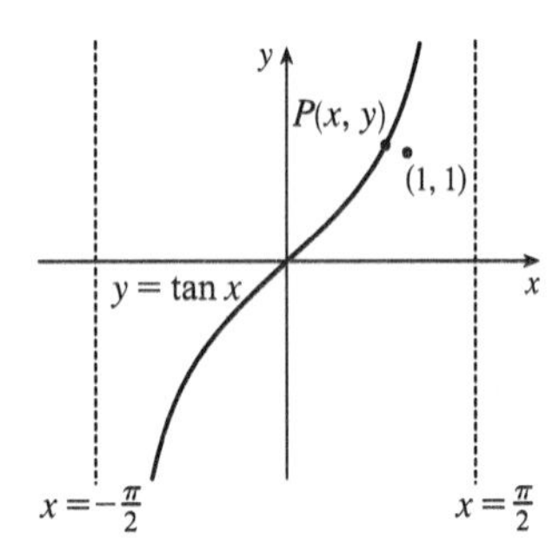

The distance d from $(1,1)$ to an arbitrary point $P(x,y)$ on the curve $y = \tan x$ is $d = \sqrt{(x-1)^2 + (y-1)^2}$ and the square of the distance is $S = d^2 = (x-1)^2 + (\tan x - 1)^2$. $S' = 2(x-1) + 2(\tan x - 1)\sec^2 x$. Graphing S' on $\left(-\frac{\pi}{2}, \frac{\pi}{2}\right)$ gives us a zero at $x \approx 0.82$, and so $\tan x \approx 1.08$. The point on $y = \tan x$ that is closest to $(1,1)$ is approximately $(0.82, 1.08)$.

15.

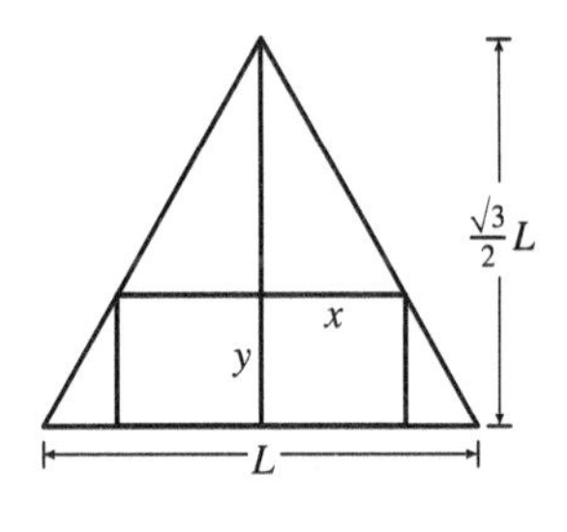

The height h of the equilateral triangle with sides of length L is $\frac{\sqrt{3}}{2}L$, since $h^2 + (L/2)^2 = L^2 \Rightarrow h^2 = L^2 - \frac{1}{4}L^2 = \frac{3}{4}L^2 \Rightarrow h = \frac{\sqrt{3}}{2}L$. Using similar triangles, $\dfrac{\frac{\sqrt{3}}{2}L - y}{x} = \dfrac{\frac{\sqrt{3}}{2}L}{L/2} = \sqrt{3} \Rightarrow \sqrt{3}\,x = \dfrac{\sqrt{3}}{2}L - y \Rightarrow$

$y = \dfrac{\sqrt{3}}{2}L - \sqrt{3}\,x \Rightarrow y = \dfrac{\sqrt{3}}{2}(L - 2x)$.

The area of the inscribed rectangle is $A(x) = (2x)y = \sqrt{3}\,x(L - 2x) = \sqrt{3}\,Lx - 2\sqrt{3}\,x^2$, where $0 \leq x \leq L/2$. Now $0 = A'(x) = \sqrt{3}\,L - 4\sqrt{3}\,x \Rightarrow x = \sqrt{3}\,L/(4\sqrt{3}) = L/4$. Since $A(0) = A(L/2) = 0$, the maximum occurs when $x = L/4$, and $y = \frac{\sqrt{3}}{2}L - \frac{\sqrt{3}}{4}L = \frac{\sqrt{3}}{4}L$, so the dimensions are $L/2$ and $\frac{\sqrt{3}}{4}L$.

16.

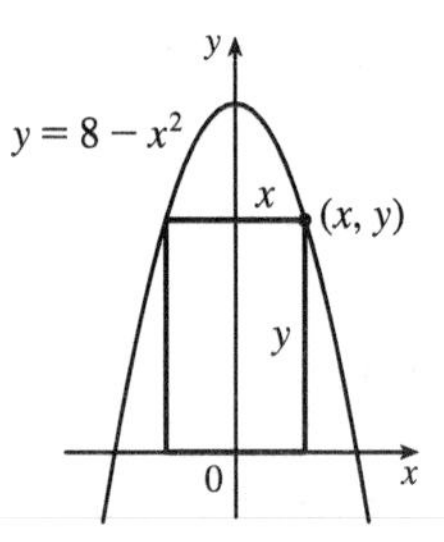

The rectangle has area $A(x) = 2xy = 2x(8 - x^2) = 16x - 2x^3$, where $0 \leq x \leq 2\sqrt{2}$. Now $A'(x) = 16 - 6x^2 = 0 \Rightarrow x = 2\sqrt{\frac{2}{3}}$. Since $A(0) = A(2\sqrt{2}) = 0$, there is a maximum when $x = 2\sqrt{\frac{2}{3}}$. Then $y = \frac{16}{3}$, so the rectangle has dimensions $4\sqrt{\frac{2}{3}}$ and $\frac{16}{3}$.

17. 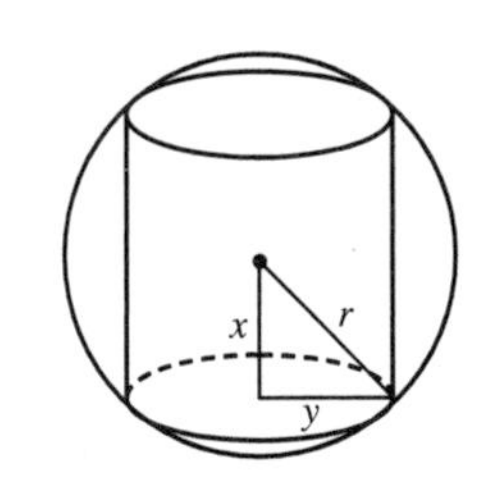

The cylinder has volume $V = \pi y^2(2x)$. Also $x^2 + y^2 = r^2 \Rightarrow y^2 = r^2 - x^2$, so $V(x) = \pi(r^2 - x^2)(2x) = 2\pi(r^2 x - x^3)$, where $0 \le x \le r$.

$V'(x) = 2\pi(r^2 - 3x^2) = 0 \Rightarrow x = r/\sqrt{3}$. Now $V(0) = V(r) = 0$, so there is a maximum when $x = r/\sqrt{3}$ and

$V(r/\sqrt{3}) = \pi(r^2 - r^2/3)(2r/\sqrt{3}) = 4\pi r^3/(3\sqrt{3})$.

18. 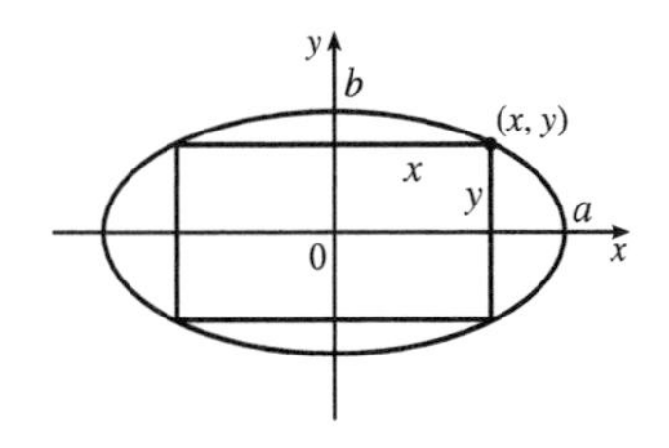

The area of the rectangle is $(2x)(2y) = 4xy$. Now $\dfrac{x^2}{a^2} + \dfrac{y^2}{b^2} = 1$ gives

$y = \dfrac{b}{a}\sqrt{a^2 - x^2}$, so we maximize $A(x) = 4\dfrac{b}{a}x\sqrt{a^2 - x^2}$.

$$A'(x) = \frac{4b}{a}\left[x \cdot \tfrac{1}{2}(a^2 - x^2)^{-1/2}(-2x) + (a^2 - x^2)^{1/2} \cdot 1\right]$$

$$= \frac{4b}{a}(a^2 - x^2)^{-1/2}\left[-x^2 + a^2 - x^2\right] = \frac{4b}{a\sqrt{a^2 - x^2}}\left[a^2 - 2x^2\right]$$

So the critical number is $x = \frac{1}{\sqrt{2}}a$, and this clearly gives a maximum. Then $y = \frac{1}{\sqrt{2}}b$, so the maximum area is $4\left(\frac{1}{\sqrt{2}}a\right)\left(\frac{1}{\sqrt{2}}b\right) = 2ab$.

19. 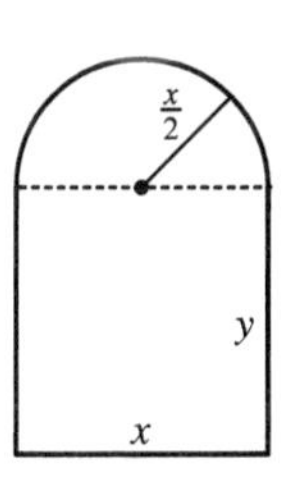

Perimeter $= 30 \Rightarrow 2y + x + \pi\left(\dfrac{x}{2}\right) = 30 \Rightarrow$

$y = \dfrac{1}{2}\left(30 - x - \dfrac{\pi x}{2}\right) = 15 - \dfrac{x}{2} - \dfrac{\pi x}{4}$. The area is the area of the rectangle plus the area of the semicircle, or $xy + \frac{1}{2}\pi\left(\dfrac{x}{2}\right)^2$,

so $A(x) = x\left(15 - \dfrac{x}{2} - \dfrac{\pi x}{4}\right) + \frac{1}{8}\pi x^2 = 15x - \frac{1}{2}x^2 - \frac{\pi}{8}x^2$.

$A'(x) = 15 - \left(1 + \frac{\pi}{4}\right)x = 0 \Rightarrow x = \dfrac{15}{1 + \pi/4} = \dfrac{60}{4 + \pi}$. $A''(x) = -\left(1 + \dfrac{\pi}{4}\right) < 0$, so this gives a maximum.

The dimensions are $x = \dfrac{60}{4 + \pi}$ ft and $y = 15 - \dfrac{30}{4 + \pi} - \dfrac{15\pi}{4 + \pi} = \dfrac{60 + 15\pi - 30 - 15\pi}{4 + \pi} = \dfrac{30}{4 + \pi}$ ft, so the height of the rectangle is half the base.

20. 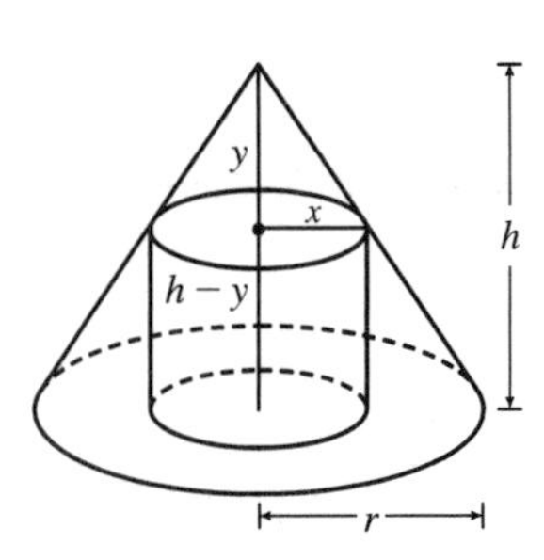

By similar triangles, $y/x = h/r$, so $y = hx/r$.

The volume of the cylinder is $\pi x^2(h - y) = \pi h x^2 - (\pi h/r)x^3 = V(x)$.

Now $V'(x) = 2\pi h x - (3\pi h/r)x^2 = \pi h x(2 - 3x/r)$.

So $V'(x) = 0 \Rightarrow x = 0$ or $x = \frac{2}{3}r$.

The maximum clearly occurs when $x = \frac{2}{3}r$ and then the volume is

$\pi h x^2 - (\pi h/r)x^3 = \pi h x^2(1 - x/r) = \pi\left(\frac{2}{3}r\right)^2 h\left(1 - \frac{2}{3}\right) = \frac{4}{27}\pi r^2 h$.

21. Let x be the length of the wire used for the square. The total area is

$$A(x) = \left(\frac{x}{4}\right)^2 + \frac{1}{2}\left(\frac{10-x}{3}\right)\frac{\sqrt{3}}{2}\left(\frac{10-x}{3}\right)$$
$$= \tfrac{1}{16}x^2 + \tfrac{\sqrt{3}}{36}(10-x)^2, \quad 0 \le x \le 10$$

$A'(x) = \frac{1}{8}x - \frac{\sqrt{3}}{18}(10-x) = 0 \quad \Leftrightarrow \quad \frac{9}{72}x + \frac{4\sqrt{3}}{72}x - \frac{40\sqrt{3}}{72} = 0 \quad \Leftrightarrow \quad x = \frac{40\sqrt{3}}{9+4\sqrt{3}}$.

Now $A(0) = \left(\frac{\sqrt{3}}{36}\right)100 \approx 4.81$, $A(10) = \frac{100}{16} = 6.25$ and $A\left(\frac{40\sqrt{3}}{9+4\sqrt{3}}\right) \approx 2.72$, so

(a) The maximum area occurs when $x = 10$ m, and all the wire is used for the square.

(b) The minimum area occurs when $x = \frac{40\sqrt{3}}{9+4\sqrt{3}} \approx 4.35$ m.

22.

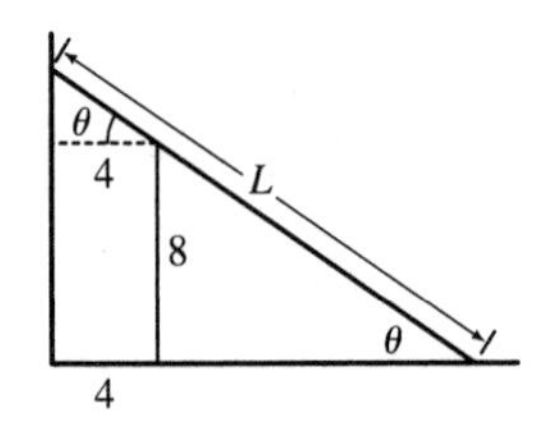

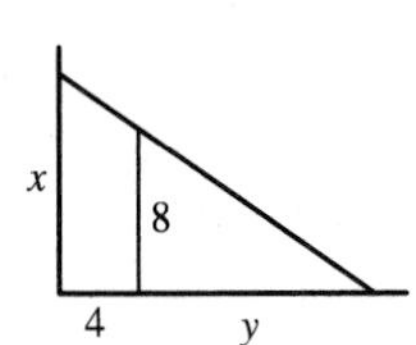

$L = 8\csc\theta + 4\sec\theta$, $0 < \theta < \frac{\pi}{2}$, $\frac{dL}{d\theta} = -8\csc\theta\,\cot\theta + 4\sec\theta\,\tan\theta = 0$ when

$\sec\theta\,\tan\theta = 2\csc\theta\,\cot\theta \quad \Leftrightarrow \quad \tan^3\theta = 2 \quad \Leftrightarrow \quad \tan\theta = \sqrt[3]{2} \quad \Leftrightarrow$

$\theta = \tan^{-1}\sqrt[3]{2}$. $dL/d\theta < 0$ when $0 < \theta < \tan^{-1}\sqrt[3]{2}$, $dL/d\theta > 0$ when

$\tan^{-1}\sqrt[3]{2} < \theta < \frac{\pi}{2}$, so L has an absolute minimum when $\theta = \tan^{-1}\sqrt[3]{2}$, and the

shortest ladder has length $L = 8\frac{\sqrt{1+2^{2/3}}}{2^{1/3}} + 4\sqrt{1+2^{2/3}} \approx 16.65$ ft.

Another method: Minimize $L^2 = x^2 + (4+y)^2$, where $\frac{x}{4+y} = \frac{8}{y}$.

23.

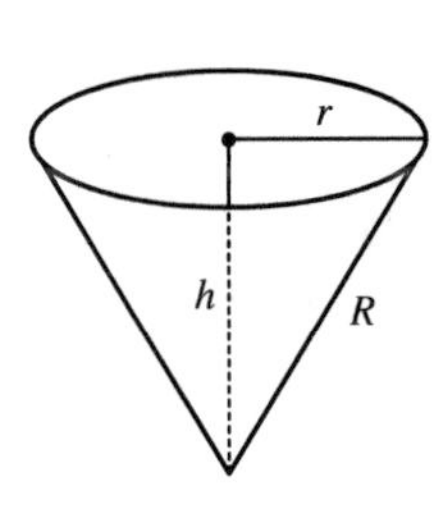

$h^2 + r^2 = R^2 \quad \Rightarrow \quad V = \frac{\pi}{3}r^2 h = \frac{\pi}{3}(R^2 - h^2)h = \frac{\pi}{3}(R^2 h - h^3)$.

$V'(h) = \frac{\pi}{3}(R^2 - 3h^2) = 0$ when $h = \frac{1}{\sqrt{3}}R$. This gives an absolute maximum, since

$V'(h) > 0$ for $0 < h < \frac{1}{\sqrt{3}}R$ and $V'(h) < 0$ for $h > \frac{1}{\sqrt{3}}R$. The maximum volume

is $V\left(\frac{1}{\sqrt{3}}R\right) = \frac{\pi}{3}\left(\frac{1}{\sqrt{3}}R^3 - \frac{1}{3\sqrt{3}}R^3\right) = \frac{2}{9\sqrt{3}}\pi R^3$.

24. The volume and surface area of a cone with radius r and height h are given by $V = \frac{1}{3}\pi r^2 h$ and $S = \pi r\sqrt{r^2+h^2}$.

We'll minimize $A = S^2$ subject to $V = 27$. $V = 27 \quad \Rightarrow \quad \frac{1}{3}\pi r^2 h = 27 \quad \Rightarrow \quad r^2 = \frac{81}{\pi h}$ **(1)**.

$A = \pi^2 r^2(r^2+h^2) = \pi^2\left(\frac{81}{\pi h}\right)\left(\frac{81}{\pi h} + h^2\right) = \frac{81^2}{h^2} + 81\pi h$, so $A' = 0 \quad \Rightarrow \quad \frac{-2\cdot 81^2}{h^3} + 81\pi = 0 \quad \Rightarrow$

$81\pi = \frac{2\cdot 81^2}{h^3} \quad \Rightarrow \quad h^3 = \frac{162}{\pi} \quad \Rightarrow \quad h = \sqrt[3]{\frac{162}{\pi}} = 3\sqrt[3]{\frac{6}{\pi}} \approx 3.722$. From **(1)**, $r^2 = \frac{81}{\pi h} = \frac{81}{\pi\cdot 3\sqrt[3]{6/\pi}} = \frac{27}{\sqrt[3]{6\pi^2}} \quad \Rightarrow$

$r = \frac{3\sqrt{3}}{\sqrt[6]{6\pi^2}} \approx 2.632$. $A'' = 6\cdot 81^2/h^4 > 0$, so A and hence S has an absolute minimum at these values of r and h.

25. 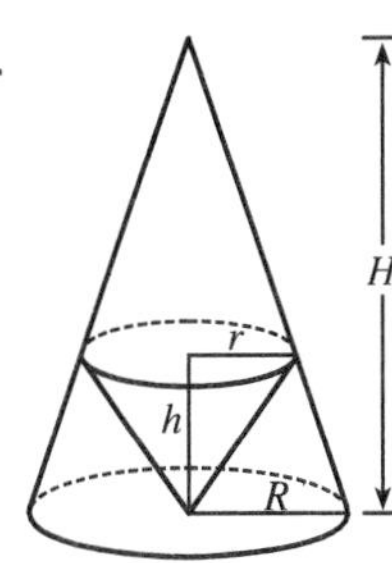

By similar triangles, $\dfrac{H}{R} = \dfrac{H-h}{r}$ **(1)**. The volume of the inner cone is $V = \frac{1}{3}\pi r^2 h$, so we'll solve **(1)** for h. $\dfrac{Hr}{R} = H - h \;\Rightarrow\; h = H - \dfrac{Hr}{R} = \dfrac{HR - Hr}{R} = \dfrac{H}{R}(R - r)$ **(2)**.

Thus, $V(r) = \dfrac{\pi}{3}r^2 \cdot \dfrac{H}{R}(R - r) = \dfrac{\pi H}{3R}(Rr^2 - r^3) \;\Rightarrow$

$V'(r) = \dfrac{\pi H}{3R}(2Rr - 3r^2) = \dfrac{\pi H}{3R}r(2R - 3r)$.

$V'(r) = 0 \;\Rightarrow\; r = 0$ or $2R = 3r \;\Rightarrow\; r = \frac{2}{3}R$ and from **(2)**, $h = \dfrac{H}{R}\left(R - \frac{2}{3}R\right) = \dfrac{H}{R}\left(\frac{1}{3}R\right) = \frac{1}{3}H$.

$V'(r)$ changes from positive to negative at $r = \frac{2}{3}R$, so the inner cone has a maximum volume of

$V = \frac{1}{3}\pi r^2 h = \frac{1}{3}\pi\left(\frac{2}{3}R\right)^2\left(\frac{1}{3}H\right) = \frac{4}{27} \cdot \frac{1}{3}\pi R^2 H$, which is approximately 15% of the volume of the larger cone.

26. We note that since c is the consumption in gallons per hour, and v is the velocity in miles per hour, then

$\dfrac{c}{v} = \dfrac{\text{gallons/hour}}{\text{miles/hour}} = \dfrac{\text{gallons}}{\text{mile}}$ gives us the consumption in gallons per mile, that is, the quantity G. To find the minimum,

we calculate $\dfrac{dG}{dv} = \dfrac{d}{dv}\left(\dfrac{c}{v}\right) = \dfrac{v\dfrac{dc}{dv} - c\dfrac{dv}{dv}}{v^2} = \dfrac{v\dfrac{dc}{dv} - c}{v^2}$. This is 0 when $v\dfrac{dc}{dv} - c = 0 \;\Leftrightarrow\; \dfrac{dc}{dv} = \dfrac{c}{v}$. This implies that the tangent line of $c(v)$ passes through the origin, and this occurs when $v \approx 53$ mi/h. Note that the slope of the secant line through the origin and a point $(v, c(v))$ on the graph is equal to $G(v)$, and it is intuitively clear that G is minimized in the case where the secant is in fact a tangent.

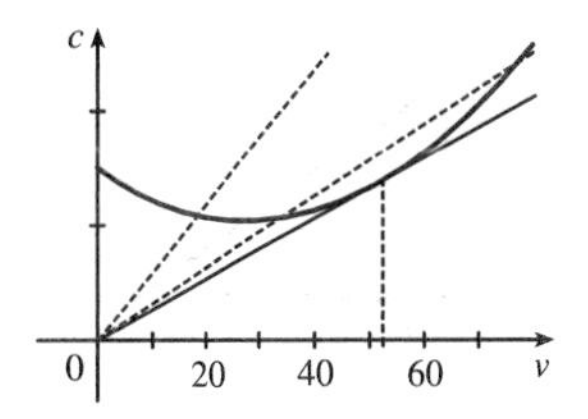

27. $P(R) = \dfrac{E^2 R}{(R+r)^2} \;\Rightarrow$

$$P'(R) = \frac{(R+r)^2 \cdot E^2 - E^2 R \cdot 2(R+r)}{[(R+r)^2]^2} = \frac{(R^2 + 2Rr + r^2)E^2 - 2E^2R^2 - 2E^2Rr}{(R+r)^4}$$

$$= \frac{E^2r^2 - E^2R^2}{(R+r)^4} = \frac{E^2(r^2 - R^2)}{(R+r)^4} = \frac{E^2(r+R)(r-R)}{(R+r)^4} = \frac{E^2(r-R)}{(R+r)^3}$$

$P'(R) = 0 \;\Rightarrow\; R = r \;\Rightarrow\; P(r) = \dfrac{E^2 r}{(r+r)^2} = \dfrac{E^2 r}{4r^2} = \dfrac{E^2}{4r}$. The expression for $P'(R)$ shows that $P'(R) > 0$ for $R < r$ and $P'(R) < 0$ for $R > r$. Thus, the maximum value of the power is $E^2/(4r)$, and this occurs when $R = r$.

28. (a) $E(v) = \dfrac{aLv^3}{v - u} \;\Rightarrow\; E'(v) = aL\dfrac{(v-u)3v^2 - v^3}{(v-u)^2} = 0$ when $2v^3 = 3uv^2 \;\Rightarrow\; 2v = 3u \;\Rightarrow\; v = \frac{3}{2}u$. The First Derivative Test shows that this value of v gives the minimum value of E.

(b)

E
0
u
$\frac{3}{2}u$
v

29. $S = 6sh - \frac{3}{2}s^2 \cot\theta + 3s^2 \frac{\sqrt{3}}{2} \csc\theta$

(a) $\dfrac{dS}{d\theta} = \frac{3}{2}s^2 \csc^2\theta - 3s^2 \frac{\sqrt{3}}{2} \csc\theta \cot\theta$ or $\frac{3}{2}s^2 \csc\theta \left(\csc\theta - \sqrt{3}\cot\theta\right)$.

(b) $\dfrac{dS}{d\theta} = 0$ when $\csc\theta - \sqrt{3}\cot\theta = 0 \;\Rightarrow\; \dfrac{1}{\sin\theta} - \sqrt{3}\dfrac{\cos\theta}{\sin\theta} = 0 \;\Rightarrow\; \cos\theta = \frac{1}{\sqrt{3}}$. The First Derivative Test shows that the minimum surface area occurs when $\theta = \cos^{-1}\left(\frac{1}{\sqrt{3}}\right) \approx 55^\circ$.

(c) If $\cos\theta = \frac{1}{\sqrt{3}}$, then $\cot\theta = \frac{1}{\sqrt{2}}$ and $\csc\theta = \frac{\sqrt{3}}{\sqrt{2}}$, so the surface area is

$$S = 6sh - \tfrac{3}{2}s^2 \tfrac{1}{\sqrt{2}} + 3s^2 \tfrac{\sqrt{3}}{2}\tfrac{\sqrt{3}}{\sqrt{2}} = 6sh - \tfrac{3}{2\sqrt{2}}s^2 + \tfrac{9}{2\sqrt{2}}s^2$$

$$= 6sh + \tfrac{6}{2\sqrt{2}}s^2 = 6s\left(h + \tfrac{1}{2\sqrt{2}}s\right)$$

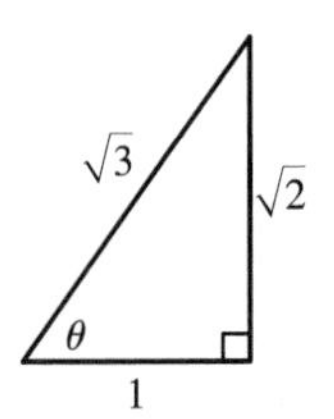

30.

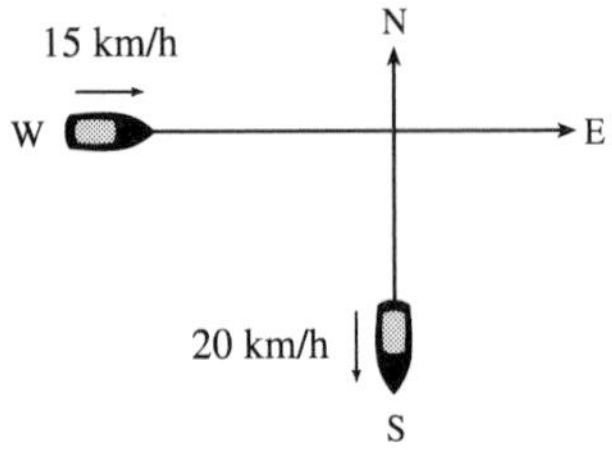

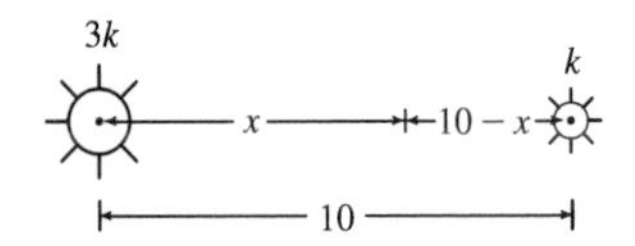

Let t be the time, in hours, after 2:00 PM. The position of the boat heading south at time t is $(0, -20t)$. The position of the boat heading east at time t is $(-15 + 15t, 0)$. If $D(t)$ is the distance between the boats at time t, we minimize

$f(t) = [D(t)]^2 = 20^2t^2 + 15^2(t-1)^2$.

$f'(t) = 800t + 450(t-1) = 1250t - 450 = 0$ when $t = \frac{450}{1250} = 0.36$ h.

$0.36 \text{ h} \times \frac{60 \text{ min}}{\text{h}} = 21.6 \text{ min} = 21 \text{ min } 36 \text{ s}$. Since $f''(t) > 0$, this gives a minimum, so the boats are closest together at 2:21:36 PM.

31.

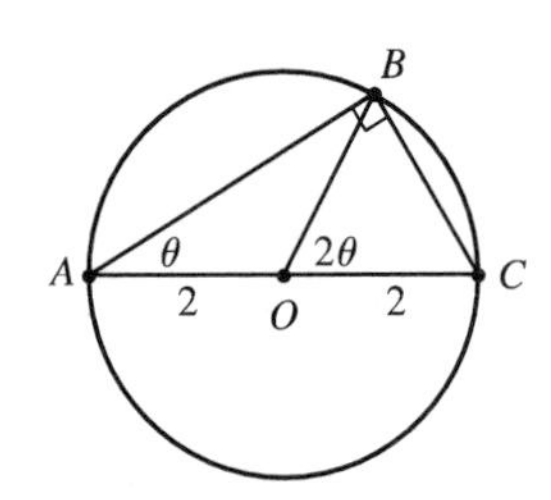

The total illumination is $I(x) = \dfrac{3k}{x^2} + \dfrac{k}{(10-x)^2}$, $0 < x < 10$. Then

$$I'(x) = \frac{-6k}{x^3} + \frac{2k}{(10-x)^3} = 0 \;\Rightarrow\; 6k(10-x)^3 = 2kx^3 \;\Rightarrow$$

$3(10-x)^3 = x^3 \;\Rightarrow\; \sqrt[3]{3}\,(10-x) = x \;\Rightarrow\; 10\sqrt[3]{3} - \sqrt[3]{3}\,x = x \;\Rightarrow\; 10\sqrt[3]{3} = x + \sqrt[3]{3}\,x \;\Rightarrow$

$10\sqrt[3]{3} = \left(1 + \sqrt[3]{3}\right)x \;\Rightarrow\; x = \dfrac{10\sqrt[3]{3}}{1 + \sqrt[3]{3}} \approx 5.9$ ft. This gives a minimum since $I''(x) > 0$ for $0 < x < 10$.

32.

In isosceles triangle AOB, $\angle O = 180^\circ - \theta - \theta$, so $\angle BOC = 2\theta$. The distance rowed is $4\cos\theta$ while the distance walked is the length of arc $BC = 2(2\theta) = 4\theta$.

The time taken is given by $T(\theta) = \dfrac{4\cos\theta}{2} + \dfrac{4\theta}{4} = 2\cos\theta + \theta, \quad 0 \le \theta \le \frac{\pi}{2}$.

$T'(\theta) = -2\sin\theta + 1 = 0 \;\Leftrightarrow\; \sin\theta = \frac{1}{2} \;\Rightarrow\; \theta = \frac{\pi}{6}$.

Check the value of T at $\theta = \frac{\pi}{6}$ and at the endpoints of the domain of T; that is, $\theta = 0$ and $\theta = \frac{\pi}{2}$. $T(0) = 2$, $T\left(\frac{\pi}{6}\right) = \sqrt{3} + \frac{\pi}{6} \approx 2.26$, and $T\left(\frac{\pi}{2}\right) = \frac{\pi}{2} \approx 1.57$. Therefore, the minimum value of T is $\frac{\pi}{2}$ when $\theta = \frac{\pi}{2}$; that is, the woman should walk all the way. Note that $T''(\theta) = -2\cos\theta < 0$ for $0 \le \theta < \frac{\pi}{2}$, so $\theta = \frac{\pi}{6}$ gives a maximum time.

33. 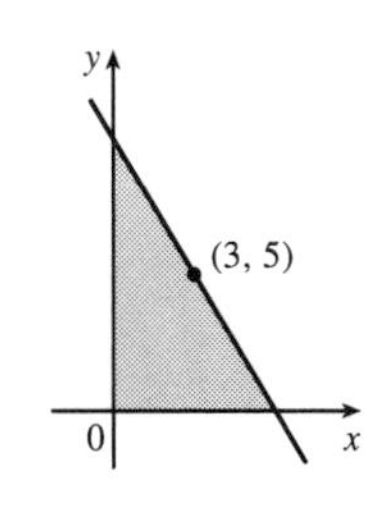

The line with slope m (where $m < 0$) through $(3, 5)$ has equation $y - 5 = m(x - 3)$ or $y = mx + (5 - 3m)$. The y-intercept is $5 - 3m$ and the x-intercept is $-5/m + 3$. So the triangle has area $A(m) = \frac{1}{2}(5 - 3m)(-5/m + 3) = 15 - 25/(2m) - \frac{9}{2}m$. Now

$A'(m) = \dfrac{25}{2m^2} - \dfrac{9}{2} = 0 \quad \Leftrightarrow \quad m^2 = \frac{25}{9} \quad \Rightarrow \quad m = -\frac{5}{3}$ (since $m < 0$).

$A''(m) = -\dfrac{25}{m^3} > 0$, so there is an absolute minimum when $m = -\frac{5}{3}$. Thus, an equation of the line is $y - 5 = -\frac{5}{3}(x - 3)$ or $y = -\frac{5}{3}x + 10$.

34. $y = 1 + 40x^3 - 3x^5 \quad \Rightarrow \quad y' = 120x^2 - 15x^4$, so the tangent line to the curve at $x = a$ has slope $m(a) = 120a^2 - 15a^4$. Now $m'(a) = 240a - 60a^3 = -60a(a^2 - 4) = -60a(a + 2)(a - 2)$, so $m'(a) > 0$ for $a < -2$, and $0 < a < 2$, and $m'(a) < 0$ for $-2 < a < 0$ and $a > 2$. Thus, m is increasing on $(-\infty, -2)$, decreasing on $(-2, 0)$, increasing on $(0, 2)$, and decreasing on $(2, \infty)$. Clearly, $m(a) \to -\infty$ as $a \to \pm\infty$, so the maximum value of $m(a)$ must be one of the two local maxima, $m(-2)$ or $m(2)$. But both $m(-2)$ and $m(2)$ equal $120 \cdot 2^2 - 15 \cdot 2^4 = 480 - 240 = 240$. So 240 is the largest slope, and it occurs at the points $(-2, -223)$ and $(2, 225)$. [*Note:* $a = 0$ corresponds to a local *minimum* of m.]

35. (a) If $c(x) = \dfrac{C(x)}{x}$, then, by Quotient Rule, we have $c'(x) = \dfrac{xC'(x) - C(x)}{x^2}$. Now $c'(x) = 0$ when $xC'(x) - C(x) = 0$ and this gives $C'(x) = \dfrac{C(x)}{x} = c(x)$. Therefore, the marginal cost equals the average cost.

(b) (i) $C(x) = 16{,}000 + 200x + 4x^{3/2}$, $C(1000) = 16{,}000 + 200{,}000 + 40{,}000\sqrt{10} \approx 216{,}000 + 126{,}491$, so $C(1000) \approx \$342{,}491$. $c(x) = C(x)/x = \dfrac{16{,}000}{x} + 200 + 4x^{1/2}$, $c(1000) \approx \$342.49/\text{unit}$.

$C'(x) = 200 + 6x^{1/2}$, $C'(1000) = 200 + 60\sqrt{10} \approx \$389.74/\text{unit}$.

(ii) We must have $C'(x) = c(x) \quad \Leftrightarrow \quad 200 + 6x^{1/2} = \dfrac{16{,}000}{x} + 200 + 4x^{1/2} \quad \Leftrightarrow \quad 2x^{3/2} = 16{,}000 \quad \Leftrightarrow$

$x = (8{,}000)^{2/3} = 400$ units. To check that this is a minimum, we calculate

$c'(x) = \dfrac{-16{,}000}{x^2} + \dfrac{2}{\sqrt{x}} = \dfrac{2}{x^2}(x^{3/2} - 8000)$. This is negative for $x < (8000)^{2/3} = 400$, zero at $x = 400$, and positive for $x > 400$, so c is decreasing on $(0, 400)$ and increasing on $(400, \infty)$. Thus, c has an absolute minimum at $x = 400$. [*Note:* $c''(x)$ is *not* positive for all $x > 0$.]

(iii) The minimum average cost is $c(400) = 40 + 200 + 80 = \$320/\text{unit}$.

36. (a) The total profit is $P(x) = R(x) - C(x)$. In order to maximize profit we look for the critical numbers of P, that is, the numbers where the marginal profit is 0. But if $P'(x) = R'(x) - C'(x) = 0$, then $R'(x) = C'(x)$. Therefore, if the profit is a maximum, then the marginal revenue equals the marginal cost.

(b) $C(x) = 16{,}000 + 500x - 1.6x^2 + 0.004x^3$, $p(x) = 1700 - 7x$. Then $R(x) = xp(x) = 1700x - 7x^2$. If the profit is maximum, then $R'(x) = C'(x) \quad \Leftrightarrow \quad 1700 - 14x = 500 - 3.2x + 0.012x^2 \quad \Leftrightarrow \quad 0.012x^2 + 10.8x - 1200 = 0 \quad \Leftrightarrow$ $x^2 + 900x - 100{,}000 = 0 \quad \Leftrightarrow \quad (x + 1000)(x - 100) = 0 \quad \Leftrightarrow \quad x = 100$ (since $x > 0$). The profit is maximized if $P''(x) < 0$, but since $P''(x) = R''(x) - C''(x)$, we can just check the condition $R''(x) < C''(x)$. Now $R''(x) = -14 < -3.2 + 0.024x = C''(x)$ for $x > 0$, so there is a maximum at $x = 100$.

37. (a) We are given that the demand function p is linear and $p(27{,}000) = 10$, $p(33{,}000) = 8$, so the slope is

$\frac{10-8}{27{,}000-33{,}000} = -\frac{1}{3000}$ and an equation of the line is $y - 10 = \left(-\frac{1}{3000}\right)(x - 27{,}000) \Rightarrow$

$y = p(x) = -\frac{1}{3000}x + 19 = 19 - (x/3000)$.

(b) The revenue is $R(x) = xp(x) = 19x - (x^2/3000) \Rightarrow R'(x) = 19 - (x/1500) = 0$ when $x = 28{,}500$. Since $R''(x) = -1/1500 < 0$, the maximum revenue occurs when $x = 28{,}500 \Rightarrow$ the price is $p(28{,}500) = \$9.50$.

38. (a) Let $p(x)$ be the demand function. Then $p(x)$ is linear and $y = p(x)$ passes through $(20, 10)$ and $(18, 11)$, so the slope is $-\frac{1}{2}$ and an equation of the line is $y - 10 = -\frac{1}{2}(x - 20) \Leftrightarrow y = -\frac{1}{2}x + 20$. Thus, the demand is $p(x) = -\frac{1}{2}x + 20$ and the revenue is $R(x) = xp(x) = -\frac{1}{2}x^2 + 20x$.

(b) The cost is $C(x) = 6x$, so the profit is $P(x) = R(x) - C(x) = -\frac{1}{2}x^2 + 14x$. Then $0 = P'(x) = -x + 14 \Rightarrow$ $x = 14$. Since $P''(x) = -1 < 0$, the selling price for maximum profit is $p(14) = -\frac{1}{2}(14) + 20 = \$13$.

39. (a) As in Example 6, we see that the demand function p is linear. We are given that $p(1000) = 450$ and deduce that $p(1100) = 440$, since a \$10 reduction in price increases sales by 100 per week. The slope for p is $\frac{440-450}{1100-1000} = -\frac{1}{10}$, so an equation is $p - 450 = -\frac{1}{10}(x - 1000)$ or $p(x) = -\frac{1}{10}x + 550$.

(b) $R(x) = xp(x) = -\frac{1}{10}x^2 + 550x$. $R'(x) = -\frac{1}{5}x + 550 = 0$ when $x = 5(550) = 2750$.

$p(2750) = 275$, so the rebate should be $450 - 275 = \$175$.

(c) $C(x) = 68{,}000 + 150x \Rightarrow P(x) = R(x) - C(x) = -\frac{1}{10}x^2 + 550x - 68{,}000 - 150x = -\frac{1}{10}x^2 + 400x - 68{,}000$,

$P'(x) = -\frac{1}{5}x + 400 = 0$ when $x = 2000$. $p(2000) = 350$. Therefore, the rebate to maximize profits should be $450 - 350 = \$100$.

40. Let x denote the number of \$10 increases in rent. Then the price is $p(x) = 800 + 10x$, and the number of units occupied is $100 - x$. Now the revenue is

$$\begin{aligned} R(x) &= (\text{rental price per unit}) \times (\text{number of units rented}) \\ &= (800 + 10x)(100 - x) = -10x^2 + 200x + 80{,}000 \text{ for } 0 \le x \le 100 \quad \Rightarrow \end{aligned}$$

$R'(x) = -20x + 200 = 0 \Leftrightarrow x = 10$. This is a maximum since $R''(x) = -20 < 0$ for all x. Now we must check the value of $R(x) = (800 + 10x)(100 - x)$ at $x = 10$ and at the endpoints of the domain to see which value of x gives the maximum value of R. $R(0) = 80{,}000$, $R(10) = (900)(90) = 81{,}000$, and $R(100) = (1800)(0) = 0$. Thus, the maximum revenue of \$81,000/week occurs when 90 units are occupied at a rent of \$900/week.

41.

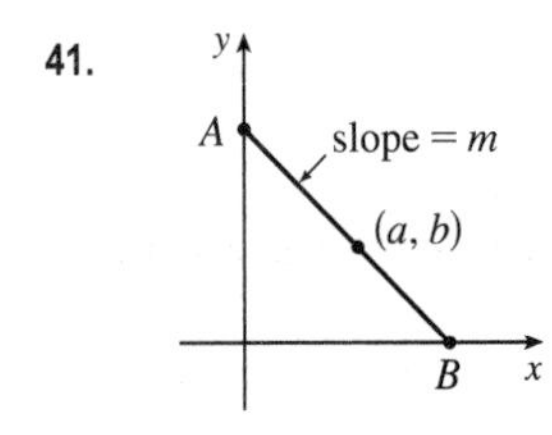

Every line segment in the first quadrant passing through (a, b) with endpoints on the x- and y-axes satisfies an equation of the form $y - b = m(x - a)$, where $m < 0$. By setting $x = 0$ and then $y = 0$, we find its endpoints, $A(0, b - am)$ and $B\left(a - \frac{b}{m}, 0\right)$. The distance d from A to B is given by $d = \sqrt{\left[\left(a - \frac{b}{m}\right) - 0\right]^2 + [0 - (b - am)]^2}$. It follows that the square of the length of the line segment, as a function of m, is given by $S(m) = \left(a - \frac{b}{m}\right)^2 + (am - b)^2 = a^2 - \frac{2ab}{m} + \frac{b^2}{m^2} + a^2m^2 - 2abm + b^2$.

Thus, $$S'(m) = \frac{2ab}{m^2} - \frac{2b^2}{m^3} + 2a^2m - 2ab = \frac{2}{m^3}(abm - b^2 + a^2m^4 - abm^3)$$
$$= \frac{2}{m^3}[b(am - b) + am^3(am - b)] = \frac{2}{m^3}(am - b)(b + am^3)$$

Thus, $S'(m) = 0 \quad \Leftrightarrow \quad m = b/a$ or $m = -\sqrt[3]{\frac{b}{a}}$. Since $b/a > 0$ and $m < 0$, m must equal $-\sqrt[3]{\frac{b}{a}}$. Since $\frac{2}{m^3} < 0$, we see that $S'(m) < 0$ for $m < -\sqrt[3]{\frac{b}{a}}$ and $S'(m) > 0$ for $m > -\sqrt[3]{\frac{b}{a}}$. Thus, S has its absolute minimum value when $m = -\sqrt[3]{\frac{b}{a}}$. That value is

$$S\left(-\sqrt[3]{\tfrac{b}{a}}\right) = \left(a + b\sqrt[3]{\tfrac{a}{b}}\right)^2 + \left(-a\sqrt[3]{\tfrac{b}{a}} - b\right)^2 = \left(a + \sqrt[3]{ab^2}\right)^2 + \left(\sqrt[3]{a^2b} + b\right)^2$$
$$= a^2 + 2a^{4/3}b^{2/3} + a^{2/3}b^{4/3} + a^{4/3}b^{2/3} + 2a^{2/3}b^{4/3} + b^2 = a^2 + 3a^{4/3}b^{2/3} + 3a^{2/3}b^{4/3} + b^2$$

The last expression is of the form $x^3 + 3x^2y + 3xy^2 + y^3 \quad [= (x + y)^3]$ with $x = a^{2/3}$ and $y = b^{2/3}$, so we can write it as $(a^{2/3} + b^{2/3})^3$ and the shortest such line segment has length $\sqrt{S} = (a^{2/3} + b^{2/3})^{3/2}$.

42. See the figure. The area is given by

$$A(x) = \tfrac{1}{2}\left(2\sqrt{a^2 - x^2}\right)x + \tfrac{1}{2}\left(2\sqrt{a^2 - x^2}\right)\left(\sqrt{x^2 + b^2 - a^2}\right)$$
$$= \sqrt{a^2 - x^2}\left(x + \sqrt{x^2 + b^2 - a^2}\right) \quad \text{for } 0 \le x \le a.$$

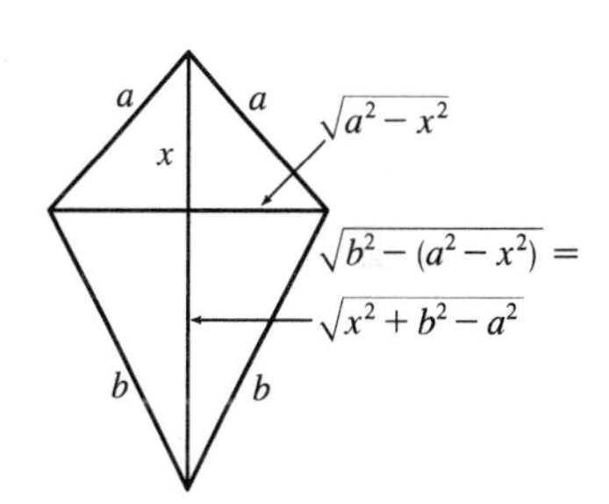

Now

$$A'(x) = \sqrt{a^2 - x^2}\left(1 + \frac{x}{\sqrt{x^2 + b^2 - a^2}}\right) + \left(x + \sqrt{x^2 + b^2 - a^2}\right)\frac{-x}{\sqrt{a^2 - x^2}} = 0$$

$$\Leftrightarrow \quad \frac{x}{\sqrt{a^2 - x^2}}\left(x + \sqrt{x^2 + b^2 - a^2}\right) = \sqrt{a^2 - x^2}\left(\frac{x + \sqrt{x^2 + b^2 - a^2}}{\sqrt{x^2 + b^2 - a^2}}\right).$$

Except for the trivial case where $x = 0$, $a = b$ and $A(x) = 0$, we have $x + \sqrt{x^2 + b^2 - a^2} > 0$. Hence, cancelling this factor gives $\frac{x}{\sqrt{a^2 - x^2}} = \frac{\sqrt{a^2 - x^2}}{\sqrt{x^2 + b^2 - a^2}} \quad \Rightarrow \quad x\sqrt{x^2 + b^2 - a^2} = a^2 - x^2 \quad \Rightarrow \quad x^2(x^2 + b^2 - a^2) = a^4 - 2a^2x^2 + x^4 \quad \Rightarrow$

$x^2(b^2 - a^2) = a^4 - 2a^2x^2 \quad \Rightarrow \quad x^2(b^2 + a^2) = a^4 \quad \Rightarrow \quad x = \frac{a^2}{\sqrt{a^2 + b^2}}$. Now we must check the value of A at this point as well as at the endpoints of the domain to see which gives the maximum value. $A(0) = a\sqrt{b^2 - a^2}$, $A(a) = 0$ and

$$A\left(\frac{a^2}{\sqrt{a^2 + b^2}}\right) = \sqrt{a^2 - \left(\frac{a^2}{\sqrt{a^2 + b^2}}\right)^2}\left[\frac{a^2}{\sqrt{a^2 + b^2}} + \sqrt{\left(\frac{a^2}{\sqrt{a^2 + b^2}}\right)^2 + b^2 - a^2}\right]$$
$$= \frac{ab}{\sqrt{a^2 + b^2}}\left[\frac{a^2}{\sqrt{a^2 + b^2}} + \frac{b^2}{\sqrt{a^2 + b^2}}\right] = \frac{ab(a^2 + b^2)}{a^2 + b^2} = ab$$

Since $b \ge \sqrt{b^2 - a^2}$, $A\left(a^2/\sqrt{a^2 + b^2}\right) \ge A(0)$. So there is an absolute maximum when $x = \frac{a^2}{\sqrt{a^2 + b^2}}$. In this case the horizontal piece should be $\frac{2ab}{\sqrt{a^2 + b^2}}$ and the vertical piece should be $\frac{a^2 + b^2}{\sqrt{a^2 + b^2}} = \sqrt{a^2 + b^2}$.

43. 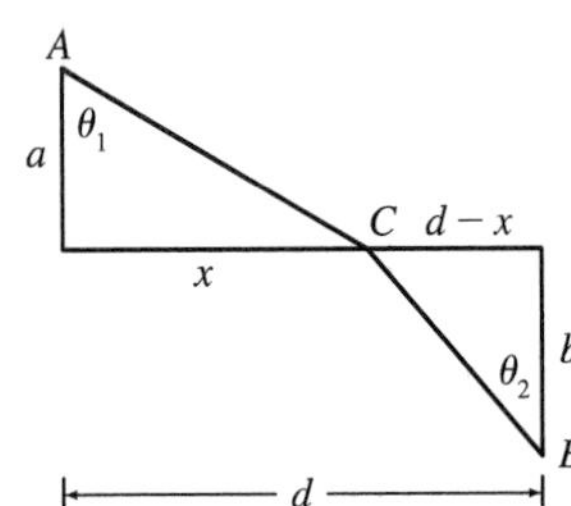

The total time is

$$(x) = \begin{pmatrix}\text{time from} \\ A \text{ to } C\end{pmatrix} + \begin{pmatrix}\text{time from} \\ C \text{ to } B\end{pmatrix} = \frac{\sqrt{a^2+x^2}}{v_1} + \frac{\sqrt{b^2+(d-x)^2}}{v_2}, \quad 0 < x < d.$$

$$T'(x) = \frac{x}{v_1\sqrt{a^2+x^2}} - \frac{d-x}{v_2\sqrt{b^2+(d-x)^2}} = \frac{\sin\theta_1}{v_1} - \frac{\sin\theta_2}{v_2}$$

The minimum occurs when $T'(x) = 0 \;\Rightarrow\; \dfrac{\sin\theta_1}{v_1} = \dfrac{\sin\theta_2}{v_2}$. [*Note:* $T''(x) > 0$]

44. If $d = |QT|$, we minimize $f(\theta_1) = |PR| + |RS| = a\csc\theta_1 + b\csc\theta_2$.

Differentiating with respect to θ_1, and setting $\dfrac{df}{d\theta_1}$ equal to 0, we get

$\dfrac{df}{d\theta_1} = 0 = -a\csc\theta_1\cot\theta_1 - b\csc\theta_2\cot\theta_2\dfrac{d\theta_2}{d\theta_1}$. So we need to find an expression for $\dfrac{d\theta_2}{d\theta_1}$. We can do this by observing that $|QT| = \text{constant} = a\cot\theta_1 + b\cot\theta_2$. Differentiating this equation implicitly with respect to θ_1, we get $-a\csc^2\theta_1 - b\csc^2\theta_2\dfrac{d\theta_2}{d\theta_1} = 0 \;\Rightarrow\; \dfrac{d\theta_2}{d\theta_1} = -\dfrac{a\csc^2\theta_1}{b\csc^2\theta_2}$. We substitute this into the expression for $\dfrac{df}{d\theta_1}$ to get $-a\csc\theta_1\cot\theta_1 - b\csc\theta_2\cot\theta_2\left(-\dfrac{a\csc^2\theta_1}{b\csc^2\theta_2}\right) = 0 \;\Leftrightarrow$

$$-a\csc\theta_1\cot\theta_1 + a\frac{\csc^2\theta_1\cot\theta_2}{\csc\theta_2} = 0 \;\Leftrightarrow\; \cot\theta_1\csc\theta_2 = \csc\theta_1\cot\theta_2 \;\Leftrightarrow\; \frac{\cot\theta_1}{\csc\theta_1} = \frac{\cot\theta_2}{\csc\theta_2} \;\Leftrightarrow\; \cos\theta_1 = \cos\theta_2.$$

Since θ_1 and θ_2 are both acute, we have $\theta_1 = \theta_2$.

45. 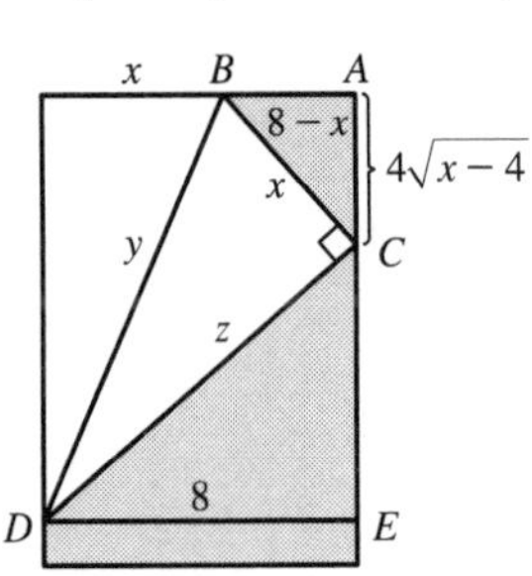

$y^2 = x^2 + z^2$, but triangles CDE and BCA are similar, so $z/8 = x/\left(4\sqrt{x-4}\right) \;\Rightarrow$ $z = 2x/\sqrt{x-4}$. Thus, we minimize $f(x) = y^2 = x^2 + 4x^2/(x-4) = x^3/(x-4)$, $4 < x \le 8$. $f'(x) = \dfrac{(x-4)(3x^2) - x^3}{(x-4)^2} = \dfrac{x^2[3(x-4) - x]}{(x-4)^2} = \dfrac{2x^2(x-6)}{(x-4)^2} = 0$ when $x = 6$. $f'(x) < 0$ when $x < 6$, $f'(x) > 0$ when $x > 6$, so the minimum occurs when $x = 6$ in.

46. 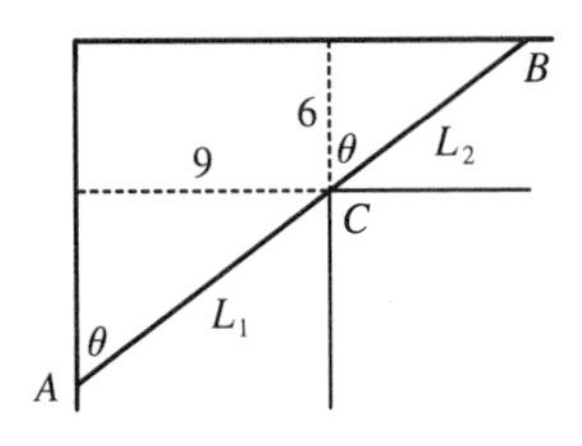

Paradoxically, we solve this maximum problem by solving a minimum problem. Let L be the length of the line ACB going from wall to wall touching the inner corner C. As $\theta \to 0$ or $\theta \to \frac{\pi}{2}$, we have $L \to \infty$ and there will be an angle that makes L a minimum. A pipe of this length will just fit around the corner. From the diagram,

$L = L_1 + L_2 = 9\csc\theta + 6\sec\theta \;\Rightarrow\; dL/d\theta = -9\csc\theta\,\cot\theta + 6\sec\theta\,\tan\theta = 0$ when $6\sec\theta\,\tan\theta = 9\csc\theta\,\cot\theta$ $\Leftrightarrow\; \tan^3\theta = \frac{9}{6} = 1.5 \;\Leftrightarrow\; \tan\theta = \sqrt[3]{1.5}$. Then $\sec^2\theta = 1 + \left(\frac{3}{2}\right)^{2/3}$ and $\csc^2\theta = 1 + \left(\frac{3}{2}\right)^{-2/3}$, so the longest pipe has length $L = 9\left[1 + \left(\frac{3}{2}\right)^{-2/3}\right]^{1/2} + 6\left[1 + \left(\frac{3}{2}\right)^{2/3}\right]^{1/2} \approx 21.07$ ft.

Or, use $\theta = \tan^{-1}\left(\sqrt[3]{1.5}\right) \approx 0.853 \;\Rightarrow\; L = 9\csc\theta + 6\sec\theta \approx 21.07$ ft.

47.

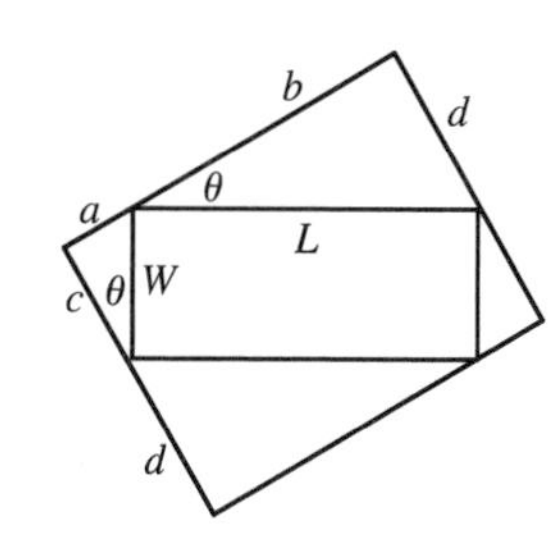

In the small triangle with sides a and c and hypotenuse W, $\sin\theta = \dfrac{a}{W}$ and $\cos\theta = \dfrac{c}{W}$. In the triangle with sides b and d and hypotenuse L, $\sin\theta = \dfrac{d}{L}$ and $\cos\theta = \dfrac{b}{L}$. Thus, $a = W\sin\theta$, $c = W\cos\theta$, $d = L\sin\theta$, and $b = L\cos\theta$, so the area of the circumscribed rectangle is

$$\begin{aligned}A(\theta) &= (a+b)(c+d) = (W\sin\theta + L\cos\theta)(W\cos\theta + L\sin\theta)\\ &= W^2\sin\theta\,\cos\theta + WL\sin^2\theta + LW\cos^2\theta + L^2\sin\theta\,\cos\theta = LW\sin^2\theta + LW\cos^2\theta + (L^2+W^2)\sin\theta\,\cos\theta\\ &= LW(\sin^2\theta + \cos^2\theta) + (L^2+W^2)\cdot\tfrac{1}{2}\cdot 2\sin\theta\,\cos\theta = LW + \tfrac{1}{2}(L^2+W^2)\sin 2\theta,\quad 0\le\theta\le\tfrac{\pi}{2}\end{aligned}$$

This expression shows, without calculus, that the maximum value of $A(\theta)$ occurs when $\sin 2\theta = 1 \Leftrightarrow 2\theta = \frac{\pi}{2} \Rightarrow \theta = \frac{\pi}{4}$. So the maximum area is $A\left(\frac{\pi}{4}\right) = LW + \frac{1}{2}(L^2+W^2) = \frac{1}{2}(L^2+2LW+W^2) = \frac{1}{2}(L+W)^2$.

48. We maximize the cross-sectional area

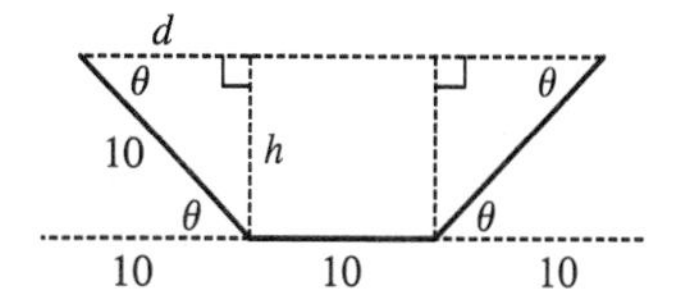

$$\begin{aligned}A(\theta) &= 10h + 2\left(\tfrac{1}{2}dh\right) = 10h + dh = 10(10\sin\theta) + (10\cos\theta)(10\sin\theta)\\ &= 100(\sin\theta + \sin\theta\,\cos\theta),\quad 0\le\theta\le\tfrac{\pi}{2}\end{aligned}$$

$$\begin{aligned}A'(\theta) &= 100(\cos\theta + \cos^2\theta - \sin^2\theta) = 100(\cos\theta + 2\cos^2\theta - 1)\\ &= 100(2\cos\theta - 1)(\cos\theta + 1) = 0 \text{ when } \cos\theta = \tfrac{1}{2} \quad\Leftrightarrow\quad \theta = \tfrac{\pi}{3} \quad [\cos\theta \ne -1 \text{ since } 0\le\theta\le\tfrac{\pi}{2}.]\end{aligned}$$

Now $A(0) = 0$, $A\left(\frac{\pi}{2}\right) = 100$ and $A\left(\frac{\pi}{3}\right) = 75\sqrt{3} \approx 129.9$, so the maximum occurs when $\theta = \frac{\pi}{3}$.

49. (a) $I(x) \propto \dfrac{\text{strength of source}}{(\text{distance from source})^2}$. Adding the intensities from the left and right lightbulbs,

$$I(x) = \frac{k}{x^2+d^2} + \frac{k}{(10-x)^2+d^2} = \frac{k}{x^2+d^2} + \frac{k}{x^2-20x+100+d^2}.$$

(b) The magnitude of the constant k won't affect the location of the point of maximum intensity, so for convenience we take $k=1$. $I'(x) = -\dfrac{2x}{(x^2+d^2)^2} - \dfrac{2(x-10)}{(x^2-20x+100+d^2)^2}$.

Substituting $d = 5$ into the equations for $I(x)$ and $I'(x)$, we get

$$I_5(x) = \frac{1}{x^2+25} + \frac{1}{x^2-20x+125} \qquad\text{and}\qquad I_5'(x) = -\frac{2x}{(x^2+25)^2} - \frac{2(x-10)}{(x^2-20x+125)^2}$$

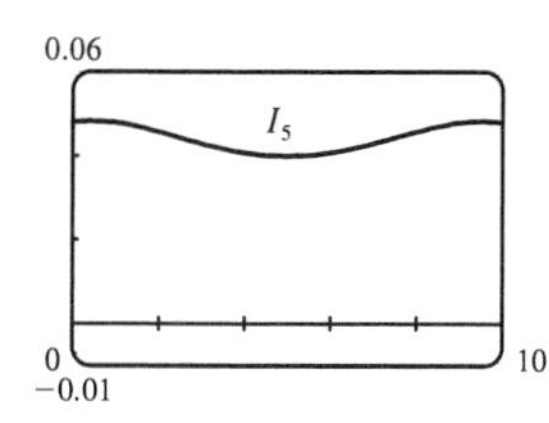

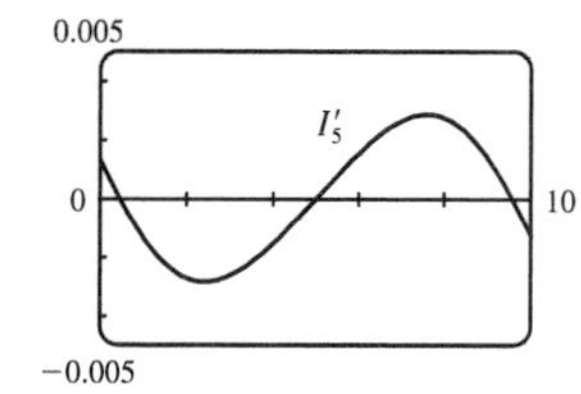

From the graphs, it appears that $I_5(x)$ has a minimum at $x = 5$ m.

(c) Substituting $d = 10$ into the equations for $I(x)$ and $I'(x)$ gives $I_{10}(x) = \dfrac{1}{x^2 + 100} + \dfrac{1}{x^2 - 20x + 200}$ and

$$I'_{10}(x) = -\frac{2x}{(x^2 + 100)^2} - \frac{2(x - 10)}{(x^2 - 20x + 200)^2}.$$

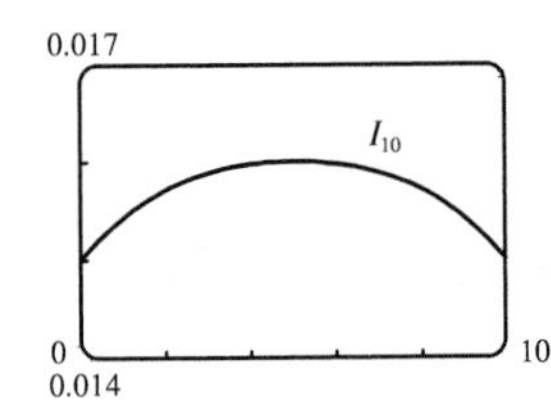

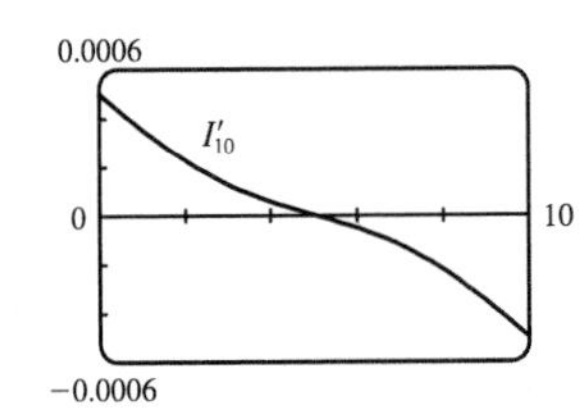

From the graphs, it seems that for $d = 10$, the intensity is minimized at the endpoints, that is, $x = 0$ and $x = 10$. The midpoint is now the most brightly lit point!

(d) From the first figures in parts (b) and (c), we see that the minimal illumination changes from the midpoint ($x = 5$ with $d = 5$) to the endpoints ($x = 0$ and $x = 10$ with $d = 10$).

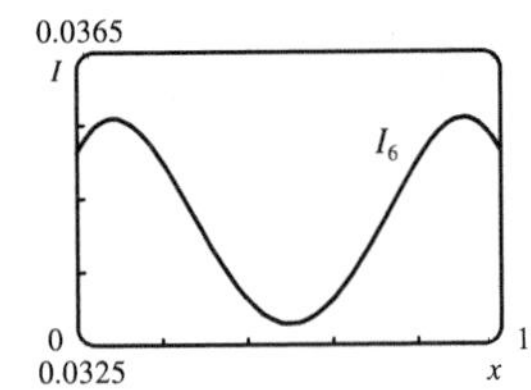

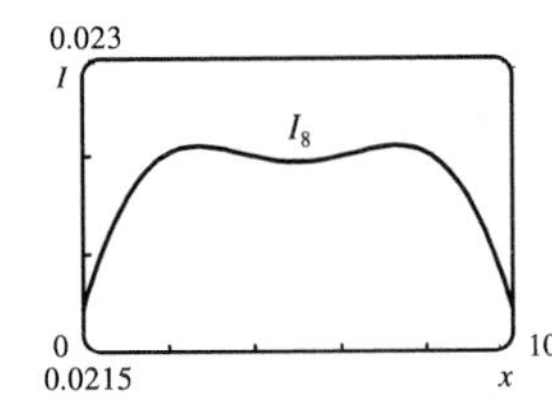

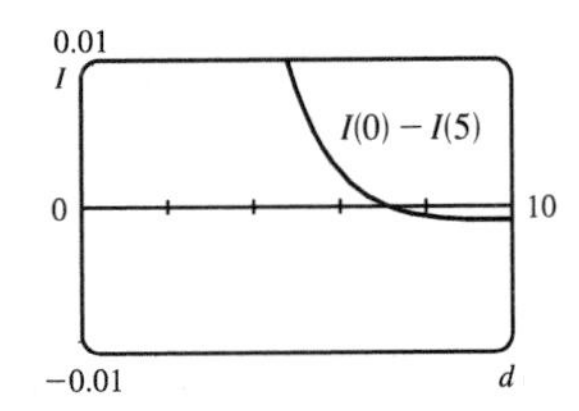

So we try $d = 6$ (see the first figure) and we see that the minimum value still occurs at $x = 5$. Next, we let $d = 8$ (see the second figure) and we see that the minimum value occurs at the endpoints. It appears that for some value of d between 6 and 8, we must have minima at both the midpoint and the endpoints, that is, $I(5)$ must equal $I(0)$. To find this value of d, we solve $I(0) = I(5)$ (with $k = 1$):

$$\frac{1}{d^2} + \frac{1}{100 + d^2} = \frac{1}{25 + d^2} + \frac{1}{25 + d^2} = \frac{2}{25 + d^2} \quad \Rightarrow$$

$$(25 + d^2)(100 + d^2) + d^2(25 + d^2) = 2d^2(100 + d^2) \quad \Rightarrow \quad 2500 + 125d^2 + d^4 + 25d^2 + d^4 = 200d^2 + 2d^4 \quad \Rightarrow$$

$2500 = 50d^2 \;\Rightarrow\; d^2 = 50 \;\Rightarrow\; d = 5\sqrt{2} \approx 7.071$ (for $0 \le d \le 10$). The third figure, a graph of $I(0) - I(5)$ with d independent, confirms that $I(0) - I(5) = 0$, that is, $I(0) = I(5)$, when $d = 5\sqrt{2}$. Thus, the point of minimal illumination changes abruptly from the midpoint to the endpoints when $d = 5\sqrt{2}$.

3.6 Newton's Method

1. (a)

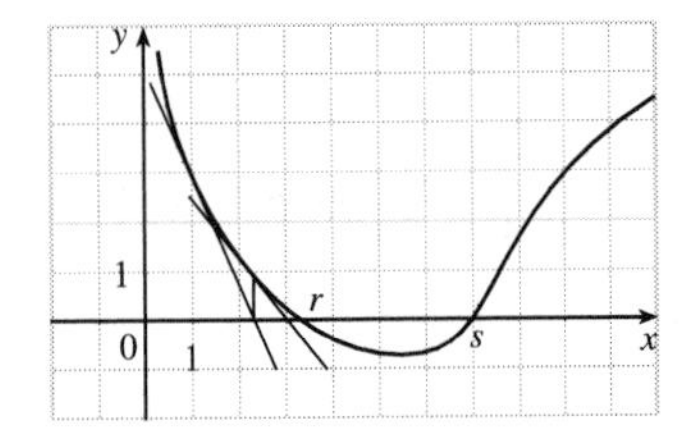

The tangent line at $x = 1$ intersects the x-axis at $x \approx 2.3$, so $x_2 \approx 2.3$. The tangent line at $x = 2.3$ intersects the x-axis at $x \approx 3$, so $x_3 \approx 3.0$.

(b) $x_1 = 5$ would *not* be a better first approximation than $x_1 = 1$ since the tangent line is nearly horizontal. In fact, the second approximation for $x_1 = 5$ appears to be to the left of $x = 1$.

2.

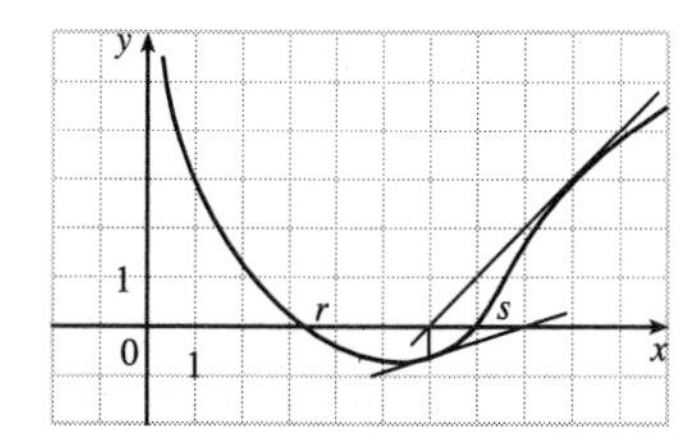

The tangent line at $x = 9$ intersects the x-axis at $x \approx 6.0$, so $x_2 \approx 6.0$. The tangent line at $x = 6.0$ intersects the x-axis at $x \approx 8.0$, so $x_3 \approx 8.0$.

3. Since $x_1 = 3$ and $y = 5x - 4$ is tangent to $y = f(x)$ at $x = 3$, we simply need to find where the tangent line intersects the x-axis. $y = 0 \Rightarrow 5x_2 - 4 = 0 \Rightarrow x_2 = \frac{4}{5}$.

4. (a)

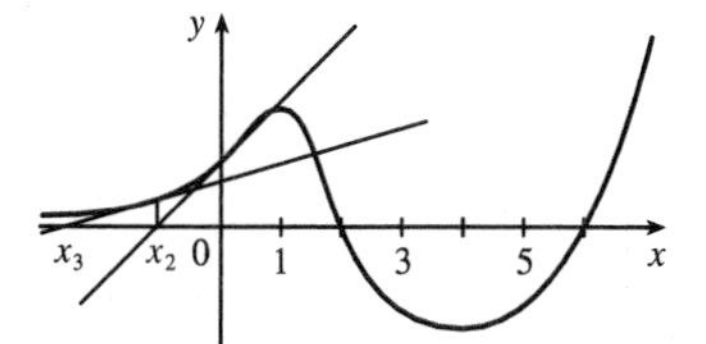

If $x_1 = 0$, then x_2 is negative, and x_3 is even more negative. The sequence of approximations does not converge, that is, Newton's method fails.

(b)

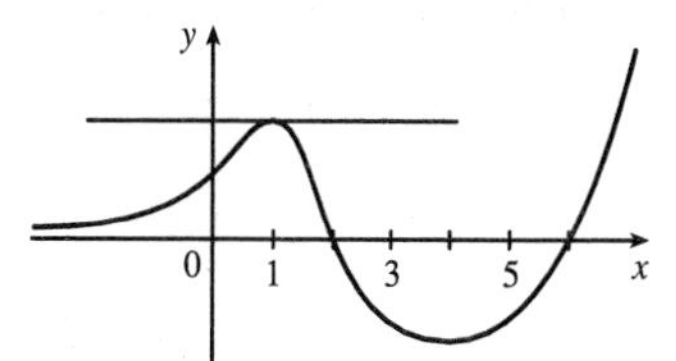

If $x_1 = 1$, the tangent line is horizontal and Newton's method fails.

(c)

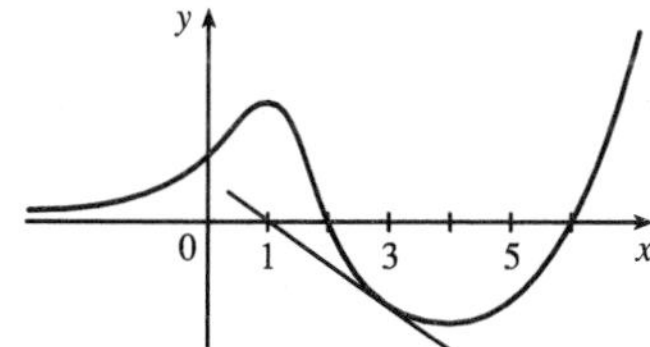

If $x_1 = 3$, then $x_2 = 1$ and we have the same situation as in part (b). Newton's method fails again.

(d)

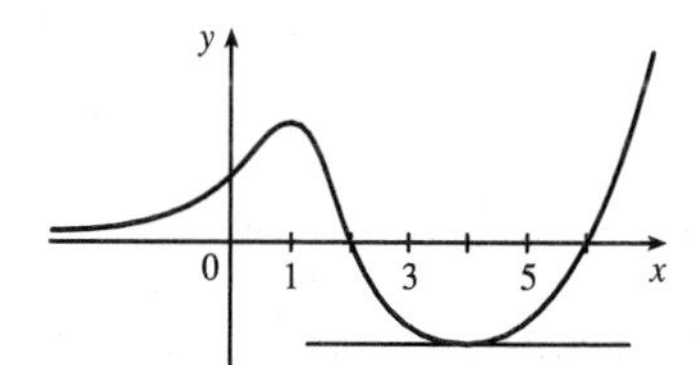

If $x_1 = 4$, the tangent line is horizontal and Newton's method fails.

(e)

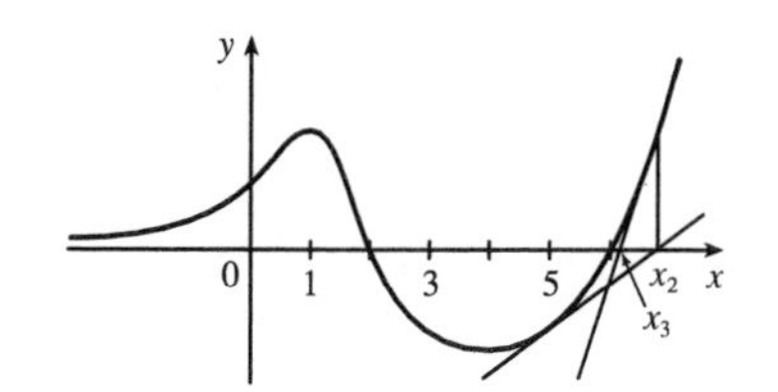

If $x_1 = 5$, then x_2 is greater than 6, x_3 gets closer to 6, and the sequence of approximations converges to 6. Newton's method succeeds!

5. $f(x) = x^3 + 2x - 4 \Rightarrow f'(x) = 3x^2 + 2$, so $x_{n+1} = x_n - \dfrac{x_n^3 + 2x_n - 4}{3x_n^2 + 2}$.

Now $x_1 = 1 \Rightarrow x_2 = 1 - \dfrac{1 + 2 - 4}{3 \cdot 1^2 + 2} = 1 - \dfrac{-1}{5} = 1.2 \Rightarrow x_3 = 1.2 - \dfrac{(1.2)^3 + 2(1.2) - 4}{3(1.2)^2 + 2} \approx 1.1797$.

6. $f(x) = x^5 + 2 \Rightarrow f'(x) = 5x^4$, so $x_{n+1} = x_n - \dfrac{x_n^5 + 2}{5x_n^4}$.

Now $x_1 = -1 \Rightarrow x_2 = -1 - \dfrac{(-1)^5 + 2}{5 \cdot (-1)^4} = -1 - \dfrac{1}{5} = -1.2 \Rightarrow x_3 = -1.2 - \dfrac{(-1.2)^5 + 2}{5(-1.2)^4} \approx -1.1529$.

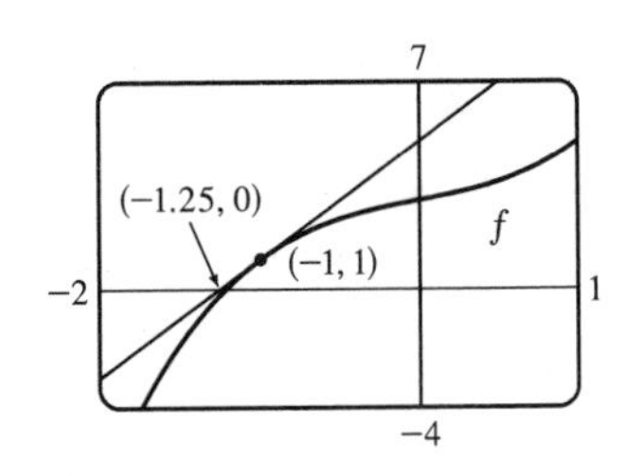

7. $f(x) = x^3 + x + 3 \quad \Rightarrow \quad f'(x) = 3x^2 + 1$, so $x_{n+1} = x_n - \dfrac{x_n^3 + x_n + 3}{3x_n^2 + 1}$.

Now $x_1 = -1 \quad \Rightarrow$

$$x_2 = -1 - \frac{(-1)^3 + (-1) + 3}{3(-1)^2 + 1} = -1 - \frac{-1 - 1 + 3}{3 + 1} = -1 - \frac{1}{4} = -1.25.$$

Newton's method follows the tangent line at $(-1, 1)$ down to its intersection with the x-axis at $(-1.25, 0)$, giving the second approximation $x_2 = -1.25$.

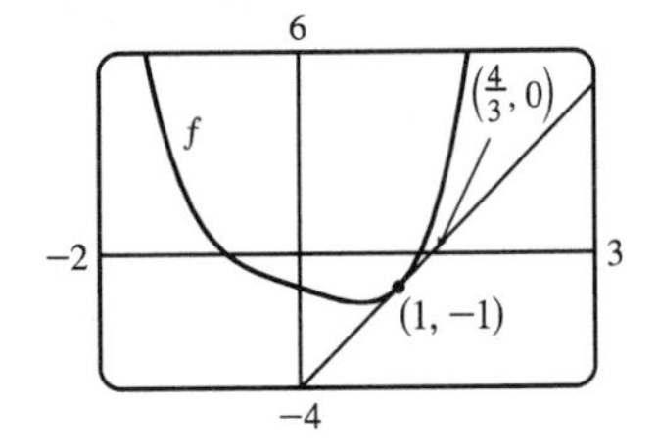

8. $f(x) = x^4 - x - 1 \quad \Rightarrow \quad f'(x) = 4x^3 - 1$, so $x_{n+1} = x_n - \dfrac{x_n^4 - x_n - 1}{4x_n^3 - 1}$.

Now $x_1 = 1 \quad \Rightarrow \quad x_2 = 1 - \dfrac{1^4 - 1 - 1}{4 \cdot 1^3 - 1} = 1 - \dfrac{-1}{3} = \dfrac{4}{3}$. Newton's method follows the tangent line at $(1, -1)$ up to its intersection with the x-axis at $\left(\frac{4}{3}, 0\right)$, giving the second approximation $x_2 = \frac{4}{3}$.

9. To approximate $x = \sqrt[3]{30}$ (so that $x^3 = 30$), we can take $f(x) = x^3 - 30$. So $f'(x) = 3x^2$, and thus, $x_{n+1} = x_n - \dfrac{x_n^3 - 30}{3x_n^2}$. Since $\sqrt[3]{27} = 3$ and 27 is close to 30, we'll use $x_1 = 3$. We need to find approximations until they agree to eight decimal places. $x_1 = 3 \quad \Rightarrow \quad x_2 \approx 3.11111111$, $x_3 \approx 3.10723734$, $x_4 \approx 3.10723251 \approx x_5$. So $\sqrt[3]{30} \approx 3.10723251$, to eight decimal places.

Here is a quick and easy method for finding the iterations for Newton's method on a programmable calculator. (The screens shown are from the TI-83 Plus, but the method is similar on other calculators.) Assign $f(x) = x^3 - 30$ to Y_1, and $f'(x) = 3x^2$ to Y_2. Now store $x_1 = 3$ in X and then enter $\text{X} - \text{Y}_1/\text{Y}_2 \rightarrow \text{X}$ to get $x_2 = 3.\overline{1}$. By successively pressing the ENTER key, you get the approximations $x_3, x_4, \ldots$.

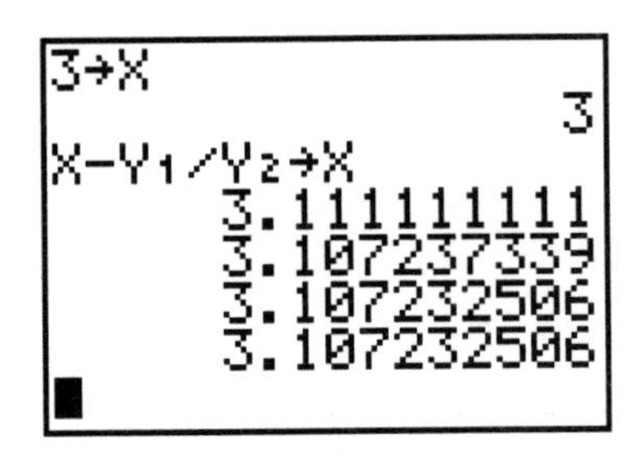

In Derive, load the utility file `SOLVE`. Enter `NEWTON(x^3-30,x,3)` and then `APPROXIMATE` to get [3, 3.11111111, 3.10723733, 3.10723250, 3.10723250]. You can request a specific iteration by adding a fourth argument. For example, `NEWTON(x^3-30,x,3,2)` gives [3, 3.11111111, 3.10723733].

In Maple, make the assignments $f := x \rightarrow x\hat{\ }3 - 30;$, $g := x \rightarrow x - f(x)/D(f)(x);$, and $x := 3.;$. Repeatedly execute the command $x := g(x);$ to generate successive approximations.

In Mathematica, make the assignments $f[x\_] := x\hat{\ }3 - 30$, $g[x\_] := x - f[x]/f'[x]$, and $x = 3$. Repeatedly execute the command $x = g[x]$ to generate successive approximations.

10. $f(x) = x^7 - 1000 \;\Rightarrow\; f'(x) = 7x^6$, so $x_{n+1} = x_n - \dfrac{x_n^7 - 1000}{7x_n^6}$. We need to find approximations until they agree to eight decimal places. $x_1 = 3 \;\Rightarrow\; x_2 \approx 2.76739173$, $x_3 \approx 2.69008741$, $x_4 \approx 2.68275645$, $x_5 \approx 2.68269580 \approx x_6$. So ≈ 2.68269580, to eight decimal places.

11. $\sin x = x^2$, so $f(x) = \sin x - x^2 \;\Rightarrow\; f'(x) = \cos x - 2x \;\Rightarrow$

$x_{n+1} = x_n - \dfrac{\sin x_n - x_n^2}{\cos x_n - 2x_n}$. From the figure, the positive root of $\sin x = x^2$ is near 1. $x_1 = 1 \;\Rightarrow\; x_2 \approx 0.891396$, $x_3 \approx 0.876985$, $x_4 \approx 0.876726 \approx x_5$. So the positive root is 0.876726, to six decimal places.

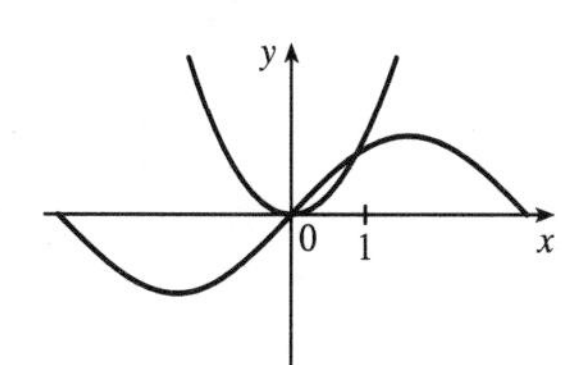

12. $2\cos x = x^4$, so $f(x) = 2\cos x - x^4 \;\Rightarrow\; f'(x) = -2\sin x - 4x^3 \;\Rightarrow$

$x_{n+1} = x_n - \dfrac{2\cos x_n - x_n^4}{-2\sin x_n - 4x_n^3}$. From the figure, the positive root of $2\cos x = x^4$ is near 1. $x_1 = 1 \;\Rightarrow\; x_2 \approx 1.014184$, $x_3 \approx 1.013958 \approx x_4$. So the positive root is 1.013958, to six decimal places.

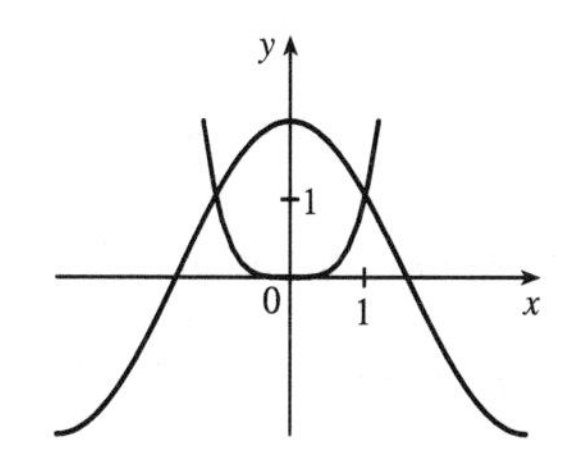

13.

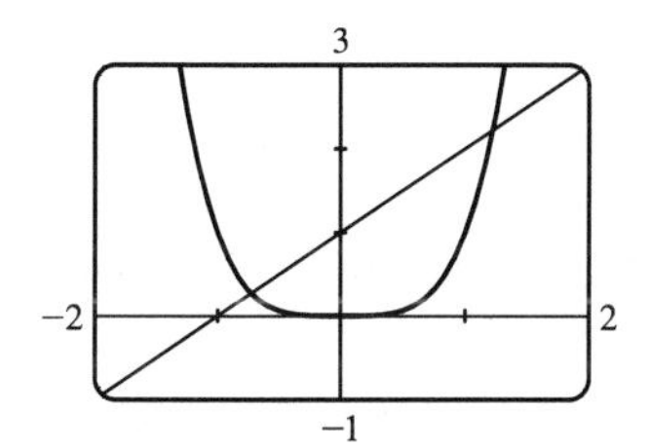

From the graph, we see that there appear to be points of intersection near $x = -0.7$ and $x = 1.2$. Solving $x^4 = 1 + x$ is the same as solving $f(x) = x^4 - x - 1 = 0$. $f(x) = x^4 - x - 1 \;\Rightarrow\; f'(x) = 4x^3 - 1$, so $x_{n+1} = x_n - \dfrac{x_n^4 - x_n - 1}{4x_n^3 - 1}$.

$x_1 = -0.7$	$x_1 = 1.2$
$x_2 \approx -0.725253$	$x_2 \approx 1.221380$
$x_3 \approx -0.724493$	$x_3 \approx 1.220745$
$x_4 \approx -0.724492 \approx x_5$	$x_4 \approx 1.220744 \approx x_5$

To six decimal places, the roots of the equation are -0.724492 and 1.220744.

14.

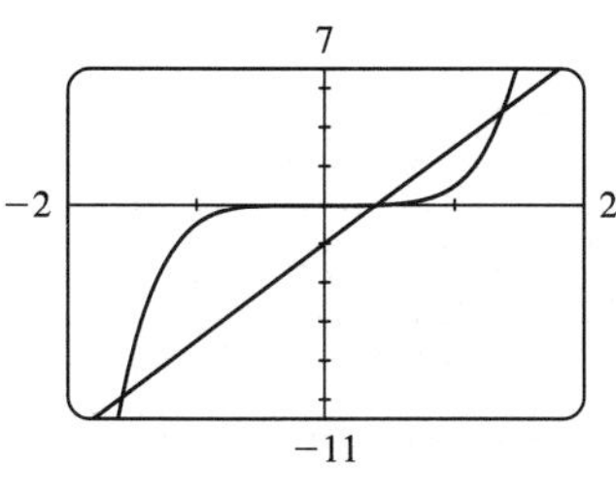

From the graph, we see that reasonable first approximations are $x = 0.5$ and $x = \pm 1.5$. $f(x) = x^5 - 5x + 2 \;\Rightarrow\; f'(x) = 5x^4 - 5$, so

$$x_{n+1} = x_n - \frac{x_n^5 - 5x_n + 2}{5x_n^4 - 5}.$$

$x_1 = -1.5$	$x_1 = 0.5$	$x_1 = 1.5$
$x_2 \approx -1.593846$	$x_2 = 0.4$	$x_2 \approx 1.396923$
$x_3 \approx -1.582241$	$x_3 \approx 0.402102 \approx x_4$	$x_3 \approx 1.373078$
$x_4 \approx -1.582036 \approx x_5$		$x_4 \approx 1.371885$
		$x_5 \approx 1.371882 \approx x_6$

To six decimal places, the roots are -1.582036, 0.402102, and 1.371882.

15. 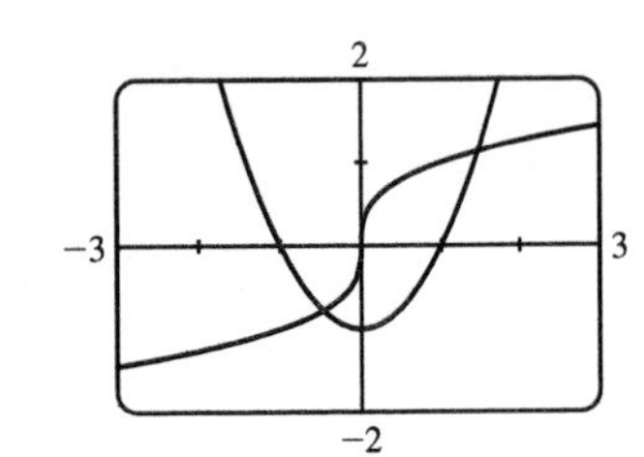

From the graph, we see that there appear to be points of intersection near $x = -0.5$ and $x = 1.5$. Solving $\sqrt[3]{x} = x^2 - 1$ is the same as solving $f(x) = \sqrt[3]{x} - x^2 + 1 = 0$. $f(x) = \sqrt[3]{x} - x^2 + 1 \Rightarrow$ $f'(x) = \frac{1}{3}x^{-2/3} - 2x$, so $x_{n+1} = x_n - \dfrac{\sqrt[3]{x_n} - x_n^2 + 1}{\frac{1}{3}x_n^{-2/3} - 2x_n}$.

$x_1 = -0.5$	$x_1 = 1.5$
$x_2 \approx -0.471421$	$x_2 \approx 1.461653$
$x_3 \approx -0.471074 \approx x_4$	$x_3 \approx 1.461070 \approx x_4$

To six decimal places, the roots are -0.471074 and 1.461070.

16. From the graph, there appears to be a point of intersection near $x = 0.7$. Solving $\tan x = \sqrt{1 - x^2}$ is the same as solving $f(x) = \tan x - \sqrt{1 - x^2} = 0$. $f(x) = \tan x - \sqrt{1 - x^2} \Rightarrow$

$f'(x) = \sec^2 x + x/\sqrt{1 - x^2}$, so $x_{n+1} = x_n - \dfrac{\tan x_n - \sqrt{1 - x_n^2}}{\sec^2 x_n + x_n/\sqrt{1 - x_n^2}}$.

$x_1 = 0.7 \Rightarrow x_2 \approx 0.652356$, $x_3 \approx 0.649895$, $x_4 \approx 0.649889 \approx x_5$.

To six decimal places, the root of the equation is 0.649889.

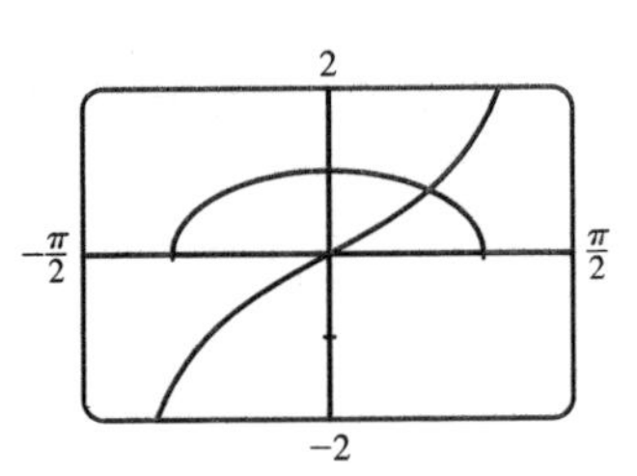

17. $f(x) = x^5 - x^4 - 5x^3 - x^2 + 4x + 3 \Rightarrow f'(x) = 5x^4 - 4x^3 - 15x^2 - 2x + 4 \Rightarrow$

$x_{n+1} = x_n - \dfrac{x_n^5 - x_n^4 - 5x_n^3 - x_n^2 + 4x_n + 3}{5x_n^4 - 4x_n^3 - 15x_n^2 - 2x_n + 4}$. From the graph of f, there appear to be roots near -1.4, 1.1, and 2.7.

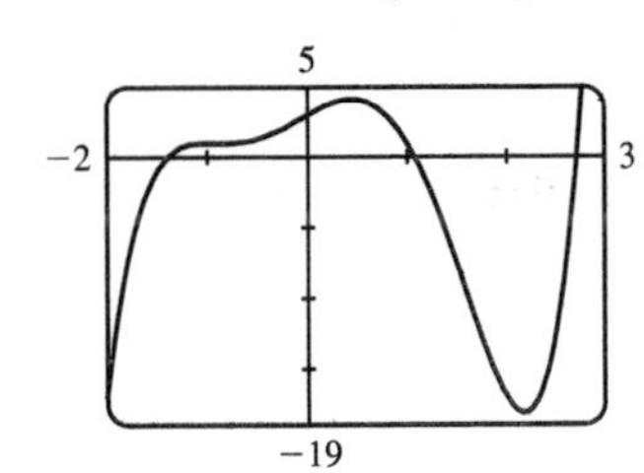

$x_1 = -1.4$	$x_1 = 1.1$	$x_1 = 2.7$
$x_2 \approx -1.39210970$	$x_2 \approx 1.07780402$	$x_2 \approx 2.72046250$
$x_3 \approx -1.39194698$	$x_3 \approx 1.07739442$	$x_3 \approx 2.71987870$
$x_4 \approx -1.39194691 \approx x_5$	$x_4 \approx 1.07739428 \approx x_5$	$x_4 \approx 2.71987822 \approx x_5$

To eight decimal places, the roots of the equation are -1.39194691, 1.07739428, and 2.71987822.

18. Solving $x^2(4 - x^2) = \dfrac{4}{x^2 + 1}$ is the same as solving $f(x) = 4x^2 - x^4 - \dfrac{4}{x^2 + 1} = 0$.

$f'(x) = 8x - 4x^3 + \dfrac{8x}{(x^2 + 1)^2} \Rightarrow x_{n+1} = x_n - \dfrac{4x_n^2 - x_n^4 - 4/(x_n^2 + 1)}{8x_n - 4x_n^3 + 8x_n/(x_n^2 + 1)^2}$. From the graph of $f(x)$, there appear to be roots near $x = \pm 1.9$ and $x = \pm 0.8$. Since f is even, we only need to find the positive roots.

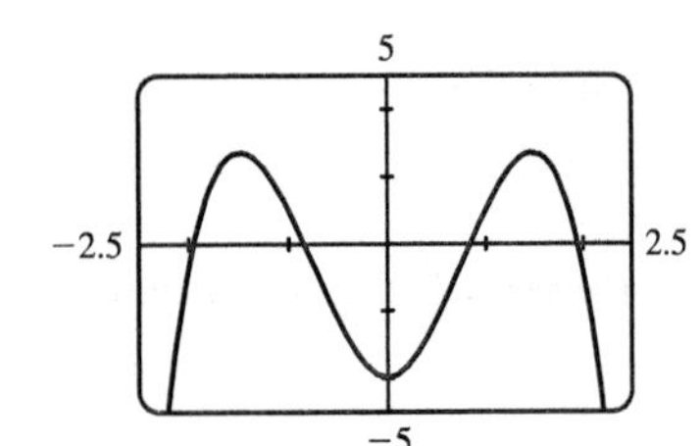

$x_1 = 0.8$	$x_1 = 1.9$
$x_2 \approx 0.84287645$	$x_2 \approx 1.94689103$
$x_3 \approx 0.84310820$	$x_3 \approx 1.94383891$
$x_4 \approx 0.84310821 \approx x_5$	$x_4 \approx 1.94382538 \approx x_5$

To eight decimal places, the roots of the equation are ± 0.84310821 and ± 1.94382538.

19. From the graph, $y = x^2\sqrt{2-x-x^2}$ and $y = 1$ intersect twice, at $x \approx -2$ and at $x \approx -1$.

$$f(x) = x^2\sqrt{2-x-x^2} - 1 \quad\Rightarrow\quad f'(x) = x^2 \cdot \tfrac{1}{2}(2-x-x^2)^{-1/2}(-1-2x) + (2-x-x^2)^{1/2} \cdot 2x$$

$$= \tfrac{1}{2}x(2-x-x^2)^{-1/2}\left[x(-1-2x) + 4(2-x-x^2)\right] = \frac{x(8-5x-6x^2)}{2\sqrt{(2+x)(1-x)}},$$

so $x_{n+1} = x_n - \dfrac{x_n^2\sqrt{2-x_n-x_n^2} - 1}{\dfrac{x_n(8-5x_n-6x_n^2)}{2\sqrt{(2+x_n)(1-x_n)}}}$. Trying $x_1 = -2$ won't work because $f'(-2)$ is undefined, so we try $x_1 = -1.95$.

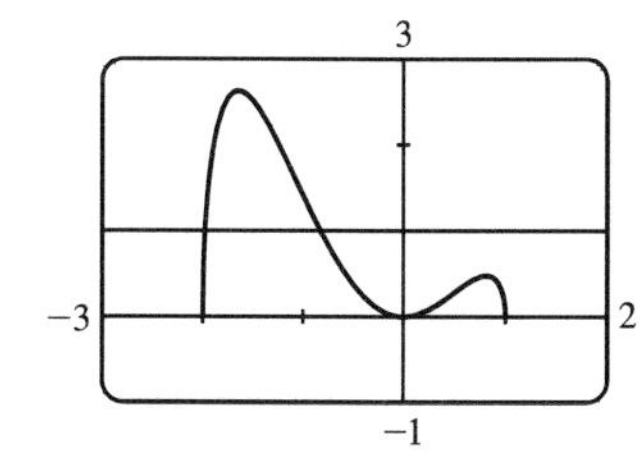

$x_1 = -1.95$
$x_2 \approx -1.98580357$
$x_3 \approx -1.97899778$
$x_4 \approx -1.97807848$
$x_5 \approx -1.97806682$
$x_6 \approx -1.97806681 \approx x_7$

$x_1 = -0.8$
$x_2 \approx -0.82674444$
$x_3 \approx -0.82646236$
$x_4 \approx -0.82646233 \approx x_5$

To eight decimal places, the roots of the equation are -1.97806681 and -0.82646233.

20. From the equations $y = 3\sin(x^2)$ and $y = 2x$ and the graph, we deduce that one root of the equation $3\sin(x^2) = 2x$ is $x = 0$. We also see that the graphs intersect at approximately $x = 0.7$ and $x = 1.4$. $f(x) = 3\sin(x^2) - 2x \quad\Rightarrow$

$f'(x) = 3\cos(x^2) \cdot 2x - 2$, so $x_{n+1} = x_n - \dfrac{3\sin(x_n^2) - 2x_n}{6x_n\cos(x_n^2) - 2}$.

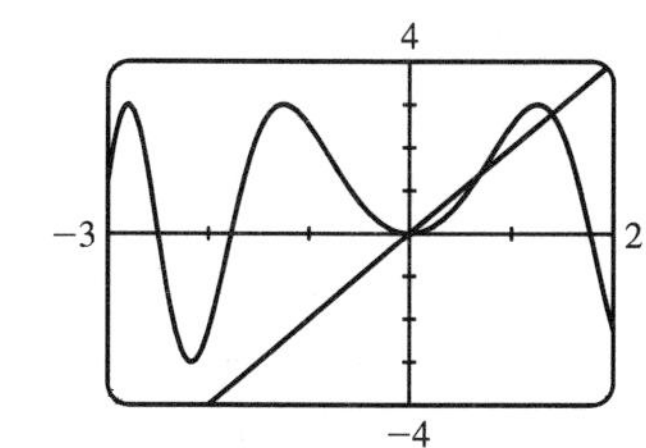

$x_1 = 0.7$
$x_2 \approx 0.69303689$
$x_3 \approx 0.69299996 \approx x_4$

$x_1 = 1.4$
$x_2 \approx 1.39530295$
$x_3 \approx 1.39525078$
$x_4 \approx 1.39525077 \approx x_5$

To eight decimal places, the other roots of the equation are 0.69299996 and 1.39525077.

21. (a) $f(x) = x^2 - a \quad\Rightarrow\quad f'(x) = 2x$, so Newton's method gives

$$x_{n+1} = x_n - \frac{x_n^2 - a}{2x_n} = x_n - \frac{1}{2}x_n + \frac{a}{2x_n} = \frac{1}{2}x_n + \frac{a}{2x_n} = \frac{1}{2}\left(x_n + \frac{a}{x_n}\right).$$

(b) Using (a) with $a = 1000$ and $x_1 = \sqrt{900} = 30$, we get $x_2 \approx 31.666667$, $x_3 \approx 31.622807$, and $x_4 \approx 31.622777 \approx x_5$. So $\sqrt{1000} \approx 31.622777$.

22. (a) $f(x) = \dfrac{1}{x} - a \quad\Rightarrow\quad f'(x) = -\dfrac{1}{x^2}$, so $x_{n+1} = x_n - \dfrac{1/x_n - a}{-1/x_n^2} = x_n + x_n - ax_n^2 = 2x_n - ax_n^2$.

(b) Using (a) with $a = 1.6894$ and $x_1 = \frac{1}{2} = 0.5$, we get $x_2 = 0.5754$, $x_3 \approx 0.588485$, and $x_4 \approx 0.588789 \approx x_5$. So $1/1.6984 \approx 0.588789$.

23. $f(x) = x^3 - 3x + 6 \quad\Rightarrow\quad f'(x) = 3x^2 - 3$. If $x_1 = 1$, then $f'(x_1) = 0$ and the tangent line used for approximating x_2 is horizontal. Attempting to find x_2 results in trying to divide by zero.

24. $x^3 - x = 1 \Leftrightarrow x^3 - x - 1 = 0$. $f(x) = x^3 - x - 1 \Rightarrow f'(x) = 3x^2 - 1$, so $x_{n+1} = x_n - \dfrac{x_n^3 - x_n - 1}{3x_n^2 - 1}$.

(a) $x_1 = 1$, $x_2 = 1.5$, $x_3 \approx 1.347826$, $x_4 \approx 1.325200$, $x_5 \approx 1.324718 \approx x_6$

(b) $x_1 = 0.6$, $x_2 = 17.9$, $x_3 \approx 11.946802$, $x_4 \approx 7.985520$, $x_5 \approx 5.356909$, $x_6 \approx 3.624996$, $x_7 \approx 2.505589$, $x_8 \approx 1.820129$, $x_9 \approx 1.461044$, $x_{10} \approx 1.339323$, $x_{11} \approx 1.324913$, $x_{12} \approx 1.324718 \approx x_{13}$

(c)

$x_1 = 0.57$	$x_2 \approx -54.165455$	$x_3 \approx -36.114293$	$x_4 \approx -24.082094$	$x_5 \approx -16.063387$
$x_6 \approx -10.721483$	$x_7 \approx -7.165534$	$x_8 \approx -4.801704$	$x_9 \approx -3.233425$	$x_{10} \approx -2.193674$
$x_{11} \approx -1.496867$	$x_{12} \approx -0.997546$	$x_{13} \approx -0.496305$	$x_{14} \approx -2.894162$	$x_{15} \approx -1.967962$
$x_{16} \approx -1.341355$	$x_{17} \approx -0.870187$	$x_{18} \approx -0.249949$	$x_{19} \approx -1.192219$	$x_{20} \approx -0.731952$
$x_{21} \approx 0.355213$	$x_{22} \approx -1.753322$	$x_{23} \approx -1.189420$	$x_{24} \approx -0.729123$	$x_{25} \approx 0.377844$
$x_{26} \approx -1.937872$	$x_{27} \approx -1.320350$	$x_{28} \approx -0.851919$	$x_{29} \approx -0.200959$	$x_{30} \approx -1.119386$
$x_{31} \approx -0.654291$	$x_{32} \approx 1.547010$	$x_{33} \approx 1.360051$	$x_{34} \approx 1.325828$	$x_{35} \approx 1.324719$
$x_{36} \approx 1.324718 \approx x_{37}$				

(d)

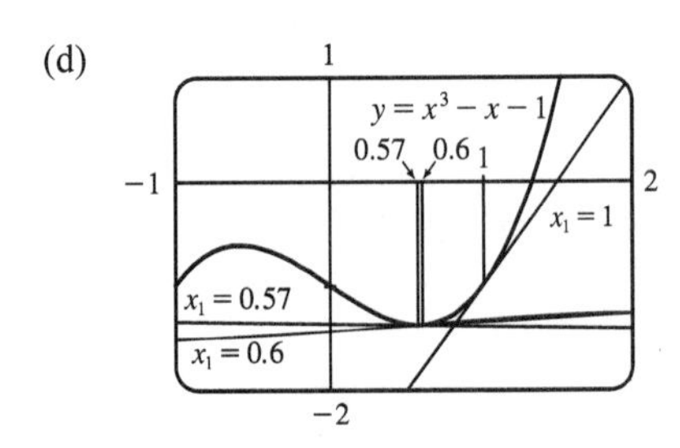

From the figure, we see that the tangent line corresponding to $x_1 = 1$ results in a sequence of approximations that converges quite quickly ($x_5 \approx x_6$). The tangent line corresponding to $x_1 = 0.6$ is close to being horizontal, so x_2 is quite far from the root. But the sequence still converges — just a little more slowly ($x_{12} \approx x_{13}$). Lastly, the tangent line corresponding to $x_1 = 0.57$ is very nearly horizontal, x_2 is farther away from the root, and the sequence takes more iterations to converge ($x_{36} \approx x_{37}$).

25. For $f(x) = x^{1/3}$, $f'(x) = \frac{1}{3}x^{-2/3}$ and

$$x_{n+1} = x_n - \frac{f(x_n)}{f'(x_n)} = x_n - \frac{x_n^{1/3}}{\frac{1}{3}x_n^{-2/3}} = x_n - 3x_n = -2x_n.$$

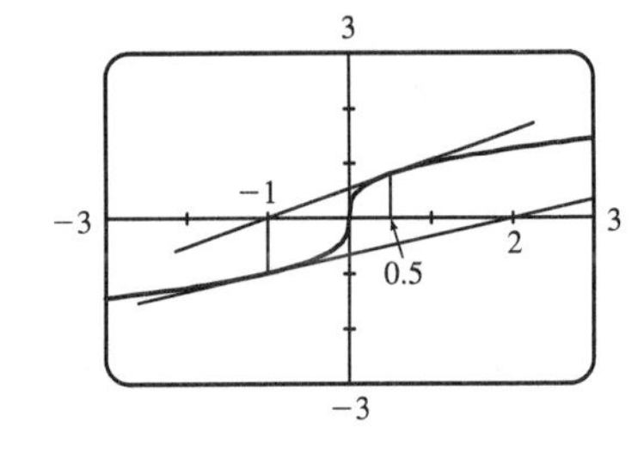

Therefore, each successive approximation becomes twice as large as the previous one in absolute value, so the sequence of approximations fails to converge to the root, which is 0. In the figure, we have $x_1 = 0.5$, $x_2 = -2(0.5) = -1$, and $x_3 = -2(-1) = 2$.

26. $f(x) = x^2 + \sin x \Rightarrow f'(x) = 2x + \cos x$. $f'(x)$ exists for all x, so to find the minimum of f, we can examine the zeros of f'. From the graph of f', we see that a good choice for x_1 is $x_1 = -0.5$. Use $g(x) = 2x + \cos x$ and $g'(x) = 2 - \sin x$ to obtain $x_2 \approx -0.450627$, $x_3 \approx -0.450184 \approx x_4$. Since $f''(x) = 2 - \sin x > 0$ for all x, $f(-0.450184) \approx -0.232466$ is the absolute minimum.

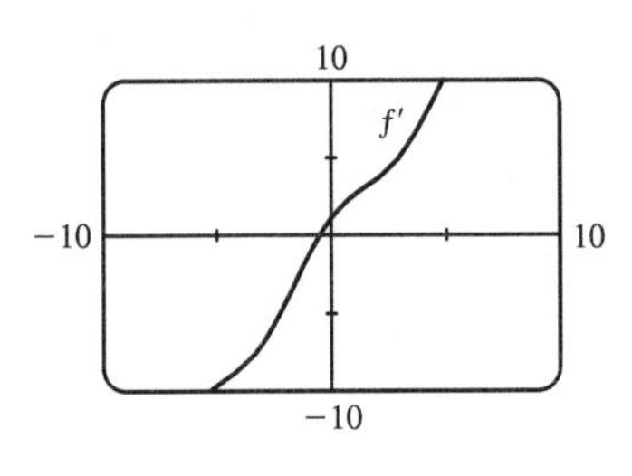

27. $y = x^3 + \cos x \Rightarrow y' = 3x^2 - \sin x \Rightarrow y'' = 6x - \cos x \Rightarrow y''' = 6 + \sin x$. Now to solve $y'' = 0$, try $x_1 = 0$, and then $x_2 = x_1 - \dfrac{y''(x_1)}{y'''(x_1)} = \dfrac{1}{6} \Rightarrow x_3 \approx 0.164419 \approx x_4$. For $x < 0.164419$, $y'' < 0$, and for $x > 0.164419$, $y'' > 0$. Therefore, the point of inflection, correct to six decimal places, is $(0.164419, y(0.164419)) \approx (0.164419, 0.990958)$.

28. 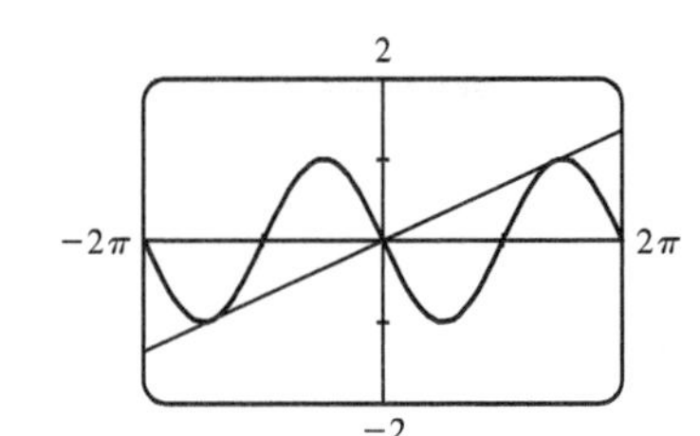

$f(x) = -\sin x \;\Rightarrow\; f'(x) = -\cos x$. At $x = a$, the slope of the tangent line is $f'(a) = -\cos a$. The line through the origin and $(a, f(a))$ is $y = \dfrac{-\sin a - 0}{a - 0}x$. If this line is to be tangent to f at $x = a$, then its slope must equal $f'(a)$. Thus, $\dfrac{-\sin a}{a} = -\cos a \;\Rightarrow\; \tan a = a$.

To solve this equation using Newton's method, let $g(x) = \tan x - x$, $g'(x) = \sec^2 x - 1$, and $x_{n+1} = x_n - \dfrac{\tan x_n - x_n}{\sec^2 x_n - 1}$ with $x_1 = 4.5$ (estimated from the figure). $x_2 \approx 4.493614$, $x_3 \approx 4.493410$, $x_4 \approx 4.493409 \approx x_5$. Thus, the slope of the line that has the largest slope is $f'(x_5) \approx 0.217234$.

29. In this case, $A = 18{,}000$, $R = 375$, and $n = 5(12) = 60$. So the formula $A = \dfrac{R}{i}\left[1 - (1+i)^{-n}\right]$ becomes

$18{,}000 = \dfrac{375}{x}\left[1 - (1+x)^{-60}\right] \;\Leftrightarrow\; 48x = 1 - (1+x)^{-60}$ [multiply each term by $(1+x)^{60}$] $\Leftrightarrow$

$48x(1+x)^{60} - (1+x)^{60} + 1 = 0$. Let the LHS be called $f(x)$, so that

$$\begin{aligned} f'(x) &= 48x(60)(1+x)^{59} + 48(1+x)^{60} - 60(1+x)^{59} \\ &= 12(1+x)^{59}\left[4x(60) + 4(1+x) - 5\right] = 12(1+x)^{59}(244x - 1) \end{aligned}$$

$x_{n+1} = x_n - \dfrac{48x_n(1+x_n)^{60} - (1+x_n)^{60} + 1}{12(1+x_n)^{59}(244x_n - 1)}$. An interest rate of 1% per month seems like a reasonable estimate for $x = i$. So let $x_1 = 1\% = 0.01$, and we get $x_2 \approx 0.0082202$, $x_3 \approx 0.0076802$, $x_4 \approx 0.0076291$, $x_5 \approx 0.0076286 \approx x_6$. Thus, the dealer is charging a monthly interest rate of 0.76286% (or 9.55% per year, compounded monthly).

30. (a) $p(x) = x^5 - (2+r)x^4 + (1+2r)x^3 - (1-r)x^2 + 2(1-r)x + r - 1 \;\Rightarrow$

$p'(x) = 5x^4 - 4(2+r)x^3 + 3(1+2r)x^2 - 2(1-r)x + 2(1-r)$. So we use

$$x_{n+1} = x_n - \frac{x_n^5 - (2+r)x_n^4 + (1+2r)x_n^3 - (1-r)x_n^2 + 2(1-r)x_n + r - 1}{5x_n^4 - 4(2+r)x_n^3 + 3(1+2r)x_n^2 - 2(1-r)x_n + 2(1-r)}.$$

We substitute in the value $r \approx 3.04042 \times 10^{-6}$ in order to evaluate the approximations numerically. The libration point L_1 is slightly less than 1 AU from the Sun, so we take $x_1 = 0.95$ as our first approximation, and get $x_2 \approx 0.96682$, $x_3 \approx 0.97770$, $x_4 \approx 0.98451$, $x_5 \approx 0.98830$, $x_6 \approx 0.98976$, $x_7 \approx 0.98998$, $x_8 \approx 0.98999 \approx x_9$. So, to five decimal places, L_1 is located 0.98999 AU from the Sun (or 0.01001 AU from the Earth).

(b) In this case we use Newton's method with the function

$p(x) - 2rx^2 = x^5 - (2+r)x^4 + (1+2r)x^3 - (1+r)x^2 + 2(1-r)x + r - 1 \;\Rightarrow$

$\left[p(x) - 2rx^2\right]' = 5x^4 - 4(2+r)x^3 + 3(1+2r)x^2 - 2(1+r)x + 2(1-r)$. So

$$x_{n+1} = x_n - \frac{x_n^5 - (2+r)x_n^4 + (1+2r)x_n^3 - (1+r)x_n^2 + 2(1-r)x_n + r - 1}{5x_n^4 - 4(2+r)x_n^3 + 3(1+2r)x_n^2 - 2(1+r)x_n + 2(1-r)}.$$

Again, we substitute $r \approx 3.04042 \times 10^{-6}$. L_2 is slightly more than 1 AU from the Sun and, judging from the result of part (a), probably less than 0.02 AU from Earth. So we take $x_1 = 1.02$ and get $x_2 \approx 1.01422$, $x_3 \approx 1.01118$, $x_4 \approx 1.01018$, $x_5 \approx 1.01008 \approx x_6$. So, to five decimal places, L_2 is located 1.01008 AU from the Sun (or 0.01008 AU from the Earth).

3.7 Antiderivatives

1. $f(x) = 6x^2 - 8x + 3 \quad \Rightarrow \quad F(x) = 6\dfrac{x^{2+1}}{2+1} - 8\dfrac{x^{1+1}}{1+1} + 3x + C = 2x^3 - 4x^2 + 3x + C$

Check: $F'(x) = 2 \cdot 3x^2 - 4 \cdot 2x + 3 + 0 = 6x^2 - 8x + 3 = f(x)$

2. $f(x) = 1 - x^3 + 12x^5 \quad \Rightarrow \quad F(x) = x - \dfrac{x^{3+1}}{3+1} + 12\dfrac{x^{5+1}}{5+1} + C = x - \frac{1}{4}x^4 + 2x^6 + C$

3. $f(x) = 5x^{1/4} - 7x^{3/4} \quad \Rightarrow \quad F(x) = 5\dfrac{x^{1/4+1}}{\frac{1}{4}+1} - 7\dfrac{x^{3/4+1}}{\frac{3}{4}+1} + C = 5\dfrac{x^{5/4}}{5/4} - 7\dfrac{x^{7/4}}{7/4} + C = 4x^{5/4} - 4x^{7/4} + C$

4. $f(x) = 2x + 3x^{1.7} \quad \Rightarrow \quad F(x) = x^2 + \frac{3}{2.7}x^{2.7} + C = x^2 + \frac{10}{9}x^{2.7} + C$

5. $f(x) = \sqrt[3]{x} + \dfrac{5}{x^6} = x^{1/3} + 5x^{-6}$ has domain $(-\infty, 0) \cup (0, \infty)$, so

$$F(x) = \begin{cases} \dfrac{x^{1/3+1}}{\frac{1}{3}+1} + 5\,\dfrac{x^{-6+1}}{-6+1} + C_1 = \frac{3}{4}x^{4/3} - x^{-5} + C_1 & \text{if } x < 0 \\ \frac{3}{4}x^{4/3} - x^{-5} + C_2 & \text{if } x > 0 \end{cases}$$

6. $f(x) = \sqrt[4]{x^3} + \sqrt[3]{x^4} = x^{3/4} + x^{4/3} \quad \Rightarrow \quad F(x) = \dfrac{x^{7/4}}{7/4} + \dfrac{x^{7/3}}{7/3} + C = \frac{4}{7}x^{7/4} + \frac{3}{7}x^{7/3} + C$

7. $f(u) = \dfrac{u^4 + 3\sqrt{u}}{u^2} = \dfrac{u^4}{u^2} + \dfrac{3u^{1/2}}{u^2} = u^2 + 3u^{-3/2} \quad \Rightarrow$

$$F(u) = \frac{u^3}{3} + 3\,\frac{u^{-3/2+1}}{-3/2+1} + C = \frac{1}{3}u^3 + 3\frac{u^{-1/2}}{-1/2} + C = \frac{1}{3}u^3 - \frac{6}{\sqrt{u}} + C$$

8. $g(x) = \dfrac{5 - 4x^3 + 2x^6}{x^6} = 5x^{-6} - 4x^{-3} + 2$ has domain $(-\infty, 0) \cup (0, \infty)$, so

$$G(x) = \begin{cases} 5\,\dfrac{x^{-5}}{-5} - 4\,\dfrac{x^{-2}}{-2} + 2x + C_1 = -\dfrac{1}{x^5} + \dfrac{2}{x^2} + 2x + C_1 & \text{if } x < 0 \\ -\dfrac{1}{x^5} + \dfrac{2}{x^2} + 2x + C_2 & \text{if } x > 0 \end{cases}$$

9. $h(x) = x^3 + 5\sin x \quad \Rightarrow \quad H(x) = \frac{1}{4}x^4 + 5(-\cos x) + C = \frac{1}{4}x^4 - 5\cos x + C$

10. $f(t) = 3\cos t - 4\sin t \quad \Rightarrow \quad F(t) = 3(\sin t) - 4(-\cos t) + C = 3\sin t + 4\cos t + C$

11. $f(t) = 4\sqrt{t} - \sec t \tan t \quad \Rightarrow \quad F(t) = \frac{4}{3/2}t^{3/2} - \sec t + C = \frac{8}{3}t^{3/2} - \sec t + C_n$ on the interval $\left(n\pi - \frac{\pi}{2}, n\pi + \frac{\pi}{2}\right)$.

12. $f(\theta) = 6\theta^2 - 7\sec^2\theta \quad \Rightarrow \quad F(\theta) = 2\theta^3 - 7\tan\theta + C_n$ on the interval $\left(n\pi - \frac{\pi}{2}, n\pi + \frac{\pi}{2}\right)$.

13. $f''(x) = 6x + 12x^2 \quad \Rightarrow \quad f'(x) = 6 \cdot \dfrac{x^2}{2} + 12 \cdot \dfrac{x^3}{3} + C = 3x^2 + 4x^3 + C \quad \Rightarrow$

$f(x) = 3 \cdot \dfrac{x^3}{3} + 4 \cdot \dfrac{x^4}{4} + Cx + D = x^3 + x^4 + Cx + D$ [C and D are just arbitrary constants]

14. $f''(x) = 2 + x^3 + x^6 \quad \Rightarrow \quad f'(x) = 2x + \frac{1}{4}x^4 + \frac{1}{7}x^7 + C \quad \Rightarrow \quad f(x) = x^2 + \frac{1}{20}x^5 + \frac{1}{56}x^8 + Cx + D$

15. $f''(x) = 1 + x^{4/5} \quad \Rightarrow \quad f'(x) = x + \frac{5}{9}x^{9/5} + C \quad \Rightarrow$

$f(x) = \frac{1}{2}x^2 + \frac{5}{9} \cdot \frac{5}{14}x^{14/5} + Cx + D = \frac{1}{2}x^2 + \frac{25}{126}x^{14/5} + Cx + D$

16. $f''(x) = \cos x \Rightarrow f'(x) = \sin x + C \Rightarrow f(x) = -\cos x + Cx + D$

17. $f'''(t) = 60t^2 \Rightarrow f''(t) = 20t^3 + C \Rightarrow f'(t) = 5t^4 + Ct + D \Rightarrow f(t) = t^5 + \frac{1}{2}Ct^2 + Dt + E$

18. $f'''(t) = t - \sqrt{t} \Rightarrow f''(t) = \frac{1}{2}t^2 - \frac{2}{3}t^{3/2} + C \Rightarrow f'(t) = \frac{1}{6}t^3 - \frac{4}{15}t^{5/2} + Ct + D \Rightarrow$

$f(t) = \frac{1}{24}t^4 - \frac{8}{105}t^{7/2} + \frac{1}{2}Ct^2 + Dt + E$

19. $f'(x) = \sqrt{x}(6 + 5x) = 6x^{1/2} + 5x^{3/2} \Rightarrow f(x) = 4x^{3/2} + 2x^{5/2} + C$.

$f(1) = 6 + C$ and $f(1) = 10 \Rightarrow C = 4$, so $f(x) = 4x^{3/2} + 2x^{5/2} + 4$.

20. $f'(x) = 2x - 3/x^4 = 2x - 3x^{-4} \Rightarrow f(x) = x^2 + x^{-3} + C$ because we're given that $x > 0$.

$f(1) = 2 + C$ and $f(1) = 3 \Rightarrow C = 1$, so $f(x) = x^2 + 1/x^3 + 1$.

21. $f'(t) = 2\cos t + \sec^2 t \Rightarrow f(t) = 2\sin t + \tan t + C$ because $-\pi/2 < t < \pi/2$.

$f\left(\frac{\pi}{3}\right) = 2\left(\sqrt{3}/2\right) + \sqrt{3} + C = 2\sqrt{3} + C$ and $f\left(\frac{\pi}{3}\right) = 4 \Rightarrow C = 4 - 2\sqrt{3}$, so $f(t) = 2\sin t + \tan t + 4 - 2\sqrt{3}$.

22. $f'(x) = 3x^{-2} \Rightarrow f(x) = \begin{cases} -3/x + C_1 & \text{if } x > 0 \\ -3/x + C_2 & \text{if } x < 0 \end{cases}$ $\quad f(1) = -3 + C_1 = 0 \Rightarrow C_1 = 3$, $f(-1) = 3 + C_2 = 0$

$\Rightarrow C_2 = -3$. So $f(x) = \begin{cases} -3/x + 3 & \text{if } x > 0 \\ -3/x - 3 & \text{if } x < 0 \end{cases}$

23. $f''(x) = 24x^2 + 2x + 10 \Rightarrow f'(x) = 8x^3 + x^2 + 10x + C$. $f'(1) = 8 + 1 + 10 + C$ and $f'(1) = -3 \Rightarrow$

$19 + C = -3 \Rightarrow C = -22$, so $f'(x) = 8x^3 + x^2 + 10x - 22$ and hence, $f(x) = 2x^4 + \frac{1}{3}x^3 + 5x^2 - 22x + D$.

$f(1) = 2 + \frac{1}{3} + 5 - 22 + D$ and $f(1) = 5 \Rightarrow D = 22 - \frac{7}{3} = \frac{59}{3}$, so $f(x) = 2x^4 + \frac{1}{3}x^3 + 5x^2 - 22x + \frac{59}{3}$.

24. $f''(x) = 4 - 6x - 40x^3 \Rightarrow f'(x) = 4x - 3x^2 - 10x^4 + C$. $f'(0) = C$ and $f'(0) = 1 \Rightarrow C = 1$, so

$f'(x) = 4x - 3x^2 - 10x^4 + 1$ and hence, $f(x) = 2x^2 - x^3 - 2x^5 + x + D$. $f(0) = D$ and $f(0) = 2 \Rightarrow D = 2$, so

$f(x) = 2x^2 - x^3 - 2x^5 + x + 2$.

25. $f''(\theta) = \sin\theta + \cos\theta \Rightarrow f'(\theta) = -\cos\theta + \sin\theta + C$. $f'(0) = -1 + C$ and $f'(0) = 4 \Rightarrow C = 5$, so

$f'(\theta) = -\cos\theta + \sin\theta + 5$ and hence, $f(\theta) = -\sin\theta - \cos\theta + 5\theta + D$. $f(0) = -1 + D$ and $f(0) = 3 \Rightarrow D = 4$, so

$f(\theta) = -\sin\theta - \cos\theta + 5\theta + 4$.

26. $f''(t) = 3/\sqrt{t} = 3t^{-1/2} \Rightarrow f'(t) = 6t^{1/2} + C$. $f'(4) = 12 + C$ and $f'(4) = 7 \Rightarrow C = -5$, so $f'(t) = 6t^{1/2} - 5$

and hence, $f(t) = 4t^{3/2} - 5t + D$. $f(4) = 32 - 20 + D$ and $f(4) = 20 \Rightarrow D = 8$, so $f(t) = 4t^{3/2} - 5t + 8$.

27. $f''(x) = 2 - 12x \Rightarrow f'(x) = 2x - 6x^2 + C \Rightarrow f(x) = x^2 - 2x^3 + Cx + D$.

$f(0) = D$ and $f(0) = 9 \Rightarrow D = 9$. $f(2) = 4 - 16 + 2C + 9 = 2C - 3$ and $f(2) = 15 \Rightarrow 2C = 18 \Rightarrow C = 9$,

so $f(x) = x^2 - 2x^3 + 9x + 9$.

28. $f''(x) = 20x^3 + 12x^2 + 4 \Rightarrow f'(x) = 5x^4 + 4x^3 + 4x + C \Rightarrow f(x) = x^5 + x^4 + 2x^2 + Cx + D$. $f(0) = D$ and

$f(0) = 8 \Rightarrow D = 8$. $f(1) = 1 + 1 + 2 + C + 8 = C + 12$ and $f(1) = 5 \Rightarrow C = -7$, so

$f(x) = x^5 + x^4 + 2x^2 - 7x + 8$.

29. Given $f'(x) = 2x + 1$, we have $f(x) = x^2 + x + C$. Since f passes through $(1, 6)$, $f(1) = 6 \Rightarrow 1^2 + 1 + C = 6 \Rightarrow$

$C = 4$. Therefore, $f(x) = x^2 + x + 4$ and $f(2) = 2^2 + 2 + 4 = 10$.

30. $f'(x) = x^3 \Rightarrow f(x) = \frac{1}{4}x^4 + C$. $x + y = 0 \Rightarrow y = -x \Rightarrow m = -1$. Now $m = f'(x) \Rightarrow -1 = x^3 \Rightarrow$ $x = -1 \Rightarrow y = 1$ (from the equation of the tangent line), so $(-1, 1)$ is a point on the graph of f. From f, $1 = \frac{1}{4}(-1)^4 + C \Rightarrow C = \frac{3}{4}$. Therefore, the function is $f(x) = \frac{1}{4}x^4 + \frac{3}{4}$.

31. b is the antiderivative of f. For small x, f is negative, so the graph of its antiderivative must be decreasing. But both a and c are increasing for small x, so only b can be f's antiderivative. Also, f is positive where b is increasing, which supports our conclusion.

32. We know right away that c cannot be f's antiderivative, since the slope of c is not zero at the x-value where $f = 0$. Now f is positive when a is increasing and negative when a is decreasing, so a is the antiderivative of f.

33. $v(t) = s'(t) = \sin t - \cos t \Rightarrow s(t) = -\cos t - \sin t + C$. $s(0) = -1 + C$ and $s(0) = 0 \Rightarrow C = 1$, so $s(t) = -\cos t - \sin t + 1$.

34. $v(t) = s'(t) = 1.5\sqrt{t} \Rightarrow s(t) = t^{3/2} + C$. $s(4) = 8 + C$ and $s(4) = 10 \Rightarrow C = 2$, so $s(t) = t^{3/2} + 2$.

35. $a(t) = v'(t) = 10 \sin t + 3 \cos t \Rightarrow v(t) = -10 \cos t + 3 \sin t + C \Rightarrow s(t) = -10 \sin t - 3 \cos t + Ct + D$. $s(0) = -3 + D = 0$ and $s(2\pi) = -3 + 2\pi C + D = 12 \Rightarrow D = 3$ and $C = \frac{6}{\pi}$. Thus, $s(t) = -10 \sin t - 3 \cos t + \frac{6}{\pi}t + 3$.

36. $a(t) = v'(t) = 10 + 3t - 3t^2 \Rightarrow v(t) = 10t + \frac{3}{2}t^2 - t^3 + C \Rightarrow s(t) = 5t^2 + \frac{1}{2}t^3 - \frac{1}{4}t^4 + Ct + D \Rightarrow$ $0 = s(0) = D$ and $10 = s(2) = 20 + 4 - 4 + 2C \Rightarrow C = -5$, so $s(t) = -5t + 5t^2 + \frac{1}{2}t^3 - \frac{1}{4}t^4$.

37. (a) We first observe that since the stone is dropped 450 m above the ground, $v(0) = 0$ and $s(0) = 450$.
$v'(t) = a(t) = -9.8 \Rightarrow v(t) = -9.8t + C$. Now $v(0) = 0 \Rightarrow C = 0$, so $v(t) = -9.8t \Rightarrow$ $s(t) = -4.9t^2 + D$. Last, $s(0) = 450 \Rightarrow D = 450 \Rightarrow s(t) = 450 - 4.9t^2$.

(b) The stone reaches the ground when $s(t) = 0$. $450 - 4.9t^2 = 0 \Rightarrow t^2 = 450/4.9 \Rightarrow t_1 = \sqrt{450/4.9} \approx 9.58$ s.

(c) The velocity with which the stone strikes the ground is $v(t_1) = -9.8\sqrt{450/4.9} \approx -93.9$ m/s.

(d) This is just reworking parts (a) and (b) with $v(0) = -5$. Using $v(t) = -9.8t + C$, $v(0) = -5 \Rightarrow 0 + C = -5 \Rightarrow$ $v(t) = -9.8t - 5$. So $s(t) = -4.9t^2 - 5t + D$ and $s(0) = 450 \Rightarrow D = 450 \Rightarrow s(t) = -4.9t^2 - 5t + 450$. Solving $s(t) = 0$ by using the quadratic formula gives us $t = (5 \pm \sqrt{8845})/(-9.8) \Rightarrow t_1 \approx 9.09$ s.

38. $v'(t) = a(t) = a \Rightarrow v(t) = at + C$ and $v_0 = v(0) = C \Rightarrow v(t) = at + v_0 \Rightarrow$ $s(t) = \frac{1}{2}at^2 + v_0 t + D \Rightarrow s_0 = s(0) = D \Rightarrow s(t) = \frac{1}{2}at^2 + v_0 t + s_0$

39. By Exercise 38 with $a = -9.8$, $s(t) = -4.9t^2 + v_0 t + s_0$ and $v(t) = s'(t) = -9.8t + v_0$. So $[v(t)]^2 = (-9.8t + v_0)^2 = (9.8)^2 t^2 - 19.6 v_0 t + v_0^2 = v_0^2 + 96.04t^2 - 19.6 v_0 t = v_0^2 - 19.6(-4.9t^2 + v_0 t)$.
But $-4.9t^2 + v_0 t$ is just $s(t)$ without the s_0 term; that is, $s(t) - s_0$. Thus, $[v(t)]^2 = v_0^2 - 19.6[s(t) - s_0]$.

40. For the first ball, $s_1(t) = -16t^2 + 48t + 432$ from Example 6. For the second ball, $a(t) = -32 \Rightarrow v(t) = -32t + C$, but $v(1) = -32(1) + C = 24 \Rightarrow C = 56$, so $v(t) = -32t + 56 \Rightarrow s(t) = -16t^2 + 56t + D$, but $s(1) = -16(1)^2 + 56(1) + D = 432 \Rightarrow D = 392$, and $s_2(t) = -16t^2 + 56t + 392$. The balls pass each other when

$s_1(t) = s_2(t) \;\Rightarrow\; -16t^2 + 48t + 432 = -16t^2 + 56t + 392 \;\Leftrightarrow\; 8t = 40 \;\Leftrightarrow\; t = 5$ s.

Another solution: From Exercise 38, we have $s_1(t) = -16t^2 + 48t + 432$ and $s_2(t) = -16t^2 + 24t + 432$.

We now want to solve $s_1(t) = s_2(t-1) \;\Rightarrow\; -16t^2 + 48t + 432 = -16(t-1)^2 + 24(t-1) + 432 \;\Rightarrow$

$48t = 32t - 16 + 24t - 24 \;\Rightarrow\; 40 = 8t \;\Rightarrow\; t = 5$ s.

41. Using Exercise 38 with $a = -32$, $v_0 = 0$, and $s_0 = h$ (the height of the cliff), we know that the height at time t is $s(t) = -16t^2 + h$. $v(t) = s'(t) = -32t$ and $v(t) = -120 \;\Rightarrow\; -32t = -120 \;\Rightarrow\; t = 3.75$, so $0 = s(3.75) = -16(3.75)^2 + h \;\Rightarrow\; h = 16(3.75)^2 = 225$ ft.

42. (a) $EIy'' = mg(L-x) + \frac{1}{2}\rho g(L-x)^2 \;\Rightarrow\; EIy' = -\frac{1}{2}mg(L-x)^2 - \frac{1}{6}\rho g(L-x)^3 + C \;\Rightarrow$

$EIy = \frac{1}{6}mg(L-x)^3 + \frac{1}{24}\rho g(L-x)^4 + Cx + D$. Since the left end of the board is fixed, we must have $y = y' = 0$ when $x = 0$. Thus, $0 = -\frac{1}{2}mgL^2 - \frac{1}{6}\rho gL^3 + C$ and $0 = \frac{1}{6}mgL^3 + \frac{1}{24}\rho gL^4 + D$. It follows that

$EIy = \frac{1}{6}mg(L-x)^3 + \frac{1}{24}\rho g(L-x)^4 + \left(\frac{1}{2}mgL^2 + \frac{1}{6}\rho gL^3\right)x - \left(\frac{1}{6}mgL^3 + \frac{1}{24}\rho gL^4\right)$ and

$$f(x) = y = \frac{1}{EI}\left[\tfrac{1}{6}mg(L-x)^3 + \tfrac{1}{24}\rho g(L-x)^4 + \left(\tfrac{1}{2}mgL^2 + \tfrac{1}{6}\rho gL^3\right)x - \left(\tfrac{1}{6}mgL^3 + \tfrac{1}{24}\rho gL^4\right)\right]$$

(b) $f(L) < 0$, so the end of the board is a *distance* approximately $-f(L)$ below the horizontal. From our result in (a), we calculate

$$-f(L) = \frac{-1}{EI}\left[\tfrac{1}{2}mgL^3 + \tfrac{1}{6}\rho gL^4 - \tfrac{1}{6}mgL^3 - \tfrac{1}{24}\rho gL^4\right] = \frac{-1}{EI}\left(\tfrac{1}{3}mgL^3 + \tfrac{1}{8}\rho gL^4\right) = -\frac{gL^3}{EI}\left(\frac{m}{3} + \frac{\rho L}{8}\right)$$

Note: This is positive because g is negative.

43. Taking the upward direction to be positive we have that for $0 \le t \le 10$ (using the subscript 1 to refer to $0 \le t \le 10$),

$a_1(t) = -(9 - 0.9t) = v_1'(t) \;\Rightarrow\; v_1(t) = -9t + 0.45t^2 + v_0$, but $v_1(0) = v_0 = -10 \;\Rightarrow$

$v_1(t) = -9t + 0.45t^2 - 10 = s_1'(t) \;\Rightarrow\; s_1(t) = -\frac{9}{2}t^2 + 0.15t^3 - 10t + s_0$. But $s_1(0) = 500 = s_0 \;\Rightarrow$

$s_1(t) = -\frac{9}{2}t^2 + 0.15t^3 - 10t + 500$. $s_1(10) = -450 + 150 - 100 + 500 = 100$, so it takes more than 10 seconds for the raindrop to fall. Now for $t > 10$, $a(t) = 0 = v'(t) \;\Rightarrow$

$v(t) = \text{constant} = v_1(10) = -9(10) + 0.45(10)^2 - 10 = -55 \;\Rightarrow\; v(t) = -55$.

At 55 m/s, it will take $100/55 \approx 1.8$ s to fall the last 100 m. Hence, the total time is $10 + \frac{100}{55} = \frac{130}{11} \approx 11.8$ s.

44. $v'(t) = a(t) = -22$. The initial velocity is 50 mi/h $= \frac{50 \cdot 5280}{3600} = \frac{220}{3}$ ft/s, so $v(t) = -22t + \frac{220}{3}$. The car stops when $v(t) = 0 \;\Leftrightarrow\; t = \frac{220}{3 \cdot 22} = \frac{10}{3}$. Since $s(t) = -11t^2 + \frac{220}{3}t$, the distance covered is

$s\left(\frac{10}{3}\right) = -11\left(\frac{10}{3}\right)^2 + \frac{220}{3} \cdot \frac{10}{3} = \frac{1100}{9} = 122.\overline{2}$ ft.

45. $a(t) = k$, the initial velocity is 30 mi/h $= 30 \cdot \frac{5280}{3600} = 44$ ft/s, and the final velocity (after 5 seconds) is 50 mi/h $= 50 \cdot \frac{5280}{3600} = \frac{220}{3}$ ft/s. So $v(t) = kt + C$ and $v(0) = 44 \;\Rightarrow\; C = 44$. Thus, $v(t) = kt + 44 \;\Rightarrow$

$v(5) = 5k + 44$. But $v(5) = \frac{220}{3}$, so $5k + 44 = \frac{220}{3} \;\Rightarrow\; 5k = \frac{88}{3} \;\Rightarrow\; k = \frac{88}{15} \approx 5.87$ ft/s^2.

46. $a(t) = -16 \;\Rightarrow\; v(t) = -16t + v_0$ where v_0 is the car's speed (in ft/s) when the brakes were applied. The car stops when $-16t + v_0 = 0 \;\Leftrightarrow\; t = \frac{1}{16}v_0$. Now $s(t) = \frac{1}{2}(-16)t^2 + v_0 t = -8t^2 + v_0 t$. The car travels 200 ft in the time that it takes to stop, so $s\left(\frac{1}{16}v_0\right) = 200 \;\Rightarrow\; 200 = -8\left(\frac{1}{16}v_0\right)^2 + v_0\left(\frac{1}{16}v_0\right) = \frac{1}{32}v_0^2 \;\Rightarrow\; v_0^2 = 32 \cdot 200 = 6400 \;\Rightarrow\; v_0 = 80$ ft/s ($54.\overline{54}$ mi/h).

47. Let the acceleration be $a(t) = k\ \mathrm{km/h^2}$. We have $v(0) = 100\ \mathrm{km/h}$ and we can take the initial position $s(0)$ to be 0. We want the time t_f for which $v(t) = 0$ to satisfy $s(t) < 0.08$ km. In general, $v'(t) = a(t) = k$, so $v(t) = kt + C$, where $C = v(0) = 100$. Now $s'(t) = v(t) = kt + 100$, so $s(t) = \frac{1}{2}kt^2 + 100t + D$, where $D = s(0) = 0$. Thus, $s(t) = \frac{1}{2}kt^2 + 100t$. Since $v(t_f) = 0$, we have $kt_f + 100 = 0$ or $t_f = -100/k$, so

$s(t_f) = \frac{1}{2}k\left(-\frac{100}{k}\right)^2 + 100\left(-\frac{100}{k}\right) = 10{,}000\left(\frac{1}{2k} - \frac{1}{k}\right) = -\frac{5{,}000}{k}$. The condition $s(t_f)$ must satisfy is

$-\frac{5{,}000}{k} < 0.08 \;\Rightarrow\; -\frac{5{,}000}{0.08} > k$ [k is negative] $\Rightarrow\; k < -62{,}500\ \mathrm{km/h^2}$, or equivalently,

$k < -\frac{3125}{648} \approx -4.82\ \mathrm{m/s^2}$. Thus, a constant deceleration of $4.82\ \mathrm{m/s^2}$ is required.

48. (a) For $0 \le t \le 3$ we have $a(t) = 60t \;\Rightarrow\; v(t) = 30t^2 + C \;\Rightarrow\; v(0) = 0 = C \;\Rightarrow\; v(t) = 30t^2$, so $s(t) = 10t^3 + C \;\Rightarrow\; s(0) = 0 = C \;\Rightarrow\; s(t) = 10t^3$. Note that $v(3) = 270$ and $s(3) = 270$.

For $3 < t \le 17$: $a(t) = -g = -32$ ft/s $\Rightarrow\; v(t) = -32(t-3) + C \;\Rightarrow\; v(3) = 270 = C \;\Rightarrow$ $v(t) = -32(t-3) + 270 \;\Rightarrow\; s(t) = -16(t-3)^2 + 270(t-3) + C \;\Rightarrow\; s(3) = 270 = C \;\Rightarrow$ $s(t) = -16(t-3)^2 + 270(t-3) + 270$. Note that $v(17) = -178$ and $s(17) = 914$.

For $17 < t \le 22$: The velocity increases linearly from -178 ft/s to -18 ft/s during this period, so

$\frac{\Delta v}{\Delta t} = \frac{-18 - (-178)}{22 - 17} = \frac{160}{5} = 32$. Thus, $v(t) = 32(t-17) - 178 \;\Rightarrow$

$s(t) = 16(t-17)^2 - 178(t-17) + 914$ and $s(22) = 424$ ft.

For $t > 22$: $v(t) = -18 \;\Rightarrow\; s(t) = -18(t-22) + C$. But $s(22) = 424 = C \;\Rightarrow\; s(t) = -18(t-22) + 424$.

Therefore, until the rocket lands, we have

$$v(t) = \begin{cases} 30t^2 & \text{if } 0 \le t \le 3 \\ -32\,(t-3) + 270 & \text{if } 3 < t \le 17 \\ 32(t-17) - 178 & \text{if } 17 < t \le 22 \\ -18 & \text{if } t > 22 \end{cases}$$

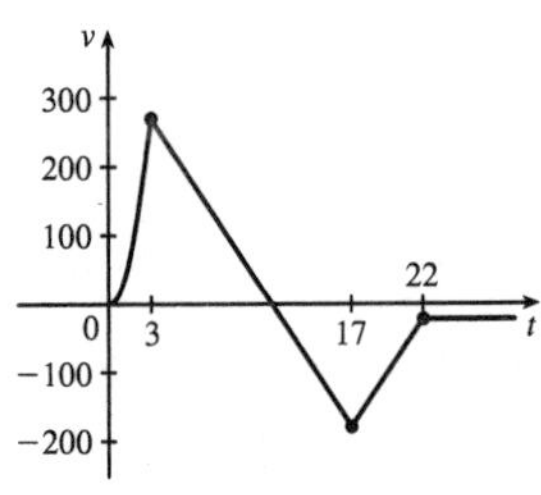

and
$$s(t) = \begin{cases} 10t^3 & \text{if } 0 \le t \le 3 \\ -16(t-3)^2 + 270(t-3) + 270 & \text{if } 3 < t \le 17 \\ 16(t-17)^2 - 178\,(t-17) + 914 & \text{if } 17 < t \le 22 \\ -18(t-22) + 424 & \text{if } t > 22 \end{cases}$$

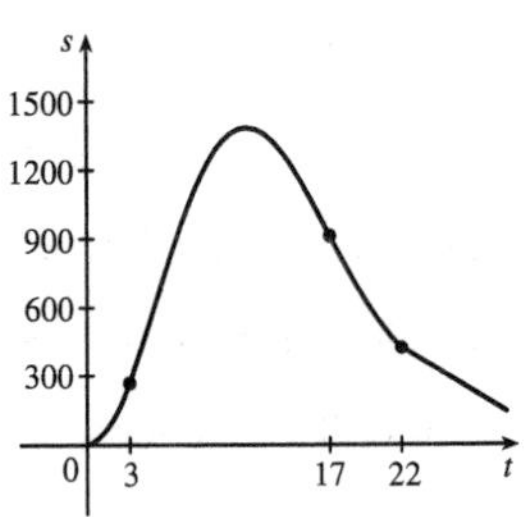

(b) To find the maximum height, set $v(t)$ on $3 < t \le 17$ equal to 0. $-32(t-3) + 270 = 0 \;\Rightarrow\; t_1 = 11.4375$ s and the maximum height is $s(t_1) = -16(t_1-3)^2 + 270(t_1-3) + 270 = 1409.0625$ ft.

(c) To find the time to land, set $s(t) = -18(t-22) + 424 = 0$. Then $t - 22 = \frac{424}{18} = 23.\overline{5}$, so $t \approx 45.6$ s.

49. (a) First note that $90 \text{ mi/h} = 90 \times \frac{5280}{3600} \text{ ft/s} = 132 \text{ ft/s}$. Then $a(t) = 4 \text{ ft/s}^2 \Rightarrow v(t) = 4t + C$, but $v(0) = 0 \Rightarrow$ $C = 0$. Now $4t = 132$ when $t = \frac{132}{4} = 33$ s, so it takes 33 s to reach 132 ft/s. Therefore, taking $s(0) = 0$, we have $s(t) = 2t^2$, $0 \le t \le 33$. So $s(33) = 2178$ ft. 15 minutes $= 15(60) = 900$ s, so for $33 < t \le 933$ we have $v(t) = 132 \text{ ft/s} \Rightarrow s(933) = 132(900) + 2178 = 120{,}978 \text{ ft} = 22.9125$ mi.

(b) As in part (a), the train accelerates for 33 s and travels 2178 ft while doing so. Similarly, it decelerates for 33 s and travels 2178 ft at the end of its trip. During the remaining $900 - 66 = 834$ s it travels at 132 ft/s, so the distance traveled is $132 \cdot 834 = 110{,}088$ ft. Thus, the total distance is $2178 + 110{,}088 + 2178 = 114{,}444 \text{ ft} = 21.675$ mi.

(c) $45 \text{ mi} = 45(5280) = 237{,}600$ ft. Subtract $2(2178)$ to take care of the speeding up and slowing down, and we have 233,244 ft at 132 ft/s for a trip of $233{,}244/132 = 1767$ s at 90 mi/h. The total time is $1767 + 2(33) = 1833 \text{ s} = 30 \text{ min } 33 \text{ s} = 30.55$ min.

(d) $37.5(60) = 2250$ s. $2250 - 2(33) = 2184$ s at maximum speed. $2184(132) + 2(2178) = 292{,}644$ total feet or $292{,}644/5280 = 55.425$ mi.

3 Review

CONCEPT CHECK

1. A function f has an **absolute maximum** at $x = c$ if $f(c)$ is the largest function value on the entire domain of f, whereas f has a **local maximum** at c if $f(c)$ is the largest function value when x is near c. See Figure 4 in Section 3.1.

2. (a) See Theorem 3.1.3.

(b) See the Closed Interval Method before Example 6 in Section 3.1.

3. (a) See Theorem 3.1.4.

(b) See Definition 3.1.6.

4. (a) See Rolle's Theorem at the beginning of Section 3.2.

(b) See the Mean Value Theorem in Section 3.2. Geometric interpretation—there is some point P on the graph of a function f [on the interval (a, b)] where the tangent line is parallel to the secant line that connects $(a, f(a))$ and $(b, f(b))$.

5. (a) See the I/D Test before Example 1 in Section 3.3.

(b) A function f is concave upward on an interval I if the graph of f lies above all of its tangent lines on I.

(c) See the Concavity Test before Example 4 in Section 3.3.

(d) An inflection point is a point where a curve changes its direction of concavity. They can be found by determining the points at which the second derivative changes sign.

6. (a) See the First Derivative Test after Example 1 in Section 3.3.

(b) See the Second Derivative Test before Example 5 in Section 3.3.

(c) See the note before Example 6 in Section 3.3.

7. Without calculus you could get misleading graphs that fail to show the most interesting features of a function. See the third paragraph in Section 3.4.

8. (a) See Figure 3 in Section 3.6.

(b) $x_2 = x_1 - \dfrac{f(x_1)}{f'(x_1)}$

(c) $x_{n+1} = x_n - \dfrac{f(x_n)}{f'(x_n)}$

(d) Newton's method is likely to fail or to work very slowly when $f'(x_1)$ is close to 0.

9. (a) See the definition at the beginning of Section 3.7.

(b) If F_1 and F_2 are both antiderivatives of f on an interval I, then they differ by a constant.

TRUE-FALSE QUIZ

1. False. For example, take $f(x) = x^3$, then $f'(x) = 3x^2$ and $f'(0) = 0$, but $f(0) = 0$ is not a maximum or minimum; $(0, 0)$ is an inflection point.

2. False. For example, $f(x) = |x|$ has an absolute minimum at 0, but $f'(0)$ does not exist.

3. False. For example, $f(x) = x$ is continuous on $(0, 1)$ but attains neither a maximum nor a minimum value on $(0, 1)$. Don't confuse this with f being continuous on the *closed* interval $[a, b]$, which would make the statement true.

4. True. By the Mean Value Theorem, $f'(c) = \dfrac{f(1) - f(-1)}{1 - (-1)} = \dfrac{0}{2} = 0$. Note that $|c| < 1 \quad \Leftrightarrow \quad c \in (-1, 1)$.

5. True. This is an example of part (b) of the I/D Test.

6. False. For example, the curve $y = f(x) = 1$ has no inflection point but $f''(c) = 0$ for all c.

7. False. $f'(x) = g'(x) \quad \Rightarrow \quad f(x) = g(x) + C$. For example, if $f(x) = x + 2$ and $g(x) = x + 1$, then $f'(x) = g'(x) = 1$, but $f(x) \neq g(x)$.

8. False. Assume there is a function f such that $f(1) = -2$ and $f(3) = 0$. Then by the Mean Value Theorem there exists a number $c \in (1, 3)$ such that $f'(c) = \dfrac{f(3) - f(1)}{3 - 1} = \dfrac{0 - (-2)}{2} = 1$. But $f'(x) > 1$ for all x, a contradiction.

9. True. The graph of one such function is sketched.

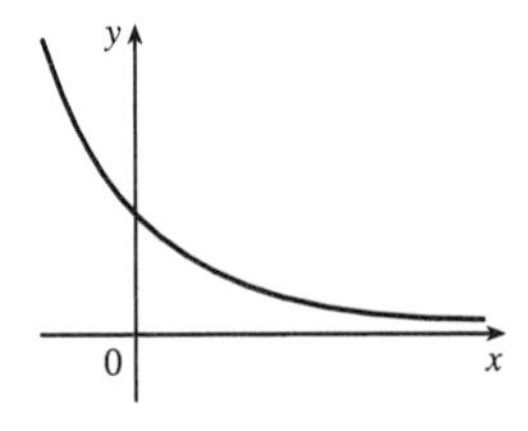

10. False. At any point $(a, f(a))$, we know that $f'(a) < 0$. So since the tangent line at $(a, f(a))$ is not horizontal, it must cross the x-axis—at $x = b$, say. But since $f''(x) > 0$ for all x, the graph of f must lie above all of its tangents; in particular, $f(b) > 0$. But this is a contradiction, since we are given that $f(x) < 0$ for all x.

11. True. Let $x_1 < x_2$ where $x_1, x_2 \in I$. Then $f(x_1) < f(x_2)$ and $g(x_1) < g(x_2)$ (since f and g are increasing on I), so $(f + g)(x_1) = f(x_1) + g(x_1) < f(x_2) + g(x_2) = (f + g)(x_2)$.

12. False. $f(x) = x$ and $g(x) = 2x$ are both increasing on $(0, 1)$, but $f(x) - g(x) = -x$ is not increasing on $(0, 1)$.

13. False. Take $f(x) = x$ and $g(x) = x - 1$. Then both f and g are increasing on $(0, 1)$. But $f(x)g(x) = x(x-1)$ is not increasing on $(0, 1)$.

14. True. Let $x_1 < x_2$ where $x_1, x_2 \in I$. Then $0 < f(x_1) < f(x_2)$ and $0 < g(x_1) < g(x_2)$ (since f and g are both positive and increasing). Hence, $f(x_1)\, g(x_1) < f(x_2)\, g(x_1) < f(x_2)\, g(x_2)$. So fg is increasing on I.

15. True. Let $x_1, x_2 \in I$ and $x_1 < x_2$. Then $f(x_1) < f(x_2)$ (f is increasing) $\Rightarrow$ $\dfrac{1}{f(x_1)} > \dfrac{1}{f(x_2)}$ (f is positive) $\Rightarrow$ $g(x_1) > g(x_2)$ $\Rightarrow$ $g(x) = 1/f(x)$ is decreasing on I.

16. False. The most general antiderivative is $F(x) = -1/x + C_1$ for $x < 0$ and $F(x) = -1/x + C_2$ for $x > 0$ (see Example 1 in Section 3.7).

17. True. By the Mean Value Theorem, there exists a number c in $(0, 1)$ such that $f(1) - f(0) = f'(c)(1 - 0) = f'(c)$. Since $f'(c)$ is nonzero, $f(1) - f(0) \neq 0$, so $f(1) \neq f(0)$.

EXERCISES

1. $f(x) = 10 + 27x - x^3$, $0 \le x \le 4$. $f'(x) = 27 - 3x^2 = -3(x^2 - 9) = -3(x+3)(x-3) = 0$ only when $x = 3$ (since -3 is not in the domain). $f'(x) > 0$ for $x < 3$ and $f'(x) < 0$ for $x > 3$, so $f(3) = 64$ is a local maximum value. Checking the endpoints, we find $f(0) = 10$ and $f(4) = 54$. Thus, $f(0) = 10$ is the absolute minimum value and $f(3) = 64$ is the absolute maximum value.

2. $f(x) = x - \sqrt{x}$, $0 \le x \le 4$. $f'(x) = 1 - 1/(2\sqrt{x}) = 0 \Leftrightarrow 2\sqrt{x} = 1 \Rightarrow x = \frac{1}{4}$. $f'(x)$ does not exist $\Leftrightarrow$ $x = 0$. $f'(x) < 0$ for $0 < x < \frac{1}{4}$ and $f'(x) > 0$ for $\frac{1}{4} < x < 4$, so $f\left(\frac{1}{4}\right) = -\frac{1}{4}$ is a local and absolute minimum value. $f(0) = 0$ and $f(4) = 2$, so $f(4) = 2$ is the absolute maximum value.

3. $f(x) = \dfrac{x}{x^2 + x + 1}$, $-2 \le x \le 0$. $f'(x) = \dfrac{(x^2 + x + 1)(1) - x(2x + 1)}{(x^2 + x + 1)^2} = \dfrac{1 - x^2}{(x^2 + x + 1)^2} = 0 \Leftrightarrow x = -1$ (since 1 is not in the domain). $f'(x) < 0$ for $-2 < x < -1$ and $f'(x) > 0$ for $-1 < x < 0$, so $f(-1) = -1$ is a local and absolute minimum value. $f(-2) = -\frac{2}{3}$ and $f(0) = 0$, so $f(0) = 0$ is an absolute maximum value.

4. $f(x) = \sin x + \cos^2 x$, $[0, \pi]$. $f'(x) = \cos x - 2\cos x \sin x = \cos x\,(1 - 2\sin x)$, so $f'(x) = 0$ for x in $(0, \pi)$ $\Leftrightarrow$ $\cos x = 0$ or $\sin x = \frac{1}{2}$ $\Leftrightarrow$ $x = \frac{\pi}{6}, \frac{\pi}{2}$, or $\frac{5\pi}{6}$. $f'(x) = \cos x - \sin 2x \Rightarrow f''(x) = -\sin x - 2\cos 2x$, so $f''\left(\frac{\pi}{6}\right) = -\frac{1}{2} - 2\left(\frac{1}{2}\right) = -\frac{3}{2}$, $f''\left(\frac{\pi}{2}\right) = -1 - 2(-1) = 1$, and $f''\left(\frac{5\pi}{6}\right) = -\frac{1}{2} - 2\left(\frac{1}{2}\right) = -\frac{3}{2}$. Thus, $f\left(\frac{\pi}{6}\right) = \frac{5}{4}$ and $f\left(\frac{5\pi}{6}\right) = \frac{5}{4}$ are local maxima and $f\left(\frac{\pi}{2}\right) = 1$ is a local minimum. $f(0) = 1$ and $f(\pi) = 1$, so f has its absolute minimum value of 1 at 0, $\frac{\pi}{2}$, and π. f attains its absolute maximum value of $\frac{5}{4}$ at $\frac{\pi}{6}$ and $\frac{5\pi}{6}$.

5. $f(0) = 0$, $f'(-2) = f'(1) = f'(9) = 0$, $\lim\limits_{x \to \infty} f(x) = 0$, $\lim\limits_{x \to 6} f(x) = -\infty$, $f'(x) < 0$ on $(-\infty, -2)$, $(1, 6)$, and $(9, \infty)$, $f'(x) > 0$ on $(-2, 1)$ and $(6, 9)$, $f''(x) > 0$ on $(-\infty, 0)$ and $(12, \infty)$, $f''(x) < 0$ on $(0, 6)$ and $(6, 12)$

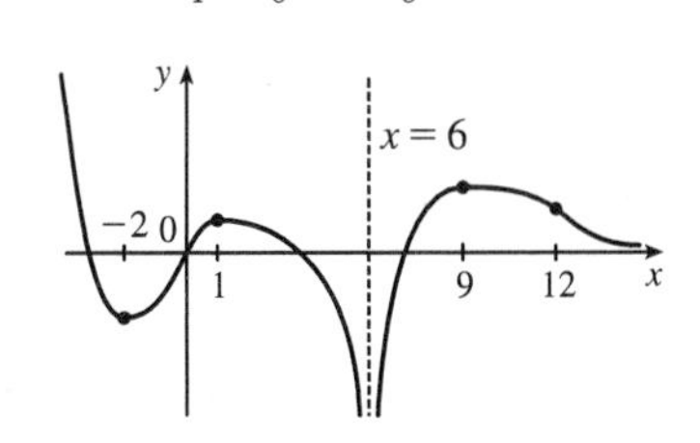

6. For $0 < x < 1$, $f'(x) = 2x$, so $f(x) = x^2 + C$. Since $f(0) = 0$, $f(x) = x^2$ on $[0, 1]$. For $1 < x < 3$, $f'(x) = -1$, so $f(x) = -x + D$. $1 = f(1) = -1 + D \Rightarrow D = 2$, so $f(x) = 2 - x$. For $x > 3$, $f'(x) = 1$, so $f(x) = x + E$. $-1 = f(3) = 3 + E \Rightarrow E = -4$, so $f(x) = x - 4$. Since f is even, its graph is symmetric about the y-axis.

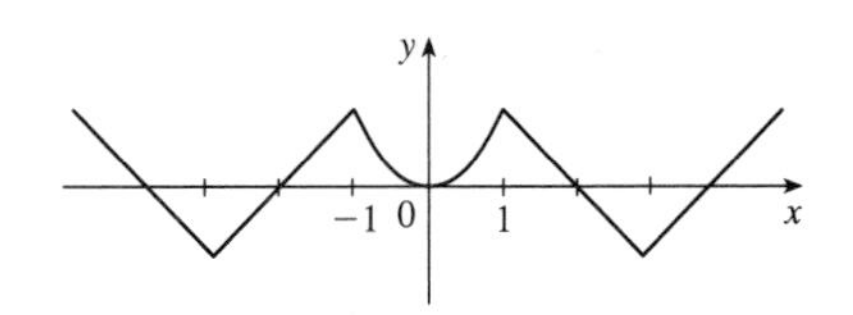

7. f is odd, $f'(x) < 0$ for $0 < x < 2$, $f'(x) > 0$ for $x > 2$, $f''(x) > 0$ for $0 < x < 3$, $f''(x) < 0$ for $x > 3$, $\lim_{x\to\infty} f(x) = -2$

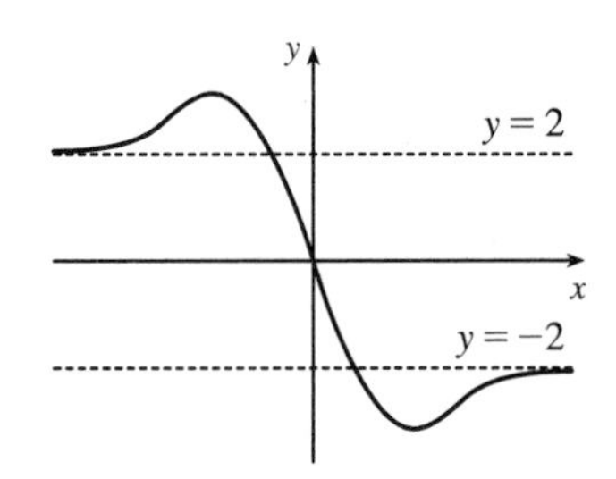

8. (a) Using the Test for Monotonic Functions we know that f is increasing on $(-2, 0)$ and $(4, \infty)$ because $f' > 0$ on $(-2, 0)$ and $(4, \infty)$, and that f is decreasing on $(-\infty, -2)$ and $(0, 4)$ because $f' < 0$ on $(-\infty, -2)$ and $(0, 4)$.

(b) Using the First Derivative Test, we know that f has a local maximum at $x = 0$ because f' changes from positive to negative at $x = 0$, and that f has a local minimum at $x = 4$ because f' changes from negative to positive at $x = 4$.

(c)

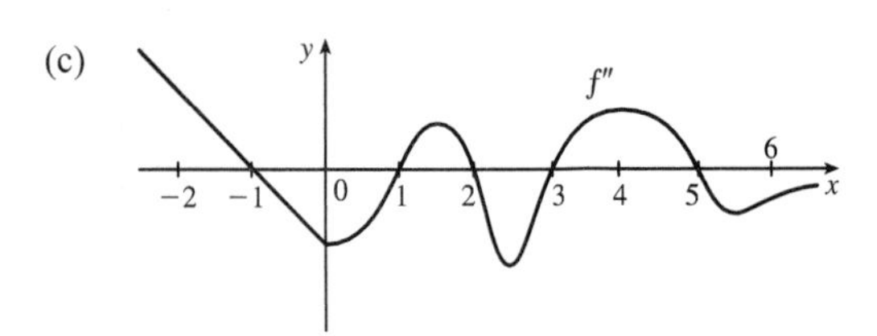

(d)

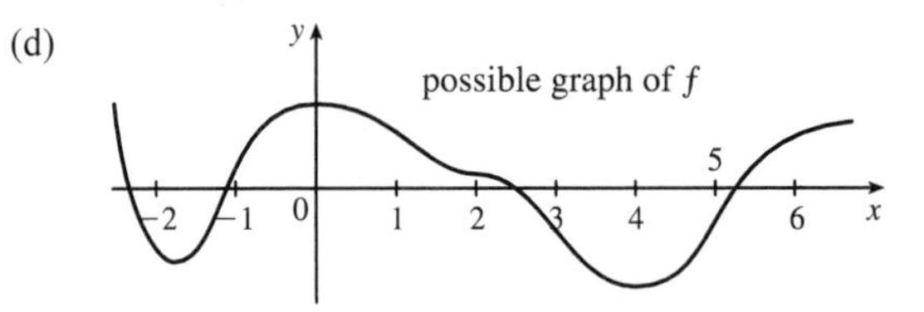

9. $y = f(x) = 2 - 2x - x^3$ **A.** $D = \mathbb{R}$ **B.** y-intercept: $f(0) = 2$. The x-intercept (approximately 0.770917) can be found using Newton's Method. **C.** No symmetry **D.** No asymptote **E.** $f'(x) = -2 - 3x^2 = -(3x^2 + 2) < 0$, so f is decreasing on $\mathbb{R}$. **F.** No extreme value **G.** $f''(x) = -6x < 0$ on $(0, \infty)$ and $f''(x) > 0$ on $(-\infty, 0)$, so f is CD on $(0, \infty)$ and CU on $(-\infty, 0)$. There is an IP at $(0, 2)$.

H.

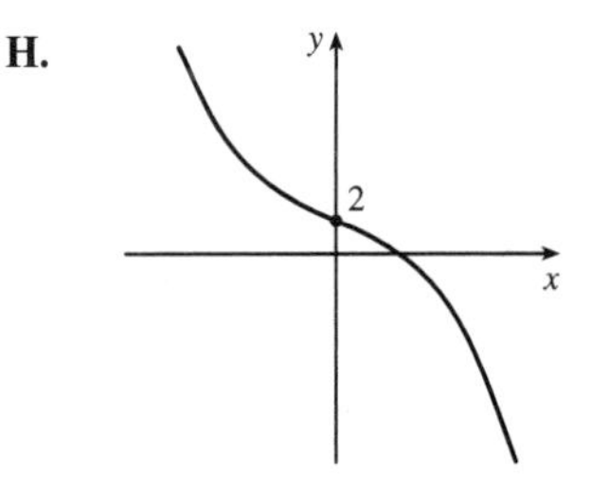

10. $y = f(x) = x^3 - 6x^2 - 15x + 4$ **A.** $D = \mathbb{R}$ **B.** y-intercept: $f(0) = 4$; x-intercepts: $f(x) = 0 \Rightarrow x \approx -2.09, 0.24, 7.85$ **C.** No symmetry **D.** No asymptote **E.** $f'(x) = 3x^2 - 12x - 15 = 3(x^2 - 4x - 5) = 3(x + 1)(x - 5)$, so f is increasing on $(-\infty, -1)$, decreasing on $(-1, 5)$, and increasing on $(5, \infty)$. **F.** Local maximum value $f(-1) = 12$, local minimum value $f(5) = -96$. **G.** $f''(x) = 6x - 12 = 6(x - 2)$, so f is CD on $(-\infty, 2)$ and CU on $(2, \infty)$. IP at $(2, -42)$

H.

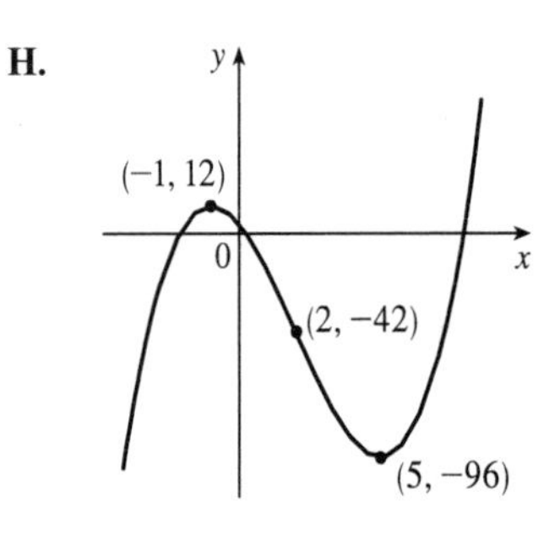

11. $y = f(x) = x^4 - 3x^3 + 3x^2 - x = x(x-1)^3$ **A.** $D = \mathbb{R}$ **B.** y-intercept: $f(0) = 0$; x-intercepts: $f(x) = 0 \Leftrightarrow$ $x = 0$ or $x = 1$ **C.** No symmetry **D.** f is a polynomial function and hence, it has no asymptote.
E. $f'(x) = 4x^3 - 9x^2 + 6x - 1$. Since the sum of the coefficients is 0, 1 is a root of f',
so $f'(x) = (x-1)(4x^2 - 5x + 1) = (x-1)^2(4x - 1)$. $f'(x) < 0 \Rightarrow x < \frac{1}{4}$, so f is decreasing on $\left(-\infty, \frac{1}{4}\right)$ and f is increasing on $\left(\frac{1}{4}, \infty\right)$. **F.** $f'(x)$ does not change sign at $x = 1$, so there

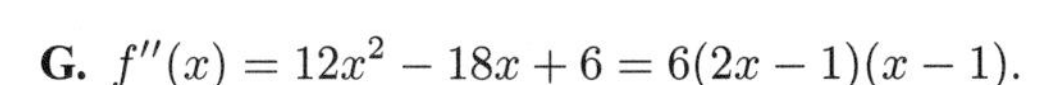

is not a local extremum there. $f\left(\frac{1}{4}\right) = -\frac{27}{256}$ is a local minimum value.
G. $f''(x) = 12x^2 - 18x + 6 = 6(2x-1)(x-1)$.
$f''(x) = 0 \Leftrightarrow x = \frac{1}{2}$ or 1. $f''(x) < 0 \Leftrightarrow \frac{1}{2} < x < 1 \Rightarrow$
f is CD on $\left(\frac{1}{2}, 1\right)$ and CU on $\left(-\infty, \frac{1}{2}\right)$ and $(1, \infty)$. IPs at $\left(\frac{1}{2}, -\frac{1}{16}\right)$ and $(1, 0)$

H.

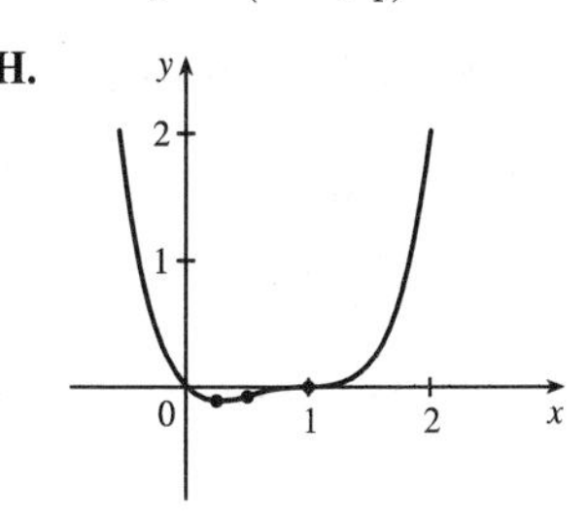

12. $y = f(x) = \dfrac{1}{1-x^2} = \dfrac{1}{(1+x)(1-x)}$ **A.** $D = \{x \mid x \neq \pm 1\}$ **B.** y-intercept: $f(0) = 1$; no x-intercept
C. $f(-x) = f(x)$, so f is even and the graph of f is symmetric about the y-axis. **D.** Vertical asymptotes: $x = \pm 1$.
Horizontal asymptote: $y = 0$ **E.** $y' = \dfrac{2x}{(1-x^2)^2} = 0 \Leftrightarrow x = 0$, so f is decreasing on $(-\infty, -1)$ and $(-1, 0)$,
and increasing on $(0, 1)$ and $(1, \infty)$.
F. Local minimum value $f(0) = 1$; no local maximum

$$\textbf{G. } f''(x) = \frac{(1-x^2)^2 \cdot 2 - 2x \cdot 2(1-x^2)(-2x)}{(1-x^2)^4}$$
$$= \frac{2(1-x^2) + 8x^2}{(1-x^2)^3} = \frac{6x^2 + 2}{(1-x^2)^3} < 0 \Leftrightarrow x^2 > 1,$$

so f is CD on $(-\infty, -1)$ and $(1, \infty)$, and CU on $(-1, 1)$. No IP

H.

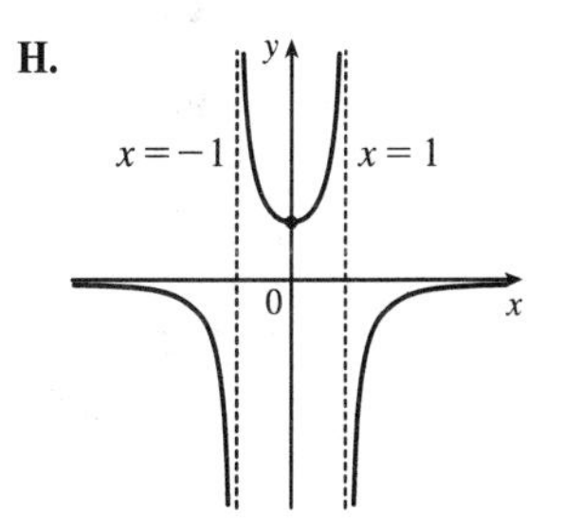

13. $y = f(x) = \dfrac{1}{x(x-3)^2}$ **A.** $D = \{x \mid x \neq 0, 3\} = (-\infty, 0) \cup (0, 3) \cup (3, \infty)$ **B.** No intercepts. **C.** No symmetry.

D. $\displaystyle\lim_{x \to \pm\infty} \frac{1}{x(x-3)^2} = 0$, so $y = 0$ is a HA. $\displaystyle\lim_{x \to 0^+} \frac{1}{x(x-3)^2} = \infty$, $\displaystyle\lim_{x \to 0^-} \frac{1}{x(x-3)^2} = -\infty$, $\displaystyle\lim_{x \to 3} \frac{1}{x(x-3)^2} = \infty$,

so $x = 0$ and $x = 3$ are VA. **E.** $f'(x) = -\dfrac{(x-3)^2 + 2x(x-3)}{x^2(x-3)^4} = \dfrac{3(1-x)}{x^2(x-3)^3} \Rightarrow f'(x) > 0 \Leftrightarrow 1 < x < 3$,

so f is increasing on $(1, 3)$ and decreasing on $(-\infty, 0)$, $(0, 1)$, and $(3, \infty)$.

F. Local minimum value $f(1) = \frac{1}{4}$ **G.** $f''(x) = \dfrac{6(2x^2 - 4x + 3)}{x^3(x-3)^4}$. Note that
$2x^2 - 4x + 3 > 0$ for all x since it has negative discriminant. So $f''(x) > 0 \Leftrightarrow$
$x > 0 \Rightarrow f$ is CU on $(0, 3)$ and $(3, \infty)$ and CD on $(-\infty, 0)$. No IP

H.

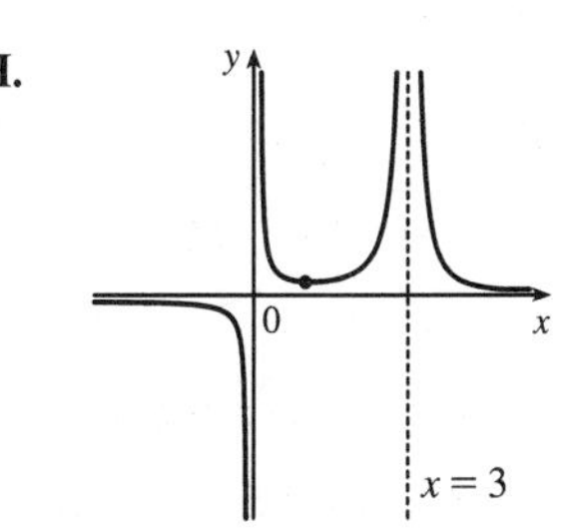

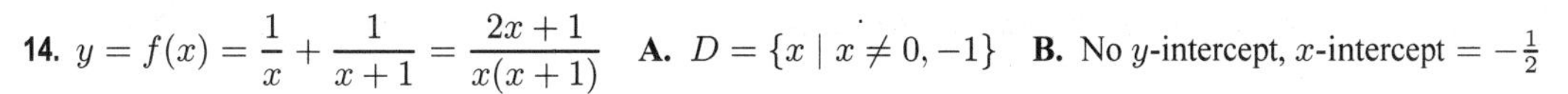

14. $y = f(x) = \dfrac{1}{x} + \dfrac{1}{x+1} = \dfrac{2x+1}{x(x+1)}$ **A.** $D = \{x \mid x \neq 0, -1\}$ **B.** No y-intercept, x-intercept $= -\frac{1}{2}$

C. No symmetry **D.** $\displaystyle\lim_{x \to \pm\infty} f(x) = 0$, so $y = 0$ is a HA. $\displaystyle\lim_{x \to 0^+} \frac{2x+1}{x(x+1)} = \infty$, $\displaystyle\lim_{x \to 0^-} \frac{2x+1}{x(x+1)} = -\infty$,

$\displaystyle\lim_{x \to -1^+} \frac{2x+1}{x(x+1)} = \infty$, $\displaystyle\lim_{x \to -1^-} \frac{2x+1}{x(x+1)} = -\infty$, so $x = 0$, $x = -1$ are VAs. **E.** $f'(x) = -\dfrac{1}{x^2} - \dfrac{1}{(x+1)^2} < 0$, so f is decreasing on $(-\infty,-1)$, $(-1,0)$ and $(0,\infty)$. **F.** No extreme values **H.**

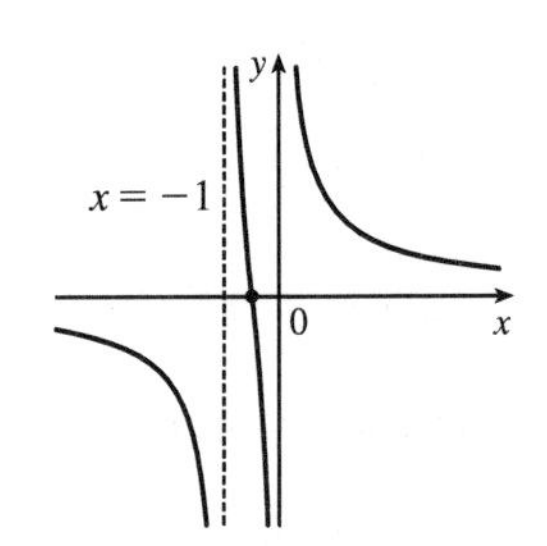

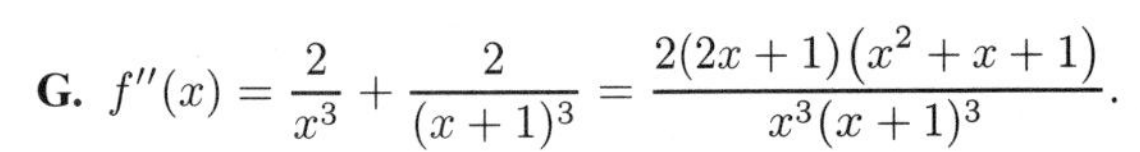

G. $f''(x) = \dfrac{2}{x^3} + \dfrac{2}{(x+1)^3} = \dfrac{2(2x+1)(x^2+x+1)}{x^3(x+1)^3}$.

$f''(x) > 0 \quad\Leftrightarrow\quad x > 0$ or $-1 < x < -\frac{1}{2}$, so f is CU on $(0,\infty)$ and $\left(-1,-\frac{1}{2}\right)$ and CD on $(-\infty,-1)$ and $\left(-\frac{1}{2},0\right)$. IP at $\left(-\frac{1}{2},0\right)$

15. $y = f(x) = x\sqrt{2+x}$ **A.** $D = [-2,\infty)$ **B.** y-intercept: $f(0)=0$; x-intercepts: -2 and 0 **C.** No symmetry

D. No asymptote **E.** $f'(x) = \dfrac{x}{2\sqrt{2+x}} + \sqrt{2+x} = \dfrac{1}{2\sqrt{2+x}}\,[x+2(2+x)] = \dfrac{3x+4}{2\sqrt{2+x}} = 0$ when $x = -\frac{4}{3}$, so f is decreasing on $\left(-2,-\frac{4}{3}\right)$ and increasing on $\left(-\frac{4}{3},\infty\right)$. **F.** Local minimum value **H.**

$f\left(-\frac{4}{3}\right) = -\frac{4}{3}\sqrt{\frac{2}{3}} = -\frac{4\sqrt{6}}{9} \approx -1.09$, no local maximum

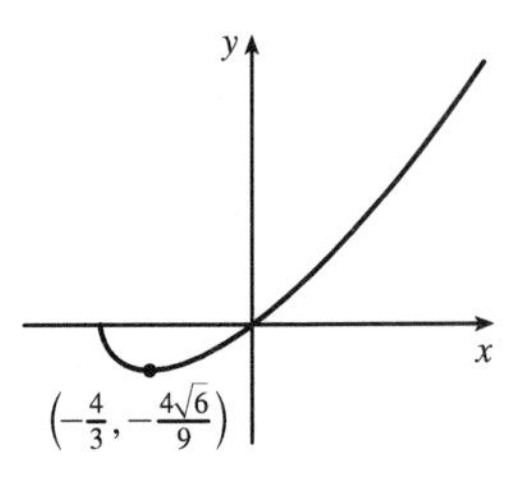

$$\textbf{G. } f''(x) = \frac{2\sqrt{2+x}\cdot 3 - (3x+4)\dfrac{1}{\sqrt{2+x}}}{4(2+x)} = \frac{6(2+x)-(3x+4)}{4(2+x)^{3/2}} = \frac{3x+8}{4(2+x)^{3/2}}$$

$f''(x) > 0$ for $x > -2$, so f is CU on $(-2,\infty)$. No IP

16. $y = f(x) = \sqrt{x} - \sqrt[3]{x}$ **A.** $D = [0,\infty)$ **B.** y-intercept 0; x-intercepts 0, 1 **C.** No symmetry

D. $\lim\limits_{x\to\infty}\left(x^{1/2}-x^{1/3}\right) = \lim\limits_{x\to\infty}\left[x^{1/3}\left(x^{1/6}-1\right)\right] = \infty$, no asymptote **E.** $f'(x) = \frac{1}{2}x^{-1/2} - \frac{1}{3}x^{-2/3} = \dfrac{3x^{1/6}-2}{6x^{2/3}} > 0$

$\Leftrightarrow \quad 3x^{1/6} > 2 \quad\Leftrightarrow\quad x > \left(\frac{2}{3}\right)^6$, so f is increasing on $\left(\left(\frac{2}{3}\right)^6,\infty\right)$ and decreasing on $\left(0,\left(\frac{2}{3}\right)^6\right)$.

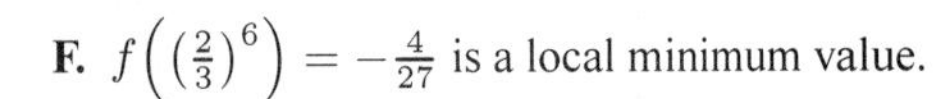

F. $f\left(\left(\frac{2}{3}\right)^6\right) = -\frac{4}{27}$ is a local minimum value. **H.**

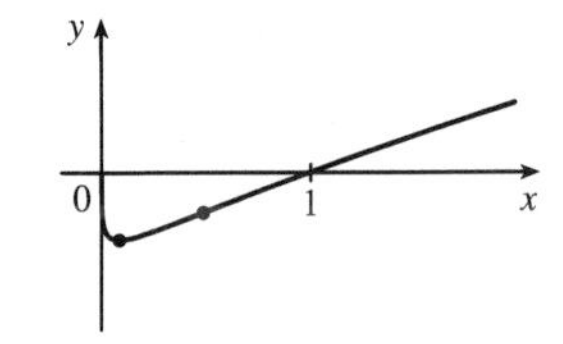

G. $f''(x) = -\frac{1}{4}x^{-3/2} + \frac{2}{9}x^{-5/3} = \dfrac{8-9x^{1/6}}{36x^{5/3}} > 0 \quad\Leftrightarrow\quad x^{1/6} < \frac{8}{9} \quad\Leftrightarrow$

$x < \left(\frac{8}{9}\right)^6$, so f is CU on $\left(0,\left(\frac{8}{9}\right)^6\right)$ and CD on $\left(\left(\frac{8}{9}\right)^6,\infty\right)$. IP at $\left(\left(\frac{8}{9}\right)^6, -\frac{64}{729}\right)$

17. (a) $y = f(x) = \sin^2 x - 2\cos x$ has no asymptote.

(b) $y' = 2\sin x\,\cos x + 2\sin x = 2\sin x\,(\cos x + 1)$. $y' = 0 \quad\Leftrightarrow\quad \sin x = 0$ or $\cos x = -1 \quad\Leftrightarrow\quad x = n\pi$ or $x = (2n+1)\pi$. $y' > 0$ when $\sin x > 0$, since $\cos x + 1 \geq 0$ for all x. Therefore, $y' > 0$ (and so f is increasing) on $(2n\pi,(2n+1)\pi)$; $y' < 0$ (and so f is decreasing) on $((2n-1)\pi, 2n\pi)$ or equivalently, $((2n+1)\pi,(2n+2)\pi)$.

(c) Local maxima are $f((2n+1)\pi) = 2$; local minima are $f(2n\pi) = -2$.

(d) $y' = \sin 2x + 2\sin x \quad\Rightarrow$ (e)

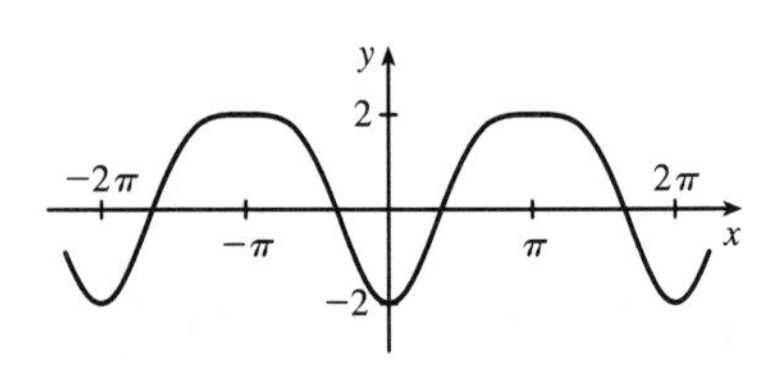

$$\begin{aligned} y'' &= 2\cos 2x + 2\cos x = 2(2\cos^2 x - 1) + 2\cos x \\ &= 4\cos^2 x + 2\cos x - 2 = 2(2\cos^2 x + \cos x - 1) \\ &= 2(2\cos x - 1)(\cos x + 1) \end{aligned}$$

$y'' = 0 \quad\Leftrightarrow\quad \cos x = \frac{1}{2}$ or $-1 \quad\Leftrightarrow\quad x = 2n\pi \pm \frac{\pi}{3}$ or $x = (2n+1)\pi$.

$y'' > 0$ (and so f is CU) on $\left(2n\pi - \frac{\pi}{3}, 2n\pi + \frac{\pi}{3}\right)$; $y'' \leq 0$ (and so f is CD) on $\left(2n\pi + \frac{\pi}{3}, 2n\pi + \frac{5\pi}{3}\right)$.

IPs at $\left(2n\pi \pm \frac{\pi}{3}, -\frac{1}{4}\right)$

18. $y = f(x) = 4x - \tan x$, $-\frac{\pi}{2} < x < \frac{\pi}{2}$ **A.** $D = \left(-\frac{\pi}{2}, \frac{\pi}{2}\right)$. **B.** y-intercept $= f(0) = 0$ **C.** $f(-x) = -f(x)$, so the curve is symmetric about $(0, 0)$. **D.** $\lim\limits_{x \to \pi/2^-} (4x - \tan x) = -\infty$, $\lim\limits_{x \to -\pi/2^+} (4x - \tan x) = \infty$, so $x = \frac{\pi}{2}$ and $x = -\frac{\pi}{2}$ are VAs. **E.** $f'(x) = 4 - \sec^2 x > 0 \Leftrightarrow \sec x < 2 \Leftrightarrow \cos x > \frac{1}{2} \Leftrightarrow -\frac{\pi}{3} < x < \frac{\pi}{3}$, so f is increasing on $\left(-\frac{\pi}{3}, \frac{\pi}{3}\right)$ and decreasing on $\left(-\frac{\pi}{2}, -\frac{\pi}{3}\right)$ and $\left(\frac{\pi}{3}, \frac{\pi}{2}\right)$. **F.** $f\left(\frac{\pi}{3}\right) = \frac{4\pi}{3} - \sqrt{3}$ is a local maximum value, $f\left(-\frac{\pi}{3}\right) = \sqrt{3} - \frac{4\pi}{3}$ is a local minimum value. **G.** $f''(x) = -2\sec^2 x \tan x > 0 \Leftrightarrow \tan x < 0 \Leftrightarrow -\frac{\pi}{2} < x < 0$, so f is CU on $\left(-\frac{\pi}{2}, 0\right)$ and CD on $\left(0, \frac{\pi}{2}\right)$. IP at $(0, 0)$

H.

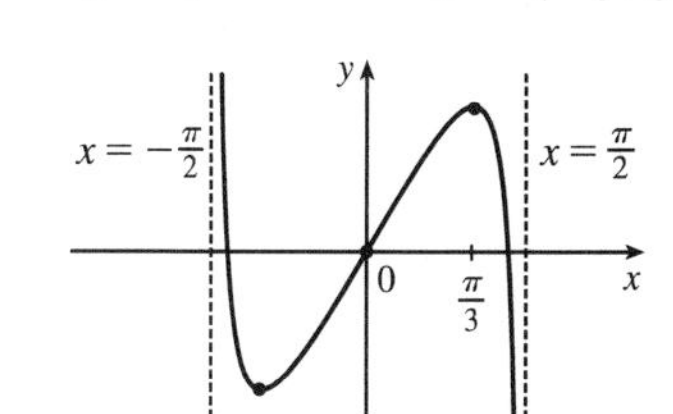

19. $f(x) = \dfrac{x^2 - 1}{x^3} \Rightarrow f'(x) = \dfrac{x^3(2x) - (x^2 - 1)3x^2}{x^6} = \dfrac{3 - x^2}{x^4} \Rightarrow$

$$f''(x) = \frac{x^4(-2x) - (3 - x^2)4x^3}{x^8} = \frac{2x^2 - 12}{x^5}$$

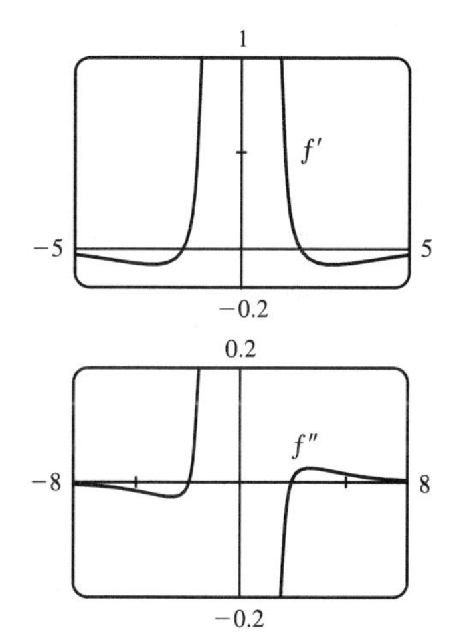

Estimates: From the graphs of f' and f'', it appears that f is increasing on $(-1.73, 0)$ and $(0, 1.73)$ and decreasing on $(-\infty, -1.73)$ and $(1.73, \infty)$; f has a local maximum of about $f(1.73) = 0.38$ and a local minimum of about $f(-1.7) = -0.38$; f is CU on $(-2.45, 0)$ and $(2.45, \infty)$, and CD on $(-\infty, -2.45)$ and $(0, 2.45)$; and f has IPs at about $(-2.45, -0.34)$ and $(2.45, 0.34)$.

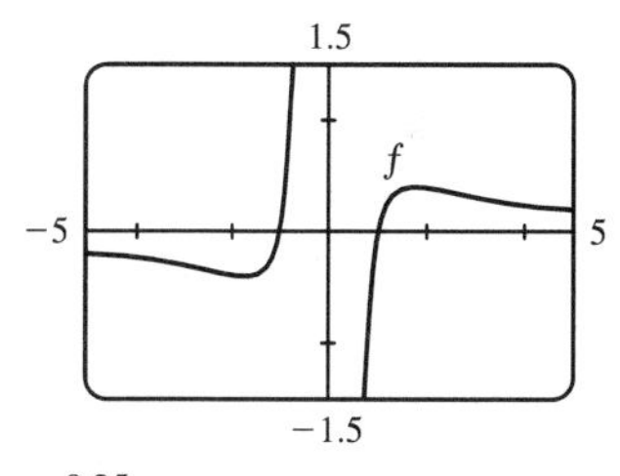

Exact: Now $f'(x) = \dfrac{3 - x^2}{x^4}$ is positive for $0 < x^2 < 3$, that is, f is increasing on $\left(-\sqrt{3}, 0\right)$ and $\left(0, \sqrt{3}\right)$; and $f'(x)$ is negative (and so f is decreasing) on $\left(-\infty, -\sqrt{3}\right)$ and $\left(\sqrt{3}, \infty\right)$. $f'(x) = 0$ when $x = \pm\sqrt{3}$.

f' goes from positive to negative at $x = \sqrt{3}$, so f has a local maximum of $f\left(\sqrt{3}\right) = \frac{\left(\sqrt{3}\right)^2 - 1}{\left(\sqrt{3}\right)^3} = \frac{2\sqrt{3}}{9}$; and since f is odd, we know that maxima on the interval $(0, \infty)$ correspond to minima on $(-\infty, 0)$, so f has a local minimum of $f\left(-\sqrt{3}\right) = -\frac{2\sqrt{3}}{9}$. Also, $f''(x) = \dfrac{2x^2 - 12}{x^5}$ is positive (so f is CU) on $\left(-\sqrt{6}, 0\right)$ and $\left(\sqrt{6}, \infty\right)$, and negative (so f is CD) on $\left(-\infty, -\sqrt{6}\right)$ and $\left(0, \sqrt{6}\right)$. There are IP at $\left(\sqrt{6}, \frac{5\sqrt{6}}{36}\right)$ and $\left(-\sqrt{6}, -\frac{5\sqrt{6}}{36}\right)$.

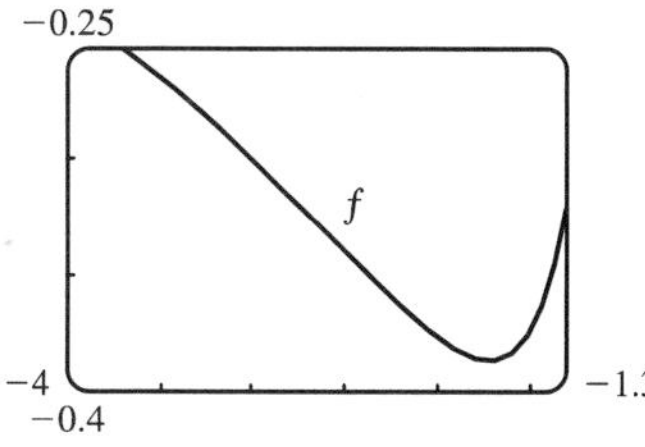

20. $f(x) = \dfrac{\sqrt[3]{x}}{1-x} = x^{1/3}(1-x)^{-1} \;\Rightarrow\; f'(x) = x^{1/3}(-1)(1-x)^{-2}(-1) + (1-x)^{-1}\left(\frac{1}{3}\right)x^{-2/3} = \dfrac{x^{-2/3}}{3}\,\dfrac{1+2x}{(x-1)^2} \;\Rightarrow$

$$f''(x) = \frac{x^{-2/3}}{3}\,\frac{(x-1)^2(2)-(1+2x)(2)(x-1)}{(x-1)^4} + \frac{1+2x}{(x-1)^2}\left(\frac{-2x^{-5/3}}{9}\right) = -\frac{2x^{-5/3}}{9}\,\frac{5x^2+5x-1}{(x-1)^3}$$

From the graphs, it appears that f is increasing on $(-0.50, 1)$ and $(1, \infty)$, with a vertical asymptote at $x = 1$, and decreasing on $(-\infty, -0.50)$; f has no local maximum, but a local minimum of about $f(-0.50) = -0.53$; f is CU on $(-1.17, 0)$ and $(0.17, 1)$ and CD on $(-\infty, -1.17)$, $(0, 0.17)$ and $(1, \infty)$; and f has inflection points at about $(-1.17, -0.49)$, $(0, 0)$ and $(0.17, 0.67)$. Note also that $\lim\limits_{x\to\pm\infty} f(x) = 0$, so $y = 0$ is a horizontal asymptote.

21. $f(x) = 3x^6 - 5x^5 + x^4 - 5x^3 - 2x^2 + 2 \;\Rightarrow\; f'(x) = 18x^5 - 25x^4 + 4x^3 - 15x^2 - 4x \;\Rightarrow$

$f''(x) = 90x^4 - 100x^3 + 12x^2 - 30x - 4$

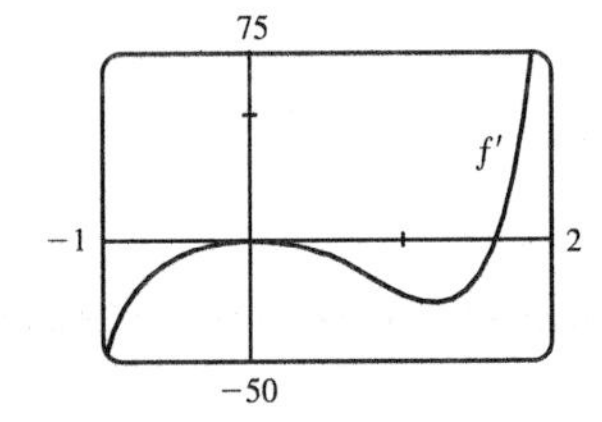

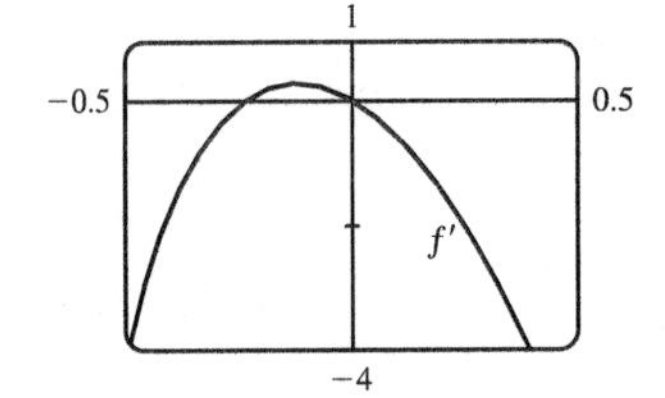

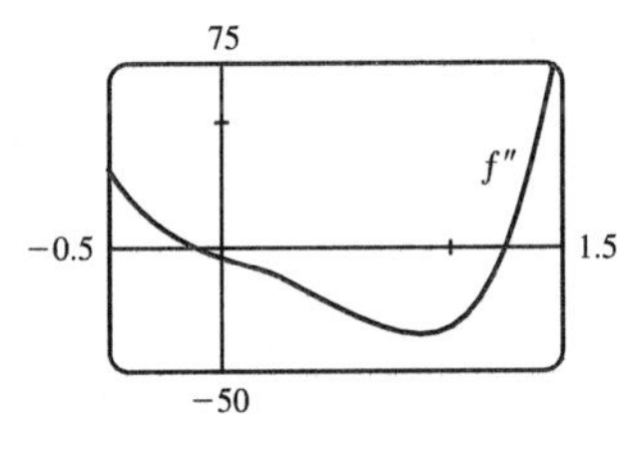

From the graphs of f' and f'', it appears that f is increasing on $(-0.23, 0)$ and $(1.62, \infty)$ and decreasing on $(-\infty, -0.23)$ and $(0, 1.62)$; f has a local maximum of about $f(0) = 2$ and local minima of about $f(-0.23) = 1.96$ and $f(1.62) = -19.2$; f is CU on $(-\infty, -0.12)$ and $(1.24, \infty)$ and CD on $(-0.12, 1.24)$; and f has inflection points at about $(-0.12, 1.98)$ and $(1.24, -12.1)$.

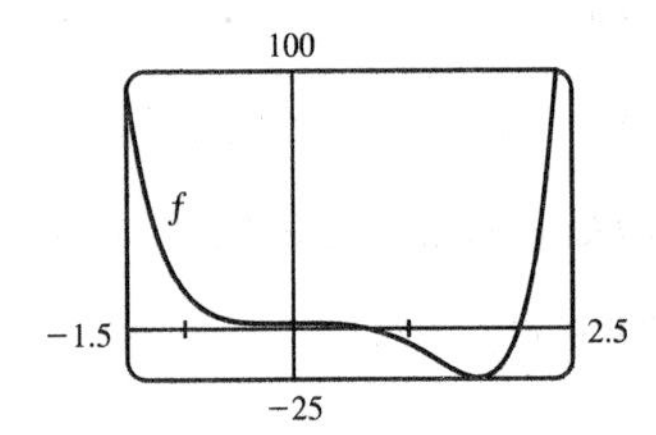

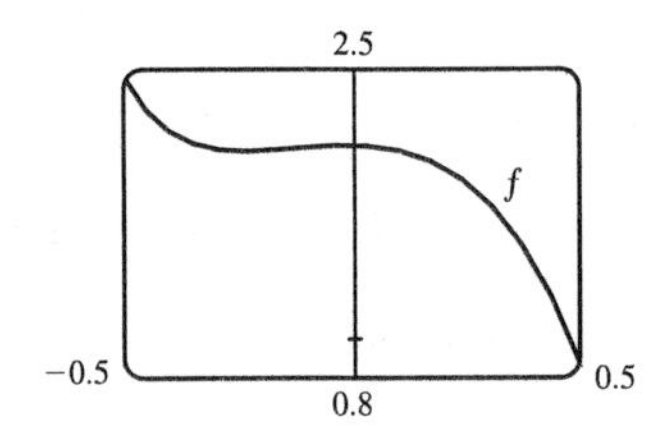

22. $f(x) = \sin x \cos^2 x \quad \Rightarrow \quad f'(x) = \cos^3 x - 2\sin^2 x \, \cos x \quad \Rightarrow \quad f''(x) = -7 \sin x \, \cos^2 x + 2 \sin^3 x$

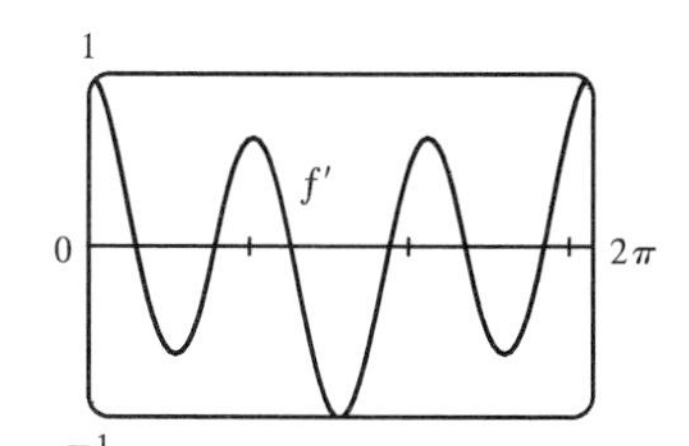

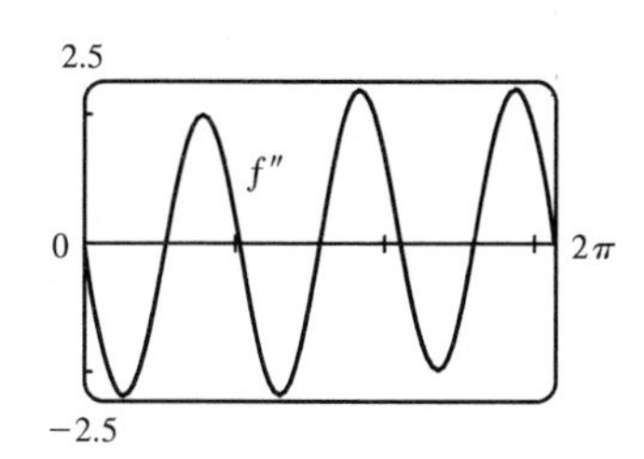

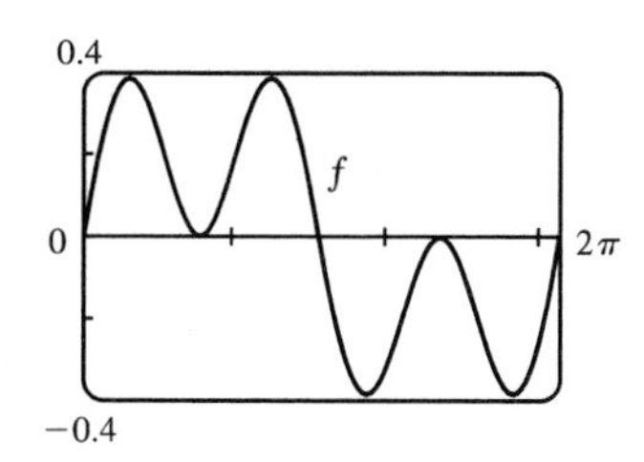

From the graphs of f' and f'', it appears that f is increasing on $(0, 0.62)$, $(1.57, 2.53)$, $(3.76, 4.71)$ and $(5.67, 2\pi)$ and decreasing on $(0.62, 1.57)$, $(2.53, 3.76)$ and $(4.71, 5.67)$; f has local maxima of about $f(0.62) = f(2.53) = 0.38$ and $f(4.71) = 0$ and local minima of about $f(1.57) = 0$ and $f(3.76) = f(5.67) = -0.38$; f is CU on $(1.08, 2.06)$, $(3.14, 4.22)$ and $(5.20, 2\pi)$ and CD on $(0, 1.08)$, $(2.06, 3.14)$ and $(4.22, 5.20)$; and f has inflection points at about $(0, 0)$, $(1.08, 0.20)$, $(2.06, 0.20)$, $(3.14, 0)$, $(4.22, -0.20)$, $(5.20, -0.20)$ and $(2\pi, 0)$.

23. $f(x) = x^{101} + x^{51} + x - 1 = 0$. Since f is continuous and $f(0) = -1$ and $f(1) = 2$, the equation has at least one root in $(0, 1)$, by the Intermediate Value Theorem. Suppose the equation has two roots, a and b, with $a < b$. Then $f(a) = 0 = f(b)$, so by the Mean Value Theorem, there is a number x in (a, b) such that

$f'(x) = \dfrac{f(b) - f(a)}{b - a} = \dfrac{0}{b - a} = 0$, so f' has a root in (a, b). But this is impossible since

$f'(x) = 101x^{100} + 51x^{50} + 1 \geq 1$ for all x.

24. By the Mean Value Theorem, $f'(c) = \dfrac{f(4) - f(0)}{4 - 0} \quad \Leftrightarrow \quad 4f'(c) = f(4) - 1$ for some c with $0 < c < 4$. Since

$2 \leq f'(c) \leq 5$, we have $4(2) \leq 4f'(c) \leq 4(5) \quad \Leftrightarrow \quad 4(2) \leq f(4) - 1 \leq 4(5) \quad \Leftrightarrow \quad 8 \leq f(4) - 1 \leq 20 \quad \Leftrightarrow$

$9 \leq f(4) \leq 21$.

25. Since f is continuous on $[32, 33]$ and differentiable on $(32, 33)$, then by the Mean Value Theorem there exists a number c in

$(32, 33)$ such that $f'(c) = \frac{1}{5}c^{-4/5} = \dfrac{\sqrt[5]{33} - \sqrt[5]{32}}{33 - 32} = \sqrt[5]{33} - 2$, but $\frac{1}{5}c^{-4/5} > 0 \quad \Rightarrow \quad \sqrt[5]{33} - 2 > 0 \quad \Rightarrow \quad \sqrt[5]{33} > 2$. Also

f' is decreasing, so that $f'(c) < f'(32) = \frac{1}{5}(32)^{-4/5} = 0.0125 \quad \Rightarrow \quad 0.0125 > f'(c) = \sqrt[5]{33} - 2 \quad \Rightarrow \quad \sqrt[5]{33} < 2.0125$.

Therefore, $2 < \sqrt[5]{33} < 2.0125$.

26. For $(1, 6)$ to be on the curve $y = x^3 + ax^2 + bx + 1$, we have that $6 = 1 + a + b + 1 \quad \Rightarrow \quad b = 4 - a$. Now

$y' = 3x^2 + 2ax + b$ and $y'' = 6x + 2a$. Also, for $(1, 6)$ to be an IP it must be true that $y''(1) = 6(1) + 2a = 0 \quad \Rightarrow$

$a = -3 \quad \Rightarrow \quad b = 4 - (-3) = 7$. Note that with $a = -3$, we have $y'' = 6x - 6 = 6(x - 1)$, so y'' changes sign at $x = 1$, proving that $(1, 6)$ is an IP. [This does not follow from the fact that $y''(1) = 0$.]

27. Call the two integers x and y. Then $x + 4y = 1000$, so $x = 1000 - 4y$. Their product is $P = xy = (1000 - 4y)y$, so our problem is to maximize the function $P(y) = 1000y - 4y^2$, where $0 < y < 250$ and y is an integer. $P'(y) = 1000 - 8y$, so $P'(y) = 0 \quad \Leftrightarrow \quad y = 125$. $\quad P''(y) = -8 < 0$, so $P(125) = 62{,}500$ is an absolute maximum. Since the optimal y turned out to be an integer, we have found the desired pair of numbers, namely $x = 1000 - 4(125) = 500$ and $y = 125$.

28. On the hyperbola $xy = 8$, if $d(x)$ is the distance from the point $(x, y) = (x, 8/x)$ to the point $(3, 0)$, then $[d(x)]^2 = (x-3)^2 + 64/x^2 = f(x)$. $\quad f'(x) = 2(x-3) - 128/x^3 = 0 \quad \Rightarrow \quad x^4 - 3x^3 - 64 = 0 \quad \Rightarrow$ $(x-4)(x^3 + x^2 + 4x + 16) = 0 \quad \Rightarrow \quad x = 4$ since the solution must have $x > 0$. Then $y = \frac{8}{4} = 2$, so the point is $(4, 2)$.

29. By similar triangles, $\dfrac{y}{x} = \dfrac{r}{\sqrt{x^2 - 2rx}}$, so the area of the triangle is

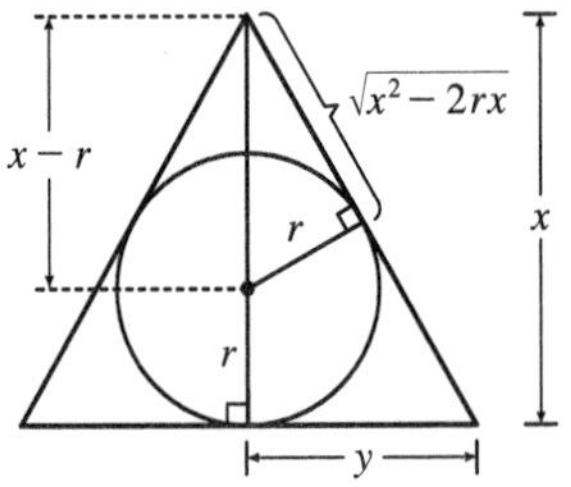

$$A(x) = \tfrac{1}{2}(2y)x = xy = \frac{rx^2}{\sqrt{x^2 - 2rx}} \quad \Rightarrow$$

$$A'(x) = \frac{2rx\sqrt{x^2 - 2rx} - rx^2(x - r)/\sqrt{x^2 - 2rx}}{x^2 - 2rx} = \frac{rx^2(x - 3r)}{(x^2 - 2rx)^{3/2}} = 0$$

when $x = 3r$. $\quad A'(x) < 0$ when $2r < x < 3r$, $A'(x) > 0$ when $x > 3r$. So $x = 3r$ gives a minimum and $A(3r) = r(9r^2)/(\sqrt{3}\,r) = 3\sqrt{3}\,r^2$.

30. The volume of the cone is $V = \frac{1}{3}\pi y^2(r + x) = \frac{1}{3}\pi(r^2 - x^2)(r + x)$, $-r \leq x \leq r$.

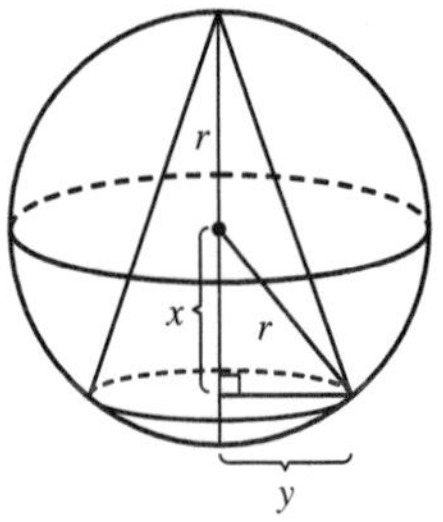

$$V'(x) = \tfrac{\pi}{3}\left[(r^2 - x^2)(1) + (r + x)(-2x)\right] = \tfrac{\pi}{3}\left[(r + x)(r - x - 2x)\right]$$

$$= \tfrac{\pi}{3}(r + x)(r - 3x) = 0 \text{ when } x = -r \text{ or } x = r/3.$$

Now $V(r) = 0 = V(-r)$, so the maximum occurs at $x = r/3$ and the volume is $V\left(\dfrac{r}{3}\right) = \dfrac{\pi}{3}\left(r^2 - \dfrac{r^2}{9}\right)\left(\dfrac{4r}{3}\right) = \dfrac{32\pi r^3}{81}$.

31. We minimize $L(x) = |PA| + |PB| + |PC| = 2\sqrt{x^2 + 16} + (5 - x)$, $0 \leq x \leq 5$. $\; L'(x) = 2x/\sqrt{x^2 + 16} - 1 = 0 \quad \Leftrightarrow \quad 2x = \sqrt{x^2 + 16} \quad \Leftrightarrow$ $4x^2 = x^2 + 16 \quad \Leftrightarrow \quad x = \frac{4}{\sqrt{3}}$. $\; L(0) = 13$, $L\left(\frac{4}{\sqrt{3}}\right) \approx 11.9$, $L(5) \approx 12.8$, so the minimum occurs when $x = \frac{4}{\sqrt{3}} \approx 2.3$.

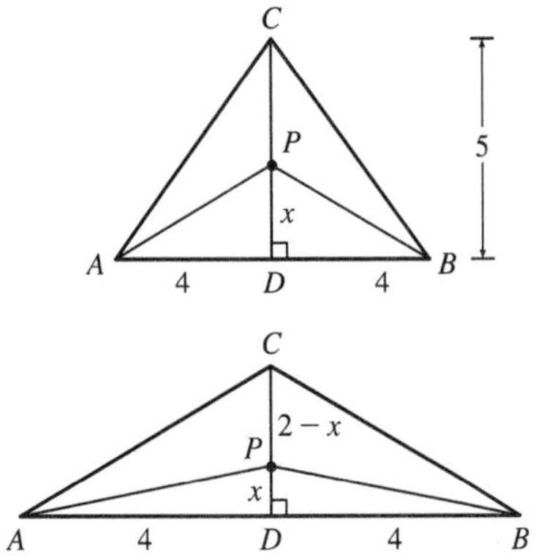

If $|CD| = 2$, $L(x)$ changes from $(5 - x)$ to $(2 - x)$ with $0 \leq x \leq 2$. But we still get $L'(x) = 0 \quad \Leftrightarrow \quad x = \frac{4}{\sqrt{3}}$, which isn't in the interval $[0, 2]$. Now $L(0) = 10$ and $L(2) = 2\sqrt{20} = 4\sqrt{5} \approx 8.9$. The minimum occurs when $P = C$.

32. 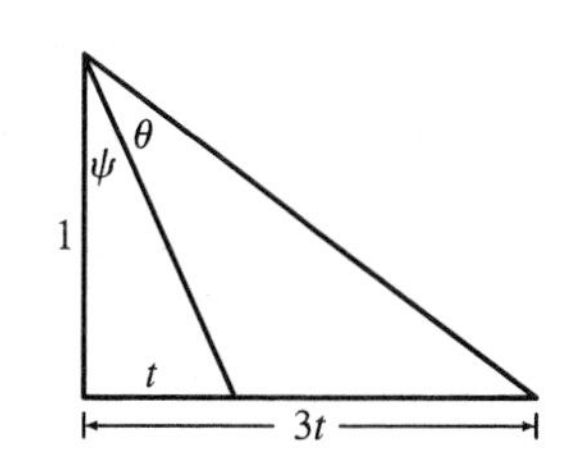

It suffices to maximize $\tan\theta$. Now

$$\frac{3t}{1} = \tan(\psi + \theta) = \frac{\tan\psi + \tan\theta}{1 - \tan\psi\,\tan\theta} = \frac{t + \tan\theta}{1 - t\tan\theta}.$$

So $3t(1 - t\tan\theta) = t + \tan\theta \;\Rightarrow\; 2t = \left(1 + 3t^2\right)\tan\theta \;\Rightarrow$

$\tan\theta = \dfrac{2t}{1+3t^2}$. Let $f(t) = \tan\theta = \dfrac{2t}{1+3t^2} \;\Rightarrow$

$f'(t) = \dfrac{2(1+3t^2) - 2t(6t)}{(1+3t^2)^2} = \dfrac{2(1-3t^2)}{(1+3t^2)^2} = 0 \;\Leftrightarrow\; 1 - 3t^2 = 0 \;\Leftrightarrow\; t = \frac{1}{\sqrt{3}}$ since $t \ge 0$. Now $f'(t) > 0$ for $0 \le t < \frac{1}{\sqrt{3}}$ and $f'(t) < 0$ for $t > \frac{1}{\sqrt{3}}$, so f has an absolute maximum when $t = \frac{1}{\sqrt{3}}$ and

$\tan\theta = \dfrac{2\left(1/\sqrt{3}\right)}{1 + 3\left(1/\sqrt{3}\right)^2} = \dfrac{1}{\sqrt{3}} \;\Rightarrow\; \theta = \frac{\pi}{6}$. Substituting for t and θ in $3t = \tan(\psi + \theta)$ gives

us $\sqrt{3} = \tan\left(\psi + \frac{\pi}{6}\right) \;\Rightarrow\; \psi = \frac{\pi}{6}$.

33. $v = K\sqrt{\dfrac{L}{C} + \dfrac{C}{L}} \;\Rightarrow\; \dfrac{dv}{dL} = \dfrac{K}{2\sqrt{(L/C) + (C/L)}}\left(\dfrac{1}{C} - \dfrac{C}{L^2}\right) = 0 \;\Leftrightarrow\; \dfrac{1}{C} = \dfrac{C}{L^2} \;\Leftrightarrow\; L^2 = C^2 \;\Leftrightarrow\; L = C.$

This gives the minimum velocity since $v' < 0$ for $0 < L < C$ and $v' > 0$ for $L > C$.

34. 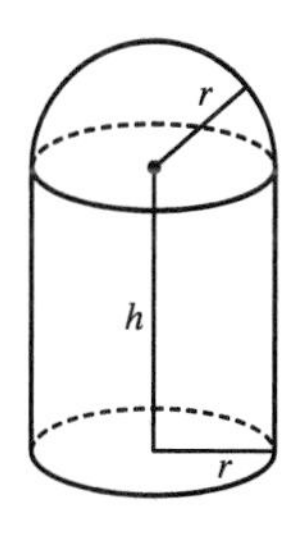

We minimize the surface area $S = \pi r^2 + 2\pi rh + \frac{1}{2}(4\pi r^2) = 3\pi r^2 + 2\pi rh$.

Solving $V = \pi r^2 h + \frac{2}{3}\pi r^3$ for h, we get $h = \dfrac{V - \frac{2}{3}\pi r^3}{\pi r^2} = \dfrac{V}{\pi r^2} - \frac{2}{3}r$, so

$S(r) = 3\pi r^2 + 2\pi r\left[\dfrac{V}{\pi r^2} - \frac{2}{3}r\right] = \frac{5}{3}\pi r^2 + \dfrac{2V}{r}$.

$S'(r) = -\dfrac{2V}{r^2} + \frac{10}{3}\pi r = \dfrac{\frac{10}{3}\pi r^3 - 2V}{r^2} = 0 \;\Leftrightarrow\; \frac{10}{3}\pi r^3 = 2V \;\Leftrightarrow r^3 = \dfrac{3V}{5\pi} \;\Leftrightarrow$

$r = \sqrt[3]{\dfrac{3V}{5\pi}}$. This gives an absolute minimum since $S'(r) < 0$ for $0 < r < \sqrt[3]{\dfrac{3V}{5\pi}}$ and $S'(r) > 0$ for $r > \sqrt[3]{\dfrac{3V}{5\pi}}$.

Thus, $h = \dfrac{V - \frac{2}{3}\pi \cdot \dfrac{3V}{5\pi}}{\pi\sqrt[3]{\dfrac{(3V)^2}{(5\pi)^2}}} = \dfrac{\left(V - \frac{2}{5}V\right)\sqrt[3]{(5\pi)^2}}{\pi\sqrt[3]{(3V)^2}} = \dfrac{3V\sqrt[3]{(5\pi)^2}}{5\pi\sqrt[3]{(3V)^2}} = \sqrt[3]{\dfrac{3V}{5\pi}} = r.$

35. Let x denote the number of \$1 decreases in ticket price. Then the ticket price is $\$12 - \$1(x)$, and the average attendance is $11{,}000 + 1000(x)$. Now the revenue per game is

$$\begin{aligned} R(x) &= (\text{price per person}) \times (\text{number of people per game}) \\ &= (12 - x)(11{,}000 + 1000x) = -1000x^2 + 1000x + 132{,}000 \end{aligned}$$

for $0 \le x \le 4$ [since the seating capacity is 15,000] $\Rightarrow$ $R'(x) = -2000x + 1000 = 0 \;\Leftrightarrow\; x = 0.5$. This is a maximum since $R''(x) = -2000 < 0$ for all x. Now we must check the value of $R(x) = (12 - x)(11{,}000 + 1000x)$ at $x = 0.5$ and at the endpoints of the domain to see which value of x gives the maximum value of R.
$R(0) = (12)(11{,}000) = 132{,}000$, $R(0.5) = (11.5)(11{,}500) = 132{,}250$, and $R(4) = (8)(15{,}000) = 120{,}000$. Thus, the maximum revenue of \$132,250 per game occurs when the average attendance is 11,500 and the ticket price is \$11.50.

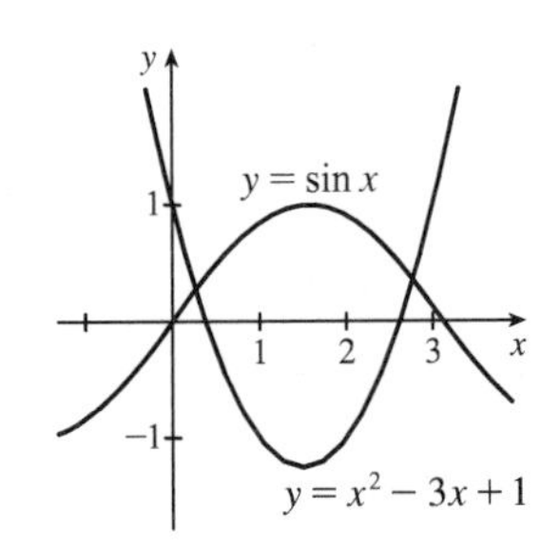

36. Graphing $y = \sin x$ and $y = x^2 - 3x + 1$ shows that there are two roots, one about 0.3 and the other about 2.8. $f(x) = \sin x - x^2 + 3x - 1 \Rightarrow$

$$f'(x) = \cos x - 2x + 3 \Rightarrow x_{n+1} = x_n - \frac{\sin x_n - x_n^2 + 3x_n - 1}{\cos x_n - 2x_n + 3}.$$

Now $x_1 = 0.3 \Rightarrow x_2 \approx 0.268552 \Rightarrow x_3 \approx 0.268881 \approx x_4$ and $x_1 = 2.8 \Rightarrow x_2 \approx 2.770354 \Rightarrow x_3 \approx 2.770058 \approx x_4$, so to six decimal places, the roots are 0.268881 and 2.770058.

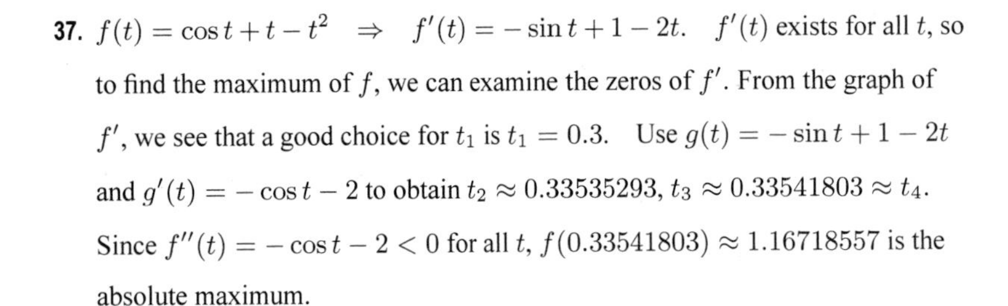

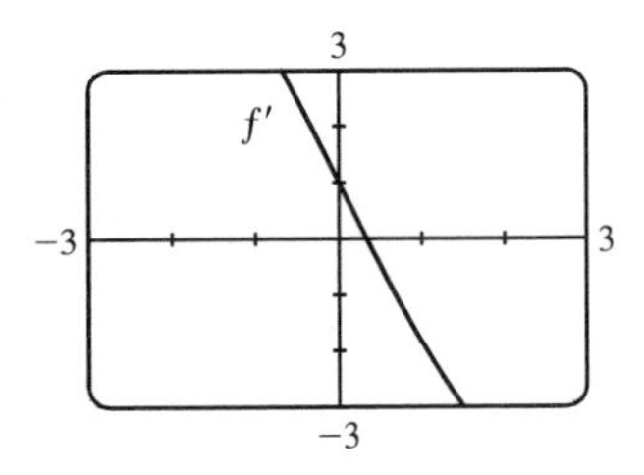

37. $f(t) = \cos t + t - t^2 \Rightarrow f'(t) = -\sin t + 1 - 2t$. $f'(t)$ exists for all t, so to find the maximum of f, we can examine the zeros of f'. From the graph of f', we see that a good choice for t_1 is $t_1 = 0.3$. Use $g(t) = -\sin t + 1 - 2t$ and $g'(t) = -\cos t - 2$ to obtain $t_2 \approx 0.33535293$, $t_3 \approx 0.33541803 \approx t_4$. Since $f''(t) = -\cos t - 2 < 0$ for all t, $f(0.33541803) \approx 1.16718557$ is the absolute maximum.

38. $y = f(x) = x \sin x$, $0 \le x \le 2\pi$. **A.** $D = [0, 2\pi]$ **B.** y-intercept: $f(0) = 0$; x-intercepts: $f(x) = 0 \Leftrightarrow x = 0$ or $\sin x = 0 \Leftrightarrow x = 0, \pi$, or 2π. **C.** There is no symmetry on D, but if f is defined for all real numbers x, then f is an even function. **D.** No asymptote **E.** $f'(x) = x \cos x + \sin x$. To find critical numbers in $(0, 2\pi)$, we graph f' and see that there are two critical numbers, about 2 and 4.9. To find them more precisely, we use Newton's method, setting $g(x) = f'(x) = x \cos x + \sin x$, so that $g'(x) = f''(x) = 2 \cos x - x \sin x$ and $x_{n+1} = x_n - \dfrac{x_n \cos x_n + \sin x_n}{2 \cos x_n - x_n \sin x_n}$.

$x_1 = 2 \Rightarrow x_2 \approx 2.029048$, $x_3 \approx 2.028758 \approx x_4$ and $x_1 = 4.9 \Rightarrow x_2 \approx 4.913214$, $x_3 \approx 4.913180 \approx x_4$, so the critical numbers, to six decimal places, are $r_1 = 2.028758$ and $r_2 = 4.913180$. By checking sample values of f' in $(0, r_1)$, (r_1, r_2), and $(r_2, 2\pi)$, we see that f is increasing on $(0, r_1)$, decreasing on (r_1, r_2), and increasing on $(r_2, 2\pi)$.

F. Local maximum value $f(r_1) \approx 1.819706$, local minimum value $f(r_2) \approx -4.814470$.

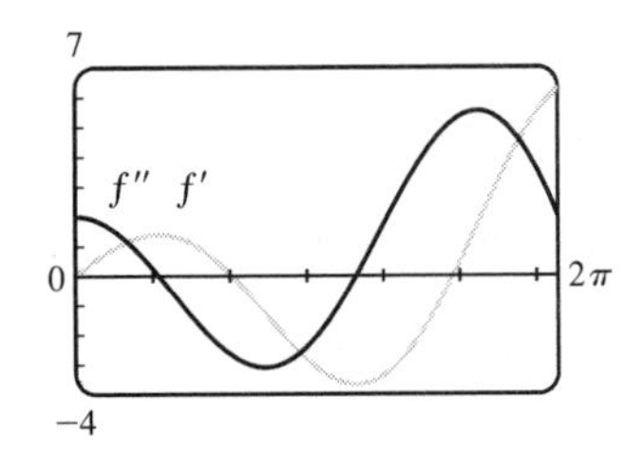

G. $f''(x) = 2 \cos x - x \sin x$. To find points where $f''(x) = 0$, we graph f'' and find that $f''(x) = 0$ at about 1 and 3.6. To find the values more precisely, we use Newton's method. Set $h(x) = f''(x) = 2 \cos x - x \sin x$. Then

$$h'(x) = -3 \sin x - x \cos x, \text{ so } x_{n+1} = x_n - \frac{2 \cos x_n - x_n \sin x_n}{-3 \sin x_n - x_n \cos x_n}.$$

$x_1 = 1 \Rightarrow x_2 \approx 1.078028$, $x_3 \approx 1.076874 \approx x_4$ and $x_1 = 3.6 \Rightarrow x_2 \approx 3.643996$, $x_3 \approx 3.643597 \approx x_4$, so the zeros of f'', to six decimal places, are $r_3 = 1.076874$ and $r_4 = 3.643597$. By checking sample values of f'' in $(0, r_3)$, (r_3, r_4), and $(r_4, 2\pi)$, we see that f is CU on $(0, r_3)$, CD on (r_3, r_4), and CU on $(r_4, 2\pi)$. f has IPs at $(r_3, f(r_3) \approx 0.948166)$ and $(r_4, f(r_4) \approx -1.753240)$.

H.

39. $f'(x) = \sqrt{x^5} - 4/\sqrt[5]{x} = x^{5/2} - 4x^{-1/5} \quad\Rightarrow\quad f(x) = \frac{2}{7}x^{7/2} - 4\left(\frac{5}{4}x^{4/5}\right) + C = \frac{2}{7}x^{7/2} - 5x^{4/5} + C$

40. $f'(x) = 8x - 3\sec^2 x \quad\Rightarrow\quad f(x) = 8\left(\frac{1}{2}x^2\right) - 3\tan x + C_n = 4x^2 - 3\tan x + C_n$ on the interval $\left(n\pi - \frac{\pi}{2}, n\pi + \frac{\pi}{2}\right)$.

41. $f'(t) = 2t - 3\sin t \quad\Rightarrow\quad f(t) = t^2 + 3\cos t + C$.

$f(0) = 3 + C$ and $f(0) = 5 \quad\Rightarrow\quad C = 2$, so $f(t) = t^2 + 3\cos t + 2$.

42. $f'(u) = \dfrac{u^2 + \sqrt{u}}{u} = u + u^{-1/2} \quad\Rightarrow\quad f(u) = \frac{1}{2}u^2 + 2u^{1/2} + C$.

$f(1) = \frac{1}{2} + 2 + C$ and $f(1) = 3 \quad\Rightarrow\quad C = \frac{1}{2}$, so $f(u) = \frac{1}{2}u^2 + 2\sqrt{u} + \frac{1}{2}$.

43. $f''(x) = 1 - 6x + 48x^2 \quad\Rightarrow\quad f'(x) = x - 3x^2 + 16x^3 + C$. $f'(0) = C$ and $f'(0) = 2 \quad\Rightarrow\quad C = 2$,

so $f'(x) = x - 3x^2 + 16x^3 + 2$ and hence, $f(x) = \frac{1}{2}x^2 - x^3 + 4x^4 + 2x + D$.

$f(0) = D$ and $f(0) = 1 \quad\Rightarrow\quad D = 1$, so $f(x) = \frac{1}{2}x^2 - x^3 + 4x^4 + 2x + 1$.

44. $f''(x) = 2x^3 + 3x^2 - 4x + 5 \quad\Rightarrow\quad f'(x) = \frac{1}{2}x^4 + x^3 - 2x^2 + 5x + C \quad\Rightarrow$

$f(x) = \frac{1}{10}x^5 + \frac{1}{4}x^4 - \frac{2}{3}x^3 + \frac{5}{2}x^2 + Cx + D$. $f(0) = D$ and $f(0) = 2 \quad\Rightarrow\quad D = 2$.

$f(1) = \frac{1}{10} + \frac{1}{4} - \frac{2}{3} + \frac{5}{2} + C + 2$ and $f(1) = 0 \quad\Rightarrow\quad C = -\frac{6}{60} - \frac{15}{60} + \frac{40}{60} - \frac{150}{60} - \frac{120}{60} = -\frac{251}{60}$, so

$f(x) = \frac{1}{10}x^5 + \frac{1}{4}x^4 - \frac{2}{3}x^3 + \frac{5}{2}x^2 - \frac{251}{60}x + 2$.

45. $a(t) = v'(t) = t - 2 \quad\Rightarrow\quad v(t) = \frac{1}{2}t^2 - 2t + C$. $v(0) = C$ and $v(0) = 3 \quad\Rightarrow\quad C = 3$, so $v(t) = \frac{1}{2}t^2 - 2t + 3$

and $s(t) = \frac{1}{6}t^3 - t^2 + 3t + D$. $s(0) = D$ and $s(0) = 1 \quad\Rightarrow\quad D = 1$, and $s(t) = \frac{1}{6}t^3 - t^2 + 3t + 1$.

46. $a(t) = v'(t) = \cos t + \sin t \quad\Rightarrow\quad v(t) = \sin t - \cos t + C \quad\Rightarrow\quad 5 = v(0) = -1 + C \quad\Rightarrow\quad C = 6$, so

$v(t) = \sin t - \cos t + 6 \quad\Rightarrow\quad s(t) = -\cos t - \sin t + 6t + D \quad\Rightarrow\quad 0 = s(0) = -1 + D \quad\Rightarrow\quad D = 1$, so

$s(t) = -\cos t - \sin t + 6t + 1$.

47. Choosing the positive direction to be upward, we have $a(t) = -9.8 \quad\Rightarrow\quad v(t) = -9.8t + v_0$, but $v(0) = 0 = v_0 \quad\Rightarrow$

$v(t) = -9.8t = s'(t) \quad\Rightarrow\quad s(t) = -4.9t^2 + s_0$, but $s(0) = s_0 = 500 \quad\Rightarrow\quad s(t) = -4.9t^2 + 500$. When $s = 0$,

$-4.9t^2 + 500 = 0 \quad\Rightarrow\quad t_1 = \sqrt{\frac{500}{4.9}} \approx 10.1 \quad\Rightarrow\quad v(t_1) = -9.8\sqrt{\frac{500}{4.9}} \approx -98.995$ m/s. Since the canister has been designed to withstand an impact velocity of 100 m/s, the canister will *not burst*.

48. $f(x) = x^4 + x^3 + cx^2 \quad\Rightarrow\quad f'(x) = 4x^3 + 3x^2 + 2cx$. This is 0 when $x(4x^2 + 3x + 2c) = 0 \quad\Leftrightarrow\quad x = 0$

or $4x^2 + 3x + 2c = 0$. Using the quadratic formula, we find that the roots of this last equation are $x = \dfrac{-3 \pm \sqrt{9 - 32c}}{8}$.

Now if $9 - 32c < 0 \quad\Leftrightarrow\quad c > \frac{9}{32}$, then $(0, 0)$ is the only critical point, a minimum. If $c = \frac{9}{32}$, then there are two critical points (a minimum at $x = 0$, and a horizontal tangent with no maximum or minimum at $x = -\frac{3}{8}$) and if $c < \frac{9}{32}$, then there are three critical points except when $c = 0$, in which case the root with the $+$ sign coincides with the critical point at $x = 0$. For

$0 < c < \frac{9}{32}$, there is a minimum at $x = -\dfrac{3}{8} - \dfrac{\sqrt{9 - 32c}}{8}$, a maximum at $x = -\dfrac{3}{8} + \dfrac{\sqrt{9 - 32c}}{8}$, and a minimum at $x = 0$.

For $c = 0$, there is a minimum at $x = -\frac{3}{4}$ and a horizontal tangent with no extremum at $x = 0$, and for $c < 0$, there is a maximum at $x = 0$, and there are minima at $x = -\frac{3}{8} \pm \frac{\sqrt{9-32c}}{8}$. Now we calculate $f''(x) = 12x^2 + 6x + 2c$.

The roots of this equation are $x = \frac{-6 \pm \sqrt{36 - 4 \cdot 12 \cdot 2c}}{24}$. So if $36 - 96c \le 0 \iff c \ge \frac{3}{8}$, then there is no IP.

If $c < \frac{3}{8}$, then there are two IPs at $x = -\frac{1}{4} \pm \frac{\sqrt{9-24c}}{12}$.

Value of c	No. of CP	No. of IP
$c < 0$	3	2
$c = 0$	2	2
$0 < c < \frac{9}{32}$	3	2
$c = \frac{9}{32}$	2	2
$\frac{9}{32} < c < \frac{3}{8}$	1	2
$c \ge \frac{3}{8}$	1	0

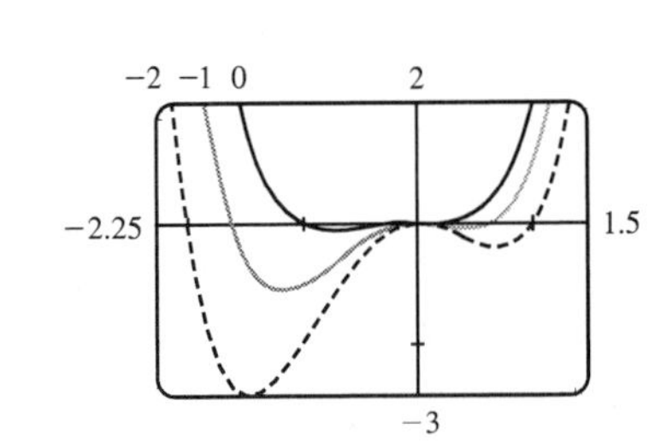

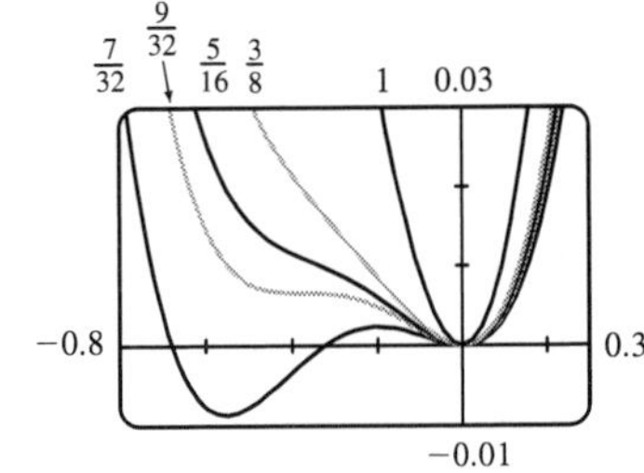

49. (a)

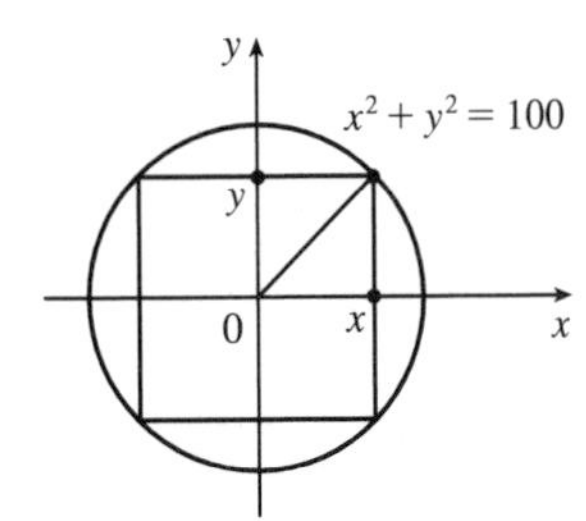

The cross-sectional area of the rectangular beam is

$A = 2x \cdot 2y = 4xy = 4x\sqrt{100 - x^2}$, $0 \le x \le 10$, so

$$\frac{dA}{dx} = 4x\left(\tfrac{1}{2}\right)\left(100 - x^2\right)^{-1/2}(-2x) + \left(100 - x^2\right)^{1/2} \cdot 4$$

$$= \frac{-4x^2}{\left(100 - x^2\right)^{1/2}} + 4(100 - x^2)^{1/2} = \frac{4[-x^2 + (100 - x^2)]}{\left(100 - x^2\right)^{1/2}}$$

$\frac{dA}{dx} = 0$ when $-x^2 + (100 - x^2) = 0 \Rightarrow x^2 = 50 \Rightarrow x = \sqrt{50} \approx 7.07 \Rightarrow y = \sqrt{100 - \left(\sqrt{50}\right)^2} = \sqrt{50}$.

Since $A(0) = A(10) = 0$, the rectangle of maximum area is a square.

(b)

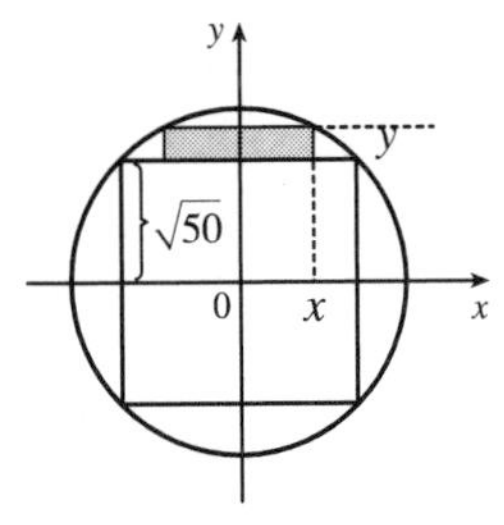

The cross-sectional area of each rectangular plank (shaded in the figure) is

$A = 2x\left(y - \sqrt{50}\right) = 2x\left[\sqrt{100 - x^2} - \sqrt{50}\right]$, $0 \le x \le \sqrt{50}$, so

$$\frac{dA}{dx} = 2\left(\sqrt{100 - x^2} - \sqrt{50}\right) + 2x\left(\tfrac{1}{2}\right)(100 - x^2)^{-1/2}(-2x)$$

$$= 2(100 - x^2)^{1/2} - 2\sqrt{50} - \frac{2x^2}{(100 - x^2)^{1/2}}$$

Set $\frac{dA}{dx} = 0$: $(100 - x^2) - \sqrt{50}\,(100 - x^2)^{1/2} - x^2 = 0 \Rightarrow 100 - 2x^2 = \sqrt{50}\,(100 - x^2)^{1/2} \Rightarrow$

$10{,}000 - 400x^2 + 4x^4 = 50(100 - x^2) \Rightarrow 4x^4 - 350x^2 + 5000 = 0 \Rightarrow 2x^4 - 175x^2 + 2500 = 0 \Rightarrow$

$x^2 = \frac{175 \pm \sqrt{10{,}625}}{4} \approx 69.52$ or $17.98 \Rightarrow x \approx 8.34$ or 4.24. But $8.34 > \sqrt{50}$, so $x_1 \approx 4.24 \Rightarrow$

$y - \sqrt{50} = \sqrt{100 - x_1^2} - \sqrt{50} \approx 1.99$. Each plank should have dimensions about $8\frac{1}{2}$ inches by 2 inches.

(c) From the figure in part (a), the width is $2x$ and the depth is $2y$, so the strength is

$S = k(2x)(2y)^2 = 8kxy^2 = 8kx(100 - x^2) = 800kx - 8kx^3$, $0 \le x \le 10$. $dS/dx = 800k - 24kx^2 = 0$ when

$24kx^2 = 800k \Rightarrow x^2 = \frac{100}{3} \Rightarrow x = \frac{10}{\sqrt{3}} \Rightarrow y = \sqrt{\frac{200}{3}} = \frac{10\sqrt{2}}{\sqrt{3}} = \sqrt{2}\,x$. Since $S(0) = S(10) = 0$, the

maximum strength occurs when $x = \frac{10}{\sqrt{3}}$. The dimensions should be $\frac{20}{\sqrt{3}} \approx 11.55$ inches by $\frac{20\sqrt{2}}{\sqrt{3}} \approx 16.33$ inches.

4 ☐ INTEGRALS

4.1 Areas and Distances

1. (a) Since f is *increasing*, we can obtain a *lower* estimate by using *left* endpoints. We are instructed to use five rectangles, so $n = 5$.

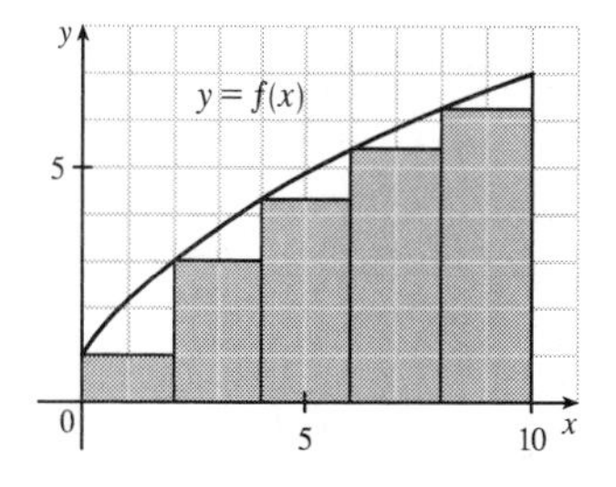

$$L_5 = \sum_{i=1}^{5} f(x_{i-1})\,\Delta x \quad \left[\Delta x = \tfrac{b-a}{n} = \tfrac{10-0}{5} = 2\right]$$

$$= f(x_0)\cdot 2 + f(x_1)\cdot 2 + f(x_2)\cdot 2 + f(x_3)\cdot 2 + f(x_4)\cdot 2$$

$$= 2\,[f(0) + f(2) + f(4) + f(6) + f(8)]$$

$$\approx 2(1 + 3 + 4.3 + 5.4 + 6.3) = 2(20) = 40$$

Since f is *increasing*, we can obtain an *upper* estimate by using *right* endpoints.

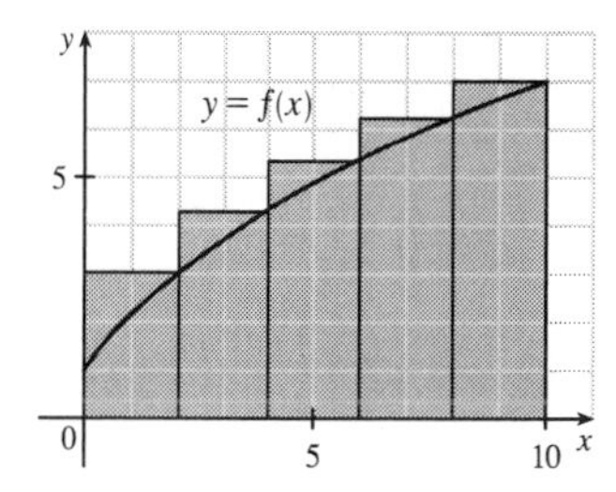

$$R_5 = \sum_{i=1}^{5} f(x_i)\,\Delta x$$

$$= 2\,[f(x_1) + f(x_2) + f(x_3) + f(x_4) + f(x_5)]$$

$$= 2\,[f(2) + f(4) + f(6) + f(8) + f(10)]$$

$$\approx 2(3 + 4.3 + 5.4 + 6.3 + 7) = 2(26) = 52$$

Comparing R_5 to L_5, we see that we have added the area of the rightmost upper rectangle, $f(10)\cdot 2$, to the sum and subtracted the area of the leftmost lower rectangle, $f(0)\cdot 2$, from the sum.

(b) $$L_{10} = \sum_{i=1}^{10} f(x_{i-1})\,\Delta x \quad \left[\Delta x = \tfrac{10-0}{10} = 1\right]$$

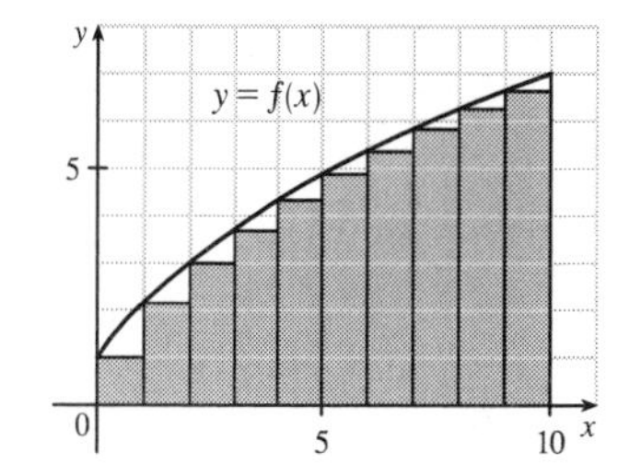

$$= 1\,[f(x_0) + f(x_1) + \cdots + f(x_9)]$$

$$= f(0) + f(1) + \cdots + f(9)$$

$$\approx 1 + 2.1 + 3 + 3.7 + 4.3 + 4.9 + 5.4 + 5.8 + 6.3 + 6.7$$

$$= 43.2$$

$$R_{10} = \sum_{i=1}^{10} f(x_i)\,\Delta x = f(1) + f(2) + \cdots + f(10)$$

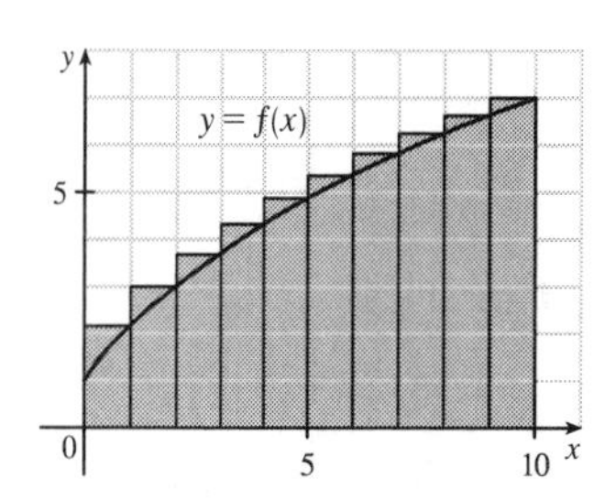

$$= L_{10} + 1\cdot f(10) - 1\cdot f(0) \quad \left[\begin{array}{c}\text{add rightmost upper rectangle,}\\ \text{subtract leftmost lower rectangle}\end{array}\right]$$

$$= 43.2 + 7 - 1 = 49.2$$

2. (a) (i) $L_6 = \sum_{i=1}^{6} f(x_{i-1})\Delta x \quad [\Delta x = \frac{12-0}{6} = 2]$

$= 2[f(x_0) + f(x_1) + f(x_2) + f(x_3) + f(x_4) + f(x_5)]$

$= 2[f(0) + f(2) + f(4) + f(6) + f(8) + f(10)]$

$\approx 2(9 + 8.8 + 8.2 + 7.3 + 5.9 + 4.1)$

$= 2(43.3) = 86.6$

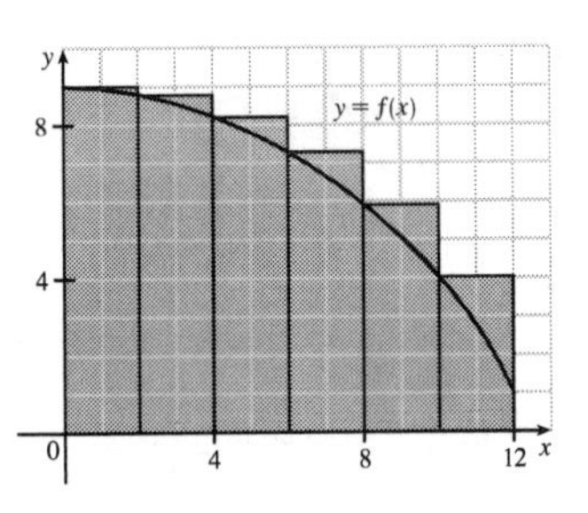

(ii) $R_6 = L_6 + 2 \cdot f(12) - 2 \cdot f(0)$

$\approx 86.6 + 2(1) - 2(9) = 70.6$

[Add area of rightmost lower rectangle and subtract area of leftmost upper rectangle.]

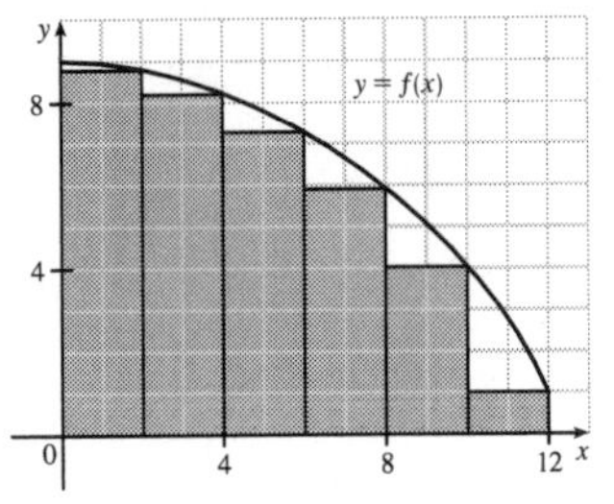

(iii) $M_6 = \sum_{i=1}^{6} f(x_i^*)\,\Delta x$

$= 2[f(1) + f(3) + f(5) + f(7) + f(9) + f(11)]$

$\approx 2(8.9 + 8.5 + 7.8 + 6.6 + 5.1 + 2.8)$

$= 2(39.7) = 79.4$

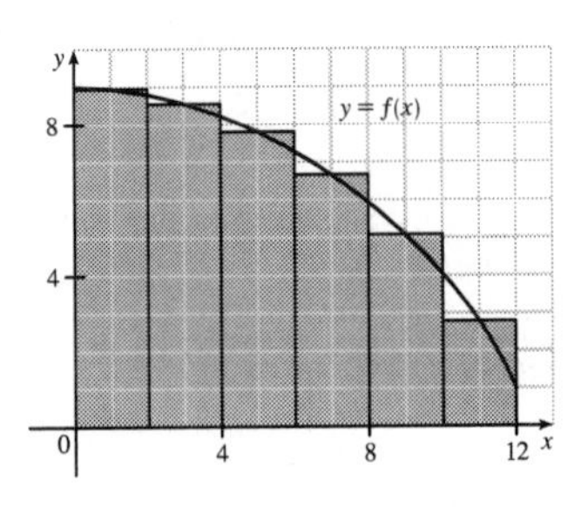

(b) Since f is *decreasing*, we obtain an *overestimate* by using *left* endpoints; that is, L_6.

(c) Since f is *decreasing*, we obtain an *underestimate* by using *right* endpoints; that is, R_6.

(d) M_6 gives the best estimate, since the area of each rectangle appears to be closer to the true area than the overestimates and underestimates in L_6 and R_6.

3. (a) $R_4 = \sum_{i=1}^{4} f(x_i)\,\Delta x \quad [\Delta x = \frac{5-1}{4} = 1]$

$= f(x_1) \cdot 1 + f(x_2) \cdot 1 + f(x_3) \cdot 1 + f(x_4) \cdot 1$

$= f(2) + f(3) + f(4) + f(5)$

$= \frac{1}{2} + \frac{1}{3} + \frac{1}{4} + \frac{1}{5} = \frac{77}{60} = 1.28\overline{3}$

Since f is *decreasing* on $[1, 5]$, an *underestimate* is obtained by using the *right* endpoint approximation, R_4.

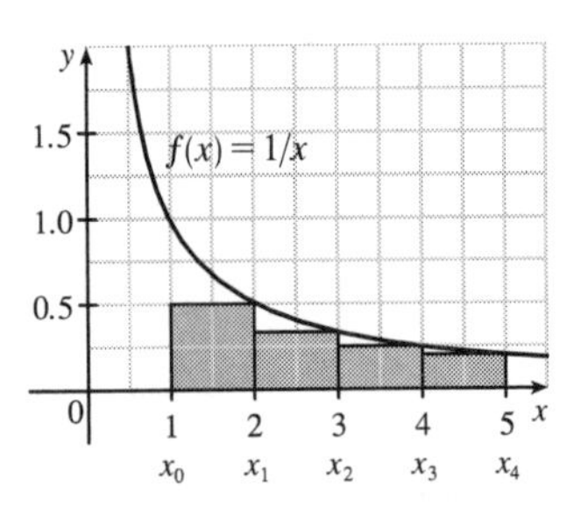

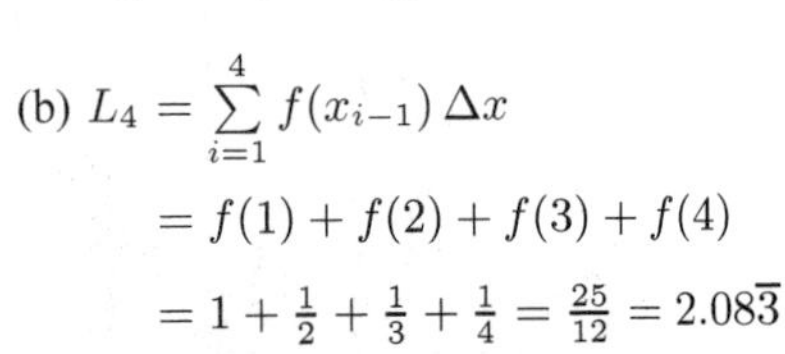

(b) $L_4 = \sum_{i=1}^{4} f(x_{i-1})\,\Delta x$

$= f(1) + f(2) + f(3) + f(4)$

$= 1 + \frac{1}{2} + \frac{1}{3} + \frac{1}{4} = \frac{25}{12} = 2.08\overline{3}$

L_4 is an overestimate. Alternatively, we could just add the area of the leftmost upper rectangle and subtract the area of the rightmost lower rectangle; that is, $L_4 = R_4 + f(1) \cdot 1 - f(5) \cdot 1$.

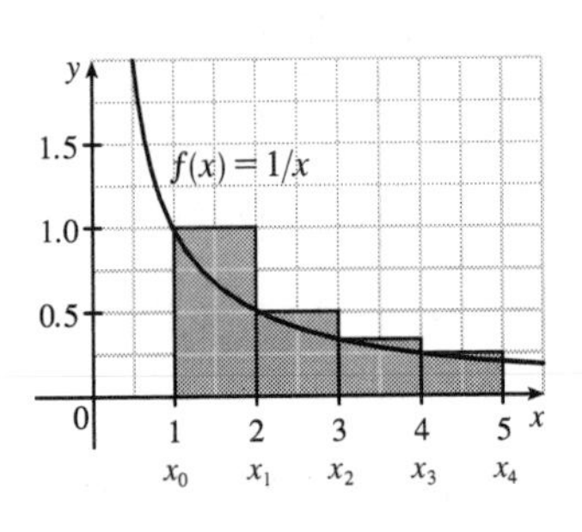

4. (a) $R_5 = \sum_{i=1}^{5} f(x_i)\,\Delta x \quad \left[\Delta x = \frac{5-0}{5} = 1\right]$

$= f(x_1)\cdot 1 + f(x_2)\cdot 1 + f(x_3)\cdot 1 + f(x_4)\cdot 1 + f(x_5)\cdot 1$

$= f(1) + f(2) + f(3) + f(4) + f(5)$

$= 24 + 21 + 16 + 9 + 0 = 70$

Since f is decreasing on $[0, 5]$, R_5 is an underestimate.

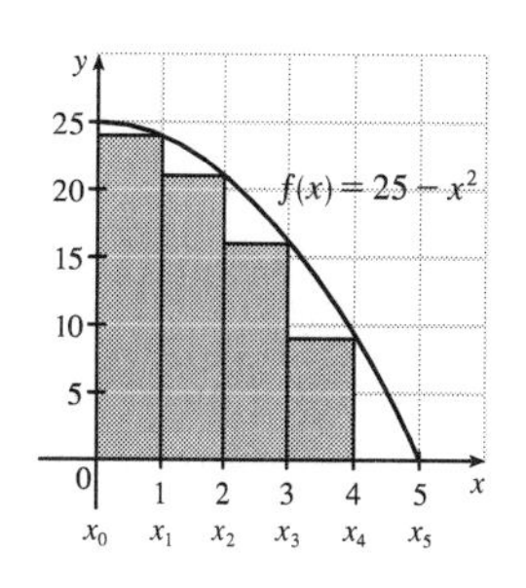

(b) $L_5 = \sum_{i=1}^{5} f(x_{i-1})\,\Delta x$

$= f(0) + f(1) + f(2) + f(3) + f(4)$

$= 25 + 24 + 21 + 16 + 9 = 95$

L_5 is an overestimate.

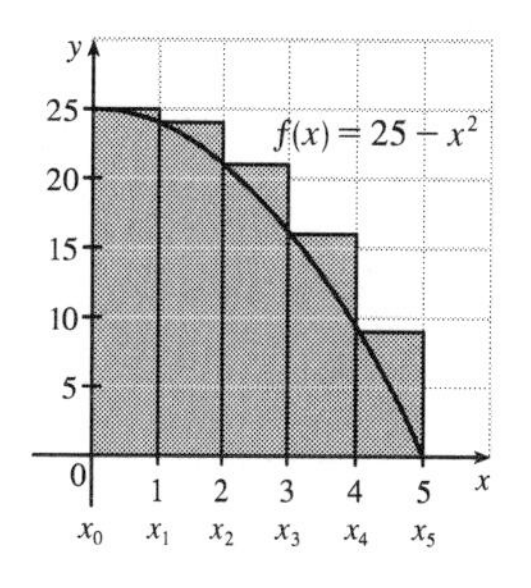

5. (a) $f(x) = 1 + x^2$ and $\Delta x = \dfrac{2-(-1)}{3} = 1 \quad\Rightarrow$

$R_3 = 1\cdot f(0) + 1\cdot f(1) + 1\cdot f(2) = 1\cdot 1 + 1\cdot 2 + 1\cdot 5 = 8.$

$\Delta x = \dfrac{2-(-1)}{6} = 0.5 \quad\Rightarrow$

$R_6 = 0.5[f(-0.5) + f(0) + f(0.5) + f(1) + f(1.5) + f(2)]$

$= 0.5(1.25 + 1 + 1.25 + 2 + 3.25 + 5)$

$= 0.5(13.75) = 6.875$

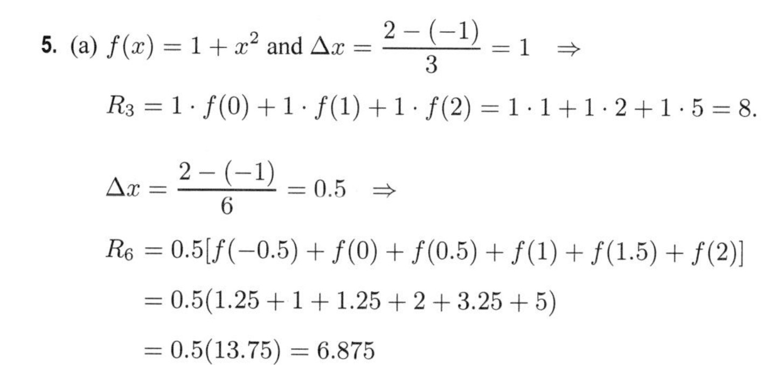

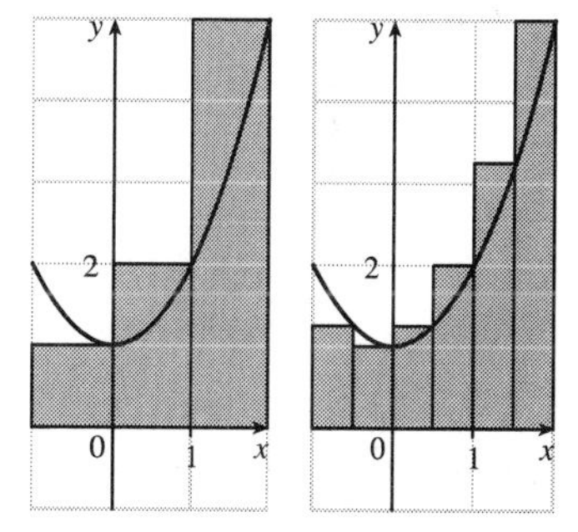

(b) $L_3 = 1\cdot f(-1) + 1\cdot f(0) + 1\cdot f(1) = 1\cdot 2 + 1\cdot 1 + 1\cdot 2 = 5$

$L_6 = 0.5[f(-1) + f(-0.5) + f(0) + f(0.5) + f(1) + f(1.5)]$

$= 0.5(2 + 1.25 + 1 + 1.25 + 2 + 3.25)$

$= 0.5(10.75) = 5.375$

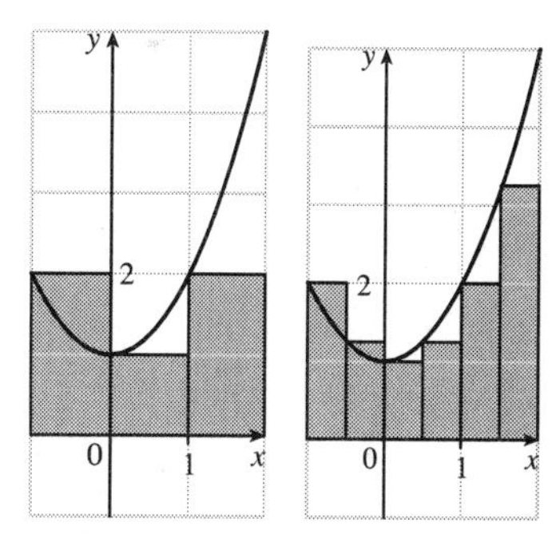

(c) $M_3 = 1\cdot f(-0.5) + 1\cdot f(0.5) + 1\cdot f(1.5)$

$= 1\cdot 1.25 + 1\cdot 1.25 + 1\cdot 3.25 = 5.75$

$M_6 = 0.5[f(-0.75) + f(-0.25) + f(0.25) + f(0.75) + f(1.25) + f(1.75)]$

$= 0.5(1.5625 + 1.0625 + 1.0625 + 1.5625 + 2.5625 + 4.0625)$

$= 0.5(11.875) = 5.9375$

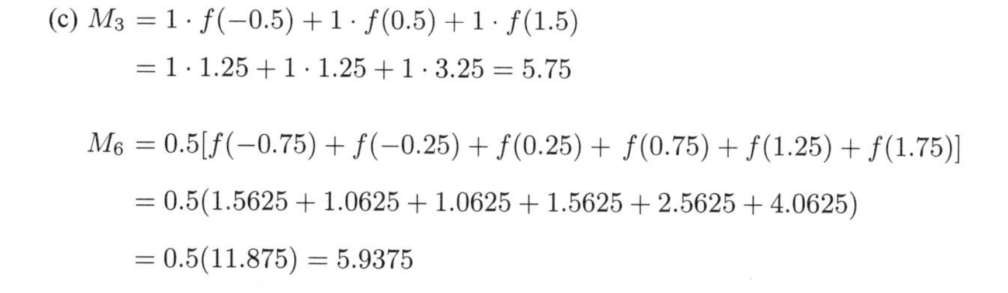

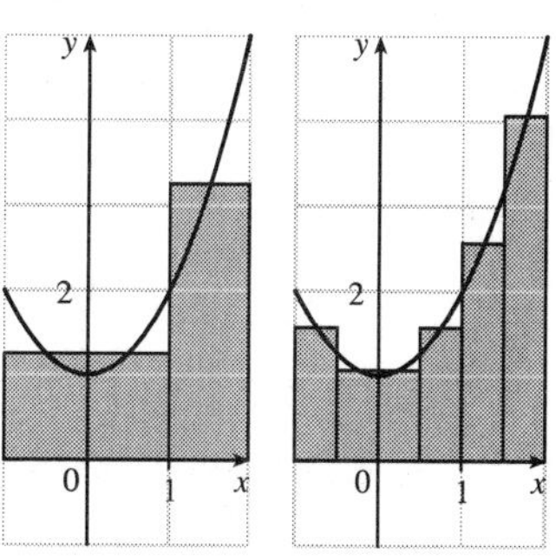

(d) M_6 appears to be the best estimate.

6. (a)

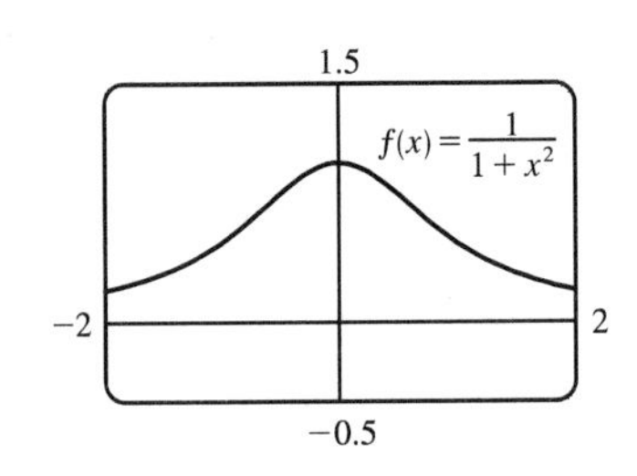

(b) $f(x) = 1/(1 + x^2)$ and $\Delta x = \frac{2-(-2)}{4} = 1 \Rightarrow$

(i) $R_4 = \sum_{i=1}^{4} f(x_i)\,\Delta x$

$= f(-1) \cdot 1 + f(0) \cdot 1$

$+ f(1) \cdot 1 + f(2) \cdot 1$

$= \frac{1}{2} + 1 + \frac{1}{2} + \frac{1}{5} = \frac{11}{5} = 2.2$

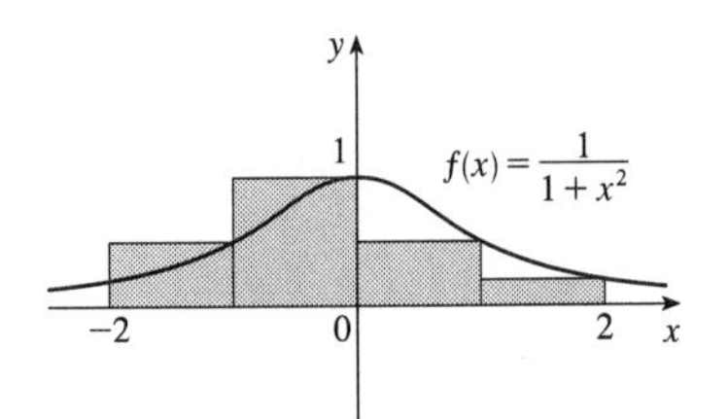

(ii) $M_4 = \sum_{i=1}^{4} f(\overline{x}_i)\,\Delta x \quad [\overline{x}_i = \frac{1}{2}(x_{i-1} + x_i)]$

$= f(-1.5) \cdot 1 + f(-0.5) \cdot 1$

$+ f(0.5) \cdot 1 + f(1.5) \cdot 1$

$= \frac{4}{13} + \frac{4}{5} + \frac{4}{5} + \frac{4}{13} = \frac{144}{65} \approx 2.2154$

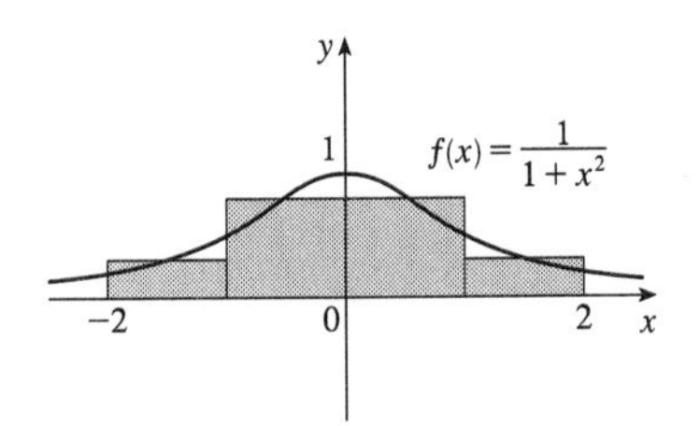

(c) $n = 8$, so $\Delta x = \frac{2-(-2)}{8} = \frac{1}{2}$.

$$R_8 = \tfrac{1}{2}[f(-1.5) + f(-1) + f(-0.5) + f(0) + f(0.5) + f(1) + f(1.5) + f(2)]$$

$$= \tfrac{1}{2}\left[\tfrac{4}{13} + \tfrac{1}{2} + \tfrac{4}{5} + 1 + \tfrac{4}{5} + \tfrac{1}{2} + \tfrac{4}{13} + \tfrac{1}{5}\right] = \tfrac{287}{130} \approx 2.2077$$

$$M_8 = \tfrac{1}{2}[f(-1.75) + f(-1.25) + f(-0.75) + f(-0.25) + f(0.25) + f(0.75) + f(1.25) + f(1.75)]$$

$$= \tfrac{1}{2}\left[2\left(\tfrac{16}{65} + \tfrac{16}{41} + \tfrac{16}{25} + \tfrac{16}{17}\right)\right] \approx 2.2176$$

7. Since v is an increasing function, L_6 will give us a lower estimate and R_6 will give us an upper estimate.

$L_6 = (0 \text{ ft/s})(0.5 \text{ s}) + (6.2)(0.5) + (10.8)(0.5) + (14.9)(0.5) + (18.1)(0.5) + (19.4)(0.5) = 0.5(69.4) = 34.7 \text{ ft}$

$R_6 = 0.5(6.2 + 10.8 + 14.9 + 18.1 + 19.4 + 20.2) = 0.5(89.6) = 44.8 \text{ ft}$

8. (a) $d \approx L_5 = (30 \text{ ft/s})(12 \text{ s}) + 28 \cdot 12 + 25 \cdot 12 + 22 \cdot 12 + 24 \cdot 12$

$= (30 + 28 + 25 + 22 + 24) \cdot 12 = 129 \cdot 12 = 1548 \text{ ft}$

(b) $d \approx R_5 = (28 + 25 + 22 + 24 + 27) \cdot 12 = 126 \cdot 12 = 1512 \text{ ft}$

(c) The estimates are neither lower nor upper estimates since v is neither an increasing nor a decreasing function of t.

9. Lower estimate for oil leakage: $R_5 = (7.6 + 6.8 + 6.2 + 5.7 + 5.3)(2) = (31.6)(2) = 63.2 \text{ L}$.

Upper estimate for oil leakage: $L_5 = (8.7 + 7.6 + 6.8 + 6.2 + 5.7)(2) = (35)(2) = 70 \text{ L}$.

10. We can find an upper estimate by using the final velocity for each time interval. Thus, the distance d traveled after 62 seconds can be approximated by

$$d = \sum_{i=1}^{6} v(t_i)\Delta t_i = (185 \text{ ft/s})(10 \text{ s}) + 319 \cdot 5 + 447 \cdot 5 + 742 \cdot 12 + 1325 \cdot 27 + 1445 \cdot 3 = 54{,}694 \text{ ft}$$

11. For a decreasing function, using left endpoints gives us an overestimate and using right endpoints results in an underestimate. We will use M_6 to get an estimate. $\Delta t = 1$, so

$$\begin{aligned} M_6 &= 1[v(0.5) + v(1.5) + v(2.5) + v(3.5) + v(4.5) + v(5.5)] \\ &\approx 55 + 40 + 28 + 18 + 10 + 4 = 155 \text{ ft} \end{aligned}$$

For a very rough check on the above calculation, we can draw a line from $(0, 70)$ to $(6, 0)$ and calculate the area of the triangle: $\frac{1}{2}(70)(6) = 210$. This is clearly an overestimate, so our midpoint estimate of 155 is reasonable.

12. For an increasing function, using left endpoints gives us an underestimate and using right endpoints results in an overestimate. We will use M_6 to get an estimate. $\Delta t = \frac{30-0}{6} = 5 \text{ s} = \frac{5}{3600} \text{ h} = \frac{1}{720} \text{ h}$.

$$\begin{aligned} M_6 &= \tfrac{1}{720}[v(2.5) + v(7.5) + v(12.5) + v(17.5) + v(22.5) + v(27.5)] \\ &= \tfrac{1}{720}(31.25 + 66 + 88 + 103.5 + 113.75 + 119.25) = \tfrac{1}{720}(521.75) \approx 0.725 \text{ km} \end{aligned}$$

For a very rough check on the above calculation, we can draw a line from $(0, 0)$ to $(30, 120)$ and calculate the area of the triangle: $\frac{1}{2}(30)(120) = 1800$. Divide by 3600 to get 0.5, which is clearly an underestimate, making our midpoint estimate of 0.725 seem reasonable. Of course, answers will vary due to different readings of the graph.

13. $f(x) = \sqrt[4]{x}$, $1 \le x \le 16$. $\Delta x = (16 - 1)/n = 15/n$ and $x_i = 1 + i\,\Delta x = 1 + 15i/n$.

$$A = \lim_{n\to\infty} R_n = \lim_{n\to\infty} \sum_{i=1}^{n} f(x_i)\,\Delta x = \lim_{n\to\infty} \sum_{i=1}^{n} \sqrt[4]{1 + \tfrac{15i}{n}} \cdot \tfrac{15}{n}.$$

14. $f(x) = 1 + x^4$, $2 \le x \le 5$. $\Delta x = (5 - 2)/n = 3/n$ and $x_i = 2 + i\,\Delta x = 2 + 3i/n$.

$$A = \lim_{n\to\infty} R_n = \lim_{n\to\infty} \sum_{i=1}^{n} f(x_i)\Delta x = \lim_{n\to\infty} \sum_{i=1}^{n} \left[1 + \left(2 + \frac{3i}{n}\right)^4\right] \cdot \frac{3}{n}.$$

15. $\displaystyle\lim_{n\to\infty} \sum_{i=1}^{n} \frac{\pi}{4n} \tan\frac{i\pi}{4n}$ can be interpreted as the area of the region lying under the graph of $y = \tan x$ on the interval $\left[0, \frac{\pi}{4}\right]$, since for $y = \tan x$ on $\left[0, \frac{\pi}{4}\right]$ with $\Delta x = \dfrac{\pi/4 - 0}{n} = \dfrac{\pi}{4n}$, $x_i = 0 + i\,\Delta x = \dfrac{i\pi}{4n}$, and $x_i^* = x_i$, the expression for the area is $\displaystyle A = \lim_{n\to\infty} \sum_{i=1}^{n} f(x_i^*)\,\Delta x = \lim_{n\to\infty} \sum_{i=1}^{n} \tan\left(\frac{i\pi}{4n}\right) \frac{\pi}{4n}$. Note that this answer is not unique, since the expression for the area is the same for the function $y = \tan(x - k\pi)$ on the interval $\left[k\pi, k\pi + \frac{\pi}{4}\right]$, where k is any integer.

16. (a) $\Delta x = \dfrac{1-0}{n} = \dfrac{1}{n}$ and $x_i = 0 + i\,\Delta x = \dfrac{i}{n}$. $\displaystyle A = \lim_{n\to\infty} R_n = \lim_{n\to\infty} \sum_{i=1}^{n} f(x_i)\,\Delta x = \lim_{n\to\infty} \sum_{i=1}^{n} \left(\frac{i}{n}\right)^3 \cdot \frac{1}{n}$.

(b) $$\lim_{n\to\infty} \sum_{i=1}^{n} \frac{i^3}{n^3} \cdot \frac{1}{n} = \lim_{n\to\infty} \frac{1}{n^4} \sum_{i=1}^{n} i^3 = \lim_{n\to\infty} \frac{1}{n^4} \left[\frac{n(n+1)}{2}\right]^2 = \lim_{n\to\infty} \frac{(n+1)^2}{4n^2} = \frac{1}{4} \lim_{n\to\infty} \left(1 + \frac{1}{n}\right)^2 = \frac{1}{4}$$

17. (a) $y = f(x) = x^5$. $\Delta x = \dfrac{2-0}{n} = \dfrac{2}{n}$ and $x_i = 0 + i\,\Delta x = \dfrac{2i}{n}$.

$$A = \lim_{n\to\infty} R_n = \lim_{n\to\infty} \sum_{i=1}^{n} f(x_i)\,\Delta x = \lim_{n\to\infty} \sum_{i=1}^{n} \left(\frac{2i}{n}\right)^5 \cdot \frac{2}{n} = \lim_{n\to\infty} \sum_{i=1}^{n} \frac{32 i^5}{n^5} \cdot \frac{2}{n} = \lim_{n\to\infty} \frac{64}{n^6} \sum_{i=1}^{n} i^5.$$

(b) $$\sum_{i=1}^{n} i^5 \overset{\text{CAS}}{=} \frac{n^2(n+1)^2(2n^2 + 2n - 1)}{12}$$

(c) $\lim\limits_{n\to\infty} \dfrac{64}{n^6} \cdot \dfrac{n^2(n+1)^2(2n^2+2n-1)}{12} = \dfrac{64}{12} \lim\limits_{n\to\infty} \dfrac{(n^2+2n+1)(2n^2+2n-1)}{n^2 \cdot n^2}$

$$= \frac{16}{3} \lim_{n\to\infty} \left(1 + \frac{2}{n} + \frac{1}{n^2}\right)\left(2 + \frac{2}{n} - \frac{1}{n^2}\right) = \tfrac{16}{3} \cdot 1 \cdot 2 = \tfrac{32}{3}$$

18. (a) $y = f(x) = x^4 + 5x^2 + x,\ 2 \le x \le 7 \ \Rightarrow\ \Delta x = \dfrac{7-2}{n} = \dfrac{5}{n},\ x_i = 2 + i\,\Delta x = 2 + \dfrac{5i}{n} \ \Rightarrow$

$$A = \lim_{n\to\infty} R_n = \lim_{n\to\infty} \frac{5}{n} \sum_{i=1}^{n} \left[\left(2 + \frac{5i}{n}\right)^4 + 5\left(2 + \frac{5i}{n}\right)^2 + \left(2 + \frac{5i}{n}\right)\right]$$

(b) $R_n = \dfrac{5}{n} \cdot \dfrac{4723n^4 + 7845n^3 + 3475n^2 - 125}{6n^3}$

(c) $A = \lim\limits_{n\to\infty} R_n = \dfrac{23{,}615}{6} = 3935.8\overline{3}$

19. $y = f(x) = \cos x.\ \ \Delta x = \dfrac{b-0}{n} = \dfrac{b}{n}$ and $x_i = 0 + i\,\Delta x = \dfrac{bi}{n}$.

$$A = \lim_{n\to\infty} R_n = \lim_{n\to\infty} \sum_{i=1}^{n} f(x_i)\,\Delta x = \lim_{n\to\infty} \sum_{i=1}^{n} \cos\left(\frac{bi}{n}\right) \cdot \frac{b}{n} \overset{\text{CAS}}{=} \lim_{n\to\infty} \left[\frac{b \sin\left(b\left(\frac{1}{2n} + 1\right)\right)}{2n \sin\left(\frac{b}{2n}\right)} - \frac{b}{2n}\right] \overset{\text{CAS}}{=} \sin b$$

If $b = \frac{\pi}{2}$, then $A = \sin\frac{\pi}{2} = 1$.

20. (a) The diagram shows one of the n congruent triangles, $\triangle AOB$, with central angle $2\pi/n$. O is the center of the circle and AB is one of the sides of the polygon. Radius OC is drawn so as to bisect $\angle AOB$. It follows that OC intersects AB at right angles and bisects AB. Thus, $\triangle AOB$ is divided into two right triangles with legs of length $\frac{1}{2}(AB) = r\sin(\pi/n)$ and $r\cos(\pi/n)$.

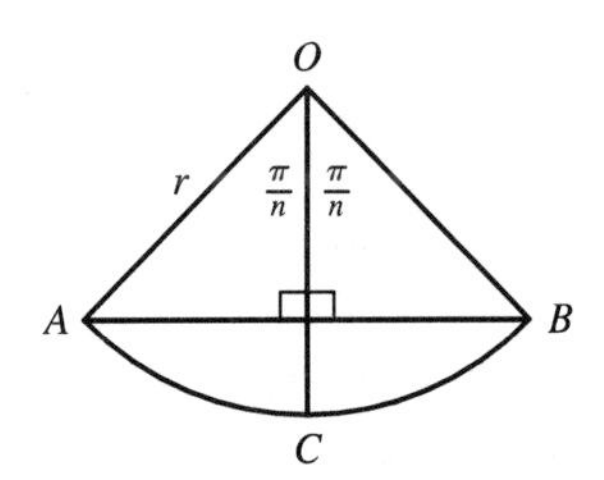

$\triangle AOB$ has area $2 \cdot \frac{1}{2}[r\sin(\pi/n)][r\cos(\pi/n)] = r^2 \sin(\pi/n)\cos(\pi/n) = \frac{1}{2}r^2 \sin(2\pi/n)$, so

$A_n = n \cdot \text{area}(\triangle AOB) = \frac{1}{2}nr^2 \sin(2\pi/n)$.

(b) To use Equation 1.4.5, $\lim\limits_{\theta\to 0} \dfrac{\sin\theta}{\theta} = 1$, we need to have the same expression in the denominator as we have in the argument of the sine function—in this case, $2\pi/n$.

$$\lim_{n\to\infty} A_n = \lim_{n\to\infty} \tfrac{1}{2}nr^2 \sin(2\pi/n) = \lim_{n\to\infty} \tfrac{1}{2}nr^2 \frac{\sin(2\pi/n)}{2\pi/n} \cdot \frac{2\pi}{n} = \lim_{n\to\infty} \frac{\sin(2\pi/n)}{2\pi/n}\pi r^2.$$

Let $\theta = \dfrac{2\pi}{n}$. Then as $n \to \infty$, $\theta \to 0$, so $\lim\limits_{n\to\infty} \dfrac{\sin(2\pi/n)}{2\pi/n}\pi r^2 = \lim\limits_{\theta\to 0} \dfrac{\sin\theta}{\theta}\pi r^2 = (1)\,\pi r^2 = \pi r^2$.

4.2 The Definite Integral

1. $R_4 = \sum_{i=1}^{4} f(x_i)\,\Delta x \quad [x_i^* = x_i \text{ is a right endpoint and } \Delta x = 0.5]$

$= 0.5\,[f(0.5) + f(1) + f(1.5) + f(2)] \quad [f(x) = 2 - x^2]$

$= 0.5\,[1.75 + 1 + (-0.25) + (-2)]$

$= 0.5(0.5) = 0.25$

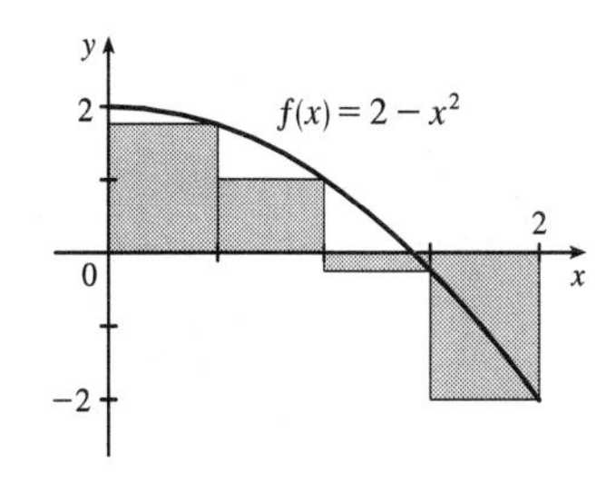

The Riemann sum represents the sum of the areas of the two rectangles above the x-axis minus the sum of the areas of the two rectangles below the x-axis; that is, the *net area* of the rectangles with respect to the x-axis.

2. 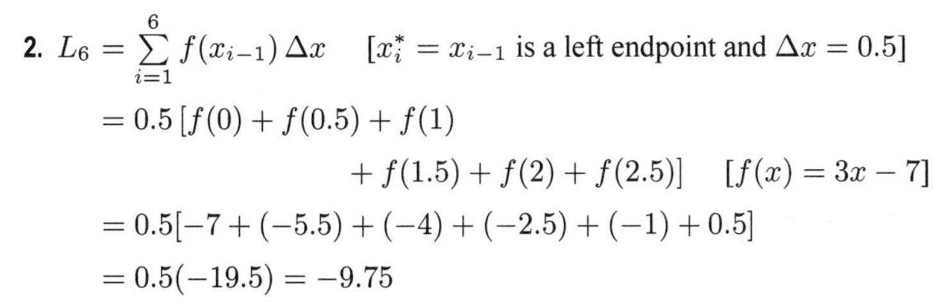

$L_6 = \sum_{i=1}^{6} f(x_{i-1})\,\Delta x \quad [x_i^* = x_{i-1} \text{ is a left endpoint and } \Delta x = 0.5]$

$= 0.5\,[f(0) + f(0.5) + f(1)$

$\qquad + f(1.5) + f(2) + f(2.5)] \quad [f(x) = 3x - 7]$

$= 0.5[-7 + (-5.5) + (-4) + (-2.5) + (-1) + 0.5]$

$= 0.5(-19.5) = -9.75$

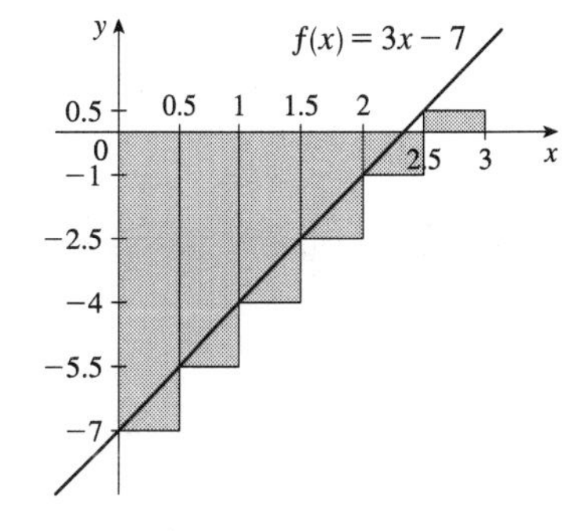

The Riemann sum represents the area of the rectangle above the x-axis minus the area of the five rectangles below the x-axis.

3. 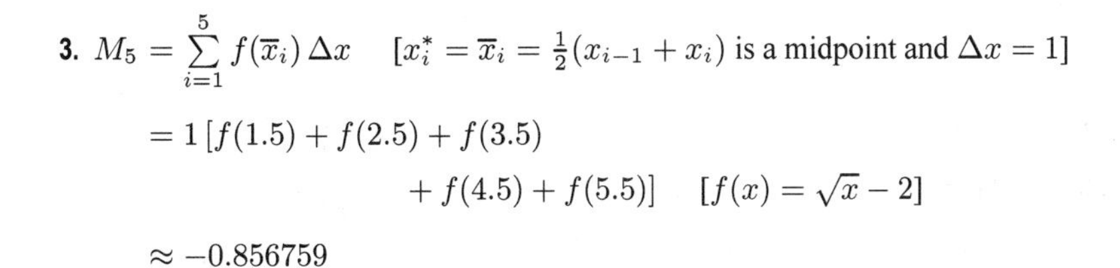

$M_5 = \sum_{i=1}^{5} f(\bar{x}_i)\,\Delta x \quad [x_i^* = \bar{x}_i = \tfrac{1}{2}(x_{i-1} + x_i) \text{ is a midpoint and } \Delta x = 1]$

$= 1\,[f(1.5) + f(2.5) + f(3.5)$

$\qquad + f(4.5) + f(5.5)] \quad [f(x) = \sqrt{x} - 2]$

≈ -0.856759

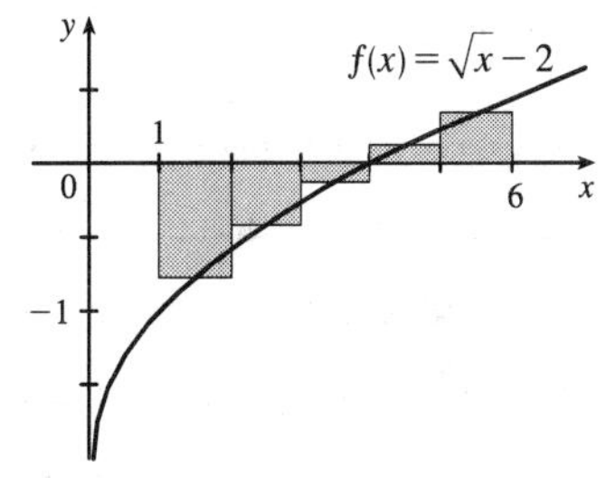

The Riemann sum represents the sum of the areas of the two rectangles above the x-axis minus the sum of the areas of the three rectangles below the x-axis.

4. 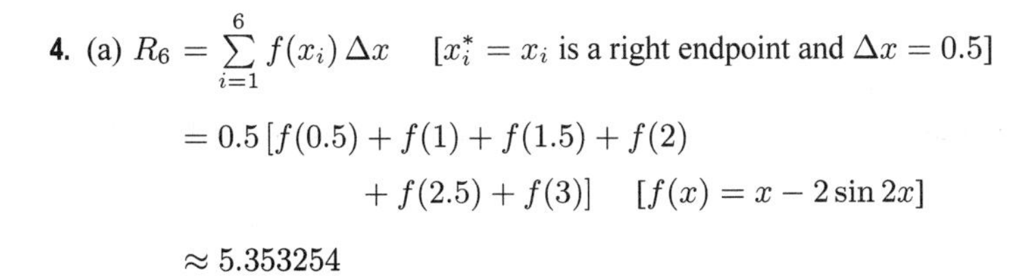

(a) $R_6 = \sum_{i=1}^{6} f(x_i)\,\Delta x \quad [x_i^* = x_i \text{ is a right endpoint and } \Delta x = 0.5]$

$= 0.5\,[f(0.5) + f(1) + f(1.5) + f(2)$

$\qquad + f(2.5) + f(3)] \quad [f(x) = x - 2\sin 2x]$

≈ 5.353254

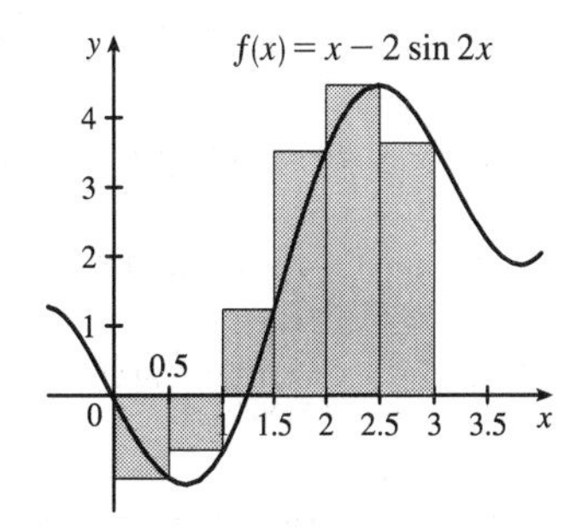

The Riemann sum represents the sum of the areas of the four rectangles above the x-axis minus the sum of the areas of the two rectangles below the x-axis.

(b) $M_6 = \sum_{i=1}^{6} f(\overline{x}_i)\,\Delta x \quad [x_i^* = \overline{x}_i \text{ is a midpoint and } \Delta x = 0.5]$

$$= 0.5[f(0.25) + f(0.75) + f(1.25) + f(1.75) + f(2.25) + f(2.75)] \quad [f(x) = x - 2\sin 2x]$$

$$\approx 4.458461$$

The Riemann sum represents the sum of the areas of the four rectangles above the x-axis minus the sum of the areas of the two rectangles below the x-axis.

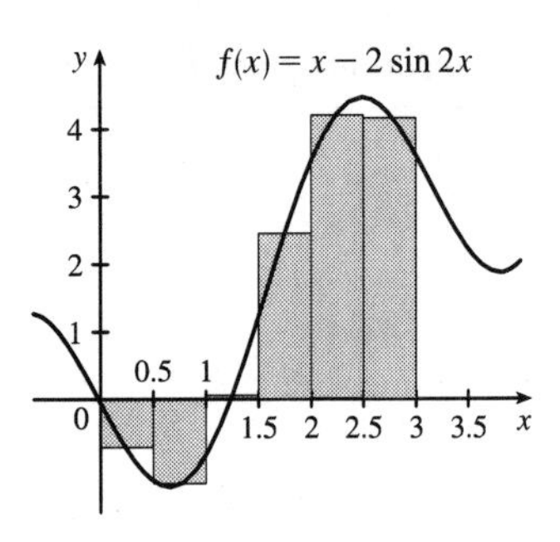

5. $f(x) = x^3$. $\sum_{i=1}^{n} f(x_i^*)\,\Delta x_i = \frac{1}{2}\sum_{i=1}^{n} f(x_i^*) = \frac{1}{2}[(-1)^3 + (-0.4)^3 + (0.2)^3 + 1^3] = -0.028$

6. $f(x) = x + x^2$

$$\sum_{i=1}^{5} f(x_i^*)\,\Delta x_i = f(-2)(0.5) + f(-1.5)(0.5) + f(-1)(0.3) + f(-0.7)(0.3) + f(-0.4)(0.4)$$
$$= 2(0.5) + (0.75)(0.5) + (0)(0.3) + (-0.21)(0.3) + (-0.24)(0.4) = 1.216$$

$\max \Delta x_i = \max\{0.5, 0.5, 0.3, 0.3, 0.4\} = 0.5$

7. $\Delta x = (b - a)/n = (8 - 0)/4 = 8/4 = 2$.

(a) Using the right endpoints to approximate $\int_0^8 f(x)\,dx$, we have

$$\sum_{i=1}^{4} f(x_i)\,\Delta x = 2[f(2) + f(4) + f(6) + f(8)] \approx 2[1 + 2 + (-2) + 1] = 4.$$

(b) Using the left endpoints to approximate $\int_0^8 f(x)\,dx$, we have

$$\sum_{i=1}^{4} f(x_{i-1})\,\Delta x = 2[f(0) + f(2) + f(4) + f(6)] \approx 2[2 + 1 + 2 + (-2)] = 6.$$

(c) Using the midpoint of each subinterval to approximate $\int_0^8 f(x)\,dx$, we have

$$\sum_{i=1}^{4} f(\overline{x}_i)\,\Delta x = 2[f(1) + f(3) + f(5) + f(7)] \approx 2[3 + 2 + 1 + (-1)] = 10.$$

8. (a) Using the right endpoints to approximate $\int_{-3}^{3} g(x)\,dx$, we have

$$\sum_{i=1}^{6} g(x_i)\,\Delta x = 1[g(-2) + g(-1) + g(0) + g(1) + g(2) + g(3)]$$
$$\approx 1 - 0.5 - 1.5 - 1.5 - 0.5 + 2.5 = -0.5$$

(b) Using the left endpoints to approximate $\int_{-3}^{3} g(x)\,dx$, we have

$$\sum_{i=1}^{6} g(x_{i-1})\,\Delta x = 1[g(-3) + g(-2) + g(-1) + g(0) + g(1) + g(2)]$$
$$\approx 2 + 1 - 0.5 - 1.5 - 1.5 - 0.5 = -1$$

(c) Using the midpoint of each subinterval to approximate $\int_{-3}^{3} g(x)\,dx$, we have

$$\sum_{i=1}^{6} g(\overline{x}_i)\,\Delta x = 1[g(-2.5) + g(-1.5) + g(-0.5) + g(0.5) + g(1.5) + g(2.5)]$$

$$\approx 1.5 + 0 - 1 - 1.75 - 1 + 0.5 = -1.75$$

9. Since f is increasing, $L_5 \leq \int_0^{25} f(x)\,dx \leq R_5$.

$$\text{Lower estimate} = L_5 = \sum_{i=1}^{5} f(x_{i-1})\,\Delta x = 5[f(0) + f(5) + f(10) + f(15) + f(20)]$$

$$= 5(-42 - 37 - 25 - 6 + 15) = 5(-95) = -475$$

$$\text{Upper estimate} = R_5 = \sum_{i=1}^{5} f(x_i)\,\Delta x = 5[f(5) + f(10) + f(15) + f(20) + f(25)]$$

$$= 5(-37 - 25 - 6 + 15 + 36) = 5(-17) = -85$$

10. (a) Using the right endpoints to approximate $\int_0^6 f(x)\,dx$, we have

$$\sum_{i=1}^{3} f(x_i)\,\Delta x = 2[f(2) + f(4) + f(6)] = 2(8.3 + 2.3 - 10.5) = 0.2.$$

(b) Using the left endpoints to approximate $\int_0^6 f(x)\,dx$, we have

$$\sum_{i=1}^{3} f(x_{i-1})\,\Delta x = 2[f(0) + f(2) + f(4)] = 2(9.3 + 8.3 + 2.3) = 39.8.$$

11. $\Delta x = (10 - 2)/4 = 2$, so the endpoints are 2, 4, 6, 8, and 10, and the midpoints are 3, 5, 7, and 9. The Midpoint Rule gives $\int_2^{10} \sqrt{x^3 + 1}\,dx \approx \sum_{i=1}^{4} f(\overline{x}_i)\,\Delta x = 2\left(\sqrt{3^3 + 1} + \sqrt{5^3 + 1} + \sqrt{7^3 + 1} + \sqrt{9^3 + 1}\right) \approx 124.1644$.

12. $\Delta x = (\pi - 0)/6 = \frac{\pi}{6}$, so the endpoints are 0, $\frac{\pi}{6}$, $\frac{2\pi}{6}$, $\frac{3\pi}{6}$, $\frac{4\pi}{6}$, $\frac{5\pi}{6}$, and $\frac{6\pi}{6}$, and the midpoints are $\frac{\pi}{12}$, $\frac{3\pi}{12}$, $\frac{5\pi}{12}$, $\frac{7\pi}{12}$, $\frac{9\pi}{12}$, and $\frac{11\pi}{12}$. The Midpoint Rule gives

$$\int_0^{\pi} \sec(x/3)\,dx \approx \sum_{i=1}^{6} f(\overline{x}_i)\,\Delta x = \tfrac{\pi}{6}\left(\sec\tfrac{\pi}{36} + \sec\tfrac{3\pi}{36} + \sec\tfrac{5\pi}{36} + \sec\tfrac{7\pi}{36} + \sec\tfrac{9\pi}{36} + \sec\tfrac{11\pi}{36}\right) \approx 3.9379.$$

13. $\Delta x = (1 - 0)/5 = 0.2$, so the endpoints are 0, 0.2, 0.4, 0.6, 0.8, and 1, and the midpoints are 0.1, 0.3, 0.5, 0.7, and 0.9. The Midpoint Rule gives

$$\int_0^1 \sin(x^2)\,dx \approx \sum_{i=1}^{5} f(\overline{x}_i)\,\Delta x = 0.2\left[\sin(0.1)^2 + \sin(0.3)^2 + \sin(0.5)^2 + \sin(0.7)^2 + \sin(0.9)^2\right] \approx 0.3084.$$

14. $\Delta x = \dfrac{5 - 1}{4} = 1$, so the endpoints are 1, 2, 3, 4, and 5, and the midpoints are 1.5, 2.5, 3.5, and 4.5. The Midpoint Rule gives

$$\int_1^5 \frac{x - 1}{x + 1}\,dx \approx \sum_{i=1}^{4} f(\overline{x}_i)\Delta x = 1\left[\frac{1.5 - 1}{1.5 + 1} + \frac{2.5 - 1}{2.5 + 1} + \frac{3.5 - 1}{3.5 + 1} + \frac{4.5 - 1}{4.5 + 1}\right] \approx 1.8205.$$

15. On $[0, \pi]$, $\lim_{n\to\infty} \sum_{i=1}^{n} x_i \sin x_i\,\Delta x = \int_0^{\pi} x \sin x\,dx$.

16. On $[1, 5]$, $\lim_{n\to\infty} \sum_{i=1}^{n} \dfrac{x_i}{1 + x_i}\,\Delta x = \displaystyle\int_1^5 \frac{x}{1 + x}\,dx$.

17. On $[1,8]$, $\lim\limits_{\max \Delta x_i \to 0} \sum\limits_{i=1}^{n} \sqrt{2x_i^* + (x_i^*)^2}\,\Delta x_i = \int_1^8 \sqrt{2x + x^2}\,dx$.

18. On $[0,2]$, $\lim\limits_{\max \Delta x_i \to 0} \sum\limits_{i=1}^{n} [4 - 3(x_i^*)^2 + 6(x_i^*)^5]\,\Delta x_i = \int_0^2 (4 - 3x^2 + 6x^5)\,dx$.

19. Note that $\Delta x = \dfrac{5-(-1)}{n} = \dfrac{6}{n}$ and $x_i = -1 + i\,\Delta x = -1 + \dfrac{6i}{n}$.

$$\begin{aligned}
\int_{-1}^{5}(1+3x)\,dx &= \lim_{n\to\infty}\sum_{i=1}^{n} f(x_i)\,\Delta x = \lim_{n\to\infty}\sum_{i=1}^{n}\left[1+3\left(-1+\frac{6i}{n}\right)\right]\frac{6}{n} = \lim_{n\to\infty}\frac{6}{n}\sum_{i=1}^{n}\left[-2+\frac{18i}{n}\right] \\
&= \lim_{n\to\infty}\frac{6}{n}\left[\sum_{i=1}^{n}(-2)+\sum_{i=1}^{n}\frac{18i}{n}\right] = \lim_{n\to\infty}\frac{6}{n}\left[-2n+\frac{18}{n}\sum_{i=1}^{n} i\right] \\
&= \lim_{n\to\infty}\frac{6}{n}\left[-2n+\frac{18}{n}\cdot\frac{n(n+1)}{2}\right] = \lim_{n\to\infty}\left[-12+\frac{108}{n^2}\cdot\frac{n(n+1)}{2}\right] \\
&= \lim_{n\to\infty}\left[-12+54\frac{n+1}{n}\right] = \lim_{n\to\infty}\left[-12+54\left(1+\frac{1}{n}\right)\right] = -12+54\cdot 1 = 42
\end{aligned}$$

20. $\displaystyle\int_1^4 (x^2+2x-5)\,dx = \lim_{n\to\infty}\sum_{i=1}^{n} f(x_i)\Delta x \qquad [\Delta x = 3/n \text{ and } x_i = 1+3i/n]$

$$\begin{aligned}
&= \lim_{n\to\infty}\sum_{i=1}^{n}\left[\left(1+\frac{3i}{n}\right)^2+2\left(1+\frac{3i}{n}\right)-5\right]\left(\frac{3}{n}\right) \\
&= \lim_{n\to\infty}\frac{3}{n}\left[\sum_{i=1}^{n}\left(1+\frac{6i}{n}+\frac{9i^2}{n^2}+2+\frac{6i}{n}-5\right)\right] \\
&= \lim_{n\to\infty}\frac{3}{n}\left[\sum_{i=1}^{n}\left(\frac{9}{n^2}\cdot i^2+\frac{12}{n}\cdot i-2\right)\right] = \lim_{n\to\infty}\frac{3}{n}\left[\frac{9}{n^2}\sum_{i=1}^{n} i^2+\frac{12}{n}\sum_{i=1}^{n} i-\sum_{i=1}^{n} 2\right] \\
&= \lim_{n\to\infty}\left(\frac{27}{n^3}\cdot\frac{n(n+1)(2n+1)}{6}+\frac{36}{n^2}\cdot\frac{n(n+1)}{2}-\frac{6}{n}\cdot n\right) \\
&= \lim_{n\to\infty}\left(\frac{9}{2}\cdot\frac{n+1}{n}\cdot\frac{2n+1}{n}+18\cdot\frac{n+1}{n}-6\right) \\
&= \lim_{n\to\infty}\left[\frac{9}{2}\left(1+\frac{1}{n}\right)\left(2+\frac{1}{n}\right)+18\left(1+\frac{1}{n}\right)-6\right] = \tfrac{9}{2}\cdot 1\cdot 2+18\cdot 1-6 = 21
\end{aligned}$$

21. Note that $\Delta x = \dfrac{2-0}{n} = \dfrac{2}{n}$ and $x_i = 0 + i\,\Delta x = \dfrac{2i}{n}$.

$$\begin{aligned}
\int_0^2 (2-x^2)\,dx &= \lim_{n\to\infty}\sum_{i=1}^{n} f(x_i)\,\Delta x = \lim_{n\to\infty}\sum_{i=1}^{n}\left(2-\frac{4i^2}{n^2}\right)\left(\frac{2}{n}\right) = \lim_{n\to\infty}\frac{2}{n}\left[\sum_{i=1}^{n} 2-\frac{4}{n^2}\sum_{i=1}^{n} i^2\right] \\
&= \lim_{n\to\infty}\frac{2}{n}\left(2n-\frac{4}{n^2}\sum_{i=1}^{n} i^2\right) = \lim_{n\to\infty}\left[4-\frac{8}{n^3}\cdot\frac{n(n+1)(2n+1)}{6}\right] \\
&= \lim_{n\to\infty}\left(4-\frac{4}{3}\cdot\frac{n+1}{n}\cdot\frac{2n+1}{n}\right) = \lim_{n\to\infty}\left[4-\frac{4}{3}\left(1+\frac{1}{n}\right)\left(2+\frac{1}{n}\right)\right] \\
&= 4-\tfrac{4}{3}\cdot 1\cdot 2 = \tfrac{4}{3}
\end{aligned}$$

22. $\displaystyle\int_0^5 (1+2x^3)\,dx = \lim_{n\to\infty} \sum_{i=1}^n f(x_i)\,\Delta x \quad [\Delta x = 5/n \text{ and } x_i = 5i/n]$

$$= \lim_{n\to\infty} \sum_{i=1}^n \left(1 + 2\cdot\frac{125i^3}{n^3}\right)\left(\frac{5}{n}\right) = \lim_{n\to\infty} \frac{5}{n}\left[\sum_{i=1}^n 1 + \frac{250}{n^3}\sum_{i=1}^n i^3\right] = \lim_{n\to\infty} \frac{5}{n}\left(1\cdot n + \frac{250}{n^3}\sum_{i=1}^n i^3\right)$$

$$= \lim_{n\to\infty}\left[5 + \frac{1250}{n^4}\cdot\frac{n^2(n+1)^2}{4}\right] = \lim_{n\to\infty}\left[5 + 312.5\cdot\frac{(n+1)^2}{n^2}\right] = \lim_{n\to\infty}\left[5 + 312.5\left(1+\frac{1}{n}\right)^2\right]$$

$$= 5 + 312.5 = 317.5$$

23. Note that $\Delta x = \dfrac{2-1}{n} = \dfrac{1}{n}$ and $x_i = 1 + i\,\Delta x = 1 + i(1/n) = 1 + i/n$.

$$\int_1^2 x^3\,dx = \lim_{n\to\infty}\sum_{i=1}^n f(x_i)\,\Delta x = \lim_{n\to\infty}\sum_{i=1}^n \left(1+\frac{i}{n}\right)^3\left(\frac{1}{n}\right) = \lim_{n\to\infty}\frac{1}{n}\sum_{i=1}^n\left(\frac{n+i}{n}\right)^3$$

$$= \lim_{n\to\infty}\frac{1}{n^4}\sum_{i=1}^n (n^3 + 3n^2 i + 3ni^2 + i^3) = \lim_{n\to\infty}\frac{1}{n^4}\left[\sum_{i=1}^n n^3 + \sum_{i=1}^n 3n^2 i + \sum_{i=1}^n 3ni^2 + \sum_{i=1}^n i^3\right]$$

$$= \lim_{n\to\infty}\frac{1}{n^4}\left[n\cdot n^3 + 3n^2\sum_{i=1}^n i + 3n\sum_{i=1}^n i^2 + \sum_{i=1}^n i^3\right]$$

$$= \lim_{n\to\infty}\left[1 + \frac{3}{n^2}\cdot\frac{n(n+1)}{2} + \frac{3}{n^3}\cdot\frac{n(n+1)(2n+1)}{6} + \frac{1}{n^4}\cdot\frac{n^2(n+1)^2}{4}\right]$$

$$= \lim_{n\to\infty}\left[1 + \frac{3}{2}\cdot\frac{n+1}{n} + \frac{1}{2}\cdot\frac{n+1}{n}\cdot\frac{2n+1}{n} + \frac{1}{4}\cdot\frac{(n+1)^2}{n^2}\right]$$

$$= \lim_{n\to\infty}\left[1 + \frac{3}{2}\left(1+\frac{1}{n}\right) + \frac{1}{2}\left(1+\frac{1}{n}\right)\left(2+\frac{1}{n}\right) + \frac{1}{4}\left(1+\frac{1}{n}\right)^2\right] = 1 + \frac{3}{2} + \tfrac{1}{2}\cdot 2 + \tfrac{1}{4} = 3.75$$

24. (a) $\Delta x = (4-0)/8 = 0.5$ and $x_i^* = x_i = 0.5i$.

$$\textstyle\int_0^4 (x^2 - 3x)\,dx \approx \sum_{i=1}^8 f(x_i^*)\,\Delta x$$

$$= 0.5\{[0.5^2 - 3(0.5)] + [1.0^2 - 3(1.0)] + \cdots + [3.5^2 - 3(3.5)] + [4.0^2 - 3(4.0)]\}$$

$$= \tfrac{1}{2}\left(-\tfrac{5}{4} - 2 - \tfrac{9}{4} - 2 - \tfrac{5}{4} + 0 + \tfrac{7}{4} + 4\right) = -1.5$$

(b)

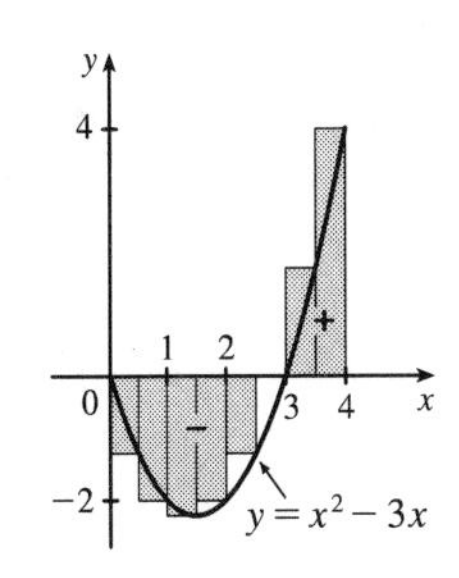

(c) $$\int_0^4 (x^2 - 3x)\,dx = \lim_{n\to\infty}\sum_{i=1}^n\left[\left(\frac{4i}{n}\right)^2 - 3\left(\frac{4i}{n}\right)\right]\left(\frac{4}{n}\right)$$

$$= \lim_{n\to\infty}\frac{4}{n}\left[\frac{16}{n^2}\sum_{i=1}^n i^2 - \frac{12}{n}\sum_{i=1}^n i\right]$$

$$= \lim_{n\to\infty}\left[\frac{64}{n^3}\cdot\frac{n(n+1)(2n+1)}{6} - \frac{48}{n^2}\cdot\frac{n(n+1)}{2}\right]$$

$$= \lim_{n\to\infty}\left[\frac{32}{3}\left(1+\frac{1}{n}\right)\left(2+\frac{1}{n}\right) - 24\left(1+\frac{1}{n}\right)\right]$$

$$= \tfrac{32}{3}\cdot 2 - 24 = -\tfrac{8}{3}$$

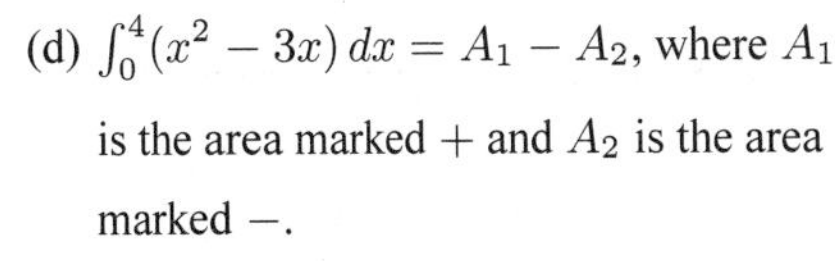
(d) $\int_0^4 (x^2 - 3x)\,dx = A_1 - A_2$, where A_1 is the area marked + and A_2 is the area marked −.

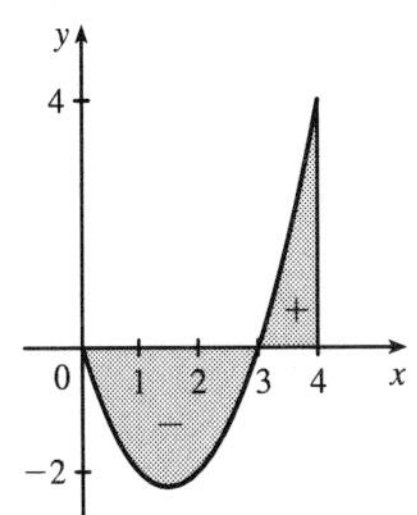

25. $f(x) = \dfrac{x}{1+x^5}$, $a = 2$, $b = 6$, and $\Delta x = \dfrac{6-2}{n} = \dfrac{4}{n}$. Using Equation 3, we get $x_i^* = x_i = 2 + i\,\Delta x = 2 + \dfrac{4i}{n}$,

so $\displaystyle\int_2^6 \frac{x}{1+x^5}\,dx = \lim_{n\to\infty} R_n = \lim_{n\to\infty}\sum_{i=1}^{n} \frac{2+\frac{4i}{n}}{1+\left(2+\frac{4i}{n}\right)^5}\cdot\frac{4}{n}$.

26. $f(x) = x^2\sin x$, $[a,b] = [0,2\pi]$, $\Delta x = \dfrac{2\pi - 0}{n} = \dfrac{2\pi}{n}$, and $x_i = a + i\,\Delta x = \dfrac{2\pi i}{n}$, so

$$\int_0^{2\pi} x^2\sin x\,dx = \lim_{n\to\infty} R_n = \lim_{n\to\infty}\sum_{i=1}^{n}\left[\left(\frac{2\pi i}{n}\right)^2\sin\!\left(\frac{2\pi i}{n}\right)\right]\cdot\frac{2\pi}{n}.$$

27. $\Delta x = (\pi - 0)/n = \pi/n$ and $x_i^* = x_i = \pi i/n$.

$$\int_0^{\pi}\sin 5x\,dx = \lim_{n\to\infty}\sum_{i=1}^{n}(\sin 5x_i)\left(\frac{\pi}{n}\right) = \lim_{n\to\infty}\sum_{i=1}^{n}\left(\sin\frac{5\pi i}{n}\right)\frac{\pi}{n} \overset{\text{CAS}}{=} \pi\lim_{n\to\infty}\frac{1}{n}\cot\left(\frac{5\pi}{2n}\right) \overset{\text{CAS}}{=} \pi\left(\frac{2}{5\pi}\right) = \frac{2}{5}$$

28. $\Delta x = (10-2)/n = 8/n$ and $x_i^* = x_i = 2 + 8i/n$.

$$\begin{aligned}\int_2^{10} x^6\,dx &= \lim_{n\to\infty}\sum_{i=1}^{n}\left(2+\frac{8i}{n}\right)^6\left(\frac{8}{n}\right) = 8\lim_{n\to\infty}\frac{1}{n}\sum_{i=1}^{n}\left(2+\frac{8i}{n}\right)^6 \\ &\overset{\text{CAS}}{=} 8\lim_{n\to\infty}\frac{1}{n}\cdot\frac{64(58{,}593n^6 + 164{,}052n^5 + 131{,}208n^4 - 27{,}776n^2 + 2048)}{21n^5} \\ &\overset{\text{CAS}}{=} 8\left(\frac{1{,}249{,}984}{7}\right) = \frac{9{,}999{,}872}{7} \approx 1{,}428{,}553.1\end{aligned}$$

29. (a) Think of $\int_0^2 f(x)\,dx$ as the area of a trapezoid with bases 1 and 3 and height 2. The area of a trapezoid is $A = \frac{1}{2}(b+B)h$, so $\int_0^2 f(x)\,dx = \frac{1}{2}(1+3)2 = 4$.

(b) $\int_0^5 f(x)\,dx = \int_0^2 f(x)\,dx + \int_2^3 f(x)\,dx + \int_3^5 f(x)\,dx$

trapezoid rectangle triangle

$= \frac{1}{2}(1+3)2 + 3\cdot 1 + \frac{1}{2}\cdot 2\cdot 3 = 4 + 3 + 3 = 10$

(c) $\int_5^7 f(x)\,dx$ is the negative of the area of the triangle with base 2 and height 3. $\int_5^7 f(x)\,dx = -\frac{1}{2}\cdot 2\cdot 3 = -3$.

(d) $\int_7^9 f(x)\,dx$ is the negative of the area of a trapezoid with bases 3 and 2 and height 2, so it equals $-\frac{1}{2}(B+b)h = -\frac{1}{2}(3+2)2 = -5$. Thus,

$\int_0^9 f(x)\,dx = \int_0^5 f(x)\,dx + \int_5^7 f(x)\,dx + \int_7^9 f(x)\,dx = 10 + (-3) + (-5) = 2$.

30. (a) $\int_0^2 g(x)\,dx = \frac{1}{2}\cdot 4\cdot 2 = 4$ (area of a triangle)

(b) $\int_2^6 g(x)\,dx = -\frac{1}{2}\pi(2)^2 = -2\pi$ (negative of the area of a semicircle)

(c) $\int_6^7 g(x)\,dx = \frac{1}{2}\cdot 1\cdot 1 = \frac{1}{2}$ (area of a triangle)

$\int_0^7 g(x)\,dx = \int_0^2 g(x)\,dx + \int_2^6 g(x)\,dx + \int_6^7 g(x)\,dx = 4 - 2\pi + \frac{1}{2} = 4.5 - 2\pi$

31. $\int_0^3 \left(\frac{1}{2}x - 1\right) dx$ can be interpreted as the area of the triangle above the x-axis minus the area of the triangle below the x-axis; that is,

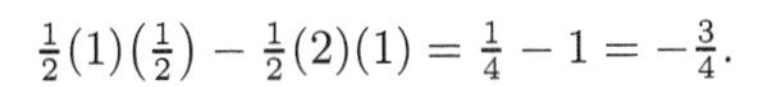

$\frac{1}{2}(1)\left(\frac{1}{2}\right) - \frac{1}{2}(2)(1) = \frac{1}{4} - 1 = -\frac{3}{4}$.

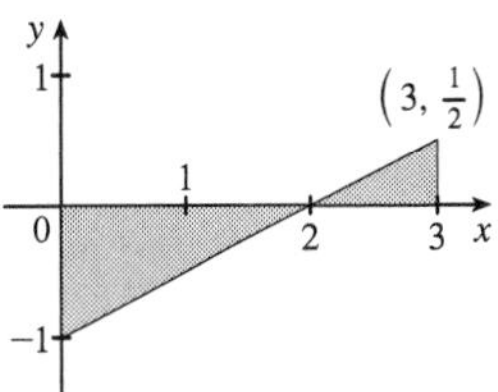

32. $\int_{-2}^{2} \sqrt{4 - x^2}\, dx$ can be interpreted as the area under the graph of $f(x) = \sqrt{4 - x^2}$ between $x = -2$ and $x = 2$. This is equal to half the area of the circle with radius 2, so $\int_{-2}^{2} \sqrt{4 - x^2}\, dx = \frac{1}{2}\pi \cdot 2^2 = 2\pi$.

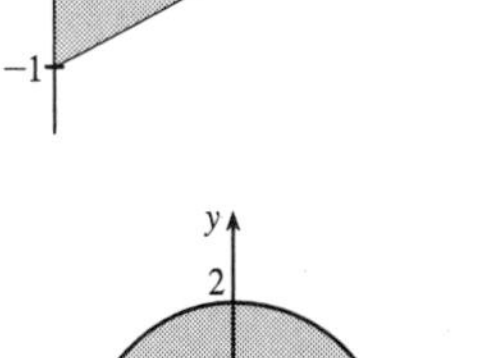

33. $\int_{-3}^{0} \left(1 + \sqrt{9 - x^2}\right) dx$ can be interpreted as the area under the graph of $f(x) = 1 + \sqrt{9 - x^2}$ between $x = -3$ and $x = 0$. This is equal to one-quarter the area of the circle with radius 3, plus the area of the rectangle, so $\int_{-3}^{0} \left(1 + \sqrt{9 - x^2}\right) dx = \frac{1}{4}\pi \cdot 3^2 + 1 \cdot 3 = 3 + \frac{9}{4}\pi$.

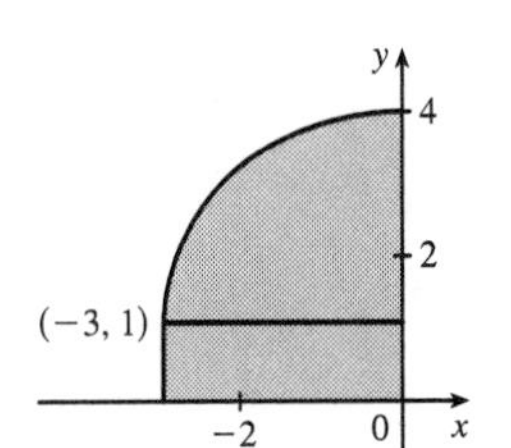

34. $\int_{-1}^{3} (3 - 2x)\, dx$ can be interpreted as the area of the triangle above the x-axis minus the area of the triangle below the x-axis; that is,

$\frac{1}{2}\left(\frac{5}{2}\right)(5) - \frac{1}{2}\left(\frac{3}{2}\right)(3) = \frac{25}{4} - \frac{9}{4} = 4$.

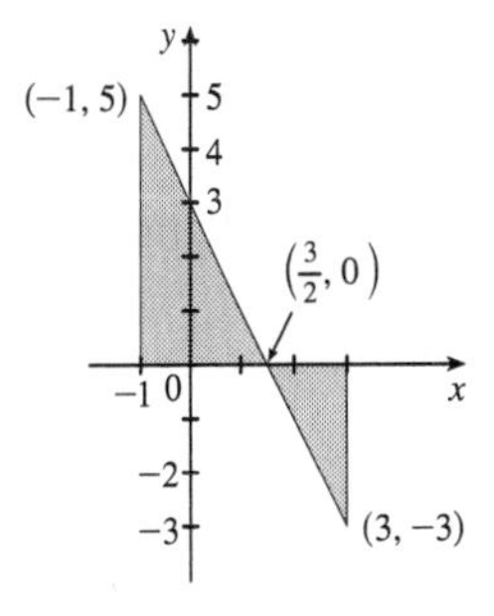

35. $\int_{-1}^{2} |x|\, dx$ can be interpreted as the sum of the areas of the two shaded triangles; that is, $\frac{1}{2}(1)(1) + \frac{1}{2}(2)(2) = \frac{1}{2} + \frac{4}{2} = \frac{5}{2}$.

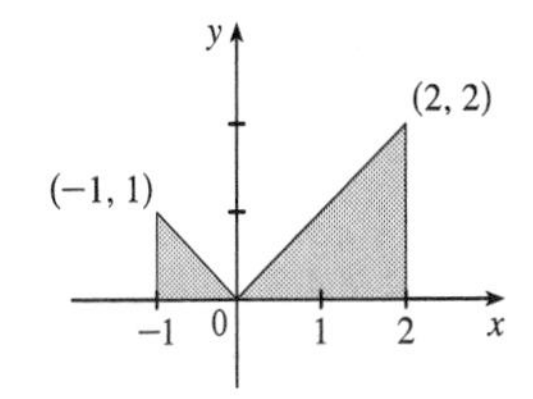

36. $\int_0^{10} |x - 5|\, dx$ can be interpreted as the sum of the areas of the two shaded triangles; that is, $2\left(\frac{1}{2}\right)(5)(5) = 25$.

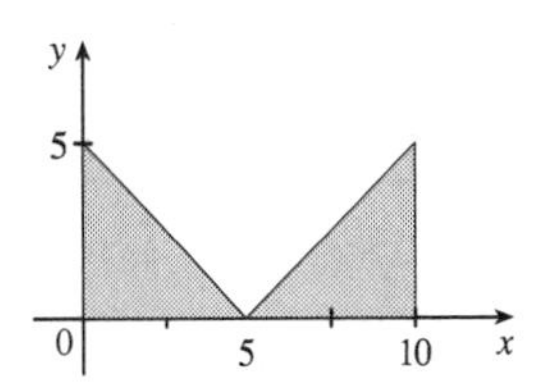

37. $\int_9^4 \sqrt{t}\, dt = -\int_4^9 \sqrt{t}\, dt$ [because we reversed the limits of integration]

$= -\int_4^9 \sqrt{x}\, dx$ [we can use any letter without changing the value of the integral]

$= -\frac{38}{3}$

38. $\int_1^1 x^2 \cos x\, dx = 0$ since the limits of integration are equal.

39. $\int_{-2}^{2} f(x)\,dx + \int_{2}^{5} f(x)\,dx - \int_{-2}^{-1} f(x)\,dx = \int_{-2}^{5} f(x)\,dx + \int_{-1}^{-2} f(x)\,dx$ [by Property 5 and reversing limits]

$= \int_{-1}^{5} f(x)\,dx$ [Property 5]

40. $\int_{1}^{4} f(x)\,dx = \int_{1}^{5} f(x)\,dx - \int_{4}^{5} f(x)\,dx = 12 - 3.6 = 8.4$

41. $\int_{0}^{9}[2f(x) + 3g(x)]\,dx = 2\int_{0}^{9} f(x)\,dx + 3\int_{0}^{9} g(x)\,dx = 2(37) + 3(16) = 122$

42. If $f(x) = \begin{cases} 3 & \text{for } x < 3 \\ x & \text{for } x \geq 3 \end{cases}$, then $\int_{0}^{5} f(x)\,dx$ can be interpreted as the area of the shaded region, which consists of a 5-by-3 rectangle surmounted by an isosceles right triangle whose legs have length 2. Thus,

$$\int_{0}^{5} f(x)\,dx = 5(3) + \tfrac{1}{2}(2)(2) = 17$$

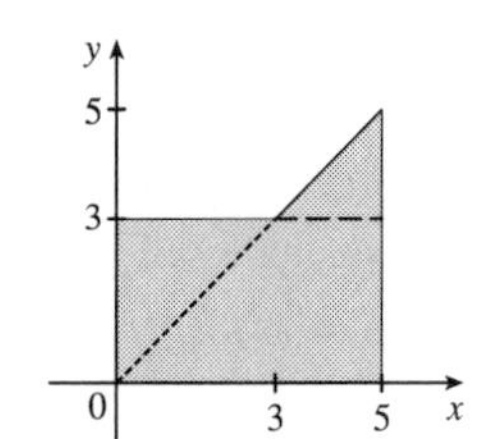

43. $\int_{0}^{1}(5 - 6x^2)\,dx = \int_{0}^{1} 5\,dx - 6\int_{0}^{1} x^2\,dx = 5(1-0) - 6\left(\tfrac{1}{3}\right) = 5 - 2 = 3$

44. Using Integral Comparison Property 8,

$m \leq f(x) \leq M \quad\Rightarrow\quad m(2-0) \leq \int_{0}^{2} f(x)\,dx \leq M(2-0) \quad\Rightarrow\quad 2m \leq \int_{0}^{2} f(x)\,dx \leq 2M.$

45. If $-1 \leq x \leq 1$, then $0 \leq x^2 \leq 1$ and $1 \leq 1 + x^2 \leq 2$, so $1 \leq \sqrt{1+x^2} \leq \sqrt{2}$ and $1[1-(-1)] \leq \int_{-1}^{1}\sqrt{1+x^2}\,dx \leq \sqrt{2}\,[1-(-1)]$ [Property 8]; that is, $2 \leq \int_{-1}^{1}\sqrt{1+x^2}\,dx \leq 2\sqrt{2}$.

46. If $0 \leq x \leq 2$, then $0 \leq x^3 \leq 8$, so $1 \leq x^3 + 1 \leq 9$ and $1 \leq \sqrt{x^3+1} \leq 3$.

Thus, $1(2-0) \leq \int_{0}^{2}\sqrt{x^3+1}\,dx \leq 3(2-0)$; that is, $2 \leq \int_{0}^{2}\sqrt{x^3+1}\,dx \leq 6$.

47. If $1 \leq x \leq 2$, then $\frac{1}{2} \leq \frac{1}{x} \leq 1$, so $\frac{1}{2}(2-1) \leq \displaystyle\int_{1}^{2}\frac{1}{x}\,dx \leq 1(2-1)$ or $\frac{1}{2} \leq \displaystyle\int_{1}^{2}\frac{1}{x}\,dx \leq 1$.

48. Let $f(x) = x^3 - 3x + 3$ for $0 \leq x \leq 2$. Then $f'(x) = 3x^2 - 3 = 3(x+1)(x-1)$, so f is decreasing on $(0,1)$ and increasing on $(1,2)$. f has the absolute minimum value $f(1) = 1$. Since $f(0) = 3$ and $f(2) = 5$, the absolute maximum value of f is $f(2) = 5$. Thus, $1 \leq x^3 - 3x + 3 \leq 5$ for x in $[0,2]$. It follows from Property 8 that $1\cdot(2-0) \leq \int_{0}^{2}(x^3 - 3x + 3)\,dx \leq 5\cdot(2-0)$; that is, $2 \leq \int_{0}^{2}(x^3 - 3x + 3)\,dx \leq 10$.

49. If $\frac{\pi}{4} \leq x \leq \frac{\pi}{3}$, then $1 \leq \tan x \leq \sqrt{3}$, so $1\left(\frac{\pi}{3} - \frac{\pi}{4}\right) \leq \int_{\pi/4}^{\pi/3}\tan x\,dx \leq \sqrt{3}\left(\frac{\pi}{3} - \frac{\pi}{4}\right)$ or $\frac{\pi}{12} \leq \int_{\pi/4}^{\pi/3}\tan x\,dx \leq \frac{\pi}{12}\sqrt{3}$.

50. If $\frac{1}{4}\pi \leq x \leq \frac{3}{4}\pi$, then $\frac{\sqrt{2}}{2} \leq \sin x \leq 1$ and $\frac{1}{2} \leq \sin^2 x \leq 1$, so $\frac{1}{2}\left(\frac{3}{4}\pi - \frac{1}{4}\pi\right) \leq \int_{\pi/4}^{3\pi/4}\sin^2 x\,dx \leq 1\left(\frac{3}{4}\pi - \frac{1}{4}\pi\right)$; that is, $\frac{1}{4}\pi \leq \int_{\pi/4}^{3\pi/4}\sin^2 x\,dx \leq \frac{1}{2}\pi$.

51. $\displaystyle\lim_{n\to\infty}\sum_{i=1}^{n}\frac{i^4}{n^5} = \lim_{n\to\infty}\sum_{i=1}^{n}\frac{i^4}{n^4}\cdot\frac{1}{n} = \lim_{n\to\infty}\sum_{i=1}^{n}\left(\frac{i}{n}\right)^4\frac{1}{n}$. At this point, we need to recognize the limit as being of the form $\displaystyle\lim_{n\to\infty}\sum_{i=1}^{n} f(x_i)\,\Delta x$, where $\Delta x = (1-0)/n = 1/n$, $x_i = 0 + i\,\Delta x = i/n$, and $f(x) = x^4$. Thus, the definite integral is $\int_{0}^{1} x^4\,dx$.

4.3 Evaluating Definite Integrals

1. $\displaystyle\int_{-1}^{3} x^5\,dx = \left[\frac{x^6}{6}\right]_{-1}^{3} = \frac{3^6}{6} - \frac{(-1)^6}{6} = \frac{729-1}{6} = \frac{364}{3}$

2. $\int_1^3 (1+2x-4x^3)\,dx = \left[x + 2\cdot\frac{1}{2}x^2 - 4\cdot\frac{1}{4}x^4\right]_1^3 = \left[x+x^2-x^4\right]_1^3 = (3+9-81)-(1+1-1) = -69-1 = -70$

3. $\int_0^2 (6x^2-4x+5)\,dx = \left[6\cdot\frac{1}{3}x^3 - 4\cdot\frac{1}{2}x^2 + 5x\right]_0^2 = \left[2x^3-2x^2+5x\right]_0^2 = (16-8+10)-0 = 18$

4. $\int_{-2}^0 (u^5-u^3+u^2)\,du = \left[\frac{1}{6}u^6 - \frac{1}{4}u^4 + \frac{1}{3}u^3\right]_{-2}^0 = 0 - \left(\frac{32}{3} - 4 - \frac{8}{3}\right) = -4$

5. $\int_0^1 x^{4/5}\,dx = \left[\frac{5}{9}x^{9/5}\right]_0^1 = \frac{5}{9} - 0 = \frac{5}{9}$

6. $\int_1^8 \sqrt[3]{x}\,dx = \int_1^8 x^{1/3}\,dx = \left[\frac{3}{4}x^{4/3}\right]_1^8 = \frac{3}{4}(8^{4/3} - 1^{4/3}) = \frac{3}{4}(2^4-1) = \frac{3}{4}(16-1) = \frac{3}{4}(15) = \frac{45}{4}$

7. $\int_0^2 x(2+x^5)\,dx = \int_0^2 (2x+x^6)\,dx = \left[x^2 + \frac{1}{7}x^7\right]_0^2 = \left(4 + \frac{128}{7}\right) - (0+0) = \frac{156}{7}$

8. $\int_\pi^{2\pi} \cos\theta\,d\theta = \left[\sin\theta\right]_\pi^{2\pi} = \sin 2\pi - \sin\pi = 0 - 0 = 0$

9. $\int_{-2}^2 (3u+1)^2\,du = \int_{-2}^2 (9u^2+6u+1)\,du = \left[9\cdot\frac{1}{3}u^3 + 6\cdot\frac{1}{2}u^2 + u\right]_{-2}^2 = \left[3u^3+3u^2+u\right]_{-2}^2$
$= (24+12+2) - (-24+12-2) = 38 - (-14) = 52$

10. $\int_0^4 (2v+5)(3v-1)\,dv = \int_0^4 (6v^2+13v-5)\,dv = \left[6\cdot\frac{1}{3}v^3 + 13\cdot\frac{1}{2}v^2 - 5v\right]_0^4 = \left[2v^3 + \frac{13}{2}v^2 - 5v\right]_0^4$
$= (128+104-20) - 0 = 212$

11. $\displaystyle\int_{-2}^{-1}\left(4y^3 + \frac{2}{y^3}\right)dy = \left[4\cdot\frac{1}{4}y^4 + 2\cdot\frac{1}{-2}y^{-2}\right]_{-2}^{-1} = \left[y^4 - \frac{1}{y^2}\right]_{-2}^{-1} = (1-1) - \left(16 - \frac{1}{4}\right) = -\frac{63}{4}$

12. $\displaystyle\int_1^2 \frac{y+5y^7}{y^3}\,dy = \int_1^2 (y^{-2}+5y^4)\,dy = \left[-y^{-1} + 5\cdot\frac{1}{5}y^5\right]_1^2 = \left[-\frac{1}{y} + y^5\right]_1^2 = \left(-\frac{1}{2}+32\right) - (-1+1) = \frac{63}{2}$

13. $\int_0^1 x(\sqrt[3]{x} + \sqrt[4]{x})\,dx = \int_0^1 (x^{4/3} + x^{5/4})\,dx = \left[\frac{3}{7}x^{7/3} + \frac{4}{9}x^{9/4}\right]_0^1 = \left(\frac{3}{7} + \frac{4}{9}\right) - 0 = \frac{55}{63}$

14. $\displaystyle\int_1^9 \frac{3x-2}{\sqrt{x}}\,dx = \int_1^9 (3x^{1/2} - 2x^{-1/2})\,dx = \left[3\cdot\frac{2}{3}x^{3/2} - 2\cdot 2x^{1/2}\right]_1^9 = \left[2x^{3/2} - 4x^{1/2}\right]_1^9 = (54-12) - (2-4) = 44$

15. $\int_0^{\pi/4} \sec^2 t\,dt = \left[\tan t\right]_0^{\pi/4} = \tan\frac{\pi}{4} - \tan 0 = 1 - 0 = 1$

16. $\int_0^1 (3 + x\sqrt{x})\,dx = \int_0^1 \left(3 + x^{3/2}\right)dx = \left[3x + \frac{2}{5}x^{5/2}\right]_0^1 = \left[\left(3+\frac{2}{5}\right) - 0\right] = \frac{17}{5}$

17. $\int_\pi^{2\pi} \csc^2\theta\,d\theta$ does not exist because the function $f(\theta) = \csc^2\theta$ has infinite discontinuities at $\theta = \pi$ and $\theta = 2\pi$; that is, f is discontinuous on the interval $[\pi, 2\pi]$.

18. $\int_0^{\pi/6} \csc\theta\cot\theta\,d\theta$ does not exist because the function $f(\theta) = \csc\theta\cot\theta$ has an infinite discontinuity at $\theta = 0$; that is, f is discontinuous on the interval $\left[0, \frac{\pi}{6}\right]$.

19. $\int_0^1 \left(\sqrt[4]{x^5} + \sqrt[5]{x^4}\right) dx = \int_0^1 \left(x^{5/4} + x^{4/5}\right) dx = \left[\dfrac{x^{9/4}}{9/4} + \dfrac{x^{9/5}}{9/5}\right]_0^1 = \left[\frac{4}{9}x^{9/4} + \frac{5}{9}x^{9/5}\right]_0^1 = \frac{4}{9} + \frac{5}{9} - 0 = 1$

20. $\int_0^9 \sqrt{2t}\, dt = \int_0^9 \sqrt{2}\, t^{1/2}\, dt = \left[\sqrt{2} \cdot \frac{2}{3} t^{3/2}\right]_0^9 = \sqrt{2} \cdot \frac{2}{3} \cdot 27 - 0 = 18\sqrt{2}$

21. $$\int_1^{64} \frac{1 + \sqrt[3]{x}}{\sqrt{x}}\, dx = \int_1^{64} \left(\frac{1}{x^{1/2}} + \frac{x^{1/3}}{x^{1/2}}\right) dx = \int_1^{64} \left(x^{-1/2} + x^{(1/3)-(1/2)}\right) dx = \int_1^{64} \left(x^{-1/2} + x^{-1/6}\right) dx$$
$$= \left[2x^{1/2} + \tfrac{6}{5}x^{5/6}\right]_1^{64} = \left(16 + \tfrac{192}{5}\right) - \left(2 + \tfrac{6}{5}\right) = 14 + \tfrac{186}{5} = \tfrac{256}{5}$$

22. $\int_{\pi/4}^{\pi/3} \sec\theta \tan\theta\, d\theta = \left[\sec\theta\right]_{\pi/4}^{\pi/3} = \sec\frac{\pi}{3} - \sec\frac{\pi}{4} = 2 - \sqrt{2}$

23. $\int_1^4 \sqrt{5/x}\, dx = \sqrt{5}\int_1^4 x^{-1/2}\, dx = \sqrt{5}\left[2\sqrt{x}\right]_1^4 = \sqrt{5}\,(2 \cdot 2 - 2 \cdot 1) = 2\sqrt{5}$

24. $\int_0^1 (1 + x^2)^3\, dx = \int_0^1 (1 + 3x^2 + 3x^4 + x^6)\, dx = \left[x + x^3 + \frac{3}{5}x^5 + \frac{1}{7}x^7\right]_0^1 = \left(1 + 1 + \frac{3}{5} + \frac{1}{7}\right) - 0 = \frac{96}{35}$

25. $$\int_0^{\pi/4} \frac{1 + \cos^2\theta}{\cos^2\theta}\, d\theta = \int_0^{\pi/4} \left(\frac{1}{\cos^2\theta} + \frac{\cos^2\theta}{\cos^2\theta}\right) d\theta = \int_0^{\pi/4} (\sec^2\theta + 1)\, d\theta$$
$$= \left[\tan\theta + \theta\right]_0^{\pi/4} = \left(\tan\tfrac{\pi}{4} + \tfrac{\pi}{4}\right) - (0 + 0) = 1 + \tfrac{\pi}{4}$$

26. $$\int_0^{\pi/3} \frac{\sin\theta + \sin\theta\tan^2\theta}{\sec^2\theta}\, d\theta = \int_0^{\pi/3} \frac{\sin\theta\,(1 + \tan^2\theta)}{\sec^2\theta}\, d\theta = \int_0^{\pi/3} \frac{\sin\theta\sec^2\theta}{\sec^2\theta}\, d\theta = \int_0^{\pi/3} \sin\theta\, d\theta$$
$$= \left[-\cos\theta\right]_0^{\pi/3} = -\tfrac{1}{2} - (-1) = \tfrac{1}{2}$$

27. $$\int_{-1}^2 (x - 2|x|)\, dx = \int_{-1}^0 [x - 2(-x)]\, dx + \int_0^2 [x - 2(x)]\, dx = \int_{-1}^0 3x\, dx + \int_0^2 (-x)\, dx = 3\left[\tfrac{1}{2}x^2\right]_{-1}^0 - \left[\tfrac{1}{2}x^2\right]_0^2$$
$$= 3\left(0 - \tfrac{1}{2}\right) - (2 - 0) = -\tfrac{7}{2} = -3.5$$

28. $$\int_0^{3\pi/2} |\sin x|\, dx = \int_0^{\pi} \sin x\, dx + \int_\pi^{3\pi/2} (-\sin x)\, dx = [-\cos x]_0^\pi + [\cos x]_\pi^{3\pi/2}$$
$$= [1 - (-1)] + [0 - (-1)] = 2 + 1 = 3$$

29. $f(x) = 1/x^2$ is not continuous on the interval $[-1, 3]$, so the Evaluation Theorem does not apply. In fact, f has an infinite discontinuity at $x = 0$, so $\int_{-1}^3 (1/x^2)\, dx$ does not exist.

30. $f(x) = \sec^2 x$ is not continuous on the interval $[0, \pi]$, so the Evaluation Theorem does not apply. In fact, f has an infinite discontinuity at $x = \pi/2$, so $\int_0^\pi \sec^2 x\, dx$ does not exist.

31. It appears that the area under the graph is about $\frac{2}{3}$ of the area of the viewing rectangle, or about $\frac{2}{3}\pi \approx 2.1$. The actual area is $\int_0^\pi \sin x\, dx = [-\cos x]_0^\pi = (-\cos\pi) - (-\cos 0) = -(-1) + 1 = 2.$

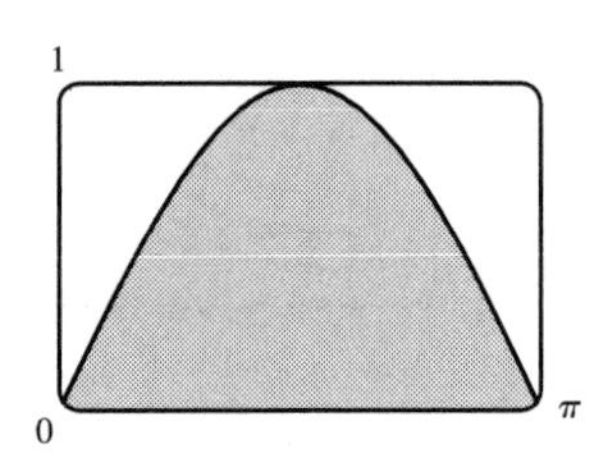

32. Splitting up the region as shown, we estimate that the area under the graph is $\frac{\pi}{3} + \frac{1}{4}\left(3 \cdot \frac{\pi}{3}\right) \approx 1.8$. The actual area is $\int_0^{\pi/3} \sec^2 x\, dx = [\tan x]_0^{\pi/3} = \sqrt{3} - 0 = \sqrt{3} \approx 1.73.$

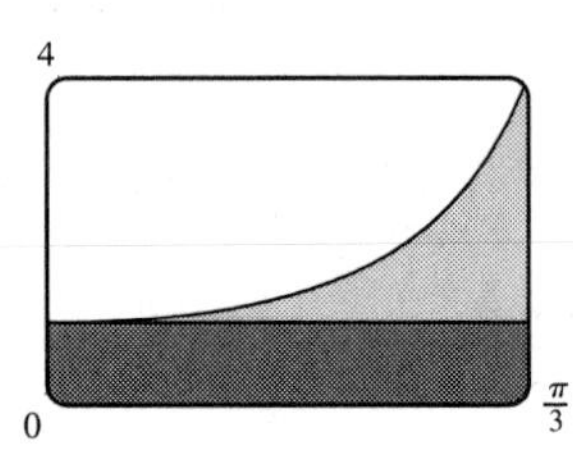

33. The graph shows that $y = x + x^2 - x^4$ has x-intercepts at $x = 0$ and at $x = a \approx 1.32$. So the area of the region that lies under the curve and above the x-axis is

$$\int_0^a (x + x^2 - x^4)\,dx = \left[\tfrac{1}{2}x^2 + \tfrac{1}{3}x^3 - \tfrac{1}{5}x^5\right]_0^a = \left(\tfrac{1}{2}a^2 + \tfrac{1}{3}a^3 - \tfrac{1}{5}a^5\right) - 0$$
$$\approx 0.84$$

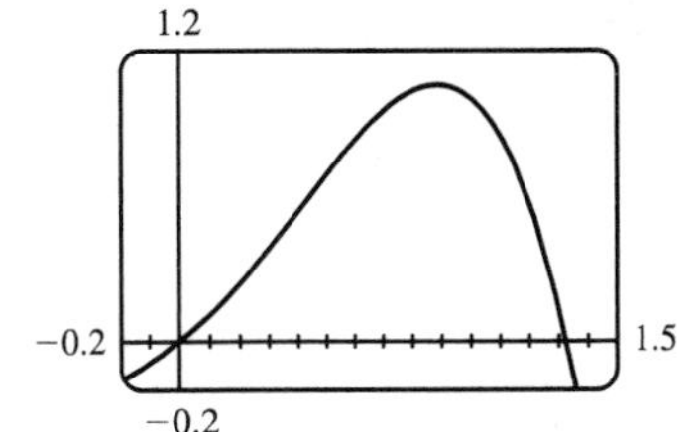

34. The graph shows that $y = 2x + 3x^4 - 2x^6$ has x-intercepts at $x = 0$ and at $x = a \approx 1.37$. So the area of the region that lies under the curve and above the x-axis is

$$\int_0^a (2x + 3x^4 - 2x^6)\,dx = \left[x^2 + \tfrac{3}{5}x^5 - \tfrac{2}{7}x^7\right]_0^a = \left(a^2 + \tfrac{3}{5}a^5 - \tfrac{2}{7}a^7\right) - 0$$
$$\approx 2.18$$

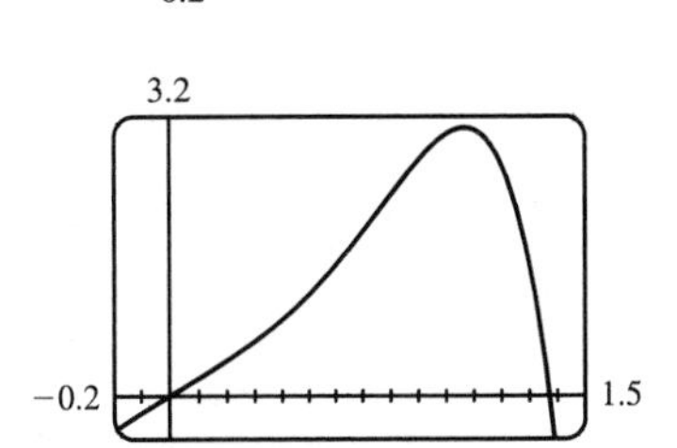

35. $\int_{-1}^{2} x^3\,dx = \left[\tfrac{1}{4}x^4\right]_{-1}^{2} = 4 - \tfrac{1}{4} = \tfrac{15}{4} = 3.75$

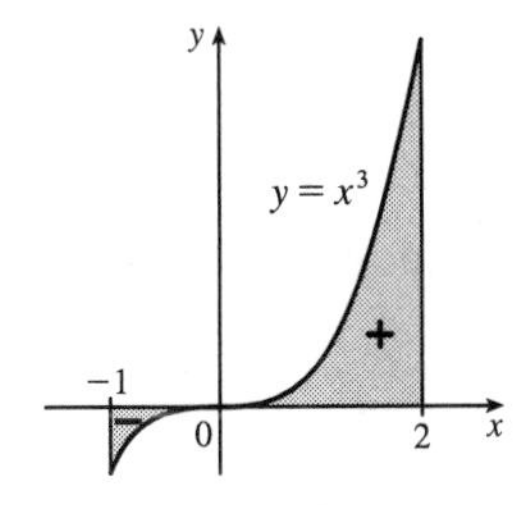

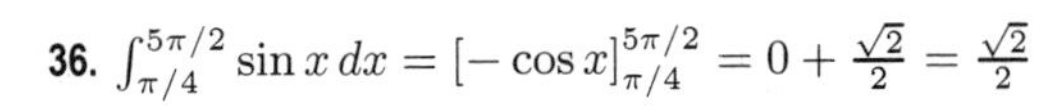

36. $\int_{\pi/4}^{5\pi/2} \sin x\,dx = [-\cos x]_{\pi/4}^{5\pi/2} = 0 + \frac{\sqrt{2}}{2} = \frac{\sqrt{2}}{2}$

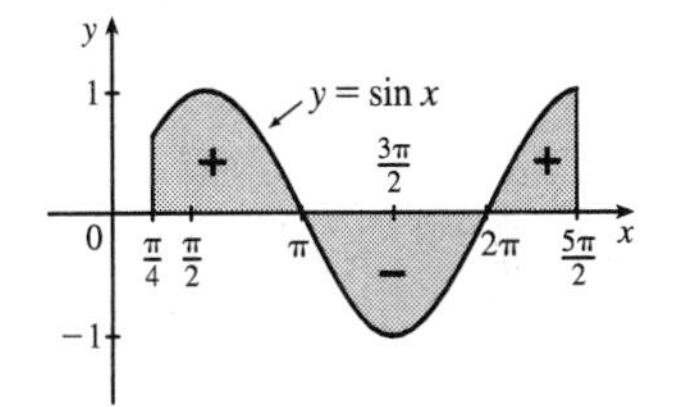

37. $\dfrac{d}{dx}\left[\sqrt{x^2+1} + C\right] = \dfrac{d}{dx}\left[(x^2+1)^{1/2} + C\right] = \tfrac{1}{2}(x^2+1)^{-1/2} \cdot 2x = \dfrac{x}{\sqrt{x^2+1}}$

38. $\dfrac{d}{dx}\left[x \sin x + \cos x + C\right] = x \cos x + (\sin x) \cdot 1 - \sin x = x \cos x$

39. $\int x\sqrt{x}\,dx = \int x^{3/2}\,dx = \tfrac{2}{5}x^{5/2} + C.$

The members of the family in the figure correspond to $C = 5, 3, 0, -2$, and -4.

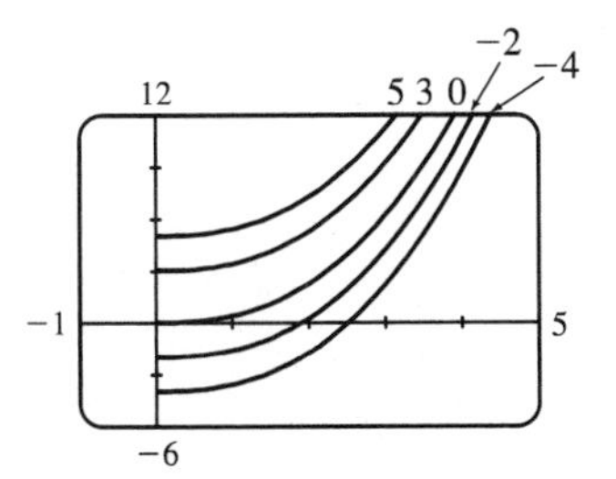

40. $\int (\cos x - 2\sin x)\,dx = \sin x + 2\cos x + C.$

The members of the family in the figure correspond to $C = 5, 3, 0, -2$, and -4.

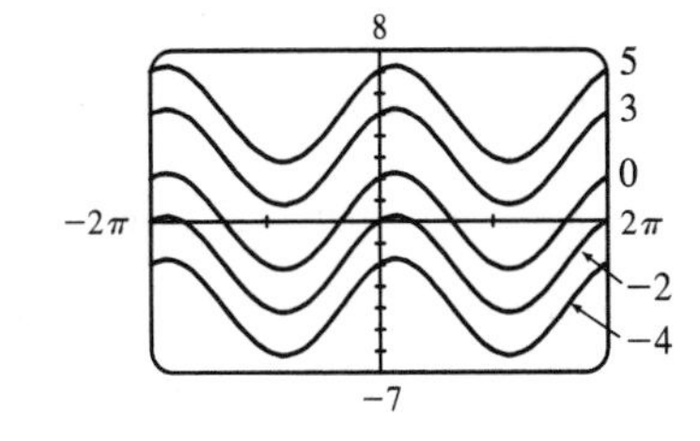

41. $\displaystyle\int (1-t)(2+t^2)\,dt = \int (2 - 2t + t^2 - t^3)\,dt = 2t - 2\frac{t^2}{2} + \frac{t^3}{3} - \frac{t^4}{4} + C = 2t - t^2 + \tfrac{1}{3}t^3 - \tfrac{1}{4}t^4 + C$

42. $\displaystyle\int x(1 + 2x^4)\,dx = \int (x + 2x^5)\,dx = \frac{x^2}{2} + 2\frac{x^6}{6} + C = \tfrac{1}{2}x^2 + \tfrac{1}{3}x^6 + C$

43. $\displaystyle\int \frac{\sin x}{1 - \sin^2 x}\,dx = \int \frac{\sin x}{\cos^2 x}\,dx = \int \frac{1}{\cos x} \cdot \frac{\sin x}{\cos x}\,dx = \int \sec x\,\tan x\,dx = \sec x + C$

44. $\displaystyle\int \frac{\sin 2x}{\sin x}\,dx = \int \frac{2\sin x\,\cos x}{\sin x}\,dx = \int 2\cos x\,dx = 2\sin x + C$

45. $A = \int_0^2 (2y - y^2)\,dy = \left[y^2 - \frac{1}{3}y^3\right]_0^2 = \left(4 - \frac{8}{3}\right) - 0 = \frac{4}{3}$

46. $y = \sqrt[4]{x} \;\Rightarrow\; x = y^4$, so $A = \int_0^1 y^4\,dy = \left[\frac{1}{5}y^5\right]_0^1 = \frac{1}{5}$.

47. If $w'(t)$ is the rate of change of weight in pounds per year, then $w(t)$ represents the weight in pounds of the child at age t. We know from the Net Change Theorem that $\int_5^{10} w'(t)\,dt = w(10) - w(5)$, so the integral represents the increase in the child's weight (in pounds) between the ages of 5 and 10.

48. $\int_a^b I(t)\,dt = \int_a^b Q'(t)\,dt = Q(b) - Q(a)$ by the Net Change Theorem, so it represents the change in the charge Q from time $t = a$ to $t = b$.

49. Since $r(t)$ is the rate at which oil leaks, we can write $r(t) = -V'(t)$, where $V(t)$ is the volume of oil at time t. [Note that the minus sign is needed because V is decreasing, so $V'(t)$ is negative, but $r(t)$ is positive.] Thus, by the Net Change Theorem, $\int_0^{120} r(t)\,dt = -\int_0^{120} V'(t)\,dt = -\left[V(120) - V(0)\right] = V(0) - V(120)$, which is the number of gallons of oil that leaked from the tank in the first two hours (120 minutes).

50. By the Net Change Theorem, $\int_0^{15} n'(t)\,dt = n(15) - n(0) = n(15) - 100$ represents the increase in the bee population in 15 weeks. So $100 + \int_0^{15} n'(t)\,dt = n(15)$ represents the total bee population after 15 weeks.

51. By the Net Change Theorem, $\int_{1000}^{5000} R'(x)\,dx = R(5000) - R(1000)$, so it represents the increase in revenue when production is increased from 1000 units to 5000 units.

52. The slope of the trail is the rate of change of the elevation E, so $f(x) = E'(x)$. By the Net Change Theorem, $\int_3^5 f(x)\,dx = \int_3^5 E'(x)\,dx = E(5) - E(3)$ is the change in the elevation E between $x = 3$ miles and $x = 5$ miles from the start of the trail.

53. In general, the unit of measurement for $\int_a^b f(x)\,dx$ is the product of the unit for $f(x)$ and the unit for x. Since $f(x)$ is measured in newtons and x is measured in meters, the units for $\int_0^{100} f(x)\,dx$ are newton-meters. (A newton-meter is abbreviated N·m.)

54. The units for $a(x)$ are pounds per foot and the units for x are feet, so the units for da/dx are pounds per foot per foot, denoted (lb/ft)/ft. The unit of measurement for $\int_2^8 a(x)\,dx$ is the product of pounds per foot and feet; that is, pounds.

55. (a) Displacement $= \int_0^3 (3t - 5)\,dt = \left[\frac{3}{2}t^2 - 5t\right]_0^3 = \frac{27}{2} - 15 = -\frac{3}{2}$ m

(b) Distance traveled $= \int_0^3 |3t - 5|\,dt = \int_0^{5/3} (5 - 3t)\,dt + \int_{5/3}^3 (3t - 5)\,dt$

$$= \left[5t - \tfrac{3}{2}t^2\right]_0^{5/3} + \left[\tfrac{3}{2}t^2 - 5t\right]_{5/3}^3 = \tfrac{25}{3} - \tfrac{3}{2}\cdot\tfrac{25}{9} + \tfrac{27}{2} - 15 - \left(\tfrac{3}{2}\cdot\tfrac{25}{9} - \tfrac{25}{3}\right) = \tfrac{41}{6}\text{ m}$$

56. (a) Displacement $= \int_1^6 (t^2 - 2t - 8)\,dt = \left[\frac{1}{3}t^3 - t^2 - 8t\right]_1^6 = (72 - 36 - 48) - \left(\frac{1}{3} - 1 - 8\right) = -\frac{10}{3}$ m

(b) Distance traveled $= \int_1^6 |t^2 - 2t - 8|\,dt = \int_1^6 |(t-4)(t+2)|\,dt$

$$= \int_1^4 (-t^2 + 2t + 8)\,dt + \int_4^6 (t^2 - 2t - 8)\,dt = \left[-\tfrac{1}{3}t^3 + t^2 + 8t\right]_1^4 + \left[\tfrac{1}{3}t^3 - t^2 - 8t\right]_4^6$$

$$= \left(-\tfrac{64}{3} + 16 + 32\right) - \left(-\tfrac{1}{3} + 1 + 8\right) + (72 - 36 - 48) - \left(\tfrac{64}{3} - 16 - 32\right) = \tfrac{98}{3}\text{ m}$$

57. (a) $v'(t) = a(t) = t + 4 \;\Rightarrow\; v(t) = \frac{1}{2}t^2 + 4t + C \;\Rightarrow\; v(0) = C = 5 \;\Rightarrow\; v(t) = \frac{1}{2}t^2 + 4t + 5$ m/s

(b) Distance traveled $= \int_0^{10} |v(t)|\,dt = \int_0^{10} \left|\frac{1}{2}t^2 + 4t + 5\right|\,dt = \int_0^{10} \left(\frac{1}{2}t^2 + 4t + 5\right)dt$

$$= \left[\tfrac{1}{6}t^3 + 2t^2 + 5t\right]_0^{10} = \tfrac{500}{3} + 200 + 50 = 416\tfrac{2}{3}\text{ m}$$

58. (a) $v'(t) = a(t) = 2t + 3 \;\Rightarrow\; v(t) = t^2 + 3t + C \;\Rightarrow\; v(0) = C = -4 \;\Rightarrow\; v(t) = t^2 + 3t - 4$

(b) Distance traveled $= \int_0^3 |t^2 + 3t - 4|\,dt = \int_0^3 |(t+4)(t-1)|\,dt = \int_0^1 (-t^2 - 3t + 4)\,dt + (t^2 + 3t - 4)\,dt$

$$= \left[-\tfrac{1}{3}t^3 - \tfrac{3}{2}t^2 + 4t\right]_0^1 + \left[\tfrac{1}{3}t^3 + \tfrac{3}{2}t^2 - 4t\right]_1^3$$

$$= \left(-\tfrac{1}{3} - \tfrac{3}{2} + 4\right) + \left(9 + \tfrac{27}{2} - 12\right) - \left(\tfrac{1}{3} + \tfrac{3}{2} - 4\right) = \tfrac{89}{6}\text{ m}$$

59. Let s be the position of the car. We know from Equation 2 that $s(100) - s(0) = \int_0^{100} v(t)\,dt$. We use the Midpoint Rule for $0 \le t \le 100$ with $n = 5$. Note that the length of each of the five time intervals is 20 seconds $= \frac{20}{3600}$ hour $= \frac{1}{180}$ hour. So the distance traveled is

$$\int_0^{100} v(t)\,dt \approx \tfrac{1}{180}[v(10) + v(30) + v(50) + v(70) + v(90)] = \tfrac{1}{180}(38 + 58 + 51 + 53 + 47) = \tfrac{247}{180} \approx 1.4\text{ miles.}$$

60. (a) By the Net Change Theorem, the total amount spewed into the atmosphere is $Q(6) - Q(0) = \int_0^6 r(t)\,dt = Q(6)$ since $Q(0) = 0$. The rate $r(t)$ is positive, so Q is an increasing function. Thus, an upper estimate for $Q(6)$ is R_6 and a lower estimate for $Q(6)$ is L_6. $\Delta t = \dfrac{b-a}{n} = \dfrac{6-0}{6} = 1$.

$$R_6 = \sum_{i=1}^{6} r(t_i)\,\Delta t = 10 + 24 + 36 + 46 + 54 + 60 = 230\text{ tonnes.}$$

$$L_6 = \sum_{i=1}^{6} r(t_{i-1})\,\Delta t = R_6 + r(0) - r(6) = 230 + 2 - 60 = 172\text{ tonnes.}$$

(b) $\Delta t = \dfrac{b-a}{n} = \dfrac{6-0}{3} = 2$. $Q(6) \approx M_3 = 2[r(1) + r(3) + r(5)] = 2(10 + 36 + 54) = 2(100) = 200$ tonnes.

61. By the Net Change Theorem, the amount of water that flows from the tank during the first 10 minutes is

$$\int_0^{10} r(t)\,dt = \int_0^{10} (200 - 4t)\,dt = \left[200t - 2t^2\right]_0^{10} = (2000 - 200) - 0 = 1800\text{ liters.}$$

62. By the Net Change Theorem, the amount of water after four days is

$$25{,}000 + \int_0^4 r(t)\,dt \approx 25{,}000 + M_4 = 25{,}000 + \tfrac{4-0}{4}\,[r(0.5) + r(1.5) + r(2.5) + r(3.5)]$$

$$\approx 25{,}000 + [1500 + 1770 + 740 + (-690)] = 28{,}320\text{ liters}$$

4.4 The Fundamental Theorem of Calculus

1. (a) $g(x) = \int_0^x f(t)\,dt$.

$g(0) = \int_0^0 f(t)\,dt = 0$

$g(1) = \int_0^1 f(t)\,dt = 1 \cdot 2 = 2$ [rectangle]

$g(2) = \int_0^2 f(t)\,dt = \int_0^1 f(t)\,dt + \int_1^2 f(t)\,dt$
$= g(1) + \int_1^2 f(t)\,dt$
$= 2 + 1 \cdot 2 + \frac{1}{2} \cdot 1 \cdot 2 = 5$ [rectangle plus triangle]

$g(3) = \int_0^3 f(t)\,dt = g(2) + \int_2^3 f(t)\,dt$
$= 5 + \frac{1}{2} \cdot 1 \cdot 4 = 7$

$g(6) = g(3) + \int_3^6 f(t)\,dt$ [the integral is negative since f lies under the x-axis]
$= 7 + \left[-\left(\frac{1}{2} \cdot 2 \cdot 2 + 1 \cdot 2\right)\right] = 7 - 4 = 3$

(b) g is increasing on $(0, 3)$ because as x increases from 0 to 3, we keep adding more area.

(c) g has a maximum value when we start subtracting area; that is, at $x = 3$.

(d)

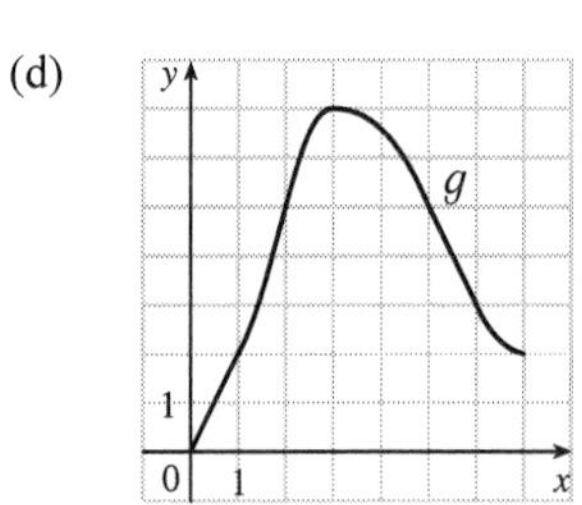

2. (a) $g(x) = \int_0^x f(t)\,dt$, so $g(0) = \int_0^0 f(t)\,dt = 0$.

$g(1) = \int_0^1 f(t)\,dt = \frac{1}{2} \cdot 1 \cdot 1$ [area of triangle]
$= \frac{1}{2}$

$g(2) = \int_0^2 f(t)\,dt = \int_0^1 f(t)\,dt + \int_1^2 f(t)\,dt$ [below the x-axis]
$= \frac{1}{2} - \frac{1}{2} \cdot 1 \cdot 1 = 0$

$g(3) = g(2) + \int_2^3 f(t)\,dt = 0 - \frac{1}{2} \cdot 1 \cdot 1 = -\frac{1}{2}$

$g(4) = g(3) + \int_3^4 f(t)\,dt = -\frac{1}{2} + \frac{1}{2} \cdot 1 \cdot 1 = 0$

$g(5) = g(4) + \int_4^5 f(t)\,dt = 0 + 1.5 = 1.5$

$g(6) = g(5) + \int_5^6 f(t)\,dt = 1.5 + 2.5 = 4$

(b) $g(7) = g(6) + \int_6^7 f(t)\,dt$
$\approx 4 + 2.2$ [estimate from the graph]
$= 6.2$

(c) The answers from part (a) and part (b) indicate that g has a minimum at $x = 3$ and a maximum at $x = 7$. This makes sense from the graph of f since we are subtracting area on $1 < x < 3$ and adding area on $3 < x < 7$.

(d)

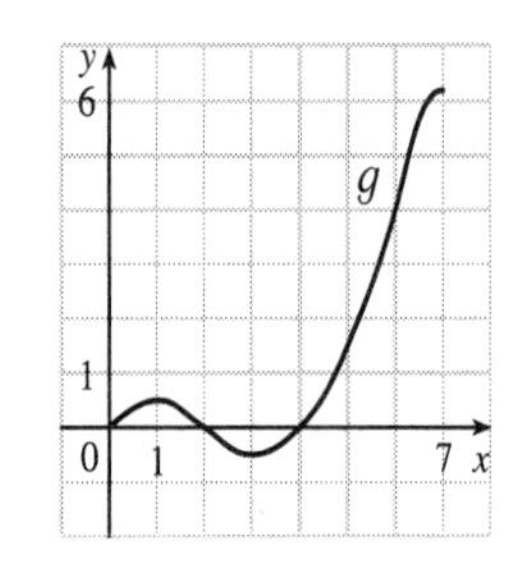

3.

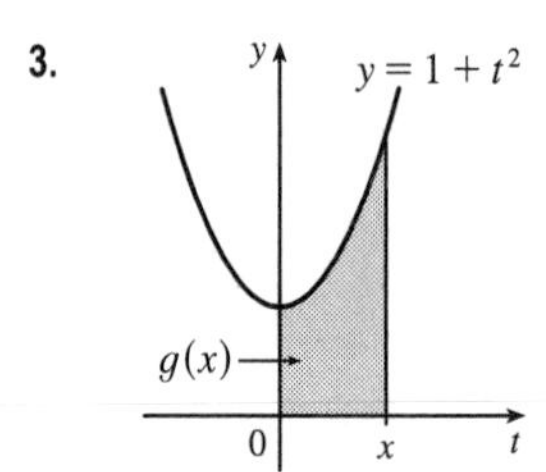

(a) By FTC1, $g(x) = \int_0^x (1 + t^2)\,dt \Rightarrow g'(x) = f(x) = 1 + x^2$.

(b) By FTC2, $g(x) = \int_0^x (1 + t^2)\,dt = \left[t + \frac{1}{3}t^3\right]_0^x = \left(x + \frac{1}{3}x^3\right) - 0 \Rightarrow g'(x) = 1 + x^2$.

4. 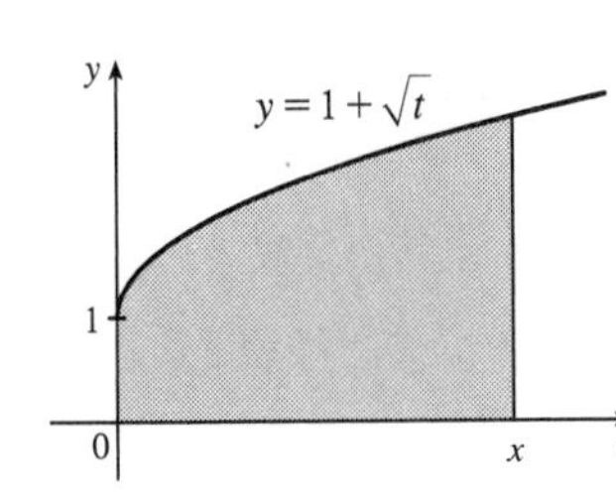

(a) By FTC1 with $f(t) = 1 + \sqrt{t}$ and $a = 0$,

$g(x) = \int_0^x (1 + \sqrt{t})\,dt \Rightarrow g'(x) = f(x) = 1 + \sqrt{x}$.

(b) Using FTC2, $g(x) = \int_0^x (1 + \sqrt{t})\,dt = \left[t + \frac{2}{3}t^{3/2}\right]_0^x = x + \frac{2}{3}x^{3/2} \Rightarrow$

$g'(x) = 1 + x^{1/2} = 1 + \sqrt{x}$.

5. $f(t) = \sqrt{1 + 2t}$ and $g(x) = \int_0^x \sqrt{1 + 2t}\,dt$, so by FTC1, $g'(x) = f(x) = \sqrt{1 + 2x}$.

6. $f(t) = (2 + t^4)^5$ and $g(x) = \int_1^x (2 + t^4)^5\,dt$, so $g'(x) = f(x) = (2 + x^4)^5$.

7. $f(t) = t^2 \sin t$ and $g(y) = \int_2^y t^2 \sin t\,dt$, so by FTC1, $g'(y) = f(y) = y^2 \sin y$.

8. $f(\theta) = \tan\theta$ and $F(x) = \int_x^{10} \tan\theta\,d\theta = -\int_{10}^x \tan\theta\,d\theta$, so by FTC1, $F'(x) = -f(x) = -\tan x$.

9. Let $u = \frac{1}{x}$. Then $\frac{du}{dx} = -\frac{1}{x^2}$. Also, $\frac{dh}{dx} = \frac{dh}{du}\frac{du}{dx}$, so

$$h'(x) = \frac{d}{dx}\int_2^{1/x} \sin^4 t\,dt = \frac{d}{du}\int_2^u \sin^4 t\,dt \cdot \frac{du}{dx} = \sin^4 u\,\frac{du}{dx} = \frac{-\sin^4(1/x)}{x^2}.$$

10. Let $u = x^2$. Then $\frac{du}{dx} = 2x$. Also, $\frac{dh}{dx} = \frac{dh}{du}\frac{du}{dx}$, so

$$h'(x) = \frac{d}{dx}\int_0^{x^2} \sqrt{1 + r^3}\,dr = \frac{d}{du}\int_0^u \sqrt{1 + r^3}\,dr \cdot \frac{du}{dx} = \sqrt{1 + u^3}(2x) = 2x\sqrt{1 + (x^2)^3} = 2x\sqrt{1 + x^6}.$$

11. Let $u = \sqrt{x}$. Then $\frac{du}{dx} = \frac{1}{2\sqrt{x}}$. Also, $\frac{dy}{dx} = \frac{dy}{du}\frac{du}{dx}$, so

$$y' = \frac{d}{dx}\int_3^{\sqrt{x}} \frac{\cos t}{t}\,dt = \frac{d}{du}\int_3^u \frac{\cos t}{t}\,dt \cdot \frac{du}{dx} = \frac{\cos u}{u} \cdot \frac{1}{2\sqrt{x}} = \frac{\cos\sqrt{x}}{\sqrt{x}} \cdot \frac{1}{2\sqrt{x}} = \frac{\cos\sqrt{x}}{2x}.$$

12. Let $u = \frac{1}{x^2}$. Then $\frac{du}{dx} = -\frac{2}{x^3}$. Also, $\frac{dy}{dx} = \frac{dy}{du}\frac{du}{dx}$, so

$$y' = \frac{d}{dx}\int_{1/x^2}^0 \sin^3 t\,dt = \frac{d}{du}\int_u^0 \sin^3 t\,dt \cdot \frac{du}{dx} = -\frac{d}{du}\int_0^u \sin^3 t\,dt \cdot \frac{du}{dx} = -\sin^3 u\left(-\frac{2}{x^3}\right) = \frac{2\sin^3(1/x^2)}{x^3}.$$

13. $$g(x) = \int_{2x}^{3x} \frac{u^2 - 1}{u^2 + 1}\,du = \int_{2x}^0 \frac{u^2 - 1}{u^2 + 1}\,du + \int_0^{3x} \frac{u^2 - 1}{u^2 + 1}\,du = -\int_0^{2x} \frac{u^2 - 1}{u^2 + 1}\,du + \int_0^{3x} \frac{u^2 - 1}{u^2 + 1}\,du \Rightarrow$$

$$g'(x) = -\frac{(2x)^2 - 1}{(2x)^2 + 1} \cdot \frac{d}{dx}(2x) + \frac{(3x)^2 - 1}{(3x)^2 + 1} \cdot \frac{d}{dx}(3x) = -2 \cdot \frac{4x^2 - 1}{4x^2 + 1} + 3 \cdot \frac{9x^2 - 1}{9x^2 + 1}$$

14. $$y = \int_{\sin x}^{\cos x} (1 + v^2)^{10}\,dv = \int_{\sin x}^0 (1 + v^2)^{10}\,dv + \int_0^{\cos x} (1 + v^2)^{10}\,dv$$

$$= -\int_0^{\sin x} (1 + v^2)^{10}\,dv + \int_0^{\cos x} (1 + v^2)^{10}\,dv \Rightarrow$$

$$y' = -(1 + (\sin x)^2)^{10} \cdot \frac{d}{dx}(\sin x) + (1 + (\cos x)^2)^{10} \cdot \frac{d}{dx}(\cos x) = -(1 + \sin^2 x)^{10}\cos x - (1 + \cos^2 x)^{10}\sin x$$

15. $$f_{\text{ave}} = \frac{1}{b - a}\int_a^b f(x)\,dx = \frac{1}{1 - (-1)}\int_{-1}^1 x^2\,dx = \frac{1}{2} \cdot 2\int_0^1 x^2\,dx = \left[\frac{1}{3}x^3\right]_0^1 = \frac{1}{3}$$

16. $f_{\text{ave}} = \frac{1}{2 - 0}\int_0^2 (x - x^2)\,dx = \frac{1}{2}\left[\frac{1}{2}x^2 - \frac{1}{3}x^3\right]_0^2 = \frac{1}{2}\left(2 - \frac{8}{3}\right) = -\frac{1}{3}$

17. $g_{\text{ave}} = \frac{1}{\frac{\pi}{2} - 0}\int_0^{\pi/2} \cos x\,dx = \frac{2}{\pi}\left[\sin x\right]_0^{\pi/2} = \frac{2}{\pi}(1 - 0) = \frac{2}{\pi}$

18. $f_{\text{ave}} = \frac{1}{\frac{\pi}{4} - 0} \int_0^{\pi/4} \sec\theta \tan\theta \, d\theta = \frac{4}{\pi} \left[\sec\theta\right]_0^{\pi/4} = \frac{4}{\pi}\left(\sqrt{2} - 1\right)$

19. (a) $f_{\text{ave}} = \frac{1}{5-2} \int_2^5 (x-3)^2 \, dx = \frac{1}{3} \int_2^5 (x^2 - 6x + 9) \, dx$

$= \frac{1}{3}\left[\frac{1}{3}x^3 - 3x^2 + 9x\right]_2^5$

$= \frac{1}{3}\left[\frac{1}{3}(125) - 3(25) + 9(5) - \frac{1}{3}(8) + 3(4) - 9(2)\right] = 1$

(b) $f(c) = f_{\text{ave}} \;\Leftrightarrow\; (c-3)^2 = 1 \;\Leftrightarrow$

$c - 3 = \pm 1 \;\Leftrightarrow\; c = 2$ or 4

(c)

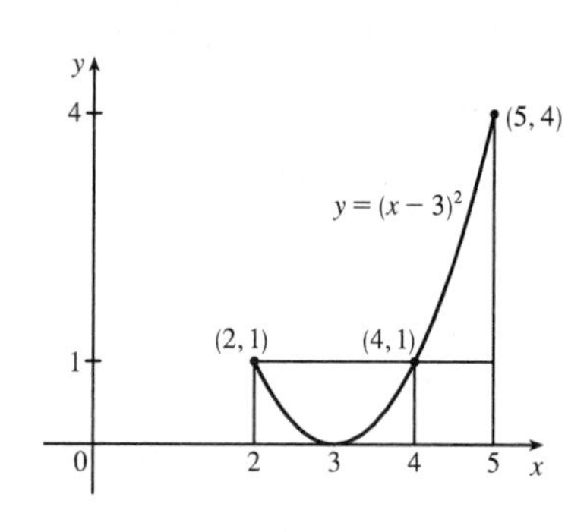

20. (a) $f_{\text{ave}} = \dfrac{1}{4-0} \displaystyle\int_0^4 \sqrt{x} \, dx = \frac{1}{4}\left[\frac{2}{3}x^{3/2}\right]_0^4$

$= \frac{1}{6}\left[x^{3/2}\right]_0^4 = \frac{1}{6}[8 - 0] = \frac{4}{3}$

(b) $f(c) = f_{\text{ave}} \;\Leftrightarrow\; \sqrt{c} = \frac{4}{3} \;\Leftrightarrow\; c = \frac{16}{9}$

(c)

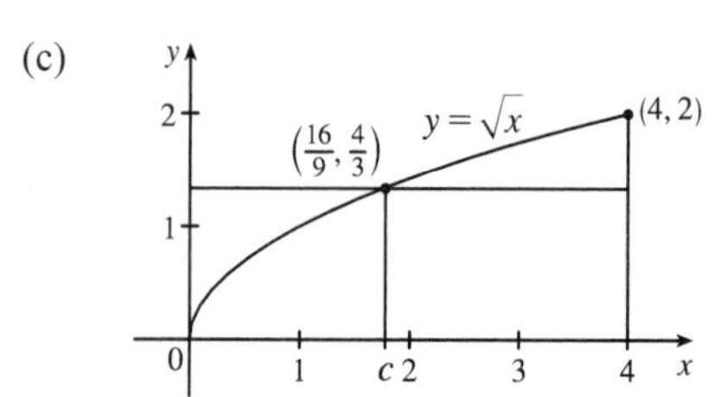

21. $f_{\text{ave}} = \dfrac{1}{50-20} \displaystyle\int_{20}^{50} f(x) \, dx \approx \frac{1}{30} M_3 = \frac{1}{30} \cdot \frac{50-20}{3}[f(25) + f(35) + f(45)] = \frac{1}{3}(38 + 29 + 48) = \frac{115}{3} = 38\frac{1}{3}$

22. (a) $v_{\text{ave}} = \frac{1}{12-0} \int_0^{12} v(t) \, dt = \frac{1}{12} I$. Use the Midpoint Rule with $n = 3$ and $\Delta t = \frac{12-0}{3} = 4$ to estimate I.

$I \approx M_3 = 4[v(2) + v(6) + v(10)] = 4[21 + 50 + 66] = 4(137) = 548$. Thus, $v_{\text{ave}} \approx \frac{1}{12}(548) = 45\frac{2}{3}$ km/h.

(b) Estimating from the graph, $v(t) = 45\frac{2}{3}$ when $t \approx 5.2$ s.

23. $F(x) = \displaystyle\int_1^x f(t) \, dt \;\Rightarrow\; F'(x) = f(x) = \int_1^{x^2} \frac{\sqrt{1+u^4}}{u} \, du \quad \left[\text{since } f(t) = \int_1^{t^2} \frac{\sqrt{1+u^4}}{u} \, du\right] \;\Rightarrow$

$F''(x) = f'(x) = \dfrac{\sqrt{1+(x^2)^4}}{x^2} \cdot \dfrac{d}{dx}(x^2) = \dfrac{\sqrt{1+x^8}}{x^2} \cdot 2x = \dfrac{2\sqrt{1+x^8}}{x}$. So $F''(2) = \sqrt{1+2^8} = \sqrt{257}$.

24. For the curve to be concave upward, we must have $y'' > 0$. $y = \displaystyle\int_0^x \frac{1}{1+t+t^2} \, dt \;\Rightarrow\; y' = \frac{1}{1+x+x^2} \;\Rightarrow$

$y'' = \dfrac{-(1+2x)}{(1+x+x^2)^2}$. For this expression to be positive, we must have $(1+2x) < 0$, since $(1+x+x^2)^2 > 0$ for all x.

$(1+2x) < 0 \;\Leftrightarrow\; x < -\frac{1}{2}$. Thus, the curve is concave upward on $\left(-\infty, -\frac{1}{2}\right)$.

25. (a) By FTC1, $g'(x) = f(x)$. So $g'(x) = f(x) = 0$ at $x = 1, 3, 5, 7$, and 9. g has local maxima at $x = 1$ and 5 (since $f = g'$ changes from positive to negative there) and local minima at $x = 3$ and 7. There is no local maximum or minimum at $x = 9$, since f is not defined for $x > 9$.

(b) We can see from the graph that $\left|\int_0^1 f \, dt\right| < \left|\int_1^3 f \, dt\right| < \left|\int_3^5 f \, dt\right| < \left|\int_5^7 f \, dt\right| < \left|\int_7^9 f \, dt\right|$. So $g(1) = \left|\int_0^1 f \, dt\right|$,

$g(5) = \int_0^5 f \, dt = g(1) - \left|\int_1^3 f \, dt\right| + \left|\int_3^5 f \, dt\right|$, and $g(9) = \int_0^9 f \, dt = g(5) - \left|\int_5^7 f \, dt\right| + \left|\int_7^9 f \, dt\right|$. Thus,

$g(1) < g(5) < g(9)$, and so the absolute maximum of $g(x)$ occurs at $x = 9$.

(c) g is concave downward on those intervals where $g'' < 0$.
But $g'(x) = f(x)$, so $g''(x) = f'(x)$, which is negative on (approximately) $\left(\frac{1}{2}, 2\right)$, $(4, 6)$ and $(8, 9)$. So g is concave downward on these intervals.

(d)

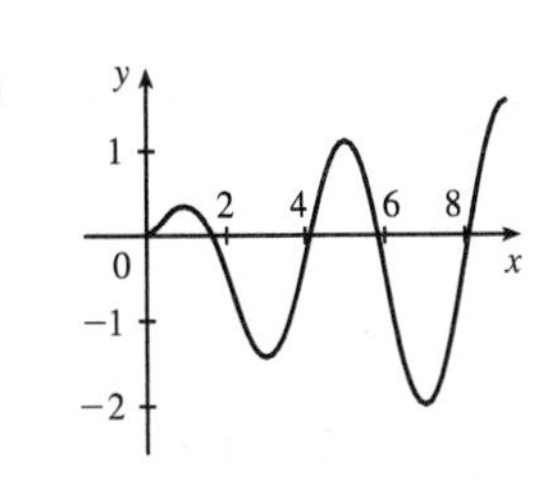

26. (a) By FTC1, $g'(x) = f(x)$. So $g'(x) = f(x) = 0$ at $x = 2, 4, 6, 8$, and 10. g has local maxima at $x = 2$ and 6 (since $f = g'$ changes from positive to negative there) and local minima at $x = 4$ and 8. There is no local maximum or minimum at $x = 10$, since f is not defined for $x > 10$.

(b) We can see from the graph that $\left|\int_0^2 f\,dt\right| > \left|\int_2^4 f\,dt\right| > \left|\int_4^6 f\,dt\right| > \left|\int_6^8 f\,dt\right| > \left|\int_8^{10} f\,dt\right|$. So $g(2) = \left|\int_0^2 f\,dt\right|$, $g(6) = \int_0^6 f\,dt = g(2) - \left|\int_2^4 f\,dt\right| + \left|\int_4^6 f\,dt\right|$, and $g(10) = \int_0^{10} f\,dt = g(6) - \left|\int_6^8 f\,dt\right| + \left|\int_8^{10} f\,dt\right|$.
Thus, $g(2) > g(6) > g(10)$, and so the absolute maximum of $g(x)$ occurs at $x = 2$.

(c) g is concave downward on those intervals where $g'' < 0$. But $g'(x) = f(x)$, so $g''(x) = f'(x)$, which is negative on $(1, 3)$, $(5, 7)$, and $(9, 10)$. So g is concave downward on these intervals.

(d)

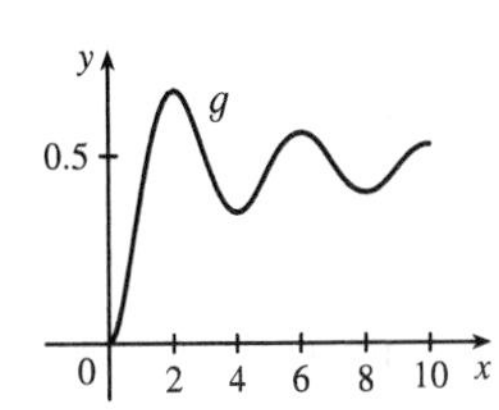

27. By FTC2, $\int_1^4 f'(x)\,dx = f(4) - f(1)$, so $17 = f(4) - 12 \;\Rightarrow\; f(4) = 17 + 12 = 29$.

28. The second derivative is the derivative of the first derivative, so we'll apply the Net Change Theorem with $F = h'$.
$\int_1^2 h''(u)\,du = \int_1^2 (h')'(u)\,du = h'(2) - h'(1) = 5 - 2 = 3$. The other information is unnecessary.

29. (a) The Fresnel function $S(x) = \int_0^x \sin\left(\frac{\pi}{2}t^2\right)dt$ has local maximum values where $0 = S'(x) = \sin\left(\frac{\pi}{2}x^2\right)$ and S' changes from positive to negative. For $x > 0$, this happens when $\frac{\pi}{2}x^2 = (2n-1)\pi$ [odd multiples of π] $\Leftrightarrow$ $x^2 = 2(2n-1) \;\Leftrightarrow\; x = \sqrt{4n-2}$, n any positive integer. For $x < 0$, S' changes from positive to negative where $\frac{\pi}{2}x^2 = 2n\pi$ [even multiples of π] $\Leftrightarrow\; x^2 = 4n \;\Leftrightarrow\; x = -2\sqrt{n}$. S' does not change sign at $x = 0$.

(b) S is concave upward on those intervals where $S''(x) > 0$. Differentiating our expression for $S'(x)$, we get $S''(x) = \cos\left(\frac{\pi}{2}x^2\right)\left(2\frac{\pi}{2}x\right) = \pi x\cos\left(\frac{\pi}{2}x^2\right)$. For $x > 0$, $S''(x) > 0$ where $\cos\left(\frac{\pi}{2}x^2\right) > 0 \;\Leftrightarrow\; 0 < \frac{\pi}{2}x^2 < \frac{\pi}{2}$ or $\left(2n - \frac{1}{2}\right)\pi < \frac{\pi}{2}x^2 < \left(2n + \frac{1}{2}\right)\pi$, n any integer $\Leftrightarrow\; 0 < x < 1$ or $\sqrt{4n-1} < x < \sqrt{4n+1}$, n any positive integer.
For $x < 0$, $S''(x) > 0$ where $\cos\left(\frac{\pi}{2}x^2\right) < 0 \;\Leftrightarrow\; \left(2n - \frac{3}{2}\right)\pi < \frac{\pi}{2}x^2 < \left(2n - \frac{1}{2}\right)\pi$, n any integer $\Leftrightarrow$ $4n - 3 < x^2 < 4n - 1 \;\Leftrightarrow\; \sqrt{4n-3} < |x| < \sqrt{4n-1} \;\Rightarrow\; \sqrt{4n-3} < -x < \sqrt{4n-1} \;\Rightarrow$ $-\sqrt{4n-3} > x > -\sqrt{4n-1}$, so the intervals of upward concavity for $x < 0$ are $\left(-\sqrt{4n-1}, -\sqrt{4n-3}\right)$, n any positive integer. To summarize: S is concave upward on the intervals $(0, 1)$, $\left(-\sqrt{3}, -1\right)$, $\left(\sqrt{3}, \sqrt{5}\right)$, $\left(-\sqrt{7}, -\sqrt{5}\right)$, $\left(\sqrt{7}, 3\right)$, ...

(c) In Maple, we use `plot({int(sin(Pi*t^2/2),t=0..x),0.2},x=0..2);`. Note that Maple recognizes the Fresnel function, calling it `FresnelS(x)`. In Mathematica, we use `Plot[{Integrate[Sin[Pi*t^2/2],{t,0,x}],0.2},{x,0,2}]`. In Derive, we load the utility file `FRESNEL` and plot `FRESNEL_SIN(x)`. From the graphs, we see that $\int_0^x \sin\left(\frac{\pi}{2}t^2\right) dt = 0.2$ at $x \approx 0.74$.

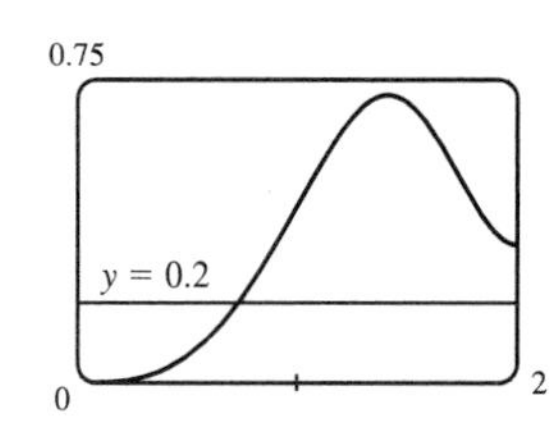

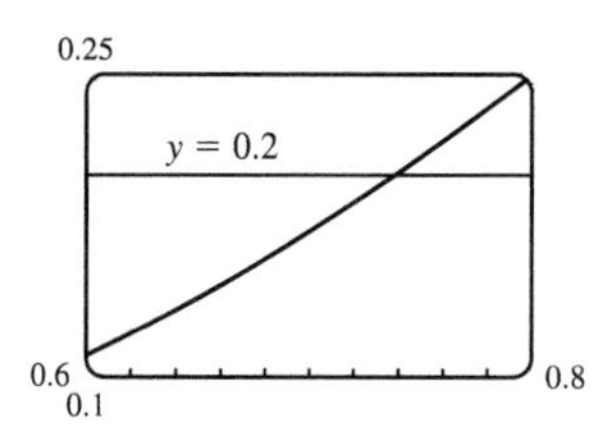

30. (a) In Maple, we should start by setting `si:=int(sin(t)/t,t=0..x);`. In Mathematica, the command is `si=Integrate[Sin[t]/t,{t,0,x}]`. Note that both systems recognize this function; Maple calls it `Si(x)` and Mathematica calls it `SinIntegral[x]`. In Maple, the command to generate the graph is `plot(si,x=-4*Pi..4*Pi);`. In Mathematica, it is `Plot[si,{x,-4*Pi,4*Pi}]`. In Derive, we load the utility file `EXP_INT` and plot `SI(x)`.

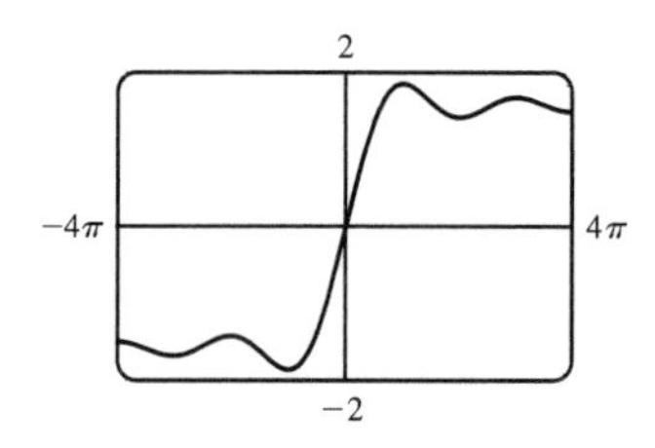

(b) $\mathrm{Si}(x)$ has local maximum values where $\mathrm{Si}'(x)$ changes from positive to negative, passing through 0. From the Fundamental Theorem we know that $\mathrm{Si}'(x) = \dfrac{d}{dx}\displaystyle\int_0^x \frac{\sin t}{t}\,dt = \frac{\sin x}{x}$, so we must have $\sin x = 0$ for a maximum, and for $x > 0$ we must have $x = (2n-1)\pi$, n any positive integer, for Si' to be changing from positive to negative at x. For $x < 0$, we must have $x = 2n\pi$, n any positive integer, for a maximum, since the denominator of $\mathrm{Si}'(x)$ is negative for $x < 0$. Thus, the local maxima occur at $x = \pi, -2\pi, 3\pi, -4\pi, 5\pi, -6\pi, \ldots$.

(c) To find the first inflection point, we solve $\mathrm{Si}''(x) = \dfrac{\cos x}{x} - \dfrac{\sin x}{x^2} = 0$. We can see from the graph that the first inflection point lies somewhere between $x = 3$ and $x = 5$. Using a rootfinder gives the value $x \approx 4.4934$. To find the y-coordinate of the inflection point, we evaluate $\mathrm{Si}(4.4934) \approx 1.6556$. So the coordinates of the first inflection point to the right of the origin are about $(4.4934, 1.6556)$. Alternatively, we could graph $S''(x)$ and estimate the first positive x-value at which it changes sign.

(d) It seems from the graph that the function has horizontal asymptotes at $y \approx 1.5$, with $\lim\limits_{x\to\pm\infty} \mathrm{Si}(x) \approx \pm 1.5$ respectively. Using the limit command, we get $\lim\limits_{x\to\infty} \mathrm{Si}(x) = \frac{\pi}{2}$. Since $\mathrm{Si}(x)$ is an odd function, $\lim\limits_{x\to-\infty} \mathrm{Si}(x) = -\frac{\pi}{2}$. So $\mathrm{Si}(x)$ has the horizontal asymptotes $y = \pm\frac{\pi}{2}$.

(e) We use the `fsolve` command in Maple (or `FindRoot` in Mathematica) to find that the solution is $x \approx 1.1$. Or, as in Exercise 29(c), we graph $y = \mathrm{Si}(x)$ and $y = 1$ on the same screen to see where they intersect.

31. Using FTC1, we differentiate both sides of $6 + \displaystyle\int_a^x \frac{f(t)}{t^2}\,dt = 2\sqrt{x}$ to get $\dfrac{f(x)}{x^2} = 2\dfrac{1}{2\sqrt{x}} \Rightarrow f(x) = x^{3/2}$.

To find a, we substitute $x = a$ in the original equation to obtain $6 + \displaystyle\int_a^a \frac{f(t)}{t^2}\,dt = 2\sqrt{a} \Rightarrow 6 + 0 = 2\sqrt{a} \Rightarrow$

$3 = \sqrt{a} \Rightarrow a = 9$.

32. (a) $C(t) = \frac{1}{t}\int_0^t [f(s)+g(s)]\,ds$. Using FTC1 and the Product Rule, we have

$C'(t) = \frac{1}{t}[f(t)+g(t)] - \frac{1}{t^2}\int_0^t [f(s)+g(s)]\,ds$. Set $C'(t)=0$: $\frac{1}{t}[f(t)+g(t)] - \frac{1}{t^2}\int_0^t [f(s)+g(s)]\,ds = 0 \;\Rightarrow$

$[f(t)+g(t)] - \frac{1}{t}\int_0^t [f(s)+g(s)]\,ds = 0 \;\Rightarrow\; [f(t)+g(t)] - C(t) = 0 \;\Rightarrow\; C(t) = f(t)+g(t)$.

(b) For $0 \le t \le 30$, we have $D(t) = \int_0^t \left(\frac{V}{15} - \frac{V}{450}s\right)ds = \left[\frac{V}{15}s - \frac{V}{900}s^2\right]_0^t = \frac{V}{15}t - \frac{V}{900}t^2$.

So $D(t) = V \;\Rightarrow\; \frac{V}{15}t - \frac{V}{900}t^2 = V \;\Rightarrow\; 60t - t^2 = 900 \;\Rightarrow\; t^2 - 60t + 900 = 0 \;\Rightarrow$

$(t-30)^2 = 0 \;\Rightarrow\; t = 30$. So the length of time T is 30 months.

(c) $C(t) = \frac{1}{t}\int_0^t \left(\frac{V}{15} - \frac{V}{450}s + \frac{V}{12{,}900}s^2\right)ds = \frac{1}{t}\left[\frac{V}{15}s - \frac{V}{900}s^2 + \frac{V}{38{,}700}s^3\right]_0^t$

$= \frac{1}{t}\left(\frac{V}{15}t - \frac{V}{900}t^2 + \frac{V}{38{,}700}t^3\right) = \frac{V}{15} - \frac{V}{900}t + \frac{V}{38{,}700}t^2 \;\Rightarrow$

$C'(t) = -\frac{V}{900} + \frac{V}{19{,}350}t = 0$ when $\frac{1}{19{,}350}t = \frac{1}{900} \;\Rightarrow\; t = 21.5$.

$C(21.5) = \frac{V}{15} - \frac{V}{900}(21.5) + \frac{V}{38{,}700}(21.5)^2 \approx 0.05472V$, $C(0) = \frac{V}{15} \approx 0.06667V$, and

$C(30) = \frac{V}{15} - \frac{V}{900}(30) + \frac{V}{38{,}700}(30)^2 \approx 0.05659V$, so the absolute minimum is $C(21.5) \approx 0.05472V$.

(d) As in part (c), we have $C(t) = \frac{V}{15} - \frac{V}{900}t + \frac{V}{38{,}700}t^2$, so $C(t) = f(t)+g(t) \;\Leftrightarrow$

$\frac{V}{15} - \frac{V}{900}t + \frac{V}{38{,}700}t^2 = \frac{V}{15} - \frac{V}{450}t + \frac{V}{12{,}900}t^2 \;\Leftrightarrow$

$t^2\left(\frac{1}{12{,}900} - \frac{1}{38{,}700}\right) = t\left(\frac{1}{450} - \frac{1}{900}\right) \;\Leftrightarrow$

$t = \frac{1/900}{2/38{,}700} = \frac{43}{2} = 21.5$. This is the value of t that we obtained as the critical number of C in part (c), so we have verified the result of (a) in this case.

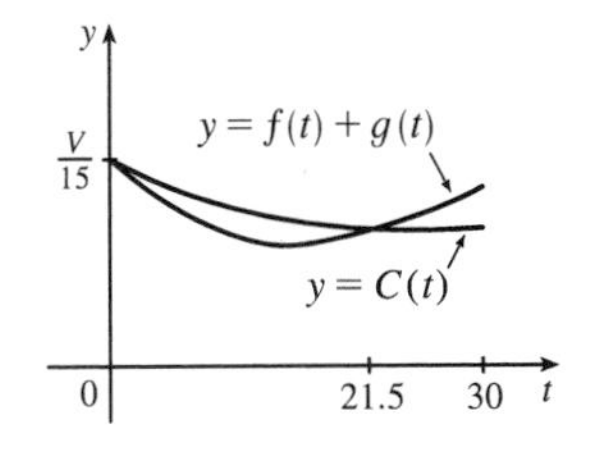

33. (a) Let $F(t) = \int_0^t f(s)\,ds$. Then, by FTC1, $F'(t) = f(t) =$ rate of depreciation, so $F(t)$ represents the loss in value over the interval $[0,t]$.

(b) $C(t) = \frac{1}{t}\left[A + \int_0^t f(s)\,ds\right] = \frac{A+F(t)}{t}$ represents the average expenditure per unit of t during the interval $[0,t]$, assuming that there has been only one overhaul during that time period. The company wants to minimize average expenditure.

(c) $C(t) = \frac{1}{t}\left[A + \int_0^t f(s)\,ds\right]$. Using FTC1, we have $C'(t) = -\frac{1}{t^2}\left[A + \int_0^t f(s)\,ds\right] + \frac{1}{t}f(t)$.

$C'(t) = 0 \;\Rightarrow\; t\,f(t) = A + \int_0^t f(s)\,ds \;\Rightarrow\; f(t) = \frac{1}{t}\left[A + \int_0^t f(s)\,ds\right] = C(t)$.

4.5 The Substitution Rule

1. Let $u = 3x$. Then $du = 3\,dx$, so $dx = \frac{1}{3}du$. Thus,

$\int \cos 3x\,dx = \int \cos u\left(\frac{1}{3}\,du\right) = \frac{1}{3}\int \cos u\,du = \frac{1}{3}\sin u + C = \frac{1}{3}\sin 3x + C$. Don't forget that it is often very easy to check an indefinite integration by differentiating your answer. In this case, $\frac{d}{dx}\left(\frac{1}{3}\sin 3x + C\right) = \frac{1}{3}(\cos 3x)\cdot 3 = \cos 3x$, the desired result.

2. Let $u = 4 + x^2$. Then $du = 2x\,dx$ and $x\,dx = \frac{1}{2}\,du$, so

$\int x\left(4+x^2\right)^{10}\,dx = \int u^{10}\left(\frac{1}{2}\,du\right) = \frac{1}{2}\cdot\frac{1}{11}u^{11} + C = \frac{1}{22}\left(4+x^2\right)^{11} + C.$

3. Let $u = x^3 + 1$. Then $du = 3x^2\,dx$ and $x^2\,dx = \frac{1}{3}\,du$, so

$$\int x^2\sqrt{x^3+1}\,dx = \int \sqrt{u}\left(\frac{1}{3}\,du\right) = \frac{1}{3}\frac{u^{3/2}}{3/2} + C = \frac{1}{3}\cdot\frac{2}{3}u^{3/2} + C = \frac{2}{9}(x^3+1)^{3/2} + C.$$

4. Let $u = \sqrt{x}$. Then $du = \dfrac{1}{2\sqrt{x}}\,dx$ and $\dfrac{1}{\sqrt{x}}\,dx = 2\,du$, so

$$\int \frac{\sin\sqrt{x}}{\sqrt{x}}\,dx = \int \sin u\,(2\,du) = 2(-\cos u) + C = -2\cos\sqrt{x} + C.$$

5. Let $u = 1 + 2x$. Then $du = 2\,dx$ and $dx = \frac{1}{2}\,du$, so

$$\int \frac{4}{(1+2x)^3}\,dx = 4\int u^{-3}\left(\tfrac{1}{2}\,du\right) = 2\frac{u^{-2}}{-2} + C = -\frac{1}{u^2} + C = -\frac{1}{(1+2x)^2} + C.$$

6. Let $u = \cos\theta$. Then $du = -\sin\theta\,d\theta$ and $\sin\theta\,d\theta = -du$, so

$\int \cos^4\theta\sin\theta\,d\theta = \int u^4\,(-du) = -\frac{1}{5}u^5 + C = -\frac{1}{5}\cos^5\theta + C.$

7. Let $u = x^2 + 3$. Then $du = 2x\,dx$, so $\int 2x(x^2+3)^4\,dx = \int u^4\,du = \frac{1}{5}u^5 + C = \frac{1}{5}(x^2+3)^5 + C$.

8. Let $u = x^3 + 5$. Then $du = 3x^2dx$ and $x^2dx = \frac{1}{3}\,du$, so

$\int x^2(x^3+5)^9\,dx = \int u^9\left(\frac{1}{3}\,du\right) = \frac{1}{3}\cdot\frac{1}{10}u^{10} + C = \frac{1}{30}(x^3+5)^{10} + C.$

9. Let $u = 3x - 2$. Then $du = 3\,dx$ and $dx = \frac{1}{3}\,du$, so $\int(3x-2)^{20}\,dx = \int u^{20}\left(\frac{1}{3}\,du\right) = \frac{1}{3}\cdot\frac{1}{21}u^{21} + C = \frac{1}{63}(3x-2)^{21} + C$.

10. Let $u = 2 - x$. Then $du = -dx$ and $dx = -du$, so $\int(2-x)^6\,dx = \int u^6(-du) = -\frac{1}{7}u^7 + C = -\frac{1}{7}(2-x)^7 + C$.

11. Let $u = 2y + 1$. Then $du = 2\,dy$ and $dy = \frac{1}{2}\,du$, so

$$\int \frac{3}{(2y+1)^5}\,dy = \int 3u^{-5}\left(\tfrac{1}{2}\,du\right) = \frac{3}{2}\cdot\frac{1}{-4}u^{-4} + C = \frac{-3}{8(2y+1)^4} + C.$$

12. Let $u = 1 + x^{3/2}$. Then $du = \frac{3}{2}x^{1/2}\,dx$ and $\sqrt{x}\,dx = \frac{2}{3}\,du$, so

$\int \sqrt{x}\sin(1+x^{3/2})\,dx = \int \sin u\left(\frac{2}{3}\,du\right) = \frac{2}{3}\cdot(-\cos u) + C = -\frac{2}{3}\cos(1+x^{3/2}) + C.$

13. Let $u = \sqrt{t}$. Then $du = \dfrac{dt}{2\sqrt{t}}$ and $\dfrac{1}{\sqrt{t}}\,dt = 2\,du$, so $\displaystyle\int \frac{\cos\sqrt{t}}{\sqrt{t}}\,dt = \int \cos u\,(2\,du) = 2\sin u + C = 2\sin\sqrt{t} + C$.

14. Let $u = x^2 + 1$. Then $du = 2x\,dx$ and $x\,dx = \frac{1}{2}\,du$, so

$$\int \frac{x}{(x^2+1)^2}\,dx = \int u^{-2}\left(\tfrac{1}{2}\,du\right) = \tfrac{1}{2}\cdot\frac{-1}{u} + C = \frac{-1}{2u} + C = \frac{-1}{2(x^2+1)} + C.$$

15. Let $u = 3ax + bx^3$. Then $du = (3a + 3bx^2)\,dx = 3(a + bx^2)\,dx$, so

$$\int \frac{a + bx^2}{\sqrt{3ax + bx^3}}\,dx = \int \frac{\frac{1}{3}\,du}{u^{1/2}} = \frac{1}{3}\int u^{-1/2}\,du = \tfrac{1}{3}\cdot 2u^2 + C = \tfrac{2}{3}\sqrt{3ax + bx^3} + C.$$

16. Let $u = 5t + 4$. Then $du = 5\,dt$ and $dt = \frac{1}{5}\,du$, so

$$\int \frac{1}{(5t+4)^{2.7}}\,dt = \int u^{-2.7}\left(\tfrac{1}{5}\,du\right) = \frac{1}{5}\cdot\frac{1}{-1.7}u^{-1.7} + C = \frac{-1}{8.5}u^{-1.7} + C = \frac{-2}{17(5t+4)^{1.7}} + C.$$

17. Let $u = \pi t$. Then $du = \pi\,dt$ and $dt = \frac{1}{\pi}\,du$, so $\int \sin \pi t\,dt = \int \sin u\left(\frac{1}{\pi}\,du\right) = \frac{1}{\pi}(-\cos u) + C = -\frac{1}{\pi}\cos \pi t + C$.

18. Let $u = 2y^4 - 1$. Then $du = 8y^3\,dy$ and $y^3\,dy = \frac{1}{8}\,du$, so

$$\int y^3\sqrt{2y^4 - 1}\,dy = \int u^{1/2}\left(\tfrac{1}{8}\,du\right) = \tfrac{1}{8}\cdot\tfrac{2}{3}u^{3/2} + C = \tfrac{1}{12}(2y^4 - 1)^{3/2} + C.$$

19. Let $u = 1 + z^3$. Then $du = 3z^2\,dz$ and $z^2\,dz = \frac{1}{3}\,du$, so

$$\int \frac{z^2}{\sqrt[3]{1 + z^3}}\,dz = \int u^{-1/3}\left(\tfrac{1}{3}\,du\right) = \tfrac{1}{3}\cdot\tfrac{3}{2}u^{2/3} + C = \tfrac{1}{2}(1 + z^3)^{2/3} + C.$$

20. Let $u = 2\theta$. Then $du = 2\,d\theta$ and $d\theta = \frac{1}{2}\,du$, so $\int \sec 2\theta\,\tan 2\theta\,d\theta = \int \sec u \tan u\left(\frac{1}{2}\,du\right) = \frac{1}{2}\sec u + C = \frac{1}{2}\sec 2\theta + C$.

21. Let $u = \sin\theta$. Then $du = \cos\theta\,d\theta$, so $\int \cos\theta\,\sin^6\theta\,d\theta = \int u^6\,du = \frac{1}{7}u^7 + C = \frac{1}{7}\sin^7\theta + C$.

22. Let $u = 1 + \tan\theta$. Then $du = \sec^2\theta\,d\theta$, so $\int(1 + \tan\theta)^5\sec^2\theta\,d\theta = \int u^5\,du = \frac{1}{6}u^6 + C = \frac{1}{6}(1 + \tan\theta)^6 + C$.

23. Let $u = \cot x$. Then $du = -\csc^2 x\,dx$ and $\csc^2 x\,dx = -du$, so

$$\int \sqrt{\cot x}\,\csc^2 x\,dx = \int \sqrt{u}\,(-du) = -\frac{u^{3/2}}{3/2} + C = -\tfrac{2}{3}(\cot x)^{3/2} + C.$$

24. Let $u = x^3 + 1$. Then $x^3 = u - 1$ and $du = 3x^2\,dx$, so

$$\begin{aligned}\int \sqrt[3]{x^3+1}\,x^5\,dx &= \int \sqrt[3]{x^3+1}\cdot x^3\cdot x^2\,dx = \int u^{1/3}(u-1)\left(\tfrac{1}{3}\,du\right) = \tfrac{1}{3}\int\left(u^{4/3} - u^{1/3}\right)du \\ &= \tfrac{1}{3}\left(\tfrac{3}{7}u^{7/3} - \tfrac{3}{4}u^{4/3}\right) + C = \tfrac{1}{7}(x^3+1)^{7/3} - \tfrac{1}{4}(x^3+1)^{4/3} + C\end{aligned}$$

25. Let $u = b + cx^{a+1}$. Then $du = (a+1)cx^a\,dx$, so

$$\int x^a\sqrt{b + cx^{a+1}}\,dx = \int u^{1/2}\frac{1}{(a+1)c}\,du = \frac{1}{(a+1)c}\left(\tfrac{2}{3}u^{3/2}\right) + C = \frac{2}{3c(a+1)}\left(b + cx^{a+1}\right)^{3/2} + C.$$

26. Let $u = \dfrac{\pi}{x}$. Then $du = -\dfrac{\pi}{x^2}\,dx$ and $\dfrac{1}{x^2}\,dx = -\dfrac{1}{\pi}\,du$, so

$$\int \frac{\cos(\pi/x)}{x^2}\,dx = \int \cos u\left(-\frac{1}{\pi}\,du\right) = -\frac{1}{\pi}\sin u + C = -\frac{1}{\pi}\sin\frac{\pi}{x} + C.$$

27. Let $u = \sec x$. Then $du = \sec x\,\tan x\,dx$, so

$$\int \sec^3 x\,\tan x\,dx = \int \sec^2 x\,(\sec x\,\tan x)\,dx = \int u^2\,du = \tfrac{1}{3}u^3 + C = \tfrac{1}{3}\sec^3 x + C.$$

28. Let $u = \cos t$. Then $du = -\sin t\, dt$ and $\sin t\, dt = -du$, so

$\int \sin t \sec^2(\cos t)\, dt = \int \sec^2 u \cdot (-du) = -\tan u + C = -\tan(\cos t) + C$.

29. Let $u = x + 2$. Then $du = dx$, so

$$\int \frac{x}{\sqrt[4]{x+2}}\, dx = \int \frac{u-2}{\sqrt[4]{u}}\, du = \int \left(u^{3/4} - 2u^{-1/4}\right) du = \tfrac{4}{7}u^{7/4} - 2 \cdot \tfrac{4}{3}u^{3/4} + C$$
$$= \tfrac{4}{7}(x+2)^{7/4} - \tfrac{8}{3}(x+2)^{3/4} + C$$

30. Let $u = 1 - x$. Then $x = 1 - u$ and $dx = -du$, so

$$\int \frac{x^2}{\sqrt{1-x}}\, dx = \int \frac{(1-u)^2}{\sqrt{u}}\,(-du) = -\int \frac{1 - 2u + u^2}{\sqrt{u}}\, du = -\int \left(u^{-1/2} - 2u^{1/2} + u^{3/2}\right) du$$
$$= -\left(2u^{1/2} - 2 \cdot \tfrac{2}{3}u^{3/2} + \tfrac{2}{5}u^{5/2}\right) + C = -2\sqrt{1-x} + \tfrac{4}{3}(1-x)^{3/2} - \tfrac{2}{5}(1-x)^{5/2} + C$$

In Exercises 31--32, let $f(x)$ denote the integrand and $F(x)$ its antiderivative (with $C = 0$).

31. Let $u = x - 1$, so $du = dx$. When $x = 0$, $u = -1$; when $x = 2$, $u = 1$. Thus, $\int_0^2 (x-1)^{25}\, dx = \int_{-1}^1 u^{25}\, du = 0$ by Theorem 6(b), since $f(u) = u^{25}$ is an odd function.

32. Let $u = 4 + 3x$, so $du = 3\, dx$. When $x = 0$, $u = 4$; when $x = 7$, $u = 25$. Thus,

$\int_0^7 \sqrt{4+3x}\, dx = \int_4^{25} \sqrt{u}\left(\tfrac{1}{3}\, du\right) = \dfrac{1}{3}\left[\dfrac{u^{3/2}}{3/2}\right]_4^{25} = \dfrac{2}{9}(25^{3/2} - 4^{3/2}) = \dfrac{2}{9}(125 - 8) = \dfrac{234}{9} = 26.$

33. Let $u = 1 + 2x^3$, so $du = 6x^2\, dx$. When $x = 0$, $u = 1$; when $x = 1$, $u = 3$. Thus,

$\int_0^1 x^2 \left(1 + 2x^3\right)^5\, dx = \int_1^3 u^5 \left(\tfrac{1}{6}\, du\right) = \tfrac{1}{6}\left[\tfrac{1}{6}u^6\right]_1^3 = \tfrac{1}{36}(3^6 - 1^6) = \tfrac{1}{36}(729 - 1) = \tfrac{728}{36} = \tfrac{182}{9}.$

34. Let $u = x^2$, so $du = 2x\, dx$. When $x = 0$, $u = 0$; when $x = \sqrt{\pi}$, $u = \pi$. Thus,

$\int_0^{\sqrt{\pi}} x\cos(x^2)\, dx = \int_0^{\pi} \cos u \left(\tfrac{1}{2}\, du\right) = \tfrac{1}{2}[\sin u]_0^{\pi} = \tfrac{1}{2}(\sin \pi - \sin 0) = \tfrac{1}{2}(0 - 0) = 0.$

35. Let $u = t/4$, so $du = \tfrac{1}{4}\, dt$. When $t = 0$, $u = 0$; when $t = \pi$, $u = \pi/4$. Thus,

$\int_0^{\pi} \sec^2(t/4)\, dt = \int_0^{\pi/4} \sec^2 u\, (4\, du) = 4\big[\tan u\big]_0^{\pi/4} = 4\left(\tan \tfrac{\pi}{4} - \tan 0\right) = 4(1 - 0) = 4.$

36. Let $u = \pi t$, so $du = \pi\, dt$. When $t = \tfrac{1}{6}$, $u = \tfrac{\pi}{6}$; when $t = \tfrac{1}{2}$, $u = \tfrac{\pi}{2}$. Thus,

$\int_{1/6}^{1/2} \csc \pi t \cot \pi t\, dt = \int_{\pi/6}^{\pi/2} \csc u \cot u \left(\tfrac{1}{\pi}\, du\right) = \tfrac{1}{\pi}\big[-\csc u\big]_{\pi/6}^{\pi/2} = -\tfrac{1}{\pi}(1 - 2) = \tfrac{1}{\pi}.$

37. Let $u = \cos\theta$, so $du = -\sin\theta\, d\theta$. When $\theta = 0$, $u = 1$; when $\theta = \tfrac{\pi}{3}$, $u = \tfrac{1}{2}$. Thus,

$$\int_0^{\pi/3} \frac{\sin\theta}{\cos^2\theta}\, d\theta = \int_1^{1/2} \frac{-du}{u^2} = \int_{1/2}^1 u^{-2}\, du = \left[-\frac{1}{u}\right]_{1/2}^1 = -1 - (-2) = 1.$$

38. Let $u = \sin x$, so $du = \cos x\, dx$. When $x = 0$, $u = 0$; when $x = \tfrac{\pi}{2}$, $u = 1$. Thus,

$\int_0^{\pi/2} \cos x \sin(\sin x)\, dx = \int_0^1 \sin u\, du = \big[-\cos u\big]_0^1 = -(\cos 1 - 1) = 1 - \cos 1.$

39. Let $u = x - 1$, so $u + 1 = x$ and $du = dx$. When $x = 1$, $u = 0$; when $x = 2$, $u = 1$. Thus,

$\int_1^2 x\sqrt{x-1}\, dx = \int_0^1 (u+1)\sqrt{u}\, du = \int_0^1 (u^{3/2} + u^{1/2})\, du = \left[\tfrac{2}{5}u^{5/2} + \tfrac{2}{3}u^{3/2}\right]_0^1 = \tfrac{2}{5} + \tfrac{2}{3} = \tfrac{16}{15}.$

40. $\displaystyle\int_{-\pi/2}^{\pi/2} \frac{x^2 \sin x}{1 + x^6}\, dx = 0$ by Theorem 6(b), since $f(x) = \dfrac{x^2 \sin x}{1 + x^6}$ is an odd function.

41. Let $u = 1 + 2x$, so $du = 2\, dx$. When $x = 0$, $u = 1$; when $x = 13$, $u = 27$. Thus,

$$\int_0^{13} \frac{dx}{\sqrt[3]{(1+2x)^2}} = \int_1^{27} u^{-2/3}\left(\tfrac{1}{2}\, du\right) = \left[\tfrac{1}{2} \cdot 3u^{1/3}\right]_1^{27} = \tfrac{3}{2}(3 - 1) = 3.$$

42. Let $u = 1 + 2x$, so $x = \frac{1}{2}(u-1)$ and $du = 2\,dx$. When $x = 0$, $u = 1$; when $x = 4$, $u = 9$. Thus,

$$\int_0^4 \frac{x\,dx}{\sqrt{1+2x}} = \int_1^9 \frac{\frac{1}{2}(u-1)}{\sqrt{u}}\frac{du}{2} = \frac{1}{4}\int_1^9 \left(u^{1/2} - u^{-1/2}\right) du = \frac{1}{4}\left[\frac{2}{3}u^{3/2} - 2u^{1/2}\right]_1^9$$

$$= \frac{1}{4}\cdot\frac{2}{3}\left[u^{3/2} - 3u^{1/2}\right]_1^9 = \frac{1}{6}[(27-9)-(1-3)] = \frac{20}{6} = \frac{10}{3}$$

43. $\int_{-\pi/6}^{\pi/6} \tan^3\theta\,d\theta = 0$ by Theorem 6(b), since $f(\theta) = \tan^3\theta$ is an odd function.

44. Assume $a > 0$. Let $u = a^2 - x^2$, so $du = -2x\,dx$. When $x = 0$, $u = a^2$; when $x = a$, $u = 0$. Thus,

$\int_0^a x\sqrt{a^2 - x^2}\,dx = \int_{a^2}^0 u^{1/2}\left(-\frac{1}{2}\,du\right) = \frac{1}{2}\int_0^{a^2} u^{1/2}\,du = \frac{1}{2}\cdot\left[\frac{2}{3}u^{3/2}\right]_0^{a^2} = \frac{1}{3}a^3.$

45. $f_{\text{ave}} = \frac{1}{5-0}\int_0^5 t\sqrt{1+t^2}\,dt = \frac{1}{5}\int_1^{26}\sqrt{u}\left(\frac{1}{2}\,du\right)$ $\quad [u = 1+t^2,\ du = 2t\,dt]$

$= \frac{1}{10}\int_1^{26} u^{1/2}\,du = \frac{1}{10}\cdot\frac{2}{3}\left[u^{3/2}\right]_1^{26} = \frac{1}{15}\left(26^{3/2} - 1\right)$

46. $g_{\text{ave}} = \frac{1}{2-0}\int_0^2 x^2\sqrt{1+x^3}\,dx = \frac{1}{2}\int_1^9 \sqrt{u}\cdot\frac{1}{3}\,du$ $\quad [u = 1+x^3,\ du = 3x^2\,dx]$ $\quad = \frac{1}{6}\left[\frac{2}{3}u^{3/2}\right]_1^9 = \frac{1}{9}(27-1) = \frac{26}{9}$

47. $h_{\text{ave}} = \frac{1}{\pi-0}\int_0^\pi \cos^4 x\,\sin x\,dx = \frac{1}{\pi}\int_1^{-1} u^4(-du)$ $\quad [u = \cos x,\ du = -\sin x\,dx]$

$= \frac{1}{\pi}\int_{-1}^1 u^4\,du = \frac{1}{\pi}\cdot 2\int_0^1 u^4 du = \frac{2}{\pi}\left[\frac{1}{5}u^5\right]_0^1 = \frac{2}{5\pi}$

48. $h_{\text{ave}} = \dfrac{1}{6-1}\displaystyle\int_1^6 \frac{3}{(1+r)^2}\,dr = \frac{1}{5}\int_2^7 3u^{-2}\,du$ $\quad [u = 1+r,\ du = dr]$

$= -\frac{3}{5}\left[u^{-1}\right]_2^7 = -\frac{3}{5}\left(\frac{1}{7} - \frac{1}{2}\right) = \frac{3}{5}\left(\frac{1}{2} - \frac{1}{7}\right) = \frac{3}{5}\cdot\frac{5}{14} = \frac{3}{14}$

49. First write the integral as a sum of two integrals:

$I = \int_{-2}^2 (x+3)\sqrt{4-x^2}\,dx = I_1 + I_2 = \int_{-2}^2 x\sqrt{4-x^2}\,dx + \int_{-2}^2 3\sqrt{4-x^2}\,dx.$ $\quad I_1 = 0$ by Theorem 6(b), since $f(x) = x\sqrt{4-x^2}$ is an odd function and we are integrating from $x = -2$ to $x = 2$. We interpret I_2 as three times the area of a semicircle with radius 2, so $I = 0 + 3\cdot\frac{1}{2}\left(\pi\cdot 2^2\right) = 6\pi$.

50. Let $u = x^2$. Then $du = 2x\,dx$ and the limits are unchanged ($0^2 = 0$ and $1^2 = 1$), so

$I = \int_0^1 x\sqrt{1-x^4}\,dx = \frac{1}{2}\int_0^1 \sqrt{1-u^2}\,du$. But this integral can be interpreted as the area of a quarter-circle with radius 1. So $I = \frac{1}{2}\cdot\frac{1}{4}(\pi\cdot 1^2) = \frac{1}{8}\pi$.

51. The volume of inhaled air in the lungs at time t is

$V(t) = \int_0^t f(u)\,du = \int_0^t \frac{1}{2}\sin\left(\frac{2\pi}{5}\,u\right)du = \int_0^{2\pi t/5} \frac{1}{2}\sin v\left(\frac{5}{2\pi}\,dv\right)$ $\quad$ [substitute $v = \frac{2\pi}{5}u$, $dv = \frac{2\pi}{5}\,du$]

$= \frac{5}{4\pi}\left[-\cos v\right]_0^{2\pi t/5} = \frac{5}{4\pi}\left[-\cos\left(\frac{2\pi}{5}t\right) + 1\right] = \frac{5}{4\pi}\left[1 - \cos\left(\frac{2\pi}{5}t\right)\right]$ liters

52. Number of calculators $= x(4) - x(2) = \int_2^4 5000[1 - 100(t+10)^{-2}]\,dt$

$= 5000\left[t + 100(t+10)^{-1}\right]_2^4 = 5000\left[\left(4 + \frac{100}{14}\right) - \left(2 + \frac{100}{12}\right)\right] \approx 4048$

53. Let $u = 2x$. Then $du = 2\,dx$, so $\int_0^2 f(2x)\,dx = \int_0^4 f(u)\left(\frac{1}{2}\,du\right) = \frac{1}{2}\int_0^4 f(u)\,du = \frac{1}{2}(10) = 5$.

54. Let $u = x^2$. Then $du = 2x\,dx$, so $\int_0^3 xf(x^2)\,dx = \int_0^9 f(u)\left(\frac{1}{2}\,du\right) = \frac{1}{2}\int_0^9 f(u)\,du = \frac{1}{2}(4) = 2$.

55. Let $u = -x$. Then $du = -dx$, so

$\int_a^b f(-x)\,dx = \int_{-a}^{-b} f(u)(-du) = \int_{-b}^{-a} f(u)\,du = \int_{-b}^{-a} f(x)\,dx.$

From the diagram, we see that the equality follows from the fact that we are reflecting the graph of f, and the limits of integration, about the y-axis.

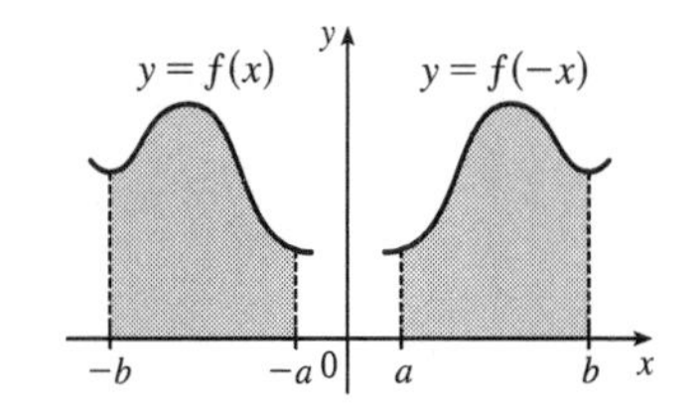

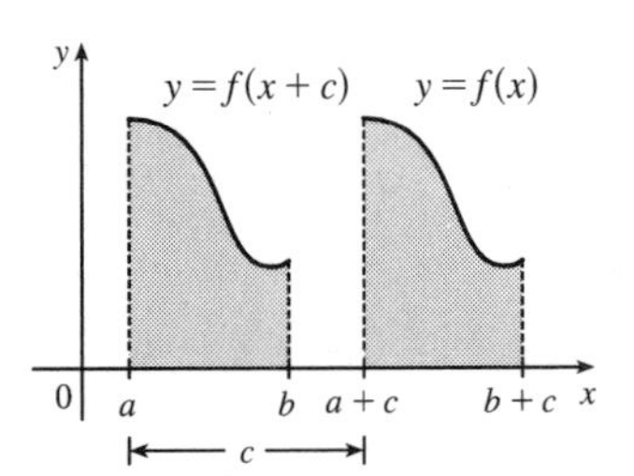

56. Let $u = x + c$. Then $du = dx$, so
$\int_a^b f(x+c)\,dx = \int_{a+c}^{b+c} f(u)\,du = \int_{a+c}^{b+c} f(x)\,dx$. From the diagram, we see that the equality follows from the fact that we are translating the graph of f, and the limits of integration, by a distance c.

57. Let $u = 1 - x$. Then $x = 1 - u$ and $dx = -du$, so
$\int_0^1 x^a(1-x)^b\,dx = \int_1^0 (1-u)^a\,u^b(-du) = \int_0^1 u^b(1-u)^a\,du = \int_0^1 x^b(1-x)^a\,dx.$

4 Review

CONCEPT CHECK

1. (a) $\sum_{i=1}^{n} f(x_i^*)\,\Delta x_i$ is an expression for a Riemann sum of a function f.
x_i^* is a point in the ith subinterval $[x_{i-1}, x_i]$ and $\Delta x_i = x_i - x_{i-1}$ is the length of the ith subinterval.

(b) See Figure 3 in Section 4.2.

(c) In Section 4.2, see Figure 5 and the paragraph before it.

2. (a) See the definition of the definite integral in Definition 4.2.2. If f is continuous and the subintervals have equal lengths, then the expression for the definite integral simplifies to the one in Theorem 4.2.4.

(b) It's the area under the curve $y = f(x)$ from a to b.

(c) In Section 4.2, see Figure 6 and the paragraph before it.

3. See the statement of the Midpoint Rule on page 211.

4. (a) See the Evaluation Theorem on page 218.

(b) See the Net Change Theorem on page 222.

(c) $\int_{t_1}^{t_2} r(t)\,dt$ represents the change in the amount of water in the reservoir between time t_1 and time t_2.

5. (a) $\int f(x)\,dx$ is an antiderivative of f (or the family of functions $\{F \mid F' = f\}$). Any two such functions differ by a constant.

(b) The connection is given by the Evaluation Theorem: $\int_a^b f(x)\,dx = \left[\int f(x)\,dx\right]_a^b$ if f is continuous.

6. See the Fundamental Theorem of Calculus on page 231.

7. (a) $\int_{60}^{120} v(t)\,dt$ represents the change in position of the particle from $t = 60$ to $t = 120$ seconds.

(b) $\int_{60}^{120} |v(t)|\,dt$ represents the total distance traveled by the particle from $t = 60$ to 120 seconds.

(c) $\int_{60}^{120} a(t)\,dt$ represents the change in the velocity of the particle from $t = 60$ to $t = 120$ seconds.

8. (a) The average value of a function f on an interval $[a, b]$ is $f_{\text{ave}} = \dfrac{1}{b-a}\displaystyle\int_a^b f(x)\,dx$.

(b) The Mean Value Theorem for Integrals says that there is a number c at which the value of f is exactly equal to the average value of the function, that is, $f(c) = f_{\text{ave}}$. For a geometric interpretation of the Mean Value Theorem for Integrals, see Figure 10 in Section 4.4 and the discussion that accompanies it.

9. The precise version of this statement is given by the Fundamental Theorem of Calculus. See the statement of this theorem and the paragraph that follows it on page 231.

10. See the Substitution Rule (4.5.4). This says that it is permissible to operate with the dx after an integral sign as if it were a differential.

TRUE-FALSE QUIZ

1. True by Property 2 of the Integral in Section 4.2.

2. False. Try $a = 0$, $b = 2$, $f(x) = g(x) = 1$ as a counterexample.

3. True by Property 3 of the Integral in Section 4.2.

4. False. You can't take a variable outside the integral sign. For example, using $f(x) = 1$ on $[0, 1]$,

$\int_0^1 x\,f(x)\,dx = \int_0^1 x\,dx = \left[\frac{1}{2}x^2\right]_0^1 = \frac{1}{2}$ (a constant) while $x\int_0^1 1\,dx = x\left[x\right]_0^1 = x \cdot 1 = x$ (a variable).

5. False. For example, let $f(x) = x^2$. Then $\int_0^1 \sqrt{x^2}\,dx = \int_0^1 x\,dx = \frac{1}{2}$, but $\sqrt{\int_0^1 x^2\,dx} = \sqrt{\frac{1}{3}} = \frac{1}{\sqrt{3}}$.

6. True by the Net Change Theorem.

7. True by Comparison Property 7 of the Integral in Section 4.2.

8. False. For example, let $a = 0$, $b = 1$, $f(x) = 3$, $g(x) = x$. $f(x) > g(x)$ for each x in $(0, 1)$, but $f'(x) = 0 < 1 = g'(x)$ for $x \in (0, 1)$.

9. True. The integrand is an odd function that is continuous on $[-1, 1]$, so the result follows from Theorem 4.5.6(b).

10. True.

$$\begin{aligned}\int_{-5}^{5}(ax^2 + bx + c)\,dx &= \int_{-5}^{5}(ax^2 + c)\,dx + \int_{-5}^{5} bx\,dx \\ &= 2\int_0^5 (ax^2 + c)\,dx \quad \text{[by 4.5.6(a)]} \; + 0 \;\; \text{[by 4.5.5(b)]}\end{aligned}$$

11. False. The function $f(x) = 1/x^4$ is not bounded on the interval $[-2, 1]$. It has an infinite discontinuity at $x = 0$, so it is not integrable on the interval. (If the integral were to exist, a positive value would be expected, by Comparison Property 6 of Integrals.)

12. False. See Figure 6 and the remarks before it in Section 4.2, and notice that $y = x - x^3 < 0$ for $1 < x \leq 2$.

13. True by FTC1.

EXERCISES

1. (a)

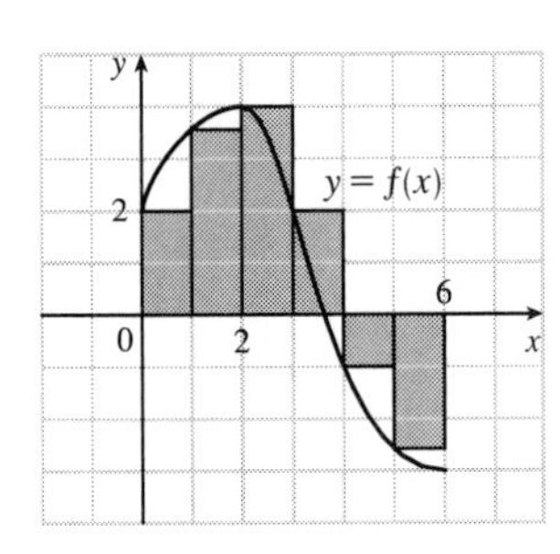

$$L_6 = \sum_{i=1}^{6} f(x_{i-1})\,\Delta x \quad [\Delta x = \tfrac{6-0}{6} = 1]$$
$$= f(x_0)\cdot 1 + f(x_1)\cdot 1 + f(x_2)\cdot 1 + f(x_3)\cdot 1 + f(x_4)\cdot 1 + f(x_5)\cdot 1$$
$$\approx 2 + 3.5 + 4 + 2 + (-1) + (-2.5) = 8$$

The Riemann sum represents the sum of the areas of the four rectangles above the x-axis minus the sum of the areas of the two rectangles below the x-axis.

(b)

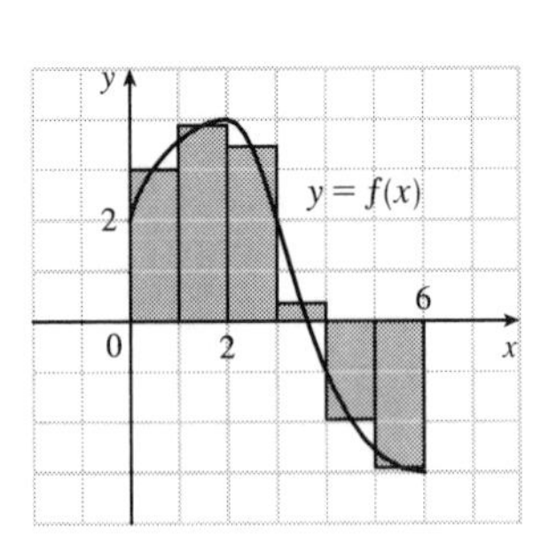

$$M_6 = \sum_{i=1}^{6} f(\overline{x}_i)\,\Delta x \quad [\Delta x = \tfrac{6-0}{6} = 1]$$
$$= f(\overline{x}_1)\cdot 1 + f(\overline{x}_2)\cdot 1 + f(\overline{x}_3)\cdot 1 + f(\overline{x}_4)\cdot 1 + f(\overline{x}_5)\cdot 1 + f(\overline{x}_6)\cdot 1$$
$$= f(0.5) + f(1.5) + f(2.5) + f(3.5) + f(4.5) + f(5.5)$$
$$\approx 3 + 3.9 + 3.4 + 0.3 + (-2) + (-2.9) = 5.7$$

The Riemann sum represents the sum of the areas of the four rectangles above the x-axis minus the sum of the areas of the two rectangles below the x-axis.

2. (a)

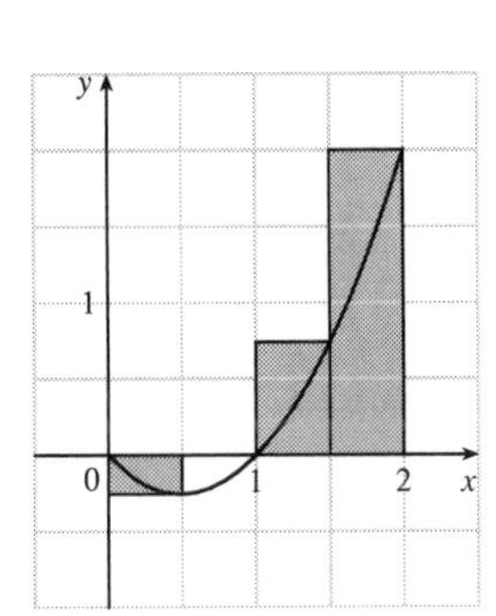

$f(x) = x^2 - x$ and $\Delta x = \frac{2-0}{4} = 0.5 \;\Rightarrow$

$$R_4 = 0.5f(0.5) + 0.5f(1) + 0.5f(1.5) + 0.5f(2)$$
$$= 0.5(-0.25 + 0 + 0.75 + 2) = 1.25$$

The Riemann sum represents the sum of the areas of the two rectangles above the x-axis minus the area of the rectangle below the x-axis. (The second rectangle vanishes.)

(b) $\int_0^2 (x^2 - x)\,dx = \lim\limits_{n\to\infty} \sum\limits_{i=1}^{n} f(x_i)\,\Delta x \qquad [\Delta x = 2/n \text{ and } x_i = 2i/n]$

$$= \lim_{n\to\infty} \sum_{i=1}^{n} \left(\frac{4i^2}{n^2} - \frac{2i}{n}\right)\left(\frac{2}{n}\right) = \lim_{n\to\infty} \frac{2}{n}\left[\frac{4}{n^2}\sum_{i=1}^{n} i^2 - \frac{2}{n}\sum_{i=1}^{n} i\right]$$
$$= \lim_{n\to\infty}\left[\frac{8}{n^3}\cdot\frac{n(n+1)(2n+1)}{6} - \frac{4}{n^2}\cdot\frac{n(n+1)}{2}\right] = \lim_{n\to\infty}\left[\frac{4}{3}\cdot\frac{n+1}{n}\cdot\frac{2n+1}{n} - 2\cdot\frac{n+1}{n}\right]$$
$$= \lim_{n\to\infty}\left[\frac{4}{3}\left(1+\frac{1}{n}\right)\left(2+\frac{1}{n}\right) - 2\left(1+\frac{1}{n}\right)\right] = \tfrac{4}{3}\cdot 1\cdot 2 - 2\cdot 1 = \tfrac{2}{3}$$

(c) $\int_0^2 (x^2 - x)\,dx = \left[\frac{1}{3}x^3 - \frac{1}{2}x^2\right]_0^2 = \left(\frac{8}{3} - 2\right) = \frac{2}{3}$

(d)

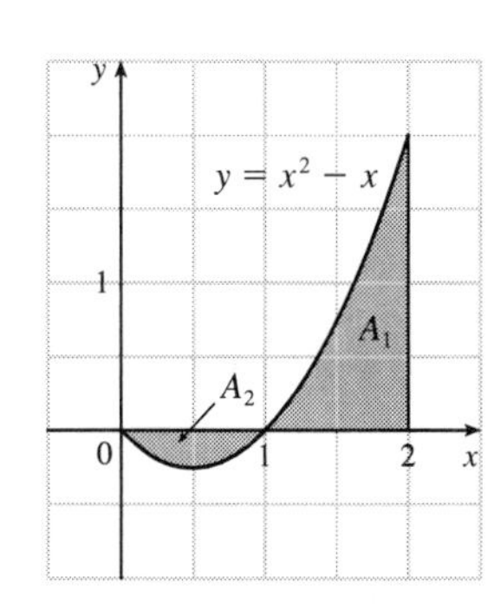

$\int_0^2 (x^2 - x)\,dx = A_1 - A_2$, where A_1 and A_2 are the areas shown in the diagram.

3. $\int_0^1 \left(x + \sqrt{1 - x^2}\,\right) dx = \int_0^1 x\,dx + \int_0^1 \sqrt{1 - x^2}\,dx = I_1 + I_2$.
I_1 can be interpreted as the area of the triangle shown in the figure and I_2 can be interpreted as the area of the quarter-circle.
Area $= \frac{1}{2}(1)(1) + \frac{1}{4}(\pi)(1)^2 = \frac{1}{2} + \frac{\pi}{4}$.

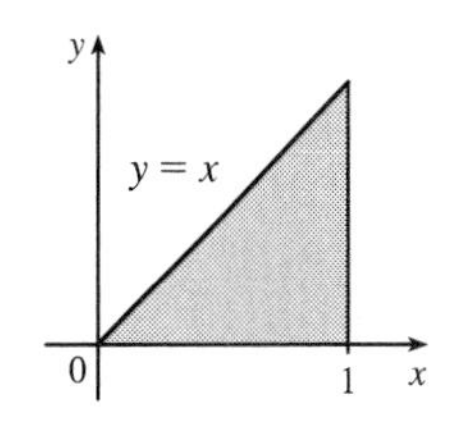

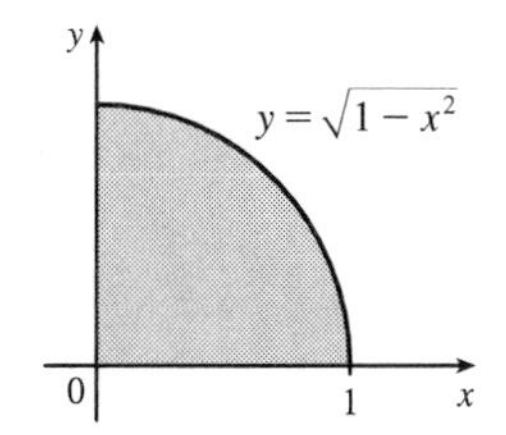

4. On $[0, \pi]$, $\lim\limits_{n\to\infty} \sum\limits_{i=1}^{n} \sin x_i\,\Delta x = \int_0^\pi \sin x\,dx = [-\cos x]_0^\pi = -(-1) - (-1) = 2$.

5. First note that either a or b must be the graph of $\int_0^x f(t)\,dt$, since $\int_0^0 f(t)\,dt = 0$, and $c(0) \neq 0$. Now notice that $b > 0$ when c is increasing, and that $c > 0$ when a is increasing. It follows that c is the graph of $f(x)$, b is the graph of $f'(x)$, and a is the graph of $\int_0^x f(t)\,dt$.

6. (a) By FTC2, we have $\displaystyle\int_0^{\pi/2} \frac{d}{dx}\left(\sin\frac{x}{2}\cos\frac{x}{3}\right) dx = \left[\sin\frac{x}{2}\cos\frac{x}{3}\right]_0^{\pi/2} = \frac{1}{\sqrt{2}}\cdot\frac{\sqrt{3}}{2} - 0\cdot 1 = \frac{\sqrt{6}}{4}$.

(b) $\displaystyle\frac{d}{dx}\int_0^{\pi/2} \sin\frac{x}{2}\cos\frac{x}{3}\,dx = 0$, since the definite integral is a constant.

(c) $\displaystyle\frac{d}{dx}\int_x^{\pi/2} \sin\frac{t}{2}\cos\frac{t}{3}\,dt = \frac{d}{dx}\left(-\int_{\pi/2}^{x} \sin\frac{t}{2}\cos\frac{t}{3}\,dt\right) = -\frac{d}{dx}\int_{\pi/2}^{x} \sin\frac{t}{2}\cos\frac{t}{2}\,dt = -\sin\frac{x}{2}\cos\frac{x}{3}$, by FTC1.

7. $\int_1^2 (8x^3 + 3x^2)\,dx = \left[8\cdot\frac{1}{4}x^4 + 3\cdot\frac{1}{3}x^3\right]_1^2 = \left[2x^4 + x^3\right]_1^2 = (2\cdot 2^4 + 2^3) - (2 + 1) = 40 - 3 = 37$

8. $\displaystyle\int_0^T (x^4 - 8x + 7)\,dx = \left[\tfrac{1}{5}x^5 - 4x^2 + 7x\right]_0^T = \left(\tfrac{1}{5}T^5 - 4T^2 + 7T\right) - 0 = \tfrac{1}{5}T^5 - 4T^2 + 7T$

9. $\int_0^1 (1 - x^9)\,dx = \left[x - \frac{1}{10}x^{10}\right]_0^1 = \left(1 - \frac{1}{10}\right) - 0 = \frac{9}{10}$

10. Let $u = 1 - x$, so $du = -dx$ and $dx = -du$. When $x = 0$, $u = 1$; when $x = 1$, $u = 0$.
Thus, $\int_0^1 (1 - x)^9\,dx = \int_1^0 u^9(-du) = \int_0^1 u^9\,du = \frac{1}{10}\left[u^{10}\right]_0^1 = \frac{1}{10}(1 - 0) = \frac{1}{10}$.

11. $\displaystyle\int_1^9 \frac{\sqrt{u} - 2u^2}{u}\,du = \int_1^9 (u^{-1/2} - 2u)\,du = \left[2u^{1/2} - u^2\right]_1^9 = (6 - 81) - (2 - 1) = -76$

12. $\int_0^1 \left(\sqrt[4]{u}+1\right)^2 du = \int_0^1 (u^{1/2}+2u^{1/4}+1)\,du = \left[\frac{2}{3}u^{3/2}+\frac{8}{5}u^{5/4}+u\right]_0^1 = \left(\frac{2}{3}+\frac{8}{5}+1\right)-0=\frac{49}{15}$

13. Let $u=y^2+1$, so $du=2y\,dy$ and $y\,dy=\frac{1}{2}\,du$. When $y=0$, $u=1$; when $y=1$, $u=2$.

Thus, $\int_0^1 y(y^2+1)^5\,dy=\int_1^2 u^5\left(\frac{1}{2}\,du\right)=\frac{1}{2}\left[\frac{1}{6}u^6\right]_1^2=\frac{1}{12}(64-1)=\frac{63}{12}=\frac{21}{4}$.

14. Let $u=1+y^3$, so $du=3y^2\,dy$ and $y^2\,dy=\frac{1}{3}\,du$. When $y=0$, $u=1$; when $y=2$, $u=9$.

Thus, $\int_0^2 y^2\sqrt{1+y^3}\,dy=\int_1^9 u^{1/2}\left(\frac{1}{3}\,du\right)=\frac{1}{3}\left[\frac{2}{3}u^{3/2}\right]_1^9=\frac{2}{9}(27-1)=\frac{52}{9}$.

15. Let $u=v^3$, so $du=3v^2\,dv$. When $v=0$, $u=0$; when $v=1$, $u=1$.

Thus, $\int_0^1 v^2\cos(v^3)\,dv=\int_0^1\cos u\left(\frac{1}{3}\,du\right)=\frac{1}{3}\left[\sin u\right]_0^1=\frac{1}{3}(\sin 1-0)=\frac{1}{3}\sin 1$.

16. Let $u=3\pi t$, so $du=3\pi\,dt$. When $t=0$, $u=1$; when $t=1$, $u=3\pi$.

Thus, $\displaystyle\int_0^1 \sin(3\pi t)\,dt=\int_0^{3\pi}\sin u\left(\frac{1}{3\pi}\,du\right)=\frac{1}{3\pi}\left[-\cos u\right]_0^{3\pi}=-\frac{1}{3\pi}(-1-1)=\frac{2}{3\pi}$.

17. Let $u=x^2+4x$. Then $du=(2x+4)\,dx=2(x+2)\,dx$, so

$$\int\frac{x+2}{\sqrt{x^2+4x}}\,dx=\int u^{-1/2}\left(\tfrac{1}{2}\,du\right)=\tfrac{1}{2}\cdot 2u^{1/2}+C=\sqrt{u}+C=\sqrt{x^2+4x}+C.$$

18. Let $u=3t$. Then $du=3\,dt$, so $\int\csc^2 3t\,dt=\int\csc^2 u\left(\frac{1}{3}\,du\right)=\frac{1}{3}(-\cot u)+C=-\frac{1}{3}\cot 3t+C$.

19. Let $u=\sin\pi t$. Then $du=\pi\cos\pi t\,dt$, so $\int\sin\pi t\cos\pi t\,dt=\int u\left(\frac{1}{\pi}\,du\right)=\frac{1}{\pi}\cdot\frac{1}{2}u^2+C=\frac{1}{2\pi}(\sin\pi t)^2+C$.

20. Let $u=\cos x$. Then $du=-\sin x\,dx$, so $\int\sin x\cos(\cos x)\,dx=-\int\cos u\,du=-\sin u+C=-\sin(\cos x)+C$.

21. Let $u=2\theta$. Then $du=2\,d\theta$, so

$$\begin{aligned}\int_0^{\pi/8}\sec 2\theta\tan 2\theta\,d\theta&=\int_0^{\pi/4}\sec u\tan u\left(\tfrac{1}{2}\,du\right)=\tfrac{1}{2}[\sec u]_0^{\pi/4}=\tfrac{1}{2}\left(\sec\tfrac{\pi}{4}-\sec 0\right)\\&=\tfrac{1}{2}\left(\sqrt{2}-1\right)=\tfrac{1}{2}\sqrt{2}-\tfrac{1}{2}\end{aligned}$$

22. Since $\sqrt{x}-1<0$ for $0\le x<1$ and $\sqrt{x}-1>0$ for $1<x\le 4$, we have $|\sqrt{x}-1|=-(\sqrt{x}-1)=1-\sqrt{x}$ for $0\le x<1$ and $|\sqrt{x}-1|=\sqrt{x}-1$ for $1<x\le 4$. Thus,

$$\begin{aligned}\int_0^4|\sqrt{x}-1|\,dx&=\int_0^1(1-\sqrt{x})\,dx+\int_1^4(\sqrt{x}-1)\,dx=\left[x-\tfrac{2}{3}x^{3/2}\right]_0^1+\left[\tfrac{2}{3}x^{3/2}-x\right]_1^4\\&=\left(1-\tfrac{2}{3}\right)-0+\left(\tfrac{16}{3}-4\right)-\left(\tfrac{2}{3}-1\right)=\tfrac{1}{3}+\tfrac{16}{3}-4+\tfrac{1}{3}=6-4=2\end{aligned}$$

23. Since $x^2-4<0$ for $0\le x<2$ and $x^2-4>0$ for $2<x\le 3$, we have $\left|x^2-4\right|=-(x^2-4)=4-x^2$ for $0\le x<2$ and $\left|x^2-4\right|=x^2-4$ for $2<x\le 3$. Thus,

$$\begin{aligned}\int_0^3\left|x^2-4\right|dx&=\int_0^2(4-x^2)\,dx+\int_2^3(x^2-4)\,dx=\left[4x-\frac{x^3}{3}\right]_0^2+\left[\frac{x^3}{3}-4x\right]_2^3\\&=\left(8-\tfrac{8}{3}\right)-0+(9-12)-\left(\tfrac{8}{3}-8\right)=\tfrac{16}{3}-3+\tfrac{16}{3}=\tfrac{32}{3}-\tfrac{9}{3}=\tfrac{23}{3}\end{aligned}$$

24. $\displaystyle\int_{-1}^1\frac{\sin x}{1+x^2}\,dx=0$ by Theorem 4.5.6(b), since $f(x)=\dfrac{\sin x}{1+x^2}$ is an odd function.

25. From the graph, it appears that the area under the curve $y = x\sqrt{x}$ between $x = 0$ and $x = 4$ is somewhat less than half the area of an 8×4 rectangle, so perhaps about 13 or 14. To find the exact value, we evaluate

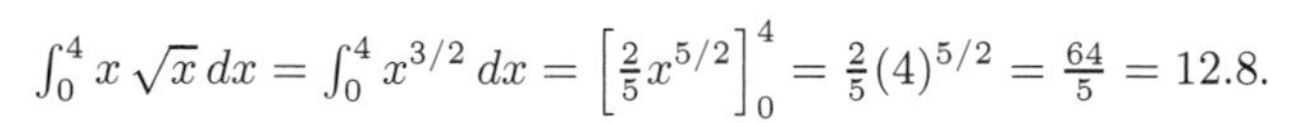

$\int_0^4 x\sqrt{x}\,dx = \int_0^4 x^{3/2}\,dx = \left[\frac{2}{5}x^{5/2}\right]_0^4 = \frac{2}{5}(4)^{5/2} = \frac{64}{5} = 12.8.$

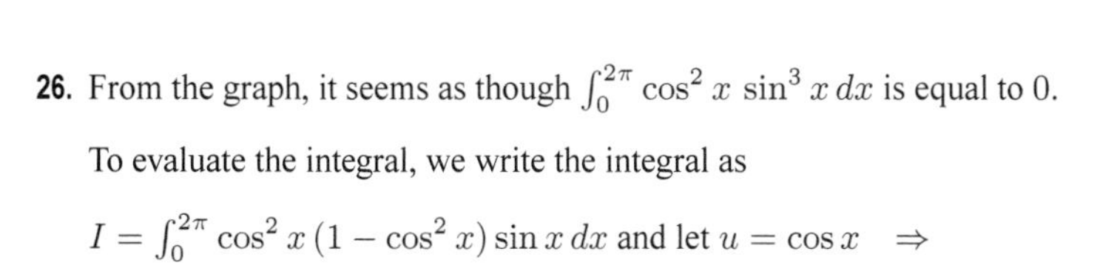

26. From the graph, it seems as though $\int_0^{2\pi} \cos^2 x \sin^3 x\,dx$ is equal to 0.

To evaluate the integral, we write the integral as

$I = \int_0^{2\pi} \cos^2 x\,(1 - \cos^2 x)\sin x\,dx$ and let $u = \cos x \Rightarrow$

$du = -\sin x\,dx$. Thus, $I = \int_1^1 u^2(1 - u^2)(-du) = 0.$

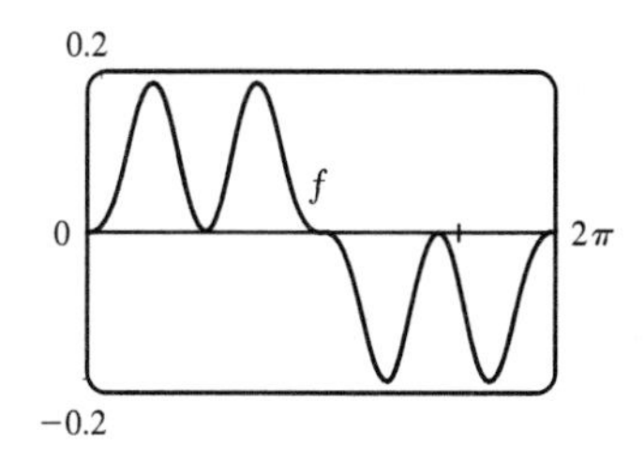

27. By FTC1, $F(x) = \int_1^x \sqrt{1 + t^4}\,dt \Rightarrow F'(x) = \sqrt{1 + x^4}.$

28. Let $u = \cos x$. Then $\dfrac{du}{dx} = -\sin x$. Also, $\dfrac{dg}{dx} = \dfrac{dg}{du}\dfrac{du}{dx}$, so

$$\frac{d}{dx}\int_1^{\cos x} \sqrt[3]{1 - t^2}\,dt = \frac{d}{du}\int_1^u \sqrt[3]{1 - t^2}\,dt \cdot \frac{du}{dx} = \sqrt[3]{1 - u^2}\,(-\sin x)$$

$$= \sqrt[3]{1 - \cos^2 x}\,(-\sin x) = -\sin x\,\sqrt[3]{\sin^2 x} = -(\sin x)^{5/3}$$

29. $y = \displaystyle\int_{\sqrt{x}}^{x} \frac{\cos\theta}{\theta}\,d\theta = \int_1^x \frac{\cos\theta}{\theta}\,d\theta + \int_{\sqrt{x}}^{1} \frac{\cos\theta}{\theta}\,d\theta = \int_1^x \frac{\cos\theta}{\theta}\,d\theta - \int_1^{\sqrt{x}} \frac{\cos\theta}{\theta}\,d\theta \Rightarrow$

$y' = \dfrac{\cos x}{x} - \dfrac{\cos\sqrt{x}}{\sqrt{x}}\,\dfrac{1}{2\sqrt{x}} = \dfrac{2\cos x - \cos\sqrt{x}}{2x}$

30. $y = \int_{2x}^{3x+1} \sin(t^4)\,dt = \int_{2x}^{0} \sin(t^4)\,dt + \int_0^{3x+1} \sin(t^4)\,dt = \int_0^{3x+1} \sin(t^4)\,dt - \int_0^{2x} \sin(t^4)\,dt \Rightarrow$

$y' = \sin[(3x+1)^4]\cdot\dfrac{d}{dx}(3x+1) - \sin[(2x)^4]\cdot\dfrac{d}{dx}(2x) = 3\sin[(3x+1)^4] - 2\sin[(2x)^4]$

31. If $1 \le x \le 3$, then $\sqrt{1^2 + 3} \le \sqrt{x^2 + 3} \le \sqrt{3^2 + 3} \Rightarrow 2 \le \sqrt{x^2 + 3} \le 2\sqrt{3}$, so
$2(3 - 1) \le \int_1^3 \sqrt{x^2 + 3}\,dx \le 2\sqrt{3}(3 - 1)$; that is, $4 \le \int_1^3 \sqrt{x^2 + 3}\,dx \le 4\sqrt{3}$.

32. On $[0, 1]$, $x^4 \ge x^4 \cos x$ (since $0 \le \cos x \le 1$), so by Property 7, $\int_0^1 x^4\,dx \ge \int_0^1 x^4 \cos x\,dx$. Also, $x^4 \cos x \ge 0$, so by Property 6, $\int_0^1 x^4 \cos x\,dx \ge 0$. But $\int_0^1 x^4\,dx = \left[\frac{1}{5}x^5\right]_0^1 = \frac{1}{5} = 0.2$, so $0 \le \int_0^1 x^4 \cos x\,dx \le 0.2$.

33. Let $f(x) = \sqrt{1 + x^3}$ on $[0, 1]$. The Midpoint Rule with $n = 5$ gives

$$\int_0^1 \sqrt{1 + x^3}\,dx \approx \tfrac{1}{5}[f(0.1) + f(0.3) + f(0.5) + f(0.7) + f(0.9)]$$

$$= \tfrac{1}{5}\left[\sqrt{1 + (0.1)^3} + \sqrt{1 + (0.3)^3} + \cdots + \sqrt{1 + (0.9)^3}\right] \approx 1.110$$

34. (a) Displacement $= \int_0^5 (t^2 - t)\,dt = \left[\frac{1}{3}t^3 - \frac{1}{2}t^2\right]_0^5 = \frac{125}{3} - \frac{25}{2} = \frac{175}{6} = 29.1\overline{6}$ meters

(b) Distance traveled $= \int_0^5 \left|t^2 - t\right|dt = \int_0^5 |t(t - 1)|\,dt = \int_0^1 (t - t^2)\,dt + \int_1^5 (t^2 - t)\,dt = \left[\frac{1}{2}t^2 - \frac{1}{3}t^3\right]_0^1 + \left[\frac{1}{3}t^3 - \frac{1}{2}t^2\right]_1^5$

$= \frac{1}{2} - \frac{1}{3} - 0 + \left(\frac{125}{3} - \frac{25}{2}\right) - \left(\frac{1}{3} - \frac{1}{2}\right) = \frac{177}{6} = 29.5$ meters

35. Note that $r(t) = b'(t)$, where $b(t) =$ the number of barrels of oil consumed up to time t. So, by the Net Change Theorem, $\int_0^3 r(t)\,dt = b(3) - b(0)$ represents the number of barrels of oil consumed from Jan. 1, 2000, through Jan. 1, 2003.

36. Distance covered $= \int_0^{5.0} v(t)\,dt \approx M_5 = \frac{5.0-0}{5}[v(0.5) + v(1.5) + v(2.5) + v(3.5) + v(4.5)]$

$$= 1(4.67 + 8.86 + 10.22 + 10.67 + 10.81) = 45.23 \text{ m}$$

37. We use the Midpoint Rule with $n = 6$ and $\Delta t = \frac{24-0}{6} = 4$. The increase in the bee population was

$$\int_0^{24} r(t)\,dt \approx M_6 = 4[r(2) + r(6) + r(10) + r(14) + r(18) + r(22)]$$

$$\approx 4[50 + 1000 + 7000 + 8550 + 1350 + 150] = 4(18{,}100) = 72{,}400$$

38. $f_{\text{ave}} = \frac{1}{2-0}\int_0^2 x^2\sqrt{1+x^3}\,dx = \frac{1}{2}\cdot\frac{1}{3}\int_1^9 \sqrt{u}\,du \qquad [u = 1 + x^3,\ du = 3x^2\,dx]$

$= \frac{1}{6}\left[\frac{2}{3}u^{3/2}\right]_1^9 = (9^{3/2} - 1^{3/2}) = \frac{1}{9}(27-1) = \frac{26}{9}$

39. $\lim\limits_{h\to 0} f_{\text{ave}} = \lim\limits_{h\to 0} \dfrac{1}{(x+h)-x}\displaystyle\int_x^{x+h} f(t)\,dt = \lim\limits_{h\to 0}\dfrac{F(x+h)-F(x)}{h}$, where $F(x) = \int_a^x f(t)\,dt$. But we recognize this limit as being $F'(x)$ by the definition of a derivative. Therefore, $\lim\limits_{h\to 0} f_{\text{ave}} = F'(x) = f(x)$ by FTC1.

40. $A_1 = \frac{1}{2}bh = \frac{1}{2}(2)(2) = 2$, $A_2 = \frac{1}{2}bh = \frac{1}{2}(1)(1) = \frac{1}{2}$, and since $y = -\sqrt{1-x^2}$ for $0 \le x \le 1$ represents a quarter-circle with radius 1, $A_3 = \frac{1}{4}\pi r^2 = \frac{1}{4}\pi(1)^2 = \frac{\pi}{4}$. So $\int_{-3}^1 f(x)\,dx = A_1 - A_2 - A_3 = 2 - \frac{1}{2} - \frac{\pi}{4} = \frac{1}{4}(6-\pi)$.

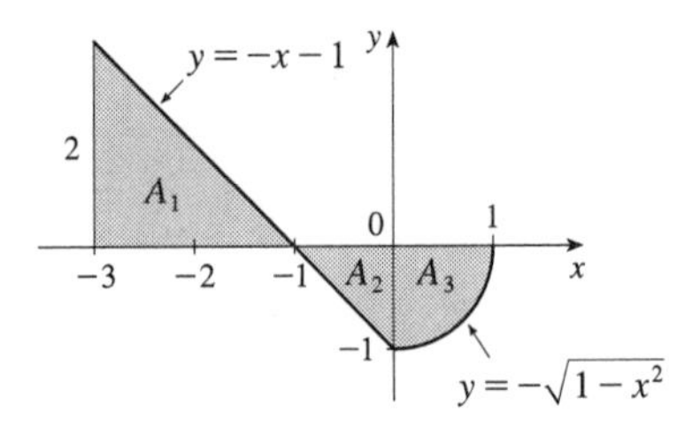

41. Let $u = 1 - x$. Then $du = -dx$, so $\int_0^1 f(1-x)\,dx = \int_1^0 f(u)(-du) = \int_0^1 f(u)\,du = \int_0^1 f(x)\,dx$.

42. $\displaystyle\int_0^x f(t)\,dt = x\sin x + \int_0^x \frac{f(t)}{1+t^2}\,dt \;\Rightarrow\; f(x) = x\cos x + \sin x + \frac{f(x)}{1+x^2}$ (by differentiation) $\Rightarrow$

$$f(x)\left(1 - \frac{1}{1+x^2}\right) = x\cos x + \sin x \;\Rightarrow\; f(x)\left(\frac{x^2}{1+x^2}\right) = x\cos x + \sin x \;\Rightarrow\; f(x) = \frac{1+x^2}{x^2}(x\cos x + \sin x)$$

43. Let $u = f(x)$ and $du = f'(x)\,dx$. So $2\int_a^b f(x)f'(x)\,dx = 2\int_{f(a)}^{f(b)} u\,du = \left[u^2\right]_{f(a)}^{f(b)} = [f(b)]^2 - [f(a)]^2$.

44. $$\lim_{n\to\infty}\frac{1}{n}\left[\left(\frac{1}{n}\right)^9 + \left(\frac{2}{n}\right)^9 + \left(\frac{3}{n}\right)^9 + \cdots + \left(\frac{n}{n}\right)^9\right] = \lim_{n\to\infty}\frac{1-0}{n}\sum_{i=1}^n\left(\frac{i}{n}\right)^9 = \int_0^1 x^9\,dx = \left[\frac{x^{10}}{10}\right]_0^1 = \frac{1}{10}.$$

The limit is based on Riemann sums using right endpoints and subintervals of equal length.

5 □ Inverse Functions: Exponential, Logarithmic, and Inverse Trigonometric Functions

5.1 Inverse Functions

1. (a) See Definition 1.

(b) It must pass the Horizontal Line Test.

2. (a) $f^{-1}(y) = x \quad \Leftrightarrow \quad f(x) = y$ for any y in B. The domain of f^{-1} is B and the range of f^{-1} is A.

(b) See the steps in (5).

(c) Reflect the graph of f about the line $y = x$.

3. f is not one-to-one because $2 \neq 6$, but $f(2) = 2.0 = f(6)$.

4. f is one-to-one since for any two different domain values, there are different range values.

5. No horizontal line intersects the graph of f more than once. Thus, by the Horizontal Line Test, f is one-to-one.

6. The horizontal line $y = 0$ (the x-axis) intersects the graph of f in more than one point. Thus, by the Horizontal Line Test, f is not one-to-one.

7. The horizontal line $y = 0$ (the x-axis) intersects the graph of f in more than one point. Thus, by the Horizontal Line Test, f is not one-to-one.

8. No horizontal line intersects the graph of f more than once. Thus, by the Horizontal Line Test, f is one-to-one.

9. The graph of $f(x) = x^2 - 2x$ is a parabola with axis of symmetry $x = -\dfrac{b}{2a} = -\dfrac{-2}{2(1)} = 1$. Pick any x-values equidistant from 1 to find two equal function values. For example, $f(0) = 0$ and $f(2) = 0$, so f is not one-to-one.

10. The graph of $f(x) = 10 - 3x$ is a line with slope -3. It passes the Horizontal Line Test, so f is one-to-one.

Algebraic solution: If $x_1 \neq x_2$, then $-3x_1 \neq -3x_2 \quad \Rightarrow \quad 10 - 3x_1 \neq 10 - 3x_2 \quad \Rightarrow \quad f(x_1) \neq f(x_2)$, so f is one-to-one.

11. $g(x) = 1/x. \quad x_1 \neq x_2 \quad \Rightarrow \quad 1/x_1 \neq 1/x_2 \quad \Rightarrow \quad g(x_1) \neq g(x_2)$, so g is one-to-one.

Geometric solution: The graph of g is the hyperbola shown in Figure 9 in Section 1.2. It passes the Horizontal Line Test, so g is one-to-one.

12. $g(x) = \cos x. \quad g(0) = 1 = g(2\pi)$, so g is not one-to-one.

13. A football will attain every height h up to its maximum height twice: once on the way up, and again on the way down. Thus, even if t_1 does not equal t_2, $f(t_1)$ may equal $f(t_2)$, so f is not 1-1.

14. f is not 1-1 because eventually we all stop growing and therefore, there are two times at which we have the same height.

15. Since $f(2) = 9$ and f is 1-1, we know that $f^{-1}(9) = 2$. Remember, if the point $(2, 9)$ is on the graph of f, then the point $(9, 2)$ is on the graph of f^{-1}.

16. $f(x) = x + \cos x \quad \Rightarrow \quad f'(x) = 1 - \sin x \geq 0$, with equality only if $x = \frac{\pi}{2} + 2n\pi$. So f is increasing on $\mathbb{R}$, and hence, 1-1. By inspection, $f(0) = 0 + \cos 0 = 1$, so $f^{-1}(1) = 0$.

17. $h(x) = x + \sqrt{x} \Rightarrow h'(x) = 1 + 1/(2\sqrt{x}) > 0$ on $(0, \infty)$. So h is increasing and hence, 1-1. By inspection, $h(4) = 4 + \sqrt{4} = 6$, so $h^{-1}(6) = 4$.

18. (a) f is 1-1 because it passes the Horizontal Line Test.

(b) Domain of $f = [-3, 3] =$ Range of f^{-1}. Range of $f = [-1, 3] =$ Domain of f^{-1}.

(c) Since $f(0) = 2$, $f^{-1}(2) = 0$.

(d) Since $f(-1.7) \approx 0$, $f^{-1}(0) = -1.7$.

19. We solve $C = \frac{5}{9}(F - 32)$ for F: $\frac{9}{5}C = F - 32 \Rightarrow F = \frac{9}{5}C + 32$. This gives us a formula for the inverse function, that is, the Fahrenheit temperature F as a function of the Celsius temperature C.

$F \geq -459.67 \Rightarrow \frac{9}{5}C + 32 \geq -459.67 \Rightarrow \frac{9}{5}C \geq -491.67 \Rightarrow C \geq -273.15$, the domain of the inverse function.

20. $m = \dfrac{m_0}{\sqrt{1 - v^2/c^2}} \Rightarrow 1 - \dfrac{v^2}{c^2} = \dfrac{m_0^2}{m^2} \Rightarrow \dfrac{v^2}{c^2} = 1 - \dfrac{m_0^2}{m^2} \Rightarrow v^2 = c^2\left(1 - \dfrac{m_0^2}{m^2}\right) \Rightarrow v = c\sqrt{1 - \dfrac{m_0^2}{m^2}}$.

This formula gives us the speed v of the particle in terms of its mass m, that is, $v = f^{-1}(m)$.

21. $y = f(x) = 3 - 2x \Rightarrow 2x = 3 - y \Rightarrow x = \dfrac{3 - y}{2}$. Interchange x and y: $y = \dfrac{3 - x}{2}$. So $f^{-1}(x) = \dfrac{3 - x}{2}$.

22. $y = f(x) = \dfrac{4x - 1}{2x + 3} \Rightarrow y(2x + 3) = 4x - 1 \Rightarrow 2xy + 3y = 4x - 1 \Rightarrow 3y + 1 = 4x - 2xy \Rightarrow$

$3y + 1 = (4 - 2y)x \Rightarrow x = \dfrac{3y + 1}{4 - 2y}$. Interchange x and y: $y = \dfrac{3x + 1}{4 - 2x}$. So $f^{-1}(x) = \dfrac{3x + 1}{4 - 2x}$.

23. $f(x) = \sqrt{10 - 3x} \Rightarrow y = \sqrt{10 - 3x}$ $(y \geq 0) \Rightarrow y^2 = 10 - 3x \Rightarrow 3x = 10 - y^2 \Rightarrow x = -\frac{1}{3}y^2 + \frac{10}{3}$.

Interchange x and y: $y = -\frac{1}{3}x^2 + \frac{10}{3}$. So $f^{-1}(x) = -\frac{1}{3}x^2 + \frac{10}{3}$. Note that the domain of f^{-1} is $x \geq 0$.

24. $y = f(x) = 2x^3 + 3 \Rightarrow y - 3 = 2x^3 \Rightarrow \dfrac{y - 3}{2} = x^3 \Rightarrow x = \sqrt[3]{\dfrac{y - 3}{2}}$. Interchange x and y: $y = \sqrt[3]{\dfrac{x - 3}{2}}$.

So $f^{-1}(x) = \sqrt[3]{\dfrac{x - 3}{2}}$.

25. For $f(x) = \dfrac{1 - \sqrt{x}}{1 + \sqrt{x}}$, the domain is $x \geq 0$. $f(0) = 1$ and as x increases, y decreases. As $x \to \infty$,

$\dfrac{1 - \sqrt{x}}{1 + \sqrt{x}} \cdot \dfrac{1/\sqrt{x}}{1/\sqrt{x}} = \dfrac{1/\sqrt{x} - 1}{1/\sqrt{x} + 1} \to \dfrac{-1}{1} = -1$, so the range of f is $-1 < y \leq 1$. Thus, the domain of f^{-1} is $-1 < x \leq 1$.

$y = \dfrac{1 - \sqrt{x}}{1 + \sqrt{x}} \Rightarrow y(1 + \sqrt{x}) = 1 - \sqrt{x} \Rightarrow y + y\sqrt{x} = 1 - \sqrt{x} \Rightarrow \sqrt{x} + y\sqrt{x} = 1 - y \Rightarrow$

$\sqrt{x}(1 + y) = 1 - y \Rightarrow \sqrt{x} = \dfrac{1 - y}{1 + y} \Rightarrow x = \left(\dfrac{1 - y}{1 + y}\right)^2$. Interchange x and y: $y = \left(\dfrac{1 - x}{1 + x}\right)^2$. So

$f^{-1}(x) = \left(\dfrac{1 - x}{1 + x}\right)^2$ with $-1 < x \leq 1$.

26. $y = f(x) = 2x^2 - 8x,\ x \geq 2 \ \Rightarrow\ 2x^2 - 8x - y = 0,\ x \geq 2 \ \Rightarrow$

$$x = \frac{8 + \sqrt{64 + 8y}}{4} \left[\begin{array}{c}\text{quadratic formula with} \\ a = 2,\ b = -8, \text{ and } c = -y\end{array}\right] = \frac{8 + 2\sqrt{16 + 2y}}{4} = 2 + \tfrac{1}{2}\sqrt{16 + 2y}.$$ Interchange x and y:

$y = 2 + \frac{1}{2}\sqrt{16 + 2x}$. So $f^{-1}(x) = 2 + \frac{1}{2}\sqrt{16 + 2x}$.

Alternate solution (by completing the square): $y = 2x^2 - 8x,\ x \geq 2 \ \Rightarrow\ x^2 - 4x = y/2,\ x \geq 2 \ \Rightarrow$

$$(x - 2)^2 = x^2 - 4x + 4 = \frac{y}{2} + 4 = \frac{y + 8}{2} = \frac{2y + 16}{4},\ x \geq 2 \ \Rightarrow\ x - 2 = +\sqrt{\frac{2y + 16}{4}} \ \Rightarrow\ x = 2 + \tfrac{1}{2}\sqrt{2y + 16}.$$

Interchange x and y: $y = 2 + \frac{1}{2}\sqrt{2x + 16}$. So $f^{-1}(x) = 2 + \frac{1}{2}\sqrt{2x + 16}$.

27. $y = f(x) = x^4 + 1 \ \Rightarrow\ y - 1 = x^4 \ \Rightarrow\ x = \sqrt[4]{y - 1}$ (not $\pm$ since $x \geq 0$). Interchange x and y: $y = \sqrt[4]{x - 1}$. So $f^{-1}(x) = \sqrt[4]{x - 1}$. The graph of $y = \sqrt[4]{x - 1}$ is just the graph of $y = \sqrt[4]{x}$ shifted right one unit. From the graph, we see that f and f^{-1} are reflections about the line $y = x$.

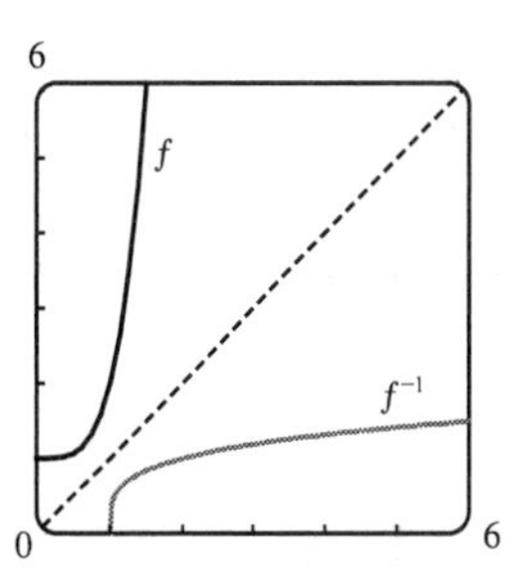

28. $y = f(x) = \sqrt{x^2 + 2x},\ x > 0 \ \Rightarrow\ y > 0$ and $y^2 = x^2 + 2x \ \Rightarrow$ $x^2 + 2x - y^2 = 0$. Now we use the quadratic formula:

$$x = \frac{-2 \pm \sqrt{2^2 - 4 \cdot 1 \cdot (-y^2)}}{2 \cdot 1} = -1 \pm \sqrt{1 + y^2}.$$ But $x > 0$, so the negative root is inadmissible. Interchange x and y: $y = -1 + \sqrt{1 + x^2}$. So $f^{-1}(x) = -1 + \sqrt{1 + x^2},\ x > 0$.

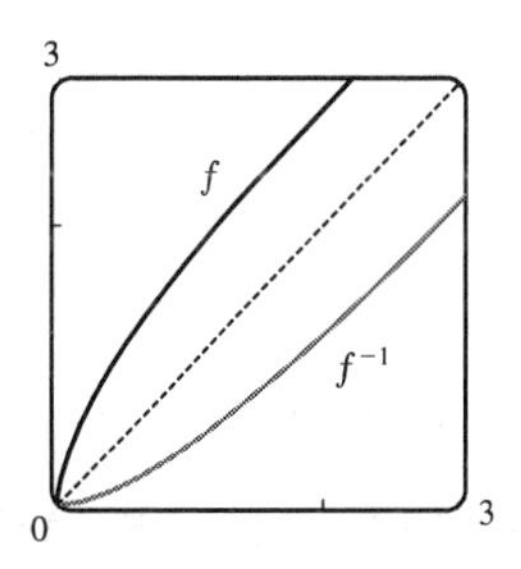

29. Reflect the graph of f about the line $y = x$. The points $(-1, -2)$, $(1, -1)$, $(2, 2)$, and $(3, 3)$ on f are reflected to $(-2, -1)$, $(-1, 1)$, $(2, 2)$, and $(3, 3)$ on f^{-1}.

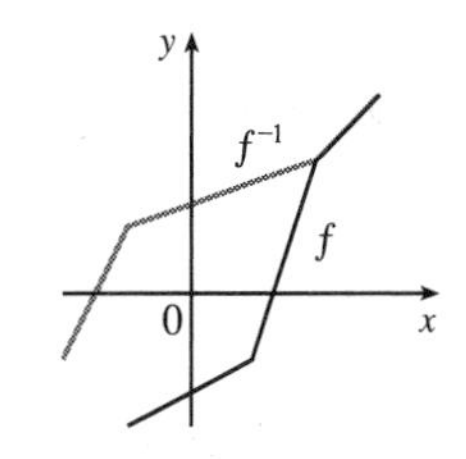

30. Reflect the graph of f about the line $y = x$.

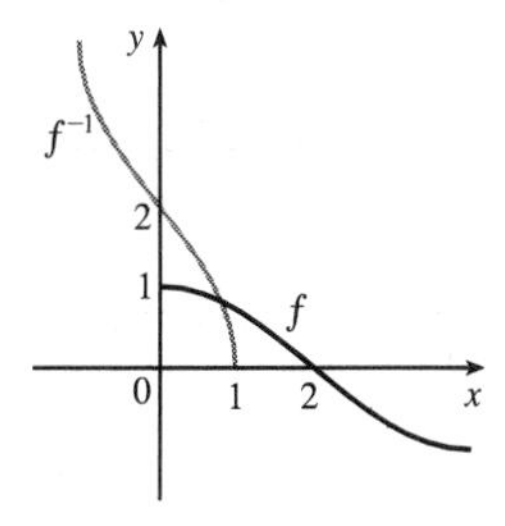

31. (a) $x_1 \neq x_2 \ \Rightarrow\ x_1^3 \neq x_2^3 \ \Rightarrow\ f(x_1) \neq f(x_2)$, so f is one-to-one.

(b) $f'(x) = 3x^2$ and $f(2) = 8 \ \Rightarrow\ f^{-1}(8) = 2$, so $(f^{-1})'(8) = 1/f'(f^{-1}(8)) = 1/f'(2) = \frac{1}{12}$.

(c) $y = x^3 \Rightarrow x = y^{1/3}$. Interchanging x and y gives $y = x^{1/3}$, so $f^{-1}(x) = x^{1/3}$. Domain(f^{-1}) = range$(f) = \mathbb{R}$.
Range(f^{-1}) = domain$(f) = \mathbb{R}$.

(d) $f^{-1}(x) = x^{1/3} \Rightarrow (f^{-1})'(x) = \frac{1}{3}x^{-2/3} \Rightarrow$
$(f^{-1})'(8) = \frac{1}{3}\left(\frac{1}{4}\right) = \frac{1}{12}$ as in part (b).

(e)

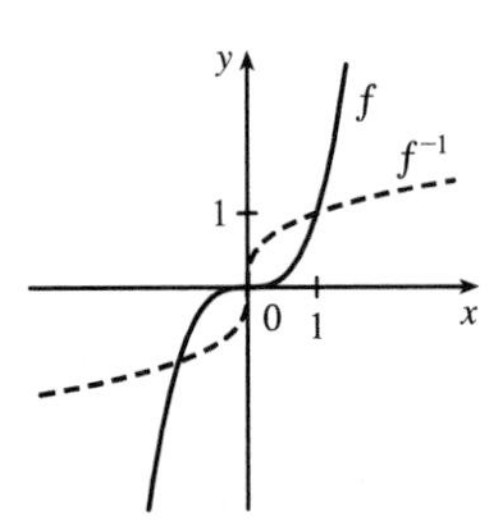

32. (a) $x_1 \neq x_2 \Rightarrow x_1 - 2 \neq x_2 - 2 \Rightarrow \sqrt{x_1 - 2} \neq \sqrt{x_2 - 2} \Rightarrow f(x_1) \neq f(x_2)$, so f is 1-1.

(b) $f(6) = 2$, so $f^{-1}(2) = 6$. Also $f'(x) = \dfrac{1}{2\sqrt{x-2}}$, so $(f^{-1})'(2) = \dfrac{1}{f'(f^{-1}(2))} = \dfrac{1}{f'(6)} = \dfrac{1}{1/4} = 4$.

(c) $y = \sqrt{x-2} \Rightarrow y^2 = x - 2 \Rightarrow x = y^2 + 2$.
Interchange x and y: $y = x^2 + 2$. So $f^{-1}(x) = x^2 + 2$.
Domain $= [0, \infty)$, range $= [2, \infty)$.

(d) $f^{-1}(x) = x^2 + 2 \Rightarrow (f^{-1})'(x) = 2x \Rightarrow (f^{-1})'(2) = 4$.

(e)

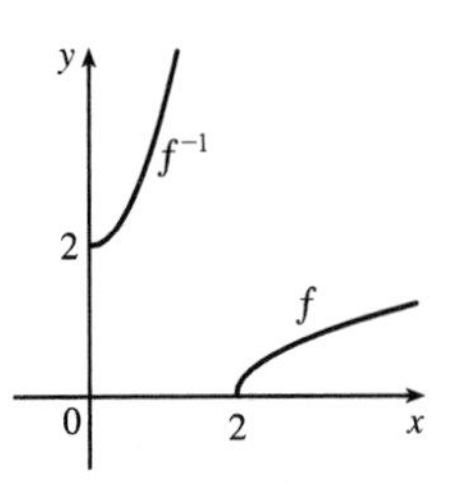

33. (a) Since $x \geq 0$, $x_1 \neq x_2 \Rightarrow x_1^2 \neq x_2^2 \Rightarrow 9 - x_1^2 \neq 9 - x_2^2 \Rightarrow f(x_1) \neq f(x_2)$, so f is 1-1.

(b) $f'(x) = -2x$ and $f(1) = 8 \Rightarrow f^{-1}(8) = 1$, so $(f^{-1})'(8) = \dfrac{1}{f'(f^{-1}(8))} = \dfrac{1}{f'(1)} = \dfrac{1}{(-2)} = -\dfrac{1}{2}$.

(c) $y = 9 - x^2 \Rightarrow x^2 = 9 - y \Rightarrow x = \sqrt{9 - y}$.
Interchange x and y: $y = \sqrt{9 - x}$, so $f^{-1}(x) = \sqrt{9 - x}$.
Domain(f^{-1}) = range $(f) = [0, 9]$.
Range(f^{-1}) = domain $(f) = [0, 3]$.

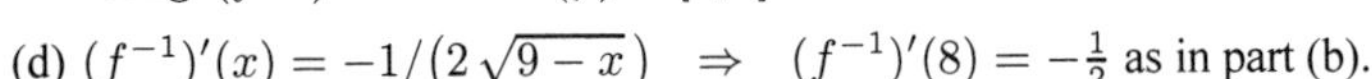
(d) $(f^{-1})'(x) = -1/(2\sqrt{9 - x}) \Rightarrow (f^{-1})'(8) = -\frac{1}{2}$ as in part (b).

(e)

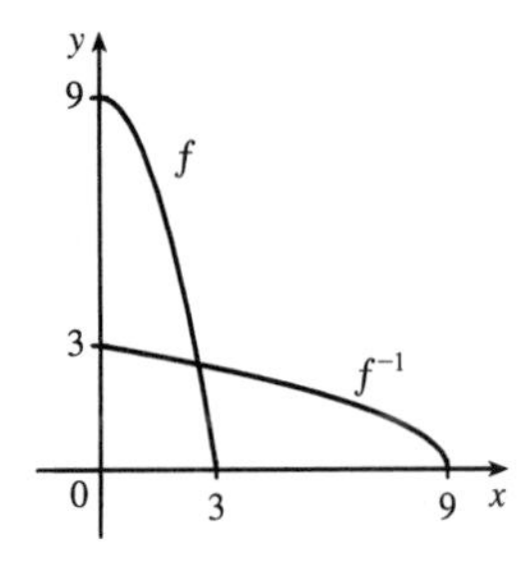

34. (a) $x_1 \neq x_2 \Rightarrow x_1 - 1 \neq x_2 - 1 \Rightarrow \dfrac{1}{x_1 - 1} \neq \dfrac{1}{x_2 - 1} \Rightarrow f(x_1) \neq f(x_2)$, so f is 1-1.

(b) $f^{-1}(2) = \frac{3}{2}$ since $f\left(\frac{3}{2}\right) = 2$. Also $f'(x) = -1/(x-1)^2$, so $(f^{-1})'(2) = 1/f'\left(\frac{3}{2}\right) = \frac{1}{-4} = -\frac{1}{4}$.

(c) $y = 1/(x - 1) \Rightarrow x - 1 = 1/y \Rightarrow x = 1 + 1/y$. Interchange x and y: $y = 1 + 1/x$. So $f^{-1}(x) = 1 + 1/x$, $x > 0$ (since $y > 1$).
Domain $= (0, \infty)$, range $= (1, \infty)$.

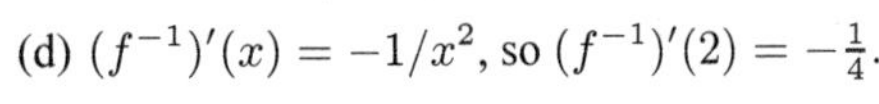
(d) $(f^{-1})'(x) = -1/x^2$, so $(f^{-1})'(2) = -\frac{1}{4}$.

(e)

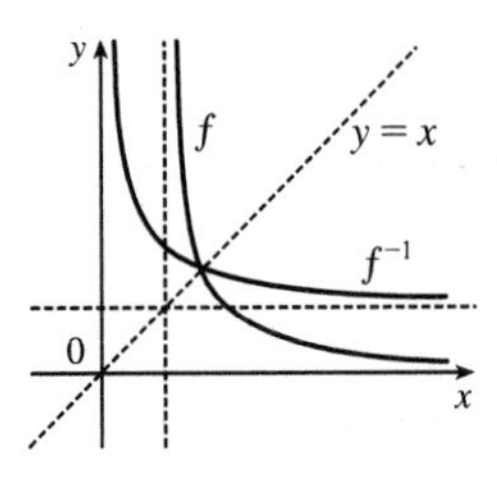

35. $f(0) = 1 \Rightarrow f^{-1}(1) = 0$, and $f(x) = x^3 + x + 1 \Rightarrow f'(x) = 3x^2 + 1$ and $f'(0) = 1$. Thus,

$$(f^{-1})'(1) = \frac{1}{f'(f^{-1}(1))} = \frac{1}{f'(0)} = \frac{1}{1} = 1.$$

36. $f(1) = 2 \Rightarrow f^{-1}(2) = 1$, and $f(x) = x^5 - x^3 + 2x \Rightarrow f'(x) = 5x^4 - 3x^2 + 2$ and $f'(1) = 4$. Thus,

$$(f^{-1})'(2) = \frac{1}{f'(f^{-1}(2))} = \frac{1}{f'(1)} = \frac{1}{4}.$$

37. $f(0) = 3 \quad \Rightarrow \quad f^{-1}(3) = 0$, and $f(x) = 3 + x^2 + \tan(\pi x/2) \quad \Rightarrow \quad f'(x) = 2x + \frac{\pi}{2}\sec^2(\pi x/2)$ and $f'(0) = \frac{\pi}{2} \cdot 1 = \frac{\pi}{2}$. Thus, $(f^{-1})'(3) = 1/f'(f^{-1}(3)) = 1/f'(0) = 2/\pi$.

38. $f(1) = 2 \quad \Rightarrow \quad f^{-1}(2) = 1$, and $f(x) = \sqrt{x^3 + x^2 + x + 1} \quad \Rightarrow \quad f'(x) = \dfrac{3x^2 + 2x + 1}{2\sqrt{x^3 + x^2 + x + 1}}$ and

$f'(1) = \dfrac{3 + 2 + 1}{2\sqrt{1 + 1 + 1 + 1}} = \dfrac{3}{2}$. Thus, $(f^{-1})'(2) = 1/f'(f^{-1}(2)) = 1/f'(1) = \frac{2}{3}$.

39. $f(4) = 5 \quad \Rightarrow \quad f^{-1}(5) = 4$. Thus, $g'(5) = \dfrac{1}{f'(f^{-1}(5))} = \dfrac{1}{f'(4)} = \dfrac{1}{2/3} = \dfrac{3}{2}$.

40. $f(3) = 2 \quad \Rightarrow \quad f^{-1}(2) = 3$. Thus, $g'(2) = \dfrac{1}{f'(f^{-1}(2))} = \dfrac{1}{f'(3)} = 9$. Hence, $G(x) = \dfrac{1}{f^{-1}(x)} \quad \Rightarrow$

$G'(x) = -\dfrac{(f^{-1})'(x)}{[f^{-1}(x)]^2} \quad \Rightarrow \quad G'(2) = -\dfrac{(f^{-1})'(2)}{[f^{-1}(2)]^2} = -\dfrac{9}{(3)^2} = -1$.

41. We see that the graph of $y = f(x) = \sqrt{x^3 + x^2 + x + 1}$ is increasing, so f is 1-1. Enter $x = \sqrt{y^3 + y^2 + y + 1}$ and use your CAS to solve the equation for y. Using Derive, we get two (irrelevant) solutions involving imaginary expressions, as well as one which can be simplified to the following:

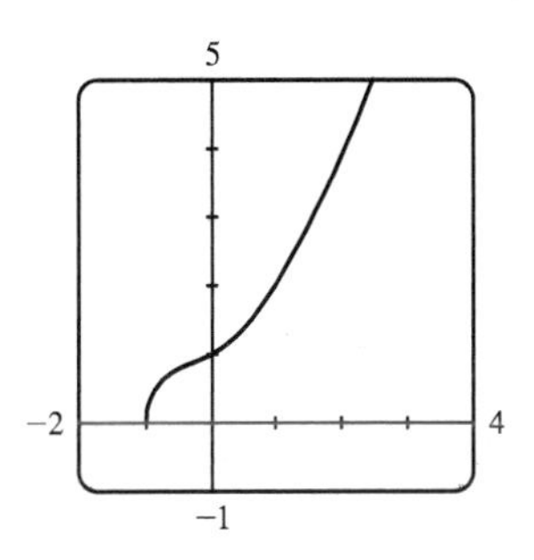

$$y = f^{-1}(x) = -\tfrac{\sqrt[3]{4}}{6}\left(\sqrt[3]{D - 27x^2 + 20} - \sqrt[3]{D + 27x^2 - 20} + \sqrt[3]{2}\right)$$

where $D = 3\sqrt{3}\sqrt{27x^4 - 40x^2 + 16}$. Maple and Mathematica each give two complex expressions and one real expression, and the real expression is equivalent to that given by Derive. For example, Maple's expression simplifies to

$\dfrac{1}{6}\dfrac{M^{2/3} - 8 - 2M^{1/3}}{2M^{1/3}}$, where $M = 108x^2 + 12\sqrt{48 - 120x^2 + 81x^4} - 80$.

42. Since $\sin(2n\pi) = 0$, $h(x) = \sin x$ is not one-to-one. $h'(x) = \cos x > 0$ on $\left(-\frac{\pi}{2}, \frac{\pi}{2}\right)$, so h is increasing and hence 1-1 on $\left[-\frac{\pi}{2}, \frac{\pi}{2}\right]$. Let $y = f^{-1}(x) = \sin^{-1} x$ so that $\sin y = x$. Differentiating $\sin y = x$ implicitly with respect to x gives us

$\cos y \dfrac{dy}{dx} = 1 \quad \Rightarrow \quad \dfrac{dy}{dx} = \dfrac{1}{\cos y}$. Now $\cos^2 y + \sin^2 y = 1 \quad \Rightarrow \quad \cos y = \pm\sqrt{1 - \sin^2 y}$, but since $\cos y > 0$ on $\left(-\frac{\pi}{2}, \frac{\pi}{2}\right)$, we have $\dfrac{dy}{dx} = \dfrac{1}{\sqrt{1 - \sin^2 y}} = \dfrac{1}{\sqrt{1 - x^2}}$.

43. (a) If the point (x, y) is on the graph of $y = f(x)$, then the point $(x - c, y)$ is that point shifted c units to the left. Since f is 1-1, the point (y, x) is on the graph of $y = f^{-1}(x)$ and the point corresponding to $(x - c, y)$ on the graph of f is $(y, x - c)$ on the graph of f^{-1}. Thus, the curve's reflection is shifted *down* the same number of units as the curve itself is shifted to the left. So an expression for the inverse function is $g^{-1}(x) = f^{-1}(x) - c$.

(b) If we compress (or stretch) a curve horizontally, the curve's reflection in the line $y = x$ is compressed (or stretched) *vertically* by the same factor. Using this geometric principle, we see that the inverse of $h(x) = f(cx)$ can be expressed as $h^{-1}(x) = (1/c)\, f^{-1}(x)$.

44. (a) We know that $g'(x) = \dfrac{1}{f'(g(x))}$. Thus,

$$g''(x) = -\frac{f''(g(x)) \cdot g'(x)}{[f'(g(x))]^2} = -\frac{f''(g(x)) \cdot [1/f'(g'(x))]}{[f'(g(x))]^2} = -\frac{f''(g(x))}{f'(g(x))[f'(g(x))]^2} = -\frac{f''(g(x))}{[f'(g(x))]^3}.$$

(b) f is increasing $\Rightarrow$ $f'(g(x)) > 0$ $\Rightarrow$ $[f'(g(x))]^3 > 0$. f is concave upward $\Rightarrow$ $f''(g(x)) > 0$. So

$g''(x) = -\dfrac{f''(g(x))}{[f'(g(x))]^3} < 0$, which implies that g (f's inverse) is concave downward.

5.2 The Natural Logarithmic Function

1. $\ln \dfrac{x^3 y}{z^2} = \ln x^3 y - \ln z^2 = \ln x^3 + \ln y - \ln z^2 = 3 \ln x + \ln y - 2 \ln z$

2. $\ln \sqrt{a(b^2+c^2)} = \ln(a(b^2+c^2))^{1/2} = \frac{1}{2}\ln(a(b^2+c^2)) = \frac{1}{2}\left[\ln a + \ln(b^2+c^2)\right]$
$= \frac{1}{2}\ln a + \frac{1}{2}\ln(b^2+c^2)$

3. $\ln(uv)^{10} = 10\ln(uv) = 10(\ln u + \ln v) = 10\ln u + 10\ln v$

4. $\ln \dfrac{3x^2}{(x+1)^5} = \ln 3x^2 - \ln(x+1)^5 = \ln 3 + \ln x^2 - 5\ln(x+1) = \ln 3 + 2\ln x - 5\ln(x+1)$

5. $2\ln 4 - \ln 2 = \ln 4^2 - \ln 2 = \ln 16 - \ln 2 = \ln \frac{16}{2} = \ln 8$

6. $\ln 3 + \frac{1}{3}\ln 8 = \ln 3 + \ln 8^{1/3} = \ln 3 + \ln 2 = \ln(3 \cdot 2) = \ln 6$

7. $\frac{1}{2}\ln x - 5\ln(x^2+1) = \ln x^{1/2} - \ln(x^2+1)^5 = \ln \dfrac{\sqrt{x}}{(x^2+1)^5}$

8. $\ln x + a\ln y - b\ln z = \ln x + \ln y^a - \ln z^b = \ln(x \cdot y^a) - \ln z^b = \ln(xy^a/z^b)$

9. Reflect the graph of $y = \ln x$ about the x-axis to obtain the graph of $y = -\ln x$.

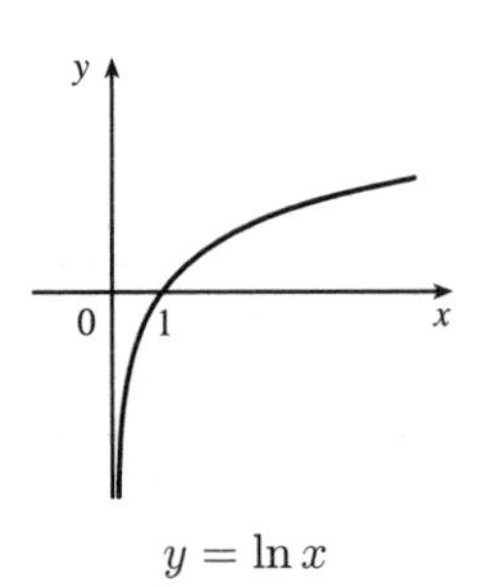

$y = \ln x$

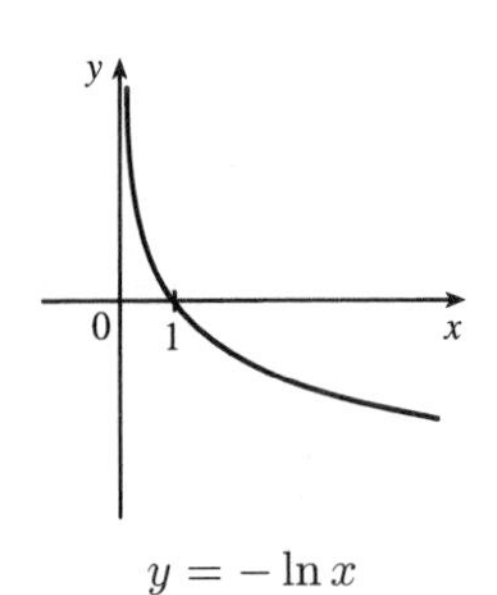

$y = -\ln x$

10. Reflect the portion of the graph of $y = \ln x$ to the right of the y-axis about the y-axis. The graph of $y = \ln|x|$ is that reflection in addition to the original portion.

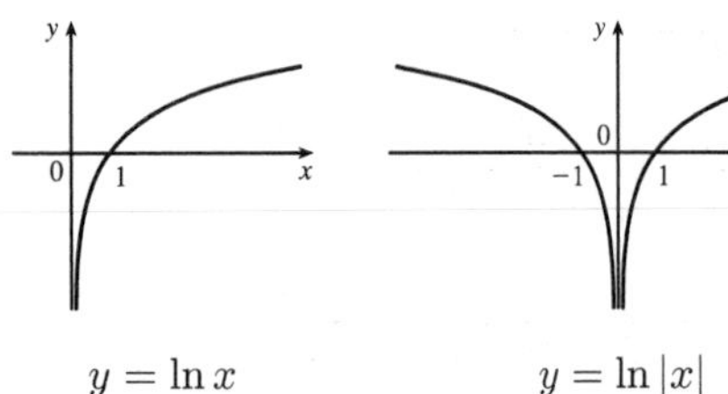

$y = \ln x$ $\qquad$ $y = \ln|x|$

11.

$y = \ln x$ $\qquad$ $y = \ln(x+3)$

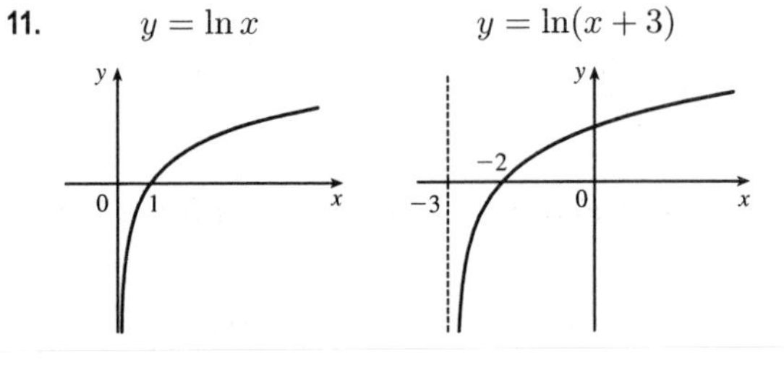

12. $y = \ln(x-2)$

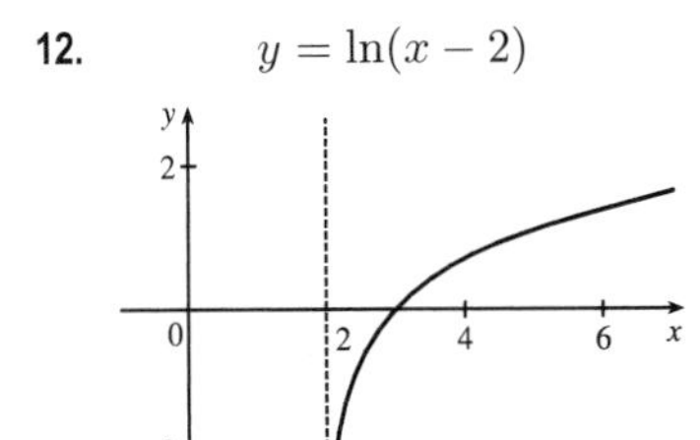

$y = 1 + \ln(x-2)$

13. $f(x) = \sqrt{x}\ln x \quad \Rightarrow \quad f'(x) = \dfrac{1}{2\sqrt{x}}\ln x + \sqrt{x}\left(\dfrac{1}{x}\right) = \dfrac{\ln x + 2}{2\sqrt{x}}$

14. $f(x) = \ln(x^2+10) \quad \Rightarrow \quad f'(x) = \dfrac{1}{x^2+10}\dfrac{d}{dx}(x^2+10) = \dfrac{2x}{x^2+10}$

15. $f(\theta) = \ln(\cos\theta) \quad \Rightarrow \quad f'(\theta) = \dfrac{1}{\cos\theta}\dfrac{d}{d\theta}(\cos\theta) = \dfrac{-\sin\theta}{\cos\theta} = -\tan\theta$

16. $f(x) = \cos(\ln x) \quad \Rightarrow \quad f'(x) = -\sin(\ln x)\cdot\dfrac{1}{x} = \dfrac{-\sin(\ln x)}{x}$

17. $f(x) = \sqrt[5]{\ln x} = (\ln x)^{1/5} \quad \Rightarrow \quad f'(x) = \frac{1}{5}(\ln x)^{-4/5}\dfrac{d}{dx}(\ln x) = \dfrac{1}{5(\ln x)^{4/5}}\cdot\dfrac{1}{x} = \dfrac{1}{5x\,\sqrt[5]{(\ln x)^4}}$

18. $f(x) = \ln\sqrt[5]{x} = \ln x^{1/5} = \frac{1}{5}\ln x \quad \Rightarrow \quad f'(x) = \dfrac{1}{5}\cdot\dfrac{1}{x} = \dfrac{1}{5x}$

19. $g(x) = \ln\dfrac{a-x}{a+x} = \ln(a-x) - \ln(a+x) \quad \Rightarrow$

$$g'(x) = \frac{1}{a-x}(-1) - \frac{1}{a+x} = \frac{-(a+x)-(a-x)}{(a-x)(a+x)} = \frac{-2a}{a^2-x^2}$$

20. $h(x) = \ln\left(x+\sqrt{x^2-1}\right) \quad \Rightarrow \quad h'(x) = \dfrac{1}{x+\sqrt{x^2-1}}\left(1+\dfrac{x}{\sqrt{x^2-1}}\right) = \dfrac{1}{x+\sqrt{x^2-1}}\cdot\dfrac{\sqrt{x^2-1}+x}{\sqrt{x^2-1}} = \dfrac{1}{\sqrt{x^2-1}}$

21. $f(u) = \dfrac{\ln u}{1+\ln(2u)} \quad \Rightarrow$

$$f'(u) = \frac{[1+\ln(2u)]\cdot\frac{1}{u} - \ln u\cdot\frac{1}{2u}\cdot 2}{[1+\ln(2u)]^2} = \frac{\frac{1}{u}[1+\ln(2u)-\ln u]}{[1+\ln(2u)]^2} = \frac{1+(\ln 2+\ln u)-\ln u}{u[1+\ln(2u)]^2} = \frac{1+\ln 2}{u[1+\ln(2u)]^2}$$

22. $f(t) = \dfrac{1+\ln t}{1-\ln t} \quad \Rightarrow \quad f'(t) = \dfrac{(1-\ln t)(1/t)-(1+\ln t)(-1/t)}{(1-\ln t)^2} = \dfrac{(1/t)[(1-\ln t)+(1+\ln t)]}{(1-\ln t)^2} = \dfrac{2}{t(1-\ln t)^2}$

23. $F(t) = \ln\dfrac{(2t+1)^3}{(3t-1)^4} = \ln(2t+1)^3 - \ln(3t-1)^4 = 3\ln(2t+1) - 4\ln(3t-1) \quad \Rightarrow$

$F'(t) = 3\cdot\dfrac{1}{2t+1}\cdot 2 - 4\cdot\dfrac{1}{3t-1}\cdot 3 = \dfrac{6}{2t+1} - \dfrac{12}{3t-1}$, or combined, $\dfrac{-6(t+3)}{(2t+1)(3t-1)}$.

24. $y = \ln(x^4\sin^2 x) = \ln x^4 + \ln(\sin x)^2 = 4\ln x + 2\ln\sin x \quad \Rightarrow \quad y' = 4\cdot\dfrac{1}{x} + 2\cdot\dfrac{1}{\sin x}\cdot\cos x = \dfrac{4}{x} + 2\cot x$

25. $y = \ln\left|2-x-5x^2\right| \quad \Rightarrow \quad y' = \dfrac{1}{2-x-5x^2}\cdot(-1-10x) = \dfrac{-10x-1}{2-x-5x^2}$ or $\dfrac{10x+1}{5x^2+x-2}$

26. $G(u) = \ln\sqrt{\dfrac{3u+2}{3u-2}} = \frac{1}{2}[\ln(3u+2) - \ln(3u-2)] \quad \Rightarrow \quad G'(u) = \dfrac{1}{2}\left(\dfrac{3}{3u+2} - \dfrac{3}{3u-2}\right) = \dfrac{-6}{9u^2-4}$

27. $y = \ln\left(\dfrac{x+1}{x-1}\right)^{3/5} = \frac{3}{5}[\ln(x+1) - \ln(x-1)] \Rightarrow y' = \dfrac{3}{5}\left(\dfrac{1}{x+1} - \dfrac{1}{x-1}\right) = \dfrac{-6}{5(x^2-1)}$

28. $y = (\ln\tan x)^2 \Rightarrow y' = 2(\ln\tan x) \cdot \dfrac{1}{\tan x} \cdot \sec^2 x = \dfrac{2(\ln\tan x)\sec^2 x}{\tan x}$

29. $y = \tan[\ln(ax+b)] \Rightarrow y' = \sec^2[\ln(ax+b)] \cdot \dfrac{1}{ax+b} \cdot a = \sec^2[\ln(ax+b)]\dfrac{a}{ax+b}$

30. $y = \ln|\tan 2x| \Rightarrow y' = \dfrac{2\sec^2 2x}{\tan 2x}$

31. $y = \ln\ln x \Rightarrow y' = \dfrac{1}{\ln x}\dfrac{d}{dx}(\ln x) = \dfrac{1}{\ln x} \cdot \dfrac{1}{x} = \dfrac{1}{x\ln x} \Rightarrow$

$$y'' = -\frac{\frac{d}{dx}(x\ln x)}{(x\ln x)^2} \quad \text{[Reciprocal Rule]} \quad = -\frac{x \cdot \frac{1}{x} + \ln x \cdot 1}{(x\ln x)^2} = -\frac{1+\ln x}{(x\ln x)^2}$$

32. $y = \dfrac{\ln x}{x^2} \Rightarrow y' = \dfrac{x^2(1/x) - (\ln x)(2x)}{(x^2)^2} = \dfrac{x(1-2\ln x)}{x^4} = \dfrac{1-2\ln x}{x^3} \Rightarrow$

$$y'' = \frac{x^3(-2/x) - (1-2\ln x)(3x^2)}{(x^3)^2} = \frac{x^2(-2-3+6\ln x)}{x^6} = \frac{6\ln x - 5}{x^4}$$

33. $f(x) = \dfrac{x}{1-\ln(x-1)} \Rightarrow$

$$f'(x) = \frac{[1-\ln(x-1)] \cdot 1 - x \cdot \frac{-1}{x-1}}{[1-\ln(x-1)]^2} = \frac{\frac{(x-1)[1-\ln(x-1)] + x}{x-1}}{[1-\ln(x-1)]^2} = \frac{x-1-(x-1)\ln(x-1)+x}{(x-1)[1-\ln(x-1)]^2}$$
$$= \frac{2x-1-(x-1)\ln(x-1)}{(x-1)[1-\ln(x-1)]^2}$$

$\text{Dom}(f) = \{x \mid x-1>0 \text{ and } 1-\ln(x-1) \neq 0\} = \{x \mid x>1 \text{ and } \ln(x-1) \neq 1\}$
$= \{x \mid x>1 \text{ and } x-1 \neq e^1\} = \{x \mid x>1 \text{ and } x \neq 1+e\} = (1, 1+e) \cup (1+e, \infty)$

34. $f(x) = \ln\ln\ln x \Rightarrow f'(x) = \dfrac{1}{\ln\ln x} \cdot \dfrac{1}{\ln x} \cdot \dfrac{1}{x}$.

$\text{Dom}(f) = \{x \mid \ln\ln x > 0\} = \{x \mid \ln x > 1\} = \{x \mid x > e\} = (e, \infty)$.

35. $f(x) = \dfrac{x}{\ln x} \Rightarrow f'(x) = \dfrac{\ln x - x(1/x)}{(\ln x)^2} = \dfrac{\ln x - 1}{(\ln x)^2} \Rightarrow f'(e) = \dfrac{1-1}{1^2} = 0$

36. $f(t) = t\ln(4+3t) \Rightarrow f'(t) = t \cdot \dfrac{1}{4+3t} \cdot 3 + \ln(4+3t) = \dfrac{3t}{4+3t} + \ln(4+3t)$,

so $f'(-1) = \frac{-3}{1} + \ln 1 = -3 + 0 = -3$.

37. $y = \sin(2\ln x) \Rightarrow y' = \cos(2\ln x) \cdot \dfrac{2}{x}$. At $(1,0)$, $y' = \cos 0 \cdot \dfrac{2}{1} = 2$, so an equation of the tangent line is

$y - 0 = 2 \cdot (x-1)$, or $y = 2x - 2$.

38. $y = \ln(x^3-7) \Rightarrow y' = \dfrac{1}{x^3-7} \cdot 3x^2 \Rightarrow y'(2) = \dfrac{12}{8-7} = 12$, so an equation of a tangent line at $(2,0)$ is

$y - 0 = 12(x-2)$ or $y = 12x - 24$.

39. $y = \ln(x^2 + y^2) \Rightarrow y' = \dfrac{1}{x^2 + y^2} \dfrac{d}{dx}(x^2 + y^2) \Rightarrow y' = \dfrac{2x + 2yy'}{x^2 + y^2} \Rightarrow x^2y' + y^2y' = 2x + 2yy' \Rightarrow$

$x^2y' + y^2y' - 2yy' = 2x \Rightarrow (x^2 + y^2 - 2y)y' = 2x \Rightarrow y' = \dfrac{2x}{x^2 + y^2 - 2y}$

40. $\ln xy = \ln x + \ln y = y \sin x \Rightarrow 1/x + y'/y = y \cos x + y' \sin x \Rightarrow y'(1/y - \sin x) = y \cos x - 1/x \Rightarrow$

$y' = \dfrac{y \cos x - 1/x}{1/y - \sin x} = \left(\dfrac{y}{x}\right) \dfrac{xy \cos x - 1}{1 - y \sin x}$

41. $f(x) = \ln(x - 1) \Rightarrow f'(x) = 1/(x - 1) = (x - 1)^{-1} \Rightarrow f''(x) = -(x - 1)^{-2} \Rightarrow f'''(x) = 2(x - 1)^{-3} \Rightarrow$

$f^{(4)}(x) = -2 \cdot 3(x - 1)^{-4} \Rightarrow \cdots \Rightarrow f^{(n)}(x) = (-1)^{n-1} \cdot 2 \cdot 3 \cdot 4 \cdot \cdots \cdot (n - 1)(x - 1)^{-n} = (-1)^{n-1} \dfrac{(n - 1)!}{(x - 1)^n}$

42. $y = x^8 \ln x$, so $D^9 y = D^8 y' = D^8 (8x^7 \ln x + x^7)$. But the eighth derivative of x^7 is 0, so we now have

$D^8 (8x^7 \ln x) = D^7 (8 \cdot 7x^6 \ln x + 8x^6) = D^7 (8 \cdot 7x^6 \ln x) = D^6 (8 \cdot 7 \cdot 6x^5 \ln x) = \cdots = D (8!\, x^0 \ln x) = 8!/x$.

43.

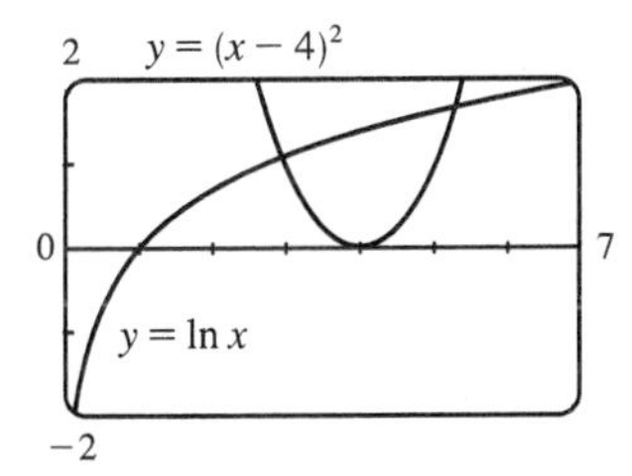

From the graph, it appears that the curves $y = (x - 4)^2$ and $y = \ln x$ intersect just to the left of $x = 3$ and to the right of $x = 5$, at about $x = 5.3$. Let $f(x) = \ln x - (x - 4)^2$. Then $f'(x) = 1/x - 2(x - 4)$, so Newton's Method says that

$x_{n+1} = x_n - f(x_n)/f'(x_n) = x_n - \dfrac{\ln x_n - (x_n - 4)^2}{1/x_n - 2(x_n - 4)}$. Taking

$x_0 = 3$, we get $x_1 \approx 2.957738$, $x_2 \approx 2.958516 \approx x_3$, so the first root is 2.958516, to six decimal places. Taking $x_0 = 5$, we get $x_1 \approx 5.290755$, $x_2 \approx 5.290718 \approx x_3$, so the second (and final) root is 5.290718, to six decimal places.

44.

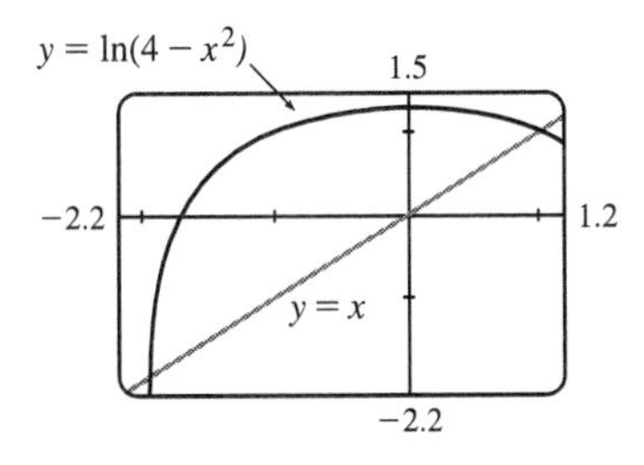

We use Newton's Method with $f(x) = \ln(4 - x^2) - x$ and

$f'(x) = \dfrac{1}{4 - x^2}(-2x) - 1 = -1 - \dfrac{2x}{4 - x^2}$. The formula is

$x_{n+1} = x_n - f(x_n)/f'(x_n)$. From the graphs it seems that the roots occur at approximately $x = -1.9$ and $x = 1.1$. However, if we use $x_1 = -1.9$ as an initial approximation to the first root, we get $x_2 \approx -2.009611$, and $f(x) = \ln(x - 2)^2 - x$ is undefined at this point, making it impossible to calculate x_3. We must use a more accurate first estimate, such as $x_1 = -1.95$. With this approximation, we get $x_1 = -1.95$, $x_2 \approx -1.1967495$, $x_3 \approx -1.964760$, $x_4 \approx x_5 \approx -1.964636$. Calculating the second root gives $x_1 = 1.1$, $x_2 \approx 1.058649$, $x_3 \approx 1.058007$, $x_4 \approx x_5 \approx 1.058006$. So, correct to six decimal places, the two roots of the equation $\ln(4 - x^2) = x$ are $x = -1.964636$ and $x = 1.058006$.

45. $y = f(x) = \ln(\sin x)$

A. $D = \{x \text{ in } \mathbb{R} \mid \sin x > 0\} = \bigcup_{n=-\infty}^{\infty} (2n\pi, (2n+1)\pi) = \cdots \cup (-4\pi, -3\pi) \cup (-2\pi, -\pi) \cup (0, \pi) \cup (2\pi, 3\pi) \cup \cdots$

B. No y-intercept; x-intercepts: $f(x) = 0 \Leftrightarrow \ln(\sin x) = 0 \Leftrightarrow \sin x = e^0 = 1 \Leftrightarrow x = 2n\pi + \frac{\pi}{2}$ for each integer n. **C.** f is periodic with period 2π. **D.** $\lim\limits_{x \to (2n\pi)^+} f(x) = -\infty$ and $\lim\limits_{x \to [(2n+1)\pi]^-} f(x) = -\infty$, so the lines $x = n\pi$ are VAs for all integers n. **E.** $f'(x) = \frac{\cos x}{\sin x} = \cot x$, so $f'(x) > 0$ when $2n\pi < x < 2n\pi + \frac{\pi}{2}$ for each integer n, and $f'(x) < 0$ when $2n\pi + \frac{\pi}{2} < x < (2n+1)\pi$. Thus, f is increasing on $\left(2n\pi, 2n\pi + \frac{\pi}{2}\right)$ and decreasing on $\left(2n\pi + \frac{\pi}{2}, (2n+1)\pi\right)$ for each integer n. **F.** Local maximum values $f\left(2n\pi + \frac{\pi}{2}\right) = 0$, no local minimum.

G. $f''(x) = -\csc^2 x < 0$, so f is CD on $(2n\pi, (2n+1)\pi)$ for each integer n. No IP

H.

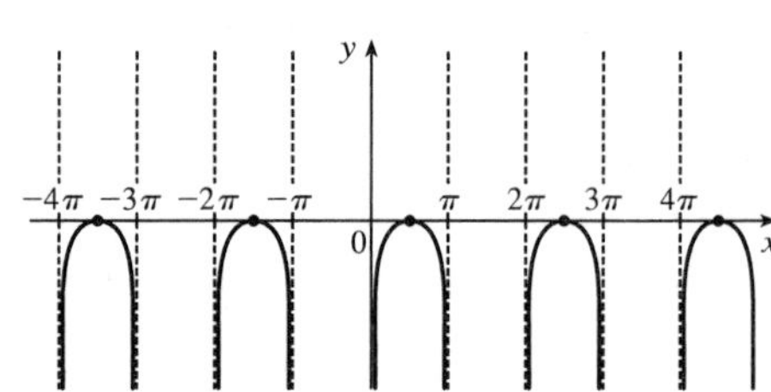

46. $y = \ln(\tan^2 x)$ **A.** $D = \{x \mid x \neq n\pi/2\}$ **B.** x-intercepts $n\pi + \frac{\pi}{4}$, no y-intercept. **C.** $f(-x) = f(x)$, so the curve is symmetric about the y-axis. Also $f(x + \pi) = f(x)$, so f is periodic with period π, and we consider parts D–G only for $-\frac{\pi}{2} < x < \frac{\pi}{2}$. **D.** $\lim\limits_{x \to 0} \ln(\tan^2 x) = -\infty$ and $\lim\limits_{x \to (\pi/2)^-} \ln(\tan^2 x) = \infty$, $\lim\limits_{x \to -(-\pi/2)^+} \ln(\tan^2 x) = \infty$, so $x = 0$, $x = \pm\frac{\pi}{2}$ are VA. **E.** $f'(x) = \dfrac{2\tan x \sec^2 x}{\tan^2 x} = 2\dfrac{\sec^2 x}{\tan x} > 0 \Leftrightarrow$

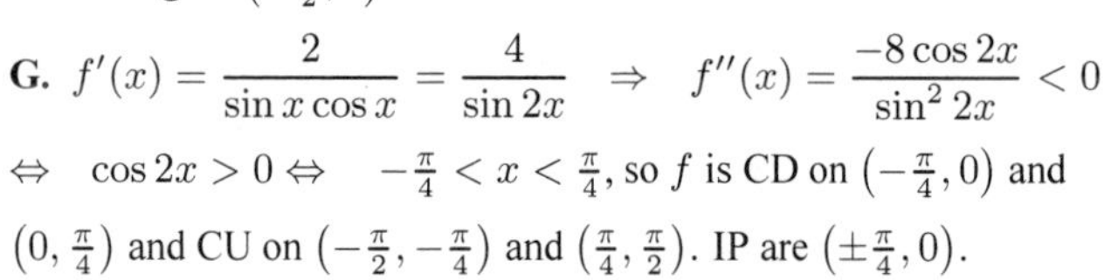

$\tan x > 0 \Leftrightarrow 0 < x < \frac{\pi}{2}$, so f is increasing on $\left(0, \frac{\pi}{2}\right)$ and decreasing on $\left(-\frac{\pi}{2}, 0\right)$. **F.** No maximum or minimum

G. $f'(x) = \dfrac{2}{\sin x \cos x} = \dfrac{4}{\sin 2x} \Rightarrow f''(x) = \dfrac{-8\cos 2x}{\sin^2 2x} < 0$

$\Leftrightarrow \cos 2x > 0 \Leftrightarrow -\frac{\pi}{4} < x < \frac{\pi}{4}$, so f is CD on $\left(-\frac{\pi}{4}, 0\right)$ and $\left(0, \frac{\pi}{4}\right)$ and CU on $\left(-\frac{\pi}{2}, -\frac{\pi}{4}\right)$ and $\left(\frac{\pi}{4}, \frac{\pi}{2}\right)$. IP are $\left(\pm\frac{\pi}{4}, 0\right)$.

H.

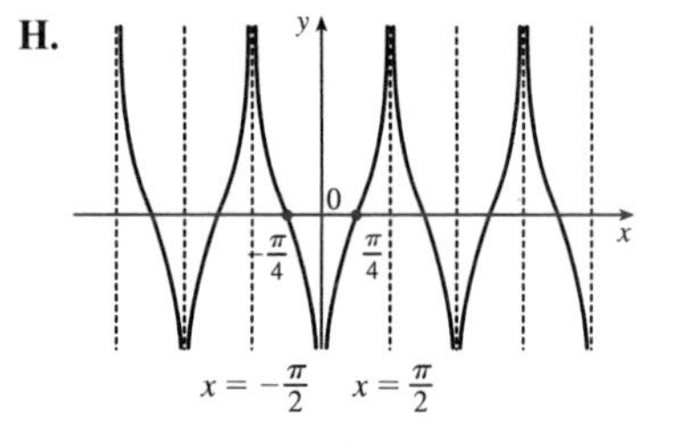

47. $y = f(x) = \ln(1 + x^2)$ **A.** $D = \mathbb{R}$ **B.** Both intercepts are 0. **C.** $f(-x) = f(x)$, so the curve is symmetric about the y-axis. **D.** $\lim\limits_{x \to \pm\infty} \ln(1 + x^2) = \infty$, no asymptotes. **E.** $f'(x) = \dfrac{2x}{1 + x^2} > 0 \Leftrightarrow$ $x > 0$, so f is increasing on $(0, \infty)$ and decreasing on $(-\infty, 0)$.

F. $f(0) = 0$ is a local and absolute minimum.

G. $f''(x) = \dfrac{2(1 + x^2) - 2x(2x)}{(1 + x^2)^2} = \dfrac{2(1 - x^2)}{(1 + x^2)^2} > 0 \Leftrightarrow$

$|x| < 1$, so f is CU on $(-1, 1)$, CD on $(-\infty, -1)$ and $(1, \infty)$. IP $(1, \ln 2)$ and $(-1, \ln 2)$.

H.

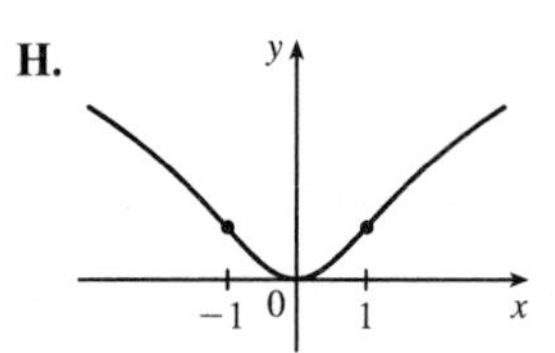

48. $y = f(x) = \ln(x^2 - 3x + 2) = \ln[(x-1)(x-2)]$

A. $D = \{x \text{ in } \mathbb{R}: x^2 - 3x + 2 > 0\} = (-\infty, 1) \cup (2, \infty)$.

B. y-intercept: $f(0) = \ln 2$; x-intercepts: $f(x) = 0 \Leftrightarrow x^2 - 3x + 2 = e^0 \Leftrightarrow$

$x^2 - 3x + 1 = 0 \Leftrightarrow x = \dfrac{3 \pm \sqrt{5}}{2} \Rightarrow x \approx 0.38, 2.62$ **C.** No symmetry **D.** $\lim\limits_{x\to 1^-} f(x) = \lim\limits_{x\to 2^+} f(x) = -\infty$,

so $x = 1$ and $x = 2$ are VAs. No HA. **E.** $f'(x) = \dfrac{2x-3}{x^2-3x+2} = \dfrac{2(x-3/2)}{(x-1)(x-2)}$, so $f'(x) < 0$ for $x < 1$ and $f'(x) > 0$

for $x > 2$. Thus, f is decreasing on $(-\infty, 1)$ and increasing on $(2, \infty)$. **F.** No extreme values

G.
$$f''(x) = \frac{(x^2-3x+2)\cdot 2 - (2x-3)^2}{(x^2-3x+2)^2} = \frac{2x^2 - 6x + 4 - 4x^2 + 12x - 9}{(x^2-3x+2)^2} = \frac{-2x^2+6x-5}{(x^2-3x+2)^2}$$

The numerator is negative for all x and the denominator is positive, so $f''(x) < 0$ for all x in the domain of f. Thus, f is CD on $(-\infty, 1)$ and $(2, \infty)$. No IP

H.

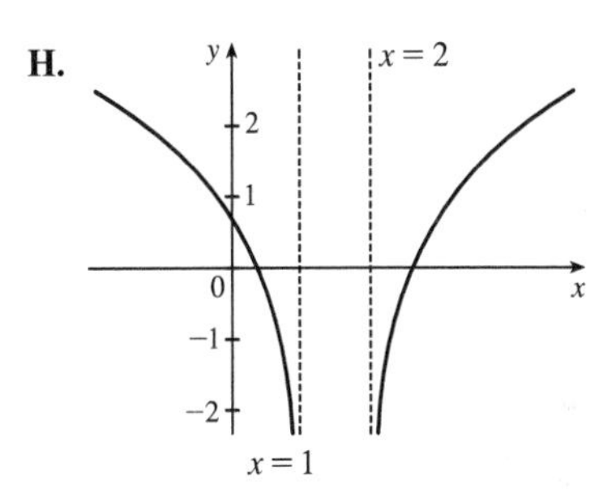

49. We use the CAS to calculate $f'(x) = \dfrac{2 + \sin x + x\cos x}{2x + x\sin x}$ and

$f''(x) = \dfrac{2x^2\sin x + 4\sin x - \cos^2 x + x^2 + 5}{x^2(\cos^2 x - 4\sin x - 5)}$. From the graphs, it seems that $f' > 0$ (and so f is increasing) on approximately the intervals $(0, 2.7)$, $(4.5, 8.2)$ and $(10.9, 14.3)$. It seems that f'' changes sign (indicating inflection points) at $x \approx 3.8, 5.7, 10.0$ and 12.0.

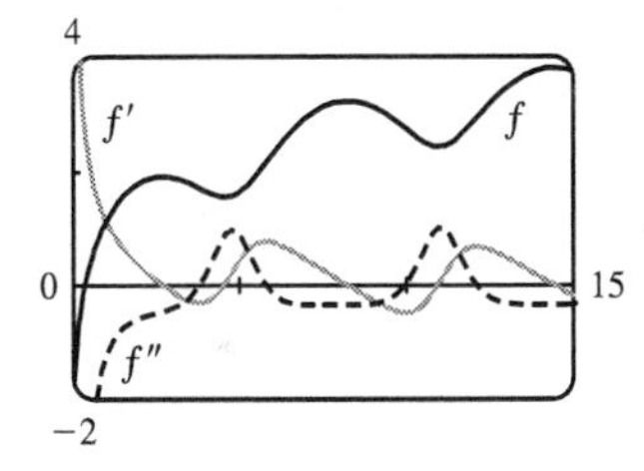

Looking back at the graph of $f(x) = \ln(2x + x\sin x)$, this implies that the inflection points have approximate coordinates $(3.8, 1.7)$, $(5.7, 2.1)$, $(10.0, 2.7)$, and $(12.0, 2.9)$.

50. We see that if $c \le 0$, $f(x) = \ln(x^2 + c)$ is only defined for $x^2 > -c \Rightarrow |x| > \sqrt{-c}$, and

$\lim\limits_{x\to\sqrt{-c}^+} f(x) = \lim\limits_{x\to-\sqrt{-c}^-} f(x) = -\infty$, since $\ln y \to -\infty$ as $y \to 0$. Thus, for $c < 0$, there are vertical asymptotes at

$x = \pm\sqrt{c}$, and as c decreases (that is, $|c|$ increases), the asymptotes get further apart. For $c = 0$,

$\lim\limits_{x\to 0} f(x) = -\infty$, so there is a vertical asymptote at $x = 0$. If $c > 0$, there is no asymptote. To find the maxima, minima, and

inflection points, we differentiate: $f(x) = \ln(x^2 + c) \Rightarrow f'(x) = \dfrac{1}{x^2 + c}(2x)$, so by the First Derivative Test there is a

local and absolute minimum at $x = 0$. Differentiating again, we get

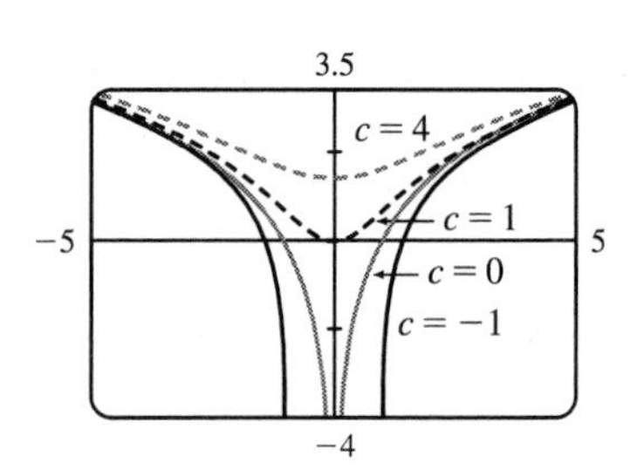

$f''(x) = \dfrac{1}{x^2+c}(2) + 2x\left[-\left(x^2+c\right)^{-2}(2x)\right] = \dfrac{2(c-x^2)}{(x^2+c)^2}$. Now if $c \le 0$, this is always negative, so f is concave down on both of the intervals on which it is defined. If $c > 0$, then f'' changes sign when $c = x^2 \Leftrightarrow x = \pm\sqrt{c}$. So for $c > 0$ there are inflection points at $\pm\sqrt{c}$, and as c increases, the inflection points get further apart.

51. $y = (2x+1)^5(x^4-3)^6 \Rightarrow \ln y = \ln\left((2x+1)^5(x^4-3)^6\right) \Rightarrow \ln y = 5\ln(2x+1) + 6\ln(x^4-3) \Rightarrow$

$\dfrac{1}{y}y' = 5\cdot\dfrac{1}{2x+1}\cdot 2 + 6\cdot\dfrac{1}{x^4-3}\cdot 4x^3 \Rightarrow$

$y' = y\left(\dfrac{10}{2x+1} + \dfrac{24x^3}{x^4-3}\right) = (2x+1)^5(x^4-3)^6\left(\dfrac{10}{2x+1} + \dfrac{24x^3}{x^4-3}\right).$

[The answer could be simplified to $y' = 2(2x+1)^4(x^4-3)^5(29x^4+12x^3-15)$, but this is unnecessary.]

52. $y = \dfrac{(x^3+1)^4\sin^2 x}{x^{1/3}} \Rightarrow \ln|y| = 4\ln\left|x^3+1\right| + 2\ln|\sin x| - \frac{1}{3}\ln|x|.$

So $\dfrac{y'}{y} = 4\dfrac{3x^2}{x^3+1} + 2\dfrac{\cos x}{\sin x} - \dfrac{1}{3x} \Rightarrow y' = \dfrac{(x^3+1)^4\sin^2 x}{x^{1/3}}\left(\dfrac{12x^2}{x^3+1} + 2\cot x - \dfrac{1}{3x}\right).$

53. $y = \dfrac{\sin^2 x\tan^4 x}{(x^2+1)^2} \Rightarrow \ln y = \ln(\sin^2 x\tan^4 x) - \ln(x^2+1)^2 \Rightarrow$

$\ln y = \ln(\sin x)^2 + \ln(\tan x)^4 - \ln(x^2+1)^2 \Rightarrow \ln y = 2\ln|\sin x| + 4\ln|\tan x| - 2\ln(x^2+1) \Rightarrow$

$\dfrac{1}{y}y' = 2\cdot\dfrac{1}{\sin x}\cdot\cos x + 4\cdot\dfrac{1}{\tan x}\cdot\sec^2 x - 2\cdot\dfrac{1}{x^2+1}\cdot 2x \Rightarrow y' = \dfrac{\sin^2 x\tan^4 x}{(x^2+1)^2}\left(2\cot x + \dfrac{4\sec^2 x}{\tan x} - \dfrac{4x}{x^2+1}\right)$

54. $y = \sqrt[4]{\dfrac{x^2+1}{x^2-1}} \Rightarrow \ln y = \frac{1}{4}\ln(x^2+1) - \frac{1}{4}\ln(x^2-1) \Rightarrow \dfrac{1}{y}y' = \dfrac{1}{4}\cdot\dfrac{1}{x^2+1}\cdot 2x - \dfrac{1}{4}\cdot\dfrac{1}{x^2-1}\cdot 2x \Rightarrow$

$y' = \sqrt[4]{\dfrac{x^2+1}{x^2-1}}\cdot\dfrac{1}{2}\left(\dfrac{x}{x^2+1} - \dfrac{x}{x^2-1}\right) = \dfrac{1}{2}\sqrt[4]{\dfrac{x^2+1}{x^2-1}}\left(\dfrac{-2x}{x^4-1}\right) = \dfrac{x}{1-x^4}\sqrt[4]{\dfrac{x^2+1}{x^2-1}}$

55. $\displaystyle\int_1^2 \frac{dt}{8-3t} = \left[-\frac{1}{3}\ln|8-3t|\right]_1^2 = -\frac{1}{3}\ln 2 - \left(-\frac{1}{3}\ln 5\right) = \frac{1}{3}(\ln 5 - \ln 2) = \frac{1}{3}\ln\frac{5}{2}$

Or: Let $u = 8-3t$. Then $du = -3\,dt$, so

$\displaystyle\int_1^2 \frac{dt}{8-3t} = \int_5^2 \frac{-\frac{1}{3}\,du}{u} = \left[-\frac{1}{3}\ln|u|\right]_5^2 = -\frac{1}{3}\ln 2 - \left(-\frac{1}{3}\ln 5\right) = \frac{1}{3}(\ln 5 - \ln 2) = \frac{1}{3}\ln\frac{5}{2}.$

56. $\displaystyle\int_1^2 \frac{4+u^2}{u^3}\,du = \int_1^2 (4u^{-3} + u^{-1})\,du = \left[\frac{4}{-2}u^{-2} + \ln|u|\right]_1^2 = \left[\frac{-2}{u^2} + \ln u\right]_1^2$

$= \left(-\frac{1}{2} + \ln 2\right) - (-2 + \ln 1) = \frac{3}{2} + \ln 2$

57. $\displaystyle\int_1^e \frac{x^2+x+1}{x}\,dx = \int_1^e \left(x + 1 + \frac{1}{x}\right)dx = \left[\tfrac{1}{2}x^2 + x + \ln x\right]_1^e = \left(\tfrac{1}{2}e^2 + e + 1\right) - \left(\tfrac{1}{2} + 1 + 0\right)$

$= \frac{1}{2}e^2 + e - \frac{1}{2}$

58. $\displaystyle\int_4^9 \left(\sqrt{x} + \frac{1}{\sqrt{x}}\right)^2 dx = \int_4^9 \left(x + 2 + \frac{1}{x}\right)dx = \left[\tfrac{1}{2}x^2 + 2x + \ln x\right]_4^9 = \frac{81}{2} + 18 + \ln 9 - (8 + 8 + \ln 4)$

$= \frac{85}{2} + \ln\frac{9}{4}$

59. Let $u = 6x - x^3$. Then $du = (6 - 3x^2)\,dx = 3(2 - x^2)\,dx$, so

$$\int \frac{2 - x^2}{6x - x^3}\,dx = \int \frac{\frac{1}{3}\,du}{u} = \frac{1}{3}\ln|u| + C = \frac{1}{3}\ln\left|6x - x^3\right| + C.$$

60. Let $u = \ln x$. Then $du = \frac{1}{x}\,dx$, so $\displaystyle\int_e^6 \frac{dx}{x\ln x} = \int_1^{\ln 6} \frac{1}{u}\,du = \Big[\ln|u|\Big]_1^{\ln 6} = \ln\ln 6 - \ln 1 = \ln\ln 6$

61. Let $u = \ln x$. Then $du = \dfrac{dx}{x} \;\Rightarrow\; \displaystyle\int \frac{(\ln x)^2}{x}\,dx = \int u^2\,du = \frac{1}{3}u^3 + C = \frac{1}{3}(\ln x)^3 + C.$

62. Let $u = 2 + \sin x$. Then $du = \cos x\,dx$, so

$$\int \frac{\cos x}{2 + \sin x}\,dx = \int \frac{1}{u}\,du = \ln|u| + C = \ln|2 + \sin x| + C = \ln(2 + \sin x) + C \quad [\text{since } 2 + \sin x > 0].$$

63. (a) $\dfrac{d}{dx}(\ln|\sin x| + C) = \dfrac{1}{\sin x}\cos x = \cot x$

(b) Let $u = \sin x$. Then $du = \cos x\,dx$, so $\displaystyle\int \cot x\,dx = \int \frac{\cos x}{\sin x}\,dx = \int \frac{du}{u} = \ln|u| + C = \ln|\sin x| + C.$

64. $f''(x) = x^{-2}, x > 0 \;\Rightarrow\; f'(x) = -1/x + C \;\Rightarrow\; f(x) = -\ln x + Cx + D.$ $0 = f(1) = C + D$ and $0 = f(2) = -\ln 2 + 2C + D = -\ln 2 + 2C - C = -\ln 2 + C \;\Rightarrow\; C = \ln 2$ and $D = -\ln 2$. So $f(x) = -\ln x + (\ln 2)x - \ln 2.$

65. $f(x) = 2x + \ln x \;\Rightarrow\; f'(x) = 2 + 1/x$. If $g = f^{-1}$, then $f(1) = 2 \;\Rightarrow\; g(2) = 1$, so $g'(2) = 1/f'(g(2)) = 1/f'(1) = \frac{1}{3}.$

66. (a) Let $f(x) = \ln x \;\Rightarrow\; f'(x) = 1/x \;\Rightarrow\; f''(x) = -1/x^2$. The linear approximation to $\ln x$ near 1 is $\ln x \approx f(1) + f'(1)(x - 1) = \ln 1 + \frac{1}{1}(x - 1) = x - 1.$

(b)

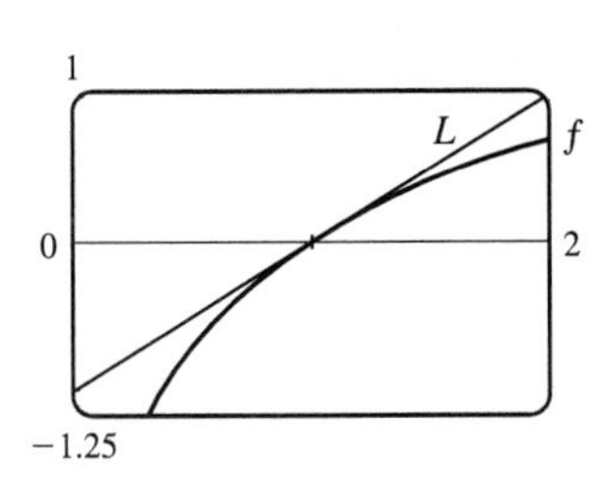

(c)

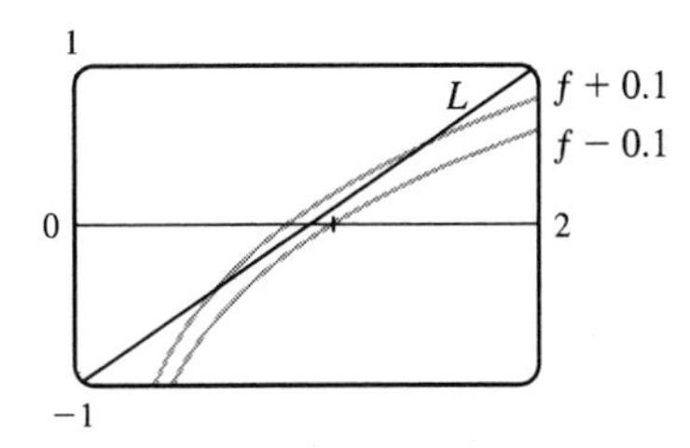

From the graph, it appears that the linear approximation is accurate to within 0.1 for x between about 0.62 and 1.51.

67. (a)

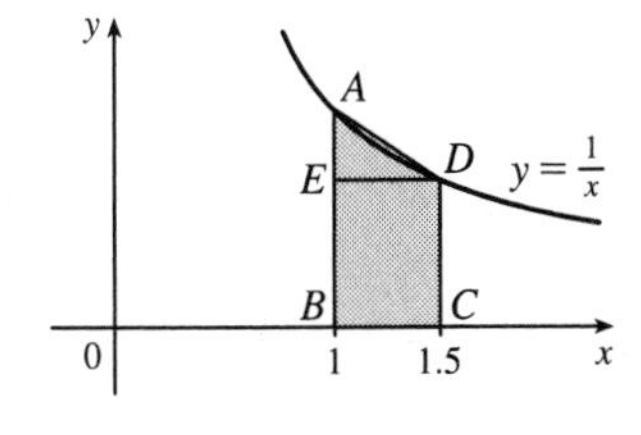

We interpret $\ln 1.5$ as the area under the curve $y = 1/x$ from $x = 1$ to $x = 1.5$. The area of the rectangle $BCDE$ is $\frac{1}{2}\cdot\frac{2}{3} = \frac{1}{3}$. The area of the trapezoid $ABCD$ is $\frac{1}{2}\cdot\frac{1}{2}\left(1 + \frac{2}{3}\right) = \frac{5}{12}$. Thus, by comparing areas, we observe that $\frac{1}{3} < \ln 1.5 < \frac{5}{12}$.

(b) With $f(t) = 1/t$, $n = 10$, and $\Delta t = 0.05$, we have

$$\begin{aligned}\ln 1.5 &= \textstyle\int_1^{1.5}(1/t)\,dt \approx (0.05)[f(1.025) + f(1.075) + \cdots + f(1.475)] \\ &= (0.05)\left[\tfrac{1}{1.025} + \tfrac{1}{1.075} + \cdots + \tfrac{1}{1.475}\right] \approx 0.4054\end{aligned}$$

68. (a) $y = \frac{1}{t}$, $y' = -\frac{1}{t^2}$. The slope of AD is $\frac{1/2 - 1}{2 - 1} = -\frac{1}{2}$. Let c be the t-coordinate of the point on $y = \frac{1}{t}$ with slope $-\frac{1}{2}$.

Then $-\frac{1}{c^2} = -\frac{1}{2} \Rightarrow c^2 = 2 \Rightarrow c = \sqrt{2}$ since $c > 0$. Therefore the tangent line is given by

$y - \frac{1}{\sqrt{2}} = -\frac{1}{2}\left(t - \sqrt{2}\right) \Rightarrow y = -\frac{1}{2}t + \sqrt{2}$.

(b)

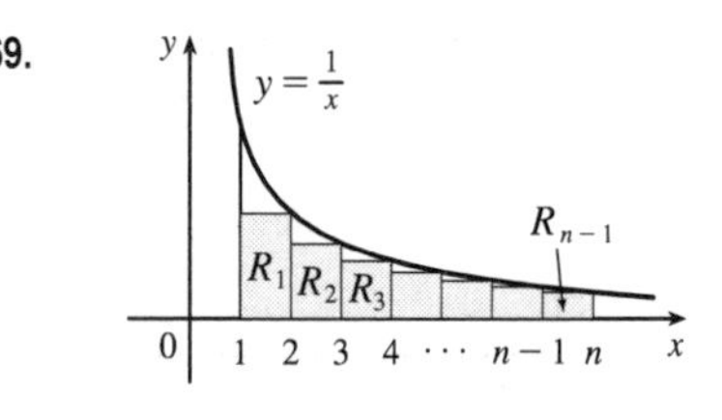

Since the graph of $y = 1/t$ is concave upward, the graph lies above the tangent line, that is, above the line segment BC. Now $|AB| = -\frac{1}{2} + \sqrt{2}$ and $|CD| = -1 + \sqrt{2}$. So the area of the trapezoid $ABCD$ is

$\frac{1}{2}\left[\left(-\frac{1}{2} + \sqrt{2}\right) + \left(-1 + \sqrt{2}\right)1\right] = -\frac{3}{4} + \sqrt{2} \approx 0.6642$. So $\ln 2 >$ area of trapezoid $ABCD > 0.66$.

69.

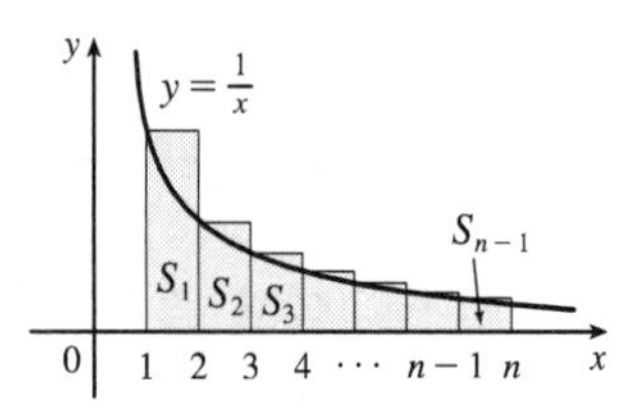

The area of R_i is $\frac{1}{i+1}$ and so $\frac{1}{2} + \frac{1}{3} + \cdots + \frac{1}{n} < \int_1^n \frac{1}{t}\,dt = \ln n$.

The area of S_i is $\frac{1}{i}$ and so $1 + \frac{1}{2} + \cdots + \frac{1}{n-1} > \int_1^n \frac{1}{t}\,dt = \ln n$.

70. If $f(x) = \ln(x^r)$, then $f'(x) = (1/x^r)(rx^{r-1}) = r/x$. But if $g(x) = r\ln x$, then $g'(x) = r/x$. So f and g must differ by a constant: $\ln(x^r) = r\ln x + C$. Put $x = 1$: $\ln(1^r) = r\ln 1 + C \Rightarrow C = 0$, so $\ln(x^r) = r\ln x$.

71. The curve and the line will determine a region when they intersect at two or more points. So we solve the equation $x/(x^2+1) = mx \Rightarrow$ $x = 0$ or $mx^2 + m - 1 = 0 \Rightarrow x = 0$ or

$x = \frac{\pm\sqrt{-4(m)(m-1)}}{2m} = \pm\sqrt{\frac{1}{m} - 1}$. Note that if $m = 1$, this has only the solution $x = 0$, and no region is determined. But if $1/m - 1 > 0 \Leftrightarrow 1/m > 1 \Leftrightarrow 0 < m < 1$, then there are two

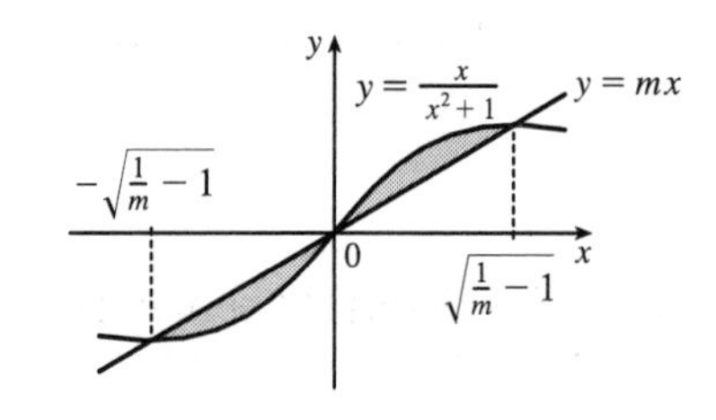

solutions. [Another way of seeing this is to observe that the slope of the tangent to $y = x/(x^2+1)$ at the origin is $y' = 1$ and therefore we must have $0 < m < 1$.] Note that we cannot just integrate between the positive and negative roots, since the curve and the line cross at the origin. Since mx and $x/(x^2+1)$ are both odd functions, the total area is twice the area between the curves on the interval $\left[0, \sqrt{1/m - 1}\right]$. So the total area enclosed is

$$2\int_0^{\sqrt{1/m-1}}\left[\frac{x}{x^2+1}-mx\right]dx = 2\left[\tfrac{1}{2}\ln(x^2+1)-\tfrac{1}{2}mx^2\right]_0^{\sqrt{1/m-1}}$$
$$= \left[\ln\left(\frac{1}{m}-1+1\right)-m\left(\frac{1}{m}-1\right)\right]-(\ln 1-0)$$
$$= \ln\left(\frac{1}{m}\right)+m-1 = m-\ln m-1$$

72. $\lim\limits_{x\to\infty}[\ln(2+x)-\ln(1+x)] = \lim\limits_{x\to\infty}\ln\left(\frac{2+x}{1+x}\right) = \lim\limits_{x\to\infty}\ln\left(\frac{2/x+1}{1/x+1}\right) = \ln\frac{1}{1} = \ln 1 = 0$

73. If $f(x)=\ln(1+x)$, then $f'(x)=\dfrac{1}{1+x}$, so $f'(0)=1$.

Thus, $\lim\limits_{x\to 0}\dfrac{\ln(1+x)}{x} = \lim\limits_{x\to 0}\dfrac{f(x)}{x} = \lim\limits_{x\to 0}\dfrac{f(x)-f(0)}{x-0} = f'(0) = 1$.

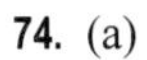

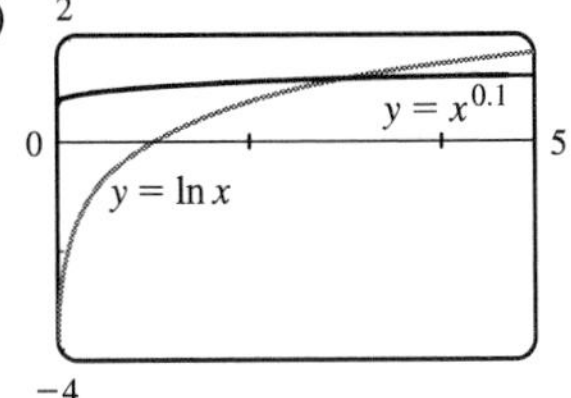

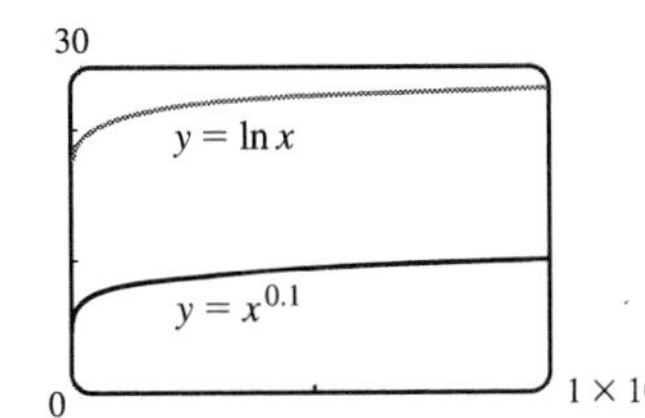

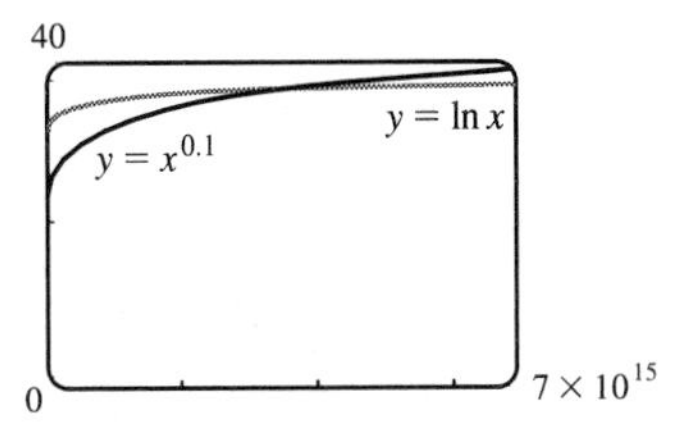

From the graphs, we see that $f(x)=x^{0.1} > g(x)=\ln x$ for approximately $0<x<3.06$, and then $g(x)>f(x)$ for $3.06<x<3.43\times 10^{15}$ (approximately). At that point, the graph of f finally surpasses the graph of g for good.

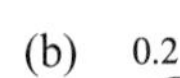

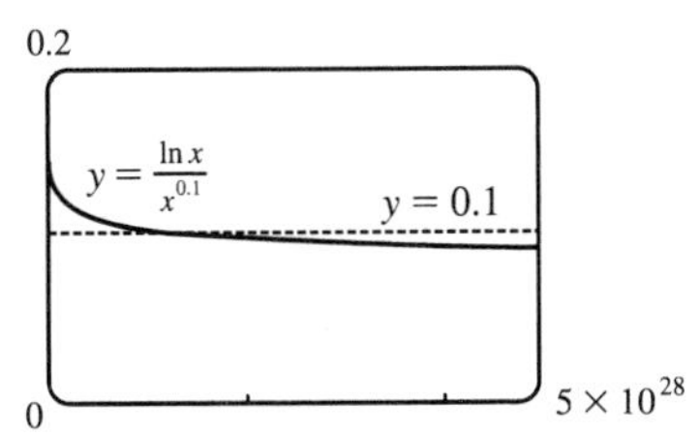

(c) From the graph at left, it seems that $\dfrac{\ln x}{x^{0.1}} < 0.1$ whenever $x > 1.3\times 10^{28}$ (approximately). So we can take $N = 1.3\times 10^{28}$, or any larger number.

5.3 The Natural Exponential Function

1. (a) e is the number such that $\ln e = 1$. (b) $e \approx 2.71828$

(c)

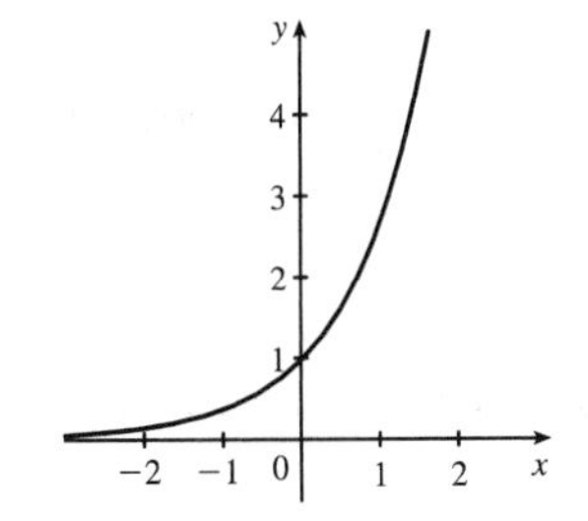

The function value at $x=0$ is 1 and the slope at $x=0$ is 1.

2. (a) $\ln\sqrt{e} = \ln(e^{1/2}) = \frac{1}{2}$ (b) $\ln\sqrt{e} = \ln(e^{1/2}) = \frac{1}{2}$

3. (a) $\ln e^{\sqrt{2}} = \sqrt{2}$ (b) $e^{3\ln 2} = \left(e^{\ln 2}\right)^3 = 2^3 = 8$

4. (a) $\ln e^{\sin x} = \sin x$ (b) $e^{x+\ln x} = e^x e^{\ln x} = xe^x$

5. (a) $2\ln x = 1 \;\Rightarrow\; \ln x = \frac{1}{2} \;\Rightarrow\; x = e^{1/2} = \sqrt{e}$

(b) $e^{-x} = 5 \;\Rightarrow\; -x = \ln 5 \;\Rightarrow\; x = -\ln 5$

6. (a) $e^{2x+3} - 7 = 0 \;\Rightarrow\; e^{2x+3} = 7 \;\Rightarrow\; 2x + 3 = \ln 7 \;\Rightarrow\; 2x = \ln 7 - 3 \;\Rightarrow\; x = \frac{1}{2}(\ln 7 - 3)$

(b) $\ln(5 - 2x) = -3 \;\Rightarrow\; 5 - 2x = e^{-3} \;\Rightarrow\; 2x = 5 - e^{-3} \;\Rightarrow\; x = \frac{1}{2}(5 - e^{-3})$

7. (a) $2^{x-5} = 3 \;\Leftrightarrow\; \log_2 3 = x - 5 \;\Leftrightarrow\; x = 5 + \log_2 3$.

Or: $2^{x-5} = 3 \;\Leftrightarrow\; \ln(2^{x-5}) = \ln 3 \;\Leftrightarrow\; (x-5)\ln 2 = \ln 3 \;\Leftrightarrow\; x - 5 = \dfrac{\ln 3}{\ln 2} \;\Leftrightarrow\; x = 5 + \dfrac{\ln 3}{\ln 2}$

(b) $\ln x + \ln(x-1) = \ln(x(x-1)) = 1 \;\Leftrightarrow\; x(x-1) = e^1 \;\Leftrightarrow\; x^2 - x - e = 0$. The quadratic formula (with $a = 1$, $b = -1$, and $c = -e$) gives $x = \frac{1}{2}\left(1 \pm \sqrt{1+4e}\right)$, but we reject the negative root since the natural logarithm is not defined for $x < 0$. So $x = \frac{1}{2}\left(1 + \sqrt{1+4e}\right)$.

8. (a) $\ln(\ln x) = 1 \;\Leftrightarrow\; e^{\ln(\ln x)} = e^1 \;\Leftrightarrow\; \ln x = e^1 = e \;\Leftrightarrow\; e^{\ln x} = e^e \;\Leftrightarrow\; x = e^e$

(b) $e^{ax} = Ce^{bx} \;\Leftrightarrow\; \ln e^{ax} = \ln[C(e^{bx})] \;\Leftrightarrow\; ax = \ln C + bx + \ln e^{bx} \;\Leftrightarrow\; ax = \ln C + bx \;\Leftrightarrow$

$ax - bx = \ln C \;\Leftrightarrow\; (a-b)x = \ln C \;\Leftrightarrow\; x = \dfrac{\ln C}{a-b}$

9. (a) $e^x < 10 \;\Rightarrow\; \ln e^x < \ln 10 \;\Rightarrow\; x < \ln 10 \;\Rightarrow\; x \in (-\infty, \ln 10)$

(b) $\ln x > -1 \;\Rightarrow\; e^{\ln x} > e^{-1} \;\Rightarrow\; x > e^{-1} \;\Rightarrow\; x \in (1/e, \infty)$

10. (a) $2 < \ln x < 9 \;\Rightarrow\; e^2 < e^{\ln x} < e^9 \;\Rightarrow\; e^2 < x < e^9 \;\Rightarrow\; x \in \left(e^2, e^9\right)$

(b) $e^{2-3x} > 4 \;\Rightarrow\; \ln e^{2-3x} > \ln 4 \;\Rightarrow\; 2 - 3x > \ln 4 \;\Rightarrow\; -3x > \ln 4 - 2 \;\Rightarrow\; x < -\frac{1}{3}(\ln 4 - 2) \;\Rightarrow$

$x \in \left(-\infty, \frac{1}{3}(2 - \ln 4)\right)$

11.

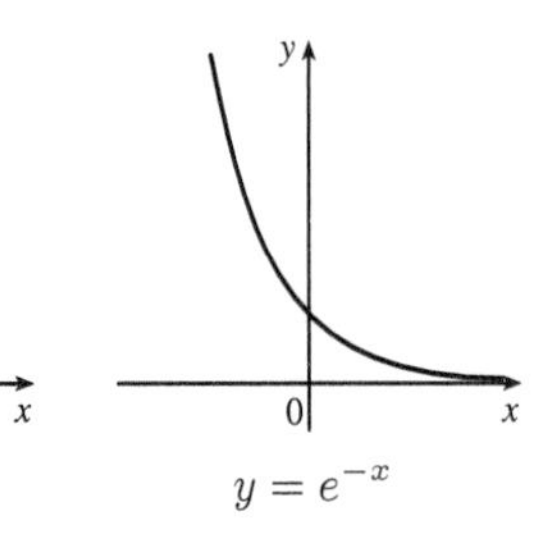

$y = e^x$ $y = e^{-x}$

12. We start with the graph of $y = e^x$ (Figure 10), vertically stretch by a factor of 2, and then shift 1 unit upward. There is a horizontal asymptote of $y = 1$.

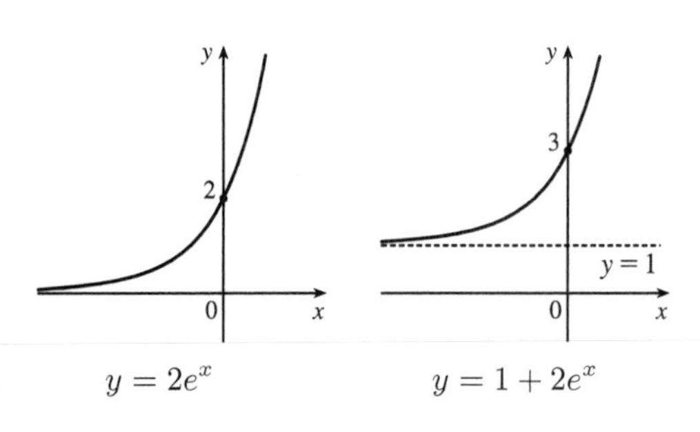

$y = 2e^x$ $y = 1 + 2e^x$

13. We start with the graph of $y = e^x$ (Figure 2), reflect it about the x-axis, and then shift 3 units upward. Note the horizontal asymptote of $y = 3$.

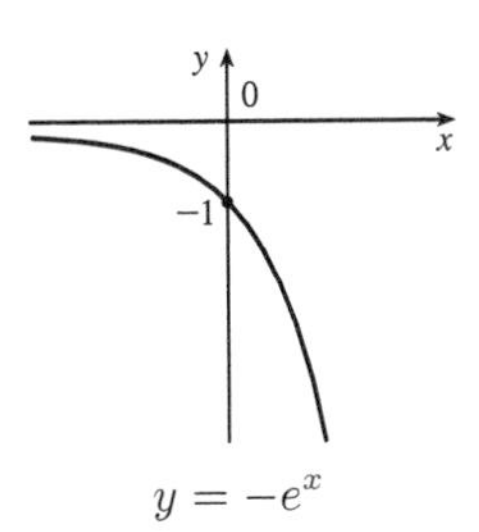

$y = -e^x$

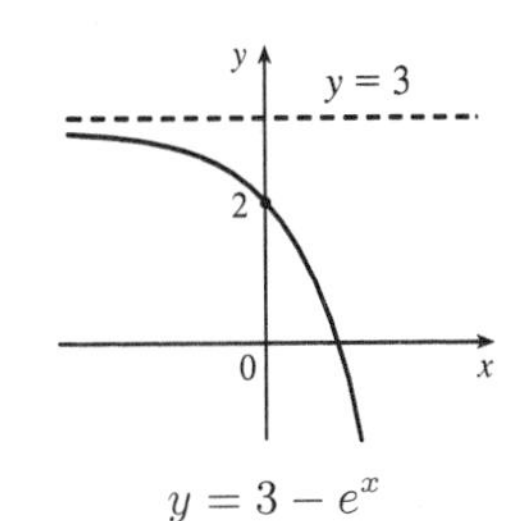

$y = 3 - e^x$

14. We start with the graph of $y = e^x$ (Figure 2), reflect it about the y-axis, and then about the x-axis (or just rotate 180° to handle both reflections) to obtain the graph of $y = -e^{-x}$. Now shift this graph 1 unit upward, vertically stretch by a factor of 5, and then shift 2 units upward.

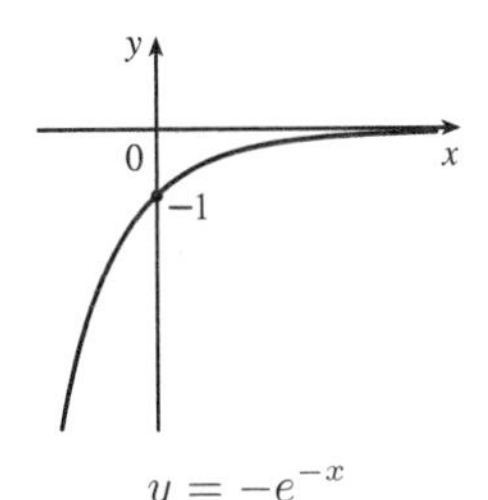

$y = -e^{-x}$

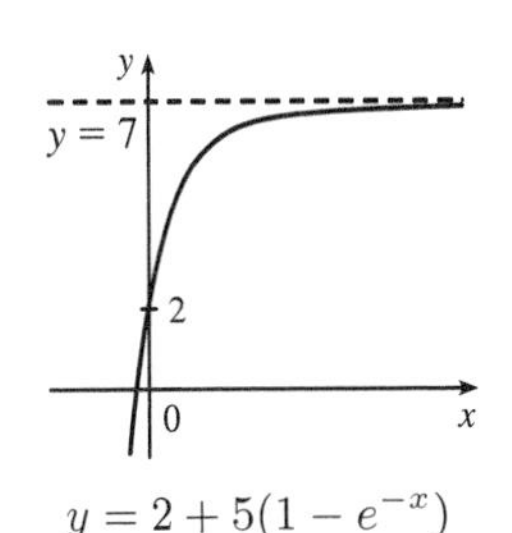

$y = 2 + 5(1 - e^{-x})$

15. $\lim\limits_{x\to\infty} e^{1-x^3} = \lim\limits_{x\to\infty} (e^1 \cdot e^{-x^3}) = e \lim\limits_{x\to\infty} \dfrac{1}{e^{x^3}} = e \cdot 0 = 0$

16. If we let $t = \tan x$, then as $x \to (\pi/2)^+$, $t \to -\infty$. Thus, $\lim\limits_{x\to(\pi/2)^+} e^{\tan x} = \lim\limits_{t\to-\infty} e^t = 0$.

17. Divide numerator and denominator by e^{3x}: $\lim\limits_{x\to\infty} \dfrac{e^{3x} - e^{-3x}}{e^{3x} + e^{-3x}} = \lim\limits_{x\to\infty} \dfrac{1 - e^{-6x}}{1 + e^{-6x}} = \dfrac{1 - 0}{1 + 0} = 1$

18. Divide numerator and denominator by e^{-3x}: $\lim\limits_{x\to-\infty} \dfrac{e^{3x} - e^{-3x}}{e^{3x} + e^{-3x}} = \lim\limits_{x\to-\infty} \dfrac{e^{6x} - 1}{e^{6x} + 1} = \dfrac{0 - 1}{0 + 1} = -1$

19. Let $t = 3/(2 - x)$. As $x \to 2^+$, $t \to -\infty$. So $\lim\limits_{x\to2^+} e^{3/(2-x)} = \lim\limits_{t\to-\infty} e^t = 0$ by (5).

20. Let $t = 3/(2 - x)$. As $x \to 2^-$, $t \to \infty$. So $\lim\limits_{x\to2^-} e^{3/(2-x)} = \lim\limits_{t\to\infty} e^t = \infty$ by (5).

21. By the Product Rule, $f(x) = x^2 e^x \Rightarrow f'(x) = x^2 \dfrac{d}{dx}(e^x) + e^x \dfrac{d}{dx}(x^2) = x^2 e^x + e^x(2x) = xe^x(x + 2)$.

22. By the Quotient Rule, $y = \dfrac{e^x}{1 + x} \Rightarrow y' = \dfrac{(1 + x)e^x - e^x(1)}{(1 + x)^2} = \dfrac{e^x + xe^x - e^x}{(x + 1)^2} = \dfrac{xe^x}{(x + 1)^2}$.

23. By (9), $y = e^{ax^3} \Rightarrow y' = e^{ax^3} \dfrac{d}{dx}(ax^3) = 3ax^2 e^{ax^3}$.

24. $y = e^u(\cos u + cu) \Rightarrow y' = e^u(-\sin u + c) + (\cos u + cu)e^u = e^u(\cos u - \sin u + cu + c)$

25. $f(u) = e^{1/u} \Rightarrow f'(u) = e^{1/u} \cdot \dfrac{d}{du}\left(\dfrac{1}{u}\right) = e^{1/u}\left(\dfrac{-1}{u^2}\right) = \left(\dfrac{-1}{u^2}\right)e^{1/u}$

26. $y = e^x \ln x \Rightarrow y' = e^x\left(\dfrac{1}{x}\right) + (\ln x)(e^x) = e^x\left(\ln x + \dfrac{1}{x}\right)$

27. $F(t) = e^{t \sin 2t} \Rightarrow F'(t) = e^{t \sin 2t}(t \sin 2t)' = e^{t \sin 2t}(t \cdot 2\cos 2t + \sin 2t \cdot 1) = e^{t \sin 2t}(2t \cos 2t + \sin 2t)$

28. $y = e^{k\tan\sqrt{x}} \Rightarrow y' = e^{k\tan\sqrt{x}} \cdot \dfrac{d}{dx}\left(k\tan\sqrt{x}\right) = e^{k\tan\sqrt{x}}\left(k\sec^2\sqrt{x}\cdot\tfrac{1}{2}x^{-1/2}\right) = \dfrac{k\sec^2\sqrt{x}}{2\sqrt{x}}\,e^{k\tan\sqrt{x}}$

29. $y = \sqrt{1+2e^{3x}} \Rightarrow y' = \dfrac{1}{2}(1+2e^{3x})^{-1/2}\dfrac{d}{dx}(1+2e^{3x}) = \dfrac{1}{2\sqrt{1+2e^{3x}}}(2e^{3x}\cdot 3) = \dfrac{3e^{3x}}{\sqrt{1+2e^{3x}}}$

30. $y = \cos(e^{\pi x}) \Rightarrow y' = -\sin(e^{\pi x})\cdot e^{\pi x}\cdot\pi = -\pi e^{\pi x}\sin(e^{\pi x})$

31. $y = e^{e^x} \Rightarrow y' = e^{e^x}\cdot\dfrac{d}{dx}(e^x) = e^{e^x}\cdot e^x$ or e^{e^x+x}

32. $y = \sqrt{1+xe^{-2x}} \Rightarrow y' = \tfrac{1}{2}\left(1+xe^{-2x}\right)^{-1/2}\left[x(-2e^{-2x})+e^{-2x}\right] = \dfrac{e^{-2x}(-2x+1)}{2\sqrt{1+xe^{-2x}}}$

33. By the Quotient Rule, $y = \dfrac{ae^x+b}{ce^x+d} \Rightarrow$

$$y' = \frac{(ce^x+d)(ae^x)-(ae^x+b)(ce^x)}{(ce^x+d)^2} = \frac{(ace^x+ad-ace^x-bc)e^x}{(ce^x+d)^2} = \frac{(ad-bc)e^x}{(ce^x+d)^2}.$$

The notations $\overset{\text{PR}}{\Rightarrow}$ and $\overset{\text{QR}}{\Rightarrow}$ indicate the use of the Product and Quotient Rules, respectively.

34. $y = \dfrac{e^x+e^{-x}}{e^x-e^{-x}} \Rightarrow y' = \dfrac{(e^x-e^{-x})(e^x-e^{-x})-(e^x+e^{-x})(e^x+e^{-x})}{(e^x-e^{-x})^2}$

$$= \frac{\left(e^{2x}-2+e^{-2x}\right)-\left(e^{2x}+2+e^{-2x}\right)}{\left(e^x-e^{-x}\right)^2} = -\frac{4}{\left(e^x-e^{-x}\right)^2}$$

35. $y = e^{2x}\cos\pi x \Rightarrow y' = e^{2x}(-\pi\sin\pi x)+(\cos\pi x)(2e^{2x}) = e^{2x}(2\cos\pi x-\pi\sin\pi x)$.

At $(0,1)$, $y' = 1(2-0) = 2$, so an equation of the tangent line is $y-1 = 2(x-0)$, or $y = 2x+1$.

36. $y = \dfrac{e^x}{x} \Rightarrow y' = \dfrac{x\cdot e^x - e^x\cdot 1}{x^2} = \dfrac{e^x(x-1)}{x^2}$. At $(1,e)$, $y' = 0$, and an equation of the tangent line is

$y - e = 0(x-1)$, or $y = e$.

37. $\dfrac{d}{dx}\left(e^{x^2y}\right) = \dfrac{d}{dx}(x+y) \Rightarrow e^{x^2y}(x^2y'+y\cdot 2x) = 1+y' \Rightarrow x^2e^{x^2y}y'+2xye^{x^2y} = 1+y' \Rightarrow$

$x^2e^{x^2y}y'-y' = 1-2xye^{x^2y} \Rightarrow y'(x^2e^{x^2y}-1) = 1-2xye^{x^2y} \Rightarrow y' = \dfrac{1-2xye^{x^2y}}{x^2e^{x^2y}-1}$

38. $y = Ae^{-x}+Bxe^{-x} \Rightarrow y' = -Ae^{-x}+Be^{-x}-Bxe^{-x} = (B-A)e^{-x}-Bxe^{-x} \Rightarrow$

$y'' = (A-B)e^{-x}-Be^{-x}+Bxe^{-x} = (A-2B)e^{-x}+Bxe^{-x}$,

so $y''+2y'+y = (A-2B)e^{-x}+Bxe^{-x}+2\left[(B-A)e^{-x}-Bxe^{-x}\right]+Ae^{-x}+Bxe^{-x} = 0$.

39. $y = e^{rx} \Rightarrow y' = re^{rx} \Rightarrow y'' = r^2e^{rx}$, so if $y = e^{rx}$ satisfies the differential equation $y''+6y'+8y = 0$,

then $r^2e^{rx}+6re^{rx}+8e^{rx} = 0$; that is, $e^{rx}(r^2+6r+8) = 0$. Since $e^{rx} > 0$ for all x, we must have $r^2+6r+8 = 0$,

or $(r+2)(r+4) = 0$, so $r = -2$ or -4.

40. $y = e^{\lambda x} \Rightarrow y' = \lambda e^{\lambda x} \Rightarrow y'' = \lambda^2e^{\lambda x}$. Thus, $y+y' = y'' \Leftrightarrow e^{\lambda x}+\lambda e^{\lambda x} = \lambda^2e^{\lambda x} \Leftrightarrow$

$e^{\lambda x}\left(\lambda^2-\lambda-1\right) = 0 \Leftrightarrow \lambda = \frac{1\pm\sqrt{5}}{2}$, since $e^{\lambda x} \neq 0$.

41. $f(x) = e^{2x} \;\Rightarrow\; f'(x) = 2e^{2x} \;\Rightarrow\; f''(x) = 2 \cdot 2e^{2x} = 2^2 e^{2x} \;\Rightarrow$

$f'''(x) = 2^2 \cdot 2e^{2x} = 2^3 e^{2x} \;\Rightarrow\; \cdots \;\Rightarrow\; f^{(n)}(x) = 2^n e^{2x}$

42. $f(x) = xe^{-x} \;\Rightarrow\; f'(x) = x(-e^{-x}) + e^{-x} = (1-x)e^{-x} \;\Rightarrow$

$f''(x) = (1-x)(-e^{-x}) + e^{-x}(-1) = (x-2)e^{-x} \;\Rightarrow\; f'''(x) = (x-2)(-e^{-x}) + e^{-x} = (3-x)e^{-x} \;\Rightarrow$

$f^{(4)}(x) = (3-x)(-e^{-x}) + e^{-x}(-1) = (x-4)e^{-x} \;\Rightarrow\; \cdots \;\Rightarrow\; f^{(n)}(x) = (-1)^n(x-n)e^{-x}$.

So $D^{1000}xe^{-x} = (x-1000)e^{-x}$.

43. (a) $f(x) = e^x + x$ is continuous on $\mathbb{R}$ and $f(-1) = e^{-1} - 1 < 0 < 1 = f(0)$, so by the Intermediate Value Theorem, $e^x + x = 0$ has a root in $(-1, 0)$.

(b) $f(x) = e^x + x \;\Rightarrow\; f'(x) = e^x + 1$, so $x_{n+1} = x_n - \dfrac{e^{x_n} + x_n}{e^{x_n} + 1}$. Using $x_1 = -0.5$, we get $x_2 \approx -0.566311$, $x_3 \approx -0.567143 \approx x_4$, so the root is -0.567143 to six decimal places.

44.

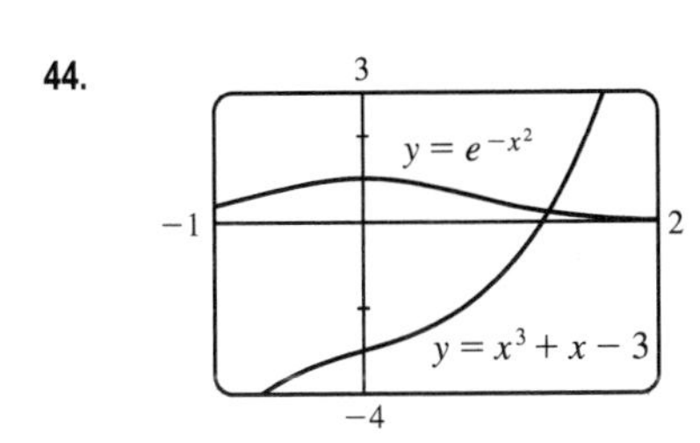

From the graph, it appears that the curves intersect at about $x \approx 1.2$ or 1.3. We use Newton's Method with $f(x) = x^3 + x - 3 - e^{-x^2}$, so $f'(x) = 3x^2 + 1 + 2xe^{-x^2}$, and the formula is $x_{n+1} = x_n - f(x_n)/f'(x_n)$. We take $x_1 = 1.2$, and the formula gives $x_2 \approx 1.252462$, $x_3 \approx 1.251045$, and $x_4 \approx x_5 \approx 1.251044$. So the root of the equation, correct to six decimal places, is $x = 1.251044$.

45. (a) $\displaystyle\lim_{t\to\infty} p(t) = \lim_{t\to\infty} \frac{1}{1 + ae^{-kt}} = \frac{1}{1 + a \cdot 0} = 1$, since $k > 0 \;\Rightarrow\; -kt \to -\infty \;\Rightarrow\; e^{-kt} \to 0$.

(b) $p(t) = (1 + ae^{-kt})^{-1} \;\Rightarrow\; \dfrac{dp}{dt} = -(1 + ae^{-kt})^{-2}(-kae^{-kt}) = \dfrac{kae^{-kt}}{(1 + ae^{-kt})^2}$

(c)

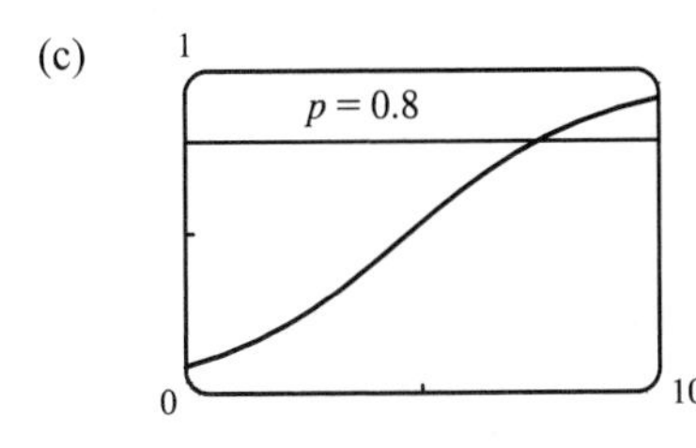

From the graph of $p(t) = (1 + 10e^{-0.5t})^{-1}$, it seems that $p(t) = 0.8$ (indicating that 80% of the population has heard the rumor) when $t \approx 7.4$ hours.

46. (a) The displacement function is squeezed between the other two functions. This is because $-1 \le \sin 4t \le 1 \;\Rightarrow$ $-8e^{-t/2} \le 8e^{-t/2} \sin 4t \le 8e^{-t/2}$.

(b) The maximum value of the displacement is about 6.6 cm, occurring at $t \approx 0.36$ s. It occurs just before the graph of the displacement function touches the graph of $8e^{-t/2}$ (when $t = \frac{\pi}{8} \approx 0.39$).

(c) The velocity of the object is the derivative of its displacement function, that is,

$\frac{d}{dt}\left(8e^{-t/2}\sin 4t\right) = 8\left[e^{-t/2}\cos 4t(4) + \sin 4t\left(-\frac{1}{2}\right)e^{-t/2}\right].$

If the displacement is zero, then we must have $\sin 4t = 0$ (since the exponential term in the displacement function is always positive). The first time that $\sin 4t = 0$ after $t = 0$ occurs at $t = \frac{\pi}{4}$. Substituting this into our expression for the velocity, and noting that the second term vanishes, we get

$v\left(\frac{\pi}{4}\right) = 8e^{-\pi/8}\cos\left(4\cdot\frac{\pi}{4}\right)\cdot 4 = -32e^{-\pi/8} \approx -21.6\text{ cm/s}.$

(d) 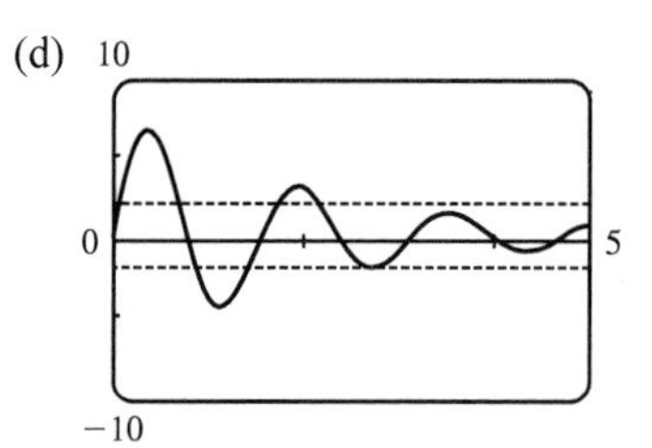

The graph indicates that the displacement is less than 2 cm from equilibrium whenever t is larger than about 2.8.

47. $f(x) = x - e^x \;\Rightarrow\; f'(x) = 1 - e^x = 0 \;\Leftrightarrow\; e^x = 1 \;\Leftrightarrow\; x = 0$. Now $f'(x) > 0$ for all $x < 0$ and $f'(x) < 0$ for all $x > 0$, so the absolute maximum value is $f(0) = 0 - 1 = -1$.

48. $g(x) = \frac{e^x}{x} \;\Rightarrow\; g'(x) = \frac{xe^x - e^x}{x^2} = 0 \;\Leftrightarrow\; e^x(x-1) = 0 \;\Rightarrow\; x = 1$. Now $g'(x) > 0 \;\Leftrightarrow\; \frac{xe^x - e^x}{x^2} > 0 \;\Leftrightarrow\; x - 1 > 0 \;\Leftrightarrow\; x > 1$ and $g'(x) < 0 \;\Leftrightarrow\; \frac{xe^x - e^x}{x^2} < 0 \;\Leftrightarrow\; x - 1 < 0 \;\Leftrightarrow\; x < 1$. Thus there is an absolute minimum value of $g(1) = e$ at $x = 1$.

49. $y = xe^{3x} \;\Rightarrow\; y' = xe^{3x}\cdot 3 + e^{3x}\cdot 1 = (3x+1)e^{3x} \;\Rightarrow\; y'' = (3x+1)e^{3x}\cdot 3 + e^{3x}\cdot 3 = (9x+6)e^{3x}$. The curve is concave upward at $x \;\Leftrightarrow\; y'' > 0$ at $x \;\Leftrightarrow\; 9x + 6 > 0 \;\Leftrightarrow\; x > -\frac{2}{3}$. Thus, the curve is concave upward on $\left(-\frac{2}{3}, \infty\right)$.

50. $f(x) = x^2e^{-x} \;\Rightarrow\; f'(x) = x^2(-e^{-x}) + e^{-x}\cdot 2x = (2x - x^2)e^{-x}$, so $f'(x) > 0 \;\Leftrightarrow\; 2x - x^2 > 0 \;\Leftrightarrow\; x(2-x) > 0 \;\Leftrightarrow\; 0 < x < 2$, so f is increasing on $(0, 2)$.

51. $y = 1/(1 + e^{-x})$ **A.** $D = \mathbb{R}$ **B.** No x-intercept; y-intercept $= f(0) = \frac{1}{2}$. **C.** No symmetry

D. $\lim_{x\to\infty} 1/(1 + e^{-x}) = \frac{1}{1+0} = 1$ and $\lim_{x\to-\infty} 1/(1 + e^{-x}) = 0$ (since $\lim_{x\to-\infty} e^{-x} = \infty$), so f has HAs $y = 0$ and $y = 1$.

E. $f'(x) = -(1 + e^{-x})^{-2}(-e^{-x}) = e^{-x}/(1 + e^{-x})^2$. This is positive for all x, so f is increasing on $\mathbb{R}$.

F. No extreme values

G. $f''(x) = \frac{(1 + e^{-x})^2(-e^{-x}) - e^{-x}(2)(1 + e^{-x})(-e^{-x})}{(1 + e^{-x})^4} = \frac{e^{-x}(e^{-x} - 1)}{(1 + e^{-x})^3}$

The second factor in the numerator is negative for $x > 0$ and positive for $x < 0$, and the other factors are always positive, so f is CU on $(-\infty, 0)$ and CD on $(0, \infty)$. f has an IP at $\left(0, \frac{1}{2}\right)$.

H. 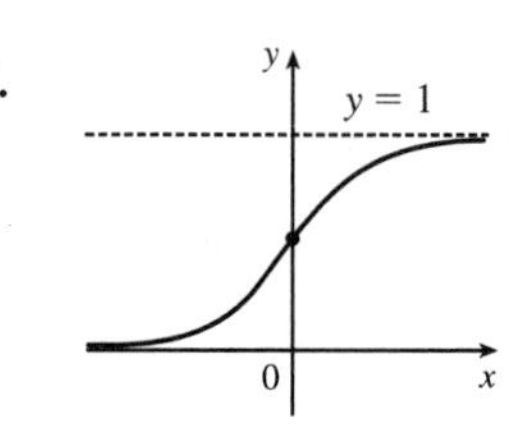

52. $y = f(x) = e^{2x} - e^x$ **A.** $D = \mathbb{R}$ **B.** y-intercept: $f(0) = 0$; x-intercepts: $f(x) = 0 \Rightarrow e^{2x} = e^x \Rightarrow$ $e^x = 1 \Rightarrow x = 0$. **C.** No symmetry **D.** $\lim\limits_{x \to -\infty} e^{2x} - e^x = 0$, so $y = 0$ is a HA. No VA.

E. $f'(x) = 2e^{2x} - e^x = e^x(2e^x - 1)$, so $f'(x) > 0 \Leftrightarrow e^x > \frac{1}{2} \Leftrightarrow$ $x > \ln\frac{1}{2} = -\ln 2$ and $f'(x) < 0 \Leftrightarrow e^x < \frac{1}{2} \Leftrightarrow x < \ln\frac{1}{2}$, so f is decreasing on $\left(-\infty, \ln\frac{1}{2}\right)$ and increasing on $\left(\ln\frac{1}{2}, \infty\right)$.

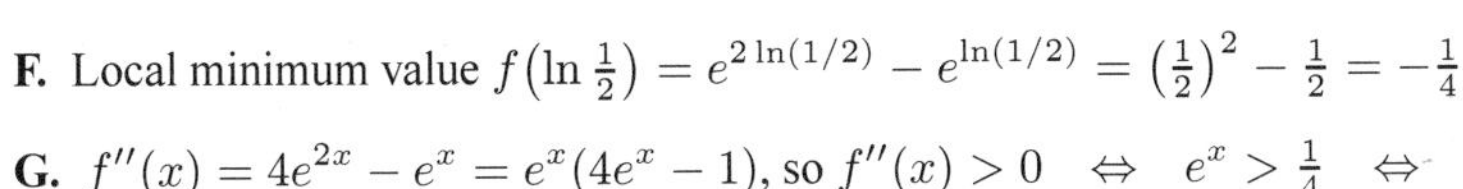

F. Local minimum value $f\left(\ln\frac{1}{2}\right) = e^{2\ln(1/2)} - e^{\ln(1/2)} = \left(\frac{1}{2}\right)^2 - \frac{1}{2} = -\frac{1}{4}$

G. $f''(x) = 4e^{2x} - e^x = e^x(4e^x - 1)$, so $f''(x) > 0 \Leftrightarrow e^x > \frac{1}{4} \Leftrightarrow$ $x > \ln\frac{1}{4}$ and $f''(x) < 0 \Leftrightarrow x < \ln\frac{1}{4}$. Thus, f is CD on $\left(-\infty, \ln\frac{1}{4}\right)$ and CU on $\left(\ln\frac{1}{4}, \infty\right)$. f has an IP at $\left(\ln\frac{1}{4}, \left(\frac{1}{4}\right)^2 - \frac{1}{4}\right) = \left(\ln\frac{1}{4}, -\frac{3}{16}\right)$.

H.

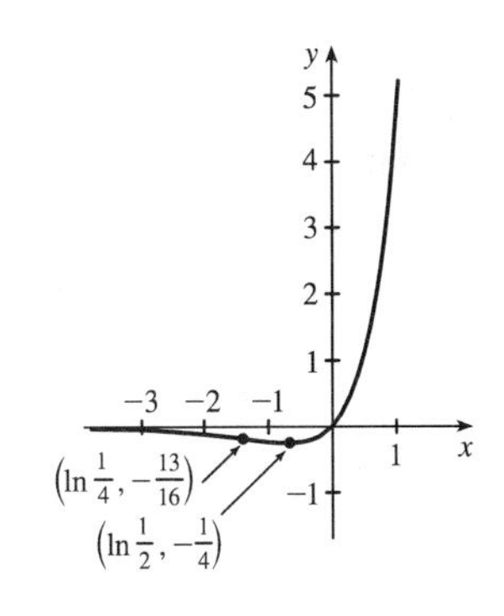

53. $y = f(x) = e^{3x} + e^{-2x}$ **A.** $D = \mathbb{R}$ **B.** y-intercept $= f(0) = 2$; no x-intercept **C.** No symmetry **D.** No asymptotes

E. $f'(x) = 3e^{3x} - 2e^{-2x}$, so $f'(x) > 0 \Leftrightarrow 3e^{3x} > 2e^{-2x}$ [multiply by e^{2x}] $\Leftrightarrow e^{5x} > \frac{2}{3} \Leftrightarrow 5x > \ln\frac{2}{3} \Leftrightarrow$ $x > \frac{1}{5}\ln\frac{2}{3} \approx -0.081$. Similarly, $f'(x) < 0 \Leftrightarrow x < \frac{1}{5}\ln\frac{2}{3}$. f is decreasing on $\left(-\infty, \frac{1}{5}\ln\frac{2}{3}\right)$ and increasing on $\left(\frac{1}{5}\ln\frac{2}{3}, \infty\right)$.

H.

F. Local minimum value $f\left(\frac{1}{5}\ln\frac{2}{3}\right) = \left(\frac{2}{3}\right)^{3/5} + \left(\frac{2}{3}\right)^{-2/5} \approx 1.96$; no local maximum.

G. $f''(x) = 9e^{3x} + 4e^{-2x}$, so $f''(x) > 0$ for all x, and f is CU on $(-\infty, \infty)$. No IP

54. The function $f(x) = e^{\cos x}$ is periodic with period 2π, so we consider it only on the interval $[0, 2\pi]$. We see that it has local maxima of about $f(0) \approx 2.72$ and $f(2\pi) \approx 2.72$, and a local minimum of about $f(3.14) \approx 0.37$. To find the

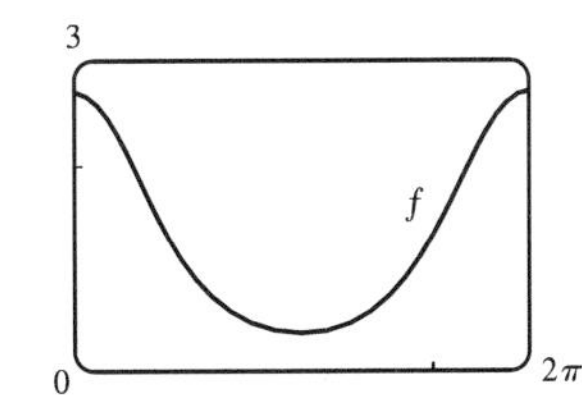

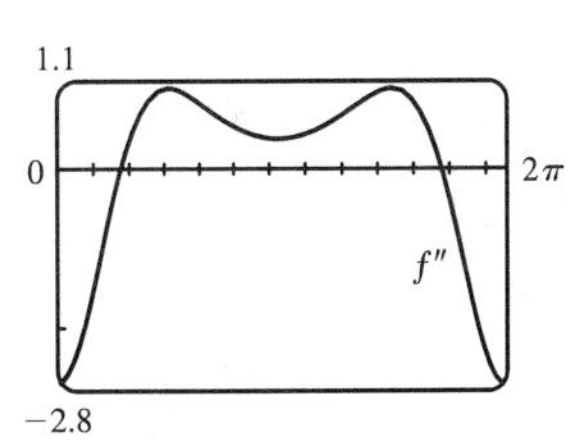

exact values, we calculate $f'(x) = -\sin x e^{\cos x}$. This is 0 when $-\sin x = 0 \Leftrightarrow x = 0, \pi$ or 2π (since we are only considering $x \in [0, 2\pi]$). Also $f'(x) > 0 \Leftrightarrow \sin x < 0 \Leftrightarrow 0 < x < \pi$. So $f(0) = f(2\pi) = e$ (both maxima) and $f(\pi) = e^{\cos \pi} = 1/e$ (minimum). To find the inflection points, we calculate and graph

$$f''(x) = \frac{d}{dx}\left(-\sin x\, e^{\cos x}\right) = -\cos x\, e^{\cos x} - \sin x\left(e^{\cos x}\right)(-\sin x) = e^{\cos x}\left(\sin^2 x - \cos x\right).$$

From the graph of $f''(x)$, we see that f has inflection points at $x \approx 0.90$ and at $x \approx 5.38$. These x-coordinates correspond to inflection points $(0.90, 1.86)$ and $(5.38, 1.86)$.

55. $f(x) = e^{x^3 - x} \to 0$ as $x \to -\infty$, and $f(x) \to \infty$ as $x \to \infty$. From the graph, it appears that f has a local minimum of about $f(0.58) = 0.68$, and a local maximum of about $f(-0.58) = 1.47$. To find the exact values, we calculate

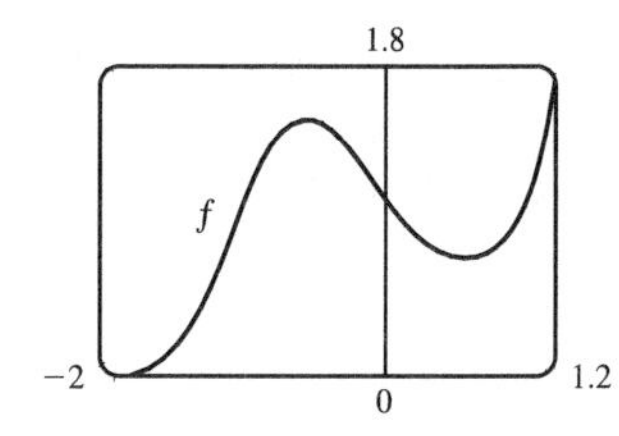

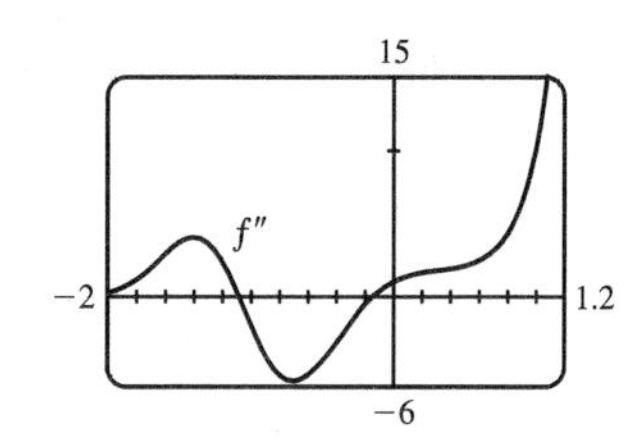

$f'(x) = (3x^2 - 1)e^{x^3 - x}$, which is 0 when $3x^2 - 1 = 0 \Leftrightarrow x = \pm\frac{1}{\sqrt{3}}$. The negative root corresponds to the local maximum $f\left(-\frac{1}{\sqrt{3}}\right) = e^{(-1/\sqrt{3})^3 - (-1/\sqrt{3})} = e^{2\sqrt{3}/9}$, and the positive root corresponds to the local minimum $f\left(\frac{1}{\sqrt{3}}\right) = e^{(1/\sqrt{3})^3 - (1/\sqrt{3})} = e^{-2\sqrt{3}/9}$. To estimate the inflection points, we calculate and graph

$$f''(x) = \frac{d}{dx}\left[(3x^2 - 1)e^{x^3 - x}\right] = (3x^2 - 1)e^{x^3 - x}(3x^2 - 1) + e^{x^3 - x}(6x) = e^{x^3 - x}(9x^4 - 6x^2 + 6x + 1).$$

From the graph, it appears that $f''(x)$ changes sign (and thus f has inflection points) at $x \approx -0.15$ and $x \approx -1.09$. From the graph of f, we see that these x-values correspond to inflection points at about $(-0.15, 1.15)$ and $(-1.09, 0.82)$.

56. (a) As $|x| \to \infty$, $t = -x^2/(2\sigma^2) \to -\infty$, and $e^t \to 0$. The HA is $y = 0$. Since t takes on its maximum value at $x = 0$, so does e^t. Showing this result using derivatives, we have $f(x) = e^{-x^2/(2\sigma^2)} \Rightarrow f'(x) = e^{-x^2/(2\sigma^2)}(-x/\sigma^2)$. $f'(x) = 0 \Leftrightarrow x = 0$. Because f' changes from positive to negative at $x = 0$, $f(0) = 1$ is a local maximum. For inflection points, we find $f''(x) = -\frac{1}{\sigma^2}\left[e^{-x^2/(2\sigma^2)} \cdot 1 + xe^{-x^2/(2\sigma^2)}(-x/\sigma^2)\right] = \frac{-1}{\sigma^2}e^{-x^2/(2\sigma^2)}(1 - x^2/\sigma^2)$.

$f''(x) = 0 \Leftrightarrow x^2 = \sigma^2 \Leftrightarrow x = \pm\sigma$. $f''(x) < 0 \Leftrightarrow x^2 < \sigma^2 \Leftrightarrow -\sigma < x < \sigma$.

So f is CD on $(-\sigma, \sigma)$ and CU on $(-\infty, -\sigma)$ and (σ, ∞). There are IPs at $(\pm\sigma, e^{-1/2})$.

(b) Since we have IP at $x = \pm\sigma$, the inflection points move away from the y-axis as σ increases.

(c)

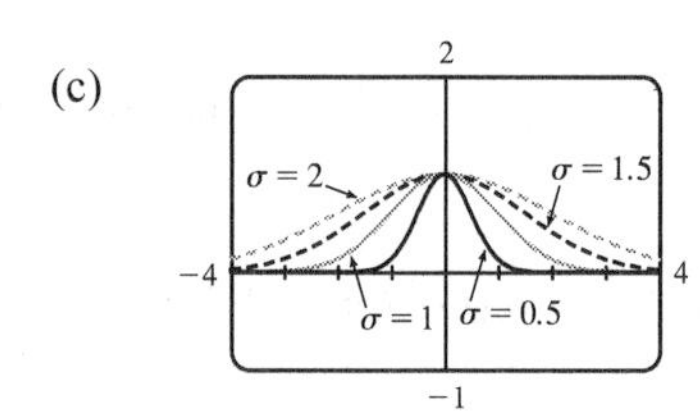

From the graph, we see that as σ increases, the graph tends to spread out and there is more area between the curve and the x-axis.

57. Let $u = -3x$. Then $du = -3\,dx$, so $\int_0^5 e^{-3x}\,dx = -\frac{1}{3}\int_0^{-15} e^u\,du = -\frac{1}{3}\left[e^u\right]_0^{-15} = -\frac{1}{3}(e^{-15} - e^0) = \frac{1}{3}(1 - e^{-15})$.

58. Let $u = -x^2$, so $du = -2x\,dx$. When $x = 0$, $u = 0$; when $x = 1$, $u = -1$. Thus,

$\int_0^1 xe^{-x^2}\,dx = \int_0^{-1} e^u\left(-\frac{1}{2}\,du\right) = -\frac{1}{2}\left[e^u\right]_0^{-1} = -\frac{1}{2}(e^{-1} - e^0) = \frac{1}{2}(1 - 1/e)$.

59. Let $u = 1 + e^x$. Then $du = e^x\,dx$, so $\int e^x\sqrt{1 + e^x}\,dx = \int \sqrt{u}\,du = \frac{2}{3}u^{3/2} + C = \frac{2}{3}(1 + e^x)^{3/2} + C$.

60. Let $u = \tan x$. Then $du = \sec^2 x\,dx$, so $\int \sec^2 x\,e^{\tan x}\,dx = \int e^u\,du = e^u + C = e^{\tan x} + C$.

61. $\displaystyle\int \frac{e^x + 1}{e^x}\,dx = \int (1 + e^{-x})\,dx = x - e^{-x} + C$

62. Let $u = \dfrac{1}{x}$. Then $du = -\dfrac{1}{x^2}\,dx$, so $\displaystyle\int \frac{e^{1/x}}{x^2}\,dx = -\int e^u\,du = -e^u + C = -e^{1/x} + C$.

63. Let $u = \sqrt{x}$. Then $du = \dfrac{1}{2\sqrt{x}}\,dx$, so $\displaystyle\int \frac{e^{\sqrt{x}}}{\sqrt{x}}\,dx = 2\int e^u\,du = 2e^u + C = 2e^{\sqrt{x}} + C$.

64. Let $u = e^x$. Then $du = e^x\,dx$, so $\int e^x \sin(e^x)\,dx = \int \sin u\,du = -\cos u + C = -\cos(e^x) + C$.

65. $y = \ln(x+3) \;\Rightarrow\; e^y = x + 3 \;\Rightarrow\; x = e^y - 3$.

Interchanging x and y, we get $y = e^x - 3$, so

$f^{-1}(x) = e^x - 3$.

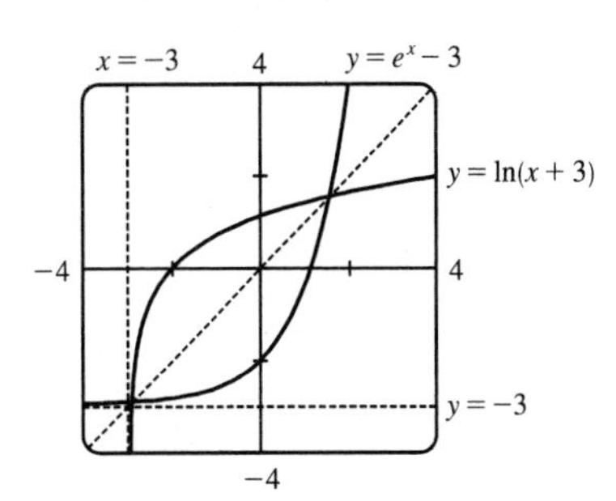

66. $y = \dfrac{1+e^x}{1-e^x} \;\Rightarrow\; y - ye^x = 1 + e^x \;\Rightarrow$

$e^x(y+1) = y - 1 \;\Rightarrow\; e^x = \dfrac{y-1}{y+1} \;\Rightarrow$

$x = \ln\left(\dfrac{y-1}{y+1}\right)$. Interchange x and y:

$y = \ln\left(\dfrac{x-1}{x+1}\right)$ is the inverse function.

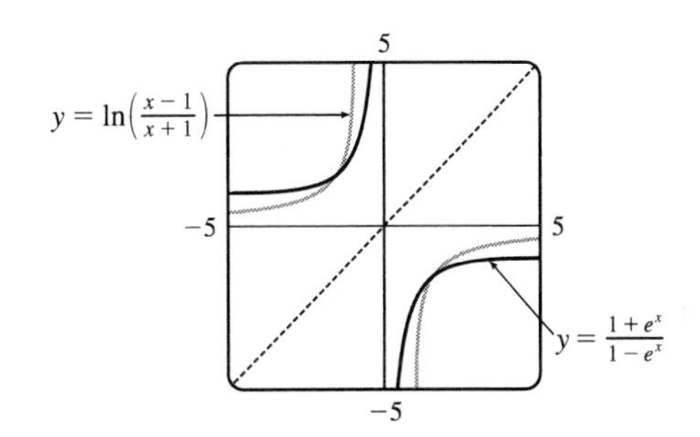

67. We use Theorem 5.1.7. Note that $f(0) = 3 + 0 + e^0 = 4$, so $f^{-1}(4) = 0$. Also $f'(x) = 1 + e^x$. Therefore,

$$(f^{-1})'(4) = \frac{1}{f'(f^{-1}(4))} = \frac{1}{f'(0)} = \frac{1}{1+e^0} = \frac{1}{2}.$$

68. We recognize this limit as the definition of the derivative of the function $f(x) = e^{\sin x}$ at $x = \pi$, since it is of the form $\displaystyle\lim_{x\to\pi} \frac{f(x) - f(\pi)}{x - \pi}$. Therefore, the limit is equal to $f'(\pi) = (\cos\pi)e^{\sin\pi} = -1 \cdot e^0 = -1$.

69. Using the second law of logarithms and Equation 5, we have $\ln(e^x/e^y) = \ln e^x - \ln e^y = x - y = \ln(e^{x-y})$. Since ln is a one-to-one function, it follows that $e^x/e^y = e^{x-y}$.

70. Using the third law of logarithms and Equation 5, we have $\ln e^{rx} = rx = r\ln e^x = \ln(e^x)^r$. Since ln is a one-to-one function, it follows that $e^{rx} = (e^x)^r$.

71. (a) Let $f(x) = e^x - 1 - x$. Now $f(0) = e^0 - 1 = 0$, and for $x \ge 0$, we have $f'(x) = e^x - 1 \ge 0$. Now, since $f(0) = 0$ and f is increasing on $[0, \infty)$, $f(x) \ge 0$ for $x \ge 0 \;\Rightarrow\; e^x - 1 - x \ge 0 \;\Rightarrow\; e^x \ge 1 + x$.

(b) For $0 \le x \le 1$, $x^2 \le x$, so $e^{x^2} \le e^x$ [since e^x is increasing]. Hence [from (a)] $1 + x^2 \le e^{x^2} \le e^x$.

So $\frac{4}{3} = \int_0^1 (1 + x^2)\,dx \le \int_0^1 e^{x^2}\,dx \le \int_0^1 e^x\,dx = e - 1 < e \;\Rightarrow\; \frac{4}{3} \le \int_0^1 e^{x^2}\,dx \le e$.

72. (a) Let $f(x) = e^x - 1 - x - \frac{1}{2}x^2$. Thus, $f'(x) = e^x - 1 - x$, which is positive for $x \ge 0$ by Exercise 71(a). Thus $f(x)$ is increasing on $(0, \infty)$, so on that interval, $0 = f(0) \le f(x) = e^x - 1 - x - \frac{1}{2}x^2 \;\Rightarrow\; e^x \ge 1 + x + \frac{1}{2}x^2$.

(b) Using the same argument as in Exercise 71(b), from part (a) we have $1 + x^2 + \frac{1}{2}x^4 \le e^{x^2} \le e^x$ [for $0 \le x \le 1$] $\Rightarrow\; \int_0^1 (1 + x^2 + \frac{1}{2}x^4)\,dx \le \int_0^1 e^{x^2}\,dx \le \int_0^1 e^x\,dx \;\Rightarrow\; \frac{43}{30} \le \int_0^1 e^{x^2}\,dx \le e - 1$.

73. (a) By Exercise 71(a), the result holds for $n = 1$. Suppose that $e^x \geq 1 + x + \frac{x^2}{2!} + \cdots + \frac{x^k}{k!}$ for $x \geq 0$.

Let $f(x) = e^x - 1 - x - \frac{x^2}{2!} - \cdots - \frac{x^k}{k!} - \frac{x^{k+1}}{(k+1)!}$. Then $f'(x) = e^x - 1 - x - \cdots - \frac{x^k}{k!} \geq 0$ by assumption. Hence

$f(x)$ is increasing on $(0, \infty)$. So $0 \leq x$ implies that $0 = f(0) \leq f(x) = e^x - 1 - x - \cdots - \frac{x^k}{k!} - \frac{x^{k+1}}{(k+1)!}$, and hence

$e^x \geq 1 + x + \cdots + \frac{x^k}{k!} + \frac{x^{k+1}}{(k+1)!}$ for $x \geq 0$. Therefore, for $x \geq 0$, $e^x \geq 1 + x + \frac{x^2}{2!} + \cdots + \frac{x^n}{n!}$ for every positive integer n, by mathematical induction.

(b) Taking $n = 4$ and $x = 1$ in (a), we have $e = e^1 \geq 1 + \frac{1}{2} + \frac{1}{6} + \frac{1}{24} = 2.708\overline{3} > 2.7$.

(c) $e^x \geq 1 + x + \cdots + \frac{x^k}{k!} + \frac{x^{k+1}}{(k+1)!} \quad \Rightarrow \quad \frac{e^x}{x^k} \geq \frac{1}{x^k} + \frac{1}{x^{k-1}} + \cdots + \frac{1}{k!} + \frac{x}{(k+1)!} \geq \frac{x}{(k+1)!}$.

But $\lim\limits_{x \to \infty} \frac{x}{(k+1)!} = \infty$, so $\lim\limits_{x \to \infty} \frac{e^x}{x^k} = \infty$.

74. (a) The graph of g finally surpasses that of f at $x \approx 35.8$.

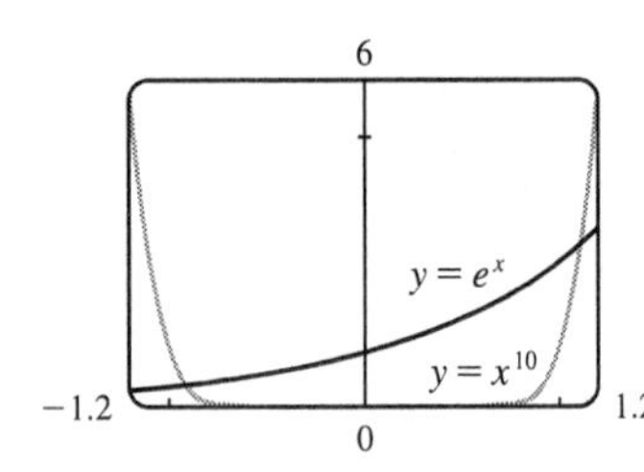

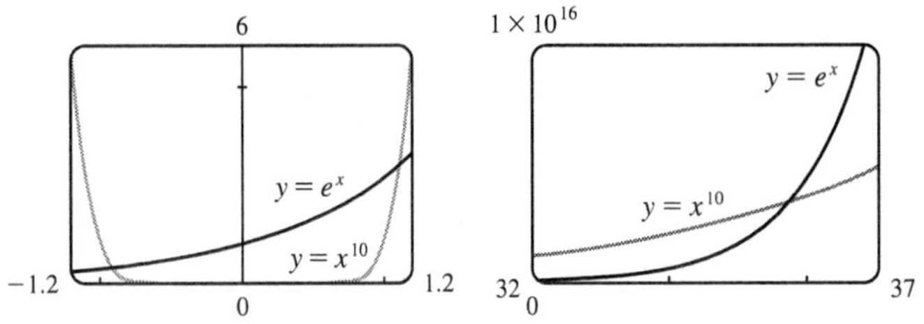

(b)

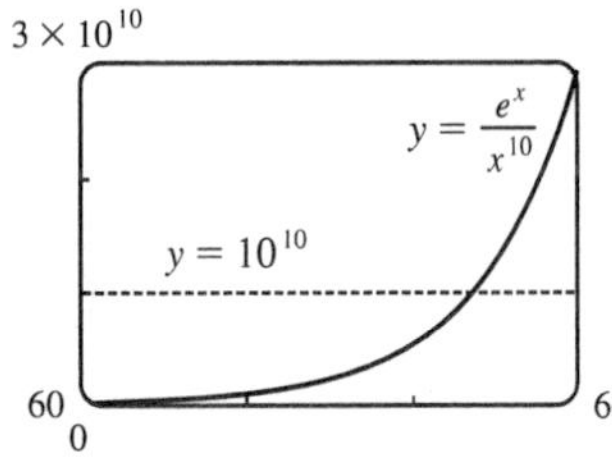

(c) From the graph in part (b), it seems that $e^x/x^{10} > 10^{10}$ whenever $x > 65$, approximately. So we can take $N \geq 65$.

5.4 General Logarithmic and Exponential Functions

1. (a) $a^x = e^{x \ln a}$

(b) The domain of $f(x) = a^x$ is $\mathbb{R}$.

(c) The range of $f(x) = a^x$ $(a \neq 1)$ is $(0, \infty)$.

(d) (i) See Figure 1. (ii) See Figure 3. (iii) See Figure 2.

2. (a) $\log_a x$ is the number y such that $a^y = x$.

(b) The domain of $f(x) = \log_a x$ is $(0, \infty)$.

(c) The range of $f(x) = \log_a x$ is $\mathbb{R}$.

(d) See Figure 7.

3. $5^{\sqrt{7}} = (e^{\ln 5})^{\sqrt{7}} = e^{\sqrt{7}\ln 5}$

4. $10^{x^2} = (e^{\ln 10})^{x^2} = e^{x^2 \ln 10}$

5. $(\cos x)^x = (e^{\ln \cos x})^x = e^{x \ln(\cos x)}$.

6. $x^{\cos x} = (e^{\ln x})^{\cos x} = e^{(\cos x)(\ln x)}$

7. (a) $\log_{10} 1000 = 3$ because $10^3 = 1000$.

(b) $\log_2 \frac{1}{16} = -4$ since $2^{-4} = \frac{1}{16}$. [*Or:* $\log_2 \frac{1}{16} = \log_2 2^{-4} = -4$]

8. (a) $\log_{10} 0.1 = -1$ since $10^{-1} = 0.1$.

(b) $\log_8 320 - \log_8 5 = \log_8 \frac{320}{5} = \log_8 64 = 2$ since $8^2 = 64$.

9. (a) $\log_{12} 3 + \log_{12} 48 = \log_{12}(3 \cdot 48) = \log_{12} 144 = 2$ since $12^2 = 144$.

(b) $\log_5 5^{\sqrt{2}} = \sqrt{2}$ by the cancellation property $\log_a a^x = x$.

[*Or:* $\log_5 5^{\sqrt{2}} = \sqrt{2}\log_5 5 = \sqrt{2} \cdot 1 = \sqrt{2}$]

10. (a) $\log_a \dfrac{1}{a} = -1$ since $a^{-1} = \dfrac{1}{a}$. [*Or:* $\log_a \dfrac{1}{a} = \log_a a^{-1} = -1$]

(b) $10^{(\log_{10} 4 + \log_{10} 7)} = 10^{\log_{10} 4} \cdot 10^{\log_{10} 7} = 4 \cdot 7 = 28$

[*Or:* $10^{(\log_{10} 4 + \log_{10} 7)} = 10^{\log_{10}(4 \cdot 7)} = 10^{\log_{10} 28} = 28$]

11. All of these graphs approach 0 as $x \to -\infty$, all of them pass through the point $(0, 1)$, and all of them are increasing and approach ∞ as $x \to \infty$. The larger the base, the faster the function increases for $x > 0$, and the faster it approaches 0 as $x \to -\infty$.

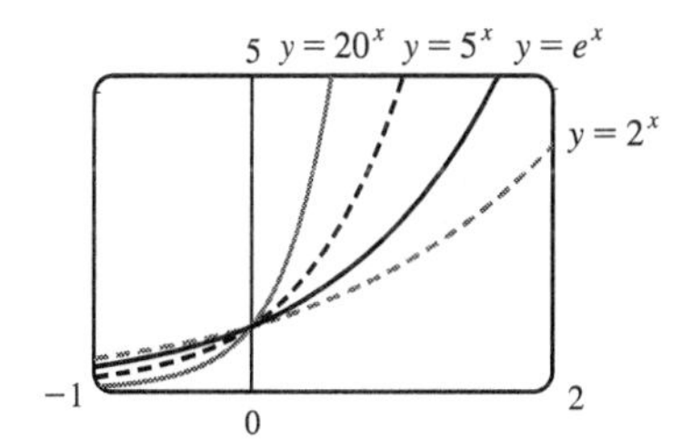

12. The functions with bases greater than 1 (3^x and 10^x) are increasing, while those with bases less than 1 $\left[\left(\frac{1}{3}\right)^x \text{ and } \left(\frac{1}{10}\right)^x\right]$ are decreasing. The graph of $\left(\frac{1}{3}\right)^x$ is the reflection of that of 3^x about the y-axis, and the graph of $\left(\frac{1}{10}\right)^x$ is the reflection of that of 10^x about the y-axis. The graph of 10^x increases more quickly than that of 3^x for $x > 0$, and approaches 0 faster as $x \to -\infty$.

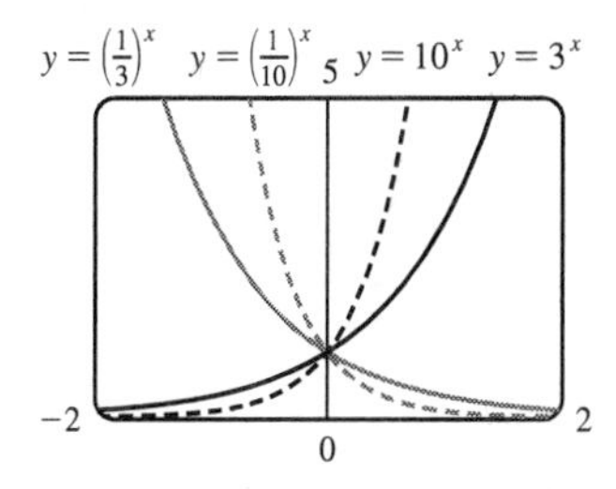

13. (a) $\log_{12} e = \dfrac{\ln e}{\ln 12} = \dfrac{1}{\ln 12} \approx 0.402430$

(b) $\log_6 13.54 = \dfrac{\ln 13.54}{\ln 6} \approx 1.454240$

(c) $\log_2 \pi = \dfrac{\ln \pi}{\ln 2} \approx 1.651496$

14. To graph the functions, we use $\log_2 x = \dfrac{\ln x}{\ln 2}$, $\log_4 x = \dfrac{\ln x}{\ln 4}$, etc. These graphs all approach $-\infty$ as $x \to 0^+$, and they all pass through the point $(1, 0)$. Also, they are all increasing, and all approach ∞ as $x \to \infty$. The smaller the base, the larger the rate of increase of the function (for $x > 1$) and the closer the approach to the y-axis (as $x \to 0^+$).

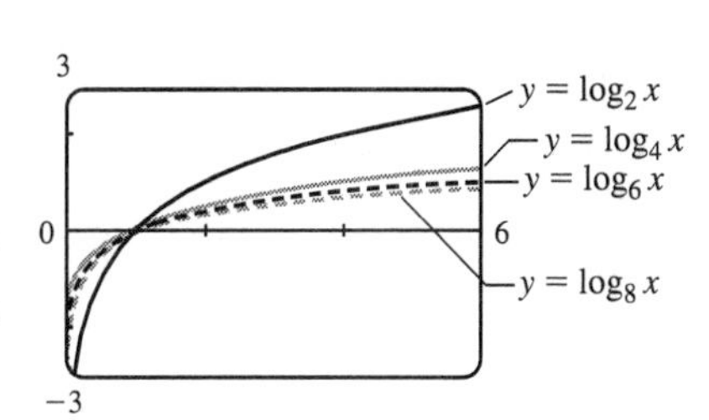

15. To graph these functions, we use $\log_{1.5} x = \dfrac{\ln x}{\ln 1.5}$ and $\log_{50} x = \dfrac{\ln x}{\ln 50}$. These graphs all approach $-\infty$ as $x \to 0^+$, and they all pass through the point $(1, 0)$. Also, they are all increasing, and all approach ∞ as $x \to \infty$. The functions with larger bases increase extremely slowly, and the ones with smaller bases do so somewhat more quickly. The functions with large bases approach the y-axis more closely as $x \to 0^+$.

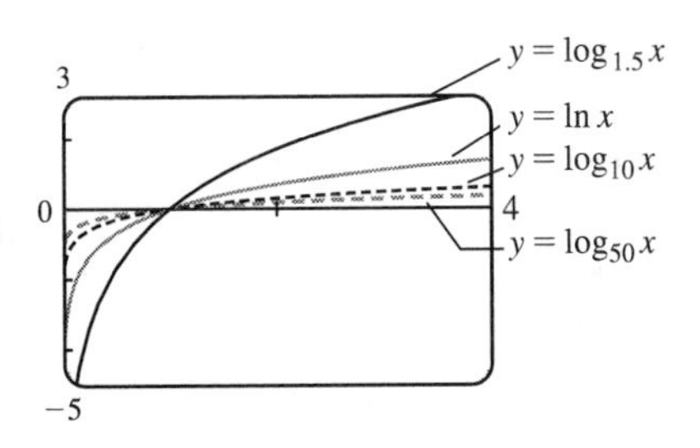

16. We see that the graph of $\ln x$ is the reflection of the graph of e^x about the line $y = x$, and that the graph of $\log_{10} x$ is the reflection of the graph of 10^x about the same line. The graph of 10^x increases more quickly than that of e^x. Also note that $\log_{10} x \to \infty$ as $x \to \infty$ more slowly than $\ln x$.

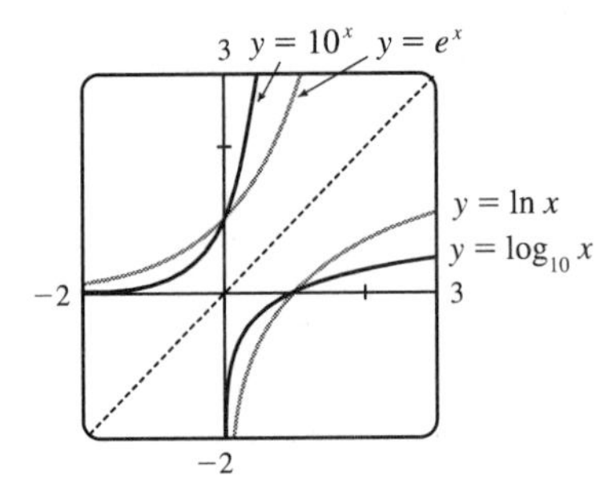

17. Use $y = Ca^x$ with the points $(1, 6)$ and $(3, 24)$. $6 = Ca^1$ $\left[C = \frac{6}{a}\right]$ and $24 = Ca^3$ $\Rightarrow$ $24 = \left(\dfrac{6}{a}\right)a^3$ $\Rightarrow$ $4 = a^2$ $\Rightarrow$ $a = 2$ [since $a > 0$] and $C = \frac{6}{2} = 3$. The function is $f(x) = 3 \cdot 2^x$.

18. Given the y-intercept $(0, 2)$, we have $y = Ca^x = 2a^x$. Using the point $\left(2, \frac{2}{9}\right)$ gives us $\frac{2}{9} = 2a^2$ $\Rightarrow$ $\frac{1}{9} = a^2$ $\Rightarrow$ $a = \frac{1}{3}$ [since $a > 0$]. The function is $f(x) = 2\left(\frac{1}{3}\right)^x$ or $f(x) = 2(3)^{-x}$.

19. (a) 2 ft $= 24$ in, $f(24) = 24^2$ in $= 576$ in $= 48$ ft. $g(24) = 2^{24}$ in $= 2^{24}/(12 \cdot 5280)$ mi ≈ 265 mi

(b) 3 ft $= 36$ in, so we need x such that $\log_2 x = 36$ $\Leftrightarrow$ $x = 2^{36} = 68{,}719{,}476{,}736$. In miles, this is

$$68{,}719{,}476{,}736 \text{ in} \cdot \frac{1 \text{ ft}}{12 \text{ in}} \cdot \frac{1 \text{ mi}}{5280 \text{ ft}} \approx 1{,}084{,}587.7 \text{ mi.}$$

20. We see from the graphs that for x less than about 1.8, $g(x) = 5^x > f(x) = x^5$, and then near the point $(1.8, 17.1)$ the curves intersect. Then $f(x) > g(x)$ from $x \approx 1.8$ until $x = 5$. At $(5, 3125)$ there is another point of intersection, and for $x > 5$ we see that $g(x) > f(x)$. In fact, g increases much more rapidly than f beyond that point.

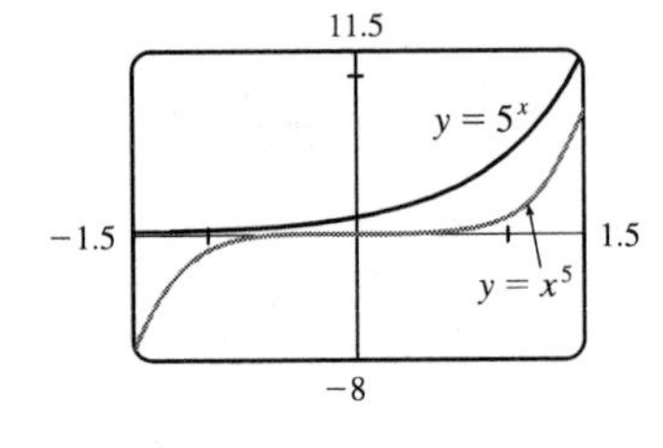

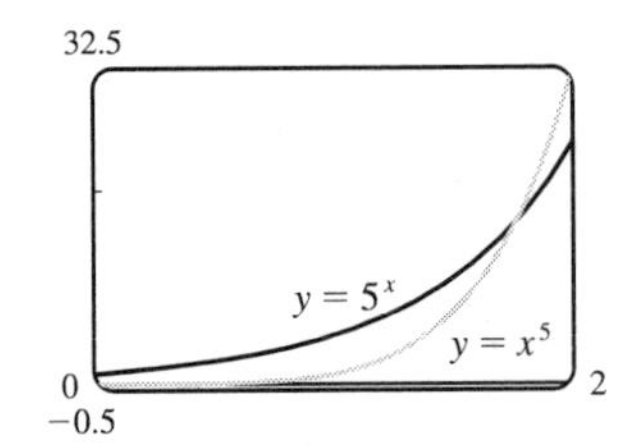

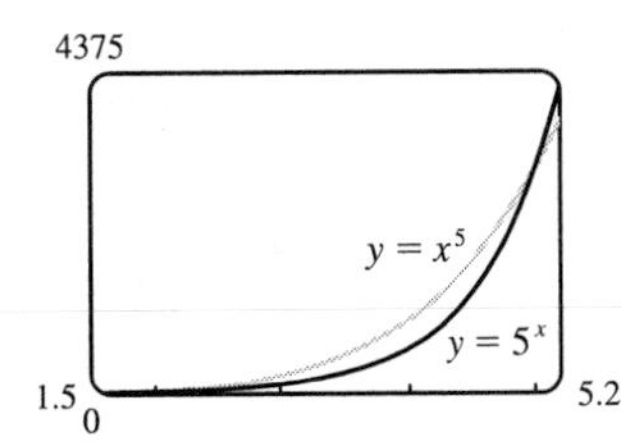

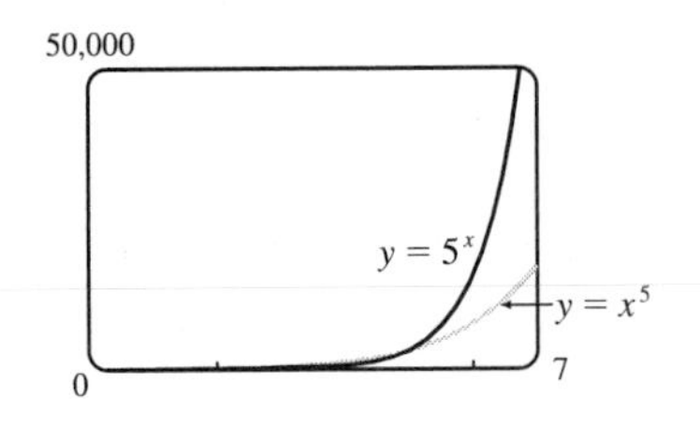

21. $\lim_{t\to\infty} 2^{-t^2} = \lim_{u\to-\infty} 2^u$ [where $u = -t^2$] $= 0$

22. Let $t = x^2 - 5x + 6$. As $x \to 3^+$, $t = (x-2)(x-3) \to 0^+$. $\lim_{x\to 3^+} \log_{10}\left(x^2 - 5x + 6\right) = \lim_{t\to 0^+} \log_{10} t = -\infty$ by (11).

23. $h(t) = t^3 - 3^t \Rightarrow h'(t) = 3t^2 - 3^t \ln 3$

24. $g(x) = x^4 4^x \Rightarrow g'(x) = x^4 4^x \ln 4 + 4^x \cdot 4x^3 = x^3 4^x (x \ln 4 + 4)$

25. Using Formula 4 and the Chain Rule, $y = 5^{-1/x} \Rightarrow y' = 5^{-1/x}(\ln 5)\left[-1 \cdot \left(-x^{-2}\right)\right] = 5^{-1/x}(\ln 5)/x^2$.

26. $y = 10^{\tan\theta} \Rightarrow y' = 10^{\tan\theta}(\ln 10)(\sec^2\theta)$

27. $f(u) = (2^u + 2^{-u})^{10} \Rightarrow$

$$f'(u) = 10(2^u + 2^{-u})^9 \frac{d}{du}(2^u + 2^{-u}) = 10(2^u + 2^{-u})^9\left[2^u \ln 2 + 2^{-u}\ln 2 \cdot (-1)\right]$$
$$= 10 \ln 2 (2^u + 2^{-u})^9 (2^u - 2^{-u})$$

28. $y = 2^{3^{x^2}} \Rightarrow y' = 2^{3^{x^2}}(\ln 2)\frac{d}{dx}\left(3^{x^2}\right) = 2^{3^{x^2}}(\ln 2) 3^{x^2}(\ln 3)(2x)$

29. $f(x) = \log_3(x^2 - 4) \Rightarrow f'(x) = \dfrac{1}{(x^2-4)\ln 3}(2x) = \dfrac{2x}{(x^2-4)\ln 3}$

30. $f(x) = \log_{10}\left(\dfrac{x}{x-1}\right) = \log_{10} x - \log_{10}(x-1) \Rightarrow f'(x) = \dfrac{1}{x\ln 10} - \dfrac{1}{(x-1)\ln 10}$ or $-\dfrac{1}{x(x-1)\ln 10}$

31. $y = x^x \Rightarrow \ln y = \ln x^x \Rightarrow \ln y = x \ln x \Rightarrow y'/y = x(1/x) + (\ln x)\cdot 1 \Rightarrow y' = y(1 + \ln x) \Rightarrow$
$y' = x^x(1 + \ln x)$

32. $y = x^{1/x} \Rightarrow \ln y = \dfrac{1}{x}\ln x \Rightarrow \dfrac{y'}{y} = -\dfrac{1}{x^2}\ln x + \dfrac{1}{x}\left(\dfrac{1}{x}\right) \Rightarrow y' = x^{1/x}\dfrac{1 - \ln x}{x^2}$

33. $y = x^{\sin x} \Rightarrow \ln y = \sin x \ln x \Rightarrow \dfrac{y'}{y} = \cos x \ln x + \dfrac{\sin x}{x} \Rightarrow y' = x^{\sin x}\left(\cos x \ln x + \dfrac{\sin x}{x}\right)$

34. $y = (\sin x)^x \Rightarrow \ln y = x\ln(\sin x) \Rightarrow y'/y = \ln(\sin x) + x(\cos x)/(\sin x) \Rightarrow y' = (\sin x)^x[\ln(\sin x) + x \cot x]$

35. $y = (\ln x)^x \Rightarrow \ln y = x \ln\ln x \Rightarrow \dfrac{y'}{y} = \ln\ln x + x \cdot \dfrac{1}{\ln x}\cdot\dfrac{1}{x} \Rightarrow y' = (\ln x)^x\left(\ln\ln x + \dfrac{1}{\ln x}\right)$

36. $y = x^{\ln x} \Rightarrow \ln y = \ln x \ln x = (\ln x)^2 \Rightarrow \dfrac{y'}{y} = 2\ln x\left(\dfrac{1}{x}\right) \Rightarrow y' = x^{\ln x}\left(\dfrac{2\ln x}{x}\right)$

37. $y = x^{e^x} \Rightarrow \ln y = e^x \ln x \Rightarrow \dfrac{y'}{y} = e^x \ln x + \dfrac{e^x}{x} \Rightarrow y' = x^{e^x} e^x\left(\ln x + \dfrac{1}{x}\right)$

38. $y = (\ln x)^{\cos x} \Rightarrow \ln y = \cos x \ln(\ln x) \Rightarrow \dfrac{y'}{y} = \cos x \cdot \dfrac{1}{\ln x}\cdot\dfrac{1}{x} + (\ln\ln x)(-\sin x) \Rightarrow$
$y' = (\ln x)^{\cos x}\left(\dfrac{\cos x}{x \ln x} - \sin x \ln\ln x\right)$

39. $y = 10^x \Rightarrow y' = 10^x \ln 10$, so at $(1, 10)$, the slope of the tangent line is $10^1 \ln 10 = 10 \ln 10$, and its equation is $y - 10 = 10\ln 10(x - 1)$, or $y = (10\ln 10)x + 10(1 - \ln 10)$.

40. $f(x) = x^{\cos x} = e^{\ln x \cos x} \Rightarrow$

$$f'(x) = e^{\ln x \cos x}\left[\ln x(-\sin x) + \cos x\left(\frac{1}{x}\right)\right]$$
$$= x^{\cos x}\left[\frac{\cos x}{x} - \sin x \ln x\right]$$

This is reasonable, because the graph shows that f increases when $f'(x)$ is positive.

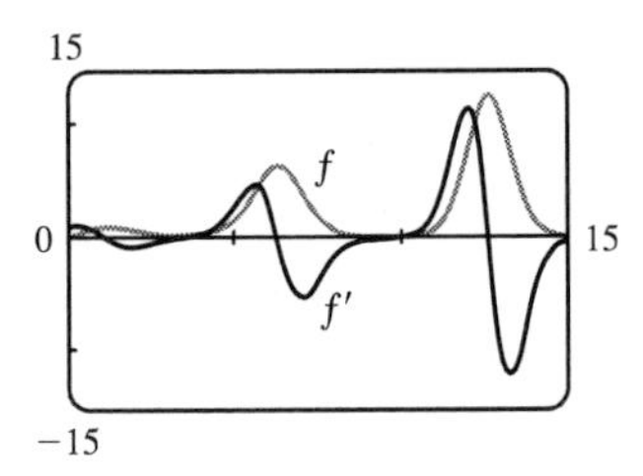

41. $\displaystyle\int_1^2 10^t\,dt = \left[\frac{10^t}{\ln 10}\right]_1^2 = \frac{10^2}{\ln 10} - \frac{10^1}{\ln 10} = \frac{100-10}{\ln 10} = \frac{90}{\ln 10}$

42. Let $v = -2u$. Then $dv = -2\,du$ and

$$\int_0^1 4^{-2u}\,du = \int_0^{-2} 4^v\left(-\frac{1}{2}\right)dv = -\frac{1}{2}\left[\frac{4^v}{\ln 4}\right]_0^{-2} = -\frac{1}{2\ln 4}\left(4^{-2} - 4^0\right) = -\frac{1}{2\ln 2^2}\left(\frac{1}{16} - 1\right)$$
$$= -\frac{1}{4\ln 2}\left(-\frac{15}{16}\right) = \frac{15}{64\ln 2}$$

43. $\displaystyle\int \frac{\log_{10} x}{x}\,dx = \int \frac{(\ln x)/(\ln 10)}{x}\,dx = \frac{1}{\ln 10}\int \frac{\ln x}{x}\,dx$. Now put $u = \ln x$, so $du = \dfrac{1}{x}\,dx$, and the expression becomes $\displaystyle\frac{1}{\ln 10}\int u\,du = \frac{1}{\ln 10}\left(\tfrac{1}{2}u^2 + C_1\right) = \frac{1}{2\ln 10}(\ln x)^2 + C.$

Or: The substitution $u = \log_{10} x$ gives $du = \dfrac{dx}{x\ln 10}$ and we get $\displaystyle\int \frac{\log_{10} x}{x}\,dx = \tfrac{1}{2}\ln 10(\log_{10} x)^2 + C.$

44. $\int \left(x^5 + 5^x\right)dx = \frac{1}{6}x^6 + \frac{1}{\ln 5}5^x + C$

45. Let $u = \sin\theta$. Then $du = \cos\theta\,d\theta$ and $\displaystyle\int 3^{\sin\theta}\cos\theta\,d\theta = \int 3^u\,du = \frac{3^u}{\ln 3} + C = \frac{1}{\ln 3}3^{\sin\theta} + C.$

46. Let $u = 2^x + 1$. Then $du = 2^x\ln 2\,dx$, so $\displaystyle\int \frac{2^x}{2^x+1}\,dx = \int \frac{1}{u}\frac{du}{\ln 2} = \frac{1}{\ln 2}\ln|u| + C = \frac{1}{\ln 2}\ln(2^x+1) + C.$

47. $\displaystyle y = \frac{10^x}{10^x+1} \Leftrightarrow (10^x+1)y = 10^x \Leftrightarrow 10^x\cdot y + y = 10^x \Leftrightarrow y = 10^x - 10^x y \Leftrightarrow$

$\displaystyle y = 10^x(1-y) \Leftrightarrow 10^x = \frac{y}{1-y} \Leftrightarrow \log_{10}10^x = \log_{10}\left(\frac{y}{1-y}\right) \Leftrightarrow x = \log_{10} y - \log_{10}(1-y).$

Interchange x and y: $y = \log_{10} x - \log_{10}(1-x)$ is the inverse function.

48. $\displaystyle x^y = y^x \Rightarrow y\ln x = x\ln y \Rightarrow y\cdot\frac{1}{x} + (\ln x)\cdot y' = x\cdot\frac{1}{y}\cdot y' + \ln y \Rightarrow y'\ln x - \frac{x}{y}y' = \ln y - \frac{y}{x} \Rightarrow$

$$y' = \frac{\ln y - y/x}{\ln x - x/y}$$

49. $\displaystyle\lim_{x\to 0+} x^{-\ln x} = \lim_{x\to 0+}\left(e^{\ln x}\right)^{-\ln x} = \lim_{x\to 0+} e^{-(\ln x)^2} = 0$ since $-(\ln x)^2 \to -\infty$ as $x \to 0^+$.

50. (a) $I(x) = I_0a^x \Rightarrow I'(x) = I_0(\ln a)a^x = (I_0a^x)\ln a = I(x)\ln a$

(b) We substitute $I_0 = 8$, $a = 0.38$ and $x = 20$ into the first expression for $I'(x)$ above:

$I'(20) = 8(\ln 0.38)(0.38)^{20} \approx -3.05 \times 10^{-8}$.

(c) The average value of the function $I(x)$ between $x = 0$ and $x = 20$ is

$$\frac{\int_0^{20} I(x)\,dx}{20 - 0} = \frac{1}{20}\int_0^{20} 8(0.38)^x\,dx = \frac{2}{5}\left[\frac{(0.38)^x}{\ln 0.38}\right]_0^{20} = \frac{2(0.38^{20} - 1)}{5\ln 0.38} \approx 0.41.$$

51. Using Definition 1 and the second law of exponents for e^x, we have $a^{x-y} = e^{(x-y)\ln a} = e^{x\ln a - y\ln a} = \dfrac{e^{x\ln a}}{e^{y\ln a}} = \dfrac{a^x}{a^y}$.

52. Using Definition 1, the first law of logarithms, and the first law of exponents for e^x, we have

$(ab)^x = e^{x\ln(ab)} = e^{x(\ln a + \ln b)} = e^{x\ln a + x\ln b} = e^{x\ln a}e^{x\ln b} = a^x b^x$.

53. Let $\log_a x = r$ and $\log_a y = s$. Then $a^r = x$ and $a^s = y$.

(a) $xy = a^r a^s = a^{r+s} \Rightarrow \log_a(xy) = r + s = \log_a x + \log_a y$

(b) $\dfrac{x}{y} = \dfrac{a^r}{a^s} = a^{r-s} \Rightarrow \log_a \dfrac{x}{y} = r - s = \log_a x - \log_a y$ $\dfrac{x}{y} = \dfrac{a^r}{a^s} = a^{r-s} \Rightarrow \log_a \dfrac{x}{y} = r - s = \log_a x - \log_a y$

(c) $x^y = (a^r)^y = a^{ry} \Rightarrow \log_a(x^y) = ry = y\log_a x$

$x^y = (a^r)^y = a^{ry} \Rightarrow \log_a(x^y) = ry = y\log_a x$

54. Let $m = n/x$. Then $n = xm$, and as $n \to \infty$, $m \to \infty$.

Therefore, $\displaystyle\lim_{n\to\infty}\left(1 + \frac{x}{n}\right)^n = \lim_{m\to\infty}\left(1 + \frac{1}{m}\right)^{mx} = \left[\lim_{m\to\infty}\left(1 + \frac{1}{m}\right)^m\right]^x = e^x$ by Equation 9.

5.5 Exponential Growth and Decay

1. The relative growth rate is $\dfrac{1}{P}\dfrac{dP}{dt} = 0.7944$, so $\dfrac{dP}{dt} = 0.7944P$ and, by Theorem 2, $P(t) = P(0)e^{0.7944t} = 2e^{0.7944t}$.

Thus, $P(6) = 2e^{0.7944(6)} \approx 234.99$ or about 235 members.

2. (a) By Theorem 2, $P(t) = P(0)e^{kt} = 60e^{kt}$. In 20 minutes ($\frac{1}{3}$ hour), there are 120 cells, so $P\left(\frac{1}{3}\right) = 60e^{k/3} = 120 \Rightarrow$

$e^{k/3} = 2 \Rightarrow k/3 = \ln 2 \Rightarrow k = 3\ln 2 = \ln(2^3) = \ln 8$.

(b) $P(t) = 60e^{(\ln 8)t} = 60 \cdot 8^t$

(c) $P(8) = 60 \cdot 8^8 = 60 \cdot 2^{24} = 1{,}006{,}632{,}960$

(d) $dP/dt = kP \Rightarrow P'(8) = kP(8) = (\ln 8)P(8) \approx 2.093$ billion cells/h

(e) $P(t) = 20{,}000 \Rightarrow 60 \cdot 8^t = 20{,}000 \Rightarrow 8^t = 1000/3 \Rightarrow t\ln 8 = \ln(1000/3) \Rightarrow$

$t = \dfrac{\ln(1000/3)}{\ln 8} \approx 2.79$ h

3. (a) By Theorem 2, $P(t) = P(0)e^{kt} = 100e^{kt}$. Now $P(1) = 100e^{k(1)} = 420 \Rightarrow e^k = \frac{420}{100} \Rightarrow k = \ln 4.2$.

So $P(t) = 100e^{(\ln 4.2)t} = 100(4.2)^t$.

(b) $P(3) = 100(4.2)^3 = 7408.8 \approx 7409$ bacteria

(c) $dP/dt = kP \Rightarrow P'(3) = k \cdot P(3) = (\ln 4.2)\left(100(4.2)^3\right)$ [from part (a)] $\approx 10{,}632$ bacteria/hour

(d) $P(t) = 100(4.2)^t = 10{,}000 \Rightarrow (4.2)^t = 100 \Rightarrow t = (\ln 100)/(\ln 4.2) \approx 3.2$ hours

4. (a) $y(t) = y(0)e^{kt} \Rightarrow y(2) = y(0)e^{2k} = 600$, $y(8) = y(0)e^{8k} = 75{,}000$. Dividing these equations, we get $e^{8k}/e^{2k} = 75{,}000/600 \Rightarrow e^{6k} = 125 \Rightarrow 6k = \ln 125 = \ln 5^3 = 3\ln 5 \Rightarrow k = \frac{3}{6}\ln 5 = \frac{1}{2}\ln 5$. Thus, $y(0) = 600/e^{2k} = 600/e^{\ln 5} = \frac{600}{5} = 120$.

(b) $y(t) = y(0)e^{kt} = 120e^{(\ln 5)t/2}$ or $y = 120 \cdot 5^{t/2}$

(c) $y(5) = 120 \cdot 5^{5/2} = 120 \cdot 25\sqrt{5} = 3000\sqrt{5} \approx 6708$ bacteria.

(d) $y(t) = 120 \cdot 5^{t/2} \Rightarrow y'(t) = 120 \cdot 5^{t/2} \cdot \ln 5 \cdot \frac{1}{2} = 60 \cdot \ln 5 \cdot 5^{t/2}$.
$y'(5) = 60 \cdot \ln 5 \cdot 5^{5/2} = 60 \cdot \ln 5 \cdot 25\sqrt{5} \approx 5398$ bacteria/hour.

(e) $y(t) = 200{,}000 \Leftrightarrow 120e^{(\ln 5)t/2} = 200{,}000 \Leftrightarrow e^{(\ln 5)t/2} = \frac{5000}{3} \Leftrightarrow (\ln 5)t/2 = \ln\frac{5000}{3} \Leftrightarrow$ $t = \left(2\ln\frac{5000}{3}\right)/\ln 5 \approx 9.2$ h.

5. (a) Let the population (in millions) in the year t be $P(t)$. Since the initial time is the year 1750, we substitute $t - 1750$ for t in Theorem 2, so the exponential model gives $P(t) = P(1750)e^{k(t-1750)}$. Then $P(1800) = 980 = 790e^{k(1800-1750)} \Rightarrow$ $\frac{980}{790} = e^{k(50)} \Rightarrow \ln\frac{980}{790} = 50k \Rightarrow k = \frac{1}{50}\ln\frac{980}{790} \approx 0.0043104$. So with this model, we have $P(1900) = 790e^{k(1900-1750)} \approx 1508$ million and $P(1950) = 790e^{k(1950-1750)} \approx 1871$ million. Both of these estimates are much too low.

(b) In this case, the exponential model gives $P(t) = P(1850)e^{k(t-1850)} \Rightarrow P(1900) = 1650 = 1260e^{k(1900-1850)} \Rightarrow$ $\ln\frac{1650}{1260} = k(50) \Rightarrow k = \frac{1}{50}\ln\frac{1650}{1260} \approx 0.005393$. So with this model, we estimate $P(1950) = 1260e^{k(1950-1850)} \approx 2161$ million. This is still too low, but closer than the estimate of $P(1950)$ in part (a).

(c) The exponential model gives $P(t) = P(1900)e^{k(t-1900)} \Rightarrow P(1950) = 2560 = 1650e^{k(1950-1900)} \Rightarrow$ $\ln\frac{2560}{1650} = k(50) \Rightarrow k = \frac{1}{50}\ln\frac{2560}{1650} \approx 0.008785$. With this model, we estimate $P(2000) = 1650e^{k(2000-1900)} \approx 3972$ million. This is much too low. The discrepancy is explained by the wars in the first part of the 20th century and the lower mortality rate in the latter part of the century due to advances in medical science. The exponential model assumes, among other things, that the birth and mortality rates will remain constant.

6. (a) Let $P(t)$ be the population (in millions) in the year t. Since the initial time is the year 1900, we substitute $t - 1900$ for t in Theorem 2, and find that the exponential model gives $P(t) = P(1900)e^{k(t-1900)} \Rightarrow$ $P(1910) = 92 = 76e^{k(1910-1900)} \Rightarrow k = \frac{1}{10}\ln\frac{92}{76} \approx 0.0191$. With this model, we estimate $P(2000) = 76e^{k(2000-1900)} \approx 514$ million. This estimate is much too high. The discrepancy is explained by the fact that, between the years 1900 and 1910, an enormous number of immigrants (compared to the total population) came to the United States. Since that time, immigration (as a proportion of total population) has been much lower. Also, the birth rate in the United States has declined since the turn of the 20th century. So our calculation of the constant k was based partly on factors which no longer exist.

(b) Substituting $t - 1980$ for t in Theorem 2, we find that the exponential model gives $P(t) = P(1980)e^{k(t-1980)} \Rightarrow$ $P(1990) = 250 = 227e^{k(1990-1980)} \Rightarrow k = \frac{1}{10}\ln\frac{250}{227} \approx 0.00965$. With this model, we estimate $P(2000) = 227e^{k(2000-1980)} \approx 275.3$ million. This is quite accurate. The further estimates are $P(2010) = 227e^{30k} \approx 303$ million and $P(2020) = 227e^{40k} \approx 334$ million.

(c) 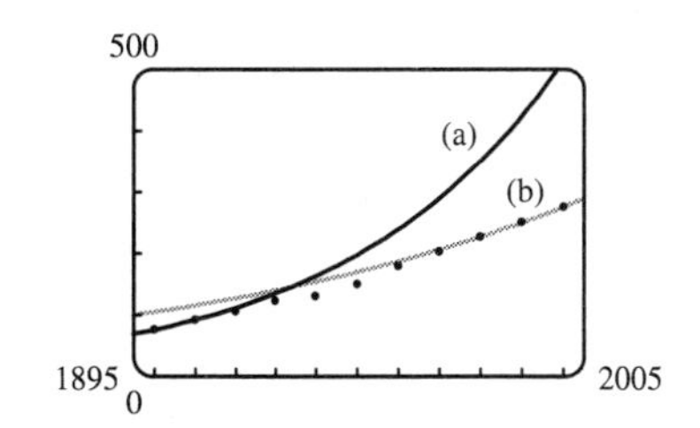

The model in part (a) is quite inaccurate after 1910 (off by 5 million in 1920 and 12 million in 1930). The model in part (b) is more accurate (which is not surprising, since it is based on more recent information).

7. (a) If $y = [N_2O_5]$ then by Theorem 2, $\dfrac{dy}{dt} = -0.0005y \;\Rightarrow\; y(t) = y(0)e^{-0.0005t} = Ce^{-0.0005t}$.

(b) $y(t) = Ce^{-0.0005t} = 0.9C \;\Rightarrow\; e^{-0.0005t} = 0.9 \;\Rightarrow\; -0.0005t = \ln 0.9 \;\Rightarrow\; t = -2000\ln 0.9 \approx 211$ s

8. (a) The mass remaining after t days is $y(t) = y(0)e^{kt} = 800e^{kt}$. Since the half-life is 5.0 days, $y(5) = 800e^{5k} = 400 \;\Rightarrow$ $e^{5k} = \frac{1}{2} \;\Rightarrow\; 5k = \ln\frac{1}{2} \;\Rightarrow\; k = -(\ln 2)/5$, so $y(t) = 800e^{-(\ln 2)t/5} = 800 \cdot 2^{-t/5}$.

(b) $y(30) = 800 \cdot 2^{-30/5} = 12.5$ mg

(c) $800e^{-(\ln 2)t/5} = 1 \;\Leftrightarrow\; -(\ln 2)\frac{t}{5} = \ln\frac{1}{800} = -\ln 800 \;\Leftrightarrow$ $t = 5\,\frac{\ln 800}{\ln 2} \approx 48$ days

(d) 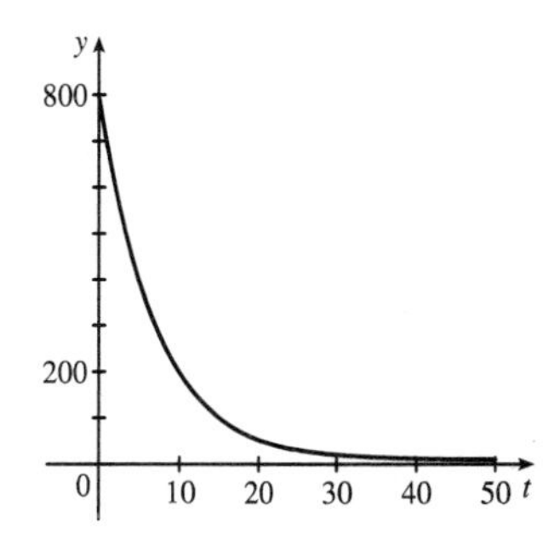

9. (a) If $y(t)$ is the mass (in mg) remaining after t years, then $y(t) = y(0)e^{kt} = 100e^{kt}$. $y(30) = 100e^{30k} = \frac{1}{2}(100) \;\Rightarrow$ $e^{30k} = \frac{1}{2} \;\Rightarrow\; k = -(\ln 2)/30 \;\Rightarrow\; y(t) = 100e^{-(\ln 2)t/30} = 100 \cdot 2^{-t/30}$

(b) $y(100) = 100 \cdot 2^{-100/30} \approx 9.92$ mg

(c) $100e^{-(\ln 2)t/30} = 1 \;\Rightarrow\; -(\ln 2)t/30 = \ln\frac{1}{100} \;\Rightarrow\; t = -30\,\frac{\ln 0.01}{\ln 2} \approx 199.3$ years

10. (a) If $y(t)$ is the mass after t days and $y(0) = A$, then $y(t) = Ae^{kt}$. $y(1) = Ae^{k} = 0.945A \;\Rightarrow$ $e^{k} = 0.945 \;\Rightarrow\; k = \ln 0.945$. Then $Ae^{(\ln 0.945)t} = \frac{1}{2}A \;\Leftrightarrow\; \ln e^{(\ln 0.945)t} = \ln\frac{1}{2} \;\Leftrightarrow\; (\ln 0.945)t = \ln\frac{1}{2} \;\Leftrightarrow$ $t = -\dfrac{\ln 2}{\ln 0.945} \approx 12.25$ years.

(b) $Ae^{(\ln 0.945)t} = 0.20A \;\Leftrightarrow\; (\ln 0.945)t = \ln\frac{1}{5} \;\Leftrightarrow\; t = -\dfrac{\ln 5}{\ln 0.945} \approx 28.45$ years

11. Let $y(t)$ be the level of radioactivity. Thus, $y(t) = y(0)e^{-kt}$ and k is determined by using the half-life: $y(5730) = \frac{1}{2}y(0) \;\Rightarrow\; y(0)e^{-k(5730)} = \frac{1}{2}y(0) \;\Rightarrow\; e^{-5730k} = \frac{1}{2} \;\Rightarrow\; -5730k = \ln\frac{1}{2} \;\Rightarrow$ $k = -\dfrac{\ln\frac{1}{2}}{5730} = \dfrac{\ln 2}{5730}$. If 74% of the ^{14}C remains, then we know that $y(t) = 0.74y(0) \;\Rightarrow\; 0.74 = e^{-t(\ln 2)/5730} \;\Rightarrow$ $\ln 0.74 = -\dfrac{t\ln 2}{5730} \;\Rightarrow\; t = -\dfrac{5730(\ln 0.74)}{\ln 2} \approx 2489 \approx 2500$ years.

12. From the information given, we know that $\dfrac{dy}{dx} = 2y \;\Rightarrow\; y = Ce^{2x}$ by Theorem 2. To calculate C we use the point $(0, 5)$: $5 = Ce^{2(0)} \;\Rightarrow\; C = 5$. Thus, the equation of the curve is $y = 5e^{2x}$.

13. (a) Using Newton's Law of Cooling, $\dfrac{dT}{dt} = k(T - T_s)$, we have $\dfrac{dT}{dt} = k(T - 75)$. Now let $y = T - 75$, so $y(0) = T(0) - 75 = 185 - 75 = 110$, so y is a solution of the initial-value problem $dy/dt = ky$ with $y(0) = 110$ and by Theorem 2 we have $y(t) = y(0)e^{kt} = 110e^{kt}$.

$y(30) = 110e^{30k} = 150 - 75 \;\Rightarrow\; e^{30k} = \frac{75}{110} = \frac{15}{22} \;\Rightarrow\; k = \frac{1}{30}\ln\frac{15}{22}$, so $y(t) = 110e^{\frac{1}{30}t\ln\left(\frac{15}{22}\right)}$ and $y(45) = 110e^{\frac{45}{30}\ln\left(\frac{15}{22}\right)} \approx 62\,^\circ\text{F}$. Thus, $T(45) \approx 62 + 75 = 137\,^\circ\text{F}$.

(b) $T(t) = 100 \;\Rightarrow\; y(t) = 25$. $y(t) = 110e^{\frac{1}{30}t\ln\left(\frac{15}{22}\right)} = 25 \;\Rightarrow\; e^{\frac{1}{30}t\ln\left(\frac{15}{22}\right)} = \frac{25}{110} \;\Rightarrow\; \frac{1}{30}t\ln\frac{15}{22} = \ln\frac{25}{110} \;\Rightarrow$

$$t = \frac{30\ln\frac{25}{110}}{\ln\frac{15}{22}} \approx 116\text{ min.}$$

14. (a) Let $T(t)$ = temperature after t minutes. Newton's Law of Cooling implies that $\dfrac{dT}{dt} = k(T - 5)$. Let $y(t) = T(t) - 5$.

Then $\dfrac{dy}{dt} = ky$, so $y(t) = y(0)e^{kt} = 15e^{kt} \;\Rightarrow\; T(t) = 5 + 15e^{kt} \;\Rightarrow\; T(1) = 5 + 15e^{k} = 12 \;\Rightarrow\; e^{k} = \frac{7}{15} \;\Rightarrow$ $k = \ln\frac{7}{15}$, so $T(t) = 5 + 15e^{\ln(7/15)t}$ and $T(2) = 5 + 15e^{2\ln(7/15)} \approx 8.3°\text{C}$.

(b) $5 + 15e^{\ln(7/15)t} = 6$ when $e^{\ln(7/15)t} = \frac{1}{15} \;\Rightarrow\; \ln\left(\frac{7}{15}\right)t = \ln\frac{1}{15} \;\Rightarrow\; t = \dfrac{\ln\frac{1}{15}}{\ln\frac{7}{15}} \approx 3.6\text{ min.}$

15. $\dfrac{dT}{dt} = k(T - 20)$. Letting $y = T - 20$, we get $\dfrac{dy}{dt} = ky$, so $y(t) = y(0)e^{kt}$. $y(0) = T(0) - 20 = 5 - 20 = -15$, so $y(25) = y(0)e^{25k} = -15e^{25k}$, and $y(25) = T(25) - 20 = 10 - 20 = -10$, so $-15e^{25k} = -10 \;\Rightarrow\; e^{25k} = \frac{2}{3}$. Thus, $25k = \ln\left(\frac{2}{3}\right)$ and $k = \frac{1}{25}\ln\left(\frac{2}{3}\right)$, so $y(t) = y(0)e^{kt} = -15e^{(1/25)\ln(2/3)t}$. More simply, $e^{25k} = \frac{2}{3} \;\Rightarrow\; e^{k} = \left(\frac{2}{3}\right)^{1/25} \;\Rightarrow$ $e^{kt} = \left(\frac{2}{3}\right)^{t/25} \;\Rightarrow\; y(t) = -15\cdot\left(\frac{2}{3}\right)^{t/25}$.

(a) $T(50) = 20 + y(50) = 20 - 15\cdot\left(\frac{2}{3}\right)^{50/25} = 20 - 15\cdot\left(\frac{2}{3}\right)^2 = 20 - \frac{20}{3} = 13.\bar{3}\,^\circ\text{C}$

(b) $15 = T(t) = 20 + y(t) = 20 - 15\cdot\left(\frac{2}{3}\right)^{t/25} \;\Rightarrow\; 15\cdot\left(\frac{2}{3}\right)^{t/25} = 5 \;\Rightarrow\; \left(\frac{2}{3}\right)^{t/25} = \frac{1}{3} \;\Rightarrow$ $(t/25)\ln\left(\frac{2}{3}\right) = \ln\left(\frac{1}{3}\right) \;\Rightarrow\; t = 25\ln\left(\frac{1}{3}\right)/\ln\left(\frac{2}{3}\right) \approx 67.74\text{ min.}$

16. $\dfrac{dT}{dt} = k(T - 20)$. Let $y = T - 20$. Then $\dfrac{dy}{dt} = ky$, so $y(t) = y(0)e^{kt}$. $y(0) = T(0) - 20 = 95 - 20 = 75$, so $y(t) = 75e^{kt}$. When $T(t) = 70$, $\dfrac{dT}{dt} = -1°\text{C/min}$. Equivalently, $\dfrac{dy}{dt} = -1$ when $y(t) = 50$. Thus, $-1 = \dfrac{dy}{dt} = ky(t) = 50k$ and $50 = y(t) = 75e^{kt}$. The first relation implies $k = -1/50$, so the second relation says $50 = 75e^{-t/50}$. Thus, $e^{-t/50} = \frac{2}{3} \;\Rightarrow\; -t/50 = \ln\left(\frac{2}{3}\right) \;\Rightarrow\; t = -50\ln\left(\frac{2}{3}\right) \approx 20.27\text{ min.}$

17. (a) Let $P(h)$ be the pressure at altitude h. Then $dP/dh = kP \;\Rightarrow\; P(h) = P(0)e^{kh} = 101.3e^{kh}$.

$P(1000) = 101.3e^{1000k} = 87.14 \;\Rightarrow\; 1000k = \ln\left(\frac{87.14}{101.3}\right) \;\Rightarrow\; k = \frac{1}{1000}\ln\left(\frac{87.14}{101.3}\right) \;\Rightarrow$

$P(h) = 101.3\,e^{\frac{1}{1000}h\ln\left(\frac{87.14}{101.3}\right)}$, so $P(3000) = 101.3e^{3\ln\left(\frac{87.14}{101.3}\right)} \approx 64.5\text{ kPa}$.

(b) $P(6187) = 101.3\,e^{\frac{6187}{1000}\ln\left(\frac{87.14}{101.3}\right)} \approx 39.9\text{ kPa}$

18. (a) Using $A = A_0\left(1 + \frac{r}{n}\right)^{nt}$ with $A_0 = 500$, $r = 0.14$, and $t = 2$, we have:

(i) Annually: $n = 1$; $A = 500\left(1 + \frac{0.14}{1}\right)^{1 \cdot 2} = \$649.80$

(ii) Quarterly: $n = 4$; $A = 500\left(1 + \frac{0.14}{4}\right)^{4 \cdot 2} = \$658.40$

(iii) Monthly: $n = 12$; $A = 500\left(1 + \frac{0.14}{12}\right)^{12 \cdot 2} = \$660.49$

(iv) Daily: $n = 365$; $A = 500\left(1 + \frac{0.14}{365}\right)^{365 \cdot 2} = \$661.53$

(v) Hourly: $n = 365 \cdot 24$; $A = 500\left(1 + \frac{0.14}{365 \cdot 24}\right)^{365 \cdot 24 \cdot 2} = \$661.56$

(vi) Continuously: $A = 500e^{(0.14)2} = \$661.56$

(b)

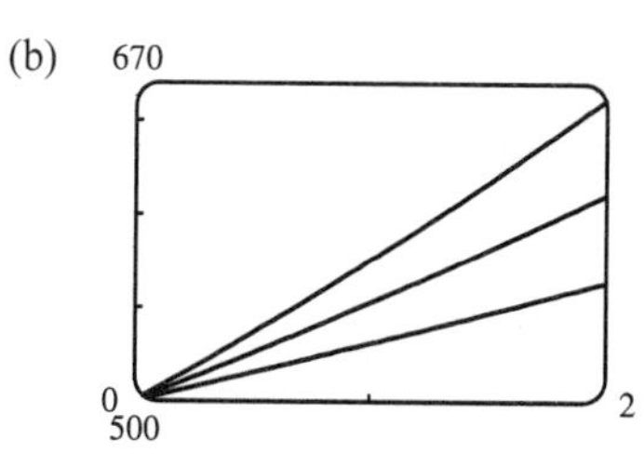

$A_{0.14}(2) = \$661.56$, $A_{0.10}(2) = \$610.70$, and $A_{0.06}(2) = \$563.75$.

19. Using $A = A_0\left(1 + \frac{r}{n}\right)^{nt}$ with $A_0 = 3000$, $r = 0.05$, and $t = 5$, we have:

(a) Annually: $n = 1$; $A = 3000\left(1 + \frac{0.05}{1}\right)^{1 \cdot 5} = \$3828.84$

(b) Semiannually: $n = 2$; $A = 3000\left(1 + \frac{0.05}{2}\right)^{2 \cdot 5} = \$3840.25$

(c) Monthly: $n = 12$; $A = 3000\left(1 + \frac{0.05}{12}\right)^{12 \cdot 5} = \$3850.08$

(d) Weekly: $n = 52$; $A = 3000\left(1 + \frac{0.05}{52}\right)^{52 \cdot 5} = \$3851.61$

(e) Daily: $n = 365$; $A = 3000\left(1 + \frac{0.05}{365}\right)^{365 \cdot 5} = \$3852.01$

(f) Continuously: $A = 3000e^{(0.05)5} = \$3852.08$

20. (a) $A_0 e^{0.06t} = 2A_0 \Leftrightarrow e^{0.06t} = 2 \Leftrightarrow 0.06t = \ln 2 \Leftrightarrow t = \frac{50}{3}\ln 2 \approx 11.55$, so the investment will double in about 11.55 years.

(b) The annual interest rate in $A = A_0(1 + r)^t$ is r. From part (a), we have $A = A_0 e^{0.06t}$. These amounts must be equal, so $(1 + r)^t = e^{0.06t} \Rightarrow 1 + r = e^{0.06} \Rightarrow r = e^{0.06} - 1 \approx 0.0618 = 6.18\%$, which is the equivalent annual interest rate.

5.6 Inverse Trigonometric Functions

1. (a) $\sin^{-1}\left(\frac{\sqrt{3}}{2}\right) = \frac{\pi}{3}$ since $\sin\frac{\pi}{3} = \frac{\sqrt{3}}{2}$ and $\frac{\pi}{3}$ is in $\left[-\frac{\pi}{2}, \frac{\pi}{2}\right]$.

(b) $\cos^{-1}(-1) = \pi$ since $\cos\pi = -1$ and π is in $[0, \pi]$.

2. (a) $\arctan(-1) = -\frac{\pi}{4}$ since $\tan\left(-\frac{\pi}{4}\right) = -1$ and $-\frac{\pi}{4}$ is in $\left(-\frac{\pi}{2}, \frac{\pi}{2}\right)$.

(b) $\csc^{-1} 2 = \frac{\pi}{6}$ since $\csc\frac{\pi}{6} = 2$ and $\frac{\pi}{6}$ is in $\left(0, \frac{\pi}{2}\right] \cup \left(\pi, \frac{3\pi}{2}\right]$.

3. (a) $\tan^{-1}\sqrt{3} = \frac{\pi}{3}$ since $\tan\frac{\pi}{3} = \sqrt{3}$ and $\frac{\pi}{3}$ is in $\left(-\frac{\pi}{2}, \frac{\pi}{2}\right)$.

(b) $\arcsin\left(-\frac{1}{\sqrt{2}}\right) = -\frac{\pi}{4}$ since $\sin\left(-\frac{\pi}{4}\right) = -\frac{1}{\sqrt{2}}$ and $-\frac{\pi}{4}$ is in $\left[-\frac{\pi}{2}, \frac{\pi}{2}\right]$.

4. (a) $\sec^{-1}\sqrt{2} = \frac{\pi}{4}$ since $\sec\frac{\pi}{4} = \sqrt{2}$ and $\frac{\pi}{4}$ is in $\left[0, \frac{\pi}{2}\right) \cup \left[\pi, \frac{3\pi}{2}\right)$.

(b) $\arcsin 1 = \frac{\pi}{2}$ since $\sin\frac{\pi}{2} = 1$ and $\frac{\pi}{2}$ is in $\left[-\frac{\pi}{2}, \frac{\pi}{2}\right]$.

5. (a) $\sin(\sin^{-1}(0.7)) = 0.7$ since 0.7 is in $[-1, 1]$.

(b) $\tan^{-1}\left(\tan\frac{4\pi}{3}\right) = \tan^{-1}\sqrt{3} = \frac{\pi}{3}$ since $\frac{\pi}{3}$ is in $\left[-\frac{\pi}{2}, \frac{\pi}{2}\right]$.

6. (a) Let $\theta = \arctan 2$, so $\tan\theta = 2 \;\Rightarrow\; \sec^2\theta = 1 + \tan^2\theta = 1 + 4 = 5 \;\Rightarrow\; \sec\theta = \sqrt{5} \;\Rightarrow$

$\sec(\arctan 2) = \sec\theta = \sqrt{5}$.

(b) Let $\theta = \sin^{-1}\left(\frac{5}{13}\right)$. Then $\sin\theta = \frac{5}{13}$, so $\cos\left(2\sin^{-1}\left(\frac{5}{13}\right)\right) = \cos 2\theta = 1 - 2\sin^2\theta = 1 - 2\left(\frac{5}{13}\right)^2 = \frac{119}{169}$.

7. Let $y = \sin^{-1}x$. Then $-\frac{\pi}{2} \le y \le \frac{\pi}{2} \;\Rightarrow\; \cos y \ge 0$, so $\cos(\sin^{-1}x) = \cos y = \sqrt{1 - \sin^2 y} = \sqrt{1 - x^2}$.

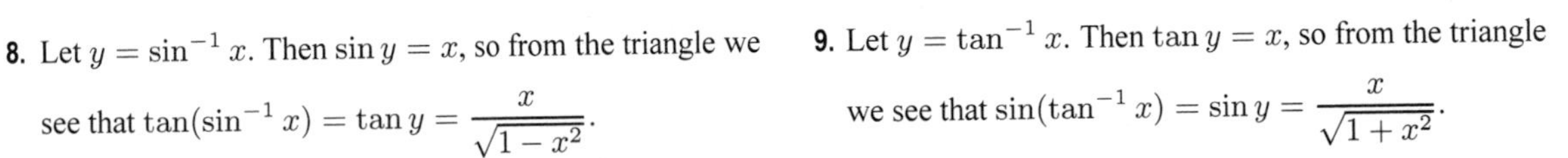

8. Let $y = \sin^{-1}x$. Then $\sin y = x$, so from the triangle we see that $\tan(\sin^{-1}x) = \tan y = \dfrac{x}{\sqrt{1 - x^2}}$.

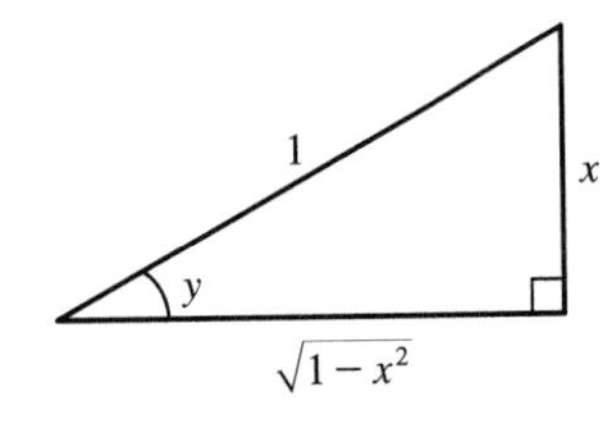

9. Let $y = \tan^{-1}x$. Then $\tan y = x$, so from the triangle we see that $\sin(\tan^{-1}x) = \sin y = \dfrac{x}{\sqrt{1 + x^2}}$.

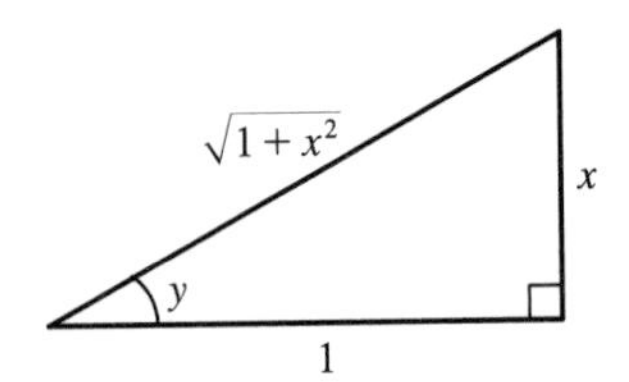

10. Let $\theta = \arctan 2x$. Then $\tan\theta = 2x$, so from the diagram we see that $\csc(\arctan 2x) = \csc\theta = \dfrac{\sqrt{4x^2 + 1}}{2x}$.

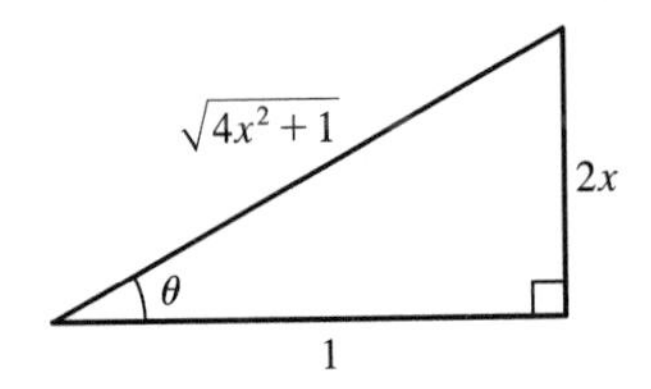

11. Let $y = \cos^{-1}x$. Then $\cos y = x$ and $0 \le y \le \pi \;\Rightarrow\; -\sin y\,\dfrac{dy}{dx} = 1 \;\Rightarrow$

$\dfrac{dy}{dx} = -\dfrac{1}{\sin y} = -\dfrac{1}{\sqrt{1 - \cos^2 y}} = -\dfrac{1}{\sqrt{1 - x^2}}$. [Note that $\sin y \ge 0$ for $0 \le y \le \pi$.]

12. (a) Let $a = \sin^{-1}x$ and $b = \cos^{-1}x$. Then $\cos a = \sqrt{1 - \sin^2 a} = \sqrt{1 - x^2}$ since $\cos a \ge 0$ for $-\frac{\pi}{2} \le a \le \frac{\pi}{2}$. Similarly, $\sin b = \sqrt{1 - x^2}$. So

$$\begin{aligned}\sin(\sin^{-1}x + \cos^{-1}x) &= \sin(a + b) = \sin a\cos b + \cos a\sin b = x \cdot x + \sqrt{1 - x^2}\,\sqrt{1 - x^2} \\ &= x^2 + (1 - x^2) = 1\end{aligned}$$

But $-\frac{\pi}{2} \le \sin^{-1}x + \cos^{-1}x \le \frac{3\pi}{2}$, and so $\sin^{-1}x + \cos^{-1}x = \frac{\pi}{2}$.

(b) We differentiate $\sin^{-1}x + \cos^{-1}x = \frac{\pi}{2}$ with respect to x, and get

$$\frac{1}{\sqrt{1 - x^2}} + \frac{d}{dx}(\cos^{-1}x) = 0 \;\Rightarrow\; \frac{d}{dx}(\cos^{-1}x) = -\frac{1}{\sqrt{1 - x^2}}.$$

13. Let $y = \cot^{-1} x$. Then $\cot y = x \Rightarrow -\csc^2 y \dfrac{dy}{dx} = 1 \Rightarrow \dfrac{dy}{dx} = -\dfrac{1}{\csc^2 y} = -\dfrac{1}{1+\cot^2 y} = -\dfrac{1}{1+x^2}$.

14. Let $y = \sec^{-1} x$. Then $\sec y = x$ and $y \in \left(0, \frac{\pi}{2}\right] \cup \left[\pi, \frac{3\pi}{2}\right)$. Differentiate with respect to x:

$\sec y \, \tan y \left(\dfrac{dy}{dx}\right) = 1 \Rightarrow \dfrac{dy}{dx} = \dfrac{1}{\sec y \, \tan y} = \dfrac{1}{\sec y \sqrt{\sec^2 y - 1}} = \dfrac{1}{x\sqrt{x^2-1}}$. Note that $\tan^2 y = \sec^2 y - 1 \Rightarrow$ $\tan y = \sqrt{\sec^2 y - 1}$ since $\tan y > 0$ when $0 < y < \frac{\pi}{2}$ or $\pi < y < \frac{3\pi}{2}$.

15. Let $y = \csc^{-1} x$. Then $\csc y = x \Rightarrow -\csc y \cot y \dfrac{dy}{dx} = 1 \Rightarrow$

$\dfrac{dy}{dx} = -\dfrac{1}{\csc y \cot y} = -\dfrac{1}{\csc y \sqrt{\csc^2 y - 1}} = -\dfrac{1}{x\sqrt{x^2-1}}$. Note that $\cot y \geq 0$ on the domain of $\csc^{-1} x$.

16. $y = \sqrt{\tan^{-1} x} = (\tan^{-1} x)^{1/2} \Rightarrow$

$$y' = \tfrac{1}{2}(\tan^{-1} x)^{-1/2} \cdot \frac{d}{dx}(\tan^{-1} x) = \frac{1}{2\sqrt{\tan^{-1} x}} \cdot \frac{1}{1+x^2} = \frac{1}{2\sqrt{\tan^{-1} x}\,(1+x^2)}$$

17. $y = \tan^{-1}\sqrt{x} \Rightarrow y' = \dfrac{1}{1+(\sqrt{x})^2} \cdot \dfrac{d}{dx}(\sqrt{x}) = \dfrac{1}{1+x}\left(\tfrac{1}{2}x^{-1/2}\right) = \dfrac{1}{2\sqrt{x}\,(1+x)}$

18. $h(x) = \sqrt{1-x^2} \arcsin x \Rightarrow h'(x) = \sqrt{1-x^2} \cdot \dfrac{1}{\sqrt{1-x^2}} + \arcsin x \left[\tfrac{1}{2}(1-x^2)^{-1/2}(-2x)\right] = 1 - \dfrac{x \arcsin x}{\sqrt{1-x^2}}$

19. $y = \sin^{-1}(2x+1) \Rightarrow$

$$y' = \frac{1}{\sqrt{1-(2x+1)^2}} \cdot \frac{d}{dx}(2x+1) = \frac{1}{\sqrt{1-(4x^2+4x+1)}} \cdot 2 = \frac{2}{\sqrt{-4x^2-4x}} = \frac{1}{\sqrt{-x^2-x}}$$

20. $f(x) = x \ln(\arctan x) \Rightarrow f'(x) = x \cdot \dfrac{1}{\arctan x} \cdot \dfrac{1}{1+x^2} + \ln(\arctan x) \cdot 1 = \dfrac{x}{(1+x^2)\arctan x} + \ln(\arctan x)$

21. $H(x) = (1+x^2)\arctan x \Rightarrow H'(x) = (1+x^2)\dfrac{1}{1+x^2} + (\arctan x)(2x) = 1 + 2x \arctan x$

22. $h(t) = e^{\sec^{-1} t} \Rightarrow h'(t) = e^{\sec^{-1} t} \dfrac{d}{dt}\left(\sec^{-1} t\right) = \dfrac{e^{\sec^{-1} t}}{t\sqrt{t^2-1}}$

23. $y = \cos^{-1}(e^{2x}) \Rightarrow y' = -\dfrac{1}{\sqrt{1-(e^{2x})^2}} \cdot \dfrac{d}{dx}\left(e^{2x}\right) = -\dfrac{2e^{2x}}{\sqrt{1-e^{4x}}}$

24. $y = x\cos^{-1} x - \sqrt{1-x^2} \Rightarrow y' = \cos^{-1} x - \dfrac{x}{\sqrt{1-x^2}} + \dfrac{x}{\sqrt{1-x^2}} = \cos^{-1} x$

25. $y = \arctan(\cos\theta) \Rightarrow y' = \dfrac{1}{1+(\cos\theta)^2}(-\sin\theta) = -\dfrac{\sin\theta}{1+\cos^2\theta}$

26. $y = \tan^{-1}\left(x - \sqrt{x^2+1}\right) \Rightarrow$

$$y' = \frac{1}{1+\left(x-\sqrt{x^2+1}\right)^2}\left(1 - \frac{x}{\sqrt{x^2+1}}\right) = \frac{1}{1+x^2-2x\sqrt{x^2+1}+x^2+1}\left(\frac{\sqrt{x^2+1}-x}{\sqrt{x^2+1}}\right)$$

$$= \frac{\sqrt{x^2+1}-x}{2\left(1+x^2-x\sqrt{x^2+1}\right)\sqrt{x^2+1}} = \frac{\sqrt{x^2+1}-x}{2\left[\sqrt{x^2+1}\,(1+x^2)-x(x^2+1)\right]} = \frac{\sqrt{x^2+1}-x}{2\left[(1+x^2)\left(\sqrt{x^2+1}-x\right)\right]}$$

$$= \frac{1}{2(1+x^2)}$$

27. $h(t) = \cot^{-1}(t) + \cot^{-1}(1/t) \Rightarrow$

$$h'(t) = -\frac{1}{1+t^2} - \frac{1}{1+(1/t)^2} \cdot \frac{d}{dt}\frac{1}{t} = -\frac{1}{1+t^2} - \frac{t^2}{t^2+1} \cdot \left(-\frac{1}{t^2}\right) = -\frac{1}{1+t^2} + \frac{1}{t^2+1} = 0.$$

Note that this makes sense because $h(t) = \frac{\pi}{2}$ for $t > 0$ and $h(t) = -\frac{\pi}{2}$ for $t < 0$.

28. $y = \tan^{-1}\left(\frac{x}{a}\right) + \ln\sqrt{\frac{x-a}{x+a}} = \tan^{-1}\left(\frac{x}{a}\right) + \frac{1}{2}\ln(x-a) - \frac{1}{2}\ln(x+a) \Rightarrow$

$$y' = \frac{a}{x^2+a^2} + \frac{1/2}{x-a} - \frac{1/2}{x+a} = \frac{a}{x^2+a^2} + \frac{a}{x^2-a^2} = \frac{2ax^2}{x^4-a^4}$$

29. $y = \arccos\left(\frac{b+a\cos x}{a+b\cos x}\right) \Rightarrow$

$$y' = -\frac{1}{\sqrt{1-\left(\frac{b+a\cos x}{a+b\cos x}\right)^2}} \frac{(a+b\cos x)(-a\sin x) - (b+a\cos x)(-b\sin x)}{(a+b\cos x)^2}$$

$$= \frac{1}{\sqrt{a^2+b^2\cos^2 x - b^2 - a^2\cos^2 x}} \frac{(a^2-b^2)\sin x}{|a+b\cos x|}$$

$$= \frac{1}{\sqrt{a^2-b^2}\sqrt{1-\cos^2 x}} \frac{(a^2-b^2)\sin x}{|a+b\cos x|} = \frac{\sqrt{a^2-b^2}}{|a+b\cos x|} \frac{\sin x}{|\sin x|}$$

But $0 \le x \le \pi$, so $|\sin x| = \sin x$. Also $a > b > 0 \Rightarrow b\cos x \ge -b > -a$, so $a + b\cos x > 0$. Thus $y' = \frac{\sqrt{a^2-b^2}}{a+b\cos x}$.

30. $f(x) = \arcsin(e^x) \Rightarrow f'(x) = \frac{1}{\sqrt{1-(e^x)^2}} \cdot e^x = \frac{e^x}{\sqrt{1-e^{2x}}}$.

$\text{Domain}(f) = \{x \mid -1 \le e^x \le 1\} = \{x \mid 0 < e^x \le 1\} = (-\infty, 0]$.

$\text{Domain}(f') = \{x \mid 1 - e^{2x} > 0\} = \{x \mid e^{2x} < 1\} = \{x \mid 2x < 0\} = (-\infty, 0)$.

31. $g(x) = \cos^{-1}(3-2x) \Rightarrow g'(x) = -\frac{1}{\sqrt{1-(3-2x)^2}}(-2) = \frac{2}{\sqrt{1-(3-2x)^2}}$.

$\text{Domain}(g) = \{x \mid -1 \le 3-2x \le 1\} = \{x \mid -4 \le -2x \le -2\} = \{x \mid 2 \ge x \ge 1\} = [1, 2]$.

$\text{Domain}(g') = \{x \mid 1-(3-2x)^2 > 0\} = \{x \mid (3-2x)^2 < 1\} = \{x \mid |3-2x| < 1\}$
$= \{x \mid -1 < 3-2x < 1\} = \{x \mid -4 < -2x < -2\} = \{x \mid 2 > x > 1\} = (1, 2)$

32. $\tan^{-1}(xy) = 1 + x^2y \Rightarrow \frac{1}{1+x^2y^2}(xy' + y \cdot 1) = 0 + x^2y' + 2xy \Rightarrow$

$$y'\left(\frac{x}{1+x^2y^2} - x^2\right) = 2xy - \frac{y}{1+x^2y^2} \Rightarrow$$

$$y' = \frac{2xy - \frac{y}{1+x^2y^2}}{\frac{x}{1+x^2y^2} - x^2} = \frac{2xy(1+x^2y^2) - y}{x - x^2(1+x^2y^2)} = \frac{y(-1-2x-2x^3y^2)}{x(1-x-x^3y^2)}.$$

33. $g(x) = x\sin^{-1}\left(\frac{x}{4}\right) + \sqrt{16-x^2} \Rightarrow g'(x) = \sin^{-1}\left(\frac{x}{4}\right) + \frac{x}{4\sqrt{1-(x/4)^2}} - \frac{x}{\sqrt{16-x^2}} = \sin^{-1}\left(\frac{x}{4}\right) \Rightarrow$

$g'(2) = \sin^{-1}\left(\frac{1}{2}\right) = \frac{\pi}{6}$

34. $y = 3\arccos\dfrac{x}{2} \;\Rightarrow\; y' = 3\left[-\dfrac{1}{\sqrt{1-(x/2)^2}}\right]\left(\dfrac{1}{2}\right)$, so at $(1,\pi)$, $y' = -\dfrac{3}{2\sqrt{1-\frac{1}{4}}} = -\sqrt{3}$. An equation of the tangent line is $y - \pi = -\sqrt{3}\,(x-1)$, or $y = -\sqrt{3}\,x + \pi + \sqrt{3}$.

35. $\displaystyle\lim_{x\to -1^+} \sin^{-1} x = \sin^{-1}(-1) = -\tfrac{\pi}{2}$

36. Let $t = \dfrac{1+x^2}{1+2x^2}$. As $x \to \infty$, $t = \dfrac{1+x^2}{1+2x^2} = \dfrac{1/x^2+1}{1/x^2+2} \to \dfrac{1}{2}$.

$$\lim_{x\to\infty} \arccos\left(\frac{1+x^2}{1+2x^2}\right) = \lim_{t\to 1/2} \arccos t = \arccos\tfrac{1}{2} = \tfrac{\pi}{3}.$$

37. Let $t = e^x$. As $x \to \infty$, $t \to \infty$. $\displaystyle\lim_{x\to\infty} \arctan(e^x) = \lim_{t\to\infty} \arctan t = \tfrac{\pi}{2}$ by (8).

38. Let $t = \ln x$. As $x \to 0^+$, $t \to -\infty$. $\displaystyle\lim_{x\to 0^+} \tan^{-1}(\ln x) = \lim_{t\to -\infty} \tan^{-1} t = -\tfrac{\pi}{2}$ by (8).

39.

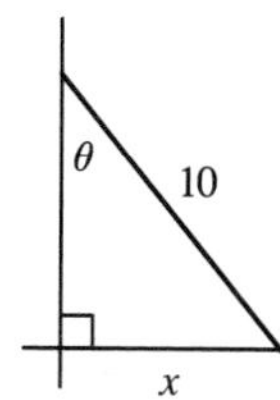

$\dfrac{dx}{dt} = 2$ ft/s, $\sin\theta = \dfrac{x}{10} \;\Rightarrow\; \theta = \sin^{-1}\left(\dfrac{x}{10}\right)$, $\dfrac{d\theta}{dx} = \dfrac{1/10}{\sqrt{1-(x/10)^2}}$,

$\dfrac{d\theta}{dt} = \dfrac{d\theta}{dx}\dfrac{dx}{dt} = \dfrac{1/10}{\sqrt{1-(x/10)^2}}\,(2)$ rad/s, $\left.\dfrac{d\theta}{dt}\right]_{x=6} = \dfrac{2/10}{\sqrt{1-(6/10)^2}}$ rad/s $= \tfrac{1}{4}$ rad/s

40.

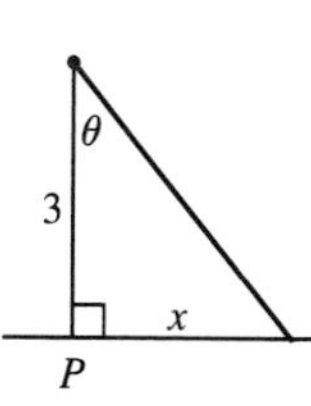

$\dfrac{d\theta}{dt} = 4$ rev/min $= 8\pi\cdot 60$ rad/h. From the diagram, we see that $\tan\theta = \dfrac{x}{3} \;\Rightarrow\; \theta = \tan^{-1}\left(\dfrac{x}{3}\right)$.

Thus, $8\pi\cdot 60 = \dfrac{d\theta}{dt} = \dfrac{d\theta}{dx}\dfrac{dx}{dt} = \dfrac{1/3}{1+(x/3)^2}\dfrac{dx}{dt}$. So $\dfrac{dx}{dt} = 8\pi\cdot 60\cdot 3\left[1+\left(\dfrac{x}{3}\right)^2\right]$ km/h, and

at $x = 1$, $\dfrac{dx}{dt} = 8\pi\cdot 60\cdot 3\left[1+\frac{1}{9}\right]$ km/h $= 1600\pi$ km/h.

41. $y = \sec^{-1} x \;\Rightarrow\; \sec y = x \;\Rightarrow\; \sec y\tan y\,\dfrac{dy}{dx} = 1 \;\Rightarrow\; \dfrac{dy}{dx} = \dfrac{1}{\sec y\tan y}$. Now $\tan^2 y = \sec^2 y - 1 = x^2 - 1$, so

$\tan y = \pm\sqrt{x^2-1}$. For $y \in \left[0, \frac{\pi}{2}\right)$, $x \ge 1$, so $\sec y = x = |x|$ and $\tan y \ge 0 \;\Rightarrow\; \dfrac{dy}{dx} = \dfrac{1}{x\sqrt{x^2-1}} = \dfrac{1}{|x|\sqrt{x^2-1}}$.

For $y \in \left(\frac{\pi}{2}, \pi\right]$, $x \le -1$, so $|x| = -x$ and $\tan y = -\sqrt{x^2-1} \;\Rightarrow$

$$\frac{dy}{dx} = \frac{1}{\sec y\,\tan y} = \frac{1}{x\left(-\sqrt{x^2-1}\right)} = \frac{1}{(-x)\sqrt{x^2-1}} = \frac{1}{|x|\sqrt{x^2-1}}$$

42. (a) $f(x) = \sin(\sin^{-1} x)$

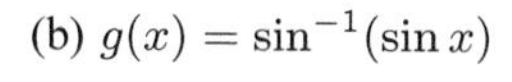

(b) $g(x) = \sin^{-1}(\sin x)$

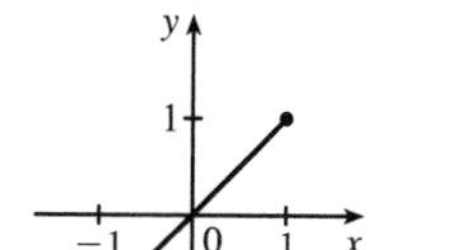

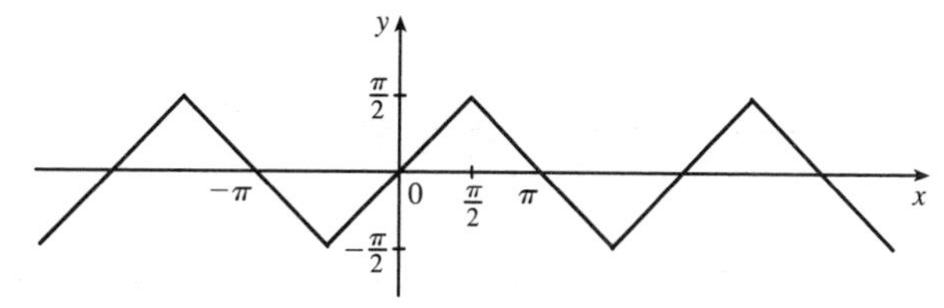

(c) $g'(x) = \frac{d}{dx}\sin^{-1}(\sin x) = \frac{1}{\sqrt{1-\sin^2 x}}\cos x = \frac{\cos x}{\sqrt{\cos^2 x}} = \frac{\cos x}{|\cos x|}$

(d) $h(x) = \cos^{-1}(\sin x)$, so

$h'(x) = -\frac{\cos x}{\sqrt{1-\sin^2 x}} = -\frac{\cos x}{|\cos x|}$.

Notice that $h(x) = \frac{\pi}{2} - g(x)$ because $\sin^{-1} t + \cos^{-1} t = \frac{\pi}{2}$ for all t.

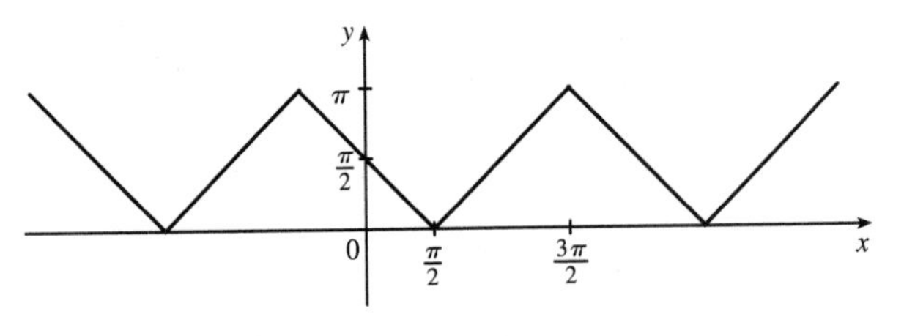

5.7 Hyperbolic Functions

1. (a) $\sinh 0 = \frac{1}{2}(e^0 - e^0) = 0$

(b) $\cosh 0 = \frac{1}{2}(e^0 + e^0) = \frac{1}{2}(1+1) = 1$

2. (a) $\tanh 0 = \frac{(e^0 - e^{-0})/2}{(e^0 + e^{-0})/2} = 0$

(b) $\tanh 1 = \frac{e^1 - e^{-1}}{e^1 + e^{-1}} = \frac{e^2 - 1}{e^2 + 1} \approx 0.76159$

3. (a) $\sinh(\ln 2) = \frac{e^{\ln 2} - e^{-\ln 2}}{2} = \frac{e^{\ln 2} - (e^{\ln 2})^{-1}}{2} = \frac{2 - 2^{-1}}{2} = \frac{2 - \frac{1}{2}}{2} = \frac{3}{4}$

(b) $\sinh 2 = \frac{1}{2}(e^2 - e^{-2}) \approx 3.62686$

4. (a) $\cosh 3 = \frac{1}{2}(e^3 + e^{-3}) \approx 10.06766$

(b) $\cosh(\ln 3) = \frac{e^{\ln 3} + e^{-\ln 3}}{2} = \frac{3 + \frac{1}{3}}{2} = \frac{5}{3}$

5. (a) $\operatorname{sech} 0 = \frac{1}{\cosh 0} = \frac{1}{1} = 1$

(b) $\cosh^{-1} 1 = 0$ because $\cosh 0 = 1$.

6. (a) $\sinh 1 = \frac{1}{2}(e^1 - e^{-1}) \approx 1.17520$

(b) Using Equation 3, we have $\sinh^{-1} 1 = \ln\left(1 + \sqrt{1^2 + 1}\right) = \ln\left(1 + \sqrt{2}\right) \approx 0.88137$.

7. $\sinh(-x) = \frac{1}{2}\left[e^{-x} - e^{-(-x)}\right] = \frac{1}{2}(e^{-x} - e^x) = -\frac{1}{2}(e^x - e^{-x}) = -\sinh x$

8. $\cosh(-x) = \frac{1}{2}\left[e^{-x} + e^{-(-x)}\right] = \frac{1}{2}(e^{-x} + e^x) = \frac{1}{2}(e^x + e^{-x}) = \cosh x$

9. $\cosh x + \sinh x = \frac{1}{2}(e^x + e^{-x}) + \frac{1}{2}(e^x - e^{-x}) = \frac{1}{2}(2e^x) = e^x$

10. $\cosh x - \sinh x = \frac{1}{2}(e^x + e^{-x}) - \frac{1}{2}(e^x - e^{-x}) = \frac{1}{2}(2e^{-x}) = e^{-x}$

11. $\sinh x \cosh y + \cosh x \sinh y = \left[\frac{1}{2}(e^x - e^{-x})\right]\left[\frac{1}{2}(e^y + e^{-y})\right] + \left[\frac{1}{2}(e^x + e^{-x})\right]\left[\frac{1}{2}(e^y - e^{-y})\right]$

$= \frac{1}{4}[(e^{x+y} + e^{x-y} - e^{-x+y} - e^{-x-y}) + (e^{x+y} - e^{x-y} + e^{-x+y} - e^{-x-y})]$

$= \frac{1}{4}(2e^{x+y} - 2e^{-x-y}) = \frac{1}{2}[e^{x+y} - e^{-(x+y)}] = \sinh(x+y)$

12. $\cosh x \cosh y + \sinh x \sinh y = \left[\frac{1}{2}(e^x + e^{-x})\right]\left[\frac{1}{2}(e^y + e^{-y})\right] + \left[\frac{1}{2}(e^x - e^{-x})\right]\left[\frac{1}{2}(e^y - e^{-y})\right]$

$= \frac{1}{4}\left[(e^{x+y} + e^{x-y} + e^{-x+y} + e^{-x-y}) + (e^{x+y} - e^{x-y} - e^{-x+y} + e^{-x-y})\right]$

$= \frac{1}{4}(2e^{x+y} + 2e^{-x-y}) = \frac{1}{2}[e^{x+y} + e^{-(x+y)}] = \cosh(x+y)$

13. Putting $y = x$ in the result from Exercise 11, we have

$\sinh 2x = \sinh(x+x) = \sinh x \cosh x + \cosh x \sinh x = 2\sinh x \cosh x$.

14. $\dfrac{1+\tanh x}{1-\tanh x} = \dfrac{1+(\sinh x)/\cosh x}{1-(\sinh x)/\cosh x} = \dfrac{\cosh x + \sinh x}{\cosh x - \sinh x} = \dfrac{\frac{1}{2}(e^x+e^{-x})+\frac{1}{2}(e^x-e^{-x})}{\frac{1}{2}(e^x+e^{-x})-\frac{1}{2}(e^x-e^{-x})}$

$= \dfrac{e^x+e^{-x}+e^x-e^{-x}}{e^x+e^{-x}-e^x+e^{-x}} = \dfrac{2e^x}{2e^{-x}} = e^{2x}$

Or: Using the results of Exercises 9 and 10, $\dfrac{\cosh x+\sinh x}{\cosh x-\sinh x} = \dfrac{e^x}{e^{-x}} = e^{2x}$

15. By Exercise 9, $(\cosh x+\sinh x)^n = (e^x)^n = e^{nx} = \cosh nx + \sinh nx$.

16. $\sinh x = \frac{3}{4} \;\Rightarrow\; \operatorname{csch} x = 1/\sinh x = \frac{4}{3}$. $\cosh^2 x = \sinh^2 x + 1 = \frac{9}{16}+1 = \frac{25}{16} \;\Rightarrow\; \cosh x = \frac{5}{4}$ (since $\cosh x > 0$).

$\operatorname{sech} x = 1/\cosh x = \frac{4}{5}$, $\tanh x = \sinh x/\cosh x = \frac{3/4}{5/4} = \frac{3}{5}$, and $\coth x = 1/\tanh x = \frac{5}{3}$.

17. $\tanh x = \frac{4}{5} > 0$, so $x > 0$. $\coth x = 1/\tanh x = \frac{5}{4}$, $\operatorname{sech}^2 x = 1-\tanh^2 x = 1-\left(\frac{4}{5}\right)^2 = \frac{9}{25} \;\Rightarrow\; \operatorname{sech} x = \frac{3}{5}$ (since $\operatorname{sech} x > 0$), $\cosh x = 1/\operatorname{sech} x = \frac{5}{3}$, $\sinh x = \tanh x \cosh x = \frac{4}{5}\cdot\frac{5}{3} = \frac{4}{3}$, and $\operatorname{csch} x = 1/\sinh x = \frac{3}{4}$.

18.

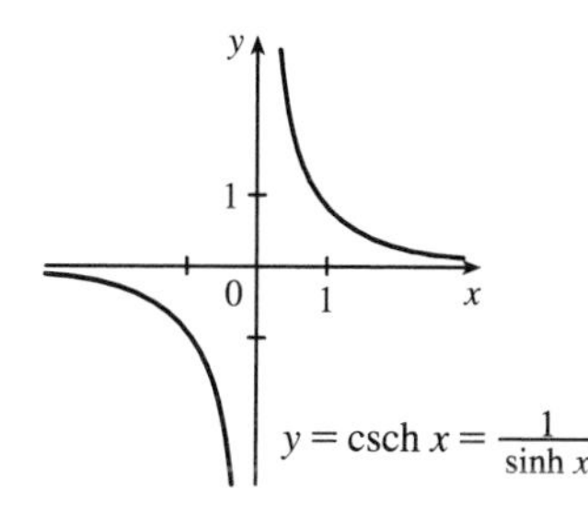

y = sech x = 1/cosh x

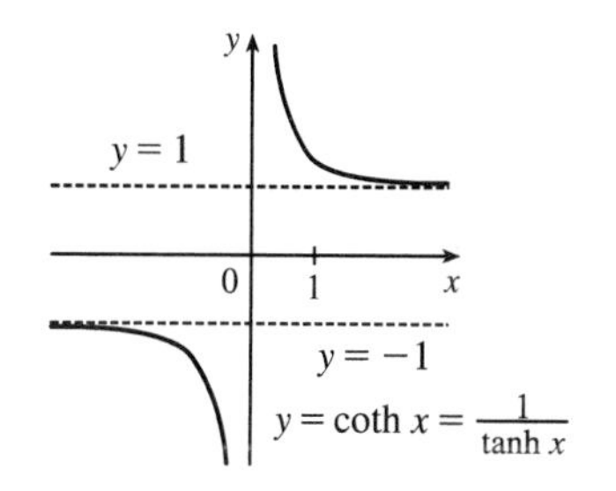

19. (a) $\lim\limits_{x\to\infty}\tanh x = \lim\limits_{x\to\infty}\dfrac{e^x-e^{-x}}{e^x+e^{-x}}\cdot\dfrac{e^{-x}}{e^{-x}} = \lim\limits_{x\to\infty}\dfrac{1-e^{-2x}}{1+e^{-2x}} = \dfrac{1-0}{1+0} = 1$

(b) $\lim\limits_{x\to-\infty}\tanh x = \lim\limits_{x\to-\infty}\dfrac{e^x-e^{-x}}{e^x+e^{-x}}\cdot\dfrac{e^x}{e^x} = \lim\limits_{x\to-\infty}\dfrac{e^{2x}-1}{e^{2x}+1} = \dfrac{0-1}{0+1} = -1$

(c) $\lim\limits_{x\to\infty}\sinh x = \lim\limits_{x\to\infty}\dfrac{e^x-e^{-x}}{2} = \infty$

(d) $\lim\limits_{x\to-\infty}\sinh x = \lim\limits_{x\to-\infty}\dfrac{e^x-e^{-x}}{2} = -\infty$

(e) $\lim\limits_{x\to\infty}\operatorname{sech} x = \lim\limits_{x\to\infty}\dfrac{2}{e^x+e^{-x}} = 0$

(f) $\lim\limits_{x\to\infty}\coth x = \lim\limits_{x\to\infty}\dfrac{e^x+e^{-x}}{e^x-e^{-x}}\cdot\dfrac{e^{-x}}{e^{-x}} = \lim\limits_{x\to\infty}\dfrac{1+e^{-2x}}{1-e^{-2x}} = \dfrac{1+0}{1-0} = 1$ [*Or:* Use part (a)]

(g) $\lim\limits_{x\to0^+}\coth x = \lim\limits_{x\to0^+}\dfrac{\cosh x}{\sinh x} = \infty$, since $\sinh x \to 0$ through positive values and $\cosh x \to 1$.

(h) $\lim\limits_{x\to0^-}\coth x = \lim\limits_{x\to0^-}\dfrac{\cosh x}{\sinh x} = -\infty$, since $\sinh x \to 0$ through negative values and $\cosh x \to 1$.

(i) $\lim\limits_{x\to-\infty}\operatorname{csch} x = \lim\limits_{x\to-\infty}\dfrac{2}{e^x-e^{-x}} = 0$

20. (a) $\dfrac{d}{dx}\cosh x = \dfrac{d}{dx}\left[\frac{1}{2}(e^x+e^{-x})\right] = \frac{1}{2}(e^x-e^{-x}) = \sinh x$

(b) $\dfrac{d}{dx}\tanh x = \dfrac{d}{dx}\left[\dfrac{\sinh x}{\cosh x}\right] = \dfrac{\cosh x\cosh x - \sinh x\sinh x}{\cosh^2 x} = \dfrac{\cosh^2 x-\sinh^2 x}{\cosh^2 x} = \dfrac{1}{\cosh^2 x} = \operatorname{sech}^2 x$

(c) $\dfrac{d}{dx}\operatorname{csch} x = \dfrac{d}{dx}\left[\dfrac{1}{\sinh x}\right] = -\dfrac{\cosh x}{\sinh^2 x} = -\dfrac{1}{\sinh x}\cdot\dfrac{\cosh x}{\sinh x} = -\operatorname{csch} x \coth x$

(d) $\dfrac{d}{dx}\operatorname{sech} x = \dfrac{d}{dx}\left[\dfrac{1}{\cosh x}\right] = -\dfrac{\sinh x}{\cosh^2 x} = -\dfrac{1}{\cosh x}\cdot\dfrac{\sinh x}{\cosh x} = -\operatorname{sech} x \tanh x$

(e) $\dfrac{d}{dx}\coth x = \dfrac{d}{dx}\left[\dfrac{\cosh x}{\sinh x}\right] = \dfrac{\sinh x \sinh x - \cosh x \cosh x}{\sinh^2 x} = \dfrac{\sinh^2 x - \cosh^2 x}{\sinh^2 x} = -\dfrac{1}{\sinh^2 x} = -\operatorname{csch}^2 x$

21. Let $y = \sinh^{-1} x$. Then $\sinh y = x$ and, by Example 1(a), $\cosh^2 y - \sinh^2 y = 1 \;\Rightarrow\;$ [with $\cosh y > 0$] $\cosh y = \sqrt{1+\sinh^2 y} = \sqrt{1+x^2}$. So by Exercise 9, $e^y = \sinh y + \cosh y = x + \sqrt{1+x^2} \;\Rightarrow\; y = \ln\left(x+\sqrt{1+x^2}\right)$.

22. Let $y = \cosh^{-1} x$. Then $\cosh y = x$ and $y \geq 0$, so $\sinh y = \sqrt{\cosh^2 y - 1} = \sqrt{x^2-1}$. So, by Exercise 9, $e^y = \cosh y + \sinh y = x + \sqrt{x^2-1} \;\Rightarrow\; y = \ln\left(x + \sqrt{x^2-1}\right)$.

Another method: Write $x = \cosh y = \frac{1}{2}\left(e^y + e^{-y}\right)$ and solve a quadratic, as in Example 3.

23. (a) Let $y = \tanh^{-1} x$. Then $x = \tanh y = \dfrac{\sinh y}{\cosh y} = \dfrac{(e^y - e^{-y})/2}{(e^y + e^{-y})/2}\cdot\dfrac{e^y}{e^y} = \dfrac{e^{2y}-1}{e^{2y}+1} \;\Rightarrow$

$xe^{2y} + x = e^{2y} - 1 \;\Rightarrow\; 1 + x = e^{2y} - xe^{2y} \;\Rightarrow\; 1 + x = e^{2y}(1-x) \;\Rightarrow$

$e^{2y} = \dfrac{1+x}{1-x} \;\Rightarrow\; 2y = \ln\left(\dfrac{1+x}{1-x}\right) \;\Rightarrow\; y = \frac{1}{2}\ln\left(\dfrac{1+x}{1-x}\right)$.

(b) Let $y = \tanh^{-1} x$. Then $x = \tanh y$, so from Exercise 14 we have

$e^{2y} = \dfrac{1+\tanh y}{1-\tanh y} = \dfrac{1+x}{1-x} \;\Rightarrow\; 2y = \ln\left(\dfrac{1+x}{1-x}\right) \;\Rightarrow\; y = \frac{1}{2}\ln\left(\dfrac{1+x}{1-x}\right)$.

24. (a) (i) $y = \operatorname{csch}^{-1} x \;\Leftrightarrow\; \operatorname{csch} y = x \quad (x \neq 0)$

(ii) We sketch the graph of csch^{-1} by reflecting the graph of csch (see Exercise 18) about the line $y = x$.

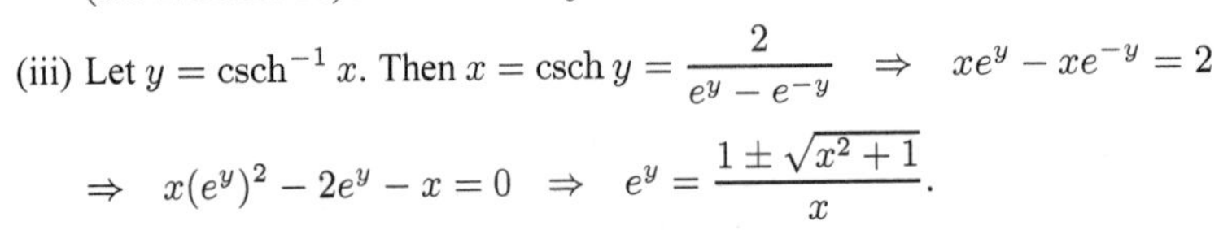

(iii) Let $y = \operatorname{csch}^{-1} x$. Then $x = \operatorname{csch} y = \dfrac{2}{e^y - e^{-y}} \;\Rightarrow\; xe^y - xe^{-y} = 2$

$\Rightarrow\; x(e^y)^2 - 2e^y - x = 0 \;\Rightarrow\; e^y = \dfrac{1 \pm \sqrt{x^2+1}}{x}$.

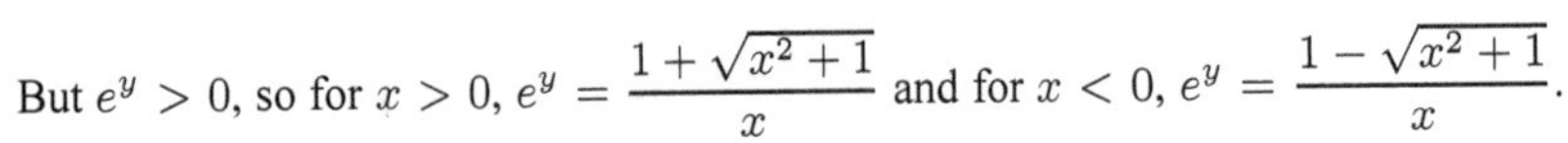

But $e^y > 0$, so for $x > 0$, $e^y = \dfrac{1+\sqrt{x^2+1}}{x}$ and for $x < 0$, $e^y = \dfrac{1-\sqrt{x^2+1}}{x}$.

Thus, $\operatorname{csch}^{-1} x = \ln\left(\dfrac{1}{x} + \dfrac{\sqrt{x^2+1}}{|x|}\right)$.

(b) (i) $y = \operatorname{sech}^{-1} x \;\Leftrightarrow\; \operatorname{sech} y = x$ and $y > 0$.

(ii) We sketch the graph of sech^{-1} by reflecting the graph of sech (see Exercise 18) about the line $y = x$.

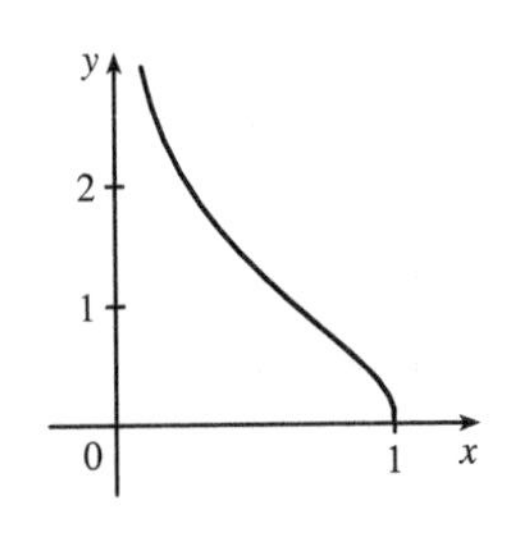

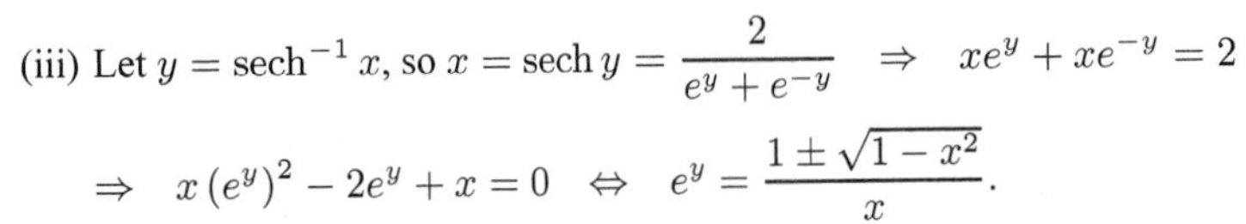

(iii) Let $y = \operatorname{sech}^{-1} x$, so $x = \operatorname{sech} y = \dfrac{2}{e^y + e^{-y}} \;\Rightarrow\; xe^y + xe^{-y} = 2$

$\Rightarrow\; x\left(e^y\right)^2 - 2e^y + x = 0 \;\Leftrightarrow\; e^y = \dfrac{1 \pm \sqrt{1-x^2}}{x}$.

But $y > 0 \;\Rightarrow\; e^y > 1$. This rules out the minus sign because $\dfrac{1-\sqrt{1-x^2}}{x} > 1 \;\Leftrightarrow\; 1 - \sqrt{1-x^2} > x \;\Leftrightarrow$

$1 - x > \sqrt{1 - x^2} \;\Leftrightarrow\; 1 - 2x + x^2 > 1 - x^2 \;\Leftrightarrow\; x^2 > x \;\Leftrightarrow\; x > 1$, but $x = \operatorname{sech} y \le 1$. Thus,

$$e^y = \frac{1 + \sqrt{1 - x^2}}{x} \;\Rightarrow\; \operatorname{sech}^{-1} x = \ln\left(\frac{1 + \sqrt{1 - x^2}}{x}\right).$$

(c) (i) $y = \coth^{-1} x \;\Leftrightarrow\; \coth y = x$

(ii) We sketch the graph of $\coth^{-1}$ by reflecting the graph of coth (see Exercise 18) about the line $y = x$.

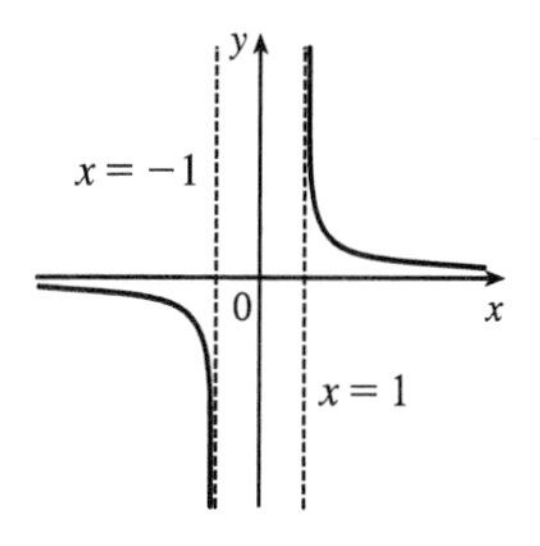

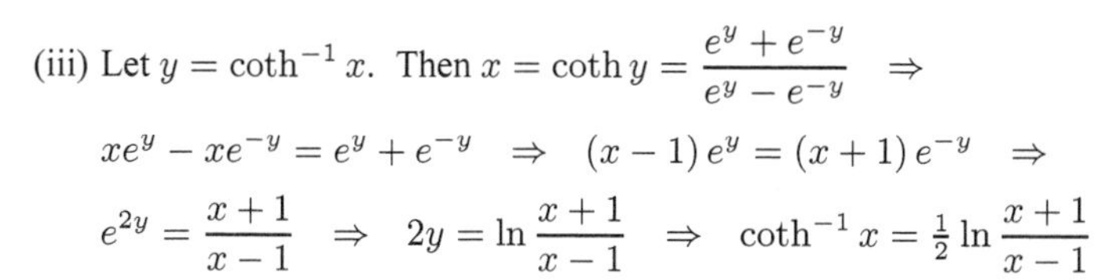

(iii) Let $y = \coth^{-1} x$. Then $x = \coth y = \dfrac{e^y + e^{-y}}{e^y - e^{-y}} \;\Rightarrow$

$xe^y - xe^{-y} = e^y + e^{-y} \;\Rightarrow\; (x - 1)\, e^y = (x + 1)\, e^{-y} \;\Rightarrow$

$$e^{2y} = \frac{x + 1}{x - 1} \;\Rightarrow\; 2y = \ln\frac{x + 1}{x - 1} \;\Rightarrow\; \coth^{-1} x = \tfrac{1}{2}\ln\frac{x + 1}{x - 1}$$

25. (a) Let $y = \cosh^{-1} x$. Then $\cosh y = x$ and $y \ge 0 \;\Rightarrow\; \sinh y \dfrac{dy}{dx} = 1 \;\Rightarrow$

$$\frac{dy}{dx} = \frac{1}{\sinh y} = \frac{1}{\sqrt{\cosh^2 y - 1}} = \frac{1}{\sqrt{x^2 - 1}}$$ (since $\sinh y \ge 0$ for $y \ge 0$). *Or:* Use Formula 4.

(b) Let $y = \tanh^{-1} x$. Then $\tanh y = x \;\Rightarrow\; \operatorname{sech}^2 y \dfrac{dy}{dx} = 1 \;\Rightarrow\; \dfrac{dy}{dx} = \dfrac{1}{\operatorname{sech}^2 y} = \dfrac{1}{1 - \tanh^2 y} = \dfrac{1}{1 - x^2}$.

Or: Use Formula 5.

(c) Let $y = \operatorname{sech}^{-1} x$. Then $\operatorname{sech} y = x \;\Rightarrow\; -\operatorname{sech} y \tanh y \dfrac{dy}{dx} = 1 \;\Rightarrow$

$$\frac{dy}{dx} = -\frac{1}{\operatorname{sech} y \tanh y} = -\frac{1}{\operatorname{sech} y\,\sqrt{1 - \operatorname{sech}^2 y}} = -\frac{1}{x\,\sqrt{1 - x^2}}.$$ (Note that $y > 0$ and so $\tanh y > 0$.)

26. $g(x) = \sinh^2 x \;\Rightarrow\; g'(x) = 2 \sinh x \cosh x$

27. $f(x) = x \cosh x \;\Rightarrow\; f'(x) = x\,(\cosh x)' + (\cosh x)(x)' = x \sinh x + \cosh x$

28. $F(x) = \sinh x \tanh x \;\Rightarrow\; F'(x) = \sinh x \operatorname{sech}^2 x + \tanh x \cosh x$

29. $h(x) = \sinh(x^2) \;\Rightarrow\; h'(x) = \cosh(x^2) \cdot 2x = 2x \cosh(x^2)$

30. $f(t) = e^t \operatorname{sech} t \;\Rightarrow\; f'(t) = e^t(-\operatorname{sech} t \tanh t) + (\operatorname{sech} t)\, e^t = e^t \operatorname{sech} t\,(1 - \tanh t)$

31. $h(t) = \coth\sqrt{1 + t^2} \;\Rightarrow\; h'(t) = -\operatorname{csch}^2\sqrt{1 + t^2} \cdot \tfrac{1}{2}(1 + t^2)^{-1/2}\,(2t) = -\dfrac{t \operatorname{csch}^2\sqrt{1 + t^2}}{\sqrt{1 + t^2}}$

32. $f(t) = \ln(\sinh t) \;\Rightarrow\; f'(t) = \dfrac{1}{\sinh t}\cosh t = \coth t$

33. $H(t) = \tanh(e^t) \;\Rightarrow\; H'(t) = \operatorname{sech}^2(e^t) \cdot e^t = e^t \operatorname{sech}^2(e^t)$

34. $y = \sinh(\cosh x) \;\Rightarrow\; y' = \cosh(\cosh x) \cdot \sinh x$

35. $y = e^{\cosh 3x} \;\Rightarrow\; y' = e^{\cosh 3x} \cdot \sinh 3x \cdot 3 = 3e^{\cosh 3x} \sinh 3x$

36. $y = x^2 \sinh^{-1}(2x) \;\Rightarrow\; y' = x^2 \cdot \dfrac{1}{\sqrt{1 + (2x)^2}} \cdot 2 + \sinh^{-1}(2x) \cdot 2x = 2x\left[\dfrac{x}{\sqrt{1 + 4x^2}} + \sinh^{-1}(2x)\right]$

37. $y = \tanh^{-1}\sqrt{x} \;\Rightarrow\; y' = \dfrac{1}{1 - (\sqrt{x})^2} \cdot \tfrac{1}{2}x^{-1/2} = \dfrac{1}{2\sqrt{x}(1 - x)}$

38. $y = x\tanh^{-1}x + \ln\sqrt{1-x^2} = x\tanh^{-1}x + \frac{1}{2}\ln(1-x^2) \Rightarrow$

$$y' = \tanh^{-1}x + \frac{x}{1-x^2} + \frac{1}{2}\left(\frac{1}{1-x^2}\right)(-2x) = \tanh^{-1}x$$

39. $y = x\sinh^{-1}(x/3) - \sqrt{9+x^2} \Rightarrow$

$$y' = \sinh^{-1}\left(\frac{x}{3}\right) + x\frac{1/3}{\sqrt{1+(x/3)^2}} - \frac{2x}{2\sqrt{9+x^2}} = \sinh^{-1}\left(\frac{x}{3}\right) + \frac{x}{\sqrt{9+x^2}} - \frac{x}{\sqrt{9+x^2}} = \sinh^{-1}\left(\frac{x}{3}\right)$$

40. $y = \operatorname{sech}^{-1}\sqrt{1-x^2} \Rightarrow y' = -\dfrac{1}{\sqrt{1-x^2}\sqrt{1-(1-x^2)}}\dfrac{-2x}{2\sqrt{1-x^2}} = \dfrac{x}{(1-x^2)|x|}$

41. $y = \coth^{-1}\sqrt{x^2+1} \Rightarrow y' = \dfrac{1}{1-(x^2+1)}\dfrac{2x}{2\sqrt{x^2+1}} = -\dfrac{1}{x\sqrt{x^2+1}}$

42.

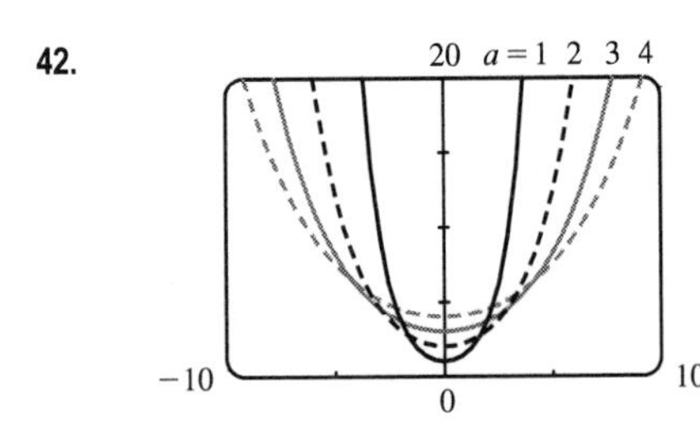

For $y = a\cosh(x/a)$ with $a > 0$, we have the y-intercept equal to a. As a increases, the graph flattens.

43. (a) $y = 20\cosh(x/20) - 15 \Rightarrow y' = 20\sinh(x/20)\cdot\frac{1}{20} = \sinh(x/20)$. Since the right pole is positioned at $x = 7$, we have $y'(7) = \sinh\frac{7}{20} \approx 0.3572$.

(b) If α is the angle between the tangent line and the x-axis, then $\tan\alpha =$ slope of the line $= \sinh\frac{7}{20}$, so $\alpha = \tan^{-1}\left(\sinh\frac{7}{20}\right) \approx 0.343 \text{ rad} \approx 19.66^\circ$. Thus, the angle between the line and the pole is $\theta = 90^\circ - \alpha \approx 70.34^\circ$.

44. We differentiate the function twice, then substitute into the differential equation: $y = \dfrac{T}{\rho g}\cosh\dfrac{\rho g x}{T} \Rightarrow$

$$\frac{dy}{dx} = \frac{T}{\rho g}\sinh\left(\frac{\rho g x}{T}\right)\frac{\rho g}{T} = \sinh\frac{\rho g x}{T} \Rightarrow \frac{d^2y}{dx^2} = \cosh\left(\frac{\rho g x}{T}\right)\frac{\rho g}{T} = \frac{\rho g}{T}\cosh\frac{\rho g x}{T}.$$

We evaluate the two sides separately: LHS $= \dfrac{d^2y}{dx^2} = \dfrac{\rho g}{T}\cosh\dfrac{\rho g x}{T}$,

RHS $= \dfrac{\rho g}{T}\sqrt{1+\left(\dfrac{dy}{dx}\right)^2} = \dfrac{\rho g}{T}\sqrt{1+\sinh^2\dfrac{\rho g x}{T}} = \dfrac{\rho g}{T}\cosh\dfrac{\rho g x}{T}$, by the identity proved in Example 1(a).

45. (a) $y = A\sinh mx + B\cosh mx \Rightarrow y' = mA\cosh mx + mB\sinh mx \Rightarrow$

$$y'' = m^2A\sinh mx + m^2B\cosh mx = m^2(A\sinh mx + B\cosh mx) = m^2y$$

(b) From part (a), a solution of $y'' = 9y$ is $y(x) = A\sinh 3x + B\cosh 3x$. So $-4 = y(0) = A\sinh 0 + B\cosh 0 = B$, so $B = -4$. Now $y'(x) = 3A\cosh 3x - 12\sinh 3x \Rightarrow 6 = y'(0) = 3A \Rightarrow A = 2$, so $y = 2\sinh 3x - 4\cosh 3x$.

46. $\displaystyle\lim_{x\to\infty}\frac{\sinh x}{e^x} = \lim_{x\to\infty}\frac{e^x - e^{-x}}{2e^x} = \lim_{x\to\infty}\frac{1-e^{-2x}}{2} = \frac{1-0}{2} = \frac{1}{2}$

47. The tangent to $y = \cosh x$ has slope 1 when $y' = \sinh x = 1 \Rightarrow x = \sinh^{-1}1 = \ln(1+\sqrt{2})$, by Equation 3.

Since $\sinh x = 1$ and $y = \cosh x = \sqrt{1+\sinh^2 x}$, we have $\cosh x = \sqrt{2}$. The point is $\left(\ln(1+\sqrt{2}), \sqrt{2}\right)$.

48. $\cosh x = \cosh[\ln(\sec\theta + \tan\theta)] = \frac{1}{2}\left[e^{\ln(\sec\theta+\tan\theta)} + e^{-\ln(\sec\theta+\tan\theta)}\right]$

$$= \frac{1}{2}\left[\sec\theta + \tan\theta + \frac{1}{\sec\theta + \tan\theta}\right] = \frac{1}{2}\left[\sec\theta + \tan\theta + \frac{\sec\theta - \tan\theta}{(\sec\theta + \tan\theta)(\sec\theta - \tan\theta)}\right]$$

$$= \frac{1}{2}\left[\sec\theta + \tan\theta + \frac{\sec\theta - \tan\theta}{\sec^2\theta - \tan^2\theta}\right] = \tfrac{1}{2}(\sec\theta + \tan\theta + \sec\theta - \tan\theta) = \sec\theta$$

49. If $ae^x + be^{-x} = \alpha\cosh(x+\beta)$ [or $\alpha\sinh(x+\beta)$], then

$ae^x + be^{-x} = \frac{\alpha}{2}\left(e^{x+\beta} \pm e^{-x-\beta}\right) = \frac{\alpha}{2}\left(e^x e^\beta \pm e^{-x}e^{-\beta}\right) = \left(\frac{\alpha}{2}e^\beta\right)e^x \pm \left(\frac{\alpha}{2}e^{-\beta}\right)e^{-x}$. Comparing coefficients of e^x and e^{-x}, we have $a = \frac{\alpha}{2}e^\beta$ **(1)** and $b = \pm\frac{\alpha}{2}e^{-\beta}$ **(2)**. We need to find α and β. Dividing equation **(1)** by equation **(2)** gives us $\frac{a}{b} = \pm e^{2\beta}$ $\Rightarrow$ $(\star)$ $2\beta = \ln\left(\pm\frac{a}{b}\right)$ $\Rightarrow$ $\beta = \frac{1}{2}\ln\left(\pm\frac{a}{b}\right)$. Solving equations **(1)** and **(2)** for e^β gives us $e^\beta = \frac{2a}{\alpha}$ and $e^\beta = \pm\frac{\alpha}{2b}$, so $\frac{2a}{\alpha} = \pm\frac{\alpha}{2b}$ $\Rightarrow$ $\alpha^2 = \pm 4ab$ $\Rightarrow$ $\alpha = 2\sqrt{\pm ab}$.

$(\star)$ If $\frac{a}{b} > 0$, we use the $+$ sign and obtain a cosh function, whereas if $\frac{a}{b} < 0$, we use the $-$ sign and obtain a sinh function.

In summary, if a and b have the same sign, we have $ae^x + be^{-x} = 2\sqrt{ab}\cosh\left(x + \frac{1}{2}\ln\frac{a}{b}\right)$, whereas, if a and b have the opposite sign, then $ae^x + be^{-x} = 2\sqrt{-ab}\sinh\left(x + \frac{1}{2}\ln\left(-\frac{a}{b}\right)\right)$.

5.8 Indeterminate Forms and l'Hospital's Rule

Note: The use of l'Hospital's Rule is indicated by an H above the equal sign: $\overset{\text{H}}{=}$

1. This limit has the form $\frac{0}{0}$. We can simply factor the numerator to evaluate this limit.

$$\lim_{x\to -1}\frac{x^2-1}{x+1} = \lim_{x\to -1}\frac{(x+1)(x-1)}{x+1} = \lim_{x\to -1}(x-1) = -2$$

2. $\displaystyle\lim_{x\to 1}\frac{x^a - 1}{x^b - 1} \overset{\text{H}}{=} \lim_{x\to 1}\frac{ax^{a-1}}{bx^{b-1}} = \frac{a}{b}$

3. This limit has the form $\frac{0}{0}$. $\displaystyle\lim_{x\to(\pi/2)^+}\frac{\cos x}{1-\sin x} \overset{\text{H}}{=} \lim_{x\to(\pi/2)^+}\frac{-\sin x}{-\cos x} = \lim_{x\to(\pi/2)^+}\tan x = -\infty$.

4. $\displaystyle\lim_{x\to 0}\frac{x + \tan x}{\sin x} \overset{\text{H}}{=} \lim_{x\to 0}\frac{1 + \sec^2 x}{\cos x} = \frac{1+1^2}{1} = 2$

5. This limit has the form $\frac{0}{0}$. $\displaystyle\lim_{t\to 0}\frac{e^t - 1}{t^3} \overset{\text{H}}{=} \lim_{t\to 0}\frac{e^t}{3t^2} = \infty$ since $e^t \to 1$ and $3t^2 \to 0^+$ as $t \to 0$.

6. $\displaystyle\lim_{t\to 0}\frac{e^{3t}-1}{t} \overset{\text{H}}{=} \lim_{t\to 0}\frac{3e^{3t}}{1} = 3$

7. This limit has the form $\frac{0}{0}$. $\displaystyle\lim_{x\to 0}\frac{\tan px}{\tan qx} \overset{\text{H}}{=} \lim_{x\to 0}\frac{p\sec^2 px}{q\sec^2 qx} = \frac{p(1)^2}{q(1)^2} = \frac{p}{q}$

8. $\displaystyle\lim_{\theta\to\pi/2}\frac{1-\sin\theta}{\csc\theta} = \frac{0}{1} = 0$. L'Hospital's Rule does not apply.

9. $\lim\limits_{x\to 0^+} [(\ln x)/x] = -\infty$ since $\ln x \to -\infty$ as $x \to 0^+$ and dividing by small values of x just increases the magnitude of the quotient $(\ln x)/x$. L'Hospital's Rule does not apply.

10. $\lim\limits_{x\to\infty} \dfrac{\ln\ln x}{x} \overset{\text{H}}{=} \lim\limits_{x\to\infty} \dfrac{\frac{1}{\ln x}\cdot\frac{1}{x}}{1} = \lim\limits_{x\to\infty} \dfrac{1}{x\ln x} = 0$

11. This limit has the form $\frac{0}{0}$. $\lim\limits_{t\to 0} \dfrac{5^t - 3^t}{t} \overset{\text{H}}{=} \lim\limits_{t\to 0} \dfrac{5^t \ln 5 - 3^t \ln 3}{1} = \ln 5 - \ln 3 = \ln \frac{5}{3}$

12. $\lim\limits_{x\to 1} \dfrac{\ln x}{\sin \pi x} \overset{\text{H}}{=} \lim\limits_{x\to 1} \dfrac{1/x}{\pi\cos \pi x} = \dfrac{1}{\pi(-1)} = -\dfrac{1}{\pi}$

13. This limit has the form $\frac{0}{0}$. $\lim\limits_{x\to 0} \dfrac{e^x - 1 - x}{x^2} \overset{\text{H}}{=} \lim\limits_{x\to 0} \dfrac{e^x - 1}{2x} \overset{\text{H}}{=} \lim\limits_{x\to 0} \dfrac{e^x}{2} = \dfrac{1}{2}$

14. This limit has the form $\frac{\infty}{\infty}$. $\lim\limits_{x\to\infty} \dfrac{e^x}{x^3} \overset{\text{H}}{=} \lim\limits_{x\to\infty} \dfrac{e^x}{3x^2} \overset{\text{H}}{=} \lim\limits_{x\to\infty} \dfrac{e^x}{6x} \overset{\text{H}}{=} \lim\limits_{x\to\infty} \dfrac{e^x}{6} = \infty$

15. This limit has the form $\frac{\infty}{\infty}$. $\lim\limits_{x\to\infty} \dfrac{x}{\ln(1+2e^x)} \overset{\text{H}}{=} \lim\limits_{x\to\infty} \dfrac{1}{\dfrac{1}{1+2e^x}\cdot 2e^x} = \lim\limits_{x\to\infty} \dfrac{1+2e^x}{2e^x} \overset{\text{H}}{=} \lim\limits_{x\to\infty} \dfrac{2e^x}{2e^x} = 1$

16. $\lim\limits_{x\to 0} \dfrac{\cos mx - \cos nx}{x^2} \overset{\text{H}}{=} \lim\limits_{x\to 0} \dfrac{-m\sin mx + n\sin nx}{2x} \overset{\text{H}}{=} \lim\limits_{x\to 0} \dfrac{-m^2\cos mx + n^2\cos nx}{2} = \frac{1}{2}(n^2 - m^2)$

17. This limit has the form $\frac{0}{0}$. $\lim\limits_{x\to 1} \dfrac{1 - x + \ln x}{1 + \cos \pi x} \overset{\text{H}}{=} \lim\limits_{x\to 1} \dfrac{-1 + 1/x}{-\pi\sin \pi x} \overset{\text{H}}{=} \lim\limits_{x\to 1} \dfrac{-1/x^2}{-\pi^2\cos \pi x} = \dfrac{-1}{-\pi^2(-1)} = -\dfrac{1}{\pi^2}$

18. $\lim\limits_{x\to 0} \dfrac{x}{\tan^{-1}(4x)} \overset{\text{H}}{=} \lim\limits_{x\to 0} \dfrac{1}{\dfrac{1}{1+(4x)^2}\cdot 4} = \lim\limits_{x\to 0} \dfrac{1+16x^2}{4} = \dfrac{1}{4}$

19. This limit has the form $\frac{0}{0}$. $\lim\limits_{x\to 1} \dfrac{x^a - ax + a - 1}{(x-1)^2} \overset{\text{H}}{=} \lim\limits_{x\to 1} \dfrac{ax^{a-1} - a}{2(x-1)} \overset{\text{H}}{=} \lim\limits_{x\to 1} \dfrac{a(a-1)x^{a-2}}{2} = \dfrac{a(a-1)}{2}$

20. $\lim\limits_{x\to 0} \dfrac{1 - e^{-2x}}{\sec x} = \dfrac{1-1}{1} = 0$. L'Hospital's Rule does not apply.

21. This limit has the form $0\cdot(-\infty)$. We need to write this product as a quotient, but keep in mind that we will have to differentiate both the numerator and the denominator. If we differentiate $\dfrac{1}{\ln x}$, we get a complicated expression that results in a more difficult limit. Instead we write the quotient as $\dfrac{\ln x}{x^{-1/2}}$.

$$\lim_{x\to 0^+} \sqrt{x}\ln x = \lim_{x\to 0^+} \frac{\ln x}{x^{-1/2}} \overset{\text{H}}{=} \lim_{x\to 0^+} \frac{1/x}{-\frac{1}{2}x^{-3/2}}\cdot\frac{-2x^{3/2}}{-2x^{3/2}} = \lim_{x\to 0^+}(-2\sqrt{x}) = 0$$

22. $\lim\limits_{x\to -\infty} x^2 e^x = \lim\limits_{x\to -\infty} \dfrac{x^2}{e^{-x}} \overset{\text{H}}{=} \lim\limits_{x\to -\infty} \dfrac{2x}{-e^{-x}} \overset{\text{H}}{=} \lim\limits_{x\to -\infty} \dfrac{2}{e^{-x}} = \lim\limits_{x\to -\infty} 2e^x = 0$

23. This limit has the form $\infty\cdot 0$. We'll change it to the form $\frac{0}{0}$.

$$\lim_{x\to 0} \cot 2x \sin 6x = \lim_{x\to 0} \frac{\sin 6x}{\tan 2x} \overset{\text{H}}{=} \lim_{x\to 0} \frac{6\cos 6x}{2\sec^2 2x} = \frac{6(1)}{2(1)^2} = 3$$

24. $\lim_{x\to 0^+} \sin x \ln x = \lim_{x\to 0^+} \frac{\ln x}{\csc x} \overset{\text{H}}{=} \lim_{x\to 0^+} \frac{1/x}{-\csc x \cot x} = -\lim_{x\to 0^+} \left(\frac{\sin x}{x} \cdot \tan x\right)$

$$= -\left(\lim_{x\to 0^+} \frac{\sin x}{x}\right)\left(\lim_{x\to 0^+} \tan x\right)$$

$$= -1 \cdot 0 = 0$$

25. This limit has the form $\infty \cdot 0$. $\lim_{x\to\infty} x^3 e^{-x^2} = \lim_{x\to\infty} \frac{x^3}{e^{x^2}} \overset{\text{H}}{=} \lim_{x\to\infty} \frac{3x^2}{2xe^{x^2}} = \lim_{x\to\infty} \frac{3x}{2e^{x^2}} \overset{\text{H}}{=} \lim_{x\to\infty} \frac{3}{4xe^{x^2}} = 0$

26. $\lim_{x\to\infty} x\tan(1/x) = \lim_{x\to\infty} \frac{\tan(1/x)}{1/x} \overset{\text{H}}{=} \lim_{x\to\infty} \frac{\sec^2(1/x)(-1/x^2)}{-1/x^2} = \lim_{x\to\infty} \sec^2(1/x) = 1^2 = 1$

27. As $x \to \infty$, $1/x \to 0$, and $e^{1/x} \to 1$. So the limit has the form $\infty - \infty$ and we will change the form to a product by factoring out x.

$$\lim_{x\to\infty} (xe^{1/x} - x) = \lim_{x\to\infty} x(e^{1/x} - 1) = \lim_{x\to\infty} \frac{e^{1/x} - 1}{1/x} \overset{\text{H}}{=} \lim_{x\to\infty} \frac{e^{1/x}(-1/x^2)}{-1/x^2} = \lim_{x\to\infty} e^{1/x} = e^0 = 1$$

28. $\lim_{x\to 0} (\csc x - \cot x) = \lim_{x\to 0} \left(\frac{1}{\sin x} - \frac{\cos x}{\sin x}\right) = \lim_{x\to 0} \frac{1-\cos x}{\sin x} \overset{\text{H}}{=} \lim_{x\to 0} \frac{\sin x}{\cos x} = 0$

29. The limit has the form $\infty - \infty$ and we will change the form to a product by factoring out x.

$$\lim_{x\to\infty} (x - \ln x) = \lim_{x\to\infty} x\left(1 - \frac{\ln x}{x}\right) = \infty \text{ since } \lim_{x\to\infty} \frac{\ln x}{x} \overset{\text{H}}{=} \lim_{x\to\infty} \frac{1/x}{1} = 0$$

30. $\lim_{x\to 1} \left(\frac{1}{\ln x} - \frac{1}{x-1}\right) = \lim_{x\to 1} \frac{x - 1 - \ln x}{(x-1)\ln x} \overset{\text{H}}{=} \lim_{x\to 1} \frac{1 - 1/x}{(x-1)(1/x) + \ln x} \cdot \frac{x}{x}$

$$= \lim_{x\to 1} \frac{x-1}{x - 1 + x\ln x} \overset{\text{H}}{=} \lim_{x\to 1} \frac{1}{1 + 1 + \ln x} = \frac{1}{2+0} = \frac{1}{2}$$

31. $y = x^{x^2} \Rightarrow \ln y = x^2 \ln x$, so $\lim_{x\to 0^+} \ln y = \lim_{x\to 0^+} x^2 \ln x = \lim_{x\to 0^+} \frac{\ln x}{1/x^2} \overset{\text{H}}{=} \lim_{x\to 0^+} \frac{1/x}{-2/x^3} = \lim_{x\to 0^+} \left(-\tfrac{1}{2}x^2\right) = 0 \Rightarrow$

$\lim_{x\to 0^+} x^{x^2} = \lim_{x\to 0^+} e^{\ln y} = e^0 = 1.$

32. $y = (\tan 2x)^x \Rightarrow \ln y = x \cdot \ln \tan 2x$, so

$$\lim_{x\to 0^+} \ln y = \lim_{x\to 0^+} x \cdot \ln \tan 2x = \lim_{x\to 0^+} \frac{\ln \tan 2x}{1/x} \overset{\text{H}}{=} \lim_{x\to 0^+} \frac{(1/\tan 2x)(2\sec^2 2x)}{-1/x^2}$$

$$= \lim_{x\to 0^+} \frac{-2x^2 \cos 2x}{\sin 2x \cos^2 2x} = \lim_{x\to 0^+} \frac{2x}{\sin 2x} \cdot \lim_{x\to 0^+} \frac{-x}{\cos 2x} = 1 \cdot 0 = 0 \Rightarrow$$

$\lim_{x\to 0^+} (\tan 2x)^x = \lim_{x\to 0^+} e^{\ln y} = e^0 = 1.$

33. $y = (1-2x)^{1/x} \Rightarrow \ln y = \frac{1}{x}\ln(1-2x)$, so $\lim_{x\to 0} \ln y = \lim_{x\to 0} \frac{\ln(1-2x)}{x} \overset{\text{H}}{=} \lim_{x\to 0} \frac{-2/(1-2x)}{1} = -2 \Rightarrow$

$\lim_{x\to 0} (1-2x)^{1/x} = \lim_{x\to 0} e^{\ln y} = e^{-2}.$

34. $y = \left(1 + \frac{a}{x}\right)^{bx} \Rightarrow \ln y = bx \ln\left(1 + \frac{a}{x}\right)$, so

$$\lim_{x\to\infty} \ln y = \lim_{x\to\infty} \frac{b \ln(1 + a/x)}{1/x} \overset{\text{H}}{=} \lim_{x\to\infty} \frac{b\left(\frac{1}{1 + a/x}\right)\left(-\frac{a}{x^2}\right)}{-1/x^2} = \lim_{x\to\infty} \frac{ab}{1 + a/x} = ab \Rightarrow$$

$$\lim_{x\to\infty} \left(1 + \frac{a}{x}\right)^{bx} = \lim_{x\to\infty} e^{\ln y} = e^{ab}.$$

35. $y = (\cos x)^{1/x^2} \Rightarrow \ln y = \frac{1}{x^2} \ln \cos x \Rightarrow$

$$\lim_{x\to 0^+} \ln y = \lim_{x\to 0^+} \frac{\ln \cos x}{x^2} \overset{\text{H}}{=} \lim_{x\to 0^+} \frac{-\tan x}{2x} \overset{\text{H}}{=} \lim_{x\to 0^+} \frac{-\sec^2 x}{2} = -\frac{1}{2} \Rightarrow$$

$$\lim_{x\to 0^+} (\cos x)^{1/x^2} = \lim_{x\to 0^+} e^{\ln y} = e^{-1/2} = 1/\sqrt{e}$$

36. $y = x^{(\ln 2)/(1 + \ln x)} \Rightarrow \ln y = \frac{\ln 2}{1 + \ln x} \ln x \Rightarrow$

$$\lim_{x\to\infty} \ln y = \lim_{x\to\infty} \frac{(\ln 2)(\ln x)}{1 + \ln x} \overset{\text{H}}{=} \lim_{x\to\infty} \frac{(\ln 2)(1/x)}{1/x} = \lim_{x\to\infty} \ln 2 = \ln 2,$$

so $\lim_{x\to\infty} x^{(\ln 2)/(1 + \ln x)} = \lim_{x\to\infty} e^{\ln y} = e^{\ln 2} = 2$.

37. From the graph, it appears that $\lim_{x\to\infty} x\,[\ln(x+5) - \ln x] = 5$. To prove this, we first note that

$\ln(x+5) - \ln x = \ln \frac{x+5}{x} = \ln\left(1 + \frac{5}{x}\right) \to \ln 1 = 0$ as $x \to \infty$. Thus,

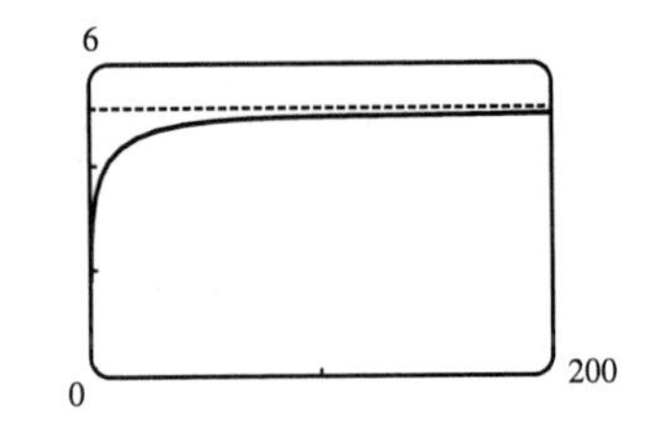

$$\lim_{x\to\infty} x\,[\ln(x+5) - \ln x] = \lim_{x\to\infty} \frac{\ln(x+5) - \ln x}{1/x} \overset{\text{H}}{=} \lim_{x\to\infty} \frac{\frac{1}{x+5} - \frac{1}{x}}{-1/x^2}$$

$$= \lim_{x\to\infty} \left[\frac{x - (x+5)}{x(x+5)} \cdot \frac{-x^2}{1}\right] = \lim_{x\to\infty} \frac{5x^2}{x^2 + 5x} = 5$$

38. From the graph, it appears that $\lim_{x\to\pi/4} (\tan x)^{\tan 2x} \approx 0.368$. The limit has the form 1^∞.

Now $y = (\tan x)^{\tan 2x} \Rightarrow \ln y = \tan 2x \ln(\tan x)$, so

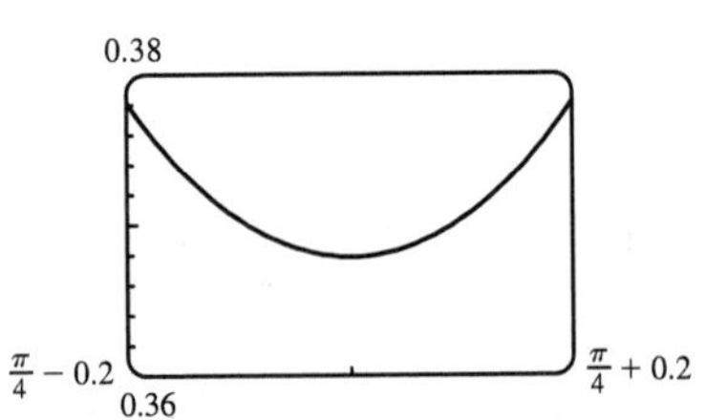

$$\lim_{x\to\pi/4} \ln y = \lim_{x\to\pi/4} \frac{\ln(\tan x)}{\cot 2x} \overset{\text{H}}{=} \lim_{x\to\pi/4} \frac{\sec^2 x/\tan x}{-2\csc^2 2x} = \frac{2/1}{-2(1)} = -1 \Rightarrow$$

$$\lim_{x\to\pi/4} (\tan x)^{\tan 2x} = \lim_{x\to\pi/4} e^{\ln y} = e^{-1} = 1/e \approx 0.3679.$$

39. $\lim_{x\to\infty} \frac{e^x}{x^n} \overset{\text{H}}{=} \lim_{x\to\infty} \frac{e^x}{nx^{n-1}} \overset{\text{H}}{=} \lim_{x\to\infty} \frac{e^x}{n(n-1)x^{n-2}} \overset{\text{H}}{=} \cdots \overset{\text{H}}{=} \lim_{x\to\infty} \frac{e^x}{n!} = \infty$

40. $\lim_{x\to\infty} \frac{\ln x}{x^p} \overset{\text{H}}{=} \lim_{x\to\infty} \frac{1/x}{px^{p-1}} = \lim_{x\to\infty} \frac{1}{px^p} = 0$ since $p > 0$.

41. First we will find $\lim\limits_{n\to\infty}\left(1+\dfrac{r}{n}\right)^{nt}$, which is of the form 1^∞. $y=\left(1+\dfrac{r}{n}\right)^{nt} \Rightarrow \ln y = nt\ln\left(1+\dfrac{r}{n}\right)$, so

$$\lim_{n\to\infty}\ln y = \lim_{n\to\infty} nt\ln\left(1+\frac{r}{n}\right) = t\lim_{n\to\infty}\frac{\ln(1+r/n)}{1/n} \overset{\text{H}}{=} t\lim_{n\to\infty}\frac{(-r/n^2)}{(1+r/n)(-1/n^2)} = t\lim_{n\to\infty}\frac{r}{1+i/n} = tr \Rightarrow$$

$\lim\limits_{n\to\infty} y = e^{rt}$. Thus, as $n\to\infty$, $A = A_0\left(1+\dfrac{r}{n}\right)^{nt}\to A_0e^{rt}$.

42. (a) $$\lim_{t\to\infty} v = \lim_{t\to\infty}\frac{mg}{c}\left(1-e^{-ct/m}\right) = \frac{mg}{c}\lim_{t\to\infty}\left(1-e^{-ct/m}\right)$$

$$= \frac{mg}{c}(1-0) \qquad [\text{because } -ct/m\to-\infty \text{ as } t\to\infty]$$

$$= \frac{mg}{c},$$ which is the speed the object approaches as time goes on, the so-called limiting velocity.

(b) $$\lim_{c\to 0^+} v = \lim_{c\to 0^+}\frac{mg}{c}\left(1-e^{-ct/m}\right) = mg\lim_{c\to 0^+}\frac{1-e^{-ct/m}}{c} \qquad [\text{form is } \tfrac{0}{0}]$$

$$\overset{\text{H}}{=} mg\lim_{c\to 0^+}\frac{\left(-e^{-ct/m}\right)\cdot(-t/m)}{1} = \frac{mgt}{m}\lim_{c\to 0^+} e^{-ct/m} = gt(1) = gt$$

The velocity of a falling object in a vacuum is directly proportional to the amount of time it falls.

43. $$\lim_{E\to 0^+} P(E) = \lim_{E\to 0^+}\left(\frac{e^E+e^{-E}}{e^E-e^{-E}}-\frac{1}{E}\right)$$

$$= \lim_{E\to 0^+}\frac{E(e^E+e^{-E})-1(e^E-e^{-E})}{(e^E-e^{-E})E} = \lim_{E\to 0^+}\frac{Ee^E+Ee^{-E}-e^E+e^{-E}}{Ee^E-Ee^{-E}} \qquad [\text{form is } \tfrac{0}{0}]$$

$$\overset{\text{H}}{=} \lim_{E\to 0^+}\frac{Ee^E+e^E\cdot 1+E(-e^{-E})+e^{-E}\cdot 1-e^E+(-e^{-E})}{Ee^E+e^E\cdot 1-[E(-e^{-E})+e^{-E}\cdot 1]}$$

$$= \lim_{E\to 0^+}\frac{Ee^E-Ee^{-E}}{Ee^E+e^E+Ee^{-E}-e^{-E}} = \lim_{E\to 0^+}\frac{e^E-e^{-E}}{e^E+\dfrac{e^E}{E}+e^{-E}-\dfrac{e^{-E}}{E}} \qquad [\text{divide by } E]$$

$$= \frac{0}{2+L}, \text{ where } L = \lim_{E\to 0^+}\frac{e^E-e^{-E}}{E} \qquad [\text{form is } \tfrac{0}{0}]$$

$$\overset{\text{H}}{=} \lim_{E\to 0^+}\frac{e^E+e^{-E}}{1} = \frac{1+1}{1} = 2$$

Thus, $\lim\limits_{E\to 0^+} P(E) = \dfrac{0}{2+2} = 0$.

44. (a) $$\lim_{R\to r^+} v = \lim_{R\to r^+}\left[-c\left(\frac{r}{R}\right)^2\ln\left(\frac{r}{R}\right)\right] = -cr^2\lim_{R\to r^+}\left[\left(\frac{1}{R}\right)^2\ln\left(\frac{r}{R}\right)\right] = -cr^2\cdot\frac{1}{r^2}\cdot\ln 1 = -c\cdot 0 = 0$$

As the insulation of a metal cable becomes thinner, the velocity of an electrical impulse in the cable approaches zero.

(b) $$\lim_{r\to 0^+} v = \lim_{r\to 0^+}\left[-c\left(\frac{r}{R}\right)^2\ln\left(\frac{r}{R}\right)\right] = -\frac{c}{R^2}\lim_{r\to 0^+}\left[r^2\ln\left(\frac{r}{R}\right)\right] \qquad [\text{form is } 0\cdot\infty]$$

$$= -\frac{c}{R^2}\lim_{r\to 0^+}\frac{\ln\left(\dfrac{r}{R}\right)}{\dfrac{1}{r^2}} \quad [\text{form is } \infty/\infty] \quad \overset{\text{H}}{=} -\frac{c}{R^2}\lim_{r\to 0^+}\frac{\dfrac{R}{r}\cdot\dfrac{1}{R}}{\dfrac{-2}{r^3}} = -\frac{c}{R^2}\lim_{r\to 0^+}\left(-\frac{r^2}{2}\right) = 0$$

As the radius of the metal cable approaches zero, the velocity of an electrical impulse in the cable approaches zero.

45. We see that both numerator and denominator approach 0, so we can use l'Hospital's Rule:

$$\lim_{x\to a}\frac{\sqrt{2a^3x-x^4}-a\sqrt[3]{aax}}{a-\sqrt[4]{ax^3}}\overset{\text{H}}{=}\lim_{x\to a}\frac{\frac12(2a^3x-x^4)^{-1/2}(2a^3-4x^3)-a\left(\frac13\right)(aax)^{-2/3}a^2}{-\frac14(ax^3)^{-3/4}(3ax^2)}$$

$$=\frac{\frac12(2a^3a-a^4)^{-1/2}(2a^3-4a^3)-\frac13a^3(a^2a)^{-2/3}}{-\frac14(aa^3)^{-3/4}(3aa^2)}$$

$$=\frac{(a^4)^{-1/2}(-a^3)-\frac13a^3(a^3)^{-2/3}}{-\frac34a^3(a^4)^{-3/4}}=\frac{-a-\frac13a}{-\frac34}=\frac43\left(\frac43a\right)=\frac{16}{9}a$$

46. Let the radius of the circle be r. We see that $A(\theta)$ is the area of the whole figure (a sector of the circle with radius 1), minus the area of $\triangle OPR$. But the area of the sector of the circle is $\frac12r^2\theta$ (see Reference Page 1), and the area of the triangle is $\frac12r\,|PQ|=\frac12r(r\sin\theta)=\frac12r^2\sin\theta$. So we have $A(\theta)=\frac12r^2\theta-\frac12r^2\sin\theta=\frac12r^2(\theta-\sin\theta)$. Now by elementary trigonometry, $B(\theta)=\frac12\,|QR|\,|PQ|=\frac12(r-|OQ|)\,|PQ|=\frac12(r-r\cos\theta)(r\sin\theta)=\frac12r^2(1-\cos\theta)\sin\theta$.

So the limit we want is

$$\lim_{\theta\to0^+}\frac{A(\theta)}{B(\theta)}\lim_{\theta\to0^+}\frac{A(\theta)}{B(\theta)}=\lim_{\theta\to0^+}\frac{\frac12r^2(\theta-\sin\theta)}{\frac12r^2(1-\cos\theta)\sin\theta}\overset{\text{H}}{=}\lim_{\theta\to0^+}\frac{1-\cos\theta}{(1-\cos\theta)\cos\theta+\sin\theta\,(\sin\theta)}$$

$$=\lim_{\theta\to0^+}\frac{1-\cos\theta}{\cos\theta-\cos^2\theta+\sin^2\theta}\overset{\text{H}}{=}\lim_{\theta\to0^+}\frac{\sin\theta}{-\sin\theta-2\cos\theta\,(-\sin\theta)+2\sin\theta\,(\cos\theta)}$$

$$=\lim_{\theta\to0^+}\frac{\sin\theta}{-\sin\theta+4\sin\theta\,\cos\theta}=\lim_{\theta\to0^+}\frac{1}{-1+4\cos\theta}=\frac{1}{-1+4\cos0}=\frac13$$

47. Since $f(2)=0$, the given limit has the form $\frac00$.

$$\lim_{x\to0}\frac{f(2+3x)+f(2+5x)}{x}\overset{\text{H}}{=}\lim_{x\to0}\frac{f'(2+3x)\cdot3+f'(2+5x)\cdot5}{1}=f'(2)\cdot3+f'(2)\cdot5=8f'(2)=8\cdot7=56$$

48. $L=\lim\limits_{x\to0}\left(\dfrac{\sin2x}{x^3}+a+\dfrac{b}{x^2}\right)=\lim\limits_{x\to0}\dfrac{\sin2x+ax^3+bx}{x^3}\overset{\text{H}}{=}\lim\limits_{x\to0}\dfrac{2\cos2x+3ax^2+b}{3x^2}$. As $x\to0$, $3x^2\to0$, and $(2\cos2x+3ax^2+b)\to b+2$, so the last limit exists only if $b+2=0$, that is, $b=-2$. Thus,

$$\lim_{x\to0}\frac{2\cos2x+3ax^2-2}{3x^2}\overset{\text{H}}{=}\lim_{x\to0}\frac{-4\sin2x+6ax}{6x}\overset{\text{H}}{=}\lim_{x\to0}\frac{-8\cos2x+6a}{6}=\frac{6a-8}{6},$$ which is equal to 0 if and only if $a=\frac43$. Hence, $L=0$ if and only if $b=-2$ and $a=\frac43$.

49. Since $\lim\limits_{h\to0}[f(x+h)-f(x-h)]=f(x)-f(x)=0$ (f is differentiable and hence continuous) and $\lim\limits_{h\to0}2h=0$, we use l'Hospital's Rule:

$$\lim_{h\to0}\frac{f(x+h)-f(x-h)}{2h}\overset{\text{H}}{=}\lim_{h\to0}\frac{f'(x+h)(1)-f'(x-h)(-1)}{2}$$

$$=\frac{f'(x)+f'(x)}{2}=\frac{2f'(x)}{2}=f'(x)$$

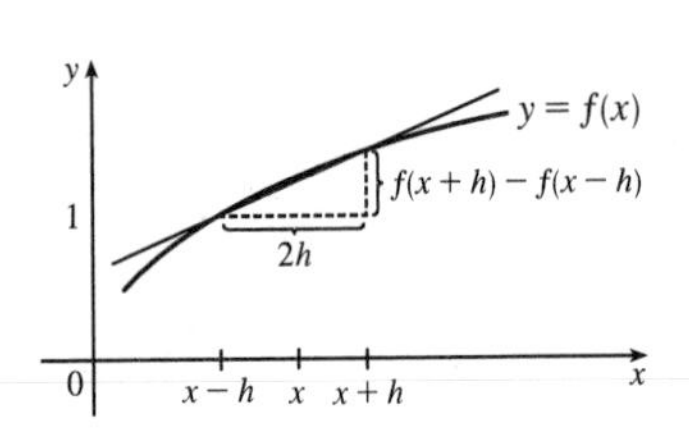

$\dfrac{f(x+h)-f(x-h)}{2h}$ is the slope of the secant line between $(x-h,f(x-h))$ and $(x+h,f(x+h))$. As $h\to0$, this line gets closer to the tangent line and its slope approaches $f'(x)$.

50. Since $\lim_{h\to 0}[f(x+h)-2f(x)+f(x-h)] = f(x)-2f(x)+f(x) = 0$ (f is differentiable and hence continuous) and $\lim_{h\to 0} h^2 = 0$, we can apply l'Hospital's Rule:

$$\lim_{h\to 0}\frac{f(x+h)-2f(x)+f(x-h)}{h^2} \overset{\text{H}}{=} \lim_{h\to 0}\frac{f'(x+h)-f'(x-h)}{2h} = f''(x)$$

At the last step, we have applied the result of Exercise 49 to $f'(x)$.

51. (a) We show that $\lim_{x\to 0}\frac{f(x)}{x^n} = 0$ for every integer $n \geq 0$. Let $y = \frac{1}{x^2}$. Then

$$\lim_{x\to 0}\frac{f(x)}{x^{2n}} = \lim_{x\to 0}\frac{e^{-1/x^2}}{(x^2)^n} = \lim_{y\to\infty}\frac{y^n}{e^y} \overset{\text{H}}{=} \lim_{y\to\infty}\frac{ny^{n-1}}{e^y} \overset{\text{H}}{=} \cdots \overset{\text{H}}{=} \lim_{y\to\infty}\frac{n!}{e^y} = 0 \;\Rightarrow$$

$$\lim_{x\to 0}\frac{f(x)}{x^n} = \lim_{x\to 0} x^n\frac{f(x)}{x^{2n}} = \lim_{x\to 0} x^n \lim_{x\to 0}\frac{f(x)}{x^{2n}} = 0. \text{ Thus, } f'(0) = \lim_{x\to 0}\frac{f(x)-f(0)}{x-0} = \lim_{x\to 0}\frac{f(x)}{x} = 0.$$

(b) Using the Chain Rule and the Quotient Rule we see that $f^{(n)}(x)$ exists for $x \neq 0$. In fact, we prove by induction that for each $n \geq 0$, there is a polynomial p_n and a non-negative integer k_n with $f^{(n)}(x) = p_n(x)f(x)/x^{k_n}$ for $x \neq 0$. This is true for $n = 0$; suppose it is true for the nth derivative. Then $f'(x) = f(x)(2/x^3)$, so

$$\begin{aligned} f^{(n+1)}(x) &= \left[x^{k_n}\left[p_n'(x)f(x) + p_n(x)f'(x)\right] - k_n x^{k_n-1}p_n(x)f(x)\right]x^{-2k_n} \\ &= \left[x^{k_n}p_n'(x) + p_n(x)\left(2/x^3\right) - k_n x^{k_n-1}p_n(x)\right]f(x)x^{-2k_n} \\ &= \left[x^{k_n+3}p_n'(x) + 2p_n(x) - k_n x^{k_n+2}p_n(x)\right]f(x)x^{-(2k_n+3)} \end{aligned}$$

which has the desired form.

Now we show by induction that $f^{(n)}(0) = 0$ for all n. By part (a), $f'(0) = 0$. Suppose that $f^{(n)}(0) = 0$. Then

$$\begin{aligned} f^{(n+1)}(0) &= \lim_{x\to 0}\frac{f^{(n)}(x) - f^{(n)}(0)}{x-0} = \lim_{x\to 0}\frac{f^{(n)}(x)}{x} = \lim_{x\to 0}\frac{p_n(x)f(x)/x^{k_n}}{x} = \lim_{x\to 0}\frac{p_n(x)f(x)}{x^{k_n+1}} \\ &= \lim_{x\to 0}p_n(x)\lim_{x\to 0}\frac{f(x)}{x^{k_n+1}} = p_n(0)\cdot 0 = 0 \end{aligned}$$

52. (a) For f to be continuous, we need $\lim_{x\to 0} f(x) = f(0) = 1$. We note that for $x \neq 0$, $\ln f(x) = \ln|x|^x = x\ln|x|$. So

$$\lim_{x\to 0}\ln f(x) = \lim_{x\to 0} x\ln|x| = \lim_{x\to 0}\frac{\ln|x|}{1/x} \overset{\text{H}}{=} \lim_{x\to 0}\frac{1/x}{-1/x^2} = 0. \text{ Therefore, } \lim_{x\to 0} f(x) = \lim_{x\to 0} e^{\ln f(x)} = e^0 = 1.$$

So f is continuous at 0.

(b) From the graphs, it appears that f is differentiable at 0.

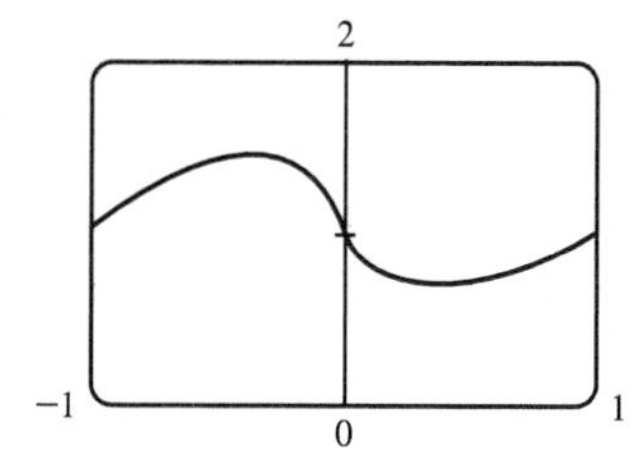

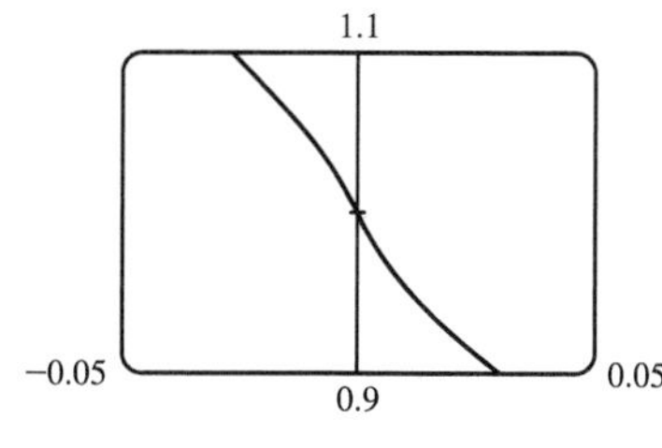

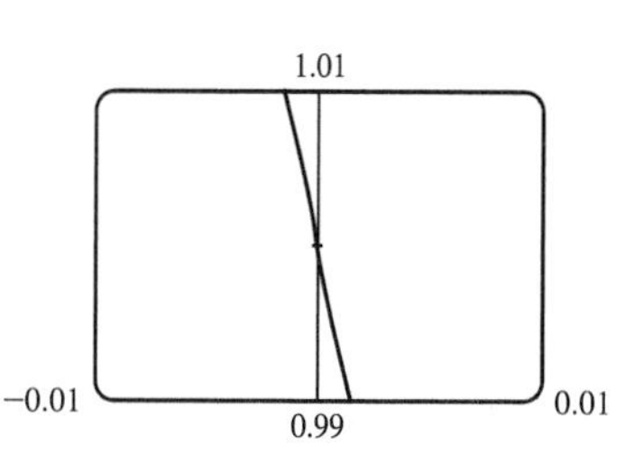

(c) To find f', we use logarithmic differentiation: $\ln f(x) = x \ln |x| \Rightarrow \dfrac{f'(x)}{f(x)} = x\left(\dfrac{1}{x}\right) + \ln|x| \Rightarrow$

$f'(x) = f(x)(1 + \ln|x|) = |x|^x (1 + \ln|x|)$, $x \neq 0$. Now $f'(x) \to -\infty$ as $x \to 0$ [since $|x|^x \to 1$ and $(1 + \ln|x|) \to -\infty$], so the curve has a vertical tangent at $(0, 1)$ and is therefore not differentiable there. The fact cannot be seen in the graphs in part (b) because $\ln|x| \to -\infty$ very slowly as $x \to 0$.

5 Review

CONCEPT CHECK

1. (a) A function f is called a *one-to-one function* if it never takes on the same value twice; that is, if $f(x_1) \neq f(x_2)$ whenever $x_1 \neq x_2$. (Or, f is 1-1 if each output corresponds to only one input.)

Use the Horizontal Line Test: A function is one-to-one if and only if no horizontal line intersects its graph more than once.

(b) If f is a one-to-one function with domain A and range B, then its *inverse function* f^{-1} has domain B and range A and is defined by $f^{-1}(y) = x \Leftrightarrow f(x) = y$ for any y in B. The graph of f^{-1} is obtained by reflecting the graph of f about the line $y = x$.

(c) $(f^{-1})'(a) = \dfrac{1}{f'(f^{-1}(a))}$

2. (a) $e = \lim\limits_{x \to 0} (1 + x)^{1/x}$

(b) $e \approx 2.71828$

(c) The differentiation formula for $y = a^x$ $[y' = a^x \ln a]$ is simplest when $a = e$ because $\ln e = 1$.

(d) The differentiation formula for $y = \log_a x$ $[y' = 1/(x \ln a)]$ is simplest when $a = e$ because $\ln e = 1$.

3. (a) The function $f(x) = e^x$ has domain $\mathbb{R}$ and range $(0, \infty)$.

(b) The function $f(x) = \ln x$ has domain $(0, \infty)$ and range $\mathbb{R}$.

(c) The graphs are reflections of one another about the line $y = x$. See Figure 5.3.1.

(d) $\log_a x = \dfrac{\ln x}{\ln a}$

4. (a) The inverse sine function $f(x) = \sin^{-1} x$ is defined as follows:

$$\sin^{-1} x = y \quad \Leftrightarrow \quad \sin y = x \quad \text{and} \quad -\frac{\pi}{2} \leq y \leq \frac{\pi}{2}$$

Its domain is $-1 \leq x \leq 1$ and its range is $-\frac{\pi}{2} \leq y \leq \frac{\pi}{2}$.

(b) The inverse cosine function $f(x) = \cos^{-1} x$ is defined as follows:

$$\cos^{-1} x = y \quad \Leftrightarrow \quad \cos y = x \quad \text{and} \quad 0 \leq y \leq \pi$$

Its domain is $-1 \leq x \leq 1$ and its range is $0 \leq y \leq \pi$.

(c) See Definition 5.6.7. Domain $= \mathbb{R}$, Range $= \left(-\frac{\pi}{2}, \frac{\pi}{2}\right)$. See Figure 10 in Section 5.6.

5. $\sinh x = \dfrac{e^x - e^{-x}}{2}$, $\cosh x = \dfrac{e^x + e^{-x}}{2}$, $\tanh x = \dfrac{\sinh x}{\cosh x} = \dfrac{e^x - e^{-x}}{e^x + e^{-x}}$

6. (a) $y = e^x \;\Rightarrow\; y' = e^x$
(b) $y = a^x \;\Rightarrow\; y' = a^x \ln a$
(c) $y = \ln x \;\Rightarrow\; y' = 1/x$
(d) $y = \log_a x \;\Rightarrow\; y' = 1/(x \ln a)$
(e) $y = \sin^{-1} x \;\Rightarrow\; y' = 1/\sqrt{1 - x^2}$
(f) $y = \cos^{-1} x \;\Rightarrow\; y' = -1/\sqrt{1 - x^2}$
(g) $y = \tan^{-1} x \;\Rightarrow\; y' = 1/(1 + x^2)$
(h) $y = \sinh x \;\Rightarrow\; y' = \cosh x$
(i) $y = \cosh x \;\Rightarrow\; y' = \sinh x$
(j) $y = \tanh x \;\Rightarrow\; y' = \operatorname{sech}^2 x$
(k) $y = \sinh^{-1} x \;\Rightarrow\; y' = 1/\sqrt{1 + x^2}$
(l) $y = \cosh^{-1} x \;\Rightarrow\; y' = 1/\sqrt{x^2 - 1}$
(m) $y = \tanh^{-1} x \;\Rightarrow\; y' = 1/(1 - x^2)$

7. (a) $\dfrac{dy}{dt} = ky$; the relative growth rate, $\dfrac{1}{y}\dfrac{dy}{dt}$, is constant.

(b) The equation in part (a) is an appropriate model for population growth if we assume that there is enough room and nutrition to support the growth.

(c) If $y(0) = y_0$, then the solution is $y(t) = y_0 e^{kt}$.

8. (a) See l'Hospital's Rule and the three notes that follow it in Section 5.8.

(b) Write fg as $\dfrac{f}{1/g}$ or $\dfrac{g}{1/f}$.

(c) Convert the difference into a quotient using a common denominator, rationalizing, factoring, or some other method.

(d) Convert the power to a product by taking the natural logarithm of both sides of $y = f^g$ or by writing f^g as $e^{g \ln f}$.

TRUE-FALSE QUIZ

1. True. If f is one-to-one, with domain $\mathbb{R}$, then $f^{-1}(f(6)) = 6$ by the first cancellation equation in (3.2.4).

2. False. By Theorem 3.2.7, $(f^{-1})'(6) = \dfrac{1}{f'(f^{-1}(6))}$, *not* $\dfrac{1}{f'(6)}$ unless $f^{-1}(6) = 6$.

3. False. For example, $\cos \frac{\pi}{2} = \cos\left(-\frac{\pi}{2}\right)$, so $\cos x$ is not 1-1.

4. False. It is true that $\tan \frac{3\pi}{4} = -1$, but since the range of $\tan^{-1}$ is $\left(-\frac{\pi}{2}, \frac{\pi}{2}\right)$, we must have $\tan^{-1}(-1) = -\frac{\pi}{4}$.

5. True, since $\ln x$ is an increasing function on $(0, \infty)$.

6. True. $\pi^{\sqrt{5}} = \left(e^{\ln \pi}\right)^{\sqrt{5}} = e^{\sqrt{5} \ln \pi}$

7. True. We can divide by e^x since $e^x \neq 0$ for every x.

8. False. For example, $\ln(1 + 1) = \ln 2$, but $\ln 1 + \ln 1 = 0$. In fact $\ln a + \ln b = \ln(ab)$.

9. False. Let $x = e$. Then $(\ln x)^6 = (\ln e)^6 = 1^6 = 1$, but $6 \ln x = 6 \ln e = 6 \cdot 1 = 6 \neq 1 = (\ln x)^6$.

10. False. $\dfrac{d}{dx} 10^x = 10^x \ln 10$

11. False. $\ln 10$ is a constant, so its derivative is 0.

12. True. $y = e^{3x} \Rightarrow \ln y = 3x \Rightarrow x = \frac{1}{3}\ln y \Rightarrow$ the inverse function is $y = \frac{1}{3}\ln x$.

13. False. The "-1" is not an exponent; it is an indication of an inverse function.

14. False. For example, $\tan^{-1} 20$ is defined; $\sin^{-1} 20$ and $\cos^{-1} 20$ are not.

15. True. See Figure 3.6.2.

16. False. L'Hospital's Rule does not apply since $\lim\limits_{x\to\pi^-} \dfrac{\tan x}{1-\cos x} = \dfrac{0}{2} = 0.$

EXERCISES

1. No. f is not 1-1 because the graph of f fails the Horizontal Line Test.

2. (a) g is one-to-one because it passes the Horizontal Line Test.

(b) When $y = 2$, $x \approx 0.2$. So $g^{-1}(2) \approx 0.2$.

(c) The range of g is $[-1, 3.5]$, which is the same as the domain of g^{-1}.

(d) We reflect the graph of g through the line $y = x$ to obtain the graph of g^{-1}.

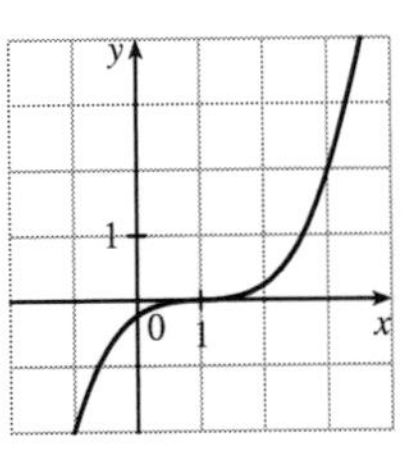

3. (a) $f^{-1}(3) = 7$ since $f(7) = 3$.

(b) $(f^{-1})'(3) = \dfrac{1}{f'(f^{-1}(3))} = \dfrac{1}{f'(7)} = \dfrac{1}{8}$

4. $y = \dfrac{x+1}{2x+1}$. Interchanging x and y gives us $x = \dfrac{y+1}{2y+1} \Rightarrow 2xy + x = y + 1 \Rightarrow 2xy - y = 1 - x \Rightarrow$

$y(2x-1) = 1 - x \Rightarrow y = \dfrac{1-x}{2x-1} = f^{-1}(x)$.

5.

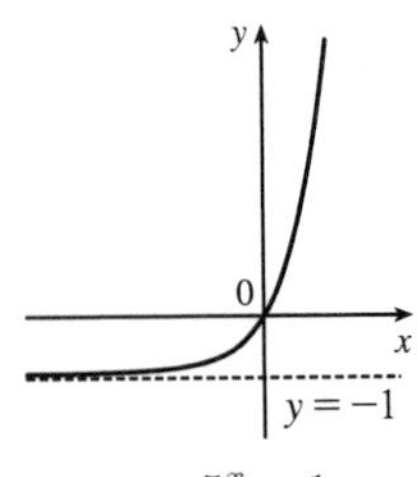

$y = 5^x - 1$

6.

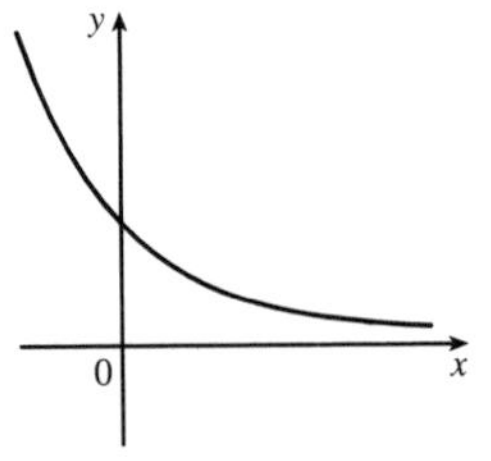

$y = e^{-x}$

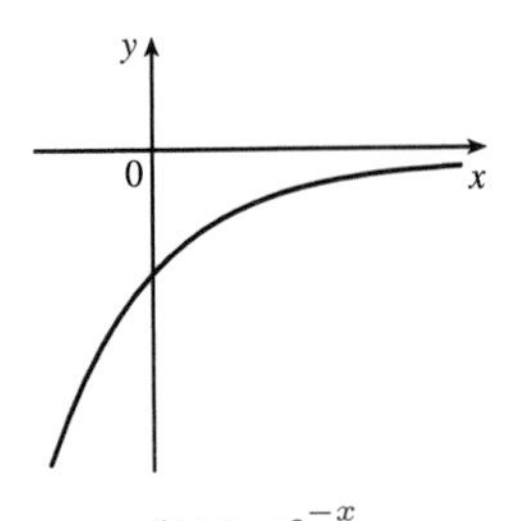

$y = -e^{-x}$

7. Reflect the graph of $y = \ln x$ about the x-axis to obtain the graph of $y = -\ln x$.

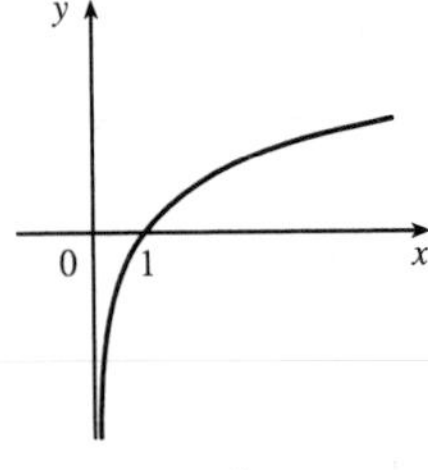

$y = \ln x$

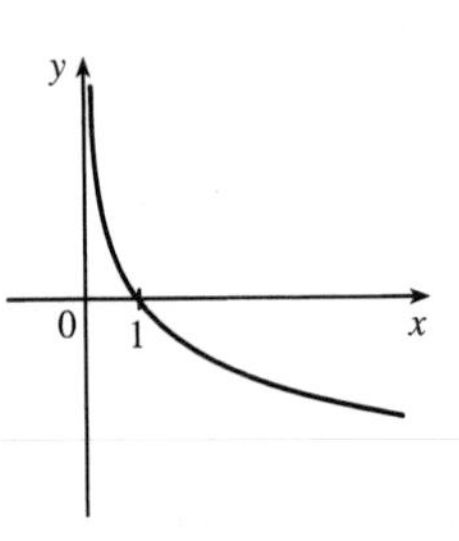

$y = -\ln x$

8. 

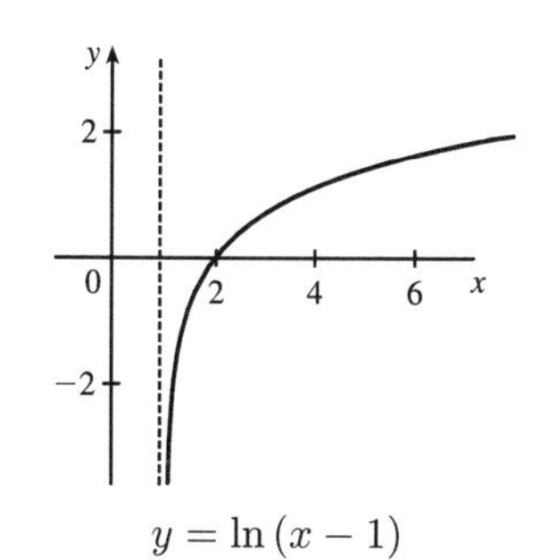

$y = \ln(x-1)$

9.

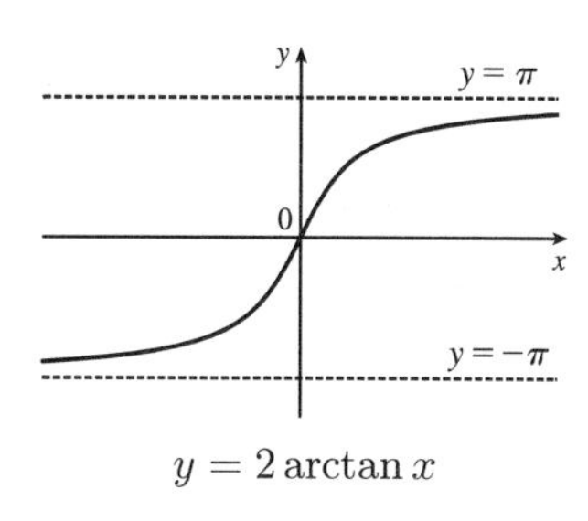

$y = 2\arctan x$

10. We have seen that if $a > 1$, then $a^x > x^a$ for sufficiently large x. (See Exercise 5.4.20.) In general, we could show that $\lim\limits_{x\to\infty}(a^x/x^a) = \infty$ by using l'Hospital's Rule repeatedly. Also, $\log_a x$ increases much more slowly than either x^a or a^x. [Compare the graph of $\log_a x$ with those of x^a and a^x, or use l'Hospital's Rule to show that $\lim\limits_{x\to\infty}[(\log_a x)/x^a] = 0$.] So for large x, $\log_a x < x^a < a^x$.

11. (a) $e^{2\ln 3} = \left(e^{\ln 3}\right)^2 = 3^2 = 9$

(b) $\log_{10} 25 + \log_{10} 4 = \log_{10}(25 \cdot 4) = \log_{10} 100 = \log_{10} 10^2 = 2$

12. (a) $\ln e^{\pi} = \pi$

(b) $\tan\left(\arcsin \frac{1}{2}\right) = \tan \frac{\pi}{6} = \frac{1}{\sqrt{3}}$

13. $\ln x = \frac{1}{3} \quad \Leftrightarrow \quad \log_e x = \frac{1}{3} \quad \Rightarrow \quad x = e^{1/3}$

14. $e^x = \frac{1}{3} \quad \Rightarrow \quad x = \ln \frac{1}{3} = \ln 1 - \ln 3 = -\ln 3$

15. $e^{e^x} = 17 \quad \Rightarrow \quad \ln e^{e^x} = \ln 17 \quad \Rightarrow \quad e^x = \ln 17 \quad \Rightarrow \quad \ln e^x = \ln(\ln 17) \quad \Rightarrow \quad x = \ln\ln 17$

16. $\ln(1+e^{-x}) = 3 \quad \Rightarrow \quad 1 + e^{-x} = e^3 \quad \Rightarrow \quad e^{-x} = e^3 - 1 \quad \Rightarrow \quad \ln e^{-x} = \ln(e^3-1) \quad \Rightarrow \quad -x = \ln(e^3-1) \quad \Rightarrow$ $x = -\ln(e^3-1)$

17. $\ln(x+1) + \ln(x-1) = 1 \quad \Rightarrow \quad \ln[(x+1)(x-1)] = 1 \quad \Rightarrow \quad \ln(x^2-1) = \ln e \quad \Rightarrow \quad x^2 - 1 = e \quad \Rightarrow \quad x^2 = e+1$ $\Rightarrow \quad x = \sqrt{e+1}$ since $\ln(x-1)$ is defined only when $x > 1$.

18. $\log_5(c^x) = d \quad \Rightarrow \quad x\log_5 c = d \quad \Rightarrow \quad x = \dfrac{d}{\log_5 c}$.

Or: $\log_5(c^x) = d \quad \Rightarrow \quad 5^d = c^x \quad \Rightarrow \quad \ln 5^d = \ln c^x \quad \Rightarrow \quad d\ln 5 = x\ln c \quad \Rightarrow \quad x = \dfrac{d\ln 5}{\ln c}$.

19. $\tan^{-1} x = 1 \quad \Rightarrow \quad \tan\tan^{-1} x = \tan 1 \quad \Rightarrow \quad x = \tan 1 \ (\approx 1.5574)$

20. $\sin x = 0.3 \quad \Rightarrow \quad x = \sin^{-1} 0.3 = \alpha$ for $-\frac{\pi}{2} \le x \le \frac{\pi}{2}$. The reference angle for α is $\pi - \alpha$, so all solutions are $x = \alpha + 2n\pi$ and $x = \pi - \alpha + 2n\pi$ [or $(2n+1)\pi - \alpha$]

21. $f(t) = t^2\ln t \quad \Rightarrow \quad f'(t) = t^2 \cdot \dfrac{1}{t} + (\ln t)(2t) = t + 2t\ln t$ or $t(1 + 2\ln t)$

22. $g(t) = \dfrac{e^t}{1+e^t} \quad \Rightarrow \quad g'(t) = \dfrac{(1+e^t)e^t - e^t(e^t)}{(1+e^t)^2} = \dfrac{e^t}{(1+e^t)^2}$

23. $h(\theta) = e^{\tan 2\theta} \quad \Rightarrow \quad h'(\theta) = e^{\tan 2\theta} \cdot \sec^2 2\theta \cdot 2 = 2\sec^2(2\theta)\, e^{\tan 2\theta}$

24. $h(u) = 10^{\sqrt{u}} \quad \Rightarrow \quad h'(u) = 10^{\sqrt{u}} \cdot \ln 10 \cdot \dfrac{1}{2\sqrt{u}} = \dfrac{(\ln 10)10^{\sqrt{u}}}{2\sqrt{u}}$

25. $y = \ln|\sec 5x + \tan 5x| \quad \Rightarrow$

$$y' = \frac{1}{\sec 5x + \tan 5x}(\sec 5x\, \tan 5x \cdot 5 + \sec^2 5x \cdot 5) = \frac{5\sec 5x\,(\tan 5x + \sec 5x)}{\sec 5x + \tan 5x} = 5\sec 5x$$

26. $y = e^{-t}(t^2 - 2t + 2) \quad \Rightarrow \quad y' = e^{-t}(2t - 2) + (t^2 - 2t + 2)(-e^{-t}) = e^{-t}(2t - 2 - t^2 + 2t - 2) = e^{-t}(-t^2 + 4t - 4)$

27. $y = e^{cx}(c\sin x - \cos x) \quad \Rightarrow \quad y' = ce^{cx}(c\sin x - \cos x) + e^{cx}(c\cos x + \sin x) = (c^2 + 1)e^{cx}\sin x$

28. $y = \sin^{-1}(e^x) \quad \Rightarrow \quad y' = 1/\sqrt{1 - (e^x)^2} \cdot e^x = e^x/\sqrt{1 - e^{2x}}$

29. $y = \ln(\sec^2 x) = 2\ln|\sec x| \quad \Rightarrow \quad y' = (2/\sec x)(\sec x \tan x) = 2\tan x$

30. $y = \ln(x^2 e^x) = 2\ln|x| + x \quad \Rightarrow \quad y' = 2/x + 1$

31. $y = xe^{-1/x} \quad \Rightarrow \quad y' = e^{-1/x} + xe^{-1/x}(1/x^2) = e^{-1/x}(1 + 1/x)$

32. $y = x^r e^{sx} \quad \Rightarrow \quad y' = rx^{r-1}e^{sx} + sx^r e^{sx}$

33. $y = 2^{-t^2} \quad \Rightarrow \quad y' = 2^{-t^2}(\ln 2)(-2t) = (-2\ln 2)t\,2^{-t^2}$

34. $y = e^{\cos x} + \cos(e^x) \quad \Rightarrow \quad y' = -\sin x\, e^{\cos x} - e^x \sin(e^x)$

35. $H(v) = v\tan^{-1} v \quad \Rightarrow \quad H'(v) = v \cdot \dfrac{1}{1 + v^2} + \tan^{-1} v \cdot 1 = \dfrac{v}{1 + v^2} + \tan^{-1} v$

36. $F(z) = \log_{10}(1 + z^2) \quad \Rightarrow \quad F'(z) = \dfrac{1}{(\ln 10)(1 + z^2)} \cdot 2z = \dfrac{2z}{(\ln 10)(1 + z^2)}$

37. $y = x\sinh(x^2) \quad \Rightarrow \quad y' = x\cosh(x^2) \cdot 2x + \sinh(x^2) \cdot 1 = 2x^2\cosh(x^2) + \sinh(x^2)$

38. $y = (\cos x)^x \quad \Rightarrow \quad \ln y = \ln(\cos x)^x = x\ln\cos x \quad \Rightarrow \quad \dfrac{y'}{y} = x \cdot \dfrac{1}{\cos x} \cdot (-\sin x) + \ln\cos x \cdot 1 \quad \Rightarrow$

$$y' = (\cos x)^x(\ln\cos x - x\tan x)$$

39. $y = \ln\sin x - \frac{1}{2}\sin^2 x \quad \Rightarrow \quad y' = \dfrac{1}{\sin x} \cdot \cos x - \frac{1}{2} \cdot 2\sin x \cdot \cos x = \cot x - \sin x\,\cos x$

40. $y = \arctan(\arcsin\sqrt{x}\,) \quad \Rightarrow \quad y' = \dfrac{1}{1 + (\arcsin\sqrt{x}\,)^2} \cdot \dfrac{1}{\sqrt{1 - x}} \cdot \dfrac{1}{2\sqrt{x}}$

41. $y = \ln\left(\dfrac{1}{x}\right) + \dfrac{1}{\ln x} = \ln x^{-1} + (\ln x)^{-1} = -\ln x + (\ln x)^{-1} \quad \Rightarrow \quad y' = -1 \cdot \dfrac{1}{x} + (-1)(\ln x)^{-2} \cdot \dfrac{1}{x} = -\dfrac{1}{x} - \dfrac{1}{x\,(\ln x)^2}$

42. $xe^y = y - 1 \quad \Rightarrow \quad e^y + xe^y y' = y' \quad \Rightarrow \quad y' = e^y/(1 - xe^y)$

43. $y = \ln(\cosh 3x) \quad \Rightarrow \quad y' = (1/\cosh 3x)(\sinh 3x)(3) = 3\tanh 3x$

44. $y = \dfrac{(x^2+1)^4}{(2x+1)^3(3x-1)^5} \quad \Rightarrow$

$$\ln y = \ln \frac{(x^2+1)^4}{(2x+1)^3(3x-1)^5} = \ln(x^2+1)^4 - \ln[(2x+1)^3(3x-1)^5]$$

$$= 4\ln(x^2+1) - [\ln(2x+1)^3 + \ln(3x-1)^5] = 4\ln(x^2+1) - 3\ln(2x+1) - 5\ln(3x-1) \quad \Rightarrow$$

$$\frac{y'}{y} = 4\cdot\frac{1}{x^2+1}\cdot 2x - 3\cdot\frac{1}{2x+1}\cdot 2 - 5\cdot\frac{1}{3x-1}\cdot 3 \quad \Rightarrow \quad y' = \frac{(x^2+1)^4}{(2x+1)^3(3x-1)^5}\left(\frac{8x}{x^2+1} - \frac{6}{2x+1} - \frac{15}{3x-1}\right).$$

[The answer could be simplified to $y' = -\dfrac{(x^2+56x+9)(x^2+1)^3}{(2x+1)^4(3x-1)^6}$, but this is unnecessary.]

45. $y = \cosh^{-1}(\sinh x) \quad \Rightarrow \quad y' = (\cosh x)/\sqrt{\sinh^2 x - 1}$

46. $y = x\tanh^{-1}\sqrt{x} \quad \Rightarrow \quad y' = \tanh^{-1}\sqrt{x} + x\dfrac{1}{1-(\sqrt{x})^2}\dfrac{1}{2\sqrt{x}} = \tanh^{-1}\sqrt{x} + \dfrac{\sqrt{x}}{2(1-x)}$

47. $f(x) = e^{\sin^3(\ln(x^2+1))} \quad \Rightarrow$

$$f'(x) = e^{\sin^3(\ln(x^2+1))}\cdot 3\sin^2(\ln(x^2+1))\cdot\cos(\ln(x^2+1))\cdot\frac{1}{x^2+1}\cdot 2x$$

$$= \frac{6x}{x^2+1}\sin^2(\ln(x^2+1))\cdot\cos(\ln(x^2+1))\cdot e^{\sin^3(\ln(x^2+1))}$$

48. $\dfrac{d}{dx}\left(\dfrac{1}{2}\tan^{-1}x + \dfrac{1}{4}\ln\dfrac{(x+1)^2}{x^2+1}\right) = \dfrac{d}{dx}\left(\dfrac{1}{2}\tan^{-1}x + \dfrac{1}{2}\ln|x+1| - \dfrac{1}{4}\ln(x^2+1)\right)$

$$= \frac{1}{2}\frac{1}{x^2+1} + \frac{1}{2}\frac{1}{x+1} - \frac{1}{4}\frac{2x}{x^2+1} = \frac{1}{2}\left(\frac{1}{x^2+1} - \frac{x}{x^2+1} + \frac{1}{x+1}\right)$$

$$= \frac{1}{2}\left(\frac{1-x}{x^2+1} + \frac{1}{x+1}\right) = \frac{1}{2}\left(\frac{1-x^2}{(x^2+1)(1+x)} + \frac{x^2+1}{(x^2+1)(1+x)}\right)$$

$$= \frac{1}{2}\frac{2}{(x^2+1)(1+x)} = \frac{1}{(1+x)(x^2+1)}$$

49. $f(x) = e^{g(x)} \quad \Rightarrow \quad f'(x) = e^{g(x)}g'(x)$

50. $f(x) = g(e^x) \quad \Rightarrow \quad f'(x) = g'(e^x)e^x$

51. $f(x) = \ln|g(x)| \quad \Rightarrow \quad f'(x) = \dfrac{1}{g(x)}g'(x) = \dfrac{g'(x)}{g(x)}$

52. $f(x) = g(\ln x) \quad \Rightarrow \quad f'(x) = g'(\ln x)\cdot\dfrac{1}{x} = \dfrac{g'(\ln x)}{x}$

53. $f(x) = 2^x \quad \Rightarrow \quad f'(x) = 2^x\ln 2 \quad \Rightarrow \quad f''(x) = 2^x(\ln 2)^2 \quad \Rightarrow \quad \cdots \quad \Rightarrow \quad f^{(n)}(x) = 2^x(\ln 2)^n$

54. $f(x) = \ln(2x) = \ln 2 + \ln x \quad \Rightarrow \quad f'(x) = x^{-1},\ f''(x) = -x^{-2},\ f'''(x) = 2x^{-3},\ f^{(4)}(x) = -2\cdot 3x^{-4}, \ldots,$

$f^{(n)}(x) = (-1)^{n-1}(n-1)!\,x^{-n}$

55. We first show it is true for $n = 1$: $f'(x) = e^x + xe^x = (x+1)e^x$. We now assume it is true for $n = k$: $f^{(k)}(x) = (x+k)e^x$. With this assumption, we must show it is true for $n = k+1$:

$$f^{(k+1)}(x) = \frac{d}{dx}\left[f^{(k)}(x)\right] = \frac{d}{dx}\left[(x+k)e^x\right] = e^x + (x+k)e^x = [x+(k+1)]\,e^x.$$

Therefore, $f^{(n)}(x) = (x+n)e^x$ by mathematical induction.

56. Using implicit differentiation, $y = x + \arctan y \;\Rightarrow\; y' = 1 + \dfrac{1}{1+y^2}y' \;\Rightarrow\; y'\left(1 - \dfrac{1}{1+y^2}\right) = 1 \;\Rightarrow$

$y'\left(\dfrac{y^2}{1+y^2}\right) = 1 \;\Rightarrow\; y' = \dfrac{1+y^2}{y^2} = \dfrac{1}{y^2} + 1.$

57. $y = (2+x)e^{-x} \;\Rightarrow\; y' = (2+x)(-e^{-x}) + e^{-x}\cdot 1 = e^{-x}[-(2+x)+1] = e^{-x}(-x-1)$. At $(0,2)$, $y' = 1(-1) = -1$, so an equation of the tangent line is $y - 2 = -1(x-0)$, or $y = -x+2$.

58. $y = f(x) = x\ln x \;\Rightarrow\; f'(x) = \ln x + 1$, so the slope of the tangent at (e,e) is $f'(e) = 2$ and an equation is $y - e = 2(x-e)$ or $y = 2x - e$.

59. $y = [\ln(x+4)]^2 \;\Rightarrow\; y' = 2[\ln(x+4)]^1 \cdot \dfrac{1}{x+4}\cdot 1 = 2\dfrac{\ln(x+4)}{x+4}$ and $y' = 0 \;\Leftrightarrow\; \ln(x+4) = 0 \;\Leftrightarrow$

$x + 4 = e^0 \;\Rightarrow\; x+4 = 1 \;\Leftrightarrow\; x = -3$, so the tangent is horizontal at the point $(-3, 0)$.

60. $f(x) = xe^{\sin x} \;\Rightarrow\; f'(x) = x[e^{\sin x}(\cos x)] + e^{\sin x}(1) = e^{\sin x}(x\cos x + 1)$. As a check on our work, we notice from the graphs that $f'(x) > 0$ when f is increasing. Also, we see in the larger viewing rectangle a certain similarity in the graphs of f and f': the sizes of the oscillations of f and f' are linked.

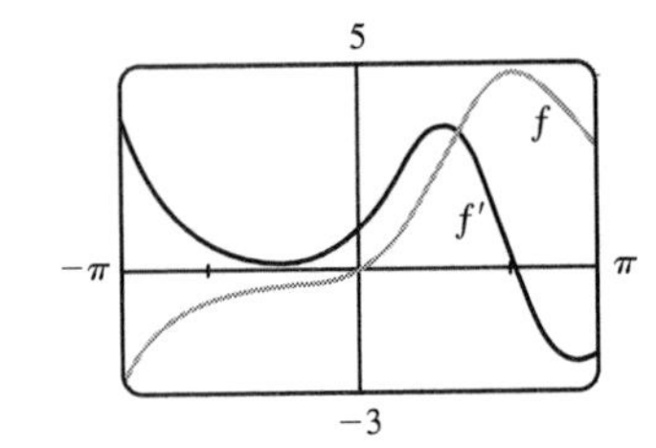

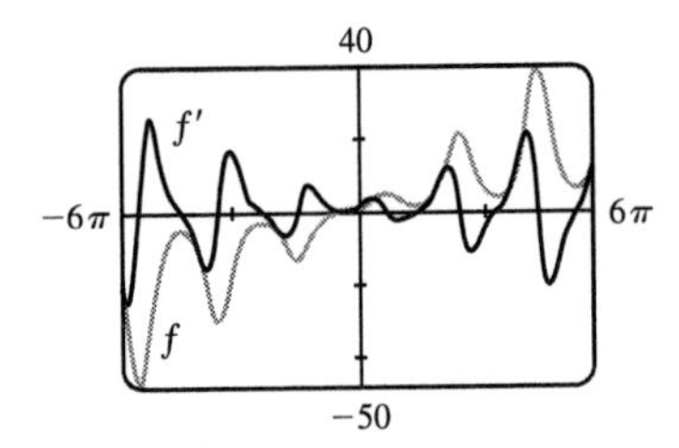

61. (a) The line $x - 4y = 1$ has slope $\frac{1}{4}$. A tangent to $y = e^x$ has slope $\frac{1}{4}$ when $y' = e^x = \frac{1}{4} \;\Rightarrow\; x = \ln\frac{1}{4} = -\ln 4$. Since $y = e^x$, the y-coordinate is $\frac{1}{4}$ and the point of tangency is $\left(-\ln 4, \frac{1}{4}\right)$. Thus, an equation of the tangent line is $y - \frac{1}{4} = \frac{1}{4}(x + \ln 4)$ or $y = \frac{1}{4}x + \frac{1}{4}(\ln 4 + 1)$.

(b) The slope of the tangent at the point (a, e^a) is $\left.\dfrac{d}{dx}e^x\right|_{x=a} = e^a$. Thus, an equation of the tangent line is $y - e^a = e^a(x-a)$. We substitute $x = 0$, $y = 0$ into this equation, since we want the line to pass through the origin: $0 - e^a = e^a(0-a) \;\Leftrightarrow\; -e^a = e^a(-a) \;\Leftrightarrow\; a = 1$. So an equation of the tangent line at the point $(a, e^a) = (1, e)$ is $y - e = e(x-1)$ or $y = ex$.

62. (a) $\lim\limits_{t\to\infty} C(t) = \lim\limits_{t\to\infty}[K(e^{-at} - e^{-bt})] = K\lim\limits_{t\to\infty}(e^{-at} - e^{-bt}) = K(0-0) = 0$ because $-at \to -\infty$ and $-bt \to -\infty$ as $t \to \infty$.

(b) $C(t) = K(e^{-at} - e^{-bt}) \;\Rightarrow\; C'(t) = K(e^{-at}(-a) - e^{-bt}(-b)) = K(-ae^{-at} + be^{-bt})$

(c) $C'(t) = 0 \;\Rightarrow\; be^{-bt} = ae^{-at} \;\Rightarrow\; \dfrac{b}{a} = e^{(-a+b)t} \;\Rightarrow\; \ln\dfrac{b}{a} = (b-a)t \;\Rightarrow\; t = \dfrac{\ln(b/a)}{b-a}$

63. (a) $y(t) = y(0)e^{kt} = 200e^{kt} \;\Rightarrow\; y(0.5) = 200e^{0.5k} = 360 \;\Rightarrow\; e^{0.5k} = 1.8 \;\Rightarrow\; 0.5k = \ln 1.8 \;\Rightarrow$

$k = 2\ln 1.8 = \ln(1.8)^2 = \ln 3.24 \;\Rightarrow\; y(t) = 200e^{(\ln 3.24)t} = 200(3.24)^t$

(b) $y(4) = 200(3.24)^4 \approx 22{,}040$ bacteria

(c) $y'(t) = 200(3.24)^t \cdot \ln 3.24$, so $y'(4) = 200(3.24)^4 \cdot \ln 3.24 \approx 25{,}910$ bacteria per hour

(d) $200(3.24)^t = 10{,}000 \;\Rightarrow\; (3.24)^t = 50 \;\Rightarrow\; t \ln 3.24 = \ln 50 \;\Rightarrow\; t = \ln 50/\ln 3.24 \approx 3.33$ hours

64. (a) If $y(t)$ is the mass remaining after t years, then $y(t) = y(0)e^{kt} = 100e^{kt}$. $y(5.24) = 100e^{5.24k} = \frac{1}{2} \cdot 100 \;\Rightarrow$ $e^{5.24k} = \frac{1}{2} \;\Rightarrow\; 5.24k = -\ln 2 \;\Rightarrow\; k = -\frac{1}{5.24}\ln 2 \;\Rightarrow\; y(t) = 100e^{-(\ln 2)t/5.24} = 100 \cdot 2^{-t/5.24}$. Thus, $y(20) = 100 \cdot 2^{-20/5.24} \approx 7.1$ mg.

(b) $100 \cdot 2^{-t/5.24} = 1 \;\Rightarrow\; 2^{-t/5.24} = \dfrac{1}{100} \;\Rightarrow\; -\dfrac{t}{5.24}\ln 2 = \ln\dfrac{1}{100} \;\Rightarrow\; t = 5.24\,\dfrac{\ln 100}{\ln 2} \approx 34.8$ years

65. (a) $C'(t) = -kC(t) \;\Rightarrow\; C(t) = C(0)e^{-kt}$ by Theorem 5.5.2. But $C(0) = C_0$, so $C(t) = C_0e^{-kt}$.

(b) $C(30) = \frac{1}{2}C_0$ since the concentration is reduced by half. Thus, $\frac{1}{2}C_0 = C_0e^{-30k} \Rightarrow \ln\frac{1}{2} = -30k \;\Rightarrow$ $k = -\frac{1}{30}\ln\frac{1}{2} = \frac{1}{30}\ln 2$. Since 10% of the original concentration remains if 90% is eliminated, we want the value of t such that $C(t) = \frac{1}{10}C_0$. Therefore, $\frac{1}{10}C_0 = C_0e^{-t(\ln 2)/30} \;\Rightarrow\; \ln 0.1 = -t(\ln 2)/30 \;\Rightarrow\; t = -\frac{30}{\ln 2}\ln 0.1 \approx 100$ h.

66. (a) If $y = u - 20$, $u(0) = 80 \;\Rightarrow\; y(0) = 80 - 20 = 60$, and the initial-value problem is $dy/dt = ky$ with $y(0) = 60$. So the solution is $y(t) = 60e^{kt}$. Now $y(0.5) = 60e^{k(0.5)} = 60 - 20 \;\Rightarrow\; e^{0.5k} = \frac{40}{60} = \frac{2}{3} \;\Rightarrow\; k = 2\ln\frac{2}{3} = \ln\frac{4}{9}$, so $y(t) = 60e^{(\ln 4/9)t} = 60(\frac{4}{9})^t$. Thus, $y(1) = 60(\frac{4}{9})^1 = \frac{80}{3} = 26\frac{2}{3}$°C and $u(1) = 46\frac{2}{3}$°C.

(b) $u(t) = 40 \;\Rightarrow\; y(t) = 20$. $y(t) = 60(\frac{4}{9})^t = 20 \;\Rightarrow\; (\frac{4}{9})^t = \frac{1}{3} \;\Rightarrow\; t\ln\frac{4}{9} = \ln\frac{1}{3} \;\Rightarrow$

$t = \dfrac{\ln\frac{1}{3}}{\ln\frac{4}{9}} \approx 1.35$ h or 81.3 min.

67. $\lim\limits_{x\to\infty} e^{-3x} = 0$ since $-3x \to -\infty$ as $x \to \infty$ and $\lim\limits_{t\to-\infty} e^t = 0$.

68. $\lim\limits_{x\to 10^-} \ln(100 - x^2) = -\infty$ since as $x \to 10^-$, $(100 - x^2) \to 0^+$.

69. Let $t = 2/(x-3)$. As $x \to 3^-$, $t \to -\infty$. $\lim\limits_{x\to 3^-} e^{2/(x-3)} = \lim\limits_{t\to-\infty} e^t = 0$

70. If $y = x^3 - x = x(x^2 - 1)$, then as $x \to \infty$, $y \to \infty$. $\lim\limits_{x\to\infty} \arctan(x^3 - x) = \lim\limits_{y\to\infty} \arctan y = \frac{\pi}{2}$ by (5.6.8).

71. Let $t = \sinh x$. As $x \to 0^+$, $t \to 0^+$. $\lim\limits_{x\to 0^+} \ln(\sinh x) = \lim\limits_{t\to 0^+} \ln t = -\infty$

72. $-1 \le \sin x \le 1 \;\Rightarrow\; -e^{-x} \le e^{-x}\sin x \le e^{-x}$. Now $\lim\limits_{x\to\infty}(\pm e^{-x}) = 0$, so by the Squeeze Theorem, $\lim\limits_{x\to\infty} e^{-x}\sin x = 0$.

73. $\lim\limits_{x\to\infty} \dfrac{(1+2^x)/2^x}{(1-2^x)/2^x} = \lim\limits_{x\to\infty} \dfrac{1/2^x + 1}{1/2^x - 1} = \dfrac{0+1}{0-1} = -1$

74. Let $t = x/4$, so $x = 4t$. As $x \to \infty$, $t \to \infty$. $\lim\limits_{x\to\infty}\left(1 + \dfrac{4}{x}\right)^x = \lim\limits_{t\to\infty}\left(1 + \dfrac{1}{t}\right)^{4t} = \left[\lim\limits_{t\to\infty}\left(1 + \dfrac{1}{t}\right)^t\right]^4 = e^4$

75. $\lim\limits_{x\to 0} \dfrac{\tan \pi x}{\ln(1+x)} \overset{\text{H}}{=} \lim\limits_{x\to 0} \dfrac{\pi \sec^2 \pi x}{1/(1+x)} = \dfrac{\pi \cdot 1^2}{1/1} = \pi$

76. $\lim\limits_{x\to 0} \dfrac{1 - \cos x}{x^2 + x} \overset{\text{H}}{=} \lim\limits_{x\to 0} \dfrac{\sin x}{2x + 1} = \dfrac{0}{1} = 0$

77. $\displaystyle\lim_{x\to 0}\frac{e^{4x}-1-4x}{x^2}\overset{\text{H}}{=}\lim_{x\to 0}\frac{4e^{4x}-4}{2x}\overset{\text{H}}{=}\lim_{x\to 0}\frac{16e^{4x}}{2}=\lim_{x\to 0}8e^{4x}=8\cdot 1=8$

78. $\displaystyle\lim_{x\to \infty}\frac{e^{4x}-1-4x}{x^2}\overset{\text{H}}{=}\lim_{x\to \infty}\frac{4e^{4x}-4}{2x}\overset{\text{H}}{=}\lim_{x\to \infty}\frac{16e^{4x}}{2}=\lim_{x\to \infty}8e^{4x}=\infty$

79. $\displaystyle\lim_{x\to\infty}x^3e^{-x}=\lim_{x\to\infty}\frac{x^3}{e^x}\overset{\text{H}}{=}\lim_{x\to\infty}\frac{3x^2}{e^x}\overset{\text{H}}{=}\lim_{x\to\infty}\frac{6x}{e^x}\overset{\text{H}}{=}\lim_{x\to\infty}\frac{6}{e^x}=0$

80. $\displaystyle\lim_{x\to 0+}x^2\ln x=\lim_{x\to 0+}\frac{\ln x}{1/x^2}\overset{\text{H}}{=}\lim_{x\to 0+}\frac{1/x}{-2/x^3}=\lim_{x\to 0+}\left(-\tfrac{1}{2}x^2\right)=0$

81. $\displaystyle\lim_{x\to 1+}\left(\frac{x}{x-1}-\frac{1}{\ln x}\right)=\lim_{x\to 1+}\left(\frac{x\ln x-x+1}{(x-1)\ln x}\right)\overset{\text{H}}{=}\lim_{x\to 1+}\frac{x\cdot(1/x)+\ln x-1}{(x-1)\cdot(1/x)+\ln x}$

$$=\lim_{x\to 1+}\frac{\ln x}{1-1/x+\ln x}\overset{\text{H}}{=}\lim_{x\to 1+}\frac{1/x}{1/x^2+1/x}=\frac{1}{1+1}=\frac{1}{2}$$

82. $y=(\tan x)^{\cos x}\quad\Rightarrow\quad \ln y=\cos x\ln\tan x$, so

$$\lim_{x\to(\pi/2)^-}\ln y=\lim_{x\to(\pi/2)^-}\frac{\ln\tan x}{\sec x}\overset{\text{H}}{=}\lim_{x\to(\pi/2)^-}\frac{(1/\tan x)\sec^2 x}{\sec x\tan x}=\lim_{x\to(\pi/2)^-}\frac{\sec x}{\tan^2 x}=\lim_{x\to(\pi/2)^-}\frac{\cos x}{\sin^2 x}=\frac{0}{1^2}=0,$$

so $\displaystyle\lim_{x\to(\pi/2)^-}(\tan x)^{\cos x}=\lim_{x\to(\pi/2)^-}e^{\ln y}=e^0=1.$

83. $f(x)=\ln x+\tan^{-1}x\quad\Rightarrow\quad f(1)=\ln 1+\tan^{-1}1=\frac{\pi}{4}\quad\Rightarrow\quad f^{-1}\left(\frac{\pi}{4}\right)=1.$

$\displaystyle f'(x)=\frac{1}{x}+\frac{1}{1+x^2}$, so $\displaystyle (f^{-1})'\left(\tfrac{\pi}{4}\right)=\frac{1}{f'(1)}=\frac{1}{3/2}=\frac{2}{3}.$

84. Let $\theta_1=\operatorname{arccot}x$, so $\cot\theta_1=x=x/1$.

So $\displaystyle\sin(\operatorname{arccot}x)=\sin\theta_1=\frac{1}{\sqrt{x^2+1}}.$

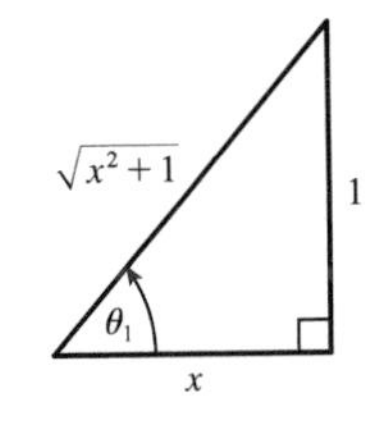

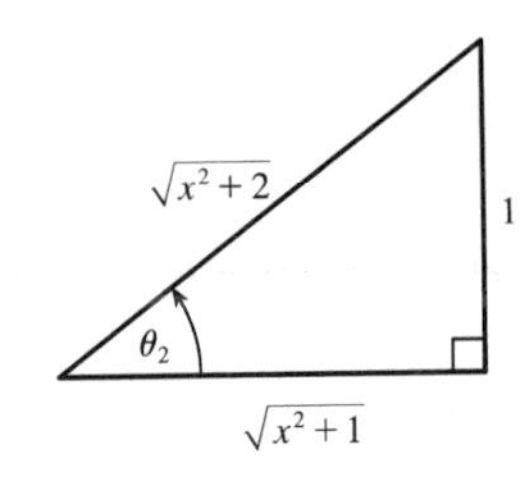

Let $\displaystyle\theta_2=\arctan\left[\frac{1}{\sqrt{x^2+1}}\right]$, so $\displaystyle\tan\theta_2=\frac{1}{\sqrt{x^2+1}}.$

Hence, $\displaystyle\cos\{\arctan[\sin(\operatorname{arccot}x)]\}=\cos\theta_2=\frac{\sqrt{x^2+1}}{\sqrt{x^2+2}}=\sqrt{\frac{x^2+1}{x^2+2}}.$

6 ☐ TECHNIQUES OF INTEGRATION

6.1 Integration by Parts

1. Let $u = \ln x$, $dv = x\,dx \Rightarrow du = dx/x$, $v = \frac{1}{2}x^2$. Then by Equation 2, $\int u\,dv = uv - \int v\,du$,

$\int x \ln x\,dx = \frac{1}{2}x^2 \ln x - \int \frac{1}{2}x^2\,(dx/x) = \frac{1}{2}x^2 \ln x - \frac{1}{2}\int x\,dx = \frac{1}{2}x^2 \ln x - \frac{1}{2} \cdot \frac{1}{2}x^2 + C = \frac{1}{2}x^2 \ln x - \frac{1}{4}x^2 + C.$

2. Let $u = \theta$, $dv = \sec^2\theta\,d\theta \Rightarrow du = d\theta$, $v = \tan\theta$. Then $\int \theta \sec^2\theta\,d\theta = \theta\tan\theta - \int \tan\theta\,d\theta = \theta\tan\theta - \ln|\sec\theta| + C$.

Note: A mnemonic device which is helpful for selecting u when using integration by parts is the LIATE principle of precedence for u:

Logarithmic
Inverse trigonometric
Algebraic
Trigonometric
Exponential

If the integrand has several factors, then we try to choose among them a u which appears as high as possible on the list. For example, in $\int xe^{2x}\,dx$ the integrand is xe^{2x}, which is the product of an algebraic function (x) and an exponential function (e^{2x}). Since Algebraic appears before Exponential, we choose $u = x$. Sometimes the integration turns out to be similar regardless of the selection of u and dv, but it is advisable to refer to LIATE when in doubt.

3. Let $u = x$, $dv = \cos 5x\,dx \Rightarrow du = dx$, $v = \frac{1}{5}\sin 5x$. Then by Equation 2,

$\int x\cos 5x\,dx = \frac{1}{5}x\sin 5x - \int \frac{1}{5}\sin 5x\,dx = \frac{1}{5}x\sin 5x + \frac{1}{25}\cos 5x + C.$

4. Let $u = x$, $dv = e^{-x}\,dx \Rightarrow du = dx$, $v = -e^{-x}$. Then $\int xe^{-x}\,dx = -xe^{-x} + \int e^{-x}\,dx = -xe^{-x} - e^{-x} + C$.

5. Let $u = r$, $dv = e^{r/2}\,dr \Rightarrow du = dr$, $v = 2e^{r/2}$. Then $\int re^{r/2}\,dr = 2re^{r/2} - \int 2e^{r/2}\,dr = 2re^{r/2} - 4e^{r/2} + C$.

6. Let $u = t$, $dv = \sin 2t\,dt \Rightarrow du = dt$, $v = -\frac{1}{2}\cos 2t$. Then

$\int t\sin 2t\,dt = -\frac{1}{2}t\cos 2t + \frac{1}{2}\int \cos 2t\,dt = -\frac{1}{2}t\cos 2t + \frac{1}{4}\sin 2t + C.$

7. Let $u = x^2$, $dv = \sin \pi x\,dx \Rightarrow du = 2x\,dx$ and $v = -\frac{1}{\pi}\cos \pi x$. Then

$I = \int x^2 \sin \pi x\,dx = -\frac{1}{\pi}x^2 \cos \pi x + \frac{2}{\pi}\int x\cos \pi x\,dx$ $(\star)$. Next let $U = x$, $dV = \cos \pi x\,dx \Rightarrow$

$dU = dx$, $V = \frac{1}{\pi}\sin \pi x$, so $\int x\cos \pi x\,dx = \frac{1}{\pi}x\sin \pi x - \frac{1}{\pi}\int \sin \pi x\,dx = \frac{1}{\pi}x\sin \pi x + \frac{1}{\pi^2}\cos \pi x + C_1$.

Substituting for $\int x\cos \pi x\,dx$ in $(\star)$, we get

$I = -\frac{1}{\pi}x^2\cos \pi x + \frac{2}{\pi}\left(\frac{1}{\pi}x\sin \pi x + \frac{1}{\pi^2}\cos \pi x + C_1\right) = -\frac{1}{\pi}x^2\cos \pi x + \frac{2}{\pi^2}x\sin \pi x + \frac{2}{\pi^3}\cos \pi x + C$, where $C = \frac{2}{\pi}C_1$.

8. Let $u = x^2$, $dv = \cos mx\,dx \Rightarrow du = 2x\,dx$, $v = \frac{1}{m}\sin mx$. Then

$I = \int x^2 \cos mx\,dx = \frac{1}{m}x^2 \sin mx - \frac{2}{m}\int x\sin mx\,dx$ $(\star)$. Next let $U = x$, $dV = \sin mx\,dx \Rightarrow$

$dU = dx$, $V = -\frac{1}{m}\cos mx$, so $\int x\sin mx\,dx = -\frac{1}{m}x\cos mx + \frac{1}{m}\int \cos mx\,dx = -\frac{1}{m}x\cos mx + \frac{1}{m^2}\sin mx + C_1$.

Substituting for $\int x\sin mx\,dx$ in $(\star)$, we get

$I = \frac{1}{m}x^2 \sin mx - \frac{2}{m}\left(-\frac{1}{m}x\cos mx + \frac{1}{m^2}\sin mx + C_1\right) = \frac{1}{m}x^2 \sin mx + \frac{2}{m^2}x\cos mx - \frac{2}{m^3}\sin mx + C,$

where $C = -\frac{2}{m}C_1$.

9. Let $u = \ln(2x+1)$, $dv = dx \;\Rightarrow\; du = \dfrac{2}{2x+1}\,dx$, $v = x$. Then

$$\begin{aligned}\int \ln(2x+1)\,dx &= x\ln(2x+1) - \int \frac{2x}{2x+1}\,dx = x\ln(2x+1) - \int \frac{(2x+1)-1}{2x+1}\,dx \\ &= x\ln(2x+1) - \int \left(1 - \frac{1}{2x+1}\right)dx = x\ln(2x+1) - x + \tfrac{1}{2}\ln(2x+1) + C \\ &= \tfrac{1}{2}(2x+1)\ln(2x+1) - x + C\end{aligned}$$

10. Let $u = \ln p$, $dv = p^5\,dp \;\Rightarrow\; du = \dfrac{1}{p}\,dp$, $v = \frac{1}{6}p^6$. Then $\int p^5 \ln p\,dp = \frac{1}{6}p^6\ln p - \frac{1}{6}\int p^5\,dp = \frac{1}{6}p^6 \ln p - \frac{1}{36}p^6 + C$.

11. Let $u = \arctan 4t$, $dv = dt \;\Rightarrow\; du = \dfrac{4}{1+(4t)^2}\,dt = \dfrac{4}{1+16t^2}\,dt$, $v = t$. Then

$$\int \arctan 4t\,dt = t\arctan 4t - \int \frac{4t}{1+16t^2}\,dt = t\arctan 4t - \frac{1}{8}\int \frac{32t}{1+16t^2}\,dt = t\arctan 4t - \tfrac{1}{8}\ln(1+16t^2) + C.$$

12. Let $u = t^3$, $dv = e^t\,dt \;\Rightarrow\; du = 3t^2\,dt$, $v = e^t$. Then $I = \int t^3 e^t\,dt = t^3e^t - \int 3t^2e^t\,dt$. Integrate by parts twice more with $dv = e^t\,dt$.

$$\begin{aligned}I &= t^3e^t - \left(3t^2e^t - \textstyle\int 6te^t\,dt\right) = t^3e^t - 3t^2e^t + 6te^t - \textstyle\int 6e^t\,dt \\ &= t^3e^t - 3t^2e^t + 6te^t - 6e^t + C = (t^3 - 3t^2 + 6t - 6)e^t + C\end{aligned}$$

More generally, if $p(t)$ is a polynomial of degree n in t, then repeated integration by parts shows that

$$\int p(t)e^t\,dt = [p(t) - p'(t) + p''(t) - p'''(t) + \cdots + (-1)^n\,p^{(n)}(t)]e^t + C$$

13. First let $u = \sin 3\theta$, $dv = e^{2\theta}\,d\theta \;\Rightarrow\; du = 3\cos 3\theta\,d\theta$, $v = \frac{1}{2}e^{2\theta}$. Then

$I = \int e^{2\theta}\sin 3\theta\,d\theta = \frac{1}{2}e^{2\theta}\sin 3\theta - \frac{3}{2}\int e^{2\theta}\cos 3\theta\,d\theta$. Next let $U = \cos 3\theta$, $dV = e^{2\theta}\,d\theta \;\Rightarrow\; dU = -3\sin 3\theta\,d\theta$,

$V = \frac{1}{2}e^{2\theta}$ to get $\int e^{2\theta}\cos 3\theta\,d\theta = \frac{1}{2}e^{2\theta}\cos 3\theta + \frac{3}{2}\int e^{2\theta}\sin 3\theta\,d\theta$. Substituting in the previous formula gives

$I = \frac{1}{2}e^{2\theta}\sin 3\theta - \frac{3}{4}e^{2\theta}\cos 3\theta - \frac{9}{4}\int e^{2\theta}\sin 3\theta\,d\theta = \frac{1}{2}e^{2\theta}\sin 3\theta - \frac{3}{4}e^{2\theta}\cos 3\theta - \frac{9}{4}I \;\Rightarrow$

$\frac{13}{4}I = \frac{1}{2}e^{2\theta}\sin 3\theta - \frac{3}{4}e^{2\theta}\cos 3\theta + C_1$. Hence, $I = \frac{1}{13}e^{2\theta}(2\sin 3\theta - 3\cos 3\theta) + C$, where $C = \frac{4}{13}C_1$.

14. First let $u = e^{-\theta}$, $dv = \cos 2\theta\,d\theta \;\Rightarrow\; du = -e^{-\theta}\,d\theta$, $v = \frac{1}{2}\sin 2\theta$. Then

$I = \int e^{-\theta}\cos 2\theta\,d\theta = \frac{1}{2}e^{-\theta}\sin 2\theta - \int \frac{1}{2}\sin 2\theta\,(-e^{-\theta}\,d\theta) = \frac{1}{2}e^{-\theta}\sin 2\theta + \frac{1}{2}\int e^{-\theta}\sin 2\theta\,d\theta$.

Next let $U = e^{-\theta}$, $dV = \sin 2\theta\,d\theta \;\Rightarrow\; dU = -e^{-\theta}\,d\theta$, $V = -\frac{1}{2}\cos 2\theta$, so

$\int e^{-\theta}\sin 2\theta\,d\theta = -\frac{1}{2}e^{-\theta}\cos 2\theta - \int\left(-\frac{1}{2}\right)\cos 2\theta\left(-e^{-\theta}\,d\theta\right) = -\frac{1}{2}e^{-\theta}\cos 2\theta - \frac{1}{2}\int e^{-\theta}\cos 2\theta\,d\theta$.

So $I = \frac{1}{2}e^{-\theta}\sin 2\theta + \frac{1}{2}\left[\left(-\frac{1}{2}e^{-\theta}\cos 2\theta\right) - \frac{1}{2}I\right] = \frac{1}{2}e^{-\theta}\sin 2\theta - \frac{1}{4}e^{-\theta}\cos 2\theta - \frac{1}{4}I \;\Rightarrow$

$\frac{5}{4}I = \frac{1}{2}e^{-\theta}\sin 2\theta - \frac{1}{4}e^{-\theta}\cos 2\theta + C_1 \;\Rightarrow\; I = \frac{4}{5}\left(\frac{1}{2}e^{-\theta}\sin 2\theta - \frac{1}{4}e^{-\theta}\cos 2\theta + C_1\right) = \frac{2}{5}e^{-\theta}\sin 2\theta - \frac{1}{5}e^{-\theta}\cos 2\theta + C.$

15. Let $u = t$, $dv = \sin 3t\,dt \Rightarrow du = dt$, $v = -\frac{1}{3}\cos 3t$. Then

$\int_0^\pi t\sin 3t\,dt = \left[-\frac{1}{3}t\cos 3t\right]_0^\pi + \frac{1}{3}\int_0^\pi \cos 3t\,dt = \left(\frac{1}{3}\pi - 0\right) + \frac{1}{9}\left[\sin 3t\right]_0^\pi = \frac{\pi}{3}$.

16. First let $u = x^2 + 1$, $dv = e^{-x}\,dx \Rightarrow du = 2x\,dx$, $v = -e^{-x}$. By (6),

$\int_0^1 (x^2+1)e^{-x}\,dx = \left[-(x^2+1)e^{-x}\right]_0^1 + \int_0^1 2xe^{-x}\,dx = -2e^{-1} + 1 + 2\int_0^1 xe^{-x}\,dx$.

Next let $U = x$, $dV = e^{-x}\,dx \Rightarrow dU = dx$, $V = -e^{-x}$. By (6) again,

$\int_0^1 xe^{-x}\,dx = \left[-xe^{-x}\right]_0^1 + \int_0^1 e^{-x}\,dx = -e^{-1} + \left[-e^{-x}\right]_0^1 = -e^{-1} - e^{-1} + 1 = -2e^{-1} + 1$.

So $\int_0^1 (x^2+1)e^{-x}\,dx = -2e^{-1} + 1 + 2(-2e^{-1}+1) = -2e^{-1} + 1 - 4e^{-1} + 2 = -6e^{-1} + 3$.

17. Let $u = \ln x$, $dv = x^{-2}\,dx \Rightarrow du = \dfrac{1}{x}\,dx$, $v = -x^{-1}$. By (6),

$$\int_1^2 \frac{\ln x}{x^2}\,dx = \left[-\frac{\ln x}{x}\right]_1^2 + \int_1^2 x^{-2}\,dx = -\tfrac{1}{2}\ln 2 + \ln 1 + \left[-\frac{1}{x}\right]_1^2 = -\tfrac{1}{2}\ln 2 + 0 - \tfrac{1}{2} + 1 = \tfrac{1}{2} - \tfrac{1}{2}\ln 2.$$

18. Let $u = \ln t$, $dv = \sqrt{t}\,dt \Rightarrow du = dt/t$, $v = \frac{2}{3}t^{3/2}$. By (6),

$\int_1^4 \sqrt{t}\ln t\,dt = \left[\frac{2}{3}t^{3/2}\ln t\right]_1^4 - \frac{2}{3}\int_1^4 \sqrt{t}\,dt = \frac{2}{3}\cdot 8\cdot\ln 4 - 0 - \left[\frac{2}{3}\cdot\frac{2}{3}t^{3/2}\right]_1^4 = \frac{16}{3}\ln 4 - \frac{4}{9}(8-1) = \frac{16}{3}\ln 4 - \frac{28}{9}$.

19. Let $u = y$, $dv = \dfrac{dy}{e^{2y}} = e^{-2y}dy \Rightarrow du = dy$, $v = -\dfrac{1}{2}e^{-2y}$. Then

$$\int_0^1 \frac{y}{e^{2y}}\,dy = \left[-\tfrac{1}{2}ye^{-2y}\right]_0^1 + \tfrac{1}{2}\int_0^1 e^{-2y}dy = \left(-\tfrac{1}{2}e^{-2} + 0\right) - \tfrac{1}{4}\left[e^{-2y}\right]_0^1 = -\tfrac{1}{2}e^{-2} - \tfrac{1}{4}e^{-2} + \tfrac{1}{4} = \tfrac{1}{4} - \tfrac{3}{4}e^{-2}.$$

20. Let $u = \arctan(1/x)$, $dv = dx \Rightarrow du = \dfrac{1}{1+(1/x)^2}\cdot\dfrac{-1}{x^2}\,dx = \dfrac{-dx}{x^2+1}$, $v = x$. Then

$$\begin{aligned}\int_1^{\sqrt{3}} \arctan(1/x)\,dx &= \left[x\arctan\left(\frac{1}{x}\right)\right]_1^{\sqrt{3}} + \int_1^{\sqrt{3}} \frac{x\,dx}{x^2+1} = \sqrt{3}\,\frac{\pi}{6} - 1\cdot\frac{\pi}{4} + \frac{1}{2}\left[\ln(x^2+1)\right]_1^{\sqrt{3}} \\ &= \frac{\pi\sqrt{3}}{6} - \frac{\pi}{4} + \frac{1}{2}(\ln 4 - \ln 2) = \frac{\pi\sqrt{3}}{6} - \frac{\pi}{4} + \frac{1}{2}\ln\frac{4}{2} = \frac{\pi\sqrt{3}}{6} - \frac{\pi}{4} + \frac{1}{2}\ln 2\end{aligned}$$

21. Let $u = \sin^{-1}x$, $dv = dx \Rightarrow du = \dfrac{dx}{\sqrt{1-x^2}}$, $v = x$. By (6),

$I = \displaystyle\int_0^{1/2} \sin^{-1}x\,dx = \left[x\sin^{-1}x\right]_0^{1/2} - \int_0^{1/2} \frac{x\,dx}{\sqrt{1-x^2}} = \frac{1}{2}\cdot\frac{\pi}{6} - \int_0^{1/2} \frac{x\,dx}{\sqrt{1-x^2}}$. To evaluate the last integral,

let $t = 1 - x^2$, so $dt = -2x\,dx$ and $x\,dx = -\frac{1}{2}\,dt$. When $x = 0$, $t = 1$; when $x = \frac{1}{2}$, $t = \frac{3}{4}$.

So $\displaystyle\int_0^{1/2} \frac{x\,dx}{\sqrt{1-x^2}} = \int_1^{3/4} \frac{1}{\sqrt{t}}\left(-\frac{1}{2}\,dt\right) = \frac{1}{2}\int_{3/4}^1 t^{-1/2}dt = \frac{1}{2}\left[2t^{1/2}\right]_{3/4}^1 = \sqrt{1} - \sqrt{\tfrac{3}{4}} = 1 - \tfrac{\sqrt{3}}{2}$.

Thus, $I = \frac{\pi}{12} - \left(1 - \frac{\sqrt{3}}{2}\right) = \frac{\pi}{12} - 1 + \frac{\sqrt{3}}{2} = \frac{1}{12}\left(\pi - 12 + 6\sqrt{3}\right)$.

22. Let $u = r^2$, $dv = \dfrac{r}{\sqrt{4+r^2}}\,dr \;\Rightarrow\; du = 2r\,dr$, $v = \sqrt{4+r^2}$. By (6),

$$\int_0^1 \frac{r^3}{\sqrt{4+r^2}}\,dr = \left[r^2\sqrt{4+r^2}\right]_0^1 - 2\int_0^1 r\sqrt{4+r^2}\,dr = \sqrt{5} - \tfrac{2}{3}\left[(4+r^2)^{3/2}\right]_0^1$$
$$= \sqrt{5} - \tfrac{2}{3}(5)^{3/2} + \tfrac{2}{3}(8) = \sqrt{5}\left(1-\tfrac{10}{3}\right) + \tfrac{16}{3} = \tfrac{16}{3} - \tfrac{7}{3}\sqrt{5}$$

23. Let $u = (\ln x)^2$, $dv = dx \;\Rightarrow\; du = \dfrac{2}{x}\ln x\,dx$, $v = x$. By (6), $I = \int_1^2 (\ln x)^2\,dx = \left[x(\ln x)^2\right]_1^2 - 2\int_1^2 \ln x\,dx$.

To evaluate the last integral, let $U = \ln x$, $dV = dx \;\Rightarrow\; dU = \dfrac{1}{x}\,dx$, $V = x$. Thus,

$$I = \left[x(\ln x)^2\right]_1^2 - 2\left(\left[x\ln x\right]_1^2 - \int_1^2 dx\right) = \left[x(\ln x)^2 - 2x\ln x + 2x\right]_1^2$$
$$= (2(\ln 2)^2 - 4\ln 2 + 4) - (0 - 0 + 2) = 2(\ln 2)^2 - 4\ln 2 + 2$$

24. Let $u = \sin(t-s)$, $dv = e^s\,ds \;\Rightarrow\; du = -\cos(t-s)\,ds$, $v = e^s$. Then

$I = \int_0^t e^s \sin(t-s)\,ds = \left[e^s\sin(t-s)\right]_0^t + \int_0^t e^s\cos(t-s)\,ds = e^t\sin 0 - e^0\sin t + I_1$. For I_1, let $U = \cos(t-s)$,

$dV = e^s\,ds \;\Rightarrow\; dU = \sin(t-s)\,ds$, $V = e^s$. So $I_1 = \left[e^s\cos(t-s)\right]_0^t - \int_0^t e^s\sin(t-s)ds = e^t\cos 0 - e^0\cos t - I$.

Thus, $I = -\sin t + e^t - \cos t - I \;\Rightarrow\; 2I = e^t - \cos t - \sin t \;\Rightarrow\; I = \frac{1}{2}(e^t - \cos t - \sin t)$.

25. Let $w = \sqrt{x}$, so that $x = w^2$ and $dx = 2w\,dw$. Thus, $\int \sin\sqrt{x}\,dx = \int 2w\sin w\,dw$. Now use parts with $u = 2w$,

$dv = \sin w\,dw$, $du = 2\,dw$, $v = -\cos w$ to get

$$\int 2w\sin w\,dw = -2w\cos w + \int 2\cos w\,dw = -2w\cos w + 2\sin w + C$$
$$= -2\sqrt{x}\cos\sqrt{x} + 2\sin\sqrt{x} + C = 2(\sin\sqrt{x} - \sqrt{x}\cos\sqrt{x}) + C$$

26. Let $t = x^3$, so that $dt = 3x^2\,dx$. Thus, $\int x^5\cos(x^3)\,dx = \frac{1}{3}\int x^3\cos(x^3)\cdot 3x^2\,dx = \frac{1}{3}\int t\cos t\,dt$.

Now use parts with $u = t$, $dv = \cos t\,dt$, $du = dt$, $v = \sin t$ to get

$\frac{1}{3}\int t\cos t\,dt = \frac{1}{3}\left(t\sin t - \int \sin t\,dt\right) = \frac{1}{3}t\sin t + \frac{1}{3}\cos t + C = \frac{1}{3}x^3\sin(x^3) + \frac{1}{3}\cos(x^3) + C$.

27. Let $x = \theta^2$, so that $dx = 2\theta\,d\theta$. Thus, $\int_{\sqrt{\pi/2}}^{\sqrt{\pi}} \theta^3\cos(\theta^2)\,d\theta = \int_{\sqrt{\pi/2}}^{\sqrt{\pi}} \theta^2\cos(\theta^2)\cdot\frac{1}{2}(2\theta\,d\theta) = \frac{1}{2}\int_{\pi/2}^{\pi} x\cos x\,dx$. Now use

parts with $u = x$, $dv = \cos x\,dx$, $du = dx$, $v = \sin x$ to get

$$\tfrac{1}{2}\int_{\pi/2}^{\pi} x\cos x\,dx = \tfrac{1}{2}\left(\left[x\sin x\right]_{\pi/2}^{\pi} - \int_{\pi/2}^{\pi}\sin x\,dx\right) = \tfrac{1}{2}\left[x\sin x + \cos x\right]_{\pi/2}^{\pi}$$
$$= \tfrac{1}{2}(\pi\sin\pi + \cos\pi) - \tfrac{1}{2}\left(\tfrac{\pi}{2}\sin\tfrac{\pi}{2} + \cos\tfrac{\pi}{2}\right) = \tfrac{1}{2}(\pi\cdot 0 - 1) - \tfrac{1}{2}\left(\tfrac{\pi}{2}\cdot 1 + 0\right) = -\tfrac{1}{2} - \tfrac{\pi}{4}$$

28. Let $w = \sqrt{x}$, so that $x = w^2$ and $dx = 2w\,dw$. Thus, $\int_1^4 e^{\sqrt{x}}\,dx = \int_1^2 e^w 2w\,dw$. Now use parts with $u = 2w$, $dv = e^w\,dw$,

$du = 2\,dw$, $v = e^w$ to get $\int_1^2 e^w 2w\,dw = \left[2we^w\right]_1^2 - 2\int_1^2 e^w\,dw = 4e^2 - 2e - 2(e^2 - e) = 2e^2$.

29. (a) Take $n = 2$ in Example 6 to get $\displaystyle\int \sin^2 x\,dx = -\frac{1}{2}\cos x\sin x + \frac{1}{2}\int 1\,dx = \frac{x}{2} - \frac{\sin 2x}{4} + C$.

(b) $\int \sin^4 x\,dx = -\frac{1}{4}\cos x\sin^3 x + \frac{3}{4}\int \sin^2 x\,dx = -\frac{1}{4}\cos x\sin^3 x + \frac{3}{8}x - \frac{3}{16}\sin 2x + C$.

30. (a) Let $u = \cos^{n-1} x$, $dv = \cos x\,dx \Rightarrow du = -(n-1)\cos^{n-2} x \sin x\,dx$, $v = \sin x$ in (2):

$$\begin{aligned}\int \cos^n x\,dx &= \cos^{n-1} x \sin x + (n-1)\int \cos^{n-2} x \sin^2 x\,dx \\ &= \cos^{n-1} x \sin x + (n-1)\int \cos^{n-2} x\,(1-\cos^2 x)dx \\ &= \cos^{n-1} x \sin x + (n-1)\int \cos^{n-2} x\,dx - (n-1)\int \cos^n x\,dx\end{aligned}$$

Rearranging terms gives $n\int \cos^n x\,dx = \cos^{n-1} x \sin x + (n-1)\int \cos^{n-2} x\,dx$ or

$$\int \cos^n x\,dx = \frac{1}{n}\cos^{n-1} x \sin x + \frac{n-1}{n}\int \cos^{n-2} x\,dx$$

(b) Take $n = 2$ in part (a) to get $\int \cos^2 x\,dx = \frac{1}{2}\cos x \sin x + \frac{1}{2}\int 1\,dx = \dfrac{x}{2} + \dfrac{\sin 2x}{4} + C$.

(c) $\int \cos^4 x\,dx = \frac{1}{4}\cos^3 x \sin x + \frac{3}{4}\int \cos^2 x\,dx = \frac{1}{4}\cos^3 x \sin x + \frac{3}{8}x + \frac{3}{16}\sin 2x + C$

31. (a) From Example 6, $\displaystyle\int \sin^n x\,dx = -\frac{1}{n}\cos x \sin^{n-1} x + \frac{n-1}{n}\int \sin^{n-2} x\,dx$. Using (6),

$$\begin{aligned}\int_0^{\pi/2} \sin^n x\,dx &= \left[-\frac{\cos x \sin^{n-1} x}{n}\right]_0^{\pi/2} + \frac{n-1}{n}\int_0^{\pi/2} \sin^{n-2} x\,dx \\ &= (0-0) + \frac{n-1}{n}\int_0^{\pi/2} \sin^{n-2} x\,dx = \frac{n-1}{n}\int_0^{\pi/2} \sin^{n-2} x\,dx\end{aligned}$$

(b) Using $n = 3$ in part (a), we have $\int_0^{\pi/2} \sin^3 x\,dx = \frac{2}{3}\int_0^{\pi/2} \sin x\,dx = \left[-\frac{2}{3}\cos x\right]_0^{\pi/2} = \frac{2}{3}$.

Using $n = 5$ in part (a), we have $\int_0^{\pi/2} \sin^5 x\,dx = \frac{4}{5}\int_0^{\pi/2} \sin^3 x\,dx = \frac{4}{5}\cdot\frac{2}{3} = \frac{8}{15}$.

(c) The formula holds for $n = 1$ (that is, $2n+1 = 3$) by (b). Assume it holds for some $k \geq 1$. Then

$\displaystyle\int_0^{\pi/2} \sin^{2k+1} x\,dx = \frac{2\cdot 4\cdot 6\cdot\cdots\cdot(2k)}{3\cdot 5\cdot 7\cdot\cdots\cdot(2k+1)}$. By Example 6,

$$\begin{aligned}\int_0^{\pi/2} \sin^{2k+3} x\,dx &= \frac{2k+2}{2k+3}\int_0^{\pi/2} \sin^{2k+1} x\,dx = \frac{2k+2}{2k+3}\cdot\frac{2\cdot 4\cdot 6\cdot\cdots\cdot(2k)}{3\cdot 5\cdot 7\cdot\cdots\cdot(2k+1)} \\ &= \frac{2\cdot 4\cdot 6\cdot\cdots\cdot(2k)[2(k+1)]}{3\cdot 5\cdot 7\cdot\cdots\cdot(2k+1)[2(k+1)+1]},\end{aligned}$$

so the formula holds for $n = k+1$. By induction, the formula holds for all $n \geq 1$.

32. Using Exercise 31(a), we see that the formula holds for $n = 1$, because $\int_0^{\pi/2} \sin^2 x\,dx = \frac{1}{2}\int_0^{\pi/2} 1\,dx = \frac{1}{2}\left[x\right]_0^{\pi/2} = \frac{1}{2}\cdot\frac{\pi}{2}$.

Now assume it holds for some $k \geq 1$. Then $\displaystyle\int_0^{\pi/2} \sin^{2k} x\,dx = \frac{1\cdot 3\cdot 5\cdot\cdots\cdot(2k-1)}{2\cdot 4\cdot 6\cdot\cdots\cdot(2k)}\frac{\pi}{2}$. By Exercise 31(a),

$$\begin{aligned}\int_0^{\pi/2} \sin^{2(k+1)} x\,dx &= \frac{2k+1}{2k+2}\int_0^{\pi/2} \sin^{2k} x\,dx = \frac{2k+1}{2k+2}\cdot\frac{1\cdot 3\cdot 5\cdot\cdots\cdot(2k-1)}{2\cdot 4\cdot 6\cdot\cdots\cdot(2k)}\frac{\pi}{2} \\ &= \frac{1\cdot 3\cdot 5\cdot\cdots\cdot(2k-1)(2k+1)}{2\cdot 4\cdot 6\cdot\cdots\cdot(2k)(2k+2)}\cdot\frac{\pi}{2},\end{aligned}$$

so the formula holds for $n = k+1$. By induction, the formula holds for all $n \geq 1$.

33. Let $u = (\ln x)^n$, $dv = dx \Rightarrow du = n(\ln x)^{n-1}(dx/x)$, $v = x$. Then

$\int (\ln x)^n\,dx = x(\ln x)^n - \int nx(\ln x)^{n-1}(dx/x) = x(\ln x)^n - n\int (\ln x)^{n-1}\,dx$.

34. Let $u = x^n$, $dv = e^x\,dx \Rightarrow du = nx^{n-1}\,dx$, $v = e^x$. Then $\int x^n e^x\,dx = x^n e^x - n\int x^{n-1}e^x\,dx$.

35. Let $u = (x^2+a^2)^n$, $dv = dx \Rightarrow du = n(x^2+a^2)^{n-1}\,2x\,dx$, $v = x$. Then

$$\begin{aligned}\int (x^2+a^2)^n\,dx &= x(x^2+a^2)^n - 2n\int x^2(x^2+a^2)^{n-1}\,dx \\ &= x(x^2+a^2)^n - 2n\left[\int (x^2+a^2)^n\,dx - a^2\int (x^2+a^2)^{n-1}\,dx\right] \quad [\text{since } x^2 = (x^2+a^2)-a^2] \quad \Rightarrow\end{aligned}$$

$(2n+1)\int (x^2+a^2)^n\,dx = x(x^2+a^2)^n + 2na^2\int (x^2+a^2)^{n-1}\,dx$, and

$$\int (x^2+a^2)^n\,dx = \frac{x(x^2+a^2)^n}{2n+1} + \frac{2na^2}{2n+1}\int (x^2+a^2)^{n-1}\,dx \quad [\text{provided } 2n+1 \neq 0].$$

36. Let $u = \sec^{n-2} x$, $dv = \sec^2 x\,dx \Rightarrow du = (n-2)\sec^{n-3} x\,\sec x\,\tan x\,dx$, $v = \tan x$. Then

$$\begin{aligned}\int \sec^n x\,dx &= \tan x\,\sec^{n-2} x - (n-2)\int \sec^{n-2} x\,\tan^2 x\,dx = \tan x\,\sec^{n-2} x - (n-2)\int \sec^{n-2} x\,(\sec^2 x - 1)\,dx \\ &= \tan x\,\sec^{n-2} x - (n-2)\int \sec^n x\,dx + (n-2)\int \sec^{n-2} x\,dx\end{aligned}$$

so $(n-1)\int \sec^n x\,dx = \tan x\,\sec^{n-2} x + (n-2)\int \sec^{n-2} x\,dx$.

If $n - 1 \neq 0$, then $\displaystyle\int \sec^n x\,dx = \frac{\tan x\,\sec^{n-2} x}{n-1} + \frac{n-2}{n-1}\int \sec^{n-2} x\,dx$.

37. Take $n = 3$ in Exercise 33 to get $\int (\ln x)^3\,dx = x\,(\ln x)^3 - 3\int (\ln x)^2\,dx = x(\ln x)^3 - 3x(\ln x)^2 + 6x\ln x - 6x + C$

[by Exercise 23].

Or: Instead of using Exercise 23, apply Exercise 33 again with $n = 2$.

38. Take $n = 4$ in Exercise 34 to get

$$\begin{aligned}\int x^4 e^x\,dx &= x^4 e^x - 4\int x^3 e^x\,dx = x^4 e^x - 4(x^3 - 3x^2 + 6x - 6)e^x + C \quad [\text{by Exercise 12}] \\ &= e^x(x^4 - 4x^3 + 12x^2 - 24x + 24) + C\end{aligned}$$

Or: Instead of using Exercise 12, apply Exercise 34 with $n = 3$, then $n = 2$, then $n = 1$.

39. The average value of $f(x) = x^2\ln x$ on the interval $[1, 3]$ is $f_{\text{ave}} = \dfrac{1}{3-1}\displaystyle\int_1^3 x^2\ln x\,dx = \tfrac{1}{2}I$.

Let $u = \ln x$, $dv = x^2\,dx \Rightarrow du = (1/x)\,dx$, $v = \frac{1}{3}x^3$.

So $I = \left[\frac{1}{3}x^3\ln x\right]_1^3 - \int_1^3 \frac{1}{3}x^2 dx = (9\ln 3 - 0) - \left[\frac{1}{9}x^3\right]_1^3 = 9\ln 3 - \left(3 - \frac{1}{9}\right) = 9\ln 3 - \frac{26}{9}$.

Thus, $f_{\text{ave}} = \frac{1}{2}I = \frac{1}{2}\left(9\ln 3 - \frac{26}{9}\right) = \frac{9}{2}\ln 3 - \frac{13}{9}$.

40. The rocket will have height $H = \int_0^{60} v(t)\,dt$ after 60 seconds.

$$\begin{aligned}H &= \int_0^{60}\left[-gt - v_e\ln\left(\frac{m-rt}{m}\right)\right]dt = -g\left[\tfrac{1}{2}t^2\right]_0^{60} - v_e\left[\int_0^{60}\ln(m-rt)\,dt - \int_0^{60}\ln m\,dt\right] \\ &= -g(1800) + v_e(\ln m)(60) - v_e\int_0^{60}\ln(m-rt)\,dt\end{aligned}$$

Let $u = \ln(m-rt)$, $dv = dt \Rightarrow du = \dfrac{1}{m-rt}(-r)\,dt$, $v = t$. Then

$$\int_0^{60} \ln(m - rt)\,dt = \left[t\ln(m-rt)\right]_0^{60} + \int_0^{60} \frac{rt}{m-rt}\,dt = 60\ln(m-60r) + \int_0^{60}\left(-1 + \frac{m}{m-rt}\right)dt$$

$$= 60\ln(m-60r) + \left[-t - \frac{m}{r}\ln(m-rt)\right]_0^{60} = 60\ln(m-60r) - 60 - \frac{m}{r}\ln(m-60r) + \frac{m}{r}\ln m$$

So $H = -1800g + 60v_e \ln m - 60v_e \ln(m-60r) + 60v_e + \frac{m}{r}v_e \ln(m-60r) - \frac{m}{r}v_e \ln m$. Substituting $g = 9.8$, $m = 30{,}000$, $r = 160$, and $v_e = 3000$ gives us $H \approx 14{,}844$ m.

41. Since $v(t) > 0$ for all t, the desired distance is $s(t) = \int_0^t v(w)\,dw = \int_0^t w^2 e^{-w}\,dw$.

First let $u = w^2$, $dv = e^{-w}\,dw \Rightarrow du = 2w\,dw$, $v = -e^{-w}$. Then $s(t) = \left[-w^2e^{-w}\right]_0^t + 2\int_0^t we^{-w}\,dw$.

Next let $U = w$, $dV = e^{-w}\,dw \Rightarrow dU = dw$, $V = -e^{-w}$. Then

$$s(t) = -t^2e^{-t} + 2\left(\left[-we^{-w}\right]_0^t + \int_0^t e^{-w}\,dw\right) = -t^2e^{-t} + 2\left(-te^{-t} + 0 + \left[-e^{-w}\right]_0^t\right)$$

$$= -t^2e^{-t} + 2(-te^{-t} - e^{-t} + 1) = -t^2e^{-t} - 2te^{-t} - 2e^{-t} + 2 = 2 - e^{-t}(t^2 + 2t + 2) \text{ meters}$$

42. Suppose $f(0) = g(0) = 0$ and let $u = f(x)$, $dv = g''(x)\,dx \Rightarrow du = f'(x)\,dx$, $v = g'(x)$.

Then $\int_0^a f(x)g''(x)\,dx = \left[f(x)g'(x)\right]_0^a - \int_0^a f'(x)g'(x)\,dx = f(a)g'(a) - \int_0^a f'(x)g'(x)\,dx$.

Now let $U = f'(x)$, $dV = g'(x)\,dx \Rightarrow dU = f''(x)\,dx$ and $V = g(x)$, so

$\int_0^a f'(x)g'(x)\,dx = \left[f'(x)g(x)\right]_0^a - \int_0^a f''(x)g(x)\,dx = f'(a)g(a) - \int_0^a f''(x)g(x)\,dx$.

Combining the two results, we get $\int_0^a f(x)g''(x)\,dx = f(a)g'(a) - f'(a)g(a) + \int_0^a f''(x)g(x)\,dx$.

43. For $I = \int_1^4 xf''(x)\,dx$, let $u = x$, $dv = f''(x)\,dx \Rightarrow du = dx$, $v = f'(x)$. Then

$I = \left[xf'(x)\right]_1^4 - \int_1^4 f'(x)\,dx = 4f'(4) - 1\cdot f'(1) - [f(4) - f(1)] = 4\cdot 3 - 1\cdot 5 - (7-2) = 12 - 5 - 5 = 2$.

We used the fact that f'' is continuous to guarantee that I exists.

44. (a) Take $g(x) = x$ and $g'(x) = 1$ in Equation 1.

(b) By part (a), $\int_a^b f(x)\,dx = bf(b) - a\,f(a) - \int_a^b x\,f'(x)\,dx$. Now let $y = f(x)$, so that $x = g(y)$ and $dy = f'(x)\,dx$. Then $\int_a^b x\,f'(x)\,dx = \int_{f(a)}^{f(b)} g(y)\,dy$. The result follows.

(c) Part (b) says that

the area of region $ABFC$ = $bf(b)$ − $af(a)$ − $\int_{f(a)}^{f(b)} g(y)\,dy$

= (area of rectangle $OBFE$) − (area of rectangle $OACD$) − (area of region $DCFE$)

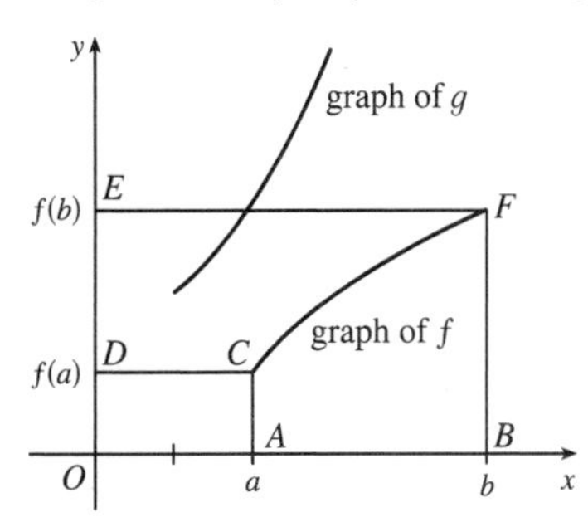

(d) We have $f(x) = \ln x$, so $f^{-1}(x) = e^x$, and since $g = f^{-1}$, we have $g(y) = e^y$. By part (b),

$\int_1^e \ln x\,dx = e\ln e - 1\ln 1 - \int_{\ln 1}^{\ln e} e^y\,dy = e - \int_0^1 e^y\,dy = e - \left[e^y\right]_0^1 = e - (e-1) = 1$.

6.2 Trigonometric Integrals and Substitutions

The symbols $\overset{s}{=}$ and $\overset{c}{=}$ indicate the use of the substitutions $\{u = \sin x, du = \cos x\,dx\}$ and $\{u = \cos x, du = -\sin x\,dx\}$, respectively.

1. $\int \sin^3 x \cos^2 x\,dx = \int \sin^2 x \cos^2 x \sin x\,dx = \int (1-\cos^2 x)\cos^2 x \sin x\,dx \overset{c}{=} \int (1-u^2)u^2\,(-du)$
$= \int (u^2-1)u^2\,du = \int (u^4-u^2)\,du = \frac{1}{5}u^5 - \frac{1}{3}u^3 + C = \frac{1}{5}\cos^5 x - \frac{1}{3}\cos^3 x + C$

2. $\int \sin^6 x \cos^3 x\,dx = \int \sin^6 x \cos^2 x \cos x\,dx = \int \sin^6 x\,(1-\sin^2 x)\cos x\,dx \overset{s}{=} \int u^6(1-u^2)\,du$
$= \int (u^6-u^8)\,du = \frac{1}{7}u^7 - \frac{1}{9}u^9 + C = \frac{1}{7}\sin^7 x - \frac{1}{9}\sin^9 x + C$

3. $\int_{\pi/2}^{3\pi/4} \sin^5 x \cos^3 x\,dx = \int_{\pi/2}^{3\pi/4} \sin^5 x \cos^2 x \cos x\,dx = \int_{\pi/2}^{3\pi/4} \sin^5 x\,(1-\sin^2 x)\cos x\,dx \overset{s}{=} \int_1^{\sqrt{2}/2} u^5(1-u^2)\,du$
$= \int_1^{\sqrt{2}/2}(u^5-u^7)\,du = \left[\frac{1}{6}u^6 - \frac{1}{8}u^8\right]_1^{\sqrt{2}/2} = \left(\frac{1/8}{6} - \frac{1/16}{8}\right) - \left(\frac{1}{6} - \frac{1}{8}\right) = -\frac{11}{384}$

4. $\int_0^{\pi/2} \cos^5 x\,dx = \int_0^{\pi/2} (\cos^2 x)^2 \cos x\,dx = \int_0^{\pi/2}(1-\sin^2 x)^2 \cos x\,dx \overset{s}{=} \int_0^1 (1-u^2)^2\,du$
$= \int_0^1 (1-2u^2+u^4)\,du = \left[u - \frac{2}{3}u^3 + \frac{1}{5}u^5\right]_0^1 = \left(1 - \frac{2}{3} + \frac{1}{5}\right) - 0 = \frac{8}{15}$

5. $\int_0^{\pi/2} \cos^2\theta\,d\theta = \int_0^{\pi/2} \frac{1}{2}(1+\cos 2\theta)\,d\theta$ [half-angle identity] $= \frac{1}{2}\left[\theta + \frac{1}{2}\sin 2\theta\right]_0^{\pi/2} = \frac{1}{2}\left[\left(\frac{\pi}{2}+0\right) - (0+0)\right] = \frac{\pi}{4}$

6. $\int \sin^3(mx)\,dx = \int (1-\cos^2 mx)\sin mx\,dx = -\frac{1}{m}\int (1-u^2)\,du$ $[u = \cos mx, du = -m\sin mx\,dx]$
$= -\frac{1}{m}\left(u - \frac{1}{3}u^3\right) + C = -\frac{1}{m}\left(\cos mx - \frac{1}{3}\cos^3 mx\right) + C = \frac{1}{3m}\cos^3 mx - \frac{1}{m}\cos mx + C$

7. $\int_0^{\pi} \sin^4(3t)\,dt = \int_0^{\pi}\left[\sin^2(3t)\right]^2 dt = \int_0^{\pi}\left[\frac{1}{2}(1-\cos 6t)\right]^2 dt = \frac{1}{4}\int_0^{\pi}(1-2\cos 6t+\cos^2 6t)\,dt$
$= \frac{1}{4}\int_0^{\pi}\left[1 - 2\cos 6t + \frac{1}{2}(1+\cos 12t)\right]dt = \frac{1}{4}\int_0^{\pi}\left(\frac{3}{2} - 2\cos 6t + \frac{1}{2}\cos 12t\right)dt$
$= \frac{1}{4}\left[\frac{3}{2}t - \frac{1}{3}\sin 6t + \frac{1}{24}\sin 12t\right]_0^{\pi} = \frac{1}{4}\left[\left(\frac{3\pi}{2} - 0 + 0\right) - (0-0+0)\right] = \frac{3\pi}{8}$

8. $\int_0^{\pi/2} \sin^2(2\theta)\,d\theta = \int_0^{\pi/2} \frac{1}{2}(1-\cos 4\theta)\,d\theta = \frac{1}{2}\left[\theta - \frac{1}{4}\sin 4\theta\right]_0^{\pi/2} = \frac{1}{2}\left[\left(\frac{\pi}{2}-0\right) - (0-0)\right] = \frac{\pi}{4}$

9. $\int (1+\cos\theta)^2\,d\theta = \int (1+2\cos\theta+\cos^2\theta)\,d\theta = \theta + 2\sin\theta + \frac{1}{2}\int(1+\cos 2\theta)\,d\theta$
$= \theta + 2\sin\theta + \frac{1}{2}\theta + \frac{1}{4}\sin 2\theta + C = \frac{3}{2}\theta + 2\sin\theta + \frac{1}{4}\sin 2\theta + C$

10. $\int_0^{\pi} \cos^6\theta\,d\theta = \int_0^{\pi}(\cos^2\theta)^3\,d\theta = \int_0^{\pi}\left[\frac{1}{2}(1+\cos 2\theta)\right]^3 d\theta = \frac{1}{8}\int_0^{\pi}(1+3\cos 2\theta+3\cos^2 2\theta+\cos^3 2\theta)\,d\theta$
$= \frac{1}{8}\left[\theta + \frac{3}{2}\sin 2\theta\right]_0^{\pi} + \frac{1}{8}\int_0^{\pi}\left[\frac{3}{2}(1+\cos 4\theta)\right]d\theta + \frac{1}{8}\int_0^{\pi}\left[(1-\sin^2 2\theta)\cos 2\theta\right]d\theta$
$= \frac{1}{8}\pi + \frac{3}{16}\left[\theta + \frac{1}{4}\sin 4\theta\right]_0^{\pi} + \frac{1}{8}\int_0^0(1-u^2)\left(\frac{1}{2}du\right)$ $[u = \sin 2\theta, du = 2\cos 2\theta\,d\theta]$
$= \frac{\pi}{8} + \frac{3\pi}{16} + 0 = \frac{5\pi}{16}$

11. $\int_0^{\pi/4} \sin^4 x \cos^2 x\,dx = \int_0^{\pi/4} \sin^2 x\,(\sin x\cos x)^2\,dx = \int_0^{\pi/4}\frac{1}{2}(1-\cos 2x)\left(\frac{1}{2}\sin 2x\right)^2 dx$
$= \frac{1}{8}\int_0^{\pi/4}(1-\cos 2x)\sin^2 2x\,dx = \frac{1}{8}\int_0^{\pi/4}\sin^2 2x\,dx - \frac{1}{8}\int_0^{\pi/4}\sin^2 2x\cos 2x\,dx$
$= \frac{1}{16}\int_0^{\pi/4}(1-\cos 4x)\,dx - \frac{1}{16}\left[\frac{1}{3}\sin^3 2x\right]_0^{\pi/4} = \frac{1}{16}\left[x - \frac{1}{4}\sin 4x - \frac{1}{3}\sin^3 2x\right]_0^{\pi/4}$
$= \frac{1}{16}\left(\frac{\pi}{4} - 0 - \frac{1}{3}\right) = \frac{1}{192}(3\pi - 4)$

12. Let $u = x$, $dv = \cos^2 x\,dx \Rightarrow du = dx$, $v = \int \cos^2 x\,dx = \int \frac{1}{2}(1 + \cos 2x)\,dx = \frac{1}{2}x + \frac{1}{4}\sin 2x$, so

$$\int x\cos^2 x\,dx = x\left(\tfrac{1}{2}x + \tfrac{1}{4}\sin 2x\right) - \int\left(\tfrac{1}{2}x + \tfrac{1}{4}\sin 2x\right)dx = \tfrac{1}{2}x^2 + \tfrac{1}{4}x\sin 2x - \tfrac{1}{4}x^2 + \tfrac{1}{8}\cos 2x + C$$
$$= \tfrac{1}{4}x^2 + \tfrac{1}{4}x\sin 2x + \tfrac{1}{8}\cos 2x + C$$

13. $$\int \cos^2 x\,\tan^3 x\,dx = \int \frac{\sin^3 x}{\cos x}\,dx \overset{c}{=} \int \frac{(1-u^2)(-du)}{u} = \int\left[\frac{-1}{u} + u\right]du$$
$$= -\ln|u| + \tfrac{1}{2}u^2 + C = \tfrac{1}{2}\cos^2 x - \ln|\cos x| + C$$

14. $$\int_0^{\pi/2}\sin^2 x\,\cos^2 x\,dx = \int_0^{\pi/2}\tfrac{1}{4}(4\sin^2 x\,\cos^2 x)\,dx = \int_0^{\pi/2}\tfrac{1}{4}(2\sin x\,\cos x)^2\,dx = \tfrac{1}{4}\int_0^{\pi/2}\sin^2 2x\,dx$$
$$= \tfrac{1}{4}\int_0^{\pi/2}\tfrac{1}{2}(1-\cos 4x)\,dx = \tfrac{1}{8}\int_0^{\pi/2}(1-\cos 4x)\,dx = \tfrac{1}{8}\left[x - \tfrac{1}{4}\sin 4x\right]_0^{\pi/2} = \tfrac{1}{8}\left(\tfrac{\pi}{2}\right) = \tfrac{\pi}{16}$$

15. $$\int \frac{1-\sin x}{\cos x}\,dx = \int(\sec x - \tan x)\,dx = \ln|\sec x + \tan x| - \ln|\sec x| + C \quad \begin{bmatrix}\text{by (1) and the boxed}\\ \text{formula above it}\end{bmatrix}$$
$$= \ln|(\sec x + \tan x)\cos x| + C = \ln|1 + \sin x| + C$$
$$= \ln(1 + \sin x) + C \quad \text{since } 1 + \sin x \geq 0$$

Or: $$\int\frac{1-\sin x}{\cos x}\,dx = \int\frac{1-\sin x}{\cos x}\cdot\frac{1+\sin x}{1+\sin x}\,dx = \int\frac{(1-\sin^2 x)dx}{\cos x\,(1+\sin x)} = \int\frac{\cos x\,dx}{1+\sin x}$$
$$= \int\frac{dw}{w} \quad [\text{where } w = 1 + \sin x,\ dw = \cos x\,dx]$$
$$= \ln|w| + C = \ln|1+\sin x| + C = \ln(1+\sin x) + C$$

16. $\int \cos^2 x\,\sin 2x\,dx = 2\int\cos^3 x\,\sin x\,dx \overset{c}{=} -2\int u^3\,du = -\frac{1}{2}u^4 + C = -\frac{1}{2}\cos^4 x + C$

17. Let $u = \tan x$, $du = \sec^2 x\,dx$. Then $\int\sec^2 x\,\tan x\,dx = \int u\,du = \frac{1}{2}u^2 + C = \frac{1}{2}\tan^2 x + C$.

Or: Let $v = \sec x$, $dv = \sec x\,\tan x\,dx$. Then $\int\sec^2 x\,\tan x\,dx = \int v\,dv = \frac{1}{2}v^2 + C = \frac{1}{2}\sec^2 x + C$.

18. $$\int_0^{\pi/2}\sec^4(t/2)\,dt = \int_0^{\pi/4}\sec^4 x\,(2\,dx) \quad [x = t/2,\ dx = \tfrac{1}{2}\,dt] \quad = 2\int_0^{\pi/4}\sec^2 x\,(1+\tan^2 x)\,dx$$
$$= 2\int_0^1(1+u^2)\,du \quad [u = \tan x,\ du = \sec^2 x\,dx] \quad = 2\left[u + \tfrac{1}{3}u^3\right]_0^1 = 2\left(1+\tfrac{1}{3}\right) = \tfrac{8}{3}$$

19. $\int\tan^2 x\,dx = \int(\sec^2 x - 1)\,dx = \tan x - x + C$

20. $\int\tan^4 x\,dx = \int\tan^2 x\,(\sec^2 x - 1)\,dx = \int\tan^2 x\,\sec^2 x\,dx - \int\tan^2 x\,dx = \frac{1}{3}\tan^3 x - \tan x + x + C$

(Set $u = \tan x$ in the first integral and use Exercise 19 for the second.)

21. $$\int\sec^6 t\,dt = \int\sec^4 t\cdot\sec^2 t\,dt = \int(\tan^2 t + 1)^2\sec^2 t\,dt = \int(u^2+1)^2\,du \quad [u = \tan t,\ du = \sec^2 t\,dt]$$
$$= \int(u^4 + 2u^2 + 1)\,du = \tfrac{1}{5}u^5 + \tfrac{2}{3}u^3 + u + C = \tfrac{1}{5}\tan^5 t + \tfrac{2}{3}\tan^3 t + \tan t + C$$

22. $$\int_0^{\pi/4}\sec^4\theta\,\tan^4\theta\,d\theta = \int_0^{\pi/4}(\tan^2\theta + 1)\,\tan^4\theta\,\sec^2\theta\,d\theta = \int_0^1(u^2+1)u^4\,du \quad [u = \tan\theta,\ du = \sec^2\theta\,d\theta]$$
$$= \int_0^1(u^6 + u^4)\,du = \left[\tfrac{1}{7}u^7 + \tfrac{1}{5}u^5\right]_0^1 = \tfrac{1}{7} + \tfrac{1}{5} = \tfrac{12}{35}$$

23. $\int_0^{\pi/3} \tan^5 x \sec^4 x\,dx = \int_0^{\pi/3} \tan^5 x\,(\tan^2 x + 1)\sec^2 x\,dx = \int_0^{\sqrt{3}} u^5(u^2+1)\,du \qquad [u = \tan x,\ du = \sec^2 x\,dx]$

$= \int_0^{\sqrt{3}} (u^7 + u^5)\,du = \left[\frac{1}{8}u^8 + \frac{1}{6}u^6\right]_0^{\sqrt{3}} = \frac{81}{8} + \frac{27}{6} = \frac{81}{8} + \frac{9}{2} = \frac{81}{8} + \frac{36}{8} = \frac{117}{8}$

Alternate solution:

$\int_0^{\pi/3} \tan^5 x \sec^4 x\,dx = \int_0^{\pi/3} \tan^4 x \sec^3 x \sec x \tan x\,dx = \int_0^{\pi/3} (\sec^2 x - 1)^2 \sec^3 x \sec x \tan x\,dx$

$= \int_1^2 (u^2-1)^2 u^3\,du \quad [u = \sec x,\ du = \sec x \tan x\,dx] \quad = \int_1^2 (u^4 - 2u^2 + 1)u^3\,du$

$= \int_1^2 (u^7 - 2u^5 + u^3)\,du = \left[\frac{1}{8}u^8 - \frac{1}{3}u^6 + \frac{1}{4}u^4\right]_1^2 = \left(32 - \frac{64}{3} + 4\right) - \left(\frac{1}{8} - \frac{1}{3} + \frac{1}{4}\right) = \frac{117}{8}$

24. $\int \tan^3(2x)\sec^5(2x)\,dx = \int \tan^2(2x)\sec^4(2x)\cdot\sec(2x)\tan(2x)\,dx$

$= \int (u^2-1)u^4\left(\frac{1}{2}\,du\right) \qquad [u = \sec(2x),\ du = 2\sec(2x)\tan(2x)\,dx]$

$= \frac{1}{2}\int (u^6 - u^4)\,du = \frac{1}{14}u^7 - \frac{1}{10}u^5 + C = \frac{1}{14}\sec^7(2x) - \frac{1}{10}\sec^5(2x) + C$

25. $\int \tan^3 x \sec x\,dx = \int \tan^2 x \sec x \tan x\,dx = \int (\sec^2 x - 1)\sec x \tan x\,dx$

$= \int (u^2 - 1)\,du \qquad [u = \sec x,\ du = \sec x \tan x\,dx]$

$= \frac{1}{3}u^3 - u + C = \frac{1}{3}\sec^3 x - \sec x + C$

26. $\int_0^{\pi/3} \tan^5 x \sec^6 x\,dx = \int_0^{\pi/3} \tan^5 x \sec^4 x \sec^2 x\,dx = \int_0^{\pi/3} \tan^5 x\,(1 + \tan^2 x)^2 \sec^2 x\,dx$

$= \int_0^{\sqrt{3}} u^5(1+u^2)^2\,du \qquad [u = \tan x,\ du = \sec^2 x\,dx] \qquad = \int_0^{\sqrt{3}} u^5(1 + 2u^2 + u^4)du$

$= \int_0^{\sqrt{3}} (u^5 + 2u^7 + u^9)\,du = \left[\frac{1}{6}u^6 + \frac{1}{4}u^8 + \frac{1}{10}u^{10}\right]_0^{\sqrt{3}} = \frac{27}{6} + \frac{81}{4} + \frac{243}{10} = \frac{981}{20}$

Alternate solution:

$\int_0^{\pi/3} \tan^5 x \sec^6 x\,dx = \int_0^{\pi/3} \tan^4 x \sec^5 x \sec x \tan x\,dx = \int_0^{\pi/3} (\sec^2 x - 1)^2 \sec^5 x \sec x \tan x\,dx$

$= \int_1^2 (u^2-1)^2 u^5\,du \quad [u = \sec x,\ du = \sec x \tan x\,dx] \quad = \int_1^2 (u^4 - 2u^2 + 1)u^5\,du$

$= \int_1^2 (u^9 - 2u^7 + u^5)\,du = \left[\frac{1}{10}u^{10} - \frac{1}{4}u^8 + \frac{1}{6}u^6\right]_1^2 = \left(\frac{512}{5} - 64 + \frac{32}{3}\right) - \left(\frac{1}{10} - \frac{1}{4} + \frac{1}{6}\right) = \frac{981}{20}$

27. $\int \tan^5 x\,dx = \int (\sec^2 x - 1)^2 \tan x\,dx = \int \sec^4 x \tan x\,dx - 2\int \sec^2 x \tan x\,dx + \int \tan x\,dx$

$= \int \sec^3 x \sec x \tan x\,dx - 2\int \tan x \sec^2 x\,dx + \int \tan x\,dx$

$= \frac{1}{4}\sec^4 x - \tan^2 x + \ln|\sec x| + C \qquad \left[\text{or } \frac{1}{4}\sec^4 x - \sec^2 x + \ln|\sec x| + C\right]$

28. $\int \tan^6(ay)\,dy = \int \tan^4 ay\,(\sec^2 ay - 1)\,dy = \int \tan^4 ay \sec^2 ay\,dy - \int \tan^4 ay\,dy$

$= \frac{1}{5a}\tan^5 ay - \int \tan^2 ay\,(\sec^2 ay - 1)\,dy = \frac{1}{5a}\tan^5 ay - \int \tan^2 ay \sec^2 ay\,dy + \int (\sec^2 ay - 1)\,dy$

$= \frac{1}{5a}\tan^5 ay - \frac{1}{3a}\tan^3 ay + \frac{1}{a}\tan ay - y + C$

29. $\int_{\pi/6}^{\pi/2} \cot^2 x\,dx = \int_{\pi/6}^{\pi/2} (\csc^2 x - 1)\,dx = \left[-\cot x - x\right]_{\pi/6}^{\pi/2} = \left(0 - \frac{\pi}{2}\right) - \left(-\sqrt{3} - \frac{\pi}{6}\right) = \sqrt{3} - \frac{\pi}{3}$

30. $$\int_{\pi/4}^{\pi/2} \cot^3 x\,dx = \int_{\pi/4}^{\pi/2} \cot x\,(\csc^2 x - 1)\,dx = \int_{\pi/4}^{\pi/2} \cot x \csc^2 x\,dx - \int_{\pi/4}^{\pi/2} \frac{\cos x}{\sin x}\,dx$$

$$= \left[-\frac{1}{2}\cot^2 x - \ln|\sin x|\right]_{\pi/4}^{\pi/2} = (0 - \ln 1) - \left[-\frac{1}{2} - \ln\frac{1}{\sqrt{2}}\right] = \frac{1}{2} + \ln\frac{1}{\sqrt{2}} = \frac{1}{2}(1 - \ln 2)$$

31. $\int \cot^3 \alpha \csc^3 \alpha\,d\alpha = \int \cot^2 \alpha \csc^2 \alpha \cdot \csc \alpha \cot \alpha\,d\alpha = \int (\csc^2 \alpha - 1)\csc^2 \alpha \cdot \csc \alpha \cot \alpha\,d\alpha$

$= \int (u^2 - 1)u^2 \cdot (-du) \qquad [u = \csc \alpha,\ du = -\csc \alpha \cot \alpha\,d\alpha]$

$= \int (u^2 - u^4)\,du = \frac{1}{3}u^3 - \frac{1}{5}u^5 + C = \frac{1}{3}\csc^3 \alpha - \frac{1}{5}\csc^5 \alpha + C$

32. $\int \csc^4 x \cot^6 x\,dx = \int \cot^6 x\,(\cot^2 x + 1)\csc^2 x\,dx = \int u^6(u^2+1)\cdot(-du) \qquad [u = \cot x,\, du = -\csc^2 x\,dx]$

$= \int(-u^8 - u^6)\,du = -\frac{1}{9}u^9 - \frac{1}{7}u^7 + C = -\frac{1}{9}\cot^9 x - \frac{1}{7}\cot^7 x + C$

33. $I = \displaystyle\int \csc x\,dx = \int \frac{\csc x\,(\csc x - \cot x)}{\csc x - \cot x}\,dx = \int \frac{-\csc x\,\cot x + \csc^2 x}{\csc x - \cot x}\,dx$. Let $u = \csc x - \cot x \;\Rightarrow$

$du = (-\csc x\,\cot x + \csc^2 x)\,dx$. Then $I = \int du/u = \ln|u| = \ln|\csc x - \cot x| + C$.

34. $\displaystyle\int \frac{1 - \tan^2 x}{\sec^2 x}\,dx = \int(\cos^2 x - \sin^2 x)dx = \int \cos 2x\,dx = \frac{1}{2}\sin 2x + C$

35. (a) $\frac{1}{2}[\cos(A-B) - \cos(A+B)] = \frac{1}{2}[(\cos A\,\cos B + \sin A\,\sin B) - (\cos A\,\cos B - \sin A\,\sin B)]$

$= \frac{1}{2}(2\sin A\,\sin B) = \sin A\,\sin B$

(b) By part (a): $\int \sin 5x\,\sin 2x\,dx = \int \frac{1}{2}[\cos(5x - 2x) - \cos(5x + 2x)]\,dx$

$= \frac{1}{2}\int(\cos 3x - \cos 7x)\,dx = \frac{1}{6}\sin 3x - \frac{1}{14}\sin 7x + C$

36. (a) $\frac{1}{2}[\sin(A-B) + \sin(A+B)] = \frac{1}{2}[(\sin A\,\cos B - \cos A\,\sin B) + (\sin A\,\cos B + \cos A\,\sin B)]$

$= \frac{1}{2}(2\sin A\,\cos B) = \sin A\,\cos B$

(b) By part (a): $\int \sin 3x\,\cos x\,dx = \int \frac{1}{2}[\sin(3x + x) + \sin(3x - x)]\,dx = \frac{1}{2}\int(\sin 4x + \sin 2x)\,dx$

$= -\frac{1}{8}\cos 4x - \frac{1}{4}\cos 2x + C$

37. Let $x = 3\sec\theta$, where $0 \le \theta < \frac{\pi}{2}$ or $\pi \le \theta < \frac{3\pi}{2}$.

Then $dx = 3\sec\theta\,\tan\theta\,d\theta$ and

$\sqrt{x^2 - 9} = \sqrt{9\sec^2\theta - 9} = \sqrt{9(\sec^2\theta - 1)} = \sqrt{9\tan^2\theta}$

$= 3\,|\tan\theta| = 3\tan\theta$ for the relevant values of θ.

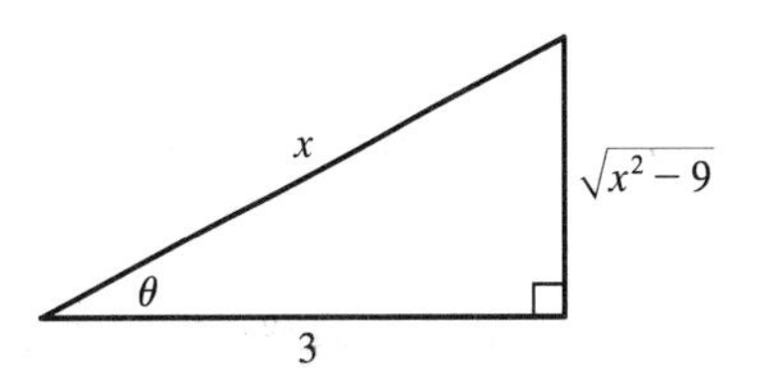

$$\int \frac{1}{x^2\sqrt{x^2 - 9}}\,dx = \int \frac{1}{9\sec^2\theta\cdot 3\tan\theta}\,3\sec\theta\,\tan\theta\,d\theta = \frac{1}{9}\int \cos\theta\,d\theta = \tfrac{1}{9}\sin\theta + C = \frac{1}{9}\frac{\sqrt{x^2 - 9}}{x} + C$$

Note that $-\sec(\theta + \pi) = \sec\theta$, so the figure is sufficient for the case $\pi \le \theta < \frac{3\pi}{2}$.

38. Let $x = 3\sin\theta$, where $-\frac{\pi}{2} \le \theta \le \frac{\pi}{2}$. Then $dx = 3\cos\theta\,d\theta$ and

$\sqrt{9 - x^2} = \sqrt{9 - 9\sin^2\theta} = \sqrt{9(1 - \sin^2\theta)} = \sqrt{9\cos^2\theta}$

$= 3\,|\cos\theta| = 3\cos\theta$ for the relevant values of θ.

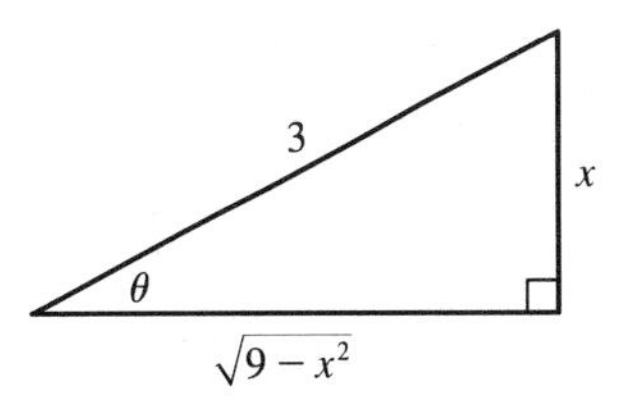

$\int x^3\sqrt{9 - x^2}\,dx = \int 3^3\sin^3\theta\cdot 3\cos\theta\cdot 3\cos\theta\,d\theta = 3^5\int\sin^3\theta\,\cos^2\theta\,d\theta = 3^5\int\sin^2\theta\,\cos^2\theta\,\sin\theta\,d\theta$

$= 3^5\int(1 - \cos^2\theta)\cos^2\theta\,\sin\theta\,d\theta = 3^5\int(1 - u^2)u^2\,(-du) \qquad [u = \cos\theta,\, du = -\sin\theta\,d\theta]$

$= 3^5\int(u^4 - u^2)\,du = 3^5\left(\frac{1}{5}u^5 - \frac{1}{3}u^3\right) + C = 3^5\left(\frac{1}{5}\cos^5\theta - \frac{1}{3}\cos^3\theta\right) + C$

$$= 3^5\left[\frac{1}{5}\frac{(9 - x^2)^{5/2}}{3^5} - \frac{1}{3}\frac{(9 - x^2)^{3/2}}{3^3}\right] + C$$

$= \frac{1}{5}(9 - x^2)^{5/2} - 3(9 - x^2)^{3/2} + C$ or $-\frac{1}{5}(x^2 + 6)(9 - x^2)^{3/2} + C$

39. Let $x = 3\tan\theta$, where $-\frac{\pi}{2} < \theta < \frac{\pi}{2}$. Then $dx = 3\sec^2\theta\,d\theta$ and

$$\sqrt{x^2+9} = \sqrt{9\tan^2\theta + 9} = \sqrt{9(\tan^2\theta+1)} = \sqrt{9\sec^2\theta}$$

$$= 3\,|\sec\theta| = 3\sec\theta \text{ for the relevant values of } \theta.$$

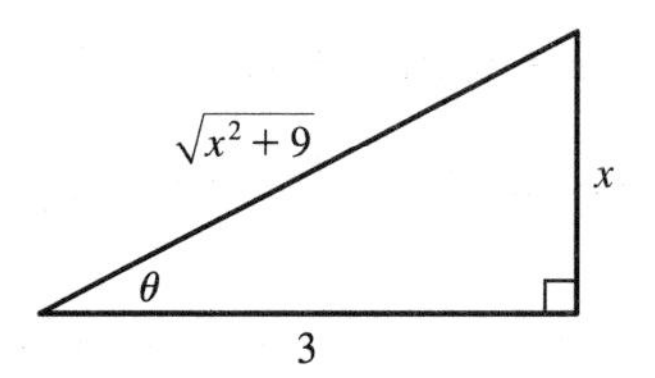

$$\int \frac{x^3}{\sqrt{x^2+9}}\,dx = \int \frac{3^3\tan^3\theta}{3\sec\theta}\,3\sec^2\theta\,d\theta = 3^3\int \tan^3\theta\,\sec\theta\,d\theta = 3^3\int \tan^2\theta\,\tan\theta\,\sec\theta\,d\theta$$

$$= 3^3\int(\sec^2\theta - 1)\tan\theta\,\sec\theta\,d\theta = 3^3\int(u^2-1)du \qquad [u = \sec\theta,\ du = \sec\theta\,\tan\theta\,d\theta]$$

$$= 3^3\left(\tfrac{1}{3}u^3 - u\right) + C = 3^3\left(\tfrac{1}{3}\sec^3\theta - \sec\theta\right) + C = 3^3\left[\frac{1}{3}\frac{(x^2+9)^{3/2}}{3^3} - \frac{\sqrt{x^2+9}}{3}\right] + C$$

$$= \tfrac{1}{3}(x^2+9)^{3/2} - 9\sqrt{x^2+9} + C \text{ or } \tfrac{1}{3}(x^2-18)\sqrt{x^2+9} + C$$

40. Let $x = 4\sin\theta$, where $-\pi/2 \le \theta \le \pi/2$. Then $dx = 4\cos\theta\,d\theta$ and

$\sqrt{16-x^2} = \sqrt{16-16\sin^2\theta} = \sqrt{16\cos^2\theta} = 4\,|\cos\theta| = 4\cos\theta$. When $x = 0$, $4\sin\theta = 0 \Rightarrow \theta = 0$,

and when $x = 2\sqrt{3}$, $4\sin\theta = 2\sqrt{3} \Rightarrow \sin\theta = \frac{\sqrt{3}}{2} \Rightarrow \theta = \frac{\pi}{3}$. Thus, substitution gives

$$\int_0^{2\sqrt{3}} \frac{x^3}{\sqrt{16-x^2}}\,dx = \int_0^{\pi/3} \frac{4^3\sin^3\theta}{4\cos\theta}\,4\cos\theta\,d\theta = 4^3\int_0^{\pi/3}\sin^3\theta\,d\theta = 4^3\int_0^{\pi/3}(1-\cos^2\theta)\sin\theta\,d\theta$$

$$\stackrel{c}{=} -4^3\int_1^{1/2}(1-u^2)\,du = -64\left[u - \tfrac{1}{3}u^3\right]_1^{1/2}$$

$$= -64\left[\left(\tfrac{1}{2} - \tfrac{1}{24}\right) - \left(1 - \tfrac{1}{3}\right)\right] = -64\left(-\tfrac{5}{24}\right) = \tfrac{40}{3}$$

Or: Let $u = 16 - x^2$, $x^2 = 16 - u$, $du = -2x\,dx$.

41. Let $t = \sec\theta$, so $dt = \sec\theta\,\tan\theta\,d\theta$, $t = \sqrt{2} \Rightarrow \theta = \frac{\pi}{4}$, and $t = 2 \Rightarrow \theta = \frac{\pi}{3}$. Then

$$\int_{\sqrt{2}}^{2} \frac{1}{t^3\sqrt{t^2-1}}\,dt = \int_{\pi/4}^{\pi/3} \frac{1}{\sec^3\theta\,\tan\theta}\sec\theta\,\tan\theta\,d\theta = \int_{\pi/4}^{\pi/3}\frac{1}{\sec^2\theta}\,d\theta$$

$$= \int_{\pi/4}^{\pi/3}\cos^2\theta\,d\theta = \int_{\pi/4}^{\pi/3}\tfrac{1}{2}(1+\cos 2\theta)\,d\theta = \tfrac{1}{2}\left[\theta + \tfrac{1}{2}\sin 2\theta\right]_{\pi/4}^{\pi/3}$$

$$= \tfrac{1}{2}\left[\left(\tfrac{\pi}{3} + \tfrac{1}{2}\tfrac{\sqrt{3}}{2}\right) - \left(\tfrac{\pi}{4} + \tfrac{1}{2}\cdot 1\right)\right] = \tfrac{1}{2}\left(\tfrac{\pi}{12} + \tfrac{\sqrt{3}}{4} - \tfrac{1}{2}\right) = \tfrac{\pi}{24} + \tfrac{\sqrt{3}}{8} - \tfrac{1}{4}$$

42. Let $x = 2\tan\theta$, so $dx = 2\sec^2\theta\,d\theta$, $x = 0 \Rightarrow \theta = 0$, and $x = 2 \Rightarrow \theta = \frac{\pi}{4}$. Then

$$\int_0^2 x^3\sqrt{x^2+4}\,dx = \int_0^{\pi/4} 2^3\tan^3\theta\cdot 2\sec\theta\cdot 2\sec^2\theta\,d\theta = 2^5\int_0^{\pi/4}\tan^2\theta\,\sec^2\theta\,\sec\theta\,\tan\theta\,d\theta$$

$$= 2^5\int_0^{\pi/4}(\sec^2\theta - 1)\,\sec^2\theta\,\sec\theta\,\tan\theta\,d\theta$$

$$= 2^5\int_1^{\sqrt{2}}(u^2-1)u^2\,du \quad [u = \sec\theta,\ du = \sec\theta\,\tan\theta\,d\theta]$$

$$= 2^5\int_1^{\sqrt{2}}(u^4-u^2)\,du = 2^5\left[\tfrac{1}{5}u^5 - \tfrac{1}{3}u^3\right]_1^{\sqrt{2}} = 2^5\left[\left(\tfrac{1}{5}\cdot 4\sqrt{2} - \tfrac{1}{3}\cdot 2\sqrt{2}\right) - \left(\tfrac{1}{5} - \tfrac{1}{3}\right)\right]$$

$$= 32\left(\tfrac{2}{15}\sqrt{2} + \tfrac{2}{15}\right) = \tfrac{64}{15}\left(\sqrt{2}+1\right)$$

Or: Let $u = x^2 + 4$, $x^2 = u - 4$, $du = 2x\,dx$.

43. Let $x = 5\sin\theta$, so $dx = 5\cos\theta\,d\theta$. Then

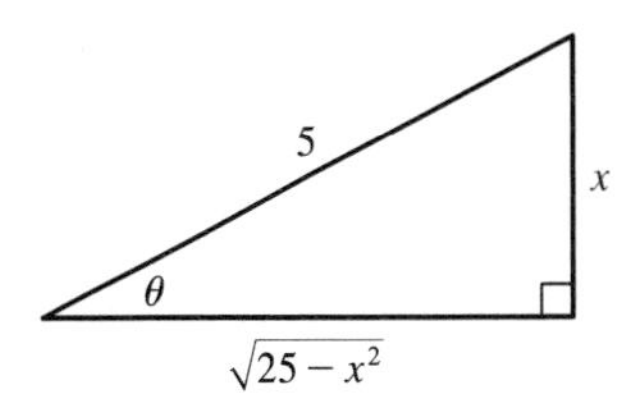

$$\int \frac{1}{x^2\sqrt{25-x^2}}\,dx = \int \frac{1}{5^2\sin^2\theta\cdot 5\cos\theta}\,5\cos\theta\,d\theta = \frac{1}{25}\int \csc^2\theta\,d\theta$$
$$= -\frac{1}{25}\cot\theta + C = -\frac{1}{25}\frac{\sqrt{25-x^2}}{x} + C$$

44. Let $x = a\sec\theta$, where $0 \le \theta < \frac{\pi}{2}$ or $\pi \le \theta < \frac{3\pi}{2}$.

Then $dx = a\sec\theta\tan\theta\,d\theta$ and $\sqrt{x^2-a^2} = a\tan\theta$, so

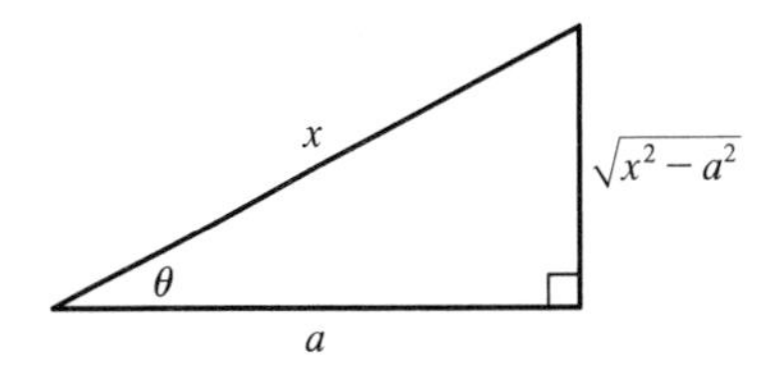

$$\int \frac{\sqrt{x^2-a^2}}{x^4}\,dx = \int \frac{a\tan\theta}{a^4\sec^4\theta}\,a\sec\theta\,\tan\theta\,d\theta = \frac{1}{a^2}\int \sin^2\theta\cos\theta\,d\theta$$
$$= \frac{1}{3a^2}\sin^3\theta + C = \frac{(x^2-a^2)^{3/2}}{3a^2x^3} + C$$

45. Let $x = 4\tan\theta$, where $-\frac{\pi}{2} < \theta < \frac{\pi}{2}$. Then $dx = 4\sec^2\theta\,d\theta$ and

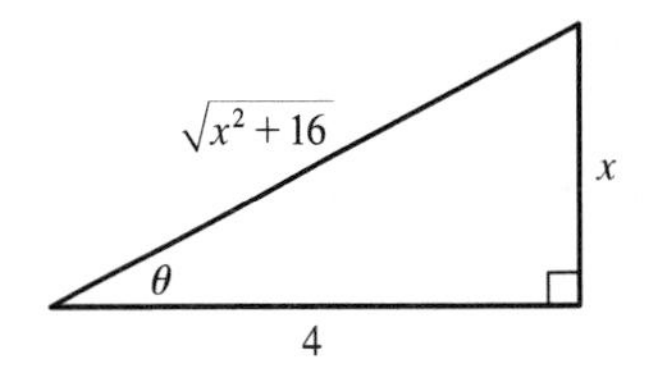

$$\sqrt{x^2+16} = \sqrt{16\tan^2\theta+16} = \sqrt{16(\tan^2\theta+1)} = \sqrt{16\sec^2\theta}$$
$$= 4\,|\sec\theta| = 4\sec\theta \text{ for the relevant values of } \theta.$$

$$\int \frac{dx}{\sqrt{x^2+16}} = \int \frac{4\sec^2\theta\,d\theta}{4\sec\theta} = \int \sec\theta\,d\theta = \ln|\sec\theta+\tan\theta| + C_1 = \ln\left|\frac{\sqrt{x^2+16}}{4} + \frac{x}{4}\right| + C_1$$
$$= \ln\left|\sqrt{x^2+16}+x\right| - \ln|4| + C_1 = \ln\left(\sqrt{x^2+16}+x\right) + C, \text{ where } C = C_1 - \ln 4.$$

(Since $\sqrt{x^2+16}+x > 0$, we don't need the absolute value.)

46. Let $t = \sqrt{2}\tan\theta$, where $-\frac{\pi}{2} < \theta < \frac{\pi}{2}$. Then $dt = \sqrt{2}\sec^2\theta\,d\theta$ and

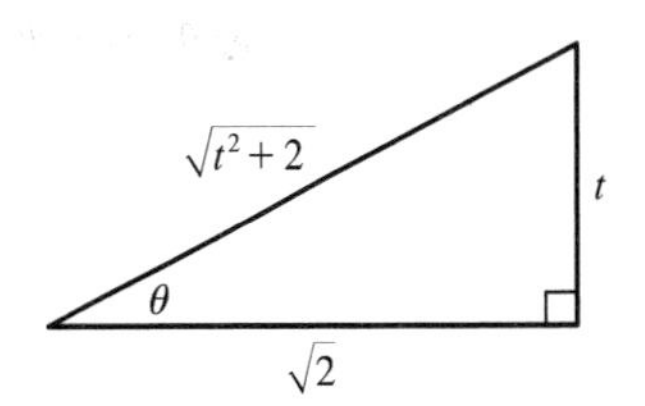

$$\sqrt{t^2+2} = \sqrt{2\tan^2\theta+2} = \sqrt{2(\tan^2\theta+1)} = \sqrt{2\sec^2\theta}$$
$$= \sqrt{2}\,|\sec\theta| = \sqrt{2}\sec\theta \text{ for the relevant values of } \theta$$

$$\int \frac{t^5}{\sqrt{t^2+2}}\,dt = \int \frac{4\sqrt{2}\tan^5\theta}{\sqrt{2}\sec\theta}\sqrt{2}\sec^2\theta\,d\theta = 4\sqrt{2}\int \tan^5\theta\,\sec\theta\,d\theta$$
$$= 4\sqrt{2}\int (\sec^2\theta-1)^2\sec\theta\,\tan\theta\,d\theta$$
$$= 4\sqrt{2}\int (u^2-1)^2\,du \qquad [u = \sec\theta,\ du = \sec\theta\,\tan\theta\,d\theta]$$
$$= 4\sqrt{2}\int (u^4-2u^2+1)\,du = 4\sqrt{2}\left(\frac{1}{5}u^5 - \frac{2}{3}u^3 + u\right) + C$$
$$= \frac{4\sqrt{2}}{15}u(3u^4-10u^2+15) + C = \frac{4\sqrt{2}}{15}\cdot\frac{\sqrt{t^2+2}}{\sqrt{2}}\left[3\cdot\frac{(t^2+2)^2}{2^2} - 10\,\frac{t^2+2}{2} + 15\right] + C$$
$$= \tfrac{4}{15}\sqrt{t^2+2}\cdot\tfrac{1}{4}\left[3(t^4+4t^2+4) - 20(t^2+2) + 60\right] + C$$
$$= \tfrac{1}{15}\sqrt{t^2+2}\,(3t^4-8t^2+32) + C$$

47. Let $2x = \sin\theta$, where $-\frac{\pi}{2} \le \theta \le \frac{\pi}{2}$.

Then $x = \frac{1}{2}\sin\theta$, $dx = \frac{1}{2}\cos\theta\, d\theta$, and $\sqrt{1-4x^2} = \sqrt{1-(2x)^2} = \cos\theta$.

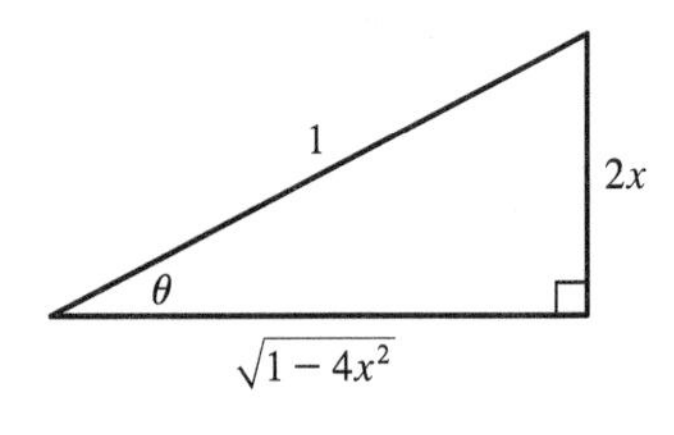

$$\int \sqrt{1-4x^2}\,dx = \int \cos\theta\left(\tfrac{1}{2}\cos\theta\right)d\theta = \tfrac{1}{4}\int(1+\cos 2\theta)\,d\theta$$
$$= \tfrac{1}{4}\left(\theta + \tfrac{1}{2}\sin 2\theta\right) + C = \tfrac{1}{4}(\theta + \sin\theta\cos\theta) + C$$
$$= \tfrac{1}{4}\left[\sin^{-1}(2x) + 2x\sqrt{1-4x^2}\,\right] + C$$

48. $\int_0^1 x\sqrt{x^2+4}\,dx = \int_4^5 \sqrt{u}\left(\frac{1}{2}\,du\right)$ $[u = x^2+4,\ du = 2x\,dx]$ $= \frac{1}{2}\cdot\frac{2}{3}\left[u^{3/2}\right]_4^5 = \frac{1}{3}\left(5\sqrt{5}-8\right)$

49. Let $x = 3\sec\theta$, where $0 \le \theta < \frac{\pi}{2}$ or $\pi \le \theta < \frac{3\pi}{2}$.

Then $dx = 3\sec\theta\tan\theta\,d\theta$ and $\sqrt{x^2-9} = 3\tan\theta$, so

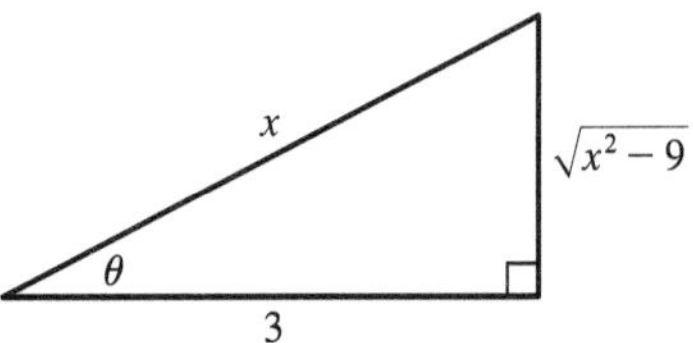

$$\int \frac{\sqrt{x^2-9}}{x^3}\,dx = \int \frac{3\tan\theta}{27\sec^3\theta}\,3\sec\theta\tan\theta\,d\theta = \frac{1}{3}\int\frac{\tan^2\theta}{\sec^2\theta}\,d\theta$$
$$= \tfrac{1}{3}\int\sin^2\theta\,d\theta = \tfrac{1}{3}\int\tfrac{1}{2}(1-\cos 2\theta)\,d\theta = \tfrac{1}{6}\theta - \tfrac{1}{12}\sin 2\theta + C = \tfrac{1}{6}\theta - \tfrac{1}{6}\sin\theta\cos\theta + C$$
$$= \frac{1}{6}\sec^{-1}\left(\frac{x}{3}\right) - \frac{1}{6}\frac{\sqrt{x^2-9}}{x}\frac{3}{x} + C = \frac{1}{6}\sec^{-1}\left(\frac{x}{3}\right) - \frac{\sqrt{x^2-9}}{2x^2} + C$$

50. Let $u = \sqrt{5}\sin\theta$, so $du = \sqrt{5}\cos\theta\,d\theta$. Then

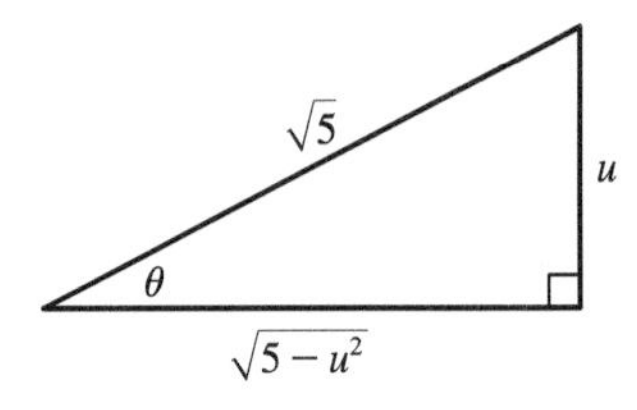

$$\int\frac{du}{u\sqrt{5-u^2}} = \int\frac{1}{\sqrt{5}\sin\theta\cdot\sqrt{5}\cos\theta}\sqrt{5}\cos\theta\,d\theta = \frac{1}{\sqrt{5}}\int\csc\theta\,d\theta$$
$$= \frac{1}{\sqrt{5}}\ln|\csc\theta - \cot\theta| + C \qquad \text{[by Exercise 33]}$$
$$= \frac{1}{\sqrt{5}}\ln\left|\frac{\sqrt{5}}{u} - \frac{\sqrt{5-u^2}}{u}\right| + C = \frac{1}{\sqrt{5}}\ln\left|\frac{\sqrt{5}-\sqrt{5-u^2}}{u}\right| + C$$

51. Let $x = a\sin\theta$, where $-\frac{\pi}{2} \le \theta \le \frac{\pi}{2}$. Then $dx = a\cos\theta\,d\theta$ and

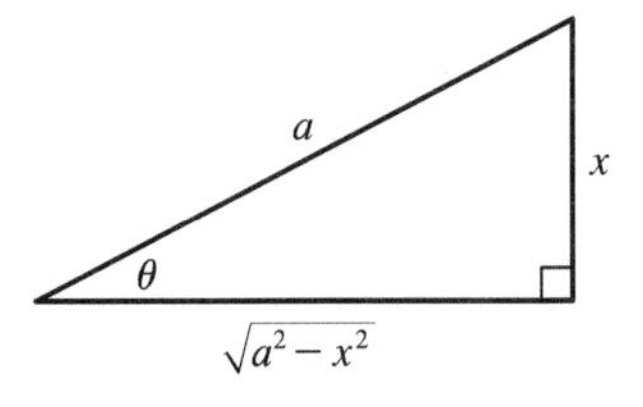

$$\int\frac{x^2\,dx}{(a^2-x^2)^{3/2}} = \int\frac{a^2\sin^2\theta\,a\cos\theta\,d\theta}{a^3\cos^3\theta} = \int\tan^2\theta\,d\theta = \int(\sec^2\theta - 1)\,d\theta$$
$$= \tan\theta - \theta + C = \frac{x}{\sqrt{a^2-x^2}} - \sin^{-1}\left(\frac{x}{a}\right) + C$$

52. Let $4x = 3\sec\theta$, where $0 \le \theta < \frac{\pi}{2}$ or $\pi \le \theta < \frac{3\pi}{2}$.

Then $dx = \frac{3}{4}\sec\theta\tan\theta\,d\theta$ and $\sqrt{16x^2-9} = 3\tan\theta$, so

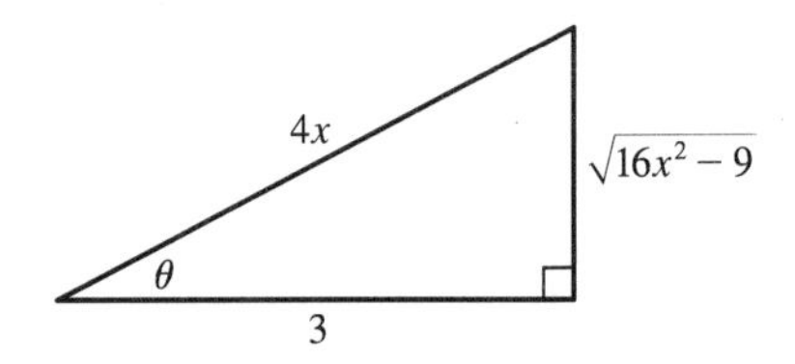

$$\int\frac{dx}{x^2\sqrt{16x^2-9}} = \int\frac{\frac{3}{4}\sec\theta\tan\theta\,d\theta}{\left(\frac{3}{4}\right)^2\sec^2\theta\;3\tan\theta} = \frac{4}{9}\int\cos\theta\,d\theta$$
$$= \frac{4}{9}\sin\theta + C = \frac{4}{9}\frac{\sqrt{16x^2-9}}{4x} + C = \frac{\sqrt{16x^2-9}}{9x} + C$$

53. Let $u = x^2 - 7$, so $du = 2x\,dx$. Then $\displaystyle\int\frac{x}{\sqrt{x^2-7}}\,dx = \frac{1}{2}\int\frac{1}{\sqrt{u}}\,du = \tfrac{1}{2}\cdot 2\sqrt{u} + C = \sqrt{x^2-7} + C.$

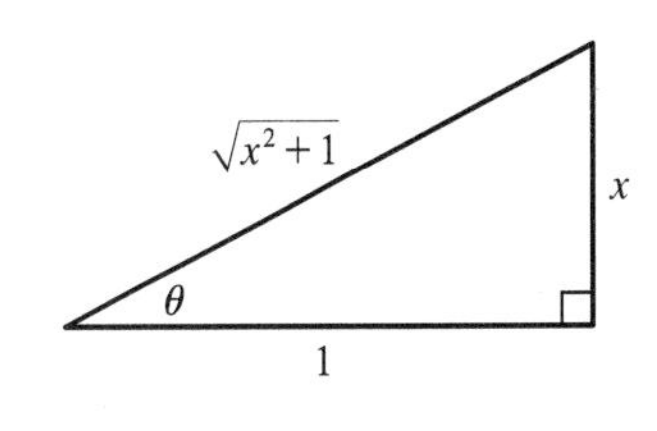

54. Let $x = \tan\theta$, where $-\frac{\pi}{2} < \theta < \frac{\pi}{2}$. Then $dx = \sec^2\theta\,d\theta$, $\sqrt{x^2+1} = \sec\theta$ and $x = 0 \Rightarrow \theta = 0$, $x = 1 \Rightarrow \theta = \frac{\pi}{4}$, so

$$\int_0^1 \sqrt{x^2+1}\,dx = \int_0^{\pi/4} \sec\theta\,\sec^2\theta\,d\theta = \int_0^{\pi/4}\sec^3\theta\,d\theta$$
$$= \tfrac{1}{2}\Big[\sec\theta\,\tan\theta + \ln|\sec\theta + \tan\theta|\Big]_0^{\pi/4} \quad \text{[by Example 8]}$$
$$= \tfrac{1}{2}\big[\sqrt{2}\cdot 1 + \ln\big(1+\sqrt{2}\big) - 0 - \ln(1+0)\big] = \tfrac{1}{2}\big[\sqrt{2} + \ln\big(1+\sqrt{2}\big)\big]$$

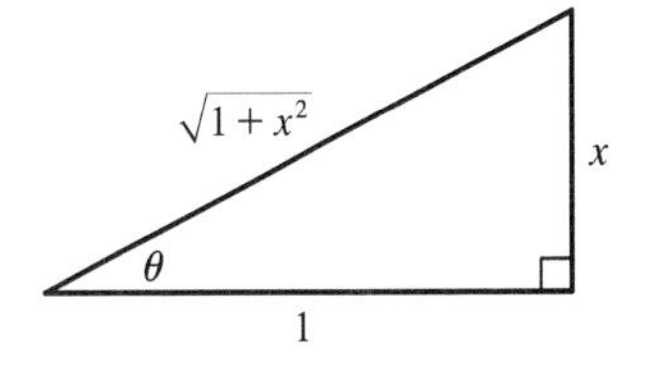

55. Let $x = \tan\theta$, where $-\frac{\pi}{2} < \theta < \frac{\pi}{2}$. Then $dx = \sec^2\theta\,d\theta$ and $\sqrt{1+x^2} = \sec\theta$, so

$$\int \frac{\sqrt{1+x^2}}{x}\,dx = \int \frac{\sec\theta}{\tan\theta}\sec^2\theta\,d\theta = \int \frac{\sec\theta}{\tan\theta}\left(1+\tan^2\theta\right)d\theta$$
$$= \int(\csc\theta + \sec\theta\,\tan\theta)\,d\theta$$
$$= \ln|\csc\theta - \cot\theta| + \sec\theta + C \quad \text{[by Exercise 33]}$$
$$= \ln\left|\frac{\sqrt{1+x^2}}{x} - \frac{1}{x}\right| + \frac{\sqrt{1+x^2}}{1} + C = \ln\left|\frac{\sqrt{1+x^2}-1}{x}\right| + \sqrt{1+x^2} + C$$

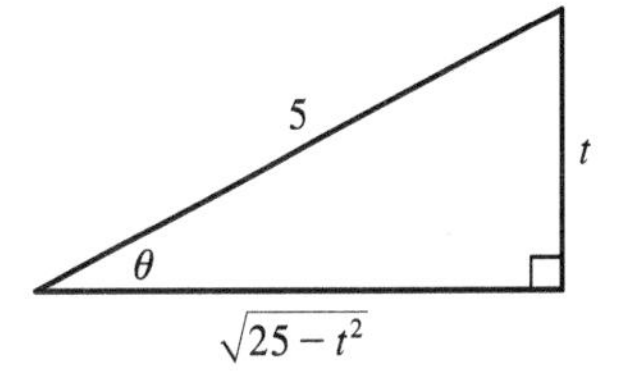

56. Let $t = 5\sin\theta$, where $-\frac{\pi}{2} \le \theta \le \frac{\pi}{2}$. Then $dt = 5\cos\theta\,d\theta$ and $\sqrt{25-t^2} = 5\cos\theta$, so

$$\int \frac{t}{\sqrt{25-t^2}}\,dt = \int \frac{5\sin\theta}{5\cos\theta}5\cos\theta\,d\theta = 5\int\sin\theta\,d\theta = -5\cos\theta + C$$
$$= -5\cdot\frac{\sqrt{25-t^2}}{5} + C = -\sqrt{25-t^2} + C$$

Or: Let $u = 25 - t^2$, so $du = -2t\,dt$.

57. Let $u = x^2$, $du = 2x\,dx$. Then

$$\int x\sqrt{1-x^4}\,dx = \int\sqrt{1-u^2}\left(\tfrac{1}{2}\,du\right) = \tfrac{1}{2}\int\cos\theta\cdot\cos\theta\,d\theta \quad \begin{bmatrix}\text{where } u = \sin\theta,\ du = \cos\theta\,d\theta \\ \text{and } \sqrt{1-u^2} = \cos\theta\end{bmatrix}$$
$$= \tfrac{1}{2}\int\tfrac{1}{2}(1+\cos 2\theta)\,d\theta = \tfrac{1}{4}\theta + \tfrac{1}{8}\sin 2\theta + C = \tfrac{1}{4}\theta + \tfrac{1}{4}\sin\theta\,\cos\theta + C$$
$$= \tfrac{1}{4}\sin^{-1}u + \tfrac{1}{4}u\sqrt{1-u^2} + C = \tfrac{1}{4}\sin^{-1}(x^2) + \tfrac{1}{4}x^2\sqrt{1-x^4} + C$$

58. Let $u = \sin t$, $du = \cos t\,dt$. Then

$$\int_0^{\pi/2}\frac{\cos t}{\sqrt{1+\sin^2 t}}\,dt = \int_0^1\frac{1}{\sqrt{1+u^2}}\,du = \int_0^{\pi/4}\frac{1}{\sec\theta}\sec^2\theta\,d\theta \quad \begin{bmatrix}\text{where } u = \tan\theta,\ du = \sec^2\theta\,d\theta, \\ \text{and } \sqrt{1+u^2} = \sec\theta\end{bmatrix}$$
$$= \int_0^{\pi/4}\sec\theta\,d\theta = \Big[\ln|\sec\theta + \tan\theta|\Big]_0^{\pi/4} \quad \text{[by (1)]}$$
$$= \ln\big(\sqrt{2}+1\big) - \ln(1+0) = \ln\big(\sqrt{2}+1\big)$$

59. $9x^2 + 6x - 8 = (3x+1)^2 - 9$, so let $u = 3x+1$, $du = 3dx$. Then $\displaystyle\int\frac{dx}{\sqrt{9x^2+6x-8}} = \int\frac{\frac{1}{3}\,du}{\sqrt{u^2-9}}$.

Now let $u = 3\sec\theta$, where $0 \le \theta < \frac{\pi}{2}$ or $\pi \le \theta < \frac{3\pi}{2}$. Then $du = 3\sec\theta\,\tan\theta\,d\theta$ and $\sqrt{u^2-9} = 3\tan\theta$, so

$$\int\frac{\frac{1}{3}\,du}{\sqrt{u^2-9}} = \int\frac{\sec\theta\,\tan\theta\,d\theta}{3\tan\theta} = \tfrac{1}{3}\int\sec\theta\,d\theta = \tfrac{1}{3}\ln|\sec\theta + \tan\theta| + C_1 = \tfrac{1}{3}\ln\left|\frac{u+\sqrt{u^2-9}}{3}\right| + C_1$$
$$= \tfrac{1}{3}\ln\left|u + \sqrt{u^2-9}\right| + C = \tfrac{1}{3}\ln\left|3x+1+\sqrt{9x^2+6x-8}\right| + C$$

60. $t^2 - 6t + 13 = (t^2 - 6t + 9) + 4 = (t-3)^2 + 2^2$.

Let $t - 3 = 2\tan\theta$, so $dt = 2\sec^2\theta\, d\theta$. Then

$$\int \frac{dt}{\sqrt{t^2-6t+13}} = \int \frac{1}{\sqrt{(2\tan\theta)^2+2^2}}\, 2\sec^2\theta\, d\theta = \int \frac{2\sec^2\theta}{2\sec\theta}\, d\theta$$

$$= \int \sec\theta\, d\theta = \ln|\sec\theta + \tan\theta| + C_1 \qquad \text{[by Formula 1]}$$

$$= \ln\left|\frac{\sqrt{t^2-6t+13}}{2} + \frac{t-3}{2}\right| + C_1 = \ln\left|\sqrt{t^2-6t+13} + t - 3\right| + C \qquad \text{where } C = C_1 - \ln 2$$

$\sqrt{(t-3)^2+2^2} = \sqrt{t^2-6t+13}$

$t-3$

θ

2

61. $x^2 + 2x + 2 = (x+1)^2 + 1$. Let $u = x + 1$, $du = dx$. Then

$$\int \frac{dx}{(x^2+2x+2)^2} = \int \frac{du}{(u^2+1)^2} = \int \frac{\sec^2\theta\, d\theta}{\sec^4\theta} \qquad \begin{bmatrix}\text{where } u = \tan\theta,\ du = \sec^2\theta\, d\theta, \\ \text{and } u^2+1 = \sec^2\theta\end{bmatrix}$$

$$= \int \cos^2\theta\, d\theta = \tfrac{1}{2}\int (1+\cos 2\theta)\, d\theta = \tfrac{1}{2}(\theta + \sin\theta\,\cos\theta) + C$$

$$= \frac{1}{2}\left[\tan^{-1} u + \frac{u}{1+u^2}\right] + C = \frac{1}{2}\left[\tan^{-1}(x+1) + \frac{x+1}{x^2+2x+2}\right] + C$$

62. $4x - x^2 = -(x^2 - 4x + 4) + 4 = 4 - (x-2)^2$, so let $u = x - 2$. Then $x = u + 2$ and $dx = du$, so

$$\int \frac{x^2\, dx}{\sqrt{4x-x^2}} = \int \frac{(u+2)^2\, du}{\sqrt{4-u^2}} = \int \frac{(2\sin\theta+2)^2}{2\cos\theta} 2\cos\theta\, d\theta \qquad [\text{Put } u = 2\sin\theta]$$

$$= 4\int(\sin^2\theta + 2\sin\theta + 1)\, d\theta = 2\int(1-\cos 2\theta)\, d\theta + 8\int\sin\theta\, d\theta + 4\int d\theta$$

$$= 2\theta - \sin 2\theta - 8\cos\theta + 4\theta + C = 6\theta - 8\cos\theta - 2\sin\theta\,\cos\theta + C$$

$$= 6\sin^{-1}\left(\tfrac{1}{2}u\right) - 4\sqrt{4-u^2} - \tfrac{1}{2}u\sqrt{4-u^2} + C$$

$$= 6\sin^{-1}\left(\frac{x-2}{2}\right) - 4\sqrt{4x-x^2} - \left(\frac{x-2}{2}\right)\sqrt{4x-x^2} + C$$

2

u

θ

$\sqrt{4-u^2}$

63. $s = f(t) = \int_0^t \sin\omega u \cos^2\omega u\, du$. Let $y = \cos\omega u \;\Rightarrow\; dy = -\omega\sin\omega u\, du$. Then

$$s = -\tfrac{1}{\omega}\int_1^{\cos\omega t} y^2 dy = -\tfrac{1}{\omega}\left[\tfrac{1}{3}y^3\right]_1^{\cos\omega t} = \tfrac{1}{3\omega}(1-\cos^3\omega t).$$

64. (a) We want to calculate the square root of the average value of $[E(t)]^2 = [155\sin(120\pi t)]^2 = 155^2\sin^2(120\pi t)$. First, we calculate the average value itself, by integrating $[E(t)]^2$ over one cycle (between $t = 0$ and $t = \frac{1}{60}$, since there are 60 cycles per second) and dividing by $\left(\frac{1}{60} - 0\right)$:

$$[E(t)]^2_{\text{ave}} = \tfrac{1}{1/60}\int_0^{1/60}\left[155^2\sin^2(120\pi t)\right] dt = 60 \cdot 155^2 \int_0^{1/60} \tfrac{1}{2}[1-\cos(240\pi t)]\, dt$$

$$= 60\cdot 155^2\left(\tfrac{1}{2}\right)\left[t - \tfrac{1}{240\pi}\sin(240\pi t)\right]_0^{1/60} = 60\cdot 155^2\left(\tfrac{1}{2}\right)\left[\left(\tfrac{1}{60}-0\right) - (0-0)\right] = \tfrac{155^2}{2}$$

The RMS value is just the square root of this quantity, which is $\frac{155}{\sqrt{2}} \approx 110$ V.

(b) $220 = \sqrt{[E(t)]^2_{\text{ave}}} \quad \Rightarrow$

$$220^2 = [E(t)]^2_{\text{ave}} = \tfrac{1}{1/60} \int_0^{1/60} A^2 \sin^2(120\pi t)\, dt = 60A^2 \int_0^{1/60} \tfrac{1}{2}[1 - \cos(240\pi t)]\, dt$$

$$= 30A^2 \left[t - \tfrac{1}{240\pi} \sin(240\pi t)\right]_0^{1/60} = 30A^2 \left[\left(\tfrac{1}{60} - 0\right) - (0 - 0)\right] = \tfrac{1}{2}A^2$$

Thus, $220^2 = \frac{1}{2}A^2 \quad \Rightarrow \quad A = 220\sqrt{2} \approx 311$ V.

65. The average value of $f(x) = \sqrt{x^2 - 1}/x$ on the interval $[1, 7]$ is

$$\frac{1}{7-1} \int_1^7 \frac{\sqrt{x^2-1}}{x}\, dx = \frac{1}{6} \int_0^\alpha \frac{\tan\theta}{\sec\theta} \cdot \sec\theta\, \tan\theta\, d\theta \qquad \left[\begin{array}{l}\text{where } x = \sec\theta,\ dx = \sec\theta\,\tan\theta\, d\theta, \\ \sqrt{x^2-1} = \tan\theta\text{, and } \alpha = \sec^{-1} 7\end{array}\right]$$

$$= \tfrac{1}{6} \int_0^\alpha \tan^2\theta\, d\theta = \tfrac{1}{6} \int_0^\alpha (\sec^2\theta - 1)\, d\theta = \tfrac{1}{6}\Big[\tan\theta - \theta\Big]_0^\alpha$$

$$= \tfrac{1}{6}(\tan\alpha - \alpha) = \tfrac{1}{6}\left(\sqrt{48} - \sec^{-1} 7\right)$$

66. $9x^2 - 4y^2 = 36 \quad \Rightarrow \quad y = \pm\frac{3}{2}\sqrt{x^2 - 4} \quad \Rightarrow$

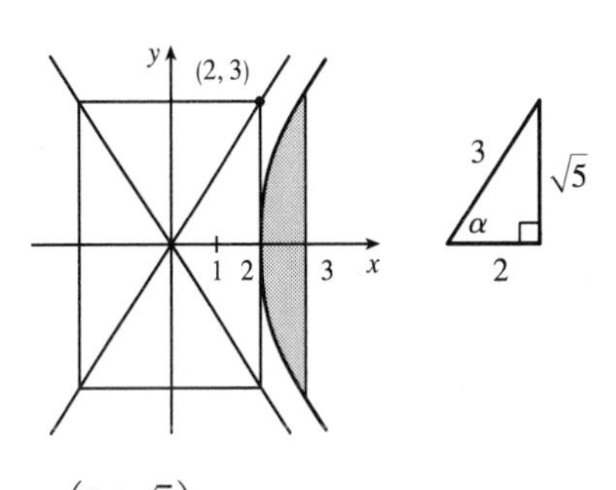

$$\text{area} = 2\int_2^3 \tfrac{3}{2}\sqrt{x^2-4}\, dx = 3\int_2^3 \sqrt{x^2-4}\, dx$$

$$= 3\int_0^\alpha 2\tan\theta\, 2\sec\theta\, \tan\theta\, d\theta \qquad \left[\begin{array}{l}\text{where } x = 2\sec\theta,\ dx = 2\sec\theta\,\tan\theta\, d\theta, \\ \text{and } \alpha = \sec^{-1}\left(\tfrac{3}{2}\right)\end{array}\right]$$

$$= 12\int_0^\alpha \left(\sec^2\theta - 1\right)\sec\theta d\theta = 12\int_0^\alpha \left(\sec^3\theta - \sec\theta\right) d\theta$$

$$= 12\left[\tfrac{1}{2}(\sec\theta\,\tan\theta + \ln|\sec\theta + \tan\theta|) - \ln|\sec\theta + \tan\theta|\right]_0^\alpha$$

$$= 6\Big[\sec\theta\,\tan\theta - \ln|\sec\theta + \tan\theta|\Big]_0^\alpha = 6\left[\tfrac{3\sqrt{5}}{4} - \ln\left(\tfrac{3}{2} + \tfrac{\sqrt{5}}{2}\right)\right] = \tfrac{9\sqrt{5}}{2} - 6\ln\left(\tfrac{3+\sqrt{5}}{2}\right)$$

67. Area of $\triangle POQ = \frac{1}{2}(r\cos\theta)(r\sin\theta) = \frac{1}{2}r^2\sin\theta\cos\theta$. Area of region $PQR = \int_{r\cos\theta}^r \sqrt{r^2 - x^2}\, dx$. Let $x = r\cos u \quad \Rightarrow$ $dx = -r\sin u\, du$ for $\theta \le u \le \frac{\pi}{2}$. Then we obtain

$$\int \sqrt{r^2 - x^2}\, dx = \int r\sin u\,(-r\sin u)\, du = -r^2 \int \sin^2 u\, du = -\tfrac{1}{2}r^2(u - \sin u\cos u) + C$$

$$= -\tfrac{1}{2}r^2\cos^{-1}(x/r) + \tfrac{1}{2}x\sqrt{r^2 - x^2} + C$$

so

$$\text{area of region } PQR = \tfrac{1}{2}\left[-r^2\cos^{-1}(x/r) + x\sqrt{r^2 - x^2}\right]_{r\cos\theta}^r$$

$$= \tfrac{1}{2}\left[0 - \left(-r^2\theta + r\cos\theta\, r\sin\theta\right)\right] = \tfrac{1}{2}r^2\theta - \tfrac{1}{2}r^2\sin\theta\cos\theta$$

and thus, (area of sector POR) = (area of $\triangle POQ$) + (area of region PQR) $= \frac{1}{2}r^2\theta$.

68. Let $x = b\tan\theta$, so that $dx = b\sec^2\theta\, d\theta$ and $\sqrt{x^2 + b^2} = b\sec\theta$.

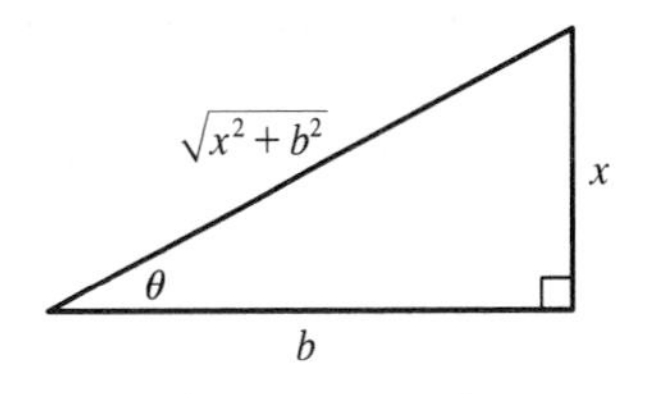

$$E(P) = \int_{-a}^{L-a} \frac{\lambda b}{4\pi\varepsilon_0(x^2 + b^2)^{3/2}}\, dx = \frac{\lambda b}{4\pi\varepsilon_0} \int_{\theta_1}^{\theta_2} \frac{1}{(b\sec\theta)^3}\, b\sec^2\theta\, d\theta$$

$$= \frac{\lambda}{4\pi\varepsilon_0 b} \int_{\theta_1}^{\theta_2} \frac{1}{\sec\theta}\, d\theta = \frac{\lambda}{4\pi\varepsilon_0 b} \int_{\theta_1}^{\theta_2} \cos\theta\, d\theta = \frac{\lambda}{4\pi\varepsilon_0 b}\left[\sin\theta\right]_{\theta_1}^{\theta_2}$$

$$= \frac{\lambda}{4\pi\varepsilon_0 b}\left[\frac{x}{\sqrt{x^2 + b^2}}\right]_{-a}^{L-a} = \frac{\lambda}{4\pi\varepsilon_0 b}\left(\frac{L-a}{\sqrt{(L-a)^2 + b^2}} + \frac{a}{\sqrt{a^2 + b^2}}\right)$$

6.3 Partial Fractions

1. (a) $\dfrac{2x}{(x+3)(3x+1)} = \dfrac{A}{x+3} + \dfrac{B}{3x+1}$

(b) $\dfrac{1}{x^3+2x^2+x} = \dfrac{1}{x(x^2+2x+1)} = \dfrac{1}{x(x+1)^2} = \dfrac{A}{x} + \dfrac{B}{x+1} + \dfrac{C}{(x+1)^2}$

2. (a) $\dfrac{x-1}{x^3+x^2} = \dfrac{x-1}{x^2(x+1)} = \dfrac{A}{x} + \dfrac{B}{x^2} + \dfrac{C}{x+1}$

(b) $\dfrac{x-1}{x^3+x} = \dfrac{x-1}{x(x^2+1)} = \dfrac{A}{x} + \dfrac{Bx+C}{x^2+1}$

3. (a) $\dfrac{2}{x^2+3x-4} = \dfrac{2}{(x+4)(x-1)} = \dfrac{A}{x+4} + \dfrac{B}{x-1}$

(b) x^2+x+1 is irreducible, so $\dfrac{x^2}{(x-1)(x^2+x+1)} = \dfrac{A}{x-1} + \dfrac{Bx+C}{x^2+x+1}$.

4. (a) $\dfrac{x^3}{x^2+4x+3} = x-4+\dfrac{13x+12}{x^2+4x+3} = x-4+\dfrac{13x+12}{(x+1)(x+3)} = x-4+\dfrac{A}{x+1}+\dfrac{B}{x+3}$

(b) $\dfrac{2x+1}{(x+1)^3(x^2+4)^2} = \dfrac{A}{x+1} + \dfrac{B}{(x+1)^2} + \dfrac{C}{(x+1)^3} + \dfrac{Dx+E}{x^2+4} + \dfrac{Fx+G}{(x^2+4)^2}$

5. (a) $\dfrac{x^4}{x^4-1} = \dfrac{(x^4-1)+1}{x^4-1} = 1+\dfrac{1}{x^4-1}$ [or use long division] $= 1+\dfrac{1}{(x^2-1)(x^2+1)}$

$= 1+\dfrac{1}{(x-1)(x+1)(x^2+1)} = 1+\dfrac{A}{x-1}+\dfrac{B}{x+1}+\dfrac{Cx+D}{x^2+1}$

(b) $\dfrac{t^4+t^2+1}{(t^2+1)(t^2+4)^2} = \dfrac{At+B}{t^2+1} + \dfrac{Ct+D}{t^2+4} + \dfrac{Et+F}{(t^2+4)^2}$

6. (a) $\dfrac{x^4}{(x^3+x)(x^2-x+3)} = \dfrac{x^4}{x(x^2+1)(x^2-x+3)} = \dfrac{x^3}{(x^2+1)(x^2-x+3)} = \dfrac{Ax+B}{x^2+1} + \dfrac{Cx+D}{x^2-x+3}$

(b) $\dfrac{1}{x^6-x^3} = \dfrac{1}{x^3(x^3-1)} = \dfrac{1}{x^3(x-1)(x^2+x+1)} = \dfrac{A}{x} + \dfrac{B}{x^2} + \dfrac{C}{x^3} + \dfrac{D}{x-1} + \dfrac{Ex+F}{x^2+x+1}$

7. $\displaystyle\int \frac{x}{x-6}\,dx = \int \frac{(x-6)+6}{x-6}\,dx = \int \left(1+\frac{6}{x-6}\right)dx = x+6\ln|x-6|+C$

8. $\displaystyle\int \frac{r^2}{r+4}\,dr = \int \left(\frac{r^2-16}{r+4}+\frac{16}{r+4}\right)dr = \int \left(r-4+\frac{16}{r+4}\right)dr$ [or use long division]

$= \frac{1}{2}r^2 - 4r + 16\ln|r+4| + C$

9. $\dfrac{x-9}{(x+5)(x-2)} = \dfrac{A}{x+5} + \dfrac{B}{x-2}$. Multiply both sides by $(x+5)(x-2)$ to get $x-9 = A(x-2)+B(x+5)$ $(\star)$, or equivalently, $x-9=(A+B)x-2A+5B$. Equating coefficients of x on each side of the equation gives us $1=A+B$ **(1)** and equating constants gives us $-9=-2A+5B$ **(2)**. Adding two times **(1)** to **(2)** gives us $-7=7B$ $\Leftrightarrow$ $B=-1$ and hence, $A = 2$. [Alternatively, to find the coefficients A and B, we may use substitution as follows: substitute 2 for x in $(\star)$ to get $-7 = 7B \Leftrightarrow B = -1$, then substitute -5 for x in $(\star)$ to get $-14 = -7A \Leftrightarrow A = 2$.]

Thus, $\displaystyle\int \frac{x-9}{(x+5)(x-2)}\,dx = \int\left(\frac{2}{x+5} + \frac{-1}{x-2}\right)dx = 2\ln|x+5| - \ln|x-2| + C.$

To find the constants in problems involving partial fractions, we may use the coefficient comparison method or the substitution method (as in the solution for Exercise 9) or a combination of both methods.

10. $\dfrac{1}{(t+4)(t-1)} = \dfrac{A}{t+4} + \dfrac{B}{t-1} \quad\Rightarrow\quad 1 = A(t-1) + B(t+4).$

$t = 1 \;\Rightarrow\; 1 = 5B \;\Rightarrow\; B = \frac{1}{5}$. $t = -4 \;\Rightarrow\; 1 = -5A \;\Rightarrow\; A = -\frac{1}{5}$. Thus,

$$\int \frac{1}{(t+4)(t-1)}\,dt = \int\left(\frac{-1/5}{t+4} + \frac{1/5}{t-1}\right)dt = -\tfrac{1}{5}\ln|t+4| + \tfrac{1}{5}\ln|t-1| + C \quad\text{or}\quad \frac{1}{5}\ln\left|\frac{t-1}{t+4}\right| + C$$

11. $\dfrac{1}{x^2-1} = \dfrac{1}{(x+1)(x-1)} = \dfrac{A}{x+1} + \dfrac{B}{x-1}$. Multiply both sides by $(x+1)(x-1)$ to get $1 = A(x-1) + B(x+1)$.

Substituting 1 for x gives $1 = 2B \Leftrightarrow B = \frac{1}{2}$. Substituting -1 for x gives $1 = -2A \Leftrightarrow A = -\frac{1}{2}$. Thus,

$$\begin{aligned}\int_2^3 \frac{1}{x^2-1}\,dx &= \int_2^3\left(\frac{-1/2}{x+1} + \frac{1/2}{x-1}\right)dx = \left[-\tfrac{1}{2}\ln|x+1| + \tfrac{1}{2}\ln|x-1|\right]_2^3 \\ &= \left(-\tfrac{1}{2}\ln 4 + \tfrac{1}{2}\ln 2\right) - \left(-\tfrac{1}{2}\ln 3 + \tfrac{1}{2}\ln 1\right) = \tfrac{1}{2}(\ln 2 + \ln 3 - \ln 4) \quad \left[\text{or } \tfrac{1}{2}\ln\tfrac{3}{2}\right]\end{aligned}$$

12. $\dfrac{x-1}{x^2+3x+2} = \dfrac{A}{x+1} + \dfrac{B}{x+2}$. Multiply both sides by $(x+1)(x+2)$ to get $x-1 = A(x+2) + B(x+1)$. Substituting -2 for x gives $-3 = -B \Leftrightarrow B = 3$. Substituting -1 for x gives $-2 = A$. Thus,

$$\begin{aligned}\int_0^1 \frac{x-1}{x^2+3x+2}\,dx &= \int_0^1\left(\frac{-2}{x+1} + \frac{3}{x+2}\right)dx = \left[-2\ln|x+1| + 3\ln|x+2|\right]_0^1 \\ &= (-2\ln 2 + 3\ln 3) - (-2\ln 1 + 3\ln 2) = 3\ln 3 - 5\ln 2 \quad \left[\text{or } \ln\tfrac{27}{32}\right]\end{aligned}$$

13. $\displaystyle\int \frac{ax}{x^2-bx}\,dx = \int \frac{ax}{x(x-b)}\,dx = \int \frac{a}{x-b}\,dx = a\ln|x-b| + C$

14. If $a \neq b$, $\dfrac{1}{(x+a)(x+b)} = \dfrac{1}{b-a}\left(\dfrac{1}{x+a} - \dfrac{1}{x+b}\right)$, so if $a \neq b$, then

$$\int \frac{dx}{(x+a)(x+b)} = \frac{1}{b-a}\left(\ln|x+a| - \ln|x+b|\right) + C = \frac{1}{b-a}\ln\left|\frac{x+a}{x+b}\right| + C$$

If $a = b$, then $\displaystyle\int \frac{dx}{(x+a)^2} = -\frac{1}{x+a} + C.$

15. $\dfrac{2x+3}{(x+1)^2} = \dfrac{A}{x+1} + \dfrac{B}{(x+1)^2} \quad\Rightarrow\quad 2x+3 = A(x+1) + B$. Take $x = -1$ to get $B = 1$, and equate coefficients of x to get $A = 2$. Now

$$\begin{aligned}\int_0^1 \frac{2x+3}{(x+1)^2}\,dx &= \int_0^1\left[\frac{2}{x+1} + \frac{1}{(x+1)^2}\right]dx = \left[2\ln(x+1) - \frac{1}{x+1}\right]_0^1 \\ &= 2\ln 2 - \tfrac{1}{2} - (2\ln 1 - 1) = 2\ln 2 + \tfrac{1}{2}\end{aligned}$$

16. $\dfrac{x^3-4x-10}{x^2-x-6} = x+1+\dfrac{3x-4}{(x-3)(x+2)}$. Write $\dfrac{3x-4}{(x-3)(x+2)} = \dfrac{A}{x-3}+\dfrac{B}{x+2}$. Then

$3x-4 = A(x+2)+B(x-3)$. Taking $x=3$ and $x=-2$, we get $5=5A \;\Leftrightarrow\; A=1$ and $-10=-5B \;\Leftrightarrow\; B=2$, so

$$\int_0^1 \frac{x^3-4x-10}{x^2-x-6}\,dx = \int_0^1\left(x+1+\frac{1}{x-3}+\frac{2}{x+2}\right)dx = \left[\frac{1}{2}x^2+x+\ln|x-3|+2\ln(x+2)\right]_0^1$$
$$= \left(\tfrac{1}{2}+1+\ln 2+2\ln 3\right)-(0+0+\ln 3+2\ln 2) = \tfrac{3}{2}+\ln 3-\ln 2 = \tfrac{3}{2}+\ln\tfrac{3}{2}$$

17. $\dfrac{4y^2-7y-12}{y(y+2)(y-3)} = \dfrac{A}{y}+\dfrac{B}{y+2}+\dfrac{C}{y-3} \;\Rightarrow\; 4y^2-7y-12 = A(y+2)(y-3)+By(y-3)+Cy(y+2)$. Setting $y=0$ gives $-12=-6A$, so $A=2$. Setting $y=-2$ gives $18=10B$, so $B=\frac{9}{5}$. Setting $y=3$ gives $3=15C$, so $C=\frac{1}{5}$.

Now

$$\int_1^2 \frac{4y^2-7y-12}{y(y+2)(y-3)}\,dy = \int_1^2\left(\frac{2}{y}+\frac{9/5}{y+2}+\frac{1/5}{y-3}\right)dy = \left[2\ln|y|+\tfrac{9}{5}\ln|y+2|+\tfrac{1}{5}\ln|y-3|\right]_1^2$$
$$= 2\ln 2+\tfrac{9}{5}\ln 4+\tfrac{1}{5}\ln 1-2\ln 1-\tfrac{9}{5}\ln 3-\tfrac{1}{5}\ln 2$$
$$= 2\ln 2+\tfrac{18}{5}\ln 2-\tfrac{1}{5}\ln 2-\tfrac{9}{5}\ln 3 = \tfrac{27}{5}\ln 2-\tfrac{9}{5}\ln 3 = \tfrac{9}{5}(3\ln 2-\ln 3) = \tfrac{9}{5}\ln\tfrac{8}{3}$$

18. $\dfrac{x^2+2x-1}{x^3-x} = \dfrac{x^2+2x-1}{x(x+1)(x-1)} = \dfrac{A}{x}+\dfrac{B}{x+1}+\dfrac{C}{x-1}$. Multiply both sides by $x(x+1)(x-1)$ to get

$x^2+2x-1 = A(x+1)(x-1)+Bx(x-1)+Cx(x+1)$. Substituting 0 for x gives $-1=-A \;\Leftrightarrow\; A=1$.

Substituting -1 for x gives $-2=2B \;\Leftrightarrow\; B=-1$. Substituting 1 for x gives $2=2C \;\Leftrightarrow\; C=1$. Thus,

$$\int\frac{x^2+2x-1}{x^3-x}\,dx = \int\left(\frac{1}{x}-\frac{1}{x+1}+\frac{1}{x-1}\right)dx = \ln|x|-\ln|x+1|+\ln|x-1|+C = \ln\left|\frac{x(x-1)}{x+1}\right|+C.$$

19. $\dfrac{1}{(x+5)^2(x-1)} = \dfrac{A}{x+5}+\dfrac{B}{(x+5)^2}+\dfrac{C}{x-1} \;\Rightarrow\; 1 = A(x+5)(x-1)+B(x-1)+C(x+5)^2$.

Setting $x=-5$ gives $1=-6B$, so $B=-\frac{1}{6}$. Setting $x=1$ gives $1=36C$, so $C=\frac{1}{36}$. Setting $x=-2$ gives

$1 = A(3)(-3)+B(-3)+C(3^2) = -9A-3B+9C = -9A+\frac{1}{2}+\frac{1}{4} = -9A+\frac{3}{4}$, so $9A=-\frac{1}{4}$ and $A=-\frac{1}{36}$.

Now $\displaystyle\int\frac{1}{(x+5)^2(x-1)}\,dx = \int\left[\frac{-1/36}{x+5}-\frac{1/6}{(x+5)^2}+\frac{1/36}{x-1}\right]dx = -\frac{1}{36}\ln|x+5|+\frac{1}{6(x+5)}+\frac{1}{36}\ln|x-1|+C.$

20. $\dfrac{x^2}{(x-3)(x+2)^2} = \dfrac{A}{x-3}+\dfrac{B}{x+2}+\dfrac{C}{(x+2)^2} \;\Rightarrow\; x^2 = A(x+2)^2+B(x-3)(x+2)+C(x-3)$.

Setting $x=3$ gives $A=\frac{9}{25}$. Take $x=-2$ to get $C=-\frac{4}{5}$, and equate the coefficients of x^2 to get $1=A+B \;\Rightarrow\; B=\frac{16}{25}$. Then

$$\int\frac{x^2}{(x-3)(x+2)^2}dx = \int\left[\frac{9/25}{x-3}+\frac{16/25}{x+2}-\frac{4/5}{(x+2)^2}\right]dx = \frac{9}{25}\ln|x-3|+\frac{16}{25}\ln|x+2|+\frac{4}{5(x+2)}+C.$$

21. $\dfrac{5x^2+3x-2}{x^3+2x^2} = \dfrac{5x^2+3x-2}{x^2(x+2)} = \dfrac{A}{x} + \dfrac{B}{x^2} + \dfrac{C}{x+2}$. Multiply by $x^2(x+2)$ to get

$5x^2+3x-2 = Ax(x+2)+B(x+2)+Cx^2$. Set $x=-2$ to get $C=3$, and take $x=0$ to get $B=-1$.

Equating the coefficients of x^2 gives $5 = A + C \Rightarrow A = 2$.

So $\displaystyle\int \frac{5x^2+3x-2}{x^3+2x^2}\,dx = \int\left(\frac{2}{x} - \frac{1}{x^2} + \frac{3}{x+2}\right)dx = 2\ln|x| + \frac{1}{x} + 3\ln|x+2| + C$.

22. $\dfrac{x^2-x+6}{x^3+3x} = \dfrac{x^2-x+6}{x(x^2+3)} = \dfrac{A}{x} + \dfrac{Bx+C}{x^2+3}$. Multiply by $x(x^2+3)$ to get $x^2-x+6 = A(x^2+3)+(Bx+C)x$.

Substituting 0 for x gives $6 = 3A \Leftrightarrow A = 2$. The coefficients of the x^2-terms must be equal, so $1 = A + B \Rightarrow$ $B = 1 - 2 = -1$. The coefficients of the x-terms must be equal, so $-1 = C$. Thus,

$$\begin{aligned}\int \frac{x^2-x+6}{x^3+3x}\,dx &= \int\left(\frac{2}{x} + \frac{-x-1}{x^2+3}\right)dx = \int\left(\frac{2}{x} - \frac{x}{x^2+3} - \frac{1}{x^2+3}\right)dx \\ &= 2\ln|x| - \frac{1}{2}\ln(x^2+3) - \frac{1}{\sqrt{3}}\tan^{-1}\frac{x}{\sqrt{3}} + C\end{aligned}$$

23. $\dfrac{10}{(x-1)(x^2+9)} = \dfrac{A}{x-1} + \dfrac{Bx+C}{x^2+9}$. Multiply both sides by $(x-1)(x^2+9)$ to get

$10 = A(x^2+9)+(Bx+C)(x-1)$ $(*)$. Substituting 1 for x gives $10 = 10A \Leftrightarrow A = 1$. Substituting 0 for x gives $10 = 9A - C \Rightarrow C = 9(1) - 10 = -1$. The coefficients of the x^2-terms in $(*)$ must be equal, so $0 = A + B \Rightarrow B = -1$. Thus,

$$\begin{aligned}\int \frac{10}{(x-1)(x^2+9)}\,dx &= \int\left(\frac{1}{x-1} + \frac{-x-1}{x^2+9}\right)dx = \int\left(\frac{1}{x-1} - \frac{x}{x^2+9} - \frac{1}{x^2+9}\right)dx \\ &= \ln|x-1| - \tfrac{1}{2}\ln(x^2+9)\ \text{[let u} = x^2+9] \ - \ \tfrac{1}{3}\tan^{-1}\left(\tfrac{x}{3}\right)\ \text{[Formula 10]} \ + \ C\end{aligned}$$

24. $\dfrac{x^2-2x-1}{(x-1)^2(x^2+1)} = \dfrac{A}{x-1} + \dfrac{B}{(x-1)^2} + \dfrac{Cx+D}{x^2+1} \Rightarrow$

$x^2-2x-1 = A(x-1)(x^2+1)+B(x^2+1)+(Cx+D)(x-1)^2$. Setting $x=1$ gives $B=-1$. Equating the coefficients of x^3 gives $A = -C$. Equating the constant terms gives $-1 = -A - 1 + D$, so $D = A$, and setting $x = 2$ gives $-1 = 5A - 5 - 2A + A$ or $A = 1$. We have

$$\int \frac{x^2-2x-1}{(x-1)^2(x^2+1)}\,dx = \int\left[\frac{1}{x-1} - \frac{1}{(x-1)^2} - \frac{x-1}{x^2+1}\right]dx = \ln|x-1| + \frac{1}{x-1} - \tfrac{1}{2}\ln(x^2+1) + \tan^{-1}x + C.$$

25. $\dfrac{x^3+x^2+2x+1}{(x^2+1)(x^2+2)} = \dfrac{Ax+B}{x^2+1} + \dfrac{Cx+D}{x^2+2}$. Multiply both sides by $(x^2+1)(x^2+2)$ to get

$x^3+x^2+2x+1 = (Ax+B)(x^2+2)+(Cx+D)(x^2+1) \Leftrightarrow$

$x^3+x^2+2x+1 = (Ax^3+Bx^2+2Ax+2B)+(Cx^3+Dx^2+Cx+D) \Leftrightarrow$

$x^3+x^2+2x+1 = (A+C)x^3+(B+D)x^2+(2A+C)x+(2B+D)$. Comparing coefficients gives us the following system of equations:

$$A + C = 1 \quad \textbf{(1)} \qquad B + D = 1 \quad \textbf{(2)}$$
$$2A + C = 2 \quad \textbf{(3)} \qquad 2B + D = 1 \quad \textbf{(4)}$$

Subtracting equation **(1)** from equation **(3)** gives us $A = 1$, so $C = 0$. Subtracting equation **(2)** from equation **(4)** gives us $B = 0$, so $D = 1$. Thus, $I = \int \frac{x^3 + x^2 + 2x + 1}{(x^2+1)(x^2+2)}\,dx = \int \left(\frac{x}{x^2+1} + \frac{1}{x^2+2}\right) dx$. For $\int \frac{x}{x^2+1}\,dx$, let $u = x^2 + 1$ so $du = 2x\,dx$ and then $\int \frac{x}{x^2+1}\,dx = \frac{1}{2}\int \frac{1}{u}\,du = \frac{1}{2}\ln|u| + C = \frac{1}{2}\ln(x^2+1) + C$. For $\int \frac{1}{x^2+2}\,dx$, use Formula 10 with $a = \sqrt{2}$. So $\int \frac{1}{x^2+2}\,dx = \int \frac{1}{x^2 + \left(\sqrt{2}\right)^2}\,dx = \frac{1}{\sqrt{2}}\tan^{-1}\frac{x}{\sqrt{2}} + C$.

Thus, $I = \frac{1}{2}\ln(x^2+1) + \frac{1}{\sqrt{2}}\tan^{-1}\frac{x}{\sqrt{2}} + C$.

26. $\frac{x^3 - 2x^2 + x + 1}{x^4 + 5x^2 + 4} = \frac{x^3 - 2x^2 + x + 1}{(x^2+1)(x^2+4)} = \frac{Ax+B}{x^2+1} + \frac{Cx+D}{x^2+4} \quad \Rightarrow$

$x^3 - 2x^2 + x + 1 = (Ax+B)(x^2+4) + (Cx+D)(x^2+1)$.

Equating coefficients gives $A + C = 1$, $B + D = -2$, $4A + C = 1$, $4B + D = 1 \quad \Rightarrow \quad A = 0$, $C = 1$, $B = 1$, $D = -3$.

Now $\int \frac{x^3 - 2x^2 + x + 1}{x^4 + 5x^2 + 4}\,dx = \int \frac{dx}{x^2+1} + \int \frac{x-3}{x^2+4}\,dx = \tan^{-1}x + \frac{1}{2}\ln(x^2+4) - \frac{3}{2}\tan^{-1}(x/2) + C$.

27.
$$\int \frac{x+4}{x^2+2x+5}\,dx = \int \frac{x+1}{x^2+2x+5}\,dx + \int \frac{3}{x^2+2x+5}\,dx = \frac{1}{2}\int \frac{(2x+2)\,dx}{x^2+2x+5} + \int \frac{3\,dx}{(x+1)^2+4}$$
$$= \frac{1}{2}\ln\left|x^2+2x+5\right| + 3\int \frac{2\,du}{4(u^2+1)} \qquad [\text{where } x+1 = 2u \text{ and } dx = 2\,du]$$
$$= \frac{1}{2}\ln(x^2+2x+5) + \frac{3}{2}\tan^{-1}u + C = \frac{1}{2}\ln(x^2+2x+5) + \frac{3}{2}\tan^{-1}\left(\frac{x+1}{2}\right) + C$$

28.
$$\int_0^1 \frac{x}{x^2+4x+13}\,dx = \int_0^1 \frac{\frac{1}{2}(2x+4)}{x^2+4x+13}\,dx - 2\int_0^1 \frac{dx}{(x+2)^2+9}$$
$$= \frac{1}{2}\int_{13}^{18} \frac{dy}{y} - 2\int_{2/3}^{1} \frac{3\,du}{9u^2+9} \qquad \left[\begin{array}{c}\text{where } y = x^2+4x+13,\ dy = (2x+4)\,dx,\\ x+2 = 3u, \text{ and } dx = 3\,du\end{array}\right]$$
$$= \tfrac{1}{2}\left[\ln y\right]_{13}^{18} - \tfrac{2}{3}\left[\tan^{-1}u\right]_{2/3}^{1} = \tfrac{1}{2}\ln\tfrac{18}{13} - \tfrac{2}{3}\left(\tfrac{\pi}{4} - \tan^{-1}\left(\tfrac{2}{3}\right)\right)$$
$$= \tfrac{1}{2}\ln\tfrac{18}{13} - \tfrac{\pi}{6} + \tfrac{2}{3}\tan^{-1}\left(\tfrac{2}{3}\right)$$

29. $\frac{1}{x^3-1} = \frac{1}{(x-1)(x^2+x+1)} = \frac{A}{x-1} + \frac{Bx+C}{x^2+x+1} \quad \Rightarrow \quad 1 = A(x^2+x+1) + (Bx+C)(x-1)$.

Take $x = 1$ to get $A = \frac{1}{3}$. Equating coefficients of x^2 and then comparing the constant terms, we get

$0 = \frac{1}{3} + B$, $1 = \frac{1}{3} - C$, so $B = -\frac{1}{3}$, $C = -\frac{2}{3} \quad \Rightarrow$

$$\int \frac{1}{x^3-1}\,dx = \int \frac{\frac{1}{3}}{x-1}\,dx + \int \frac{-\frac{1}{3}x-\frac{2}{3}}{x^2+x+1}\,dx = \tfrac{1}{3}\ln|x-1| - \frac{1}{3}\int \frac{x+2}{x^2+x+1}dx$$
$$= \tfrac{1}{3}\ln|x-1| - \frac{1}{3}\int \frac{x+1/2}{x^2+x+1}dx - \frac{1}{3}\int \frac{(3/2)dx}{(x+1/2)^2+3/4}$$
$$= \tfrac{1}{3}\ln|x-1| - \tfrac{1}{6}\ln(x^2+x+1) - \tfrac{1}{2}\left(\frac{2}{\sqrt{3}}\right)\tan^{-1}\left(\frac{x+\frac{1}{2}}{\sqrt{3}/2}\right) + K$$
$$= \tfrac{1}{3}\ln|x-1| - \tfrac{1}{6}\ln(x^2+x+1) - \tfrac{1}{\sqrt{3}}\tan^{-1}\left(\tfrac{1}{\sqrt{3}}(2x+1)\right) + K$$

30. $\dfrac{x^3}{x^3+1} = \dfrac{(x^3+1)-1}{x^3+1} = 1 - \dfrac{1}{x^3+1} = 1 - \left(\dfrac{A}{x+1} + \dfrac{Bx+C}{x^2-x+1}\right) \Rightarrow 1 = A(x^2-x+1) + (Bx+C)(x+1).$

Equate the terms of degree 2, 1 and 0 to get $0 = A+B$, $0 = -A+B+C$, $1 = A+C$. Solve the three equations to get $A = \frac{1}{3}$, $B = -\frac{1}{3}$, and $C = \frac{2}{3}$. So

$$\int \frac{x^3}{x^3+1}\,dx = \int \left[1 - \frac{\frac{1}{3}}{x+1} + \frac{\frac{1}{3}x-\frac{2}{3}}{x^2-x+1}\right]dx = x - \tfrac{1}{3}\ln|x+1| + \frac{1}{6}\int \frac{2x-1}{x^2-x+1}\,dx - \frac{1}{2}\int \frac{dx}{\left(x-\frac{1}{2}\right)^2+\frac{3}{4}}$$
$$= x - \tfrac{1}{3}\ln|x+1| + \tfrac{1}{6}\ln(x^2-x+1) - \tfrac{1}{\sqrt{3}}\tan^{-1}\left(\tfrac{1}{\sqrt{3}}(2x-1)\right) + K$$

31. $\dfrac{1}{x^4-x^2} = \dfrac{1}{x^2(x-1)(x+1)} = \dfrac{A}{x} + \dfrac{B}{x^2} + \dfrac{C}{x-1} + \dfrac{D}{x+1}$. Multiply by $x^2(x-1)(x+1)$ to get

$1 = Ax(x-1)(x+1) + B(x-1)(x+1) + Cx^2(x+1) + Dx^2(x-1)$. Setting $x=1$ gives $C = \frac{1}{2}$,

taking $x=-1$ gives $D = -\frac{1}{2}$. Equating the coefficients of x^3 gives $0 = A+C+D = A$. Finally, setting $x=0$

yields $B=-1$. Now $\displaystyle\int \frac{dx}{x^4-x^2} = \int \left[\frac{-1}{x^2} + \frac{1/2}{x-1} - \frac{1/2}{x+1}\right]dx = \frac{1}{x} + \tfrac{1}{2}\ln\left|\frac{x-1}{x+1}\right| + C.$

32. Let $u = x^4+5x^2+4 \Rightarrow du = (4x^3+10x)\,dx = 2(2x^3+5x)\,dx$, so

$$\int_0^1 \frac{2x^3+5x}{x^4+5x^2+4}\,dx = \frac{1}{2}\int_4^{10}\frac{du}{u} = \tfrac{1}{2}\Big[\ln|u|\Big]_4^{10} = \tfrac{1}{2}(\ln 10 - \ln 4) = \tfrac{1}{2}\ln\tfrac{5}{2}.$$

33. $$\int \frac{x-3}{(x^2+2x+4)^2}\,dx = \int \frac{x-3}{(x^2+2x+4)^2}\,dx = \int \frac{x-3}{[(x+1)^2+3]^2}\,dx = \int \frac{u-4}{(u^2+3)^2}\,du \qquad [\text{with } u = x+1]$$
$$= \int \frac{u\,du}{(u^2+3)^2} - 4\int \frac{du}{(u^2+3)^2} = \frac{1}{2}\int \frac{dv}{v^2} - 4\int \frac{\sqrt{3}\sec^2\theta\,d\theta}{9\sec^4\theta} \qquad \left[\begin{array}{l} v = u^2+3 \text{ in the first integral;} \\ u = \sqrt{3}\tan\theta \text{ in the second} \end{array}\right]$$
$$= \frac{-1}{(2v)} - \frac{4\sqrt{3}}{9}\int \cos^2\theta\,d\theta = \frac{-1}{2(u^2+3)} - \frac{2\sqrt{3}}{9}(\theta + \sin\theta\cos\theta) + C$$
$$= \frac{-1}{2(x^2+2x+4)} - \frac{2\sqrt{3}}{9}\left[\tan^{-1}\left(\frac{x+1}{\sqrt{3}}\right) + \frac{\sqrt{3}(x+1)}{x^2+2x+4}\right] + C$$
$$= \frac{-1}{2(x^2+2x+4)} - \frac{2\sqrt{3}}{9}\tan^{-1}\left(\frac{x+1}{\sqrt{3}}\right) - \frac{2(x+1)}{3(x^2+2x+4)} + C$$

34. $\dfrac{x^4+1}{x(x^2+1)^2} = \dfrac{A}{x} + \dfrac{Bx+C}{x^2+1} + \dfrac{Dx+E}{(x^2+1)^2} \quad \Rightarrow \quad x^4+1 = A(x^2+1)^2 + (Bx+C)x(x^2+1) + (Dx+E)x.$

Setting $x=0$ gives $A=1$, and equating the coefficients of x^4 gives $1 = A+B$, so $B=0$. Now

$$\frac{C}{x^2+1} + \frac{Dx+E}{(x^2+1)^2} = \frac{x^4+1}{x(x^2+1)^2} - \frac{1}{x} = \frac{1}{x}\left[\frac{x^4+1-(x^4+2x^2+1)}{(x^2+1)^2}\right] = \frac{-2x}{(x^2+1)^2},$$

so we can take $C=0$, $D=-2$, and $E=0$. Hence,

$$\int \frac{x^4+1}{x(x^2+1)^2}\,dx = \int \left[\frac{1}{x} - \frac{2x}{(x^2+1)^2}\right] dx = \ln|x| + \frac{1}{x^2+1} + C$$

35. Let $u=\sqrt{x}$, so $u^2=x$ and $dx = 2u\,du$. Thus,

$$\int_9^{16} \frac{\sqrt{x}}{x-4}\,dx = \int_3^4 \frac{u}{u^2-4}\,2u\,du = 2\int_3^4 \frac{u^2}{u^2-4}\,du = 2\int_3^4 \left(1+\frac{4}{u^2-4}\right) du \qquad \text{[by long division]}$$

$$= 2 + 8\int_3^4 \frac{du}{(u+2)(u-2)} \quad (\star)$$

Multiply $\dfrac{1}{(u+2)(u-2)} = \dfrac{A}{u+2} + \dfrac{B}{u-2}$ by $(u+2)(u-2)$ to get $1 = A(u-2)+B(u+2)$. Equating coefficients we get $A+B=0$ and $-2A+2B=1$. Solving gives us $B=\frac{1}{4}$ and $A=-\frac{1}{4}$, so $\dfrac{1}{(u+2)(u-2)} = \dfrac{-1/4}{u+2} + \dfrac{1/4}{u-2}$ and $(\star)$ is

$$2+8\int_3^4 \left(\frac{-1/4}{u+2}+\frac{1/4}{u-2}\right) du = 2+8\Big[-\tfrac{1}{4}\ln|u+2| + \tfrac{1}{4}\ln|u-2|\Big]_3^4 = 2 + \Big[2\ln|u-2| - 2\ln|u+2|\Big]_3^4$$

$$= 2+2\left[\ln\left|\frac{u-2}{u+2}\right|\right]_3^4 = 2+2\left(\ln\tfrac{2}{6} - \ln\tfrac{1}{5}\right) = 2+2\ln\tfrac{2/6}{1/5}$$

$$= 2+2\ln\tfrac{5}{3} \;\text{ or }\; 2+\ln\left(\tfrac{5}{3}\right)^2 = 2+\ln\tfrac{25}{9}$$

36. Let $u=\sqrt[3]{x}$. Then $x=u^3$, $dx=3u^2\,du \quad \Rightarrow$

$$\int_0^1 \frac{1}{1+\sqrt[3]{x}}\,dx = \int_0^1 \frac{3u^2\,du}{1+u} = \int_0^1 \left(3u-3+\frac{3}{1+u}\right) du = \left[\tfrac{3}{2}u^2 - 3u + 3\ln(1+u)\right]_0^1 = 3\left(\ln 2 - \tfrac{1}{2}\right).$$

37. Let $u=\sqrt[3]{x^2+1}$. Then $x^2=u^3-1$, $2x\,dx = 3u^2\,du \quad \Rightarrow$

$$\int \frac{x^3\,dx}{\sqrt[3]{x^2+1}} = \int \frac{(u^3-1)\frac{3}{2}u^2\,du}{u} = \tfrac{3}{2}\int (u^4-u)\,du = \tfrac{3}{10}u^5 - \tfrac{3}{4}u^2 + C = \tfrac{3}{10}(x^2+1)^{5/3} - \tfrac{3}{4}(x^2+1)^{2/3} + C.$$

38. Let $u=\sqrt{x}$. Then $x=u^2$, $dx=2u\,du \quad \Rightarrow$

$$\int_{1/3}^3 \frac{\sqrt{x}}{x^2+x}\,dx = \int_{1/\sqrt{3}}^{\sqrt{3}} \frac{u\cdot 2u\,du}{u^4+u^2} = 2\int_{1/\sqrt{3}}^{\sqrt{3}} \frac{du}{u^2+1} = 2\left[\tan^{-1}u\right]_{1/\sqrt{3}}^{\sqrt{3}} = 2\left(\tfrac{\pi}{3}-\tfrac{\pi}{6}\right) = \tfrac{\pi}{3}.$$

39. Let $u = e^x$. Then $x = \ln u$, $dx = \dfrac{du}{u}$ $\Rightarrow$

$$\int \frac{e^{2x}\,dx}{e^{2x} + 3e^x + 2} = \int \frac{u^2\,(du/u)}{u^2 + 3u + 2} = \int \frac{u\,du}{(u+1)(u+2)} = \int \left[\frac{-1}{u+1} + \frac{2}{u+2}\right] du$$

$$= 2\ln|u+2| - \ln|u+1| + C = \ln\left[\frac{(e^x+2)^2}{e^x+1}\right] + C$$

40. Let $u = \sin x$. Then $du = \cos x\,dx$ $\Rightarrow$

$$\int \frac{\cos x\,dx}{\sin^2 x + \sin x} = \int \frac{du}{u^2 + u} = \int \frac{du}{u(u+1)} = \int \left[\frac{1}{u} - \frac{1}{u+1}\right] du = \ln\left|\frac{u}{u+1}\right| + C = \ln\left|\frac{\sin x}{1 + \sin x}\right| + C.$$

41. Let $u = \ln(x^2 - x + 2)$, $dv = dx$. Then $du = \dfrac{2x-1}{x^2 - x + 2}\,dx$, $v = x$, and (by integration by parts)

$$\int \ln(x^2 - x + 2)\,dx = x\ln(x^2 - x + 2) - \int \frac{2x^2 - x}{x^2 - x + 2}\,dx = x\ln(x^2 - x + 2) - \int \left(2 + \frac{x-4}{x^2 - x + 2}\right) dx$$

$$= x\ln(x^2 - x + 2) - 2x - \int \frac{\frac{1}{2}(2x-1)}{x^2 - x + 2}\,dx + \frac{7}{2}\int \frac{dx}{(x - \frac{1}{2})^2 + \frac{7}{4}}$$

$$= x\ln(x^2 - x + 2) - 2x - \frac{1}{2}\ln(x^2 - x + 2) + \frac{7}{2}\int \frac{\frac{\sqrt{7}}{2}\,du}{\frac{7}{4}(u^2+1)} \qquad \left[\begin{array}{c}\text{where } x - \frac{1}{2} = \frac{\sqrt{7}}{2}\,u, \\ dx = \frac{\sqrt{7}}{2}\,du, \\ (x - \frac{1}{2})^2 + \frac{7}{4} = \frac{7}{4}(u^2+1)\end{array}\right]$$

$$= (x - \tfrac{1}{2})\ln(x^2 - x + 2) - 2x + \sqrt{7}\tan^{-1} u + C$$

$$= (x - \tfrac{1}{2})\ln(x^2 - x + 2) - 2x + \sqrt{7}\tan^{-1}\left(\frac{2x-1}{\sqrt{7}}\right) + C$$

42. Let $u = \tan^{-1} x$, $dv = x\,dx$ $\Rightarrow$ $du = dx/(1+x^2)$, $v = \frac{1}{2}x^2$.

Then $\displaystyle\int x\tan^{-1} x\,dx = \frac{1}{2}x^2\tan^{-1} x - \frac{1}{2}\int \frac{x^2}{1+x^2}\,dx$. To evaluate the last integral, use long division or

observe that $\displaystyle\int \frac{x^2}{1+x^2}\,dx = \int \frac{(1+x^2) - 1}{1+x^2}\,dx = \int 1\,dx - \int \frac{1}{1+x^2}\,dx = x - \tan^{-1} x + C_1$.

So $\int x\tan^{-1} x\,dx = \frac{1}{2}x^2\tan^{-1} x - \frac{1}{2}(x - \tan^{-1} x + C_1) = \frac{1}{2}(x^2\tan^{-1} x + \tan^{-1} x - x) + C$.

43. $\dfrac{P+S}{P\,[(r-1)P - S]} = \dfrac{A}{P} + \dfrac{B}{(r-1)P - S}$ $\Rightarrow$ $P + S = A\,[(r-1)P - S] + BP = [(r-1)A + B]\,P - AS$ $\Rightarrow$

$(r-1)A + B = 1$, $-A = 1$ $\Rightarrow$ $A = -1$, $B = r$. Now

$$t = \int \frac{P+S}{P\,[(r-1)P - S]}\,dP = \int \left[\frac{-1}{P} + \frac{r}{(r-1)P - S}\right] dP = -\int \frac{dP}{P} + \frac{r}{r-1}\int \frac{r-1}{(r-1)P - S}\,dP$$

so $t = -\ln P + \dfrac{r}{r-1}\ln|(r-1)P - S| + C$. Here $r = 0.10$ and $S = 900$, so

$t = -\ln P + \frac{0.1}{-0.9}\ln|-0.9P - 900| + C = -\ln P - \frac{1}{9}\ln(|-1|\,|0.9P + 900|\,) = -\ln P - \frac{1}{9}\ln(0.9P + 900) + C$.

When $t = 0$, $P = 10{,}000$, so $0 = -\ln 10{,}000 - \frac{1}{9}\ln(9900) + C$. Thus, $C = \ln 10{,}000 + \frac{1}{9}\ln 9900$ $[\approx 10.2326]$,

so our equation becomes

$$t = \ln 10{,}000 - \ln P + \tfrac{1}{9}\ln 9900 - \tfrac{1}{9}\ln(0.9P + 900) = \ln\frac{10{,}000}{P} + \frac{1}{9}\ln\frac{9900}{0.9P + 900}$$

$$= \ln\frac{10{,}000}{P} + \frac{1}{9}\ln\frac{1100}{0.1P + 100} = \ln\frac{10{,}000}{P} + \frac{1}{9}\ln\frac{11{,}000}{P + 1000}$$

44. If we subtract and add $2x^2$, we get

$$\begin{aligned} x^4 + 1 &= x^4 + 2x^2 + 1 - 2x^2 = (x^2+1)^2 - 2x^2 = (x^2+1)^2 - \left(\sqrt{2}\,x\right)^2 \\ &= \left[(x^2+1) - \sqrt{2}\,x\right]\left[(x^2+1) + \sqrt{2}\,x\right] = \left(x^2 - \sqrt{2}\,x + 1\right)\left(x^2 + \sqrt{2}\,x + 1\right) \end{aligned}$$

So we can decompose $\dfrac{1}{x^4+1} = \dfrac{Ax+B}{x^2+\sqrt{2}x+1} + \dfrac{Cx+D}{x^2-\sqrt{2}x+1} \quad \Rightarrow$

$1 = (Ax+B)\left(x^2 - \sqrt{2}x + 1\right) + (Cx+D)\left(x^2 + \sqrt{2}x + 1\right)$. Setting the constant terms equal gives $B + D = 1$, then from the coefficients of x^3 we get $A + C = 0$. Now from the coefficients of x we get $A + C + (B - D)\sqrt{2} = 0 \quad \Leftrightarrow$ $[(1-D) - D]\sqrt{2} = 0 \quad \Rightarrow \quad D = \frac{1}{2} \quad \Rightarrow \quad B = \frac{1}{2}$, and finally, from the coefficients of x^2 we get $\sqrt{2}\,(C - A) + B + D = 0 \quad \Rightarrow \quad C - A = -\frac{1}{\sqrt{2}} \quad \Rightarrow \quad C = -\frac{\sqrt{2}}{4}$ and $A = \frac{\sqrt{2}}{4}$.

So we rewrite the integrand, splitting the terms into forms which we know how to integrate:

$$\begin{aligned} \frac{1}{x^4+1} &= \frac{\frac{\sqrt{2}}{4}x + \frac{1}{2}}{x^2+\sqrt{2}\,x+1} + \frac{-\frac{\sqrt{2}}{4}x + \frac{1}{2}}{x^2-\sqrt{2}\,x+1} = \frac{1}{4\sqrt{2}}\left[\frac{2x+2\sqrt{2}}{x^2+\sqrt{2}\,x+1} - \frac{2x-2\sqrt{2}}{x^2-\sqrt{2}\,x+1}\right] \\ &= \frac{\sqrt{2}}{8}\left[\frac{2x+\sqrt{2}}{x^2+\sqrt{2}\,x+1} - \frac{2x-\sqrt{2}}{x^2-\sqrt{2}\,x+1}\right] + \frac{1}{4}\left[\frac{1}{\left(x+\frac{1}{\sqrt{2}}\right)^2 + \frac{1}{2}} + \frac{1}{\left(x-\frac{1}{\sqrt{2}}\right)^2 + \frac{1}{2}}\right] \end{aligned}$$

Now we integrate: $\displaystyle\int \frac{dx}{x^4+1} = \frac{\sqrt{2}}{8}\ln\left(\frac{x^2+\sqrt{2}\,x+1}{x^2-\sqrt{2}\,x+1}\right) + \frac{\sqrt{2}}{4}\left[\tan^{-1}\left(\sqrt{2}\,x+1\right) + \tan^{-1}\left(\sqrt{2}\,x-1\right)\right] + C.$

45. There are only finitely many values of x where $Q(x) = 0$ (assuming that Q is not the zero polynomial). At all other values of x, $F(x)/Q(x) = G(x)/Q(x)$, so $F(x) = G(x)$. In other words, the values of F and G agree at all except perhaps finitely many values of x. By continuity of F and G, the polynomials F and G must agree at those values of x too.

More explicitly: If a is a value of x such that $Q(a) = 0$, then $Q(x) \neq 0$ for all x sufficiently close to a. Thus,

$$\begin{aligned} F(a) &= \lim_{x\to a} F(x) && \text{[by continuity of } F\text{]} \\ &= \lim_{x\to a} G(x) && \text{[whenever } Q(x) \neq 0\text{]} \\ &= G(a) && \text{[by continuity of } G\text{]} \end{aligned}$$

46. Let $f(x) = ax^2 + bx + c$. We calculate the partial fraction decomposition of $\dfrac{f(x)}{x^2(x+1)^3}$. Since $f(0) = 1$, we must have $c = 1$, so $\dfrac{f(x)}{x^2(x+1)^3} = \dfrac{ax^2+bx+1}{x^2(x+1)^3} = \dfrac{A}{x} + \dfrac{B}{x^2} + \dfrac{C}{x+1} + \dfrac{D}{(x+1)^2} + \dfrac{E}{(x+1)^3}$. Now in order for the integral not to contain any logarithms (that is, in order for it to be a rational function), we must have $A = C = 0$, so $ax^2 + bx + 1 = B(x+1)^3 + Dx^2(x+1) + Ex^2$. Equating constant terms gives $B = 1$, then equating coefficients of x gives $3B = b \quad \Rightarrow \quad b = 3$. This is the quantity we are looking for, since $f'(0) = b$.

6.4 Integration with Tables and Computer Algebra Systems

Keep in mind that there are several ways to approach many of these exercises, and different methods can lead to different forms of the answer.

1. $\displaystyle\int_0^1 2x\cos^{-1}x\,dx \overset{91}{=} 2\left[\frac{2x^2-1}{4}\cos^{-1}x - \frac{x\sqrt{1-x^2}}{4}\right]_0^1 = 2\left[\left(\tfrac{1}{4}\cdot 0 - 0\right) - \left(-\tfrac{1}{4}\cdot\tfrac{\pi}{2} - 0\right)\right] = 2\left(\tfrac{\pi}{8}\right) = \tfrac{\pi}{4}$

2. $\displaystyle\int e^{2\theta}\sin 3\theta\,d\theta \overset{98}{=} \frac{e^{2\theta}}{2^2+3^2}(2\sin 3\theta - 3\cos 3\theta) + C = \tfrac{2}{13}e^{2\theta}\sin 3\theta - \tfrac{3}{13}e^{2\theta}\cos 3\theta + C$

3. Let $u = \pi x \;\Rightarrow\; du = \pi\,dx$, so

$$\begin{aligned}\textstyle\int \sec^3(\pi x)\,dx &= \tfrac{1}{\pi}\textstyle\int \sec^3 u\,du \overset{71}{=} \tfrac{1}{\pi}\left(\tfrac{1}{2}\sec u\,\tan u + \tfrac{1}{2}\ln|\sec u + \tan u|\right) + C\\ &= \tfrac{1}{2\pi}\sec\pi x\,\tan\pi x + \tfrac{1}{2\pi}\ln|\sec\pi x + \tan\pi x| + C\end{aligned}$$

4. $$\begin{aligned}\int_2^3 \frac{1}{x^2\sqrt{4x^2-7}}\,dx &= \int_4^6 \frac{1}{\left(\frac{1}{2}u\right)^2\sqrt{u^2-7}}\left(\frac{1}{2}\,du\right) \qquad [u = 2x,\ du = 2\,dx]\\ &= 2\int_4^6 \frac{du}{u^2\sqrt{u^2-7}} \overset{45}{=} 2\left[\frac{\sqrt{u^2-7}}{7u}\right]_4^6 = 2\left(\frac{\sqrt{29}}{42} - \frac{3}{28}\right) = \frac{\sqrt{29}}{21} - \frac{3}{14}\end{aligned}$$

5. Let $u = 2x$ and $a = 3$. Then $du = 2\,dx$ and

$$\begin{aligned}\int \frac{dx}{x^2\sqrt{4x^2+9}} &= \int \frac{\frac{1}{2}\,du}{\frac{u^2}{4}\sqrt{u^2+a^2}} = 2\int \frac{du}{u^2\sqrt{a^2+u^2}} \overset{28}{=} -2\,\frac{\sqrt{a^2+u^2}}{a^2u} + C\\ &= -2\,\frac{\sqrt{4x^2+9}}{9\cdot 2x} + C = -\frac{\sqrt{4x^2+9}}{9x} + C\end{aligned}$$

6. Let $u = \sqrt{2}\,y$ and $a = \sqrt{3}$. Then $du = \sqrt{2}\,dy$ and

$$\begin{aligned}\int \frac{\sqrt{2y^2-3}}{y^2}\,dy &= \int \frac{\sqrt{u^2-a^2}}{\frac{1}{2}u^2}\,\frac{du}{\sqrt{2}} = \sqrt{2}\int \frac{\sqrt{u^2-a^2}}{u^2}\,du\\ &\overset{42}{=} \sqrt{2}\left(-\frac{\sqrt{u^2-a^2}}{u} + \ln\left|u + \sqrt{u^2-a^2}\right|\right) + C\\ &= \sqrt{2}\left(-\frac{\sqrt{2y^2-3}}{\sqrt{2}\,y} + \ln\left|\sqrt{2}\,y + \sqrt{2y^2-3}\right|\right) + C\\ &= -\frac{\sqrt{2y^2-3}}{y} + \sqrt{2}\ln\left|\sqrt{2}\,y + \sqrt{2y^2-3}\right| + C\end{aligned}$$

7. $\int x^3\sin x\,dx \overset{84}{=} -x^3\cos x + 3\int x^2\cos x\,dx$, $\int x^2\cos x\,dx \overset{85}{=} x^2\sin x - 2\int x\sin x\,dx$, and

$\int x\sin x\,dx \overset{82}{=} \sin x - x\cos x + C$. Substituting, we get

$\int x^3\sin x\,dx = -x^3\cos x + 3\left[x^2\sin x - 2(\sin x - x\cos x)\right] + C = -x^3\cos x + 3x^2\sin x - 6\sin x + 6x\cos x + C.$

So $\int_0^\pi x^3\sin x\,dx = \left[-x^3\cos x + 3x^2\sin x - 6\sin x + 6x\cos x\right]_0^\pi = \left(-\pi^3\cdot -1 + 6\pi\cdot -1\right) - (0) = \pi^3 - 6\pi.$

8. Let $u = e^x$. Then $du = e^x\,dx$, so

$$\int \frac{e^{2x}}{\sqrt{2+e^x}}\,dx = \int \frac{e^x}{\sqrt{2+e^x}}(e^x\,dx) = \int \frac{u}{\sqrt{2+u}}\,du \overset{55}{=} \tfrac{2}{3}(u-2\cdot 2)\sqrt{2+u} + C = \tfrac{2}{3}(e^x-4)\sqrt{2+e^x} + C.$$

Another method: Let $u = 2 + e^x$. Then $du = e^x\,dx$, so

$$\begin{aligned}\int \frac{e^{2x}}{\sqrt{2+e^x}}\,dx &= \int \frac{e^x}{\sqrt{2+e^x}}(e^x\,dx) = \int \frac{u-2}{\sqrt{u}}\,du = \int (u^{1/2} - 2u^{-1/2})\,du = \tfrac{2}{3}u^{3/2} - 2(2u^{1/2}) + C \\ &= \tfrac{2}{3}(2+e^x)^{3/2} - 4(2+e^x)^{1/2} + C\end{aligned}$$

9. $$\int \frac{\tan^3(1/z)}{z^2}\,dz \quad \begin{bmatrix} u = 1/z, \\ du = -dz/z^2 \end{bmatrix} \quad = -\int \tan^3 u\,du \overset{69}{=} -\tfrac{1}{2}\tan^2 u - \ln|\cos u| + C = -\tfrac{1}{2}\tan^2\!\left(\frac{1}{z}\right) - \ln\left|\cos\!\left(\frac{1}{z}\right)\right| + C$$

10. Let $u = \sqrt{x}$. Then $u^2 = x$ and $2u\,du = dx$, so

$$\int \sin^{-1}\sqrt{x}\,dx = 2\int u\sin^{-1}u\,du \overset{90}{=} \frac{2u^2-1}{2}\sin^{-1}u + \frac{u\sqrt{1-u^2}}{2} + C = \frac{2x-1}{2}\sin^{-1}\sqrt{x} + \frac{\sqrt{x(1-x)}}{2} + C.$$

11. Let $z = 6 + 4y - 4y^2 = 6 - (4y^2 - 4y + 1) + 1 = 7 - (2y-1)^2$, $u = 2y - 1$, and $a = \sqrt{7}$. Then $z = a^2 - u^2$, $du = 2\,dy$, and

$$\begin{aligned}\int y\sqrt{6+4y-4y^2}\,dy &= \int y\sqrt{z}\,dy = \int \tfrac{1}{2}(u+1)\sqrt{a^2-u^2}\,\tfrac{1}{2}\,du = \tfrac{1}{4}\int u\sqrt{a^2-u^2}\,du + \tfrac{1}{4}\int \sqrt{a^2-u^2}\,du \\ &= \tfrac{1}{4}\int \sqrt{a^2-u^2}\,du - \tfrac{1}{8}\int (-2u)\sqrt{a^2-u^2}\,du \\ &\overset{30}{=} \frac{u}{8}\sqrt{a^2-u^2} + \frac{a^2}{8}\sin^{-1}\!\left(\frac{u}{a}\right) - \frac{1}{8}\int \sqrt{w}\,dw \qquad [w = a^2 - u^2,\ dw = -2u\,du] \\ &= \frac{2y-1}{8}\sqrt{6+4y-4y^2} + \frac{7}{8}\sin^{-1}\!\left(\frac{2y-1}{\sqrt{7}}\right) - \frac{1}{8}\cdot\frac{2}{3}w^{3/2} + C \\ &= \frac{2y-1}{8}\sqrt{6+4y-4y^2} + \frac{7}{8}\sin^{-1}\!\left(\frac{2y-1}{\sqrt{7}}\right) - \frac{1}{12}(6+4y-4y^2)^{3/2} + C\end{aligned}$$

This can be rewritten as

$$\begin{aligned}&\sqrt{6+4y-4y^2}\left[\frac{1}{8}(2y-1) - \frac{1}{12}(6+4y-4y^2)\right] + \frac{7}{8}\sin^{-1}\!\left(\frac{2y-1}{\sqrt{7}}\right) + C \\ &\qquad = \left(\frac{1}{3}y^2 - \frac{1}{12}y - \frac{5}{8}\right)\sqrt{6+4y-4y^2} + \frac{7}{8}\sin^{-1}\!\left(\frac{2y-1}{\sqrt{7}}\right) + C \\ &\qquad = \frac{1}{24}(8y^2 - 2y - 15)\sqrt{6+4y-4y^2} + \frac{7}{8}\sin^{-1}\!\left(\frac{2y-1}{\sqrt{7}}\right) + C\end{aligned}$$

12. Let $u = x^2$, so that $du = 2x\,dx$. Then

$$\begin{aligned}\int x\sin(x^2)\cos(3x^2)\,dx &= \frac{1}{2}\int \sin u\cos 3u\,du \overset{81}{=} -\frac{1}{2}\,\frac{\cos(1-3)u}{2(1-3)} - \frac{1}{2}\,\frac{\cos(1+3)u}{2(1+3)} + C \\ &= \tfrac{1}{8}\cos 2u - \tfrac{1}{16}\cos 4u + C = \tfrac{1}{8}\cos(2x^2) - \tfrac{1}{16}\cos(4x^2) + C\end{aligned}$$

13. Let $u = \sin x$. Then $du = \cos x\,dx$, so

$$\begin{aligned}\int \sin^2 x\cos x\ln(\sin x)\,dx &= \int u^2\ln u\,du \overset{101}{=} \frac{u^{2+1}}{(2+1)^2}[(2+1)\ln u - 1] + C \\ &= \tfrac{1}{9}u^3(3\ln u - 1) + C = \tfrac{1}{9}\sin^3 x\,[3\ln(\sin x) - 1] + C\end{aligned}$$

14. Let $u = 3\theta$. Then $du = 3\,d\theta$, so $I = \int_0^\pi \cos^4(3\theta)\,d\theta = \frac{1}{3}\int_0^{3\pi} \cos^4 u\,du$. Now

$\int \cos^4 u\,du \overset{74}{=} \frac{1}{4}\cos^3 u\,\sin u + \frac{3}{4}\int \cos^2 u\,du$ and $\int \cos^2 u\,du \overset{64}{=} \frac{1}{2}u + \frac{1}{4}\sin 2u + C$. So

$$I = \tfrac{1}{3}\left[\tfrac{1}{4}\cos^3 u\,\sin u + \tfrac{3}{4}\left(\tfrac{1}{2}u + \tfrac{1}{4}\sin 2u\right)\right]_0^{3\pi} = \left[\tfrac{1}{12}\cos^3 u\,\sin u + \tfrac{1}{8}u + \tfrac{1}{16}\sin 2u\right]_0^{3\pi} = \tfrac{3\pi}{8} - 0 = \tfrac{3\pi}{8}.$$

15. Let $u = e^x$ and $a = \sqrt{3}$. Then $du = e^x\,dx$ and

$$\int \frac{e^x}{3 - e^{2x}}\,dx = \int \frac{du}{a^2 - u^2} \overset{19}{=} \frac{1}{2a}\ln\left|\frac{u + a}{u - a}\right| + C = \frac{1}{2\sqrt{3}}\ln\left|\frac{e^x + \sqrt{3}}{e^x - \sqrt{3}}\right| + C.$$

16. Let $u = x^2$ and $a = 2$. Then $du = 2x\,dx$ and

$$\begin{aligned}
\int_0^2 x^3\sqrt{4x^2 - x^4}\,dx &= \frac{1}{2}\int_0^2 x^2\sqrt{2\cdot 2\cdot x^2 - (x^2)^2}\cdot 2x\,dx = \frac{1}{2}\int_0^4 u\sqrt{2au - u^2}\,du \\
&\overset{114}{=} \left[\frac{2u^2 - au - 3a^2}{12}\sqrt{2au - u^2} + \frac{a^3}{4}\cos^{-1}\left(\frac{a - u}{a}\right)\right]_0^4 \\
&= \left[\frac{2u^2 - 2u - 12}{12}\sqrt{4u - u^2} + \frac{8}{4}\cos^{-1}\left(\frac{2 - u}{2}\right)\right]_0^4 \\
&= \left[\frac{u^2 - u - 6}{6}\sqrt{4u - u^2} + 2\cos^{-1}\left(\frac{2 - u}{2}\right)\right]_0^4 \\
&= [0 + 2\cos^{-1}(-1)] - (0 + 2\cos^{-1}1) = 2\cdot\pi - 2\cdot 0 = 2\pi
\end{aligned}$$

17. $\displaystyle\int \frac{x^4\,dx}{\sqrt{x^{10} - 2}} = \int \frac{x^4\,dx}{\sqrt{(x^5)^2 - 2}} = \frac{1}{5}\int \frac{du}{\sqrt{u^2 - 2}}$ $\quad [u = x^5,\ du = 5x^4\,dx]$

$\overset{43}{=} \frac{1}{5}\ln\left|u + \sqrt{u^2 - 2}\right| + C = \frac{1}{5}\ln\left|x^5 + \sqrt{x^{10} - 2}\right| + C$

18. $\int x^4 e^{-x}\,dx \overset{97}{=} -x^4 e^{-x} + 4\int x^3 e^{-x}\,dx \overset{97}{=} -x^4 e^{-x} + 4\left(-x^3 e^{-x} + 3\int x^2 e^{-x}\,dx\right)$

$\overset{97}{=} -(x^4 + 4x^3)e^{-x} + 12\left(-x^2 e^{-x} + 2\int x e^{-x}\,dx\right)$

$\overset{96}{=} -(x^4 + 4x^3 + 12x^2)e^{-x} + 24[(-x - 1)e^{-x}] + C = -(x^4 + 4x^3 + 12x^2 + 24x + 24)e^{-x} + C$

So $\int_0^1 x^4 e^{-x}\,dx = \left[-(x^4 + 4x^3 + 12x^2 + 24x + 24)e^{-x}\right]_0^1 = -(1 + 4 + 12 + 24 + 24)e^{-1} + 24e^0 = 24 - 65e^{-1}$.

19. Let $u = \ln x$ and $a = 2$. Then $du = \dfrac{dx}{x}$ and

$$\begin{aligned}
\int \frac{\sqrt{4 + (\ln x)^2}}{x}\,dx &= \int \sqrt{a^2 + u^2}\,du \overset{21}{=} \frac{u}{2}\sqrt{a^2 + u^2} + \frac{a^2}{2}\ln\left(u + \sqrt{a^2 + u^2}\right) + C \\
&= \tfrac{1}{2}(\ln x)\sqrt{4 + (\ln x)^2} + 2\ln\left[\ln x + \sqrt{4 + (\ln x)^2}\right] + C
\end{aligned}$$

20. Let $u = \tan\theta$ and $a = 3$. Then $du = \sec^2\theta\,d\theta$ and

$$\begin{aligned}
\int \frac{\sec^2\theta\,\tan^2\theta}{\sqrt{9 - \tan^2\theta}}\,d\theta &= \int \frac{u^2}{\sqrt{a^2 - u^2}}\,du \overset{34}{=} -\frac{u}{2}\sqrt{a^2 - u^2} + \frac{a^2}{2}\sin^{-1}\left(\frac{u}{a}\right) + C \\
&= -\frac{1}{2}\tan\theta\sqrt{9 - \tan^2\theta} + \frac{9}{2}\sin^{-1}\left(\frac{\tan\theta}{3}\right) + C
\end{aligned}$$

21. Let $u = e^x$. Then $x = \ln u$, $dx = du/u$, so

$$\int \sqrt{e^{2x}-1}\,dx = \int \frac{\sqrt{u^2-1}}{u}\,du \overset{41}{=} \sqrt{u^2-1} - \cos^{-1}(1/u) + C = \sqrt{e^{2x}-1} - \cos^{-1}(e^{-x}) + C.$$

22. Let $u = \alpha t - 3$ and assume that $\alpha \neq 0$. Then $du = \alpha\,dt$ and

$$\begin{aligned}\int e^t \sin(\alpha t - 3)\,dt &= \frac{1}{\alpha}\int e^{(u+3)/\alpha}\sin u\,du = \frac{1}{\alpha}e^{3/\alpha}\int e^{(1/\alpha)u}\sin u\,du \overset{98}{=} \frac{1}{\alpha}e^{3/\alpha}\frac{e^{(1/\alpha)u}}{(1/\alpha)^2+1^2}\left(\frac{1}{\alpha}\sin u - \cos u\right) + C \\ &= \frac{1}{\alpha}e^{3/\alpha}e^{(1/\alpha)u}\frac{\alpha^2}{1+\alpha^2}\left(\frac{1}{\alpha}\sin u - \cos u\right) + C = \frac{1}{1+\alpha^2}e^{(u+3)/\alpha}(\sin u - \alpha\cos u) + C \\ &= \frac{1}{1+\alpha^2}e^t\left[\sin(\alpha t - 3) - \alpha\cos(\alpha t - 3)\right] + C\end{aligned}$$

23. (a)
$$\begin{aligned}\frac{d}{du}\left[\frac{1}{b^3}\left(a + bu - \frac{a^2}{a+bu} - 2a\ln|a+bu|\right) + C\right] &= \frac{1}{b^3}\left[b + \frac{ba^2}{(a+bu)^2} - \frac{2ab}{(a+bu)}\right] \\ &= \frac{1}{b^3}\left[\frac{b(a+bu)^2 + ba^2 - (a+bu)2ab}{(a+bu)^2}\right] \\ &= \frac{1}{b^3}\left[\frac{b^3u^2}{(a+bu)^2}\right] = \frac{u^2}{(a+bu)^2}\end{aligned}$$

(b) Let $t = a + bu \Rightarrow dt = b\,du$. Note that $u = \dfrac{t-a}{b}$ and $du = \dfrac{1}{b}\,dt$.

$$\begin{aligned}\int \frac{u^2\,du}{(a+bu)^2} &= \frac{1}{b^3}\int \frac{(t-a)^2}{t^2}\,dt = \frac{1}{b^3}\int \frac{t^2 - 2at + a^2}{t^2}\,dt = \frac{1}{b^3}\int\left(1 - \frac{2a}{t} + \frac{a^2}{t^2}\right)dt \\ &= \frac{1}{b^3}\left(t - 2a\ln|t| - \tfrac{a^2}{t}\right) + C = \frac{1}{b^3}\left(a + bu - \frac{a^2}{a+bu} - 2a\ln|a+bu|\right) + C\end{aligned}$$

24. (a)
$$\begin{aligned}&\frac{d}{du}\left[\frac{u}{8}(2u^2 - a^2)\sqrt{a^2-u^2} + \frac{a^4}{8}\sin^{-1}\frac{u}{a} + C\right] \\ &= \frac{u}{8}(2u^2-a^2)\frac{-u}{\sqrt{a^2-u^2}} + \sqrt{a^2-u^2}\left[\frac{u}{8}(4u) + (2u^2-a^2)\tfrac{1}{8}\right] + \frac{a^4}{8}\frac{1/a}{\sqrt{1-u^2/a^2}} \\ &= -\frac{u^2(2u^2-a^2)}{8\sqrt{a^2-u^2}} + \sqrt{a^2-u^2}\left[\frac{u^2}{2} + \frac{2u^2-a^2}{8}\right] + \frac{a^4}{8\sqrt{a^2-u^2}} \\ &= \tfrac{1}{2}(a^2-u^2)^{-1/2}\left[-\frac{u^2}{4}(2u^2-a^2) + u^2(a^2-u^2) + \tfrac{1}{4}(a^2-u^2)(2u^2-a^2) + \frac{a^4}{4}\right] \\ &= \tfrac{1}{2}(a^2-u^2)^{-1/2}\left[2u^2a^2 - 2u^4\right] = \frac{u^2(a^2-u^2)}{\sqrt{a^2-u^2}} = u^2\sqrt{a^2-u^2}\end{aligned}$$

(b) Let $u = a\sin\theta \Rightarrow du = a\cos\theta\,d\theta$. Then

$$\begin{aligned}\textstyle\int u^2\sqrt{a^2-u^2}\,du &= \textstyle\int a^2\sin^2\theta\, a\sqrt{1-\sin^2\theta}\, a\cos\theta\,d\theta = a^4\int\sin^2\theta\cos^2\theta\,d\theta \\ &= \textstyle a^4\int\frac{1}{2}(1+\cos 2\theta)\frac{1}{2}(1-\cos 2\theta)\,d\theta = \frac{1}{4}a^4\int(1-\cos^2 2\theta)\,d\theta \\ &= \textstyle\frac{1}{4}a^4\int\left[1 - \frac{1}{2}(1+\cos 4\theta)\right]d\theta = \frac{1}{4}a^4\left(\frac{1}{2}\theta - \frac{1}{8}\sin 4\theta\right) + C \\ &= \textstyle\frac{1}{4}a^4\left(\frac{1}{2}\theta - \frac{1}{8}\cdot 2\sin 2\theta\,\cos 2\theta\right) + C = \frac{1}{4}a^4\left[\frac{1}{2}\theta - \frac{1}{2}\sin\theta\,\cos\theta(1-2\sin^2\theta)\right] + C \\ &= \frac{a^4}{8}\left[\sin^{-1}\frac{u}{a} - \frac{u}{a}\frac{\sqrt{a^2-u^2}}{a}\left(1 - \frac{2u^2}{a^2}\right)\right] + C = \frac{a^4}{8}\left[\sin^{-1}\frac{u}{a} - \frac{u}{a}\frac{\sqrt{a^2-u^2}}{a}\frac{a^2-2u^2}{a^2}\right] + C \\ &= \frac{u}{8}(2u^2-a^2)\sqrt{a^2-u^2} + \frac{a^4}{8}\sin^{-1}\frac{u}{a} + C\end{aligned}$$

25. Maple, Mathematica and Derive all give $\int x^2\sqrt{5-x^2}\,dx = -\frac{1}{4}x(5-x^2)^{3/2} + \frac{5}{8}x\sqrt{5-x^2} + \frac{25}{8}\sin^{-1}\!\left(\frac{1}{\sqrt{5}}x\right)$.

Using Formula 31, we get $\int x^2\sqrt{5-x^2}\,dx = \frac{1}{8}x(2x^2-5)\sqrt{5-x^2} + \frac{1}{8}(5^2)\sin^{-1}\!\left(\frac{1}{\sqrt{5}}x\right) + C$.

But $-\frac{1}{4}x(5-x^2)^{3/2} + \frac{5}{8}x\sqrt{5-x^2} = \frac{1}{8}x\sqrt{5-x^2}\left[5-2(5-x^2)\right] = \frac{1}{8}x(2x^2-5)\sqrt{5-x^2}$, and the $\sin^{-1}$ terms are the same in each expression, so the answers are equivalent.

26. Maple and Mathematica both give $\int x^2(1+x^3)^4\,dx = \frac{1}{15}x^{15} + \frac{1}{3}x^{12} + \frac{2}{3}x^9 + \frac{2}{3}x^6 + \frac{1}{3}x^3$, while Derive gives $\int x^2(1+x^3)^4\,dx = \frac{1}{15}(x^3+1)^5$. Using the substitution $u = 1+x^3 \;\Rightarrow\; du = 3x^2\,dx$, we get $\int x^2(1+x^3)^4\,dx = \int u^4\left(\frac{1}{3}\,du\right) = \frac{1}{15}u^5 + C = \frac{1}{15}(1+x^3)^5 + C$. We can use the Binomial Theorem or a CAS to expand this expression, and we get $\frac{1}{15}(1+x^3)^5 + C = \frac{1}{15} + \frac{1}{3}x^3 + \frac{2}{3}x^6 + \frac{2}{3}x^9 + \frac{1}{3}x^{12} + \frac{1}{15}x^{15} + C$.

27. Maple and Derive both give $\int \sin^3 x\,\cos^2 x\,dx = -\frac{1}{5}\sin^2 x\,\cos^3 x - \frac{2}{15}\cos^3 x$ (although Derive factors the expression), and Mathematica gives $\int \sin^3 x\,\cos^2 x\,dx = -\frac{1}{8}\cos x - \frac{1}{48}\cos 3x + \frac{1}{80}\cos 5x$. We can use a CAS to show that both of these expressions are equal to $-\frac{1}{3}\cos^3 x + \frac{1}{5}\cos^5 x$. Using Formula 86, we write

$$\begin{aligned}\int \sin^3 x\,\cos^2 x\,dx &= -\tfrac{1}{5}\sin^2 x\,\cos^3 x + \tfrac{2}{5}\int \sin x\,\cos^2 x\,dx = -\tfrac{1}{5}\sin^2 x\,\cos^3 x + \tfrac{2}{5}\left(-\tfrac{1}{3}\cos^3 x\right) + C \\ &= -\tfrac{1}{5}\sin^2 x\,\cos^3 x - \tfrac{2}{15}\cos^3 x + C\end{aligned}$$

28. Maple gives $\displaystyle\int \tan^2 x\,\sec^4\,dx = \frac{1}{5}\frac{\sin^3 x}{\cos^5 x} + \frac{2}{15}\frac{\sin^3 x}{\cos^3 x}$,

Mathematica gives $\int \tan^2 x\,\sec^4\,dx = -\frac{1}{120}\sec^5 x\,(-20\sin x + 5\sin 3x + \sin 5x)$, and

Derive gives $\displaystyle\int \tan^2 x\,\sec^4\,dx = -\frac{2}{15}\tan x - \frac{\sin x}{15\cos^3 x} + \frac{\sin x}{5\cos^5 x}$. All of these expressions can be "simplified" to $\displaystyle -\frac{1}{15}\frac{\sin x(\cos^2 x - 2\cos^4 x - 3)}{\cos^5 x}$ using Maple. Using the identity $1+\tan^2 x = \sec^2 x$, we write

$\int \tan^2 x\,\sec^4 x\,dx = \int \tan^2 x\,(1+\tan^2 x)\,\sec^2 x\,dx = \int (\tan^2 x + \tan^4 x)\,\sec^2 x\,dx$. Now we substitute $u = \tan x \;\Rightarrow\; du = \sec^2 x\,dx$, and the integral becomes $\int (u^2+u^4)\,du = \frac{1}{3}u^3 + \frac{1}{5}u^5 + C = \frac{1}{3}\tan^3 x + \frac{1}{5}\tan^5 x + C$.

If we write $\sin^5 x = \sin^3 x\,(1-\cos^2 x)$ and substitute into the numerator of the $\tan^5 x$ term, this becomes

$$\frac{1}{3}\frac{\sin^3 x}{\cos^3 x} + \frac{1}{5}\frac{\sin^3 x\,(1-\cos^2 x)}{\cos^5 x} + C = \frac{1}{5}\frac{\sin^3 x}{\cos^5 x} + \left(\frac{1}{3}-\frac{1}{5}\right)\frac{\sin^3 x}{\cos^3 x} + C = \frac{1}{5}\frac{\sin^3 x}{\cos^5 x} + \frac{2}{15}\frac{\sin^3 x}{\cos^3 x} + C,$$

which is the same as Maple's expression.

29. Maple gives $\int x\sqrt{1+2x}\,dx = \frac{1}{10}(1+2x)^{5/2} - \frac{1}{6}(1+2x)^{3/2}$, Mathematica gives $\sqrt{1+2x}\left(\frac{2}{5}x^2 + \frac{1}{15}x - \frac{1}{15}\right)$, and Derive gives $\frac{1}{15}(1+2x)^{3/2}(3x-1)$. The first two expressions can be simplified to Derive's result. If we use Formula 54, we get $\int x\sqrt{1+2x}\,dx = \frac{2}{15(2)^2}(3\cdot 2x - 2\cdot 1)(1+2x)^{3/2} + C = \frac{1}{30}(6x-2)(1+2x)^{3/2} + C = \frac{1}{15}(3x-1)(1+2x)^{3/2}$.

30. Maple and Derive both give $\int \sin^4 x\,dx = -\frac{1}{4}\sin^3 x\,\cos x - \frac{3}{8}\cos x\,\sin x + \frac{3}{8}x$, while Mathematica gives $\frac{1}{32}(12x - 8\sin 2x + \sin 4x)$, which can be expanded and simplified to give the other expression. Now

$$\begin{aligned}\int \sin^4 x\,dx &\overset{73}{=} -\tfrac{1}{4}\sin^3 x\,\cos x + \tfrac{3}{4}\int \sin^2 x\,dx\\ &\overset{63}{=} -\tfrac{1}{4}\sin^3 x\,\cos x + \tfrac{3}{4}\left(\tfrac{1}{2}x - \tfrac{1}{4}\sin 2x\right) + C\\ &= -\tfrac{1}{4}\sin^3 x\,\cos x - \tfrac{3}{8}\sin x\,\cos x + \tfrac{3}{8}x + C \text{ since } \sin 2x = 2\sin x\,\cos x\end{aligned}$$

31. Maple gives $\int \tan^5 x\,dx = \frac{1}{4}\tan^4 x - \frac{1}{2}\tan^2 x + \frac{1}{2}\ln(1+\tan^2 x)$, Mathematica gives $\int \tan^5 x\,dx = \frac{1}{4}[-1 - 2\cos(2x)]\sec^4 x - \ln(\cos x)$, and Derive gives $\int \tan^5 x\,dx = \frac{1}{4}\tan^4 x - \frac{1}{2}\tan^2 x - \ln(\cos x)$. These expressions are equivalent, and none includes absolute value bars or a constant of integration. Note that Mathematica's and Derive's expressions suggest that the integral is undefined where $\cos x < 0$, which is not the case.

Using Formula 75, $\int \tan^5 x\,dx = \frac{1}{5-1}\tan^{5-1} x - \int \tan^{5-2} x\,dx = \frac{1}{4}\tan^4 x - \int \tan^3 x\,dx$.

Using Formula 69, $\int \tan^3 x\,dx = \frac{1}{2}\tan^2 x + \ln|\cos x| + C$, so $\int \tan^5 x\,dx = \frac{1}{4}\tan^4 x - \frac{1}{2}\tan^2 x - \ln|\cos x| + C$.

32. Maple gives $\int x^5\sqrt{x^2+1}\,dx = \frac{1}{35}x^4\sqrt{1+x^2} - \frac{4}{105}x^2\sqrt{1+x^2} + \frac{8}{105}\sqrt{1+x^2} + \frac{1}{7}x^6\sqrt{1+x^2}$.

When we use the `factor` command on this expression, it becomes $\frac{1}{105}(1+x^2)^{3/2}(15x^4 - 12x^2 + 8)$.

Mathematica gives $\sqrt{1+x^2}\left(\frac{8}{105} - \frac{4}{105}x^2 + \frac{1}{35}x^4 + \frac{1}{7}x^6\right)$, which again factors to give the above expression, and Derive gives the factored form immediately. If we substitute $u = \sqrt{x^2+1} \;\Rightarrow\; x^4 = (u^2-1)^2$, $x\,dx = u\,du$, then the integral becomes

$$\begin{aligned}\int (u^2-1)^2 u(u\,du) &= \int (u^4 - 2u^2 + 1)u^2\,du = \tfrac{1}{7}u^7 - \tfrac{2}{5}u^5 + \tfrac{1}{3}u^3 + C\\ &= (x^2+1)^{3/2}\left[\tfrac{1}{7}(x^2+1)^2 - \tfrac{2}{5}(x^2+1) + \tfrac{1}{3}\right] + C\\ &= \tfrac{1}{105}(x^2+1)^{3/2}\left[15(x^2+1)^2 - 42(x^2+1) + 35\right] + C\\ &= \tfrac{1}{105}(x^2+1)^{3/2}(15x^4 - 12x^2 + 8) + C\end{aligned}$$

33. Derive gives $I = \displaystyle\int 2^x\sqrt{4^x - 1}\,dx = \frac{2^{x-1}\sqrt{2^{2x}-1}}{\ln 2} - \frac{\ln\left(\sqrt{2^{2x}-1} + 2^x\right)}{2\ln 2}$ immediately. Neither Maple nor Mathematica is able to evaluate I in its given form. However, if we instead write I as $\int 2^x\sqrt{(2^x)^2 - 1}\,dx$, both systems give the same answer as Derive (after minor simplification). Our trick works because the CAS now recognizes 2^x as a promising substitution.

34. None of Maple, Mathematica and Derive is able to evaluate $\int (1+\ln x)\sqrt{1+(x\ln x)^2}\,dx$. However, if we let $u = x\ln x$, then $du = (1+\ln x)\,dx$ and the integral is simply $\int \sqrt{1+u^2}\,du$, which any CAS can evaluate. The antiderivative is $\frac{1}{2}\ln\left(x\ln x + \sqrt{1+(x\ln x)^2}\right) + \frac{1}{2}x\ln x\sqrt{1+(x\ln x)^2} + C$.

6.5 Approximate Integration

1. (a) $\Delta x = (b - a)/n = (4 - 0)/2 = 2$

$$L_2 = \sum_{i=1}^{2} f(x_{i-1})\,\Delta x = f(x_0)\cdot 2 + f(x_1)\cdot 2 = 2\,[f(0) + f(2)] = 2(0.5 + 2.5) = 6$$

$$R_2 = \sum_{i=1}^{2} f(x_i)\,\Delta x = f(x_1)\cdot 2 + f(x_2)\cdot 2 = 2\,[f(2) + f(4)] = 2(2.5 + 3.5) = 12$$

$$M_2 = \sum_{i=1}^{2} f(\overline{x}_i)\Delta x = f(\overline{x}_1)\cdot 2 + f(\overline{x}_2)\cdot 2 = 2\,[f(1) + f(3)] \approx 2(1.6 + 3.2) = 9.6$$

(b)

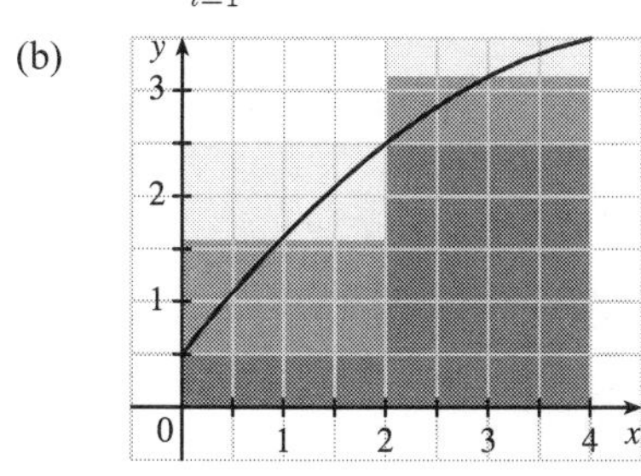

L_2 is an underestimate, since the area under the small rectangles is less than the area under the curve, and R_2 is an overestimate, since the area under the large rectangles is greater than the area under the curve. It appears that M_2 is an overestimate, though it is fairly close to I. See the solution to Exercise 37 for a proof of the fact that if f is concave down on $[a, b]$, then the Midpoint Rule is an overestimate of $\int_a^b f(x)\,dx$.

(c) $T_2 = \left(\frac{1}{2}\,\Delta x\right)[f(x_0) + 2f(x_1) + f(x_2)] = \frac{2}{2}[f(0) + 2f(2) + f(4)] = 0.5 + 2(2.5) + 3.5 = 9.$

This approximation is an underestimate, since the graph is concave down. Thus, $T_2 = 9 < I$. See the solution to Exercise 37 for a general proof of this conclusion.

(d) For any n, we will have $L_n < T_n < I < M_n < R_n$.

2.

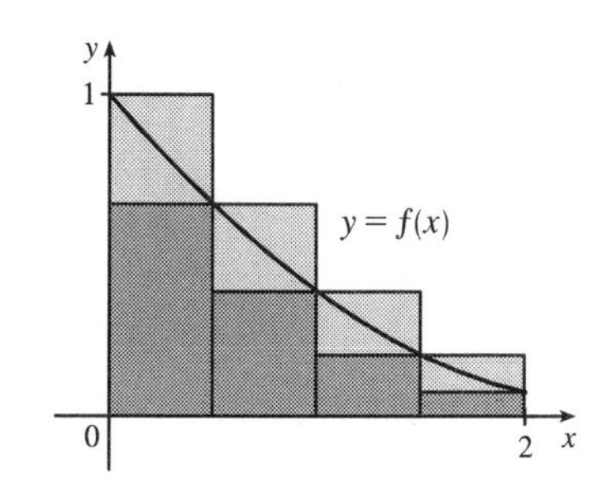

The diagram shows that $L_4 > T_4 > \int_0^2 f(x)\,dx > R_4$, and it appears that M_4 is a bit less than $\int_0^2 f(x)\,dx$. In fact, for any function that is concave upward, it can be shown that

$L_n > T_n > \int_0^2 f(x)\,dx > M_n > R_n$.

(a) Since $0.9540 > 0.8675 > 0.8632 > 0.7811$, it follows that $L_n = 0.9540$, $T_n = 0.8675$, $M_n = 0.8632$, and $R_n = 0.7811$.

(b) Since $M_n < \int_0^2 f(x)\,dx < T_n$, we have $0.8632 < \int_0^2 f(x)\,dx < 0.8675$.

3. $f(x) = \cos(x^2)$, $\Delta x = \frac{1-0}{4} = \frac{1}{4}$

(a) $T_4 = \frac{1}{4\cdot 2}\left[f(0) + 2f\left(\frac{1}{4}\right) + 2f\left(\frac{2}{4}\right) + 2f\left(\frac{3}{4}\right) + f(1)\right] \approx 0.895759$

(b) $M_4 = \frac{1}{4}\left[f\left(\frac{1}{8}\right) + f\left(\frac{3}{8}\right) + f\left(\frac{5}{8}\right) + f\left(\frac{7}{8}\right)\right] \approx 0.908907$

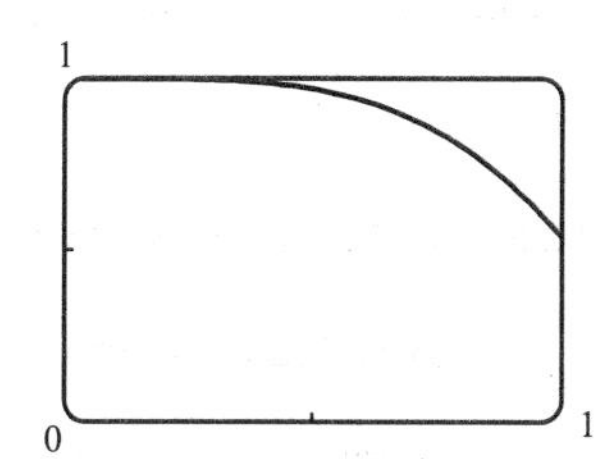

The graph shows that f is concave down on $[0, 1]$. So T_4 is an underestimate and M_4 is an overestimate. We can conclude that $0.895759 < \int_0^1 \cos(x^2)\,dx < 0.908907$.

4. 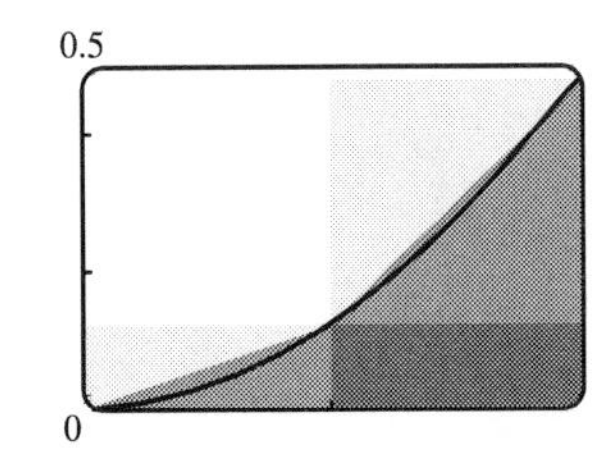

(a) Since f is increasing on $[0, 1]$, L_2 will underestimate I (since the area of the darkest rectangle is less than the area under the curve), and R_2 will overestimate I. Since f is concave upward on $[0, 1]$, M_2 will underestimate I and T_2 will overestimate I (the area under the straight line segments is greater than the area under the curve).

(b) For any n, we will have $L_n < M_n < I < T_n < R_n$.

(c) $L_5 = \sum_{i=1}^{5} f(x_{i-1})\,\Delta x = \frac{1}{5}[f(0.0) + f(0.2) + f(0.4) + f(0.6) + f(0.8)] \approx 0.1187$

$R_5 = \sum_{i=1}^{5} f(x_i)\,\Delta x = \frac{1}{5}[f(0.2) + f(0.4) + f(0.6) + f(0.8) + f(1)] \approx 0.2146$

$M_5 = \sum_{i=1}^{5} f(\overline{x}_i)\,\Delta x = \frac{1}{5}[f(0.1) + f(0.3) + f(0.5) + f(0.7) + f(0.9)] \approx 0.1622$

$T_5 = \left(\frac{1}{2}\,\Delta x\right)[f(0) + 2f(0.2) + 2f(0.4) + 2f(0.6) + 2f(0.8) + f(1)] \approx 0.1666$

From the graph, it appears that the Midpoint Rule gives the best approximation. (This is in fact the case, since $I \approx 0.16371405$.)

5. $f(x) = x^2 \sin x$, $\Delta x = \dfrac{b-a}{n} = \dfrac{\pi - 0}{8} = \dfrac{\pi}{8}$

(a) $M_8 = \frac{\pi}{8}\left[f\left(\frac{\pi}{16}\right) + f\left(\frac{3\pi}{16}\right) + f\left(\frac{5\pi}{16}\right) + \cdots + f\left(\frac{15\pi}{16}\right)\right] \approx 5.932957$

(b) $S_8 = \frac{\pi}{8 \cdot 3}\left[f(0) + 4f\left(\frac{\pi}{8}\right) + 2f\left(\frac{2\pi}{8}\right) + 4f\left(\frac{3\pi}{8}\right) + 2f\left(\frac{4\pi}{8}\right) + 4f\left(\frac{5\pi}{8}\right) + 2f\left(\frac{6\pi}{8}\right) + 4f\left(\frac{7\pi}{8}\right) + f(\pi)\right]$

≈ 5.869247

Actual: $\int_0^\pi x^2 \sin x\,dx \overset{84}{=} \left[-x^2 \cos x\right]_0^\pi + 2\int_0^\pi x \cos x\,dx \overset{83}{=} \left[-\pi^2(-1) - 0\right] + 2[\cos x + x \sin x]_0^\pi$

$= \pi^2 + 2[(-1+0) - (1+0)] = \pi^2 - 4 \approx 5.869604$

Errors: $E_M = \text{actual} - M_8 = \int_0^\pi x^2 \sin x\,dx - M_8 \approx -0.063353$

$E_S = \text{actual} - S_8 = \int_0^\pi x^2 \sin x\,dx - S_8 \approx 0.000357$

6. $f(x) = e^{-\sqrt{x}}$, $\Delta x = \dfrac{b-a}{n} = \dfrac{1-0}{6} = \dfrac{1}{6}$

(a) $M_6 = \frac{1}{6}\left[f\left(\frac{1}{12}\right) + f\left(\frac{3}{12}\right) + f\left(\frac{5}{12}\right) + f\left(\frac{7}{12}\right) + f\left(\frac{9}{12}\right) + f\left(\frac{11}{12}\right)\right] \approx 0.525100$

(b) $S_6 = \frac{1}{6 \cdot 3}\left[f(0) + 4f\left(\frac{1}{6}\right) + 2f\left(\frac{2}{6}\right) + 4f\left(\frac{3}{6}\right) + 2f\left(\frac{4}{6}\right) + 4f\left(\frac{5}{6}\right) + f(1)\right] \approx 0.533979$

Actual: $\int_0^1 e^{-\sqrt{x}}\,dx = \int_0^{-1} e^u 2u\,du \qquad [u = -\sqrt{x},\ u^2 = x,\ 2u\,du = dx]$

$\overset{96}{=} 2[(u-1)e^u]_0^{-1} = 2\left[-2e^{-1} - (-1e^0)\right] = 2 - 4e^{-1} \approx 0.528482$

Errors: $E_M = \text{actual} - M_6 = \int_0^1 e^{-\sqrt{x}}\,dx - M_6 \approx 0.003382$

$E_S = \text{actual} - S_6 = \int_0^1 e^{-\sqrt{x}}\,dx - S_6 \approx -0.005497$

7. $f(x) = \sqrt[4]{1+x^2}$, $\Delta x = \dfrac{2-0}{8} = \dfrac{1}{4}$

(a) $T_8 = \frac{1}{4 \cdot 2}\left[f(0) + 2f\left(\frac{1}{4}\right) + 2f\left(\frac{1}{2}\right) + \cdots + 2f\left(\frac{3}{2}\right) + 2f\left(\frac{7}{4}\right) + f(2)\right] \approx 2.413790$

(b) $M_8 = \frac{1}{4}\left[f\left(\frac{1}{8}\right) + f\left(\frac{3}{8}\right) + \cdots + f\left(\frac{13}{8}\right) + f\left(\frac{15}{8}\right)\right] \approx 2.411453$

(c) $S_8 = \frac{1}{4 \cdot 3}\left[f(0) + 4f\left(\frac{1}{4}\right) + 2f\left(\frac{1}{2}\right) + 4f\left(\frac{3}{4}\right) + 2f(1) + 4f\left(\frac{5}{4}\right) + 2f\left(\frac{3}{2}\right) + 4f\left(\frac{7}{4}\right) + f(2)\right] \approx 2.412232$

8. $f(x) = \sin(x^2)$, $\Delta x = \dfrac{\frac{1}{2} - 0}{4} = \dfrac{1}{8}$

(a) $T_4 = \frac{1}{8 \cdot 2}\left[f(0) + 2f\left(\frac{1}{8}\right) + 2f\left(\frac{2}{8}\right) + 2f\left(\frac{3}{8}\right) + f\left(\frac{1}{2}\right)\right] \approx 0.042743$

(b) $M_4 = \frac{1}{8}\left[f\left(\frac{1}{16}\right) + f\left(\frac{3}{16}\right) + f\left(\frac{5}{16}\right) + f\left(\frac{7}{16}\right)\right] \approx 0.040850$

(c) $S_4 = \frac{1}{8 \cdot 3}\left[f(0) + 4f\left(\frac{1}{8}\right) + 2f\left(\frac{2}{8}\right) + 4f\left(\frac{3}{8}\right) + f\left(\frac{1}{2}\right)\right] \approx 0.041478$

9. $f(x) = \dfrac{\ln x}{1 + x}$, $\Delta x = \dfrac{2 - 1}{10} = \dfrac{1}{10}$

(a) $T_{10} = \frac{1}{10 \cdot 2}[f(1) + 2f(1.1) + 2f(1.2) + \cdots + 2f(1.8) + 2f(1.9) + f(2)] \approx 0.146879$

(b) $M_{10} = \frac{1}{10}[f(1.05) + f(1.15) + \cdots + f(1.85) + f(1.95)] \approx 0.147391$

(c) $S_{10} = \frac{1}{10 \cdot 3}[f(1) + 4f(1.1) + 2f(1.2) + 4f(1.3) + 2f(1.4) + 4f(1.5) + 2f(1.6) + 4f(1.7)$
$+ 2f(1.8) + 4f(1.9) + f(2)]$
≈ 0.147219

10. $f(t) = \dfrac{1}{1 + t^2 + t^4}$, $\Delta t = \dfrac{3 - 0}{6} = \dfrac{1}{2}$

(a) $T_6 = \frac{1}{2 \cdot 2}\left[f(0) + 2f\left(\frac{1}{2}\right) + 2f(1) + 2f\left(\frac{3}{2}\right) + 2f(2) + 2f\left(\frac{5}{2}\right) + f(3)\right] \approx 0.895122$

(b) $M_6 = \frac{1}{2}\left[f\left(\frac{1}{4}\right) + f\left(\frac{3}{4}\right) + f\left(\frac{5}{4}\right) + f\left(\frac{7}{4}\right) + f\left(\frac{9}{4}\right) + f\left(\frac{11}{4}\right)\right] \approx 0.895478$

(c) $S_6 = \frac{1}{2 \cdot 3}\left[f(0) + 4f\left(\frac{1}{2}\right) + 2f(1) + 4f\left(\frac{3}{2}\right) + 2f(2) + 4f\left(\frac{5}{2}\right) + f(3)\right] \approx 0.898014$

11. $f(t) = e^{\sqrt{t}} \sin t$, $\Delta t = \dfrac{4 - 0}{8} = \dfrac{1}{2}$

(a) $T_8 = \frac{1}{2 \cdot 2}\left[f(0) + 2f\left(\frac{1}{2}\right) + 2f(1) + 2f\left(\frac{3}{2}\right) + 2f(2) + 2f\left(\frac{5}{2}\right) + 2f(3) + 2f\left(\frac{7}{2}\right) + f(4)\right] \approx 4.513618$

(b) $M_8 = \frac{1}{2}\left[f\left(\frac{1}{4}\right) + f\left(\frac{3}{4}\right) + f\left(\frac{5}{4}\right) + f\left(\frac{7}{4}\right) + f\left(\frac{9}{4}\right) + f\left(\frac{11}{4}\right) + f\left(\frac{13}{4}\right) + f\left(\frac{15}{4}\right)\right] \approx 4.748256$

(c) $S_8 = \frac{1}{2 \cdot 3}\left[f(0) + 4f\left(\frac{1}{2}\right) + 2f(1) + 4f\left(\frac{3}{2}\right) + 2f(2) + 4f\left(\frac{5}{2}\right) + 2f(3) + 4f\left(\frac{7}{2}\right) + f(4)\right] \approx 4.675111$

12. $f(x) = \sqrt{1 + \sqrt{x}}$, $\Delta x = \dfrac{4 - 0}{8} = \dfrac{1}{2}$

(a) $T_8 = \frac{1}{2 \cdot 2}\left[f(0) + 2f\left(\frac{1}{2}\right) + 2f(1) + \cdots + 2f(3) + 2f\left(\frac{7}{2}\right) + f(4)\right] \approx 6.042985$

(b) $M_8 = \frac{1}{2}\left[f\left(\frac{1}{4}\right) + f\left(\frac{3}{4}\right) + \cdots + f\left(\frac{13}{4}\right) + f\left(\frac{15}{4}\right)\right] \approx 6.084778$

(c) $S_8 = \frac{1}{2 \cdot 3}\left[f(0) + 4f\left(\frac{1}{2}\right) + 2f(1) + 4f\left(\frac{3}{2}\right) + 2f(2) + 4f\left(\frac{5}{2}\right) + 2f(3) + 4f\left(\frac{7}{2}\right) + f(4)\right] \approx 6.061678$

13. $f(x) = \dfrac{\cos x}{x}$, $\Delta x = \dfrac{5 - 1}{8} = \dfrac{1}{2}$

(a) $T_8 = \frac{1}{2 \cdot 2}\left[f(1) + 2f\left(\frac{3}{2}\right) + 2f(2) + \cdots + 2f(4) + 2f\left(\frac{9}{2}\right) + f(5)\right] \approx -0.495333$

(b) $M_8 = \frac{1}{2}\left[f\left(\frac{5}{4}\right) + f\left(\frac{7}{4}\right) + f\left(\frac{9}{4}\right) + f\left(\frac{11}{4}\right) + f\left(\frac{13}{4}\right) + f\left(\frac{15}{4}\right) + f\left(\frac{17}{4}\right) + f\left(\frac{19}{4}\right)\right] \approx -0.543321$

(c) $S_8 = \frac{1}{2 \cdot 3}\left[f(1) + 4f\left(\frac{3}{2}\right) + 2f(2) + 4f\left(\frac{5}{2}\right) + 2f(3) + 4f\left(\frac{7}{2}\right) + 2f(4) + 4f\left(\frac{9}{2}\right) + f(5)\right] \approx -0.526123$

14. $f(x) = \ln(x^3 + 2)$, $\Delta x = \dfrac{6 - 4}{10} = \dfrac{1}{5}$

(a) $T_{10} = \frac{1}{5 \cdot 2}\left[f(4) + 2f(4.2) + 2f(4.4) + \cdots + 2f(5.6) + 2f(5.8) + f(6)\right] \approx 9.649753$

(b) $M_{10} = \frac{1}{5}[f(4.1) + f(4.3) + \cdots + f(5.7) + f(5.9)] \approx 9.650912$

(c) $S_{10} = \frac{1}{5 \cdot 3}[f(4) + 4f(4.2) + 2f(4.4) + 4f(4.6) + 2f(4.8) + 4f(5) + 2f(5.2) + 4f(5.4)$
$+2f(5.6) + 4f(5.8) + f(6)]$

≈ 9.650526

15. $f(y) = \dfrac{1}{1+y^5}$, $\Delta y = \dfrac{3-0}{6} = \dfrac{1}{2}$

(a) $T_6 = \frac{1}{2 \cdot 2}\left[f(0) + 2f\left(\frac{1}{2}\right) + 2f\left(\frac{2}{2}\right) + 2f\left(\frac{3}{2}\right) + 2f\left(\frac{4}{2}\right) + 2f\left(\frac{5}{2}\right) + f(3)\right] \approx 1.064275$

(b) $M_6 = \frac{1}{2}\left[f\left(\frac{1}{4}\right) + f\left(\frac{3}{4}\right) + f\left(\frac{5}{4}\right) + f\left(\frac{7}{4}\right) + f\left(\frac{9}{4}\right) + f\left(\frac{11}{4}\right)\right] \approx 1.067416$

(c) $S_6 = \frac{1}{2 \cdot 3}\left[f(0) + 4f\left(\frac{1}{2}\right) + 2f\left(\frac{2}{2}\right) + 4f\left(\frac{3}{2}\right) + 2f\left(\frac{4}{2}\right) + 4f\left(\frac{5}{2}\right) + f(3)\right] \approx 1.074915$

16. $f(z) = \sqrt{z}e^{-z}$, $\Delta z = \dfrac{1-0}{10} = \dfrac{1}{10}$

(a) $T_{10} = \frac{1}{10 \cdot 2}\left\{f(0) + 2\left[f(0.1) + f(0.2) + \cdots + f(0.9)\right] + f(1)\right\} \approx 0.372299$

(b) $M_{10} = \frac{1}{10}\left[f(0.05) + f(0.15) + f(0.25) + \cdots + f(0.95)\right] \approx 0.380894$

(c) $S_{10} = \frac{1}{10 \cdot 3}[f(0) + 4f(0.1) + 2f(0.2) + 4f(0.3) + 2f(0.4) + 4f(0.5) + 2f(0.6)$
$+ 4f(0.7) + 2f(0.8) + 4f(0.9) + f(1)]$

≈ 0.376330

17. (a) $f(x) = e^{-x^2}$, $\Delta x = \dfrac{2-0}{10} = \dfrac{1}{5}$

$T_{10} = \frac{1}{5 \cdot 2}\{f(0) + 2[f(0.2) + f(0.4) + \cdots + f(1.8)] + f(2)\} \approx 0.881839$

$M_{10} = \frac{1}{5}[f(0.1) + f(0.3) + f(0.5) + \cdots + f(1.7) + f(1.9)] \approx 0.882202$

(b) $f(x) = e^{-x^2}$, $f'(x) = -2xe^{-x^2}$, $f''(x) = (4x^2 - 2)e^{-x^2}$, $f'''(x) = 4x(3 - 2x^2)e^{-x^2}$.

$f'''(x) = 0 \quad \Leftrightarrow \quad x = 0$ or $x = \pm\sqrt{\frac{3}{2}}$. So to find the maximum value of $|f''(x)|$ on $[0, 2]$, we need only consider its values at $x = 0$, $x = 2$, and $x = \sqrt{\frac{3}{2}}$. $|f''(0)| = 2$, $|f''(2)| \approx 0.2564$ and $\left|f''\left(\sqrt{\frac{3}{2}}\right)\right| \approx 0.8925$. Thus, taking $K = 2$, $a = 0$, $b = 2$, and $n = 10$ in Theorem 3, we get $|E_T| \leq 2 \cdot 2^3/(12 \cdot 10^2) = \frac{1}{75} = 0.01\overline{3}$, and $|E_M| \leq |E_T|/2 \leq 0.00\overline{6}$.

(c) Take $K = 2$ [as in part (b)] in Theorem 3. $|E_T| \leq \dfrac{K(b-a)^3}{12n^2} \leq 10^{-5} \quad \Leftrightarrow \quad \dfrac{2(2-0)^3}{12n^2} \leq 10^{-5} \quad \Leftrightarrow \quad \frac{3}{4}n^2 \geq 10^5 \quad \Leftrightarrow$ $n \geq 365.1\ldots \quad \Leftrightarrow \quad n \geq 366$. Take $n = 366$ for T_n. For E_M, again take $K = 2$ in Theorem 3 to get $|E_M| \leq 10^{-5} \quad \Leftrightarrow$ $\frac{3}{2}n^2 \geq 10^5 \quad \Leftrightarrow \quad n \geq 258.2 \quad \Rightarrow \quad n \geq 259$. Take $n = 259$ for M_n.

18. (a) $T_8 = \frac{1}{8 \cdot 2}\left\{f(0) + 2\left[f\left(\frac{1}{8}\right) + f\left(\frac{2}{8}\right) + \cdots + f\left(\frac{7}{8}\right)\right] + f(1)\right\} \approx 0.902333$

$M_8 = \frac{1}{8}\left[f\left(\frac{1}{16}\right) + f\left(\frac{3}{16}\right) + f\left(\frac{5}{16}\right) + \cdots + f\left(\frac{15}{16}\right)\right] = 0.905620$

(b) $f(x) = \cos(x^2)$, $f'(x) = -2x\sin(x^2)$, $f''(x) = -2\sin(x^2) - 4x^2\cos(x^2)$. For $0 \leq x \leq 1$, sin and cos are positive, so $|f''(x)| = 2\sin(x^2) + 4x^2\cos(x^2) \leq 2 \cdot 1 + 4 \cdot 1 \cdot 1 = 6$ since $\sin(x^2) \leq 1$ and $\cos(x^2) \leq 1$ for all x, and $x^2 \leq 1$ for $0 \leq x \leq 1$. So for $n = 8$, we take $K = 6$, $a = 0$, and $b = 1$ in Theorem 3, to get $|E_T| \leq 6 \cdot 1^3/(12 \cdot 8^2) = \frac{1}{128} = 0.0078125$ and $|E_M| \leq \frac{1}{256} = 0.00390625$. [A better estimate is obtained by noting from a graph of f'' that $|f''(x)| \leq 4$ for $0 \leq x \leq 1$.]

(c) Using $K = 6$ as in part (b), we have $|E_T| \leq 6 \cdot 1^3/(12n^2) = 1/(2n^2) \leq 10^{-5} \Rightarrow 2n^2 \geq 10^5 \Rightarrow$

$n \geq \sqrt{\frac{1}{2} \cdot 10^5}$ or $n \geq 224$. To guarantee that $|E_M| \leq 0.00001$, we need $6 \cdot 1^3/(24n^2) \leq 10^{-5} \Rightarrow$

$4n^2 \geq 10^5 \Rightarrow n \geq \sqrt{\frac{1}{4} \cdot 10^5}$ or $n \geq 159$.

19. (a) $T_{10} = \frac{1}{10 \cdot 2}\{f(0) + 2[f(0.1) + f(0.2) + \cdots + f(0.9)] + f(1)\} \approx 1.71971349$

$S_{10} = \frac{1}{10 \cdot 3}[f(0) + 4f(0.1) + 2f(0.2) + 4f(0.3) + \cdots + 4f(0.9) + f(1)] \approx 1.71828278$

Since $I = \int_0^1 e^x dx = [e^x]_0^1 = e - 1 \approx 1.71828183$, $E_T = I - T_{10} \approx -0.00143166$ and

$E_S = I - S_{10} \approx -0.00000095$.

(b) $f(x) = e^x \Rightarrow f''(x) = e^x \leq e$ for $0 \leq x \leq 1$. Taking $K = e$, $a = 0$, $b = 1$, and $n = 10$ in Theorem 3, we get

$|E_T| \leq e(1)^3/(12 \cdot 10^2) \approx 0.002265 > 0.00143166$ [actual $|E_T|$ from (a)]. $f^{(4)}(x) = e^x < e$ for $0 \leq x \leq 1$.

Using Theorem 4, we have $|E_S| \leq e(1)^5/(180 \cdot 10^4) \approx 0.0000015 > 0.00000095$ [actual $|E_S|$ from (a)].

We see that the actual errors are about two-thirds the size of the error estimates.

(c) From part (b), we take $K = e$ to get $|E_T| \leq \dfrac{K(b-a)^3}{12n^2} \leq 0.00001 \Rightarrow n^2 \geq \dfrac{e(1^3)}{12(0.00001)} \Rightarrow n \geq 150.5$.

Take $n = 151$ for T_n. Now $|E_M| \leq \dfrac{K(b-a)^3}{24n^2} \leq 0.00001 \Rightarrow n \geq 106.4$. Take $n = 107$ for M_n.

Finally, $|E_S| \leq \dfrac{K(b-a)^5}{180n^4} \leq 0.00001 \Rightarrow n^4 \geq \dfrac{e(1^5)}{180(0.00001)} \Rightarrow n \geq 6.23$.

Take $n = 8$ for S_n (since n has to be even for Simpson's Rule).

20. From Example 7(b), we take $K = 76e$ to get $|E_S| \leq 76e(1)^5/(180n^4) \leq 0.00001 \Rightarrow n^4 \geq 76e/[180(0.00001)] \Rightarrow$

$n \geq 18.4$. Take $n = 20$ (since n must be even).

21. (a) Using a CAS, we differentiate $f(x) = e^{\cos x}$ twice, and find that $f''(x) = e^{\cos x}(\sin^2 x - \cos x)$. From the graph, we see that the maximum value of $|f''(x)|$ occurs at the endpoints of the interval $[0, 2\pi]$. Since $f''(0) = -e$, we can use $K = e$ or $K = 2.8$.

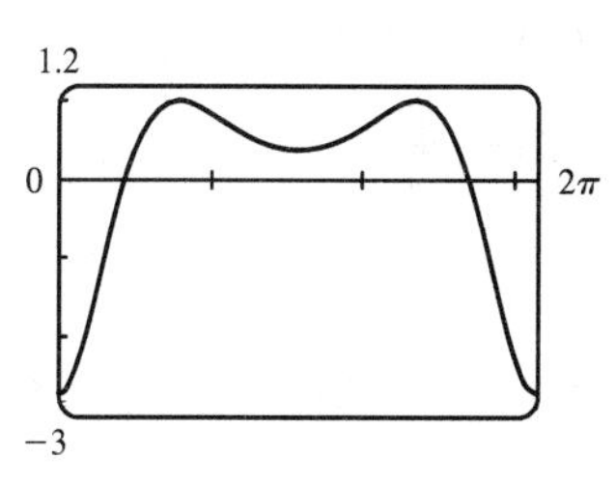

(b) A CAS gives $M_{10} \approx 7.954926518$. (In Maple, use `student[middlesum]`.)

(c) Using Theorem 3 for the Midpoint Rule, with $K = e$, we get $|E_M| \leq \dfrac{e(2\pi - 0)^3}{24 \cdot 10^2} \approx 0.280945995$.

With $K = 2.8$, we get $|E_M| \leq \dfrac{2.8(2\pi - 0)^3}{24 \cdot 10^2} = 0.289391916$.

(d) A CAS gives $I \approx 7.954926521$.

(e) The actual error is only about 3×10^{-9}, much less than the estimate in part (c).

(f) We use the CAS to differentiate twice more, and then graph

$f^{(4)}(x) = e^{\cos x}(\sin^4 x - 6\sin^2 x \cos x + 3 - 7\sin^2 x + \cos x)$.

From the graph, we see that the maximum value of $\left|f^{(4)}(x)\right|$ occurs at the endpoints of the interval $[0, 2\pi]$. Since $f^{(4)}(0) = 4e$, we can use $K = 4e$ or $K = 10.9$.

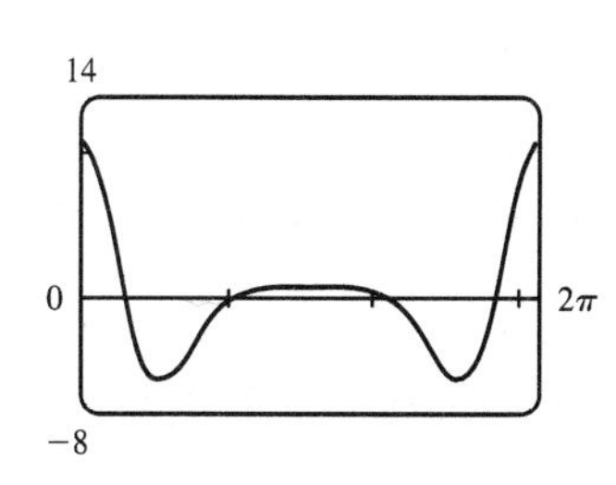

(g) A CAS gives $S_{10} \approx 7.953789422$. (In Maple, use `student[simpson]`.)

(h) Using Theorem 4 with $K = 4e$, we get $|E_S| \le \dfrac{4e(2\pi - 0)^5}{180 \cdot 10^4} \approx 0.059153618$.

With $K = 10.9$, we get $|E_S| \le \dfrac{10.9(2\pi - 0)^5}{180 \cdot 10^4} \approx 0.059299814$.

(i) The actual error is about $7.954926521 - 7.953789422 \approx 0.00114$. This is quite a bit smaller than the estimate in part (h), though the difference is not nearly as great as it was in the case of the Midpoint Rule.

(j) To ensure that $|E_S| \le 0.0001$, we use Theorem 4: $|E_S| \le \dfrac{4e(2\pi)^5}{180 \cdot n^4} \le 0.0001 \Rightarrow \dfrac{4e(2\pi)^5}{180 \cdot 0.0001} \le n^4 \Rightarrow$

$n^4 \ge 5{,}915{,}362 \Leftrightarrow n \ge 49.3$. So we must take $n \ge 50$ to ensure that $|I - S_n| \le 0.0001$. ($K = 10.9$ leads to the same value of n.)

22. (a) Using the CAS, we differentiate $f(x) = \sqrt{4 - x^3}$ twice,

and find that $f''(x) = -\dfrac{9x^4}{4(4 - x^3)^{3/2}} - \dfrac{3x}{(4 - x^3)^{1/2}}$.

From the graph, we see that $|f''(x)| < 2.2$ on $[-1, 1]$.

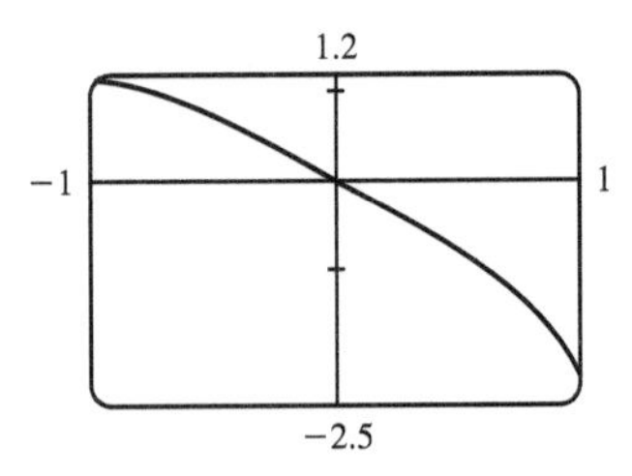

(b) A CAS gives $M_{10} \approx 3.995804152$. (In Maple, use `student[middlesum]`.)

(c) Using Theorem 3 for the Midpoint Rule, with $K = 2.2$, we get $|E_M| \le \dfrac{2.2[1 - (-1)]^3}{24 \cdot 10^2} \approx 0.00733$.

(d) A CAS gives $I \approx 3.995487677$.

(e) The actual error is about -0.0003165, much less than the estimate in part (c).

(f) We use the CAS to differentiate twice more,

and then graph $f^{(4)}(x) = \dfrac{9}{16} \dfrac{x^2(x^6 - 224x^3 - 1280)}{(4 - x^3)^{7/2}}$.

From the graph, we see that $\left|f^{(4)}(x)\right| < 18.1$ on $[-1, 1]$.

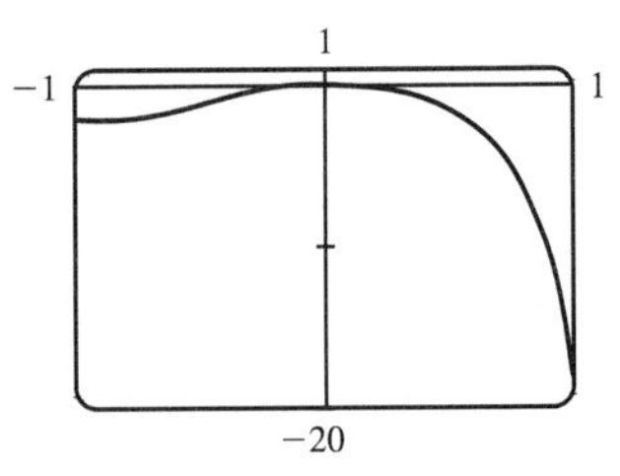

(g) A CAS gives $S_{10} \approx 3.995449790$. (In Maple, use `student[simpson]`.)

(h) Using Theorem 4 with $K = 18.1$, we get $|E_S| \le \dfrac{18.1[1 - (-1)]^5}{180 \cdot 10^4} \approx 0.000322$.

(i) The actual error is about $3.995487677 - 3.995449790 \approx 0.0000379$. This is quite a bit smaller than the estimate in part (h).

(j) To ensure that $|E_S| \le 0.0001$, we use Theorem 4: $|E_S| \le \dfrac{18.1(2)^5}{180 \cdot n^4} \le 0.0001 \Rightarrow \dfrac{18.1(2)^5}{180 \cdot 0.0001} \le n^4 \Rightarrow$

$n^4 \ge 32{,}178 \Rightarrow n \ge 13.4$. So we must take $n \ge 14$ to ensure that $|I - S_n| \le 0.0001$.

23. $I = \int_0^1 x^3\,dx = \left[\frac{1}{4}x^4\right]_0^1 = 0.25. \quad f(x) = x^3.$

$n = 4$: $L_4 = \frac{1}{4}\left[0^3 + \left(\frac{1}{4}\right)^3 + \left(\frac{2}{4}\right)^3 + \left(\frac{3}{4}\right)^3\right] = 0.140625$

$R_4 = \frac{1}{4}\left[\left(\frac{1}{4}\right)^3 + \left(\frac{2}{4}\right)^3 + \left(\frac{3}{4}\right)^3 + 1^3\right] = 0.390625$

$T_4 = \frac{1}{4\cdot 2}\left[0^3 + 2\left(\frac{1}{4}\right)^3 + 2\left(\frac{2}{4}\right)^3 + 2\left(\frac{3}{4}\right)^3 + 1^3\right] = 0.265625,$

$M_4 = \frac{1}{4}\left[\left(\frac{1}{8}\right)^3 + \left(\frac{3}{8}\right)^3 + \left(\frac{5}{8}\right)^3 + \left(\frac{7}{8}\right)^3\right] = 0.2421875,$

$E_L = I - L_4 = \frac{1}{4} - 0.140625 = 0.109375,\ E_R = \frac{1}{4} - 0.390625 = -0.140625,$

$E_T = \frac{1}{4} - 0.265625 = -0.015625,\ E_M = \frac{1}{4} - 0.2421875 = 0.0078125$

$n = 8$: $L_8 = \frac{1}{8}\left[f(0) + f\left(\frac{1}{8}\right) + f\left(\frac{2}{8}\right) + \cdots + f\left(\frac{7}{8}\right)\right] \approx 0.191406$

$R_8 = \frac{1}{8}\left[f\left(\frac{1}{8}\right) + f\left(\frac{2}{8}\right) + \cdots + f\left(\frac{7}{8}\right) + f(1)\right] \approx 0.316406$

$T_8 = \frac{1}{8\cdot 2}\left\{f(0) + 2\left[f\left(\frac{1}{8}\right) + f\left(\frac{2}{8}\right) + \cdots + f\left(\frac{7}{8}\right)\right] + f(1)\right\} \approx 0.253906$

$M_8 = \frac{1}{8}\left[f\left(\frac{1}{16}\right) + f\left(\frac{3}{16}\right) + \cdots + f\left(\frac{13}{16}\right) + f\left(\frac{15}{16}\right)\right] = 0.248047$

$E_L \approx \frac{1}{4} - 0.191406 \approx 0.058594,\ E_R \approx \frac{1}{4} - 0.316406 \approx -0.066406,$

$E_T \approx \frac{1}{4} - 0.253906 \approx -0.003906,\ E_M \approx \frac{1}{4} - 0.248047 \approx 0.001953.$

$n = 16$: $L_{16} = \frac{1}{16}\left[f(0) + f\left(\frac{1}{16}\right) + f\left(\frac{2}{16}\right) + \cdots + f\left(\frac{15}{16}\right)\right] \approx 0.219727$

$R_{16} = \frac{1}{16}\left[f\left(\frac{1}{16}\right) + f\left(\frac{2}{16}\right) + \cdots + f\left(\frac{15}{16}\right) + f(1)\right] \approx 0.282227$

$T_{16} = \frac{1}{16\cdot 2}\left\{f(0) + 2\left[f\left(\frac{1}{16}\right) + f\left(\frac{2}{16}\right) + \cdots + f\left(\frac{15}{16}\right)\right] + f(1)\right\} \approx 0.250977$

$M_{16} = \frac{1}{16}\left[f\left(\frac{1}{32}\right) + f\left(\frac{3}{32}\right) + \cdots + f\left(\frac{31}{32}\right)\right] \approx 0.249512$

$E_L \approx \frac{1}{4} - 0.219727 \approx 0.030273,\ E_R \approx \frac{1}{4} - 0.282227 \approx -0.032227,$

$E_T \approx \frac{1}{4} - 0.250977 \approx -0.000977,\ E_M \approx \frac{1}{4} - 0.249512 \approx 0.000488.$

n	L_n	R_n	T_n	M_n
4	0.140625	0.390625	0.265625	0.242188
8	0.191406	0.316406	0.253906	0.248047
16	0.219727	0.282227	0.250977	0.249512

n	E_L	E_R	E_T	E_M
4	0.109375	−0.140625	−0.015625	0.007813
8	0.058594	−0.066406	−0.003906	0.001953
16	0.030273	−0.032227	−0.000977	0.000488

Observations:

1. E_L and E_R are always opposite in sign, as are E_T and E_M.
2. As n is doubled, E_L and E_R are decreased by about a factor of 2, and E_T and E_M are decreased by a factor of about 4.
3. The Midpoint approximation is about twice as accurate as the Trapezoidal approximation.
4. All the approximations become more accurate as the value of n increases.
5. The Midpoint and Trapezoidal approximations are much more accurate than the endpoint approximations.

24. $I = \int_{-1}^{2} xe^x\,dx = [xe^x - e^x]_{-1}^{2} = e^2 + 2/e \approx 8.124815$. $f(x) = xe^x$.

$n = 6$: $\Delta x = [2 - (-1)]/6 = \frac{1}{2}$

$T_6 = \frac{1}{2 \cdot 2}\{f(-1) + 2[f(-0.5) + f(0) + \cdots + f(1.5)] + f(2)\} \approx 8.583514$

$M_6 = \frac{1}{2}[f(-0.75) + f(-0.25) + \cdots + f(1.75)] \approx 7.896632$

$S_6 = \frac{1}{2 \cdot 3}[f(-1) + 4f(-0.5) + 2f(0) + 4f(0.5) + 2f(1) + 4f(1.5) + f(2)] \approx 8.136885$

$E_T \approx I - 8.583514 \approx -0.458699$, $E_M \approx I - 7.896632 \approx 0.228183$,

$E_S \approx I - 8.136885 \approx -0.012070$.

$n = 12$: $\Delta x = [2 - (-1)]/12 = \frac{1}{4}$

$T_{12} = \frac{1}{4 \cdot 2}\{f(-1) + 2[f(-0.75) + f(-0.5) + \cdots + f(1.75)] + f(2)\} \approx 8.240073$

$M_{12} = \frac{1}{4}\left[f\left(-\frac{7}{8}\right) + f\left(-\frac{5}{8}\right) + \cdots + f\left(\frac{13}{8}\right) + f\left(\frac{15}{8}\right)\right] \approx 8.067259$

$S_{12} = \frac{1}{4 \cdot 3}[f(-1) + 4f(-0.75) + 2f(-0.5) + \cdots + 2f(1.5) + 4f(1.75) + f(2)] \approx 8.125593$

$E_T \approx I - 8.240073 \approx -0.115258$, $E_M \approx I - 8.067259 \approx 0.057556$,

$E_S \approx I - 8.125593 \approx -0.000778$

n	T_n	M_n	S_n
6	8.583514	7.896632	8.136885
12	8.240073	8.067259	8.125593

n	E_T	E_M	E_S
6	−0.458699	0.228183	−0.012070
12	−0.115258	0.057556	−0.000778

Observations:

1. E_T and E_M are opposite in sign and decrease by a factor of about 4 as n is doubled.
2. The Simpson's approximation is much more accurate than the Midpoint and Trapezoidal approximations, and seems to decrease by a factor of about 16 as n is doubled.

25. $\Delta x = (4 - 0)/4 = 1$

(a) $T_4 = \frac{1}{2}[f(0) + 2f(1) + 2f(2) + 2f(3) + f(4)] \approx \frac{1}{2}[0 + 2(3) + 2(5) + 2(3) + 1] = 11.5$

(b) $M_4 = 1 \cdot [f(0.5) + f(1.5) + f(2.5) + f(3.5)] \approx 1 + 4.5 + 4.5 + 2 = 12$

(c) $S_4 = \frac{1}{3}[f(0) + 4f(1) + 2f(2) + 4f(3) + f(4)] \approx \frac{1}{3}[0 + 4(3) + 2(5) + 4(3) + 1] = 11.\overline{6}$

26. We use Simpson's Rule with $n = 10$ and $\Delta x = \frac{1}{2}$:

$$\begin{aligned}\text{distance} = \int_0^5 v(t)\,dt \approx S_{10} &= \frac{1}{2 \cdot 3}[f(0) + 4f(0.5) + 2f(1) + \cdots + 4f(4.5) + f(5)] \\ &= \frac{1}{6}[0 + 4(4.67) + 2(7.34) + 4(8.86) + 2(9.73) + 4(10.22) \\ &\qquad + 2(10.51) + 4(10.67) + 2(10.76) + 4(10.81) + 10.81] \\ &= \frac{1}{6}(268.41) = 44.735 \text{ m}\end{aligned}$$

27. By the Net Change Theorem, the increase in velocity is equal to $\int_0^6 a(t)\,dt$. We use Simpson's Rule with $n = 6$ and $\Delta t = (6 - 0)/6 = 1$ to estimate this integral:

$$\begin{aligned}\int_0^6 a(t)\,dt \approx S_6 &= \frac{1}{3}[a(0) + 4a(1) + 2a(2) + 4a(3) + 2a(4) + 4a(5) + a(6)] \\ &\approx \frac{1}{3}[0 + 4(0.5) + 2(4.1) + 4(9.8) + 2(12.9) + 4(9.5) + 0] = \frac{1}{3}(113.2) = 37.7\overline{3} \text{ ft/s}\end{aligned}$$

28. By the Net Change Theorem, the total amount of water that leaked out during the first six hours is equal to $\int_0^6 r(t)\,dt$.

We use Simpson's Rule with $n = 6$ and $\Delta t = (6-0)/6 = 1$ to estimate this integral:

$$\begin{aligned}\int_0^6 r(t)\,dt \approx S_6 &= \tfrac{1}{3}[r(0) + 4r(1) + 2r(2) + 4r(3) + 2r(4) + 4r(5) + r(6)] \\ &\approx \tfrac{1}{3}[4 + 4(3) + 2(2.4) + 4(1.9) + 2(1.4) + 4(1.1) + 1] = \tfrac{1}{3}(36.6) = 12.2 \text{ liters}\end{aligned}$$

The function values were obtained from a high-resolution graph.

29. By the Net Change Theorem, the energy used is equal to $\int_0^6 P(t)\,dt$. We use Simpson's Rule with $n = 12$ and $\Delta t = (6-0)/12 = \frac{1}{2}$ to estimate this integral:

$$\begin{aligned}\int_0^6 P(t)\,dt \approx S_{12} &= \tfrac{1/2}{3}[P(0) + 4P(0.5) + 2P(1) + 4P(1.5) + 2P(2) + 4P(2.5) \\ &\qquad + 2P(3) + 4P(3.5) + 2P(4) + 4P(4.5) + 2P(5) + 4P(5.5) + P(6)] \\ &= \tfrac{1}{6}[1814 + 4(1735) + 2(1686) + 4(1646) + 2(1637) + 4(1609) + 2(1604) \\ &\qquad + 4(1611) + 2(1621) + 4(1666) + 2(1745) + 4(1886) + 2052] \\ &= \tfrac{1}{6}(61{,}064) = 10{,}177.\overline{3} \text{ megawatt-hours}\end{aligned}$$

30. By the Net Change Theorem, the total amount of data transmitted is equal to $\int_0^8 D(t)\,dt \times 3600$ [since $D(t)$ is measured in megabits per second and t is in hours]. We use Simpson's Rule with $n = 8$ and $\Delta t = (8-0)/8 = 1$ to estimate this integral:

$$\begin{aligned}\int_0^8 D(t)\,dt \approx S_8 &= \tfrac{1}{3}[D(0) + 4D(1) + 2D(2) + 4D(3) + 2D(4) + 4D(5) + 2D(6) + 4D(7) + D(8)] \\ &\approx \tfrac{1}{3}[0.35 + 4(0.32) + 2(0.41) + 4(0.50) + 2(0.51) + 4(0.56) + 2(0.56) + 4(0.83) + 0.88] \\ &= \tfrac{1}{3}(13.03) = 4.34\overline{3}\end{aligned}$$

Now multiply by 3600 to obtain 15,636 megabits.

31. (a) We are given the function values at the endpoints of 8 intervals of length 0.4, so we'll use the Midpoint Rule with $n = 8/2 = 4$ and $\Delta x = (3.2-0)/4 = 0.8$.

$$\int_0^{3.2} f(x)\,dx \approx M_4 = 0.8[f(0.4) + f(1.2) + f(2.0) + f(2.8)] = 0.8[6.5 + 6.4 + 7.6 + 8.8] = 0.8(29.3) = 23.44$$

(b) $-4 \le f''(x) \le 1 \;\Rightarrow\; |f''(x)| \le 4$, so use $K = 4$, $a = 0$, $b = 3.2$, and $n = 4$ in Theorem 3.

So $|E_M| \le \dfrac{4(3.2-0)^3}{24(4)^2} = \dfrac{128}{375} = 0.341\overline{3}$.

32. Using Simpson's Rule with $n = 10$, $\Delta x = \frac{\pi/2}{10}$, $L = 1$, $\theta_0 = \frac{42\pi}{180}$ radians, $g = 9.8\text{ m/s}^2$, $k^2 = \sin^2\left(\frac{1}{2}\theta_0\right)$, and $f(x) = 1/\sqrt{1 - k^2\sin^2 x}$, we get

$$T = 4\sqrt{\frac{L}{g}}\int_0^{\pi/2}\frac{dx}{\sqrt{1-k^2\sin^2 x}} \approx 4\sqrt{\frac{L}{g}}\,S_{10} = 4\sqrt{\tfrac{1}{9.8}}\left(\tfrac{\pi/2}{10\cdot 3}\right)\left[f(0) + 4f\left(\tfrac{\pi}{20}\right) + 2f\left(\tfrac{2\pi}{20}\right) + \cdots + 4f\left(\tfrac{9\pi}{20}\right) + f\left(\tfrac{\pi}{2}\right)\right]$$

≈ 2.07665

33. $I(\theta) = \dfrac{N^2\sin^2 k}{k^2}$, where $k = \dfrac{\pi N d\sin\theta}{\lambda}$, $N = 10{,}000$, $d = 10^{-4}$, and $\lambda = 632.8 \times 10^{-9}$.

So $I(\theta) = \dfrac{(10^4)^2\sin^2 k}{k^2}$, where $k = \dfrac{\pi(10^4)(10^{-4})\sin\theta}{632.8 \times 10^{-9}}$. Now $n = 10$ and $\Delta\theta = \dfrac{10^{-6} - (-10^{-6})}{10} = 2 \times 10^{-7}$,

so $M_{10} = 2 \times 10^{-7}[I(-0.0000009) + I(-0.0000007) + \cdots + I(0.0000009)] \approx 59.4$.

34. Consider the function $f(x) = |x - 1|$, $0 \le x \le 2$. The area $\int_0^2 f(x)\,dx$ is exactly 1. So is the right endpoint approximation: $R_2 = f(1)\,\Delta x + f(2)\,\Delta x = 0 \cdot 1 + 1 \cdot 1 = 1$. But Simpson's Rule approximates f with the parabola $y = (x-1)^2$, shown dashed, and

$$S_2 = \frac{\Delta x}{3}\,[f(0) + 4f(1) + f(2)] = \frac{1}{3}\,[1 + 4 \cdot 0 + 1] = \frac{2}{3}.$$

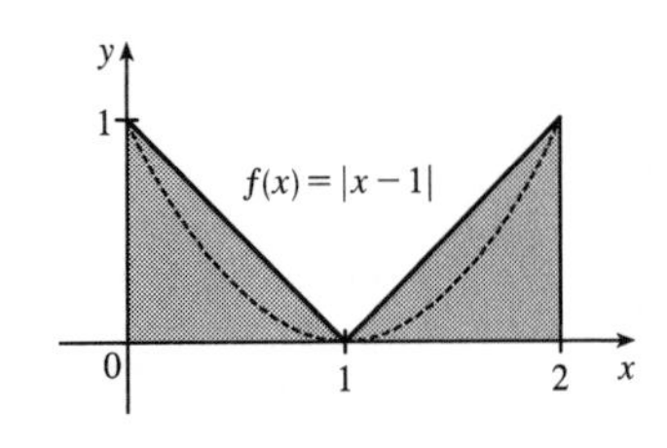

35. Consider the function f whose graph is shown.

The area $\int_0^2 f(x)\,dx$ is close to 2. The Trapezoidal Rule gives

$T_2 = \frac{2-0}{2 \cdot 2}\,[f(0) + 2f(1) + f(2)] = \frac{1}{2}\,[1 + 2 \cdot 1 + 1] = 2$.

The Midpoint Rule gives $M_2 = \frac{2-0}{2}\,[f(0.5) + f(1.5)] = 1[0 + 0] = 0$, so the Trapezoidal Rule is more accurate.

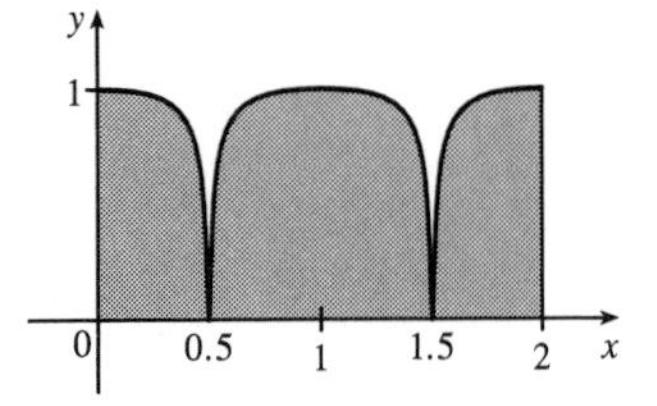

36. $f(x) = \cos(\pi x)$, $\Delta x = \frac{20-0}{10} = 2 \;\Rightarrow$

$$\begin{aligned} T_{10} &= \tfrac{2}{2}\{f(0) + 2[f(2) + f(4) + \cdots + f(18)] + f(20)\} \\ &= 1[\cos 0 + 2(\cos 2\pi + \cos 4\pi + \cdots + \cos 18\pi) + \cos 20\pi] \\ &= 1 + 2(1+1+1+1+1+1+1+1+1) + 1 = 20 \end{aligned}$$

The actual value is $\int_0^{20} \cos(\pi x)\,dx = \frac{1}{\pi}\,[\sin \pi x]_0^{20} = \frac{1}{\pi}(\sin 20\pi - \sin 0) = 0$. The discrepancy is due to the fact that the function is sampled only at points of the form $2n$, where its value is $f(2n) = \cos(2n\pi) = 1$.

37. Since the Trapezoidal and Midpoint approximations on the interval $[a,b]$ are the sums of the Trapezoidal and Midpoint approximations on the subintervals $[x_{i-1}, x_i]$, $i = 1, 2, \ldots, n$, we can focus our attention on one such interval. The condition $f''(x) < 0$ for $a \le x \le b$ means that the graph of f is concave down as in Figure 5. In that figure, T_n is the area of the trapezoid $AQRD$, $\int_a^b f(x)\,dx$ is the area of the region $AQPRD$, and M_n is the area of the trapezoid $ABCD$, so $T_n < \int_a^b f(x)\,dx < M_n$. In general, the condition $f'' < 0$ implies that the graph of f on $[a,b]$ lies above the chord joining the points $(a, f(a))$ and $(b, f(b))$. Thus, $\int_a^b f(x)\,dx > T_n$. Since M_n is the area under a tangent to the graph, and since $f'' < 0$ implies that the tangent lies above the graph, we also have $M_n > \int_a^b f(x)\,dx$. Thus, $T_n < \int_a^b f(x)\,dx < M_n$.

38. Let f be a polynomial of degree ≤ 3; say $f(x) = Ax^3 + Bx^2 + Cx + D$. It will suffice to show that Simpson's estimate is exact when there are two subintervals ($n = 2$), because for a larger even number of subintervals the sum of exact estimates is exact. As in the derivation of Simpson's Rule, we can assume that $x_0 = -h$, $x_1 = 0$, and $x_2 = h$. Then Simpson's approximation is

$$\begin{aligned} \int_{-h}^{h} f(x)\,dx &\approx \tfrac{1}{3}h[f(-h) + 4f(0) + f(h)] = \tfrac{1}{3}h[(-Ah^3 + Bh^2 - Ch + D) + 4D + (Ah^3 + Bh^2 + Ch + D)] \\ &= \tfrac{1}{3}h[2Bh^2 + 6D] = \tfrac{2}{3}Bh^3 + 2Dh \end{aligned}$$

The exact value of the integral is

$$\begin{aligned} \int_{-h}^{h} (Ax^3 + Bx^2 + Cx + D)\,dx &= 2\int_0^h (Bx^2 + D)\,dx \quad \text{[by Theorem 5.5.7(a) and (b)]} \\ &= 2\left[\tfrac{1}{3}Bx^3 + Dx\right]_0^h = \tfrac{2}{3}Bh^3 + 2Dh \end{aligned}$$

Thus, Simpson's Rule is exact.

39. $T_n = \frac{1}{2}\,\Delta x\,[f(x_0) + 2f(x_1) + \cdots + 2f(x_{n-1}) + f(x_n)]$ and

$M_n = \Delta x\,[f(\overline{x}_1) + f(\overline{x}_2) + \cdots + f(\overline{x}_{n-1}) + f(\overline{x}_n)]$, where $\overline{x}_i = \frac{1}{2}(x_{i-1} + x_i)$. Now

$T_{2n} = \frac{1}{2}\left(\frac{1}{2}\Delta x\right)[f(x_0) + 2f(\overline{x}_1) + 2f(x_1) + 2f(\overline{x}_2) + 2f(x_2) + \cdots + 2f(\overline{x}_{n-1}) + 2f(x_{n-1}) + 2f(\overline{x}_n) + f(x_n)]$, so

$$\begin{aligned}\tfrac{1}{2}(T_n + M_n) &= \tfrac{1}{2}T_n + \tfrac{1}{2}M_n \\ &= \tfrac{1}{4}\,\Delta x[f(x_0) + 2f(x_1) + \cdots + 2f(x_{n-1}) + f(x_n)] \\ &\qquad + \tfrac{1}{4}\Delta x 2f(\overline{x}_1) + 2f(\overline{x}_2) + \cdots + 2f(\overline{x}_{n-1}) + 2f(\overline{x}_n) \\ &= T_{2n}\end{aligned}$$

40. $T_n = \dfrac{\Delta x}{2}\left[f(x_0) + 2\sum\limits_{i=1}^{n-1} f(x_i) + f(x_n)\right]$ and $M_n = \Delta x \sum\limits_{i=1}^{n} f\left(x_i - \dfrac{\Delta x}{2}\right)$, so

$$\tfrac{1}{3}T_n + \tfrac{2}{3}M_n = \tfrac{1}{3}(T_n + 2M_n) = \frac{\Delta x}{3\cdot 2}\left[f(x_0) + 2\sum_{i=1}^{n-1} f(x_i) + f(x_n) + 4\sum_{i=1}^{n} f\left(x_i - \frac{\Delta x}{2}\right)\right]$$

where $\Delta x = \dfrac{b-a}{n}$. Let $\delta x = \dfrac{b-a}{2n}$. Then $\Delta x = 2\delta x$, so

$$\begin{aligned}\tfrac{1}{3}T_n + \tfrac{2}{3}M_n &= \frac{\delta x}{3}\left[f(x_0) + 2\sum_{i=1}^{n-1} f(x_i) + f(x_n) + 4\sum_{i=1}^{n} f(x_i - \delta x)\right] \\ &= \tfrac{1}{3}\delta x[f(x_0) + 4f(x_1 - \delta x) + 2f(x_1) + 4f(x_2 - \delta x) \\ &\qquad + 2f(x_2) + \cdots + 2f(x_{n-1}) + 4f(x_n - \delta x) + f(x_n)]\end{aligned}$$

Since $x_0, x_1 - \delta x, x_1, x_2 - \delta x, x_2, \ldots, x_{n-1}, x_n - \delta x, x_n$ are the subinterval endpoints for S_{2n}, and since $\delta x = \dfrac{b-a}{2n}$ is the width of the subintervals for S_{2n}, the last expression for $\frac{1}{3}T_n + \frac{2}{3}M_n$ is the usual expression for S_{2n}. Therefore, $\frac{1}{3}T_n + \frac{2}{3}M_n = S_{2n}$.

6.6 Improper Integrals

1. (a) Since $\int_1^\infty x^4 e^{-x^4}\,dx$ has an infinite interval of integration, it is an improper integral of Type I.

(b) Since $y = \sec x$ has an infinite discontinuity at $x = \frac{\pi}{2}$, $\int_0^{\pi/2} \sec x\,dx$ is a Type II improper integral.

(c) Since $y = \dfrac{x}{(x-2)(x-3)}$ has an infinite discontinuity at $x = 2$, $\displaystyle\int_0^2 \frac{x}{x^2 - 5x + 6}\,dx$ is a Type II improper integral.

(d) Since $\displaystyle\int_{-\infty}^0 \frac{1}{x^2+5}\,dx$ has an infinite interval of integration, it is an improper integral of Type I.

2. (a) Since $y = \dfrac{1}{2x-1}$ is defined and continuous on $[1, 2]$, $\displaystyle\int_1^2 \frac{1}{2x-1}\,dx$ is proper.

(b) Since $y = \dfrac{1}{2x-1}$ has an infinite discontinuity at $x = \frac{1}{2}$, $\displaystyle\int_0^1 \frac{1}{2x-1}\,dx$ is a Type II improper integral.

(c) Since $\displaystyle\int_{-\infty}^\infty \frac{\sin x}{1+x^2}\,dx$ has an infinite interval of integration, it is an improper integral of Type I.

(d) Since $y = \ln(x-1)$ has an infinite discontinuity at $x = 1$, $\int_1^2 \ln(x-1)\,dx$ is a Type II improper integral.

3. The area under the graph of $y = 1/x^3 = x^{-3}$ between $x = 1$ and $x = t$ is

$A(t) = \int_1^t x^{-3}\,dx = \left[-\frac{1}{2}x^{-2}\right]_1^t = -\frac{1}{2}t^{-2} - \left(-\frac{1}{2}\right) = \frac{1}{2} - 1/(2t^2)$. So the area for $1 \le x \le 10$ is

$A(10) = 0.5 - 0.005 = 0.495$, the area for $1 \le x \le 100$ is $A(100) = 0.5 - 0.00005 = 0.49995$, and the area for

$1 \le x \le 1000$ is $A(1000) = 0.5 - 0.0000005 = 0.4999995$. The total area under the curve for $x \ge 1$ is

$\lim_{t\to\infty} A(t) = \lim_{t\to\infty} \left[\frac{1}{2} - 1/(2t^2)\right] = \frac{1}{2}$.

4. (a)

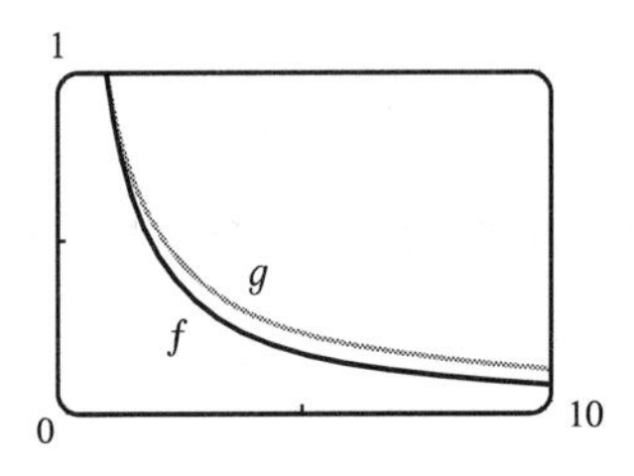

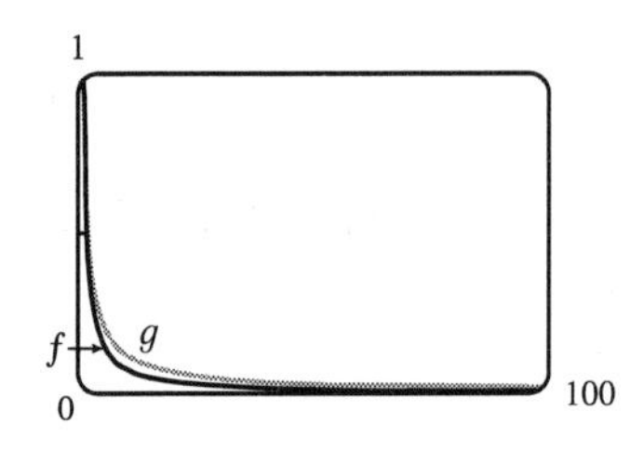

(b) The area under the graph of f from $x = 1$ to $x = t$ is

$$F(t) = \int_1^t f(x)\,dx = \int_1^t x^{-1.1}dx = \left[-\frac{1}{0.1}x^{-0.1}\right]_1^t$$

$$= -10(t^{-0.1} - 1) = 10(1 - t^{-0.1})$$

and the area under the graph of g is

$$G(t) = \int_1^t g(x)dx = \int_1^t x^{-0.9}dx = \left[\frac{1}{0.1}x^{0.1}\right]_1^t = 10(t^{0.1} - 1)$$

t	$F(t)$	$G(t)$
10	2.06	2.59
100	3.69	5.85
10^4	6.02	15.12
10^6	7.49	29.81
10^{10}	9	90
10^{20}	9.9	990

(c) The total area under the graph of f is $\lim_{t\to\infty} F(t) = \lim_{t\to\infty} 10(1 - t^{-0.1}) = 10$.

The total area under the graph of g does not exist, since $\lim_{t\to\infty} G(t) = \lim_{t\to\infty} 10(t^{0.1} - 1) = \infty$.

5. $I = \int_1^\infty \frac{1}{(3x+1)^2}\,dx = \lim_{t\to\infty} \int_1^t \frac{1}{(3x+1)^2}\,dx$. Now

$$\int \frac{1}{(3x+1)^2}\,dx = \frac{1}{3}\int \frac{1}{u^2}\,du \quad [u = 3x+1,\, du = 3\,dx] \quad = -\frac{1}{3u} + C = -\frac{1}{3(3x+1)} + C,$$

so $I = \lim_{t\to\infty} \left[-\frac{1}{3(3x+1)}\right]_1^t = \lim_{t\to\infty} \left[-\frac{1}{3(3t+1)} + \frac{1}{12}\right] = 0 + \frac{1}{12} = \frac{1}{12}$. Convergent

6. $\int_{-\infty}^0 \frac{1}{2x-5}\,dx = \lim_{t\to-\infty} \int_t^0 \frac{1}{2x-5}\,dx = \lim_{t\to-\infty} \left[\frac{1}{2}\ln|2x-5|\right]_t^0 = \lim_{t\to\infty} \left[\frac{1}{2}\ln 5 - \frac{1}{2}\ln|2t-5|\right] = -\infty$. Divergent

7. $\int_{-\infty}^{-1} \frac{1}{\sqrt{2-w}}\,dw = \lim_{t\to-\infty} \int_t^{-1} \frac{1}{\sqrt{2-w}}\,dw = \lim_{t\to-\infty} \left[-2\sqrt{2-w}\right]_t^{-1}$ $\quad [u = 2-w,\, du = -dw]$

$= \lim_{t\to-\infty} \left[-2\sqrt{3} + 2\sqrt{2-t}\right] = \infty$. Divergent

8. $\int_0^\infty \frac{x}{(x^2+2)^2}\,dx = \lim_{t\to\infty} \int_0^t \frac{x}{(x^2+2)^2}\,dx = \lim_{t\to\infty} \frac{1}{2}\left[\frac{-1}{x^2+2}\right]_0^t = \frac{1}{2}\lim_{t\to\infty} \left(\frac{-1}{t^2+2} + \frac{1}{2}\right) = \frac{1}{2}\left(0 + \frac{1}{2}\right) = \frac{1}{4}$.

Convergent

9. $\int_4^\infty e^{-y/2}\,dy = \lim\limits_{t\to\infty} \int_4^t e^{-y/2}\,dy = \lim\limits_{t\to\infty} \left[-2e^{-y/2}\right]_4^t = \lim\limits_{t\to\infty} (-2e^{-t/2} + 2e^{-2}) = 0 + 2e^{-2} = 2e^{-2}$. Convergent

10. $\int_{-\infty}^{-1} e^{-2t}\,dt = \lim\limits_{x\to-\infty} \int_x^{-1} e^{-2t}\,dt = \lim\limits_{x\to-\infty} \left[-\frac{1}{2}e^{-2t}\right]_x^{-1} = \lim\limits_{x\to-\infty} \left[-\frac{1}{2}e^2 + \frac{1}{2}e^{-2x}\right] = \infty$. Divergent

11. $\int_{2\pi}^\infty \sin\theta\,d\theta = \lim\limits_{t\to\infty} \int_{2\pi}^t \sin\theta\,d\theta = \lim\limits_{t\to\infty} \left[-\cos\theta\right]_{2\pi}^t = \lim\limits_{t\to\infty} (-\cos t + 1)$.

This limit does not exist, so the integral is divergent. Divergent

12. $I = \int_{-\infty}^\infty (2 - v^4)\,dv = I_1 + I_2 = \int_{-\infty}^0 (2 - v^4)\,dv + \int_0^\infty (2 - v^4)\,dv$, but

$I_1 = \lim\limits_{t\to-\infty} \left[2v - \frac{1}{5}v^5\right]_t^0 = \lim\limits_{t\to-\infty} \left(-2t + \frac{1}{5}t^5\right) = -\infty$. Since I_1 is divergent, I is divergent, and there is no need to evaluate I_2. Divergent

13. $\int_{-\infty}^\infty xe^{-x^2}\,dx = \int_{-\infty}^0 xe^{-x^2}\,dx + \int_0^\infty xe^{-x^2}\,dx$.

$\int_{-\infty}^0 xe^{-x^2}\,dx = \lim\limits_{t\to-\infty} \left(-\frac{1}{2}\right)\left[e^{-x^2}\right]_t^0 = \lim\limits_{t\to-\infty} \left(-\frac{1}{2}\right)\left(1 - e^{-t^2}\right) = -\frac{1}{2}\cdot 1 = -\frac{1}{2}$, and

$\int_0^\infty xe^{-x^2}\,dx = \lim\limits_{t\to\infty} \left(-\frac{1}{2}\right)\left[e^{-x^2}\right]_0^t = \lim\limits_{t\to\infty} \left(-\frac{1}{2}\right)\left(e^{-t^2} - 1\right) = -\frac{1}{2}\cdot(-1) = \frac{1}{2}$. Therefore,

$\int_{-\infty}^\infty xe^{-x^2}\,dx = -\frac{1}{2} + \frac{1}{2} = 0$. Convergent

14. $\int_{-\infty}^\infty x^2e^{-x^3}\,dx = \int_{-\infty}^0 x^2e^{-x^3}\,dx + \int_0^\infty x^2e^{-x^3}\,dx$, and

$\int_{-\infty}^0 x^2e^{-x^3}\,dx = \lim\limits_{t\to-\infty} \left[-\frac{1}{3}e^{-x^3}\right]_t^0 = -\frac{1}{3} + \frac{1}{3}\left(\lim\limits_{t\to-\infty} e^{-t^3}\right) = \infty$. Divergent

15. $$\int_0^\infty se^{-5s}\,ds = \lim_{t\to\infty} \int_0^t se^{-5s}\,ds = \lim_{t\to\infty} \left[-\frac{1}{5}se^{-5s} - \frac{1}{25}e^{-5s}\right]_0^t \quad \left[\begin{array}{l}\text{by integration by}\\ \text{parts with } u = s\end{array}\right]$$

$$= \lim_{t\to\infty} \left(-\tfrac{1}{5}te^{-5t} - \tfrac{1}{25}e^{-5t} + \tfrac{1}{25}\right) = 0 - 0 + \tfrac{1}{25} \quad \text{[by l'Hospital's Rule]}$$

$$= \tfrac{1}{25}$$ Convergent

16. $I = \int_{-\infty}^\infty \cos\pi t\,dt = I_1 + I_2 = \int_{-\infty}^0 \cos\pi t\,dt + \int_0^\infty \cos\pi t\,dt$, but $I_1 = \lim\limits_{s\to-\infty} \left[\frac{1}{\pi}\sin\pi t\right]_s^0 = \lim\limits_{s\to-\infty} \left(-\frac{1}{\pi}\sin\pi t\right)$

and this limit does not exist. Since I_1 is divergent, I is divergent, and there is no need to evaluate I_2. Divergent

17. $\displaystyle\int_1^\infty \frac{\ln x}{x}\,dx = \lim_{t\to\infty} \left[\frac{(\ln x)^2}{2}\right]_1^t$ [by substitution with $u = \ln x$, $du = dx/x$] $\displaystyle= \lim_{t\to\infty} \frac{(\ln t)^2}{2} = \infty$. Divergent

18. $$\int_{-\infty}^6 re^{r/3}\,dr = \lim_{t\to-\infty} \int_t^6 re^{r/3}\,dr = \lim_{t\to-\infty} \left[3re^{r/3} - 9e^{r/3}\right]_t^6 \quad \left[\begin{array}{l}\text{by integration by}\\ \text{parts with } u = r\end{array}\right]$$

$$= \lim_{t\to-\infty} (18e^2 - 9e^2 - 3te^{t/3} + 9e^{t/3}) = 9e^2 - 0 + 0 \quad \text{[by l'Hospital's Rule]}$$

$$= 9e^2$$ Convergent

19. Integrate by parts with $u = \ln x$, $dv = dx/x^2 \Rightarrow du = dx/x$, $v = -1/x$.

$$\int_1^\infty \frac{\ln x}{x^2}\,dx = \lim_{t\to\infty}\int_1^t \frac{\ln x}{x^2}\,dx = \lim_{t\to\infty}\left[-\frac{\ln x}{x} - \frac{1}{x}\right]_1^t = \lim_{t\to\infty}\left(-\frac{\ln t}{t} - \frac{1}{t} + 0 + 1\right) = -0 - 0 + 0 + 1 = 1$$

since $\displaystyle\lim_{t\to\infty}\frac{\ln t}{t} \overset{\text{H}}{=} \lim_{t\to\infty}\frac{1/t}{1} = 0.$ Convergent

20. Integrate by parts with $u = \ln x$, $dv = dx/x^3 \Rightarrow du = dx/x$, $v = -1/(2x^2)$.

$$\int_1^\infty \frac{\ln x}{x^3}\,dx = \lim_{t\to\infty}\int_1^t \frac{\ln x}{x^3}\,dx = \lim_{t\to\infty}\left(\left[-\frac{1}{2x^2}\ln x\right]_1^t + \frac{1}{2}\int_1^t \frac{1}{x^3}\,dx\right) = \lim_{t\to\infty}\left(-\frac{1}{2}\frac{\ln t}{t^2} + 0 - \frac{1}{4t^2} + \frac{1}{4}\right) = \frac{1}{4}$$

since $\displaystyle\lim_{t\to\infty}\frac{\ln t}{t^2} \overset{\text{H}}{=} \lim_{t\to\infty}\frac{1/t}{2t} = \lim_{t\to\infty}\frac{1}{2t^2} = 0.$ Convergent

21. $\displaystyle\int_{-\infty}^\infty \frac{x^2}{9+x^6}\,dx = \int_{-\infty}^0 \frac{x^2}{9+x^6}\,dx + \int_0^\infty \frac{x^2}{9+x^6}\,dx = 2\int_0^\infty \frac{x^2}{9+x^6}\,dx$ [since the integrand is even].

Now $\displaystyle\int \frac{x^2\,dx}{9+x^6}\ \left[\begin{matrix} u = x^3 \\ du = 3x^2\,dx\end{matrix}\right] = \int \frac{\frac{1}{3}\,du}{9+u^2}\ \left[\begin{matrix} u = 3v \\ du = 3\,dv\end{matrix}\right] = \int \frac{\frac{1}{3}(3\,dv)}{9+9v^2} = \frac{1}{9}\int \frac{dv}{1+v^2}$

$$= \frac{1}{9}\tan^{-1}v + C = \frac{1}{9}\tan^{-1}\left(\frac{u}{3}\right) + C = \frac{1}{9}\tan^{-1}\left(\frac{x^3}{3}\right) + C,$$

so $\displaystyle 2\int_0^\infty \frac{x^2}{9+x^6}\,dx = 2\lim_{t\to\infty}\int_0^t \frac{x^2}{9+x^6}\,dx = 2\lim_{t\to\infty}\left[\frac{1}{9}\tan^{-1}\left(\frac{x^3}{3}\right)\right]_0^t = 2\lim_{t\to\infty}\frac{1}{9}\tan^{-1}\left(\frac{t^3}{3}\right) = \frac{2}{9}\cdot\frac{\pi}{2} = \frac{\pi}{9}.$

Convergent

22. $\displaystyle\int_0^\infty \frac{e^x}{e^{2x}+3}\,dx = \lim_{t\to\infty}\int_0^t \frac{e^x}{(e^x)^2 + (\sqrt{3})^2}\,dx = \lim_{t\to\infty}\left[\frac{1}{\sqrt{3}}\arctan\frac{e^x}{\sqrt{3}}\right]_0^t = \frac{1}{\sqrt{3}}\lim_{t\to\infty}\left(\arctan\frac{e^t}{\sqrt{3}} - \arctan\frac{1}{\sqrt{3}}\right)$

$$= \frac{1}{\sqrt{3}}\left(\frac{\pi}{2} - \frac{\pi}{6}\right) = \frac{1}{\sqrt{3}}\left(\frac{\pi}{3}\right) = \frac{\pi\sqrt{3}}{9}.$$ Convergent

23. $\displaystyle\int_0^1 \frac{3}{x^5}\,dx = \lim_{t\to 0^+}\int_t^1 3x^{-5}\,dx = \lim_{t\to 0^+}\left[-\frac{3}{4x^4}\right]_t^1 = -\frac{3}{4}\lim_{t\to 0^+}\left(1 - \frac{1}{t^4}\right) = \infty.$ Divergent

24. $\displaystyle\int_2^3 \frac{1}{\sqrt{3-x}}\,dx = \lim_{t\to 3^-}\int_2^t (3-x)^{-1/2}\,dx = \lim_{t\to 3^-}\left[-2(3-x)^{1/2}\right]_2^t = -2\lim_{t\to 3^-}\left(\sqrt{3-t} - \sqrt{1}\right) = -2(0-1) = 2.$

Convergent

25. $\displaystyle\int_{-2}^{14} \frac{dx}{\sqrt[4]{x+2}} = \lim_{t\to -2^+}\int_t^{14} (x+2)^{-1/4}\,dx = \lim_{t\to -2^+}\left[\frac{4}{3}(x+2)^{3/4}\right]_t^{14} = \frac{4}{3}\lim_{t\to -2^+}\left[16^{3/4} - (t+2)^{3/4}\right]$

$= \frac{4}{3}(8-0) = \frac{32}{3}$ Convergent

26. $\displaystyle\int_6^8 \frac{4}{(x-6)^3}\,dx = \lim_{t\to 6^+}\int_t^8 4(x-6)^{-3}\,dx = \lim_{t\to 6^+}\left[-2(x-6)^{-2}\right]_t^8 = -2\lim_{t\to 6^+}\left[\frac{1}{2^2} - \frac{1}{(t-6)^2}\right] = \infty.$ Divergent

27. There is an infinite discontinuity at $x = 1$. $\int_0^{33} (x-1)^{-1/5}\,dx = \int_0^1 (x-1)^{-1/5}\,dx + \int_1^{33} (x-1)^{-1/5}\,dx$.

Here $\int_0^1 (x-1)^{-1/5}\,dx = \lim\limits_{t\to 1^-} \int_0^t (x-1)^{-1/5}\,dx = \lim\limits_{t\to 1^-} \left[\frac{5}{4}(x-1)^{4/5}\right]_0^t = \lim\limits_{t\to 1^-} \left[\frac{5}{4}(t-1)^{4/5} - \frac{5}{4}\right] = -\frac{5}{4}$

and $\int_1^{33} (x-1)^{-1/5}\,dx = \lim\limits_{t\to 1^+} \int_t^{33} (x-1)^{-1/5}\,dx = \lim\limits_{t\to 1^+} \left[\frac{5}{4}(x-1)^{4/5}\right]_t^{33} = \lim\limits_{t\to 1^+} \left[\frac{5}{4}\cdot 16 - \frac{5}{4}(t-1)^{4/5}\right] = 20.$

Thus, $\int_0^{33} (x-1)^{-1/5}\,dx = -\frac{5}{4} + 20 = \frac{75}{4}$. Convergent

28. $f(y) = 1/(4y-1)$ has an infinite discontinuity at $y = \frac{1}{4}$.

$$\int_{1/4}^1 \frac{1}{4y-1}\,dy = \lim_{t\to(1/4)^+} \int_t^1 \frac{1}{4y-1}\,dy = \lim_{t\to(1/4)^+} \left[\tfrac{1}{4}\ln|4y-1|\right]_t^1 = \lim_{t\to(1/4)^+} \left[\tfrac{1}{4}\ln 3 - \tfrac{1}{4}\ln(4t-1)\right] = \infty$$

so $\displaystyle\int_{1/4}^1 \frac{1}{4y-1}\,dy$ diverges, and hence, $\displaystyle\int_0^1 \frac{1}{4y-1}\,dy$ diverges. Divergent

29. There is an infinite discontinuity at $x = 0$. $\displaystyle\int_{-1}^1 \frac{e^x}{e^x-1}\,dx = \int_{-1}^0 \frac{e^x}{e^x-1}\,dx + \int_0^1 \frac{e^x}{e^x-1}\,dx$.

$$\int_{-1}^0 \frac{e^x}{e^x-1}\,dx = \lim_{t\to 0^-} \int_{-1}^t \frac{e^x}{e^x-1}\,dx = \lim_{t\to 0^-} \Big[\ln|e^x-1|\Big]_{-1}^t = \lim_{t\to 0^-} \Big[\ln|e^t-1| - \ln|e^{-1}-1|\Big] = -\infty,$$

so $\displaystyle\int_{-1}^1 \frac{e^x}{e^x-1}\,dx$ is divergent. The integral $\displaystyle\int_0^1 \frac{e^x}{e^x-1}\,dx$ also diverges since

$$\int_0^1 \frac{e^x}{e^x-1}\,dx = \lim_{t\to 0^+} \int_t^1 \frac{e^x}{e^x-1}\,dx = \lim_{t\to 0^+} \Big[\ln|e^x-1|\Big]_t^1 = \lim_{t\to 0^+} \Big[\ln|e-1| - \ln|e^t-1|\Big] = \infty.$$ Divergent

30. $\displaystyle\int_0^1 \frac{dx}{\sqrt{1-x^2}} = \lim_{t\to 1^-} \int_0^t \frac{dx}{\sqrt{1-x^2}} = \lim_{t\to 1^-} \left[\sin^{-1} x\right]_0^t = \lim_{t\to 1^-} \sin^{-1} t = \frac{\pi}{2}$. Convergent

31. $I = \int_0^2 z^2 \ln z\,dz = \lim\limits_{t\to 0^+} \int_t^2 z^2 \ln z\,dz \overset{101}{=} \lim\limits_{t\to 0^+} \left[\dfrac{z^3}{3^2}(3\ln z - 1)\right]_t^2 = \lim\limits_{t\to 0^+} \left[\frac{8}{9}(3\ln 2 - 1) - \frac{1}{9}t^3(3\ln t - 1)\right]$

$= \frac{8}{3}\ln 2 - \frac{8}{9} - \frac{1}{9}\lim\limits_{t\to 0^+} \left[t^3(3\ln t - 1)\right] = \frac{8}{3}\ln 2 - \frac{8}{9} - \frac{1}{9}L$

Now $L = \lim\limits_{t\to 0^+} \left[t^3(3\ln t - 1)\right] = \lim\limits_{t\to 0^+} \dfrac{3\ln t - 1}{t^{-3}} \overset{\text{H}}{=} \lim\limits_{t\to 0^+} \dfrac{3/t}{-3/t^4} = \lim\limits_{t\to 0^+} (-t^3) = 0.$

Thus, $L = 0$ and $I = \frac{8}{3}\ln 2 - \frac{8}{9}$. Convergent

32. Integrate by parts with $u = \ln x$, $dv = dx/\sqrt{x}$ $\Rightarrow$ $du = dx/x$, $v = 2\sqrt{x}$.

$$\int_0^1 \frac{\ln x}{\sqrt{x}}\,dx = \lim_{t\to 0^+} \int_t^1 \frac{\ln x}{\sqrt{x}}\,dx = \lim_{t\to 0^+} \left(\left[2\sqrt{x}\ln x\right]_t^1 - 2\int_t^1 \frac{dx}{\sqrt{x}}\right) = \lim_{t\to 0^+} \left(-2\sqrt{t}\ln t - 4\left[\sqrt{x}\right]_t^1\right)$$

$$= \lim_{t\to 0^+} \left(-2\sqrt{t}\ln t - 4 + 4\sqrt{t}\right) = -4$$

since $\lim\limits_{t\to 0^+} \sqrt{t}\ln t = \lim\limits_{t\to 0^+} \dfrac{\ln t}{t^{-1/2}} \overset{\text{H}}{=} \lim\limits_{t\to 0^+} \dfrac{1/t}{-t^{-3/2}/2} = \lim\limits_{t\to 0^+} \left(-2\sqrt{t}\right) = 0.$ Convergent

33.

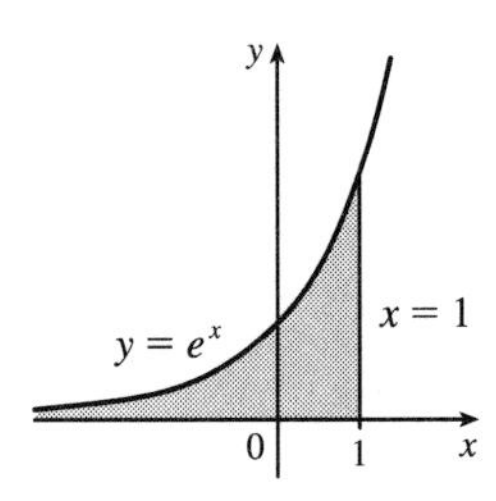

$\text{Area} = \int_{-\infty}^{1} e^x\,dx = \lim_{t\to-\infty}\left[e^x\right]_t^1 = e - \lim_{t\to-\infty} e^t = e$

34.

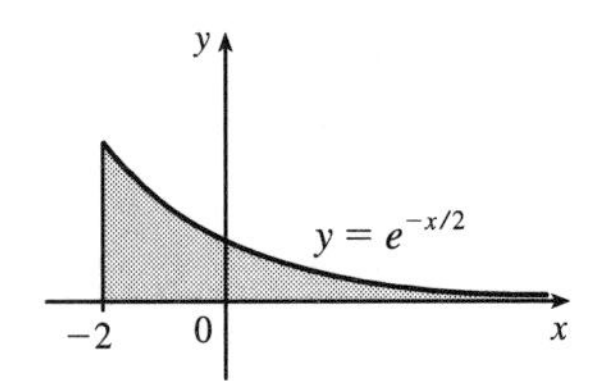

$\text{Area} = \int_{-2}^{\infty} e^{-x/2}\,dx = -2\lim_{t\to\infty}\left[e^{-x/2}\right]_{-2}^{t} = -2\lim_{t\to\infty} e^{-t/2} + 2e = 2e$

35.

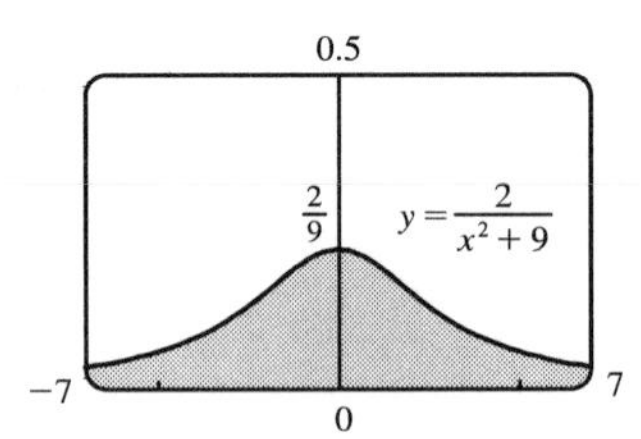

$$\text{Area} = \int_{-\infty}^{\infty} \frac{2}{x^2+9}\,dx = 2\cdot 2\int_0^{\infty}\frac{1}{x^2+9}\,dx$$

$$= 4\lim_{t\to\infty}\int_0^t \frac{1}{x^2+9}\,dx = 4\lim_{t\to\infty}\left[\frac{1}{3}\tan^{-1}\frac{x}{3}\right]_0^t$$

$$= \frac{4}{3}\lim_{t\to\infty}\left[\tan^{-1}\frac{t}{3} - 0\right] = \frac{4}{3}\cdot\frac{\pi}{2} = \frac{2\pi}{3}$$

36.

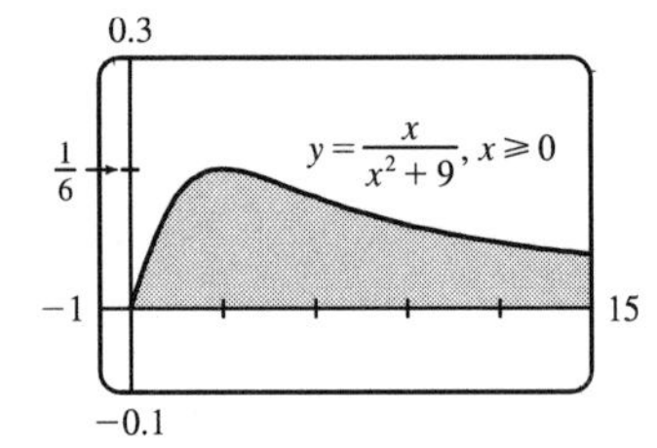

$$\text{Area} = \int_0^{\infty}\frac{x}{x^2+9}\,dx = \lim_{t\to\infty}\int_0^t \frac{x}{x^2+9}\,dx$$

$$= \lim_{t\to\infty}\left[\tfrac{1}{2}\ln(x^2+9)\right]_0^t$$

$$= \tfrac{1}{2}\lim_{t\to\infty}\left[\ln(t^2+9) - \ln 9\right] = \infty$$

Infinite area

37.

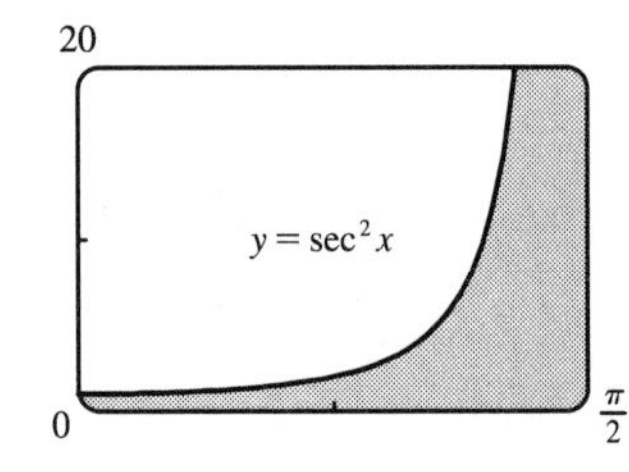

$$\text{Area} = \int_0^{\pi/2}\sec^2 x\,dx = \lim_{t\to(\pi/2)^-}\int_0^t \sec^2 x\,dx$$

$$= \lim_{t\to(\pi/2)^-}\left[\tan x\right]_0^t = \lim_{t\to(\pi/2)^-}(\tan t - 0)$$

$$= \infty$$

Infinite area

38.

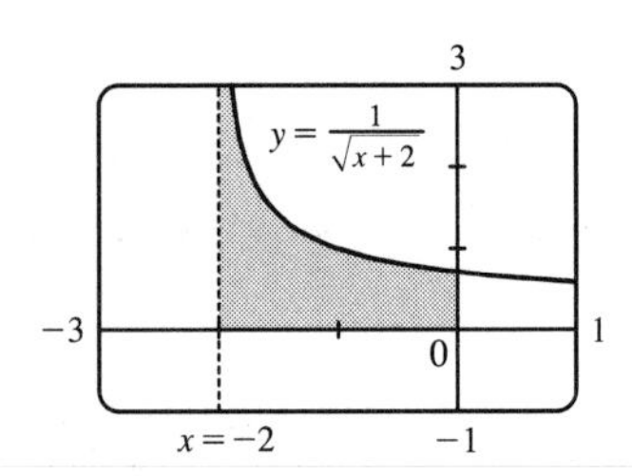

$$\text{Area} = \int_{-2}^{0}\frac{1}{\sqrt{x+2}}\,dx = \lim_{t\to-2^+}\int_t^0 \frac{1}{\sqrt{x+2}}\,dx$$

$$= \lim_{t\to-2^+}\left[2\sqrt{x+2}\right]_t^0 = \lim_{t\to-2^+}\left(2\sqrt{2} - 2\sqrt{t+2}\right)$$

$$= 2\sqrt{2} - 0 = 2\sqrt{2}$$

39. (a)

t	$\int_1^t g(x)\,dx$
2	0.447453
5	0.577101
10	0.621306
100	0.668479
1000	0.672957
10,000	0.673407

$g(x) = \dfrac{\sin^2 x}{x^2}$.

It appears that the integral is convergent.

(b) $-1 \le \sin x \le 1 \;\Rightarrow\; 0 \le \sin^2 x \le 1 \;\Rightarrow\; 0 \le \dfrac{\sin^2 x}{x^2} \le \dfrac{1}{x^2}$. Since $\displaystyle\int_1^\infty \frac{1}{x^2}\,dx$ is convergent (Equation 2 with $p = 2 > 1$), $\displaystyle\int_1^\infty \frac{\sin^2 x}{x^2}\,dx$ is convergent by the Comparison Theorem.

(c)

1

$f(x) = \frac{1}{x^2}$

$g(x) = \frac{\sin^2 x}{x^2}$

1 10

-0.1

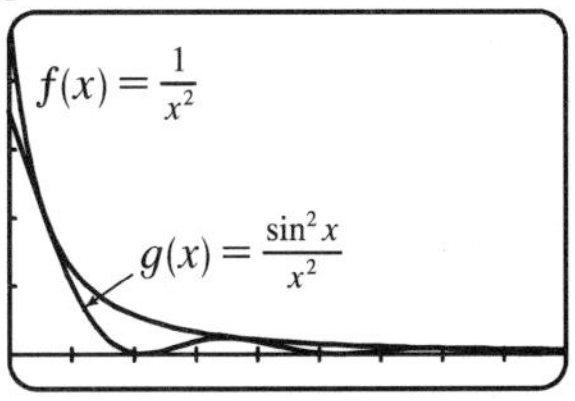

Since $\int_1^\infty f(x)\,dx$ is finite and the area under $g(x)$ is less than the area under $f(x)$ on any interval $[1, t]$, $\int_1^\infty g(x)\,dx$ must be finite; that is, the integral is convergent.

40. (a)

t	$\int_2^t g(x)dx$
5	3.830327
10	6.801200
100	23.328769
1000	69.023361
10,000	208.124560

$g(x) = \dfrac{1}{\sqrt{x} - 1}$.

It appears that the integral is divergent.

(b) For $x \ge 2$, $\sqrt{x} > \sqrt{x} - 1 \;\Rightarrow\; \dfrac{1}{\sqrt{x}} < \dfrac{1}{\sqrt{x} - 1}$. Since $\displaystyle\int_2^\infty \frac{1}{\sqrt{x}}\,dx$ is divergent (Equation 2 with $p = \frac{1}{2} \le 1$), $\displaystyle\int_2^\infty \frac{1}{\sqrt{x} - 1}\,dx$ is divergent by the Comparison Theorem.

(c)

2.5

$g(x) = \frac{1}{\sqrt{x} - 1}$

$f(x) = \frac{1}{\sqrt{x}}$

2 20

-0.5

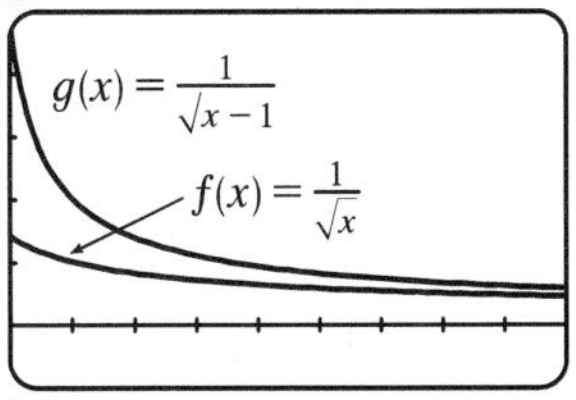

Since $\int_2^\infty f(x)\,dx$ is infinite and the area under $g(x)$ is greater than the area under $f(x)$ on any interval $[2, t]$, $\int_2^\infty g(x)\,dx$ must be infinite; that is, the integral is divergent.

41. For $x \ge 1$, $\dfrac{\cos^2 x}{1 + x^2} \le \dfrac{1}{1 + x^2} < \dfrac{1}{x^2}$. $\displaystyle\int_1^\infty \frac{1}{x^2}\,dx$ is convergent by (2) with $p = 2 > 1$, so $\displaystyle\int_1^\infty \frac{\cos^2 x}{1 + x^2}\,dx$ is convergent by the Comparison Theorem.

42. For $x \ge 1$, $\dfrac{2 + e^{-x}}{x} > \dfrac{2}{x}$ [since $e^{-x} > 0$] $> \dfrac{1}{x}$. $\displaystyle\int_1^\infty \frac{1}{x}\,dx$ is divergent by (2) with $p = 1 \le 1$, so $\displaystyle\int_1^\infty \frac{2 + e^{-x}}{x}\,dx$ is divergent by the Comparison Theorem.

43. For $x \geq 1$, $x + e^{2x} > e^{2x} > 0 \Rightarrow \dfrac{1}{x + e^{2x}} \leq \dfrac{1}{e^{2x}} = e^{-2x}$ on $[1, \infty)$.

$\int_1^\infty e^{-2x}\,dx = \lim\limits_{t\to\infty}\left[-\frac{1}{2}e^{-2x}\right]_1^t = \lim\limits_{t\to\infty}\left[-\frac{1}{2}e^{-2t} + \frac{1}{2}e^{-2}\right] = \frac{1}{2}e^{-2}$. Therefore, $\int_1^\infty e^{-2x}\,dx$ is convergent, and by the Comparison Theorem, $\displaystyle\int_1^\infty \frac{dx}{x + e^{2x}}$ is also convergent.

44. For $x \geq 1$, $0 < \dfrac{x}{\sqrt{1+x^6}} < \dfrac{x}{\sqrt{x^6}} = \dfrac{x}{x^3} = \dfrac{1}{x^2}$. $\displaystyle\int_1^\infty \frac{1}{x^2}\,dx$ is convergent by (2) with $p = 2 > 1$, so $\displaystyle\int_1^\infty \frac{x}{\sqrt{1+x^6}}\,dx$ is convergent by the Comparison Theorem.

45. $\dfrac{1}{x\sin x} \geq \dfrac{1}{x}$ on $\left(0, \frac{\pi}{2}\right]$ since $0 \leq \sin x \leq 1$. $\displaystyle\int_0^{\pi/2}\frac{dx}{x} = \lim_{t\to 0^+}\int_t^{\pi/2}\frac{dx}{x} = \lim_{t\to 0^+}\left[\ln x\right]_t^{\pi/2}$.

But $\ln t \to -\infty$ as $t \to 0^+$, so $\displaystyle\int_0^{\pi/2}\frac{dx}{x}$ is divergent, and by the Comparison Theorem, $\displaystyle\int_0^{\pi/2}\frac{dx}{x\sin x}$ is also divergent.

46. For $0 \leq x \leq 1$, $e^{-x} \leq 1 \Rightarrow \dfrac{e^{-x}}{\sqrt{x}} \leq \dfrac{1}{\sqrt{x}}$.

$$\int_0^1 \frac{1}{\sqrt{x}}\,dx = \lim_{t\to 0^+}\int_t^1 \frac{1}{\sqrt{x}}\,dx = \lim_{t\to 0^+}\left[2\sqrt{x}\right]_t^1 = \lim_{t\to 0^+}\left(2 - 2\sqrt{t}\right) = 2 \text{ is convergent.}$$

Therefore, $\displaystyle\int_0^1 \frac{e^{-x}}{\sqrt{x}}\,dx$ is convergent by the Comparison Theorem.

47. $$\int_0^\infty \frac{dx}{\sqrt{x}\,(1+x)} = \int_0^1 \frac{dx}{\sqrt{x}\,(1+x)} + \int_1^\infty \frac{dx}{\sqrt{x}\,(1+x)} = \lim_{t\to 0^+}\int_t^1 \frac{dx}{\sqrt{x}\,(1+x)} + \lim_{t\to\infty}\int_1^t \frac{dx}{\sqrt{x}\,(1+x)}.$$

Now $$\int \frac{dx}{\sqrt{x}\,(1+x)} = \int \frac{2u\,du}{u(1+u^2)} \qquad [u = \sqrt{x}, x = u^2, dx = 2u\,du]$$
$$= 2\int \frac{du}{1+u^2} = 2\tan^{-1}u + C = 2\tan^{-1}\sqrt{x} + C,$$

so $$\int_0^\infty \frac{dx}{\sqrt{x}\,(1+x)} = \lim_{t\to 0^+}\left[2\tan^{-1}\sqrt{x}\right]_t^1 + \lim_{t\to\infty}\left[2\tan^{-1}\sqrt{x}\right]_1^t$$
$$= \lim_{t\to 0^+}\left[2\left(\tfrac{\pi}{4}\right) - 2\tan^{-1}\sqrt{t}\right] + \lim_{t\to\infty}\left[2\tan^{-1}\sqrt{t} - 2\left(\tfrac{\pi}{4}\right)\right] = \tfrac{\pi}{2} - 0 + 2\left(\tfrac{\pi}{2}\right) - \tfrac{\pi}{2} = \pi.$$

48. Let $u = \ln x$. Then $du = dx/x \Rightarrow \displaystyle\int_e^\infty \frac{dx}{x(\ln x)^p} = \int_1^\infty \frac{du}{u^p}$. By Example 4, this converges to $\dfrac{1}{p-1}$ if $p > 1$ and diverges otherwise.

49. If $p = 1$, then $\displaystyle\int_0^1 \frac{dx}{x^p} = \lim_{t\to 0^+}\int_t^1 \frac{dx}{x} = \lim_{t\to 0^+}\left[\ln x\right]_t^1 = \infty$. Divergent.

If $p \neq 1$, then $\displaystyle\int_0^1 \frac{dx}{x^p} = \lim_{t\to 0^+}\int_t^1 \frac{dx}{x^p}$ [note that the integral is not improper if $p < 0$] $\displaystyle = \lim_{t\to 0^+}\left[\frac{x^{-p+1}}{-p+1}\right]_t^1 = \lim_{t\to 0^+}\frac{1}{1-p}\left[1 - \frac{1}{t^{p-1}}\right]$

If $p > 1$, then $p - 1 > 0$, so $\dfrac{1}{t^{p-1}} \to \infty$ as $t \to 0^+$, and the integral diverges.

If $p < 1$, then $p - 1 < 0$, so $\dfrac{1}{t^{p-1}} \to 0$ as $t \to 0^+$ and $\displaystyle\int_0^1 \frac{dx}{x^p} = \frac{1}{1-p}\left[\lim_{t\to 0^+}\left(1 - t^{1-p}\right)\right] = \frac{1}{1-p}$.

Thus, the integral converges if and only if $p < 1$, and in that case its value is $\dfrac{1}{1-p}$.

50. (a) $\boldsymbol{n=0}$: $\displaystyle\int_0^\infty x^n e^{-x}\,dx = \lim_{t\to\infty}\int_0^t e^{-x}\,dx = \lim_{t\to\infty}\left[-e^{-x}\right] = \lim_{t\to\infty}\left[-e^{-t}+1\right] = 0+1 = 1$

$\boldsymbol{n=1}$: $\displaystyle\int_0^\infty x^n e^{-x}\,dx = \lim_{t\to\infty}\int_0^t xe^{-x}\,dx.$

To evaluate $\int xe^{-x}\,dx$, we'll use integration by parts with $u = x$, $dv = e^{-x}\,dx \Rightarrow du = dx$, $v = -e^{-x}$.

So $\int xe^{-x}\,dx = -xe^{-x} - \int -e^{-x}\,dx = -xe^{-x} - e^{-x} + C = (-x-1)e^{-x} + C$ and

$$\lim_{t\to\infty}\int_0^t xe^{-x}\,dx = \lim_{t\to\infty}\left[(-x-1)e^{-x}\right]_0^t = \lim_{t\to\infty}\left[(-t-1)e^{-t}+1\right] = \lim_{t\to\infty}\left[-te^{-t}-e^{-t}+1\right]$$

$$= 0 - 0 + 1 \quad \text{[use l'Hospital's Rule]} \quad = 1$$

$\boldsymbol{n=2}$: $\displaystyle\int_0^\infty x^n e^{-x}\,dx = \lim_{t\to\infty}\int_0^t x^2e^{-x}\,dx.$

To evaluate $\int x^2e^{-x}\,dx$, we could use integration by parts again or Formula 97. Thus,

$$\lim_{t\to\infty}\int_0^t x^2e^{-x}\,dx = \lim_{t\to\infty}\left[-x^2e^{-x}\right]_0^t + 2\lim_{t\to\infty}\int_0^t xe^{-x}\,dx$$

$$= 0 + 0 + 2(1) \quad \text{[use l'Hospital's Rule and the result for } n=1\text{]} \quad = 2$$

$\boldsymbol{n=3}$: $\displaystyle\int_0^\infty x^n e^{-x}\,dx = \lim_{t\to\infty}\int_0^t x^3e^{-x}\,dx \overset{97}{=} \lim_{t\to\infty}\left[-x^3e^{-x}\right]_0^t + 3\lim_{t\to\infty}\int_0^t x^2e^{-x}\,dx$

$$= 0 + 0 + 3(2) \quad \text{[use l'Hospital's Rule and the result for } n=2\text{]} \quad = 6$$

(b) For $n = 1$, 2, and 3, we have $\int_0^\infty x^n e^{-x}\,dx = 1$, 2, and 6. The values for the integral are equal to the factorials for n, so we guess $\int_0^\infty x^n e^{-x}\,dx = n!$.

(c) Suppose that $\int_0^\infty x^k e^{-x}\,dx = k!$ for some positive integer k. Then $\int_0^\infty x^{k+1}e^{-x}\,dx = \lim_{t\to\infty}\int_0^t x^{k+1}e^{-x}\,dx.$

To evaluate $\int x^{k+1}e^{-x}\,dx$, we use parts with $u = x^{k+1}$, $dv = e^{-x}\,dx \Rightarrow du = (k+1)x^k\,dx$, $v = -e^{-x}$.

So $\int x^{k+1}e^{-x}\,dx = -x^{k+1}e^{-x} - \int -(k+1)x^k e^{-x}\,dx = -x^{k+1}e^{-x} + (k+1)\int x^k e^{-x}\,dx$ and

$$\lim_{t\to\infty}\int_0^t x^{k+1}e^{-x}\,dx = \lim_{t\to\infty}\left[-x^{k+1}e^{-x}\right]_0^t + (k+1)\lim_{t\to\infty}\int_0^t x^k e^{-x}\,dx = \lim_{t\to\infty}\left[-t^{k+1}e^{-t}+0\right] + (k+1)k!$$

$$= 0 + 0 + (k+1)! = (k+1)!,$$

so the formula holds for $k+1$. By induction, the formula holds for all positive integers. (Since $0! = 1$, the formula holds for $n = 0$, too.)

51. (a) $I = \int_{-\infty}^\infty x\,dx = \int_{-\infty}^0 x\,dx + \int_0^\infty x\,dx$, and $\int_0^\infty x\,dx = \lim_{t\to\infty}\int_0^t x\,dx = \lim_{t\to\infty}\left[\frac{1}{2}x^2\right]_0^t = \lim_{t\to\infty}\left[\frac{1}{2}t^2 - 0\right] = \infty$, so I is divergent.

(b) $\int_{-t}^t x\,dx = \left[\frac{1}{2}x^2\right]_{-t}^t = \frac{1}{2}t^2 - \frac{1}{2}t^2 = 0$, so $\lim_{t\to\infty}\int_{-t}^t x\,dx = 0$. Therefore, $\int_{-\infty}^\infty x\,dx \neq \lim_{t\to\infty}\int_{-t}^t x\,dx$.

52. Assume without loss of generality that $a < b$. Then

$$\begin{aligned}\int_{-\infty}^{a} f(x)\,dx + \int_{a}^{\infty} f(x)\,dx &= \lim_{t\to-\infty}\int_t^a f(x)\,dx + \lim_{u\to\infty}\int_a^u f(x)\,dx \\ &= \lim_{t\to-\infty}\int_t^a f(x)\,dx + \lim_{u\to\infty}\left[\int_a^b f(x)\,dx + \int_b^u f(x)\,dx\right] \\ &= \lim_{t\to-\infty}\int_t^a f(x)\,dx + \int_a^b f(x)\,dx + \lim_{u\to\infty}\int_b^u f(x)\,dx \\ &= \lim_{t\to-\infty}\left[\int_t^a f(x)\,dx + \int_a^b f(x)\,dx\right] + \int_b^\infty f(x)\,dx \\ &= \lim_{t\to-\infty}\int_t^b f(x)\,dx + \int_b^\infty f(x)\,dx = \int_{-\infty}^b f(x)\,dx + \int_b^\infty f(x)\,dx\end{aligned}$$

53. We would expect a small percentage of bulbs to burn out in the first few hundred hours, most of the bulbs to burn out after close to 700 hours, and a few overachievers to burn on and on.

(a)

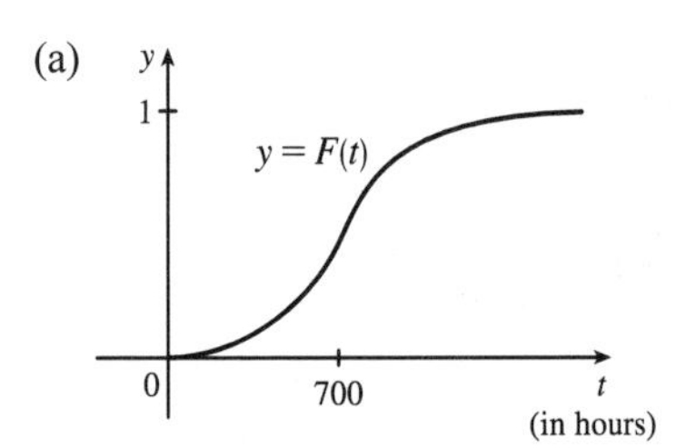

(b) $r(t) = F'(t)$ is the rate at which the fraction $F(t)$ of burnt-out bulbs increases as t increases. This could be interpreted as a fractional burnout rate.

(c) $\int_0^\infty r(t)\,dt = \lim_{x\to\infty} F(x) = 1$, since all of the bulbs will eventually burn out.

54. Let $k = \dfrac{M}{2RT}$ so that $\overline{v} = \dfrac{4}{\sqrt{\pi}}\,k^{3/2}\displaystyle\int_0^\infty v^3 e^{-kv^2}\,dv$. Let I denote the integral and use parts to integrate I.

Let $\alpha = v^2$, $d\beta = ve^{-kv^2}\,dv \quad\Rightarrow\quad d\alpha = 2v\,dv$, $\beta = -\dfrac{1}{2k}e^{-kv^2}$:

$$\begin{aligned}I &= \lim_{t\to\infty}\left[-\frac{1}{2k}v^2e^{-kv^2}\right]_0^t + \frac{1}{k}\int_0^\infty ve^{-kv^2}\,dv_0^t \\ &= -\frac{1}{2k}\lim_{t\to\infty}\left(t^2e^{-kt^2}\right) + \frac{1}{k}\lim_{t\to\infty}\left[-\frac{1}{2k}e^{-kv^2}\right] \\ &\overset{\text{H}}{=} -\frac{1}{2k}\cdot 0 - \frac{1}{2k^2}(0-1) = \frac{1}{2k^2}\end{aligned}$$

Thus, $\overline{v} = \dfrac{4}{\sqrt{\pi}}k^{3/2}\cdot\dfrac{1}{2k^2} = \dfrac{2}{(k\pi)^{1/2}} = \dfrac{2}{[\pi M/(2RT)]^{1/2}} = \dfrac{2\sqrt{2}\sqrt{RT}}{\sqrt{\pi M}} = \sqrt{\dfrac{8RT}{\pi M}}$.

55. $I = \displaystyle\int_0^\infty te^{kt}\,dt = \lim_{s\to\infty}\left[\frac{1}{k^2}(kt-1)\,e^{kt}\right]_0^s$ [Formula 96, or parts] $= \displaystyle\lim_{s\to\infty}\left[\left(\frac{1}{k}se^{ks} - \frac{1}{k^2}e^{ks}\right) - \left(-\frac{1}{k^2}\right)\right]$.

Since $k < 0$ the first two terms approach 0 (you can verify that the first term does so with l'Hospital's Rule), so the limit is equal to $1/k^2$. Thus, $M = -kI = -k(1/k^2) = -1/k = -1/(-0.000121) \approx 8264.5$ years.

56. $y(s) = \int_s^R \frac{2r}{\sqrt{r^2 - s^2}}\, x(r)\, dr$ and $x(r) = \frac{1}{2}(R - r)^2 \Rightarrow$

$$y(s) = \lim_{t \to s^+} \int_t^R \frac{r(R-r)^2}{\sqrt{r^2 - s^2}}\, dr = \lim_{t \to s^+} \int_t^R \frac{r^3 - 2Rr^2 + R^2 r}{\sqrt{r^2 - s^2}}\, dr$$

$$= \lim_{t \to s^+} \left[\int_t^R \frac{r^3\, dr}{\sqrt{r^2 - s^2}} - 2R \int_t^R \frac{r^2\, dr}{\sqrt{r^2 - s^2}} + R^2 \int_t^R \frac{r\, dr}{\sqrt{r^2 - s^2}} \right] = \lim_{t \to s^+} \left(I_1 - 2RI_2 + R^2 I_3\right) = L$$

For I_1: Let $u = \sqrt{r^2 - s^2} \Rightarrow u^2 = r^2 - s^2$, $r^2 = u^2 + s^2$, $2r\, dr = 2u\, du$, so, omitting limits and constant of integration,

$$I_1 = \int \frac{(u^2 + s^2)u}{u}\, du = \int (u^2 + s^2)\, du = \tfrac{1}{3}u^3 + s^2 u = \tfrac{1}{3}u(u^2 + 3s^2)$$

$$= \tfrac{1}{3}\sqrt{r^2 - s^2}\,(r^2 - s^2 + 3s^2) = \tfrac{1}{3}\sqrt{r^2 - s^2}\,(r^2 + 2s^2)$$

For I_2: Using Formula 44, $I_2 = \frac{r}{2}\sqrt{r^2 - s^2} + \frac{s^2}{2}\ln\left|r + \sqrt{r^2 - s^2}\,\right|$.

For I_3: Let $u = r^2 - s^2 \Rightarrow du = 2r\, dr$. Then $I_3 = \frac{1}{2}\int \frac{du}{\sqrt{u}} = \frac{1}{2} \cdot 2\sqrt{u} = \sqrt{r^2 - s^2}$. Thus,

$$L = \lim_{t \to s^+} \left[\tfrac{1}{3}\sqrt{r^2 - s^2}\,(r^2 + 2s^2) - 2R\left(\frac{r}{2}\sqrt{r^2 - s^2} + \frac{s^2}{2}\ln\left|r + \sqrt{r^2 - s^2}\,\right|\right) + R^2\sqrt{r^2 - s^2} \right]_t^R$$

$$= \lim_{t \to s^+} \left[\tfrac{1}{3}\sqrt{R^2 - s^2}\,(R^2 + 2s^2) - 2R\left(\frac{R}{2}\sqrt{R^2 - s^2} + \frac{s^2}{2}\ln\left|R + \sqrt{R^2 - s^2}\,\right|\right) + R^2\sqrt{R^2 - s^2} \right]$$

$$- \lim_{t \to s^+} \left[\tfrac{1}{3}\sqrt{t^2 - s^2}\,(t^2 + 2s^2) - 2R\left(\frac{t}{2}\sqrt{t^2 - s^2} + \frac{s^2}{2}\ln\left|t + \sqrt{t^2 - s^2}\,\right|\right) + R^2\sqrt{t^2 - s^2} \right]$$

$$= \left[\tfrac{1}{3}\sqrt{R^2 - s^2}\,(R^2 + 2s^2) - Rs^2\ln\left|R + \sqrt{R^2 - s^2}\,\right|\right] - \left[-Rs^2\ln|s|\right]$$

$$= \tfrac{1}{3}\sqrt{R^2 - s^2}\,(R^2 + 2s^2) - Rs^2\ln\left(\frac{R + \sqrt{R^2 - s^2}}{s}\right)$$

57. $I = \int_a^\infty \frac{1}{x^2 + 1}\, dx = \lim_{t \to \infty} \int_a^t \frac{1}{x^2 + 1}\, dx = \lim_{t \to \infty} \left[\tan^{-1} x\right]_a^t = \lim_{t \to \infty} (\tan^{-1} t - \tan^{-1} a) = \frac{\pi}{2} - \tan^{-1} a$.

$I < 0.001 \Rightarrow \frac{\pi}{2} - \tan^{-1} a < 0.001 \Rightarrow \tan^{-1} a > \frac{\pi}{2} - 0.001 \Rightarrow a > \tan\left(\frac{\pi}{2} - 0.001\right) \approx 1000$.

58. $f(x) = e^{-x^2}$ and $\Delta x = \frac{4 - 0}{8} = \frac{1}{2}$.

$\int_0^4 f(x)\, dx \approx S_8 = \frac{1}{2 \cdot 3}[f(0) + 4f(0.5) + 2f(1) + \cdots + 2f(3) + 4f(3.5) + f(4)] \approx \frac{1}{6}(5.31717808) \approx 0.8862$.

Now $x > 4 \Rightarrow -x \cdot x < -x \cdot 4 \Rightarrow e^{-x^2} < e^{-4x} \Rightarrow \int_4^\infty e^{-x^2}\, dx < \int_4^\infty e^{-4x}\, dx$.

$\int_4^\infty e^{-4x}\, dx = \lim_{t \to \infty} \left[-\frac{1}{4}e^{-4x}\right]_4^t = -\frac{1}{4}(0 - e^{-16}) = 1/(4e^{16}) \approx 0.0000000281 < 0.0000001$, as desired.

59. We use integration by parts: Let $u = x$, $dv = xe^{-x^2}\,dx \quad \Rightarrow \quad du = dx$, $v = -\frac{1}{2}e^{-x^2}$. So

$$\int_0^\infty x^2 e^{-x^2}\,dx = \lim_{t\to\infty}\left[-\tfrac{1}{2}xe^{-x^2}\right]_0^t + \tfrac{1}{2}\int_0^\infty e^{-x^2}\,dx = \lim_{t\to\infty}\left[-t\big/\left(2e^{t^2}\right)\right] + \tfrac{1}{2}\int_0^\infty e^{-x^2}\,dx = \tfrac{1}{2}\int_0^\infty e^{-x^2}\,dx$$

(The limit is 0 by l'Hospital's Rule.)

60. $\int_0^\infty e^{-x^2}\,dx$ is the area under the curve $y = e^{-x^2}$ for $0 \le x < \infty$ and $0 < y \le 1$. Solving $y = e^{-x^2}$ for x, we get $y = e^{-x^2} \quad \Rightarrow \quad \ln y = -x^2 \quad \Rightarrow \quad -\ln y = x^2 \quad \Rightarrow \quad x = \pm\sqrt{-\ln y}$. Since x is positive, choose $x = \sqrt{-\ln y}$, and the area is represented by $\int_0^1 \sqrt{-\ln y}\,dy$. Therefore, each integral represents the same area, so the integrals are equal.

61. For the first part of the integral, let $x = 2\tan\theta \quad \Rightarrow \quad dx = 2\sec^2\theta\,d\theta$.

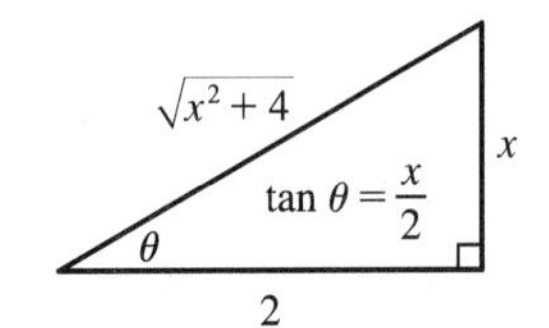

$$\int \frac{1}{\sqrt{x^2+4}}\,dx = \int \frac{2\sec^2\theta}{2\sec\theta}\,d\theta = \int \sec\theta\,d\theta = \ln|\sec\theta + \tan\theta|.$$

From the figure, $\tan\theta = \dfrac{x}{2}$, and $\sec\theta = \dfrac{\sqrt{x^2+4}}{2}$. So

$$\begin{aligned} I = \int_0^\infty \left(\frac{1}{\sqrt{x^2+4}} - \frac{C}{x+2}\right)dx &= \lim_{t\to\infty}\left[\ln\left|\frac{\sqrt{x^2+4}}{2} + \frac{x}{2}\right| - C\ln|x+2|\right]_0^t \\ &= \lim_{t\to\infty}\left[\ln\frac{\sqrt{t^2+4}+t}{2} - C\ln(t+2) - (\ln 1 - C\ln 2)\right] \\ &= \lim_{t\to\infty}\left[\ln\left(\frac{\sqrt{t^2+4}+t}{2\,(t+2)^C}\right) + \ln 2^C\right] = \ln\left(\lim_{t\to\infty}\frac{t+\sqrt{t^2+4}}{(t+2)^C}\right) + \ln 2^{C-1} \end{aligned}$$

Now $L = \displaystyle\lim_{t\to\infty}\frac{t+\sqrt{t^2+4}}{(t+2)^C} \overset{\text{H}}{=} \lim_{t\to\infty}\frac{1+t/\sqrt{t^2+4}}{C\,(t+2)^{C-1}} = \frac{2}{C\displaystyle\lim_{t\to\infty}(t+2)^{C-1}}$.

If $C < 1$, $L = \infty$ and I diverges.

If $C = 1$, $L = 2$ and I converges to $\ln 2 + \ln 2^0 = \ln 2$.

If $C > 1$, $L = 0$ and I diverges to $-\infty$.

62.
$$\begin{aligned} I = \int_0^\infty \left(\frac{x}{x^2+1} - \frac{C}{3x+1}\right)dx &= \lim_{t\to\infty}\left[\tfrac{1}{2}\ln(x^2+1) - \tfrac{1}{3}C\ln(3x+1)\right]_0^t = \lim_{t\to\infty}\left[\ln(t^2+1)^{1/2} - \ln(3t+1)^{C/3}\right] \\ &= \lim_{t\to\infty}\left(\ln\frac{(t^2+1)^{1/2}}{(3t+1)^{C/3}}\right) = \ln\left(\lim_{t\to\infty}\frac{\sqrt{t^2+1}}{(3t+1)^{C/3}}\right) \end{aligned}$$

For $C \le 0$, the integral diverges. For $C > 0$, we have

$$L = \lim_{t\to\infty}\frac{\sqrt{t^2+1}}{(3t+1)^{C/3}} \overset{\text{H}}{=} \lim_{t\to\infty}\frac{t\big/\sqrt{t^2+1}}{C(3t+1)^{(C/3)-1}} = \frac{1}{C}\lim_{t\to\infty}\frac{1}{(3t+1)^{(C/3)-1}}.$$

For $C/3 < 1 \quad \Leftrightarrow \quad C < 3$, $L = \infty$ and I diverges.

For $C = 3$, $L = \frac{1}{3}$ and $I = \ln\frac{1}{3}$.

For $C > 3$, $L = 0$ and I diverges to $-\infty$.

6 Review

CONCEPT CHECK

1. See Formula 6.1.1 or 6.1.2. We try to choose $u = f(x)$ to be a function that becomes simpler when differentiated (or at least not more complicated) as long as $dv = g'(x)\,dx$ can be readily integrated to give v.

2. See the margin note on page 314.

3. If $\sqrt{a^2 - x^2}$ occurs, try $x = a\sin\theta$; if $\sqrt{a^2 + x^2}$ occurs, try $x = a\tan\theta$, and if $\sqrt{x^2 - a^2}$ occurs, try $x = a\sec\theta$. See the Table of Trigonometric Substitutions on page 317.

4. See Equation 2 and Expressions 6, 8, and 10 in Section 6.3.

5. See the Midpoint Rule, the Trapezoidal Rule, and Simpson's Rule, as well as their associated error bounds, all in Section 6.6. We would expect the best estimate to be given by Simpson's Rule.

6. See Definitions 1(a), (b), and (c) in Section 6.6.

7. See Definitions 3(b), (a), and (c) in Section 6.6.

8. See the Comparison Theorem on page 353.

TRUE-FALSE QUIZ

1. False. Since the numerator has a higher degree than the denominator, $\dfrac{x(x^2+4)}{x^2-4} = x + \dfrac{8x}{x^2-4} = x + \dfrac{A}{x+2} + \dfrac{B}{x-2}$.

2. True. In fact, $A = -1$, $B = C = 1$.

3. False. It can be put in the form $\dfrac{A}{x} + \dfrac{B}{x^2} + \dfrac{C}{x-4}$.

4. False. The form is $\dfrac{A}{x} + \dfrac{Bx + C}{x^2+4}$.

5. False. This is an improper integral, since the denominator vanishes at $x = 1$.

$$\int_0^4 \frac{x}{x^2-1}\,dx = \int_0^1 \frac{x}{x^2-1}\,dx + \int_1^4 \frac{x}{x^2-1}\,dx \text{ and}$$

$$\int_0^1 \frac{x}{x^2-1}\,dx = \lim_{t\to 1^-}\int_0^t \frac{x}{x^2-1}\,dx = \lim_{t\to 1^-}\left[\tfrac{1}{2}\ln\left|x^2-1\right|\right]_0^t = \lim_{t\to 1^-}\tfrac{1}{2}\ln\left|t^2-1\right| = \infty$$

So the integral diverges.

6. True by Theorem 6.6.2 with $p = \sqrt{2} > 1$.

7. False. See Exercise 51 in Section 6.6.

8. False. For example, with $n = 1$ the Trapezoidal Rule is much more accurate than the Midpoint Rule for the function in the diagram.

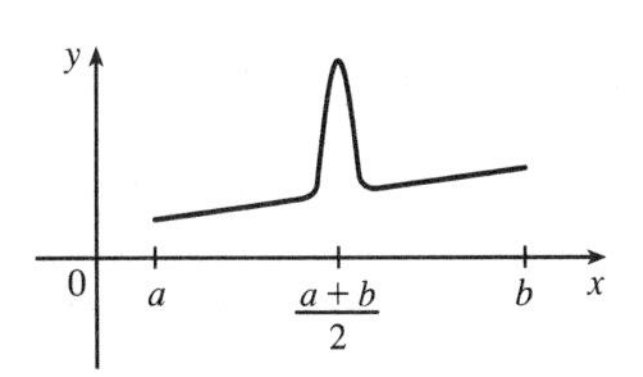

9. (a) True. See the end of Section 6.4.

(b) False. Examples include the functions $f(x) = e^{x^2}$, $g(x) = \sin(x^2)$, and $h(x) = \dfrac{\sin x}{x}$.

10. True. If f is continuous on $[0, \infty)$, then $\int_0^1 f(x)\,dx$ is finite. Since $\int_1^\infty f(x)\,dx$ is finite, so is $\int_0^\infty f(x)\,dx = \int_0^1 f(x)\,dx + \int_1^\infty f(x)\,dx$.

11. False. If $f(x) = 1/x$, then f is continuous and decreasing on $[1, \infty)$ with $\lim\limits_{x\to\infty} f(x) = 0$, but $\int_1^\infty f(x)\,dx$ is divergent.

12. True.
$$\begin{aligned}\int_a^\infty [f(x) + g(x)]\,dx &= \lim_{t\to\infty} \int_a^t [f(x) + g(x)]\,dx = \lim_{t\to\infty} \left(\int_a^t f(x)\,dx + \int_a^t g(x)\,dx\right)\\ &= \lim_{t\to\infty} \int_a^t f(x)\,dx + \lim_{t\to\infty} \int_a^t g(x)\,dx \quad \begin{bmatrix}\text{since both limits}\\ \text{in the sum exist}\end{bmatrix}\\ &= \int_a^\infty f(x)\,dx + \int_a^\infty g(x)\,dx\end{aligned}$$

Since the two integrals are finite, so is their sum.

13. False. Take $f(x) = 1$ for all x and $g(x) = -1$ for all x. Then $\int_a^\infty f(x)\,dx = \infty$ [divergent] and $\int_a^\infty g(x)\,dx = -\infty$ [divergent], but $\int_a^\infty [f(x) + g(x)]\,dx = 0$ [convergent].

14. False. $\int_0^\infty f(x)\,dx$ could converge or diverge. For example, if $g(x) = 1$, then $\int_0^\infty f(x)\,dx$ diverges if $f(x) = 1$ and converges if $f(x) = 0$.

EXERCISES

1. $$\int_0^5 \frac{x}{x+10}\,dx = \int_0^5 \left(1 - \frac{10}{x+10}\right)dx = \Big[x - 10\ln(x+10)\Big]_0^5 = 5 - 10\ln 15 + 10\ln 10 = 5 + 10\ln\tfrac{10}{15} = 5 + 10\ln\tfrac{2}{3}$$

2. $$\int_0^5 ye^{-0.6y}\,dy \quad \begin{bmatrix}u = y & dv = e^{-0.6y}\,dy\\ du = dy & v = -\frac{5}{3}e^{-0.6y}\end{bmatrix} = \left[-\tfrac{5}{3}ye^{-0.6y}\right]_0^5 - \int_0^5 \left(-\tfrac{5}{3}e^{-0.6y}\right)dy = -\tfrac{25}{3}e^{-3} - \tfrac{25}{9}\left[e^{-0.6y}\right]_0^5$$
$$= -\tfrac{25}{3}e^{-3} - \tfrac{25}{9}(e^{-3} - 1) = -\tfrac{25}{3}e^{-3} - \tfrac{25}{9}e^{-3} + \tfrac{25}{9} = \tfrac{25}{9} - \tfrac{100}{9}e^{-3}$$

3. $$\int_0^{\pi/2} \frac{\cos\theta}{1+\sin\theta}\,d\theta = \Big[\ln(1+\sin\theta)\Big]_0^{\pi/2} = \ln 2 - \ln 1 = \ln 2$$

4. $$\int_1^4 \frac{dt}{(2t+1)^3} \quad \begin{bmatrix}u = 2t+1\\ du = 2\,dt\end{bmatrix} = \int_3^9 \frac{\frac{1}{2}\,du}{u^3} = \frac{-1}{4}\left[\frac{1}{u^2}\right]_3^9 = -\frac{1}{4}\left(\frac{1}{81} - \frac{1}{9}\right) = -\frac{1}{4}\left(-\frac{8}{81}\right) = \frac{2}{81}$$

5. Let $u = \sec x$. Then $du = \sec x\,\tan x\,dx$, so
$$\begin{aligned}\int \tan^7 x\,\sec^3 x\,dx &= \int \tan^6 x\,\sec^2 x\,\sec x\,\tan x\,dx = \int (u^2-1)^3 u^2\,du = \int (u^8 - 3u^6 + 3u^4 - u^2)\,du\\ &= \tfrac{1}{9}u^9 - \tfrac{3}{7}u^7 + \tfrac{3}{5}u^5 - \tfrac{1}{3}u^3 + C = \tfrac{1}{9}\sec^9 x - \tfrac{3}{7}\sec^7 x + \tfrac{3}{5}\sec^5 x - \tfrac{1}{3}\sec^3 x + C\end{aligned}$$

6. $\dfrac{1}{y^2 - 4y - 12} = \dfrac{1}{(y-6)(y+2)} = \dfrac{A}{y-6} + \dfrac{B}{y+2} \;\Rightarrow\; 1 = A(y+2) + B(y-6)$. Letting $y = -2 \;\Rightarrow\; B = -\frac{1}{8}$ and letting $y = 6 \;\Rightarrow\; A = \frac{1}{8}$. So $\displaystyle\int \frac{1}{y^2-4y-12}\,dy = \int \left(\frac{1/8}{y-6} + \frac{-1/8}{y+2}\right)dy = \tfrac{1}{8}\ln|y-6| - \tfrac{1}{8}\ln|y+2| + C.$

7. Let $u = \ln t$, $du = dt/t$. Then $\displaystyle\int \frac{\sin(\ln t)}{t}\,dt = \int \sin u\,du = -\cos u + C = -\cos(\ln t) + C$.

8. Let $x = \tan\theta$, $-\frac{\pi}{2} < \theta < \frac{\pi}{2}$. Then

$$\int \frac{dx}{x^2\sqrt{1+x^2}} = \int \frac{\sec^2\theta\,d\theta}{\tan^2\theta\sec\theta} = \int \frac{\sec\theta\,d\theta}{\tan^2\theta} = \int \frac{\cos\theta\,d\theta}{\sin^2\theta} = \int \frac{du}{u^2} \qquad [\text{put } u = \sin\theta]$$

$$= -\frac{1}{u} + C = -\frac{1}{\sin\theta} + C = -\frac{\sqrt{1+x^2}}{x} + C$$

9. $\displaystyle\int_1^4 x^{3/2}\ln x\,dx \quad \begin{bmatrix} u = \ln x, & dv = x^{3/2}\,dx, \\ du = dx/x & v = \frac{2}{5}x^{5/2} \end{bmatrix} \quad = \frac{2}{5}\Big[x^{5/2}\ln x\Big]_1^4 - \frac{2}{5}\int_1^4 x^{3/2}\,dx = \frac{2}{5}(32\ln 4 - \ln 1) - \frac{2}{5}\Big[\frac{2}{5}x^{5/2}\Big]_1^4$

$$= \tfrac{2}{5}(64\ln 2) - \tfrac{4}{25}(32-1) = \tfrac{128}{5}\ln 2 - \tfrac{124}{25} \quad \left[\text{or } \tfrac{64}{5}\ln 4 - \tfrac{124}{25}\right]$$

10. Let $u = \arctan x$, $du = dx/(1+x^2)$. Then

$$\int_0^1 \frac{\sqrt{\arctan x}}{1+x^2}\,dx = \int_0^{\pi/4} \sqrt{u}\,du = \frac{2}{3}\Big[u^{3/2}\Big]_0^{\pi/4} = \frac{2}{3}\left[\frac{\pi^{3/2}}{4^{3/2}} - 0\right] = \frac{2}{3}\cdot\frac{1}{8}\pi^{3/2} = \frac{1}{12}\pi^{3/2}.$$

11. Let $x = \sec\theta$. Then

$$\int_1^2 \frac{\sqrt{x^2-1}}{x}\,dx = \int_0^{\pi/3} \frac{\tan\theta}{\sec\theta}\sec\theta\,\tan\theta\,d\theta = \int_0^{\pi/3}\tan^2\theta\,d\theta = \int_0^{\pi/3}(\sec^2\theta - 1)\,d\theta = \Big[\tan\theta - \theta\Big]_0^{\pi/3} = \sqrt{3} - \tfrac{\pi}{3}.$$

12. $\displaystyle\int_{-1}^1 \frac{\sin x}{1+x^2}\,dx = 0$ by Theorem 4.5.6(b), since $f(x) = \dfrac{\sin x}{1+x^2}$ is an odd function.

13. $\displaystyle\int \frac{dx}{x^3+x} = \int\left(\frac{1}{x} - \frac{x}{x^2+1}\right)dx = \ln|x| - \tfrac{1}{2}\ln(x^2+1) + C$

14. $\displaystyle\int \frac{x^2+2}{x+2}\,dx = \int\left(x - 2 + \frac{6}{x+2}\right)dx = \tfrac{1}{2}x^2 - 2x + 6\ln|x+2| + C$

15. $\int \sin^2\theta\,\cos^5\theta\,d\theta = \int \sin^2\theta\,(\cos^2\theta)^2\cos\theta\,d\theta = \int \sin^2\theta\,(1-\sin^2\theta)^2\cos\theta\,d\theta$

$$= \int u^2\left(1-u^2\right)^2 du \quad [u = \sin\theta,\ du = \cos\theta\,d\theta] \quad = \int u^2(1 - 2u^2 + u^4)\,du$$

$$= \int (u^2 - 2u^4 + u^6)\,du = \tfrac{1}{3}u^3 - \tfrac{2}{5}u^5 + \tfrac{1}{7}u^7 + C = \tfrac{1}{3}\sin^3\theta - \tfrac{2}{5}\sin^5\theta + \tfrac{1}{7}\sin^7\theta + C$$

16. $\displaystyle\int \frac{\sec^6\theta}{\tan^2\theta}\,d\theta = \int \frac{(\tan^2\theta+1)^2\sec^2\theta}{\tan^2\theta}\,d\theta \quad \begin{bmatrix} u = \tan\theta, \\ du = \sec^2\theta\,d\theta \end{bmatrix} \quad = \int \frac{(u^2+1)^2}{u^2}\,du = \int \frac{u^4+2u^2+1}{u^2}\,du$

$$= \int\left(u^2 + 2 + \frac{1}{u^2}\right)du = \frac{u^3}{3} + 2u - \frac{1}{u} + C = \frac{1}{3}\tan^3\theta + 2\tan\theta - \cot\theta + C$$

17. Integrate by parts with $u = x$, $dv = \sec x\,\tan x\,dx \quad\Rightarrow\quad du = dx$, $v = \sec x$:

$\int x\sec x\,\tan x\,dx = x\sec x - \int \sec x\,dx \overset{14}{=} x\sec x - \ln|\sec x + \tan x| + C$.

18. $\displaystyle\frac{x^2+8x-3}{x^3+3x^2} = \frac{x^2+8x-3}{x^2(x+3)} = \frac{A}{x} + \frac{B}{x^2} + \frac{C}{x+3} \quad\Rightarrow\quad x^2 + 8x - 3 = Ax(x+3) + B(x+3) + Cx^2.$

Taking $x = 0$, we get $-3 = 3B$, so $B = -1$. Taking $x = -3$, we get $-18 = 9C$, so $C = -2$.

Taking $x = 1$, we get $6 = 4A + 4B + C = 4A - 4 - 2$, so $4A = 12$ and $A = 3$. Now

$$\int \frac{x^2+8x-3}{x^3+3x^2}\,dx = \int\left(\frac{3}{x} - \frac{1}{x^2} - \frac{2}{x+3}\right)dx = 3\ln|x| + \frac{1}{x} - 2\ln|x+3| + C.$$

19. $\displaystyle\int \frac{x+1}{9x^2+6x+5}\,dx = \int \frac{x+1}{(9x^2+6x+1)+4}\,dx = \int \frac{x+1}{(3x+1)^2+4}\,dx \qquad [u = 3x+1,\ u = 3\,dx]$

$$= \int \frac{\left[\frac{1}{3}(u-1)\right]+1}{u^2+4}\left(\frac{1}{3}\,du\right) = \frac{1}{3}\cdot\frac{1}{3}\int \frac{(u-1)+3}{u^2+4}\,du$$

$$= \frac{1}{9}\int \frac{u}{u^2+4}\,du + \frac{1}{9}\int \frac{2}{u^2+2^2}\,du = \frac{1}{9}\cdot\frac{1}{2}\ln(u^2+4) + \frac{2}{9}\cdot\frac{1}{2}\tan^{-1}\left(\frac{1}{2}u\right) + C$$

$$= \tfrac{1}{18}\ln(9x^2+6x+5) + \tfrac{1}{9}\tan^{-1}\left[\tfrac{1}{2}(3x+1)\right] + C$$

20. $\displaystyle\int \frac{dt}{\sin^2 t + \cos 2t} = \int \frac{dt}{\sin^2 t + (\cos^2 t - \sin^2 t)} = \int \frac{dt}{\cos^2 t} = \int \sec^2 t\,dt = \tan t + C$

21.

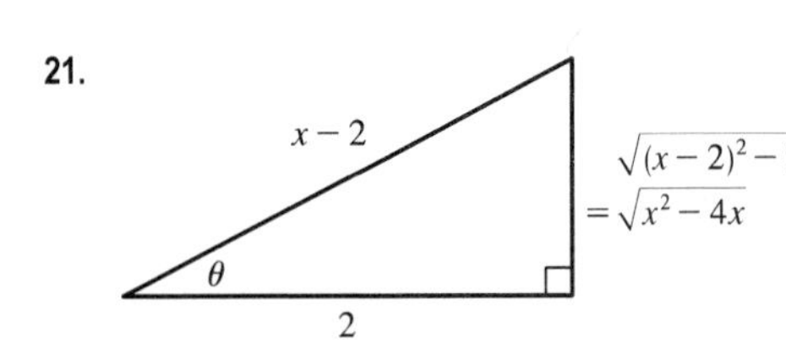

$$\int \frac{dx}{\sqrt{x^2-4x}} = \int \frac{dx}{\sqrt{(x^2-4x+4)-4}} = \int \frac{dx}{\sqrt{(x-2)^2-2^2}}$$

$$= \int \frac{2\sec\theta\tan\theta\,d\theta}{2\tan\theta} \qquad \begin{bmatrix} x-2 = 2\sec\theta, \\ dx = 2\sec\theta\,\tan\theta\,d\theta \end{bmatrix}$$

$$= \int \sec\theta\,d\theta = \ln|\sec\theta + \tan\theta| + C_1$$

$$= \ln\left|\frac{x-2}{2} + \frac{\sqrt{x^2-4x}}{2}\right| + C_1$$

$$= \ln\left|x-2+\sqrt{x^2-4x}\right| + C, \text{ where } C = C_1 - \ln 2$$

22. Let $u = x+1$. Then

$$\int \frac{x^3}{(x+1)^{10}}\,dx = \int \frac{(u-1)^3}{u^{10}}\,dx = \int \frac{u^3-3u^2+3u-1}{u^{10}}\,du$$

$$= \int (u^{-7} - 3u^{-8} + 3u^{-9} - u^{-10})\,du = -\tfrac{1}{6}u^{-6} + \tfrac{3}{7}u^{-7} - \tfrac{3}{8}u^{-8} + \tfrac{1}{9}u^{-9} + C$$

$$= \frac{-1}{6(x+1)^6} + \frac{3}{7(x+1)^7} - \frac{3}{8(x+1)^8} + \frac{1}{9(x+1)^9} + C$$

23. Let $u = \cot 4x$. Then $du = -4\csc^2 4x\,dx \quad \Rightarrow$

$\int \csc^4 4x\,dx = \int (\cot^2 4x + 1)\csc^2 4x\,dx = \int (u^2+1)\left(-\frac{1}{4}\,du\right) = -\frac{1}{4}\left(\frac{1}{3}u^3 + u\right) + C = -\frac{1}{12}(\cot^3 4x + 3\cot 4x) + C.$

24. Let $u = \cos x$, $dv = e^x\,dx \quad \Rightarrow \quad du = -\sin x\,dx$, $v = e^x$: $I = \int e^x \cos x\,dx = e^x\cos x + \int e^x \sin x\,dx$ $(\star)$.

To integrate $\int e^x \sin x\,dx$, let $U = \sin x$, $dV = e^x\,dx \quad \Rightarrow \quad dU = \cos x\,dx$, $V = e^x$.

Then $\int e^x \sin x\,dx = e^x \sin x - \int e^x \cos x\,dx = e^x\sin x - I$. By substitution in $(\star)$, $I = e^x\cos x + e^x \sin x - I \quad \Rightarrow$

$2I = e^x(\cos x + \sin x) \quad \Rightarrow \quad I = \frac{1}{2}e^x(\cos x + \sin x) + C.$

25. $\displaystyle\frac{3x^3 - x^2 + 6x - 4}{(x^2+1)(x^2+2)} = \frac{Ax+B}{x^2+1} + \frac{Cx+D}{x^2+2} \quad \Rightarrow \quad 3x^3 - x^2 + 6x - 4 = (Ax+B)(x^2+2) + (Cx+D)(x^2+1).$

Equating the coefficients gives $A + C = 3$, $B + D = -1$, $2A + C = 6$, and $2B + D = -4 \quad \Rightarrow$

$A = 3$, $C = 0$, $B = -3$, and $D = 2$. Now

$$\int \frac{3x^3 - x^2 + 6x - 4}{(x^2+1)(x^2+2)}\,dx = 3\int \frac{x-1}{x^2+1}\,dx + 2\int \frac{dx}{x^2+2} = \tfrac{3}{2}\ln(x^2+1) - 3\tan^{-1}x + \sqrt{2}\tan^{-1}\left(\frac{1}{\sqrt{2}}x\right) + C.$$

26. Let $u = e^x$. Then $x = \ln u$, $dx = \dfrac{du}{u}$, so

$$\int \frac{dx}{1+e^x} = \int \frac{du/u}{1+u} = \int \left[\frac{1}{u} - \frac{1}{u+1}\right] du = \ln u - \ln(u+1) + C = \ln e^x - \ln(1+e^x) + C = x - \ln(1+e^x) + C.$$

27. $\int_0^{\pi/2} \cos^3 x \sin 2x\, dx = \int_0^{\pi/2} \cos^3 x\,(2\sin x \cos x)\, dx = \int_0^{\pi/2} 2\cos^4 x \sin x\, dx = \left[-\frac{2}{5}\cos^5 x\right]_0^{\pi/2} = \frac{2}{5}$

28. Let $u = \sqrt[3]{x}$. Then $x = u^3$, $dx = 3u^2 du \Rightarrow$

$$\int \frac{\sqrt[3]{x}+1}{\sqrt[3]{x}-1}\, dx = \int \frac{u+1}{u-1} 3u^2\, du = 3\int \left(u^2 + 2u + 2 + \frac{2}{u-1}\right) du$$

$$= u^3 + 3u^2 + 6u + 6\ln|u-1| + C = x + 3x^{2/3} + 6\sqrt[3]{x} + 6\ln|\sqrt[3]{x} - 1| + C$$

29. The product of an odd function and an even function is an odd function, so $f(x) = x^5 \sec x$ is an odd function.

By Theorem 4.5.6(b), $\int_{-1}^{1} x^5 \sec x\, dx = 0$.

30. Let $u = e^{-x}$, $du = -e^{-x}\, dx$. Then

$$\int \frac{dx}{e^x\sqrt{1-e^{-2x}}} = \int \frac{e^{-x}\, dx}{\sqrt{1-(e^{-x})^2}} = \int \frac{-du}{\sqrt{1-u^2}} = -\sin^{-1} u + C = -\sin^{-1}(e^{-x}) + C.$$

31. Let $u = \sqrt{e^x - 1}$. Then $u^2 = e^x - 1$ and $2u\, du = e^x\, dx$. Also, $e^x + 8 = u^2 + 9$. Thus,

$$\int_0^{\ln 10} \frac{e^x\sqrt{e^x-1}}{e^x+8}\, dx = \int_0^3 \frac{u \cdot 2u\, du}{u^2+9} = 2\int_0^3 \frac{u^2}{u^2+9}\, du = 2\int_0^3 \left(1 - \frac{9}{u^2+9}\right) du = 2\left[u - \frac{9}{3}\tan^{-1}\left(\frac{u}{3}\right)\right]_0^3$$

$$= 2\left[(3 - 3\tan^{-1} 1) - 0\right] = 2\left(3 - 3\cdot\frac{\pi}{4}\right) = 6 - \frac{3\pi}{2}$$

32. $$\int_0^{\pi/4} \frac{x\sin x}{\cos^3 x}\, dx = \int_0^{\pi/4} x\tan x \sec^2 x\, dx \qquad \left[\begin{matrix} u = x, & dv = \tan x \sec^2 x\, dx, \\ du = dx & v = \frac{1}{2}\tan^2 x \end{matrix}\right]$$

$$= \left[\frac{x}{2}\tan^2 x\right]_0^{\pi/4} - \frac{1}{2}\int_0^{\pi/4} \tan^2 x\, dx = \frac{\pi}{8}\cdot 1^2 - 0 - \frac{1}{2}\int_0^{\pi/4} (\sec^2 x - 1)\, dx$$

$$= \frac{\pi}{8} - \frac{1}{2}\left[\tan x - x\right]_0^{\pi/4} = \frac{\pi}{8} - \frac{1}{2}\left(1 - \frac{\pi}{4}\right) = \frac{\pi}{4} - \frac{1}{2}$$

33. Let $x = 2\sin\theta \Rightarrow (4 - x^2)^{3/2} = (2\cos\theta)^3$, $dx = 2\cos\theta\, d\theta$, so

$$\int \frac{x^2}{(4-x^2)^{3/2}}\, dx = \int \frac{4\sin^2\theta}{8\cos^3\theta} 2\cos\theta\, d\theta = \int \tan^2\theta\, d\theta = \int (\sec^2\theta - 1) d\theta$$

$$= \tan\theta - \theta + C = \frac{x}{\sqrt{4-x^2}} - \sin^{-1}\left(\frac{x}{2}\right) + C$$

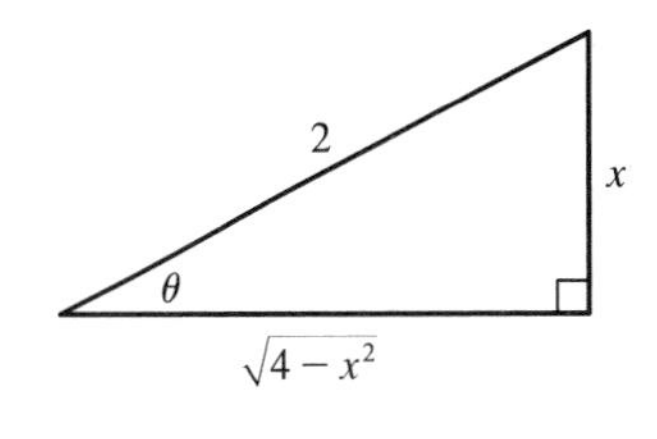

34. Integrate by parts twice, first with $u = (\arcsin x)^2$, $dv = dx$:

$$I = \int (\arcsin x)^2\, dx = x(\arcsin x)^2 - \int 2x \arcsin x \left(\frac{dx}{\sqrt{1-x^2}}\right)$$

Now let $U = \arcsin x$, $dV = \dfrac{x}{\sqrt{1-x^2}}\, dx \Rightarrow dU = \dfrac{1}{\sqrt{1-x^2}}\, dx$, $V = -\sqrt{1-x^2}$. So

$$I = x(\arcsin x)^2 - 2\left[\arcsin x\,\left(-\sqrt{1-x^2}\right) + \int dx\right] = x(\arcsin x)^2 + 2\sqrt{1-x^2}\arcsin x - 2x + C$$

35. $\displaystyle\int \frac{1}{\sqrt{x+x^{3/2}}}\,dx = \int \frac{dx}{\sqrt{x\,(1+\sqrt{x}\,)}} = \int \frac{dx}{\sqrt{x}\sqrt{1+\sqrt{x}}} \quad \left[\begin{array}{l} u = 1+\sqrt{x}, \\ du = \dfrac{dx}{2\sqrt{x}} \end{array}\right] \quad = \int \frac{2\,du}{\sqrt{u}} = \int 2u^{-1/2}\,du$

$\displaystyle = 4\sqrt{u} + C = 4\sqrt{1+\sqrt{x}} + C$

36. $\displaystyle\int \frac{1-\tan\theta}{1+\tan\theta}\,d\theta = \int \frac{\frac{\cos\theta}{\cos\theta} - \frac{\sin\theta}{\cos\theta}}{\frac{\cos\theta}{\cos\theta} + \frac{\sin\theta}{\cos\theta}}\,d\theta = \int \frac{\cos\theta - \sin\theta}{\cos\theta + \sin\theta}\,d\theta = \ln|\cos\theta + \sin\theta| + C$

37. $\int (\cos x + \sin x)^2 \cos 2x\,dx = \int (\cos^2 x + 2\sin x\,\cos x + \sin^2 x)\cos 2x\,dx = \int (1+\sin 2x)\cos 2x\,dx$

$= \int \cos 2x\,dx + \frac{1}{2}\int \sin 4x\,dx = \frac{1}{2}\sin 2x - \frac{1}{8}\cos 4x + C$

Or: $\int (\cos x + \sin x)^2 \cos 2x\,dx = \int (\cos x + \sin x)^2(\cos^2 x - \sin^2 x)\,dx$

$= \int (\cos x + \sin x)^3(\cos x - \sin x)\,dx = \frac{1}{4}(\cos x + \sin x)^4 + C_1$

38. Let $u = (\tan^{-1} x)^2$, $dv = x\,dx \;\Rightarrow\; du = 2(\tan^{-1} x)/(1+x^2)\,dx$, $v = \frac{1}{2}x^2$. Then

$$I = \int x(\tan^{-1} x)^2\,dx = \tfrac{1}{2}x^2(\tan^{-1} x)^2 - \int \frac{x^2\tan^{-1} x}{1+x^2}\,dx$$

Now let $w = \tan^{-1} x$, $dw = 1/(1+x^2)\,dx$, and $x^2 = \tan^2 w$. So

$I = \frac{1}{2}x^2(\tan^{-1} x)^2 - \int w\tan^2 w\,dw = \frac{1}{2}x^2(\tan^{-1} x)^2 - \int w\sec^2 w\,dw + \int w\,dw$

$= \frac{1}{2}x^2(\tan^{-1} x)^2 - \left(x\tan^{-1} x - \ln\sqrt{x^2+1}\,\right) + \frac{1}{2}(\tan^{-1} x)^2$ [by parts with $u = w$, $dv = \sec^2 w\,dw$]

$= \frac{1}{2}(x^2+1)(\tan^{-1} x)^2 - x\tan^{-1} x + \ln\sqrt{x^2+1} + C$

or $\frac{1}{2}(x^2+1)(\tan^{-1} x)^2 - x\tan^{-1} x + \frac{1}{2}\ln(x^2+1) + C$

39. We'll integrate $I = \displaystyle\int \frac{xe^{2x}}{(1+2x)^2}\,dx$ by parts with $u = xe^{2x}$ and $dv = \dfrac{dx}{(1+2x)^2}$. Then $du = (x\cdot 2e^{2x} + e^{2x}\cdot 1)\,dx$

and $v = -\dfrac{1}{2}\cdot\dfrac{1}{1+2x}$, so

$$I = -\frac{1}{2}\cdot\frac{xe^{2x}}{1+2x} - \int\left[-\frac{1}{2}\cdot\frac{e^{2x}(2x+1)}{1+2x}\right]dx = -\frac{xe^{2x}}{4x+2} + \frac{1}{2}\cdot\frac{1}{2}e^{2x} + C = e^{2x}\left(\frac{1}{4} - \frac{x}{4x+2}\right) + C.$$

Thus, $\displaystyle\int_0^{1/2} \frac{xe^{2x}}{(1+2x)^2}\,dx = \left[e^{2x}\left(\frac{1}{4} - \frac{x}{4x+2}\right)\right]_0^{1/2} = e\left(\frac{1}{4} - \frac{1}{8}\right) - 1\left(\frac{1}{4} - 0\right) = \frac{1}{8}e - \frac{1}{4}.$

40. $\displaystyle\int_{\pi/4}^{\pi/3} \frac{\sqrt{\tan\theta}}{\sin 2\theta}\,d\theta = \int_{\pi/4}^{\pi/3} \frac{\sqrt{\frac{\sin\theta}{\cos\theta}}}{2\sin\theta\cos\theta}\,d\theta = \int_{\pi/4}^{\pi/3} \frac{1}{2}(\sin\theta)^{-1/2}(\cos\theta)^{-3/2}d\theta$

$\displaystyle = \int_{\pi/4}^{\pi/3} \frac{1}{2}\left(\frac{\sin\theta}{\cos\theta}\right)^{-1/2}(\cos\theta)^{-2}\,d\theta = \int_{\pi/4}^{\pi/3} \frac{1}{2}(\tan\theta)^{-1/2}\sec^2\theta\,d\theta$

$\displaystyle = \left[\sqrt{\tan\theta}\,\right]_{\pi/4}^{\pi/3} = \sqrt{\sqrt{3}} - \sqrt{1} = \sqrt[4]{3} - 1$

41. $\displaystyle\int_1^{\infty} \frac{1}{(2x+1)^3}\,dx = \lim_{t\to\infty}\int_1^t \frac{1}{(2x+1)^3}\,dx = \lim_{t\to\infty}\int_1^t \tfrac{1}{2}(2x+1)^{-3}\,2\,dx$

$\displaystyle = \lim_{t\to\infty}\left[-\frac{1}{4(2x+1)^2}\right]_1^t = -\frac{1}{4}\lim_{t\to\infty}\left[\frac{1}{(2t+1)^2} - \frac{1}{9}\right] = -\frac{1}{4}\left(0 - \frac{1}{9}\right) = \frac{1}{36}$

42. $\dfrac{t^2+1}{t^2-1} = \dfrac{(t^2-1)+2}{t^2-1} = 1 + \dfrac{2}{(t+1)(t-1)}$. Now $\dfrac{2}{(t+1)(t-1)} = \dfrac{A}{t+1} + \dfrac{B}{t-1} \;\Rightarrow\; 2 = A(t-1) + B(t+1)$.

Letting $t = 1 \;\Rightarrow\; B = 1$ and letting $t = -1 \;\Rightarrow\; A = -1$. So

$$\begin{aligned}\int_0^1 \frac{t^2+1}{t^2-1}\,dt &= \lim_{b\to 1^-}\int_0^b \left(1 + \frac{-1}{t+1} + \frac{1}{t-1}\right)dt = \lim_{b\to 1^-}\Big[t - \ln|t+1| + \ln|t-1|\Big]_0^b \\ &= \lim_{b\to 1^-}\left[t + \ln\left|\frac{t-1}{t+1}\right|\right]_0^b = \lim_{b\to 1^-}\left[\left(b + \ln\left|\frac{b-1}{b+1}\right|\right) - (0+0)\right] = -\infty. \qquad \text{Divergent}\end{aligned}$$

43. $\displaystyle\int \frac{dx}{x\ln x} \quad \begin{bmatrix} u = \ln x, \\ du = dx/x \end{bmatrix} \quad = \int \frac{du}{u} = \ln|u| + C = \ln|\ln x| + C$, so

$\displaystyle\int_2^\infty \frac{dx}{x\ln x} = \lim_{t\to\infty}\int_2^t \frac{dx}{x\ln x} = \lim_{t\to\infty}\Big[\ln|\ln x|\Big]_2^t = \lim_{t\to\infty}\left[\ln(\ln t) - \ln(\ln 2)\right] = \infty$, so the integral is divergent.

44. Let $u = \sqrt{y-2}$. Then $y = u^2 + 2$ and $dy = 2u\,du$, so

$$\int \frac{y\,dy}{\sqrt{y-2}} = \int \frac{(u^2+2)2u\,du}{u} = 2\int (u^2+2)\,du = 2\left[\tfrac{1}{3}u^3 + 2u\right] + C$$

Thus,

$$\begin{aligned}\int_2^6 \frac{y\,dy}{\sqrt{y-2}} &= \lim_{t\to 2^+}\int_t^6 \frac{y\,dy}{\sqrt{y-2}} = \lim_{t\to 2^+}\left[\tfrac{2}{3}(y-2)^{3/2} + 4\sqrt{y-2}\right]_t^6 \\ &= \lim_{t\to 2^+}\left[\tfrac{16}{3} + 8 - \tfrac{2}{3}(t-2)^{3/2} - 4\sqrt{t-2}\right] = \tfrac{40}{3}\end{aligned}$$

45.
$$\begin{aligned}\int_0^4 \frac{\ln x}{\sqrt{x}}\,dx &= \lim_{t\to 0^+}\int_t^4 \frac{\ln x}{\sqrt{x}}\,dx \overset{\star}{=} \lim_{t\to 0^+}\left[2\sqrt{x}\,\ln x - 4\sqrt{x}\right]_t^4 \\ &= \lim_{t\to 0^+}\left[(2\cdot 2\ln 4 - 4\cdot 2) - \left(2\sqrt{t}\,\ln t - 4\sqrt{t}\right)\right] \overset{\star\star}{=} (4\ln 4 - 8) - (0-0) = 4\ln 4 - 8\end{aligned}$$

$(\star)$ Let $u = \ln x$, $dv = \dfrac{1}{\sqrt{x}}\,dx \;\Rightarrow\; du = \dfrac{1}{x}\,dx$, $v = 2\sqrt{x}$. Then

$$\int \frac{\ln x}{\sqrt{x}}\,dx = 2\sqrt{x}\,\ln x - 2\int \frac{dx}{\sqrt{x}} = 2\sqrt{x}\,\ln x - 4\sqrt{x} + C$$

$(\star\star)$ $\displaystyle\lim_{t\to 0^+}\left(2\sqrt{t}\,\ln t\right) = \lim_{t\to 0^+}\frac{2\ln t}{t^{-1/2}} \overset{\text{H}}{=} \lim_{t\to 0^+}\frac{2/t}{-\frac{1}{2}t^{-3/2}} = \lim_{t\to 0^+}\left(-4\sqrt{t}\right) = 0$

46. Note that $f(x) = 1/(2-3x)$ has an infinite discontinuity at $x = \frac{2}{3}$. Now

$$\int_0^{2/3} \frac{1}{2-3x}\,dx = \lim_{t\to(2/3)^-}\int_0^t \frac{1}{2-3x}\,dx = \lim_{t\to(2/3)^-}\left[-\tfrac{1}{3}\ln|2-3x|\right]_0^t = -\tfrac{1}{3}\lim_{t\to(2/3)^-}\left[\ln|2-3t| - \ln 2\right] = \infty.$$

Since $\displaystyle\int_0^{2/3} \frac{1}{2-3x}\,dx$ diverges, so does $\displaystyle\int_0^1 \frac{1}{2-3x}\,dx$.

47. $\displaystyle\int_0^3 \frac{dx}{x^2-x-2} = \int_0^3 \frac{dx}{(x+1)(x-2)} = \int_0^2 \frac{dx}{(x+1)(x-2)} + \int_2^3 \frac{dx}{(x+1)(x-2)}$, and

$$\int_2^3 \frac{dx}{x^2-x-2} = \lim_{t\to 2^+}\int_t^3 \left[\frac{-1/3}{x+1} + \frac{1/3}{x-2}\right]dx = \lim_{t\to 2^+}\left[\frac{1}{3}\ln\left|\frac{x-2}{x+1}\right|\right]_t^3 = \lim_{t\to 2^+}\left[\frac{1}{3}\ln\frac{1}{4} - \frac{1}{3}\ln\left|\frac{t-2}{t+1}\right|\right] = \infty,$$

so $\displaystyle\int_0^3 \frac{dx}{x^2-x-2}$ diverges.

48. $\int_{-1}^{1} \frac{x+1}{\sqrt[3]{x^4}}\,dx = \int_{-1}^{1} (x^{-1/3} + x^{-4/3})\,dx = \int_{-1}^{0} (x^{-1/3} + x^{-4/3})\,dx + \int_{0}^{1} (x^{-1/3} + x^{-4/3})\,dx$. But

$$\int_0^1 (x^{-1/3} + x^{-4/3})\,dx = \lim_{t\to 0^+} \int_t^1 (x^{-1/3} + x^{-4/3})\,dx = \lim_{t\to 0^+} \left[\tfrac{3}{2}x^{2/3} - 3x^{-1/3}\right]_t^1$$
$$= \lim_{t\to 0^+} \left[\tfrac{3}{2} - 3 - \tfrac{3}{2}t^{2/3} + 3t^{-1/3}\right] = \infty. \qquad \text{Divergent}$$

49. Let $u = 2x + 1$. Then

$$\int_{-\infty}^{\infty} \frac{dx}{4x^2+4x+5} = \int_{-\infty}^{\infty} \frac{\frac{1}{2}\,du}{u^2+4} = \frac{1}{2}\int_{-\infty}^{0} \frac{du}{u^2+4} + \frac{1}{2}\int_{0}^{\infty} \frac{du}{u^2+4}$$
$$= \tfrac{1}{2}\lim_{t\to-\infty}\left[\tfrac{1}{2}\tan^{-1}\left(\tfrac{1}{2}u\right)\right]_t^0 + \tfrac{1}{2}\lim_{t\to\infty}\left[\tfrac{1}{2}\tan^{-1}\left(\tfrac{1}{2}u\right)\right]_0^t = \tfrac{1}{4}\left[0 - \left(-\tfrac{\pi}{2}\right)\right] + \tfrac{1}{4}\left[\tfrac{\pi}{2} - 0\right] = \tfrac{\pi}{4}$$

50. $\int_1^{\infty} \frac{\tan^{-1}x}{x^2}\,dx = \lim_{t\to\infty} \int_1^t \frac{\tan^{-1}x}{x^2}\,dx$. Integrate by parts:

$$\int \frac{\tan^{-1}x}{x^2}\,dx = \frac{-\tan^{-1}x}{x} + \int \frac{1}{x}\frac{dx}{1+x^2} = \frac{-\tan^{-1}x}{x} + \int \left[\frac{1}{x} - \frac{x}{x^2+1}\right]dx$$
$$= \frac{-\tan^{-1}x}{x} + \ln|x| - \frac{1}{2}\ln(x^2+1) + C = \frac{-\tan^{-1}x}{x} + \frac{1}{2}\ln\frac{x^2}{x^2+1} + C$$

Thus,

$$\int_1^{\infty} \frac{\tan^{-1}x}{x^2}\,dx = \lim_{t\to\infty}\left[-\frac{\tan^{-1}x}{x} + \frac{1}{2}\ln\frac{x^2}{x^2+1}\right]_1^t = \lim_{t\to\infty}\left[-\frac{\tan^{-1}t}{t} + \frac{1}{2}\ln\frac{t^2}{t^2+1} + \frac{\pi}{4} - \frac{1}{2}\ln\frac{1}{2}\right]$$
$$= 0 + \tfrac{1}{2}\ln 1 + \tfrac{\pi}{4} + \tfrac{1}{2}\ln 2 = \tfrac{\pi}{4} + \tfrac{1}{2}\ln 2$$

51. $u = e^x \;\Rightarrow\; du = e^x\,dx$, so

$$\int e^x\sqrt{1-e^{2x}}\,dx = \int \sqrt{1-u^2}\,du \overset{30}{=} \tfrac{1}{2}u\sqrt{1-u^2} + \tfrac{1}{2}\sin^{-1}u + C = \tfrac{1}{2}\left[e^x\sqrt{1-e^{2x}} + \sin^{-1}(e^x)\right] + C.$$

52. $\int \csc^5 t\,dt \overset{78}{=} -\tfrac{1}{4}\cot t\,\csc^3 t + \tfrac{3}{4}\int \csc^3 t\,dt$

$$\overset{72}{=} -\tfrac{1}{4}\cot t\,\csc^3 t + \tfrac{3}{4}\left[-\tfrac{1}{2}\csc t\,\cot t + \tfrac{1}{2}\ln|\csc t - \cot t|\right] + C$$
$$= -\tfrac{1}{4}\cot t\,\csc^3 t - \tfrac{3}{8}\csc t\,\cot t + \tfrac{3}{8}\ln|\csc t - \cot t| + C$$

53. $\int \sqrt{x^2+x+1}\,dx = \int \sqrt{x^2+x+\frac{1}{4}+\frac{3}{4}}\,dx = \int \sqrt{\left(x+\frac{1}{2}\right)^2+\frac{3}{4}}\,dx$

$$= \int \sqrt{u^2 + \left(\tfrac{\sqrt{3}}{2}\right)^2}\,du \qquad \left[u = x + \tfrac{1}{2},\ du = dx\right]$$
$$\overset{21}{=} \tfrac{1}{2}u\sqrt{u^2+\tfrac{3}{4}} + \tfrac{3}{8}\ln\left(u + \sqrt{u^2+\tfrac{3}{4}}\right) + C$$
$$= \frac{2x+1}{4}\sqrt{x^2+x+1} + \tfrac{3}{8}\ln\left(x + \tfrac{1}{2} + \sqrt{x^2+x+1}\right) + C$$

54. Let $u = \sin x$. Then $du = \cos x\,dx$, so

$$\int \frac{\cot x\,dx}{\sqrt{1+2\sin x}} = \int \frac{du}{u\sqrt{1+2u}} \overset{57 \text{ with } a=1,\,b=2}{=} \ln\left|\frac{\sqrt{1+2u}-1}{\sqrt{1+2u}+1}\right| + C = \ln\left|\frac{\sqrt{1+2\sin x}-1}{\sqrt{1+2\sin x}+1}\right| + C$$

55. For $n \geq 0$, $\int_0^\infty x^n\,dx = \lim\limits_{t\to\infty}\left[x^{n+1}/(n+1)\right]_0^t = \infty$. For $n < 0$, $\int_0^\infty x^n\,dx = \int_0^1 x^n\,dx + \int_1^\infty x^n\,dx$. Both integrals are improper. By (6.6.2), the second integral diverges if $-1 \leq n < 0$. By Exercise 6.6.49, the first integral diverges if $n \leq -1$. Thus, $\int_0^\infty x^n\,dx$ is divergent for all values of n.

56. $$I = \int_0^\infty e^{ax}\cos x\,dx = \lim_{t\to\infty}\int_0^t e^{ax}\cos x\,dx \overset{\substack{99 \text{ with} \\ b=1}}{=} \lim_{t\to\infty}\left[\frac{e^{ax}}{a^2+1}(a\cos x + \sin x)\right]_0^t$$

$$= \lim_{t\to\infty}\left[\frac{e^{at}}{a^2+1}(a\cos t + \sin t) - \frac{1}{a^2+1}(a)\right] = \frac{1}{a^2+1}\lim_{t\to\infty}\left[e^{at}(a\cos t + \sin t) - a\right].$$

For $a \geq 0$, the limit does not exist due to oscillation. For $a < 0$, $\lim\limits_{t\to\infty}\left[e^{at}(a\cos t + \sin t)\right] = 0$ by the Squeeze Theorem, because $\left|e^{at}(a\cos t + \sin t)\right| \leq e^{at}(|a|+1)$, so $I = \dfrac{1}{a^2+1}(-a) = -\dfrac{a}{a^2+1}$.

57. $f(x) = \sqrt{1+x^4}$, $\Delta x = \dfrac{b-a}{n} = \dfrac{1-0}{10} = \dfrac{1}{10}$.

(a) $T_{10} = \frac{1}{10\cdot 2}\{f(0) + 2[f(0.1) + f(0.2) + \cdots + f(0.9)] + f(1)\} \approx 1.090608$

(b) $M_{10} = \frac{1}{10}\left[f\left(\frac{1}{20}\right) + f\left(\frac{3}{20}\right) + f\left(\frac{5}{20}\right) + \cdots + f\left(\frac{19}{20}\right)\right] \approx 1.088840$

(c) $S_{10} = \frac{1}{10\cdot 3}[f(0) + 4f(0.1) + 2f(0.2) + \cdots + 4f(0.9) + f(1)] \approx 1.089429$

f is concave upward, so the Trapezoidal Rule gives us an overestimate, the Midpoint Rule gives an underestimate, and we cannot tell whether Simpson's Rule gives us an overestimate or an underestimate.

58. $f(x) = \sqrt{\sin x}$, $\Delta x = \dfrac{\frac{\pi}{2} - 0}{10} = \dfrac{\pi}{20}$.

(a) $T_{10} = \frac{\pi}{20\cdot 2}\left\{f(0) + 2\left[f\left(\frac{\pi}{20}\right) + f\left(\frac{2\pi}{20}\right) + \cdots + f\left(\frac{9\pi}{20}\right)\right] + f\left(\frac{\pi}{2}\right)\right\} \approx 1.185197$

(b) $M_{10} = \frac{\pi}{20}\left[f\left(\frac{\pi}{40}\right) + f\left(\frac{3\pi}{40}\right) + f\left(\frac{5\pi}{40}\right) + \cdots + f\left(\frac{17\pi}{40}\right) + f\left(\frac{19\pi}{40}\right)\right] \approx 1.201932$

(c) $S_{10} = \frac{\pi}{20\cdot 3}\left[f(0) + 4f\left(\frac{\pi}{20}\right) + 2f\left(\frac{2\pi}{20}\right) + \cdots + 4f\left(\frac{9\pi}{20}\right) + f\left(\frac{\pi}{2}\right)\right] \approx 1.193089$

f is concave downward, so the Trapezoidal Rule gives us an underestimate, the Midpoint Rule gives an overestimate, and we cannot tell whether Simpson's Rule gives us an overestimate or an underestimate.

59. $f(x) = (1+x^4)^{1/2}$, $f'(x) = \frac{1}{2}(1+x^4)^{-1/2}(4x^3) = 2x^3(1+x^4)^{-1/2}$, $f''(x) = (2x^6 + 6x^2)(1+x^4)^{-3/2}$.

A graph of f'' on $[0,1]$ shows that it has its maximum at $x = 1$, so $|f''(x)| \leq f''(1) = \sqrt{8}$ on $[0,1]$. By taking $K = \sqrt{8}$, we find that the error in Exercise 57(a) is bounded by $\dfrac{K(b-a)^3}{12n^2} = \dfrac{\sqrt{8}}{1200} \approx 0.0024$, and in (b) by about $\frac{1}{2}(0.0024) = 0.0012$.

Note: Another way to estimate K is to let $x = 1$ in the factor $2x^6 + 6x^2$ (maximizing the numerator) and let $x = 0$ in the factor $(1+x^4)^{-3/2}$ (minimizing the denominator). Doing so gives us $K = 8$ and errors of $0.00\overline{6}$ and $0.00\overline{3}$. Using $K = 8$ for the Trapezoidal Rule, we have $|E_T| \leq \dfrac{K(b-a)^3}{12n^2} \leq 0.00001 \iff \dfrac{8(1-0)^3}{12n^2} \leq \dfrac{1}{100{,}000} \iff n^2 \geq \dfrac{800{,}000}{12} \iff$ $n \gtrsim 258.2$, so we should take $n = 259$. For the Midpoint Rule, $|E_M| \leq \dfrac{K(b-a)^3}{24n^2} \leq 0.00001 \iff n^2 \geq \dfrac{800{,}000}{24} \iff$ $n \gtrsim 182.6$, so we should take $n = 183$.

60. $\displaystyle\int_1^4 \frac{e^x}{x}\,dx \approx S_6 = \frac{(4-1)/6}{3}\,[f(1)+4f(1.5)+2f(2)+4f(2.5)+2f(3)+4f(3.5)+f(4)] \approx 17.739438$

61. $\Delta t = \left(\frac{10}{60}-0\right)/10 = \frac{1}{60}$.

$$\begin{aligned}\text{Distance traveled} &= \textstyle\int_0^{10} v\,dt \approx S_{10}\\ &= \tfrac{1}{60\cdot 3}[40+4(42)+2(45)+4(49)+2(52)+4(54)+2(56)+4(57)+2(57)+4(55)+56]\\ &= \tfrac{1}{180}(1544) = 8.5\overline{7}\text{ mi}\end{aligned}$$

62. We use Simpson's Rule with $n = 6$ and $\Delta t = \frac{24-0}{6} = 4$:

$$\begin{aligned}\text{Increase in bee population} &= \textstyle\int_0^{24} r(t)\,dt \approx S_6\\ &= \tfrac{4}{3}[r(0)+4r(4)+2r(8)+4r(12)+2r(16)+4r(20)+r(24)]\\ &= \tfrac{4}{3}[0+4(300)+2(3000)+4(11{,}000)+2(4000)+4(400)+0]\\ &= \tfrac{4}{3}(60{,}800) \approx 81{,}067 \text{ bees}\end{aligned}$$

63. (a) $f(x) = \sin(\sin x)$. A CAS gives

$$\begin{aligned}f^{(4)}(x) &= \sin(\sin x)\,[\cos^4 x + 7\cos^2 x - 3]\\ &\quad + \cos(\sin x)\,[6\cos^2 x\,\sin x + \sin x]\end{aligned}$$

From the graph, we see that $\left|f^{(4)}(x)\right| < 3.8$ for $x \in [0,\pi]$.

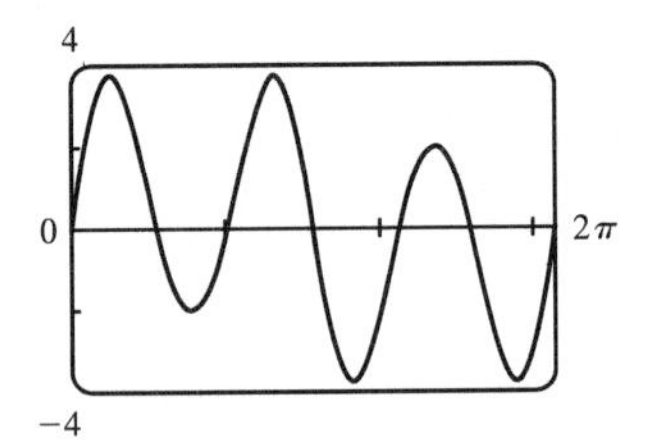

(b) We use Simpson's Rule with $f(x) = \sin(\sin x)$ and $\Delta x = \frac{\pi}{10}$:

$$\textstyle\int_0^\pi f(x)\,dx \approx \frac{\pi}{10\cdot 3}\left[f(0)+4f\left(\frac{\pi}{10}\right)+2f\left(\frac{2\pi}{10}\right)+\cdots+4f\left(\frac{9\pi}{10}\right)+f(\pi)\right] \approx 1.786721$$

From part (a), we know that $\left|f^{(4)}(x)\right| < 3.8$ on $[0,\pi]$, so we use Theorem 6.5.4 with $K = 3.8$, and estimate the error as $|E_S| \le \dfrac{3.8(\pi-0)^5}{180(10)^4} \approx 0.000646$.

(c) If we want the error to be less than 0.00001, we must have $|E_S| \le \dfrac{3.8\pi^5}{180n^4} \le 0.00001$, so

$n^4 \ge \dfrac{3.8\pi^5}{180(0.00001)} \approx 646{,}041.6 \;\Rightarrow\; n \ge 28.35$. Since n must be even for Simpson's Rule, we must have $n \ge 30$ to ensure the desired accuracy.

64. $\dfrac{x^3}{x^5+2} \le \dfrac{x^3}{x^5} = \dfrac{1}{x^2}$ for x in $[1,\infty)$. $\displaystyle\int_1^\infty \frac{1}{x^2}\,dx$ is convergent by (6.6.2) with $p = 2 > 1$. Therefore, $\displaystyle\int_1^\infty \frac{x^3}{x^5+2}\,dx$ is convergent by the Comparison Theorem.

65. By the Fundamental Theorem of Calculus,

$$\textstyle\int_0^\infty f'(x)\,dx = \lim\limits_{t\to\infty}\int_0^t f'(x)\,dx = \lim\limits_{t\to\infty}[f(t)-f(0)] = \lim\limits_{t\to\infty} f(t) - f(0) = 0 - f(0) = -f(0)$$

66. If the distance between P and the point charge is d, then the potential V at P is

$$V = W = \int_\infty^d F\,dr = \int_\infty^d \frac{q}{4\pi\varepsilon_0 r^2}\,dr = \lim_{t\to\infty}\frac{q}{4\pi\varepsilon_0}\left[-\frac{1}{r}\right]_t^d = \frac{q}{4\pi\varepsilon_0}\lim_{t\to\infty}\left(-\frac{1}{d}+\frac{1}{t}\right) = -\frac{q}{4\pi\varepsilon_0 d}$$

7 □ APPLICATIONS OF INTEGRATION

7.1 Areas Between Curves

1. $A = \int_{x=0}^{x=4} (y_T - y_B)\,dx = \int_0^4 [(5x - x^2) - x]\,dx = \int_0^4 (4x - x^2)\,dx = \left[2x^2 - \frac{1}{3}x^3\right]_0^4 = \left(32 - \frac{64}{3}\right) - (0) = \frac{32}{3}$

2. $A = \int_0^2 \left(\sqrt{x+2} - \frac{1}{x+1}\right)dx = \left[\frac{2}{3}(x+2)^{3/2} - \ln(x+1)\right]_0^2 = \left[\frac{2}{3}(4)^{3/2} - \ln 3\right] - \left[\frac{2}{3}(2)^{3/2} - \ln 1\right]$
$= \frac{16}{3} - \ln 3 - \frac{4}{3}\sqrt{2}$

3. $A = \int_{y=-1}^{y=1} (x_R - x_L)\,dy = \int_{-1}^1 [e^y - (y^2 - 2)]\,dy = \int_{-1}^1 (e^y - y^2 + 2)\,dy = \left[e^y - \frac{1}{3}y^3 + 2y\right]_{-1}^1$
$= \left(e^1 - \frac{1}{3} + 2\right) - \left(e^{-1} + \frac{1}{3} - 2\right) = e - \frac{1}{e} + \frac{10}{3}$

4. $A = \int_0^3 [(2y - y^2) - (y^2 - 4y)]\,dy = \int_0^3 (-2y^2 + 6y)\,dy = \left[-\frac{2}{3}y^3 + 3y^2\right]_0^3 = (-18 + 27) - 0 = 9$

5. $A = \int_{-1}^2 [(9 - x^2) - (x+1)]\,dx$
$= \int_{-1}^2 (8 - x - x^2)\,dx$
$= \left[8x - \frac{x^2}{2} - \frac{x^3}{3}\right]_{-1}^2$
$= \left(16 - 2 - \frac{8}{3}\right) - \left(-8 - \frac{1}{2} + \frac{1}{3}\right)$
$= 22 - 3 + \frac{1}{2} = \frac{39}{2}$

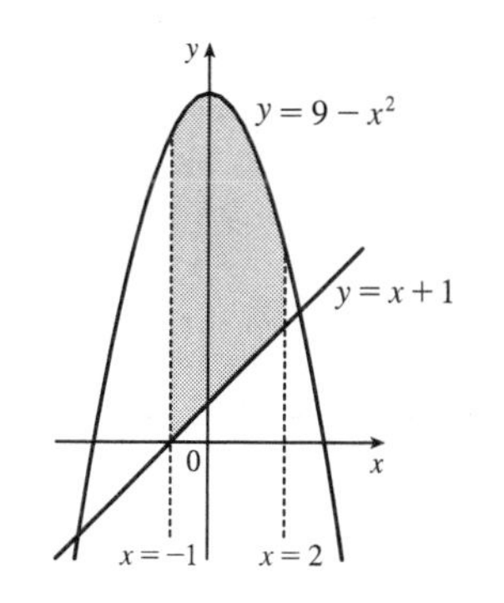

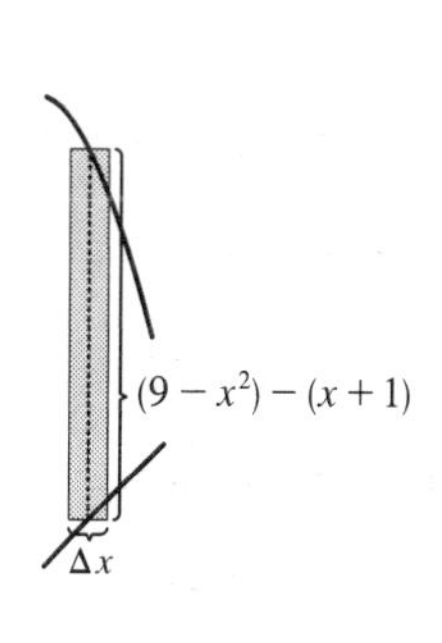

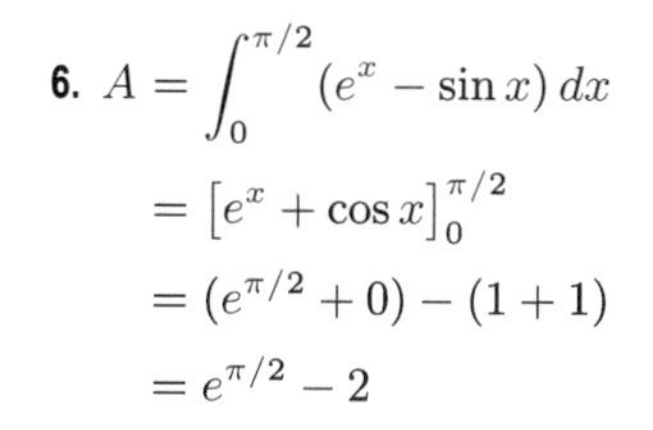

6. $A = \int_0^{\pi/2} (e^x - \sin x)\,dx$
$= \left[e^x + \cos x\right]_0^{\pi/2}$
$= (e^{\pi/2} + 0) - (1 + 1)$
$= e^{\pi/2} - 2$

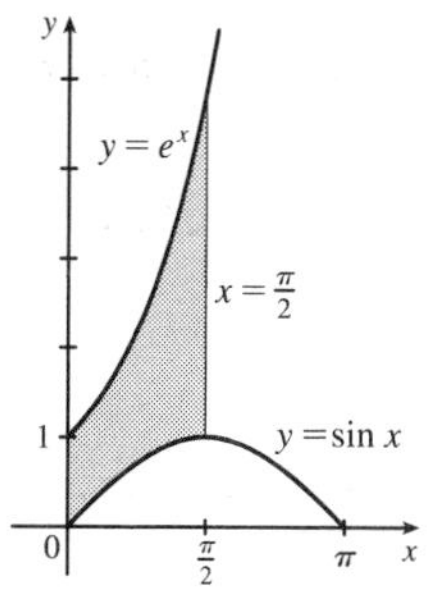

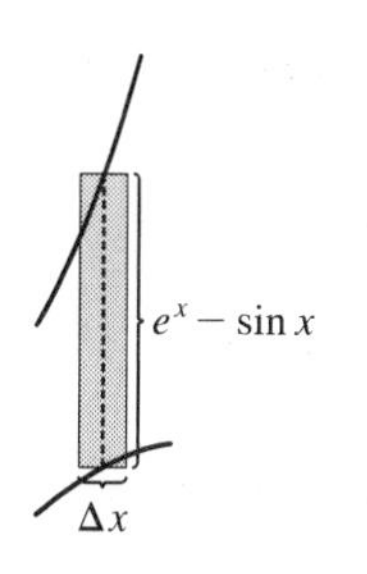

7. The curves intersect when $x = x^2 \;\Rightarrow\; x^2 - x = 0 \;\Leftrightarrow\; x(x-1) = 0 \;\Leftrightarrow\; x = 0, 1.$

$A = \int_0^1 (x - x^2)\,dx = \left[\frac{1}{2}x^2 - \frac{1}{3}x^3\right]_0^1$
$= \frac{1}{2} - \frac{1}{3} = \frac{1}{6}$

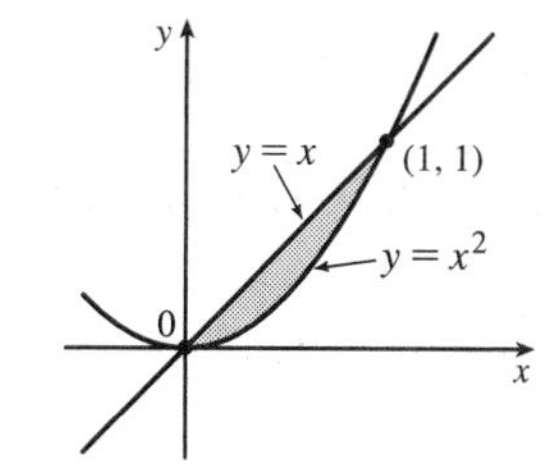

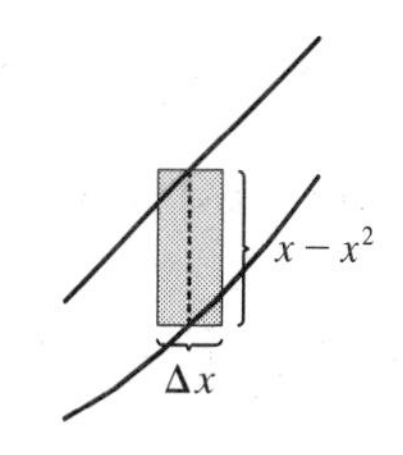

8. $1+\sqrt{x} = 1+\frac{1}{3}x \;\Rightarrow\; \sqrt{x} = \frac{1}{3}x \;\Rightarrow\; x = \dfrac{x^2}{9} \;\Rightarrow\; 9x - x^2 = 0 \;\Rightarrow\; x(9-x) = 0 \;\Rightarrow$

$x = 0$ or 9, so

$$A = \int_0^9 \left[(1+\sqrt{x}) - (1+\tfrac{1}{3}x)\right] dx = \int_0^9 \left(\sqrt{x} - \tfrac{1}{3}x\right) dx = \left[\tfrac{2}{3}x^{3/2} - \tfrac{1}{6}x^2\right]_0^9 = 18 - \tfrac{27}{2} = \tfrac{9}{2}$$

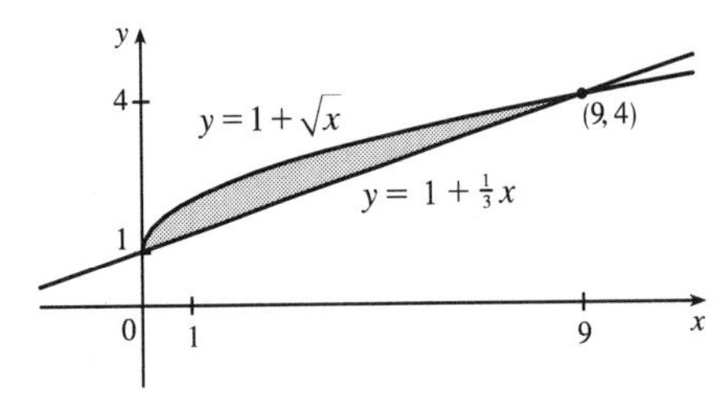

9. $12 - x^2 = x^2 - 6 \;\Leftrightarrow\; 2x^2 = 18 \;\Leftrightarrow\; x^2 = 9 \;\Leftrightarrow\; x = \pm 3$, so

$$A = \int_{-3}^{3} \left[(12-x^2) - (x^2-6)\right] dx = 2\int_0^3 (18 - 2x^2)\, dx \qquad \text{[by symmetry]}$$

$$= 2\left[18x - \tfrac{2}{3}x^3\right]_0^3 = 2[(54-18) - 0] = 2(36) = 72$$

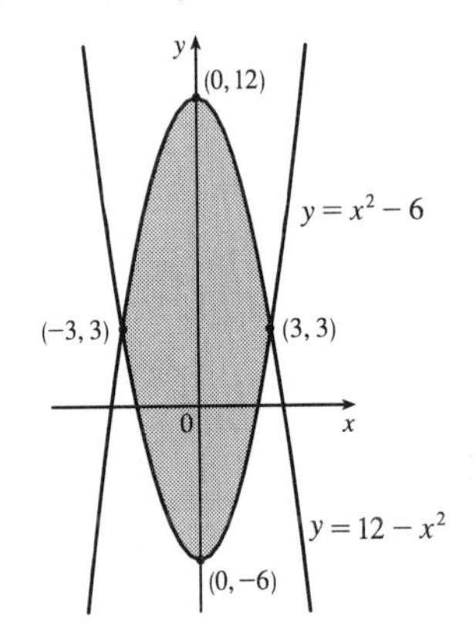

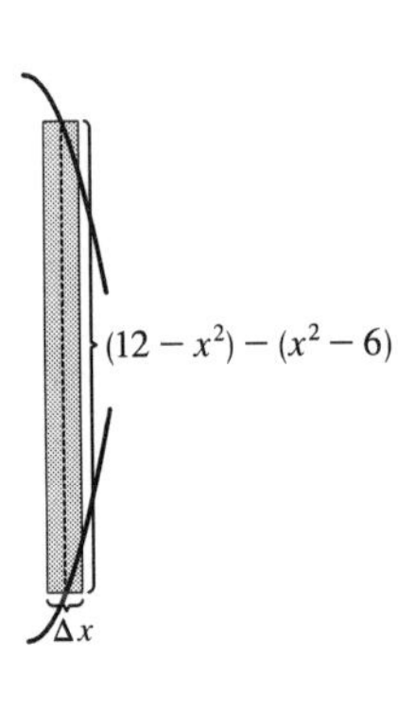

10. $x^2 = 4x - x^2 \;\Leftrightarrow\; 2x^2 - 4x = 0 \;\Leftrightarrow\; 2x(x-2) = 0 \;\Leftrightarrow\; x = 0$ or 2, so

$$A = \int_0^2 \left[(4x - x^2) - x^2\right] dx = \int_0^2 (4x - 2x^2)\, dx = \left[2x^2 - \tfrac{2}{3}x^3\right]_0^2 = 8 - \tfrac{16}{3} = \tfrac{8}{3}$$

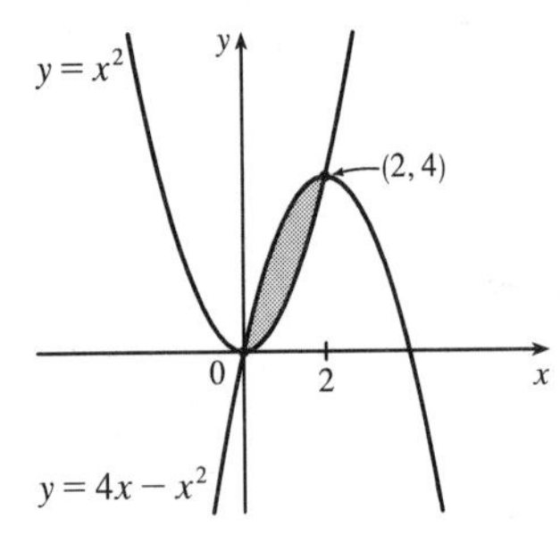

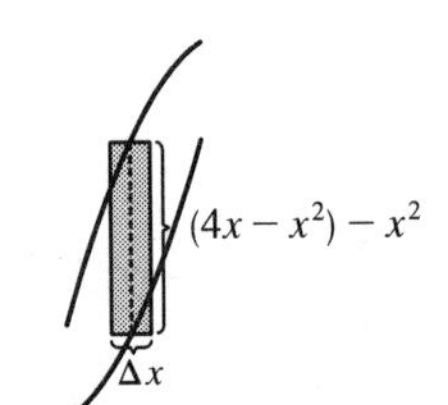

11. $2y^2 = 1 - y \quad \Leftrightarrow \quad 2y^2 + y - 1 = 0 \quad \Leftrightarrow \quad (2y-1)(y+1) = 0 \quad \Leftrightarrow \quad y = \frac{1}{2}$ or -1, so $x = \frac{1}{2}$ or 2 and

$$A = \int_{-1}^{1/2} [(1-y) - 2y^2]\,dy = \int_{-1}^{1/2} (1 - y - 2y^2)\,dy = \left[y - \tfrac{1}{2}y^2 - \tfrac{2}{3}y^3\right]_{-1}^{1/2}$$

$$= \left(\tfrac{1}{2} - \tfrac{1}{8} - \tfrac{1}{12}\right) - \left(-1 - \tfrac{1}{2} + \tfrac{2}{3}\right) = \tfrac{7}{24} - \left(-\tfrac{5}{6}\right) = \tfrac{7}{24} + \tfrac{20}{24} = \tfrac{27}{24} = \tfrac{9}{8}$$

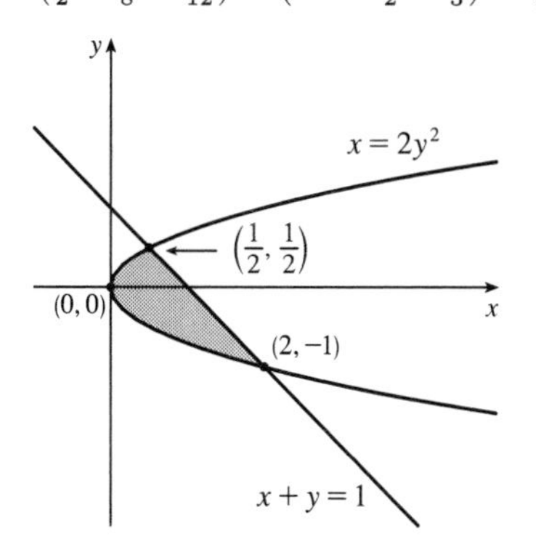

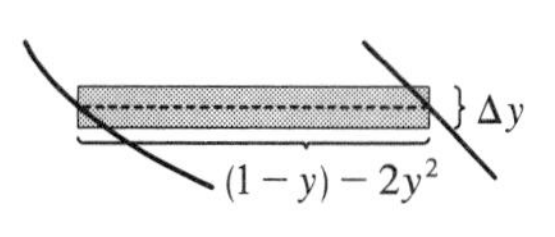

12. $4x + x^2 = 12 \quad \Leftrightarrow \quad (x+6)(x-2) = 0 \quad \Leftrightarrow \quad x = -6$ or $x = 2$, so $y = -6$ or $y = 2$ and

$$A = \int_{-6}^{2} \left[\left(-\tfrac{1}{4}y^2 + 3\right) - y\right] dy = \left[-\tfrac{1}{12}y^3 - \tfrac{1}{2}y^2 + 3y\right]_{-6}^{2}$$

$$= \left(-\tfrac{2}{3} - 2 + 6\right) - (18 - 18 - 18) = 22 - \tfrac{2}{3} = \tfrac{64}{3}$$

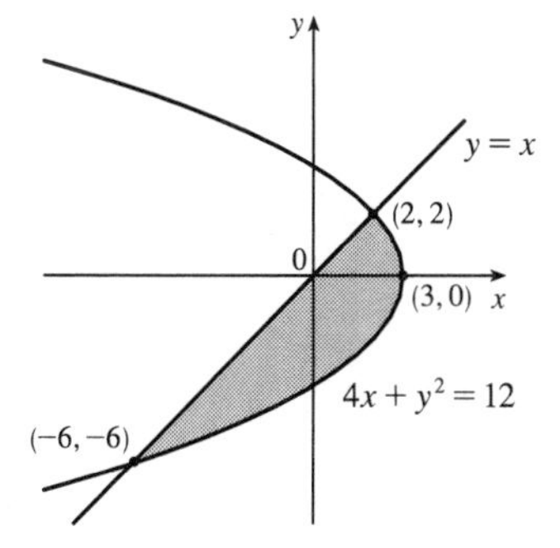

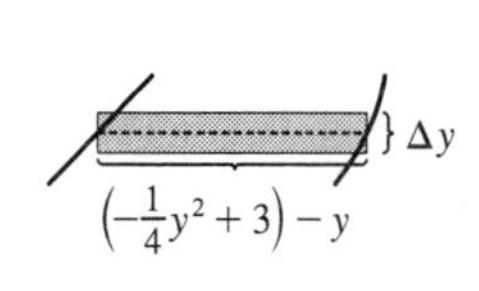

13. $2y^2 = 4 + y^2 \quad \Leftrightarrow \quad y^2 = 4 \quad \Leftrightarrow \quad y = \pm 2$, so

$$A = \int_{-2}^{2} [(4 + y^2) - 2y^2]\,dy = 2\int_0^2 (4 - y^2)\,dy \qquad \text{[by symmetry]}$$

$$= 2\left[4y - \tfrac{1}{3}y^3\right]_0^2 = 2\left(8 - \tfrac{8}{3}\right) = \tfrac{32}{3}$$

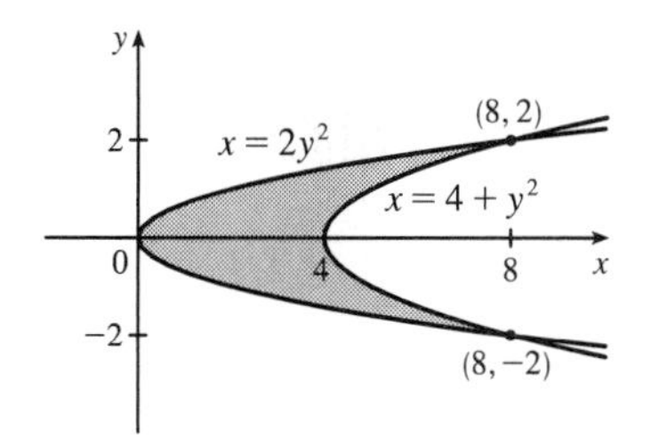

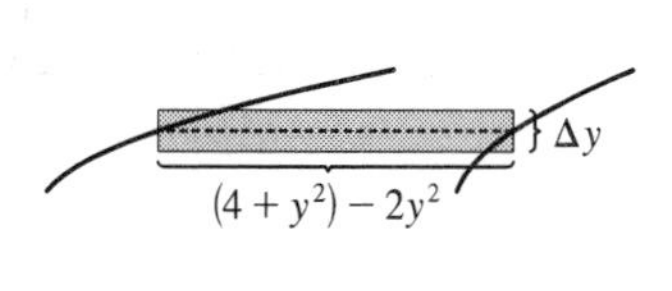

14. By observation, $y = \sin x$ and $y = 2x/\pi$ intersect at $(0, 0)$ and $(\pi/2, 1)$ for $x \geq 0$.

$$A = \int_0^{\pi/2} \left(\sin x - \frac{2x}{\pi} \right) dx = \left[-\cos x - \frac{1}{\pi} x^2 \right]_0^{\pi/2} = \left(0 - \frac{\pi}{4} \right) - (-1) = 1 - \frac{\pi}{4}$$

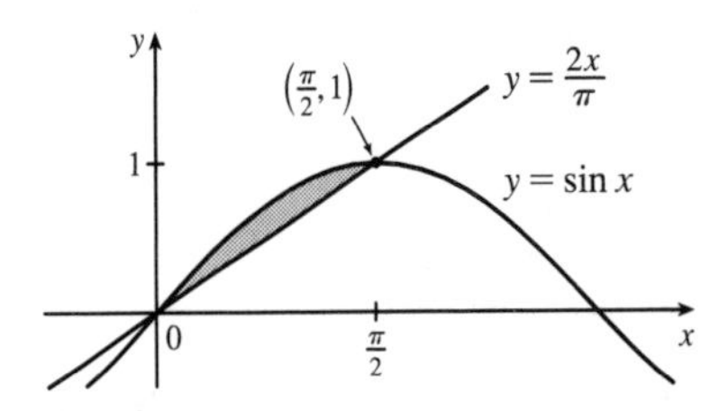

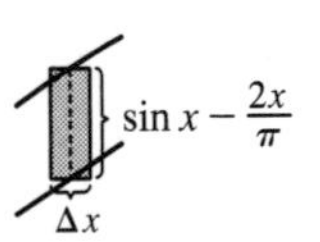

15. $1/x = x \quad \Leftrightarrow \quad 1 = x^2 \quad \Leftrightarrow \quad x = \pm 1$ and $1/x = \frac{1}{4}x \quad \Leftrightarrow \quad 4 = x^2 \quad \Leftrightarrow \quad x = \pm 2$, so for $x > 0$,

$$A = \int_0^1 \left(x - \frac{1}{4}x \right) dx + \int_1^2 \left(\frac{1}{x} - \frac{1}{4}x \right) dx = \int_0^1 \left(\frac{3}{4}x \right) dx + \int_1^2 \left(\frac{1}{x} - \frac{1}{4}x \right) dx$$

$$= \left[\frac{3}{8}x^2 \right]_0^1 + \left[\ln|x| - \frac{1}{8}x^2 \right]_1^2 = \tfrac{3}{8} + \left(\ln 2 - \tfrac{1}{2} \right) - \left(0 - \tfrac{1}{8} \right) = \ln 2$$

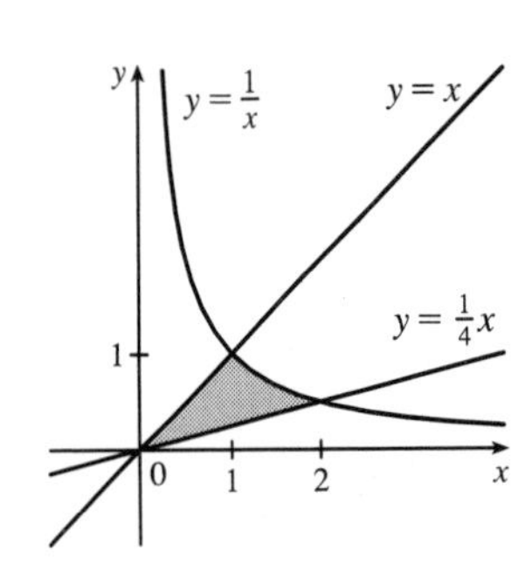

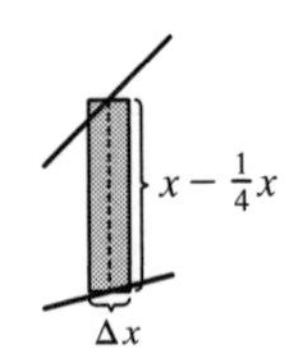

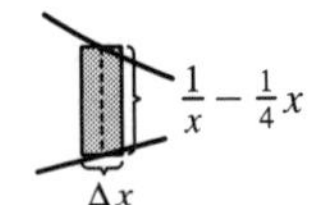

16. For $x > 0$, $x = x^2 - 2 \quad \Rightarrow \quad 0 = x^2 - x - 2 \quad \Rightarrow \quad 0 = (x - 2)(x + 1) \quad \Rightarrow \quad x = 2$. By symmetry,

$$\int_{-2}^{2} \left[|x| - (x^2 - 2) \right] dx = 2 \int_0^2 \left[x - (x^2 - 2) \right] dx = 2 \int_0^2 (x - x^2 + 2)\, dx = 2\left[\tfrac{1}{2}x^2 - \tfrac{1}{3}x^3 + 2x \right]_0^2$$

$$= 2\left(2 - \tfrac{8}{3} + 4 \right) = \tfrac{20}{3}$$

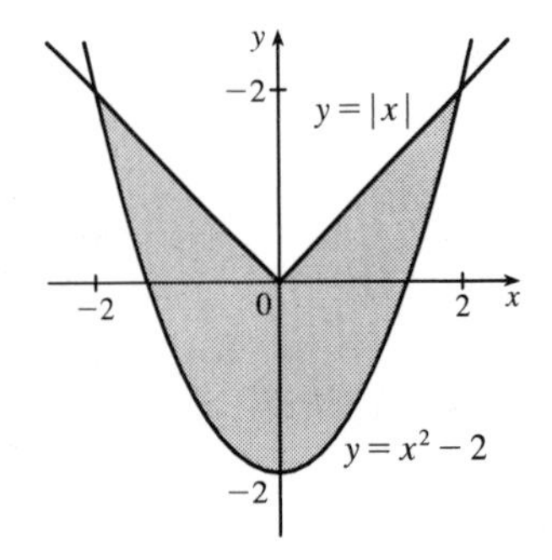

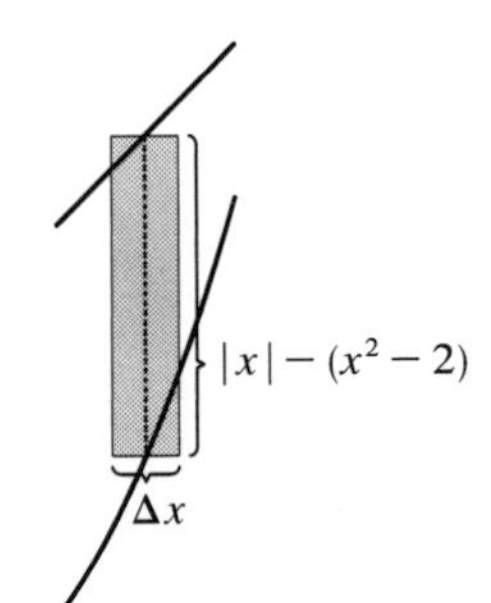

17.

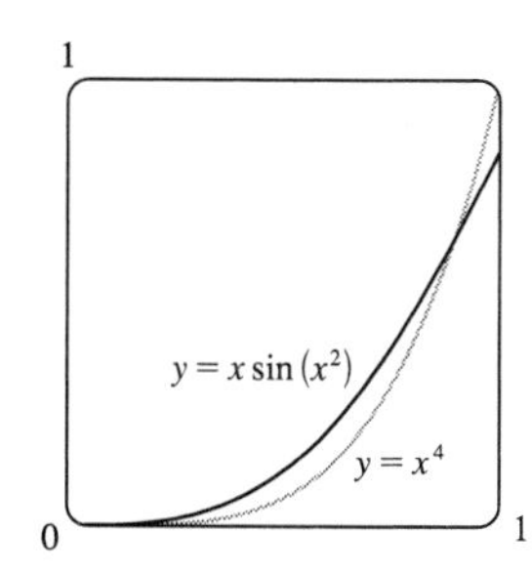

From the graph, we see that the curves intersect at $x = 0$ and $x = a \approx 0.896$, with $x\sin(x^2) > x^4$ on $(0, a)$. So the area A of the region bounded by the curves is

$$A = \int_0^a \left[x\sin(x^2) - x^4\right] dx = \left[-\tfrac{1}{2}\cos(x^2) - \tfrac{1}{5}x^5\right]_0^a$$
$$= -\tfrac{1}{2}\cos(a^2) - \tfrac{1}{5}a^5 + \tfrac{1}{2} \approx 0.037$$

18.

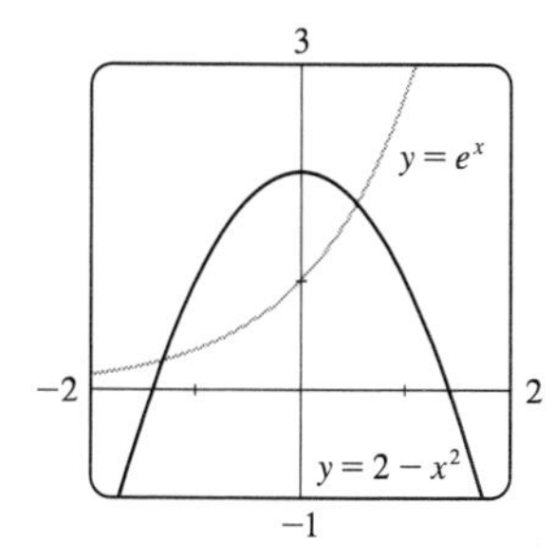

From the graph, we see that the curves intersect at $x = a \approx -1.32$ and $x = b \approx 0.54$, with $2 - x^2 > e^x$ on (a, b). So the area A of the region bounded by the curves is

$$A = \int_a^b \left[(2 - x^2) - e^x\right] dx = \left[2x - \tfrac{1}{3}x^3 - e^x\right]_a^b \approx 1.45$$

19.

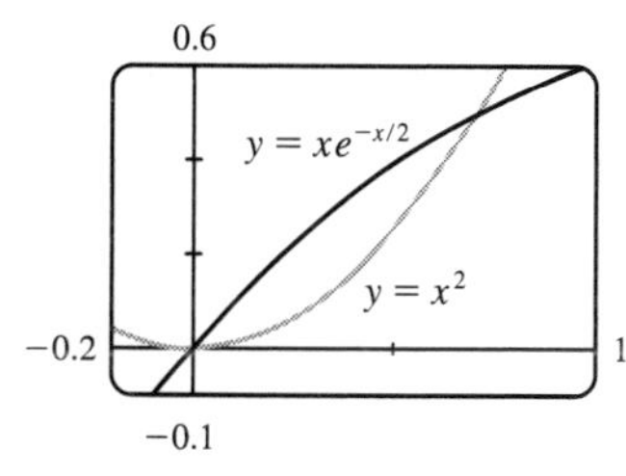

From the graph, we see that the curves intersect at $x = 0$ and $x = a \approx 0.70$, with $xe^{-x/2} > x^2$ on $(0, a)$. So the area A of the region bounded by the curves is

$$A = \int_0^a \left(xe^{-x/2} - x^2\right) dx$$
$$= \left[4\left(-\tfrac{1}{2}x - 1\right)e^{-x/2} - \tfrac{1}{3}x^3\right]_0^a \qquad \left[\text{Formula 96 with } a = -\tfrac{1}{2}\right]$$
$$\approx 0.08$$

20.

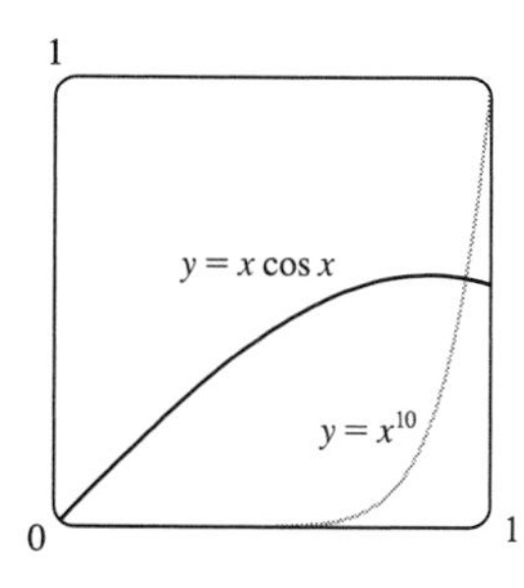

From the graph, we see that the curves intersect at $x = 0$ and $x = a \approx 0.94$, with $x\cos x > x^{10}$ on $(0, a)$. So the area A of the region bounded by the curves is

$$A = \int_0^a \left(x\cos x - x^{10}\right) dx$$
$$= \left[x\sin x + \cos x - \tfrac{1}{11}x^{11}\right]_0^a \qquad \left[\begin{array}{ll} u = x, & dv = \cos x\,dx, \\ du = dx, & v = \sin x \end{array}\right]$$
$$\approx 0.030$$

21. $\cos x = \sin 2x = 2\sin x\cos x \iff 2\sin x\cos x - \cos x = 0 \iff \cos x\,(2\sin x - 1) = 0 \iff$ $2\sin x = 1$ or $\cos x = 0 \iff x = \frac{\pi}{6}$ or $\frac{\pi}{2}$.

$$A = \int_0^{\pi/6}(\cos x - \sin 2x)\,dx + \int_{\pi/6}^{\pi/2}(\sin 2x - \cos x)\,dx$$
$$= \left[\sin x + \tfrac{1}{2}\cos 2x\right]_0^{\pi/6} + \left[-\tfrac{1}{2}\cos 2x - \sin x\right]_{\pi/6}^{\pi/2}$$
$$= \left(\tfrac{1}{2} + \tfrac{1}{2}\cdot\tfrac{1}{2}\right) - \left(0 + \tfrac{1}{2}\cdot 1\right) + \left[-\tfrac{1}{2}\cdot(-1) - 1\right] - \left(-\tfrac{1}{2}\cdot\tfrac{1}{2} - \tfrac{1}{2}\right)$$
$$= \tfrac{3}{4} - \tfrac{1}{2} - \tfrac{1}{2} + \tfrac{3}{4} = \tfrac{1}{2}$$

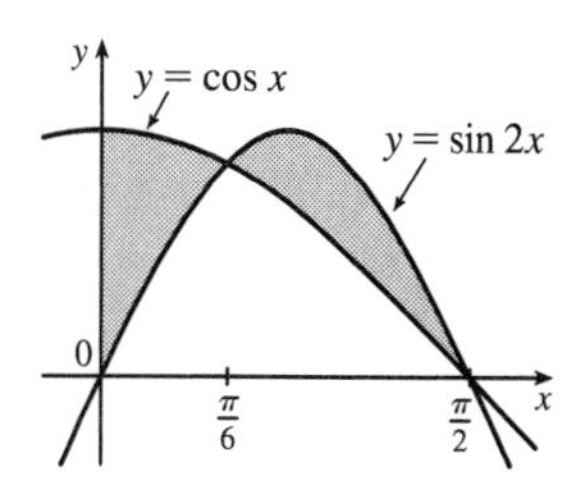

22. $A = \int_0^1 \left[(x^3 - 4x^2 + 3x) - (x^2 - x)\right] dx$

$\quad + \int_1^4 \left[(x^2 - x) - (x^3 - 4x^2 + 3x)\right] dx$

$= \int_0^1 (x^3 - 5x^2 + 4x)\, dx + \int_1^4 (-x^3 + 5x^2 - 4x)\, dx$

$= \left[\frac{1}{4}x^4 - \frac{5}{3}x^3 + 2x^2\right]_0^1 + \left[-\frac{1}{4}x^4 + \frac{5}{3}x^3 - 2x^2\right]_1^4$

$= \left(\frac{1}{4} - \frac{5}{3} + 2\right) - 0 + \left(-64 + \frac{320}{3} - 32\right) - \left(-\frac{1}{4} + \frac{5}{3} - 2\right) = \frac{71}{6}$

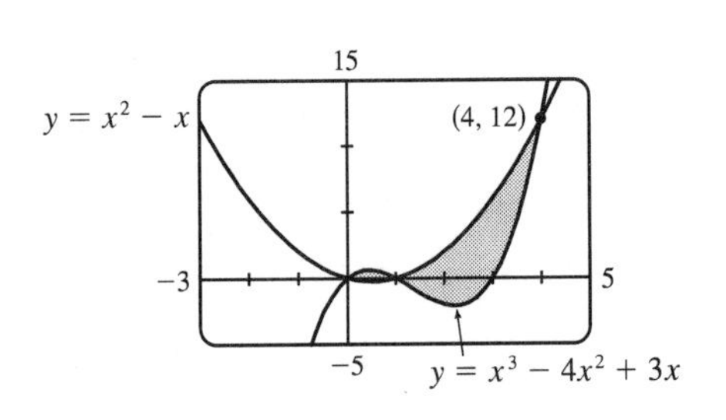

23. As in Example 4, we approximate the distance between the two cars after ten seconds using Simpson's Rule with $\Delta t = 1\text{ s} = \frac{1}{3600}$ h.

$$\text{distance}_{\text{Kelly}} - \text{distance}_{\text{Chris}} = \int_0^{10} v_K\, dt - \int_0^{10} v_C\, dt = \int_0^{10} (v_K - v_C)\, dt \approx S_{10}$$

$$= \frac{1}{3 \cdot 3600}[(0-0) + 4(22-20) + 2(37-32) + 4(52-46) + 2(61-54) + 4(71-62)$$
$$+ 2(80-69) + 4(86-75) + 2(93-81) + 4(98-86) + (102-90)]$$

$$= \frac{1}{10{,}800}(242) = \frac{121}{5400}\text{ mi}$$

So after 10 seconds, Kelly's car is about $\dfrac{121}{5400}\text{ mi}\left(5280\dfrac{\text{ft}}{\text{mi}}\right) \approx 118$ ft ahead of Chris's.

24. We know that the area under curve A between $t = 0$ and $t = x$ is $\int_0^x v_A(t)\, dt = s_A(x)$, where $v_A(t)$ is the velocity of car A and s_A is its displacement. Similarly, the area under curve B between $t = 0$ and $t = x$ is $\int_0^x v_B(t)\, dt = s_B(x)$.

(a) After one minute, the area under curve A is greater than the area under curve B. So car A is ahead after one minute.

(b) The area of the shaded region has numerical value $s_A(1) - s_B(1)$, which is the distance by which A is ahead of B after 1 minute.

(c) After two minutes, car B is traveling faster than car A and has gained some ground, but the area under curve A from $t = 0$ to $t = 2$ is still greater than the corresponding area for curve B, so car A is still ahead.

(d) From the graph, it appears that the area between curves A and B for $0 \le t \le 1$ (when car A is going faster), which corresponds to the distance by which car A is ahead, seems to be about 3 squares. Therefore, the cars will be side by side at the time x where the area between the curves for $1 \le t \le x$ (when car B is going faster) is the same as the area for $0 \le t \le 1$. From the graph, it appears that this time is $x \approx 2.2$. So the cars are side by side when $t \approx 2.2$ minutes.

25. If x = distance from left end of pool and $w = w(x)$ = width at x, then Simpson's Rule with $n = 8$ and $\Delta x = 2$ gives

$$\text{Area} = \int_0^{16} w\, dx \approx \tfrac{2}{3}[0 + 4(6.2) + 2(7.2) + 4(6.8) + 2(5.6) + 4(5.0) + 2(4.8) + 4(4.8) + 0] = \tfrac{2}{3}(126.4) \approx 84\text{ m}^2.$$

26. Let $h(x)$ denote the height of the wing at x cm from the left end.

$$A \approx S_{10} = \frac{200 - 0}{3(10)}[h(0) + 4h(20) + 2h(40) + \cdots + 4h(180) + h(200)]$$

$$= \tfrac{20}{3}[5.8 + 4(20.3) + 2(26.7) + 4(29.0) + 2(27.6) + 4(27.3) + 2(23.8) + 4(20.5) + 2(15.1) + 4(8.7) + 2.8]$$

$$= \tfrac{20}{3}(618.2) \approx 4121\text{ cm}^2$$

27. For $0 \le t \le 10$, $b(t) > d(t)$, so the area between the curves is given by

$$\int_0^{10} [b(t) - d(t)]\, dt = \int_0^{10} (2200e^{0.024t} - 1460e^{0.018t})\, dt = \left[\frac{2200}{0.024}e^{0.024t} - \frac{1460}{0.018}e^{0.018t}\right]_0^{10}$$

$$= \left(\frac{275{,}000}{3}e^{0.24} - \frac{730{,}000}{9}e^{0.18}\right) - \left(\frac{275{,}000}{3} - \frac{730{,}000}{9}\right) \approx 8868\text{ people}$$

This area represents the increase in population over a 10-year period.

28. Note that the circular cross-sections of the tank are the same everywhere, so the percentage of the total capacity that is being used is equal to the percentage of any cross-section that is under water. The underwater area is

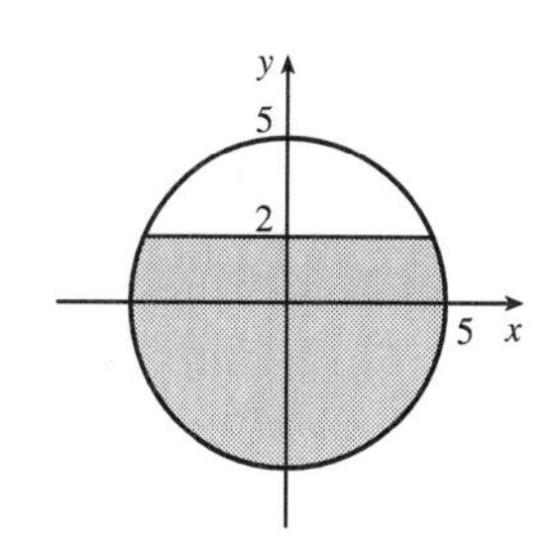

$$A = 2\int_{-5}^{2} \sqrt{25 - y^2}\,dy$$
$$= \left[25\arcsin(y/5) + y\sqrt{25 - y^2}\right]_{-5}^{2} \qquad \text{[substitute } y = 5\sin\theta\text{]}$$
$$= 25\arcsin\tfrac{2}{5} + 2\sqrt{21} + \tfrac{25}{2}\pi \approx 58.72 \text{ ft}^2$$

so the fraction of the total capacity in use is $\dfrac{A}{\pi(5)^2} \approx \dfrac{58.72}{25\pi} \approx 0.748$ or 74.8%.

29. Let the equation of the large circle be $x^2 + y^2 = R^2$. Then the equation of the small circle is $x^2 + (y - b)^2 = r^2$, where $b = \sqrt{R^2 - r^2}$ is the distance between the centers of the circles. The desired area is

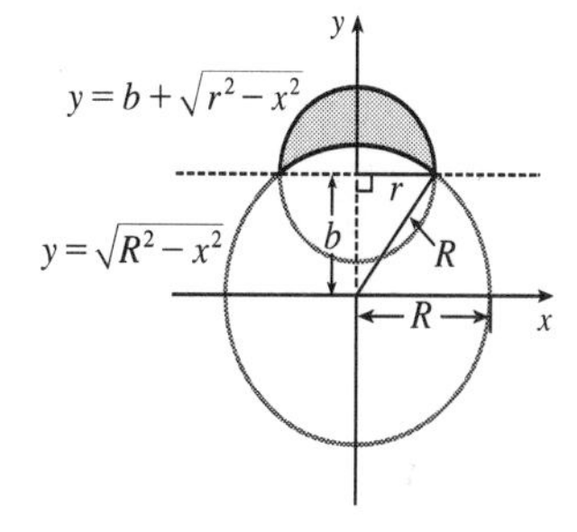

$$A = \int_{-r}^{r} \left[\left(b + \sqrt{r^2 - x^2}\right) - \sqrt{R^2 - x^2}\right] dx$$
$$= 2\int_0^r \left(b + \sqrt{r^2 - x^2} - \sqrt{R^2 - x^2}\right) dx$$
$$= 2\int_0^r b\,dx + 2\int_0^r \sqrt{r^2 - x^2}\,dx - 2\int_0^r \sqrt{R^2 - x^2}\,dx$$

The first integral is just $2br = 2r\sqrt{R^2 - r^2}$. The second integral represents the area of a quarter-circle of radius r, so its value is $\frac{1}{4}\pi r^2$. To evaluate the other integral, note that

$$\int \sqrt{a^2 - x^2}\,dx = \int a^2\cos^2\theta\,d\theta \;\; [x = a\sin\theta,\, dx = a\cos\theta\,d\theta] \;\; = \left(\tfrac{1}{2}a^2\right)\int(1 + \cos 2\theta)\,d\theta$$
$$= \tfrac{1}{2}a^2\left(\theta + \tfrac{1}{2}\sin 2\theta\right) + C = \tfrac{1}{2}a^2(\theta + \sin\theta\,\cos\theta) + C$$
$$= \frac{a^2}{2}\arcsin\left(\frac{x}{a}\right) + \frac{a^2}{2}\left(\frac{x}{a}\right)\frac{\sqrt{a^2 - x^2}}{a} + C = \frac{a^2}{2}\arcsin\left(\frac{x}{a}\right) + \frac{x}{2}\sqrt{a^2 - x^2} + C$$

Thus, the desired area is

$$A = 2r\sqrt{R^2 - r^2} + 2\left(\tfrac{1}{4}\pi r^2\right) - \left[R^2\arcsin(x/R) + x\sqrt{R^2 - x^2}\right]_0^r$$
$$= 2r\sqrt{R^2 - r^2} + \tfrac{1}{2}\pi r^2 - \left[R^2\arcsin(r/R) + r\sqrt{R^2 - r^2}\right] = r\sqrt{R^2 - r^2} + \tfrac{\pi}{2}r^2 - R^2\arcsin(r/R)$$

30. The inequality $x \geq 2y^2$ describes the region that lies on, or to the right of, the parabola $x = 2y^2$. The inequality $x \leq 1 - |y|$ describes the region that lies on, or to the left of, the curve $x = 1 - |y| = \begin{cases} 1 - y & \text{if } y \geq 0 \\ 1 + y & \text{if } y < 0 \end{cases}$.

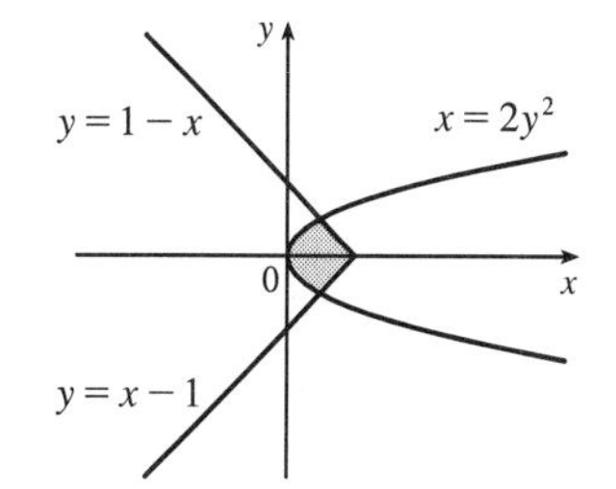

So the given region is the shaded region that lies between the curves. The graphs of $x = 1 - y$ and $x = 2y^2$ intersect when $1 - y = 2y^2 \;\Leftrightarrow\; 2y^2 + y - 1 = 0 \;\Leftrightarrow\; (2y - 1)(y + 1) = 0 \;\Rightarrow\; y = \frac{1}{2}$ (for $y \geq 0$).

By symmetry, $A = 2\int_0^{1/2}\left[(1 - y) - 2y^2\right] dy = 2\left[-\frac{2}{3}y^3 - \frac{1}{2}y^2 + y\right]_0^{1/2} = 2\left[\left(-\frac{1}{12} - \frac{1}{8} + \frac{1}{2}\right) - 0\right] = 2\left(\frac{7}{24}\right) = \frac{7}{12}$.

31. We first assume that $c > 0$, since c can be replaced by $-c$ in both equations without changing the graphs, and if $c = 0$ the curves do not enclose a region. We see from the graph that the enclosed area A lies between $x = -c$ and $x = c$, and by symmetry, it is equal to four times the area in the first quadrant. The enclosed area is

$A = 4\int_0^c (c^2 - x^2)\,dx = 4\left[c^2x - \frac{1}{3}x^3\right]_0^c = 4\left(c^3 - \frac{1}{3}c^3\right) = 4\left(\frac{2}{3}c^3\right) = \frac{8}{3}c^3.$

So $A = 576 \quad\Leftrightarrow\quad \frac{8}{3}c^3 = 576 \quad\Leftrightarrow\quad c^3 = 216 \quad\Leftrightarrow\quad c = \sqrt[3]{216} = 6$. Note that $c = -6$ is another solution, since the graphs are the same.

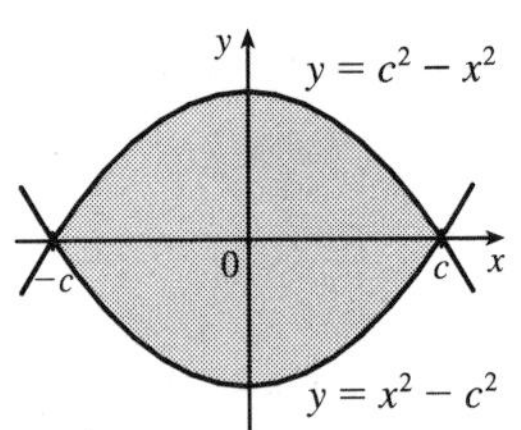

32.

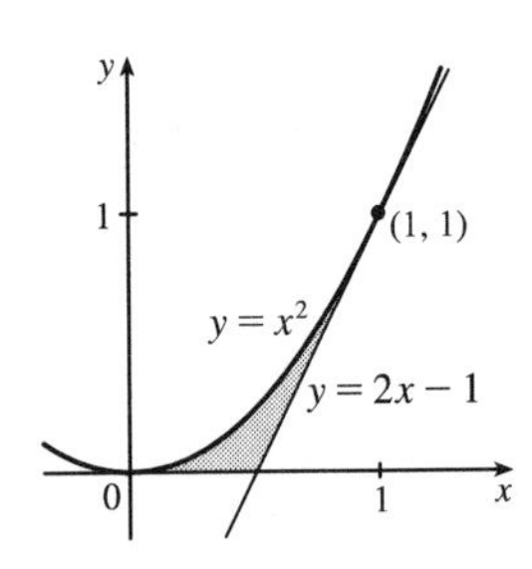

We start by finding the equation of the tangent line to $y = x^2$ at the point $(1, 1)$: $y' = 2x$, so the slope of the tangent is $2(1) = 2$, and its equation is $y - 1 = 2(x - 1)$, or $y = 2x - 1$. We would need two integrals to integrate with respect to x, but only one to integrate with respect to y.

$$A = \int_0^1 \left[\tfrac{1}{2}(y+1) - \sqrt{y}\right] dy = \left[\tfrac{1}{4}y^2 + \tfrac{1}{2}y - \tfrac{2}{3}y^{3/2}\right]_0^1 = \tfrac{1}{4} + \tfrac{1}{2} - \tfrac{2}{3} = \tfrac{1}{12}$$

33.

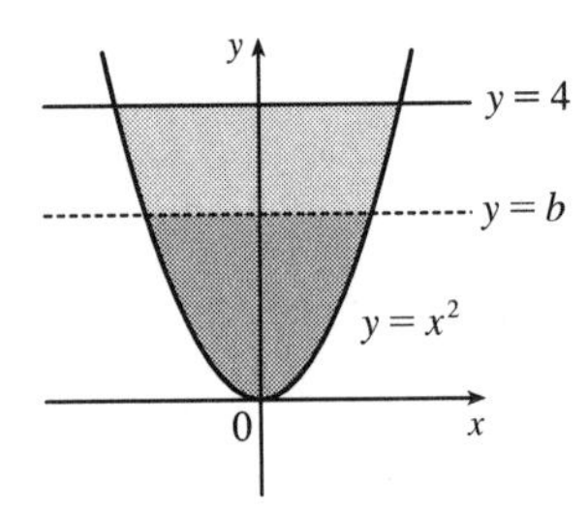

By the symmetry of the problem, we consider only the first quadrant, where $y = x^2 \quad\Rightarrow\quad x = \sqrt{y}$. We are looking for a number b such that

$\int_0^b \sqrt{y}\,dy = \int_b^4 \sqrt{y}\,dy \quad\Rightarrow\quad \frac{2}{3}\left[y^{3/2}\right]_0^b = \frac{2}{3}\left[y^{3/2}\right]_b^4 \quad\Rightarrow$

$b^{3/2} = 4^{3/2} - b^{3/2} \quad\Rightarrow\quad 2b^{3/2} = 8 \quad\Rightarrow\quad b^{3/2} = 4 \quad\Rightarrow\quad b = 4^{2/3} \approx 2.52.$

34. (a) We want to choose a so that

$$\int_1^a \frac{1}{x^2}\,dx = \int_a^4 \frac{1}{x^2}\,dx \quad\Rightarrow\quad \left[\frac{-1}{x}\right]_1^a = \left[\frac{-1}{x}\right]_a^4 \quad\Rightarrow\quad -\frac{1}{a} + 1 = -\frac{1}{4} + \frac{1}{a} \quad\Rightarrow\quad \frac{5}{4} = \frac{2}{a} \quad\Rightarrow\quad a = \frac{8}{5}.$$

(b) The area under the curve $y = 1/x^2$ from $x = 1$ to $x = 4$ is $\frac{3}{4}$ [take $a = 4$ in the first integral in part (a)]. Now the line $y = b$ must intersect the curve $x = 1/\sqrt{y}$ and not the line $x = 4$, since the area under the line $y = 1/4^2$ from $x = 1$ to $x = 4$ is only $\frac{3}{16}$, which is less than half of $\frac{3}{4}$. We want to choose b so that the upper area in the diagram is half of the total area under the curve $y = 1/x^2$ from $x = 1$ to $x = 4$. This implies that $\int_b^1 \left(1/\sqrt{y} - 1\right) dy = \frac{1}{2} \cdot \frac{3}{4} \quad\Rightarrow$

$\left[2\sqrt{y} - y\right]_b^1 = \frac{3}{8} \quad\Rightarrow\quad 1 - 2\sqrt{b} + b = \frac{3}{8} \quad\Rightarrow\quad b - 2\sqrt{b} + \frac{5}{8} = 0.$

Letting $c = \sqrt{b}$, we get $c^2 - 2c + \frac{5}{8} = 0 \quad\Rightarrow\quad 8c^2 - 16c + 5 = 0$. Thus,

$c = \frac{16 \pm \sqrt{256 - 160}}{16} = 1 \pm \frac{\sqrt{6}}{4}$. But $c = \sqrt{b} < 1 \quad\Rightarrow\quad c = 1 - \frac{\sqrt{6}}{4} \quad\Rightarrow$

$b = c^2 = 1 + \frac{3}{8} - \frac{\sqrt{6}}{2} = \frac{1}{8}\left(11 - 4\sqrt{6}\right) \approx 0.1503.$

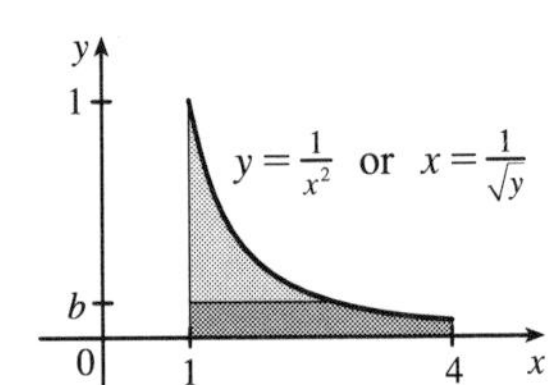

35. The area under the graph of f from 0 to t is equal to $\int_0^t f(x)\,dx$, so the requirement is that $\int_0^t f(x)\,dx = t^3$ for all t. We differentiate both sides of this equation with respect to t (with the help of FTC1) to get $f(t) = 3t^2$. This function is positive and continuous, as required.

36. 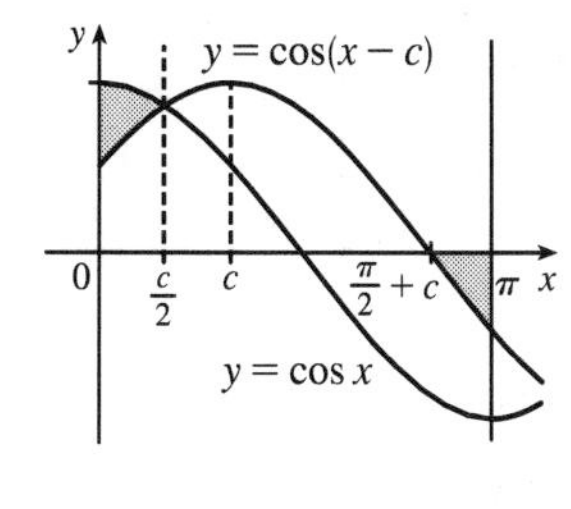

It appears from the diagram that the curves $y = \cos x$ and $y = \cos(x - c)$ intersect halfway between 0 and c, namely, when $x = c/2$. We can verify that this is indeed true by noting that $\cos(c/2 - c) = \cos(-c/2) = \cos(c/2)$. The point where $\cos(x - c)$ crosses the x-axis is $x = \frac{\pi}{2} + c$. So we require that $\int_0^{c/2} [\cos x - \cos(x - c)]\,dx = -\int_{\pi/2+c}^{\pi} \cos(x - c)\,dx$ [the negative sign on the RHS is needed since the second area is beneath the x-axis] $\Leftrightarrow$ $[\sin x - \sin(x - c)]_0^{c/2} = -[\sin(x - c)]_{\pi/2+c}^{\pi}$ $\Rightarrow$ $[\sin(c/2) - \sin(-c/2)] - [-\sin(-c)] = -\sin(\pi - c) + \sin\left[\left(\frac{\pi}{2} + c\right) - c\right]$ $\Leftrightarrow$ $2\sin(c/2) - \sin c = -\sin c + 1$. [Here we have used the oddness of the sine function, and the fact that $\sin(\pi - c) = \sin c$]. So $2\sin(c/2) = 1$ $\Leftrightarrow$ $\sin(c/2) = \frac{1}{2}$ $\Leftrightarrow$ $c/2 = \frac{\pi}{6}$ $\Leftrightarrow$ $c = \frac{\pi}{3}$.

37. 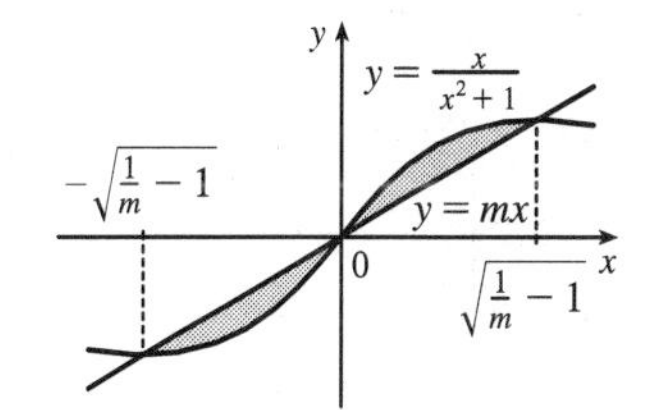

The curve and the line will determine a region when they intersect at two or more points. So we solve the equation $x/(x^2 + 1) = mx$ $\Rightarrow$ $x = x(mx^2 + m)$ $\Rightarrow$ $x(mx^2 + m) - x = 0$ $\Rightarrow$ $x(mx^2 + m - 1) = 0$ $\Rightarrow$ $x = 0$ or $mx^2 + m - 1 = 0$ $\Rightarrow$ $x = 0$ or $x^2 = \dfrac{1 - m}{m}$ $\Rightarrow$ $x = 0$ or $x = \pm\sqrt{\dfrac{1}{m} - 1}$.

Note that if $m = 1$, this has only the solution $x = 0$, and no region is determined. But if $1/m - 1 > 0$ $\Leftrightarrow$ $1/m > 1$ $\Leftrightarrow$ $0 < m < 1$, then there are two solutions. [Another way of seeing this is to observe that the slope of the tangent to $y = x/(x^2 + 1)$ at the origin is $y' = 1$ and therefore we must have $0 < m < 1$.] Note that we cannot just integrate between the positive and negative roots, since the curve and the line cross at the origin. Since mx and $x/(x^2 + 1)$ are both odd functions, the total area is twice the area between the curves on the interval $\left[0, \sqrt{1/m - 1}\right]$. So the total area enclosed is

$$\begin{aligned} 2\int_0^{\sqrt{1/m-1}} \left[\frac{x}{x^2 + 1} - mx\right] dx &= 2\left[\frac{1}{2}\ln(x^2 + 1) - \frac{1}{2}mx^2\right]_0^{\sqrt{1/m-1}} \\ &= [\ln(1/m - 1 + 1) - m(1/m - 1)] - (\ln 1 - 0) \\ &= \ln(1/m) - 1 + m = m - \ln m - 1 \end{aligned}$$

7.2 Volumes

1. A cross-section is a disk with radius $1/x$, so its area is $A(x) = \pi(1/x)^2$.

$$V = \int_1^2 A(x)\,dx = \int_1^2 \pi\left(\frac{1}{x}\right)^2 dx$$
$$= \pi\int_1^2 \frac{1}{x^2}\,dx = \pi\left[-\frac{1}{x}\right]_1^2$$
$$= \pi\left[-\tfrac{1}{2} - (-1)\right] = \frac{\pi}{2}$$

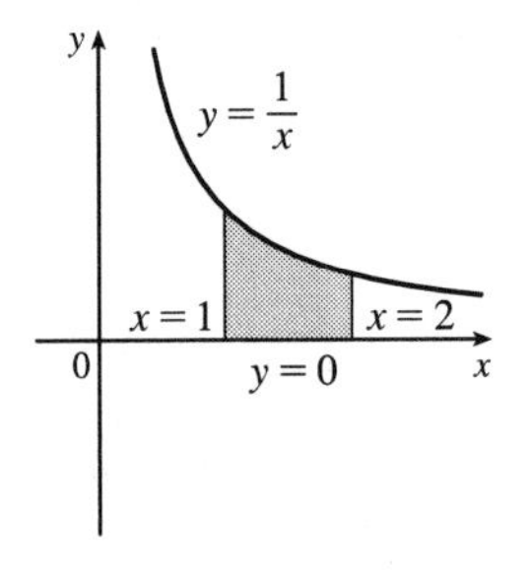

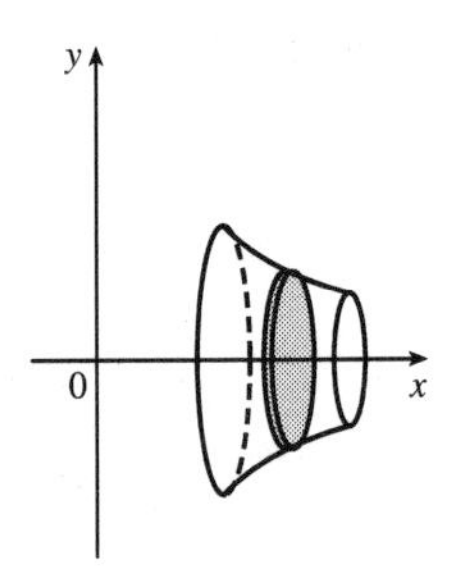

2. A cross-section is a disk with radius $1 - x^2$, so its area is $A(x) = \pi(1 - x^2)^2$.

$$V = \int_{-1}^{1} A(x)\,dx = \int_{-1}^{1} \pi(1 - x^2)^2\,dx$$
$$= 2\pi\int_0^1 (1 - 2x^2 + x^4)\,dx = 2\pi\left[x - \tfrac{2}{3}x^3 + \tfrac{1}{5}x^5\right]_0^1$$
$$= 2\pi\left(1 - \tfrac{2}{3} + \tfrac{1}{5}\right) = 2\pi\left(\tfrac{8}{15}\right) = \tfrac{16\pi}{15}$$

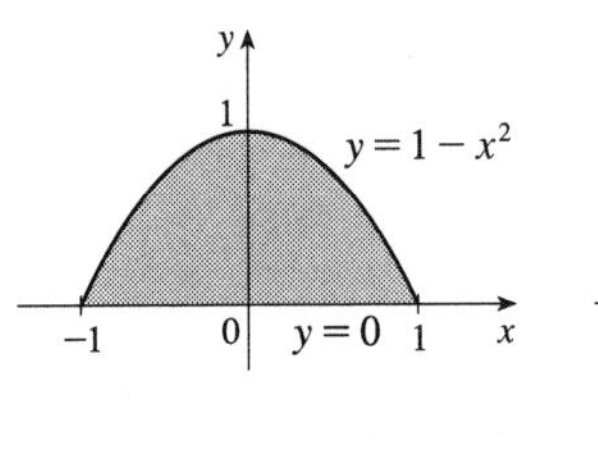

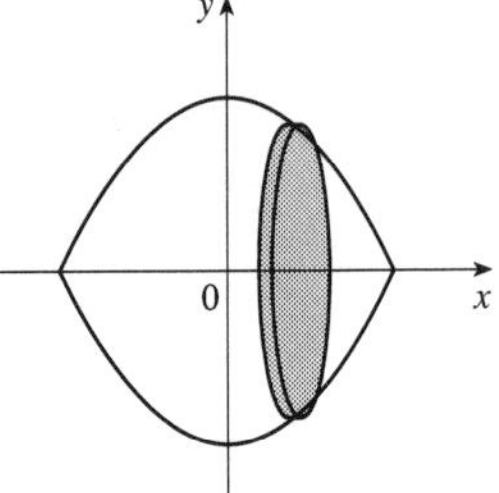

3. A cross-section is a disk with radius $2\sqrt{y}$, so its area is $A(y) = \pi\left(2\sqrt{y}\right)^2$.

$$V = \int_0^9 A(y)\,dy = \int_0^9 \pi\left(2\sqrt{y}\right)^2 dy$$
$$= 4\pi\int_0^9 y\,dy = 4\pi\left[\tfrac{1}{2}y^2\right]_0^9$$
$$= 2\pi(81) = 162\pi$$

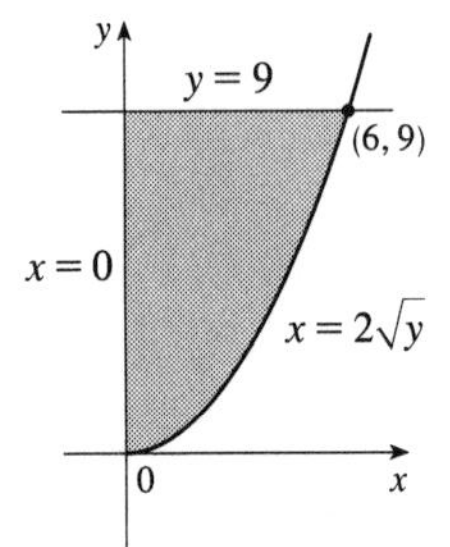

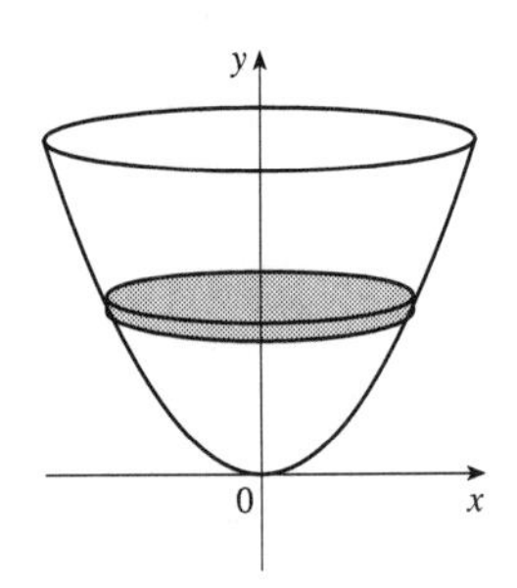

4. A cross-section is a disk with radius e^y, so its area is $A(y) = \pi(e^y)^2$.

$$V = \int_1^2 \pi(e^y)^2\,dy = \pi\int_1^2 e^{2y}\,dy = \pi\left[\frac{1}{2}e^{2y}\right]_1^2 = \frac{\pi}{2}(e^4 - e^2)$$

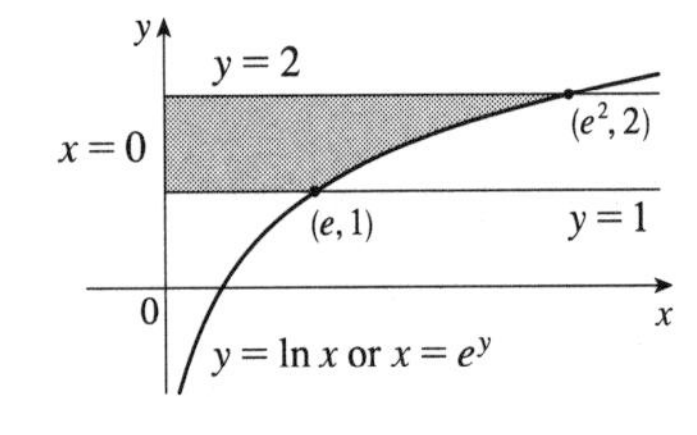

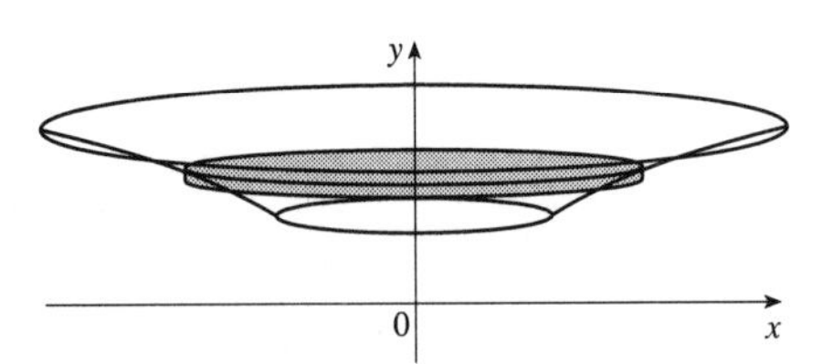

5. A cross-section is a washer (annulus) with inner radius x^3 and outer radius x, so its area is $A(x) = \pi(x)^2 - \pi(x^3)^2 = \pi(x^2 - x^6)$.

$$V = \int_0^1 A(x)\,dx = \int_0^1 \pi(x^2 - x^6)\,dx$$
$$= \pi\left[\tfrac{1}{3}x^3 - \tfrac{1}{7}x^7\right]_0^1 = \pi\left(\tfrac{1}{3} - \tfrac{1}{7}\right) = \tfrac{4\pi}{21}$$

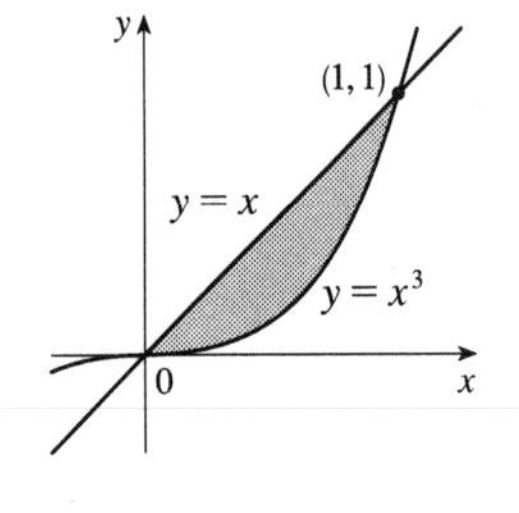

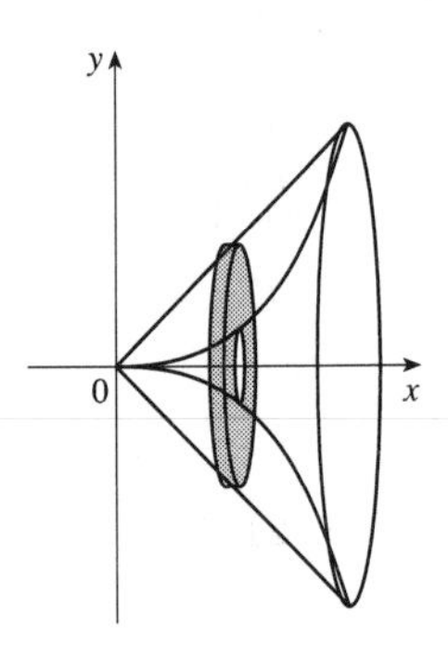

6. A cross-section is a washer with inner radius $\frac{1}{4}x^2$ and outer radius $5 - x^2$, so its area is

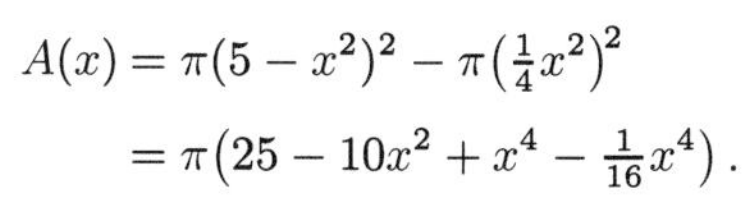

$$A(x) = \pi(5 - x^2)^2 - \pi\left(\tfrac{1}{4}x^2\right)^2 = \pi\left(25 - 10x^2 + x^4 - \tfrac{1}{16}x^4\right).$$

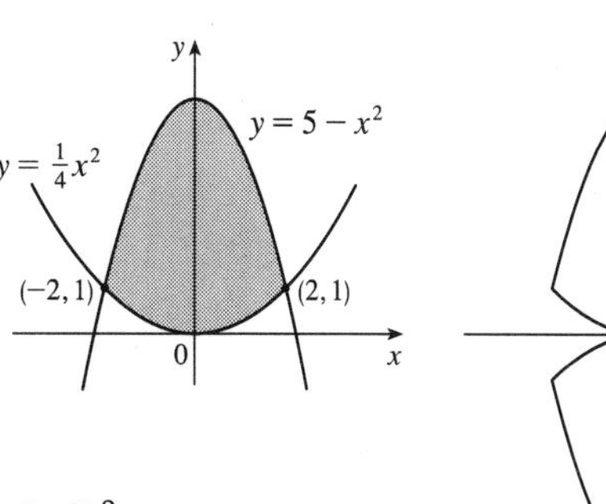

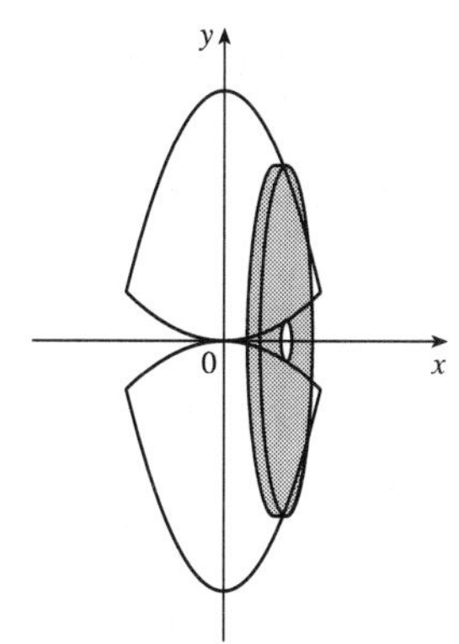

$$V = \int_{-2}^{2} A(x)\,dx = \int_{-2}^{2} \pi\left(25 - 10x^2 + \tfrac{15}{16}x^4\right)dx$$
$$= 2\pi\int_0^2 \left(25 - 10x^2 + \tfrac{15}{16}x^4\right)dx = 2\pi\left[25x - \tfrac{10}{3}x^3 + \tfrac{3}{16}x^5\right]_0^2$$
$$= 2\pi\left(50 - \tfrac{80}{3} + 6\right) = \tfrac{176\pi}{3}$$

7. A cross-section is a washer with inner radius y^2 and outer radius $2y$, so its area is

$$A(y) = \pi(2y)^2 - \pi(y^2)^2 = \pi(4y^2 - y^4).$$

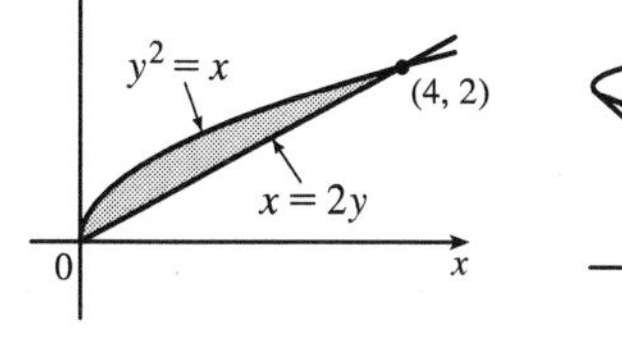

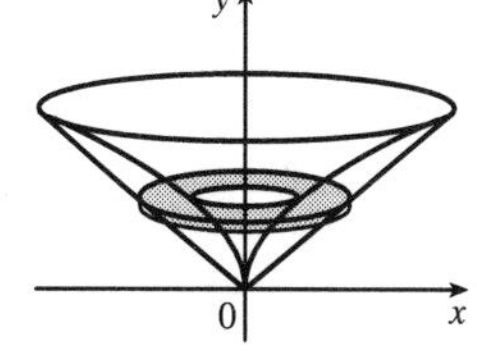

$$V = \int_0^2 A(y)\,dy = \pi\int_0^2 (4y^2 - y^4)\,dy$$
$$= \pi\left[\tfrac{4}{3}y^3 - \tfrac{1}{5}y^5\right]_0^2 = \pi\left(\tfrac{32}{3} - \tfrac{32}{5}\right) = \tfrac{64\pi}{15}$$

8. $y = x^{2/3} \quad\Leftrightarrow\quad x = y^{3/2}$, so a cross-section is a washer with inner radius $y^{3/2}$ and outer radius 1, and its area is

$$A(y) = \pi(1)^2 - \pi(y^{3/2})^2 = \pi(1 - y^3).$$

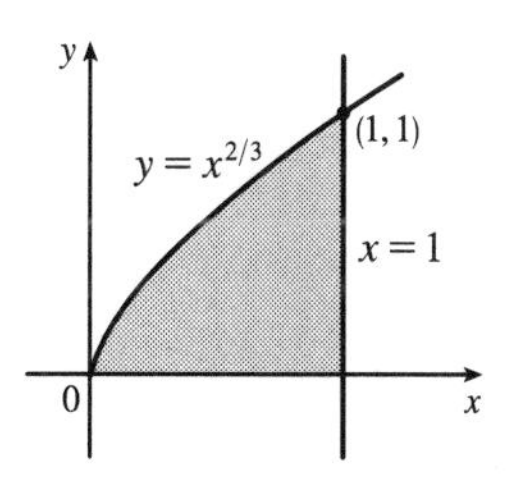

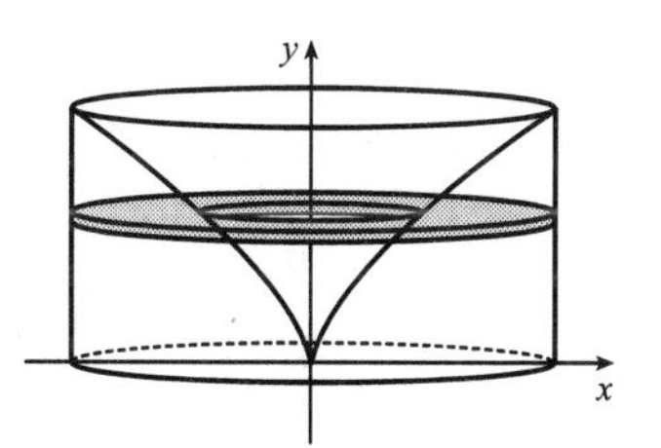

$$V = \int_0^1 A(y)\,dy = \pi\int_0^1 (1 - y^3)\,dy$$
$$= \pi\left[y - \tfrac{1}{4}y^4\right]_0^1 = \tfrac{3}{4}\pi$$

9. A cross-section is a washer with inner radius $1 - \sqrt{x}$ and outer radius $1 - x$, so its area is

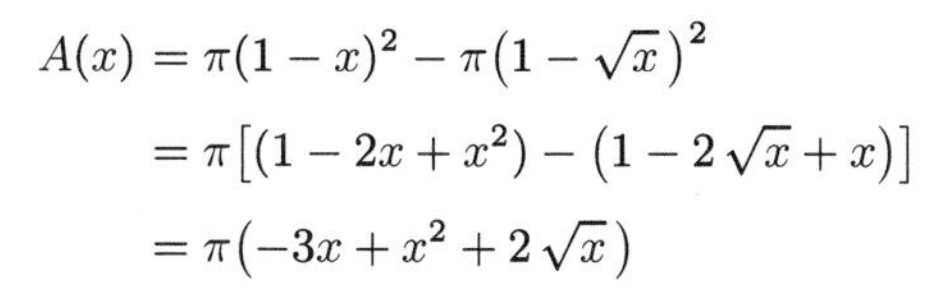

$$A(x) = \pi(1 - x)^2 - \pi\left(1 - \sqrt{x}\right)^2$$
$$= \pi\left[(1 - 2x + x^2) - \left(1 - 2\sqrt{x} + x\right)\right]$$
$$= \pi\left(-3x + x^2 + 2\sqrt{x}\right)$$

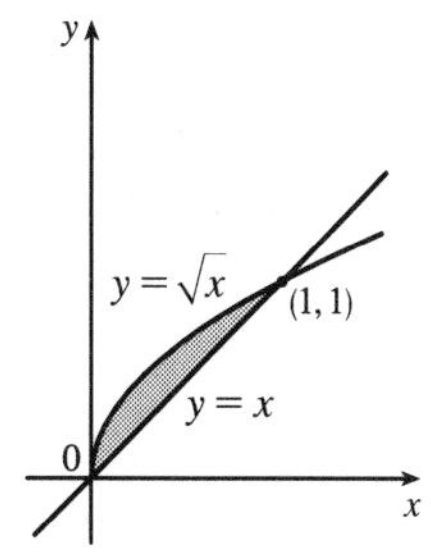

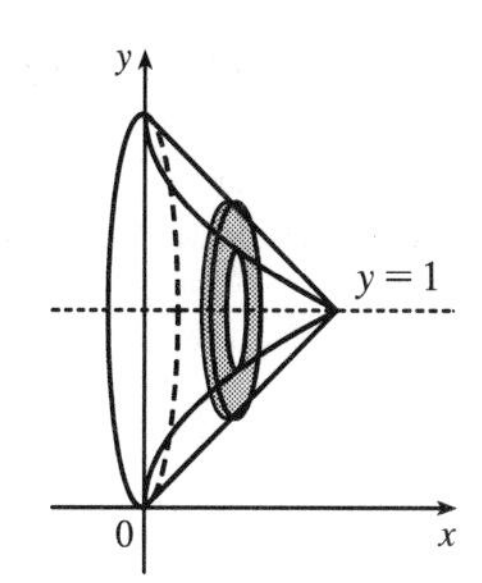

$$V = \int_0^1 A(x)\,dx = \pi\int_0^1 \left(-3x + x^2 + 2\sqrt{x}\right)dx = \pi\left[-\tfrac{3}{2}x^2 + \tfrac{1}{3}x^3 + \tfrac{4}{3}x^{3/2}\right]_0^1 = \pi\left(-\tfrac{3}{2} + \tfrac{5}{3}\right) = \tfrac{\pi}{6}$$

10. $V = \displaystyle\int_1^3 \pi\left\{\left[\frac{1}{x} - (-1)\right]^2 - [0 - (-1)]^2\right\}dx$

$$= \pi\int_1^3 \left[\left(\frac{1}{x} + 1\right)^2 - 1^2\right]dx$$
$$= \pi\int_1^3 \left(\frac{1}{x^2} + \frac{2}{x}\right)dx = \pi\left[-\frac{1}{x} + 2\ln x\right]_1^3$$
$$= \pi\left[\left(-\tfrac{1}{3} + 2\ln 3\right) - (-1 + 0)\right]$$
$$= \pi\left(2\ln 3 + \tfrac{2}{3}\right) = 2\pi\left(\ln 3 + \tfrac{1}{3}\right)$$

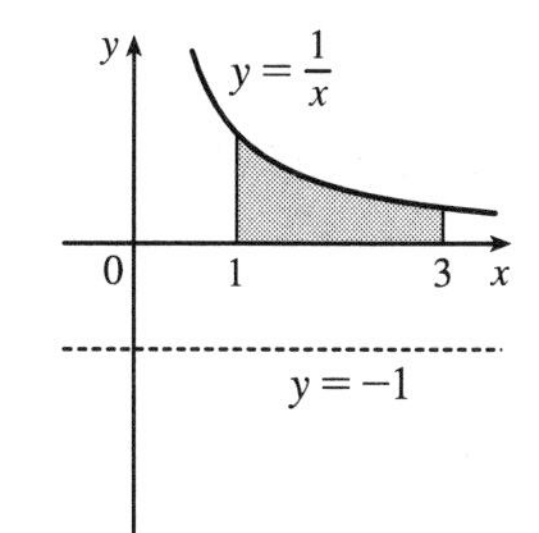

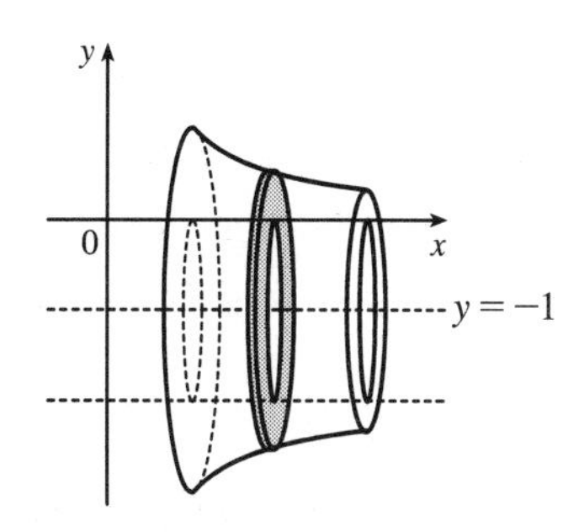

11. $y = x^2 \;\Rightarrow\; x = \sqrt{y}$ for $x \geq 0$. The outer radius is the distance from $x = -1$ to $x = \sqrt{y}$ and the inner radius is the distance from $x = -1$ to $x = y^2$.

$$V = \int_0^1 \pi \left\{ \left[\sqrt{y} - (-1)\right]^2 - \left[y^2 - (-1)\right]^2 \right\} dy = \pi \int_0^1 \left[\left(\sqrt{y} + 1\right)^2 - \left(y^2 + 1\right)^2 \right] dy$$

$$= \pi \int_0^1 \left(y + 2\sqrt{y} + 1 - y^4 - 2y^2 - 1 \right) dy = \pi \int_0^1 \left(y + 2\sqrt{y} - y^4 - 2y^2 \right) dy$$

$$= \pi \left[\tfrac{1}{2}y^2 + \tfrac{4}{3}y^{3/2} - \tfrac{1}{5}y^5 - \tfrac{2}{3}y^3 \right]_0^1 = \pi \left(\tfrac{1}{2} + \tfrac{4}{3} - \tfrac{1}{5} - \tfrac{2}{3} \right) = \tfrac{29}{30}\pi$$

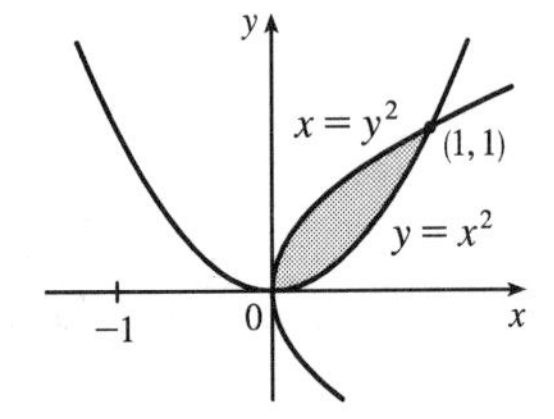

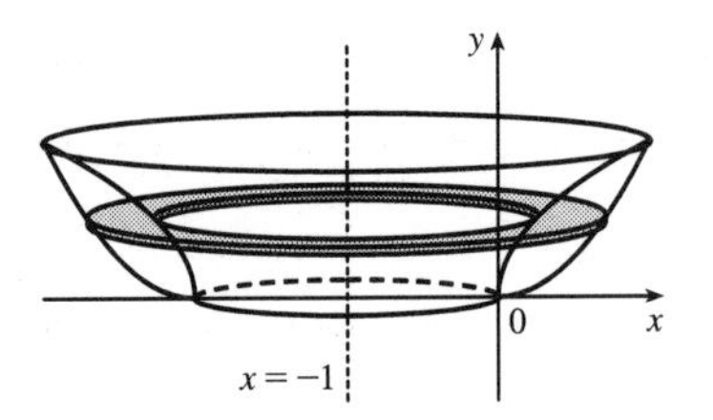

12. $y = \sqrt{x} \;\Rightarrow\; x = y^2$, so the outer radius is $2 - y^2$.

$$V = \int_0^1 \pi \left[(2 - y^2)^2 - (2 - y)^2 \right] dy = \pi \int_0^1 \left[(4 - 4y^2 + y^4) - (4 - 4y + y^2) \right] dy$$

$$= \pi \int_0^1 (y^4 - 5y^2 + 4y)\, dy = \pi \left[\tfrac{1}{5}y^5 - \tfrac{5}{3}y^3 + 2y^2 \right]_0^1 = \pi \left(\tfrac{1}{5} - \tfrac{5}{3} + 2 \right) = \tfrac{8}{15}\pi$$

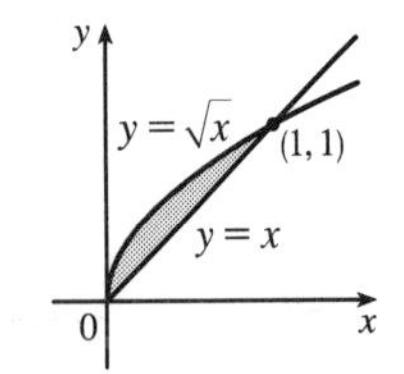

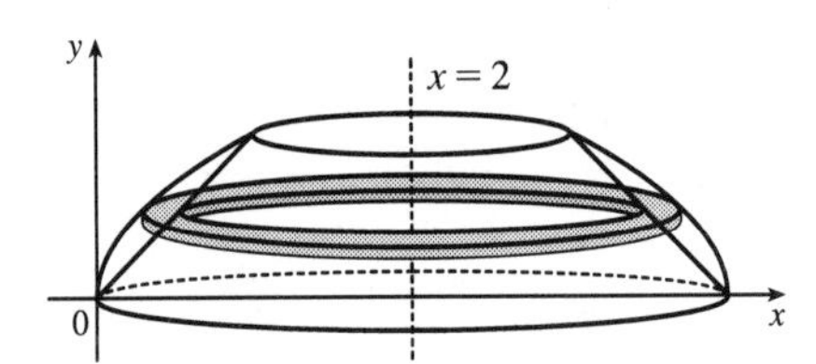

13. $y = \sqrt{x} \;\Rightarrow\; x = y^2$ and $y = x^3 \;\Rightarrow\; x = \sqrt[3]{y}$. A cross-section is a washer with inner radius $1 - \sqrt[3]{y}$ and outer radius $1 - y^2$, so its area is $A(y) = \pi(1 - y^2)^2 - \pi\left(1 - \sqrt[3]{y}\right)^2$.

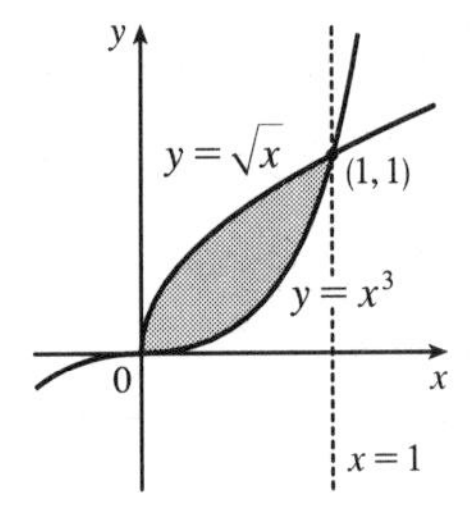

$$V = \int_0^1 A(y)\, dy = \int_0^1 \left[\pi(1 - y^2)^2 - \pi\left(1 - \sqrt[3]{y}\right)^2 \right] dy$$

$$= \pi \int_0^1 \left[(1 - 2y^2 + y^4) - (1 - 2y^{1/3} + y^{2/3}) \right] dy$$

$$= \pi \int_0^1 (-2y^2 + y^4 + 2y^{1/3} - y^{2/3})\, dy = \pi \left[-\tfrac{2}{3}y^3 + \tfrac{1}{5}y^5 + \tfrac{3}{2}y^{4/3} - \tfrac{3}{5}y^{5/3} \right]_0^1 = \pi \left(-\tfrac{2}{3} + \tfrac{1}{5} + \tfrac{3}{2} - \tfrac{3}{5} \right) = \tfrac{13\pi}{30}$$

14. A cross-section is a washer with inner radius $1 - \sqrt{x}$ and outer radius $1 - x^3$, so its area is $A(x) = \pi(1 - x^3)^2 - \pi\left(1 - \sqrt{x}\right)^2$.

$$V = \int_0^1 A(x)\, dx = \int_0^1 \left[\pi(1 - x^3)^2 - \pi\left(1 - \sqrt{x}\right)^2 \right] dx = \pi \int_0^1 \left[(1 - 2x^3 + x^6) - (1 - 2x^{1/2} + x) \right] dx$$

$$= \pi \int_0^1 (-2x^3 + x^6 + 2x^{1/2} - x)\, dx = \pi \left[-\tfrac{1}{2}x^4 + \tfrac{1}{7}x^7 + \tfrac{4}{3}x^{3/2} - \tfrac{1}{2}x^2 \right]_0^1 = \pi \left(-\tfrac{1}{2} + \tfrac{1}{7} + \tfrac{4}{3} - \tfrac{1}{2} \right) = \tfrac{10\pi}{21}$$

15. $V = \pi \int_0^{\pi/4} (1 - \tan^3 x)^2\, dx$

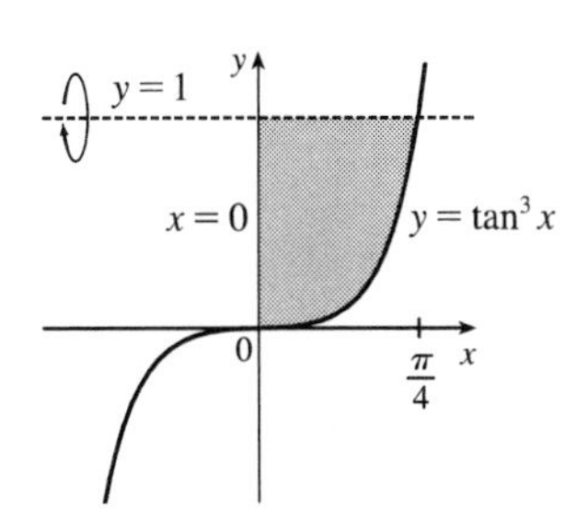

16. $y = (x-2)^4$ and $8x - y = 16$ intersect when

$(x-2)^4 = 8x - 16 = 8(x-2) \iff (x-2)^4 - 8(x-2) = 0 \iff$

$(x-2)\left[(x-2)^3 - 8\right] = 0 \iff x - 2 = 0$ or $x - 2 = 2 \iff$

$x = 2$ or 4. $y = (x-2)^4 \Rightarrow x - 2 = \pm\sqrt[4]{y} \Rightarrow x = 2 + \sqrt[4]{y}$

[since $x \geq 2$]. $8x - y = 16 \Rightarrow 8x = y + 16 \Rightarrow x = \frac{1}{8}y + 2$.

$$V = \pi \int_0^{16} \left\{ \left[10 - \left(\tfrac{1}{8}y + 2\right)\right]^2 - \left[10 - \left(2 + \sqrt[4]{y}\right)\right]^2 \right\} dy$$

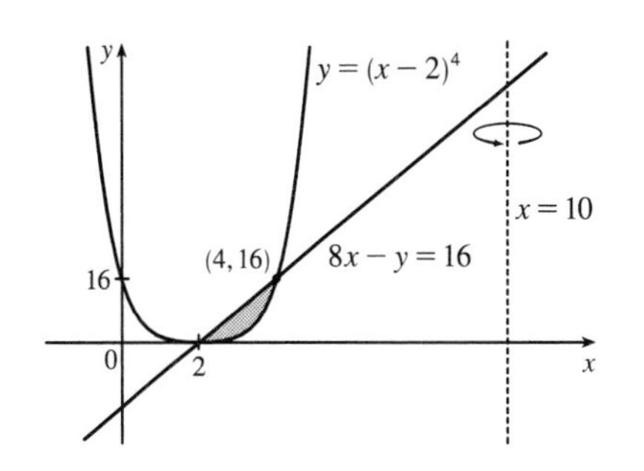

17.

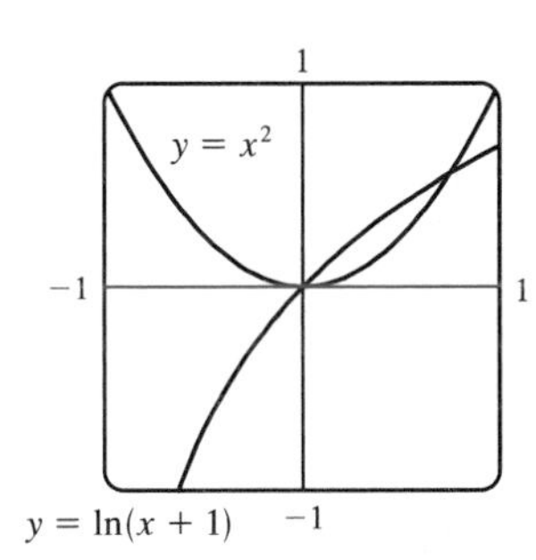

$y = x^2$ and $y = \ln(x+1)$ intersect at $x = 0$ and at $x = a \approx 0.747$.

$$V = \pi \int_0^a \left\{ [\ln(x+1)]^2 - (x^2)^2 \right\} dx \approx 0.132$$

18.

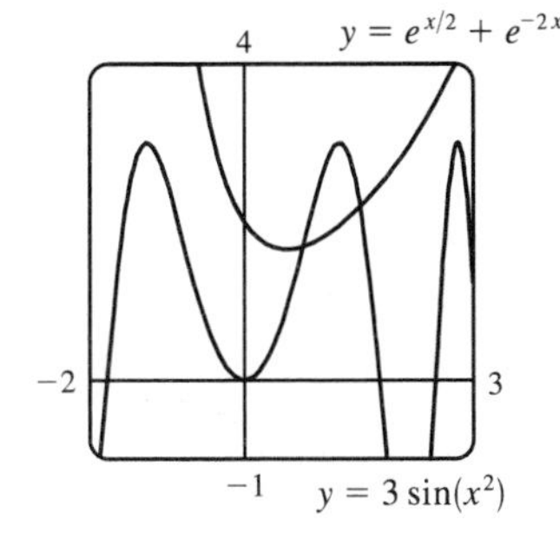

$y = 3\sin(x^2)$ and $y = e^{x/2} + e^{-2x}$ intersect at $x = a \approx 0.772$ and at $x = b \approx 1.524$.

$$V = \pi \int_a^b \left\{ [3\sin(x^2)]^2 - (e^{x/2} + e^{-2x})^2 \right\} dx \approx 7.519$$

19. $V = \pi \int_0^{\pi} \left\{ [\sin^2 x - (-1)]^2 - [0 - (-1)]^2 \right\} dx$

$\overset{\text{CAS}}{=} \dfrac{11}{8}\pi^2$

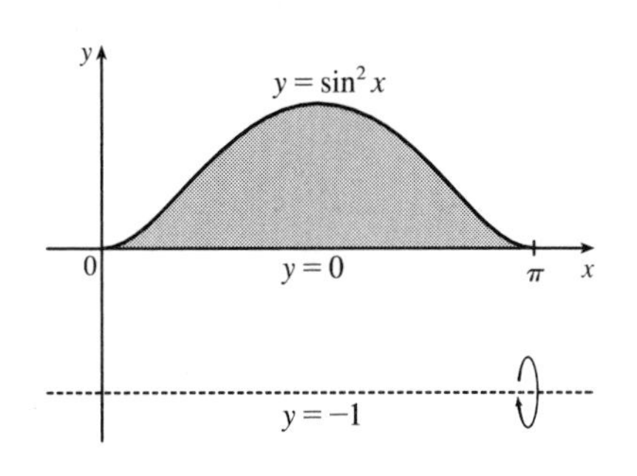

20. $V = \pi \int_0^2 \left[(3-x)^2 - (3 - xe^{1-x/2})^2\right] dx$

$\overset{\text{CAS}}{=} \pi\left(-2e^2 + 24e - \frac{142}{3}\right)$

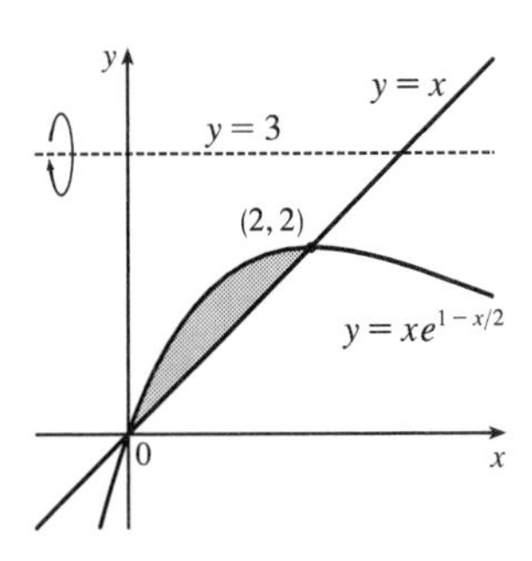

21. (a) $\pi \int_0^{\pi/2} \cos^2 x\, dx$ describes the volume of the solid obtained by rotating the region $\mathcal{R} = \left\{(x,y) \mid 0 \le x \le \frac{\pi}{2}, 0 \le y \le \cos x\right\}$ of the xy-plane about the x-axis.

(b) $\pi \int_0^1 (y^4 - y^8)\, dy = \pi \int_0^1 \left[(y^2)^2 - (y^4)^2\right] dy$ describes the volume of the solid obtained by rotating the region $\mathcal{R} = \left\{(x,y) \mid 0 \le y \le 1, y^4 \le x \le y^2\right\}$ of the xy-plane about the y-axis.

22. (a) $\pi \int_2^5 y\, dy = \pi \int_2^5 \left(\sqrt{y}\right)^2 dy$ describes the volume of the solid obtained by rotating the region $\mathcal{R} = \left\{(x,y) \mid 2 \le y \le 5, 0 \le x \le \sqrt{y}\right\}$ of the xy-plane about the y-axis.

(b) $\pi \int_0^{\pi/2} \left[(1+\cos x)^2 - 1^2\right] dx$ describes the volume of the solid obtained by rotating the region $\mathcal{R} = \left\{(x,y) \mid 0 \le x \le \frac{\pi}{2}, 1 \le y \le 1 + \cos x\right\}$ of the xy-plane about the x-axis.

Or: The solid could be obtained by rotating the region $\mathcal{R}' = \left\{(x,y) \mid 0 \le x \le \frac{\pi}{2}, 0 \le y \le \cos x\right\}$ about the line $y = -1$.

23. There are 10 subintervals over the 15-cm length, so we'll use $n = 10/2 = 5$ for the Midpoint Rule.

$$V = \int_0^{15} A(x)\, dx \approx M_5 = \tfrac{15-0}{5}[A(1.5) + A(4.5) + A(7.5) + A(10.5) + A(13.5)]$$
$$= 3(18 + 79 + 106 + 128 + 39) = 3 \cdot 370 = 1110 \text{ cm}^3$$

24. $V = \int_0^{10} A(x)\, dx \approx M_5 = \frac{10-0}{5}[A(1) + A(3) + A(5) + A(7) + A(9)]$
$$= 2(0.65 + 0.61 + 0.59 + 0.55 + 0.50) = 2(2.90) = 5.80 \text{ m}^3$$

25. We'll form a right circular cone with height h and base radius r by revolving the line $y = \frac{r}{h}x$ about the x-axis.

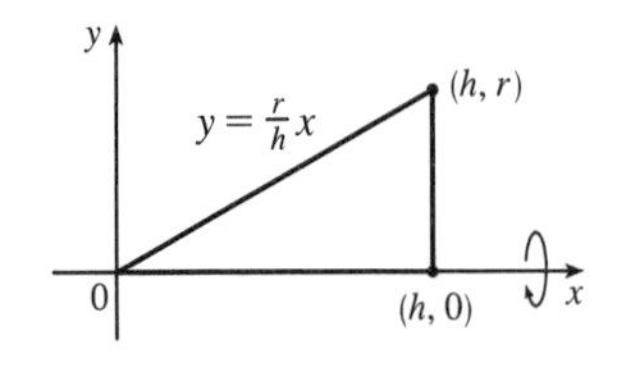

$$V = \pi \int_0^h \left(\frac{r}{h}x\right)^2 dx = \pi \int_0^h \frac{r^2}{h^2} x^2\, dx = \pi \frac{r^2}{h^2}\left[\frac{1}{3}x^3\right]_0^h$$
$$= \pi \frac{r^2}{h^2}\left(\frac{1}{3}h^3\right) = \frac{1}{3}\pi r^2 h$$

Another solution: Revolve $x = -\dfrac{r}{h}y + r$ about the y-axis.

$$V = \pi \int_0^h \left(-\frac{r}{h}y + r\right)^2 dy \overset{*}{=} \pi \int_0^h \left[\frac{r^2}{h^2}y^2 - \frac{2r^2}{h}y + r^2\right] dy$$
$$= \pi\left[\frac{r^2}{3h^2}y^3 - \frac{r^2}{h}y^2 + r^2 y\right]_0^h = \pi\left(\tfrac{1}{3}r^2h - r^2h + r^2h\right) = \tfrac{1}{3}\pi r^2 h$$

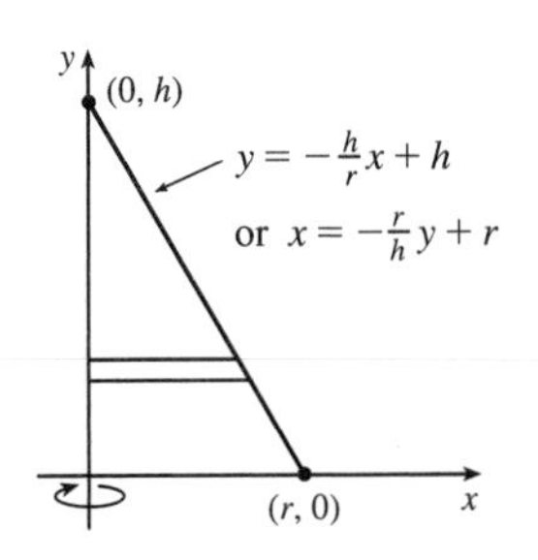

* Or use substitution with $u = r - \dfrac{r}{h}y$ and $du = -\dfrac{r}{h}dy$ to get

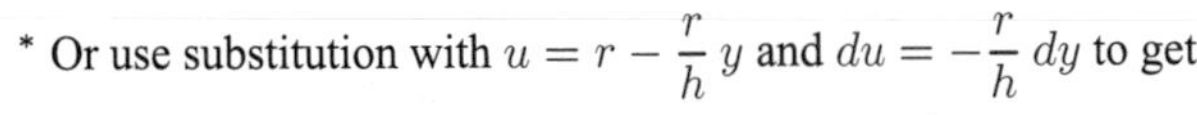

$$\pi \int_r^0 u^2\left(-\frac{h}{r}du\right) = -\pi\frac{h}{r}\left[\frac{1}{3}u^3\right]_r^0 = -\pi\frac{h}{r}\left(-\frac{1}{3}r^3\right) = \frac{1}{3}\pi r^2 h.$$

26. $V = \pi \int_0^h \left(R - \frac{R-r}{h}y\right)^2 dy = \pi \int_0^h \left[R^2 - \frac{2R(R-r)}{h}y + \left(\frac{R-r}{h}\right)^2 y^2\right] dy$

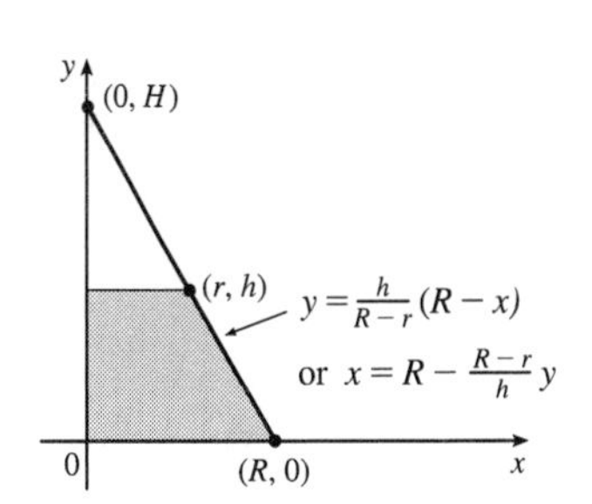

$= \pi \left[R^2 y - \frac{R(R-r)}{h}y^2 + \frac{1}{3}\left(\frac{R-r}{h}\right)^2 y^3\right]_0^h$

$= \pi\left[R^2 h - R(R-r)h + \frac{1}{3}(R-r)^2 h\right]$

$= \frac{1}{3}\pi h\left[3Rr + (R^2 - 2Rr + r^2)\right] = \frac{1}{3}\pi h(R^2 + Rr + r^2)$

Another solution: $\frac{H}{R} = \frac{H-h}{r}$ by similar triangles. Therefore,

$Hr = HR - hR \;\Rightarrow\; hR = H(R-r) \;\Rightarrow\; H = \frac{hR}{R-r}$. Now

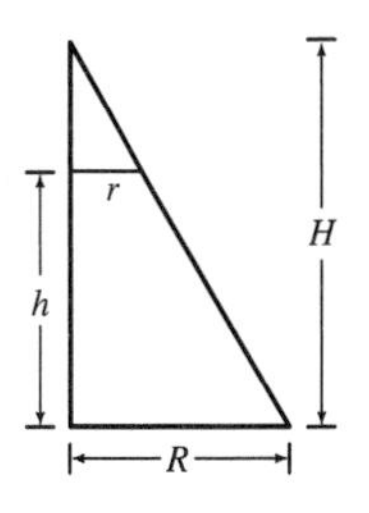

$V = \frac{1}{3}\pi R^2 H - \frac{1}{3}\pi r^2 (H-h)$ [by Exercise 25]

$= \frac{1}{3}\pi R^2 \frac{hR}{R-r} - \frac{1}{3}\pi r^2 \frac{rh}{R-r}$ $\left[H - h = \frac{rH}{R} = \frac{rhR}{R(R-r)}\right]$

$= \frac{1}{3}\pi h \frac{R^3 - r^3}{R-r} = \frac{1}{3}\pi h(R^2 + Rr + r^2) = \frac{1}{3}\left[\pi R^2 + \pi r^2 + \sqrt{(\pi R^2)(\pi r^2)}\right]h = \frac{1}{3}\left(A_1 + A_2 + \sqrt{A_1 A_2}\,\right)h$

where A_1 and A_2 are the areas of the bases of the frustum. (See Exercise 28 for a related result.)

27. $x^2 + y^2 = r^2 \;\Leftrightarrow\; x^2 = r^2 - y^2$

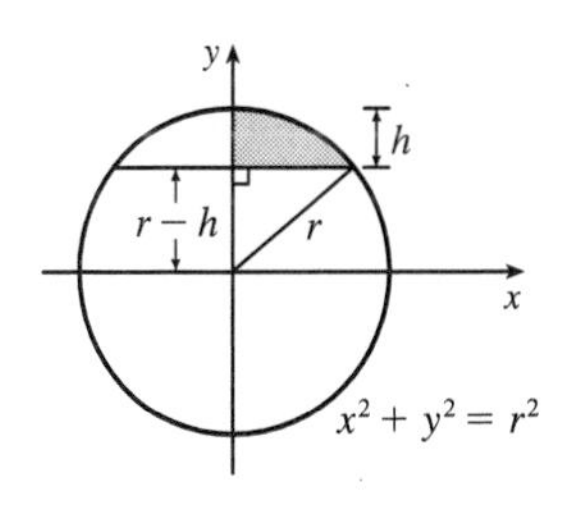

$V = \pi \int_{r-h}^{r} (r^2 - y^2)\,dy = \pi \left[r^2 y - \frac{y^3}{3}\right]_{r-h}^{r}$

$= \pi \left\{\left[r^3 - \frac{r^3}{3}\right] - \left[r^2(r-h) - \frac{(r-h)^3}{3}\right]\right\}$

$= \pi\left\{\frac{2}{3}r^3 - \frac{1}{3}(r-h)[3r^2 - (r-h)^2]\right\}$

$= \frac{1}{3}\pi\left\{2r^3 - (r-h)[3r^2 - (r^2 - 2rh + h^2)]\right\}$

$= \frac{1}{3}\pi\left\{2r^3 - (r-h)[2r^2 + 2rh - h^2]\right\} = \frac{1}{3}\pi(2r^3 - 2r^3 - 2r^2 h + rh^2 + 2r^2 h + 2rh^2 - h^3)$

$= \frac{1}{3}\pi(3rh^2 - h^3) = \frac{1}{3}\pi h^2(3r - h)$, or, equivalently, $\pi h^2\left(r - \frac{h}{3}\right)$

28. An equation of the line is $x = \frac{\Delta x}{\Delta y}y + (x\text{-intercept}) = \frac{a/2 - b/2}{h - 0}y + \frac{b}{2} = \frac{a-b}{2h}y + \frac{b}{2}$.

$V = \int_0^h A(y)\,dy = \int_0^h (2x)^2\,dy = \int_0^h \left[2\left(\frac{a-b}{2h}y + \frac{b}{2}\right)\right]^2 dy$

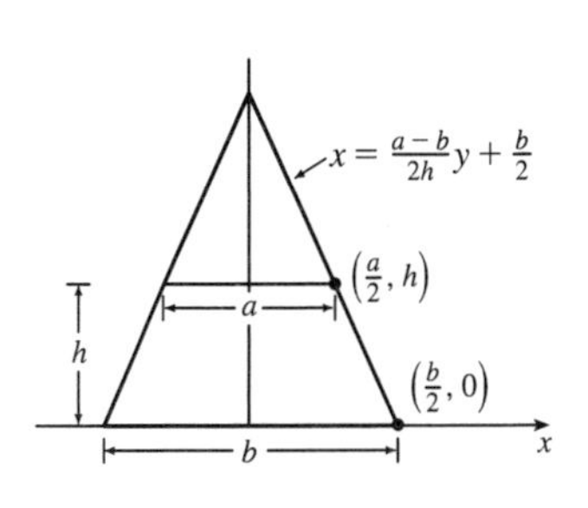

$= \int_0^h \left[\frac{a-b}{h}y + b\right]^2 dy = \int_0^h \left[\frac{(a-b)^2}{h^2}y^2 + \frac{2b(a-b)}{h}y + b^2\right] dy$

$= \left[\frac{(a-b)^2}{3h^2}y^3 + \frac{b(a-b)}{h}y^2 + b^2 y\right]_0^h = \frac{1}{3}(a-b)^2 h + b(a-b)h + b^2 h$

$= \frac{1}{3}(a^2 - 2ab + b^2 + 3ab)h = \frac{1}{3}(a^2 + ab + b^2)h$

[Note that this can be written as $\frac{1}{3}\left(A_1 + A_2 + \sqrt{A_1 A_2}\,\right)h$, as in Exercise 26.]

If $a = b$, we get a rectangular solid with volume $b^2 h$. If $a = 0$, we get a square pyramid with volume $\frac{1}{3}b^2 h$.

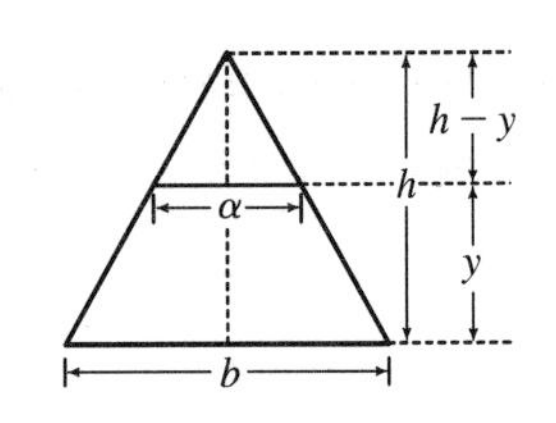

29. For a cross-section at height y, we see from similar triangles that $\dfrac{\alpha/2}{b/2} = \dfrac{h-y}{h}$, so

$\alpha = b\left(1 - \dfrac{y}{h}\right)$. Similarly, for cross-sections having $2b$ as their base and β replacing

α, $\beta = 2b\left(1 - \dfrac{y}{h}\right)$. So

$$V = \int_0^h A(y)\,dy = \int_0^h \left[b\left(1-\frac{y}{h}\right)\right]\left[2b\left(1-\frac{y}{h}\right)\right]dy = \int_0^h 2b^2\left(1-\frac{y}{h}\right)^2 dy$$

$$= 2b^2 \int_0^h \left(1 - \frac{2y}{h} + \frac{y^2}{h^2}\right)dy = 2b^2\left[y - \frac{y^2}{h} + \frac{y^3}{3h^2}\right]_0^h = 2b^2\left[h - h + \tfrac{1}{3}h\right]$$

$$= \tfrac{2}{3}b^2 h \quad \left[= \tfrac{1}{3}Bh \text{ where } B \text{ is the area of the base, as with any pyramid.}\right]$$

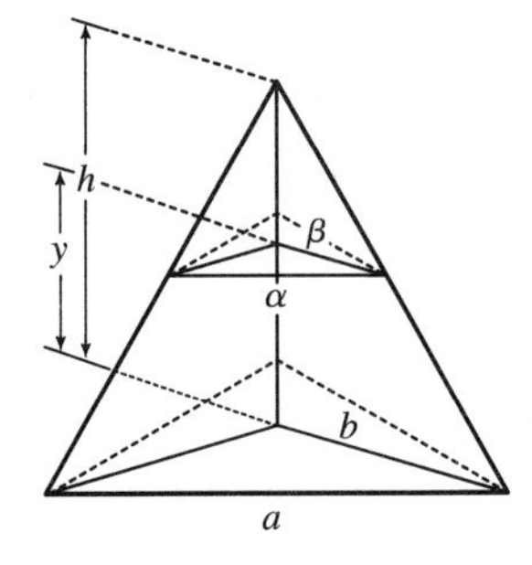

30. Consider the triangle consisting of two vertices of the base and the center of the base. This triangle is similar to the corresponding triangle at a height y, so $a/b = \alpha/\beta \;\Rightarrow\; \alpha = a\beta/b$. Also by similar triangles, $b/h = \beta/(h-y) \;\Rightarrow\; \beta = b(h-y)/h$. These two equations imply that $\alpha = a(1 - y/h)$, and since the cross-section is an equilateral triangle, it has area

$$A(y) = \frac{1}{2}\cdot\alpha\cdot\frac{\sqrt{3}}{2}\alpha = \frac{a^2(1-y/h)^2}{4}\sqrt{3}$$

so $V = \displaystyle\int_0^h A(y)\,dy = \frac{a^2\sqrt{3}}{4}\int_0^h \left(1-\frac{y}{h}\right)^2 dy = \frac{a^2\sqrt{3}}{4}\left[-\frac{h}{3}\left(1-\frac{y}{h}\right)^3\right]_0^h = -\frac{\sqrt{3}}{12}a^2h(-1) = \frac{\sqrt{3}}{12}a^2h.$

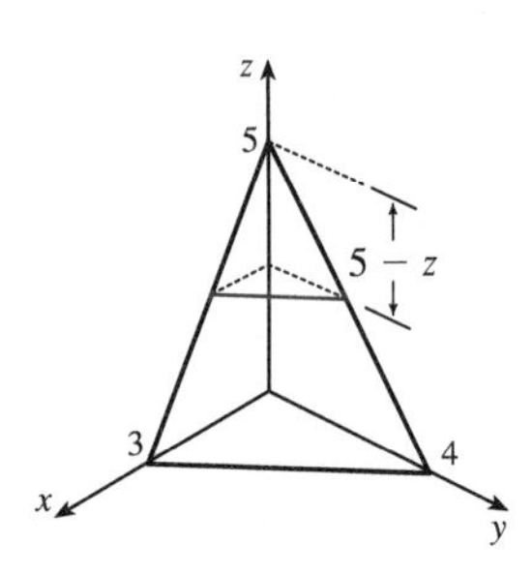

31. A cross-section at height z is a triangle similar to the base, so we'll multiply the legs of the base triangle, 3 and 4, by a proportionality factor of $(5-z)/5$. Thus, the triangle at height z has area $A(z) = \dfrac{1}{2}\cdot 3\left(\dfrac{5-z}{5}\right)\cdot 4\left(\dfrac{5-z}{5}\right) = 6\left(1 - \dfrac{z}{5}\right)^2$, so

$$V = \int_0^5 A(z)\,dz = 6\int_0^5 (1 - z/5)^2\,dz = 6\int_1^0 u^2(-5\,du) \qquad \begin{bmatrix} u = 1 - z/5, \\ du = -\frac{1}{5}dz \end{bmatrix}$$

$$= -30\left[\tfrac{1}{3}u^3\right]_1^0 = -30\left(-\tfrac{1}{3}\right) = 10 \text{ cm}^3$$

32. A cross-section is shaded in the diagram.

$A(x) = (2y)^2 = \left(2\sqrt{r^2 - x^2}\right)^2$, so

$$V = \int_{-r}^r A(x)\,dx = 2\int_0^r 4(r^2 - x^2)\,dx$$

$$= 8\left[r^2x - \tfrac{1}{3}x^3\right]_0^r = 8\left(\tfrac{2}{3}r^3\right) = \tfrac{16}{3}r^3$$

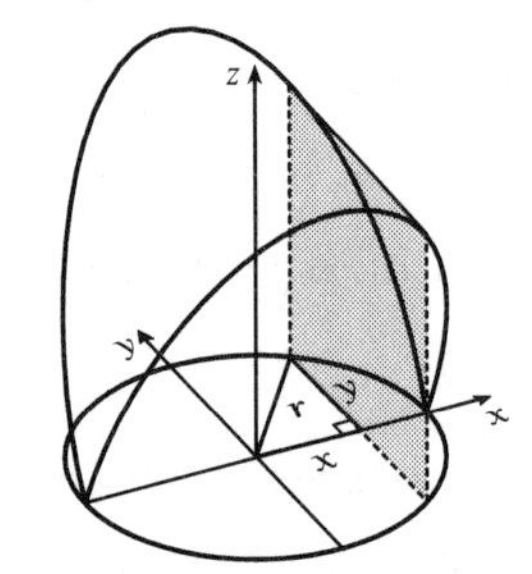

33. If l is a leg of the isosceles right triangle and $2y$ is the hypotenuse, then $l^2 + l^2 = (2y)^2 \;\Rightarrow\; 2l^2 = 4y^2 \;\Rightarrow\; l^2 = 2y^2$.

$$V = \int_{-2}^2 A(x)\,dx = 2\int_0^2 A(x)\,dx = 2\int_0^2 \tfrac{1}{2}(l)(l)\,dx = 2\int_0^2 y^2\,dx$$

$$= 2\int_0^2 \tfrac{1}{4}(36 - 9x^2)\,dx = \tfrac{9}{2}(4 - x^2)\,dx = \tfrac{9}{2}\left[4x - \tfrac{1}{3}x^3\right]_0^2 = \tfrac{9}{2}\left(8 - \tfrac{8}{3}\right) = 24$$

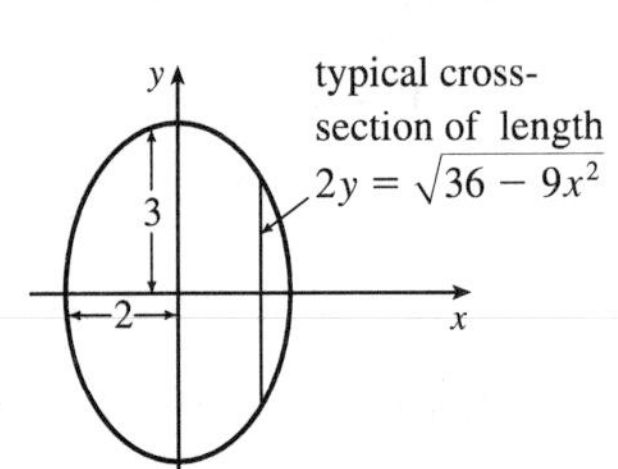

34. The cross-section of the base corresponding to the coordinate y has length $2x = 2\sqrt{y}$. The corresponding equilateral triangle with side s has area

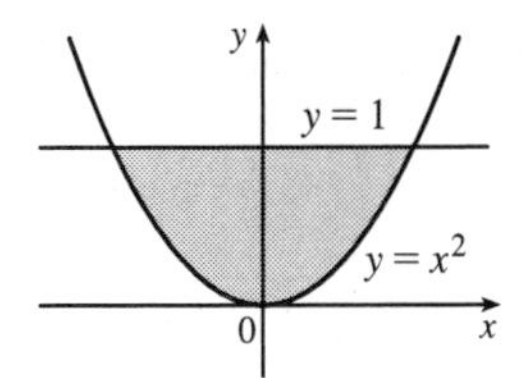

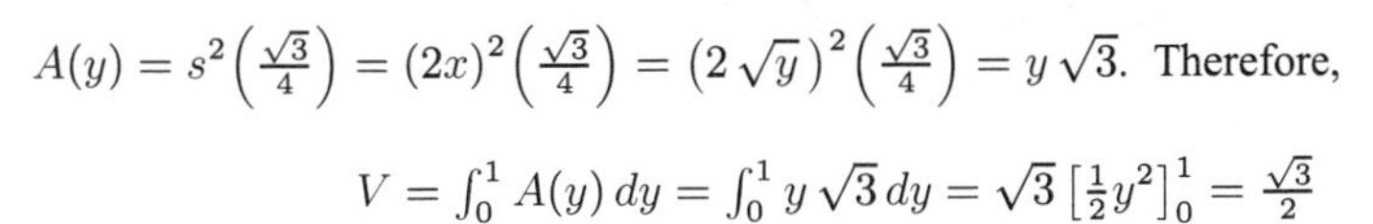

$A(y) = s^2\left(\frac{\sqrt{3}}{4}\right) = (2x)^2\left(\frac{\sqrt{3}}{4}\right) = \left(2\sqrt{y}\right)^2\left(\frac{\sqrt{3}}{4}\right) = y\sqrt{3}$. Therefore,

$$V = \int_0^1 A(y)\,dy = \int_0^1 y\sqrt{3}\,dy = \sqrt{3}\left[\tfrac{1}{2}y^2\right]_0^1 = \tfrac{\sqrt{3}}{2}$$

35. The cross-section of the base corresponding to the coordinate y has length $2x = 2\sqrt{y}$. The square has area $A(y) = \left(2\sqrt{y}\right)^2 = 4y$, so

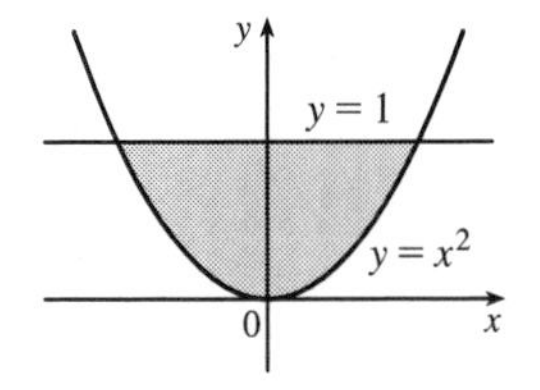

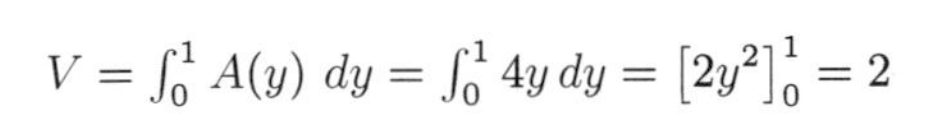

$$V = \int_0^1 A(y)\,dy = \int_0^1 4y\,dy = \left[2y^2\right]_0^1 = 2$$

36. A typical cross-section perpendicular to the y-axis in the base has length $\ell(y) = 3 - \frac{3}{2}y$. This length is the diameter of a cross-sectional semicircle in S, so

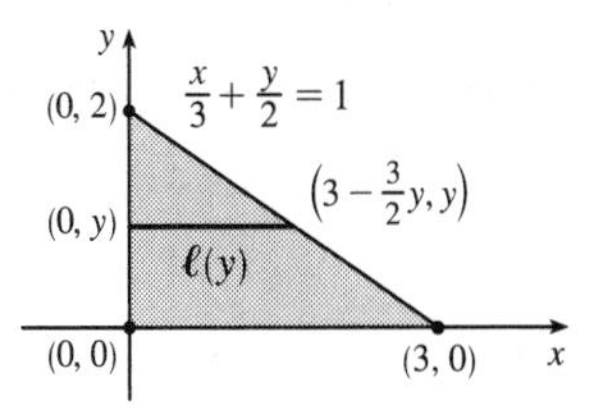

$$\begin{aligned} V &= \int_0^2 A(y)\,dy = \int_0^2 \frac{\pi}{2}\left[\frac{\ell(y)}{2}\right]^2 dy = \frac{\pi}{8}\int_0^2 \left(3 - \tfrac{3}{2}y\right)^2 dy \\ &= \tfrac{\pi}{8}\int_3^0 u^2\left(-\tfrac{2}{3}\,du\right) \quad \left[u = 3 - \tfrac{3}{2}y,\ du = -\tfrac{3}{2}\,dy\right] \\ &= -\tfrac{\pi}{12}\left[\tfrac{1}{3}u^3\right]_3^0 = -\tfrac{\pi}{12}(-9) = \tfrac{3\pi}{4} \end{aligned}$$

37. A typical cross-section perpendicular to the y-axis in the base has length $\ell(y) = 3 - \frac{3}{2}y$. This length is the leg of an isosceles right triangle, so

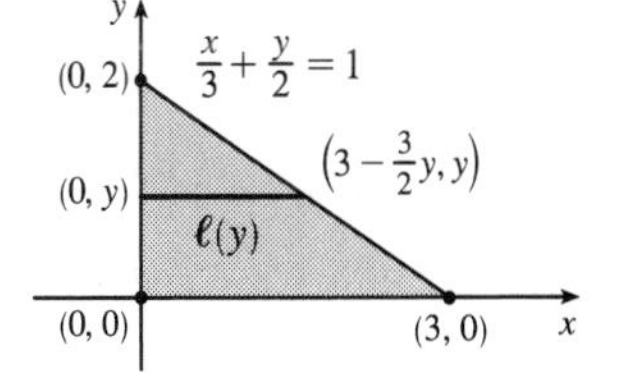

$A(y) = \frac{1}{2}\left[\ell(y)\right]^2 \quad \left[\frac{1}{2}bh \text{ with base} = \text{height}\right] \quad = \frac{1}{2}\left[3\left(1 - \frac{1}{2}y\right)\right]^2 = \frac{9}{2}\left(1 - \frac{1}{2}y\right)^2$.

Thus,

$$V = \int_0^2 A(y)\,dy = \frac{9}{2}\int_1^0 u^2(-2\,du) \left[\begin{matrix} u = 1 - \frac{1}{2}y, \\ du = -\frac{1}{2}y \end{matrix}\right] = -9\left[\tfrac{1}{3}u^3\right]_1^0 = -9\left(-\tfrac{1}{3}\right) = 3.$$

38. (a) $V = \displaystyle\int_{-r}^{r} A(x)\,dx = 2\int_0^r A(x)\,dx = 2\int_0^r \tfrac{1}{2}h\left(2\sqrt{r^2 - x^2}\right)dx = 2h\int_0^r \sqrt{r^2 - x^2}\,dx$

(b) Observe that the integral represents one quarter of the area of a circle of radius r, so $V = 2h \cdot \frac{1}{4}\pi r^2 = \frac{1}{2}\pi h r^2$.

39. (a) The radius of the barrel is the same at each end by symmetry, since the function $y = R - cx^2$ is even. Since the barrel is obtained by rotating the graph of the function y about the x-axis, this radius is equal to the value of y at $x = \frac{1}{2}h$, which is $R - c\left(\frac{1}{2}h\right)^2 = R - d = r$.

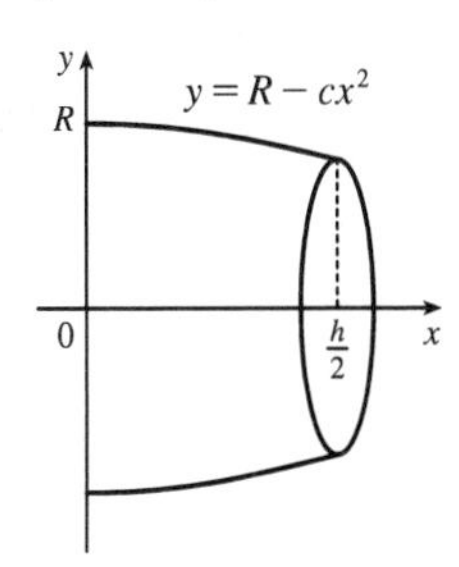

(b) The barrel is symmetric about the y-axis, so its volume is twice the volume of that part of the barrel for $x > 0$. Also, the barrel is a volume of rotation, so

$$\begin{aligned} V &= 2\int_0^{h/2} \pi y^2\,dx = 2\pi\int_0^{h/2}(R - cx^2)^2 dx \\ &= 2\pi\left[R^2x - \tfrac{2}{3}Rcx^3 + \tfrac{1}{5}c^2x^5\right]_0^{h/2} = 2\pi\left(\tfrac{1}{2}R^2h - \tfrac{1}{12}Rch^3 + \tfrac{1}{160}c^2h^5\right) \end{aligned}$$

Trying to make this look more like the expression we want, we rewrite it as $V = \frac{1}{3}\pi h\left[2R^2 + \left(R^2 - \frac{1}{2}Rch^2 + \frac{3}{80}c^2h^4\right)\right]$.

But $R^2 - \frac{1}{2}Rch^2 + \frac{3}{80}c^2h^4 = \left(R - \frac{1}{4}ch^2\right)^2 - \frac{1}{40}c^2h^4 = (R - d)^2 - \frac{2}{5}\left(\frac{1}{4}ch^2\right)^2 = r^2 - \frac{2}{5}d^2$.

Substituting this back into V, we see that $V = \frac{1}{3}\pi h\left(2R^2 + r^2 - \frac{2}{5}d^2\right)$, as required.

40. (a) $V = \int_{-1}^{1} \pi \left[(ax^3 + bx^2 + cx + d)\sqrt{1 - x^2}\right]^2 dx \overset{\text{CAS}}{=} \dfrac{4\{5a^2 + 18ac + 3[3b^2 + 14bd + 7(c^2 + 5d^2)]\}\pi}{315}$

(b) $y = (-0.06x^3 + 0.04x^2 + 0.1x + 0.54)\sqrt{1 - x^2}$ is graphed in the figure.

Substitute $a = -0.06$, $b = 0.04$, $c = 0.1$, and $d = 0.54$ in the answer for part (a)

to get $V \overset{\text{CAS}}{=} \dfrac{3769\pi}{9375} \approx 1.263$.

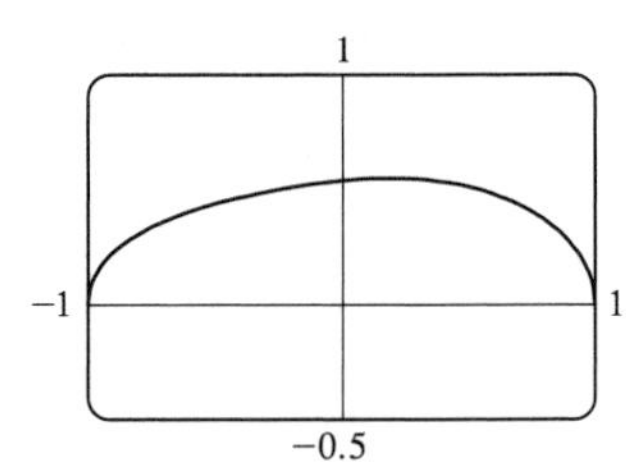

41. (a) The torus is obtained by rotating the circle $(x - R)^2 + y^2 = r^2$ about the y-axis. Solving for x, we see that the right half of the circle is given by $x = R + \sqrt{r^2 - y^2} = f(y)$ and the left half by $x = R - \sqrt{r^2 - y^2} = g(y)$. So

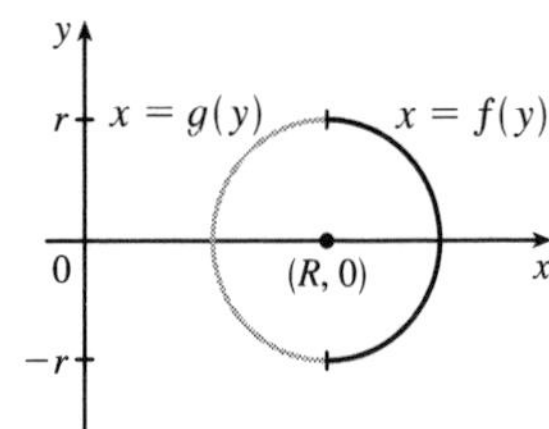

$$\begin{aligned} V &= \pi \textstyle\int_{-r}^{r} \{[f(y)]^2 - [g(y)]^2\}\, dy \\ &= 2\pi \textstyle\int_0^r \left[\left(R^2 + 2R\sqrt{r^2 - y^2} + r^2 - y^2\right) - \left(R^2 - 2R\sqrt{r^2 - y^2} + r^2 - y^2\right)\right] dy \\ &= 2\pi \textstyle\int_0^r 4R\sqrt{r^2 - y^2}\, dy = 8\pi R \int_0^r \sqrt{r^2 - y^2}\, dy \end{aligned}$$

(b) Observe that the integral represents a quarter of the area of a circle with radius r, so

$$8\pi R \int_0^r \sqrt{r^2 - y^2}\, dy = 8\pi R \cdot \tfrac{1}{4}\pi r^2 = 2\pi^2 r^2 R$$

42. If we place the x-axis along the diameter where the planes meet, then the base of the solid is a semicircle with equation $y = \sqrt{16 - x^2}$, $-4 \le x \le 4$. A cross-section perpendicular to the x-axis at a distance x from the origin is a triangle ABC, as shown in the figure, whose base is $y = \sqrt{16 - x^2}$ and whose height is $|BC| = y \tan 30° = \sqrt{16 - x^2}/\sqrt{3}$. Thus, the cross-sectional area is

$A(x) = \frac{1}{2}\sqrt{16 - x^2} \cdot \dfrac{1}{\sqrt{3}}\sqrt{16 - x^2} = \dfrac{16 - x^2}{2\sqrt{3}}$ and the volume is

$$\begin{aligned} V &= \int_{-4}^{4} A(x)\, dx = \int_{-4}^{4} \frac{16 - x^2}{2\sqrt{3}}\, dx = \frac{1}{\sqrt{3}} \int_0^4 (16 - x^2)\, dx \\ &= \frac{1}{\sqrt{3}} \left[16x - \frac{1}{3}x^3\right]_0^4 = \frac{128}{3\sqrt{3}} \end{aligned}$$

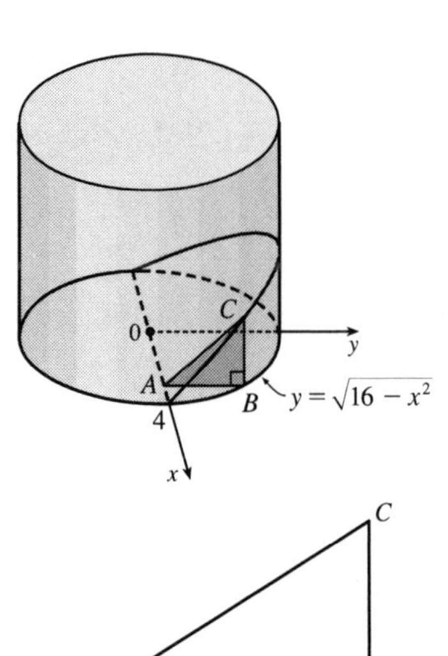

Another method: The cross-sections perpendicular to the y-axis in the figure are rectangles. The rectangle corresponding to the coordinate y has a base of length $2\sqrt{16 - y^2}$ in the xy-plane and a height of $\frac{1}{\sqrt{3}}y$, since $\angle BAC = 30°$ and $|BC| = \frac{1}{\sqrt{3}}|AB|$. Thus, $A(y) = \frac{2}{\sqrt{3}}\, y\sqrt{16 - y^2}$ and

$$\begin{aligned} V &= \int_0^4 A(y)\, dy = \frac{2}{\sqrt{3}} \int_0^4 \sqrt{16 - y^2}\, y\, dy = \frac{2}{\sqrt{3}} \int_{16}^{0} u^{1/2}\left(-\frac{1}{2}\, du\right) \quad \left[\begin{array}{l} u = 16 - y^2, \\ du = -2y\, dy \end{array}\right] \\ &= \frac{1}{\sqrt{3}} \int_0^{16} u^{1/2}\, du = \frac{1}{\sqrt{3}} \frac{2}{3}\left[u^{3/2}\right]_0^{16} = \frac{2}{3\sqrt{3}}(64) = \frac{128}{3\sqrt{3}} \end{aligned}$$

43. (a) Volume$(S_1) = \int_0^h A(z)\,dz =$ Volume(S_2) since the cross-sectional area $A(z)$ at height z is the same for both solids.

(b) By Cavalieri's Principle, the volume of the cylinder in the figure is the same as that of a right circular cylinder with radius r and height h, that is, $\pi r^2 h$.

44. Each cross-section of the solid S in a plane perpendicular to the x-axis is a square (since the edges of the cut lie on the cylinders, which are perpendicular). One-quarter of this square and one-eighth of S are shown.

The area of this quarter-square is $|PQ|^2 = r^2 - x^2$.

Therefore, $A(x) = 4(r^2 - x^2)$ and the volume of S is

$$V = \int_{-r}^{r} A(x)\,dx = 4\int_{-r}^{r}(r^2 - x^2)\,dx$$

$$= 8\int_0^r (r^2 - x^2)\,dx = 8\left[r^2x - \tfrac{1}{3}x^3\right]_0^r = \tfrac{16}{3}r^3$$

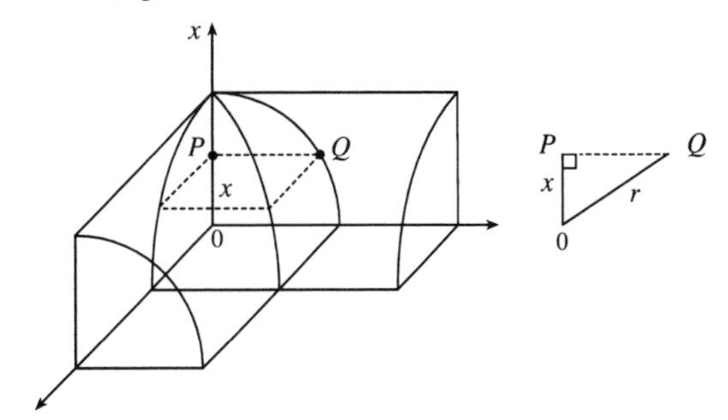

45. The volume is obtained by rotating the area common to two circles of radius r, as shown. The volume of the right half is

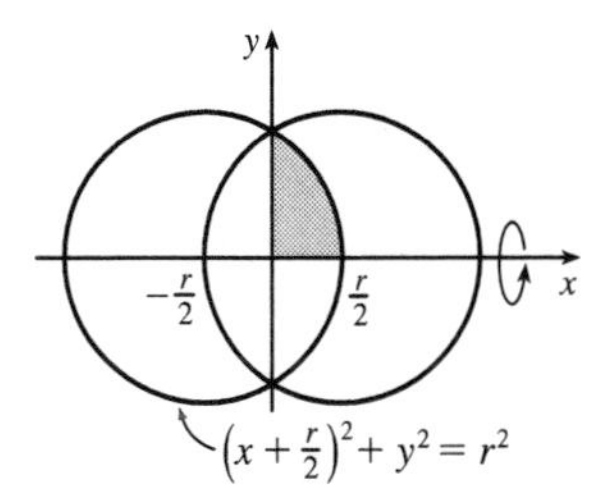

$$V_{\text{right}} = \pi\int_0^{r/2} y^2\,dx = \pi\int_0^{r/2}\left[r^2 - \left(\tfrac{1}{2}r + x\right)^2\right]dx$$

$$= \pi\left[r^2x - \tfrac{1}{3}\left(\tfrac{1}{2}r + x\right)^3\right]_0^{r/2} = \pi\left[\left(\tfrac{1}{2}r^3 - \tfrac{1}{3}r^3\right) - \left(0 - \tfrac{1}{24}r^3\right)\right] = \tfrac{5}{24}\pi r^3$$

So by symmetry, the total volume is twice this, or $\frac{5}{12}\pi r^3$.

Another solution: We observe that the volume is the twice the volume of a cap of a sphere, so we can use the formula from Exercise 27 with $h = \frac{1}{2}r$: $V = 2\cdot\frac{1}{3}\pi h^2(3r - h) = \frac{2}{3}\pi\left(\frac{1}{2}r\right)^2\left(3r - \frac{1}{2}r\right) = \frac{5}{12}\pi r^3$.

46. We consider two cases: one in which the ball is not completely submerged and the other in which it is.

Case 1: $0 \le h \le 10$ The ball will not be completely submerged, and so a cross-section of the water parallel to the surface will be the shaded area shown in the first diagram. We can find the area of the cross-section at height x above the bottom of the bowl by using the Pythagorean Theorem: $R^2 = 15^2 - (15 - x)^2$ and $r^2 = 5^2 - (x - 5)^2$, so $A(x) = \pi(R^2 - r^2) = 20\pi x$. The volume of water when it has depth h is then $V(h) = \int_0^h A(x)\,dx = \int_0^h 20\pi x\,dx = \left[10\pi x^2\right]_0^h = 10\pi h^2\ \text{cm}^3$, $0 \le h \le 10$.

Case 2: $10 < h \le 15$ In this case we can find the volume by simply subtracting the volume displaced by the ball from the total volume inside the bowl underneath the surface of the water. The total volume underneath the surface is just the volume of a cap of the bowl, so we use the formula from Exercise 27: $V_{\text{cap}}(h) = \frac{1}{3}\pi h^2(45 - h)$. The volume of the small sphere is $V_{\text{ball}} = \frac{4}{3}\pi(5)^3 = \frac{500}{3}\pi$, so the total volume is $V_{\text{cap}} - V_{\text{ball}} = \frac{1}{3}\pi(45h^2 - h^3 - 500)\ \text{cm}^3$.

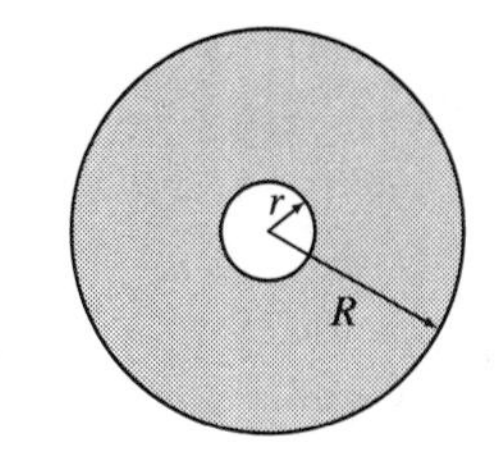

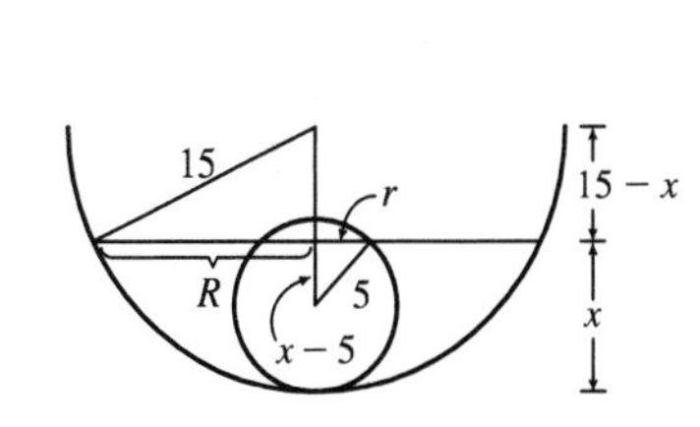

47. Take the x-axis to be the axis of the cylindrical hole of radius r. A quarter of the cross-section through y, perpendicular to the y-axis, is the rectangle shown. Using the Pythagorean Theorem twice, we see that the dimensions of this rectangle are $x = \sqrt{R^2 - y^2}$ and $z = \sqrt{r^2 - y^2}$, so $\frac{1}{4}A(y) = xz = \sqrt{r^2 - y^2}\,\sqrt{R^2 - y^2}$, and

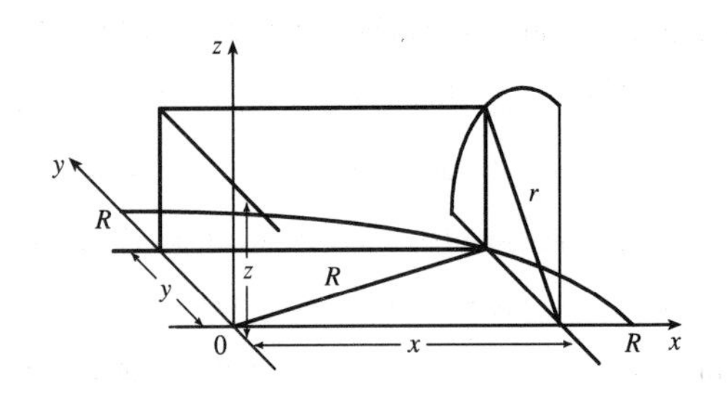

$$V = \int_{-r}^{r} A(y)\,dy = \int_{-r}^{r} 4\sqrt{r^2 - y^2}\,\sqrt{R^2 - y^2}\,dy = 8\int_0^r \sqrt{r^2 - y^2}\,\sqrt{R^2 - y^2}\,dy$$

48. The line $y = r$ intersects the semicircle $y = \sqrt{R^2 - x^2}$ when $r = \sqrt{R^2 - x^2} \;\Rightarrow\; r^2 = R^2 - x^2 \;\Rightarrow\; x^2 = R^2 - r^2$ $\Rightarrow\; x = \pm\sqrt{R^2 - r^2}$. Rotating the shaded region about the x-axis gives us

$$V = \int_{-\sqrt{R^2-r^2}}^{\sqrt{R^2-r^2}} \pi\left[\left(\sqrt{R^2 - x^2}\right)^2 - r^2\right] dx = 2\pi \int_0^{\sqrt{R^2-r^2}} (R^2 - x^2 - r^2)\,dx \qquad \text{[by symmetry]}$$

$$= 2\pi \int_0^{\sqrt{R^2-r^2}} \left[(R^2 - r^2) - x^2\right] dx = 2\pi\left[(R^2 - r^2)x - \tfrac{1}{3}x^3\right]_0^{\sqrt{R^2-r^2}}$$

$$= 2\pi\left[(R^2 - r^2)^{3/2} - \tfrac{1}{3}(R^2 - r^2)^{3/2}\right] = 2\pi \cdot \tfrac{2}{3}(R^2 - r^2)^{3/2} = \tfrac{4\pi}{3}(R^2 - r^2)^{3/2}$$

Our answer makes sense in limiting cases. As $r \to 0$, $V \to \frac{4}{3}\pi R^3$, which is the volume of the full sphere. As $r \to R$, $V \to 0$, which makes sense because the hole's radius is approaching that of the sphere.

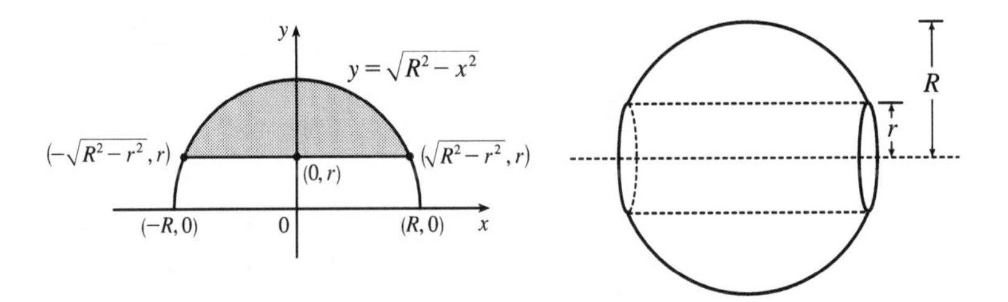

7.3 Volumes by Cylindrical Shells

1. If we were to use the "washer" method, we would first have to locate the local maximum point (a, b) of $y = x(x - 1)^2$ using the methods of Chapter 3. Then we would have to solve the equation $y = x(x - 1)^2$ for x in terms of y to obtain the functions $x = g_1(y)$ and $x = g_2(y)$ shown in the first figure. This step would be difficult because it involves the cubic formula. Finally we would find the volume using $V = \pi \int_0^b \left\{[g_1(y)]^2 - [g_2(y)]^2\right\} dy$. Using shells, we find that a typical approximating shell has radius x, so its circumference is $2\pi x$. Its height is y, that is, $x(x - 1)^2$. So the total volume is

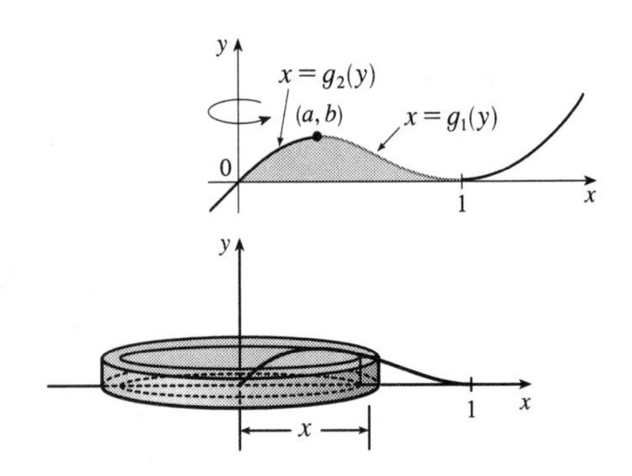

$$V = \int_0^1 2\pi x\left[x(x - 1)^2\right] dx = 2\pi \int_0^1 (x^4 - 2x^3 + x^2)\,dx = 2\pi\left[\frac{x^5}{5} - 2\frac{x^4}{4} + \frac{x^3}{3}\right]_0^1 = \frac{\pi}{15}$$

2. A typical cylindrical shell has circumference $2\pi x$ and height $\sin(x^2)$.

$V = \int_0^{\sqrt{\pi}} 2\pi x \sin(x^2)\,dx$. Let $u = x^2$. Then $du = 2x\,dx$, so

$V = \pi \int_0^{\pi} \sin u\,du = \pi\left[-\cos u\right]_0^{\pi} = \pi[1 - (-1)] = 2\pi$.

For slicing, we would first have to locate the local maximum point (a, b) of $y = \sin(x^2)$ using the methods of Chapter 3. Then we would have to solve the equation $y = \sin(x^2)$ for x in terms of y to obtain the functions $x = g_1(y)$ and $x = g_2(y)$ shown in the second figure. Finally we would find the volume using $V = \pi \int_0^b \{[g_1(y)]^2 - [g_2(y)]^2\}\,dy$. Using shells is definitely preferable to slicing.

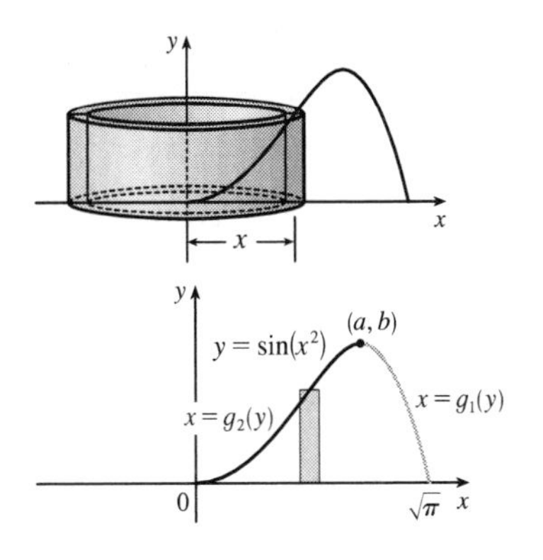

3. $V = \displaystyle\int_1^2 2\pi x \cdot \frac{1}{x}\,dx = 2\pi \int_1^2 1\,dx$

$= 2\pi\left[x\right]_1^2 = 2\pi(2-1) = 2\pi$

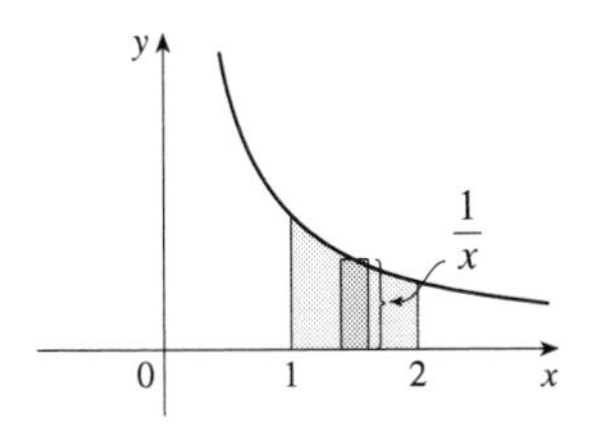

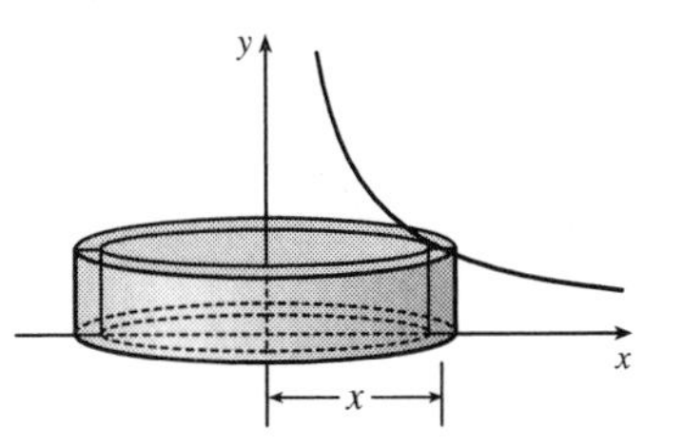

4. $V = \displaystyle\int_0^1 2\pi x \cdot x^2\,dx = 2\pi \int_0^1 x^3\,dx$

$= 2\pi\left[\frac{1}{4}x^4\right]_0^1 = 2\pi \cdot \frac{1}{4} = \frac{\pi}{2}$

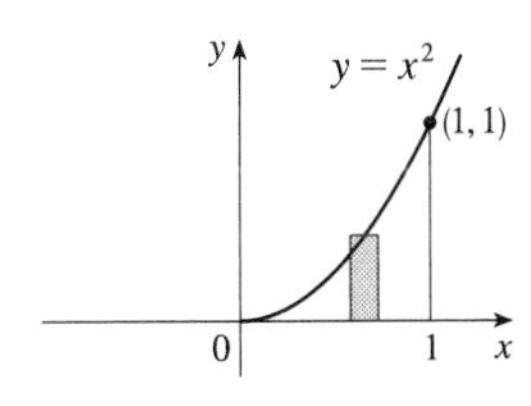

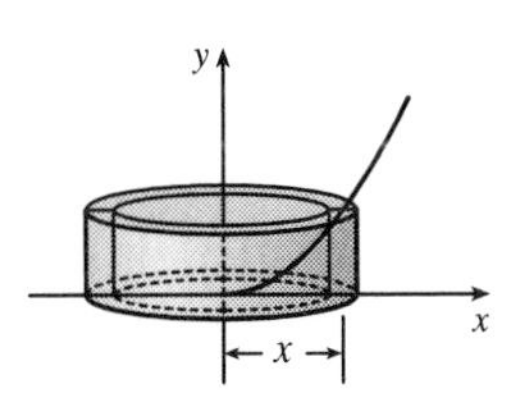

5. $V = \int_0^1 2\pi x e^{-x^2}\,dx$. Let $u = x^2$. Thus, $du = 2x\,dx$, so $V = \pi \int_0^1 e^{-u}\,du = \pi\left[-e^{-u}\right]_0^1 = \pi(1 - 1/e)$.

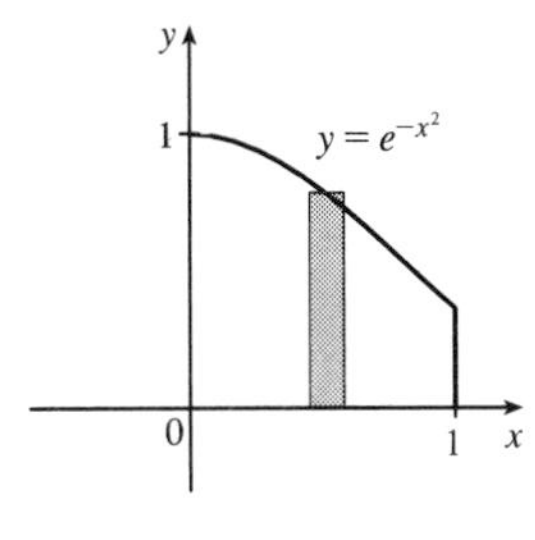

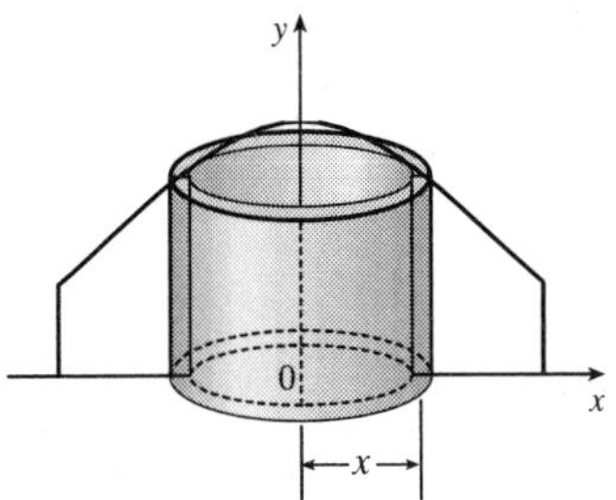

6. $V = 2\pi \int_0^3 \{x[(3 + 2x - x^2) - (3 - x)]\}\,dx = 2\pi \int_0^3 [x(3x - x^2)]\,dx$

$= 2\pi \int_0^3 (3x^2 - x^3)\,dx = 2\pi\left[x^3 - \frac{1}{4}x^4\right]_0^3 = 2\pi\left(27 - \frac{81}{4}\right) = 2\pi\left(\frac{27}{4}\right) = \frac{27\pi}{2}$

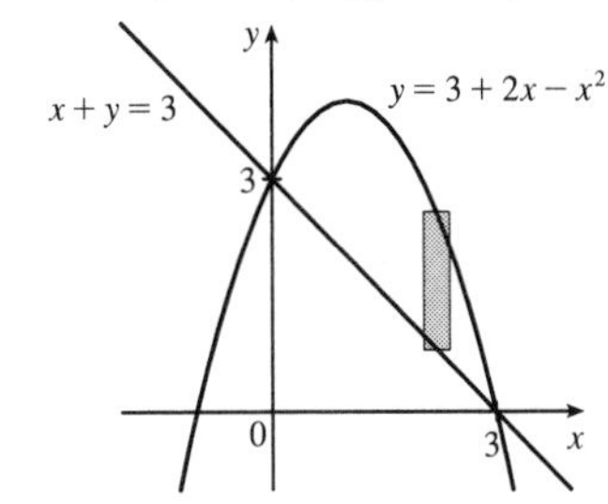

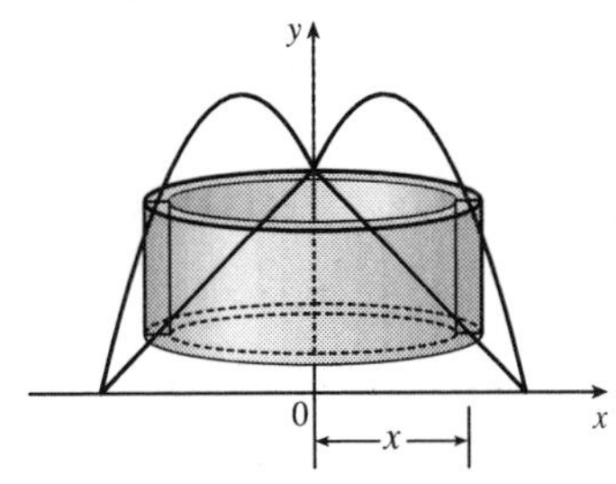

7. The curves intersect when $4(x-2)^2 = x^2 - 4x + 7 \iff 4x^2 - 16x + 16 = x^2 - 4x + 7 \iff$ $3x^2 - 12x + 9 = 0 \iff 3(x^2 - 4x + 3) = 0 \iff 3(x-1)(x-3) = 0$, so $x = 1$ or 3.

$$V = 2\pi \int_1^3 \{x[(x^2 - 4x + 7) - 4(x-2)^2]\}\,dx = 2\pi \int_1^3 [x(x^2 - 4x + 7 - 4x^2 + 16x - 16)]\,dx$$
$$= 2\pi \int_1^3 [x(-3x^2 + 12x - 9)]\,dx = 2\pi(-3)\int_1^3 (x^3 - 4x^2 + 3x)\,dx = -6\pi\left[\tfrac{1}{4}x^4 - \tfrac{4}{3}x^3 + \tfrac{3}{2}x^2\right]_1^3$$
$$= -6\pi\left[\left(\tfrac{81}{4} - 36 + \tfrac{27}{2}\right) - \left(\tfrac{1}{4} - \tfrac{4}{3} + \tfrac{3}{2}\right)\right] = -6\pi\left(20 - 36 + 12 + \tfrac{4}{3}\right) = -6\pi\left(-\tfrac{8}{3}\right) = 16\pi$$

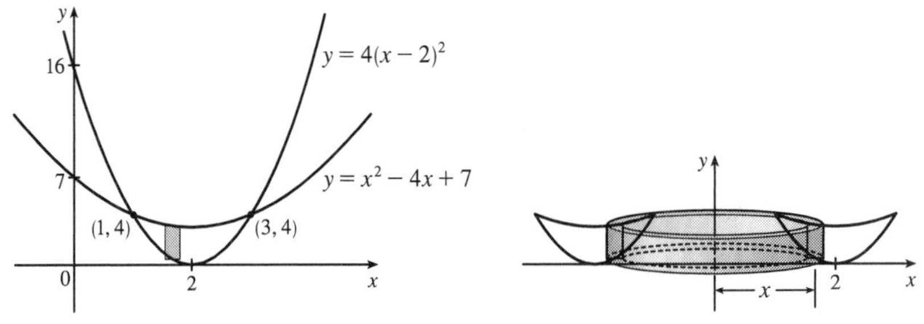

8. By slicing: $V = \int_0^1 \pi\left[\left(\sqrt{y}\right)^2 - (y^2)^2\right] dy = \pi \int_0^1 (y - y^4)\,dy = \pi\left[\tfrac{1}{2}y^2 - \tfrac{1}{5}y^5\right]_0^1 = \pi\left(\tfrac{1}{2} - \tfrac{1}{5}\right) = \tfrac{3\pi}{10}$

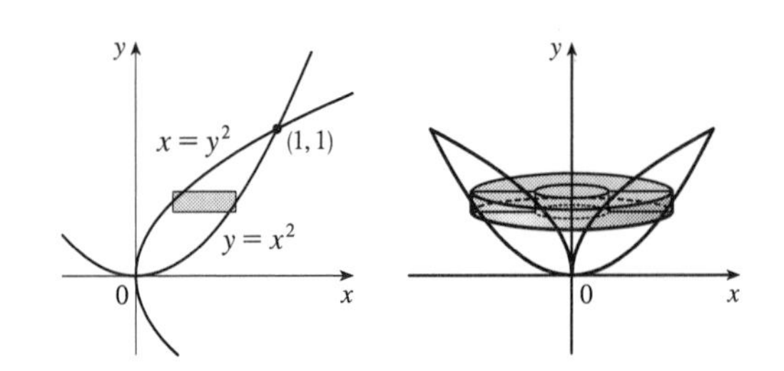

By cylindrical shells:

$$V = \int_0^1 2\pi x\left(\sqrt{x} - x^2\right) dx = 2\pi \int_0^1 (x^{3/2} - x^3)\,dx$$
$$= 2\pi\left[\tfrac{2}{5}x^{5/2} - \tfrac{1}{4}x^4\right]_0^1 = 2\pi\left(\tfrac{2}{5} - \tfrac{1}{4}\right)$$
$$= 2\pi\left(\tfrac{3}{20}\right) = \tfrac{3\pi}{10}$$

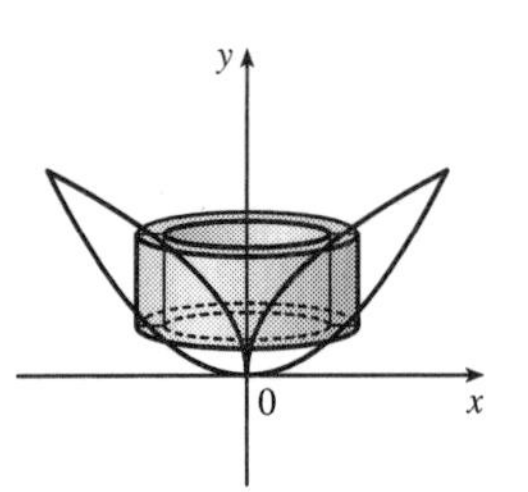

9. $V = \int_1^2 2\pi y(1 + y^2)\,dy = 2\pi \int_1^2 (y + y^3)\,dy = 2\pi\left[\tfrac{1}{2}y^2 + \tfrac{1}{4}y^4\right]_1^2 = 2\pi\left[(2 + 4) - \left(\tfrac{1}{2} + \tfrac{1}{4}\right)\right] = 2\pi\left(\tfrac{21}{4}\right) = \tfrac{21\pi}{2}$

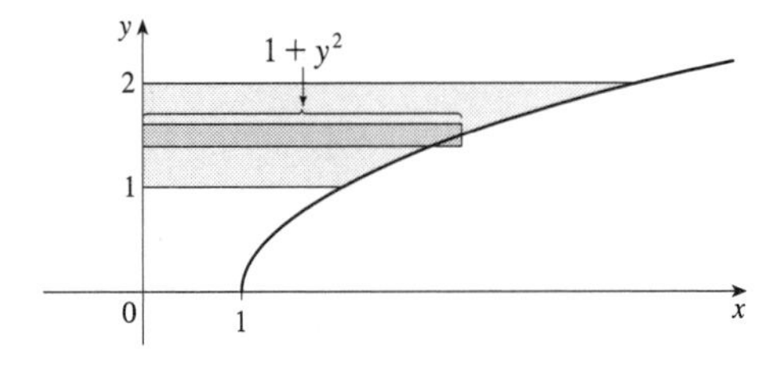

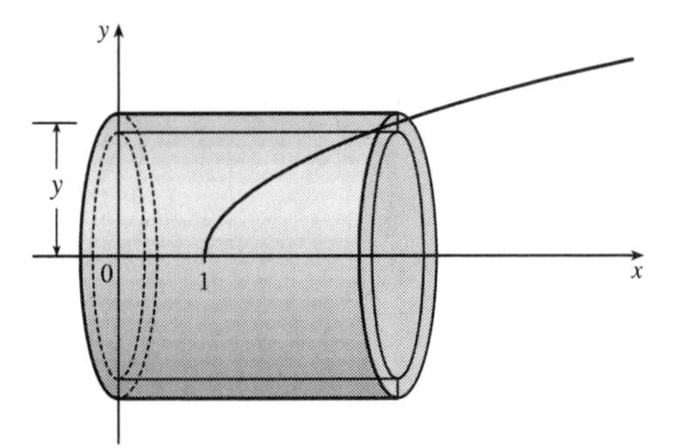

10. $V = \int_0^1 2\pi y\,\sqrt{y}\,dy = 2\pi\int_0^1 y^{3/2}\,dy$

$= 2\pi\left[\frac{2}{5}y^{5/2}\right]_0^1 = \frac{4\pi}{5}$

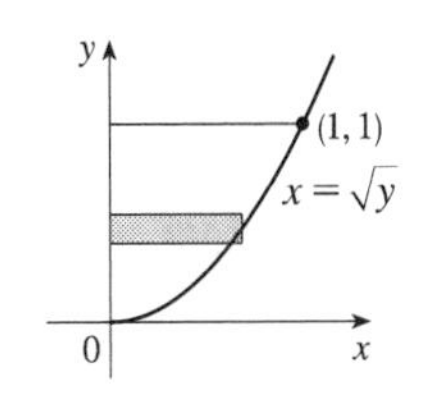

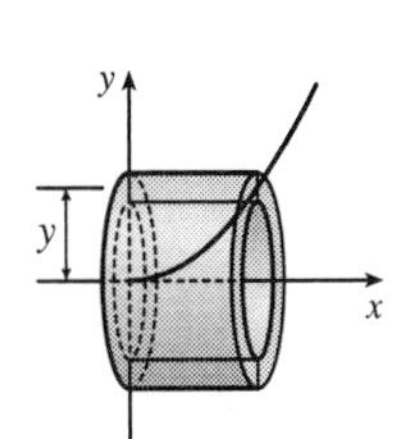

11. $V = 2\pi\displaystyle\int_0^8 [y(\sqrt[3]{y} - 0)]\,dy$

$= 2\pi\displaystyle\int_0^8 y^{4/3}\,dy = 2\pi\left[\frac{3}{7}y^{7/3}\right]_0^8$

$= \dfrac{6\pi}{7}(8^{7/3}) = \dfrac{6\pi}{7}(2^7) = \dfrac{768\pi}{7}$

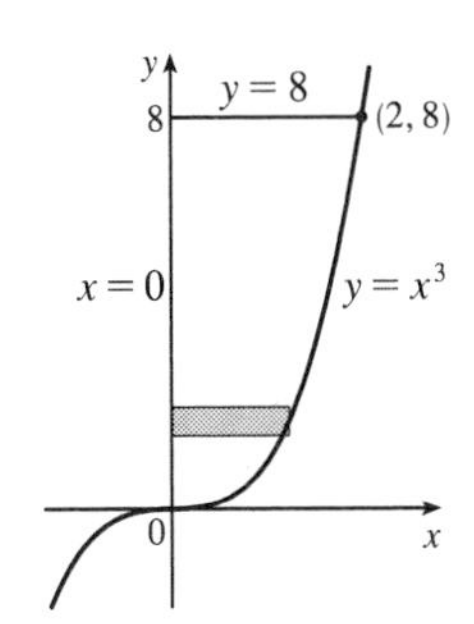

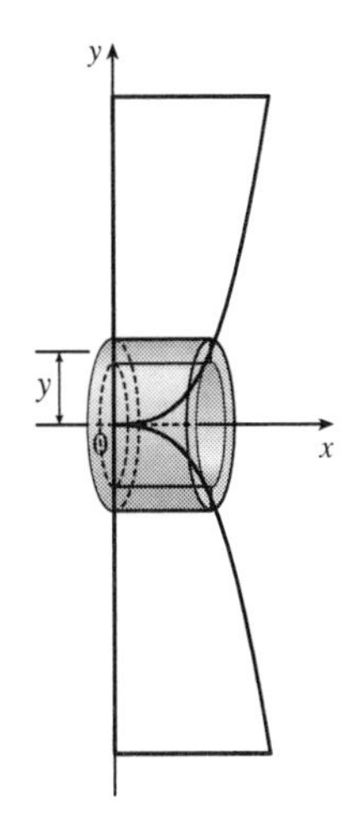

12. $V = 2\pi\displaystyle\int_0^4 [y(4y^2 - y^3)]\,dy$

$= 2\pi\displaystyle\int_0^4 (4y^3 - y^4)\,dy$

$= 2\pi\left[y^4 - \frac{1}{5}y^5\right]_0^4 = 2\pi\left(256 - \frac{1024}{5}\right)$

$= 2\pi\left(\frac{256}{5}\right) = \frac{512\pi}{5}$

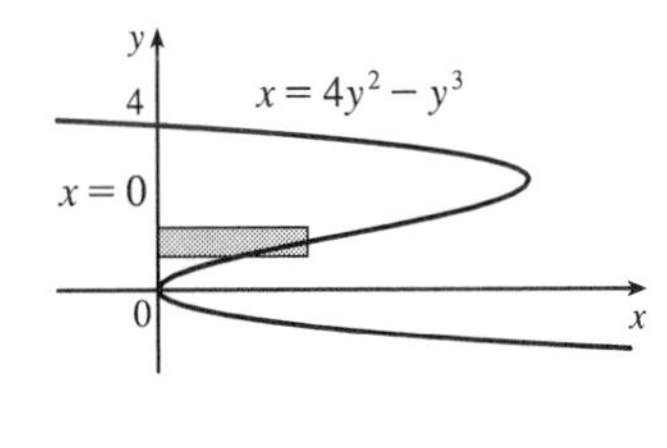

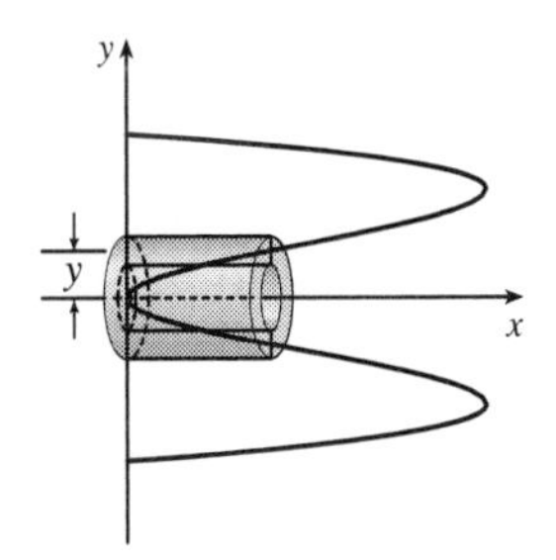

13. The curves intersect when $4x^2 = 6 - 2x \;\Leftrightarrow\; 2x^2 + x - 3 = 0 \;\Leftrightarrow\; (2x+3)(x-1) = 0 \;\Leftrightarrow\; x = -\frac{3}{2}$ or 1. Solving the equations for x gives us $y = 4x^2 \;\Rightarrow\; x = \pm\frac{1}{2}\sqrt{y}$ and $2x + y = 6 \;\Rightarrow\; x = -\frac{1}{2}y + 3$.

$$V = 2\pi\int_0^4 \left\{y\left[\left(\tfrac{1}{2}\sqrt{y}\right) - \left(-\tfrac{1}{2}\sqrt{y}\right)\right]\right\}dy + 2\pi\int_4^9 \left\{y\left[\left(-\tfrac{1}{2}y + 3\right) - \left(-\tfrac{1}{2}\sqrt{y}\right)\right]\right\}dy$$

$$= 2\pi\int_0^4 \left(y\sqrt{y}\right)dy + 2\pi\int_4^9 \left(-\tfrac{1}{2}y^2 + 3y + \tfrac{1}{2}y^{3/2}\right)dy = 2\pi\left[\tfrac{2}{5}y^{5/2}\right]_0^4 + 2\pi\left[-\tfrac{1}{6}y^3 + \tfrac{3}{2}y^2 + \tfrac{1}{5}y^{5/2}\right]_4^9$$

$$= 2\pi\left(\tfrac{2}{5}\cdot 32\right) + 2\pi\left[\left(-\tfrac{243}{2} + \tfrac{243}{2} + \tfrac{243}{5}\right) - \left(-\tfrac{32}{3} + 24 + \tfrac{32}{5}\right)\right] = \tfrac{128}{5}\pi + 2\pi\left(\tfrac{433}{15}\right) = \tfrac{1250}{15}\pi = \tfrac{250}{3}\pi$$

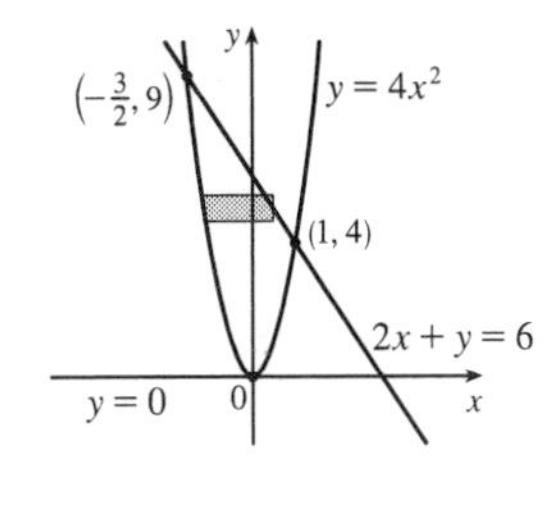

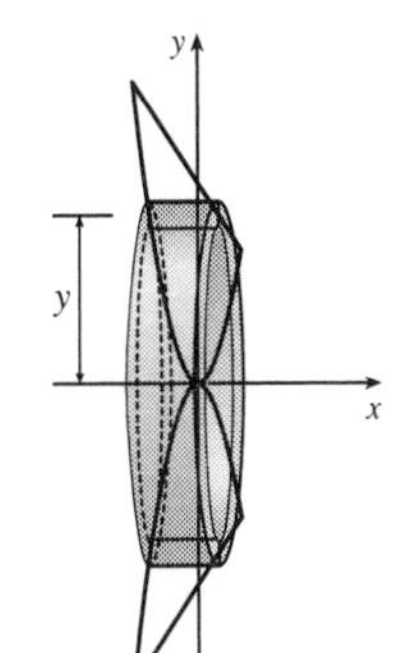

14. $V = \int_0^3 2\pi y\left[4 - (y-1)^2 - (3-y)\right] dy$

$= 2\pi \int_0^3 y\left(-y^2 + 3y\right) dy$

$= 2\pi \int_0^3 \left(-y^3 + 3y^2\right) dy = 2\pi\left[-\frac{1}{4}y^4 + y^3\right]_0^3$

$= 2\pi\left(-\frac{81}{4} + 27\right) = 2\pi\left(\frac{27}{4}\right) = \frac{27\pi}{2}$

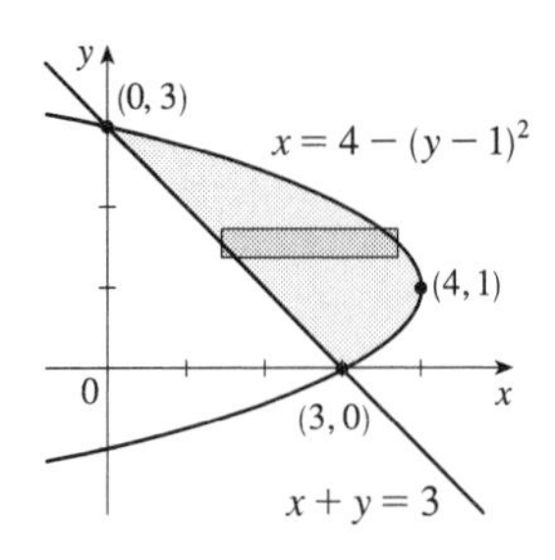

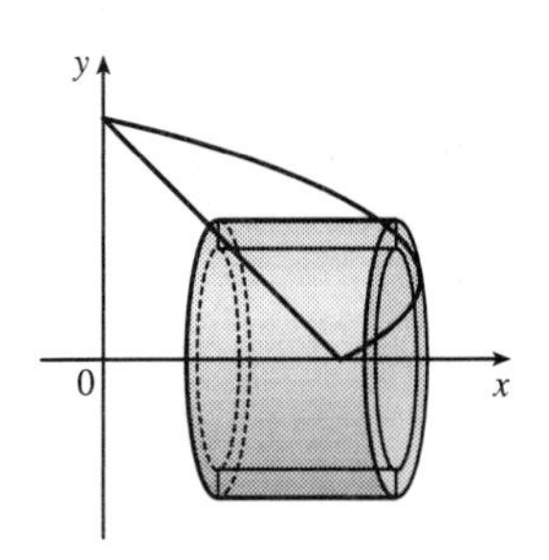

15. $V = \int_1^2 2\pi(x-1)x^2\, dx = 2\pi\left[\frac{1}{4}x^4 - \frac{1}{3}x^3\right]_1^2$

$= 2\pi\left[\left(4 - \frac{8}{3}\right) - \left(\frac{1}{4} - \frac{1}{3}\right)\right] = \frac{17}{6}\pi$

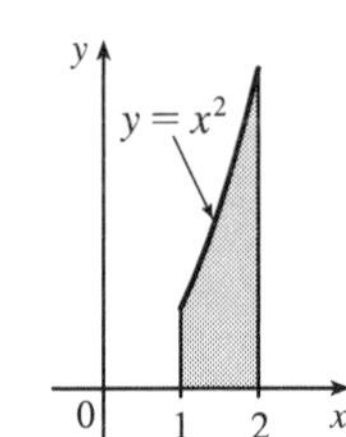

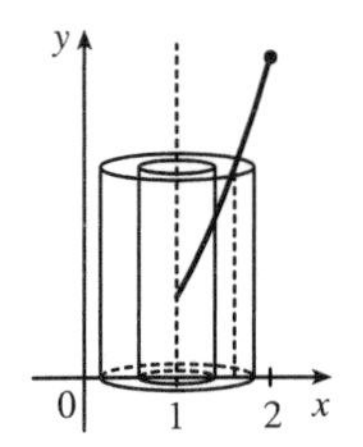

16. $V = \int_{-2}^{-1} 2\pi(-x)\cdot x^2\, dx = 2\pi\left[-\frac{1}{4}x^4\right]_{-2}^{-1}$

$= 2\pi\left[\left(-\frac{1}{4}\right) - (-4)\right] = \frac{15}{2}\pi$

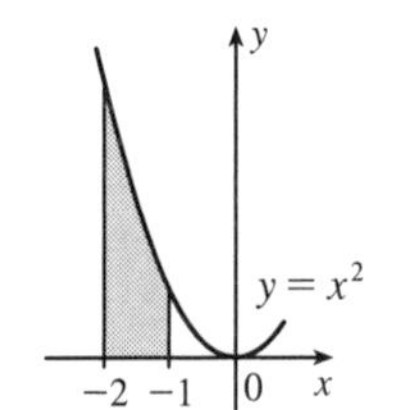

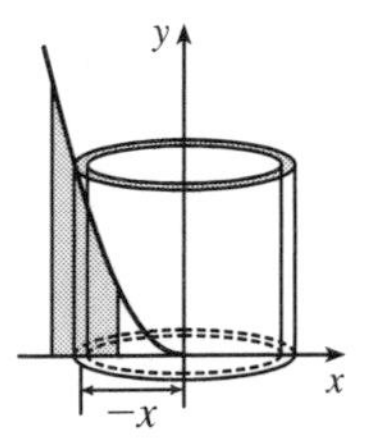

17. $V = \int_1^2 2\pi(4-x)x^2\, dx = 2\pi\left[\frac{4}{3}x^3 - \frac{1}{4}x^4\right]_1^2$

$= 2\pi\left[\left(\frac{32}{3} - 4\right) - \left(\frac{4}{3} - \frac{1}{4}\right)\right] = \frac{67}{6}\pi$

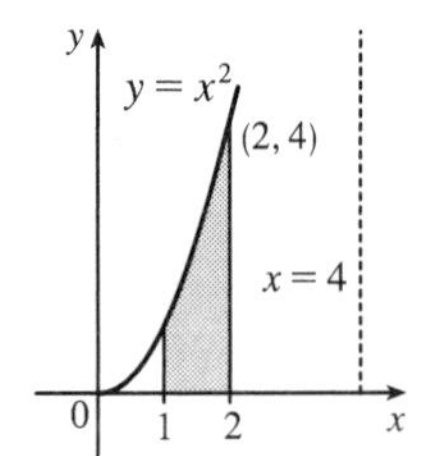

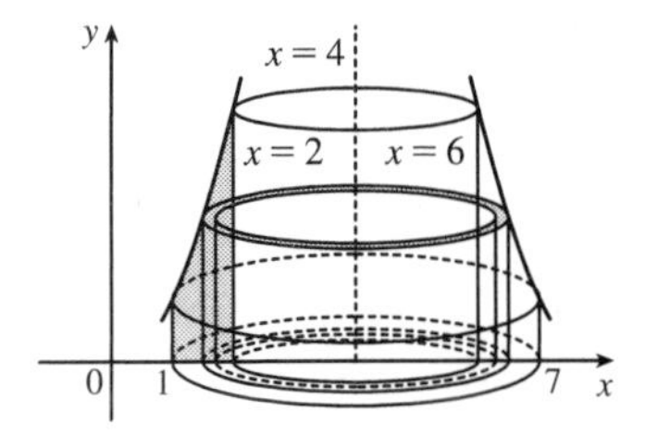

18. $V = \int_0^4 2\pi[x - (-2)][(8x - 2x^2) - (4x - x^2)]\, dx$

$= \int_0^4 2\pi(2 + x)(4x - x^2)\, dx$

$= 2\pi \int_0^4 (8x + 2x^2 - x^3)\, dx$

$= 2\pi\left[4x^2 + \frac{2}{3}x^3 - \frac{1}{4}x^4\right]_0^4$

$= 2\pi\left(64 + \frac{128}{3} - 64\right) = \frac{256}{3}\pi$

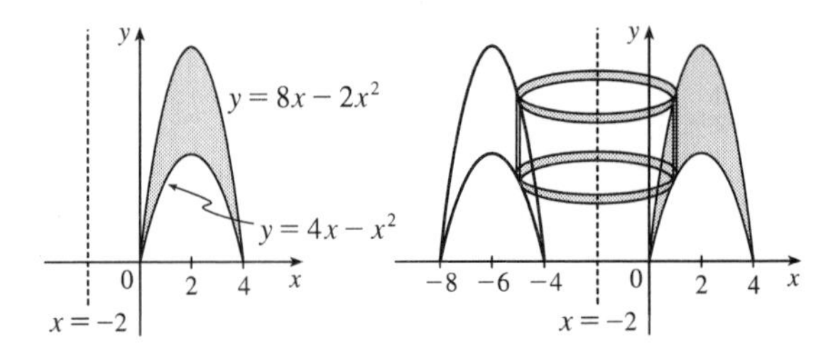

19. $V = \int_0^2 2\pi(3-y)(5-x)dy$

$= \int_0^2 2\pi(3-y)\left(5 - y^2 - 1\right) dy$

$= \int_0^2 2\pi\left(12 - 4y - 3y^2 + y^3\right) dy$

$= 2\pi\left[12y - 2y^2 - y^3 + \frac{1}{4}y^4\right]_0^2$

$= 2\pi(24 - 8 - 8 + 4) = 24\pi$

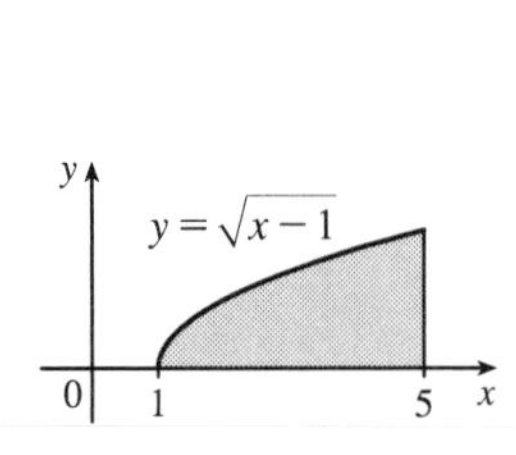

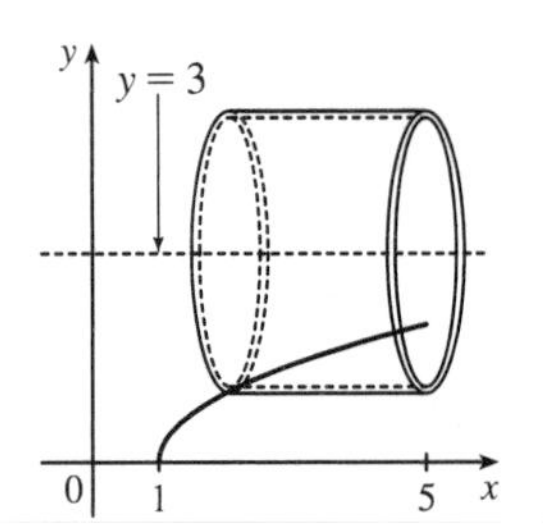

20. $V = \int_0^1 2\pi(y+1)\left(\sqrt{y} - y^2\right) dy$

$= 2\pi \int_0^1 (y^{3/2} + y^{1/2} - y^3 - y^2)\, dy$

$= 2\pi \left[\frac{2}{5}y^{5/2} + \frac{2}{3}y^{3/2} - \frac{1}{4}y^4 - \frac{1}{3}y^3\right]_0^1$

$= 2\pi\left(\frac{2}{5} + \frac{2}{3} - \frac{1}{4} - \frac{1}{3}\right) = 2\pi\left(\frac{29}{60}\right) = \frac{29\pi}{30}$

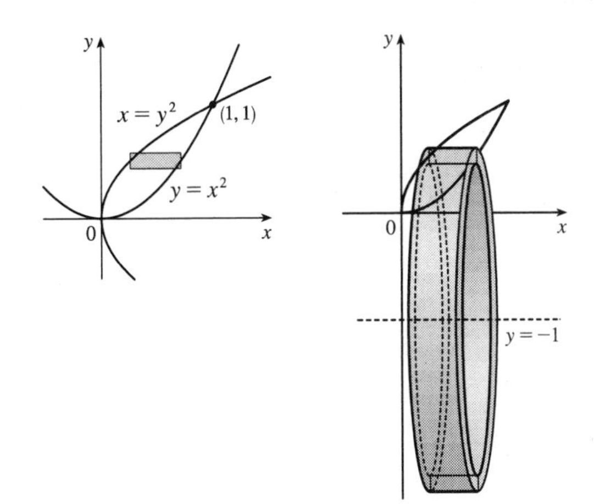

21. $V = \int_1^2 2\pi x \ln x\, dx$

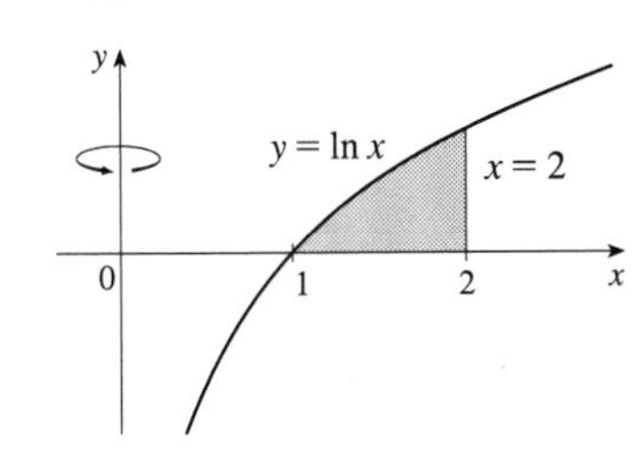

22. $V = \int_0^3 2\pi(7-x)[(4x - x^2) - x]\, dx$

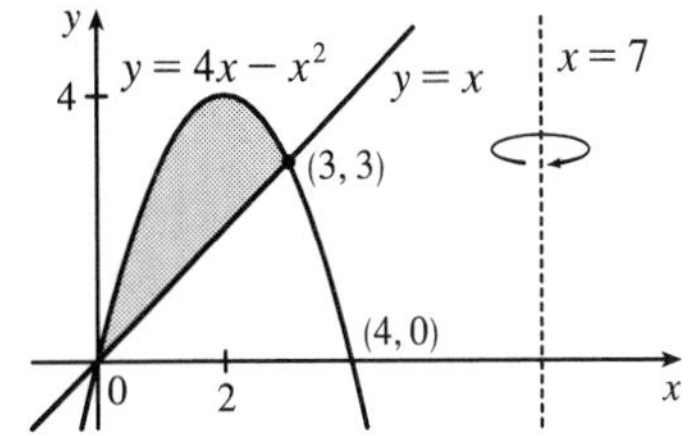

23. $V = \int_0^1 2\pi[x - (-1)]\left(\sin \frac{\pi}{2}x - x^4\right) dx$

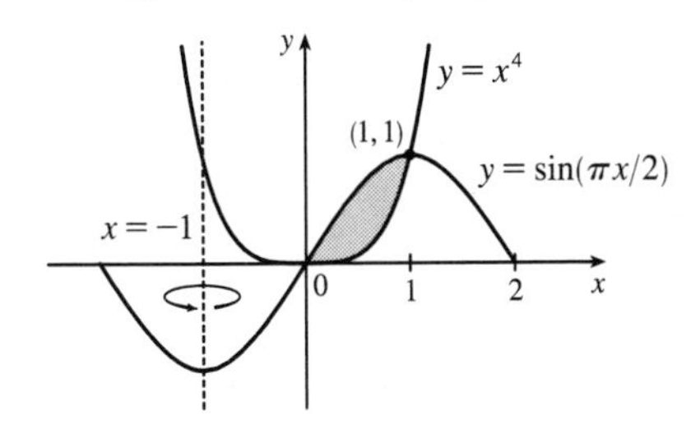

24. $V = \displaystyle\int_0^2 2\pi(2-x)\left(\frac{1}{1+x^2}\right) dx$

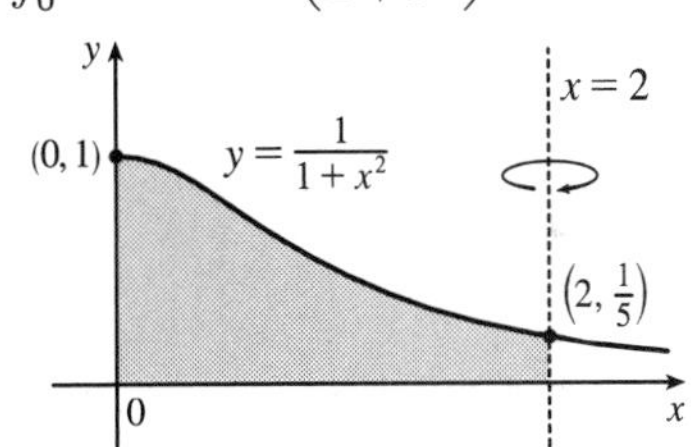

25. $V = \int_0^{\pi} 2\pi(4-y)\sqrt{\sin y}\, dy$

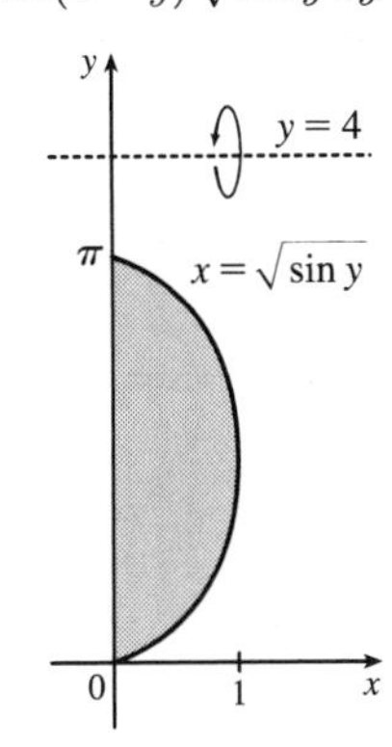

26. $V = \int_{-3}^{3} 2\pi(5-y)\left(4 - \sqrt{y^2+7}\right) dy$

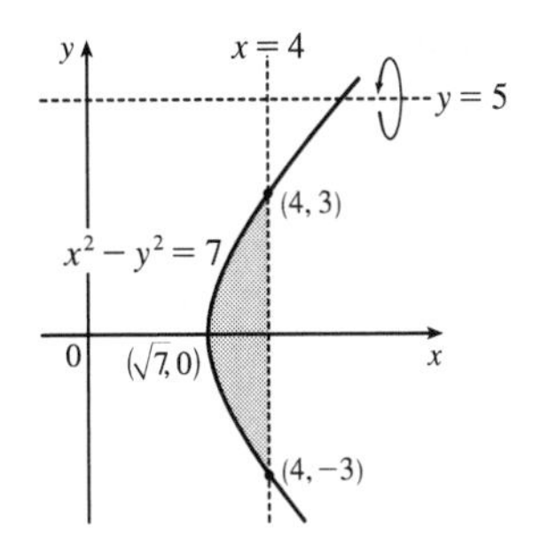

27. $\Delta x = \dfrac{\pi/4 - 0}{4} = \dfrac{\pi}{16}$.

$V = \int_0^{\pi/4} 2\pi x \tan x\, dx \approx 2\pi \cdot \frac{\pi}{16}\left(\frac{\pi}{32}\tan\frac{\pi}{32} + \frac{3\pi}{32}\tan\frac{3\pi}{32} + \frac{5\pi}{32}\tan\frac{5\pi}{32} + \frac{7\pi}{32}\tan\frac{7\pi}{32}\right) \approx 1.142$

28. (a) Let $y = f(x)$ denote the curve. Using cylindrical shells, $V = \int_2^{10} 2\pi x f(x)\,dx = 2\pi \int_2^{10} x f(x)\,dx = 2\pi I_1$.

Now use Simpson's Rule to approximate I_1:

$$\begin{aligned} I_1 \approx S_8 &= \tfrac{10-2}{3(8)}\,[2f(2) + 4\cdot 3f(3) + 2\cdot 4f(4) + 4\cdot 5f(5) + 2\cdot 6f(6) \\ &\qquad + 4\cdot 7f(7) + 2\cdot 8f(8) + 4\cdot 9f(9) + 10f(10)] \\ &\approx \tfrac{1}{3}[2(0) + 12(1.5) + 8(1.9) + 20(2.2) + 12(3.0) + 28(3.8) + 16(4.0) + 36(3.1) + 10(0)] \\ &= \tfrac{1}{3}(395.2) \end{aligned}$$

Thus, $V \approx 2\pi \cdot \frac{1}{3}(395.2) \approx 827.7$ or 828 cubic units.

(b) Using disks, $V = \int_2^{10} \pi [f(x)]^2\,dx = \pi \int_2^{10} [f(x)]^2\,dx = \pi I_2$. Now use Simpson's Rule to approximate I_2:

$$\begin{aligned} I_2 \approx S_8 &= \tfrac{10-2}{3(8)}\,\{[f(2)]^2 + 4\,[f(3)]^2 + 2\,[f(4)]^2 + 4\,[f(5)]^2 + 2\,[f(6)]^2 \\ &\qquad + 4\,[f(7)]^2 + 2\,[f(8)]^2 + 4\,[f(9)]^2 + [f(10)]^2\} \\ &\approx \tfrac{1}{3}\left[(0)^2 + 4(1.5)^2 + 2(1.9)^2 + 4(2.2)^2 + 2(3.0)^2 + 4(3.8)^2 + 2(4.0)^2 + 4(3.1)^2 + (0)^2\right] \\ &= \tfrac{1}{3}(181.78) \end{aligned}$$

Thus, $V \approx \pi \cdot \frac{1}{3}(181.78) \approx 190.4$ or 190 cubic units.

29. $\int_0^3 2\pi x^5\,dx = 2\pi \int_0^3 x(x^4)\,dx$. The solid is obtained by rotating the region $0 \le y \le x^4$, $0 \le x \le 3$ about the y-axis using cylindrical shells.

30. $2\pi \displaystyle\int_0^2 \frac{y}{1+y^2}\,dy = 2\pi \int_0^2 y\left(\frac{1}{1+y^2}\right)dy$. The solid is obtained by rotating the region $0 \le x \le \dfrac{1}{1+y^2}$, $0 \le y \le 2$ about the x-axis using cylindrical shells.

31. $\int_0^1 2\pi(3-y)(1-y^2)\,dy$. The solid is obtained by rotating the region bounded by (i) $x = 1 - y^2$, $x = 0$, and $y = 0$ or (ii) $x = y^2$, $x = 1$, and $y = 0$ about the line $y = 3$ using cylindrical shells.

32. $\int_0^{\pi/4} 2\pi(\pi - x)(\cos x - \sin x)\,dx$. The solid is obtained by rotating the region bounded by (i) $0 \le y \le \cos x - \sin x$, $0 \le x \le \frac{\pi}{4}$ or (ii) $\sin x \le y \le \cos x$, $0 \le x \le \frac{\pi}{4}$ about the line $x = \pi$ using cylindrical shells.

33. Use disks:

$$\begin{aligned} V &= \int_{-2}^{1} \pi(x^2 + x - 2)^2\,dx = \pi \int_{-2}^{1}(x^4 + 2x^3 - 3x^2 - 4x + 4)\,dx = \pi\left[\tfrac{1}{5}x^5 + \tfrac{1}{2}x^4 - x^3 - 2x^2 + 4x\right]_{-2}^{1} \\ &= \pi\left[\left(\tfrac{1}{5} + \tfrac{1}{2} - 1 - 2 + 4\right) - \left(-\tfrac{32}{5} + 8 + 8 - 8 - 8\right)\right] = \pi\left(\tfrac{33}{5} + \tfrac{3}{2}\right) = \tfrac{81}{10}\pi \end{aligned}$$

34. Use shells:

$$\begin{aligned} V &= \int_1^2 2\pi x(-x^2 + 3x - 2)\,dx = 2\pi \int_1^2 (-x^3 + 3x^2 - 2x)\,dx = 2\pi\left[-\tfrac{1}{4}x^4 + x^3 - x^2\right]_1^2 \\ &= 2\pi\left[(-4 + 8 - 4) - \left(-\tfrac{1}{4} + 1 - 1\right)\right] = \tfrac{\pi}{2} \end{aligned}$$

35. Use shells:

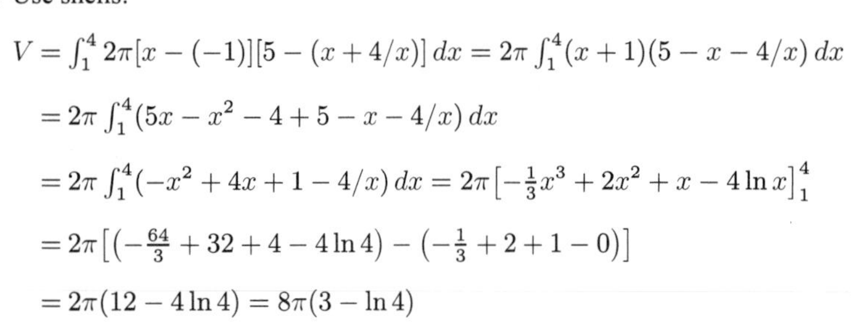

$$\begin{aligned} V &= \int_1^4 2\pi[x - (-1)][5 - (x + 4/x)]\,dx = 2\pi \int_1^4 (x+1)(5 - x - 4/x)\,dx \\ &= 2\pi \int_1^4 (5x - x^2 - 4 + 5 - x - 4/x)\,dx \\ &= 2\pi \int_1^4 (-x^2 + 4x + 1 - 4/x)\,dx = 2\pi\left[-\tfrac{1}{3}x^3 + 2x^2 + x - 4\ln x\right]_1^4 \\ &= 2\pi\left[\left(-\tfrac{64}{3} + 32 + 4 - 4\ln 4\right) - \left(-\tfrac{1}{3} + 2 + 1 - 0\right)\right] \\ &= 2\pi(12 - 4\ln 4) = 8\pi(3 - \ln 4) \end{aligned}$$

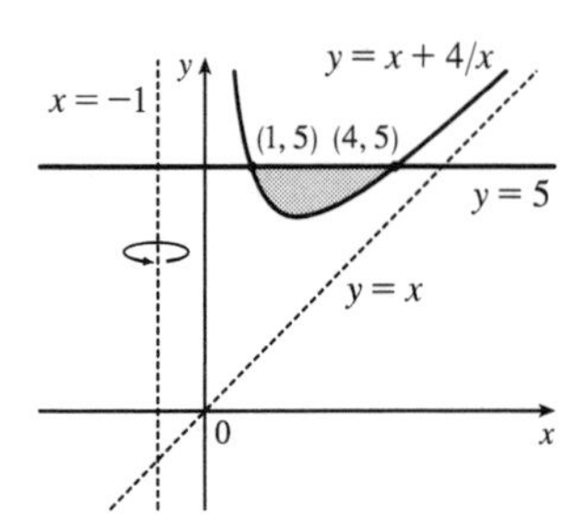

36. Use washers:

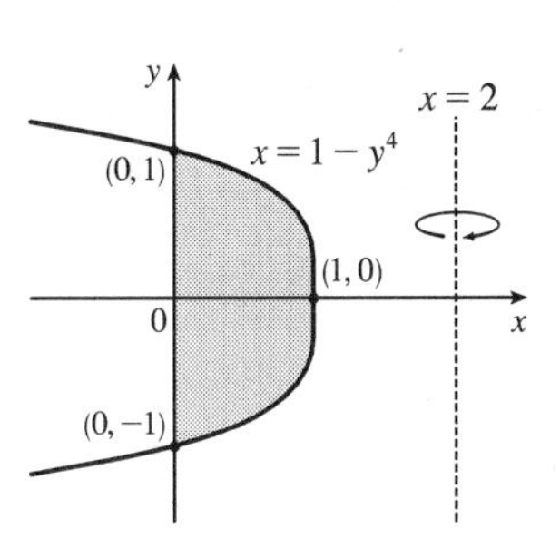

$$V = \int_{-1}^{1} \pi \left\{[2-0]^2 - [2-(1-y^4)]^2\right\} dy$$
$$= 2\pi \int_0^1 \left[4-(1+y^4)^2\right] dy \qquad \text{[by symmetry]}$$
$$= 2\pi \int_0^1 [4-(1+2y^4+y^8)]\,dy = 2\pi \int_0^1 (3-2y^4-y^8)\,dy$$
$$= 2\pi\left[3y - \tfrac{2}{5}y^5 - \tfrac{1}{9}y^9\right]_0^1 = 2\pi\left(3-\tfrac{2}{5}-\tfrac{1}{9}\right) = 2\pi\left(\tfrac{112}{45}\right) = \tfrac{224\pi}{45}$$

37. Use disks: $V = \pi \int_0^2 \left[\sqrt{1-(y-1)^2}\right]^2 dy = \pi \int_0^2 (2y-y^2)\,dy = \pi\left[y^2 - \tfrac{1}{3}y^3\right]_0^2 = \pi\left(4-\tfrac{8}{3}\right) = \tfrac{4}{3}\pi$

38. Using shells, we have

$$V = \int_0^2 2\pi y\left[\sqrt{1-(y-1)^2} - \left(-\sqrt{1-(y-1)^2}\right)\right] dy = 2\pi \int_0^2 y \cdot 2\sqrt{1-(y-1)^2}\,dy$$
$$= 4\pi \int_{-1}^{1} (u+1)\sqrt{1-u^2}\,du \quad [\text{let } u = y-1] \quad = 4\pi \int_{-1}^{1} u\sqrt{1-u^2}\,du + 4\pi \int_{-1}^{1} \sqrt{1-u^2}\,du$$

The first definite integral equals zero because its integrand is an odd function. The second is the area of a semicircle of radius 1, that is, $\frac{\pi}{2}$. Thus, $V = 4\pi \cdot 0 + 4\pi \cdot \frac{\pi}{2} = 2\pi^2$.

39. $V = 2\int_0^r 2\pi x\sqrt{r^2-x^2}\,dx = -2\pi \int_0^r (r^2-x^2)^{1/2}(-2x)\,dx = \left[-2\pi \cdot \tfrac{2}{3}(r^2-x^2)^{3/2}\right]_0^r = -\tfrac{4}{3}\pi(0-r^3) = \tfrac{4}{3}\pi r^3$

40. $V = \int_{R-r}^{R+r} 2\pi x \cdot 2\sqrt{r^2-(x-R)^2}\,dx$

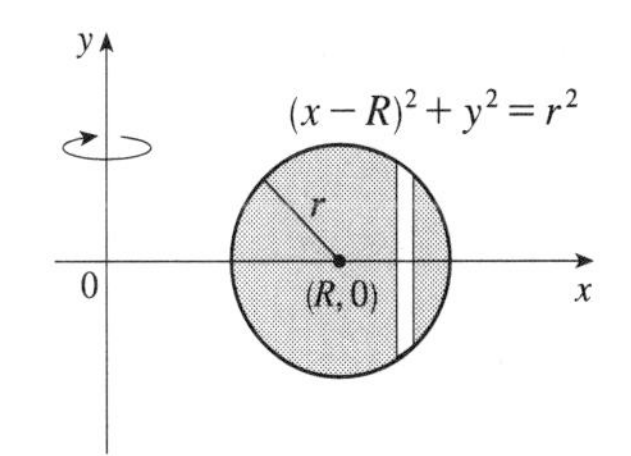

$$= \int_{-r}^{r} 4\pi(u+R)\sqrt{r^2-u^2}\,du \qquad [\text{let } u = x-R]$$
$$= 4\pi R\int_{-r}^{r} \sqrt{r^2-u^2}du + 4\pi \int_{-r}^{r} u\sqrt{r^2-u^2}du$$

The first integral is the area of a semicircle of radius r, that is, $\frac{1}{2}\pi r^2$, and the second is zero since the integrand is an odd function. Thus,

$V = 4\pi R\left(\frac{1}{2}\pi r^2\right) + 4\pi \cdot 0 = 2\pi R r^2$.

41. $V = 2\pi \displaystyle\int_0^r x\left(-\frac{h}{r}x+h\right) dx = 2\pi h \int_0^r \left(-\frac{x^2}{r}+x\right) dx = 2\pi h\left[-\frac{x^3}{3r}+\frac{x^2}{2}\right]_0^r = 2\pi h\,\frac{r^2}{6} = \frac{\pi r^2 h}{3}$

42. By symmetry, the volume of a napkin ring obtained by drilling a hole of radius r through a sphere with radius R is twice the volume obtained by rotating the area above the x-axis and below the curve $y = \sqrt{R^2-x^2}$ (the equation of the top half of the cross-section of the sphere), between $x = r$ and $x = R$, about the y-axis. This volume is equal to

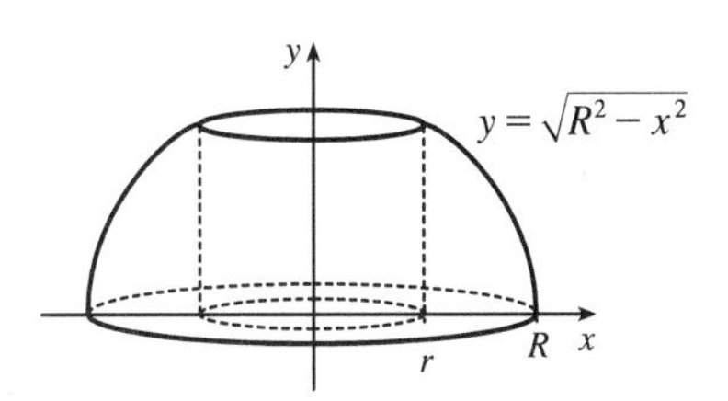

$$2\int_{\text{inner radius}}^{\text{outer radius}} 2\pi rh\,dx = 2 \cdot 2\pi \int_r^R x\sqrt{R^2-x^2}\,dx = 4\pi\left[-\tfrac{1}{3}(R^2-x^2)^{3/2}\right]_r^R = \tfrac{4}{3}\pi(R^2-r^2)^{3/2}$$

But by the Pythagorean Theorem, $R^2 - r^2 = \left(\frac{1}{2}h\right)^2$, so the volume of the napkin ring is $\frac{4}{3}\pi\left(\frac{1}{2}h\right)^3 = \frac{1}{6}\pi h^3$, which is independent of both R and r; that is, the amount of wood in a napkin ring of height h is the same regardless of the size of the sphere used. Note that most of this calculation has been done already, but with more difficulty, in Exercise 6.2.47.

Another solution: The height of the missing cap is the radius of the sphere minus half the height of the cut-out cylinder, that is, $R - \frac{1}{2}h$. Using Exercise 6.2.27,

$$V_{\text{napkin ring}} = V_{\text{sphere}} - V_{\text{cylinder}} - 2V_{\text{cap}} = \tfrac{4}{3}\pi R^3 - \pi r^2 h - 2 \cdot \tfrac{\pi}{3}\left(R-\tfrac{1}{2}h\right)^2\left[3R - \left(R-\tfrac{1}{2}h\right)\right] = \tfrac{1}{6}\pi h^3$$

43. Using the formula for volumes of rotation and the figure, we see that

Volume $= \int_0^d \pi b^2\,dy - \int_0^c \pi a^2\,dy - \int_c^d \pi\left[f^{-1}(y)\right]^2\,dy = \pi b^2 d - \pi a^2 c - \int_c^d \pi\left[f^{-1}(y)\right]^2\,dy$. Let $y = f(x)$,

which gives $dy = f'(x)\,dx$ and $f^{-1}(y) = x$, so that $V = \pi b^2 d - \pi a^2 c - \pi\int_a^b x^2 f'(x)\,dx$.

Now integrate by parts with $u = x^2$, and $dv = f'(x)\,dx \;\Rightarrow\; du = 2x\,dx$, $v = f(x)$, and

$\int_a^b x^2 f'(x)\,dx = \left[x^2 f(x)\right]_a^b - \int_a^b 2x\,f(x)\,dx = b^2 f(b) - a^2 f(a) - \int_a^b 2x\,f(x)\,dx$, but $f(a) = c$ and $f(b) = d \;\Rightarrow$

$V = \pi b^2 d - \pi a^2 c - \pi\left[b^2 d - a^2 c - \int_a^b 2x f(x)\,dx\right] = \int_a^b 2\pi x f(x)\,dx$.

7.4 Arc Length

1. $y = 2 - 3x \;\Rightarrow\; L = \int_{-2}^{1}\sqrt{1 + (dy/dx)^2}\,dx = \int_{-2}^{1}\sqrt{1 + (-3)^2}\,dx = \sqrt{10}\,[1 - (-2)] = 3\sqrt{10}$.

The arc length can be calculated using the distance formula, since the curve is a line segment, so

$L = [\text{distance from } (-2, 8) \text{ to } (1, -1)] = \sqrt{[1 - (-2)]^2 + [(-1) - 8]^2} = \sqrt{90} = 3\sqrt{10}$.

2. Using the arc length formula with $y = \sqrt{4 - x^2} \;\Rightarrow\; \dfrac{dy}{dx} = -\dfrac{x}{\sqrt{4 - x^2}}$, we get

$$L = \int_0^2 \sqrt{1 + \left(\frac{dy}{dx}\right)^2}\,dx = \int_0^2 \sqrt{1 + \frac{x^2}{4 - x^2}}\,dx = \int_0^2 \frac{2\,dx}{\sqrt{4 - x^2}} = 2\lim_{t\to 2^-}\int_0^t \frac{dx}{\sqrt{2^2 - x^2}}$$

$$= 2\lim_{t\to 2^-}\left[\sin^{-1}(x/2)\right]_0^t = 2\lim_{t\to 2^-}\left[\sin^{-1}(t/2) - \sin^{-1}0\right] = 2\left(\tfrac{\pi}{2} - 0\right) = \pi$$

The curve is a quarter of a circle with radius 2, so the length of the arc is $\frac{1}{4}(2\pi\cdot 2) = \pi$, as above.

3. $y = 1 + 6x^{3/2} \;\Rightarrow\; dy/dx = 9x^{1/2} \;\Rightarrow\; 1 + (dy/dx)^2 = 1 + 81x$. So

$L = \int_0^1 \sqrt{1 + 81x}\,dx = \int_1^{82} u^{1/2}\left(\frac{1}{81}\,du\right)$ [where $u = 1 + 81x$ and $du = 81\,dx$]

$= \frac{1}{81}\cdot\frac{2}{3}\left[u^{3/2}\right]_1^{82} = \frac{2}{243}\left(82\sqrt{82} - 1\right)$

4. $y^2 = 4(x + 4)^3,\; y > 0 \;\Rightarrow\; y = 2(x + 4)^{3/2} \;\Rightarrow\; dy/dx = 3(x + 4)^{1/2} \;\Rightarrow$

$1 + (dy/dx)^2 = 1 + 9(x + 4) = 9x + 37$. So

$$L = \int_0^2 \sqrt{9x + 37}\,dx \quad \begin{bmatrix} u = 9x + 37, \\ du = 9\,dx \end{bmatrix} \quad = \int_{37}^{55} u^{1/2}\left(\frac{1}{9}\,du\right) = \frac{1}{9}\cdot\frac{2}{3}\left[u^{3/2}\right]_{37}^{55} = \frac{2}{27}\left(55\sqrt{55} - 37\sqrt{37}\right)$$

5. $y = \dfrac{x^5}{6} + \dfrac{1}{10x^3} \;\Rightarrow\; \dfrac{dy}{dx} = \dfrac{5}{6}x^4 - \dfrac{3}{10}x^{-4} \;\Rightarrow$

$1 + (dy/dx)^2 = 1 + \frac{25}{36}x^8 - \frac{1}{2} + \frac{9}{100}x^{-8} = \frac{25}{36}x^8 + \frac{1}{2} + \frac{9}{100}x^{-8} = \left(\frac{5}{6}x^4 + \frac{3}{10}x^{-4}\right)^2$. So

$$L = \int_1^2 \sqrt{\left(\tfrac{5}{6}x^4 + \tfrac{3}{10}x^{-4}\right)^2}\,dx = \int_1^2 \left(\tfrac{5}{6}x^4 + \tfrac{3}{10}x^{-4}\right)dx = \left[\tfrac{1}{6}x^5 - \tfrac{1}{10}x^{-3}\right]_1^2$$

$$= \left(\tfrac{32}{6} - \tfrac{1}{80}\right) - \left(\tfrac{1}{6} - \tfrac{1}{10}\right) = \tfrac{31}{6} + \tfrac{7}{80} = \tfrac{1261}{240}$$

6. 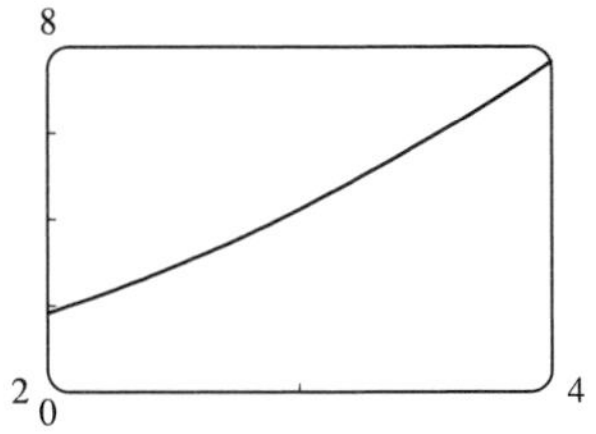 $y = \dfrac{x^2}{2} - \dfrac{\ln x}{4} \;\Rightarrow\; \dfrac{dy}{dx} = x - \dfrac{1}{4x} \;\Rightarrow\; 1 + \left(\dfrac{dy}{dx}\right)^2 = x^2 + \dfrac{1}{2} + \dfrac{1}{16x^2}$. So

$$L = \int_2^4 \left(x + \frac{1}{4x}\right) dx = \left[\frac{x^2}{2} + \frac{\ln x}{4}\right]_2^4 = \left(8 + \frac{2\ln 2}{4}\right) - \left(2 + \frac{\ln 2}{4}\right)$$
$$= 6 + \frac{\ln 2}{4}$$

7. $x = \frac{1}{3}\sqrt{y}\,(y-3) = \frac{1}{3}y^{3/2} - y^{1/2} \;\Rightarrow\; dx/dy = \frac{1}{2}y^{1/2} - \frac{1}{2}y^{-1/2} \;\Rightarrow$

$1 + (dx/dy)^2 = 1 + \frac{1}{4}y - \frac{1}{2} + \frac{1}{4}y^{-1} = \frac{1}{4}y + \frac{1}{2} + \frac{1}{4}y^{-1} = \left(\frac{1}{2}y^{1/2} + \frac{1}{2}y^{-1/2}\right)^2$. So

$L = \int_1^9 \left(\frac{1}{2}y^{1/2} + \frac{1}{2}y^{-1/2}\right) dy = \frac{1}{2}\left[\frac{2}{3}y^{3/2} + 2y^{1/2}\right]_1^9 = \frac{1}{2}\left[\left(\frac{2}{3}\cdot 27 + 2\cdot 3\right) - \left(\frac{2}{3}\cdot 1 + 2\cdot 1\right)\right]$

$= \frac{1}{2}\left(24 - \frac{8}{3}\right) = \frac{1}{2}\left(\frac{64}{3}\right) = \frac{32}{3}$

8. $y = \ln(\cos x) \;\Rightarrow\; dy/dx = -\tan x \;\Rightarrow\; 1 + (dy/dx)^2 = 1 + \tan^2 x = \sec^2 x$. So

$L = \int_0^{\pi/3} \sqrt{\sec^2 x}\, dx = \int_0^{\pi/3} \sec x\, dx = \left[\ln|\sec x + \tan x|\right]_0^{\pi/3} = \ln\left(2 + \sqrt{3}\right) - \ln(1 + 0) = \ln\left(2 + \sqrt{3}\right)$.

9. $y = \ln(\sec x) \;\Rightarrow\; \dfrac{dy}{dx} = \dfrac{\sec x \tan x}{\sec x} = \tan x \;\Rightarrow\; 1 + \left(\dfrac{dy}{dx}\right)^2 = 1 + \tan^2 x = \sec^2 x$, so

$L = \int_0^{\pi/4} \sqrt{\sec^2 x}\, dx = \int_0^{\pi/4} |\sec x|\, dx = \int_0^{\pi/4} \sec x\, dx = \left[\ln(\sec x + \tan x)\right]_0^{\pi/4}$

$= \ln\left(\sqrt{2} + 1\right) - \ln(1 + 0) = \ln\left(\sqrt{2} + 1\right)$

10. $y = \ln x \;\Rightarrow\; \dfrac{dy}{dx} = \dfrac{1}{x} \;\Rightarrow\; \sqrt{1 + \left(\dfrac{dy}{dx}\right)^2} = \sqrt{1 + \left(\dfrac{1}{x}\right)^2} = \dfrac{\sqrt{1+x^2}}{x}$. So $L = \displaystyle\int_1^{\sqrt{3}} \frac{\sqrt{1+x^2}}{x}\, dx$.

Now let $v = \sqrt{1+x^2}$, so $v^2 = 1 + x^2$ and $v\, dv = x\, dx$. Thus

$$L = \int_{\sqrt{2}}^2 \frac{v}{v^2 - 1} v\, dv = \int_{\sqrt{2}}^2 \left(1 + \frac{1/2}{v-1} - \frac{1/2}{v+1}\right) dv = \left[v + \tfrac{1}{2}\ln|v-1| - \tfrac{1}{2}\ln|v+1|\right]_{\sqrt{2}}^2$$
$$= \left[v - \tfrac{1}{2}\ln\left|\frac{v+1}{v-1}\right|\right]_{\sqrt{2}}^2 = 2 - \tfrac{1}{2}\ln 3 - \sqrt{2} + \frac{1}{2}\ln\left(\frac{\sqrt{2}+1}{\sqrt{2}-1}\right) = 2 - \sqrt{2} + \ln\left(\sqrt{2}+1\right) - \tfrac{1}{2}\ln 3$$

Or: Use Formula 23 in the Table of Integrals.

11. $y = \cosh x \;\Rightarrow\; y' = \sinh x \;\Rightarrow\; 1 + (y')^2 = 1 + \sinh^2 x = \cosh^2 x$.

So $L = \int_0^1 \cosh x\, dx = [\sinh x]_0^1 = \sinh 1 = \frac{1}{2}(e - 1/e)$.

12. $y^2 = 4x$, $x = \frac{1}{4}y^2 \;\Rightarrow\; dx/dy = \frac{1}{2}y \;\Rightarrow\; 1 + (dx/dy)^2 = 1 + \frac{1}{4}y^2$. So

$$L = \int_0^2 \sqrt{1 + \tfrac{1}{4}y^2}\, dy = \int_0^1 \sqrt{1 + u^2}\cdot 2\, du \qquad \left[u = \tfrac{1}{2}y,\ dy = 2\, du\right]$$
$$\stackrel{21}{=} \left[u\sqrt{1+u^2} + \ln\left|u + \sqrt{1+u^2}\right|\right]_0^1 = \sqrt{2} + \ln\left(1 + \sqrt{2}\right)$$

13. $y = e^x \Rightarrow y' = e^x \Rightarrow 1 + (y')^2 = 1 + e^{2x}$. So

$$L = \int_0^1 \sqrt{1+e^{2x}}\,dx = \int_1^e \sqrt{1+u^2}\,\frac{du}{u} \qquad [u = e^x, \text{ so } x = \ln u, dx = du/u]$$

$$= \int_1^e \frac{\sqrt{1+u^2}}{u^2}\,u\,du = \int_{\sqrt{2}}^{\sqrt{1+e^2}} \frac{v}{v^2-1}\,v\,dv \qquad \left[v = \sqrt{1+u^2}, \text{ so } v^2 = 1+u^2, v\,dv = u\,du\right]$$

$$= \int_{\sqrt{2}}^{\sqrt{1+e^2}} \left(1 + \frac{1/2}{v-1} - \frac{1/2}{v+1}\right) dv = \left[v + \frac{1}{2}\ln\frac{v-1}{v+1}\right]_{\sqrt{2}}^{\sqrt{1+e^2}}$$

$$= \sqrt{1+e^2} + \tfrac{1}{2}\ln\frac{\sqrt{1+e^2}-1}{\sqrt{1+e^2}+1} - \sqrt{2} - \frac{1}{2}\ln\frac{\sqrt{2}-1}{\sqrt{2}+1}$$

$$= \sqrt{1+e^2} - \sqrt{2} + \ln\left(\sqrt{1+e^2}-1\right) - 1 - \ln\left(\sqrt{2}-1\right)$$

Or: Use Formula 23 for $\int \left(\sqrt{1+u^2}/u\right) du$, or substitute $u = \tan\theta$.

14. $y = \ln\left(\dfrac{e^x+1}{e^x-1}\right) = \ln(e^x+1) - \ln(e^x-1) \Rightarrow y' = \dfrac{e^x}{e^x+1} - \dfrac{e^x}{e^x-1} = \dfrac{-2e^x}{e^{2x}-1} \Rightarrow$

$$1 + (y')^2 = 1 + \frac{4e^{2x}}{(e^{2x}-1)^2} = \frac{(e^{2x}+1)^2}{(e^{2x}-1)^2} \Rightarrow \sqrt{1+(y')^2} = \frac{e^{2x}+1}{e^{2x}-1} = \frac{e^x+e^{-x}}{e^x-e^{-x}} = \frac{\cosh x}{\sinh x}.$$

So $L = \displaystyle\int_a^b \frac{\cosh x}{\sinh x}\,dx = \Big[\ln\sinh x\Big]_a^b = \ln\sinh b - \ln\sinh a = \ln\left(\frac{\sinh b}{\sinh a}\right) = \ln\left(\frac{e^b - e^{-b}}{e^a - e^{-a}}\right).$

15. $y = \cos x \Rightarrow dy/dx = -\sin x \Rightarrow 1 + (dy/dx)^2 = 1 + \sin^2 x$. So $L = \int_0^{2\pi} \sqrt{1+\sin^2 x}\,dx$.

16. $y = 2^x \Rightarrow dy/dx = (2^x)\ln 2 \Rightarrow L = \int_0^3 \sqrt{1 + (\ln 2)^2\,2^{2x}}\,dx$

17. $x = y + y^3 \Rightarrow dx/dy = 1 + 3y^2 \Rightarrow 1 + (dx/dy)^2 = 1 + (1+3y^2)^2 = 9y^4 + 6y^2 + 2$.

So $L = \int_1^4 \sqrt{9y^4 + 6y^2 + 2}\,dy$.

18. $\dfrac{x^2}{a^2} + \dfrac{y^2}{b^2} = 1$, $y = \pm b\sqrt{1 - x^2/a^2} = \pm\dfrac{b}{a}\sqrt{a^2-x^2}$ [assume $a > 0$].

$$y = \frac{b}{a}\sqrt{a^2-x^2} \Rightarrow \frac{dy}{dx} = \frac{-bx}{a\sqrt{a^2-x^2}} \Rightarrow \left(\frac{dy}{dx}\right)^2 = \frac{b^2x^2}{a^2(a^2-x^2)}.$$

$$\text{So } L = 2\int_{-a}^{a}\left[1 + \frac{b^2x^2}{a^2(a^2-x^2)}\right]^{1/2} dx = \frac{4}{a}\int_0^a \left[\frac{(b^2-a^2)x^2 + a^4}{a^2-x^2}\right]^{1/2} dx.$$

19. $y = xe^{-x} \Rightarrow dy/dx = e^{-x} - xe^{-x} = e^{-x}(1-x) \Rightarrow 1 + (dy/dx)^2 = 1 + e^{-2x}(1-x)^2$. Let $f(x) = \sqrt{1 + (dy/dx)^2} = \sqrt{1 + e^{-2x}(1-x)^2}$. Then $L = \int_0^5 f(x)\,dx$. Since $n = 10$, $\Delta x = \frac{5-0}{10} = \frac{1}{2}$. Now

$$L \approx S_{10} = \tfrac{1/2}{3}\left[f(0) + 4f(\tfrac{1}{2}) + 2f(1) + 4f(\tfrac{3}{2}) + 2f(2) + 4f(\tfrac{5}{2}) + 2f(3) + 4f(\tfrac{7}{2}) + 2f(4) + 4f(\tfrac{9}{2}) + f(5)\right]$$

$$\approx 5.115840$$

The value of the integral produced by a calculator is 5.113568 (to six decimal places).

20. $x = y + \sqrt{y} \;\Rightarrow\; dx/dy = 1 + \dfrac{1}{2\sqrt{y}} \;\Rightarrow\; 1 + (dx/dy)^2 = 1 + \left(1 + \dfrac{1}{2\sqrt{y}}\right)^2 = 2 + \dfrac{1}{\sqrt{y}} + \dfrac{1}{4y}.$

Let $g(y) = \sqrt{1 + (dx/dy)^2}$. Then $L = \int_1^2 g(y)\,dy$. Since $n = 10$, $\Delta y = \frac{2-1}{10} = \frac{1}{10}$. Now

$$L \approx S_{10} = \tfrac{1/10}{3}[g(1) + 4g(1.1) + 2g(1.2) + 4g(1.3) + 2g(1.4) + 4g(1.5) + 2g(1.6) + 4g(1.7) + 2g(1.8) + 4g(1.9) + g(2)] \approx 1.732215,$$

which is the same value of the integral produced by a calculator to six decimal places.

21. $y = \sec x \;\Rightarrow\; dy/dx = \sec x \tan x \;\Rightarrow\; L = \int_0^{\pi/3} f(x)\,dx$, where $f(x) = \sqrt{1 + \sec^2 x \tan^2 x}$.

Since $n = 10$, $\Delta x = \dfrac{\pi/3 - 0}{10} = \dfrac{\pi}{30}$. Now

$$L \approx S_{10} = \frac{\pi/30}{3}\left[f(0) + 4f\left(\frac{\pi}{30}\right) + 2f\left(\frac{2\pi}{30}\right) + 4f\left(\frac{3\pi}{30}\right) + 2f\left(\frac{4\pi}{30}\right) + 4f\left(\frac{5\pi}{30}\right) + 2f\left(\frac{6\pi}{30}\right) + 4f\left(\frac{7\pi}{30}\right) + 2f\left(\frac{8\pi}{30}\right) + 4f\left(\frac{9\pi}{30}\right) + f\left(\frac{\pi}{3}\right)\right] \approx 1.569619$$

The value of the integral produced by a calculator is 1.569259 (to six decimal places).

22. $y = x \ln x \;\Rightarrow\; dy/dx = 1 + \ln x$. Let $f(x) = \sqrt{1 + (dy/dx)^2} = \sqrt{1 + (1 + \ln x)^2}$.

Then $L = \int_1^3 f(x)\,dx$. Since $n = 10$, $\Delta x = \frac{3-1}{10} = \frac{1}{5}$. Now

$$L \approx S_{10} = \tfrac{1/5}{3}[f(1) + 4f(1.2) + 2f(1.4) + 4f(1.6) + 2f(1.8) + 4f(2) + 2f(2.2) + 4f(2.4) + 2f(2.6) + 4f(2.8) + f(3)] \approx 3.869618$$

The value of the integral produced by a calculator is 3.869617 (to six decimal places).

23. $x = \ln(1 - y^2) \;\Rightarrow\; \dfrac{dx}{dy} = \dfrac{-2y}{1 - y^2} \;\Rightarrow\; 1 + \left(\dfrac{dx}{dy}\right)^2 = 1 + \dfrac{4y^2}{(1 - y^2)^2} = \dfrac{(1 + y^2)^2}{(1 - y^2)^2}$. So

$$L = \int_0^{1/2} \sqrt{\frac{(1 + y^2)^2}{(1 - y^2)^2}}\,dy = \int_0^{1/2} \frac{1 + y^2}{1 - y^2}\,dy = \ln 3 - \tfrac{1}{2} \quad \text{[from a CAS]}$$

24. $y = x^{4/3} \;\Rightarrow\; dy/dx = \frac{4}{3}x^{1/3} \;\Rightarrow\; 1 + (dy/dx)^2 = 1 + \frac{16}{9}x^{2/3} \;\Rightarrow$

$$L = \int_0^1 \sqrt{1 + \tfrac{16}{9}x^{2/3}}\,dx = \int_0^{4/3} \sqrt{1 + u^2}\,\tfrac{81}{64}u^2\,du \quad \left[\begin{array}{c} u = \frac{4}{3}x^{1/3},\ du = \frac{4}{9}x^{-2/3}\,dx, \\ dx = \frac{9}{4}x^{2/3}\,du = \frac{9}{4}\cdot\frac{9}{16}u^2\,du = \frac{81}{64}u^2\,du \end{array}\right]$$

$$\overset{22}{=} \tfrac{81}{64}\left[\tfrac{1}{8}u(1 + 2u^2)\sqrt{1 + u^2} - \tfrac{1}{8}\ln\left(u + \sqrt{1 + u^2}\right)\right]_0^{4/3} = \tfrac{81}{64}\left[\tfrac{1}{6}\left(1 + \tfrac{32}{9}\right)\sqrt{\tfrac{25}{9}} - \tfrac{1}{8}\ln\left(\tfrac{4}{3} + \sqrt{\tfrac{25}{9}}\right)\right]$$

$$= \tfrac{81}{64}\left(\tfrac{1}{6}\cdot\tfrac{41}{9}\cdot\tfrac{5}{3} - \tfrac{1}{8}\ln 3\right) = \tfrac{205}{128} - \tfrac{81}{512}\ln 3$$

25. $y^{2/3} = 1 - x^{2/3} \;\Rightarrow\; y = (1 - x^{2/3})^{3/2} \;\Rightarrow$

$$\frac{dy}{dx} = \tfrac{3}{2}(1 - x^{2/3})^{1/2}\left(-\tfrac{2}{3}x^{-1/3}\right) = -x^{-1/3}(1 - x^{2/3})^{1/2} \;\Rightarrow$$

$$\left(\frac{dy}{dx}\right)^2 = x^{-2/3}(1 - x^{2/3}) = x^{-2/3} - 1. \text{ Thus}$$

$$L = 4\int_0^1 \sqrt{1 + (x^{-2/3} - 1)}\,dx = 4\int_0^1 x^{-1/3}\,dx = 4\lim_{t\to 0^+}\left[\tfrac{3}{2}x^{2/3}\right]_t^1 = 6.$$

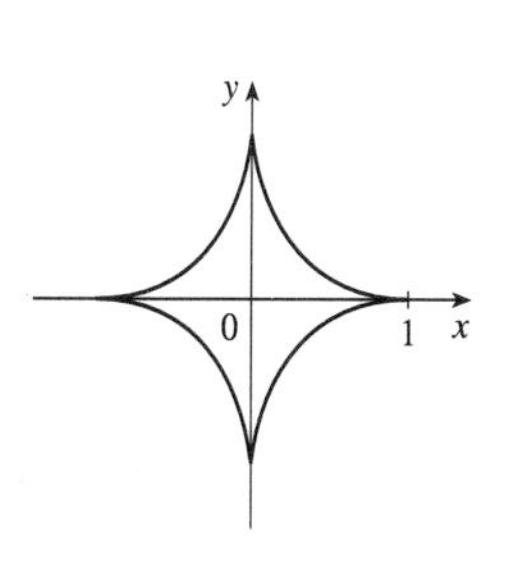

26. (a)

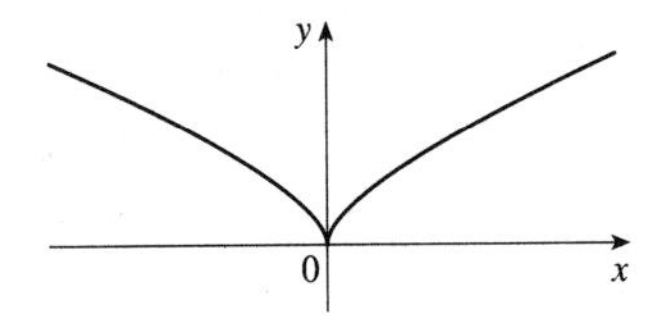

(b) $y = x^{2/3} \Rightarrow 1 + (dy/dx)^2 = 1 + \left(\frac{2}{3}x^{-1/3}\right)^2 = 1 + \frac{4}{9}x^{-2/3}$.

So $L = \int_0^1 \sqrt{1 + \frac{4}{9}x^{-2/3}}\,dx$ [an improper integral]. $x = y^{3/2} \Rightarrow 1 + (dx/dy)^2 = 1 + \left(\frac{3}{2}y^{1/2}\right)^2 = 1 + \frac{9}{4}y$.

So $L = \int_0^1 \sqrt{1 + \frac{9}{4}y}\,dy$. The second integral equals $\frac{4}{9} \cdot \frac{2}{3}\left[\left(1 + \frac{9}{4}y\right)^{3/2}\right]_0^1 = \frac{8}{27}\left(\frac{13\sqrt{13}}{8} - 1\right) = \frac{13\sqrt{13} - 8}{27}$.

The first integral can be evaluated as follows:

$$\int_0^1 \sqrt{1 + \tfrac{4}{9}x^{-2/3}}\,dx = \lim_{t \to 0^+} \int_t^1 \frac{\sqrt{9x^{2/3} + 4}}{3x^{1/3}}\,dx = \lim_{t \to 0^+} \int_{9t^{2/3}}^9 \frac{\sqrt{u + 4}}{18}\,du \qquad \left[\begin{array}{l} u = 9x^{2/3}, \\ du = 6x^{-1/3}\,dx \end{array}\right]$$

$$= \int_0^9 \frac{\sqrt{u + 4}}{18}\,du = \frac{1}{18} \cdot \left[\frac{2}{3}(u + 4)^{3/2}\right]_0^9 = \frac{1}{27}(13^{3/2} - 4^{3/2}) = \frac{13\sqrt{13} - 8}{27}$$

(c) L = length of the arc of this curve from $(-1, 1)$ to $(8, 4)$

$= \int_0^1 \sqrt{1 + \frac{9}{4}y}\,dy + \int_0^4 \sqrt{1 + \frac{9}{4}y}\,dy = \frac{13\sqrt{13} - 8}{27} + \frac{8}{27}\left[\left(1 + \frac{9}{4}y\right)^{3/2}\right]_0^4$ [from part (b)]

$= \frac{13\sqrt{13} - 8}{27} + \frac{8}{27}(10\sqrt{10} - 1) = \frac{13\sqrt{13} + 80\sqrt{10} - 16}{27}$

27. $y = 2x^{3/2} \Rightarrow y' = 3x^{1/2} \Rightarrow 1 + (y')^2 = 1 + 9x$. The arc length function with starting point $P_0(1, 2)$ is

$s(x) = \int_1^x \sqrt{1 + 9t}\,dt = \left[\frac{2}{27}(1 + 9t)^{3/2}\right]_1^x = \frac{2}{27}\left[(1 + 9x)^{3/2} - 10\sqrt{10}\right]$.

28. (a)

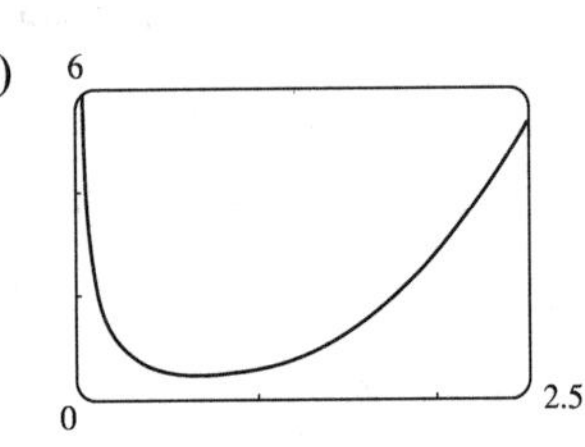

(b) $1 + \left(\frac{dy}{dx}\right)^2 = x^4 + \frac{1}{2} + \frac{1}{16x^4}$,

$$s(x) = \int_1^x \left[t^2 + \frac{1}{4t^2}\right] dt = \left[\tfrac{1}{3}t^3 - \frac{1}{4t}\right]_1^x$$

$$= \frac{1}{3}x^3 - \frac{1}{4x} - \left(\frac{1}{3} - \frac{1}{4}\right) = \frac{1}{3}x^3 - \frac{1}{4x} - \frac{1}{12} \quad \text{for } x \geq 1$$

(c)

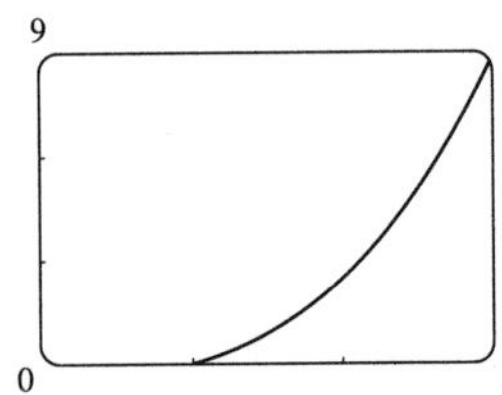

29. The prey hits the ground when $y = 0 \Leftrightarrow 180 - \frac{1}{45}x^2 = 0 \Leftrightarrow x^2 = 45 \cdot 180 \Rightarrow x = \sqrt{8100} = 90$, since x must be positive. $y' = -\frac{2}{45}x \Rightarrow 1 + (y')^2 = 1 + \frac{4}{45^2}x^2$, so the distance traveled by the prey is

$L = \int_0^{90} \sqrt{1 + \frac{4}{45^2}x^2}\,dx = \int_0^4 \sqrt{1 + u^2}\left(\frac{45}{2}\,du\right)$ $\quad \left[u = \frac{2}{45}x,\ du = \frac{2}{45}\,dx\right]$

$\overset{21}{=} \frac{45}{2}\left[\frac{1}{2}u\sqrt{1 + u^2} + \frac{1}{2}\ln\left(u + \sqrt{1 + u^2}\right)\right]_0^4 = \frac{45}{2}\left[2\sqrt{17} + \frac{1}{2}\ln(4 + \sqrt{17})\right] = 45\sqrt{17} + \frac{45}{4}\ln(4 + \sqrt{17}) \approx 209.1$ m

30. $y = 150 - \frac{1}{40}(x-50)^2 \Rightarrow y' = -\frac{1}{20}(x-50) \Rightarrow 1 + (y')^2 = 1 + \frac{1}{20^2}(x-50)^2$, so the distance traveled by the kite is

$$L = \int_0^{80} \sqrt{1 + \tfrac{1}{20^2}(x-50)^2}\, dx = \int_{-5/2}^{3/2} \sqrt{1+u^2}\,(20\,du) \qquad \left[u = \tfrac{1}{20}(x-50),\, du = \tfrac{1}{20}\,dx\right]$$

$$\overset{21}{=} 20\left[\tfrac{1}{2}u\sqrt{1+u^2} + \tfrac{1}{2}\ln\left(u+\sqrt{1+u^2}\right)\right]_{-5/2}^{3/2} = 10\left[\tfrac{3}{2}\sqrt{\tfrac{13}{4}} + \ln\left(\tfrac{3}{2}+\sqrt{\tfrac{13}{4}}\right) + \tfrac{5}{2}\sqrt{\tfrac{29}{4}} - \ln\left(-\tfrac{5}{2}+\sqrt{\tfrac{29}{4}}\right)\right]$$

$$= \tfrac{15}{2}\sqrt{13} + \tfrac{25}{2}\sqrt{29} + 10\ln\left(\frac{3+\sqrt{13}}{-5+\sqrt{29}}\right) \approx 122.8 \text{ ft}$$

31. The sine wave has amplitude 1 and period 14, since it goes through two periods in a distance of 28 in., so its equation is $y = 1\sin\left(\frac{2\pi}{14}x\right) = \sin\left(\frac{\pi}{7}x\right)$. The width w of the flat metal sheet needed to make the panel is the arc length of the sine curve from $x = 0$ to $x = 28$. We set up the integral to evaluate w using the arc length formula with $\frac{dy}{dx} = \frac{\pi}{7}\cos\left(\frac{\pi}{7}x\right)$:

$L = \int_0^{28} \sqrt{1 + \left[\frac{\pi}{7}\cos\left(\frac{\pi}{7}x\right)\right]^2}\, dx = 2\int_0^{14} \sqrt{1 + \left[\frac{\pi}{7}\cos\left(\frac{\pi}{7}x\right)\right]^2}\, dx$. This integral would be very difficult to evaluate exactly, so we use a CAS, and find that $L \approx 29.36$ inches.

32. By symmetry, the length of the curve in each quadrant is the same, so we'll find the length in the first quadrant and multiply by 4.

$x^{2k} + y^{2k} = 1 \Rightarrow y^{2k} = 1 - x^{2k} \Rightarrow y = (1 - x^{2k})^{1/(2k)}$

(in the first quadrant), so we use the arc length formula with

$$\frac{dy}{dx} = \frac{1}{2k}(1-x^{2k})^{1/(2k)-1}(-2kx^{2k-1}) = -x^{2k-1}(1-x^{2k})^{1/(2k)-1}.$$

The total length is therefore

$$L_{2k} = 4\int_0^1 \sqrt{1 + \left[-x^{2k-1}(1-x^{2k})^{1/(2k)-1}\right]^2}\, dx$$
$$= 4\int_0^1 \sqrt{1 + x^{2(2k-1)}(1-x^{2k})^{1/k-2}}\, dx$$

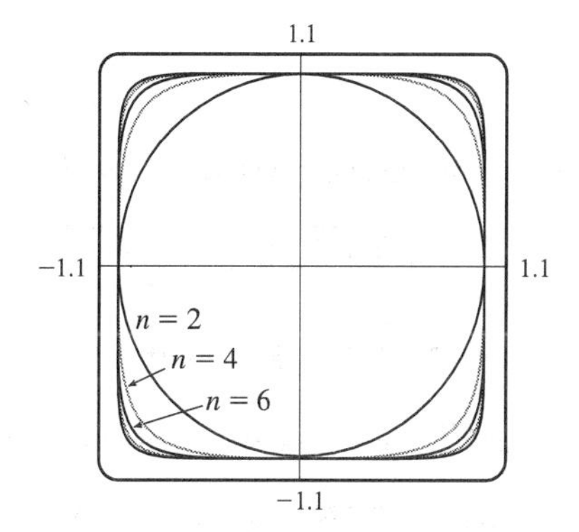

Now from the graph, we see that as k increases, the "corners" of these fat circles get closer to the points $(\pm1, \pm1)$ and $(\pm1, \mp1)$, and the "edges" of the fat circles approach the lines joining these four points. It seems plausible that as $k \to \infty$, the total length of the fat circle with $n = 2k$ will approach the length of the perimeter of the square with sides of length 2. This is supported by taking the limit as $k \to \infty$ of the equation of the fat circle in the first quadrant: $\lim_{k\to\infty}(1-x^{2k})^{1/(2k)} = 1$ for $0 \le x < 1$. So we guess that $\lim_{k\to\infty} L_{2k} = 4 \cdot 2 = 8$.

7.5 Applications to Physics and Engineering

1. $W = \displaystyle\int_a^b f(x)\,dx = \int_0^9 \frac{10}{(1+x)^2}\,dx = 10\int_1^{10}\frac{1}{u^2}\,du \quad [u = 1+x,\; du = dx] \;= 10\left[-\frac{1}{u}\right]_1^{10} = 10\left(-\tfrac{1}{10}+1\right) = 9$ ft-lb

2. $W = \int_1^2 \cos\left(\frac{1}{3}\pi x\right)dx = \frac{3}{\pi}\left[\sin\left(\frac{1}{3}\pi x\right)\right]_1^2 = \frac{3}{\pi}\left(\frac{\sqrt{3}}{2} - \frac{\sqrt{3}}{2}\right) = 0$ N·m $= 0$ J.

Interpretation: From $x = 1$ to $x = \frac{3}{2}$, the force does work equal to $\int_1^{3/2}\cos\left(\frac{1}{3}\pi x\right)dx = \frac{3}{\pi}\left(1 - \frac{\sqrt{3}}{2}\right)$ J in accelerating the particle and increasing its kinetic energy. From $x = \frac{3}{2}$ to $x = 2$, the force opposes the motion of the particle, decreasing its kinetic energy. This is negative work, equal in magnitude but opposite in sign to the work done from $x = 1$ to $x = \frac{3}{2}$.

3. The force function is given by $F(x)$ (in newtons) and the work (in joules) is the area under the curve, given by

$\int_0^8 F(x)\,dx = \int_0^4 F(x)\,dx + \int_4^8 F(x)\,dx = \frac{1}{2}(4)(30) + (4)(30) = 180$ J.

4. Work $= \int_0^{18} f(x)\,dx \approx S_6 = \frac{18-0}{6\cdot 3}\,[f(0) + 4f(3) + 2f(6) + 4f(9) + 2f(12) + 4f(15) + f(18)]$

$= 1\cdot[9.8 + 4(9.1) + 2(8.5) + 4(8.0) + 2(7.7) + 4(7.5) + 7.4] = 148$ J

5. $10 = f(x) = kx = \frac{1}{3}k$ [4 inches $= \frac{1}{3}$ foot], so $k = 30$ lb/ft and $f(x) = 30x$. Now 6 inches $= \frac{1}{2}$ foot, so

$W = \int_0^{1/2} 30x\,dx = \left[15x^2\right]_0^{1/2} = \frac{15}{4}$ ft-lb.

6. $25 = f(x) = kx = k(0.1)$ [10 cm $= 0.1$ m], so $k = 250$ N/m and $f(x) = 250x$. Now 5 cm $= 0.05$ m, so

$W = \int_0^{0.05} 250x\,dx = \left[125x^2\right]_0^{0.05} = 125(0.0025) = 0.3125 \approx 0.31$ J.

7. (a) If $\int_0^{0.12} kx\,dx = 2$ J, then $2 = \left[\frac{1}{2}kx^2\right]_0^{0.12} = \frac{1}{2}k(0.0144) = 0.0072k$ and $k = \frac{2}{0.0072} = \frac{2500}{9} \approx 277.78$ N/m.

Thus, the work needed to stretch the spring from 35 cm to 40 cm is

$\int_{0.05}^{0.10} \frac{2500}{9}x\,dx = \left[\frac{1250}{9}x^2\right]_{1/20}^{1/10} = \frac{1250}{9}\left(\frac{1}{100} - \frac{1}{400}\right) = \frac{25}{24} \approx 1.04$ J.

(b) $f(x) = kx$, so $30 = \frac{2500}{9}x$ and $x = \frac{270}{2500}$ m $= 10.8$ cm

8. Let L be the natural length of the spring in meters. Then

$6 = \int_{0.10-L}^{0.12-L} kx\,dx = \left[\frac{1}{2}kx^2\right]_{0.10-L}^{0.12-L} = \frac{1}{2}k\left[(0.12-L)^2 - (0.10-L)^2\right]$ and

$10 = \int_{0.12-L}^{0.14-L} kx\,dx = \left[\frac{1}{2}kx^2\right]_{0.12-L}^{0.14-L} = \frac{1}{2}k\left[(0.14-L)^2 - (0.12-L)^2\right]$. Simplifying gives us $12 = k(0.0044 - 0.04L)$ and $20 = k(0.0052 - 0.04L)$. Subtracting the first equation from the second gives $8 = 0.0008k$, so $k = 10{,}000$. Now the second equation becomes $20 = 52 - 400L$, so $L = \frac{32}{400}$ m $= 8$ cm.

In Exercises 9–16, n is the number of subintervals of length Δx, and x_i^* is a sample point in the ith subinterval $[x_{i-1}, x_i]$.

9. (a) The portion of the rope from x ft to $(x + \Delta x)$ ft below the top of the building weighs $\frac{1}{2}\,\Delta x$ lb and must be lifted x_i^* ft, so its contribution to the total work is $\frac{1}{2}x_i^*\,\Delta x$ ft-lb. The total work is

$$W = \lim_{n\to\infty}\sum_{i=1}^{n}\frac{1}{2}x_i^*\,\Delta x = \int_0^{50}\frac{1}{2}x\,dx = \left[\frac{1}{4}x^2\right]_0^{50} = \frac{2500}{4} = 625 \text{ ft-lb}$$

Notice that the exact height of the building does not matter (as long as it is more than 50 ft).

(b) When half the rope is pulled to the top of the building, the work to lift the top half of the rope is $W_1 = \int_0^{25}\frac{1}{2}x\,dx = \left[\frac{1}{4}x^2\right]_0^{25} = \frac{625}{4}$ ft-lb. The bottom half of the rope is lifted 25 ft and the work needed to accomplish that is $W_2 = \int_{25}^{50}\frac{1}{2}\cdot 25\,dx = \frac{25}{2}\left[x\right]_{25}^{50} = \frac{625}{2}$ ft-lb. The total work done in pulling half the rope to the top of the building is $W = W_1 + W_2 = \frac{625}{2} + \frac{625}{4} = \frac{3}{4}\cdot 625 = \frac{1875}{4}$ ft-lb.

10. *Assumptions*:

1. After lifting, the chain is L-shaped, with 4 m of the chain lying along the ground.
2. The chain slides effortlessly and without friction along the ground while its end is lifted.
3. The weight density of the chain is constant throughout its length and therefore equals $(8\text{ kg/m})(9.8\text{ m/s}^2) = 78.4$ N/m.

The part of the chain x m from the lifted end is raised $6 - x$ m if $0 \le x \le 6$ m, and it is lifted 0 m if $x > 6$ m.

Thus, the work needed is

$$W = \lim_{n\to\infty}\sum_{i=1}^{n}(6 - x_i^*)\cdot 78.4\,\Delta x = \int_0^6 (6-x)78.4\,dx = 78.4\left[6x - \frac{1}{2}x^2\right]_0^6 = (78.4)(18) = 1411.2 \text{ J}$$

11. The work needed to lift the cable is $\lim_{n\to\infty}\sum_{i=1}^{n} 2x_i^*\,\Delta x = \int_0^{500} 2x\,dx = \left[x^2\right]_0^{500} = 250{,}000$ ft-lb. The work needed to lift the coal is 800 lb · 500 ft = 400,000 ft-lb. Thus, the total work required is 250,000 + 400,000 = 650,000 ft-lb.

12. The work needed to lift the bucket itself is 4 lb · 80 ft = 320 ft-lb. At time t (in seconds) the bucket is $x_i^* = 2t$ ft above its original 80 ft depth, but it now holds only $(40 - 0.2t)$ lb of water. In terms of distance, the bucket holds $\left[40 - 0.2\left(\frac{1}{2}x_i^*\right)\right]$ lb of water when it is x_i^* ft above its original 80 ft depth. Moving this amount of water a distance Δx requires $\left(40 - \frac{1}{10}x_i^*\right)\Delta x$ ft-lb of work. Thus, the work needed to lift the water is

$$W = \lim_{n\to\infty}\sum_{i=1}^{n}\left(40 - \tfrac{1}{10}x_i^*\right)\Delta x = \int_0^{80}\left(40 - \tfrac{1}{10}x\right)dx = \left[40x - \tfrac{1}{20}x^2\right]_0^{80} = (3200 - 320)\text{ ft-lb}$$

Adding the work of lifting the bucket gives a total of 3200 ft-lb of work.

13. At a height of x meters ($0 \le x \le 12$), the mass of the rope is $(0.8\text{ kg/m})(12 - x\text{ m}) = (9.6 - 0.8x)$ kg and the mass of the water is $\left(\frac{36}{12}\text{ kg/m}\right)(12 - x\text{ m}) = (36 - 3x)$ kg. The mass of the bucket is 10 kg, so the total mass is $(9.6 - 0.8x) + (36 - 3x) + 10 = (55.6 - 3.8x)$ kg, and hence, the total force is $9.8(55.6 - 3.8x)$ N. The work needed to lift the bucket Δx m through the ith subinterval of $[0, 12]$ is $9.8(55.6 - 3.8x_i^*)\Delta x$, so the total work is

$$W = \lim_{n\to\infty}\sum_{i=1}^{n} 9.8(55.6 - 3.8x_i^*)\,\Delta x = \int_0^{12}(9.8)(55.6 - 3.8x)\,dx = 9.8\left[55.6x - 1.9x^2\right]_0^{12} = 9.8(393.6) \approx 3857\text{ J}$$

14. The chain's weight density is $\dfrac{25\text{ lb}}{10\text{ ft}} = 2.5$ lb/ft. The part of the chain x ft below the ceiling (for $5 \le x \le 10$) has to be lifted $2(x - 5)$ ft, so the work needed to lift the ith subinterval of the chain is $2(x_i^* - 5)(2.5\,\Delta x)$. The total work needed is

$$\begin{aligned} W &= \lim_{n\to\infty}\sum_{i=1}^{n} 2(x_i^* - 5)(2.5)\,\Delta x = \int_5^{10}[2(x - 5)(2.5)]\,dx = 5\int_5^{10}(x - 5)\,dx \\ &= 5\left[\tfrac{1}{2}x^2 - 5x\right]_5^{10} = 5\left[(50 - 50) - \left(\tfrac{25}{2} - 25\right)\right] = 5\left(\tfrac{25}{2}\right) = 62.5\text{ ft-lb} \end{aligned}$$

15. A "slice" of water Δx m thick and lying at a depth of x_i^* m (where $0 \le x_i^* \le \frac{1}{2}$) has volume $(2 \times 1 \times \Delta x)$ m^3, a mass of $2000\,\Delta x$ kg, weighs about $(9.8)(2000\,\Delta x) = 19{,}600\,\Delta x$ N, and thus requires about $19{,}600x_i^*\,\Delta x$ J of work for its removal.

So $W = \lim_{n\to\infty}\sum_{i=1}^{n} 19{,}600x_i^*\,\Delta x = \int_0^{1/2} 19{,}600x\,dx = \left[9800x^2\right]_0^{1/2} = 2450$ J.

16. A horizontal cylindrical slice of water Δx ft thick has a volume of $\pi r^2 h = \pi \cdot 12^2 \cdot \Delta x$ ft^3 and weighs about $\left(62.5\text{ lb/ft}^3\right)\left(144\pi\,\Delta x\text{ ft}^3\right) = 9000\pi\,\Delta x$ lb. If the slice lies x_i^* ft below the edge of the pool (where $1 \le x_i^* \le 5$), then the work needed to pump it out is about $9000\pi x_i^*\,\Delta x$. Thus,

$$W = \lim_{n\to\infty}\sum_{i=1}^{n} 9000\pi x_i^*\,\Delta x = \int_1^5 9000\pi x\,dx = \left[4500\pi x^2\right]_1^5 = 4500\pi(25 - 1) = 108{,}000\pi\text{ ft-lb}$$

17. (a) A rectangular "slice" of water Δx m thick and lying x m above the bottom has width x m and volume $8x\,\Delta x$ m^3. It weighs about $(9.8 \times 1000)(8x\,\Delta x)$ N, and must be lifted $(5 - x)$ m by the pump, so the work needed is about $(9.8 \times 10^3)(5 - x)(8x\,\Delta x)$ J. The total work required is

$$\begin{aligned} W &\approx \int_0^3 (9.8 \times 10^3)(5 - x)8x\,dx = (9.8 \times 10^3)\int_0^3 (40x - 8x^2)\,dx = (9.8 \times 10^3)\left[20x^2 - \tfrac{8}{3}x^3\right]_0^3 \\ &= (9.8 \times 10^3)(180 - 72) = (9.8 \times 10^3)(108) = 1058.4 \times 10^3 \approx 1.06 \times 10^6\text{ J} \end{aligned}$$

(b) If only 4.7×10^5 J of work is done, then only the water above a certain level (call it h) will be pumped out. So we use the same formula as in par (a), except that the work is fixed, and we are trying to find the lower limit of integration:

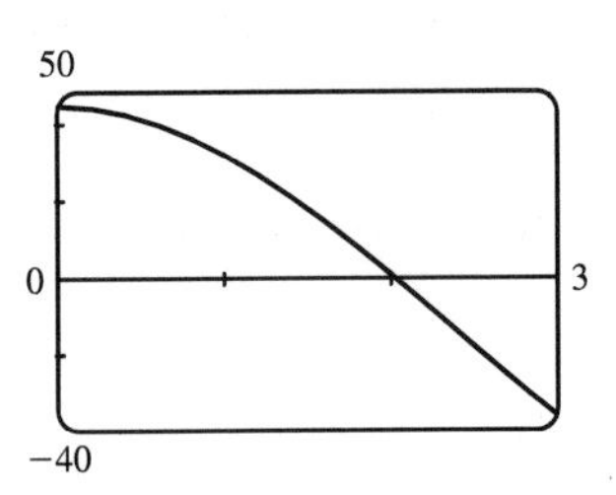

$4.7 \times 10^5 \approx \int_h^3 (9.8 \times 10^3)(5 - x)8x\,dx = (9.8 \times 10^3)\left[20x^2 - \frac{8}{3}x^3\right]_h^3 \Leftrightarrow$

$\frac{4.7}{9.8} \times 10^2 \approx 48 = \left(20 \cdot 3^2 - \frac{8}{3} \cdot 3^3\right) - \left(20h^2 - \frac{8}{3}h^3\right) \Leftrightarrow$

$2h^3 - 15h^2 + 45 = 0$. To find the solution of this equation, we plot $2h^3 - 15h^2 + 45$ between $h = 0$ and $h = 3$. We see that the equation is satisfied for $h \approx 2.0$. So the depth of water remaining in the tank is about 2.0 m.

18. Let x be depth in feet, so that $0 \le x \le 5$. Then $\Delta W = (62.5)\pi\left(\sqrt{5^2 - x^2}\right)^2 \Delta x \cdot x$ ft-lb and

$W \approx 62.5\pi \int_0^5 x(25 - x^2)\,dx = 62.5\pi\left[\frac{25}{2}x^2 - \frac{1}{4}x^4\right]_0^5 = 62.5\pi\left(\frac{625}{2} - \frac{625}{4}\right) = 62.5\pi\left(\frac{625}{4}\right) \approx 3.07 \times 10^4$ ft-lb.

19. $V = \pi r^2 x$, so V is a function of x and P can also be regarded as a function of x. If $V_1 = \pi r^2 x_1$ and $V_2 = \pi r^2 x_2$, then

$$W = \int_{x_1}^{x_2} F(x)\,dx = \int_{x_1}^{x_2} \pi r^2 P(V(x))\,dx = \int_{x_1}^{x_2} P(V(x))\,dV(x) \qquad \text{[Let } V(x) = \pi r^2 x\text{, so } dV(x) = \pi r^2\,dx.]$$

$$= \int_{V_1}^{V_2} P(V)\,dV \quad \text{by the Substitution Rule.}$$

20. $160 \text{ lb/in}^2 = 160 \cdot 144 \text{ lb/ft}^2$, $100 \text{ in}^3 = \frac{100}{1728} \text{ ft}^3$, and $800 \text{ in}^3 = \frac{800}{1728} \text{ ft}^3$.

$k = PV^{1.4} = (160 \cdot 144)\left(\frac{100}{1728}\right)^{1.4} = 23{,}040\left(\frac{25}{432}\right)^{1.4} \approx 426.5$. Therefore, $P \approx 426.5V^{-1.4}$ and

$$W = \int_{100/1728}^{800/1728} 426.5V^{-1.4}\,dV = 426.5\left[\frac{1}{-0.4}V^{-0.4}\right]_{25/432}^{25/54} = (426.5)(2.5)\left[\left(\frac{432}{25}\right)^{0.4} - \left(\frac{54}{25}\right)^{0.4}\right] \approx 1.88 \times 10^3 \text{ ft-lb.}$$

21. (a) $W = \displaystyle\int_a^b F(r)\,dr = \int_a^b G\frac{m_1 m_2}{r^2}\,dr = Gm_1m_2\left[\frac{-1}{r}\right]_a^b = Gm_1m_2\left(\frac{1}{a} - \frac{1}{b}\right)$

(b) By part (a), $W = GMm\left(\dfrac{1}{R} - \dfrac{1}{R + 1{,}000{,}000}\right)$ where M = mass of the Earth in kg, R = radius of the Earth in m, and m = mass of the satellite in kg. (Note that 1000 km = 1,000,000 m.) Thus,

$$W = (6.67 \times 10^{-11})(5.98 \times 10^{24})(1000) \times \left(\frac{1}{6.37 \times 10^6} - \frac{1}{7.37 \times 10^6}\right) \approx 8.50 \times 10^9 \text{ J}$$

22. (a) $W = \displaystyle\int_R^\infty \frac{GMm}{r^2}\,dr = \lim_{t\to\infty}\int_R^t \frac{GMm}{r^2}\,dr = \lim_{t\to\infty} GMm\left[\frac{-1}{r}\right]_R^t = GMm\lim_{t\to\infty}\left(\frac{-1}{t} + \frac{1}{R}\right) = \frac{GMm}{R}$,

where M = mass of the Earth = 5.98×10^{24} kg, m = mass of the satellite = 10^3 kg,

R = radius of the Earth = 6.37×10^6 m, and G = gravitational constant = 6.67×10^{-11} N·m²/kg².

Therefore, work $= \dfrac{6.67 \times 10^{-11} \cdot 5.98 \times 10^{24} \cdot 10^3}{6.37 \times 10^6} \approx 6.26 \times 10^{10}$ J.

(b) From part (a), $W = \dfrac{GMm}{R}$. The initial kinetic energy supplies the needed work, so $\dfrac{1}{2}mv_0^2 = \dfrac{GMm}{R} \Rightarrow$

$v_0 = \sqrt{\dfrac{2GM}{R}}$.

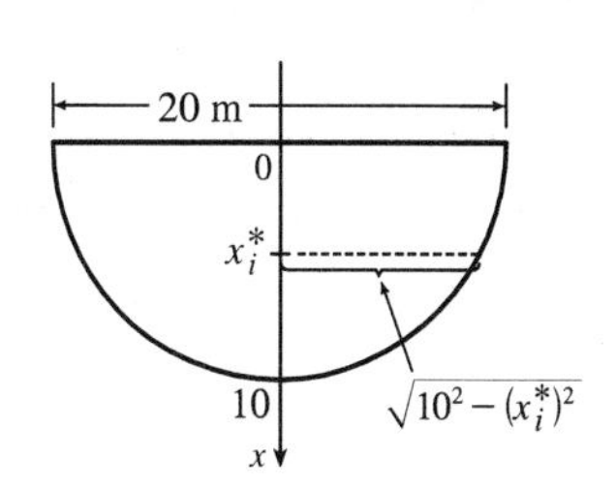

23. Since an equation for the shape is $x^2 + y^2 = 10^2$ $(x \geq 0)$, we have $y = \sqrt{100 - x^2}$. Thus, the area of the ith strip is $2\sqrt{100 - (x_i^*)^2}\,\Delta x$ and the pressure on the strip is $\rho g x_i^*$, so the hydrostatic force on the strip is $\rho g x_i^* \cdot 2\sqrt{100 - (x_i^*)^2}\,\Delta x$ and the total force on the plate $\approx \sum_{i=1}^{n} \rho g x_i^* \cdot 2\sqrt{100 - (x_i^*)^2}\,\Delta x$. The total force

$$F = \lim_{n\to\infty} \sum_{i=1}^{n} \rho g x_i^* \cdot 2\sqrt{100 - (x_i^*)^2}\,\Delta x = \int_0^{10} 2\rho g x\sqrt{100 - x^2}\,dx$$
$$= -\rho g \int_0^{10} (100 - x^2)^{1/2}(-2x)\,dx = -\rho g\left[\tfrac{2}{3}(100 - x^2)^{3/2}\right]_0^{10} = -\tfrac{2}{3}\rho g(0 - 1000)$$
$$= \tfrac{2000}{3}\rho g \approx \tfrac{2000}{3} \cdot 1000 \cdot 9.8 \approx 6.5 \times 10^6 \text{ N} \qquad \left[\rho \approx 1000 \text{ kg/m}^3 \text{ and } g \approx 9.8 \text{ m/s}^2\right]$$

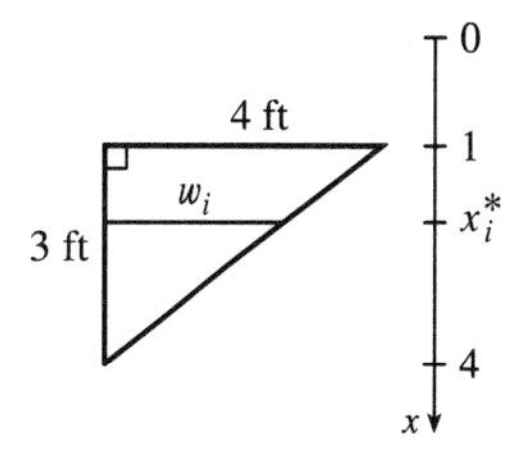

24. Set up a vertical x-axis as shown. Then the area of the ith rectangular strip is $\frac{4}{3}(4 - x_i^*)\,\Delta x$. [By similar triangles, $\dfrac{w_i}{4 - x_i^*} = \dfrac{4}{3}$, so $w_i = \frac{4}{3}(4 - x_i^*)$.]

The pressure on the strip is δx_i^*, so the hydrostatic force on the strip is $\delta x_i^* \cdot \frac{4}{3}(4 - x_i^*)\,\Delta x$ and the total force on the plate $\approx \sum_{i=1}^{n} \delta x_i^* \cdot \frac{4}{3}(4 - x_i^*)\,\Delta x$.

The total force

$$F = \lim_{n\to\infty} \sum_{i=1}^{n} \delta x_i^* \cdot \tfrac{4}{3}(4 - x_i^*)\,\Delta x = \int_1^4 \delta x \cdot \tfrac{4}{3}(4 - x)\,dx = \tfrac{4}{3}\delta \int_1^4 (4x - x^2)\,dx$$
$$= \tfrac{4}{3}\delta\left[2x^2 - \tfrac{1}{3}x^3\right]_1^4 = \tfrac{4}{3}\delta\left[\left(32 - \tfrac{64}{3}\right) - \left(2 - \tfrac{1}{3}\right)\right] = \tfrac{4}{3}\delta(9) = 12\delta \approx 750 \text{ lb}$$

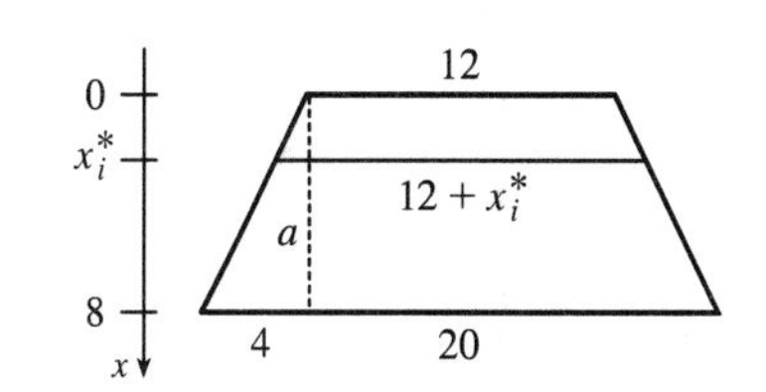

25. Using similar triangles, $\dfrac{4 \text{ ft wide}}{8 \text{ ft high}} = \dfrac{a \text{ ft wide}}{x_i^* \text{ ft high}}$, so $a = \frac{1}{2}x_i^*$ and the width of the ith rectangular strip is $12 + 2a = 12 + x_i^*$. The area of the strip is $(12 + x_i^*)\,\Delta x$. The pressure on the strip is δx_i^*.

$$F = \lim_{n\to\infty} \sum_{i=1}^{n} \delta x_i^*(12 + x_i^*)\,\Delta x = \int_0^8 \delta x \cdot (12 + x)\,dx = \delta \int_0^8 (12x + x^2)\,dx$$
$$= \delta\left[6x^2 + \tfrac{x^3}{3}\right]_0^8 = \delta\left(384 + \tfrac{512}{3}\right) = (62.5)\tfrac{1664}{3} \approx 3.47 \times 10^4 \text{ lb}$$

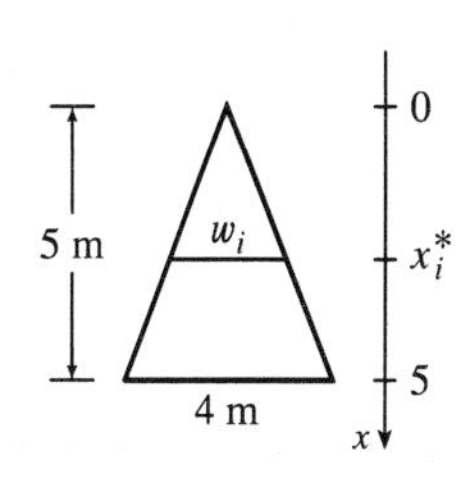

26. By similar triangles, $w_i/4 = x_i^*/5$, so $w_i = \frac{4}{5}x_i^*$ and the area of the ith strip is $\frac{4}{5}x_i^*\,\Delta x$. The pressure on the strip is $\rho g x_i^*$, so the hydrostatic force on the strip is $\rho g x_i^* \cdot \frac{4}{5}x_i^*\,\Delta x$ and the total force on the plate $\approx \sum_{i=1}^{n} \rho g x_i^* \cdot \frac{4}{5}x_i^*\,\Delta x$. The total force

$$F = \lim_{n\to\infty} \sum_{i=1}^{n} \rho g x_i^* \cdot \tfrac{4}{5}x_i^*\,\Delta x = \int_0^5 \rho g x \cdot \tfrac{4}{5}x\,dx = \tfrac{4}{5}\rho g\left[\tfrac{1}{3}x^3\right]_0^5 = \tfrac{4}{5}\rho g \cdot \tfrac{125}{3} = \tfrac{100}{3}\rho g$$
$$\approx \tfrac{100}{3} \cdot 1000 \cdot 9.8 \approx 3.3 \times 10^5 \text{ N}$$

27. By similar triangles, $\dfrac{8}{4\sqrt{3}} = \dfrac{w_i}{x_i^*} \Rightarrow w_i = \dfrac{2x_i^*}{\sqrt{3}}$. The area of the ith rectangular strip is $\dfrac{2x_i^*}{\sqrt{3}}\,\Delta x$ and the pressure on it is $\rho g\left(4\sqrt{3} - x_i^*\right)$.

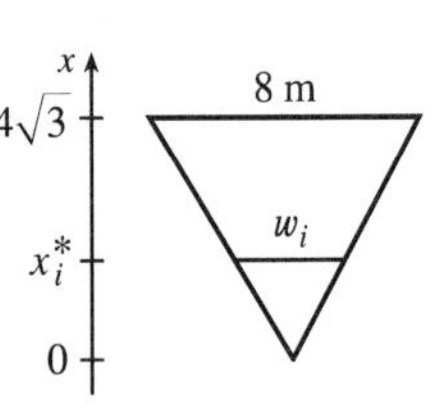

$$\begin{aligned} F &= \int_0^{4\sqrt{3}} \rho g\left(4\sqrt{3} - x\right)\frac{2x}{\sqrt{3}}\,dx = 8\rho g\int_0^{4\sqrt{3}} x\,dx - \frac{2\rho g}{\sqrt{3}}\int_0^{4\sqrt{3}} x^2\,dx \\ &= 4\rho g\left[x^2\right]_0^{4\sqrt{3}} - \frac{2\rho g}{3\sqrt{3}}\left[x^3\right]_0^{4\sqrt{3}} = 192\rho g - \frac{2\rho g}{3\sqrt{3}}\,64\cdot 3\sqrt{3} = 192\rho g - 128\rho g = 64\rho g \\ &\approx 64(840)(9.8) \approx 5.27\times 10^5\text{ N} \end{aligned}$$

28. The area of the ith rectangular strip is $2\sqrt{2y_i^*}\,\Delta y$ and the pressure on it is $\delta d_i = \delta(8 - y_i^*)$.

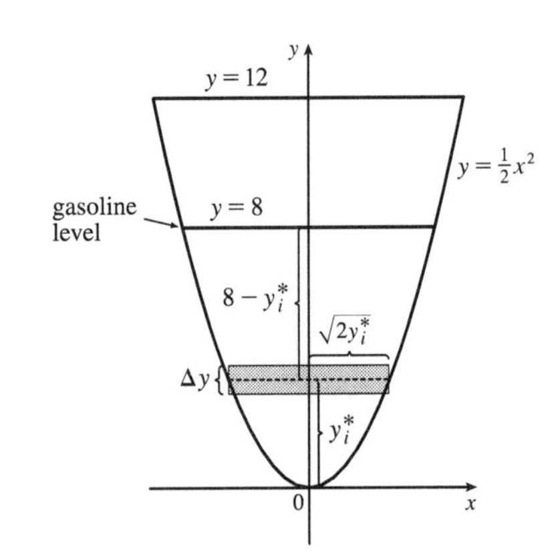

$$\begin{aligned} F &= \textstyle\int_0^8 \delta(8-y)2\sqrt{2y}\,dy = 42\cdot 2\cdot\sqrt{2}\int_0^8 (8-y)y^{1/2}\,dy \\ &= 84\sqrt{2}\textstyle\int_0^8 (8y^{1/2} - y^{3/2})\,dy = 84\sqrt{2}\left[8\cdot\frac{2}{3}y^{3/2} - \frac{2}{5}y^{5/2}\right]_0^8 \\ &= 84\sqrt{2}\left[8\cdot\frac{2}{3}\cdot 16\sqrt{2} - \frac{2}{5}\cdot 128\sqrt{2}\right] \\ &= 84\sqrt{2}\cdot 256\sqrt{2}\left(\frac{1}{3} - \frac{1}{5}\right) = 43{,}008\cdot\frac{2}{15} = 5734.4\text{ lb} \end{aligned}$$

29. (a) The area of a strip is $20\,\Delta x$ and the pressure on it is δx_i.

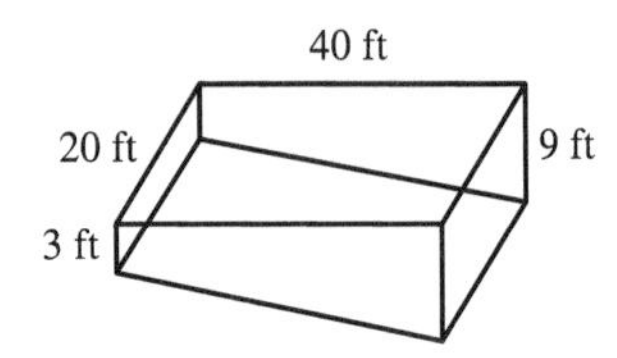

$$\begin{aligned} F &= \textstyle\int_0^3 \delta x 20\,dx = 20\delta\left[\frac{1}{2}x^2\right]_0^3 = 20\delta\cdot\frac{9}{2} = 90\delta \\ &= 90(62.5) = 5625\text{ lb} \approx 5.63\times 10^3\text{ lb} \end{aligned}$$

(b) $F = \int_0^9 \delta x 20\,dx = 20\delta\left[\frac{1}{2}x^2\right]_0^9 = 20\delta\cdot\frac{81}{2} = 810\delta = 810(62.5) = 50{,}625\text{ lb} \approx 5.06\times 10^4$ lb.

(c) For the first 3 ft, the length of the side is constant at 40 ft. For $3 < x \le 9$, we can use similar triangles to find the length a:

$\dfrac{a}{40} = \dfrac{9-x}{6} \Rightarrow a = 40\cdot\dfrac{9-x}{6}$.

$$\begin{aligned} F &= \textstyle\int_0^3 \delta x 40\,dx + \int_3^9 \delta x(40)\frac{9-x}{6}\,dx = 40\delta\left[\frac{1}{2}x^2\right]_0^3 + \frac{20}{3}\delta\int_3^9 (9x - x^2)\,dx_3^9 = 180\delta + \frac{20}{3}\delta\left[\frac{9}{2}x^2 - \frac{1}{3}x^3\right]_3^9 \\ &= 180\delta + \tfrac{20}{3}\delta\left[\left(\tfrac{729}{2} - 243\right) - \left(\tfrac{81}{2} - 9\right)\right] = 180\delta + 600\delta = 780\delta = 780(62.5) = 48{,}750\text{ lb} \approx 4.88\times 10^4\text{ lb} \end{aligned}$$

(d) For any right triangle with hypotenuse on the bottom,

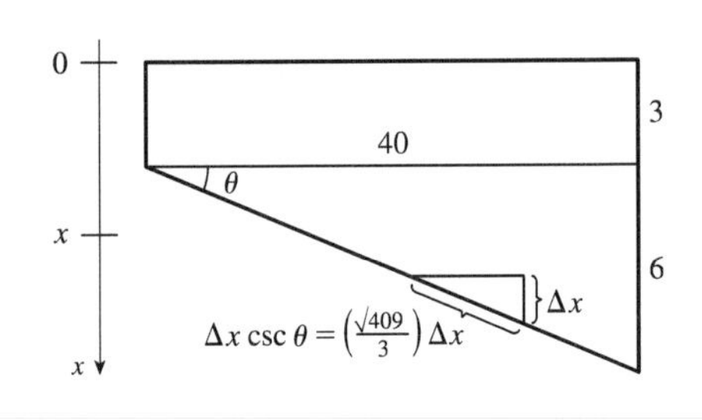

$\sin\theta = \dfrac{\Delta x}{\text{hypotenuse}} \Rightarrow$

$\text{hypotenuse} = \Delta x\csc\theta = \Delta x\,\dfrac{\sqrt{40^2 + 6^2}}{6} = \dfrac{\sqrt{409}}{3}\,\Delta x.$

$$\begin{aligned} F &= \textstyle\int_3^9 \delta x 20\,\frac{\sqrt{409}}{3}\,dx = \frac{1}{3}\left(20\sqrt{409}\right)\delta\left[\frac{1}{2}x^2\right]_3^9 \\ &= \tfrac{1}{3}\cdot 10\sqrt{409}\,\delta(81 - 9) \approx 303{,}356\text{ lb} \approx 3.03\times 10^5\text{ lb} \end{aligned}$$

30. $F = \int_0^2 \rho g(10 - x)2\sqrt{4 - x^2}\,dx$

$= 20\rho g \int_0^2 \sqrt{4 - x^2}\,dx - \rho g \int_0^2 \sqrt{4 - x^2}\,2x\,dx$

$= 20\rho g \frac{1}{4}\pi(2^2) - \rho g \int_0^4 u^{1/2}\,du \qquad [u = 4 - x^2,\ du = -2x\,dx]$

$= 20\pi\rho g - \frac{2}{3}\rho g\left[u^{3/2}\right]_0^4 = 20\pi\rho g - \frac{16}{3}\rho g = \rho g\left(20\pi - \frac{16}{3}\right)$

$= (1000)(9.8)\left(20\pi - \frac{16}{3}\right) \approx 5.63 \times 10^5$ N

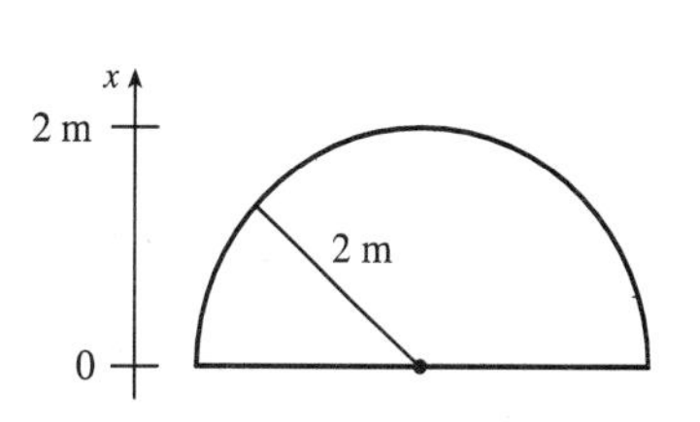

31. $F = \int_2^5 \rho g x \cdot w(x)\,dx$, where $w(x)$ is the width of the plate at depth x. Since $n = 6$, $\Delta x = \frac{5-2}{6} = \frac{1}{2}$, and

$$F \approx S_6 = \rho g \cdot \frac{1/2}{3}[2 \cdot w(2) + 4 \cdot 2.5 \cdot w(2.5) + 2 \cdot 3 \cdot w(3) + 4 \cdot 3.5 \cdot w(3.5) + 2 \cdot 4 \cdot w(4) + 4 \cdot 4.5 \cdot w(4.5) + 5 \cdot w(5)]$$

$$= \tfrac{1}{6}\rho g(2 \cdot 0 + 10 \cdot 0.8 + 6 \cdot 1.7 + 14 \cdot 2.4 + 8 \cdot 2.9 + 18 \cdot 3.3 + 5 \cdot 3.6)$$

$$= \tfrac{1}{6}(1000)(9.8)(152.4) \approx 2.5 \times 10^5 \text{ N}$$

32. $M = m_1x_1 + m_2x_2 + m_3x_3 = 25(-2) + 20(3) + 10(7) = 80$; $\overline{x} = M/(m_1 + m_2 + m_3) = \frac{80}{55} = \frac{16}{11}$.

33. $m = \sum_{i=1}^{3} m_i = 6 + 5 + 10 = 21$.

$M_x = \sum_{i=1}^{3} m_i y_i = 6(5) + 5(-2) + 10(-1) = 10$; $M_y = \sum_{i=1}^{3} m_i x_i = 6(1) + 5(3) + 10(-2) = 1$.

$\overline{x} = \dfrac{M_y}{m} = \dfrac{1}{21}$ and $\overline{y} = \dfrac{M_x}{m} = \dfrac{10}{21}$, so the center of mass of the system is $\left(\frac{1}{21}, \frac{10}{21}\right)$.

34. $M_x = \sum_{i=1}^{4} m_i y_i = 6(-2) + 5(4) + 1(-7) + 4(-1) = -3$, $M_y = \sum_{i=1}^{4} m_i x_i = 6(1) + 5(3) + 1(-3) + 4(6) = 42$,

and $m = \sum_{i=1}^{4} m_i = 16$, so $\overline{x} = \dfrac{M_y}{m} = \dfrac{42}{16} = \dfrac{21}{8}$ and $\overline{y} = \dfrac{M_x}{m} = -\dfrac{3}{16}$; the center of mass is $(\overline{x}, \overline{y}) = \left(\frac{21}{8}, -\frac{3}{16}\right)$.

35. Since the region in the figure is symmetric about the y-axis, we know that $\overline{x} = 0$. The region is "bottom-heavy," so we know that $\overline{y} < 2$, and we might guess that $\overline{y} = 1.5$.

$A = \int_{-2}^{2}(4 - x^2)\,dx = 2\int_0^2(4 - x^2)\,dx = 2\left[4x - \frac{1}{3}x^3\right]_0^2$

$= 2\left(8 - \frac{8}{3}\right) = \frac{32}{3}$

$\overline{x} = \frac{1}{A}\int_{-2}^{2} x(4 - x^2)\,dx = 0$ since $f(x) = x(4 - x^2)$ is an odd function (or since the region is symmetric about the y-axis).

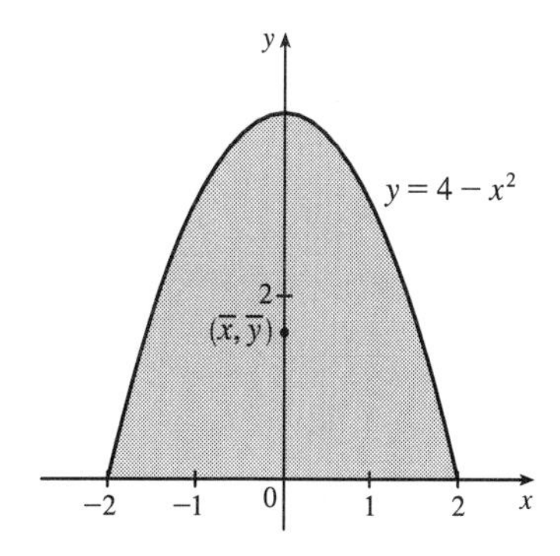

$$\overline{y} = \frac{1}{A}\int_{-2}^{2} \tfrac{1}{2}(4 - x^2)^2\,dx = \tfrac{3}{32} \cdot \tfrac{1}{2} \cdot 2\int_0^2 (16 - 8x^2 + x^4)\,dx = \tfrac{3}{32}\left[16x - \tfrac{8}{3}x^3 + \tfrac{1}{5}x^5\right]_0^2$$

$$= \tfrac{3}{32}\left(32 - \tfrac{64}{3} + \tfrac{32}{5}\right) = 3\left(1 - \tfrac{2}{3} + \tfrac{1}{5}\right) = 3\left(\tfrac{8}{15}\right) = \tfrac{8}{5}$$

Thus, the centroid is $(\overline{x}, \overline{y}) = \left(0, \frac{8}{5}\right)$.

36. The region in the figure is "left-heavy" and "bottom-heavy," so we know $\overline{x} < 1$ and $\overline{y} < 1.5$, and we might guess that $\overline{x} = 0.7$ and $\overline{y} = 1.2$.

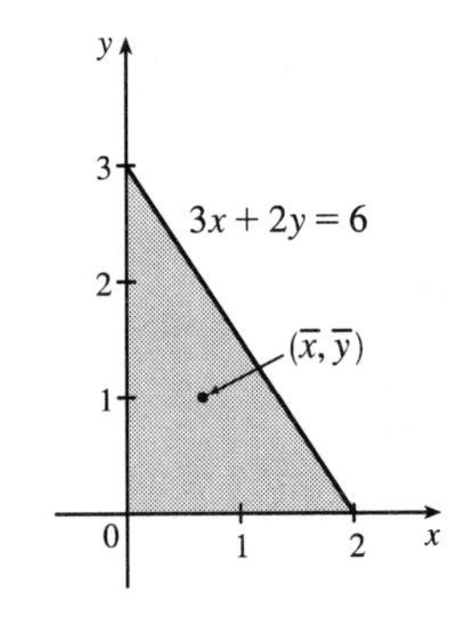

$3x + 2y = 6 \quad \Leftrightarrow \quad 2y = 6 - 3x \quad \Leftrightarrow \quad y = 3 - \frac{3}{2}x.$

$A = \int_0^2 \left(3 - \frac{3}{2}x\right) dx = \left[3x - \frac{3}{4}x^2\right]_0^2 = 6 - 3 = 3.$

$\overline{x} = \frac{1}{A}\int_0^2 x\left(3 - \frac{3}{2}x\right) dx = \frac{1}{3}\int_0^2 \left(3x - \frac{3}{2}x^2\right) dx = \frac{1}{3}\left[\frac{3}{2}x^2 - \frac{1}{2}x^3\right]_0^2$

$= \frac{1}{3}(6 - 4) = \frac{2}{3},$

$\overline{y} = \frac{1}{A}\int_0^2 \frac{1}{2}\left(3 - \frac{3}{2}x\right)^2 dx = \frac{1}{3} \cdot \frac{1}{2}\int_0^2 \left(9 - 9x + \frac{9}{4}x^2\right) dx = \frac{1}{6}\left[9x - \frac{9}{2}x^2 + \frac{3}{4}x^3\right]_0^2$

$= \frac{1}{6}(18 - 18 + 6) = 1.$

Thus, the centroid is $(\overline{x}, \overline{y}) = \left(\frac{2}{3}, 1\right)$.

37. The region in the figure is "right-heavy" and "bottom-heavy," so we know $\overline{x} > 0.5$ and $\overline{y} < 1$, and we might guess that $\overline{x} = 0.6$ and $\overline{y} = 0.9$.

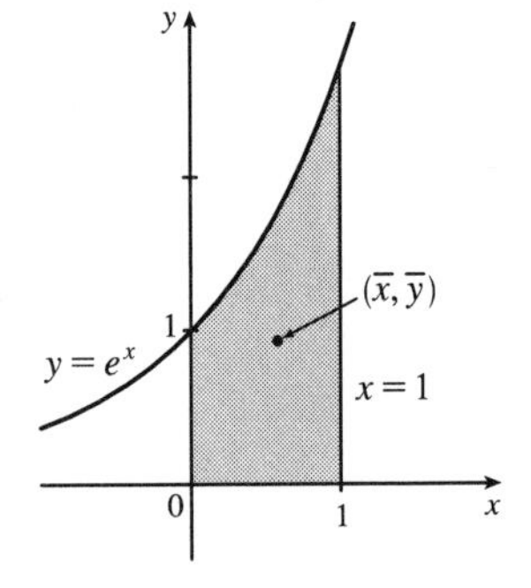

$A = \int_0^1 e^x\, dx = \left[e^x\right]_0^1 = e - 1,$

$$\overline{x} = \frac{1}{A}\int_0^1 xe^x\, dx = \frac{1}{e - 1}\left[xe^x - e^x\right]_0^1 \quad \text{[by parts]}$$

$$= \frac{1}{e - 1}[0 - (-1)] = \frac{1}{e - 1},$$

$$\overline{y} = \frac{1}{A}\int_0^1 \tfrac{1}{2}(e^x)^2\, dx = \frac{1}{e - 1} \cdot \tfrac{1}{4}\left[e^{2x}\right]_0^1 = \frac{1}{4(e - 1)}(e^2 - 1) = \frac{e + 1}{4}.$$

Thus, the centroid is $(\overline{x}, \overline{y}) = \left(\dfrac{1}{e - 1}, \dfrac{e + 1}{4}\right) \approx (0.58, 0.93)$.

38. The region in the figure is "left-heavy" and "bottom-heavy," so we know $\overline{x} < 1.5$ and $\overline{y} < 0.5$, and we might guess that $\overline{x} = 1.4$ and $\overline{y} = 0.4$.

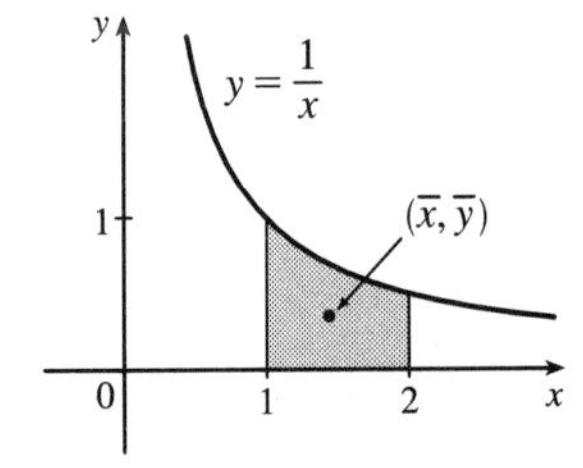

$A = \int_1^2 \frac{1}{x}\, dx = \left[\ln x\right]_1^2 = \ln 2,$

$\overline{x} = \frac{1}{A}\int_1^2 x \cdot \frac{1}{x}\, dx = \frac{1}{A}\left[x\right]_1^2 = \frac{1}{A} = \frac{1}{\ln 2},$

$\overline{y} = \frac{1}{A}\int_1^2 \frac{1}{2}\left(\frac{1}{x}\right)^2 dx = \frac{1}{2A}\int_1^2 x^{-2}\, dx = \frac{1}{2A}\left[-\frac{1}{x}\right]_1^2 = \frac{1}{2\ln 2}\left(-\frac{1}{2} + 1\right) = \frac{1}{4\ln 2}.$

Thus, the centroid is $(\overline{x}, \overline{y}) = \left(\dfrac{1}{\ln 2}, \dfrac{1}{4\ln 2}\right) \approx (1.44, 0.36)$.

39. $A = \int_0^1 (\sqrt{x} - x)\, dx = \left[\frac{2}{3}x^{3/2} - \frac{1}{2}x^2\right]_0^1 = \frac{2}{3} - \frac{1}{2} = \frac{1}{6}.$

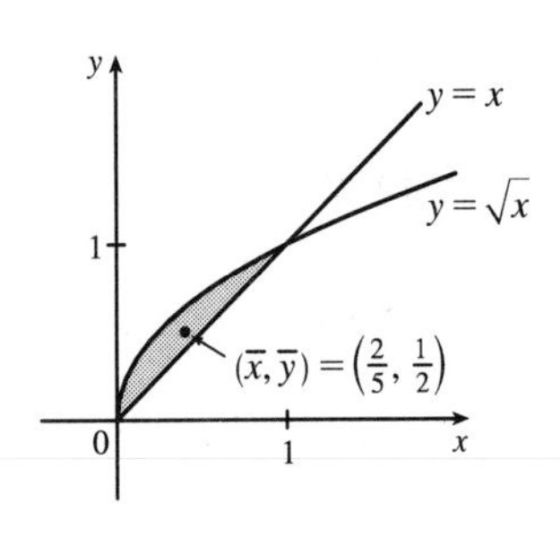

$\overline{x} = \frac{1}{A}\int_0^1 x(\sqrt{x} - x)\, dx = 6\int_0^1 (x^{3/2} - x^2)\, dx = 6\left[\frac{2}{5}x^{5/2} - \frac{1}{3}x^3\right]_0^1$

$= 6\left(\frac{2}{5} - \frac{1}{3}\right) = 6\left(\frac{1}{15}\right) = \frac{2}{5},$

$\overline{y} = \frac{1}{A}\int_0^1 \frac{1}{2}\left[(\sqrt{x})^2 - x^2\right] dx = 6 \cdot \frac{1}{2}\int_0^1 (x - x^2)\, dx = 3\left[\frac{1}{2}x^2 - \frac{1}{3}x^3\right]_0^1$

$= 3\left(\frac{1}{2} - \frac{1}{3}\right) = \frac{1}{2}.$

Thus, the centroid is $(\overline{x}, \overline{y}) = \left(\frac{2}{5}, \frac{1}{2}\right)$.

40. $A = \int_{-1}^{2}(x+2-x^2)\,dx = \left[\frac{1}{2}x^2 + 2x - \frac{1}{3}x^3\right]_{-1}^{2}$
$= \left(2+4-\frac{8}{3}\right) - \left(\frac{1}{2} - 2 + \frac{1}{3}\right) = \frac{9}{2}.$

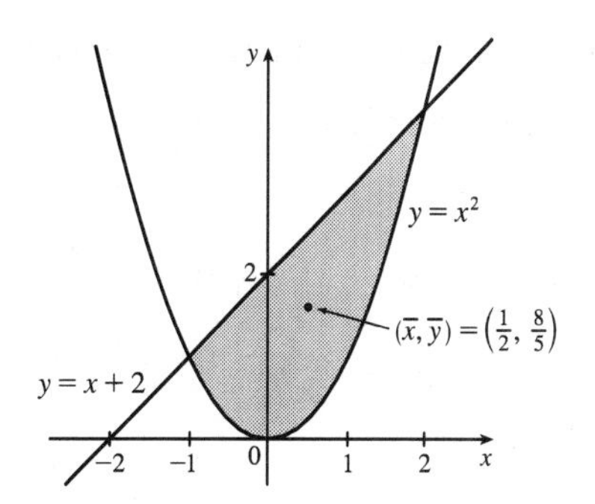

$\overline{x} = \frac{1}{A}\int_{-1}^{2} x(x+2-x^2)\,dx = \frac{2}{9}\int_{-1}^{2}(x^2+2x-x^3)\,dx$

$= \frac{2}{9}\left[\frac{1}{3}x^3 + x^2 - \frac{1}{4}x^4\right]_{-1}^{2}$

$= \frac{2}{9}\left[\left(\frac{8}{3}+4-4\right) - \left(-\frac{1}{3}+1-\frac{1}{4}\right)\right] = \frac{2}{9}\cdot\frac{9}{4} = \frac{1}{2},$

$\overline{y} = \frac{1}{A}\int_{-1}^{2}\frac{1}{2}\left[(x+2)^2 - (x^2)^2\right]dx = \frac{2}{9}\cdot\frac{1}{2}\int_{-1}^{2}(x^2+4x+4-x^4)\,dx = \frac{1}{9}\left[\frac{1}{3}x^3 + 2x^2 + 4x - \frac{1}{5}x^5\right]_{-1}^{2}$
$= \frac{1}{9}\left[\left(\frac{8}{3}+8+8-\frac{32}{5}\right) - \left(-\frac{1}{3}+2-4+\frac{1}{5}\right)\right] = \frac{1}{9}\left(18+\frac{9}{3}-\frac{33}{5}\right) = \frac{1}{9}\cdot\frac{72}{5} = \frac{8}{5}.$

Thus, the centroid is $(\overline{x},\overline{y}) = \left(\frac{1}{2},\frac{8}{5}\right)$.

41. $A = \int_0^{\pi/4}(\cos x - \sin x)\,dx = \left[\sin x + \cos x\right]_0^{\pi/4} = \sqrt{2} - 1,$

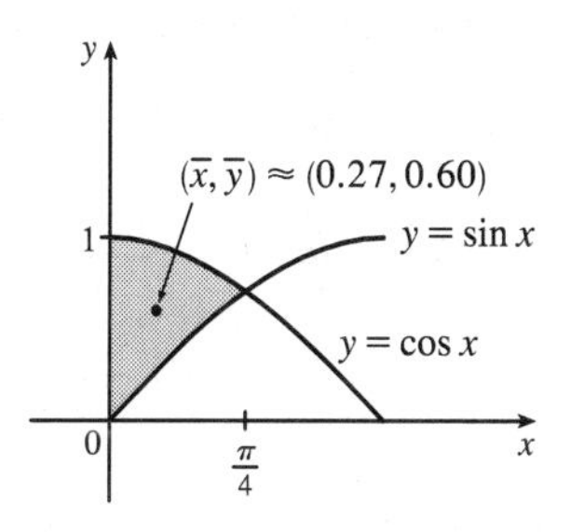

$\overline{x} = A^{-1}\int_0^{\pi/4} x(\cos x - \sin x)\,dx$

$= A^{-1}\left[x(\sin x + \cos x) + \cos x - \sin x\right]_0^{\pi/4}$ [integration by parts]

$= A^{-1}\left(\frac{\pi}{4}\sqrt{2} - 1\right) = \frac{\frac{1}{4}\pi\sqrt{2} - 1}{\sqrt{2} - 1},$

$\overline{y} = A^{-1}\int_0^{\pi/4}\frac{1}{2}(\cos^2 x - \sin^2 x)\,dx = \frac{1}{2A}\int_0^{\pi/4}\cos 2x\,dx = \frac{1}{4A}\left[\sin 2x\right]_0^{\pi/4} = \frac{1}{4A} = \frac{1}{4\left(\sqrt{2}-1\right)}$

Thus, the centroid is $(\overline{x},\overline{y}) = \left(\frac{\pi\sqrt{2}-4}{4(\sqrt{2}-1)}, \frac{1}{4(\sqrt{2}-1)}\right) \approx (0.27, 0.60)$.

42. $A = \int_0^1 x\,dx + \int_1^2 \frac{1}{x}\,dx = \left[\frac{1}{2}x^2\right]_0^1 + [\ln x]_1^2 = \frac{1}{2} + \ln 2,$

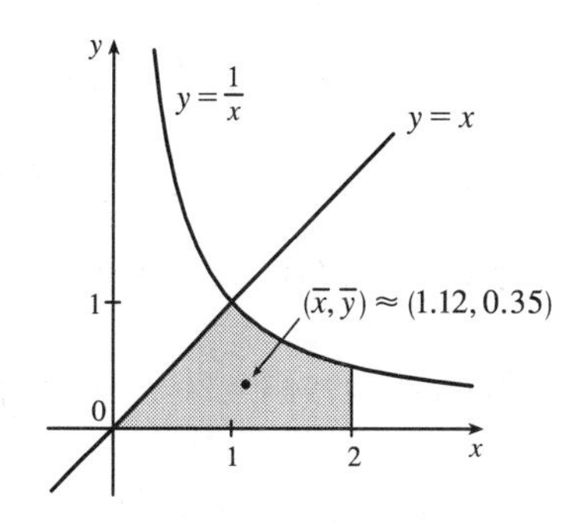

$\overline{x} = \frac{1}{A}\left[\int_0^1 x^2\,dx + \int_1^2 1\,dx\right] = \frac{1}{A}\left(\left[\frac{1}{3}x^3\right]_0^1 + [x]_1^2\right)$

$= \frac{1}{A}\left(\frac{1}{3}+1\right) = \frac{2}{1+2\ln 2}\cdot\frac{4}{3} = \frac{8}{3(1+2\ln 2)},$

$\overline{y} = \frac{1}{A}\left[\int_0^1 \frac{1}{2}x^2\,dx + \int_1^2 \frac{1}{2x^2}\,dx\right] = \frac{1}{2A}\left(\left[\frac{1}{3}x^3\right]_0^1 + \left[-\frac{1}{x}\right]_1^2\right)$

$= \frac{1}{2A}\left(\frac{1}{3}+\frac{1}{2}\right) = \frac{5}{12A} = \frac{5}{6+12\ln 2}.$

Thus, the centroid is $(\overline{x},\overline{y}) = \left(\frac{8}{3(1+2\ln 2)}, \frac{5}{6(1+2\ln 2)}\right) \approx (1.12, 0.35)$. The principle used in this problem is stated after Example 7: the moment of the union of two nonoverlapping regions is the sum of the moments of the individual regions.

43. By symmetry, $M_y = 0$ and $\overline{x} = 0$. $A = \frac{1}{2}bh = \frac{1}{2} \cdot 2 \cdot 2 = 2$.

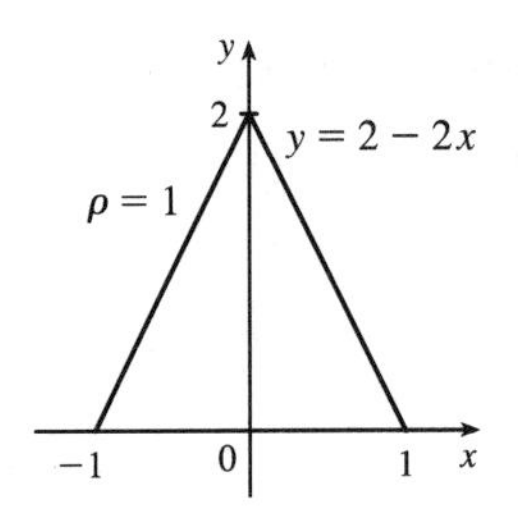

$$M_x = \rho \int_{-1}^{1} \tfrac{1}{2}(2-2x)^2\,dx = 2\rho \int_0^1 \tfrac{1}{2}(2-2x)^2\,dx$$
$$= \left(2 \cdot 1 \cdot \tfrac{1}{2} \cdot 2^2\right) \int_0^1 (1-x)^2\,dx = 4\int_1^0 u^2(-du) \quad [u = 1-x,\ du = -dx]$$
$$= -4\left[\tfrac{1}{3}u^3\right]_1^0 = -4\left(-\tfrac{1}{3}\right) = \tfrac{4}{3}$$

$\overline{y} = \frac{1}{m}M_x = \frac{1}{\rho A}M_x = \frac{1}{1 \cdot 2} \cdot \frac{4}{3} = \frac{2}{3}$. Thus, the centroid is $(\overline{x}, \overline{y}) = \left(0, \frac{2}{3}\right)$.

44. By symmetry about the line $y = x$, we expect that $\overline{x} = \overline{y}$. $A = \frac{1}{4}\pi r^2$, so $m = \rho A = 2A = \frac{1}{2}\pi r^2$.

$$M_x = \rho \int_0^r \tfrac{1}{2}\left(\sqrt{r^2 - x^2}\right)^2 dx = 2 \cdot \tfrac{1}{2}\int_0^r (r^2 - x^2)\,dx = \left[r^2 x - \tfrac{1}{3}x^3\right]_0^r = \tfrac{2}{3}r^3.$$

$$M_y = \rho \int_0^r x\sqrt{r^2 - x^2}\,dx = \int_0^r (r^2 - x^2)^{1/2} 2x\,dx = \int_0^{r^2} u^{1/2}\,du \quad [u = r^2 - x^2] = \left[\tfrac{2}{3}u^{3/2}\right]_0^{r^2} = \tfrac{2}{3}r^3.$$

$$\overline{x} = \frac{1}{m}M_y = \frac{2}{\pi r^2}\left(\tfrac{2}{3}r^3\right) = \tfrac{4}{3\pi}r, \quad \overline{y} = \frac{1}{m}M_x = \frac{2}{\pi r^2}\left(\tfrac{2}{3}r^3\right) = \tfrac{4}{3\pi}r. \text{ Thus, the centroid is } (\overline{x}, \overline{y}) = \left(\frac{4}{3\pi}r, \frac{4}{3\pi}r\right).$$

45. Choose x- and y-axes so that the base (one side of the triangle) lies along the x-axis with the other vertex along the positive y-axis as shown. From geometry, we know the medians intersect at a point $\frac{2}{3}$ of the way from each vertex (along the median) to the opposite side. The median from B goes to the midpoint $\left(\frac{1}{2}(a + c), 0\right)$ of side AC, so the point of intersection of the medians is $\left(\frac{2}{3} \cdot \frac{1}{2}(a + c), \frac{1}{3}b\right) = \left(\frac{1}{3}(a + c), \frac{1}{3}b\right)$.

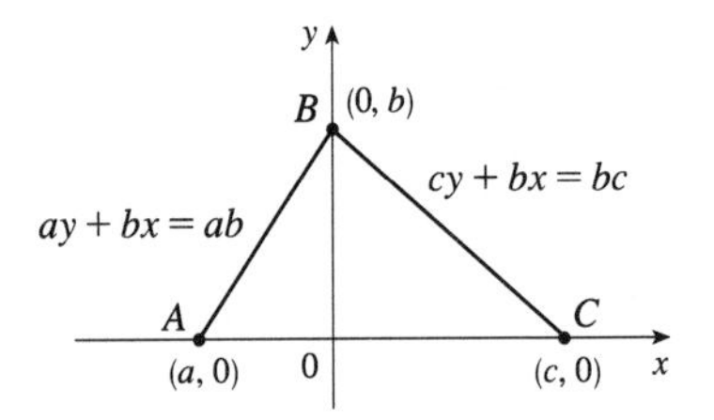

This can also be verified by finding the equations of two medians, and solving them simultaneously to find their point of intersection. Now let us compute the location of the centroid of the triangle. The area is $A = \frac{1}{2}(c - a)b$.

$$\overline{x} = \tfrac{1}{A}\left[\int_a^0 x \cdot \tfrac{b}{a}(a - x)\,dx + \int_0^c x \cdot \tfrac{b}{c}(c - x)\,dx\right] = \tfrac{1}{A}\left[\tfrac{b}{a}\int_a^0 (ax - x^2)\,dx + \tfrac{b}{c}\int_0^c (cx - x^2)\,dx\right]$$
$$= \tfrac{b}{Aa}\left[\tfrac{1}{2}ax^2 - \tfrac{1}{3}x^3\right]_a^0 + \tfrac{b}{Ac}\left[\tfrac{1}{2}cx^2 - \tfrac{1}{3}x^3\right]_0^c = \tfrac{b}{Aa}\left[-\tfrac{1}{2}a^3 + \tfrac{1}{3}a^3\right] + \tfrac{b}{Ac}\left[\tfrac{1}{2}c^3 - \tfrac{1}{3}c^3\right]$$
$$= \tfrac{2}{a(c-a)} \cdot \tfrac{-a^3}{6} + \tfrac{2}{c(c-a)} \cdot \tfrac{c^3}{6} = \tfrac{1}{3(c-a)}(c^2 - a^2) = \tfrac{a+c}{3}$$

and

$$\overline{y} = \tfrac{1}{A}\left[\int_a^0 \tfrac{1}{2}\left(\tfrac{b}{a}(a - x)\right)^2 dx + \int_0^c \tfrac{1}{2}\left(\tfrac{b}{c}(c - x)\right)^2 dx\right]$$
$$= \tfrac{1}{A}\left[\tfrac{b^2}{2a^2}\int_a^0 (a^2 - 2ax + x^2)\,dx + \tfrac{b^2}{2c^2}\int_0^c (c^2 - 2cx + x^2)\,dx\right]$$
$$= \tfrac{1}{A}\left[\tfrac{b^2}{2a^2}\left[a^2x - ax^2 + \tfrac{1}{3}x^3\right]_a^0 + \tfrac{b^2}{2c^2}\left[c^2x - cx^2 + \tfrac{1}{3}x^3\right]_0^c\right]$$
$$= \tfrac{1}{A}\left[\tfrac{b^2}{2a^2}\left(-a^3 + a^3 - \tfrac{1}{3}a^3\right) + \tfrac{b^2}{2c^2}\left(c^3 - c^3 + \tfrac{1}{3}c^3\right)\right] = \tfrac{1}{A}\left[\tfrac{b^2}{6}(-a + c)\right] = \tfrac{2}{(c-a)b} \cdot \tfrac{(c-a)b^2}{6} = \tfrac{b}{3}$$

Thus, the centroid is $(\overline{x}, \overline{y}) = \left(\dfrac{a+c}{3}, \dfrac{b}{3}\right)$, as claimed.

Remarks: Actually the computation of $\overline{y}$ is all that is needed. By considering each side of the triangle in turn to be the base, we see that the centroid is $\frac{1}{3}$ of the way from each side to the opposite vertex and must therefore be the intersection of the medians.

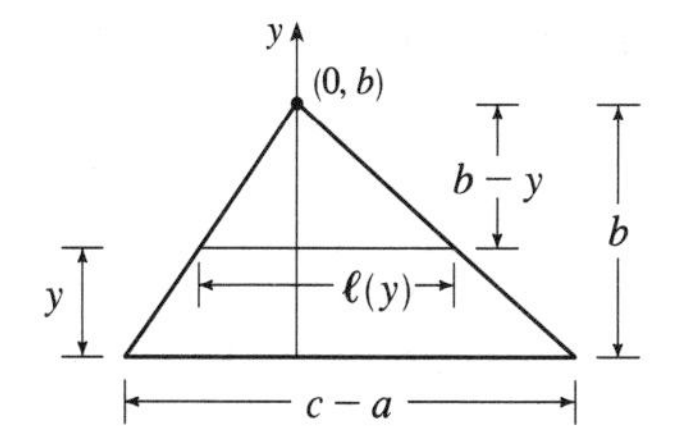

The computation of $\overline{y}$ in this problem (and many others) can be simplified by using horizontal rather than vertical approximating rectangles. If the length of a thin rectangle at coordinate y is $\ell(y)$, then its area is $\ell(y)\,\Delta y$, its mass is $\rho\ell(y)\,\Delta y$, and its moment about the x-axis is $\Delta M_x = \rho y\ell(y)\,\Delta y$. Thus,

$$M_x = \int \rho y\ell(y)\,dy \text{ and } \overline{y} = \frac{\int \rho y\ell(y)\,dy}{\rho A} = \tfrac{1}{A}\int y\ell(y)\,dy.$$

In this problem, $\ell(y) = \dfrac{c-a}{b}(b-y)$ by similar triangles, so

$$\overline{y} = \tfrac{1}{A}\int_0^b \tfrac{c-a}{b}\,y(b-y)dy = \tfrac{2}{b^2}\int_0^b (by-y^2)\,dy = \tfrac{2}{b^2}\left[\tfrac{1}{2}by^2 - \tfrac{1}{3}y^3\right]_0^b = \tfrac{2}{b^2}\cdot\tfrac{b^3}{6} = \tfrac{b}{3}$$

Notice that only one integral is needed when this method is used.

46. Divide the lamina into three rectangles with masses 2, 2 and 6, with centroids $\left(-\frac{3}{2},1\right)$, $\left(0,\frac{1}{2}\right)$ and $\left(2,\frac{3}{2}\right)$, respectively. The total mass of the lamina is 10. So, using Formulas 9, 10, and 11, we have

$$\overline{x} = \frac{M_y}{m} = \frac{1}{m}\sum_{i=1}^{3} m_i x_i = \tfrac{1}{10}\left[2\left(-\tfrac{3}{2}\right) + 2(0) + 6(2)\right] = \tfrac{1}{10}(9), \text{ and}$$

$$\overline{y} = \frac{M_x}{m} = \frac{1}{m}\sum_{i=1}^{3} m_i y_i = \tfrac{1}{10}\left[2(1) + 2\left(\tfrac{1}{2}\right) + 6\left(\tfrac{3}{2}\right)\right] = \tfrac{1}{10}(12).$$

Thus, the centroid is $(\overline{x},\overline{y}) = \left(\frac{9}{10},\frac{6}{5}\right)$.

47. Divide the lamina into two triangles and one rectangle with respective masses of 2, 2 and 4, so that the total mass is 8. Using the result of Exercise 45, the triangles have centroids $\left(-1,\frac{2}{3}\right)$ and $\left(1,\frac{2}{3}\right)$. The centroid of the rectangle (its center) is $\left(0,-\frac{1}{2}\right)$. So, using Formulas 9 and 11, we have $\overline{y} = \dfrac{M_x}{m} = \dfrac{1}{m}\displaystyle\sum_{i=1}^{3} m_i y_i = \tfrac{1}{8}\left[2\left(\tfrac{2}{3}\right) + 2\left(\tfrac{2}{3}\right) + 4\left(-\tfrac{1}{2}\right)\right] = \tfrac{1}{8}\left(\tfrac{2}{3}\right) = \tfrac{1}{12}$, and $\overline{x} = 0$, since the lamina is symmetric about the line $x = 0$. Thus, the centroid is $(\overline{x},\overline{y}) = \left(0,\frac{1}{12}\right)$.

48. A sphere can be generated by rotating a semicircle about its diameter. By Example 8, the center of mass travels a distance $2\pi\overline{y} = 2\pi\left(\frac{4r}{3\pi}\right) = \frac{8r}{3}$, so by the Theorem of Pappus, the volume of the sphere is $V = Ad = \dfrac{\pi r^2}{2}\cdot\dfrac{8r}{3} = \frac{4}{3}\pi r^3$.

49. A cone of height h and radius r can be generated by rotating a right triangle about one of its legs as shown. By Exercise 45, $\overline{x} = \frac{1}{3}r$, so by the Theorem of Pappus, the volume of the cone is

$$V = Ad = \left(\tfrac{1}{2}\cdot\text{base}\cdot\text{height}\right)\cdot(2\pi\overline{x}) = \tfrac{1}{2}rh\cdot 2\pi\left(\tfrac{1}{3}r\right) = \tfrac{1}{3}\pi r^2 h.$$

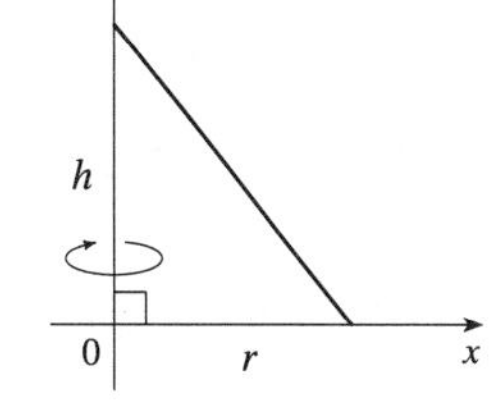

50. From the symmetry in the figure, $\overline{y} = 4$. So the distance traveled by the centroid when rotating the triangle about the x-axis is $d = 2\pi\cdot 4 = 8\pi$. The area of the triangle is $A = \frac{1}{2}bh = \frac{1}{2}(2)(3) = 3$. By the Theorem of Pappus, the volume of the resulting solid is $Ad = 3(8\pi) = 24\pi$.

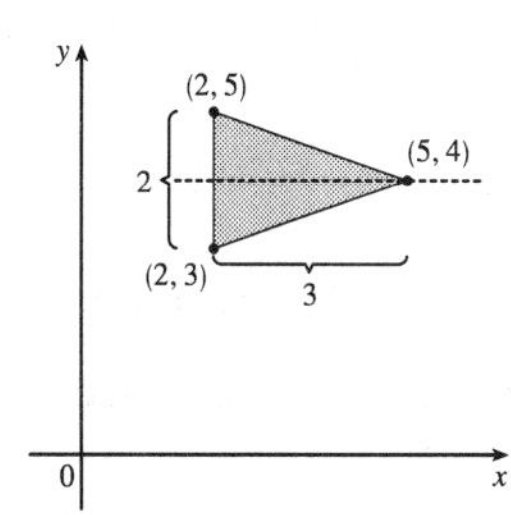

51. Suppose the region lies between two curves $y = f(x)$ and $y = g(x)$ where $f(x) \geq g(x)$, as illustrated in Figure 16. Choose points x_i with $a = x_0 < x_1 < \cdots < x_n = b$ and choose x_i^* to be the midpoint of the ith subinterval; that is, $x_i^* = \overline{x}_i = \frac{1}{2}(x_{i-1} + x_i)$. Then the centroid of the ith approximating rectangle R_i is its center $C_i = \left(\overline{x}_i, \frac{1}{2}[f(\overline{x}_i) + g(\overline{x}_i)]\right)$. Its area is $[f(\overline{x}_i) - g(\overline{x}_i)]\,\Delta x$, so its mass is $\rho[f(\overline{x}_i) - g(\overline{x}_i)]\,\Delta x$. Thus,

$M_y(R_i) = \rho[f(\overline{x}_i) - g(\overline{x}_i)]\,\Delta x \cdot \overline{x}_i = \rho\overline{x}_i[f(\overline{x}_i) - g(\overline{x}_i)]\,\Delta x$ and

$M_x(R_i) = \rho[f(\overline{x}_i) - g(\overline{x}_i)]\,\Delta x \cdot \frac{1}{2}[f(\overline{x}_i) + g(\overline{x}_i)] = \rho \cdot \frac{1}{2}\left[f(\overline{x}_i)^2 - g(\overline{x}_i)^2\right]\Delta x$. Summing over i and taking the limit as $n \to \infty$, we get $M_y = \lim\limits_{n\to\infty} \sum_i \rho\overline{x}_i\,[f(\overline{x}_i) - g(\overline{x}_i)]\,\Delta x = \rho \int_a^b x[f(x) - g(x)]\,dx$ and

$M_x = \lim\limits_{n\to\infty} \sum_i \rho \cdot \frac{1}{2}\left[f(\overline{x}_i)^2 - g(\overline{x}_i)^2\right]\Delta x = \rho \int_a^b \frac{1}{2}[f(x)^2 - g(x)^2]\,dx$. Thus,

$$\overline{x} = \frac{M_y}{m} = \frac{M_y}{\rho A} = \frac{1}{A}\int_a^b x[f(x) - g(x)]\,dx \quad \text{and} \quad \overline{y} = \frac{M_x}{m} = \frac{M_x}{\rho A} = \frac{1}{A}\int_a^b \tfrac{1}{2}[f(x)^2 - g(x)^2]\,dx$$

52. (a) Let $0 \leq x \leq 1$. If $n < m$, then $x^n > x^m$; that is, raising x to a larger power produces a smaller number.

(b) Using Formulas 13 and the fact that the area of $\mathfrak{R}$ is

$$A = \int_0^1 (x^n - x^m)\,dx = \frac{1}{n+1} - \frac{1}{m+1} = \frac{m-n}{(n+1)(m+1)}, \text{ we get}$$

$$\overline{x} = \frac{(n+1)(m+1)}{m-n}\int_0^1 x[x^n - x^m]\,dx$$

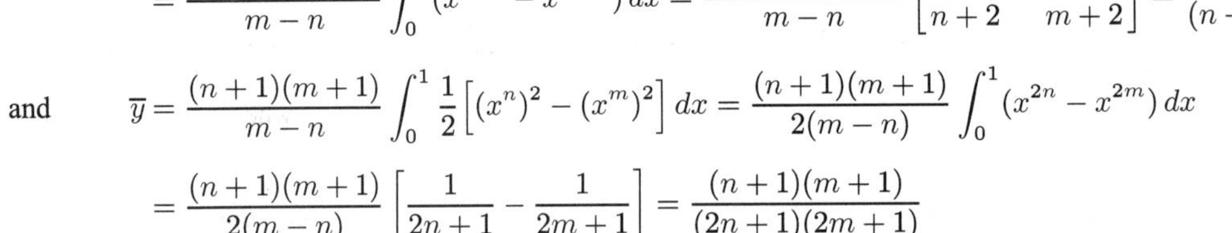

$$= \frac{(n+1)(m+1)}{m-n}\int_0^1 (x^{n+1} - x^{m+1})\,dx = \frac{(n+1)(m+1)}{m-n}\left[\frac{1}{n+2} - \frac{1}{m+2}\right] = \frac{(n+1)(m+1)}{(n+2)(m+2)}$$

and $$\overline{y} = \frac{(n+1)(m+1)}{m-n}\int_0^1 \frac{1}{2}\left[(x^n)^2 - (x^m)^2\right]dx = \frac{(n+1)(m+1)}{2(m-n)}\int_0^1 (x^{2n} - x^{2m})\,dx$$

$$= \frac{(n+1)(m+1)}{2(m-n)}\left[\frac{1}{2n+1} - \frac{1}{2m+1}\right] = \frac{(n+1)(m+1)}{(2n+1)(2m+1)}$$

(c) If we take $n = 3$ and $m = 4$, then

$$(\overline{x}, \overline{y}) = \left(\frac{4\cdot 5}{5\cdot 6}, \frac{4\cdot 5}{7\cdot 9}\right) = \left(\frac{2}{3}, \frac{20}{63}\right)$$

which lies outside $\mathfrak{R}$ since $\left(\frac{2}{3}\right)^3 = \frac{8}{27} < \frac{20}{63}$. This is the simplest of many possibilities.

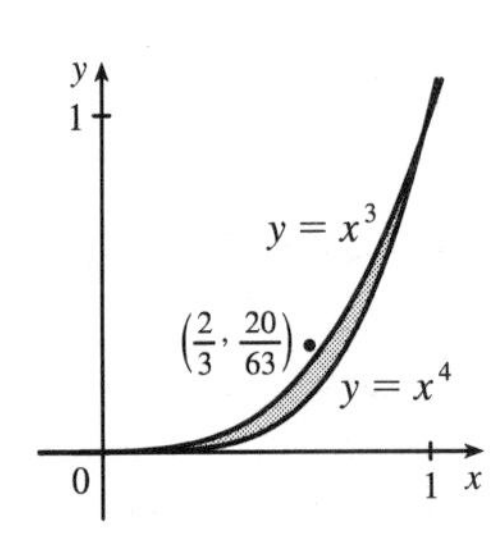

7.6 Differential Equations

1. $\dfrac{dy}{dx} = \dfrac{y}{x} \;\Rightarrow\; \dfrac{dy}{y} = \dfrac{dx}{x}\;\; [y \neq 0] \;\Rightarrow\; \displaystyle\int \frac{dy}{y} = \int \frac{dx}{x} \;\Rightarrow\; \ln|y| = \ln|x| + C \;\Rightarrow$

$|y| = e^{\ln|x| + C} = e^{\ln|x|}e^C = e^C\,|x| \;\Rightarrow\; y = Kx$, where $K = \pm e^C$ is a constant. (In our derivation, K was nonzero, but we can restore the excluded case $y = 0$ by allowing K to be zero.)

2. $\dfrac{dy}{dx} = \dfrac{\sqrt{x}}{e^y} \;\Rightarrow\; e^y\,dy = \sqrt{x}\,dx \;\Rightarrow\; \int e^y\,dy = \int x^{1/2}\,dx \;\Rightarrow\; e^y = \frac{2}{3}x^{3/2} + C \;\Rightarrow\; y = \ln\!\left(\frac{2}{3}x^{3/2} + C\right)$

3. $(x^2+1)y' = xy \;\Rightarrow\; \dfrac{dy}{dx} = \dfrac{xy}{x^2+1} \;\Rightarrow\; \dfrac{dy}{y} = \dfrac{x\,dx}{x^2+1}\;\; [y \neq 0] \;\Rightarrow\; \displaystyle\int \frac{dy}{y} = \int \frac{x\,dx}{x^2+1} \;\Rightarrow$

$\ln|y| = \frac{1}{2}\ln(x^2+1) + C\;\; [u = x^2+1,\, du = 2x\,dx] \;= \ln(x^2+1)^{1/2} + \ln e^C = \ln\!\left(e^C\sqrt{x^2+1}\right) \;\Rightarrow$

$|y| = e^C\sqrt{x^2+1} \;\Rightarrow\; y = K\sqrt{x^2+1}$, where $K = \pm e^C$ is a constant. (In our derivation, K was nonzero, but we can restore the excluded case $y = 0$ by allowing K to be zero.)

4. $y' = y^2\sin x \;\Rightarrow\; \dfrac{dy}{dx} = y^2\sin x \;\Rightarrow\; \dfrac{dy}{y^2} = \sin x\,dx\;\; [y \neq 0] \;\Rightarrow\; \displaystyle\int \frac{dy}{y^2} = \int \sin x\,dx \;\Rightarrow$

$-\dfrac{1}{y} = -\cos x + C \;\Rightarrow\; \dfrac{1}{y} = \cos x - C \;\Rightarrow\; y = \dfrac{1}{\cos x + K}$, where $K = -C$. $y = 0$ is also a solution.

5. $(1+\tan y)\,y' = x^2+1 \;\Rightarrow\; (1+\tan y)\dfrac{dy}{dx} = x^2+1 \;\Rightarrow\; \left(1 + \dfrac{\sin y}{\cos y}\right)dy = (x^2+1)\,dx \;\Rightarrow$

$\displaystyle\int\left(1 - \frac{-\sin y}{\cos y}\right)dy = \int (x^2+1)\,dx \;\Rightarrow\; y - \ln|\cos y| = \tfrac{1}{3}x^3 + x + C$. Note: The left side is equivalent to $y + \ln|\sec y|$.

6. $\dfrac{dy}{d\theta} = \dfrac{e^y\sin^2\theta}{y\sec\theta} \;\Rightarrow\; \dfrac{y}{e^y}\,dy = \dfrac{\sin^2\theta}{\sec\theta}\,d\theta \;\Rightarrow\; \displaystyle\int ye^{-y}\,dy = \int \sin^2\theta\,\cos\theta\,d\theta$. Integrating the left side by parts with $u = y$, $dv = e^{-y}\,dy$ and the right side by the substitution $u = \sin\theta$, we obtain $-ye^{-y} - e^{-y} = \frac{1}{3}\sin^3\theta + C$. We cannot solve explicitly for y.

7. $\dfrac{du}{dt} = 2 + 2u + t + tu \;\Rightarrow\; \dfrac{du}{dt} = (1+u)(2+t) \;\Rightarrow\; \displaystyle\int \frac{du}{1+u} = \int (2+t)\,dt\;\; [u \neq -1] \;\Rightarrow$

$\ln|1+u| = \frac{1}{2}t^2 + 2t + C \;\Rightarrow\; |1+u| = e^{t^2/2+2t+C} = Ke^{t^2/2+2t}$, where $K = e^C \;\Rightarrow\; 1 + u = \pm Ke^{t^2/2+2t} \;\Rightarrow$

$u = -1 \pm Ke^{t^2/2+2t}$ where $K > 0$. $u = -1$ is also a solution, so $u = -1 + Ae^{t^2/2+2t}$, where A is an arbitrary constant.

8. $\dfrac{dz}{dt} + e^{t+z} = 0 \;\Rightarrow\; \dfrac{dz}{dt} = -e^te^z \;\Rightarrow\; \displaystyle\int e^{-z}\,dz = -\int e^t\,dt \;\Rightarrow\; -e^{-z} = -e^t + C \;\Rightarrow\; e^{-z} = e^t - C \;\Rightarrow$

$\dfrac{1}{e^z} = e^t - C \;\Rightarrow\; e^z = \dfrac{1}{e^t - C} \;\Rightarrow\; z = \ln\!\left(\dfrac{1}{e^t - C}\right) \;\Rightarrow\; z = -\ln(e^t - C)$

9. $\dfrac{du}{dt} = \dfrac{2t + \sec^2 t}{2u}$, $u(0) = -5$. $\int 2u\,du = \int (2t + \sec^2 t)\,dt \;\Rightarrow\; u^2 = t^2 + \tan t + C$, where $[u(0)]^2 = 0^2 + \tan 0 + C \;\Rightarrow\; C = (-5)^2 = 25$. Therefore, $u^2 = t^2 + \tan t + 25$, so $u = \pm\sqrt{t^2 + \tan t + 25}$. Since $u(0) = -5$, we must have $u = -\sqrt{t^2 + \tan t + 25}$.

10. $\dfrac{dy}{dx} = \dfrac{y\cos x}{1+y^2}$, $(0) = 1$. $dy = y\cos x\,dx \;\Rightarrow\; \dfrac{1+y^2}{y}\,dy = \cos x\,dx \;\Rightarrow\; \displaystyle\int\left(\frac{1}{y} + y\right)dy = \int \cos x\,dx \;\Rightarrow$

$\ln|y| + \frac{1}{2}y^2 = \sin x + C$. $y(0) = 1 \;\Rightarrow\; \ln 1 + \frac{1}{2} = \sin 0 + C \;\Rightarrow\; C = \frac{1}{2}$, so $\ln|y| + \frac{1}{2}y^2 = \sin x + \frac{1}{2}$. We cannot solve explicitly for y.

11. $x\cos x = (2y + e^{3y})\,y' \;\Rightarrow\; x\cos x\,dx = (2y + e^{3y})\,dy \;\Rightarrow\; \int (2y + e^{3y})\,dy = \int x\cos x\,dx \;\Rightarrow$

$y^2 + \frac{1}{3}e^{3y} = x\sin x + \cos x + C$ [where the second integral is evaluated using integration by parts]. Now $y(0) = 0 \;\Rightarrow$ $0 + \frac{1}{3} = 0 + 1 + C \;\Rightarrow\; C = -\frac{2}{3}$. Thus, a solution is $y^2 + \frac{1}{3}e^{3y} = x\sin x + \cos x - \frac{2}{3}$. We cannot solve explicitly for y.

12. $\dfrac{dP}{dt} = \sqrt{Pt} \;\Rightarrow\; dP/\sqrt{P} = \sqrt{t}\,dt \;\Rightarrow\; \int P^{-1/2}\,dP = \int t^{1/2}\,dt \;\Rightarrow\; 2P^{1/2} = \frac{2}{3}t^{3/2} + C.$

$P(1) = 2 \;\Rightarrow\; 2\sqrt{2} = \frac{2}{3} + C \;\Rightarrow\; C = 2\sqrt{2} - \frac{2}{3}$, so $2P^{1/2} = \frac{2}{3}t^{3/2} + 2\sqrt{2} - \frac{2}{3} \;\Rightarrow\; \sqrt{P} = \frac{1}{3}t^{3/2} + \sqrt{2} - \frac{1}{3} \;\Rightarrow$

$P = \left(\frac{1}{3}t^{3/2} + \sqrt{2} - \frac{1}{3}\right)^2.$

13. $y' \tan x = a + y,\; 0 < x < \pi/2 \;\Rightarrow\; \dfrac{dy}{dx} = \dfrac{a+y}{\tan x} \;\Rightarrow\; \dfrac{dy}{a+y} = \cot x\,dx \quad [a + y \neq 0] \;\Rightarrow$

$\displaystyle\int \frac{dy}{a+y} = \int \frac{\cos x}{\sin x}\,dx \;\Rightarrow\; \ln|a+y| = \ln|\sin x| + C \;\Rightarrow\; |a+y| = e^{\ln|\sin x| + C} = e^{\ln|\sin x|} \cdot e^C = e^C\,|\sin x| \;\Rightarrow$

$a + y = K \sin x$, where $K = \pm e^C$. (In our derivation, K was nonzero, but we can restore the excluded case $y = -a$ by allowing K to be zero.) $y(\pi/3) = a \;\Rightarrow\; a + a = K \sin(\pi/3) \;\Rightarrow\; 2a = K\dfrac{\sqrt{3}}{2} \;\Rightarrow\; K = \dfrac{4a}{\sqrt{3}}.$

Thus, $a + y = \dfrac{4a}{\sqrt{3}}\sin x$ and so $y = \dfrac{4a}{\sqrt{3}}\sin x - a.$

14. $\dfrac{dL}{dt} = kL^2 \ln t \;\Rightarrow\; \dfrac{dL}{L^2} = k \ln t\,dt \;\Rightarrow\; \displaystyle\int \frac{dL}{L^2} = \int k \ln t\,dt \;\Rightarrow\; -\frac{1}{L} = kt \ln t - \int k\,dt$ [by parts with $u = \ln t$, $dv = k\,dt$] $\Rightarrow\; -\dfrac{1}{L} = kt \ln t - kt + C \;\Rightarrow\; L = \dfrac{1}{kt - kt \ln t - C}.$

$L(1) = -1 \;\Rightarrow\; -1 = \dfrac{1}{k - k \ln 1 - C} \;\Rightarrow\; C - k = 1 \;\Rightarrow\; C = k + 1$. Thus, $L = \dfrac{1}{kt - kt \ln t - k - 1}.$

15. $\dfrac{dy}{dx} = 4x^3 y,\; y(0) = 7.\quad \dfrac{dy}{y} = 4x^3\,dx \quad [\text{if } y \neq 0] \;\Rightarrow\; \displaystyle\int \frac{dy}{y} = \int 4x^3\,dx \;\Rightarrow\; \ln|y| = x^4 + C \;\Rightarrow$

$e^{\ln|y|} = e^{x^4 + C} \;\Rightarrow\; |y| = e^{x^4} e^C \;\Rightarrow\; y = Ae^{x^4};\quad y(0) = 7 \;\Rightarrow\; A = 7 \;\Rightarrow\; y = 7e^{x^4}.$

16. $\dfrac{dy}{dx} = \dfrac{y^2}{x^3},\; y(1) = 1.\quad \displaystyle\int \frac{dy}{y^2} = \int \frac{dx}{x^3} \;\Rightarrow\; -\frac{1}{y} = -\frac{1}{2x^2} + C.\quad y(1) = 1 \;\Rightarrow\; -1 = -\frac{1}{2} + C \;\Rightarrow\; C = -\frac{1}{2}.$

So $\dfrac{1}{y} = \dfrac{1}{2x^2} + \dfrac{1}{2} = \dfrac{2 + 2x^2}{2 \cdot 2x^2} \;\Rightarrow\; y = \dfrac{2x^2}{x^2 + 1}.$

17. (a) $y' = 2x\sqrt{1 - y^2} \;\Rightarrow\; \dfrac{dy}{dx} = 2x\sqrt{1 - y^2} \;\Rightarrow\; \dfrac{dy}{\sqrt{1 - y^2}} = 2x\,dx \;\Rightarrow\; \displaystyle\int \frac{dy}{\sqrt{1 - y^2}} = \int 2x\,dx \;\Rightarrow$

$\sin^{-1} y = x^2 + C$ for $-\frac{\pi}{2} \le x^2 + C \le \frac{\pi}{2}.$

(b) $y(0) = 0 \;\Rightarrow\; \sin^{-1} 0 = 0^2 + C \;\Rightarrow\; C = 0$, so $\sin^{-1} y = x^2$ and $y = \sin(x^2)$ for $-\sqrt{\pi/2} \le x \le \sqrt{\pi/2}.$

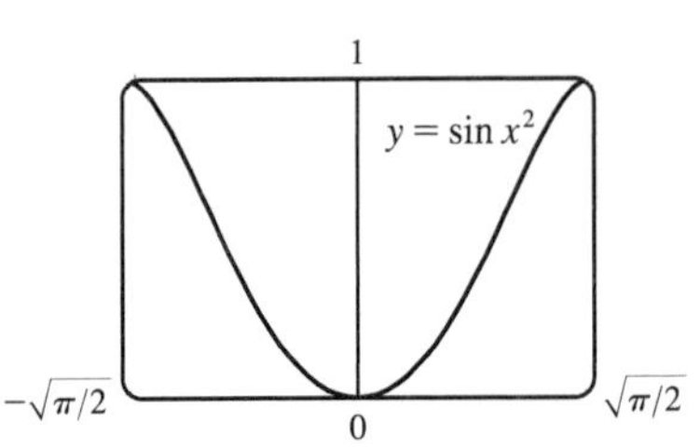

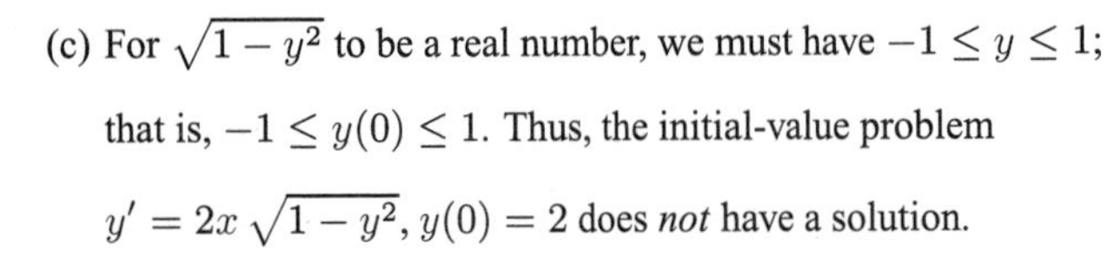

(c) For $\sqrt{1 - y^2}$ to be a real number, we must have $-1 \le y \le 1$; that is, $-1 \le y(0) \le 1$. Thus, the initial-value problem $y' = 2x\sqrt{1 - y^2}$, $y(0) = 2$ does *not* have a solution.

18. $e^{-y}y' + \cos x = 0 \quad\Leftrightarrow\quad \int e^{-y}\,dy = -\int \cos x\,dx \quad\Leftrightarrow\quad -e^{-y} = -\sin x + C_1 \quad\Leftrightarrow\quad y = -\ln(\sin x + C)$. The solution is periodic, with period 2π. Note that for $C > 1$, the domain of the solution is $\mathbb{R}$, but for $-1 < C \leq 1$ it is only defined on the intervals where $\sin x + C > 0$, and it is meaningless for $C \leq -1$, since then $\sin x + C \leq 0$, and the logarithm is undefined.

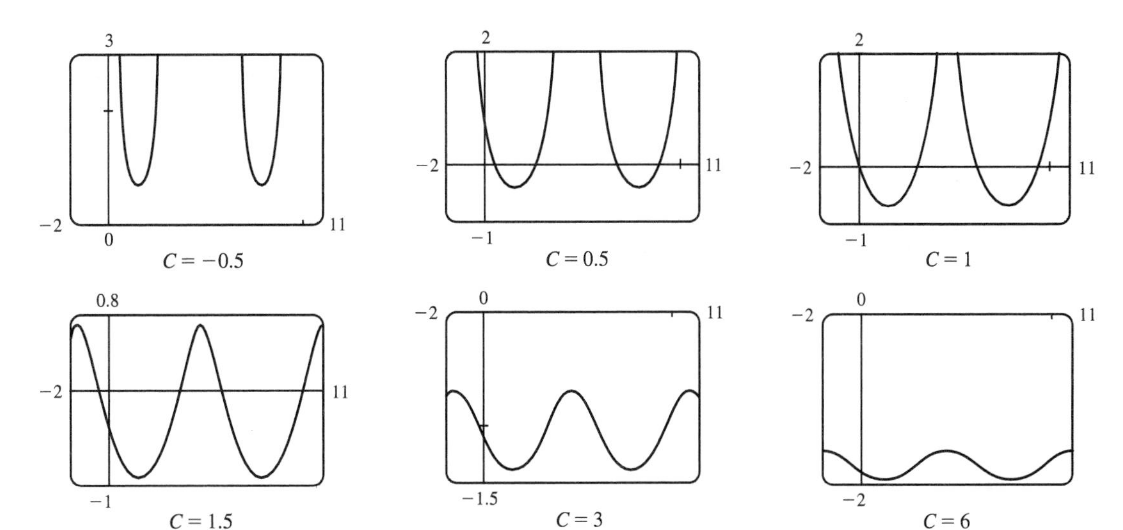

For $-1 < C < 1$, the solution curve consists of concave-up pieces separated by intervals on which the solution is not defined (where $\sin x + C \leq 0$). For $C = 1$, the solution curve consists of concave-up pieces separated by vertical asymptotes at the points where $\sin x + C = 0 \quad\Leftrightarrow\quad \sin x = -1$. For $C > 1$, the curve is continuous, and as C increases, the graph moves downward, and the amplitude of the oscillations decreases.

19. $\dfrac{dy}{dx} = \dfrac{\sin x}{\sin y}$, $y(0) = \dfrac{\pi}{2}$. So $\int \sin y\,dy = \int \sin x\,dx \quad\Leftrightarrow$

$-\cos y = -\cos x + C \quad\Leftrightarrow\quad \cos y = \cos x - C$. From the initial condition, we need $\cos\frac{\pi}{2} = \cos 0 - C \quad\Rightarrow\quad 0 = 1 - C \quad\Rightarrow\quad C = 1$, so the solution is $\cos y = \cos x - 1$. Note that we cannot take $\cos^{-1}$ of both sides, since that would unnecessarily restrict the solution to the case where $-1 \leq \cos x - 1 \quad\Leftrightarrow\quad 0 \leq \cos x$, as $\cos^{-1}$ is defined only on $[-1, 1]$. Instead we plot the graph using Maple's `plots[implicitplot]` or Mathematica's `Plot[Evaluate[···]]`.

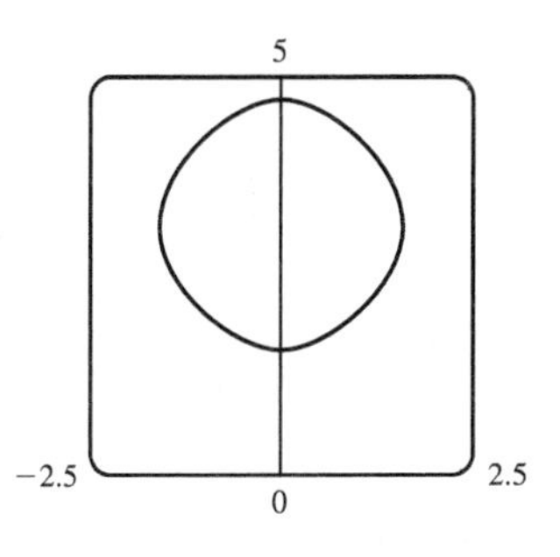

20. $\dfrac{dy}{dx} = \dfrac{x\sqrt{x^2+1}}{ye^y} \quad\Leftrightarrow\quad \int ye^y\,dy = \int x\sqrt{x^2+1}\,dx$. We use parts on the LHS with $u = y$, $dv = e^y\,dy$, and on the RHS we use the substitution $z = x^2 + 1$, so $dz = 2x\,dx$. The equation becomes $ye^y - \int e^y\,dy = \frac{1}{2}\int \sqrt{z}\,dz \quad\Leftrightarrow$

$e^y(y-1) = \frac{1}{3}(x^2+1)^{3/2} + C$, so we see that the curves are symmetric about the y-axis. Every point (x, y) in the plane lies on one of the curves, namely the one for which $C = (y-1)e^y - \frac{1}{3}(x^2+1)^{3/2}$. For example, along the y-axis, $C = (y-1)e^y - \frac{1}{3}$, so the origin lies on the curve with $C = -\frac{4}{3}$. We use Maple's `plots[implicitplot]` command or `Plot[Evaluate[···]]` in Mathematica to plot the solution curves for various values of C.

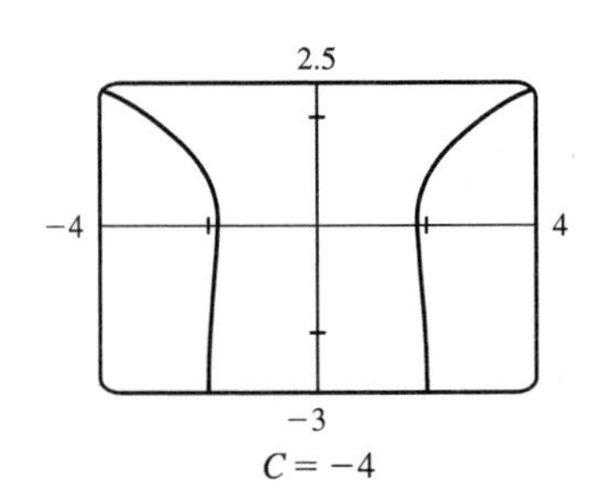

$C = -4$

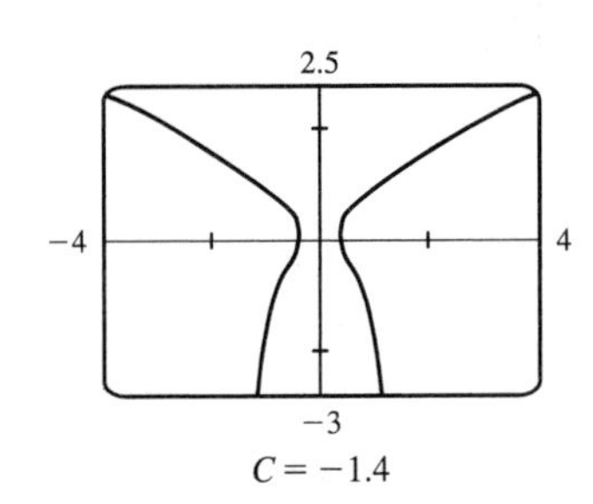

$C = -1.4$

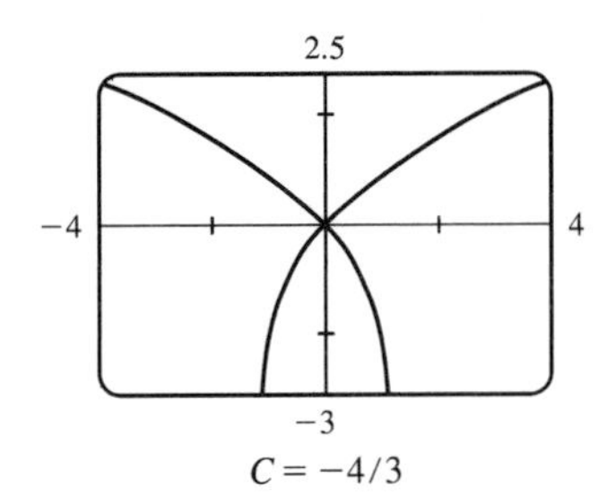

$C = -4/3$

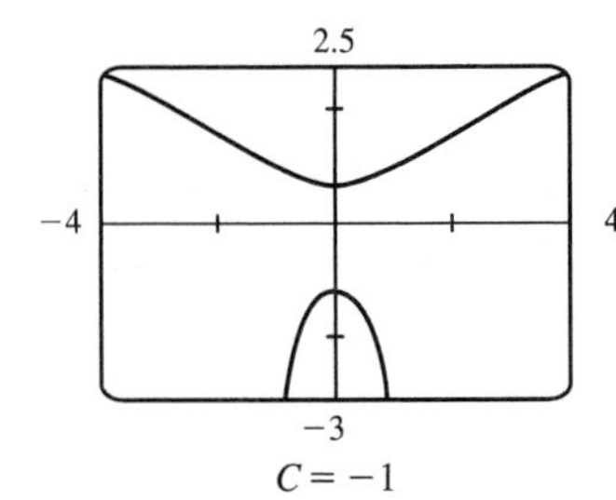

$C = -1$

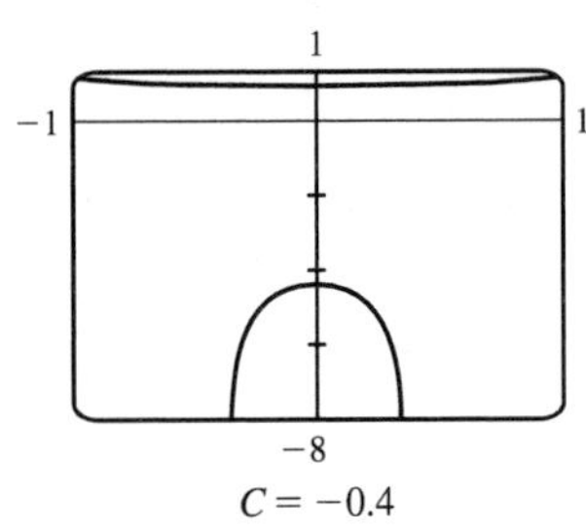

$C = -0.4$

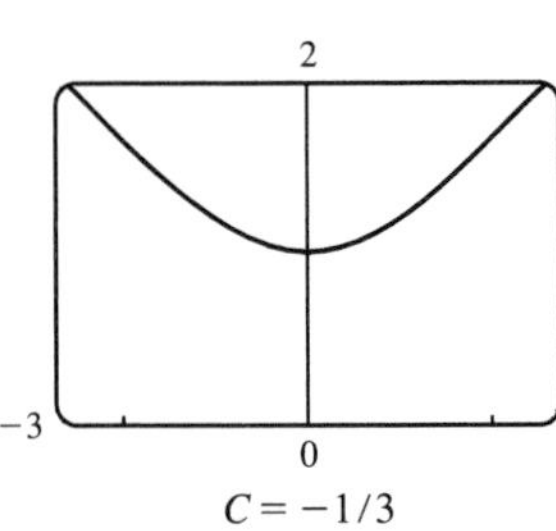

$C = -1/3$

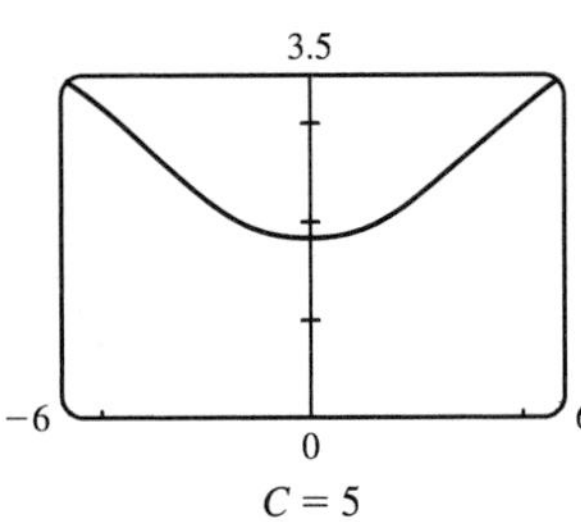

$C = 5$

It seems that the transitional values of C are $-\frac{4}{3}$ and $-\frac{1}{3}$. For $C < -\frac{4}{3}$, the graph consists of left and right branches. At $C = -\frac{4}{3}$, the two branches become connected at the origin, and as C increases, the graph splits into top and bottom branches. At $C = -\frac{1}{3}$, the bottom half disappears. As C increases further, the graph moves upward, but doesn't change shape much.

21. $y' = 2 - y$. The slopes at each point are independent of x, so the slopes are the same along each line parallel to the x-axis. Thus, III is the direction field for this equation. Note that for $y = 2$, $y' = 0$.

22. $y' = x(2 - y) = 0$ on the lines $x = 0$ and $y = 2$. Direction field I satisfies these conditions.

23. $y' = x + y - 1 = 0$ on the line $y = -x + 1$. Direction field IV satisfies this condition. Notice also that on the line $y = -x$ we have $y' = -1$, which is true in IV.

24. $y' = \sin x \, \sin y = 0$ on the lines $x = 0$ and $y = 0$, and $y' > 0$ for $0 < x < \pi$, $0 < y < \pi$. Direction field II satisfies these conditions.

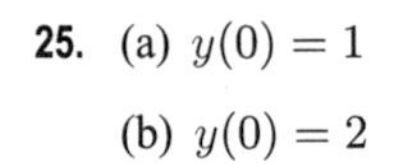

25. (a) $y(0) = 1$
(b) $y(0) = 2$
(c) $y(0) = -1$

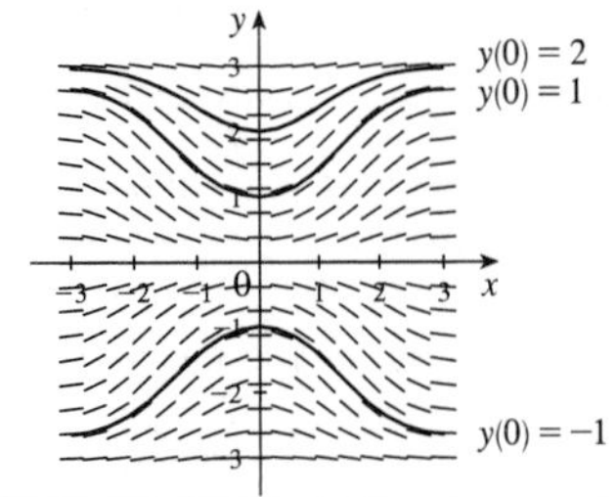

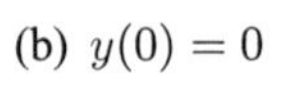

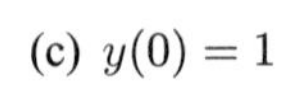

26. (a) $y(0) = -1$
(b) $y(0) = 0$
(c) $y(0) = 1$

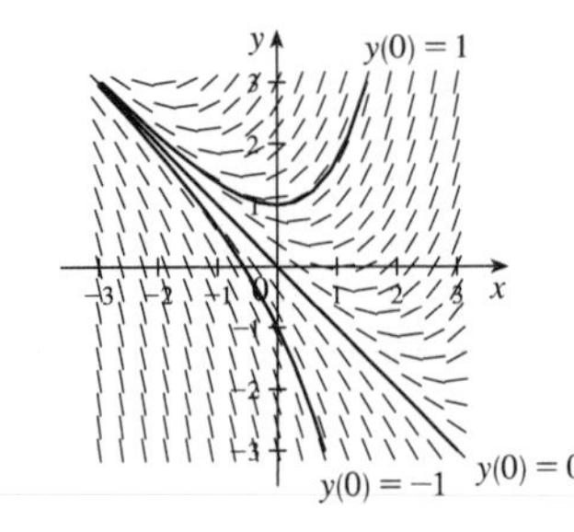

27.

x	y	$y' = 1 + y$
0	0	1
0	1	2
0	2	3
0	-3	-2
0	-2	-1

Note that for $y = -1$, $y' = 0$. The three solution curves sketched go through $(0, 0)$, $(0, -1)$, and $(0, -2)$.

28.

x	y	$y' = x^2 - y^2$
± 1	± 3	-8
± 3	± 1	8
± 1	± 0.5	0.75
± 0.5	± 1	-0.75

Note that $y' = 0$ for $y = \pm x$. If $|x| < |y|$, then $y' < 0$; that is, the slopes are negative for all points in quadrants I and II above both of the lines $y = x$ and $y = -x$, and all points in quadrants III and IV below both of the lines $y = -x$ and $y = x$. A similar statement holds for positive slopes.

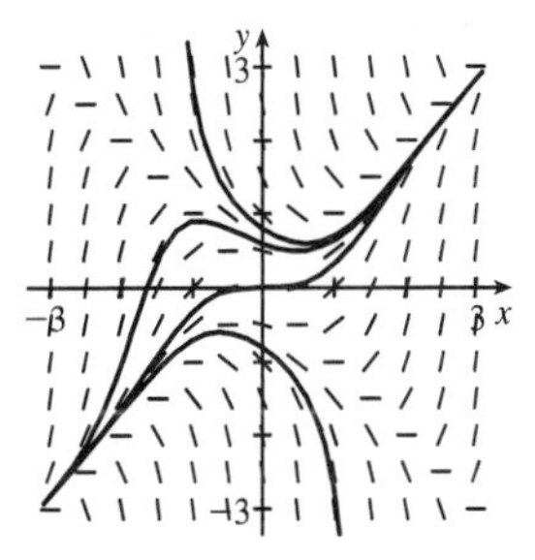

29.

x	y	$y' = y - 2x$
-2	-2	2
-2	2	6
2	2	-2
2	-2	-6

Note that $y' = 0$ for any point on the line $y = 2x$. The slopes are positive to the left of the line and negative to the right of the line. The solution curve in the graph passes through $(1, 0)$.

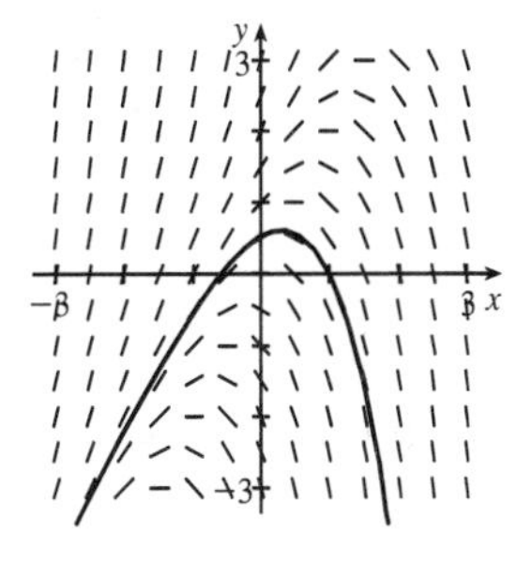

30.

x	y	$y' = 1 - xy$
± 1	± 1	0
± 2	± 2	-3
± 2	∓ 2	5

Note that $y' = 0$ for any point on the hyperbola $xy = 1$ (or $y = 1/x$). The slopes are negative at points "inside" the branches and positive at points everywhere else. The solution curve in the graph passes through $(0, 0)$.

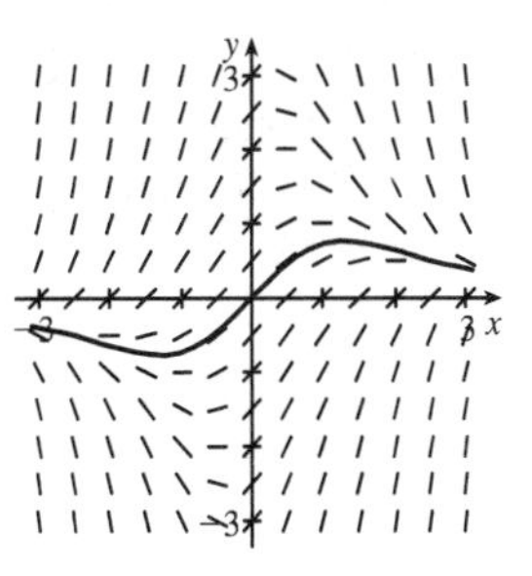

31.

x	y	$y' = y + xy$
0	± 2	± 2
1	± 2	± 4
-3	± 2	∓ 4

Note that $y' = y(x+1) = 0$ for any point on $y = 0$ or on $x = -1$. The slopes are positive when the factors y and $x+1$ have the same sign and negative when they have opposite signs. The solution curve in the graph passes through $(0, 1)$.

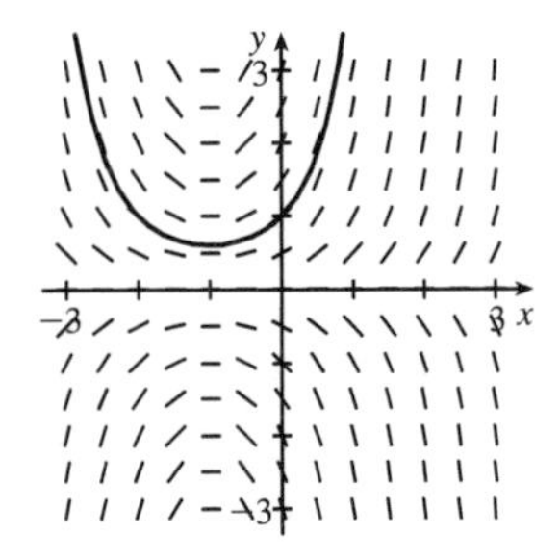

32.

x	y	$y' = x - xy$
± 2	0	± 2
± 2	3	∓ 4
± 2	-1	± 4

Note that $y' = x(1-y) = 0$ for any point on $x = 0$ or on $y = 1$. The slopes are positive when the factors x and $1-y$ have the same sign and negative when they have opposite signs. The solution curve in the graph passes through $(1, 0)$.

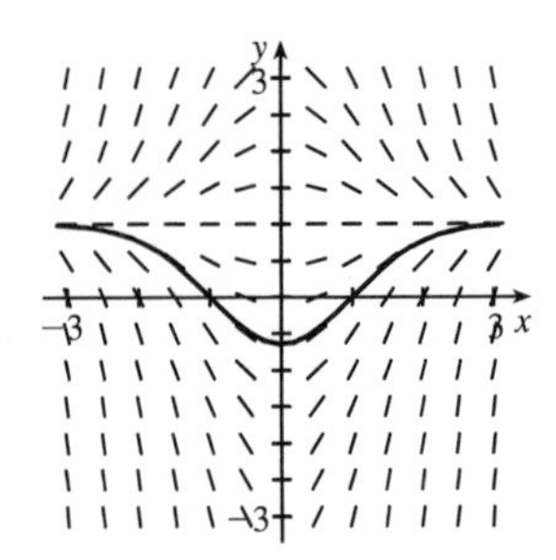

33. (a) $\dfrac{dP}{dt} = k(M-P)$ is always positive, so the level of performance P is increasing. As P gets close to M, dP/dt gets close to 0; that is, the performance levels off, as explained in part (a).

(b) $\dfrac{dP}{dt} = k(M-P) \;\Leftrightarrow\; \displaystyle\int \frac{dP}{P-M} = \int (-k)\,dt \;\Leftrightarrow\; \ln|P-M| = -kt + C \;\Leftrightarrow\; |P-M| = e^{-kt+C} \;\Leftrightarrow$

$P - M = Ae^{-kt}\ \left[A = \pm e^C\right] \;\Leftrightarrow\; P = M + Ae^{-kt}$. If we assume that performance is at level 0 when $t = 0$, then

$P(0) = 0 \;\Leftrightarrow\; 0 = M + A \;\Leftrightarrow\; A = -M \;\Leftrightarrow\; P(t) = M - Me^{-kt}$. $\displaystyle\lim_{t\to\infty} P(t) = M - M \cdot 0 = M$.

34. If $S = \dfrac{dT}{dr}$, then $\dfrac{dS}{dr} = \dfrac{d^2T}{dr^2}$. The differential equation $\dfrac{d^2T}{dr^2} + \dfrac{2}{r}\dfrac{dT}{dr} = 0$ can be written as $\dfrac{dS}{dr} + \dfrac{2}{r}S = 0$. Thus,

$\dfrac{dS}{dr} = \dfrac{-2S}{r} \;\Rightarrow\; \dfrac{dS}{S} = -\dfrac{2}{r}\,dr \;\Rightarrow\; \displaystyle\int \frac{1}{S}\,dS = \int -\frac{2}{r}\,dr \;\Rightarrow\; \ln|S| = -2\ln|r| + C$. Assuming $S = dT/dr > 0$

and $r > 0$, we have $S = e^{-2\ln r + C} = e^{\ln r^{-2}}e^C = r^{-2}k\ \left[k = e^C\right] \;\Rightarrow\; S = \dfrac{1}{r^2}k \;\Rightarrow\; \dfrac{dT}{dr} = \dfrac{1}{r^2}k \;\Rightarrow$

$dT = \dfrac{1}{r^2}k\,dr \;\Rightarrow\; \displaystyle\int dT = \int \frac{1}{r^2}k\,dr \;\Rightarrow\; T(r) = -\frac{k}{r} + A.$

$$T(1) = 15 \;\Rightarrow\; 15 = -k + A \ \textbf{(1)} \quad \text{and} \quad T(2) = 25 \;\Rightarrow\; 25 = -\tfrac{1}{2}k + A \ \textbf{(2)}$$

Now solve for k and A: $-2\textbf{(2)} + \textbf{(1)} \;\Rightarrow\; -35 = -A$, so $A = 35$ and $k = 20$, and $T(r) = -20/r + 35$.

35. (a) $\frac{dC}{dt} = r - kC \quad \Rightarrow \quad \frac{dC}{dt} = -(kC - r) \quad \Rightarrow \quad \int \frac{dC}{kC - r} = \int -dt \quad \Rightarrow \quad (1/k)\ln|kC - r| = -t + M_1 \quad \Rightarrow$

$\ln|kC - r| = -kt + M_2 \quad \Rightarrow \quad |kC - r| = e^{-kt + M_2} \quad \Rightarrow \quad kC - r = M_3 e^{-kt} \quad \Rightarrow \quad kC = M_3 e^{-kt} + r \quad \Rightarrow$

$C(t) = M_4 e^{-kt} + r/k. \quad C(0) = C_0 \quad \Rightarrow \quad C_0 = M_4 + r/k \quad \Rightarrow \quad M_4 = C_0 - r/k \quad \Rightarrow$

$C(t) = (C_0 - r/k)e^{-kt} + r/k.$

(b) If $C_0 < r/k$, then $C_0 - r/k < 0$ and the formula for $C(t)$ shows that $C(t)$ increases and $\lim_{t \to \infty} C(t) = r/k$.

As t increases, the formula for $C(t)$ shows how the role of C_0 steadily diminishes as that of r/k increases.

36. (a) Use 1 billion dollars as the x-unit and 1 day as the t-unit. Initially, there is \$10 billion of old currency in circulation, so all of the \$50 million returned to the banks is old. At time t, the amount of new currency is $x(t)$ billion dollars, so $10 - x(t)$ billion dollars of currency is old. The fraction of circulating money that is old is $[10 - x(t)]/10$, and the amount of old currency being returned to the banks each day is $\frac{10 - x(t)}{10}\,0.05$ billion dollars. This amount of new currency per day is introduced into circulation, so $\frac{dx}{dt} = \frac{10 - x}{10} \cdot 0.05 = 0.005(10 - x)$ billion dollars per day.

(b) $\frac{dx}{10 - x} = 0.005\,dt \quad \Rightarrow \quad \frac{-dx}{10 - x} = -0.005\,dt \quad \Rightarrow \quad \ln(10 - x) = -0.005t + c \quad \Rightarrow \quad 10 - x = Ce^{-0.005t}$,

where $C = e^c \quad \Rightarrow \quad x(t) = 10 - Ce^{-0.005t}$. From $x(0) = 0$, we get $C = 10$, so $x(t) = 10(1 - e^{-0.005t})$.

(c) The new bills make up 90% of the circulating currency when $x(t) = 0.9 \cdot 10 = 9$ billion dollars.

$9 = 10(1 - e^{-0.005t}) \quad \Rightarrow \quad 0.9 = 1 - e^{-0.005t} \quad \Rightarrow \quad e^{-0.005t} = 0.1 \quad \Rightarrow \quad -0.005t = -\ln 10 \quad \Rightarrow$

$t = 200 \ln 10 \approx 460.517$ days ≈ 1.26 years.

37. The differential equation is a logistic equation with $k = 0.00008$, carrying capacity $M = 1000$, and initial population $y_0 = P(0) = 100$. So Equation 8 gives the population at time t as

$$P(t) = \frac{100 \cdot 1000}{100 + (1000 - 100)e^{-0.08t}} = \frac{100{,}000}{100 + 900e^{-0.08t}} = \frac{1000}{1 + 9e^{-0.08t}}$$

So the population sizes when $t = 40$ and 80 are

$$P(40) = \frac{1000}{1 + 9e^{-3.2}} \approx 731.6 \qquad P(80) = \frac{1000}{1 + 9e^{-6.4}} \approx 985.3$$

The population reaches 900 when $\frac{1000}{1 + 9e^{-0.08t}} = 900$. Solving this equation for t, we get $1 + 9e^{-0.08t} = \frac{10}{9} \quad \Rightarrow$

$e^{-0.08t} = \frac{1}{81} \quad \Rightarrow \quad -0.08t = \ln \frac{1}{81} = -\ln 81 \quad \Rightarrow \quad t = \frac{\ln 81}{0.08} \approx 54.9$. So the population reaches 900 when t is approximately 55.

38. (a) $\frac{dy}{dt} = ky(M - y) \quad \Rightarrow \quad y(t) = \frac{y_0 M}{y_0 + (M - y_0)e^{-kMt}}$ by (8). With $M = 8 \times 10^7$, $k = 8.875 \times 10^{-9}$, and

$y_0 = 2 \times 10^7$, we get the model $y(t) = \frac{(2 \times 10^7)(8 \times 10^7)}{2 \times 10^7 + (6 \times 10^7)e^{-0.71t}} = \frac{8 \times 10^7}{1 + 3e^{-0.71t}}$, so

$y(1) = \frac{8 \times 10^7}{1 + 3e^{-0.71}} \approx 3.23 \times 10^7$ kg.

(b) $y(t) = 4 \times 10^7 \Rightarrow \dfrac{8 \times 10^7}{1 + 3e^{-0.71t}} = 4 \times 10^7 \Rightarrow 2 = 1 + 3e^{-0.71t} \Rightarrow e^{-0.71t} = \frac{1}{3} \Rightarrow -0.71t = \ln\frac{1}{3} \Rightarrow$

$t = \dfrac{\ln 3}{0.71} \approx 1.55$ years

39. (a) Our assumption is that $\dfrac{dy}{dt} = ky(1 - y)$, where y is the fraction of the population that has heard the rumor.

(b) The equation in part (a) is the logistic differential equation (7) with $M = 1$, so the solution is given by (8):

$$y = \frac{y_0}{y_0 + (1 - y_0)e^{-kt}}.$$

(c) Let t be the number of hours since 8 AM. Then $y_0 = y(0) = \frac{80}{1000} = 0.08$ and $y(4) = \frac{1}{2}$, so

$\dfrac{1}{2} = y(4) = \dfrac{0.08}{0.08 + 0.92e^{-4k}}$. Thus, $0.08 + 0.92e^{-4k} = 0.16$, $e^{-4k} = \frac{0.08}{0.92} = \frac{2}{23}$, and $e^{-k} = \left(\frac{2}{23}\right)^{1/4}$,

so $y = \dfrac{0.08}{0.08 + 0.92(2/23)^{t/4}} = \dfrac{2}{2 + 23(2/23)^{t/4}}$. Solving this equation for t, we get

$$2y + 23y\left(\frac{2}{23}\right)^{t/4} = 2 \Rightarrow \left(\frac{2}{23}\right)^{t/4} = \frac{2 - 2y}{23y} \Rightarrow \left(\frac{2}{23}\right)^{t/4} = \frac{2}{23} \cdot \frac{1 - y}{y} \Rightarrow \left(\frac{2}{23}\right)^{t/4 - 1} = \frac{1 - y}{y}.$$

It follows that $\dfrac{t}{4} - 1 = \dfrac{\ln[(1 - y)/y]}{\ln\frac{2}{23}}$, so $t = 4\left[1 + \dfrac{\ln((1 - y)/y)}{\ln\frac{2}{23}}\right]$.

When $y = 0.9$, $\dfrac{1 - y}{y} = \frac{1}{9}$, so $t = 4\left(1 - \dfrac{\ln 9}{\ln\frac{2}{23}}\right) \approx 7.6$ h or 7 h 36 min. Thus, 90% of the population will have heard the rumor by 3:36 PM.

40. (a) $P(0) = P_0 = 400$, $P(1) = 1200$ and $M = 10{,}000$. From the solution (8) to the logistic differential equation

$P(t) = \dfrac{P_0 M}{P_0 + (M - P_0)e^{-kMt}}$, we get $P = \dfrac{400\,(10{,}000)}{400 + (9600)e^{-kMt}} = \dfrac{10{,}000}{1 + 24e^{-kMt}}$. $P(1) = 1200 \Rightarrow$

$1 + 24e^{-kM} = \frac{100}{12} \Rightarrow e^{kM} = \frac{288}{88} \Rightarrow kM = \ln\frac{36}{11}$. So $P = \dfrac{10{,}000}{1 + 24e^{-t\ln(36/11)}} = \dfrac{10{,}000}{1 + 24 \cdot (11/36)^t}$.

(b) $5000 = \dfrac{10{,}000}{1 + 24(11/36)^t} \Rightarrow 24\left(\frac{11}{36}\right)^t = 1 \Rightarrow t\ln\frac{11}{36} = \ln\frac{1}{24} \Rightarrow t \approx 2.68$ years.

41. (a) $\dfrac{dy}{dt} = ky(M - y) \Rightarrow$

$$\frac{d^2y}{dt^2} = ky\left(-\frac{dy}{dt}\right) + k(M - y)\frac{dy}{dt} = k\frac{dy}{dt}(M - 2y) = k[ky(M - y)](M - 2y) = k^2y(M - y)(M - 2y)$$

(b) y grows fastest when y' has a maximum, that is, when $y'' = 0$. From part (a), $y'' = 0 \Leftrightarrow y = 0, y = M$, or $y = M/2$. Since $0 < y < M$, we see that $y'' = 0 \Leftrightarrow y = M/2$.

42. First we keep k constant (at 0.1, say) and change y_0 in the function $y = \dfrac{10y_0}{y_0 + (10 - y_0)e^{-t}}$. (Notice that y_0 is the y-intercept.) If $y_0 = 0$, the function is 0 everywhere. For $0 < y_0 < 5$, the curve has an inflection point, which moves to the right as y_0 decreases. If $5 < y_0 < 10$, the graph is concave down everywhere. (We are considering only $t \geq 0$.) If $y_0 = 10$, the function is the constant function $y = 10$, and if $y_0 > 10$, the function decreases. For all $y_0 \neq 0$, $\lim\limits_{t\to\infty} y = 10$.

Now we instead keep y_0 constant (at $y_0 = 1$) and change k in the function $y = \dfrac{10}{1 + 9e^{-10kt}}$. It seems that as k increases, the graph approaches the line $y = 10$ more and more quickly. (Note that the only difference in the shape of the curves is in the horizontal scaling; if we choose suitable x-scales, the graphs all look the same.)

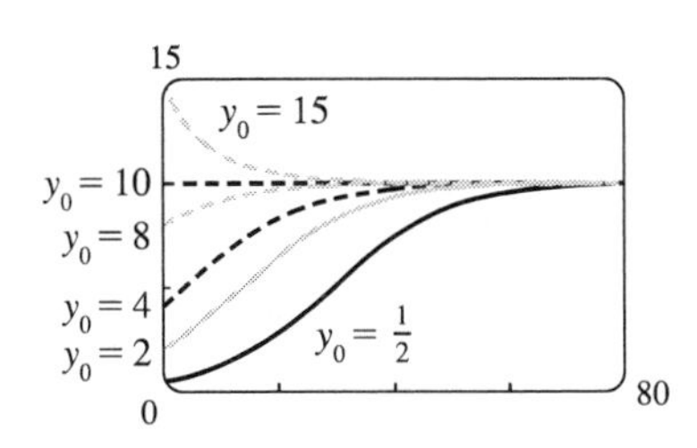

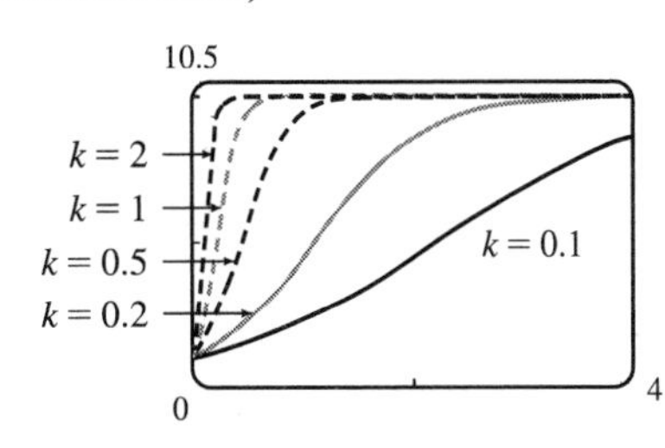

43. (a) Let $y(t)$ be the amount of salt (in kg) after t minutes. Then $y(0) = 15$. The amount of liquid in the tank is 1000 L at all times, so the concentration at time t (in minutes) is $y(t)/1000$ kg/L and $\frac{dy}{dt} = -\left[\frac{y(t)}{1000}\frac{\text{kg}}{\text{L}}\right]\left(10\,\frac{\text{L}}{\text{min}}\right) = -\frac{y(t)}{100}\frac{\text{kg}}{\text{min}}$.

$\int \frac{dy}{y} = -\frac{1}{100}\int dt \;\Rightarrow\; \ln y = -\frac{t}{100} + C$, and $y(0) = 15 \;\Rightarrow\; \ln 15 = C$, so $\ln y = \ln 15 - \frac{t}{100}$.

It follows that $\ln\left(\frac{y}{15}\right) = -\frac{t}{100}$ and $\frac{y}{15} = e^{-t/100}$, so $y = 15e^{-t/100}$ kg.

(b) After 20 minutes, $y = 15e^{-20/100} = 15e^{-0.2} \approx 12.3$ kg.

44. Let $y(t)$ be the amount of carbon dioxide in the room after t minutes. Then $y(0) = 0.0015(180) = 0.27\text{ m}^3$. The amount of air in the room is 180 m^3 at all times, so the percentage at time t (in mimutes) is $y(t)/180 \times 100$, and the change in the amount of carbon dioxide with respect to time is

$$\frac{dy}{dt} = (0.0005)\left(2\,\frac{\text{m}^3}{\text{min}}\right) - \frac{y(t)}{180}\left(2\,\frac{\text{m}^3}{\text{min}}\right) = 0.001 - \frac{y}{90} = \frac{9 - 100y}{9000}\,\frac{\text{m}^3}{\text{min}}$$

Hence, $\int \frac{dy}{9 - 100y} = \int \frac{dt}{9000}$ and $-\frac{1}{100}\ln|9 - 100y| = \frac{1}{9000}t + C$. Because $y(0) = 0.27$, we have $-\frac{1}{100}\ln 18 = C$, so $-\frac{1}{100}\ln|9 - 100y| = \frac{1}{9000}t - \frac{1}{100}\ln 18 \;\Rightarrow\; \ln|9 - 100y| = -\frac{1}{90}t + \ln 18 \;\Rightarrow$ $\ln|9 - 100y| = \ln e^{-t/90} + \ln 18 \;\Rightarrow\; \ln|9 - 100y| = \ln(18e^{-t/90})$, and $|9 - 100y| = 18e^{-t/90}$. Since y is continuous, $y(0) = 0.27$, and the right-hand side is never zero, we deduce that $9 - 100y$ is always negative. Thus, $|9 - 100y| = 100y - 9$ and we have $100y - 9 = 18e^{-t/90} \;\Rightarrow\; 100y = 9 + 18e^{-t/90} \;\Rightarrow\; y = 0.09 + 0.18e^{-t/90}$. The percentage of carbon dioxide in the room is

$$p(t) = \frac{y}{180} \times 100 = \frac{0.09 + 0.18e^{-t/90}}{180} \times 100 = (0.0005 + 0.001e^{-t/90}) \times 100 = 0.05 + 0.1e^{-t/90}$$

In the long run, we have $\lim_{t\to\infty} p(t) = 0.05 + 0.1(0) = 0.05$; that is, the amount of carbon dioxide approaches 0.05% as time goes on.

45. Let $y(t)$ be the amount of alcohol in the vat after t minutes. Then $y(0) = 0.04(500) = 20$ gal. The amount of beer in the vat is 500 gallons at all times, so the percentage at time t (in minutes) is $y(t)/500 \times 100$, and the change in the amount of alcohol with respect to time t is

$$\frac{dy}{dt} = \text{rate in} - \text{rate out} = 0.06\left(5\,\frac{\text{gal}}{\text{min}}\right) - \frac{y(t)}{500}\left(5\,\frac{\text{gal}}{\text{min}}\right) = 0.3 - \frac{y}{100} = \frac{30 - y}{100}\,\frac{\text{gal}}{\text{min}}$$

Hence, $\int \frac{dy}{30 - y} = \int \frac{dt}{100}$ and $-\ln|30 - y| = \frac{1}{100}t + C$. Because $y(0) = 20$, we have $-\ln 10 = C$, so

$-\ln|30-y| = \frac{1}{100}t - \ln 10 \Rightarrow \ln|30-y| = -t/100 + \ln 10 \Rightarrow \ln|30-y| = \ln e^{-t/100} + \ln 10 \Rightarrow$ $\ln|30-y| = \ln(10e^{-t/100}) \Rightarrow |30-y| = 10e^{-t/100}$. Since y is continuous, $y(0) = 20$, and the right-hand side is never zero, we deduce that $30 - y$ is always positive. Thus, $30 - y = 10e^{-t/100} \Rightarrow y = 30 - 10e^{-t/100}$.

The percentage of alcohol is $p(t) = y(t)/500 \times 100 = y(t)/5 = 6 - 2e^{-t/100}$. The percentage of alcohol after one hour is $p(60) = 6 - 2e^{-60/100} \approx 4.9$.

46. (a) If $y(t)$ is the amount of salt (in kg) after t minutes, then $y(0) = 0$ and the total amount of liquid in the tank remains constant at 1000 L.

$$\frac{dy}{dt} = \left(0.05\ \frac{\text{kg}}{\text{L}}\right)\left(5\ \frac{\text{L}}{\text{min}}\right) + \left(0.04\ \frac{\text{kg}}{\text{L}}\right)\left(10\ \frac{\text{L}}{\text{min}}\right) - \left(\frac{y(t)}{1000}\ \frac{\text{kg}}{\text{L}}\right)\left(15\ \frac{\text{L}}{\text{min}}\right)$$

$$= 0.25 + 0.40 - 0.015y = 0.65 - 0.015y = \frac{130 - 3y}{200}\ \frac{\text{kg}}{\text{min}}$$

Hence, $\displaystyle\int \frac{dy}{130 - 3y} = \int \frac{dt}{200}$ and $-\frac{1}{3}\ln|130 - 3y| = \frac{1}{200}t + C$. Because $y(0) = 0$, we have $-\frac{1}{3}\ln 130 = C$, so $-\frac{1}{3}\ln|130 - 3y| = \frac{1}{200}t - \frac{1}{3}\ln 130 \Rightarrow \ln|130 - 3y| = -\frac{3}{200}t + \ln 130 = \ln\left(130e^{-3t/200}\right)$, and $|130 - 3y| = 130e^{-3t/200}$. Since y is continuous, $y(0) = 0$, and the right-hand side is never zero, we deduce that $130 - 3y$ is always positive. Thus, $130 - 3y = 130e^{-3t/200}$ and $y = \frac{130}{3}\left(1 - e^{-3t/200}\right)$ kg.

(b) After one hour, $y = \frac{130}{3}\left(1 - e^{-3\cdot 60/200}\right) = \frac{130}{3}\left(1 - e^{-0.9}\right) \approx 25.7$ kg. *Note:* As $t \to \infty$, $y(t) \to \frac{130}{3} = 43\frac{1}{3}$ kg.

47. Assume that the raindrop begins at rest, so that $v(0) = 0$. $dm/dt = km$ and $(mv)' = gm \Rightarrow mv' + vm' = gm \Rightarrow$ $mv' + v(km) = gm \Rightarrow v' + vk = g \Rightarrow \dfrac{dv}{dt} = g - kv \Rightarrow \displaystyle\int \frac{dv}{g - kv} = \int dt \Rightarrow$ $-(1/k)\ln|g - kv| = t + C \Rightarrow \ln|g - kv| = -kt - kC \Rightarrow g - kv = Ae^{-kt}$. $v(0) = 0 \Rightarrow A = g$. So $kv = g - ge^{-kt} \Rightarrow v = (g/k)(1 - e^{-kt})$. Since $k > 0$, as $t \to \infty$, $e^{-kt} \to 0$ and therefore, $\lim\limits_{t\to\infty} v(t) = g/k$.

48. (a) $m\dfrac{dv}{dt} = -kv \Rightarrow \dfrac{dv}{v} = -\dfrac{k}{m}\,dt \Rightarrow \ln|v| = -\dfrac{k}{m}t + C$. Since $v(0) = v_0$, $\ln|v_0| = C$.

Therefore, $\ln\left|\dfrac{v}{v_0}\right| = -\dfrac{k}{m}t \Rightarrow \left|\dfrac{v}{v_0}\right| = e^{-kt/m} \Rightarrow v(t) = \pm v_0e^{-kt/m}$. The sign is $+$ when $t = 0$, and we assume v is continuous, so that the sign is $+$ for all t. Thus, $v(t) = v_0e^{-kt/m}$.

$ds/dt = v_0e^{-kt/m} \Rightarrow s(t) = -\dfrac{mv_0}{k}e^{-kt/m} + C'$.

From $s(0) = s_0$, we get $s_0 = -\dfrac{mv_0}{k} + C'$, so $C' = s_0 + \dfrac{mv_0}{k}$ and $s(t) = s_0 + \dfrac{mv_0}{k}\left(1 - e^{-kt/m}\right)$. The distance traveled from time 0 to time t is $s(t) - s_0$, so the total distance traveled is $\lim\limits_{t\to\infty}[s(t) - s_0] = \dfrac{mv_0}{k}$.

Note: In finding the limit, we use the fact that $k > 0$ to conclude that $\lim\limits_{t\to\infty} e^{-kt/m} = 0$.

(b) $m\dfrac{dv}{dt} = -kv^2 \Rightarrow \dfrac{dv}{v^2} = -\dfrac{k}{m}\,dt \Rightarrow \dfrac{-1}{v} = -\dfrac{kt}{m} + C \Rightarrow \dfrac{1}{v} = \dfrac{kt}{m} - C$.

Since $v(0) = v_0$, $C = -\dfrac{1}{v_0}$ and $\dfrac{1}{v} = \dfrac{kt}{m} + \dfrac{1}{v_0}$. Therefore, $v(t) = \dfrac{1}{kt/m + 1/v_0} = \dfrac{mv_0}{kv_0t + m}$.

$\dfrac{ds}{dt} = \dfrac{mv_0}{kv_0t + m} \Rightarrow s(t) = \dfrac{m}{k}\displaystyle\int \frac{kv_0\,dt}{kv_0t + m} = \dfrac{m}{k}\ln|kv_0t + m| + C'$. Since $s(0) = s_0$, we get

$$s_0 = \frac{m}{k}\ln m + C' \;\Rightarrow\; C' = s_0 - \frac{m}{k}\ln m \;\Rightarrow\; s(t) = s_0 + \frac{m}{k}(\ln|kv_0t + m| - \ln m) = s_0 + \frac{m}{k}\ln\left|\frac{kv_0t + m}{m}\right|.$$

We can rewrite the formulas for $v(t)$ and $s(t)$ as $v(t) = \dfrac{v_0}{1 + (kv_0/m)t}$ and $s(t) = s_0 + \dfrac{m}{k}\ln\left|1 + \dfrac{kv_0}{m}t\right|$.

Remarks: This model of horizontal motion through a resistive medium was designed to handle the case in which $v_0 > 0$. Then the term $-kv^2$ representing the resisting force causes the object to decelerate. The absolute value in the expression for $s(t)$ is unnecessary (since k, v_0, and m are all positive), and $\lim\limits_{t\to\infty} s(t) = \infty$. In other words, the object travels infinitely far. However, $\lim\limits_{t\to\infty} v(t) = 0$. When $v_0 < 0$, the term $-kv^2$ increases the magnitude of the object's negative velocity. According to the formula for $s(t)$, the position of the object approaches $-\infty$ as t approaches $m/k(-v_0)$: $\lim\limits_{t\to -m/(kv_0)} s(t) = -\infty$. Again the object travels infinitely far, but this time the feat is accomplished in a finite amount of time. Notice also that $\lim\limits_{t\to -m/(kv_0)} v(t) = -\infty$ when $v_0 < 0$, showing that the speed of the object increases without limit.

49. (a) The rate of growth of the area is jointly proportional to $\sqrt{A(t)}$ and $M - A(t)$; that is, the rate is proportional to the product of those two quantities. So for some constant k, $dA/dt = k\sqrt{A}\,(M - A)$. We are interested in the maximum of the function dA/dt (when the tissue grows the fastest), so we differentiate, using the Chain Rule and then substituting for dA/dt from the differential equation:

$$\frac{d}{dt}\left(\frac{dA}{dt}\right) = k\left[\sqrt{A}\,(-1)\frac{dA}{dt} + (M - A)\cdot\tfrac{1}{2}A^{-1/2}\frac{dA}{dt}\right] = \tfrac{1}{2}kA^{-1/2}\frac{dA}{dt}\,[-2A + (M - A)]$$

$$= \tfrac{1}{2}kA^{-1/2}\left[k\sqrt{A}(M - A)\right][M - 3A] = \tfrac{1}{2}k^2(M - A)(M - 3A)$$

This is 0 when $M - A = 0$ [this situation never actually occurs, since the graph of $A(t)$ is asymptotic to the line $y = M$, as in the logistic model] and when $M - 3A = 0 \;\Leftrightarrow\; A(t) = M/3$. This represents a maximum by the First Derivative Test, since $\dfrac{d}{dt}\left(\dfrac{dA}{dt}\right)$ goes from positive to negative when $A(t) = M/3$.

(b) From the CAS, we get $A(t) = M\left(\dfrac{Ce^{\sqrt{M}kt} - 1}{Ce^{\sqrt{M}kt} + 1}\right)^2$. To get C in terms of the initial area A_0 and the maximum area M, we substitute $t = 0$ and $A = A_0 = A(0)$: $A_0 = M\left(\dfrac{C - 1}{C + 1}\right)^2 \;\Leftrightarrow\; (C + 1)\sqrt{A_0} = (C - 1)\sqrt{M} \;\Leftrightarrow$

$C\sqrt{A_0} + \sqrt{A_0} = C\sqrt{M} - \sqrt{M} \;\Leftrightarrow\; \sqrt{M} + \sqrt{A_0} = C\sqrt{M} - C\sqrt{A_0} \;\Leftrightarrow$

$\sqrt{M} + \sqrt{A_0} = C\left(\sqrt{M} - \sqrt{A_0}\right) \;\Leftrightarrow\; C = \dfrac{\sqrt{M} + \sqrt{A_0}}{\sqrt{M} - \sqrt{A_0}}$. (Notice that if $A_0 = 0$, then $C = 1$.)

50. (a) According to the hint we use the Chain Rule: $m\dfrac{dv}{dt} = m\dfrac{dv}{dx}\cdot\dfrac{dx}{dt} = mv\dfrac{dv}{dx} = -\dfrac{mgR^2}{(x + R)^2} \;\Rightarrow$

$\displaystyle\int v\,dv = \int\frac{-gR^2\,dx}{(x + R)^2} \;\Rightarrow\; \frac{v^2}{2} = \frac{gR^2}{x + R} + C$. When $x = 0$, $v = v_0$, so $\dfrac{v_0^2}{2} = \dfrac{gR^2}{0 + R} + C \;\Rightarrow\; C = \tfrac{1}{2}v_0^2 - gR$

$\Rightarrow\; \tfrac{1}{2}v^2 - \tfrac{1}{2}v_0^2 = \dfrac{gR^2}{x + R} - gR$. Now at the top of its flight, the rocket's velocity will be 0, and its height will be $x = h$.

Solving for v_0: $-\tfrac{1}{2}v_0^2 = \dfrac{gR^2}{h + R} - gR \;\Rightarrow\; \dfrac{v_0^2}{2} = g\left[-\dfrac{R^2}{R + h} + \dfrac{R(R + h)}{R + h}\right] = \dfrac{gRh}{R + h} \;\Rightarrow\; v_0 = \sqrt{\dfrac{2gRh}{R + h}}$.

(b) $v_e = \lim_{h\to\infty} v_0 = \lim_{h\to\infty} \sqrt{\dfrac{2gRh}{R+h}} = \lim_{h\to\infty} \sqrt{\dfrac{2gR}{(R/h)+1}} = \sqrt{2gR}$

(c) $v_e = \sqrt{2 \cdot 32 \text{ ft/s}^2 \cdot 3960 \text{ mi} \cdot 5280 \text{ ft/mi}} \approx 36{,}581 \text{ ft/s} \approx 6.93 \text{ mi/s}$

7 Review

CONCEPT CHECK

1. (a) See Section 7.1, Figure 2 and Equations 7.1.1 and 7.1.2.

(b) Instead of using "top minus bottom" and integrating from left to right, we use "right minus left" and integrate from bottom to top. See Figures 8 and 9 in Section 7.1.

2. The numerical value of the area represents the number of meters by which Sue is ahead of Kathy after 1 minute.

3. (a) See the discussion on pages 365–66.

(b) See the discussion between Examples 5 and 6 in Section 7.2. If the cross-section is a disk, find the radius in terms of x or y and use $A = \pi(\text{radius})^2$. If the cross-section is a washer, find the inner radius r_{in} and outer radius r_{out} and use $A = \pi\left(r_{\text{out}}^2\right) - \pi\left(r_{\text{in}}^2\right)$.

4. (a) $V = 2\pi rh\,\Delta r = (\text{circumference})\,(\text{height})\,(\text{thickness})$

(b) For a typical shell, find the circumference and height in terms of x or y and calculate $V = \int_a^b (\text{circumference})\,(\text{height})\,(dx \text{ or } dy)$, where a and b are the limits on x or y.

(c) Sometimes slicing produces washers or disks whose radii are difficult (or impossible) to find explicitly. On other occasions, the cylindrical shell method leads to an easier integral than slicing does.

5. (a) The length of a curve is defined to be the limit of the lengths of the inscribed polygons, as described near Figure 3 in Section 7.4.

(b) See Equation 7.4.2.

(c) See Equation 7.4.4.

6. $\int_0^6 f(x)\,dx$ represents the amount of work done. Its units are newton-meters, or joules.

7. Let $c(x)$ be the cross-sectional length of the wall (measured parallel to the surface of the fluid) at depth x. Then the hydrostatic force against the wall is given by $F = \int_a^b \delta x c(x)\,dx$, where a and b are the lower and upper limits for x at points of the wall and δ is the weight density of the fluid.

8. (a) The center of mass is the point at which the plate balances horizontally.

(b) See Equations 7.5.12.

9. If a plane region $\mathcal{R}$ that lies entirely on one side of a line ℓ in its plane is rotated about ℓ, then the volume of the resulting solid is the product of the area of $\mathcal{R}$ and the distance traveled by the centroid of $\mathcal{R}$.

10. (a) A differential equation is an equation that contains an unknown function and one or more of its derivatives.

(b) The order of a differential equation is the order of the highest derivative that occurs in the equation.

(c) An initial condition is a condition of the form $y(t_0) = y_0$.

11. See the paragraph preceding Example 6 in Section 7.6.

12. A separable equation is a first-order differential equation in which the expression for dy/dx can be factored as a function of x times a function of y, that is, $dy/dx = g(x)f(y)$. We can solve the equation by integrating both sides of the equation $dy/f(y) = g(x)\,dx$ and solving for y.

EXERCISES

1. $0 = x^2 - x - 6 = (x-3)(x+2) \quad \Leftrightarrow \quad x = 3$ or -2. So

$$A = \int_{-2}^{3} \left[0 - (x^2 - x - 6)\right] dx = \int_{-2}^{3} (-x^2 + x + 6)\, dx$$
$$= \left[-\tfrac{1}{3}x^3 + \tfrac{1}{2}x^2 + 6x\right]_{-2}^{3}$$
$$= \left(-9 + \tfrac{9}{2} + 18\right) - \left(\tfrac{8}{3} + 2 - 12\right)$$
$$= \tfrac{125}{6}$$

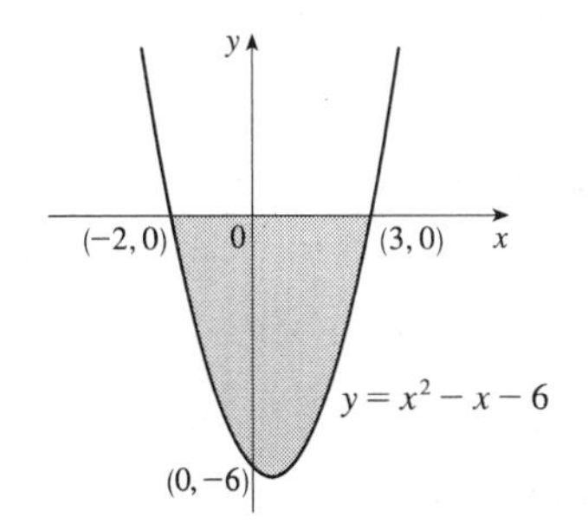

2. $20 - x^2 = x^2 - 12 \quad \Leftrightarrow \quad 32 = 2x^2 \quad \Leftrightarrow \quad x^2 = 16 \quad \Leftrightarrow \quad x = \pm 4$. So

$$A = \int_{-4}^{4} \left[(20 - x^2) - (x^2 - 12)\right] dx = \int_{-4}^{4} (32 - 2x^2)\, dx$$
$$= 2\int_{0}^{4} (32 - 2x^2)\, dx \qquad \text{[even function]}$$
$$= 2\left[32x - \tfrac{2}{3}x^3\right]_{0}^{4}$$
$$= 2\left(128 - \tfrac{128}{3}\right) = \tfrac{512}{3}$$

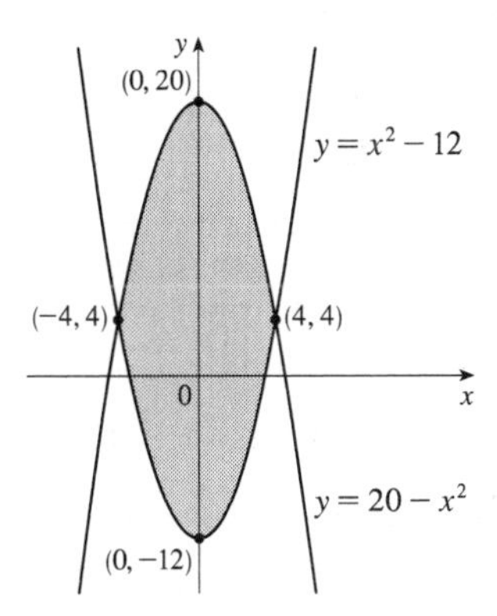

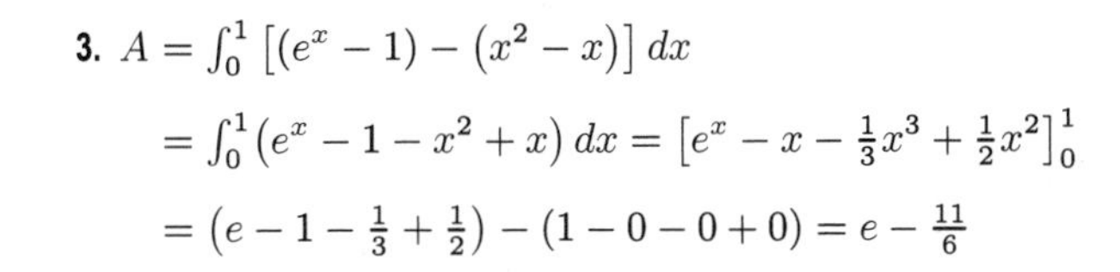

3. $A = \int_{0}^{1} \left[(e^x - 1) - (x^2 - x)\right] dx$

$$= \int_{0}^{1} (e^x - 1 - x^2 + x)\, dx = \left[e^x - x - \tfrac{1}{3}x^3 + \tfrac{1}{2}x^2\right]_{0}^{1}$$
$$= \left(e - 1 - \tfrac{1}{3} + \tfrac{1}{2}\right) - (1 - 0 - 0 + 0) = e - \tfrac{11}{6}$$

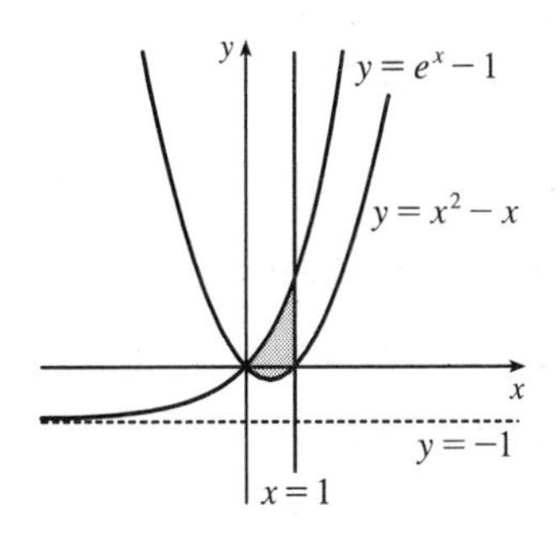

4. $y^2 + 3y = -y \quad \Leftrightarrow \quad y^2 + 4y = 0 \quad \Leftrightarrow \quad y(y+4) = 0 \quad \Leftrightarrow$
$y = 0$ or -4.

$$A = \int_{-4}^{0} \left[-y - (y^2 + 3y)\right] dy = \int_{-4}^{0} (-y^2 - 4y)\, dy$$
$$= \left[-\tfrac{1}{3}y^3 - 2y^2\right]_{-4}^{0} = 0 - \left(\tfrac{64}{3} - 32\right) = \tfrac{32}{3}$$

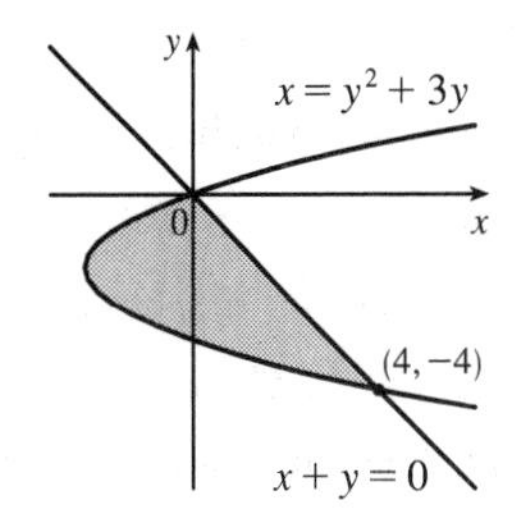

5. Using washers with inner radius x^2 and outer radius $2x$, we have

$$V = \pi \int_0^2 \left[(2x)^2 - (x^2)^2\right] dx = \pi \int_0^2 (4x^2 - x^4)\, dx$$
$$= \pi\left[\tfrac{4}{3}x^3 - \tfrac{1}{5}x^5\right]_0^2 = \pi\left(\tfrac{32}{3} - \tfrac{32}{5}\right) = 32\pi \cdot \tfrac{2}{15}$$
$$= \tfrac{64\pi}{15}$$

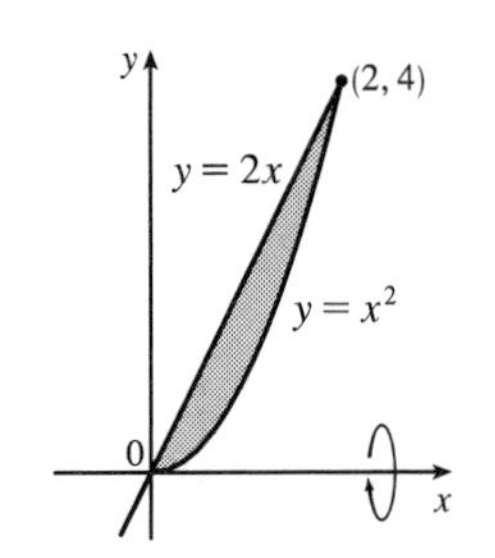

6. $1 + y^2 = y + 3 \;\Leftrightarrow\; y^2 - y - 2 = 0 \;\Leftrightarrow\; (y-2)(y+1) = 0 \;\Leftrightarrow$ $y = 2$ or -1.

$$V = \pi \int_{-1}^2 \left[(y+3)^2 - (1+y^2)^2\right] dy$$
$$= \pi \int_{-1}^2 (y^2 + 6y + 9 - 1 - 2y^2 - y^4)\, dy$$
$$= \pi \int_{-1}^2 (8 + 6y - y^2 - y^4)\, dy$$
$$= \pi\left[8y + 3y^2 - \tfrac{1}{3}y^3 - \tfrac{1}{5}y^5\right]_{-1}^2$$
$$= \pi\left[\left(16 + 12 - \tfrac{8}{3} - \tfrac{32}{5}\right) - \left(-8 + 3 + \tfrac{1}{3} + \tfrac{1}{5}\right)\right]$$
$$= \pi\left(33 - \tfrac{9}{3} - \tfrac{33}{5}\right) = \tfrac{117\pi}{5}$$

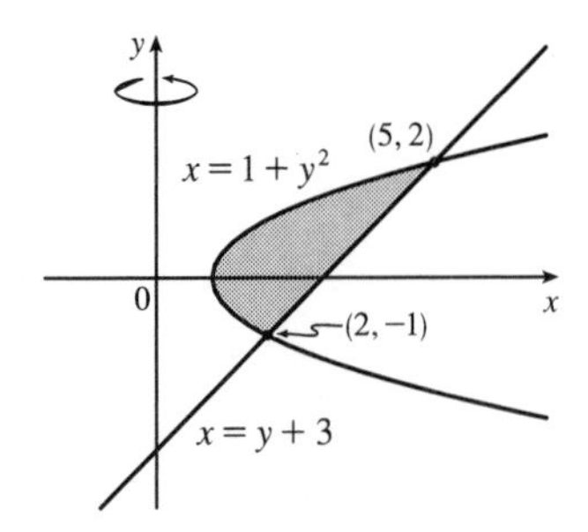

7. $V = \pi \int_{-3}^3 \left\{[(9 - y^2) - (-1)]^2 - [0 - (-1)]^2\right\} dy$

$$= 2\pi \int_0^3 \left[(10 - y^2)^2 - 1\right] dy$$
$$= 2\pi \int_0^3 (100 - 20y^2 + y^4 - 1)\, dy$$
$$= 2\pi \int_0^3 (99 - 20y^2 + y^4)\, dy = 2\pi\left[99y - \tfrac{20}{3}y^3 + \tfrac{1}{5}y^5\right]_0^3$$
$$= 2\pi\left(297 - 180 + \tfrac{243}{5}\right) = \tfrac{1656\pi}{5}$$

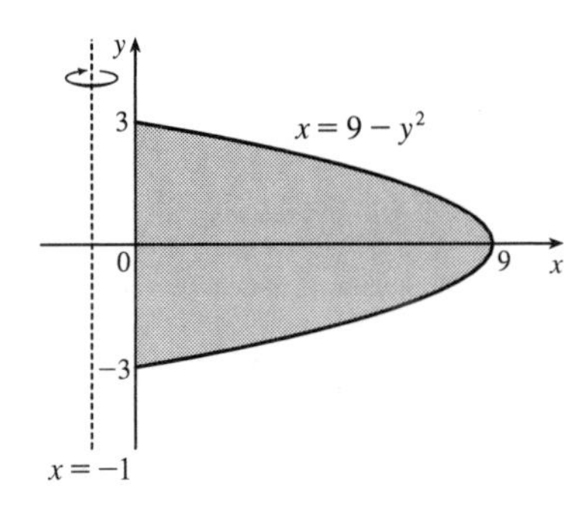

8. $V = \pi \int_{-2}^2 \left\{[(9 - x^2) - (-1)]^2 - [(x^2 + 1) - (-1)]^2\right\} dx$

$$= \pi \int_{-2}^2 \left[(10 - x^2)^2 - (x^2 + 2)^2\right] dx$$
$$= 2\pi \int_0^2 (96 - 24x^2)\, dx = 48\pi \int_0^2 (4 - x^2)\, dx$$
$$= 48\pi\left[4x - \tfrac{1}{3}x^3\right]_0^2 = 48\pi\left(8 - \tfrac{8}{3}\right) = 256\pi$$

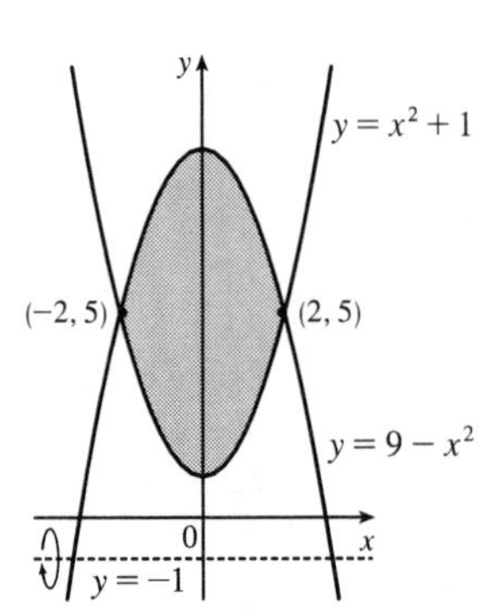

9. The graph of $x^2 - y^2 = a^2$ is a hyperbola with right and left branches. Solving for y gives us $y^2 = x^2 - a^2 \Rightarrow y = \pm\sqrt{x^2 - a^2}$. We'll use shells and the height of each shell is $\sqrt{x^2 - a^2} - \left(-\sqrt{x^2 - a^2}\right) = 2\sqrt{x^2 - a^2}$.

The volume is $V = \int_a^{a+h} 2\pi x \cdot 2\sqrt{x^2 - a^2}\,dx$. To evaluate, let $u = x^2 - a^2$, so $du = 2x\,dx$ and $x\,dx = \frac{1}{2}\,du$. When $x = a$, $u = 0$, and when $x = a + h$, $u = (a+h)^2 - a^2 = a^2 + 2ah + h^2 - a^2 = 2ah + h^2$.

Thus, $V = 4\pi \int_0^{2ah+h^2} \sqrt{u}\left(\frac{1}{2}\,du\right) = 2\pi\left[\frac{2}{3}u^{3/2}\right]_0^{2ah+h^2} = \frac{4}{3}\pi(2ah + h^2)^{3/2}$.

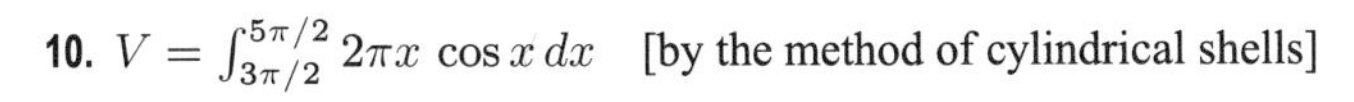

10. $V = \int_{3\pi/2}^{5\pi/2} 2\pi x \cos x\,dx$ [by the method of cylindrical shells]

11. $V = \int_0^1 \pi[(1 - x^3)^2 - (1 - x^2)^2]\,dx$

12. $V = \int_0^2 2\pi(8 - x^3)(2 - x)\,dx$

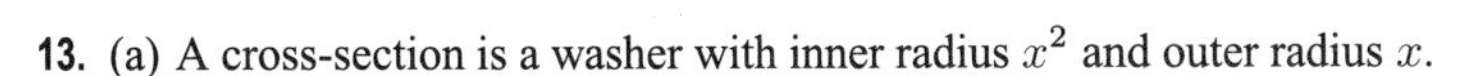

13. (a) A cross-section is a washer with inner radius x^2 and outer radius x.

$$V = \int_0^1 \pi\left[(x)^2 - (x^2)^2\right]dx = \int_0^1 \pi(x^2 - x^4)\,dx = \pi\left[\tfrac{1}{3}x^3 - \tfrac{1}{5}x^5\right]_0^1 = \pi\left[\tfrac{1}{3} - \tfrac{1}{5}\right] = \tfrac{2\pi}{15}$$

(b) A cross-section is a washer with inner radius y and outer radius $\sqrt{y}$.

$$V = \int_0^1 \pi\left[\left(\sqrt{y}\right)^2 - y^2\right]dy = \int_0^1 \pi(y - y^2)\,dy = \pi\left[\tfrac{1}{2}y^2 - \tfrac{1}{3}y^3\right]_0^1 = \pi\left[\tfrac{1}{2} - \tfrac{1}{3}\right] = \tfrac{\pi}{6}$$

(c) A cross-section is a washer with inner radius $2 - x$ and outer radius $2 - x^2$.

$$V = \int_0^1 \pi\left[(2 - x^2)^2 - (2 - x)^2\right]dx = \int_0^1 \pi(x^4 - 5x^2 + 4x)\,dx = \pi\left[\tfrac{1}{5}x^5 - \tfrac{5}{3}x^3 + 2x^2\right]_0^1 = \pi\left[\tfrac{1}{5} - \tfrac{5}{3} + 2\right] = \tfrac{8\pi}{15}$$

14. (a) $A = \int_0^1 (2x - x^2 - x^3)\,dx = \left[x^2 - \frac{1}{3}x^3 - \frac{1}{4}x^4\right]_0^1 = 1 - \frac{1}{3} - \frac{1}{4} = \frac{5}{12}$

(b) A cross-section is a washer with inner radius x^3 and outer radius $2x - x^2$, so its area is $\pi(2x - x^2)^2 - \pi(x^3)^2$.

$$\begin{aligned} V &= \int_0^1 A(x)\,dx = \int_0^1 \pi\left[(2x - x^2)^2 - (x^3)^2\right]dx = \int_0^1 \pi(4x^2 - 4x^3 + x^4 - x^6)\,dx \\ &= \pi\left[\tfrac{4}{3}x^3 - x^4 + \tfrac{1}{5}x^5 - \tfrac{1}{7}x^7\right]_0^1 = \pi\left(\tfrac{4}{3} - 1 + \tfrac{1}{5} - \tfrac{1}{7}\right) = \tfrac{41\pi}{105} \end{aligned}$$

(c) Using the method of cylindrical shells,

$$V = \int_0^1 2\pi x(2x - x^2 - x^3)\,dx = \int_0^1 2\pi(2x^2 - x^3 - x^4)\,dx = 2\pi\left[\tfrac{2}{3}x^3 - \tfrac{1}{4}x^4 - \tfrac{1}{5}x^5\right]_0^1 = 2\pi\left(\tfrac{2}{3} - \tfrac{1}{4} - \tfrac{1}{5}\right) = \tfrac{13\pi}{30}.$$

15. (a) Using the Midpoint Rule on $[0, 1]$ with $f(x) = \tan(x^2)$ and $n = 4$, we estimate

$$A = \int_0^1 \tan(x^2)\,dx \approx \tfrac{1}{4}\left[\tan\left(\left(\tfrac{1}{8}\right)^2\right) + \tan\left(\left(\tfrac{3}{8}\right)^2\right) + \tan\left(\left(\tfrac{5}{8}\right)^2\right) + \tan\left(\left(\tfrac{7}{8}\right)^2\right)\right] \approx \tfrac{1}{4}(1.53) \approx 0.38$$

(b) Using the Midpoint Rule on $[0, 1]$ with $f(x) = \pi\tan^2(x^2)$ (for disks) and $n = 4$, we estimate

$$V = \int_0^1 f(x)\,dx \approx \tfrac{1}{4}\pi\left[\tan^2\left(\left(\tfrac{1}{8}\right)^2\right) + \tan^2\left(\left(\tfrac{3}{8}\right)^2\right) + \tan^2\left(\left(\tfrac{5}{8}\right)^2\right) + \tan^2\left(\left(\tfrac{7}{8}\right)^2\right)\right] \approx \tfrac{\pi}{4}(1.114) \approx 0.87$$

16. (a)

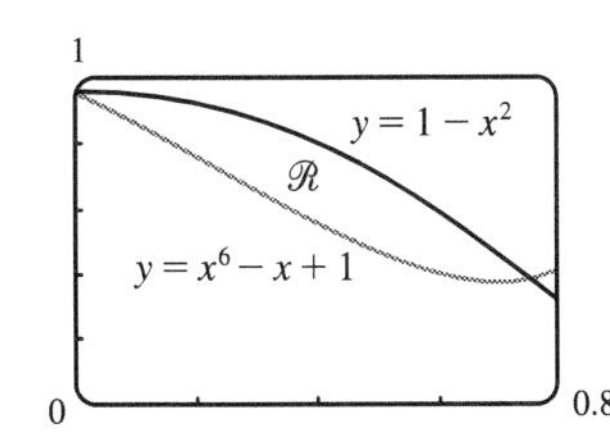

From the graph, we see that the curves intersect at $x = 0$ and at $x = a \approx 0.75$, with $1 - x^2 > x^6 - x + 1$ on $(0, a)$.

(b) The area of $\mathcal{R}$ is $A = \int_0^a [(1-x^2)-(x^6-x+1)]\,dx = \left[-\frac{1}{3}x^3 - \frac{1}{7}x^7 + \frac{1}{2}x^2\right]_0^a \approx 0.12$.

(c) Using washers, the volume generated when $\mathcal{R}$ is rotated about the x-axis is

$$V = \pi \int_0^a [(1-x^2)^2 - (x^6-x+1)^2]\,dx = \pi \int_0^a (-x^{12} + 2x^7 - 2x^6 + x^4 - 3x^2 + 2x)\,dx$$
$$= \pi\left[-\tfrac{1}{13}x^{13} + \tfrac{1}{4}x^8 - \tfrac{2}{7}x^7 + \tfrac{1}{5}x^5 - x^3 + x^2\right]_0^a \approx 0.54$$

(d) Using shells, the volume generated when $\mathcal{R}$ is rotated about the y-axis is

$$V = \int_0^a 2\pi x[(1-x^2)-(x^6-x+1)]\,dx = 2\pi\int_0^a (-x^3 - x^7 + x^2)\,dx = 2\pi\left[-\tfrac{1}{4}x^4 - \tfrac{1}{8}x^8 + \tfrac{1}{3}x^3\right]_0^a \approx 0.31.$$

17. The solid is obtained by rotating the region $\mathcal{R} = \left\{(x,y) \mid 0 \le x \le \frac{\pi}{2}, 0 \le y \le \cos x\right\}$ about the y-axis.

18. The solid is obtained by rotating the region $\mathcal{R} = \left\{(x,y) \mid 0 \le x \le \frac{\pi}{2}, 0 \le y \le \sqrt{2}\cos x\right\}$ about the x-axis.

19. The solid is obtained by rotating the region $\mathcal{R} = \left\{(x,y) \mid 0 \le y \le 2, 0 \le x \le 4 - y^2\right\}$ about the x-axis.

20. The solid is obtained by rotating the region $\mathcal{R} = \left\{(x,y) \mid 0 \le x \le 1, 2 - \sqrt{x} \le y \le 2 - x^2\right\}$ about the x-axis.

Or: The solid is obtained by rotating the region $\mathcal{R} = \left\{(x,y) \mid 0 \le x \le 1, x^2 \le y \le \sqrt{x}\right\}$ about the line $y = 2$.

21. Take the base to be the disk $x^2 + y^2 \le 9$. Then $V = \int_{-3}^3 A(x)\,dx$, where $A(x_0)$ is the area of the isosceles right triangle whose hypotenuse lies along the line $x = x_0$ in the xy-plane. The length of the hypotenuse is $2\sqrt{9-x^2}$ and the length of each leg is $\sqrt{2}\sqrt{9-x^2}$. $A(x) = \frac{1}{2}\left(\sqrt{2}\sqrt{9-x^2}\right)^2 = 9 - x^2$, so

$$V = 2\int_0^3 A(x)\,dx = 2\int_0^3 (9-x^2)\,dx = 2\left[9x - \tfrac{1}{3}x^3\right]_0^3 = 2(27-9) = 36$$

22. $V = \int_{-1}^1 A(x)\,dx = 2\int_0^1 A(x)\,dx = 2\int_0^1 \left[(2-x^2) - x^2\right]^2 dx = 2\int_0^1 [2(1-x^2)]^2 dx$

$= 8\int_0^1 (1 - 2x^2 + x^4)\,dx = 8\left[x - \frac{2}{3}x^3 + \frac{1}{5}x^5\right]_0^1 = 8\left(1 - \frac{2}{3} + \frac{1}{5}\right) = \frac{64}{15}$

23. Equilateral triangles with sides measuring $\frac{1}{4}x$ meters have height $\frac{1}{4}x \sin 60^\circ = \frac{\sqrt{3}}{8}x$. Therefore,

$$A(x) = \tfrac{1}{2}\cdot\tfrac{1}{4}x\cdot\tfrac{\sqrt{3}}{8}x = \tfrac{\sqrt{3}}{64}x^2. \quad V = \int_0^{20} A(x)\,dx = \frac{\sqrt{3}}{64}\int_0^{20} x^2\,dx = \frac{\sqrt{3}}{64}\left[\frac{1}{3}x^3\right]_0^{20} = \frac{8000\sqrt{3}}{64\cdot 3} = \frac{125\sqrt{3}}{3}\ \text{m}^3.$$

24. (a) By the symmetry of the problem, we consider only the solid to the right of the origin. The semicircular cross-sections perpendicular to the x-axis have radius $1 - x$, so $A(x) = \frac{1}{2}\pi(1-x)^2$. Now we can calculate

$V = 2\int_0^1 A(x)\,dx = 2\int_0^1 \frac{1}{2}\pi(1-x)^2\,dx = \int_0^1 \pi(1-x)^2\,dx = -\frac{\pi}{3}\left[(1-x)^3\right]_0^1 = \frac{\pi}{3}$.

(b) Cut the solid with a plane perpendicular to the x-axis and passing through the y-axis. Fold the half of the solid in the region $x \le 0$ under the xy-plane so that the point $(-1,0)$ comes around and touches the point $(1,0)$. The resulting solid is a right circular cone of radius 1 with vertex at $(x,y,z) = (1,0,0)$ and with its base in the yz-plane, centered at the origin. The volume of this cone is $\frac{1}{3}\pi r^2 h = \frac{1}{3}\pi\cdot 1^2\cdot 1 = \frac{\pi}{3}$.

25. $y = \frac{1}{6}(x^2+4)^{3/2} \;\Rightarrow\; dy/dx = \frac{1}{4}(x^2+4)^{1/2}(2x) \;\Rightarrow$

$1 + (dy/dx)^2 = 1 + \left[\frac{1}{2}x(x^2+4)^{1/2}\right]^2 = 1 + \frac{1}{4}x^2(x^2+4) = \frac{1}{4}x^4 + x^2 + 1 = \left(\frac{1}{2}x^2+1\right)^2$.

Thus, $L = \int_0^3 \sqrt{\left(\frac{1}{2}x^2+1\right)^2}\,dx = \int_0^3 \left(\frac{1}{2}x^2+1\right)dx = \left[\frac{1}{6}x^3 + x\right]_0^3 = \frac{15}{2}$.

26. $y = 2\ln(\sin\frac{1}{2}x) \Rightarrow \dfrac{dy}{dx} = 2\cdot\dfrac{1}{\sin(\frac{1}{2}x)}\cdot\cos(\frac{1}{2}x)\cdot\frac{1}{2} = \cot(\frac{1}{2}x) \Rightarrow 1+\left(\dfrac{dy}{dx}\right)^2 = 1+\cot^2(\frac{1}{2}x) = \csc^2(\frac{1}{2}x)$.

Thus,

$$L = \int_{\pi/3}^{\pi}\sqrt{\csc^2(\tfrac{1}{2}x)}\,dx = \int_{\pi/3}^{\pi}\left|\csc(\tfrac{1}{2}x)\right|dx = \int_{\pi/3}^{\pi}\csc(\tfrac{1}{2}x)\,dx = \int_{\pi/6}^{\pi/2}\csc u\,(2\,du) \qquad \left[u = \tfrac{1}{2}x,\, du = \tfrac{1}{2}\,dx\right]$$

$$= 2\Big[\ln|\csc u - \cot u|\Big]_{\pi/6}^{\pi/2} = 2\Big[\ln\left|\csc\tfrac{\pi}{2} - \cot\tfrac{\pi}{2}\right| - \ln\left|\csc\tfrac{\pi}{6} - \cot\tfrac{\pi}{6}\right|\Big]$$

$$= 2\Big[\ln|1-0| - \ln\big|2-\sqrt{3}\big|\Big] = -2\ln(2-\sqrt{3}) \approx 2.63$$

27. $y = e^{-x^2} \Rightarrow dy/dx = -2xe^{-x^2} \Rightarrow 1+(dy/dx)^2 = 1+4x^2e^{-2x^2}$. Let $f(x) = \sqrt{1+4x^2e^{-2x^2}}$. Then

$L = \int_0^3 f(x)\,dx \approx S_6 = \frac{(3-0)/6}{3}[f(0)+4f(0.5)+2f(1)+4f(1.5)+2f(2)+4f(2.5)+f(3)] \approx 3.292287$.

28. $y = \int_1^x \sqrt{\sqrt{t}-1}\,dt \Rightarrow dy/dx = \sqrt{\sqrt{x}-1} \Rightarrow 1+(dy/dx)^2 = 1+(\sqrt{x}-1) = \sqrt{x}$.

Thus, $L = \int_1^{16}\sqrt{\sqrt{x}}\,dx = \int_1^{16} x^{1/4}\,dx = \frac{4}{5}\left[x^{5/4}\right]_1^{16} = \frac{4}{5}(32-1) = \frac{124}{5}$.

29. $f(x) = kx \Rightarrow 30\text{ N} = k(15-12)\text{ cm} \Rightarrow k = 10\text{ N/cm} = 1000\text{ N/m}$. $20\text{ cm} - 12\text{ cm} = 0.08\text{ m} \Rightarrow$

$W = \int_0^{0.08} kx\,dx = 1000\int_0^{0.08} x\,dx = 500\left[x^2\right]_0^{0.08} = 500(0.08)^2 = 3.2\text{ N·m} = 3.2\text{ J}$.

30. The work needed to raise the elevator alone is 1600 lb × 30 ft = 48,000 ft-lb. The work needed to raise the bottom 170 ft of cable is 170 ft × 10 lb/ft × 30 ft = 51,000 ft-lb. The work needed to raise the top 30 ft of cable is $\int_0^{30} 10x\,dx = \left[5x^2\right]_0^{30} = 5\cdot 900 = 4500$ ft-lb. Adding these, we see that the total work needed is 48,000 + 51,000 + 4,500 = 103,500 ft-lb.

31. (a) The parabola has equation $y = ax^2$ with vertex at the origin and passing through $(4,4)$. $4 = a\cdot 4^2 \Rightarrow a = \frac{1}{4} \Rightarrow y = \frac{1}{4}x^2 \Rightarrow x^2 = 4y \Rightarrow x = 2\sqrt{y}$. Each circular disk has radius $2\sqrt{y}$ and is moved $4-y$ ft.

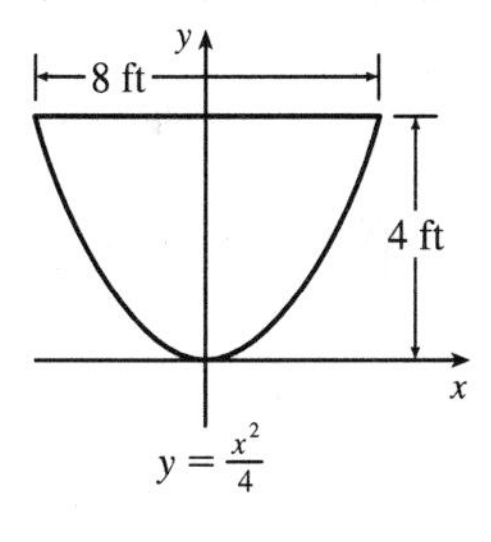

$$W = \int_0^4 \pi\left(2\sqrt{y}\right)^2 62.5\,(4-y)\,dy = 250\pi\int_0^4 y(4-y)\,dy$$

$$= 250\pi\left[2y^2 - \tfrac{1}{3}y^3\right]_0^4 = 250\pi\left(32-\tfrac{64}{3}\right) = \tfrac{8000\pi}{3} \approx 8378\text{ ft-lb}$$

(b) In part (a) we knew the final water level (0) but not the amount of work done. Here we use the same equation, except with the work fixed, and the lower limit of integration (that is, the final water level — call it h) unknown: $W = 4000 \Leftrightarrow$

$250\pi\left[2y^2 - \frac{1}{3}y^3\right]_h^4 = 4000 \Leftrightarrow \frac{16}{\pi} = \left[\left(32-\frac{64}{3}\right) - \left(2h^2 - \frac{1}{3}h^3\right)\right] \Leftrightarrow$

$h^3 - 6h^2 + 32 - \frac{48}{\pi} = 0$. We graph the function $f(h) = h^3 - 6h^2 + 32 - \frac{48}{\pi}$ on the interval $[0,4]$ to see where it is 0. From the graph, $f(h) = 0$ for $h \approx 2.1$. So the depth of water remaining is about 2.1 ft.

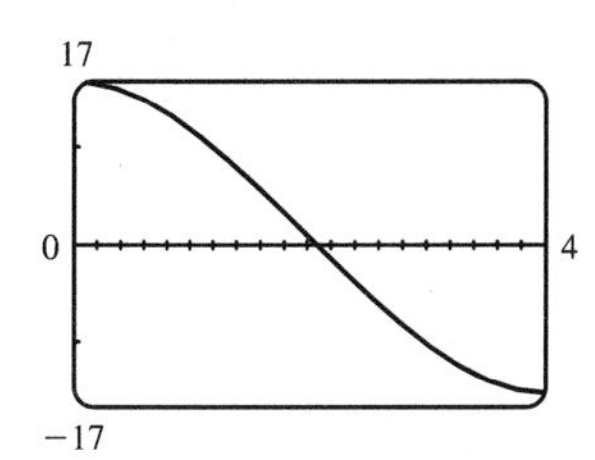

32. $F = \int_0^4 \delta(4-y)2(2\sqrt{y})\,dy = 4\delta\int_0^4 (4y^{1/2} - y^{3/2})\,dy$

$= 4\delta\left[\frac{8}{3}y^{3/2} - \frac{2}{5}y^{5/2}\right]_0^4$

$= 4\delta\left(\frac{64}{3} - \frac{64}{5}\right) = 256\delta\left(\frac{1}{3} - \frac{1}{5}\right)$

$= \frac{512}{15}\delta \approx 2133.3 \text{ lb} \qquad [\delta \approx 62.5 \text{ lb/ft}^3]$

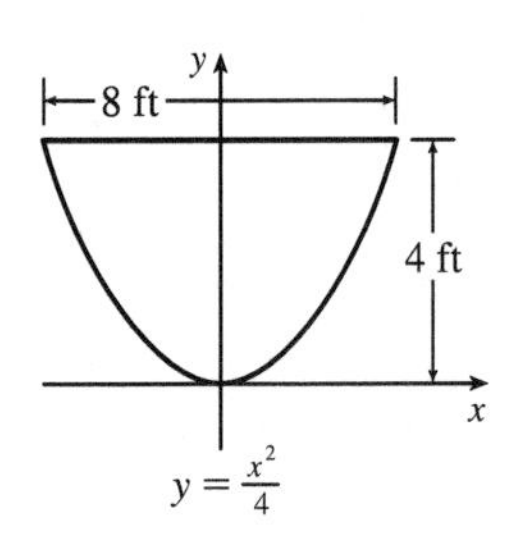

33. As in Example 5 of Section 7.5, $\dfrac{a}{2-x} = \dfrac{1}{2} \;\Rightarrow\; 2a = 2 - x$ and $w = 2(1.5 + a) = 3 + 2a = 3 + 2 - x = 5 - x$.

Thus, $F = \int_0^2 \rho g x(5-x)\,dx = \rho g\left[\frac{5}{2}x^2 - \frac{1}{3}x^3\right]_0^2 = \rho g\left(10 - \frac{8}{3}\right) = \frac{22}{3}\delta \;\; [\rho g = \delta] \;\; \approx \frac{22}{3}\cdot 62.5 \approx 458$ lb.

34. An equation of the line passing through $(0, 0)$ and $(3, 2)$ is $y = \frac{2}{3}x$. $A = \frac{1}{2}\cdot 3\cdot 2 = 3$. Therefore, using Equations 7.5.12, $\overline{x} = \frac{1}{3}\int_0^3 x\left(\frac{2}{3}x\right)dx = \frac{2}{27}\left[x^3\right]_0^3 = 2$ and $\overline{y} = \frac{1}{3}\int_0^3 \frac{1}{2}\left(\frac{2}{3}x\right)^2 dx = \frac{2}{81}\left[x^3\right]_0^3 = \frac{2}{3}$. Thus, the centroid is $(\overline{x}, \overline{y}) = \left(2, \frac{2}{3}\right)$.

35. $A = \int_{-2}^1 [(4 - x^2) - (x + 2)]\,dx = \int_{-2}^1 (2 - x - x^2)\,dx = \left[2x - \frac{1}{2}x^2 - \frac{1}{3}x^3\right]_{-2}^1$

$= \left(2 - \frac{1}{2} - \frac{1}{3}\right) - \left(-4 - 2 + \frac{8}{3}\right) = \frac{9}{2} \;\Rightarrow$

$\overline{x} = A^{-1}\int_{-2}^1 x(2 - x - x^2)\,dx = \frac{2}{9}\int_{-2}^1 (2x - x^2 - x^3)\,dx = \frac{2}{9}\left[x^2 - \frac{1}{3}x^3 - \frac{1}{4}x^4\right]_{-2}^1$

$= \frac{2}{9}\left[\left(1 - \frac{1}{3} - \frac{1}{4}\right) - \left(4 + \frac{8}{3} - 4\right)\right] = -\frac{1}{2}$

and $\quad \overline{y} = A^{-1}\int_{-2}^1 \frac{1}{2}\left[(4 - x^2)^2 - (x + 2)^2\right]dx = \frac{1}{9}\int_{-2}^1 (x^4 - 9x^2 - 4x + 12)\,dx$

$= \frac{1}{9}\left[\frac{1}{5}x^5 - 3x^3 - 2x^2 + 12x\right]_{-2}^1 = \frac{1}{9}\left[\left(\frac{1}{5} - 3 - 2 + 12\right) - \left(-\frac{32}{5} + 24 - 8 - 24\right)\right] = \frac{12}{5}$

Thus, the centroid is $(\overline{x}, \overline{y}) = \left(-\frac{1}{2}, \frac{12}{5}\right)$.

36. From the symmetry of the region, $\overline{x} = \frac{\pi}{2}$. $\quad A = \int_{\pi/4}^{3\pi/4} \sin x\,dx = \left[-\cos x\right]_{\pi/4}^{3\pi/4} = \frac{1}{\sqrt{2}} - \left(-\frac{1}{\sqrt{2}}\right) = \sqrt{2}$

$\overline{y} = \frac{1}{A}\int_{\pi/4}^{3\pi/4} \frac{1}{2}\sin^2 x\,dx = \frac{1}{A}\int_{\pi/4}^{3\pi/4} \frac{1}{4}(1 - \cos 2x)\,dx$

$= \frac{1}{4\sqrt{2}}\left[x - \frac{1}{2}\sin 2x\right]_{\pi/4}^{3\pi/4}$

$= \frac{1}{4\sqrt{2}}\left[\frac{3\pi}{4} - \frac{1}{2}(-1) - \frac{\pi}{4} + \frac{1}{2}\cdot 1\right] = \frac{1}{4\sqrt{2}}\left(\frac{\pi}{2} + 1\right)$

Thus, the centroid is $(\overline{x}, \overline{y}) = \left(\frac{\pi}{2}, \frac{1}{4\sqrt{2}}\left(\frac{\pi}{2} + 1\right)\right) \approx (1.57, 0.45)$.

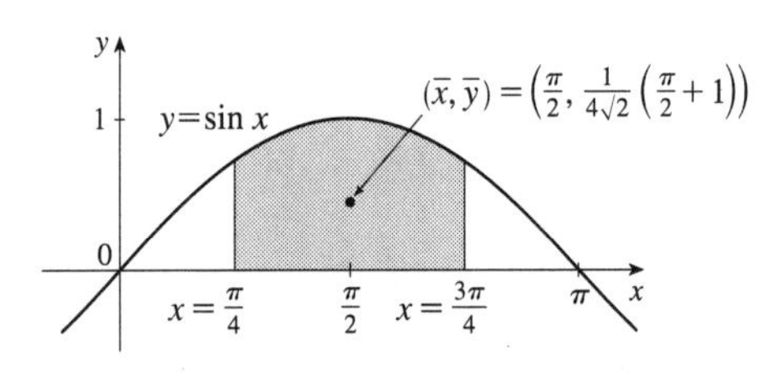

37. The centroid of this circle, $(1, 0)$, travels a distance $2\pi(1)$ when the lamina is rotated about the y-axis. The area of the circle is $\pi(1)^2$. So by the Theorem of Pappus, $V = A(2\pi\overline{x}) = \pi(1)^2 2\pi(1) = 2\pi^2$.

38. The semicircular region has an area of $\frac{1}{2}\pi r^2$, and sweeps out a sphere of radius r when rotated about the x-axis. $\overline{x} = 0$ because of symmetry about the line $x = 0$. And by the Theorem of Pappus, $V = A(2\pi\overline{y}) \;\Rightarrow\; \frac{4}{3}\pi r^3 = \frac{1}{2}\pi r^2(2\pi\overline{y}) \;\Rightarrow$ $\overline{y} = \frac{4}{3\pi}r$. Thus, the centroid is $(\overline{x}, \overline{y}) = \left(0, \frac{4}{3\pi}r\right)$.

39. $(3y^2 + 2y)\,y' = x\cos x \;\Rightarrow\; (3y^2 + 2y)\,dy = (x\cos x)\,dx \;\Rightarrow\; \int(3y^2 + 2y)\,dy = \int(x\cos x)\,dx \;\Rightarrow$

$y^3 + y^2 = \cos x + x\sin x + C$. For the last step, use integration by parts or Formula 83 in the Table of Integrals.

40. $\dfrac{dx}{dt} = 1 - t + x - tx = 1(1-t) + x(1-t) = (1+x)(1-t) \;\Rightarrow\; \dfrac{dx}{1+x} = (1-t)\,dt \;\Rightarrow$

$\displaystyle\int \frac{dx}{1+x} = \int (1-t)\,dt \;\Rightarrow\; \ln|1+x| = t - \tfrac{1}{2}t^2 + C \;\Rightarrow\; |1+x| = e^{t-t^2/2+C} \;\Rightarrow\; 1 + x = \pm e^{t-t^2/2}\cdot e^C \;\Rightarrow$

$x = -1 + Ke^{t-t^2/2}$, where K is any nonzero constant.

41. $\dfrac{dr}{dt} + 2tr = r \;\Rightarrow\; \dfrac{dr}{dt} = r - 2tr = r(1-2t) \;\Rightarrow\; \displaystyle\int \frac{dr}{r} = \int (1-2t)\,dt \;\Rightarrow\; \ln|r| = t - t^2 + C \;\Rightarrow$

$|r| = e^{t-t^2+C} = ke^{t-t^2}$. Since $r(0) = 5$, $5 = ke^0 = k$. Thus, $r(t) = 5e^{t-t^2}$.

42. $(1+\cos x)y' = (1+e^{-y})\sin x \;\Rightarrow\; \dfrac{dy}{1+e^{-y}} = \dfrac{\sin x\,dx}{1+\cos x} \;\Rightarrow\; \displaystyle\int \frac{dy}{1+1/e^y} = \int \frac{\sin x\,dx}{1+\cos x} \;\Rightarrow$

$\displaystyle\int \frac{e^y\,dy}{1+e^y} = \int \frac{\sin x\,dx}{1+\cos x} \;\Rightarrow\; \ln|1+e^y| = -\ln|1+\cos x| + C \;\Rightarrow\; \ln(1+e^y) = -\ln(1+\cos x) + C \;\Rightarrow$

$1 + e^y = e^{-\ln(1+\cos x)}\cdot e^C \;\Rightarrow\; e^y = ke^{-\ln(1+\cos x)} - 1 \;\Rightarrow\; y = \ln[ke^{-\ln(1+\cos x)} - 1]$. Since $y(0) = 0$,

$0 = \ln[ke^{-\ln 2} - 1] \;\Rightarrow\; e^0 = k(\tfrac{1}{2}) - 1 \;\Rightarrow\; k = 4$. Thus, $y(x) = \ln[4e^{-\ln(1+\cos x)} - 1]$. An equavalent form

is $y(x) = \ln\dfrac{3-\cos x}{1+\cos x}$.

43. (a)

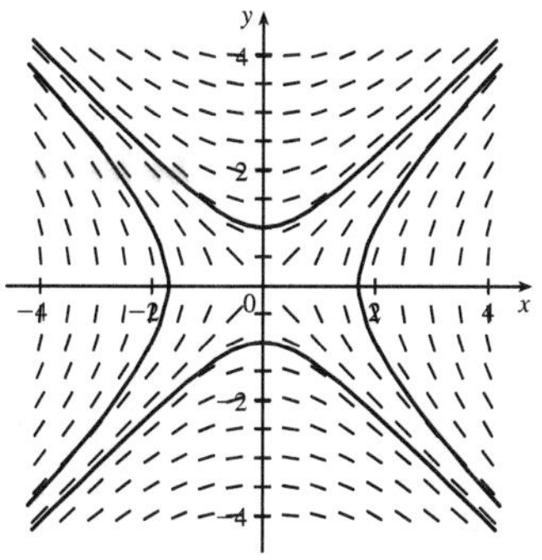

We sketch the direction field and four solution curves, as shown. Note that the slope $y' = x/y$ is not defined on the line $y = 0$.

(b) $y' = x/y \;\Leftrightarrow\; y\,dy = x\,dx \;\Leftrightarrow\; y^2 = x^2 + C$. For $C = 0$, this is the pair of lines $y = \pm x$. For $C \neq 0$, it is the hyperbola $x^2 - y^2 = -C$.

44. (a) $\mathscr{R}_1$ is the region below the graph of $y = x^2$ and above the x-axis between $x = 0$ and $x = b$, and $\mathscr{R}_2$ is the region to the left of the graph of $x = \sqrt{y}$ and to the right of the y-axis between $y = 0$ and $y = b^2$. So the area of $\mathscr{R}_1$ is

$A_1 = \int_0^b x^2\,dx = \left[\tfrac{1}{3}x^3\right]_0^b = \tfrac{1}{3}b^3$, and the area of $\mathscr{R}_2$ is $A_2 = \int_0^{b^2} \sqrt{y}\,dy = \left[\tfrac{2}{3}y^{3/2}\right]_0^{b^2} = \tfrac{2}{3}b^3$. So there is no solution to

$A_1 = A_2$ for $b \neq 0$.

(b) Using disks, we calculate the volume of rotation of $\mathcal{R}_1$ about the x-axis to be $V_{1,x} = \pi \int_0^b \left(x^2\right)^2 dx = \frac{1}{5}\pi b^5$.

Using cylindrical shells, we calculate the volume of rotation of $\mathcal{R}_1$ about the y-axis to be

$V_{1,y} = 2\pi \int_0^b x\left(x^2\right) dx = 2\pi\left[\frac{1}{4}x^4\right]_0^b = \frac{1}{2}\pi b^4$. So $V_{1,x} = V_{1,y} \iff \frac{1}{5}\pi b^5 = \frac{1}{2}\pi b^4 \iff 2b = 5 \iff b = \frac{5}{2}$.

So the volumes of rotation about the x- and y-axes are the same for $b = \frac{5}{2}$.

(c) We use cylindrical shells to calculate the volume of rotation of $\mathcal{R}_2$ about the x-axis:

$\mathcal{R}_{2,x} = 2\pi \int_0^{b^2} y\left(\sqrt{y}\right) dy = 2\pi\left[\frac{2}{5}y^{5/2}\right]_0^{b^2} = \frac{4}{5}\pi b^5$. We already know the volume of rotation of $\mathcal{R}_1$ about the x-axis from part (b), and $\mathcal{R}_{1,x} = \mathcal{R}_{2,x} \iff \frac{1}{5}\pi b^5 = \frac{4}{5}\pi b^5$, which has no solution for $b \neq 0$.

(d) We use disks to calculate the volume of rotation of $\mathcal{R}_2$ about the y-axis: $\mathcal{R}_{2,y} = \pi \int_0^{b^2} \left(\sqrt{y}\right)^2 dy = \pi\left[\frac{1}{2}y^2\right]_0^{b^2} = \frac{1}{2}\pi b^4$.

We know the volume of rotation of $\mathcal{R}_1$ about the y-axis from part (b), and $\mathcal{R}_{1,y} = \mathcal{R}_{2,y} \iff \frac{1}{2}\pi b^4 = \frac{1}{2}\pi b^4$. But this equation is true for all b, so the volumes of rotation about the y-axis are equal for all values of b.

8 □ SERIES

8.1 Sequences

1. (a) A sequence is an ordered list of numbers. It can also be defined as a function whose domain is the set of positive integers.

(b) The terms a_n approach 8 as n becomes large. In fact, we can make a_n as close to 8 as we like by taking n sufficiently large.

(c) The terms a_n become large as n becomes large. In fact, we can make a_n as large as we like by taking n sufficiently large.

2. (a) From Definition 1, a convergent sequence is a sequence for which $\lim\limits_{n\to\infty} a_n$ exists. Examples: $\{1/n\}$, $\{1/2^n\}$

(b) A divergent sequence is a sequence for which $\lim\limits_{n\to\infty} a_n$ *does not* exist. Examples: $\{n\}$, $\{\sin n\}$

3. The first six terms of $a_n = \dfrac{n}{2n+1}$ are $\dfrac{1}{3}, \dfrac{2}{5}, \dfrac{3}{7}, \dfrac{4}{9}, \dfrac{5}{11}, \dfrac{6}{13}$. It appears that the sequence is approaching $\dfrac{1}{2}$.

$$\lim_{n\to\infty} \frac{n}{2n+1} = \lim_{n\to\infty} \frac{1}{2+1/n} = \frac{1}{2}$$

4. $\{\cos(n\pi/3)\}_{n=1}^{9} = \left\{\frac{1}{2}, -\frac{1}{2}, -1, -\frac{1}{2}, \frac{1}{2}, 1, \frac{1}{2}, -\frac{1}{2}, -1\right\}$. The sequence does not appear to have a limit. The values will cycle through the first six numbers in the sequence—never approaching a particular number.

5. $\left\{1, -\frac{2}{3}, \frac{4}{9}, -\frac{8}{27}, \ldots\right\}$. Each term is $-\frac{2}{3}$ times the preceding one, so $a_n = \left(-\frac{2}{3}\right)^{n-1}$.

6. $\left\{-\frac{1}{4}, \frac{2}{9}, -\frac{3}{16}, \frac{4}{25}, \ldots\right\}$. The numerator of the nth term is n and its denominator is $(n+1)^2$. Including the alternating signs, we get $a_n = (-1)^n \dfrac{n}{(n+1)^2}$.

7. $\{2, 7, 12, 17, \ldots\}$. Each term is larger than the preceding one by 5, so $a_n = a_1 + d(n-1) = 2 + 5(n-1) = 5n - 3$.

8. $\{5, 1, 5, 1, 5, 1, \ldots\}$. The average of 5 and 1 is 3, so we can think of the sequence as alternately adding 2 and -2 to 3. Thus, $a_n = 3 + (-1)^{n+1} \cdot 2$.

9. $a_n = \dfrac{3+5n^2}{n+n^2} = \dfrac{(3+5n^2)/n^2}{(n+n^2)/n^2} = \dfrac{5+3/n^2}{1+1/n}$, so $a_n \to \dfrac{5+0}{1+0} = 5$ as $n \to \infty$. Converges

10. $a_n = \dfrac{n+1}{3n-1} = \dfrac{1+1/n}{3-1/n}$, so $a_n \to \dfrac{1+0}{3-0} = \dfrac{1}{3}$ as $n \to \infty$. Converges

11. $a_n = \dfrac{2^n}{3^{n+1}} = \dfrac{1}{3}\left(\dfrac{2}{3}\right)^n$, so $\lim\limits_{n\to\infty} a_n = \frac{1}{3} \lim\limits_{n\to\infty} \left(\frac{2}{3}\right)^n = \frac{1}{3} \cdot 0 = 0$ by (8) with $r = \frac{2}{3}$. Converges

12. $a_n = \dfrac{\sqrt{n}}{1+\sqrt{n}} = \dfrac{1}{1/\sqrt{n}+1}$, so $a_n \to \dfrac{1}{0+1} = 1$ as $n \to \infty$. Converges

13. $a_n = \dfrac{(n+2)!}{n!} = \dfrac{(n+2)(n+1)n!}{n!} = (n+2)(n+1)$, so $a_n \to \infty$ as $n \to \infty$ and the sequence diverges.

14. $a_n = \dfrac{n}{1+\sqrt{n}} = \dfrac{\sqrt{n}}{1/\sqrt{n}+1}$. The numerator approaches ∞ and the denominator approaches $0 + 1 = 1$ as $n \to \infty$, so $a_n \to \infty$ as $n \to \infty$ and the sequence diverges.

15. $a_n = \dfrac{(-1)^{n-1}n}{n^2+1} = \dfrac{(-1)^{n-1}}{n+1/n}$, so $0 \le |a_n| = \dfrac{1}{n+1/n} \le \dfrac{1}{n} \to 0$ as $n \to \infty$, so $a_n \to 0$ by the Squeeze Theorem and Theorem 6. Converges

16. $a_n = \dfrac{(-1)^n n^3}{n^3+2n^2+1}$. Now $|a_n| = \dfrac{n^3}{n^3+2n^2+1} = \dfrac{1}{1+\frac{2}{n}+\frac{1}{n^3}} \to 1$ as $n \to \infty$, but the terms of the sequence $\{a_n\}$ alternate in sign, so the sequence $a_1, a_3, a_5, \ldots$ converges to -1 and the sequence $a_2, a_4, a_6, \ldots$ converges to $+1$. This shows that the given sequence diverges since its terms don't approach a single real number.

17. $a_n = \dfrac{e^n + e^{-n}}{e^{2n}-1} \cdot \dfrac{e^{-n}}{e^{-n}} = \dfrac{1+e^{-2n}}{e^n - e^{-n}} \to \dfrac{1+0}{e^n - 0} \to 0$ as $n \to \infty$. Converges

18. $a_n = \cos(2/n)$. As $n \to \infty$, $2/n \to 0$, so $\cos(2/n) \to \cos 0 = 1$. Converges

19. $a_n = n^2 e^{-n} = \dfrac{n^2}{e^n}$. Since $\lim\limits_{x\to\infty} \dfrac{x^2}{e^x} \overset{H}{=} \lim\limits_{x\to\infty} \dfrac{2x}{e^x} \overset{H}{=} \lim\limits_{x\to\infty} \dfrac{2}{e^x} = 0$, it follows from Theorem 3 that $\lim\limits_{n\to\infty} a_n = 0$. Converges

20. $2n \to \infty$ as $n \to \infty$, so since $\lim\limits_{x\to\infty} \arctan x = \frac{\pi}{2}$, we have $\lim\limits_{n\to\infty} \arctan 2n = \frac{\pi}{2}$. Converges

21. $0 \le \dfrac{\cos^2 n}{2^n} \le \dfrac{1}{2^n}$ [since $0 \le \cos^2 n \le 1$], so since $\lim\limits_{n\to\infty} \dfrac{1}{2^n} = 0$, $\left\{\dfrac{\cos^2 n}{2^n}\right\}$ converges to 0 by the Squeeze Theorem.

22. $a_n = n\cos n\pi = n(-1)^n$. Since $|a_n| = n \to \infty$ as $n \to \infty$, the given sequence diverges.

23. $y = \left(1+\dfrac{2}{x}\right)^x \Rightarrow \ln y = x\ln\left(1+\dfrac{2}{x}\right)$, so

$$\lim_{x\to\infty} \ln y = \lim_{x\to\infty} \frac{\ln(1+2/x)}{1/x} \overset{H}{=} \lim_{x\to\infty} \frac{\left(\dfrac{1}{1+2/x}\right)\left(-\dfrac{2}{x^2}\right)}{-1/x^2} = \lim_{x\to\infty} \frac{2}{1+2/x} = 2 \Rightarrow$$

$\lim\limits_{x\to\infty} \left(1+\dfrac{2}{x}\right)^x = \lim\limits_{x\to\infty} e^{\ln y} = e^2$, so by Theorem 2, $\lim\limits_{n\to\infty} \left(1+\dfrac{2}{n}\right)^n = e^2$. Convergent

24. $a_n = \dfrac{\sin 2n}{1+\sqrt{n}}$. $|a_n| \le \dfrac{1}{1+\sqrt{n}}$ and $\lim\limits_{n\to\infty} \dfrac{1}{1+\sqrt{n}} = 0$, so $\dfrac{-1}{1+\sqrt{n}} \le a_n \le \dfrac{1}{1+\sqrt{n}} \Rightarrow \lim\limits_{n\to\infty} a_n = 0$ by the Squeeze Theorem. Converges

25. $\{0, 1, 0, 0, 1, 0, 0, 0, 1, \ldots\}$ diverges since the sequence takes on only two values, 0 and 1, and never stays arbitrarily close to either one (or any other value) for n sufficiently large.

26. $\lim\limits_{x\to\infty} \dfrac{(\ln x)^2}{x} \overset{H}{=} \lim\limits_{x\to\infty} \dfrac{2(\ln x)(1/x)}{1} = 2\lim\limits_{x\to\infty} \dfrac{\ln x}{x} \overset{H}{=} 2\lim\limits_{x\to\infty} \dfrac{1/x}{1} = 0$, so by Theorem 3, $\lim\limits_{n\to\infty} \dfrac{(\ln n)^2}{n} = 0$. Convergent

27. $a_n = \ln(2n^2+1) - \ln(n^2+1) = \ln\left(\dfrac{2n^2+1}{n^2+1}\right) = \ln\left(\dfrac{2+1/n^2}{1+1/n^2}\right) \to \ln 2$ as $n \to \infty$. Convergent

28. $0 < |a_n| = \dfrac{3^n}{n!} = \dfrac{3}{1}\cdot\dfrac{3}{2}\cdot\dfrac{3}{3}\cdot\cdots\cdot\dfrac{3}{(n-1)}\cdot\dfrac{3}{n} \le \dfrac{3}{1}\cdot\dfrac{3}{2}\cdot\dfrac{3}{n}$ [for $n > 2$] $= \dfrac{27}{2n} \to 0$ as $n \to \infty$, so by the Squeeze Theorem and Theorem 6, $\{(-3)^n/n\}$ converges to 0.

29. (a) $a_n = 1000(1.06)^n \Rightarrow a_1 = 1060$, $a_2 = 1123.60$, $a_3 = 1191.02$, $a_4 = 1262.48$, and $a_5 = 1338.23$.

(b) $\lim\limits_{n\to\infty} a_n = 1000 \lim\limits_{n\to\infty} (1.06)^n$, so the sequence diverges by (8) with $r = 1.06 > 1$.

30. $a_{n+1} = \begin{cases} \frac{1}{2}a_n & \text{if } a_n \text{ is an even number} \\ 3a_n + 1 & \text{if } a_n \text{ is an odd number} \end{cases}$ When $a_1 = 11$, the first 40 terms are 11, 34, 17, 52, 26, 13, 40, 20, 10, 5, 16, 8, 4, 2, 1, 4, 2, 1, 4, 2, 1, 4, 2, 1, 4, 2, 1, 4, 2, 1, 4, 2, 1, 4, 2, 1, 4, 2, 1, 4. When $a_1 = 25$, the first 40 terms are 25, 76, 38, 19, 58, 29, 88, 44, 22, 11, 34, 17, 52, 26, 13, 40, 20, 10, 5, 16, 8, 4, 2, 1, 4, 2, 1, 4, 2, 1, 4, 2, 1, 4, 2, 1, 4, 2, 1, 4. The famous Collatz conjecture is that this sequence always reaches 1, regardless of the starting point a_1.

31. Since $\{a_n\}$ is a decreasing sequence, $a_n > a_{n+1}$ for all $n \geq 1$. Because all of its terms lie between 5 and 8, $\{a_n\}$ is a bounded sequence. By the Monotonic Sequence Theorem, $\{a_n\}$ is convergent; that is, $\{a_n\}$ has a limit L. L must be less than 8 since $\{a_n\}$ is decreasing, so $5 \leq L < 8$.

32. (a) Let $\lim\limits_{n\to\infty} a_n = L$. By Definition 2, this means that for every $\varepsilon > 0$ there is an integer N such that $|a_n - L| < \varepsilon$ whenever $n > N$. Thus, $|a_{n+1} - L| < \varepsilon$ whenever $n + 1 > N \quad \Leftrightarrow \quad n > N - 1$. It follows that $\lim\limits_{n\to\infty} a_{n+1} = L$ and so $\lim\limits_{n\to\infty} a_n = \lim\limits_{n\to\infty} a_{n+1}$.

(b) If $L = \lim\limits_{n\to\infty} a_n$ then $\lim\limits_{n\to\infty} a_{n+1} = L$ also, so L must satisfy $L = 1/(1+L) \quad \Rightarrow \quad L^2 + L - 1 = 0 \quad \Rightarrow \quad L = \frac{-1+\sqrt{5}}{2}$ (since L has to be nonnegative if it exists).

33. $a_n = \dfrac{1}{2n+3}$ is decreasing since $a_{n+1} = \dfrac{1}{2(n+1)+3} = \dfrac{1}{2n+5} < \dfrac{1}{2n+3} = a_n$ for each $n \geq 1$. The sequence is bounded since $0 < a_n \leq \frac{1}{5}$ for all $n \geq 1$. Note that $a_1 = \frac{1}{5}$.

34. $a_n = \dfrac{2n-3}{3n+4}$ defines an increasing sequence since for $f(x) = \dfrac{2x-3}{3x+4}$,

$f'(x) = \dfrac{(3x+4)(2) - (2x-3)(3)}{(3x+4)^2} = \dfrac{17}{(3x+4)^2} > 0$. The sequence is bounded since $a_n \geq a_1 = -\frac{1}{7}$ for $n \geq 1$, and

$a_n < \dfrac{2n-3}{3n} < \dfrac{2n}{3n} = \dfrac{2}{3}$ for $n \geq 1$.

35. $a_n = \cos(n\pi/2)$ is not monotonic. The first few terms are $0, -1, 0, 1, 0, -1, 0, 1, \ldots$. In fact, the sequence consists of the terms $0, -1, 0, 1$ repeated over and over again in that order. The sequence is bounded since $|a_n| \leq 1$ for all $n \geq 1$.

36. $a_n = n + \dfrac{1}{n}$ defines an increasing sequence since the function $g(x) = x + \dfrac{1}{x}$ is increasing for $x > 1$. [$g'(x) = 1 - 1/x^2 > 0$ for $x > 1$.] The sequence is unbounded since $a_n \to \infty$ as $n \to \infty$. (It is, however, bounded below by $a_1 = 2$.)

37. $a_1 = 2^{1/2}$, $a_2 = 2^{3/4}$, $a_3 = 2^{7/8}$, ..., so $a_n = 2^{(2^n-1)/2^n} = 2^{1-(1/2^n)}$. $\lim\limits_{n\to\infty} a_n = \lim\limits_{n\to\infty} 2^{1-(1/2^n)} = 2^1 = 2$.

Alternate solution: Let $L = \lim\limits_{n\to\infty} a_n$. (We could show the limit exists by showing that $\{a_n\}$ is bounded and increasing.) Then L must satisfy $L = \sqrt{2 \cdot L} \quad \Rightarrow \quad L^2 = 2L \quad \Rightarrow \quad L(L-2) = 0$. $L \neq 0$ since the sequence increases, so $L = 2$.

38. (a) Let P_n be the statement that $a_{n+1} \geq a_n$ and $a_n \leq 3$. P_1 is obviously true. We will assume that P_n is true and then show that as a consequence P_{n+1} must also be true. $a_{n+2} \geq a_{n+1} \quad \Leftrightarrow \quad \sqrt{2 + a_{n+1}} \geq \sqrt{2 + a_n} \quad \Leftrightarrow$ $2 + a_{n+1} \geq 2 + a_n \quad \Leftrightarrow \quad a_{n+1} \geq a_n$, which is the induction hypothesis. $a_{n+1} \leq 3 \quad \Leftrightarrow \quad \sqrt{2 + a_n} \leq 3 \quad \Leftrightarrow$ $2 + a_n \leq 9 \quad \Leftrightarrow \quad a_n \leq 7$, which is certainly true because we are assuming that $a_n \leq 3$. So P_n is true for all n, and so $a_1 \leq a_n \leq 3$ (showing that the sequence is bounded), and hence by the Monotonic Sequence Theorem, $\lim\limits_{n\to\infty} a_n$ exists.

(b) If $L = \lim\limits_{n\to\infty} a_n$, then $\lim\limits_{n\to\infty} a_{n+1} = L$ also, so $L = \sqrt{2 + L} \quad \Rightarrow \quad L^2 = 2 + L \quad \Leftrightarrow \quad L^2 - L - 2 = 0 \quad \Leftrightarrow$ $(L+1)(L-2) = 0 \quad \Leftrightarrow \quad L = 2$ (since L can't be negative).

39. We show by induction that $\{a_n\}$ is increasing and bounded above by 3. Let P_n be the proposition that $a_{n+1} > a_n$ and $0 < a_n < 3$. Clearly P_1 is true. Assume that P_n is true.

Then $a_{n+1} > a_n \;\Rightarrow\; \dfrac{1}{a_{n+1}} < \dfrac{1}{a_n} \;\Rightarrow\; -\dfrac{1}{a_{n+1}} > -\dfrac{1}{a_n}$. Now $a_{n+2} = 3 - \dfrac{1}{a_{n+1}} > 3 - \dfrac{1}{a_n} = a_{n+1} \;\Leftrightarrow\; P_{n+1}$.

This proves that $\{a_n\}$ is increasing and bounded above by 3, so $1 = a_1 < a_n < 3$, that is, $\{a_n\}$ is bounded, and hence convergent by the Monotonic Sequence Theorem. If $L = \lim_{n\to\infty} a_n$, then $\lim_{n\to\infty} a_{n+1} = L$ also, so L must satisfy

$L = 3 - 1/L \;\Rightarrow\; L^2 - 3L + 1 = 0 \;\Rightarrow\; L = \frac{3 \pm \sqrt{5}}{2}$. But $L > 1$, so $L = \frac{3 + \sqrt{5}}{2}$.

40. We use induction. Let P_n be the statement that $0 < a_{n+1} \le a_n \le 2$. Clearly P_1 is true, since $a_2 = 1/(3-2) = 1$. Now assume that P_n is true. Then $a_{n+1} \le a_n \;\Rightarrow\; -a_{n+1} \ge -a_n \;\Rightarrow\; 3 - a_{n+1} \ge 3 - a_n \;\Rightarrow$

$a_{n+2} = \dfrac{1}{3 - a_{n+1}} \le \dfrac{1}{3 - a_n} = a_{n+1}$. Also $a_{n+2} > 0$ (since $3 - a_{n+1}$ is positive) and $a_{n+1} \le 2$ by the induction hypothesis, so P_{n+1} is true. To find the limit, we use the fact that $\lim_{n\to\infty} a_n = \lim_{n\to\infty} a_{n+1} \;\Rightarrow\; L = \frac{1}{3 - L} \;\Rightarrow$

$L^2 - 3L + 1 = 0 \;\Rightarrow\; L = \frac{3 \pm \sqrt{5}}{2}$. But $L \le 2$, so we must have $L = \frac{3 - \sqrt{5}}{2}$.

41. (a) Let a_n be the number of rabbit pairs in the nth month. Clearly $a_1 = 1 = a_2$. In the nth month, each pair that is 2 or more months old (that is, a_{n-2} pairs) will produce a new pair to add to the a_{n-1} pairs already present. Thus, $a_n = a_{n-1} + a_{n-2}$, so that $\{a_n\} = \{f_n\}$, the Fibonacci sequence.

(b) $a_n = \dfrac{f_{n+1}}{f_n} \;\Rightarrow\; a_{n-1} = \dfrac{f_n}{f_{n-1}} = \dfrac{f_{n-1} + f_{n-2}}{f_{n-1}} = 1 + \dfrac{f_{n-2}}{f_{n-1}} = 1 + \dfrac{1}{f_{n-1}/f_{n-2}} = 1 + \dfrac{1}{a_{n-2}}$. If $L = \lim_{n\to\infty} a_n$, then $L = \lim_{n\to\infty} a_{n-1}$ and $L = \lim_{n\to\infty} a_{n-2}$, so L must satisfy $L = 1 + \dfrac{1}{L} \;\Rightarrow\; L^2 - L - 1 = 0 \;\Rightarrow\; L = \frac{1 + \sqrt{5}}{2}$ (since L must be positive).

42. (a) If f is continuous, then $f(L) = f\left(\lim_{n\to\infty} a_n\right) = \lim_{n\to\infty} f(a_n) = \lim_{n\to\infty} a_{n+1} = L$ by Exercise 38(a).

(b) By repeatedly pressing the cosine key on the calculator (that is, taking cosine of the previous answer) until the displayed value stabilizes, we see that $L \approx 0.73909$.

43. $(0.8)^n < 0.000001 \;\Rightarrow\; \ln(0.8)^n < \ln(0.000001) \;\Rightarrow\; n\ln(0.8) < \ln(0.000001) \;\Rightarrow\; n > \dfrac{\ln(0.000001)}{\ln(0.8)} \;\Rightarrow$

$n > 61.9$, so n must be at least 62 to satisfy the given inequality.

44. Let $\varepsilon > 0$ and let N be any positive integer larger than $\ln(\varepsilon)/\ln|r|$. If $n > N$ then $n > \ln(\varepsilon)/\ln|r| \;\Rightarrow$ $n\ln|r| < \ln\varepsilon$ [since $|r| < 1 \;\Rightarrow\; \ln|r| < 0$] $\Rightarrow\; \ln(|r|^n) < \ln\varepsilon \;\Rightarrow\; |r|^n < \varepsilon \;\Rightarrow\; |r^n - 0| < \varepsilon$, and so by Definition 2, $\lim_{n\to\infty} r^n = 0$.

45. If $\lim_{n\to\infty} |a_n| = 0$ then $\lim_{n\to\infty} -|a_n| = 0$, and since $-|a_n| \le a_n \le |a_n|$, we have that $\lim_{n\to\infty} a_n = 0$ by the Squeeze Theorem.

46. (a) Let $\varepsilon > 0$. Since $\lim_{n\to\infty} a_{2n} = L$, there exists N_1 such that $|a_{2n} - L| < \varepsilon$ for $n > N_1$. Since $\lim_{n\to\infty} a_{2n+1} = L$, there exists N_2 such that $|a_{2n+1} - L| < \varepsilon$ for $n > N_2$. Let $N = \max\{2N_1, 2N_2 + 1\}$ and let $n > N$. If n is even, then $n = 2m$ where $m > N_1$, so $|a_n - L| = |a_{2m} - L| < \varepsilon$. If n is odd, then $n = 2m + 1$, where $m > N_2$, so $|a_n - L| = |a_{2m+1} - L| < \varepsilon$. Therefore $\lim_{n\to\infty} a_n = L$.

(b) $a_1 = 1$, $a_2 = 1 + \frac{1}{1+1} = \frac{3}{2} = 1.5$, $a_3 = 1 + \frac{1}{5/2} = \frac{7}{5} = 1.4$, $a_4 = 1 + \frac{1}{12/5} = \frac{17}{12} = 1.41\overline{6}$,

$a_5 = 1 + \frac{1}{29/12} = \frac{41}{29} \approx 1.413793$, $a_6 = 1 + \frac{1}{70/29} = \frac{99}{70} \approx 1.414286$, $a_7 = 1 + \frac{1}{169/70} = \frac{239}{169} \approx 1.414201$,

$a_8 = 1 + \frac{1}{408/169} = \frac{577}{408} \approx 1.414216$. Notice that $a_1 < a_3 < a_5 < a_7$ and $a_2 > a_4 > a_6 > a_8$. It appears that the odd terms are increasing and the even terms are decreasing. Let's prove that $a_{2n-2} > a_{2n}$ and $a_{2n-1} < a_{2n+1}$ by mathematical induction.

Suppose that $a_{2k-2} > a_{2k}$. Then $1 + a_{2k-2} > 1 + a_{2k} \Rightarrow \dfrac{1}{1+a_{2k-2}} < \dfrac{1}{1+a_{2k}} \Rightarrow$

$1 + \dfrac{1}{1+a_{2k-2}} < 1 + \dfrac{1}{1+a_{2k}} \Rightarrow a_{2k-1} < a_{2k+1} \Rightarrow 1 + a_{2k-1} < 1 + a_{2k+1} \Rightarrow$

$\dfrac{1}{1+a_{2k-1}} > \dfrac{1}{1+a_{2k+1}} \Rightarrow 1 + \dfrac{1}{1+a_{2k-1}} > 1 + \dfrac{1}{1+a_{2k+1}} \Rightarrow a_{2k} > a_{2k+2}$. We have thus shown, by induction, that the odd terms are increasing and the even terms are decreasing. Also all terms lie between 1 and 2, so both $\{a_n\}$ and $\{b_n\}$ are bounded monotonic sequences and are therefore convergent by the Monotonic Sequence Theorem.

Let $\lim\limits_{n\to\infty} a_{2n} = L$. Then $\lim\limits_{n\to\infty} a_{2n+2} = L$ also. We have

$a_{n+2} = 1 + \dfrac{1}{1 + 1 + 1/(1+a_n)} = 1 + \dfrac{1}{(3+2a_n)/(1+a_n)} = \dfrac{4+3a_n}{3+2a_n}$, so $a_{2n+2} = \dfrac{4+3a_{2n}}{3+2a_{2n}}$. Taking limits of both sides, we get $L = \dfrac{4+3L}{3+2L} \Rightarrow 3L + 2L^2 = 4 + 3L \Rightarrow L^2 = 2 \Rightarrow L = \sqrt{2}$ [since $L > 0$].

Thus, $\lim\limits_{n\to\infty} a_{2n} = \sqrt{2}$. Similarly we find that $\lim\limits_{n\to\infty} a_{2n+1} = \sqrt{2}$. So, by part (a), $\lim\limits_{n\to\infty} a_n = \sqrt{2}$.

47. (a) Suppose $\{p_n\}$ converges to p. Then $p_{n+1} = \dfrac{bp_n}{a+p_n} \Rightarrow \lim\limits_{n\to\infty} p_{n+1} = \dfrac{b \lim\limits_{n\to\infty} p_n}{a + \lim\limits_{n\to\infty} p_n} \Rightarrow p = \dfrac{bp}{a+p} \Rightarrow$

$p^2 + ap = bp \Rightarrow p(p + a - b) = 0 \Rightarrow p = 0$ or $p = b - a$.

(b) $p_{n+1} = \dfrac{bp_n}{a+p_n} = \dfrac{\left(\frac{b}{a}\right)p_n}{1 + \frac{p_n}{a}} < \left(\dfrac{b}{a}\right)p_n$ since $1 + \dfrac{p_n}{a} > 1$.

(c) By part (b), $p_1 < \left(\frac{b}{a}\right)p_0$, $p_2 < \left(\frac{b}{a}\right)p_1 < \left(\frac{b}{a}\right)^2 p_0$, $p_3 < \left(\frac{b}{a}\right)p_2 < \left(\frac{b}{a}\right)^3 p_0$, etc. In general, $p_n < \left(\frac{b}{a}\right)^n p_0$,

so $\lim\limits_{n\to\infty} p_n \le \lim\limits_{n\to\infty} \left(\frac{b}{a}\right)^n \cdot p_0 = 0$ since $b < a$. $\left[\text{By result 8, } \lim\limits_{n\to\infty} r^n = 0 \text{ if } -1 < r < 1. \text{ Here } r = \frac{b}{a} \in (0,1).\right]$

(d) Let $a < b$. We first show, by induction, that if $p_0 < b - a$, then $p_n < b - a$ and $p_{n+1} > p_n$.

For $n = 0$, we have $p_1 - p_0 = \dfrac{bp_0}{a+p_0} - p_0 = \dfrac{p_0(b-a-p_0)}{a+p_0} > 0$ since $p_0 < b - a$. So $p_1 > p_0$.

Now we suppose the assertion is true for $n = k$, that is, $p_k < b - a$ and $p_{k+1} > p_k$. Then

$b - a - p_{k+1} = b - a - \dfrac{bp_k}{a+p_k} = \dfrac{a(b-a) + bp_k - ap_k - bp_k}{a+p_k} = \dfrac{a(b-a-p_k)}{a+p_k} > 0$ because $p_k < b - a$.

So $p_{k+1} < b - a$. And $p_{k+2} - p_{k+1} = \dfrac{bp_{k+1}}{a+p_{k+1}} - p_{k+1} = \dfrac{p_{k+1}(b-a-p_{k+1})}{a+p_{k+1}} > 0$ since $p_{k+1} < b - a$. Therefore, $p_{k+2} > p_{k+1}$. Thus, the assertion is true for $n = k+1$. It is therefore true for all n by mathematical induction. A similar proof by induction shows that if $p_0 > b - a$, then $p_n > b - a$ and $\{p_n\}$ is decreasing.

In either case the sequence $\{p_n\}$ is bounded and monotonic, so it is convergent by the Monotonic Sequence Theorem. It then follows from part (a) that $\lim\limits_{n\to\infty} p_n = b - a$.

8.2 Series

1. (a) A sequence is an ordered list of numbers whereas a series is the *sum* of a list of numbers.

(b) A series is convergent if the sequence of partial sums is a convergent sequence. A series is divergent if it is not convergent.

2. $\sum\limits_{n=1}^{\infty} a_n = 5$ means that by adding sufficiently many terms of the series we can get as close as we like to the number 5.

In other words, it means that $\lim_{n\to\infty} s_n = 5$, where s_n is the nth partial sum, that is, $\sum\limits_{i=1}^{n} a_i$.

3. $5 - \frac{10}{3} + \frac{20}{9} - \frac{40}{27} + \cdots$ is a geometric series with $a = 5$ and $r = -\frac{2}{3}$. Since $|r| = \frac{2}{3} < 1$, the series converges to

$$\frac{a}{1-r} = \frac{5}{1-(-2/3)} = \frac{5}{5/3} = 3.$$

4. $1 + 0.4 + 0.16 + 0.064 + \cdots$ is a geometric series with ratio $0.4 = \frac{2}{5}$. Since $|r| = \frac{2}{5} < 1$, the series converges to

$$\frac{a}{1-r} = \frac{1}{1-2/5} = \frac{5}{3}.$$

5. $\sum\limits_{n=1}^{\infty} 5\left(\frac{2}{3}\right)^{n-1}$ is a geometric series with $a = 5$ and $r = \frac{2}{3}$. Since $|r| = \frac{2}{3} < 1$, the series converges to

$$\frac{a}{1-r} = \frac{5}{1-2/3} = \frac{5}{1/3} = 15.$$

6. $\sum\limits_{n=1}^{\infty} \frac{(-6)^{n-1}}{5^{n-1}}$ is a geometric series with $a = 1$ and $r = -\frac{6}{5}$. The series diverges since $|r| = \frac{6}{5} > 1$.

7. $\sum\limits_{n=0}^{\infty} \frac{\pi^n}{3^{n+1}} = \frac{1}{3}\sum\limits_{n=0}^{\infty}\left(\frac{\pi}{3}\right)^n$ is a geometric series with ratio $r = \frac{\pi}{3}$. Since $|r| > 1$, the series diverges.

8. $\sum\limits_{n=0}^{\infty} \frac{1}{\left(\sqrt{2}\right)^n}$ is a geometric series with ratio $r = \frac{1}{\sqrt{2}}$. Since $|r| = \frac{1}{\sqrt{2}} < 1$, the series converges. Its sum is

$$\frac{1}{1-1/\sqrt{2}} = \frac{\sqrt{2}}{\sqrt{2}-1} = \frac{\sqrt{2}}{\sqrt{2}-1}\cdot\frac{\sqrt{2}+1}{\sqrt{2}+1} = \sqrt{2}\left(\sqrt{2}+1\right) = 2+\sqrt{2}.$$

9. $\sum\limits_{n=1}^{\infty} \frac{1}{2n} = \frac{1}{2}\sum\limits_{n=1}^{\infty}\frac{1}{n}$ diverges since each of its partial sums is $\frac{1}{2}$ times the corresponding partial sum of the harmonic series

$\sum\limits_{n=1}^{\infty}\frac{1}{n}$, which diverges. $\left[\text{If } \sum\limits_{n=1}^{\infty}\frac{1}{2n} \text{ were to converge, then } \sum\limits_{n=1}^{\infty}\frac{1}{n} \text{ would also have to converge by Theorem 8(i).}\right]$

In general, constant multiples of divergent series are divergent.

10. $\sum\limits_{n=1}^{\infty} \frac{n+1}{2n-3}$ diverges since $\lim\limits_{n\to\infty} a_n = \lim\limits_{n\to\infty}\frac{n+1}{2n-3} = \frac{1}{2} \neq 0$. [Use (7), the Test for Divergence.]

11. $\sum\limits_{k=2}^{\infty} \frac{k^2}{k^2-1}$ diverges by the Test for Divergence since $\lim\limits_{k\to\infty} a_k = \lim\limits_{k\to\infty}\frac{k^2}{k^2-1} = 1 \neq 0$.

12. $\sum\limits_{k=1}^{\infty} \frac{k(k+2)}{(k+3)^2}$ diverges by (7), the Test for Divergence, since $\lim\limits_{k\to\infty} a_k = \lim\limits_{k\to\infty}\frac{k(k+2)}{(k+3)^2} = \lim\limits_{k\to\infty}\frac{1\cdot(1+2/k)}{(1+3/k)^2} = 1 \neq 0$.

13. Converges.

$$\sum_{n=1}^{\infty}\frac{1+2^n}{3^n}=\sum_{n=1}^{\infty}\left(\frac{1}{3^n}+\frac{2^n}{3^n}\right)=\sum_{n=1}^{\infty}\left[\left(\frac{1}{3}\right)^n+\left(\frac{2}{3}\right)^n\right] \quad \text{[sum of two convergent geometric series]}$$

$$=\frac{1/3}{1-1/3}+\frac{2/3}{1-2/3}=\frac{1}{2}+2=\frac{5}{2}$$

14. $\sum_{n=1}^{\infty}\frac{1+3^n}{2^n}=\sum_{n=1}^{\infty}\left(\frac{1}{2^n}+\frac{3^n}{2^n}\right)=\sum_{n=1}^{\infty}\left[\left(\frac{1}{2}\right)^n+\left(\frac{3}{2}\right)^n\right]=\sum_{n=1}^{\infty}\left(\frac{1}{2}\right)^n+\sum_{n=1}^{\infty}\left(\frac{3}{2}\right)^n$. The first series is a convergent geometric series ($|r|=\frac{1}{2}<1$), but the second series is a divergent geometric series ($|r|=\frac{3}{2}\geq 1$), so the original series is divergent.

15. $\sum_{n=1}^{\infty}\sqrt[n]{2}=2+\sqrt{2}+\sqrt[3]{2}+\sqrt[4]{2}+\cdots$ diverges by the Test for Divergence since

$\lim_{n\to\infty}a_n=\lim_{n\to\infty}\sqrt[n]{2}=\lim_{n\to\infty}2^{1/n}=2^0=1\neq 0.$

16. $$\sum_{n=1}^{\infty}\left[(0.8)^{n-1}-(0.3)^n\right]=\sum_{n=1}^{\infty}(0.8)^{n-1}-\sum_{n=1}^{\infty}(0.3)^n \quad \text{[difference of two convergent geometric series]}$$

$$=\frac{1}{1-0.8}-\frac{0.3}{1-0.3}=5-\frac{3}{7}=\frac{32}{7}$$

17. $\lim_{n\to\infty}a_n=\lim_{n\to\infty}\arctan n=\frac{\pi}{2}\neq 0$, so the series diverges by the Test for Divergence.

18. $\sum_{k=1}^{\infty}(\cos 1)^k$ is a geometric series with ratio $r=\cos 1\approx 0.540302$. It converges because $|r|<1$.

Its sum is $\dfrac{\cos 1}{1-\cos 1}\approx 1.175343$.

19. Using partial fractions, the partial sums of the series $\sum_{n=2}^{\infty}\frac{2}{n^2-1}$ are

$$s_n=\sum_{i=2}^{n}\frac{2}{(i-1)(i+1)}=\sum_{i=2}^{n}\left(\frac{1}{i-1}-\frac{1}{i+1}\right)$$

$$=\left(1-\frac{1}{3}\right)+\left(\frac{1}{2}-\frac{1}{4}\right)+\left(\frac{1}{3}-\frac{1}{5}\right)+\cdots+\left(\frac{1}{n-3}-\frac{1}{n-1}\right)+\left(\frac{1}{n-2}-\frac{1}{n}\right)$$

This sum is a telescoping series and $s_n=1+\frac{1}{2}-\frac{1}{n-1}-\frac{1}{n}$.

Thus, $\sum_{n=2}^{\infty}\frac{2}{n^2-1}=\lim_{n\to\infty}s_n=\lim_{n\to\infty}\left(1+\frac{1}{2}-\frac{1}{n-1}-\frac{1}{n}\right)=\frac{3}{2}$.

20. For the series $\sum_{n=1}^{\infty}\frac{2}{n^2+4n+3}$, $s_n=\sum_{i=1}^{n}\frac{2}{i^2+4i+3}=\sum_{i=1}^{n}\left(\frac{1}{i+1}-\frac{1}{i+3}\right)$ [using partial fractions]. The latter sum is

$$\left(\tfrac{1}{2}-\tfrac{1}{4}\right)+\left(\tfrac{1}{3}-\tfrac{1}{5}\right)+\left(\tfrac{1}{4}-\tfrac{1}{6}\right)+\left(\tfrac{1}{5}-\tfrac{1}{7}\right)+\cdots+\left(\tfrac{1}{n}-\tfrac{1}{n+2}\right)+\left(\tfrac{1}{n+1}-\tfrac{1}{n+3}\right)$$

$$=\tfrac{1}{2}+\tfrac{1}{3}-\tfrac{1}{n+2}-\tfrac{1}{n+3} \quad \text{[telescoping series]}$$

Thus, $\sum_{n=1}^{\infty}\frac{2}{n^2+4n+3}=\lim_{n\to\infty}s_n=\lim_{n\to\infty}\left(\frac{1}{2}+\frac{1}{3}-\frac{1}{n+2}-\frac{1}{n+3}\right)=\frac{1}{2}+\frac{1}{3}=\frac{5}{6}$. Converges

21. For the series $\sum_{n=1}^{\infty} \frac{3}{n(n+3)}$, $s_n = \sum_{i=1}^{n} \frac{3}{i(i+3)} = \sum_{i=1}^{n} \left(\frac{1}{i} - \frac{1}{i+3}\right)$ [using partial fractions]. The latter sum is

$$\left(1-\tfrac{1}{4}\right)+\left(\tfrac{1}{2}-\tfrac{1}{5}\right)+\left(\tfrac{1}{3}-\tfrac{1}{6}\right)+\left(\tfrac{1}{4}-\tfrac{1}{7}\right)+\cdots+\left(\tfrac{1}{n-3}-\tfrac{1}{n}\right)+\left(\tfrac{1}{n-2}-\tfrac{1}{n+1}\right)+\left(\tfrac{1}{n-1}-\tfrac{1}{n+2}\right)+\left(\tfrac{1}{n}-\tfrac{1}{n+3}\right)$$

$$= 1+\tfrac{1}{2}+\tfrac{1}{3}-\tfrac{1}{n+1}-\tfrac{1}{n+2}-\tfrac{1}{n+3} \quad \text{[telescoping series]}$$

Thus, $\sum_{n=1}^{\infty} \frac{3}{n(n+3)} = \lim_{n\to\infty} s_n = \lim_{n\to\infty}\left(1+\tfrac{1}{2}+\tfrac{1}{3}-\tfrac{1}{n+1}-\tfrac{1}{n+2}-\tfrac{1}{n+3}\right) = 1+\tfrac{1}{2}+\tfrac{1}{3} = \tfrac{11}{6}$. Converges

22. For the series $\sum_{n=1}^{\infty} \ln \frac{n}{n+1}$,

$$s_n = (\ln 1 - \ln 2) + (\ln 2 - \ln 3) + (\ln 3 - \ln 4) + \cdots + [\ln n - \ln(n+1)]$$

$$= \ln 1 - \ln(n+1) = -\ln(n+1) \quad \text{[telescoping series]}$$

Thus, $\lim_{n\to\infty} s_n = -\infty$, so the series is divergent.

23. $0.\overline{2} = \frac{2}{10} + \frac{2}{10^2} + \cdots$ is a geometric series with $a = \frac{2}{10}$ and $r = \frac{1}{10}$. It converges to $\frac{a}{1-r} = \frac{2/10}{1-1/10} = \frac{2}{9}$.

24. $0.\overline{73} = \frac{73}{10^2} + \frac{73}{10^4} + \cdots = \frac{73/10^2}{1-1/10^2} = \frac{73/100}{99/100} = \frac{73}{99}$

25. $3.\overline{417} = 3 + \frac{417}{10^3} + \frac{417}{10^6} + \cdots$. Now $\frac{417}{10^3} + \frac{417}{10^6} + \cdots$ is a geometric series with $a = \frac{417}{10^3}$ and $r = \frac{1}{10^3}$.

It converges to $\frac{a}{1-r} = \frac{417/10^3}{1-1/10^3} = \frac{417/10^3}{999/10^3} = \frac{417}{999}$. Thus, $3.\overline{417} = 3 + \frac{417}{999} = \frac{3414}{999} = \frac{1138}{333}$.

26. $6.2\overline{54} = 6.2 + \frac{54}{10^3} + \frac{54}{10^5} + \cdots = 6.2 + \frac{54/10^3}{1-1/10^2} = \frac{62}{10} + \frac{54}{990} = \frac{6192}{990} = \frac{344}{55}$

27. $\sum_{n=1}^{\infty} \frac{x^n}{3^n} = \sum_{n=1}^{\infty} \left(\frac{x}{3}\right)^n$ is a geometric series with $r = \frac{x}{3}$, so the series converges $\Leftrightarrow$ $|r| < 1$ $\Leftrightarrow$ $\frac{|x|}{3} < 1$ $\Leftrightarrow$ $|x| < 3$;

that is, $-3 < x < 3$. In that case, the sum of the series is $\frac{a}{1-r} = \frac{x/3}{1-x/3} = \frac{x/3}{1-x/3} \cdot \frac{3}{3} = \frac{x}{3-x}$.

28. $\sum_{n=0}^{\infty} 2^n (x+1)^n = \sum_{n=0}^{\infty} [2(x+1)]^n = \sum_{n=1}^{\infty} [2(x+1)]^{n-1}$ is a geometric series with $r = 2(x+1)$, so the series

converges $\Leftrightarrow$ $|r| < 1$ $\Leftrightarrow$ $|2(x+1)| < 1$ $\Leftrightarrow$ $|x+1| < \frac{1}{2}$ $\Leftrightarrow$ $-\frac{1}{2} < x+1 < \frac{1}{2}$ $\Leftrightarrow$ $-\frac{3}{2} < x < -\frac{1}{2}$.

In that case, the sum of the series is $\frac{a}{1-r} = \frac{1}{1-2(x+1)} = \frac{1}{-1-2x}$ or $\frac{-1}{2x+1}$.

29. $\sum_{n=0}^{\infty} \frac{\cos^n x}{2^n}$ is a geometric series with first term 1 and ratio $r = \frac{\cos x}{2}$, so it converges $\Leftrightarrow$ $|r| < 1$. But $|r| = \frac{|\cos x|}{2} \leq \frac{1}{2}$

for all x. Thus, the series converges for all real values of x and the sum of the series is $\frac{1}{1-(\cos x)/2} = \frac{2}{2-\cos x}$.

30. Because $\frac{1}{n} \to 0$ and ln is continuous, we have $\lim_{n\to\infty} \ln\left(1+\frac{1}{n}\right) = \ln 1 = 0$.

We now show that the series $\sum_{n=1}^{\infty} \ln\left(1+\frac{1}{n}\right) = \sum_{n=1}^{\infty} \ln\left(\frac{n+1}{n}\right) = \sum_{n=1}^{\infty} [\ln(n+1) - \ln n]$ diverges:

$s_n = (\ln 2 - \ln 1) + (\ln 3 - \ln 2) + \cdots + (\ln(n+1) - \ln n) = \ln(n+1) - \ln 1 = \ln(n+1)$. As $n \to \infty$,

$s_n = \ln(n+1) \to \infty$, so the series diverges.

31. For $n = 1$, $a_1 = 0$ since $s_1 = 0$. For $n > 1$,

$$a_n = s_n - s_{n-1} = \frac{n-1}{n+1} - \frac{(n-1)-1}{(n-1)+1} = \frac{(n-1)n - (n+1)(n-2)}{(n+1)n} = \frac{2}{n(n+1)}$$

Also, $\sum_{n=1}^{\infty} a_n = \lim_{n\to\infty} s_n = \lim_{n\to\infty} \frac{1 - 1/n}{1 + 1/n} = 1$.

32. $a_1 = s_1 = 3 - \frac{1}{2} = \frac{5}{2}$. For $n \neq 1$,

$$a_n = s_n - s_{n-1} = \left(3 - n2^{-n}\right) - \left[3 - (n-1)2^{-(n-1)}\right] = -\frac{n}{2^n} + \frac{n-1}{2^{n-1}} \cdot \frac{2}{2} = \frac{2(n-1)}{2^n} - \frac{n}{2^n} = \frac{n-2}{2^n}$$

Also, $\sum_{n=1}^{\infty} a_n = \lim_{n\to\infty} s_n = \lim_{n\to\infty} \left(3 - \frac{n}{2^n}\right) = 3$ because $\lim_{x\to\infty} \frac{x}{2^x} \overset{\text{H}}{=} \lim_{x\to\infty} \frac{1}{2^x \ln 2} = 0$.

33. (a) The first step in the chain occurs when the local government spends D dollars. The people who receive it spend a fraction c of those D dollars, that is, Dc dollars. Those who receive the Dc dollars spend a fraction c of it, that is, Dc^2 dollars. Continuing in this way, we see that the total spending after n transactions is

$$S_n = D + Dc + Dc^2 + \cdots + Dc^{n-1} = \frac{D(1-c^n)}{1-c} \text{ by (3).}$$

(b) $$\lim_{n\to\infty} S_n = \lim_{n\to\infty} \frac{D(1-c^n)}{1-c} = \frac{D}{1-c} \lim_{n\to\infty} (1-c^n) = \frac{D}{1-c} \quad [\text{since } 0 < c < 1 \;\Rightarrow\; \lim_{n\to\infty} c^n = 0]$$
$$= \frac{D}{s} \quad [\text{since } c + s = 1] = kD \quad [\text{since } k = 1/s]$$

If $c = 0.8$, then $s = 1 - c = 0.2$ and the multiplier is $k = 1/s = 5$.

34. (a) Initially, the ball falls a distance H, then rebounds a distance rH, falls rH, rebounds r^2H, falls r^2H, etc. The total distance it travels is

$$H + 2rH + 2r^2H + 2r^3H + \cdots = H\left(1 + 2r + 2r^2 + 2r^3 + \cdots\right) = H\left[1 + 2r\left(1 + r + r^2 + \cdots\right)\right]$$
$$= H\left[1 + 2r\left(\frac{1}{1-r}\right)\right] = H\left(\frac{1+r}{1-r}\right) \text{ meters}$$

(b) From Example 3 in Section 2.1, we know that a ball falls $\frac{1}{2}gt^2$ meters in t seconds, where g is the gravitational acceleration. Thus, a ball falls h meters in $t = \sqrt{2h/g}$ seconds. The total travel time in seconds is

$$\sqrt{\frac{2H}{g}} + 2\sqrt{\frac{2H}{g}r} + 2\sqrt{\frac{2H}{g}r^2} + 2\sqrt{\frac{2H}{g}r^3} + \cdots$$
$$= \sqrt{\frac{2H}{g}}\left[1 + 2\sqrt{r} + 2\sqrt{r}^2 + 2\sqrt{r}^3 + \cdots\right]$$
$$= \sqrt{\frac{2H}{g}}\left(1 + 2\sqrt{r}\left[1 + \sqrt{r} + \sqrt{r}^2 + \cdots\right]\right) = \sqrt{\frac{2H}{g}}\left[1 + 2\sqrt{r}\left(\frac{1}{1-\sqrt{r}}\right)\right] = \sqrt{\frac{2H}{g}}\frac{1+\sqrt{r}}{1-\sqrt{r}}$$

(c) It will help to make a chart of the time for each descent and each rebound of the ball, together with the velocity just before and just after each bounce. Recall that the time in seconds needed to fall h meters is $\sqrt{2h/g}$. The ball hits the ground with velocity $-g\sqrt{2h/g} = -\sqrt{2hg}$ (taking the upward direction to be positive) and rebounds with velocity $kg\sqrt{2h/g} = k\sqrt{2hg}$, taking time $k\sqrt{2h/g}$ to reach the top of its bounce, where its velocity is 0. At that point, its height is k^2h. All these results follow from the formulas for vertical motion with gravitational acceleration $-g$:

$$\frac{d^2y}{dt^2} = -g \Rightarrow v = \frac{dy}{dt} = v_0 - gt \Rightarrow y = y_0 + v_0t - \tfrac{1}{2}gt^2.$$

number of descent	time of descent	speed before bounce	speed after bounce	time of ascent	peak height
1	$\sqrt{2H/g}$	$\sqrt{2Hg}$	$k\sqrt{2Hg}$	$k\sqrt{2H/g}$	k^2H
2	$\sqrt{2k^2H/g}$	$\sqrt{2k^2Hg}$	$k\sqrt{2k^2Hg}$	$k\sqrt{2k^2H/g}$	k^4H
3	$\sqrt{2k^4H/g}$	$\sqrt{2k^4Hg}$	$k\sqrt{2k^4Hg}$	$k\sqrt{2k^4H/g}$	k^6H
...	...	...	...	...	...

The total travel time in seconds is

$$\sqrt{\frac{2H}{g}} + k\sqrt{\frac{2H}{g}} + k\sqrt{\frac{2H}{g}} + k^2\sqrt{\frac{2H}{g}} + k^2\sqrt{\frac{2H}{g}} + \cdots = \sqrt{\frac{2H}{g}}\left(1 + 2k + 2k^2 + 2k^3 + \cdots\right)$$

$$= \sqrt{\frac{2H}{g}}\left[1 + 2k(1 + k + k^2 + \cdots)\right] = \sqrt{\frac{2H}{g}}\left[1 + 2k\left(\frac{1}{1-k}\right)\right] = \sqrt{\frac{2H}{g}}\frac{1+k}{1-k}$$

Another method: We could use part (b). At the top of the bounce, the height is $k^2h = rh$, so $\sqrt{r} = k$ and the result follows from part (b).

35. $\sum\limits_{n=2}^{\infty}(1+c)^{-n}$ is a geometric series with $a = (1+c)^{-2}$ and $r = (1+c)^{-1}$, so the series converges when $\left|(1+c)^{-1}\right| < 1 \Leftrightarrow |1+c| > 1 \Leftrightarrow 1+c > 1$ or $1+c < -1 \Leftrightarrow c > 0$ or $c < -2$. We calculate the sum of the series and set it equal to 2: $\dfrac{(1+c)^{-2}}{1-(1+c)^{-1}} = 2 \Leftrightarrow \left(\dfrac{1}{1+c}\right)^2 = 2 - 2\left(\dfrac{1}{1+c}\right) \Leftrightarrow 1 = 2(1+c)^2 - 2(1+c) \Leftrightarrow$ $2c^2 + 2c - 1 = 0 \Leftrightarrow c = \frac{-2\pm\sqrt{12}}{4} = \frac{\pm\sqrt{3}-1}{2}$. However, the negative root is inadmissible because $-2 < \frac{-\sqrt{3}-1}{2} < 0$. So $c = \frac{\sqrt{3}-1}{2}$.

36. The area between $y = x^{n-1}$ and $y = x^n$ for $0 \le x \le 1$ is

$$\int_0^1 (x^{n-1} - x^n)\,dx = \left[\frac{x^n}{n} - \frac{x^{n+1}}{n+1}\right]_0^1 = \frac{1}{n} - \frac{1}{n+1}$$

$$= \frac{(n+1) - n}{n(n+1)} = \frac{1}{n(n+1)}$$

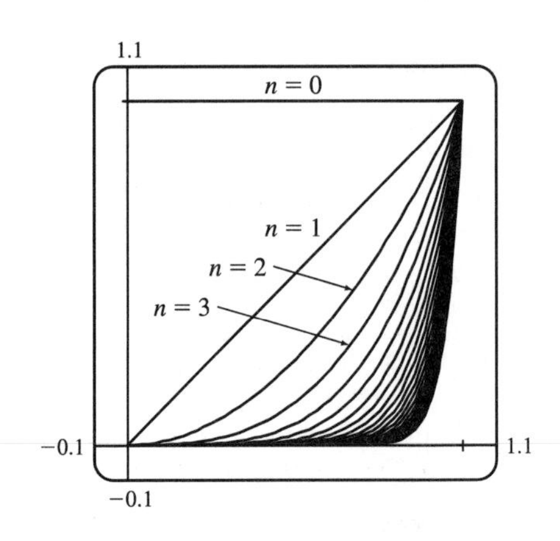

We can see from the diagram that as $n \to \infty$, the sum of the areas between the successive curves approaches the area of the unit square, that is, 1.

So $\sum\limits_{n=1}^{\infty}\dfrac{1}{n(n+1)} = 1$.

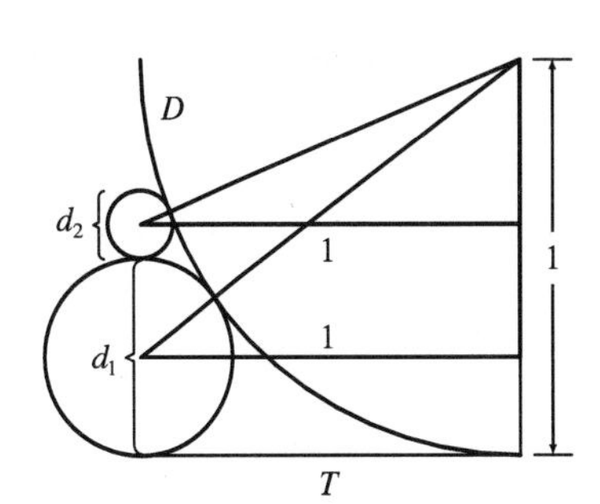

37. Let d_n be the diameter of C_n. We draw lines from the centers of the C_i to the center of D (or C), and using the Pythagorean Theorem, we can write

$1^2 + \left(1 - \frac{1}{2}d_1\right)^2 = \left(1 + \frac{1}{2}d_1\right)^2 \quad \Leftrightarrow$

$1 = \left(1 + \frac{1}{2}d_1\right)^2 - \left(1 - \frac{1}{2}d_1\right)^2 = 2d_1$ [difference of squares] $\Rightarrow \quad d_1 = \frac{1}{2}$.

Similarly,

$$\begin{aligned} 1 &= \left(1 + \tfrac{1}{2}d_2\right)^2 - \left(1 - d_1 - \tfrac{1}{2}d_2\right)^2 \\ &= 2d_2 + 2d_1 - d_1^2 - d_1 d_2 = (2 - d_1)(d_1 + d_2) \quad \Leftrightarrow \end{aligned}$$

$d_2 = \dfrac{1}{2 - d_1} - d_1 = \dfrac{(1 - d_1)^2}{2 - d_1}$, $1 = \left(1 + \frac{1}{2}d_3\right)^2 - \left(1 - d_1 - d_2 - \frac{1}{2}d_3\right)^2 \quad \Leftrightarrow \quad d_3 = \dfrac{[1 - (d_1 + d_2)]^2}{2 - (d_1 + d_2)}$, and in general,

$d_{n+1} = \dfrac{\left(1 - \sum_{i=1}^n d_i\right)^2}{2 - \sum_{i=1}^n d_i}$. If we actually calculate d_2 and d_3 from the formulas above, we find that they are $\dfrac{1}{6} = \dfrac{1}{2 \cdot 3}$ and

$\dfrac{1}{12} = \dfrac{1}{3 \cdot 4}$ respectively, so we suspect that in general, $d_n = \dfrac{1}{n(n+1)}$. To prove this, we use induction: Assume that for all

$k \le n$, $d_k = \dfrac{1}{k(k+1)} = \dfrac{1}{k} - \dfrac{1}{k+1}$. Then $\sum\limits_{i=1}^n d_i = 1 - \dfrac{1}{n+1} = \dfrac{n}{n+1}$ [telescoping sum]. Substituting this into our

formula for d_{n+1}, we get $d_{n+1} = \dfrac{\left[1 - \dfrac{n}{n+1}\right]^2}{2 - \left(\dfrac{n}{n+1}\right)} = \dfrac{\dfrac{1}{(n+1)^2}}{\dfrac{n+2}{n+1}} = \dfrac{1}{(n+1)(n+2)}$, and the induction is complete.

Now, we observe that the partial sums $\sum\limits_{i=1}^n d_i$ of the diameters of the circles approach 1 as $n \to \infty$; that is,

$\sum\limits_{n=1}^\infty a_n = \sum\limits_{n=1}^\infty \dfrac{1}{n(n+1)} = 1$, which is what we wanted to prove.

38. $|CD| = b\sin\theta$, $|DE| = |CD|\sin\theta = b\sin^2\theta$, $|EF| = |DE|\sin\theta = b\sin^3\theta$, ... Therefore,

$|CD| + |DE| + |EF| + |FG| + \cdots = b\sum\limits_{n=1}^\infty \sin^n\theta = b\left(\dfrac{\sin\theta}{1 - \sin\theta}\right)$ since this is a geometric series with $r = \sin\theta$ and

$|\sin\theta| < 1$ $\left[\text{because } 0 < \theta < \frac{\pi}{2}\right]$.

39. The series $1 - 1 + 1 - 1 + 1 - 1 + \cdots$ diverges (geometric series with $r = -1$) so we cannot say that $0 = 1 - 1 + 1 - 1 + 1 - 1 + \cdots$.

40. If $\sum\limits_{n=1}^\infty a_n$ is convergent, then $\lim\limits_{n\to\infty} a_n = 0$ by Theorem 6, so $\lim\limits_{n\to\infty} \dfrac{1}{a_n} \ne 0$, and so $\sum\limits_{n=1}^\infty \dfrac{1}{a_n}$ is divergent by the Test for Divergence.

41. $\sum\limits_{n=1}^\infty ca_n = \lim\limits_{n\to\infty} \sum\limits_{i=1}^n ca_i = \lim\limits_{n\to\infty} c\sum\limits_{i=1}^n a_i = c\lim\limits_{n\to\infty} \sum\limits_{i=1}^n a_i = c\sum\limits_{n=1}^\infty a_n$, which exists by hypothesis.

42. If $\sum ca_n$ were convergent, then $\sum(1/c)(ca_n) = \sum a_n$ would be also, by Theorem 8. But this is not the case, so $\sum ca_n$ must diverge.

43. Suppose on the contrary that $\sum(a_n + b_n)$ converges. Then $\sum(a_n + b_n)$ and $\sum a_n$ are convergent series. So by Theorem 8, $\sum[(a_n + b_n) - a_n]$ would also be convergent. But $\sum[(a_n + b_n) - a_n] = \sum b_n$, a contradiction, since $\sum b_n$ is given to be divergent.

44. No. For example, take $\sum a_n = \sum n$ and $\sum b_n = \sum(-n)$, which both diverge, yet $\sum(a_n + b_n) = \sum 0$, which converges with sum 0.

45. The partial sums $\{s_n\}$ form an increasing sequence, since $s_n - s_{n-1} = a_n > 0$ for all n. Also, the sequence $\{s_n\}$ is bounded since $s_n \leq 1000$ for all n. So by the Monotonic Sequence Theorem, the sequence of partial sums converges, that is, the series $\sum a_n$ is convergent.

46. (a) $\text{RHS} = \dfrac{1}{f_{n-1}f_n} - \dfrac{1}{f_n f_{n+1}} = \dfrac{f_n f_{n+1} - f_n f_{n-1}}{f_n^2 f_{n-1} f_{n+1}} = \dfrac{f_{n+1} - f_{n-1}}{f_n f_{n-1} f_{n+1}} = \dfrac{(f_{n-1} + f_n) - f_{n-1}}{f_n f_{n-1} f_{n+1}} = \dfrac{1}{f_{n-1} f_{n+1}} = \text{LHS}$

(b)
$$\sum_{n=2}^{\infty} \frac{1}{f_{n-1}f_{n+1}} = \sum_{n=2}^{\infty}\left(\frac{1}{f_{n-1}f_n} - \frac{1}{f_n f_{n+1}}\right) \quad \text{[from part (a)]}$$
$$= \lim_{n\to\infty}\left[\left(\frac{1}{f_1 f_2} - \frac{1}{f_2 f_3}\right) + \left(\frac{1}{f_2 f_3} - \frac{1}{f_3 f_4}\right) + \left(\frac{1}{f_3 f_4} - \frac{1}{f_4 f_5}\right) + \cdots + \left(\frac{1}{f_{n-1} f_n} - \frac{1}{f_n f_{n+1}}\right)\right]$$
$$= \lim_{n\to\infty}\left(\frac{1}{f_1 f_2} - \frac{1}{f_n f_{n+1}}\right) = \frac{1}{f_1 f_2} - 0 = \frac{1}{1\cdot 1} = 1 \quad \text{because } f_n \to \infty \text{ as } n \to \infty.$$

(c)
$$\sum_{n=2}^{\infty} \frac{f_n}{f_{n-1}f_{n+1}} = \sum_{n=2}^{\infty}\left(\frac{f_n}{f_{n-1}f_n} - \frac{f_n}{f_n f_{n+1}}\right) \quad \text{[as above]}$$
$$= \sum_{n=2}^{\infty}\left(\frac{1}{f_{n-1}} - \frac{1}{f_{n+1}}\right)$$
$$= \lim_{n\to\infty}\left[\left(\frac{1}{f_1} - \frac{1}{f_3}\right) + \left(\frac{1}{f_2} - \frac{1}{f_4}\right) + \left(\frac{1}{f_3} - \frac{1}{f_5}\right) + \left(\frac{1}{f_4} - \frac{1}{f_6}\right) + \cdots + \left(\frac{1}{f_{n-1}} - \frac{1}{f_{n+1}}\right)\right]$$
$$= \lim_{n\to\infty}\left(\frac{1}{f_1} + \frac{1}{f_2} - \frac{1}{f_n} - \frac{1}{f_{n+1}}\right) = 1 + 1 - 0 - 0 = 2 \quad \text{because } f_n \to \infty \text{ as } n \to \infty.$$

47. (a) At the first step, only the interval $\left(\frac{1}{3}, \frac{2}{3}\right)$ (length $\frac{1}{3}$) is removed. At the second step, we remove the intervals $\left(\frac{1}{9}, \frac{2}{9}\right)$ and $\left(\frac{7}{9}, \frac{8}{9}\right)$, which have a total length of $2 \cdot \left(\frac{1}{3}\right)^2$. At the third step, we remove 2^2 intervals, each of length $\left(\frac{1}{3}\right)^3$. In general, at the nth step we remove 2^{n-1} intervals, each of length $\left(\frac{1}{3}\right)^n$, for a length of $2^{n-1} \cdot \left(\frac{1}{3}\right)^n = \frac{1}{3}\left(\frac{2}{3}\right)^{n-1}$. Thus, the total length of all removed intervals is $\sum\limits_{n=1}^{\infty} \frac{1}{3}\left(\frac{2}{3}\right)^{n-1} = \frac{1/3}{1 - 2/3} = 1$ $\left[\text{geometric series with } a = \frac{1}{3} \text{ and } r = \frac{2}{3}\right]$. Notice that at the nth step, the leftmost interval that is removed is $\left(\left(\frac{1}{3}\right)^n, \left(\frac{2}{3}\right)^n\right)$, so we never remove 0, and 0 is in the Cantor set. Also, the rightmost interval removed is $\left(1 - \left(\frac{2}{3}\right)^n, 1 - \left(\frac{1}{3}\right)^n\right)$, so 1 is never removed. Some other numbers in the Cantor set are $\frac{1}{3}$, $\frac{2}{3}$, $\frac{1}{9}$, $\frac{2}{9}$, $\frac{7}{9}$, and $\frac{8}{9}$.

(b) The area removed at the first step is $\frac{1}{9}$; at the second step, $8 \cdot \left(\frac{1}{9}\right)^2$; at the third step, $(8)^2 \cdot \left(\frac{1}{9}\right)^3$. In general, the area removed at the nth step is $(8)^{n-1}\left(\frac{1}{9}\right)^n = \frac{1}{9}\left(\frac{8}{9}\right)^{n-1}$, so the total area of all removed squares is

$$\sum_{n=1}^{\infty} \frac{1}{9}\left(\frac{8}{9}\right)^{n-1} = \frac{\frac{1}{9}}{1 - \frac{8}{9}} = 1.$$

48. (a)

a_1	1	2	4	1	1	1000
a_2	2	3	1	4	1000	1
a_3	1.5	2.5	2.5	2.5	500.5	500.5
a_4	1.75	2.75	1.75	3.25	750.25	250.75
a_5	1.625	2.625	2.125	2.875	625.375	375.625
a_6	1.6875	2.6875	1.9375	3.0625	687.813	313.188
a_7	1.65625	2.65625	2.03125	2.96875	656.594	344.406
a_8	1.67188	2.67188	1.98438	3.01563	672.203	328.797
a_9	1.66406	2.66406	2.00781	2.99219	664.398	336.602
a_{10}	1.66797	2.66797	1.99609	3.00391	668.301	332.699
a_{11}	1.66602	2.66602	2.00195	2.99805	666.350	334.650
a_{12}	1.66699	2.66699	1.99902	3.00098	667.325	333.675

The limits seem to be $\frac{5}{3}, \frac{8}{3}$, 2, 3, 667, and 334. Note that the limits appear to be "weighted" more toward a_2.

In general, we guess that the limit is $\dfrac{a_1 + 2a_2}{3}$.

(b) $a_{n+1} - a_n = \frac{1}{2}(a_n + a_{n-1}) - a_n = -\frac{1}{2}(a_n - a_{n-1}) = -\frac{1}{2}\left[\frac{1}{2}(a_{n-1} + a_{n-2}) - a_{n-1}\right]$

$$= -\tfrac{1}{2}\left[-\tfrac{1}{2}(a_{n-1} - a_{n-2})\right] = \cdots = \left(-\tfrac{1}{2}\right)^{n-1}(a_2 - a_1)$$

Note that we have used the formula $a_k = \frac{1}{2}(a_{k-1} + a_{k-2})$ a total of $n - 1$ times in this calculation, once for each k between 3 and $n + 1$. Now we can write

$$a_n = a_1 + (a_2 - a_1) + (a_3 - a_2) + \cdots + (a_{n-1} - a_{n-2}) + (a_n - a_{n-1})$$
$$= a_1 + \sum_{k=1}^{n-1}(a_{k+1} - a_k) = a_1 + \sum_{k=1}^{n-1}\left(-\tfrac{1}{2}\right)^{k-1}(a_2 - a_1)$$

and so

$$\lim_{n\to\infty} a_n = a_1 + (a_2 - a_1)\sum_{k=1}^{\infty}\left(-\tfrac{1}{2}\right)^{k-1} = a_1 + (a_2 - a_1)\left[\frac{1}{1 - (-1/2)}\right] = a_1 + \tfrac{2}{3}(a_2 - a_1) = \frac{a_1 + 2a_2}{3}.$$

49. (a) For $\displaystyle\sum_{n=1}^{\infty} \frac{n}{(n+1)!}$, $s_1 = \dfrac{1}{1 \cdot 2} = \dfrac{1}{2}$, $s_2 = \dfrac{1}{2} + \dfrac{2}{1 \cdot 2 \cdot 3} = \dfrac{5}{6}$, $s_3 = \dfrac{5}{6} + \dfrac{3}{1 \cdot 2 \cdot 3 \cdot 4} = \dfrac{23}{24}$,

$s_4 = \dfrac{23}{24} + \dfrac{4}{1 \cdot 2 \cdot 3 \cdot 4 \cdot 5} = \dfrac{119}{120}$. The denominators are $(n+1)!$, so a guess would be $s_n = \dfrac{(n+1)! - 1}{(n+1)!}$.

(b) For $n = 1$, $s_1 = \dfrac{1}{2} = \dfrac{2! - 1}{2!}$, so the formula holds for $n = 1$. Assume $s_k = \dfrac{(k+1)! - 1}{(k+1)!}$. Then

$$s_{k+1} = \frac{(k+1)! - 1}{(k+1)!} + \frac{k+1}{(k+2)!} = \frac{(k+1)! - 1}{(k+1)!} + \frac{k+1}{(k+1)!(k+2)}$$
$$= \frac{(k+2)! - (k+2) + k + 1}{(k+2)!} = \frac{(k+2)! - 1}{(k+2)!}$$

Thus, the formula is true for $n = k + 1$. So by induction, the guess is correct.

(c) $\displaystyle\lim_{n\to\infty} s_n = \lim_{n\to\infty} \frac{(n+1)! - 1}{(n+1)!} = \lim_{n\to\infty}\left[1 - \frac{1}{(n+1)!}\right] = 1$ and so $\displaystyle\sum_{n=1}^{\infty} \frac{n}{(n+1)!} = 1$.

50.

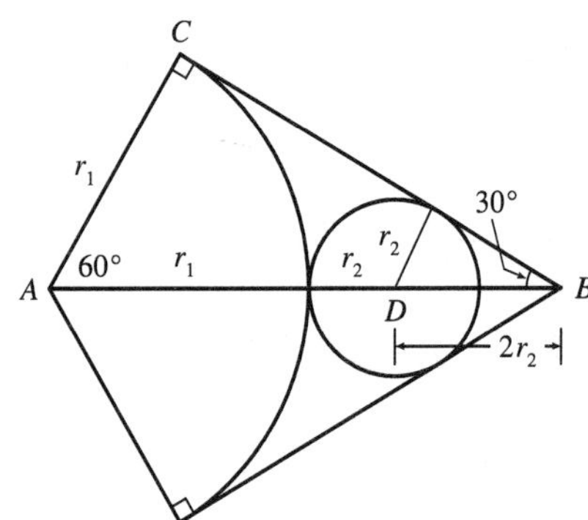

Let r_1 = radius of the large circle, r_2 = radius of next circle, and so on. From the figure we have $\angle BAC = 60°$ and $\cos 60° = r_1/|AB|$, so $|AB| = 2r_1$ and $|DB| = 2r_2$. Therefore, $2r_1 = r_1 + r_2 + 2r_2 \Rightarrow$ $r_1 = 3r_2$. In general, we have $r_{n+1} = \frac{1}{3}r_n$, so the total area is

$$A = \pi r_1^2 + 3\pi r_2^2 + 3\pi r_3^2 + \cdots = \pi r_1^2 + 3\pi r_2^2\left(1 + \frac{1}{3^2} + \frac{1}{3^4} + \frac{1}{3^6} + \cdots\right)$$

$$= \pi r_1^2 + 3\pi r_2^2 \cdot \frac{1}{1 - 1/9} = \pi r_1^2 + \tfrac{27}{8}\pi r_2^2$$

Since the sides of the triangle have length 1, $|BC| = \frac{1}{2}$ and $\tan 30° = \dfrac{r_1}{1/2}$. Thus, $r_1 = \dfrac{\tan 30°}{2} = \dfrac{1}{2\sqrt{3}} \Rightarrow r_2 = \dfrac{1}{6\sqrt{3}}$,

so $A = \pi\left(\dfrac{1}{2\sqrt{3}}\right)^2 + \dfrac{27\pi}{8}\left(\dfrac{1}{6\sqrt{3}}\right)^2 = \dfrac{\pi}{12} + \dfrac{\pi}{32} = \dfrac{11\pi}{96}$. The area of the triangle is $\dfrac{\sqrt{3}}{4}$, so the circles occupy about 83.1% of the area of the triangle.

8.3 The Integral and Comparison Tests

1. The picture shows that $a_2 = \dfrac{1}{2^{1.3}} < \displaystyle\int_1^2 \frac{1}{x^{1.3}}\,dx$,

$a_3 = \dfrac{1}{3^{1.3}} < \displaystyle\int_2^3 \frac{1}{x^{1.3}}\,dx$, and so on, so $\displaystyle\sum_{n=2}^{\infty} \frac{1}{n^{1.3}} < \int_1^{\infty} \frac{1}{x^{1.3}}\,dx$. The integral converges by (6.6.2) with $p = 1.3 > 1$, so the series converges.

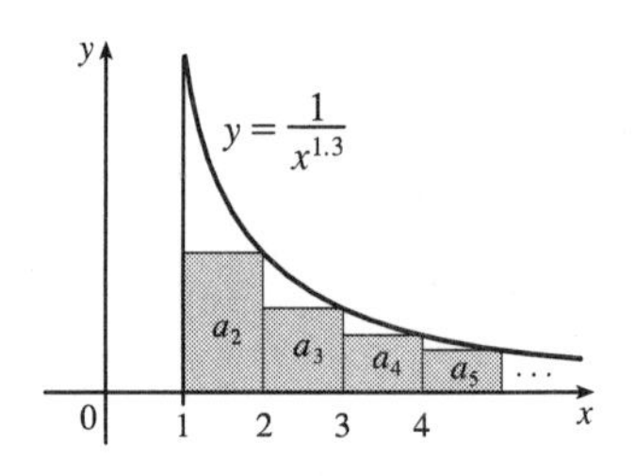

2. From the first figure, we see that $\int_1^6 f(x)dx < \sum_{i=1}^5 a_i$. From the second figure, we see that $\sum_{i=2}^6 a_i < \int_1^6 f(x)\,dx$. Thus, we have $\sum_{i=2}^6 a_i < \int_1^6 f(x)\,dx < \sum_{i=1}^5 a_i$.

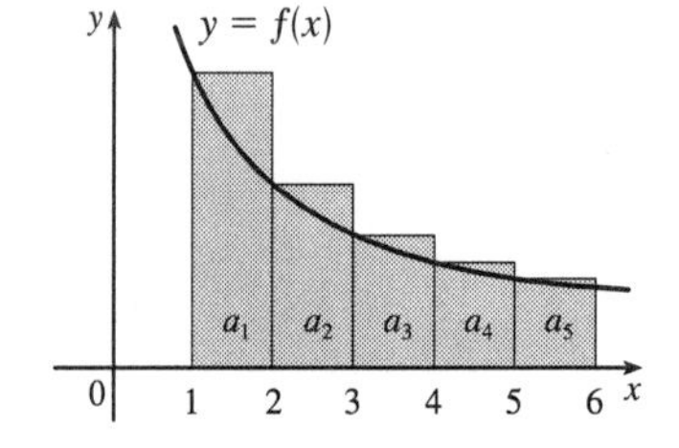

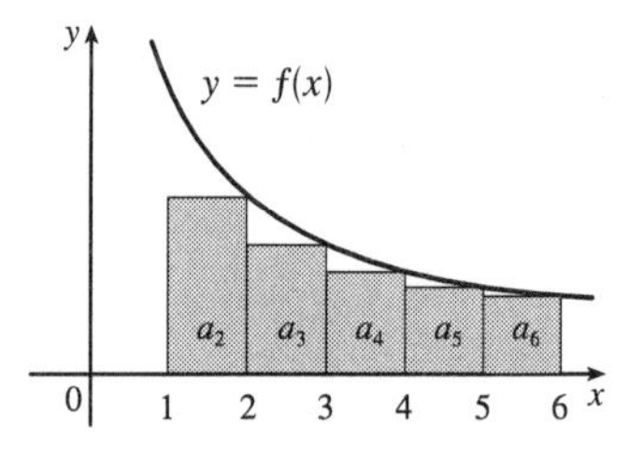

3. (a) We cannot say anything about $\sum a_n$. If $a_n > b_n$ for all n and $\sum b_n$ is convergent, then $\sum a_n$ could be convergent or divergent. (See the note after Example 4.)

(b) If $a_n < b_n$ for all n, then $\sum a_n$ is convergent. [This is part (i) of the Comparison Test.]

4. (a) If $a_n > b_n$ for all n, then $\sum a_n$ is divergent. [This is part (ii) of the Comparison Test.]

(b) We cannot say anything about $\sum a_n$. If $a_n < b_n$ for all n and $\sum b_n$ is divergent, then $\sum a_n$ could be convergent or divergent.

5. $\displaystyle\sum_{n=1}^{\infty} n^b$ is a p-series with $p = -b$. $\displaystyle\sum_{n=1}^{\infty} b^n$ is a geometric series. By (1), the p-series is convergent if $p > 1$. In this case, $\displaystyle\sum_{n=1}^{\infty} n^b = \sum_{n=1}^{\infty} (1/n^{-b})$, so $-b > 1 \Leftrightarrow b < -1$ are the values for which the series converge. A geometric series $\displaystyle\sum_{n=1}^{\infty} ar^{n-1}$ converges if $|r| < 1$, so $\displaystyle\sum_{n=1}^{\infty} b^n$ converges if $|b| < 1 \Leftrightarrow -1 < b < 1$.

6. The function $f(x) = 1/\sqrt[4]{x} = x^{-1/4}$ is continuous, positive, and decreasing on $[1, \infty)$, so the Integral Test applies.

$\int_1^\infty x^{-1/4}\,dx = \lim\limits_{t\to\infty} \int_1^t x^{-1/4}\,dx = \lim\limits_{t\to\infty} \left[\frac{4}{3}x^{3/4}\right]_1^t = \lim\limits_{t\to\infty} \left(\frac{4}{3}t^{3/4} - \frac{4}{3}\right) = \infty$, so $\sum_{n=1}^\infty 1/\sqrt[4]{n}$ diverges.

7. The function $f(x) = 1/x^4$ is continuous, positive, and decreasing on $[1, \infty)$, so the Integral Test applies.

$\displaystyle\int_1^\infty \frac{1}{x^4}\,dx = \lim_{t\to\infty} \int_1^t x^{-4}\,dx = \lim_{t\to\infty} \left[\frac{x^{-3}}{-3}\right]_1^t = \lim_{t\to\infty} \left(-\frac{1}{3t^3} + \frac{1}{3}\right) = \frac{1}{3}$. Since this improper integral is convergent, the series $\displaystyle\sum_{n=1}^\infty \frac{1}{n^4}$ is also convergent by the Integral Test.

8. The function $f(x) = 1/(x^2 + 1)$ is continuous, positive, and decreasing on $[1, \infty)$, so the Integral Test applies.

$\displaystyle\int_1^\infty \frac{1}{x^2+1}\,dx = \lim_{t\to\infty} \int_1^t \frac{1}{x^2+1}\,dx = \lim_{t\to\infty} \left[\tan^{-1} x\right]_1^t = \lim_{t\to\infty} \left(\tan^{-1} t - \tan^{-1} 1\right) = \frac{\pi}{2} - \frac{\pi}{4} = \frac{\pi}{4}$, so $\displaystyle\sum_{n=1}^\infty \frac{1}{n^2+1}$ converges.

9. $\displaystyle\frac{1}{n^2+n+1} < \frac{1}{n^2}$ for all $n \geq 1$, so $\displaystyle\sum_{n=1}^\infty \frac{1}{n^2+n+1}$ converges by comparison with $\displaystyle\sum_{n=1}^\infty \frac{1}{n^2}$, which converges because it is a p-series with $p = 2 > 1$.

10. $\displaystyle\frac{\sqrt{n}}{n-1} > \frac{\sqrt{n}}{n} = \frac{1}{\sqrt{n}}$, so $\displaystyle\sum_{n=2}^\infty \frac{\sqrt{n}}{n-1}$ diverges by comparison with the divergent (partial) p-series $\displaystyle\sum_{n=2}^\infty \frac{1}{\sqrt{n}}$ $\left[p = \frac{1}{2} \leq 1\right]$.

11. $\displaystyle 1 + \frac{1}{8} + \frac{1}{27} + \frac{1}{64} + \frac{1}{125} + \cdots = \sum_{n=1}^\infty \frac{1}{n^3}$. This is a p-series with $p = 3 > 1$, so it converges by (1).

12. $\displaystyle\sum_{n=1}^\infty \frac{1}{n^4}$ and $\displaystyle\sum_{n=1}^\infty \frac{1}{n^{3/2}}$ are convergent p-series with $p = 4 > 1$ and $p = \frac{3}{2} > 1$, respectively. Thus,

$\displaystyle\sum_{n=1}^\infty \left(\frac{5}{n^4} + \frac{4}{n\sqrt{n}}\right) = 5\sum_{n=1}^\infty \frac{1}{n^4} + 4\sum_{n=1}^\infty \frac{1}{n^{3/2}}$ is convergent by Theorems 8.2.8(i) and 8.2.8(ii).

13. $f(x) = xe^{-x}$ is continuous and positive on $[1, \infty)$. $f'(x) = -xe^{-x} + e^{-x} = e^{-x}(1 - x) < 0$ for $x > 1$, so f is decreasing on $[1, \infty)$. Thus, the Integral Test applies.

$$\begin{aligned} \int_1^\infty xe^{-x}\,dx &= \lim_{b\to\infty} \int_1^b xe^{-x}\,dx = \lim_{b\to\infty} \left[-xe^{-x} - e^{-x}\right]_1^b \qquad \text{[by parts]} \\ &= \lim_{b\to\infty} \left[-be^{-b} - e^{-b} + e^{-1} + e^{-1}\right] = 2/e \end{aligned}$$

since $\displaystyle\lim_{b\to\infty} be^{-b} = \lim_{b\to\infty} (b/e^b) \overset{\text{H}}{=} \lim_{b\to\infty} (1/e^b) = 0$ and $\displaystyle\lim_{b\to\infty} e^{-b} = 0$. Thus, $\displaystyle\sum_{n=1}^\infty ne^{-n}$ converges.

14. $\displaystyle f(x) = \frac{x^2}{x^3+1}$ is continuous and positive on $[2, \infty)$, and also decreasing since $\displaystyle f'(x) = \frac{x(2 - x^3)}{(x^3+1)^2} < 0$ for $x \geq 2$, so we can use the Integral Test [note that f is *not* decreasing on $[1, \infty)$].

$\displaystyle\int_2^\infty \frac{x^2}{x^3+1}\,dx = \lim_{t\to\infty} \left[\tfrac{1}{3}\ln(x^3+1)\right]_2^t = \tfrac{1}{3}\lim_{t\to\infty} \left[\ln(t^3+1) - \ln 9\right] = \infty$, so the series $\displaystyle\sum_{n=2}^\infty \frac{n^2}{n^3+1}$ diverges, and so does the given series, $\displaystyle\sum_{n=1}^\infty \frac{n^2}{n^3+1}$.

Another solution: Use the Limit Comparison Test with $\displaystyle a_n = \frac{n^2}{n^3+1}$ and $\displaystyle b_n = \frac{1}{n}$:

$\displaystyle\lim_{n\to\infty} \frac{a_n}{b_n} = \lim_{n\to\infty} \frac{n^2 \cdot n}{n^3+1} = \lim_{n\to\infty} \frac{1}{1 + 1/n^3} = 1 > 0$. Since the harmonic series $\displaystyle\sum_{n=1}^\infty \frac{1}{n}$ diverges, so does $\displaystyle\sum_{n=1}^\infty \frac{n^2}{n^3+1}$.

15. $f(x) = \dfrac{1}{x\ln x}$ is continuous and positive on $[2, \infty)$, and also decreasing since $f'(x) = -\dfrac{1+\ln x}{x^2(\ln x)^2} < 0$ for $x > 2$, so we can use the Integral Test. $\displaystyle\int_2^\infty \frac{1}{x\ln x}\,dx = \lim_{t\to\infty}\left[\ln(\ln x)\right]_2^t = \lim_{t\to\infty}\left[\ln(\ln t) - \ln(\ln 2)\right] = \infty$, so the series $\displaystyle\sum_{n=2}^\infty \frac{1}{n\ln n}$ diverges.

16. $\dfrac{n^2-1}{3n^4+1} < \dfrac{n^2}{3n^4+1} < \dfrac{n^2}{3n^4} = \dfrac{1}{3}\dfrac{1}{n^2}$. $\displaystyle\sum_{n=1}^\infty \frac{n^2-1}{3n^4+1}$ converges by comparison with $\displaystyle\sum_{n=1}^\infty \frac{1}{3n^2}$, which converges because it is a constant multiple of a convergent p-series $[p = 2 > 1]$. The terms of the given series are positive for $n > 1$, which is good enough.

17. $\dfrac{\cos^2 n}{n^2+1} \le \dfrac{1}{n^2+1} < \dfrac{1}{n^2}$, so the series $\displaystyle\sum_{n=1}^\infty \frac{\cos^2 n}{n^2+1}$ converges by comparison with the p-series $\displaystyle\sum_{n=1}^\infty \frac{1}{n^2}$ $[p = 2 > 1]$.

18. $\dfrac{4+3^n}{2^n} > \dfrac{3^n}{2^n} = \left(\dfrac{3}{2}\right)^n$ for all $n \ge 1$, so $\displaystyle\sum_{n=1}^\infty \frac{4+3^n}{2^n}$ diverges by comparison with the divergent geometric series $\displaystyle\sum_{n=1}^\infty \left(\tfrac{3}{2}\right)^n$.

19. $\dfrac{n-1}{n\,4^n}$ is positive for $n > 1$ and $\dfrac{n-1}{n\,4^n} < \dfrac{n}{n\,4^n} = \dfrac{1}{4^n} = \left(\dfrac{1}{4}\right)^n$, so $\displaystyle\sum_{n=1}^\infty \frac{n-1}{n\,4^n}$ converges by comparison with the convergent geometric series $\displaystyle\sum_{n=1}^\infty \left(\frac{1}{4}\right)^n$.

20. $\dfrac{1}{\sqrt{n^3+1}} < \dfrac{1}{\sqrt{n^3}} = \dfrac{1}{n^{3/2}}$, so $\displaystyle\sum_{n=1}^\infty \frac{1}{\sqrt{n^3+1}}$ converges by comparison with the convergent p-series $\displaystyle\sum_{n=1}^\infty \frac{1}{n^{3/2}}$ $\left[p = \tfrac{3}{2} > 1\right]$.

21. Use the Limit Comparison Test with $a_n = \dfrac{1}{\sqrt{n^2+1}}$ and $b_n = \dfrac{1}{n}$:

$\displaystyle\lim_{n\to\infty}\frac{a_n}{b_n} = \lim_{n\to\infty}\frac{n}{\sqrt{n^2+1}} = \lim_{n\to\infty}\frac{1}{\sqrt{1+(1/n^2)}} = 1 > 0$. Since the harmonic series $\displaystyle\sum_{n=1}^\infty \frac{1}{n}$ diverges, so does $\displaystyle\sum_{n=1}^\infty \frac{1}{\sqrt{n^2+1}}$.

22. Use the Limit Comparison Test with $a_n = \dfrac{1}{2n+3}$ and $b_n = \dfrac{1}{n}$: $\displaystyle\lim_{n\to\infty}\frac{a_n}{b_n} = \lim_{n\to\infty}\frac{n}{2n+3} = \lim_{n\to\infty}\frac{1}{2+(3/n)} = \frac{1}{2} > 0$.

Since the harmonic series $\displaystyle\sum_{n=1}^\infty \frac{1}{n}$ diverges, so does $\displaystyle\sum_{n=1}^\infty \frac{1}{2n+3}$.

23. $\dfrac{2+(-1)^n}{n\sqrt{n}} \le \dfrac{3}{n\sqrt{n}}$, and $\displaystyle\sum_{n=1}^\infty \frac{3}{n\sqrt{n}}$ converges because it is a constant multiple of the convergent p-series $\displaystyle\sum_{n=1}^\infty \frac{1}{n\sqrt{n}}$ $\left[p = \tfrac{3}{2} > 1\right]$, so the given series converges by the Comparison Test.

24. $\dfrac{1+\sin n}{10^n} \le \dfrac{2}{10^n}$ and $\displaystyle\sum_{n=0}^\infty \frac{2}{10^n} = 2\sum_{n=0}^\infty \left(\frac{1}{10}\right)^n$, so the given series converges by comparison with a constant multiple of a convergent geometric series.

25. Use the Limit Comparison Test with $a_n = \sin\left(\frac{1}{n}\right)$ and $b_n = \frac{1}{n}$. Then $\sum a_n$ and $\sum b_n$ are series with positive terms and

$\lim\limits_{n\to\infty} \frac{a_n}{b_n} = \lim\limits_{n\to\infty} \frac{\sin(1/n)}{1/n} = \lim\limits_{\theta\to 0} \frac{\sin\theta}{\theta} = 1 > 0$. Since $\sum\limits_{n=1}^{\infty} b_n$ is the divergent harmonic series, $\sum\limits_{n=1}^{\infty} \sin(1/n)$ also diverges.

Note: We could also use l'Hospital's Rule to evaluate the limit:

$$\lim_{x\to\infty} \frac{\sin(1/x)}{1/x} \overset{\text{H}}{=} \lim_{x\to\infty} \frac{\cos(1/x)\cdot(-1/x^2)}{-1/x^2} = \lim_{x\to\infty} \cos\frac{1}{x} = \cos 0 = 1$$

26. If $a_n = \dfrac{n+5}{\sqrt[3]{n^7+n^2}}$ and $b_n = \dfrac{n}{\sqrt[3]{n^7}} = \dfrac{n}{n^{7/3}} = \dfrac{1}{n^{4/3}}$, then

$$\lim_{n\to\infty} \frac{a_n}{b_n} = \lim_{n\to\infty} \frac{n^{7/3}+5n^{4/3}}{(n^7+n^2)^{1/3}}\cdot\frac{n^{-7/3}}{n^{-7/3}} = \lim_{n\to\infty} \frac{1+5/n}{[(n^7+n^2)/n^7]^{1/3}} = \lim_{n\to\infty} \frac{1+5/n}{(1+1/n^5)^{1/3}} = \frac{1+0}{(1+0)^{1/3}} = 1 > 0,$$

so $\sum\limits_{n=1}^{\infty} \dfrac{n+5}{\sqrt[3]{n^7+n^2}}$ converges by the Limit Comparison Test with the convergent p-series $\sum\limits_{n=1}^{\infty} \dfrac{1}{n^{4/3}}$.

27. We have already shown (in Exercise 15) that when $p = 1$ the series $\sum\limits_{n=2}^{\infty} \dfrac{1}{n(\ln n)^p}$ diverges, so assume that $p \neq 1$.

$f(x) = \dfrac{1}{x(\ln x)^p}$ is continuous and positive on $[2,\infty)$, and $f'(x) = -\dfrac{p+\ln x}{x^2(\ln x)^{p+1}} < 0$ if $x > e^{-p}$, so that f is eventually decreasing and we can use the Integral Test.

$$\int_2^{\infty} \frac{1}{x(\ln x)^p}\,dx = \lim_{t\to\infty}\left[\frac{(\ln x)^{1-p}}{1-p}\right]_2^t \quad [\text{for } p \neq 1] = \lim_{t\to\infty}\left[\frac{(\ln t)^{1-p}}{1-p}\right] - \frac{(\ln 2)^{1-p}}{1-p}$$

This limit exists whenever $1-p<0 \iff p>1$, so the series converges for $p>1$.

28. If $p \le 0$, $\lim\limits_{n\to\infty} \dfrac{\ln n}{n^p} = \infty$ and the series diverges, so assume $p > 0$. $f(x) = \dfrac{\ln x}{x^p}$ is positive and continuous and $f'(x) < 0$ for $x > e^{1/p}$, so f is eventually decreasing and we can use the Integral Test. Integration by parts gives

$$\int_1^{\infty} \frac{\ln x}{x^p}\,dx = \lim_{t\to\infty}\left[\frac{x^{1-p}\,[(1-p)\ln x - 1]}{(1-p)^2}\right]_1^t \quad [\text{for } p \neq 1] = \frac{1}{(1-p)^2}\left[\lim_{t\to\infty} t^{1-p}\,[(1-p)\ln t - 1] + 1\right]$$

which exists whenever $1-p<0 \iff p>1$. Since we have already done the case $p=1$ in Exercise 27 (set $p=-1$ in that exercise), $\sum\limits_{n=1}^{\infty} \dfrac{\ln n}{n^p}$ converges $\iff p > 1$.

29. (a)

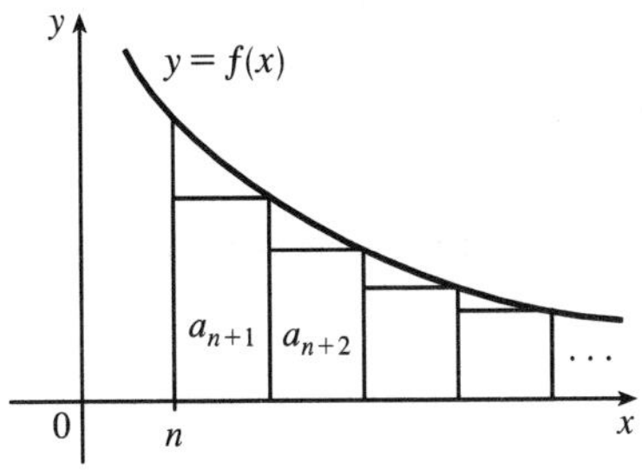

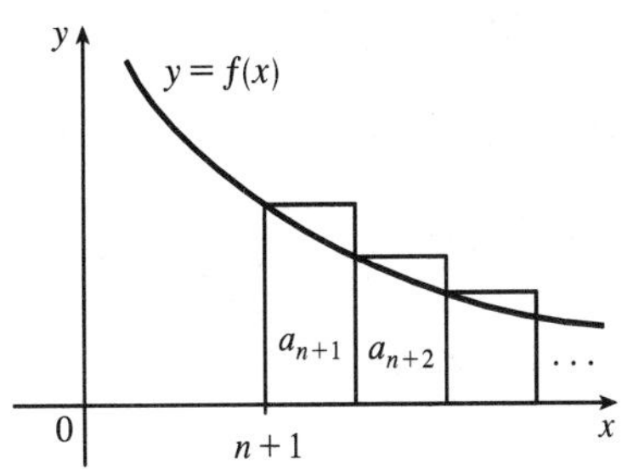

We use the same notation and ideas as in the Integral Test, assuming that f is decreasing on $[n,\infty)$. Comparing the areas of the rectangles with the area under $y = f(x)$ for $x > n$ in the first figure, we see that

$$R_n = a_{n+1} + a_{n+2} + \cdots \le \int_n^\infty f(x)\,dx$$

Similarly, we see from the second figure that

$$R_n = a_{n+1} + a_{n+2} + \cdots \ge \int_{n+1}^\infty f(x)\,dx$$

So we have proved that $\int_{n+1}^\infty f(x)\,dx \le R_n \le \int_n^\infty f(x)\,dx$.

(b) If we add s_n to each side of the inequalities in part (a), we get $s_n + \int_{n+1}^\infty f(x)\,dx \le s \le s_n + \int_n^\infty f(x)\,dx$ because $s_n + R_n = s$.

30. (a) $f(x) = 1/x^4$ is positive and continuous and $f'(x) = -4/x^5$ is negative for $x > 0$, and so the Integral Test applies.

$\sum\limits_{n=1}^\infty \dfrac{1}{n^4} \approx s_{10} = \dfrac{1}{1^4} + \dfrac{1}{2^4} + \dfrac{1}{3^4} + \cdots + \dfrac{1}{10^4} \approx 1.082037$. From Exercise 29(a) we have

$R_{10} \le \displaystyle\int_{10}^\infty \frac{1}{x^4}\,dx = \lim_{t\to\infty}\left[\frac{1}{-3x^3}\right]_{10}^t = \lim_{t\to\infty}\left(-\frac{1}{3t^3} + \frac{1}{3\,(10)^3}\right) = \frac{1}{3000}$, so the error is at most $0.000\overline{3}$.

(b) $s_{10} + \displaystyle\int_{11}^\infty \frac{1}{x^4}\,dx \le s \le s_{10} + \int_{10}^\infty \frac{1}{x^4}\,dx \;\Rightarrow\; s_{10} + \frac{1}{3(11)^3} \le s \le s_{10} + \frac{1}{3(10)^3} \;\Rightarrow$

$1.082037 + 0.000250 = 1.082287 \le s \le 1.082037 + 0.000333 = 1.082370$, so we get $s \approx 1.08233$ with error ≤ 0.00005.

(c) $R_n \le \displaystyle\int_n^\infty \frac{1}{x^4}\,dx = \frac{1}{3n^3}$. So $R_n < 0.00001 \;\Rightarrow\; \dfrac{1}{3n^3} < \dfrac{1}{10^5} \;\Rightarrow\; 3n^3 > 10^5 \;\Rightarrow\; n > \sqrt[3]{(10)^5/3} \approx 32.2$, that is, for $n > 32$.

31. (a) $f(x) = \dfrac{1}{x^2}$ is positive and continuous and $f'(x) = -\dfrac{2}{x^3}$ is negative for $x > 0$, and so the Integral Test applies.

$\sum\limits_{n=1}^\infty \dfrac{1}{n^2} \approx s_{10} = \dfrac{1}{1^2} + \dfrac{1}{2^2} + \dfrac{1}{3^2} + \cdots + \dfrac{1}{10^2} \approx 1.549768$. From Exercise 29(a) we have

$R_{10} \le \displaystyle\int_{10}^\infty \frac{1}{x^2}\,dx = \lim_{t\to\infty}\left[\frac{-1}{x}\right]_{10}^t = \lim_{t\to\infty}\left(-\frac{1}{t} + \frac{1}{10}\right) = \frac{1}{10}$, so the error is at most 0.1.

(b) $s_{10} + \displaystyle\int_{11}^\infty \frac{1}{x^2}\,dx \le s \le s_{10} + \int_{10}^\infty \frac{1}{x^2}\,dx \;\Rightarrow\; s_{10} + \tfrac{1}{11} \le s \le s_{10} + \tfrac{1}{10} \;\Rightarrow$

$1.549768 + 0.090909 = 1.640677 \le s \le 1.549768 + 0.1 = 1.649768$, so we get $s \approx 1.64522$ (the average of 1.640677 and 1.649768) with error ≤ 0.005 (the maximum of $1.649768 - 1.64522$ and $1.64522 - 1.640677$, rounded up).

(c) $R_n \le \displaystyle\int_n^\infty \frac{1}{x^2}\,dx = \frac{1}{n}$. So $R_n < 0.001$ if $\dfrac{1}{n} < \dfrac{1}{1000} \;\Leftrightarrow\; n > 1000$.

32. $f(x) = 1/x^5$ is positive and continuous and $f'(x) = -5/x^6$ is negative for $x > 0$, and so the Integral Test applies.

Using Exercise 29(a), $R_n \le \displaystyle\int_n^\infty x^{-5}\,dx = \lim_{t\to\infty}\left[\frac{-1}{4x^4}\right]_n^t = \frac{1}{4n^4}$. If we take $n = 5$, then $s_5 \approx 1.036662$ and $R_5 \le 0.0004$.

So $s \approx s_5 \approx 1.037$.

33. (a) From the figure, $a_2 + a_3 + \cdots + a_n \leq \int_1^n f(x)\,dx$, so with

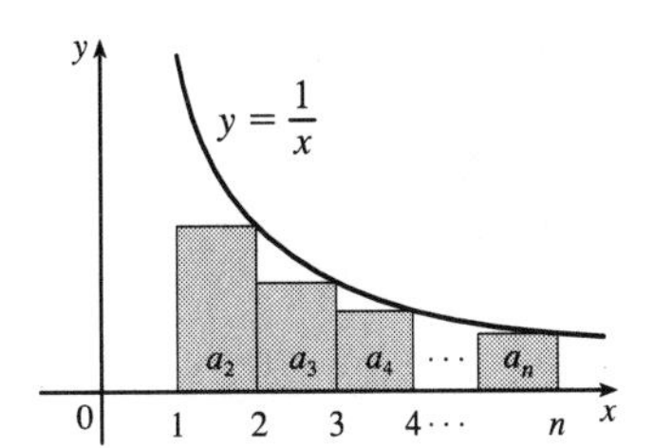

$$f(x) = \frac{1}{x},\ \frac{1}{2} + \frac{1}{3} + \frac{1}{4} + \cdots + \frac{1}{n} \leq \int_1^n \frac{1}{x}\,dx = \ln n. \text{ Thus,}$$

$$s_n = 1 + \frac{1}{2} + \frac{1}{3} + \frac{1}{4} + \cdots + \frac{1}{n} \leq 1 + \ln n.$$

(b) By part (a), $s_{10^6} \leq 1 + \ln 10^6 \approx 14.82 < 15$ and $s_{10^9} \leq 1 + \ln 10^9 \approx 21.72 < 22$.

34. $\sum\limits_{n=1}^{\infty} n^{-1.001} = \sum\limits_{n=1}^{\infty} \frac{1}{n^{1.001}}$ is a convergent p-series with $p = 1.001 > 1$. Using Exercise 29(a), we get

$$R_n \leq \int_n^{\infty} x^{-1.001}\,dx = \lim_{t\to\infty}\left[\frac{x^{-0.001}}{-0.001}\right]_n^t = -1000 \lim_{t\to\infty}\left[\frac{1}{x^{0.001}}\right]_n^t = -1000\left(-\frac{1}{n^{0.001}}\right) = \frac{1000}{n^{0.001}}.$$

We want $R_n < 0.000\,000\,005 \quad \Leftrightarrow \quad \frac{1000}{n^{0.001}} < 5 \times 10^{-9} \quad \Leftrightarrow \quad n^{0.001} > \frac{1000}{5 \times 10^{-9}} \quad \Leftrightarrow$

$n > \left(2 \times 10^{11}\right)^{1000} = 2^{1000} \times 10^{11,000} \approx 1.07 \times 10^{301} \times 10^{11,000} = 1.07 \times 10^{11,301}$.

35. Since $\frac{d_n}{10^n} \leq \frac{9}{10^n}$ for each n, and since $\sum\limits_{n=1}^{\infty} \frac{9}{10^n}$ is a convergent geometric series $\left(|r| = \frac{1}{10} < 1\right)$, $0.d_1d_2d_3 \ldots = \sum\limits_{n=1}^{\infty} \frac{d_n}{10^n}$ will always converge by the Comparison Test.

36. First we observe that, by l'Hospital's Rule, $\lim\limits_{x\to 0} \frac{\ln(1+x)}{x} = \lim\limits_{x\to 0} \frac{1}{1+x} = 1$. Also, if $\sum a_n$ converges, then $\lim\limits_{n\to\infty} a_n = 0$ by Theorem 8.2.6. Therefore, $\lim\limits_{n\to\infty} \frac{\ln(1+a_n)}{a_n} = \lim\limits_{x\to 0} \frac{\ln(1+x)}{x} = 1 > 0$. We are given that $\sum a_n$ is convergent and $a_n > 0$. Thus, $\sum \ln(1+a_n)$ is convergent by the Limit Comparison Test.

37. Yes. Since $\sum a_n$ is a convergent series with positive terms, $\lim\limits_{n\to\infty} a_n = 0$ by Theorem 8.2.6, and $\sum b_n = \sum \sin(a_n)$ is a series with positive terms (for large enough n). We have $\lim\limits_{n\to\infty} \frac{b_n}{a_n} = \lim\limits_{n\to\infty} \frac{\sin(a_n)}{a_n} = 1 > 0$ by Theorem 1.4.5. Thus, $\sum b_n$ is also convergent by the Limit Comparison Test.

38. (a) Since $\lim\limits_{n\to\infty} (a_n/b_n) = 0$, there is a number $N > 0$ such that $|a_n/b_n - 0| < 1$ for all $n > N$, and so $a_n < b_n$ since a_n and b_n are positive. Thus, since $\sum b_n$ converges, so does $\sum a_n$ by the Comparison Test.

(b) (i) If $a_n = \frac{\ln n}{n^3}$ and $b_n = \frac{1}{n^2}$, then $\lim\limits_{n\to\infty} \frac{a_n}{b_n} = \lim\limits_{n\to\infty} \frac{\ln n}{n} = \lim\limits_{x\to\infty} \frac{\ln x}{x} \overset{H}{=} \lim\limits_{x\to\infty} \frac{1/x}{1} = 0$, so $\sum\limits_{n=1}^{\infty} \frac{\ln n}{n^3}$ converges by part (a).

(ii) If $a_n = \frac{\ln n}{\sqrt{n}\,e^n}$ and $b_n = \frac{1}{e^n}$, then

$$\lim_{n\to\infty} \frac{a_n}{b_n} = \lim_{n\to\infty} \frac{\ln n}{\sqrt{n}} = \lim_{x\to\infty} \frac{\ln x}{\sqrt{x}} \overset{H}{=} \lim_{x\to\infty} \frac{1/x}{1/\left(2\sqrt{x}\right)} = \lim_{x\to\infty} \frac{2}{\sqrt{x}} = 0$$

Now $\sum b_n$ is a convergent geometric series with ratio $r = 1/e$ ($|r| < 1$), so $\sum a_n$ converges by part (a).

39. (a) Since $\lim\limits_{n\to\infty} \frac{a_n}{b_n} = \infty$, there is an integer N such that $\frac{a_n}{b_n} > 1$ whenever $n > N$. (Take $M = 1$ in Definition 8.1.5.) Then $a_n > b_n$ whenever $n > N$ and since $\sum b_n$ is divergent, $\sum a_n$ is also divergent by the Comparison Test.

(b) (i) If $a_n = \frac{1}{\ln n}$ and $b_n = \frac{1}{n}$ for $n \geq 2$, then $\lim\limits_{n\to\infty} \frac{a_n}{b_n} = \lim\limits_{n\to\infty} \frac{n}{\ln n} = \lim\limits_{x\to\infty} \frac{x}{\ln x} \overset{H}{=} \lim\limits_{x\to\infty} \frac{1}{1/x} = \lim\limits_{x\to\infty} x = \infty$, so by part (a), $\sum\limits_{n=2}^{\infty} \frac{1}{\ln n}$ is divergent.

(ii) If $a_n = \dfrac{\ln n}{n}$ and $b_n = \dfrac{1}{n}$, then $\sum\limits_{n=1}^{\infty} b_n$ is the divergent harmonic series and $\lim\limits_{n\to\infty} \dfrac{a_n}{b_n} = \lim\limits_{n\to\infty} \ln n = \lim\limits_{x\to\infty} \ln x = \infty$, so $\sum\limits_{n=1}^{\infty} a_n$ diverges by part (a).

40. Let $a_n = \dfrac{1}{n^2}$ and $b_n = \dfrac{1}{n}$. Then $\lim\limits_{n\to\infty} \dfrac{a_n}{b_n} = \lim\limits_{n\to\infty} \dfrac{1}{n} = 0$, but $\sum b_n$ diverges while $\sum a_n$ converges.

41. Since $\sum a_n$ converges, $\lim\limits_{n\to\infty} a_n = 0$, so there exists N such that $|a_n - 0| < 1$ for all $n > N \Rightarrow 0 \le a_n < 1$ for all $n > N \Rightarrow 0 \le a_n^2 \le a_n$. Since $\sum a_n$ converges, so does $\sum a_n^2$ by the Comparison Test.

42. $b^{\ln n} = \left(e^{\ln b}\right)^{\ln n} = \left(e^{\ln n}\right)^{\ln b} = n^{\ln b} = \dfrac{1}{n^{-\ln b}}$. This is a p-series, which converges for all b such that $-\ln b > 1 \Leftrightarrow \ln b < -1 \Leftrightarrow b < e^{-1} \Leftrightarrow b < 1/e$ [with $b > 0$].

8.4 Other Convergence Tests

1. (a) An alternating series is a series whose terms are alternately positive and negative.

(b) An alternating series $\sum\limits_{n=1}^{\infty} (-1)^{n-1} b_n$ converges if $0 < b_{n+1} \le b_n$ for all n and $\lim\limits_{n\to\infty} b_n = 0$. (This is the Alternating Series Test.)

(c) The error involved in using the partial sum s_n as an approximation to the total sum s is the remainder $R_n = s - s_n$ and the size of the error is smaller than b_{n+1}; that is, $|R_n| \le b_{n+1}$. (This is the Alternating Series Estimation Theorem.)

2. (a) Since $\lim\limits_{n\to\infty} \left|\dfrac{a_{n+1}}{a_n}\right| = 8 > 1$, part (b) of the Ratio Test tells us that the series $\sum a_n$ is divergent.

(b) Since $\lim\limits_{n\to\infty} \left|\dfrac{a_{n+1}}{a_n}\right| = 0.8 < 1$, part (a) of the Ratio Test tells us that the series $\sum a_n$ is absolutely convergent (and therefore convergent).

(c) Since $\lim\limits_{n\to\infty} \left|\dfrac{a_{n+1}}{a_n}\right| = 1$, the Ratio Test fails and the series $\sum a_n$ might converge or it might diverge.

3. $\dfrac{4}{7} - \dfrac{4}{8} + \dfrac{4}{9} - \dfrac{4}{10} + \dfrac{4}{11} - \cdots = \sum\limits_{n=1}^{\infty} (-1)^{n-1} \dfrac{4}{n+6}$. Now $b_n = \dfrac{4}{n+6} > 0$, $\{b_n\}$ is decreasing, and $\lim\limits_{n\to\infty} b_n = 0$, so the series converges by the Alternating Series Test.

4. $-\dfrac{1}{3} + \dfrac{2}{4} - \dfrac{3}{5} + \dfrac{4}{6} - \dfrac{5}{7} + \cdots = \sum\limits_{n=1}^{\infty} (-1)^n \dfrac{n}{n+2}$. Here $a_n = (-1)^n \dfrac{n}{n+2}$. Since $\lim\limits_{n\to\infty} a_n \ne 0$ (in fact the limit does not exist), the series diverges by the Test for Divergence.

5. $b_n = \dfrac{1}{\sqrt{n}} > 0$, $\{b_n\}$ is decreasing, and $\lim\limits_{n\to\infty} b_n = 0$, so the series $\sum\limits_{n=1}^{\infty} \dfrac{(-1)^{n-1}}{\sqrt{n}}$ converges by the Alternating Series Test.

6. $\sum\limits_{n=1}^{\infty} a_n = \sum\limits_{n=1}^{\infty} (-1)^n \dfrac{\sqrt{n}}{1 + 2\sqrt{n}} = \sum\limits_{n=1}^{\infty} (-1)^n b_n$. Now $\lim\limits_{n\to\infty} b_n = \lim\limits_{n\to\infty} \dfrac{1}{2 + 1/\sqrt{n}} = \dfrac{1}{2} \ne 0$. Since $\lim\limits_{n\to\infty} a_n \ne 0$ (in fact the limit does not exist), the series diverges by the Test for Divergence.

7. $\sum\limits_{n=1}^{\infty} a_n = \sum\limits_{n=1}^{\infty} (-1)^n \dfrac{3n-1}{2n+1} = \sum\limits_{n=1}^{\infty} (-1)^n b_n$. Now $\lim\limits_{n\to\infty} b_n = \lim\limits_{n\to\infty} \dfrac{3 - 1/n}{2 + 1/n} = \dfrac{3}{2} \ne 0$. Since $\lim\limits_{n\to\infty} a_n \ne 0$ (in fact the limit does not exist), the series diverges by the Test for Divergence.

8. $\sum_{n=1}^{\infty}(-1)^{n-1}\left(\frac{\ln n}{n}\right) = 0 + \sum_{n=2}^{\infty}(-1)^{n-1}\left(\frac{\ln n}{n}\right)$. $b_n = \frac{\ln n}{n} > 0$ for $n \geq 2$, and if $f(x) = \frac{\ln x}{x}$, then

$f'(x) = \frac{1 - \ln x}{x^2} < 0$ for $x > e$, so $\{b_n\}$ is eventually decreasing. Also,

$\lim\limits_{n\to\infty} b_n = \lim\limits_{n\to\infty} \frac{\ln n}{n} = \lim\limits_{x\to\infty} \frac{\ln x}{x} \overset{\text{H}}{=} \lim\limits_{x\to\infty} \frac{1/x}{1} = 0$, so the series converges by the Alternating Series Test.

9. The series $\sum_{n=1}^{\infty} \frac{(-2)^n}{n!} = \sum_{n=1}^{\infty}(-1)^n \frac{2^n}{n!}$ satisfies (i) of the Alternating Series Test because

$b_{n+1} = \frac{2^{n+1}}{(n+1)!} = \frac{2 \cdot 2^n}{(n+1)n!} = \frac{2}{n+1} \cdot \frac{2^n}{n!} = \frac{2}{n+1} \cdot b_n \leq b_n$ and (ii) $\lim\limits_{n\to\infty} \frac{2^n}{n!} = \frac{2}{n} \cdot \frac{2}{n-1} \cdot \cdots \cdot \frac{2}{2} \cdot \frac{2}{1} = 0$, so the series is convergent. Now $b_7 = 2^7/7! \approx 0.025 > 0.01$ and $b_8 = 2^8/8! \approx 0.006 < 0.01$, so by the Alternating Series Estimation Theorem, $n = 7$. (That is, since the 8th term is less than the desired error, we need to add the first 7 terms to get the sum to the desired accuracy.)

10. The series $\sum_{n=1}^{\infty} \frac{(-1)^n}{n\,5^n}$ satisfies (i) of the Alternating Series Test because $\frac{1}{(n+1)5^{n+1}} < \frac{1}{n\,5^n}$ and (ii) $\lim\limits_{n\to\infty} \frac{1}{n\,5^n} = 0$, so the series is convergent. Now $b_4 = \frac{1}{4 \cdot 5^4} = 0.0004 > 0.0001$ and $b_5 = \frac{1}{5 \cdot 5^5} = 0.000064 < 0.0001$, so by the Alternating Series Estimation Theorem, $n = 4$. (That is, since the 5th term is less than the desired error, we need to add the first 4 terms to get the sum to the desired accuracy.)

11. The series $\sum_{n=1}^{\infty} \frac{(-1)^{n+1}}{n^6}$ satisfies (i) of the Alternating Series Test because $\frac{1}{(n+1)^6} < \frac{1}{n^6}$ and (ii) $\lim\limits_{n\to\infty} \frac{1}{n^6} = 0$, so the series is convergent. Now $b_5 = \frac{1}{5^6} = 0.000064 > 0.00005$ and $b_6 = \frac{1}{6^6} \approx 0.00002 < 0.00005$, so by the Alternating Series Estimation Theorem, $n = 5$. (That is, since the 6th term is less than the desired error, we need to add the first 5 terms to get the sum to the desired accuracy.)

12. Using the Ratio Test with the series $\sum_{n=1}^{\infty}(-1)^{n-1}ne^{-n} = \sum_{n=1}^{\infty}(-1)^{n-1}\frac{n}{e^n}$,

$$\lim_{n\to\infty}\left|\frac{a_{n+1}}{a_n}\right| = \lim_{n\to\infty}\left|\frac{(-1)^n(n+1)}{e^{n+1}} \cdot \frac{e^n}{(-1)^{n-1}n}\right| = \lim_{n\to\infty}\left|\frac{(-1)^1(n+1)}{en}\right| = \frac{1}{e}\lim_{n\to\infty}\frac{n+1}{n} = \frac{1}{e}(1) = \frac{1}{e} < 1,$$

so the series is absolutely convergent (and therefore convergent). Now $b_6 = 6/e^6 \approx 0.015 > 0.01$ and $b_7 = 7/e^7 \approx 0.006 < 0.01$, so by the Alternating Series Estimation Theorem, $n = 6$. (That is, since the 7th term is less than the desired error, we need to add the first 6 terms to get the sum to the desired accuracy.)

13. $b_7 = \frac{1}{7^5} = \frac{1}{16{,}807} \approx 0.000\,059\,5$, so

$\sum_{n=1}^{\infty} \frac{(-1)^{n+1}}{n^5} \approx s_6 = \sum_{n=1}^{6} \frac{(-1)^{n+1}}{n^5} = 1 - \frac{1}{32} + \frac{1}{243} - \frac{1}{1024} + \frac{1}{3125} - \frac{1}{7776} \approx 0.972\,080$. Adding b_7 to s_6 does not change the fourth decimal place of s_6, so the sum of the series, correct to four decimal places, is 0.9721.

14. $b_6 = \frac{6}{8^6} = \frac{6}{262{,}144} \approx 0.000\,023$, so $\sum_{n=1}^{\infty} \frac{(-1)^n n}{8^n} \approx s_5 = \sum_{n=1}^{5} \frac{(-1)^n n}{8^n} = -\frac{1}{8} + \frac{2}{64} - \frac{3}{512} + \frac{4}{4096} - \frac{5}{32{,}768} \approx -0.098\,785$.

Adding b_6 to s_5 does not change the fourth decimal place of s_5, so the sum of the series, correct to four decimal places, is -0.0988.

15. $b_7 = \dfrac{7^2}{10^7} = 0.000\,004\,9$, so

$\displaystyle\sum_{n=1}^{\infty} \frac{(-1)^{n-1}n^2}{10^n} \approx s_6 = \sum_{n=1}^{6} \frac{(-1)^{n-1}n^2}{10^n} = \tfrac{1}{10} - \tfrac{4}{100} + \tfrac{9}{1000} - \tfrac{16}{10{,}000} + \tfrac{25}{100{,}000} - \tfrac{36}{1{,}000{,}000} = 0.067\,614$. Adding b_7 to s_6 does not change the fourth decimal place of s_6, so the sum of the series, correct to four decimal places, is 0.0676.

16. $b_6 = \dfrac{1}{3^6 \cdot 6!} = \dfrac{1}{524{,}880} \approx 0.000\,001\,9$, so

$\displaystyle\sum_{n=1}^{\infty} \frac{(-1)^n}{3^n n!} \approx s_5 = \sum_{n=1}^{5} \frac{(-1)^n}{3^n n!} = -\tfrac{1}{3} + \tfrac{1}{18} - \tfrac{1}{162} + \tfrac{1}{1944} - \tfrac{1}{29{,}160} \approx -0.283\,471$. Adding b_6 to s_5 does not change the fourth decimal place of s_5, so the sum of the series, correct to four decimal places, is -0.2835.

17. $\displaystyle\sum_{n=1}^{\infty} \frac{(-1)^{n-1}}{n} = 1 - \frac{1}{2} + \frac{1}{3} - \frac{1}{4} + \cdots + \frac{1}{49} - \frac{1}{50} + \frac{1}{51} - \frac{1}{52} + \cdots$. The 50th partial sum of this series is an underestimate, since $\displaystyle\sum_{n=1}^{\infty} \frac{(-1)^{n-1}}{n} = s_{50} + \left(\frac{1}{51} - \frac{1}{52}\right) + \left(\frac{1}{53} - \frac{1}{54}\right) + \cdots$, and the terms in parentheses are all positive. The result can be seen geometrically in Figure 1.

18. If $p > 0$, $\dfrac{1}{(n+1)^p} \le \dfrac{1}{n^p}$ $[\{1/n^p\}$ is decreasing$]$ and $\displaystyle\lim_{n\to\infty} \frac{1}{n^p} = 0$, so the series converges by the Alternating Series Test. If $p \le 0$, $\displaystyle\lim_{n\to\infty} \frac{(-1)^{n-1}}{n^p}$ does not exist, so the series diverges by the Test for Divergence. Thus,

$\displaystyle\sum_{n=1}^{\infty} \frac{(-1)^{n-1}}{n^p}$ converges $\Leftrightarrow$ $p > 0$.

19. Using the Ratio Test, $\displaystyle\lim_{n\to\infty} \left|\frac{a_{n+1}}{a_n}\right| = \lim_{n\to\infty} \left|\frac{(-3)^{n+1}/(n+1)^3}{(-3)^n/n^3}\right| = \lim_{n\to\infty} \left|\frac{(-3)n^3}{(n+1)^3}\right| = 3 \lim_{n\to\infty} \left(\frac{n}{n+1}\right)^3 = 3 > 1$,

so the series diverges.

20. The series $\displaystyle\sum_{n=1}^{\infty} \frac{n^2}{2^n}$ has positive terms and $\displaystyle\lim_{n\to\infty} \frac{a_{n+1}}{a_n} = \lim_{n\to\infty} \left[\frac{(n+1)^2}{2^{n+1}} \cdot \frac{2^n}{n^2}\right] = \lim_{n\to\infty} \left(1 + \frac{1}{n}\right)^2 \cdot \frac{1}{2} = \frac{1}{2} < 1$,

so the series is absolutely convergent by the Ratio Test.

21. $\displaystyle\sum_{n=0}^{\infty} \frac{(-10)^n}{n!}$. Using the Ratio Test, $\displaystyle\lim_{n\to\infty} \left|\frac{a_{n+1}}{a_n}\right| = \lim_{n\to\infty} \left|\frac{(-10)^{n+1}}{(n+1)!} \cdot \frac{n!}{(-10)^n}\right| = \lim_{n\to\infty} \left|\frac{-10}{n+1}\right| = 0 < 1$, so the series is absolutely convergent.

22. $\displaystyle\sum_{n=1}^{\infty} \frac{n}{n^2+1}$ diverges by the Limit Comparison Test with the harmonic series: $\displaystyle\lim_{n\to\infty} \frac{n/(n^2+1)}{1/n} = \lim_{n\to\infty} \frac{n^2}{n^2+1} = 1$. But

$\displaystyle\sum_{n=1}^{\infty} (-1)^{n-1} \frac{n}{n^2+1}$ converges by the Alternating Series Test: $\left\{\dfrac{n}{n^2+1}\right\}$ has positive terms, is decreasing since

$\displaystyle\left(\frac{x}{x^2+1}\right)' = \frac{1-x^2}{(x^2+1)^2} \le 0$ for $x \ge 1$, and $\displaystyle\lim_{n\to\infty} \frac{n}{n^2+1} = 0$. Thus, $\displaystyle\sum_{n=1}^{\infty} (-1)^{n-1} \frac{n}{n^2+1}$ is conditionally convergent.

23. $\displaystyle\sum_{n=1}^{\infty} \frac{(-1)^{n+1}}{\sqrt[4]{n}}$ converges by the Alternating Series Test, but $\displaystyle\sum_{n=1}^{\infty} \frac{1}{\sqrt[4]{n}}$ is a divergent p-series $\left[p = \frac{1}{4} \le 1\right]$,

so the given series is conditionally convergent.

24. $\sum_{n=1}^{\infty}(-1)^{n-1}\frac{2^n}{n^4}$ diverges by the Test for Divergence. $\lim_{n\to\infty}\frac{2^n}{n^4}=\infty$, so $\lim_{n\to\infty}(-1)^{n-1}\frac{2^n}{n^4}$ does not exist.

25. $\lim_{n\to\infty}\left|\frac{a_{n+1}}{a_n}\right|=\lim_{n\to\infty}\left[\frac{10^{n+1}}{(n+2)4^{2(n+1)+1}}\cdot\frac{(n+1)4^{2n+1}}{10^n}\right]=\lim_{n\to\infty}\left[\frac{10^{n+1}}{(n+2)4^{2n+3}}\cdot\frac{(n+1)4^{2n+1}}{10^n}\right]$

$$=\lim_{n\to\infty}\left(\frac{10}{4^2}\cdot\frac{n+1}{n+2}\right)=\frac{5}{8}<1,$$

so the series is absolutely convergent by the Ratio Test. Since the terms of this series are positive, absolute convergence is the same as convergence.

26. $\left|\frac{\sin 4n}{4^n}\right|\le\frac{1}{4^n}$, so $\sum_{n=1}^{\infty}\left|\frac{\sin 4n}{4^n}\right|$ converges by comparison with the convergent geometric series $\sum_{n=1}^{\infty}\frac{1}{4^n}$ $\left[|r|=\frac{1}{4}<1\right]$. Thus, $\sum_{n=1}^{\infty}\frac{\sin 4n}{4^n}$ is absolutely convergent.

27. $\frac{|\cos(n\pi/3)|}{n!}\le\frac{1}{n!}$ and $\sum_{n=1}^{\infty}\frac{1}{n!}$ converges (use the Ratio Test), so the series $\sum_{n=1}^{\infty}\frac{\cos(n\pi/3)}{n!}$ converges absolutely by the Comparison Test.

28. $\lim_{n\to\infty}\left|\frac{a_{n+1}}{a_n}\right|=\lim_{n\to\infty}\frac{5^n/\left[(n+2)^2\,4^{n+3}\right]}{5^{n-1}/\left[(n+1)^2\,4^{n+2}\right]}=\frac{5}{4}\lim_{n\to\infty}\left(\frac{n+1}{n+2}\right)^2=\frac{5}{4}>1$, so the series diverges by the Ratio Test.

29. $\left|\frac{(-1)^n\arctan n}{n^2}\right|<\frac{\pi/2}{n^2}$, so since $\sum_{n=1}^{\infty}\frac{\pi/2}{n^2}=\frac{\pi}{2}\sum_{n=1}^{\infty}\frac{1}{n^2}$ converges $[p=2>1]$, the given series $\sum_{n=1}^{\infty}\frac{(-1)^n\arctan n}{n^2}$ converges absolutely by the Comparison Test.

30. $\lim_{n\to\infty}\sqrt[n]{|a_n|}=\lim_{n\to\infty}\frac{1}{\ln n}=0<1$, so the series $\sum_{n=2}^{\infty}\frac{(-1)^n}{(\ln n)^n}$ converges absolutely by the Root Test.

31. $\lim_{n\to\infty}\sqrt[n]{|a_n|}=\lim_{n\to\infty}\left(\frac{n^n}{3^{1+3n}}\right)^{1/n}=\lim_{n\to\infty}\frac{n}{\sqrt[n]{3}\cdot 3^3}=\infty$, so the series $\sum_{n=1}^{\infty}\frac{n^n}{3^{1+3n}}$ is divergent by the Root Test.

Or: $\lim_{n\to\infty}\left|\frac{a_{n+1}}{a_n}\right|=\lim_{n\to\infty}\left[\frac{(n+1)^{n+1}}{3^{4+3n}}\cdot\frac{3^{1+3n}}{n^n}\right]=\lim_{n\to\infty}\left[\frac{1}{3^3}\cdot\left(\frac{n+1}{n}\right)^n(n+1)\right]$

$$=\frac{1}{27}\lim_{n\to\infty}\left(1+\frac{1}{n}\right)^n\lim_{n\to\infty}(n+1)=\tfrac{1}{27}e\lim_{n\to\infty}(n+1)=\infty,$$

so the series is divergent by the Ratio Test.

32. Since $\left\{\frac{1}{n\ln n}\right\}$ is decreasing and $\lim_{n\to\infty}\frac{1}{n\ln n}=0$, the series $\sum_{n=2}^{\infty}\frac{(-1)^n}{n\ln n}$ converges by the Alternating Series Test. Since $\sum_{n=2}^{\infty}\frac{1}{n\ln n}$ diverges by the Integral Test (Exercise 8.3.15), the series $\sum_{n=2}^{\infty}\frac{(-1)^n}{n\ln n}$ is conditionally convergent.

33. $\lim_{n\to\infty}\sqrt[n]{|a_n|}=\lim_{n\to\infty}\frac{n^2+1}{2n^2+1}=\lim_{n\to\infty}\frac{1+1/n^2}{2+1/n^2}=\frac{1}{2}<1$, so the series $\sum_{n=1}^{\infty}\left(\frac{n^2+1}{2n^2+1}\right)^n$ is absolutely convergent by the Root Test.

34. $\lim_{n\to\infty}\sqrt[n]{|a_n|}=\lim_{n\to\infty}\frac{1}{\arctan n}=\frac{1}{\pi/2}=\frac{2}{\pi}<1$, so the series $\sum_{n=1}^{\infty}\frac{(-1)^n}{(\arctan n)^n}$ is absolutely convergent by the Root Test.

35. Use the Ratio Test with the series

$$1-\frac{1\cdot 3}{3!}+\frac{1\cdot 3\cdot 5}{5!}-\frac{1\cdot 3\cdot 5\cdot 7}{7!}+\cdots+(-1)^{n-1}\frac{1\cdot 3\cdot 5\cdot\cdots\cdot(2n-1)}{(2n-1)!}+\cdots=\sum_{n=1}^{\infty}(-1)^{n-1}\frac{1\cdot 3\cdot 5\cdot\cdots\cdot(2n-1)}{(2n-1)!}.$$

$$\lim_{n\to\infty}\left|\frac{a_{n+1}}{a_n}\right|=\lim_{n\to\infty}\left|\frac{(-1)^n\cdot 1\cdot 3\cdot 5\cdot\cdots\cdot(2n-1)[2(n+1)-1]}{[2(n+1)-1]!}\cdot\frac{(2n-1)!}{(-1)^{n-1}\cdot 1\cdot 3\cdot 5\cdot\cdots\cdot(2n-1)}\right|$$

$$=\lim_{n\to\infty}\left|\frac{(-1)(2n+1)(2n-1)!}{(2n+1)(2n)(2n-1)!}\right|=\lim_{n\to\infty}\frac{1}{2n}=0<1,$$

so the given series is absolutely convergent and therefore convergent.

36. Use the Ratio Test with the series $\frac{2}{5}+\frac{2\cdot 6}{5\cdot 8}+\frac{2\cdot 6\cdot 10}{5\cdot 8\cdot 11}+\frac{2\cdot 6\cdot 10\cdot 14}{5\cdot 8\cdot 11\cdot 14}+\cdots=\sum\limits_{n=1}^{\infty}\frac{2\cdot 6\cdot 10\cdot 14\cdot\cdots\cdot(4n-2)}{5\cdot 8\cdot 11\cdot 14\cdot\cdots\cdot(3n+2)}$.

$$\lim_{n\to\infty}\left|\frac{a_{n+1}}{a_n}\right|=\lim_{n\to\infty}\left|\frac{2\cdot 6\cdot 10\cdot\cdots\cdot(4n-2)[4(n+1)-2]}{5\cdot 8\cdot 11\cdot\cdots\cdot(3n+2)[3(n+1)+2]}\cdot\frac{5\cdot 8\cdot 11\cdot\cdots\cdot(3n+2)}{2\cdot 6\cdot 10\cdot\cdots\cdot(4n-2)}\right|=\lim_{n\to\infty}\frac{4n+2}{3n+5}=\frac{4}{3}>1,$$

so the given series is divergent.

37. $\sum\limits_{n=1}^{\infty}\frac{2\cdot 4\cdot 6\cdot\cdots\cdot(2n)}{n!}=\sum\limits_{n=1}^{\infty}\frac{(2\cdot 1)\cdot(2\cdot 2)\cdot(2\cdot 3)\cdot\cdots\cdot(2\cdot n)}{n!}=\sum\limits_{n=1}^{\infty}\frac{2^n n!}{n!}=\sum\limits_{n=1}^{\infty}2^n$, which diverges by the Test for Divergence since $\lim\limits_{n\to\infty}2^n=\infty$.

38. $\lim\limits_{n\to\infty}\left|\frac{a_{n+1}}{a_n}\right|=\lim\limits_{n\to\infty}\left|\dfrac{\dfrac{2^{n+1}(n+1)!}{5\cdot 8\cdot 11\cdot\cdots\cdot(3n+5)}}{\dfrac{2^n n!}{5\cdot 8\cdot 11\cdot\cdots\cdot(3n+2)}}\right|=\lim\limits_{n\to\infty}\frac{2(n+1)}{3n+5}=\frac{2}{3}<1$, so the series converges absolutely by the Ratio Test.

39. (a) $\lim\limits_{n\to\infty}\left|\frac{1/(n+1)^3}{1/n^3}\right|=\lim\limits_{n\to\infty}\frac{n^3}{(n+1)^3}=\lim\limits_{n\to\infty}\frac{1}{(1+1/n)^3}=1.$ Inconclusive

(b) $\lim\limits_{n\to\infty}\left|\frac{(n+1)}{2^{n+1}}\cdot\frac{2^n}{n}\right|=\lim\limits_{n\to\infty}\frac{n+1}{2n}=\lim\limits_{n\to\infty}\left(\frac{1}{2}+\frac{1}{2n}\right)=\frac{1}{2}.$ Conclusive (convergent)

(c) $\lim\limits_{n\to\infty}\left|\frac{(-3)^n}{\sqrt{n+1}}\cdot\frac{\sqrt{n}}{(-3)^{n-1}}\right|=3\lim\limits_{n\to\infty}\sqrt{\frac{n}{n+1}}=3\lim\limits_{n\to\infty}\sqrt{\frac{1}{1+1/n}}=3.$ Conclusive (divergent)

(d) $\lim\limits_{n\to\infty}\left|\frac{\sqrt{n+1}}{1+(n+1)^2}\cdot\frac{1+n^2}{\sqrt{n}}\right|=\lim\limits_{n\to\infty}\left[\sqrt{1+\frac{1}{n}}\cdot\frac{1/n^2+1}{1/n^2+(1+1/n)^2}\right]=1.$ Inconclusive

40. We use the Ratio Test:

$$\lim_{n\to\infty}\left|\frac{a_{n+1}}{a_n}\right|=\lim_{n\to\infty}\left|\frac{[(n+1)!]^2/[k(n+1)]!}{(n!)^2/(kn)!}\right|=\lim_{n\to\infty}\left|\frac{(n+1)^2}{[k(n+1)]\,[k(n+1)-1]\cdots[kn+1]}\right|$$

Now if $k=1$, then this is equal to $\lim\limits_{n\to\infty}\left|\frac{(n+1)^2}{(n+1)}\right|=\infty$, so the series diverges; if $k=2$, the limit is $\lim\limits_{n\to\infty}\left|\frac{(n+1)^2}{(2n+2)(2n+1)}\right|=\frac{1}{4}<1$, so the series converges, and if $k>2$, then the highest power of n in the denominator is larger than 2, and so the limit is 0, indicating convergence. So the series converges for $k\geq 2$.

41. (a) $\lim_{n\to\infty}\left|\frac{a_{n+1}}{a_n}\right| = \lim_{n\to\infty}\left|\frac{x^{n+1}}{(n+1)!}\cdot\frac{n!}{x^n}\right| = \lim_{n\to\infty}\left|\frac{x}{n+1}\right| = |x|\lim_{n\to\infty}\frac{1}{n+1} = |x|\cdot 0 = 0 < 1$, so by the Ratio Test the series $\sum_{n=0}^{\infty}\frac{x^n}{n!}$ converges for all x.

(b) Since the series of part (a) always converges, we must have $\lim_{n\to\infty}\frac{x^n}{n!} = 0$ by Theorem 8.2.6.

42. (a) $\lim_{n\to\infty}\left|\frac{a_{n+1}}{a_n}\right| = \lim_{n\to\infty}\left|\frac{[4(n+1)]!\,[1103+26{,}390(n+1)]}{[(n+1)!]^4\,396^{4(n+1)}}\cdot\frac{(n!)^4\,396^{4n}}{(4n)!\,(1103+26{,}390n)}\right|$

$$= \lim_{n\to\infty}\frac{(4n+4)(4n+3)(4n+2)(4n+1)(26{,}390n+27{,}493)}{(n+1)^4\,396^4\,(26{,}390n+1103)} = \frac{4^4}{396^4} = \frac{1}{99^4} < 1,$$

so by the Ratio Test, the series $\sum_{n=0}^{\infty}\frac{(4n)!\,(1103+26{,}390n)}{(n!)^4\,396^{4n}}$ converges.

(b) $\frac{1}{\pi} = \frac{2\sqrt{2}}{9801}\sum_{n=0}^{\infty}\frac{(4n)!\,(1103+26{,}390n)}{(n!)^4\,396^{4n}}$

With the first term $(n=0)$, $\frac{1}{\pi}\approx\frac{2\sqrt{2}}{9801}\cdot\frac{1103}{1} \quad\Rightarrow\quad \pi\approx 3.141\,592\,73$, so we get 6 correct decimal places of π, which is 3.141 592 653 589 793 238 to 18 decimal places.

With the second term $(n=1)$, $\frac{1}{\pi}\approx\frac{2\sqrt{2}}{9801}\left(\frac{1103}{1}+\frac{4!\,(1103+26{,}390)}{396^4}\right) \quad\Rightarrow\quad \pi\approx 3.141\,592\,653\,589\,793\,878$, so we get 15 correct decimal places of π.

43. Following the hint, we get that $|a_n| < r^n$ for $n\geq N$, and so since the geometric series $\sum_{n=1}^{\infty} r^n$ converges $(0<r<1)$, the series $\sum_{n=N}^{\infty}|a_n|$ converges as well by the Comparison Test, and hence so does $\sum_{n=1}^{\infty}|a_n|$, so $\sum_{n=1}^{\infty}a_n$ is absolutely convergent.

8.5 Power Series

1. A power series is a series of the form $\sum_{n=0}^{\infty}c_nx^n = c_0+c_1x+c_2x^2+c_3x^3+\cdots$, where x is a variable and the c_n's are constants called the coefficients of the series. More generally, a series of the form $\sum_{n=0}^{\infty}c_n(x-a)^n = c_0+c_1(x-a)+c_2(x-a)^2+\cdots$ is called a power series in $(x-a)$ or a power series centered at a or a power series about a, where a is a constant.

2. (a) Given the power series $\sum_{n=0}^{\infty}c_n(x-a)^n$, the radius of convergence is:

(i) 0 if the series converges only when $x=a$

(ii) ∞ if the series converges for all x, or

(iii) a positive number R such that the series converges if $|x-a|<R$ and diverges if $|x-a|>R$.

In most cases, R can be found by using the Ratio Test.

(b) The interval of convergence of a power series is the interval that consists of all values of x for which the series converges. Corresponding to the cases in part (a), the interval of convergence is: (i) the single point $\{a\}$, (ii) all real numbers; that is, the real number line $(-\infty, \infty)$, or (iii) an interval with endpoints $a - R$ and $a + R$ which can contain neither, either, or both of the endpoints. In this case, we must test the series for convergence at each endpoint to determine the interval of convergence.

3. If $a_n = \dfrac{x^n}{\sqrt{n}}$, then $\displaystyle\lim_{n\to\infty}\left|\frac{a_{n+1}}{a_n}\right| = \lim_{n\to\infty}\left|\frac{x^{n+1}}{\sqrt{n+1}}\cdot\frac{\sqrt{n}}{x^n}\right| = \lim_{n\to\infty}\left|\frac{x}{\sqrt{n+1}/\sqrt{n}}\right| = \lim_{n\to\infty}\frac{|x|}{\sqrt{1+1/n}} = |x|$.

By the Ratio Test, the series $\displaystyle\sum_{n=1}^{\infty}\frac{x^n}{\sqrt{n}}$ converges when $|x| < 1$, so the radius of convergence $R = 1$. Now we'll check the endpoints, that is, $x = \pm 1$. When $x = 1$, the series $\displaystyle\sum_{n=1}^{\infty}\frac{1}{\sqrt{n}}$ diverges because it is a p-series with $p = \frac{1}{2} \le 1$. When $x = -1$, the series $\displaystyle\sum_{n=1}^{\infty}\frac{(-1)^n}{\sqrt{n}}$ converges by the Alternating Series Test. Thus, the interval of convergence is $I = [-1, 1)$.

4. If $a_n = \dfrac{(-1)^n x^n}{n+1}$, then $\displaystyle\lim_{n\to\infty}\left|\frac{a_{n+1}}{a_n}\right| = \lim_{n\to\infty}\left|\frac{x^{n+1}}{n+2}\cdot\frac{n+1}{x^n}\right| = \lim_{n\to\infty}\frac{|x|}{1+1/(n+1)} = |x|$. By the Ratio Test, the series $\displaystyle\sum_{n=0}^{\infty}\frac{(-1)^n x^n}{n+1}$ converges when $|x| < 1$, so $R = 1$. When $x = -1$, the series diverges because it is the harmonic series; when $x = 1$, it is the alternating harmonic series, which converges by the Alternating Series Test. Thus, $I = (-1, 1]$.

5. If $a_n = \dfrac{(-1)^{n-1}x^n}{n^3}$, then

$$\lim_{n\to\infty}\left|\frac{a_{n+1}}{a_n}\right| = \lim_{n\to\infty}\left|\frac{(-1)^n x^{n+1}}{(n+1)^3}\cdot\frac{n^3}{(-1)^{n-1}x^n}\right| = \lim_{n\to\infty}\left|\frac{(-1)xn^3}{(n+1)^3}\right| = \lim_{n\to\infty}\left[\left(\frac{n}{n+1}\right)^3|x|\right] = 1^3\cdot|x| = |x|$$

converges by the Alternating Series Test. When $x = -1$, the series $\displaystyle\sum_{n=1}^{\infty}\frac{(-1)^{n-1}(-1)^n}{n^3} = -\sum_{n=1}^{\infty}\frac{1}{n^3}$ converges because it is a constant multiple of a convergent p-series $[p = 3 > 1]$. Thus, the interval of convergence is $I = [-1, 1]$.

6. $a_n = \sqrt{n}\,x^n$, so we need $\displaystyle\lim_{n\to\infty}\left|\frac{a_{n+1}}{a_n}\right| = \lim_{n\to\infty}\frac{\sqrt{n+1}\,|x|^{n+1}}{\sqrt{n}\,|x|^n} = \lim_{n\to\infty}\sqrt{1+\frac{1}{n}}\,|x| = |x| < 1$ for convergence (by the Ratio Test), so $R = 1$. When $x = \pm 1$, $\displaystyle\lim_{n\to\infty}|a_n| = \lim_{n\to\infty}\sqrt{n} = \infty$, so the series diverges by the Test for Divergence. Thus, $I = (-1, 1)$.

7. If $a_n = \dfrac{x^n}{n!}$, then $\displaystyle\lim_{n\to\infty}\left|\frac{a_{n+1}}{a_n}\right| = \lim_{n\to\infty}\left|\frac{x^{n+1}}{(n+1)!}\cdot\frac{n!}{x^n}\right| = \lim_{n\to\infty}\left|\frac{x}{n+1}\right| = |x|\lim_{n\to\infty}\frac{1}{n+1} = |x|\cdot 0 = 0 < 1$ for *all* real x. So, by the Ratio Test, $R = \infty$, and $I = (-\infty, \infty)$.

8. If $a_n = \dfrac{x^n}{n3^n}$, then $\displaystyle\lim_{n\to\infty}\left|\frac{a_{n+1}}{a_n}\right| = \lim_{n\to\infty}\left|\frac{x^{n+1}}{(n+1)3^{n+1}}\cdot\frac{n3^n}{x^n}\right| = \lim_{n\to\infty}\left|\frac{xn}{(n+1)3}\right| = \frac{|x|}{3}\lim_{n\to\infty}\frac{n}{n+1} = \frac{|x|}{3}$. By the Ratio Test, the series converges when $\dfrac{|x|}{3} < 1 \iff |x| < 3$, so $R = 3$. When $x = -3$, the series is the alternating harmonic series, which converges by the Alternating Series Test. When $x = 3$, it is the harmonic series, which diverges. Thus, $I = [-3, 3)$.

9. $a_n = \dfrac{(-2)^n x^n}{\sqrt[4]{n}}$, so $\displaystyle\lim_{n\to\infty}\left|\frac{a_{n+1}}{a_n}\right| = \lim_{n\to\infty}\frac{2^{n+1}|x|^{n+1}}{\sqrt[4]{n+1}}\cdot\frac{\sqrt[4]{n}}{2^n|x|^n} = \lim_{n\to\infty} 2|x|\sqrt[4]{\frac{n}{n+1}} = 2|x|$, so by the Ratio Test, the series converges when $2|x| < 1 \Leftrightarrow |x| < \frac{1}{2}$, so $R = \frac{1}{2}$. When $x = -\frac{1}{2}$, we get the divergent p-series $\displaystyle\sum_{n=1}^{\infty}\frac{1}{\sqrt[4]{n}}$ $\left[p = \frac{1}{4} \le 1\right]$. When $x = \frac{1}{2}$, we get the series $\displaystyle\sum_{n=1}^{\infty}\frac{(-1)^n}{\sqrt[4]{n}}$, which converges by the Alternating Series Test. Thus, $I = \left(-\frac{1}{2}, \frac{1}{2}\right]$.

10. $a_n = \dfrac{x^n}{5^n n^5}$, so $\displaystyle\lim_{n\to\infty}\left|\frac{a_{n+1}}{a_n}\right| = \lim_{n\to\infty}\left|\frac{x^{n+1}}{5^{n+1}(n+1)^5}\cdot\frac{5^n n^5}{x^n}\right| = \lim_{n\to\infty}\frac{|x|}{5}\left(\frac{n}{n+1}\right)^5 = \frac{|x|}{5}$. By the Ratio Test, the series $\displaystyle\sum_{n=0}^{\infty}\frac{x^n}{5^n n^5}$ converges when $\frac{|x|}{5} < 1 \Leftrightarrow |x| < 5$, so $R = 5$. When $x = -5$, we get the series $\displaystyle\sum_{n=1}^{\infty}\frac{(-1)^n}{n^5}$, which converges by the Alternating Series Test. When $x = 5$, we get the convergent p-series $\displaystyle\sum_{n=1}^{\infty}\frac{1}{n^5}$ $[p = 5 > 1]$. Thus, $I = [-5, 5]$.

11. If $a_n = (-1)^n\dfrac{x^n}{4^n \ln n}$, then $\displaystyle\lim_{n\to\infty}\left|\frac{a_{n+1}}{a_n}\right| = \lim_{n\to\infty}\left|\frac{x^{n+1}}{4^{n+1}\ln(n+1)}\cdot\frac{4^n\ln n}{x^n}\right| = \frac{|x|}{4}\lim_{n\to\infty}\frac{\ln n}{\ln(n+1)} = \frac{|x|}{4}\cdot 1$ [by l'Hospital's Rule] $= \frac{|x|}{4}$. By the Ratio Test, the series converges when $\frac{|x|}{4} < 1 \Leftrightarrow |x| < 4$, so $R = 4$. When $x = -4$, $\displaystyle\sum_{n=2}^{\infty}(-1)^n\frac{x^n}{4^n\ln n} = \sum_{n=2}^{\infty}\frac{[(-1)(-4)]^n}{4^n\ln n} = \sum_{n=2}^{\infty}\frac{1}{\ln n}$. Since $\ln n < n$ for $n \ge 2$, $\frac{1}{\ln n} > \frac{1}{n}$ and $\displaystyle\sum_{n=2}^{\infty}\frac{1}{n}$ is the divergent harmonic series (without the $n = 1$ term), $\displaystyle\sum_{n=2}^{\infty}\frac{1}{\ln n}$ is divergent by the Comparison Test. When $x = 4$, $\displaystyle\sum_{n=2}^{\infty}(-1)^n\frac{x^n}{4^n\ln n} = \sum_{n=2}^{\infty}(-1)^n\frac{1}{\ln n}$, which converges by the Alternating Series Test. Thus, $I = (-4, 4]$.

12. $a_n = (-1)^n\dfrac{x^{2n}}{(2n)!}$, so $\displaystyle\lim_{n\to\infty}\left|\frac{a_{n+1}}{a_n}\right| = \lim_{n\to\infty}\frac{|x|^{2n+2}}{(2n+2)!}\cdot\frac{(2n)!}{|x|^{2n}} = \lim_{n\to\infty}\frac{|x|^2}{(2n+1)(2n+2)} = 0$. Thus, by the Ratio Test, the series converges for *all* real x and we have $R = \infty$ and $I = (-\infty, \infty)$.

13. If $a_n = (-1)^n\dfrac{(x+2)^n}{n\,2^n}$, then $\displaystyle\lim_{n\to\infty}\left|\frac{a_{n+1}}{a_n}\right| = \lim_{n\to\infty}\left[\frac{|x+2|^{n+1}}{(n+1)2^{n+1}}\cdot\frac{n2^n}{|x+2|^n}\right] = \lim_{n\to\infty}\frac{n}{n+1}\cdot\frac{|x+2|}{2} = \frac{|x+2|}{2}$.

By the Ratio Test, the series converges when $\frac{|x+2|}{2} < 1 \Leftrightarrow |x+2| < 2$ [so $R = 2$] $\Leftrightarrow -2 < x+2 < 2 \Leftrightarrow -4 < x < 0$. When $x = -4$, the series becomes $\displaystyle\sum_{n=1}^{\infty}(-1)^n\frac{(-2)^n}{n2^n} = \sum_{n=1}^{\infty}\frac{2^n}{n\,2^n} = \sum_{n=1}^{\infty}\frac{1}{n}$, which is the divergent harmonic series. When $x = 0$, the series is $\displaystyle\sum_{n=1}^{\infty}\frac{(-1)^n}{n}$, the alternating harmonic series, which converges by the Alternating Series Test. Thus, $I = (-4, 0]$.

14. If $a_n = \dfrac{(-2)^n}{\sqrt{n}}(x+3)^n$, then

$$\lim_{n\to\infty}\left|\frac{a_{n+1}}{a_n}\right| = \lim_{n\to\infty}\left|\frac{(-2)^{n+1}(x+3)^{n+1}}{\sqrt{n+1}}\cdot\frac{\sqrt{n}}{(-2)^n(x+3)^n}\right| = \lim_{n\to\infty}\frac{2|x+3|}{\sqrt{1+1/n}} = 2|x+3| < 1 \Leftrightarrow |x+3| < \tfrac{1}{2}$$

$\left[\text{so } R = \frac{1}{2}\right] \Leftrightarrow -\frac{7}{2} < x < -\frac{5}{2}$. When $x = -\frac{7}{2}$, the series becomes $\sum_{n=1}^{\infty} \frac{1}{\sqrt{n}}$, which diverges because it is a p-series with $p = \frac{1}{2} \leq 1$. When $x = -\frac{5}{2}$, the series becomes $\sum_{n=1}^{\infty} \frac{(-1)^n}{\sqrt{n}}$, which converges by the Alternating Series Test.

Thus, $I = \left(-\frac{7}{2}, -\frac{5}{2}\right]$.

15. $a_n = \dfrac{n}{b^n}(x-a)^n$, where $b > 0$.

$$\lim_{n\to\infty}\left|\frac{a_{n+1}}{a_n}\right| = \lim_{n\to\infty}\frac{(n+1)\,|x-a|^{n+1}}{b^{n+1}} \cdot \frac{b^n}{n\,|x-a|^n} = \lim_{n\to\infty}\left(1+\frac{1}{n}\right)\frac{|x-a|}{b} = \frac{|x-a|}{b}.$$

By the Ratio Test, the series converges when $\dfrac{|x-a|}{b} < 1 \Leftrightarrow |x-a| < b$ $\left[\text{so } R = b\right] \Leftrightarrow -b < x-a < b \Leftrightarrow$ $a-b < x < a+b$. When $|x-a| = b$, $\lim_{n\to\infty}|a_n| = \lim_{n\to\infty} n = \infty$, so the series diverges. Thus, $I = (a-b, a+b)$.

16. $a_n = \dfrac{n(x-4)^n}{n^3+1}$, so

$$\lim_{n\to\infty}\left|\frac{a_{n+1}}{a_n}\right| = \lim_{n\to\infty}\frac{(n+1)\,|x-4|^{n+1}}{(n+1)^3+1} \cdot \frac{n^3+1}{n\,|x-4|^n} = \lim_{n\to\infty}\left(1+\frac{1}{n}\right)\frac{n^3+1}{n^3+3n^2+3n+2}\,|x-4| = |x-4|.$$

By the Ratio Test, the series converges when $|x-4| < 1$ $\left[\text{so } R = 1\right] \Leftrightarrow -1 < x-4 < 1 \Leftrightarrow 3 < x < 5$.

When $|x-4| = 1$, $\sum_{n=1}^{\infty}|a_n| = \sum_{n=1}^{\infty}\frac{n}{n^3+1}$, which converges by comparison with the convergent p-series $\sum_{n=1}^{\infty}\frac{1}{n^2}$ $\left[p = 2 > 1\right]$. Thus, $I = [3, 5]$.

17. If $a_n = n!(2x-1)^n$, then $\lim_{n\to\infty}\left|\frac{a_{n+1}}{a_n}\right| = \lim_{n\to\infty}\left|\frac{(n+1)!(2x-1)^{n+1}}{n!(2x-1)^n}\right| = \lim_{n\to\infty}(n+1)\,|2x-1| \to \infty$ as $n \to \infty$ for all $x \neq \frac{1}{2}$. Since the series diverges for all $x \neq \frac{1}{2}$, $R = 0$ and $I = \left\{\frac{1}{2}\right\}$.

18. $a_n = \dfrac{n^2x^n}{2\cdot 4\cdot 6\cdot\cdots\cdot(2n)} = \dfrac{n^2x^n}{2^n n!} = \dfrac{nx^n}{2^n(n-1)!}$, so

$$\lim_{n\to\infty}\left|\frac{a_{n+1}}{a_n}\right| = \lim_{n\to\infty}\frac{(n+1)\,|x|^{n+1}}{2^{n+1}n!} \cdot \frac{2^n(n-1)!}{n\,|x|^n} = \lim_{n\to\infty}\frac{n+1}{n^2}\frac{|x|}{2} = 0.$$ Thus, by the Ratio Test, the series converges for *all* real x and we have $R = \infty$ and $I = (-\infty, \infty)$.

19. (a) We are given that the power series $\sum_{n=0}^{\infty} c_n x^n$ is convergent for $x = 4$. So by Theorem 3, it must converge for at least $-4 < x \leq 4$. In particular, it converges when $x = -2$; that is, $\sum_{n=0}^{\infty} c_n(-2)^n$ is convergent.

(b) It does not follow that $\sum_{n=0}^{\infty} c_n(-4)^n$ is necessarily convergent. [See the comments after Theorem 3 about convergence at the endpoint of an interval. An example is $c_n = (-1)^n/(n4^n)$.]

20. We are given that the power series $\sum_{n=0}^{\infty} c_n x^n$ is convergent for $x = -4$ and divergent when $x = 6$. So by Theorem 3 it converges for at least $-4 \leq x < 4$ and diverges for at least $x \geq 6$ and $x < -6$. Therefore:

(a) It converges when $x = 1$; that is, $\sum c_n$ is convergent.

(b) It diverges when $x = 8$; that is, $\sum c_n 8^n$ is divergent.

(c) It converges when $x = -3$; that is, $\sum c_n(-3^n)$ is convergent.

(d) It diverges when $x = -9$; that is, $\sum c_n(-9)^n = \sum(-1)^n c_n 9^n$ is divergent.

21. If $a_n = \dfrac{(n!)^k}{(kn)!}x^n$, then

$$\lim_{n\to\infty}\left|\frac{a_{n+1}}{a_n}\right| = \lim_{n\to\infty}\frac{[(n+1)!]^k\,(kn)!}{(n!)^k\,[k(n+1)]!}|x| = \lim_{n\to\infty}\frac{(n+1)^k}{(kn+k)(kn+k-1)\cdots(kn+2)(kn+1)}|x|$$

$$= \lim_{n\to\infty}\left[\frac{(n+1)}{(kn+1)}\frac{(n+1)}{(kn+2)}\cdots\frac{(n+1)}{(kn+k)}\right]|x| = \lim_{n\to\infty}\left[\frac{n+1}{kn+1}\right]\lim_{n\to\infty}\left[\frac{n+1}{kn+2}\right]\cdots\lim_{n\to\infty}\left[\frac{n+1}{kn+k}\right]|x|$$

$$= \left(\frac{1}{k}\right)^k|x| < 1 \quad\Leftrightarrow\quad |x| < k^k \text{ for convergence, and the radius of convergence is } R = k^k.$$

22. The partial sums of the series $\sum_{n=0}^{\infty} x^n$ definitely do not converge to $f(x) = 1/(1-x)$ for $x \geq 1$, since f is undefined at $x = 1$ and negative on $(1, \infty)$, while all the partial sums are positive on this interval. The partial sums also fail to converge to f for $x \leq -1$, since $0 < f(x) < 1$ on this interval, while the partial sums are either larger than 1 or less than 0. The partial sums seem to converge to f on $(-1, 1)$. This graphical evidence is consistent with what we know about geometric series: convergence for $|x| < 1$, divergence for $|x| \geq 1$ (see Example 8.2.5).

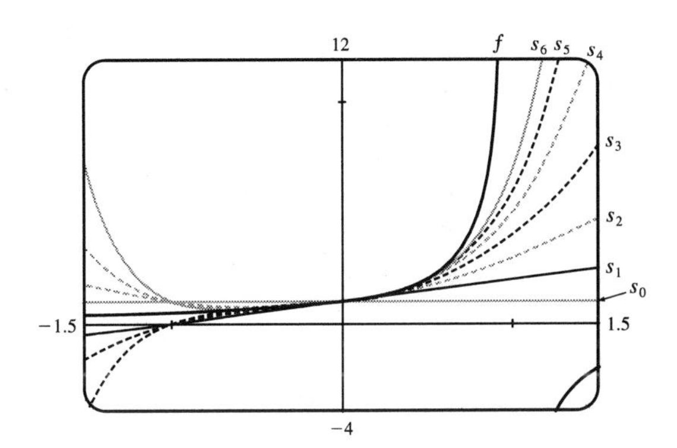

23. (a) If $a_n = \dfrac{(-1)^n\,x^{2n+1}}{n!(n+1)!\,2^{2n+1}}$, then

$$\lim_{n\to\infty}\left|\frac{a_{n+1}}{a_n}\right| = \lim_{n\to\infty}\left|\frac{x^{2n+3}}{(n+1)!(n+2)!\,2^{2n+3}}\cdot\frac{n!(n+1)!\,2^{2n+1}}{x^{2n+1}}\right| = \left(\frac{x}{2}\right)^2\lim_{n\to\infty}\frac{1}{(n+1)(n+2)} = 0 \text{ for all } x.$$

So $J_1(x)$ converges for all x and its domain is $(-\infty, \infty)$.

(b), (c) The initial terms of $J_1(x)$ up to $n = 5$ are $a_0 = \dfrac{x}{2}$,

$$a_1 = -\frac{x^3}{16},\ a_2 = \frac{x^5}{384},\ a_3 = -\frac{x^7}{18{,}432},\ a_4 = \frac{x^9}{1{,}474{,}560},$$

and $a_5 = -\dfrac{x^{11}}{176{,}947{,}200}$. The partial sums seem to approximate $J_1(x)$ well near the origin, but as $|x|$ increases, we need to take a large number of terms to get a good approximation.

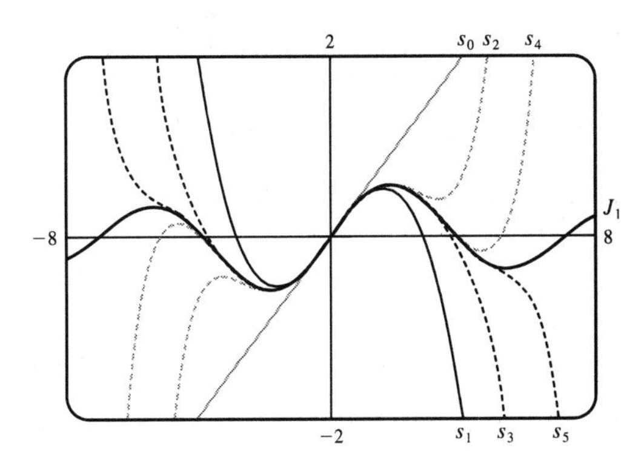

24. (a) $A(x) = 1 + \sum\limits_{n=1}^{\infty} a_n$, where $a_n = \dfrac{x^{3n}}{2\cdot 3\cdot 5\cdot 6\cdot\cdots\cdot(3n-1)(3n)}$, so $\lim\limits_{n\to\infty}\left|\dfrac{a_{n+1}}{a_n}\right| = |x|^3\lim\limits_{n\to\infty}\dfrac{1}{(3n+2)(3n+3)} = 0$

for all x, so the domain is $\mathbb{R}$.

(b), (c)

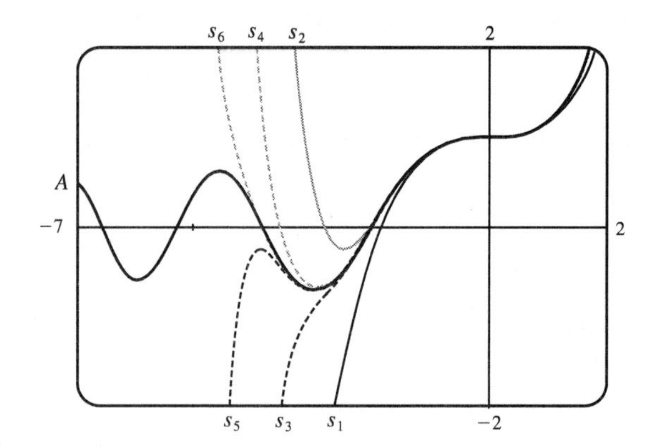

$s_0 = 1$ has been omitted from the graph. The partial sums seem to approximate $A(x)$ well near the origin, but as $|x|$ increases, we need to take a large number of terms to get a good approximation.

To plot A, we must first define $A(x)$ for the CAS. Note that for $n \geq 1$, the denominator of a_n is

$2 \cdot 3 \cdot 5 \cdot 6 \cdot \cdots \cdot (3n-1) \cdot 3n = \dfrac{(3n)!}{1 \cdot 4 \cdot 7 \cdot \cdots \cdot (3n-2)} = \dfrac{(3n)!}{\prod_{k=1}^{n}(3k-2)}$, so $a_n = \dfrac{\prod_{k=1}^{n}(3k-2)}{(3n)!}x^{3n}$ and thus

$A(x) = 1 + \sum\limits_{n=1}^{\infty} \dfrac{\prod_{k=1}^{n}(3k-2)}{(3n)!}x^{3n}$. Both Maple and Mathematica are able to plot A if we define it this way, and Derive is able to produce a similar graph using a suitable partial sum of $A(x)$.

Derive, Maple and Mathematica all have two initially known Airy functions, called `AI·SERIES(z,m)` and `BI·SERIES(z,m)` from `BESSEL.MTH` in Derive and `AiryAi` and `AiryBi` in Maple and Mathematica (just `Ai` and `Bi` in older versions of Maple). However, it is very difficult to solve for A in terms of the CAS's Airy functions, although in fact $A(x) = \dfrac{\sqrt{3}\,\texttt{AiryAi}(x) + \texttt{AiryBi}(x)}{\sqrt{3}\,\texttt{AiryAi}(0) + \texttt{AiryBi}(0)}$.

25. $s_{2n-1} = 1 + 2x + x^2 + 2x^3 + x^4 + 2x^5 + \cdots + x^{2n-2} + 2x^{2n-1}$

$= 1(1+2x) + x^2(1+2x) + x^4(1+2x) + \cdots + x^{2n-2}(1+2x) = (1+2x)(1 + x^2 + x^4 + \cdots + x^{2n-2})$

$= (1+2x)\dfrac{1-x^{2n}}{1-x^2}$ [by (8.2.3)] with $r = x^2$] $\to \dfrac{1+2x}{1-x^2}$ as $n \to \infty$ [by (8.2.4)], when $|x| < 1$.

Also $s_{2n} = s_{2n-1} + x^{2n} \to \dfrac{1+2x}{1-x^2}$ since $x^{2n} \to 0$ for $|x| < 1$. Therefore, $s_n \to \dfrac{1+2x}{1-x^2}$ since s_{2n} and s_{2n-1} both approach $\dfrac{1+2x}{1-x^2}$ as $n \to \infty$. Thus, the interval of convergence is $(-1, 1)$ and $f(x) = \dfrac{1+2x}{1-x^2}$.

26. $s_{4n-1} = c_0 + c_1 x + c_2 x^2 + c_3 x^3 + c_0 x^4 + c_1 x^5 + c_2 x^6 + c_3 x^7 + \cdots + c_3 x^{4n-1}$

$= \left(c_0 + c_1 x + c_2 x^2 + c_3 x^3\right)\left(1 + x^4 + x^8 + \cdots + x^{4n-4}\right) \to \dfrac{c_0 + c_1 x + c_2 x^2 + c_3 x^3}{1 - x^4}$ as $n \to \infty$

[by (8.2.4) with $r = x^4$] for $|x^4| < 1 \Leftrightarrow |x| < 1$. Also s_{4n}, s_{4n+1}, s_{4n+2} have the same limits (for example, $s_{4n} = s_{4n-1} + c_0 x^{4n}$ and $x^{4n} \to 0$ for $|x| < 1$). So if at least one of c_0, c_1, c_2, and c_3 is nonzero, then the interval of convergence is $(-1, 1)$ and $f(x) = \dfrac{c_0 + c_1 x + c_2 x^2 + c_3 x^3}{1 - x^4}$.

27. We use the Root Test on the series $\sum c_n x^n$. We need $\lim\limits_{n\to\infty} \sqrt[n]{|c_n x^n|} = |x| \lim\limits_{n\to\infty} \sqrt[n]{|c_n|} = c\,|x| < 1$ for convergence, or $|x| < 1/c$, so $R = 1/c$.

28. Suppose $c_n \neq 0$. Applying the Ratio Test to the series $\sum c_n (x-a)^n$, we find that

$$L = \lim_{n\to\infty} \left| \frac{a_{n+1}}{a_n} \right| = \lim_{n\to\infty} \left| \frac{c_{n+1}(x-a)^{n+1}}{c_n (x-a)^n} \right| = \lim_{n\to\infty} \frac{|x-a|}{|c_n/c_{n+1}|} \quad (*) \quad = \frac{|x-a|}{\lim\limits_{n\to\infty} |c_n/c_{n+1}|} \quad \left[\textit{if} \lim_{n\to\infty} |c_n/c_{n+1}| \neq 0 \right],$$

so the series converges when $\dfrac{|x-a|}{\lim\limits_{n\to\infty} |c_n/c_{n+1}|} < 1 \quad \Leftrightarrow \quad |x-a| < \lim\limits_{n\to\infty} \left| \dfrac{c_n}{c_{n+1}} \right|$. Thus, $R = \lim\limits_{n\to\infty} \left| \dfrac{c_n}{c_{n+1}} \right|$. If $\lim\limits_{n\to\infty} \left| \dfrac{c_n}{c_{n+1}} \right| = 0$ and $|x-a| \neq 0$, then $(*)$ shows that $L = \infty$ and so the series diverges, and hence, $R = 0$. Thus, in all cases, $R = \lim\limits_{n\to\infty} \left| \dfrac{c_n}{c_{n+1}} \right|$.

29. For $2 < x < 3$, $\sum c_n x^n$ diverges and $\sum d_n x^n$ converges. By Exercise 8.2.43, $\sum (c_n + d_n)\, x^n$ diverges. Since both series converge for $|x| < 2$, the radius of convergence of $\sum (c_n + d_n)\, x^n$ is 2.

30. Since $\sum c_n x^n$ converges whenever $|x| < R$, $\sum c_n x^{2n} = \sum c_n \left(x^2\right)^n$ converges whenever $\left|x^2\right| < R \quad \Leftrightarrow \quad |x| < \sqrt{R}$, so the second series has radius of convergence $\sqrt{R}$.

8.6 Representing Functions as Power Series

1. If $f(x) = \sum\limits_{n=0}^{\infty} c_n x^n$ has radius of convergence 10, then $f'(x) = \sum\limits_{n=1}^{\infty} n c_n x^{n-1}$ also has radius of convergence 10 by Theorem 2.

2. If $f(x) = \sum\limits_{n=0}^{\infty} b_n x^n$ converges on $(-2, 2)$, then $\int f(x)dx = C + \sum\limits_{n=0}^{\infty} \dfrac{b_n}{n+1} x^{n+1}$ has the same radius of convergence (by Theorem 2), but may not have the same interval of convergence—it may happen that the integrated series converges at an endpoint (or both endpoints).

3. Our goal is to write the function in the form $\dfrac{1}{1-r}$, and then use Equation (1) to represent the function as a sum of a power series. $f(x) = \dfrac{1}{1+x} = \dfrac{1}{1-(-x)} = \sum\limits_{n=0}^{\infty} (-x)^n = \sum\limits_{n=0}^{\infty} (-1)^n x^n$ with $|-x| < 1 \quad \Leftrightarrow \quad |x| < 1$, so $R = 1$ and $I = (-1, 1)$.

4. $f(x) = \dfrac{3}{1-x^4} = 3\left(\dfrac{1}{1-x^4}\right) = 3(1 + x^4 + x^8 + x^{12} + \cdots) = 3\sum\limits_{n=0}^{\infty} (x^4)^n = \sum\limits_{n=0}^{\infty} 3x^{4n}$ with $\left|x^4\right| < 1 \quad \Leftrightarrow$ $|x| < 1$, so $R = 1$ and $I = (-1, 1)$.

$\left[\text{Note that } 3\sum\limits_{n=0}^{\infty} (x^4)^n \text{ converges} \quad \Leftrightarrow \quad \sum\limits_{n=0}^{\infty} (x^4)^n \text{ converges, so the appropriate condition [from Equation (1)] is } \left|x^4\right| < 1.\right]$

5. Replacing x with x^3 in (1) gives $f(x) = \dfrac{1}{1-x^3} = \sum\limits_{n=0}^{\infty} (x^3)^n = \sum\limits_{n=0}^{\infty} x^{3n}$. The series converges when $\left|x^3\right| < 1 \quad \Leftrightarrow$ $|x|^3 < 1 \quad \Leftrightarrow \quad |x| < \sqrt[3]{1} \quad \Leftrightarrow \quad |x| < 1$. Thus, $R = 1$ and $I = (-1, 1)$.

6. $f(x) = \dfrac{1}{1+9x^2} = \dfrac{1}{1-(-9x^2)} = \sum\limits_{n=0}^{\infty}(-9x^2)^n = \sum\limits_{n=0}^{\infty}(-1)^n\, 3^{2n}\, x^{2n}$. The series converges when $\left|-9x^2\right| < 1$; that is, when $|x| < \frac{1}{3}$, so $I = \left(-\frac{1}{3}, \frac{1}{3}\right)$.

7. $f(x) = \dfrac{1}{x-5} = -\dfrac{1}{5}\left(\dfrac{1}{1-x/5}\right) = -\dfrac{1}{5}\sum\limits_{n=0}^{\infty}\left(\dfrac{x}{5}\right)^n$ or equivalently, $-\sum\limits_{n=0}^{\infty}\dfrac{1}{5^{n+1}}\,x^n$. The series converges when $\left|\dfrac{x}{5}\right| < 1$; that is, when $|x| < 5$, so $I = (-5, 5)$.

8. $f(x) = \dfrac{x}{4x+1} = x\cdot\dfrac{1}{1-(-4x)} = x\sum\limits_{n=0}^{\infty}(-4x)^n = \sum\limits_{n=0}^{\infty}(-1)^n 2^{2n} x^{n+1}$. The series converges when $|-4x| < 1$; that is, when $|x| < \frac{1}{4}$, so $I = \left(-\frac{1}{4}, \frac{1}{4}\right)$.

9. $f(x) = \dfrac{x}{9+x^2} = \dfrac{x}{9}\left[\dfrac{1}{1+(x/3)^2}\right] = \dfrac{x}{9}\left[\dfrac{1}{1-\{-(x/3)^2\}}\right] = \dfrac{x}{9}\sum\limits_{n=0}^{\infty}\left[-\left(\dfrac{x}{3}\right)^2\right]^n = \dfrac{x}{9}\sum\limits_{n=0}^{\infty}(-1)^n\dfrac{x^{2n}}{9^n}$

$$= \sum_{n=0}^{\infty}(-1)^n\frac{x^{2n+1}}{9^{n+1}}$$

The geometric series $\sum\limits_{n=0}^{\infty}\left[-\left(\dfrac{x}{3}\right)^2\right]^n$ converges when $\left|-\left(\dfrac{x}{3}\right)^2\right| < 1 \;\Leftrightarrow\; \dfrac{|x^2|}{9} < 1 \;\Leftrightarrow\; |x|^2 < 9 \;\Leftrightarrow\; |x| < 3$, so $R = 3$ and $I = (-3, 3)$.

10. $f(x) = \dfrac{x^2}{a^3 - x^3} = \dfrac{x^2}{a^3}\cdot\dfrac{1}{1-x^3/a^3} = \dfrac{x^2}{a^3}\sum\limits_{n=0}^{\infty}\left(\dfrac{x^3}{a^3}\right)^n = \sum\limits_{n=0}^{\infty}\dfrac{x^{3n+2}}{a^{3n+3}}$. The series converges when $\left|x^3/a^3\right| < 1 \;\Leftrightarrow$ $\left|x^3\right| < \left|a^3\right| \;\Leftrightarrow\; |x| < |a|$, so $R = |a|$ and $I = (-|a|, |a|)$.

11. $f(x) = \dfrac{3}{x^2+x-2} = \dfrac{3}{(x+2)(x-1)} = \dfrac{A}{x+2} + \dfrac{B}{x-1} \;\Rightarrow\; 3 = A(x-1) + B(x+2)$. Taking $x = -2$, we get $A = -1$. Taking $x = 1$, we get $B = 1$. Thus,

$$\frac{3}{x^2+x-2} = \frac{1}{x-1} - \frac{1}{x+2} = -\frac{1}{1-x} - \frac{1}{2}\frac{1}{1+x/2} = -\sum_{n=0}^{\infty}x^n - \frac{1}{2}\sum_{n=0}^{\infty}\left(-\frac{x}{2}\right)^n$$

$$= \sum_{n=0}^{\infty}\left[-1 - \tfrac{1}{2}\left(-\tfrac{1}{2}\right)^n\right]x^n = \sum_{n=0}^{\infty}\left[-1 + \left(-\tfrac{1}{2}\right)^{n+1}\right]x^n = \sum_{n=0}^{\infty}\left[\frac{(-1)^{n+1}}{2^{n+1}} - 1\right]x^n$$

We represented the given function as the sum of two geometric series; the first converges for $x \in (-1, 1)$ and the second converges for $x \in (-2, 2)$. Thus, the sum converges for $x \in (-1, 1) = I$.

12. $f(x) = \dfrac{7x-1}{3x^2+2x-1} = \dfrac{7x-1}{(3x-1)(x+1)} = \dfrac{A}{3x-1} + \dfrac{B}{x+1} = \dfrac{1}{3x-1} + \dfrac{2}{x+1} = 2\cdot\dfrac{1}{1-(-x)} - \dfrac{1}{1-3x}$

$$= 2\sum_{n=0}^{\infty}(-x)^n - \sum_{n=0}^{\infty}(3x)^n = \sum_{n=0}^{\infty}\left[2(-1)^n - 3^n\right]x^n$$

The series $\sum(-x)^n$ converges for $x \in (-1, 1)$ and the series $\sum(3x)^n$ converges for $x \in \left(-\frac{1}{3}, \frac{1}{3}\right)$, so their sum converges for $x \in \left(-\frac{1}{3}, \frac{1}{3}\right) = I$.

13. (a) $f(x) = \dfrac{1}{(1+x)^2} = \dfrac{d}{dx}\left(\dfrac{-1}{1+x}\right) = -\dfrac{d}{dx}\left[\sum\limits_{n=0}^{\infty}(-1)^n x^n\right]$ [from Exercise 3]

$= \sum\limits_{n=1}^{\infty}(-1)^{n+1}nx^{n-1}$ [from Theorem 2(i)] $= \sum\limits_{n=0}^{\infty}(-1)^n(n+1)x^n$ with $R = 1$.

In the last step, note that we *decreased* the initial value of the summation variable n by 1, and then *increased* each occurrence of n in the term by 1 [also note that $(-1)^{n+2} = (-1)^n$].

(b) $f(x) = \dfrac{1}{(1+x)^3} = -\dfrac{1}{2}\dfrac{d}{dx}\left[\dfrac{1}{(1+x)^2}\right] = -\dfrac{1}{2}\dfrac{d}{dx}\left[\sum_{n=0}^{\infty}(-1)^n(n+1)x^n\right]$ [from part (a)]

$= -\frac{1}{2}\sum_{n=1}^{\infty}(-1)^n(n+1)nx^{n-1} = \frac{1}{2}\sum_{n=0}^{\infty}(-1)^n(n+2)(n+1)x^n$ with $R = 1$.

(c) $f(x) = \dfrac{x^2}{(1+x)^3} = x^2 \cdot \dfrac{1}{(1+x)^3} = x^2 \cdot \dfrac{1}{2}\sum_{n=0}^{\infty}(-1)^n(n+2)(n+1)x^n$ [from part (b)]

$= \frac{1}{2}\sum_{n=0}^{\infty}(-1)^n(n+2)(n+1)x^{n+2}$

To write the power series with x^n rather than x^{n+2}, we will *decrease* each occurrence of n in the term by 2 and *increase* the initial value of the summation variable by 2. This gives us $\dfrac{1}{2}\sum_{n=2}^{\infty}(-1)^n(n)(n-1)x^n$ with $R = 1$.

14. (a) $\dfrac{1}{1+x} = \dfrac{1}{1-(-x)} = \sum_{n=0}^{\infty}(-1)^n x^n$ [geometric series with $R = 1$], so

$f(x) = \ln(1+x) = \int \dfrac{dx}{1+x} = \int\left[\sum_{n=0}^{\infty}(-1)^n x^n\right]dx = C + \sum_{n=0}^{\infty}(-1)^n\dfrac{x^{n+1}}{n+1} = \sum_{n=1}^{\infty}\dfrac{(-1)^{n-1}x^n}{n}$

$[C = 0$ since $f(0) = \ln 1 = 0]$, with $R = 1$

(b) $f(x) = x\ln(1+x) = x\left[\sum_{n=1}^{\infty}\dfrac{(-1)^{n-1}x^n}{n}\right]$ [by part (a)] $= \sum_{n=1}^{\infty}\dfrac{(-1)^{n-1}x^{n+1}}{n} = \sum_{n=2}^{\infty}\dfrac{(-1)^n x^n}{n-1}$ with $R = 1$.

(c) $f(x) = \ln(x^2+1) = \sum_{n=1}^{\infty}\dfrac{(-1)^{n-1}(x^2)^n}{n}$ [by part (a)] $= \sum_{n=1}^{\infty}\dfrac{(-1)^{n-1}x^{2n}}{n}$ with $R = 1$.

15. $f(x) = \ln(5-x) = -\displaystyle\int\dfrac{dx}{5-x} = -\dfrac{1}{5}\int\dfrac{dx}{1-x/5} = -\dfrac{1}{5}\int\left[\sum_{n=0}^{\infty}\left(\dfrac{x}{5}\right)^n\right]dx = C - \dfrac{1}{5}\sum_{n=0}^{\infty}\dfrac{x^{n+1}}{5^n(n+1)} = C - \sum_{n=1}^{\infty}\dfrac{x^n}{n\,5^n}$.

Putting $x = 0$, we get $C = \ln 5$ The series converges for $|x/5| < 1 \Leftrightarrow |x| < 5$, so $R = 5$.

16. We know that $\dfrac{1}{1-2x} = \sum_{n=0}^{\infty}(2x)^n$. Differentiating, we get $\dfrac{2}{(1-2x)^2} = \sum_{n=1}^{\infty}2^n nx^{n-1} = \sum_{n=0}^{\infty}2^{n+1}(n+1)x^n$, so

$f(x) = \dfrac{x^2}{(1-2x)^2} = \dfrac{x^2}{2}\cdot\dfrac{2}{(1-2x)^2} = \dfrac{x^2}{2}\sum_{n=0}^{\infty}2^{n+1}(n+1)x^n = \sum_{n=0}^{\infty}2^n(n+1)x^{n+2}$ or $\sum_{n=2}^{\infty}2^{n-2}(n-1)x^n$,

with $R = \frac{1}{2}$.

17. $\dfrac{1}{2-x} = \dfrac{1}{2(1-x/2)} = \dfrac{1}{2}\sum_{n=0}^{\infty}\left(\dfrac{x}{2}\right)^n = \sum_{n=0}^{\infty}\dfrac{1}{2^{n+1}}x^n$ for $\left|\dfrac{x}{2}\right| < 1 \Leftrightarrow |x| < 2$. Now

$\dfrac{1}{(x-2)^2} = \dfrac{d}{dx}\left(\dfrac{1}{2-x}\right) = \dfrac{d}{dx}\left(\sum_{n=0}^{\infty}\dfrac{1}{2^{n+1}}x^n\right) = \sum_{n=1}^{\infty}\dfrac{n}{2^{n+1}}x^{n-1} = \sum_{n=0}^{\infty}\dfrac{n+1}{2^{n+2}}x^n$. So

$f(x) = \dfrac{x^3}{(x-2)^2} = x^3\sum_{n=0}^{\infty}\dfrac{n+1}{2^{n+2}}x^n = \sum_{n=0}^{\infty}\dfrac{n+1}{2^{n+2}}x^{n+3}$ or $\sum_{n=3}^{\infty}\dfrac{n-2}{2^{n-1}}x^n$ for $|x| < 2$. Thus, $R = 2$ and $I = (-2, 2)$.

18. From Example 7, $g(x) = \arctan x = \sum_{n=0}^{\infty}(-1)^n\dfrac{x^{2n+1}}{2n+1}$. Thus,

$f(x) = \arctan(x/3) = \sum_{n=0}^{\infty}(-1)^n\dfrac{(x/3)^{2n+1}}{2n+1} = \sum_{n=0}^{\infty}(-1)^n\dfrac{1}{3^{2n+1}(2n+1)}x^{2n+1}$ for $\left|\dfrac{x}{3}\right| < 1 \Leftrightarrow |x| < 3$, so $R = 3$.

19. $f(x) = \ln(3+x) = \int \frac{dx}{3+x} = \frac{1}{3}\int \frac{dx}{1+x/3} = \frac{1}{3}\int \frac{dx}{1-(-x/3)} = \frac{1}{3}\int \sum_{n=0}^{\infty}\left(-\frac{x}{3}\right)^n dx$

$$= C + \frac{1}{3}\sum_{n=0}^{\infty}\frac{(-1)^n}{(n+1)3^n}x^{n+1} = \ln 3 + \frac{1}{3}\sum_{n=1}^{\infty}\frac{(-1)^{n-1}}{n3^{n-1}}x^n \quad [C = f(0) = \ln 3] \quad = \ln 3 + \sum_{n=1}^{\infty}\frac{(-1)^{n-1}}{n\,3^n}x^n.$$

The series converges when $|-x/3| < 1 \Leftrightarrow |x| < 3$, so $R = 3$. The terms of the series are $a_0 = \ln 3$, $a_1 = \frac{x}{3}$, $a_2 = -\frac{x^2}{18}$, $a_3 = \frac{x^3}{81}$, $a_4 = -\frac{x^4}{324}$, $a_5 = \frac{x^5}{1215}, \ldots$. As n increases, $s_n(x)$ approximates f better on the interval of convergence, which is $(-3, 3)$.

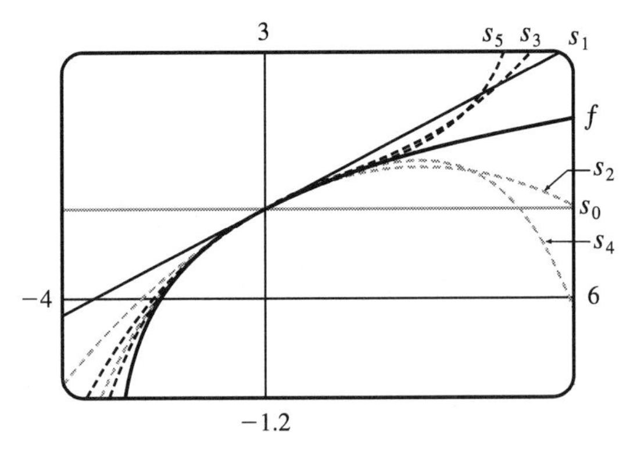

20. $f(x) = \frac{1}{x^2+25} = \frac{1}{25}\left(\frac{1}{1+x^2/25}\right) = \frac{1}{25}\left(\frac{1}{1-(-x^2/25)}\right) = \frac{1}{25}\sum_{n=0}^{\infty}\left(-\frac{x^2}{25}\right)^n = \frac{1}{25}\sum_{n=0}^{\infty}(-1)^n\left(\frac{x}{5}\right)^{2n}.$

The series converges when $\left|-x^2/25\right| < 1 \Leftrightarrow x^2 < 25 \Leftrightarrow |x| < 5$, so $R = 5$. The terms of the series are $a_0 = \frac{1}{25}$, $a_1 = -\frac{x^2}{625}$, $a_2 = \frac{x^4}{15{,}625}, \ldots$. As n increases, $s_n(x)$ approximates f better on the interval of convergence, which is $(-5, 5)$.

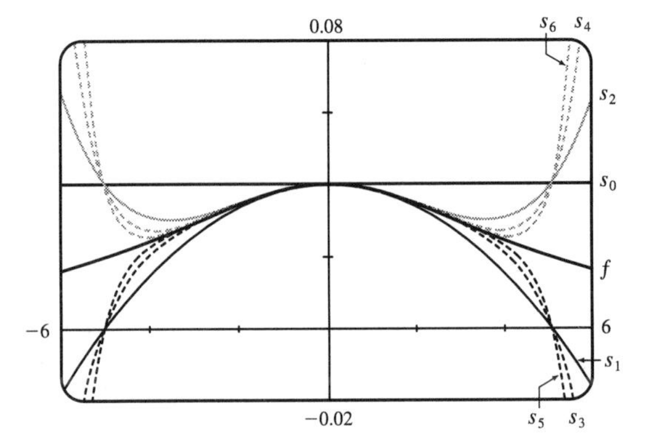

21. $f(x) = \ln\left(\frac{1+x}{1-x}\right) = \ln(1+x) - \ln(1-x) = \int \frac{dx}{1+x} + \int \frac{dx}{1-x} = \int \frac{dx}{1-(-x)} + \int \frac{dx}{1-x}$

$$= \int\left[\sum_{n=0}^{\infty}(-1)^n x^n + \sum_{n=0}^{\infty}x^n\right]dx = \int\left[(1 - x + x^2 - x^3 + x^4 - \cdots) + (1 + x + x^2 + x^3 + x^4 + \cdots)\right]dx$$

$$= \int (2 + 2x^2 + 2x^4 + \cdots)\,dx = \int \sum_{n=0}^{\infty} 2x^{2n}\,dx = C + \sum_{n=0}^{\infty}\frac{2x^{2n+1}}{2n+1}$$

But $f(0) = \ln\frac{1}{1} = 0$, so $C = 0$ and we have $f(x) = \sum_{n=0}^{\infty}\frac{2x^{2n+1}}{2n+1}$ with $R = 1$. If $x = \pm 1$, then $f(x) = \pm 2\sum_{n=0}^{\infty}\frac{1}{2n+1}$, which both diverge by the Limit Comparison Test with $b_n = \frac{1}{n}$. As n increases, $s_n(x)$ approximates f better on the interval of convergence, which is $(-1, 1)$.

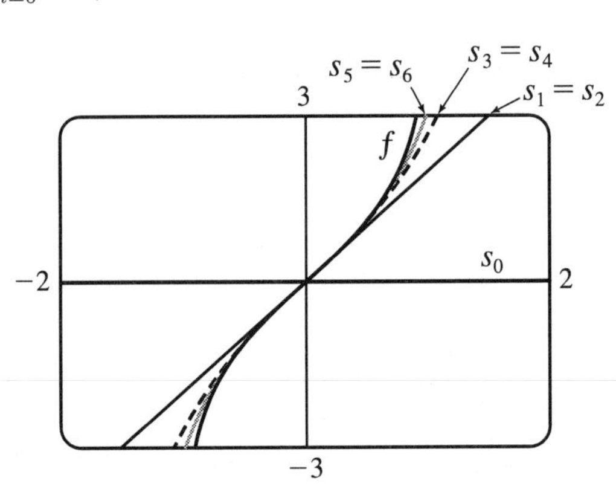

22. $f(x) = \tan^{-1}(2x) = 2\int \dfrac{dx}{1+4x^2} = 2\int \sum\limits_{n=0}^{\infty}(-1)^n\left(4x^2\right)^n dx = 2\int \sum\limits_{n=0}^{\infty}(-1)^n 4^n x^{2n}\,dx$

$= C + 2\sum\limits_{n=0}^{\infty}\dfrac{(-1)^n 4^n x^{2n+1}}{2n+1} = \sum\limits_{n=0}^{\infty}\dfrac{(-1)^n 2^{2n+1} x^{2n+1}}{2n+1}$ $\quad [f(0) = \tan^{-1} 0 = 0$, so $C = 0]$.

The series converges when $\left|4x^2\right| < 1 \;\Leftrightarrow\; |x| < \frac{1}{2}$, so $R = \frac{1}{2}$. If $x = \pm\frac{1}{2}$, then $f(x) = \sum\limits_{n=0}^{\infty}(-1)^n\dfrac{1}{2n+1}$ and $f(x) = \sum\limits_{n=0}^{\infty}(-1)^{n+1}\dfrac{1}{2n+1}$, respectively. Both series converge by the Alternating Series Test. As n increases, $s_n(x)$ approximates f better on the interval of convergence, which is $\left[-\frac{1}{2}, \frac{1}{2}\right]$.

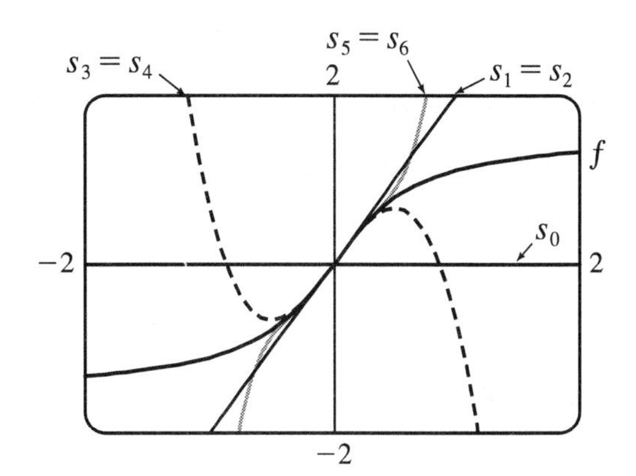

23. $\dfrac{t}{1-t^8} = t\cdot\dfrac{1}{1-t^8} = t\sum\limits_{n=0}^{\infty}(t^8)^n = \sum\limits_{n=0}^{\infty}t^{8n+1} \;\Rightarrow\; \displaystyle\int\dfrac{t}{1-t^8}\,dt = C + \sum_{n=0}^{\infty}\dfrac{t^{8n+2}}{8n+2}$. The series for $\dfrac{1}{1-t^8}$ converges when $\left|t^8\right| < 1 \;\Leftrightarrow\; |t| < 1$, so $R = 1$ for that series and also the series for $t/(1-t^8)$. By Theorem 2, the series for $\displaystyle\int\dfrac{t}{1-t^8}\,dt$ also has $R = 1$.

24. By Example 6, $\ln(1-t) = -\sum\limits_{n=1}^{\infty}\dfrac{t^n}{n}$ for $|t| < 1$, so $\dfrac{\ln(1-t)}{t} = -\sum\limits_{n=1}^{\infty}\dfrac{t^{n-1}}{n}$ and $\displaystyle\int\dfrac{\ln(1-t)}{t}\,dt = C - \sum_{n=1}^{\infty}\dfrac{t^n}{n^2}$.

By Theorem 2, $R = 1$.

25. By Example 7, $\tan^{-1}x = \sum\limits_{n=0}^{\infty}(-1)^n\dfrac{x^{2n+1}}{2n+1}$ with $R = 1$, so

$x - \tan^{-1}x = x - \left(x - \dfrac{x^3}{3} + \dfrac{x^5}{5} - \dfrac{x^7}{7} + \cdots\right) = \dfrac{x^3}{3} - \dfrac{x^5}{5} + \dfrac{x^7}{7} - \cdots = \sum\limits_{n=1}^{\infty}(-1)^{n+1}\dfrac{x^{2n+1}}{2n+1}$ and

$\dfrac{x - \tan^{-1}x}{x^3} = \sum\limits_{n=1}^{\infty}(-1)^{n+1}\dfrac{x^{2n-2}}{2n+1}$, so

$\displaystyle\int\dfrac{x-\tan^{-1}x}{x^3}\,dx = C + \sum_{n=1}^{\infty}(-1)^{n+1}\dfrac{x^{2n-1}}{(2n+1)(2n-1)} = C + \sum_{n=1}^{\infty}(-1)^{n+1}\dfrac{x^{2n-1}}{4n^2-1}$. By Theorem 2, $R = 1$.

26. By Example 7, $\displaystyle\int\tan^{-1}(x^2)\,dx = \int\sum_{n=0}^{\infty}(-1)^n\dfrac{(x^2)^{2n+1}}{2n+1}\,dx = C + \sum_{n=0}^{\infty}(-1)^n\dfrac{x^{4n+3}}{(2n+1)(4n+3)}$ with $R = 1$.

27. $\dfrac{1}{1+x^5} = \dfrac{1}{1-(-x^5)} = \sum\limits_{n=0}^{\infty}\left(-x^5\right)^n = \sum\limits_{n=0}^{\infty}(-1)^n x^{5n} \;\Rightarrow$

$\displaystyle\int\dfrac{1}{1+x^5}\,dx = \int\sum_{n=0}^{\infty}(-1)^n x^{5n}\,dx = C + \sum_{n=0}^{\infty}(-1)^n\dfrac{x^{5n+1}}{5n+1}$. Thus,

$I = \displaystyle\int_0^{0.2}\dfrac{1}{1+x^5}\,dx = \left[x - \dfrac{x^6}{6} + \dfrac{x^{11}}{11} - \cdots\right]_0^{0.2} = 0.2 - \dfrac{(0.2)^6}{6} + \dfrac{(0.2)^{11}}{11} - \cdots$. The series is alternating, so if we use the first two terms, the error is at most $(0.2)^{11}/11 \approx 1.9\times 10^{-9}$. So $I \approx 0.2 - (0.2)^6/6 \approx 0.199989$ to six decimal places.

SERIES

28. From Example 6, we know $\ln(1-x) = -\sum\limits_{n=1}^{\infty} \frac{x^n}{n}$, so

$$\ln(1+x^4) = \ln[1-(-x^4)] = -\sum_{n=1}^{\infty} \frac{(-x^4)^n}{n} = \sum_{n=1}^{\infty} (-1)^{n+1} \frac{x^{4n}}{n} \quad \Rightarrow$$

$$\int \ln(1+x^4)\,dx = \int \sum_{n=1}^{\infty} (-1)^{n+1} \frac{x^{4n}}{n}\,dx = C + \sum_{n=1}^{\infty} (-1)^{n+1} \frac{x^{4n+1}}{n(4n+1)}. \text{ Thus,}$$

$$I = \int_0^{0.4} \ln(1+x^4)\,dx = \left[\frac{x^5}{5} - \frac{x^9}{18} + \frac{x^{13}}{39} - \frac{x^{17}}{68} + \cdots\right]_0^{0.4} = \frac{(0.4)^5}{5} - \frac{(0.4)^9}{18} + \frac{(0.4)^{13}}{39} - \frac{(0.4)^{17}}{68} + \cdots.$$

The series is alternating, so if we use the first three terms, the error is at most $(0.4)^{17}/68 \approx 2.5 \times 10^{-9}$.

So $I \approx (0.4)^5/5 - (0.4)^9/18 + (0.9)^{13}/39 \approx 0.002034$ to six decimal places.

29. We substitute $3x$ for x in Example 7, and find that

$$\int x\arctan(3x)\,dx = \int x \sum_{n=0}^{\infty} (-1)^n \frac{(3x)^{2n+1}}{2n+1}\,dx = \int \sum_{n=0}^{\infty} (-1)^n \frac{3^{2n+1}\,x^{2n+2}}{2n+1}\,dx$$
$$= C + \sum_{n=0}^{\infty} (-1)^n \frac{3^{2n+1}\,x^{2n+3}}{(2n+1)(2n+3)}$$

So
$$\int_0^{0.1} x\arctan(3x)\,dx = \left[\frac{3x^3}{1\cdot 3} - \frac{3^3x^5}{3\cdot 5} + \frac{3^5x^7}{5\cdot 7} - \frac{3^7x^9}{7\cdot 9} + \cdots\right]_0^{0.1}$$
$$= \frac{1}{10^3} - \frac{9}{5\times 10^5} + \frac{243}{35\times 10^7} - \frac{2187}{63\times 10^9} + \cdots.$$

The series is alternating, so if we use three terms, the error is at most $\frac{2187}{63\times 10^9} \approx 3.5\times 10^{-8}$. So

$$\int_0^{0.1} x\arctan(3x)\,dx \approx \frac{1}{10^3} - \frac{9}{5\times 10^5} + \frac{243}{35\times 10^7} \approx 0.000\,983 \text{ to six decimal places.}$$

30. $$\int_0^{0.3} \frac{x^2}{1+x^4}\,dx = \int_0^{0.3} x^2 \sum_{n=0}^{\infty} (-1)^n x^{4n}\,dx = \sum_{n=0}^{\infty} \left[\frac{(-1)^n x^{4n+3}}{4n+3}\right]_0^{0.3} = \sum_{n=0}^{\infty} \frac{(-1)^n\, 3^{4n+3}}{(4n+3)10^{4n+3}}$$
$$= \frac{3^3}{3\times 10^3} - \frac{3^7}{7\times 10^7} + \frac{3^{11}}{11\times 10^{11}} - \cdots$$

The series is alternating, so if we use only two terms, the error is at most $\frac{3^{11}}{11\times 10^{11}} \approx 0.000\,000\,16$. So, to six decimal places, $\int_0^{0.3} \frac{x^2}{1+x^4}\,dx \approx \frac{3^3}{3\times 10^3} - \frac{3^7}{7\times 10^7} \approx 0.008\,969$.

31. Using the result of Example 6, $\ln(1-x) = -\sum\limits_{n=1}^{\infty} \frac{x^n}{n}$, with $x = -0.1$, we have

$\ln 1.1 = \ln[1-(-0.1)] = 0.1 - \frac{0.01}{2} + \frac{0.001}{3} - \frac{0.0001}{4} + \frac{0.00001}{5} - \cdots$. The series is alternating, so if we use only the first four terms, the error is at most $\frac{0.00001}{5} = 0.000\,002$. So $\ln 1.1 \approx 0.1 - \frac{0.01}{2} + \frac{0.001}{3} - \frac{0.0001}{4} \approx 0.09531$.

32. $f(x) = \sum\limits_{n=0}^{\infty} \dfrac{(-1)^n x^{2n}}{(2n)!} \quad \Rightarrow \quad f'(x) = \sum\limits_{n=1}^{\infty} \dfrac{(-1)^n 2n x^{2n-1}}{(2n)!}$ [the first term disappears], so

$$f''(x) = \sum_{n=1}^{\infty} \frac{(-1)^n (2n)(2n-1) x^{2n-2}}{(2n)!} = \sum_{n=1}^{\infty} \frac{(-1)^n x^{2(n-1)}}{[2(n-1)]!} = \sum_{n=0}^{\infty} \frac{(-1)^{n+1} x^{2n}}{(2n)!} \quad \text{[substituting } n+1 \text{ for } n\text{]}$$

$$= -\sum_{n=0}^{\infty} \frac{(-1)^n x^{2n}}{(2n)!} = -f(x) \quad \Rightarrow \quad f''(x) + f(x) = 0.$$

33. (a) $J_0(x) = \sum\limits_{n=0}^{\infty} \dfrac{(-1)^n x^{2n}}{2^{2n}(n!)^2}$, $J_0'(x) = \sum\limits_{n=1}^{\infty} \dfrac{(-1)^n 2n x^{2n-1}}{2^{2n}(n!)^2}$, and $J_0''(x) = \sum\limits_{n=1}^{\infty} \dfrac{(-1)^n 2n(2n-1) x^{2n-2}}{2^{2n}(n!)^2}$, so

$$x^2 J_0''(x) + x J_0'(x) + x^2 J_0(x) = \sum_{n=1}^{\infty} \frac{(-1)^n 2n(2n-1) x^{2n}}{2^{2n}(n!)^2} + \sum_{n=1}^{\infty} \frac{(-1)^n 2n x^{2n}}{2^{2n}(n!)^2} + \sum_{n=0}^{\infty} \frac{(-1)^n x^{2n+2}}{2^{2n}(n!)^2}$$

$$= \sum_{n=1}^{\infty} \frac{(-1)^n 2n(2n-1) x^{2n}}{2^{2n}(n!)^2} + \sum_{n=1}^{\infty} \frac{(-1)^n 2n x^{2n}}{2^{2n}(n!)^2} + \sum_{n=1}^{\infty} \frac{(-1)^{n-1} x^{2n}}{2^{2n-2}[(n-1)!]^2}$$

$$= \sum_{n=1}^{\infty} \frac{(-1)^n 2n(2n-1) x^{2n}}{2^{2n}(n!)^2} + \sum_{n=1}^{\infty} \frac{(-1)^n 2n x^{2n}}{2^{2n}(n!)^2} + \sum_{n=1}^{\infty} \frac{(-1)^n (-1)^{-1} 2^2 n^2 x^{2n}}{2^{2n}(n!)^2}$$

$$= \sum_{n=1}^{\infty} (-1)^n \left[\frac{2n(2n-1) + 2n - 2^2 n^2}{2^{2n}(n!)^2} \right] x^{2n}$$

$$= \sum_{n=1}^{\infty} (-1)^n \left[\frac{4n^2 - 2n + 2n - 4n^2}{2^{2n}(n!)^2} \right] x^{2n} = 0$$

(b) $$\int_0^1 J_0(x)\,dx = \int_0^1 \left[\sum_{n=0}^{\infty} \frac{(-1)^n x^{2n}}{2^{2n}(n!)^2} \right] dx = \int_0^1 \left(1 - \frac{x^2}{4} + \frac{x^4}{64} - \frac{x^6}{2304} + \cdots \right) dx$$

$$= \left[x - \frac{x^3}{3 \cdot 4} + \frac{x^5}{5 \cdot 64} - \frac{x^7}{7 \cdot 2304} + \cdots \right]_0^1 = 1 - \frac{1}{12} + \frac{1}{320} - \frac{1}{16{,}128} + \cdots$$

Since $\frac{1}{16{,}128} \approx 0.000062$, it follows from The Alternating Series Estimation Theorem that, correct to three decimal places, $\int_0^1 J_0(x)\,dx \approx 1 - \frac{1}{12} + \frac{1}{320} \approx 0.920$.

34. (a) $J_1(x) = \sum\limits_{n=0}^{\infty} \dfrac{(-1)^n x^{2n+1}}{n!\,(n+1)!\,2^{2n+1}}$, $J_1'(x) = \sum\limits_{n=0}^{\infty} \dfrac{(-1)^n (2n+1) x^{2n}}{n!\,(n+1)!\,2^{2n+1}}$, and $J_1''(x) = \sum\limits_{n=1}^{\infty} \dfrac{(-1)^n (2n+1)(2n) x^{2n-1}}{n!\,(n+1)!\,2^{2n+1}}$.

$$x^2 J_1''(x) + x J_1'(x) + (x^2 - 1) J_1(x)$$

$$= \sum_{n=1}^{\infty} \frac{(-1)^n (2n+1)(2n) x^{2n+1}}{n!\,(n+1)!\,2^{2n+1}} + \sum_{n=0}^{\infty} \frac{(-1)^n (2n+1) x^{2n+1}}{n!\,(n+1)!\,2^{2n+1}}$$

$$+ \sum_{n=0}^{\infty} \frac{(-1)^n x^{2n+3}}{n!\,(n+1)!\,2^{2n+1}} - \sum_{n=0}^{\infty} \frac{(-1)^n x^{2n+1}}{n!\,(n+1)!\,2^{2n+1}}$$

$$= \sum_{n=1}^{\infty} \frac{(-1)^n (2n+1)(2n) x^{2n+1}}{n!\,(n+1)!\,2^{2n+1}} + \sum_{n=0}^{\infty} \frac{(-1)^n (2n+1) x^{2n+1}}{n!\,(n+1)!\,2^{2n+1}}$$

$$- \sum_{n=1}^{\infty} \frac{(-1)^n x^{2n+1}}{(n-1)!\,n!\,2^{2n-1}} - \sum_{n=0}^{\infty} \frac{(-1)^n x^{2n+1}}{n!\,(n+1)!\,2^{2n+1}} \quad \left[\begin{array}{c} \text{Replace } n \text{ with } n-1 \\ \text{in the third term} \end{array} \right]$$

$$= \frac{x}{2} - \frac{x}{2} + \sum_{n=1}^{\infty} (-1)^n \left[\frac{(2n+1)(2n) + (2n+1) - (n)(n+1)2^2 - 1}{n!\,(n+1)!\,2^{2n+1}} \right] x^{2n+1} = 0$$

(b) $J_0(x) = \sum\limits_{n=0}^{\infty} \dfrac{(-1)^n\, x^{2n}}{2^{2n}\,(n!)^2} \quad \Rightarrow$

$$J_0'(x) = \sum_{n=1}^{\infty} \frac{(-1)^n\,(2n)x^{2n-1}}{2^{2n}\,(n!)^2} = \sum_{n=0}^{\infty} \frac{(-1)^{n+1}\,2(n+1)x^{2n+1}}{2^{2n+2}\,[(n+1)!]^2} \qquad \text{[Replace } n \text{ with } n+1]$$

$$= -\sum_{n=0}^{\infty} \frac{(-1)^n\,x^{2n+1}}{2^{2n+1}(n+1)!\,n!} \qquad [\text{cancel 2 and } n+1; \text{ take } -1 \text{ outside sum}] \quad = -J_1(x)$$

35. (a) $f(x) = \sum\limits_{n=0}^{\infty} \dfrac{x^n}{n!} \quad \Rightarrow \quad f'(x) = \sum\limits_{n=1}^{\infty} \dfrac{nx^{n-1}}{n!} = \sum\limits_{n=1}^{\infty} \dfrac{x^{n-1}}{(n-1)!} = \sum\limits_{n=0}^{\infty} \dfrac{x^n}{n!} = f(x)$

(b) By Theorem 5.5.2, the only solution to the differential equation $df(x)/dx = f(x)$ is $f(x) = Ke^x$, but $f(0) = 1$, so $K = 1$ and $f(x) = e^x$.

Or: We could solve the equation $df(x)/dx = f(x)$ as a separable differential equation.

36. $\dfrac{|\sin nx|}{n^2} \le \dfrac{1}{n^2}$, so $\sum\limits_{n=1}^{\infty} \dfrac{\sin nx}{n^2}$ converges by the Comparison Test. $\dfrac{d}{dx}\left(\dfrac{\sin nx}{n^2}\right) = \dfrac{\cos nx}{n}$, so when $x = 2k\pi$ (k an integer), $\sum\limits_{n=1}^{\infty} f_n'(x) = \sum\limits_{n=1}^{\infty} \dfrac{\cos(2kn\pi)}{n} = \sum\limits_{n=1}^{\infty} \dfrac{1}{n}$, which diverges (harmonic series). $f_n''(x) = -\sin nx$, so $\sum\limits_{n=1}^{\infty} f_n''(x) = -\sum\limits_{n=1}^{\infty} \sin nx$, which converges only if $\sin nx = 0$, or $x = k\pi$ (k an integer).

37. If $a_n = \dfrac{x^n}{n^2}$, then by the Ratio Test, $\lim\limits_{n\to\infty} \left|\dfrac{a_{n+1}}{a_n}\right| = \lim\limits_{n\to\infty} \left|\dfrac{x^{n+1}}{(n+1)^2} \cdot \dfrac{n^2}{x^n}\right| = |x| \lim\limits_{n\to\infty} \left(\dfrac{n}{n+1}\right)^2 = |x| < 1$ for convergence, so $R = 1$. When $x = \pm 1$, $\sum\limits_{n=1}^{\infty} \left|\dfrac{x^n}{n^2}\right| = \sum\limits_{n=1}^{\infty} \dfrac{1}{n^2}$ which is a convergent p-series ($p = 2 > 1$), so the interval of convergence for f is $[-1, 1]$. By Theorem 2, the radii of convergence of f' and f'' are both 1, so we need only check the endpoints. $f(x) = \sum\limits_{n=1}^{\infty} \dfrac{x^n}{n^2} \quad \Rightarrow \quad f'(x) = \sum\limits_{n=1}^{\infty} \dfrac{nx^{n-1}}{n^2} = \sum\limits_{n=0}^{\infty} \dfrac{x^n}{n+1}$, and this series diverges for $x = 1$ (harmonic series) and converges for $x = -1$ (Alternating Series Test), so the interval of convergence is $[-1, 1)$. $f''(x) = \sum\limits_{n=1}^{\infty} \dfrac{nx^{n-1}}{n+1}$ diverges at both 1 and -1 (Test for Divergence) since $\lim\limits_{n\to\infty} \dfrac{n}{n+1} = 1 \ne 0$, so its interval of convergence is $(-1, 1)$.

38. (a) $\sum\limits_{n=1}^{\infty} nx^{n-1} = \sum\limits_{n=0}^{\infty} \dfrac{d}{dx} x^n = \dfrac{d}{dx}\left[\sum\limits_{n=0}^{\infty} x^n\right] = \dfrac{d}{dx}\left[\dfrac{1}{1-x}\right] = -\dfrac{1}{(1-x)^2}(-1) = \dfrac{1}{(1-x)^2}$, $|x| < 1$.

(b) (i) $\sum\limits_{n=1}^{\infty} nx^n = x \sum\limits_{n=1}^{\infty} nx^{n-1} = x\left[\dfrac{1}{(1-x)^2}\right]$ [from part (a)] $= \dfrac{x}{(1-x)^2}$ for $|x| < 1$.

(ii) Put $x = \frac{1}{2}$ in (i): $\sum\limits_{n=1}^{\infty} \dfrac{n}{2^n} = \sum\limits_{n=1}^{\infty} n\left(\frac{1}{2}\right)^n = \dfrac{1/2}{(1-1/2)^2} = 2$.

(c) (i) $\sum\limits_{n=2}^{\infty} n(n-1)x^n = x^2 \sum\limits_{n=2}^{\infty} n(n-1)x^{n-2} = x^2 \dfrac{d}{dx}\left[\sum\limits_{n=1}^{\infty} nx^{n-1}\right] = x^2 \dfrac{d}{dx} \dfrac{1}{(1-x)^2}$

$$= x^2 \frac{2}{(1-x)^3} = \frac{2x^2}{(1-x)^3} \text{ for } |x| < 1.$$

(ii) Put $x = \frac{1}{2}$ in (i): $\sum_{n=2}^{\infty} \frac{n^2 - n}{2^n} = \sum_{n=2}^{\infty} n(n-1)\left(\frac{1}{2}\right)^n = \frac{2(1/2)^2}{(1-1/2)^3} = 4.$

(iii) From (b)(ii) and (c)(ii), we have $\sum_{n=1}^{\infty} \frac{n^2}{2^n} = \sum_{n=1}^{\infty} \frac{n^2 - n}{2^n} + \sum_{n=1}^{\infty} \frac{n}{2^n} = 4 + 2 = 6.$

39. By Example 7, $\tan^{-1} x = \sum_{n=0}^{\infty} (-1)^n \frac{x^{2n+1}}{2n+1}$ for $|x| < 1$. In particular, for $x = \frac{1}{\sqrt{3}}$,

we have $\frac{\pi}{6} = \tan^{-1}\left(\frac{1}{\sqrt{3}}\right) = \sum_{n=0}^{\infty} (-1)^n \frac{\left(1/\sqrt{3}\right)^{2n+1}}{2n+1} = \sum_{n=0}^{\infty} (-1)^n \left(\frac{1}{3}\right)^n \frac{1}{\sqrt{3}} \frac{1}{2n+1}$,

so $\pi = \frac{6}{\sqrt{3}} \sum_{n=0}^{\infty} \frac{(-1)^n}{(2n+1)3^n} = 2\sqrt{3} \sum_{n=0}^{\infty} \frac{(-1)^n}{(2n+1)3^n}$.

8.7 Taylor and Maclaurin Series

1. Using Theorem 5 with $\sum_{n=0}^{\infty} b_n (x-5)^n$, $b_n = \frac{f^{(n)}(a)}{n!}$, so $b_8 = \frac{f^{(8)}(5)}{8!}$.

2. (a) Using Formula 6, a power series expansion of f at 1 must have the form $f(1) + f'(1)(x-1) + \cdots$. Comparing to the given series, $1.6 - 0.8(x-1) + \cdots$, we must have $f'(1) = -0.8$. But from the graph, $f'(1)$ is positive. Hence, the given series is *not* the Taylor series of f centered at 1.

(b) A power series expansion of f at 2 must have the form $f(2) + f'(2)(x-2) + \frac{1}{2}f''(2)(x-2)^2 + \cdots$. Comparing to the given series, $2.8 + 0.5(x-2) + 1.5(x-2)^2 - 0.1(x-2)^3 + \cdots$, we must have $\frac{1}{2}f''(2) = 1.5$; that is, $f''(2)$ is positive. But from the graph, f is concave downward near $x = 2$, so $f''(2)$ must be negative. Hence, the given series is *not* the Taylor series of f centered at 2.

3. Since $f^{(n)}(0) = (n+1)!$, Equation 7 gives the Maclaurin series

$\sum_{n=0}^{\infty} \frac{f^{(n)}(0)}{n!} x^n = \sum_{n=0}^{\infty} \frac{(n+1)!}{n!} x^n = \sum_{n=0}^{\infty} (n+1)x^n$. Applying the Ratio Test with $a_n = (n+1)x^n$ gives us

$\lim_{n\to\infty} \left|\frac{a_{n+1}}{a_n}\right| = \lim_{n\to\infty} \left|\frac{(n+2)x^{n+1}}{(n+1)x^n}\right| = |x| \lim_{n\to\infty} \frac{n+2}{n+1} = |x| \cdot 1 = |x|$. For convergence, we must have $|x| < 1$, so the radius of convergence $R = 1$.

4. Since $f^{(n)}(4) = \frac{(-1)^n n!}{3^n (n+1)}$, Equation 6 gives the Taylor series

$\sum_{n=0}^{\infty} \frac{f^{(n)}(4)}{n!}(x-4)^n = \sum_{n=0}^{\infty} \frac{(-1)^n n!}{3^n (n+1)\, n!}(x-4)^n = \sum_{n=0}^{\infty} \frac{(-1)^n}{3^n (n+1)}(x-4)^n$, which is the Taylor series for f centered at 4. Apply the Ratio Test to find the radius of convergence R.

$$\lim_{n\to\infty} \left|\frac{a_{n+1}}{a_n}\right| = \lim_{n\to\infty} \left|\frac{(-1)^{n+1}(x-4)^{n+1}}{3^{n+1}(n+2)} \cdot \frac{3^n(n+1)}{(-1)^n(x-4)^n}\right| = \lim_{n\to\infty} \left|\frac{(-1)(x-4)(n+1)}{3(n+2)}\right|$$

$$= \frac{1}{3}|x-4| \lim_{n\to\infty} \frac{n+1}{n+2} = \frac{1}{3}|x-4|$$

For convergence, $\frac{1}{3}|x-4| < 1 \quad \Leftrightarrow \quad |x-4| < 3$, so $R = 3$.

5.

n	$f^{(n)}(x)$	$f^{(n)}(0)$
0	$\cos x$	1
1	$-\sin x$	0
2	$-\cos x$	-1
3	$\sin x$	0
4	$\cos x$	1
$\vdots$	$\vdots$	$\vdots$

We use Equation 7 with $f(x) = \cos x$.

$$\cos x = f(0) + f'(0)x + \frac{f''(0)}{2!}x^2 + \frac{f^{(3)}(0)}{3!}x^3 + \frac{f^{(4)}(0)}{4!}x^4 + \cdots$$

$$= 1 - \frac{x^2}{2!} + \frac{x^4}{4!} - \cdots = \sum_{n=0}^{\infty} \frac{(-1)^n x^{2n}}{(2n)!}$$

If $a_n = \dfrac{(-1)^n x^{2n}}{(2n)!}$, then

$$\lim_{n\to\infty} \left|\frac{a_{n+1}}{a_n}\right| = \lim_{n\to\infty} \left|\frac{x^{2n+2}}{(2n+2)!} \cdot \frac{(2n)!}{x^{2n}}\right| = x^2 \lim_{n\to\infty} \frac{1}{(2n+2)(2n+1)}$$

$$= 0 < 1 \text{ for all } x.$$

So $R = \infty$ (Ratio Test).

6.

n	$f^{(n)}(x)$	$f^{(n)}(0)$
0	$\sin 2x$	0
1	$2\cos 2x$	2
2	$-2^2 \sin 2x$	0
3	$-2^3 \cos 2x$	-2^3
4	$2^4 \sin 2x$	0
$\vdots$	$\vdots$	$\vdots$

$f^{(n)}(0) = 0$ if n is even and $f^{(2n+1)}(0) = (-1)^n 2^{2n+1}$, so

$$\sin 2x = \sum_{n=0}^{\infty} \frac{f^{(n)}(0)}{n!}x^n = \sum_{n=0}^{\infty} \frac{f^{(2n+1)}(0)}{(2n+1)!}x^{2n+1} = \sum_{n=0}^{\infty} \frac{(-1)^n 2^{2n+1} x^{2n+1}}{(2n+1)!}$$

$$\lim_{n\to\infty} \left|\frac{a_{n+1}}{a_n}\right| = \lim_{n\to\infty} \frac{2^2 |x|^2}{(2n+3)(2n+2)} = 0 < 1 \text{ for all } x.$$

So $R = \infty$ (Ratio Test).

7.

n	$f^{(n)}(x)$	$f^{(n)}(0)$
0	e^{5x}	1
1	$5e^{5x}$	5
2	$5^2 e^{5x}$	25
3	$5^3 e^{5x}$	125
4	$5^4 e^{5x}$	625
$\vdots$	$\vdots$	$\vdots$

$$e^{5x} = \sum_{n=0}^{\infty} \frac{f^{(n)}(0)}{n!}x^n = \sum_{n=0}^{\infty} \frac{5^n}{n!}x^n$$

$$\lim_{n\to\infty} \left|\frac{a_{n+1}}{a_n}\right| = \lim_{n\to\infty} \left[\frac{5^{n+1}|x|^{n+1}}{(n+1)!} \cdot \frac{n!}{5^n |x|^n}\right] = \lim_{n\to\infty} \frac{5|x|}{n+1}$$

$$= 0 < 1 \text{ for all } x$$

So $R = \infty$.

8.

n	$f^{(n)}(x)$	$f^{(n)}(0)$
0	xe^x	0
1	$(x+1)e^x$	1
2	$(x+2)e^x$	2
3	$(x+3)e^x$	3
$\vdots$	$\vdots$	$\vdots$

$$xe^x = \sum_{n=0}^{\infty} \frac{f^{(n)}(0)}{n!}x^n = \sum_{n=0}^{\infty} \frac{n}{n!}x^n = \sum_{n=1}^{\infty} \frac{n}{n!}x^n = \sum_{n=1}^{\infty} \frac{x^n}{(n-1)!}.$$

$$\lim_{n\to\infty} \left|\frac{a_{n+1}}{a_n}\right| = \lim_{n\to\infty} \left[\frac{|x|^{n+1}}{n!} \cdot \frac{(n-1)!}{|x|^n}\right] = \lim_{n\to\infty} \frac{|x|}{n}$$

$$= 0 < 1 \text{ for all } x$$

So $R = \infty$.

9.

n	$f^{(n)}(x)$	$f^{(n)}(0)$
0	$\sinh x$	0
1	$\cosh x$	1
2	$\sinh x$	0
3	$\cosh x$	1
4	$\sinh x$	0
⋮	⋮	⋮

$$f^{(n)}(0) = \begin{cases} 0 & \text{if } n \text{ is even} \\ 1 & \text{if } n \text{ is odd} \end{cases} \quad \text{so } \sinh x = \sum_{n=0}^{\infty} \frac{x^{2n+1}}{(2n+1)!}.$$

Use the Ratio Test to find R. If $a_n = \dfrac{x^{2n+1}}{(2n+1)!}$, then

$$\lim_{n\to\infty}\left|\frac{a_{n+1}}{a_n}\right| = \lim_{n\to\infty}\left|\frac{x^{2n+3}}{(2n+3)!}\cdot\frac{(2n+1)!}{x^{2n+1}}\right| = x^2\cdot\lim_{n\to\infty}\frac{1}{(2n+3)(2n+2)}$$
$$= 0 < 1 \text{ for all } x$$

So $R = \infty$.

10.

n	$f^{(n)}(x)$	$f^{(n)}(0)$
0	$\cosh x$	1
1	$\sinh x$	0
2	$\cosh x$	1
3	$\sinh x$	0
⋮	⋮	⋮

$$f^{(n)}(0) = \begin{cases} 1 & \text{if } n \text{ is even} \\ 0 & \text{if } n \text{ is odd} \end{cases} \quad \text{so } \cosh x = \sum_{n=0}^{\infty} \frac{x^{2n}}{(2n)!}.$$

Use the Ratio Test to find R. If $a_n = \dfrac{x^{2n}}{(2n)!}$, then

$$\lim_{n\to\infty}\left|\frac{a_{n+1}}{a_n}\right| = \lim_{n\to\infty}\left|\frac{x^{2n+2}}{(2n+2)!}\cdot\frac{(2n)!}{x^{2n}}\right| = x^2\cdot\lim_{n\to\infty}\frac{1}{(2n+2)(2n+1)}$$
$$= 0 < 1 \text{ for all } x$$

So $R = \infty$.

11.

n	$f^{(n)}(x)$	$f^{(n)}(2)$
0	$1+x+x^2$	7
1	$1+2x$	5
2	2	2
3	0	0
4	0	0
⋮	⋮	⋮

$$f(x) = 7 + 5(x-2) + \frac{2}{2!}(x-2)^2 + \sum_{n=3}^{\infty}\frac{0}{n!}(x-2)^n$$
$$= 7 + 5(x-2) + (x-2)^2$$

Since $a_n = 0$ for large n, $R = \infty$.

12.

n	$f^{(n)}(x)$	$f^{(n)}(-1)$
0	x^3	-1
1	$3x^2$	3
2	$6x$	-6
3	6	6
4	0	0
5	0	0
⋮	⋮	⋮

$$f(x) = -1 + 3(x+1) - \frac{6}{2!}(x+1)^2 + \frac{6}{3!}(x+1)^3$$
$$= -1 + 3(x+1) - 3(x+1)^2 + (x+1)^3$$

Since $a_n = 0$ for large n, $R = \infty$.

13. Clearly, $f^{(n)}(x) = e^x$, so $f^{(n)}(3) = e^3$ and $e^x = \sum\limits_{n=0}^{\infty}\dfrac{e^3}{n!}(x-3)^n$. If $a_n = \dfrac{e^3}{n!}(x-3)^n$, then

$$\lim_{n\to\infty}\left|\frac{a_{n+1}}{a_n}\right| = \lim_{n\to\infty}\left|\frac{e^3(x-3)^{n+1}}{(n+1)!}\cdot\frac{n!}{e^3(x-3)^n}\right| = \lim_{n\to\infty}\frac{|x-3|}{n+1} = 0 < 1 \text{ for all } x, \text{ so } R = \infty.$$

14.

n	$f^{(n)}(x)$	$f^{(n)}(2)$
0	$\ln x$	$\ln 2$
1	x^{-1}	$\frac{1}{2}$
2	$-x^{-2}$	$-\frac{1}{4}$
3	$2x^{-3}$	$\frac{2}{8}$
4	$-3\cdot 2x^{-4}$	$-\frac{3\cdot 2}{16}$
$\vdots$	$\vdots$	$\vdots$

$f^{(n)}(2) = \dfrac{(-1)^{n-1}(n-1)!}{2^n}$ for $n \geq 1$, so

$$\ln x = \ln 2 + \sum_{n=1}^{\infty} \frac{(-1)^{n-1}(x-2)^n}{n \cdot 2^n}.$$

$$\lim_{n\to\infty}\left|\frac{a_{n+1}}{a_n}\right| = \frac{|x-2|}{2}\lim_{n\to\infty}\frac{n}{n+1} = \frac{|x-2|}{2} < 1 \text{ for convergence,}$$

so $|x-2| < 2 \;\Rightarrow\; R = 2$.

15.

n	$f^{(n)}(x)$	$f^{(n)}(\pi)$
0	$\cos x$	-1
1	$-\sin x$	0
2	$-\cos x$	1
3	$\sin x$	0
4	$\cos x$	-1
$\vdots$	$\vdots$	$\vdots$

$$\cos x = \sum_{k=0}^{\infty}\frac{f^{(k)}(\pi)}{k!}(x-\pi)^k = -1 + \frac{(x-\pi)^2}{2!} - \frac{(x-\pi)^4}{4!} + \frac{(x-\pi)^6}{6!} - \cdots$$

$$= \sum_{n=0}^{\infty}(-1)^{n+1}\frac{(x-\pi)^{2n}}{(2n)!}$$

$$\lim_{n\to\infty}\left|\frac{a_{n+1}}{a_n}\right| = \lim_{n\to\infty}\left[\frac{|x-\pi|^{2n+2}}{(2n+2)!}\cdot\frac{(2n)!}{|x-\pi|^{2n}}\right] = \lim_{n\to\infty}\frac{|x-\pi|^2}{(2n+2)(2n+1)}$$

$$= 0 < 1 \text{ for all } x$$

So $R = \infty$.

16.

n	$f^{(n)}(x)$	$f^{(n)}(\pi/2)$
0	$\sin x$	1
1	$\cos x$	0
2	$-\sin x$	-1
3	$-\cos x$	0
4	$\sin x$	1
$\vdots$	$\vdots$	$\vdots$

$$\sin x = \sum_{k=0}^{\infty}\frac{f^{(k)}(\pi/2)}{k!}\left(x-\frac{\pi}{2}\right)^k$$

$$= 1 - \frac{(x-\pi/2)^2}{2!} + \frac{(x-\pi/2)^4}{4!} - \frac{(x-\pi/2)^6}{6!} + \cdots$$

$$= \sum_{n=0}^{\infty}(-1)^n\frac{(x-\pi/2)^{2n}}{(2n)!}$$

$$\lim_{n\to\infty}\left|\frac{a_{n+1}}{a_n}\right| = \lim_{n\to\infty}\left[\frac{|x-\pi/2|^{2n+2}}{(2n+2)!}\cdot\frac{(2n)!}{|x-\pi/2|^{2n}}\right] = \lim_{n\to\infty}\frac{|x-\pi/2|^2}{(2n+2)(2n+1)}$$

$$= 0 < 1 \text{ for all } x$$

So $R = \infty$.

17.

n	$f^{(n)}(x)$	$f^{(n)}(9)$
0	$x^{-1/2}$	$\frac{1}{3}$
1	$-\frac{1}{2}x^{-3/2}$	$-\frac{1}{2}\cdot\frac{1}{3^3}$
2	$\frac{3}{4}x^{-5/2}$	$-\frac{1}{2}\cdot\left(-\frac{3}{2}\right)\cdot\frac{1}{3^5}$
3	$-\frac{15}{8}x^{-7/2}$	$-\frac{1}{2}\cdot\left(-\frac{3}{2}\right)\cdot\left(-\frac{5}{2}\right)\cdot\frac{1}{3^7}$
$\vdots$	$\vdots$	$\vdots$

$$\frac{1}{\sqrt{x}} = \frac{1}{3} - \frac{1}{2\cdot 3^3}(x-9) + \frac{3}{2^2\cdot 3^5}\frac{(x-9)^2}{2!} - \frac{3\cdot 5}{2^3\cdot 3^7}\frac{(x-9)^3}{3!} + \cdots$$

$$= \frac{1}{3} + \sum_{n=0}^{\infty}(-1)^n\frac{1\cdot 3\cdot 5\cdot\cdots\cdot(2n-1)}{2^n\cdot 3^{2n+1}\cdot n!}(x-9)^n.$$

$$\lim_{n\to\infty}\left|\frac{a_{n+1}}{a_n}\right| = \lim_{n\to\infty}\left[\frac{1\cdot 3\cdot 5\cdot\cdots\cdot(2n-1)[2(n+1)-1]\,|x-9|^{n+1}}{2^{n+1}\cdot 3^{[2(n+1)+1]}\cdot(n+1)!}\cdot\frac{2^n\cdot 3^{2n+1}\cdot n!}{1\cdot 3\cdot 5\cdot\cdots\cdot(2n-1)\,|x-9|^n}\right]$$

$$= \lim_{n\to\infty}\left[\frac{(2n+1)\,|x-9|}{2\cdot 3^2(n+1)}\right] = \frac{1}{9}\,|x-9| < 1 \text{ for convergence, so } |x-9| < 9 \text{ and } R = 9.$$

18.

n	$f^{(n)}(x)$	$f^{(n)}(1)$
0	x^{-2}	1
1	$-2x^{-3}$	-2
2	$6x^{-4}$	6
3	$-24x^{-5}$	-24
4	$120x^{-6}$	120
$\vdots$	$\vdots$	$\vdots$

$$x^{-2} = 1 - 2(x-1) + 6\cdot\frac{(x-1)^2}{2!} - 24\cdot\frac{(x-1)^3}{3!} + 120\cdot\frac{(x-1)^4}{4!} - \cdots$$

$$= 1 - 2(x-1) + 3(x-1)^2 - 4(x-1)^3 + 5(x-1)^4 - \cdots$$

$$= \sum_{n=0}^{\infty}(-1)^n(n+1)(x-1)^n.$$

$$\lim_{n\to\infty}\left|\frac{a_{n+1}}{a_n}\right| = \lim_{n\to\infty}\frac{(n+2)\,|x-1|^{n+1}}{(n+1)\,|x-1|^n} = \lim_{n\to\infty}\left[\frac{n+2}{n+1}\cdot|x-1|\right]$$

$$= |x-1| < 1 \text{ for convergence, so } R = 1.$$

19. If $f(x) = \cos x$, then by Taylor's Formula $R_n(x) = \dfrac{f^{(n+1)}(z)}{(n+1)!}x^{n+1}$, where $0 < |z| < |x|$. But $f^{(n+1)}(z) = \pm\sin z$ or $\pm\cos z$. In each case, $\left|f^{(n+1)}(z)\right| \le 1$, so $|R_n(x)| \le \dfrac{1}{(n+1)!}\,|x|^{n+1}$. Thus, $|R_n(x)| \to 0$ as $n\to\infty$ by Equation 11. So $\lim\limits_{n\to\infty} R_n(x) = 0$ and, by Theorem 8, the series in Exercise 5 represents $\cos x$ for all x.

20. If $f(x) = \sin x$, then by Taylor's Formula $R_n(x) = \dfrac{f^{(n+1)}(z)}{(n+1)!}\left(x - \dfrac{\pi}{2}\right)^{n+1}$, where z lies between x and $\pi/2$. But $f^{(n+1)}(z) = \pm\sin z$ or $\pm\cos z$. In each case, $\left|f^{(n+1)}(z)\right| \le 1$, so $|R_n(x)| \le \dfrac{1}{(n+1)!}\left|x - \dfrac{\pi}{2}\right|^{n+1}$. Thus, $|R_n(x)| \to 0$ as $n\to\infty$ by Equation 11. So $\lim\limits_{n\to\infty} R_n(x) = 0$ and, by Theorem 8, the series in Exercise 16 represents $\sin x$ for all x.

21. If $f(x) = \sinh x$, then by Taylor's Formula $R_n(x) = \dfrac{f^{(n+1)}(z)}{(n+1)!}x^{n+1}$, where $0 < |z| < |x|$. But for all n, $\left|f^{(n+1)}(z)\right| \le \cosh z \le \cosh x$ (because all derivatives are either sinh or cosh, $|\sinh z| < |\cosh z|$ for all z, and $|z| < |x| \Rightarrow \cosh z < \cosh x$). So $|R_n(x)| \le \dfrac{\cosh x}{(n+1)!}x^{n+1} \to 0$ as $n\to\infty$ by Equation 11. So by Theorem 8, the series represents $\sinh x$ for all x.

22. If $f(x) = \cosh x$, then by Taylor's Formula $R_n(x) = \dfrac{f^{(n+1)}(z)}{(n+1)!}x^{n+1}$, where $0 < |z| < |x|$. But for all n, $\left|f^{(n+1)}(z)\right| \le \cosh z \le \cosh x$ (because all derivatives are either sinh or cosh, $|\sinh z| < |\cosh z|$ for all z, and $|z| < |x| \Rightarrow \cosh z < \cosh x$). So $|R_n(x)| \le \dfrac{\cosh x}{(n+1)!}x^{n+1} \to 0$ as $n\to\infty$ by Equation 11. So by Theorem 8, the series represents $\cosh x$ for all x.

23. The general binomial series in (18) is

$$(1+x)^k = \sum_{n=0}^{\infty} \binom{k}{n} x^n = 1 + kx + \frac{k(k-1)}{2!}x^2 + \frac{k(k-1)(k-2)}{3!}x^3 + \cdots.$$

$$\begin{aligned}(1+x)^{1/2} &= \sum_{n=0}^{\infty} \binom{\frac{1}{2}}{n} x^n = 1 + \left(\tfrac{1}{2}\right)x + \frac{\left(\frac{1}{2}\right)\left(-\frac{1}{2}\right)}{2!}x^2 + \frac{\left(\frac{1}{2}\right)\left(-\frac{1}{2}\right)\left(-\frac{3}{2}\right)}{3!}x^3 + \cdots \\ &= 1 + \frac{x}{2} - \frac{x^2}{2^2 \cdot 2!} + \frac{1 \cdot 3 \cdot x^3}{2^3 \cdot 3!} - \frac{1 \cdot 3 \cdot 5 \cdot x^4}{2^4 \cdot 4!} + \cdots \\ &= 1 + \frac{x}{2} + \sum_{n=2}^{\infty} \frac{(-1)^{n-1}\, 1 \cdot 3 \cdot 5 \cdot \cdots \cdot (2n-3)x^n}{2^n \cdot n!} \text{ for } |x| < 1, \quad \text{so } R = 1.\end{aligned}$$

24. $\dfrac{1}{(1+x)^4} = (1+x)^{-4} = \displaystyle\sum_{n=0}^{\infty} \binom{-4}{n} x^n$. The binomial coefficient is

$$\begin{aligned}\binom{-4}{n} &= \frac{(-4)(-5)(-6)\cdot \cdots \cdot (-4-n+1)}{n!} = \frac{(-4)(-5)(-6)\cdot \cdots \cdot [-(n+3)]}{n!} \\ &= \frac{(-1)^n \cdot 2 \cdot 3 \cdot 4 \cdot 5 \cdot 6 \cdot \cdots \cdot (n+1)(n+2)(n+3)}{2 \cdot 3 \cdot n!} = \frac{(-1)^n (n+1)(n+2)(n+3)}{6}\end{aligned}$$

Thus, $\dfrac{1}{(1+x)^4} = \displaystyle\sum_{n=0}^{\infty} \frac{(-1)^n (n+1)(n+2)(n+3)}{6} x^n$ for $|x| < 1$, so $R = 1$.

25. $\dfrac{1}{(2+x)^3} = \dfrac{1}{[2(1+x/2)]^3} = \dfrac{1}{8}\left(1 + \dfrac{x}{2}\right)^{-3} = \dfrac{1}{8} \displaystyle\sum_{n=0}^{\infty} \binom{-3}{n} \left(\frac{x}{2}\right)^n$. The binomial coefficient is

$$\begin{aligned}\binom{-3}{n} &= \frac{(-3)(-4)(-5)\cdot \cdots \cdot (-3-n+1)}{n!} = \frac{(-3)(-4)(-5)\cdot \cdots \cdot [-(n+2)]}{n!} \\ &= \frac{(-1)^n \cdot 2 \cdot 3 \cdot 4 \cdot 5 \cdot \cdots \cdot (n+1)(n+2)}{2 \cdot n!} = \frac{(-1)^n (n+1)(n+2)}{2}\end{aligned}$$

Thus, $\dfrac{1}{(2+x)^3} = \dfrac{1}{8} \displaystyle\sum_{n=0}^{\infty} \frac{(-1)^n (n+1)(n+2)}{2} \frac{x^n}{2^n} = \sum_{n=0}^{\infty} \frac{(-1)^n (n+1)(n+2)x^n}{2^{n+4}}$ for $\left|\dfrac{x}{2}\right| < 1 \;\Leftrightarrow\; |x| < 2$, so $R = 2$.

26.
$$\begin{aligned}(1-x)^{2/3} &= \sum_{n=0}^{\infty} \binom{\frac{2}{3}}{n} (-x)^n = 1 + \tfrac{2}{3}(-x) + \frac{\frac{2}{3}\left(-\frac{1}{3}\right)}{2!}(-x)^2 + \frac{\frac{2}{3}\left(-\frac{1}{3}\right)\left(-\frac{4}{3}\right)}{3!}(-x)^3 + \cdots \\ &= 1 - \tfrac{2}{3}x + \sum_{n=2}^{\infty} \frac{(-1)^{n-1}(-1)^n \cdot 2 \cdot [1 \cdot 4 \cdot 7 \cdot \cdots \cdot (3n-5)]}{3^n \cdot n!} x^n \\ &= 1 - \tfrac{2}{3}x - 2\sum_{n=2}^{\infty} \frac{1 \cdot 4 \cdot 7 \cdot \cdots \cdot (3n-5)}{3^n \cdot n!} x^n\end{aligned}$$

and $|-x| < 1 \;\Leftrightarrow\; |x| < 1$, so $R = 1$.

27. $\cos x = \displaystyle\sum_{n=0}^{\infty} (-1)^n \frac{x^{2n}}{(2n)!} \;\Rightarrow\; f(x) = \cos(\pi x) = \sum_{n=0}^{\infty} \frac{(-1)^n (\pi x)^{2n}}{(2n)!} = \sum_{n=0}^{\infty} \frac{(-1)^n \pi^{2n} x^{2n}}{(2n)!}, \quad R = \infty$

28. $e^x = \displaystyle\sum_{n=0}^{\infty} \frac{x^n}{n!} \;\Rightarrow\; f(x) = e^{-x/2} = \sum_{n=0}^{\infty} \frac{(-x/2)^n}{n!} = \sum_{n=0}^{\infty} \frac{(-1)^n}{2^n\, n!} x^n, \quad R = \infty$

29. $\tan^{-1} x = \sum\limits_{n=0}^{\infty} (-1)^n \dfrac{x^{2n+1}}{2n+1} \Rightarrow f(x) = x \tan^{-1} x = x \sum\limits_{n=0}^{\infty} (-1)^n \dfrac{x^{2n+1}}{2n+1} = \sum\limits_{n=0}^{\infty} (-1)^n \dfrac{x^{2n+2}}{2n+1}, \quad R = 1$

30. $\sin x = \sum\limits_{n=0}^{\infty} (-1)^n \dfrac{x^{2n+1}}{(2n+1)!} \Rightarrow f(x) = \sin(x^4) = \sum\limits_{n=0}^{\infty} (-1)^n \dfrac{(x^4)^{2n+1}}{(2n+1)!} = \sum\limits_{n=0}^{\infty} \dfrac{(-1)^n}{(2n+1)!} x^{8n+4}, \quad R = \infty$

31. $e^x = \sum\limits_{n=0}^{\infty} \dfrac{x^n}{n!} \Rightarrow f(x) = x^2 e^{-x} = x^2 \sum\limits_{n=0}^{\infty} \dfrac{(-x)^n}{n!} = \sum\limits_{n=0}^{\infty} \dfrac{(-1)^n x^{n+2}}{n!}, \quad R = \infty$

32. $\cos x = \sum\limits_{n=0}^{\infty} (-1)^n \dfrac{x^{2n}}{(2n)!} \Rightarrow \cos 2x = \sum\limits_{n=0}^{\infty} (-1)^n \dfrac{(2x)^{2n}}{(2n)!} = \sum\limits_{n=0}^{\infty} \dfrac{(-1)^n 2^{2n}}{(2n)!} x^{2n} \Rightarrow$

$f(x) = x \cos 2x = \sum\limits_{n=0}^{\infty} \dfrac{(-1)^n 2^{2n}}{(2n)!} x^{2n+1}, \quad R = \infty$

33. We must write the binomial in the form (1+ expression), so we'll factor out a 4.

$$\frac{x}{\sqrt{4+x^2}} = \frac{x}{\sqrt{4(1+x^2/4)}} = \frac{x}{2\sqrt{1+x^2/4}} = \frac{x}{2}\left(1+\frac{x^2}{4}\right)^{-1/2} = \frac{x}{2} \sum_{n=0}^{\infty} \binom{-\frac{1}{2}}{n} \left(\frac{x^2}{4}\right)^n$$

$$= \frac{x}{2}\left[1 + \left(-\tfrac{1}{2}\right)\frac{x^2}{4} + \frac{\left(-\frac{1}{2}\right)\left(-\frac{3}{2}\right)}{2!}\left(\frac{x^2}{4}\right)^2 + \frac{\left(-\frac{1}{2}\right)\left(-\frac{3}{2}\right)\left(-\frac{5}{2}\right)}{3!}\left(\frac{x^2}{4}\right)^3 + \cdots\right]$$

$$= \frac{x}{2} + \frac{x}{2} \sum_{n=1}^{\infty} (-1)^n \frac{1 \cdot 3 \cdot 5 \cdot \cdots \cdot (2n-1)}{2^n \cdot 4^n \cdot n!} x^{2n}$$

$$= \frac{x}{2} + \sum_{n=1}^{\infty} (-1)^n \frac{1 \cdot 3 \cdot 5 \cdot \cdots \cdot (2n-1)}{n!\, 2^{3n+1}} x^{2n+1} \text{ and } \frac{x^2}{4} < 1 \Leftrightarrow \frac{|x|}{2} < 1 \Leftrightarrow |x| < 2, \text{ so } R = 2.$$

34. $\dfrac{x^2}{\sqrt{2+x}} = \dfrac{x^2}{\sqrt{2(1+x/2)}} = \dfrac{x^2}{\sqrt{2}}\left(1+\dfrac{x}{2}\right)^{-1/2} = \dfrac{x^2}{\sqrt{2}} \sum\limits_{n=0}^{\infty} \dbinom{-\frac{1}{2}}{n} \left(\dfrac{x}{2}\right)^n$

$$= \frac{x^2}{\sqrt{2}}\left[1 + \left(-\tfrac{1}{2}\right)\left(\frac{x}{2}\right) + \frac{\left(-\frac{1}{2}\right)\left(-\frac{3}{2}\right)}{2!}\left(\frac{x}{2}\right)^2 + \frac{\left(-\frac{1}{2}\right)\left(-\frac{3}{2}\right)\left(-\frac{5}{2}\right)}{3!}\left(\frac{x}{2}\right)^3 + \cdots\right]$$

$$= \frac{x^2}{\sqrt{2}} + \frac{x^2}{\sqrt{2}} \sum_{n=1}^{\infty} (-1)^n \frac{1 \cdot 3 \cdot 5 \cdot \cdots \cdot (2n-1)}{n!\, 2^{2n}} x^n$$

$$= \frac{x^2}{\sqrt{2}} + \sum_{n=1}^{\infty} (-1)^n \frac{1 \cdot 3 \cdot 5 \cdot \cdots \cdot (2n-1)}{n!\, 2^{2n+1/2}} x^{n+2} \text{ and } \left|\frac{x}{2}\right| < 1 \Leftrightarrow |x| < 2, \text{ so } R = 2.$$

35. $\sin^2 x = \dfrac{1}{2}(1 - \cos 2x) = \dfrac{1}{2}\left[1 - \sum\limits_{n=0}^{\infty} \dfrac{(-1)^n (2x)^{2n}}{(2n)!}\right] = \dfrac{1}{2}\left[1 - 1 - \sum\limits_{n=1}^{\infty} \dfrac{(-1)^n (2x)^{2n}}{(2n)!}\right]$

$= \sum\limits_{n=1}^{\infty} \dfrac{(-1)^{n+1} 2^{2n-1} x^{2n}}{(2n)!}, \quad R = \infty$

36. $\dfrac{x - \sin x}{x^3} = \dfrac{1}{x^3}\left[x - \sum\limits_{n=0}^{\infty} \dfrac{(-1)^n x^{2n+1}}{(2n+1)!}\right] = \dfrac{1}{x^3}\left[x - x - \sum\limits_{n=1}^{\infty} \dfrac{(-1)^n x^{2n+1}}{(2n+1)!}\right] = \dfrac{1}{x^3}\left[-\sum\limits_{n=0}^{\infty} \dfrac{(-1)^{n+1} x^{2n+3}}{(2n+3)!}\right]$

$= \dfrac{1}{x^3} \sum\limits_{n=0}^{\infty} \dfrac{(-1)^n x^{2n+3}}{(2n+3)!} = \sum\limits_{n=0}^{\infty} \dfrac{(-1)^n x^{2n}}{(2n+3)!}$

and this series also gives the required value at $x = 0$ (namely $1/6$); $R = \infty$.

37. $\cos x = \sum_{n=0}^{\infty} (-1)^n \frac{x^{2n}}{(2n)!} \Rightarrow$

$$f(x) = \cos(x^2) = \sum_{n=0}^{\infty} \frac{(-1)^n (x^2)^{2n}}{(2n)!} = \sum_{n=0}^{\infty} \frac{(-1)^n x^{4n}}{(2n)!},\ R = \infty$$

Notice that, as n increases, $T_n(x)$ becomes a better approximation to $f(x)$.

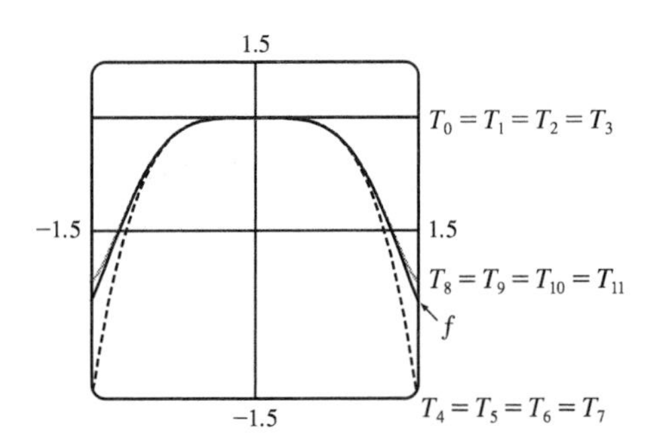

38. $e^x \overset{(12)}{=} \sum_{n=0}^{\infty} \frac{x^n}{n!}$, so $e^{-x^2} = \sum_{n=0}^{\infty} \frac{(-x^2)^n}{n!} = \sum_{n=0}^{\infty} (-1)^n \frac{x^{2n}}{n!}$.

Also, $\cos x \overset{(17)}{=} \sum_{n=0}^{\infty} (-1)^n \frac{x^{2n}}{(2n)!}$, so

$$f(x) = e^{-x^2} + \cos x = \sum_{n=0}^{\infty} (-1)^n \left(\frac{1}{n!} + \frac{1}{(2n)!} \right) x^{2n}$$

$$= 2 - \frac{3}{2}x^2 + \frac{13}{24}x^4 - \frac{121}{720}x^6 + \cdots.$$

The series for e^x and $\cos x$ converge for all x, so the same is true of the series for $f(x)$; that is, $R = \infty$. From the graphs of f and the first few Taylor polynomials, we see that $T_n(x)$ provides a closer fit to $f(x)$ near 0 as n increases.

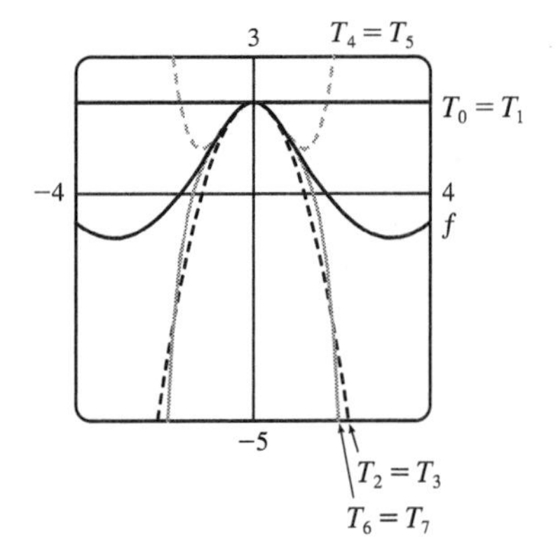

39. $e^x = \sum_{n=0}^{\infty} \frac{x^n}{n!}$, so $e^{-0.2} = \sum_{n=0}^{\infty} \frac{(-0.2)^n}{n!} = 1 - 0.2 + \frac{1}{2!}(0.2)^2 - \frac{1}{3!}(0.2)^3 + \frac{1}{4!}(0.2)^4 - \frac{1}{5!}(0.2)^5 + \frac{1}{6!}(0.2)^6 - \cdots$.

But $\frac{1}{6!}(0.2)^6 = 8.\overline{8} \times 10^{-8}$, so by the Alternating Series Estimation Theorem, $e^{-0.2} \approx \sum_{n=0}^{5} \frac{(-0.2)^n}{n!} \approx 0.81873$, correct to five decimal places.

40. $3^\circ = \frac{\pi}{60}$ radians and $\sin x = \sum_{n=0}^{\infty} \frac{(-1)^n x^{2n+1}}{(2n+1)!}$, so

$\sin \frac{\pi}{60} = \frac{\pi}{60} - \frac{\left(\frac{\pi}{60}\right)^3}{3!} + \frac{\left(\frac{\pi}{60}\right)^5}{5!} - \cdots = \frac{\pi}{60} - \frac{\pi^3}{1{,}296{,}000} + \frac{\pi^5}{93{,}312{,}000{,}000} - \cdots$. But $\frac{\pi^5}{93{,}312{,}000{,}000} < 10^{-8}$, so by the Alternating Series Estimation Theorem, $\sin \frac{\pi}{60} \approx \frac{\pi}{60} - \frac{\pi^3}{1{,}296{,}000} \approx 0.05234$.

41. (a) $1/\sqrt{1-x^2} = \left[1 + (-x^2)\right]^{-1/2} = 1 + \left(-\frac{1}{2}\right)(-x^2) + \frac{\left(-\frac{1}{2}\right)\left(-\frac{3}{2}\right)}{2!}(-x^2)^2 + \frac{\left(-\frac{1}{2}\right)\left(-\frac{3}{2}\right)\left(-\frac{5}{2}\right)}{3!}(-x^2)^3 + \cdots$

$$= 1 + \sum_{n=1}^{\infty} \frac{1 \cdot 3 \cdot 5 \cdot \cdots \cdot (2n-1)}{2^n \cdot n!} x^{2n}$$

(b) $\sin^{-1} x = \int \frac{1}{\sqrt{1-x^2}}\,dx = C + x + \sum_{n=1}^{\infty} \frac{1 \cdot 3 \cdot 5 \cdot \cdots \cdot (2n-1)}{(2n+1)2^n \cdot n!} x^{2n+1}$

$$= x + \sum_{n=1}^{\infty} \frac{1 \cdot 3 \cdot 5 \cdot \cdots \cdot (2n-1)}{(2n+1)2^n \cdot n!} x^{2n+1} \quad \text{since } 0 = \sin^{-1} 0 = C.$$

42. (a) $1/\sqrt[4]{1+x} = (1+x)^{-1/4} = \sum_{n=0}^{\infty} \binom{-\frac{1}{4}}{n} x^n = 1 - \frac{1}{4}x + \frac{\left(-\frac{1}{4}\right)\left(-\frac{5}{4}\right)}{2!}x^2 + \frac{\left(-\frac{1}{4}\right)\left(-\frac{5}{4}\right)\left(-\frac{9}{4}\right)}{3!}x^3 + \cdots$

$$= 1 - \frac{1}{4}x + \sum_{n=2}^{\infty} (-1)^n \frac{1 \cdot 5 \cdot 9 \cdot \cdots \cdot (4n-3)}{4^n \cdot n!} x^n$$

(b) $1/\sqrt[4]{1+x} = 1 - \frac{1}{4}x + \frac{5}{32}x^2 - \frac{15}{128}x^3 + \frac{195}{2048}x^4 - \cdots$. $1/\sqrt[4]{1.1} = 1/\sqrt[4]{1+0.1}$, so let $x = 0.1$. The sum of the first four terms is then $1 - \frac{1}{4}(0.1) + \frac{5}{32}(0.1)^2 - \frac{15}{128}(0.1)^3 \approx 0.976$. The fifth term is $\frac{195}{2048}(0.1)^4 \approx 0.000\,009\,5$, which does not affect the third decimal place of the sum, so we have $1/\sqrt[4]{1.1} \approx 0.976$. (Note that the third decimal place of the sum of the first three terms is affected by the fourth term, so we need to use more than three terms for the sum.)

43. $\cos x \overset{(17)}{=} \sum_{n=0}^{\infty} (-1)^n \frac{x^{2n}}{(2n)!} \quad \Rightarrow \quad \cos(x^3) = \sum_{n=0}^{\infty} (-1)^n \frac{(x^3)^{2n}}{(2n)!} = \sum_{n=0}^{\infty} (-1)^n \frac{x^{6n}}{(2n)!} \quad \Rightarrow$

$x\cos(x^3) = \sum_{n=0}^{\infty} (-1)^n \frac{x^{6n+1}}{(2n)!} \quad \Rightarrow \quad \int x\cos(x^3)\,dx = C + \sum_{n=0}^{\infty} (-1)^n \frac{x^{6n+2}}{(6n+2)(2n)!}$, with $R = \infty$.

44. $\frac{\sin x}{x} = \frac{1}{x}\sum_{n=0}^{\infty} \frac{(-1)^n x^{2n+1}}{(2n+1)!} = \sum_{n=0}^{\infty} \frac{(-1)^n x^{2n}}{(2n+1)!}$, so $\int \frac{\sin x}{x}\,dx = \int \sum_{n=0}^{\infty} \frac{(-1)^n x^{2n}}{(2n+1)!}\,dx = C + \sum_{n=0}^{\infty} \frac{(-1)^n x^{2n+1}}{(2n+1)(2n+1)!}$

45. Using the series from Exercise 23 and substituting x^3 for x, we get

$$\int \sqrt{x^3+1}\,dx = \int \left[1 + \frac{x^3}{2} + \sum_{n=2}^{\infty} \frac{(-1)^{n-1}\, 1 \cdot 3 \cdot 5 \cdot \cdots \cdot (2n-3)}{2^n n!} x^{3n}\right] dx$$

$$= C + x + \frac{x^4}{8} + \sum_{n=2}^{\infty} \frac{(-1)^{n-1}\, 1 \cdot 3 \cdot 5 \cdot \cdots \cdot (2n-3)}{2^n n!(3n+1)} x^{3n+1}$$

46. $e^x \overset{(12)}{=} \sum_{n=0}^{\infty} \frac{x^n}{n!} \quad \Rightarrow \quad e^x - 1 = \sum_{n=1}^{\infty} \frac{x^n}{n!} \quad \Rightarrow \quad \frac{e^x - 1}{x} = \sum_{n=1}^{\infty} \frac{x^{n-1}}{n!} \quad \Rightarrow \quad \int \frac{e^x - 1}{x}\,dx = C + \sum_{n=1}^{\infty} \frac{x^n}{n \cdot n!}$,

with $R = \infty$.

47. By Exercise 43, $\int x\cos(x^3)\,dx = C + \sum_{n=0}^{\infty} (-1)^n \frac{x^{6n+2}}{(6n+2)(2n)!}$, so

$\int_0^1 x\cos(x^3)\,dx = \left[\sum_{n=0}^{\infty} (-1)^n \frac{x^{6n+2}}{(6n+2)(2n)!}\right]_0^1 = \sum_{n=0}^{\infty} \frac{(-1)^n}{(6n+2)(2n)!} = \frac{1}{2} - \frac{1}{8 \cdot 2!} + \frac{1}{14 \cdot 4!} - \frac{1}{20 \cdot 6!} + \cdots$, but

$\frac{1}{20 \cdot 6!} = \frac{1}{14{,}400} \approx 0.000\,069$, so $\int_0^1 x\cos(x^3)\,dx \approx \frac{1}{2} - \frac{1}{16} + \frac{1}{336} \approx 0.440$ (correct to three decimal places) by the Alternating Series Estimation Theorem.

48. From the table of Maclaurin series in this section, we see that

$\tan^{-1} x = \sum_{n=0}^{\infty} (-1)^n \frac{x^{2n+1}}{2n+1}$ for x in $[-1, 1]$ and $\sin x = \sum_{n=0}^{\infty} (-1)^n \frac{x^{2n+1}}{(2n+1)!}$ for all real numbers x, so

$\tan^{-1}(x^3) + \sin(x^3) = \sum_{n=0}^{\infty} (-1)^n \frac{x^{6n+3}}{2n+1} + \sum_{n=0}^{\infty} (-1)^n \frac{x^{6n+3}}{(2n+1)!}$ for x^3 in $[-1, 1] \quad \Leftrightarrow \quad x$ in $[-1, 1]$. Thus,

$$I = \int_0^{0.2} [\tan^{-1}(x^3) + \sin(x^3)]\,dx = \int_0^{0.2} \sum_{n=0}^{\infty} (-1)^n x^{6n+3}\left(\frac{1}{2n+1} + \frac{1}{(2n+1)!}\right) dx$$
$$= \left[\sum_{n=0}^{\infty} (-1)^n \frac{x^{6n+4}}{6n+4}\left(\frac{1}{2n+1} + \frac{1}{(2n+1)!}\right)\right]_0^{0.2} = \sum_{n=0}^{\infty} (-1)^n \frac{(0.2)^{6n+4}}{6n+4}\left(\frac{1}{2n+1} + \frac{1}{(2n+1)!}\right)$$
$$= \frac{(0.2)^4}{4}(1+1) - \frac{(0.2)^{10}}{10}\left(\frac{1}{3} + \frac{1}{3!}\right) + \cdots$$

But $\frac{(0.2)^{10}}{10}\left(\frac{1}{3} + \frac{1}{3!}\right) = \frac{(0.2)^{10}}{20} = 5.12 \times 10^{-9}$, so by the Alternating Series Estimation Theorem,

$I \approx \frac{(0.2)^4}{2} = 0.000\,80$ (correct to five decimal places). [Actually, the value is $0.000\,800\,0$, correct to seven decimal places.]

49. We first find a series representation for $f(x) = (1+x)^{-1/2}$, and then substitute.

n	$f^{(n)}(x)$	$f^{(n)}(0)$
0	$(1+x)^{-1/2}$	1
1	$-\frac{1}{2}(1+x)^{-3/2}$	$-\frac{1}{2}$
2	$\frac{3}{4}(1+x)^{-5/2}$	$\frac{3}{4}$
3	$-\frac{15}{8}(1+x)^{-7/2}$	$-\frac{15}{8}$
$\vdots$	$\vdots$	$\vdots$

$$\frac{1}{\sqrt{1+x}} = 1 - \frac{x}{2} + \frac{3}{4}\left(\frac{x^2}{2!}\right) - \frac{15}{8}\left(\frac{x^3}{3!}\right) + \cdots \quad \Rightarrow$$
$$\frac{1}{\sqrt{1+x^3}} = 1 - \frac{1}{2}x^3 + \frac{3}{8}x^6 - \frac{5}{16}x^9 + \cdots \quad \Rightarrow$$

$\int_0^{0.1} \frac{dx}{\sqrt{1+x^3}} = \left[x - \frac{1}{8}x^4 + \frac{3}{56}x^7 - \frac{1}{32}x^{10} + \cdots\right]_0^{0.1} \approx (0.1) - \frac{1}{8}(0.1)^4$, by the Alternating Series Estimation Theorem, since $\frac{3}{56}(0.1)^7 \approx 0.000\,000\,005\,4 < 10^{-8}$, which is the maximum desired error. Therefore,

$$\int_0^{0.1} \frac{dx}{\sqrt{1+x^3}} \approx 0.099\,987\,50.$$

50. $\int_0^{0.5} x^2 e^{-x^2}\,dx = \int_0^{0.5} \sum_{n=0}^{\infty} \frac{(-1)^n x^{2n+2}}{n!}\,dx = \sum_{n=0}^{\infty}\left[\frac{(-1)^n x^{2n+3}}{n!(2n+3)}\right]_0^{1/2} = \sum_{n=0}^{\infty} \frac{(-1)^n}{n!(2n+3)2^{2n+3}}$ and since the term with $n=2$ is $\frac{1}{1792} < 0.001$, we use $\sum_{n=0}^{1} \frac{(-1)^n}{n!(2n+3)2^{2n+3}} = \frac{1}{24} - \frac{1}{160} \approx 0.0354$.

51.
$$\lim_{x\to 0} \frac{x - \tan^{-1} x}{x^3} = \lim_{x\to 0} \frac{x - \left(x - \frac{1}{3}x^3 + \frac{1}{5}x^5 - \frac{1}{7}x^7 + \cdots\right)}{x^3} = \lim_{x\to 0} \frac{\frac{1}{3}x^3 - \frac{1}{5}x^5 + \frac{1}{7}x^7 - \cdots}{x^3}$$
$$= \lim_{x\to 0}\left(\tfrac{1}{3} - \tfrac{1}{5}x^2 + \tfrac{1}{7}x^4 - \cdots\right) = \tfrac{1}{3}$$

since power series are continuous functions.

52.
$$\lim_{x\to 0} \frac{1-\cos x}{1+x-e^x} = \lim_{x\to 0} \frac{1 - \left(1 - \frac{1}{2!}x^2 + \frac{1}{4!}x^4 - \frac{1}{6!}x^6 + \cdots\right)}{1 + x - \left(1 + x + \frac{1}{2!}x^2 + \frac{1}{3!}x^3 + \frac{1}{4!}x^4 + \frac{1}{5!}x^5 + \frac{1}{6!}x^6 + \cdots\right)}$$
$$= \lim_{x\to 0} \frac{\frac{1}{2!}x^2 - \frac{1}{4!}x^4 + \frac{1}{6!}x^6 - \cdots}{-\frac{1}{2!}x^2 - \frac{1}{3!}x^3 - \frac{1}{4!}x^4 - \frac{1}{5!}x^5 - \frac{1}{6!}x^6 - \cdots}$$
$$= \lim_{x\to 0} \frac{\frac{1}{2!} - \frac{1}{4!}x^2 + \frac{1}{6!}x^4 - \cdots}{-\frac{1}{2!} - \frac{1}{3!}x - \frac{1}{4!}x^2 - \frac{1}{5!}x^3 - \frac{1}{6!}x^4 - \cdots} = \frac{\frac{1}{2} - 0}{-\frac{1}{2} - 0} = -1$$

since power series are continuous functions.

53. $$\lim_{x\to 0}\frac{\sin x - x + \frac{1}{6}x^3}{x^5} = \lim_{x\to 0}\frac{\left(x - \frac{1}{3!}x^3 + \frac{1}{5!}x^5 - \frac{1}{7!}x^7 + \cdots\right) - x + \frac{1}{6}x^3}{x^5}$$
$$= \lim_{x\to 0}\frac{\frac{1}{5!}x^5 - \frac{1}{7!}x^7 + \cdots}{x^5} = \lim_{x\to 0}\left(\frac{1}{5!} - \frac{x^2}{7!} + \frac{x^4}{9!} - \cdots\right) = \frac{1}{5!} = \frac{1}{120}$$

since power series are continuous functions.

54. $$\lim_{x\to 0}\frac{\tan x - x}{x^3} = \lim_{x\to 0}\frac{\left(x + \frac{1}{3}x^3 + \frac{2}{15}x^5 + \cdots\right) - x}{x^3} = \lim_{x\to 0}\frac{\frac{1}{3}x^3 + \frac{2}{15}x^5 + \cdots}{x^3} = \lim_{x\to 0}\left(\tfrac{1}{3} + \tfrac{2}{15}x^2 + \cdots\right) = \tfrac{1}{3}$$

since power series are continuous functions.

55. As in Example 9(a), we have $e^{-x^2} = 1 - \dfrac{x^2}{1!} + \dfrac{x^4}{2!} - \dfrac{x^6}{3!} + \cdots$ and we know that $\cos x = 1 - \dfrac{x^2}{2!} + \dfrac{x^4}{4!} - \cdots$ from Equation 17. Therefore, $e^{-x^2}\cos x = \left(1 - x^2 + \frac{1}{2}x^4 - \cdots\right)\left(1 - \frac{1}{2}x^2 + \frac{1}{24}x^4 - \cdots\right)$. Writing only the terms with degree ≤ 4, we get $e^{-x^2}\cos x = 1 - \frac{1}{2}x^2 + \frac{1}{24}x^4 - x^2 + \frac{1}{2}x^4 + \frac{1}{2}x^4 + \cdots = 1 - \frac{3}{2}x^2 + \frac{25}{24}x^4 + \cdots$.

56. $$\sec x = \frac{1}{\cos x} \overset{(17)}{=} \frac{1}{1 - \frac{1}{2}x^2 + \frac{1}{24}x^4 - \cdots}.$$

$$\begin{array}{r|l}
 & 1 + \frac{1}{2}x^2 + \frac{5}{24}x^4 + \cdots \\ \hline
1 - \frac{1}{2}x^2 + \frac{1}{24}x^4 - \cdots & 1 \\
 & 1 - \frac{1}{2}x^2 + \frac{1}{24}x^4 - \cdots \\ \cline{2-2}
 & \quad \frac{1}{2}x^2 - \frac{1}{24}x^4 + \cdots \\
 & \quad \frac{1}{2}x^2 - \frac{1}{4}x^4 + \cdots \\ \cline{2-2}
 & \qquad \frac{5}{24}x^4 + \cdots \\
 & \qquad \frac{5}{24}x^4 + \cdots \\ \cline{2-2}
 & \qquad \cdots
\end{array}$$

From the long division, $\sec x = 1 + \frac{1}{2}x^2 + \frac{5}{24}x^4 + \cdots$.

57. $$\frac{x}{\sin x} \overset{(16)}{=} \frac{x}{x - \frac{1}{6}x^3 + \frac{1}{120}x^5 - \cdots}.$$

$$\begin{array}{r|l}
 & 1 + \frac{1}{6}x^2 + \frac{7}{360}x^4 + \cdots \\ \hline
x - \frac{1}{6}x^3 + \frac{1}{120}x^5 - \cdots & x \\
 & x - \frac{1}{6}x^3 + \frac{1}{120}x^5 - \cdots \\ \cline{2-2}
 & \quad \frac{1}{6}x^3 - \frac{1}{120}x^5 + \cdots \\
 & \quad \frac{1}{6}x^3 - \frac{1}{36}x^5 + \cdots \\ \cline{2-2}
 & \qquad \frac{7}{360}x^5 + \cdots \\
 & \qquad \frac{7}{360}x^5 + \cdots \\ \cline{2-2}
 & \qquad \cdots
\end{array}$$

From the long division, $\dfrac{x}{\sin x} = 1 + \frac{1}{6}x^2 + \frac{7}{360}x^4 + \cdots$.

58. From Example 6 in Section 8.6, we have $\ln(1-x) = -x - \frac{1}{2}x^2 - \frac{1}{3}x^3 - \cdots$, $|x| < 1$. Therefore,

$$e^x \ln(1-x) = \left(1 + x + \tfrac{1}{2}x^2 + \cdots\right)\left(-x - \tfrac{1}{2}x^2 - \tfrac{1}{3}x^3 - \cdots\right)$$
$$= -x - \tfrac{1}{2}x^2 - \tfrac{1}{3}x^3 - x^2 - \tfrac{1}{2}x^3 - \tfrac{1}{2}x^3 - \cdots = -x - \tfrac{3}{2}x^2 - \tfrac{4}{3}x^3 - \cdots,\ |x| < 1$$

59. $\sum\limits_{n=0}^{\infty} (-1)^n \dfrac{x^{4n}}{n!} = \sum\limits_{n=0}^{\infty} \dfrac{\left(-x^4\right)^n}{n!} = e^{-x^4}$, by (12).

60. $\sum\limits_{n=0}^{\infty} \dfrac{(-1)^n \pi^{2n}}{6^{2n}(2n)!} = \sum\limits_{n=0}^{\infty} (-1)^n \dfrac{\left(\frac{\pi}{6}\right)^{2n}}{(2n)!} = \cos\frac{\pi}{6} = \frac{\sqrt{3}}{2}$, by (17).

61. $\sum\limits_{n=0}^{\infty} \dfrac{(-1)^n \pi^{2n+1}}{4^{2n+1}(2n+1)!} = \sum\limits_{n=0}^{\infty} \dfrac{(-1)^n \left(\frac{\pi}{4}\right)^{2n+1}}{(2n+1)!} = \sin\frac{\pi}{4} = \frac{1}{\sqrt{2}}$, by (16).

62. $\sum\limits_{n=0}^{\infty} \dfrac{3^n}{5^n\, n!} = \sum\limits_{n=0}^{\infty} \dfrac{(3/5)^n}{n!} = e^{3/5}$, by (12).

63. $3 + \dfrac{9}{2!} + \dfrac{27}{3!} + \dfrac{81}{4!} + \cdots = \dfrac{3^1}{1!} + \dfrac{3^2}{2!} + \dfrac{3^3}{3!} + \dfrac{3^4}{4!} + \cdots = \sum\limits_{n=1}^{\infty} \dfrac{3^n}{n!} = \sum\limits_{n=0}^{\infty} \dfrac{3^n}{n!} - 1 = e^3 - 1$, by (12).

64. $1 - \ln 2 + \dfrac{(\ln 2)^2}{2!} - \dfrac{(\ln 2)^3}{3!} + \cdots = \sum\limits_{n=0}^{\infty} \dfrac{(-\ln 2)^n}{n!} = e^{-\ln 2} = \left(e^{\ln 2}\right)^{-1} = 2^{-1} = \frac{1}{2}$, by (12).

65. (a) $[1 + (-x)]^{-2} = 1 + (-2)(-x) + \dfrac{(-2)(-3)}{2!}(-x)^2 + \dfrac{(-2)(-3)(-4)}{3!}(-x)^3 + \cdots$

$$= 1 + 2x + 3x^2 + 4x^3 + \cdots = \sum_{n=0}^{\infty} (n+1)x^n,$$

so $\dfrac{x}{(1-x)^2} = x \sum\limits_{n=0}^{\infty} (n+1)x^n = \sum\limits_{n=0}^{\infty} (n+1)x^{n+1} = \sum\limits_{n=1}^{\infty} nx^n$.

(b) With $x = \frac{1}{2}$ in part (a), we have $\sum\limits_{n=1}^{\infty} n\left(\frac{1}{2}\right)^n = \sum\limits_{n=1}^{\infty} \dfrac{n}{2^n} = \dfrac{\frac{1}{2}}{\left(1-\frac{1}{2}\right)^2} = \dfrac{\frac{1}{2}}{\frac{1}{4}} = 2$.

66. (a) $[1 + (-x)]^{-3} = \sum\limits_{n=0}^{\infty} \dbinom{-3}{n} (-x)^n = 1 + (-3)(-x) + \dfrac{(-3)(-4)}{2!}(-x)^2 + \dfrac{(-3)(-4)(-5)}{3!}(-x)^3 + \cdots$

$$= 1 + \sum_{n=1}^{\infty} \frac{3 \cdot 4 \cdot 5 \cdot \cdots \cdot (n+2)}{n!} x^n = \sum_{n=0}^{\infty} \frac{2 \cdot 3 \cdot 4 \cdot 5 \cdot \cdots \cdot (n+2)}{2 \cdot n!} x^n = \sum_{n=0}^{\infty} \frac{(n+1)(n+2)}{2} x^n \quad \Rightarrow$$

$$\left(x + x^2\right)[1 + (-x)]^{-3} = x\,[1 + (-x)]^{-3} + x^2\,[1 + (-x)]^{-3}$$
$$= \sum_{n=0}^{\infty} \frac{(n+1)(n+2)}{2} x^{n+1} + \sum_{n=0}^{\infty} \frac{(n+1)(n+2)}{2} x^{n+2}$$
$$= \sum_{n=1}^{\infty} \frac{n(n+1)}{2} x^n + \sum_{n=1}^{\infty} \frac{n(n+1)}{2} x^{n+1} = x + \sum_{n=2}^{\infty} \frac{n(n+1)}{2} x^n + \sum_{n=2}^{\infty} \frac{(n-1)n}{2} x^n$$
$$= x + \sum_{n=2}^{\infty} \left[\frac{n(n+1)}{2} + \frac{(n-1)n}{2}\right] x^n = x + \sum_{n=2}^{\infty} n^2 x^n = \sum_{n=1}^{\infty} n^2 x^n, \quad -1 < x < 1$$

(b) Setting $x = \frac{1}{2}$ in the last series above gives the required series, so $\sum\limits_{n=1}^{\infty} \dfrac{n^2}{2^n} = \dfrac{\frac{1}{2} + \left(\frac{1}{2}\right)^2}{\left(1 - \frac{1}{2}\right)^3} = \dfrac{\frac{3}{4}}{\frac{1}{8}} = 6$.

67. (a) $\left(1+x^2\right)^{1/2} = 1 + \left(\frac{1}{2}\right)x^2 + \frac{\left(\frac{1}{2}\right)\left(-\frac{1}{2}\right)}{2!}\left(x^2\right)^2 + \frac{\left(\frac{1}{2}\right)\left(-\frac{1}{2}\right)\left(-\frac{3}{2}\right)}{3!}\left(x^2\right)^3 + \cdots$

$$= 1 + \frac{x^2}{2} + \sum_{n=2}^{\infty} \frac{(-1)^{n-1}\, 1 \cdot 3 \cdot 5 \cdot \cdots \cdot (2n-3)}{2^n \cdot n!} x^{2n}$$

(b) The coefficient of x^{10} (corresponding to $n = 5$) in the above Maclaurin series is $\frac{f^{(10)}(0)}{10!}$, so

$$\frac{f^{(10)}(0)}{10!} = \frac{(-1)^4 \cdot 1 \cdot 3 \cdot 5 \cdot 7}{2^5 \cdot 5!} \quad \Rightarrow \quad f^{(10)}(0) = 10!\left(\frac{1 \cdot 3 \cdot 5 \cdot 7}{2^5 \cdot 5!}\right) = 99{,}225.$$

68. (a) $\left(1+x^3\right)^{-1/2} = \sum_{n=0}^{\infty} \binom{-\frac{1}{2}}{n} \left(x^3\right)^n = 1 + \left(-\frac{1}{2}\right)\left(x^3\right) + \frac{\left(-\frac{1}{2}\right)\left(-\frac{3}{2}\right)}{2!}\left(x^3\right)^2 + \frac{\left(-\frac{1}{2}\right)\left(-\frac{3}{2}\right)\left(-\frac{5}{2}\right)}{3!}\left(x^3\right)^3 + \cdots$

$$= 1 + \sum_{n=1}^{\infty} \frac{(-1)^n\, 1 \cdot 3 \cdot 5 \cdot \cdots \cdot (2n-1)\, x^{3n}}{2^n \cdot n!}$$

(b) The coefficient of x^9 (corresponding to $n = 3$) in the preceding series is $\frac{f^{(9)}(0)}{9!}$, so

$$\frac{f^{(9)}(0)}{9!} = \frac{(-1)^3\, 1 \cdot 3 \cdot 5}{2^3 \cdot 3!} \quad \Rightarrow \quad f^{(9)}(0) = -\frac{9! \cdot 5}{8 \cdot 2} = -113{,}400.$$

69. (a) $g(x) = \sum_{n=0}^{\infty} \binom{k}{n} x^n \quad \Rightarrow \quad g'(x) = \sum_{n=1}^{\infty} \binom{k}{n} n x^{n-1}$, so

$$(1+x)g'(x) = (1+x)\sum_{n=1}^{\infty} \binom{k}{n} n x^{n-1} = \sum_{n=1}^{\infty} \binom{k}{n} n x^{n-1} + \sum_{n=1}^{\infty} \binom{k}{n} n x^n$$

$$= \sum_{n=0}^{\infty} \binom{k}{n+1} (n+1) x^n + \sum_{n=0}^{\infty} \binom{k}{n} n x^n \qquad \left[\begin{array}{c}\text{Replace } n \text{ with } n+1 \\ \text{in the first series}\end{array}\right]$$

$$= \sum_{n=0}^{\infty} (n+1) \frac{k(k-1)(k-2)\cdots(k-n+1)(k-n)}{(n+1)!} x^n + \sum_{n=0}^{\infty} \left[(n)\frac{k(k-1)(k-2)\cdots(k-n+1)}{n!}\right] x^n$$

$$= \sum_{n=0}^{\infty} \frac{(n+1)k(k-1)(k-2)\cdots(k-n+1)}{(n+1)!} \left[(k-n)+n\right] x^n$$

$$= k \sum_{n=0}^{\infty} \frac{k(k-1)(k-2)\cdots(k-n+1)}{n!} x^n = k \sum_{n=0}^{\infty} \binom{k}{n} x^n = kg(x)$$

Thus, $g'(x) = \frac{kg(x)}{1+x}$.

(b) $h(x) = (1+x)^{-k} g(x) \quad \Rightarrow$

$$h'(x) = -k(1+x)^{-k-1} g(x) + (1+x)^{-k} g'(x) \qquad \text{[Product Rule]}$$

$$= -k(1+x)^{-k-1} g(x) + (1+x)^{-k} \frac{kg(x)}{1+x} \qquad \text{[from part (a)]}$$

$$= -k(1+x)^{-k-1} g(x) + k(1+x)^{-k-1} g(x) = 0$$

(c) From part (b) we see that $h(x)$ must be constant for $x \in (-1, 1)$, so $h(x) = h(0) = 1$ for $x \in (-1, 1)$.

Thus, $h(x) = 1 = (1+x)^{-k} g(x) \quad \Leftrightarrow \quad g(x) = (1+x)^k$ for $x \in (-1, 1)$.

70. (a) $f(x) = \begin{cases} e^{-1/x^2} & \text{if } x \neq 0 \\ 0 & \text{if } x = 0 \end{cases}$ so $f'(0) = \lim_{x \to 0} \frac{f(x) - f(0)}{x - 0} = \lim_{x \to 0} \frac{e^{-1/x^2}}{x} = \lim_{x \to 0} \frac{1/x}{e^{1/x^2}} = \lim_{x \to 0} \frac{x}{2e^{1/x^2}} = 0$

(using l'Hospital's Rule and simplifying in the penultimate step). Similarly, we can use the definition of the derivative and l'Hospital's Rule to show that $f''(0) = 0$, $f^{(3)}(0) = 0, \ldots, f^{(n)}(0) = 0$, so that the Maclaurin series for f consists entirely of zero terms. But since $f(x) \neq 0$ except for $x = 0$, we see that f cannot equal its Maclaurin series except at $x = 0$.

(b)

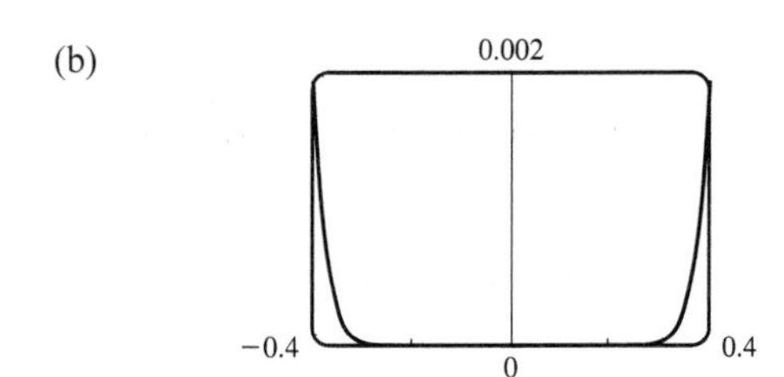

From the graph, it seems that the function is extremely flat at the origin. In fact, it could be said to be "infinitely flat" at $x = 0$, since all of its derivatives are 0 there.

8.8 Applications of Taylor Polynomials

1. (a)

n	$f^{(n)}(x)$	$f^{(n)}(0)$	$T_n(x)$
0	$\cos x$	1	1
1	$-\sin x$	0	1
2	$-\cos x$	-1	$1 - \frac{1}{2}x^2$
3	$\sin x$	0	$1 - \frac{1}{2}x^2$
4	$\cos x$	1	$1 - \frac{1}{2}x^2 + \frac{1}{24}x^4$
5	$-\sin x$	0	$1 - \frac{1}{2}x^2 + \frac{1}{24}x^4$
6	$-\cos x$	-1	$1 - \frac{1}{2}x^2 + \frac{1}{24}x^4 - \frac{1}{720}x^6$

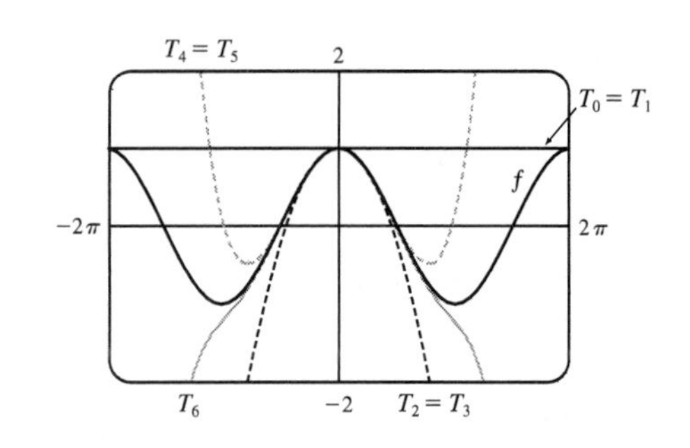

(b)

x	f	$T_0 = T_1$	$T_2 = T_3$	$T_4 = T_5$	T_6
$\frac{\pi}{4}$	0.7071	1	0.6916	0.7074	0.7071
$\frac{\pi}{2}$	0	1	-0.2337	0.0200	-0.0009
π	-1	1	-3.9348	0.1239	-1.2114

(c) As n increases, $T_n(x)$ is a good approximation to $f(x)$ on a larger and larger interval.

2. (a)

n	$f^{(n)}(x)$	$f^{(n)}(1)$	$T_n(x)$
0	x^{-1}	1	1
1	$-x^{-2}$	-1	$1 - (x - 1) = 2 - x$
2	$2x^{-3}$	2	$1 - (x - 1) + (x - 1)^2 = x^2 - 3x + 3$
3	$-6x^{-4}$	-6	$1 - (x - 1) + (x - 1)^2 - (x - 1)^3 = -x^3 + 4x^2 - 6x + 4$

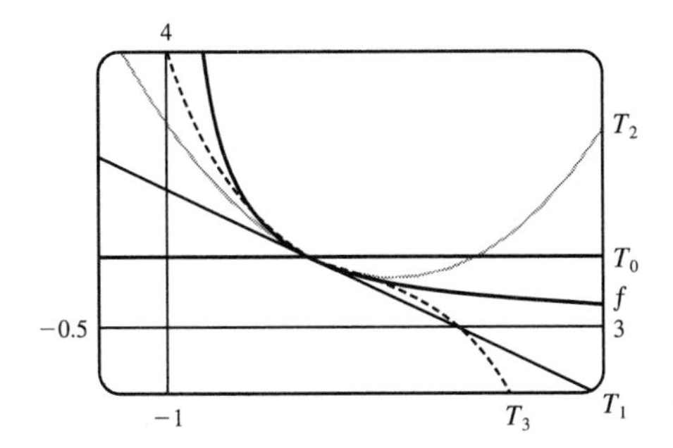

(b)

x	f	T_0	T_1	T_2	T_3
0.9	$1.\overline{1}$	1	1.1	1.11	1.111
1.3	0.7692	1	0.7	0.79	0.763

(c) As n increases, $T_n(x)$ is a good approximation to $f(x)$ on a larger and larger interval.

3.

n	$f^{(n)}(x)$	$f^{(n)}(\frac{\pi}{6})$
0	$\sin x$	$\frac{1}{2}$
1	$\cos x$	$\frac{\sqrt{3}}{2}$
2	$-\sin x$	$-\frac{1}{2}$
3	$-\cos x$	$-\frac{\sqrt{3}}{2}$

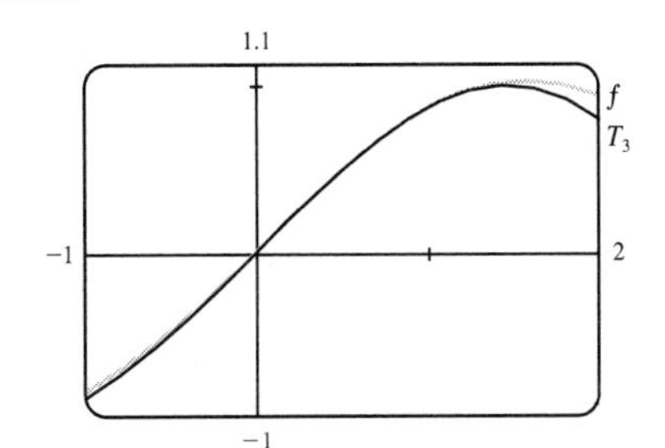

$$T_3(x) = \sum_{n=0}^{3} \frac{f^{(n)}(\frac{\pi}{6})}{n!}\left(x - \frac{\pi}{6}\right)^n = \frac{1}{2} + \frac{\sqrt{3}}{2}\left(x - \frac{\pi}{6}\right) - \frac{1}{4}\left(x - \frac{\pi}{6}\right)^2 - \frac{\sqrt{3}}{12}\left(x - \frac{\pi}{6}\right)^3$$

4.

n	$f^{(n)}(x)$	$f^{(n)}(2)$
0	e^x	e^2
1	e^x	e^2
2	e^x	e^2
3	e^x	e^2

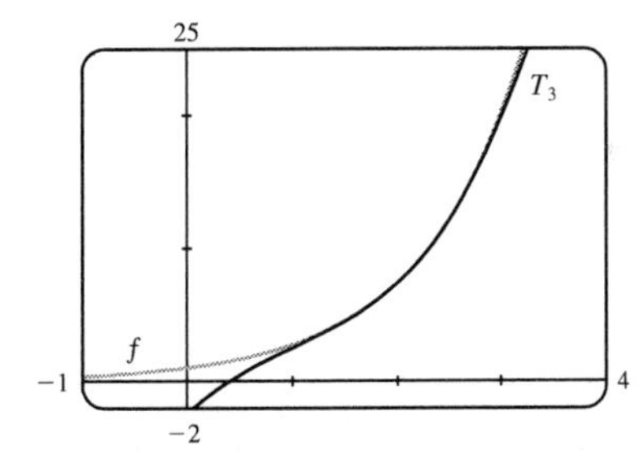

$$T_3(x) = \sum_{n=0}^{3} \frac{f^{(n)}(2)}{n!}(x-2)^n = e^2 + e^2(x-2) + \frac{e^2}{2}(x-2)^2 + \frac{e^2}{6}(x-2)^3$$

5.

n	$f^{(n)}(x)$	$f^{(n)}(0)$
0	$\arcsin x$	0
1	$1/\sqrt{1-x^2}$	1
2	$x/(1-x^2)^{3/2}$	0
3	$(2x^2+1)/(1-x^2)^{5/2}$	1

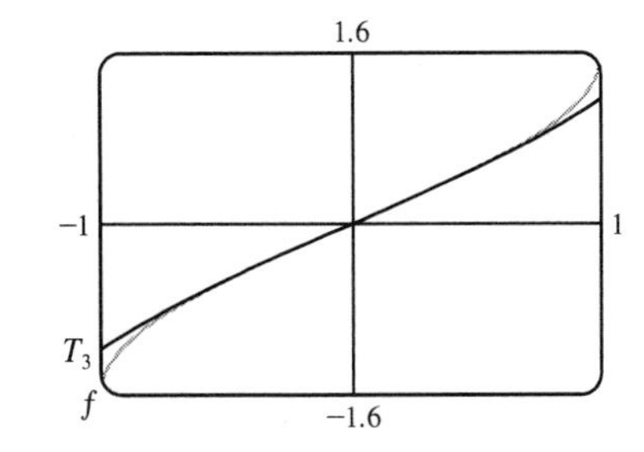

$$T_3(x) = \sum_{n=0}^{3} \frac{f^{(n)}(0)}{n!}x^n = x + \frac{x^3}{6}$$

6.

n	$f^{(n)}(x)$	$f^{(n)}(1)$
0	$(\ln x)/x$	0
1	$(1-\ln x)/x^2$	1
2	$(-3+2\ln x)/x^3$	-3
3	$(11-6\ln x)/x^4$	11

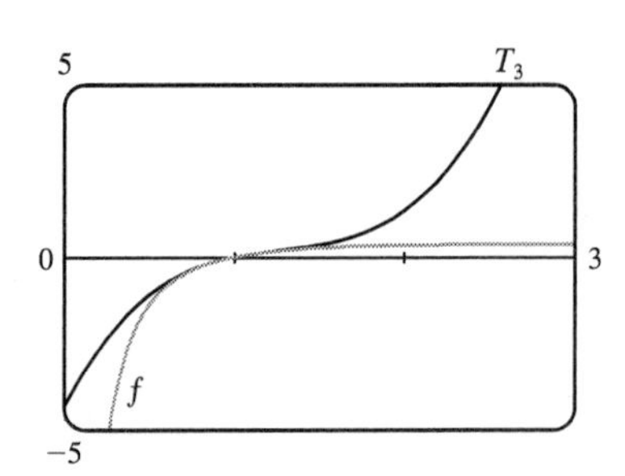

$$T_3(x) = \sum_{n=0}^{3} \frac{f^{(n)}(1)}{n!}(x-1)^n = (x-1) - \tfrac{3}{2}(x-1)^2 + \tfrac{11}{6}(x-1)^3$$

7.

n	$f^{(n)}(x)$	$f^{(n)}(0)$
0	xe^{-2x}	0
1	$(1-2x)e^{-2x}$	1
2	$4(x-1)e^{-2x}$	-4
3	$4(3-2x)e^{-2x}$	12

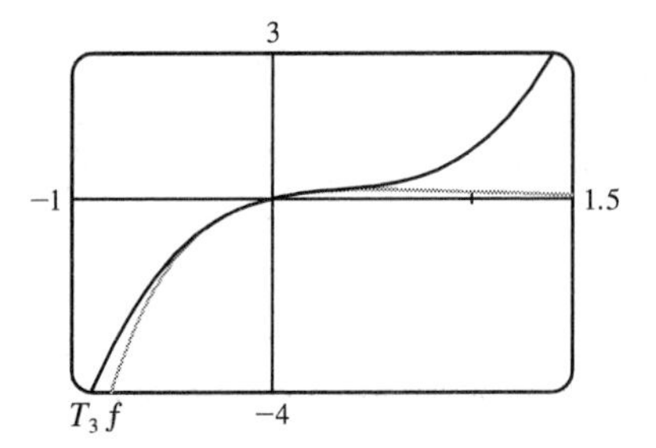

$$T_3(x) = \sum_{n=0}^{3} \frac{f^{(n)}(0)}{n!}x^n = \tfrac{0}{1}\cdot 1 + \tfrac{1}{1}x^1 + \tfrac{-4}{2}x^2 + \tfrac{12}{6}x^3 = x - 2x^2 + 2x^3$$

8.

n	$f^{(n)}(x)$	$f^{(n)}(1)$
0	$(3+x^2)^{1/2}$	2
1	$x(3+x^2)^{-1/2}$	$\frac{1}{2}$
2	$3(3+x^2)^{-3/2}$	$\frac{3}{8}$

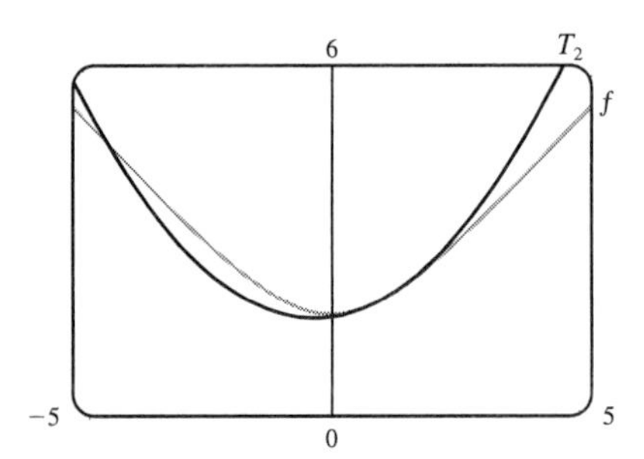

$$T_2(x) = \sum_{n=0}^{2} \frac{f^{(n)}(1)}{n!}(x-1)^n = 2 + \tfrac{1}{2}(x-1) + \tfrac{3/8}{2}(x-1)^2 = 2 + \tfrac{1}{2}(x-1) + \tfrac{3}{16}(x-1)^2$$

9. $f(x) = (1+x)^{1/2}$ $\qquad f(0) = 1$

$f'(x) = \frac{1}{2}(1+x)^{-1/2}$ $\qquad f'(0) = \frac{1}{2}$

$f''(x) = -\frac{1}{4}(1+x)^{-3/2}$

(a) $(1+x)^{1/2} \approx T_1(x) = 1 + \frac{1}{2}x$

(b) By Taylor's Formula, the remainder is $R_1(x) = \dfrac{f''(z)}{2!}x^2 = -\dfrac{1}{8(1+z)^{3/2}}x^2$, where z lies between 0 and x. Now

$0 \le x \le 0.1 \;\Rightarrow\; 0 \le x^2 \le 0.01$, and $0 < z < 0.1 \;\Rightarrow\; 1 < 1+z < 1.1$ so $|R_1(x)| < \frac{0.01}{8\cdot 1} = 0.00125$.

(c) From the graph of $|R_1(x)| = \left|\sqrt{1+x} - \left(1+\frac{1}{2}x\right)\right|$, it seems that the error is at most 0.0013 on $(0, 0.1)$.

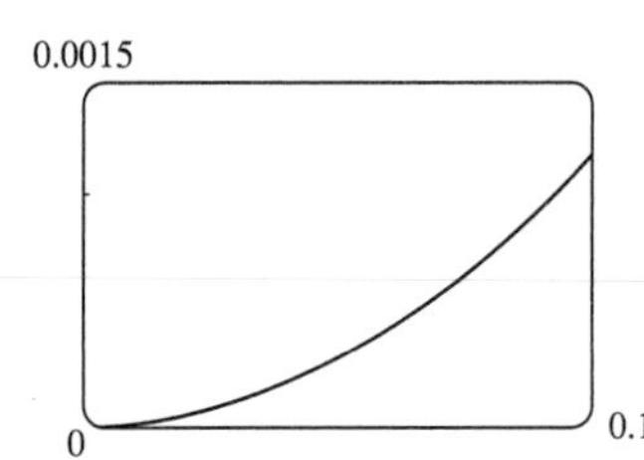

10. $f(x) = x^{-1}$ $\quad f(1) = 1$

$f'(x) = -x^{-2}$ $\quad f''(1) = -1$

$f''(x) = 2x^{-3}$ $\quad f''(1) = 2$

$f'''(x) = -6x^{-4}$ $\quad f'''(1) = -6$

(a) From Exercise 2, $1/x \approx T_3(x) = 1 - (x-1) + (x-1)^2 - (x-1)^3$.

(b) $f^{(4)}(x) = 24x^{-5}$, so $R_3(x) = \dfrac{24z^{-5}}{4!}(x-1)^4 = -\dfrac{(x-1)^4}{z^5}$, where z is between 0 and x. Now $0.8 \leq x \leq 01.2 \Rightarrow$

$|x-1| \leq 0.2$ and $z^5 > (0.8)^5 \Rightarrow |R_3(x)| < \dfrac{(0.2)^4}{(0.8)^5} < 0.005$.

(c) From the graph of $|R_3(x)| = |\tan x - T_3(3)|$, it seems that the error is less than 0.006 on $[0, \pi/6]$.

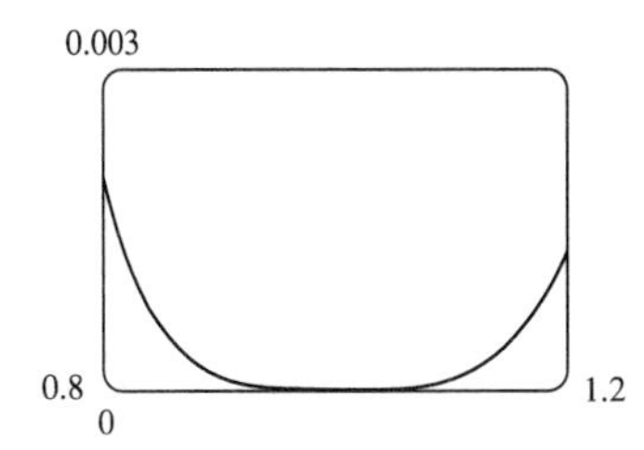

11.

n	$f^{(n)}(x)$	$f^{(n)}(0)$
0	$\tan x$	0
1	$\sec^2 x$	1
2	$2\sec^2 x \tan x$	0
3	$4\sec^2 x \tan^2 x + 2\sec^4 x$	2
4	$8\sec^2 x \tan^3 x + 16\sec^4 x \tan x$	

(a) $\tan x \approx T_3(x) = x + \frac{1}{3}x^3$

(b) The remainder is $R_3(x) = \dfrac{f^{(4)}(z)}{4!}x^4 = \dfrac{8\sec^2 z\,\tan^3 z + 16\sec^4 z\,\tan z}{4!}x^4 = \dfrac{\sec^2 z\,\tan^3 z + 2\sec^4 z\,\tan z}{3}x^4$

where z lies between 0 and x. Now $0 \leq x^4 \leq (\pi/6)^4$ and $0 < z < \pi/6 \Rightarrow \sec^2 z < \frac{4}{3}$ and $\tan z < \sqrt{3}/3$, so

$$|R_3(x)| \leq \frac{\frac{4}{3}\left(\frac{1}{\sqrt{3}}\right)^3 + 2\left(\frac{2}{\sqrt{3}}\right)^4\left(\frac{1}{\sqrt{3}}\right)}{3}\left(\frac{\pi}{6}\right)^4 = \frac{4\sqrt{3}}{9}\left(\frac{\pi}{6}\right)^4 \approx 0.057859 < 0.06$$

(c) From the graph, it seems that the error is less than 0.006 on $(0, \pi)$.

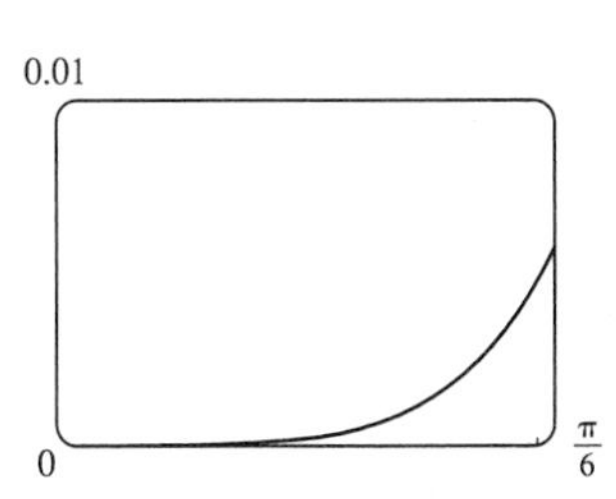

12.

n	$f^{(n)}(x)$	$f^{(n)}\left(\frac{\pi}{3}\right)$
0	$\cos x$	$\frac{1}{2}$
1	$-\sin x$	$-\frac{\sqrt{3}}{2}$
2	$-\cos x$	$-\frac{1}{2}$
3	$\sin x$	$\frac{\sqrt{3}}{2}$
4	$\cos x$	$\frac{1}{2}$
5	$-\sin x$	

(a) $f(x) = \cos x \approx T_4(x) = \frac{1}{2} - \frac{\sqrt{3}}{2}\left(x - \frac{\pi}{3}\right) - \frac{1}{4}\left(x - \frac{\pi}{3}\right)^2 + \frac{\sqrt{3}}{12}\left(x - \frac{\pi}{3}\right)^3 + \frac{1}{48}\left(x - \frac{\pi}{3}\right)^4$

(b) The remainder is $R_4(x) = \frac{1}{5!}(-\sin z)\left(x - \frac{\pi}{3}\right)^5$ where z lies between x and $\frac{\pi}{3}$. Now $0 \le x \le \frac{2\pi}{3} \Rightarrow$ $\left|x - \frac{\pi}{3}\right| \le \frac{\pi}{3} \Rightarrow |R_4(x)| \le \frac{1}{5!}\left(\frac{\pi}{3}\right)^5 \approx 0.0105$.

(c) From the graph of $|R_4(x)| = |\cos x - T_4(x)|$, it seems that the error is less than 0.01 on $\left[0, \frac{2\pi}{3}\right]$.

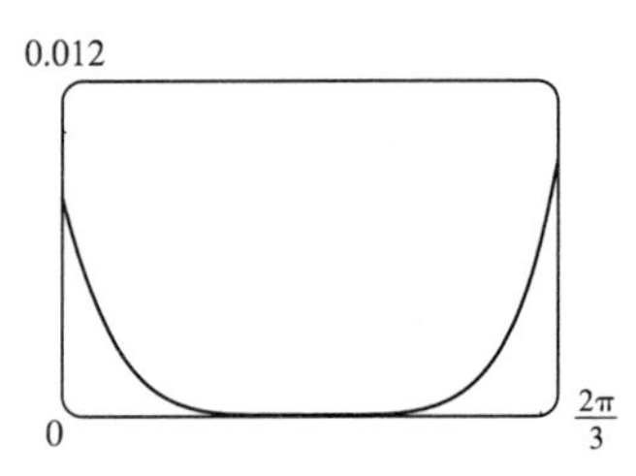

13.

n	$f^{(n)}(x)$	$f^{(n)}(0)$
0	e^{x^2}	1
1	$e^{x^2}(2x)$	0
2	$e^{x^2}(2 + 4x^2)$	2
3	$e^{x^2}(12x + 8x^3)$	0
4	$e^{x^2}(12 + 48x^2 + 16x^4)$	

(a) $f(x) = e^{x^2} \approx T_3(x) = 1 + \frac{2}{2!}x^2 = 1 + x^2$

(b) $R_3(x) = \dfrac{f^{(4)}(z)}{4!}x^4 = \dfrac{e^{z^2}(3 + 12z^2 + 4z^4)}{6}$, where z lies between 0 and x. Now $0 \le x \le 0.1 \Rightarrow$
$|R_3(x)| \le \dfrac{e^{0.01}(3 + 0.12 + 0.0004)}{6}(0.1)^4 < 0.00006$.

(c) From the graph of $|R_3(x)| = \left|e^{x^2} - (1 + x^2)\right|$, it appears that the error is less than 0.000051 on $[0, 0.1]$.

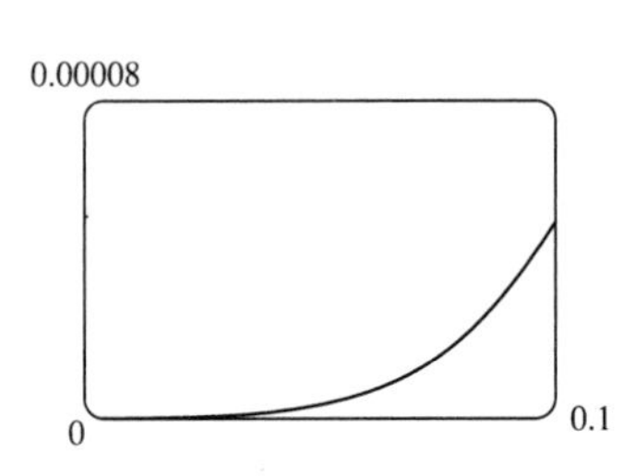

14. (a) Clearly $f^{(2n)}(0) = 1$ and $f^{(2n+1)}(0) = 0$, so $\cosh x \approx T_5(x) = 1 + \dfrac{x^2}{2} + \dfrac{x^4}{24}$.

(b) $|R_5(x)| \le \dfrac{f^{(6)}(z)}{6!}x^6 = \dfrac{\cosh z}{6!}x^6 \le \dfrac{\cosh(1)}{6!} \approx 0.00214$.

(c) It appears that the error is less than 0.0015 on $(-1, 1)$.

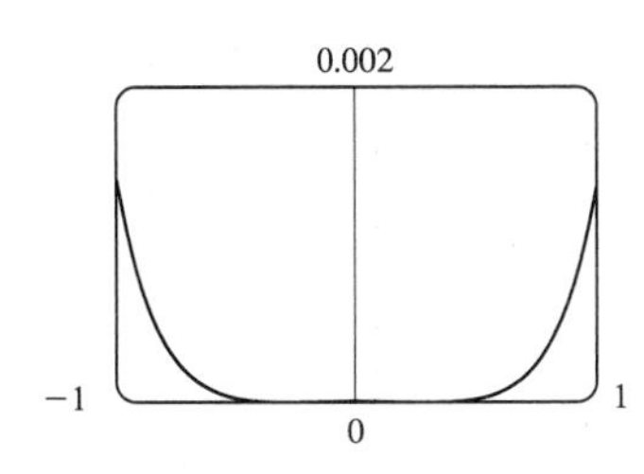

15.

$f(x) = x^{3/4}$ $\qquad f(16) = 8$

$f'(x) = \frac{3}{4}x^{-1/4}$ $\qquad f'(16) = \frac{3}{8}$

$f''(x) = -\frac{3}{16}x^{-5/4}$ $\qquad f''(16) = -\frac{3}{512}$

$f'''(x) = \frac{15}{64}x^{-9/4}$ $\qquad f'''(16) = \frac{15}{32{,}768}$

$f^{(4)}(x) = -\frac{135}{256}x^{-13/4}$

(a) $x^{3/4} \approx T_3(x) = 8 + \frac{3}{8}(x-16) - \frac{3}{1024}(x-16)^2 + \frac{5}{65{,}536}(x-16)^3$

(b) $R_3(x) \le \dfrac{-6z^{-4}}{4!}(x-4)^4$, where z lies between x and 4. Now $3 \le x \le 5 \;\Rightarrow\; |x-4| \le 1$ and $z > 3 \;\Rightarrow$ $\dfrac{1}{z^4} < \dfrac{1}{3^4}$, so $|R_3(x)| < \dfrac{6}{4!\,3^4}3 \le x \le 5 \;\Rightarrow\; |x-4| \le 1$ and $z > 3 \;\Rightarrow\; \dfrac{1}{z^4} < \dfrac{1}{3^4}$, so $|R_3(x)| < \dfrac{6}{4!\,3^4} = \dfrac{1}{324} \approx 0.0031$.

(c) It appears that the error is less than 3×10^{-6} on $(15, 17)$.

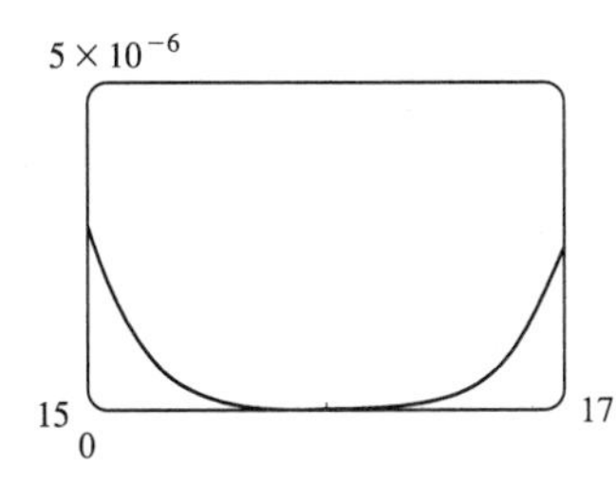

16.

$f(x) = \ln x$ $\qquad f(4) = \ln 4$

$f'(x) = x^{-1}$ $\qquad f'(4) = \frac{1}{4}$

$f''(x) = -x^{-2}$ $\qquad f''(4) = -\frac{1}{16}$

$f'''(x) = 2x^{-3}$ $\qquad f'''(4) = \frac{1}{32}$

$f^{(4)}(x) = -6x^{-4}$

(a) $\ln x \approx T_3(x) = \ln 4 + \frac{1}{4}(x-4) - \frac{1}{32}(x-4)^2 + \frac{1}{192}(x-4)^3$

(b) $R_3(x) = \dfrac{-6z^{-4}}{4!}(x-4)^4$, where z lies between x and 4. $3 \le x \le 5 \;\Rightarrow\; |x-4| \le 1$ and $z > 3 \;\Rightarrow\; \dfrac{1}{z^4} < \dfrac{1}{3^4}$, so $|R_3(x)| < \dfrac{6}{4!\,3^4} = \dfrac{1}{324} \approx 0.0031$.

(c) It appears that the error is less than 0.0013 on $(3, 5)$.

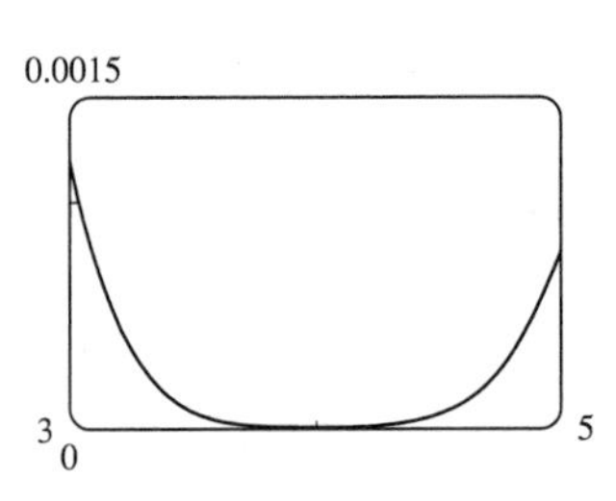

17. From Exercise 3, $\sin x = \frac{1}{2} + \frac{\sqrt{3}}{2}\left(x - \frac{\pi}{6}\right) - \frac{1}{4}\left(x - \frac{\pi}{6}\right)^2 - \frac{\sqrt{3}}{12}\left(x - \frac{\pi}{6}\right)^3 + R_3(x)$, where $|R_3(x)| \le \dfrac{1}{4!}\left|x - \frac{\pi}{6}\right|^4$ because $\left|f^{(4)}(z)\right| = |\sin z| \le 1$. Now $x = 35^\circ = (30^\circ + 5^\circ) = \left(\frac{\pi}{6} + \frac{\pi}{36}\right)$ radians, so the error is $\left|R_3\left(\frac{\pi}{36}\right)\right| \le \dfrac{\left(\frac{\pi}{36}\right)^4}{4!} < 0.000003$.

Therefore, to five decimal places, $\sin 35^\circ \approx \frac{1}{2} + \frac{\sqrt{3}}{2}\left(\frac{\pi}{36}\right) - \frac{1}{4}\left(\frac{\pi}{36}\right)^2 - \frac{\sqrt{3}}{12}\left(\frac{\pi}{36}\right)^3 \approx 0.57358$.

18. From Exercise 12, $\cos x = \frac{1}{2} - \frac{\sqrt{3}}{2}\left(x - \frac{\pi}{3}\right) - \frac{1}{4}\left(x - \frac{\pi}{3}\right)^2 + \frac{\sqrt{3}}{12}\left(x - \frac{\pi}{3}\right)^3 + \frac{1}{48}\left(x - \frac{\pi}{3}\right)^4 + R_4(x)$, where $|R_4(x)| \le \frac{1}{5!}\left|x - \frac{\pi}{3}\right|^5$ because $\left|f^{(5)}(z)\right| = |-\sin z| \le 1$. Now $x = 69^\circ = (60^\circ + 9^\circ) = \left(\frac{\pi}{3} + \frac{\pi}{20}\right)$ radians,

so the error is $|R_4(x)| \le \dfrac{\left(\frac{\pi}{20}\right)^5}{5!} < 8 \times 10^{-7}$. Therefore, to five decimal places,

$\cos 69^\circ \approx \frac{1}{2} - \frac{\sqrt{3}}{2}\left(\frac{\pi}{20}\right) - \frac{1}{4}\left(\frac{\pi}{20}\right)^2 + \frac{\sqrt{3}}{12}\left(\frac{\pi}{20}\right)^3 + \frac{1}{48}\left(\frac{\pi}{20}\right)^4 \approx 0.35837$.

19. All derivatives of e^x are e^x, so the remainder term is $R_n(x) = \dfrac{e^z}{(n+1)!}x^{n+1}$, where $0 < z < 0.1$. So we want

$R_n(0.1) \le \dfrac{e^{0.1}}{(n+1)!}(0.1)^{n+1} < 0.00001$, and we find that $n = 3$ satisfies this inequality. [In fact $R_3(0.1) < 0.0000046$.]

20. $f(x) = \sum\limits_{n=0}^{\infty} \dfrac{f^{(n)}(4)}{n!}(x-4)^n = \sum\limits_{n=0}^{\infty} \dfrac{(-1)^n\, n!}{3^n(n+1)\, n!}(x-4)^n = \sum\limits_{n=0}^{\infty} \dfrac{(-1)^n}{3^n(n+1)}(x-4)^n$. Now

$f(5) = \sum\limits_{n=0}^{\infty} \dfrac{(-1)^n}{3^n(n+1)} = \sum\limits_{n=0}^{\infty} (-1)^n b_n$ is the sum of an alternating series that satisfies (i) $b_{n+1} \le b_n$ and

(ii) $\lim\limits_{n\to\infty} b_n = 0$, so by the Alternating Series Estimation Theorem, $|R_5(5)| = |f(5) - T_5(5)| \le b_6$, and

$b_6 = \dfrac{1}{3^6(7)} = \dfrac{1}{5103} \approx 0.000196 < 0.0002$; that is, the fifth-degree Taylor polynomial approximates $f(5)$ with error less than 0.0002.

21. $\sin x = x - \dfrac{1}{3!}x^3 + \dfrac{1}{5!}x^5 - \cdots$. By the Alternating Series Estimation Theorem, the error in the approximation

$\sin x = x - \dfrac{1}{3!}x^3$ is less than $\left|\dfrac{1}{5!}x^5\right| < 0.01 \iff$

$\left|x^5\right| < 120(0.01) \iff |x| < (1.2)^{1/5} \approx 1.037$. The curves $y = x - \frac{1}{6}x^3$ and $y = \sin x - 0.01$ intersect at $x \approx 1.043$, so the graph confirms our estimate. Since both the sine function and the given approximation are odd functions, we need to check the estimate only for $x > 0$. Thus, the desired range of values for x is $-1.037 < x < 1.037$.

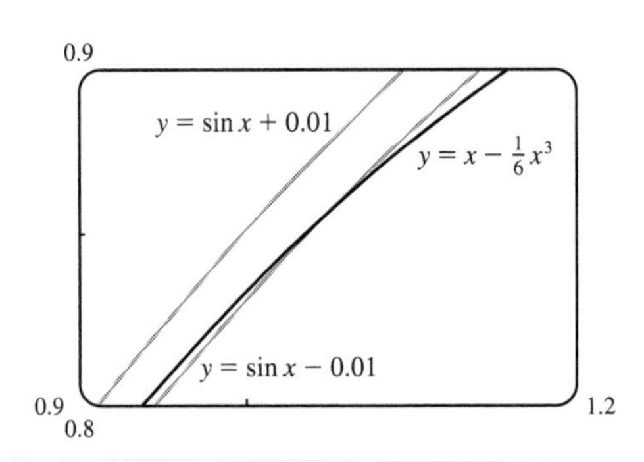

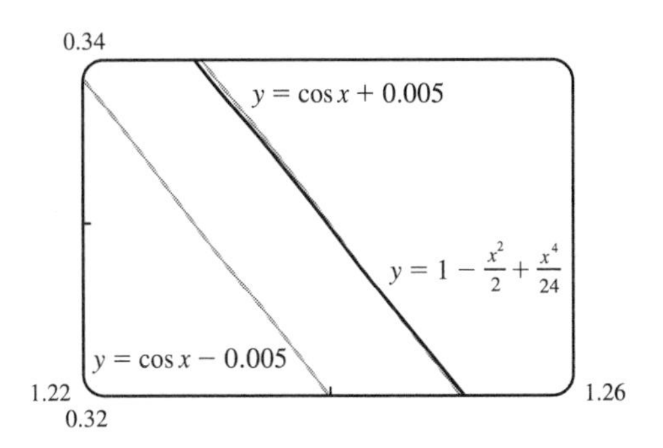

22. $\cos x = 1 - \frac{1}{2!}x^2 + \frac{1}{4!}x^4 - \frac{1}{6!}x^6 + \cdots$. By the Alternating Series Estimation Theorem, the error is less than $\left|-\frac{1}{6!}x^6\right| < 0.005 \Leftrightarrow x^6 < 720(0.005) \Leftrightarrow |x| < (3.6)^{1/6} \approx 1.238$. The curves $y = 1 - \frac{1}{2}x^2 + \frac{1}{24}x^4$ and $y = \cos x + 0.005$ intersect at $x \approx 1.244$, so the graph confirms our estimate. Since both the cosine function and the given approximation are even functions, we need to check the estimate only for $x > 0$. Thus, the desired range of values for x is $-1.238 < x < 1.238$.

23. Let $s(t)$ be the position function of the car, and for convenience set $s(0) = 0$. The velocity of the car is $v(t) = s'(t)$ and the acceleration is $a(t) = s''(t)$, so the second degree Taylor polynomial is $T_2(t) = s(0) + v(0)t + \frac{a(0)}{2}t^2 = 20t + t^2$. We estimate the distance traveled during the next second to be $s(1) \approx T_2(1) = 20 + 1 = 21$ m. The function $T_2(t)$ would not be accurate over a full minute, since the car could not possibly maintain an acceleration of 2 m/s^2 for that long (if it did, its final speed would be 140 m/s $\approx$ 313 mi/h!)

24. (a)

n	$\rho^{(n)}(t)$	$\rho^{(n)}(20)$
0	$\rho_{20}e^{\alpha(t-20)}$	ρ_{20}
1	$\alpha\rho_{20}e^{\alpha(t-20)}$	$\alpha\rho_{20}$
2	$\alpha^2\rho_{20}e^{\alpha(t-20)}$	$\alpha^2\rho_{20}$

The linear approximation is

$$T_1(t) = \rho(20) + \rho'(20)(t-20) = \rho_{20}\left[1 + \alpha(t-20)\right]$$

The quadratic approximation is

$$T_2(t) = \rho(20) + \rho'(20)(t-20) + \frac{\rho''(20)}{2}(t-20)^2 = \rho_{20}\left[1 + \alpha(t-20) + \tfrac{1}{2}\alpha^2(t-20)^2\right]$$

(b)

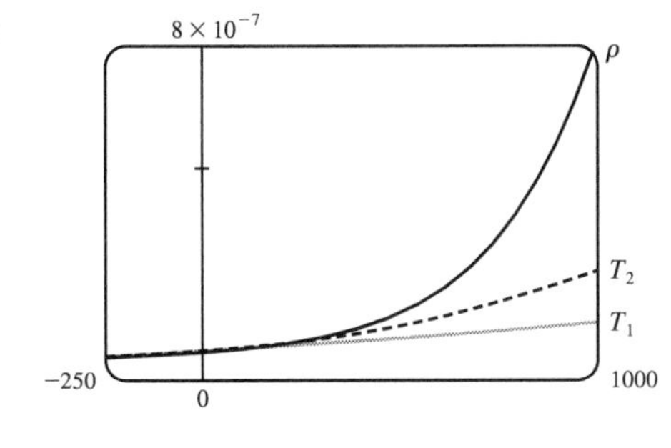

(c)

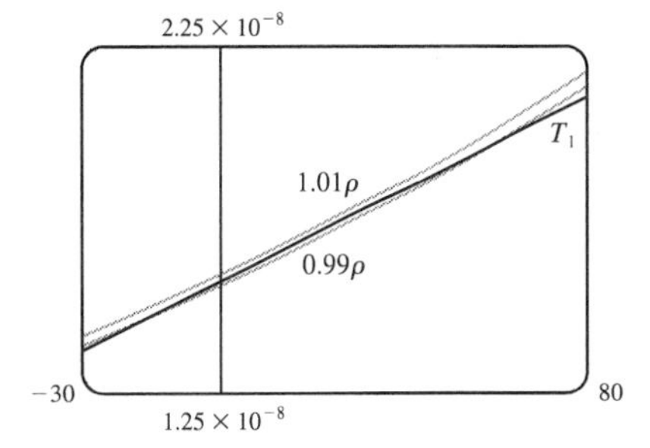

From the graph, it seems that $T_1(t)$ is within 1% of $\rho(t)$, that is, $0.99\rho(t) \le T_1(t) \le 1.01\rho(t)$, for $-14°\text{C} \le t \le 58°\text{C}$.

25. $E = \dfrac{q}{D^2} - \dfrac{q}{(D+d)^2} = \dfrac{q}{D^2} - \dfrac{q}{D^2(1+d/D)^2} = \dfrac{q}{D^2}\left[1 - \left(1 + \dfrac{d}{D}\right)^{-2}\right]$.

We use the Binomial Series to expand $(1 + d/D)^{-2}$:

$$E = \frac{q}{D^2}\left[1 - \left(1 - 2\left(\frac{d}{D}\right) + \frac{2\cdot 3}{2!}\left(\frac{d}{D}\right)^2 - \frac{2\cdot 3\cdot 4}{3!}\left(\frac{d}{D}\right)^3 + \cdots\right)\right] = \frac{q}{D^2}\left[2\left(\frac{d}{D}\right) - 3\left(\frac{d}{D}\right)^2 + 4\left(\frac{d}{D}\right)^3 - \cdots\right]$$

$$\approx \frac{q}{D^2}\cdot 2\left(\frac{d}{D}\right) = 2qd\cdot\frac{1}{D^3}$$

when D is much larger than d; that is, when P is far away from the dipole.

26. (a) If the water is deep, then $2\pi d/L$ is large, and we know that $\tanh x \to 1$ as $x \to \infty$. So we can approximate $\tanh(2\pi d/L) \approx 1$, and so $v^2 \approx gL/(2\pi) \quad \Leftrightarrow \quad v \approx \sqrt{gL/(2\pi)}$.

(b) From the table, the first term in the Maclaurin series of $\tanh x$ is x, so if the water is shallow, we can approximate $\tanh \frac{2\pi d}{L} \approx \frac{2\pi d}{L}$, and so $v^2 \approx \frac{gL}{2\pi} \cdot \frac{2\pi d}{L} \quad \Leftrightarrow \quad v \approx \sqrt{gd}$.

n	$f^{(n)}(x)$	$f^{(n)}(0)$
0	$\tanh x$	0
1	$\operatorname{sech}^2 x$	1
2	$-2\operatorname{sech}^2 x \tanh x$	0
3	$2\operatorname{sech}^2 x\,(3\tanh^2 x - 1)$	-2

(c) Since $\tanh x$ is an odd function, its Maclaurin series is alternating, so the error in the approximation $\tanh \frac{2\pi d}{L} \approx \frac{2\pi d}{L}$ is less than the first neglected term, which is $\frac{|f'''(0)|}{3!}\left(\frac{2\pi d}{L}\right)^3 = \frac{1}{3}\left(\frac{2\pi d}{L}\right)^3$. If $L > 10d$, then $\frac{1}{3}\left(\frac{2\pi d}{L}\right)^3 < \frac{1}{3}\left(2\pi \cdot \frac{1}{10}\right)^3 = \frac{\pi^3}{375}$, so the error in the approximation $v^2 = gd$ is less than $\frac{gL}{2\pi} \cdot \frac{\pi^3}{375} \approx 0.0132gL$.

27. (a) L is the length of the arc subtended by the angle θ, so $L = R\theta \quad \Rightarrow$

$\theta = L/R$. Now $\sec\theta = (R + C)/R \quad \Rightarrow \quad R\sec\theta = R + C \quad \Rightarrow$

$C = R\sec\theta - R = R\sec(L/R) - R$.

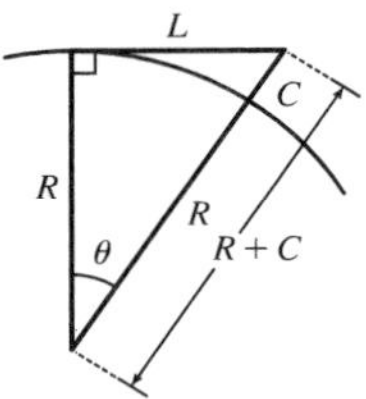

(b) If $f(x) = \sec x$, then $f'(x) = \sec x \tan x$,

$f''(x) = \sec^3 x + \sec x \tan^2 x$, $f'''(x) = 5\sec^3 x \tan x + \sec x \tan^3 x$,

$f^{(4)}(x) = 5\sec^5 x + 18\sec^3 x \tan^2 x + \sec x \tan^4 x$. So $f(0) = 1$, $f'(0) = 0$, $f''(0) = 1$, $f'''(0) = 0$,

$f^{(4)}(0) = 5$, and $\sec x \approx T_4(x) = 1 + \frac{1}{2}x^2 + \frac{5}{24}x^4$. By part (a),

$$C \approx R\left[1 + \frac{1}{2}\left(\frac{L}{R}\right)^2 + \frac{5}{24}\left(\frac{L}{R}\right)^4\right] - R = R + \frac{1}{2}R \cdot \frac{L^2}{R^2} + \frac{5}{24}R \cdot \frac{L^4}{R^4} - R = \frac{L^2}{2R} + \frac{5L^4}{24R^3}.$$

(c) Taking $L = 100$ km and $R = 6370$ km, the formula in part (a) says that

$C = R\sec(L/R) - R = 6370\sec(100/6370) - 6370 \approx 0.785\,009\,965\,44$ km. The formula in part (b) says that

$$C \approx \frac{L^2}{2R} + \frac{5L^4}{24R^3} = \frac{100^2}{2 \cdot 6370} + \frac{5 \cdot 100^4}{24 \cdot 6370^3} \approx 0.785\,009\,957\,36 \text{ km.}$$

The difference between these two results is only 0.000 000 008 08 km, or 0.000 008 08 m!

28. (a) $$4\sqrt{\frac{L}{g}}\int_0^{\pi/2} \frac{dx}{\sqrt{1 - k^2\sin^2 x}} = 4\sqrt{\frac{L}{g}}\int_0^{\pi/2} \left[1 + (-k^2\sin^2 x)\right]^{-1/2} dx$$

$$= 4\sqrt{\frac{L}{g}}\int_0^{\pi/2} \left[1 - \frac{1}{2}(-k^2\sin^2 x) + \frac{\frac{1}{2}\cdot\frac{3}{2}}{2!}(-k^2\sin^2 x)^2 - \frac{\frac{1}{2}\cdot\frac{3}{2}\cdot\frac{5}{2}}{3!}(-k^2\sin^2 x)^3 + \cdots\right] dx$$

$$= 4\sqrt{\frac{L}{g}}\int_0^{\pi/2} \left[1 + \left(\frac{1}{2}\right)k^2\sin^2 x + \left(\frac{1\cdot 3}{2\cdot 4}\right)k^4\sin^4 x + \left(\frac{1\cdot 3\cdot 5}{2\cdot 4\cdot 6}\right)k^6\sin^6 x + \cdots\right] dx$$

$$= 4\sqrt{\frac{L}{g}}\left[\frac{\pi}{2} + \left(\frac{1}{2}\right)\left(\frac{1}{2}\cdot\frac{\pi}{2}\right)k^2 + \left(\frac{1\cdot 3}{2\cdot 4}\right)\left(\frac{1\cdot 3}{2\cdot 4}\cdot\frac{\pi}{2}\right)k^4 + \left(\frac{1\cdot 3\cdot 5}{2\cdot 4\cdot 6}\right)\left(\frac{1\cdot 3\cdot 5}{2\cdot 4\cdot 6}\cdot\frac{\pi}{2}\right)k^6 + \cdots\right]$$

[split up the integral and use the result from Exercise 6.1.32]

$$= 2\pi\sqrt{\frac{L}{g}}\left[1 + \frac{1^2}{2^2}k^2 + \frac{1^2\cdot 3^2}{2^2\cdot 4^2}k^4 + \frac{1^2\cdot 3^2\cdot 5^2}{2^2\cdot 4^2\cdot 6^2}k^6 + \cdots\right]$$

(b) The first of the two inequalities is true because all of the terms in the series are positive. For the second,

$$T = 2\pi\sqrt{\frac{L}{g}}\left[1 + \frac{1^2}{2^2}k^2 + \frac{1^2 \cdot 3^2}{2^2 \cdot 4^2}k^4 + \frac{1^2 \cdot 3^2 \cdot 5^2}{2^2 \cdot 4^2 \cdot 6^2}k^6 + \frac{1^2 \cdot 3^2 \cdot 5^2 \cdot 7^2}{2^2 \cdot 4^2 \cdot 6^2 \cdot 8^2}k^8 + \cdots\right]$$

$$\leq 2\pi\sqrt{\frac{L}{g}}\left[1 + \tfrac{1}{4}k^2 + \tfrac{1}{4}k^4 + \tfrac{1}{4}k^6 + \tfrac{1}{4}k^8 + \cdots\right]$$

The terms in brackets (after the first) form a geometric series with $a = \frac{1}{4}k^2$ and $r = k^2 = \sin^2\left(\frac{1}{2}\theta_0\right) < 1$.

So $T \leq 2\pi\sqrt{\frac{L}{g}}\left[1 + \frac{k^2/4}{1-k^2}\right] = 2\pi\sqrt{\frac{L}{g}}\,\frac{4-3k^2}{4-4k^2}$.

(c) We substitute $L = 1$, $g = 9.8$, and $k = \sin(10°/2) \approx 0.08716$, and the inequality from part (b) becomes $2.01090 \leq T \leq 2.01093$, so $T \approx 2.0109$. The estimate $T \approx 2\pi\sqrt{L/g} \approx 2.0071$ differs by about 0.2%.

If $\theta_0 = 42°$, then $k \approx 0.35837$ and the inequality becomes $2.07153 \leq T \leq 2.08103$, so $T \approx 2.0763$. The one-term estimate is the same, and the discrepancy between the two estimates increases to about 3.4%.

29. Using Taylor's Formula with $n = 1$, $a = x_n$, $x = r$, we get $f(r) = f(x_n) + f'(x_n)(r - x_n) + R_1(x)$, where $R_1(x) = \frac{1}{2}f''(z)(r - x_n)^2$ and z lies between x_n and r. But r is a root, so $f(r) = 0$ and Taylor's Formula becomes $0 = f(x_n) + f'(x_n)(r - x_n) + \frac{1}{2}f''(z)(r - x_n)^2$. Taking the first two terms to the left side and dividing by $f'(x_n)$, we have

$x_n - r - \dfrac{f(x_n)}{f'(x_n)} = \dfrac{1}{2}\dfrac{f''(z)}{f'(x_n)}\,|x_n - r|^2$. By the formula for Newton's Method, we have

$$|x_{n+1} - r| = \left|x_n - \frac{f(x_n)}{f'(x_n)} - r\right| = \frac{1}{2}\frac{|f''(z)|}{|f'(x_n)|}\,|x_n - r|^2 \leq \frac{M}{2K}\,|x_n - r|^2 \text{ since } |f''(z)| \leq M \text{ and } |f'(x_n)| \geq K.$$

30. $q!(e - s_q) = q!\left(\dfrac{p}{q} - 1 - \dfrac{1}{1!} - \dfrac{1}{2!} - \cdots - \dfrac{1}{q!}\right) = p(q-1)! - q! - q! - \dfrac{q!}{2!} - \cdots - 1$, which is clearly an integer, and

$q!(e - s_q) = q!\left[\dfrac{e^z}{(q+1)!}\right] = \dfrac{e^z}{q+1}$. We have $0 < \dfrac{e^z}{q+1} < \dfrac{e}{q+1} < \dfrac{e}{3} < 1$ since $0 < z < 1$ and $q > 2$, and so $0 < q!(e - s_q) < 1$, which is a contradiction since we have already shown $q!(e - s_q)$ must be an integer. So e cannot be rational.

8 Review

CONCEPT CHECK

1. (a) See Definition 8.1.1.

(b) See Definition 8.2.2.

(c) The terms of the sequence $\{a_n\}$ approach 3 as n becomes large.

(d) By adding sufficiently many terms of the series, we can make the partial sums as close to 3 as we like.

2. (a) A sequence $\{a_n\}$ is bounded if there are numbers m and M such that $m \leq a_n \leq M$ for all $n \geq 1$.

(b) A sequence is monotonic if it is either increasing or decreasing.

(c) By Theorem 8.1.11, every bounded, monotonic sequence is convergent.

3. (a) See (4) in Section 8.2.

(b) The p-series $\sum_{n=1}^{\infty} \frac{1}{n^p}$ is convergent if $p > 1$.

4. If $\sum a_n = 3$, then $\lim_{n\to\infty} a_n = 0$ and $\lim_{n\to\infty} s_n = 3$.

5. (a) See the Test for Divergence on page 427.

(b) See the Integral Test on page 433.

(c) See the Comparison Test on page 435.

(d) See the Limit Comparison Test on page 436.

(e) See the Alternating Series Test on page 440.

(f) See the Ratio Test on page 445.

(g) See the Root Test on page 447.

6. (a) A series $\sum a_n$ is called *absolutely convergent* if the series of absolute values $\sum |a_n|$ is convergent.

(b) If a series $\sum a_n$ is absolutely convergent, then it is convergent.

(c) A series $\sum a_n$ is called *conditionally convergent* if it is convergent but not absolutely convergent.

7. By adding terms until you reach the desired accuracy given by the Alternating Series Estimation Theorem on page 440.

8. (a) $\sum_{n=0}^{\infty} c_n(x-a)^n$

(b) Given the power series $\sum_{n=0}^{\infty} c_n(x-a)^n$, the radius of convergence is:

(i) 0 if the series converges only when $x = a$

(ii) ∞ if the series converges for all x, or

(iii) a positive number R such that the series converges if $|x-a| < R$ and diverges if $|x-a| > R$.

(c) The interval of convergence of a power series is the interval that consists of all values of x for which the series converges. Corresponding to the cases in part (b), the interval of convergence is: (i) the single point $\{a\}$, (ii) all real numbers, that is, the real number line $(-\infty, \infty)$, or (iii) an interval with endpoints $a - R$ and $a + R$ which can contain neither, either, or both of the endpoints. In this case, we must test the series for convergence at each endpoint to determine the interval of convergence.

9. (a), (b) See Theorem 8.6.2.

10. (a) $T_n(x) = \sum_{i=0}^{n} \frac{f^{(i)}(a)}{i!}(x-a)^i$

(b) $\sum_{n=0}^{\infty} \frac{f^{(n)}(a)}{n!}(x-a)^n$

(c) $\sum_{n=0}^{\infty} \frac{f^{(n)}(0)}{n!}x^n \quad [a = 0 \text{ in part (b)}]$

(d) See Theorem 8.7.8.

(e) See Taylor's Formula.

11. (a) – (e) See the table on page 468.

12. See the Binomial Series (8.7.18) for the expansion. The radius of convergence for the binomial series is 1.

TRUE-FALSE QUIZ

1. False. See Note 2 on page 427.

2. False. The series $\sum_{n=1}^{\infty} n^{-\sin 1} = \sum_{n=1}^{\infty} \frac{1}{n^{\sin 1}}$ is a p-series with $p = \sin 1 \approx 0.84 \leq 1$, so the series diverges.

3. True. If $\lim_{n\to\infty} a_n = L$, then given any $\varepsilon > 0$, we can find a positive integer N such that $|a_n - L| < \varepsilon$ whenever $n > N$. If $n > N$, then $2n + 1 > N$ and $|a_{2n+1} - L| < \varepsilon$. Thus, $\lim_{n\to\infty} a_{2n+1} = L$.

4. True by Theorem 8.5.3.

Or: Use the Comparison Test to show that $\sum c_n(-2)^n$ converges absolutely.

5. False. For example, take $c_n = (-1)^n/(n6^n)$.

6. True by Theorem 8.5.3.

7. False, since $\lim_{n\to\infty} \left|\frac{a_{n+1}}{a_n}\right| = \lim_{n\to\infty} \left|\frac{1}{(n+1)^3} \cdot \frac{n^3}{1}\right| = \lim_{n\to\infty} \left|\frac{n^3}{(n+1)^3} \cdot \frac{1/n^3}{1/n^3}\right| = \lim_{n\to\infty} \frac{1}{(1+1/n)^3} = 1.$

8. True, since $\lim_{n\to\infty} \left|\frac{a_{n+1}}{a_n}\right| = \lim_{n\to\infty} \left|\frac{1}{(n+1)!} \cdot \frac{n!}{1}\right| = \lim_{n\to\infty} \frac{1}{n+1} = 0 < 1.$

9. False. See the note after Example 4 in Section 8.3.

10. True, since $\frac{1}{e} = e^{-1}$ and $e^x = \sum_{n=0}^{\infty} \frac{x^n}{n!}$, so $e^{-1} = \sum_{n=0}^{\infty} \frac{(-1)^n}{n!}$.

11. True. See (8) in Section 8.1.

12. True, because if $\sum |a_n|$ is convergent, then so is $\sum a_n$ by Theorem 8.4.1.

13. True. By Theorem 8.7.5 the coefficient of x^3 is $\frac{f'''(0)}{3!} = \frac{1}{3} \Rightarrow f'''(0) = 2$.

Or: Use Theorem 8.6.2 to differentiate f three times.

14. False. Let $a_n = n$ and $b_n = -n$. Then $\{a_n\}$ and $\{b_n\}$ are divergent, but $a_n + b_n = 0$, so $\{a_n + b_n\}$ is convergent.

15. False. For example, let $a_n = b_n = (-1)^n$. Then $\{a_n\}$ and $\{b_n\}$ are divergent, but $a_n b_n = 1$, so $\{a_n b_n\}$ is convergent.

16. True by the Monotonic Sequence Theorem, since $\{a_n\}$ is decreasing and $0 < a_n \leq a_1$ for all $n \Rightarrow \{a_n\}$ is bounded.

17. True by Theorem 8.4.1. [$\sum (-1)^n a_n$ is absolutely convergent and hence convergent.]

18. True. $\lim_{n\to\infty} \frac{a_{n+1}}{a_n} < 1 \Rightarrow \sum a_n$ converges (Ratio Test) $\Rightarrow \lim_{n\to\infty} a_n = 0$ [Theorem 8.2.6].

EXERCISES

1. $\left\{\frac{2+n^3}{1+2n^3}\right\}$ converges since $\lim_{n\to\infty} \frac{2+n^3}{1+2n^3} = \lim_{n\to\infty} \frac{2/n^3+1}{1/n^3+2} = \frac{1}{2}$.

2. $a_n = \frac{9^{n+1}}{10^n} = 9 \cdot \left(\frac{9}{10}\right)^n$, so $\lim_{n\to\infty} a_n = 9 \lim_{n\to\infty} \left(\frac{9}{10}\right)^n = 9 \cdot 0 = 0$ by (8.1.8).

3. $\lim\limits_{n\to\infty} a_n = \lim\limits_{n\to\infty} \dfrac{n^3}{1+n^2} = \lim\limits_{n\to\infty} \dfrac{n}{1/n^2+1} = \infty$, so the sequence diverges.

4. $a_n = \cos(n\pi/2)$, so $a_n = 0$ if n is odd and $a_n = \pm 1$ if n is even. As n increases, a_n keeps cycling through the values 0, 1, 0, -1, so the sequence $\{a_n\}$ is divergent.

5. $|a_n| = \left|\dfrac{n \sin n}{n^2+1}\right| \le \dfrac{n}{n^2+1} < \dfrac{1}{n}$, so $|a_n| \to 0$ as $n \to \infty$. Thus, $\lim\limits_{n\to\infty} a_n = 0$. The sequence $\{a_n\}$ is convergent.

6. $a_n = \dfrac{\ln n}{\sqrt{n}}$. Let $f(x) = \dfrac{\ln x}{\sqrt{x}}$ for $x > 0$. Then $\lim\limits_{x\to\infty} f(x) = \lim\limits_{x\to\infty} \dfrac{\ln x}{\sqrt{x}} \overset{\text{H}}{=} \lim\limits_{x\to\infty} \dfrac{1/x}{1/(2\sqrt{x})} = \lim\limits_{x\to\infty} \dfrac{2}{\sqrt{x}} = 0$. Thus, by Theorem 3 in Section 8.1, $\{a_n\}$ converges and $\lim\limits_{n\to\infty} a_n = 0$.

7. $\left\{\left(1+\dfrac{3}{n}\right)^{4n}\right\}$ is convergent. Let $y = \left(1+\dfrac{3}{x}\right)^{4x}$. Then

$$\lim_{x\to\infty} \ln y = \lim_{x\to\infty} 4x \ln(1+3/x) = \lim_{x\to\infty} \frac{\ln(1+3/x)}{1/(4x)} \overset{\text{H}}{=} \lim_{x\to\infty} \frac{\dfrac{1}{1+3/x}\left(-\dfrac{3}{x^2}\right)}{-1/(4x^2)} = \lim_{x\to\infty} \frac{12}{1+3/x} = 12$$

so $\lim\limits_{x\to\infty} y = \lim\limits_{n\to\infty} \left(1+\dfrac{3}{n}\right)^{4n} = e^{12}$.

8. $\left\{\dfrac{(-10)^n}{n!}\right\}$ converges, since $\dfrac{10^n}{n!} = \dfrac{10\cdot 10\cdot 10\cdot\cdots\cdot 10}{1\cdot 2\cdot 3\cdot\cdots\cdot 10} \cdot \dfrac{10\cdot 10\cdot\cdots\cdot 10}{11\cdot 12\cdot\cdots\cdot n} \le 10^{10}\left(\dfrac{10}{11}\right)^{n-10} \to 0$ as $n \to \infty$,

so $\lim\limits_{n\to\infty} \dfrac{(-10)^n}{n!} = 0$ (Squeeze Theorem).

9. $\dfrac{n}{n^3+1} < \dfrac{n}{n^3} = \dfrac{1}{n^2}$, so $\sum\limits_{n=1}^{\infty} \dfrac{n}{n^3+1}$ converges by the Comparison Test with the convergent p-series $\sum\limits_{n=1}^{\infty} \dfrac{1}{n^2}$ $[p = 2 > 1]$.

10. Let $a_n = \dfrac{n^2+1}{n^3+1}$ and $b_n = \dfrac{1}{n}$, so $\lim\limits_{n\to\infty} \dfrac{a_n}{b_n} = \lim\limits_{n\to\infty} \dfrac{n^3+n}{n^3+1} = \lim\limits_{n\to\infty} \dfrac{1+1/n^2}{1+1/n^3} = 1 > 0$. Since $\sum_{n=1}^{\infty} b_n$ is the divergent harmonic series, $\sum_{n=1}^{\infty} a_n$ also diverges by the Limit Comparison Test.

11. $\lim\limits_{n\to\infty} \left|\dfrac{a_{n+1}}{a_n}\right| = \lim\limits_{n\to\infty} \left[\dfrac{(n+1)^3}{5^{n+1}} \cdot \dfrac{5^n}{n^3}\right] = \lim\limits_{n\to\infty} \left(1+\dfrac{1}{n}\right)^3 \cdot \dfrac{1}{5} = \dfrac{1}{5} < 1$, so $\sum\limits_{n=1}^{\infty} \dfrac{n^3}{5^n}$ converges by the Ratio Test.

12. Let $b_n = \dfrac{1}{\sqrt{n+1}}$. Then b_n is positive for $n \ge 1$, the sequence $\{b_n\}$ is decreasing, and $\lim\limits_{n\to\infty} b_n = 0$, so the series $\sum\limits_{n=1}^{\infty} \dfrac{(-1)^n}{\sqrt{n+1}}$ converges by the Alternating Series Test.

13. Let $f(x) = \dfrac{1}{x\sqrt{\ln x}}$. Then f is continuous, positive, and decreasing on $[2, \infty)$, so the Integral Test applies.

$$\int_2^{\infty} f(x)\,dx = \lim_{t\to\infty} \int_2^t \frac{1}{x\sqrt{\ln x}}\,dx \quad \left[\begin{array}{l} u = \ln x, \\ du = \frac{1}{x}\,dx \end{array}\right] \quad = \lim_{t\to\infty} \int_{\ln 2}^{\ln t} u^{-1/2}\,du$$

$$= \lim_{t\to\infty} \left[2\sqrt{u}\right]_{\ln 2}^{\ln t} = \lim_{t\to\infty} \left(2\sqrt{\ln t} - 2\sqrt{\ln 2}\right) = \infty, \text{ so the series } \sum_{n=2}^{\infty} \frac{1}{n\sqrt{\ln n}} \text{ diverges.}$$

14. $\lim\limits_{n\to\infty} \dfrac{n}{3n+1} = \dfrac{1}{3}$, so $\lim\limits_{n\to\infty} \ln\left(\dfrac{n}{3n+1}\right) = \ln \frac{1}{3} \neq 0$. Thus, the series $\sum\limits_{n=1}^{\infty} \ln\left(\dfrac{n}{3n+1}\right)$ diverges by the Test for Divergence.

15. $|a_n| = \left|\dfrac{\cos 3n}{1+(1.2)^n}\right| \leq \dfrac{1}{1+(1.2)^n} < \dfrac{1}{(1.2)^n} = \left(\dfrac{5}{6}\right)^n$, so $\sum\limits_{n=1}^{\infty} |a_n|$ converges by comparison with the convergent geometric series $\sum\limits_{n=1}^{\infty} \left(\dfrac{5}{6}\right)^n$ $\left[r = \frac{5}{6} < 1\right]$. It follows that $\sum\limits_{n=1}^{\infty} a_n$ converges (by Theorem 1 in Section 8.4).

16. $\lim\limits_{n\to\infty} \sqrt[n]{|a_n|} = \lim\limits_{n\to\infty} \sqrt[n]{\left|\dfrac{n^{2n}}{(1+2n^2)^n}\right|} = \lim\limits_{n\to\infty} \dfrac{n^2}{1+2n^2} = \lim\limits_{n\to\infty} \dfrac{1}{1/n^2+2} = \dfrac{1}{2} < 1$, so $\sum\limits_{n=1}^{\infty} \dfrac{n^{2n}}{(1+2n^2)^n}$ converges by the Root Test.

17. $\lim\limits_{n\to\infty} \left|\dfrac{a_{n+1}}{a_n}\right| = \lim\limits_{n\to\infty} \dfrac{1\cdot 3\cdot 5\cdot\cdots\cdot(2n-1)(2n+1)}{5^{n+1}(n+1)!} \cdot \dfrac{5^n n!}{1\cdot 3\cdot 5\cdot\cdots\cdot(2n-1)} = \lim\limits_{n\to\infty} \dfrac{2n+1}{5(n+1)} = \dfrac{2}{5} < 1$, so the series converges by the Ratio Test.

18. $\sum\limits_{n=1}^{\infty} \dfrac{(-5)^{2n}}{n^2\, 9^n} = \sum\limits_{n=1}^{\infty} \dfrac{1}{n^2}\left(\dfrac{25}{9}\right)^n$. Now $\lim\limits_{n\to\infty} \left|\dfrac{a_{n+1}}{a_n}\right| = \lim\limits_{n\to\infty} \dfrac{25^{n+1}}{(n+1)^2\cdot 9^{n+1}} \cdot \dfrac{n^2\cdot 9^n}{25^n} = \lim\limits_{n\to\infty} \dfrac{25n^2}{9(n+1)^2} = \dfrac{25}{9} > 1$, so the series diverges by the Ratio Test.

19. $b_n = \dfrac{\sqrt{n}}{n+1} > 0$, $\{b_n\}$ is decreasing, and $\lim\limits_{n\to\infty} b_n = 0$, so the series $\sum\limits_{n=1}^{\infty} (-1)^{n-1}\dfrac{\sqrt{n}}{n+1}$ converges by the Alternating Series Test.

20. Use the Limit Comparison Test with $a_n = \dfrac{\sqrt{n+1}-\sqrt{n-1}}{n} = \dfrac{2}{n\left(\sqrt{n+1}+\sqrt{n-1}\right)}$ (rationalizing the numerator) and $b_n = \dfrac{1}{n^{3/2}}$. $\lim\limits_{n\to\infty} \dfrac{a_n}{b_n} = \lim\limits_{n\to\infty} \dfrac{2\sqrt{n}}{\sqrt{n+1}+\sqrt{n-1}} = 1$, so since $\sum\limits_{n=1}^{\infty} b_n$ converges $\left[p = \frac{3}{2} > 1\right]$, $\sum\limits_{n=1}^{\infty} a_n$ converges also.

21. Consider the series of absolute values: $\sum\limits_{n=1}^{\infty} n^{-1/3}$ is a p-series with $p = \frac{1}{3} \leq 1$ and is therefore divergent. But if we apply the Alternating Series Test, we see that $b_n = \dfrac{1}{\sqrt[3]{n}} > 0$, $\{b_n\}$ is decreasing, and $\lim\limits_{n\to\infty} b_n = 0$, so the series $\sum\limits_{n=1}^{\infty} (-1)^{n-1} n^{-1/3}$ converges. Thus, $\sum\limits_{n=1}^{\infty} (-1)^{n-1} n^{-1/3}$ is conditionally convergent.

22. $\sum\limits_{n=1}^{\infty} \left|(-1)^{n-1}\, n^{-3}\right| = \sum\limits_{n=1}^{\infty} n^{-3}$ is a convergent p-series $[p = 3 > 1]$. Therefore, $\sum\limits_{n=1}^{\infty} (-1)^{n-1}\, n^{-3}$ is absolutely convergent.

23. $\left|\dfrac{a_{n+1}}{a_n}\right| = \left|\dfrac{(-1)^{n+1}(n+2)\,3^{n+1}}{2^{2n+3}} \cdot \dfrac{2^{2n+1}}{(-1)^n(n+1)\,3^n}\right| = \dfrac{n+2}{n+1}\cdot\dfrac{3}{4} = \dfrac{1+(2/n)}{1+(1/n)}\cdot\dfrac{3}{4} \to \dfrac{3}{4} < 1$ as $n\to\infty$, so by the Ratio Test, $\sum\limits_{n=1}^{\infty} \dfrac{(-1)^n(n+1)\,3^n}{2^{2n+1}}$ is absolutely convergent.

24. $\lim\limits_{x\to\infty} \dfrac{\sqrt{x}}{\ln x} \overset{\text{H}}{=} \lim\limits_{x\to\infty} \dfrac{1/(2\sqrt{x})}{1/x} = \lim\limits_{x\to\infty} \dfrac{\sqrt{x}}{2} = \infty$. Therefore, $\lim\limits_{n\to\infty} \dfrac{(-1)^n\sqrt{n}}{\ln n} \neq 0$, so the given series is divergent by the Test for Divergence.

25. $\dfrac{2^{2n+1}}{5^n} = \dfrac{2^{2n} \cdot 2^1}{5^n} = \dfrac{(2^2)^n \cdot 2}{5^n} = 2\left(\dfrac{4}{5}\right)^n$, so $\sum\limits_{n=1}^{\infty} \dfrac{2^{2n+1}}{5^n} = 2\sum\limits_{n=1}^{\infty}\left(\dfrac{4}{5}\right)^n$ is a geometric series with $a = \dfrac{8}{5}$ and $r = \dfrac{4}{5}$.

Since $|r| = \dfrac{4}{5} < 1$, the series converges to $\dfrac{a}{1-r} = \dfrac{8/5}{1-4/5} = \dfrac{8/5}{1/5} = 8$.

26. $\sum\limits_{n=1}^{\infty} \dfrac{1}{n(n+3)} = \sum\limits_{n=1}^{\infty}\left[\dfrac{1}{3n} - \dfrac{1}{3(n+3)}\right]$ (partial fractions).

$s_n = \sum\limits_{i=1}^{n}\left[\dfrac{1}{3i} - \dfrac{1}{3(i+3)}\right] = \dfrac{1}{3} + \dfrac{1}{6} + \dfrac{1}{9} - \dfrac{1}{3(n+1)} - \dfrac{1}{3(n+2)} - \dfrac{1}{3(n+3)}$ (telescoping sum),

so $\sum\limits_{n=1}^{\infty} \dfrac{1}{n(n+3)} = \lim\limits_{n\to\infty} s_n = \dfrac{1}{3} + \dfrac{1}{6} + \dfrac{1}{9} = \dfrac{11}{18}$.

27.
$$\begin{aligned}\sum_{n=1}^{\infty}\left[\tan^{-1}(n+1) - \tan^{-1} n\right] &= \lim_{n\to\infty} s_n \\ &= \lim_{n\to\infty}\left[\left(\tan^{-1}2 - \tan^{-1}1\right) + \left(\tan^{-1}3 - \tan^{-1}2\right) + \cdots \right.\\ &\qquad \left. + \left(\tan^{-1}(n+1) - \tan^{-1}n\right)\right] \\ &= \lim_{n\to\infty}\left[\tan^{-1}(n+1) - \tan^{-1}1\right] = \tfrac{\pi}{2} - \tfrac{\pi}{4} = \tfrac{\pi}{4}\end{aligned}$$

28. $\sum\limits_{n=0}^{\infty} \dfrac{(-1)^n \pi^n}{3^{2n}(2n)!} = \sum\limits_{n=0}^{\infty}(-1)^n \dfrac{1}{(2n)!}\cdot\dfrac{\pi^n}{3^{2n}} = \sum\limits_{n=0}^{\infty}(-1)^n\dfrac{1}{(2n)!}\cdot\left(\dfrac{\sqrt{\pi}}{3}\right)^{2n} = \cos\left(\dfrac{\sqrt{\pi}}{3}\right)$

since $\cos x = \sum\limits_{n=0}^{\infty}(-1)^n \dfrac{x^{2n}}{(2n)!}$ for all x.

29. $1 - e + \dfrac{e^2}{2!} - \dfrac{e^3}{3!} + \dfrac{e^4}{4!} - \cdots = \sum\limits_{n=0}^{\infty}(-1)^n\dfrac{e^n}{n!} = \sum\limits_{n=0}^{\infty}\dfrac{(-e)^n}{n!} = e^{-e}$ since $e^x = \sum\limits_{n=0}^{\infty}\dfrac{x^n}{n!}$ for all x.

30. $4.17\overline{326} = 4.17 + \dfrac{326}{10^5} + \dfrac{326}{10^8} + \cdots = 4.17 + \dfrac{326/10^5}{1 - 1/10^3} = \dfrac{417}{100} + \dfrac{326}{99{,}900} = \dfrac{416{,}909}{99{,}900}$

31.
$$\begin{aligned}\cosh x &= \frac{1}{2}(e^x + e^{-x}) = \frac{1}{2}\left(\sum_{n=0}^{\infty}\frac{x^n}{n!} + \sum_{n=0}^{\infty}\frac{(-x)^n}{n!}\right) \\ &= \frac{1}{2}\left[\left(1 + x + \frac{x^2}{2!} + \frac{x^3}{3!} + \frac{x^4}{4!} + \cdots\right) + \left(1 - x + \frac{x^2}{2!} - \frac{x^3}{3!} + \frac{x^4}{4!} - \cdots\right)\right] \\ &= \frac{1}{2}\left(2 + 2\cdot\frac{x^2}{2!} + 2\cdot\frac{x^4}{4!} + \cdots\right) = 1 + \frac{1}{2}x^2 + \sum_{n=2}^{\infty}\frac{x^{2n}}{(2n)!} \\ &\geq 1 + \frac{1}{2}x^2 \quad \text{for all } x\end{aligned}$$

32. $\sum\limits_{n=1}^{\infty}(\ln x)^n$ is a geometric series which converges whenever $|\ln x| < 1 \Rightarrow -1 < \ln x < 1 \Rightarrow e^{-1} < x < e$.

33. $\sum\limits_{n=1}^{\infty}\dfrac{(-1)^{n+1}}{n^5} = 1 - \dfrac{1}{32} + \dfrac{1}{243} - \dfrac{1}{1024} + \dfrac{1}{3125} - \dfrac{1}{7776} + \dfrac{1}{16{,}807} - \dfrac{1}{32{,}768} + \cdots$.

Since $b_8 = \dfrac{1}{8^5} = \dfrac{1}{32{,}768} < 0.000031$, $\sum\limits_{n=1}^{\infty}\dfrac{(-1)^{n+1}}{n^5} \approx \sum\limits_{n=1}^{7}\dfrac{(-1)^{n+1}}{n^5} \approx 0.9721$.

34. (a) $\displaystyle\lim_{n\to\infty}\left|\frac{a_{n+1}}{a_n}\right| = \lim_{n\to\infty}\left|\frac{(n+1)^{n+1}}{[2(n+1)]!}\cdot\frac{(2n)!}{n^n}\right| = \lim_{n\to\infty}\frac{(n+1)^n\,(n+1)^1}{(2n+2)(2n+1)n^n} = \lim_{n\to\infty}\left(\frac{n+1}{n}\right)^n\frac{1}{2(2n+1)}$

$$= \lim_{n\to\infty}\left(1+\frac{1}{n}\right)^n\frac{1}{2(2n+1)} = e\cdot 0 = 0 < 1$$

so the series converges by the Ratio Test.

(b) The series in part (a) is convergent, so $\lim\limits_{n\to\infty} a_n = 0$ by Theorem 8.2.6.

35. Use the Limit Comparison Test. $\displaystyle\lim_{n\to\infty}\left|\frac{\left(\frac{n+1}{n}\right)a_n}{a_n}\right| = \lim_{n\to\infty}\frac{n+1}{n} = \lim_{n\to\infty}\left(1+\frac{1}{n}\right) = 1 > 0.$

Since $\sum |a_n|$ is convergent, so is $\sum\left|\left(\dfrac{n+1}{n}\right)a_n\right|$, by the Limit Comparison Test.

36. $\displaystyle\lim_{n\to\infty}\left|\frac{a_{n+1}}{a_n}\right| = \lim_{n\to\infty}\left|\frac{x^{n+1}}{(n+1)^2\,5^{n+1}}\cdot\frac{n^2 5^n}{x^n}\right| = \lim_{n\to\infty}\frac{1}{(1+1/n)^2}\frac{|x|}{5} = \frac{|x|}{5}$, so by the Ratio Test, $\displaystyle\sum_{n=1}^{\infty}(-1)^n\frac{x^n}{n^2\,5^n}$

converges when $\dfrac{|x|}{5} < 1 \Leftrightarrow |x| < 5$, so $R = 5$. When $x = -5$, the series becomes the convergent p-series $\displaystyle\sum_{n=1}^{\infty}\frac{1}{n^2}$ with

$p = 2 > 1$. When $x = 5$, the series becomes $\displaystyle\sum_{n=1}^{\infty}\frac{(-1)^n}{n^2}$, which converges by the Alternating Series Test. Thus, $I = [-5, 5]$.

37. $\displaystyle\lim_{n\to\infty}\left|\frac{a_{n+1}}{a_n}\right| = \lim_{n\to\infty}\left[\frac{|x+2|^{n+1}}{(n+1)\,4^{n+1}}\cdot\frac{n\,4^n}{|x+2|^n}\right] = \lim_{n\to\infty}\left[\frac{n}{n+1}\frac{|x+2|}{4}\right] = \frac{|x+2|}{4} < 1 \Leftrightarrow |x+2| < 4$, so $R = 4$.

$|x+2| < 4 \Leftrightarrow -4 < x+2 < 4 \Leftrightarrow -6 < x < 2$. If $x = -6$, then the series $\displaystyle\sum_{n=1}^{\infty}\frac{(x+2)^n}{n4^n}$ becomes

$\displaystyle\sum_{n=1}^{\infty}\frac{(-4)^n}{n4^n} = \sum_{n=1}^{\infty}\frac{(-1)^n}{n}$, the alternating harmonic series, which converges by the Alternating Series Test. When $x = 2$, the

series becomes the harmonic series $\displaystyle\sum_{n=1}^{\infty}\frac{1}{n}$, which diverges. Thus, $I = [-6, 2)$.

38. $\displaystyle\lim_{n\to\infty}\left|\frac{a_{n+1}}{a_n}\right| = \lim_{n\to\infty}\left|\frac{2^{n+1}\,(x-2)^{n+1}}{(n+3)!}\cdot\frac{(n+2)!}{2^n(x-2)^n}\right| = \lim_{n\to\infty}\frac{2}{n+3}\,|x-2| = 0 < 1$, so the series $\displaystyle\sum_{n=1}^{\infty}\frac{2^n\,(x-2)^n}{(n+2)!}$

converges for all x. $R = \infty$ and $I = (-\infty, \infty)$.

39. $\displaystyle\lim_{n\to\infty}\left|\frac{a_{n+1}}{a_n}\right| = \lim_{n\to\infty}\left|\frac{2^{n+1}(x-3)^{n+1}}{\sqrt{n+4}}\cdot\frac{\sqrt{n+3}}{2^n(x-3)^n}\right| = 2\,|x-3|\lim_{n\to\infty}\sqrt{\frac{n+3}{n+4}} = 2\,|x-3| < 1 \Leftrightarrow |x-3| < \tfrac{1}{2}$, so

$R = \frac{1}{2}$. $|x-3| < \frac{1}{2} \Leftrightarrow -\frac{1}{2} < x-3 < \frac{1}{2} \Leftrightarrow \frac{5}{2} < x < \frac{7}{2}$. For $x = \frac{7}{2}$, the series $\displaystyle\sum_{n=1}^{\infty}\frac{2^n(x-3)^n}{\sqrt{n+3}}$ becomes

$\displaystyle\sum_{n=0}^{\infty}\frac{1}{\sqrt{n+3}} = \sum_{n=3}^{\infty}\frac{1}{n^{1/2}}$, which diverges ($p = \frac{1}{2} \le 1$), but for $x = \frac{5}{2}$, we get $\displaystyle\sum_{n=0}^{\infty}\frac{(-1)^n}{\sqrt{n+3}}$, which is a convergent

alternating series, so $I = \left[\frac{5}{2}, \frac{7}{2}\right)$.

40. $\displaystyle\lim_{n\to\infty}\left|\frac{a_{n+1}}{a_n}\right| = \lim_{n\to\infty}\left|\frac{(2n+2)!\,x^{n+1}}{[(n+1)!]^2}\cdot\frac{(n!)^2}{(2n)!\,x^n}\right| = \lim_{n\to\infty}\frac{(2n+2)(2n+1)}{(n+1)(n+1)}\,|x| = 4\,|x|.$

To converge, we must have $4\,|x| < 1 \Leftrightarrow |x| < \frac{1}{4}$, so $R = \frac{1}{4}$.

41.

n	$f^{(n)}(x)$	$f^{(n)}\left(\frac{\pi}{6}\right)$
0	$\sin x$	$\frac{1}{2}$
1	$\cos x$	$\frac{\sqrt{3}}{2}$
2	$-\sin x$	$-\frac{1}{2}$
3	$-\cos x$	$-\frac{\sqrt{3}}{2}$
4	$\sin x$	$\frac{1}{2}$
$\vdots$	$\vdots$	$\vdots$

$$\sin x = f\left(\frac{\pi}{6}\right) + f'\left(\frac{\pi}{6}\right)\left(x-\frac{\pi}{6}\right) + \frac{f''\left(\frac{\pi}{6}\right)}{2!}\left(x-\frac{\pi}{6}\right)^2 + \frac{f^{(3)}\left(\frac{\pi}{6}\right)}{3!}\left(x-\frac{\pi}{6}\right)^3 + \frac{f^{(4)}\left(\frac{\pi}{6}\right)}{4!}\left(x-\frac{\pi}{6}\right)^4 +$$

$$= \frac{1}{2}\left[1 - \frac{1}{2!}\left(x-\frac{\pi}{6}\right)^2 + \frac{1}{4!}\left(x-\frac{\pi}{6}\right)^4 - \cdots\right] + \frac{\sqrt{3}}{2}\left[\left(x-\frac{\pi}{6}\right) - \frac{1}{3!}\left(x-\frac{\pi}{6}\right)^3 + \cdots\right]$$

$$= \frac{1}{2}\sum_{n=0}^{\infty}(-1)^n\frac{1}{(2n)!}\left(x-\frac{\pi}{6}\right)^{2n} + \frac{\sqrt{3}}{2}\sum_{n=0}^{\infty}(-1)^n\frac{1}{(2n+1)!}\left(x-\frac{\pi}{6}\right)^{2n+1}$$

42.

n	$f^{(n)}(x)$	$f^{(n)}\left(\frac{\pi}{3}\right)$
0	$\cos x$	$\frac{1}{2}$
1	$-\sin x$	$-\frac{\sqrt{3}}{2}$
2	$-\cos x$	$-\frac{1}{2}$
3	$\sin x$	$\frac{\sqrt{3}}{2}$
4	$\cos x$	$\frac{1}{2}$
$\vdots$	$\vdots$	$\vdots$

$$\cos x = f\left(\frac{\pi}{3}\right) + f'\left(\frac{\pi}{3}\right)\left(x-\frac{\pi}{3}\right) + \frac{f''\left(\frac{\pi}{3}\right)}{2!}\left(x-\frac{\pi}{3}\right)^2 + \frac{f^{(3)}\left(\frac{\pi}{3}\right)}{3!}\left(x-\frac{\pi}{3}\right)^3 + \frac{f^{(4)}\left(\frac{\pi}{3}\right)}{4!}\left(x-\frac{\pi}{3}\right)^4 + \cdots$$

$$= \frac{1}{2}\left[1 - \frac{1}{2!}\left(x-\frac{\pi}{3}\right)^2 + \frac{1}{4!}\left(x-\frac{\pi}{3}\right)^4 - \cdots\right] + \frac{\sqrt{3}}{2}\left[-\left(x-\frac{\pi}{3}\right) + \frac{1}{3!}\left(x-\frac{\pi}{3}\right)^3 - \cdots\right]$$

$$= \frac{1}{2}\sum_{n=0}^{\infty}(-1)^n\frac{1}{(2n)!}\left(x-\frac{\pi}{3}\right)^{2n} + \frac{\sqrt{3}}{2}\sum_{n=0}^{\infty}(-1)^{n+1}\frac{1}{(2n+1)!}\left(x-\frac{\pi}{3}\right)^{2n+1}$$

43. $\dfrac{1}{1+x} = \dfrac{1}{1-(-x)} = \sum_{n=0}^{\infty}(-x)^n = \sum_{n=0}^{\infty}(-1)^n x^n$ for $|x| < 1$ $\Rightarrow$ $\dfrac{x^2}{1+x} = \sum_{n=0}^{\infty}(-1)^n x^{n+2}$ with $R = 1$.

44. $\tan^{-1} x = \sum_{n=0}^{\infty}(-1)^n \dfrac{x^{2n+1}}{2n+1}$ with interval of convergence $[-1,1]$, so

$\tan^{-1}(x^2) = \sum_{n=0}^{\infty}(-1)^n \dfrac{(x^2)^{2n+1}}{2n+1} = \sum_{n=0}^{\infty}(-1)^n \dfrac{x^{4n+2}}{2n+1}$, which converges when $x^2 \in [-1,1]$ $\Leftrightarrow$ $x \in [-1,1]$.

Therefore, $R = 1$.

45. $\dfrac{1}{1-x} = \sum\limits_{n=0}^{\infty} x^n$ for $|x| < 1 \;\Rightarrow\; \ln(1-x) = -\displaystyle\int \frac{dx}{1-x} = -\int \sum_{n=0}^{\infty} x^n\,dx = C - \sum_{n=0}^{\infty} \frac{x^{n+1}}{n+1}$.

$\ln(1-0) = C - 0 \;\Rightarrow\; C = 0 \;\Rightarrow\; \ln(1-x) = -\displaystyle\sum_{n=0}^{\infty} \frac{x^{n+1}}{n+1} = \sum_{n=1}^{\infty} \frac{-x^n}{n}$ with $R = 1$.

46. $e^x = \displaystyle\sum_{n=0}^{\infty} \frac{x^n}{n!} \;\Rightarrow\; e^{2x} = \sum_{n=0}^{\infty} \frac{(2x)^n}{n!} \;\Rightarrow\; xe^{2x} = x\sum_{n=0}^{\infty} \frac{2^n x^n}{n!} = \sum_{n=0}^{\infty} \frac{2^n x^{n+1}}{n!}$, $R = \infty$

47. $\sin x = \displaystyle\sum_{n=0}^{\infty} \frac{(-1)^n x^{2n+1}}{(2n+1)!} \;\Rightarrow\; \sin(x^4) = \sum_{n=0}^{\infty} \frac{(-1)^n (x^4)^{2n+1}}{(2n+1)!} = \sum_{n=0}^{\infty} \frac{(-1)^n x^{8n+4}}{(2n+1)!}$ for all x, so the radius of convergence is ∞.

48. $e^x = \displaystyle\sum_{n=0}^{\infty} \frac{x^n}{n!} \;\Rightarrow\; 10^x = e^{(\ln 10)x} = \sum_{n=0}^{\infty} \frac{[(\ln 10)x]^n}{n!} = \sum_{n=0}^{\infty} \frac{(\ln 10)^n x^n}{n!}$, $R = \infty$

49. $f(x) = \dfrac{1}{\sqrt[4]{16-x}} = \dfrac{1}{\sqrt[4]{16(1-x/16)}} = \dfrac{1}{\sqrt[4]{16}\left(1-\frac{1}{16}x\right)^{1/4}} = \frac{1}{2}\left(1-\frac{1}{16}x\right)^{-1/4}$

$$= \frac{1}{2}\left[1 + \left(-\frac{1}{4}\right)\left(-\frac{x}{16}\right) + \frac{\left(-\frac{1}{4}\right)\left(-\frac{5}{4}\right)}{2!}\left(-\frac{x}{16}\right)^2 + \frac{\left(-\frac{1}{4}\right)\left(-\frac{5}{4}\right)\left(-\frac{9}{4}\right)}{3!}\left(-\frac{x}{16}\right)^3 + \cdots\right]$$

$$= \frac{1}{2} + \sum_{n=1}^{\infty} \frac{1 \cdot 5 \cdot 9 \cdot \cdots \cdot (4n-3)}{2 \cdot 4^n \cdot n! \cdot 16^n} x^n = \frac{1}{2} + \sum_{n=1}^{\infty} \frac{1 \cdot 5 \cdot 9 \cdot \cdots \cdot (4n-3)}{2^{6n+1}\, n!} x^n$$

for $\left|-\dfrac{x}{16}\right| < 1 \;\Leftrightarrow\; |x| < 16$, so $R = 16$.

50. $(1-3x)^{-5} = \displaystyle\sum_{n=0}^{\infty} \binom{-5}{n} (-3x)^n = 1 + (-5)(-3x) + \frac{(-5)(-6)}{2!}(-3x)^2 + \frac{(-5)(-6)(-7)}{3!}(-3x)^3 + \cdots$

$= 1 + \displaystyle\sum_{n=1}^{\infty} \frac{5 \cdot 6 \cdot 7 \cdot \cdots \cdot (n+4) \cdot 3^n x^n}{n!}$ for $|-3x| < 1 \;\Leftrightarrow\; |x| < \frac{1}{3}$, so $R = \frac{1}{3}$.

51. $e^x = \displaystyle\sum_{n=0}^{\infty} \frac{x^n}{n!}$, so $\dfrac{e^x}{x} = \dfrac{1}{x}\displaystyle\sum_{n=0}^{\infty} \frac{x^n}{n!} = \sum_{n=0}^{\infty} \frac{x^{n-1}}{n!} = x^{-1} + \sum_{n=1}^{\infty} \frac{x^{n-1}}{n!} = \frac{1}{x} + \sum_{n=1}^{\infty} \frac{x^{n-1}}{n!}$ and

$\displaystyle\int \frac{e^x}{x}\,dx = C + \ln|x| + \sum_{n=1}^{\infty} \frac{x^n}{n \cdot n!}$.

52. $(1+x^4)^{1/2} = \displaystyle\sum_{n=0}^{\infty} \binom{\frac{1}{2}}{n} (x^4)^n = 1 + \left(\tfrac{1}{2}\right)x^4 + \frac{\left(\frac{1}{2}\right)\left(-\frac{1}{2}\right)}{2!}(x^4)^2 + \frac{\left(\frac{1}{2}\right)\left(-\frac{1}{2}\right)\left(-\frac{3}{2}\right)}{3!}(x^4)^3 + \cdots$

$= 1 + \frac{1}{2}x^4 - \frac{1}{8}x^8 + \frac{1}{16}x^{12} - \cdots$

so $\int_0^1 (1+x^4)^{1/2}\,dx = \left[x + \frac{1}{10}x^5 - \frac{1}{72}x^9 + \frac{1}{208}x^{13} - \cdots\right]_0^1 = 1 + \frac{1}{10} - \frac{1}{72} + \frac{1}{208} - \cdots$.

This is an alternating series, so by the Alternating Series Test, the error in the approximation $\int_0^1 (1+x^4)^{1/2}\,dx \approx 1 + \frac{1}{10} - \frac{1}{72} \approx 1.086$ is less than $\frac{1}{208}$, sufficient for the desired accuracy.

Thus, correct to two decimal places, $\int_0^1 (1+x^4)^{1/2}\,dx \approx 1.09$.

53. (a)

n	$f^{(n)}(x)$	$f^{(n)}(1)$
0	$x^{1/2}$	1
1	$\frac{1}{2}x^{-1/2}$	$\frac{1}{2}$
2	$-\frac{1}{4}x^{-3/2}$	$-\frac{1}{4}$
3	$\frac{3}{8}x^{-5/2}$	$\frac{3}{8}$
4	$-\frac{15}{16}x^{-7/2}$	$-\frac{15}{16}$
⋮	⋮	⋮

$$\sqrt{x} \approx T_3(x) = 1 + \frac{1/2}{1!}(x-1) - \frac{1/4}{2!}(x-1)^2 + \frac{3/8}{3!}(x-1)^3$$
$$= 1 + \tfrac{1}{2}(x-1) - \tfrac{1}{8}(x-1)^2 + \tfrac{1}{16}(x-1)^3$$

(b)

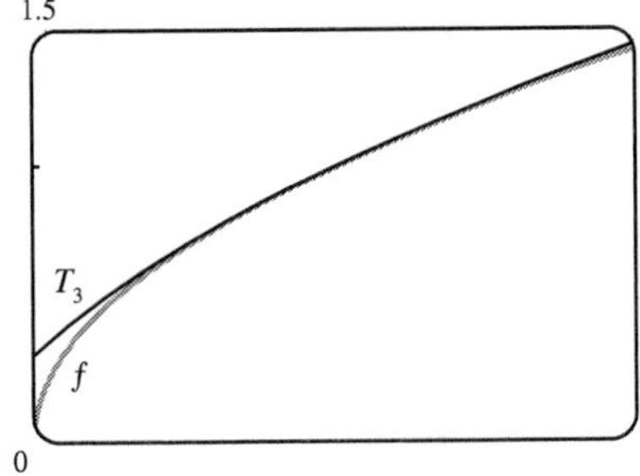

(c) By Taylor's Formula, $R_3(x) = \dfrac{f^{(4)}(z)}{4!}(x-1)^4 = -\dfrac{5(x-1)^4}{128\,z^{7/2}}$, with z between x and 1. If $0.9 \leq x \leq 1.1$, then $0 \leq |x-1| \leq 0.1$ and $z^{7/2} > (0.9)^{7/2}$ so $|R_3(x)| < \dfrac{5(0.1)^4}{128(0.9)^{7/2}} < 0.000006$.

(d) From the graph of $|R_3(x)| = |\sqrt{x} - T_3(x)|$, it appears that the error is less than 5×10^{-6} on $[0.9, 1.1]$.

54. (a)

n	$f^{(n)}(x)$	$f^{(n)}(0)$
0	$\sec x$	1
1	$\sec x \tan x$	0
2	$\sec x \tan^2 x + \sec^3 x$	1
3	$\sec x \tan^3 x + 5\sec^3 x \tan x$	0
⋮	⋮	⋮

$$\sec x \approx T_2(x) = 1 + \tfrac{1}{2}x^2$$

(b)

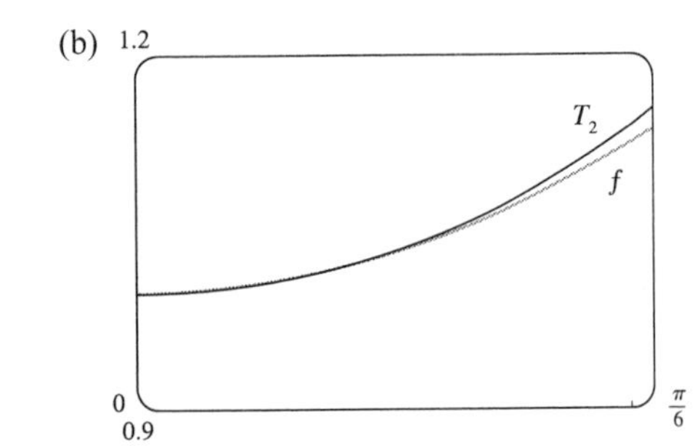

(c) $R_2(x) = \dfrac{\sec x \tan^3 x + 5\sec^3 x \tan x}{3!}$ with z between 0 and x.

If $0 < z < \frac{\pi}{6}$, $1 < \sec z < \frac{2}{\sqrt{3}}$ and $1 < \tan z < \frac{1}{\sqrt{3}}$, so

$$|R_2(x)| \leq \frac{\frac{2}{9} + \frac{40}{9}}{6}\left(\frac{\pi}{6}\right)^3 \approx 0.1117.$$

(d) 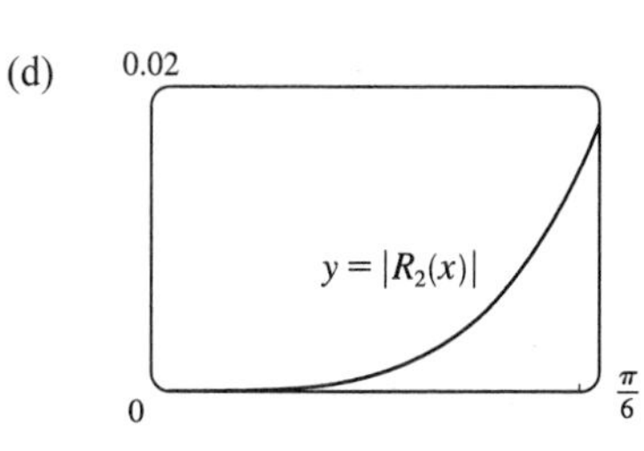

From the graph of $|R_2(x)| = |\sec x - T_2(x)|$, it appears that the error is less than 0.02 on $\left[0, \frac{\pi}{6}\right]$.

55. $\sin x = \sum\limits_{n=0}^{\infty} (-1)^n \dfrac{x^{2n+1}}{(2n+1)!} = x - \dfrac{x^3}{3!} + \dfrac{x^5}{5!} - \dfrac{x^7}{7!} + \cdots$, so $\sin x - x = -\dfrac{x^3}{3!} + \dfrac{x^5}{5!} - \dfrac{x^7}{7!} + \cdots$ and

$\dfrac{\sin x - x}{x^3} = -\dfrac{1}{3!} + \dfrac{x^2}{5!} - \dfrac{x^4}{7!} + \cdots$. Thus, $\lim\limits_{x\to 0} \dfrac{\sin x - x}{x^3} = \lim\limits_{x\to 0}\left(-\dfrac{1}{6} + \dfrac{x^2}{120} - \dfrac{x^4}{5040} + \cdots\right) = -\dfrac{1}{6}$.

56. (a) $F = \dfrac{mgR^2}{(R+h)^2} = \dfrac{mg}{(1+h/R)^2} = mg \sum\limits_{n=0}^{\infty} \dbinom{-2}{n} \left(\dfrac{h}{R}\right)^n$ [Binomial Series]

(b) We expand $F = mg\left[1 - 2(h/R) + 3(h/R)^2 - \cdots\right]$. This is an alternating series, so by the Alternating Series Estimation Theorem, the error in the approximation $F = mg$ is less than $2mgh/R$, so for accuracy within 1% we want $\left|\dfrac{2mgh/R}{mgR^2/(R+h)^2}\right| < 0.01 \Leftrightarrow$

$\dfrac{2h(R+h)^2}{R^3} < 0.01$. This inequality would be difficult to solve for h,

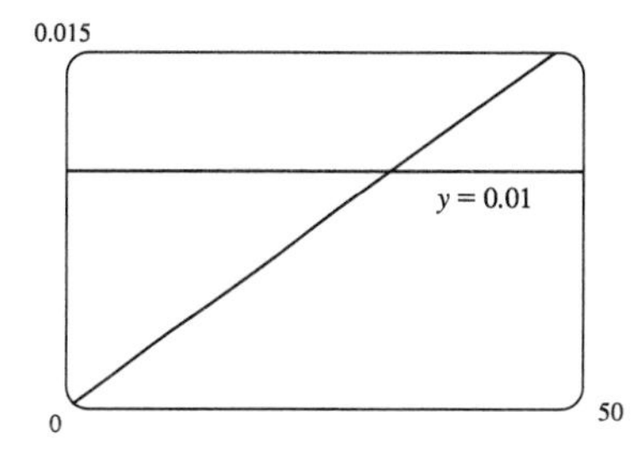

so we substitute $R = 6{,}400$ km and plot both sides of the inequality. It appears that the approximation is accurate to within 1% for $h < 31$ km.

57. $f(x) = \sum\limits_{n=0}^{\infty} c_n x^n \Rightarrow f(-x) = \sum\limits_{n=0}^{\infty} c_n (-x)^n = \sum\limits_{n=0}^{\infty} (-1)^n c_n x^n$

(a) If f is an odd function, then $f(-x) = -f(x) \Rightarrow \sum\limits_{n=0}^{\infty} (-1)^n c_n x^n = \sum\limits_{n=0}^{\infty} -c_n x^n$. The coefficients of any power series are uniquely determined (by Theorem 8.7.5), so $(-1)^n c_n = -c_n$. If n is even, then $(-1)^n = 1$, so $c_n = -c_n \Rightarrow 2c_n = 0 \Rightarrow c_n = 0$. Thus, all even coefficients are 0, that is, $c_0 = c_2 = c_4 = \cdots = 0$.

(b) If f is even, then $f(-x) = f(x) \Rightarrow \sum\limits_{n=0}^{\infty} (-1)^n c_n x^n = \sum\limits_{n=0}^{\infty} c_n x^n \Rightarrow (-1)^n c_n = c_n$. If n is odd, then $(-1)^n = -1$, so $-c_n = c_n \Rightarrow 2c_n = 0 \Rightarrow c_n = 0$. Thus, all odd coefficients are 0, that is, $c_1 = c_3 = c_5 = \cdots = 0$.

58. $e^x = \sum\limits_{n=0}^{\infty} \dfrac{x^n}{n!} \Rightarrow f(x) = e^{x^2} = \sum\limits_{n=0}^{\infty} \dfrac{\left(x^2\right)^n}{n!} = \sum\limits_{n=0}^{\infty} \dfrac{x^{2n}}{n!} = \sum\limits_{n=0}^{\infty} \dfrac{1}{n!} x^{2n}$. By Theorem 8.7.6 with $a = 0$, we also have $f(x) = \sum\limits_{k=0}^{\infty} \dfrac{f^{(k)}(0)}{k!} x^k$. Comparing coefficients for $k = 2n$, we have $\dfrac{f^{(2n)}(0)}{(2n)!} = \dfrac{1}{n!} \Rightarrow f^{(2n)}(0) = \dfrac{(2n)!}{n!}$.

9 □ PARAMETRIC EQUATIONS AND POLAR COORDINATES

9.1 Parametric Curves

1. $x = 1 + \sqrt{t},\ y = t^2 - 4t,\ 0 \le t \le 5$

t	0	1	2	3	4	5
x	1	2	$1+\sqrt{2}$ 2.41	$1+\sqrt{3}$ 2.73	3	$1+\sqrt{5}$ 3.24
y	0	-3	-4	-3	0	5

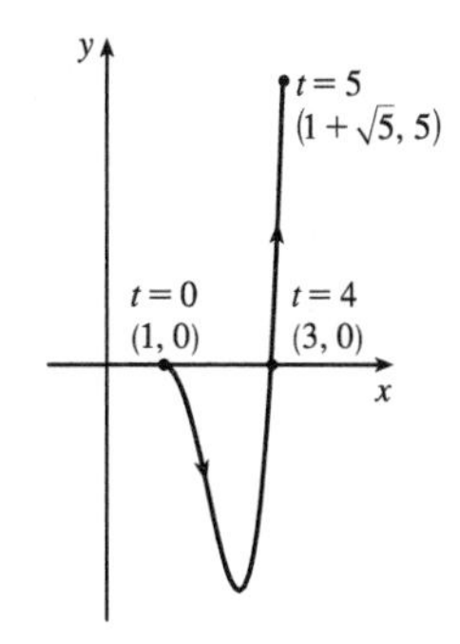

2. $x = 2\cos t,\ y = t - \cos t,\ 0 \le t \le 2\pi$

t	0	$\pi/2$	π	$3\pi/2$	2π
x	2	0	-2	0	2
y	-1	$\pi/2$ 1.57	$\pi+1$ 4.14	$3\pi/2$ 4.71	$2\pi-1$ 5.28

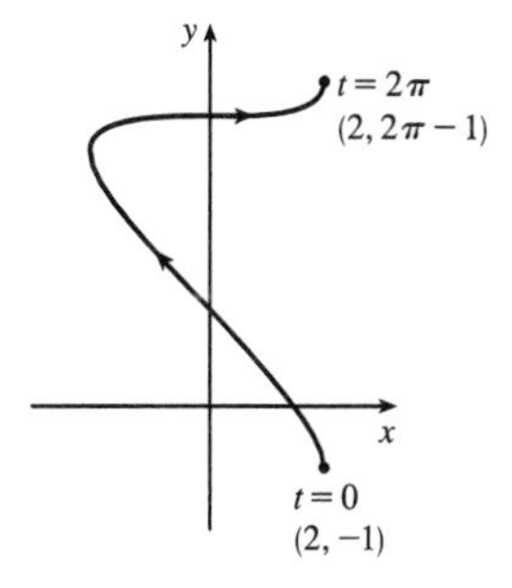

3. $x = 5\sin t,\ y = t^2,\ -\pi \le t \le \pi$

t	$-\pi$	$-\pi/2$	0	$\pi/2$	π
x	0	-5	0	5	0
y	π^2 9.87	$\pi^2/4$ 2.47	0	$\pi^2/4$ 2.47	π^2 9.87

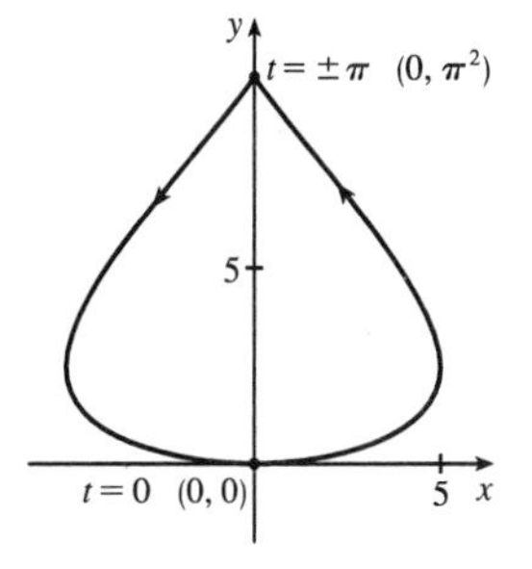

4. $x = e^{-t} + t,\ y = e^t - t,\ -2 \le t \le 2$

t	-2	-1	0	1	2
x	e^2-2 5.39	$e-1$ 1.72	1	$e^{-1}+1$ 1.37	$e^{-2}+2$ 2.14
y	$e^{-2}+2$ 2.14	$e^{-1}+1$ 1.37	1	$e-1$ 1.72	e^2-2 5.39

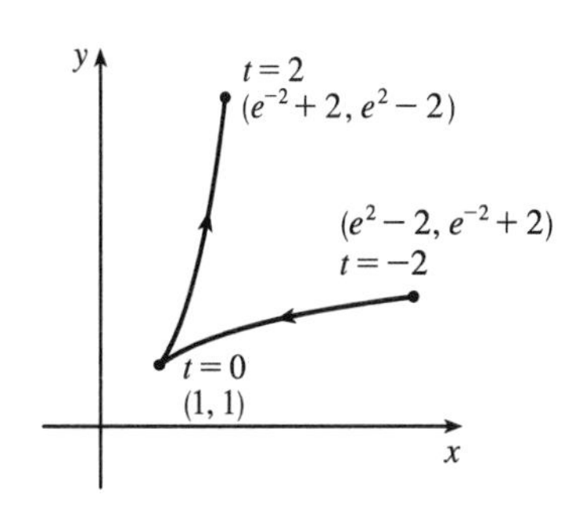

5. $x = 3t - 5,\ y = 2t + 1$

(a)

t	-2	-1	0	1	2	3	4
x	-11	-8	-5	-2	1	4	7
y	-3	-1	1	3	5	7	9

(b) $x = 3t - 5 \ \Rightarrow\ 3t = x + 5 \ \Rightarrow\ t = \frac{1}{3}(x + 5) \ \Rightarrow$

$y = 2 \cdot \frac{1}{3}(x + 5) + 1$, so $y = \frac{2}{3}x + \frac{13}{3}$.

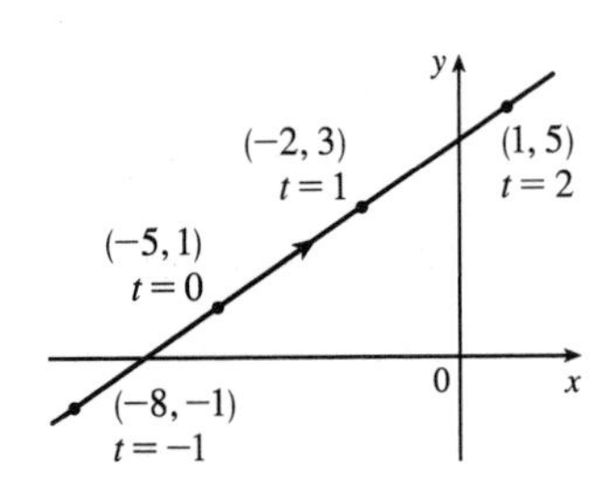

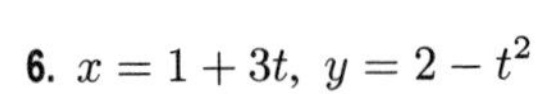

6. $x = 1 + 3t,\ y = 2 - t^2$

(a)

t	-3	-2	-1	0	1	2	3
x	-8	-5	-2	1	4	7	10
y	-7	-2	1	2	1	-2	-7

(b) $x = 1 + 3t \ \Rightarrow\ t = \frac{1}{3}(x - 1) \ \Rightarrow\ y = 2 - \left[\frac{1}{3}(x - 1)\right]^2$,

so $y = -\frac{1}{9}(x - 1)^2 + 2$.

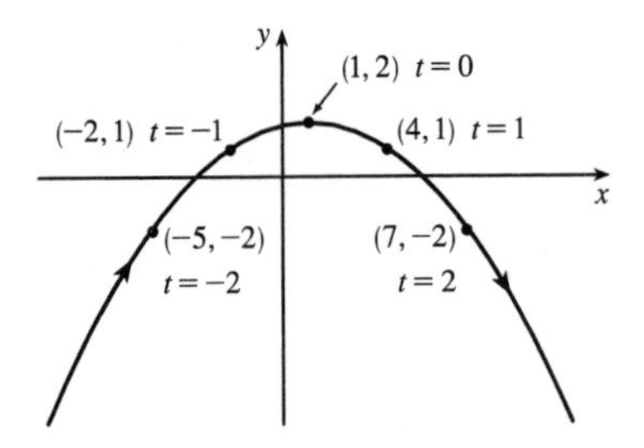

7. $x = \sqrt{t},\ y = 1 - t$

(a)

t	0	1	2	3	4
x	0	1	1.414	1.732	2
y	1	0	-1	-2	-3

(b) $x = \sqrt{t} \ \Rightarrow\ t = x^2 \ \Rightarrow\ y = 1 - t = 1 - x^2$.

Since $t \ge 0$, $x \ge 0$.

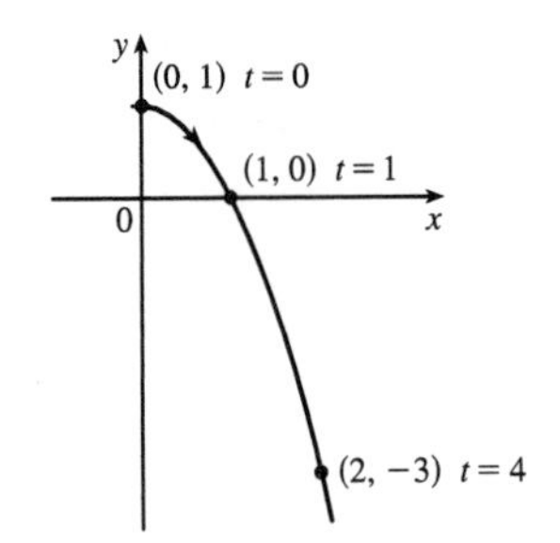

8. $x = t^2,\ y = t^3$

(a)

t	-2	-1	0	1	2
x	4	1	0	1	4
y	-8	-1	0	1	8

(b) $y = t^3 \ \Rightarrow\ t = \sqrt[3]{y} \ \Rightarrow\ x = t^2 = \left(\sqrt[3]{y}\right)^2 = y^{2/3}$.

$t \in \mathbb{R}$, $y \in \mathbb{R}$, $x \ge 0$.

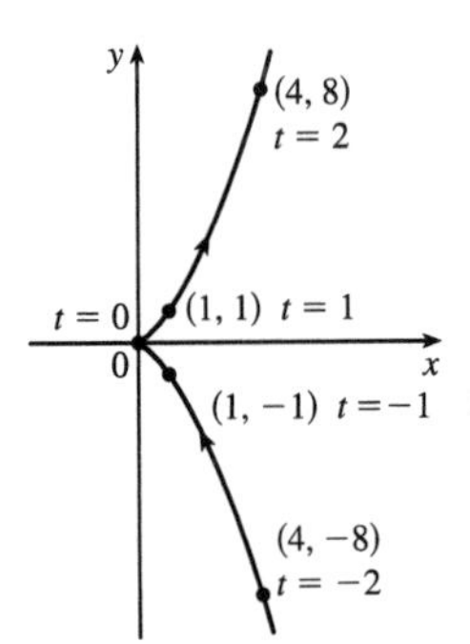

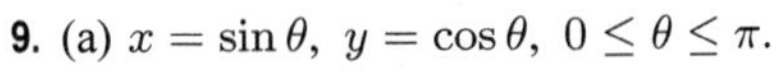

9. (a) $x = \sin\theta,\ y = \cos\theta,\ 0 \le \theta \le \pi$.

$x^2 + y^2 = \sin^2\theta + \cos^2\theta = 1$.

Since $0 \le \theta \le \pi$, we have $\sin\theta \ge 0$, so $x \ge 0$. Thus, the curve is the right half of the circle $x^2 + y^2 = 1$.

(b)

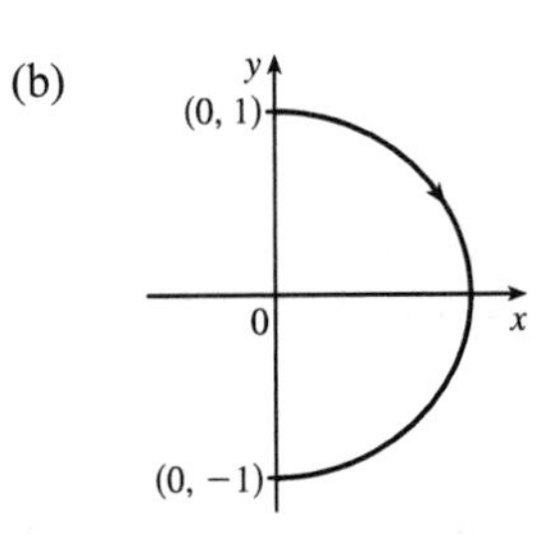

10. (a) $x = 4\cos\theta,\ y = 5\sin\theta,\ -\pi/2 \le \theta \le \pi/2$.

$\left(\frac{x}{4}\right)^2 + \left(\frac{y}{5}\right)^2 = \cos^2\theta + \sin^2\theta = 1$, which is an ellipse with x-intercepts $(\pm 4, 0)$ and y-intercepts $(0, \pm 5)$. We obtain the portion of the ellipse with $x \ge 0$ since $4\cos\theta \ge 0$ for $-\pi/2 \le \theta \le \pi/2$.

(b)

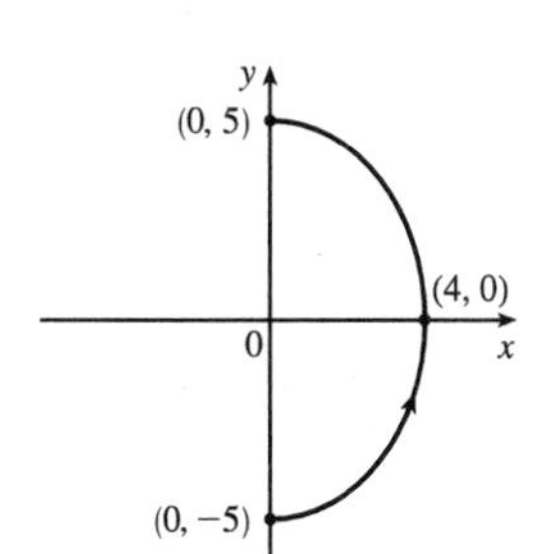

11. (a) $x = \sin t,\ y = \csc t,\ 0 < t < \frac{\pi}{2}$.

$y = \csc t = \dfrac{1}{\sin t} = \dfrac{1}{x}$. For $0 < t < \frac{\pi}{2}$, we have $0 < x < 1$ and $y > 1$.

Thus, the curve is the portion of the hyperbola $y = 1/x$ with $y > 1$.

(b)

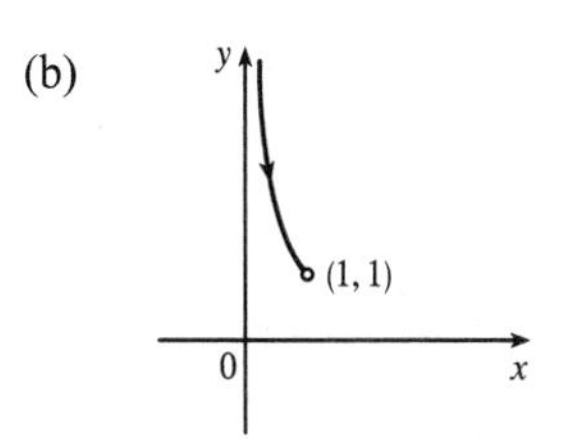

12. (a) $x = \sec\theta,\ y = \tan\theta,\ -\frac{\pi}{2} < \theta < \frac{\pi}{2}$.

$x^2 - y^2 = \sec^2\theta - \tan^2\theta = 1,\ x \ge 1$, or $x = \sqrt{y^2 + 1}$.

Thus, the curve is the right branch of the hyperbola $x^2 - y^2 = 1$.

(b)

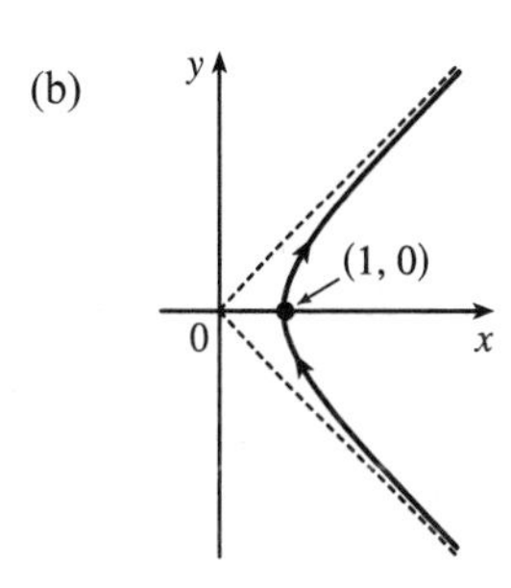

13. (a) $x = e^{2t} \Rightarrow 2t = \ln x \Rightarrow t = \frac{1}{2}\ln x$.

$y = t + 1 = \frac{1}{2}\ln x + 1$.

(b)

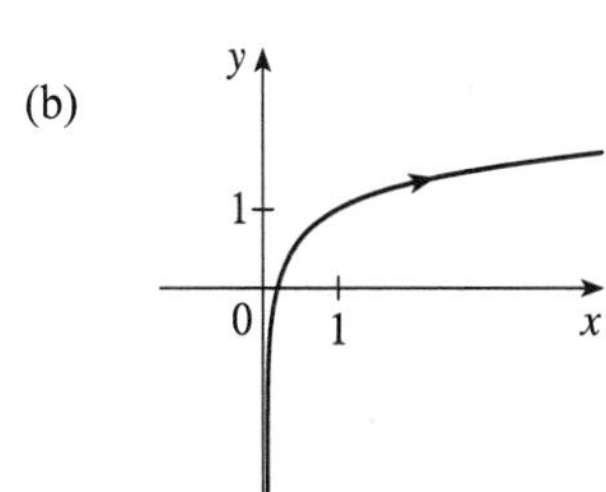

14. (a) $x = 1 + \cos\theta \Rightarrow \cos\theta = x - 1$.

$y = 2\cos\theta - 1 = 2(x - 1) - 1 = 2x - 3,\ 0 \le x \le 2$.

(b)

15. $x = 3 + 2\cos t,\ y = 1 + 2\sin t,\ \pi/2 \le t \le 3\pi/2$. By Example 4 with $r = 2$, $h = 3$, and $k = 1$, the motion of the particle takes place on a circle centered at $(3, 1)$ with a radius of 2. As t goes from $\frac{\pi}{2}$ to $\frac{3\pi}{2}$, the particle starts at the point $(3, 3)$ and moves counterclockwise to $(3, -1)$ [one-half of a circle].

16. $x = 2\sin t,\ y = 4 + \cos t \ \Rightarrow\ \sin t = \dfrac{x}{2},\ \cos t = y - 4.\ \ \sin^2 t + \cos^2 t = 1 \ \Rightarrow\ \left(\dfrac{x}{2}\right)^2 + (y-4)^2 = 1$. The motion of the particle takes place on an ellipse centered at $(0, 4)$. As t goes from 0 to $\frac{3\pi}{2}$, the particle starts at the point $(0, 5)$ and moves clockwise to $(-2, 4)$ [three-quarters of an ellipse].

17. $x = 5\sin t,\ y = 2\cos t \ \Rightarrow\ \sin t = \dfrac{x}{5},\ \cos t = \dfrac{y}{2}.\ \ \sin^2 t + \cos^2 t = 1 \ \Rightarrow\ \left(\dfrac{x}{5}\right)^2 + \left(\dfrac{y}{2}\right)^2 = 1$. The motion of the particle takes place on an ellipse centered at $(0, 0)$. As t goes from $-\pi$ to 5π, the particle starts at the point $(0, -2)$ and moves clockwise around the ellipse 3 times.

18. $x = \sin t,\ y = \cos^2 t,\ -2\pi \le t \le 2\pi.\ \ y = \cos^2 t = 1 - \sin^2 t = 1 - x^2$. The motion of the particle takes place on the parabola $y = 1 - x^2$. As t goes from -2π to $-\pi$, the particle starts at the point $(0, 1)$, moves to $(1, 0)$, and goes back to $(0, 1)$. As t goes from $-\pi$ to 0, the particle moves to $(-1, 0)$ and goes back to $(0, 1)$. The particle repeats this motion as t goes from 0 to 2π.

19. When $t = -1$, $(x, y) = (0, -1)$. As t increases to 0, x decreases to -1 and y increases to 0. As t increases from 0 to 1, x increases to 0 and y increases to 1. As t increases beyond 1, both x and y increase. For $t < -1$, x is positive and decreasing and y is negative and increasing. We could achieve greater accuracy by estimating x- and y-values for selected values of t from the given graphs and plotting the corresponding points.

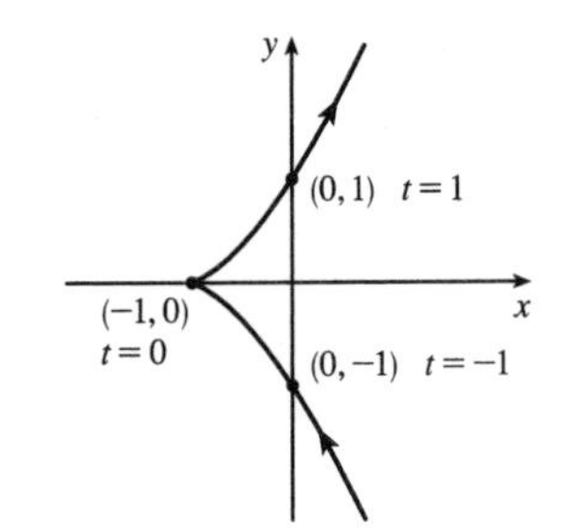

20. For $t < -1$, x is positive and decreasing, while y is negative and increasing (these points are in Quadrant IV). When $t = -1$, $(x, y) = (0, 0)$ and, as t increases from -1 to 0, x becomes negative and y increases from 0 to 1. At $t = 0$, $(x, y) = (0, 1)$ and, as t increases from 0 to 1, y decreases from 1 to 0 and x is positive. At $t = 1$, $(x, y) = (0, 0)$ again, so the loop is completed. For $t > 1$, x and y both become large negative. This enables us to draw a rough sketch. We could achieve greater accuracy by estimating x- and y-values for selected values of t from the given graphs and plotting the corresponding points.

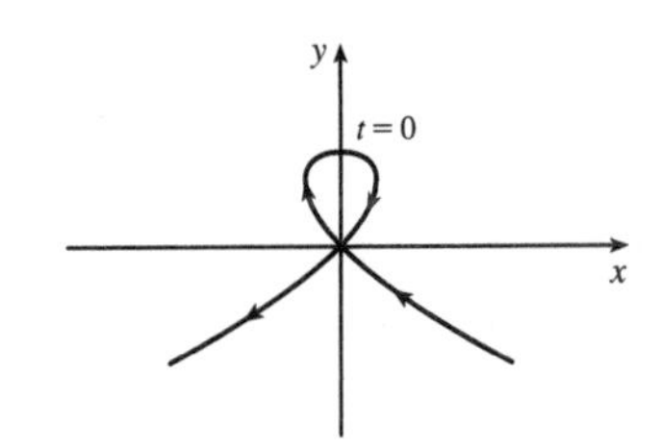

21. When $t = 0$ we see that $x = 0$ and $y = 0$, so the curve starts at the origin. As t increases from 0 to $\frac{1}{2}$, the graphs show that y increases from 0 to 1 while x increases from 0 to 1, decreases to 0 and to -1, then increases back to 0, so we arrive at the point $(0, 1)$. Similarly, as t increases from $\frac{1}{2}$ to 1, y decreases from 1 to 0 while x repeats its pattern, and we arrive back at the origin. We could achieve greater accuracy by estimating x- and y-values for selected values of t from the given graphs and plotting the corresponding points.

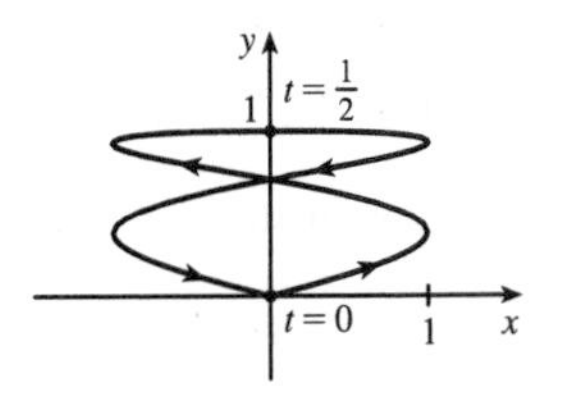

22. (a) Note that as $t \to -\infty$, we have $x \to -\infty$ and $y \to \infty$, whereas when $t \to \infty$, both x and $y \to \infty$. This description fits only IV. [But also note that $x(t)$ increases, then decreases, then increases again.]

(b) Note that as $t \to \pm\infty$, $y \to -\infty$. This is only the case with VI.

(c) If $t = 0$, then $(x, y) = (\sin 0, \sin 0) = (0, 0)$. Also, $|x| = |\sin 3t| \le 1$ for all t, and $|y| = |\sin 4t| \le 1$ for all t. The only graph which includes the point $(0, 0)$ and which has $|x| \le 1$ and $|y| \le 1$, is V.

(d) Note that as $t \to -\infty$, both x and $y \to -\infty$, and as $t \to \infty$, both x and $y \to \infty$. This description fits only III. (Also note that, since $\sin 2t$ and $\sin 3t$ lie between -1 and 1, the curve never strays very far from the line $y = x$.)

(e) Note that both $x(t)$ and $y(t)$ are periodic with period 2π and satisfy $|x| \le 1$ and $|y| \le 1$. Now the only y-intercepts occur when $x = \sin(t + \sin t) = 0 \quad \Leftrightarrow \quad t = 0$ or π. So there should be two y-intercepts: $y(0) = \cos 1 \approx 0.54$ and $y(\pi) = \cos(\pi - 1) \approx -0.54$. Similarly, there should be two x-intercepts: $x\left(\frac{\pi}{2}\right) = \sin\left(\frac{\pi}{2} + 1\right) \approx 0.54$ and $x\left(\frac{3\pi}{2}\right) = \sin\left(\frac{3\pi}{2} - 1\right) \approx -0.54$. The only curve with these x- and y-intercepts is I.

(f) Note that $x(t)$ is periodic with period 2π, so the only y-intercepts occur when $x = \cos t = 0 \quad \Leftrightarrow \quad t = \frac{\pi}{2}$ or $\frac{3\pi}{2}$.

Also, the graph is symmetric about the x-axis, since

$y(-t) = \sin(-t + \sin 5(-t)) = \sin(-t - \sin 5t) = -\sin(t + \sin 5t) = -y(t)$, and $x(-t) = \cos(-t) = \cos t = x(t)$.

The only graph which has only two y-intercepts, and is symmetric about the x-axis, is II.

23. As in Example 6, we let $y = t$ and $x = t - 3t^3 + t^5$ and use a t-interval of $[-2\pi, 2\pi]$.

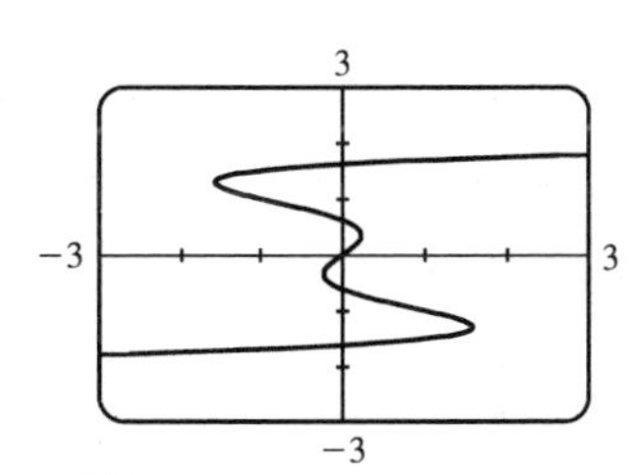

24. We use $x_1 = t,\ y_1 = t^5$ and $x_2 = t\,(t-1)^2,\ y_2 = t$ with $-2\pi \le t \le 2\pi$.

There are 3 points of intersection; $(0, 0)$ is fairly obvious. The point in quadrant III is approximately $(-0.8, -0.4)$ and the point in quadrant I is approximately $(1.1, 1.8)$.

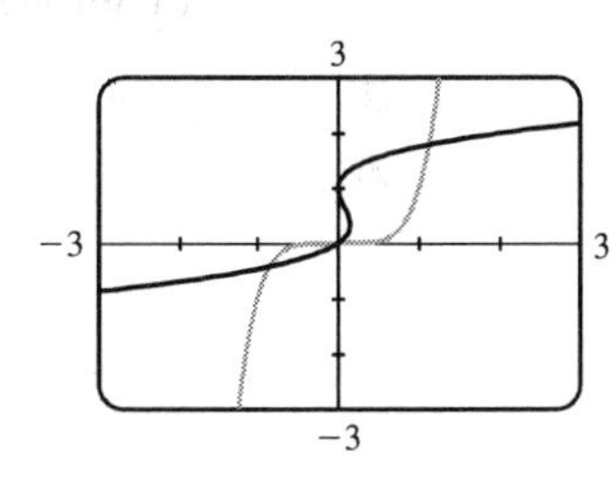

25. (a) $x = x_1 + (x_2 - x_1)t,\ y = y_1 + (y_2 - y_1)t\ , 0 \le t \le 1$. Clearly the curve passes through $P_1(x_1, y_1)$ when $t = 0$ and through $P_2(x_2, y_2)$ when $t = 1$. For $0 < t < 1$, x is strictly between x_1 and x_2 and y is strictly between y_1 and y_2. For every value of t, x and y satisfy the relation $y - y_1 = \dfrac{y_2 - y_1}{x_2 - x_1}(x - x_1)$, which is the equation of the line through $P_1(x_1, y_1)$ and $P_2(x_2, y_2)$.

Finally, any point (x, y) on that line satisfies $\dfrac{y - y_1}{y_2 - y_1} = \dfrac{x - x_1}{x_2 - x_1}$; if we call that common value t, then the given parametric equations yield the point (x, y); and any (x, y) on the line between $P_1(x_1, y_1)$ and $P_2(x_2, y_2)$ yields a value of t in $[0, 1]$. So the given parametric equations exactly specify the line segment from $P_1(x_1, y_1)$ to $P_2(x_2, y_2)$.

(b) $x = -2 + [3 - (-2)]t = -2 + 5t$ and $y = 7 + (-1 - 7)t = 7 - 8t$ for $0 \le t \le 1$.

26. For the side of the triangle from A to B, use $(x_1, y_1) = (1, 1)$ and $(x_2, y_2) = (4, 2)$. Hence, the equations are

$$x = x_1 + (x_2 - x_1)\,t = 1 + (4 - 1)\,t = 1 + 3t$$

$$y = y_1 + (y_2 - y_1)\,t = 1 + (2 - 1)\,t = 1 + t$$

Graphing $x = 1 + 3t$ and $y = 1 + t$ with $0 \le t \le 1$ gives us the side of the triangle from A to B. Similarly, for the side BC we use $x = 4 - 3t$ and $y = 2 + 3t$, and for the side AC we use $x = 1$ and $y = 1 + 4t$.

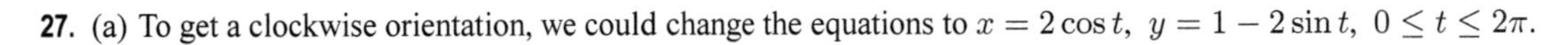

27. (a) To get a clockwise orientation, we could change the equations to $x = 2\cos t,\ y = 1 - 2\sin t,\ 0 \le t \le 2\pi$.

(b) To get three times around in the counterclockwise direction, we use the original equations $x = 2\cos t,\ y = 1 + 2\sin t$ with the domain expanded to $0 \le t \le 6\pi$.

(c) To start at $(0, 3)$ using the original equations, we must have $x_1 = 0$; that is, $2\cos t = 0$. Hence, $t = \frac{\pi}{2}$. So we use $x = 2\cos t,\ y = 1 + 2\sin t,\ \frac{\pi}{2} \le t \le \frac{3\pi}{2}$.

Alternatively, if we want t to start at 0, we could change the equations of the curve. For example, we could use $x = -2\sin t,\ y = 1 + 2\cos t,\ 0 \le t \le \pi$.

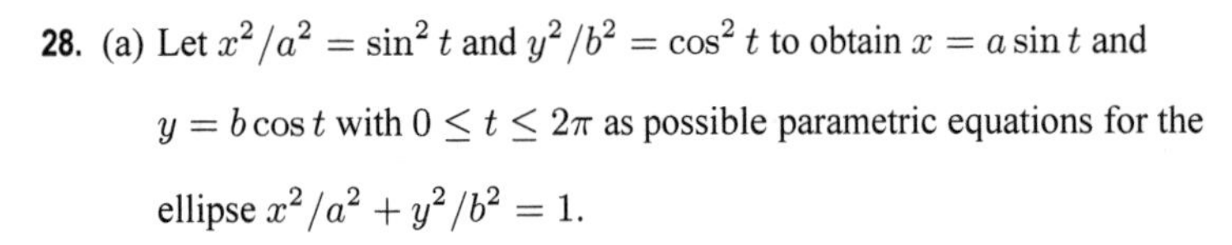

28. (a) Let $x^2/a^2 = \sin^2 t$ and $y^2/b^2 = \cos^2 t$ to obtain $x = a\sin t$ and $y = b\cos t$ with $0 \le t \le 2\pi$ as possible parametric equations for the ellipse $x^2/a^2 + y^2/b^2 = 1$.

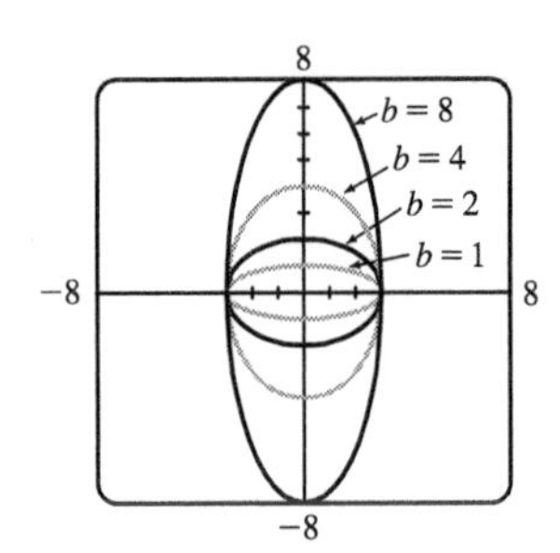

(b) The equations are $x = 3\sin t$ and $y = b\cos t$ for $b \in \{1, 2, 4, 8\}$.

(c) As b increases, the ellipse stretches vertically.

29. *Big circle:* It's centered at $(2, 2)$ with a radius of 2, so by Example 4, parametric equations are

$$x = 2 + 2\cos t, \qquad y = 2 + 2\sin t, \qquad 0 \le t \le 2\pi$$

Small circles: They are centered at $(1, 3)$ and $(3, 3)$ with a radius of 0.1. By Example 4, parametric equations are

(left) $\quad x = 1 + 0.1\cos t, \quad y = 3 + 0.1\sin t, \quad 0 \le t \le 2\pi$

and *(right)* $\quad x = 3 + 0.1\cos t, \quad y = 3 + 0.1\sin t, \quad 0 \le t \le 2\pi$

Semicircle: It's the lower half of a circle centered at $(2, 2)$ with radius 1. By Example 4, parametric equations are

$$x = 2 + 1\cos t, \qquad y = 2 + 1\sin t, \qquad \pi \le t \le 2\pi$$

To get all four graphs on the same screen with a typical graphing calculator, we need to change the last t-interval to $[0, 2\pi]$ in order to match the others. We can do this by changing t to $0.5t$. This change gives us the upper half. There are several ways to get the lower half—one is to change the "+" to a "−" in the y-assignment, giving us

$$x = 2 + 1\cos(0.5t), \qquad y = 2 - 1\sin(0.5t), \qquad 0 \le t \le 2\pi$$

30. If you are using a calculator or computer that can overlay graphs (using multiple t-intervals), the following is appropriate.

Left side: $x = 1$ and y goes from 1.5 to 4, so use

$$x = 1, \qquad y = t, \qquad 1.5 \le t \le 4$$

Right side: $x = 10$ and y goes from 1.5 to 4, so use

$$x = 10, \qquad y = t, \qquad 1.5 \le t \le 4$$

Bottom: x goes from 1 to 10 and $y = 1.5$, so use

$$x = t, \qquad y = 1.5, \qquad 1 \le t \le 10$$

Handle: It starts at $(10, 4)$ and ends at $(13, 7)$, so use

$$x = 10 + t, \qquad y = 4 + t, \qquad 0 \le t \le 3$$

Left wheel: It's centered at $(3, 1)$, has a radius of 1, and appears to go about $30°$ above the horizontal, so use

$$x = 3 + 1\cos t, \qquad y = 1 + 1\sin t, \qquad \tfrac{5\pi}{6} \le t \le \tfrac{13\pi}{6}$$

Right wheel: Similar to the left wheel with center $(8, 1)$, so use

$$x = 8 + 1\cos t, \qquad y = 1 + 1\sin t, \qquad \tfrac{5\pi}{6} \le t \le \tfrac{13\pi}{6}$$

If you are using a calculator or computer that cannot overlay graphs (using one t-interval), the following is appropriate. We'll start by picking the t-interval $[0, 2.5]$ since it easily matches the t-values for the two sides. We now need to find parametric equations for all graphs with $0 \le t \le 2.5$.

Left side: $x = 1$ and y goes from 1.5 to 4, so use

$$x = 1, \qquad y = 1.5 + t, \qquad 0 \le t \le 2.5$$

Right side: $x = 10$ and y goes from 1.5 to 4, so use

$$x = 10, \qquad y = 1.5 + t, \qquad 0 \le t \le 2.5$$

Bottom: x goes from 1 to 10 and $y = 1.5$, so use

$$x = 1 + 3.6t, \qquad y = 1.5, \qquad 0 \le t \le 2.5$$

To get the x-assignment, think of creating a linear function such that when $t = 0$, $x = 1$ and when $t = 2.5$, $x = 10$. We can use the point-slope form of a line with $(t_1, x_1) = (0, 1)$ and $(t_2, x_2) = (2.5, 10)$.

$$x - 1 = \frac{10 - 1}{2.5 - 0}(t - 0) \quad \Rightarrow \quad x = 1 + 3.6t.$$

Handle: It starts at $(10, 4)$ and ends at $(13, 7)$, so use

$$x = 10 + 1.2t, \qquad y = 4 + 1.2t, \qquad 0 \le t \le 2.5$$

$(t_1, x_1) = (0, 10)$ and $(t_2, x_2) = (2.5, 13)$ gives us $x - 10 = \dfrac{13 - 10}{2.5 - 0}(t - 0) \quad \Rightarrow \quad x = 10 + 1.2t.$

$(t_1, y_1) = (0, 4)$ and $(t_2, y_2) = (2.5, 7)$ gives us $y - 4 = \dfrac{7 - 4}{2.5 - 0}(t - 0) \quad \Rightarrow \quad y = 4 + 1.2t.$

Left wheel: It's centered at $(3, 1)$, has a radius of 1, and appears to go about $30°$ above the horizontal, so use

$$x = 3 + 1\cos\left(\tfrac{8\pi}{15}t + \tfrac{5\pi}{6}\right), \qquad y = 1 + 1\sin\left(\tfrac{8\pi}{15}t + \tfrac{5\pi}{6}\right), \qquad 0 \le t \le 2.5$$

$(t_1, \theta_1) = \left(0, \frac{5\pi}{6}\right)$ and $(t_2, \theta_2) = \left(\frac{5}{2}, \frac{13\pi}{6}\right)$ gives us $\theta - \frac{5\pi}{6} = \dfrac{\frac{13\pi}{6} - \frac{5\pi}{6}}{\frac{5}{2} - 0}(t - 0) \quad \Rightarrow \quad \theta = \frac{5\pi}{6} + \frac{8\pi}{15}t.$

Right wheel: Similar to the left wheel with center $(8, 1)$, so use

$$x = 8 + 1\cos\left(\tfrac{8\pi}{15}t + \tfrac{5\pi}{6}\right), \qquad y = 1 + 1\sin\left(\tfrac{8\pi}{15}t + \tfrac{5\pi}{6}\right), \qquad 0 \le t \le 2.5$$

31. (a) $x = t^3 \quad\Rightarrow\quad t = x^{1/3}$, so $y = t^2 = x^{2/3}$.

We get the entire curve $y = x^{2/3}$ traversed in a left to right direction.

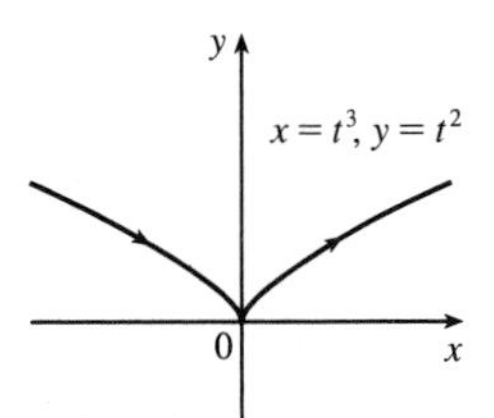

(b) $x = t^6 \quad\Rightarrow\quad t = x^{1/6}$, so $y = t^4 = x^{4/6} = x^{2/3}$.

Since $x = t^6 \geq 0$, we only get the right half of the curve $y = x^{2/3}$.

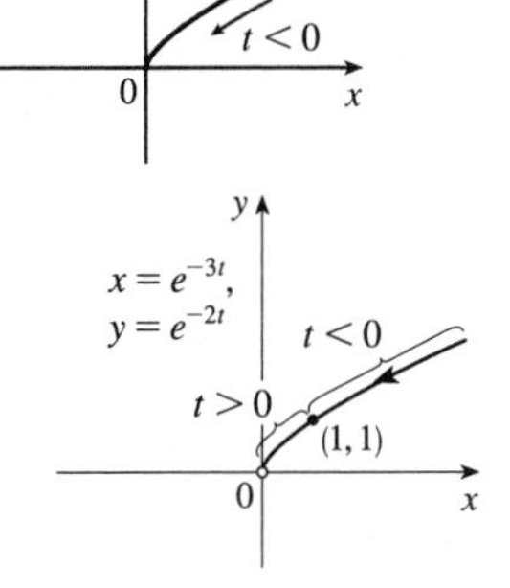

(c) $x = e^{-3t} = (e^{-t})^3$ [so $e^{-t} = x^{1/3}$],

$y = e^{-2t} = (e^{-t})^2 = (x^{1/3})^2 = x^{2/3}$.

If $t < 0$, then x and y are both larger than 1. If $t > 0$, then x and y are between 0 and 1. Since $x > 0$ and $y > 0$, the curve never quite reaches the origin.

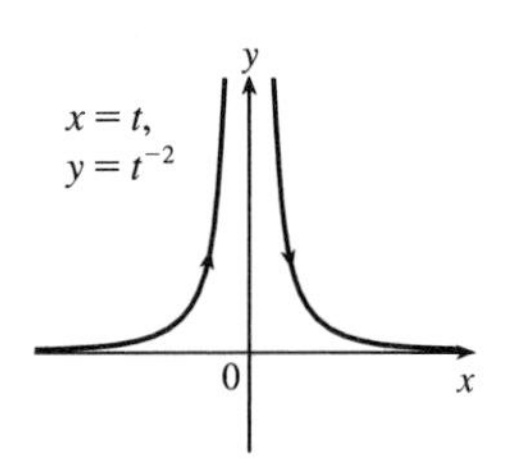

32. (a) $x = t$, so $y = t^{-2} = x^{-2}$.

We get the entire curve $y = 1/x^2$ traversed in a left-to-right direction.

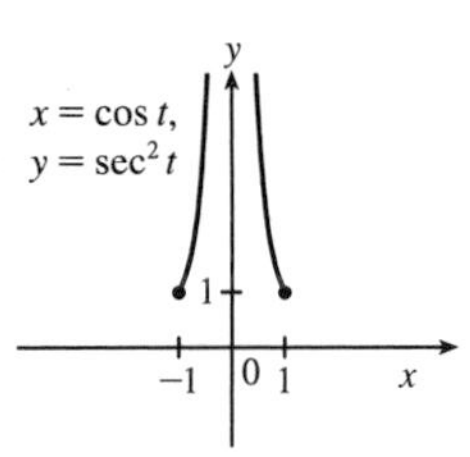

(b) $x = \cos t,\ y = \sec^2 t = \dfrac{1}{\cos^2 t} = \dfrac{1}{x^2}$.

Since $\sec t \geq 1$, we only get the parts of the curve $y = 1/x^2$ with $y \geq 1$. We get the first quadrant portion of the curve when $x > 0$, that is, $\cos t > 0$, and we get the second quadrant portion of the curve when $x < 0$, that is, $\cos t < 0$.

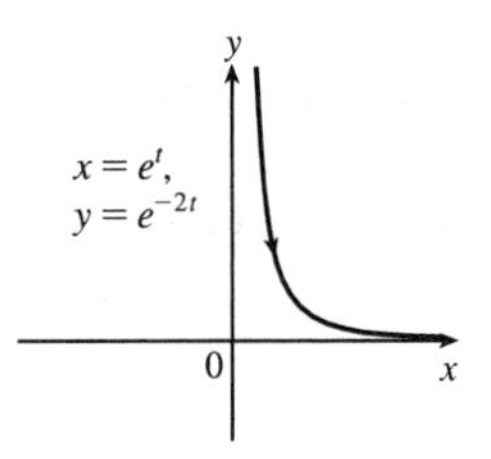

(c) $x = e^t,\ y = e^{-2t} = (e^t)^{-2} = x^{-2}$.

Since e^t and e^{-2t} are both positive, we only get the first quadrant portion of the curve $y = 1/x^2$.

33. The case $\frac{\pi}{2} < \theta < \pi$ is illustrated. C has coordinates $(r\theta, r)$ as in Example 6, and Q has coordinates $(r\theta, r + r\cos(\pi - \theta)) = (r\theta, r(1 - \cos\theta))$ [since $\cos(\pi - \alpha) = \cos\pi\cos\alpha + \sin\pi\sin\alpha = -\cos\alpha$], so P has coordinates $(r\theta - r\sin(\pi - \theta), r(1 - \cos\theta)) = (r(\theta - \sin\theta), r(1 - \cos\theta))$ [since $\sin(\pi - \alpha) = \sin\pi\cos\alpha - \cos\pi\sin\alpha = \sin\alpha$]. Again we have the parametric equations $x = r(\theta - \sin\theta)$, $y = r(1 - \cos\theta)$.

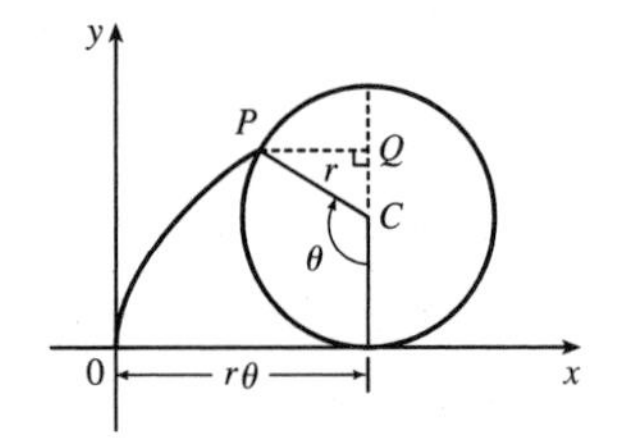

34. The first two diagrams depict the case $\pi < \theta < \frac{3\pi}{2}$, $d < r$. As in Example 6, C has coordinates $(r\theta, r)$. Now Q (in the second diagram) has coordinates $(r\theta, r + d\cos(\theta - \pi)) = (r\theta, r - d\cos\theta)$, so a typical point P of the trochoid has coordinates $(r\theta + d\sin(\theta - \pi), r - d\cos\theta)$. That is, P has coordinates (x, y), where $x = r\theta - d\sin\theta$ and $y = r - d\cos\theta$. When $d = r$, these equations agree with those of the cycloid.

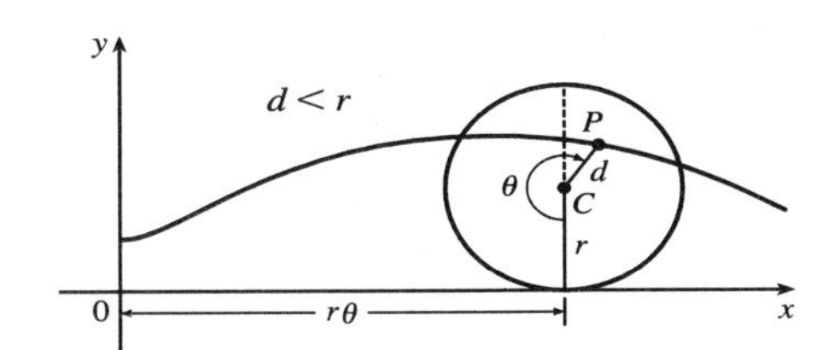

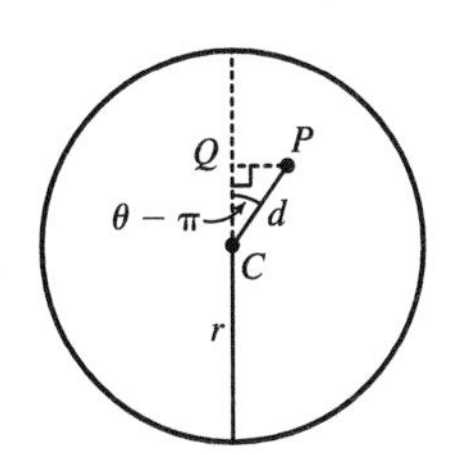

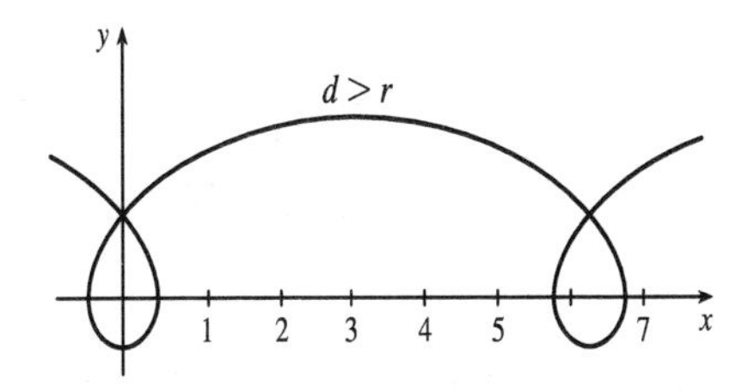

35. It is apparent that $x = |OQ|$ and $y = |QP| = |ST|$. From the diagram, $x = |OQ| = a\cos\theta$ and $y = |ST| = b\sin\theta$. Thus, the parametric equations are $x = a\cos\theta$ and $y = b\sin\theta$. To eliminate θ we rearrange: $\sin\theta = y/b \Rightarrow \sin^2\theta = (y/b)^2$ and $\cos\theta = x/a \Rightarrow \cos^2\theta = (x/a)^2$. Adding the two equations: $\sin^2\theta + \cos^2\theta = 1 = x^2/a^2 + y^2/b^2$. Thus, we have an ellipse.

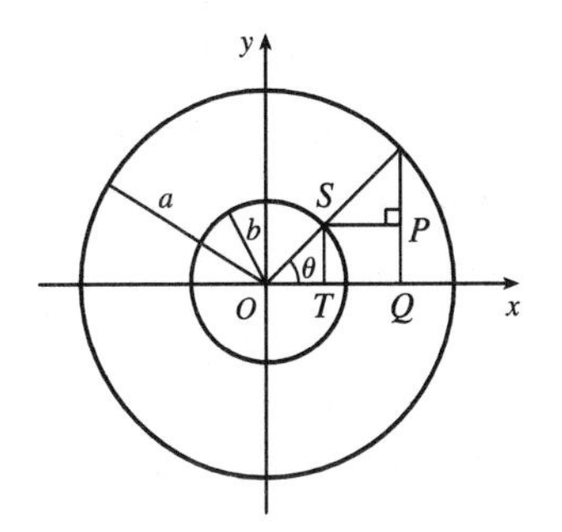

36. $C = (2a\cot\theta, 2a)$, so the x-coordinate of P is $x = 2a\cot\theta$. Let $B = (0, 2a)$. Then $\angle OAB$ is a right angle and $\angle OBA = \theta$, so $|OA| = 2a\sin\theta$ and $A = ((2a\sin\theta)\cos\theta, (2a\sin\theta)\sin\theta)$. Thus, the y-coordinate of P is $y = 2a\sin^2\theta$.

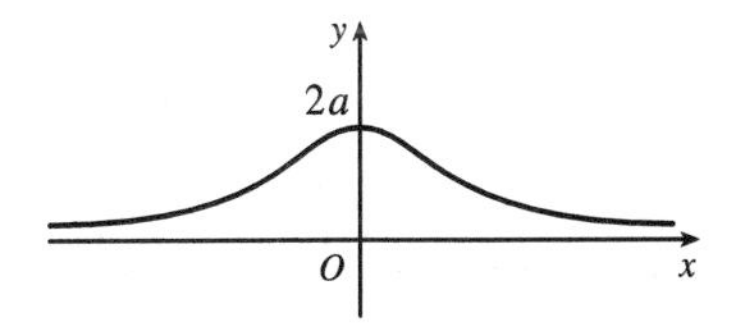

37. (a)

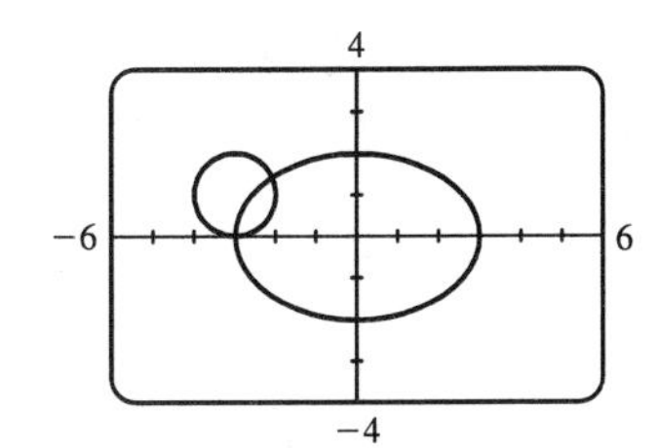

There are 2 points of intersection: $(-3, 0)$ and approximately $(-2.1, 1.4)$.

(b) A collision point occurs when $x_1 = x_2$ and $y_1 = y_2$ for the same t. So solve the equations:

$$3\sin t = -3 + \cos t \quad \textbf{(1)}$$
$$2\cos t = 1 + \sin t \quad \textbf{(2)}$$

From **(2)**, $\sin t = 2\cos t - 1$. Substituting into **(1)**, we get $3(2\cos t - 1) = -3 + \cos t \Rightarrow 5\cos t = 0$ $(\star)$ $\Rightarrow \cos t = 0 \Rightarrow t = \frac{\pi}{2}$ or $\frac{3\pi}{2}$. We check that $t = \frac{3\pi}{2}$ satisfies **(1)** and **(2)** but $t = \frac{\pi}{2}$ does not. So the only collision point occurs when $t = \frac{3\pi}{2}$, and this gives the point $(-3, 0)$. [We could check our work by graphing x_1 and x_2 together as functions of t and, on another plot, y_1 and y_2 as functions of t. If we do so, we see that the only value of t for which *both* pairs of graphs intersect is $t = \frac{3\pi}{2}$.]

(c) The circle is centered at $(3, 1)$ instead of $(-3, 1)$. There are still 2 intersection points: $(3, 0)$ and $(2.1, 1.4)$, but there are no collision points, since $(\star)$ in part (b) becomes $5\cos t = 6 \Rightarrow \cos t = \frac{6}{5} > 1$.

38. (a) If $\alpha = 30^\circ$ and $v_0 = 500$ m/s, then the equations become $x = (500 \cos 30^\circ)t = 250\sqrt{3}\,t$ and $y = (500 \sin 30^\circ)t - \frac{1}{2}(9.8)t^2 = 250t - 4.9t^2$. $y = 0$ when $t = 0$ (when the gun is fired) and again when $t = \frac{250}{4.9} \approx 51$ s. Then $x = \left(250\sqrt{3}\right)\left(\frac{250}{4.9}\right) \approx 22{,}092$ m, so the bullet hits the ground about 22 km from the gun.

The formula for y is quadratic in t. To find the maximum y-value, we will complete the square:

$$y = -4.9\left(t^2 - \tfrac{250}{4.9}t\right) = -4.9\left[t^2 - \tfrac{250}{4.9}t + \left(\tfrac{125}{4.9}\right)^2\right] + \tfrac{125^2}{4.9} = -4.9\left(t - \tfrac{125}{4.9}\right)^2 + \tfrac{125^2}{4.9} \le \tfrac{125^2}{4.9}$$

with equality when $t = \frac{125}{4.9}$ s, so the maximum height attained is $\frac{125^2}{4.9} \approx 3189$ m.

(b)

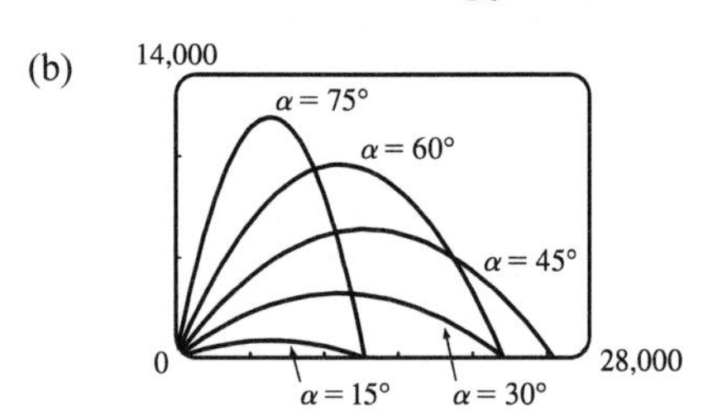

As α $(0^\circ < \alpha < 90^\circ)$ increases up to 45°, the projectile attains a greater height and a greater range. As α increases past 45°, the projectile attains a greater height, but its range decreases.

(c) $x = (v_0 \cos\alpha)t \;\Rightarrow\; t = \dfrac{x}{v_0 \cos\alpha}$.

$$y = (v_0 \sin\alpha)t - \tfrac{1}{2}gt^2 \;\Rightarrow\; y = (v_0 \sin\alpha)\frac{x}{v_0\cos\alpha} - \frac{g}{2}\left(\frac{x}{v_0\cos\alpha}\right)^2 = (\tan\alpha)x - \left(\frac{g}{2v_0^2\cos^2\alpha}\right)x^2,$$

which is the equation of a parabola (quadratic in x).

39. $x = t^2$, $y = t^3 - ct$. We use a graphing device to produce the graphs for various values of c with $-\pi \le t \le \pi$. Note that all the members of the family are symmetric about the x-axis. For $c < 0$, the graph does not cross itself, but for $c = 0$ it has a cusp at $(0, 0)$ and for $c > 0$ the graph crosses itself at $x = c$, so the loop grows larger as c increases.

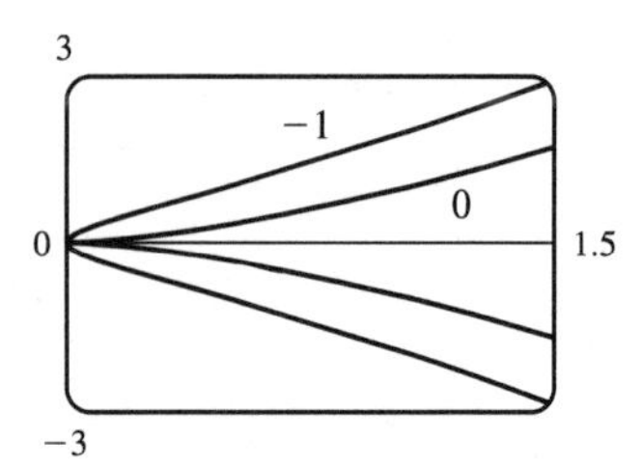

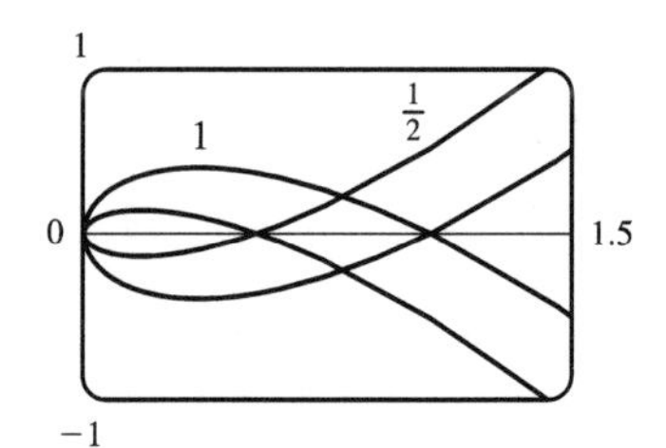

40. $x = 2ct - 4t^3$, $y = -ct^2 + 3t^4$. We use a graphing device to produce the graphs for various values of c with $-\pi \le t \le \pi$. Note that all the members of the family are symmetric about the y-axis. When $c < 0$, the graph resembles that of a polynomial of even degree, but when $c = 0$ there is a corner at the origin, and when $c > 0$, the graph crosses itself at the origin, and has two cusps below the x-axis. The size of the "swallowtail" increases as c increases.

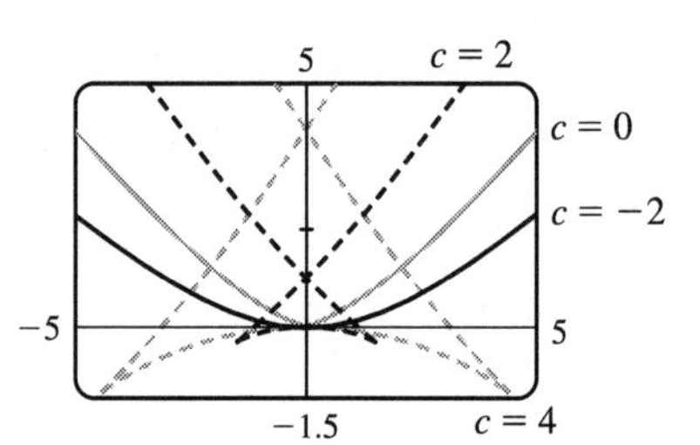

41. Note that all the Lissajous figures are symmetric about the x-axis. The parameters a and b simply stretch the graph in the x- and y-directions respectively. For $a = b = n = 1$ the graph is simply a circle with radius 1. For $n = 2$ the graph crosses itself at the origin and there are loops above and below the x-axis. In general, the figures have $n - 1$ points of intersection, all of which are on the y-axis, and a total of n closed loops.

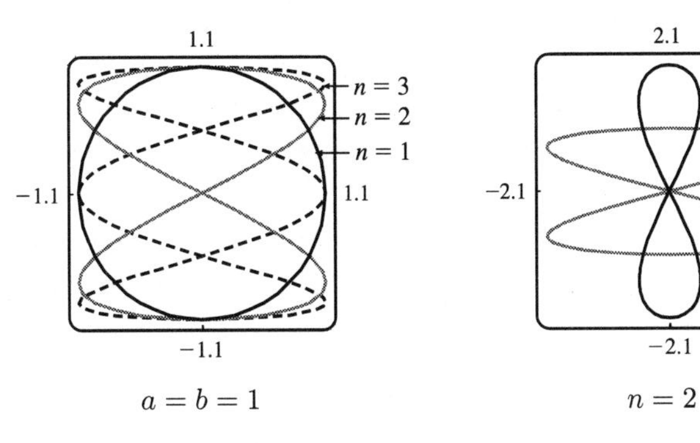

$a = b = 1$ $n = 2$

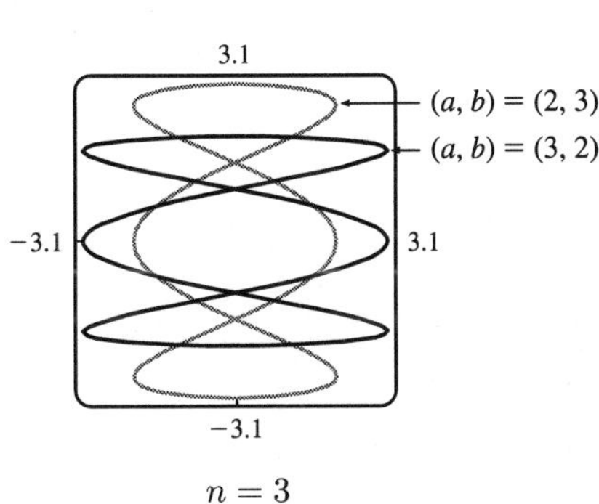

$n = 3$

42. We use $-\pi \leq t \leq \pi$ in the viewing rectangle $[-4, 2] \times [-3, 3]$. We first observe that for $c = 0$, we obtain a circle with center $\left(-\frac{1}{2}, 0\right)$ and radius $\frac{1}{2}$. As the value of c increases, there is a larger outer loop and a smaller inner loop until $c = 1$, when we obtain a curve with a dent (called a **cardioid**). As c increases, we get a curve with a dimple (called a **limaçon**) until $c = 2$. For $c > 2$, we have convex limaçons. For negative values of c, we obtain the same graphs as for positive c, but with different values of t corresponding to the points on the curve.

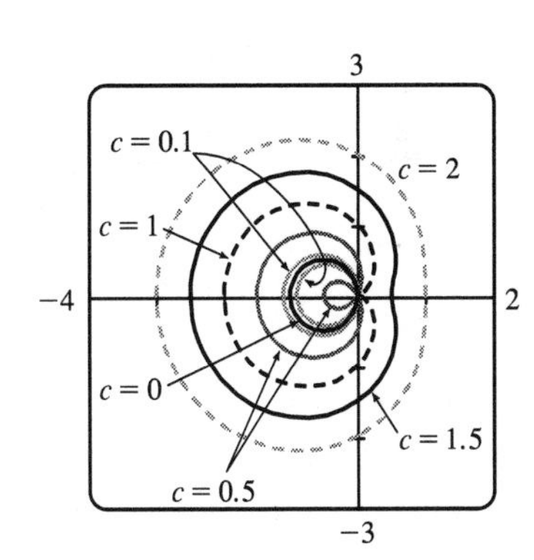

9.2 Calculus with Parametric Curves

1. $x = t - t^3,\ y = 2 - 5t \quad \Rightarrow \quad \dfrac{dy}{dt} = -5,\ \dfrac{dx}{dt} = 1 - 3t^2$, and $\dfrac{dy}{dx} = \dfrac{dy/dt}{dx/dt} = \dfrac{-5}{1-3t^2}$ or $\dfrac{5}{3t^2-1}$.

2. $x = te^t,\ y = t + e^t \quad \Rightarrow \quad \dfrac{dy}{dt} = 1 + e^t,\ \dfrac{dx}{dt} = te^t + e^t$, and $\dfrac{dy}{dx} = \dfrac{dy/dt}{dx/dt} = \dfrac{1+e^t}{te^t + e^t}$.

3. $x = t^4 + 1,\ y = t^3 + t;\ t = -1.\quad \dfrac{dy}{dt} = 3t^2 + 1,\ \dfrac{dx}{dt} = 4t^3$, and $\dfrac{dy}{dx} = \dfrac{dy/dt}{dx/dt} = \dfrac{3t^2+1}{4t^3}$. When $t = -1$, $(x, y) = (2, -2)$ and $dy/dx = \frac{4}{-4} = -1$, so an equation of the tangent to the curve at the point corresponding to $t = -1$ is $y - (-2) = (-1)(x - 2)$, or $y = -x$.

4. $x = 2t^2 + 1,\ y = \frac{1}{3}t^3 - t;\ t = 3.\quad \dfrac{dy}{dt} = t^2 - 1,\ \dfrac{dx}{dt} = 4t$, and $\dfrac{dy}{dx} = \dfrac{dy/dt}{dx/dt} = \dfrac{t^2-1}{4t}$. When $t = 3$, $(x, y) = (19, 6)$ and $dy/dx = \frac{8}{12} = \frac{2}{3}$, so an equation of the tangent line is $y - 6 = \frac{2}{3}(x - 19)$, or $y = \frac{2}{3}x - \frac{20}{3}$.

5. $x = e^{\sqrt{t}},\ y = t - \ln t^2;\ t = 1.\quad \dfrac{dy}{dt} = 1 - \dfrac{2t}{t^2} = 1 - \dfrac{2}{t},\ \dfrac{dx}{dt} = \dfrac{e^{\sqrt{t}}}{2\sqrt{t}}$, and $\dfrac{dy}{dx} = \dfrac{dy/dt}{dx/dt} = \dfrac{1 - 2/t}{e^{\sqrt{t}}/(2\sqrt{t})} \cdot \dfrac{2t}{2t} = \dfrac{2t-4}{\sqrt{t}\,e^{\sqrt{t}}}$. When $t = 1$, $(x, y) = (e, 1)$ and $\dfrac{dy}{dx} = -\dfrac{2}{e}$, so an equation of the tangent line is $y - 1 = -\frac{2}{e}(x - e)$, or $y = -\frac{2}{e}x + 3$.

6. $x = \cos\theta + \sin 2\theta,\ y = \sin\theta + \cos 2\theta;\ \theta = 0.\quad \dfrac{dy}{dx} = \dfrac{dy/d\theta}{dx/d\theta} = \dfrac{\cos\theta - 2\sin 2\theta}{-\sin\theta + 2\cos 2\theta}$. When $\theta = 0$, $(x, y) = (1, 1)$ and $dy/dx = \frac{1}{2}$, so an equation of the tangent to the curve is $y - 1 = \frac{1}{2}(x - 1)$, or $y = \frac{1}{2}x + \frac{1}{2}$.

7. (a) $x = e^t,\ y = (t-1)^2;\ (1, 1).\quad \dfrac{dy}{dt} = 2(t-1),\ \dfrac{dx}{dt} = e^t$, and $\dfrac{dy}{dx} = \dfrac{dy/dt}{dx/dt} = \dfrac{2(t-1)}{e^t}$.

At $(1, 1)$, $t = 0$ and $\dfrac{dy}{dx} = -2$, so an equation of the tangent is $y - 1 = -2(x - 1)$, or $y = -2x + 3$.

(b) $x = e^t \quad \Rightarrow \quad t = \ln x$, so $y = (t-1)^2 = (\ln x - 1)^2$ and $\dfrac{dy}{dx} = 2(\ln x - 1)\left(\dfrac{1}{x}\right)$. When $x = 1$, $\dfrac{dy}{dx} = 2(-1)(1) = -2$, so an equation of the tangent is $y = -2x + 3$, as in part (a).

8. $x = \sin t,\ y = \sin(t + \sin t);\ (0, 0).$

$$\frac{dy}{dx} = \frac{dy/dt}{dx/dt} = \frac{\cos(t + \sin t)(1 + \cos t)}{\cos t} = (\sec t + 1)\cos(t + \sin t)$$

Note that there are two tangents at the point $(0, 0)$, since both $t = 0$ and $t = \pi$ correspond to the origin. The tangent corresponding to $t = 0$ has slope $(\sec 0 + 1)\cos(0 + \sin 0) = 2\cos 0 = 2$, and its equation is $y = 2x$. The tangent corresponding to $t = \pi$ has slope $(\sec\pi + 1)\cos(\pi + \sin\pi) = 0$, so it is the x-axis.

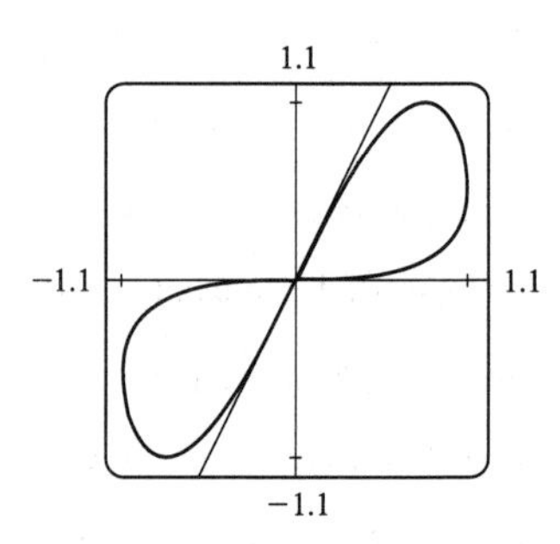

9. $x = 4 + t^2,\ y = t^2 + t^3 \quad \Rightarrow \quad \dfrac{dy}{dx} = \dfrac{dy/dt}{dx/dt} = \dfrac{2t + 3t^2}{2t} = 1 + \dfrac{3}{2}t \quad \Rightarrow$

$\dfrac{d^2y}{dx^2} = \dfrac{d}{dx}\left(\dfrac{dy}{dx}\right) = \dfrac{d(dy/dx)/dt}{dx/dt} = \dfrac{(d/dt)\left(1 + \frac{3}{2}t\right)}{2t} = \dfrac{3/2}{2t} = \dfrac{3}{4t}$. The curve is CU when $\dfrac{d^2y}{dx^2} > 0$, that is, when $t > 0$.

10. $x = t^3 - 12t,\ y = t^2 - 1 \ \Rightarrow\ \dfrac{dy}{dx} = \dfrac{dy/dt}{dx/dt} = \dfrac{2t}{3t^2 - 12} \ \Rightarrow$

$$\frac{d^2y}{dx^2} = \frac{\dfrac{d}{dt}\left(\dfrac{dy}{dx}\right)}{dx/dt} = \frac{\dfrac{(3t^2-12)\cdot 2 - 2t(6t)}{(3t^2-12)^2}}{3t^2-12} = \frac{-6t^2-24}{(3t^2-12)^3} = \frac{-6(t^2+4)}{3^3(t^2-4)^3} = \frac{-2(t^2+4)}{9(t^2-4)^3}.$$

Thus, the curve is CU when $t^2 - 4 < 0 \ \Rightarrow\ |t| < 2 \ \Rightarrow\ -2 < t < 2$.

11. $x = t - e^t,\ y = t + e^{-t} \ \Rightarrow$

$$\frac{dy}{dx} = \frac{dy/dt}{dx/dt} = \frac{1-e^{-t}}{1-e^t} = \frac{1-\dfrac{1}{e^t}}{1-e^t} = \frac{\dfrac{e^t-1}{e^t}}{1-e^t} = -e^{-t} \ \Rightarrow\ \frac{d^2y}{dx^2} = \frac{\dfrac{d}{dt}\left(\dfrac{dy}{dx}\right)}{dx/dt} = \frac{\dfrac{d}{dt}\left(-e^{-t}\right)}{dx/dt} = \frac{e^{-t}}{1-e^t}.$$

The curve is CU when $e^t < 1$ [since $e^{-t} > 0$] $\Rightarrow\ t < 0$.

12. $x = t + \ln t,\ y = t - \ln t \ \Rightarrow\ \dfrac{dy}{dx} = \dfrac{dy/dt}{dx/dt} = \dfrac{1 - 1/t}{1 + 1/t} = \dfrac{t-1}{t+1} = 1 - \dfrac{2}{t+1} \ \Rightarrow$

$$\frac{d^2y}{dx^2} = \frac{\dfrac{d}{dt}\left(\dfrac{dy}{dx}\right)}{dx/dt} = \frac{\dfrac{d}{dt}\left(1 - \dfrac{2}{t+1}\right)}{1+1/t} = \frac{2/(t+1)^2}{(t+1)/t} = \frac{2t}{(t+1)^3},$$ so the curve is CU for all t in its domain, that is, $t > 0$ [$t < -1$ not in domain].

13. $x = 10 - t^2,\ y = t^3 - 12t$.

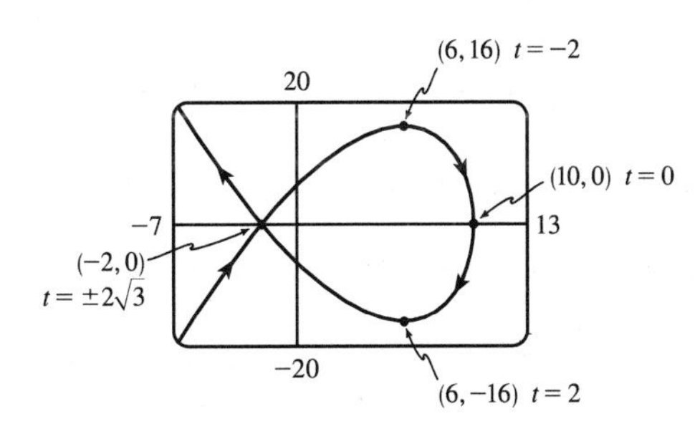

$dy/dt = 3t^2 - 12 = 3(t+2)(t-2)$, so $dy/dt = 0 \ \Leftrightarrow$ $t = \pm 2 \ \Leftrightarrow\ (x,y) = (6, \mp 16)$. $dx/dt = -2t$, so $dx/dt = 0 \ \Leftrightarrow\ t = 0 \ \Leftrightarrow\ (x,y) = (10,0)$. The curve has horizontal tangents at $(6, \pm 16)$ and a vertical tangent at $(10, 0)$.

14. $x = 2t^3 + 3t^2 - 12t,\ y = 2t^3 + 3t^2 + 1$.

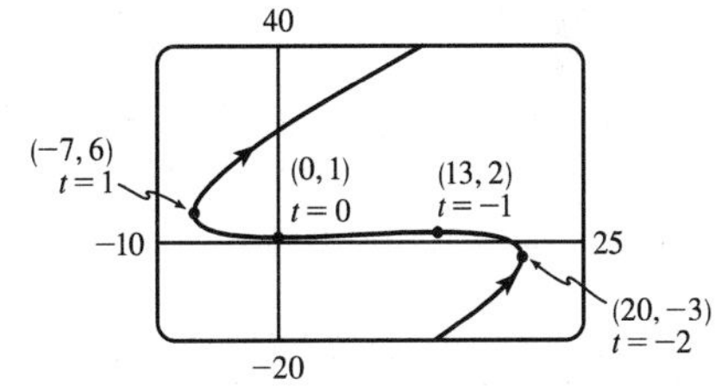

$dy/dt = 6t^2 + 6t = 6t(t+1)$, so $dy/dt = 0 \ \Leftrightarrow$ $t = 0$ or $-1 \ \Leftrightarrow\ (x,y) = (0,1)$ or $(13,2)$.

$dx/dt = 6t^2 + 6t - 12 = 6(t+2)(t-1)$, so $dx/dt = 0 \ \Leftrightarrow\ t = -2$ or $1 \ \Leftrightarrow\ (x,y) = (20,-3)$ or $(-7,6)$. The curve has horizontal tangents at $(0,1)$ and $(13,2)$, and vertical tangents at $(20,-3)$ and $(-7,6)$.

15. $x = 2\cos\theta,\ y = \sin 2\theta$.

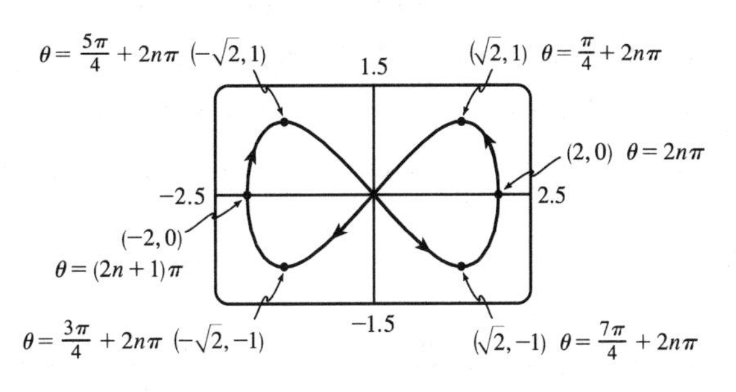

$dy/d\theta = 2\cos 2\theta$, so $dy/d\theta = 0 \ \Leftrightarrow\ 2\theta = \frac{\pi}{2} + n\pi$ (n an integer) $\Leftrightarrow\ \theta = \frac{\pi}{4} + \frac{\pi}{2}n \ \Leftrightarrow$ $(x,y) = \left(\pm\sqrt{2}, \pm 1\right)$. Also, $dx/d\theta = -2\sin\theta$, so $dx/d\theta = 0 \ \Leftrightarrow\ \theta = n\pi \ \Leftrightarrow\ (x,y) = (\pm 2, 0)$. The curve has horizontal tangents at $\left(\pm\sqrt{2}, \pm 1\right)$ (four points), and vertical tangents at $(\pm 2, 0)$.

16. $x = \cos 3\theta,\ y = 2\sin\theta.\ \ dy/d\theta = 2\cos\theta$, so $dy/d\theta = 0 \ \Leftrightarrow\ \theta = \frac{\pi}{2} + n\pi$ (n an integer) $\Leftrightarrow\ (x,y) = (0,\pm 2)$.

Also, $dx/d\theta = -3\sin 3\theta$, so $dx/d\theta = 0 \ \Leftrightarrow\ 3\theta = n\pi \ \Leftrightarrow\ \theta = \frac{\pi}{3}n \ \Leftrightarrow\ (x,y) = (\pm 1, 0)$ or $\left(\pm 1, \pm\sqrt{3}\right)$.

The curve has horizontal tangents at $(0,\pm 2)$, and vertical tangents at $(\pm 1, 0)$, $\left(\pm 1, -\sqrt{3}\right)$ and $\left(\pm 1, \sqrt{3}\right)$.

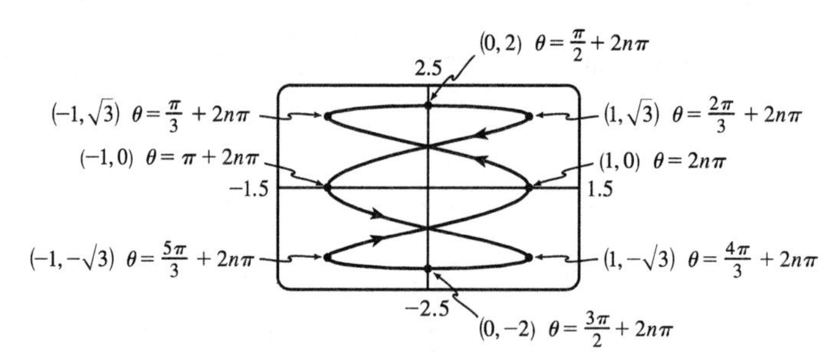

17. From the graph, it appears that the leftmost point on the curve $x = t^4 - t^2$, $y = t + \ln t$ is about $(-0.25, 0.36)$. To find the exact coordinates, we find the value of t for which the graph has a vertical tangent, that is, $0 = dx/dt = 4t^3 - 2t \ \Leftrightarrow\ 2t(2t^2 - 1) = 0 \ \Leftrightarrow$ $2t\left(\sqrt{2}\,t + 1\right)\left(\sqrt{2}\,t - 1\right) = 0 \ \Leftrightarrow\ t = 0$ or $\pm\frac{1}{\sqrt{2}}$. The negative and 0 roots are inadmissible since $y(t)$ is only defined for $t > 0$, so the leftmost point must be

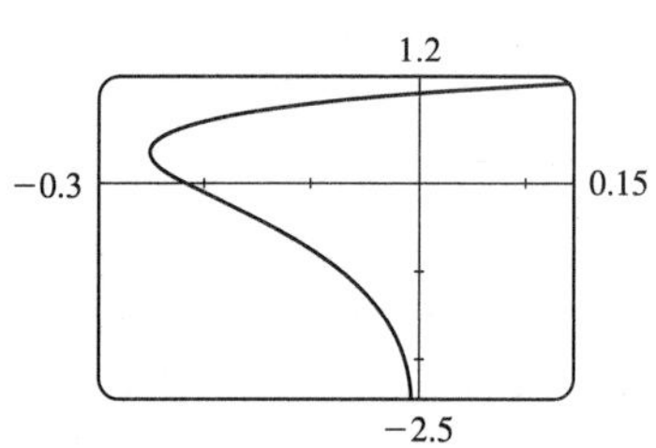

$$\left(x\left(\frac{1}{\sqrt{2}}\right), y\left(\frac{1}{\sqrt{2}}\right)\right) = \left(\left(\frac{1}{\sqrt{2}}\right)^4 - \left(\frac{1}{\sqrt{2}}\right)^2, \frac{1}{\sqrt{2}} + \ln\frac{1}{\sqrt{2}}\right) = \left(-\frac{1}{4}, \frac{1}{\sqrt{2}} - \frac{1}{2}\ln 2\right)$$

18. The curve is symmetric about the line $y = -x$ since replacing t with $-t$ has the effect of replacing (x,y) with $(-y,-x)$, so if we can find the highest point (x_h, y_h), then the leftmost point is $(x_l, y_l) = (-y_h, -x_h)$. After carefully zooming in, we estimate that the highest point on the curve $x = te^t$, $y = te^{-t}$ is about $(2.7, 0.37)$.

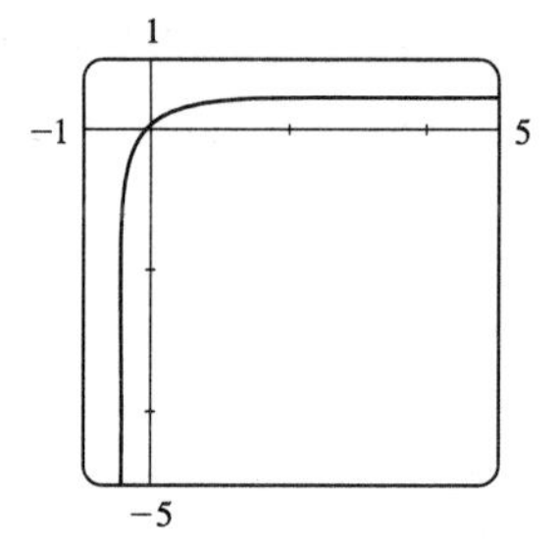

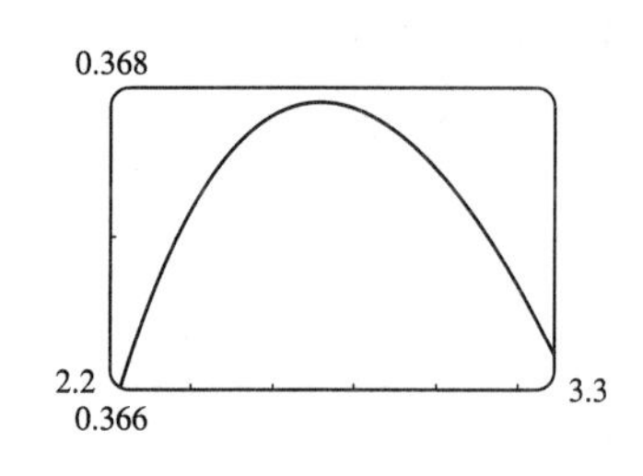

To find the exact coordinates of the highest point, we find the value of t for which the curve has a horizontal tangent, that is, $dy/dt = 0 \ \Leftrightarrow\ t(-e^{-t}) + e^{-t} = 0 \ \Leftrightarrow\ (1-t)e^{-t} = 0 \ \Leftrightarrow\ t = 1$. This corresponds to the point $(x(1), y(1)) = (e, 1/e)$. To find the leftmost point, we find the value of t for which $0 = dx/dt = te^t + e^t \ \Leftrightarrow$ $(1+t)e^t = 0 \ \Leftrightarrow\ t = -1$. This corresponds to the point $(x(-1), y(-1)) = (-1/e, -e)$.

As $t \to -\infty$, $x(t) = te^t \to 0^-$ by l'Hospital's Rule and $y(t) = te^{-t} \to -\infty$, so the y-axis is an asymptote. As $t \to \infty$, $x(t) \to \infty$ and $y(t) \to 0^+$, so the x-axis is the other asymptote. The asymptotes can also be determined from the graph, if we use a larger t-interval.

19. We graph the curve $x = t^4 - 2t^3 - 2t^2$, $y = t^3 - t$ in the viewing rectangle $[-2, 1.1]$ by $[-0.5, 0.5]$. This rectangle corresponds approximately to $t \in [-1, 0.8]$. We estimate that the curve has horizontal tangents at about $(-1, -0.4)$ and $(-0.17, 0.39)$ and vertical tangents at about $(0, 0)$ and $(-0.19, 0.37)$. We calculate

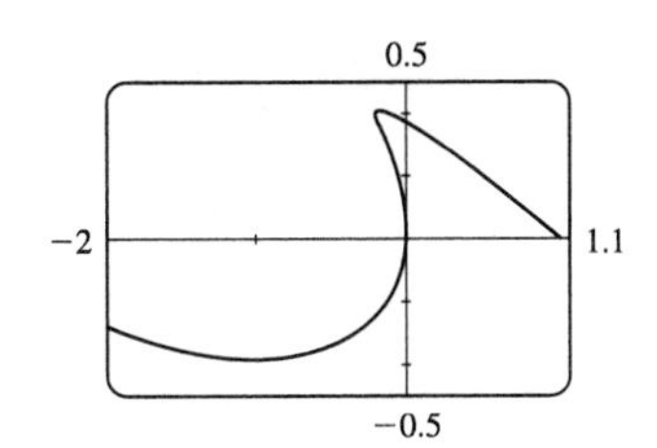

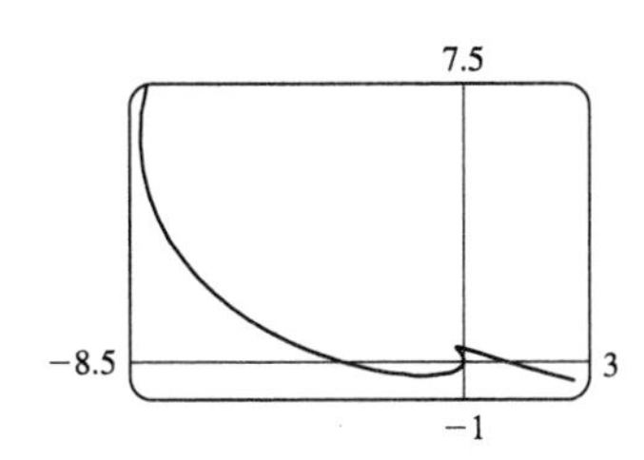

$\dfrac{dy}{dx} = \dfrac{dy/dt}{dx/dt} = \dfrac{3t^2 - 1}{4t^3 - 6t^2 - 4t}$. The horizontal tangents occur when $dy/dt = 3t^2 - 1 = 0 \Leftrightarrow t = \pm\frac{1}{\sqrt{3}}$, so both horizontal tangents are shown in our graph. The vertical tangents occur when $dx/dt = 2t(2t^2 - 3t - 2) = 0 \Leftrightarrow 2t(2t + 1)(t - 2) = 0 \Leftrightarrow t = 0, -\frac{1}{2}$ or 2. It seems that we have missed one vertical tangent, and indeed if we plot the curve on the t-interval $[-1.2, 2.2]$ we see that there is another vertical tangent at $(-8, 6)$.

20. We graph the curve $x = t^4 + 4t^3 - 8t^2$, $y = 2t^2 - t$ in the viewing rectangle $[-3.7, 0.2]$ by $[-0.2, 1.4]$. It appears that there is a horizontal tangent at about $(-0.4, -0.1)$, and vertical tangents at about $(-3, 1)$ and $(0, 0)$. We calculate

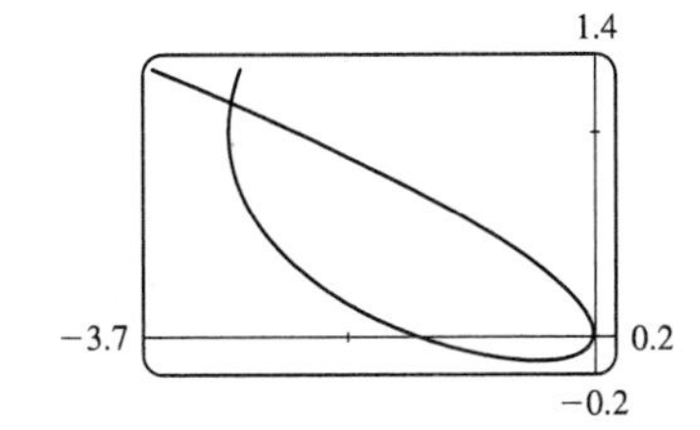

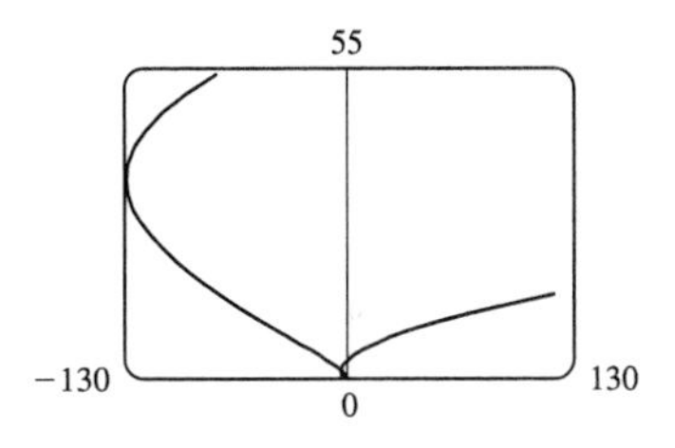

$\dfrac{dy}{dx} = \dfrac{dy/dt}{dx/dt} = \dfrac{4t - 1}{4t^3 + 12t^2 - 16t}$, so there is a horizontal tangent where $dy/dt = 4t - 1 = 0 \Leftrightarrow t = \frac{1}{4}$. This point (the lowest point) is shown in the first graph. There are vertical tangents where $dx/dt = 4t^3 + 12t^2 - 16t = 0 \Leftrightarrow 4t(t^2 + 3t - 4) = 0 \Leftrightarrow 4t(t + 4)(t - 1) = 0$. We have missed one vertical tangent corresponding to $t = -4$, and if we plot the graph for $t \in [-5, 3]$, we see that the curve has another vertical tangent line at approximately $(-128, 36)$.

21. $x = \cos t$, $y = \sin t \cos t$. $\dfrac{dx}{dt} = -\sin t$, $\dfrac{dy}{dt} = -\sin^2 t + \cos^2 t = \cos 2t$. $(x, y) = (0, 0) \Leftrightarrow \cos t = 0 \Leftrightarrow t$ is an odd multiple of $\dfrac{\pi}{2}$. When $t = \dfrac{\pi}{2}$, $\dfrac{dx}{dt} = -1$ and $\dfrac{dy}{dt} = -1$, so $\dfrac{dy}{dx} = 1$. When $t = \dfrac{3\pi}{2}$, $\dfrac{dx}{dt} = 1$ and $\dfrac{dy}{dt} = -1$. So $\dfrac{dy}{dx} = -1$. Thus, $y = x$ and $y = -x$ are both tangent to the curve at $(0, 0)$.

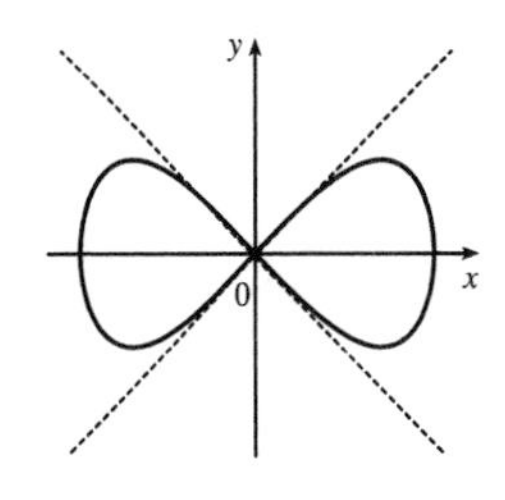

22. $x = 1 - 2\cos^2 t = -\cos 2t$, $y = (\tan t)(1 - 2\cos^2 t) = -(\tan t)\cos 2t$. To find a point where the curve crosses itself, we look for two values of t that give the same point (x, y). Call these values t_1 and t_2. Then $\cos^2 t_1 = \cos^2 t_2$ (from the equation for x) and either $\tan t_1 = \tan t_2$ or $\cos^2 t_1 = \cos^2 t_2 = \frac{1}{2}$ (from the equation for y). We can satisfy $\cos^2 t_1 = \cos^2 t_2$ and $\tan t_1 = \tan t_2$ by choosing t_1 arbitrarily and taking $t_2 = t_1 + \pi$, so evidently the whole curve is retraced every time t traverses an interval of length π. Thus, we can restrict our attention to the interval $\left(-\frac{\pi}{2}, \frac{\pi}{2}\right)$. If $t_2 = -t_1$, then $\cos^2 t_2 = \cos^2 t_1$, but $\tan t_2 = -\tan t_1$. This suggests that we try to satisfy the condition $\cos^2 t_1 = \cos^2 t_2 = \frac{1}{2}$.

Taking $t_1 = \frac{\pi}{4}$ and $t_2 = -\frac{\pi}{4}$ gives $(x, y) = (0, 0)$ for both values of t.
$dx/dt = 2\sin 2t$, and $dy/dt = 2\sin 2t \tan t - \cos 2t \sec^2 t$. When $t = \frac{\pi}{4}$, $dx/dt = 2$ and $dy/dt = 2$, so $dy/dx = 1$. When $t = -\frac{\pi}{4}$, $dx/dt = -2$ and $dy/dt = 2$, so $dy/dx = -1$. Thus, the equations of the two tangents at $(0, 0)$ are $y = x$ and $y = -x$.

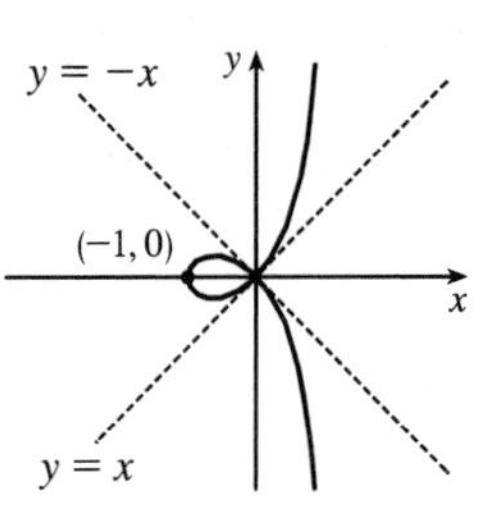

23. (a) $x = r\theta - d\sin\theta,\ y = r - d\cos\theta$. $\dfrac{dx}{d\theta} = r - d\cos\theta$, $\dfrac{dy}{d\theta} = d\sin\theta$. So $\dfrac{dy}{dx} = \dfrac{d\sin\theta}{r - d\cos\theta}$.

(b) If $0 < d < r$, then $|d\cos\theta| \le d < r$, so $r - d\cos\theta \ge r - d > 0$. This shows that $dx/d\theta$ never vanishes, so the trochoid can have no vertical tangent if $d < r$.

24. $x = a\cos^3\theta,\ y = a\sin^3\theta$.

(a) $\dfrac{dx}{d\theta} = -3a\cos^2\theta\sin\theta$, $\dfrac{dy}{d\theta} = 3a\sin^2\theta\cos\theta$, so $\dfrac{dy}{dx} = -\dfrac{\sin\theta}{\cos\theta} = -\tan\theta$.

(b) The tangent is horizontal $\Leftrightarrow$ $dy/dx = 0$ $\Leftrightarrow$ $\tan\theta = 0$ $\Leftrightarrow$ $\theta = n\pi$ $\Leftrightarrow$ $(x, y) = (\pm a, 0)$.

The tangent is vertical $\Leftrightarrow$ $\cos\theta = 0$ $\Leftrightarrow$ θ is an odd multiple of $\frac{\pi}{2}$ $\Leftrightarrow$ $(x, y) = (0, \pm a)$.

(c) $dy/dx = \pm 1$ $\Leftrightarrow$ $\tan\theta = \pm 1$ $\Leftrightarrow$ θ is an odd multiple of $\frac{\pi}{4}$ $\Leftrightarrow$ $(x, y) = \left(\pm\frac{\sqrt{2}}{4}a, \pm\frac{\sqrt{2}}{4}a\right)$

(All sign choices are valid.)

25. The line with parametric equations $x = -7t,\ y = 12t - 5$ is $y = 12\left(-\frac{1}{7}x\right) - 5$, which has slope $-\frac{12}{7}$.

The curve $x = t^3 + 4t,\ y = 6t^2$ has slope $\dfrac{dy}{dx} = \dfrac{dy/dt}{dx/dt} = \dfrac{12t}{3t^2 + 4}$. This equals $-\frac{12}{7}$ $\Leftrightarrow$ $3t^2 + 4 = -7t$ $\Leftrightarrow$ $(3t + 4)(t + 1) = 0$ $\Leftrightarrow$ $t = -1$ or $t = -\frac{4}{3}$ $\Leftrightarrow$ $(x, y) = (-5, 6)$ or $\left(-\frac{208}{27}, \frac{32}{3}\right)$.

26. $x = 3t^2 + 1,\ y = 2t^3 + 1$. $\dfrac{dx}{dt} = 6t$, $\dfrac{dy}{dt} = 6t^2$, so $\dfrac{dy}{dx} = \dfrac{6t^2}{6t} = t$ (even where $t = 0$).

So at the point corresponding to parameter value t, an equation of the tangent line is $y - (2t^3 + 1) = t\left[x - (3t^2 + 1)\right]$.
If this line is to pass through $(4, 3)$, we must have $3 - (2t^3 + 1) = t\left[4 - (3t^2 + 1)\right]$ $\Leftrightarrow$ $2t^3 - 2 = 3t^3 - 3t$ $\Leftrightarrow$ $t^3 - 3t + 2 = 0$ $\Leftrightarrow$ $(t - 1)^2(t + 2) = 0$ $\Leftrightarrow$ $t = 1$ or -2. Hence, the desired equations are $y - 3 = x - 4$, or $y = x - 1$, tangent to the curve at $(4, 3)$, and $y - (-15) = -2(x - 13)$, or $y = -2x + 11$, tangent to the curve at $(13, -15)$.

27. By symmetry of the ellipse about the x- and y-axes,

$$A = 4\int_0^a y\,dx = 4\int_{\pi/2}^0 b\sin\theta\,(-a\sin\theta)\,d\theta = 4ab\int_0^{\pi/2}\sin^2\theta\,d\theta = 4ab\int_0^{\pi/2}\tfrac{1}{2}(1 - \cos 2\theta)\,d\theta$$
$$= 2ab\left[\theta - \tfrac{1}{2}\sin 2\theta\right]_0^{\pi/2} = 2ab\left(\tfrac{\pi}{2}\right) = \pi ab$$

28. $t + 1/t = 2.5$ $\Leftrightarrow$ $t = \frac{1}{2}$ or 2, and for $\frac{1}{2} < t < 2$, we have $t + 1/t < 2.5$. $x = -\frac{3}{2}$ when $t = \frac{1}{2}$ and $x = \frac{3}{2}$ when $t = 2$.

$$A = \int_{-3/2}^{3/2}(2.5 - y)\,dx = \int_{1/2}^{2}\left(\tfrac{5}{2} - t - 1/t\right)(1 + 1/t^2)\,dt \qquad [x = t - 1/t,\ dx = (1 + 1/t^2)\,dt]$$
$$= \int_{1/2}^{2}\left(-t + \tfrac{5}{2} - 2t^{-1} + \tfrac{5}{2}t^{-2} - t^{-3}\right)dt = \left[\frac{-t^2}{2} + \frac{5t}{2} - 2\ln|t| - \frac{5}{2t} + \frac{1}{2t^2}\right]_{1/2}^{2}$$
$$= \left(-2 + 5 - 2\ln 2 - \tfrac{5}{4} + \tfrac{1}{8}\right) - \left(-\tfrac{1}{8} + \tfrac{5}{4} + 2\ln 2 - 5 + 2\right) = \tfrac{15}{4} - 4\ln 2$$

29. $A = \int_0^1 (y - 1)\,dx = \int_{\pi/2}^0 (e^t - 1)(-\sin t)\,dt = \int_0^{\pi/2}(e^t\sin t - \sin t)\,dt \overset{98}{=} \left[\frac{1}{2}e^t(\sin t - \cos t) + \cos t\right]_0^{\pi/2}$
$= \frac{1}{2}\left(e^{\pi/2} - 1\right)$

30. By symmetry, $A = 4\int_0^a y\,dx = 4\int_{\pi/2}^0 a\sin^3\theta(-3a\cos^2\theta\sin\theta)\,d\theta = 12a^2\int_0^{\pi/2}\sin^4\theta\cos^2\theta\,d\theta$. Now

$$\begin{aligned}\int \sin^4\theta\cos^2\theta\,d\theta &= \int \sin^2\theta\left(\tfrac{1}{4}\sin^2 2\theta\right)d\theta = \tfrac{1}{8}\int(1-\cos 2\theta)\sin^2 2\theta\,d\theta \\ &= \tfrac{1}{8}\int\left[\tfrac{1}{2}(1-\cos 4\theta) - \sin^2 2\theta\cos 2\theta\right]d\theta = \tfrac{1}{16}\theta - \tfrac{1}{64}\sin 4\theta - \tfrac{1}{48}\sin^3 2\theta + C\end{aligned}$$

so $\int_0^{\pi/2}\sin^4\theta\cos^2\theta\,d\theta = \left[\tfrac{1}{16}\theta - \tfrac{1}{64}\sin 4\theta - \tfrac{1}{48}\sin^3 2\theta\right]_0^{\pi/2} = \frac{\pi}{32}$. Thus, $A = 12a^2\left(\frac{\pi}{32}\right) = \frac{3}{8}\pi a^2$.

31. $A = \int_0^{2\pi r} y\,dx = \int_0^{2\pi}(r - d\cos\theta)(r - d\cos\theta)\,d\theta = \int_0^{2\pi}(r^2 - 2dr\cos\theta + d^2\cos^2\theta)\,d\theta$

$= \left[r^2\theta - 2dr\sin\theta + \tfrac{1}{2}d^2\left(\theta + \tfrac{1}{2}\sin 2\theta\right)\right]_0^{2\pi} = 2\pi r^2 + \pi d^2$

32. (a) By symmetry, the area of $\mathcal{R}$ is twice the area inside $\mathcal{R}$ above the x-axis. The top half of the loop is described by $x = t^2$, $y = t^3 - 3t$, $-\sqrt{3} \le t \le 0$, so, using the Substitution Rule with $y = t^3 - 3t$ and $dx = 2t\,dt$, we find that

$$\begin{aligned}\text{area} &= 2\int_0^3 y\,dx = 2\int_0^{-\sqrt{3}}(t^3 - 3t)2t\,dt = 2\int_0^{-\sqrt{3}}(2t^4 - 6t^2)\,dt = 2\left[\tfrac{2}{5}t^5 - 2t^3\right]_0^{-\sqrt{3}} \\ &= 2\left[\tfrac{2}{5}(-3^{1/2})^5 - 2(-3^{1/2})^3\right] = 2\left[\tfrac{2}{5}(-9\sqrt{3}) - 2(-3\sqrt{3})\right] = \tfrac{24}{5}\sqrt{3}\end{aligned}$$

(b) Here we use the formula for disks and use the Substitution Rule as in part (a):

$$\begin{aligned}\text{volume} &= \pi\int_0^3 y^2\,dx = \pi\int_0^{-\sqrt{3}}(t^3 - 3t)^2 2t\,dt = 2\pi\int_0^{-\sqrt{3}}(t^6 - 6t^4 + 9t^2)t\,dt \\ &= 2\pi\left[\tfrac{1}{8}t^8 - t^6 + \tfrac{9}{4}t^4\right]_0^{-\sqrt{3}} = 2\pi\left[\tfrac{1}{8}(-3^{1/2})^8 - (-3^{1/2})^6 + \tfrac{9}{4}(-3^{1/2})^4\right] \\ &= 2\pi\left[\tfrac{81}{8} - 27 + \tfrac{81}{4}\right] = \tfrac{27}{4}\pi\end{aligned}$$

(c) By symmetry, the y-coordinate of the centroid is 0. To find the x-coordinate, we note that it is the same as the x-coordinate of the centroid of the top half of $\mathcal{R}$, the area of which is $\frac{1}{2}\cdot\frac{24}{5}\sqrt{3} = \frac{12}{5}\sqrt{3}$. So, using Formula 7.5.12 with $A = \frac{12}{5}\sqrt{3}$, we get

$$\begin{aligned}\overline{x} &= \tfrac{5}{12\sqrt{3}}\int_0^3 xy\,dx = \tfrac{5}{12\sqrt{3}}\int_0^{-\sqrt{3}} t^2(t^3 - 3t)2t\,dt = \tfrac{5}{6\sqrt{3}}\left[\tfrac{1}{7}t^7 - \tfrac{3}{5}t^5\right]_0^{-\sqrt{3}} \\ &= \tfrac{5}{6\sqrt{3}}\left[\tfrac{1}{7}(-3^{1/2})^7 - \tfrac{3}{5}(-3^{1/2})^5\right] = \tfrac{5}{6\sqrt{3}}\left[-\tfrac{27}{7}\sqrt{3} + \tfrac{27}{5}\sqrt{3}\right] = \tfrac{9}{7}\end{aligned}$$

So the coordinates of the centroid of $\mathcal{R}$ are $(x, y) = \left(\frac{9}{7}, 0\right)$.

33. $x = t - t^2,\ y = \frac{4}{3}t^{3/2},\ 1 \le t \le 2.\ dx/dt = 1 - 2t$ and $dy/dt = 2t^{1/2}$, so

$(dx/dt)^2 + (dy/dt)^2 = (1 - 2t)^2 + (2t^{1/2})^2 = 1 - 4t + 4t^2 + 4t = 1 + 4t^2$. Thus,

$L = \int_a^b \sqrt{(dx/dt)^2 + (dy/dt)^2}\,dt = \int_1^2 \sqrt{1 + 4t^2}\,dt.$

34. $x = 1 + e^t,\ y = t^2,\ -3 \le t \le 3.\ dx/dt = e^t$ and $dy/dt = 2t$, so $(dx/dt)^2 + (dy/dt)^2 = e^{2t} + 4t^2$. Thus,

$L = \int_a^b \sqrt{(dx/dt)^2 + (dy/dt)^2}\,dt = \int_{-3}^3 \sqrt{e^{2t} + 4t^2}\,dt.$

35. $x = t + \cos t,\ y = t - \sin t,\ 0 \le t \le 2\pi.\ dx/dt = 1 - \sin t$ and $dy/dt = 1 - \cos t$, so

$$\begin{aligned}(dx/dt)^2 + (dy/dt)^2 &= (1 - \sin t)^2 + (1 - \cos t)^2 = (1 - 2\sin t + \sin^2 t) + (1 - 2\cos t + \cos^2 t) \\ &= 3 - 2\sin t - 2\cos t\end{aligned}$$

Thus, $L = \int_a^b \sqrt{(dx/dt)^2 + (dy/dt)^2}\,dt = \int_0^{2\pi}\sqrt{3 - 2\sin t - 2\cos t}\,dt.$

36. $x = \ln t,\ y = \sqrt{t+1},\ \ 1 \le t \le 5.\ \ \dfrac{dx}{dt} = \dfrac{1}{t}$ and $\dfrac{dy}{dt} = \dfrac{1}{2\sqrt{t+1}}$, so $\left(\dfrac{dx}{dt}\right)^2 + \left(\dfrac{dy}{dt}\right)^2 = \dfrac{1}{t^2} + \dfrac{1}{4\,(t+1)} = \dfrac{t^2+4t+4}{4t^2\,(t+1)}$.

Thus, $L = \displaystyle\int_a^b \sqrt{\left(\frac{dx}{dt}\right)^2 + \left(\frac{dy}{dt}\right)^2}\,dt = \int_1^5 \sqrt{\frac{t^2+4t+4}{4t^2\,(t+1)}}\,dt = \int_1^5 \sqrt{\frac{(t+2)^2}{(2t)^2\,(t+1)}}\,dt = \int_1^5 \frac{t+2}{2t\,\sqrt{t+1}}\,dt.$

37.

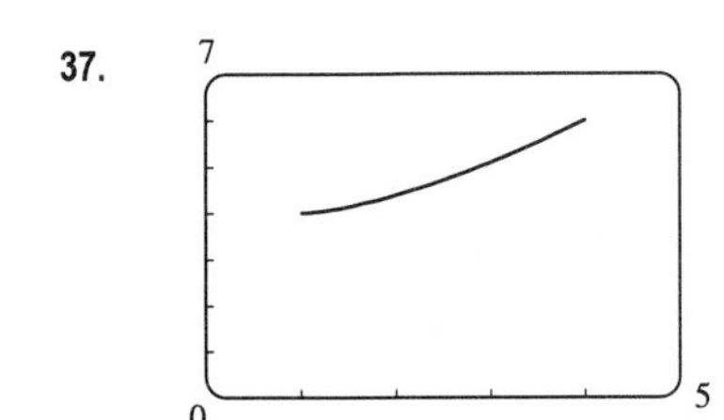

$x = 1 + 3t^2,\ \ y = 4 + 2t^3,\ \ 0 \le t \le 1.$

$dx/dt = 6t$ and $dy/dt = 6t^2$, so $(dx/dt)^2 + (dy/dt)^2 = 36t^2 + 36t^4$.

Thus,
$$\begin{aligned} L &= \textstyle\int_0^1 \sqrt{36t^2 + 36t^4}\,dt \\ &= \textstyle\int_0^1 6t\,\sqrt{1+t^2}\,dt = 6\int_1^2 \sqrt{u}\,\left(\frac{1}{2}du\right) \quad [u = 1+t^2,\ du = 2t\,dt] \\ &= 3\left[\tfrac{2}{3}u^{3/2}\right]_1^2 = 2(2^{3/2} - 1) = 2\left(2\sqrt{2} - 1\right) \end{aligned}$$

38.

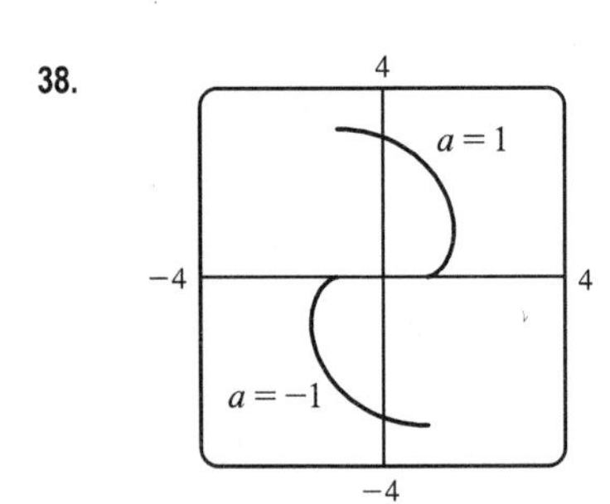

$x = a(\cos\theta + \theta\sin\theta),\ \ y = a(\sin\theta - \theta\cos\theta),\ \ 0 \le \theta \le \pi$

$$\begin{aligned} \left(\frac{dx}{d\theta}\right)^2 + \left(\frac{dy}{d\theta}\right)^2 &= a^2[(-\sin\theta + \theta\cos\theta + \sin\theta)^2 + (\cos\theta + \theta\sin\theta - \cos\theta)^2] \\ &= a^2\theta^2(\cos^2\theta + \sin^2\theta) = (a\theta)^2 \end{aligned}$$

$L = \int_0^\pi a\theta\,d\theta = a\left[\frac{1}{2}\theta^2\right]_0^\pi = \frac{1}{2}\pi^2 a$

39. $x = \dfrac{t}{1+t},\ \ y = \ln(1+t),\ 0 \le t \le 2.\ \ \dfrac{dx}{dt} = \dfrac{(1+t)\cdot 1 - t\cdot 1}{(1+t)^2} = \dfrac{1}{(1+t)^2}$ and $\dfrac{dy}{dt} = \dfrac{1}{1+t}$, so

$\left(\dfrac{dx}{dt}\right)^2 + \left(\dfrac{dy}{dt}\right)^2 = \dfrac{1}{(1+t)^4} + \dfrac{1}{(1+t)^2} = \dfrac{1}{(1+t)^4}\left[1 + (1+t)^2\right] = \dfrac{t^2+2t+2}{(1+t)^4}$. Thus,

$$L = \int_0^2 \frac{\sqrt{t^2+2t+2}}{(1+t)^2}\,dt = \int_1^3 \frac{\sqrt{u^2+1}}{u^2}\,du\ \ [u = t+1,\ du = dt]\ \overset{24}{=} \left[-\frac{\sqrt{u^2+1}}{u} + \ln\left(u + \sqrt{u^2+1}\right)\right]_1^3$$
$$= -\tfrac{\sqrt{10}}{3} + \ln\left(3 + \sqrt{10}\right) + \sqrt{2} - \ln\left(1 + \sqrt{2}\right)$$

40.

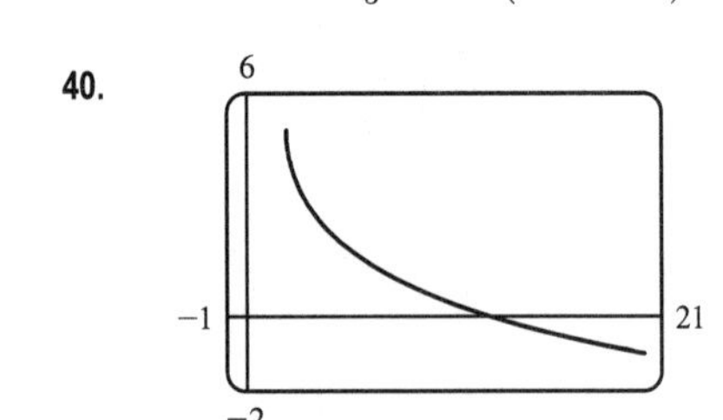

$x = e^t + e^{-t},\ \ y = 5 - 2t,\ \ 0 \le t \le 3.$

$dx/dt = e^t - e^{-t}$ and $dy/dt = -2$, so

$\left(\dfrac{dx}{dt}\right)^2 + \left(\dfrac{dy}{dt}\right)^2 = e^{2t} - 2 + e^{-2t} + 4 = e^{2t} + 2 + e^{-2t} = (e^t + e^{-t})^2$

and $L = \int_0^3 (e^t + e^{-t})\,dt = \left[e^t - e^{-t}\right]_0^3 = e^3 - e^{-3} - (1 - 1) = e^3 - e^{-3}$.

41. $x = e^t\cos t,\ \ y = e^t\sin t,\ \ 0 \le t \le \pi.$

$$\begin{aligned} \left(\frac{dx}{dt}\right)^2 + \left(\frac{dy}{dt}\right)^2 &= \left[e^t(\cos t - \sin t)\right]^2 + \left[e^t(\sin t + \cos t)\right]^2 \\ &= \left(e^t\right)^2\left(\cos^2 t - 2\cos t\,\sin t + \sin^2 t\right) \\ &\quad + \left(e^t\right)^2\left(\sin^2 t + 2\sin t\,\cos t + \cos^2 t\right) \\ &= e^{2t}\left(2\cos^2 t + 2\sin^2 t\right) = 2e^{2t} \end{aligned}$$

Thus, $L = \int_0^\pi \sqrt{2e^{2t}}\,dt = \int_0^\pi \sqrt{2}\,e^t\,dt = \sqrt{2}\left[e^t\right]_0^\pi = \sqrt{2}\left(e^\pi - 1\right)$.

42. $x = \cos t + \ln\left(\tan \frac{1}{2}t\right),\ y = \sin t,\ \pi/4 \le t \le 3\pi/4.$

$$\frac{dx}{dt} = -\sin t + \frac{\frac{1}{2}\sec^2(t/2)}{\tan(t/2)} = -\sin t + \frac{1}{2\sin(t/2)\cos(t/2)} = -\sin t + \frac{1}{\sin t}$$

$$\frac{dx}{dt} = -\sin t + \frac{\frac{1}{2}\sec^2(t/2)}{\tan(t/2)} = -\sin t + \frac{1}{2\sin(t/2)\cos(t/2)} = -\sin t + \frac{1}{\sin t}$$

and $\frac{dy}{dt} = \cos t$, so $\left(\frac{dx}{dt}\right)^2 + \left(\frac{dy}{dt}\right)^2 = \sin^2 t - 2 + \frac{1}{\sin^2 t} + \cos^2 t = 1 - 2 + \csc^2 t = \cot^2 t$. Thus,

$$L = \int_{\pi/4}^{3\pi/4} \left|\cot t\right| dt = 2\int_{\pi/4}^{\pi/2} \cot t\,dt = 2\left[\ln|\sin t|\right]_{\pi/4}^{\pi/2} = 2\left(\ln 1 - \ln\frac{1}{\sqrt{2}}\right)$$
$$= 2\left(0 + \ln\sqrt{2}\right) = 2\left(\tfrac{1}{2}\ln 2\right) = \ln 2$$

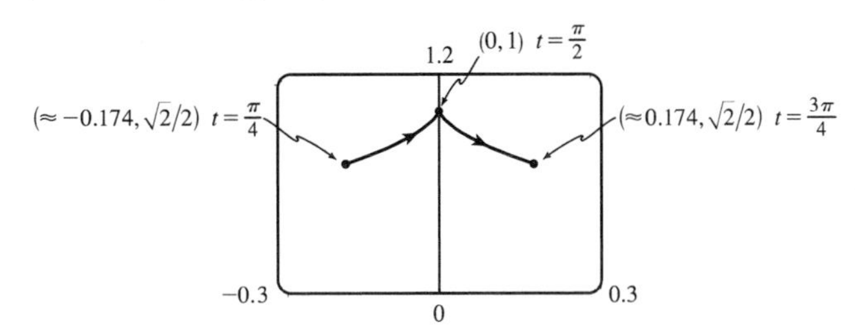

43.

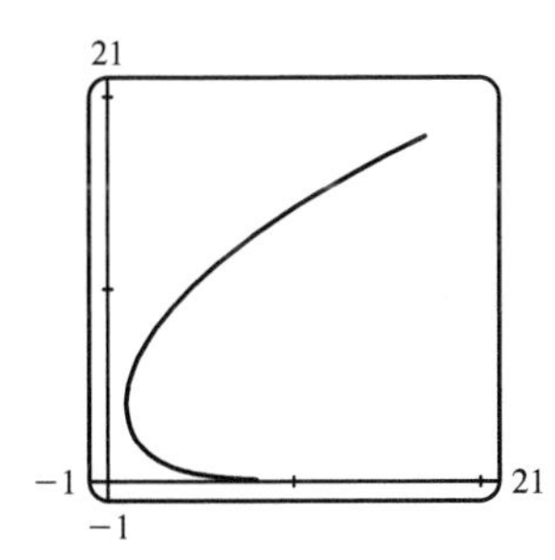

$x = e^t - t,\ y = 4e^{t/2},\ -8 \le t \le 3.$

$$(dx/dt)^2 + (dy/dt)^2 = (e^t - 1)^2 + (2e^{t/2})^2 = e^{2t} - 2e^t + 1 + 4e^t$$
$$= e^{2t} + 2e^t + 1 = (e^t + 1)^2$$

$$L = \int_{-8}^{3} \sqrt{(e^t+1)^2}\,dt = \int_{-8}^{3}(e^t + 1)\,dt = \left[e^t + t\right]_{-8}^{3t}$$
$$= (e^3 + 3) - (e^{-8} - 8) = e^3 - e^{-8} + 11$$

44. $x = 3t - t^3,\ y = 3t^2.\quad dx/dt = 3 - 3t^2$ and $dy/dt = 6t$, so

$(dx/dt)^2 + (dy/dt)^2 = (3 - 3t^2)^2 + (6t)^2 = (3 + 3t^2)^2$ and the length of the loop is given by

$$L = \int_{-\sqrt{3}}^{\sqrt{3}} (3 + 3t^2)\,dt = 2\int_0^{\sqrt{3}} (3 + 3t^2)\,dt = 2[3t + t^3]_0^{\sqrt{3}}$$
$$= 2\left(3\sqrt{3} + 3\sqrt{3}\right) = 12\sqrt{3}$$

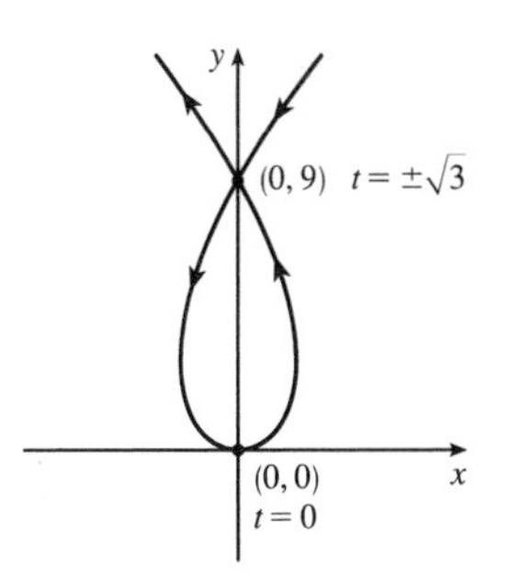

45. $x = t - e^t,\ y = t + e^t,\ -6 \le t \le 6.$

$\left(\frac{dx}{dt}\right)^2 + \left(\frac{dy}{dt}\right)^2 = (1 - e^t)^2 + (1 + e^t)^2 = \left(1 - 2e^t + e^{2t}\right) + \left(1 + 2e^t + e^{2t}\right) = 2 + 2e^{2t}$, so $L = \int_{-6}^{6} \sqrt{2 + 2e^{2t}}\,dt$.

Set $f(t) = \sqrt{2 + 2e^{2t}}$. Then by Simpson's Rule with $n = 6$ and $\Delta t = \frac{6-(-6)}{6} = 2$, we get

$L \approx \frac{2}{3}[f(-6) + 4f(-4) + 2f(-2) + 4f(0) + 2f(2) + 4f(4) + f(6)] \approx 612.3053.$

46. $x = 2a\cot\theta \;\Rightarrow\; dx/dt = -2a\csc^2\theta$ and $y = 2a\sin^2\theta \;\Rightarrow\; dy/dt = 4a\sin\theta\,\cos\theta = 2a\sin 2\theta$.

So $L = \int_{\pi/4}^{\pi/2}\sqrt{4a^2\csc^4\theta + 4a^2\sin^2 2\theta}\,d\theta = 2a\int_{\pi/4}^{\pi/2}\sqrt{\csc^4\theta + \sin^2 2\theta}\,d\theta$. Using Simpson's Rule with $n = 4$,

$\Delta\theta = \frac{\pi/2 - \pi/4}{4} = \frac{\pi}{16}$, and $f(\theta) = \sqrt{\csc^4\theta + \sin^2 2\theta}$, we get

$L \approx 2a\cdot S_4 = (2a)\frac{\pi}{16\cdot 3}\left[f\left(\frac{\pi}{4}\right) + 4f\left(\frac{5\pi}{16}\right) + 2f\left(\frac{3\pi}{8}\right) + 4f\left(\frac{7\pi}{16}\right) + f\left(\frac{\pi}{2}\right)\right] \approx 2.2605a$.

47. $x = \sin^2 t,\ y = \cos^2 t,\ 0 \le t \le 3\pi$.

$(dx/dt)^2 + (dy/dt)^2 = (2\sin t\cos t)^2 + (-2\cos t\sin t)^2 = 8\sin^2 t\cos^2 t = 2\sin^2 2t \;\Rightarrow$

$$\begin{aligned}\text{Distance} &= \int_0^{3\pi}\sqrt{2}\,|\sin 2t|\,dt = 6\sqrt{2}\int_0^{\pi/2}\sin 2t\,dt \quad [\text{by symmetry}] \; = -3\sqrt{2}\left[\cos 2t\right]_0^{\pi/2}\\ &= -3\sqrt{2}\,(-1-1) = 6\sqrt{2}\end{aligned}$$

The full curve is traversed as t goes from 0 to $\frac{\pi}{2}$, because the curve is the segment of $x + y = 1$ that lies in the first quadrant (since $x, y \ge 0$), and this segment is completely traversed as t goes from 0 to $\frac{\pi}{2}$. Thus, $L = \int_0^{\pi/2}\sin 2t\,dt = \sqrt{2}$, as above.

48. $x = \cos^2 t,\ y = \cos t,\ 0 \le t \le 4\pi$. $(dx/dt)^2 + (dy/dt)^2 = (-2\cos t\sin t)^2 + (-\sin t)^2 = \sin^2 t(4\cos^2 t + 1)$

$$\begin{aligned}\text{Distance} &= \int_0^{4\pi}|\sin t|\,\sqrt{4\cos^2 t + 1}\,dt = 4\int_0^{\pi}\sin t\sqrt{4\cos^2 t + 1}\,dt\\ &= -4\int_1^{-1}\sqrt{4u^2+1}\,du \quad [u = \cos t,\ du = -\sin t\,dt] \; = 4\int_{-1}^{1}\sqrt{4u^2+1}\,du = 8\int_0^1\sqrt{4u^2+1}\,du\\ &= 8\int_0^{\tan^{-1}2}\sec\theta\cdot\tfrac{1}{2}\sec^2\theta\,d\theta \qquad [2u = \tan\theta,\ 2\,du = \sec^2\theta\,d\theta]\\ &= 4\int_0^{\tan^{-1}2}\sec^3\theta\,d\theta \overset{71}{=} \left[2\sec\theta\tan\theta + 2\ln|\sec\theta + \tan\theta|\right]_0^{\tan^{-1}2} = 4\sqrt{5} + 2\ln\left(\sqrt{5}+2\right)\end{aligned}$$

Thus, $L = \int_0^{\pi}|\sin t|\,\sqrt{4\cos^2 t + 1}\,dt = \sqrt{5} + \frac{1}{2}\ln\left(\sqrt{5}+2\right)$.

49. $x = a\sin\theta,\ y = b\cos\theta,\ 0 \le \theta \le 2\pi$.

$$\begin{aligned}(dx/d\theta)^2 + (dy/d\theta)^2 &= (a\cos\theta)^2 + (-b\sin\theta)^2 = a^2\cos^2\theta + b^2\sin^2\theta = a^2\left(1 - \sin^2\theta\right) + b^2\sin^2\theta\\ &= a^2 - \left(a^2 - b^2\right)\sin^2\theta = a^2 - c^2\sin^2\theta = a^2\left(1 - \frac{c^2}{a^2}\sin^2\theta\right) = a^2\left(1 - e^2\sin^2\theta\right)\end{aligned}$$

So $L = 4\int_0^{\pi/2}\sqrt{a^2\left(1 - e^2\sin^2\theta\right)}\,d\theta$ [by symmetry] $= 4a\int_0^{\pi/2}\sqrt{1 - e^2\sin^2\theta}\,d\theta$.

50. $x = a\cos^3\theta,\ y = a\sin^3\theta$.

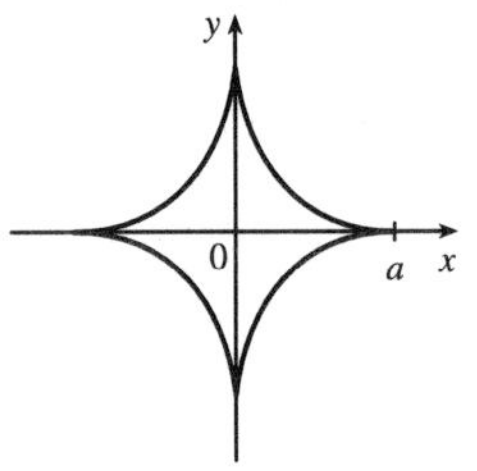

$$\begin{aligned}(dx/d\theta)^2 + (dy/d\theta)^2 &= (-3a\cos^2\theta\,\sin\theta)^2 + (3a\sin^2\theta\,\cos\theta)^2\\ &= 9a^2\cos^4\theta\,\sin^2\theta + 9a^2\sin^4\theta\,\cos^2\theta\\ &= 9a^2\sin^2\theta\,\cos^2\theta(\cos^2\theta + \sin^2\theta) = 9a^2\sin^2\theta\,\cos^2\theta\end{aligned}$$

The graph has four-fold symmetry and the curve in the first quadrant corresponds to $0 \le \theta \le \pi/2$. Thus,

$$\begin{aligned}L &= 4\int_0^{\pi/2}3a\sin\theta\,\cos\theta\,d\theta \qquad [\text{since } a > 0 \text{ and } \sin\theta \text{ and } \cos\theta \text{ are positive for } 0 \le \theta \le \pi/2]\\ &= 12a\left[\tfrac{1}{2}\sin^2\theta\right]_0^{\pi/2} = 12a\left(\tfrac{1}{2} - 0\right) = 6a\end{aligned}$$

51. (a) Notice that $0 \le t \le 2\pi$ does not give the complete curve because $x(0) \ne x(2\pi)$. In fact, we must take $t \in [0, 4\pi]$ in order to obtain the complete curve, since the first term in each of the parametric equations has period 2π and the second has period $\frac{2\pi}{11/2} = \frac{4\pi}{11}$, and the least common integer multiple of these two numbers is 4π.

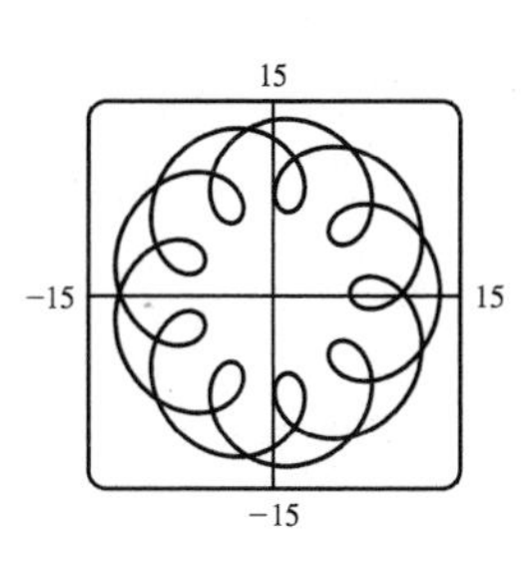

(b) We use the CAS to find the derivatives dx/dt and dy/dt, and then use Formula 1 to find the arc length. Recent versions of Maple express the integral $\int_0^{4\pi} \sqrt{(dx/dt)^2 + (dy/dt)^2}\,dt$ as $88E(2\sqrt{2}\,i)$, where $E(x)$ is the elliptic integral $\displaystyle\int_0^1 \frac{\sqrt{1 - x^2t^2}}{\sqrt{1 - t^2}}\,dt$ and i is the imaginary number $\sqrt{-1}$. Some earlier versions of Maple (as well as Mathematica) cannot do the integral exactly, so we use the command `evalf(Int(sqrt(diff(x,t)^2+diff(y,t)^2),t=0..4*Pi));` to estimate the length, and find that the arc length is approximately 294.03. Derive's `Para_arc_length` function in the utility file `Int_apps` simplifies the integral to $11\int_0^{4\pi} \sqrt{-4\cos t\,\cos\left(\frac{11t}{2}\right) - 4\sin t\,\sin\left(\frac{11t}{2}\right) + 5}\,dt$.

52. (a) It appears that as $t \to \infty$, $(x, y) \to \left(\frac{1}{2}, \frac{1}{2}\right)$, and as $t \to -\infty$, $(x, y) \to \left(-\frac{1}{2}, -\frac{1}{2}\right)$.

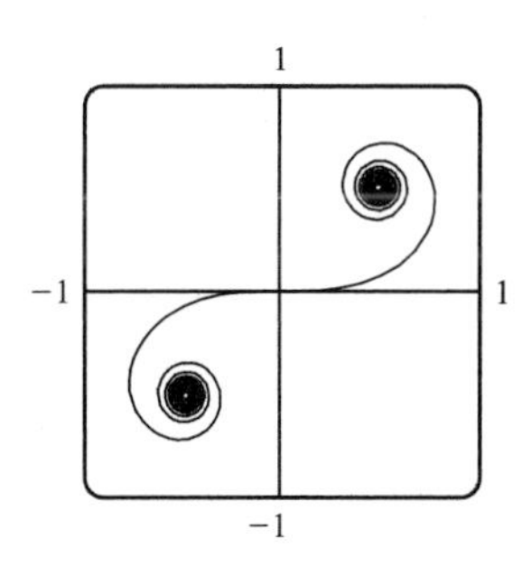

(b) By the Fundamental Theorem of Calculus, $dx/dt = \cos\left(\frac{\pi}{2}t^2\right)$ and $dy/dt = \sin\left(\frac{\pi}{2}t^2\right)$, so by Formula 1, the length of the curve from the origin to the point with parameter value t is

$$L = \int_0^t \sqrt{\left(\tfrac{dx}{du}\right)^2 + \left(\tfrac{dy}{du}\right)^2}\,du = \int_0^t \sqrt{\cos^2\left(\tfrac{\pi}{2}u^2\right) + \sin^2\left(\tfrac{\pi}{2}u^2\right)}\,du$$
$$= \int_0^t 1\,du = t \qquad [\text{or } -t \text{ if } t < 0]$$

We have used u as the dummy variable so as not to confuse it with the upper limit of integration.

53. The coordinates of T are $(r\cos\theta, r\sin\theta)$. Since TP was unwound from arc TA, TP has length $r\theta$. Also $\angle PTQ = \angle PTR - \angle QTR = \frac{1}{2}\pi - \theta$, so P has coordinates

$$x = r\cos\theta + r\theta\cos\left(\tfrac{1}{2}\pi - \theta\right) = r(\cos\theta + \theta\sin\theta),$$
$$y = r\sin\theta - r\theta\sin\left(\tfrac{1}{2}\pi - \theta\right) = r(\sin\theta - \theta\cos\theta).$$

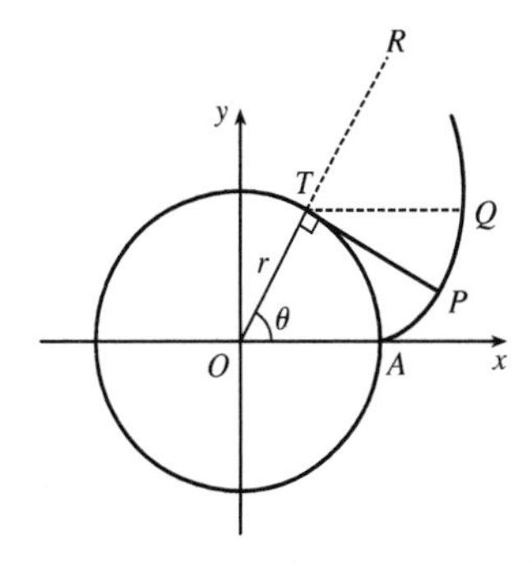

54. If the cow walks with the rope taut, it traces out the portion of the involute in Exercise 53 corresponding to the range $0 \le \theta \le \pi$, arriving at the point $(-r, \pi r)$ when $\theta = \pi$. With the rope now fully extended, the cow walks in a semicircle of radius πr, arriving at $(-r, -\pi r)$. Finally, the cow traces out another portion of the involute, namely the reflection about the

x-axis of the initial involute path. (This corresponds to the range $-\pi \leq \theta \leq 0$.) Referring to the figure, we see that the total grazing area is $2(A_1 + A_3)$. A_3 is one-quarter of the area of a circle of radius πr, so $A_3 = \frac{1}{4}\pi(\pi r)^2 = \frac{1}{4}\pi^3 r^2$. We will compute $A_1 + A_2$ and then subtract $A_2 = \frac{1}{2}\pi r^2$ to obtain A_1.

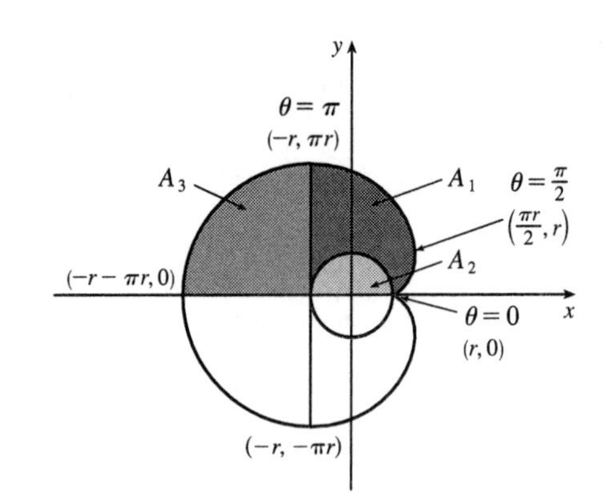

To find $A_1 + A_2$, first note that the rightmost point of the involute is $\left(\frac{\pi}{2}r, r\right)$. [To see this, note that $dx/d\theta = 0$ when $\theta = 0$ or $\frac{\pi}{2}$. $\theta = 0$ corresponds to the cusp at $(r, 0)$ and $\theta = \frac{\pi}{2}$ corresponds to $\left(\frac{\pi}{2}r, r\right)$.]

The leftmost point of the involute is $(-r, \pi r)$. Thus, $A_1 + A_2 = \int_{\theta=\pi}^{\pi/2} y\,dx - \int_{\theta=0}^{\pi/2} y\,dx = \int_{\theta=\pi}^{0} y\,dx$.

Now $y\,dx = r(\sin\theta - \theta\cos\theta)r\theta\cos\theta\,d\theta = r^2(\theta\sin\theta\cos\theta - \theta^2\cos^2\theta)d\theta$. Integrate:

$(1/r^2)\int y\,dx = -\theta\cos^2\theta - \frac{1}{2}(\theta^2 - 1)\sin\theta\cos\theta - \frac{1}{6}\theta^3 + \frac{1}{2}\theta + C$. This enables us to compute

$$A_1 + A_2 = r^2\left[-\theta\cos^2\theta - \tfrac{1}{2}(\theta^2 - 1)\sin\theta\cos\theta - \tfrac{1}{6}\theta^3 + \tfrac{1}{2}\theta\right]_\pi^0 = r^2\left[0 - \left(-\pi - \tfrac{\pi^3}{6} + \tfrac{\pi}{2}\right)\right] = r^2\left(\tfrac{\pi}{2} + \tfrac{\pi^3}{6}\right)$$

Therefore, $A_1 = (A_1 + A_2) - A_2 = \frac{1}{6}\pi^3 r^2$, so the grazing area is $2(A_1 + A_3) = 2\left(\frac{1}{6}\pi^3 r^2 + \frac{1}{4}\pi^3 r^2\right) = \frac{5}{6}\pi^3 r^2$.

9.3 Polar Coordinates

1. (a) By adding 2π to $\frac{\pi}{2}$, we obtain the point $\left(1, \frac{5\pi}{2}\right)$. The direction opposite $\frac{\pi}{2}$ is $\frac{3\pi}{2}$, so $\left(-1, \frac{3\pi}{2}\right)$ is a point that satisfies the $r < 0$ requirement.

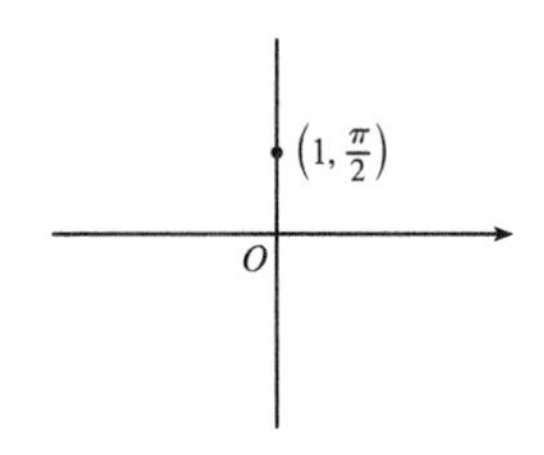

(b) $\left(-2, \frac{\pi}{4}\right)$

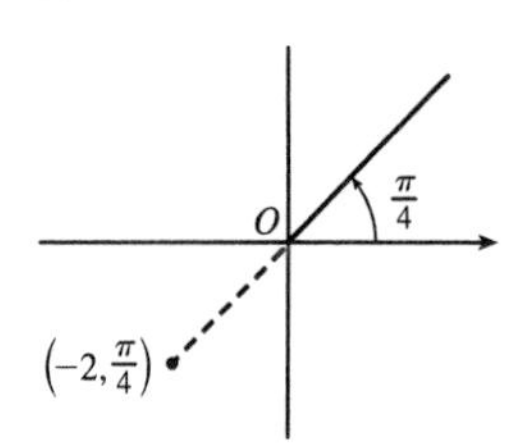

$\left(2, \frac{5\pi}{4}\right), \left(-2, \frac{9\pi}{4}\right)$

(c) $(3, 2)$

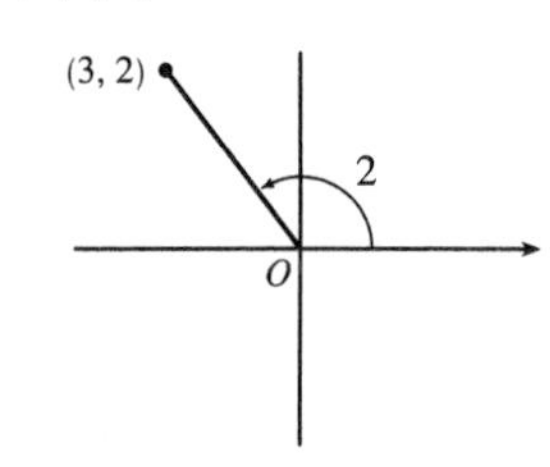

$(3, 2 + 2\pi), (-3, 2 + \pi)$

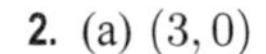

2. (a) $(3, 0)$

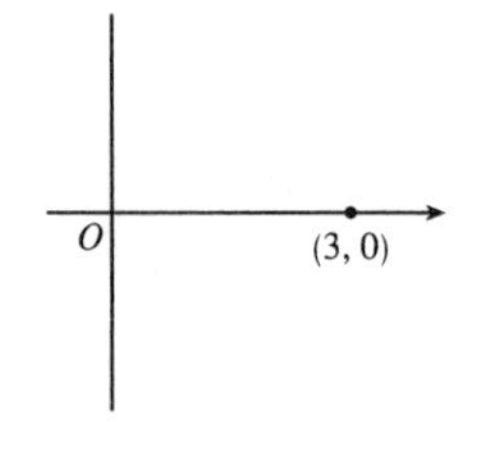

$(3, 2\pi), (-3, \pi)$

(b) $\left(2, -\frac{\pi}{7}\right)$

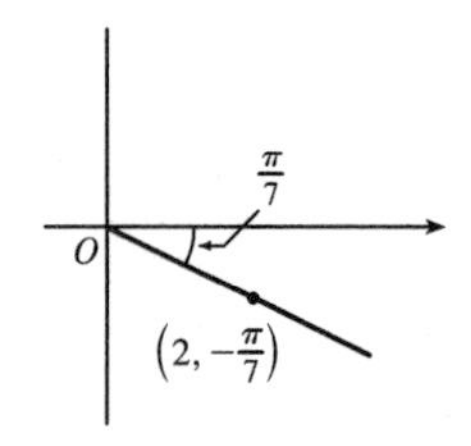

$\left(2, \frac{13\pi}{7}\right), \left(-2, \frac{6\pi}{7}\right)$

(c) $\left(-1, -\frac{\pi}{2}\right)$

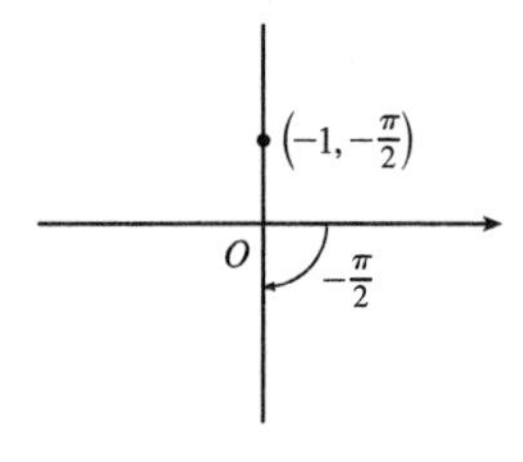

$\left(1, \frac{\pi}{2}\right), \left(-1, \frac{3\pi}{2}\right)$

3. (a)

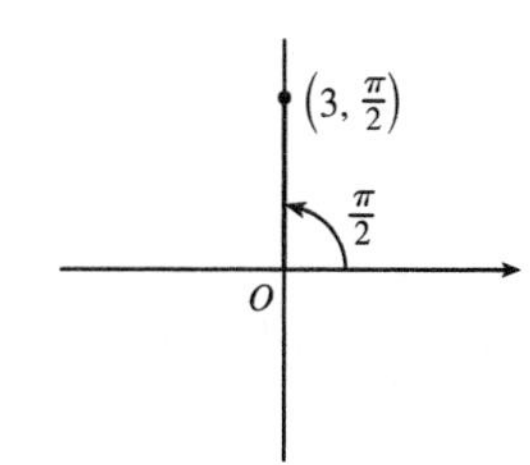

$x = 3\cos\frac{\pi}{2} = 3(0) = 0$ and

$y = 3\sin\frac{\pi}{2} = 3(1) = 3$ give us

the Cartesian coordinates $(0, 3)$.

(b)

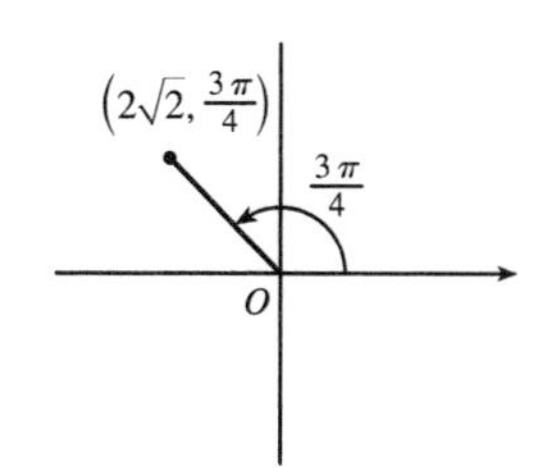

$x = 2\sqrt{2}\cos\frac{3\pi}{4}$

$= 2\sqrt{2}\left(-\frac{1}{\sqrt{2}}\right) = -2$ and

$y = 2\sqrt{2}\sin\frac{3\pi}{4} = 2\sqrt{2}\left(\frac{1}{\sqrt{2}}\right) = 2$

give us $(-2, 2)$.

(c)

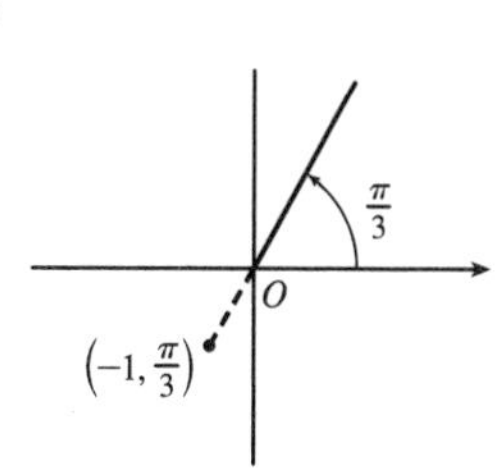

$x = -1\cos\frac{\pi}{3} = -\frac{1}{2}$ and

$y = -1\sin\frac{\pi}{3} = -\frac{\sqrt{3}}{2}$ give

us $\left(-\frac{1}{2}, -\frac{\sqrt{3}}{2}\right)$.

4. (a)

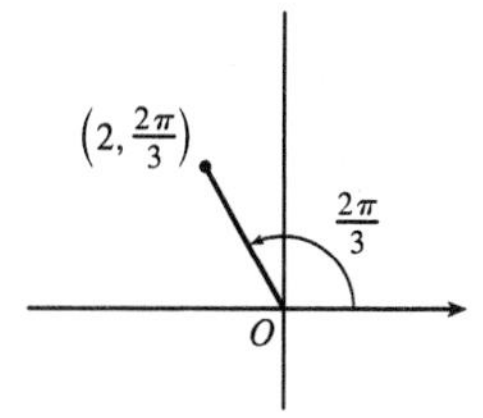

$x = 2\cos\frac{2\pi}{3} = -1$ and

$y = 2\sin\frac{2\pi}{3} = \sqrt{3}$ give

us $\left(-1, \sqrt{3}\right)$.

(b)

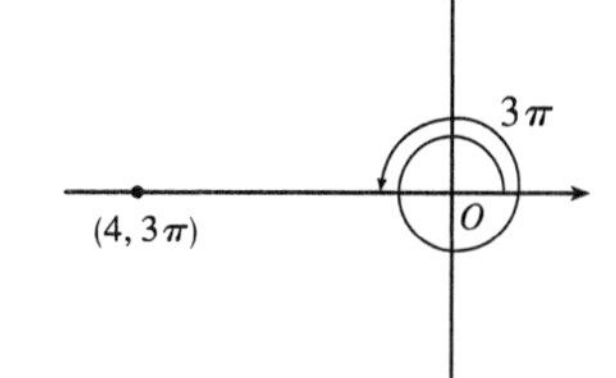

$x = 4\cos 3\pi = -4$ and

$y = 4\sin 3\pi = 0$ give

us $(-4, 0)$.

(c)

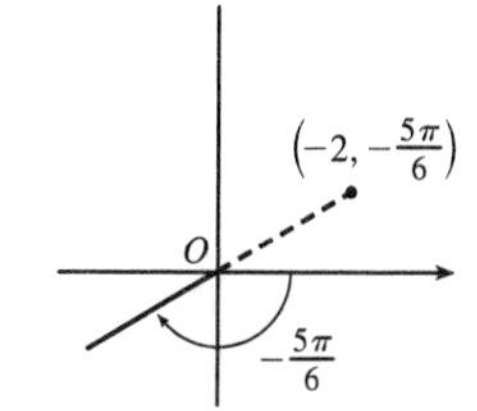

$x = -2\cos\left(-\frac{5\pi}{6}\right) = \sqrt{3}$

and $y = -2\sin\left(-\frac{5\pi}{6}\right) = 1$

give us $\left(\sqrt{3}, 1\right)$.

5. (a) $x = 1$ and $y = 1 \quad\Rightarrow\quad r = \sqrt{1^2 + 1^2} = \sqrt{2}$ and $\theta = \tan^{-1}\left(\frac{1}{1}\right) = \frac{\pi}{4}$. Since $(1, 1)$ is in the first quadrant, the polar coordinates are (i) $\left(\sqrt{2}, \frac{\pi}{4}\right)$ and (ii) $\left(-\sqrt{2}, \frac{5\pi}{4}\right)$.

(b) $x = 2\sqrt{3}$ and $y = -2 \quad\Rightarrow\quad r = \sqrt{\left(2\sqrt{3}\right)^2 + (-2)^2} = \sqrt{12 + 4} = \sqrt{16} = 4$ and

$\theta = \tan^{-1}\left(-\frac{2}{2\sqrt{3}}\right) = \tan^{-1}\left(-\frac{1}{\sqrt{3}}\right) = -\frac{\pi}{6}$. Since $\left(2\sqrt{3}, -2\right)$ is in the fourth quadrant and $0 \le \theta \le 2\pi$, the polar coordinates are (i) $\left(4, \frac{11\pi}{6}\right)$ and (ii) $\left(-4, \frac{5\pi}{6}\right)$.

6. (a) $(x, y) = \left(-1, -\sqrt{3}\right)$, $r = \sqrt{1 + 3} = 2$, $\tan\theta = y/x = \sqrt{3}$ and (x, y) is in the third quadrant, so $\theta = \frac{4\pi}{3}$. The polar coordinates are (i) $\left(2, \frac{4\pi}{3}\right)$ and (ii) $\left(-2, \frac{\pi}{3}\right)$.

(b) $(x, y) = (-2, 3)$, $r = \sqrt{4 + 9} = \sqrt{13}$, $\tan\theta = y/x = -\frac{3}{2}$ and (x, y) is in the second quadrant, so $\theta = \tan^{-1}\left(-\frac{3}{2}\right) + \pi$. The polar coordinates are (i) $\left(\sqrt{13}, \theta\right)$ and (ii) $\left(-\sqrt{13}, \theta + \pi\right)$.

7. The curves $r = 1$ and $r = 2$ represent circles with center O and radii 1 and 2. The region in the plane satisfying $1 \le r \le 2$ consists of both circles and the shaded region between them in the figure.

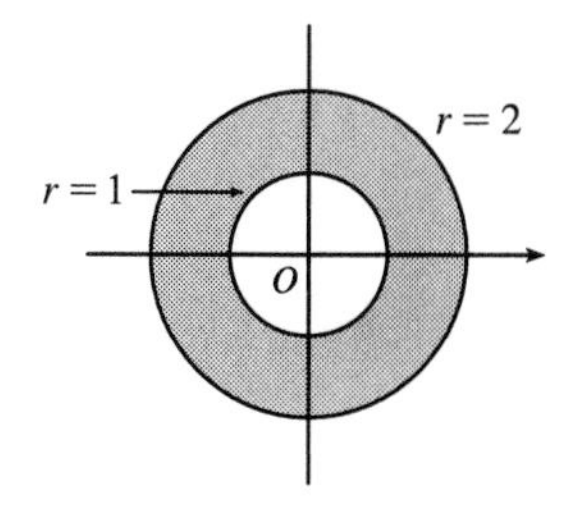

8. $r \ge 0, \quad \pi/3 \le \theta \le 2\pi/3$

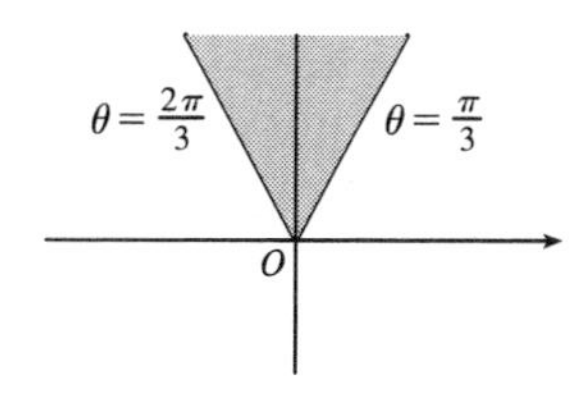

9. The region satisfying $0 \le r < 4$ and $-\pi/2 \le \theta < \pi/6$ does not include the circle $r = 4$ nor the line $\theta = \frac{\pi}{6}$.

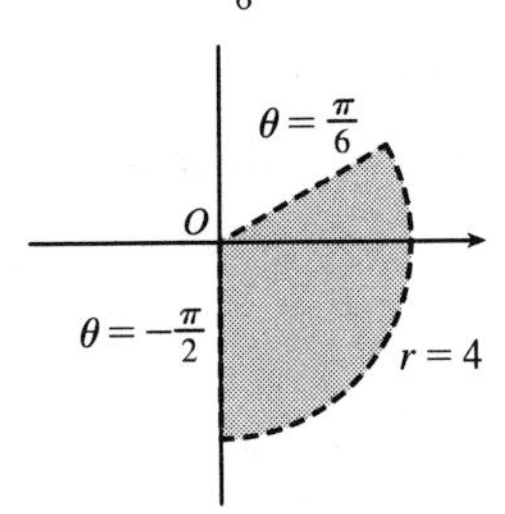

10. $2 < r \le 5$, $3\pi/4 < \theta < 5\pi/4$

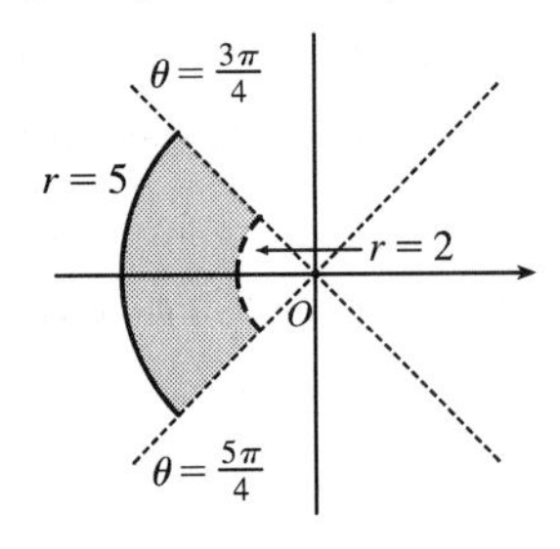

11. $2 < r < 3$, $\frac{5\pi}{3} \le \theta \le \frac{7\pi}{3}$

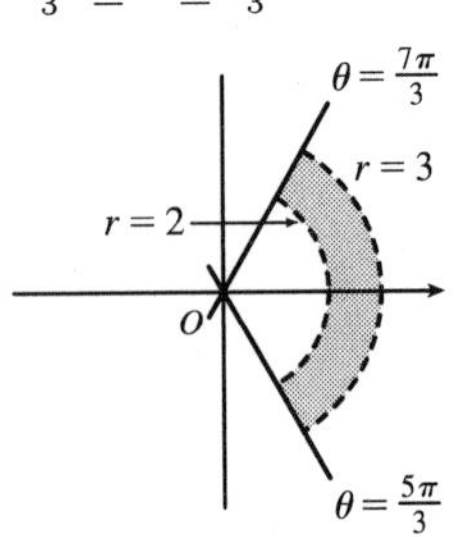

12. $-1 \le r \le 1$, $\frac{\pi}{4} \le \theta \le \frac{3\pi}{4}$

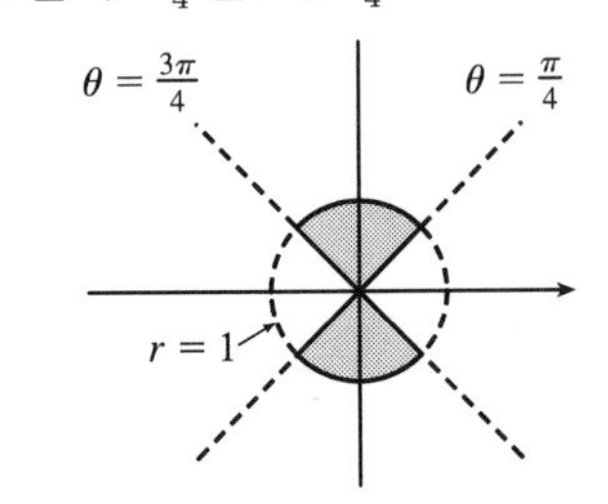

13. $r = 3\sin\theta \;\Rightarrow\; r^2 = 3r\sin\theta \;\Leftrightarrow\; x^2 + y^2 = 3y \;\Leftrightarrow\; x^2 + \left(y - \frac{3}{2}\right)^2 = \left(\frac{3}{2}\right)^2$, a circle of radius $\frac{3}{2}$ centered at $\left(0, \frac{3}{2}\right)$. The first two equations are actually equivalent since $r^2 = 3r\sin\theta \;\Rightarrow\; r(r - 3\sin\theta) = 0 \;\Rightarrow\; r = 0$ or $r = 3\sin\theta$. But $r = 3\sin\theta$ gives the point $r = 0$ (the pole) when $\theta = 0$. Thus, the single equation $r = 3\sin\theta$ is equivalent to the compound condition ($r = 0$ or $r = 3\sin\theta$).

14. $r = 2\sin\theta + 2\cos\theta \;\Rightarrow\; r^2 = 2r\sin\theta + 2r\cos\theta \;\Leftrightarrow\; x^2 + y^2 = 2y + 2x \;\Leftrightarrow\; (x^2 - 2x + 1) + (y^2 - 2y + 1) = 2$ $\Leftrightarrow\; (x-1)^2 + (y-1)^2 = 2$. The first implication is reversible since $r^2 = 2r\sin\theta + 2r\cos\theta \;\Rightarrow\; r = 0$ or $r = 2\sin\theta + 2\cos\theta$, but the curve $r = 2\sin\theta + 2\cos\theta$ passes through the pole $(r = 0)$ when $\theta = -\frac{\pi}{4}$, so $r = 2\sin\theta + 2\cos\theta$ includes the single point of $r = 0$. The curve is a circle of radius $\sqrt{2}$, centered at $(1, 1)$.

15. $r = \csc\theta \;\Leftrightarrow\; r = \dfrac{1}{\sin\theta} \;\Leftrightarrow\; r\sin\theta = 1 \;\Leftrightarrow\; y = 1$, a horizontal line 1 unit above the x-axis.

16. $r = \tan\theta\sec\theta = \dfrac{\sin\theta}{\cos^2\theta} \;\Rightarrow\; r\cos^2\theta = \sin\theta \;\Leftrightarrow\; (r\cos\theta)^2 = r\sin\theta \;\Leftrightarrow\; x^2 = y$, a parabola with vertex at the origin opening upward. The first implication is reversible since $\cos\theta = 0$ would imply $\sin\theta = r\cos^2\theta = 0$, contradicting the fact that $\cos^2\theta + \sin^2\theta = 1$.

17. $x = -y^2 \;\Leftrightarrow\; r\cos\theta = -r^2\sin^2\theta \;\Leftrightarrow\; \cos\theta = -r\sin^2\theta \;\Leftrightarrow\; r = -\dfrac{\cos\theta}{\sin^2\theta} = -\cot\theta\csc\theta$.

18. $x + y = 9 \;\Leftrightarrow\; r\cos\theta + r\sin\theta = 9 \;\Leftrightarrow\; r = 9/(\cos\theta + \sin\theta)$.

19. $x^2 + y^2 = 2cx \;\Leftrightarrow\; r^2 = 2cr\cos\theta \;\Leftrightarrow\; r^2 - 2cr\cos\theta = 0 \;\Leftrightarrow\; r(r - 2c\cos\theta) = 0 \;\Leftrightarrow\; r = 0$ or $r = 2c\cos\theta$. $r = 0$ is included in $r = 2c\cos\theta$ when $\theta = \frac{\pi}{2} + n\pi$, so the curve is represented by the single equation $r = 2c\cos\theta$.

20. $x^2 - y^2 = 1 \;\Leftrightarrow\; (r\cos\theta)^2 - (r\sin\theta)^2 = 1 \;\Leftrightarrow\; r^2(\cos^2\theta - \sin^2\theta) = 1 \;\Leftrightarrow\; r^2\cos 2\theta = 1 \;\Rightarrow\; r^2 = \sec 2\theta$

21. (a) The description leads immediately to the polar equation $\theta = \frac{\pi}{6}$, and the Cartesian equation $\tan\theta = y/x \Rightarrow$ $y = \left(\tan\frac{\pi}{6}\right)x = \frac{1}{\sqrt{3}}x$ is slightly more difficult to derive.

(b) The easier description here is the Cartesian equation $x = 3$.

22. (a) Because its center is not at the origin, it is more easily described by its Cartesian equation, $(x-2)^2 + (y-3)^2 = 5^2$.

(b) This circle is more easily given in polar coordinates: $r = 4$. The Cartesian equation is also simple: $x^2 + y^2 = 16$.

23. $\theta = -\pi/6$

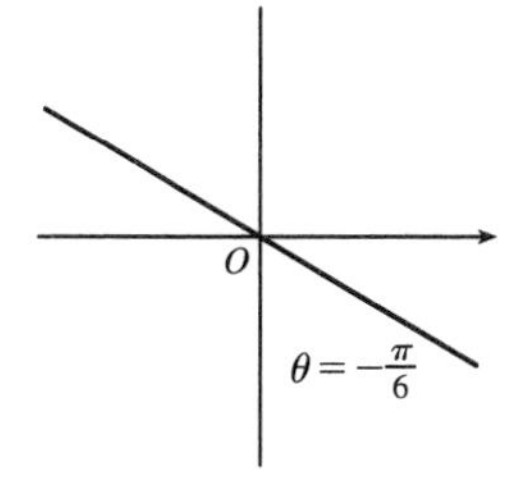

24. $r^2 - 3r + 2 = 0 \Leftrightarrow (r-1)(r-2) = 0 \Leftrightarrow$ $r = 1$ or $r = 2$

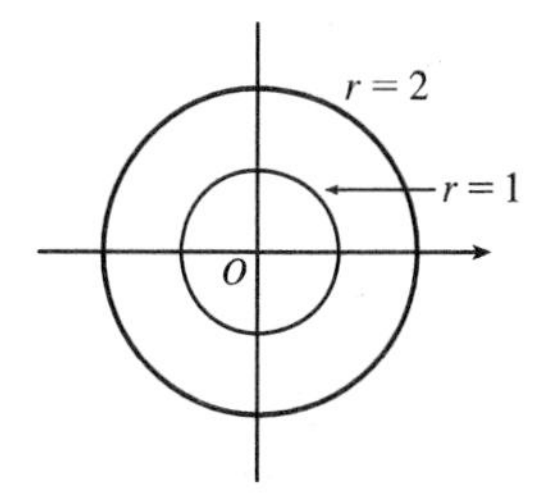

25. $r = \sin\theta \Leftrightarrow r^2 = r\sin\theta \Leftrightarrow$ $x^2 + y^2 = y \Leftrightarrow x^2 + \left(y - \frac{1}{2}\right)^2 = \left(\frac{1}{2}\right)^2$.

The reasoning here is the same as in Exercise 13. This is a circle of radius $\frac{1}{2}$ centered at $\left(0, \frac{1}{2}\right)$.

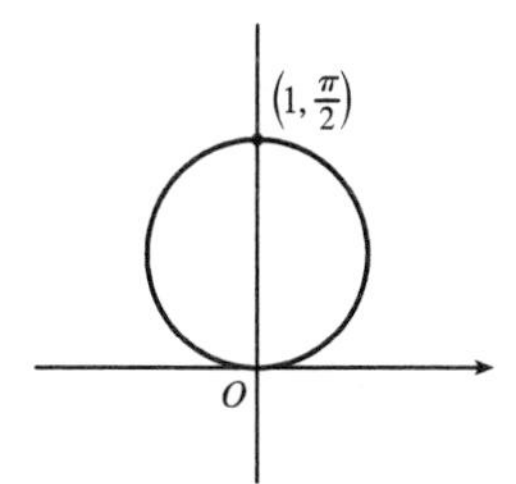

26. $r = -3\cos\theta \Leftrightarrow r^2 = -3r\cos\theta \Leftrightarrow$ $x^2 + y^2 = -3x \Leftrightarrow \left(x + \frac{3}{2}\right)^2 + y^2 = \left(\frac{3}{2}\right)^2$.

This curve is a circle of radius $\frac{3}{2}$ centered at $\left(-\frac{3}{2}, 0\right)$.

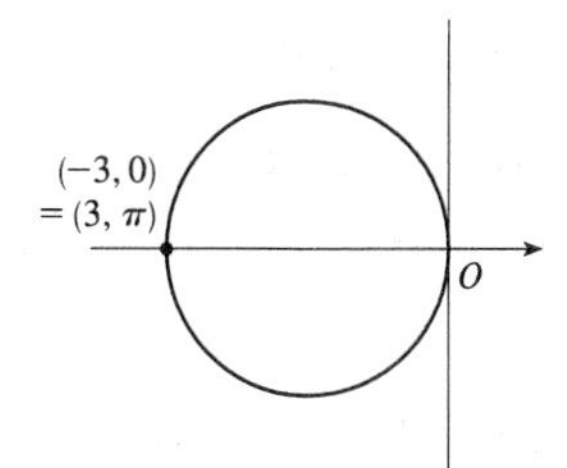

27. $r = 2(1 - \sin\theta)$. This curve is a cardioid.

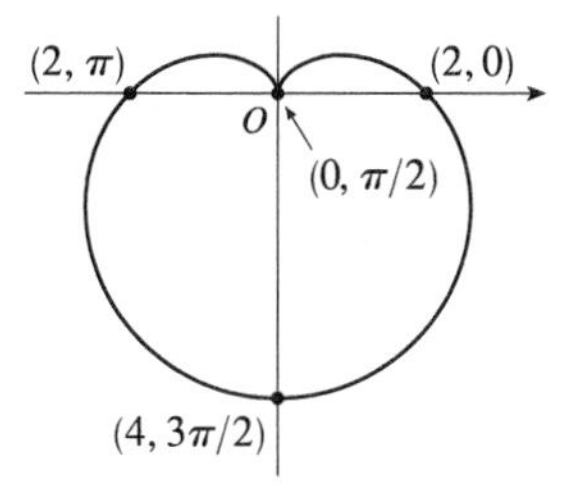

28. $r = 1 - 3\cos\theta$. This is a limaçon.

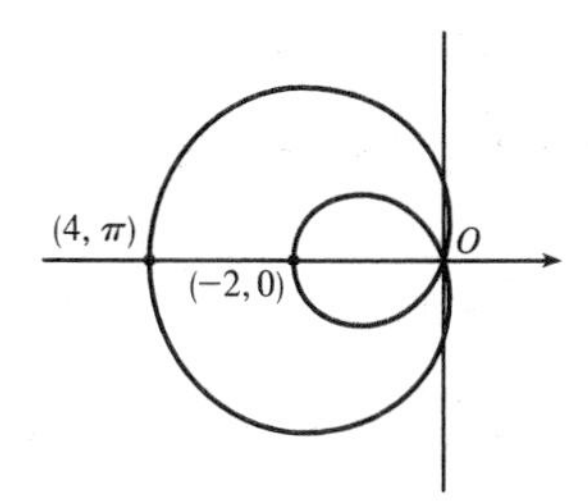

29. $r = \theta, \quad \theta \geq 0$

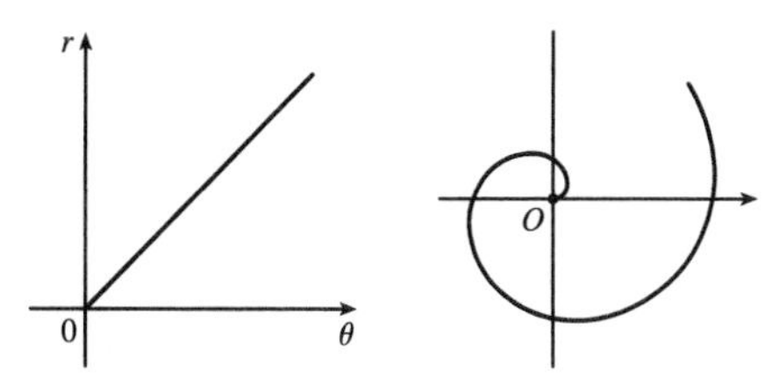

30. $r = \ln\theta, \quad \theta \geq 1$

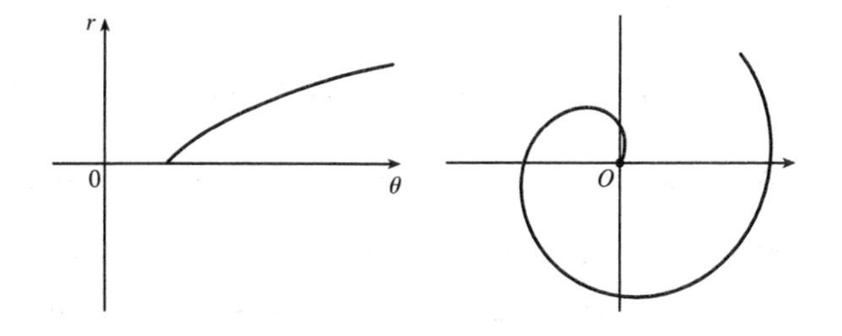

31. $r = \sin 2\theta$

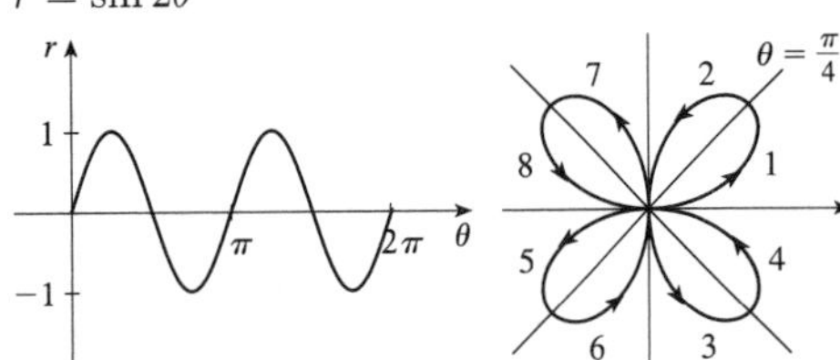

32. $r = 2\cos 3\theta$

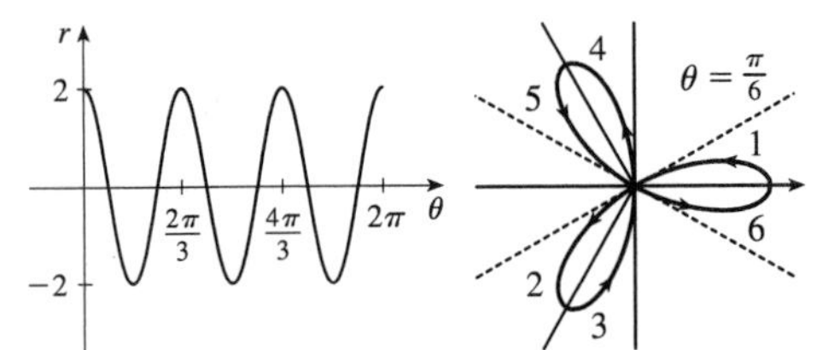

33. $r = 2\cos 4\theta$

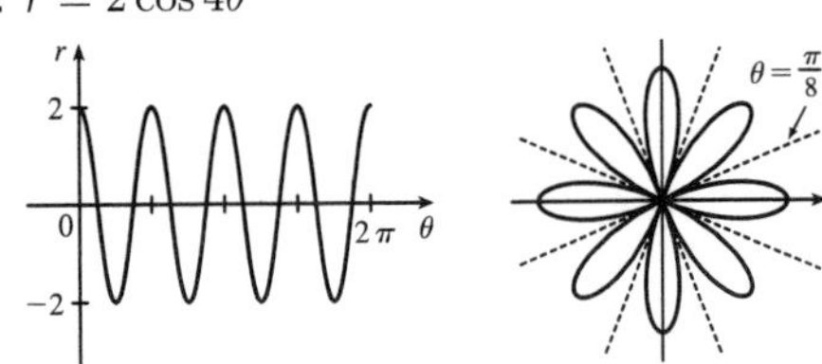

34. $r = \sin 5\theta$

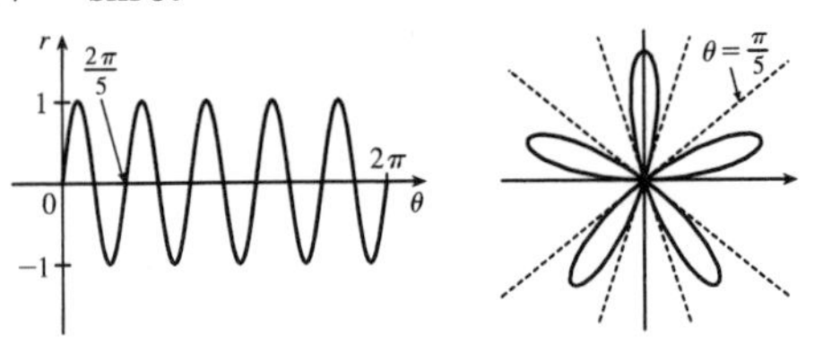

35. $r^2 = 4\cos 2\theta$

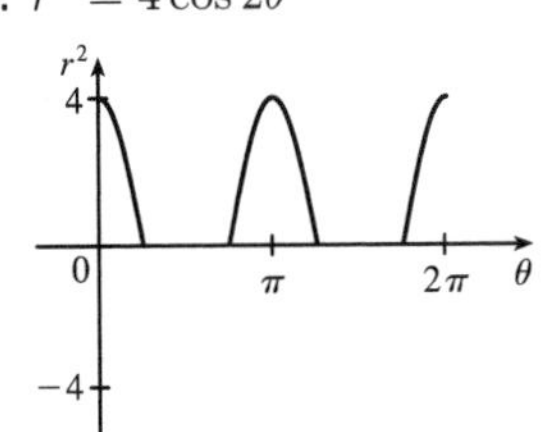

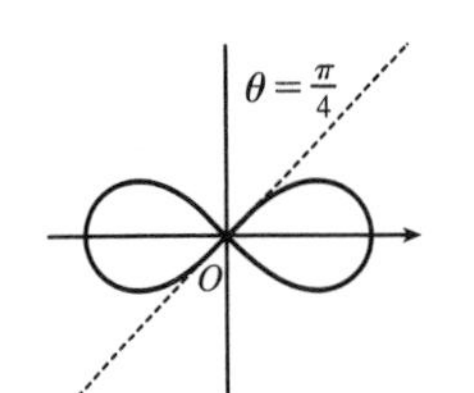

36. $r^2 = \sin 2\theta$

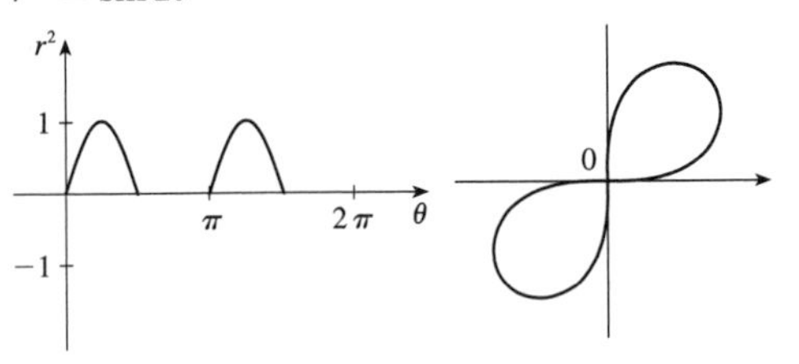

37. $r = 2\cos\left(\frac{3}{2}\theta\right)$

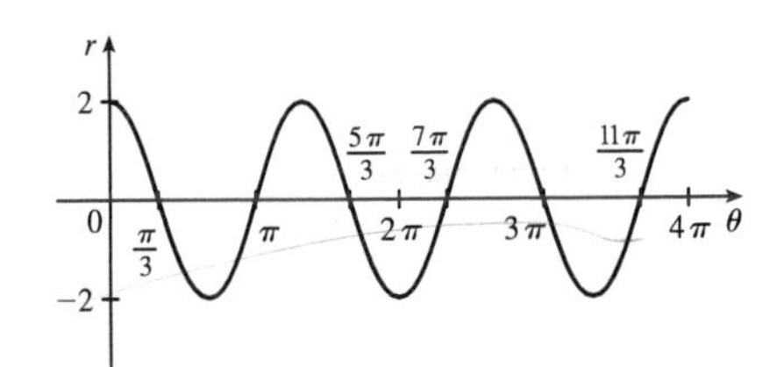

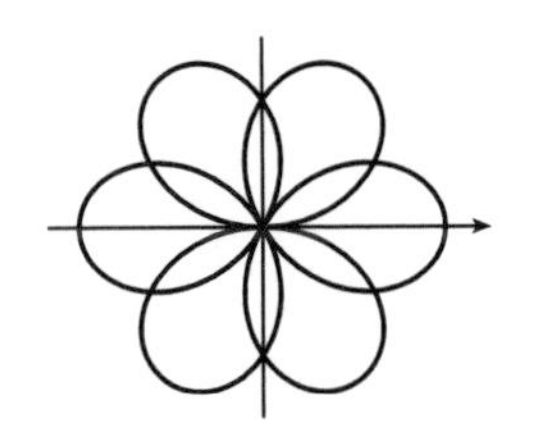

38. $r^2\theta = 1 \quad \Leftrightarrow \quad r = \pm 1/\sqrt{\theta}$ for $\theta > 0$

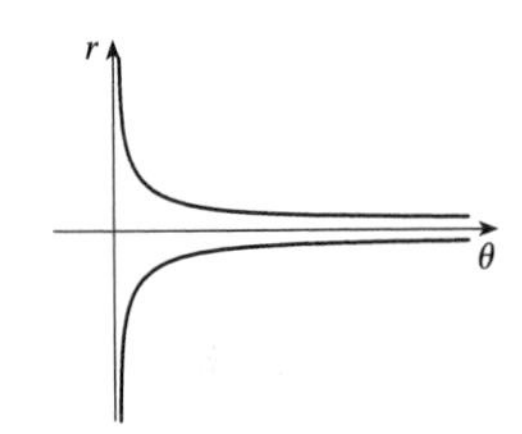

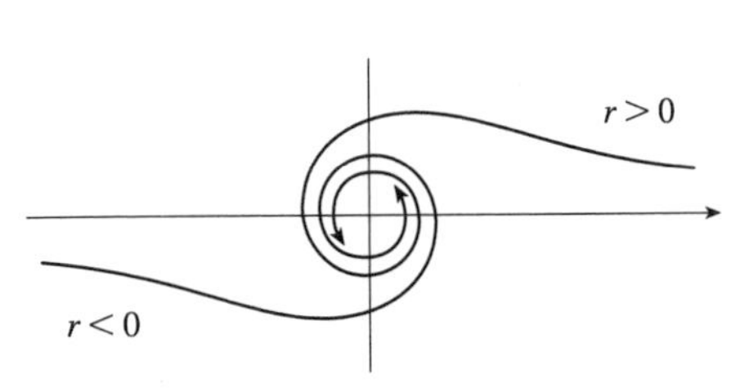

39. $r = 1 + 2\cos 2\theta$

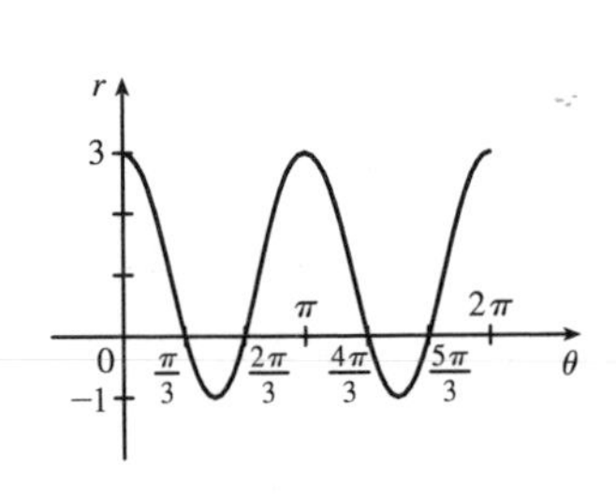

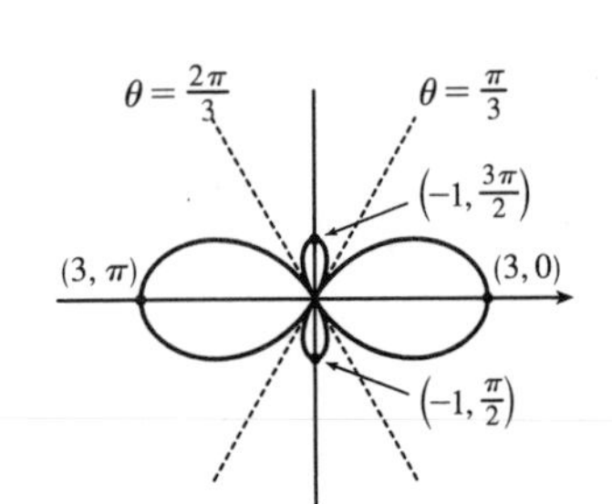

40. $r = 1 + 2\cos(\theta/2)$

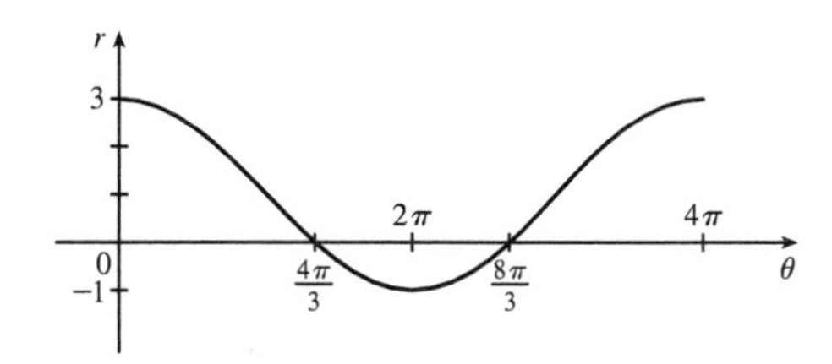

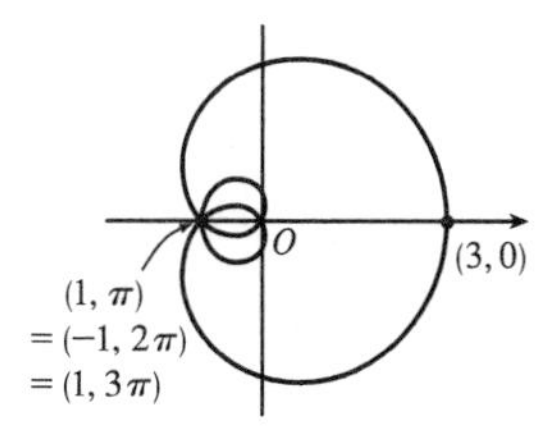

41. For $\theta = 0$, π, and 2π, r has its minimum value of about 0.5. For $\theta = \frac{\pi}{2}$ and $\frac{3\pi}{2}$, r attains its maximum value of 2. We see that the graph has a similar shape for $0 \le \theta \le \pi$ and $\pi \le \theta \le 2\pi$.

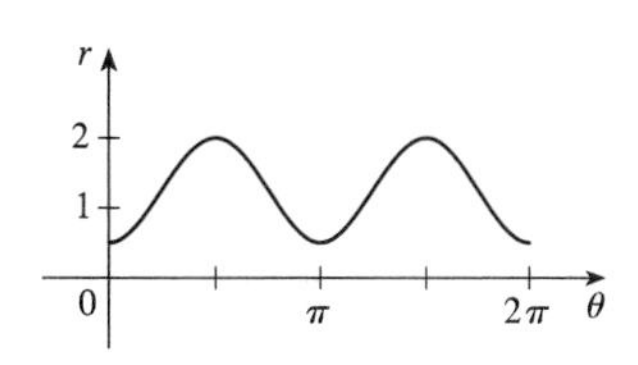

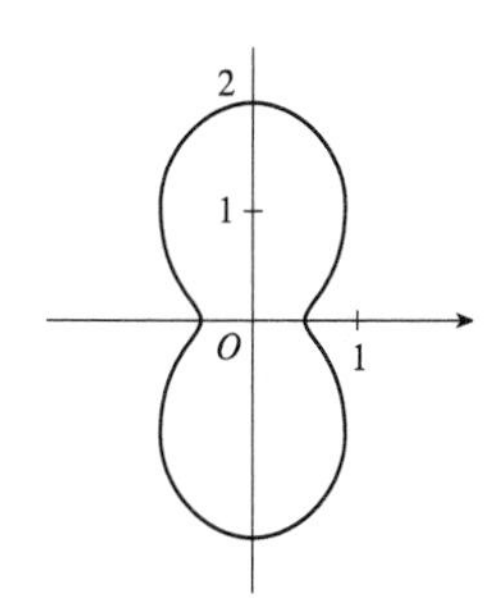

42.

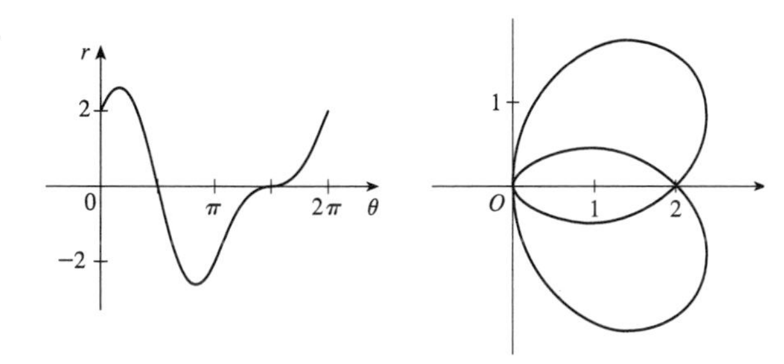

43. $x = r\cos\theta = (4 + 2\sec\theta)\cos\theta = 4\cos\theta + 2$. Now, $r \to \infty \;\Rightarrow$ $(4 + 2\sec\theta) \to \infty \;\Rightarrow\; \theta \to \left(\frac{\pi}{2}\right)^-$ or $\theta \to \left(\frac{3\pi}{2}\right)^+$ (since we need only consider $0 \le \theta < 2\pi$), so $\lim\limits_{r\to\infty} x = \lim\limits_{\theta\to\pi/2^-} (4\cos\theta + 2) = 2$. Also, $r \to -\infty \;\Rightarrow\; (4 + 2\sec\theta) \to -\infty \;\Rightarrow\; \theta \to \left(\frac{\pi}{2}\right)^+$ or $\theta \to \left(\frac{3\pi}{2}\right)^-$, so $\lim\limits_{r\to-\infty} x = \lim\limits_{\theta\to\pi/2^+} (4\cos\theta + 2) = 2$. Therefore, $\lim\limits_{r\to\pm\infty} x = 2 \;\Rightarrow$ $x = 2$ is a vertical asymptote.

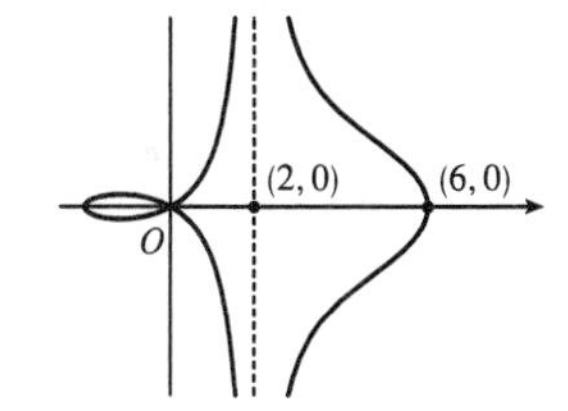

44. The equation is $(x^2 + y^2)^3 = 4x^2y^2$, but using polar coordinates we know that $x^2 + y^2 = r^2$ and $x = r\cos\theta$ and $y = r\sin\theta$. Substituting into the given equation: $r^6 = 4r^2\cos^2\theta\, r^2\sin^2\theta \;\Rightarrow\; r^2 = 4\cos^2\theta\,\sin^2\theta \;\Rightarrow$ $r = \pm 2\cos\theta\sin\theta = \pm\sin 2\theta$. $r = \pm\sin 2\theta$ is sketched at right.

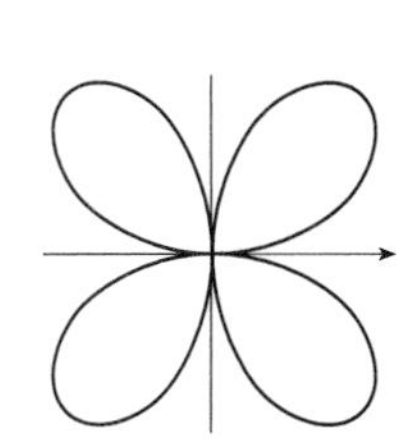

45. To show that $x = 1$ is an asymptote we must prove $\lim\limits_{r\to\pm\infty} x = 1$.

$x = r\cos\theta = (\sin\theta\,\tan\theta)\cos\theta = \sin^2\theta$. Now, $r \to \infty \;\Rightarrow\; \sin\theta\,\tan\theta \to \infty \;\Rightarrow$ $\theta \to \left(\frac{\pi}{2}\right)^-$, so $\lim\limits_{r\to\infty} x = \lim\limits_{\theta\to\pi/2^-} \sin^2\theta = 1$. Also, $r \to -\infty \;\Rightarrow\; \sin\theta\,\tan\theta \to -\infty \;\Rightarrow$

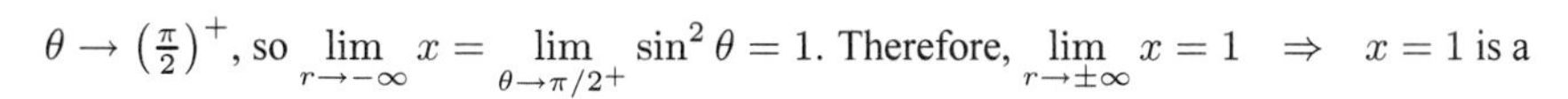

$\theta \to \left(\frac{\pi}{2}\right)^+$, so $\lim\limits_{r\to-\infty} x = \lim\limits_{\theta\to\pi/2^+} \sin^2\theta = 1$. Therefore, $\lim\limits_{r\to\pm\infty} x = 1 \;\Rightarrow\; x = 1$ is a

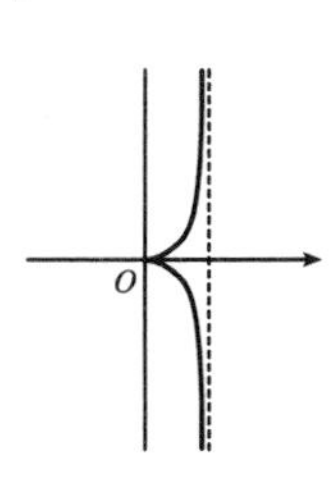

vertical asymptote. Also notice that $x = \sin^2\theta \geq 0$ for all θ, and $x = \sin^2\theta \leq 1$ for all θ. And $x \neq 1$, since the curve is not defined at odd multiples of $\frac{\pi}{2}$. Therefore, the curve lies entirely within the vertical strip $0 \leq x < 1$.

46. (a) $r = \sin(\theta/2)$. This equation must correspond to one of II, III or VI, since these are the only graphs which are bounded. In fact it must be VI, since this is the only graph which is completed after a rotation of exactly 4π.

(b) $r = \sin(\theta/4)$. This equation must correspond to III, since this is the only graph which is completed after a rotation of exactly 8π.

(c) $r = \sec(3\theta)$. This must correspond to IV, since the graph is unbounded at $\theta = \frac{\pi}{6}, \frac{\pi}{2}, \frac{2\pi}{3}$, and so on.

(d) $r = \theta\sin\theta$. This must correspond to V. Note that $r = 0$ whenever θ is a multiple of π. This graph is unbounded, and each time θ moves through an interval of 2π, the same basic shape is repeated (because of the periodic $\sin\theta$ factor) but it gets larger each time (since θ increases each time we go around.)

(e) $r = 1 + 4\cos 5\theta$. This corresponds to II, since it is bounded, has fivefold rotational symmetry, and takes only one takes only one rotation through 2π to be complete.

(f) $r = 1/\sqrt{\theta}$. This corresponds to I, since it is unbounded at $\theta = 0$, and r decreases as θ increases; in fact $r \to 0$ as $\theta \to \infty$.

47. $r = 2\sin\theta \quad\Rightarrow\quad x = r\cos\theta = 2\sin\theta\,\cos\theta = \sin 2\theta,\ y = r\sin\theta = 2\sin^2\theta \quad\Rightarrow$

$$\frac{dy}{dx} = \frac{dy/d\theta}{dx/d\theta} = \frac{2\cdot 2\sin\theta\,\cos\theta}{\cos 2\theta\cdot 2} = \frac{\sin 2\theta}{\cos 2\theta} = \tan 2\theta. \text{ When } \theta = \frac{\pi}{6}, \frac{dy}{dx} = \tan\left(2\cdot\frac{\pi}{6}\right) = \tan\frac{\pi}{3} = \sqrt{3}.$$

48. $r = 2 - \sin\theta \quad\Rightarrow\quad x = r\cos\theta = (2-\sin\theta)\cos\theta,\ y = r\sin\theta = (2-\sin\theta)\sin\theta \quad\Rightarrow$

$$\frac{dy}{dx} = \frac{dy/d\theta}{dx/d\theta} = \frac{(2-\sin\theta)\cos\theta + \sin\theta(-\cos\theta)}{(2-\sin\theta)(-\sin\theta)+\cos\theta(-\cos\theta)} = \frac{2\cos\theta - 2\sin\theta\,\cos\theta}{-2\sin\theta+\sin^2\theta-\cos^2\theta} = \frac{2\cos\theta-\sin 2\theta}{-2\sin\theta-\cos 2\theta}$$

$$\text{When } \theta = \frac{\pi}{3}, \frac{dy}{dx} = \frac{2(1/2)-(\sqrt{3}/2)}{-2(\sqrt{3}/2)-(-1/2)} = \frac{1-\sqrt{3}/2}{-\sqrt{3}+1/2}\cdot\frac{2}{2} = \frac{2-\sqrt{3}}{1-2\sqrt{3}}.$$

49. $r = 1/\theta \quad\Rightarrow\quad x = r\cos\theta = (\cos\theta)/\theta,\ y = r\sin\theta = (\sin\theta)/\theta \quad\Rightarrow$

$$\frac{dy}{dx} = \frac{dy/d\theta}{dx/d\theta} = \frac{\sin\theta(-1/\theta^2)+(1/\theta)\cos\theta}{\cos\theta(-1/\theta^2)-(1/\theta)\sin\theta}\cdot\frac{\theta^2}{\theta^2} = \frac{-\sin\theta+\theta\cos\theta}{-\cos\theta-\theta\sin\theta}$$

$$\text{When } \theta = \pi, \frac{dy}{dx} = \frac{-0+\pi(-1)}{-(-1)-\pi(0)} = \frac{-\pi}{1} = -\pi.$$

50. $r = \sin 3\theta \quad\Rightarrow\quad x = r\cos\theta = \sin 3\theta\cos\theta,\ y = r\sin\theta = \sin 3\theta\sin\theta \quad\Rightarrow$

$$\frac{dy}{dx} = \frac{dy/d\theta}{dx/d\theta} = \frac{3\cos 3\theta\sin\theta + \sin 3\theta\cos\theta}{3\cos 3\theta\cos\theta - \sin 3\theta\sin\theta}$$

$$\text{When } \theta = \frac{\pi}{6}, \frac{dy}{dx} = \frac{3(0)(1/2)+1(\sqrt{3}/2)}{3(0)(\sqrt{3}/2)-1(1/2)} = \frac{\sqrt{3}/2}{-1/2} = -\sqrt{3}.$$

51. $r = 3\cos\theta \quad\Rightarrow\quad x = r\cos\theta = 3\cos\theta\,\cos\theta,\ y = r\sin\theta = 3\cos\theta\,\sin\theta \quad\Rightarrow$

$dy/d\theta = -3\sin^2\theta + 3\cos^2\theta = 3\cos 2\theta = 0 \quad\Rightarrow\quad 2\theta = \frac{\pi}{2}$ or $\frac{3\pi}{2} \quad\Leftrightarrow\quad \theta = \frac{\pi}{4}$ or $\frac{3\pi}{4}$. So the tangent is horizontal at $\left(\frac{3}{\sqrt{2}}, \frac{\pi}{4}\right)$ and $\left(-\frac{3}{\sqrt{2}}, \frac{3\pi}{4}\right)$ $\left[\text{same as } \left(\frac{3}{\sqrt{2}}, -\frac{\pi}{4}\right)\right]$. $dx/d\theta = -6\sin\theta\cos\theta = -3\sin 2\theta = 0 \;\Rightarrow\; 2\theta = 0$ or $\pi \;\Leftrightarrow$ $\theta = 0$ or $\frac{\pi}{2}$. So the tangent is vertical at $(3, 0)$ and $\left(0, \frac{\pi}{2}\right)$.

52. $dy/d\theta = e^\theta \sin\theta + e^\theta \cos\theta = e^\theta(\sin\theta + \cos\theta) = 0 \;\Rightarrow\; \sin\theta = -\cos\theta \;\Rightarrow\; \tan\theta = -1 \;\Rightarrow$

$\theta = -\frac{1}{4}\pi + n\pi$ [n any integer] $\Rightarrow$ horizontal tangents at $\left(e^{\pi(n-1/4)}, \pi\left(n - \frac{1}{4}\right)\right)$.

$dx/d\theta = e^\theta \cos\theta - e^\theta \sin\theta = e^\theta(\cos\theta - \sin\theta) = 0 \;\Rightarrow\; \sin\theta = \cos\theta \;\Rightarrow\; \tan\theta = 1 \;\Rightarrow\; \theta = \frac{1}{4}\pi + n\pi$

[n any integer] $\Rightarrow$ vertical tangents at $\left(e^{\pi(n+1/4)}, \pi\left(n + \frac{1}{4}\right)\right)$.

53. $r = 1 + \cos\theta \;\Rightarrow\; x = r\cos\theta = \cos\theta(1 + \cos\theta),\; y = r\sin\theta = \sin\theta(1 + \cos\theta) \;\Rightarrow$

$dy/d\theta = (1 + \cos\theta)\cos\theta - \sin^2\theta = 2\cos^2\theta + \cos\theta - 1 = (2\cos\theta - 1)(\cos\theta + 1) = 0 \;\Rightarrow$

$\cos\theta = \frac{1}{2}$ or $-1 \;\Rightarrow\; \theta = \frac{\pi}{3}, \pi$, or $\frac{5\pi}{3} \;\Rightarrow$ horizontal tangent at $\left(\frac{3}{2}, \frac{\pi}{3}\right)$, $(0, \pi)$ [the pole], and $\left(\frac{3}{2}, \frac{5\pi}{3}\right)$.

$dx/d\theta = -(1 + \cos\theta)\sin\theta - \cos\theta\sin\theta = -\sin\theta(1 + 2\cos\theta) = 0 \;\Rightarrow\; \sin\theta = 0$ or $\cos\theta = -\frac{1}{2} \;\Rightarrow$

$\theta = 0, \pi, \frac{2\pi}{3}$, or $\frac{4\pi}{3} \;\Rightarrow$ vertical tangent at $(2, 0)$, $\left(\frac{1}{2}, \frac{2\pi}{3}\right)$, and $\left(\frac{1}{2}, \frac{4\pi}{3}\right)$. Note that the tangent is horizontal, not vertical when $\theta = \pi$, since $\displaystyle\lim_{\theta\to\pi} \frac{dy/d\theta}{dx/d\theta} = 0$.

54. By differentiating implicitly, $r^2 = \sin 2\theta \;\Rightarrow\; 2r\,(dr/d\theta) = 2\cos 2\theta \;\Rightarrow$ $dr/d\theta = (1/r)\cos 2\theta$, so

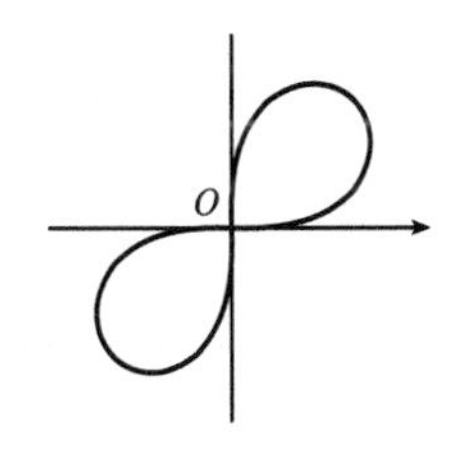

$$\frac{dy}{d\theta} = \frac{1}{r}\cos 2\theta\,\sin\theta + r\cos\theta = \frac{1}{r}\left(\cos 2\theta\,\sin\theta + r^2\cos\theta\right)$$
$$= \frac{1}{r}(\cos 2\theta\,\sin\theta + \sin 2\theta\,\cos\theta) = \frac{1}{r}\sin 3\theta$$

This is 0 when $\sin 3\theta = 0 \;\Rightarrow\; \theta = 0, \frac{\pi}{3}$ or $\frac{4\pi}{3}$ (restricting θ to the domain of the lemniscate), so there are horizontal tangents at $\left(\sqrt[4]{\frac{3}{4}}, \frac{\pi}{3}\right)$, $\left(\sqrt[4]{\frac{3}{4}}, \frac{4\pi}{3}\right)$ and $(0, 0)$. Similarly, $dx/d\theta = (1/r)\cos 3\theta = 0$ when $\theta = \frac{\pi}{6}$ or $\frac{7\pi}{6}$, so there are vertical tangents at $\left(\sqrt[4]{\frac{3}{4}}, \frac{\pi}{6}\right)$ and $\left(\sqrt[4]{\frac{3}{4}}, \frac{7\pi}{6}\right)$ [and $(0, 0)$].

55. $r = a\sin\theta + b\cos\theta \;\Rightarrow\; r^2 = ar\sin\theta + br\cos\theta \;\Rightarrow\; x^2 + y^2 = ay + bx \;\Rightarrow$

$x^2 - bx + \left(\frac{1}{2}b\right)^2 + y^2 - ay + \left(\frac{1}{2}a\right)^2 = \left(\frac{1}{2}b\right)^2 + \left(\frac{1}{2}a\right)^2 \;\Rightarrow\; \left(x - \frac{1}{2}b\right)^2 + \left(y - \frac{1}{2}a\right)^2 = \frac{1}{4}\left(a^2 + b^2\right)$, and this is a circle with center $\left(\frac{1}{2}b, \frac{1}{2}a\right)$ and radius $\frac{1}{2}\sqrt{a^2 + b^2}$.

56. These curves are circles which intersect at the origin and at $\left(\frac{1}{\sqrt{2}}a, \frac{\pi}{4}\right)$. At the origin, the first circle has a horizontal tangent and the second a vertical one, so the tangents are perpendicular here. For the first circle ($r = a\sin\theta$), $dy/d\theta = a\cos\theta\,\sin\theta + a\sin\theta\,\cos\theta = a\sin 2\theta = a$ at $\theta = \frac{\pi}{4}$ and $dx/d\theta = a\cos^2\theta - a\sin^2\theta = a\cos 2\theta = 0$ at $\theta = \frac{\pi}{4}$, so the tangent here is vertical. Similarly, for the second circle ($r = a\cos\theta$), $dy/d\theta = a\cos 2\theta = 0$ and $dx/d\theta = -a\sin 2\theta = -a$ at $\theta = \frac{\pi}{4}$, so the tangent is horizontal, and again the tangents are perpendicular.

Note for Exercises 57–60: Maple is able to plot polar curves using the `polarplot` command, or using the `coords=polar` option in a regular `plot` command. In Mathematica, use `PolarPlot`. In Derive, change to `Polar` under `Options State`. If your graphing device cannot plot polar equations, you must convert to parametric equations. For example, in Exercise 59, $x = r\cos\theta = [2 - 5\sin(\theta/6)]\cos\theta$, $y = r\sin\theta = [2 - 5\sin(\theta/6)]\sin\theta$.

57. $r = e^{\sin\theta} - 2\cos(4\theta)$.

The parameter interval is $[0, 2\pi]$.

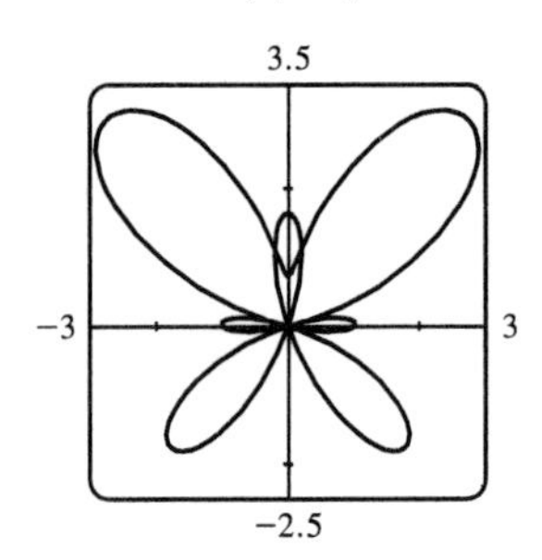

58. $r = \sin^2(4\theta) + \cos(4\theta)$.

The parameter interval is $[0, 2\pi]$.

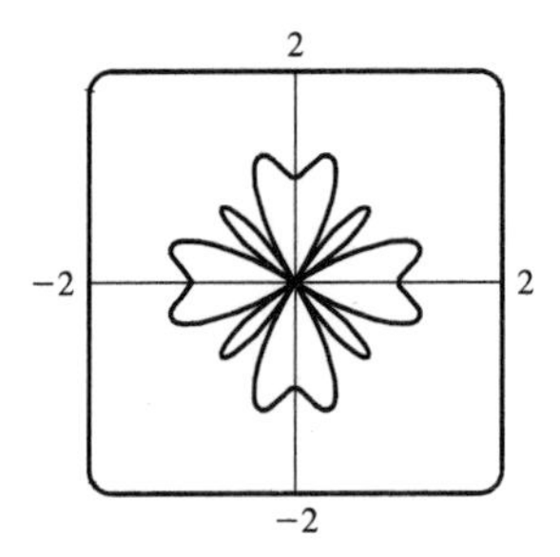

59. $r = 2 - 5\sin(\theta/6)$.

The parameter interval is $[-6\pi, 6\pi]$.

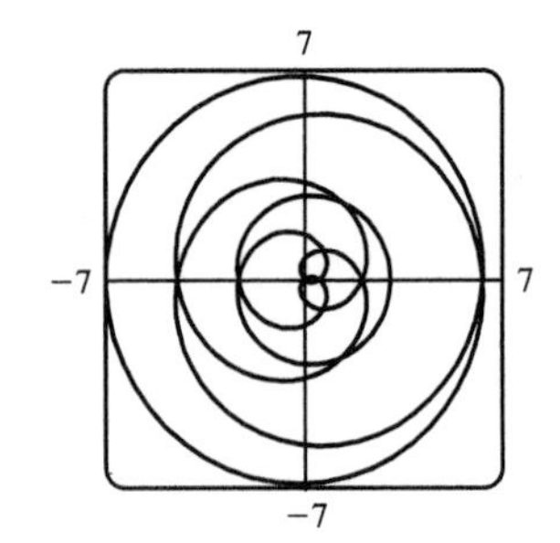

60. $r = \cos(\theta/2) + \cos(\theta/3)$.

The parameter interval is $[-6\pi, 6\pi]$.

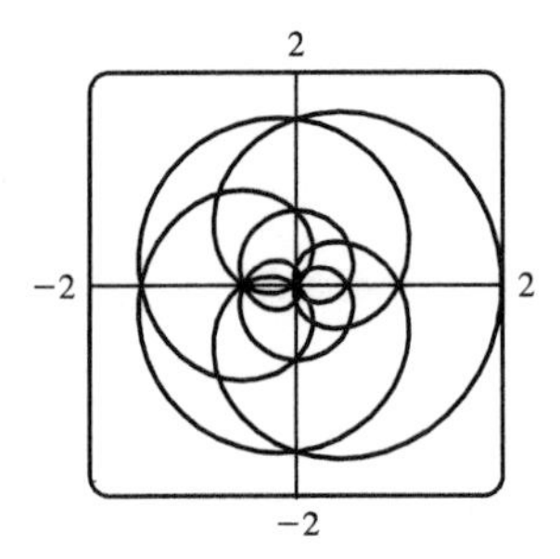

61.

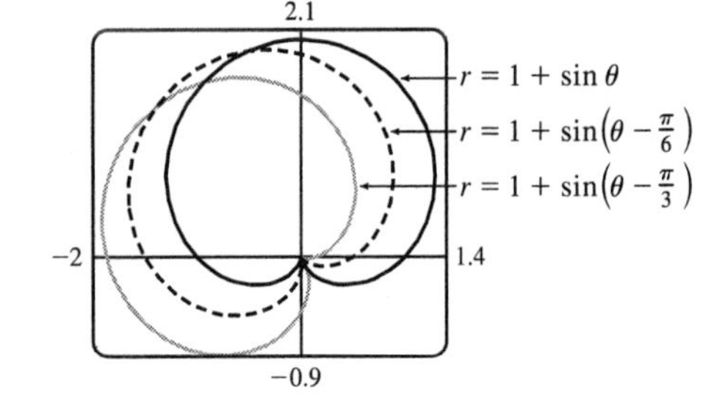

It appears that the graph of $r = 1 + \sin\left(\theta - \frac{\pi}{6}\right)$ is the same shape as the graph of $r = 1 + \sin\theta$, but rotated counterclockwise about the origin by $\frac{\pi}{6}$. Similarly, the graph of $r = 1 + \sin\left(\theta - \frac{\pi}{3}\right)$ is rotated by $\frac{\pi}{3}$. In general, the graph of $r = f(\theta - \alpha)$ is the same shape as that of $r = f(\theta)$, but rotated counterclockwise through α about the origin. That is, for any point (r_0, θ_0) on the curve $r = f(\theta)$, the point $(r_0, \theta_0 + \alpha)$ is on the curve $r = f(\theta - \alpha)$, since $r_0 = f(\theta_0) = f((\theta_0 + \alpha) - \alpha)$.

62.

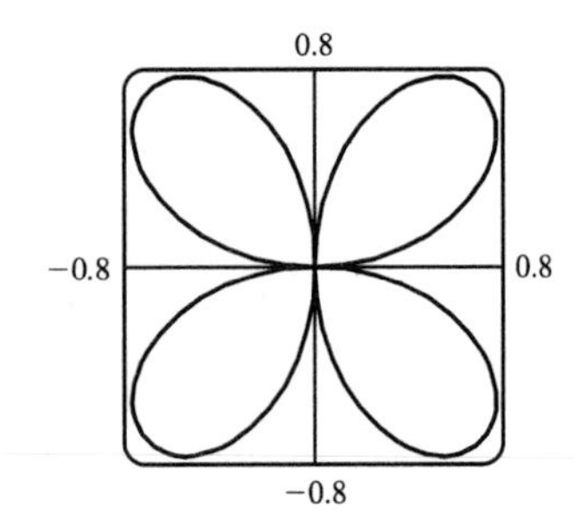

From the graph, the highest points seem to have $y \approx 0.77$. To find the exact value, we solve $dy/d\theta = 0$. $y = r\sin\theta = \sin\theta\,\sin 2\theta \quad\Rightarrow$

$$dy/d\theta = 2\sin\theta\,\cos 2\theta + \cos\theta\,\sin 2\theta$$

$$= 2\sin\theta\,(2\cos^2\theta - 1) + \cos\theta\,(2\sin\theta\,\cos\theta) = 2\sin\theta\,(3\cos^2\theta - 1)$$

In the first quadrant, this is 0 when $\cos\theta = \frac{1}{\sqrt{3}} \quad\Leftrightarrow\quad \sin\theta = \sqrt{\frac{2}{3}} \quad\Leftrightarrow$

$y = 2\sin^2\theta\cos\theta = 2\cdot\frac{2}{3}\cdot\frac{1}{\sqrt{3}} = \frac{4\sqrt{3}}{9} \approx 0.77.$

63. (a) $r = \sin n\theta$. From the graphs, it seems that when n is even, the number of loops in the curve (called a rose) is $2n$, and when n is odd, the number of loops is simply n. This is because in the case of n odd, every point on the graph is traversed twice, due to the fact that

$$r(\theta + \pi) = \sin\left[n(\theta + \pi)\right] = \sin n\theta \, \cos n\pi + \cos n\theta \, \sin n\pi = \begin{cases} \sin n\theta & \text{if } n \text{ is even} \\ -\sin n\theta & \text{if } n \text{ is odd} \end{cases}$$

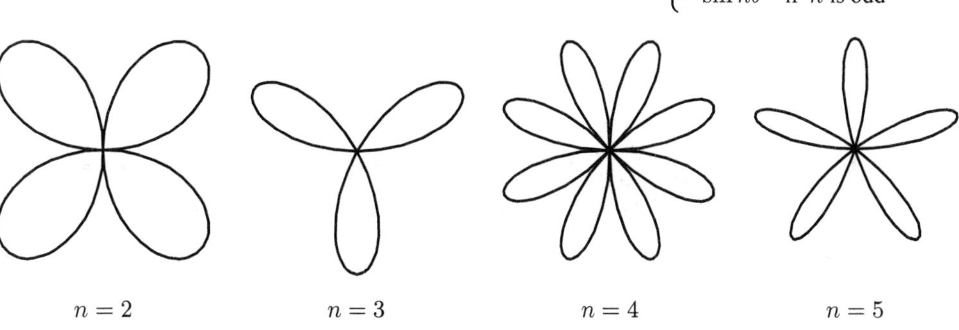

(b) The graph of $r = |\sin n\theta|$ has $2n$ loops whether n is odd or even, since $r(\theta + \pi) = r(\theta)$.

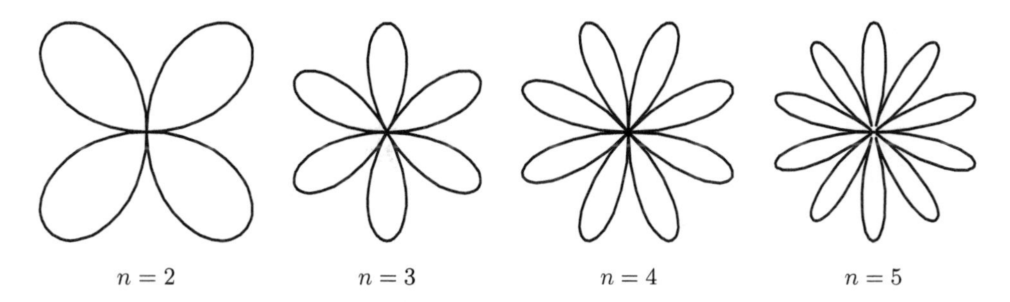

64. $r = 1 + c\sin n\theta$. We vary n while keeping c constant at 2. As n changes, the curves change in the same way as those in Exercise 63: the number of loops increases. Note that if n is even, the smaller loops are outside the larger ones; if n is odd, they are inside.

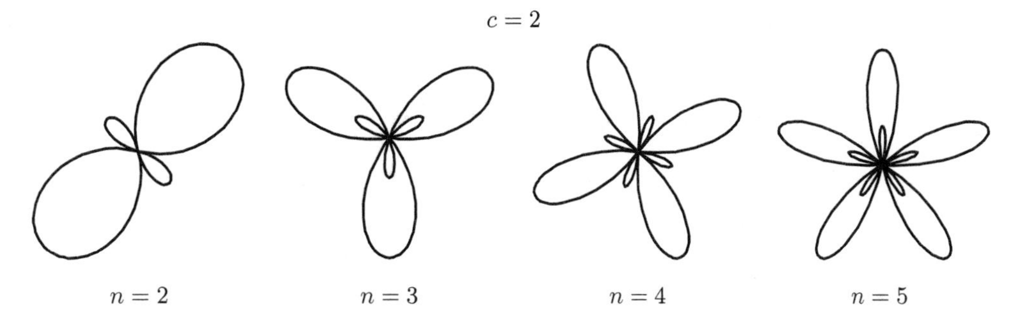

Now we vary c while keeping $n = 3$. As c increases toward 0, the entire graph gets smaller (the graphs below are not to scale) and the smaller loops shrink in relation to the large ones. At $c = -1$, the small loops disappear entirely, and for $-1 < c < 1$, the graph is a simple, closed curve (at $c = 0$ it is a circle). As c continues to increase, the same changes are seen, but in reverse order, since $1 + (-c)\sin n\theta = 1 + c\sin n(\theta + \pi)$, so the graph for $c = c_0$ is the same as that for $c = -c_0$, with a rotation through π. As $c \to \infty$, the smaller loops get relatively closer in size to the large ones. Note that the distance between the outermost points of corresponding inner and outer loops is always 2. Maple's `animate` command (or Mathematica's `Animate`) is very useful for seeing the changes that occur as c varies.

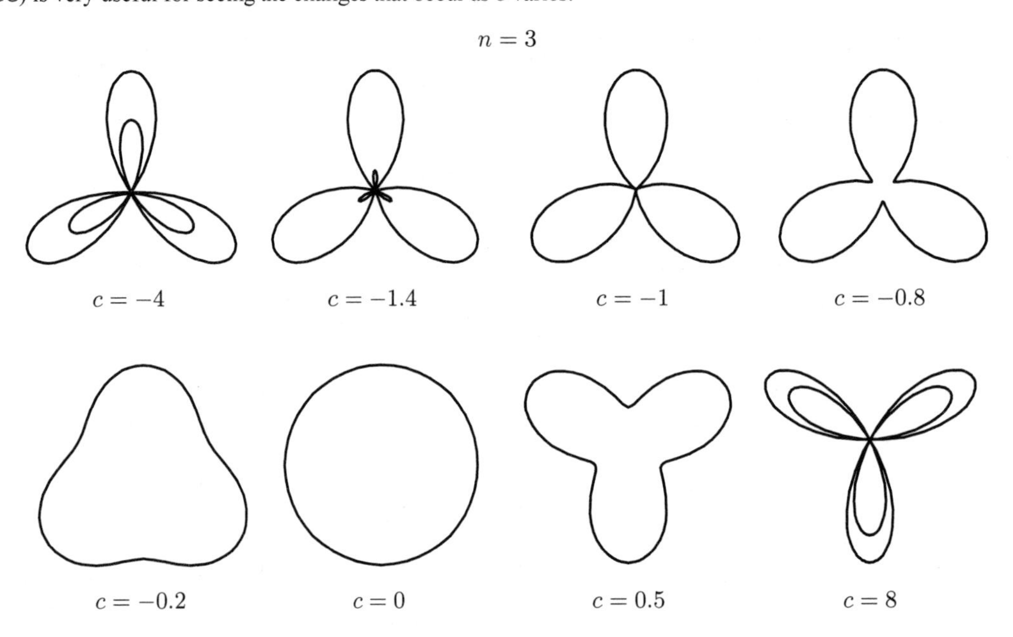

65. $r = \dfrac{1 - a\cos\theta}{1 + a\cos\theta}$. We start with $a = 0$, since in this case the curve is simply the circle $r = 1$.

As a increases, the graph moves to the left, and its right side becomes flattened. As a increases through about 0.4, the right side seems to grow a dimple, which upon closer investigation (with narrower θ-ranges) seems to appear at $a \approx 0.42$ (the actual value is $\sqrt{2} - 1$). As $a \to 1$, this dimple becomes more pronounced, and the curve begins to stretch out horizontally, until at $a = 1$ the denominator vanishes at $\theta = \pi$, and the dimple becomes an actual cusp. For $a > 1$ we must choose our parameter interval carefully, since $r \to \infty$ as $1 + a\cos\theta \to 0 \quad \Leftrightarrow \quad \theta \to \pm\cos^{-1}(-1/a)$. As a increases from 1, the curve splits into two parts. The left part has a loop, which grows larger as a increases, and the right part grows broader vertically, and its left tip develops a dimple when $a \approx 2.42$ (actually, $\sqrt{2} + 1$). As a increases, the dimple grows more and more pronounced. If $a < 0$, we get the same graph as we do for the corresponding positive a-value, but with a rotation through π about the pole, as happened when c was replaced with $-c$ in Exercise 64.

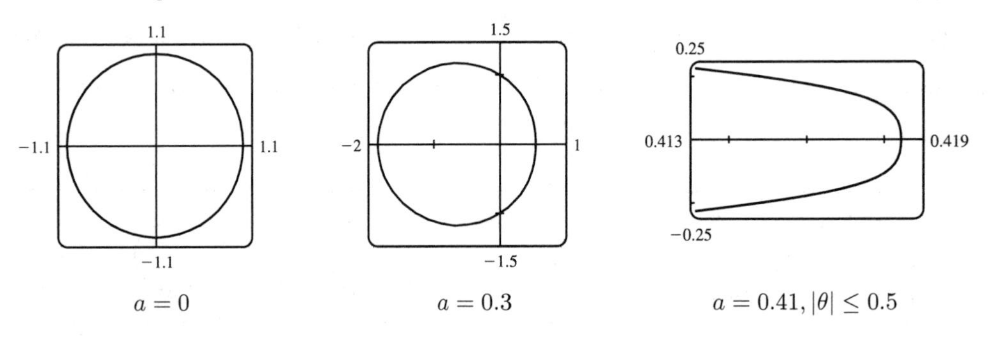

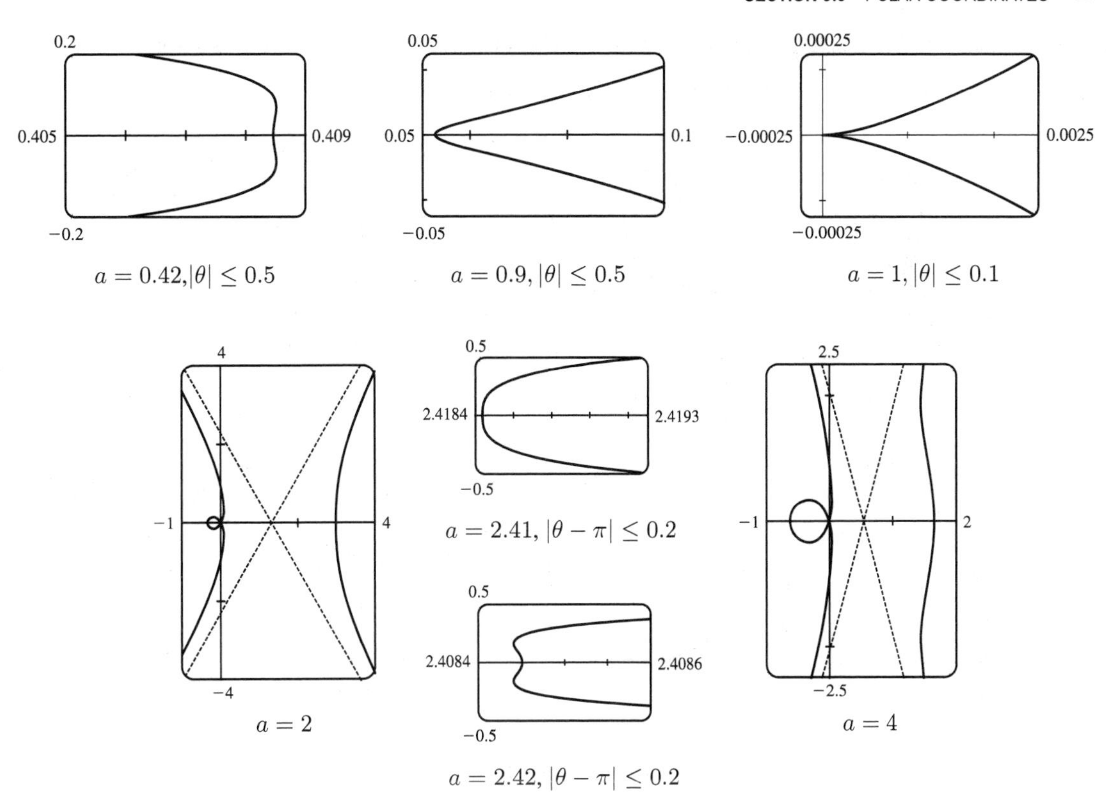

$a = 0.42, |\theta| \leq 0.5$ $a = 0.9, |\theta| \leq 0.5$ $a = 1, |\theta| \leq 0.1$

$a = 2$ $a = 2.41, |\theta - \pi| \leq 0.2$ $a = 2.42, |\theta - \pi| \leq 0.2$ $a = 4$

66. Most graphing devices cannot plot implicit polar equations, so we must first find an explicit expression (or expressions) for r in terms of θ, a, and c. We note that the given equation is a quadratic in r^2, so we use the quadratic formula and find that

$$r^2 = \frac{2c^2 \cos 2\theta \pm \sqrt{4c^4 \cos^2 2\theta - 4(c^4 - a^4)}}{2} = c^2 \cos 2\theta \pm \sqrt{a^4 - c^4 \sin^2 2\theta}$$

so $r = \pm\sqrt{c^2 \cos 2\theta \pm \sqrt{a^4 - c^4 \sin^2 2\theta}}$. So for each graph, we must plot four curves to be sure of plotting all the points which satisfy the given equation. Note that all four functions have period π.

We start with the case $a = c = 1$, and the resulting curve resembles the symbol for infinity. If we let a decrease, the curve splits into two symmetric parts, and as a decreases further, the parts become smaller, further apart, and rounder. If instead we let a increase from 1, the two lobes of the curve join together, and as a increases further they continue to merge, until at $a \approx 1.4$, the graph no longer has dimples, and has an oval shape. As $a \to \infty$, the oval becomes larger and rounder, since the c^2 and c^4 terms lose their significance. Note that the shape of the graph seems to depend only on the ratio c/a, while the size of the graph varies as c and a jointly increase.

0.75 0.75 0.75

−1.5 1.5 −1.5 1.5 −1.5 1.5

−0.75 −0.75 −0.75

$(a, c) = (1, 1)$ $(a, c) = (0.99, 1)$ $(a, c) = (0.9, 1)$

[continued]

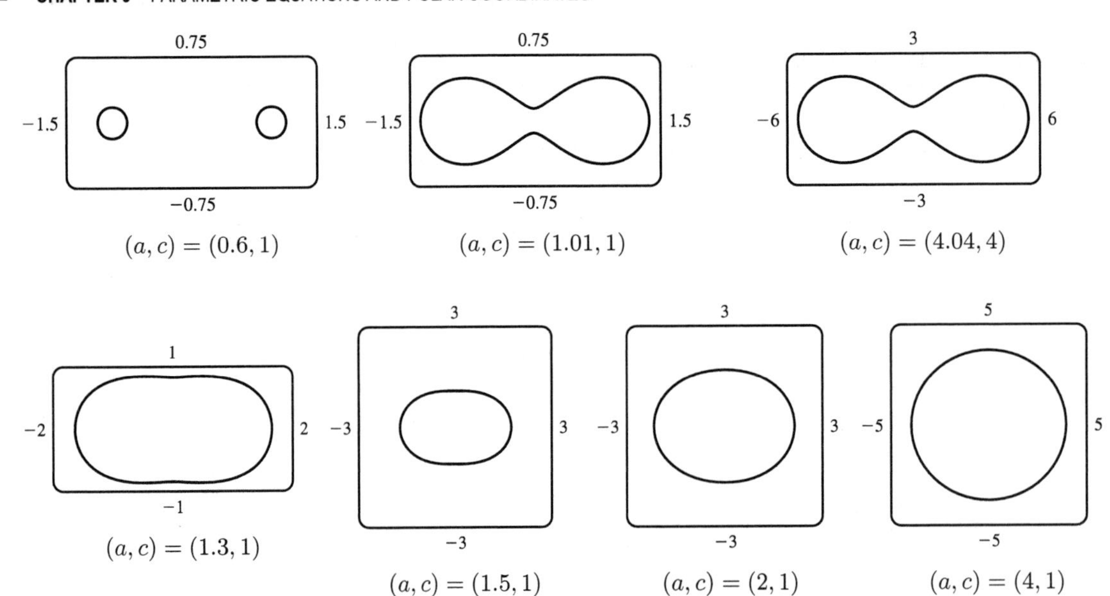

67. $\tan\psi = \tan(\phi - \theta) = \dfrac{\tan\phi - \tan\theta}{1 + \tan\phi\,\tan\theta} = \dfrac{\dfrac{dy}{dx} - \tan\theta}{1 + \dfrac{dy}{dx}\tan\theta} = \dfrac{\dfrac{dy/d\theta}{dx/d\theta} - \tan\theta}{1 + \dfrac{dy/d\theta}{dx/d\theta}\tan\theta} = \dfrac{\dfrac{dy}{d\theta} - \dfrac{dx}{d\theta}\tan\theta}{\dfrac{dx}{d\theta} + \dfrac{dy}{d\theta}\tan\theta}$

$$= \frac{\left(\dfrac{dr}{d\theta}\sin\theta + r\cos\theta\right) - \tan\theta\left(\dfrac{dr}{d\theta}\cos\theta - r\sin\theta\right)}{\left(\dfrac{dr}{d\theta}\cos\theta - r\sin\theta\right) + \tan\theta\left(\dfrac{dr}{d\theta}\sin\theta + r\cos\theta\right)} = \frac{r\cos\theta + r\cdot\dfrac{\sin^2\theta}{\cos\theta}}{\dfrac{dr}{d\theta}\cos\theta + \dfrac{dr}{d\theta}\cdot\dfrac{\sin^2\theta}{\cos\theta}}$$

$$= \frac{r\cos^2\theta + r\sin^2\theta}{\dfrac{dr}{d\theta}\cos^2\theta + \dfrac{dr}{d\theta}\sin^2\theta} = \frac{r}{dr/d\theta}$$

68. (a) $r = e^\theta \;\Rightarrow\; dr/d\theta = e^\theta$, so by Exercise 67, $\tan\psi = r/e^\theta = 1 \;\Rightarrow\; \psi = \arctan 1 = \frac{\pi}{4}$.

(b) The Cartesian equation of the tangent line at $(1, 0)$ is $y = x - 1$,

and that of the tangent line at $(0, e^{\pi/2})$ is $y = e^{\pi/2} - x$.

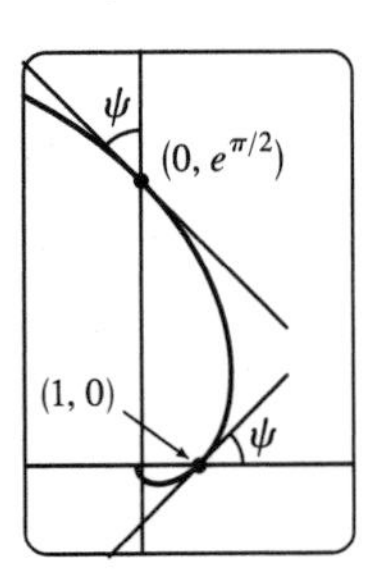

(c) Let a be the tangent of the angle between the tangent and radial lines, that is, $a = \tan\psi$. Then, by Exercise 67,

$a = \dfrac{r}{dr/d\theta} \;\Rightarrow\; \dfrac{dr}{d\theta} = \dfrac{1}{a}r \;\Rightarrow\; r = Ce^{\theta/a}$ (by Theorem 5.5.2).

9.4 Areas and Lengths in Polar Coordinates

1. $r = \sqrt{\theta},\ 0 \le \theta \le \frac{\pi}{4}$. $A = \int_0^{\pi/4} \frac{1}{2} r^2\, d\theta = \int_0^{\pi/4} \frac{1}{2}\left(\sqrt{\theta}\right)^2 d\theta = \int_0^{\pi/4} \frac{1}{2}\theta\, d\theta = \left[\frac{1}{4}\theta^2\right]_0^{\pi/4} = \frac{1}{64}\pi^2$

2. $r = e^{\theta/2}$, $\pi \le \theta \le 2\pi$. $A = \int_\pi^{2\pi} \frac{1}{2}\left(e^{\theta/2}\right)^2 d\theta = \int_\pi^{2\pi} \frac{1}{2} e^\theta\, d\theta = \frac{1}{2}\left[e^\theta\right]_\pi^{2\pi} = \frac{1}{2}\left(e^{2\pi} - e^\pi\right)$

3. $r = \sin\theta,\ \frac{\pi}{3} \le \theta \le \frac{2\pi}{3}$.

$$A = \int_{\pi/3}^{2\pi/3} \tfrac{1}{2}\sin^2\theta\, d\theta = \tfrac{1}{4}\int_{\pi/3}^{2\pi/3}(1 - \cos 2\theta)\, d\theta = \tfrac{1}{4}\left[\theta - \tfrac{1}{2}\sin 2\theta\right]_{\pi/3}^{2\pi/3}$$

$$= \tfrac{1}{4}\left[\tfrac{2\pi}{3} - \tfrac{1}{2}\sin\tfrac{4\pi}{3} - \tfrac{\pi}{3} + \tfrac{1}{2}\sin\tfrac{2\pi}{3}\right] = \tfrac{1}{4}\left[\tfrac{2\pi}{3} - \tfrac{1}{2}\left(-\tfrac{\sqrt{3}}{2}\right) - \tfrac{\pi}{3} + \tfrac{1}{2}\left(\tfrac{\sqrt{3}}{2}\right)\right] = \tfrac{1}{4}\left(\tfrac{\pi}{3} + \tfrac{\sqrt{3}}{2}\right) = \tfrac{\pi}{12} + \tfrac{\sqrt{3}}{8}$$

4. $r = \sqrt{\sin\theta},\ 0 \le \theta \le \pi$. $A = \int_0^\pi \frac{1}{2}\left(\sqrt{\sin\theta}\right)^2 d\theta = \int_0^\pi \frac{1}{2}\sin\theta\, d\theta = \left[-\frac{1}{2}\cos\theta\right]_0^\pi = \frac{1}{2} + \frac{1}{2} = 1$

5. $r = \theta$, $0 \le \theta \le \pi$. $A = \int_0^\pi \frac{1}{2}\theta^2 d\theta = \left[\frac{1}{6}\theta^3\right]_0^\pi = \frac{1}{6}\pi^3$

6. $r = 1 + \sin\theta,\ \frac{\pi}{2} \le \theta \le \pi$.

$$A = \int_{\pi/2}^{\pi} \tfrac{1}{2}(1+\sin\theta)^2\, d\theta = \tfrac{1}{2}\int_{\pi/2}^{\pi}\left(1 + 2\sin\theta + \sin^2\theta\right) d\theta = \tfrac{1}{2}\int_{\pi/2}^{\pi}\left[1 + 2\sin\theta + \tfrac{1}{2}(1 - \cos 2\theta)\right] d\theta$$

$$= \tfrac{1}{2}\left[\theta - 2\cos\theta + \tfrac{1}{2}\theta - \tfrac{1}{4}\sin 2\theta\right]_{\pi/2}^{\pi} = \tfrac{1}{2}\left[\pi + 2 + \tfrac{\pi}{2} - 0 - \left(\tfrac{\pi}{2} - 0 + \tfrac{\pi}{4} - 0\right)\right] = \tfrac{1}{2}\left(\tfrac{3\pi}{4} + 2\right) = \tfrac{3\pi}{8} + 1$$

7. $r = 4 + 3\sin\theta,\ -\frac{\pi}{2} \le \theta \le \frac{\pi}{2}$.

$$A = \int_{-\pi/2}^{\pi/2} \tfrac{1}{2}(4 + 3\sin\theta)^2 d\theta = \tfrac{1}{2}\int_{-\pi/2}^{\pi/2}\left(16 + 24\sin\theta + 9\sin^2\theta\right) d\theta$$

$$= \tfrac{1}{2}\int_{-\pi/2}^{\pi/2}\left(16 + 9\sin^2\theta\right) d\theta \quad \text{[by Theorem 4.5.6(b)]}$$

$$= \tfrac{1}{2}\cdot 2\int_0^{\pi/2}\left[16 + 9\cdot\tfrac{1}{2}(1 - \cos 2\theta)\right] d\theta \quad \text{[by Theorem 4.5.6(a)]}$$

$$= \int_0^{\pi/2}\left(\tfrac{41}{2} - \tfrac{9}{2}\cos 2\theta\right) d\theta = \left[\tfrac{41}{2}\theta - \tfrac{9}{4}\sin 2\theta\right]_0^{\pi/2} = \left(\tfrac{41\pi}{4} - 0\right) - (0 - 0) = \tfrac{41\pi}{4}$$

8. $r = \sin 4\theta,\ 0 \le \theta \le \frac{\pi}{4}$. $A = \int_0^{\pi/4} \frac{1}{2}\sin^2 4\theta\, d\theta = \int_0^{\pi/4} \frac{1}{4}(1 - \cos 8\theta)\, d\theta = \left[\frac{1}{4}\theta - \frac{1}{32}\sin 8\theta\right]_0^{\pi/4} = \frac{\pi}{16}$

9. The curve $r^2 = 4\cos 2\theta$ goes through the pole when $\theta = \pi/4$, so we'll find the area for $0 \le \theta \le \pi/4$ and multiply it by 4.

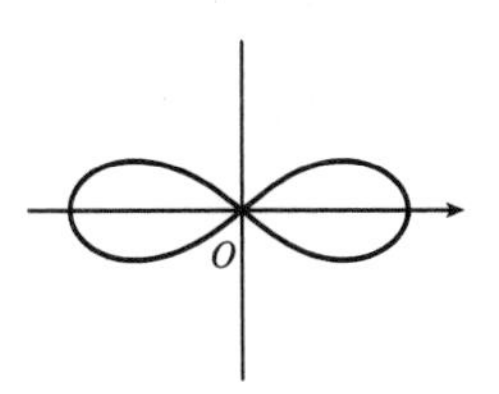

$$A = 4\int_0^{\pi/4} \tfrac{1}{2} r^2\, d\theta = 2\int_0^{\pi/4}(4\cos 2\theta)\, d\theta = 8\int_0^{\pi/4}\cos 2\theta\, d\theta$$

$$= 4\left[\sin 2\theta\right]_0^{\pi/4} = 4(1 - 0) = 4$$

10.

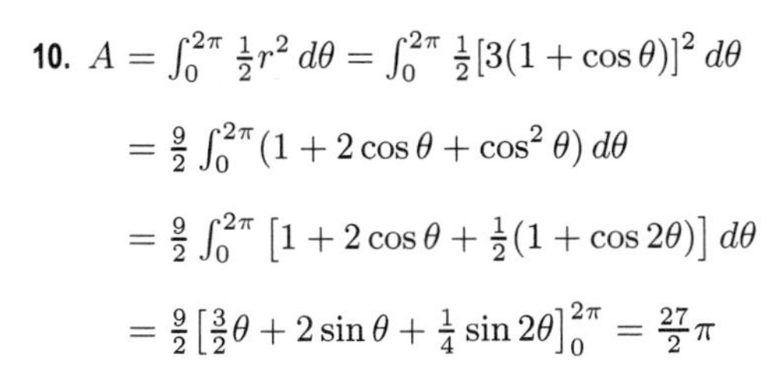

$$A = \int_0^{2\pi} \tfrac{1}{2} r^2\, d\theta = \int_0^{2\pi} \tfrac{1}{2}[3(1 + \cos\theta)]^2\, d\theta$$

$$= \tfrac{9}{2}\int_0^{2\pi}\left(1 + 2\cos\theta + \cos^2\theta\right) d\theta$$

$$= \tfrac{9}{2}\int_0^{2\pi}\left[1 + 2\cos\theta + \tfrac{1}{2}(1 + \cos 2\theta)\right] d\theta$$

$$= \tfrac{9}{2}\left[\tfrac{3}{2}\theta + 2\sin\theta + \tfrac{1}{4}\sin 2\theta\right]_0^{2\pi} = \tfrac{27}{2}\pi$$

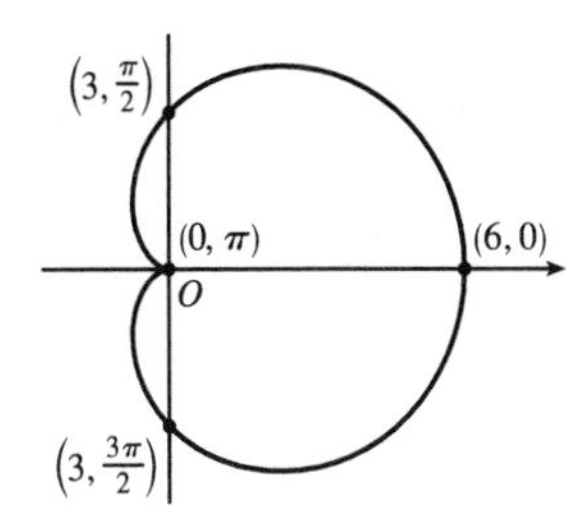

11. One-sixth of the area lies above the polar axis and is bounded by the curve $r = 2\cos 3\theta$ for $\theta = 0$ to $\theta = \pi/6$.

$A = 6\int_0^{\pi/6} \frac{1}{2}(2\cos 3\theta)^2\, d\theta = 12\int_0^{\pi/6} \cos^2 3\theta\, d\theta$

$= \frac{12}{2}\int_0^{\pi/6}(1 + \cos 6\theta)d\theta$

$= 6\left[\theta + \frac{1}{6}\sin 6\theta\right]_0^{\pi/6} = 6\left(\frac{\pi}{6}\right) = \pi$

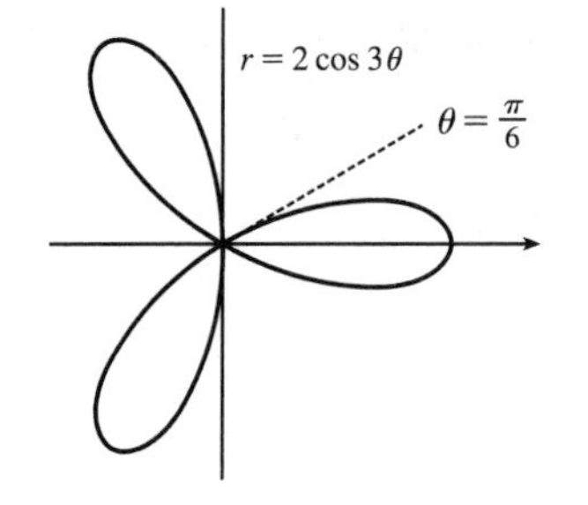

12. $A = \int_0^{2\pi} \frac{1}{2}(2 + \cos 2\theta)^2\, d\theta = \frac{1}{2}\int_0^{2\pi}(4 + 4\cos 2\theta + \cos^2 2\theta)\, d\theta$

$= \frac{1}{2}\int_0^{2\pi}\left(4 + 4\cos 2\theta + \frac{1}{2} + \frac{1}{2}\cos 4\theta\right)\, d\theta$

$= \frac{1}{2}\left[\frac{9}{2}\theta + 2\sin 2\theta + \frac{1}{8}\sin 4\theta\right]_0^{2\pi}$

$= \frac{1}{2}(9\pi) = \frac{9\pi}{2}$

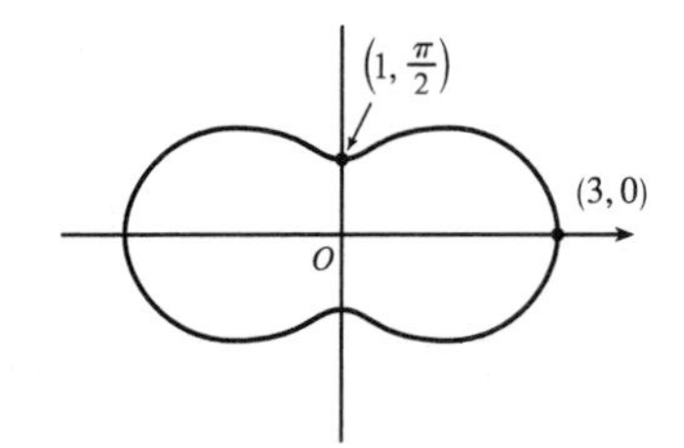

13. $A = \int_0^{2\pi} \frac{1}{2}(1 + 2\sin 6\theta)^2\, d\theta = \frac{1}{2}\int_0^{2\pi}(1 + 4\sin 6\theta + 4\sin^2 6\theta)d\theta$

$= \frac{1}{2}\int_0^{2\pi}\left[1 + 4\sin 6\theta + 4\cdot\frac{1}{2}(1 - \cos 12\theta)\right]\, d\theta$

$= \frac{1}{2}\int_0^{2\pi}(3 + 4\sin 6\theta - 2\cos 12\theta)\, d\theta$

$= \frac{1}{2}\left[3\theta - \frac{2}{3}\cos 6\theta - \frac{1}{6}\sin 12\theta\right]_0^{2\pi}$

$= \frac{1}{2}\left[\left(6\pi - \frac{2}{3} - 0\right) - \left(0 - \frac{2}{3} - 0\right)\right] = 3\pi.$

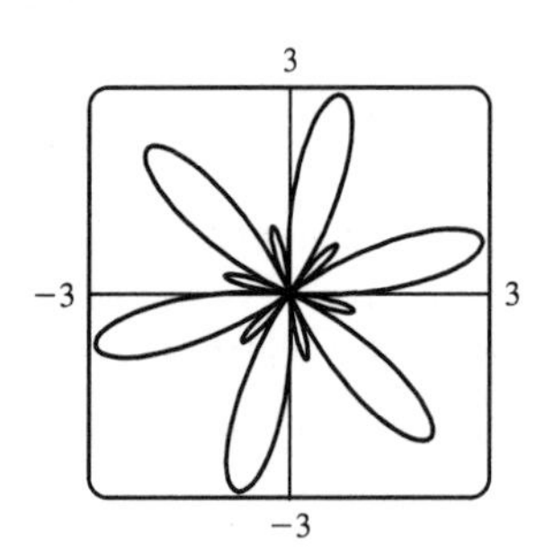

14. $A = \int_0^{\pi} \frac{1}{2}(2\sin\theta + 3\sin 9\theta)^2\, d\theta = 2\int_0^{\pi/2} \frac{1}{2}(2\sin\theta + 3\sin 9\theta)^2\, d\theta$

$= \int_0^{\pi/2}(4\sin^2\theta + 12\sin\theta\,\sin 9\theta + 9\sin^2 9\theta)\, d\theta$

$= \int_0^{\pi/2}\left[2(1 - \cos 2\theta) + 12\cdot\frac{1}{2}(\cos(\theta - 9\theta) - \cos(\theta + 9\theta)) + \frac{9}{2}(1 - \cos 18\theta)\right]\, d\theta$

[integration by parts could be used for $\int \sin\theta\,\sin 9\theta\, d\theta$]

$= \int_0^{\pi/2}(2 - 2\cos 2\theta + 6\cos 8\theta - 6\cos 10\theta + \frac{9}{2} - \frac{9}{2}\cos 18\theta)\, d\theta$

$= \left[\frac{13}{2}\theta - \sin 2\theta + \frac{3}{4}\sin 8\theta - \frac{3}{5}\sin 10\theta - \frac{1}{4}\sin 18\theta\right]_0^{\pi/2} = \frac{13}{4}\pi$

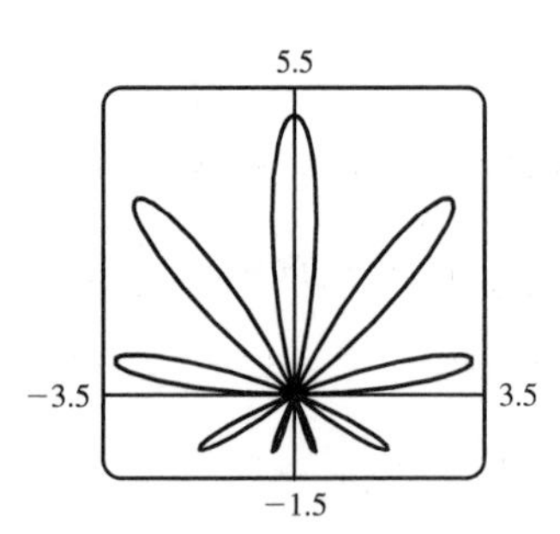

15. The shaded loop is traced out from $\theta = 0$ to $\theta = \pi/2$.

$A = \int_0^{\pi/2} \frac{1}{2}r^2\, d\theta = \frac{1}{2}\int_0^{\pi/2} \sin^2 2\theta\, d\theta$

$= \frac{1}{2}\int_0^{\pi/2} \frac{1}{2}(1 - \cos 4\theta)\, d\theta$

$= \frac{1}{4}\left[\theta - \frac{1}{4}\sin 4\theta\right]_0^{\pi/2} = \frac{1}{4}\left(\frac{\pi}{2}\right) = \frac{\pi}{8}$

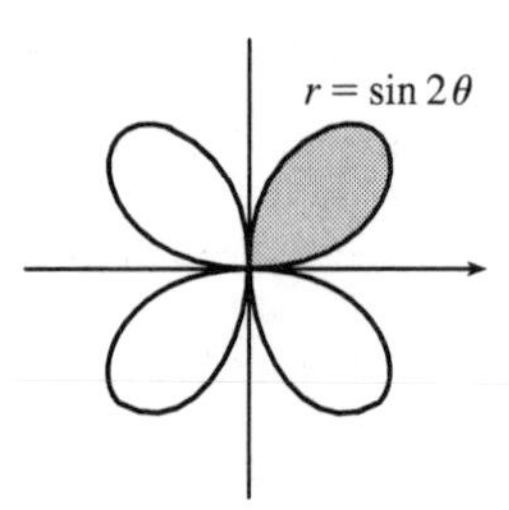

16. $A = \int_0^{\pi/3} \frac{1}{2}(4\sin 3\theta)^2\,d\theta = 8\int_0^{\pi/3} \sin^2 3\theta\,d\theta$

$= 4\int_0^{\pi/3} (1 - \cos 6\theta)\,d\theta$

$= 4\left[\theta - \frac{1}{6}\sin 6\theta\right]_0^{\pi/3} = \frac{4\pi}{3}$

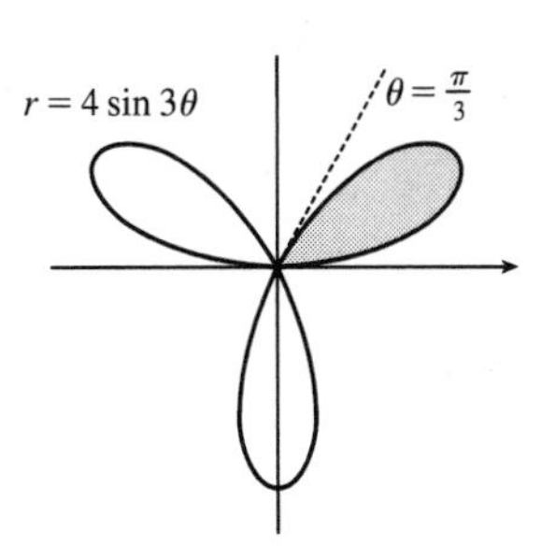

17.

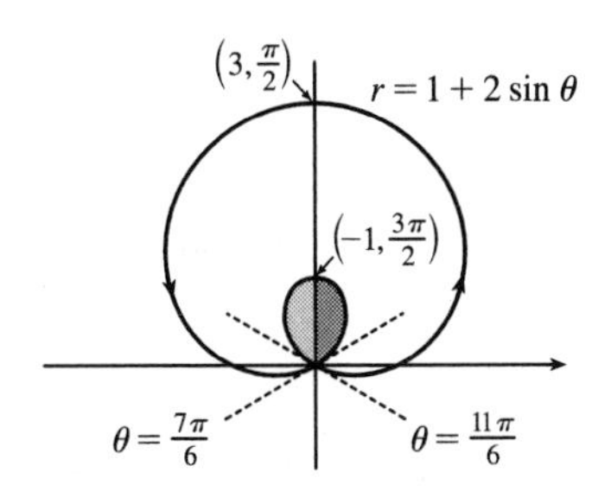

This is a limaçon, with inner loop traced out between $\theta = \frac{7\pi}{6}$ and $\frac{11\pi}{6}$ [found by solving $r = 0$].

$$A = 2\int_{7\pi/6}^{3\pi/2} \frac{1}{2}(1 + 2\sin\theta)^2\,d\theta = \int_{7\pi/6}^{3\pi/2}(1 + 4\sin\theta + 4\sin^2\theta)\,d\theta = \int_{7\pi/6}^{3\pi/2}\left[1 + 4\sin\theta + 4\cdot\tfrac{1}{2}(1 - \cos 2\theta)\right]d\theta$$

$$= \left[\theta - 4\cos\theta + 2\theta - \sin 2\theta\right]_{7\pi/6}^{3\pi/2} = \left(\tfrac{9\pi}{2}\right) - \left(\tfrac{7\pi}{2} + 2\sqrt{3} - \tfrac{\sqrt{3}}{2}\right) = \pi - \tfrac{3\sqrt{3}}{2}$$

18. To determine when the strophoid $r = 2\cos\theta - \sec\theta$ passes through the pole, we solve $r = 0 \;\Rightarrow\; 2\cos\theta - \dfrac{1}{\cos\theta} = 0 \;\Rightarrow$ $2\cos^2\theta - 1 = 0 \;\Rightarrow\; \cos^2\theta = \frac{1}{2} \;\Rightarrow\; \cos\theta = \pm\frac{1}{\sqrt{2}} \;\Rightarrow$ $\theta = \frac{\pi}{4}$ or $\theta = \frac{3\pi}{4}$ for $0 \le \theta \le \pi$ with $\theta \ne \frac{\pi}{2}$.

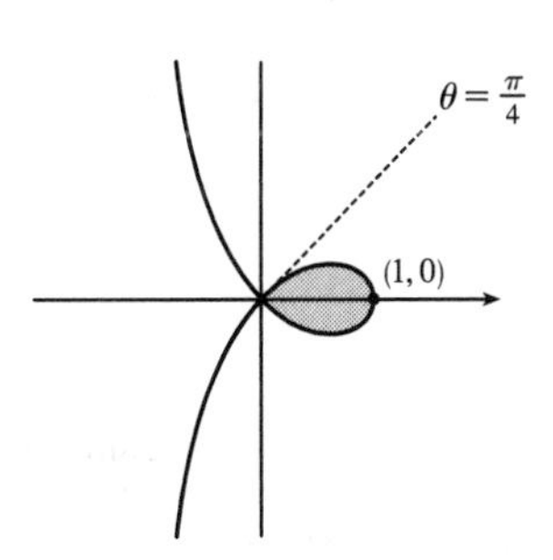

$$A = 2\int_0^{\pi/4} \tfrac{1}{2}(2\cos\theta - \sec\theta)^2\,d\theta = \int_0^{\pi/4}(4\cos^2\theta - 4 + \sec^2\theta)\,d\theta$$

$$= \int_0^{\pi/4}\left[4\cdot\tfrac{1}{2}(1 + \cos 2\theta) - 4 + \sec^2\theta\right]d\theta = \int_0^{\pi/4}(-2 + 2\cos 2\theta + \sec^2\theta)\,d\theta$$

$$= \left[-2\theta + \sin 2\theta + \tan\theta\right]_0^{\pi/4} = \left(-\tfrac{\pi}{2} + 1 + 1\right) - 0 = 2 - \tfrac{\pi}{2}$$

19. $4\sin\theta = 2 \;\Leftrightarrow\; \sin\theta = \frac{1}{2} \;\Leftrightarrow\; \theta = \frac{\pi}{6}$ or $\frac{5\pi}{6}$ (for $0 \le \theta \le 2\pi$). We'll subtract the unshaded area from the shaded area for $\pi/6 \le \theta \le \pi/2$ and double that value.

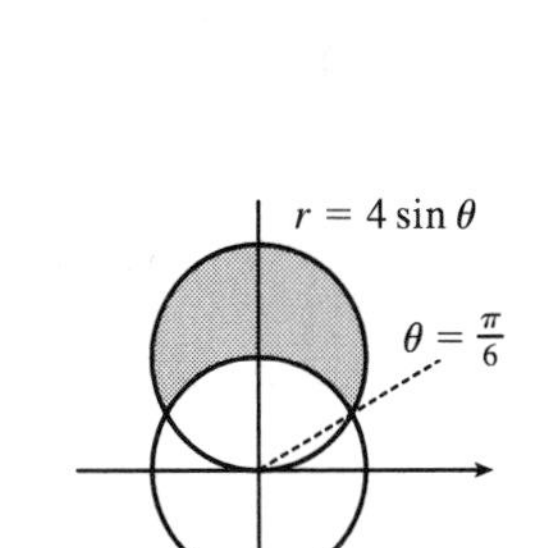

$$A = 2\int_{\pi/6}^{\pi/2} \tfrac{1}{2}(4\sin\theta)^2\,d\theta - 2\int_{\pi/6}^{\pi/2}\tfrac{1}{2}(2)^2\,d\theta = 2\int_{\pi/6}^{\pi/2}\tfrac{1}{2}\left[(4\sin\theta)^2 - 2^2\right]d\theta$$

$$= \int_{\pi/6}^{\pi/2}(16\sin^2\theta - 4)\,d\theta = \int_{\pi/6}^{\pi/2}\left[8(1 - \cos 2\theta) - 4\right]d\theta = \int_{\pi/6}^{\pi/2}(4 - 8\cos 2\theta)\,d\theta$$

$$= \left[4\theta - 4\sin 2\theta\right]_{\pi/6}^{\pi/2} = (2\pi - 0) - \left(\tfrac{2\pi}{3} - 4\cdot\tfrac{\sqrt{3}}{2}\right) = \tfrac{4}{3}\pi + 2\sqrt{3}$$

20. $1 - \sin\theta = 1 \;\Rightarrow\; \sin\theta = 0 \;\Rightarrow\; \theta = 0$ or $\pi \;\Rightarrow$

$$A = \int_\pi^{2\pi}\tfrac{1}{2}[(1 - \sin\theta)^2 - 1]\,d\theta = \tfrac{1}{2}\int_\pi^{2\pi}(\sin^2\theta - 2\sin\theta)\,d\theta$$

$$= \tfrac{1}{4}\int_\pi^{2\pi}(1 - \cos 2\theta - 4\sin\theta)\,d\theta$$

$$= \tfrac{1}{4}\left[\theta - \tfrac{1}{2}\sin 2\theta + 4\cos\theta\right]_\pi^{2\pi} = \tfrac{1}{4}\pi + 2$$

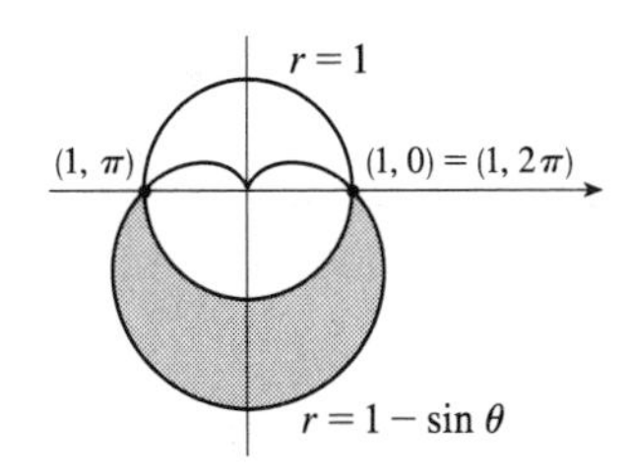

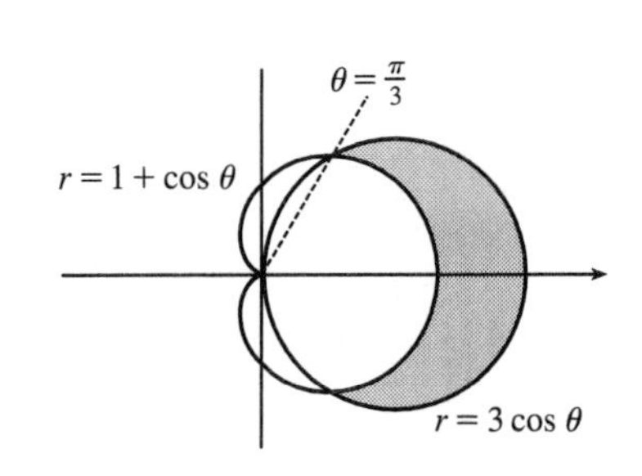

21. $3\cos\theta = 1 + \cos\theta \;\Leftrightarrow\; \cos\theta = \frac{1}{2} \;\Rightarrow\; \theta = \frac{\pi}{3}$ or $-\frac{\pi}{3}$.

$$\begin{aligned} A &= 2\int_0^{\pi/3} \tfrac{1}{2}[(3\cos\theta)^2 - (1+\cos\theta)^2]\,d\theta \\ &= \int_0^{\pi/3}(8\cos^2\theta - 2\cos\theta - 1)\,d\theta \\ &= \int_0^{\pi/3}[4(1+\cos 2\theta) - 2\cos\theta - 1]\,d\theta \\ &= \int_0^{\pi/3}(3 + 4\cos 2\theta - 2\cos\theta)\,d\theta = \big[3\theta + 2\sin 2\theta - 2\sin\theta\big]_0^{\pi/3} \\ &= \pi + \sqrt{3} - \sqrt{3} = \pi \end{aligned}$$

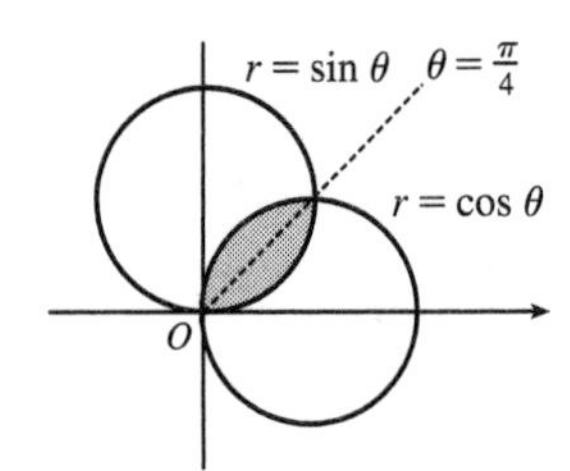

22. To find the shaded area A, we'll find the area A_1 inside the curve $r = 2 + \sin\theta$ and subtract $\pi\left(\frac{3}{2}\right)^2$ since $r = 3\sin\theta$ is a circle with radius $\frac{3}{2}$.

$$\begin{aligned} A_1 &= \int_0^{2\pi} \tfrac{1}{2}(2+\sin\theta)^2\,d\theta = \tfrac{1}{2}\int_0^{2\pi}(4 + 4\sin\theta + \sin^2\theta)\,d\theta \\ &= \tfrac{1}{2}\int_0^{2\pi}\left[4 + 4\sin\theta + \tfrac{1}{2}\cdot(1-\cos 2\theta)\right]d\theta = \tfrac{1}{2}\int_0^{2\pi}\left(\tfrac{9}{2} + 4\sin\theta - \tfrac{1}{2}\cos 2\theta\right)d \\ &= \tfrac{1}{2}\left[\tfrac{9}{2}\theta - 4\cos\theta - \tfrac{1}{4}\sin 2\theta\right]_0^{2\pi} = \tfrac{1}{2}[(9\pi - 4) - (-4)] = \tfrac{9\pi}{2} \end{aligned}$$

So $A = A_1 - \frac{9\pi}{4} = \frac{9\pi}{2} - \frac{9\pi}{4} = \frac{9\pi}{4}$.

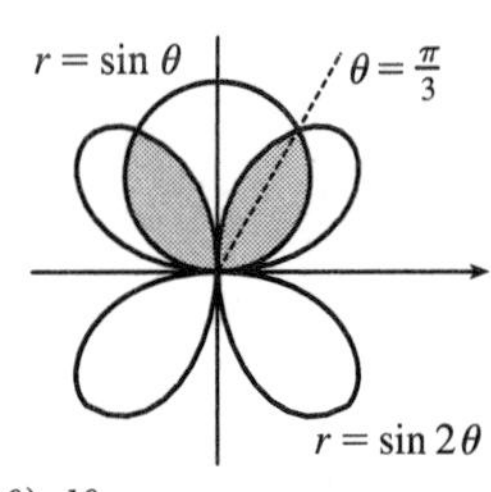

23. $A = 2\int_0^{\pi/4} \frac{1}{2}\sin^2\theta\,d\theta = \int_0^{\pi/4}\frac{1}{2}(1 - \cos 2\theta)\,d\theta$

$$\begin{aligned} &= \tfrac{1}{2}\left[\theta - \tfrac{1}{2}\sin 2\theta\right]_0^{\pi/4} = \tfrac{1}{2}\left[\left(\tfrac{\pi}{4} - \tfrac{1}{2}\cdot 1\right) - (0 - 0)\right] \\ &= \tfrac{1}{8}\pi - \tfrac{1}{4} \end{aligned}$$

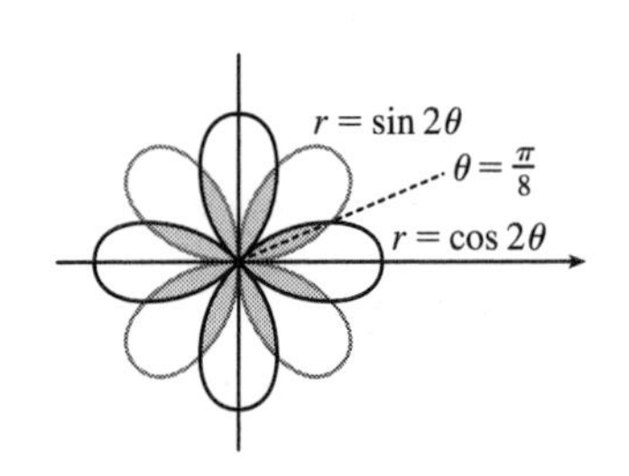

24. $r = \sin 2\theta$ takes on both positive and negative values.

$\sin\theta = \pm\sin 2\theta = \pm 2\sin\theta\cos\theta \;\Rightarrow\; \sin\theta\,(1 \pm 2\cos\theta) = 0$.

From the figure we can see that the intersections occur where $\cos\theta = \pm\frac{1}{2}$, or $\theta = \frac{\pi}{3}$ and $\frac{2\pi}{3}$.

$$\begin{aligned} A &= 2\left[\int_0^{\pi/3}\tfrac{1}{2}\sin^2\theta\,d\theta + \int_{\pi/3}^{\pi/2}\tfrac{1}{2}\sin^2 2\theta\,d\theta\right] = \int_0^{\pi/3}\tfrac{1}{2}(1-\cos 2\theta)\,d\theta + \int_{\pi/3}^{\pi/2}\tfrac{1}{2}(1 - \cos 4\theta)\,d\theta \\ &= \tfrac{1}{2}\left[\theta - \tfrac{1}{2}\sin 2\theta\right]_0^{\pi/3} + \tfrac{1}{2}\left[\theta - \tfrac{1}{4}\sin 4\theta\right]_{\pi/3}^{\pi/2} = \tfrac{4\pi - 3\sqrt{3}}{16} \end{aligned}$$

25. $\sin 2\theta = \cos 2\theta \;\Rightarrow\; \dfrac{\sin 2\theta}{\cos 2\theta} = 1 \;\Rightarrow\; \tan 2\theta = 1 \;\Rightarrow$

$2\theta = \frac{\pi}{4} \;\Rightarrow\; \theta = \frac{\pi}{8} \;\Rightarrow$

$$\begin{aligned} A &= 8\cdot 2\int_0^{\pi/8}\tfrac{1}{2}\sin^2 2\theta\,d\theta = 8\int_0^{\pi/8}\tfrac{1}{2}(1 - \cos 4\theta)\,d \\ &= 4\left[\theta - \tfrac{1}{4}\sin 4\theta\right]_0^{\pi/8} = 4\left(\tfrac{\pi}{8} - \tfrac{1}{4}\cdot 1\right) = \tfrac{1}{2}\pi - 1 \end{aligned}$$

26. $2\sin 2\theta = 1^2 \quad\Rightarrow\quad \sin 2\theta = \frac{1}{2} \quad\Rightarrow\quad 2\theta = \frac{\pi}{6}$ or $\frac{5\pi}{6} \quad\Rightarrow\quad \theta = \frac{\pi}{12}$ or $\frac{5\pi}{12}$.

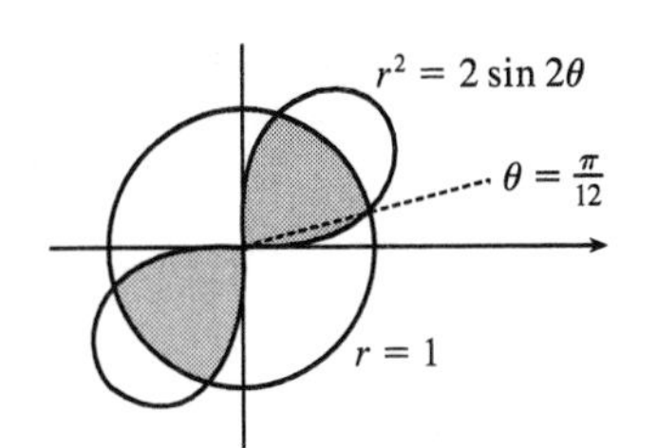

$$A = 4\left[\int_0^{\pi/12} \tfrac{1}{2}\cdot 2\sin 2\theta\, d\theta + \int_{\pi/12}^{\pi/4} \tfrac{1}{2}(1^2)\, d\theta\right]$$

$$= \left[-2\cos 2\theta\right]_0^{\pi/12} + \left[2\theta\right]_{\pi/12}^{\pi/4}$$

$$= -2\left(\tfrac{\sqrt{3}}{2} - 1\right) + 2\left(\tfrac{1}{4}\pi - \tfrac{1}{12}\pi\right) = 2 - \sqrt{3} + \tfrac{\pi}{3}$$

27. The darker shaded region (from $\theta = 0$ to $\theta = 2\pi/3$) represents $\frac{1}{2}$ of the desired area plus $\frac{1}{2}$ of the area of the inner loop. From this area, we'll subtract $\frac{1}{2}$ of the area of the inner loop (the lighter shaded region from $\theta = 2\pi/3$ to $\theta = \pi$), and then double that difference to obtain the desired area.

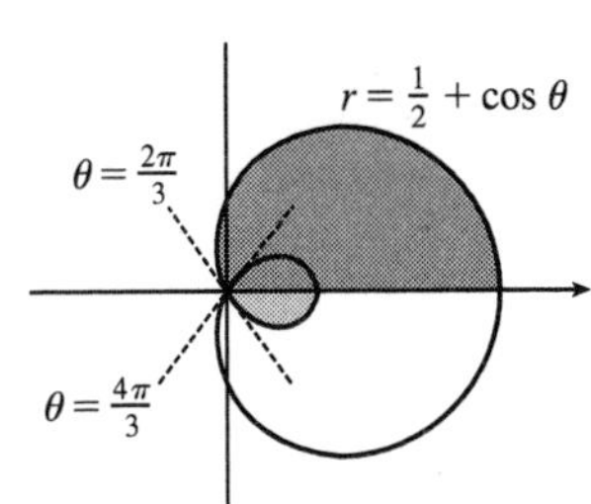

$$A = 2\left[\int_0^{2\pi/3} \tfrac{1}{2}\left(\tfrac{1}{2} + \cos\theta\right)^2 d\theta - \int_{2\pi/3}^{\pi} \tfrac{1}{2}\left(\tfrac{1}{2} + \cos\theta\right)^2 d\theta\right]$$

$$= \int_0^{2\pi/3} \left(\tfrac{1}{4} + \cos\theta + \cos^2\theta\right) d\theta - \int_{2\pi/3}^{\pi} \left(\tfrac{1}{4} + \cos\theta + \cos^2\theta\right) d\theta$$

$$= \int_0^{2\pi/3} \left[\tfrac{1}{4} + \cos\theta + \tfrac{1}{2}(1 + \cos 2\theta)\right] d\theta - \int_{2\pi/3}^{\pi} \left[\tfrac{1}{4} + \cos\theta + \tfrac{1}{2}(1 + \cos 2\theta)\right] d\theta$$

$$= \left[\frac{\theta}{4} + \sin\theta + \frac{\theta}{2} + \frac{\sin 2\theta}{4}\right]_0^{2\pi/3} - \left[\frac{\theta}{4} + \sin\theta + \frac{\theta}{2} + \frac{\sin 2\theta}{4}\right]_{2\pi/3}^{\pi}$$

$$= \left(\tfrac{\pi}{6} + \tfrac{\sqrt{3}}{2} + \tfrac{\pi}{3} - \tfrac{\sqrt{3}}{8}\right) - \left(\tfrac{\pi}{4} + \tfrac{\pi}{2}\right) + \left(\tfrac{\pi}{6} + \tfrac{\sqrt{3}}{2} + \tfrac{\pi}{3} - \tfrac{\sqrt{3}}{8}\right) = \tfrac{\pi}{4} + \tfrac{3}{4}\sqrt{3} = \tfrac{1}{4}\left(\pi + 3\sqrt{3}\right)$$

28. We need to find the shaded area A in the figure. The horizontal line representing the front of the stage has equation $y = 4 \quad\Leftrightarrow$ $r\sin\theta = 4 \quad\Rightarrow\quad r = 4/\sin\theta$. This line intersects the curve $r = 8 + 8\sin\theta$ when $8 + 8\sin\theta = \dfrac{4}{\sin\theta} \quad\Rightarrow$ $8\sin\theta + 8\sin^2\theta = 4 \quad\Rightarrow\quad 2\sin^2\theta + 2\sin\theta - 1 = 0 \quad\Rightarrow$

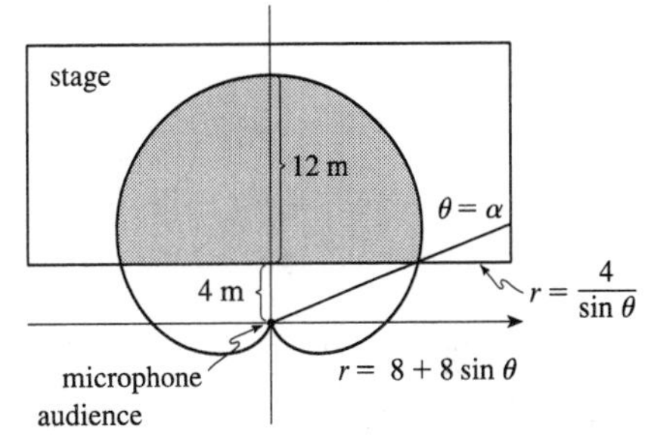

$\sin\theta = \dfrac{-2 \pm \sqrt{4+8}}{4} = \dfrac{-2 \pm 2\sqrt{3}}{4} = \dfrac{-1 + \sqrt{3}}{2}$ [the other value is less than -1] $\Rightarrow\quad \theta = \sin^{-1}\left(\dfrac{\sqrt{3} - 1}{2}\right)$.

This angle is about 21.5° and is denoted by α in the figure.

$$A = 2\int_\alpha^{\pi/2} \tfrac{1}{2}(8 + 8\sin\theta)^2\, d\theta - 2\int_\alpha^{\pi/2} \tfrac{1}{2}(4\csc\theta)^2\, d\theta = 64\int_\alpha^{\pi/2}(1 + 2\sin\theta + \sin^2\theta)\, d\theta - 16\int_\alpha^{\pi/2} \csc^2\theta\, d\theta$$

$$= 64\int_\alpha^{\pi/2} \left(1 + 2\sin\theta + \tfrac{1}{2} - \tfrac{1}{2}\cos 2\theta\right) d\theta + 16\int_\alpha^{\pi/2}(-\csc^2\theta)\, d\theta$$

$$= 64\left[\tfrac{3}{2}\theta - 2\cos\theta - \tfrac{1}{4}\sin 2\theta\right]_\alpha^{\pi/2} + 16\left[\cot\theta\right]_\alpha^{\pi/2} = 16\left[6\theta - 8\cos\theta - \sin 2\theta + \cot\theta\right]_\alpha^{\pi/a}$$

$$= 16[(3\pi - 0 - 0 + 0) - (6\alpha - 8\cos\alpha - \sin 2\alpha + \cot\alpha)] = 48\pi - 96\alpha + 128\cos\alpha + 16\sin 2\alpha - 16\cot\alpha$$

From the figure, $x^2 + \left(\sqrt{3} - 1\right)^2 = 2^2 \quad\Rightarrow\quad x^2 = 4 - \left(3 - 2\sqrt{3} + 1\right)$

$\Rightarrow\quad x^2 = 2\sqrt{3} = \sqrt{12}$, so $x = \sqrt{2\sqrt{3}} = \sqrt[4]{12}$. Using the trigonometric

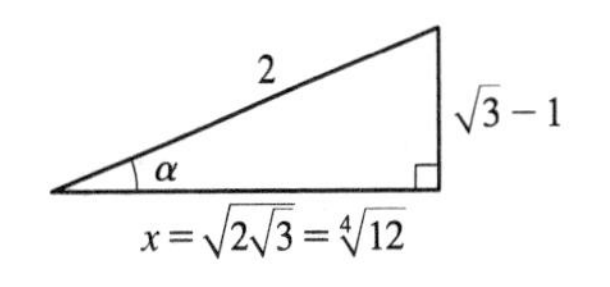

relationships for a right triangle and the identity $\sin 2\alpha = 2\sin\alpha\cos\alpha$, we continue:

$$A = 48\pi - 96\alpha + 128 \cdot \frac{\sqrt[4]{12}}{2} + 16 \cdot 2 \cdot \frac{\sqrt{3}-1}{2} \cdot \frac{\sqrt[4]{12}}{2} - 16 \cdot \frac{\sqrt[4]{12}}{\sqrt{3}-1} \cdot \frac{\sqrt{3}+1}{\sqrt{3}+1}$$

$$= 48\pi - 96\alpha + 64\sqrt[4]{12} + 8\sqrt[4]{12}\,(\sqrt{3}-1) - 8\sqrt[4]{12}\,(\sqrt{3}+1)$$

$$= 48\pi + 48\sqrt[4]{12} - 96\sin^{-1}\left(\frac{\sqrt{3}-1}{2}\right) \approx 204.16 \text{ m}^2$$

29. The curves intersect at the pole since $\left(0, \frac{\pi}{2}\right)$ satisfies $r = \cos\theta$ and $(0, 0)$ satisfies $r = 1 - \cos\theta$. Now $\cos\theta = 1 - \cos\theta \;\Rightarrow\; 2\cos\theta = 1 \;\Rightarrow\; \cos\theta = \frac{1}{2} \;\Rightarrow\; \theta = \frac{\pi}{3}$ or $\frac{5\pi}{3} \;\Rightarrow$ the other intersection points are $\left(\frac{1}{2}, \frac{\pi}{3}\right)$ and $\left(\frac{1}{2}, \frac{5\pi}{3}\right)$.

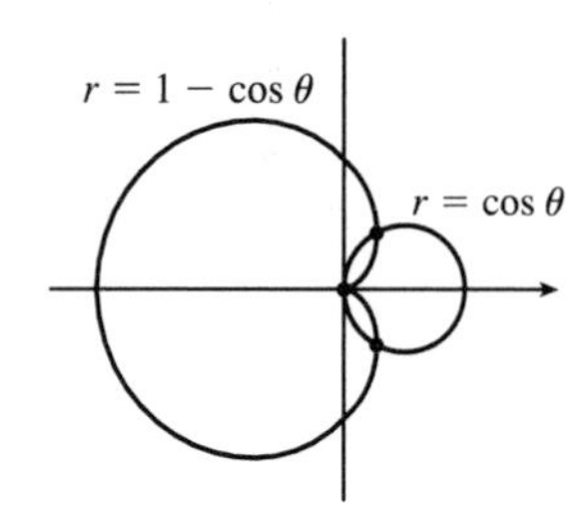

30. Clearly the pole lies on both curves.

$\sin 3\theta = \cos 3\theta \;\Rightarrow\; \tan 3\theta = 1 \;\Rightarrow\; 3\theta = \frac{\pi}{4} + n\pi$ [n any integer] $\Rightarrow\; \theta = \frac{\pi}{12} + \frac{\pi}{3}n \;\Rightarrow\; \theta = \frac{\pi}{12}, \frac{5\pi}{12}$, or $\frac{3\pi}{4}$, so the three remaining intersection points are $\left(\frac{1}{\sqrt{2}}, \frac{\pi}{12}\right)$, $\left(-\frac{1}{\sqrt{2}}, \frac{5\pi}{12}\right)$, and $\left(\frac{1}{\sqrt{2}}, \frac{3\pi}{4}\right)$.

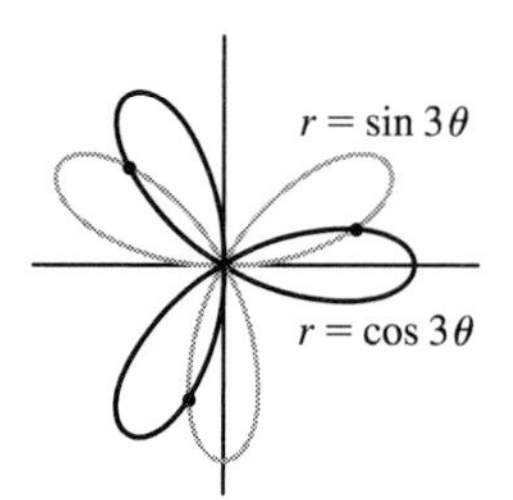

31. The pole is a point of intersection.

$\sin\theta = \sin 2\theta = 2\sin\theta\cos\theta \;\Leftrightarrow\; \sin\theta\,(1 - 2\cos\theta) = 0 \;\Leftrightarrow\; \sin\theta = 0$ or $\cos\theta = \frac{1}{2} \;\Rightarrow\; \theta = 0, \pi, \frac{\pi}{3}, -\frac{\pi}{3} \;\Rightarrow\; \left(\frac{\sqrt{3}}{2}, \frac{\pi}{3}\right)$ and $\left(\frac{\sqrt{3}}{2}, \frac{2\pi}{3}\right)$ (by symmetry) are the other intersection points.

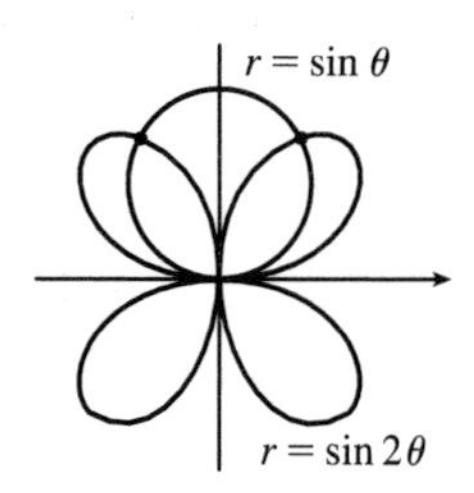

32. Clearly the pole is a point of intersection.

$\sin 2\theta = \cos 2\theta \;\Rightarrow\; \tan 2\theta = 1 \;\Rightarrow\; 2\theta = \frac{\pi}{4} + 2n\pi$ [since $\sin 2\theta$ and $\cos 2\theta$ must be positive in the equations] $\Rightarrow\; \theta = \frac{\pi}{8} + n\pi \;\Rightarrow\; \theta = \frac{\pi}{8}$ or $\frac{9\pi}{8}$. So the curves also intersect at $\left(\frac{1}{\sqrt[4]{2}}, \frac{\pi}{8}\right)$ and $\left(\frac{1}{\sqrt[4]{2}}, \frac{9\pi}{8}\right)$.

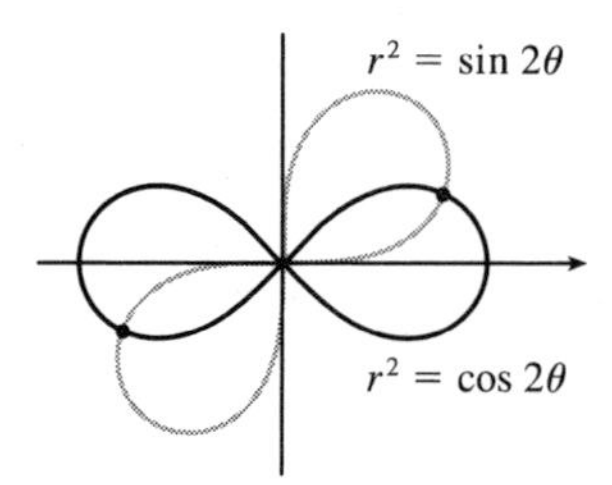

33. $L = \int_a^b \sqrt{r^2 + (dr/d\theta)^2}\,d\theta = \int_0^{\pi/3} \sqrt{(3\sin\theta)^2 + (3\cos\theta)^2}\,d\theta = \int_0^{\pi/3} \sqrt{9(\sin^2\theta + \cos^2\theta)}\,d\theta$
$= 3\int_0^{\pi/3} d\theta = 3[\theta]_0^{\pi/3} = 3\left(\frac{\pi}{3}\right) = \pi.$

As a check, note that the circumference of a circle with radius $\frac{3}{2}$ is $2\pi\left(\frac{3}{2}\right) = 3\pi$, and since $\theta = 0$ to $\pi = \frac{\pi}{3}$ traces out $\frac{1}{3}$ of the circle (from $\theta = 0$ to $\theta = \pi$), $\frac{1}{3}(3\pi) = \pi$.

34. $L = \int_a^b \sqrt{r^2 + (dr/d\theta)^2}\, d\theta = \int_0^{2\pi} \sqrt{(e^{2\theta})^2 + (2e^{2\theta})^2}\, d\theta = \int_0^{2\pi} \sqrt{e^{4\theta} + 4e^{4\theta}}\, d\theta = \int_0^{2\pi} \sqrt{5e^{4\theta}}\, d\theta$

$= \sqrt{5}\int_0^{2\pi} e^{2\theta}\, d\theta = \frac{\sqrt{5}}{2}\left[e^{2\theta}\right]_0^{2\pi} = \frac{\sqrt{5}}{2}\left(e^{4\pi} - 1\right)$

35. $L = \int_a^b \sqrt{r^2 + (dr/d\theta)^2}\, d\theta = \int_0^{2\pi} \sqrt{\left(\theta^2\right)^2 + (2\theta)^2}\, d\theta = \int_0^{2\pi} \sqrt{\theta^4 + 4\theta^2}\, d\theta$

$= \int_0^{2\pi} \sqrt{\theta^2\left(\theta^2 + 4\right)}\, d\theta = \int_0^{2\pi} \theta\sqrt{\theta^2 + 4}\, d\theta$

Now let $u = \theta^2 + 4$, so that $du = 2\theta\, d\theta$ $\left[\theta\, d\theta = \frac{1}{2}\, du\right]$ and

$\int_0^{2\pi} \theta\sqrt{\theta^2 + 4}\, d\theta = \int_4^{4\pi^2+4} \frac{1}{2}\sqrt{u}\, du = \frac{1}{2}\cdot\frac{2}{3}\left[u^{3/2}\right]_4^{4(\pi^2+1)} = \frac{1}{3}\left[4^{3/2}(\pi^2+1)^{3/2} - 4^{3/2}\right] = \frac{8}{3}\left[(\pi^2+1)^{3/2} - 1\right]$

36. $L = \int_a^b \sqrt{r^2 + (dr/d\theta)^2}\, d\theta = \int_0^{2\pi} \sqrt{\theta^2 + 1}\, d\theta \overset{21}{=} \left[\frac{\theta}{2}\sqrt{\theta^2+1} + \frac{1}{2}\ln\left(\theta + \sqrt{\theta^2+1}\right)\right]_0^{2\pi}$

$= \pi\sqrt{4\pi^2+1} + \frac{1}{2}\ln\left(2\pi + \sqrt{4\pi^2+1}\right)$

37. The curve $r = 3\sin 2\theta$ is completely traced with $0 \le \theta \le 2\pi$.

$r^2 + (dr/d\theta)^2 = (3\sin 2\theta)^2 + (6\cos 2\theta)^2 \Rightarrow$

$L = \int_0^{2\pi} \sqrt{9\sin^2 2\theta + 36\cos^2 2\theta}\, d\theta \approx 29.0653$

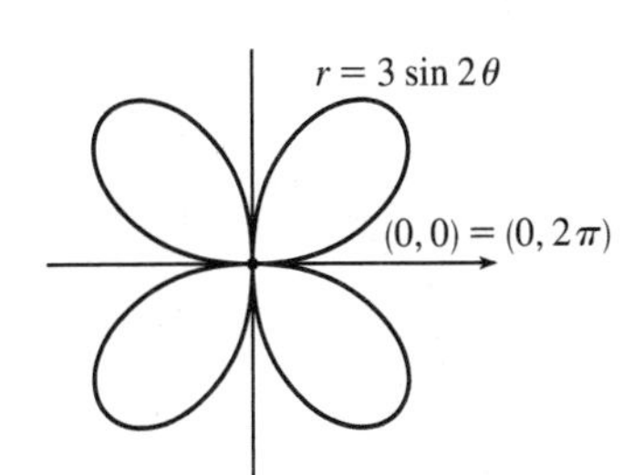

38. The curve $r = 4\sin 3\theta$ is completely traced with $0 \le \theta \le \pi$.

$r^2 + (dr/d\theta)^2 = (4\sin 3\theta)^2 + (12\cos 3\theta)^2 \Rightarrow$

$L = \int_0^{\pi} \sqrt{16\sin^2 3\theta + 144\cos^2 3\theta}\, d\theta \approx 26.7298$

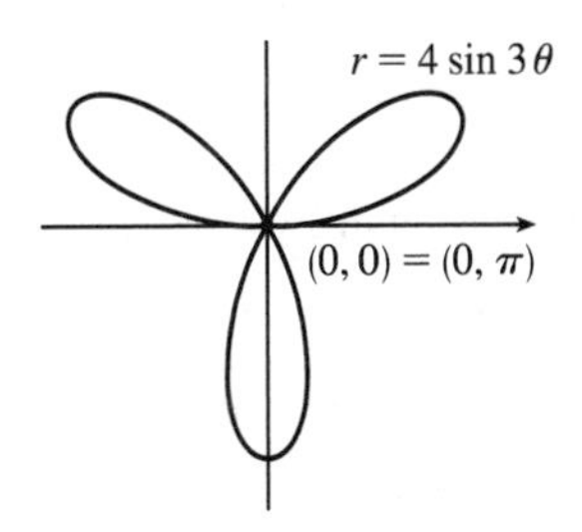

9.5 Conic Sections in Polar Coordinates

1. The directrix $y = 6$ is above the focus at the origin, so we use the form with "$+\, e\sin\theta$" in the denominator. (See Theorem 8 and Figure 8.) $r = \dfrac{ed}{1 + e\sin\theta} = \dfrac{\frac{7}{4}\cdot 6}{1 + \frac{7}{4}\sin\theta} = \dfrac{42}{4 + 7\sin\theta}$

2. The directrix $x = 4$ is to the right of the focus at the origin, so we use the form with "$+\, e\cos\theta$" in the denominator. $e = 1$ for a parabola, so an equation is $r = \dfrac{ed}{1 + e\cos\theta} = \dfrac{1\cdot 4}{1 + 1\cos\theta} = \dfrac{4}{1 + \cos\theta}$

3. The directrix $x = -5$ is to the left of the focus at the origin, so we use the form with "$-\, e\cos\theta$" in the denominator.

$$r = \frac{ed}{1 - e\cos\theta} = \frac{\frac{3}{4}\cdot 5}{1 - \frac{3}{4}\cos\theta} = \frac{15}{4 - 3\cos\theta}$$

4. The directrix $y = -2$ is below the focus at the origin, so we use the form with "$-\, e\sin\theta$" in the denominator.

$$r = \frac{ed}{1 - e\sin\theta} = \frac{2\cdot 2}{1 - 2\sin\theta} = \frac{4}{1 - 2\sin\theta}$$

5. The vertex $(4, 3\pi/2)$ is 4 units below the focus at the origin, so the directrix is 8 units below the focus ($d = 8$), and we use the form with "$-e\sin\theta$" in the denominator.

$e = 1$ for a parabola, so an equation is $r = \dfrac{ed}{1 - e\sin\theta} = \dfrac{1(8)}{1 - 1\sin\theta} = \dfrac{8}{1 - \sin\theta}$.

6. The vertex $P(1, \pi/2)$ is 1 unit above the focus F at the origin, so $|PF| = 1$ and we use the form with "$+e\sin\theta$" in the denominator. The distance from the focus to the directrix l is d, so $e = \dfrac{|PF|}{|Pl|} \Rightarrow 0.8 = \dfrac{1}{d-1} \Rightarrow 0.8d - 0.8 = 1 \Rightarrow$

$0.8d = 1.8 \Rightarrow d = 2.25$. An equation is $r = \dfrac{ed}{1 + e\sin\theta} = \dfrac{0.8(2.25)}{1 + 0.8\sin\theta} \cdot \dfrac{5}{5} = \dfrac{9}{5 + 4\sin\theta}$.

7. The directrix $r = 4\sec\theta$ (equivalent to $r\cos\theta = 4$ or $x = 4$) is to the right of the focus at the origin, so we will use the form with "$+e\cos\theta$" in the denominator. The distance from the focus to the directrix is $d = 4$, so an equation is

$r = \dfrac{ed}{1 + e\cos\theta} = \dfrac{0.5(4)}{1 + 0.5\cos\theta} \cdot \dfrac{2}{2} = \dfrac{4}{2 + \cos\theta}$.

8. The directrix $r = -6\csc\theta$ (equivalent to $r\sin\theta = -6$ or $y = -6$) is below the focus at the origin, so we will use the form with "$-e\sin\theta$" in the denominator. The distance from the focus to the directrix is $d = 6$, so an equation is

$r = \dfrac{ed}{1 - e\sin\theta} = \dfrac{3(6)}{1 - 3\sin\theta} = \dfrac{18}{1 - 3\sin\theta}$.

9. $r = \dfrac{1}{1 + \sin\theta} = \dfrac{ed}{1 + e\sin\theta}$, where $d = e = 1$.

(a) Eccentricity $= e = 1$

(b) Since $e = 1$, the conic is a parabola.

(c) Since "$+e\sin\theta$" appears in the denominator, the directrix is above the focus at the origin. $d = |Fl| = 1$, so an equation of the directrix is $y = 1$.

(d) The vertex is at $\left(\frac{1}{2}, \frac{\pi}{2}\right)$, midway between the focus and the directrix.

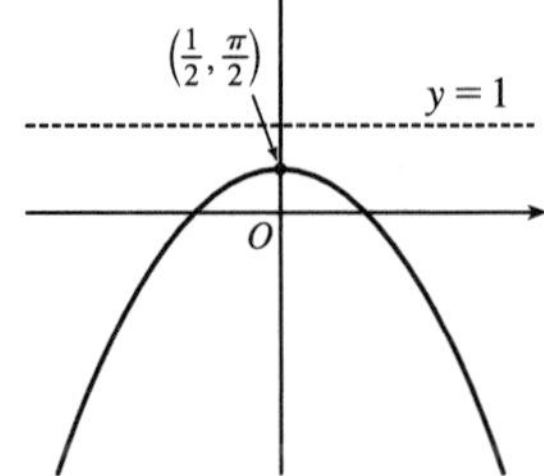

10. $r = \dfrac{6}{3 + 2\sin\theta} = \dfrac{2}{1 + \frac{2}{3}\sin\theta} = \dfrac{\frac{2}{3} \cdot 3}{1 + \frac{2}{3}\sin\theta}$

(a) $e = \frac{2}{3}$

(b) Ellipse

(c) $y = 3$

(d) Vertices $\left(\frac{6}{5}, \frac{\pi}{2}\right)$ and $\left(6, \frac{3\pi}{2}\right)$; center $\left(\frac{12}{5}, \frac{3\pi}{2}\right)$

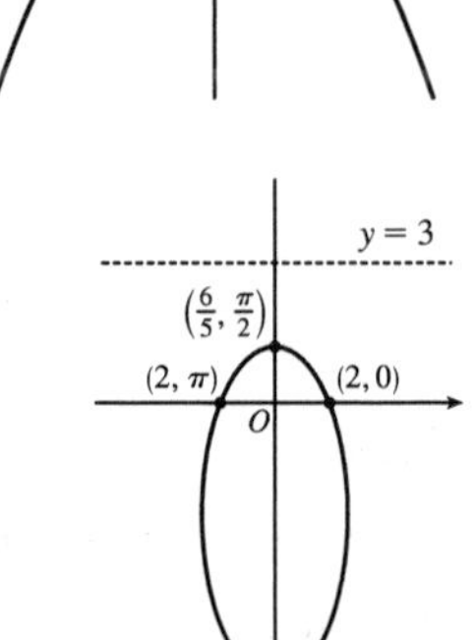

11. $r = \dfrac{12}{4 - \sin\theta} \cdot \dfrac{1/4}{1/4} = \dfrac{3}{1 - \frac{1}{4}\sin\theta}$, where $e = \frac{1}{4}$ and $ed = 3 \Rightarrow d = 12$.

(a) Eccentricity $= e = \frac{1}{4}$

(b) Since $e = \frac{1}{4} < 1$, the conic is an ellipse.

(c) Since "$-e\sin\theta$" appears in the denominator, the directrix is below the focus at the origin. $d = |Fl| = 12$, so an equation of the directrix is $y = -12$.

(d) The vertices are $\left(4, \frac{\pi}{2}\right)$ and $\left(\frac{12}{5}, \frac{3\pi}{2}\right)$, so the center is midway between them, that is, $\left(\frac{4}{5}, \frac{\pi}{2}\right)$.

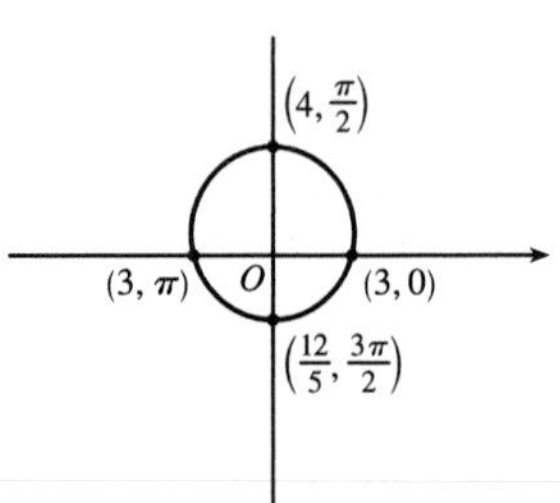

12. $r = \dfrac{4}{2 - 3\cos\theta} = \dfrac{2}{1 - \frac{3}{2}\cos\theta} = \dfrac{\frac{3}{2}\cdot\frac{4}{3}}{1 - \frac{3}{2}\cos\theta}$

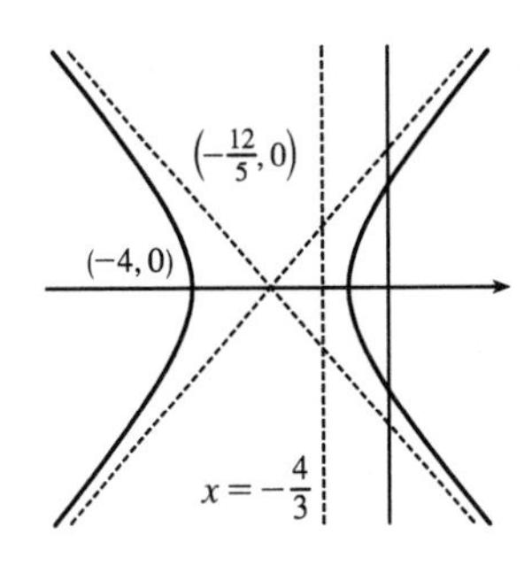

(a) $e = \frac{3}{2}$

(b) Hyperbola

(c) $x = -\frac{4}{3}$

(d) The vertices are $(-4, 0)$ and $\left(\frac{4}{5}, \pi\right) = \left(-\frac{4}{5}, 0\right)$, so the center is $\left(-\frac{12}{5}, 0\right)$.

The asymptotes are parallel to $\theta = \pm\cos^{-1}\frac{2}{3}$. [Their slopes are

$\pm\tan\left(\cos^{-1}\frac{2}{3}\right) = \pm\frac{\sqrt{5}}{2}$.]

13. $r = \dfrac{9}{6 + 2\cos\theta}\cdot\dfrac{1/6}{1/6} = \dfrac{3/2}{1 + \frac{1}{3}\cos\theta}$, where $e = \frac{1}{3}$ and $ed = \frac{3}{2} \Rightarrow d = \frac{9}{2}$.

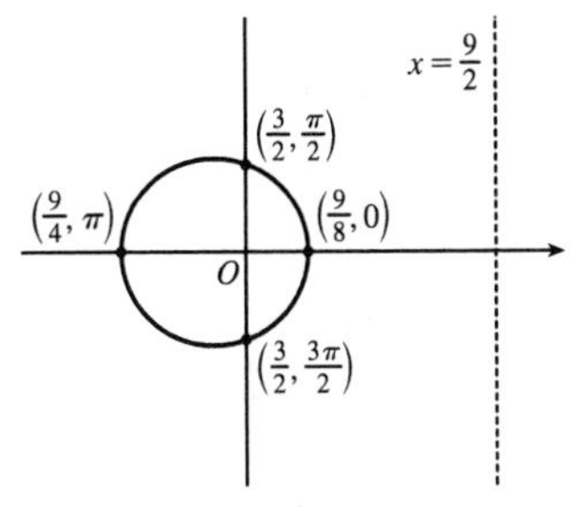

(a) Eccentricity $= e = \frac{1}{3}$

(b) Since $e = \frac{1}{3} < 1$, the conic is an ellipse.

(c) Since "$+e\cos\theta$" appears in the denominator, the directrix is to the right of the focus at the origin. $d = |Fl| = \frac{9}{2}$, so an equation of the directrix is $x = \frac{9}{2}$.

(d) The vertices are $\left(\frac{9}{8}, 0\right)$ and $\left(\frac{9}{4}, \pi\right)$, so the center is midway between them, that is, $\left(\frac{9}{16}, \pi\right)$.

14. $r = \dfrac{5}{2 - 2\sin\theta} = \dfrac{\frac{5}{2}}{1 - \sin\theta}$

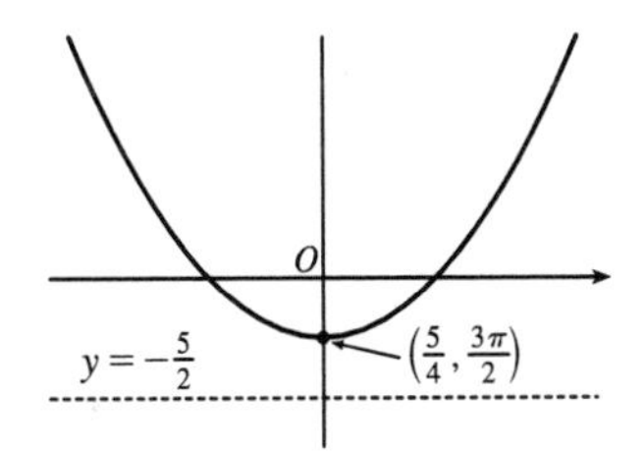

(a) $e = 1$

(b) Parabola

(c) $y = -\frac{5}{2}$

(d) The focus is $(0, 0)$, so the vertex is $\left(\frac{5}{4}, \frac{3\pi}{2}\right)$ and the parabola opens up.

15. $r = \dfrac{3}{4 - 8\cos\theta}\cdot\dfrac{1/4}{1/4} = \dfrac{3/4}{1 - 2\cos\theta}$, where $e = 2$ and $ed = \frac{3}{4} \Rightarrow d = \frac{3}{8}$.

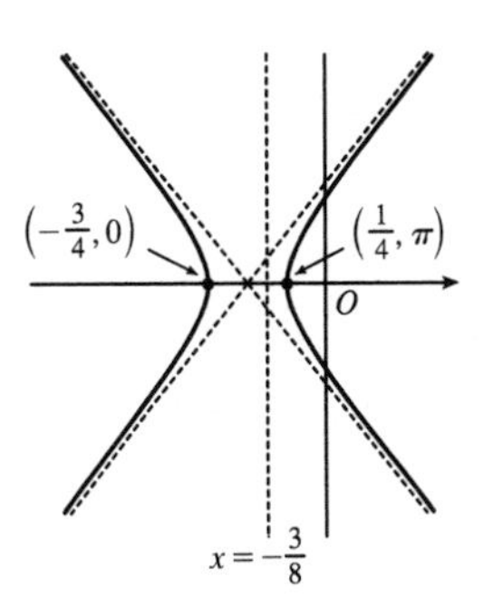

(a) Eccentricity $= e = 2$

(b) Since $e = 2 > 1$, the conic is a hyperbola.

(c) Since "$-e\cos\theta$" appears in the denominator, the directrix is to the left of the focus at the origin. $d = |Fl| = \frac{3}{8}$, so an equation of the directrix is $x = -\frac{3}{8}$.

(d) The vertices are $\left(-\frac{3}{4}, 0\right)$ and $\left(\frac{1}{4}, \pi\right)$, so the center is midway between them, that is, $\left(\frac{1}{2}, \pi\right)$.

16. $r = \dfrac{4}{2 + \cos\theta} = \dfrac{2}{1 + \frac{1}{2}\cos\theta} = \dfrac{\frac{1}{2}\cdot 4}{1 + \frac{1}{2}\cos\theta}$

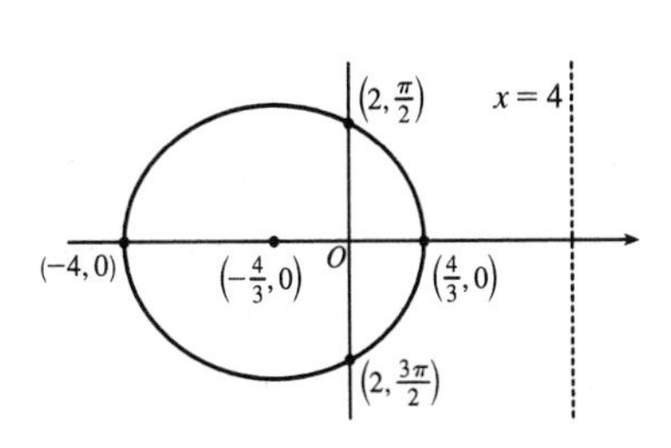

(a) $e = \frac{1}{2}$

(b) Ellipse

(c) $x = 4$

(d) The vertices are $\left(\frac{4}{3}, 0\right)$ and $(4, \pi) = (-4, 0)$, so the center is $\left(-\frac{4}{3}, 0\right)$.

17. For $e < 1$ the curve is an ellipse. It is nearly circular when e is close to 0. As e increases, the graph is stretched out to the right, and grows larger (that is, its right-hand focus moves to the right while its left-hand focus remains at the origin.) At $e = 1$, the curve becomes a parabola with focus at the origin.

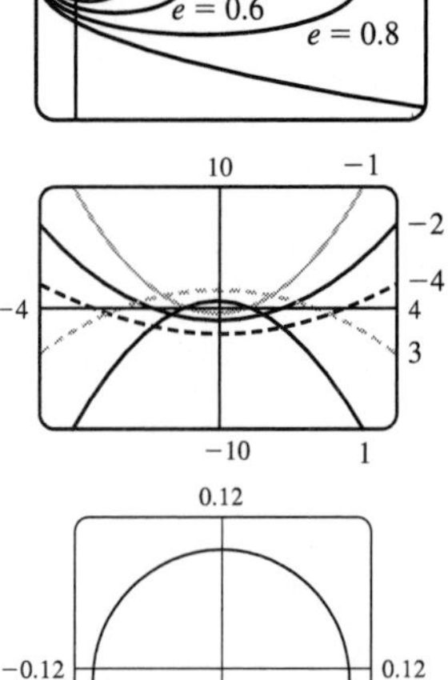

18. (a) The value of d does not seem to affect the shape of the conic (a parabola) at all, just its size, position, and orientation (for $d < 0$ it opens upward, for $d > 0$ it opens downward).

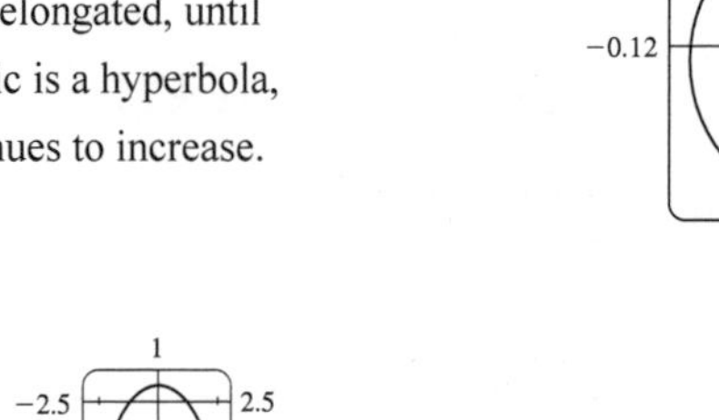

(b) We consider only positive values of e. When $0 < e < 1$, the conic is an ellipse. As $e \to 0^+$, the graph approaches perfect roundness and zero size. As e increases, the ellipse becomes more elongated, until at $e = 1$ it turns into a parabola. For $e > 1$, the conic is a hyperbola, which moves downward and gets broader as e continues to increase.

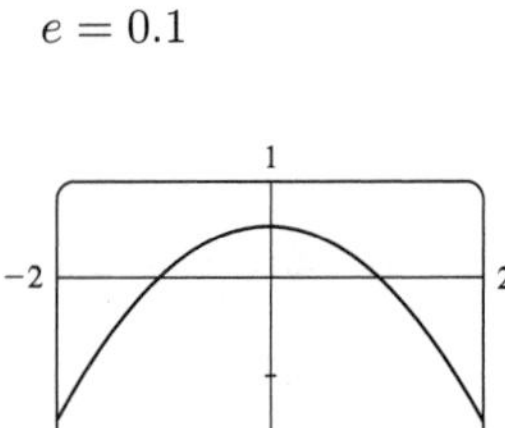

$e = 0.1$

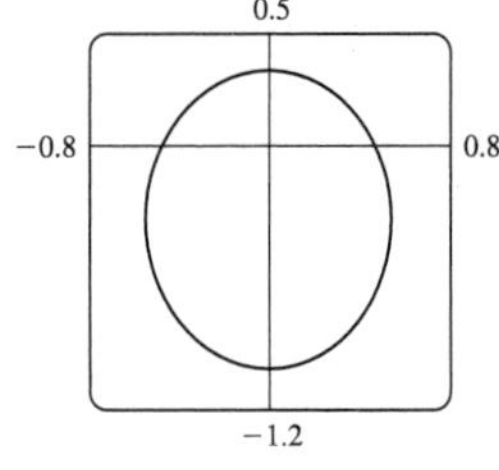

$e = 0.5$

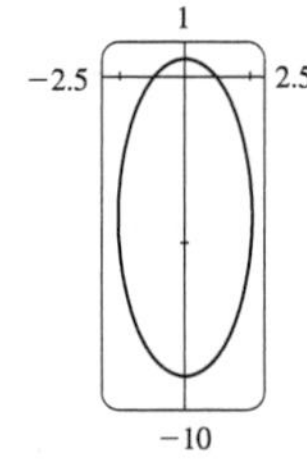

$e = 0.9$

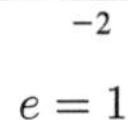

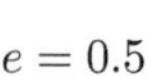

$e = 1$

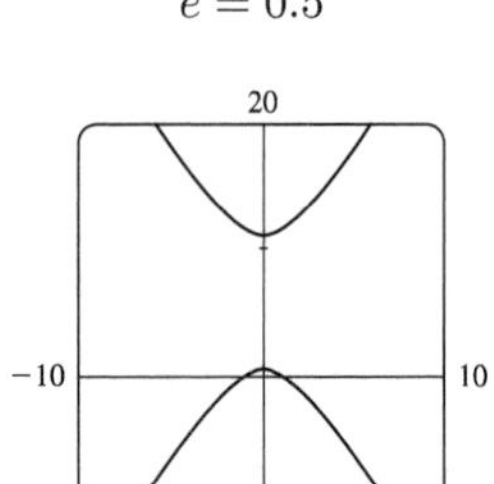

$e = 1.1$

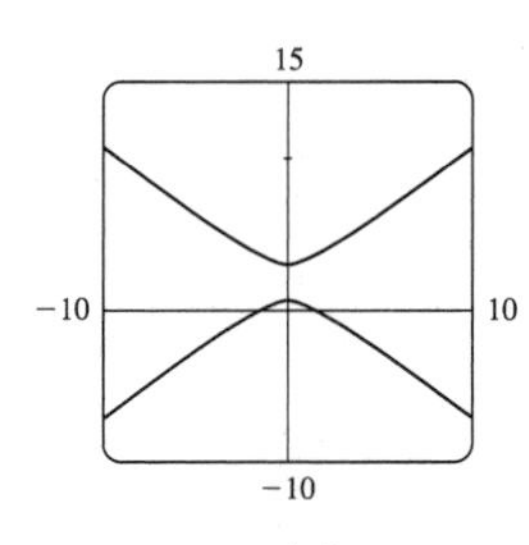

$e = 1.5$

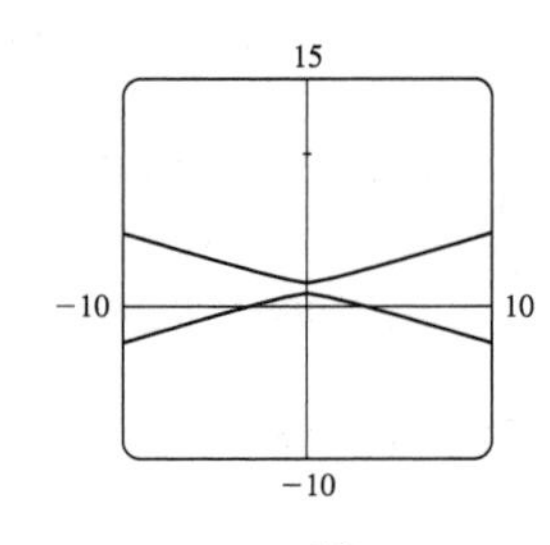

$e = 10$

19. 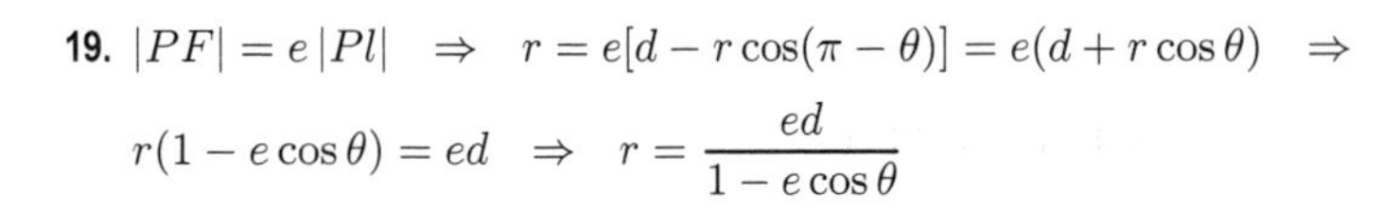

$|PF| = e\,|Pl| \;\Rightarrow\; r = e[d - r\cos(\pi - \theta)] = e(d + r\cos\theta) \;\Rightarrow$

$r(1 - e\cos\theta) = ed \;\Rightarrow\; r = \dfrac{ed}{1 - e\cos\theta}$

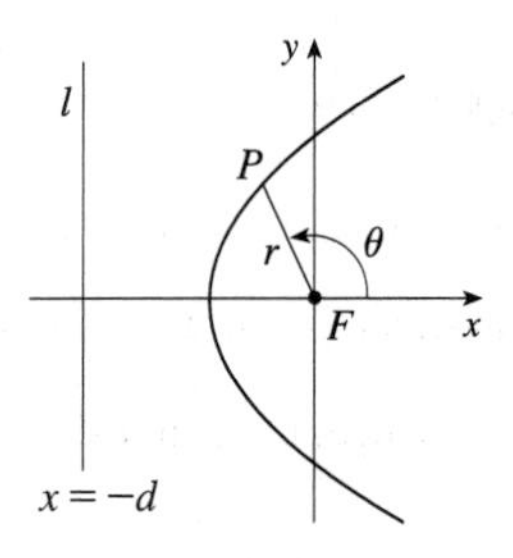

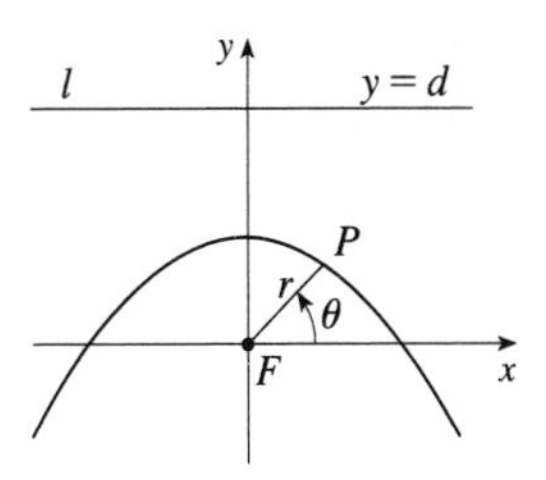

20. $|PF| = e\,|Pl| \;\Rightarrow\; r = e[d - r\sin\theta] \;\Rightarrow\; r(1 + e\sin\theta) = ed \;\Rightarrow$

$$r = \frac{ed}{1 + e\sin\theta}$$

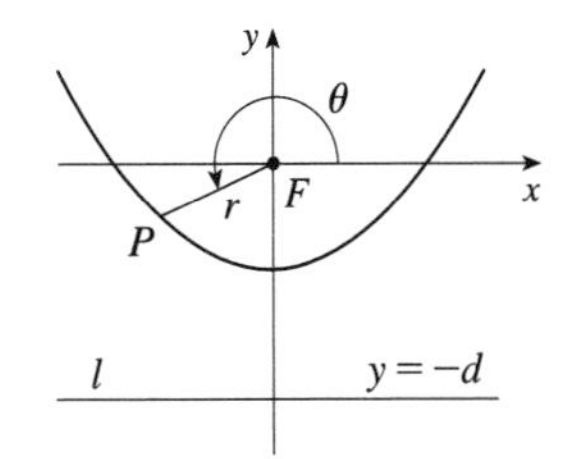

21. $|PF| = e\,|Pl| \;\Rightarrow\; r = e[d - r\sin(\theta - \pi)] = e(d + r\sin\theta) \;\Rightarrow$

$r(1 - e\sin\theta) = ed \;\Rightarrow\; r = \dfrac{ed}{1 - e\sin\theta}$

22. The parabolas intersect at the two points where $\dfrac{c}{1+\cos\theta} = \dfrac{d}{1-\cos\theta} \;\Rightarrow\; \cos\theta = \dfrac{c-d}{c+d} \;\Rightarrow\; r = \dfrac{c+d}{2}$.

For the first parabola, $\dfrac{dr}{d\theta} = \dfrac{c\sin\theta}{(1+\cos\theta)^2}$, so

$$\frac{dy}{dx} = \frac{(dr/d\theta)\sin\theta + r\cos\theta}{(dr/d\theta)\cos\theta - r\sin\theta} = \frac{c\sin^2\theta + c\cos\theta(1+\cos\theta)}{c\sin\theta\cos\theta - c\sin\theta(1+\cos\theta)} = \frac{1+\cos\theta}{-\sin\theta}$$

and similarly for the second, $\dfrac{dy}{dx} = \dfrac{1-\cos\theta}{\sin\theta} = \dfrac{\sin\theta}{1+\cos\theta}$. Since the product of these slopes is -1, the parabolas intersect at right angles.

23. (a) If the directrix is $x = -d$, then $r = \dfrac{ed}{1 - e\cos\theta}$ [see Figure 8(b)], and, from (6), $a^2 = \dfrac{e^2d^2}{(1-e^2)^2} \;\Rightarrow$

$ed = a(1-e^2)$. Therefore, $r = \dfrac{a(1-e^2)}{1 - e\cos\theta}$.

(b) $e = 0.017$ and the major axis $= 2a = 2.99 \times 10^8 \;\Rightarrow\; a = 1.495 \times 10^8$.

Therefore $r = \dfrac{1.495 \times 10^8\left[1 - (0.017)^2\right]}{1 - 0.017\cos\theta} \approx \dfrac{1.49 \times 10^8}{1 - 0.017\cos\theta}$.

24. (a) The Sun is at point F in Figure 7 so that perihelion is in the positive x-direction and aphelion is in the negative x-direction.

At perihelion, $\theta = 0$, so $r = \dfrac{a(1-e^2)}{1 + e\cos 0} = \dfrac{a(1-e)(1+e)}{1+e} = a(1-e)$.

At aphelion, $\theta = \pi$, so $r = \dfrac{a(1-e^2)}{1 + e\cos\pi} = \dfrac{a(1-e)(1+e)}{1-e} = a(1+e)$.

(b) At perihelion, $r = a(1-e) \approx (1.495 \times 10^8)(1 - 0.017) \approx 1.47 \times 10^8$ km.

At aphelion, $r = a(1+e) \approx (1.495 \times 10^8)(1 + 0.017) \approx 1.52 \times 10^8$ km.

25. Here $2a =$ length of major axis $= 36.18$ AU $\;\Rightarrow\; a = 18.09$ AU and $e = 0.97$. By Exercise 23(a), the equation of the orbit is $r = \dfrac{18.09\left[1 - (0.97)^2\right]}{1 - 0.97\cos\theta} \approx \dfrac{1.07}{1 - 0.97\cos\theta}$. By Exercise 24(a), the maximum distance from the comet to the sun is $18.09(1 + 0.97) \approx 35.64$ AU or about 3.314 billion miles.

26. Here $2a =$ length of major axis $= 356.5$ AU $\Rightarrow$ $a = 178.25$ AU and $e = 0.9951$. By Exercise 23(a), the equation of the orbit is $r = \dfrac{178.25\left[1-(0.9951)^2\right]}{1-0.9951\cos\theta} \approx \dfrac{1.7426}{1-0.9951\cos\theta}$. By Exercise 24(a), the minimum distance from the comet to the sun is $178.25(1-0.9951) \approx 0.8734$ AU or about 81 million miles.

27. The minimum distance is at perihelion, where $4.6\times10^7 = r = a(1-e) = a(1-0.206) = a(0.794)$ $\Rightarrow$ $a = 4.6\times10^7/0.794$. So the maximum distance, which is at aphelion, is $r = a(1+e) = \left(4.6\times10^7/0.794\right)(1.206) \approx 7.0\times10^7$ km.

28. At perihelion, $r = a(1-e) = 4.43\times10^9$, and at aphelion, $r = a(1+e) = 7.37\times10^9$. Adding, we get $2a = 11.80\times10^9$, so $a = 5.90\times10^9$ km. Therefore $1+e = a(1+e)/a = \frac{7.37}{5.90} \approx 1.249$ and $e \approx 0.249$.

29. From Exercise 27, we have $e = 0.206$ and $a(1-e) = 4.6\times10^7$ km. Thus, $a = 4.6\times10^7/0.794$. From Exercise 23, we can write the equation of Mercury's orbit as $r = a\,\dfrac{1-e^2}{1-e\cos\theta}$. So since $\dfrac{dr}{d\theta} = \dfrac{-a(1-e^2)e\sin\theta}{(1-e\cos\theta)^2}$ $\Rightarrow$

$$r^2 + \left(\frac{dr}{d\theta}\right)^2 = \frac{a^2(1-e^2)^2}{(1-e\cos\theta)^2} + \frac{a^2(1-e^2)^2e^2\sin^2\theta}{(1-e\cos\theta)^4} = \frac{a^2(1-e^2)^2}{(1-e\cos\theta)^4}(1-2e\cos\theta+e^2)$$

the length of the orbit is $L = \displaystyle\int_0^{2\pi}\sqrt{r^2+(dr/d\theta)^2}\,d\theta = a\left(1-e^2\right)\int_0^{2\pi}\frac{\sqrt{1+e^2-2e\cos\theta}}{(1-e\cos\theta)^2}\,d\theta \approx 3.6\times10^8$ km.

This seems reasonable, since Mercury's orbit is nearly circular, and the circumference of a circle of radius a is $2\pi a \approx 3.6\times10^8$ km.

9 Review

CONCEPT CHECK

1. (a) A parametric curve is a set of points of the form $(x,y) = (f(t), g(t))$, where f and g are continuous functions of a variable t.

(b) Sketching a parametric curve, like sketching the graph of a function, is difficult to do in general. We can plot points on the curve by finding $f(t)$ and $g(t)$ for various values of t, either by hand or with a calculator or computer. Sometimes, when f and g are given by formulas, we can eliminate t from the equations $x = f(t)$ and $y = g(t)$ to get a Cartesian equation relating x and y. It may be easier to graph that equation than to work with the original formulas for x and y in terms of t.

2. (a) You can find $\dfrac{dy}{dx}$ as a function of t by calculating $\dfrac{dy}{dx} = \dfrac{dy/dt}{dx/dt}$ (if $dx/dt \neq 0$).

(b) Calculate the area as $\int_a^b y\,dx = \int_\alpha^\beta g(t)f'(t)dt$ [or $\int_\beta^\alpha g(t)f'(t)dt$ if the leftmost point is $(f(\beta), g(\beta))$ rather than $(f(\alpha), g(\alpha))$].

3. $L = \int_\alpha^\beta \sqrt{(dx/dt)^2+(dy/dt)^2}\,dt = \int_\alpha^\beta \sqrt{[f'(t)]^2+[g'(t)]^2}\,dt$

4. (a) See Figure 5 in Section 9.3.

(b) $x = r\cos\theta,\ y = r\sin\theta$

(c) To find a polar representation (r,θ) with $r \geq 0$ and $0 \leq \theta < 2\pi$, first calculate $r = \sqrt{x^2+y^2}$. Then θ is specified by $\cos\theta = x/r$ and $\sin\theta = y/r$.

5. (a) Calculate $\dfrac{dy}{dx} = \dfrac{\dfrac{dy}{d\theta}}{\dfrac{dx}{d\theta}} = \dfrac{\dfrac{d}{d\theta}(y)}{\dfrac{d}{d\theta}(x)} = \dfrac{\dfrac{d}{d\theta}(r\sin\theta)}{\dfrac{d}{d\theta}(r\cos\theta)} = \dfrac{\left(\dfrac{dr}{d\theta}\right)\sin\theta + r\cos\theta}{\left(\dfrac{dr}{d\theta}\right)\cos\theta - r\sin\theta}$, where $r = f(\theta)$.

(b) Calculate $A = \int_a^b \frac{1}{2}r^2\,d\theta = \int_a^b \frac{1}{2}[f(\theta)]^2\,d\theta$.

(c) $L = \int_a^b \sqrt{(dx/d\theta)^2 + (dy/d\theta)^2}\,d\theta = \int_a^b \sqrt{r^2 + (dr/d\theta)^2}\,d\theta = \int_a^b \sqrt{[f(\theta)]^2 + [f'(\theta)]^2}\,d\theta$

6. (a) If a conic section has focus F and corresponding directrix l, then the eccentricity e is the fixed ratio $|PF|\,/\,|Pl|$ for points P of the conic section.

(b) $e < 1$ for an ellipse; $e > 1$ for a hyperbola; $e = 1$ for a parabola.

(c) $x = d$: $r = \dfrac{ed}{1 + e\cos\theta}$; $x = -d$: $r = \dfrac{ed}{1 - e\cos\theta}$; $y = d$: $r = \dfrac{ed}{1 + e\sin\theta}$; $y = -d$: $r = \dfrac{ed}{1 - e\sin\theta}$.

TRUE-FALSE QUIZ

1. False. Consider the curve defined by $x = f(t) = (t-1)^3$ and $y = g(t) = (t-1)^2$. Then $g'(t) = 2(t-1)$, so $g'(1) = 0$, but its graph has a *vertical* tangent when $t = 1$. *Note:* The statement is true if $f'(1) \neq 0$ when $g'(1) = 0$.

2. False. If $x = f(t)$ and $y = g(t)$ are twice differentiable, then $\dfrac{d^2y}{dx^2} = \dfrac{d}{dx}\left(\dfrac{dy}{dx}\right) = \dfrac{\dfrac{d}{dt}\left(\dfrac{dy}{dx}\right)}{\dfrac{dx}{dt}}$.

3. False. For example, if $f(t) = \cos t$ and $g(t) = \sin t$ for $0 \le t \le 4\pi$, then the curve is a circle of radius 1, hence its length is 2π, but $\int_0^{4\pi} \sqrt{[f'(t)]^2 + [g'(t)]^2}\,dt = \int_0^{4\pi} \sqrt{(-\sin t)^2 + (\cos t)^2}\,dt = \int_0^{4\pi} 1\,dt = 4\pi$, since as t increases from 0 to 4π, the circle is traversed twice.

4. False. If $(r,\theta) = (1,\pi)$, then $(x,y) = (-1,0)$, so $\tan^{-1}(y/x) = \tan^{-1}0 = 0 \neq \theta$. The statement is true for points in quadrants I and IV.

5. True. The curve $r = 1 - \sin 2\theta$ is unchanged if we rotate it through $180°$ about O because $1 - \sin 2(\theta + \pi) = 1 - \sin(2\theta + 2\pi) = 1 - \sin 2\theta$. So it's unchanged if we replace r by $-r$. In other words, it's the same curve as $r = -(1 - \sin 2\theta) = \sin 2\theta - 1$.

6. True. The polar equation $r = 2$, the Cartesian equation $x^2 + y^2 = 4$, and the parametric equations $x = 2\sin 3t$, $y = 2\cos 3t$ $(0 \le t \le 2\pi)$ all describe the circle of radius 2 centered at the origin.

7. False. The first pair of equations yields the portion of the parabola $y = x^2$ with $x \ge 0$, whereas the second pair of equations traces out the whole parabola $y = x^2$.

8. True. Consider a hyperbola with focus at the origin, oriented so that its polar equation is $r = \dfrac{ed}{1 + e\cos\theta}$, where $e > 1$.

The directrix is $x = d$, but along the hyperbola we have $x = r\cos\theta = \dfrac{ed\cos\theta}{1 + e\cos\theta} = d\left(\dfrac{e\cos\theta}{1 + e\cos\theta}\right) \neq d$.

EXERCISES

1. $x = t^2 + 4t,\ y = 2 - t,\ -4 \le t \le 1.\quad t = 2 - y$, so

$x = (2 - y)^2 + 4(2 - y) = 4 - 4y + y^2 + 8 - 4y = y^2 - 8y + 12 \quad \Leftrightarrow$

$x + 4 = y^2 - 8y + 16 = (y - 4)^2$. This is part of a parabola with vertex $(-4, 4)$, opening to the right.

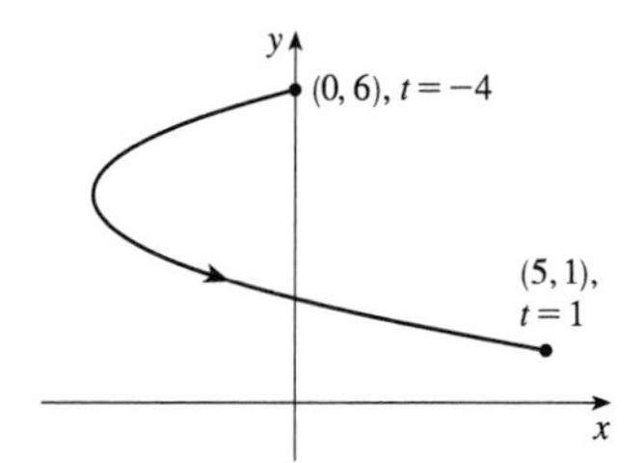

2. $x = 1 + e^{2t},\ y = e^t$.

$x = 1 + e^{2t} = 1 + (e^t)^2 = 1 + y^2,\ y > 0$.

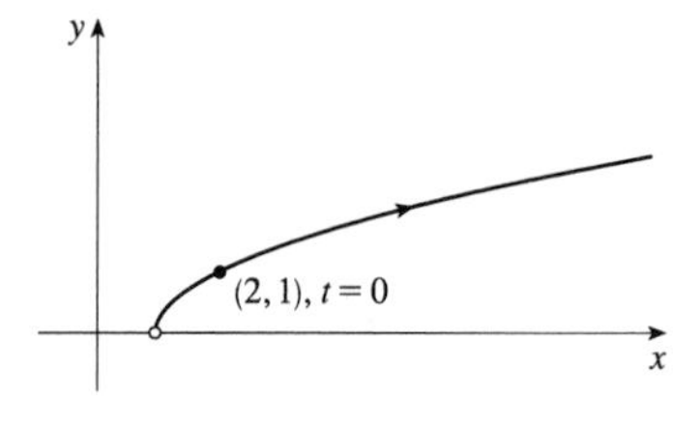

3. $x = \tan\theta,\ y = \cot\theta.\quad y = 1/\tan\theta = 1/x$. The whole curve is traced out as θ ranges over the open interval $\left(-\frac{\pi}{2}, \frac{\pi}{2}\right)$ [or any open interval of the form $\left(-\frac{\pi}{2} + n\pi, \frac{\pi}{2} + n\pi\right)$, where n is an integer].

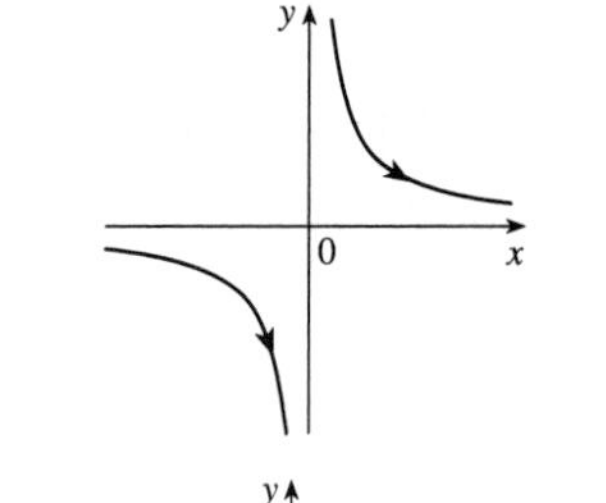

4. $x = 2\cos\theta,\ y = 1 + \sin\theta.\quad \cos^2\theta + \sin^2\theta = 1 \quad \Rightarrow$

$\left(\frac{x}{2}\right)^2 + (y - 1)^2 = 1 \quad \Rightarrow \quad \frac{x^2}{4} + (y - 1)^2 = 1$. This is an ellipse, centered at $(0, 1)$, with semimajor axis of length 2 and semiminor axis of length 1.

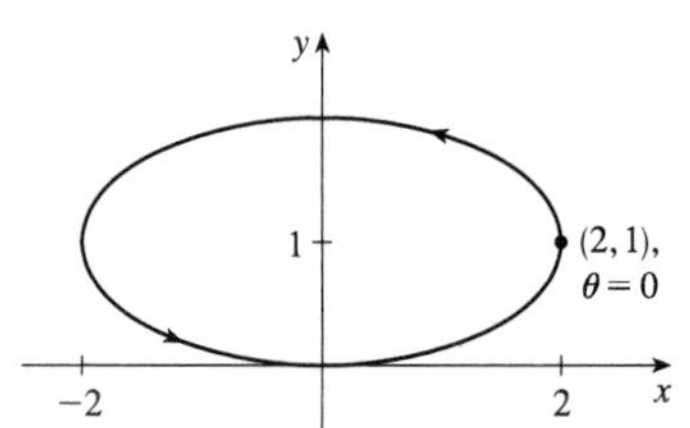

5. Three different sets of parametric equations for the curve $y = \sqrt{x}$ are

(i) $x = t,\ \ y = \sqrt{t},\ \ t \ge 0$

(ii) $x = t^4,\ \ y = t^2$

(iii) $x = \tan^2 t,\ \ y = \tan t,\ \ 0 \le t < \pi/2$

There are many other sets of equations that also give this curve.

6. For $t < -1$, $x > 0$ and $y < 0$ with x decreasing and y increasing. When $t = -1$, $(x, y) = (0, 0)$. When $-1 < t < 0$, we have $-1 < x < 0$ and $0 < y < 1/2$. When $t = 0$, $(x, y) = (-1, 0)$. When $0 < t < 1$, $-1 < x < 0$ and $-\frac{1}{2} < y < 0$. When $t = 1$, $(x, y) = (0, 0)$ again. When $t > 1$, both x and y are positive and increasing.

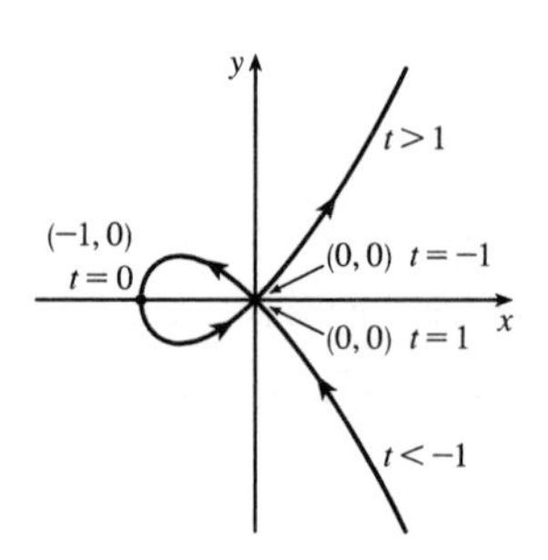

7. $r = 1 - \cos\theta$. This cardiod is symmetric about the polar axis.

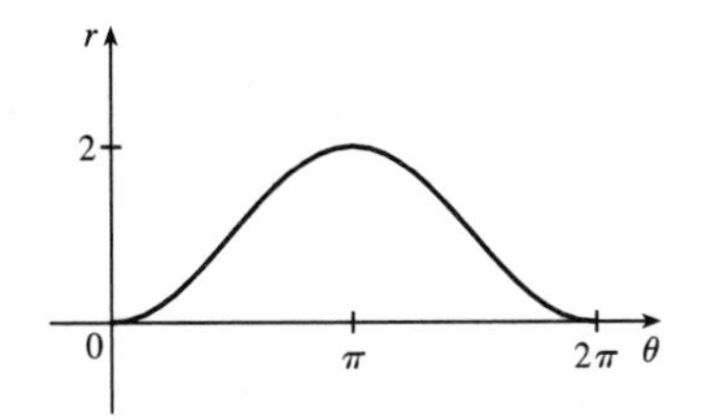

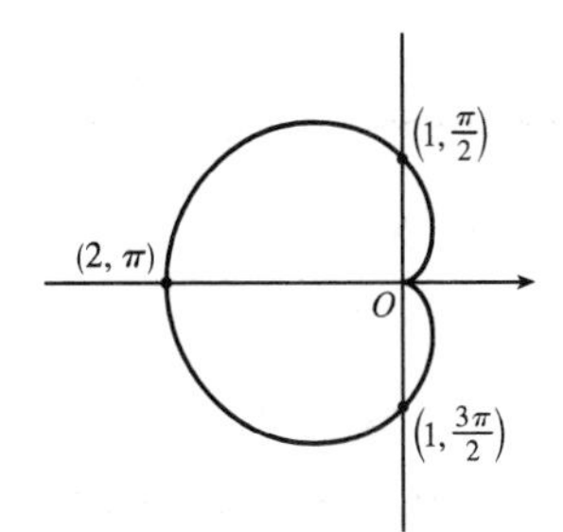

8. $r = \sin 4\theta$. This is an eight-leafed rose.

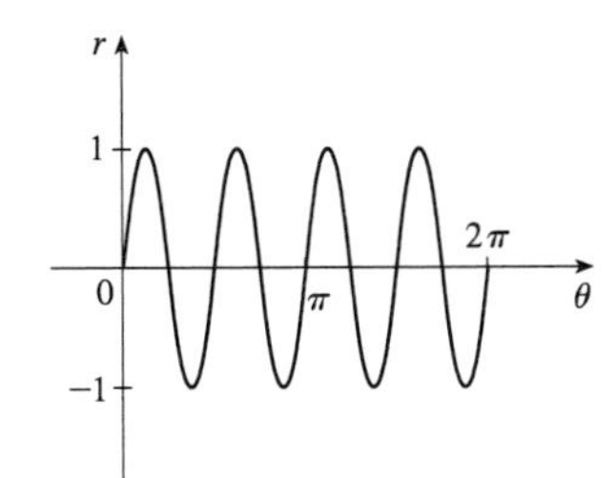

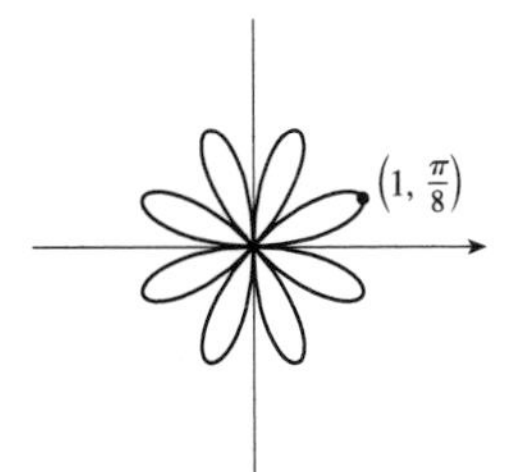

9. $r = 1 + \cos 2\theta$. The curve is symmetric about the pole and both the horizontal and vertical axes.

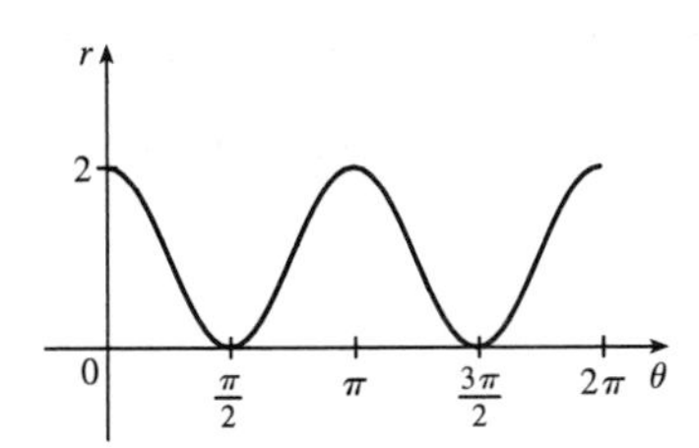

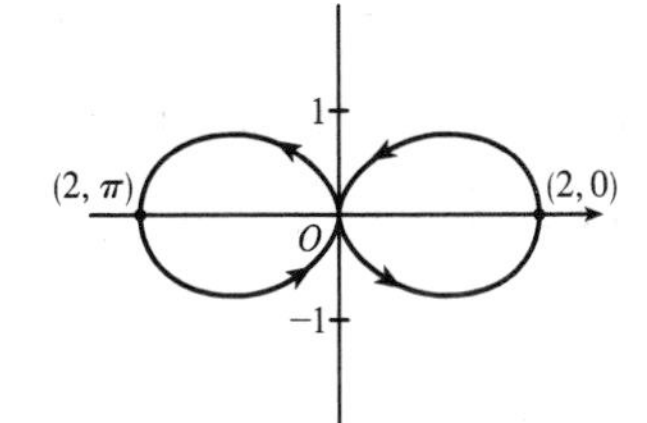

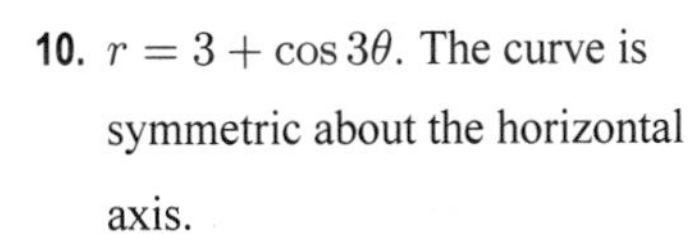
10. $r = 3 + \cos 3\theta$. The curve is symmetric about the horizontal axis.

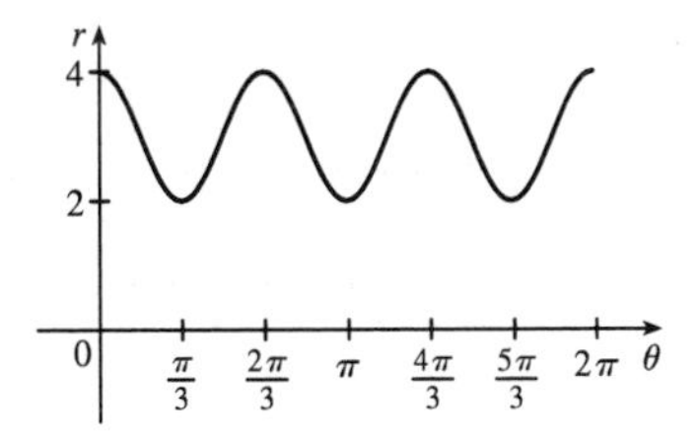

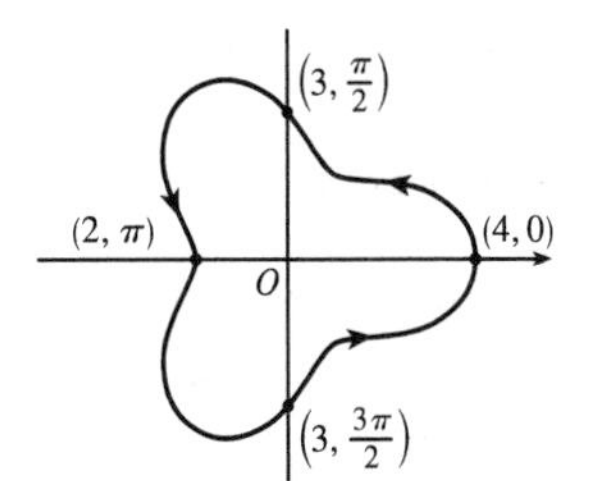

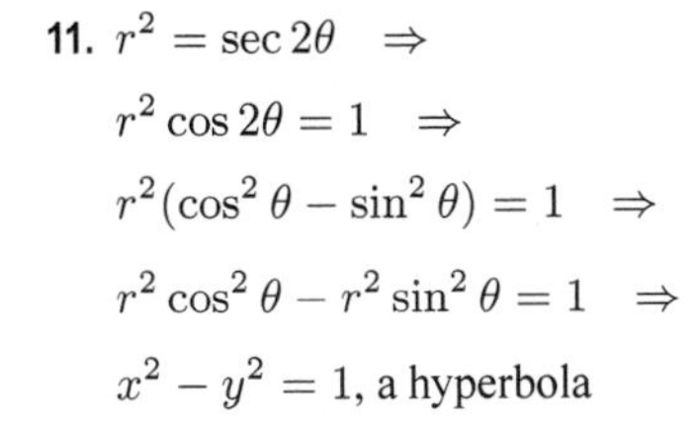
11. $r^2 = \sec 2\theta \quad\Rightarrow$

$r^2 \cos 2\theta = 1 \quad\Rightarrow$

$r^2(\cos^2\theta - \sin^2\theta) = 1 \quad\Rightarrow$

$r^2\cos^2\theta - r^2\sin^2\theta = 1 \quad\Rightarrow$

$x^2 - y^2 = 1$, a hyperbola

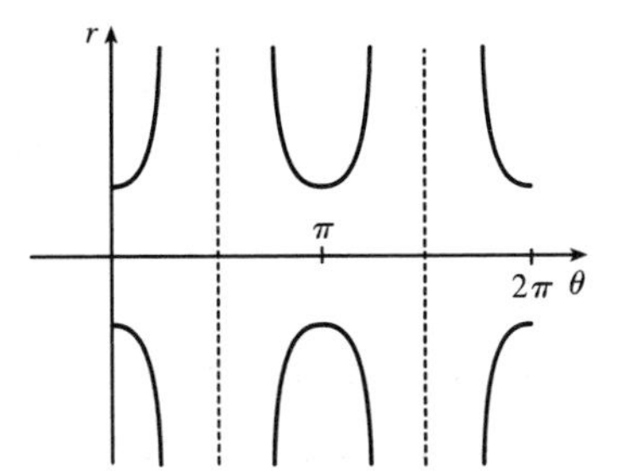

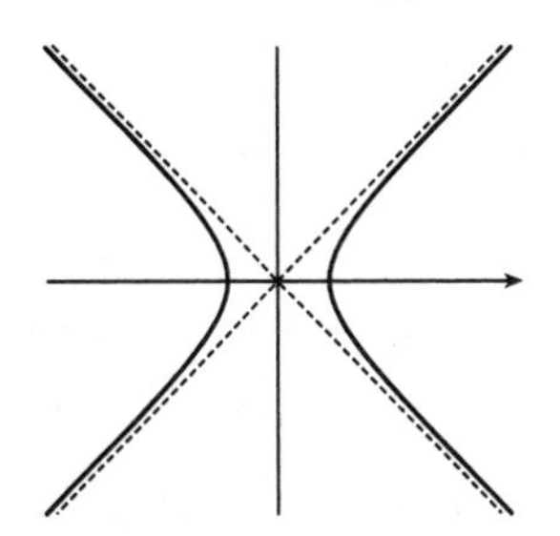

12. $r = 2\cos(\theta/2)$. The curve is symmetric about the pole and both the horizontal and vertical axes.

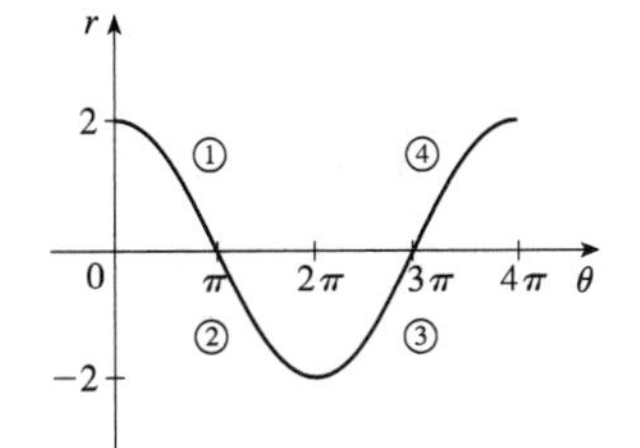

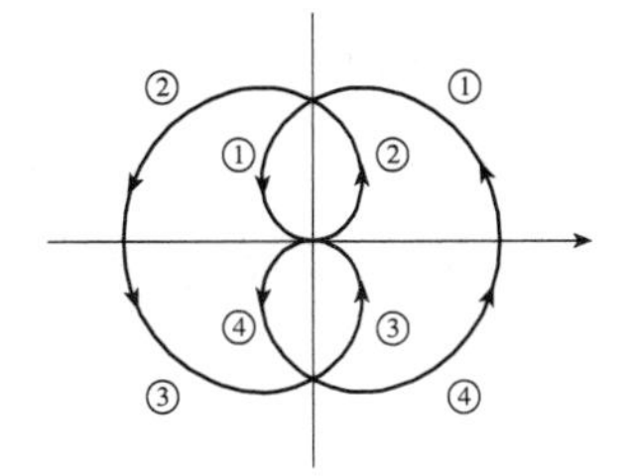

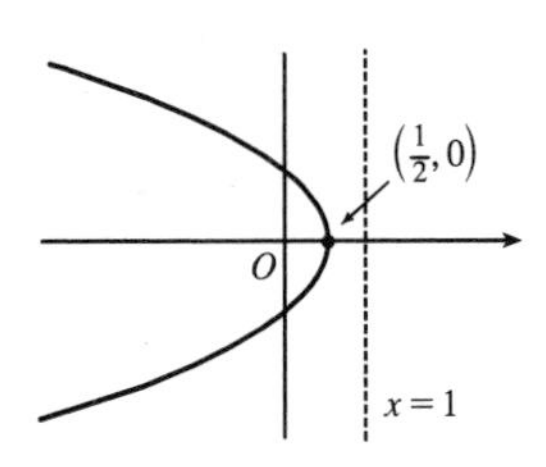

13. $r = \dfrac{1}{1 + \cos\theta} \Rightarrow e = 1 \Rightarrow$ parabola; $d = 1 \Rightarrow$ directrix $x = 1$ and vertex $\left(\frac{1}{2}, 0\right)$; y-intercepts are $\left(1, \frac{\pi}{2}\right)$ and $\left(1, \frac{3\pi}{2}\right)$.

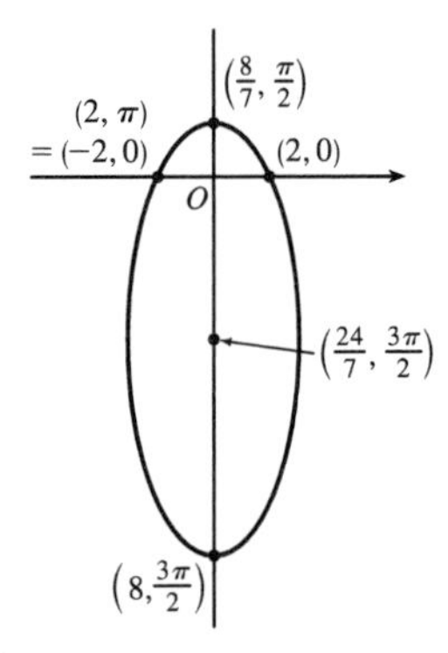

14. $r = \dfrac{8}{4 + 3\sin\theta} = \dfrac{\frac{3}{4} \cdot \frac{8}{3}}{1 + \frac{3}{4}\sin\theta}$. This is an ellipse with focus at the pole, eccentricity $\frac{3}{4}$, and directrix $y = \frac{8}{3}$. The center is $\left(\frac{24}{7}, \frac{3\pi}{2}\right)$.

15. $x + y = 2 \Leftrightarrow r\cos\theta + r\sin\theta = 2 \Leftrightarrow r(\cos\theta + \sin\theta) = 2 \Leftrightarrow r = \dfrac{2}{\cos\theta + \sin\theta}$

16. $x^2 + y^2 = 2 \Rightarrow r^2 = 2 \Rightarrow r = \sqrt{2}$. $\left[r = -\sqrt{2}\text{ gives the same curve.}\right]$

17. $r = (\sin\theta)/\theta$. As $\theta \to \pm\infty$, $r \to 0$. As $\theta \to 0$, $r \to 1$. In the first figure, there are an infinite number of x-intercepts at $x = \pi n$, n a nonzero integer. These correspond to pole points in the second figure.

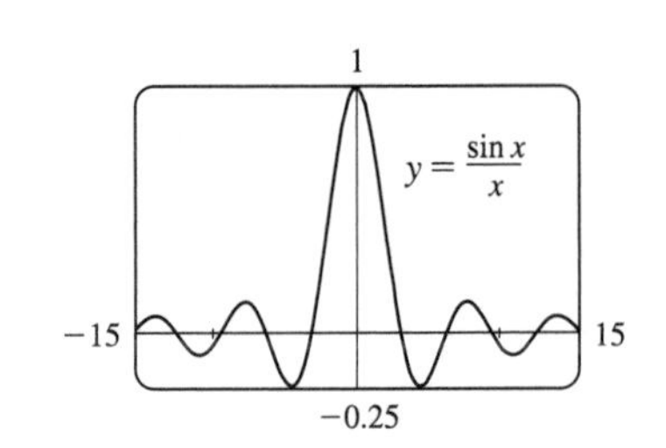

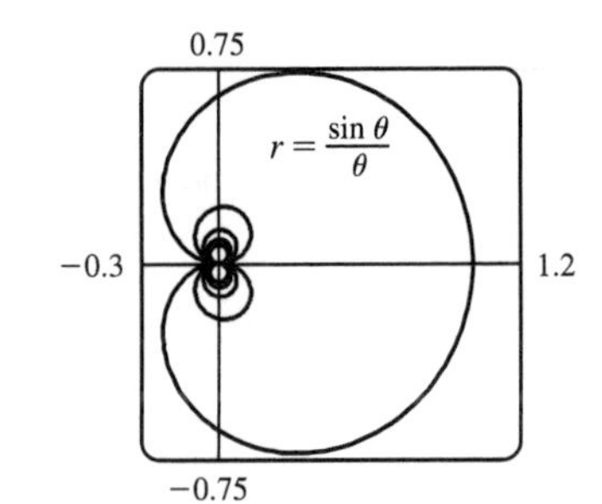

18. $r = \dfrac{2}{4 - 3\cos\theta} = \dfrac{1/2}{1 - \frac{3}{4}\cos\theta} \Rightarrow e = \frac{3}{4}$ and $d = \frac{2}{3}$. The equation of the directrix is $x = -\frac{2}{3} \Rightarrow r = -2/(3\cos\theta)$. To obtain the equation of the rotated ellipse, we replace θ in the original equation with $\theta - \frac{2\pi}{3}$, and get $r = \dfrac{2}{4 - 3\cos\left(\theta - \frac{2\pi}{3}\right)}$.

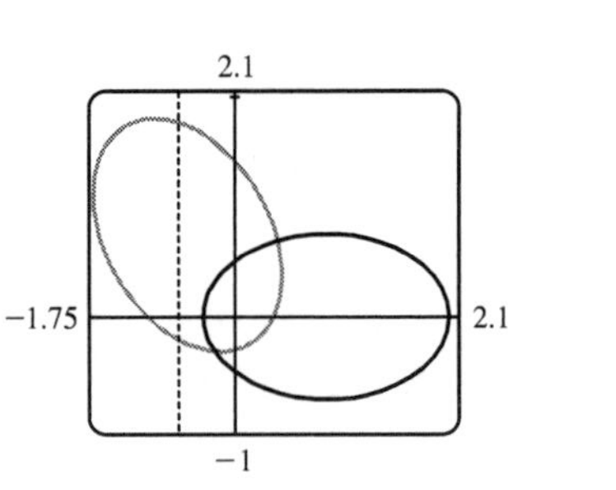

19. $x = \ln t$, $y = 1 + t^2$; $t = 1$. $\dfrac{dy}{dt} = 2t$ and $\dfrac{dx}{dt} = \dfrac{1}{t}$, so $\dfrac{dy}{dx} = \dfrac{dy/dt}{dx/dt} = \dfrac{2t}{1/t} = 2t^2$.

When $t = 1$, $(x, y) = (0, 2)$ and $dy/dx = 2$.

20. $x = t^3 + 6t + 1,\ y = 2t - t^2;\ t = -1.\quad \dfrac{dy}{dx} = \dfrac{dy/dt}{dx/dt} = \dfrac{2 - 2t}{3t^2 + 6}.$

When $t = -1$, $(x, y) = (-6, -3)$ and $dy/dx = 4/9$.

21. $r = e^{-\theta} \quad \Rightarrow \quad y = r\sin\theta = e^{-\theta}\sin\theta$ and $x = r\cos\theta = e^{-\theta}\cos\theta \quad \Rightarrow$

$$\frac{dy}{dx} = \frac{dy/d\theta}{dx/d\theta} = \frac{\frac{dr}{d\theta}\sin\theta + r\cos\theta}{\frac{dr}{d\theta}\cos\theta - r\sin\theta} = \frac{-e^{-\theta}\sin\theta + e^{-\theta}\cos\theta}{-e^{-\theta}\cos\theta - e^{-\theta}\sin\theta} \cdot \frac{-e^{\theta}}{-e^{\theta}} = \frac{\sin\theta - \cos\theta}{\cos\theta + \sin\theta}.$$

When $\theta = \pi$, $\dfrac{dy}{dx} = \dfrac{0 - (-1)}{-1 + 0} = \dfrac{1}{-1} = -1.$

22. $r = 3 + \cos 3\theta \quad \Rightarrow \quad \dfrac{dy}{dx} = \dfrac{dy/d\theta}{dx/d\theta} = \dfrac{\frac{dr}{d\theta}\sin\theta + r\cos\theta}{\frac{dr}{d\theta}\cos\theta - r\sin\theta} = \dfrac{-3\sin 3\theta\sin\theta + (3 + \cos 3\theta)\cos\theta}{-3\sin 3\theta\cos\theta - (3 + \cos 3\theta)\sin\theta}.$

When $\theta = \pi/2$, $\dfrac{dy}{dx} = \dfrac{(-3)(-1)(1) + (3 + 0)\cdot 0}{(-3)(-1)(0) - (3 + 0)\cdot 1} = \dfrac{3}{-3} = -1.$

23. $x = t\cos t,\ y = t\sin t.\quad \dfrac{dy}{dx} = \dfrac{dy/dt}{dx/dt} = \dfrac{t\cos t + \sin t}{-t\sin t + \cos t}.\quad \dfrac{d^2y}{dx^2} = \dfrac{\frac{d}{dt}\left(\frac{dy}{dx}\right)}{dx/dt}$, where

$$\frac{d}{dt}\left(\frac{dy}{dx}\right) = \frac{(-t\sin t + \cos t)(-t\sin t + 2\cos t) - (t\cos t + \sin t)(-t\cos t - 2\sin t)}{(-t\sin t + \cos t)^2} = \frac{t^2 + 2}{(-t\sin t + \cos t)^2} \quad \Rightarrow$$

$$\frac{d^2y}{dx^2} = \frac{t^2 + 2}{(-t\sin t + \cos t)^3}.$$

24. $x = 1 + t^2,\ y = t - t^3.\quad \dfrac{dy}{dt} = 1 - 3t^2$ and $\dfrac{dx}{dt} = 2t$, so $\dfrac{dy}{dx} = \dfrac{dy/dt}{dx/dt} = \dfrac{1 - 3t^2}{2t} = \frac{1}{2}t^{-1} - \frac{3}{2}t.$

$$\frac{d^2y}{dx^2} = \frac{d(dy/dx)/dt}{dx/dt} = \frac{-\frac{1}{2}t^{-2} - \frac{3}{2}}{2t} = -\tfrac{1}{4}t^{-3} - \tfrac{3}{4}t^{-1} = -\frac{1}{4t^3}(1 + 3t^2) = -\frac{3t^2 + 1}{4t^3}.$$

25. We graph the curve $x = t^3 - 3t$, $y = t^2 + t + 1$ for $-2.2 \le t \le 1.2$. By zooming in or using a cursor, we find that the lowest point is about $(1.4, 0.75)$. To find the exact values, we find the t-value at which $dy/dt = 2t + 1 = 0 \;\Leftrightarrow\; t = -\frac{1}{2} \;\Leftrightarrow\; (x, y) = \left(\frac{11}{8}, \frac{3}{4}\right)$.

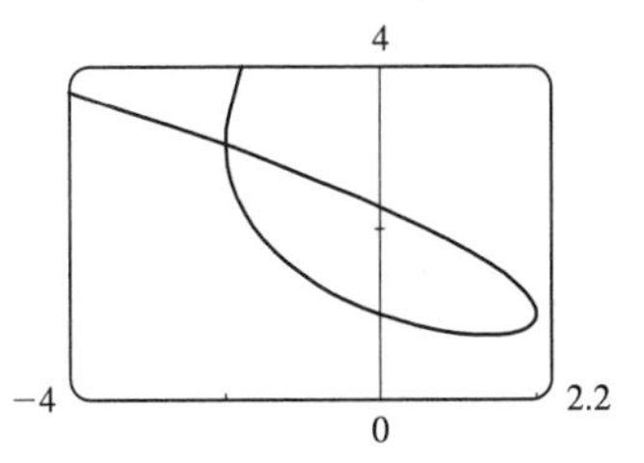

26. We estimate the coordinates of the point of intersection to be $(-2, 3)$. In fact this is exact, since both $t = -2$ and $t = 1$ give the point $(-2, 3)$. So the area enclosed by the loop is

$$\int_{t=-2}^{t=1} y\,dx = \int_{-2}^{1}(t^2 + t + 1)(3t^2 - 3)\,dt = \int_{-2}^{1}(3t^4 + 3t^3 - 3t - 3)\,dt$$

$$= \left[\tfrac{3}{5}t^5 + \tfrac{3}{4}t^4 - \tfrac{3}{2}t^2 - 3t\right]_{-2}^{1} = \left(\tfrac{3}{5} + \tfrac{3}{4} - \tfrac{3}{2} - 3\right) - \left[-\tfrac{96}{5} + 12 - 6 - (-6)\right] = \tfrac{81}{20}$$

27. $x = 2a\cos t - a\cos 2t \quad \Rightarrow \quad \dfrac{dx}{dt} = -2a\sin t + 2a\sin 2t = 2a\sin t(2\cos t - 1) = 0 \quad \Leftrightarrow \quad \sin t = 0 \text{ or } \cos t = \frac{1}{2} \quad \Rightarrow$

$t = 0, \frac{\pi}{3}, \pi, \text{ or } \frac{5\pi}{3}.$

$y = 2a\sin t - a\sin 2t \quad \Rightarrow \quad \dfrac{dy}{dt} = 2a\cos t - 2a\cos 2t = 2a\left(1 + \cos t - 2\cos^2 t\right) = 2a(1 - \cos t)(1 + 2\cos t) = 0 \quad \Rightarrow$

$t = 0, \frac{2\pi}{3}, \text{ or } \frac{4\pi}{3}.$

Thus, the graph has vertical tangents where $t = \frac{\pi}{3}$, π and $\frac{5\pi}{3}$, and horizontal tangents where $t = \frac{2\pi}{3}$ and $\frac{4\pi}{3}$. To determine what the slope is where $t = 0$, we use l'Hospital's Rule to evaluate $\lim\limits_{t\to 0} \dfrac{dy/dt}{dx/dt} = 0$, so there is a horizontal tangent there.

t	x	y
0	a	0
$\frac{\pi}{3}$	$\frac{3}{2}a$	$\frac{\sqrt{3}}{2}a$
$\frac{2\pi}{3}$	$-\frac{1}{2}a$	$\frac{3\sqrt{3}}{2}a$
π	$-3a$	0
$\frac{4\pi}{3}$	$-\frac{1}{2}a$	$-\frac{3\sqrt{3}}{2}a$
$\frac{5\pi}{3}$	$\frac{3}{2}a$	$-\frac{\sqrt{3}}{2}a$

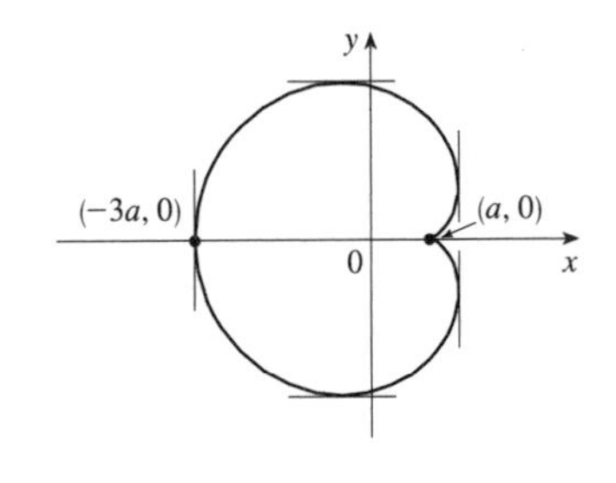

28. From Exercise 27, $x = 2a\cos t - a\cos 2t$, $y = 2a\sin t - a\sin 2t \quad \Rightarrow$

$A = 2\int_{\pi}^{0} (2a\sin t - a\sin 2t)(-2a\sin t + 2a\sin 2t)\,dt = 4a^2 \int_0^{\pi} (2\sin^2 t + \sin^2 2t - 3\sin t\sin 2t)\,dt$

$= 4a^2 \int_0^{\pi} \left[(1 - \cos 2t) + \frac{1}{2}(1 - \cos 4t) - 6\sin^2 t\cos t\right] dt = 4a^2\left[t - \frac{1}{2}\sin 2t + \frac{1}{2}t - \frac{1}{8}\sin 4t - 2\sin^3 t\right]_0^{\pi}$

$= 4a^2\left(\frac{3}{2}\right)\pi = 6\pi a^2$

29. The curve $r^2 = 9\cos 5\theta$ has 10 "petals." For instance, for $-\frac{\pi}{10} \le \theta \le \frac{\pi}{10}$, there are two petals, one with $r > 0$ and one with $r < 0$. $A = 10\displaystyle\int_{-\pi/10}^{\pi/10} \frac{1}{2}r^2\,d\theta = 5\int_{-\pi/10}^{\pi/10} 9\cos 5\theta\,d\theta = 5 \cdot 9 \cdot 2\int_0^{\pi/10} \cos 5\theta\,d\theta = 18\left[\sin 5\theta\right]_0^{\pi/10} = 18.$

30. $r = 1 - 3\sin\theta$. The inner loop is traced out as θ goes from $\alpha = \sin^{-1}\left(\frac{1}{3}\right)$ to $\pi - \alpha$, so

$$A = \int_{\alpha}^{\pi-\alpha} \tfrac{1}{2}r^2\,d\theta = \int_{\alpha}^{\pi/2}(1 - 3\sin\theta)^2\,d\theta = \int_{\alpha}^{\pi/2}\left[1 - 6\sin\theta + \tfrac{9}{2}(1 - \cos 2\theta)\right] d\theta$$

$$= \left[\tfrac{11}{2}\theta + 6\cos\theta - \tfrac{9}{4}\sin 2\theta\right]_{\alpha}^{\pi/2} = \tfrac{11}{4}\pi - \tfrac{11}{2}\sin^{-1}\left(\tfrac{1}{3}\right) - 3\sqrt{2}$$

31. The curves intersect when $4\cos\theta = 2 \quad \Rightarrow \quad \cos\theta = \frac{1}{2} \quad \Rightarrow$ $\theta = \pm\frac{\pi}{3}$ for $-\pi \le \theta \le \pi$. The points of intersection are $\left(2, \frac{\pi}{3}\right)$ and $\left(2, -\frac{\pi}{3}\right)$.

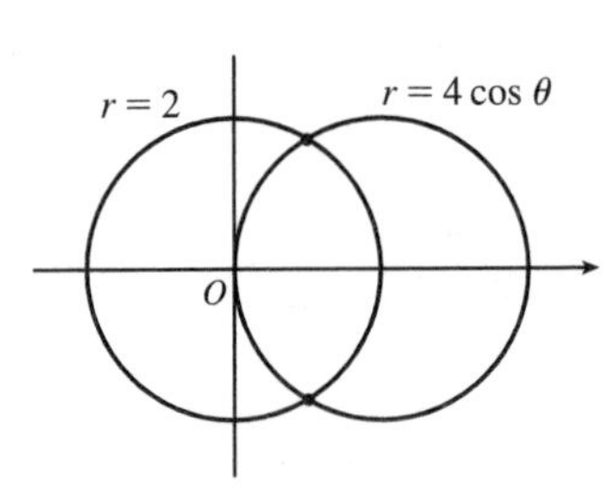

32. The two curves clearly both contain the pole. For other points of intersection, $\cot\theta = 2\cos(\theta + 2n\pi)$ or $-2\cos(\theta + \pi + 2n\pi)$, both of which reduce to $\cot\theta = 2\cos\theta \;\Leftrightarrow\; \cos\theta = 2\sin\theta\cos\theta \;\Leftrightarrow\; \cos\theta(1 - 2\sin\theta) = 0 \;\Rightarrow$ $\cos\theta = 0$ or $\sin\theta = \frac{1}{2} \;\Rightarrow\; \theta = \frac{\pi}{6}, \frac{\pi}{2}, \frac{5\pi}{6}$ or $\frac{3\pi}{2} \;\Rightarrow$ intersection points are $\left(0, \frac{\pi}{2}\right)$, $\left(\sqrt{3}, \frac{\pi}{6}\right)$, and $\left(\sqrt{3}, \frac{11\pi}{6}\right)$.

33. The curves intersect where $2\sin\theta = \sin\theta + \cos\theta \;\Rightarrow$ $\sin\theta = \cos\theta \;\Rightarrow \theta = \frac{\pi}{4}$, and also at the origin (at which $\theta = \frac{3\pi}{4}$ on the second curve).

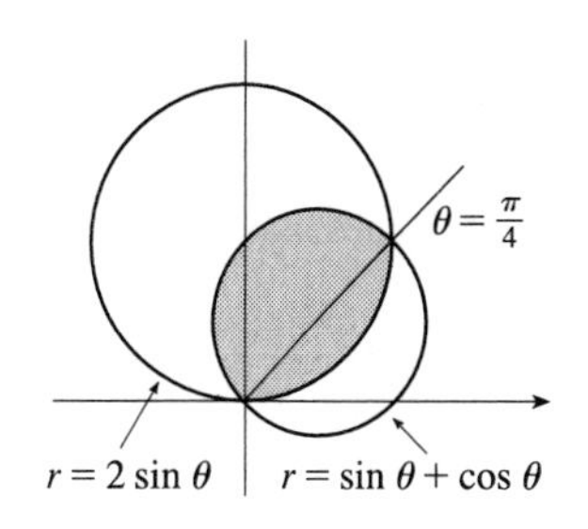

$$A = \int_0^{\pi/4} \tfrac{1}{2}(2\sin\theta)^2\,d\theta + \int_{\pi/4}^{3\pi/4} \tfrac{1}{2}(\sin\theta + \cos\theta)^2\,d\theta$$
$$= \int_0^{\pi/4}(1 - \cos 2\theta)sd\theta + \tfrac{1}{2}\int_{\pi/4}^{3\pi/4}(1 + \sin 2\theta)\,d\theta$$
$$= \left[\theta - \tfrac{1}{2}\sin 2\theta\right]_0^{\pi/4} + \left[\tfrac{1}{2}\theta - \tfrac{1}{4}\cos 2\theta\right]_{\pi/4}^{3\pi/4} = \tfrac{1}{2}(\pi - 1)$$

34. $A = 2\int_{-\pi/2}^{\pi/6} \frac{1}{2}\left[(2 + \cos 2\theta)^2 - (2 + \sin\theta)^2\right]d\theta$

$$= \int_{-\pi/2}^{\pi/6}[4\cos 2\theta + \cos^2 2\theta - 4\sin\theta - \sin^2\theta]\,d\theta$$
$$= \left[2\sin 2\theta + \tfrac{1}{2}\theta + \tfrac{1}{8}\sin 4\theta + 4\cos\theta - \tfrac{1}{2}\theta + \tfrac{1}{4}\sin 2\theta\right]_{-\pi/2}^{\pi/6}$$
$$= \tfrac{51}{16}\sqrt{3}$$

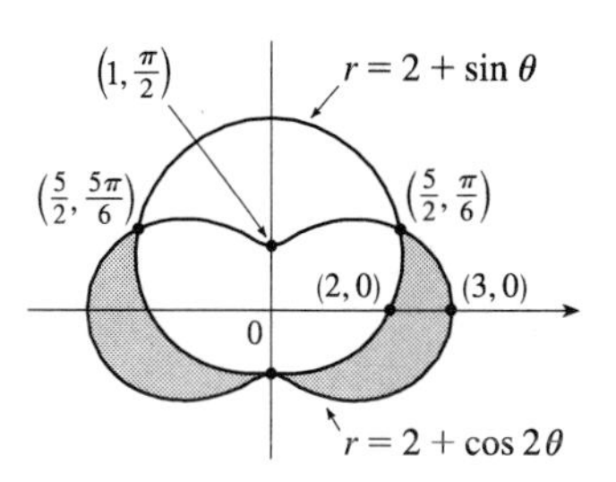

35. $x = 3t^2,\; y = 2t^3$.

$$L = \int_0^2 \sqrt{(dx/dt)^2 + (dy/dt)^2}\,dt = \int_0^2 \sqrt{(6t)^2 + (6t^2)^2}\,dt$$
$$= \int_0^2 \sqrt{36t^2 + 36t^4}\,dt = \int_0^2 \sqrt{36t^2}\,\sqrt{1 + t^2}\,dt$$
$$= \int_0^2 6\,|t|\,\sqrt{1 + t^2}\,dt = 6\int_0^2 t\sqrt{1 + t^2}\,dt = 6\int_1^5 u^{1/2}\left(\tfrac{1}{2}du\right) \qquad \left[u = 1 + t^2,\, du = 2t\,dt\right]$$
$$= 6 \cdot \tfrac{1}{2} \cdot \tfrac{2}{3}\left[u^{3/2}\right]_1^5 = 2(5^{3/2} - 1) = 2(5\sqrt{5} - 1)$$

36. $x = 2 + 3t,\; y = \cosh 3t \;\Rightarrow\; (dx/dt)^2 + (dy/dt)^2 = 3^2 + (3\sinh 3t)^2 = 9(1 + \sinh^2 3t) = 9\cosh^2 3t$, so

$$L = \int_0^1 \sqrt{9\cosh^2 3t}\,dt = \int_0^1 |3\cosh 3t|\,dt = \int_0^1 3\cosh 3t\,dt$$
$$= \left[\sinh 3t\right]_0^1 = \sinh 3 - \sinh 0 = \sinh 3$$

37. $L = \int_\pi^{2\pi} \sqrt{r^2 + (dr/d\theta)^2}\,d\theta = \int_\pi^{2\pi} \sqrt{(1/\theta)^2 + \left(-1/\theta^2\right)^2}\,d\theta = \displaystyle\int_\pi^{2\pi} \frac{\sqrt{\theta^2 + 1}}{\theta^2}\,d\theta$

$$\overset{24}{=} \left[-\frac{\sqrt{\theta^2 + 1}}{\theta} + \ln\left(\theta + \sqrt{\theta^2 + 1}\right)\right]_\pi^{2\pi} = \frac{\sqrt{\pi^2 + 1}}{\pi} - \frac{\sqrt{4\pi^2 + 1}}{2\pi} + \ln\left(\frac{2\pi + \sqrt{4\pi^2 + 1}}{\pi + \sqrt{\pi^2 + 1}}\right)$$
$$= \frac{2\sqrt{\pi^2 + 1} - \sqrt{4\pi^2 + 1}}{2\pi} + \ln\left(\frac{2\pi + \sqrt{4\pi^2 + 1}}{\pi + \sqrt{\pi^2 + 1}}\right)$$

38. $L = \int_0^\pi \sqrt{r^2 + (dr/d\theta)^2}\, d\theta = \int_0^\pi \sqrt{\sin^6\left(\frac{1}{3}\theta\right) + \sin^4\left(\frac{1}{3}\theta\right)\cos^2\left(\frac{1}{3}\theta\right)}\, d\theta = \int_0^\pi \sin^2\left(\frac{1}{3}\theta\right) d\theta$

$= \left[\frac{1}{2}\left(\theta - \frac{3}{2}\sin\left(\frac{2}{3}\theta\right)\right)\right]_0^\pi = \frac{1}{2}\pi - \frac{3}{8}\sqrt{3}$

39. For all c except -1, the curve is asymptotic to the line $x = 1$. For $c < -1$, the curve bulges to the right near $y = 0$. As c increases, the bulge becomes smaller, until at $c = -1$ the curve is the straight line $x = 1$. As c continues to increase, the curve bulges to the left, until at $c = 0$ there is a cusp at the origin. For $c > 0$, there is a loop to the left of the origin, whose size and roundness increase as c increases. Note that the x-intercept of the curve is always $-c$.

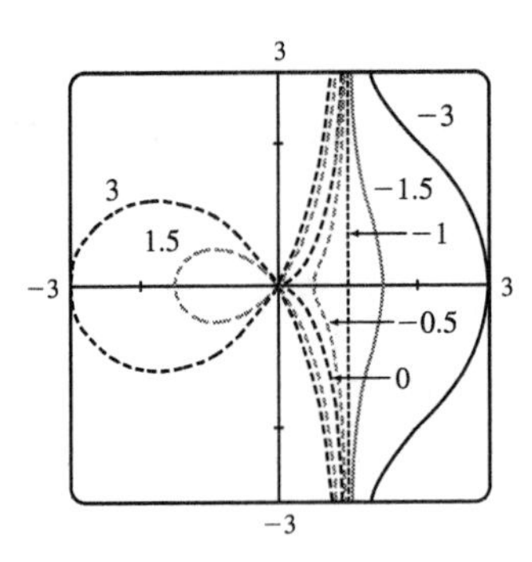

40. For a close to 0, the graph consists of four thin petals. As a increases, the petals get fatter, until as $a \to \infty$, each petal occupies almost its entire quarter-circle.

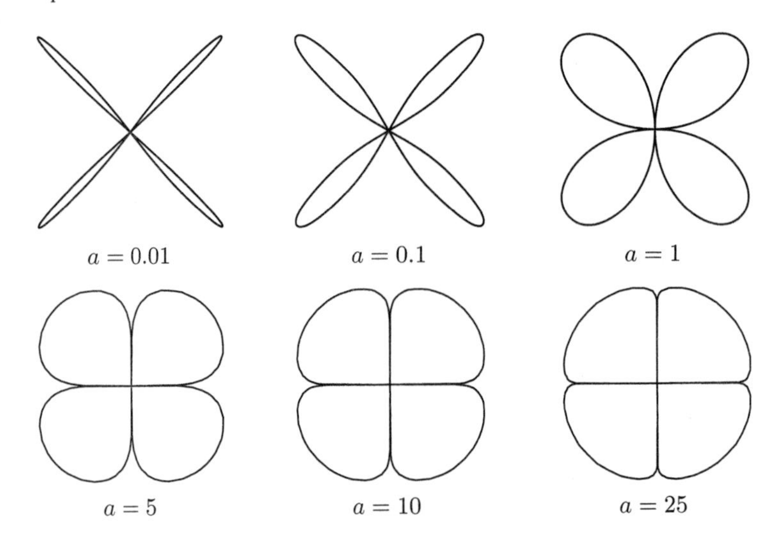

41. Directrix $x = 4 \Rightarrow d = 4$, so $e = \frac{1}{3} \Rightarrow r = \dfrac{ed}{1 + e\cos\theta} = \dfrac{4}{3 + \cos\theta}$.

42. See the end of the proof of Theorem 9.5.8. If $e > 1$, then $1 - e^2 < 0$ and Equations 9.5.6 become $a^2 = \dfrac{e^2 d^2}{(e^2 - 1)^2}$ and $b^2 = \dfrac{e^2 d^2}{e^2 - 1}$, so $\dfrac{b^2}{a^2} = e^2 - 1$. The asymptotes $y = \pm\dfrac{b}{a}x$ have slopes $\pm\dfrac{b}{a} = \pm\sqrt{e^2 - 1}$, so the angles they make with the polar axis are $\pm\tan^{-1}\left[\sqrt{e^2 - 1}\right] = \cos^{-1}(\pm 1/e)$.

43. In polar coordinates, an equation for the circle is $r = 2a\sin\theta$. Thus, the coordinates of Q are $x = r\cos\theta = 2a\sin\theta\cos\theta$ and $y = r\sin\theta = 2a\sin^2\theta$. The coordinates of R are $x = 2a\cot\theta$ and $y = 2a$. Since P is the midpoint of QR, we use the midpoint formula to get $x = a(\sin\theta\cos\theta + \cot\theta)$ and $y = a(1 + \sin^2\theta)$.

10 □ VECTORS AND THE GEOMETRY OF SPACE

10.1 Three-Dimensional Coordinate Systems

1. We start at the origin, which has coordinates $(0, 0, 0)$. First we move 4 units along the positive x-axis, affecting only the x-coordinate, bringing us to the point $(4, 0, 0)$. We then move 3 units straight downward, in the negative z-direction. Thus only the z-coordinate is affected, and we arrive at $(4, 0, -3)$.

2.

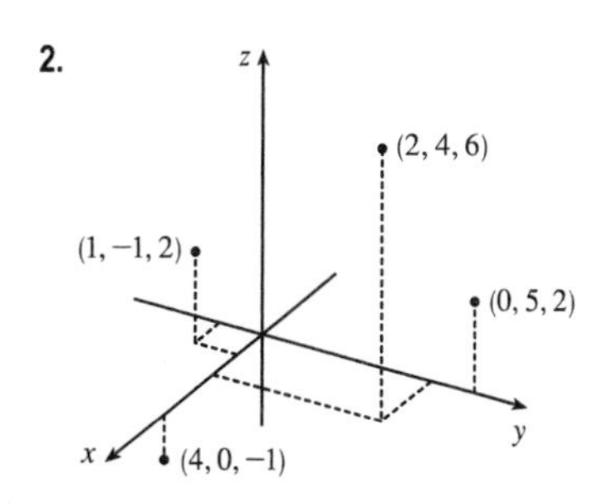

3. The distance from a point to the xz-plane is the absolute value of the y-coordinate of the point. $Q(-5, -1, 4)$ has the y-coordinate with the smallest absolute value, so Q is the point closest to the xz-plane. $R(0, 3, 8)$ must lie in the yz-plane since the distance from R to the yz-plane, given by the x-coordinate of R, is 0.

4. The projection of $(2, 3, 5)$ on the xy-plane is $(2, 3, 0)$; on the yz-plane, $(0, 3, 5)$; on the xz-plane, $(2, 0, 5)$.

The length of the diagonal of the box is the distance between the origin and $(2, 3, 5)$, given by

$$\sqrt{(2-0)^2 + (3-0)^2 + (5-0)^2} = \sqrt{38} \approx 6.16$$

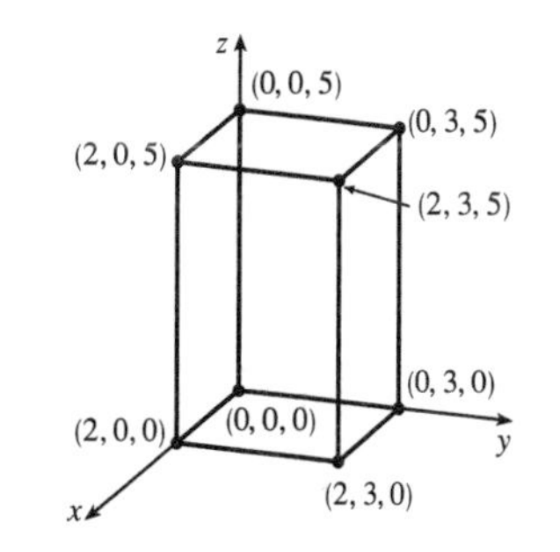

5. The equation $x + y = 2$ represents the set of all points in $\mathbb{R}^3$ whose x- and y-coordinates have a sum of 2, or equivalently where $y = 2 - x$. This is the set $\{(x, 2 - x, z) \mid x \in \mathbb{R}, z \in \mathbb{R}\}$ which is a vertical plane that intersects the xy-plane in the line $y = 2 - x$, $z = 0$.

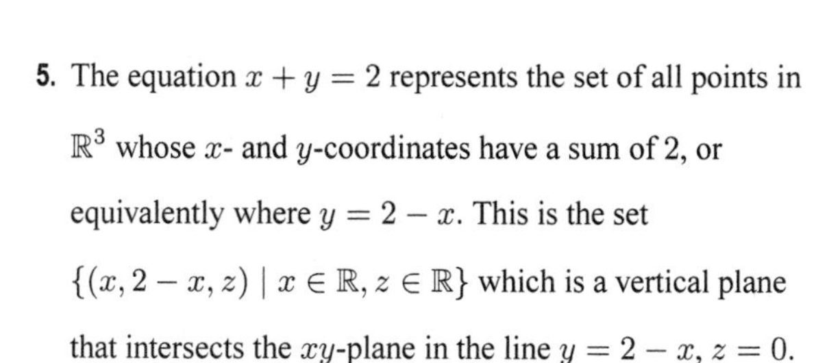

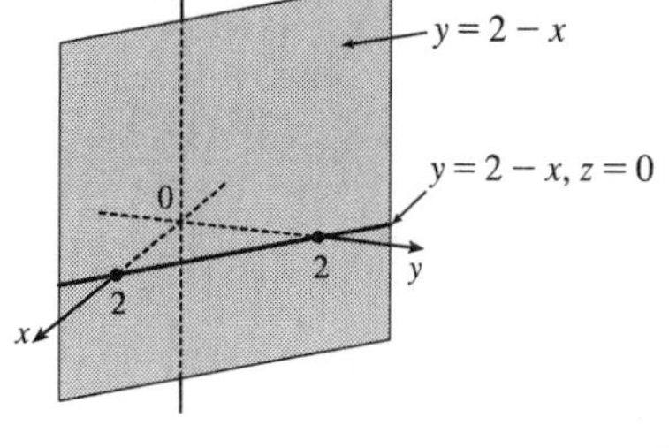

6. (a) In $\mathbb{R}^2$, the equation $x = 4$ represents a line parallel to the y-axis. In $\mathbb{R}^3$, the equation $x = 4$ represents the set $\{(x, y, z) \mid x = 4\}$, the set of all points whose x-coordinate is 4. This is the vertical plane that is parallel to the yz-plane and 4 units in front of it.

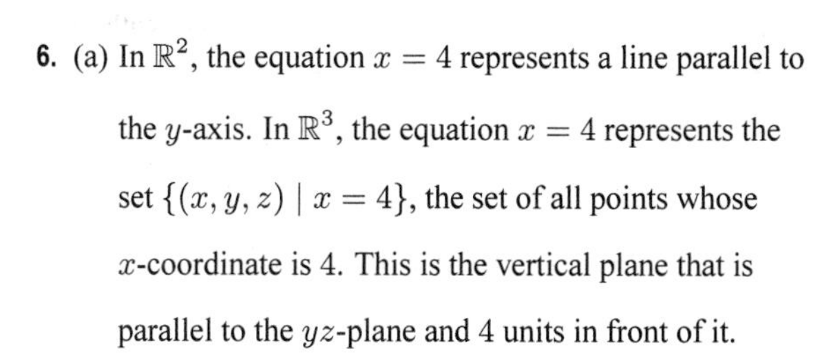

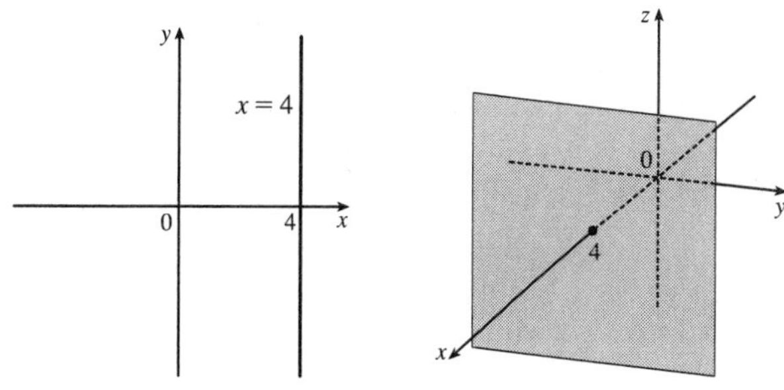

(b) In $\mathbb{R}^3$, the equation $y = 3$ represents a vertical plane that is parallel to the xz-plane and 3 units to the right of it. The equation $z = 5$ represents a horizontal plane parallel to the xy-plane and 5 units above it. The pair of equations $y = 3$, $z = 5$ represents the set of points that are simultaneously on both planes, or in other words, the line of intersection of the planes $y = 3$, $z = 5$. This line can also be described as the set $\{(x, 3, 5) \mid x \in \mathbb{R}\}$, which is the set of all points in $\mathbb{R}^3$ whose x-coordinate may vary but whose y- and z-coordinates are fixed at 3 and 5, respectively. Thus the line is parallel to the x-axis and intersects the yz-plane in the point $(0, 3, 5)$.

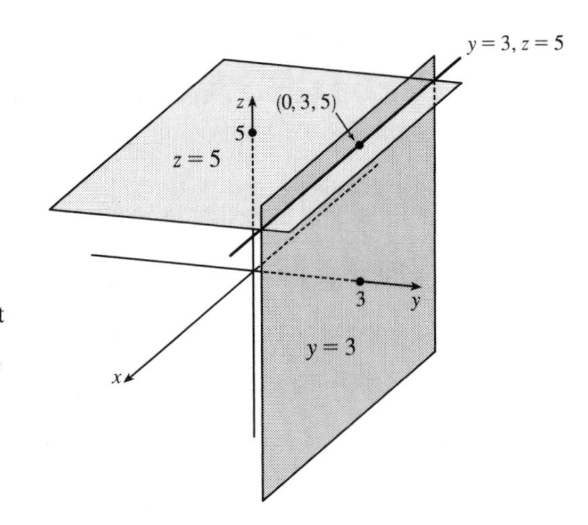

7. (a) We can find the lengths of the sides of the triangle by using the distance formula between pairs of vertices:

$$|PQ| = \sqrt{(7-3)^2 + [0-(-2)]^2 + [1-(-3)]^2} = \sqrt{16+4+16} = 6$$

$$|QR| = \sqrt{(1-7)^2 + (2-0)^2 + (1-1)^2} = \sqrt{36+4+0} = \sqrt{40} = 2\sqrt{10}$$

$$|RP| = \sqrt{(3-1)^2 + (-2-2)^2 + (-3-1)^2} = \sqrt{4+16+16} = 6$$

The longest side is QR, but the Pythagorean Theorem is not satisfied: $|PQ|^2 + |RP|^2 \neq |QR|^2$. Thus PQR is not a right triangle. PQR is isosceles, as two sides have the same length.

(b) Compute the lengths of the sides of the triangle by using the distance formula between pairs of vertices:

$$|PQ| = \sqrt{(4-2)^2 + [1-(-1)]^2 + (1-0)^2} = \sqrt{4+4+1} = 3$$

$$|QR| = \sqrt{(4-4)^2 + (-5-1)^2 + (4-1)^2} = \sqrt{0+36+9} = \sqrt{45} = 3\sqrt{5}$$

$$|RP| = \sqrt{(2-4)^2 + [-1-(-5)]^2 + (0-4)^2} = \sqrt{4+16+16} = 6$$

Since the Pythagorean Theorem is satisfied by $|PQ|^2 + |RP|^2 = |QR|^2$, PQR is a right triangle. PQR is not isosceles, as no two sides have the same length.

8. (a) The distance from a point to the xy-plane is the absolute value of the z-coordinate of the point. Thus, the distance is $|-5| = 5$.

(b) Similarly, the distance to the yz-plane is the absolute value of the x-coordinate of the point: $|3| = 3$.

(c) The distance to the xz-plane is the absolute value of the y-coordinate of the point: $|7| = 7$.

(d) The point on the x-axis closest to $(3, 7, -5)$ is the point $(3, 0, 0)$. (Approach the x-axis perpendicularly.) The distance from $(3, 7, -5)$ to the x-axis is the distance between these two points:

$\sqrt{(3-3)^2 + (7-0)^2 + (-5-0)^2} = \sqrt{74} \approx 8.60.$

(e) The point on the y-axis closest to $(3, 7, -5)$ is $(0, 7, 0)$. The distance between these points is

$\sqrt{(3-0)^2 + (7-7)^2 + (-5-0)^2} = \sqrt{34} \approx 5.83.$

(f) The point on the z-axis closest to $(3, 7, -5)$ is $(0, 0, -5)$. The distance between these points is

$\sqrt{(3-0)^2 + (7-0)^2 + [-5-(-5)]^2} = \sqrt{58} \approx 7.62.$

9. (a) First we find the distances between points:

$$|AB| = \sqrt{(3-2)^2 + (7-4)^2 + (-2-2)^2} = \sqrt{26}$$
$$|BC| = \sqrt{(1-3)^2 + (3-7)^2 + [3-(-2)]^2} = \sqrt{45} = 3\sqrt{5}$$
$$|AC| = \sqrt{(1-2)^2 + (3-4)^2 + (3-2)^2} = \sqrt{3}$$

In order for the points to lie on a straight line, the sum of the two shortest distances must be equal to the longest distance. Since $\sqrt{26} + \sqrt{3} \neq 3\sqrt{5}$, the three points do not lie on a straight line.

(b) First we find the distances between points:

$$|DE| = \sqrt{(1-0)^2 + [-2-(-5)]^2 + (4-5)^2} = \sqrt{11}$$
$$|EF| = \sqrt{(3-1)^2 + [4-(-2)]^2 + (2-4)^2} = \sqrt{44} = 2\sqrt{11}$$
$$|DF| = \sqrt{(3-0)^2 + [4-(-5)]^2 + (2-5)^2} = \sqrt{99} = 3\sqrt{11}$$

Since $|DE| + |EF| = |DF|$, the three points lie on a straight line.

10. An equation of the sphere with center $(2, -6, 4)$ and radius 5 is $(x-2)^2 + [y-(-6)]^2 + (z-4)^2 = 5^2$ or $(x-2)^2 + (y+6)^2 + (z-4)^2 = 25$. The intersection of this sphere with the xy-plane is the set of points on the sphere whose z-coordinate is 0. Putting $z = 0$ into the equation, we have $(x-2)^2 + (y+6)^2 = 9$, $z = 0$ which represents a circle in the xy-plane with center $(2, -6, 0)$ and radius 3. To find the intersection with the xz-plane, we set $y = 0$: $(x-2)^2 + (z-4)^2 = -11$. Since no points satisfy this equation, the sphere does not intersect the xz-plane. (Also note that the distance from the center of the sphere to the xz-plane is greater than the radius of the sphere.) To find the intersection with the yz-plane, we set $x = 0$: $(y+6)^2 + (z-4)^2 = 21$, $x = 0$, a circle in the yz-plane with center $(0, -6, 4)$ and radius $\sqrt{21}$.

11. The radius of the sphere is the distance between $(4, 3, -1)$ and $(3, 8, 1)$: $r = \sqrt{(3-4)^2 + (8-3)^2 + [1-(-1)]^2} = \sqrt{30}$. Thus, an equation of the sphere is $(x-3)^2 + (y-8)^2 + (z-1)^2 = 30$.

12. If the sphere passes through the origin, the radius of the sphere must be the distance from the origin to the point $(1, 2, 3)$: $r = \sqrt{(1-0)^2 + (2-0)^2 + (3-0)^2} = \sqrt{14}$. Then an equation of the sphere is $(x-1)^2 + (y-2)^2 + (z-3)^2 = 14$.

13. Completing squares in the equation $x^2 + y^2 + z^2 - 6x + 4y - 2z = 11$ gives $(x^2 - 6x + 9) + (y^2 + 4y + 4) + (z^2 - 2z + 1) = 11 + 9 + 4 + 1 \Rightarrow (x-3)^2 + (y+2)^2 + (z-1)^2 = 25$, which we recognize as an equation of a sphere with center $(3, -2, 1)$ and radius 5.

14. Completing squares in the equation gives $(x^2 - 4x + 4) + (y^2 + 2y + 1) + z^2 = 0 + 4 + 1 \Rightarrow$ $(x-2)^2 + (y+1)^2 + z^2 = 5$ which we recognize as an equation of a sphere with center $(2, -1, 0)$ and radius $\sqrt{5}$.

15. Completing squares in the equation gives $\left(x^2 - x + \frac{1}{4}\right) + \left(y^2 - y + \frac{1}{4}\right) + \left(z^2 - z + \frac{1}{4}\right) = \frac{1}{4} + \frac{1}{4} + \frac{1}{4} \Rightarrow$ $\left(x - \frac{1}{2}\right)^2 + \left(y - \frac{1}{2}\right)^2 + \left(z - \frac{1}{2}\right)^2 = \frac{3}{4}$ which we recognize as an equation of a sphere with center $\left(\frac{1}{2}, \frac{1}{2}, \frac{1}{2}\right)$ and radius $\sqrt{\frac{3}{4}} = \frac{\sqrt{3}}{2}$.

16. Completing squares in the equation gives $4(x^2 - 2x + 1) + 4(y^2 + 4y + 4) + 4z^2 = 1 + 4 + 16 \Rightarrow$

$4(x-1)^2 + 4(y+2)^2 + 4z^2 = 21 \Rightarrow (x-1)^2 + (y+2)^2 + z^2 = \frac{21}{4}$, which we recognize as an equation of a sphere with center $(1, -2, 0)$ and radius $\sqrt{\frac{21}{4}} = \frac{\sqrt{21}}{2}$.

17. (a) If the midpoint of the line segment from $P_1(x_1, y_1, z_1)$ to $P_2(x_2, y_2, z_2)$ is $Q = \left(\frac{x_1 + x_2}{2}, \frac{y_1 + y_2}{2}, \frac{z_1 + z_2}{2}\right)$, then the distances $|P_1Q|$ and $|QP_2|$ are equal, and each is half of $|P_1P_2|$. We verify that this is the case:

$$|P_1P_2| = \sqrt{(x_2 - x_1)^2 + (y_2 - y_1)^2 + (z_2 - z_1)^2}$$

$$\begin{aligned}|P_1Q| &= \sqrt{\left[\tfrac{1}{2}(x_1 + x_2) - x_1\right]^2 + \left[\tfrac{1}{2}(y_1 + y_2) - y_1\right]^2 + \left[\tfrac{1}{2}(z_1 + z_2) - z_1\right]^2} \\ &= \sqrt{\left(\tfrac{1}{2}x_2 - \tfrac{1}{2}x_1\right)^2 + \left(\tfrac{1}{2}y_2 - \tfrac{1}{2}y_1\right)^2 + \left(\tfrac{1}{2}z_2 - \tfrac{1}{2}z_1\right)^2} \\ &= \sqrt{\left(\tfrac{1}{2}\right)^2\left[(x_2 - x_1)^2 + (y_2 - y_1)^2 + (z_2 - z_1)^2\right]} = \tfrac{1}{2}\sqrt{(x_2 - x_1)^2 + (y_2 - y_1)^2 + (z_2 - z_1)^2} \\ &= \tfrac{1}{2}\,|P_1P_2|\end{aligned}$$

$$\begin{aligned}|QP_2| &= \sqrt{\left[x_2 - \tfrac{1}{2}(x_1 + x_2)\right]^2 + \left[y_2 - \tfrac{1}{2}(y_1 + y_2)\right]^2 + \left[z_2 - \tfrac{1}{2}(z_1 + z_2)\right]^2} \\ &= \sqrt{\left(\tfrac{1}{2}x_2 - \tfrac{1}{2}x_1\right)^2 + \left(\tfrac{1}{2}y_2 - \tfrac{1}{2}y_1\right)^2 + \left(\tfrac{1}{2}z_2 - \tfrac{1}{2}z_1\right)^2} = \tfrac{1}{2}\sqrt{(x_2 - x_1)^2 + (y_2 - y_1)^2 + (z_2 - z_1)^2} \\ &= \sqrt{\left(\tfrac{1}{2}\right)^2\left[(x_2 - x_1)^2 + (y_2 - y_1)^2 + (z_2 - z_1)^2\right]} = \tfrac{1}{2}\,|P_1P_2|\end{aligned}$$

So Q is indeed the midpoint of P_1P_2.

(b) By part (a), the midpoints of sides AB, BC and CA are $P_1\left(-\frac{1}{2}, 1, 4\right)$, $P_2\left(1, \frac{1}{2}, 5\right)$ and $P_3\left(\frac{5}{2}, \frac{3}{2}, 4\right)$. (Recall that a median of a triangle is a line segment from a vertex to the midpoint of the opposite side.) Then the lengths of the medians are:

$$|AP_2| = \sqrt{0^2 + \left(\tfrac{1}{2} - 2\right)^2 + (5 - 3)^2} = \sqrt{\tfrac{9}{4} + 4} = \sqrt{\tfrac{25}{4}} = \tfrac{5}{2}$$

$$|BP_3| = \sqrt{\left(\tfrac{5}{2} + 2\right)^2 + \left(\tfrac{3}{2}\right)^2 + (4 - 5)^2} = \sqrt{\tfrac{81}{4} + \tfrac{9}{4} + 1} = \sqrt{\tfrac{94}{4}} = \tfrac{1}{2}\sqrt{94}$$

$$|CP_1| = \sqrt{\left(-\tfrac{1}{2} - 4\right)^2 + (1 - 1)^2 + (4 - 5)^2} = \sqrt{\tfrac{81}{4} + 1} = \tfrac{1}{2}\sqrt{85}$$

18. By Exercise 17(a), the midpoint of the diameter (and thus the center of the sphere) is $C(3, 2, 7)$. The radius is half the diameter, so $r = \frac{1}{2}\sqrt{(4-2)^2 + (3-1)^2 + (10-4)^2} = \frac{1}{2}\sqrt{44} = \sqrt{11}$. Therefore an equation of the sphere is $(x-3)^2 + (y-2)^2 + (z-7)^2 = 11$.

19. (a) Since the sphere touches the xy-plane, its radius is the distance from its center, $(2, -3, 6)$, to the xy-plane, namely 6. Therefore $r = 6$ and an equation of the sphere is $(x-2)^2 + (y+3)^2 + (z-6)^2 = 6^2 = 36$.

(b) The radius of this sphere is the distance from its center $(2, -3, 6)$ to the yz-plane, which is 2. Therefore, an equation is $(x-2)^2 + (y+3)^2 + (z-6)^2 = 4$.

(c) Here the radius is the distance from the center $(2, -3, 6)$ to the xz-plane, which is 3. Therefore, an equation is $(x-2)^2 + (y+3)^2 + (z-6)^2 = 9$.

20. The largest sphere contained in the first octant must have a radius equal to the minimum distance from the center $(5, 4, 9)$ to any of the three coordinate planes. The shortest such distance is to the xz-plane, a distance of 4. Thus an equation of the sphere is $(x-5)^2 + (y-4)^2 + (z-9)^2 = 16$.

21. The equation $y = -4$ represents a plane parallel to the xz-plane and 4 units to the left of it.

22. The equation $x = 10$ represents a plane parallel to the yz-plane and 10 units in front of it.

23. The inequality $x > 3$ represents a half-space consisting of all points in front of the plane $x = 3$.

24. The inequality $y \geq 0$ represents a half-space consisting of all points on or to the right of the xz-plane.

25. The inequality $0 \leq z \leq 6$ represents all points on or between the horizontal planes $z = 0$ (the xy-plane) and $z = 6$.

26. The equation $z^2 = 1 \quad \Leftrightarrow \quad z = \pm 1$ represents two horizontal planes; $z = 1$ is parallel to the xy-plane, one unit above it, and $z = -1$ is one unit below it.

27. The inequality $x^2 + y^2 + z^2 \leq 3$ is equivalent to $\sqrt{x^2 + y^2 + z^2} \leq \sqrt{3}$, so the region consists of those points whose distance from the origin is at most $\sqrt{3}$. This is the set of all points on or inside the sphere with radius $\sqrt{3}$ and center $(0, 0, 0)$.

28. The equation $x = z$ represents a plane perpendicular to the xz-plane and intersecting the xz-plane in the line $x = z$, $y = 0$.

29. Here $x^2 + z^2 \leq 9$ or equivalently $\sqrt{x^2 + z^2} \leq 3$ which describes the set of all points in $\mathbb{R}^3$ whose distance from the y-axis is at most 3. Thus, the inequality represents the region consisting of all points on or inside a circular cylinder of radius 3 with axis the y-axis.

30. The inequality $x^2 + y^2 + z^2 > 2z \quad \Leftrightarrow \quad x^2 + y^2 + (z-1)^2 > 1$ is equivalent to $\sqrt{x^2 + y^2 + (z-1)^2} > 1$, so the region consists of those points whose distance from the point $(0, 0, 1)$ is greater than 1. This is the set of all points outside the sphere with radius 1 and center $(0, 0, 1)$.

31. This describes all points with negative y-coordinates, that is, $y < 0$.

32. Because the box lies in the first quadrant, each point must comprise only nonnegative coordinates. So inequalities describing the region are $0 \leq x \leq 1$, $0 \leq y \leq 2$, $0 \leq z \leq 3$.

33. This describes a region all of whose points have a distance to the origin which is greater than r, but smaller than R. So inequalities describing the region are $r < \sqrt{x^2 + y^2 + z^2} < R$, or $r^2 < x^2 + y^2 + z^2 < R^2$.

34. The solid sphere itself is represented by $\sqrt{x^2 + y^2 + z^2} \leq 2$. Since we want only the upper hemisphere, we restrict the z-coordinate to nonnegative values. Then inequalities describing the region are $\sqrt{x^2 + y^2 + z^2} \leq 2$, $z \geq 0$, or $x^2 + y^2 + z^2 \leq 4$, $z \geq 0$.

35. We need to find a set of points $\{P(x, y, z) \mid |AP| = |BP|\}$.

$\sqrt{(x+1)^2 + (y-5)^2 + (z-3)^2} = \sqrt{(x-6)^2 + (y-2)^2 + (z+2)^2} \quad \Rightarrow$

$(x+1)^2 + (y-5) + (z-3)^2 = (x-6)^2 + (y-2)^2 + (z+2)^2 \quad \Rightarrow$

$x^2 + 2x + 1 + y^2 - 10y + 25 + z^2 - 6z + 9 = x^2 - 12x + 36 + y^2 - 4y + 4 + z^2 + 4z + 4 \quad \Rightarrow \quad 14x - 6y - 10z = 9.$

Thus the set of points is a plane perpendicular to the line segment joining A and B (since this plane must contain the perpendicular bisector of the line segment AB).

36. Completing the square three times in the first equation gives $(x+2)^2+(y-1)^2+(z+2)^2=2^2$, a sphere with center $(-2,1,2)$ and radius 2. The second equation is that of a sphere with center $(0,0,0)$ and radius 2. The distance between the centers of the spheres is $\sqrt{(-2-0)^2+(1-0)^2+(-2-0)^2}=\sqrt{4+1+4}=3$. Since the spheres have the same radius, the volume inside both spheres is symmetrical about the plane containing the circle of intersection of the spheres. The distance from this plane to the center of the circles is $\frac{3}{2}$. So the region inside both spheres consists of two caps of spheres of height $h=2-\frac{3}{2}=\frac{1}{2}$. From Exercise 7.2.27, the volume of a cap of a sphere is $V=\frac{1}{3}\pi h^2(3r-h)=\frac{1}{3}\pi\left(\frac{1}{2}\right)^2\left(3\cdot 2-\frac{1}{2}\right)=\frac{11\pi}{24}$. So the total volume is $2\cdot\frac{11\pi}{24}=\frac{11\pi}{12}$.

10.2 Vectors

1. Vectors are equal when they share the same length and direction (but not necessarily location). Using the symmetry of the parallelogram as a guide, we see that $\overrightarrow{AB}=\overrightarrow{DC}$, $\overrightarrow{DA}=\overrightarrow{CB}$, $\overrightarrow{DE}=\overrightarrow{EB}$, and $\overrightarrow{EA}=\overrightarrow{CE}$.

2. (a) The initial point of $\overrightarrow{QR}$ is positioned at the terminal point of $\overrightarrow{PQ}$, so by the Triangle Law the sum $\overrightarrow{PQ}+\overrightarrow{QR}$ is the vector with initial point P and terminal point R, namely $\overrightarrow{PR}$.

(b) By the Triangle Law, $\overrightarrow{RP}+\overrightarrow{PS}$ is the vector with initial point R and terminal point S, namely $\overrightarrow{RS}$.

(c) First we consider $\overrightarrow{QS}-\overrightarrow{PS}$ as $\overrightarrow{QS}+\left(-\overrightarrow{PS}\right)$. Then since $-\overrightarrow{PS}$ has the same length as $\overrightarrow{PS}$ but points in the opposite direction, we have $-\overrightarrow{PS}=\overrightarrow{SP}$ and so $\overrightarrow{QS}-\overrightarrow{PS}=\overrightarrow{QS}+\overrightarrow{SP}=\overrightarrow{QP}$.

(d) We use the Triangle Law twice: $\overrightarrow{RS}+\overrightarrow{SP}+\overrightarrow{PQ}=\left(\overrightarrow{RS}+\overrightarrow{SP}\right)+\overrightarrow{PQ}=\overrightarrow{RP}+\overrightarrow{PQ}=\overrightarrow{RQ}$

3. (a) (b) (c) (d)

4. (a)

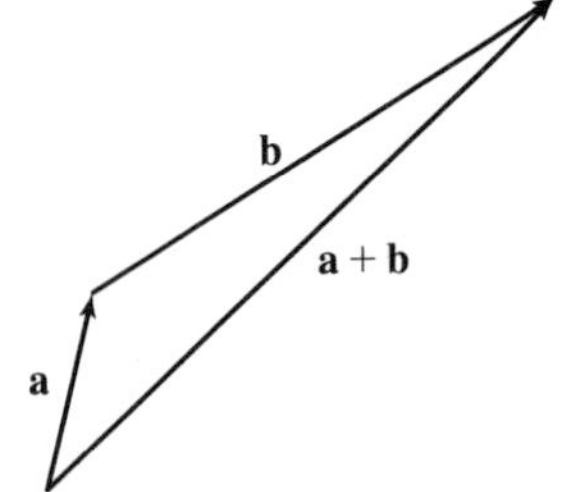

(b)

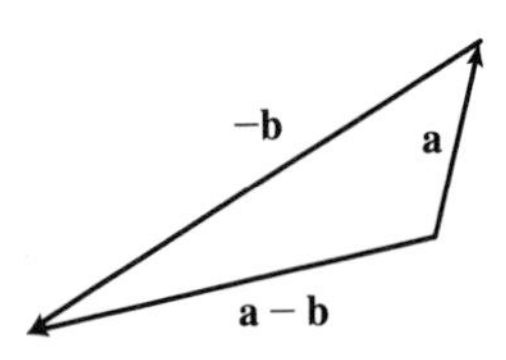

(c)

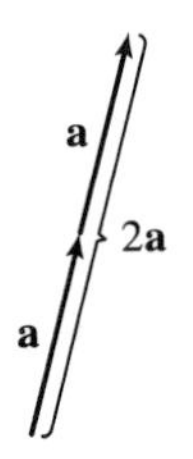

(d)

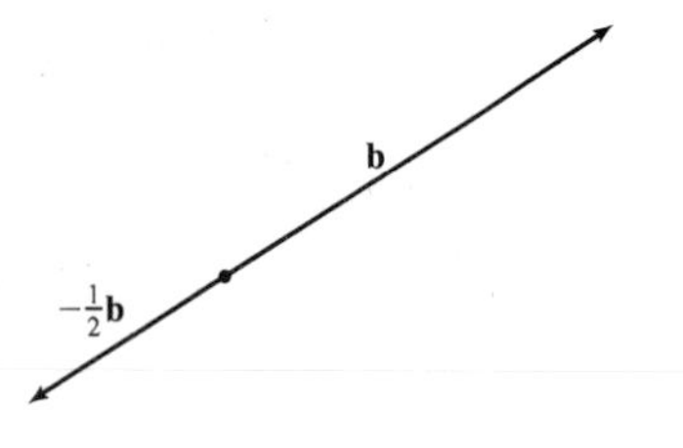

(e)

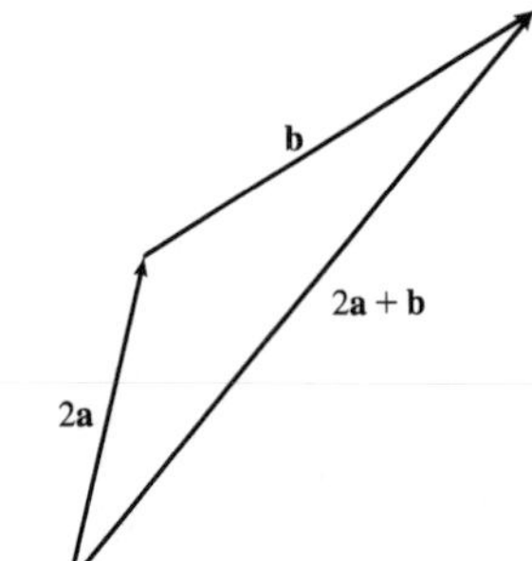

(f)

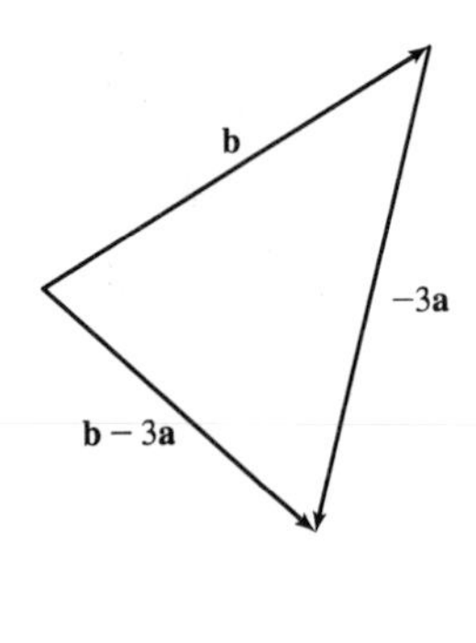

5. $\mathbf{a} = \langle -2 - 2, 1 - 3 \rangle = \langle -4, -2 \rangle$

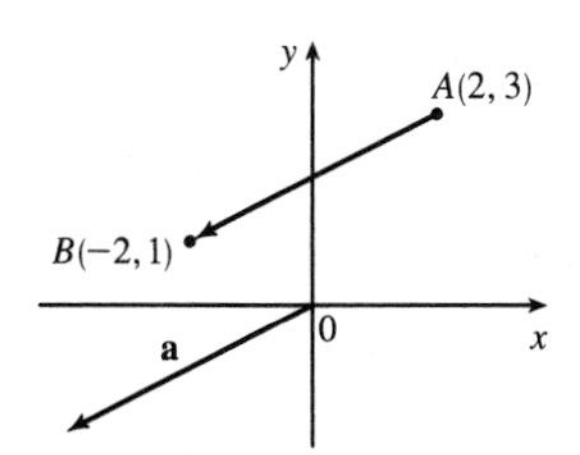

6. $\mathbf{a} = \langle 5 - (-2), 3 - (-2) \rangle = \langle 7, 5 \rangle$

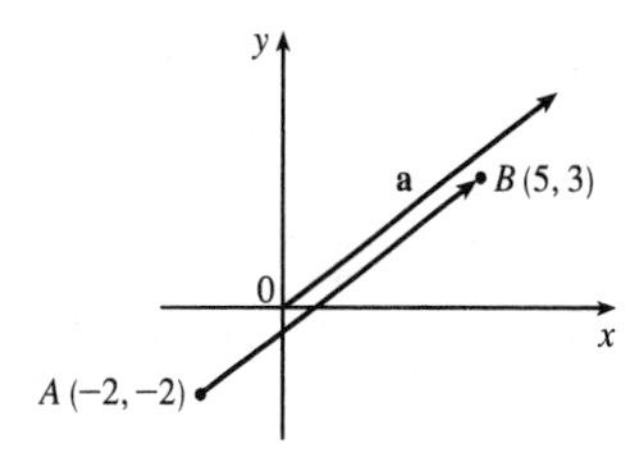

7. $\mathbf{a} = \langle 2 - 0, 3 - 3, -1 - 1 \rangle = \langle 2, 0, -2 \rangle$

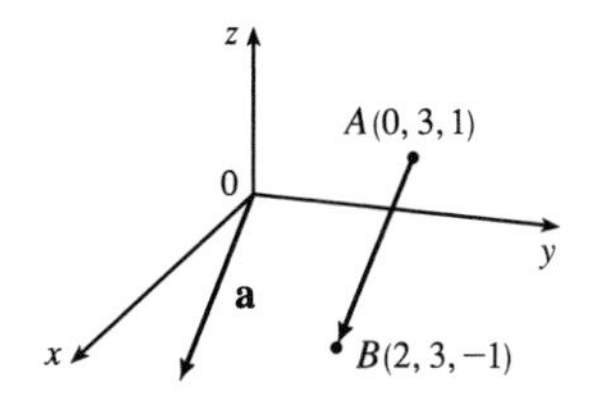

8. $\mathbf{a} = \langle 4 - 4, 2 - 0, 1 - (-2) \rangle = \langle 0, 2, 3 \rangle$

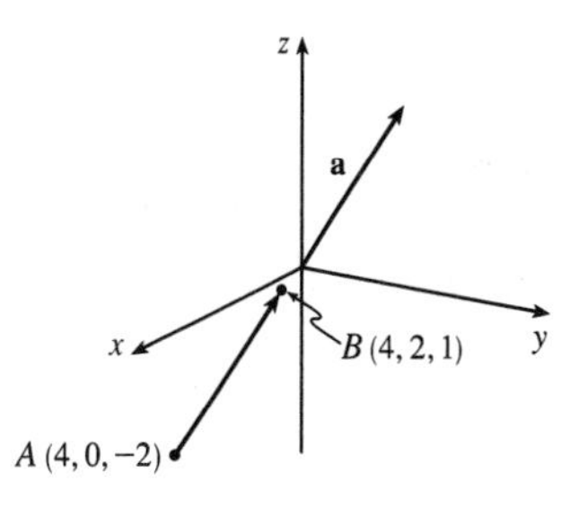

9. $\langle 3, -1 \rangle + \langle -2, 4 \rangle = \langle 3 + (-2), -1 + 4 \rangle$
$= \langle 1, 3 \rangle$

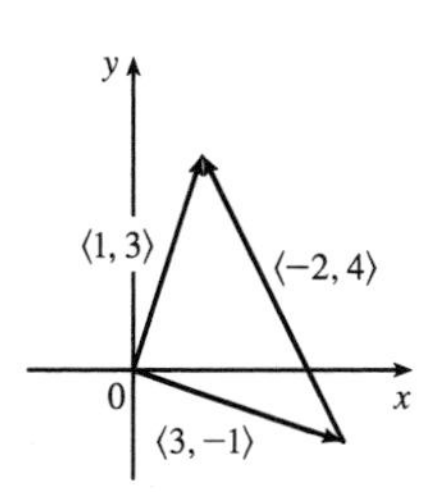

10. $\langle -2, -1 \rangle + \langle 5, 7 \rangle = \langle -2 + 5, -1 + 7 \rangle$
$= \langle 3, 6 \rangle$

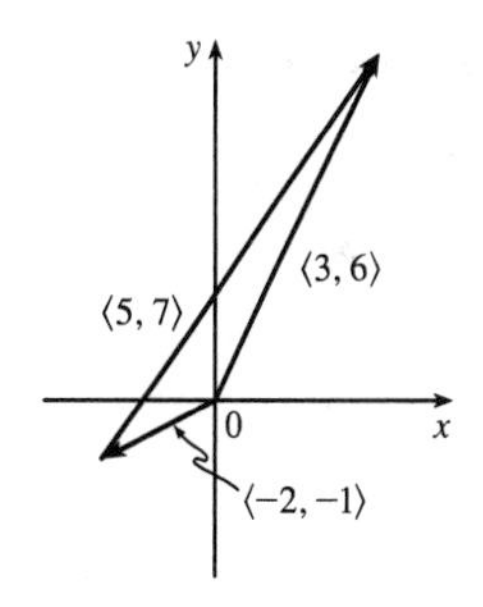

11. $\langle 0, 1, 2 \rangle + \langle 0, 0, -3 \rangle = \langle 0 + 0, 1 + 0, 2 + (-3) \rangle$
$= \langle 0, 1, -1 \rangle$

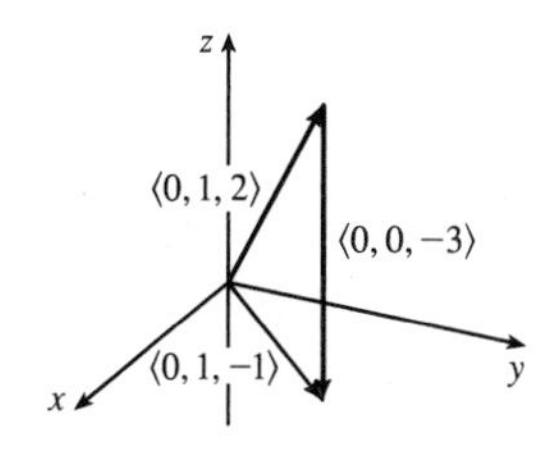

12. $\langle -1, 0, 2 \rangle + \langle 0, 4, 0 \rangle = \langle -1 + 0, 0 + 4, 2 + 0 \rangle$
$= \langle -1, 4, 2 \rangle$

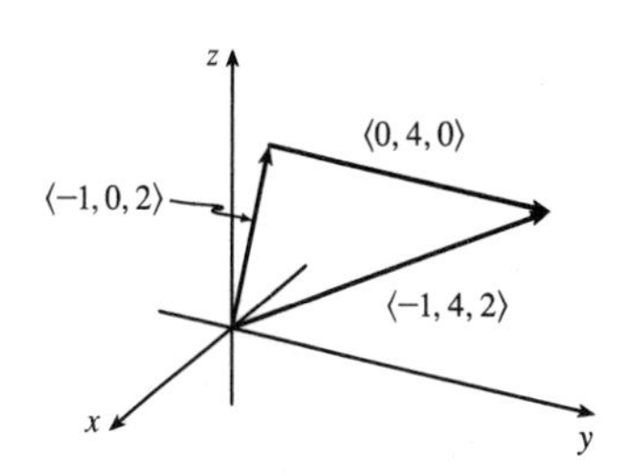

13. $\mathbf{a}+\mathbf{b} = \langle 5+(-3), -12+(-6)\rangle = \langle 2,-18\rangle$

$2\mathbf{a}+3\mathbf{b} = \langle 10,-24\rangle + \langle -9,-18\rangle = \langle 1,-42\rangle$

$|\mathbf{a}| = \sqrt{5^2+(-12)^2} = \sqrt{169} = 13$

$|\mathbf{a}-\mathbf{b}| = |\langle 5-(-3), -12-(-6)\rangle| = |\langle 8,-6\rangle| = \sqrt{8^2+(-6)^2} = \sqrt{100} = 10$

14. $\mathbf{a}+\mathbf{b} = (4\,\mathbf{i}+\mathbf{j}) + (\mathbf{i}-2\,\mathbf{j}) = 5\,\mathbf{i}-\mathbf{j}$

$2\mathbf{a}+3\mathbf{b} = 2\,(4\,\mathbf{i}+\mathbf{j}) + 3\,(\mathbf{i}-2\,\mathbf{j}) = 8\,\mathbf{i}+2\,\mathbf{j}+3\,\mathbf{i}-6\,\mathbf{j} = 11\,\mathbf{i}-4\,\mathbf{j}$

$|\mathbf{a}| = \sqrt{4^2+1^2} = \sqrt{17}$

$|\mathbf{a}-\mathbf{b}| = |(4\,\mathbf{i}+\mathbf{j}) - (\mathbf{i}-2\,\mathbf{j})| = |3\,\mathbf{i}+3\,\mathbf{j}| = \sqrt{3^2+3^2} = \sqrt{18} = 3\sqrt{2}$

15. $\mathbf{a}+\mathbf{b} = (\mathbf{i}+2\,\mathbf{j}-3\,\mathbf{k}) + (-2\,\mathbf{i}-\mathbf{j}+5\,\mathbf{k}) = -\,\mathbf{i}+\mathbf{j}+2\mathbf{k}$

$2\mathbf{a}+3\mathbf{b} = 2\,(\mathbf{i}+2\,\mathbf{j}-3\,\mathbf{k}) + 3\,(-2\,\mathbf{i}-\mathbf{j}+5\,\mathbf{k}) = 2\,\mathbf{i}+4\,\mathbf{j}-6\,\mathbf{k}-6\,\mathbf{i}-3\,\mathbf{j}+15\,\mathbf{k} = -\,4\,\mathbf{i}+\mathbf{j}+9\mathbf{k}$

$|\mathbf{a}| = \sqrt{1^2+2^2+(-3)^2} = \sqrt{14}$

$|\mathbf{a}-\mathbf{b}| = |(\mathbf{i}+2\,\mathbf{j}-3\,\mathbf{k}) - (-2\,\mathbf{i}-\mathbf{j}+5\,\mathbf{k})| = |3\,\mathbf{i}+3\,\mathbf{j}-8\,\mathbf{k}| = \sqrt{3^2+3^2+(-8)^2} = \sqrt{82}$

16. $\mathbf{a}+\mathbf{b} = (2\,\mathbf{i}-4\,\mathbf{j}+4\,\mathbf{k}) + (2\,\mathbf{j}-\mathbf{k}) = 2\,\mathbf{i}-2\,\mathbf{j}+3\,\mathbf{k}$

$2\mathbf{a}+3\mathbf{b} = 2\,(2\,\mathbf{i}-4\,\mathbf{j}+4\,\mathbf{k}) + 3\,(2\,\mathbf{j}-\mathbf{k}) = 4\,\mathbf{i}-8\,\mathbf{j}+8\,\mathbf{k}+6\,\mathbf{j}-3\,\mathbf{k} = 4\,\mathbf{i}-2\,\mathbf{j}+5\,\mathbf{k}$

$|\mathbf{a}| = \sqrt{2^2+(-4)^2+4^2} = \sqrt{36} = 6$

$|\mathbf{a}-\mathbf{b}| = |(2\,\mathbf{i}-4\,\mathbf{j}+4\,\mathbf{k}) - (2\,\mathbf{j}-\mathbf{k})| = |2\,\mathbf{i}-6\,\mathbf{j}+5\,\mathbf{k}| = \sqrt{2^2+(-6)^2+5^2} = \sqrt{65}$

17. The vector $8\,\mathbf{i}-\mathbf{j}+4\,\mathbf{k}$ has length $|8\,\mathbf{i}-\mathbf{j}+4\,\mathbf{k}| = \sqrt{8^2+(-1)^2+4^2} = \sqrt{81} = 9$, so by Equation 4 the unit vector with the same direction is $\frac{1}{9}(8\,\mathbf{i}-\mathbf{j}+4\,\mathbf{k}) = \frac{8}{9}\,\mathbf{i}-\frac{1}{9}\,\mathbf{j}+\frac{4}{9}\,\mathbf{k}$.

18. $|\langle -2,4,2\rangle| = \sqrt{(-2)^2+4^2+2^2} = \sqrt{24} = 2\sqrt{6}$, so a unit vector in the direction of $\langle -2,4,2\rangle$ is $\mathbf{u} = \dfrac{1}{2\sqrt{6}}\langle -2,4,2\rangle$.

A vector in the same direction but with length 6 is $6\mathbf{u} = 6\cdot\dfrac{1}{2\sqrt{6}}\langle -2,4,2\rangle = \left\langle -\dfrac{6}{\sqrt{6}}, \dfrac{12}{\sqrt{6}}, \dfrac{6}{\sqrt{6}}\right\rangle$ or $\langle -\sqrt{6}, 2\sqrt{6}, \sqrt{6}\,\rangle$.

19. From the figure, we see that the x-component of $\mathbf{v}$ is $v_1 = |\mathbf{v}|\cos(\pi/3) = 4\cdot\frac{1}{2} = 2$ and the y-component is $v_2 = |\mathbf{v}|\sin(\pi/3) = 4\cdot\frac{\sqrt{3}}{2} = 2\sqrt{3}$. Thus $\mathbf{v} = \langle v_1, v_2\rangle = \langle 2, 2\sqrt{3}\,\rangle$.

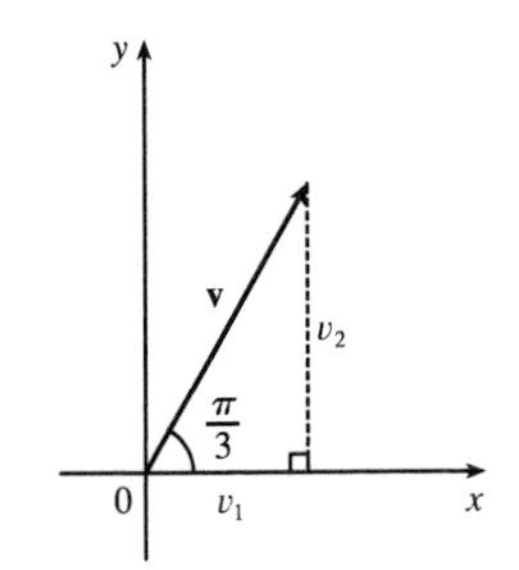

20. From the figure, we see that the horizontal component of the force $\mathbf{F}$ is $|\mathbf{F}|\cos 38^\circ = 50\cos 38^\circ \approx 39.4$ N, and the vertical component is $|\mathbf{F}|\sin 38^\circ = 50\sin 38^\circ \approx 30.8$ N.

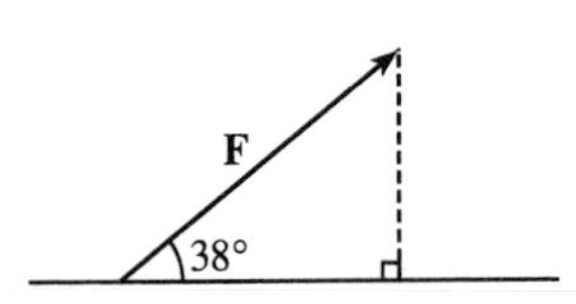

21. $|\mathbf{F}_1| = 10$ lb and $|\mathbf{F}_2| = 12$ lb.

$\mathbf{F}_1 = -|\mathbf{F}_1|\cos 45°\,\mathbf{i} + |\mathbf{F}_1|\sin 45°\,\mathbf{j} = -10\cos 45°\,\mathbf{i} + 10\sin 45°\,\mathbf{j} = -5\sqrt{2}\,\mathbf{i} + 5\sqrt{2}\,\mathbf{j}$

$\mathbf{F}_2 = |\mathbf{F}_2|\cos 30°\,\mathbf{i} + |\mathbf{F}_2|\sin 30°\,\mathbf{j} = 12\cos 30°\,\mathbf{i} + 12\sin 30°\,\mathbf{j} = 6\sqrt{3}\,\mathbf{i} + 6\,\mathbf{j}$

$\mathbf{F} = \mathbf{F}_1 + \mathbf{F}_2 = \left(6\sqrt{3} - 5\sqrt{2}\right)\mathbf{i} + \left(6 + 5\sqrt{2}\right)\mathbf{j} \approx 3.32\,\mathbf{i} + 13.07\,\mathbf{j}$

$|\mathbf{F}| \approx \sqrt{(3.32)^2 + (13.07)^2} \approx 13.5$ lb. $\tan\theta = \dfrac{6 + 5\sqrt{2}}{6\sqrt{3} - 5\sqrt{2}} \quad\Rightarrow\quad \theta = \tan^{-1}\dfrac{6 + 5\sqrt{2}}{6\sqrt{3} - 5\sqrt{2}} \approx 76°$.

22. Set up the coordinate axes so that north is the positive y-direction, and east is the positive x-direction. The wind is blowing at 50 km/h from the direction N45°W, so that its velocity vector is 50 km/h S45°E, which can be written as $\mathbf{v}_{\text{wind}} = 50(\cos 45°\,\mathbf{i} - \sin 45°\,\mathbf{j})$. With respect to the still air, the velocity vector of the plane is 250 km/h N 60°E, or equivalently $\mathbf{v}_{\text{plane}} = 250(\cos 30°\,\mathbf{i} + \sin 30°\,\mathbf{j})$. The velocity of the plane relative to the ground is

$$\begin{aligned}\mathbf{v} = \mathbf{v}_{\text{wind}} + \mathbf{v}_{\text{plane}} &= (50\cos 45° + 250\cos 30°)\,\mathbf{i} + (-50\sin 45° + 250\sin 30°)\,\mathbf{j}\\ &= \left(25\sqrt{2} + 125\sqrt{3}\right)\mathbf{i} + \left(125 - 25\sqrt{2}\right)\mathbf{j} \approx 251.9\,\mathbf{i} + 89.6\,\mathbf{j}\end{aligned}$$

The ground speed is $|\mathbf{v}| \approx \sqrt{(251.9)^2 + (89.6)^2} \approx 267$ km/h. The angle the velocity vector makes with the x-axis is $\theta \approx \tan^{-1}\left(\frac{89.6}{251.9}\right) \approx 20°$. Therefore, the true course of the plane is about N$(90 - 20)°$E = N 70°E.

23. With respect to the water's surface, the woman's velocity is the vector sum of the velocity of the ship with respect to the water, and the woman's velocity with respect to the ship. If we let north be the positive y-direction, then $\mathbf{v} = \langle 0, 22\rangle + \langle -3, 0\rangle = \langle -3, 22\rangle$. The woman's speed is $|\mathbf{v}| = \sqrt{9 + 484} \approx 22.2$ mi/h. The vector $\mathbf{v}$ makes an angle θ with the east, where $\theta = \tan^{-1}\left(\frac{22}{-3}\right) \approx 98°$. Therefore, the woman's direction is about N$(98 - 90)°$W = N8°W.

24. Call the two tensile forces $\mathbf{T}_3$ and $\mathbf{T}_5$, corresponding to the ropes of length 3 m and 5 m. In terms of vertical and horizontal components,

$$\mathbf{T}_3 = -|\mathbf{T}_3|\cos 52°\,\mathbf{i} + |\mathbf{T}_3|\sin 52°\,\mathbf{j} \quad \textbf{(1)} \qquad \text{and} \qquad \mathbf{T}_5 = |\mathbf{T}_5|\cos 40°\,\mathbf{i} + |\mathbf{T}_5|\sin 40°\,\mathbf{j} \quad \textbf{(2)}$$

The resultant of these forces, $\mathbf{T}_3 + \mathbf{T}_5$, counterbalances the force of gravity acting on the decoration [which is $-5g\,\mathbf{j} \approx -5(9.8)\,\mathbf{j} = -49\,\mathbf{j}$]. So $\mathbf{T}_3 + \mathbf{T}_5 = 49\,\mathbf{j}$. Hence

$\mathbf{T}_3 + \mathbf{T}_5 = \left(-|\mathbf{T}_3|\cos 52° + |\mathbf{T}_5|\cos 40°\right)\mathbf{i} + \left(|\mathbf{T}_3|\sin 52° + |\mathbf{T}_5|\sin 40°\right)\mathbf{j} = 49\,\mathbf{j}$.

Thus $-|\mathbf{T}_3|\cos 52° + |\mathbf{T}_5|\cos 40° = 0$ and $|\mathbf{T}_3|\sin 52° + |\mathbf{T}_5|\sin 40° = 49$.

From the first of these two equations $|\mathbf{T}_3| = |\mathbf{T}_5|\dfrac{\cos 40°}{\cos 52°}$. Substituting this into the second equation gives

$|\mathbf{T}_5| = \dfrac{49}{\cos 40°\tan 52° + \sin 40°} \approx 30$ N. Therefore, $|\mathbf{T}_3| = |\mathbf{T}_5|\dfrac{\cos 40°}{\cos 52°} \approx 38$ N. Finally, from **(1)** and **(2)**, $\mathbf{T}_3 \approx -23\,\mathbf{i} + 30\,\mathbf{j}$, and $\mathbf{T}_5 \approx 23\,\mathbf{i} + 19\,\mathbf{j}$.

25. Let $\mathbf{T}_1$ and $\mathbf{T}_2$ represent the tension vectors in each side of the clothesline as shown in the figure. $\mathbf{T}_1$ and $\mathbf{T}_2$ have equal vertical components and opposite horizontal components, so we can write

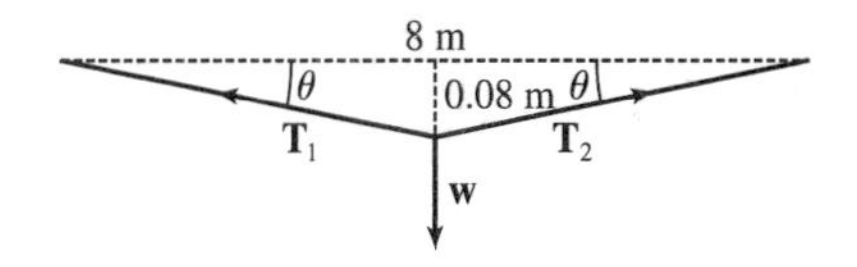

$\mathbf{T}_1 = -a\,\mathbf{i} + b\,\mathbf{j}$ and $\mathbf{T}_2 = a\,\mathbf{i} + b\,\mathbf{j}$ $[a, b > 0]$. By similar triangles, $\dfrac{b}{a} = \dfrac{0.08}{4} \quad\Rightarrow\quad a = 50b$. The force due to gravity acting on the shirt has magnitude $0.8g \approx (0.8)(9.8) = 7.84$ N, hence we have $\mathbf{w} = -7.84\,\mathbf{j}$. The resultant $\mathbf{T}_1 + \mathbf{T}_2$

of the tensile forces counterbalances $\mathbf{w}$, so $\mathbf{T}_1 + \mathbf{T}_2 = -\mathbf{w} \;\Rightarrow\; (-a\,\mathbf{i} + b\,\mathbf{j}) + (a\,\mathbf{i} + b\,\mathbf{j}) = 7.84\,\mathbf{j} \;\Rightarrow$

$(-50b\,\mathbf{i} + b\,\mathbf{j}) + (50b\,\mathbf{i} + b\,\mathbf{j}) = 2b\,\mathbf{j} = 7.84\,\mathbf{j} \;\Rightarrow\; b = \frac{7.84}{2} = 3.92$ and $a = 50b = 196$. Thus the tensions are

$\mathbf{T}_1 = -a\,\mathbf{i} + b\,\mathbf{j} = -196\,\mathbf{i} + 3.92\,\mathbf{j}$ and $\mathbf{T}_2 = a\,\mathbf{i} + b\,\mathbf{j} = 196\,\mathbf{i} + 3.92\,\mathbf{j}$.

Alternatively, we can find the value of θ and proceed as in Example 7.

26. We can consider the weight of the chain to be concentrated at its midpoint. The forces acting on the chain then are the tension vectors $\mathbf{T}_1$, $\mathbf{T}_2$ in each end of the chain and the weight $\mathbf{w}$, as shown in the figure. We know $|\mathbf{T}_1| = |\mathbf{T}_2| = 25$ N so, in terms of vertical and horizontal components, we have

$$\mathbf{T}_1 = -25\cos 37°\,\mathbf{i} + 25\sin 37°\,\mathbf{j}$$

$$\mathbf{T}_2 = 25\cos 37°\,\mathbf{i} + 25\sin 37°\,\mathbf{j}$$

The resultant vector $\mathbf{T}_1 + \mathbf{T}_2$ of the tensions counterbalances the weight $\mathbf{w}$, giving $\mathbf{T}_1 + \mathbf{T}_2 = -\mathbf{w}$. Since $\mathbf{w} = -|\mathbf{w}|\,\mathbf{j}$, we have $(-25\cos 37°\mathbf{i} + 25\sin 37°\mathbf{j}) + (25\cos 37°\mathbf{i} + 25\sin 37°\mathbf{j}) = |\mathbf{w}|\,\mathbf{j} \;\Rightarrow\; 50\sin 37°\mathbf{j} = |\mathbf{w}|\,\mathbf{j} \;\Rightarrow$

$|\mathbf{w}| = 50\sin 37° \approx 30.1$. So the weight is 30.1 N, and since $w = mg$, the mass is $\frac{30.1}{9.8} \approx 3.07$ kg.

27. (a), (b)

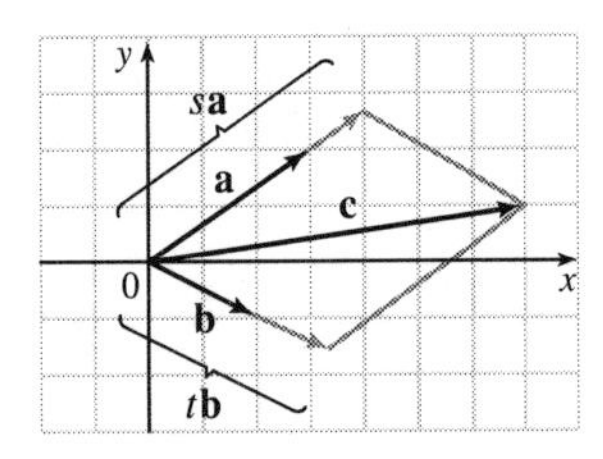

(c) From the sketch, we estimate that $s \approx 1.3$ and $t \approx 1.6$.

(d) $\mathbf{c} = s\,\mathbf{a} + t\,\mathbf{b} \;\Leftrightarrow\; 7 = 3s + 2t$ and $1 = 2s - t$.

Solving these equations gives $s = \frac{9}{7}$ and $t = \frac{11}{7}$.

28. Draw $\mathbf{a}$, $\mathbf{b}$, and $\mathbf{c}$ emanating from the origin. Extend $\mathbf{a}$ and $\mathbf{b}$ to form lines A and B, and draw lines A' and B' parallel to these two lines through the terminal point of $\mathbf{c}$. Since $\mathbf{a}$ and $\mathbf{b}$ are not parallel, A and B' must meet (at P), and A' and B must also meet (at Q). Now we see that $\overrightarrow{OP} + \overrightarrow{OQ} = \mathbf{c}$, so if

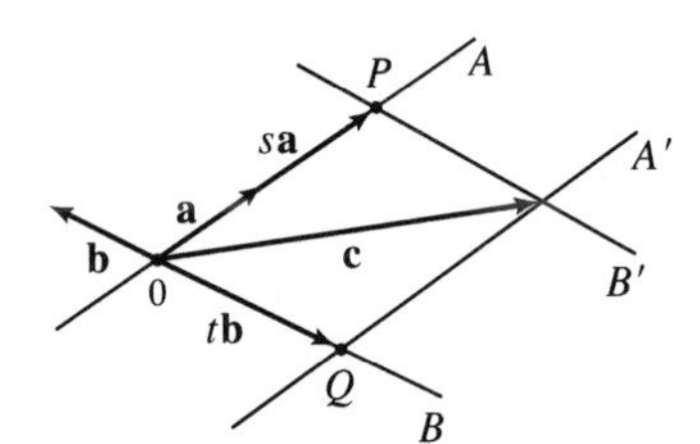

$s = \dfrac{|\overrightarrow{OP}|}{|\mathbf{a}|}$ (or its negative, if $\mathbf{a}$ points in the direction opposite $\overrightarrow{OP}$) and $t = \dfrac{|\overrightarrow{OQ}|}{|\mathbf{b}|}$ (or its negative, as in the diagram), then $\mathbf{c} = s\mathbf{a} + t\mathbf{b}$, as required.

Argument using components: Since $\mathbf{a}$, $\mathbf{b}$, and $\mathbf{c}$ all lie in the same plane, we can consider them to be vectors in two dimensions. Let $\mathbf{a} = \langle a_1, a_2 \rangle$, $\mathbf{b} = \langle b_1, b_2 \rangle$, and $\mathbf{c} = \langle c_1, c_2 \rangle$. We need $sa_1 + tb_1 = c_1$ and $sa_2 + tb_2 = c_2$. Multiplying the first equation by a_2 and the second by a_1 and subtracting, we get $t = \dfrac{c_2 a_1 - c_1 a_2}{b_2 a_1 - b_1 a_2}$. Similarly $s = \dfrac{b_2 c_1 - b_1 c_2}{b_2 a_1 - b_1 a_2}$. Since $\mathbf{a} \neq \mathbf{0}$ and $\mathbf{b} \neq \mathbf{0}$ and $\mathbf{a}$ is not a scalar multiple of $\mathbf{b}$, the denominator is not zero.

29. $|\mathbf{r} - \mathbf{r}_0|$ is the distance between the points (x, y, z) and (x_0, y_0, z_0), so the set of points is a sphere with radius 1 and center (x_0, y_0, z_0).

Alternate method: $|\mathbf{r} - \mathbf{r}_0| = 1 \;\Leftrightarrow\; \sqrt{(x - x_0)^2 + (y - y_0)^2 + (z - z_0)^2} = 1 \;\Leftrightarrow$

$(x - x_0)^2 + (y - y_0)^2 + (z - z_0)^2 = 1$, which is the equation of a sphere with radius 1 and center (x_0, y_0, z_0).

30. Let P_1 and P_2 be the points with position vectors $\mathbf{r}_1$ and $\mathbf{r}_2$ respectively. Then $|\mathbf{r}-\mathbf{r}_1|+|\mathbf{r}-\mathbf{r}_2|$ is the sum of the distances from (x,y) to P_1 and P_2. Since this sum is constant, the set of points (x,y) represents an ellipse with foci P_1 and P_2. The condition $k>|\mathbf{r}_1-\mathbf{r}_2|$ assures us that the ellipse is not degenerate.

31.
$$\begin{aligned}\mathbf{a}+(\mathbf{b}+\mathbf{c}) &= \langle a_1,a_2\rangle+(\langle b_1,b_2\rangle+\langle c_1,c_2\rangle)=\langle a_1,a_2\rangle+\langle b_1+c_1,b_2+c_2\rangle\\ &=\langle a_1+b_1+c_1,a_2+b_2+c_2\rangle=\langle (a_1+b_1)+c_1,(a_2+b_2)+c_2\rangle\\ &=\langle a_1+b_1,a_2+b_2\rangle+\langle c_1,c_2\rangle=(\langle a_1,a_2\rangle+\langle b_1,b_2\rangle)+\langle c_1,c_2\rangle\\ &=(\mathbf{a}+\mathbf{b})+\mathbf{c}\end{aligned}$$

32. *Algebraically:*

$$\begin{aligned}c(\mathbf{a}+\mathbf{b}) &= c\left(\langle a_1,a_2,a_3\rangle+\langle b_1,b_2,b_3\rangle\right)=c\,\langle a_1+b_1,a_2+b_2,a_3+b_3\rangle\\ &=\langle c\,(a_1+b_1),c\,(a_2+b_2),c\,(a_3+b_3)\rangle=\langle ca_1+cb_1,ca_2+cb_2,ca_3+cb_3\rangle\\ &=\langle ca_1,ca_2,ca_3\rangle+\langle cb_1,cb_2,cb_3\rangle=c\,\mathbf{a}+c\,\mathbf{b}\end{aligned}$$

Geometrically:

According to the Triangle Law, if $\mathbf{a}=\overrightarrow{PQ}$ and $\mathbf{b}=\overrightarrow{QR}$, then $\mathbf{a}+\mathbf{b}=\overrightarrow{PR}$. Construct triangle PST as shown so that $\overrightarrow{PS}=c\,\mathbf{a}$ and $\overrightarrow{ST}=c\,\mathbf{b}$. (We have drawn the case where $c>1$.) By the Triangle Law, $\overrightarrow{PT}=c\,\mathbf{a}+c\,\mathbf{b}$. But triangle PQR and triangle PST are similar triangles because $c\,\mathbf{b}$ is parallel to $\mathbf{b}$. Therefore, $\overrightarrow{PR}$ and $\overrightarrow{PT}$ are parallel and, in fact, $\overrightarrow{PT}=c\overrightarrow{PR}$. Thus, $c\,\mathbf{a}+c\,\mathbf{b}=c(\mathbf{a}+\mathbf{b})$.

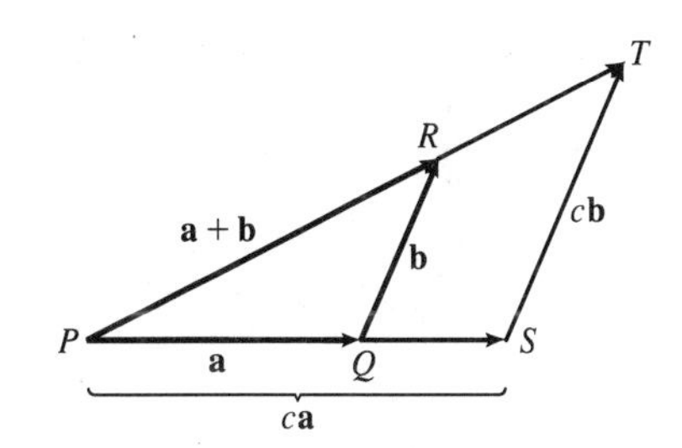

33. Consider triangle ABC, where D and E are the midpoints of AB and BC. We know that $\overrightarrow{AB}+\overrightarrow{BC}=\overrightarrow{AC}$ **(1)** and $\overrightarrow{DB}+\overrightarrow{BE}=\overrightarrow{DE}$ **(2)**. However, $\overrightarrow{DB}=\frac{1}{2}\overrightarrow{AB}$, and $\overrightarrow{BE}=\frac{1}{2}\overrightarrow{BC}$. Substituting these expressions for $\overrightarrow{DB}$ and $\overrightarrow{BE}$ into **(2)** gives $\frac{1}{2}\overrightarrow{AB}+\frac{1}{2}\overrightarrow{BC}=\overrightarrow{DE}$. Comparing this with **(1)** gives $\overrightarrow{DE}=\frac{1}{2}\overrightarrow{AC}$. Therefore $\overrightarrow{AC}$ and $\overrightarrow{DE}$ are parallel and $\left|\overrightarrow{DE}\right|=\frac{1}{2}\left|\overrightarrow{AC}\right|$.

10.3 The Dot Product

1. (a) $\mathbf{a}\cdot\mathbf{b}$ is a scalar, and the dot product is defined only for vectors, so $(\mathbf{a}\cdot\mathbf{b})\cdot\mathbf{c}$ has no meaning.

(b) $(\mathbf{a}\cdot\mathbf{b})\,\mathbf{c}$ is a scalar multiple of a vector, so it does have meaning.

(c) Both $|\mathbf{a}|$ and $\mathbf{b}\cdot\mathbf{c}$ are scalars, so $|\mathbf{a}|\,(\mathbf{b}\cdot\mathbf{c})$ is an ordinary product of real numbers, and has meaning.

(d) Both $\mathbf{a}$ and $\mathbf{b}+\mathbf{c}$ are vectors, so the dot product $\mathbf{a}\cdot(\mathbf{b}+\mathbf{c})$ has meaning.

(e) $\mathbf{a}\cdot\mathbf{b}$ is a scalar, but $\mathbf{c}$ is a vector, and so the two quantities cannot be added and $\mathbf{a}\cdot\mathbf{b}+\mathbf{c}$ has no meaning.

(f) $|\mathbf{a}|$ is a scalar, and the dot product is defined only for vectors, so $|\mathbf{a}|\cdot(\mathbf{b}+\mathbf{c})$ has no meaning.

2. Let the vectors be $\mathbf{a}$ and $\mathbf{b}$. Then by the definition of the dot product, $\mathbf{a}\cdot\mathbf{b}=|\mathbf{a}|\,|\mathbf{b}|\cos\theta=(6)\left(\frac{1}{3}\right)\cos\frac{\pi}{4}=\frac{6}{3\sqrt{2}}=\sqrt{2}$.

3. $\mathbf{a}\cdot\mathbf{b}=|\mathbf{a}|\,|\mathbf{b}|\cos\theta=(6)(5)\cos\frac{2\pi}{3}=30\left(-\frac{1}{2}\right)=-15$

4. $\mathbf{a}\cdot\mathbf{b}=\langle -2,3\rangle\cdot\langle 0.7,1.2\rangle=(-2)(0.7)+(3)(1.2)=2.2$

5. $\mathbf{a}\cdot\mathbf{b}=\left\langle 4,1,\frac{1}{4}\right\rangle\cdot\langle 6,-3,-8\rangle=(4)(6)+(1)(-3)+\left(\frac{1}{4}\right)(-8)=19$

6. $\mathbf{a}\cdot\mathbf{b}=\langle s,2s,3s\rangle\cdot\langle t,-t,5t\rangle=(s)(t)+(2s)(-t)+(3s)(5t)=st-2st+15st=14st$

7. $\mathbf{a}\cdot\mathbf{b}=(\mathbf{i}-2\mathbf{j}+3\mathbf{k})\cdot(5\,\mathbf{i}+9\,\mathbf{k})=(1)(5)+(-2)(0)+(3)(9)=32$

8. $\mathbf{a}\cdot\mathbf{b}=(4\mathbf{j}-3\mathbf{k})\cdot(2\,\mathbf{i}+4\,\mathbf{j}+6\,\mathbf{k})=(0)(2)+(4)(4)+(-3)(6)=-2$

9. $\mathbf{u}$, $\mathbf{v}$, and $\mathbf{w}$ are all unit vectors, so the triangle is an equilateral triangle. Thus the angle between $\mathbf{u}$ and $\mathbf{v}$ is 60° and $\mathbf{u}\cdot\mathbf{v}=|\mathbf{u}|\,|\mathbf{v}|\cos 60^\circ=(1)(1)\left(\frac{1}{2}\right)=\frac{1}{2}$. If $\mathbf{w}$ is moved so it has the same initial point as $\mathbf{u}$, we can see that the angle between them is 120° and we have $\mathbf{u}\cdot\mathbf{w}=|\mathbf{u}|\,|\mathbf{w}|\cos 120^\circ=(1)(1)\left(-\frac{1}{2}\right)=-\frac{1}{2}$.

10. $\mathbf{u}$ is a unit vector, so $\mathbf{w}$ is also a unit vector, and $|\mathbf{v}|$ can be determined by examining the right triangle formed by $\mathbf{u}$ and $\mathbf{v}$. Since the angle between $\mathbf{u}$ and $\mathbf{v}$ is 45°, we have $|\mathbf{v}|=|\mathbf{u}|\cos 45^\circ=\frac{\sqrt{2}}{2}$. Then $\mathbf{u}\cdot\mathbf{v}=|\mathbf{u}|\,|\mathbf{v}|\cos 45^\circ=(1)\left(\frac{\sqrt{2}}{2}\right)\frac{\sqrt{2}}{2}=\frac{1}{2}$. Since $\mathbf{u}$ and $\mathbf{w}$ are orthogonal, $\mathbf{u}\cdot\mathbf{w}=0$.

11. (a) $\mathbf{i}\cdot\mathbf{j}=\langle 1,0,0\rangle\cdot\langle 0,1,0\rangle=(1)(0)+(0)(1)+(0)(0)=0$. Similarly, $\mathbf{j}\cdot\mathbf{k}=(0)(0)+(1)(0)+(0)(1)=0$ and $\mathbf{k}\cdot\mathbf{i}=(0)(1)+(0)(0)+(1)(0)=0$.

Another method: Because **i**, **j**, and **k** are mutually perpendicular, the cosine factor in each dot product is $\cos\frac{\pi}{2}=0$.

(b) By Property 1 of the dot product, $\mathbf{i}\cdot\mathbf{i}=|\mathbf{i}|^2=1^2=1$ since **i** is a unit vector. Similarly, $\mathbf{j}\cdot\mathbf{j}=|\mathbf{j}|^2=1$ and $\mathbf{k}\cdot\mathbf{k}=|\mathbf{k}|^2=1$.

12. The dot product $\mathbf{A}\cdot\mathbf{P}$ is

$$\begin{aligned}\langle a,b,c\rangle\cdot\langle 2,1.5,1\rangle &= a(2)+b(1.5)+c(1)\\ &= (\text{number of hamburgers sold})(\text{price per hamburger})\\ &\quad+(\text{number of hot dogs sold})(\text{price per hot dog})\\ &\quad+(\text{number of soft drinks sold})(\text{price per soft drink})\end{aligned}$$

so it is equal to the vendor's total revenue for that day.

13. $|\mathbf{a}|=\sqrt{(-8)^2+6^2}=10$, $|\mathbf{b}|=\sqrt{\left(\sqrt{7}\right)^2+3^2}=4$, and $\mathbf{a}\cdot\mathbf{b}=(-8)\left(\sqrt{7}\right)+(6)(3)=18-8\sqrt{7}$. From the definition of the dot product, we have $\cos\theta=\dfrac{\mathbf{a}\cdot\mathbf{b}}{|\mathbf{a}|\,|\mathbf{b}|}=\dfrac{18-8\sqrt{7}}{10\cdot 4}=\dfrac{9-4\sqrt{7}}{20}$. So the angle between **a** and **b** is $\theta=\cos^{-1}\left(\dfrac{9-4\sqrt{7}}{20}\right)\approx 95^\circ$.

14. $|\mathbf{a}|=\sqrt{4^2+0^2+2^2}=\sqrt{20}$, $|\mathbf{b}|=\sqrt{2^2+(-1)^2+0^2}=\sqrt{5}$, and $\mathbf{a}\cdot\mathbf{b}=(4)(2)+(0)(-1)+(2)(0)=8$.

Then $\cos\theta=\dfrac{\mathbf{a}\cdot\mathbf{b}}{|\mathbf{a}|\,|\mathbf{b}|}=\dfrac{8}{\sqrt{20}\cdot\sqrt{5}}=\dfrac{4}{5}$ and $\theta=\cos^{-1}\left(\frac{4}{5}\right)\approx 37^\circ$.

15. $|\mathbf{a}|=\sqrt{0^2+1^2+1^2}=\sqrt{2}$, $|\mathbf{b}|=\sqrt{1^2+2^2+(-3)^2}=\sqrt{14}$, and $\mathbf{a}\cdot\mathbf{b}=(0)(1)+(1)(2)+(1)(-3)=-1$.

Then $\cos\theta=\dfrac{\mathbf{a}\cdot\mathbf{b}}{|\mathbf{a}|\,|\mathbf{b}|}=\dfrac{-1}{\sqrt{2}\cdot\sqrt{14}}=\dfrac{-1}{2\sqrt{7}}$ and $\theta=\cos^{-1}\left(-\dfrac{1}{2\sqrt{7}}\right)\approx 101^\circ$.

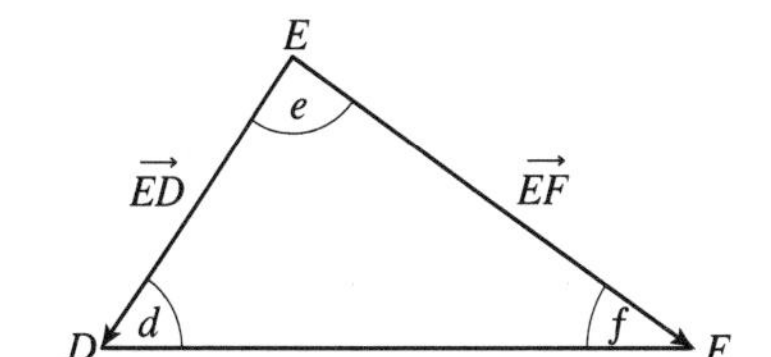

16. Let d, e, and f be the angles at vertices D, E, and F. Then d is the angle between vectors $\overrightarrow{DE}$ and $\overrightarrow{DF}$, e is the angle between vectors $\overrightarrow{ED}$ and $\overrightarrow{EF}$, and f is the angle between vectors $\overrightarrow{FD}$ and $\overrightarrow{FE}$.

Thus $\cos d = \dfrac{\overrightarrow{DE}\cdot\overrightarrow{DF}}{\left|\overrightarrow{DE}\right|\left|\overrightarrow{DF}\right|} = \dfrac{\langle -2,3,2\rangle\cdot\langle 1,1,-2\rangle}{\sqrt{(-2)^2+3^2+2^2}\sqrt{1^2+1^2+(-2)^2}} = \dfrac{1}{\sqrt{17}\sqrt{6}}(-2+3-4) = -\dfrac{3}{\sqrt{102}}$

and $d = \cos^{-1}\left(-\dfrac{3}{\sqrt{102}}\right) \approx 107^\circ$. Similarly,

$\cos e = \dfrac{\overrightarrow{ED}\cdot\overrightarrow{EF}}{\left|\overrightarrow{ED}\right|\left|\overrightarrow{EF}\right|} = \dfrac{\langle 2,-3,-2\rangle\cdot\langle 3,-2,-4\rangle}{\sqrt{4+9+4}\sqrt{9+4+16}} = \dfrac{1}{\sqrt{17}\sqrt{29}}(6+6+8) = \dfrac{20}{\sqrt{493}}$ so $e = \cos^{-1}\left(\dfrac{20}{\sqrt{493}}\right) \approx 26^\circ$

and $f \approx 180^\circ - (107^\circ + 26^\circ) = 47^\circ$.

Alternate solution: Apply the Law of Cosines three times as follows:

$$\cos d = \frac{\left|\overrightarrow{EF}\right|^2 - \left|\overrightarrow{DE}\right|^2 - \left|\overrightarrow{DF}\right|^2}{2\left|\overrightarrow{DE}\right|\left|\overrightarrow{DF}\right|} \qquad \cos e = \frac{\left|\overrightarrow{DF}\right|^2 - \left|\overrightarrow{DE}\right|^2 - \left|\overrightarrow{EF}\right|^2}{2\left|\overrightarrow{DE}\right|\left|\overrightarrow{EF}\right|} \qquad \cos f = \frac{\left|\overrightarrow{DE}\right|^2 - \left|\overrightarrow{DF}\right|^2 - \left|\overrightarrow{EF}\right|^2}{2\left|\overrightarrow{DF}\right|\left|\overrightarrow{EF}\right|}$$

17. (a) $\mathbf{a}\cdot\mathbf{b} = (-5)(6)+(3)(-8)+(7)(2) = -40 \neq 0$, so $\mathbf{a}$ and $\mathbf{b}$ are not orthogonal. Also, since $\mathbf{a}$ is not a scalar multiple of $\mathbf{b}$, $\mathbf{a}$ and $\mathbf{b}$ are not parallel.

(b) $\mathbf{a}\cdot\mathbf{b} = (4)(-3)+(6)(2) = 0$, so $\mathbf{a}$ and $\mathbf{b}$ are orthogonal (and not parallel).

(c) $\mathbf{a}\cdot\mathbf{b} = (-1)(3)+(2)(4)+(5)(-1) = 0$, so $\mathbf{a}$ and $\mathbf{b}$ are orthogonal (and not parallel).

(d) Because $\mathbf{a} = -\frac{2}{3}\,\mathbf{b}$, $\mathbf{a}$ and $\mathbf{b}$ are parallel.

18. (a) Because $\mathbf{u} = -\frac{3}{4}\,\mathbf{v}$, $\mathbf{u}$ and $\mathbf{v}$ are parallel vectors (and thus not orthogonal).

(b) $\mathbf{u}\cdot\mathbf{v} = (1)(2)+(-1)(-1)+(2)(1) = 5 \neq 0$, so $\mathbf{u}$ and $\mathbf{v}$ are not orthogonal. Also, $\mathbf{u}$ is not a scalar multiple of $\mathbf{v}$, so $\mathbf{u}$ and $\mathbf{v}$ are not parallel.

(c) $\mathbf{u}\cdot\mathbf{v} = (a)(-b)+(b)(a)+(c)(0) = -ab+ab+0 = 0$, so $\mathbf{u}$ and $\mathbf{v}$ are orthogonal (and not parallel).

19. $\overrightarrow{QP} = \langle -1,-3,2\rangle$, $\overrightarrow{QR} = \langle 4,-2,-1\rangle$, and $\overrightarrow{QP}\cdot\overrightarrow{QR} = -4+6-2 = 0$. Thus $\overrightarrow{QP}$ and $\overrightarrow{QR}$ are orthogonal, so the angle of the triangle at vertex Q is a right angle.

20. $\langle -6,b,2\rangle$ and $\langle b,b^2,b\rangle$ are orthogonal when $\langle -6,b,2\rangle\cdot\langle b,b^2,b\rangle = 0 \;\Leftrightarrow\; (-6)(b)+(b)(b^2)+(2)(b) = 0 \;\Leftrightarrow\; b^3-4b = 0 \;\Leftrightarrow\; b(b+2)(b-2) = 0 \;\Leftrightarrow\; b = 0$ or $b = \pm 2$.

21. Let $\mathbf{a} = a_1\,\mathbf{i} + a_2\,\mathbf{j} + a_3\,\mathbf{k}$ be a vector orthogonal to both $\mathbf{i}+\mathbf{j}$ and $\mathbf{i}+\mathbf{k}$. Then $\mathbf{a}\cdot(\mathbf{i}+\mathbf{j}) = 0 \;\Leftrightarrow\; a_1+a_2 = 0$ and $\mathbf{a}\cdot(\mathbf{i}+\mathbf{k}) = 0 \;\Leftrightarrow\; a_1+a_3 = 0$, so $a_1 = -a_2 = -a_3$. Furthermore $\mathbf{a}$ is to be a unit vector, so $1 = a_1^2+a_2^2+a_3^2 = 3a_1^2$ implies $a_1 = \pm\frac{1}{\sqrt{3}}$. Thus $\mathbf{a} = \frac{1}{\sqrt{3}}\,\mathbf{i} - \frac{1}{\sqrt{3}}\,\mathbf{j} - \frac{1}{\sqrt{3}}\,\mathbf{k}$ and $\mathbf{a} = -\frac{1}{\sqrt{3}}\,\mathbf{i} + \frac{1}{\sqrt{3}}\,\mathbf{j} + \frac{1}{\sqrt{3}}\,\mathbf{k}$ are two such unit vectors.

22. Let $\mathbf{u} = \langle a,b\rangle$ be a unit vector. By the definition of the dot product we need $\mathbf{u}\cdot\mathbf{v} = |\mathbf{u}|\;|\mathbf{v}|\cos 60^\circ \;\Leftrightarrow\; 3a+4b = (1)(5)\frac{1}{2} \;\Leftrightarrow\; b = \frac{5}{8}-\frac{3}{4}a$. Since $\mathbf{u}$ is a unit vector, $|\mathbf{u}| = \sqrt{a^2+b^2} = 1 \;\Leftrightarrow\; a^2+b^2 = 1 \;\Leftrightarrow$

$a^2 + \left(\frac{5}{8} - \frac{3}{4}a\right)^2 = 1 \;\Leftrightarrow\; \frac{25}{16}a^2 - \frac{15}{16}a + \frac{25}{64} = 1 \;\Leftrightarrow\; 100a^2 - 60a - 39 = 0$. By the quadratic formula,

$a = \dfrac{-(-60) \pm \sqrt{(-60)^2 - 4(100)(-39)}}{2(100)} = \dfrac{60 \pm \sqrt{19{,}200}}{200} = \dfrac{3 \pm 4\sqrt{3}}{10}$. If $a = \dfrac{3+4\sqrt{3}}{10}$ then

$b = \dfrac{5}{8} - \dfrac{3}{4}\left(\dfrac{3+4\sqrt{3}}{10}\right) = \dfrac{4-3\sqrt{3}}{10}$, and if $a = \dfrac{3-4\sqrt{3}}{10}$ then $b = \dfrac{5}{8} - \dfrac{3}{4}\left(\dfrac{3-4\sqrt{3}}{10}\right) = \dfrac{4+3\sqrt{3}}{10}$. Thus the two unit

vectors are $\left\langle \dfrac{3+4\sqrt{3}}{10}, \dfrac{4-3\sqrt{3}}{10} \right\rangle \approx \langle 0.9928, -0.1196\rangle$ and $\left\langle \dfrac{3-4\sqrt{3}}{10}, \dfrac{4+3\sqrt{3}}{10} \right\rangle \approx \langle -0.3928, 0.9196\rangle$.

23. $|\mathbf{a}| = \sqrt{3^2 + (-4)^2} = 5$. The scalar projection of $\mathbf{b}$ onto $\mathbf{a}$ is $\operatorname{comp}_{\mathbf{a}} \mathbf{b} = \dfrac{\mathbf{a}\cdot\mathbf{b}}{|\mathbf{a}|} = \dfrac{3\cdot 5 + (-4)\cdot 0}{5} = 3$ and the vector

projection of $\mathbf{b}$ onto $\mathbf{a}$ is $\operatorname{proj}_{\mathbf{a}} \mathbf{b} = \left(\dfrac{\mathbf{a}\cdot\mathbf{b}}{|\mathbf{a}|}\right)\dfrac{\mathbf{a}}{|\mathbf{a}|} = 3 \cdot \frac{1}{5}\langle 3, -4\rangle = \left\langle \frac{9}{5}, -\frac{12}{5}\right\rangle$.

24. $|\mathbf{a}| = \sqrt{1^2 + 2^2} = \sqrt{5}$, so the scalar projection of $\mathbf{b}$ onto $\mathbf{a}$ is $\operatorname{comp}_{\mathbf{a}} \mathbf{b} = \dfrac{\mathbf{a}\cdot\mathbf{b}}{|\mathbf{a}|} = \dfrac{1(-4) + 2\cdot 1}{\sqrt{5}} = -\dfrac{2}{\sqrt{5}}$ and the vector

projection of $\mathbf{b}$ onto $\mathbf{a}$ is $\operatorname{proj}_{\mathbf{a}} \mathbf{b} = \left(\dfrac{\mathbf{a}\cdot\mathbf{b}}{|\mathbf{a}|}\right)\dfrac{\mathbf{a}}{|\mathbf{a}|} = -\frac{2}{\sqrt{5}} \cdot \frac{1}{\sqrt{5}}\langle 1, 2\rangle = \left\langle -\frac{2}{5}, -\frac{4}{5}\right\rangle$.

25. $|\mathbf{a}| = \sqrt{9 + 36 + 4} = 7$ so the scalar projection of $\mathbf{b}$ onto $\mathbf{a}$ is $\operatorname{comp}_{\mathbf{a}}\mathbf{b} = \dfrac{\mathbf{a}\cdot\mathbf{b}}{|\mathbf{a}|} = \frac{1}{7}(3 + 12 - 6) = \frac{9}{7}$. The vector

projection of $\mathbf{b}$ onto $\mathbf{a}$ is $\operatorname{proj}_{\mathbf{a}}\mathbf{b} = \dfrac{9}{7}\dfrac{\mathbf{a}}{|\mathbf{a}|} = \frac{9}{7} \cdot \frac{1}{7}\langle 3, 6, -2\rangle = \frac{9}{49}\langle 3, 6, -2\rangle = \left\langle \frac{27}{49}, \frac{54}{49}, -\frac{18}{49}\right\rangle$.

26. $|\mathbf{a}| = \sqrt{1 + 1 + 1} = \sqrt{3}$, so the scalar projection of $\mathbf{b}$ onto $\mathbf{a}$ is $\operatorname{comp}_{\mathbf{a}} \mathbf{b} = \dfrac{\mathbf{a}\cdot\mathbf{b}}{|\mathbf{a}|} = \dfrac{1 - 1 + 1}{\sqrt{3}} = \dfrac{1}{\sqrt{3}}$ while the vector

projection of $\mathbf{b}$ onto $\mathbf{a}$ is $\operatorname{proj}_{\mathbf{a}} \mathbf{b} = \dfrac{1}{\sqrt{3}}\dfrac{\mathbf{a}}{|\mathbf{a}|} = \dfrac{1}{\sqrt{3}} \cdot \dfrac{\mathbf{i} + \mathbf{j} + \mathbf{k}}{\sqrt{3}} = \frac{1}{3}(\mathbf{i} + \mathbf{j} + \mathbf{k})$.

27. $(\operatorname{orth}_{\mathbf{a}} \mathbf{b}) \cdot \mathbf{a} = (\mathbf{b} - \operatorname{proj}_{\mathbf{a}} \mathbf{b}) \cdot \mathbf{a} = \mathbf{b}\cdot\mathbf{a} - (\operatorname{proj}_{\mathbf{a}} \mathbf{b}) \cdot \mathbf{a} = \mathbf{b}\cdot\mathbf{a} - \dfrac{\mathbf{a}\cdot\mathbf{b}}{|\mathbf{a}|^2}\mathbf{a}\cdot\mathbf{a} = \mathbf{b}\cdot\mathbf{a} - \dfrac{\mathbf{a}\cdot\mathbf{b}}{|\mathbf{a}|^2}|\mathbf{a}|^2 = \mathbf{b}\cdot\mathbf{a} - \mathbf{a}\cdot\mathbf{b} = 0$.

So they are orthogonal by (7).

28. Using the formula in Exercise 27 and the result of Exercise 24, we have

$\operatorname{orth}_{\mathbf{a}} \mathbf{b} = \mathbf{b} - \operatorname{proj}_{\mathbf{a}} \mathbf{b} = \langle -4, 1\rangle - \left\langle -\frac{2}{5}, -\frac{4}{5}\right\rangle = \left\langle -\frac{18}{5}, \frac{9}{5}\right\rangle$.

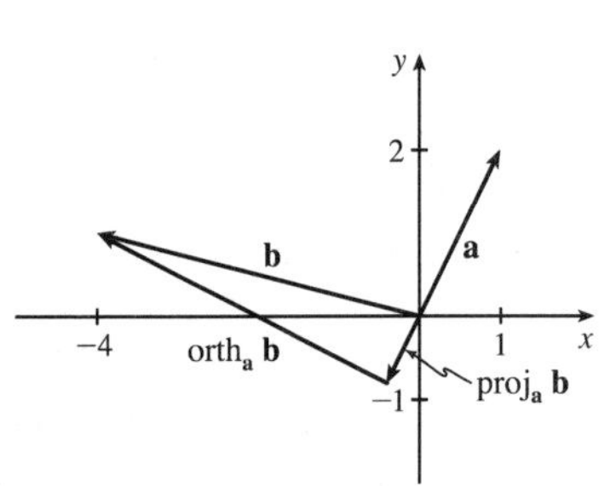

29. $\operatorname{comp}_{\mathbf{a}} \mathbf{b} = \dfrac{\mathbf{a}\cdot\mathbf{b}}{|\mathbf{a}|} = 2 \;\Leftrightarrow\; \mathbf{a}\cdot\mathbf{b} = 2|\mathbf{a}| = 2\sqrt{10}$. If $\mathbf{b} = \langle b_1, b_2, b_3\rangle$, then we need $3b_1 + 0b_2 - 1b_3 = 2\sqrt{10}$.

One possible solution is obtained by taking $b_1 = 0$, $b_2 = 0$, $b_3 = -2\sqrt{10}$. In general, $\mathbf{b} = \left\langle s, t, 3s - 2\sqrt{10}\right\rangle$, $s, t \in \mathbb{R}$.

30. (a) $\operatorname{comp}_{\mathbf{a}} \mathbf{b} = \operatorname{comp}_{\mathbf{b}} \mathbf{a} \;\Leftrightarrow\; \dfrac{\mathbf{a}\cdot\mathbf{b}}{|\mathbf{a}|} = \dfrac{\mathbf{b}\cdot\mathbf{a}}{|\mathbf{b}|} \;\Leftrightarrow\; \dfrac{1}{|\mathbf{a}|} = \dfrac{1}{|\mathbf{b}|}$ or $\mathbf{a}\cdot\mathbf{b} = 0 \;\Leftrightarrow\; |\mathbf{b}| = |\mathbf{a}|$ or $\mathbf{a}\cdot\mathbf{b} = 0$.

That is, if $\mathbf{a}$ and $\mathbf{b}$ are orthogonal or if they have the same length.

(b) $\text{proj}_{\mathbf{a}}\,\mathbf{b} = \text{proj}_{\mathbf{b}}\,\mathbf{a} \quad\Leftrightarrow\quad \dfrac{\mathbf{a}\cdot\mathbf{b}}{|\mathbf{a}|^2}\,\mathbf{a} = \dfrac{\mathbf{b}\cdot\mathbf{a}}{|\mathbf{b}|^2}\,\mathbf{b} \quad\Leftrightarrow\quad \mathbf{a}\cdot\mathbf{b}=0$ or $\dfrac{\mathbf{a}}{|\mathbf{a}|^2} = \dfrac{\mathbf{b}}{|\mathbf{b}|^2}$.

But $\dfrac{\mathbf{a}}{|\mathbf{a}|^2} = \dfrac{\mathbf{b}}{|\mathbf{b}|^2} \quad\Rightarrow\quad \dfrac{|\mathbf{a}|}{|\mathbf{a}|^2} = \dfrac{|\mathbf{b}|}{|\mathbf{b}|^2} \quad\Rightarrow\quad |\mathbf{a}| = |\mathbf{b}|$. Substituting this into the previous equation gives $\mathbf{a} = \mathbf{b}$.

So $\text{proj}_{\mathbf{a}}\,\mathbf{b} = \text{proj}_{\mathbf{b}}\,\mathbf{a} \quad\Leftrightarrow\quad \mathbf{a}$ and $\mathbf{b}$ are orthogonal, or they are equal.

31. Here $\mathbf{D} = (4-2)\,\mathbf{i} + (9-3)\,\mathbf{j} + (15-0)\,\mathbf{k} = 2\,\mathbf{i} + 6\,\mathbf{j} + 15\,\mathbf{k}$ so $W = \mathbf{F}\cdot\mathbf{D} = 20 + 108 - 90 = 38$ J.

32. $W = |\mathbf{F}|\,|\mathbf{D}|\cos\theta = (20)(4)\cos 40^\circ \approx 61$ ft-lb

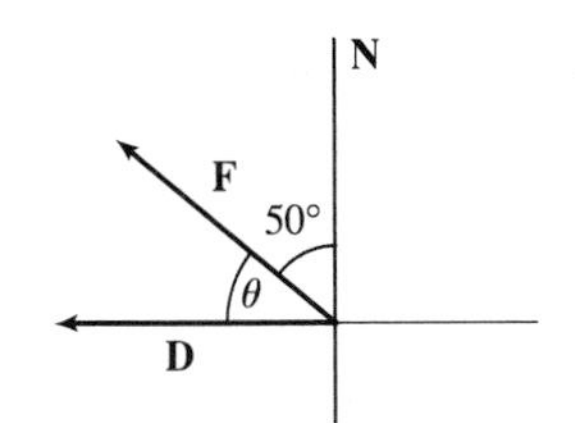

33. $W = |\mathbf{F}|\,|\mathbf{D}|\cos\theta = (25)(10)\cos 20^\circ \approx 235$ ft-lb

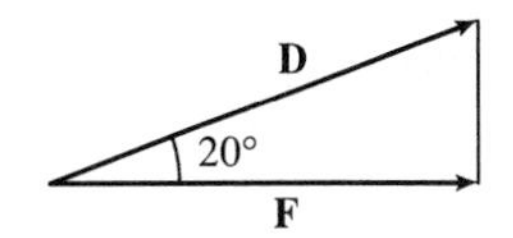

34. Here $|\mathbf{D}| = 100$ m, $|\mathbf{F}| = 50$ N, and $\theta = 30^\circ$. Thus $W = |\mathbf{F}|\,|\mathbf{D}|\cos\theta = (50)(100)\left(\frac{\sqrt{3}}{2}\right) = 2500\sqrt{3}$ J.

35. First note that $\mathbf{n} = \langle a, b\rangle$ is perpendicular to the line, because if $Q_1 = (a_1, b_1)$ and $Q_2 = (a_2, b_2)$ lie on the line, then

$\mathbf{n}\cdot\overrightarrow{Q_1Q_2} = aa_2 - aa_1 + bb_2 - bb_1 = 0$, since $aa_2 + bb_2 = -c = aa_1 + bb_1$ from the equation of the line.

Let $P_2 = (x_2, y_2)$ lie on the line. Then the distance from P_1 to the line is the absolute value of the scalar projection

of $\overrightarrow{P_1P_2}$ onto $\mathbf{n}$. $\text{comp}_{\mathbf{n}}\left(\overrightarrow{P_1P_2}\right) = \dfrac{|\mathbf{n}\cdot\langle x_2 - x_1, y_2 - y_1\rangle|}{|\mathbf{n}|} = \dfrac{|ax_2 - ax_1 + by_2 - by_1|}{\sqrt{a^2+b^2}} = \dfrac{|ax_1 + by_1 + c|}{\sqrt{a^2+b^2}}$

since $ax_2 + by_2 = -c$. The required distance is $\dfrac{|3\cdot -2 + -4\cdot 3 + 5|}{\sqrt{3^2+4^2}} = \dfrac{13}{5}$.

36. $(\mathbf{r}-\mathbf{a})\cdot(\mathbf{r}-\mathbf{b}) = 0$ implies that the vectors $\mathbf{r}-\mathbf{a}$ and $\mathbf{r}-\mathbf{b}$ are orthogonal.

From the diagram (in which A, B and R are the terminal points of the vectors), we see that this implies that R lies on a sphere whose diameter is the line from A to B.

The center of this circle is the midpoint of AB, that is,

$\frac{1}{2}(\mathbf{a}+\mathbf{b}) = \left\langle \frac{1}{2}(a_1+b_1), \frac{1}{2}(a_2+b_2), \frac{1}{2}(a_3+b_3)\right\rangle$, and its radius is

$\frac{1}{2}\,|\mathbf{a}-\mathbf{b}| = \frac{1}{2}\sqrt{(a_1-b_1)^2 + (a_2-b_2)^2 + (a_3-b_3)^2}$.

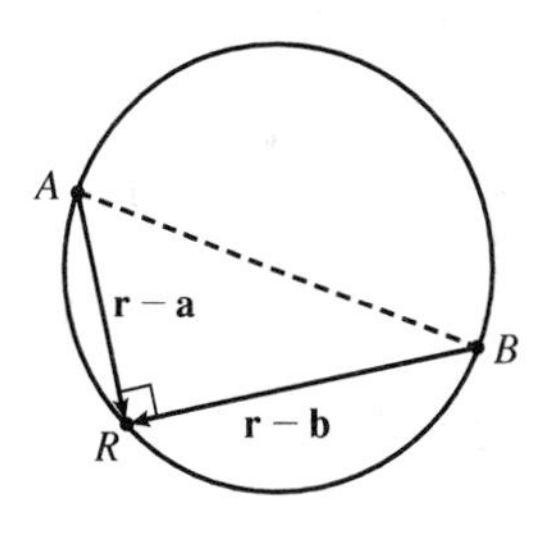

Or: Expand the given equation, substitute $\mathbf{r}\cdot\mathbf{r} = x^2 + y^2 + z^2$ and complete the squares.

37. For convenience, consider the unit cube positioned so that its back left corner is at the origin, and its edges lie along the coordinate axes. The diagonal of the cube that begins at the origin and ends at $(1,1,1)$ has vector representation $\langle 1,1,1\rangle$. The angle θ between this vector and the vector of the edge which also begins at the origin and runs along the x-axis [that is, $\langle 1,0,0\rangle$] is given by $\cos\theta = \dfrac{\langle 1,1,1\rangle\cdot\langle 1,0,0\rangle}{|\langle 1,1,1\rangle|\,|\langle 1,0,0\rangle|} = \dfrac{1}{\sqrt{3}} \quad\Rightarrow\quad \theta = \cos^{-1}\left(\frac{1}{\sqrt{3}}\right) \approx 55^\circ$.

38. Consider a cube with sides of unit length, wholly within the first octant and with edges along each of the three coordinate axes. $\mathbf{i}+\mathbf{j}+\mathbf{k}$ and $\mathbf{i}+\mathbf{j}$ are vector representations of a diagonal of the cube and a diagonal of one of its faces. If θ is the angle between these diagonals, then $\cos\theta = \dfrac{(\mathbf{i}+\mathbf{j}+\mathbf{k})\cdot(\mathbf{i}+\mathbf{j})}{|\mathbf{i}+\mathbf{j}+\mathbf{k}|\,|\mathbf{i}+\mathbf{j}|} = \dfrac{1+1}{\sqrt{3}\sqrt{2}} = \sqrt{\dfrac{2}{3}} \quad\Rightarrow\quad \theta = \cos^{-1}\sqrt{\frac{2}{3}} \approx 35^\circ$.

39. Consider the H—C—H combination consisting of the sole carbon atom and the two hydrogen atoms that are at $(1, 0, 0)$ and $(0, 1, 0)$ (or any H—C—H combination, for that matter). Vector representations of the line segments emanating from the carbon atom and extending to these two hydrogen atoms are $\langle 1 - \frac{1}{2}, 0 - \frac{1}{2}, 0 - \frac{1}{2} \rangle = \langle \frac{1}{2}, -\frac{1}{2}, -\frac{1}{2} \rangle$ and $\langle 0 - \frac{1}{2}, 1 - \frac{1}{2}, 0 - \frac{1}{2} \rangle = \langle -\frac{1}{2}, \frac{1}{2}, -\frac{1}{2} \rangle$. The bond angle, θ, is therefore given by

$$\cos\theta = \frac{\langle \frac{1}{2}, -\frac{1}{2}, -\frac{1}{2} \rangle \cdot \langle -\frac{1}{2}, \frac{1}{2}, -\frac{1}{2} \rangle}{\left|\langle \frac{1}{2}, -\frac{1}{2}, -\frac{1}{2} \rangle\right| \left|\langle -\frac{1}{2}, \frac{1}{2}, -\frac{1}{2} \rangle\right|} = \frac{-\frac{1}{4} - \frac{1}{4} + \frac{1}{4}}{\sqrt{\frac{3}{4}}\sqrt{\frac{3}{4}}} = -\frac{1}{3} \quad \Rightarrow \quad \theta = \cos^{-1}\left(-\frac{1}{3}\right) \approx 109.5^\circ.$$

40. Let α be the angle between $\mathbf{a}$ and $\mathbf{c}$ and β be the angle between $\mathbf{c}$ and $\mathbf{b}$. We need to show that $\alpha = \beta$. Now

$$\cos\alpha = \frac{\mathbf{a} \cdot \mathbf{c}}{|\mathbf{a}|\,|\mathbf{c}|} = \frac{\mathbf{a} \cdot |\mathbf{a}|\,\mathbf{b} + \mathbf{a} \cdot |\mathbf{b}|\,\mathbf{a}}{|\mathbf{a}|\,|\mathbf{c}|} = \frac{|\mathbf{a}|\,\mathbf{a} \cdot \mathbf{b} + |\mathbf{a}|^2\,|\mathbf{b}|}{|\mathbf{a}|\,|\mathbf{c}|} = \frac{\mathbf{a} \cdot \mathbf{b} + |\mathbf{a}|\,|\mathbf{b}|}{|\mathbf{c}|}.$$ Similarly,

$$\cos\beta = \frac{\mathbf{b} \cdot \mathbf{c}}{|\mathbf{b}|\,|\mathbf{c}|} = \frac{|\mathbf{a}|\,|\mathbf{b}| + \mathbf{b} \cdot \mathbf{a}}{|\mathbf{c}|}.$$ Thus $\cos\alpha = \cos\beta$. However $0^\circ \le \alpha \le 180^\circ$ and $0^\circ \le \beta \le 180^\circ$, so $\alpha = \beta$ and $\mathbf{c}$ bisects the angle between $\mathbf{a}$ and $\mathbf{b}$.

41. Let $\mathbf{a} = \langle a_1, a_2, a_3 \rangle$ and $= \langle b_1, b_2, b_3 \rangle$.

Property 2: $\mathbf{a} \cdot \mathbf{b} = \langle a_1, a_2, a_3 \rangle \cdot \langle b_1, b_2, b_3 \rangle = a_1 b_1 + a_2 b_2 + a_3 b_3$

$$= b_1 a_1 + b_2 a_2 + b_3 a_3 = \langle b_1, b_2, b_3 \rangle \cdot \langle a_1, a_2, a_3 \rangle = \mathbf{b} \cdot \mathbf{a}$$

Property 4: $(c\,\mathbf{a}) \cdot \mathbf{b} = \langle ca_1, ca_2, ca_3 \rangle \cdot \langle b_1, b_2, b_3 \rangle = (ca_1)b_1 + (ca_2)b_2 + (ca_3)b_3$

$$= c\,(a_1 b_1 + a_2 b_2 + a_3 b_3) = c\,(\mathbf{a} \cdot \mathbf{b}) = a_1(cb_1) + a_2(cb_2) + a_3(cb_3)$$

$$= \langle a_1, a_2, a_3 \rangle \cdot \langle cb_1, cb_2, cb_3 \rangle = \mathbf{a} \cdot (c\,\mathbf{b})$$

Property 5: $\mathbf{0} \cdot \mathbf{a} = \langle 0, 0, 0 \rangle \cdot \langle a_1, a_2, a_3 \rangle = (0)(a_1) + (0)(a_2) + (0)(a_3) = 0$

42. Let the figure be called quadrilateral $ABCD$. The diagonals can be represented by $\overrightarrow{AC}$ and $\overrightarrow{BD}$. $\overrightarrow{AC} = \overrightarrow{AB} + \overrightarrow{BC}$ and $\overrightarrow{BD} = \overrightarrow{BC} + \overrightarrow{CD} = \overrightarrow{BC} - \overrightarrow{DC} = \overrightarrow{BC} - \overrightarrow{AB}$ (Since opposite sides of the object are of the same length and parallel, $\overrightarrow{AB} = \overrightarrow{DC}$.) Thus

$$\overrightarrow{AC} \cdot \overrightarrow{BD} = \left(\overrightarrow{AB} + \overrightarrow{BC}\right) \cdot \left(\overrightarrow{BC} - \overrightarrow{AB}\right) = \overrightarrow{AB} \cdot \left(\overrightarrow{BC} - \overrightarrow{AB}\right) + \overrightarrow{BC} \cdot \left(\overrightarrow{BC} - \overrightarrow{AB}\right)$$

$$= \overrightarrow{AB} \cdot \overrightarrow{BC} - \left|\overrightarrow{AB}\right|^2 + \left|\overrightarrow{BC}\right|^2 - \overrightarrow{AB} \cdot \overrightarrow{BC} = \left|\overrightarrow{BC}\right|^2 - \left|\overrightarrow{AB}\right|^2$$

But $\left|\overrightarrow{AB}\right|^2 = \left|\overrightarrow{BC}\right|^2$ because all sides of the quadrilateral are equal in length. Therefore $\overrightarrow{AC} \cdot \overrightarrow{BD} = 0$, and since both of these vectors are nonzero this tells us that the diagonals of the quadrilateral are perpendicular.

43. $|\mathbf{a} \cdot \mathbf{b}| = \big|\,|\mathbf{a}|\,|\mathbf{b}| \cos\theta\big| = |\mathbf{a}|\,|\mathbf{b}|\,|\cos\theta|$. Since $|\cos\theta| \le 1$, $|\mathbf{a} \cdot \mathbf{b}| = |\mathbf{a}|\,|\mathbf{b}|\,|\cos\theta| \le |\mathbf{a}|\,|\mathbf{b}|$.

Note: We have equality in the case of $\cos\theta = \pm 1$, so $\theta = 0$ or $\theta = \pi$, thus equality when $\mathbf{a}$ and $\mathbf{b}$ are parallel.

44. (a)

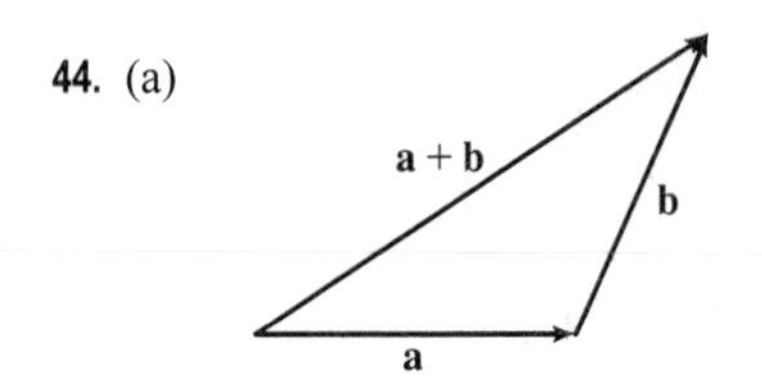

The Triangle Inequality states that the length of the longest side of a triangle is less than or equal to the sum of the lengths of the two shortest sides.

(b) $|\mathbf{a}+\mathbf{b}|^2 = (\mathbf{a}+\mathbf{b})\cdot(\mathbf{a}+\mathbf{b}) = (\mathbf{a}\cdot\mathbf{a}) + 2(\mathbf{a}\cdot\mathbf{b}) + (\mathbf{b}\cdot\mathbf{b}) = |\mathbf{a}|^2 + 2(\mathbf{a}\cdot\mathbf{b}) + |\mathbf{b}|^2$

$\leq |\mathbf{a}|^2 + 2\,|\mathbf{a}|\,|\mathbf{b}| + |\mathbf{b}|^2$ [by the Cauchy-Schwartz Inequality]

$= (|\mathbf{a}| + |\mathbf{b}|)^2$

Thus, taking the square root of both sides, $|\mathbf{a}+\mathbf{b}| \leq |\mathbf{a}| + |\mathbf{b}|$.

45. (a)

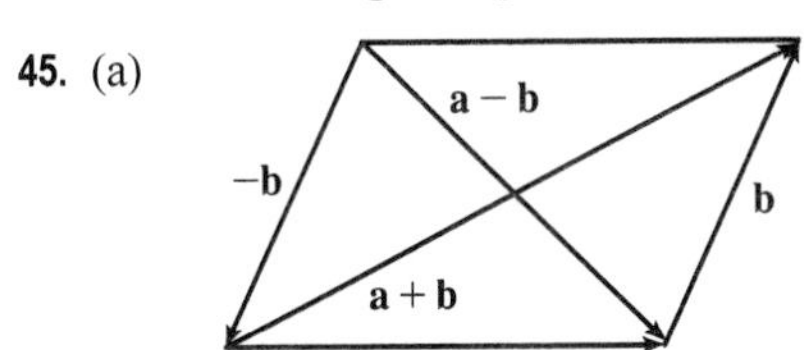

The Parallelogram Law states that the sum of the squares of the lengths of the diagonals of a parallelogram equals the sum of the squares of its (four) sides.

(b) $|\mathbf{a}+\mathbf{b}|^2 = (\mathbf{a}+\mathbf{b})\cdot(\mathbf{a}+\mathbf{b}) = |\mathbf{a}|^2 + 2(\mathbf{a}\cdot\mathbf{b}) + |\mathbf{b}|^2$ and $|\mathbf{a}-\mathbf{b}|^2 = (\mathbf{a}-\mathbf{b})\cdot(\mathbf{a}-\mathbf{b}) = |\mathbf{a}|^2 - 2(\mathbf{a}\cdot\mathbf{b}) + |\mathbf{b}|^2$.
Adding these two equations gives $|\mathbf{a}+\mathbf{b}|^2 + |\mathbf{a}-\mathbf{b}|^2 = 2\,|\mathbf{a}|^2 + 2\,|\mathbf{b}|^2$.

10.4 The Cross Product

1. $\mathbf{a}\times\mathbf{b} = \begin{vmatrix} \mathbf{i} & \mathbf{j} & \mathbf{k} \\ 1 & 2 & 0 \\ 0 & 3 & 1 \end{vmatrix} = \begin{vmatrix} 2 & 0 \\ 3 & 1 \end{vmatrix}\mathbf{i} - \begin{vmatrix} 1 & 0 \\ 0 & 1 \end{vmatrix}\mathbf{j} + \begin{vmatrix} 1 & 2 \\ 0 & 3 \end{vmatrix}\mathbf{k} = (2-0)\,\mathbf{i} - (1-0)\,\mathbf{j} + (3-0)\,\mathbf{k} = 2\,\mathbf{i} - \mathbf{j} + 3\,\mathbf{k}$

Now $(\mathbf{a}\times\mathbf{b})\cdot\mathbf{a} = \langle 2,-1,3\rangle\cdot\langle 1,2,0\rangle = 2-2+0 = 0$ and $(\mathbf{a}\times\mathbf{b})\cdot\mathbf{b} = \langle 2,-1,3\rangle\cdot\langle 0,3,1\rangle = 0-3+3 = 0$,
so $\mathbf{a}\times\mathbf{b}$ is orthogonal to both $\mathbf{a}$ and $\mathbf{b}$.

2. $\mathbf{a}\times\mathbf{b} = \begin{vmatrix} \mathbf{i} & \mathbf{j} & \mathbf{k} \\ 5 & 1 & 4 \\ -1 & 0 & 2 \end{vmatrix} = \begin{vmatrix} 1 & 4 \\ 0 & 2 \end{vmatrix}\mathbf{i} - \begin{vmatrix} 5 & 4 \\ -1 & 2 \end{vmatrix}\mathbf{j} + \begin{vmatrix} 5 & 1 \\ -1 & 0 \end{vmatrix}\mathbf{k}$

$= (2-0)\,\mathbf{i} - [10-(-4)]\,\mathbf{j} + [0-(-1)]\,\mathbf{k} = 2\,\mathbf{i} - 14\,\mathbf{j} + \mathbf{k}$

Now $(\mathbf{a}\times\mathbf{b})\cdot\mathbf{a} = \langle 2,-14,1\rangle\cdot\langle 5,1,4\rangle = 10-14+4 = 0$ and $(\mathbf{a}\times\mathbf{b})\cdot\mathbf{b} = \langle 2,-14,1\rangle\cdot\langle -1,0,2\rangle = -2+0+2 = 0$,
so $\mathbf{a}\times\mathbf{b}$ is orthogonal to both $\mathbf{a}$ and $\mathbf{b}$.

3. $\mathbf{a}\times\mathbf{b} = \begin{vmatrix} \mathbf{i} & \mathbf{j} & \mathbf{k} \\ 2 & 1 & -1 \\ 0 & 1 & 2 \end{vmatrix} = \begin{vmatrix} 1 & -1 \\ 1 & 2 \end{vmatrix}\mathbf{i} - \begin{vmatrix} 2 & -1 \\ 0 & 2 \end{vmatrix}\mathbf{j} + \begin{vmatrix} 2 & 1 \\ 0 & 1 \end{vmatrix}\mathbf{k}$

$= [2-(-1)]\,\mathbf{i} - (4-0)\,\mathbf{j} + (2-0)\,\mathbf{k} = 3\,\mathbf{i} - 4\,\mathbf{j} + 2\,\mathbf{k}$

Now $(\mathbf{a}\times\mathbf{b})\cdot\mathbf{a} = (3\,\mathbf{i} - 4\,\mathbf{j} + 2\,\mathbf{k})\cdot(2\,\mathbf{i} + \mathbf{j} - \mathbf{k}) = 6-4-2 = 0$ and
$(\mathbf{a}\times\mathbf{b})\cdot\mathbf{b} = (3\,\mathbf{i} - 4\,\mathbf{j} + 2\,\mathbf{k})\cdot(\mathbf{j} + 2\,\mathbf{k}) = 0-4+4 = 0$, so $\mathbf{a}\times\mathbf{b}$ is orthogonal to both $\mathbf{a}$ and $\mathbf{b}$.

4. $\mathbf{a}\times\mathbf{b} = \begin{vmatrix} \mathbf{i} & \mathbf{j} & \mathbf{k} \\ 1 & -1 & 1 \\ 1 & 1 & 1 \end{vmatrix} = \begin{vmatrix} -1 & 1 \\ 1 & 1 \end{vmatrix}\mathbf{i} - \begin{vmatrix} 1 & 1 \\ 1 & 1 \end{vmatrix}\mathbf{j} + \begin{vmatrix} 1 & -1 \\ 1 & 1 \end{vmatrix}\mathbf{k}$

$= (-1-1)\,\mathbf{i} - (1-1)\,\mathbf{j} + [1-(-1)]\,\mathbf{k} = -2\,\mathbf{i} + 2\,\mathbf{k}$

Now $(\mathbf{a}\times\mathbf{b})\cdot\mathbf{a} = (-2\,\mathbf{i} + 2\,\mathbf{k})\cdot(\mathbf{i} - \mathbf{j} + \mathbf{k}) = -2+0+2 = 0$ and
$(\mathbf{a}\times\mathbf{b})\cdot\mathbf{b} = (-2\,\mathbf{i} + 2\,\mathbf{k})\cdot(\mathbf{i} + \mathbf{j} + \mathbf{k}) = -2+0+2 = 0$, so $\mathbf{a}\times\mathbf{b}$ is orthogonal to both $\mathbf{a}$ and $\mathbf{b}$.

5. $\mathbf{a}\times\mathbf{b}=\begin{vmatrix}\mathbf{i}&\mathbf{j}&\mathbf{k}\\3&2&4\\1&-2&-3\end{vmatrix}=\begin{vmatrix}2&4\\-2&-3\end{vmatrix}\mathbf{i}-\begin{vmatrix}3&4\\1&-3\end{vmatrix}\mathbf{j}+\begin{vmatrix}3&2\\1&-2\end{vmatrix}\mathbf{k}$

$=[-6-(-8)]\,\mathbf{i}-(-9-4)\,\mathbf{j}+(-6-2)\,\mathbf{k}=2\,\mathbf{i}+13\,\mathbf{j}-8\,\mathbf{k}$

Since $(\mathbf{a}\times\mathbf{b})\cdot\mathbf{a}=(2\,\mathbf{i}+13\,\mathbf{j}-8\,\mathbf{k})\cdot(3\,\mathbf{i}+2\,\mathbf{j}+4\,\mathbf{k})=6+26-32=0$, $\mathbf{a}\times\mathbf{b}$ is orthogonal to $\mathbf{a}$.

Since $(\mathbf{a}\times\mathbf{b})\cdot\mathbf{b}=(2\,\mathbf{i}+13\,\mathbf{j}-8\,\mathbf{k})\cdot(\mathbf{i}-2\,\mathbf{j}-3\,\mathbf{k})=2-26+24=0$, $\mathbf{a}\times\mathbf{b}$ is orthogonal to $\mathbf{b}$.

6. $\mathbf{a}\times\mathbf{b}=\begin{vmatrix}\mathbf{i}&\mathbf{j}&\mathbf{k}\\1&e^t&e^{-t}\\2&e^t&-e^{-t}\end{vmatrix}=\begin{vmatrix}e^t&e^{-t}\\e^t&-e^{-t}\end{vmatrix}\mathbf{i}-\begin{vmatrix}1&e^{-t}\\2&-e^{-t}\end{vmatrix}\mathbf{j}+\begin{vmatrix}1&e^t\\2&e^t\end{vmatrix}\mathbf{k}$

$=(-1-1)\,\mathbf{i}-(-e^{-t}-2e^{-t})\,\mathbf{j}+(e^t-2e^t)\,\mathbf{k}=-2\,\mathbf{i}+3e^{-t}\,\mathbf{j}-e^t\,\mathbf{k}$

Since $(\mathbf{a}\times\mathbf{b})\cdot\mathbf{a}=(-2\,\mathbf{i}+3e^{-t}\,\mathbf{j}-e^t\,\mathbf{k})\cdot(\mathbf{i}+e^t\,\mathbf{j}+e^{-t}\,\mathbf{k})=-2+3-1=0$, $\mathbf{a}\times\mathbf{b}$ is orthogonal to $\mathbf{a}$.

Since $(\mathbf{a}\times\mathbf{b})\cdot\mathbf{b}=(-2\,\mathbf{i}+3e^{-t}\,\mathbf{j}-e^t\,\mathbf{k})\cdot(2\,\mathbf{i}+e^t\,\mathbf{j}-e^{-t}\,\mathbf{k})=-4+3+1=0$, $\mathbf{a}\times\mathbf{b}$ is orthogonal to $\mathbf{b}$.

7. $\mathbf{a}\times\mathbf{b}=\begin{vmatrix}\mathbf{i}&\mathbf{j}&\mathbf{k}\\t&t^2&t^3\\1&2t&3t^2\end{vmatrix}=\begin{vmatrix}t^2&t^3\\2t&3t^2\end{vmatrix}\mathbf{i}-\begin{vmatrix}t&t^3\\1&3t^2\end{vmatrix}\mathbf{j}+\begin{vmatrix}t&t^2\\1&2t\end{vmatrix}\mathbf{k}$

$=(3t^4-2t^4)\,\mathbf{i}-(3t^3-t^3)\,\mathbf{j}+(2t^2-t^2)\,\mathbf{k}=t^4\,\mathbf{i}-2t^3\,\mathbf{j}+t^2\,\mathbf{k}$

Since $(\mathbf{a}\times\mathbf{b})\cdot\mathbf{a}=\langle t^4,-2t^3,t^2\rangle\cdot\langle t,t^2,t^3\rangle=t^5-2t^5+t^5=0$, $\mathbf{a}\times\mathbf{b}$ is orthogonal to $\mathbf{a}$.

Since $(\mathbf{a}\times\mathbf{b})\cdot\mathbf{b}=\langle t^4,-2t^3,t^2\rangle\cdot\langle 1,2t,3t^2\rangle=t^4-4t^4+3t^4=0$, $\mathbf{a}\times\mathbf{b}$ is orthogonal to $\mathbf{b}$.

8. $\mathbf{a}\times\mathbf{b}=\begin{vmatrix}\mathbf{i}&\mathbf{j}&\mathbf{k}\\1&0&-2\\0&1&1\end{vmatrix}$

$=\begin{vmatrix}0&-2\\1&1\end{vmatrix}\mathbf{i}-\begin{vmatrix}1&-2\\0&1\end{vmatrix}\mathbf{j}+\begin{vmatrix}1&0\\0&1\end{vmatrix}\mathbf{k}$

$=2\,\mathbf{i}-\mathbf{j}+\mathbf{k}$

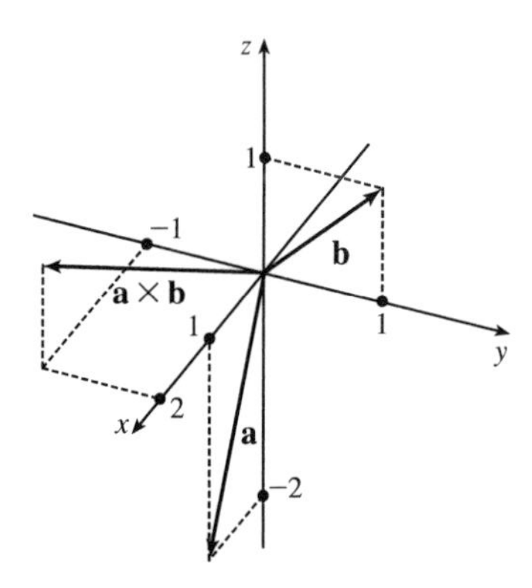

9. (a) Since $\mathbf{b}\times\mathbf{c}$ is a vector, the dot product $\mathbf{a}\cdot(\mathbf{b}\times\mathbf{c})$ is meaningful and is a scalar.

(b) $\mathbf{b}\cdot\mathbf{c}$ is a scalar, so $\mathbf{a}\times(\mathbf{b}\cdot\mathbf{c})$ is meaningless, as the cross product is defined only for two *vectors*.

(c) Since $\mathbf{b}\times\mathbf{c}$ is a vector, the cross product $\mathbf{a}\times(\mathbf{b}\times\mathbf{c})$ is meaningful and results in another vector.

(d) $\mathbf{a}\cdot\mathbf{b}$ is a scalar, so the cross product $(\mathbf{a}\cdot\mathbf{b})\times\mathbf{c}$ is meaningless.

(e) Since $(\mathbf{a}\cdot\mathbf{b})$ and $(\mathbf{c}\cdot\mathbf{d})$ are both scalars, the cross product $(\mathbf{a}\cdot\mathbf{b})\times(\mathbf{c}\cdot\mathbf{d})$ is meaningless.

(f) $\mathbf{a}\times\mathbf{b}$ and $\mathbf{c}\times\mathbf{d}$ are both vectors, so the dot product $(\mathbf{a}\times\mathbf{b})\cdot(\mathbf{c}\times\mathbf{d})$ is meaningful and is a scalar.

10. $|\mathbf{u}\times\mathbf{v}|=|\mathbf{u}|\,|\mathbf{v}|\sin\theta=(5)(10)\sin 60^\circ=25\sqrt{3}$. By the right-hand rule, $\mathbf{u}\times\mathbf{v}$ is directed into the page.

11. If we sketch $\mathbf{u}$ and $\mathbf{v}$ starting from the same initial point, we see that the angle between them is 30°, so $|\mathbf{u}\times\mathbf{v}| = |\mathbf{u}|\,|\mathbf{v}|\sin 30^\circ = (6)(8)\left(\frac{1}{2}\right) = 24$. By the right-hand rule, $\mathbf{u}\times\mathbf{v}$ is directed into the page.

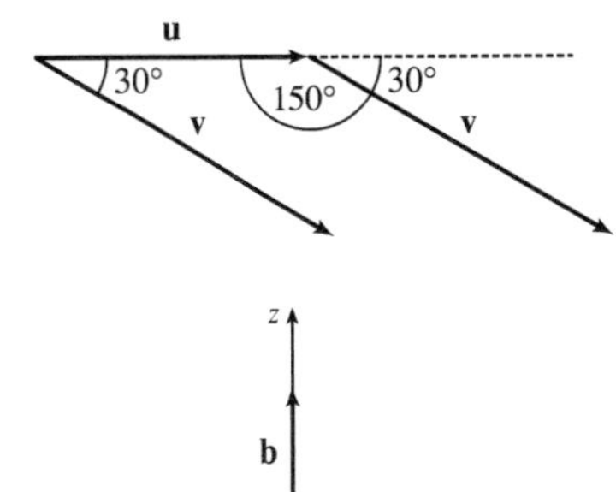

12. (a) $|\mathbf{a}\times\mathbf{b}| = |\mathbf{a}|\,|\mathbf{b}|\sin\theta = 3\cdot 2\cdot \sin\frac{\pi}{2} = 6$

(b) $\mathbf{a}\times\mathbf{b}$ is orthogonal to $\mathbf{k}$, so it lies in the xy-plane, and its z-coordinate is 0. By the right-hand rule, its y-component is negative and its x-component is positive.

13. $$\mathbf{a}\times\mathbf{b} = \begin{vmatrix} \mathbf{i} & \mathbf{j} & \mathbf{k} \\ 1 & 2 & 1 \\ 0 & 1 & 3 \end{vmatrix} = \begin{vmatrix} 2 & 1 \\ 1 & 3 \end{vmatrix}\mathbf{i} - \begin{vmatrix} 1 & 1 \\ 0 & 3 \end{vmatrix}\mathbf{j} + \begin{vmatrix} 1 & 2 \\ 0 & 1 \end{vmatrix}\mathbf{k} = (6-1)\,\mathbf{i} - (3-0)\,\mathbf{j} + (1-0)\,\mathbf{k} = 5\,\mathbf{i} - 3\mathbf{j} + \mathbf{k}$$

$$\mathbf{b}\times\mathbf{a} = \begin{vmatrix} \mathbf{i} & \mathbf{j} & \mathbf{k} \\ 0 & 1 & 3 \\ 1 & 2 & 1 \end{vmatrix} = \begin{vmatrix} 1 & 3 \\ 2 & 1 \end{vmatrix}\mathbf{i} - \begin{vmatrix} 0 & 3 \\ 1 & 1 \end{vmatrix}\mathbf{j} + \begin{vmatrix} 0 & 1 \\ 1 & 2 \end{vmatrix}\mathbf{k} = (1-6)\,\mathbf{i} - (0-3)\,\mathbf{j} + (0-1)\,\mathbf{k} = -5\,\mathbf{i} + 3\,\mathbf{j} - \mathbf{k}$$

Notice $\mathbf{a}\times\mathbf{b} = -\mathbf{b}\times\mathbf{a}$ here, as we know is always true by Theorem 8.

14. $$\mathbf{b}\times\mathbf{c} = \begin{vmatrix} \mathbf{i} & \mathbf{j} & \mathbf{k} \\ -1 & 1 & 0 \\ 0 & 0 & -4 \end{vmatrix} = \begin{vmatrix} 1 & 0 \\ 0 & -4 \end{vmatrix}\mathbf{i} - \begin{vmatrix} -1 & 0 \\ 0 & -4 \end{vmatrix}\mathbf{j} + \begin{vmatrix} -1 & 1 \\ 0 & 0 \end{vmatrix}\mathbf{k} = -4\,\mathbf{i} - 4\,\mathbf{j} \text{ so}$$

$$\mathbf{a}\times(\mathbf{b}\times\mathbf{c}) = \begin{vmatrix} \mathbf{i} & \mathbf{j} & \mathbf{k} \\ 3 & 1 & 2 \\ -4 & -4 & 0 \end{vmatrix} = \begin{vmatrix} 1 & 2 \\ -4 & 0 \end{vmatrix}\mathbf{i} - \begin{vmatrix} 3 & 2 \\ -4 & 0 \end{vmatrix}\mathbf{j} + \begin{vmatrix} 3 & 1 \\ -4 & -4 \end{vmatrix}\mathbf{k} = 8\,\mathbf{i} - 8\,\mathbf{j} - 8\,\mathbf{k}.$$

$$\mathbf{a}\times\mathbf{b} = \begin{vmatrix} \mathbf{i} & \mathbf{j} & \mathbf{k} \\ 3 & 1 & 2 \\ -1 & 1 & 0 \end{vmatrix} = \begin{vmatrix} 1 & 2 \\ 1 & 0 \end{vmatrix}\mathbf{i} - \begin{vmatrix} 3 & 2 \\ -1 & 0 \end{vmatrix}\mathbf{j} + \begin{vmatrix} 3 & 1 \\ -1 & 1 \end{vmatrix}\mathbf{k} = -2\,\mathbf{i} - 2\,\mathbf{j} + 4\,\mathbf{k} \text{ so}$$

$$(\mathbf{a}\times\mathbf{b})\times\mathbf{c} = \begin{vmatrix} \mathbf{i} & \mathbf{j} & \mathbf{k} \\ -2 & -2 & 4 \\ 0 & 0 & -4 \end{vmatrix} = \begin{vmatrix} -2 & 4 \\ 0 & -4 \end{vmatrix}\mathbf{i} - \begin{vmatrix} -2 & 4 \\ 0 & -4 \end{vmatrix}\mathbf{j} + \begin{vmatrix} -2 & -2 \\ 0 & 0 \end{vmatrix}\mathbf{k} = 8\,\mathbf{i} - 8\,\mathbf{j}.$$

Thus $\mathbf{a}\times(\mathbf{b}\times\mathbf{c}) \neq (\mathbf{a}\times\mathbf{b})\times\mathbf{c}$.

15. We know that the cross product of two vectors is orthogonal to both. So we calculate

$$\langle 2, 0, -3\rangle \times \langle -1, 4, 2\rangle = \begin{vmatrix} \mathbf{i} & \mathbf{j} & \mathbf{k} \\ 2 & 0 & -3 \\ -1 & 4 & 2 \end{vmatrix} = \begin{vmatrix} 0 & -3 \\ 4 & 2 \end{vmatrix}\mathbf{i} - \begin{vmatrix} 2 & -3 \\ -1 & 2 \end{vmatrix}\mathbf{j} + \begin{vmatrix} 2 & 0 \\ -1 & 4 \end{vmatrix}\mathbf{k} = 12\,\mathbf{i} - \mathbf{j} + 8\,\mathbf{k}$$

So two unit vectors orthogonal to both are $\pm\dfrac{\langle 12, -1, 8\rangle}{\sqrt{144+1+64}} = \pm\dfrac{\langle 12, -1, 8\rangle}{\sqrt{209}}$, that is, $\left\langle \frac{12}{\sqrt{209}}, -\frac{1}{\sqrt{209}}, \frac{8}{\sqrt{209}} \right\rangle$ and $\left\langle -\frac{12}{\sqrt{209}}, \frac{1}{\sqrt{209}}, -\frac{8}{\sqrt{209}} \right\rangle$.

16. We know that the cross product of two vectors is orthogonal to both. So we calculate

$$\begin{vmatrix} \mathbf{i} & \mathbf{j} & \mathbf{k} \\ 1 & 1 & 1 \\ 2 & 0 & 1 \end{vmatrix} = \begin{vmatrix} 1 & 1 \\ 0 & 1 \end{vmatrix} \mathbf{i} - \begin{vmatrix} 1 & 1 \\ 2 & 1 \end{vmatrix} \mathbf{j} + \begin{vmatrix} 1 & 1 \\ 2 & 0 \end{vmatrix} \mathbf{k} = \mathbf{i} + \mathbf{j} - 2\mathbf{k}$$

Thus, two unit vectors orthogonal to both are $\pm\frac{1}{\sqrt{6}}\langle 1, 1, -2\rangle$, that is, $\left\langle \frac{1}{\sqrt{6}}, \frac{1}{\sqrt{6}}, -\frac{2}{\sqrt{6}} \right\rangle$ and $\left\langle -\frac{1}{\sqrt{6}}, -\frac{1}{\sqrt{6}}, \frac{2}{\sqrt{6}} \right\rangle$.

17. Let $\mathbf{a} = \langle a_1, a_2, a_3 \rangle$. Then

$$\mathbf{0} \times \mathbf{a} = \begin{vmatrix} \mathbf{i} & \mathbf{j} & \mathbf{k} \\ 0 & 0 & 0 \\ a_1 & a_2 & a_3 \end{vmatrix} = \begin{vmatrix} 0 & 0 \\ a_2 & a_3 \end{vmatrix} \mathbf{i} - \begin{vmatrix} 0 & 0 \\ a_1 & a_3 \end{vmatrix} \mathbf{j} + \begin{vmatrix} 0 & 0 \\ a_1 & a_2 \end{vmatrix} \mathbf{k} = \mathbf{0},$$

$$\mathbf{a} \times \mathbf{0} = \begin{vmatrix} \mathbf{i} & \mathbf{j} & \mathbf{k} \\ a_1 & a_2 & a_3 \\ 0 & 0 & 0 \end{vmatrix} = \begin{vmatrix} a_2 & a_3 \\ 0 & 0 \end{vmatrix} \mathbf{i} - \begin{vmatrix} a_1 & a_3 \\ 0 & 0 \end{vmatrix} \mathbf{j} + \begin{vmatrix} a_1 & a_2 \\ 0 & 0 \end{vmatrix} \mathbf{k} = \mathbf{0}.$$

18. Let $\mathbf{a} = \langle a_1, a_2, a_3 \rangle$ and $\mathbf{b} = \langle b_1, b_2, b_3 \rangle$.

$$(\mathbf{a} \times \mathbf{b}) \cdot \mathbf{b} = \left\langle \begin{vmatrix} a_2 & a_3 \\ b_2 & b_3 \end{vmatrix}, \begin{vmatrix} a_1 & a_3 \\ b_1 & b_3 \end{vmatrix}, \begin{vmatrix} a_1 & a_2 \\ b_1 & b_2 \end{vmatrix} \right\rangle \cdot \langle b_1, b_2, b_3 \rangle = \begin{vmatrix} a_2 & a_3 \\ b_2 & b_3 \end{vmatrix} b_1 - \begin{vmatrix} a_1 & a_3 \\ b_1 & b_3 \end{vmatrix} b_2 + \begin{vmatrix} a_1 & a_2 \\ b_1 & b_2 \end{vmatrix} b_3$$

$$= (a_2b_3b_1 - a_3b_2b_1) - (a_1b_3b_2 - a_3b_1b_2) + (a_1b_2b_3 - a_2b_1b_3) = 0$$

19.
$$\begin{aligned} \mathbf{a} \times \mathbf{b} &= \langle a_2b_3 - a_3b_2, a_3b_1 - a_1b_3, a_1b_2 - a_2b_1 \rangle \\ &= \langle (-1)(b_2a_3 - b_3a_2), (-1)(b_3a_1 - b_1a_3), (-1)(b_1a_2 - b_2a_1) \rangle \\ &= -\langle b_2a_3 - b_3a_2, b_3a_1 - b_1a_3, b_1a_2 - b_2a_1 \rangle = -\mathbf{b} \times \mathbf{a} \end{aligned}$$

20. $c\mathbf{a} = \langle ca_1, ca_2, ca_3 \rangle$, so

$$\begin{aligned} (c\mathbf{a}) \times \mathbf{b} &= \langle ca_2b_3 - ca_3b_2, ca_3b_1 - ca_1b_3, ca_1b_2 - ca_2b_1 \rangle \\ &= c\langle a_2b_3 - a_3b_2, a_3b_1 - a_1b_3, a_1b_2 - a_2b_1 \rangle = c(\mathbf{a} \times \mathbf{b}) \\ &= \langle ca_2b_3 - ca_3b_2, ca_3b_1 - ca_1b_3, ca_1b_2 - ca_2b_1 \rangle \\ &= \langle a_2(cb_3) - a_3(cb_2), a_3(cb_1) - a_1(cb_3), a_1(cb_2) - a_2(cb_1) \rangle \\ &= \mathbf{a} \times c\mathbf{b} \end{aligned}$$

21.
$$\begin{aligned} \mathbf{a} \times (\mathbf{b} + \mathbf{c}) &= \mathbf{a} \times \langle b_1 + c_1, b_2 + c_2, b_3 + c_3 \rangle \\ &= \langle a_2(b_3 + c_3) - a_3(b_2 + c_2), a_3(b_1 + c_1) - a_1(b_3 + c_3), a_1(b_2 + c_2) - a_2(b_1 + c_1) \rangle \\ &= \langle a_2b_3 + a_2c_3 - a_3b_2 - a_3c_2, a_3b_1 + a_3c_1 - a_1b_3 - a_1c_3, a_1b_2 + a_1c_2 - a_2b_1 - a_2c_1 \rangle \\ &= \langle (a_2b_3 - a_3b_2) + (a_2c_3 - a_3c_2), (a_3b_1 - a_1b_3) + (a_3c_1 - a_1c_3), (a_1b_2 - a_2b_1) + (a_1c_2 - a_2c_1) \rangle \\ &= \langle a_2b_3 - a_3b_2, a_3b_1 - a_1b_3, a_1b_2 - a_2b_1 \rangle + \langle a_2c_3 - a_3c_2, a_3c_1 - a_1c_3, a_1c_2 - a_2c_1 \rangle \\ &= (\mathbf{a} \times \mathbf{b}) + (\mathbf{a} \times \mathbf{c}) \end{aligned}$$

22. $(\mathbf{a}+\mathbf{b})\times\mathbf{c} = -\mathbf{c}\times(\mathbf{a}+\mathbf{b})$ by Property 1 of Theorem 8

$= -(\mathbf{c}\times\mathbf{a}+\mathbf{c}\times\mathbf{b})$ by Property 3 of Theorem 8

$= -(-\mathbf{a}\times\mathbf{c}+(-\mathbf{b}\times\mathbf{c}))$ by Property 1 of Theorem 8

$= \mathbf{a}\times\mathbf{c}+\mathbf{b}\times\mathbf{c}$ by Property 2 of Theorem 8

23. By plotting the vertices, we can see that the parallelogram is determined by the vectors $\overrightarrow{AB} = \langle 2, 3\rangle$ and $\overrightarrow{AD} = \langle 4, -2\rangle$. We know that the area of the parallelogram determined by two vectors is equal to the length of the cross product of these vectors. In order to compute the cross product, we consider the vector $\overrightarrow{AB}$ as the three-dimensional vector $\langle 2, 3, 0\rangle$ (and similarly for $\overrightarrow{AD}$), and then the area of parallelogram $ABCD$ is

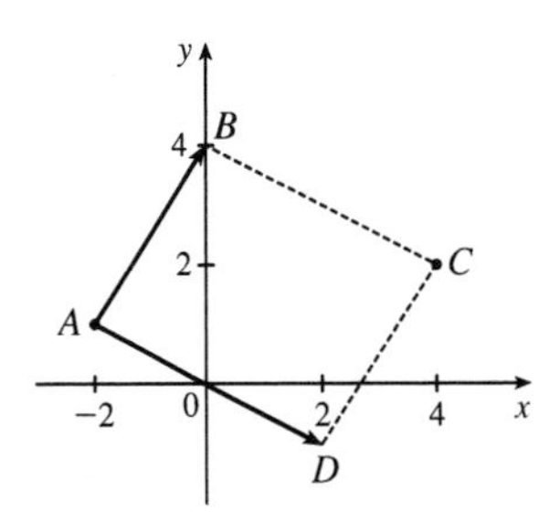

$$\left|\overrightarrow{AB}\times\overrightarrow{AD}\right| = \left\|\begin{matrix}\mathbf{i} & \mathbf{j} & \mathbf{k}\\ 2 & 3 & 0\\ 4 & -2 & 0\end{matrix}\right\| = |(0)\,\mathbf{i}-(0)\,\mathbf{j}+(-4-12)\,\mathbf{k}| = |-16\,\mathbf{k}| = 16$$

24. The parallelogram is determined by the vectors $\overrightarrow{KL} = \langle 0, 1, 3\rangle$ and $\overrightarrow{KN} = \langle 2, 5, 0\rangle$, so the area of parallelogram $KLMN$ is

$$\left|\overrightarrow{KL}\times\overrightarrow{KN}\right| = \left\|\begin{matrix}\mathbf{i} & \mathbf{j} & \mathbf{k}\\ 0 & 1 & 3\\ 2 & 5 & 0\end{matrix}\right\| = |(-15)\,\mathbf{i}-(-6)\,\mathbf{j}+(-2)\,\mathbf{k}| = |-15\,\mathbf{i}+6\,\mathbf{j}-2\,\mathbf{k}| = \sqrt{265}\approx 16.28$$

25. (a) Because the plane through P, Q, and R contains the vectors $\overrightarrow{PQ}$ and $\overrightarrow{PR}$, a vector orthogonal to both of these vectors (such as their cross product) is also orthogonal to the plane. Here $\overrightarrow{PQ} = \langle -1, 2, 0\rangle$ and $\overrightarrow{PR} = \langle -1, 0, 3\rangle$, so

$$\overrightarrow{PQ}\times\overrightarrow{PR} = \langle (2)(3)-(0)(0), (0)(-1)-(-1)(3), (-1)(0)-(2)(-1)\rangle = \langle 6, 3, 2\rangle$$

Therefore, $\langle 6, 3, 2\rangle$ (or any scalar multiple thereof) is orthogonal to the plane through P, Q, and R.

(b) Note that the area of the triangle determined by P, Q, and R is equal to half of the area of the parallelogram determined by the three points. From part (a), the area of the parallelogram is $\left|\overrightarrow{PQ}\times\overrightarrow{PR}\right| = |\langle 6, 3, 2\rangle| = \sqrt{36+9+4} = 7$, so the area of the triangle is $\frac{1}{2}(7) = \frac{7}{2}$.

26. (a) $\overrightarrow{PQ} = \langle -3, 2, -1\rangle$ and $\overrightarrow{PR} = \langle 1, -1, 1\rangle$, so a vector orthogonal to the plane through P, Q, and R is $\overrightarrow{PQ}\times\overrightarrow{PR} = \langle (2)(1)-(-1)(-1), (-1)(1)-(-3)(1), (-3)(-1)-(2)(1)\rangle = \langle 1, 2, 1\rangle$ (or any scalar mutiple thereof).

(b) The area of the parallelogram determined by $\overrightarrow{PQ}$ and $\overrightarrow{PR}$ is $\left|\overrightarrow{PQ}\times\overrightarrow{PR}\right| = |\langle 1, 2, 1\rangle| = \sqrt{1^2+2^2+1^2} = \sqrt{6}$, so the area of triangle PQR is $\frac{1}{2}\sqrt{6}$.

27. (a) $\overrightarrow{PQ} = \langle 4, 3, -2\rangle$ and $\overrightarrow{PR} = \langle 5, 5, 1\rangle$, so a vector orthogonal to the plane through P, Q, and R is $\overrightarrow{PQ}\times\overrightarrow{PR} = \langle (3)(1)-(-2)(5), (-2)(5)-(4)(1), (4)(5)-(3)(5)\rangle = \langle 13, -14, 5\rangle$ (or any scalar mutiple thereof).

(b) The area of the parallelogram determined by $\overrightarrow{PQ}$ and $\overrightarrow{PR}$ is

$\left|\overrightarrow{PQ} \times \overrightarrow{PR}\right| = |\langle 13, -14, 5\rangle| = \sqrt{13^2 + (-14)^2 + 5^2} = \sqrt{390}$, so the area of triangle PQR is $\frac{1}{2}\sqrt{390}$.

28. (a) $\overrightarrow{PQ} = \langle 1, 1, 3\rangle$ and $\overrightarrow{PR} = \langle 3, 2, 5\rangle$, so a vector orthogonal to the plane through P, Q, and R is

$\overrightarrow{PQ} \times \overrightarrow{PR} = \langle (1)(5) - (3)(2), (3)(3) - (1)(5), (1)(2) - (1)(3)\rangle = \langle -1, 4, -1\rangle$ (or any scalar mutiple thereof).

(b) The area of the parallelogram determined by $\overrightarrow{PQ}$ and $\overrightarrow{PR}$ is

$\left|\overrightarrow{PQ} \times \overrightarrow{PR}\right| = |\langle -1, 4, -1\rangle| = \sqrt{1 + 16 + 1} = \sqrt{18} = 3\sqrt{2}$, so the area of triangle PQR is $\frac{1}{2} \cdot 3\sqrt{2} = \frac{3}{2}\sqrt{2}$.

29. We know that the volume of the parallelepiped determined by **a**, **b**, and **c** is the magnitude of their scalar triple product, which

is $\mathbf{a} \cdot (\mathbf{b} \times \mathbf{c}) = \begin{vmatrix} 6 & 3 & -1 \\ 0 & 1 & 2 \\ 4 & -2 & 5 \end{vmatrix} = 6\begin{vmatrix} 1 & 2 \\ -2 & 5 \end{vmatrix} - 3\begin{vmatrix} 0 & 2 \\ 4 & 5 \end{vmatrix} + (-1)\begin{vmatrix} 0 & 1 \\ 4 & -2 \end{vmatrix} = 6(5+4) - 3(0-8) - (0-4) = 82.$

Thus the volume of the parallelepiped is 82 cubic units.

30. $\mathbf{a} \cdot (\mathbf{b} \times \mathbf{c}) = \begin{vmatrix} 1 & 1 & -1 \\ 1 & -1 & 1 \\ -1 & 1 & 1 \end{vmatrix} = 1\begin{vmatrix} -1 & 1 \\ 1 & 1 \end{vmatrix} - 1\begin{vmatrix} 1 & 1 \\ -1 & 1 \end{vmatrix} + (-1)\begin{vmatrix} 1 & -1 \\ -1 & 1 \end{vmatrix} = -2 - 2 + 0 = -4.$

So the volume of the parallelepiped determined by **a**, **b**, and **c** is $|-4| = 4$ cubic units.

31. $\mathbf{a} = \overrightarrow{PQ} = \langle 2, 1, 1\rangle$, $\mathbf{b} = \overrightarrow{PR} = \langle 1, -1, 2\rangle$, and $\mathbf{c} = \overrightarrow{PS} = \langle 0, -2, 3\rangle$.

$\mathbf{a} \cdot (\mathbf{b} \times \mathbf{c}) = \begin{vmatrix} 2 & 1 & 1 \\ 1 & -1 & 2 \\ 0 & -2 & 3 \end{vmatrix} = 2\begin{vmatrix} -1 & 2 \\ -2 & 3 \end{vmatrix} - 1\begin{vmatrix} 1 & 2 \\ 0 & 3 \end{vmatrix} + 1\begin{vmatrix} 1 & -1 \\ 0 & -2 \end{vmatrix} = 2 - 3 - 2 = -3,$

so the volume of the parallelepiped is 3 cubic units.

32. $\mathbf{a} = \overrightarrow{PQ} = \langle -4, 2, 4\rangle$, $\mathbf{b} = \overrightarrow{PR} = \langle 2, 1, -2\rangle$ and $\mathbf{c} = \overrightarrow{PS} = \langle -3, 4, 1\rangle$.

$\mathbf{a} \cdot (\mathbf{b} \times \mathbf{c}) = \begin{vmatrix} -4 & 2 & 4 \\ 2 & 1 & -2 \\ -3 & 4 & 1 \end{vmatrix} = -4\begin{vmatrix} 1 & -2 \\ 4 & 1 \end{vmatrix} - 2\begin{vmatrix} 2 & -2 \\ -3 & 1 \end{vmatrix} + 4\begin{vmatrix} 2 & 1 \\ -3 & 4 \end{vmatrix} = -36 + 8 + 44 = 16$, so the volume of the

parallelepiped is 16 cubic units.

33. $\mathbf{u} \cdot (\mathbf{v} \times \mathbf{w}) = \begin{vmatrix} 1 & 5 & -2 \\ 3 & -1 & 0 \\ 5 & 9 & -4 \end{vmatrix} = 1\begin{vmatrix} -1 & 0 \\ 9 & -4 \end{vmatrix} - 5\begin{vmatrix} 3 & 0 \\ 5 & -4 \end{vmatrix} + (-2)\begin{vmatrix} 3 & -1 \\ 5 & 9 \end{vmatrix} = 4 + 60 - 64 = 0$, which says that the volume

of the parallelepiped determined by **u**, **v** and **w** is 0, and thus these three vectors are coplanar.

34. $\mathbf{u} = \overrightarrow{AB} = \langle 2, -4, 4\rangle$, $\mathbf{v} = \overrightarrow{AC} = \langle 4, -1, -2\rangle$ and $\mathbf{w} = \overrightarrow{AD} = \langle 2, 3, -6\rangle$.

$\mathbf{u} \cdot (\mathbf{v} \times \mathbf{w}) = \begin{vmatrix} 2 & -4 & 4 \\ 4 & -1 & -2 \\ 2 & 3 & -6 \end{vmatrix} = 2\begin{vmatrix} -1 & -2 \\ 3 & -6 \end{vmatrix} - (-4)\begin{vmatrix} 4 & -2 \\ 2 & -6 \end{vmatrix} + 4\begin{vmatrix} 4 & -1 \\ 2 & 3 \end{vmatrix} = 24 - 80 + 56 = 0$, so the volume of the

parallelepiped determined by **u**, **v** and **w** is 0, which says these vectors lie in the same plane. Therefore, their initial and terminal points A, B, C and D also lie in the same plane.

35. The magnitude of the torque is $|\boldsymbol{\tau}| = |\mathbf{r} \times \mathbf{F}| = |\mathbf{r}|\,|\mathbf{F}| \sin\theta = (0.18\text{ m})(60\text{ N}) \sin(70+10)^\circ = 10.8 \sin 80^\circ \approx 10.6\text{ N·m}$.

36. $|\mathbf{r}| = \sqrt{4^2+4^2} = 4\sqrt{2}$ ft. A line drawn from the point P to the point of application of the force makes an angle of $180^\circ - (45+30)^\circ = 105^\circ$ with the force vector. Therefore,

$|\boldsymbol{\tau}| = |\mathbf{r} \times \mathbf{F}| = |\mathbf{r}|\,|\mathbf{F}| \sin\theta = \left(4\sqrt{2}\right)(36) \sin 105^\circ \approx 197$ ft-lb.

37. Using the notation of (1), $\mathbf{r} = \langle 0, 0.3, 0\rangle$ and $\mathbf{F}$ has direction $\langle 0, 3, -4\rangle$. The angle θ between them can be determined by

$$\cos\theta = \frac{\langle 0, 0.3, 0\rangle \cdot \langle 0, 3, -4\rangle}{|\langle 0, 0.3, 0\rangle|\,|\langle 0, 3, -4\rangle|} \quad\Rightarrow\quad \cos\theta = \frac{0.9}{(0.3)(5)} \quad\Rightarrow\quad \cos\theta = 0.6 \quad\Rightarrow\quad \theta \approx 53.1^\circ. \text{ Then } |\boldsymbol{\tau}| = |\mathbf{r}|\,|\mathbf{F}| \sin\theta \quad\Rightarrow$$

$100 = 0.3\,|\mathbf{F}| \sin 53.1^\circ \quad\Rightarrow\quad |\mathbf{F}| \approx 417$ N.

38. Since $|\mathbf{u} \times \mathbf{v}| = |\mathbf{u}|\,|\mathbf{v}| \sin\theta$, $0 \le \theta \le \pi$, $|\mathbf{u} \times \mathbf{v}|$ achieves its maximum value for $\sin\theta = 1 \;\Rightarrow\; \theta = \frac{\pi}{2}$, in which case $|\mathbf{u} \times \mathbf{v}| = |\mathbf{u}|\,|\mathbf{v}| = 15$. The minimum value is zero, which occurs when $\sin\theta = 0 \;\Rightarrow\; \theta = 0$ or π, so when $\mathbf{u}$, $\mathbf{v}$ are parallel. Thus, when $\mathbf{u}$ points in the same direction as $\mathbf{v}$, so $\mathbf{u} = 3\,\mathbf{j}$, $|\mathbf{u} \times \mathbf{v}| = 0$. As $\mathbf{u}$ rotates counterclockwise, $\mathbf{u} \times \mathbf{v}$ is directed in the negative z-direction (by the right-hand rule) and the length increases until $\theta = \frac{\pi}{2}$, in which case $\mathbf{u} = -3\,\mathbf{i}$ and $|\mathbf{u} \times \mathbf{v}| = 15$. As $\mathbf{u}$ rotates to the negative y-axis, $\mathbf{u} \times \mathbf{v}$ remains pointed in the negative z-direction and the length of $\mathbf{u} \times \mathbf{v}$ decreases to 0, after which the direction of $\mathbf{u} \times \mathbf{v}$ reverses to point in the positive z-direction and $|\mathbf{u} \times \mathbf{v}|$ increases. When $\mathbf{u} = 3\,\mathbf{i}$ (so $\theta = \frac{\pi}{2}$), $|\mathbf{u} \times \mathbf{v}|$ again reaches its maximum of 15, after which $|\mathbf{u} \times \mathbf{v}|$ decreases to 0 as $\mathbf{u}$ rotates to the positive y-axis.

39. (a)

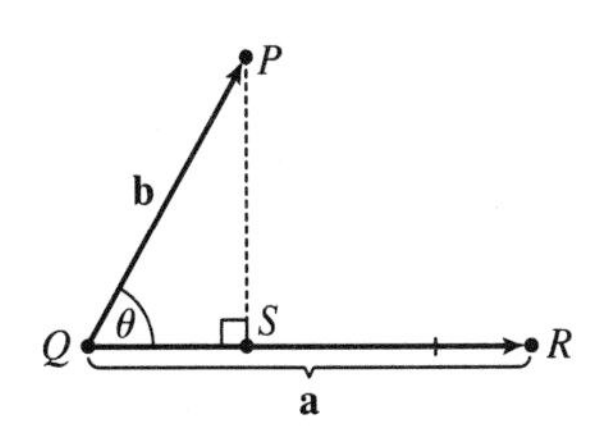

The distance between a point and a line is the length of the perpendicular from the point to the line, here $\left|\overrightarrow{PS}\right| = d$. But referring to triangle PQS, $d = \left|\overrightarrow{PS}\right| = \left|\overrightarrow{QP}\right| \sin\theta = |\mathbf{b}| \sin\theta$. But θ is the angle between $\overrightarrow{QP} = \mathbf{b}$ and $\overrightarrow{QR} = \mathbf{a}$. Thus by the definition of the cross product, $\sin\theta = \dfrac{|\mathbf{a} \times \mathbf{b}|}{|\mathbf{a}|\,|\mathbf{b}|}$ and so

$$d = |\mathbf{b}| \sin\theta = \frac{|\mathbf{b}|\,|\mathbf{a} \times \mathbf{b}|}{|\mathbf{a}|\,|\mathbf{b}|} = \frac{|\mathbf{a} \times \mathbf{b}|}{|\mathbf{a}|}.$$

(b) $\mathbf{a} = \overrightarrow{QR} = \langle -1, -2, -1\rangle$ and $\mathbf{b} = \overrightarrow{QP} = \langle 1, -5, -7\rangle$. Then

$\mathbf{a} \times \mathbf{b} = \langle (-2)(-7) - (-1)(-5), (-1)(1) - (-1)(-7), (-1)(-5) - (-2)(1)\rangle = \langle 9, -8, 7\rangle$.

Thus the distance is $d = \dfrac{|\mathbf{a} \times \mathbf{b}|}{|\mathbf{a}|} = \frac{1}{\sqrt{6}}\sqrt{81+64+49} = \sqrt{\frac{194}{6}} = \sqrt{\frac{97}{3}}$.

40. (a) The distance between a point and a plane is the length of the perpendicular from the point to the plane, here $\left|\overrightarrow{TP}\right| = d$. But $\overrightarrow{TP}$ is parallel to $\mathbf{b} \times \mathbf{a}$ (because $\mathbf{b} \times \mathbf{a}$ is perpendicular to $\mathbf{b}$ and $\mathbf{a}$) and $d = \left|\overrightarrow{TP}\right| =$ the absolute value of the scalar projection of $\mathbf{c}$ along

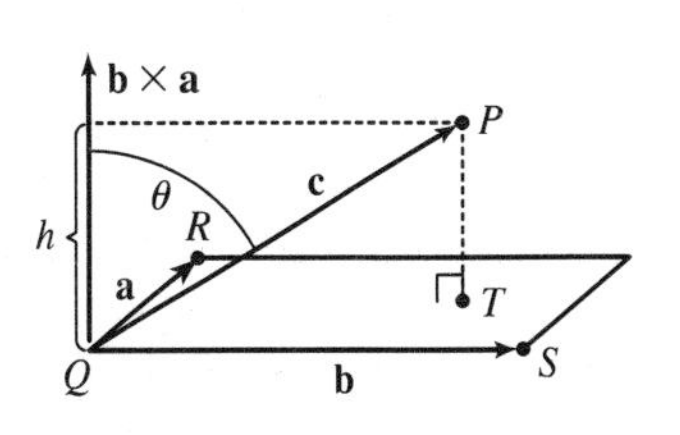

$\mathbf{b} \times \mathbf{a}$, which is $|\mathbf{c}|\,|\cos\theta|$. (Notice that this is the same setup as the development of the volume of a parallelepiped with $h = |\mathbf{c}|\,|\cos\theta|$). Thus $d = |\mathbf{c}|\,|\cos\theta| = h = V/A$ where $A = |\mathbf{a} \times \mathbf{b}|$, the area of the base. So finally

$$d = \frac{V}{A} = \frac{|\mathbf{a} \cdot (\mathbf{b} \times \mathbf{c})|}{|\mathbf{a} \times \mathbf{b}|} = \frac{|(\mathbf{a} \times \mathbf{b}) \cdot \mathbf{c}|}{|\mathbf{a} \times \mathbf{b}|} \text{ by Property 5 of Theorem 8.}$$

(b) $\mathbf{a} = \overrightarrow{QR} = \langle -1, 2, 0 \rangle$, $\mathbf{b} = \overrightarrow{QS} = \langle -1, 0, 3 \rangle$ and $\mathbf{c} = \overrightarrow{QP} = \langle 1, 1, 4 \rangle$. Then

$$\mathbf{a} \cdot (\mathbf{b} \times \mathbf{c}) = \begin{vmatrix} -1 & 2 & 0 \\ -1 & 0 & 3 \\ 1 & 1 & 4 \end{vmatrix} = (-1) \begin{vmatrix} 0 & 3 \\ 1 & 4 \end{vmatrix} - 2 \begin{vmatrix} -1 & 3 \\ 1 & 4 \end{vmatrix} + 0 = 17$$

and
$$\mathbf{a} \times \mathbf{b} = \begin{vmatrix} \mathbf{i} & \mathbf{j} & \mathbf{k} \\ -1 & 2 & 0 \\ -1 & 0 & 3 \end{vmatrix} = \begin{vmatrix} 2 & 0 \\ 0 & 3 \end{vmatrix} \mathbf{i} - \begin{vmatrix} -1 & 0 \\ -1 & 3 \end{vmatrix} \mathbf{j} + \begin{vmatrix} -1 & 2 \\ -1 & 0 \end{vmatrix} \mathbf{k} = 6\,\mathbf{i} + 3\,\mathbf{j} + 2\,\mathbf{k}$$

Thus $d = \dfrac{|\mathbf{a} \cdot (\mathbf{b} \times \mathbf{c})|}{|\mathbf{a} \times \mathbf{b}|} = \dfrac{17}{\sqrt{36 + 9 + 4}} = \dfrac{17}{7}$.

41. $(\mathbf{a} - \mathbf{b}) \times (\mathbf{a} + \mathbf{b}) = (\mathbf{a} - \mathbf{b}) \times \mathbf{a} + (\mathbf{a} - \mathbf{b}) \times \mathbf{b}$ by Property 3 of Theorem 8

$= \mathbf{a} \times \mathbf{a} + (-\mathbf{b}) \times \mathbf{a} + \mathbf{a} \times \mathbf{b} + (-\mathbf{b}) \times \mathbf{b}$ by Property 4 of Theorem 8

$= (\mathbf{a} \times \mathbf{a}) - (\mathbf{b} \times \mathbf{a}) + (\mathbf{a} \times \mathbf{b}) - (\mathbf{b} \times \mathbf{b})$ by Property 2 of Theorem 8 (with $c = -1$)

$= \mathbf{0} - (\mathbf{b} \times \mathbf{a}) + (\mathbf{a} \times \mathbf{b}) - \mathbf{0}$ by Example 2

$= (\mathbf{a} \times \mathbf{b}) + (\mathbf{a} \times \mathbf{b})$ by Property 1 of Theorem 8

$= 2(\mathbf{a} \times \mathbf{b})$

42. Let $\mathbf{a} = \langle a_1, a_2, a_3 \rangle$, $\mathbf{b} = \langle b_1, b_2, b_3 \rangle$ and $\mathbf{c} = \langle c_1, c_2, c_3 \rangle$, so $\mathbf{b} \times \mathbf{c} = \langle b_2c_3 - b_3c_2, b_3c_1 - b_1c_3, b_1c_2 - b_2c_1 \rangle$ and

$$\begin{aligned}
\mathbf{a} \times (\mathbf{b} \times \mathbf{c}) &= \langle a_2(b_1c_2 - b_2c_1) - a_3(b_3c_1 - b_1c_3), a_3(b_2c_3 - b_3c_2) - a_1(b_1c_2 - b_2c_1), \\
&\qquad a_1(b_3c_1 - b_1c_3) - a_2(b_2c_3 - b_3c_2) \rangle \\
&= \langle a_2b_1c_2 - a_2b_2c_1 - a_3b_3c_1 + a_3b_1c_3, a_3b_2c_3 - a_3b_3c_2 - a_1b_1c_2 + a_1b_2c_1, \\
&\qquad a_1b_3c_1 - a_1b_1c_3 - a_2b_2c_3 + a_2b_3c_2 \rangle \\
&= \langle (a_2c_2 + a_3c_3)b_1 - (a_2b_2 + a_3b_3)c_1, (a_1c_1 + a_3c_3)b_2 - (a_1b_1 + a_3b_3)c_2, \\
&\qquad (a_1c_1 + a_2c_2)b_3 - (a_1b_1 + a_2b_2)c_3 \rangle \\
(\star) \quad &= \langle (a_2c_2 + a_3c_3)b_1 - (a_2b_2 + a_3b_3)c_1 + a_1b_1c_1 - a_1b_1c_1, \\
&\qquad (a_1c_1 + a_3c_3)b_2 - (a_1b_1 + a_3b_3)c_2 + a_2b_2c_2 - a_2b_2c_2, \\
&\qquad (a_1c_1 + a_2c_2)b_3 - (a_1b_1 + a_2b_2)c_3 + a_3b_3c_3 - a_3b_3c_3 \rangle \\
&= \langle (a_1c_1 + a_2c_2 + a_3c_3)b_1 - (a_1b_1 + a_2b_2 + a_3b_3)c_1, \\
&\qquad (a_1c_1 + a_2c_2 + a_3c_3)b_2 - (a_1b_1 + a_2b_2 + a_3b_3)c_2, \\
&\qquad (a_1c_1 + a_2c_2 + a_3c_3)b_3 - (a_1b_1 + a_2b_2 + a_3b_3)c_3 \rangle \\
&= (a_1c_1 + a_2c_2 + a_3c_3) \langle b_1, b_2, b_3 \rangle - (a_1b_1 + a_2b_2 + a_3b_3) \langle c_1, c_2, c_3 \rangle \\
&= (\mathbf{a} \cdot \mathbf{c})\mathbf{b} - (\mathbf{a} \cdot \mathbf{b})\mathbf{c}
\end{aligned}$$

($\star$) Here we look ahead to see what terms are still needed to arrive at the desired equation. By adding and subtracting the same terms, we don't change the value of the component.

43. $\mathbf{a} \times (\mathbf{b} \times \mathbf{c}) + \mathbf{b} \times (\mathbf{c} \times \mathbf{a}) + \mathbf{c} \times (\mathbf{a} \times \mathbf{b})$

$= [(\mathbf{a} \cdot \mathbf{c})\mathbf{b} - (\mathbf{a} \cdot \mathbf{b})\mathbf{c}] + [(\mathbf{b} \cdot \mathbf{a})\mathbf{c} - (\mathbf{b} \cdot \mathbf{c})\mathbf{a}] + [(\mathbf{c} \cdot \mathbf{b})\mathbf{a} - (\mathbf{c} \cdot \mathbf{a})\mathbf{b}]$ by Exercise 42

$= (\mathbf{a} \cdot \mathbf{c})\mathbf{b} - (\mathbf{a} \cdot \mathbf{b})\mathbf{c} + (\mathbf{a} \cdot \mathbf{b})\mathbf{c} - (\mathbf{b} \cdot \mathbf{c})\mathbf{a} + (\mathbf{b} \cdot \mathbf{c})\mathbf{a} - (\mathbf{a} \cdot \mathbf{c})\mathbf{b} = \mathbf{0}$

44. Let $\mathbf{c} \times \mathbf{d} = \mathbf{v}$. Then

$$\begin{aligned}
(\mathbf{a} \times \mathbf{b}) \cdot (\mathbf{c} \times \mathbf{d}) &= (\mathbf{a} \times \mathbf{b}) \cdot \mathbf{v} = \mathbf{a} \cdot (\mathbf{b} \times \mathbf{v}) && \text{by Property 5 of Theorem 8} \\
&= \mathbf{a} \cdot [\mathbf{b} \times (\mathbf{c} \times \mathbf{d})] = \mathbf{a} \cdot [(\mathbf{b} \cdot \mathbf{d})\mathbf{c} - (\mathbf{b} \cdot \mathbf{c})\mathbf{d}] && \text{by Exercise 42} \\
&= (\mathbf{b} \cdot \mathbf{d})(\mathbf{a} \cdot \mathbf{c}) - (\mathbf{b} \cdot \mathbf{c})(\mathbf{a} \cdot \mathbf{d}) && \text{by Properties 3 and 4 of the dot product} \\
&= \begin{vmatrix} \mathbf{a} \cdot \mathbf{c} & \mathbf{b} \cdot \mathbf{c} \\ \mathbf{a} \cdot \mathbf{d} & \mathbf{b} \cdot \mathbf{d} \end{vmatrix}
\end{aligned}$$

45. (a) No. If $\mathbf{a} \cdot \mathbf{b} = \mathbf{a} \cdot \mathbf{c}$, then $\mathbf{a} \cdot (\mathbf{b} - \mathbf{c}) = 0$, so $\mathbf{a}$ is perpendicular to $\mathbf{b} - \mathbf{c}$, which can happen if $\mathbf{b} \neq \mathbf{c}$. For example, let $\mathbf{a} = \langle 1, 1, 1 \rangle$, $\mathbf{b} = \langle 1, 0, 0 \rangle$ and $\mathbf{c} = \langle 0, 1, 0 \rangle$.

(b) No. If $\mathbf{a} \times \mathbf{b} = \mathbf{a} \times \mathbf{c}$ then $\mathbf{a} \times (\mathbf{b} - \mathbf{c}) = \mathbf{0}$, which implies that $\mathbf{a}$ is parallel to $\mathbf{b} - \mathbf{c}$, which of course can happen if $\mathbf{b} \neq \mathbf{c}$.

(c) Yes. Since $\mathbf{a} \cdot \mathbf{c} = \mathbf{a} \cdot \mathbf{b}$, $\mathbf{a}$ is perpendicular to $\mathbf{b} - \mathbf{c}$, by part (a). From part (b), $\mathbf{a}$ is also parallel to $\mathbf{b} - \mathbf{c}$. Thus since $\mathbf{a} \neq \mathbf{0}$ but is both parallel and perpendicular to $\mathbf{b} - \mathbf{c}$, we have $\mathbf{b} - \mathbf{c} = \mathbf{0}$, so $\mathbf{b} = \mathbf{c}$.

46. (a) $\mathbf{k}_i$ is perpendicular to $\mathbf{v}_i$ if $i \neq j$ by the definition of $\mathbf{k}_i$ and Theorem 5.

(b) $\mathbf{k}_1 \cdot \mathbf{v}_1 = \dfrac{\mathbf{v}_2 \times \mathbf{v}_3}{\mathbf{v}_1 \cdot (\mathbf{v}_2 \times \mathbf{v}_3)} \cdot \mathbf{v}_1 = \dfrac{\mathbf{v}_1 \cdot (\mathbf{v}_2 \times \mathbf{v}_3)}{\mathbf{v}_1 \cdot (\mathbf{v}_2 \times \mathbf{v}_3)} = 1$

$$\mathbf{k}_2 \cdot \mathbf{v}_2 = \frac{\mathbf{v}_3 \times \mathbf{v}_1}{\mathbf{v}_1 \cdot (\mathbf{v}_2 \times \mathbf{v}_3)} \cdot \mathbf{v}_2 = \frac{\mathbf{v}_2 \cdot (\mathbf{v}_3 \times \mathbf{v}_1)}{\mathbf{v}_1 \cdot (\mathbf{v}_2 \times \mathbf{v}_3)} = \frac{(\mathbf{v}_2 \times \mathbf{v}_3) \cdot \mathbf{v}_1}{\mathbf{v}_1 \cdot (\mathbf{v}_2 \times \mathbf{v}_3)} = 1 \quad \text{[by Property 5 of Theorem 8]}$$

$$\mathbf{k}_3 \cdot \mathbf{v}_3 = \frac{(\mathbf{v}_1 \times \mathbf{v}_2) \cdot \mathbf{v}_3}{\mathbf{v}_1 \cdot (\mathbf{v}_2 \times \mathbf{v}_3)} = \frac{\mathbf{v}_1 \cdot (\mathbf{v}_2 \times \mathbf{v}_3)}{\mathbf{v}_1 \cdot (\mathbf{v}_2 \times \mathbf{v}_3)} = 1 \quad \text{[by Property 5 of Theorem 8]}$$

(c) $\mathbf{k}_1 \cdot (\mathbf{k}_2 \times \mathbf{k}_3) = \mathbf{k}_1 \cdot \left(\dfrac{\mathbf{v}_3 \times \mathbf{v}_1}{\mathbf{v}_1 \cdot (\mathbf{v}_2 \times \mathbf{v}_3)} \times \dfrac{\mathbf{v}_1 \times \mathbf{v}_2}{\mathbf{v}_1 \cdot (\mathbf{v}_2 \times \mathbf{v}_3)} \right) = \dfrac{\mathbf{k}_1}{[\mathbf{v}_1 \cdot (\mathbf{v}_2 \times \mathbf{v}_3)]^2} \cdot [(\mathbf{v}_3 \times \mathbf{v}_1) \times (\mathbf{v}_1 \times \mathbf{v}_2)]$

$$= \frac{\mathbf{k}_1}{[\mathbf{v}_1 \cdot (\mathbf{v}_2 \times \mathbf{v}_3)]^2} \cdot ([(\mathbf{v}_3 \times \mathbf{v}_1) \cdot \mathbf{v}_2]\, \mathbf{v}_1 - [(\mathbf{v}_3 \times \mathbf{v}_1) \cdot \mathbf{v}_1]\, \mathbf{v}_2) \quad \text{[by Exercise 42]}$$

But $(\mathbf{v}_3 \times \mathbf{v}_1) \cdot \mathbf{v}_1 = 0$ since $\mathbf{v}_3 \times \mathbf{v}_1$ is orthogonal to $\mathbf{v}_1$, and $(\mathbf{v}_3 \times \mathbf{v}_1) \cdot \mathbf{v}_2 = \mathbf{v}_2 \cdot (\mathbf{v}_3 \times \mathbf{v}_1) = (\mathbf{v}_2 \times \mathbf{v}_3) \cdot \mathbf{v}_1 = \mathbf{v}_1 \cdot (\mathbf{v}_2 \times \mathbf{v}_3)$. Thus

$$\mathbf{k}_1 \cdot (\mathbf{k}_2 \times \mathbf{k}_3) = \frac{\mathbf{k}_1}{[\mathbf{v}_1 \cdot (\mathbf{v}_2 \times \mathbf{v}_3)]^2} \cdot [\mathbf{v}_1 \cdot (\mathbf{v}_2 \times \mathbf{v}_3)]\, \mathbf{v}_1 = \frac{\mathbf{k}_1 \cdot \mathbf{v}_1}{\mathbf{v}_1 \cdot (\mathbf{v}_2 \times \mathbf{v}_3)} = \frac{1}{\mathbf{v}_1 \cdot (\mathbf{v}_2 \times \mathbf{v}_3)} \quad \text{[by part (b)]}$$

10.5 Equations of Lines and Planes

1. (a) True; each of the first two lines has a direction vector parallel to the direction vector of the third line, so these vectors are each scalar multiples of the third direction vector. Then the first two direction vectors are also scalar multiples of each other, so these vectors, and hence the two lines, are parallel.

(b) False; for example, the x- and y-axes are both perpendicular to the z-axis, yet the x- and y-axes are not parallel.

(c) True; each of the first two planes has a normal vector parallel to the normal vector of the third plane, so these two normal vectors are parallel to each other and the planes are parallel.

(d) False; for example, the xy- and yz-planes are not parallel, yet they are both perpendicular to the xz-plane.

(e) False; the x- and y-axes are not parallel, yet they are both parallel to the plane $z = 1$.

(f) True; if each line is perpendicular to a plane, then the lines' direction vectors are both parallel to a normal vector for the plane. Thus, the direction vectors are parallel to each other and the lines are parallel.

(g) False; the planes $y = 1$ and $z = 1$ are not parallel, yet they are both parallel to the x-axis.

(h) True; if each plane is perpendicular to a line, then any normal vector for each plane is parallel to a direction vector for the line. Thus, the normal vectors are parallel to each other and the planes are parallel.

(i) True; see Figure 9 and the accompanying discussion.

(j) False; they can be skew, as in Example 3.

(k) True. Consider any normal vector for the plane and any direction vector for the line. If the normal vector is perpendicular to the direction vector, the line and plane are parallel. Otherwise, the vectors meet at an angle θ, $0^\circ \leq \theta < 90^\circ$, and the line will intersect the plane at an angle $90^\circ - \theta$.

2. For this line, we have $\mathbf{r}_0 = \mathbf{i} - 3\,\mathbf{k}$ and $\mathbf{v} = 2\,\mathbf{i} - 4\,\mathbf{j} + 5\,\mathbf{k}$, so a vector equation is $\mathbf{r} = \mathbf{r}_0 + t\,\mathbf{v} = (\mathbf{i} - 3\,\mathbf{k}) + t(2\,\mathbf{i} - 4\,\mathbf{j} + 5\,\mathbf{k}) = (1 + 2t)\,\mathbf{i} - 4t\,\mathbf{j} + (-3 + 5t)\,\mathbf{k}$ and parametric equations are $x = 1 + 2t$, $y = -4t$, $z = -3 + 5t$.

3. For this line, we have $\mathbf{r}_0 = -2\,\mathbf{i} + 4\,\mathbf{j} + 10\,\mathbf{k}$ and $\mathbf{v} = 3\,\mathbf{i} + \mathbf{j} - 8\,\mathbf{k}$, so a vector equation is $\mathbf{r} = \mathbf{r}_0 + t\,\mathbf{v} = (-2\,\mathbf{i} + 4\,\mathbf{j} + 10\,\mathbf{k}) + t(3\,\mathbf{i} + \mathbf{j} - 8\,\mathbf{k}) = (-2 + 3t)\,\mathbf{i} + (4 + t)\,\mathbf{j} + (10 - 8t)\,\mathbf{k}$ and parametric equations are $x = -2 + 3t$, $y = 4 + t$, $z = 10 - 8t$.

4. This line has the same direction as the given line, $\mathbf{v} = 2\,\mathbf{i} - \mathbf{j} + 3\,\mathbf{k}$. Here $\mathbf{r}_0 = 0\,\mathbf{i} + 0\,\mathbf{j} + 0\,\mathbf{k}$, so a vector equation is $\mathbf{r} = (0\,\mathbf{i} + 0\,\mathbf{j} + 0\,\mathbf{k}) + t(2\,\mathbf{i} - \mathbf{j} + 3\,\mathbf{k}) = 2t\,\mathbf{i} - t\,\mathbf{j} + 3t\,\mathbf{k}$ and parametric equations are $x = 2t$, $y = -t$, $z = 3t$.

5. A line perpendicular to the given plane has the same direction as a normal vector to the plane, such as $\mathbf{n} = \langle 1, 3, 1 \rangle$. So $\mathbf{r}_0 = \mathbf{i} + 6\,\mathbf{k}$, and we can take $\mathbf{v} = \mathbf{i} + 3\,\mathbf{j} + \mathbf{k}$. Then a vector equation is $\mathbf{r} = (\mathbf{i} + 6\,\mathbf{k}) + t(\mathbf{i} + 3\,\mathbf{j} + \mathbf{k}) = (1 + t)\,\mathbf{i} + 3t\,\mathbf{j} + (6 + t)\,\mathbf{k}$, and parametric equations are $x = 1 + t$, $y = 3t$, $z = 6 + t$.

6. $\mathbf{v} = \langle 2 - 6, 4 - 1, 5 - (-3) \rangle = \langle -4, 3, 8 \rangle$, and letting $P_0 = (6, 1, -3)$, parametric equations are $x = 6 - 4t$, $y = 1 + 3t$, $z = -3 + 8t$, while symmetric equations are $\dfrac{x - 6}{-4} = \dfrac{y - 1}{3} = \dfrac{z + 3}{8}$.

7. $\mathbf{v} = \left\langle 2 - 0, 1 - \frac{1}{2}, -3 - 1 \right\rangle = \left\langle 2, \frac{1}{2}, -4 \right\rangle$, and letting $P_0 = (2, 1, -3)$, parametric equations are $x = 2 + 2t$, $y = 1 + \frac{1}{2}t$, $z = -3 - 4t$, while symmetric equations are $\dfrac{x - 2}{2} = \dfrac{y - 1}{1/2} = \dfrac{z + 3}{-4}$ or $\dfrac{x - 2}{2} = 2y - 2 = \dfrac{z + 3}{-4}$.

8. $\mathbf{v} = (\mathbf{i} + \mathbf{j}) \times (\mathbf{j} + \mathbf{k}) = \begin{vmatrix} \mathbf{i} & \mathbf{j} & \mathbf{k} \\ 1 & 1 & 0 \\ 0 & 1 & 1 \end{vmatrix} = \mathbf{i} - \mathbf{j} + \mathbf{k}$ is the direction of the line perpendicular to both $\mathbf{i} + \mathbf{j}$ and $\mathbf{j} + \mathbf{k}$.

With $P_0 = (2, 1, 0)$, parametric equations are $x = 2 + t$, $y = 1 - t$, $z = t$ and symmetric equations are $x - 2 = \dfrac{y - 1}{-1} = z$ or $x - 2 = 1 - y = z$.

9. The line has direction $\mathbf{v} = \langle 1, 2, 1 \rangle$. Letting $P_0 = (1, -1, 1)$, parametric equations are $x = 1 + t$, $y = -1 + 2t$, $z = 1 + t$ and symmetric equations are $x - 1 = \dfrac{y + 1}{2} = z - 1$.

10. Setting $x = 0$, we see that $(0, 1, 0)$ satisfies the equations of both planes, so they do in fact have a line of intersection. $\mathbf{v} = \mathbf{n}_1 \times \mathbf{n}_2 = \langle 1, 1, 1 \rangle \times \langle 1, 0, 1 \rangle = \langle 1, 0, -1 \rangle$ is the direction of this line. Taking the point $(0, 1, 0)$ as P_0, parametric equations are $x = t$, $y = 1$, $z = -t$, and symmetric equations are $x = -z$, $y = 1$.

11. Direction vectors of the lines are $\mathbf{v}_1 = \langle -2 - (-4), 0 - (-6), -3 - 1 \rangle = \langle 2, 6, -4 \rangle$ and $\mathbf{v}_2 = \langle 5 - 10, 3 - 18, 14 - 4 \rangle = \langle -5, -15, 10 \rangle$, and since $\mathbf{v}_2 = -\frac{5}{2}\mathbf{v}_1$, the direction vectors and thus the lines are parallel.

12. Direction vectors of the lines are $\mathbf{v}_1 = \langle -2, 4, 4 \rangle$ and $\mathbf{v}_2 = \langle 8, -1, 4 \rangle$. Since $\mathbf{v}_1 \cdot \mathbf{v}_2 = -16 - 4 + 16 \neq 0$, the vectors and thus the lines are not perpendicular.

13. (a) A direction vector of the line with parametric equations $x = 1 + 2t$, $y = 3t$, $z = 5 - 7t$ is $\mathbf{v} = \langle 2, 3, -7 \rangle$ and the desired parallel line must also have $\mathbf{v}$ as a direction vector. Here $P_0 = (0, 2, -1)$, so symmetric equations for the line are $\dfrac{x}{2} = \dfrac{y-2}{3} = \dfrac{z+1}{-7}$.

(b) The line intersects the xy-plane when $z = 0$, so we need $\dfrac{x}{2} = \dfrac{y-2}{3} = \dfrac{1}{-7}$ or $x = -\frac{2}{7}$, $y = \frac{11}{7}$. Thus the point of intersection with the xy-plane is $\left(-\frac{2}{7}, \frac{11}{7}, 0\right)$. Similarly for the yz-plane, we need $x = 0 \Leftrightarrow 0 = \dfrac{y-2}{3} = \dfrac{z+1}{-7} \Leftrightarrow y = 2$, $z = -1$. Thus the line intersects the yz-plane at $(0, 2, -1)$. For the xz-plane, we need $y = 0 \Leftrightarrow \dfrac{x}{2} = -\dfrac{2}{3} = \dfrac{z+1}{-7} \Leftrightarrow x = -\frac{4}{3}$, $z = \frac{11}{3}$. So the line intersects the xz-plane at $\left(-\frac{4}{3}, 0, \frac{11}{3}\right)$.

14. (a) A vector normal to the plane $2x - y + z = 1$ is $\mathbf{n} = \langle 2, -1, 1 \rangle$, and since the line is to be perpendicular to the plane, $\mathbf{n}$ is also a direction vector for the line. Thus parametric equations of the line are $x = 5 + 2t$, $y = 1 - t$, $z = t$.

(b) On the xy-plane, $z = 0$. So $z = t = 0$ in the parametric equations of the line, and therefore $x = 5$ and $y = 1$, giving the point of intersection $(5, 1, 0)$. For the yz-plane, $x = 0$ which implies $t = -\frac{5}{2}$, so $y = \frac{7}{2}$ and $z = -\frac{5}{2}$ and the point is $\left(0, \frac{7}{2}, -\frac{5}{2}\right)$. For the xz-plane, $y = 0$ which implies $t = 1$, so $x = 7$ and $z = 1$ and the point of intersection is $(7, 0, 1)$.

15. From Equation 4, the line segment from $\mathbf{r}_0 = 2\,\mathbf{i} - \mathbf{j} + 4\,\mathbf{k}$ to $\mathbf{r}_1 = 4\,\mathbf{i} + 6\,\mathbf{j} + \mathbf{k}$ is $\mathbf{r}(t) = (1-t)\,\mathbf{r}_0 + t\,\mathbf{r}_1 = (1-t)(2\,\mathbf{i} - \mathbf{j} + 4\,\mathbf{k}) + t(4\,\mathbf{i} + 6\,\mathbf{j} + \mathbf{k}) = (2\,\mathbf{i} - \mathbf{j} + 4\,\mathbf{k}) + t(2\,\mathbf{i} + 7\,\mathbf{j} - 3\,\mathbf{k})$, $0 \leq t \leq 1$.

16. From Equation 4, the line segment from $\mathbf{r}_0 = 10\,\mathbf{i} + 3\,\mathbf{j} + \mathbf{k}$ to $\mathbf{r}_1 = 5\,\mathbf{i} + 6\,\mathbf{j} - 3\,\mathbf{k}$ is

$$\begin{aligned}\mathbf{r}(t) &= (1-t)\,\mathbf{r}_0 + t\,\mathbf{r}_1 = (1-t)(10\,\mathbf{i} + 3\,\mathbf{j} + \mathbf{k}) + t(5\,\mathbf{i} + 6\,\mathbf{j} - 3\,\mathbf{k}) \\ &= (10\,\mathbf{i} + 3\,\mathbf{j} + \mathbf{k}) + t(-5\,\mathbf{i} + 3\,\mathbf{j} - 4\,\mathbf{k}), \quad 0 \leq t \leq 1.\end{aligned}$$

The corresponding parametric equations are $x = 10 - 5t$, $y = 3 + 3t$, $z = 1 - 4t$, $0 \leq t \leq 1$.

17. Since the direction vectors are $\mathbf{v}_1 = \langle -6, 9, -3 \rangle$ and $\mathbf{v}_2 = \langle 2, -3, 1 \rangle$, we have $\mathbf{v}_1 = -3\mathbf{v}_2$ so the lines are parallel.

18. The lines aren't parallel since the direction vectors $\langle 2, 3, -1 \rangle$ and $\langle 1, 1, 3 \rangle$ aren't parallel. For the lines to intersect we must be able to find one value of t and one value of s that produce the same point from the respective parametric equations. Thus we need to satisfy the following three equations: $1 + 2t = -1 + s$, $3t = 4 + s$, $2 - t = 1 + 3s$. Solving the first two equations we get $t = 6$, $s = 14$ and checking, we see that these values don't satisfy the third equation. Thus L_1 and L_2 aren't parallel and don't intersect, so they must be skew lines.

19. Since the direction vectors $\langle 1,2,3\rangle$ and $\langle -4,-3,2\rangle$ are not scalar multiples of each other, the lines are not parallel, so we check to see if the lines intersect. The parametric equations of the lines are L_1: $x=t$, $y=1+2t$, $z=2+3t$ and L_2: $x=3-4s$, $y=2-3s$, $z=1+2s$. For the lines to intersect, we must be able to find one value of t and one value of s that produce the same point from the respective parametric equations. Thus we need to satisfy the following three equations: $t=3-4s$, $1+2t=2-3s$, $2+3t=1+2s$. Solving the first two equations we get $t=-1$, $s=1$ and checking, we see that these values don't satisfy the third equation. Thus the lines aren't parallel and don't intersect, so they must be skew lines.

20. Since the direction vectors $\langle 2,2,-1\rangle$ and $\langle 1,-1,3\rangle$ aren't parallel, the lines aren't parallel. Here the parametric equations are L_1: $x=1+2t$, $y=3+2t$, $z=2-t$ and L_2: $x=2+s$, $y=6-s$, $z=-2+3s$. Thus, for the lines to intersect, the three equations $1+2t=2+s$, $3+2t=6-s$, and $2-t=-2+3s$ must be satisfied simultaneously. Solving the first two equations gives $t=1$, $s=1$ and, checking, we see that these values do satisfy the third equation, so the lines intersect when $t=1$ and $s=1$, that is, at the point $(3,5,1)$.

21. Since the plane is perpendicular to the vector $\langle -2,1,5\rangle$, we can take $\langle -2,1,5\rangle$ as a normal vector to the plane. $(6,3,2)$ is a point on the plane, so setting $a=-2$, $b=1$, $c=5$ and $x_0=6$, $y_0=3$, $z_0=2$ in Equation 6 gives $-2(x-6)+1(y-3)+5(z-2)=0$ or $-2x+y+5z=1$ to be an equation of the plane.

22. $\mathbf{j}+2\,\mathbf{k}=\langle 0,1,2\rangle$ is a normal vector to the plane and $(4,0,-3)$ is a point on the plane, so setting $a=0$, $b=1$, $c=2$, $x_0=4$, $y_0=0$, $z_0=-3$ in Equation 7 gives $0(x-4)+1(y-0)+2[z-(-3)]=0$ or $y+2z=-6$ to be an equation of the plane.

23. Since the two planes are parallel, they will have the same normal vectors. So we can take $\mathbf{n}=\langle 2,-1,3\rangle$, and an equation of the plane is $2(x-0)-1(y-0)+3(z-0)=0$ or $2x-y+3z=0$.

24. First, a normal vector for the plane $2x+4y+8z=17$ is $\mathbf{n}=\langle 2,4,8\rangle$. A direction vector for the line is $\mathbf{v}=\langle 2,1,-1\rangle$, and since $\mathbf{n}\cdot\mathbf{v}=0$ we know the line is perpendicular to $\mathbf{n}$ and hence parallel to the plane. Thus, there is a parallel plane which contains the line. By putting $t=0$, we know the point $(3,0,8)$ is on the line and hence the new plane. We can use the same normal vector $\mathbf{n}=\langle 2,4,8\rangle$, so an equation of the plane is $2(x-3)+4(y-0)+8(z-8)=0$ or $x+2y+4z=35$.

25. Here the vectors $\mathbf{a}=\langle 1-0,0-1,1-1\rangle=\langle 1,-1,0\rangle$ and $\mathbf{b}=\langle 1-0,1-1,0-1\rangle=\langle 1,0,-1\rangle$ lie in the plane, so $\mathbf{a}\times\mathbf{b}$ is a normal vector to the plane. Thus, we can take $\mathbf{n}=\mathbf{a}\times\mathbf{b}=\langle 1-0,0+1,0+1\rangle=\langle 1,1,1\rangle$. If P_0 is the point $(0,1,1)$, an equation of the plane is $1(x-0)+1(y-1)+1(z-1)=0$ or $x+y+z=2$.

26. Here the vectors $\mathbf{a}=\langle 2,-4,6\rangle$ and $\mathbf{b}=\langle 5,1,3\rangle$ lie in the plane, so $\mathbf{n}=\mathbf{a}\times\mathbf{b}=\langle -12-6,30-6,2+20\rangle=\langle -18,24,22\rangle$ is a normal vector to the plane and an equation of the plane is $-18(x-0)+24(y-0)+22(z-0)=0$ or $-18x+24y+22z=0$.

27. If we first find two nonparallel vectors in the plane, their cross product will be a normal vector to the plane. Since the given line lies in the plane, its direction vector $\mathbf{a}=\langle -2,5,4\rangle$ is one vector in the plane. We can verify that the given point $(6,0,-2)$ does not lie on this line, so to find another nonparallel vector $\mathbf{b}$ which lies in the plane, we can pick any point on the line and find a vector connecting the points. If we put $t=0$, we see that $(4,3,7)$ is on the line, so $\mathbf{b}=\langle 6-4,0-3,-2-7\rangle=\langle 2,-3,-9\rangle$ and $\mathbf{n}=\mathbf{a}\times\mathbf{b}=\langle -45+12,8-18,6-10\rangle=\langle -33,-10,-4\rangle$. Thus, an equation of the plane is $-33(x-6)-10(y-0)-4[z-(-2)]=0$ or $33x+10y+4z=190$.

28. Since the line $x = 2y = 3z$, or $x = \frac{y}{1/2} = \frac{z}{1/3}$, lies in the plane, its direction vector $\mathbf{a} = \left\langle 1, \frac{1}{2}, \frac{1}{3} \right\rangle$ is parallel to the plane. The point $(0, 0, 0)$ is on the line (put $t = 0$), and we can verify that the given point $(1, -1, 1)$ in the plane is not on the line. The vector connecting these two points, $\mathbf{b} = \langle 1, -1, 1 \rangle$, is therefore parallel to the plane, but not parallel to $\langle 1, 2, 3 \rangle$. Then $\mathbf{a} \times \mathbf{b} = \left\langle \frac{1}{2} + \frac{1}{3}, \frac{1}{3} - 1, -1 - \frac{1}{2} \right\rangle = \left\langle \frac{5}{6}, -\frac{2}{3}, -\frac{3}{2} \right\rangle$ is a normal vector to the plane, and an equation of the plane is $\frac{5}{6}(x - 0) - \frac{2}{3}(y - 0) - \frac{3}{2}(z - 0) = 0$ or $5x - 4y - 9z = 0$.

29. A direction vector for the line of intersection is $\mathbf{a} = \mathbf{n}_1 \times \mathbf{n}_2 = \langle 1, 1, -1 \rangle \times \langle 2, -1, 3 \rangle = \langle 2, -5, -3 \rangle$, and $\mathbf{a}$ is parallel to the desired plane. Another vector parallel to the plane is the vector connecting any point on the line of intersection to the given point $(-1, 2, 1)$ in the plane. Setting $x = 0$, the equations of the planes reduce to $y - z = 2$ and $-y + 3z = 1$ with simultaneous solution $y = \frac{7}{2}$ and $z = \frac{3}{2}$. So a point on the line is $\left(0, \frac{7}{2}, \frac{3}{2}\right)$ and another vector parallel to the plane is $\left\langle -1, -\frac{3}{2}, -\frac{1}{2} \right\rangle$. Then a normal vector to the plane is $\mathbf{n} = \langle 2, -5, -3 \rangle \times \left\langle -1, -\frac{3}{2}, -\frac{1}{2} \right\rangle = \langle -2, 4, -8 \rangle$ and an equation of the plane is $-2(x + 1) + 4(y - 2) - 8(z - 1) = 0$ or $x - 2y + 4z = -1$.

30. $\mathbf{n}_1 = \langle 1, 0, -1 \rangle$ and $\mathbf{n}_2 = \langle 0, 1, 2 \rangle$. Setting $z = 0$, it is easy to see that $(1, 3, 0)$ is a point on the line of intersection of $x - z = 1$ and $y + 2z = 3$. The direction of this line is $\mathbf{v}_1 = \mathbf{n}_1 \times \mathbf{n}_2 = \langle 1, -2, 1 \rangle$. A second vector parallel to the desired plane is $\mathbf{v}_2 = \langle 1, 1, -2 \rangle$, since it is perpendicular to $x + y - 2z = 1$. Therefore, a normal of the plane in question is $\mathbf{n} = \mathbf{v}_1 \times \mathbf{v}_2 = \langle 4 - 1, 1 + 2, 1 + 2 \rangle = \langle 3, 3, 3 \rangle$, or we can use $\langle 1, 1, 1 \rangle$. Taking $(x_0, y_0, z_0) = (1, 3, 0)$, the equation we are looking for is $(x - 1) + (y - 3) + z = 0 \quad \Leftrightarrow \quad x + y + z = 4$.

31. Substitute the parametric equations of the line into the equation of the plane: $(3 - t) - (2 + t) + 2(5t) = 9 \quad \Rightarrow$ $8t = 8 \quad \Rightarrow \quad t = 1$. Therefore, the point of intersection of the line and the plane is given by $x = 3 - 1 = 2$, $y = 2 + 1 = 3$, and $z = 5(1) = 5$, that is, the point $(2, 3, 5)$.

32. A direction vector for the line through $(1, 0, 1)$ and $(4, -2, 2)$ is $\mathbf{v} = \langle 3, -2, 1 \rangle$ and, taking $P_0 = (1, 0, 1)$, parametric equations for the line are $x = 1 + 3t$, $y = -2t$, $z = 1 + t$. Substitution of the parametric equations into the equation of the plane gives $1 + 3t - 2t + 1 + t = 6 \quad \Rightarrow \quad t = 2$. Then $x = 1 + 3(2) = 7$, $y = -2(2) = -4$, and $z = 1 + 2 = 3$ so the point of intersection is $(7, -4, 3)$.

33. Normal vectors for the planes are $\mathbf{n}_1 = \langle 1, 1, 1 \rangle$ and $\mathbf{n}_2 = \langle 1, -1, 1 \rangle$. The normals are not parallel, so neither are the planes. Furthermore, $\mathbf{n}_1 \cdot \mathbf{n}_2 = 1 - 1 + 1 = 1 \neq 0$, so the planes aren't perpendicular. The angle between them is given by

$$\cos\theta = \frac{\mathbf{n}_1 \cdot \mathbf{n}_2}{|\mathbf{n}_1|\,|\mathbf{n}_2|} = \frac{1}{\sqrt{3}\,\sqrt{3}} = \frac{1}{3} \quad \Rightarrow \quad \theta = \cos^{-1}\left(\tfrac{1}{3}\right) \approx 70.5^\circ.$$

34. The normals are $\mathbf{n}_1 = \langle 2, -3, 4 \rangle$ and $\mathbf{n}_2 = \langle 1, 6, 4 \rangle$ so the planes aren't parallel. Since $\mathbf{n}_1 \cdot \mathbf{n}_2 = 2 - 18 + 16 = 0$, the normals (and thus the planes) are perpendicular.

35. The normals are $\mathbf{n}_1 = \langle 1, -4, 2 \rangle$ and $\mathbf{n}_2 = \langle 2, -8, 4 \rangle$. Since $\mathbf{n}_2 = 2\mathbf{n}_1$, the normals (and thus the planes) are parallel.

36. The normal vectors are $\mathbf{n}_1 = \langle 1, 2, 2 \rangle$ and $\mathbf{n}_2 = \langle 2, -1, 2 \rangle$. The normals are not parallel, so neither are the planes. Furthermore, $\mathbf{n}_1 \cdot \mathbf{n}_2 = 2 - 2 + 4 = 4 \neq 0$, so the planes aren't perpendicular. The angle between them is given by

$$\cos\theta = \frac{\mathbf{n}_1 \cdot \mathbf{n}_2}{|\mathbf{n}_1|\,|\mathbf{n}_2|} = \frac{4}{\sqrt{9}\,\sqrt{9}} = \frac{4}{9} \quad \Rightarrow \quad \theta = \cos^{-1}\left(\tfrac{4}{9}\right) \approx 63.6^\circ.$$

37. (a) To find a point on the line of intersection, set one of the variables equal to a constant, say $z = 0$. (This will only work if the line of intersection crosses the xy-plane; otherwise, try setting x or y equal to 0.) Then the equations of the planes reduce to $x + y = 2$ and $3x - 4y = 6$. Solving these two equations gives $x = 2$, $y = 0$. So a point on the line of intersection is $(2, 0, 0)$. The direction of the line is $\mathbf{v} = \mathbf{n}_1 \times \mathbf{n}_2 = \langle 5 - 4, -3 - 5, -4 - 3 \rangle = \langle 1, -8, -7 \rangle$, and symmetric equations for the line are $x - 2 = \dfrac{y}{-8} = \dfrac{z}{-7}$.

(b) The angle between the planes satisfies $\cos\theta = \dfrac{\mathbf{n}_1 \cdot \mathbf{n}_2}{|\mathbf{n}_1|\,|\mathbf{n}_2|} = \dfrac{3 - 4 - 5}{\sqrt{3}\,\sqrt{50}} = -\dfrac{\sqrt{6}}{5}$. Therefore $\theta = \cos^{-1}\left(-\frac{\sqrt{6}}{5}\right) \approx 119^\circ$ (or 61°).

38. The plane will contain all perpendicular bisectors of the line segment joining the two points. Thus, a point in the plane is $P_0 = (-1, -1, 2)$, the midpoint of the line segment joining the two given points, and a normal to the plane is $\mathbf{n} = \langle 6, -6, 2 \rangle$, the vector connecting the two points. So an equation of the plane is $6(x + 1) - 6(y + 1) + 2(z - 2) = 0$ or $3x - 3y + z = 2$.

39. The plane contains the points $(a, 0, 0)$, $(0, b, 0)$ and $(0, 0, c)$. Thus the vectors $\mathbf{a} = \langle -a, b, 0 \rangle$ and $\mathbf{b} = \langle -a, 0, c \rangle$ lie in the plane, and $\mathbf{n} = \mathbf{a} \times \mathbf{b} = \langle bc - 0, 0 + ac, 0 + ab \rangle = \langle bc, ac, ab \rangle$ is a normal vector to the plane. The equation of the plane is therefore $bcx + acy + abz = abc + 0 + 0$ or $bcx + acy + abz = abc$. Notice that if $a \neq 0$, $b \neq 0$ and $c \neq 0$ then we can rewrite the equation as $\dfrac{x}{a} + \dfrac{y}{b} + \dfrac{z}{c} = 1$. This is a good equation to remember!

40. (a) For the lines to intersect, we must be able to find one value of t and one value of s satisfying the three equations $1 + t = 2 - s$, $1 - t = s$ and $2t = 2$. From the third we get $t = 1$, and putting this in the second gives $s = 0$. These values of s and t do satisfy the first equation, so the lines intersect at the point $P_0 = (1 + 1, 1 - 1, 2(1)) = (2, 0, 2)$.

(b) The direction vectors of the lines are $\langle 1, -1, 2 \rangle$ and $\langle -1, 1, 0 \rangle$, so a normal vector for the plane is $\langle -1, 1, 0 \rangle \times \langle 1, -1, 2 \rangle = \langle 2, 2, 0 \rangle$ and it contains the point $(2, 0, 2)$. Then the equation of the plane is $2(x - 2) + 2(y - 0) + 0(z - 2) = 0 \quad \Leftrightarrow \quad x + y = 2$.

41. Two vectors which are perpendicular to the required line are the normal of the given plane, $\langle 1, 1, 1 \rangle$, and a direction vector for the given line, $\langle 1, -1, 2 \rangle$. So a direction vector for the required line is $\langle 1, 1, 1 \rangle \times \langle 1, -1, 2 \rangle = \langle 3, -1, -2 \rangle$. Thus L is given by $\langle x, y, z \rangle = \langle 0, 1, 2 \rangle + t\langle 3, -1, -2 \rangle$, or in parametric form, $x = 3t$, $y = 1 - t$, $z = 2 - 2t$.

42. Let L be the given line. Then $(1, 1, 0)$ is the point on L corresponding to $t = 0$. L is in the direction of $\mathbf{a} = \langle 1, -1, 2 \rangle$ and $\mathbf{b} = \langle -1, 0, 2 \rangle$ is the vector joining $(1, 1, 0)$ and $(0, 1, 2)$. Then

$\mathbf{b} - \operatorname{proj}_{\mathbf{a}} \mathbf{b} = \langle -1, 0, 2 \rangle - \dfrac{\langle 1, -1, 2 \rangle \cdot \langle -1, 0, 2 \rangle}{1^2 + (-1)^2 + 2^2} \langle 1, -1, 2 \rangle = \langle -1, 0, 2 \rangle - \frac{1}{2}\langle 1, -1, 2 \rangle = \left\langle -\frac{3}{2}, \frac{1}{2}, 1 \right\rangle$ is a direction vector for the required line. Thus $2\left\langle -\frac{3}{2}, \frac{1}{2}, 1 \right\rangle = \langle -3, 1, 2 \rangle$ is also a direction vector, and the line has parametric equations $x = -3t$, $y = 1 + t$, $z = 2 + 2t$. (Notice that this is the same line as in Exercise 41.)

43. Let P_i have normal vector $\mathbf{n}_i$. Then $\mathbf{n}_1 = \langle 4, -2, 6 \rangle$, $\mathbf{n}_2 = \langle 4, -2, -2 \rangle$, $\mathbf{n}_3 = \langle -6, 3, -9 \rangle$, $\mathbf{n}_4 = \langle 2, -1, -1 \rangle$. Now $\mathbf{n}_1 = -\frac{2}{3}\mathbf{n}_3$, so $\mathbf{n}_1$ and $\mathbf{n}_3$ are parallel, and hence P_1 and P_3 are parallel; similarly P_2 and P_4 are parallel because $\mathbf{n}_2 = 2\mathbf{n}_4$. However, $\mathbf{n}_1$ and $\mathbf{n}_2$ are not parallel. $\left(0, 0, \frac{1}{2}\right)$ lies on P_1, but not on P_3, so they are not the same plane, but both P_2 and P_4 contain the point $(0, 0, -3)$, so these two planes are identical.

44. Let L_i have direction vector $\mathbf{v}_i$. Then $\mathbf{v}_1 = \langle 1, 1, -5 \rangle$, $\mathbf{v}_2 = \langle 1, 1, -1 \rangle$, $\mathbf{v}_3 = \langle 1, 1, -1 \rangle$, $\mathbf{v}_4 = \langle 2, 2, -10 \rangle$. $\mathbf{v}_2$ and $\mathbf{v}_3$ are equal so they're parallel. $\mathbf{v}_4 = 2\mathbf{v}_1$, so L_4 and L_1 are parallel. L_3 contains the point $(1, 4, 1)$, but this point does not lie on L_2, so they're not equal. $(2, 1, -3)$ lies on L_4, and on L_1, with $t = 1$. So L_1 and L_4 are identical.

45. Let $Q = (2, 2, 0)$ and $R = (3, -1, 5)$, points on the line corresponding to $t = 0$ and $t = 1$.

Let $P = (1, 2, 3)$. Then $\mathbf{a} = \overrightarrow{QR} = \langle 1, -3, 5 \rangle$, $\mathbf{b} = \overrightarrow{QP} = \langle -1, 0, 3 \rangle$. The distance is

$$d = \frac{|\mathbf{a} \times \mathbf{b}|}{|\mathbf{a}|} = \frac{|\langle 1, -3, 5 \rangle \times \langle -1, 0, 3 \rangle|}{|\langle 1, -3, 5 \rangle|} = \frac{|\langle -9, -8, -3 \rangle|}{|\langle 1, -3, 5 \rangle|} = \frac{\sqrt{9^2 + 8^2 + 3^2}}{\sqrt{1^2 + 3^2 + 5^2}} = \frac{\sqrt{154}}{\sqrt{35}} = \sqrt{\frac{22}{5}}.$$

46. Let $Q = (5, 0, 1)$ and $R = (4, 3, 3)$, points on the line corresponding to $t = 0$ and $t = 1$. Let $P = (1, 0, -1)$.

Then $\mathbf{a} = \overrightarrow{QR} = \langle -1, 3, 2 \rangle$ and $\mathbf{b} = \overrightarrow{QP} = \langle -4, 0, -2 \rangle$. The distance is

$$d = \frac{|\mathbf{a} \times \mathbf{b}|}{|\mathbf{a}|} = \frac{|\langle -1, 3, 2 \rangle \times \langle -4, 0, -2 \rangle|}{|\langle -1, 3, 2 \rangle|} = \frac{|\langle -6, -10, 12 \rangle|}{|\langle -1, 3, 2 \rangle|} = \frac{2\sqrt{3^2 + 5^2 + 6^2}}{\sqrt{1^2 + 3^2 + 2^2}} = \frac{2\sqrt{70}}{\sqrt{14}} = 2\sqrt{5}.$$

47. By Equation 9, the distance is $D = \dfrac{1}{\sqrt{1 + 4 + 4}} [(1)(2) + (-2)(8) + (-2)(5) - 1] = \dfrac{25}{3}$.

48. By Equation 9, the distance is $D = \dfrac{1}{\sqrt{16 + 36 + 1}} [4(3) + (-6)(-2) + 1(7) - 5] = \dfrac{26}{\sqrt{53}}$.

49. Put $y = z = 0$ in the equation of the first plane to get the point $(-1, 0, 0)$ on the plane. Because the planes are parallel, the distance D between them is the distance from $(-1, 0, 0)$ to the second plane. By Equation 9,

$$D = \frac{|3(-1) + 6(0) - 3(0) - 4|}{\sqrt{3^2 + 6^2 + (-3)^2}} = \frac{7}{3\sqrt{6}} \text{ or } \frac{7\sqrt{6}}{18}.$$

50. Put $y = z = 0$ in the equation of the first plane to get the point $\left(\frac{4}{3}, 0, 0\right)$ on the plane. Because the planes are parallel, the distance D between them is the distance from $\left(\frac{4}{3}, 0, 0\right)$ to the second plane.

By Equation 9, $D = \dfrac{\left|1\left(\frac{4}{3}\right) + 2(0) - 3(0) - 1\right|}{\sqrt{1^2 + 2^2 + (-3)^2}} = \dfrac{1}{3\sqrt{14}}$.

51. The distance between two parallel planes is the same as the distance between a point on one of the planes and the other plane. Let $P_0 = (x_0, y_0, z_0)$ be a point on the plane given by $ax + by + cz + d_1 = 0$. Then $ax_0 + by_0 + cz_0 + d_1 = 0$ and the distance between P_0 and the plane given by $ax + by + cz + d_2 = 0$ is, from Equation 9,

$$D = \frac{|ax_0 + by_0 + cz_0 + d_2|}{\sqrt{a^2 + b^2 + c^2}} = \frac{|-d_1 + d_2|}{\sqrt{a^2 + b^2 + c^2}} = \frac{|d_1 - d_2|}{\sqrt{a^2 + b^2 + c^2}}.$$

52. The planes must have parallel normal vectors, so if $ax + by + cz + d = 0$ is such a plane, then for some $t \neq 0$, $\langle a, b, c \rangle = t\langle 1, 2, -2 \rangle = \langle t, 2t, -2t \rangle$. So this plane is given by the equation $x + 2y - 2z + k = 0$, where $k = d/t$. By Exercise 51, the distance between the planes is $2 = \dfrac{|1 - k|}{\sqrt{1^2 + 2^2 + (-2)^2}} \Leftrightarrow 6 = |1 - k| \Leftrightarrow k = 7$ or -5. So the desired planes have equations $x + 2y - 2z = 7$ and $x + 2y - 2z = -5$.

53. L_1: $x = y = z \Rightarrow x = y$ **(1)**. L_2: $x + 1 = y/2 = z/3 \Rightarrow x + 1 = y/2$ **(2)**. The solution of **(1)** and **(2)** is $x = y = -2$. However, when $x = -2$, $x = z \Rightarrow z = -2$, but $x + 1 = z/3 \Rightarrow z = -3$, a contradiction. Hence the lines do not intersect. For L_1, $\mathbf{v}_1 = \langle 1, 1, 1 \rangle$, and for L_2, $\mathbf{v}_2 = \langle 1, 2, 3 \rangle$, so the lines are not parallel. Thus the lines are skew lines. If two lines are skew, they can be viewed as lying in two parallel planes and so the distance between the skew lines would be the same as the distance between these parallel planes. The common normal vector to the planes must be perpendicular to both $\langle 1,1,1\rangle$ and $\langle 1,2,3\rangle$, the direction vectors of the two lines. So set $\mathbf{n} = \langle 1,1,1\rangle \times \langle 1,2,3\rangle = \langle 3-2,-3+1,2-1\rangle = \langle 1,-2,1\rangle$. From above, we know that $(-2,-2,-2)$ and $(-2,-2,-3)$ are points of L_1 and L_2 respectively. So in the notation of Equation 8, $1(-2)-2(-2)+1(-2)+d_1=0 \Rightarrow d_1=0$ and $1(-2)-2(-2)+1(-3)+d_2=0 \Rightarrow d_2=1$.

By Exercise 51, the distance between these two skew lines is $D = \dfrac{|0-1|}{\sqrt{1+4+1}} = \dfrac{1}{\sqrt{6}}$.

Alternate solution (without reference to planes): A vector which is perpendicular to both of the lines is $\mathbf{n} = \langle 1,1,1\rangle \times \langle 1,2,3\rangle = \langle 1,-2,1\rangle$. Pick any point on each of the lines, say $(-2,-2,-2)$ and $(-2,-2,-3)$, and form the vector $\mathbf{b} = \langle 0,0,1\rangle$ connecting the two points. The distance between the two skew lines is the absolute value of the scalar projection of $\mathbf{b}$ along $\mathbf{n}$, that is, $D = \dfrac{|\mathbf{n}\cdot\mathbf{b}|}{|\mathbf{n}|} = \dfrac{|1\cdot 0 - 2\cdot 0 + 1\cdot 1|}{\sqrt{1+4+1}} = \dfrac{1}{\sqrt{6}}$.

54. First notice that if two lines are skew, they can be viewed as lying in two parallel planes and so the distance between the skew lines would be the same as the distance between these parallel planes. The common normal vector to the planes must be perpendicular to both $\mathbf{v}_1 = \langle 1,6,2\rangle$ and $\mathbf{v}_2 = \langle 2,15,6\rangle$, the direction vectors of the two lines respectively. Thus set $\mathbf{n} = \mathbf{v}_1 \times \mathbf{v}_2 = \langle 36-30, 4-6, 15-12\rangle = \langle 6,-2,3\rangle$. Setting $t=0$ and $s=0$ gives the points $(1,1,0)$ and $(1,5,-2)$. So in the notation of Equation 8, $6-2+0+d_1=0 \Rightarrow d_1=-4$ and $6-10-6+d_2=0 \Rightarrow d_2=10$.

Then by Exercise 51, the distance between the two skew lines is given by $D = \dfrac{|-4-10|}{\sqrt{36+4+9}} = \dfrac{14}{7} = 2$.

Alternate solution (without reference to planes): We already know that the direction vectors of the two lines are $\mathbf{v}_1 = \langle 1,6,2\rangle$ and $\mathbf{v}_2 = \langle 2,15,6\rangle$. Then $\mathbf{n} = \mathbf{v}_1 \times \mathbf{v}_2 = \langle 6,-2,3\rangle$ is perpendicular to both lines. Pick any point on each of the lines, say $(1,1,0)$ and $(1,5,-2)$, and form the vector $\mathbf{b} = \langle 0,4,-2\rangle$ connecting the two points. Then the distance between the two skew lines is the absolute value of the scalar projection of $\mathbf{b}$ along $\mathbf{n}$, that is,

$$D = \frac{|\mathbf{n}\cdot\mathbf{b}|}{|\mathbf{n}|} = \frac{1}{\sqrt{36+4+9}}|0-8-6| = \frac{14}{7} = 2.$$

55. If $a \neq 0$, then $ax+by+cz+d=0 \Rightarrow a(x+d/a)+b(y-0)+c(z-0)=0$ which by (7) is the scalar equation of the plane through the point $(-d/a,0,0)$ with normal vector $\langle a,b,c\rangle$. Similarly, if $b\neq 0$ (or if $c \neq 0$) the equation of the plane can be rewritten as $a(x-0)+b(y+d/b)+c(z-0)=0$ [or as $a(x-0)+b(y-0)+c(z+d/c)=0$] which by (7) is the scalar equation of a plane through the point $(0,-d/b,0)$ [or the point $(0,0,-d/c)$] with normal vector $\langle a,b,c\rangle$.

56. (a) The planes $x+y+z=c$ have normal vector $\langle 1,1,1\rangle$, so they are all parallel. Their x-, y-, and z-intercepts are all c. When $c>0$ their intersection with the first octant is an equilateral triangle and when $c<0$ their intersection with the octant diagonally opposite the first is an equilateral triangle.

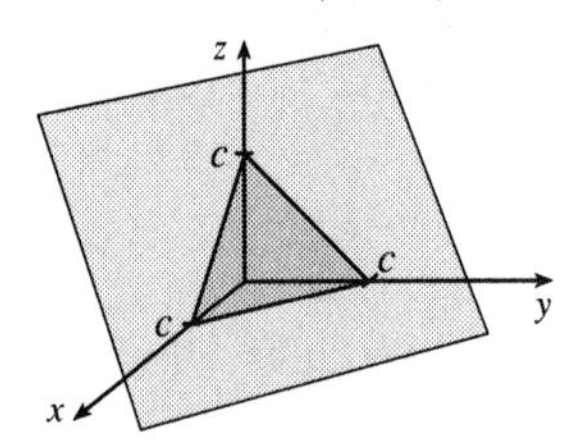

(b) The planes $x+y+cz=1$ have x-intercept 1, y-intercept 1, and z-intercept $1/c$. The plane with $c=0$ is parallel to the z-axis. As c gets larger, the planes get closer to the xy-plane.

(c) The planes $y\cos\theta + z\cos\theta = 1$ have normal vectors $\langle 0, \cos\theta, \sin\theta\rangle$, which are perpendicular to the x-axis, and so the planes are parallel to the x-axis. We look at their intersection with the yz-plane. These are lines that are perpendicular to $\langle \cos\theta, \sin\theta\rangle$ and pass through $(\cos\theta, \sin\theta)$, since $\cos^2\theta + \sin^2\theta = 1$. So these are the tangent lines to the unit circle. Thus the family consists of all planes tangent to the circular cylinder with radius 1 and axis the x-axis.

10.6 Cylinders and Quadric Surfaces

1. (a) In $\mathbb{R}^2$, the equation $y = x^2$ represents a parabola.

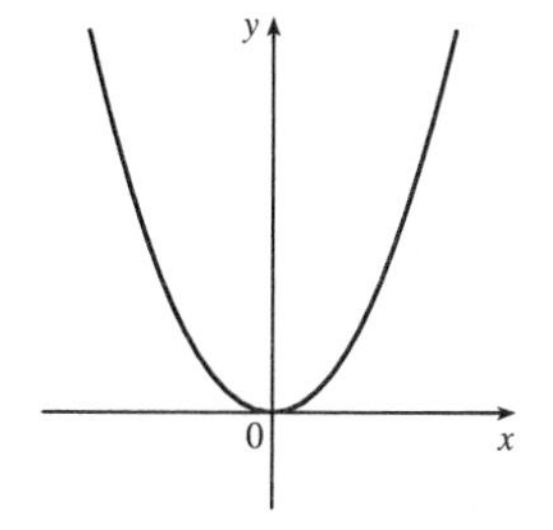

(b) In $\mathbb{R}^3$, the equation $y = x^2$ doesn't involve z, so any horizontal plane with equation $z = k$ intersects the graph in a curve with equation $y = x^2$. Thus, the surface is a parabolic cylinder, made up of infinitely many shifted copies of the same parabola. The rulings are parallel to the z-axis.

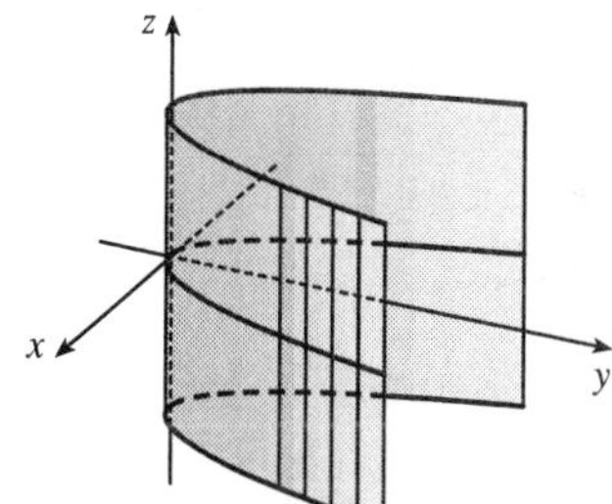

(c) In $\mathbb{R}^3$, the equation $z = y^2$ also represents a parabolic cylinder. Since x doesn't appear, the graph is formed by moving the parabola $z = y^2$ in the direction of the x-axis. Thus, the rulings of the cylinder are parallel to the x-axis.

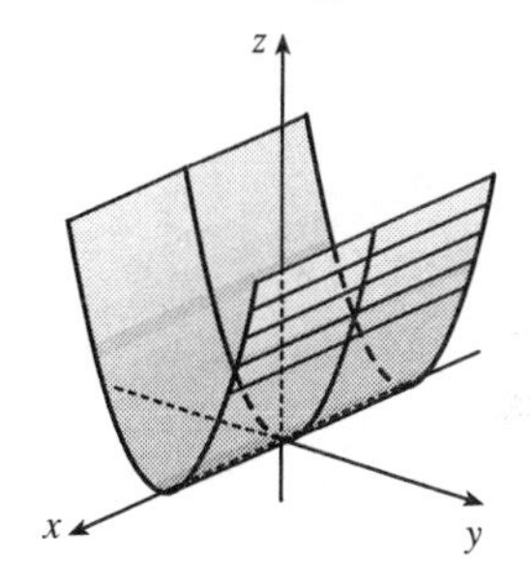

2. (a)

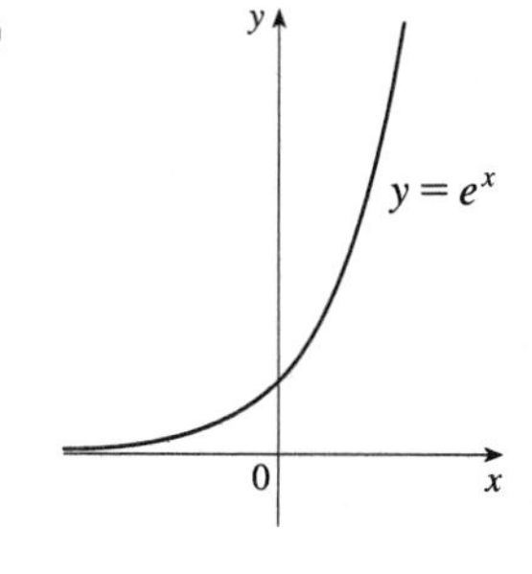

(b) Since the equation $y = e^x$ doesn't involve z, horizontal traces are copies of the curve $y = e^x$. The rulings are parallel to the z-axis.

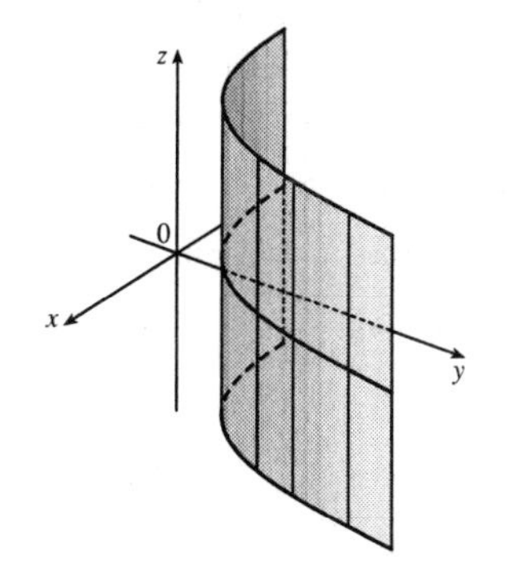

(c) The equation $z = e^y$ doesn't involve x, so vertical traces in $x = k$ (parallel to the yz-plane) are copies of the curve $z = e^y$. The rulings are parallel to the x-axis.

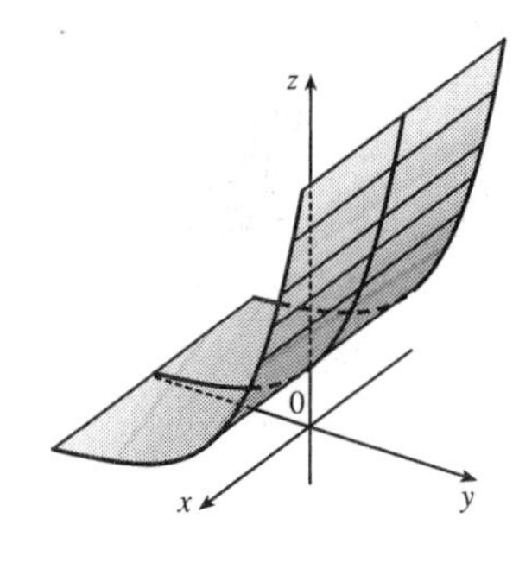

3. Since x is missing from the equation, the vertical traces $y^2 + 4z^2 = 4$, $x = k$, are copies of the same ellipse in the plane $x = k$. Thus, the surface $y^2 + 4z^2 = 4$ is an elliptic cylinder with rulings parallel to the x-axis.

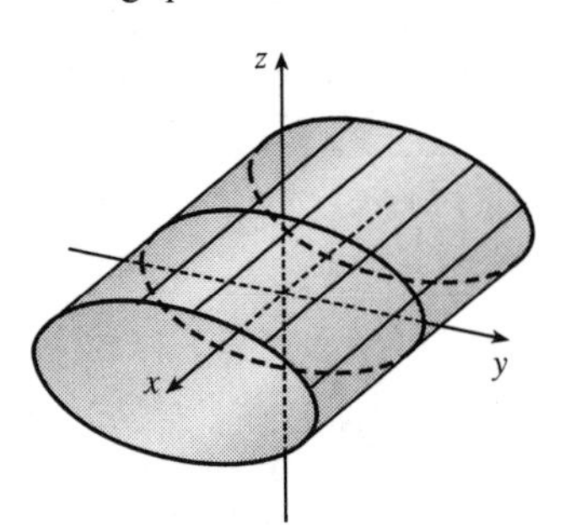

4. Since y is missing from the equation, each vertical trace $z = 4 - x^2$, $y = k$, is a copy of the same parabola in the plane $y = k$. Thus, the surface $z = 4 - x^2$ is a parabolic cylinder with rulings parallel to the y-axis.

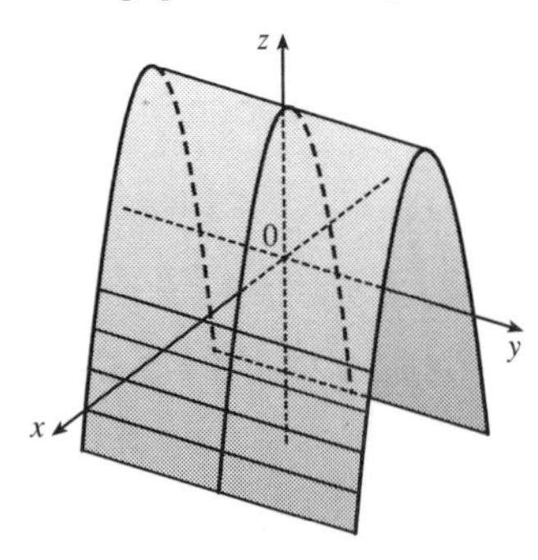

5. Since z is missing, each horizontal trace $x = y^2$, $z = k$, is a copy of the same parabola in the plane $z = k$. Thus, the surface $x - y^2 = 0$ is a parabolic cylinder with rulings parallel to the z-axis.

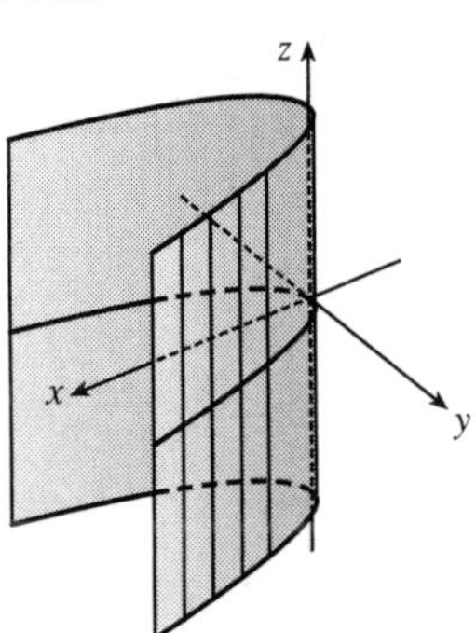

6. Since x is missing, each vertical trace $yz = 4$, $x = k$ is a copy of the same hyperbola in the plane $x = k$. Thus, the surface $yz = 4$ is a hyperbolic cylinder with rulings parallel to the x-axis.

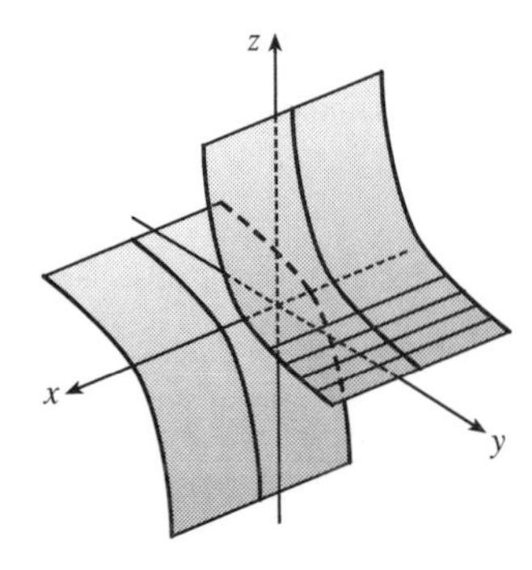

7. Since y is missing, each vertical trace $z = \cos x$, $y = k$ is a copy of a cosine curve in the plane $y = k$. Thus, the surface $z = \cos x$ is a cylindrical surface with rulings parallel to the y-axis.

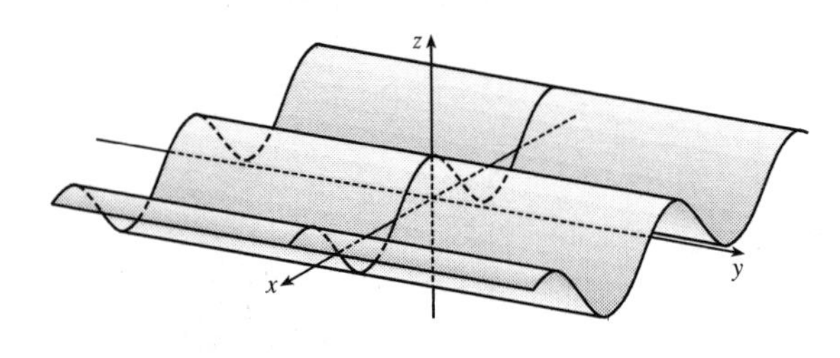

8. Since z is missing, each horizontal trace $x^2 - y^2 = 1$, $z = k$ is a copy of the same hyperbola in the plane $z = k$. Thus, the surface $x^2 - y^2 = 1$ is a hyperbolic cylinder with rulings parallel to the z-axis.

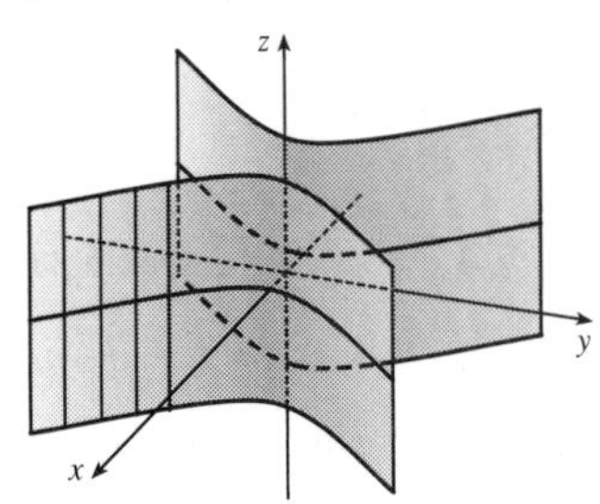

9. (a) The traces of $x^2 + y^2 - z^2 = 1$ in $x = k$ are $y^2 - z^2 = 1 - k^2$, a family of hyperbolas. (Note that the hyperbolas are oriented differently for $-1 < k < 1$ than for $k < -1$ or $k > 1$.) The traces in $y = k$ are $x^2 - z^2 = 1 - k^2$, a similar family of hyperbolas. The traces in $z = k$ are $x^2 + y^2 = 1 + k^2$, a family of circles. For $k = 0$, the trace in the

xy-plane, the circle is of radius 1. As $|k|$ increases, so does the radius of the circle. This behavior, combined with the hyperbolic vertical traces, gives the graph of the hyperboloid of one sheet in Table 1.

(b) The shape of the surface is unchanged, but the hyperboloid is rotated so that its axis is the y-axis. Traces in $y = k$ are circles, while traces in $x = k$ and $z = k$ are hyperbolas.

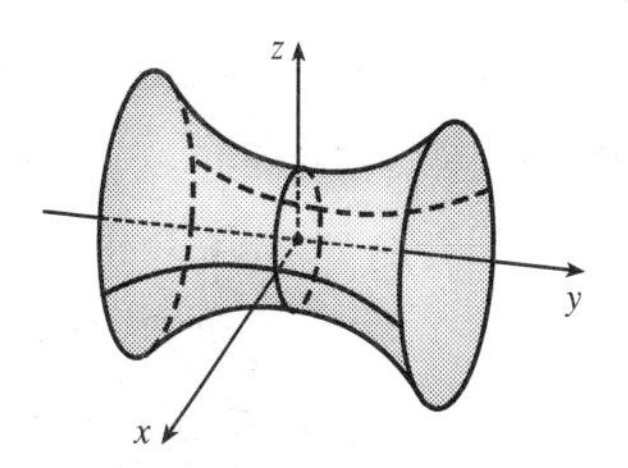

(c) Completing the square in y gives $x^2 + (y+1)^2 - z^2 = 1$. The surface is a hyperboloid identical to the one in part (a) but shifted one unit in the negative y-direction.

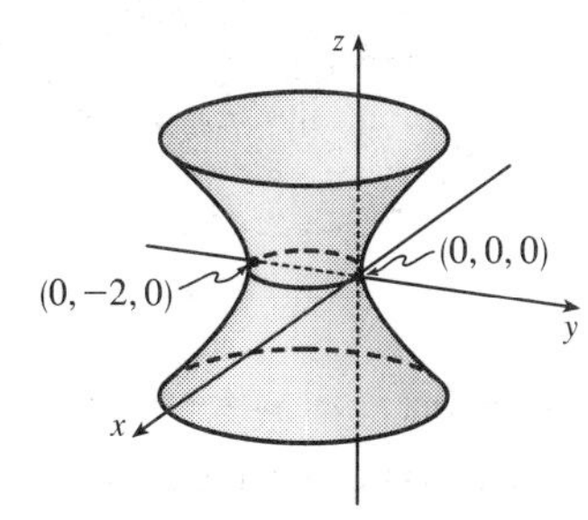

10. (a) The traces of $-x^2 - y^2 + z^2 = 1$ in $x = k$ are $-y^2 + z^2 = 1 + k^2$, a family of hyperbolas, as are the traces in $y = k$, $-x^2 + z^2 = 1 + k^2$. The traces in $z = k$ are $x^2 + y^2 = k^2 - 1$, a family of circles for $|k| > 1$. As $|k|$ increases, the radii of the circles increase; the traces are empty for $|k| < 1$. This behavior, combined with the vertical traces, gives the graph of the hyperboloid of two sheets in Table 1.

(b) The graph has the same shape as the hyperboloid in part (a) but is rotated so that its axis is the x-axis. Traces in $x = k$, $|k| > 1$, are circles, while traces in $y = k$ and $z = k$ are hyperbolas.

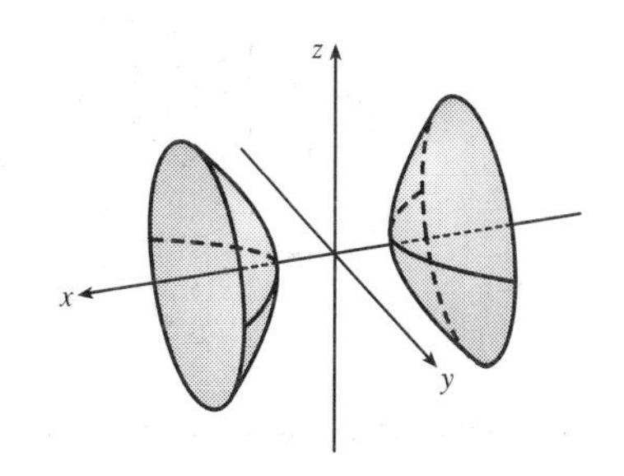

11. Traces: $x = k$, $9y^2 + 36z^2 = 36 - 4k^2$, an ellipse for $|k| < 3$; $y = k$, $4x^2 + 36z^2 = 36 - 9k^2$, an ellipse for $|k| < 2$; $z = k$, $4x^2 + 9y^2 = 36(1 - k^2)$, an ellipse for $|k| < 1$. Thus the surface is an ellipsoid with center at the origin and axes along the x-, y- and z-axes.

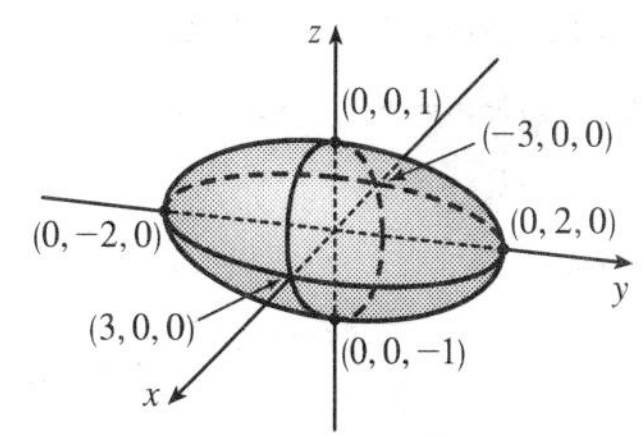

12. Traces: $x = k$, $4y = k^2 + z^2$, a parabola; $y = k$, $4k = x^2 + z^2$, a circle for $k > 0$; $z = k$, $4y = x^2 + k^2$ a parabola. Thus the surface is a circular paraboloid with axis the y-axis and vertex at $(0, 0, 0)$.

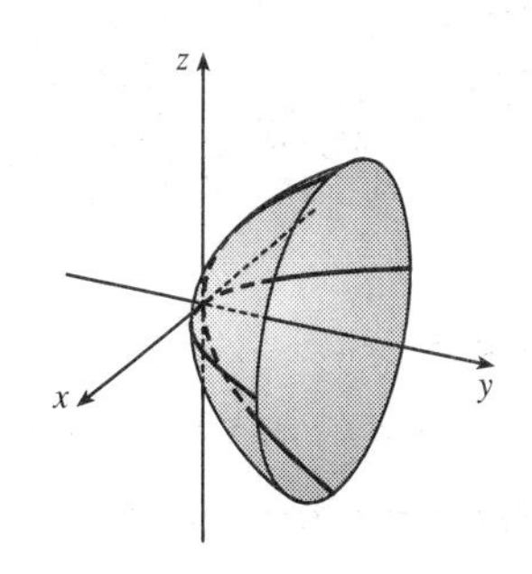

13. Traces: $x = k$, $y^2 = k^2 + z^2$ or $y^2 - z^2 = k^2$, a hyperbola for $k \neq 0$ and two intersecting lines for $k = 0$; $y = k$, $x^2 + z^2 = k^2$, a circle for $k \neq 0$; $z = k$, $y^2 = x^2 + k^2$ or $y^2 - x^2 = k^2$, a hyperbola for $k \neq 0$ and two intersecting lines for $k = 0$. Thus the surface is a cone (right circular) with axis the y-axis and vertex the origin.

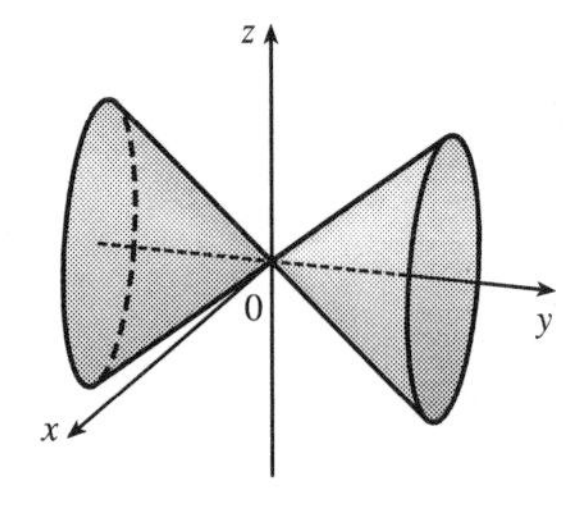

14. Traces: $x = k$, $z - k^2 = -y^2$, a parabola; $y = k$, $z + k^2 = x^2$, a parabola; $z = k$, $x^2 - y^2 = k$, a hyperbola. Thus the surface is a hyperbolic paraboloid with saddle point $(0, 0, 0)$ (and since $c > 0$, the saddle is upside down).

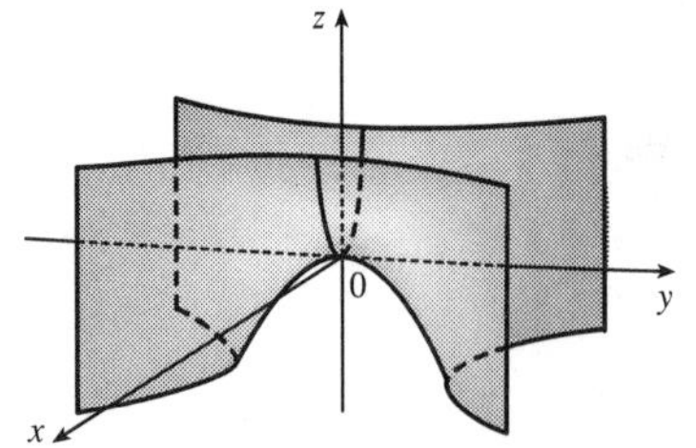

15. Traces: $x = k$, $4y^2 - z^2 = 4 + k^2$, a hyperbola; $y = k$, $x^2 + z^2 = 4k^2 - 4$, a circle for $|k| > 1$; $z = k$, $4y^2 - x^2 = 4 + k^2$, a hyperbola. Thus the surface is a hyperboloid of two sheets with axis the y-axis.

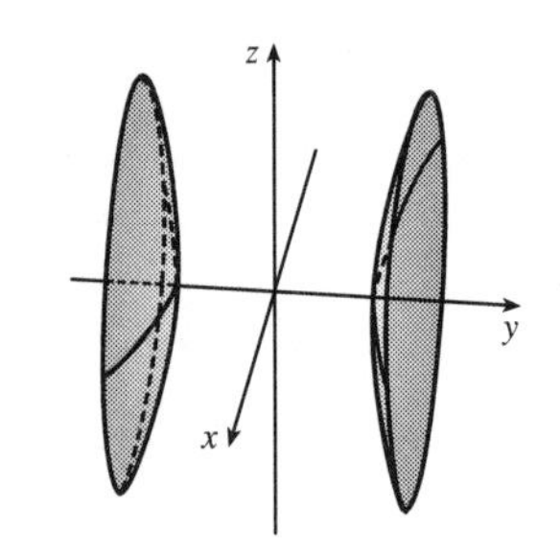

16. Traces: $x = k$, $25y^2 + z^2 = 100 + 4k^2$, an ellipse; $y = k$, $25k^2 + z^2 = 100 + 4x^2$ or $z^2 - 4x^2 = 100 - 25k^2$, a hyperbola for $|k| < 2$; $z = k$, $25y^2 + k^2 = 100 + 4x^2$ or $25y^2 - 4x^2 = 100 - k^2$, a hyperbola for $|k| < 10$. Thus the surface is a hyperboloid of one sheet with axis the x-axis.

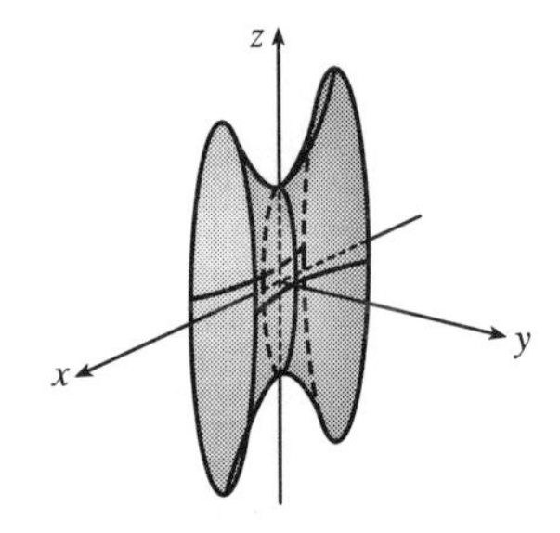

17. Traces: $x = k$, $k^2 + 4z^2 - y = 0$ or $y - k^2 = 4z^2$, a parabola; $y = k$, $x^2 + 4z^2 = k$, an ellipse for $k > 0$; $z = k$, $x^2 + 4k^2 - y = 0$ or $y - 4k^2 = x^2$, a parabola. Thus the surface is an elliptic paraboloid with axis the y-axis and vertex the origin.

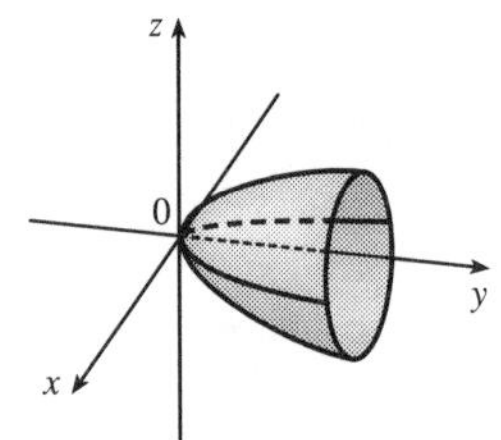

18. Traces: $x = k$, $|k| \leq 2 \;\Rightarrow\; y^2 + \frac{z^2}{4} = 1 - \frac{k^2}{4}$, ellipses;

$y = k$, $|k| \leq 1 \;\Rightarrow\; x^2 + z^2 = 4\left(1 - k^2\right)$, circles;

$z = k$, $|k| \leq 2 \;\Rightarrow\; \frac{x^2}{4} + y^2 = 1 - \frac{k^2}{4}$, ellipses.

$x^2 + 4y^2 + z^2 = 4 \;\Leftrightarrow\; \frac{x^2}{2^2} + \frac{y^2}{1^2} + \frac{z^2}{2^2} = 1$, which is the equation of an ellipsoid.

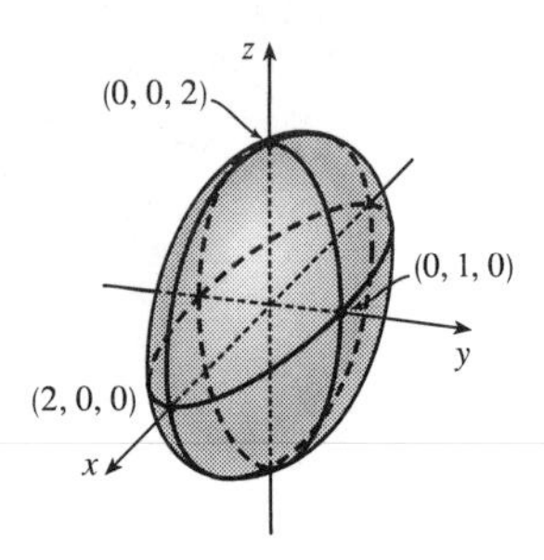

19. Traces: $x = k$, $y = z^2 - k^2$, parabolas; $y = k$, $k = z^2 - x^2$, hyperbolas (note the hyperbolas are oriented differently for $k > 0$ than for $k < 0$); $z = k$, $y = k^2 - x^2$, parabolas. Thus $\frac{y}{1} = \frac{z^2}{1^2} - \frac{x^2}{1^2}$ is a hyperbolic paraboloid.

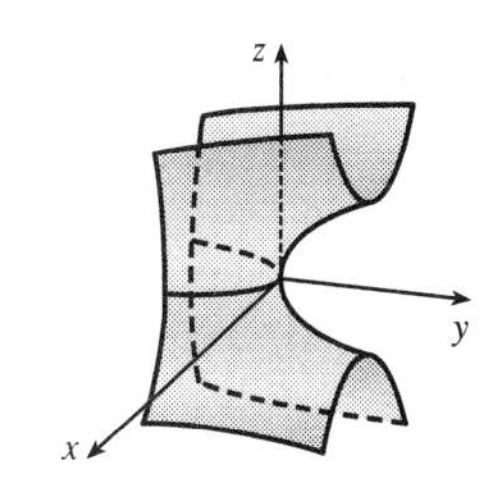

20. Traces: $x = k \Rightarrow y^2 + 4z^2 = 16k^2$, ellipses; $y = k \Rightarrow 16x^2 - 4z^2 = k^2$, hyperbolas if $k \neq 0$ and two intersecting lines if $k = 0$; $z = k \Rightarrow 16x^2 - y^2 = 4k^2$, hyperbolas if $k \neq 0$ and two intersecting lines if $k = 0$.

$16x^2 = y^2 + 4z^2 \Leftrightarrow x^2 = \frac{y^2}{4^2} + \frac{z^2}{2^2}$ is an elliptic cone with axis the x-axis and vertex the origin.

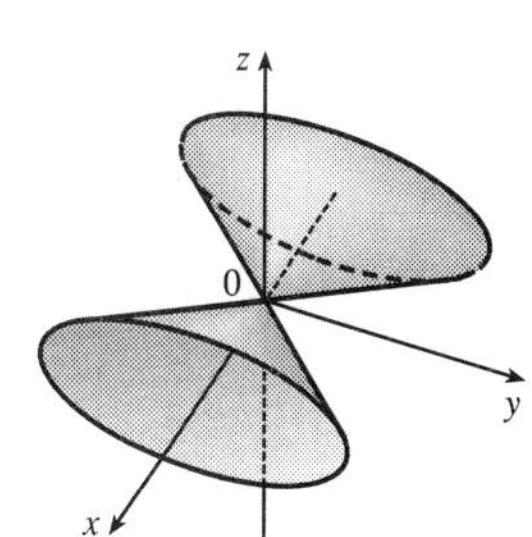

21. $z^2 = 4x^2 + 9y^2 + 36$ or $-4x^2 - 9y^2 + z^2 = 36$ or $-\frac{x^2}{9} - \frac{y^2}{4} + \frac{z^2}{36} = 1$ represents a hyperboloid of two sheets with axis the z-axis.

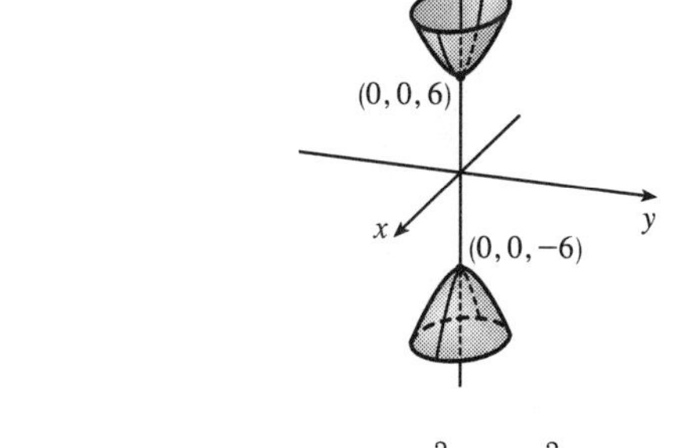

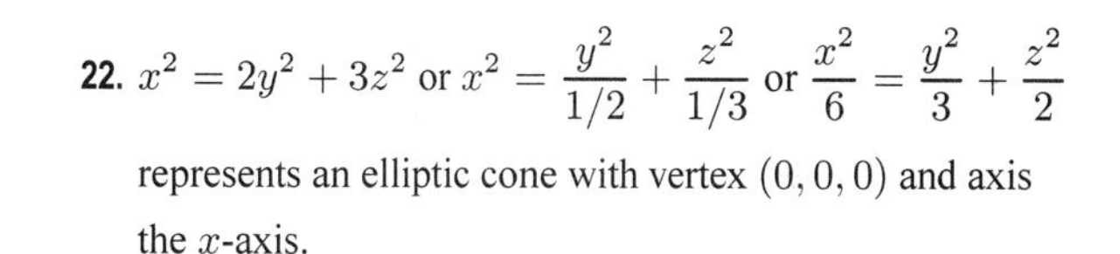

22. $x^2 = 2y^2 + 3z^2$ or $x^2 = \frac{y^2}{1/2} + \frac{z^2}{1/3}$ or $\frac{x^2}{6} = \frac{y^2}{3} + \frac{z^2}{2}$ represents an elliptic cone with vertex $(0, 0, 0)$ and axis the x-axis.

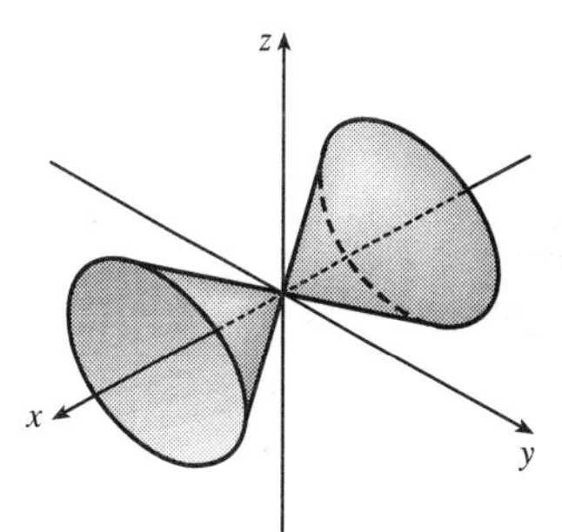

23. $x = 2y^2 + 3z^2$ or $x = \frac{y^2}{1/2} + \frac{z^2}{1/3}$ or $\frac{x}{6} = \frac{y^2}{3} + \frac{z^2}{2}$ represents an elliptic paraboloid with vertex $(0, 0, 0)$ and axis the x-axis.

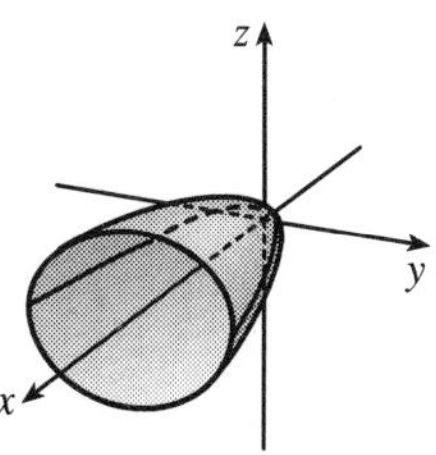

24. $4x - y^2 + 4z^2 = 0$ or $4x = y^2 - 4z^2$ or $x = \frac{y^2}{4} - z^2$ represents a hyperbolic paraboloid with center $(0, 0, 0)$.

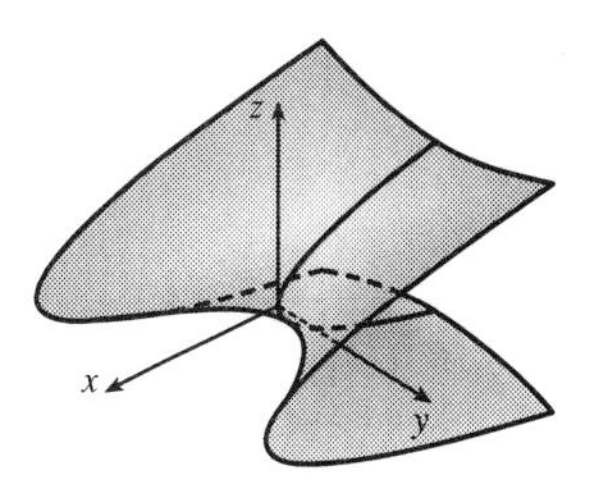

25. Completing squares in y and z gives

$4x^2 + (y-2)^2 + 4(z-3)^2 = 4$ or

$x^2 + \dfrac{(y-2)^2}{4} + (z-3)^2 = 1$, an ellipsoid with

center $(0, 2, 3)$.

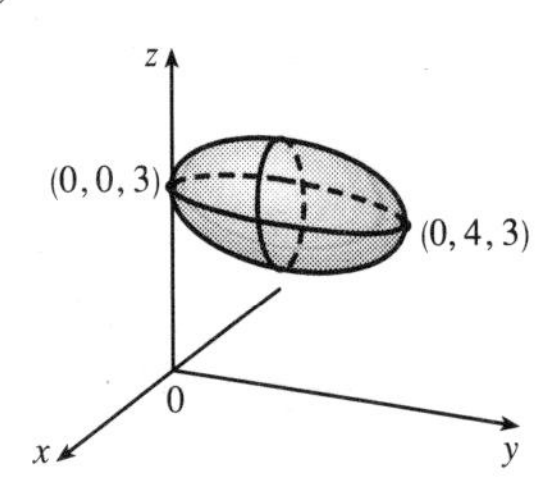

26. Completing squares in y and z gives

$4(y-2)^2 + (z-2)^2 - x = 0$ or

$\dfrac{x}{4} = (y-2)^2 + \dfrac{(z-2)^2}{4}$, an elliptic paraboloid with

vertex $(0, 2, 2)$ and axis the horizontal line $y = 2$, $z = 2$.

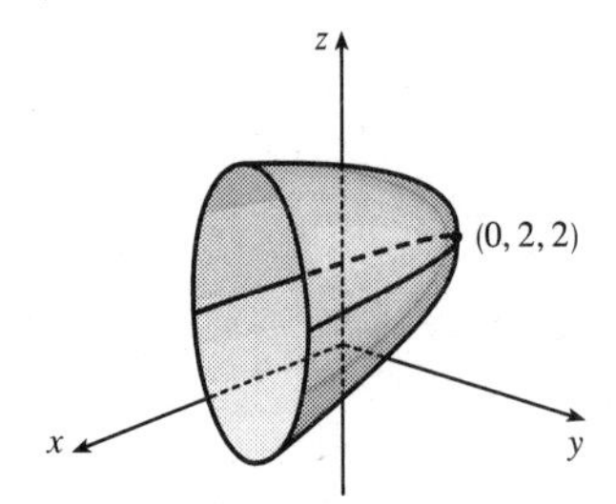

27. Completing squares in all three variables gives

$(x-2)^2 - (y+1)^2 + (z-1)^2 = 0$ or

$(y+1)^2 = (x-2)^2 + (z-1)^2$, a circular cone with

center $(2, -1, 1)$ and axis the horizontal line $x = 2$,

$z = 1$.

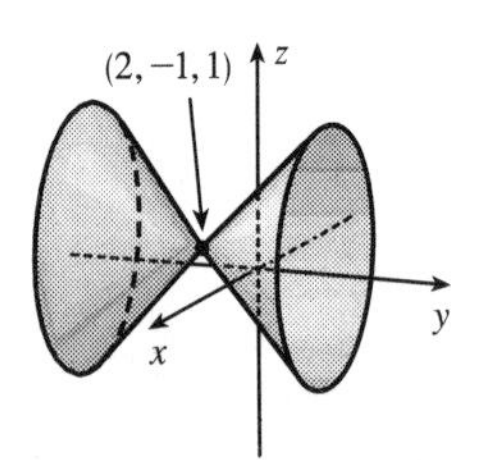

28. Completing squares in all three variables gives

$(x-1)^2 - (y-1)^2 + (z+2)^2 = 2$ or

$\dfrac{(x-1)^2}{2} - \dfrac{(y-1)^2}{2} + \dfrac{(z+2)^2}{2} = 1$, a hyperboloid of

one sheet with center $(1, 1, -2)$ and axis the horizontal

line $x = 1$, $z = -2$.

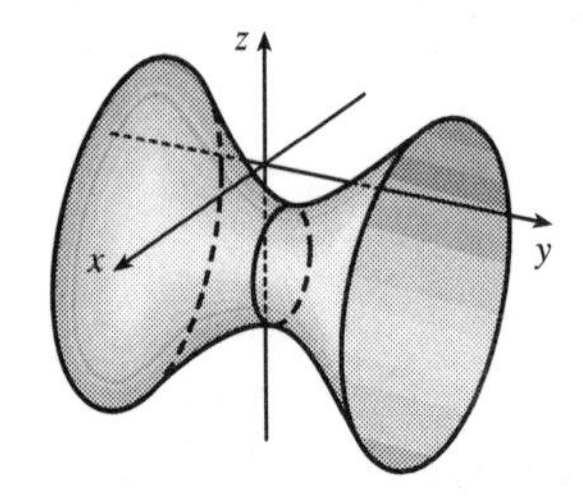

29.

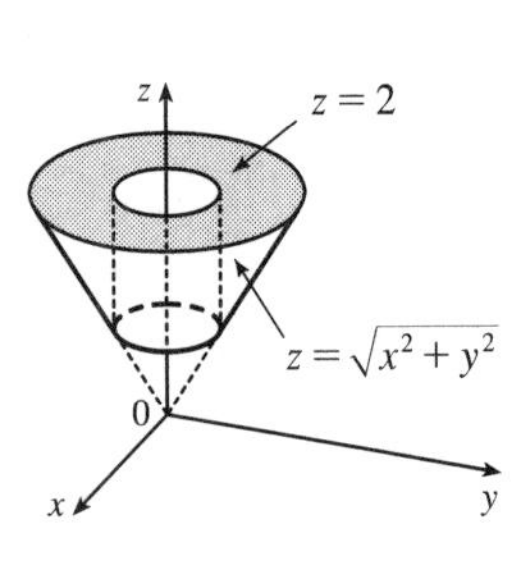

30.

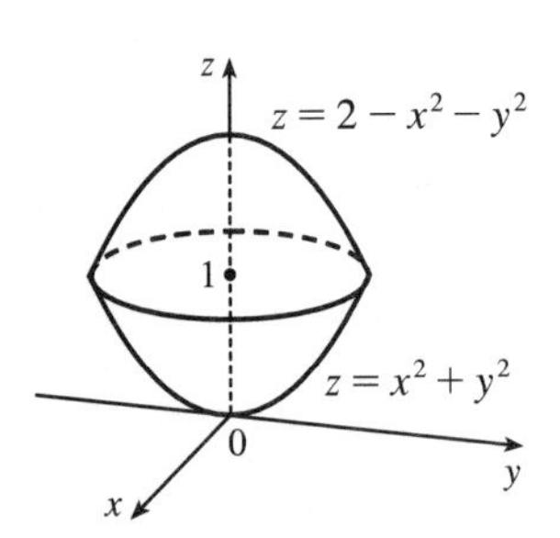

31. Let $P = (x, y, z)$ be an arbitrary point equidistant from $(-1, 0, 0)$ and the plane $x = 1$. Then the distance from P to $(-1, 0, 0)$ is $\sqrt{(x+1)^2 + y^2 + z^2}$ and the distance from P to the plane $x = 1$ is $|x - 1| / \sqrt{1^2} = |x - 1|$ (by Equation 10.5.7). So $|x - 1| = \sqrt{(x+1)^2 + y^2 + z^2} \quad \Leftrightarrow \quad (x-1)^2 = (x+1)^2 + y^2 + z^2 \quad \Leftrightarrow$ $x^2 - 2x + 1 = x^2 + 2x + 1 + y^2 + z^2 \quad \Leftrightarrow \quad -4x = y^2 + z^2$. Thus the collection of all such points P is a circular paraboloid with vertex at the origin, axis the x-axis, which opens in the negative direction.

32. Let $P = (x, y, z)$ be an arbitrary point whose distance from the x-axis is twice its distance from the yz-plane. The distance from P to the x-axis is $\sqrt{(x-x)^2 + y^2 + z^2} = \sqrt{y^2 + z^2}$ and the distance from P to the yz-plane ($x = 0$) is $|x|/1 = |x|$. Thus $\sqrt{y^2 + z^2} = 2|x| \Leftrightarrow y^2 + z^2 = 4x^2 \Leftrightarrow x^2 = (y^2/2^2) + (z^2/2^2)$. So the surface is a right circular cone with vertex the origin and axis the x-axis.

33.

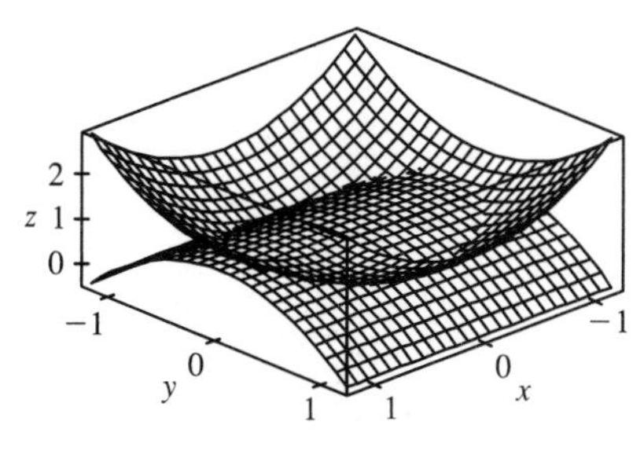

The curve of intersection looks like a bent ellipse. The projection of this curve onto the xy-plane is the set of points $(x, y, 0)$ which satisfy $x^2 + y^2 = 1 - y^2 \Leftrightarrow x^2 + 2y^2 = 1 \Leftrightarrow$ $x^2 + \dfrac{y^2}{\left(1/\sqrt{2}\right)^2} = 1$. This is an equation of an ellipse.

34. Any point on the curve of intersection must satisfy both $2x^2 + 4y^2 - 2z^2 + 6x = 2$ and $2x^2 + 4y^2 - 2z^2 - 5y = 0$. Subtracting, we get $6x + 5y = 2$, which is linear and therefore the equation of a plane. Thus the curve of intersection lies in this plane.

10.7 Vector Functions and Space Curves

1. The component functions t^2, $\sqrt{t-1}$, and $\sqrt{5-t}$ are all defined when $t - 1 \geq 0 \Rightarrow t \geq 1$ and $5 - t \geq 0 \Rightarrow t \leq 5$, so the domain of $\mathbf{r}$ is $[1, 5]$.

2. The component functions $\dfrac{t-2}{t+2}$, $\sin t$, and $\ln(9 - t^2)$ are all defined when $t \neq -2$ and $9 - t^2 > 0 \Rightarrow -3 < t < 3$, so the domain of $\mathbf{r}$ is $(-3, -2) \cup (-2, 3)$.

3. $\lim\limits_{t\to0^+} \cos t = \cos 0 = 1$, $\lim\limits_{t\to0^+} \sin t = \sin 0 = 0$, $\lim\limits_{t\to0^+} t\ln t = \lim\limits_{t\to0^+} \dfrac{\ln t}{1/t} = \lim\limits_{t\to0^+} \dfrac{1/t}{-1/t^2} = \lim\limits_{t\to0^+} -t = 0$ [by l'Hospital's Rule]. Thus $\lim\limits_{t\to0^+} \langle \cos t, \sin t, t\ln t\rangle = \left\langle \lim\limits_{t\to0^+} \cos t, \lim\limits_{t\to0^+} \sin t, \lim\limits_{t\to0^+} t\ln t \right\rangle = \langle 1, 0, 0\rangle$.

4. $\lim\limits_{t\to\infty} \arctan t = \frac{\pi}{2}$, $\lim\limits_{t\to\infty} e^{-2t} = 0$, $\lim\limits_{t\to\infty} \dfrac{\ln t}{t} = \lim\limits_{t\to\infty} \dfrac{1/t}{1} = 0$ [by l'Hospital's Rule].

Thus $\lim\limits_{t\to\infty} \left\langle \arctan t, e^{-2t}, \dfrac{\ln t}{t} \right\rangle = \left\langle \frac{\pi}{2}, 0, 0\right\rangle$.

5. The corresponding parametric equations for this curve are $x = \sin t$, $y = t$. We can make a table of values, or we can eliminate the parameter: $t = y \Rightarrow$ $x = \sin y$, with $y \in \mathbb{R}$. By comparing different values of t, we find the direction in which t increases as indicated in the graph.

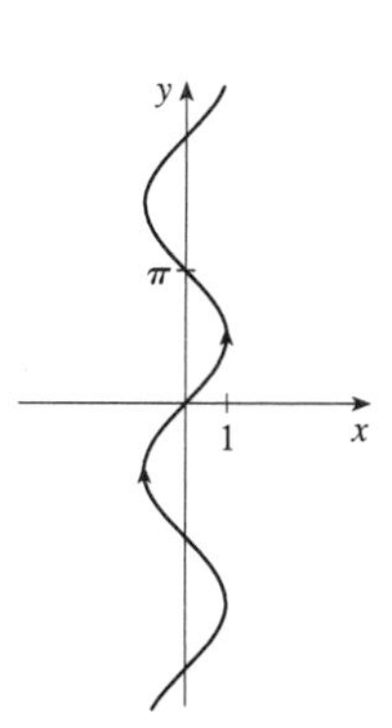

6. The corresponding parametric equations for this curve are $x = t^3$, $y = t^2$.
We can make a table of values, or we can eliminate the parameter:
$x = t^3 \;\Rightarrow\; t = \sqrt[3]{x} \;\Rightarrow\; y = t^2 = \left(\sqrt[3]{x}\right)^2 = x^{2/3}$,
with $t \in \mathbb{R} \;\Rightarrow\; x \in \mathbb{R}$. By comparing different values of t, we find the direction in which t increases as indicated in the graph.

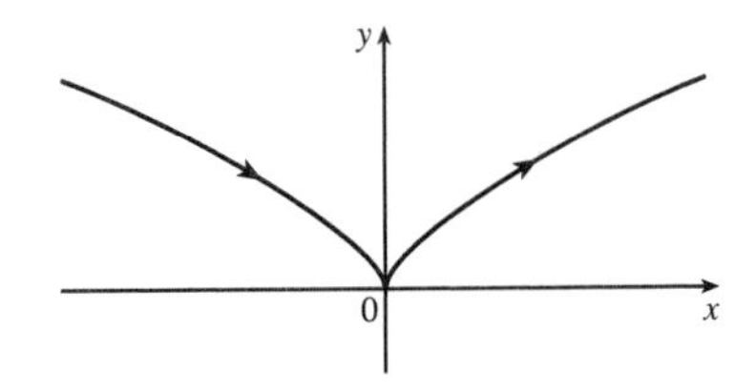

7. The corresponding parametric equations are $x = t$, $y = \cos 2t$, $z = \sin 2t$.
Note that $y^2 + z^2 = \cos^2 2t + \sin^2 2t = 1$, so the curve lies on the circular cylinder $y^2 + z^2 = 1$. Since $x = t$, the curve is a helix.

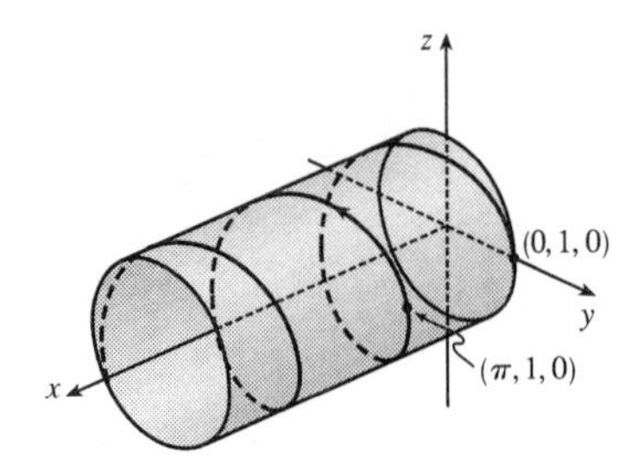

8. The corresponding parametric equations are $x = 1 + t$, $y = 3t$, $z = -t$,
which are parametric equations of a line through the point $(1, 0, 0)$ and with direction vector $\langle 1, 3, -1 \rangle$.

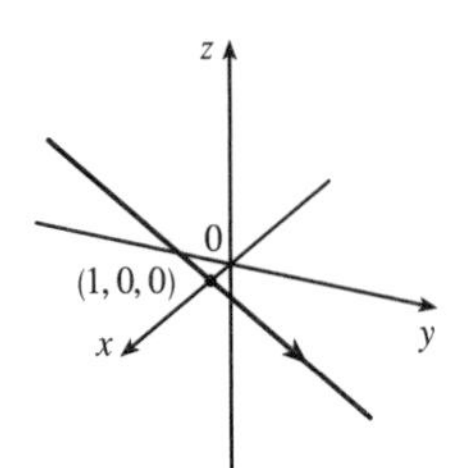

9. The corresponding parametric equations are $x = 1$, $y = \cos t$, $z = 2\sin t$.
Eliminating the parameter in y and z gives $y^2 + (z/2)^2 = \cos^2 t + \sin^2 t = 1$
or $y^2 + z^2/4 = 1$. Since $x = 1$, the curve is an ellipse centered at $(1, 0, 0)$ in the plane $x = 1$.

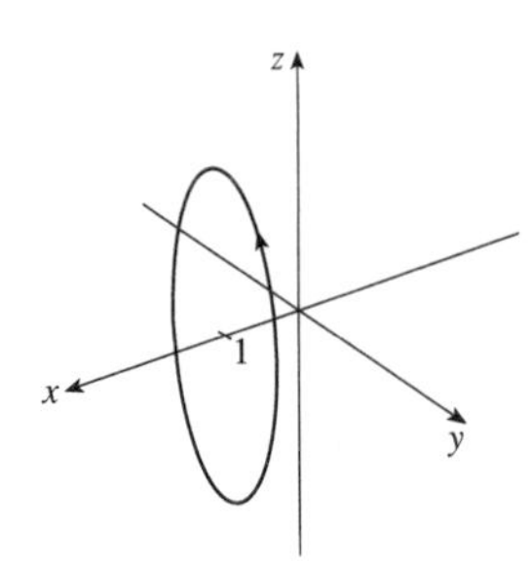

10. The parametric equations are $x = t^2$, $y = t$, $z = 2$, so we have $x = y^2$ with $z = 2$.
Thus the curve is a parabola in the plane $z = 2$ with vertex $(0, 0, 2)$.

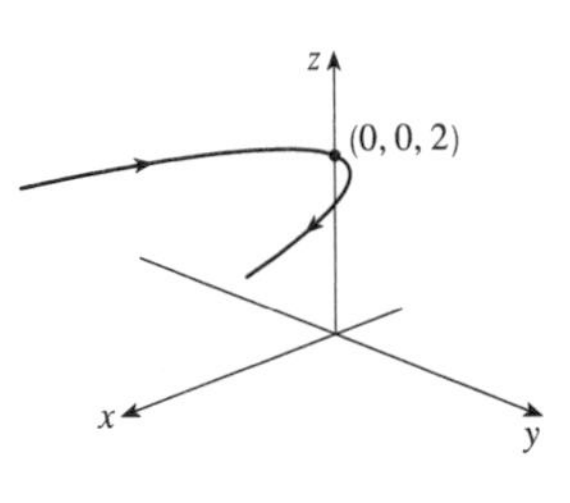

11. The parametric equations are $x = t^2$, $y = t^4$, $z = t^6$. These are positive for $t \neq 0$ and 0 when $t = 0$. So the curve lies entirely in the first quadrant.
The projection of the graph onto the xy-plane is $y = x^2$, $y > 0$, a half parabola.
On the xz-plane $z = x^3$, $z > 0$, a half cubic, and the yz-plane, $y^3 = z^2$.

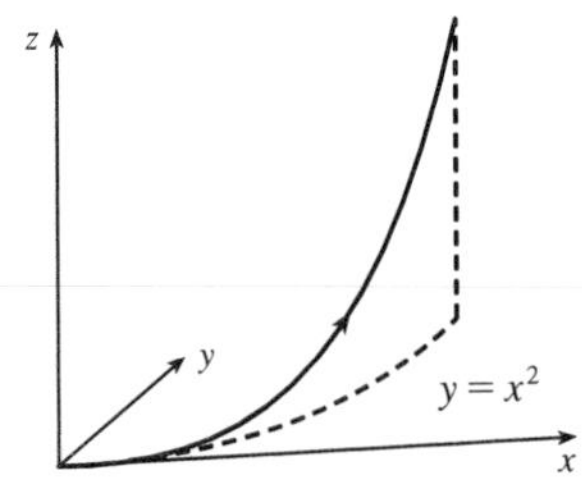

12. If $x = \cos t,\ y = -\cos t,\ z = \sin t$, then $x^2 + z^2 = 1$ and $y^2 + z^2 = 1$, so the curve is contained in the intersection of circular cylinders along the x- and y-axes. Furthermore, $y = -x$, so the curve is an ellipse in the plane $y = -x$, centered at the origin.

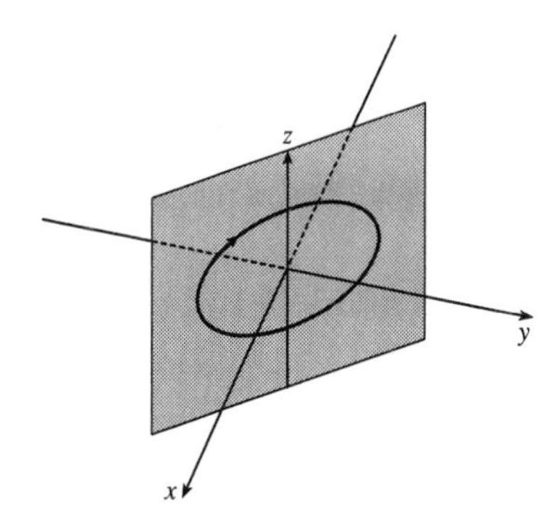

13. Taking $\mathbf{r}_0 = \langle 0, 0, 0 \rangle$ and $\mathbf{r}_1 = \langle 1, 2, 3 \rangle$, we have from Equation 10.5.4

$\mathbf{r}(t) = (1 - t)\,\mathbf{r}_0 + t\,\mathbf{r}_1 = (1 - t)\,\langle 0, 0, 0 \rangle + t\,\langle 1, 2, 3 \rangle, 0 \le t \le 1$ or $\mathbf{r}(t) = \langle t, 2t, 3t \rangle, 0 \le t \le 1$.

Parametric equations are $x = t,\ y = 2t,\ z = 3t,\ 0 \le t \le 1$.

14. Taking $\mathbf{r}_0 = \langle 1, 0, 1 \rangle$ and $\mathbf{r}_1 = \langle 2, 3, 1 \rangle$, we have from Equation 10.5.4

$\mathbf{r}(t) = (1 - t)\,\mathbf{r}_0 + t\,\mathbf{r}_1 = (1 - t)\,\langle 1, 0, 1 \rangle + t\,\langle 2, 3, 1 \rangle, 0 \le t \le 1$ or $\mathbf{r}(t) = \langle 1 + t, 3t, 1 \rangle, 0 \le t \le 1$.

Parametric equations are $x = 1 + t,\ y = 3t,\ z = 1,\ 0 \le t \le 1$.

15. Taking $\mathbf{r}_0 = \langle 1, -1, 2 \rangle$ and $\mathbf{r}_1 = \langle 4, 1, 7 \rangle$, we have

$\mathbf{r}(t) = (1 - t)\,\mathbf{r}_0 + t\,\mathbf{r}_1 = (1 - t)\,\langle 1, -1, 2 \rangle + t\,\langle 4, 1, 7 \rangle, 0 \le t \le 1$ or $\mathbf{r}(t) = \langle 1 + 3t, -1 + 2t, 2 + 5t \rangle, 0 \le t \le 1$.

Parametric equations are $x = 1 + 3t,\ y = -1 + 2t,\ z = 2 + 5t,\ 0 \le t \le 1$.

16. Taking $\mathbf{r}_0 = \langle -2, 4, 0 \rangle$ and $\mathbf{r}_1 = \langle 6, -1, 2 \rangle$, we have

$\mathbf{r}(t) = (1 - t)\,\mathbf{r}_0 + t\,\mathbf{r}_1 = (1 - t)\,\langle -2, 4, 0 \rangle + t\,\langle 6, -1, 2 \rangle, 0 \le t \le 1$ or $\mathbf{r}(t) = \langle -2 + 8t, 4 - 5t, 2t \rangle, 0 \le t \le 1$.

Parametric equations are $x = -2 + 8t,\ y = 4 - 5t,\ z = 2t,\ 0 \le t \le 1$.

17. $x = \cos 4t,\ y = t,\ z = \sin 4t$. At any point (x, y, z) on the curve, $x^2 + z^2 = \cos^2 4t + \sin^2 4t = 1$. So the curve lies on a circular cylinder with axis the y-axis. Since $y = t$, this is a helix. So the graph is VI.

18. $x = t,\ y = t^2,\ z = e^{-t}$. At any point on the curve, $y = x^2$. So the curve lies on the parabolic cylinder $y = x^2$. Note that y and z are positive for all t, and the point $(0, 0, 1)$ is on the curve (when $t = 0$). As $t \to \infty$, $(x, y, z) \to (\infty, \infty, 0)$, while as $t \to -\infty$, $(x, y, z) \to (-\infty, \infty, \infty)$, so the graph must be II.

19. $x = t,\ y = 1/(1 + t^2),\ z = t^2$. Note that y and z are positive for all t. The curve passes through $(0, 1, 0)$ when $t = 0$. As $t \to \infty$, $(x, y, z) \to (\infty, 0, \infty)$, and as $t \to -\infty$, $(x, y, z) \to (-\infty, 0, \infty)$. So the graph is IV.

20. $x = e^{-t} \cos 10t,\ y = e^{-t} \sin 10t,\ z = e^{-t}$.

$x^2 + y^2 = e^{-2t} \cos^2 10t + e^{-2t} \sin^2 10t = e^{-2t}(\cos^2 10t + \sin^2 10t) = e^{-2t} = z^2$, so the curve lies on the cone $x^2 + y^2 = z^2$. Also, z is always positive; the graph must be I.

21. $x = \cos t,\ y = \sin t,\ z = \sin 5t$. $x^2 + y^2 = \cos^2 t + \sin^2 t = 1$, so the curve lies on a circular cylinder with axis the z-axis. Each of x, y and z is periodic, and at $t = 0$ and $t = 2\pi$ the curve passes through the same point, so the curve repeats itself and the graph is V.

22. $x = \cos t,\ y = \sin t,\ z = \ln t$. $x^2 + y^2 = \cos^2 t + \sin^2 t = 1$, so the curve lies on a circular cylinder with axis the z-axis. As $t \to 0$, $z \to -\infty$, so the graph is III.

23. If $x = t\cos t,\ y = t\sin t,\ z = t$, then $x^2 + y^2 = t^2\cos^2 t + t^2\sin^2 t = t^2 = z^2$, so the curve lies on the cone $z^2 = x^2 + y^2$. Since $z = t$, the curve is a spiral on this cone.

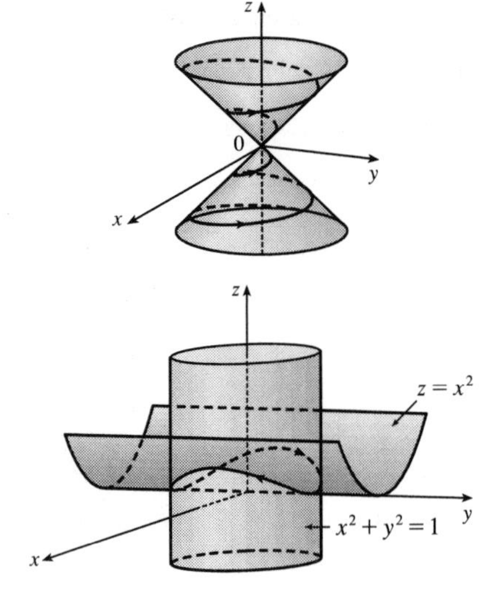

24. Here $x^2 = \sin^2 t = z$ and $x^2 + y^2 = \sin^2 t + \cos^2 t = 1$, so the curve is the intersection of the parabolic cylinder $z = x^2$ with the circular cylinder $x^2 + y^2 = 1$.

25. Parametric equations for the curve are $x = t,\ y = 0,\ z = 2t - t^2$. Substituting into the equation of the paraboloid gives $2t - t^2 = t^2 \ \Rightarrow\ 2t = 2t^2 \ \Rightarrow\ t = 0, 1$. Since $\mathbf{r}(0) = \mathbf{0}$ and $\mathbf{r}(1) = \mathbf{i} + \mathbf{k}$, the points of intersection are $(0, 0, 0)$ and $(1, 0, 1)$.

26.

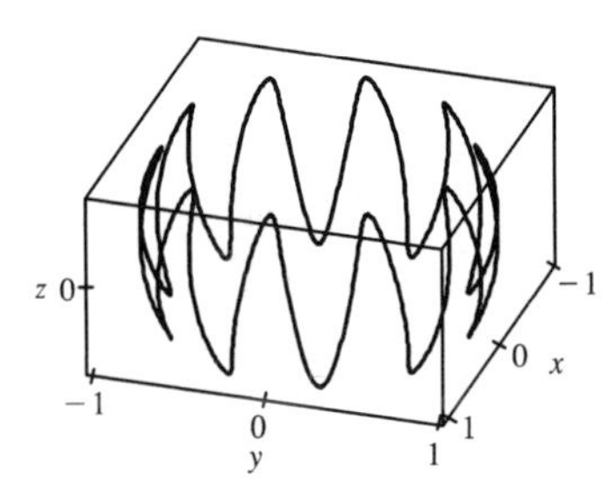

$x = \sqrt{1 - 0.25\cos^2 10t}\,\cos t,\ y = \sqrt{1 - 0.25\cos^2 10t}\,\sin t,\ z = 0.5\cos 10t$.

At any point on the graph,

$$\begin{aligned} x^2 + y^2 + z^2 &= (1 - 0.25\cos^2 10t)\cos^2 t \\ &\quad + (1 - 0.25\cos^2 10t)\sin^2 t + 0.25\cos^2 t \\ &= 1 - 0.25\cos^2 10t + 0.25\cos^2 10t = 1, \end{aligned}$$

so the graph lies on the sphere $x^2 + y^2 + z^2 = 1$, and since $z = 0.5\cos 10t$ the graph resembles a trigonometric curve with ten peaks projected onto the sphere. The graph is generated by $t \in [0, 2\pi]$.

27. If $t = -1$, then $x = 1,\ y = 4,\ z = 0$, so the curve passes through the point $(1, 4, 0)$. If $t = 3$, then $x = 9,\ y = -8,\ z = 28$, so the curve passes through the point $(9, -8, 28)$. For the point $(4, 7, -6)$ to be on the curve, we require $y = 1 - 3t = 7 \ \Rightarrow\ t = -2$. But then $z = 1 + (-2)^3 = -7 \neq -6$, so $(4, 7, -6)$ is not on the curve.

28. The projection of the curve C of intersection onto the xy-plane is the circle $x^2 + y^2 = 4, z = 0$. Then we can write $x = 2\cos t,\ y = 2\sin t,\ 0 \le t \le 2\pi$. Since C also lies on the surface $z = xy$, we have $z = xy = (2\cos t)(2\sin t) = 4\cos t\sin t$, or $2\sin(2t)$. Then parametric equations for C are $x = 2\cos t, y = 2\sin t$, $z = 2\sin(2t),\ 0 \le t \le 2\pi$, and the corresponding vector function is $\mathbf{r}(t) = 2\cos t\,\mathbf{i} + 2\sin t\,\mathbf{j} + 2\sin(2t)\,\mathbf{k}, 0 \le t \le 2\pi$.

29. Both equations are solved for z, so we can substitute to eliminate z: $\sqrt{x^2 + y^2} = 1 + y \ \Rightarrow\ x^2 + y^2 = 1 + 2y + y^2 \ \Rightarrow\ x^2 = 1 + 2y \ \Rightarrow\ y = \frac{1}{2}(x^2 - 1)$. We can form parametric equations for the curve C of intersection by choosing a parameter $x = t$, then $y = \frac{1}{2}(t^2 - 1)$ and $z = 1 + y = 1 + \frac{1}{2}(t^2 - 1) = \frac{1}{2}(t^2 + 1)$. Thus a vector function representing C is $\mathbf{r}(t) = t\,\mathbf{i} + \frac{1}{2}(t^2 - 1)\,\mathbf{j} + \frac{1}{2}(t^2 + 1)\,\mathbf{k}$.

30. The projection of the curve C of intersection onto the xy-plane is the parabola $y = x^2$, $z = 0$. Then we can choose the parameter $x = t \;\Rightarrow\; y = t^2$. Since C also lies on the surface $z = 4x^2 + y^2$, we have $z = 4x^2 + y^2 = 4t^2 + (t^2)^2$. Then parametric equations for C are $x = t,\; y = t^2,\; z = 4t^2 + t^4$, and the corresponding vector function is $\mathbf{r}(t) = t\,\mathbf{i} + t^2\,\mathbf{j} + (4t^2 + t^4)\,\mathbf{k}$.

31.

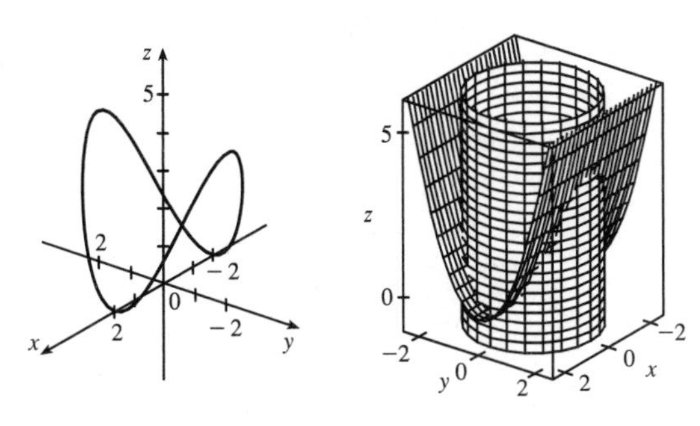

The projection of the curve C of intersection onto the xy-plane is the circle $x^2 + y^2 = 4$, $z = 0$. Then we can write $x = 2\cos t,\; y = 2\sin t,\; 0 \le t \le 2\pi$. Since C also lies on the surface $z = x^2$, we have $z = x^2 = (2\cos t)^2 = 4\cos^2 t$. Then parametric equations for C are $x = 2\cos t,\; y = 2\sin t,\; z = 4\cos^2 t,\; 0 \le t \le 2\pi$.

32.

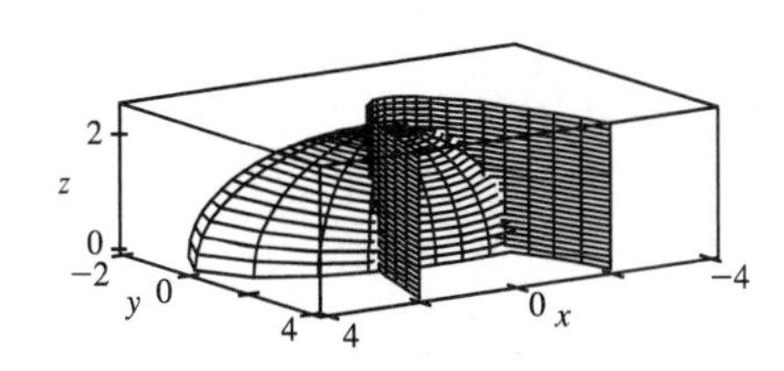

$x = t \;\Rightarrow\; y = t^2 \;\Rightarrow\; 4z^2 = 16 - x^2 - 4y^2 = 16 - t^2 - 4t^4 \;\Rightarrow\; z = \sqrt{4 - \left(\frac{1}{2}t\right)^2 - t^4}$.

Note that z is positive because the intersection is with the top half of the ellipsoid. Hence the curve is given by $x = t,\; y = t^2,\; z = \sqrt{4 - \frac{1}{4}t^2 - t^4}$.

33. (a), (c)

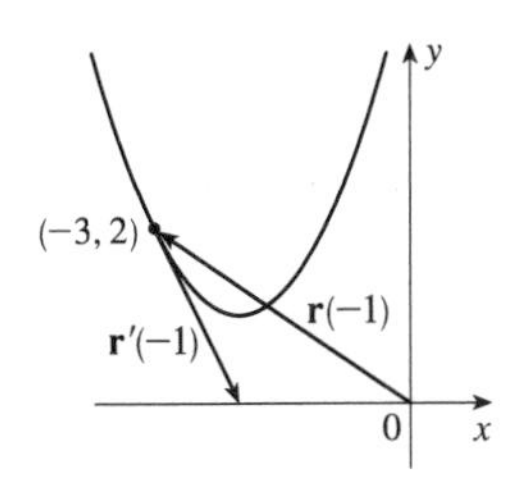

(b) $\mathbf{r}'(t) = \langle 1, 2t \rangle$

34. (a), (c)

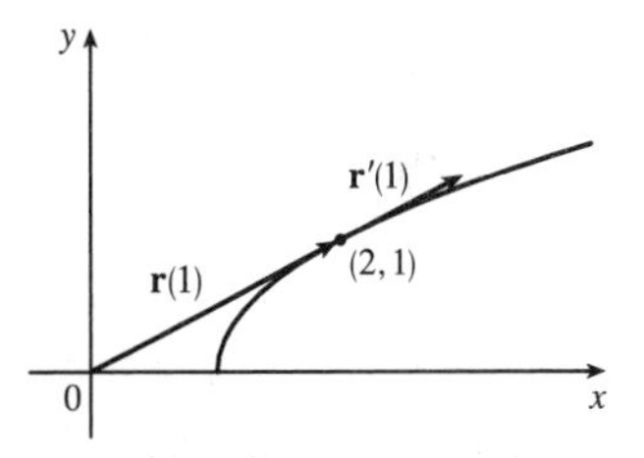

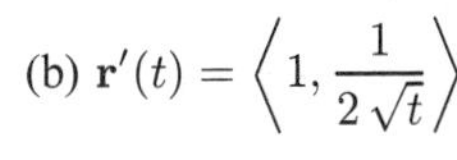

(b) $\mathbf{r}'(t) = \left\langle 1, \dfrac{1}{2\sqrt{t}} \right\rangle$

35. (a), (c)

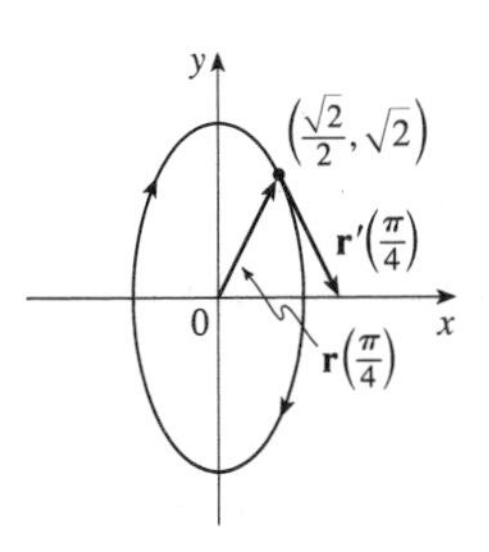

(b) $\mathbf{r}'(t) = \cos t\,\mathbf{i} - 2\sin t\,\mathbf{j}$

36. (a), (c)

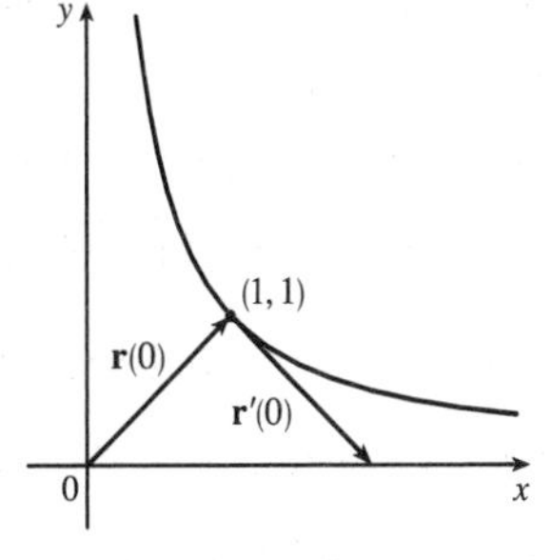

(b) $\mathbf{r}'(t) = e^t\,\mathbf{i} - e^{-t}\,\mathbf{j}$

37. (a), (c)

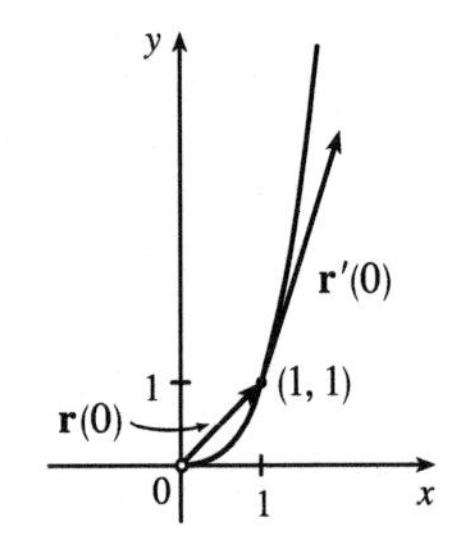

(b) $\mathbf{r}'(t) = e^t\,\mathbf{i} + 3e^{3t}\,\mathbf{j}$

38. (a), (c)

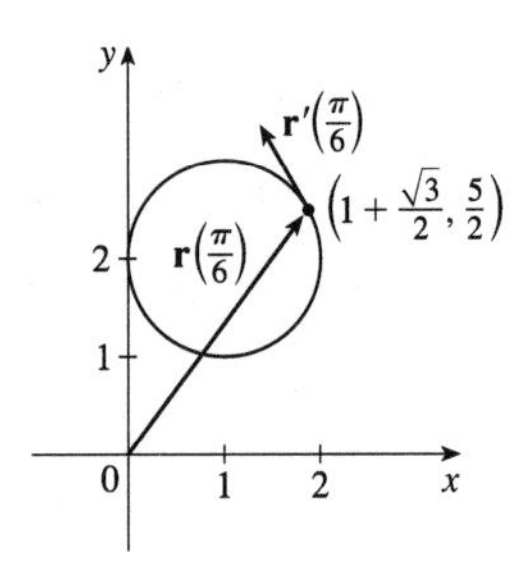

(b) $\mathbf{r}'(t) = -\sin t\,\mathbf{i} + \cos t\,\mathbf{j}$

39. $\mathbf{r}'(t) = \left\langle \dfrac{d}{dt}\left[t^2\right], \dfrac{d}{dt}\left[1-t\right], \dfrac{d}{dt}\left[\sqrt{t}\,\right] \right\rangle = \left\langle 2t, -1, \dfrac{1}{2\sqrt{t}} \right\rangle$

40. $\mathbf{r}(t) = \langle \cos 3t, t, \sin 3t\rangle \;\Rightarrow\; \mathbf{r}'(t) = \langle -3\sin 3t, 1, 3\cos 3t\rangle$

41. $\mathbf{r}(t) = e^{t^2}\,\mathbf{i} - \mathbf{j} + \ln(1+3t)\,\mathbf{k} \;\Rightarrow\; \mathbf{r}'(t) = 2te^{t^2}\,\mathbf{i} + \dfrac{3}{1+3t}\,\mathbf{k}$

42. $\mathbf{r}'(t) = [at(-3\sin 3t) + a\cos 3t]\,\mathbf{i} + b\cdot 3\sin^2 t\cos t\,\mathbf{j} + c\cdot 3\cos^2 t(-\sin t)\,\mathbf{k}$

$= (a\cos 3t - 3at\sin 3t)\,\mathbf{i} + 3b\sin^2 t\cos t\,\mathbf{j} - 3c\cos^2 t\sin t\,\mathbf{k}$

43. $\mathbf{r}'(t) = \mathbf{0} + \mathbf{b} + 2t\,\mathbf{c} = \mathbf{b} + 2t\,\mathbf{c}$ by Formulas 1 and 3 of Theorem 5.

44. To find $\mathbf{r}'(t)$, we first expand $\mathbf{r}(t) = t\,\mathbf{a}\times(\mathbf{b} + t\,\mathbf{c}) = t(\mathbf{a}\times\mathbf{b}) + t^2(\mathbf{a}\times\mathbf{c})$, so $\mathbf{r}'(t) = \mathbf{a}\times\mathbf{b} + 2t(\mathbf{a}\times\mathbf{c})$.

45. $\mathbf{r}'(t) = -\sin t\,\mathbf{i} + 3\,\mathbf{j} + 4\cos 2t\,\mathbf{k} \;\Rightarrow\; \mathbf{r}'(0) = 3\,\mathbf{j} + 4\,\mathbf{k}$. Thus

$$\mathbf{T}(0) = \frac{\mathbf{r}'(0)}{|\mathbf{r}'(0)|} = \frac{1}{\sqrt{0^2+3^2+4^2}}\,(3\,\mathbf{j} + 4\,\mathbf{k}) = \tfrac{1}{5}(3\,\mathbf{j} + 4\,\mathbf{k}) = \tfrac{3}{5}\,\mathbf{j} + \tfrac{4}{5}\,\mathbf{k}.$$

46. $\mathbf{r}'(t) = 2\cos t\,\mathbf{i} - 2\sin t\,\mathbf{j} + \sec^2 t\,\mathbf{k} \;\Rightarrow\; \mathbf{r}'\left(\frac{\pi}{4}\right) = \sqrt{2}\,\mathbf{i} - \sqrt{2}\,\mathbf{j} + 2\,\mathbf{k}$ and $\left|\mathbf{r}'\left(\frac{\pi}{4}\right)\right| = \sqrt{2+2+4} = 2\sqrt{2}$. Thus

$$\mathbf{T}\left(\tfrac{\pi}{4}\right) = \frac{\mathbf{r}'\left(\frac{\pi}{4}\right)}{\left|\mathbf{r}'\left(\frac{\pi}{4}\right)\right|} = \frac{1}{2\sqrt{2}}\left(\sqrt{2}\,\mathbf{i} - \sqrt{2}\,\mathbf{j} + 2\,\mathbf{k}\right) = \frac{1}{2}\,\mathbf{i} - \frac{1}{2}\,\mathbf{j} + \frac{1}{\sqrt{2}}\,\mathbf{k}.$$

47. $\mathbf{r}(t) = \langle t, t^2, t^3\rangle \;\Rightarrow\; \mathbf{r}'(t) = \langle 1, 2t, 3t^2\rangle$. Then $\mathbf{r}'(1) = \langle 1, 2, 3\rangle$ and $|\mathbf{r}'(1)| = \sqrt{1^2+2^2+3^2} = \sqrt{14}$, so

$\mathbf{T}(1) = \dfrac{\mathbf{r}'(1)}{|\mathbf{r}'(1)|} = \frac{1}{\sqrt{14}}\langle 1, 2, 3\rangle = \left\langle \frac{1}{\sqrt{14}}, \frac{2}{\sqrt{14}}, \frac{3}{\sqrt{14}} \right\rangle$. $\mathbf{r}''(t) = \langle 0, 2, 6t\rangle$, so

$$\mathbf{r}'(t)\times\mathbf{r}''(t) = \begin{vmatrix} \mathbf{i} & \mathbf{j} & \mathbf{k} \\ 1 & 2t & 3t^2 \\ 0 & 2 & 6t \end{vmatrix} = \begin{vmatrix} 2t & 3t^2 \\ 2 & 6t \end{vmatrix}\mathbf{i} - \begin{vmatrix} 1 & 3t^2 \\ 0 & 6t \end{vmatrix}\mathbf{j} + \begin{vmatrix} 1 & 2t \\ 0 & 2 \end{vmatrix}\mathbf{k}$$

$$= (12t^2 - 6t^2)\,\mathbf{i} - (6t - 0)\,\mathbf{j} + (2-0)\,\mathbf{k} = \left\langle 6t^2, -6t, 2\right\rangle.$$

48. $\mathbf{r}(t) = \left\langle e^{2t}, e^{-2t}, te^{2t}\right\rangle \;\Rightarrow\; \mathbf{r}'(t) = \left\langle 2e^{2t}, -2e^{-2t}, (2t+1)e^{2t}\right\rangle \;\Rightarrow\; \mathbf{r}'(0) = \left\langle 2e^0, -2e^0, (0+1)e^0\right\rangle = \langle 2, -2, 1\rangle$

and $|\mathbf{r}'(0)| = \sqrt{2^2+(-2)^2+1^2} = 3$. Then $\mathbf{T}(0) = \dfrac{\mathbf{r}'(0)}{|\mathbf{r}'(0)|} = \frac{1}{3}\langle 2, -2, 1\rangle = \left\langle \frac{2}{3}, -\frac{2}{3}, \frac{1}{3}\right\rangle$.

$\mathbf{r}''(t) = \left\langle 4e^{2t}, 4e^{-2t}, (4t+4)e^{2t}\right\rangle \;\Rightarrow\; \mathbf{r}''(0) = \left\langle 4e^0, 4e^0, (0+4)e^0\right\rangle = \langle 4, 4, 4\rangle$.

$$\mathbf{r}'(t)\cdot\mathbf{r}''(t) = \left\langle 2e^{2t}, -2e^{-2t}, (2t+1)e^{2t}\right\rangle \cdot \left\langle 4e^{2t}, 4e^{-2t}, (4t+4)e^{2t}\right\rangle$$

$$= (2e^{2t})(4e^{2t}) + (-2e^{-2t})(4e^{-2t}) + ((2t+1)e^{2t})((4t+4)e^{2t})$$

$$= 8e^{4t} - 8e^{-4t} + (8t^2 + 12t + 4)e^{4t} = (8t^2 + 12t + 12)e^{4t} - 8e^{-4t}$$

49. The vector equation for the curve is $\mathbf{r}(t) = \langle t^5, t^4, t^3 \rangle$, so $\mathbf{r}'(t) = \langle 5t^4, 4t^3, 3t^2 \rangle$. The point $(1,1,1)$ corresponds to $t = 1$, so the tangent vector there is $\mathbf{r}'(1) = \langle 5,4,3 \rangle$. Thus, the tangent line goes through the point $(1,1,1)$ and is parallel to the vector $\langle 5,4,3 \rangle$. Parametric equations are $x = 1 + 5t$, $y = 1 + 4t$, $z = 1 + 3t$.

50. The vector equation for the curve is $\mathbf{r}(t) = \langle t^2 - 1, t^2 + 1, t + 1 \rangle$, so $\mathbf{r}'(t) = \langle 2t, 2t, 1 \rangle$. The point $(-1,1,1)$ corresponds to $t = 0$, so the tangent vector there is $\mathbf{r}'(0) = \langle 0,0,1 \rangle$. Thus, the tangent line is parallel to the vector $\langle 0,0,1 \rangle$ and parametric equations are $x = -1 + 0 \cdot t = -1$, $y = 1 + 0 \cdot t = 1$, $z = 1 + 1 \cdot t = 1 + t$.

51. The vector equation for the curve is $\mathbf{r}(t) = \langle e^{-t}\cos t, e^{-t}\sin t, e^{-t} \rangle$, so

$$\begin{aligned}\mathbf{r}'(t) &= \langle e^{-t}(-\sin t) + (\cos t)(-e^{-t}), e^{-t}\cos t + (\sin t)(-e^{-t}), (-e^{-t}) \rangle \\ &= \langle -e^{-t}(\cos t + \sin t), e^{-t}(\cos t - \sin t), -e^{-t} \rangle\end{aligned}$$

The point $(1,0,1)$ corresponds to $t = 0$, so the tangent vector there is

$\mathbf{r}'(0) = \langle -e^0(\cos 0 + \sin 0), e^0(\cos 0 - \sin 0), -e^0 \rangle = \langle -1, 1, -1 \rangle$. Thus, the tangent line is parallel to the vector $\langle -1,1,-1 \rangle$ and parametric equations are $x = 1 + (-1)t = 1 - t$, $y = 0 + 1 \cdot t = t$, $z = 1 + (-1)t = 1 - t$.

52. $\mathbf{r}(t) = \langle \ln t, 2\sqrt{t}, t^2 \rangle$, $\mathbf{r}'(t) = \langle 1/t, 1/\sqrt{t}, 2t \rangle$. At $(0,2,1)$, $t = 1$ and $\mathbf{r}'(1) = \langle 1,1,2 \rangle$. Thus, parametric equations of the tangent line are $x = t$, $y = 2 + t$, $z = 1 + 2t$.

53. (a) $\mathbf{r}(t) = \langle t^3, t^4, t^5 \rangle \;\Rightarrow\; \mathbf{r}'(t) = \langle 3t^2, 4t^3, 5t^4 \rangle$, and since $\mathbf{r}'(0) = \langle 0,0,0 \rangle = \mathbf{0}$, the curve is not smooth.

(b) $\mathbf{r}(t) = \langle t^3 + t, t^4, t^5 \rangle \;\Rightarrow\; \mathbf{r}'(t) = \langle 3t^2 + 1, 4t^3, 5t^4 \rangle$. $\mathbf{r}'(t)$ is continuous since its component functions are continuous. Also, $\mathbf{r}'(t) \neq \mathbf{0}$, as the y- and z-components are 0 only for $t = 0$, but $\mathbf{r}'(0) = \langle 1,0,0 \rangle \neq \mathbf{0}$. Thus, the curve is smooth.

(c) $\mathbf{r}(t) = \langle \cos^3 t, \sin^3 t \rangle \;\Rightarrow\; \mathbf{r}'(t) = \langle -3\cos^2 t \sin t, 3\sin^2 t \cos t \rangle$. Since $\mathbf{r}'(0) = \langle -3\cos^2 0 \sin 0, 3\sin^2 0 \cos 0 \rangle = \langle 0,0 \rangle = \mathbf{0}$, the curve is not smooth.

54. (a) The tangent line at $t = 0$ is the line through the point with position vector $\mathbf{r}(0) = \langle \sin 0, 2\sin 0, \cos 0 \rangle = \langle 0,0,1 \rangle$, and in the direction of the tangent vector, $\mathbf{r}'(0) = \langle \pi\cos 0, 2\pi\cos 0, -\pi\sin 0 \rangle = \langle \pi, 2\pi, 0 \rangle$. So an equation of the line is $\langle x,y,z \rangle = \mathbf{r}(0) + u\,\mathbf{r}'(0) = \langle 0 + \pi u, 0 + 2\pi u, 1 \rangle = \langle \pi u, 2\pi u, 1 \rangle$.

$$\mathbf{r}\left(\tfrac{1}{2}\right) = \left\langle \sin\tfrac{\pi}{2}, 2\sin\tfrac{\pi}{2}, \cos\tfrac{\pi}{2} \right\rangle = \langle 1,2,0 \rangle,$$

$$\mathbf{r}'\left(\tfrac{1}{2}\right) = \left\langle \pi\cos\tfrac{\pi}{2}, 2\pi\cos\tfrac{\pi}{2}, -\pi\sin\tfrac{\pi}{2} \right\rangle = \langle 0,0,-\pi \rangle.$$

So the equation of the second line is $\langle x,y,z \rangle = \langle 1,2,0 \rangle + v\,\langle 0,0,-\pi \rangle = \langle 1,2,-\pi v \rangle$.

The lines intersect where $\langle \pi u, 2\pi u, 1 \rangle = \langle 1,2,-\pi v \rangle$, so the point of intersection is $(1,2,1)$.

(b)

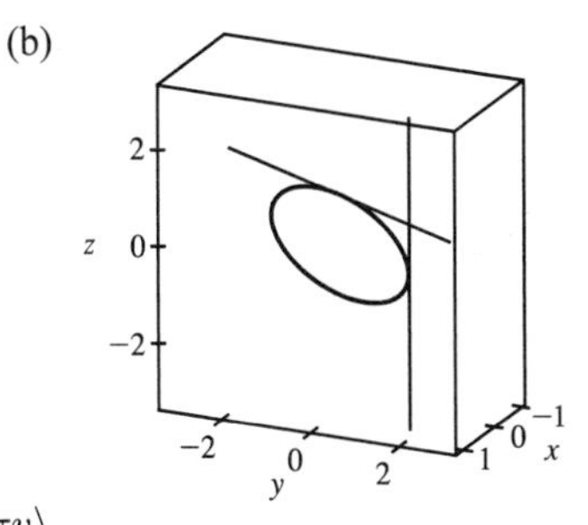

55. The angle of intersection of the two curves is the angle between the two tangent vectors to the curves at the point of intersection. Since $\mathbf{r}_1'(t) = \langle 1, 2t, 3t^2 \rangle$ and $t = 0$ at $(0,0,0)$, $\mathbf{r}_1'(0) = \langle 1,0,0 \rangle$ is a tangent vector to $\mathbf{r}_1$ at $(0,0,0)$. Similarly, $\mathbf{r}_2'(t) = \langle \cos t, 2\cos 2t, 1 \rangle$ and since $\mathbf{r}_2(0) = \langle 0,0,0 \rangle$, $\mathbf{r}_2'(0) = \langle 1,2,1 \rangle$ is a tangent vector to $\mathbf{r}_2$ at $(0,0,0)$. If θ is the angle between these two tangent vectors, then $\cos\theta = \frac{1}{\sqrt{1}\sqrt{6}}\langle 1,0,0 \rangle \cdot \langle 1,2,1 \rangle = \frac{1}{\sqrt{6}}$ and $\theta = \cos^{-1}\left(\frac{1}{\sqrt{6}}\right) \approx 66^\circ$.

56. To find the point of intersection, we must find the values of t and s which satisfy the following three equations simultaneously: $t = 3 - s$, $1 - t = s - 2$, $3 + t^2 = s^2$. Solving the last two equations gives $t = 1$, $s = 2$ (check these in the first equation). Thus the point of intersection is $(1, 0, 4)$. To find the angle θ of intersection, we proceed as in Exercise 55. The tangent vectors to the respective curves at $(1, 0, 4)$ are $\mathbf{r}_1'(1) = \langle 1, -1, 2\rangle$ and $\mathbf{r}_2'(2) = \langle -1, 1, 4\rangle$. So

$\cos\theta = \frac{1}{\sqrt{6}\sqrt{18}}(-1 - 1 + 8) = \frac{6}{6\sqrt{3}} = \frac{1}{\sqrt{3}}$ and $\theta = \cos^{-1}\left(\frac{1}{\sqrt{3}}\right) \approx 55°$.

Note: In Exercise 55, the curves intersect when the value of both parameters is zero. However, as seen in this exercise, it is not necessary for the parameters to be of equal value at the point of intersection.

57. $\int_0^1 (16t^3\,\mathbf{i} - 9t^2\,\mathbf{j} + 25t^4\,\mathbf{k})\,dt = \left(\int_0^1 16t^3\,dt\right)\mathbf{i} - \left(\int_0^1 9t^2\,dt\right)\mathbf{j} + \left(\int_0^1 25t^4\,dt\right)\mathbf{k}$

$= \left[4t^4\right]_0^1\mathbf{i} - \left[3t^3\right]_0^1\mathbf{j} + \left[5t^5\right]_0^1\mathbf{k} = 4\,\mathbf{i} - 3\,\mathbf{j} + 5\,\mathbf{k}$

58. $\displaystyle\int_0^1 \left(\frac{4}{1+t^2}\,\mathbf{j} + \frac{2t}{1+t^2}\,\mathbf{k}\right)dt = \left[4\tan^{-1}t\,\mathbf{j} + \ln(1+t^2)\,\mathbf{k}\right]_0^1 = \left[4\tan^{-1}1\,\mathbf{j} + \ln 2\,\mathbf{k}\right] - \left[4\tan^{-1}0\,\mathbf{j} + \ln 1\,\mathbf{k}\right]$

$= 4\left(\frac{\pi}{4}\right)\mathbf{j} + \ln 2\,\mathbf{k} - 0\,\mathbf{j} - 0\,\mathbf{k} = \pi\,\mathbf{j} + \ln 2\,\mathbf{k}$

59. $\int_0^{\pi/2}(3\sin^2 t\,\cos t\,\mathbf{i} + 3\sin t\,\cos^2 t\,\mathbf{j} + 2\sin t\,\cos t\,\mathbf{k})\,dt$

$= \left(\int_0^{\pi/2} 3\sin^2 t\,\cos t\,dt\right)\mathbf{i} + \left(\int_0^{\pi/2} 3\sin t\,\cos^2 t\,dt\right)\mathbf{j} + \left(\int_0^{\pi/2} 2\sin t\,\cos t\,dt\right)\mathbf{k}$

$= \left[\sin^3 t\right]_0^{\pi/2}\mathbf{i} + \left[-\cos^3 t\right]_0^{\pi/2}\mathbf{j} + \left[\sin^2 t\right]_0^{\pi/2}\mathbf{k} = (1-0)\,\mathbf{i} + (0+1)\,\mathbf{j} + (1-0)\,\mathbf{k} = \mathbf{i} + \mathbf{j} + \mathbf{k}$

60. $\int_1^2 (t^2\,\mathbf{i} + t\sqrt{t-1}\,\mathbf{j} + t\sin\pi t\,\mathbf{k})\,dt = \left[\frac{1}{3}t^3\,\mathbf{i} + \left(\frac{2}{5}(t-1)^{5/2} + \frac{2}{3}(t-1)^{3/2}\right)\mathbf{j}\right]_1^2 + \left(\left[-\frac{1}{\pi}t\cos\pi t\right]_1^2 + \int_1^2 \frac{1}{\pi}\cos\pi t\,dt\right)\mathbf{k}$

$= \frac{7}{3}\,\mathbf{i} + \frac{16}{15}\,\mathbf{j} + \left(-\frac{3}{\pi} + \left[\frac{1}{\pi^2}\sin\pi t\right]_1^2\right)\mathbf{k} = \frac{7}{3}\,\mathbf{i} + \frac{16}{15}\,\mathbf{j} - \frac{3}{\pi}\,\mathbf{k}$

61. $\int (e^t\,\mathbf{i} + 2t\,\mathbf{j} + \ln t\,\mathbf{k})\,dt = \left(\int e^t\,dt\right)\mathbf{i} + \left(\int 2t\,dt\right)\mathbf{j} + \left(\int \ln t\,dt\right)\mathbf{k}$

$= e^t\,\mathbf{i} + t^2\,\mathbf{j} + (t\ln t - t)\,\mathbf{k} + \mathbf{C}$, where $\mathbf{C}$ is a vector constant of integration.

62. $\int(\cos\pi t\,\mathbf{i} + \sin\pi t\,\mathbf{j} + t\,\mathbf{k})\,dt = \left(\int\cos\pi t\,dt\right)\mathbf{i} + \left(\int\sin\pi t\,dt\right)\mathbf{j} + \left(\int t\,dt\right)\mathbf{k} = \frac{1}{\pi}\sin\pi t\,\mathbf{i} - \frac{1}{\pi}\cos\pi t\,\mathbf{j} + \frac{1}{2}t^2\,\mathbf{k} + \mathbf{C}$

63. $\mathbf{r}'(t) = 2t\,\mathbf{i} + 3t^2\,\mathbf{j} + \sqrt{t}\,\mathbf{k} \;\Rightarrow\; \mathbf{r}(t) = t^2\,\mathbf{i} + t^3\,\mathbf{j} + \frac{2}{3}t^{3/2}\,\mathbf{k} + \mathbf{C}$, where $\mathbf{C}$ is a constant vector.

But $\mathbf{i} + \mathbf{j} = \mathbf{r}(1) = \mathbf{i} + \mathbf{j} + \frac{2}{3}\mathbf{k} + \mathbf{C}$. Thus $\mathbf{C} = -\frac{2}{3}\mathbf{k}$ and $\mathbf{r}(t) = t^2\,\mathbf{i} + t^3\,\mathbf{j} + \left(\frac{2}{3}t^{3/2} - \frac{2}{3}\right)\mathbf{k}$.

64. $\mathbf{r}'(t) = t\,\mathbf{i} + e^t\,\mathbf{j} + te^t\,\mathbf{k} \;\Rightarrow\; \mathbf{r}(t) = \frac{1}{2}t^2\,\mathbf{i} + e^t\,\mathbf{j} + \left(te^t - e^t\right)\mathbf{k} + \mathbf{C}$. But $\mathbf{i} + \mathbf{j} + \mathbf{k} = \mathbf{r}(0) = \mathbf{j} - \mathbf{k} + \mathbf{C}$. Thus $\mathbf{C} = \mathbf{i} + 2\,\mathbf{k}$ and $\mathbf{r}(t) = \left(\frac{1}{2}t^2 + 1\right)\mathbf{i} + e^t\,\mathbf{j} + (te^t - e^t + 2)\,\mathbf{k}$.

65. For the particles to collide, we require $\mathbf{r}_1(t) = \mathbf{r}_2(t) \;\Leftrightarrow\; \left\langle t^2, 7t - 12, t^2\right\rangle = \left\langle 4t - 3, t^2, 5t - 6\right\rangle$. Equating components gives $t^2 = 4t - 3$, $7t - 12 = t^2$, and $t^2 = 5t - 6$. From the first equation, $t^2 - 4t + 3 = 0 \;\Leftrightarrow\; (t-3)(t-1) = 0$ so $t = 1$ or $t = 3$. $t = 1$ does not satisfy the other two equations, but $t = 3$ does. The particles collide when $t = 3$, at the point $(9, 9, 9)$.

66. The particles collide provided $\mathbf{r}_1(t) = \mathbf{r}_2(t) \;\Leftrightarrow\; \left\langle t, t^2, t^3\right\rangle = \langle 1 + 2t, 1 + 6t, 1 + 14t\rangle$. Equating components gives $t = 1 + 2t$, $t^2 = 1 + 6t$, and $t^3 = 1 + 14t$. The first equation gives $t = -1$, but this does not satisfy the other equations, so the particles do not collide. For the paths to intersect, we need to find a value for t and a value for s where $\mathbf{r}_1(t) = \mathbf{r}_2(s) \;\Leftrightarrow\; \left\langle t, t^2, t^3\right\rangle = \langle 1 + 2s, 1 + 6s, 1 + 14s\rangle$. Equating components, $t = 1 + 2s$, $t^2 = 1 + 6s$, and $t^3 = 1 + 14s$. Substituting the first equation into the second gives $(1+2s)^2 = 1+6s \Rightarrow 4s^2 - 2s = 0 \Rightarrow 2s(2s-1) = 0 \Rightarrow s = 0$ or $s = \frac{1}{2}$. From the first equation, $s = 0 \Rightarrow t = 1$ and $s = \frac{1}{2} \Rightarrow t = 2$. Checking, we see that both pairs of values satisfy the third equation. Thus the paths intersect twice, at the point $(1,1,1)$ when $s = 0$ and $t = 1$, and at $(2,4,8)$ when $s = \frac{1}{2}$ and $t = 2$.

67. (a) $\lim_{t\to a} \mathbf{u}(t) + \lim_{t\to a} \mathbf{v}(t) = \left\langle \lim_{t\to a} u_1(t), \lim_{t\to a} u_2(t), \lim_{t\to a} u_3(t)\right\rangle + \left\langle \lim_{t\to a} v_1(t), \lim_{t\to a} v_2(t), \lim_{t\to a} v_3(t)\right\rangle$ and the limits of these component functions must each exist since the vector functions both possess limits as $t \to a$. Then adding the two vectors and using the addition property of limits for real-valued functions, we have that

$$\begin{aligned}\lim_{t\to a} \mathbf{u}(t) + \lim_{t\to a} \mathbf{v}(t) &= \left\langle \lim_{t\to a} u_1(t) + \lim_{t\to a} v_1(t), \lim_{t\to a} u_2(t) + \lim_{t\to a} v_2(t), \lim_{t\to a} u_3(t) + \lim_{t\to a} v_3(t)\right\rangle \\ &= \left\langle \lim_{t\to a}[u_1(t)+v_1(t)], \lim_{t\to a}[u_2(t)+v_2(t)], \lim_{t\to a}[u_3(t)+v_3(t)]\right\rangle \\ &= \lim_{t\to a}\langle u_1(t)+v_1(t), u_2(t)+v_2(t), u_3(t)+v_3(t)\rangle \qquad \text{[using (1) backward]} \\ &= \lim_{t\to a}[\mathbf{u}(t)+\mathbf{v}(t)]\end{aligned}$$

(b)
$$\begin{aligned}\lim_{t\to a} c\mathbf{u}(t) &= \lim_{t\to a}\langle cu_1(t), cu_2(t), cu_3(t)\rangle = \left\langle \lim_{t\to a} cu_1(t), \lim_{t\to a} cu_2(t), \lim_{t\to a} cu_3(t)\right\rangle \\ &= \left\langle c\lim_{t\to a} u_1(t), c\lim_{t\to a} u_2(t), c\lim_{t\to a} u_3(t)\right\rangle = c\left\langle \lim_{t\to a} u_1(t), \lim_{t\to a} u_2(t), \lim_{t\to a} u_3(t)\right\rangle \\ &= c\lim_{t\to a}\langle u_1(t), u_2(t), u_3(t)\rangle = c\lim_{t\to a}\mathbf{u}(t)\end{aligned}$$

(c)
$$\begin{aligned}\lim_{t\to a} \mathbf{u}(t)\cdot\lim_{t\to a} \mathbf{v}(t) &= \left\langle \lim_{t\to a} u_1(t), \lim_{t\to a} u_2(t), \lim_{t\to a} u_3(t)\right\rangle \cdot \left\langle \lim_{t\to a} v_1(t), \lim_{t\to a} v_2(t), \lim_{t\to a} v_3(t)\right\rangle \\ &= \left[\lim_{t\to a} u_1(t)\right]\left[\lim_{t\to a} v_1(t)\right] + \left[\lim_{t\to a} u_2(t)\right]\left[\lim_{t\to a} v_2(t)\right] + \left[\lim_{t\to a} u_3(t)\right]\left[\lim_{t\to a} v_3(t)\right] \\ &= \lim_{t\to a} u_1(t)v_1(t) + \lim_{t\to a} u_2(t)v_2(t) + \lim_{t\to a} u_3(t)v_3(t) \\ &= \lim_{t\to a}[u_1(t)v_1(t) + u_2(t)v_2(t) + u_3(t)v_3(t)] = \lim_{t\to a}[\mathbf{u}(t)\cdot\mathbf{v}(t)]\end{aligned}$$

(d)
$$\begin{aligned}\lim_{t\to a} \mathbf{u}(t)\times\lim_{t\to a} \mathbf{v}(t) &= \left\langle \lim_{t\to a} u_1(t), \lim_{t\to a} u_2(t), \lim_{t\to a} u_3(t)\right\rangle \times \left\langle \lim_{t\to a} v_1(t), \lim_{t\to a} v_2(t), \lim_{t\to a} v_3(t)\right\rangle \\ &= \left\langle \left[\lim_{t\to a} u_2(t)\right]\left[\lim_{t\to a} v_3(t)\right] - \left[\lim_{t\to a} u_3(t)\right]\left[\lim_{t\to a} v_2(t)\right],\right. \\ &\qquad \left[\lim_{t\to a} u_3(t)\right]\left[\lim_{t\to a} v_1(t)\right] - \left[\lim_{t\to a} u_1(t)\right]\left[\lim_{t\to a} v_3(t)\right], \\ &\qquad \left.\left[\lim_{t\to a} u_1(t)\right]\left[\lim_{t\to a} v_2(t)\right] - \left[\lim_{t\to a} u_2(t)\right]\left[\lim_{t\to a} v_1(t)\right]\right\rangle \\ &= \left\langle \lim_{t\to a}[u_2(t)v_3(t) - u_3(t)v_2(t)], \lim_{t\to a}[u_3(t)v_1(t) - u_1(t)v_3(t)],\right. \\ &\qquad \left.\lim_{t\to a}[u_1(t)v_2(t) - u_2(t)v_1(t)]\right\rangle \\ &= \lim_{t\to a}\langle u_2(t)v_3(t) - u_3(t)v_2(t), u_3(t)v_1(t) - u_1(t)v_3(t), u_1(t)v_2(t) - u_2(t)v_1(t)\rangle \\ &= \lim_{t\to a}[\mathbf{u}(t)\times\mathbf{v}(t)]\end{aligned}$$

68. Let $\mathbf{r}(t) = \langle f(t), g(t), h(t)\rangle$ and $\mathbf{b} = \langle b_1, b_2, b_3\rangle$. If $\lim_{t\to a}\mathbf{r}(t) = \mathbf{b}$, then $\lim_{t\to a}\mathbf{r}(t)$ exists, so by (1),

$\mathbf{b} = \lim_{t\to a}\mathbf{r}(t) = \left\langle \lim_{t\to a} f(t), \lim_{t\to a} g(t), \lim_{t\to a} h(t)\right\rangle$. By the definition of equal vectors we have $\lim_{t\to a} f(t) = b_1$, $\lim_{t\to a} g(t) = b_2$ and $\lim_{t\to a} h(t) = b_3$. But these are limits of real-valued functions, so by the definition of limits, for every $\varepsilon > 0$ there exists

$\delta_1 > 0$, $\delta_2 > 0$, $\delta_3 > 0$ so $|f(t) - b_1| < \varepsilon/3$ whenever $0 < |t - a| < \delta_1$, $|g(t) - b_2| < \varepsilon/3$ whenever $0 < |t - a| < \delta_2$, and $|h(t) - b_3| < \varepsilon/3$ whenever $0 < |t - a| < \delta_3$. Letting $\delta =$ minimum of $\{\delta_1, \delta_2, \delta_3\}$, we have $|f(t) - b_1| + |g(t) - b_2| + |h(t) - b_3| < \varepsilon/3 + \varepsilon/3 + \varepsilon/3 = \varepsilon$ whenever $0 < |t - a| < \delta$. But

$$\begin{aligned} |\mathbf{r}(t) - \mathbf{b}| &= |\langle f(t) - b_1, g(t) - b_2, h(t) - b_3 \rangle| = \sqrt{(f(t) - b_1)^2 + (g(t) - b_2)^2 + (h(t) - b_3)^2} \\ &\leq \sqrt{[f(t) - b_1]^2} + \sqrt{[g(t) - b_2]^2} + \sqrt{[h(t) - b_3]^2} = |f(t) - b_1| + |g(t) - b_2| + |h(t) - b_3| \end{aligned}$$

Thus for every $\varepsilon > 0$ there exists $\delta > 0$ such that $|\mathbf{r}(t) - b| \leq |f(t) - b_1| + |g(t) - b_2| + |h(t) - b_3| < \varepsilon$ whenever $0 < |t - a| < \delta$. Conversely, if for every $\varepsilon > 0$, there exists $\delta > 0$ such that $|\mathbf{r}(t) - \mathbf{b}| < \varepsilon$ whenever $0 < |t - a| < \delta$, then $|\langle f(t) - b_1, g(t) - b_2, h(t) - b_3 \rangle| < \varepsilon \;\Leftrightarrow\; \sqrt{[f(t) - b_1]^2 + [g(t) - b_2]^2 + [h(t) - b_3]^2} < \varepsilon \;\Leftrightarrow$ $[f(t) - b_1]^2 + [g(t) - b_2]^2 + [h(t) - b_3]^2 < \varepsilon^2$ whenever $0 < |t - a| < \delta$. But each term on the left side of this inequality is positive so $[f(t) - b_1]^2 < \varepsilon^2$, $[g(t) - b_2]^2 < \varepsilon^2$ and $[h(t) - b_3]^2 < \varepsilon^2$ whenever $0 < |t - a| < \delta$, or taking the square root of both sides in each of the above we have $|f(t) - b_1| < \varepsilon$, $|g(t) - b_2| < \varepsilon$ and $|h(t) - b_3| < \varepsilon$ whenever $0 < |t - a| < \delta$. And by definition of limits of real-valued functions we have $\lim\limits_{t \to a} f(t) = b_1$, $\lim\limits_{t \to a} g(t) = b_2$ and $\lim\limits_{t \to a} h(t) = b_3$. But by (1), $\lim\limits_{t \to a} \mathbf{r}(t) = \left\langle \lim\limits_{t \to a} f(t), \lim\limits_{t \to a} g(t), \lim\limits_{t \to a} h(t) \right\rangle$, so $\lim\limits_{t \to a} \mathbf{r}(t) = \langle b_1, b_2, b_3 \rangle = \mathbf{b}$.

For Exercises 69–72, let $\mathbf{u}(t) = \langle u_1(t), u_2(t), u_3(t) \rangle$ and $\mathbf{v}(t) = \langle v_1(t), v_2(t), v_3(t) \rangle$. In each of these exercises, the procedure is to apply Theorem 3 so that the corresponding properties of derivatives of real-valued functions can be used.

69. $$\begin{aligned} \frac{d}{dt}[\mathbf{u}(t) + \mathbf{v}(t)] &= \frac{d}{dt}\langle u_1(t) + v_1(t), u_2(t) + v_2(t), u_3(t) + v_3(t) \rangle \\ &= \left\langle \frac{d}{dt}[u_1(t) + v_1(t)], \frac{d}{dt}[u_2(t) + v_2(t)], \frac{d}{dt}[u_3(t) + v_3(t)] \right\rangle \\ &= \langle u_1'(t) + v_1'(t), u_2'(t) + v_2'(t), u_3'(t) + v_3'(t) \rangle \\ &= \langle u_1'(t), u_2'(t), u_3'(t) \rangle + \langle v_1'(t), v_2'(t), v_3'(t) \rangle = \mathbf{u}'(t) + \mathbf{v}'(t) \end{aligned}$$

70. $$\begin{aligned} \frac{d}{dt}[f(t)\,\mathbf{u}(t)] &= \frac{d}{dt}\langle f(t)u_1(t), f(t)u_2(t), f(t)u_3(t) \rangle \\ &= \left\langle \frac{d}{dt}[f(t)u_1(t)], \frac{d}{dt}[f(t)u_2(t)], \frac{d}{dt}[f(t)u_3(t)] \right\rangle \\ &= \langle f'(t)u_1(t) + f(t)u_1'(t), f'(t)u_2(t) + f(t)u_2'(t), f'(t)u_3(t) + f(t)u_3'(t) \rangle \\ &= f'(t)\,\langle u_1(t), u_2(t), u_3(t) \rangle + f(t)\,\langle u_1'(t), u_2'(t), u_3'(t) \rangle = f'(t)\,\mathbf{u}(t) + f(t)\,\mathbf{u}'(t) \end{aligned}$$

71. $$\begin{aligned} \frac{d}{dt}[\mathbf{u}(t) \times \mathbf{v}(t)] &= \frac{d}{dt}\langle u_2(t)v_3(t) - u_3(t)v_2(t), u_3(t)v_1(t) - u_1(t)v_3(t), u_1(t)v_2(t) - u_2(t)v_1(t) \rangle \\ &= \langle u_2'v_3(t) + u_2(t)v_3'(t) - u_3'(t)v_2(t) - u_3(t)v_2'(t), \\ &\qquad u_3'(t)v_1(t) + u_3(t)v_1'(t) - u_1'(t)v_3(t) - u_1(t)v_3'(t), \\ &\qquad\quad u_1'(t)v_2(t) + u_1(t)v_2'(t) - u_2'(t)v_1(t) - u_2(t)v_1'(t) \rangle \\ &= \langle u_2'(t)v_3(t) - u_3'(t)v_2(t), u_3'(t)v_1(t) - u_1'(t)v_3(t), u_1'(t)v_2(t) - u_2'(t)v_1(t) \rangle \\ &\qquad + \langle u_2(t)v_3'(t) - u_3(t)v_2'(t), u_3(t)v_1'(t) - u_1(t)v_3'(t), u_1(t)v_2'(t) - u_2(t)v_1'(t) \rangle \\ &= \mathbf{u}'(t) \times \mathbf{v}(t) + \mathbf{u}(t) \times \mathbf{v}'(t) \end{aligned}$$

Alternate solution: Let $\mathbf{r}(t) = \mathbf{u}(t) \times \mathbf{v}(t)$. Then

$$\begin{aligned}\mathbf{r}(t+h) - \mathbf{r}(t) &= [\mathbf{u}(t+h) \times \mathbf{v}(t+h)] - [\mathbf{u}(t) \times \mathbf{v}(t)] \\ &= [\mathbf{u}(t+h) \times \mathbf{v}(t+h)] - [\mathbf{u}(t) \times \mathbf{v}(t)] + [\mathbf{u}(t+h) \times \mathbf{v}(t)] - [\mathbf{u}(t+h) \times \mathbf{v}(t)] \\ &= \mathbf{u}(t+h) \times [\mathbf{v}(t+h) - \mathbf{v}(t)] + [\mathbf{u}(t+h) - \mathbf{u}(t)] \times \mathbf{v}(t)\end{aligned}$$

(Be careful of the order of the cross product.) Dividing through by h and taking the limit as $h \to 0$ we have

$$\mathbf{r}'(t) = \lim_{h\to 0} \frac{\mathbf{u}(t+h) \times [\mathbf{v}(t+h) - \mathbf{v}(t)]}{h} + \lim_{h\to 0} \frac{[\mathbf{u}(t+h) - \mathbf{u}(t)] \times \mathbf{v}(t)}{h} = \mathbf{u}(t) \times \mathbf{v}'(t) + \mathbf{u}'(t) \times \mathbf{v}(t)$$

by Exercise 10.7.67(a) and Definition 3.

72. $\displaystyle \frac{d}{dt}[\mathbf{u}(f(t))] = \frac{d}{dt}\langle u_1(f(t)), u_2(f(t)), u_3(f(t))\rangle = \left\langle \frac{d}{dt}[u_1(f(t))], \frac{d}{dt}[u_2(f(t))], \frac{d}{dt}[u_3(f(t))]\right\rangle$

$$= \langle f'(t)u_1'(f(t)), f'(t)u_2'(f(t)), f'(t)u_3'(f(t))\rangle = f'(t)\,\mathbf{u}'(t)$$

73. $\displaystyle \frac{d}{dt}[\mathbf{u}(t)\cdot\mathbf{v}(t)] = \mathbf{u}'(t)\cdot\mathbf{v}(t) + \mathbf{u}(t)\cdot\mathbf{v}'(t)$ [by Formula 4 of Theorem 5]

$$\begin{aligned}&= (-4t\,\mathbf{j} + 9t^2\,\mathbf{k})\cdot(t\,\mathbf{i} + \cos t\,\mathbf{j} + \sin t\,\mathbf{k}) + (\mathbf{i} - 2t^2\,\mathbf{j} + 3t^3\,\mathbf{k})\cdot(\mathbf{i} - \sin t\,\mathbf{j} + \cos t\,\mathbf{k}) \\ &= -4t\cos t + 9t^2\sin t + 1 + 2t^2\sin t + 3t^3\cos t \\ &= 1 - 4t\cos t + 11t^2\sin t + 3t^3\cos t\end{aligned}$$

74. $\displaystyle \frac{d}{dt}[\mathbf{u}(t)\times\mathbf{v}(t)] = \mathbf{u}'(t)\times\mathbf{v}(t) + \mathbf{u}(t)\times\mathbf{v}'(t)$ [by Formula 5 of Theorem 5]

$$\begin{aligned}&= (-4t\,\mathbf{j} + 9t^2\,\mathbf{k})\times(t\,\mathbf{i} + \cos t\,\mathbf{j} + \sin t\,\mathbf{k}) + (\mathbf{i} - 2t^2\,\mathbf{j} + 3t^3\,\mathbf{k})\times(\mathbf{i} - \sin t\,\mathbf{j} + \cos t\,\mathbf{k}) \\ &= (-4t\sin t - 9t^2\cos t)\,\mathbf{i} + (9t^3 - 0)\,\mathbf{j} + (0 + 4t^2)\,\mathbf{k} \\ &\qquad + (-2t^2\cos t + 3t^3\sin t)\,\mathbf{i} + (3t^3 - \cos t)\,\mathbf{j} + (-\sin t + 2t^2)\,\mathbf{k} \\ &= [(\sin t)(3t^3 - 4t) - 11t^2\cos t]\,\mathbf{i} + (12t^3 - \cos t)\,\mathbf{j} + (6t^2 - \sin t)\,\mathbf{k}\end{aligned}$$

75. $\displaystyle \frac{d}{dt}[\mathbf{r}(t)\times\mathbf{r}'(t)] = \mathbf{r}'(t)\times\mathbf{r}'(t) + \mathbf{r}(t)\times\mathbf{r}''(t)$ by Formula 5 of Theorem 5. But $\mathbf{r}'(t)\times\mathbf{r}'(t) = \mathbf{0}$ [by Example 2 in Section 10.4]. Thus, $\displaystyle \frac{d}{dt}[\mathbf{r}(t)\times\mathbf{r}'(t)] = \mathbf{r}(t)\times\mathbf{r}''(t)$.

76. $\displaystyle \frac{d}{dt}(\mathbf{u}(t)\cdot[\mathbf{v}(t)\times\mathbf{w}(t)]) = \mathbf{u}'(t)\cdot[\mathbf{v}(t)\times\mathbf{w}(t)] + \mathbf{u}(t)\cdot\frac{d}{dt}[\mathbf{v}(t)\times\mathbf{w}(t)]$

$$\begin{aligned}&= \mathbf{u}'(t)\cdot[\mathbf{v}(t)\times\mathbf{w}(t)] + \mathbf{u}(t)\cdot[\mathbf{v}'(t)\times\mathbf{w}(t) + \mathbf{v}(t)\times\mathbf{w}'(t)] \\ &= \mathbf{u}'(t)\cdot[\mathbf{v}(t)\times\mathbf{w}(t)] + \mathbf{u}(t)\cdot[\mathbf{v}'(t)\times\mathbf{w}(t)] + \mathbf{u}(t)\cdot[\mathbf{v}(t)\times\mathbf{w}'(t)] \\ &= \mathbf{u}'(t)\cdot[\mathbf{v}(t)\times\mathbf{w}(t)] - \mathbf{v}'(t)\cdot[\mathbf{u}(t)\times\mathbf{w}(t)] + \mathbf{w}'(t)\cdot[\mathbf{u}(t)\times\mathbf{v}(t)]\end{aligned}$$

77. $\displaystyle \frac{d}{dt}|\mathbf{r}(t)| = \frac{d}{dt}[\mathbf{r}(t)\cdot\mathbf{r}(t)]^{1/2} = \tfrac{1}{2}[\mathbf{r}(t)\cdot\mathbf{r}(t)]^{-1/2}\,[2\mathbf{r}(t)\cdot\mathbf{r}'(t)] = \frac{1}{|\mathbf{r}(t)|}\,\mathbf{r}(t)\cdot\mathbf{r}'(t)$

78. Since $\mathbf{r}(t)\cdot\mathbf{r}'(t) = 0$, we have $0 = 2\mathbf{r}(t)\cdot\mathbf{r}'(t) = \dfrac{d}{dt}[\mathbf{r}(t)\cdot\mathbf{r}(t)] = \dfrac{d}{dt}|\mathbf{r}(t)|^2$. Thus $|\mathbf{r}(t)|^2$, and so $|\mathbf{r}(t)|$, is a constant, and hence the curve lies on a sphere with center the origin.

79. Since $\mathbf{u}(t) = \mathbf{r}(t)\cdot[\mathbf{r}'(t)\times\mathbf{r}''(t)]$,

$$\begin{aligned}\mathbf{u}'(t) &= \mathbf{r}'(t)\cdot[\mathbf{r}'(t)\times\mathbf{r}''(t)] + \mathbf{r}(t)\cdot\frac{d}{dt}[\mathbf{r}'(t)\times\mathbf{r}''(t)] \\ &= 0 + \mathbf{r}(t)\cdot[\mathbf{r}''(t)\times\mathbf{r}''(t) + \mathbf{r}'(t)\times\mathbf{r}'''(t)] && [\text{since } \mathbf{r}'(t)\perp\mathbf{r}'(t)\times\mathbf{r}''(t)] \\ &= \mathbf{r}(t)\cdot[\mathbf{r}'(t)\times\mathbf{r}'''(t)] && [\text{since } \mathbf{r}''(t)\times\mathbf{r}''(t) = \mathbf{0}]\end{aligned}$$

10.8 Arc Length and Curvature

1. $\mathbf{r}'(t) = \langle 2\cos t, 5, -2\sin t\rangle \Rightarrow |\mathbf{r}'(t)| = \sqrt{(2\cos t)^2 + 5^2 + (-2\sin t)^2} = \sqrt{29}$.

Then using Formula 3, we have $L = \int_{-10}^{10} |\mathbf{r}'(t)|\,dt = \int_{-10}^{10} \sqrt{29}\,dt = \sqrt{29}\,t\big]_{-10}^{10} = 20\sqrt{29}$.

2. $\mathbf{r}'(t) = \langle 2t, \cos t + t\sin t - \cos t, -\sin t + t\cos t + \sin t\rangle = \langle 2t, t\sin t, t\cos t\rangle \Rightarrow$

$|\mathbf{r}'(t)| = \sqrt{(2t)^2 + (t\sin t)^2 + (t\cos t)^2} = \sqrt{4t^2 + t^2(\sin^2 t + \cos^2 t)} = \sqrt{5}\,|t| = \sqrt{5}\,t$ for $0 \le t \le \pi$.

Then using Formula 3, we have $L = \int_0^\pi |\mathbf{r}'(t)|\,dt = \int_0^\pi \sqrt{5}\,t\,dt = \sqrt{5}\,\frac{t^2}{2}\Big]_0^\pi = \frac{\sqrt{5}}{2}\pi^2$.

3. $\mathbf{r}'(t) = 2t\,\mathbf{j} + 3t^2\,\mathbf{k} \Rightarrow |\mathbf{r}'(t)| = \sqrt{4t^2 + 9t^4} = t\sqrt{4 + 9t^2}$ [since $t \ge 0$).

Then $L = \int_0^1 |\mathbf{r}'(t)|\,dt = \int_0^1 t\sqrt{4 + 9t^2}\,dt = \frac{1}{18}\cdot\frac{2}{3}(4 + 9t^2)^{3/2}\Big]_0^1 = \frac{1}{27}(13^{3/2} - 4^{3/2}) = \frac{1}{27}(13^{3/2} - 8)$.

4. $\mathbf{r}'(t) = 12\,\mathbf{i} + 12\sqrt{t}\,\mathbf{j} + 6t\,\mathbf{k} \Rightarrow |\mathbf{r}'(t)| = \sqrt{144 + 144t + 36t^2} = \sqrt{36(t+2)^2} = 6\,|t+2| = 6(t+2)$ for $0 \le t \le 1$.

Then $L = \int_0^1 |\mathbf{r}'(t)|\,dt = \int_0^1 6(t+2)\,dt = \left[3t^2 + 12t\right]_0^1 = 15$.

5. The point $(2, 4, 8)$ corresponds to $t = 2$, so by Equation 2, $L = \int_0^2 \sqrt{(1)^2 + (2t)^2 + (3t^2)^2}\,dt$. If $f(t) = \sqrt{1 + 4t^2 + 9t^4}$,

then Simpson's Rule gives $L \approx \dfrac{2-0}{10\cdot 3}\left[f(0) + 4f(0.2) + 2f(0.4) + \cdots + 4f(1.8) + f(2)\right] \approx 9.5706$.

6. Here are two views of the curve with parametric equations $x = \cos t,\ y = \sin 3t,\ z = \sin t$:

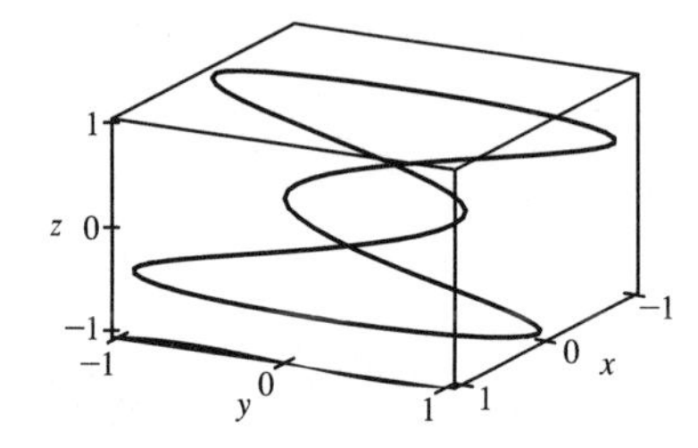

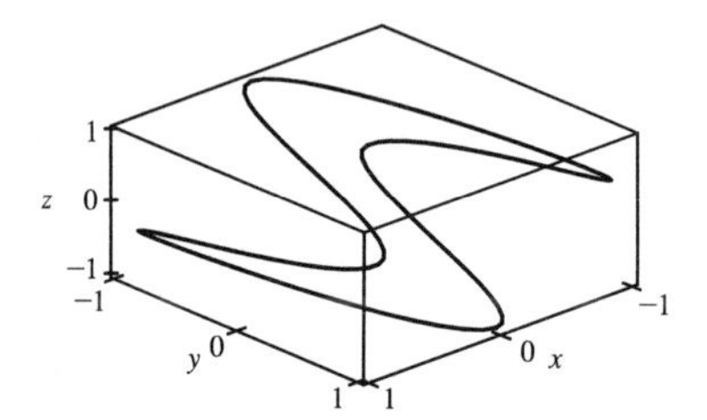

The complete curve is given by the parameter interval $[0, 2\pi]$, so

$L = \int_0^{2\pi} \sqrt{(-\sin t)^2 + (3\cos 3t)^2 + (\cos t)^2}\,dt = \int_0^{2\pi} \sqrt{1 + 9\cos^2 3t}\,dt \approx 13.9744$.

7. $\mathbf{r}'(t) = 2\,\mathbf{i} - 3\,\mathbf{j} + 4\,\mathbf{k}$ and $\frac{ds}{dt} = |\mathbf{r}'(t)| = \sqrt{4 + 9 + 16} = \sqrt{29}$. Then $s = s(t) = \int_0^t |\mathbf{r}'(u)|\,du = \int_0^t \sqrt{29}\,du = \sqrt{29}\,t$.

Therefore, $t = \frac{1}{\sqrt{29}}s$, and substituting for t in the original equation, we have

$\mathbf{r}(t(s)) = \frac{2}{\sqrt{29}}s\,\mathbf{i} + \left(1 - \frac{3}{\sqrt{29}}s\right)\mathbf{j} + \left(5 + \frac{4}{\sqrt{29}}s\right)\mathbf{k}$.

8. $\mathbf{r}'(t) = 2e^{2t}(\cos 2t - \sin 2t)\,\mathbf{i} + 2e^{2t}(\cos 2t + \sin 2t)\,\mathbf{k}$,

$\frac{ds}{dt} = |\mathbf{r}'(t)| = 2e^{2t}\sqrt{(\cos 2t - \sin 2t)^2 + (\cos 2t + \sin 2t)^2} = 2e^{2t}\sqrt{2\cos^2 2t + 2\sin^2 2t} = 2\sqrt{2}\,e^{2t}$.

$s = s(t) = \int_0^t |\mathbf{r}'(u)|\,du = \int_0^t 2\sqrt{2}\,e^{2u}\,du = \sqrt{2}\,e^{2u}\big]_0^t = \sqrt{2}\,(e^{2t} - 1) \Rightarrow$

$\frac{s}{\sqrt{2}} + 1 = e^{2t} \Rightarrow t = \frac{1}{2}\ln\left(\frac{s}{\sqrt{2}} + 1\right)$. Substituting, we have

$$\begin{aligned}\mathbf{r}(t(s)) &= e^{2\left(\frac{1}{2}\ln\left(\frac{s}{\sqrt{2}}+1\right)\right)}\cos 2\left(\tfrac{1}{2}\ln\left(\tfrac{s}{\sqrt{2}} + 1\right)\right)\mathbf{i} + 2\,\mathbf{j} + e^{2\left(\frac{1}{2}\ln\left(\frac{s}{\sqrt{2}}+1\right)\right)}\sin 2\left(\tfrac{1}{2}\ln\left(\tfrac{s}{\sqrt{2}} + 1\right)\right)\mathbf{k} \\ &= \left(\tfrac{s}{\sqrt{2}} + 1\right)\cos\left(\ln\left(\tfrac{s}{\sqrt{2}} + 1\right)\right)\mathbf{i} + 2\,\mathbf{j} + \left(\tfrac{s}{\sqrt{2}} + 1\right)\sin\left(\ln\left(\tfrac{s}{\sqrt{2}} + 1\right)\right)\mathbf{k}\end{aligned}$$

9. Here $\mathbf{r}(t) = \langle 3\sin t, 4t, 3\cos t\rangle$, so $\mathbf{r}'(t) = \langle 3\cos t, 4, -3\sin t\rangle$ and $|\mathbf{r}'(t)| = \sqrt{9\cos^2 t + 16 + 9\sin^2 t} = \sqrt{25} = 5$. The point $(0,0,3)$ corresponds to $t = 0$, so the arc length function beginning at $(0,0,3)$ and measuring in the positive direction is given by $s(t) = \int_0^t |\mathbf{r}'(u)|\,du = \int_0^t 5\,du = 5t$. $s(t) = 5 \;\Rightarrow\; 5t = 5 \;\Rightarrow\; t = 1$, thus your location after moving 5 units along the curve is $(3\sin 1, 4, 3\cos 1)$.

10. $\mathbf{r}'(t) = \dfrac{-4t}{(t^2+1)^2}\,\mathbf{i} + \dfrac{-2t^2+2}{(t^2+1)^2}\,\mathbf{j}$,

$$\frac{ds}{dt} = |\mathbf{r}'(t)| = \sqrt{\left[\frac{-4t}{(t^2+1)^2}\right]^2 + \left[\frac{-2t^2+2}{(t^2+1)^2}\right]^2} = \sqrt{\frac{4t^4+8t^2+4}{(t^2+1)^4}} = \sqrt{\frac{4(t^2+1)^2}{(t^2+1)^4}} = \sqrt{\frac{4}{(t^2+1)^2}} = \frac{2}{t^2+1}.$$

Since the initial point $(1,0)$ corresponds to $t = 0$, the arc length function

$s(t) = \displaystyle\int_0^t |\mathbf{r}'(u)|\,du = \int_0^t \frac{2}{u^2+1}\,du = 2\arctan t$. Then $\arctan t = \frac{1}{2}s \;\Rightarrow\; t = \tan\frac{1}{2}s$. Substituting, we have

$$\begin{aligned}\mathbf{r}(t(s)) &= \left[\frac{2}{\tan^2\left(\frac{1}{2}s\right)+1} - 1\right]\mathbf{i} + \frac{2\tan\left(\frac{1}{2}s\right)}{\tan^2\left(\frac{1}{2}s\right)+1}\,\mathbf{j} = \frac{1-\tan^2\left(\frac{1}{2}s\right)}{1+\tan^2\left(\frac{1}{2}s\right)}\,\mathbf{i} + \frac{2\tan\left(\frac{1}{2}s\right)}{\sec^2\left(\frac{1}{2}s\right)}\,\mathbf{j} \\ &= \frac{1-\tan^2\left(\frac{1}{2}s\right)}{\sec^2\left(\frac{1}{2}s\right)}\,\mathbf{i} + 2\tan\left(\tfrac{1}{2}s\right)\cos^2\left(\tfrac{1}{2}s\right)\mathbf{j} \\ &= \left[\cos^2\left(\tfrac{1}{2}s\right) - \sin^2\left(\tfrac{1}{2}s\right)\right]\mathbf{i} + 2\sin\left(\tfrac{1}{2}s\right)\cos\left(\tfrac{1}{2}s\right)\mathbf{j} = \cos s\,\mathbf{i} + \sin s\,\mathbf{j}\end{aligned}$$

With this parametrization, we recognize the function as representing the unit circle. Note here that the curve approaches, but does not include, the point $(-1,0)$, since $\cos s = -1$ for $s = \pi + 2k\pi$ (k an integer) but then $t = \tan\left(\frac{1}{2}s\right)$ is undefined.

11. (a) $\mathbf{r}'(t) = \langle 2\cos t, 5, -2\sin t\rangle \;\Rightarrow\; |\mathbf{r}'(t)| = \sqrt{4\cos^2 t + 25 + 4\sin^2 t} = \sqrt{29}$. Then

$\mathbf{T}(t) = \dfrac{\mathbf{r}'(t)}{|\mathbf{r}'(t)|} = \frac{1}{\sqrt{29}}\langle 2\cos t, 5, -2\sin t\rangle$ or $\left\langle \frac{2}{\sqrt{29}}\cos t, \frac{5}{\sqrt{29}}, -\frac{2}{\sqrt{29}}\sin t\right\rangle$.

$\mathbf{T}'(t) = \frac{1}{\sqrt{29}}\langle -2\sin t, 0, -2\cos t\rangle \;\Rightarrow\; |\mathbf{T}'(t)| = \frac{1}{\sqrt{29}}\sqrt{4\sin^2 t + 0 + 4\cos^2 t} = \frac{2}{\sqrt{29}}$. Thus

$\mathbf{N}(t) = \dfrac{\mathbf{T}'(t)}{|\mathbf{T}'(t)|} = \dfrac{1/\sqrt{29}}{2/\sqrt{29}}\langle -2\sin t, 0, -2\cos t\rangle = \langle -\sin t, 0, -\cos t\rangle$.

(b) $\kappa(t) = \dfrac{|\mathbf{T}'(t)|}{|\mathbf{r}'(t)|} = \dfrac{2/\sqrt{29}}{\sqrt{29}} = \dfrac{2}{29}$.

12. (a) $\mathbf{r}'(t) = \langle 2t, t\sin t, t\cos t\rangle \;\Rightarrow\; |\mathbf{r}'(t)| = \sqrt{4t^2 + t^2\sin^2 t + t^2\cos^2 t} = \sqrt{5t^2} = \sqrt{5}\,t$ (since $t > 0$). Then

$\mathbf{T}(t) = \dfrac{\mathbf{r}'(t)}{|\mathbf{r}'(t)|} = \dfrac{1}{\sqrt{5}\,t}\langle 2t, t\sin t, t\cos t\rangle = \frac{1}{\sqrt{5}}\langle 2, \sin t, \cos t\rangle$. $\mathbf{T}'(t) = \frac{1}{\sqrt{5}}\langle 0, \cos t, -\sin t\rangle \;\Rightarrow$

$|\mathbf{T}'(t)| = \frac{1}{\sqrt{5}}\sqrt{0 + \cos^2 t + \sin^2 t} = \frac{1}{\sqrt{5}}$. Thus $\mathbf{N}(t) = \dfrac{\mathbf{T}'(t)}{|\mathbf{T}'(t)|} = \dfrac{1/\sqrt{5}}{1/\sqrt{5}}\langle 0, \cos t, -\sin t\rangle = \langle 0, \cos t, -\sin t\rangle$.

(b) $\kappa(t) = \dfrac{|\mathbf{T}'(t)|}{|\mathbf{r}'(t)|} = \dfrac{1/\sqrt{5}}{\sqrt{5}\,t} = \dfrac{1}{5t}$.

13. (a) $\mathbf{r}'(t) = \left\langle \sqrt{2}, e^t, -e^{-t}\right\rangle \;\Rightarrow\; |\mathbf{r}'(t)| = \sqrt{2 + e^{2t} + e^{-2t}} = \sqrt{(e^t + e^{-t})^2} = e^t + e^{-t}$. Then

$\mathbf{T}(t) = \dfrac{\mathbf{r}'(t)}{|\mathbf{r}'(t)|} = \dfrac{1}{e^t + e^{-t}}\left\langle \sqrt{2}, e^t, -e^{-t}\right\rangle = \dfrac{1}{e^{2t}+1}\left\langle \sqrt{2}e^t, e^{2t}, -1\right\rangle$ $\left[\text{after multiplying by } \dfrac{e^t}{e^t}\right]$ and

$$\mathbf{T}'(t) = \frac{1}{e^{2t}+1}\left\langle \sqrt{2}e^t, 2e^{2t}, 0\right\rangle - \frac{2e^{2t}}{(e^{2t}+1)^2}\left\langle \sqrt{2}e^t, e^{2t}, -1\right\rangle$$

$$= \frac{1}{(e^{2t}+1)^2}\left[(e^{2t}+1)\left\langle \sqrt{2}e^t, 2e^{2t}, 0\right\rangle - 2e^{2t}\left\langle \sqrt{2}e^t, e^{2t}, -1\right\rangle\right] = \frac{1}{(e^{2t}+1)^2}\left\langle \sqrt{2}e^t\left(1-e^{2t}\right), 2e^{2t}, 2e^{2t}\right\rangle$$

Then

$$|\mathbf{T}'(t)| = \frac{1}{(e^{2t}+1)^2}\sqrt{2e^{2t}(1-2e^{2t}+e^{4t})+4e^{4t}+4e^{4t}} = \frac{1}{(e^{2t}+1)^2}\sqrt{2e^{2t}(1+2e^{2t}+e^{4t})}$$

$$= \frac{1}{(e^{2t}+1)^2}\sqrt{2e^{2t}\left(1+e^{2t}\right)^2} = \frac{\sqrt{2}e^t(1+e^{2t})}{(e^{2t}+1)^2} = \frac{\sqrt{2}\,e^t}{e^{2t}+1}$$

Therefore

$$\mathbf{N}(t) = \frac{\mathbf{T}'(t)}{|\mathbf{T}'(t)|} = \frac{e^{2t}+1}{\sqrt{2}\,e^t}\frac{1}{(e^{2t}+1)^2}\left\langle \sqrt{2}\,e^t(1-e^{2t}), 2e^{2t}, 2e^{2t}\right\rangle$$

$$= \frac{1}{\sqrt{2}\,e^t(e^{2t}+1)}\left\langle \sqrt{2}\,e^t(1-e^{2t}), 2e^{2t}, 2e^{2t}\right\rangle = \frac{1}{e^{2t}+1}\left\langle 1-e^{2t}, \sqrt{2}\,e^t, \sqrt{2}\,e^t\right\rangle$$

(b) $\kappa(t) = \dfrac{|\mathbf{T}'(t)|}{|\mathbf{r}'(t)|} = \dfrac{\sqrt{2}\,e^t}{e^{2t}+1}\cdot\dfrac{1}{e^t+e^{-t}} = \dfrac{\sqrt{2}\,e^t}{e^{3t}+2e^t+e^{-t}} = \dfrac{\sqrt{2}\,e^{2t}}{e^{4t}+2e^{2t}+1} = \dfrac{\sqrt{2}\,e^{2t}}{(e^{2t}+1)^2}.$

14. (a) $\mathbf{r}'(t) = \langle 1, t, 2t\rangle \quad\Rightarrow\quad |\mathbf{r}'(t)| = \sqrt{1+t^2+4t^2} = \sqrt{1+5t^2}$. Then $\mathbf{T}(t) = \dfrac{\mathbf{r}'(t)}{|\mathbf{r}'(t)|} = \dfrac{1}{\sqrt{1+5t^2}}\langle 1, t, 2t\rangle.$

$$\mathbf{T}'(t) = \frac{-5t}{(1+5t^2)^{3/2}}\langle 1, t, 2t\rangle + \frac{1}{\sqrt{1+5t^2}}\langle 0, 1, 2\rangle \qquad \text{[by Formula 3 of Theorem 10.7.5]}$$

$$= \frac{1}{(1+5t^2)^{3/2}}\left(\left\langle -5t, -5t^2, -10t^2\right\rangle + \left\langle 0, 1+5t^2, 2+10t^2\right\rangle\right) = \frac{1}{(1+5t^2)^{3/2}}\langle -5t, 1, 2\rangle$$

$$|\mathbf{T}'(t)| = \frac{1}{(1+5t^2)^{3/2}}\sqrt{25t^2+1+4} = \frac{1}{(1+5t^2)^{3/2}}\sqrt{25t^2+5} = \frac{\sqrt{5}\sqrt{5t^2+1}}{(1+5t^2)^{3/2}} = \frac{\sqrt{5}}{1+5t^2}$$

Thus $\mathbf{N}(t) = \dfrac{\mathbf{T}'(t)}{|\mathbf{T}'(t)|} = \dfrac{1+5t^2}{\sqrt{5}}\cdot\dfrac{1}{(1+5t^2)^{3/2}}\langle -5t, 1, 2\rangle = \dfrac{1}{\sqrt{5+25t^2}}\langle -5t, 1, 2\rangle.$

(b) $\kappa(t) = \dfrac{|\mathbf{T}'(t)|}{|\mathbf{r}'(t)|} = \dfrac{\sqrt{5}/(1+5t^2)}{\sqrt{1+5t^2}} = \dfrac{\sqrt{5}}{(1+5t^2)^{3/2}}$

15. $\mathbf{r}'(t) = 2t\,\mathbf{i}+\mathbf{k}$, $\mathbf{r}''(t) = 2\,\mathbf{i}$, $|\mathbf{r}'(t)| = \sqrt{(2t)^2+0^2+1^2} = \sqrt{4t^2+1}$, $\mathbf{r}'(t)\times\mathbf{r}''(t) = 2\,\mathbf{j}$, $|\mathbf{r}'(t)\times\mathbf{r}''(t)| = 2$.

Then $\kappa(t) = \dfrac{|\mathbf{r}'(t)\times\mathbf{r}''(t)|}{|\mathbf{r}'(t)|^3} = \dfrac{2}{\left(\sqrt{4t^2+1}\right)^3} = \dfrac{2}{(4t^2+1)^{3/2}}.$

16. $\mathbf{r}'(t) = \mathbf{i}+\mathbf{j}+2t\,\mathbf{k}$, $\mathbf{r}''(t) = 2\,\mathbf{k}$, $|\mathbf{r}'(t)| = \sqrt{1^2+1^2+(2t)^2} = \sqrt{4t^2+2}$, $\mathbf{r}'(t)\times\mathbf{r}''(t) = 2\,\mathbf{i}-2\,\mathbf{j}$,

$|\mathbf{r}'(t)\times\mathbf{r}''(t)| = \sqrt{2^2+2^2+0^2} = \sqrt{8} = 2\sqrt{2}.$

Then $\kappa(t) = \dfrac{|\mathbf{r}'(t)\times\mathbf{r}''(t)|}{|\mathbf{r}'(t)|^3} = \dfrac{2\sqrt{2}}{\left(\sqrt{4t^2+2}\right)^3} = \dfrac{2\sqrt{2}}{\left(\sqrt{2}\sqrt{2t^2+1}\right)^3} = \dfrac{1}{(2t^2+1)^{3/2}}.$

17. $\mathbf{r}'(t) = 3\,\mathbf{i}+4\cos t\,\mathbf{j}-4\sin t\,\mathbf{k}$, $\mathbf{r}''(t) = -4\sin t\,\mathbf{j}-4\cos t\,\mathbf{k}$, $|\mathbf{r}'(t)| = \sqrt{9+16\cos^2 t+16\sin^2 t} = \sqrt{9+16} = 5$,

$\mathbf{r}'(t)\times\mathbf{r}''(t) = -16\,\mathbf{i}+12\cos t\,\mathbf{j}-12\sin t\,\mathbf{k}$, $|\mathbf{r}'(t)\times\mathbf{r}''(t)| = \sqrt{256+144\cos^2 t+144\sin^2 t} = \sqrt{400} = 20.$

Then $\kappa(t) = \dfrac{|\mathbf{r}'(t)\times\mathbf{r}''(t)|}{|\mathbf{r}'(t)|^3} = \dfrac{20}{5^3} = \dfrac{4}{25}.$

18. $\mathbf{r}'(t) = \langle e^t \cos t - e^t \sin t, e^t \cos t + e^t \sin t, 1 \rangle$. The point $(1, 0, 0)$ corresponds to $t = 0$, and

$\mathbf{r}'(0) = \langle 1, 1, 1 \rangle \Rightarrow |\mathbf{r}'(0)| = \sqrt{1^2 + 1^2 + 1^2} = \sqrt{3}$.

$\mathbf{r}''(t) = \langle e^t \cos t - e^t \sin t - e^t \cos t - e^t \sin t, e^t \cos t - e^t \sin t + e^t \cos t + e^t \sin t, 0 \rangle = \langle -2e^t \sin t, 2e^t \cos t, 0 \rangle \Rightarrow$

$\mathbf{r}''(0) = \langle 0, 2, 0 \rangle$. $\mathbf{r}'(0) \times \mathbf{r}''(0) = \langle -2, 0, 2 \rangle$. $|\mathbf{r}'(0) \times \mathbf{r}''(0)| = \sqrt{(-2)^2 + 0^2 + 2^2} = \sqrt{8} = 2\sqrt{2}$.

Then $\kappa(0) = \dfrac{|\mathbf{r}'(0) \times \mathbf{r}''(0)|}{|\mathbf{r}'(0)|^3} = \dfrac{2\sqrt{2}}{(\sqrt{3})^3} = \dfrac{2\sqrt{2}}{3\sqrt{3}}$ or $\dfrac{2\sqrt{6}}{9}$.

19. $\mathbf{r}'(t) = \langle 1, 2t, 3t^2 \rangle$. The point $(1, 1, 1)$ corresponds to $t = 1$, and $\mathbf{r}'(1) = \langle 1, 2, 3 \rangle \Rightarrow |\mathbf{r}'(1)| = \sqrt{1 + 4 + 9} = \sqrt{14}$.

$\mathbf{r}''(t) = \langle 0, 2, 6t \rangle \Rightarrow \mathbf{r}''(1) = \langle 0, 2, 6 \rangle$. $\mathbf{r}'(1) \times \mathbf{r}''(1) = \langle 6, -6, 2 \rangle$, so $|\mathbf{r}'(1) \times \mathbf{r}''(1)| = \sqrt{36 + 36 + 4} = \sqrt{76}$.

Then $\kappa(1) = \dfrac{|\mathbf{r}'(1) \times \mathbf{r}''(1)|}{|\mathbf{r}'(1)|^3} = \dfrac{\sqrt{76}}{\sqrt{14}^3} = \dfrac{1}{7}\sqrt{\dfrac{19}{14}}$.

20.

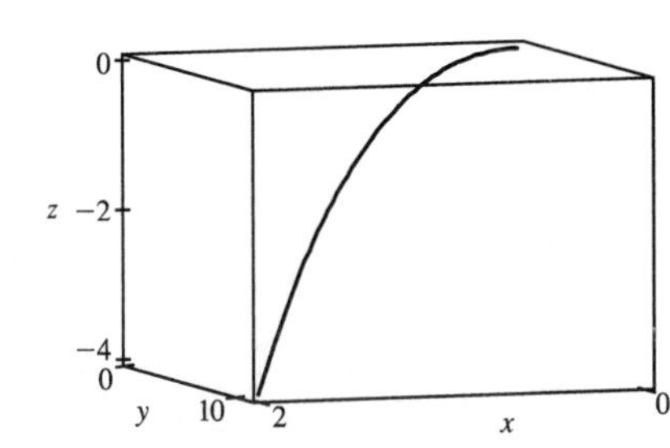

$\mathbf{r}(t) = \left\langle t, 4t^{3/2}, -t^2 \right\rangle \Rightarrow \mathbf{r}'(t) = \left\langle 1, 6t^{1/2}, -2t \right\rangle$,

$\mathbf{r}''(t) = \left\langle 0, 3t^{-1/2}, -2 \right\rangle$, $|\mathbf{r}'(t)|^3 = (1 + 36t + 4t^2)^{3/2}$,

$\mathbf{r}'(t) \times \mathbf{r}''(t) = \left\langle -12t^{1/2} + 6t^{1/2}, 2, 3t^{-1/2} \right\rangle \Rightarrow$

$|\mathbf{r}'(t) \times \mathbf{r}''(t)| = \sqrt{36t + 4 + 9t^{-1}} = \left[\dfrac{36t^2 + 4t + 9}{t}\right]^{1/2}$

$$\kappa(t) = \frac{|\mathbf{r}'(t) \times \mathbf{r}''(t)|}{|\mathbf{r}'(t)|^3} = \left(\frac{36t^2 + 4t + 9}{t}\right)^{1/2} \frac{1}{(1 + 36t + 4t^2)^{3/2}} = \frac{\sqrt{36t^2 + 4t + 9}}{t^{1/2}(1 + 36t + 4t^2)^{3/2}}.$$

The point $(1, 4, -1)$ corresponds to $t = 1$, so the curvature at this point is $\kappa(1) = \dfrac{\sqrt{36 + 4 + 9}}{(1 + 36 + 4)^{3/2}} = \dfrac{7}{41\sqrt{41}}$.

21. $f(x) = xe^x$, $f'(x) = xe^x + e^x$, $f''(x) = xe^x + 2e^x$,

$$\kappa(x) = \frac{|f''(x)|}{[1 + (f'(x))^2]^{3/2}} = \frac{|xe^x + 2e^x|}{[1 + (xe^x + e^x)^2]^{3/2}} = \frac{|x + 2|\, e^x}{[1 + (xe^x + e^x)^2]^{3/2}}$$

22. $f(x) = \cos x$, $f'(x) = -\sin x$, $f''(x) = -\cos x$,

$$\kappa(x) = \frac{|f''(x)|}{[1 + (f'(x))^2]^{3/2}} = \frac{|-\cos x|}{[1 + (-\sin x)^2]^{3/2}} = \frac{|\cos x|}{(1 + \sin^2 x)^{3/2}}$$

23. $f(x) = 4x^{5/2}$, $f'(x) = 10x^{3/2}$, $f''(x) = 15x^{1/2}$,

$$\kappa(x) = \frac{|f''(x)|}{[1 + (f'(x))^2]^{3/2}} = \frac{\left|15x^{1/2}\right|}{[1 + (10x^{3/2})^2]^{3/2}} = \frac{15\sqrt{x}}{(1 + 100x^3)^{3/2}}$$

24. $y' = \dfrac{1}{x}$, $y'' = -\dfrac{1}{x^2}$,

$$\kappa(x) = \frac{|y''(x)|}{\left[1 + (y'(x))^2\right]^{3/2}} = \left|\frac{-1}{x^2}\right| \frac{1}{(1 + 1/x^2)^{3/2}} = \frac{1}{x^2}\frac{(x^2)^{3/2}}{(x^2 + 1)^{3/2}} = \frac{|x|}{(x^2 + 1)^{3/2}} = \frac{x}{(x^2 + 1)^{3/2}} \quad [\text{since } x > 0].$$

To find the maximum curvature, we first find the critical numbers of $\kappa(x)$:

$$\kappa'(x) = \frac{(x^2 + 1)^{3/2} - x\left(\frac{3}{2}\right)(x^2 + 1)^{1/2}(2x)}{[(x^2 + 1)^{3/2}]^2} = \frac{(x^2 + 1)^{1/2}[(x^2 + 1) - 3x^2]}{(x^2 + 1)^3} = \frac{1 - 2x^2}{(x^2 + 1)^{5/2}};$$

$\kappa'(x) = 0 \Rightarrow 1 - 2x^2 = 0$, so the only critical number in the domain is $x = \frac{1}{\sqrt{2}}$. Since $\kappa'(x) > 0$ for $0 < x < \frac{1}{\sqrt{2}}$ and $\kappa'(x) < 0$ for $x > \frac{1}{\sqrt{2}}$, $\kappa(x)$ attains its maximum at $x = \frac{1}{\sqrt{2}}$. Thus, the maximum curvature occurs at $\left(\frac{1}{\sqrt{2}}, \ln \frac{1}{\sqrt{2}}\right)$.

Since $\lim\limits_{x\to\infty} \dfrac{x}{(x^2+1)^{3/2}} = 0$, $\kappa(x)$ approaches 0 as $x \to \infty$.

25. Since $y' = y'' = e^x$, the curvature is $\kappa(x) = \dfrac{|y''(x)|}{[1 + (y'(x))^2]^{3/2}} = \dfrac{e^x}{(1+e^{2x})^{3/2}} = e^x(1+e^{2x})^{-3/2}$.

To find the maximum curvature, we first find the critical numbers of $\kappa(x)$:

$$\kappa'(x) = e^x(1+e^{2x})^{-3/2} + e^x\left(-\tfrac{3}{2}\right)(1+e^{2x})^{-5/2}(2e^{2x}) = e^x\frac{1+e^{2x}-3e^{2x}}{(1+e^{2x})^{5/2}} = e^x\frac{1-2e^{2x}}{(1+e^{2x})^{5/2}}.$$

$\kappa'(x) = 0$ when $1 - 2e^{2x} = 0$, so $e^{2x} = \frac{1}{2}$ or $x = -\frac{1}{2}\ln 2$. And since $1 - 2e^{2x} > 0$ for $x < -\frac{1}{2}\ln 2$ and $1 - 2e^{2x} < 0$ for $x > -\frac{1}{2}\ln 2$, the maximum curvature is attained at the point $\left(-\frac{1}{2}\ln 2, e^{(-\ln 2)/2}\right) = \left(-\frac{1}{2}\ln 2, \frac{1}{\sqrt{2}}\right)$.

Since $\lim\limits_{x\to\infty} e^x(1+e^{2x})^{-3/2} = 0$, $\kappa(x)$ approaches 0 as $x \to \infty$.

26. We can take the parabola as having its vertex at the origin and opening upward, so the equation is $f(x) = ax^2$, $a > 0$. Then by Equation 11, $\kappa(x) = \dfrac{|f''(x)|}{[1+(f'(x))^2]^{3/2}} = \dfrac{|2a|}{[1+(2ax)^2]^{3/2}} = \dfrac{2a}{(1+4a^2x^2)^{3/2}}$, thus $\kappa(0) = 2a$. We want $\kappa(0) = 4$, so $a = 2$ and the equation is $y = 2x^2$.

27. (a) C appears to be changing direction more quickly at P than Q, so we would expect the curvature to be greater at P.

(b) First we sketch approximate osculating circles at P and Q. Using the axes scale as a guide, we measure the radius of the osculating circle at P to be approximately 0.8 units, thus $\rho = \dfrac{1}{\kappa} \Rightarrow \kappa = \dfrac{1}{\rho} \approx \dfrac{1}{0.8} \approx 1.3$. Similarly, we estimate the radius of the osculating circle at Q to be 1.4 units, so $\kappa = \dfrac{1}{\rho} \approx \dfrac{1}{1.4} \approx 0.7$.

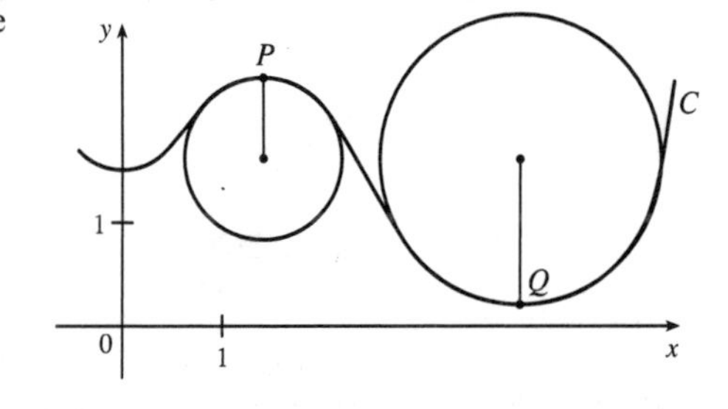

28. $y = x^4 - 2x^2 \Rightarrow y' = 4x^3 - 4x$, $y'' = 12x^2 - 4$, and

$\kappa(x) = \dfrac{|y''|}{\left[1+(y')^2\right]^{3/2}} = \dfrac{|12x^2-4|}{\left[1+(4x^3-4x)^2\right]^{3/2}}$. The graph of the curvature here is what we would expect. The graph of $y = x^4 - 2x^2$ appears to be bending most sharply at the origin and near $x = \pm 1$.

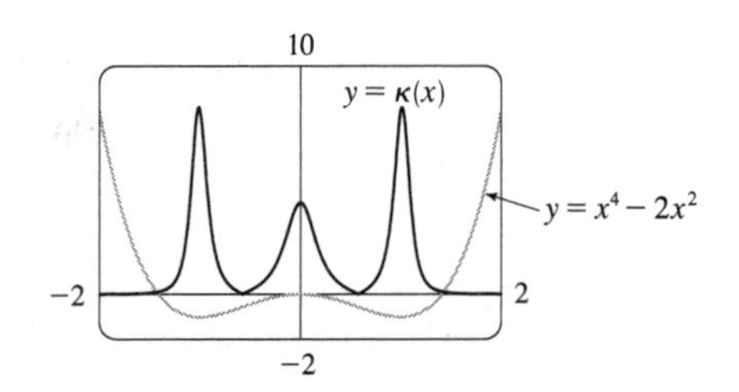

29. $y = x^{-2} \Rightarrow y' = -2x^{-3}$, $y'' = 6x^{-4}$, and

$\kappa(x) = \dfrac{|y''|}{\left[1+(y')^2\right]^{3/2}} = \dfrac{|6x^{-4}|}{\left[1+(-2x^{-3})^2\right]^{3/2}} = \dfrac{6}{x^4\left(1+4x^{-6}\right)^{3/2}}$.

The appearance of the two humps in this graph is perhaps a little surprising, but it is explained by the fact that $y = x^{-2}$ increases asymptotically at the origin from both directions, and so its graph has very little bend there. (Note that $\kappa(0)$ is undefined.)

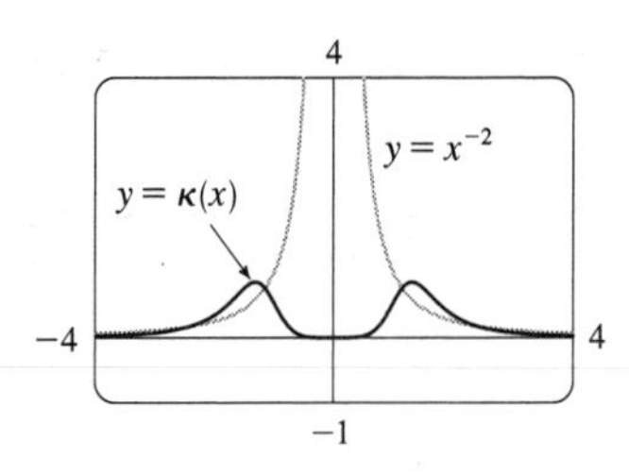

30. Notice that the curve a is highest for the same x-values at which curve b is turning more sharply, and a is 0 or near 0 where b is nearly straight. So, a must be the graph of $y = \kappa(x)$, and b is the graph of $y = f(x)$.

31. Notice that the curve b has two inflection points at which the graph appears almost straight. We would expect the curvature to be 0 or nearly 0 at these values, but the curve a isn't near 0 there. Thus, a must be the graph of $y = f(x)$ rather than the graph of curvature, and b is the graph of $y = \kappa(x)$.

32. Here $\mathbf{r}(t) = \langle f(t), g(t)\rangle$, $\mathbf{r}'(t) = \langle f'(t), g'(t)\rangle$, $\mathbf{r}''(t) = \langle f''(t), g''(t)\rangle$,

$|\mathbf{r}'(t)|^3 = \left[\sqrt{(f'(t))^2 + (g'(t))^2}\right]^3 = [(f'(t))^2 + (g'(t))^2]^{3/2} = (\dot{x}^2 + \dot{y}^2)^{3/2}$, and

$|\mathbf{r}'(t) \times \mathbf{r}''(t)| = |\langle 0, 0, f'(t)\,g''(t) - f''(t)\,g'(t)\rangle| = \left[(\dot{x}\ddot{y} - \ddot{x}\dot{y})^2\right]^{1/2} = |\dot{x}\ddot{y} - \dot{y}\ddot{x}|$. Thus $\kappa(t) = \dfrac{|\dot{x}\ddot{y} - \dot{y}\ddot{x}|}{[\dot{x}^2 + \dot{y}^2]^{3/2}}$.

33. $x = e^t \cos t \;\Rightarrow\; \dot{x} = e^t(\cos t - \sin t) \;\Rightarrow\; \ddot{x} = e^t(-\sin t - \cos t) + e^t(\cos t - \sin t) = -2e^t \sin t$,

$y = e^t \sin t \;\Rightarrow\; \dot{y} = e^t(\cos t + \sin t) \;\Rightarrow\; \ddot{y} = e^t(-\sin t + \cos t) + e^t(\cos t + \sin t) = 2e^t \cos t$. Then

$$\kappa(t) = \frac{|\dot{x}\ddot{y} - \dot{y}\ddot{x}|}{[\dot{x}^2 + \dot{y}^2]^{3/2}} = \frac{\left|e^t(\cos t - \sin t)(2e^t\cos t) - e^t(\cos t + \sin t)(-2e^t \sin t)\right|}{\left([e^t(\cos t - \sin t)]^2 + [e^t(\cos t + \sin t)]^2\right)^{3/2}}$$

$$= \frac{\left|2e^{2t}(\cos^2 t - \sin t\cos t + \sin t \cos t + \sin^2 t)\right|}{\left[e^{2t}(\cos^2 t - 2\cos t \sin t + \sin^2 t + \cos^2 t + 2\cos t \sin t + \sin^2 t)\right]^{3/2}} = \frac{\left|2e^{2t}(1)\right|}{\left[e^{2t}(1+1)\right]^{3/2}} = \frac{2e^{2t}}{e^{3t}(2)^{3/2}} = \frac{1}{\sqrt{2}\,e^t}$$

34. $x = 1 + t^3 \;\Rightarrow\; \dot{x} = 3t^2 \;\Rightarrow\; \ddot{x} = 6t$, $\;y = t + t^2 \;\Rightarrow\; \dot{y} = 1 + 2t \;\Rightarrow\; \ddot{y} = 2$.

Then $\kappa(t) = \dfrac{|\dot{x}\ddot{y} - \dot{y}\ddot{x}|}{[\dot{x}^2 + \dot{y}^2]^{3/2}} = \dfrac{\left|(3t^2)(2) - (1+2t)(6t)\right|}{[(3t^2)^2 + (1+2t)^2]^{3/2}} = \dfrac{\left|-6t^2 - 6t\right|}{(9t^4 + 4t^2 + 4t + 1)^{3/2}} = \dfrac{6\left|t^2 + t\right|}{(9t^4 + 4t^2 + 4t + 1)^{3/2}}$.

35. $\left(1, \frac{2}{3}, 1\right)$ corresponds to $t = 1$. $\mathbf{T}(t) = \dfrac{\mathbf{r}'(t)}{|\mathbf{r}'(t)|} = \dfrac{\langle 2t, 2t^2, 1\rangle}{\sqrt{4t^2 + 4t^4 + 1}} = \dfrac{\langle 2t, 2t^2, 1\rangle}{2t^2 + 1}$, so $\mathbf{T}(1) = \left\langle \frac{2}{3}, \frac{2}{3}, \frac{1}{3}\right\rangle$.

$\mathbf{T}'(t) = -4t(2t^2+1)^{-2}\langle 2t, 2t^2, 1\rangle + (2t^2+1)^{-1}\langle 2, 4t, 0\rangle$ [by Formula 3 of Theorem 10.7.5]

$= (2t^2+1)^{-2}\langle -8t^2 + 4t^2 + 2, -8t^3 + 8t^3 + 4t, -4t\rangle = 2(2t^2+1)^{-2}\langle 1 - 2t^2, 2t, -2t\rangle$

$$\mathbf{N}(t) = \frac{\mathbf{T}'(t)}{|\mathbf{T}'(t)|} = \frac{2(2t^2+1)^{-2}\langle 1 - 2t^2, 2t, -2t\rangle}{2(2t^2+1)^{-2}\sqrt{(1-2t^2)^2 + (2t)^2 + (-2t)^2}} = \frac{\langle 1 - 2t^2, 2t, -2t\rangle}{\sqrt{1 - 4t^2 + 4t^4 + 8t^2}} = \frac{\langle 1 - 2t^2, 2t, -2t\rangle}{1 + 2t^2}$$

$\mathbf{N}(1) = \left\langle -\frac{1}{3}, \frac{2}{3}, -\frac{2}{3}\right\rangle$ and $\mathbf{B}(1) = \mathbf{T}(1) \times \mathbf{N}(1) = \left\langle -\frac{4}{9} - \frac{2}{9}, -\left(-\frac{4}{9} + \frac{1}{9}\right), \frac{4}{9} + \frac{2}{9}\right\rangle = \left\langle -\frac{2}{3}, \frac{1}{3}, \frac{2}{3}\right\rangle$.

36. $(1, 0, 1)$ corresponds to $t = 0$. $\mathbf{r}(t) = e^t\langle 1, \sin t, \cos t\rangle$, so

$\mathbf{r}'(t) = e^t\langle 1, \sin t, \cos t\rangle + e^t\langle 0, \cos t, -\sin t\rangle = e^t\langle 1, \sin t + \cos t, \cos t - \sin t\rangle$ and

$$\mathbf{T}(t) = \frac{\mathbf{r}'(t)}{|\mathbf{r}'(t)|} = \frac{e^t\langle 1, \sin t + \cos t, \cos t - \sin t\rangle}{e^t\sqrt{1 + \sin^2 t + 2\sin t\,\cos t + \cos^2 t + \cos^2 t - 2\sin t\,\cos t + \sin^2 t}} = \frac{\langle 1, \sin t + \cos t, \cos t - \sin t\rangle}{\sqrt{3}}$$

$\mathbf{T}(0) = \left\langle \frac{1}{\sqrt{3}}, \frac{1}{\sqrt{3}}, \frac{1}{\sqrt{3}}\right\rangle$. $\mathbf{T}'(t) = \frac{1}{\sqrt{3}}\langle 0, \cos t - \sin t, -\sin t - \cos t\rangle$, so

$$\mathbf{N}(t) = \frac{\mathbf{T}'(t)}{|\mathbf{T}'(t)|} = \frac{\frac{1}{\sqrt{3}}\langle 0, \cos t - \sin t, -\sin t - \cos t\rangle}{\frac{1}{\sqrt{3}}\sqrt{0^2 + \cos^2 t - 2\cos t\,\sin t + \sin^2 t + \sin^2 t + 2\sin t\,\cos t + \cos^2 t}}$$

$$= \frac{1}{\sqrt{2}}\langle 0, \cos t - \sin t, -\sin t - \cos t\rangle.$$

$\mathbf{N}(0) = \left\langle 0, \frac{1}{\sqrt{2}}, -\frac{1}{\sqrt{2}}\right\rangle$ and $\mathbf{B}(0) = \mathbf{T}(0) \times \mathbf{N}(0) = \left\langle -\frac{2}{\sqrt{6}}, \frac{1}{\sqrt{6}}, \frac{1}{\sqrt{6}}\right\rangle$.

37. $(0, \pi, -2)$ corresponds to $t = \pi$. $\mathbf{r}(t) = \langle 2\sin 3t, t, 2\cos 3t \rangle \Rightarrow$

$$\mathbf{T}(t) = \frac{\mathbf{r}'(t)}{|\mathbf{r}'(t)|} = \frac{\langle 6\cos 3t, 1, -6\sin 3t \rangle}{\sqrt{36\cos^2 3t + 1 + 36\sin^2 3t}} = \frac{1}{\sqrt{37}} \langle 6\cos 3t, 1, -6\sin 3t \rangle.$$

$\mathbf{T}(\pi) = \frac{1}{\sqrt{37}} \langle -6, 1, 0 \rangle$ is a normal vector for the normal plane, and so $\langle -6, 1, 0 \rangle$ is also normal. Thus an equation for the plane is $-6(x - 0) + 1(y - \pi) + 0(z + 2) = 0$ or $y - 6x = \pi$.

$$\mathbf{T}'(t) = \frac{1}{\sqrt{37}} \langle -18\sin 3t, 0, -18\cos 3t \rangle \Rightarrow |\mathbf{T}'(t)| = \frac{\sqrt{18^2 \sin^2 3t + 18^2 \cos^2 3t}}{\sqrt{37}} = \frac{18}{\sqrt{37}} \Rightarrow$$

$$\mathbf{N}(t) = \frac{\mathbf{T}'(t)}{|\mathbf{T}'(t)|} = \langle -\sin 3t, 0, -\cos 3t \rangle. \text{ So } \mathbf{N}(\pi) = \langle 0, 0, 1 \rangle \text{ and } \mathbf{B}(\pi) = \frac{1}{\sqrt{37}} \langle -6, 1, 0 \rangle \times \langle 0, 0, 1 \rangle = \frac{1}{\sqrt{37}} \langle 1, 6, 0 \rangle.$$

Since $\mathbf{B}(\pi)$ is a normal to the osculating plane, so is $\langle 1, 6, 0 \rangle$.

An equation for the plane is $1(x - 0) + 6(y - \pi) + 0(z + 2) = 0$ or $x + 6y = 6\pi$.

38. $t = 1$ at $(1, 1, 1)$. $\mathbf{r}'(t) = \langle 1, 2t, 3t^2 \rangle$. $\mathbf{r}'(1) = \langle 1, 2, 3 \rangle$ is normal to the normal plane, so an equation for this plane is $1(x - 1) + 2(y - 1) + 3(z - 1) = 0$, or $x + 2y + 3z = 6$.

$\mathbf{T}(t) = \frac{\mathbf{r}'(t)}{|\mathbf{r}'(t)|} = \frac{1}{\sqrt{1 + 4t^2 + 9t^4}} \langle 1, 2t, 3t^2 \rangle$. Using the product rule on each term of $\mathbf{T}(t)$ gives

$$\begin{aligned}\mathbf{T}'(t) &= \frac{1}{(1 + 4t^2 + 9t^4)^{3/2}} \Big\langle -\tfrac{1}{2}(8t + 36t^3), 2(1 + 4t^2 + 9t^4) - \tfrac{1}{2}(8t + 36t^3)2t, \\ &\qquad\qquad 6t(1 + 4t^2 + 9t^4) - \tfrac{1}{2}(8t + 36t^3)3t^2 \Big\rangle \\ &= \frac{1}{(1 + 4t^2 + 9t^4)^{3/2}} \left\langle -4t - 18t^3, 2 - 18t^4, 6t + 12t^3 \right\rangle = \frac{-2}{(14)^{3/2}} \langle 11, 8, -9 \rangle \text{ when } t = 1.\end{aligned}$$

$\mathbf{N}(1) \parallel \mathbf{T}'(1) \parallel \langle 11, 8, -9 \rangle$ and $\mathbf{T}(1) \parallel \mathbf{r}'(1) = \langle 1, 2, 3 \rangle \Rightarrow$ a normal vector to the osculating plane is $\langle 11, 8, -9 \rangle \times \langle 1, 2, 3 \rangle = \langle 42, -42, 14 \rangle$ or equivalently $\langle 3, -3, 1 \rangle$.

An equation for the plane is $3(x - 1) - 3(y - 1) + (z - 1) = 0$ or $3x - 3y + z = 1$.

39. The ellipse is given by the parametric equations $x = 2\cos t$, $y = 3\sin t$, so using the result from Exercise 32,

$$\kappa(t) = \frac{|\dot{x}\ddot{y} - \ddot{x}\dot{y}|}{[\dot{x}^2 + \dot{y}^2]^{3/2}} = \frac{|(-2\sin t)(-3\sin t) - (3\cos t)(-2\cos t)|}{(4\sin^2 t + 9\cos^2 t)^{3/2}} = \frac{6}{(4\sin^2 t + 9\cos^2 t)^{3/2}}.$$

At $(2, 0)$, $t = 0$. Now $\kappa(0) = \frac{6}{27} = \frac{2}{9}$, so the radius of the osculating circle is $1/\kappa(0) = \frac{9}{2}$ and its center is $\left(-\frac{5}{2}, 0\right)$. Its equation is therefore $\left(x + \frac{5}{2}\right)^2 + y^2 = \frac{81}{4}$. At $(0, 3)$, $t = \frac{\pi}{2}$, and $\kappa\left(\frac{\pi}{2}\right) = \frac{6}{8} = \frac{3}{4}$. So the radius of the osculating circle is $\frac{4}{3}$ and its center is $\left(0, \frac{5}{3}\right)$. Hence its equation is $x^2 + \left(y - \frac{5}{3}\right)^2 = \frac{16}{9}$.

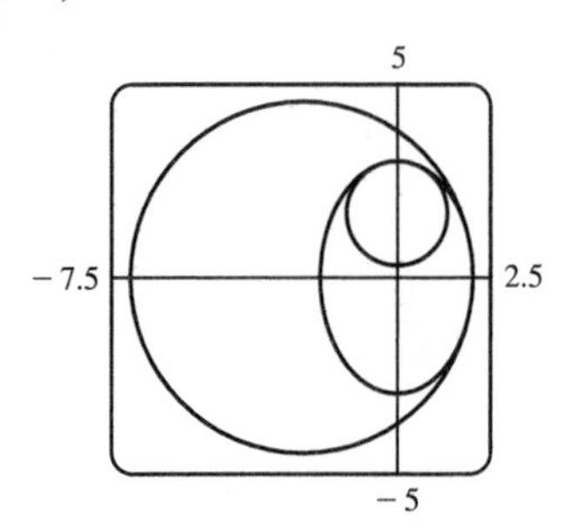

40. $y = \frac{1}{2}x^2 \Rightarrow y' = x$ and $y'' = 1$, so Formula 11 gives $\kappa(x) = \frac{1}{(1 + x^2)^{3/2}}$. So the curvature at $(0, 0)$ is $\kappa(0) = 1$ and the osculating circle has radius 1 and center $(0, 1)$, and hence equation $x^2 + (y - 1)^2 = 1$. The curvature at $\left(1, \frac{1}{2}\right)$

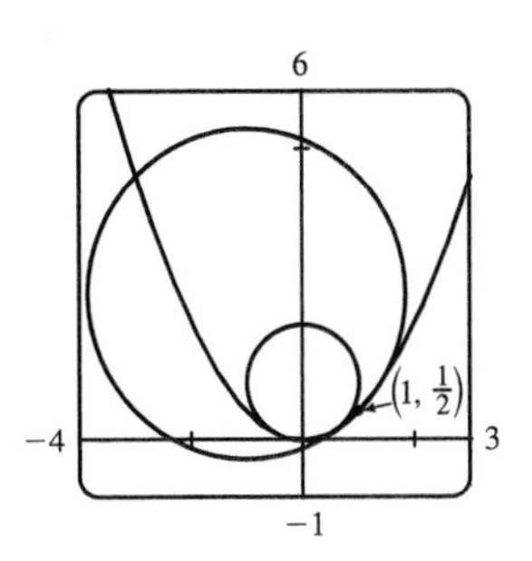

is $\kappa(1) = \dfrac{1}{(1+1^2)^{3/2}} = \dfrac{1}{2\sqrt{2}}$. The tangent line to the parabola at $\left(1, \frac{1}{2}\right)$ has slope 1, so the normal line has slope -1. Thus the center of the osculating circle lies in the direction of the unit vector $\left\langle -\frac{1}{\sqrt{2}}, \frac{1}{\sqrt{2}} \right\rangle$. The circle has radius $2\sqrt{2}$, so its center has position vector $\left\langle 1, \frac{1}{2} \right\rangle + 2\sqrt{2}\left\langle -\frac{1}{\sqrt{2}}, \frac{1}{\sqrt{2}} \right\rangle = \left\langle -1, \frac{5}{2} \right\rangle$. So the equation of the circle is $(x+1)^2 + \left(y - \frac{5}{2}\right)^2 = 8$.

41. The tangent vector is normal to the normal plane, and the vector $\langle 6, 6, -8 \rangle$ is normal to the given plane. But $\mathbf{T}(t) \parallel \mathbf{r}'(t)$ and $\langle 6, 6, -8 \rangle \parallel \langle 3, 3, -4 \rangle$, so we need to find t such that $\mathbf{r}'(t) \parallel \langle 3, 3, -4 \rangle$. $\mathbf{r}(t) = \langle t^3, 3t, t^4 \rangle \;\Rightarrow\; \mathbf{r}'(t) = \langle 3t^2, 3, 4t^3 \rangle \parallel \langle 3, 3, -4 \rangle$ when $t = -1$. So the planes are parallel at the point $(-1, -3, 1)$.

42. To find the osculating plane, we first calculate the tangent and normal vectors.

In Maple, we set `x:=t^3; y:=3*t;` and `z:=t^4;` and then calculate the components of the tangent vector $\mathbf{T}(t)$ using the `diff` command. We find that $\mathbf{T}(t) = \dfrac{\langle 3t^2, 3, 4t^3 \rangle}{\sqrt{16t^6 + 9t^4 + 9}}$. Differentiating the components of $\mathbf{T}(t)$, we find that

$$\mathbf{N}(t) = \frac{\mathbf{T}'(t)}{|\mathbf{T}'(t)|} = \frac{\left\langle -6t(8t^6 - 9), 3(48t^5 + 18t^3), 36t^2(t^4 + 3) \right\rangle}{\sqrt{144t(8t^6 - 9)^2 + 9(96t^5 + 36t^3)^2 + 5{,}184t^{12} + 31{,}104t^8 + 46{,}656t^4}}$$

In Maple, we can calculate $\mathbf{B}(t) = \mathbf{T}(t) \times \mathbf{N}(t)$ using the `linalg` package. First we define $\mathbf{T}$ and $\mathbf{N}$ using `T:=array([f,g,h]);` and `N:=array([F,G,H]);` where `f`, `g`, `h`, `F`, `G`, and `H` are the components of $\mathbf{T}$ and $\mathbf{N}$. Then we use the command `B:=crossprod(T,N);`. After normalization and simplification, we find that $\mathbf{B}(t) = b\left\langle 6t, -2t^3, -3 \right\rangle$, where $b = \dfrac{t\sqrt{16t^6 + 9t^4 + 9}}{\sqrt{16t^2(8t^6 - 9)^2 + (96t^5 + 36t^3)^2 + 576t^{12} + 3456t^8 + 5184t^4}}$. In Mathematica, we use the command `Dt` to differentiate the components of $\mathbf{r}(t)$ and subsequently $\mathbf{T}(t)$, and then load the vector analysis package with the command `<<Calculus`VectorAnalysis``. After setting `T={f,g,h}` and `N={F,G,H}`, we use `CrossProduct[T,N]` to find $\mathbf{B}$ (before normalization). Now $\mathbf{B}(t)$ is parallel to $\left\langle 6t, -2t^3, -3 \right\rangle$, so if $\mathbf{B}(t)$ is parallel to $\langle 1, 1, 1 \rangle$ for some t, then $6t = 1 \;\Rightarrow\; t = \frac{1}{6}$, but $-2\left(\frac{1}{6}\right)^3 \neq 1$. So there is no such osculating plane.

43. $\kappa = \left|\dfrac{d\mathbf{T}}{ds}\right| = \left|\dfrac{d\mathbf{T}/dt}{ds/dt}\right| = \dfrac{|d\mathbf{T}/dt|}{ds/dt}$ and $\mathbf{N} = \dfrac{d\mathbf{T}/dt}{|d\mathbf{T}/dt|}$, so $\kappa\mathbf{N} = \dfrac{\left|\dfrac{d\mathbf{T}}{dt}\right| \dfrac{d\mathbf{T}}{dt}}{\left|\dfrac{d\mathbf{T}}{dt}\right| \dfrac{ds}{dt}} = \dfrac{d\mathbf{T}/dt}{ds/dt} = \dfrac{d\mathbf{T}}{ds}$ by the Chain Rule.

44. For a plane curve, $\mathbf{T} = |\mathbf{T}| \cos\phi\, \mathbf{i} + |\mathbf{T}| \sin\phi\, \mathbf{j} = \cos\phi\, \mathbf{i} + \sin\phi\, \mathbf{j}$. Then

$\dfrac{d\mathbf{T}}{ds} = \left(\dfrac{d\mathbf{T}}{d\phi}\right)\left(\dfrac{d\phi}{ds}\right) = (-\sin\phi\, \mathbf{i} + \cos\phi\, \mathbf{j})\left(\dfrac{d\phi}{ds}\right)$ and $\left|\dfrac{d\mathbf{T}}{ds}\right| = |-\sin\phi\, \mathbf{i} + \cos\phi\, \mathbf{j}| \left|\dfrac{d\phi}{ds}\right| = \left|\dfrac{d\phi}{ds}\right|$. Hence for a plane curve, the curvature is $\kappa = |d\phi/ds|$.

45. (a) $|\mathbf{B}| = 1 \Rightarrow \mathbf{B}\cdot\mathbf{B} = 1 \Rightarrow \dfrac{d}{ds}(\mathbf{B}\cdot\mathbf{B}) = 0 \Rightarrow 2\dfrac{d\mathbf{B}}{ds}\cdot\mathbf{B} = 0 \Rightarrow \dfrac{d\mathbf{B}}{ds}\perp\mathbf{B}$

(b) $\mathbf{B} = \mathbf{T}\times\mathbf{N} \Rightarrow$

$$\frac{d\mathbf{B}}{ds} = \frac{d}{ds}(\mathbf{T}\times\mathbf{N}) = \frac{d}{dt}(\mathbf{T}\times\mathbf{N})\frac{1}{ds/dt} = \frac{d}{dt}(\mathbf{T}\times\mathbf{N})\frac{1}{|\mathbf{r}'(t)|} = [(\mathbf{T}'\times\mathbf{N}) + (\mathbf{T}\times\mathbf{N}')]\frac{1}{|\mathbf{r}'(t)|}$$

$$= \left[\left(\mathbf{T}'\times\frac{\mathbf{T}'}{|\mathbf{T}'|}\right) + (\mathbf{T}\times\mathbf{N}')\right]\frac{1}{|\mathbf{r}'(t)|} = \frac{\mathbf{T}\times\mathbf{N}'}{|\mathbf{r}'(t)|} \Rightarrow \frac{d\mathbf{B}}{ds}\perp\mathbf{T}$$

(c) $\mathbf{B} = \mathbf{T}\times\mathbf{N} \Rightarrow \mathbf{T}\perp\mathbf{N}$, $\mathbf{B}\perp\mathbf{T}$ and $\mathbf{B}\perp\mathbf{N}$. So $\mathbf{B}$, $\mathbf{T}$ and $\mathbf{N}$ form an orthogonal set of vectors in the three-dimensional space $\mathbb{R}^3$. From parts (a) and (b), $d\mathbf{B}/ds$ is perpendicular to both $\mathbf{B}$ and $\mathbf{T}$, so $d\mathbf{B}/ds$ is parallel to $\mathbf{N}$. Therefore, $d\mathbf{B}/ds = -\tau(s)\mathbf{N}$, where $\tau(s)$ is a scalar.

(d) Since $\mathbf{B} = \mathbf{T}\times\mathbf{N}$, $\mathbf{T}\perp\mathbf{N}$ and both $\mathbf{T}$ and $\mathbf{N}$ are unit vectors, $\mathbf{B}$ is a unit vector mutually perpendicular to both $\mathbf{T}$ and $\mathbf{N}$. For a plane curve, $\mathbf{T}$ and $\mathbf{N}$ always lie in the plane of the curve, so that $\mathbf{B}$ is a constant unit vector always perpendicular to the plane. Thus $d\mathbf{B}/ds = \mathbf{0}$, but $d\mathbf{B}/ds = -\tau(s)\mathbf{N}$ and $\mathbf{N}\neq\mathbf{0}$, so $\tau(s) = 0$.

46. $\mathbf{N} = \mathbf{B}\times\mathbf{T} \Rightarrow$

$$\begin{aligned}\frac{d\mathbf{N}}{ds} &= \frac{d}{ds}(\mathbf{B}\times\mathbf{T}) = \frac{d\mathbf{B}}{ds}\times\mathbf{T} + \mathbf{B}\times\frac{d\mathbf{T}}{ds} && \text{[by Fomula 5 of Theorem 10.7.5]}\\ &= -\tau\mathbf{N}\times\mathbf{T} + \mathbf{B}\times\kappa\mathbf{N} && \text{[by Formulas 3 and 1]}\\ &= -\tau(\mathbf{N}\times\mathbf{T}) + \kappa(\mathbf{B}\times\mathbf{N}) && \text{[by Property 2 of the cross product]}\end{aligned}$$

But $\mathbf{B}\times\mathbf{N} = \mathbf{B}\times(\mathbf{B}\times\mathbf{T}) = (\mathbf{B}\cdot\mathbf{T})\mathbf{B} - (\mathbf{B}\cdot\mathbf{B})\mathbf{T}$ [by Property 6 of Theorem 10.4.8] $= -\mathbf{T} \Rightarrow$ $d\mathbf{N}/ds = \tau(\mathbf{T}\times\mathbf{N}) - \kappa\mathbf{T} = -\kappa\,\mathbf{T} + \tau\,\mathbf{B}$.

47. (a) $\mathbf{r}' = s'\,\mathbf{T} \Rightarrow \mathbf{r}'' = s''\,\mathbf{T} + s'\,\mathbf{T}' = s''\,\mathbf{T} + s'\,\dfrac{d\mathbf{T}}{ds}s' = s''\,\mathbf{T} + \kappa(s')^2\,\mathbf{N}$ by the first Serret-Frenet formula.

(b) Using part (a), we have

$$\begin{aligned}\mathbf{r}'\times\mathbf{r}'' &= (s'\,\mathbf{T})\times[s''\,\mathbf{T} + \kappa(s')^2\,\mathbf{N}]\\ &= [(s'\,\mathbf{T})\times(s''\,\mathbf{T})] + [(s'\mathbf{T})\times(\kappa(s')^2\,\mathbf{N})] && \text{[by Property 3 of the cross product]}\\ &= (s's'')(\mathbf{T}\times\mathbf{T}) + \kappa(s')^3(\mathbf{T}\times\mathbf{N}) = \mathbf{0} + \kappa(s')^3\,\mathbf{B} = \kappa(s')^3\,\mathbf{B}\end{aligned}$$

(c) Using part (a), we have

$$\begin{aligned}\mathbf{r}''' &= [s''\,\mathbf{T} + \kappa(s')^2\,\mathbf{N}]' = s'''\,\mathbf{T} + s''\,\mathbf{T}' + \kappa'(s')^2\,\mathbf{N} + 2\kappa s's''\,\mathbf{N} + \kappa(s')^2\,\mathbf{N}'\\ &= s'''\,\mathbf{T} + s''\frac{d\mathbf{T}}{ds}\,s' + \kappa'(s')^2\,\mathbf{N} + 2\kappa s's''\,\mathbf{N} + \kappa(s')^2\,\frac{d\mathbf{N}}{ds}\,s'\\ &= s'''\,\mathbf{T} + s''s'\kappa\,\mathbf{N} + \kappa'(s')^2\,\mathbf{N} + 2\kappa s's''\,\mathbf{N} + \kappa(s')^3(-\kappa\,\mathbf{T} + \tau\,\mathbf{B}) && \text{[by the second formula]}\\ &= [s''' - \kappa^2(s')^3]\,\mathbf{T} + [3\kappa s's'' + \kappa'(s')^2]\,\mathbf{N} + \kappa\tau(s')^3\,\mathbf{B}\end{aligned}$$

(d) Using parts (b) and (c) and the facts that $\mathbf{B}\cdot\mathbf{T} = 0$, $\mathbf{B}\cdot\mathbf{N} = 0$, and $\mathbf{B}\cdot\mathbf{B} = 1$, we get

$$\frac{(\mathbf{r}'\times\mathbf{r}'')\cdot\mathbf{r}'''}{|\mathbf{r}'\times\mathbf{r}''|^2} = \frac{\kappa(s')^3\,\mathbf{B}\cdot\{[s''' - \kappa^2(s')^3]\,\mathbf{T} + [3\kappa s's'' + \kappa'(s')^2]\,\mathbf{N} + \kappa\tau(s')^3\,\mathbf{B}\}}{|\kappa(s')^3\,\mathbf{B}|^2} = \frac{\kappa(s')^3\kappa\tau(s')^3}{[\kappa(s')^3]^2} = \tau.$$

48. First we find the quantities required to compute κ:

$$\mathbf{r}'(t) = \langle -a\sin t, a\cos t, b\rangle \quad\Rightarrow\quad \mathbf{r}''(t) = \langle -a\cos t, -a\sin t, 0\rangle \quad\Rightarrow\quad \mathbf{r}'''(t) = \langle a\sin t, -a\cos t, 0\rangle$$

$$|\mathbf{r}'(t)| = \sqrt{(-a\sin t)^2 + (a\cos t)^2 + b^2} = \sqrt{a^2+b^2}$$

$$\mathbf{r}'(t)\times\mathbf{r}''(t) = \begin{vmatrix} \mathbf{i} & \mathbf{j} & \mathbf{k} \\ -a\sin t & a\cos t & b \\ -a\cos t & -a\sin t & 0 \end{vmatrix} = ab\sin t\,\mathbf{i} - ab\cos t\,\mathbf{j} + a^2\,\mathbf{k}$$

$$|\mathbf{r}'(t)\times\mathbf{r}''(t)| = \sqrt{(ab\sin t)^2 + (-ab\cos t)^2 + (a^2)^2} = \sqrt{a^2b^2+a^4}$$

$$(\mathbf{r}'(t)\times\mathbf{r}''(t))\cdot\mathbf{r}'''(t) = (ab\sin t)(a\sin t) + (-ab\cos t)(-a\cos t) + (a^2)(0) = a^2b$$

Then by Theorem 10, $\kappa(t) = \dfrac{|\mathbf{r}'(t)\times\mathbf{r}''(t)|}{|\mathbf{r}'(t)|^3} = \dfrac{\sqrt{a^2b^2+a^4}}{\left(\sqrt{a^2+b^2}\right)^3} = \dfrac{a\sqrt{a^2+b^2}}{\left(\sqrt{a^2+b^2}\right)^3} = \dfrac{a}{a^2+b^2}$ which is a constant.

From Exercise 47(d), the torsion τ is given by $\tau = \dfrac{(\mathbf{r}'\times\mathbf{r}'')\cdot\mathbf{r}'''}{|\mathbf{r}'\times\mathbf{r}''|^2} = \dfrac{a^2b}{\left(\sqrt{a^2b^2+a^4}\right)^2} = \dfrac{b}{a^2+b^2}$ which is also a constant.

49. For one helix, the vector equation is $\mathbf{r}(t) = \langle 10\cos t, 10\sin t, 34t/(2\pi)\rangle$ (measuring in angstroms), because the radius of each helix is 10 angstroms, and z increases by 34 angstroms for each increase of 2π in t. Using the arc length formula, letting t go from 0 to $2.9\times 10^8\times 2\pi$, we find the approximate length of each helix to be

$$L = \int_0^{2.9\times10^8\times2\pi} |\mathbf{r}'(t)|\,dt = \int_0^{2.9\times10^8\times2\pi} \sqrt{(-10\sin t)^2 + (10\cos t)^2 + \left(\tfrac{34}{2\pi}\right)^2}\,dt = \sqrt{100 + \left(\tfrac{34}{2\pi}\right)^2}\,\Big]_0^{2.9\times10^8\times2\pi}$$

$$= 2.9\times10^8\times2\pi\sqrt{100+\left(\tfrac{34}{2\pi}\right)^2} \approx 2.07\times10^{10}\text{ Å — more than two meters!}$$

50. (a) For the function $F(x) = \begin{cases} 0 & \text{if } x<0 \\ P(x) & \text{if } 0<x<1 \\ 1 & \text{if } x\geq 1 \end{cases}$ to be continuous, we must have $P(0)=0$ and $P(1)=1$.

For F' to be continuous, we must have $P'(0) = P'(1) = 0$. The curvature of the curve $y = F(x)$ at the point $(x, F(x))$ is $\kappa(x) = \dfrac{|F''(x)|}{\left(1+[F'(x)]^2\right)^{3/2}}$. For $\kappa(x)$ to be continuous, we must have $P''(0) = P''(1) = 0$.

Write $P(x) = ax^5 + bx^4 + cx^3 + dx^2 + ex + f$. Then $P'(x) = 5ax^4 + 4bx^3 + 3cx^2 + 2dx + e$ and $P''(x) = 20ax^3 + 12bx^2 + 6cx + 2d$. Our six conditions are:

$$P(0)=0 \;\Rightarrow\; f=0 \quad (1) \qquad\qquad P(1)=1 \;\Rightarrow\; a+b+c+d+e+f=1 \quad (2)$$

$$P'(0)=0 \;\Rightarrow\; e=0 \quad (3) \qquad\qquad P'(1)=0 \;\Rightarrow\; 5a+4b+3c+2d+e=0 \quad (4)$$

$$P''(0)=0 \;\Rightarrow\; d=0 \quad (5) \qquad\qquad P''(1)=0 \;\Rightarrow\; 20a+12b+6c+2d=0 \quad (6)$$

From (1), (3), and (5), we have $d=e=f=0$. Thus (2), (4) and (6) become (7) $a+b+c=1$, (8) $5a+4b+3c=0$, and (9) $10a+6b+3c=0$. Subtracting (8) from (9) gives (10) $5a+2b=0$. Multiplying (7) by 3 and subtracting from (8) gives (11) $2a+b=-3$. Multiplying (11) by 2 and subtracting from (10) gives $a=6$. By (10), $b=-15$. By (7), $c=10$. Thus, $P(x) = 6x^5 - 15x^4 + 10x^3$.

(b)

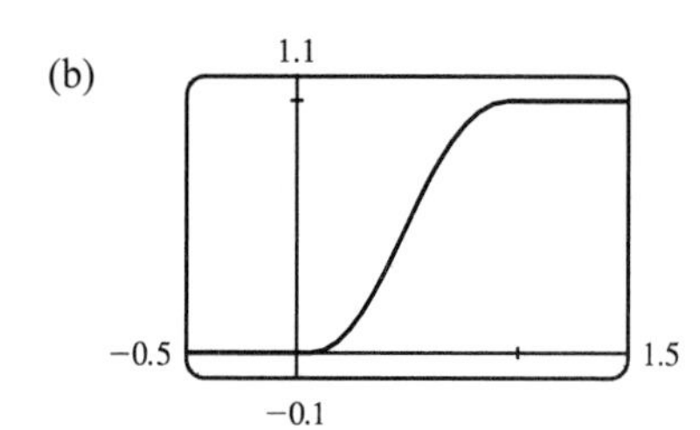

10.9 Motion in Space: Velocity and Acceleration

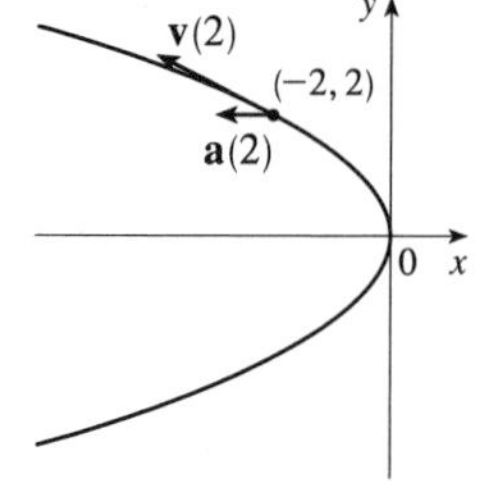

1. $\mathbf{r}(t) = \langle -\frac{1}{2}t^2, t\rangle \Rightarrow$ At $t = 2$:

$\mathbf{v}(t) = \mathbf{r}'(t) = \langle -t, 1\rangle$ $\quad\mathbf{v}(2) = \langle -2, 1\rangle$

$\mathbf{a}(t) = \mathbf{r}''(t) = \langle -1, 0\rangle$ $\quad\mathbf{a}(2) = \langle -1, 0\rangle$

$|\mathbf{v}(t)| = \sqrt{t^2 + 1}$

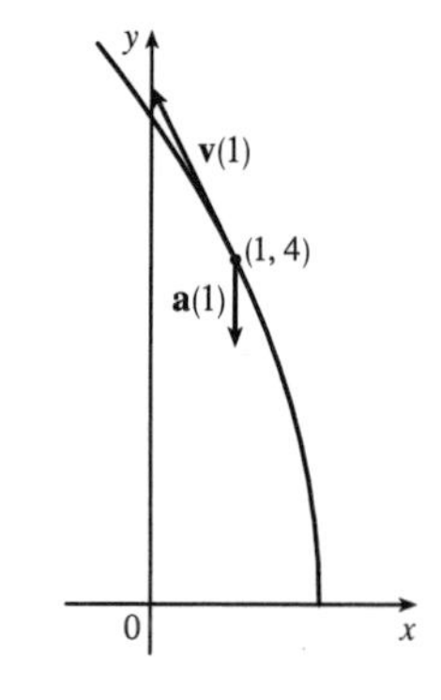

2. $\mathbf{r}(t) = \langle 2 - t, 4\sqrt{t}\rangle \Rightarrow$ At $t = 1$:

$\mathbf{v}(t) = \mathbf{r}'(t) = \langle -1, 2/\sqrt{t}\rangle$ $\quad\mathbf{v}(1) = \langle -1, 2\rangle$

$\mathbf{a}(t) = \mathbf{r}''(t) = \left\langle 0, -1/t^{3/2}\right\rangle$ $\quad\mathbf{a}(1) = \langle 0, -1\rangle$

$|\mathbf{v}(t)| = \sqrt{1 + 4/t}$

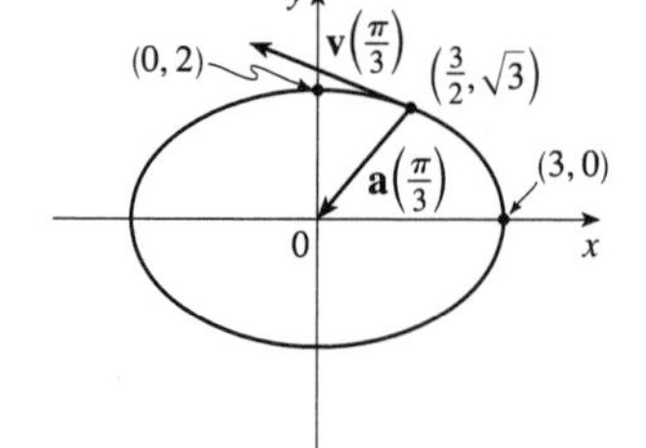

3. $(t) = 3\cos t\,\mathbf{i} + 2\sin t\,\mathbf{j} \Rightarrow$ At $t = \pi/3$:

$\mathbf{v}(t) = -3\sin t\,\mathbf{i} + 2\cos t\,\mathbf{j}$ $\quad\mathbf{v}\left(\frac{\pi}{3}\right) = -\frac{3\sqrt{3}}{2}\,\mathbf{i} + \mathbf{j}$

$\mathbf{a}(t) = -3\cos t\,\mathbf{i} - 2\sin t\,\mathbf{j}$ $\quad\mathbf{a}\left(\frac{\pi}{3}\right) = -\frac{3}{2}\,\mathbf{i} - \sqrt{3}\,\mathbf{j}$

$|\mathbf{v}(t)| = \sqrt{9\sin^2 t + 4\cos^2 t} = \sqrt{4 + 5\sin^2 t}$

Notice that $x^2/9 + y^2/4 = \sin^2 t + \cos^2 t = 1$, so the path is an ellipse.

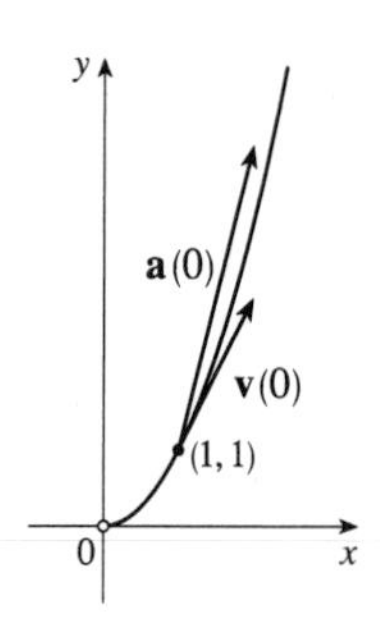

4. $\mathbf{r}(t) = e^t\,\mathbf{i} + e^{2t}\,\mathbf{j} \Rightarrow$ At $t = 0$:

$\mathbf{v}(t) = e^t\,\mathbf{i} + 2e^{2t}\,\mathbf{j}$ $\quad\mathbf{v}(0) = \mathbf{i} + 2\,\mathbf{j}$,

$\mathbf{a}(t) = e^t\,\mathbf{i} + 4e^{2t}\,\mathbf{j}$ $\quad\mathbf{a}(0) = \mathbf{i} + 4\,\mathbf{j}$

$|\mathbf{v}(t)| = \sqrt{e^{2t} + 4e^{4t}} = e^t\sqrt{1 + 4e^{2t}}$

Notice that $y = e^{2t} = \left(e^t\right)^2 = x^2$, so the particle travels along a parabola, but $x = e^t$, so $x > 0$.

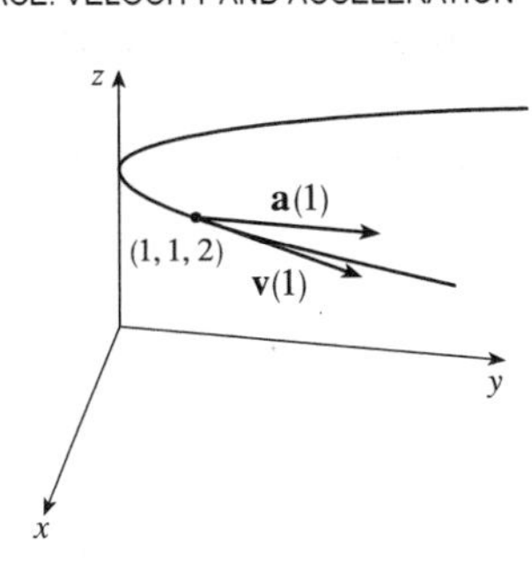

5. $\mathbf{r}(t) = t\,\mathbf{i} + t^2\,\mathbf{j} + 2\,\mathbf{k} \;\Rightarrow$ At $t = 1$:

$\mathbf{v}(t) = \mathbf{i} + 2t\,\mathbf{j}$ $\qquad$ $\mathbf{v}(1) = \mathbf{i} + 2\,\mathbf{j}$

$\mathbf{a}(t) = 2\,\mathbf{j}$ $\qquad$ $\mathbf{a}(1) = 2\,\mathbf{j}$

$|\mathbf{v}(t)| = \sqrt{1 + 4t^2}$

Here $x = t$, $y = t^2 \;\Rightarrow\; y = x^2$ and $z = 2$, so the path of the particle is a parabola in the plane $z = 2$.

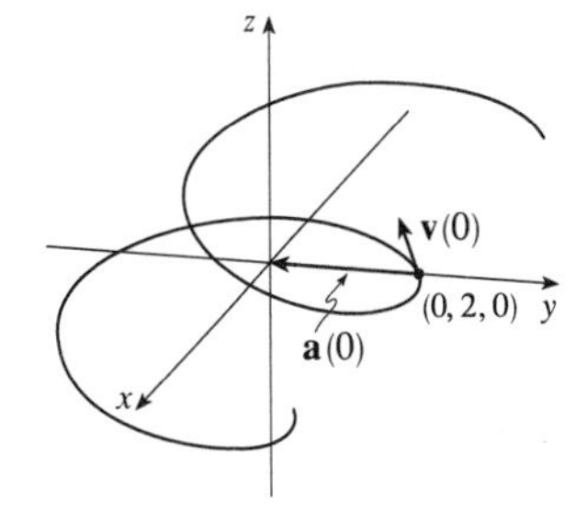

6. $\mathbf{r}(t) = t\,\mathbf{i} + 2\cos t\,\mathbf{j} + \sin t\,\mathbf{k} \;\Rightarrow$ At $t = 0$:

$\mathbf{v}(t) = \mathbf{i} - 2\sin t\,\mathbf{j} + \cos t\,\mathbf{k}$ $\qquad$ $\mathbf{v}(0) = \mathbf{i} + \mathbf{k}$

$\mathbf{a}(t) = -2\cos t\,\mathbf{j} - \sin t\,\mathbf{k}$ $\qquad$ $\mathbf{a}(0) = -2\,\mathbf{j}$

$|\mathbf{v}(t)| = \sqrt{1 + 4\sin^2 t + \cos^2 t} = \sqrt{2 + 3\sin^2 t}$

Since $y^2/4 + z^2 = 1$, $x = t$, the path of the particle is an elliptical helix about the x-axis.

7. $\mathbf{r}(t) = \langle t^2 + 1, t^3, t^2 - 1\rangle \;\Rightarrow\; \mathbf{v}(t) = \mathbf{r}'(t) = \langle 2t, 3t^2, 2t\rangle,\;\; \mathbf{a}(t) = \mathbf{v}'(t) = \langle 2, 6t, 2\rangle,$

$|\mathbf{v}(t)| = \sqrt{(2t)^2 + (3t^2)^2 + (2t)^2} = \sqrt{9t^4 + 8t^2} = |t|\,\sqrt{9t^2 + 8}.$

8. $\mathbf{r}(t) = \langle 2\cos t, 3t, 2\sin t\rangle \;\Rightarrow\; \mathbf{v}(t) = \mathbf{r}'(t) = \langle -2\sin t, 3, 2\cos t\rangle,\;\; \mathbf{a}(t) = \mathbf{v}'(t) = \langle -2\cos t, 0, -2\sin t\rangle,$

$|\mathbf{v}(t)| = \sqrt{4\sin^2 t + 9 + 4\cos^2 t} = \sqrt{13}.$

9. $\mathbf{r}(t) = \sqrt{2}\,t\,\mathbf{i} + e^t\,\mathbf{j} + e^{-t}\,\mathbf{k} \;\Rightarrow\; \mathbf{v}(t) = \mathbf{r}'(t) = \sqrt{2}\,\mathbf{i} + e^t\,\mathbf{j} - e^{-t}\,\mathbf{k},\;\; \mathbf{a}(t) = \mathbf{v}'(t) = e^t\,\mathbf{j} + e^{-t}\,\mathbf{k},$

$|\mathbf{v}(t)| = \sqrt{2 + e^{2t} + e^{-2t}} = \sqrt{(e^t + e^{-t})^2} = e^t + e^{-t}.$

10. $\mathbf{r}(t) = t\sin t\,\mathbf{i} + t\cos t\,\mathbf{j} + t^2\,\mathbf{k} \;\Rightarrow\; \mathbf{v}(t) = \mathbf{r}'(t) = (\sin t + t\cos t)\,\mathbf{i} + (\cos t - t\sin t)\,\mathbf{j} + 2t\,\mathbf{k},$

$\mathbf{a}(t) = \mathbf{v}'(t) = (2\cos t - t\sin t)\,\mathbf{i} + (-2\sin t - t\cos t)\,\mathbf{j} + 2\,\mathbf{k},$

$|\mathbf{v}(t)| = \sqrt{(\sin^2 t + 2t\sin t\,\cos t + t^2\cos^2 t) + (\cos^2 t - 2t\sin t\,\cos t + t^2\sin^2 t) + 4t^2} = \sqrt{5t^2 + 1}.$

11. $\mathbf{a}(t) = \mathbf{i} + 2\,\mathbf{j} \;\Rightarrow\; \mathbf{v}(t) = \int \mathbf{a}(t)\,dt = \int(\mathbf{i} + 2\mathbf{j})\,dt = t\,\mathbf{i} + 2t\,\mathbf{j} + \mathbf{C}$ and $\mathbf{k} = \mathbf{v}\,(0) = \mathbf{C}$,

so $\mathbf{C} = \mathbf{k}$ and $\mathbf{v}(t) = t\,\mathbf{i} + 2t\,\mathbf{j} + \mathbf{k}$. $\;\mathbf{r}(t) = \int \mathbf{v}(t)\,dt = \int (t\,\mathbf{i} + 2t\,\mathbf{j} + \mathbf{k})\,dt = \frac{1}{2}t^2\,\mathbf{i} + t^2\,\mathbf{j} + t\,\mathbf{k} + \mathbf{D}.$

But $\mathbf{i} = \mathbf{r}\,(0) = \mathbf{D}$, so $\mathbf{D} = \mathbf{i}$ and $\mathbf{r}(t) = \left(\frac{1}{2}t^2 + 1\right)\mathbf{i} + t^2\,\mathbf{j} + t\,\mathbf{k}.$

12. $\mathbf{a}(t) = 2\,\mathbf{i} + 6t\,\mathbf{j} + 12t^2\,\mathbf{k} \;\Rightarrow\; \mathbf{v}(t) = \int(2\,\mathbf{i} + 6t\,\mathbf{j} + 12t^2\,\mathbf{k})\,dt = 2t\,\mathbf{i} + 3t^2\,\mathbf{j} + 4t^3\,\mathbf{k} + \mathbf{C}$, and $\mathbf{i} = \mathbf{v}(0) = \mathbf{C}$,

so $\mathbf{C} = \mathbf{i}$ and $\mathbf{v}(t) = (2t + 1)\mathbf{i} + 3t^2\,\mathbf{j} + 4t^3\,\mathbf{k}$. $\;\mathbf{r}(t) = \int \left[(2t + 1)\mathbf{i} + 3t^2\,\mathbf{j} + 4t^3\,\mathbf{k}\right] dt = (t^2 + t)\,\mathbf{i} + t^3\,\mathbf{j} + t^4\,\mathbf{k} + \mathbf{D}.$

But $\mathbf{j} - \mathbf{k} = \mathbf{r}(0) = \mathbf{D}$, so $\mathbf{D} = \mathbf{j} - \mathbf{k}$ and $\mathbf{r}(t) = (t^2 + t)\,\mathbf{i} + (t^3 + 1)\,\mathbf{j} + (t^4 - 1)\,\mathbf{k}.$

13. (a) $\mathbf{a}(t) = 2t\,\mathbf{i} + \sin t\,\mathbf{j} + \cos 2t\,\mathbf{k} \;\Rightarrow$

$\mathbf{v}(t) = \int (2t\,\mathbf{i} + \sin t\,\mathbf{j} + \cos 2t\,\mathbf{k})\,dt = t^2\,\mathbf{i} - \cos t\,\mathbf{j} + \frac{1}{2}\sin 2t\,\mathbf{k} + \mathbf{C}$

and $\mathbf{i} = \mathbf{v}\,(0) = -\mathbf{j} + \mathbf{C}$, so $\mathbf{C} = \mathbf{i} + \mathbf{j}$

and $\mathbf{v}(t) = (t^2+1)\,\mathbf{i} + (1-\cos t)\,\mathbf{j} + \frac{1}{2}\sin 2t\,\mathbf{k}$.

$\mathbf{r}(t) = \int [(t^2+1)\;\mathbf{i} + (1-\cos t)\,\mathbf{j} + \frac{1}{2}\sin 2t\,\mathbf{k}]dt$

$= \left(\frac{1}{3}t^3 + t\right)\,\mathbf{i} + (t - \sin t)\,\mathbf{j} - \frac{1}{4}\cos 2t\,\mathbf{k} + \mathbf{D}$

(b)

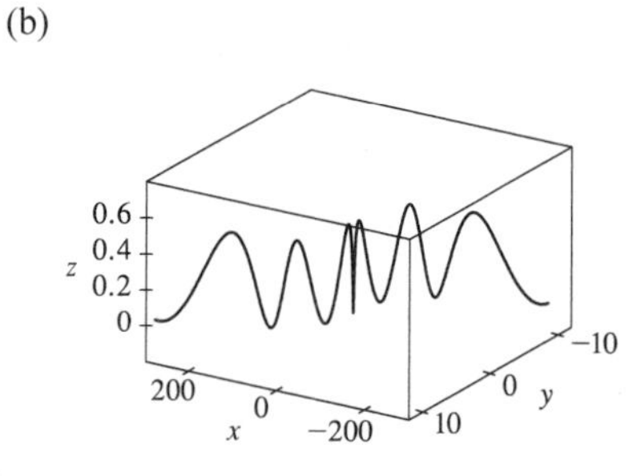

But $\mathbf{j} = \mathbf{r}\,(0) = -\frac{1}{4}\,\mathbf{k} + \mathbf{D}$, so $\mathbf{D} = \mathbf{j} + \frac{1}{4}\,\mathbf{k}$ and $\mathbf{r}(t) = \left(\frac{1}{3}t^3 + t\right)\mathbf{i} + (t - \sin t + 1)\,\mathbf{j} + \left(\frac{1}{4} - \frac{1}{4}\cos 2t\right)\mathbf{k}$.

14. (a) $\mathbf{a}(t) = t\,\mathbf{i} + e^t\,\mathbf{j} + e^{-t}\,\mathbf{k} \;\Rightarrow$

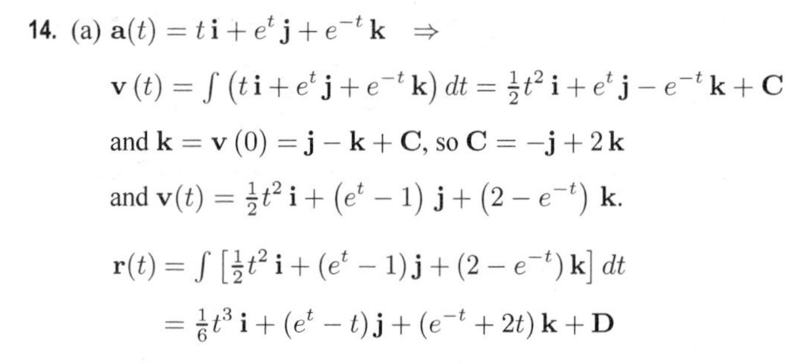

$\mathbf{v}\,(t) = \int \left(t\,\mathbf{i} + e^t\,\mathbf{j} + e^{-t}\,\mathbf{k}\right)dt = \frac{1}{2}t^2\,\mathbf{i} + e^t\,\mathbf{j} - e^{-t}\,\mathbf{k} + \mathbf{C}$

and $\mathbf{k} = \mathbf{v}\,(0) = \mathbf{j} - \mathbf{k} + \mathbf{C}$, so $\mathbf{C} = -\mathbf{j} + 2\,\mathbf{k}$

and $\mathbf{v}(t) = \frac{1}{2}t^2\,\mathbf{i} + \left(e^t - 1\right)\,\mathbf{j} + \left(2 - e^{-t}\right)\,\mathbf{k}$.

$\mathbf{r}(t) = \int \left[\frac{1}{2}t^2\,\mathbf{i} + (e^t - 1)\,\mathbf{j} + (2 - e^{-t})\,\mathbf{k}\right]dt$

$= \frac{1}{6}t^3\,\mathbf{i} + (e^t - t)\,\mathbf{j} + (e^{-t} + 2t)\,\mathbf{k} + \mathbf{D}$

(b)

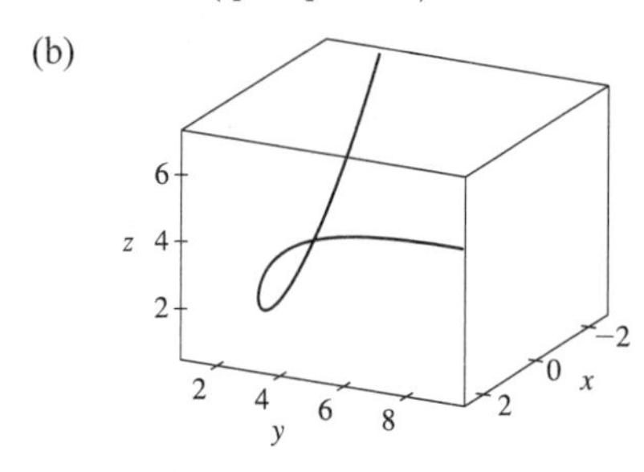

But $\mathbf{j} + \mathbf{k} = \mathbf{r}(0) = \mathbf{j} + \mathbf{k} + \mathbf{D}$, so $\mathbf{D} = \mathbf{0}$ and $\mathbf{r}(t) = \frac{1}{6}t^3\,\mathbf{i} + (e^t - t)\,\mathbf{j} + (e^{-t} + 2t)\,\mathbf{k}$.

15. $\mathbf{r}(t) = \langle t^2, 5t, t^2 - 16t\rangle \;\Rightarrow\; \mathbf{v}(t) = \langle 2t, 5, 2t - 16\rangle$, $|\mathbf{v}(t)| = \sqrt{4t^2 + 25 + 4t^2 - 64t + 256} = \sqrt{8t^2 - 64t + 281}$

and $\dfrac{d}{dt}\,|\mathbf{v}(t)| = \frac{1}{2}(8t^2 - 64t + 281)^{-1/2}(16t - 64)$. This is zero if and only if the numerator is zero, that is,

$16t - 64 = 0$ or $t = 4$. Since $\dfrac{d}{dt}\,|\mathbf{v}(t)| < 0$ for $t < 4$ and $\dfrac{d}{dt}\,|\mathbf{v}(t)| > 0$ for $t > 4$, the minimum speed of $\sqrt{153}$ is attained

at $t = 4$ units of time.

16. Since $\mathbf{r}(t) = t^3\,\mathbf{i} + t^2\,\mathbf{j} + t^3\,\mathbf{k}$, $\mathbf{a}(t) = \mathbf{r}''(t) = 6t\,\mathbf{i} + 2\,\mathbf{j} + 6t\,\mathbf{k}$. By Newton's Second Law,

$\mathbf{F}(t) = m\,\mathbf{a}(t) = 6mt\,\mathbf{i} + 2m\,\mathbf{j} + 6mt\,\mathbf{k}$ is the required force.

17. $|\mathbf{F}(t)| = 20$ N in the direction of the positive z-axis, so $\mathbf{F}(t) = 20\,\mathbf{k}$. Also $m = 4$ kg, $\mathbf{r}(0) = \mathbf{0}$ and $\mathbf{v}(0) = \mathbf{i} - \mathbf{j}$.

Since $20\mathbf{k} = \mathbf{F}(t) = 4\,\mathbf{a}(t)$, $\mathbf{a}(t) = 5\,\mathbf{k}$. Then $\mathbf{v}(t) = 5t\,\mathbf{k} + \mathbf{c}_1$ where $\mathbf{c}_1 = \mathbf{i} - \mathbf{j}$ so $\mathbf{v}(t) = \mathbf{i} - \mathbf{j} + 5t\,\mathbf{k}$ and the

speed is $|\mathbf{v}(t)| = \sqrt{1 + 1 + 25t^2} = \sqrt{25t^2 + 2}$. Also $\mathbf{r}(t) = t\,\mathbf{i} - t\,\mathbf{j} + \frac{5}{2}t^2\,\mathbf{k} + \mathbf{c}_2$ and $\mathbf{0} = \mathbf{r}(0)$, so $\mathbf{c}_2 = \mathbf{0}$

and $\mathbf{r}(t) = t\,\mathbf{i} - t\,\mathbf{j} + \frac{5}{2}t^2\,\mathbf{k}$.

18. The argument here is the same as that in Example 10.7.12 with $\mathbf{r}(t)$ replaced by $\mathbf{v}(t)$ and $\mathbf{r}'(t)$ replaced by $\mathbf{a}(t)$.

19. $|\mathbf{v}(0)| = 500$ m/s and since the angle of elevation is 30°, the direction of the velocity is $\frac{1}{2}\left(\sqrt{3}\,\mathbf{i} + \mathbf{j}\right)$. Thus

$\mathbf{v}(0) = 250\left(\sqrt{3}\,\mathbf{i} + \mathbf{j}\right)$ and if we set up the axes so the projectile starts at the origin, then $\mathbf{r}(0) = \mathbf{0}$. Ignoring air resistance, the

only force is that due to gravity, so $\mathbf{F}(t) = -mg\,\mathbf{j}$ where $g \approx 9.8$ m/s^2. Thus $\mathbf{a}(t) = -g\,\mathbf{j}$ and $\mathbf{v}(t) = -gt\,\mathbf{j} + \mathbf{c}_1$. But

$250\left(\sqrt{3}\,\mathbf{i} + \mathbf{j}\right) = \mathbf{v}(0) = \mathbf{c}_1$, so $\mathbf{v}(t) = 250\sqrt{3}\,\mathbf{i} + (250 - gt)\,\mathbf{j}$ and $\mathbf{r}(t) = 250\sqrt{3}\,t\,\mathbf{i} + \left(250t - \frac{1}{2}gt^2\right)\mathbf{j} + \mathbf{c}_2$ where

$\mathbf{0} = \mathbf{r}(0) = \mathbf{c}_2$. Thus $\mathbf{r}(t) = 250\sqrt{3}\,t\,\mathbf{i} + \left(250t - \frac{1}{2}gt^2\right)\mathbf{j}$.

(a) Setting $250t - \frac{1}{2}gt^2 = 0$ gives $t = 0$ or $t = \frac{500}{g} \approx 51.0$ s. So the range is $250\sqrt{3} \cdot \frac{500}{g} \approx 22$ km.

(b) $0 = \dfrac{d}{dt}\left(250t - \frac{1}{2}gt^2\right) = 250 - gt$ implies that the maximum height is attained when $t = 250/g \approx 25.5$ s.

Thus, the maximum height is $(250)(250/g) - g(250/g)^2\frac{1}{2} = (250)^2/(2g) \approx 3.2$ km.

(c) From part (a), impact occurs at $t = 500/g \approx 51.0$. Thus, the velocity at impact is

$\mathbf{v}(500/g) = 250\sqrt{3}\,\mathbf{i} + [250 - g(500/g)]\,\mathbf{j} = 250\sqrt{3}\,\mathbf{i} - 250\,\mathbf{j}$ and the speed is $|\mathbf{v}(500/g)| = 250\sqrt{3+1} = 500$ m/s.

20. As in Exercise 19, $\mathbf{v}(t) = 250\sqrt{3}\,\mathbf{i} + (250 - gt)\,\mathbf{j}$ and $\mathbf{r}(t) = 250\sqrt{3}\,t\,\mathbf{i} + \left(250t - \frac{1}{2}gt^2\right)\mathbf{j} + \mathbf{c}_2$.

But $\mathbf{r}(0) = 200\,\mathbf{j}$, so $\mathbf{c}_2 = 200\,\mathbf{j}$ and $\mathbf{r}(t) = 250\sqrt{3}\,t\,\mathbf{i} + \left(200 + 250t - \frac{1}{2}gt^2\right)\mathbf{j}$.

(a) $200 + 250t - \frac{1}{2}gt^2 = 0$ implies that $gt^2 - 500t - 400 = 0$ or $t = \dfrac{500 \pm \sqrt{500^2 + 1600g}}{2g}$. Taking the positive t-value

gives $t = \dfrac{500 + \sqrt{250{,}000 + 1600g}}{2g} \approx 51.8$ s. Thus the range is $\left(250\sqrt{3}\right)\dfrac{500 + \sqrt{250{,}000 + 1600g}}{2g} \approx 22.4$ km.

(b) $0 = \dfrac{d}{dt}\left(200 + 250t - \frac{1}{2}gt^2\right) = 250 - gt$ implies that the maximum height is attained when $t = 250/g \approx 25.5$ s and

thus the maximum height is $\left[200 + (250)\left(\dfrac{250}{g}\right) - \dfrac{g}{2}\left(\dfrac{250}{g}\right)^2\right] = 200 + \dfrac{(250)^2}{2g} \approx 3.4$ km.

Alternate solution: Because the projectile is fired in the same direction and with the same velocity as in Exercise 19, but from a point 200 m higher, the maximum height reached is 200 m higher than that found in Exercise 19, that is, 3.2 km + 200 m = 3.4 km.

(c) From part (a), impact occurs at $t = \dfrac{500 + \sqrt{250{,}000 + 1600g}}{2g}$. Thus the velocity at impact is

$250\sqrt{3}\,\mathbf{i} + \left[250 - g\,\dfrac{500 + \sqrt{250{,}000 + 1600g}}{2g}\right]\mathbf{j}$, so $|\mathbf{v}| \approx \sqrt{(250)^2(3) + (250 - 51.8g)^2} \approx 504$ m/s.

21. As in Example 5, $\mathbf{r}(t) = (v_0 \cos 45^\circ)t\,\mathbf{i} + \left[(v_0 \sin 45^\circ)t - \frac{1}{2}gt^2\right]\mathbf{j} = \frac{1}{2}\left[v_0\sqrt{2}\,t\,\mathbf{i} + \left(v_0\sqrt{2}\,t - gt^2\right)\mathbf{j}\right]$. Then the ball

lands at $t = \dfrac{v_0\sqrt{2}}{g}$ s. Now since it lands 90 m away, $90 = \frac{1}{2}v_0\sqrt{2}\,\dfrac{v_0\sqrt{2}}{g}$ or $v_0^2 = 90g$ and the initial velocity

is $v_0 = \sqrt{90g} \approx 30$ m/s.

22. As in Example 5, $\mathbf{r}(t) = (v_0 \cos 30^\circ)t\,\mathbf{i} + \left[(v_0 \sin 30^\circ)t - \frac{1}{2}gt^2\right]\mathbf{j} = \frac{1}{2}\left[v_0\sqrt{3}\,t\,\mathbf{i} + (v_0\,t - gt^2)\,\mathbf{j}\right]$ and then

$\mathbf{v}(t) = \mathbf{r}'(t) = \frac{1}{2}\left[v_0\sqrt{3}\,\mathbf{i} + (v_0 - 2gt)\,\mathbf{j}\right]$. The shell reaches its maximum height when the vertical component of velocity

is zero, so $\frac{1}{2}(v_0 - 2gt) = 0 \;\Rightarrow\; t = \dfrac{v_0}{2g}$. The vertical height of the shell at that time is 500 m,

$$\text{so } \frac{1}{2}\left[v_0\left(\frac{v_0}{2g}\right) - g\left(\frac{v_0}{2g}\right)^2\right] = 500 \;\Rightarrow\; \frac{v_0^2}{8g} = 500 \;\Rightarrow\; v_0 = \sqrt{4000g} = \sqrt{4000(9.8)} \approx 198 \text{ m/s}.$$

23. Let α be the angle of elevation. Then $v_0 = 150$ m/s and from Example 5, the horizontal distance traveled by the projectile is

$d = \dfrac{v_0^2 \sin 2\alpha}{g}$. Thus $\dfrac{150^2 \sin 2\alpha}{g} = 800 \;\Rightarrow\; \sin 2\alpha = \dfrac{800g}{150^2} \approx 0.3484 \;\Rightarrow\; 2\alpha \approx 20.4^\circ$ or $180 - 20.4 = 159.6^\circ$.

Two angles of elevation then are $\alpha \approx 10.2^\circ$ and $\alpha \approx 79.8^\circ$.

24. Here $v_0 = 115$ ft/s, the angle of elevation is $\alpha = 50^\circ$, and if we place the origin at home plate, then $\mathbf{r}(0) = 3\,\mathbf{j}$.

As in Example 5, we have $\mathbf{r}(t) = -\frac{1}{2}gt^2\,\mathbf{j} + t\,\mathbf{v}_0 + \mathbf{D}$ where $\mathbf{D} = \mathbf{r}(0) = 3\,\mathbf{j}$ and $\mathbf{v}_0 = v_0 \cos\alpha\,\mathbf{i} + v_0 \sin\alpha\,\mathbf{j}$, so $\mathbf{r}(t) = (v_0 \cos\alpha)t\,\mathbf{i} + \left[(v_0 \sin\alpha)t - \frac{1}{2}gt^2 + 3\right]\mathbf{j}$. Thus, parametric equations for the trajectory of the ball are $x = (v_0 \cos\alpha)t$, $y = (v_0 \sin\alpha)t - \frac{1}{2}gt^2 + 3$. The ball reaches the fence when $x = 400 \Rightarrow$ $(v_0 \cos\alpha)t = 400 \Rightarrow t = \dfrac{400}{v_0 \cos\alpha} = \dfrac{400}{115 \cos 50^\circ} \approx 5.41$ s. At this time, the height of the ball is $y = (v_0 \sin\alpha)t - \frac{1}{2}gt^2 + 3 \approx (115 \sin 50^\circ)(5.41) - \frac{1}{2}(32)(5.41)^2 + 3 \approx 11.2$ ft. Since the fence is 10 ft high, the ball clears the fence.

25. Place the catapult at the origin and assume the catapult is 100 meters from the city, so the city lies between $(100, 0)$ and $(600, 0)$. The initial speed is $v_0 = 80$ m/s and let θ be the angle the catapult is set at. As in Example 5, the trajectory of the catapulted rock is given by $\mathbf{r}(t) = (80\cos\theta)t\,\mathbf{i} + \left[(80\sin\theta)t - 4.9t^2\right]\mathbf{j}$. The top of the near city wall is at $(100, 15)$, which the rock will hit when $(80\cos\theta)\,t = 100 \Rightarrow t = \dfrac{5}{4\cos\theta}$ and $(80\sin\theta)t - 4.9t^2 = 15 \Rightarrow$

$80\sin\theta \cdot \dfrac{5}{4\cos\theta} - 4.9\left(\dfrac{5}{4\cos\theta}\right)^2 = 15 \Rightarrow 100\tan\theta - 7.65625\sec^2\theta = 15$. Replacing $\sec^2\theta$ with $\tan^2\theta + 1$ gives $7.65625\tan^2\theta - 100\tan\theta + 22.62625 = 0$. Using the quadratic formula, we have $\tan\theta \approx 0.230324, 12.8309 \Rightarrow$ $\theta \approx 13.0^\circ, 85.5^\circ$. So for $13.0^\circ < \theta < 85.5^\circ$, the rock will land beyond the near city wall. The base of the far wall is located at $(600, 0)$ which the rock hits if $(80\cos\theta)t = 600 \Rightarrow t = \dfrac{15}{2\cos\theta}$ and $(80\sin\theta)t - 4.9t^2 = 0 \Rightarrow$

$80\sin\theta \cdot \dfrac{15}{2\cos\theta} - 4.9\left(\dfrac{15}{2\cos\theta}\right)^2 = 0 \Rightarrow 600\tan\theta - 275.625\sec^2\theta = 0 \Rightarrow$

$275.625\tan^2\theta - 600\tan\theta + 275.625 = 0$. Solutions are $\tan\theta \approx 0.658678, 1.51819 \Rightarrow \theta \approx 33.4^\circ, 56.6^\circ$. Thus the rock lands beyond the enclosed city ground for $33.4^\circ < \theta < 56.6^\circ$, and the angles that allow the rock to land on city ground are $13.0^\circ < \theta < 33.4^\circ$, $56.6^\circ < \theta < 85.5^\circ$. If you consider that the rock can hit the far wall and bounce back into the city, we calculate the angles that cause the rock to hit the top of the wall at $(600, 15)$: $(80\cos\theta)t = 600 \Rightarrow t = \dfrac{15}{2\cos\theta}$ and $(80\sin\theta)t - 4.9t^2 = 15 \Rightarrow 600\tan\theta - 275.625\sec^2\theta = 15 \Rightarrow 275.625\tan^2\theta - 600\tan\theta + 290.625 = 0$. Solutions are $\tan\theta \approx 0.727506, 1.44936 \Rightarrow \theta \approx 36.0^\circ, 55.4^\circ$, so the catapult should be set with angle θ where $13.0^\circ < \theta < 36.0^\circ$, $55.4^\circ < \theta < 85.5^\circ$.

26. Place the ball at the origin and consider $\mathbf{j}$ to be pointing in the northward direction with $\mathbf{i}$ pointing east and $\mathbf{k}$ pointing upward. Force = mass × acceleration $\Rightarrow$ acceleration = force/mass, so the wind applies a constant acceleration of $4\text{ N}/0.8\text{ kg} = 5\text{ m/s}^2$ in the easterly direction. Combined with the acceleration due to gravity, the acceleration acting on the ball is $\mathbf{a}(t) = 5\,\mathbf{i} - 9.8\,\mathbf{k}$. Then $\mathbf{v}(t) = \int \mathbf{a}(t)\,dt = 5t\,\mathbf{i} - 9.8t\,\mathbf{k} + \mathbf{C}$ where $\mathbf{C}$ is a constant vector.

We know $\mathbf{v}(0) = \mathbf{C} = -30\cos 30^\circ\,\mathbf{j} + 30\sin 30^\circ\,\mathbf{k} = -15\sqrt{3}\,\mathbf{j} + 15\,\mathbf{k} \Rightarrow \mathbf{C} = -15\sqrt{3}\,\mathbf{j} + 15\,\mathbf{k}$ and $\mathbf{v}(t) = 5t\,\mathbf{i} - 15\sqrt{3}\,\mathbf{j} + (15 - 9.8t)\,\mathbf{k}$. $\mathbf{r}(t) = \int \mathbf{v}(t)\,dt = 2.5t^2\,\mathbf{i} - 15\sqrt{3}\,t\,\mathbf{j} + (15t - 4.9t^2)\,\mathbf{k} + \mathbf{D}$ but $\mathbf{r}(0) = \mathbf{D} = \mathbf{0}$ so $\mathbf{r}(t) = 2.5t^2\,\mathbf{i} - 15\sqrt{3}\,t\,\mathbf{j} + (15t - 4.9t^2)\,\mathbf{k}$. The ball lands when $15t - 4.9t^2 = 0 \Rightarrow t = 0, t = 15/4.9 \approx 3.0612$ s, so the ball lands at approximately $\mathbf{r}(3.0612) \approx 23.43\,\mathbf{i} - 79.53\,\mathbf{j}$ which is 82.9 m away in the direction S 16.4°E. Its speed is approximately $|\mathbf{v}(3.0612)| \approx |15.306\,\mathbf{i} - 15\sqrt{3}\,\mathbf{j} - 15\,\mathbf{k}| \approx 33.68$ m/s.

27. (a) After t seconds, the boat will be $5t$ meters west of point A. The velocity of the water at that location is $\frac{3}{400}(5t)(40-5t)\,\mathbf{j}$. The velocity of the boat in still water is $5\,\mathbf{i}$, so the resultant velocity of the boat is $\mathbf{v}(t) = 5\,\mathbf{i} + \frac{3}{400}(5t)(40-5t)\,\mathbf{j} = 5\mathbf{i} + \left(\frac{3}{2}t - \frac{3}{16}t^2\right)\mathbf{j}$. Integrating, we obtain $\mathbf{r}(t) = 5t\,\mathbf{i} + \left(\frac{3}{4}t^2 - \frac{1}{16}t^3\right)\mathbf{j} + \mathbf{C}$. If we place the origin at A (and consider $\mathbf{j}$ to coincide with the northern direction) then $\mathbf{r}(0) = \mathbf{0} \Rightarrow \mathbf{C} = \mathbf{0}$ and we have $\mathbf{r}(t) = 5t\,\mathbf{i} + \left(\frac{3}{4}t^2 - \frac{1}{16}t^3\right)\mathbf{j}$. The boat reaches the east bank after 8 s, and it is located at $\mathbf{r}(8) = 5(8)\mathbf{i} + \left(\frac{3}{4}(8)^2 - \frac{1}{16}(8)^3\right)\mathbf{j} = 40\,\mathbf{i} + 16\,\mathbf{j}$. Thus the boat is 16 m downstream.

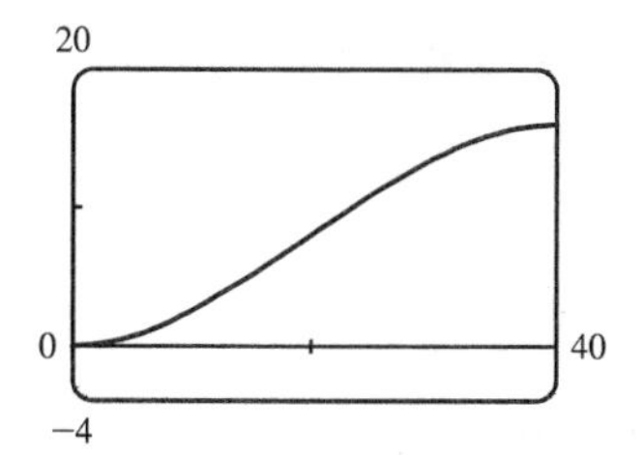

(b) Let α be the angle north of east that the boat heads. Then the velocity of the boat in still water is given by $5(\cos\alpha)\,\mathbf{i} + 5(\sin\alpha)\,\mathbf{j}$. At t seconds, the boat is $5(\cos\alpha)t$ meters from the west bank, at which point the velocity of the water is $\frac{3}{400}[5(\cos\alpha)t][40 - 5(\cos\alpha)t]\,\mathbf{j}$. The resultant velocity of the boat is given by $\mathbf{v}(t) = 5(\cos\alpha)\,\mathbf{i} + \left[5\sin\alpha + \frac{3}{400}(5t\cos\alpha)(40 - 5t\cos\alpha)\right]\mathbf{j} = (5\cos\alpha)\,\mathbf{i} + \left(5\sin\alpha + \frac{3}{2}t\cos\alpha - \frac{3}{16}t^2\cos^2\alpha\right)\mathbf{j}$.

Integrating, $\mathbf{r}(t) = (5t\cos\alpha)\,\mathbf{i} + \left(5t\sin\alpha + \frac{3}{4}t^2\cos\alpha - \frac{1}{16}t^3\cos^2\alpha\right)\mathbf{j}$ (where we have again placed the origin at A). The boat will reach the east bank when $5t\cos\alpha = 40 \Rightarrow t = \dfrac{40}{5\cos\alpha} = \dfrac{8}{\cos\alpha}$.

In order to land at point $B(40, 0)$ we need $5t\sin\alpha + \frac{3}{4}t^2\cos\alpha - \frac{1}{16}t^3\cos^2\alpha = 0 \Rightarrow$

$$5\left(\frac{8}{\cos\alpha}\right)\sin\alpha + \frac{3}{4}\left(\frac{8}{\cos\alpha}\right)^2\cos\alpha - \frac{1}{16}\left(\frac{8}{\cos\alpha}\right)^3\cos^2\alpha = 0 \quad\Rightarrow\quad \frac{1}{\cos\alpha}(40\sin\alpha + 48 - 32) = 0 \quad\Rightarrow$$

$40\sin\alpha + 16 = 0 \Rightarrow \sin\alpha = -\frac{2}{5}$. Thus $\alpha = \sin^{-1}\left(-\frac{2}{5}\right) \approx -23.6^\circ$, so the boat should head 23.6° south of east (upstream). The path does seem realistic. The boat initially heads upstream to counteract the effect of the current. Near the center of the river, the current is stronger and the boat is pushed downstream. When the boat nears the eastern bank, the current is slower and the boat is able to progress upstream to arrive at point B.

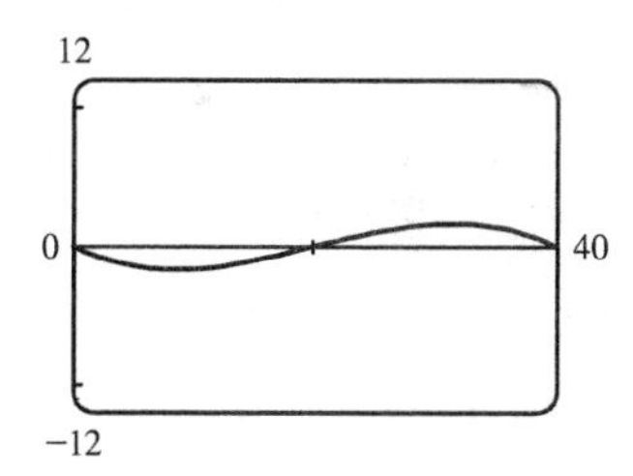

28. $\mathbf{r}(t) = (1+t)\,\mathbf{i} + (t^2 - 2t)\,\mathbf{j} \Rightarrow \mathbf{r}'(t) = \mathbf{i} + (2t-2)\,\mathbf{j}$, $|\mathbf{r}'(t)| = \sqrt{1^2 + (2t-2)^2} = \sqrt{4t^2 - 8t + 5}$,

$\mathbf{r}''(t) = 2\,\mathbf{j}$, $\mathbf{r}'(t) \times \mathbf{r}''(t) = 2\,\mathbf{k}$. Then Equation 9 gives $a_T = \dfrac{\mathbf{r}'(t)\cdot\mathbf{r}''(t)}{|\mathbf{r}'(t)|} = \dfrac{2(2t-2)}{\sqrt{4t^2-8t+5}}$ and Equation 10

gives $a_N = \dfrac{|\mathbf{r}'(t)\times\mathbf{r}''(t)|}{|\mathbf{r}'(t)|} = \dfrac{2}{\sqrt{4t^2-8t+5}}$.

29. $\mathbf{r}(t) = \cos t\,\mathbf{i} + \sin t\,\mathbf{j} + t\,\mathbf{k} \Rightarrow \mathbf{r}'(t) = -\sin t\,\mathbf{i} + \cos t\,\mathbf{j} + \mathbf{k}$, $|\mathbf{r}'(t)| = \sqrt{\sin^2 t + \cos^2 t + 1} = \sqrt{2}$,

$\mathbf{r}''(t) = -\cos t\,\mathbf{i} - \sin t\,\mathbf{j}$, $\mathbf{r}'(t)\times\mathbf{r}''(t) = \sin t\,\mathbf{i} - \cos t\,\mathbf{j} + \mathbf{k}$.

Then $a_T = \dfrac{\mathbf{r}'(t)\cdot\mathbf{r}''(t)}{|\mathbf{r}'(t)|} = \dfrac{\sin t\,\cos t - \sin t\,\cos t}{\sqrt{2}} = 0$ and $a_N = \dfrac{|\mathbf{r}'(t)\times\mathbf{r}''(t)|}{|\mathbf{r}'(t)|} = \dfrac{\sqrt{\sin^2 t + \cos^2 t + 1}}{\sqrt{2}} = \dfrac{\sqrt{2}}{\sqrt{2}} = 1$.

30. $\mathbf{r}(t) = t\,\mathbf{i} + t^2\,\mathbf{j} + 3t\,\mathbf{k} \;\Rightarrow\; \mathbf{r}'(t) = \mathbf{i} + 2t\,\mathbf{j} + 3\,\mathbf{k},\;\; |\mathbf{r}'(t)| = \sqrt{1^2 + (2t)^2 + 3^2} = \sqrt{4t^2 + 10},$

$\mathbf{r}''(t) = 2\mathbf{j},\;\; \mathbf{r}'(t) \times \mathbf{r}''(t) = -6\mathbf{i} + 2\mathbf{k}.$

Then $a_T = \dfrac{\mathbf{r}'(t) \cdot \mathbf{r}''(t)}{|\mathbf{r}'(t)|} = \dfrac{4t}{\sqrt{4t^2 + 10}}$ and $a_N = \dfrac{|\mathbf{r}'(t) \times \mathbf{r}''(t)|}{|\mathbf{r}'(t)|} = \dfrac{2\sqrt{10}}{\sqrt{4t^2 + 10}}.$

31. $\mathbf{r}(t) = e^t\,\mathbf{i} + \sqrt{2}\,t\,\mathbf{j} + e^{-t}\,\mathbf{k} \;\Rightarrow\; \mathbf{r}'(t) = e^t\,\mathbf{i} + \sqrt{2}\,\mathbf{j} - e^{-t}\,\mathbf{k},\;\; |\mathbf{r}(t)| = \sqrt{e^{2t} + 2 + e^{-2t}} = \sqrt{(e^t + e^{-t})^2} = e^t + e^{-t},$

$\mathbf{r}''(t) = e^t\,\mathbf{i} + e^{-t}\,\mathbf{k}.$ Then $a_T = \dfrac{e^{2t} - e^{-2t}}{e^t + e^{-t}} = \dfrac{(e^t + e^{-t})(e^t - e^{-t})}{e^t + e^{-t}} = e^t - e^{-t} = 2\sinh t$

and $a_N = \dfrac{\left|\sqrt{2}e^{-t}\,\mathbf{i} - 2\,\mathbf{j} - \sqrt{2}e^t\,\mathbf{k}\right|}{e^t + e^{-t}} = \dfrac{\sqrt{2(e^{-2t} + 2 + e^{2t})}}{e^t + e^{-t}} = \sqrt{2}\,\dfrac{e^t + e^{-t}}{e^t + e^{-t}} = \sqrt{2}.$

32. $\mathbf{L}(t) = m\,\mathbf{r}(t) \times \mathbf{v}(t) \;\Rightarrow$

$$\begin{aligned} \mathbf{L}'(t) &= m[\mathbf{r}'(t) \times \mathbf{v}(t) + \mathbf{r}(t) \times \mathbf{v}'(t)] \qquad \text{[by Formula 5 of Theorem 10.7.5]} \\ &= m[\mathbf{v}(t) \times \mathbf{v}(t) + \mathbf{r}(t) \times \mathbf{v}'(t)] = m[\mathbf{0} + \mathbf{r}(t) \times \mathbf{a}(t)] = \boldsymbol{\tau}(t) \end{aligned}$$

So if the torque is always $\mathbf{0}$, then $\mathbf{L}'(t) = \mathbf{0}$ for all t, and so $\mathbf{L}(t)$ is constant.

33. If the engines are turned off at time t, then the spacecraft will continue to travel in the direction of $\mathbf{v}(t)$, so we need a t such that for some scalar $s > 0$, $\mathbf{r}(t) + s\,\mathbf{v}(t) = \langle 6, 4, 9\rangle$. $\mathbf{v}(t) = \mathbf{r}'(t) = \mathbf{i} + \dfrac{1}{t}\,\mathbf{j} + \dfrac{8t}{(t^2+1)^2}\,\mathbf{k} \;\Rightarrow$

$$\mathbf{r}(t) + s\,\mathbf{v}(t) = \left\langle 3 + t + s, 2 + \ln t + \frac{s}{t}, 7 - \frac{4}{t^2+1} + \frac{8st}{(t^2+1)^2} \right\rangle \;\Rightarrow\; 3 + t + s = 6 \;\Rightarrow\; s = 3 - t,$$

so $7 - \dfrac{4}{t^2+1} + \dfrac{8(3-t)t}{(t^2+1)^2} = 9 \;\Leftrightarrow\; \dfrac{24t - 12t^2 - 4}{(t^2+1)^2} = 2 \;\Leftrightarrow\; t^4 + 8t^2 - 12t + 3 = 0.$

It is easily seen that $t = 1$ is a root of this polynomial. Also $2 + \ln 1 + \dfrac{3-1}{1} = 4$, so $t = 1$ is the desired solution.

34. (a) $m\dfrac{d\mathbf{v}}{dt} = \dfrac{dm}{dt}\mathbf{v}_e \;\Leftrightarrow\; \dfrac{d\mathbf{v}}{dt} = \dfrac{1}{m}\dfrac{dm}{dt}\mathbf{v}_e$. Integrating both sides of this equation with respect to t gives

$$\int_0^t \frac{d\mathbf{v}}{du}\,du = \mathbf{v}_e \int_0^t \frac{1}{m}\frac{dm}{du}\,du \;\Rightarrow\; \int_{\mathbf{v}(0)}^{\mathbf{v}(t)} d\mathbf{v} = \mathbf{v}_e \int_{m(0)}^{m(t)} \frac{dm}{m} \quad \text{[Substitution Rule]} \;\Rightarrow$$

$$\mathbf{v}(t) - \mathbf{v}(0) = \ln\left(\frac{m(t)}{m(0)}\right)\mathbf{v}_e \;\Rightarrow\; \mathbf{v}(t) = \mathbf{v}(0) - \ln\left(\frac{m(0)}{m(t)}\right)\mathbf{v}_e.$$

(b) $|\mathbf{v}(t)| = 2\,|\mathbf{v}_e|$, and $|\mathbf{v}(0)| = 0$. Therefore, by part (a), $2\,|\mathbf{v}_e| = \left|-\ln\left(\dfrac{m(0)}{m(t)}\right)\mathbf{v}_e\right| \;\Rightarrow$

$$2\,|\mathbf{v}_e| = \ln\left(\frac{m(0)}{m(t)}\right)|\mathbf{v}_e|. \quad \left[\textit{Note: } m(0) > m(t) \text{ so that } \ln\left(\frac{m(0)}{m(t)}\right) > 0\right] \;\Rightarrow\; m(t) = e^{-2}m(0).$$

Thus $\dfrac{m(0) - e^{-2}m(0)}{m(0)} = 1 - e^{-2}$ is the fraction of the initial mass that is burned as fuel.

10 Review

CONCEPT CHECK

1. A scalar is a real number, while a vector is a quantity that has both a real-valued magnitude and a direction.

2. To add two vectors geometrically, we can use either the Triangle Law or the Parallelogram Law, as illustrated in Figures 3 and 4 in Section 10.2. (See also the definition of vector addition on page 525.) Algebraically, we add the corresponding components of the vectors.

3. For $c > 0$, $c\,\mathbf{a}$ is a vector with the same direction as $\mathbf{a}$ and length c times the length of $\mathbf{a}$. If $c < 0$, $c\mathbf{a}$ points in the opposite direction as $\mathbf{a}$ and has length $|c|$ times the length of $\mathbf{a}$. (See Figures 7 and 15 in Section 10.2.) Algebraically, to find $c\,\mathbf{a}$ we multiply each component of $\mathbf{a}$ by c.

4. See (1) in Section 10.2.

5. See Theorem 10.3.3 and Definition 10.3.1.

6. The dot product can be used to determine the work done moving an object given the force and displacement vectors. The dot product can also be used to find the angle between two vectors and the scalar projection of one vector onto another. In particular, the dot product can determine if two vectors are orthogonal.

7. See the boxed equations on page 536 as well as Figures 3 and 4 and the accompanying discussion on pages 535–36.

8. See Theorem 10.4.6 and the preceding discussion; use either (1) or (4) in Section 10.4.

9. The cross product can be used to determine torque if the force and position vectors are known. In addition, the cross product can be used to create a vector orthogonal to two given vectors as well as to determine if two vectors are parallel. The cross product can also be used to find the area of a parallelogram determined by two vectors.

10. (a) The area of the parallelogram determined by $\mathbf{a}$ and $\mathbf{b}$ is the length of the cross product: $|\mathbf{a} \times \mathbf{b}|$.

    (b) The volume of the parallelepiped determined by $\mathbf{a}$, $\mathbf{b}$, and $\mathbf{c}$ is the magnitude of their scalar triple product: $|\mathbf{a} \cdot (\mathbf{b} \times \mathbf{c})|$.

11. If an equation of the plane is known, it can be written as $ax + by + cz + d = 0$. A normal vector, which is perpendicular to the plane, is $\langle a, b, c \rangle$ (or any scalar multiple of $\langle a, b, c \rangle$). If an equation is not known, we can use points on the plane to find two non-parallel vectors which lie in the plane. The cross product of these vectors is a vector perpendicular to the plane.

12. The angle between two intersecting planes is defined as the acute angle between their normal vectors. We can find this angle using Corollary 10.3.6.

13. See (1), (2), and (3) in Section 10.5.

14. See (5), (6), and (7) in Section 10.5.

15. (a) Two (nonzero) vectors are parallel if and only if one is a scalar multiple of the other. In addition, two nonzero vectors are parallel if and only if their cross product is $\mathbf{0}$.

    (b) Two vectors are perpendicular if and only if their dot product is 0.

    (c) Two planes are parallel if and only if their normal vectors are parallel.

16. (a) Determine the vectors $\overrightarrow{PQ} = \langle a_1, a_2, a_3 \rangle$ and $\overrightarrow{PR} = \langle b_1, b_2, b_3 \rangle$. If there is a scalar t such that $\langle a_1, a_2, a_3 \rangle = t \langle b_1, b_2, b_3 \rangle$, then the vectors are parallel and the points must all lie on the same line.

Alternatively, if $\overrightarrow{PQ} \times \overrightarrow{PR} = \mathbf{0}$, then $\overrightarrow{PQ}$ and $\overrightarrow{PR}$ are parallel, so P, Q, and R are collinear.

Thirdly, an algebraic method is to determine an equation of the line joining two of the points, and then check whether or not the third point satisfies this equation.

(b) Find the vectors $\overrightarrow{PQ} = \mathbf{a}$, $\overrightarrow{PR} = \mathbf{b}$, $\overrightarrow{PS} = \mathbf{c}$. $\mathbf{a} \times \mathbf{b}$ is normal to the plane formed by P, Q and R, and so S lies on this plane if $\mathbf{a} \times \mathbf{b}$ and $\mathbf{c}$ are orthogonal, that is, if $(\mathbf{a} \times \mathbf{b}) \cdot \mathbf{c} = 0$. (Or use the reasoning in Example 5 in Section 10.4.) Alternatively, find an equation for the plane determined by three of the points and check whether or not the fourth point satisfies this equation.

17. (a) See Exercise 10.4.39.

(b) See Example 7 in Section 10.5.

18. The traces of a surface are the curves of intersection of the surface with planes parallel to the coordinate planes. We can find the trace in the plane $x = k$ (parallel to the yz-plane) by setting $x = k$ and determining the curve represented by the resulting equation. Traces in the planes $y = k$ (parallel to the xz-plane) and $z = k$ (parallel to the xy-plane) are found similarly.

19. See Table 1 in Section 10.6.

20. A vector function is a function whose domain is a set of real numbers and whose range is a set of vectors. To find the derivative or integral, we can differentiate or integrate each component of the vector function.

21. The tip of the moving vector $\mathbf{r}(t)$ of a continuous vector function traces out a space curve.

22. (a) A curve represented by the vector function $\mathbf{r}(t)$ is smooth if $\mathbf{r}'(t)$ is continuous and $\mathbf{r}'(t) \neq \mathbf{0}$ on its parametric domain (except possibly at the endpoints).

(b) The tangent vector to a smooth curve at a point P with position vector $\mathbf{r}(t)$ is the vector $\mathbf{r}'(t)$. The tangent line at P is the line through P parallel to the tangent vector $\mathbf{r}'(t)$. The unit tangent vector is $\mathbf{T}(t) = \dfrac{\mathbf{r}'(t)}{|\mathbf{r}'(t)|}$.

23. (a)–(f) See Theorem 10.7.5.

24. Use Formula 10.8.2, or equivalently, 10.8.3.

25. (a) The curvature of a curve is $\kappa = \left|\dfrac{d\mathbf{T}}{ds}\right|$ where $\mathbf{T}$ is the unit tangent vector.

(b) $\kappa(t) = \left|\dfrac{\mathbf{T}'(t)}{\mathbf{r}'(t)}\right|$ (c) $\kappa(t) = \dfrac{|\mathbf{r}'(t) \times \mathbf{r}''(t)|}{|\mathbf{r}'(t)|^3}$ (d) $\kappa(x) = \dfrac{|f''(x)|}{[1 + (f'(x))^2]^{3/2}}$

26. The unit normal vector: $\mathbf{N}(t) = \dfrac{\mathbf{T}'(t)}{|\mathbf{T}'(t)|}$. The binormal vector: $\mathbf{B}(t) = \mathbf{T}(t) \times \mathbf{N}(t)$.

27. (a) If $\mathbf{r}(t)$ is the position vector of the particle on the space curve, the velocity $\mathbf{v}(t) = \mathbf{r}'(t)$, the speed is given by $|\mathbf{v}(t)|$, and the acceleration $\mathbf{a}(t) = \mathbf{v}'(t) = \mathbf{r}''(t)$.

(b) $\mathbf{a} = a_T\mathbf{T} + a_N\mathbf{N}$ where $a_T = v'$ and $a_N = \kappa v^2$.

28. See the statement of Kepler's Laws on page 586.

TRUE-FALSE QUIZ

1. True, by Property 2 of the dot product. (See page 533.)

2. False. Theorem 10.4.8 says that $\mathbf{u} \times \mathbf{v} = -\mathbf{v} \times \mathbf{u}$. (See page 543.)

3. True. If θ is the angle between $\mathbf{u}$ and $\mathbf{v}$, then by Theorem 10.4.6, $|\mathbf{u} \times \mathbf{v}| = |\mathbf{u}|\,|\mathbf{v}|\sin\theta = |\mathbf{v}|\,|\mathbf{u}|\sin\theta = |\mathbf{v} \times \mathbf{u}|$.

(Or, by Theorem 10.4.8, $|\mathbf{u} \times \mathbf{v}| = |-\mathbf{v} \times \mathbf{u}| = |-1|\,|\mathbf{v} \times \mathbf{u}| = |\mathbf{v} \times \mathbf{u}|$.)

4. This is true by Property 4 of the dot product.

5. Property 2 of the cross product tells us that this is true.

6. This is true by Theorem 10.4.8.

7. This is true by Theorem 10.4.8 #5.

8. In general, this assertion is false; a counterexample is $\mathbf{i} \times (\mathbf{i} \times \mathbf{j}) \neq (\mathbf{i} \times \mathbf{i}) \times \mathbf{j}$. (See the paragraph preceding Theorem 10.4.8.)

9. This is true because $\mathbf{u} \times \mathbf{v}$ is orthogonal to $\mathbf{u}$ (see Theorem 10.4.5), and the dot product of two orthogonal vectors is 0.

10.
$$\begin{aligned}(\mathbf{u}+\mathbf{v}) \times \mathbf{v} &= \mathbf{u} \times \mathbf{v} + \mathbf{v} \times \mathbf{v} && \text{[by Theorem 10.4.8]}\\ &= \mathbf{u} \times \mathbf{v} + \mathbf{0} && \text{[by Example 10.4.2]}\\ &= \mathbf{u} \times \mathbf{v}, \text{ so this is true.}\end{aligned}$$

11. If $|\mathbf{u}| = 1$, $|\mathbf{v}| = 1$ and θ is the angle between these two vectors (so $0 \leq \theta \leq \pi$), then by Theorem 10.4.6, $|\mathbf{u} \times \mathbf{v}| = |\mathbf{u}|\,|\mathbf{v}|\sin\theta = \sin\theta$, which is equal to 1 if and only if $\theta = \frac{\pi}{2}$ (that is, if and only if the two vectors are orthogonal). Therefore, the assertion that the cross product of two unit vectors is a unit vector is false.

12. This is false, because according to Equation 10.5.8, $ax + by + cz + d = 0$ is the general equation of a plane.

13. This is false. In $\mathbb{R}^2$, $x^2 + y^2 = 1$ represents a circle, but $\{(x, y, z) \mid x^2 + y^2 = 1\}$ represents a *three-dimensional surface*, namely, a circular cylinder with axis the z-axis.

14. This is false, as the dot product of two vectors is a scalar, not a vector.

15. False. For example, $\mathbf{i} \cdot \mathbf{j} = \mathbf{0}$ but $\mathbf{i} \neq \mathbf{0}$ and $\mathbf{j} \neq \mathbf{0}$.

16. This is true. We know $\mathbf{u} \cdot \mathbf{v} = |\mathbf{u}|\;|\mathbf{v}|\;\cos\theta$ where $|\mathbf{u}| \geq 0$, $|\mathbf{v}| \geq 0$, and $|\cos\theta| \leq 1$, so $|\mathbf{u} \cdot \mathbf{v}| = |\mathbf{u}|\;|\mathbf{v}|\;|\cos\theta| \leq |\mathbf{u}|\;|\mathbf{v}|$.

17. True. If we reparametrize the curve by replacing $u = t^3$, we have $\mathbf{r}(u) = u\,\mathbf{i} + 2u\,\mathbf{j} + 3u\,\mathbf{k}$, which is a line through the origin with direction vector $\mathbf{i} + 2\,\mathbf{j} + 3\,\mathbf{k}$.

18. True. $\mathbf{r}'(t) = \langle 1, 3t^2, 5t^4 \rangle$ is continuous for all t (since its component functions are each continuous) and since $x'(t) = 1$, we have $\mathbf{r}'(t) \neq \mathbf{0}$ for all t.

19. False. $\mathbf{r}'(t) = \langle -\sin t, 2t, 4t^3 \rangle$, and since $\mathbf{r}'(0) = \langle 0, 0, 0 \rangle = \mathbf{0}$, the curve is not smooth.

20. True. See Theorem 10.7.4.

21. False. By Formula 5 of Theorem 10.7.5, $\dfrac{d}{dt}\,[\mathbf{u}(t) \times \mathbf{v}(t)] = \mathbf{u}'(t) \times \mathbf{v}(t) + \mathbf{u}(t) \times \mathbf{v}'(t)$.

22. False. For example, let $\mathbf{r}(t) = \langle \cos t, \sin t \rangle$. Then $|\mathbf{r}(t)| = \sqrt{\cos^2 t + \sin^2 t} = 1 \;\Rightarrow\; \dfrac{d}{dt}\,|\mathbf{r}(t)| = 0$, but $|\mathbf{r}'(t)| = |\langle -\sin t, \cos t \rangle| = \sqrt{(-\sin t)^2 + \cos^2 t} = 1$.

23. False. κ is the magnitude of the rate of change of the unit tangent vector $\mathbf{T}$ with respect to arc length s, not with respect to t.

24. False. The binormal vector, by the definition given in Section 10.8, is $\mathbf{B}(t) = \mathbf{T}(t) \times \mathbf{N}(t) = -\,[\mathbf{N}(t) \times \mathbf{T}(t)]$.

EXERCISES

1. (a) The radius of the sphere is the distance between the points $(-1, 2, 1)$ and $(6, -2, 3)$, namely, $\sqrt{[6-(-1)]^2 + (-2-2)^2 + (3-1)^2} = \sqrt{69}$. By the formula for an equation of a sphere (see page 522), an equation of the sphere with center $(-1, 2, 1)$ and radius $\sqrt{69}$ is $(x+1)^2 + (y-2)^2 + (z-1)^2 = 69$.

(b) The intersection of this sphere with the yz-plane is the set of points on the sphere whose x-coordinate is 0. Putting $x = 0$ into the equation, we have $(y-2)^2 + (z-1)^2 = 68$, $x = 0$ which represents a circle in the yz-plane with center $(0, 2, 1)$ and radius $\sqrt{68}$.

(c) Completing squares gives $(x-4)^2 + (y+1)^2 + (z+3)^2 = -1 + 16 + 1 + 9 = 25$. Thus the sphere is centered at $(4, -1, -3)$ and has radius 5.

2. (a)

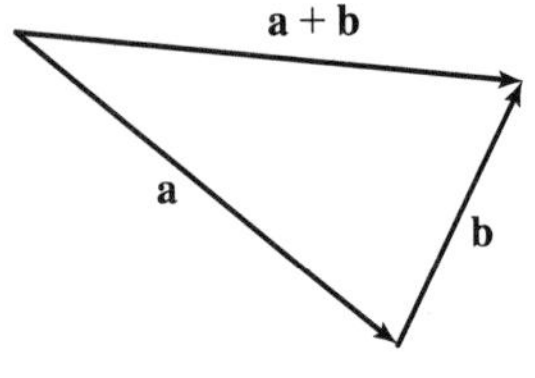

(b)

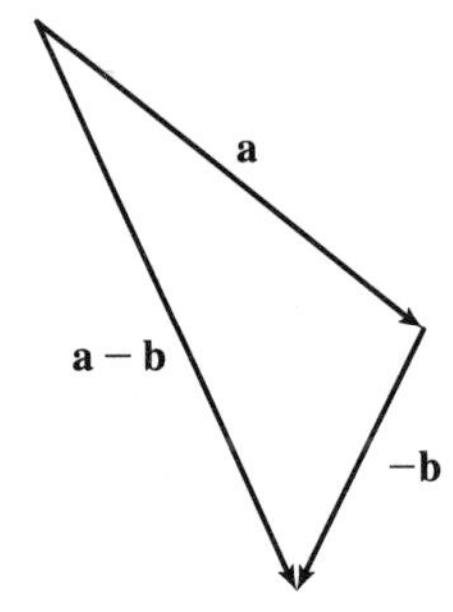

(c)

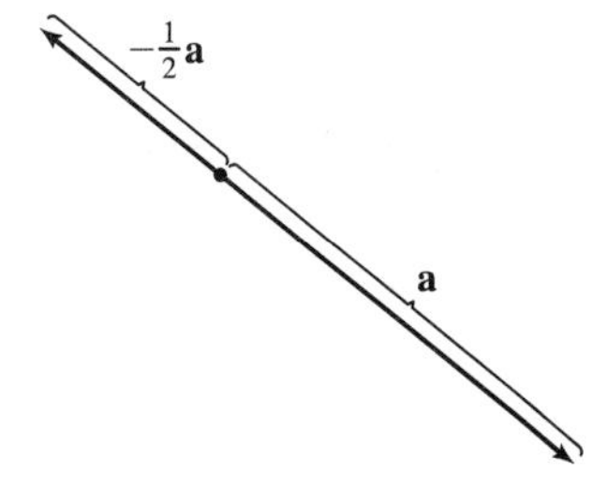

(d)

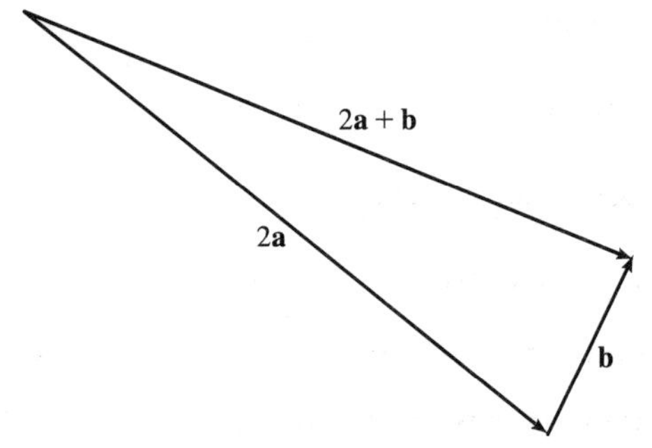

3. $\mathbf{u} \cdot \mathbf{v} = |\mathbf{u}|\,|\mathbf{v}| \cos 45^\circ = (2)(3)\frac{\sqrt{2}}{2} = 3\sqrt{2}$. $|\mathbf{u} \times \mathbf{v}| = |\mathbf{u}|\,|\mathbf{v}| \sin 45^\circ = (2)(3)\frac{\sqrt{2}}{2} = 3\sqrt{2}$.

By the right-hand rule, $\mathbf{u} \times \mathbf{v}$ is directed out of the page.

4. (a) $2\,\mathbf{a} + 3\,\mathbf{b} = 2\,\mathbf{i} + 2\,\mathbf{j} - 4\,\mathbf{k} + 9\,\mathbf{i} - 6\,\mathbf{j} + 3\,\mathbf{k} = 11\,\mathbf{i} - 4\,\mathbf{j} - \mathbf{k}$

(b) $|\mathbf{b}| = \sqrt{9 + 4 + 1} = \sqrt{14}$

(c) $\mathbf{a} \cdot \mathbf{b} = (1)(3) + (1)(-2) + (-2)(1) = -1$

(d) $\mathbf{a}\times\mathbf{b} = \begin{vmatrix} \mathbf{i} & \mathbf{j} & \mathbf{k} \\ 1 & 1 & -2 \\ 3 & -2 & 1 \end{vmatrix} = (1-4)\,\mathbf{i} - (1+6)\,\mathbf{j} + (-2-3)\,\mathbf{k} = -3\,\mathbf{i} - 7\mathbf{j} - 5\,\mathbf{k}$

(e) $\mathbf{b}\times\mathbf{c} = \begin{vmatrix} \mathbf{i} & \mathbf{j} & \mathbf{k} \\ 3 & -2 & 1 \\ 0 & 1 & -5 \end{vmatrix} = 9\,\mathbf{i} + 15\,\mathbf{j} + 3\,\mathbf{k}, \quad |\mathbf{b}\times\mathbf{c}| = 3\sqrt{9+25+1} = 3\sqrt{35}$

(f) $\mathbf{a}\cdot(\mathbf{b}\times\mathbf{c}) = \begin{vmatrix} 1 & 1 & -2 \\ 3 & -2 & 1 \\ 0 & 1 & -5 \end{vmatrix} = \begin{vmatrix} -2 & 1 \\ 1 & -5 \end{vmatrix} - \begin{vmatrix} 3 & 1 \\ 0 & -5 \end{vmatrix} - 2\begin{vmatrix} 3 & -2 \\ 0 & 1 \end{vmatrix} = 9 + 15 - 6 = 18$

(g) $\mathbf{c}\times\mathbf{c} = \mathbf{0}$ for any $\mathbf{c}$.

(h) From part (e),

$$\mathbf{a}\times(\mathbf{b}\times\mathbf{c}) = \mathbf{a}\times(9\,\mathbf{i} + 15\,\mathbf{j} + 3\,\mathbf{k}) = \begin{vmatrix} \mathbf{i} & \mathbf{j} & \mathbf{k} \\ 1 & 1 & -2 \\ 9 & 15 & 3 \end{vmatrix}$$

$$= (3+30)\,\mathbf{i} - (3+18)\,\mathbf{j} + (15-9)\,\mathbf{k} = 33\,\mathbf{i} - 21\,\mathbf{j} + 6\,\mathbf{k}$$

(i) The scalar projection is $\operatorname{comp}_{\mathbf{a}}\mathbf{b} = |\mathbf{b}|\cos\theta = \mathbf{a}\cdot\mathbf{b}/\,|\mathbf{a}| = -\frac{1}{\sqrt{6}}$.

(j) The vector projection is $\operatorname{proj}_{\mathbf{a}}\mathbf{b} = -\dfrac{1}{\sqrt{6}}\left(\dfrac{\mathbf{a}}{|\mathbf{a}|}\right) = -\frac{1}{6}(\mathbf{i} + \mathbf{j} - 2\,\mathbf{k})$.

(k) $\cos\theta = \dfrac{\mathbf{a}\cdot\mathbf{b}}{|\mathbf{a}|\,|\mathbf{b}|} = \dfrac{-1}{\sqrt{6}\sqrt{14}} = \dfrac{-1}{2\sqrt{21}}$ and $\theta = \cos^{-1}\left(\dfrac{-1}{2\sqrt{21}}\right) \approx 96^\circ$.

5. For the two vectors to be orthogonal, we need $\langle 3, 2, x\rangle \cdot \langle 2x, 4, x\rangle = 0 \quad\Leftrightarrow\quad (3)(2x) + (2)(4) + (x)(x) = 0 \quad\Leftrightarrow$
$x^2 + 6x + 8 = 0 \quad\Leftrightarrow\quad (x+2)(x+4) = 0 \quad\Leftrightarrow\quad x = -2$ or $x = -4$.

6. We know that the cross product of two vectors is orthogonal to both. So we calculate
$(\mathbf{j} + 2\,\mathbf{k}) \times (\mathbf{i} - 2\,\mathbf{j} + 3\,\mathbf{k}) = [3 - (-4)]\,\mathbf{i} - (0-2)\,\mathbf{j} + (0-1)\,\mathbf{k} = 7\,\mathbf{i} + 2\,\mathbf{j} - \mathbf{k}$.

Then two unit vectors orthogonal to both given vectors are $\pm\dfrac{7\,\mathbf{i} + 2\,\mathbf{j} - \mathbf{k}}{\sqrt{7^2 + 2^2 + (-1)^2}} = \pm\dfrac{1}{3\sqrt{6}}(7\,\mathbf{i} + 2\,\mathbf{j} - \mathbf{k})$,

that is, $\frac{7}{3\sqrt{6}}\,\mathbf{i} + \frac{2}{3\sqrt{6}}\,\mathbf{j} - \frac{1}{3\sqrt{6}}\,\mathbf{k}$ and $-\frac{7}{3\sqrt{6}}\,\mathbf{i} - \frac{2}{3\sqrt{6}}\,\mathbf{j} + \frac{1}{3\sqrt{6}}\,\mathbf{k}$.

7. (a) $(\mathbf{u}\times\mathbf{v})\cdot\mathbf{w} = \mathbf{u}\cdot(\mathbf{v}\times\mathbf{w}) = 2$

(b) $\mathbf{u}\cdot(\mathbf{w}\times\mathbf{v}) = \mathbf{u}\cdot[-(\mathbf{v}\times\mathbf{w})] = -\mathbf{u}\cdot(\mathbf{v}\times\mathbf{w}) = -2$

(c) $\mathbf{v}\cdot(\mathbf{u}\times\mathbf{w}) = (\mathbf{v}\times\mathbf{u})\cdot\mathbf{w} = -(\mathbf{u}\times\mathbf{v})\cdot\mathbf{w} = -2$

(d) $(\mathbf{u}\times\mathbf{v})\cdot\mathbf{v} = \mathbf{u}\cdot(\mathbf{v}\times\mathbf{v}) = \mathbf{u}\cdot\mathbf{0} = 0$

8. $(\mathbf{a}\times\mathbf{b})\cdot[(\mathbf{b}\times\mathbf{c})\times(\mathbf{c}\times\mathbf{a})] = (\mathbf{a}\times\mathbf{b})\cdot([(\mathbf{b}\times\mathbf{c})\cdot\mathbf{a}]\,\mathbf{c} - [(\mathbf{b}\times\mathbf{c})\cdot\mathbf{c}]\,\mathbf{a})$ [by Theorem 10.4.8, Property 6]

$$= (\mathbf{a}\times\mathbf{b})\cdot[(\mathbf{b}\times\mathbf{c})\cdot\mathbf{a}]\,\mathbf{c} = [\mathbf{a}\cdot(\mathbf{b}\times\mathbf{c})]\,(\mathbf{a}\times\mathbf{b})\cdot\mathbf{c}$$

$$= [\mathbf{a}\cdot(\mathbf{b}\times\mathbf{c})]\,[\mathbf{a}\cdot(\mathbf{b}\times\mathbf{c})] = [\mathbf{a}\cdot(\mathbf{b}\times\mathbf{c})]^2$$

9. For simplicity, consider a unit cube positioned with its back left corner at the origin. Vector representations of the diagonals joining the points $(0,0,0)$ to $(1,1,1)$ and $(1,0,0)$ to $(0,1,1)$ are $\langle 1,1,1\rangle$ and $\langle -1,1,1\rangle$. Let θ be the angle between these two vectors. $\langle 1,1,1\rangle \cdot \langle -1,1,1\rangle = -1+1+1 = 1 = |\langle 1,1,1\rangle|\,|\langle -1,1,1\rangle| \cos\theta = 3\cos\theta \quad\Rightarrow\quad \cos\theta = \frac{1}{3} \quad\Rightarrow$ $\theta = \cos^{-1}\left(\frac{1}{3}\right) \approx 71^\circ$.

10. $\overrightarrow{AB} = \langle 1,3,-1\rangle$, $\overrightarrow{AC} = \langle -2,1,3\rangle$ and $\overrightarrow{AD} = \langle -1,3,1\rangle$. By Equation 10.4.10,

$$\overrightarrow{AB}\cdot\left(\overrightarrow{AC}\times\overrightarrow{AD}\right) = \begin{vmatrix} 1 & 3 & -1 \\ -2 & 1 & 3 \\ -1 & 3 & 1 \end{vmatrix} = \begin{vmatrix} 1 & 3 \\ 3 & 1 \end{vmatrix} - 3\begin{vmatrix} -2 & 3 \\ -1 & 1 \end{vmatrix} - \begin{vmatrix} -2 & 1 \\ -1 & 3 \end{vmatrix} = -8-3+5 = -6.$$

The volume is $\left|\overrightarrow{AB}\cdot\left(\overrightarrow{AC}\times\overrightarrow{AD}\right)\right| = 6$ cubic units.

11. $\overrightarrow{AB} = \langle 1,0,-1\rangle$, $\overrightarrow{AC} = \langle 0,4,3\rangle$, so

(a) a vector perpendicular to the plane is $\overrightarrow{AB}\times\overrightarrow{AC} = \langle 0+4, -(3+0), 4-0\rangle = \langle 4,-3,4\rangle$.

(b) $\frac{1}{2}\left|\overrightarrow{AB}\times\overrightarrow{AC}\right| = \frac{1}{2}\sqrt{16+9+16} = \frac{\sqrt{41}}{2}$.

12. $\mathbf{D} = 4\,\mathbf{i} + 3\,\mathbf{j} + 6\,\mathbf{k}$, $\quad W = \mathbf{F}\cdot\mathbf{D} = 12+15+60 = 87$ J

13. Let F_1 be the magnitude of the force directed 20° away from the direction of shore, and let F_2 be the magnitude of the other force. Separating these forces into components parallel to the direction of the resultant force and perpendicular to it gives

$F_1\cos 20^\circ + F_2\cos 30^\circ = 255$ **(1)**, and $F_1\sin 20^\circ - F_2\sin 30^\circ = 0 \quad\Rightarrow\quad F_1 = F_2\dfrac{\sin 30^\circ}{\sin 20^\circ}$ **(2)**. Substituting **(2)** into **(1)** gives $F_2(\sin 30^\circ\cot 20^\circ + \cos 30^\circ) = 255 \quad\Rightarrow\quad F_2 \approx 114$ N. Substituting this into **(2)** gives $F_1 \approx 166$ N.

14. $|\tau| = |\mathbf{r}|\,|\mathbf{F}|\sin\theta = (0.40)(50)\sin(90^\circ - 30^\circ) \approx 17.3\ \mathrm{N\cdot m}$.

15. The line has direction $\mathbf{v} = \langle -3,2,3\rangle$. Letting $P_0 = (4,-1,2)$, parametric equations are $x = 4-3t,\ y = -1+2t,\ z = 2+3t$.

16. A direction vector for the line is $\mathbf{v} = \langle 3,2,1\rangle$, so parametric equations for the line are $x = 1+3t,\ y = 2t,\ z = -1+t$.

17. A direction vector for the line is a normal vector for the plane, $\mathbf{n} = \langle 2,-1,5\rangle$, and parametric equations for the line are $x = -2+2t,\ y = 2-t,\ z = 4+5t$.

18. Since the two planes are parallel, they will have the same normal vectors. Then we can take $\mathbf{n} = \langle 1,4,-3\rangle$ and an equation of the plane is $1(x-2) + 4(y-1) - 3(z-0) = 0$ or $x + 4y - 3z = 6$.

19. Here the vectors $\mathbf{a} = \langle 4-3, 0-(-1), 2-1\rangle = \langle 1,1,1\rangle$ and $\mathbf{b} = \langle 6-3, 3-(-1), 1-1\rangle = \langle 3,4,0\rangle$ lie in the plane, so $\mathbf{n} = \mathbf{a}\times\mathbf{b} = \langle -4,3,1\rangle$ is a normal vector to the plane and an equation of the plane is $-4(x-3) + 3(y-(-1)) + 1(z-1) = 0$ or $-4x + 3y + z = -14$.

20. If we first find two nonparallel vectors in the plane, their cross product will be a normal vector to the plane. Since the given line lies in the plane, its direction vector $\mathbf{a} = \langle 2,-1,3\rangle$ is one vector in the plane. We can verify that the given point $(1,2,-2)$ does not lie on this line. The point $(0,3,1)$ is on the line (obtained by putting $t = 0$) and hence in the plane, so the vector $\mathbf{b} = \langle 0-1, 3-2, 1-(-2)\rangle = \langle -1,1,3\rangle$ lies in the plane, and a normal vector is $\mathbf{n} = \mathbf{a}\times\mathbf{b} = \langle -6,-9,1\rangle$. Thus an equation of the plane is $-6(x-1) - 9(y-2) + (z+2) = 0$ or $6x + 9y - z = 26$.

21. Substitution of the parametric equations into the equation of the plane gives $2x - y + z = 2(2-t) - (1+3t) + 4t = 2 \Rightarrow -t + 3 = 2 \Rightarrow t = 1$. When $t = 1$, the parametric equations give $x = 2 - 1 = 1$, $y = 1 + 3 = 4$ and $z = 4$. Therefore, the point of intersection is $(1, 4, 4)$.

22. Use the formula proven in Exercise 10.4.39(a). In the notation used in that exercise, $\mathbf{a}$ is just the direction of the line; that is, $\mathbf{a} = \langle 1, -1, 2 \rangle$. A point on the line is $(1, 2, -1)$ (setting $t = 0$), and therefore $\mathbf{b} = \langle 1 - 0, 2 - 0, -1 - 0 \rangle = \langle 1, 2, -1 \rangle$.

Hence $d = \dfrac{|\mathbf{a} \times \mathbf{b}|}{|\mathbf{a}|} = \dfrac{|\langle 1, -1, 2 \rangle \times \langle 1, 2, -1 \rangle|}{\sqrt{1+1+4}} = \dfrac{|\langle -3, 3, 3 \rangle|}{\sqrt{6}} = \sqrt{\dfrac{27}{6}} = \dfrac{3}{\sqrt{2}}$.

23. Since the direction vectors $\langle 2, 3, 4 \rangle$ and $\langle 6, -1, 2 \rangle$ aren't parallel, neither are the lines. For the lines to intersect, the three equations $1 + 2t = -1 + 6s$, $2 + 3t = 3 - s$, $3 + 4t = -5 + 2s$ must be satisfied simultaneously. Solving the first two equations gives $t = \frac{1}{5}$, $s = \frac{2}{5}$ and checking we see these values don't satisfy the third equation. Thus the lines aren't parallel and they don't intersect, so they must be skew.

24. (a) The normal vectors are $\langle 1, 1, -1 \rangle$ and $\langle 2, -3, 4 \rangle$. Since these vectors aren't parallel, neither are the planes parallel. Also $\langle 1, 1, -1 \rangle \cdot \langle 2, -3, 4 \rangle = 2 - 3 - 4 = -5 \neq 0$ so the normal vectors, and thus the planes, are not perpendicular.

(b) $\cos\theta = \dfrac{\langle 1, 1, -1 \rangle \cdot \langle 2, -3, 4 \rangle}{\sqrt{3}\,\sqrt{29}} = -\dfrac{5}{\sqrt{87}}$ and $\theta = \cos^{-1}\left(-\frac{5}{\sqrt{87}}\right) \approx 122^\circ$ [or we can say $\approx 58^\circ$].

25. By Exercise 10.5.51, $D = \dfrac{|2 - 24|}{\sqrt{26}} = \dfrac{22}{\sqrt{26}}$.

26. The equation $x = 3$ represents a plane parallel to the yz-plane and 3 units in front of it.

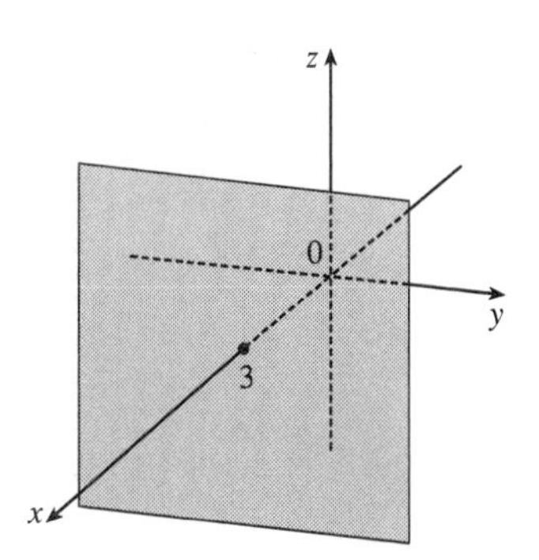

27. The equation $x = z$ represents a plane perpendicular to the xz-plane and intersecting the xz-plane in the line $x = z$, $y = 0$.

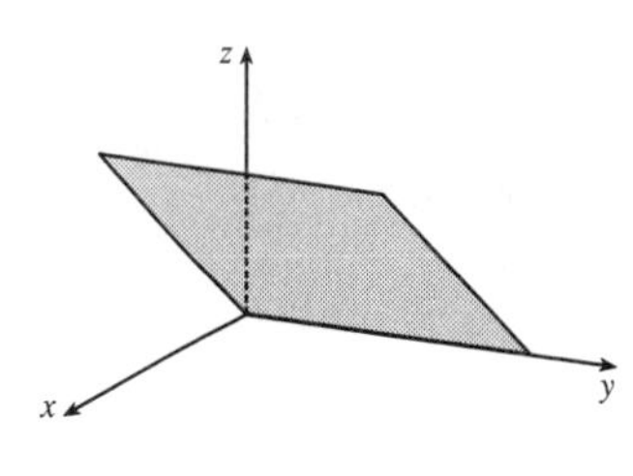

28. The equation $y = z^2$ represents a parabolic cylinder whose trace in the xz-plane is the x-axis and which opens to the right.

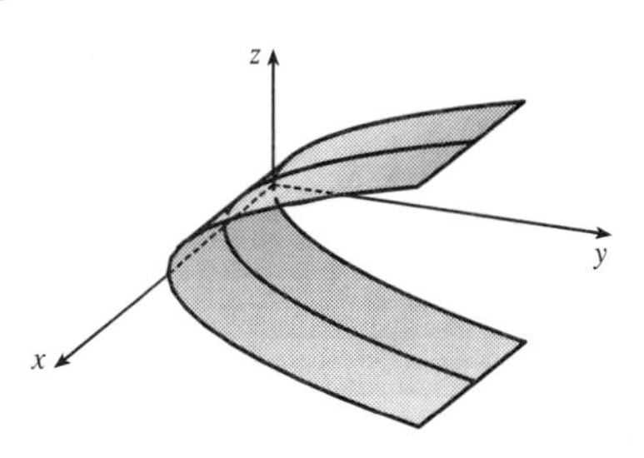

29. The equation $x^2 = y^2 + 4z^2$ represents a (right elliptical) cone with vertex at the origin and axis the x-axis.

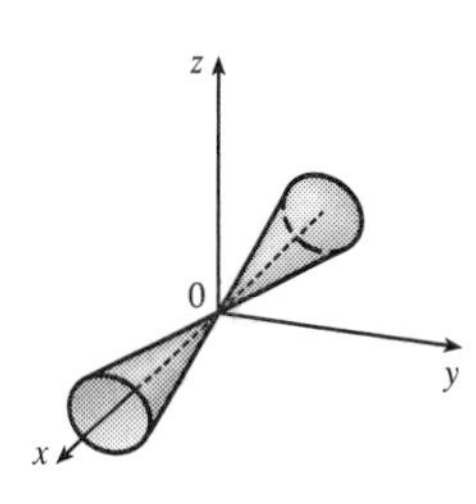

30. $4x - y + 2z = 4$ is a plane with intercepts $(1, 0, 0)$, $(0, -4, 0)$, and $(0, 0, 2)$.

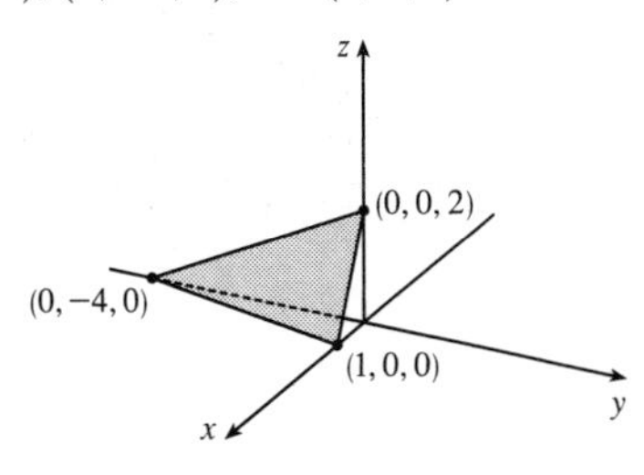

31. An equivalent equation is $-x^2 + \frac{y^2}{4} - z^2 = 1$, a hyperboloid of two sheets with axis the y-axis. For $|y| > 2$, traces parallel to the xz-plane are circles.

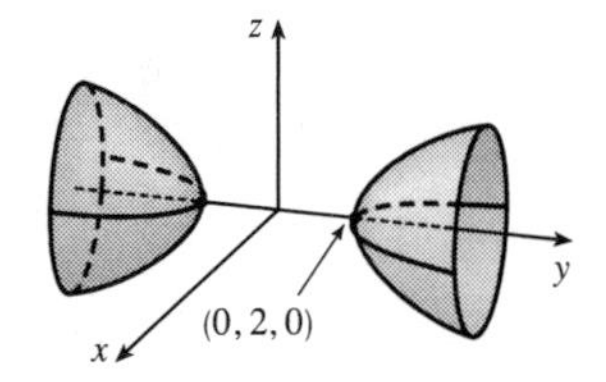

32. An equivalent equation is $-x^2 + y^2 + z^2 = 1$, a hyperboloid of one sheet with axis the x-axis.

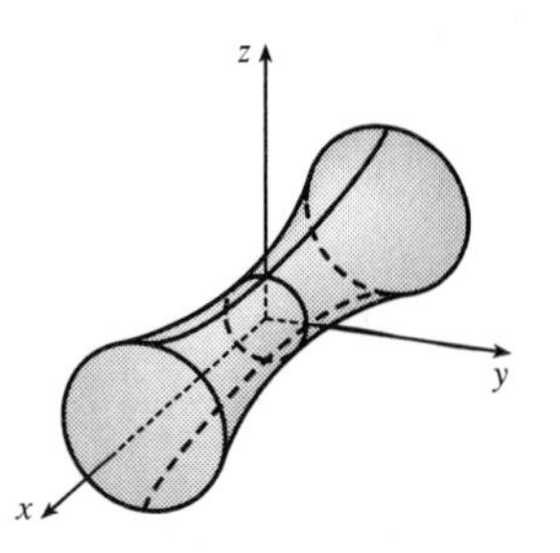

33. Completing the square in y gives $4x^2 + 4(y-1)^2 + z^2 = 4$ or $x^2 + (y-1)^2 + \frac{z^2}{4} = 1$, an ellipsoid centered at $(0, 1, 0)$.

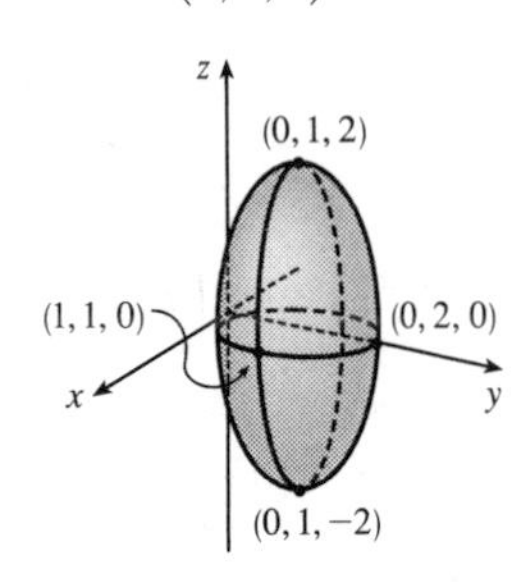

34. Completing the square in y and z gives $x = (y-1)^2 + (z-2)^2$, a circular paraboloid with vertex $(0, 1, 2)$ and axis the horizontal line $y = 1$, $z = 2$.

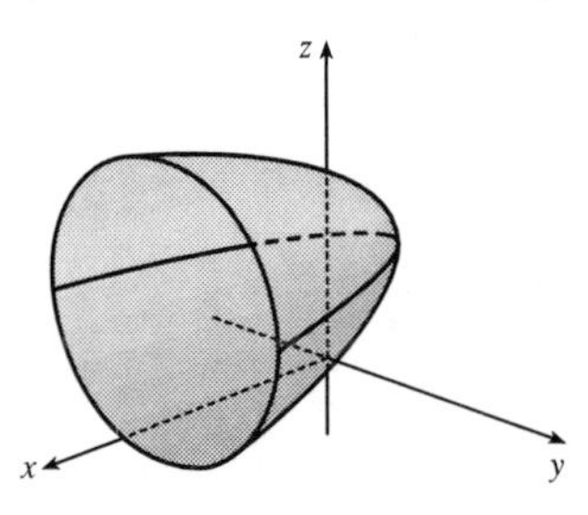

35. $4x^2 + y^2 = 16 \Leftrightarrow \frac{x^2}{4} + \frac{y^2}{16} = 1$. The equation of the ellipsoid is $\frac{x^2}{4} + \frac{y^2}{16} + \frac{z^2}{c^2} = 1$, since the horizontal trace in the plane $z = 0$ must be the original ellipse. The traces of the ellipsoid in the yz-plane must be circles since the surface is obtained by rotation about the x-axis. Therefore, $c^2 = 16$ and the equation of the ellipsoid is $\frac{x^2}{4} + \frac{y^2}{16} + \frac{z^2}{16} = 1 \Leftrightarrow$ $4x^2 + y^2 + z^2 = 16$.

36. The distance from a point $P(x, y, z)$ to the plane $y = 1$ is $|y - 1|$, so the given condition becomes

$$|y-1| = 2\sqrt{(x-0)^2 + (y+1)^2 + (z-0)^2} \Rightarrow |y-1| = 2\sqrt{x^2 + (y+1)^2 + z^2} \Rightarrow$$

$$(y-1)^2 = 4x^2 + 4(y+1)^2 + 4z^2 \Leftrightarrow -3 = 4x^2 + (3y^2 + 10y) + 4z^2 \Leftrightarrow$$

$$\tfrac{16}{3} = 4x^2 + 3\left(y + \tfrac{5}{3}\right)^2 + 4z^2 \Rightarrow \tfrac{3}{4}x^2 + \tfrac{9}{16}\left(y + \tfrac{5}{3}\right)^2 + \tfrac{3}{4}z^2 = 1.$$

This is the equation of an ellipsoid whose center is $\left(0, -\frac{5}{3}, 0\right)$.

37. (a) The corresponding parametric equations for the curve are $x = t$, $y = \cos \pi t$, $z = \sin \pi t$. Since $y^2 + z^2 = 1$, the curve is contained in a circular cylinder with axis the x-axis. Since $x = t$, the curve is a helix.

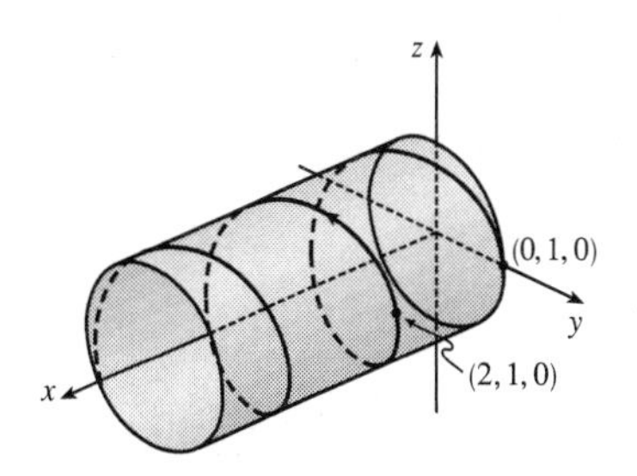

(b) $\mathbf{r}(t) = t\,\mathbf{i} + \cos \pi t\,\mathbf{j} + \sin \pi t\,\mathbf{k} \;\Rightarrow$

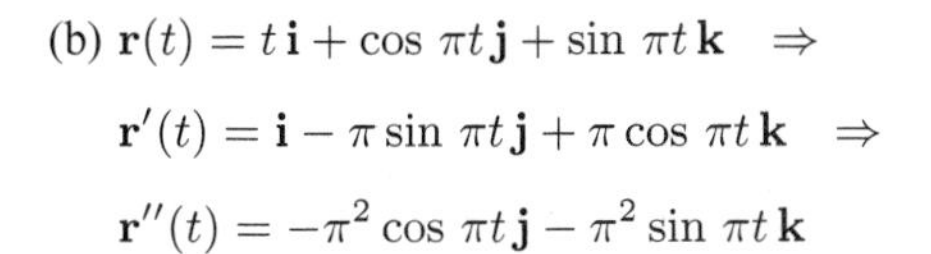

$\mathbf{r}'(t) = \mathbf{i} - \pi \sin \pi t\,\mathbf{j} + \pi \cos \pi t\,\mathbf{k} \;\Rightarrow$

$\mathbf{r}''(t) = -\pi^2 \cos \pi t\,\mathbf{j} - \pi^2 \sin \pi t\,\mathbf{k}$

38. (a) The expressions $\sqrt{2-t}$, $(e^t - 1)/t$, and $\ln(t+1)$ are all defined when $2 - t \geq 0 \;\Rightarrow\; t \leq 2, t \neq 0$, and $t + 1 > 0 \;\Rightarrow\; t > -1$. Thus the domain of $\mathbf{r}$ is $(-1, 0) \cup (0, 2]$.

(b) $\displaystyle \lim_{t\to 0} \mathbf{r}(t) = \left\langle \lim_{t\to 0} \sqrt{2-t}, \lim_{t\to 0} \frac{e^t - 1}{t}, \lim_{t\to 0} \ln(t+1) \right\rangle = \left\langle \sqrt{2-0}, \lim_{t\to 0} \frac{e^t}{1}, \ln(0+1) \right\rangle$

$= \langle \sqrt{2}, 1, 0 \rangle$ [using l'Hospital's Rule in the y-component]

(c) $\displaystyle \mathbf{r}'(t) = \left\langle \frac{d}{dt}\sqrt{2-t}, \frac{d}{dt}\frac{e^t-1}{t}, \frac{d}{dt}\ln(t+1) \right\rangle = \left\langle -\frac{1}{2\sqrt{2-t}}, \frac{te^t - e^t + 1}{t^2}, \frac{1}{t+1} \right\rangle$

39. The projection of the curve C of intersection onto the xy-plane is the circle $x^2 + y^2 = 16$, $z = 0$. So we can write $x = 4\cos t$, $y = 4\sin t$, $0 \leq t \leq 2\pi$. From the equation of the plane, we have $z = 5 - x = 5 - 4\cos t$, so parametric equations for C are $x = 4\cos t$, $y = 4\sin t$, $z = 5 - 4\cos t$, $0 \leq t \leq 2\pi$, and the corresponding vector function is $\mathbf{r}(t) = 4\cos t\,\mathbf{i} + 4\sin t\,\mathbf{j} + (5 - 4\cos t)\,\mathbf{k}$, $0 \leq t \leq 2\pi$.

40. The curve is given by $\mathbf{r}(t) = \langle 2\sin t, 2\sin 2t, 2\sin 3t \rangle$, so $\mathbf{r}'(t) = \langle 2\cos t, 4\cos 2t, 6\cos 3t \rangle$. The point $(1, \sqrt{3}, 2)$ corresponds to $t = \frac{\pi}{6}$ (or $\frac{\pi}{6} + 2k\pi$, k an integer), so the tangent vector there is $\mathbf{r}'(\frac{\pi}{6}) = \langle \sqrt{3}, 2, 0 \rangle$. Then the tangent line has direction vector $\langle \sqrt{3}, 2, 0 \rangle$ and includes the point $(1, \sqrt{3}, 2)$, so parametric equations are $x = 1 + \sqrt{3}\,t$, $y = \sqrt{3} + 2t$, $z = 2$.

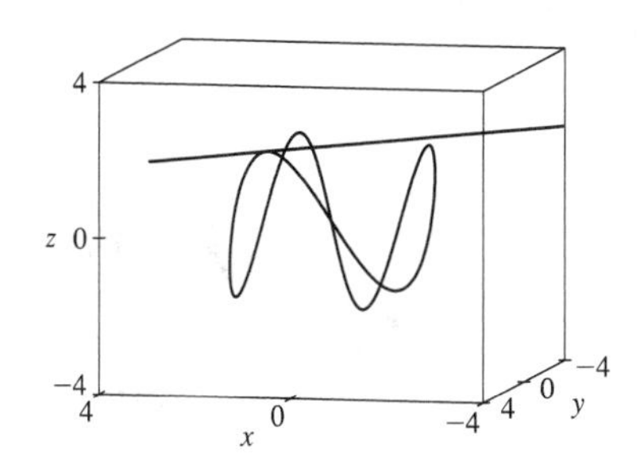

41. $\int_0^1 (t^2\,\mathbf{i} + t\cos \pi t\,\mathbf{j} + \sin \pi t\,\mathbf{k})\,dt = \left(\int_0^1 t^2\,dt\right)\mathbf{i} + \left(\int_0^1 t\cos \pi t\,dt\right)\mathbf{j} + \left(\int_0^1 \sin \pi t\,dt\right)\mathbf{k}$

$= \left[\frac{1}{3}t^3\right]_0^1 \mathbf{i} + \left(\left[\frac{t}{\pi}\sin \pi t\right]_0^1 - \int_0^1 \frac{1}{\pi}\sin \pi t\,dt\right)\mathbf{j} + \left[-\frac{1}{\pi}\cos \pi t\right]_0^1 \mathbf{k}$

$= \frac{1}{3}\mathbf{i} + \left[\frac{1}{\pi^2}\cos \pi t\right]_0^1 \mathbf{j} + \frac{2}{\pi}\mathbf{k} = \frac{1}{3}\mathbf{i} - \frac{2}{\pi^2}\mathbf{j} + \frac{2}{\pi}\mathbf{k}$

where we integrated by parts in the y-component.

42. (a) C intersects the xz-plane where $y = 0 \;\Rightarrow\; 2t - 1 = 0 \;\Rightarrow\; t = \frac{1}{2}$, so the point is $\left(2 - \left(\frac{1}{2}\right)^3, 0, \ln \frac{1}{2}\right) = \left(\frac{15}{8}, 0, -\ln 2\right)$.

(b) The curve is given by $\mathbf{r}(t) = \langle 2 - t^3, 2t - 1, \ln t \rangle$, so $\mathbf{r}'(t) = \langle -3t^2, 2, 1/t \rangle$. The point $(1, 1, 0)$ corresponds to $t = 1$, so the tangent vector there is $\mathbf{r}'(1) = \langle -3, 2, 1 \rangle$. Then the tangent line has direction vector $\langle -3, 2, 1 \rangle$ and includes the point $(1, 1, 0)$, so parametric equations are $x = 1 - 3t$, $y = 1 + 2t$, $z = t$.

(c) The normal plane has normal vector $\mathbf{r}'(1) = \langle -3, 2, 1 \rangle$ and equation $-3(x - 1) + 2(y - 1) + z = 0$ or $3x - 2y - z = 1$.

43. $\mathbf{r}(t)=\langle t^2,t^3,t^4\rangle \;\Rightarrow\; \mathbf{r}'(t)=\langle 2t,3t^2,4t^3\rangle \;\Rightarrow\; |\mathbf{r}'(t)|=\sqrt{4t^2+9t^4+16t^6}$ and

$L=\int_0^3|\mathbf{r}'(t)|\,dt=\int_0^3\sqrt{4t^2+9t^4+16t^6}\,dt$. Using Simpson's Rule with $f(t)=\sqrt{4t^2+9t^4+16t^6}$ and $n=6$ we have $\Delta t=\frac{3-0}{6}=\frac{1}{2}$ and

$$\begin{aligned}L&\approx\tfrac{\Delta t}{3}\left[f(0)+4f\left(\tfrac12\right)+2f(1)+4f\left(\tfrac32\right)+2f(2)+4f\left(\tfrac52\right)+f(3)\right]\\&=\tfrac16\left[\sqrt{0+0+0}+4\cdot\sqrt{4\left(\tfrac12\right)^2+9\left(\tfrac12\right)^4+16\left(\tfrac12\right)^6}+2\cdot\sqrt{4(1)^2+9(1)^4+16(1)^6}\right.\\&\qquad+4\cdot\sqrt{4\left(\tfrac32\right)^2+9\left(\tfrac32\right)^4+16\left(\tfrac32\right)^6}+2\cdot\sqrt{4(2)^2+9(2)^4+16(2)^6}\\&\qquad\left.+4\cdot\sqrt{4\left(\tfrac52\right)^2+9\left(\tfrac52\right)^4+16\left(\tfrac52\right)^6}+\sqrt{4(3)^2+9(3)^4+16(3)^6}\right]\\&\approx 86.631\end{aligned}$$

44. $\mathbf{r}'(t)=\left\langle 3t^{1/2},-2\sin 2t,2\cos 2t\right\rangle,\quad |\mathbf{r}'(t)|=\sqrt{9t+4(\sin^2 2t+\cos^2 2t)}=\sqrt{9t+4}.$

Thus $L=\int_0^1\sqrt{9t+4}\,dt=\int_4^{13}\frac19 u^{1/2}\,du=\frac19\cdot\frac23 u^{3/2}\Big]_4^{13}=\frac{2}{27}(13^{3/2}-8).$

45. The angle of intersection of the two curves, θ, is the angle between their respective tangents at the point of intersection. For both curves the point $(1,0,0)$ occurs when $t=0$.

$\mathbf{r}_1'(t)=-\sin t\,\mathbf{i}+\cos t\,\mathbf{j}+\mathbf{k}\;\Rightarrow\;\mathbf{r}_1'(0)=\mathbf{j}+\mathbf{k}$ and $\mathbf{r}_2'(t)=\mathbf{i}+2t\,\mathbf{j}+3t^2\,\mathbf{k}\;\Rightarrow\;\mathbf{r}_2'(0)=\mathbf{i}$.

$\mathbf{r}_1'(0)\cdot\mathbf{r}_2'(0)=(\mathbf{j}+\mathbf{k})\cdot\mathbf{i}=0.$ Therefore, the curves intersect in a right angle, that is, $\theta=\frac{\pi}{2}$.

46. The parametric value corresponding to the point $(1,0,1)$ is $t=0$.

$\mathbf{r}'(t)=e^t\,\mathbf{i}+e^t(\cos t+\sin t)\,\mathbf{j}+e^t(\cos t-\sin t)\,\mathbf{k}\;\Rightarrow\;|\mathbf{r}'(t)|=e^t\sqrt{1+(\cos t+\sin t)^2+(\cos t-\sin t)^2}=\sqrt3\,e^t$

and $s(t)=\int_0^t e^u\sqrt3\,du=\sqrt3(e^t-1)\;\Rightarrow\;t=\ln\left(1+\frac{1}{\sqrt3}s\right)$.

Therefore, $\mathbf{r}(t(s))=\left(1+\frac{1}{\sqrt3}s\right)\mathbf{i}+\left(1+\frac{1}{\sqrt3}s\right)\sin\ln\left(1+\frac{1}{\sqrt3}s\right)\mathbf{j}+\left(1+\frac{1}{\sqrt3}s\right)\cos\ln\left(1+\frac{1}{\sqrt3}s\right)\mathbf{k}.$

47. (a) $\mathbf{T}(t)=\dfrac{\mathbf{r}'(t)}{|\mathbf{r}'(t)|}=\dfrac{\langle t^2,t,1\rangle}{|\langle t^2,t,1\rangle|}=\dfrac{\langle t^2,t,1\rangle}{\sqrt{t^4+t^2+1}}$

(b) $$\begin{aligned}\mathbf{T}'(t)&=-\tfrac12(t^4+t^2+1)^{-3/2}(4t^3+2t)\langle t^2,t,1\rangle+(t^4+t^2+1)^{-1/2}\langle 2t,1,0\rangle\\&=\frac{-2t^3-t}{(t^4+t^2+1)^{3/2}}\langle t^2,t,1\rangle+\frac{1}{(t^4+t^2+1)^{1/2}}\langle 2t,1,0\rangle\\&=\frac{\langle -2t^5-t^3,-2t^4-t^2,-2t^3-t\rangle+\langle 2t^5+2t^3+2t,t^4+t^2+1,0\rangle}{(t^4+t^2+1)^{3/2}}=\frac{\langle 2t,-t^4+1,-2t^3-t\rangle}{(t^4+t^2+1)^{3/2}}\end{aligned}$$

$|\mathbf{T}'(t)|=\dfrac{\sqrt{4t^2+t^8-2t^4+1+4t^6+4t^4+t^2}}{(t^4+t^2+1)^{3/2}}=\dfrac{\sqrt{t^8+4t^6+2t^4+5t^2}}{(t^4+t^2+1)^{3/2}}$ and $\mathbf{N}(t)=\dfrac{\langle 2t,1-t^4,-2t^3-t\rangle}{\sqrt{t^8+4t^6+2t^4+5t^2}}$.

(c) $\kappa(t)=\dfrac{|\mathbf{T}'(t)|}{|\mathbf{r}'(t)|}=\dfrac{\sqrt{t^8+4t^6+2t^4+5t^2}}{(t^4+t^2+1)^2}$

48. Using Exercise 10.8.32, we have $\mathbf{r}'(t)=\langle -3\sin t,4\cos t\rangle,\quad \mathbf{r}''(t)=\langle -3\cos t,-4\sin t\rangle$,

$|\mathbf{r}'(t)|^3=\left(\sqrt{9\sin^2 t+4\cos^2 t}\right)^3$ and then

$$\kappa(t) = \frac{|(-3\sin t)(-4\sin t) - (4\cos t)(-3\cos t)|}{(9\sin^2 t + 16\cos^2 t)^{3/2}} = \frac{12}{(9\sin^2 t + 16\cos^2 t)^{3/2}}.$$

At $(3,0)$, $t = 0$ and $\kappa(0) = 12/(16)^{3/2} = \frac{12}{64} = \frac{3}{16}$. At $(0,4)$, $t = \frac{\pi}{2}$ and $\kappa\left(\frac{\pi}{2}\right) = 12/9^{3/2} = \frac{12}{27} = \frac{4}{9}$.

49. $y' = 4x^3$, $y'' = 12x^2$ and $\kappa(x) = \dfrac{|y''|}{[1+(y')^2]^{3/2}} = \dfrac{|12x^2|}{(1+16x^6)^{3/2}}$, so $\kappa(1) = \dfrac{12}{17^{3/2}}$.

50. $\kappa(x) = \dfrac{|12x^2 - 2|}{[1+(4x^3-2x)^2]^{3/2}} \quad \Rightarrow \quad \kappa(0) = 2.$

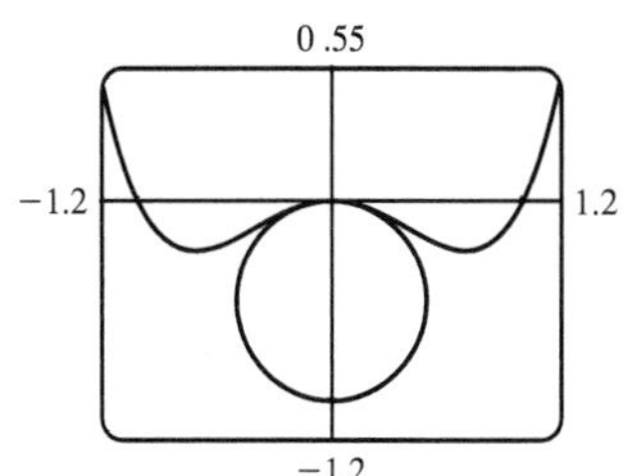

So the osculating circle has radius $\frac{1}{2}$ and center $\left(0, -\frac{1}{2}\right)$.

Thus its equation is $x^2 + \left(y + \frac{1}{2}\right)^2 = \frac{1}{4}$.

51. $\mathbf{r}(t) = t\ln t\,\mathbf{i} + t\,\mathbf{j} + e^{-t}\,\mathbf{k}$, $\quad \mathbf{v}(t) = \mathbf{r}'(t) = (1+\ln t)\,\mathbf{i} + \mathbf{j} - e^{-t}\,\mathbf{k}$,

$|\mathbf{v}(t)| = \sqrt{(1+\ln t)^2 + 1^2 + (-e^{-t})^2} = \sqrt{2 + 2\ln t + (\ln t)^2 + e^{-2t}}$, $\quad \mathbf{a}(t) = \mathbf{v}'(t) = \frac{1}{t}\,\mathbf{i} + e^{-t}\,\mathbf{k}$

52. $\mathbf{v}(t) = \int \mathbf{a}(t)\,dt = \int (6t\,\mathbf{i} + 12t^2\,\mathbf{j} - 6t\,\mathbf{k})\,dt = 3t^2\,\mathbf{i} + 4t^3\,\mathbf{j} - 3t^2\,\mathbf{k} + \mathbf{C}$, but $\mathbf{i} - \mathbf{j} + 3\,\mathbf{k} = \mathbf{v}(0) = \mathbf{0} + \mathbf{C}$,

so $\mathbf{C} = \mathbf{i} - \mathbf{j} + 3\,\mathbf{k}$ and $\mathbf{v}(t) = (3t^2+1)\,\mathbf{i} + (4t^3-1)\,\mathbf{j} + (3-3t^2)\,\mathbf{k}$.

$\mathbf{r}(t) = \int \mathbf{v}(t)\,dt = (t^3+t)\,\mathbf{i} + (t^4-t)\,\mathbf{j} + (3t-t^3)\,\mathbf{k} + \mathbf{D}$.

But $\mathbf{r}(0) = \mathbf{0}$, so $\mathbf{D} = \mathbf{0}$ and $\mathbf{r}(t) = (t^3+t)\,\mathbf{i} + (t^4-t)\,\mathbf{j} + (3t-t^3)\,\mathbf{k}$.

53. We set up the axes so that the shot leaves the athlete's hand 7 ft above the origin. Then we are given $\mathbf{r}(0) = 7\mathbf{j}$, $|\mathbf{v}(0)| = 43$ ft/s, and $\mathbf{v}(0)$ has direction given by a 45° angle of elevation. Then a unit vector in the direction of $\mathbf{v}(0)$ is $\frac{1}{\sqrt{2}}(\mathbf{i}+\mathbf{j}) \quad \Rightarrow \quad \mathbf{v}(0) = \frac{43}{\sqrt{2}}(\mathbf{i}+\mathbf{j})$. Assuming air resistance is negligible, the only external force is due to gravity, so as in Example 10.4.5 we have $\mathbf{a} = -g\,\mathbf{j}$ where here $g \approx 32$ ft/s^2. Since $\mathbf{v}'(t) = \mathbf{a}(t)$, we integrate, giving $\mathbf{v}(t) = -gt\,\mathbf{j} + \mathbf{C}$ where $\mathbf{C} = \mathbf{v}(0) = \frac{43}{\sqrt{2}}(\mathbf{i}+\mathbf{j}) \quad \Rightarrow \quad \mathbf{v}(t) = \frac{43}{\sqrt{2}}\,\mathbf{i} + \left(\frac{43}{\sqrt{2}} - gt\right)\mathbf{j}$. Since $\mathbf{r}'(t) = \mathbf{v}(t)$ we integrate again, so

$\mathbf{r}(t) = \frac{43}{\sqrt{2}}t\,\mathbf{i} + \left(\frac{43}{\sqrt{2}}t - \frac{1}{2}gt^2\right)\mathbf{j} + \mathbf{D}$. But $\mathbf{D} = \mathbf{r}(0) = 7\,\mathbf{j} \quad \Rightarrow \quad \mathbf{r}(t) = \frac{43}{\sqrt{2}}t\,\mathbf{i} + \left(\frac{43}{\sqrt{2}}t - \frac{1}{2}gt^2 + 7\right)\mathbf{j}$.

(a) At 2 seconds, the shot is at $\mathbf{r}(2) = \frac{43}{\sqrt{2}}(2)\,\mathbf{i} + \left(\frac{43}{\sqrt{2}}(2) - \frac{1}{2}g(2)^2 + 7\right)\mathbf{j} \approx 60.8\,\mathbf{i} + 3.8\,\mathbf{j}$, so the shot is about 3.8 ft above the ground, at a horizontal distance of 60.8 ft from the athlete.

(b) The shot reaches its maximum height when the vertical component of velocity is 0: $\frac{43}{\sqrt{2}} - gt = 0 \quad \Rightarrow$

$t = \dfrac{43}{\sqrt{2}\,g} \approx 0.95$ s. Then $\mathbf{r}(0.95) \approx 28.9\,\mathbf{i} + 21.4\,\mathbf{j}$, so the maximum height is approximately 21.4 ft.

(c) The shot hits the ground when the vertical component of $\mathbf{r}(t)$ is 0, so $\frac{43}{\sqrt{2}}t - \frac{1}{2}gt^2 + 7 = 0 \quad \Rightarrow$

$-16t^2 + \frac{43}{\sqrt{2}}t + 7 = 0 \quad \Rightarrow \quad t \approx 2.11$ s. $\quad \mathbf{r}(2.11) \approx 64.2\,\mathbf{i} - 0.08\,\mathbf{j}$, thus the shot lands approximately 64.2 ft from the athlete.

54. $\mathbf{r}'(t) = \mathbf{i} + 2\,\mathbf{j} + 2t\,\mathbf{k}$, $\quad \mathbf{r}''(t) = 2\,\mathbf{k}$, $\quad |\mathbf{r}'(t)| = \sqrt{1+4+4t^2} = \sqrt{4t^2+5}$.

Then $a_T = \dfrac{\mathbf{r}'(t)\cdot\mathbf{r}''(t)}{|\mathbf{r}'(t)|} = \dfrac{4t}{\sqrt{4t^2+5}}$ and $a_N = \dfrac{|\mathbf{r}'(t)\times\mathbf{r}''(t)|}{|\mathbf{r}'(t)|} = \dfrac{|4\,\mathbf{i} - 2\,\mathbf{j}|}{\sqrt{4t^2+5}} = \dfrac{2\sqrt{5}}{\sqrt{4t^2+5}}$.

55. By the Fundamental Theorem of Calculus, $\mathbf{r}'(t) = \left\langle \sin\left(\frac{1}{2}\pi t^2\right), \cos\left(\frac{1}{2}\pi t^2\right)\right\rangle$, $|\mathbf{r}'(t)| = 1$ and so $\mathbf{T}(t) = \mathbf{r}'(t)$.

Thus $\mathbf{T}'(t) = \pi t\left\langle \cos\left(\frac{1}{2}\pi t^2\right), -\sin\left(\frac{1}{2}\pi t^2\right)\right\rangle$ and the curvature is $\kappa = |\mathbf{T}'(t)| = \sqrt{(\pi t)^2(1)} = \pi\,|t|$.

11 □ PARTIAL DERIVATIVES

11.1 Functions of Several Variables

1. (a) $f(2,0) = 2^2 e^{3(2)(0)} = 4(1) = 4$

(b) Since both x^2 and the exponential function are defined everywhere, $x^2 e^{3xy}$ is defined for all choices of values for x and y. Thus the domain of f is $\mathbb{R}^2$.

(c) Because the range of $g(x,y) = 3xy$ is $\mathbb{R}$, and the range of e^x is $(0,\infty)$, the range of $e^{g(x,y)} = e^{3xy}$ is $(0,\infty)$. The range of x^2 is $[0,\infty)$, so the range of the product $x^2 e^{3xy}$ is $[0,\infty)$.

2. (a) $f(1,1) = \ln(1+1-1) = \ln 1 = 0$

(b) $f(e,1) = \ln(e+1-1) = \ln e = 1$

(c) $\ln(x+y-1)$ is defined only when $x+y-1>0$, that is, $y > 1-x$. So the domain of f is $\{(x,y) \mid y > 1-x\}$.

(d) Since $\ln(x+y-1)$ can be any real number, the range is $\mathbb{R}$.

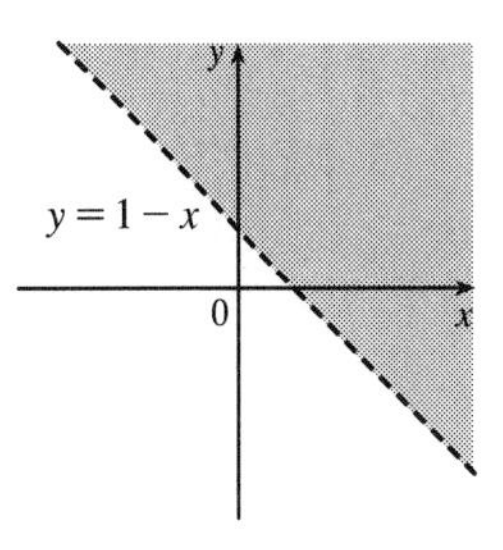

3. (a) $f(2,-1,6) = e^{\sqrt{6-2^2-(-1)^2}} = e^{\sqrt{1}} = e$.

(b) $e^{\sqrt{z-x^2-y^2}}$ is defined when $z - x^2 - y^2 \geq 0 \quad \Rightarrow \quad z \geq x^2 + y^2$. Thus the domain of f is $\{(x,y,z) \mid z \geq x^2 + y^2\}$.

(c) Since $\sqrt{z-x^2-y^2} \geq 0$, we have $e^{\sqrt{z-x^2-y^2}} \geq 1$. Thus the range of f is $[1,\infty)$.

4. (a) $g(2,-2,4) = \ln(25 - 2^2 - (-2)^2 - 4^2) = \ln 1 = 0$.

(b) For the logarithmic function to be defined, we need $25 - x^2 - y^2 - z^2 > 0$. Thus the domain of g is $\{(x,y,z) \mid x^2 + y^2 + z^2 < 25\}$, the interior of the sphere $x^2 + y^2 + z^2 = 25$.

(c) Since $0 < 25 - x^2 - y^2 - z^2 \leq 25$ for (x,y,z) in the domain of g, $\ln(25 - x^2 - y^2 - z^2) \leq \ln 25$. Thus the range of g is $(-\infty, \ln 25]$.

5. $\sqrt{x+y}$ is defined only when $x+y \geq 0$, or $y \geq -x$. So the domain of f is $\{(x,y) \mid y \geq -x\}$.

6. We need $xy \geq 0$, so $D = \{(x,y) \mid xy \geq 0\}$, the first and third quadrants.

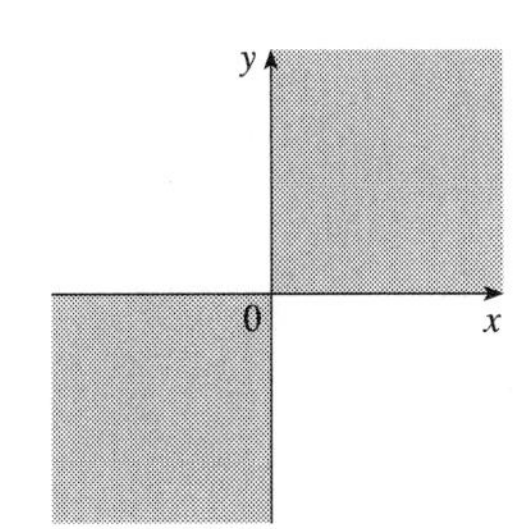

7. $\ln(9 - x^2 - 9y^2)$ is defined only when $9 - x^2 - 9y^2 > 0$, or $\frac{1}{9}x^2 + y^2 < 1$. So the domain of f is $\left\{(x, y) \mid \frac{1}{9}x^2 + y^2 < 1\right\}$, the interior of an ellipse.

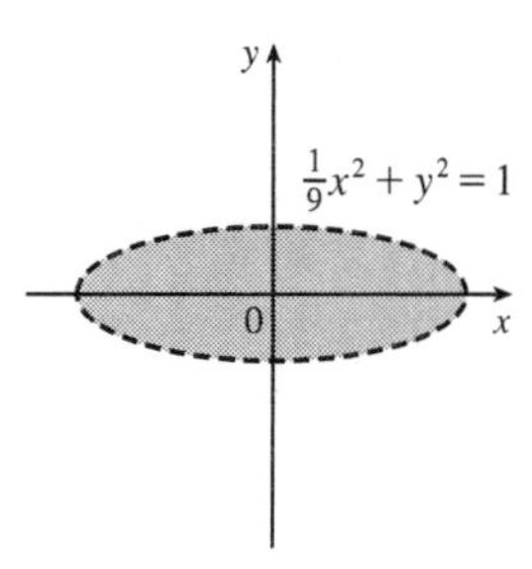

8. We need $y - x \geq 0$ or $y \geq x$ and $y + x > 0$ or $x > -y$. Thus $D = \{(x, y) \mid -y < x \leq y, y > 0\}$.

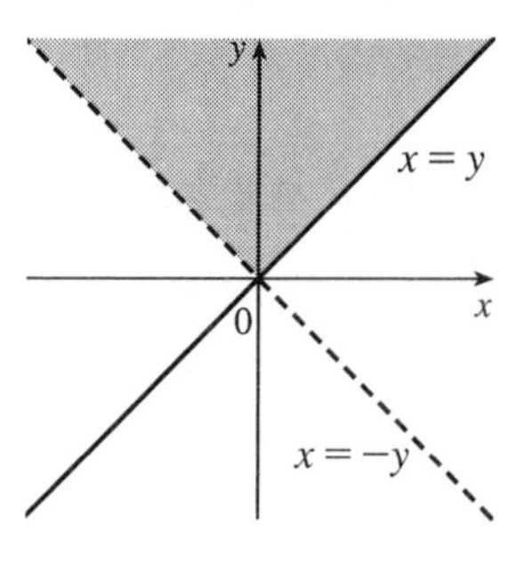

9. $\sqrt{y - x^2}$ is defined only when $y - x^2 \geq 0$, or $y \geq x^2$. In addition, f is not defined if $1 - x^2 = 0 \quad \Rightarrow \quad x = \pm 1$. Thus the domain of f is $\left\{(x, y) \mid y \geq x^2, x \neq \pm 1\right\}$.

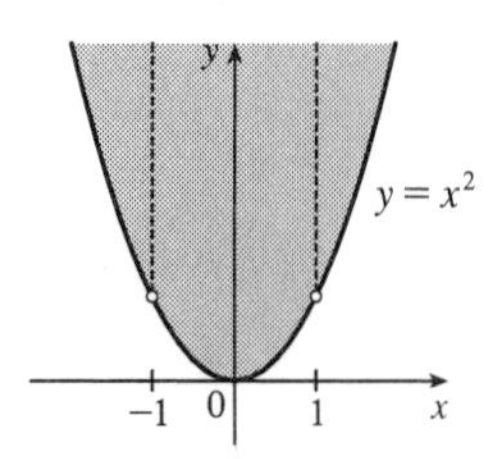

10. f is defined only when
$x^2 + y^2 - 1 \geq 0 \quad \Rightarrow \quad x^2 + y^2 \geq 1$ and
$4 - x^2 - y^2 > 0 \quad \Rightarrow \quad x^2 + y^2 < 4$.
Thus $D = \left\{(x, y) \mid 1 \leq x^2 + y^2 < 4\right\}$.

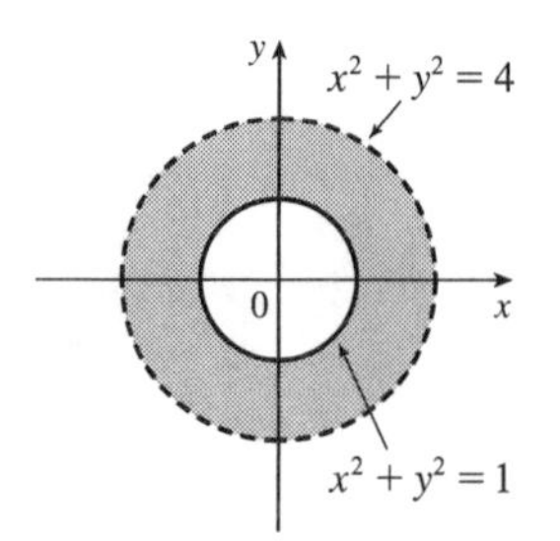

11. We need $1 - x^2 - y^2 - z^2 \geq 0$ or $x^2 + y^2 + z^2 \leq 1$, so $D = \left\{(x, y, z) \mid x^2 + y^2 + z^2 \leq 1\right\}$ (the points inside or on the sphere of radius 1, center the origin).

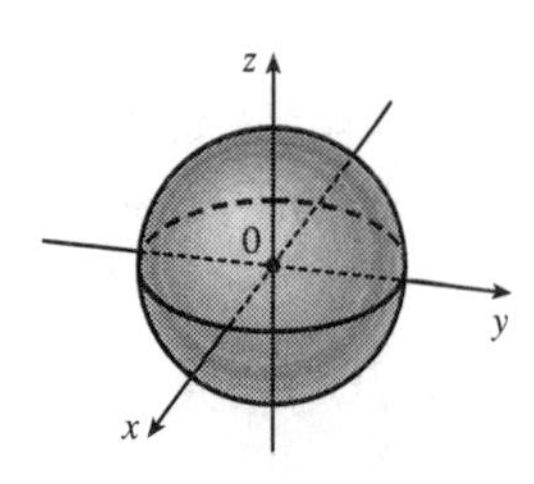

12. f is defined only when $16 - 4x^2 - 4y^2 - z^2 > 0 \quad \Rightarrow$
$\dfrac{x^2}{4} + \dfrac{y^2}{4} + \dfrac{z^2}{16} < 1$. Thus,
$D = \left\{(x, y, z) \,\middle|\, \dfrac{x^2}{4} + \dfrac{y^2}{4} + \dfrac{z^2}{16} < 1\right\}$, that is, the points inside the ellipsoid $\dfrac{x^2}{4} + \dfrac{y^2}{4} + \dfrac{z^2}{16} = 1$.

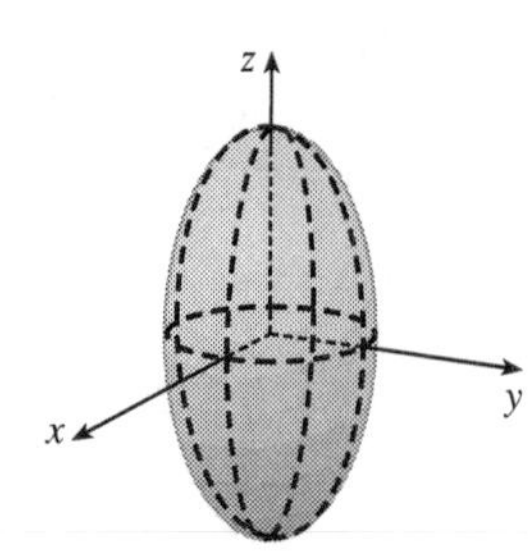

13. $z = 6 - 3x - 2y$ or $3x + 2y + z = 6$, a plane with intercepts 2, 3, and 6.

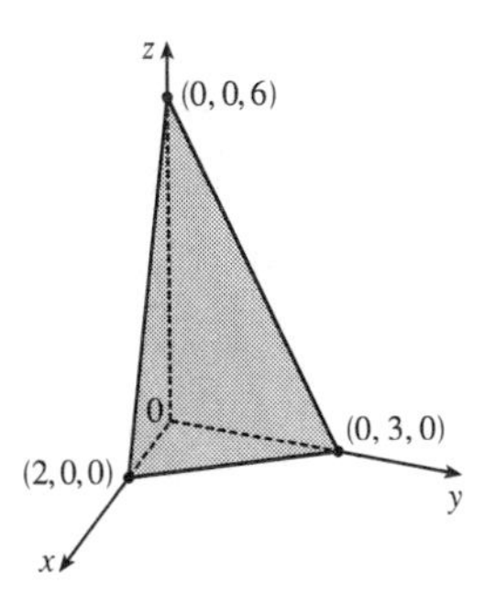

14. $z = y$, a plane which intersects the yz-plane in the line $z = y$, $x = 0$. The portion of this plane that lies in the first octant is shown.

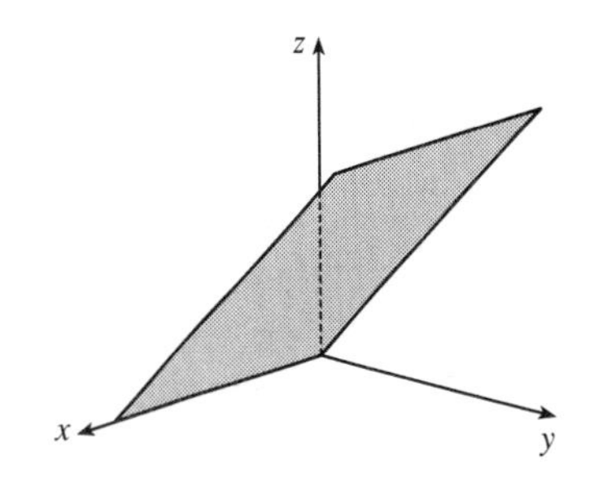

15. $z = y^2 + 1$, a parabolic cylinder

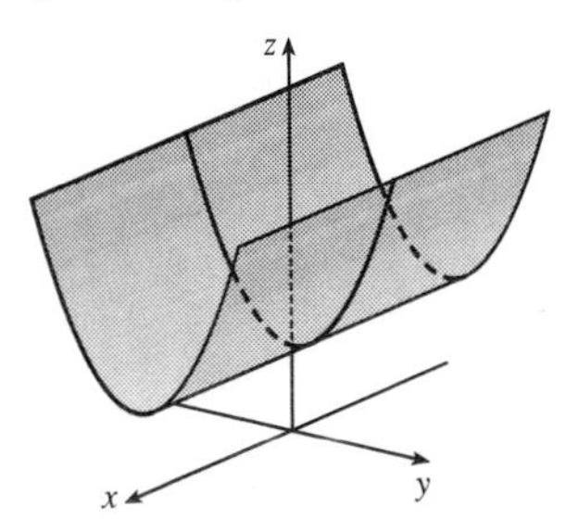

16. $z = \cos x$, a "wave."

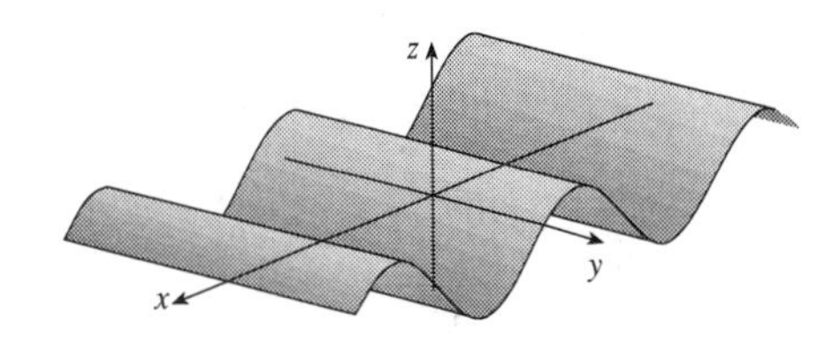

17. $z = 4x^2 + y^2 + 1$, an elliptic paraboloid with vertex at $(0, 0, 1)$.

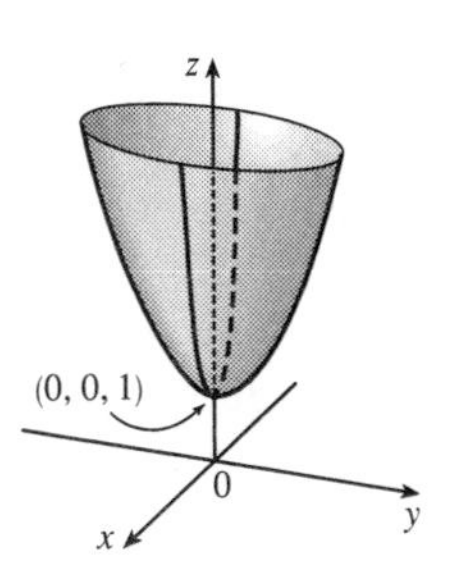

18. $z = 3 - x^2 - y^2$, a circular paraboloid with vertex at $(0, 0, 3)$.

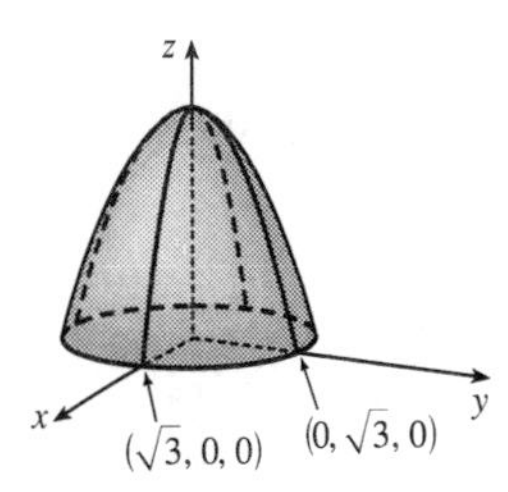

19. $z = \sqrt{x^2 + y^2}$ so $x^2 + y^2 = z^2$ and $z \geq 0$, the top half of a right circular cone.

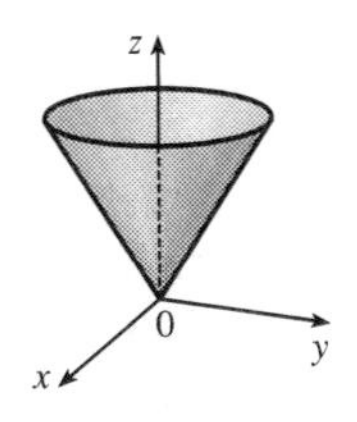

20. $z = \sqrt{16 - x^2 - 16y^2}$ so $z \geq 0$ and $z^2 + x^2 + 16y^2 = 16$, the top half of an ellipsoid.

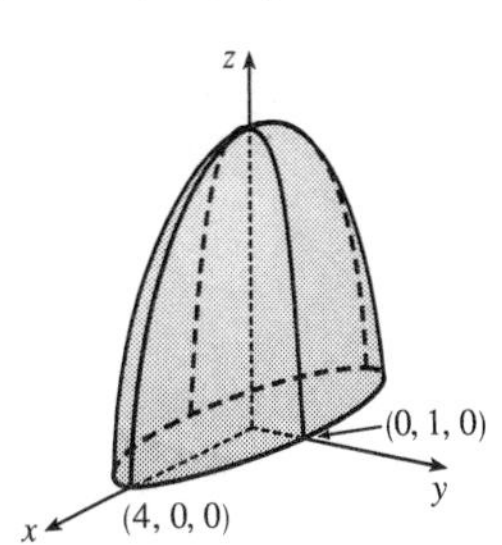

21. The point $(-3, 3)$ lies between the level curves with z-values 50 and 60. Since the point is a little closer to the level curve with $z = 60$, we estimate that $f(-3, 3) \approx 56$. The point $(3, -2)$ appears to be just about halfway between the level curves with z-values 30 and 40, so we estimate $f(3, -2) \approx 35$. The graph rises as we approach the origin, gradually from above, steeply from below.

22. If we start at the origin and move along the x-axis, for example, the z-values of a cone centered at the origin increase at a constant rate, so we would expect its level curves to be equally spaced. A paraboloid with vertex the origin, on the other hand, has z-values which change slowly near the origin and more quickly as we move farther away. Thus, we would expect its level curves near the origin to be spaced more widely apart than those farther from the origin. Therefore contour map I must correspond to the paraboloid, and contour map II the cone.

23. Near A, the level curves are very close together, indicating that the terrain is quite steep. At B, the level curves are much farther apart, so we would expect the terrain to be much less steep than near A, perhaps almost flat.

24.

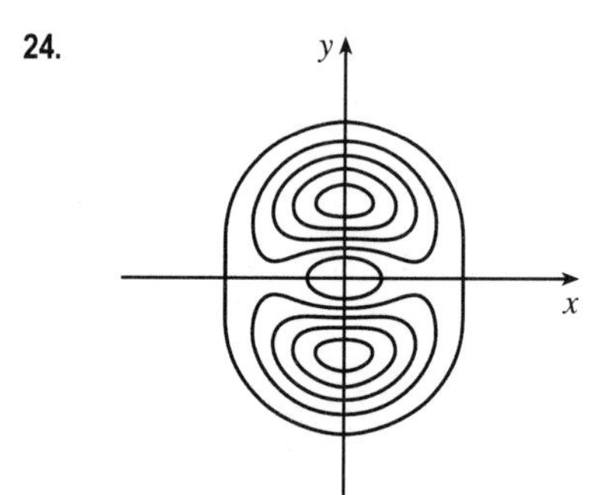

25. The level curves are $(y - 2x)^2 = k$ or $y = 2x \pm \sqrt{k}$, $k \geq 0$, a family of pairs of parallel lines.

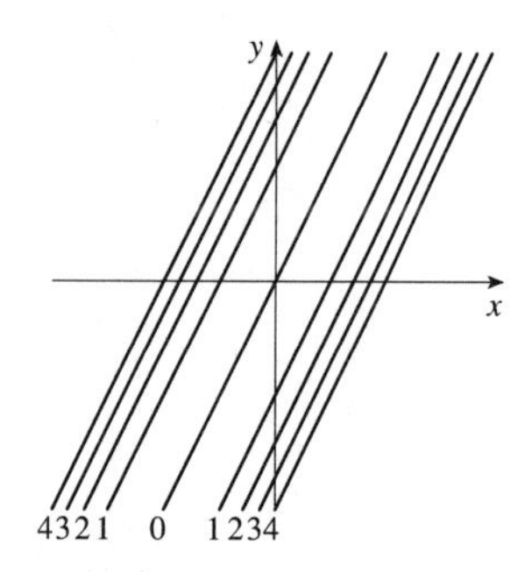

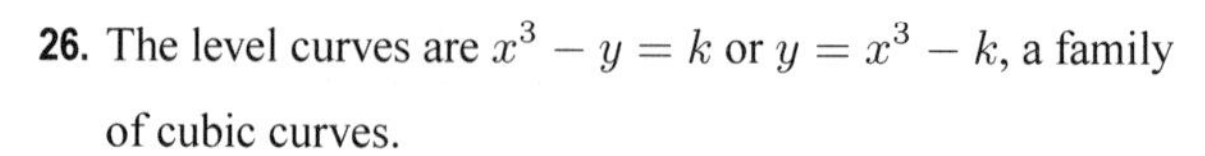

26. The level curves are $x^3 - y = k$ or $y = x^3 - k$, a family of cubic curves.

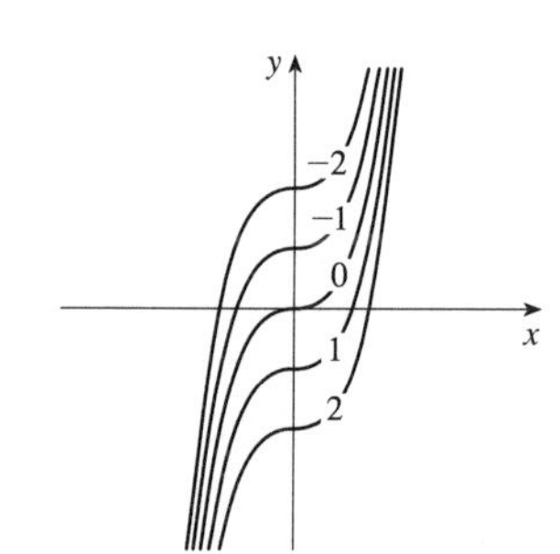

27. The level curves are $y - \ln x = k$ or $y = \ln x + k$.

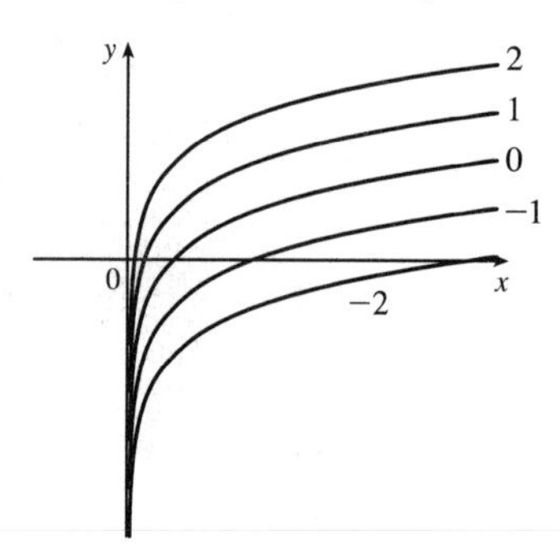

28. The level curves are $e^{y/x} = k$ or equivalently $y = x \ln k$ $(x \neq 0)$, a family of lines with slope $\ln k$ $(k > 0)$ without the origin.

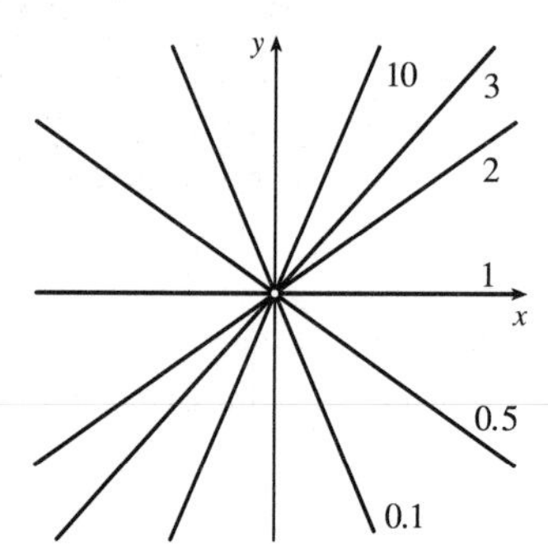

29. The level curves are $ye^x = k$ or $y = ke^{-x}$, a family of exponential curves.

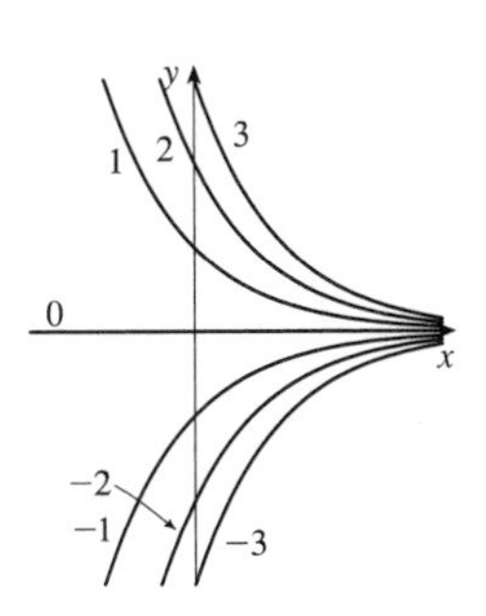

30. $k = y \sec x$ or $y = k \cos x$, $x \neq \frac{\pi}{2} + n\pi$ (n an integer).

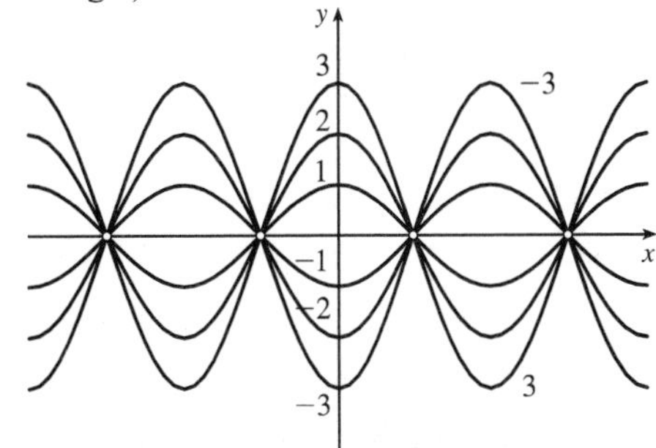

31. The level curves are $\sqrt{y^2 - x^2} = k$ or $y^2 - x^2 = k^2$, $k \geq 0$. When $k = 0$ the level curve is the pair of lines $y = \pm x$. For $k > 0$, the level curves are hyperbolas with axis the y-axis.

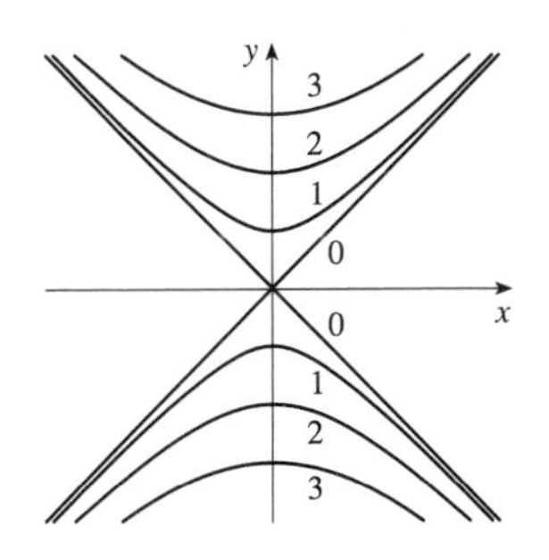

32. For $k \neq 0$ and $(x, y) \neq (0, 0)$, $k = \dfrac{y}{x^2 + y^2} \quad \Leftrightarrow$

$$x^2 + y^2 - \frac{y}{k} = 0 \quad \Leftrightarrow \quad x^2 + \left(y - \tfrac{1}{2k}\right)^2 = \frac{1}{4k^2},$$

a family of circles with center $\left(0, \frac{1}{2k}\right)$ and radius $\frac{1}{2k}$ (without the origin). If $k = 0$, the level curve is the x-axis.

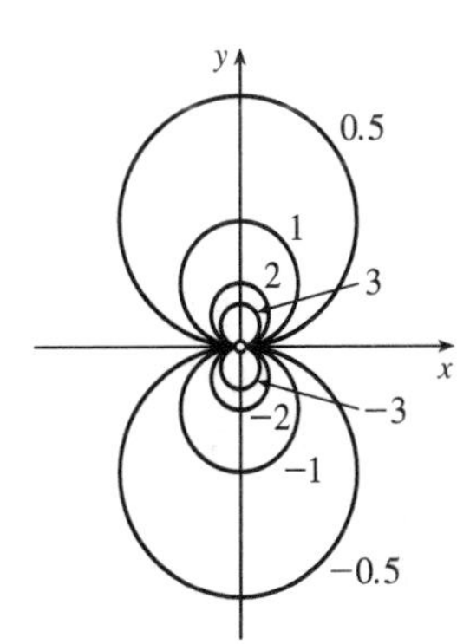

33. The contour map consists of the level curves $k = x^2 + 9y^2$, a family of ellipses with major axis the x-axis. (Or, if $k = 0$, the origin.)
The graph of $f(x, y)$ is the surface $z = x^2 + 9y^2$, an elliptic paraboloid.

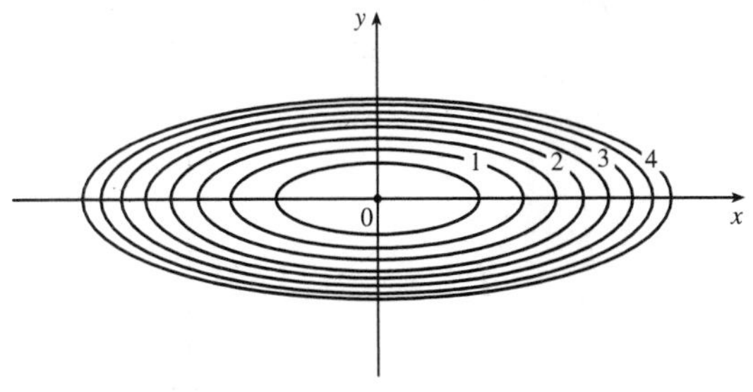

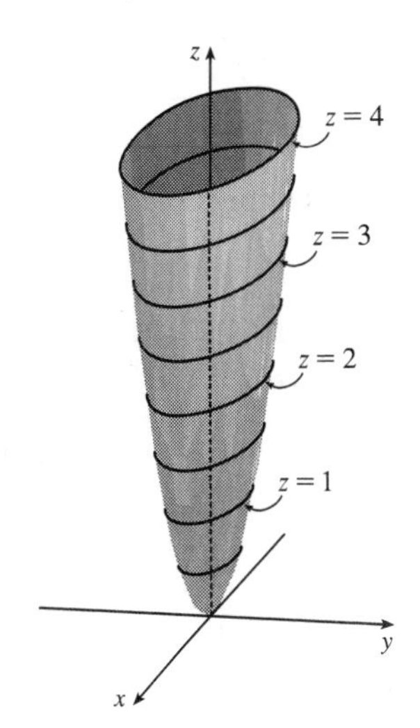

If we visualize lifting each ellipse $k = x^2 + 9y^2$ of the contour map to the plane $z = k$, we have horizontal traces that indicate the shape of the graph of f.

34.

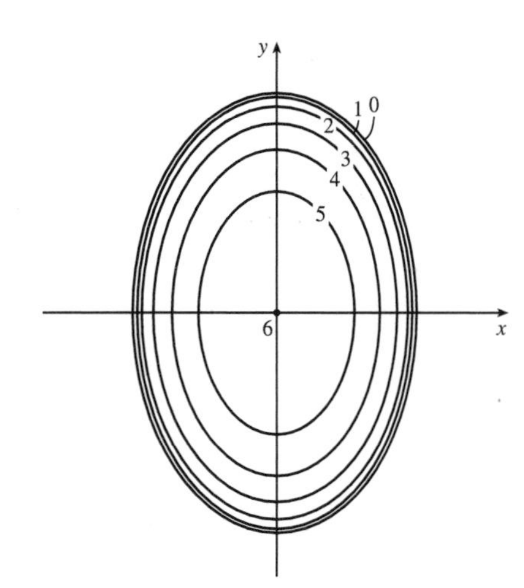

The contour map consists of the level curves $k = \sqrt{36 - 9x^2 - 4y^2} \Rightarrow$ $9x^2 + 4y^2 = 36 - k^2$, $k \geq 0$, a family of ellipses with major axis the y-axis. (Or, if $k = 6$, the origin.)

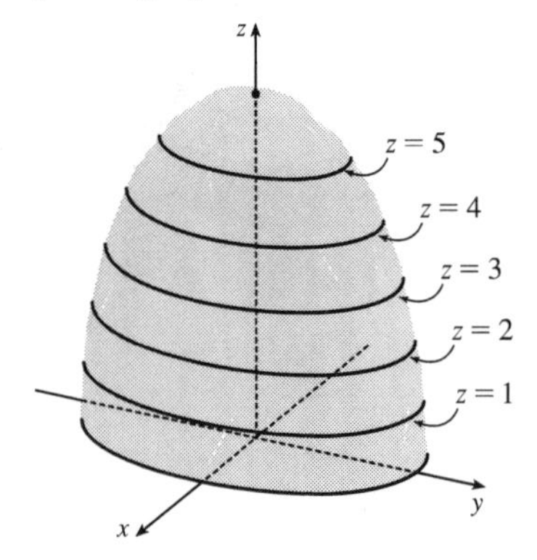

The graph of $f(x, y)$ is the surface $z = \sqrt{36 - 9x^2 - 4y^2}$, or equivalently the upper half of the ellipsoid $9x^2 + 4y^2 + z^2 = 36$. If we visualize lifting each ellipse $k = \sqrt{36 - 9x^2 - 4y^2}$ of the contour map to the plane $z = k$, we have horizontal traces that indicate the shape of the graph of f.

35. The isothermals are given by $k = 100/(1 + x^2 + 2y^2)$ or

$x^2 + 2y^2 = (100 - k)/k$ $[0 < k \leq 100]$, a family of ellipses.

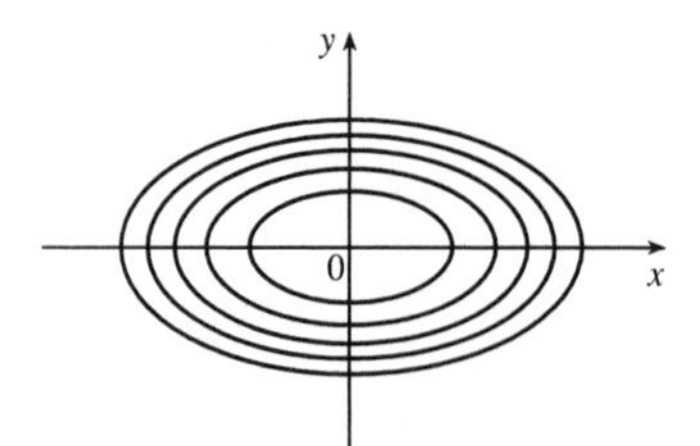

36. The equipotential curves are $k = \dfrac{c}{\sqrt{r^2 - x^2 - y^2}}$ or

$x^2 + y^2 = r^2 - \left(\dfrac{c}{k}\right)^2$, a family of circles ($k \geq c/r$).

Note: As $k \to \infty$, the radius of the circle approaches r.

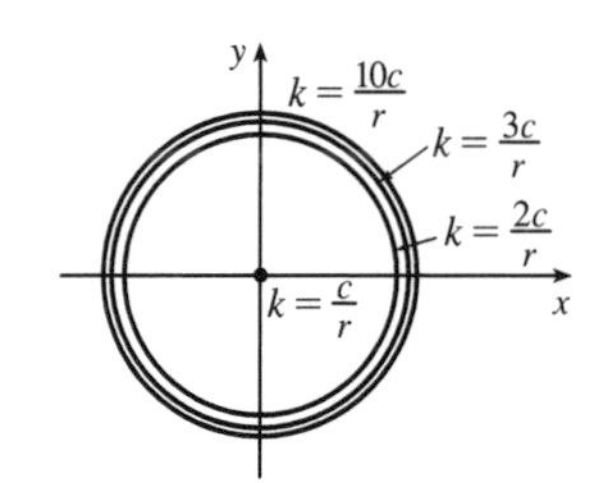

37. $f(x, y) = e^x \cos y$

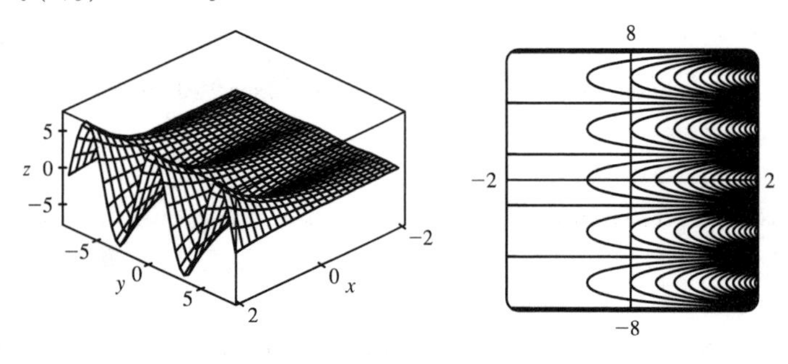

Traces parallel to the yz-plane (such as the left-front trace in the first graph above) are cosine curves. The amplitudes of these curves decrease as x decreases.

38. $f(x,y) = (1 - 3x^2 + y^2)e^{1-x^2-y^2}$

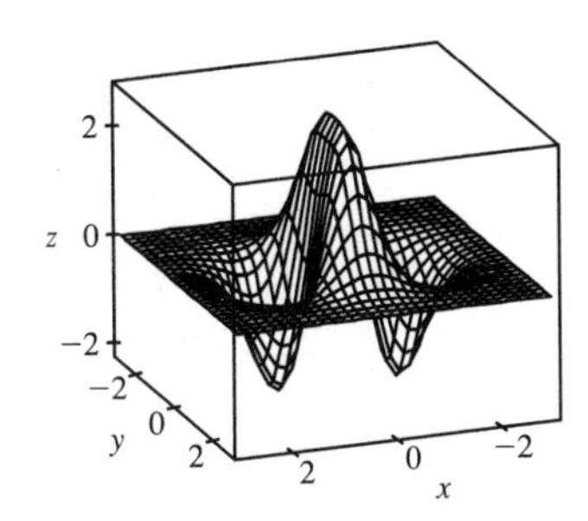

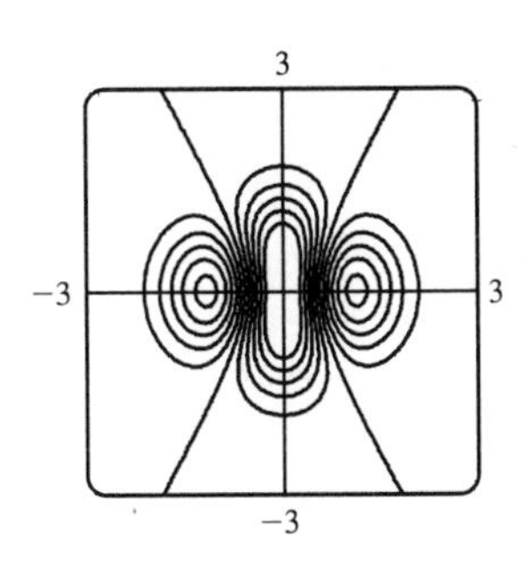

39. $f(x,y) = xy^2 - x^3$

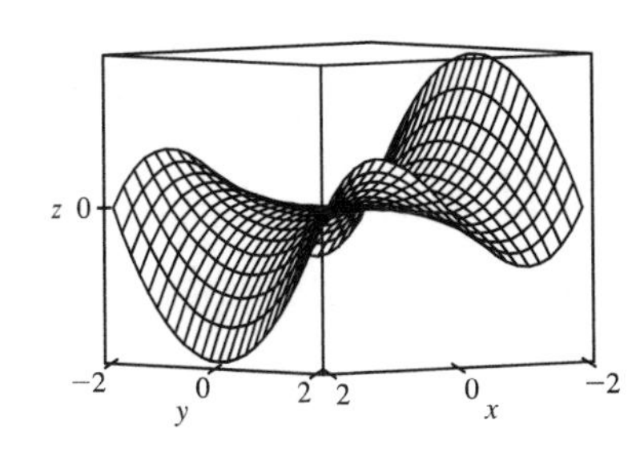

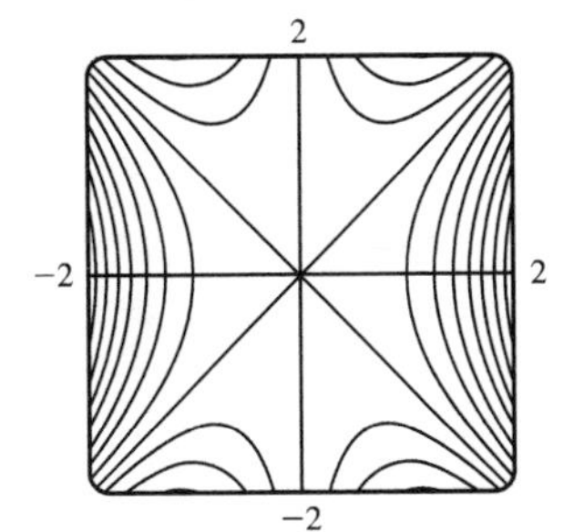

The traces parallel to the yz-plane (such as the left-front trace in the graph above) are parabolas; those parallel to the xz-plane (such as the right-front trace) are cubic curves. The surface is called a monkey saddle because a monkey sitting on the surface near the origin has places for both legs and tail to rest.

40. $f(x,y) = xy^3 - yx^3$

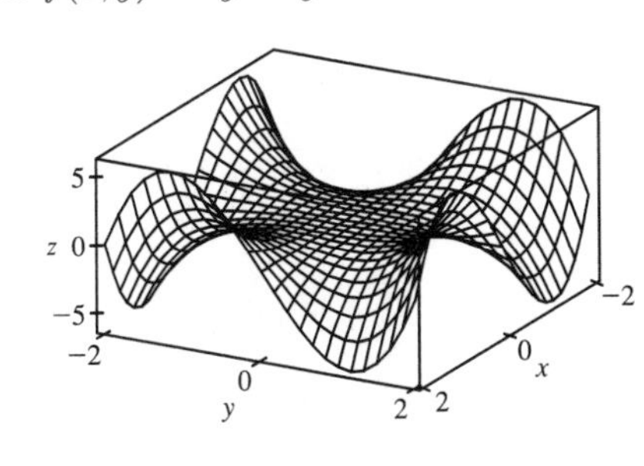

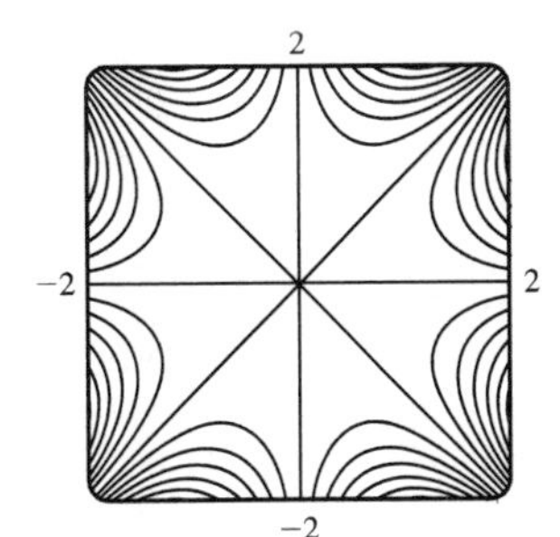

The traces parallel to either the yz-plane or the xz-plane are cubic curves.

41. (a) C (b) II

Reasons: This function is periodic in both x and y, and the function is the same when x is interchanged with y, so its graph is symmetric about the plane $y = x$. In addition, the function is 0 along the x- and y-axes. These conditions are satisfied only by C and II.

42. (a) A (b) IV

Reasons: This function is periodic in y but not x, a condition satisfied only by A and IV. Also, note that traces in $x = k$ are cosine curves with amplitude that increases as x increases.

43. (a) F (b) I

Reasons: This function is periodic in both x and y but is constant along the lines $y = x + k$, a condition satisfied only by F and I.

44. (a) E (b) III

Reasons: This function is periodic in both x and y, but unlike the function in Exercise 43, it is not constant along lines such as $y = x + \pi$, so the contour map is III. Also notice that traces in $y = k$ are vertically shifted copies of the sine wave $z = \sin x$, so the graph must be E.

45. (a) B (b) VI

Reasons: This function is 0 along the lines $x = \pm 1$ and $y = \pm 1$. The only contour map in which this could occur is VI. Also note that the trace in the xz-plane is the parabola $z = 1 - x^2$ and the trace in the yz-plane is the parabola $z = 1 - y^2$, so the graph is B.

46. (a) D (b) V

Reasons: This function is not periodic, ruling out the graphs in A, C, E, and F. Also, the values of z approach 0 as we use points farther from the origin. The only graph that shows this behavior is D, which corresponds to V.

47. $k = x + 3y + 5z$ is a family of parallel planes with normal vector $\langle 1, 3, 5 \rangle$.

48. $k = x^2 + 3y^2 + 5z^2$ is a family of ellipsoids for $k > 0$ and the origin for $k = 0$.

49. $k = x^2 - y^2 + z^2$ are the equations of the level surfaces. For $k = 0$, the surface is a right circular cone with vertex the origin and axis the y-axis. For $k > 0$, we have a family of hyperboloids of one sheet with axis the y-axis. For $k < 0$, we have a family of hyperboloids of two sheets with axis the y-axis.

50. $k = x^2 - y^2$ is a family of hyperbolic cylinders oriented vertically. The cross section of each level surface in the xy-plane is a hyperbola with axis the x-axis when $k > 0$ and y-axis when $k < 0$. (When $k = 0$, the level surface is two intersecting vertical planes.)

51. (a) The graph of g is the graph of f shifted upward 2 units.

(b) The graph of g is the graph of f stretched vertically by a factor of 2.

(c) The graph of g is the graph of f reflected about the xy-plane.

(d) The graph of $g(x, y) = -f(x, y) + 2$ is the graph of f reflected about the xy-plane and then shifted upward 2 units.

52. (a) The graph of g is the graph of f shifted 2 units in the positive x-direction.

(b) The graph of g is the graph of f shifted 2 units in the negative y-direction.

(c) The graph of g is the graph of f shifted 3 units in the negative x-direction and 4 units in the positive y-direction.

53. $f(x, y) = e^{cx^2 + y^2}$. First, if $c = 0$, the graph is the cylindrical surface $z = e^{y^2}$ (whose level curves are parallel lines). When $c > 0$, the vertical trace above the y-axis remains fixed while the sides of the surface in the x-direction "curl" upward, giving the graph a shape resembling an elliptic paraboloid. The level curves of the surface are ellipses centered at the origin.

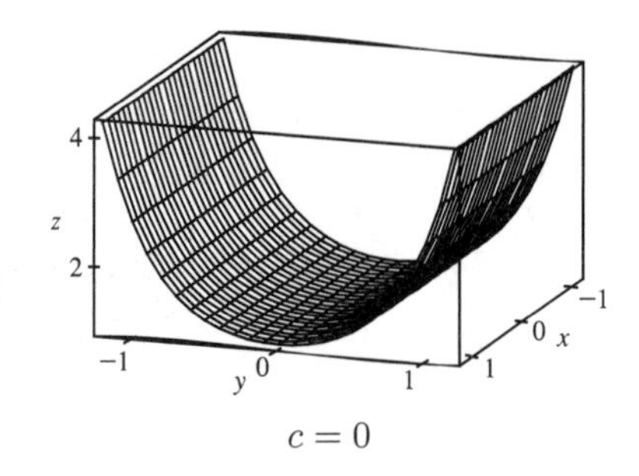

$c = 0$

For $0 < c < 1$, the ellipses have major axis the x-axis and the eccentricity increases as $c \to 0$.

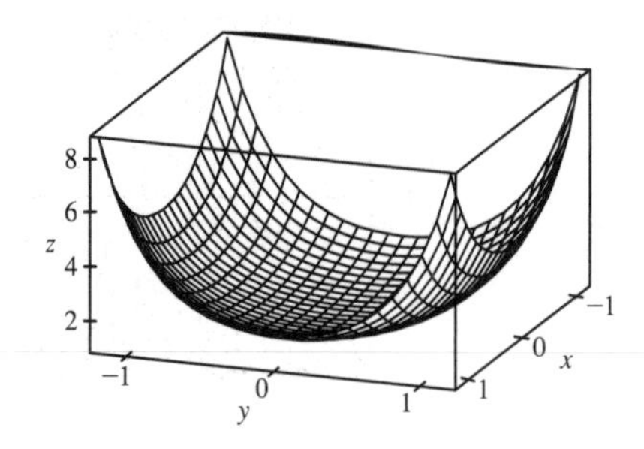

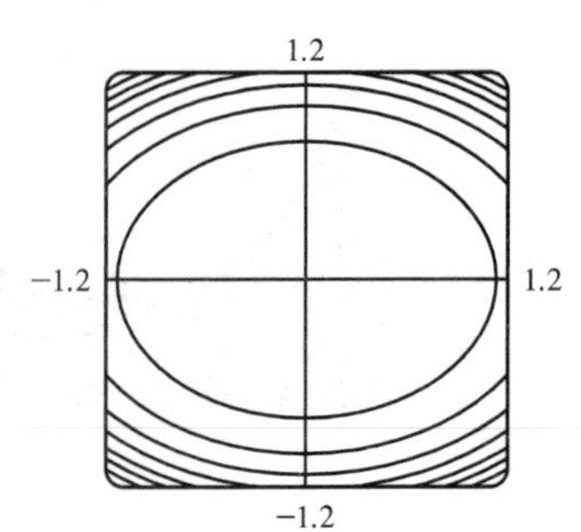

$c = 0.5$ (level curves in increments of 1)

For $c = 1$ the level curves are circles centered at the origin.

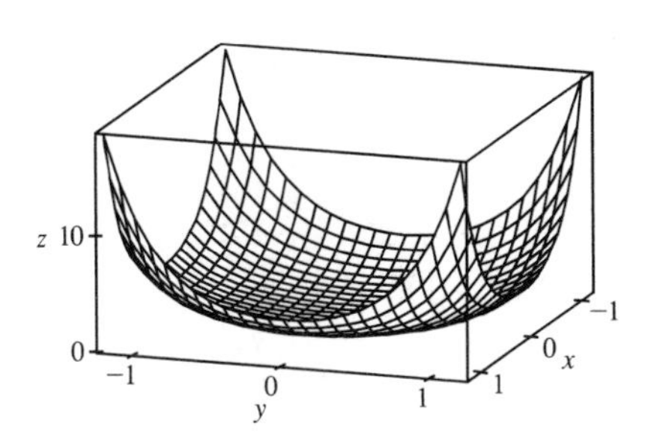

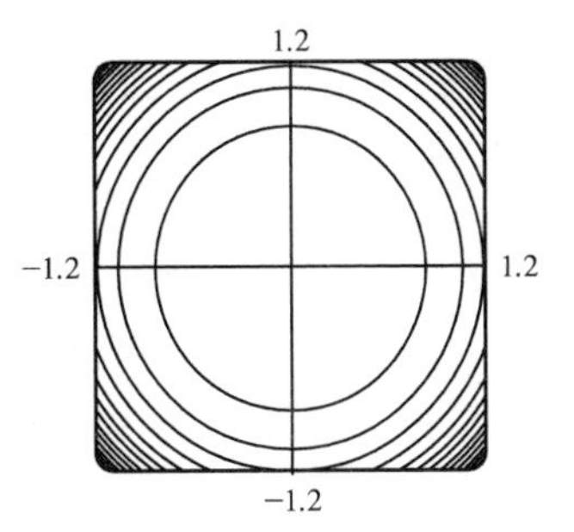

$c = 1$ (level curves in increments of 1)

When $c > 1$, the level curves are ellipses with major axis the y-axis, and the eccentricity increases as c increases.

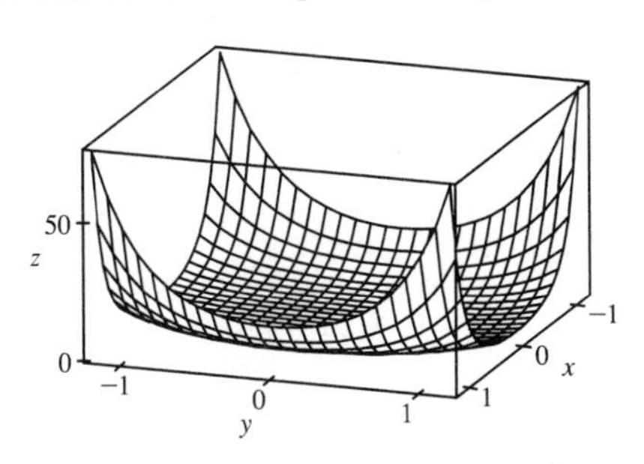

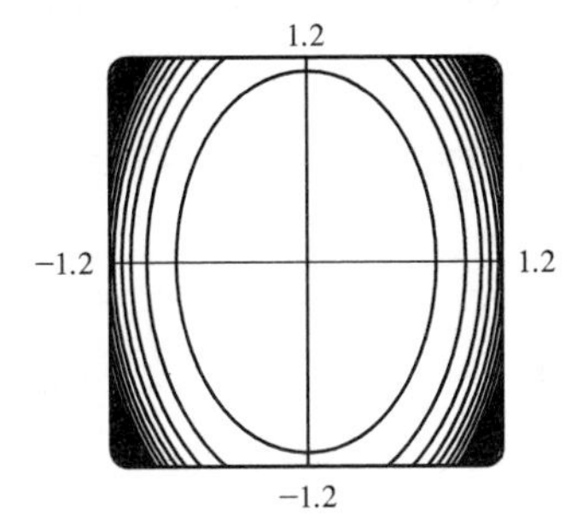

$c = 2$ (level curves in increments of 4)

For values of $c < 0$, the sides of the surface in the x-direction curl downward and approach the xy-plane (while the vertical trace $x = 0$ remains fixed), giving a saddle-shaped appearance to the graph near the point $(0, 0, 1)$. The level curves consist of a family of hyperbolas. As c decreases, the surface becomes flatter in the x-direction and the surface's approach to the curve in the trace $x = 0$ becomes steeper, as the graphs demonstrate.

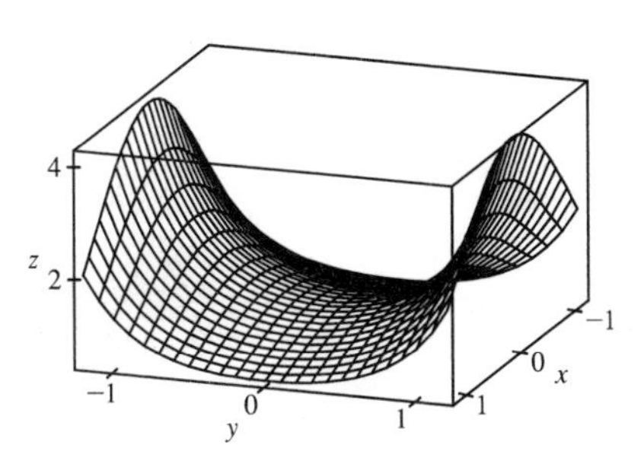

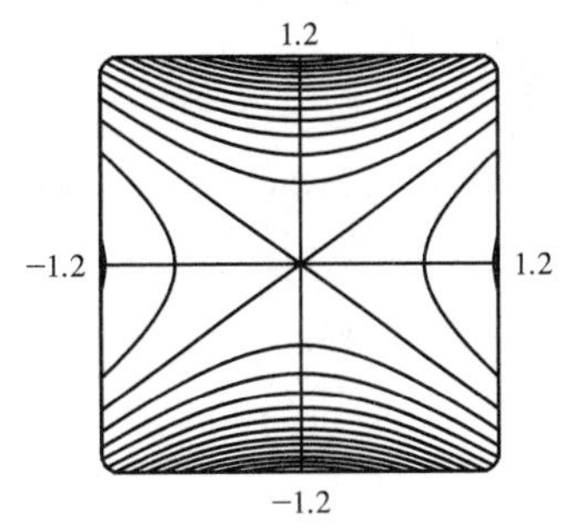

$c = -0.5$ (level curves in increments of 0.25)

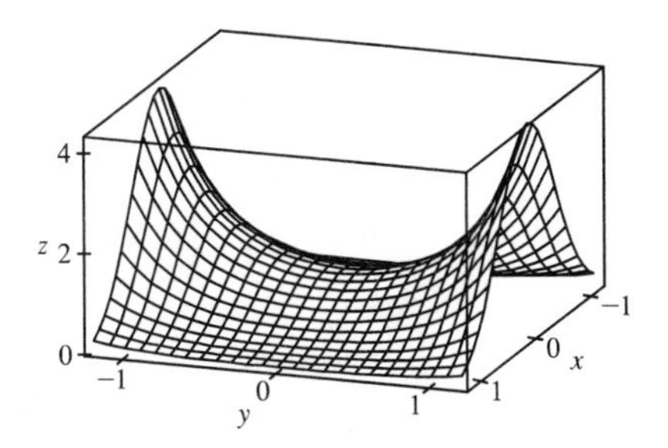

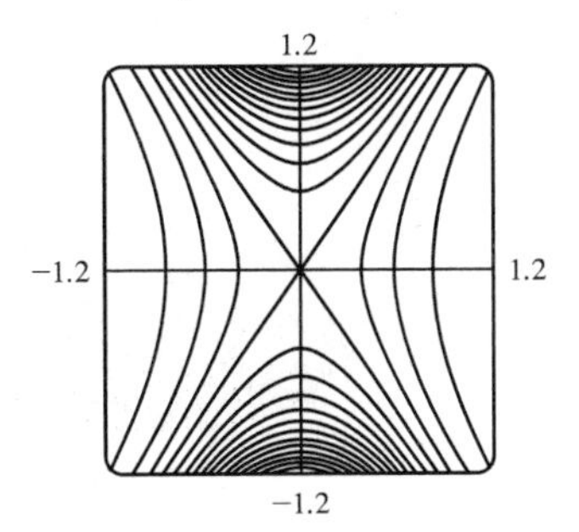

$c = -2$ (level curves in increments of 0.25)

54. First, we graph $f(x, y) = \sqrt{x^2 + y^2}$. As an alternative, the $x^2 + y^2$ expression suggests that cylindrical coordinates may be appropriate, giving the equivalent equation $z = \sqrt{r^2} = r$, $r \geq 0$ which we graph as well. Notice that the graph in cylindrical coordinates better demonstrates the symmetry of the surface.

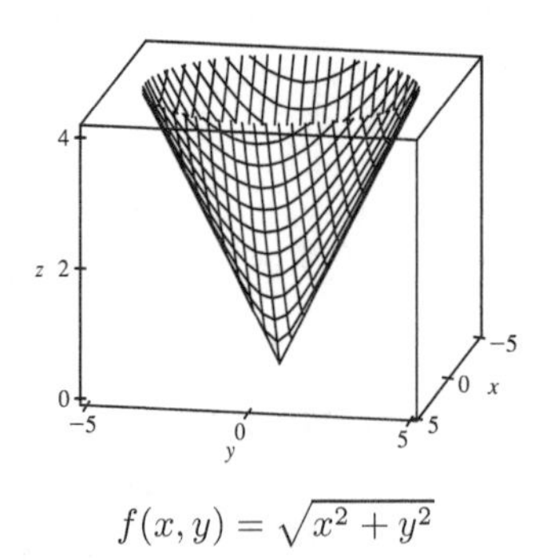

$f(x, y) = \sqrt{x^2 + y^2}$

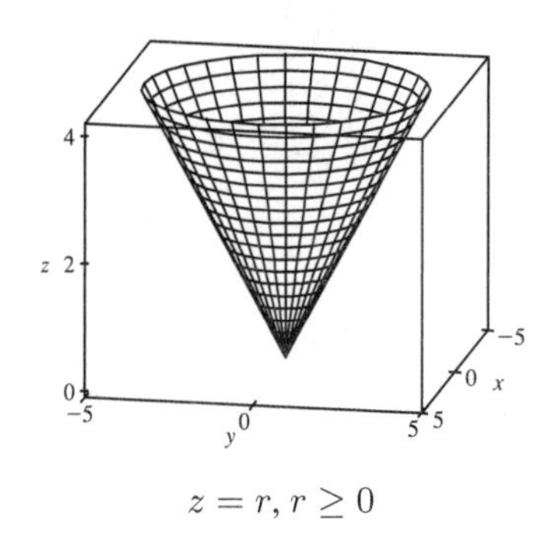

$z = r,\ r \geq 0$

Graphs of the other four functions follow.

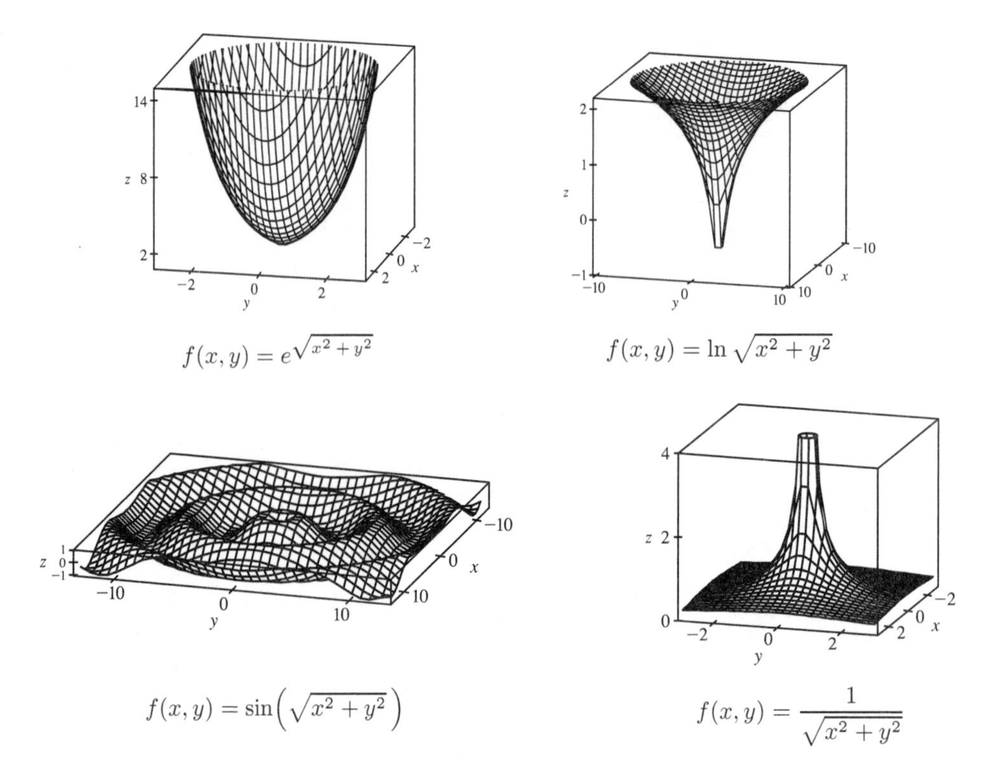

$f(x, y) = e^{\sqrt{x^2 + y^2}}$

$f(x, y) = \ln \sqrt{x^2 + y^2}$

$f(x, y) = \sin\left(\sqrt{x^2 + y^2}\right)$

$f(x, y) = \dfrac{1}{\sqrt{x^2 + y^2}}$

Notice that each graph $f(x, y) = g\left(\sqrt{x^2 + y^2}\right)$ exhibits radial symmetry about the z-axis and the trace in the xz-plane for $x \geq 0$ is the graph of $z = g(x)$, $x \geq 0$. This suggests that the graph of $f(x, y) = g\left(\sqrt{x^2 + y^2}\right)$ is obtained from the graph of g by graphing $z = g(x)$ in the xz-plane and rotating the curve about the z-axis.

11.2 Limits and Continuity

1. In general, we can't say anything about $f(3,1)$! $\lim_{(x,y)\to(3,1)} f(x,y) = 6$ means that the values of $f(x,y)$ approach 6 as (x,y) approaches, but is not equal to, $(3,1)$. If f is continuous, we know that $\lim_{(x,y)\to(a,b)} f(x,y) = f(a,b)$, so $\lim_{(x,y)\to(3,1)} f(x,y) = f(3,1) = 6$.

2. (a) The outdoor temperature as a function of longitude, latitude, and time is continuous. Small changes in longitude, latitude, or time can produce only small changes in temperature, as the temperature doesn't jump abruptly from one value to another.

(b) Elevation is not necessarily continuous. If we think of a cliff with a sudden drop-off, a very small change in longitude or latitude can produce a comparatively large change in elevation, without all the intermediate values being attained. Elevation *can* jump from one value to another.

(c) The cost of a taxi ride is usually discontinuous. The cost normally increases in jumps, so small changes in distance traveled or time can produce a jump in cost. A graph of the function would show breaks in the surface.

3. $f(x,y) = x^5 + 4x^3y - 5xy^2$ is a polynomial, and hence continuous, so $\lim_{(x,y)\to(5,-2)} f(x,y) = f(5,-2) = 5^5 + 4(5)^3(-2) - 5(5)(-2)^2 = 2025$.

4. $x - 2y$ is a polynomial and therefore continuous. Since $\cos t$ is a continuous function, the composition $\cos(x-2y)$ is also continuous. xy is also a polynomial, and hence continuous, so the product $f(x,y) = xy\cos(x-2y)$ is a continuous function. Then $\lim_{(x,y)\to(6,3)} f(x,y) = f(6,3) = (6)(3)\cos(6 - 2\cdot 3) = 18$.

5. $f(x,y) = y^4/(x^4+3y^4)$. First approach $(0,0)$ along the x-axis. Then $f(x,0) = 0/x^4 = 0$ for $x \neq 0$, so $f(x,y) \to 0$. Now approach $(0,0)$ along the y-axis. Then for $y \neq 0$, $f(0,y) = y^4/3y^4 = 1/3$, so $f(x,y) \to 1/3$. Since f has two different limits along two different lines, the limit does not exist.

6. $f(x,y) = (x^2 + \sin^2 y)/(2x^2 + y^2)$. First approach $(0,0)$ along the x-axis. Then $f(x,0) = x^2/2x^2 = \frac{1}{2}$ for $x \neq 0$, so $f(x,y) \to \frac{1}{2}$. Next approach $(0,0)$ along the y-axis. For $y \neq 0$, $f(0,y) = \dfrac{\sin^2 y}{y^2} = \left(\dfrac{\sin y}{y}\right)^2$ and $\lim_{y\to 0} \dfrac{\sin y}{y} = 1$, so $f(x,y) \to 1$. Since f has two different limits along two different lines, the limit does not exist.

7. $f(x,y) = (xy\cos y)/(3x^2+y^2)$. On the x-axis, $f(x,0) = 0$ for $x \neq 0$, so $f(x,y) \to 0$ as $(x,y) \to (0,0)$ along the x-axis. Approaching $(0,0)$ along the line $y = x$, $f(x,x) = (x^2\cos x)/4x^2 = \frac{1}{4}\cos x$ for $x \neq 0$, so $f(x,y) \to \frac{1}{4}$ along this line. Thus the limit does not exist.

8. $f(x,y) = 6x^3y/(2x^4+y^4)$. On the x-axis, $f(x,0) = 0$ for $x \neq 0$, so $f(x,y) \to 0$ as $(x,y) \to (0,0)$ along the x-axis. Approaching $(0,0)$ along the line $y = x$ gives $f(x,x) = 6x^4/(3x^4) = 2$ for $x \neq 0$, so along this line $f(x,y) \to 2$ as $(x,y) \to (0,0)$. Thus the limit does not exist.

9. $f(x,y) = \dfrac{xy}{\sqrt{x^2+y^2}}$. We can see that the limit along any line through $(0,0)$ is 0, as well as along other paths through $(0,0)$ such as $x = y^2$ and $y = x^2$. So we suspect that the limit exists and equals 0; we use the Squeeze Theorem to prove our assertion. $0 \le \left|\dfrac{xy}{\sqrt{x^2+y^2}}\right| \le |x|$ since $|y| \le \sqrt{x^2+y^2}$, and $|x| \to 0$ as $(x,y) \to (0,0)$. So $\displaystyle\lim_{(x,y)\to(0,0)} f(x,y) = 0$.

10. We can use the Squeeze Theorem to show that $\displaystyle\lim_{(x,y)\to(0,0)} \frac{x^2 \sin^2 y}{x^2+2y^2} = 0$:

$0 \le \dfrac{x^2 \sin^2 y}{x^2+2y^2} \le \sin^2 y$ since $\dfrac{x^2}{x^2+2y^2} \le 1$, and $\sin^2 y \to 0$ as $(x,y) \to (0,0)$, so $\displaystyle\lim_{(x,y)\to(0,0)} \frac{x^2 \sin^2 y}{x^2+2y^2} = 0$.

11. Let $f(x,y) = 2x^2y/(x^4+y^2)$. Then $f(x,0) = 0$ for $x \ne 0$, so $f(x,y) \to 0$ as $(x,y) \to (0,0)$ along the x-axis. But $f(x,x^2) = \dfrac{2x^4}{2x^4} = 1$ for $x \ne 0$, so $f(x,y) \to 1$ as $(x,y) \to (0,0)$ along the parabola $y = x^2$. Thus the limit doesn't exist.

12. $f(x,y) = xy^4/(x^2+y^8)$. On the x-axis, $f(x,0) = 0$ for $x \ne 0$, so $f(x,y) \to 0$ as $(x,y) \to (0,0)$ along the x-axis. Approaching $(0,0)$ along the curve $x = y^4$ gives $f(y^4,y) = y^8/2y^8 = \frac{1}{2}$ for $y \ne 0$, so along this path $f(x,y) \to \frac{1}{2}$ as $(x,y) \to (0,0)$. Thus the limit does not exist.

13.
$$\begin{aligned}\lim_{(x,y)\to(0,0)} \frac{x^2+y^2}{\sqrt{x^2+y^2+1}-1} &= \lim_{(x,y)\to(0,0)} \frac{x^2+y^2}{\sqrt{x^2+y^2+1}-1} \cdot \frac{\sqrt{x^2+y^2+1}+1}{\sqrt{x^2+y^2+1}+1} \\ &= \lim_{(x,y)\to(0,0)} \frac{(x^2+y^2)\left(\sqrt{x^2+y^2+1}+1\right)}{x^2+y^2} = \lim_{(x,y)\to(0,0)} \left(\sqrt{x^2+y^2+1}+1\right) = 2\end{aligned}$$

14. $f(x,y) = \dfrac{x^4-y^4}{x^2+y^2} = \dfrac{(x^2+y^2)(x^2-y^2)}{x^2+y^2} = x^2-y^2$ for $(x,y) \ne (0,0)$. Thus the limit as $(x,y) \to (0,0)$ is 0.

15. $f(x,y,z) = \dfrac{xy+yz^2+xz^2}{x^2+y^2+z^4}$. Then $f(x,0,0) = 0/x^2 = 0$ for $x \ne 0$, so as $(x,y,z) \to (0,0,0)$ along the x-axis, $f(x,y,z) \to 0$. But $f(x,x,0) = x^2/(2x^2) = \frac{1}{2}$ for $x \ne 0$, so as $(x,y,z) \to (0,0,0)$ along the line $y = x$, $z = 0$, $f(x,y,z) \to \frac{1}{2}$. Thus the limit doesn't exist.

16. $f(x,y,z) = \dfrac{x^2+2y^2+3z^2}{x^2+y^2+z^2}$. Then $f(x,0,0) = \dfrac{x^2+0+0}{x^2+0+0} = 1$ for $x \ne 0$, so $f(x,y,z) \to 1$ as $(x,y,z) \to (0,0,0)$ along the x-axis. But $f(0,y,0) = \dfrac{0+2y^2+0}{0+y^2+0} = 2$ for $y \ne 0$, so $f(x,y,z) \to 2$ as $(x,y,z) \to (0,0,0)$ along the y-axis. Thus, the limit doesn't exist.

17.

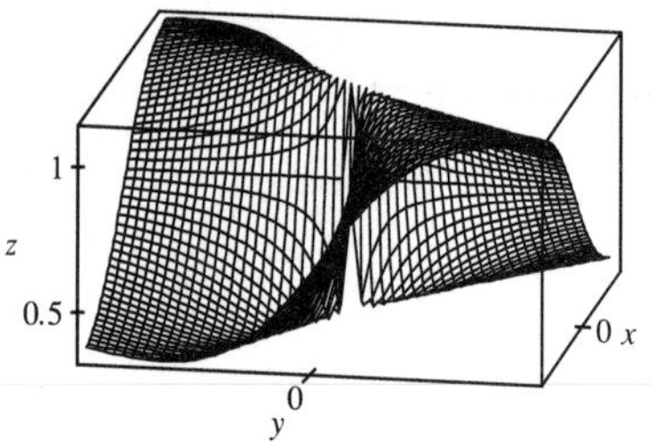

From the ridges on the graph, we see that as $(x,y) \to (0,0)$ along the lines under the two ridges, $f(x,y)$ approaches different values. So the limit does not exist.

18.

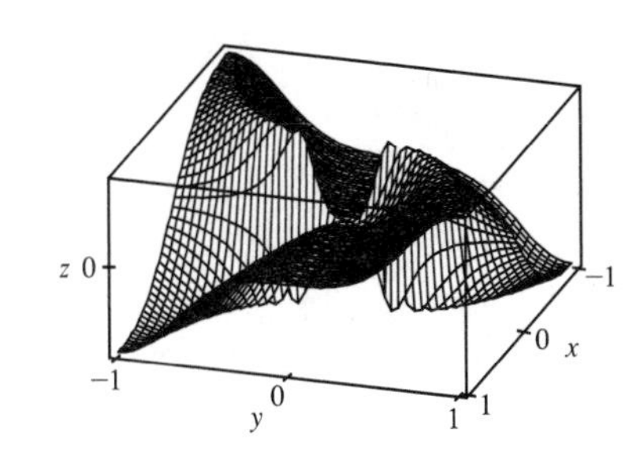

From the graph, it appears that as we approach the origin along the lines $x = 0$ or $y = 0$, the function is everywhere 0, whereas if we approach the origin along a certain curve it has a constant value of about $\frac{1}{2}$. [In fact, $f(y^3, y) = y^6/(2y^6) = \frac{1}{2}$ for $y \neq 0$, so $f(x, y) \to \frac{1}{2}$ as $(x, y) \to (0, 0)$ along the curve $x = y^3$.] Since the function approaches different values depending on the path of approach, the limit does not exist.

19. $h(x, y) = g(f(x, y)) = (2x + 3y - 6)^2 + \sqrt{2x + 3y - 6}$. Since f is a polynomial, it is continuous on $\mathbb{R}^2$ and g is continuous on its domain $\{t \mid t \geq 0\}$. Thus h is continuous on its domain.

$D = \{(x, y) \mid 2x + 3y - 6 \geq 0\} = \left\{(x, y) \mid y \geq -\frac{2}{3}x + 2\right\}$, which consists of all points on or above the line $y = -\frac{2}{3}x + 2$.

20. $h(x, y) = g(f(x, y)) = \left(\sqrt{x^2 - y} - 1\right) \Big/ \left(\sqrt{x^2 - y} + 1\right)$. Since f is a polynomial, it is continuous on $\mathbb{R}^2$ and g is continuous on its domain $\{t \mid t \geq 0\}$. Thus h is continuous on its domain $D = \left\{(x, y) \mid x^2 - y \geq 0\right\} = \left\{(x, y) \mid y \leq x^2\right\}$ which consists of all points below or on the parabola $y = x^2$.

21. The functions $\sin(xy)$ and $e^x - y^2$ are continuous everywhere, so $F(x, y) = \dfrac{\sin(xy)}{e^x - y^2}$ is continuous except where $e^x - y^2 = 0 \;\Rightarrow\; y^2 = e^x \;\Rightarrow\; y = \pm\sqrt{e^x} = \pm e^{\frac{1}{2}x}$. Thus F is continuous on its domain $\left\{(x, y) \mid y \neq \pm e^{x/2}\right\}$.

22. $F(x, y) = \dfrac{x - y}{1 + x^2 + y^2}$ is a rational function and thus is continuous on its domain $\mathbb{R}^2$ (since the denominator is never zero).

23. $G(x, y) = \ln\left(x^2 + y^2 - 4\right) = g(f(x, y))$ where $f(x, y) = x^2 + y^2 - 4$, continuous on $\mathbb{R}^2$, and $g(t) = \ln t$, continuous on its domain $\{t \mid t > 0\}$. Thus G is continuous on its domain $\left\{(x, y) \mid x^2 + y^2 - 4 > 0\right\} = \left\{(x, y) \mid x^2 + y^2 > 4\right\}$, the exterior of the circle $x^2 + y^2 = 4$.

24. $e^{x^2 y}$ is continuous on $\mathbb{R}^2$ and $\sqrt{x + y^2}$ is continuous on its domain $\left\{(x, y) \mid x + y^2 \geq 0\right\} = \left\{(x, y) \mid x \geq -y^2\right\}$, so $F(x, y) = e^{x^2 y} + \sqrt{x + y^2}$ is continuous on the set $\left\{(x, y) \mid x \geq -y^2\right\}$.

25. $\sqrt{y}$ is continuous on its domain $\{y \mid y \geq 0\}$ and $x^2 - y^2 + z^2$ is continuous everywhere, so $f(x, y, z) = \dfrac{\sqrt{y}}{x^2 - y^2 + z^2}$ is continuous for $y \geq 0$ and $x^2 - y^2 + z^2 \neq 0 \;\Rightarrow\; y^2 \neq x^2 + z^2$, that is, $\left\{(x, y, z) \mid y \geq 0, y \neq \sqrt{x^2 + z^2}\,\right\}$.

26. $f(x, y, z) = \sqrt{x + y + z} = h(g(x, y, z))$ where $g(x, y, z) = x + y + z$, continuous everywhere, and $h(t) = \sqrt{t}$ is continuous on its domain $\{t \mid t \geq 0\}$. Thus f is continuous on its domain $\{(x, y, z) \mid x + y + z \geq 0\}$, so f is continuous on and above the plane $z = -x - y$.

27. $f(x,y)=\begin{cases}\dfrac{x^2y^3}{2x^2+y^2} & \text{if } (x,y)\neq(0,0)\\ 1 & \text{if } (x,y)=(0,0)\end{cases}$ The first piece of f is a rational function defined everywhere except at the origin, so f is continuous on $\mathbb{R}^2$ except possibly at the origin. Since $x^2 \leq 2x^2+y^2$, we have $\left|x^2y^3/(2x^2+y^2)\right| \leq \left|y^3\right|$. We know that $\left|y^3\right| \to 0$ as $(x,y)\to(0,0)$. So, by the Squeeze Theorem, $\lim\limits_{(x,y)\to(0,0)} f(x,y) = \lim\limits_{(x,y)\to(0,0)} \dfrac{x^2y^3}{2x^2+y^2} = 0$. But $f(0,0)=1$, so f is discontinuous at $(0,0)$. Therefore, f is continuous on the set $\{(x,y)\mid(x,y)\neq(0,0)\}$.

28. $f(x,y)=\begin{cases}\dfrac{xy}{x^2+xy+y^2} & \text{if } (x,y)\neq(0,0)\\ 0 & \text{if } (x,y)=(0,0)\end{cases}$ The first piece of f is a rational function defined everywhere except at the origin, so f is continuous on $\mathbb{R}^2$ except possibly at the origin. $f(x,0)=0/x^2=0$ for $x\neq0$, so $f(x,y)\to0$ as $(x,y)\to(0,0)$ along the x-axis. But $f(x,x)=x^2/(3x^2)=\frac{1}{3}$ for $x\neq0$, so $f(x,y)\to\frac{1}{3}$ as $(x,y)\to(0,0)$ along the line $y=x$. Thus $\lim\limits_{(x,y)\to(0,0)} f(x,y)$ doesn't exist, so f is not continuous at $(0,0)$ and the largest set on which f is continuous is $\{(x,y)\mid(x,y)\neq(0,0)\}$.

29. $\displaystyle\lim_{(x,y)\to(0,0)}\frac{x^3+y^3}{x^2+y^2}=\lim_{r\to0^+}\frac{(r\cos\theta)^3+(r\sin\theta)^3}{r^2}=\lim_{r\to0^+}(r\cos^3\theta+r\sin^3\theta)=0$

30. $$\lim_{(x,y)\to(0,0)}(x^2+y^2)\ln(x^2+y^2)=\lim_{r\to0^+}r^2\ln r^2=\lim_{r\to0^+}\frac{\ln r^2}{1/r^2}=\lim_{r\to0^+}\frac{(1/r^2)(2r)}{-2/r^3}\quad\text{[using l'Hospital's Rule]}$$
$$=\lim_{r\to0^+}(-r^2)=0$$

31. Since $|\mathbf{x}-\mathbf{a}|^2=|\mathbf{x}|^2+|\mathbf{a}|^2-2\,|\mathbf{x}|\,|\mathbf{a}|\cos\theta\geq|\mathbf{x}|^2+|\mathbf{a}|^2-2\,|\mathbf{x}|\,|\mathbf{a}|=(|\mathbf{x}|-|\mathbf{a}|)^2$, we have $\big||\mathbf{x}|-|\mathbf{a}|\big|\leq|\mathbf{x}-\mathbf{a}|$. Let $\epsilon>0$ be given and set $\delta=\epsilon$. Then whenever $0<|\mathbf{x}-\mathbf{a}|<\delta$, $\big||\mathbf{x}|-|\mathbf{a}|\big|\leq|\mathbf{x}-\mathbf{a}|<\delta=\epsilon$. Hence $\lim_{\mathbf{x}\to\mathbf{a}}|\mathbf{x}|=|\mathbf{a}|$ and $f(\mathbf{x})=|\mathbf{x}|$ is continuous on $\mathbb{R}^n$.

32. Let $\epsilon>0$ be given. We need to find $\delta>0$ such that $|f(\mathbf{x})-f(\mathbf{a})|<\epsilon$ whenever $|\mathbf{x}-\mathbf{a}|<\delta$ or $|\mathbf{c}\cdot\mathbf{x}-\mathbf{c}\cdot\mathbf{a}|<\epsilon$ whenever $|\mathbf{x}-\mathbf{a}|<\delta$. But $|\mathbf{c}\cdot\mathbf{x}-\mathbf{c}\cdot\mathbf{a}|=|\mathbf{c}\cdot(\mathbf{x}-\mathbf{a})|$ and $|\mathbf{c}\cdot(\mathbf{x}-\mathbf{a})|\leq|\mathbf{c}|\,|\mathbf{x}-\mathbf{a}|$ by Exercise 10.3.43 (the Cauchy-Schwartz Inequality). Let $\epsilon>0$ be given and set $\delta=\epsilon/|\mathbf{c}|$. Then whenever $0<|\mathbf{x}-\mathbf{a}|<\delta$, $|f(\mathbf{x})-f(\mathbf{a})|=|\mathbf{c}\cdot\mathbf{x}-\mathbf{c}\cdot\mathbf{a}|\leq|\mathbf{c}|\,|\mathbf{x}-\mathbf{a}|<|\mathbf{c}|\,\delta=|\mathbf{c}|\,(\epsilon/|\mathbf{c}|)=\epsilon$. So f is continuous on $\mathbb{R}^{\kappa}$.

11.3 Partial Derivatives

1. (a) $\partial T/\partial x$ represents the rate of change of T when we fix y and t and consider T as a function of the single variable x, which describes how quickly the temperature changes when longitude changes but latitude and time are constant. $\partial T/\partial y$ represents the rate of change of T when we fix x and t and consider T as a function of y, which describes how quickly the temperature changes when latitude changes but longitude and time are constant. $\partial T/\partial t$ represents the rate of change of T when we fix x and y and consider T as a function of t, which describes how quickly the temperature changes over time for a constant longitude and latitude.

(b) $f_x(158, 21, 9)$ represents the rate of change of temperature at longitude 158° W, latitude 21°N at 9:00 AM when only longitude varies. Since the air is warmer to the west than to the east, increasing longitude results in an increased air temperature, so we would expect $f_x(158, 21, 9)$ to be positive. $f_y(158, 21, 9)$ represents the rate of change of temperature at the same time and location when only latitude varies. Since the air is warmer to the south and cooler to the north, increasing latitude results in a decreased air temperature, so we would expect $f_y(158, 21, 9)$ to be negative. $f_t(158, 21, 9)$ represents the rate of change of temperature at the same time and location when only time varies. Since typically air temperature increases from the morning to the afternoon as the sun warms it, we would expect $f_t(158, 21, 9)$ to be positive.

2. $f_x(2, 1)$ is the rate of change of f at $(2, 1)$ in the x-direction. If we start at $(2, 1)$, where $f(2, 1) = 10$, and move in the positive x-direction, we reach the next contour line (where $f(x, y) = 12$) after approximately 0.6 units. This represents an average rate of change of about $\frac{2}{0.6}$. If we approach the point $(2, 1)$ from the left (moving in the positive x-direction) the output values increase from 8 to 10 with an increase in x of approximately 0.9 units, corresponding to an average rate of change of $\frac{2}{0.9}$. A good estimate for $f_x(2, 1)$ would be the average of these two, so $f_x(2, 1) \approx 2.8$. Similarly, $f_y(2, 1)$ is the rate of change of f at $(2, 1)$ in the y-direction. If we approach $(2, 1)$ from below, the output values decrease from 12 to 10 with a change in y of approximately 1 unit, corresponding to an average rate of change of -2. If we start at $(2, 1)$ and move in the positive y-direction, the output values decrease from 10 to 8 after approximately 0.9 units, a rate of change of $\frac{-2}{0.9}$. Averaging these two results, we estimate $f_y(2, 1) \approx -2.1$.

3. (a) If we start at $(1, 2)$ and move in the positive x-direction, the graph of f increases. Thus $f_x(1, 2)$ is positive.

(b) If we start at $(1, 2)$ and move in the positive y-direction, the graph of f decreases. Thus $f_y(1, 2)$ is negative.

4. (a) The graph of f decreases if we start at $(-1, 2)$ and move in the positive x-direction, so $f_x(-1, 2)$ is negative.

(b) The graph of f decreases if we start at $(-1, 2)$ and move in the positive y-direction, so $f_y(-1, 2)$ is negative.

(c) $f_{xx} = \frac{\partial}{\partial x}(f_x)$, so f_{xx} is the rate of change of f_x in the x-direction. f_x is negative at $(-1, 2)$ and if we move in the positive x-direction, the surface becomes less steep. Thus the values of f_x are increasing and $f_{xx}(-1, 2)$ is positive.

(d) f_{yy} is the rate of change of f_y in the y-direction. f_y is negative at $(-1, 2)$ and if we move in the positive y-direction, the surface becomes steeper. Thus the values of f_y are decreasing, and $f_{yy}(-1, 2)$ is negative.

5. $f(x, y) = 16 - 4x^2 - y^2 \;\Rightarrow\; f_x(x, y) = -8x$ and $f_y(x, y) = -2y \;\Rightarrow\; f_x(1, 2) = -8$ and $f_y(1, 2) = -4$. The graph of f is the paraboloid $z = 16 - 4x^2 - y^2$ and the vertical plane $y = 2$ intersects it in the parabola $z = 12 - 4x^2$, $y = 2$ (the curve C_1 in the first figure). The slope of the tangent line to this parabola at $(1, 2, 8)$ is $f_x(1, 2) = -8$. Similarly the plane $x = 1$ intersects the paraboloid in the parabola $z = 12 - y^2$, $x = 1$ (the curve C_2 in the second figure) and the slope of the tangent line at $(1, 2, 8)$ is $f_y(1, 2) = -4$.

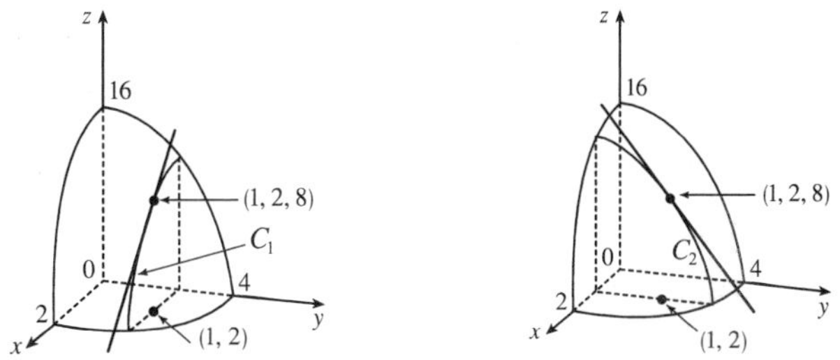

6. $f(x,y) = (4 - x^2 - 4y^2)^{1/2} \Rightarrow f_x(x,y) = -x(4 - x^2 - 4y^2)^{-1/2}$ and $f_y(x,y) = -4y(4 - x^2 - 4y^2)^{-1/2} \Rightarrow$ $f_x(1,0) = -\frac{1}{\sqrt{3}}$, $f_y(1,0) = 0$. The graph of f is the upper half of the ellipsoid $z^2 + x^2 + 4y^2 = 4$ and the plane $y = 0$ intersects the graph in the semicircle $x^2 + z^2 = 4$, $z \geq 0$ and the slope of the tangent line T_1 to this semicircle at $(1, 0, \sqrt{3})$ is $f_x(1,0) = -\frac{1}{\sqrt{3}}$. Similarly the plane $x = 1$ intersects the graph in the semi-ellipse $z^2 + 4y^2 = 3$, $z \geq 0$ and the slope of the tangent line T_2 to this semi-ellipse at $(1, 0, \sqrt{3})$ is $f_y(1,0) = 0$.

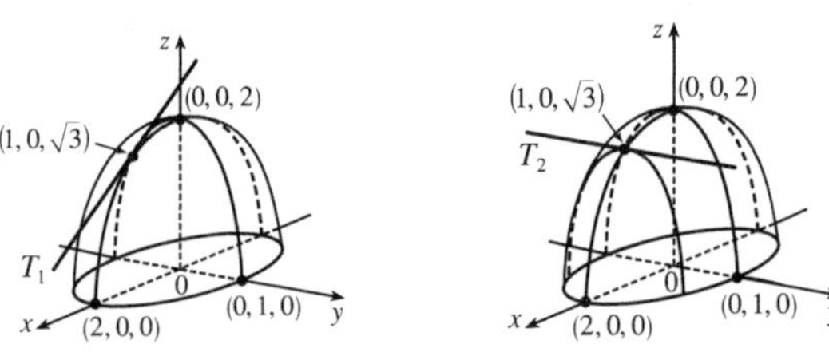

7. $f(x,y) = 3x - 2y^4 \Rightarrow f_x(x,y) = 3 - 0 = 3$, $f_y(x,y) = 0 - 8y^3 = -8y^3$

8. $f(x,y) = x^5 + 3x^3y^2 + 3xy^4 \Rightarrow f_x(x,y) = 5x^4 + 3 \cdot 3x^2 \cdot y^2 + 3 \cdot 1 \cdot y^4 = 5x^4 + 9x^2y^2 + 3y^4$,

$f_y(x,y) = 0 + 3x^3 \cdot 2y + 3x \cdot 4y^3 = 6x^3y + 12xy^3$.

9. $z = xe^{3y} \Rightarrow \dfrac{\partial z}{\partial x} = e^{3y}$, $\dfrac{\partial z}{\partial y} = 3xe^{3y}$

10. $z = y \ln x \Rightarrow \dfrac{\partial z}{\partial x} = \dfrac{y}{x}$, $\dfrac{\partial z}{\partial y} = \ln x$

11. $f(x,y) = \dfrac{x-y}{x+y} \Rightarrow f_x(x,y) = \dfrac{(1)(x+y) - (x-y)(1)}{(x+y)^2} = \dfrac{2y}{(x+y)^2}$,

$f_y(x,y) = \dfrac{(-1)(x+y) - (x-y)(1)}{(x+y)^2} = -\dfrac{2x}{(x+y)^2}$

12. $f(x,y) = x^y \Rightarrow f_x(x,y) = yx^{y-1}$, $f_y(x,y) = x^y \ln x$

13. $w = \sin\alpha \cos\beta \Rightarrow \dfrac{\partial w}{\partial \alpha} = \cos\alpha \cos\beta$, $\dfrac{\partial w}{\partial \beta} = -\sin\alpha \sin\beta$

14. $f(s,t) = \dfrac{st^2}{s^2+t^2} \Rightarrow f_s(s,t) = \dfrac{t^2(s^2+t^2) - st^2(2s)}{(s^2+t^2)^2} = \dfrac{t^4 - s^2t^2}{(s^2+t^2)^2}$,

$f_t(s,t) = \dfrac{2st(s^2+t^2) - st^2(2t)}{(s^2+t^2)^2} = \dfrac{2s^3t}{(s^2+t^2)^2}$

15. $f(r,s) = r\ln(r^2+s^2) \Rightarrow f_r(r,s) = r \cdot \dfrac{2r}{r^2+s^2} + \ln(r^2+s^2) \cdot 1 = \dfrac{2r^2}{r^2+s^2} + \ln(r^2+s^2)$,

$f_s(r,s) = r \cdot \dfrac{2s}{r^2+s^2} + 0 = \dfrac{2rs}{r^2+s^2}$

16. $f(x,t) = \arctan(x\sqrt{t}) \Rightarrow f_x(x,t) = \dfrac{1}{1+(x\sqrt{t})^2} \cdot \sqrt{t} = \dfrac{\sqrt{t}}{1+x^2t}$,

$f_t(x,t) = \dfrac{1}{1+(x\sqrt{t})^2} \cdot x\left(\dfrac{1}{2}t^{-1/2}\right) = \dfrac{x}{2\sqrt{t}\,(1+x^2t)}$

17. $u = te^{w/t} \quad \Rightarrow \quad \dfrac{\partial u}{\partial t} = t \cdot e^{w/t}(-wt^{-2}) + e^{w/t} \cdot 1 = e^{w/t} - \dfrac{w}{t}e^{w/t} = e^{w/t}\left(1 - \dfrac{w}{t}\right), \; \dfrac{\partial u}{\partial w} = te^{w/t} \cdot \dfrac{1}{t} = e^{w/t}$

18. $f(x, y) = \displaystyle\int_y^x \cos(t^2)\,dt \quad \Rightarrow \quad f_x(x, y) = \dfrac{\partial}{\partial x}\displaystyle\int_y^x \cos(t^2)\;dt = \cos(x^2)$ by the Fundamental Theorem of Calculus,

Part 1; $f_y(x, y) = \dfrac{\partial}{\partial y}\displaystyle\int_y^x \cos(t^2)\;dt = -\dfrac{\partial}{\partial y}\cos(t^2)\,dt = -\cos(y^2)$.

19. $f(x, y, z) = xy^2z^3 + 3yz \quad \Rightarrow \quad f_x(x, y, z) = y^2z^3, \; f_y(x, y, z) = 2xyz^3 + 3z \;, f_z(x, y, z) = 3xy^2z^2 + 3y$

20. $f(x, y, z) = x^2e^{yz} \quad \Rightarrow \quad f_x(x, y, z) = 2xe^{yz}, \; f_y(x, y, z) = x^2e^{yz}(z) = x^2ze^{yz}, \; f_z(x, y, z) = x^2e^{yz}(y) = x^2ye^{yz}$

21. $w = \ln(x + 2y + 3z) \quad \Rightarrow \quad \dfrac{\partial w}{\partial x} = \dfrac{1}{x + 2y + 3z}, \; \dfrac{\partial w}{\partial y} = \dfrac{2}{x + 2y + 3z}, \; \dfrac{\partial w}{\partial z} = \dfrac{3}{x + 2y + 3z}$

22. $w = \sqrt{r^2 + s^2 + t^2} \quad \Rightarrow$

$$\frac{\partial w}{\partial r} = \frac{1}{2}(r^2 + s^2 + t^2)^{-1/2}(2r) = \frac{r}{\sqrt{r^2 + s^2 + t^2}}, \; \frac{\partial w}{\partial s} = \frac{s}{\sqrt{r^2 + s^2 + t^2}}, \; \frac{\partial w}{\partial t} = \frac{t}{\sqrt{r^2 + s^2 + t^2}}$$

23. $u = xe^{-t}\sin\theta \quad \Rightarrow \quad \dfrac{\partial u}{\partial x} = e^{-t}\sin\theta, \; \dfrac{\partial u}{\partial t} = -xe^{-t}\sin\theta, \; \dfrac{\partial u}{\partial \theta} = xe^{-t}\cos\theta$

24. $u = x^{y/z} \quad \Rightarrow \quad u_x = \dfrac{y}{z}x^{(y/z)-1}, \; u_y = x^{y/z}\ln x \cdot \dfrac{1}{z} = \dfrac{x^{y/z}}{z}\ln x, \; u_z = x^{y/z}\ln x \cdot \dfrac{-y}{z^2} = -\dfrac{yx^{y/z}}{z^2}\ln x$

25. $f(x, y, z, t) = xyz^2\tan(yt) \quad \Rightarrow \quad f_x(x, y, z, t) = yz^2\tan(yt)$,

$f_y(x, y, z, t) = xyz^2 \cdot \sec^2(yt) \cdot t + xz^2\tan(yt) = xyz^2t\sec^2(yt) + xz^2\tan(yt)$,

$f_z(x, y, z, t) = 2xyz\tan(yt), \; f_t(x, y, z, t) = xyz^2\sec^2(yt) \cdot y = xy^2z^2\sec^2(yt)$

26. $f(x, y, z, t) = \dfrac{xy^2}{t + 2z} \quad \Rightarrow \quad f_x(x, y, z, t) = \dfrac{y^2}{t + 2z}, \; f_y(x, y, z, t) = \dfrac{2xy}{t + 2z}$,

$f_z(x, y, z, t) = xy^2(-1)(t + 2z)^{-2}(2) = -\dfrac{2xy^2}{(t + 2z)^2}, \; f_t(x, y, z, t) = xy^2(-1)(t + 2z)^{-2}(1) = -\dfrac{xy^2}{(t + 2z)^2}$.

27. $u = \sqrt{x_1^2 + x_2^2 + \cdots + x_n^2}$. For each $i = 1, \ldots, n$, $u_{x_i} = \frac{1}{2}\left(x_1^2 + x_2^2 + \cdots + x_n^2\right)^{-1/2}(2x_i) = \dfrac{x_i}{\sqrt{x_1^2 + x_2^2 + \cdots + x_n^2}}$.

28. $u = \sin(x_1 + 2x_2 + \cdots + nx_n)$. For each $i = 1, \ldots, n$, $u_{x_i} = i\cos(x_1 + 2x_2 + \cdots + nx_n)$.

29. $f(x, y) = \sqrt{x^2 + y^2} \quad \Rightarrow \quad f_x(x, y) = \frac{1}{2}(x^2 + y^2)^{-1/2}(2x) = \dfrac{x}{\sqrt{x^2 + y^2}}$, so $f_x(3, 4) = \dfrac{3}{\sqrt{3^2 + 4^2}} = \dfrac{3}{5}$.

30. $f(x, y) = \sin(2x + 3y) \quad \Rightarrow \quad f_y(x, y) = \cos(2x + 3y) \cdot 3 = 3\cos(2x + 3y)$, so

$f_y(-6, 4) = 3\cos[2(-6) + 3(4)] = 3\cos 0 = 3$.

31. $f(x, y, z) = \dfrac{x}{y + z} = x(y + z)^{-1} \quad \Rightarrow \quad f_z(x, y, z) = x(-1)(y + z)^{-2} = -\dfrac{x}{(y + z)^2}$,

so $f_z(3, 2, 1) = -\dfrac{3}{(2 + 1)^2} = -\dfrac{1}{3}$.

32. $f(u,v,w) = w\tan(uv) \Rightarrow f_v(u,v,w) = w\sec^2(uv)\cdot u = uw\sec^2(uv)$, so $f_v(2,0,3) = (2)(3)\sec^2(2\cdot 0) = 6$.

33. $f(x,y) = xy^2 - x^3y \Rightarrow$

$$f_x(x,y) = \lim_{h\to 0}\frac{f(x+h,y)-f(x,y)}{h} = \lim_{h\to 0}\frac{(x+h)y^2-(x+h)^3y-(xy^2-x^3y)}{h}$$

$$= \lim_{h\to 0}\frac{h(y^2-3x^2y-3xyh-yh^2)}{h} = \lim_{h\to 0}(y^2-3x^2y-3xyh-yh^2) = y^2-3x^2y$$

$$f_y(x,y) = \lim_{h\to 0}\frac{f(x,y+h)-f(x,y)}{h} = \lim_{h\to 0}\frac{x(y+h)^2-x^3(y+h)-(xy^2-x^3y)}{h} = \lim_{h\to 0}\frac{h(2xy+xh-x^3)}{h}$$

$$= \lim_{h\to 0}(2xy+xh-x^3) = 2xy-x^3$$

34. $f(x,y) = \dfrac{x}{x+y^2} \Rightarrow$

$$f_x(x,y) = \lim_{h\to 0}\frac{f(x+h,y)-f(x,y)}{h} = \lim_{h\to 0}\frac{\frac{x+h}{x+h+y^2}-\frac{x}{x+y^2}}{h}\cdot\frac{(x+h+y^2)(x+y^2)}{(x+h+y^2)(x+y^2)}$$

$$= \lim_{h\to 0}\frac{(x+h)(x+y^2)-x(x+h+y^2)}{h(x+h+y^2)(x+y^2)} = \lim_{h\to 0}\frac{y^2h}{h(x+h+y^2)(x+y^2)} = \lim_{h\to 0}\frac{y^2}{(x+h+y^2)(x+y^2)} = \frac{y^2}{(x+y^2)^2}$$

$$f_y(x,y) = \lim_{h\to 0}\frac{f(x,y+h)-f(x,y)}{h} = \lim_{h\to 0}\frac{\frac{x}{x+(y+h)^2}-\frac{x}{x+y^2}}{h}\cdot\frac{\left[x+(y+h)^2\right](x+y^2)}{\left[x+(y+h)^2\right](x+y^2)}$$

$$= \lim_{h\to 0}\frac{x(x+y^2)-x\left[x+(y+h)^2\right]}{h[x+(y+h)^2](x+y^2)} = \lim_{h\to 0}\frac{h(-2xy-xh)}{h[x+(y+h)^2](x+y^2)} = \lim_{h\to 0}\frac{-2xy-xh}{[x+(y+h)^2](x+y^2)} = \frac{-2xy}{(x+y^2)^2}$$

35. $f(x,y) = x^2+y^2+x^2y \Rightarrow f_x = 2x+2xy,\ f_y = 2y+x^2$

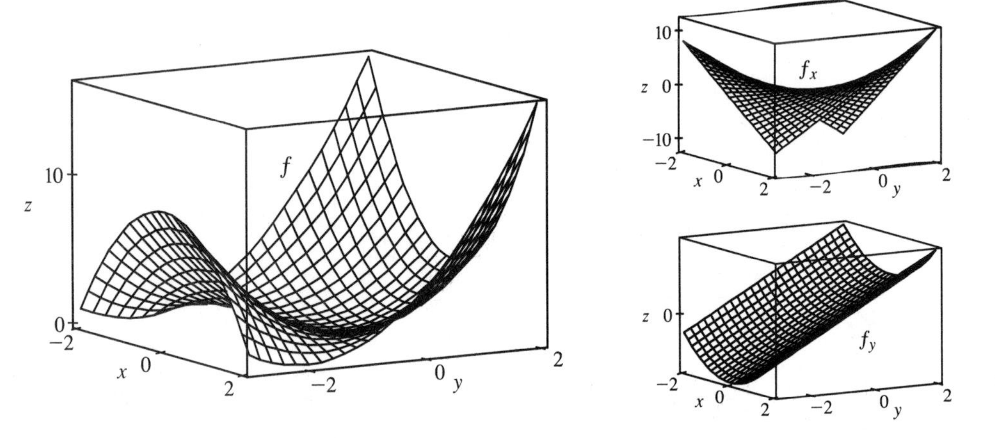

Note that the traces of f in planes parallel to the xz-plane are parabolas which open downward for $y < -1$ and upward for $y > -1$, and the traces of f_x in these planes are straight lines, which have negative slopes for $y < -1$ and positive slopes for $y > -1$. The traces of f in planes parallel to the yz-plane are parabolas which always open upward, and the traces of f_y in these planes are straight lines with positive slopes.

36. $f(x,y) = xe^{-x^2-y^2} \;\Rightarrow\; f_x = x\left(-2xe^{-x^2-y^2}\right) + e^{-x^2-y^2} = e^{-x^2-y^2}(1-2x^2),\quad f_y = -2xye^{-x^2-y^2}$

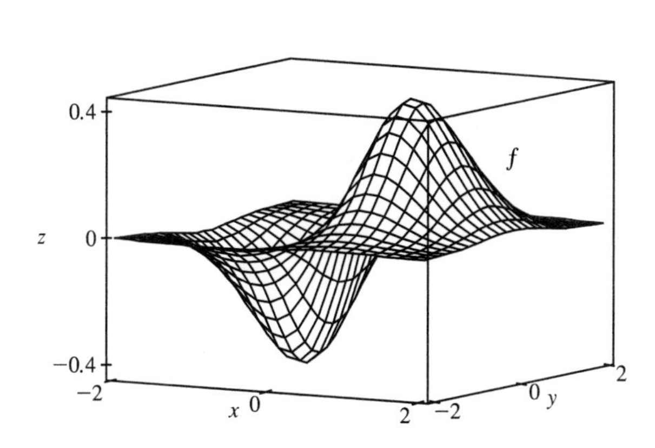

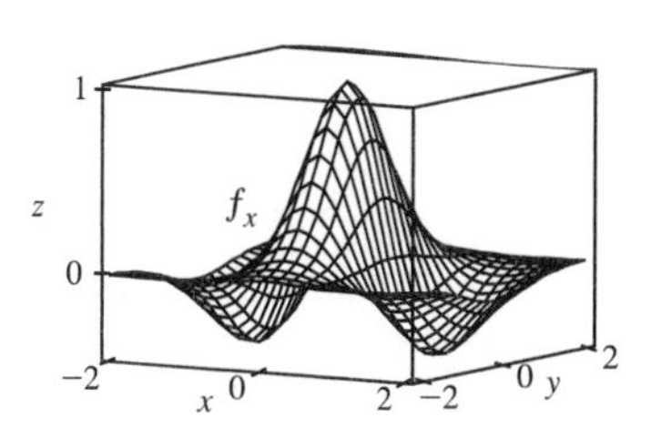

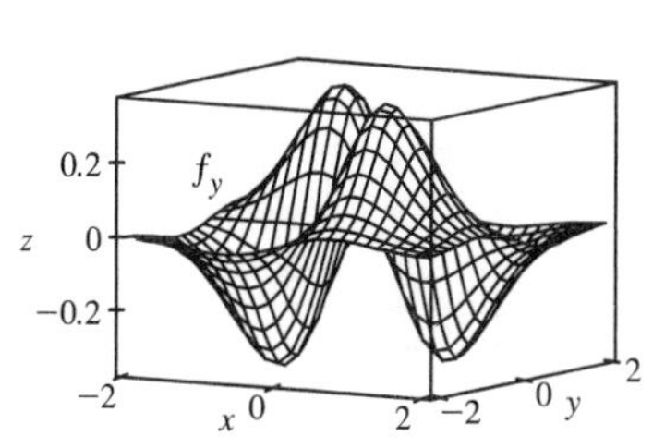

Note that traces of f in planes parallel to the xz-plane have two extreme values, while traces of f_x in these planes have two zeros. Traces of f in planes parallel to the yz-plane have only one extreme value (a minimum if $x < 0$, a maximum if $x > 0$), and traces of f_y in these planes have only one zero (going from negative to positive if $x < 0$ and from positive to negative if $x > 0$).

37. $x^2 + y^2 + z^2 = 3xyz \;\Rightarrow\; \dfrac{\partial}{\partial x}(x^2+y^2+z^2) = \dfrac{\partial}{\partial x}(3xyz) \;\Rightarrow\; 2x + 0 + 2z\,\dfrac{\partial z}{\partial x} = 3y\left(x\,\dfrac{\partial z}{\partial x} + z\cdot 1\right) \;\Leftrightarrow$

$2z\,\dfrac{\partial z}{\partial x} - 3xy\,\dfrac{\partial z}{\partial x} = 3yz - 2x \;\Leftrightarrow\; (2z - 3xy)\,\dfrac{\partial z}{\partial x} = 3yz - 2x$, so $\dfrac{\partial z}{\partial x} = \dfrac{3yz - 2x}{2z - 3xy}$.

$\dfrac{\partial}{\partial y}(x^2+y^2+z^2) = \dfrac{\partial}{\partial y}(3xyz) \;\Rightarrow\; 0 + 2y + 2z\,\dfrac{\partial z}{\partial y} = 3x\left(y\,\dfrac{\partial z}{\partial y} + z\cdot 1\right) \;\Leftrightarrow\; 2z\,\dfrac{\partial z}{\partial y} - 3xy\,\dfrac{\partial z}{\partial y} = 3xz - 2y \;\Leftrightarrow$

$(2z - 3xy)\,\dfrac{\partial z}{\partial y} = 3xz - 2y$, so $\dfrac{\partial z}{\partial y} = \dfrac{3xz - 2y}{2z - 3xy}$.

38. $yz = \ln(x+z) \;\Rightarrow\; \dfrac{\partial}{\partial x}(yz) = \dfrac{\partial}{\partial x}(\ln(x+z)) \;\Rightarrow\; y\,\dfrac{\partial z}{\partial x} = \dfrac{1}{x+z}\left(1 + \dfrac{\partial z}{\partial x}\right) \;\Leftrightarrow\; \left(y - \dfrac{1}{x+z}\right)\dfrac{\partial z}{\partial x} = \dfrac{1}{x+z}$,

so $\dfrac{\partial z}{\partial x} = \dfrac{1/(x+z)}{y - 1/(x+z)} = \dfrac{1}{y(x+z) - 1}$.

$\dfrac{\partial}{\partial y}(yz) = \dfrac{\partial}{\partial y}(\ln(x+z)) \;\Rightarrow\; y\,\dfrac{\partial z}{\partial y} + z\cdot 1 = \dfrac{1}{x+z}\left(0 + \dfrac{\partial z}{\partial y}\right) \;\Leftrightarrow\; \left(y - \dfrac{1}{x+z}\right)\dfrac{\partial z}{\partial y} = -z$,

so $\dfrac{\partial z}{\partial y} = \dfrac{-z}{y - 1/(x+z)} = \dfrac{z(x+z)}{1 - y(x+z)}$.

39. $x - z = \arctan(yz) \Rightarrow \dfrac{\partial}{\partial x}(x - z) = \dfrac{\partial}{\partial x}(\arctan(yz)) \Rightarrow 1 - \dfrac{\partial z}{\partial x} = \dfrac{1}{1+(yz)^2} \cdot y\,\dfrac{\partial z}{\partial x} \Leftrightarrow$

$1 = \left(\dfrac{y}{1+y^2z^2} + 1\right)\dfrac{\partial z}{\partial x} \Leftrightarrow 1 = \left(\dfrac{y+1+y^2z^2}{1+y^2z^2}\right)\dfrac{\partial z}{\partial x}$, so $\dfrac{\partial z}{\partial x} = \dfrac{1+y^2z^2}{1+y+y^2z^2}$.

$\dfrac{\partial}{\partial y}(x - z) = \dfrac{\partial}{\partial y}(\arctan(yz)) \Rightarrow 0 - \dfrac{\partial z}{\partial y} = \dfrac{1}{1+(yz)^2} \cdot \left(y\,\dfrac{\partial z}{\partial y} + z \cdot 1\right) \Leftrightarrow$

$-\dfrac{z}{1+y^2z^2} = \left(\dfrac{y}{1+y^2z^2} + 1\right)\dfrac{\partial z}{\partial y} \Leftrightarrow -\dfrac{z}{1+y^2z^2} = \left(\dfrac{y+1+y^2z^2}{1+y^2z^2}\right)\dfrac{\partial z}{\partial y} \Leftrightarrow \dfrac{\partial z}{\partial y} = -\dfrac{z}{1+y+y^2z^2}$.

40. $\sin(xyz) = x + 2y + 3z \Rightarrow \dfrac{\partial}{\partial x}(\sin(xyz)) = \dfrac{\partial}{\partial x}(x + 2y + 3z) \Rightarrow \cos(xyz) \cdot y\left(x\,\dfrac{\partial z}{\partial x} + z\right) = 1 + 3\,\dfrac{\partial z}{\partial x} \Leftrightarrow$

$(xy\cos(xyz) - 3)\,\dfrac{\partial z}{\partial x} = 1 - yz\cos(xyz)$, so $\dfrac{\partial z}{\partial x} = \dfrac{1 - yz\cos(xyz)}{xy\cos(xyz) - 3}$.

$\dfrac{\partial}{\partial y}(\sin(xyz)) = \dfrac{\partial}{\partial y}(x + 2y + 3z) \Rightarrow \cos(xyz) \cdot x\left(y\,\dfrac{\partial z}{\partial y} + z\right) = 2 + 3\,\dfrac{\partial z}{\partial y} \Leftrightarrow$

$(xy\cos(xyz) - 3)\,\dfrac{\partial z}{\partial y} = 2 - xz\cos(xyz)$, so $\dfrac{\partial z}{\partial y} = \dfrac{2 - xz\cos(xyz)}{xy\cos(xyz) - 3}$.

41. (a) $z = f(x) + g(y) \Rightarrow \dfrac{\partial z}{\partial x} = f'(x)$, $\dfrac{\partial z}{\partial y} = g'(y)$

(b) $z = f(x + y)$. Let $u = x + y$. Then $\dfrac{\partial z}{\partial x} = \dfrac{df}{du}\dfrac{\partial u}{\partial x} = \dfrac{df}{du}(1) = f'(u) = f'(x + y)$,

$\dfrac{\partial z}{\partial y} = \dfrac{df}{du}\dfrac{\partial u}{\partial y} = \dfrac{df}{du}(1) = f'(u) = f'(x + y)$.

42. (a) $z = f(x)g(y) \Rightarrow \dfrac{\partial z}{\partial x} = f'(x)g(y)$, $\dfrac{\partial z}{\partial y} = f(x)g'(y)$

(b) $z = f(xy)$. Let $u = xy$. Then $\dfrac{\partial u}{\partial x} = y$ and $\dfrac{\partial u}{\partial y} = x$. Hence $\dfrac{\partial z}{\partial x} = \dfrac{df}{du}\dfrac{\partial u}{\partial x} = \dfrac{df}{du} \cdot y = yf'(u) = yf'(xy)$

and $\dfrac{\partial z}{\partial y} = \dfrac{df}{du}\dfrac{\partial u}{\partial y} = \dfrac{df}{du} \cdot x = xf'(u) = xf'(xy)$.

(c) $z = f\left(\dfrac{x}{y}\right)$. Let $u = \dfrac{x}{y}$. Then $\dfrac{\partial u}{\partial x} = \dfrac{1}{y}$ and $\dfrac{\partial u}{\partial y} = -\dfrac{x}{y^2}$. Hence $\dfrac{\partial z}{\partial x} = \dfrac{df}{du}\dfrac{\partial u}{\partial x} = f'(u)\dfrac{1}{y} = \dfrac{f'(x/y)}{y}$

and $\dfrac{\partial z}{\partial y} = \dfrac{df}{du}\dfrac{\partial u}{\partial y} = f'(u)\left(-\dfrac{x}{y^2}\right) = -\dfrac{xf'(x/y)}{y^2}$.

43. $f(x, y) = x^4 - 3x^2y^3 \Rightarrow f_x(x, y) = 4x^3 - 6xy^3$, $f_y(x, y) = -9x^2y^2$.

Then $f_{xx}(x, y) = 12x^2 - 6y^3$, $f_{xy}(x, y) = -18xy^2$, $f_{yx}(x, y) = -18xy^2$, $f_{yy}(x, y) = -18x^2y$.

44. $f(x, y) = \ln(3x + 5y) \Rightarrow f_x(x, y) = \dfrac{3}{3x + 5y}$, $f_y(x, y) = \dfrac{5}{3x + 5y}$. Then

$f_{xx}(x, y) = 3(-1)(3x + 5y)^{-2}(3) = -\dfrac{9}{(3x + 5y)^2}$, $f_{xy}(x, y) = -\dfrac{15}{(3x + 5y)^2}$, $f_{yx}(x, y) = -\dfrac{15}{(3x + 5y)^2}$,

$f_{yy}(x, y) = -\dfrac{25}{(3x + 5y)^2}$.

45. $z = \dfrac{x}{x+y} = x(x+y)^{-1} \Rightarrow z_x = \dfrac{1(x+y) - 1(x)}{(x+y)^2} = \dfrac{y}{(x+y)^2}$, $z_y = x(-1)(x+y)^{-2} = -\dfrac{x}{(x+y)^2}$. Then

$z_{xx} = y(-2)(x+y)^{-3} = -\dfrac{2y}{(x+y)^3}$, $z_{xy} = \dfrac{1(x+y)^2 - y(2)(x+y)}{[(x+y)^2]^2} = \dfrac{x+y-2y}{(x+y)^3} = \dfrac{x-y}{(x+y)^3}$,

$z_{yx} = -\dfrac{1(x+y)^2 - x(2)(x+y)}{[(x+y)^2]^2} = -\dfrac{-x^2+xy+y^2}{(x+y)^2} = \dfrac{(x+y)(x-y)}{(x+y)^2} = \dfrac{x-y}{(x+y)^3}$,

$z_{yy} = -x(-2)(x+y)^{-3} = \dfrac{2x}{(x+y)^3}$.

46. $z = y\tan 2x \Rightarrow z_x = y\sec^2(2x)\cdot 2 = 2y\sec^2(2x)$, $z_y = \tan 2x$. Then

$z_{xx} = 2y(2)\sec(2x)\cdot\sec(2x)\tan(2x)\cdot 2 = 8y\sec^2(2x)\tan(2x)$, $z_{xy} = 2\sec^2(2x)$, $z_{yx} = \sec^2(2x)\cdot 2 = 2\sec^2(2x)$,

$z_{yy} = 0$.

47. $u = e^{-s}\sin t \Rightarrow u_s = -e^{-s}\sin t$, $u_t = e^{-s}\cos t$. Then $u_{ss} = e^{-s}\sin t$, $u_{st} = -e^{-s}\cos t$, $u_{ts} = -e^{-s}\cos t$,

$u_{tt} = -e^{-s}\sin t$.

48. $v = \sqrt{x+y^2} \Rightarrow v_x = \frac{1}{2}(x+y^2)^{-1/2} = \dfrac{1}{2\sqrt{x+y^2}}$, $v_y = \frac{1}{2}(x+y^2)^{-1/2}(2y) = \dfrac{y}{\sqrt{x+y^2}}$.

Then $v_{xx} = \frac{1}{2}\left(-\frac{1}{2}\right)(x+y^2)^{-3/2} = -\dfrac{1}{4(x+y^2)^{3/2}}$,

$v_{xy} = \frac{1}{2}\left(-\frac{1}{2}\right)(x+y^2)^{-3/2}(2y) = -\dfrac{y}{2(x+y^2)^{3/2}}$, $v_{yx} = y\left(-\frac{1}{2}\right)(x+y^2)^{-3/2} = -\dfrac{y}{2(x+y^2)^{3/2}}$,

$v_{yy} = \dfrac{1\sqrt{x+y^2} - y\left(\frac{1}{2}\right)(x+y^2)^{-1/2}(2y)}{\left(\sqrt{x+y^2}\right)^2} = \dfrac{(x+y^2)-y^2}{(x+y^2)^{3/2}} = \dfrac{x}{(x+y^2)^{3/2}}$.

49. $u = x\sin(x+2y) \Rightarrow u_x = x\cdot\cos(x+2y)(1) + \sin(x+2y)\cdot 1 = x\cos(x+2y) + \sin(x+2y)$,

$u_{xy} = x(-\sin(x+2y)(2)) + \cos(x+2y)(2) = 2\cos(x+2y) - 2x\sin(x+2y)$, $u_y = x\cos(x+2y)(2) = 2x\cos(x+2y)$,

$u_{yx} = 2x\cdot(-\sin(x+2y)(1)) + \cos(x+2y)\cdot 2 = 2\cos(x+2y) - 2x\sin(x+2y)$. Thus $u_{xy} = u_{yx}$.

50. $u = x^4y^2 - 2xy^5 \Rightarrow u_x = 4x^3y^2 - 2y^5$, $u_{xy} = 8x^3y - 10y^4$ and $u_y = 2x^4y - 10xy^4$, $u_{yx} = 8x^3y - 10y^4$.

Thus $u_{xy} = u_{yx}$.

51. $f(x,y) = 3xy^4 + x^3y^2 \Rightarrow f_x = 3y^4 + 3x^2y^2$, $f_{xx} = 6xy^2$, $f_{xxy} = 12xy$ and

$f_y = 12xy^3 + 2x^3y$, $f_{yy} = 36xy^2 + 2x^3$, $f_{yyy} = 72xy$.

52. $f(x,t) = x^2e^{-ct} \Rightarrow f_t = x^2(-ce^{-ct})$, $f_{tt} = x^2(c^2e^{-ct})$, $f_{ttt} = x^2(-c^3e^{-ct}) = -c^3x^2e^{-ct}$ and

$f_{tx} = 2x(-ce^{-ct})$, $f_{txx} = 2(-ce^{-ct}) = -2ce^{-ct}$.

53. $f(x,y,z) = \cos(4x+3y+2z) \Rightarrow$

$f_x = -\sin(4x+3y+2z)(4) = -4\sin(4x+3y+2z)$,

$f_{xy} = -4\cos(4x+3y+2z)(3) = -12\cos(4x+3y+2z)$,

$f_{xyz} = -12(-\sin(4x+3y+2z))(2) = 24\sin(4x+3y+2z)$ and

$f_y = -\sin(4x + 3y + 2z)(3) = -3\sin(4x + 3y + 2z)$,

$f_{yz} = -3\cos(4x + 3y + 2z)(2) = -6\cos(4x + 3y + 2z)$,

$f_{yzz} = -6(-\sin(4x + 3y + 2z))(2) = 12\sin(4x + 3y + 2z)$.

54. $f(r, s, t) = r\ln(rs^2t^3) \Rightarrow f_r = r \cdot \dfrac{1}{rs^2t^3}(s^2t^3) + \ln(rs^2t^3) \cdot 1 = \dfrac{rs^2t^3}{rs^2t^3} + \ln(rs^2t^3) = 1 + \ln(rs^2t^3)$,

$f_{rs} = \dfrac{1}{rs^2t^3}(2rst^3) = \dfrac{2}{s} = 2s^{-1}$, $f_{rss} = -2s^{-2} = -\dfrac{2}{s^2}$ and $f_{rst} = 0$.

55. $u = e^{r\theta}\sin\theta \Rightarrow \dfrac{\partial u}{\partial\theta} = e^{r\theta}\cos\theta + \sin\theta \cdot e^{r\theta}(r) = e^{r\theta}(\cos\theta + r\sin\theta)$,

$\dfrac{\partial^2 u}{\partial r\,\partial\theta} = e^{r\theta}(\sin\theta) + (\cos\theta + r\sin\theta)\,e^{r\theta}(\theta) = e^{r\theta}(\sin\theta + \theta\cos\theta + r\theta\sin\theta)$,

$\dfrac{\partial^3 u}{\partial r^2\,\partial\theta} = e^{r\theta}(\theta\sin\theta) + (\sin\theta + \theta\cos\theta + r\theta\sin\theta) \cdot e^{r\theta}(\theta) = \theta e^{r\theta}(2\sin\theta + \theta\cos\theta + r\theta\sin\theta)$.

56. $u = x^a y^b z^c$. If $a = 0$, or if $b = 0$ or 1, or if $c = 0$, 1, or 2, then $\dfrac{\partial^6 u}{\partial x\,\partial y^2\,\partial z^3} = 0$. Otherwise $\dfrac{\partial u}{\partial z} = cx^a y^b z^{c-1}$,

$\dfrac{\partial^2 u}{\partial z^2} = c(c-1)x^a y^b z^{c-2}$, $\dfrac{\partial^3 u}{\partial z^3} = c(c-1)(c-2)x^a y^b z^{c-3}$, $\dfrac{\partial^4 u}{\partial y\,\partial z^3} = bc(c-1)(c-2)x^a y^{b-1} z^{c-3}$,

$\dfrac{\partial^5 u}{\partial y^2\,\partial z^3} = b(b-1)c(c-1)(c-2)x^a y^{b-2} z^{c-3}$, and $\dfrac{\partial^6 u}{\partial x\,\partial y^2\,\partial z^3} = ab(b-1)c(c-1)(c-2)x^{a-1} y^{b-2} z^{c-3}$.

57. $u = e^{-\alpha^2 k^2 t}\sin kx \Rightarrow u_x = ke^{-\alpha^2 k^2 t}\cos kx$, $u_{xx} = -k^2 e^{-\alpha^2 k^2 t}\sin kx$, and $u_t = -\alpha^2 k^2 e^{-\alpha^2 k^2 t}\sin kx$.

Thus $\alpha^2 u_{xx} = u_t$.

58. (a) $u = x^2 + y^2 \Rightarrow u_x = 2x$, $u_{xx} = 2$; $u_y = 2y$, $u_{yy} = 2$. Thus $u_{xx} + u_{yy} \neq 0$ and $u = x^2 + y^2$ does not satisfy Laplace's Equation.

(b) $u = x^2 - y^2$ is a solution: $u_{xx} = 2$, $u_{yy} = -2$ so $u_{xx} + u_{yy} = 0$.

(c) $u = x^3 + 3xy^2$ is not a solution: $u_x = 3x^2 + 3y^2$, $u_{xx} = 6x$; $u_y = 6xy$, $u_{yy} = 6x$.

(d) $u = \ln\sqrt{x^2 + y^2}$ is a solution: $u_x = \dfrac{1}{\sqrt{x^2 + y^2}}\left(\dfrac{1}{2}\right)(x^2 + y^2)^{-1/2}(2x) = \dfrac{x}{x^2 + y^2}$,

$u_{xx} = \dfrac{(x^2 + y^2) - x(2x)}{(x^2 + y^2)^2} = \dfrac{y^2 - x^2}{(x^2 + y^2)^2}$. By symmetry, $u_{yy} = \dfrac{x^2 - y^2}{(x^2 + y^2)^2}$, so $u_{xx} + u_{yy} = 0$.

(e) $u = \sin x\cosh y + \cos x\sinh y$ is a solution: $u_x = \cos x\cosh y - \sin x\sinh y$, $u_{xx} = -\sin x\cosh y - \cos x\sinh y$, and $u_y = \sin x\sinh y + \cos x\cosh y$, $u_{yy} = \sin x\cosh y + \cos x\sinh y$.

(f) $u = e^{-x}\cos y - e^{-y}\cos x$ is a solution: $u_x = -e^{-x}\cos y + e^{-y}\sin x$, $u_{xx} = e^{-x}\cos y + e^{-y}\cos x$, and $u_y = -e^{-x}\sin y + e^{-y}\cos x$, $u_{yy} = -e^{-x}\cos y - e^{-y}\cos x$.

59. $u = \dfrac{1}{\sqrt{x^2+y^2+z^2}} \quad\Rightarrow\quad u_x = \left(-\frac{1}{2}\right)(x^2+y^2+z^2)^{-3/2}(2x) = -x(x^2+y^2+z^2)^{-3/2}$ and

$$u_{xx} = -(x^2+y^2+z^2)^{-3/2} - x\left(-\tfrac{3}{2}\right)(x^2+y^2+z^2)^{-5/2}(2x) = \frac{2x^2-y^2-z^2}{(x^2+y^2+z^2)^{5/2}}.$$

By symmetry, $u_{yy} = \dfrac{2y^2-x^2-z^2}{(x^2+y^2+z^2)^{5/2}}$ and $u_{zz} = \dfrac{2z^2-x^2-y^2}{(x^2+y^2+z^2)^{5/2}}$.

Thus $u_{xx}+u_{yy}+u_{zz} = \dfrac{2x^2-y^2-z^2+2y^2-x^2-z^2+2z^2-x^2-y^2}{(x^2+y^2+z^2)^{5/2}} = 0.$

60. (a) $u = \sin(kx)\,\sin(akt) \quad\Rightarrow\quad u_t = ak\sin(kx)\,\cos(akt)$, $u_{tt} = -a^2k^2\sin(kx)\,\sin(akt)$, $u_x = k\cos(kx)\,\sin(akt)$,

$u_{xx} = -k^2\sin(kx)\,\sin(akt)$. Thus $u_{tt} = a^2u_{xx}$.

(b) $u = \dfrac{t}{a^2t^2-x^2} \quad\Rightarrow\quad u_t = \dfrac{(a^2t^2-x^2)-t(2a^2t)}{(a^2t^2-x^2)^2} = -\dfrac{a^2t^2+x^2}{(a^2t^2-x^2)^2}$,

$$u_{tt} = \frac{-2a^2t(a^2t^2-x^2)^2+(a^2t^2-x^2)(2)(a^2t^2-x^2)(2a^2t)}{(a^2t^2-x^2)^4} = \frac{2a^4t^3+6a^2tx^2}{(a^2t^2-x^2)^3},$$

$$u_x = t(-1)(a^2t^2-x^2)^{-2}(2x) = \frac{2tx}{(a^2t^2-x^2)^2},$$

$$u_{xx} = \frac{2t(a^2t^2-x^2)^2-2tx\,(2)(a^2t^2-x^2)(-2x)}{(a^2t^2-x^2)^4} = \frac{2a^2t^3-2tx^2+8tx^2}{(a^2t^2-x^2)^3} = \frac{2a^2t^3+6tx^2}{(a^2t^2-x^2)^3}.$$

Thus $u_{tt} = a^2u_{xx}$.

(c) $u = (x-at)^6+(x+at)^6 \quad\Rightarrow\quad u_t = -6a(x-at)^5+6a(x+at)^5$, $\quad u_{tt} = 30a^2(x-at)^4+30a^2(x+at)^4$,

$u_x = 6(x-at)^5+6(x+at)^5$, $\quad u_{xx} = 30(x-at)^4+30(x+at)^4$. Thus $u_{tt} = a^2u_{xx}$.

(d) $u = \sin(x-at)+\ln(x+at) \quad\Rightarrow\quad u_t = -a\cos(x-at)+\dfrac{a}{x+at}$, $\quad u_{tt} = -a^2\sin(x-at)-\dfrac{a^2}{(x+at)^2}$,

$u_x = \cos(x-at)+\dfrac{1}{x+at}$, $\quad u_{xx} = -\sin(x-at)-\dfrac{1}{(x+at)^2}$. Thus $u_{tt} = a^2u_{xx}$.

61. Let $v = x+at$, $w = x-at$. Then $u_t = \dfrac{\partial[f(v)+g(w)]}{\partial t} = \dfrac{df(v)}{dv}\dfrac{\partial v}{\partial t}+\dfrac{dg(w)}{dw}\dfrac{\partial w}{\partial t} = af'(v)-ag'(w)$ and

$u_{tt} = \dfrac{\partial[af'(v)-ag'(w)]}{\partial t} = a[af''(v)+ag''(w)] = a^2[f''(v)+g''(w)]$. Similarly, by using the Chain Rule we have

$u_x = f'(v)+g'(w)$ and $u_{xx} = f''(v)+g''(w)$. Thus $u_{tt} = a^2u_{xx}$.

62. For each i, $i = 1,\ldots,n$, $\partial u/\partial x_i = a_ie^{a_1x_1+a_2x_2+\cdots+a_nx_n}$ and $\partial^2u/\partial x_i^2 = a_i^2e^{a_1x_1+a_2x_2+\cdots+a_nx_n}$.

Then $\dfrac{\partial^2u}{\partial x_1^2}+\dfrac{\partial^2u}{\partial x_2^2}+\cdots+\dfrac{\partial^2u}{\partial x_n^2} = \left(a_1^2+a_2^2+\cdots+a_n^2\right)e^{a_1x_1+a_2x_2+\cdots+a_nx_n} = e^{a_1x_1+a_2x_2+\cdots+a_nx_n} = u$

since $a_1^2+a_2^2+\cdots+a_n^2 = 1$.

63. $z_x = e^y + ye^x$, $z_{xx} = ye^x$, $\partial^3 z/\partial x^3 = ye^x$. By symmetry $z_y = xe^y + e^x$, $z_{yy} = xe^y$, $\partial^3 z/\partial y^3 = xe^y$.

Then $\partial^3 z/\partial x \partial y^2 = e^y$ and $\partial^3 z/\partial x^2 \partial y = e^x$. Thus $z = xe^y + ye^x$ satisfies the given partial differential equation.

64. (a) $\partial T/\partial x = -60(2x)/(1 + x^2 + y^2)^2$, so at $(2, 1)$, $T_x = -240/(1 + 4 + 1)^2 = -\frac{20}{3}$.

(b) $\partial T/\partial y = -60(2y)/(1 + x^2 + y^2)^2$, so at $(2, 1)$, $T_y = -120/36 = -\frac{10}{3}$. Thus from the point $(2, 1)$ the temperature is decreasing at a rate of $\frac{20}{3}$ °C/m in the x-direction and is decreasing at a rate of $\frac{10}{3}$ °C/m in the y-direction.

65. By the Chain Rule, taking the partial derivative of both sides with respect to R_1 gives

$$\frac{\partial R^{-1}}{\partial R}\frac{\partial R}{\partial R_1} = \frac{\partial\left[(1/R_1) + (1/R_2) + (1/R_3)\right]}{\partial R_1} \text{ or } -R^{-2}\frac{\partial R}{\partial R_1} = -R_1^{-2}. \text{ Thus } \frac{\partial R}{\partial R_1} = \frac{R^2}{R_1^2}.$$

66. $P = \dfrac{mRT}{V}$ so $\dfrac{\partial P}{\partial V} = \dfrac{-mRT}{V^2}$; $V = \dfrac{mRT}{P}$, so $\dfrac{\partial V}{\partial T} = \dfrac{mR}{P}$; $T = \dfrac{PV}{mR}$, so $\dfrac{\partial T}{\partial P} = \dfrac{V}{mR}$.

Thus $\dfrac{\partial P}{\partial V}\dfrac{\partial V}{\partial T}\dfrac{\partial T}{\partial P} = \dfrac{-mRT}{V^2}\dfrac{mR}{P}\dfrac{V}{mR} = \dfrac{-mRT}{PV} = -1$, since $PV = mRT$.

67. By Exercise 66, $PV = mRT \Rightarrow P = \dfrac{mRT}{V}$, so $\dfrac{\partial P}{\partial T} = \dfrac{mR}{V}$. Also, $PV = mRT \Rightarrow V = \dfrac{mRT}{P}$

and $\dfrac{\partial V}{\partial T} = \dfrac{mR}{P}$. Since $T = \dfrac{PV}{mR}$, we have $T\dfrac{\partial P}{\partial T}\dfrac{\partial V}{\partial T} = \dfrac{PV}{mR}\cdot\dfrac{mR}{V}\cdot\dfrac{mR}{P} = mR$.

68. $\dfrac{\partial W}{\partial T} = 0.6215 + 0.3965v^{0.16}$. When $T = -15$°C and $v = 30$ km/h, $\dfrac{\partial W}{\partial T} = 0.6215 + 0.3965(30)^{0.16} \approx 1.3048$, so we would expect the apparent temperature to drop by approximately 1.3°C if the actual temperature decreases by 1°C.

$\dfrac{\partial W}{\partial v} = -11.37(0.16)v^{-0.84} + 0.3965T(0.16)v^{-0.84}$ and when $T = -15$°C and $v = 30$ km/h,

$\dfrac{\partial W}{\partial v} = -11.37(0.16)(30)^{-0.84} + 0.3965(-15)(0.16)(30)^{-0.84} \approx -0.1592$, so we would expect the apparent temperature to drop by approximately 0.16°C if the wind speed increases by 1 km/h.

69. $\dfrac{\partial K}{\partial m} = \frac{1}{2}v^2$, $\dfrac{\partial K}{\partial v} = mv$, $\dfrac{\partial^2 K}{\partial v^2} = m$. Thus $\dfrac{\partial K}{\partial m}\cdot\dfrac{\partial^2 K}{\partial v^2} = \frac{1}{2}v^2 m = K$.

70. The Law of Cosines says that $a^2 = b^2 + c^2 - 2bc\cos A$. Thus $\dfrac{\partial(a^2)}{\partial a} = \dfrac{\partial(b^2 + c^2 - 2ab\cos A)}{\partial a}$ or

$2a = -2bc(-\sin A)\dfrac{\partial A}{\partial a}$, implying that $\dfrac{\partial A}{\partial a} = \dfrac{a}{bc\sin A}$. Taking the partial derivative of both sides with respect to b gives

$0 = 2b - 2c(\cos A) - 2bc(-\sin A)\dfrac{\partial A}{\partial b}$. Thus $\dfrac{\partial A}{\partial b} = \dfrac{c\cos A - b}{bc\sin A}$. By symmetry $\dfrac{\partial A}{\partial c} = \dfrac{b\cos A - c}{bc\sin A}$.

71. $f_x(x, y) = x + 4y \Rightarrow f_{xy}(x, y) = 4$ and $f_y(x, y) = 3x - y \Rightarrow f_{yx}(x, y) = 3$. Since f_{xy} and f_{yx} are continuous everywhere but $f_{xy}(x, y) \neq f_{yx}(x, y)$, Clairaut's Theorem implies that such a function $f(x, y)$ does not exist.

72. Setting $x = 1$, the equation of the parabola of intersection is $z = 6 - 1 - 1 - 2y^2 = 4 - 2y^2$. The slope of the tangent is $\partial z/\partial y = -4y$, so at $(1, 2, -4)$ the slope is -8. Parametric equations for the line are therefore $x = 1$, $y = 2 + t$, $z = -4 - 8t$.

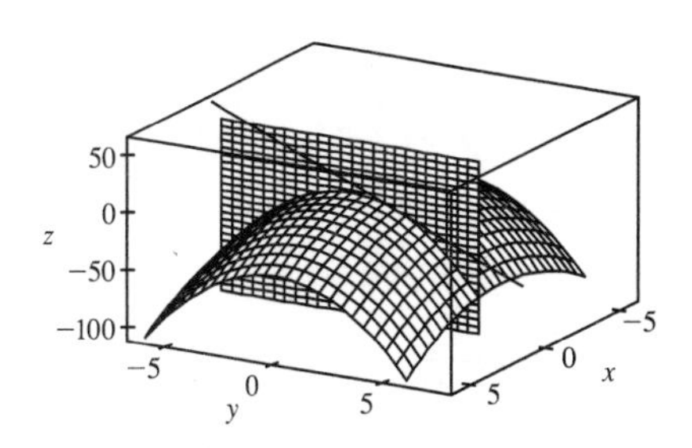

73. By the geometry of partial derivatives, the slope of the tangent line is $f_x(1, 2)$. By implicit differentiation of $4x^2 + 2y^2 + z^2 = 16$, we get $8x + 2z\,(\partial z/\partial x) = 0 \;\Rightarrow\; \partial z/\partial x = -4x/z$, so when $x = 1$ and $z = 2$ we have $\partial z/\partial x = -2$. So the slope is $f_x(1, 2) = -2$. Thus the tangent line is given by $z - 2 = -2(x - 1)$, $y = 2$. Taking the parameter to be $t = x - 1$, we can write parametric equations for this line: $x = 1 + t$, $y = 2$, $z = 2 - 2t$.

74. $T(x, t) = T_0 + T_1 e^{-\lambda x} \sin(\omega t - \lambda x)$

(a) $\partial T/\partial x = T_1 e^{-\lambda x}\left[\cos(\omega t - \lambda x)(-\lambda)\right] + T_1(-\lambda e^{-\lambda x})\sin(\omega t - \lambda x) = -\lambda T_1 e^{-\lambda x}\left[\sin(\omega t - \lambda x) + \cos(\omega t - \lambda x)\right]$.

This quantity represents the rate of change of temperature with respect to depth below the surface, at a given time t.

(b) $\partial T/\partial t = T_1 e^{-\lambda x}\left[\cos(\omega t - \lambda x)(\omega)\right] = \omega T_1 e^{-\lambda x}\cos(\omega t - \lambda x)$. This quantity represents the rate of change of temperature with respect to time at a fixed depth x.

(c)
$$\begin{aligned} T_{xx} &= \frac{\partial}{\partial x}\left(\frac{\partial T}{\partial x}\right) \\ &= -\lambda T_1\left(e^{-\lambda x}\left[\cos(\omega t - \lambda x)(-\lambda) - \sin(\omega t - \lambda x)(-\lambda)\right] + e^{-\lambda x}(-\lambda)\left[\sin(\omega t - \lambda x) + \cos(\omega t - \lambda x)\right]\right) \\ &= 2\lambda^2 T_1 e^{-\lambda x}\cos(\omega t - \lambda x) \end{aligned}$$

But from part (b), $T_t = \omega T_1 e^{-\lambda x}\cos(\omega t - \lambda x) = \dfrac{\omega}{2\lambda^2} T_{xx}$. So with $k = \dfrac{\omega}{2\lambda^2}$, the function T satisfies the heat equation.

(d)

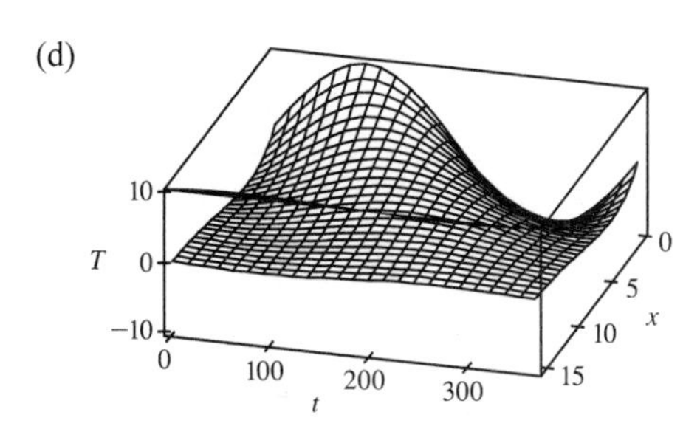

Note that near the surface (that is, for small x) the temperature varies greatly as t changes, but deeper (for large x) the temperature is more stable.

(e) The term $-\lambda x$ is a phase shift: it represents the fact that since heat diffuses slowly through soil, it takes time for changes in the surface temperature to affect the temperature at deeper points. As x increases, the phase shift also increases. For example, at the surface the highest temperature is reached at $t \approx 100$, whereas at a depth of 5 feet the peak temperature is attained at $t \approx 150$, and at a depth of 10 feet, at $t \approx 220$.

75. By Clairaut's Theorem, $f_{xyy} = (f_{xy})_y = (f_{yx})_y = f_{yxy} = (f_y)_{xy} = (f_y)_{yx} = f_{yyx}$.

76. (a) Since we are differentiating n times, with two choices of variable at each differentiation, there are 2^n nth order partial derivatives.

(b) If these partial derivatives are all continuous, then the order in which the partials are taken doesn't affect the value of the result, that is, all nth order partial derivatives with p partials with respect to x and $n-p$ partials with respect to y are equal. Since the number of partials taken with respect to x for an nth order partial derivative can range from 0 to n, a function of two variables has $n+1$ distinct partial derivatives of order n if these partial derivatives are all continuous.

(c) Since n differentiations are to be performed with three choices of variable at each differentiation, there are 3^n nth order partial derivatives of a function of three variables.

77. Let $g(x) = f(x,0) = x(x^2)^{-3/2}e^0 = x\,|x|^{-3}$. But we are using the point $(1,0)$, so near $(1,0)$, $g(x) = x^{-2}$. Then $g'(x) = -2x^{-3}$ and $g'(1) = -2$, so using (1) we have $f_x(1,0) = g'(1) = -2$.

78. $f_x(0,0) = \lim\limits_{h\to 0} \dfrac{f(0+h,0) - f(0,0)}{h} = \lim\limits_{h\to 0} \dfrac{(h^3+0)^{1/3} - 0}{h} = \lim\limits_{h\to 0} \dfrac{h}{h} = 1.$

Or: Let $g(x) = f(x,0) = \sqrt[3]{x^3+0} = x$. Then $g'(x) = 1$ and $g'(0) = 1$ so, by (1), $f_x(0,0) = g'(0) = 1$.

79. (a)

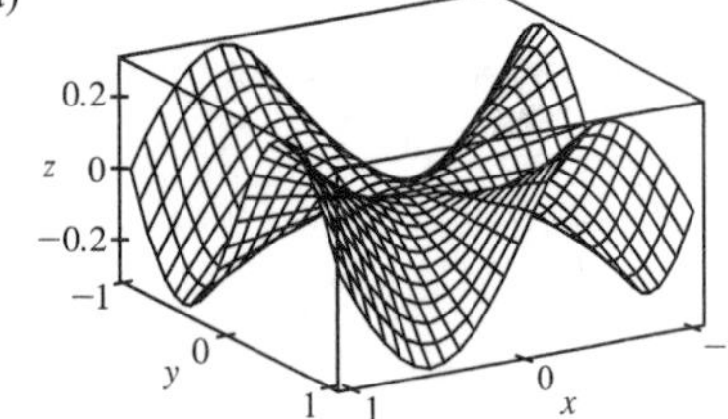

(b) For $(x,y) \neq (0,0)$, $f_x(x,y) = \dfrac{(3x^2y - y^3)(x^2+y^2) - (x^3y - xy^3)(2x)}{(x^2+y^2)^2} = \dfrac{x^4y + 4x^2y^3 - y^5}{(x^2+y^2)^2}$, and by symmetry

$$f_y(x,y) = \frac{x^5 - 4x^3y^2 - xy^4}{(x^2+y^2)^2}.$$

(c) $f_x(0,0) = \lim\limits_{h\to 0} \dfrac{f(h,0) - f(0,0)}{h} = \lim\limits_{h\to 0} \dfrac{(0/h^2) - 0}{h} = 0$ and $f_y(0,0) = \lim\limits_{h\to 0} \dfrac{f(0,h) - f(0,0)}{h} = 0.$

(d) By (3), $f_{xy}(0,0) = \dfrac{\partial f_x}{\partial y} = \lim\limits_{h\to 0} \dfrac{f_x(0,h) - f_x(0,0)}{h} = \lim\limits_{h\to 0} \dfrac{(-h^5 - 0)/h^4}{h} = -1$ while by (2),

$$f_{yx}(0,0) = \frac{\partial f_y}{\partial x} = \lim_{h\to 0} \frac{f_y(h,0) - f_y(0,0)}{h} = \lim_{h\to 0} \frac{h^5/h^4}{h} = 1.$$

(e) For $(x,y) \neq (0,0)$, we use a CAS to compute

$$f_{xy}(x,y) = \frac{x^6 + 9x^4y^2 - 9x^2y^4 - y^6}{(x^2+y^2)^3}.$$

Now as $(x,y) \to (0,0)$ along the x-axis, $f_{xy}(x,y) \to 1$ while as $(x,y) \to (0,0)$ along the y-axis, $f_{xy}(x,y) \to -1$. Thus f_{xy} isn't continuous at $(0,0)$ and Clairaut's Theorem doesn't apply, so there is no contradiction. The graphs of f_{xy} and f_{yx} are identical except at the origin, where we observe the discontinuity.

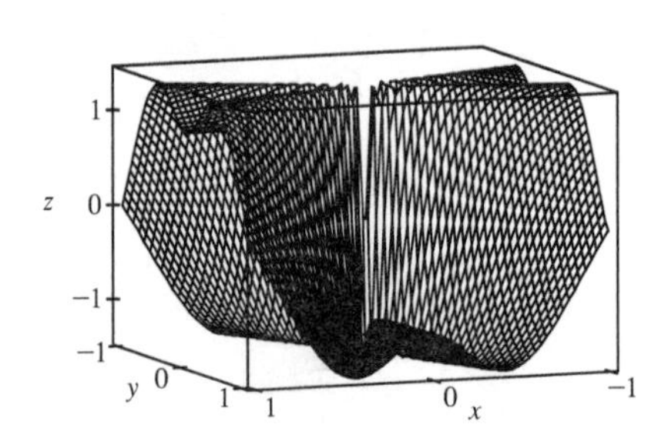

11.4 Tangent Planes and Linear Approximations

1. $z = f(x, y) = 4x^2 - y^2 + 2y \Rightarrow f_x(x, y) = 8x$, $f_y(x, y) = -2y + 2$, so $f_x(-1, 2) = -8$, $f_y(-1, 2) = -2$. By Equation 2, an equation of the tangent plane is $z - 4 = f_x(-1, 2)[x - (-1)] + f_y(-1, 2)(y - 2) \Rightarrow$ $z - 4 = -8(x + 1) - 2(y - 2)$ or $z = -8x - 2y$.

2. $z = f(x, y) = 9x^2 + y^2 + 6x - 3y + 5 \Rightarrow f_x(x, y) = 18x + 6$, $f_y(x, y) = 2y - 3$, so $f_x(1, 2) = 24$ and $f_y(1, 2) = 1$. By Equation 2, an equation of the tangent plane is $z - 18 = f_x(1, 2)(x - 1) + f_y(1, 2)(y - 2) \Rightarrow$ $z - 18 = 24(x - 1) + 1(y - 2)$ or $z = 24x + y - 8$.

3. $z = f(x, y) = \sqrt{4 - x^2 - 2y^2} \Rightarrow f_x(x, y) = \frac{1}{2}(4 - x^2 - 2y^2)^{-1/2}(-2x) = -\dfrac{x}{\sqrt{4 - x^2 - 2y^2}}$,
$f_y(x, y) = \frac{1}{2}(4 - x^2 - 2y^2)^{-1/2}(-4y) = -\dfrac{2y}{\sqrt{4 - x^2 - 2y^2}}$, so $f_x(1, -1) = -1$ and $f_y(1, -1) = 2$. Thus, an equation of the tangent plane is $z - 1 = f_x(1, -1)(x - 1) + f_y(1, -1)[y - (-1)] \Rightarrow z - 1 = -1(x - 1) + 2(y + 1)$ or $x - 2y + z = 4$.

4. $z = f(x, y) = y \ln x \Rightarrow f_x(x, y) = y/x$, $f_y(x, y) = \ln x$, so $f_x(1, 4) = 4$, $f_y(1, 4) = 0$, and an equation of the tangent plane is $z - 0 = f_x(1, 4)(x - 1) + f_y(1, 4)(y - 4) \Rightarrow z = 4(x - 1) + 0(y - 4)$ or $z = 4x - 4$.

5. $z = f(x, y) = y\cos(x - y) \Rightarrow f_x = y(-\sin(x - y)(1)) = -y\sin(x - y)$,
$f_y = y(-\sin(x - y)(-1)) + \cos(x - y) = y\sin(x - y) + \cos(x - y)$, so $f_x(2, 2) = -2\sin(0) = 0$,
$f_y(2, 2) = 2\sin(0) + \cos(0) = 1$ and an equation of the tangent plane is $z - 2 = 0(x - 2) + 1(y - 2)$ or $z = y$.

6. $z = f(x, y) = e^{x^2 - y^2} \Rightarrow f_x(x, y) = 2xe^{x^2 - y^2}$, $f_y(x, y) = -2ye^{x^2 - y^2}$, so $f_x(1, -1) = 2$, $f_y(1, -1) = 2$. By Equation 2, an equation of the tangent plane is $z - 1 = f_x(1, -1)(x - 1) + f_y(1, -1)[y - (-1)] \Rightarrow$ $z - 1 = 2(x - 1) + 2(y + 1)$ or $z = 2x + 2y + 1$.

7. $z = f(x, y) = x^2 + xy + 3y^2$, so $f_x(x, y) = 2x + y \Rightarrow f_x(1, 1) = 3$, $f_y(x, y) = x + 6y \Rightarrow f_y(1, 1) = 7$ and an equation of the tangent plane is $z - 5 = 3(x - 1) + 7(y - 1)$ or $z = 3x + 7y - 5$. After zooming in, the surface and the tangent plane become almost indistinguishable. (Here, the tangent plane is below the surface.) If we zoom in farther, the surface and the tangent plane will appear to coincide.

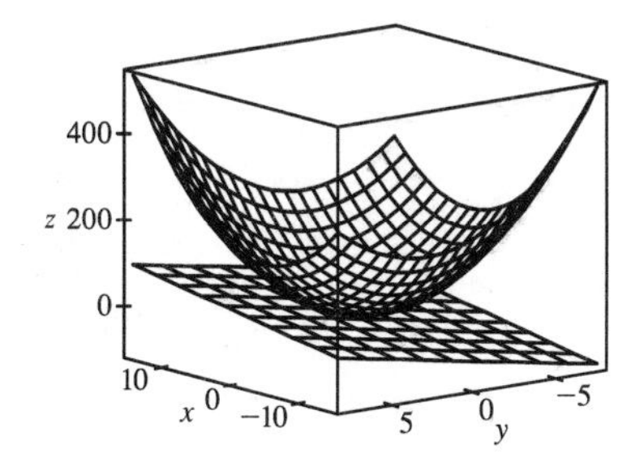

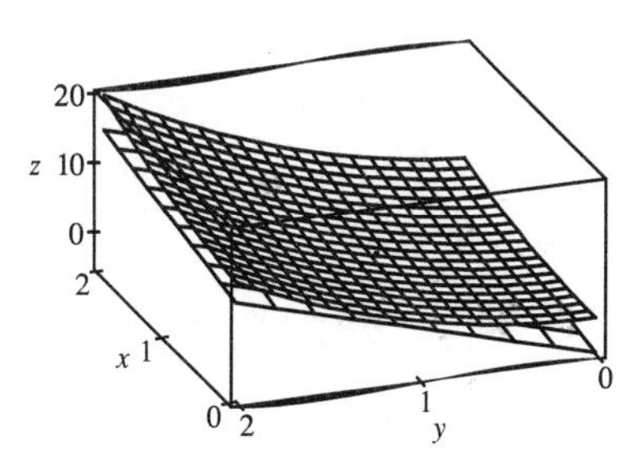

8. $z = f(x, y) = \arctan(xy^2) \Rightarrow f_x = \dfrac{1}{1 + (xy^2)^2}(y^2) = \dfrac{y^2}{1 + x^2y^4}$, $f_y = \dfrac{1}{1 + (xy^2)^2}(2xy) = \dfrac{2xy}{1 + x^2y^4}$,
$f_x(1, 1) = \frac{1}{1+1} = \frac{1}{2}$, $f_y(1, 1) = \frac{2}{1+1} = 1$, so an equation of the tangent plane is $z - \frac{\pi}{4} = \frac{1}{2}(x - 1) + 1(y - 1)$ or

$z = \frac{1}{2}x + y - \frac{3}{2} + \frac{\pi}{4}$. After zooming in, the surface and the tangent plane become almost indistinguishable. (Here the tangent plane is above the surface.) If we zoom in farther, the surface and the tangent plane will appear to coincide.

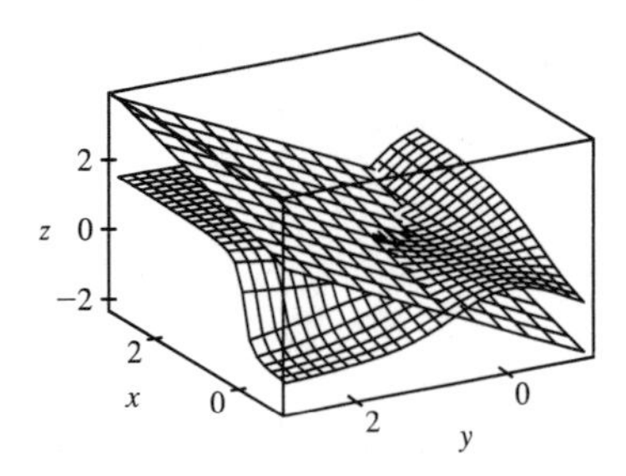

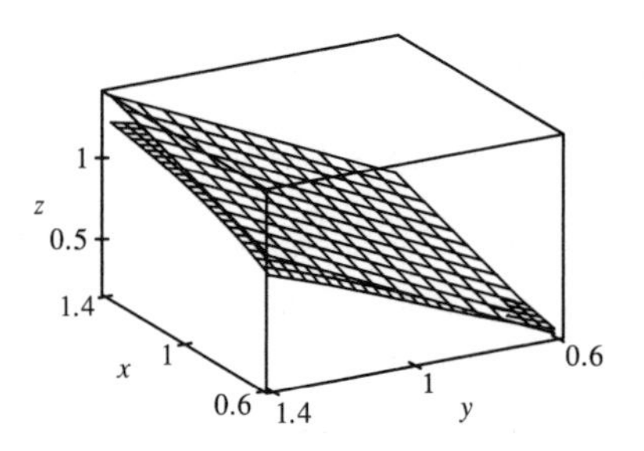

9. $f(x,y) = \dfrac{xy\sin(x-y)}{1+x^2+y^2}$. A CAS gives $f_x(x,y) = \dfrac{y\sin(x-y) + xy\cos(x-y)}{1+x^2+y^2} - \dfrac{2x^2y\sin(x-y)}{(1+x^2+y^2)^2}$ and

$f_y(x,y) = \dfrac{x\sin(x-y) - xy\cos(x-y)}{1+x^2+y^2} - \dfrac{2xy^2\sin(x-y)}{(1+x^2+y^2)^2}$. We use the CAS to evaluate these at $(1,1)$, and then substitute the results into Equation 2 to compute an equation of the tangent plane: $z = \frac{1}{3}x - \frac{1}{3}y$. The surface and tangent plane are shown in the first graph below. After zooming in, the surface and the tangent plane become almost indistinguishable, as shown in the second graph. (Here, the tangent plane is shown with fewer traces than the surface.) If we zoom in farther, the surface and the tangent plane will appear to coincide.

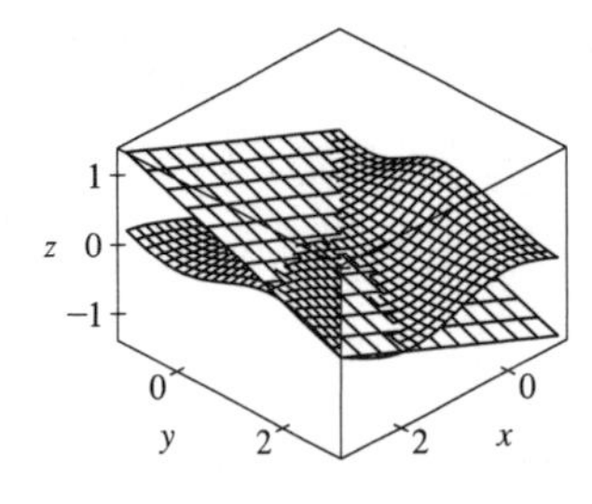

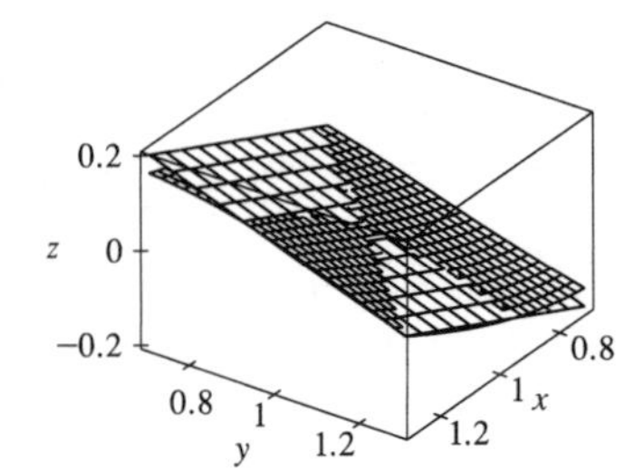

10. $f(x,y) = e^{-xy/10}\left(\sqrt{x} + \sqrt{y} + \sqrt{xy}\right)$. A CAS gives

$f_x(x,y) = -\frac{1}{10}ye^{-xy/10}\left(\sqrt{x} + \sqrt{y} + \sqrt{xy}\right) + e^{-xy/10}\left(\frac{1}{2\sqrt{x}} + \frac{y}{2\sqrt{xy}}\right)$ and

$f_y = -\frac{1}{10}xe^{-xy/10}\left(\sqrt{x} + \sqrt{y} + \sqrt{xy}\right) + e^{-xy/10}\left(\frac{1}{2\sqrt{y}} + \frac{x}{2\sqrt{xy}}\right)$. We use the CAS to evaluate these at $(1,1)$, and then substitute the results into Equation 2 to get an equation of the tangent plane: $z = 0.7e^{-0.1}x + 0.7e^{-0.1}y + 1.6e^{-0.1}$. The surface and tangent plane are shown in the first graph below. After zooming in, the surface and the tangent plane become almost indistinguishable, as shown in the second graph. (Here, the tangent plane is above the surface.) If we zoom in farther, the surface and the tangent plane will appear to coincide.

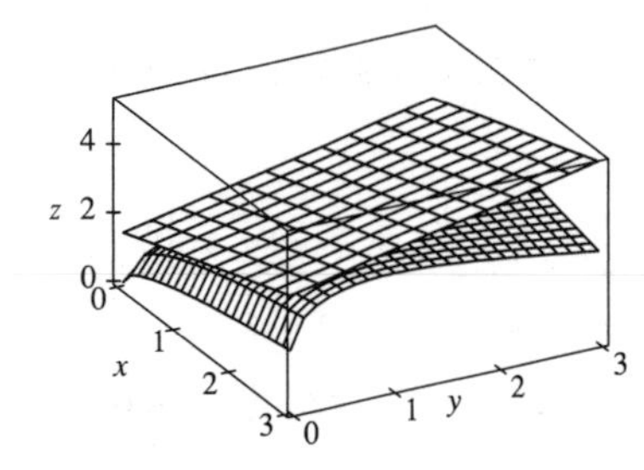

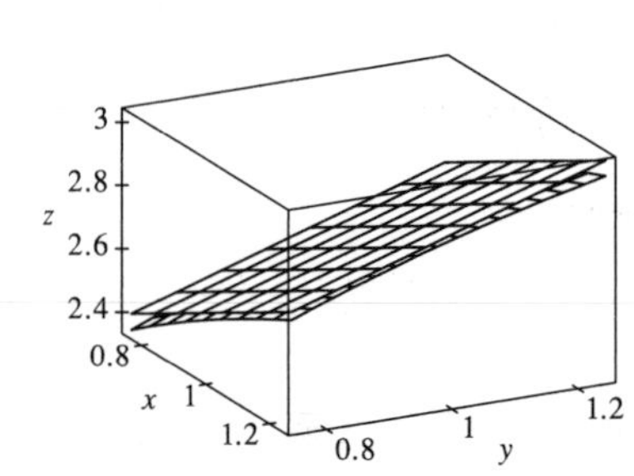

11. $f(x,y)=x\sqrt{y}$. The partial derivatives are $f_x(x,y)=\sqrt{y}$ and $f_y(x,y)=\dfrac{x}{2\sqrt{y}}$, so $f_x(1,4)=2$ and $f_y(1,4)=\frac{1}{4}$. Both f_x and f_y are continuous functions for $y>0$, so by Theorem 8, f is differentiable at $(1,4)$. By Equation 3, the linearization of f at $(1,4)$ is given by $L(x,y)=f(1,4)+f_x(1,4)(x-1)+f_y(1,4)(y-4)=2+2(x-1)+\frac{1}{4}(y-4)=2x+\frac{1}{4}y-1$.

12. $f(x,y)=\dfrac{x}{y}$. The partial derivatives are $f_x(x,y)=\dfrac{1}{y}$ and $f_y(x,y)=-\dfrac{x}{y^2}$, so $f_x(6,3)=\frac{1}{3}$ and $f_y(6,3)=-\frac{2}{3}$. Both f_x and f_y are continuous functions for $y\neq 0$, so f is differentiable at $(6,3)$ by Theorem 8. The linearization of f at $(6,3)$ is given by $L(x,y)=f(6,3)+f_x(6,3)(x-6)+f_y(6,3)(y-3)=2+\frac{1}{3}(x-6)-\frac{2}{3}(y-3)=\frac{1}{3}x-\frac{2}{3}y+2$.

13. $f(x,y)=\tan^{-1}(x+2y)$. The partial derivatives are $f_x(x,y)=\dfrac{1}{1+(x+2y)^2}$ and $f_y(x,y)=\dfrac{2}{1+(x+2y)^2}$, so $f_x(1,0)=\frac{1}{2}$ and $f_y(1,0)=1$. Both f_x and f_y are continuous functions, so f is differentiable at $(1,0)$, and the linearization of f at $(1,0)$ is $L(x,y)=f(1,0)+f_x(1,0)(x-1)+f_y(1,0)(y-0)=\frac{\pi}{4}+\frac{1}{2}(x-1)+1(y)=\frac{1}{2}x+y+\frac{\pi}{4}-\frac{1}{2}$.

14. $f(x,y)=\sqrt{x+e^{4y}}=(x+e^{4y})^{1/2}$. The partial derivatives are $f_x(x,y)=\frac{1}{2}(x+e^{4y})^{-1/2}$ and $f_y(x,y)=\frac{1}{2}(x+e^{4y})^{-1/2}(4e^{4y})=2e^{4y}(x+e^{4y})^{-1/2}$, so $f_x(3,0)=\frac{1}{2}(3+e^0)^{-1/2}=\frac{1}{4}$ and $f_y(3,0)=2e^0(3+e^0)^{-1/2}=1$. Both f_x and f_y are continuous functions near $(3,0)$, so f is differentiable at $(3,0)$ by Theorem 8. The linearization of f at $(3,0)$ is $L(x,y)=f(3,0)+f_x(3,0)(x-3)+f_y(3,0)(y-0)=2+\frac{1}{4}(x-3)+1(y-0)=\frac{1}{4}x+y+\frac{5}{4}$.

15. $f(x,y)=\sqrt{20-x^2-7y^2}\;\Rightarrow\; f_x(x,y)=-\dfrac{x}{\sqrt{20-x^2-7y^2}}$ and $f_y(x,y)=-\dfrac{7y}{\sqrt{20-x^2-7y^2}}$, so $f_x(2,1)=-\frac{2}{3}$ and $f_y(2,1)=-\frac{7}{3}$. Then the linear approximation of f at $(2,1)$ is given by $f(x,y)\approx f(2,1)+f_x(2,1)(x-2)+f_y(2,1)(y-1)=3-\frac{2}{3}(x-2)-\frac{7}{3}(y-1)=-\frac{2}{3}x-\frac{7}{3}y+\frac{20}{3}$. Thus $f(1.95,1.08)\approx-\frac{2}{3}(1.95)-\frac{7}{3}(1.08)+\frac{20}{3}=2.84\bar{6}$.

16. $f(x,y)=\ln(x-3y)\;\Rightarrow\; f_x(x,y)=\dfrac{1}{x-3y}$ and $f_y(x,y)=-\dfrac{3}{x-3y}$, so $f_x(7,2)=1$ and $f_y(7,2)=-3$.

Then the linear approximation of f at $(7,2)$ is given by

$$\begin{aligned} f(x,y)&\approx f(7,2)+f_x(7,2)(x-7)+f_y(7,2)(y-2)\\ &=0+1(x-7)-3(y-2)=x-3y-1\end{aligned}$$

Thus $f(6.9,2.06)\approx 6.9-3(2.06)-1=-0.28$. The graph shows that our approximated value is slightly greater than the actual value.

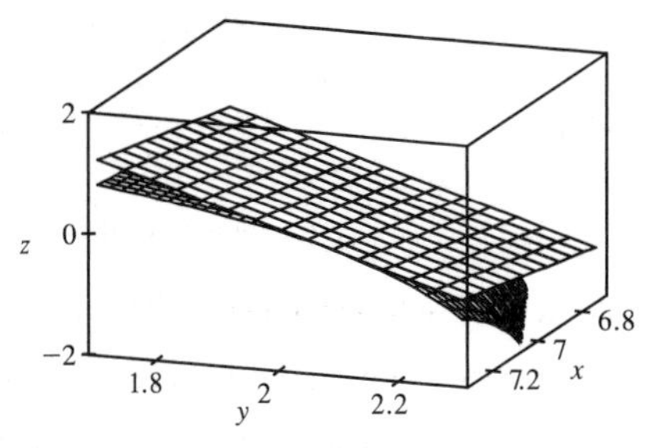

17. $f(x,y,z)=\sqrt{x^2+y^2+z^2}\;\Rightarrow\; f_x(x,y,z)=\dfrac{x}{\sqrt{x^2+y^2+z^2}}$, $f_y(x,y,z)=\dfrac{y}{\sqrt{x^2+y^2+z^2}}$, and $f_z(x,y,z)=\dfrac{z}{\sqrt{x^2+y^2+z^2}}$, so $f_x(3,2,6)=\frac{3}{7}$, $f_y(3,2,6)=\frac{2}{7}$, $f_z(3,2,6)=\frac{6}{7}$. Then the linear approximation of f at $(3,2,6)$ is given by

$$\begin{aligned} f(x,y,z)&\approx f(3,2,6)+f_x(3,2,6)(x-3)+f_y(3,2,6)(y-2)+f_z(3,2,6)(z-6)\\ &=7+\tfrac{3}{7}(x-3)+\tfrac{2}{7}(y-2)+\tfrac{6}{7}(z-6)=\tfrac{3}{7}x+\tfrac{2}{7}y+\tfrac{6}{7}z\end{aligned}$$

Thus $\sqrt{(3.02)^2+(1.97)^2+(5.99)^2}=f(3.02,1.97,5.99)\approx\frac{3}{7}(3.02)+\frac{2}{7}(1.97)+\frac{6}{7}(5.99)\approx 6.9914$.

18. $v = y\cos xy \;\Rightarrow$

$$dv = \frac{\partial v}{\partial x}dx + \frac{\partial v}{\partial y}dy = y(-\sin xy)y\,dx + [y(-\sin xy)x + \cos xy]\,dy = -y^2\sin xy\,dx + (\cos xy - xy\sin xy)\,dy$$

19. $z = x^3\ln(y^2) \;\Rightarrow\; dz = \dfrac{\partial z}{\partial x}dx + \dfrac{\partial z}{\partial y}dy = 3x^2\ln(y^2)\,dx + x^3\cdot\dfrac{1}{y^2}(2y)\,dy = 3x^2\ln(y^2)\,dx + \dfrac{2x^3}{y}dy$

20. $u = e^{-t}\sin(s+2t) \;\Rightarrow$

$$\begin{aligned} du &= \frac{\partial u}{\partial s}\,ds + \frac{\partial u}{\partial t}\,dt = e^{-t}\cos(s+2t)\,ds + \left[e^{-t}\cos(s+2t)\cdot 2 + \left(-e^{-t}\right)\sin(s+2t)\right]dt \\ &= e^{-t}\cos(s+2t)\,ds + e^{-t}\left[2\cos(s+2t) - \sin(s+2t)\right]dt \end{aligned}$$

21. $R = \alpha\beta^2\cos\gamma \;\Rightarrow\; dR = \dfrac{\partial R}{\partial\alpha}\,d\alpha + \dfrac{\partial R}{\partial\beta}\,d\beta + \dfrac{\partial R}{\partial\gamma}\,d\gamma = \beta^2\cos\gamma\,d\alpha + 2\alpha\beta\cos\gamma\,d\beta - \alpha\beta^2\sin\gamma\,d\gamma$

22. $w = xye^{xz} \;\Rightarrow$

$$\begin{aligned} dw &= \frac{\partial w}{\partial x}\,dx + \frac{\partial w}{\partial y}\,dy + \frac{\partial w}{\partial z}\,dz = (xyze^{xz} + ye^{xz})\,dx + xe^{xz}\,dy + x^2ye^{xz}\,dz \\ &= (xz+1)ye^{xz}\,dx + xe^{xz}\,dy + x^2ye^{xz}\,dz \end{aligned}$$

23. $dx = \Delta x = 0.05$, $dy = \Delta y = 0.1$, $z = 5x^2 + y^2$, $z_x = 10x$, $z_y = 2y$. Thus when $x = 1$ and $y = 2$,

$dz = z_x(1,2)\,dx + z_y(1,2)\,dy = (10)(0.05) + (4)(0.1) = 0.9$ while

$\Delta z = f(1.05, 2.1) - f(1,2) = 5(1.05)^2 + (2.1)^2 - 5 - 4 = 0.9225$.

24. $dx = \Delta x = -0.04$, $dy = \Delta y = 0.05$, $z = x^2 - xy + 3y^2$, $z_x = 2x - y$, $z_y = 6y - x$. Thus when $x = 3$ and $y = -1$,

$dz = (7)(-0.04) + (-9)(0.05) = -0.73$ while $\Delta z = (2.96)^2 - (2.96)(-0.95) + 3(-0.95)^2 - (9+3+3) = -0.7189$.

25. $dA = \dfrac{\partial A}{\partial x}\,dx + \dfrac{\partial A}{\partial y}\,dy = y\,dx + x\,dy$ and $|\Delta x| \le 0.1$, $|\Delta y| \le 0.1$. We use $dx = 0.1$, $dy = 0.1$ with $x = 30$, $y = 24$; then the maximum error in the area is about $dA = 24(0.1) + 30(0.1) = 5.4\text{ cm}^2$.

26. Let S be surface area. Then $S = 2(xy + xz + yz)$ and $dS = 2(y+z)\,dx + 2(x+z)\,dy + 2(x+y)\,dz$. The maximum error occurs with $\Delta x = \Delta y = \Delta z = 0.2$. Using $dx = \Delta x$, $dy = \Delta y$, $dz = \Delta z$ we find the maximum error in calculated surface area to be about $dS = (220)(0.2) + (260)(0.2) + (280)(0.2) = 152\text{ cm}^2$.

27. The volume of a can is $V = \pi r^2 h$ and $\Delta V \approx dV$ is an estimate of the amount of tin. Here $dV = 2\pi rh\,dr + \pi r^2\,dh$, so put $dr = 0.04$, $dh = 0.08$ (0.04 on top, 0.04 on bottom) and then $\Delta V \approx dV = 2\pi(48)(0.04) + \pi(16)(0.08) \approx 16.08\text{ cm}^3$. Thus the amount of tin is about 16 cm^3.

28. Let V be the volume. Then $V = \pi r^2 h$ and $\Delta V \approx dV = 2\pi rh\,dr + \pi r^2\,dh$ is an estimate of the amount of metal. With $dr = 0.05$ and $dh = 0.2$ we get $dV = 2\pi(2)(10)(0.05) + \pi(2)^2(0.2) = 2.80\pi \approx 8.8\text{ cm}^3$.

29. The errors in measurement are at most 2%, so $\left|\dfrac{\Delta w}{w}\right| \le 0.02$ and $\left|\dfrac{\Delta h}{h}\right| \le 0.02$. The relative error in the calculated surface area is

$$\frac{\Delta S}{S} \approx \frac{dS}{S} = \frac{0.1091(0.425w^{0.425-1})h^{0.725}\,dw + 0.1091w^{0.425}(0.725h^{0.725-1})\,dh}{0.1091w^{0.425}h^{0.725}} = 0.425\frac{dw}{w} + 0.725\frac{dh}{h}$$

To estimate the maximum relative error, we use $\dfrac{dw}{w} = \left|\dfrac{\Delta w}{w}\right| = 0.02$ and $\dfrac{dh}{h} = \left|\dfrac{\Delta h}{h}\right| = 0.02 \quad\Rightarrow$

$\dfrac{dS}{S} = 0.425\,(0.02) + 0.725\,(0.02) = 0.023$. Thus the maximum percentage error is approximately 2.3%.

30. Here $dV = \Delta V = 0.3$, $dT = \Delta T = -5$, $P = 8.31\,\dfrac{T}{V}$, so

$dP = \left(\dfrac{8.31}{V}\right) dT - \dfrac{8.31 \cdot T}{V^2}\, dV = 8.31\left[-\dfrac{5}{12} - \dfrac{310}{144} \cdot \dfrac{3}{10}\right] \approx -8.83$. Thus the pressure will drop by about 8.83 kPa.

31. First we find $\dfrac{\partial R}{\partial R_1}$ implicitly by taking partial derivatives of both sides with respect to R_1:

$\dfrac{\partial}{\partial R_1}\left(\dfrac{1}{R}\right) = \dfrac{\partial\left[(1/R_1) + (1/R_2) + (1/R_3)\right]}{\partial R_1} \quad\Rightarrow\quad -R^{-2}\dfrac{\partial R}{\partial R_1} = -R_1^{-2} \quad\Rightarrow\quad \dfrac{\partial R}{\partial R_1} = \dfrac{R^2}{R_1^2}$. Then by symmetry,

$\dfrac{\partial R}{\partial R_2} = \dfrac{R^2}{R_2^2}$, $\dfrac{\partial R}{\partial R_3} = \dfrac{R^2}{R_3^2}$. When $R_1 = 25$, $R_2 = 40$ and $R_3 = 50$, $\dfrac{1}{R} = \dfrac{17}{200} \quad\Leftrightarrow\quad R = \frac{200}{17}\ \Omega$. Since the possible error for each R_i is 0.5%, the maximum error of R is attained by setting $\Delta R_i = 0.005R_i$. So

$\Delta R \approx dR = \dfrac{\partial R}{\partial R_1}\Delta R_1 + \dfrac{\partial R}{\partial R_2}\Delta R_2 + \dfrac{\partial R}{\partial R_3}\Delta R_3 = (0.005)R^2\left(\dfrac{1}{R_1} + \dfrac{1}{R_2} + \dfrac{1}{R_3}\right) = (0.005)R = \frac{1}{17} \approx 0.059\ \Omega$.

32. $\mathbf{r}_1(t) = \langle 2 + 3t, 1 - t^2, 3 - 4t + t^2\rangle \quad\Rightarrow\quad \mathbf{r}_1'(t) = \langle 3, -2t, -4 + 2t\rangle$, $\mathbf{r}_2(u) = \langle 1 + u^2, 2u^3 - 1, 2u + 1\rangle \quad\Rightarrow$

$\mathbf{r}_2'(u) = \langle 2u, 6u^2, 2\rangle$. Both curves pass through P since $\mathbf{r}_1(0) = \mathbf{r}_2(1) = \langle 2, 1, 3\rangle$, so the tangent vectors $\mathbf{r}_1'(0) = \langle 3, 0, -4\rangle$ and $\mathbf{r}_2'(1) = \langle 2, 6, 2\rangle$ are both parallel to the tangent plane to S at P. A normal vector for the tangent plane is

$\mathbf{r}_1'(0) \times \mathbf{r}_2'(1) = \langle 3, 0, -4\rangle \times \langle 2, 6, 2\rangle = \langle 24, -14, 18\rangle$, so an equation of the tangent plane is

$24(x - 2) - 14(y - 1) + 18(z - 3) = 0$ or $12x - 7y + 9z = 44$.

33.
$$\begin{aligned}\Delta z &= f(a + \Delta x, b + \Delta y) - f(a, b) = (a + \Delta x)^2 + (b + \Delta y)^2 - (a^2 + b^2)\\ &= a^2 + 2a\,\Delta x + (\Delta x)^2 + b^2 + 2b\,\Delta y + (\Delta y)^2 - a^2 - b^2 = 2a\,\Delta x + (\Delta x)^2 + 2b\,\Delta y + (\Delta y)^2\end{aligned}$$

But $f_x(a, b) = 2a$ and $f_y(a, b) = 2b$ and so $\Delta z = f_x(a, b)\,\Delta x + f_y(a, b)\,\Delta y + \Delta x\,\Delta x + \Delta y\,\Delta y$, which is Definition 7 with $\varepsilon_1 = \Delta x$ and $\varepsilon_2 = \Delta y$. Hence f is differentiable.

34.
$$\begin{aligned}\Delta z &= f(a + \Delta x, b + \Delta y) - f(a, b) = (a + \Delta x)(b + \Delta y) - 5(b + \Delta y)^2 - (ab - 5b^2)\\ &= ab + a\,\Delta y + b\,\Delta x + \Delta x\,\Delta y - 5b^2 - 10b\,\Delta y - 5(\Delta y)^2 - ab + 5b^2\\ &= (a - 10b)\,\Delta y + b\,\Delta x + \Delta x\,\Delta y - 5\,\Delta y\,\Delta y,\end{aligned}$$

but $f_x(a, b) = b$ and $f_y(a, b) = a - 10b$ and so $\Delta z = f_x(a, b)\,\Delta x + f_y(a, b)\,\Delta y + \Delta x\,\Delta y - 5\Delta y\,\Delta y$, which is Definition 7 with $\varepsilon_1 = \Delta y$ and $\varepsilon_2 = -5\,\Delta y$. Hence f is differentiable.

35. To show that f is continuous at (a, b) we need to show that $\lim_{(x,y)\to(a,b)} f(x, y) = f(a, b)$ or equivalently $\lim_{(\Delta x,\Delta y)\to(0,0)} f(a + \Delta x, b + \Delta y) = f(a, b)$. Since f is differentiable at (a, b), $f(a + \Delta x, b + \Delta y) - f(a, b) = \Delta z = f_x(a, b)\,\Delta x + f_y(a, b)\,\Delta y + \varepsilon_1\,\Delta x + \varepsilon_2\,\Delta y$, where ϵ_1 and $\epsilon_2 \to 0$ as $(\Delta x, \Delta y) \to (0, 0)$. Thus $f(a + \Delta x, b + \Delta y) = f(a, b) + f_x(a, b)\,\Delta x + f_y(a, b)\,\Delta y + \varepsilon_1\,\Delta x + \varepsilon_2\,\Delta y$. Taking the limit of both sides as $(\Delta x, \Delta y) \to (0, 0)$ gives $\lim_{(\Delta x,\Delta y)\to(0,0)} f(a + \Delta x, b + \Delta y) = f(a, b)$. Thus f is continuous at (a, b).

36. (a) $\lim_{h\to 0} \dfrac{f(h, 0) - f(0, 0)}{h} = \lim_{h\to 0} \dfrac{0 - 0}{h} = 0$ and $\lim_{h\to 0} \dfrac{f(0, h) - f(0, 0)}{h} = \lim_{h\to 0} \dfrac{0 - 0}{h} = 0$. Thus $f_x(0, 0) = f_y(0, 0) = 0$.

To show that f isn't differentiable at $(0, 0)$ we need only show that f is not continuous at $(0, 0)$ and apply Exercise 35. As $(x, y) \to (0, 0)$ along the x-axis $f(x, y) = 0/x^2 = 0$ for $x \neq 0$ so $f(x, y) \to 0$ as $(x, y) \to (0, 0)$ along the x-axis. But as $(x, y) \to (0, 0)$ along the line $y = x$, $f(x, x) = x^2/(2x^2) = \frac{1}{2}$ for $x \neq 0$ so $f(x, y) \to \frac{1}{2}$ as $(x, y) \to (0, 0)$ along this line. Thus $\lim_{(x,y)\to(0,0)} f(x, y)$ doesn't exist, so f is discontinuous at $(0, 0)$ and thus not differentiable there.

(b) For $(x, y) \neq (0, 0)$, $f_x(x, y) = \dfrac{(x^2 + y^2)y - xy(2x)}{(x^2 + y^2)^2} = \dfrac{y(y^2 - x^2)}{(x^2 + y^2)^2}$. If we approach $(0, 0)$ along the y-axis, then $f_x(x, y) = f_x(0, y) = \dfrac{y^3}{y^4} = \dfrac{1}{y}$, so $f_x(x, y) \to \pm\infty$ as $(x, y) \to (0, 0)$. Thus $\lim_{(x,y)\to(0,0)} f_x(x, y)$ does not exist and $f_x(x, y)$ is not continuous at $(0, 0)$. Similarly, $f_y(x, y) = \dfrac{(x^2 + y^2)x - xy(2y)}{(x^2 + y^2)^2} = \dfrac{x(x^2 - y^2)}{(x^2 + y^2)^2}$ for $(x, y) \neq (0, 0)$, and if we approach $(0, 0)$ along the x-axis, then $f_y(x, y) = f_x(x, 0) = \dfrac{x^3}{x^4} = \dfrac{1}{x}$. Thus $\lim_{(x,y)\to(0,0)} f_y(x, y)$ does not exist and $f_y(x, y)$ is not continuous at $(0, 0)$.

11.5 The Chain Rule

1. $z = \sin x \cos y,\ x = \pi t,\ y = \sqrt{t} \ \Rightarrow$

$$\frac{dz}{dt} = \frac{\partial z}{\partial x}\frac{dx}{dt} + \frac{\partial z}{\partial y}\frac{dy}{dt} = \cos x\,\cos y \cdot \pi + \sin x\,(-\sin y) \cdot \tfrac{1}{2}t^{-1/2} = \pi \cos x\,\cos y - \frac{1}{2\sqrt{t}} \sin x\,\sin y$$

2. $z = x\ln(x + 2y),\ x = \sin t,\ y = \cos t \ \Rightarrow$

$$\frac{dz}{dt} = \frac{\partial z}{\partial x}\frac{dx}{dt} + \frac{\partial z}{\partial y}\frac{dy}{dt} = \left[x \cdot \frac{1}{x + 2y} + 1 \cdot \ln(x + 2y)\right]\cos t + x \cdot \frac{1}{x + 2y}\,(2) \cdot (-\sin t)$$

$$= \left[\frac{x}{x + 2y} + \ln(x + 2y)\right]\cos t - \frac{2x}{x + 2y}\,(\sin t)$$

3. $w = xe^{y/z},\ x = t^2,\ y = 1 - t,\ z = 1 + 2t \ \Rightarrow$

$$\frac{dw}{dt} = \frac{\partial w}{\partial x}\frac{dx}{dt} + \frac{\partial w}{\partial y}\frac{dy}{dt} + \frac{\partial w}{\partial z}\frac{dz}{dt} = e^{y/z} \cdot 2t + xe^{y/z}\left(\frac{1}{z}\right) \cdot (-1) + xe^{y/z}\left(-\frac{y}{z^2}\right) \cdot 2 = e^{y/z}\left(2t - \frac{x}{z} - \frac{2xy}{z^2}\right)$$

4. $w = xy + yz^2,\ x = e^t,\ y = e^t \sin t,\ z = e^t \cos t \ \Rightarrow$

$$\frac{dw}{dt} = \frac{\partial w}{\partial x}\frac{dx}{dt} + \frac{\partial w}{\partial y}\frac{dy}{dt} + \frac{\partial w}{\partial z}\frac{dz}{dt} = y \cdot e^t + (x + z^2) \cdot (e^t \cos t + e^t \sin t) + 2yz \cdot (-e^t \sin t + e^t \cos t)$$
$$= e^t\left[y + (x + z^2)(\cos t + \sin t) + 2yz(\cos t - \sin t)\right]$$

5. $z = x^2 + xy + y^2,\ x = s + t,\ y = st \ \Rightarrow$

$$\frac{\partial z}{\partial s} = \frac{\partial z}{\partial x}\frac{\partial x}{\partial s} + \frac{\partial z}{\partial y}\frac{\partial y}{\partial s} = (2x + y)(1) + (x + 2y)(t) = 2x + y + xt + 2yt$$

$$\frac{\partial z}{\partial t} = \frac{\partial z}{\partial x}\frac{\partial x}{\partial t} + \frac{\partial z}{\partial y}\frac{\partial y}{\partial t} = (2x + y)(1) + (x + 2y)(s) = 2x + y + xs + 2ys$$

6. $z = x/y,\ x = se^t,\ y = 1 + se^{-t} \ \Rightarrow$

$$\frac{\partial z}{\partial s} = \frac{\partial z}{\partial x}\frac{\partial x}{\partial s} + \frac{\partial z}{\partial y}\frac{\partial y}{\partial s} = \frac{1}{y}(e^t) + \left(-\frac{x}{y^2}\right)(e^{-t}) = \frac{1}{y}\,e^t - \frac{x}{y^2}\,e^{-t}$$

$$\frac{\partial z}{\partial t} = \frac{\partial z}{\partial x}\frac{\partial x}{\partial t} + \frac{\partial z}{\partial y}\frac{\partial y}{\partial t} = \frac{1}{y}(se^t) + \left(-\frac{x}{y^2}\right)(-se^{-t}) = \frac{s}{y}\,e^t + \frac{xs}{y^2}\,e^{-t}$$

7. $z = e^r \cos\theta,\ r = st,\ \theta = \sqrt{s^2 + t^2} \ \Rightarrow$

$$\frac{\partial z}{\partial s} = \frac{\partial z}{\partial r}\frac{\partial r}{\partial s} + \frac{\partial z}{\partial \theta}\frac{\partial \theta}{\partial s} = e^r \cos\theta \cdot t + e^r(-\sin\theta) \cdot \tfrac{1}{2}(s^2 + t^2)^{-1/2}(2s) = te^r \cos\theta - e^r \sin\theta \cdot \frac{s}{\sqrt{s^2 + t^2}}$$
$$= e^r\left(t\cos\theta - \frac{s}{\sqrt{s^2 + t^2}}\sin\theta\right)$$

$$\frac{\partial z}{\partial t} = \frac{\partial z}{\partial r}\frac{\partial r}{\partial t} + \frac{\partial z}{\partial \theta}\frac{\partial \theta}{\partial t} = e^r \cos\theta \cdot s + e^r(-\sin\theta) \cdot \tfrac{1}{2}(s^2 + t^2)^{-1/2}(2t) = se^r \cos\theta - e^r \sin\theta \cdot \frac{t}{\sqrt{s^2 + t^2}}$$
$$= e^r\left(s\cos\theta - \frac{t}{\sqrt{s^2 + t^2}}\sin\theta\right)$$

8. $z = \sin\alpha \tan\beta,\ \alpha = 3s + t,\ \beta = s - t \ \Rightarrow$

$$\frac{\partial z}{\partial s} = \frac{\partial z}{\partial \alpha}\frac{\partial \alpha}{\partial s} + \frac{\partial z}{\partial \beta}\frac{\partial \beta}{\partial s} = \cos\alpha\,\tan\beta \cdot 3 + \sin\alpha\,\sec^2\beta \cdot 1 = 3\cos\alpha\,\tan\beta + \sin\alpha\,\sec^2\beta$$

$$\frac{\partial z}{\partial t} = \frac{\partial z}{\partial \alpha}\frac{\partial \alpha}{\partial t} + \frac{\partial z}{\partial \beta}\frac{\partial \beta}{\partial t} = \cos\alpha\,\tan\beta \cdot 1 + \sin\alpha\,\sec^2\beta \cdot (-1) = \cos\alpha\,\tan\beta - \sin\alpha\,\sec^2\beta$$

9. When $t = 3$, $x = g(3) = 2$ and $y = h(3) = 7$. By the Chain Rule (2),

$$\frac{dz}{dt} = \frac{\partial f}{\partial x}\frac{dx}{dt} + \frac{\partial f}{\partial y}\frac{dy}{dt} = f_x(2,7)g'(3) + f_y(2,7)\,h'(3) = (6)(5) + (-8)(-4) = 62.$$

10. By the Chain Rule (3), $\dfrac{\partial W}{\partial s} = \dfrac{\partial W}{\partial u}\dfrac{\partial u}{\partial s} + \dfrac{\partial W}{\partial v}\dfrac{\partial v}{\partial s}$. Then

$$W_s(1,0) = F_u(u(1,0), v(1,0))\,u_s(1,0) + F_v(u(1,0), v(1,0))\,v_s(1,0) = F_u(2,3)u_s(1,0) + F_v(2,3)v_s(1,0)$$
$$= (-1)(-2) + (10)(5) = 52$$

Similarly, $\dfrac{\partial W}{\partial t} = \dfrac{\partial W}{\partial u}\dfrac{\partial u}{\partial t} + \dfrac{\partial W}{\partial v}\dfrac{\partial v}{\partial t} \ \Rightarrow$

$$W_t(1,0) = F_u(u(1,0), v(1,0))\,u_t(1,0) + F_v(u(1,0), v(1,0))\,v_t(1,0) = F_u(2,3)u_t(1,0) + F_v(2,3)v_t(1,0)$$
$$= (-1)(6) + (10)(4) = 34$$

11. $g(u,v) = f(x(u,v), y(u,v))$ where $x = e^u + \sin v,\ y = e^u + \cos v \Rightarrow$

$\frac{\partial x}{\partial u} = e^u,\ \frac{\partial x}{\partial v} = \cos v,\ \frac{\partial y}{\partial u} = e^u,\ \frac{\partial y}{\partial v} = -\sin v$. By the Chain Rule (3), $\frac{\partial g}{\partial u} = \frac{\partial f}{\partial x}\frac{\partial x}{\partial u} + \frac{\partial f}{\partial y}\frac{\partial y}{\partial u}$. Then

$$g_u(0,0) = f_x(x(0,0), y(0,0))\, x_u(0,0) + f_y(x(0,0), y(0,0))\, y_u(0,0) = f_x(1,2)(e^0) + f_y(1,2)(e^0) = 2(1) + 5(1) = 7.$$

Similarly, $\frac{\partial g}{\partial v} = \frac{\partial f}{\partial x}\frac{\partial x}{\partial v} + \frac{\partial f}{\partial y}\frac{\partial y}{\partial v}$. Then

$$\begin{aligned} g_v(0,0) &= f_x(x(0,0), y(0,0))\, x_v(0,0) + f_y(x(0,0), y(0,0))\, y_v(0,0) = f_x(1,2)(\cos 0) + f_y(1,2)(-\sin 0) \\ &= 2(1) + 5(0) = 2 \end{aligned}$$

12. $g(r,s) = f(x(r,s), y(r,s))$ where $x = 2r - s,\ y = s^2 - 4r \Rightarrow \frac{\partial x}{\partial r} = 2,\ \frac{\partial x}{\partial s} = -1,\ \frac{\partial y}{\partial r} = -4,\ \frac{\partial y}{\partial s} = 2s.$

By the Chain Rule (3) $\frac{\partial g}{\partial r} = \frac{\partial f}{\partial x}\frac{\partial x}{\partial r} + \frac{\partial f}{\partial y}\frac{\partial y}{\partial r}$. Then

$$\begin{aligned} g_r(1,2) &= f_x(x(1,2), y(1,2))\, x_r(1,2) + f_y(x(1,2), y(1,2))\, y_r(1,2) = f_x(0,0)(2) + f_y(0,0)(-4) \\ &= 4(2) + 8(-4) = -24 \end{aligned}$$

Similarly, $\frac{\partial g}{\partial s} = \frac{\partial f}{\partial x}\frac{\partial x}{\partial s} + \frac{\partial f}{\partial y}\frac{\partial y}{\partial s}$. Then

$$\begin{aligned} g_s(1,2) &= f_x(x(1,2), y(1,2))\, x_s(1,2) + f_y(x(1,2), y(1,2))\, y_s(1,2) = f_x(0,0)(-1) + f_y(0,0)(4) \\ &= 4(-1) + 8(4) = 28 \end{aligned}$$

13.

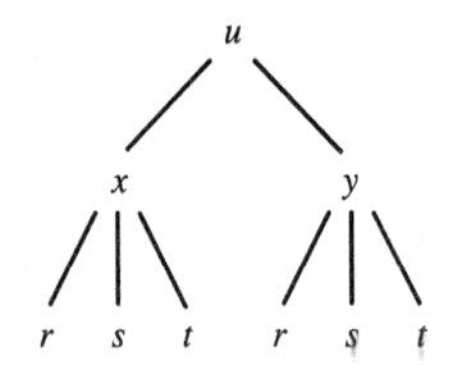

$u = f(x,y),\ x = x(r,s,t),\ y = y(r,s,t) \Rightarrow$

$$\frac{\partial u}{\partial r} = \frac{\partial u}{\partial x}\frac{\partial x}{\partial r} + \frac{\partial u}{\partial y}\frac{\partial y}{\partial r},\ \frac{\partial u}{\partial s} = \frac{\partial u}{\partial x}\frac{\partial x}{\partial s} + \frac{\partial u}{\partial y}\frac{\partial y}{\partial s},$$

$$\frac{\partial u}{\partial t} = \frac{\partial u}{\partial x}\frac{\partial x}{\partial t} + \frac{\partial u}{\partial y}\frac{\partial y}{\partial t}$$

14.

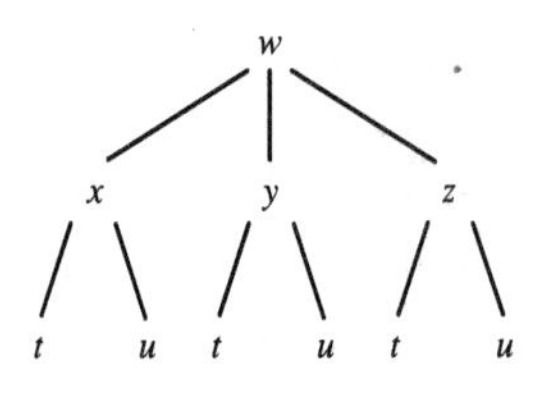

$w = f(x,y,z),\ x = x(t,u),\ y = y(t,u),\ z = z(t,u) \Rightarrow$

$$\frac{\partial w}{\partial t} = \frac{\partial w}{\partial x}\frac{\partial x}{\partial t} + \frac{\partial w}{\partial y}\frac{\partial y}{\partial t} + \frac{\partial w}{\partial z}\frac{\partial z}{\partial t},$$

$$\frac{\partial w}{\partial u} = \frac{\partial w}{\partial x}\frac{\partial x}{\partial u} + \frac{\partial w}{\partial y}\frac{\partial y}{\partial u} + \frac{\partial w}{\partial z}\frac{\partial z}{\partial u}$$

15.

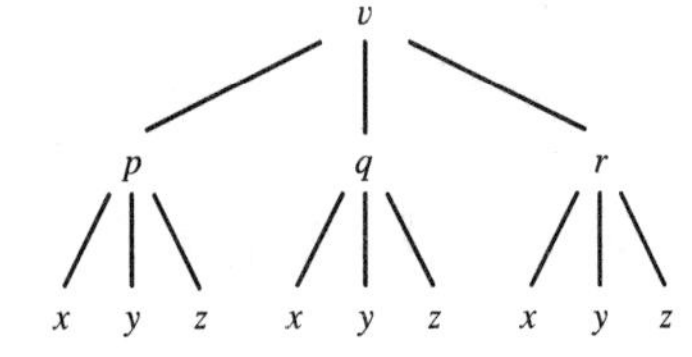

$v = f(p,q,r),\ p = p(x,y,z),\ q = q(x,y,z),\ r = r(x,y,z) \Rightarrow$

$$\frac{\partial v}{\partial x} = \frac{\partial v}{\partial p}\frac{\partial p}{\partial x} + \frac{\partial v}{\partial q}\frac{\partial q}{\partial x} + \frac{\partial v}{\partial r}\frac{\partial r}{\partial x},\ \frac{\partial v}{\partial y} = \frac{\partial v}{\partial p}\frac{\partial p}{\partial y} + \frac{\partial v}{\partial q}\frac{\partial q}{\partial y} + \frac{\partial v}{\partial r}\frac{\partial r}{\partial y},$$

$$\frac{\partial v}{\partial z} = \frac{\partial v}{\partial p}\frac{\partial p}{\partial z} + \frac{\partial v}{\partial q}\frac{\partial q}{\partial z} + \frac{\partial v}{\partial r}\frac{\partial r}{\partial z}$$

16.

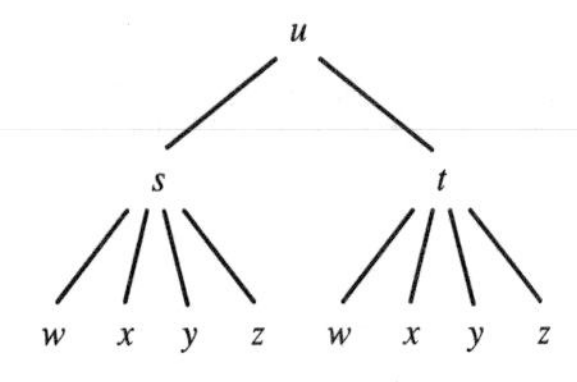

$u = f(s,t),\ s = s(w,x,y,z),\ t = t(w,x,y,z) \Rightarrow$

$$\frac{\partial u}{\partial w} = \frac{\partial u}{\partial s}\frac{\partial s}{\partial w} + \frac{\partial u}{\partial t}\frac{\partial t}{\partial w},\ \frac{\partial u}{\partial x} = \frac{\partial u}{\partial s}\frac{\partial s}{\partial x} + \frac{\partial u}{\partial t}\frac{\partial t}{\partial x},$$

$$\frac{\partial u}{\partial y} = \frac{\partial u}{\partial s}\frac{\partial s}{\partial y} + \frac{\partial u}{\partial t}\frac{\partial t}{\partial y},\ \frac{\partial u}{\partial z} = \frac{\partial u}{\partial s}\frac{\partial s}{\partial z} + \frac{\partial u}{\partial t}\frac{\partial t}{\partial z}$$

17. $z = x^2 + xy^3,\ x = uv^2 + w^3,\ y = u + ve^w \ \Rightarrow\ \dfrac{\partial z}{\partial u} = \dfrac{\partial z}{\partial x}\dfrac{\partial x}{\partial u} + \dfrac{\partial z}{\partial y}\dfrac{\partial y}{\partial u} = (2x + y^3)(v^2) + (3xy^2)(1),$

$\dfrac{\partial z}{\partial v} = \dfrac{\partial z}{\partial x}\dfrac{\partial x}{\partial v} + \dfrac{\partial z}{\partial y}\dfrac{\partial y}{\partial v} = (2x + y^3)(2uv) + (3xy^2)(e^w),\ \dfrac{\partial z}{\partial w} = \dfrac{\partial z}{\partial x}\dfrac{\partial x}{\partial w} + \dfrac{\partial z}{\partial y}\dfrac{\partial y}{\partial w} = (2x + y^3)(3w^2) + (3xy^2)(ve^w).$

When $u = 2$, $v = 1$, and $w = 0$, we have $x = 2$, $y = 3$, so $\dfrac{\partial z}{\partial u} = (31)(1) + (54)(1) = 85,$

$\dfrac{\partial z}{\partial v} = (31)(4) + (54)(1) = 178,\ \dfrac{\partial z}{\partial w} = (31)(0) + (54)(1) = 54.$

18. $u = (r^2 + s^2)^{1/2},\ r = y + x\cos t,\ s = x + y\sin t \ \Rightarrow$

$\dfrac{\partial u}{\partial x} = \dfrac{\partial u}{\partial r}\dfrac{\partial r}{\partial x} + \dfrac{\partial u}{\partial s}\dfrac{\partial s}{\partial x} = \frac{1}{2}(r^2 + s^2)^{-1/2}(2r)(\cos t) + \frac{1}{2}(r^2 + s^2)^{-1/2}(2s)(1) = (r\cos t + s)/\sqrt{r^2 + s^2},$

$\dfrac{\partial u}{\partial y} = \dfrac{\partial u}{\partial r}\dfrac{\partial r}{\partial y} + \dfrac{\partial u}{\partial s}\dfrac{\partial s}{\partial y} = \frac{1}{2}(r^2 + s^2)^{-1/2}(2r)(1) + \frac{1}{2}(r^2 + s^2)^{-1/2}(2s)(\sin t) = (r + s\sin t)/\sqrt{r^2 + s^2},$

$\dfrac{\partial u}{\partial t} = \dfrac{\partial u}{\partial r}\dfrac{\partial r}{\partial t} + \dfrac{\partial u}{\partial s}\dfrac{\partial s}{\partial t} = \frac{1}{2}(r^2 + s^2)^{-1/2}(2r)(-x\sin t) + \frac{1}{2}(r^2 + s^2)^{-1/2}(2s)(y\cos t) = \dfrac{-rx\sin t + sy\cos t}{\sqrt{r^2 + s^2}}.$

When $x = 1$, $y = 2$, and $t = 0$ we have $r = 3$ and $s = 1$, so $\dfrac{\partial u}{\partial x} = \dfrac{4}{\sqrt{10}},\ \dfrac{\partial u}{\partial y} = \dfrac{3}{\sqrt{10}},$ and $\dfrac{\partial u}{\partial t} = \dfrac{2}{\sqrt{10}}.$

19. $R = \ln(u^2 + v^2 + w^2),\ u = x + 2y,\ v = 2x - y,\ w = 2xy \ \Rightarrow$

$$\begin{aligned}\frac{\partial R}{\partial x} &= \frac{\partial R}{\partial u}\frac{\partial u}{\partial x} + \frac{\partial R}{\partial v}\frac{\partial v}{\partial x} + \frac{\partial R}{\partial w}\frac{\partial w}{\partial x} = \frac{2u}{u^2 + v^2 + w^2}(1) + \frac{2v}{u^2 + v^2 + w^2}(2) + \frac{2w}{u^2 + v^2 + w^2}(2y)\\ &= \frac{2u + 4v + 4wy}{u^2 + v^2 + w^2},\end{aligned}$$

$$\begin{aligned}\frac{\partial R}{\partial y} &= \frac{\partial R}{\partial u}\frac{\partial u}{\partial y} + \frac{\partial R}{\partial v}\frac{\partial v}{\partial y} + \frac{\partial R}{\partial w}\frac{\partial w}{\partial y} = \frac{2u}{u^2 + v^2 + w^2}(2) + \frac{2v}{u^2 + v^2 + w^2}(-1) + \frac{2w}{u^2 + v^2 + w^2}(2x)\\ &= \frac{4u - 2v + 4wx}{u^2 + v^2 + w^2}.\end{aligned}$$

When $x = y = 1$ we have $u = 3$, $v = 1$, and $w = 2$, so $\dfrac{\partial R}{\partial x} = \dfrac{9}{7}$ and $\dfrac{\partial R}{\partial y} = \dfrac{9}{7}.$

20. $M = xe^{y - z^2},\ x = 2uv,\ y = u - v,\ z = u + v \ \Rightarrow$

$\dfrac{\partial M}{\partial u} = \dfrac{\partial M}{\partial x}\dfrac{\partial x}{\partial u} + \dfrac{\partial M}{\partial y}\dfrac{\partial y}{\partial u} + \dfrac{\partial M}{\partial z}\dfrac{\partial z}{\partial u} = e^{y-z^2}(2v) + xe^{y-z^2}(1) + x(-2z)e^{y-z^2}(1) = e^{y-z^2}(2v + x - 2xz),$

$\dfrac{\partial M}{\partial v} = \dfrac{\partial M}{\partial x}\dfrac{\partial x}{\partial v} + \dfrac{\partial M}{\partial y}\dfrac{\partial y}{\partial v} + \dfrac{\partial M}{\partial z}\dfrac{\partial z}{\partial v} = e^{y-z^2}(2u) + xe^{y-z^2}(-1) + x(-2z)e^{y-z^2}(1) = e^{y-z^2}(2u - x - 2xz).$

When $u = 3$, $v = -1$ we have $x = -6$, $y = 4$, and $z = 2$, so $\dfrac{\partial M}{\partial u} = 16$ and $\dfrac{\partial M}{\partial v} = 36.$

21. $u = x^2 + yz,\ x = pr\cos\theta,\ y = pr\sin\theta,\ z = p + r \ \Rightarrow$

$\dfrac{\partial u}{\partial p} = \dfrac{\partial u}{\partial x}\dfrac{\partial x}{\partial p} + \dfrac{\partial u}{\partial y}\dfrac{\partial y}{\partial p} + \dfrac{\partial u}{\partial z}\dfrac{\partial z}{\partial p} = (2x)(r\cos\theta) + (z)(r\sin\theta) + (y)(1) = 2xr\cos\theta + zr\sin\theta + y,$

$\dfrac{\partial u}{\partial r} = \dfrac{\partial u}{\partial x}\dfrac{\partial x}{\partial r} + \dfrac{\partial u}{\partial y}\dfrac{\partial y}{\partial r} + \dfrac{\partial u}{\partial z}\dfrac{\partial z}{\partial r} = (2x)(p\cos\theta) + (z)(p\sin\theta) + (y)(1) = 2xp\cos\theta + zp\sin\theta + y,$

$$\frac{\partial u}{\partial \theta} = \frac{\partial u}{\partial x}\frac{\partial x}{\partial \theta} + \frac{\partial u}{\partial y}\frac{\partial y}{\partial \theta} + \frac{\partial u}{\partial z}\frac{\partial z}{\partial \theta} = (2x)(-pr\sin\theta) + (z)(pr\cos\theta) + (y)(0) = -2xpr\sin\theta + zpr\cos\theta.$$

When $p = 2$, $r = 3$, and $\theta = 0$ we have $x = 6$, $y = 0$, and $z = 5$, so $\dfrac{\partial u}{\partial p} = 36$, $\dfrac{\partial u}{\partial r} = 24$, and $\dfrac{\partial u}{\partial \theta} = 30$.

22. $y^5 + x^2y^3 = 1 + ye^{x^2}$, so let $F(x, y) = y^5 + x^2y^3 - 1 - ye^{x^2} = 0$. Then

$$\frac{dy}{dx} = -\frac{F_x}{F_y} = -\frac{2xy^3 - 2xye^{x^2}}{5y^4 + 3x^2y^2 - e^{x^2}} = \frac{2xye^{x^2} - 2xy^3}{5y^4 + 3x^2y^2 - e^{x^2}}.$$

23. $\sqrt{xy} = 1 + x^2y$, so let $F(x, y) = (xy)^{1/2} - 1 - x^2y = 0$. Then by Equation 6

$$\frac{dy}{dx} = -\frac{F_x}{F_y} = -\frac{\frac{1}{2}(xy)^{-1/2}(y) - 2xy}{\frac{1}{2}(xy)^{-1/2}(x) - x^2} = -\frac{y - 4xy\sqrt{xy}}{x - 2x^2\sqrt{xy}} = \frac{4(xy)^{3/2} - y}{x - 2x^2\sqrt{xy}}.$$

24. $\sin x + \cos y = \sin x \cos y$, so let $F(x, y) = \sin x + \cos y - \sin x \cos y = 0$. Then

$$\frac{dy}{dx} = -\frac{F_x}{F_y} = -\frac{\cos x - \cos x \cos y}{-\sin y + \sin x \sin y} = \frac{\cos x(\cos y - 1)}{\sin y(\sin x - 1)}.$$

25. $x^2 + y^2 + z^2 = 3xyz$, so let $F(x, y, z) = x^2 + y^2 + z^2 - 3xyz = 0$. Then by Equations 7

$$\frac{\partial z}{\partial x} = -\frac{F_x}{F_z} = -\frac{2x - 3yz}{2z - 3xy} = \frac{3yz - 2x}{2z - 3xy} \quad \text{and} \quad \frac{\partial z}{\partial y} = -\frac{F_y}{F_z} = -\frac{2y - 3xz}{2z - 3xy} = \frac{3xz - 2y}{2z - 3xy}.$$

26. $xyz = \cos(x + y + z)$. Let $F(x, y, z) = xyz - \cos(x + y + z) = 0$, so

$$\frac{\partial z}{\partial x} = -\frac{F_x}{F_z} = -\frac{yz + \sin(x + y + z)}{xy + \sin(x + y + z)} \quad \text{and} \quad \frac{\partial z}{\partial y} = -\frac{F_y}{F_z} = -\frac{xz + \sin(x + y + z)}{xy + \sin(x + y + z)}.$$

27. $x - z = \arctan(yz)$, so let $F(x, y, z) = x - z - \arctan(yz) = 0$. Then

$$\frac{\partial z}{\partial x} = -\frac{F_x}{F_z} = -\frac{1}{-1 - \dfrac{1}{1 + (yz)^2}(y)} = \frac{1 + y^2z^2}{1 + y + y^2z^2} \quad \text{and}$$

$$\frac{\partial z}{\partial y} = -\frac{F_y}{F_z} = -\frac{-\dfrac{1}{1 + (yz)^2}(z)}{-1 - \dfrac{1}{1 + (yz)^2}(y)} = -\frac{\dfrac{z}{1 + y^2z^2}}{\dfrac{1 + y^2z^2 + y}{1 + y^2z^2}} = -\frac{z}{1 + y + y^2z^2}.$$

28. $yz = \ln(x + z)$, so let $F(x, y, z) = yz - \ln(x + z) = 0$.

Then $$\frac{\partial z}{\partial x} = -\frac{F_x}{F_z} = -\frac{-\dfrac{1}{x + z}(1)}{y - \dfrac{1}{x + z}(1)} = \frac{1}{y(x + z) - 1} \quad \text{and} \quad \frac{\partial z}{\partial y} = -\frac{F_y}{F_z} = -\frac{z}{y - \dfrac{1}{x + z}} = -\frac{z(x + z)}{y(x + z) - 1}.$$

29. Since x and y are each functions of t, $T(x, y)$ is a function of t, so by the Chain Rule, $\dfrac{dT}{dt} = \dfrac{\partial T}{\partial x}\dfrac{dx}{dt} + \dfrac{\partial T}{\partial y}\dfrac{dy}{dt}$. After

3 seconds, $x = \sqrt{1 + t} = \sqrt{1 + 3} = 2$, $y = 2 + \frac{1}{3}t = 2 + \frac{1}{3}(3) = 3$, $\dfrac{dx}{dt} = \dfrac{1}{2\sqrt{1 + t}} = \dfrac{1}{2\sqrt{1 + 3}} = \dfrac{1}{4}$, and $\dfrac{dy}{dt} = \dfrac{1}{3}$.

Then $\dfrac{dT}{dt} = T_x(2, 3)\dfrac{dx}{dt} + T_y(2, 3)\dfrac{dy}{dt} = 4\left(\frac{1}{4}\right) + 3\left(\frac{1}{3}\right) = 2$. Thus the temperature is rising at a rate of $2°\mathrm{C/s}$.

30. (a) Since $\partial W/\partial T$ is negative, a rise in average temperature (while annual rainfall remains constant) causes a decrease in wheat production at the current production levels. Since $\partial W/\partial R$ is positive, an increase in annual rainfall (while the average temperature remains constant) causes an increase in wheat production.

(b) Since the average temperature is rising at a rate of $0.15°$C/year, we know that $dT/dt = 0.15$. Since rainfall is decreasing at a rate of 0.1 cm/year, we know $dR/dt = -0.1$. Then, by the Chain Rule,

$\dfrac{dW}{dt} = \dfrac{\partial W}{\partial T}\dfrac{dT}{dt} + \dfrac{\partial W}{\partial R}\dfrac{dR}{dt} = (-2)(0.15) + (8)(-0.1) = -1.1$. Thus we estimate that wheat production will decrease at a rate of 1.1 units/year.

31. $C = 1449.2 + 4.6T - 0.055T^2 + 0.00029T^3 + 0.016D$, so $\dfrac{\partial C}{\partial T} = 4.6 - 0.11T + 0.00087T^2$ and $\dfrac{\partial C}{\partial D} = 0.016$.

According to the graph, the diver is experiencing a temperature of approximately $12.5°$C at $t = 20$ minutes, so

$\dfrac{\partial C}{\partial T} = 4.6 - 0.11(12.5) + 0.00087(12.5)^2 \approx 3.36$. By sketching tangent lines at $t = 20$ to the graphs given, we estimate

$\dfrac{dD}{dt} \approx \dfrac{1}{2}$ and $\dfrac{dT}{dt} \approx -\dfrac{1}{10}$. Then, by the Chain Rule, $\dfrac{dC}{dt} = \dfrac{\partial C}{\partial T}\dfrac{dT}{dt} + \dfrac{\partial C}{\partial D}\dfrac{dD}{dt} \approx (3.36)\left(-\frac{1}{10}\right) + (0.016)\left(\frac{1}{2}\right) \approx -0.33$.

Thus the speed of sound experienced by the diver is decreasing at a rate of approximately 0.33 m/s per minute.

32. $V = \pi r^2 h/3$, so $\dfrac{dV}{dt} = \dfrac{\partial V}{\partial r}\dfrac{dr}{dt} + \dfrac{\partial V}{\partial h}\dfrac{dh}{dt} = \dfrac{2\pi rh}{3}\,1.8 + \dfrac{\pi r^2}{3}(-2.5) = 20{,}160\pi - 12{,}000\pi = 8160\pi \text{ in}^3/\text{s}$.

33. (a) $V = \ell wh$, so by the Chain Rule,

$$\frac{dV}{dt} = \frac{\partial V}{\partial \ell}\frac{d\ell}{dt} + \frac{\partial V}{\partial w}\frac{dw}{dt} + \frac{\partial V}{\partial h}\frac{dh}{dt} = wh\frac{d\ell}{dt} + \ell h\frac{dw}{dt} + \ell w\frac{dh}{dt} = 2\cdot 2\cdot 2 + 1\cdot 2\cdot 2 + 1\cdot 2\cdot(-3) = 6 \text{ m}^3/\text{s}.$$

(b) $S = 2(\ell w + \ell h + wh)$, so by the Chain Rule,

$$\begin{aligned}\frac{dS}{dt} &= \frac{\partial S}{\partial \ell}\frac{d\ell}{dt} + \frac{\partial S}{\partial w}\frac{dw}{dt} + \frac{\partial S}{\partial h}\frac{dh}{dt} = 2(w+h)\frac{d\ell}{dt} + 2(\ell+h)\frac{dw}{dt} + 2(\ell+w)\frac{dh}{dt}\\ &= 2(2+2)2 + 2(1+2)2 + 2(1+2)(-3) = 10 \text{ m}^2/\text{s}\end{aligned}$$

(c) $L^2 = \ell^2 + w^2 + h^2 \Rightarrow 2L\dfrac{dL}{dt} = 2\ell\dfrac{d\ell}{dt} + 2w\dfrac{dw}{dt} + 2h\dfrac{dh}{dt} = 2(1)(2) + 2(2)(2) + 2(2)(-3) = 0 \Rightarrow$

$dL/dt = 0$ m/s.

34. $I = \dfrac{V}{R} \Rightarrow$

$$\frac{dI}{dt} = \frac{\partial I}{\partial V}\frac{dV}{dt} + \frac{\partial I}{\partial R}\frac{dR}{dt} = \frac{1}{R}\frac{dV}{dt} - \frac{V}{R^2}\frac{dR}{dt} = \frac{1}{R}\frac{dV}{dt} - \frac{I}{R}\frac{dR}{dt} = \frac{1}{400}(-0.01) - \frac{0.08}{400}(0.03) = -0.000031 \text{ A/s}$$

35. $\dfrac{dP}{dt} = 0.05$, $\dfrac{dT}{dt} = 0.15$, $V = 8.31\dfrac{T}{P}$ and $\dfrac{dV}{dt} = \dfrac{8.31}{P}\dfrac{dT}{dt} - 8.31\dfrac{T}{P^2}\dfrac{dP}{dt}$. Thus when $P = 20$ and $T = 320$,

$$\frac{dV}{dt} = 8.31\left[\frac{0.15}{20} - \frac{(0.05)(320)}{400}\right] \approx -0.27 \text{ L/s}.$$

36. $f_o = \left(\dfrac{c + v_o}{c - v_s}\right) f_s = \left(\frac{332+34}{332-40}\right) 460 \approx 576.6$ Hz. v_o and v_s are functions of time t, so

$$\begin{aligned}\frac{df_o}{dt} &= \frac{\partial f_o}{\partial v_o}\frac{dv_o}{dt} + \frac{\partial f_o}{\partial v_s}\frac{dv_s}{dt} = \left(\frac{1}{c - v_s}\right) f_s \cdot \frac{dv_o}{dt} + \frac{c + v_o}{(c - v_s)^2} f_s \cdot \frac{dv_s}{dt}\\ &= \left(\tfrac{1}{332-40}\right)(460)(1.2) + \tfrac{332+34}{(332-40)^2}(460)(1.4) \approx 4.65 \text{ Hz/s}\end{aligned}$$

37. (a) By the Chain Rule, $\dfrac{\partial z}{\partial r} = \dfrac{\partial z}{\partial x}\cos\theta + \dfrac{\partial z}{\partial y}\sin\theta$, $\dfrac{\partial z}{\partial \theta} = \dfrac{\partial z}{\partial x}(-r\sin\theta) + \dfrac{\partial z}{\partial y}r\cos\theta$.

(b) $\left(\dfrac{\partial z}{\partial r}\right)^2 = \left(\dfrac{\partial z}{\partial x}\right)^2\cos^2\theta + 2\dfrac{\partial z}{\partial x}\dfrac{\partial z}{\partial y}\cos\theta\,\sin\theta + \left(\dfrac{\partial z}{\partial y}\right)^2\sin^2\theta$,

$\left(\dfrac{\partial z}{\partial \theta}\right)^2 = \left(\dfrac{\partial z}{\partial x}\right)^2 r^2\sin^2\theta - 2\dfrac{\partial z}{\partial x}\dfrac{\partial z}{\partial y}r^2\cos\theta\,\sin\theta + \left(\dfrac{\partial z}{\partial y}\right)^2 r^2\cos^2\theta$. Thus

$$\left(\frac{\partial z}{\partial r}\right)^2 + \frac{1}{r^2}\left(\frac{\partial z}{\partial \theta}\right)^2 = \left[\left(\frac{\partial z}{\partial x}\right)^2 + \left(\frac{\partial z}{\partial y}\right)^2\right](\cos^2\theta + \sin^2\theta) = \left(\frac{\partial z}{\partial x}\right)^2 + \left(\frac{\partial z}{\partial y}\right)^2.$$

38. By the Chain Rule, $\dfrac{\partial u}{\partial s} = \dfrac{\partial u}{\partial x}e^s\cos t + \dfrac{\partial u}{\partial y}e^s\sin t$, $\dfrac{\partial u}{\partial t} = \dfrac{\partial u}{\partial x}(-e^s\sin t) + \dfrac{\partial u}{\partial y}e^s\cos t$. Then

$\left(\dfrac{\partial u}{\partial s}\right)^2 = \left(\dfrac{\partial u}{\partial x}\right)^2 e^{2s}\cos^2 t + 2\dfrac{\partial u}{\partial x}\dfrac{\partial u}{\partial y}e^{2s}\cos t\,\sin t + \left(\dfrac{\partial u}{\partial y}\right)^2 e^{2s}\sin^2 t$ and

$\left(\dfrac{\partial u}{\partial t}\right)^2 = \left(\dfrac{\partial u}{\partial x}\right)^2 e^{2s}\sin^2 t - 2\dfrac{\partial u}{\partial x}\dfrac{\partial u}{\partial y}e^{2s}\cos t\,\sin t + \left(\dfrac{\partial u}{\partial y}\right)^2 e^{2s}\sin^2 t$. Thus

$$\left[\left(\frac{\partial u}{\partial s}\right)^2 + \left(\frac{\partial u}{\partial t}\right)^2\right]e^{-2s} = \left(\frac{\partial u}{\partial x}\right)^2 + \left(\frac{\partial u}{\partial y}\right)^2.$$

39. Let $u = x - y$. Then $\dfrac{\partial z}{\partial x} = \dfrac{dz}{du}\dfrac{\partial u}{\partial x} = \dfrac{dz}{du}$ and $\dfrac{\partial z}{\partial y} = \dfrac{dz}{du}(-1)$. Thus $\dfrac{\partial z}{\partial x} + \dfrac{\partial z}{\partial y} = 0$.

40. $\dfrac{\partial z}{\partial s} = \dfrac{\partial z}{\partial x} + \dfrac{\partial z}{\partial y}$ and $\dfrac{\partial z}{\partial t} = \dfrac{\partial z}{\partial x} - \dfrac{\partial z}{\partial y}$. Thus $\dfrac{\partial z}{\partial s}\dfrac{\partial z}{\partial t} = \left(\dfrac{\partial z}{\partial x}\right)^2 - \left(\dfrac{\partial z}{\partial y}\right)^2$.

41. Let $u = x + at$, $v = x - at$. Then $z = f(u) + g(v)$, so $\partial z/\partial u = f'(u)$ and $\partial z/\partial v = g'(v)$.

Thus $\dfrac{\partial z}{\partial t} = \dfrac{\partial z}{\partial u}\dfrac{\partial u}{\partial t} + \dfrac{\partial z}{\partial v}\dfrac{\partial v}{\partial t} = af'(u) - ag'(v)$ and

$$\frac{\partial^2 z}{\partial t^2} = a\frac{\partial}{\partial t}[f'(u) - g'(v)] = a\left(\frac{df'(u)}{du}\frac{\partial u}{\partial t} - \frac{dg'(v)}{dv}\frac{\partial v}{\partial t}\right) = a^2 f''(u) + a^2 g''(v).$$

Similarly $\dfrac{\partial z}{\partial x} = f'(u) + g'(v)$ and $\dfrac{\partial^2 z}{\partial x^2} = f''(u) + g''(v)$. Thus $\dfrac{\partial^2 z}{\partial t^2} = a^2\dfrac{\partial^2 z}{\partial x^2}$.

42. By the Chain Rule, $\dfrac{\partial u}{\partial s} = e^s\cos t\dfrac{\partial u}{\partial x} + e^s\sin t\dfrac{\partial u}{\partial y}$ and $\dfrac{\partial u}{\partial t} = -e^s\sin t\dfrac{\partial u}{\partial x} + e^s\cos t\dfrac{\partial u}{\partial y}$.

Then $\dfrac{\partial^2 u}{\partial s^2} = e^s\cos t\dfrac{\partial u}{\partial x} + e^s\cos t\dfrac{\partial}{\partial s}\left(\dfrac{\partial u}{\partial x}\right) + e^s\sin t\dfrac{\partial u}{\partial y} + e^s\sin t\dfrac{\partial}{\partial s}\left(\dfrac{\partial u}{\partial y}\right)$. But

$\dfrac{\partial}{\partial s}\left(\dfrac{\partial u}{\partial x}\right) = \dfrac{\partial^2 u}{\partial x^2}\dfrac{\partial x}{\partial s} + \dfrac{\partial^2 u}{\partial y\,\partial x}\dfrac{\partial y}{\partial s} = e^s\cos t\dfrac{\partial^2 u}{\partial x^2} + e^s\sin t\dfrac{\partial^2 u}{\partial y\,\partial x}$ and

$\dfrac{\partial}{\partial s}\left(\dfrac{\partial u}{\partial y}\right) = \dfrac{\partial^2 u}{\partial y^2}\dfrac{\partial y}{\partial s} + \dfrac{\partial^2 u}{\partial x\,\partial y}\dfrac{\partial x}{\partial s} = e^s\sin t\dfrac{\partial^2 u}{\partial y^2} + e^s\cos t\dfrac{\partial^2 u}{\partial x\,\partial y}$.

Also, by continuity of the partials, $\dfrac{\partial^2 u}{\partial x\,\partial y} = \dfrac{\partial^2 u}{\partial y\,\partial x}$. Thus

$$\frac{\partial^2 u}{\partial s^2} = e^s \cos t \frac{\partial u}{\partial x} + e^s \cos t\left(e^s \cos t \frac{\partial^2 u}{\partial x^2} + e^s \sin t \frac{\partial^2 u}{\partial x\,\partial y}\right) + e^s \sin t \frac{\partial u}{\partial y} + e^s \sin t\left(e^s \sin t \frac{\partial^2 u}{\partial y^2} + e^s \cos t \frac{\partial^2 u}{\partial x\,\partial y}\right)$$

$$= e^s \cos t \frac{\partial u}{\partial x} + e^s \sin t \frac{\partial u}{\partial y} + e^{2s} \cos^2 t \frac{\partial^2 u}{\partial x^2} + 2e^{2s} \cos t \sin t \frac{\partial^2 u}{\partial x\,\partial y} + e^{2s} \sin^2 t \frac{\partial^2 u}{\partial y^2}$$

Similarly

$$\frac{\partial^2 u}{\partial t^2} = -e^s \cos t \frac{\partial u}{\partial x} - e^s \sin t \frac{\partial}{\partial t}\left(\frac{\partial u}{\partial x}\right) - e^s \sin t \frac{\partial u}{\partial y} + e^s \cos t \frac{\partial}{\partial t}\left(\frac{\partial u}{\partial y}\right)$$

$$= -e^s \cos t \frac{\partial u}{\partial x} - e^s \sin t\left(-e^s \sin t \frac{\partial^2 u}{\partial x^2} + e^s \cos t \frac{\partial^2 u}{\partial x\,\partial y}\right)$$

$$-e^s \sin t \frac{\partial u}{\partial y} + e^s \cos t\left(e^s \cos t \frac{\partial^2 u}{\partial y^2} - e^s \sin t \frac{\partial^2 u}{\partial x\,\partial y}\right)$$

$$= -e^s \cos t \frac{\partial u}{\partial x} - e^s \sin t \frac{\partial u}{\partial y} + e^{2s} \sin^2 t \frac{\partial^2 u}{\partial x^2} - 2e^{2s} \cos t \sin t \frac{\partial^2 u}{\partial x\,\partial y} + e^{2s} \cos^2 t \frac{\partial^2 u}{\partial y^2}$$

Thus $e^{-2s}\left(\frac{\partial^2 u}{\partial s^2} + \frac{\partial^2 u}{\partial t^2}\right) = (\cos^2 t + \sin^2 t)\left(\frac{\partial^2 u}{\partial x^2} + \frac{\partial^2 u}{\partial y^2}\right) = \frac{\partial^2 u}{\partial x^2} + \frac{\partial^2 u}{\partial y^2}$, as desired.

43. $\frac{\partial z}{\partial s} = \frac{\partial z}{\partial x} 2s + \frac{\partial z}{\partial y} 2r$. Then

$$\frac{\partial^2 z}{\partial r\,\partial s} = \frac{\partial}{\partial r}\left(\frac{\partial z}{\partial x} 2s\right) + \frac{\partial}{\partial r}\left(\frac{\partial z}{\partial y} 2r\right)$$

$$= \frac{\partial^2 z}{\partial x^2}\frac{\partial x}{\partial r} 2s + \frac{\partial}{\partial y}\left(\frac{\partial z}{\partial x}\right)\frac{\partial y}{\partial r} 2s + \frac{\partial z}{\partial x}\frac{\partial}{\partial r} 2s + \frac{\partial^2 z}{\partial y^2}\frac{\partial y}{\partial r} 2r + \frac{\partial}{\partial x}\left(\frac{\partial z}{\partial y}\right)\frac{\partial x}{\partial r} 2r + \frac{\partial z}{\partial y} 2$$

$$= 4rs \frac{\partial^2 z}{\partial x^2} + \frac{\partial^2 z}{\partial y\,\partial x} 4s^2 + 0 + 4rs \frac{\partial^2 z}{\partial y^2} + \frac{\partial^2 z}{\partial x\,\partial y} 4r^2 + 2\frac{\partial z}{\partial y}$$

By the continuity of the partials, $\frac{\partial^2 z}{\partial r \partial s} = 4rs \frac{\partial^2 z}{\partial x^2} + 4rs \frac{\partial^2 z}{\partial y^2} + (4r^2 + 4s^2)\frac{\partial^2 z}{\partial x\,\partial y} + 2\frac{\partial z}{\partial y}$.

44. By the Chain Rule,

(a) $\frac{\partial z}{\partial r} = \frac{\partial z}{\partial x}\cos\theta + \frac{\partial z}{\partial y}\sin\theta$

(b) $\frac{\partial z}{\partial \theta} = -\frac{\partial z}{\partial x} r\sin\theta + \frac{\partial z}{\partial y} r\cos\theta$

(c) $$\frac{\partial^2 z}{\partial r\,\partial\theta} = \frac{\partial^2 z}{\partial\theta\,\partial r} = \frac{\partial}{\partial\theta}\left(\frac{\partial z}{\partial x}\cos\theta + \frac{\partial z}{\partial y}\sin\theta\right) = -\sin\theta\frac{\partial z}{\partial x} + \cos\theta\frac{\partial}{\partial\theta}\left(\frac{\partial z}{\partial x}\right) + \cos\theta\frac{\partial z}{\partial y} + \sin\theta\frac{\partial}{\partial\theta}\left(\frac{\partial z}{\partial y}\right)$$

$$= -\sin\theta\frac{\partial z}{\partial x} + \cos\theta\left(\frac{\partial^2 z}{\partial x^2}\frac{\partial x}{\partial\theta} + \frac{\partial^2 z}{\partial y\,\partial x}\frac{\partial y}{\partial\theta}\right) + \cos\theta\frac{\partial z}{\partial y} + \sin\theta\frac{\partial^2 z}{\partial y^2}\frac{\partial y}{\partial\theta} + \frac{\partial^2 z}{\partial x\,\partial y}\frac{\partial x}{\partial\theta}$$

$$= -\sin\theta\frac{\partial z}{\partial x} + \cos\theta\left(-r\sin\theta\frac{\partial^2 z}{\partial x^2} + r\cos\theta\frac{\partial^2 z}{\partial y\,\partial x}\right) + \cos\theta\frac{\partial z}{\partial y} + \sin\theta\left(r\cos\theta\frac{\partial^2 z}{\partial y^2} - r\sin\theta\frac{\partial^2 z}{\partial x\,\partial y}\right)$$

$$= -\sin\theta\frac{\partial z}{\partial x} - r\cos\theta\sin\theta\frac{\partial^2 z}{\partial x^2} + r\cos^2\theta\frac{\partial^2 z}{\partial y\,\partial x} + \cos\theta\frac{\partial z}{\partial y} + r\cos\theta\sin\theta\frac{\partial^2 z}{\partial y^2} - r\sin^2\theta\frac{\partial^2 z}{\partial y\,\partial x}$$

$$= \cos\theta\frac{\partial z}{\partial y} - \sin\theta\frac{\partial z}{\partial x} + r\cos\theta\sin\theta\left(\frac{\partial^2 z}{\partial y^2} - \frac{\partial^2 z}{\partial x^2}\right) + r(\cos^2\theta - \sin^2\theta)\frac{\partial^2 z}{\partial y\,\partial x}$$

45. $\dfrac{\partial z}{\partial r} = \dfrac{\partial z}{\partial x}\cos\theta + \dfrac{\partial z}{\partial y}\sin\theta$ and $\dfrac{\partial z}{\partial \theta} = -\dfrac{\partial z}{\partial x} r\sin\theta + \dfrac{\partial z}{\partial y} r\cos\theta$. Then

$$\begin{aligned}\frac{\partial^2 z}{\partial r^2} &= \cos\theta\left(\frac{\partial^2 z}{\partial x^2}\cos\theta + \frac{\partial^2 z}{\partial y\,\partial x}\sin\theta\right) + \sin\theta\left(\frac{\partial^2 z}{\partial y^2}\sin\theta + \frac{\partial^2 z}{\partial x\,\partial y}\cos\theta\right)\\ &= \cos^2\theta\,\frac{\partial^2 z}{\partial x^2} + 2\cos\theta\,\sin\theta\,\frac{\partial^2 z}{\partial x\,\partial y} + \sin^2\theta\,\frac{\partial^2 z}{\partial y^2}\end{aligned}$$

and

$$\begin{aligned}\frac{\partial^2 z}{\partial \theta^2} &= -r\cos\theta\,\frac{\partial z}{\partial x} + (-r\sin\theta)\left(\frac{\partial^2 z}{\partial x^2}(-r\sin\theta) + \frac{\partial^2 z}{\partial y\,\partial x}\,r\cos\theta\right)\\ &\qquad -r\sin\theta\,\frac{\partial z}{\partial y} + r\cos\theta\left(\frac{\partial^2 z}{\partial y^2}\,r\cos\theta + \frac{\partial^2 z}{\partial x\,\partial y}(-r\sin\theta)\right)\\ &= -r\cos\theta\,\frac{\partial z}{\partial x} - r\sin\theta\,\frac{\partial z}{\partial y} + r^2\sin^2\theta\,\frac{\partial^2 z}{\partial x^2} - 2r^2\cos\theta\,\sin\theta\,\frac{\partial^2 z}{\partial x\,\partial y} + r^2\cos^2\theta\,\frac{\partial^2 z}{\partial y^2}\end{aligned}$$

Thus

$$\begin{aligned}\frac{\partial^2 z}{\partial r^2} + \frac{1}{r^2}\frac{\partial^2 z}{\partial \theta^2} + \frac{1}{r}\frac{\partial z}{\partial r} &= (\cos^2\theta + \sin^2\theta)\,\frac{\partial^2 z}{\partial x^2} + (\sin^2\theta + \cos^2\theta)\,\frac{\partial^2 z}{\partial y^2}\\ &\qquad -\frac{1}{r}\cos\theta\,\frac{\partial z}{\partial x} - \frac{1}{r}\sin\theta\,\frac{\partial z}{\partial y} + \frac{1}{r}\left(\cos\theta\,\frac{\partial z}{\partial x} + \sin\theta\,\frac{\partial z}{\partial y}\right)\\ &= \frac{\partial^2 z}{\partial x^2} + \frac{\partial^2 z}{\partial y^2} \text{ as desired.}\end{aligned}$$

46. (a) $\dfrac{\partial z}{\partial t} = \dfrac{\partial z}{\partial x}\dfrac{\partial x}{\partial t} + \dfrac{\partial z}{\partial y}\dfrac{\partial y}{\partial t}$. Then

$$\begin{aligned}\frac{\partial^2 z}{\partial t^2} &= \frac{\partial}{\partial t}\left(\frac{\partial z}{\partial x}\frac{\partial x}{\partial t}\right) + \frac{\partial}{\partial t}\left(\frac{\partial z}{\partial y}\frac{\partial y}{\partial t}\right) = \frac{\partial}{\partial t}\left(\frac{\partial z}{\partial x}\right)\frac{\partial x}{\partial t} + \frac{\partial^2 x}{\partial t^2}\frac{\partial z}{\partial x} + \frac{\partial}{\partial t}\left(\frac{\partial z}{\partial y}\right)\frac{\partial y}{\partial t} + \frac{\partial^2 y}{\partial t^2}\frac{\partial z}{\partial y}\\ &= \frac{\partial^2 z}{\partial x^2}\left(\frac{\partial x}{\partial t}\right)^2 + \frac{\partial^2 z}{\partial y\,\partial x}\frac{\partial x}{\partial t}\frac{\partial y}{\partial t} + \frac{\partial^2 x}{\partial t^2}\frac{\partial z}{\partial x} + \frac{\partial^2 z}{\partial y^2}\left(\frac{\partial y}{\partial t}\right)^2 + \frac{\partial^2 z}{\partial x\,\partial y}\frac{\partial y}{\partial t}\frac{\partial x}{\partial t} + \frac{\partial^2 y}{\partial t^2}\frac{\partial z}{\partial y}\\ &= \frac{\partial^2 z}{\partial x^2}\left(\frac{\partial x}{\partial t}\right)^2 + 2\frac{\partial^2 z}{\partial x\,\partial y}\frac{\partial x}{\partial t}\frac{\partial y}{\partial t} + \frac{\partial^2 z}{\partial y^2}\left(\frac{\partial y}{\partial t}\right)^2 + \frac{\partial^2 x}{\partial t^2}\frac{\partial z}{\partial x} + \frac{\partial^2 y}{\partial t^2}\frac{\partial z}{\partial y}\end{aligned}$$

(b)

$$\begin{aligned}\frac{\partial^2 z}{\partial s\,\partial t} &= \frac{\partial}{\partial s}\left(\frac{\partial z}{\partial x}\frac{\partial x}{\partial t} + \frac{\partial z}{\partial y}\frac{\partial y}{\partial t}\right)\\ &= \left(\frac{\partial^2 z}{\partial x^2}\frac{\partial x}{\partial s} + \frac{\partial^2 z}{\partial y\,\partial x}\frac{\partial y}{\partial s}\right)\frac{\partial x}{\partial t} + \frac{\partial z}{\partial x}\frac{\partial^2 x}{\partial s\,\partial t} + \left(\frac{\partial^2 z}{\partial y^2}\frac{\partial y}{\partial s} + \frac{\partial^2 z}{\partial x\,\partial y}\frac{\partial x}{\partial s}\right)\frac{\partial y}{\partial t} + \frac{\partial z}{\partial y}\frac{\partial^2 y}{\partial s\,\partial t}\\ &= \frac{\partial^2 z}{\partial x^2}\frac{\partial x}{\partial s}\frac{\partial x}{\partial t} + \frac{\partial^2 z}{\partial x\,\partial y}\left(\frac{\partial y}{\partial s}\frac{\partial x}{\partial t} + \frac{\partial y}{\partial t}\frac{\partial x}{\partial s}\right) + \frac{\partial z}{\partial x}\frac{\partial^2 x}{\partial s\,\partial t} + \frac{\partial z}{\partial y}\frac{\partial^2 y}{\partial s\,\partial t} + \frac{\partial^2 z}{\partial y^2}\frac{\partial y}{\partial s}\frac{\partial y}{\partial t}\end{aligned}$$

47. $F(x,y,z) = 0$ is assumed to define z as a function of x and y, that is, $z = f(x,y)$. So by (7), $\dfrac{\partial z}{\partial x} = -\dfrac{F_x}{F_z}$ since $F_z \neq 0$. Similarly, it is assumed that $F(x,y,z) = 0$ defines x as a function of y and z, that is $x = h(x,z)$. Then $F(h(y,z),y,z) = 0$ and by the Chain Rule, $F_x\dfrac{\partial x}{\partial y} + F_y\dfrac{\partial y}{\partial y} + F_z\dfrac{\partial z}{\partial y} = 0$. But $\dfrac{\partial z}{\partial y} = 0$ and $\dfrac{\partial y}{\partial y} = 1$, so $F_x\dfrac{\partial x}{\partial y} + F_y = 0 \;\Rightarrow\; \dfrac{\partial x}{\partial y} = -\dfrac{F_y}{F_x}$. A similar calculation shows that $\dfrac{\partial y}{\partial z} = -\dfrac{F_z}{F_y}$. Thus $\dfrac{\partial z}{\partial x}\dfrac{\partial x}{\partial y}\dfrac{\partial y}{\partial z} = \left(-\dfrac{F_x}{F_z}\right)\left(-\dfrac{F_y}{F_x}\right)\left(-\dfrac{F_z}{F_y}\right) = -1$.

11.6 Directional Derivatives and the Gradient Vector

1. $f(x,y) = \sqrt{5x-4y} \;\Rightarrow\; f_x(x,y) = \frac{1}{2}(5x-4y)^{-1/2}(5) = \dfrac{5}{2\sqrt{5x-4y}}$ and

$f_y(x,y) = \frac{1}{2}(5x-4y)^{-1/2}(-4) = -\dfrac{2}{\sqrt{5x-4y}}$. If $\mathbf{u}$ is a unit vector in the direction of $\theta = -\frac{\pi}{6}$, then from Equation 6,

$D_{\mathbf{u}}\,f(4,1) = f_x(4,1)\cos\left(-\frac{\pi}{6}\right) + f_y(4,1)\sin\left(-\frac{\pi}{6}\right) = \frac{5}{8}\cdot\frac{\sqrt{3}}{2} + \left(-\frac{1}{2}\right)\left(-\frac{1}{2}\right) = \frac{5\sqrt{3}}{16} + \frac{1}{4}$.

2. $f(x,y) = x\sin(xy) \;\Rightarrow\; f_x(x,y) = x\cos(xy)\cdot y + \sin(xy) = xy\cos(xy) + \sin(xy)$ and

$f_y(x,y) = x\cos(xy)\cdot x = x^2\cos(xy)$. If $\mathbf{u}$ is a unit vector in the direction of $\theta = \frac{\pi}{3}$, then from Equation 6,

$D_{\mathbf{u}}\,f(2,0) = f_x(2,0)\cos\frac{\pi}{3} + f_y(2,0)\sin\frac{\pi}{3} = 0 + 4\left(\frac{\sqrt{3}}{2}\right) = 2\sqrt{3}$.

3. $f(x,y) = 5xy^2 - 4x^3y$

(a) $\nabla f(x,y) = \langle f_x(x,y), f_y(x,y)\rangle = \langle 5y^2 - 12x^2y, 10xy - 4x^3\rangle$

(b) $\nabla f(1,2) = \langle 5(2)^2 - 12(1)^2(2), 10(1)(2) - 4(1)^3\rangle = \langle -4, 16\rangle$

(c) By Equation 9, $D_{\mathbf{u}}\,f(1,2) = \nabla f(1,2)\cdot\mathbf{u} = \langle -4, 16\rangle\cdot\left\langle \frac{5}{13}, \frac{12}{13}\right\rangle = (-4)\left(\frac{5}{13}\right) + (16)\left(\frac{12}{13}\right) = \frac{172}{13}$.

4. $f(x,y) = y\ln x$

(a) $\nabla f(x,y) = \langle f_x(x,y), f_y(x,y)\rangle = \langle y/x, \ln x\rangle$

(b) $\nabla f(1,-3) = \left\langle \frac{-3}{1}, \ln 1\right\rangle = \langle -3, 0\rangle$

(c) By Equation 9, $D_{\mathbf{u}}\,f(1,-3) = \nabla f(1,-3)\cdot\mathbf{u} = \langle -3, 0\rangle\cdot\left\langle -\frac{4}{5}, \frac{3}{5}\right\rangle = \frac{12}{5}$.

5. $f(x,y,z) = xe^{2yz}$

(a) $\nabla f(x,y,z) = \langle f_x(x,y,z), f_y(x,y,z), f_z(x,y,z)\rangle = \left\langle e^{2yz}, 2xze^{2yz}, 2xye^{2yz}\right\rangle$

(b) $\nabla f(3,0,2) = \langle 1, 12, 0\rangle$

(c) By Equation 14, $D_{\mathbf{u}}f(3,0,2) = \nabla f(3,0,2)\cdot\mathbf{u} = \langle 1, 12, 0\rangle\cdot\left\langle \frac{2}{3}, -\frac{2}{3}, \frac{1}{3}\right\rangle = \frac{2}{3} - \frac{24}{3} + 0 = -\frac{22}{3}$.

6. $f(x,y,z) = \sqrt{x+yz} = (x+yz)^{1/2}$

(a) $\nabla f(x,y,z) = \left\langle \frac{1}{2}(x+yz)^{-1/2}(1), \frac{1}{2}(x+yz)^{-1/2}(z), \frac{1}{2}(x+yz)^{-1/2}(y)\right\rangle$

$= \left\langle 1/(2\sqrt{x+yz}\,), z/(2\sqrt{x+yz}\,), y/(2\sqrt{x+yz}\,)\right\rangle$

(b) $\nabla f(1,3,1) = \left\langle \frac{1}{4}, \frac{1}{4}, \frac{3}{4}\right\rangle$

(c) $D_{\mathbf{u}}f(1,3,1) = \nabla f(1,3,1)\cdot\mathbf{u} = \left\langle \frac{1}{4}, \frac{1}{4}, \frac{3}{4}\right\rangle\cdot\left\langle \frac{2}{7}, \frac{3}{7}, \frac{6}{7}\right\rangle = \frac{2}{28} + \frac{3}{28} + \frac{18}{28} = \frac{23}{28}$

7. $f(x,y) = 1 + 2x\sqrt{y} \;\Rightarrow\; \nabla f(x,y) = \left\langle 2\sqrt{y}, 2x\cdot\frac{1}{2}y^{-1/2}\right\rangle = \left\langle 2\sqrt{y}, x/\sqrt{y}\,\right\rangle$, $\nabla f(3,4) = \left\langle 4, \frac{3}{2}\right\rangle$, and a unit vector in

the direction of $\mathbf{v}$ is $\mathbf{u} = \dfrac{1}{\sqrt{4^2 + (-3)^2}}\langle 4, -3\rangle = \left\langle \frac{4}{5}, -\frac{3}{5}\right\rangle$, so $D_{\mathbf{u}}\,f(3,4) = \nabla f(3,4)\cdot\mathbf{u} = \left\langle 4, \frac{3}{2}\right\rangle\cdot\left\langle \frac{4}{5}, -\frac{3}{5}\right\rangle = \frac{23}{10}$.

8. $f(x,y) = \ln(x^2+y^2) \;\Rightarrow\; \nabla f(x,y) = \left\langle \dfrac{2x}{x^2+y^2}, \dfrac{2y}{x^2+y^2}\right\rangle$, $\nabla f(2,1) = \left\langle \frac{4}{5}, \frac{2}{5}\right\rangle$, and

a unit vector in the direction of $\mathbf{v} = \langle -1, 2\rangle$ is $\mathbf{u} = \dfrac{1}{\sqrt{1+4}}\langle -1, 2\rangle = \left\langle -\frac{1}{\sqrt{5}}, \frac{2}{\sqrt{5}}\right\rangle$, so

$D_{\mathbf{u}}\,f(2,1) = \nabla f(2,1)\cdot\mathbf{u} = \left\langle \frac{4}{5}, \frac{2}{5}\right\rangle\cdot\left\langle -\frac{1}{\sqrt{5}}, \frac{2}{\sqrt{5}}\right\rangle = -\frac{4}{5\sqrt{5}} + \frac{4}{5\sqrt{5}} = 0$.

9. $g(s,t)=s^2e^t \Rightarrow \nabla g(s,t)=2se^t\,\mathbf{i}+s^2e^t\,\mathbf{j}$, $\nabla g(2,0)=4\,\mathbf{i}+4\,\mathbf{j}$, and a unit vector in the direction of $\mathbf{v}$ is

$\mathbf{u}=\frac{1}{\sqrt{2}}(\mathbf{i}+\mathbf{j})$, so $D_{\mathbf{u}}\,g(2,0)=\nabla g(2,0)\cdot\mathbf{u}=(4\,\mathbf{i}+4\,\mathbf{j})\cdot\frac{1}{\sqrt{2}}(\mathbf{i}+\mathbf{j})=\frac{8}{\sqrt{2}}=4\sqrt{2}$.

10. $f(x,y,z)=\dfrac{x}{y+z} \Rightarrow \nabla f(x,y,z)=\left\langle\dfrac{1}{y+z},-\dfrac{x}{(y+z)^2},-\dfrac{x}{(y+z)^2}\right\rangle$, $\nabla f(4,1,1)=\left\langle\frac{1}{2},-1,-1\right\rangle$, and a unit vector in the direction of $\mathbf{v}$ is $\mathbf{u}=\frac{1}{\sqrt{14}}\langle 1,2,3\rangle$, so $D_{\mathbf{u}}\,f(4,1,1)=\nabla f(4,1,1)\cdot\mathbf{u}=\left\langle\frac{1}{2},-1,-1\right\rangle\cdot\frac{1}{\sqrt{14}}\langle 1,2,3\rangle=-\frac{9}{2\sqrt{14}}$.

11. $g(x,y,z)=(x+2y+3z)^{3/2} \Rightarrow$

$$\begin{aligned}\nabla g(x,y,z)&=\left\langle\tfrac{3}{2}(x+2y+3z)^{1/2}(1),\tfrac{3}{2}(x+2y+3z)^{1/2}(2),\tfrac{3}{2}(x+2y+3z)^{1/2}(3)\right\rangle\\&=\left\langle\tfrac{3}{2}\sqrt{x+2y+3z},3\sqrt{x+2y+3z},\tfrac{9}{2}\sqrt{x+2y+3z}\right\rangle,\ \nabla g(1,1,2)=\left\langle\tfrac{9}{2},9,\tfrac{27}{2}\right\rangle,\end{aligned}$$

and a unit vector in the direction of $\mathbf{v}=2\,\mathbf{j}-\mathbf{k}$ is $\mathbf{u}=\frac{2}{\sqrt{5}}\,\mathbf{j}-\frac{1}{\sqrt{5}}\,\mathbf{k}$, so

$D_{\mathbf{u}}\,g(1,1,2)=\left\langle\frac{9}{2},9,\frac{27}{2}\right\rangle\cdot\left\langle 0,\frac{2}{\sqrt{5}},-\frac{1}{\sqrt{5}}\right\rangle=\frac{18}{\sqrt{5}}-\frac{27}{2\sqrt{5}}=\frac{9}{2\sqrt{5}}$.

12. $D_{\mathbf{u}}f(2,2)=\nabla f(2,2)\cdot\mathbf{u}$, the scalar projection of $\nabla f(2,2)$ onto $\mathbf{u}$, so we draw a perpendicular from the tip of $\nabla f(2,2)$ to the line containing $\mathbf{u}$. We can use the point $(2,2)$ to determine the scale of the axes, and we estimate the length of the projection to be approximately 3.0 units. Since the angle between $\nabla f(2,2)$ and $\mathbf{u}$ is greater than $90°$, the scalar projection is negative. Thus $D_{\mathbf{u}}\,f(2,2)\approx -3$.

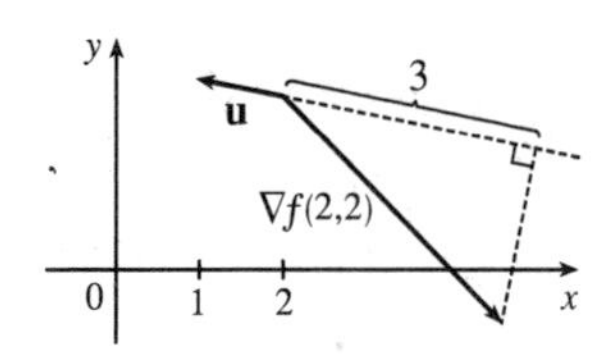

13. $f(x,y)=\sqrt{xy} \Rightarrow \nabla f(x,y)=\left\langle\frac{1}{2}(xy)^{-1/2}(y),\frac{1}{2}(xy)^{-1/2}(x)\right\rangle=\left\langle\dfrac{y}{2\sqrt{xy}},\dfrac{x}{2\sqrt{xy}}\right\rangle$, so $\nabla f(2,8)=\left\langle 1,\frac{1}{4}\right\rangle$.

The unit vector in the direction of $\overrightarrow{PQ}=\langle 5-2,4-8\rangle=\langle 3,-4\rangle$ is $\mathbf{u}=\left\langle\frac{3}{5},-\frac{4}{5}\right\rangle$, so

$D_{\mathbf{u}}\,f(2,8)=\nabla f(2,8)\cdot\mathbf{u}=\left\langle 1,\frac{1}{4}\right\rangle\cdot\left\langle\frac{3}{5},-\frac{4}{5}\right\rangle=\frac{2}{5}$.

14. $f(x,y,z)=x^2+y^2+z^2 \Rightarrow \nabla f(x,y,z)=\langle 2x,2y,2z\rangle$, so $\nabla f(2,1,3)=\langle 4,2,6\rangle$. The unit vector in the direction of $\overrightarrow{PO}=\langle -2,-1,-3\rangle$ is $\mathbf{u}=\frac{1}{\sqrt{14}}\langle -2,-1,-3\rangle$, so

$D_{\mathbf{u}}\,f(2,1,3)=\nabla f(2,1,3)\cdot\mathbf{u}=\langle 4,2,6\rangle\cdot\frac{1}{\sqrt{14}}\langle -2,-1,-3\rangle=-\frac{28}{\sqrt{14}}=-2\sqrt{14}$.

15. $f(x,y)=y^2/x=y^2x^{-1} \Rightarrow \nabla f(x,y)=\left\langle -y^2x^{-2},2yx^{-1}\right\rangle=\left\langle -y^2/x^2,2y/x\right\rangle$.

$\nabla f(2,4)=\langle -4,4\rangle$, or equivalently $\langle -1,1\rangle$, is the direction of maximum rate of change, and the maximum rate is $|\nabla f(2,4)|=\sqrt{16+16}=4\sqrt{2}$.

16. $f(p,q)=qe^{-p}+pe^{-q} \Rightarrow \nabla f(p,q)=\left\langle -qe^{-p}+e^{-q},e^{-p}-pe^{-q}\right\rangle$.

$\nabla f(0,0)=\langle 1,1\rangle$ is the direction of maximum rate of change and the maximum rate is $|\nabla f(0,0)|=\sqrt{2}$.

17. $f(x,y,z)=\ln(xy^2z^3) \Rightarrow \nabla f(x,y,z)=\left\langle\dfrac{y^2z^3}{xy^2z^3},\dfrac{2xyz^3}{xy^2z^3},\dfrac{3xy^2z^2}{xy^2z^3}\right\rangle=\left\langle\dfrac{1}{x},\dfrac{2}{y},\dfrac{3}{z}\right\rangle$.

$\nabla f(1,-2,-3)=\langle 1,-1,-1\rangle$ is the direction of maximum rate of change and the maximum rate is $|\nabla f(1,-2,-3)|=\sqrt{3}$.

18. $f(x,y,z) = \tan(x+2y+3z) \Rightarrow$

$\nabla f(x,y,z) = \langle \sec^2(x+2y+3z)(1), \sec^2(x+2y+3z)(2), \sec^2(x+2y+3z)(3) \rangle$.

$\nabla f(-5,1,1) = \langle \sec^2(0), 2\sec^2(0), 3\sec^2(0) \rangle = \langle 1,2,3 \rangle$ is the direction of maximum rate of change and the maximum rate is $|\nabla f(-5,1,1)| = \sqrt{14}$.

19. (a) As in the proof of Theorem 15, $D_{\mathbf{u}}\, f = |\nabla f| \cos\theta$. Since the minimum value of $\cos\theta$ is -1 occurring when $\theta = \pi$, the minimum value of $D_{\mathbf{u}}\, f$ is $-|\nabla f|$ occurring when $\theta = \pi$, that is when $\mathbf{u}$ is in the opposite direction of ∇f (assuming $\nabla f \neq \mathbf{0}$).

(b) $f(x,y) = x^4 y - x^2 y^3 \Rightarrow \nabla f(x,y) = \langle 4x^3 y - 2xy^3, x^4 - 3x^2 y^2 \rangle$, so f decreases fastest at the point $(2,-3)$ in the direction $-\nabla f(2,-3) = -\langle 12,-92 \rangle = \langle -12, 92 \rangle$.

20. $f(x,y) = x^2 + \sin xy \Rightarrow f_x(x,y) = 2x + y\cos xy$, $f_y(x,y) = x\cos xy$ and $f_x(1,0) = 2(1) + (0)\cos 0 = 2$, $f_y(1,0) = (1)\cos 0 = 1$. If $\mathbf{u}$ is a unit vector which makes an angle θ with the positive x-axis, then $D_{\mathbf{u}} f(1,0) = f_x(1,0)\cos\theta + f_y(1,0)\sin\theta = 2\cos\theta + \sin\theta$. We want $D_{\mathbf{u}} f(1,0) = 1$, so $2\cos\theta + \sin\theta = 1 \Rightarrow$ $\sin\theta = 1 - 2\cos\theta \Rightarrow \sin^2\theta = (1-2\cos\theta)^2 \Rightarrow 1 - \cos^2\theta = 1 - 4\cos\theta + 4\cos^2\theta \Rightarrow$ $5\cos^2\theta - 4\cos\theta = 0 \Rightarrow \cos\theta(5\cos\theta - 4) = 0 \Rightarrow \cos\theta = 0$ or $\cos\theta = \frac{4}{5} \Rightarrow \theta = \frac{\pi}{2}$ or $\theta = 2\pi - \cos^{-1}\left(\frac{4}{5}\right) \approx 5.64$.

21. The direction of fastest change is $\nabla f(x,y) = (2x-2)\,\mathbf{i} + (2y-4)\,\mathbf{j}$, so we need to find all points (x,y) where $\nabla f(x,y)$ is parallel to $\mathbf{i} + \mathbf{j} \Leftrightarrow (2x-2)\,\mathbf{i} + (2y-4)\,\mathbf{j} = k\,(\mathbf{i}+\mathbf{j}) \Leftrightarrow k = 2x-2$ and $k = 2y-4$. Then $2x - 2 = 2y - 4 \Rightarrow$ $y = x+1$, so the direction of fastest change is $\mathbf{i}+\mathbf{j}$ at all points on the line $y = x+1$.

22. The fisherman is traveling in the direction $\langle -80,-60 \rangle$. A unit vector in this direction is $\mathbf{u} = \frac{1}{100}\langle -80,-60 \rangle = \langle -\frac{4}{5}, -\frac{3}{5} \rangle$, and if the depth of the lake is given by $f(x,y) = 200 + 0.02x^2 - 0.001y^3$, then $\nabla f(x,y) = \langle 0.04x, -0.003y^2 \rangle$. $D_{\mathbf{u}}\, f(80,60) = \nabla f(80,60) \cdot \mathbf{u} = \langle 3.2, -10.8 \rangle \cdot \langle -\frac{4}{5}, -\frac{3}{5} \rangle = 3.92$. Since $D_{\mathbf{u}}\, f(80,60)$ is positive, the depth of the lake is increasing near $(80,60)$ in the direction toward the buoy.

23. $T = \dfrac{k}{\sqrt{x^2+y^2+z^2}}$ and $120 = T(1,2,2) = \dfrac{k}{3}$ so $k = 360$.

(a) $\mathbf{u} = \dfrac{\langle 1,-1,1 \rangle}{\sqrt{3}}$,

$$D_{\mathbf{u}}T(1,2,2) = \nabla T(1,2,2) \cdot \mathbf{u} = \left[-360\left(x^2+y^2+z^2\right)^{-3/2} \langle x,y,z \rangle\right]_{(1,2,2)} \cdot \mathbf{u} = -\tfrac{40}{3}\langle 1,2,2 \rangle \cdot \tfrac{1}{\sqrt{3}}\langle 1,-1,1 \rangle = -\tfrac{40}{3\sqrt{3}}$$

(b) From (a), $\nabla T = -360\left(x^2+y^2+z^2\right)^{-3/2} \langle x,y,z \rangle$, and since $\langle x,y,z \rangle$ is the position vector of the point (x,y,z), the vector $-\langle x,y,z \rangle$, and thus ∇T, always points toward the origin.

24. $\nabla T = -400e^{-x^2-3y^2-9z^2}\langle x, 3y, 9z \rangle$

(a) $\mathbf{u} = \frac{1}{\sqrt{6}}\langle 1,-2,1 \rangle$, $\nabla T(2,-1,2) = -400e^{-43}\langle 2,-3,18 \rangle$ and

$$D_{\mathbf{u}}\, T(2,-1,2) = \left(-\frac{400e^{-43}}{\sqrt{6}}\right)(26) = -\frac{5200\sqrt{6}}{3e^{43}}\ {}^\circ\mathrm{C/m}.$$

(b) $\nabla T(2,-1,2) = 400e^{-43}\langle -2, 3, -18\rangle$ or equivalently $\langle -2, 3, -18\rangle$.

(c) $|\nabla T| = 400e^{-x^2-3y^2-9z^2}\sqrt{x^2+9y^2+81z^2}$ °C/m is the maximum rate of increase. At $(2,-1,2)$ the maximum rate of increase is $400e^{-43}\sqrt{337}$°C/m.

25. $\nabla V(x,y,z) = \langle 10x - 3y + yz, xz - 3x, xy\rangle$, $\nabla V(3,4,5) = \langle 38, 6, 12\rangle$

(a) $D_{\mathbf{u}}\, V(3,4,5) = \langle 38, 6, 12\rangle \cdot \frac{1}{\sqrt{3}}\langle 1, 1, -1\rangle = \frac{32}{\sqrt{3}}$

(b) $\nabla V(3,4,5) = \langle 38, 6, 12\rangle$, or equivalently, $\langle 19, 3, 6\rangle$.

(c) $|\nabla V(3,4,5)| = \sqrt{38^2+6^2+12^2} = \sqrt{1624} = 2\sqrt{406}$

26. $z = f(x,y) = 1000 - 0.005x^2 - 0.01y^2 \quad\Rightarrow\quad \nabla f(x,y) = \langle -0.01x, -0.02y\rangle$ and $\nabla f(60,40) = \langle -0.6, -0.8\rangle$.

(a) Due south is in the direction of the unit vector $\mathbf{u} = -\mathbf{j}$ and $D_{\mathbf{u}} f(60,40) = \nabla f\,(60,40)\cdot\langle 0,-1\rangle = \langle -0.6,-0.8\rangle\cdot\langle 0,-1\rangle = 0.8$. Thus, if you walk due south from $(60,40,966)$ you will ascend at a rate of 0.8 vertical meters per horizontal meter.

(b) Northwest is in the direction of the unit vector $\mathbf{u} = \frac{1}{\sqrt{2}}\,\langle -1,1\rangle$ and

$D_{\mathbf{u}} f(60,40) = \nabla f\,(60,40)\cdot\frac{1}{\sqrt{2}}\,\langle -1,1\rangle = \langle -0.6,-0.8\rangle\cdot\frac{1}{\sqrt{2}}\,\langle -1,1\rangle = -\frac{0.2}{\sqrt{2}} \approx -0.14$. Thus, if you walk northwest from $(60,40,966)$ you will descend at a rate of approximately 0.14 vertical meters per horizontal meter.

(c) $\nabla f(60,40) = \langle -0.6,-0.8\rangle$ is the direction of largest slope with a rate of ascent given by $|\nabla f(60,40)| = \sqrt{(-0.6)^2+(-0.8)^2} = 1$. The angle above the horizontal in which the path begins is given by $\tan\theta = 1 \quad\Rightarrow\quad \theta = 45\,°$.

27. A unit vector in the direction of $\overrightarrow{AB}$ is $\mathbf{i}$ and a unit vector in the direction of $\overrightarrow{AC}$ is $\mathbf{j}$. Thus $D_{\overrightarrow{AB}}\, f(1,3) = f_x(1,3) = 3$ and $D_{\overrightarrow{AC}}\, f(1,3) = f_y(1,3) = 26$. Therefore $\nabla f(1,3) = \langle f_x(1,3), f_y(1,3)\rangle = \langle 3, 26\rangle$, and by definition,

$D_{\overrightarrow{AD}}\, f(1,3) = \nabla f\cdot\mathbf{u}$ where $\mathbf{u}$ is a unit vector in the direction of $\overrightarrow{AD}$, which is $\left\langle \frac{5}{13}, \frac{12}{13}\right\rangle$. Therefore,

$D_{\overrightarrow{AD}}\, f\,(1,3) = \langle 3, 26\rangle\cdot\left\langle \frac{5}{13}, \frac{12}{13}\right\rangle = 3\cdot\frac{5}{13} + 26\cdot\frac{12}{13} = \frac{327}{13}$.

28. The curve of steepest ascent is perpendicular to all of the contour lines.

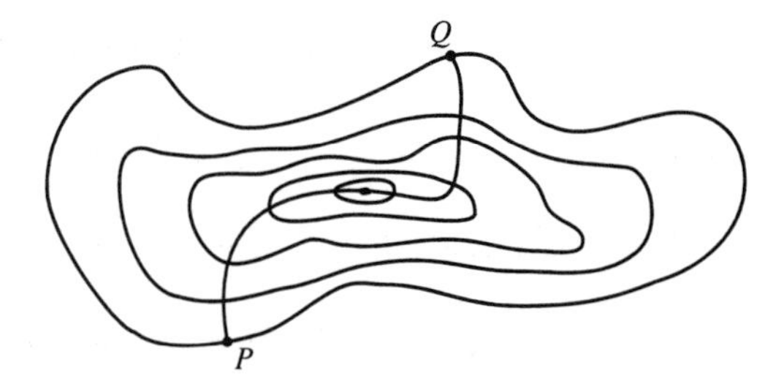

29. (a) $\nabla(au+bv) = \left\langle \dfrac{\partial(au+bv)}{\partial x}, \dfrac{\partial(au+bv)}{\partial y}\right\rangle = \left\langle a\,\dfrac{\partial u}{\partial x} + b\,\dfrac{\partial v}{\partial x}, a\,\dfrac{\partial u}{\partial y} + b\,\dfrac{\partial v}{\partial y}\right\rangle = a\left\langle \dfrac{\partial u}{\partial x}, \dfrac{\partial u}{\partial y}\right\rangle + b\left\langle \dfrac{\partial v}{\partial x}, \dfrac{\partial v}{\partial y}\right\rangle$

$= a\,\nabla u + b\,\nabla v$

(b) $\nabla(uv) = \left\langle v\,\dfrac{\partial u}{\partial x} + u\,\dfrac{\partial v}{\partial x}, v\,\dfrac{\partial u}{\partial y} + u\,\dfrac{\partial v}{\partial y}\right\rangle = v\left\langle \dfrac{\partial u}{\partial x}, \dfrac{\partial u}{\partial y}\right\rangle + u\left\langle \dfrac{\partial v}{\partial x}, \dfrac{\partial v}{\partial y}\right\rangle = v\,\nabla u + u\,\nabla v$

(c) $\nabla\left(\frac{u}{v}\right) = \left\langle \dfrac{v\,\frac{\partial u}{\partial x} - u\,\frac{\partial v}{\partial x}}{v^2}, \dfrac{v\,\frac{\partial u}{\partial y} - u\,\frac{\partial v}{\partial y}}{v^2} \right\rangle = \dfrac{v\left\langle \frac{\partial u}{\partial x}, \frac{\partial u}{\partial y}\right\rangle - u\left\langle \frac{\partial v}{\partial x}, \frac{\partial v}{\partial y}\right\rangle}{v^2} = \dfrac{v\,\nabla u - u\,\nabla v}{v^2}$

(d) $\nabla u^n = \left\langle \dfrac{\partial(u^n)}{\partial x}, \dfrac{\partial(u^n)}{\partial y} \right\rangle = \left\langle nu^{n-1}\dfrac{\partial u}{\partial x}, nu^{n-1}\dfrac{\partial u}{\partial y} \right\rangle = nu^{n-1}\,\nabla u$

30. If we place the initial point of the gradient vector $\nabla f(4,6)$ at $(4,6)$, the vector is perpendicular to the level curve of f that includes $(4,6)$, so we sketch a portion of the level curve through $(4,6)$ (using the nearby level curves as a guideline) and draw a line perpendicular to the curve at $(4,6)$. The gradient vector is parallel to this line, pointing in the direction of increasing function values, and with length equal to the maximum value of the directional derivative of f at $(4,6)$. We can estimate this length by finding the average rate of change in the direction of the gradient. The line intersects the contour lines corresponding to -2 and -3 with an estimated distance of 0.5 units. Thus the rate of change is approximately $\dfrac{-2-(-3)}{0.5} = 2$, and we sketch the gradient vector with length 2.

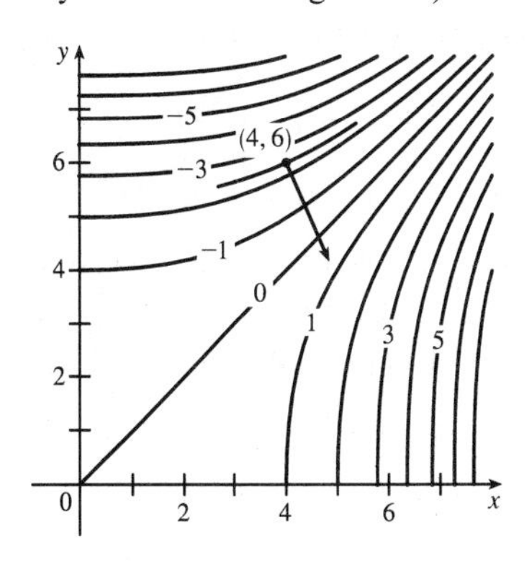

31. Let $F(x,y,z) = x^2 - 2y^2 + z^2 + yz$. Then $x^2 - 2y^2 + z^2 + yz = 2$ is a level surface of F and $\nabla F(x,y,z) = \langle 2x, -4y+z, 2z+y\rangle$.

(a) $\nabla F(2,1,-1) = \langle 4,-5,-1\rangle$ is a normal vector for the tangent plane at $(2,1,-1)$, so an equation of the tangent plane is $4(x-2) - 5(y-1) - 1(z+1) = 0$ or $4x - 5y - z = 4$.

(b) The normal line has direction $\langle 4,-5,-1\rangle$, so parametric equations are $x = 2+4t,\ y = 1-5t,\ z = -1-t$, and symmetric equations are $\dfrac{x-2}{4} = \dfrac{y-1}{-5} = \dfrac{z+1}{-1}$.

32. Let $F(x,y,z) = x - z - 4\arctan(yz)$. Then $x - z = 4\arctan(yz)$ is the level surface $F(x,y,z) = 0$, and $\nabla F(x,y,z) = \left\langle 1, -\dfrac{4z}{1+y^2z^2}, -1-\dfrac{4y}{1+y^2z^2}\right\rangle$.

(a) $\nabla F(1+\pi,1,1) = \langle 1,-2,-3\rangle$ and an equation of the tangent plane is $1(x-(1+\pi)) - 2(y-1) - 3(z-1) = 0$ or $x - 2y - 3z = -4+\pi$.

(b) The normal line has direction $\langle 1,-2,-3\rangle$, so parametric equations are $x = 1+\pi+t,\ y = 1-2t,\ z = 1-3t$, and symmetric equations are $x - 1 - \pi = \dfrac{y-1}{-2} = \dfrac{z-1}{-3}$.

33. $F(x,y,z) = -z + xe^y\cos z \quad\Rightarrow\quad \nabla F(x,y,z) = \langle e^y\cos z, xe^y\cos z, -1-xe^y\sin z\rangle$ and $\nabla F(1,0,0) = \langle 1,1,-1\rangle$.

(a) $1(x-1) + 1(y-0) - 1(z-0) = 0$ or $x + y - z = 1$

(b) $x - 1 = y = -z$

34. $F(x,y,z) = yz - \ln(x+z) \quad\Rightarrow\quad \nabla F(x,y,z) = \left\langle -\dfrac{1}{x+z}, z, y - \dfrac{1}{x+z}\right\rangle$ and $\nabla F(0,0,1) = \langle -1,1,-1\rangle$.

(a) $(-1)(x-0) + (1)(y-0) - 1(z-1) = 0$ or $x - y + z = 1$

(b) Parametric equations are $x = -t,\ y = t,\ z = 1-t$ and symmetric equations are $\dfrac{x}{-1} = \dfrac{y}{1} = \dfrac{z-1}{-1}$ or $-x = y = 1-z$.

35. $F(x, y, z) = xy + yz + zx$,
$\nabla F(x, y, z) = \langle y + z, x + z, y + x \rangle$,
$\nabla F(1, 1, 1) = \langle 2, 2, 2 \rangle$, so an equation of the tangent plane is $2x + 2y + 2z = 6$ or $x + y + z = 3$, and the normal line is given by $x - 1 = y - 1 = z - 1$ or $x = y = z$.

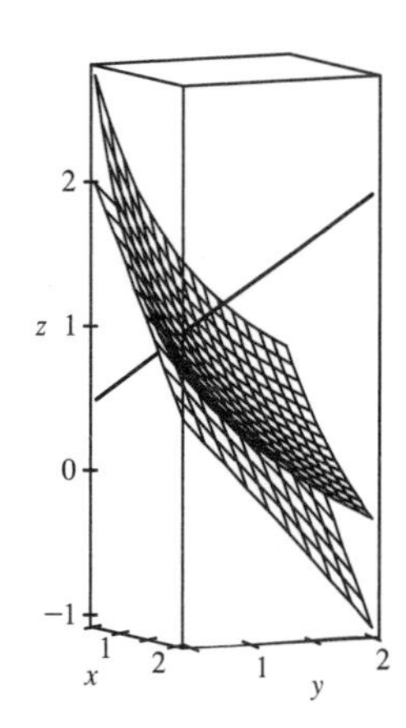

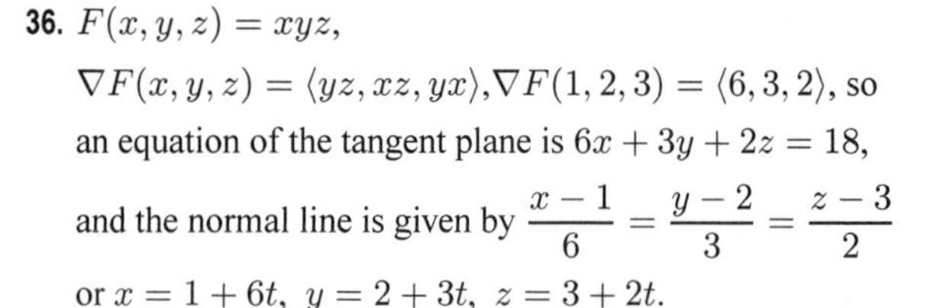

36. $F(x, y, z) = xyz$,
$\nabla F(x, y, z) = \langle yz, xz, yx \rangle, \nabla F(1, 2, 3) = \langle 6, 3, 2 \rangle$, so an equation of the tangent plane is $6x + 3y + 2z = 18$, and the normal line is given by $\dfrac{x-1}{6} = \dfrac{y-2}{3} = \dfrac{z-3}{2}$ or $x = 1 + 6t,\ y = 2 + 3t,\ z = 3 + 2t$.

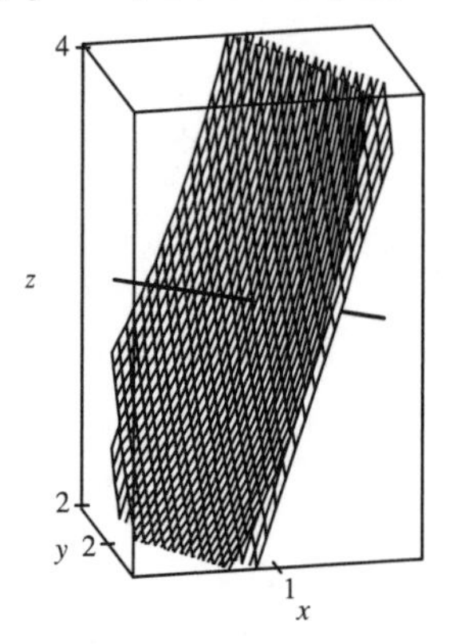

37. $\nabla f(x, y) = \langle 2x, 8y \rangle,\ \nabla f(2, 1) = \langle 4, 8 \rangle$.
The tangent line has equation $\nabla f(2, 1) \cdot \langle x - 2, y - 1 \rangle = 0 \ \Rightarrow$
$4(x - 2) + 8(y - 1) = 0$, which simplifies to $x + 2y = 4$.

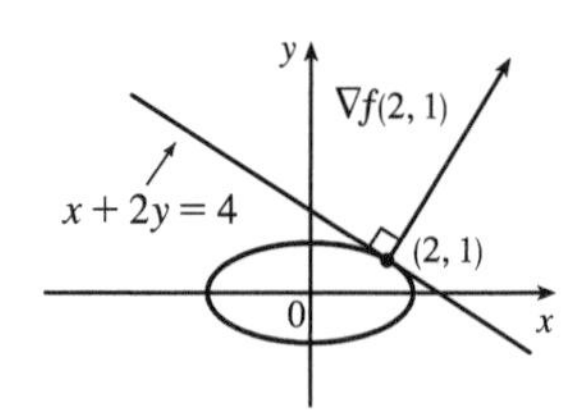

38. $\nabla g(x, y) = \langle 1, -2y \rangle,\ \nabla g(3, -1) = \langle 1, 2 \rangle$.
The tangent line has equation $\nabla g(3, -1) \cdot \langle x - 3, y + 1 \rangle = 0 \ \Rightarrow$
$1(x - 3) + 2(y + 1) = 0$, which simplifies to $x + 2y = 1$.

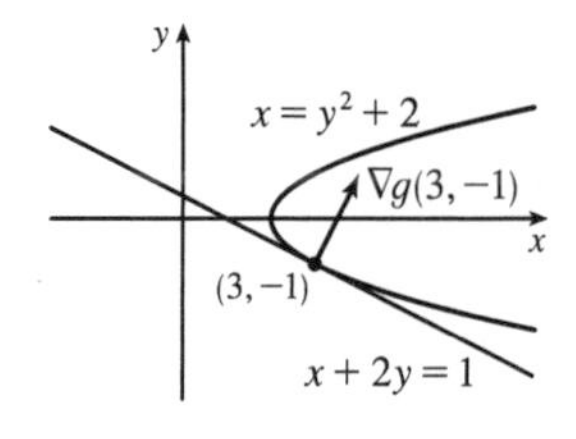

39. $\nabla F(x_0, y_0, z_0) = \left\langle \dfrac{2x_0}{a^2}, \dfrac{2y_0}{b^2}, \dfrac{2z_0}{c^2} \right\rangle$. Thus an equation of the tangent plane at (x_0, y_0, z_0) is

$\dfrac{2x_0}{a^2}\,x + \dfrac{2y_0}{b^2}\,y + \dfrac{2z_0}{c^2}\,z = 2\left(\dfrac{x_0^2}{a^2} + \dfrac{y_0^2}{b^2} + \dfrac{z_0^2}{c^2}\right) = 2(1) = 2$ since (x_0, y_0, z_0) is a point on the ellipsoid. Hence

$\dfrac{x_0}{a^2}\,x + \dfrac{y_0}{b^2}\,y + \dfrac{z_0}{c^2}\,z = 1$ is an equation of the tangent plane.

40. Since $\nabla f(x_0, y_0, z_0) = \langle 2x_0, 4y_0, 6z_0 \rangle$ and $\langle 3, -1, 3 \rangle$ are both normal vectors to the surface at (x_0, y_0, z_0), we need $\langle 2x_0, 4y_0, 6z_0 \rangle = c\,\langle 3, -1, 3 \rangle$ or $\langle x_0, 2y_0, 3z_0 \rangle = k\langle 3, -1, 3 \rangle$. Thus $x_0 = 3k$, $y_0 = -\frac{1}{2}k$ and $z_0 = k$. But $x_0^2 + 2y_0^2 + 3z_0^2 = 1$ or $\left(9 + \frac{1}{2} + 3\right)k^2 = 1$, so $k = \pm\frac{\sqrt{2}}{5}$ and there are two such points: $\left(\pm\frac{3\sqrt{2}}{5}, \mp\frac{1}{5\sqrt{2}}, \pm\frac{\sqrt{2}}{5}\right)$.

41. $\nabla f(x_0, y_0, z_0) = \langle 2x_0, -2y_0, 4z_0 \rangle$ and the given line has direction numbers $2, 4, 6$, so $\langle 2x_0, -2y_0, 4z_0 \rangle = k\langle 2, 4, 6 \rangle$ or $x_0 = k$, $y_0 = -2k$ and $z_0 = \frac{3}{2}k$. But $x_0^2 - y_0^2 + 2z_0^2 = 1$ or $\left(1 - 4 + \frac{9}{2}\right)k^2 = 1$, so $k = \pm\sqrt{\frac{2}{3}} = \pm\frac{\sqrt{6}}{3}$ and there are two such points: $\left(\pm\frac{\sqrt{6}}{3}, \mp\frac{2\sqrt{6}}{3}, \pm\frac{\sqrt{6}}{2}\right)$.

42. First note that the point $(1, 1, 2)$ is on both surfaces. For the ellipsoid, an equation of the tangent plane at $(1, 1, 2)$ is $6x + 4y + 4z = 18$ or $3x + 2y + 2z = 9$, and for the sphere, an equation of the tangent plane at $(1, 1, 2)$ is $(2 - 8)x + (2 - 6)y + (4 - 8)z = -18$ or $-6x - 4y - 4z = -18$ or $3x + 2y + 2z = 9$. Since these tangent planes are the same, the surfaces are tangent to each other at the point $(1, 1, 2)$.

43. Let (x_0, y_0, z_0) be a point on the surface. Then an equation of the tangent plane at the point is $\dfrac{x}{2\sqrt{x_0}} + \dfrac{y}{2\sqrt{y_0}} + \dfrac{z}{2\sqrt{z_0}} = \dfrac{\sqrt{x_0} + \sqrt{y_0} + \sqrt{z_0}}{2}$. But $\sqrt{x_0} + \sqrt{y_0} + \sqrt{z_0} = \sqrt{c}$, so the equation is $\dfrac{x}{\sqrt{x_0}} + \dfrac{y}{\sqrt{y_0}} + \dfrac{z}{\sqrt{z_0}} = \sqrt{c}$. The x-, y-, and z-intercepts are $\sqrt{cx_0}$, $\sqrt{cy_0}$ and $\sqrt{cz_0}$ respectively. (The x-intercept is found by setting $y = z = 0$ and solving the resulting equation for x, and the y- and z-intercepts are found similarly.) So the sum of the intercepts is $\sqrt{c}\left(\sqrt{x_0} + \sqrt{y_0} + \sqrt{z_0}\right) = c$, a constant.

44. Let (x_0, y_0, z_0) be a point on the sphere. Then the normal line is given by $\dfrac{x - x_0}{2x_0} = \dfrac{y - y_0}{2y_0} = \dfrac{z - z_0}{2z_0}$. For the center $(0, 0, 0)$ to be on the line, we need $-\dfrac{x_0}{2x_0} = -\dfrac{y_0}{2y_0} = -\dfrac{z_0}{2z_0}$ or equivalently $1 = 1 = 1$, which is true.

45. If $f(x, y, z) = z - x^2 - y^2$ and $g(x, y, z) = 4x^2 + y^2 + z^2$, then the tangent line is perpendicular to both ∇f and ∇g at $(-1, 1, 2)$. The vector $\mathbf{v} = \nabla f \times \nabla g$ will therefore be parallel to the tangent line.

We have $\nabla f(x, y, z) = \langle -2x, -2y, 1\rangle \;\Rightarrow\; \nabla f(-1, 1, 2) = \langle 2, -2, 1\rangle$, and $\nabla g(x, y, z) = \langle 8x, 2y, 2z\rangle \;\Rightarrow$

$\nabla g(-1, 1, 2) = \langle -8, 2, 4\rangle$. Hence $\mathbf{v} = \nabla f \times \nabla g = \begin{vmatrix} \mathbf{i} & \mathbf{j} & \mathbf{k} \\ 2 & -2 & 1 \\ -8 & 2 & 4 \end{vmatrix} = -10\,\mathbf{i} - 16\,\mathbf{j} - 12\,\mathbf{k}$.

Parametric equations are: $x = -1 - 10t,\; y = 1 - 16t,\; z = 2 - 12t$.

46. (a) Let $f(x, y, z) = y + z$ and $g(x, y, z) = x^2 + y^2$. Then the required tangent line is perpendicular to both ∇f and ∇g at $(1, 2, 1)$ and the vector $\mathbf{v} = \nabla f \times \nabla g$ is parallel to the tangent line. We have

$\nabla f(x, y, z) = \langle 0, 1, 1\rangle \;\Rightarrow\; \nabla f(1, 2, 1) = \langle 0, 1, 1\rangle$, and

$\nabla g(x, y, z) = \langle 2x, 2y, 0\rangle \;\Rightarrow\; \nabla g(1, 2, 1) = \langle 2, 4, 0\rangle$. Hence

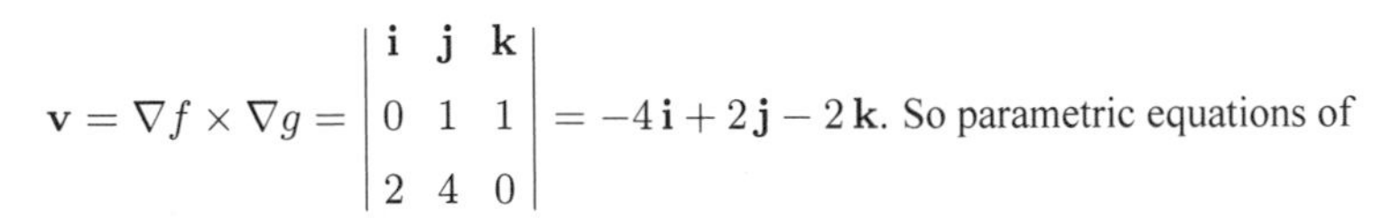

$\mathbf{v} = \nabla f \times \nabla g = \begin{vmatrix} \mathbf{i} & \mathbf{j} & \mathbf{k} \\ 0 & 1 & 1 \\ 2 & 4 & 0 \end{vmatrix} = -4\,\mathbf{i} + 2\,\mathbf{j} - 2\,\mathbf{k}$. So parametric equations of the desired tangent line are $x = 1 - 4t,\; y = 2 + 2t,\; z = 1 - 2t$.

(b)

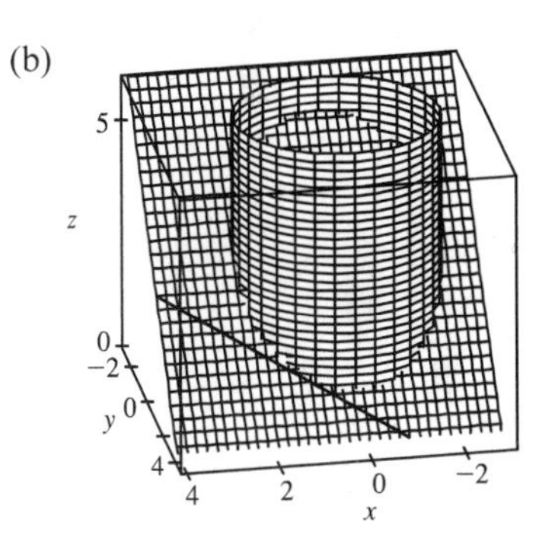

47. (a) The direction of the normal line of F is given by ∇F, and that of G by ∇G. Assuming that $\nabla F \neq 0 \neq \nabla G$, the two normal lines are perpendicular at P if $\nabla F \cdot \nabla G = 0$ at $P \;\Leftrightarrow$

$\langle \partial F/\partial x, \partial F/\partial y, \partial F/\partial z\rangle \cdot \langle \partial G/\partial x, \partial G/\partial y, \partial G/\partial z\rangle = 0$ at $P \;\Leftrightarrow\; F_x G_x + F_y G_y + F_z G_z = 0$ at P.

(b) Here $F = x^2 + y^2 - z^2$ and $G = x^2 + y^2 + z^2 - r^2$, so

$\nabla F \cdot \nabla G = \langle 2x, 2y, -2z\rangle \cdot \langle 2x, 2y, 2z\rangle = 4x^2 + 4y^2 - 4z^2 = 4F = 0$, since the point (x, y, z) lies on the graph of

$F = 0$. To see that this is true without using calculus, note that $G = 0$ is the equation of a sphere centered at the origin and $F = 0$ is the equation of a right circular cone with vertex at the origin (which is generated by lines through the origin). At any point of intersection, the sphere's normal line (which passes through the origin) lies on the cone, and thus is perpendicular to the cone's normal line. So the surfaces with equations $F = 0$ and $G = 0$ are everywhere orthogonal.

48. (a) The function $f(x, y) = (xy)^{1/3}$ is continuous on $\mathbb{R}^2$ since it is a composition of a polynomial and the cube root function, both of which are continuous. (See the text just after Example 11.2.8.)

$$f_x(0,0) = \lim_{h\to 0} \frac{f(0+h,0) - f(0,0)}{h} = \lim_{h\to 0} \frac{(h\cdot 0)^{1/3} - 0}{h} = 0,$$

$$f_y(0,0) = \lim_{h\to 0} \frac{f(0,0+h) - f(0,0)}{h} = \lim_{h\to 0} \frac{(0\cdot h)^{1/3} - 0}{h} = 0.$$

Therefore, $f_x(0,0)$ and $f_y(0,0)$ do exist and are equal to 0. Now let $\mathbf{u}$ be any unit vector other than $\mathbf{i}$ and $\mathbf{j}$ (these correspond to f_x and f_y respectively.) Then $\mathbf{u} = a\,\mathbf{i} + b\,\mathbf{j}$ where $a \neq 0$ and $b \neq 0$. Thus

$$D_{\mathbf{u}}\, f(0,0) = \lim_{h\to 0} \frac{f(0+ha, 0+hb) - f(0,0)}{h} = \lim_{h\to 0} \frac{\sqrt[3]{(ha)(hb)}}{h} = \lim_{h\to 0} \frac{\sqrt[3]{ab}}{h^{1/3}}$$ and this limit does not exist, so $D_{\mathbf{u}}\, f(0,0)$ does not exist.

(b)

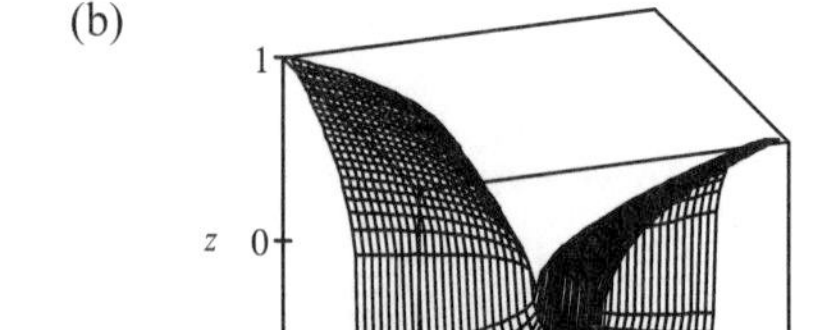

Notice that if we start at the origin and proceed in the direction of the x- or y-axis, then the graph is flat. But if we proceed in any other direction, then the graph is extremely steep.

49. Let $\mathbf{u} = \langle a, b\rangle$ and $\mathbf{v} = \langle c, d\rangle$. Then we know that at the given point, $D_{\mathbf{u}}\, f = \nabla f \cdot \mathbf{u} = af_x + bf_y$ and $D_{\mathbf{v}}\, f = \nabla f \cdot \mathbf{v} = cf_x + df_y$. But these are just two linear equations in the two unknowns f_x and f_y, and since $\mathbf{u}$ and $\mathbf{v}$ are not parallel, we can solve the equations to find $\nabla f = \langle f_x, f_y\rangle$ at the given point. In fact,

$$\nabla f = \left\langle \frac{d\,D_{\mathbf{u}}\, f - b\,D_{\mathbf{v}}\, f}{ad - bc}, \frac{a\,D_{\mathbf{v}}\, f - c\,D_{\mathbf{u}}\, f}{ad - bc} \right\rangle.$$

50. Since $z = f(x, y)$ is differentiable at $\mathbf{x}_0 = (x_0, y_0)$, by Definition 11.4.7 we have $\Delta z = f_x(x_0, y_0)\,\Delta x + f_y(x_0, y_0)\,\Delta y + \varepsilon_1\,\Delta x + \varepsilon_2\,\Delta y$ where $\varepsilon_1, \varepsilon_2 \to 0$ as $(\Delta x, \Delta y) \to (0,0)$. Now $\Delta z = f(\mathbf{x}) - f(\mathbf{x}_0)$, $\langle \Delta x, \Delta y\rangle = \mathbf{x} - \mathbf{x}_0$ so $(\Delta x, \Delta y) \to (0,0)$ is equivalent to $\mathbf{x} \to \mathbf{x}_0$ and $\langle f_x(x_0, y_0), f_y(x_0, y_0)\rangle = \nabla f(\mathbf{x}_0)$. Substituting into (11.4.7) gives $f(\mathbf{x}) - f(\mathbf{x}_0) = \nabla f(\mathbf{x}_0)\cdot(\mathbf{x} - \mathbf{x}_0) + \langle \varepsilon_1, \varepsilon_2\rangle \cdot \langle \Delta x, \Delta y\rangle$ or $\langle \varepsilon_1, \varepsilon_2\rangle \cdot (\mathbf{x} - \mathbf{x}_0) = f(\mathbf{x}) - f(\mathbf{x}_0) - \nabla f(\mathbf{x}_0)\cdot(\mathbf{x} - \mathbf{x}_0)$,

and so $\dfrac{f(\mathbf{x}) - f(\mathbf{x}_0) - \nabla f(\mathbf{x}_0)\cdot(\mathbf{x} - \mathbf{x}_0)}{|\mathbf{x} - \mathbf{x}_0|} = \dfrac{\langle \varepsilon_1, \varepsilon_2\rangle \cdot (\mathbf{x} - \mathbf{x}_0)}{|\mathbf{x} - \mathbf{x}_0|}$. But $\dfrac{\mathbf{x} - \mathbf{x}_0}{|\mathbf{x} - \mathbf{x}_0|}$ is a unit vector so

$\displaystyle\lim_{\mathbf{x}\to\mathbf{x}_0} \frac{\langle \varepsilon_1, \varepsilon_2\rangle \cdot (\mathbf{x} - \mathbf{x}_0)}{|\mathbf{x} - \mathbf{x}_0|} = 0$ since $\varepsilon_1, \varepsilon_2 \to 0$ as $\mathbf{x} \to \mathbf{x}_0$. Hence $\displaystyle\lim_{\mathbf{x}\to\mathbf{x}_0} \frac{f(\mathbf{x}) - f(\mathbf{x}_0) - \nabla f(\mathbf{x}_0)\cdot(\mathbf{x} - \mathbf{x}_0)}{|\mathbf{x} - \mathbf{x}_0|} = 0.$

11.7 Maximum and Minimum Values

1. (a) First we compute $D(1,1) = f_{xx}(1,1)\, f_{yy}(1,1) - [f_{xy}(1,1)]^2 = (4)(2) - (1)^2 = 7$. Since $D(1,1) > 0$ and $f_{xx}(1,1) > 0$, f has a local minimum at $(1,1)$ by the Second Derivatives Test.

(b) $D(1,1) = f_{xx}(1,1)\, f_{yy}(1,1) - [f_{xy}(1,1)]^2 = (4)(2) - (3)^2 = -1$. Since $D(1,1) < 0$, f has a saddle point at $(1,1)$ by the Second Derivatives Test.

2. In the figure, points at approximately $(-1,1)$ and $(-1,-1)$ are enclosed by oval-shaped level curves which indicate that as we move away from either point in any direction, the values of f are increasing. Hence we would expect local minima at or near $(-1,\pm 1)$. Similarly, the point $(1,0)$ appears to be enclosed by oval-shaped level curves which indicate that as we move away from the point in any direction the values of f are decreasing, so we should have a local maximum there. We also show hyperbola-shaped level curves near the points $(-1,0)$, $(1,1)$, and $(1,-1)$. The values of f increase along some paths leaving these points and decrease in others, so we should have a saddle point at each of these points.

To confirm our predictions, we have $f(x,y) = 3x - x^3 - 2y^2 + y^4 \Rightarrow f_x(x,y) = 3 - 3x^2$, $f_y(x,y) = -4y + 4y^3$. Setting these partial derivatives equal to 0, we have $3 - 3x^2 = 0 \Rightarrow x = \pm 1$ and $-4y + 4y^3 = 0 \Rightarrow y(y^2-1) = 0 \Rightarrow y = 0, \pm 1$. So our critical points are $(\pm 1, 0)$, $(\pm 1, \pm 1)$.

The second partial derivatives are $f_{xx}(x,y) = -6x$, $f_{xy}(x,y) = 0$, and $f_{yy}(x,y) = 12y^2 - 4$, so $D(x,y) = f_{xx}(x,y)\, f_{yy}(x,y) - [f_{xy}(x,y)]^2 = (-6x)(12y^2 - 4) - (0)^2 = -72xy^2 + 24x$.

We use the Second Derivatives Test to classify the 6 critical points:

Critical Point	D	f_{xx}	Conclusion
$(1,0)$	24	-6	$D > 0,\ f_{xx} < 0 \Rightarrow f$ has a local maximum at $(1,0)$
$(1,1)$	-48		$D < 0 \Rightarrow f$ has a saddle point at $(1,1)$
$(1,-1)$	-48		$D < 0 \Rightarrow f$ has a saddle point at $(1,-1)$
$(-1,0)$	-24		$D < 0 \Rightarrow f$ has a saddle point at $(-1,0)$
$(-1,1)$	48	6	$D > 0,\ f_{xx} > 0 \Rightarrow f$ has a local minimum at $(-1,1)$
$(-1,-1)$	48	6	$D > 0,\ f_{xx} > 0 \Rightarrow f$ has a local minimum at $(-1,-1)$

3. $f(x,y) = 9 - 2x + 4y - x^2 - 4y^2 \Rightarrow f_x = -2 - 2x$, $f_y = 4 - 8y$, $f_{xx} = -2$, $f_{xy} = 0$, $f_{yy} = -8$. Then $f_x = 0$ and $f_y = 0$ imply $x = -1$ and $y = \frac{1}{2}$, and the only critical point is $\left(-1, \frac{1}{2}\right)$.

$D(x,y) = f_{xx} f_{yy} - (f_{xy})^2 = (-2)(-8) - 0^2 = 16$, and since $D\left(-1, \frac{1}{2}\right) = 16 > 0$ and $f_{xx}\left(-1, \frac{1}{2}\right) = -2 < 0$, $f\left(-1, \frac{1}{2}\right) = 11$ is a local maximum by the Second Derivatives Test.

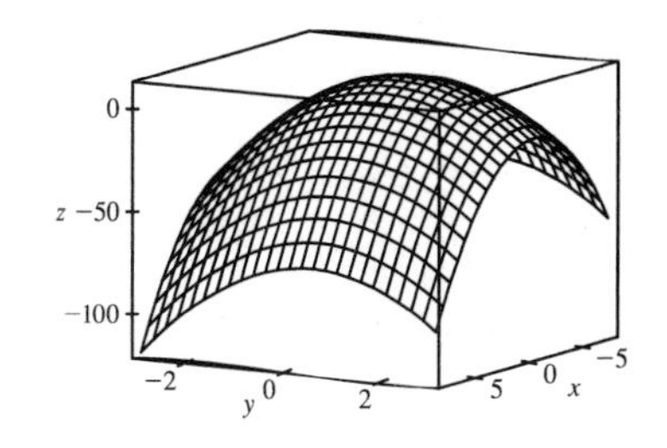

4. $f(x, y) = x^3y + 12x^2 - 8y \Rightarrow f_x = 3x^2y + 24x$,

$f_y = x^3 - 8,\ f_{xx} = 6xy + 24,\ f_{xy} = 3x^2,\ f_{yy} = 0$.

Then $f_y = 0$ implies $x = 2$, and substitution into $f_x = 0$ gives

$12y + 48 = 0 \Rightarrow y = -4$. Thus, the only critical point is $(2, -4)$.

$D(2, -4) = (-24)(0) - 12^2 = -144 < 0$, so $(2, -4)$ is a saddle point.

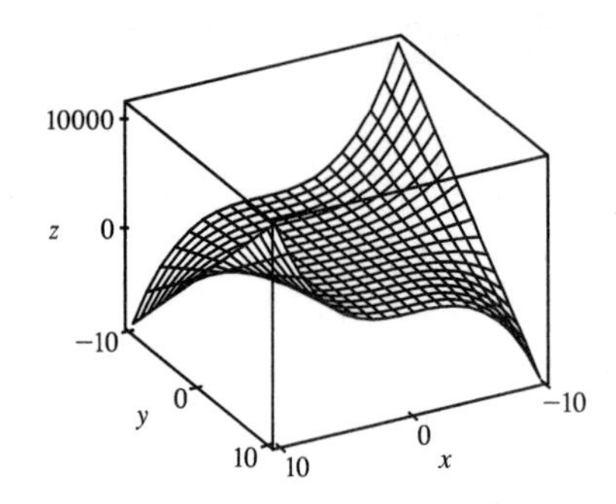

5. $f(x, y) = x^4 + y^4 - 4xy + 2 \Rightarrow f_x = 4x^3 - 4y, f_y = 4y^3 - 4x$,

$f_{xx} = 12x^2,\ f_{xy} = -4, f_{yy} = 12y^2$. Then $f_x = 0$ implies $y = x^3$,

and substitution into $f_y = 0 \Rightarrow x = y^3$ gives $x^9 - x = 0 \Rightarrow$

$x(x^8 - 1) = 0 \Rightarrow x = 0$ or $x = \pm 1$. Thus the critical points are $(0, 0)$,

$(1, 1)$, and $(-1, -1)$. Now $D(0, 0) = 0 \cdot 0 - (-4)^2 = -16 < 0$,

so $(0, 0)$ is a saddle point. $D(1, 1) = (12)(12) - (-4)^2 > 0$ and

$f_{xx}(1, 1) = 12 > 0$, so $f(1, 1) = 0$ is a local minimum. $D(-1, -1) = (12)(12) - (-4)^2 > 0$ and

$f_{xx} = (-1, -1) = 12 > 0$, so $f(-1, -1) = 0$ is also a local minimum.

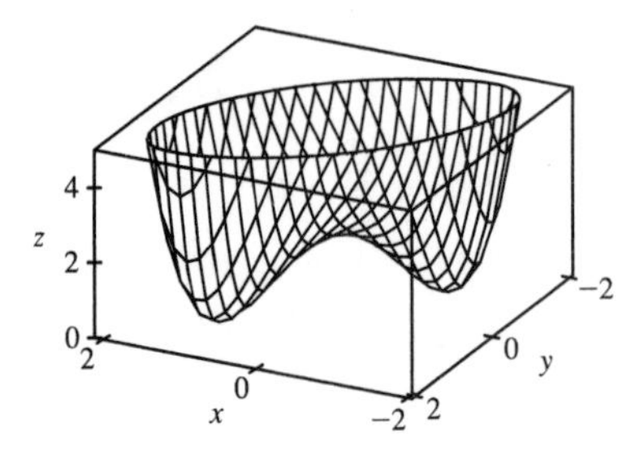

6. $f(x, y) = e^{4y - x^2 - y^2} \Rightarrow f_x = -2xe^{4y - x^2 - y^2}$,

$f_y = (4 - 2y)e^{4y - x^2 - y^2},\ f_{xx} = (4x^2 - 2)e^{4y - x^2 - y^2}$,

$f_{xy} = -2x(4 - 2y)e^{4y - x^2 - y^2}, f_{yy} = (4y^2 - 16y + 14)e^{4y - x^2 - y^2}$.

Then $f_x = 0$ and $f_y = 0$ implies $x = 0$ and $y = 2$, so the only critical

point is $(0, 2)$. Now $D(0, 2) = (-2e^4)(-2e^4) - 0^2 = 4e^8 > 0$ and

$f_{xx}(0, 2) = -2e^4 < 0$, so $f(0, 2) = e^4$ is a local maximum.

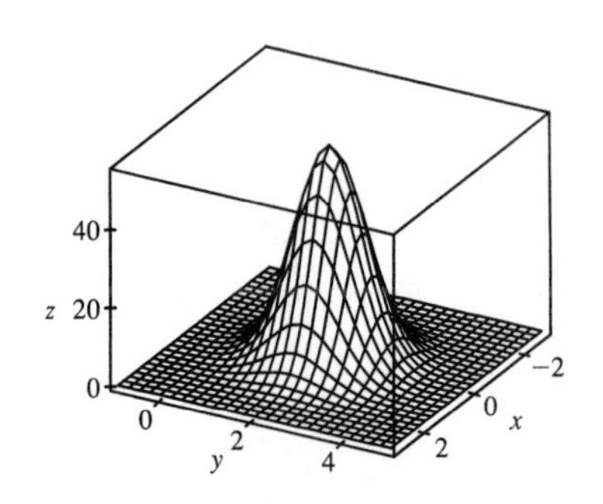

7. $f(x, y) = (1 + xy)(x + y) = x + y + x^2y + xy^2 \Rightarrow$

$f_x = 1 + 2xy + y^2,\ f_y = 1 + x^2 + 2xy,\ f_{xx} = 2y,\ f_{xy} = 2x + 2y$,

$f_{yy} = 2x$. Then $f_x = 0$ implies $1 + 2xy + y^2 = 0$ and $f_y = 0$ implies

$1 + x^2 + 2xy = 0$. Subtracting the second equation from the first gives

$y^2 - x^2 = 0 \Rightarrow y = \pm x$, but if $y = x$ then $1 + 2xy + y^2 = 0 \Rightarrow$

$1 + 3x^2 = 0$ which has no real solution. If $y = -x$ then

$1 + 2xy + y^2 = 0 \Rightarrow 1 - x^2 = 0 \Rightarrow x = \pm 1$, so critical points

are $(1, -1)$ and $(-1, 1)$. $D(1, -1) = (-2)(2) - 0 < 0$ and $D(-1, 1) = (2)(-2) - 0 < 0$, so $(-1, 1)$ and $(1, -1)$ are saddle points.

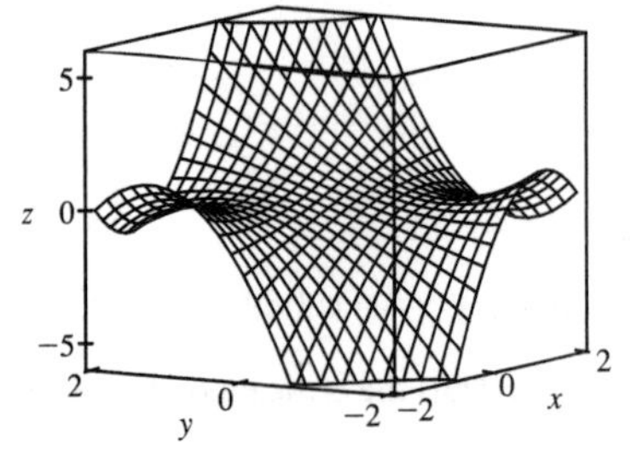

8. $f(x, y) = 2x^3 + xy^2 + 5x^2 + y^2 \Rightarrow \quad f_x = 6x^2 + y^2 + 10x$,

$f_y = 2xy + 2y,\ f_{xx} = 12x + 10,\ f_{yy} = 2x + 2,\ f_{xy} = 2y$. Then

$f_y = 0$ implies $y = 0$ or $x = -1$. Substituting into $f_x = 0$ gives the

critical points $(0, 0)$, $\left(-\frac{5}{3}, 0\right)$, $(-1, \pm 2)$. Now $D(0, 0) = 20 > 0$ and

$f_{xx}(0, 0) = 10 > 0$, so $f(0, 0) = 0$ is a local minimum. Also

$f_{xx}\left(-\frac{5}{3}, 0\right) < 0$, $D\left(-\frac{5}{3}, 0\right) > 0$, and $D(-1, \pm 2) < 0$. Hence

$f\left(-\frac{5}{3}, 0\right) = \frac{125}{27}$ is a local maximum while $(-1, \pm 2)$ are saddle points.

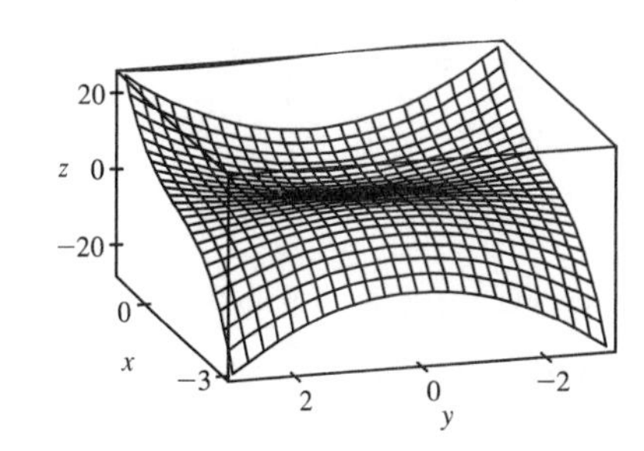

9. $f(x, y) = e^x \cos y \quad \Rightarrow \quad f_x = e^x \cos y,\ f_y = -e^x \sin y$.

Now $f_x = 0$ implies $\cos y = 0$ or $y = \frac{\pi}{2} + n\pi$ for n an integer.

But $\sin\left(\frac{\pi}{2} + n\pi\right) \neq 0$, so there are no critical points.

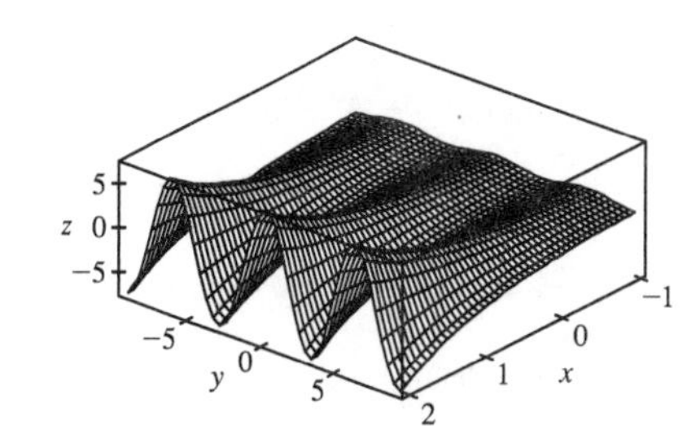

10. $f(x, y) = x^2 + y^2 + \dfrac{1}{x^2 y^2} \quad \Rightarrow \quad f_x = 2x - 2x^{-3}y^{-2}$,

$f_y = 2y - 2x^{-2}y^{-3},\ f_{xx} = 2 + 6x^{-4}y^{-2},\ f_{yy} = 2 + 6x^{-2}y^{-4}$,

$f_{xy} = 4x^{-3}y^{-3}$. Then $f_x = 0$ implies $2x^4y^2 - 2 = 0$ or $x^4y^2 = 1$ or

$y^2 = x^{-4}$. Note that neither x nor y can be zero. Now $f_y = 0$ implies

$2x^2y^4 - 2 = 0$, and with $y^2 = x^{-4}$ this implies $2x^{-6} - 2 = 0$

or $x^6 = 1$. Thus $x = \pm 1$ and if $x = 1$, $y = \pm 1$; if $x = -1$, $y = \pm 1$.

So the critical points are $(1, 1)$, $(1, -1)$,$(-1, 1)$ and $(-1, -1)$. Now $D(\pm 1, \pm 1) = D(\pm 1, \mp 1) = 64 - 16 > 0$ and $f_{xx} > 0$ always, so $f(\pm 1, \pm 1) = f(\pm 1, \mp 1) = 3$ are local minima.

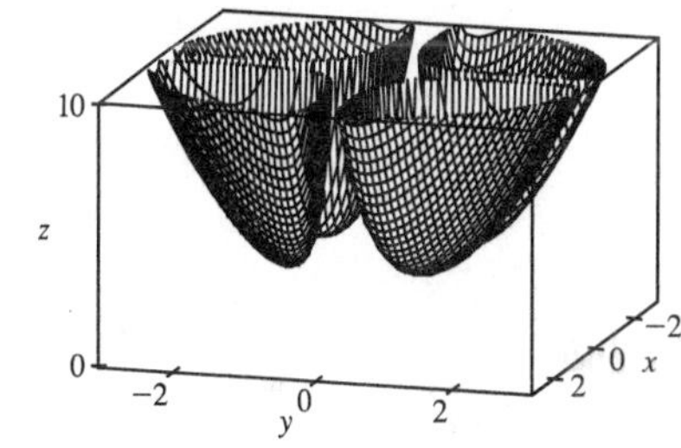

11. $f(x, y) = x \sin y \quad \Rightarrow \quad f_x = \sin y,\ f_y = x \cos y,\ f_{xx} = 0$,

$f_{yy} = -x \sin y,\ f_{xy} = \cos y$. Then $f_x = 0$ if and only if $y = n\pi$,

n an integer, and substituting into $f_y = 0$ requires $x = 0$ for each of these

y-values. Thus the critical points are $(0, n\pi)$, n an integer. But

$D(0, n\pi) = -\cos^2(n\pi) < 0$ so each critical point is a saddle point.

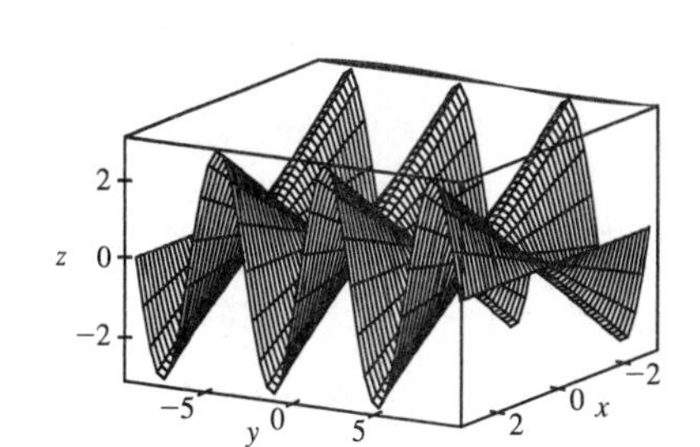

12. $f(x,y) = (2x - x^2)(2y - y^2) \Rightarrow \ f_x = (2 - 2x)(2y - y^2)$,

$f_y = (2x - x^2)(2 - 2y)$, $f_{xx} = -2(2y - y^2)$, $f_{yy} = -2(2x - x^2)$,

$f_{xy} = (2 - 2x)(2 - 2y)$. Then $f_x = 0$ implies $x = 1$ or $y = 0$ or $y = 2$ and when $x = 1$, $f_y = 0$ implies $y = 1$, when $y = 0$, $f_y = 0$ implies $x = 0$ or $x = 2$ and when $y = 2$, $f_y = 0$ implies $x = 0$ or $x = 2$. Thus the critical points are $(1,1)$, $(0,0)$, $(2,0)$, $(0,2)$ and $(2,2)$. Now $D(0,0) = D(2,0) = D(0,2) = D(2,2) = -16$ so these critical points are saddle points, and $D(1,1) = 4$ with $f_{xx}(1,1) = -2$, so $f(1,1) = 1$ is a local maximum.

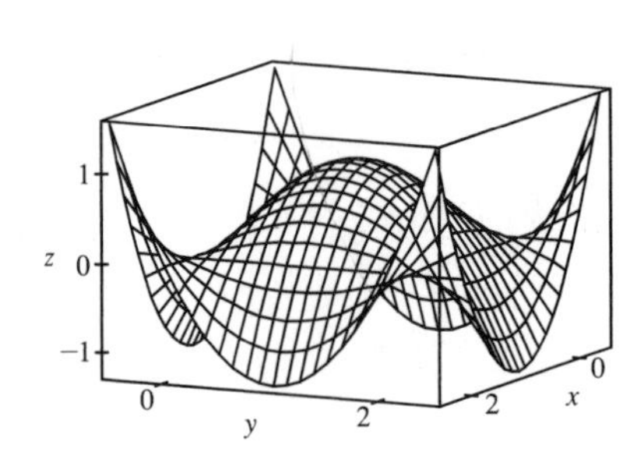

13. $f(x,y) = (x^2 + y^2)e^{y^2 - x^2} \quad \Rightarrow$

$f_x = (x^2 + y^2)e^{y^2-x^2}(-2x) + 2xe^{y^2-x^2} = 2xe^{y^2-x^2}(1 - x^2 - y^2)$,

$f_y = (x^2 + y^2)e^{y^2-x^2}(2y) + 2ye^{y^2-x^2} = 2ye^{y^2-x^2}(1 + x^2 + y^2)$,

$f_{xx} = 2xe^{y^2-x^2}(-2x) + (1 - x^2 - y^2)\left(2x\left(-2xe^{y^2-x^2}\right) + 2e^{y^2-x^2}\right) = 2e^{y^2-x^2}((1 - x^2 - y^2)(1 - 2x^2) - 2x^2)$,

$f_{xy} = 2xe^{y^2-x^2}(-2y) + 2x(2y)e^{y^2-x^2}(1 - x^2 - y^2) = -4xye^{y^2-x^2}(x^2 + y^2)$,

$f_{yy} = 2ye^{y^2-x^2}(2y) + (1 + x^2 + y^2)\left(2y\left(2ye^{y^2-x^2}\right) + 2e^{y^2-x^2}\right) = 2e^{y^2-x^2}((1 + x^2 + y^2)(1 + 2y^2) + 2y^2)$.

$f_y = 0$ implies $y = 0$, and substituting into $f_x = 0$ gives $2xe^{-x^2}(1 - x^2) = 0 \quad \Rightarrow \quad x = 0$ or $x = \pm 1$. Thus the critical points are $(0,0)$ and $(\pm 1, 0)$. Now $D(0,0) = (2)(2) - 0 > 0$ and $f_{xx}(0,0) = 2 > 0$, so $f(0,0) = 0$ is a local minimum. $D(\pm 1, 0) = (-4e^{-1})(4e^{-1}) - 0 < 0$ so $(\pm 1, 0)$ are saddle points.

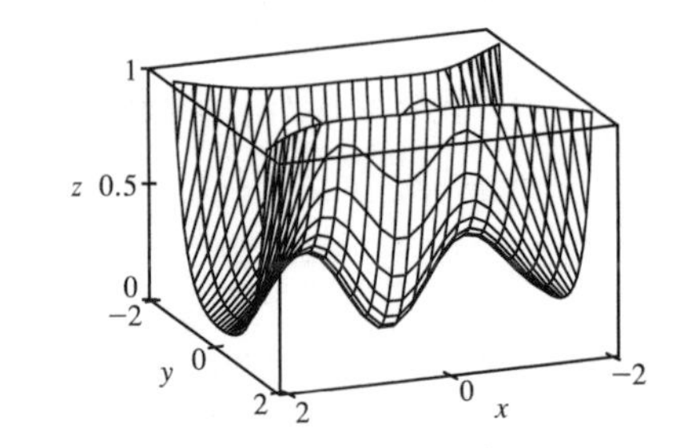

14. $f(x,y) = x^2ye^{-x^2-y^2} \quad \Rightarrow$

$f_x = x^2ye^{-x^2-y^2}(-2x) + 2xye^{-x^2-y^2} = 2xy(1 - x^2)e^{-x^2-y^2}$,

$f_y = x^2ye^{-x^2-y^2}(-2y) + x^2e^{-x^2-y^2} = x^2(1 - 2y^2)e^{-x^2-y^2}$,

$f_{xx} = 2y(2x^4 - 5x^2 + 1)e^{-x^2-y^2}$,

$f_{xy} = 2x(1 - x^2)(1 - 2y^2)e^{-x^2-y^2}$, $f_{yy} = 2x^2y(2y^2 - 3)e^{-x^2-y^2}$.

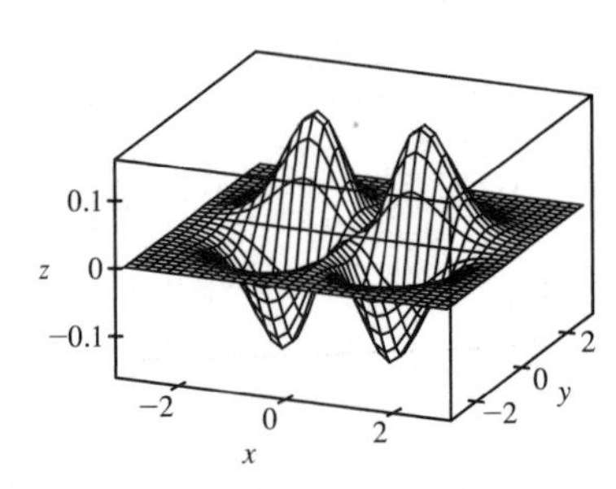

$f_x = 0$ implies $x = 0$, $y = 0$, or $x = \pm 1$. If $x = 0$ then $f_y = 0$ for any y-value, so all points of the form $(0, y)$ are critical points. If $y = 0$ then $f_y = 0 \quad \Rightarrow \quad x^2e^{-x^2} = 0 \quad \Rightarrow \quad x = 0$, so $(0,0)$ (already included above) is a critical point. If $x = \pm 1$ then $(1 - 2y^2)e^{-1-y^2} = 0 \quad \Rightarrow \quad y = \pm\frac{1}{\sqrt{2}}$, so $\left(1, \pm\frac{1}{\sqrt{2}}\right)$ and $\left(-1, \pm\frac{1}{\sqrt{2}}\right)$ are critical points. $D(0,y) = 0$, so the Second Derivatives Test gives no information. However, if $y > 0$ then $x^2ye^{-x^2-y^2} \geq 0$ with equality only when $x = 0$, so we have local minimum values $f(0,y)=0$, $y>0$. Similarly, if $y<0$ then $x^2ye^{-x^2-y^2}\le 0$ with equality when $x=0$ so $f(0,y)=0$, $y<0$ are local maximum values, and $(0,0)$ is a saddle point.

$D\left(\pm1,\frac{1}{\sqrt{2}}\right)=8e^{-3}>0$, $f_{xx}\left(\pm1,\frac{1}{\sqrt{2}}\right)=-2\sqrt{2}e^{-3/2}<0$ and

$D\left(\pm1,-\frac{1}{\sqrt{2}}\right)=8e^{-3}>0$, $f_{xx}\left(\pm1,-\frac{1}{\sqrt{2}}\right)=2\sqrt{2}e^{-3/2}>0$, so $f\left(\pm1,\frac{1}{\sqrt{2}}\right)=\frac{1}{\sqrt{2}}e^{-3/2}$ are local maximum points

while $f\left(\pm1,-\frac{1}{\sqrt{2}}\right)=-\frac{1}{\sqrt{2}}e^{-3/2}$ are local minimum points.

15. $f(x,y)=3x^2y+y^3-3x^2-3y^2+2$

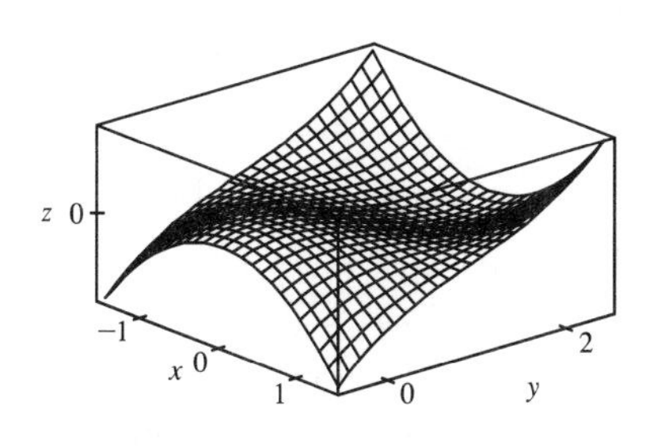

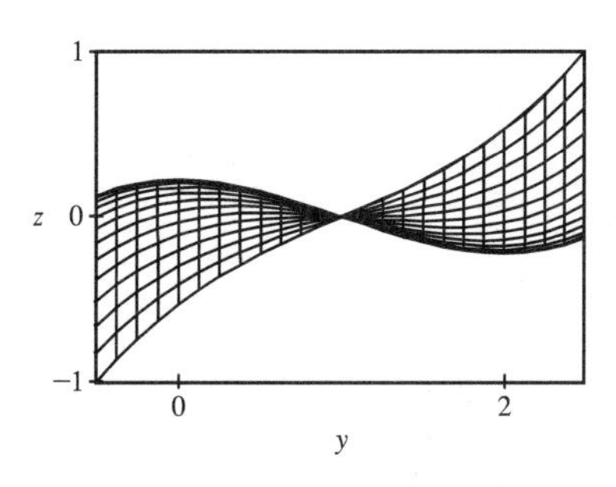

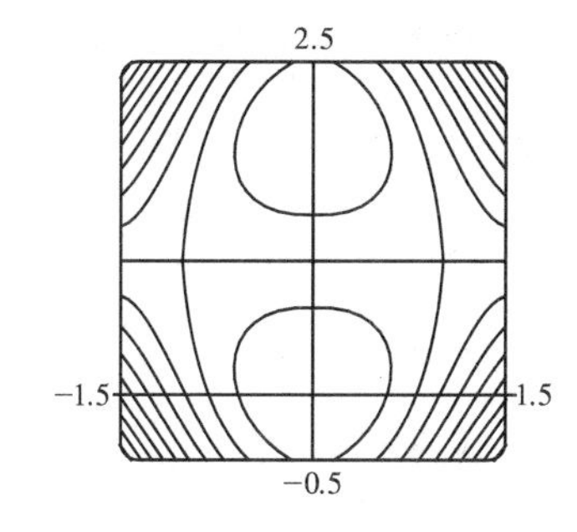

From the graphs, it appears that f has a local maximum $f(0,0)\approx 2$ and a local minimum $f(0,2)\approx -2$. There appear to be saddle points near $(\pm1,1)$.

$f_x=6xy-6x$, $f_y=3x^2+3y^2-6y$. Then $f_x=0$ implies $x=0$ or $y=1$ and when $x=0$, $f_y=0$ implies $y=0$ or $y=2$; when $y=1$, $f_y=0$ implies $x^2=1$ or $x=\pm1$. Thus the critical points are $(0,0)$, $(0,2)$, $(\pm1,1)$. Now $f_{xx}=6y-6$, $f_{yy}=6y-6$, $f_{xy}=6x$, so $D(0,0)=D(0,2)=36>0$ while $D(\pm1,1)=-36<0$ and $f_{xx}(0,0)=-6$, $f_{xx}(0,2)=6$. Hence $(\pm1,1)$ are saddle points while $f(0,0)=2$ is a local maximum and $f(0,2)=-2$ is a local minimum.

16. $f(x,y)=xye^{-x^2-y^2}$

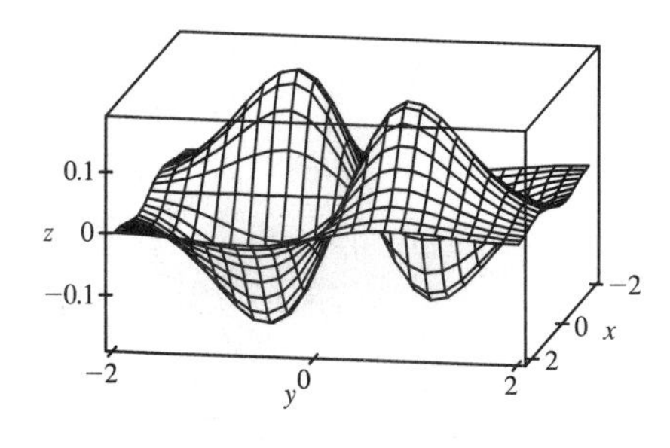

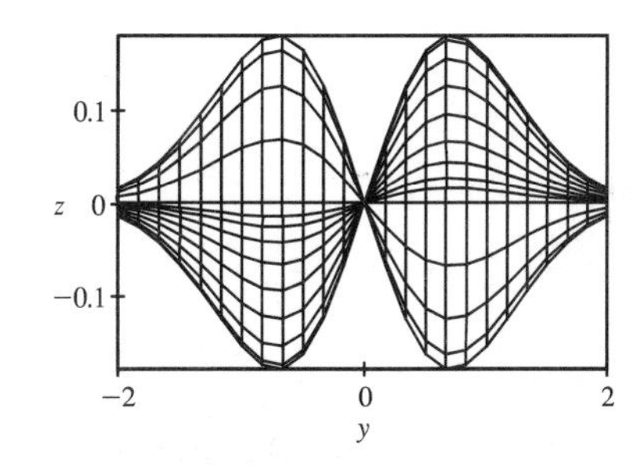

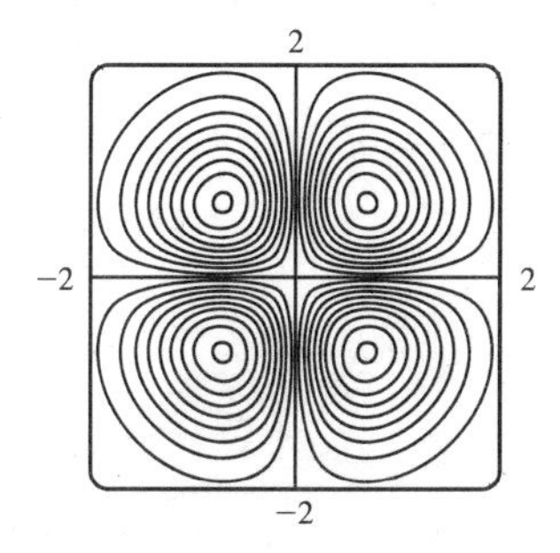

There appear to be local maxima of about $f(\pm0.7,\pm0.7)\approx 0.18$ and local minima of about $f(\pm0.7,\mp0.7)\approx -0.18$. Also, there seems to be a saddle point at the origin.

$f_x=ye^{-x^2-y^2}(1-2x^2)$, $f_y=xe^{-x^2-y^2}(1-2y^2)$, $f_{xx}=2xye^{-x^2-y^2}(2x^2-3)$, $f_{yy}=2xye^{-x^2-y^2}(2y^2-3)$, $f_{xy}=(1-2x^2)e^{-x^2-y^2}(1-2y^2)$. Then $f_x=0$ implies $y=0$ or $x=\pm\frac{1}{\sqrt{2}}$.

Substituting these values into $f_y=0$ gives the critical points $(0,0)$, $\left(\frac{1}{\sqrt{2}},\pm\frac{1}{\sqrt{2}}\right)$, $\left(-\frac{1}{\sqrt{2}},\pm\frac{1}{\sqrt{2}}\right)$. Then

$D(x, y) = e^{2(-x^2-y^2)}\left[4x^2y^2(2x^2-3)(2y^2-3) - (1-2x^2)^2(1-2y^2)^2\right]$, so $D(0,0) = -1$, while $D\left(\frac{1}{\sqrt{2}}, \pm\frac{1}{\sqrt{2}}\right) > 0$ and $D\left(-\frac{1}{\sqrt{2}}, \pm\frac{1}{\sqrt{2}}\right) > 0$. But $f_{xx}\left(\frac{1}{\sqrt{2}}, \frac{1}{\sqrt{2}}\right) < 0, f_{xx}\left(\frac{1}{\sqrt{2}}, -\frac{1}{\sqrt{2}}\right) > 0,\ f_{xx}\left(-\frac{1}{\sqrt{2}}, \frac{1}{\sqrt{2}}\right) > 0,\ f_{xx}\left(-\frac{1}{\sqrt{2}}, -\frac{1}{\sqrt{2}}\right) < 0$.

Hence $(0,0)$ is a saddle point; $f\left(\frac{1}{\sqrt{2}}, -\frac{1}{\sqrt{2}}\right) = f\left(-\frac{1}{\sqrt{2}}, \frac{1}{\sqrt{2}}\right) = -\frac{1}{2e}$ are local minima and

$f\left(\frac{1}{\sqrt{2}}, \frac{1}{\sqrt{2}}\right) = f\left(-\frac{1}{\sqrt{2}}, -\frac{1}{\sqrt{2}}\right) = \frac{1}{2e}$ are local maxima.

17. $f(x, y) = \sin x + \sin y + \sin(x+y),\ 0 \le x \le 2\pi,\ 0 \le y \le 2\pi$

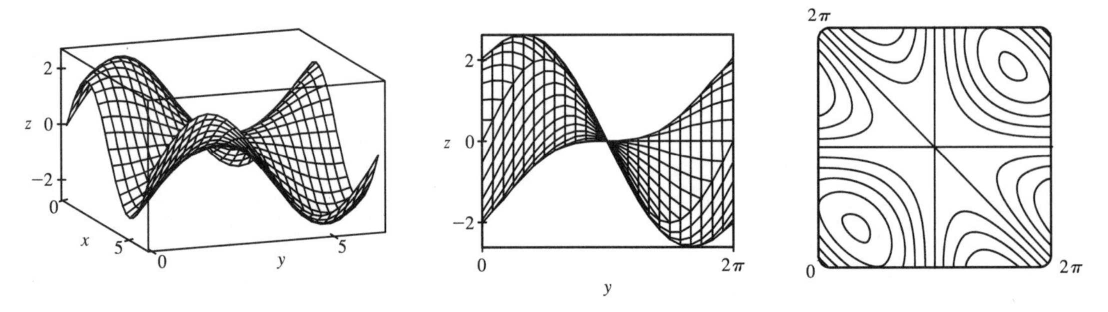

From the graphs it appears that f has a local maximum at about $(1,1)$ with value approximately 2.6, a local minimum at about $(5,5)$ with value approximately -2.6, and a saddle point at about $(3,3)$.

$f_x = \cos x + \cos(x+y),\ f_y = \cos y + \cos(x+y),\ f_{xx} = -\sin x - \sin(x+y),\ f_{yy} = -\sin y - \sin(x+y)$, $f_{xy} = -\sin(x+y)$. Setting $f_x = 0$ and $f_y = 0$ and subtracting gives $\cos x - \cos y = 0$ or $\cos x = \cos y$. Thus $x = y$ or $x = 2\pi - y$. If $x = y$, $f_x = 0$ becomes $\cos x + \cos 2x = 0$ or $2\cos^2 x + \cos x - 1 = 0$, a quadratic in $\cos x$. Thus $\cos x = -1$ or $\frac{1}{2}$ and $x = \pi, \frac{\pi}{3}$, or $\frac{5\pi}{3}$, yielding the critical points (π, π), $\left(\frac{\pi}{3}, \frac{\pi}{3}\right)$ and $\left(\frac{5\pi}{3}, \frac{5\pi}{3}\right)$. Similarly if $x = 2\pi - y$, $f_x = 0$ becomes $(\cos x) + 1 = 0$ and the resulting critical point is (π, π).

Now $D(x, y) = \sin x\ \sin y + \sin x\ \sin(x+y) + \sin y\ \sin(x+y)$. So $D(\pi, \pi) = 0$ and the Second Derivatives Test doesn't apply. $D\left(\frac{\pi}{3}, \frac{\pi}{3}\right) = \frac{9}{4} > 0$ and $f_{xx}\left(\frac{\pi}{3}, \frac{\pi}{3}\right) < 0$ so $f\left(\frac{\pi}{3}, \frac{\pi}{3}\right) = \frac{3\sqrt{3}}{2}$ is a local maximum while $D\left(\frac{5\pi}{3}, \frac{5\pi}{3}\right) = \frac{9}{4} > 0$ and $f_{xx}\left(\frac{5\pi}{3}, \frac{5\pi}{3}\right) > 0$, so $f\left(\frac{5\pi}{3}, \frac{5\pi}{3}\right) = -\frac{3\sqrt{3}}{2}$ is a local minimum.

18. $f(x, y) = \sin x + \sin y + \cos(x+y),\ 0 \le x \le \frac{\pi}{4},\ 0 \le y \le \frac{\pi}{4}$

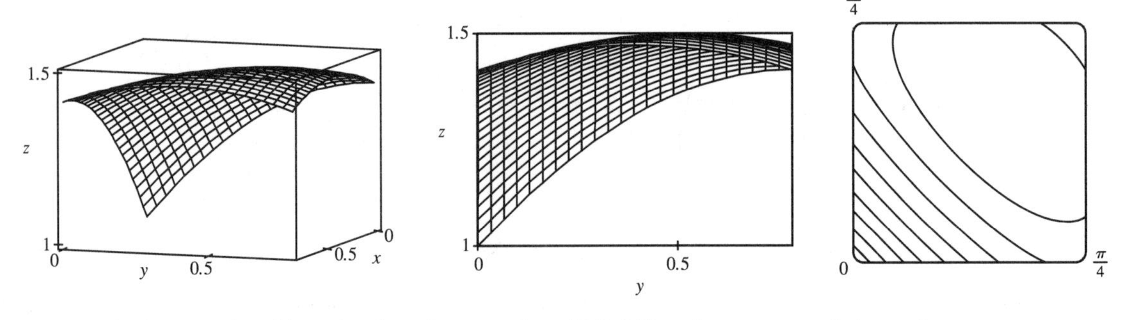

From the graphs, it seems that f has a local maximum at about $(0.5, 0.5)$. $f_x = \cos x - \sin(x+y)$, $f_y = \cos y - \sin(x+y),\ f_{xx} = -\sin x - \cos(x+y),\ f_{yy} = -\sin y - \cos(x+y),\ f_{xy} = -\cos(x+y)$. Setting $f_x = 0$ and $f_y = 0$ and subtracting gives $\cos x = \cos y$. Thus $x = y$. Substituting $x = y$ into $f_x = 0$ gives $\cos x - \sin 2x = 0$ or

$\cos x(1 - 2\sin x) = 0$. But $\cos x \neq 0$ for $0 \leq x \leq \frac{\pi}{4}$ and $1 - 2\sin x = 0$ implies $x = \frac{\pi}{6}$, so the only critical point is $\left(\frac{\pi}{6}, \frac{\pi}{6}\right)$. Here $f_{xx}\left(\frac{\pi}{6}, \frac{\pi}{6}\right) = -1 < 0$ and $D\left(\frac{\pi}{6}, \frac{\pi}{6}\right) = (-1)^2 - \frac{1}{4} > 0$. Thus $f\left(\frac{\pi}{6}, \frac{\pi}{6}\right) = \frac{3}{2}$ is a local maximum.

19. $f(x, y) = x^4 - 5x^2 + y^2 + 3x + 2 \Rightarrow f_x(x, y) = 4x^3 - 10x + 3$ and $f_y(x, y) = 2y$. $f_y = 0 \Rightarrow y = 0$, and the graph of f_x shows that the roots of $f_x = 0$ are approximately $x = -1.714$, 0.312 and 1.402. (Alternatively, we could have used a calculator or a CAS to find these roots.) So to three decimal places, the critical points are $(-1.714, 0)$, $(1.402, 0)$, and $(0.312, 0)$. Now since $f_{xx} = 12x^2 - 10$, $f_{xy} = 0$, $f_{yy} = 2$, and $D = 24x^2 - 20$, we have $D(-1.714, 0) > 0$, $f_{xx}(-1.714, 0) > 0$, $D(1.402, 0) > 0$, $f_{xx}(1.402, 0) > 0$, and $D(0.312, 0) < 0$. Therefore $f(-1.714, 0) \approx -9.200$ and $f(1.402, 0) \approx 0.242$ are local minima, and $(0.312, 0)$ is a saddle point. The lowest point on the graph is approximately $(-1.714, 0, -9.200)$.

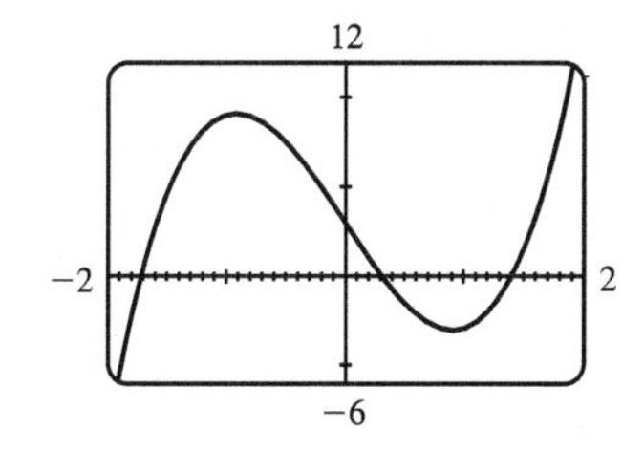

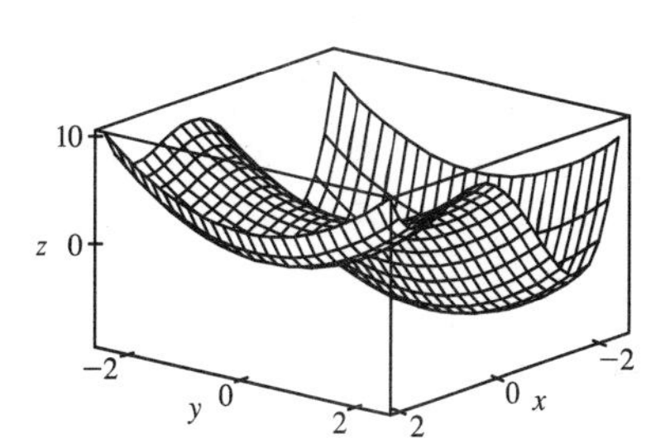

20. $f(x, y) = 5 - 10xy - 4x^2 + 3y - y^4 \Rightarrow f_x(x, y) = -10y - 8x$, $f_y(x, y) = -10x + 3 - 4y^3$.
Now $f_x = 0 \Rightarrow x = -\frac{5}{4}y$, so using a graph, we find solutions to
$0 = f_y\left(-\frac{5}{4}y, y\right) = -10\left(-\frac{5}{4}y\right) + 3 - 4y^3 = -4y^3 + \frac{25}{2}y + 3$. (Alternatively, we could have found the roots of $f_x = f_y = 0$ directly, using a calculator or a CAS.) To three decimal places, the solutions are $y \approx 1.877$, -0.245, and -1.633, so f has critical points at approximately $(-2.347, 1.877)$, $(0.306, -0.245)$, and $(2.041, -1.633)$. Now since $f_{xx} = -8$, $f_{xy} = -10$, $f_{yy} = -12y^2$, and $D = 96y^2 - 100$, we have $D(-2.347, 1.877) > 0$, $D(0.306, -0.245) < 0$, and $D(2.041, -1.633) > 0$. Therefore, since $f_{xx} < 0$ everywhere, $f(-2.347, 1.877) \approx 20.238$ and $f(2.041, -1.633) \approx 9.657$ are local maxima, and $(0.306, -0.245)$ is a saddle point. The highest point on the graph is approximately $(-2.347, 1.877, 20.238)$.

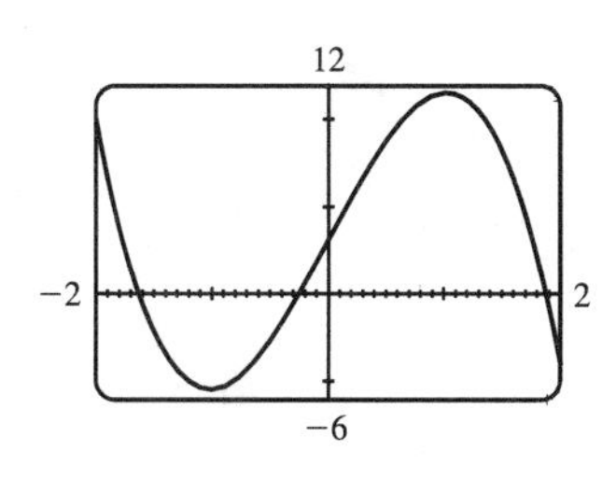

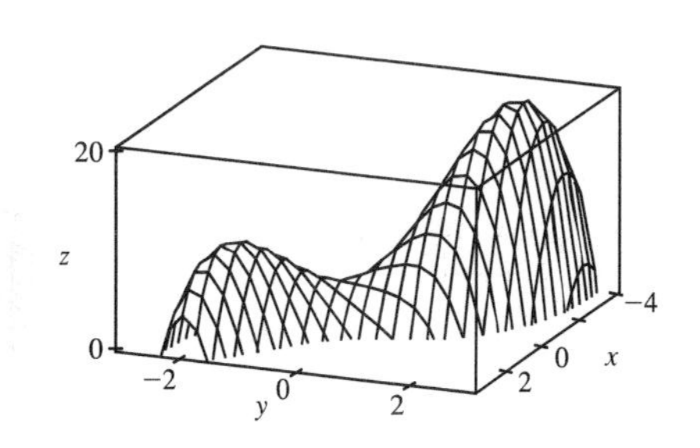

21. $f(x, y) = 2x + 4x^2 - y^2 + 2xy^2 - x^4 - y^4 \Rightarrow f_x(x, y) = 2 + 8x + 2y^2 - 4x^3$, $f_y(x, y) = -2y + 4xy - 4y^3$.
Now $f_y = 0 \Leftrightarrow 2y(2y^2 - 2x + 1) = 0 \Leftrightarrow y = 0$ or $y^2 = x - \frac{1}{2}$. The first of these implies that $f_x = -4x^3 + 8x + 2$, and the second implies that $f_x = 2 + 8x + 2\left(x - \frac{1}{2}\right) - 4x^3 = -4x^3 + 10x + 1$. From the graphs, we see that the first possibility for f_x has roots at approximately -1.267, -0.259, and 1.526, and the second has a root at

approximately 1.629 (the negative roots do not give critical points, since $y^2 = x - \frac{1}{2}$ must be positive). So to three decimal places, f has critical points at $(-1.267, 0)$, $(-0.259, 0)$, $(1.526, 0)$, and $(1.629, \pm 1.063)$. Now since $f_{xx} = 8 - 12x^2$, $f_{xy} = 4y$, $f_{yy} = 4x - 12y^2$, and $D = (8 - 12x^2)(4x - 12y^2) - 16y^2$, we have $D(-1.267, 0) > 0$, $f_{xx}(-1.267, 0) > 0$, $D(-0.259, 0) < 0$, $D(1.526, 0) < 0$, $D(1.629, \pm 1.063) > 0$, and $f_{xx}(1.629, \pm 1.063) < 0$. Therefore, to three decimal places, $f(-1.267, 0) \approx 1.310$ and $f(1.629, \pm 1.063) \approx 8.105$ are local maxima, and $(-0.259, 0)$ and $(1.526, 0)$ are saddle points. The highest points on the graph are approximately $(1.629, \pm 1.063, 8.105)$.

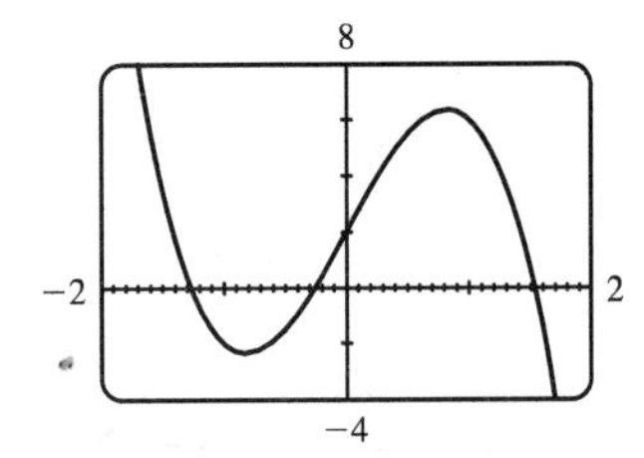

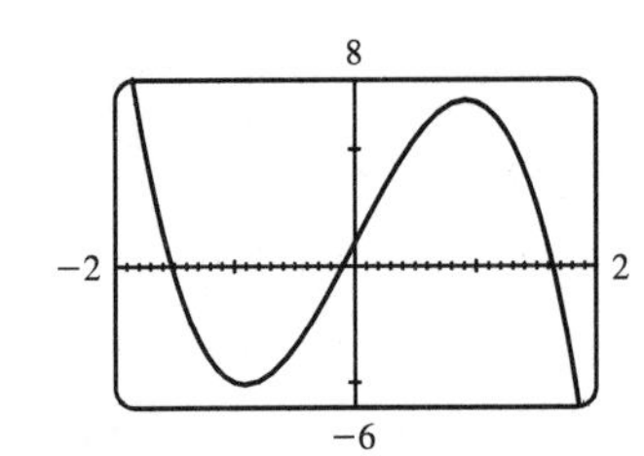

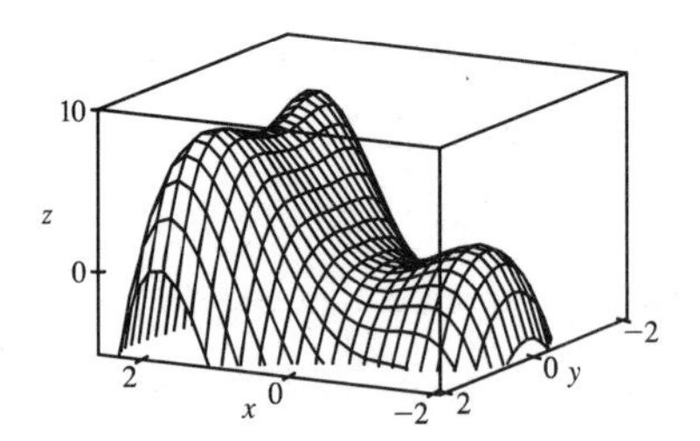

22. $f(x, y) = e^x + y^4 - x^3 + 4\cos y \Rightarrow f_x(x, y) = e^x - 3x^2$ and $f_y(x, y) = 4y^3 - 4\sin y$. From the graphs, we see that to three decimal places, $f_x = 0$ when $x \approx -0.459$, 0.910, or 3.733, and $f_y = 0$ when $y \approx 0$ or ± 0.929. (Alternatively, we could have used a calculator or a CAS to find the roots of $f_x = 0$ and $f_y = 0$.) So, to three decimal places, f has critical points at $(-0.459, 0)$, $(-0.459, \pm 0.929)$, $(0.910, 0)$, $(0.910, \pm 0.929)$, $(3.733, 0)$, and $(3.733, \pm 0.929)$. Now $f_{xx} = e^x - 6x$, $f_{xy} = 0$, $f_{yy} = 12y^2 - 4\cos y$, and $D = (e^x - 6x)(12y^2 - 4\cos y)$. Therefore $D(-0.459, 0) < 0$, $D(-0.459, \pm 0.929) > 0$, $f_{xx}(-0.459, \pm 0.929) > 0$, $D(0.910, 0) > 0$, $f_{xx}(0.910, 0) < 0$, $D(0.910, \pm 0.929) < 0$, $D(3.733, 0) < 0$, $D(3.733, \pm 0.929) > 0$, and $f_{xx}(3.733, \pm 0.929) > 0$. So $f(-0.459, \pm 0.929) \approx 3.868$ and $f(3.733, \pm 0.929) \approx -7.077$ are local minima, $f(0.910, 0) \approx 5.731$ is a local maximum, and $(-0.459, 0)$, $(0.910, \pm 0.929)$, and $(3.733, 0)$ are saddle points. The lowest points on the graph are approximately $(3.733, \pm 0.929, -7.077)$.

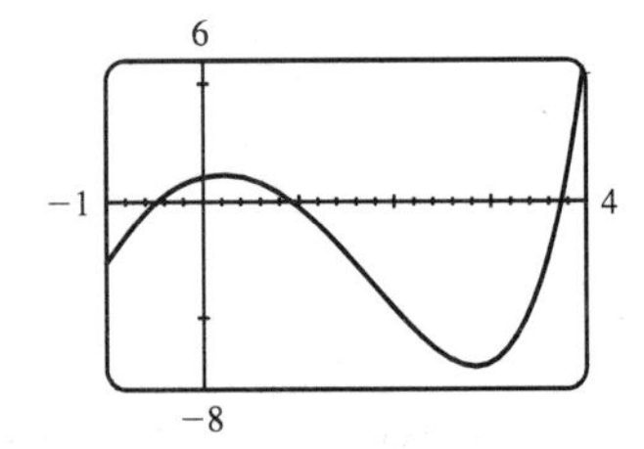

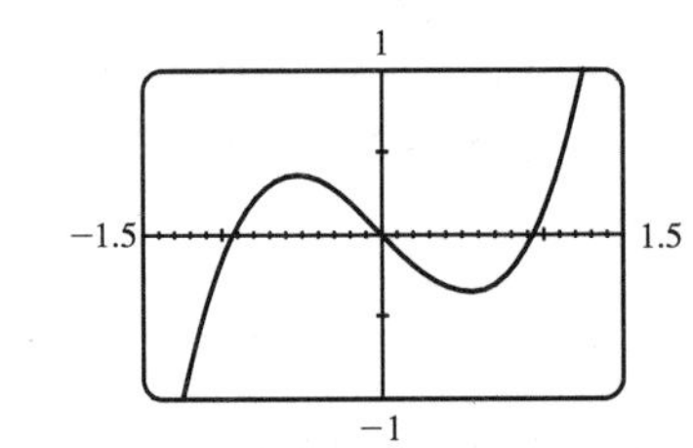

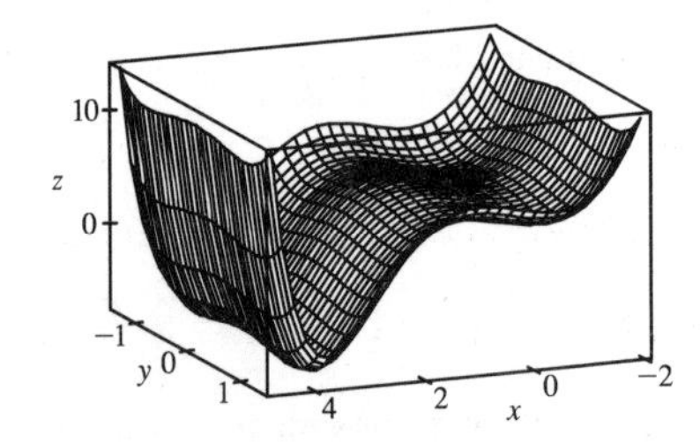

23. Since f is a polynomial it is continuous on D, so an absolute maximum and minimum exist. Here $f_x = 4$, $f_y = -5$ so there are no critical points inside D. Thus the absolute extrema must both occur on the boundary. Along L_1: $x = 0$ and $f(0, y) = 1 - 5y$ for $0 \le y \le 3$, a decreasing function in y, so the maximum value is $f(0, 0) = 1$ and the minimum value

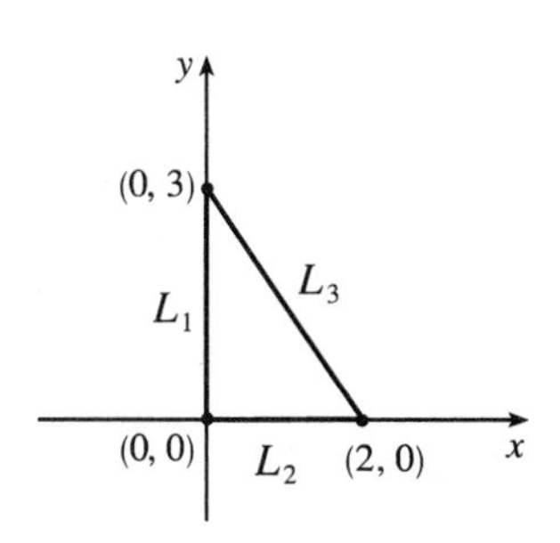

is $f(0,3) = -14$. Along L_2: $y = 0$ and $f(x,0) = 1 + 4x$ for $0 \le x \le 2$, an increasing function in x, so the minimum value is $f(0,0) = 1$ and the maximum value is $f(2,0) = 9$. Along L_3: $y = -\frac{3}{2}x + 3$ and $f\left(x, -\frac{3}{2}x + 3\right) = \frac{23}{2}x - 14$ for $0 \le x \le 2$, an increasing function in x, so the minimum value is $f(0,3) = -14$ and the maximum value is $f(2,0) = 9$. Thus the absolute maximum of f on D is $f(2,0) = 9$ and the absolute minimum is $f(0,3) = -14$.

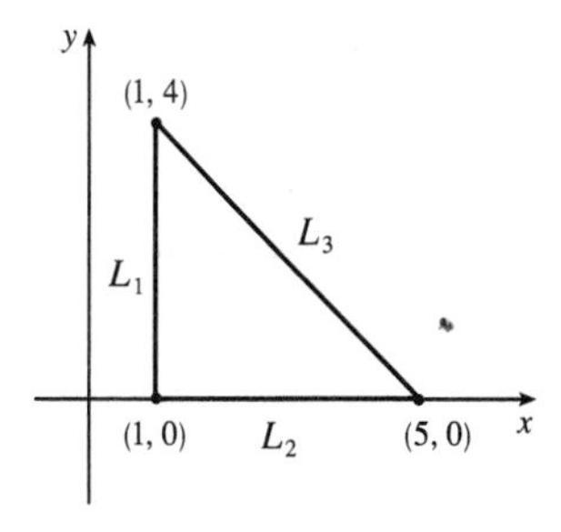

24. Since f is a polynomial it is continuous on D, so an absolute maximum and minimum exist. $f_x = y - 1$, $f_y = x - 2$, and setting $f_x = f_y = 0$ gives $(2,1)$ as the only critical point, where $f(2,1) = 1$. Along L_1: $x = 1$ and $f(1,y) = 2 - y$ for $0 \le y \le 4$, a decreasing function in y, so the maximum value is $f(1,0) = 2$ and the minimum value is $f(1,4) = -2$. Along L_2: $y = 0$ and $f(x,0) = 3 - x$ for $1 \le x \le 5$, a decreasing function in x, so the maximum value is $f(1,0) = 2$ and the minimum value is $f(5,0) = -2$. Along L_3: $y = 5 - x$ and $f(x, 5-x) = -x^2 + 6x - 7 = -(x-3)^2 + 2$ for $1 \le x \le 5$, which has a maximum at $x = 3$ where $f(3,2) = 2$ and a minimum at both $x = 1$ and $x = 5$, where $f(1,4) = f(5,0) = -2$. Thus the absolute maximum of f on D is $f(1,0) = f(3,2) = 2$ and the absolute minimum is $f(1,4) = f(5,0) = -2$.

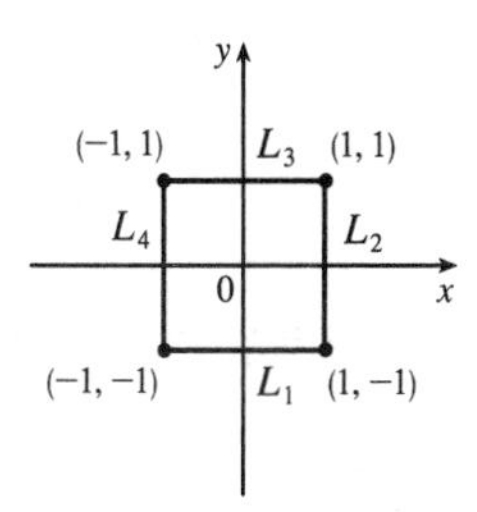

25. $f_x(x,y) = 2x + 2xy$, $f_y(x,y) = 2y + x^2$, and setting $f_x = f_y = 0$ gives $(0,0)$ as the only critical point in D, with $f(0,0) = 4$.

On L_1: $y = -1$, $f(x,-1) = 5$, a constant.

On L_2: $x = 1$, $f(1,y) = y^2 + y + 5$, a quadratic in y which attains its maximum at $(1,1)$, $f(1,1) = 7$ and its minimum at $\left(1, -\frac{1}{2}\right)$, $f\left(1, -\frac{1}{2}\right) = \frac{19}{4}$.

On L_3: $f(x,1) = 2x^2 + 5$ which attains its maximum at $(-1,1)$ and $(1,1)$ with $f(\pm 1, 1) = 7$ and its minimum at $(0,1)$, $f(0,1) = 5$.

On L_4: $f(-1,y) = y^2 + y + 5$ with maximum at $(-1,1)$, $f(-1,1) = 7$ and minimum at $\left(-1, -\frac{1}{2}\right)$, $f\left(-1, -\frac{1}{2}\right) = \frac{19}{4}$.

Thus the absolute maximum is attained at both $(\pm 1, 1)$ with $f(\pm 1, 1) = 7$ and the absolute minimum on D is attained at $(0,0)$ with $f(0,0) = 4$.

26. $f_x(x,y) = 4 - 2x$ and $f_y(x,y) = 6 - 2y$, so the only critical point is $(2,3)$ (which is in D) where $f(2,3) = 13$. Along L_1: $y = 0$, so $f(x,0) = 4x - x^2 = -(x-2)^2 + 4$, $0 \le x \le 4$, which has a maximum value when $x = 2$ where $f(2,0) = 4$ and a minimum value both when $x = 0$ and $x = 4$, where $f(0,0) = f(4,0) = 0$. Along L_2: $x = 4$, so $f(4,y) = 6y - y^2 = -(y-3)^2 + 9$, $0 \le y \le 5$, which has a maximum value when $y = 3$ where $f(4,3) = 9$ and a minimum value when $y = 0$ where $f(4,0) = 0$. Along L_3: $y = 5$, so $f(x,5) = -x^2 + 4x + 5 = -(x-2)^2 + 9$,

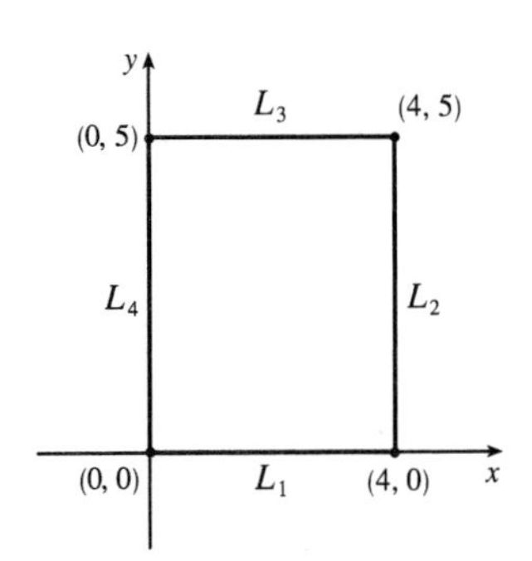

$0 \leq x \leq 4$, which has a maximum value when $x = 2$ where $f(2,5) = 9$ and a minimum value both when $x = 0$ and $x = 4$, where $f(0,5) = f(4,5) = 5$. Along L_4: $x = 0$, so $f(0, y) = 6y - y^2 = -(y-3)^2 + 9$, $0 \leq y \leq 5$, which has a maximum value when $y = 3$ where $f(0,3) = 9$ and a minimum value when $y = 0$ where $f(0,0) = 0$. Thus the absolute maximum is $f(2,3) = 13$ and the absolute minimum is attained at both $(0,0)$ and $(4,0)$, where $f(0,0) = f(4,0) = 0$.

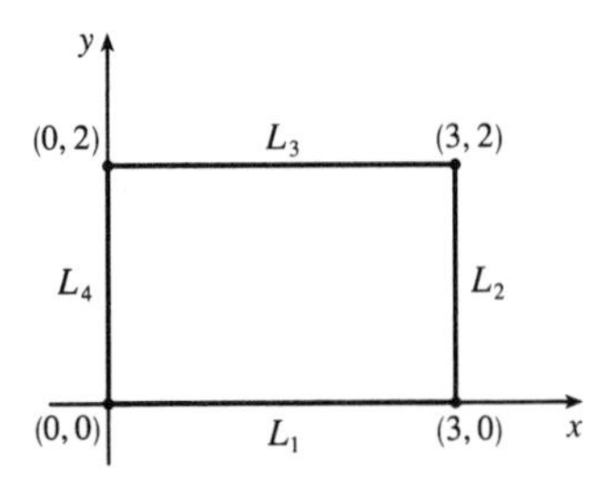

27. $f(x,y) = x^4 + y^4 - 4xy + 2$ is a polynomial and hence continuous on D, so it has an absolute maximum and minimum on D. In Exercise 5, we found the critical points of f; only $(1,1)$ with $f(1,1) = 0$ is inside D. On L_1: $y = 0$, $f(x,0) = x^4 + 2$, $0 \leq x \leq 3$, a polynomial in x which attains its maximum at $x = 3$, $f(3,0) = 83$, and its minimum at $x = 0$, $f(0,0) = 2$. On L_2: $x = 3$, $f(3,y) = y^4 - 12y + 83$, $0 \leq y \leq 2$, a polynomial in y which attains its minimum at $y = \sqrt[3]{3}$, $f\left(3, \sqrt[3]{3}\right) = 83 - 9\sqrt[3]{3} \approx 70.0$, and its maximum at $y = 0$, $f(3,0) = 83$. On L_3: $y = 2$, $f(x,2) = x^4 - 8x + 18$, $0 \leq x \leq 3$, a polynomial in x which attains its minimum at $x = \sqrt[3]{2}$, $f\left(\sqrt[3]{2}, 2\right) = 18 - 6\sqrt[3]{2} \approx 10.4$, and its maximum at $x = 3$, $f(3,2) = 75$. On L_4: $x = 0$, $f(0,y) = y^4 + 2$, $0 \leq y \leq 2$, a polynomial in y which attains its maximum at $y = 2$, $f(0,2) = 18$, and its minimum at $y = 0$, $f(0,0) = 2$. Thus the absolute maximum of f on D is $f(3,0) = 83$ and the absolute minimum is $f(1,1) = 0$.

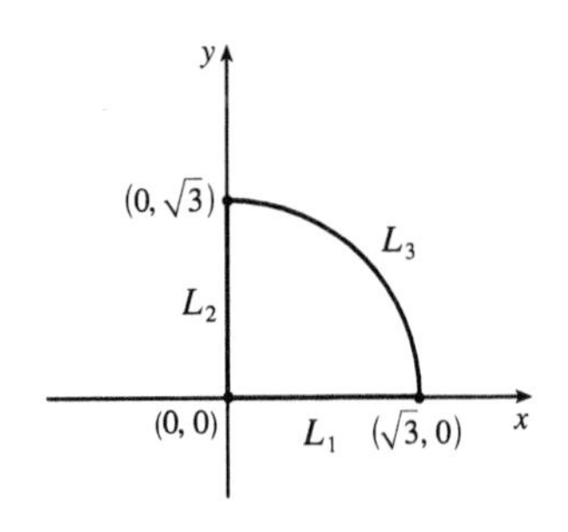

28. $f_x = y^2$ and $f_y = 2xy$, and since $f_x = 0 \Leftrightarrow y = 0$, there are no critical points in the interior of D. Along L_1: $y = 0$ and $f(x,0) = 0$. Along L_2: $x = 0$ and $f(0,y) = 0$. Along L_3: $y = \sqrt{3 - x^2}$, so let $g(x) = f\left(x, \sqrt{3 - x^2}\right) = 3x - x^3$ for $0 \leq x \leq \sqrt{3}$. Then $g'(x) = 3 - 3x^2 = 0 \Leftrightarrow x = 1$. The maximum value is $f\left(1, \sqrt{2}\right) = 2$ and the minimum occurs both at $x = 0$ and $x = \sqrt{3}$ where $f\left(0, \sqrt{3}\right) = f\left(\sqrt{3}, 0\right) = 0$. Thus the absolute maximum of f on D is $f\left(1, \sqrt{2}\right) = 2$, and the absolute minimum is 0 which occurs at all points along L_1 and L_2.

29. $f(x,y) = -(x^2 - 1)^2 - (x^2 y - x - 1)^2 \Rightarrow f_x(x,y) = -2(x^2 - 1)(2x) - 2(x^2 y - x - 1)(2xy - 1)$ and $f_y(x,y) = -2(x^2 y - x - 1)x^2$. Setting $f_y(x,y) = 0$ gives either $x = 0$ or $x^2 y - x - 1 = 0$. There are no critical points for $x = 0$, since $f_x(0,y) = -2$, so we set $x^2 y - x - 1 = 0 \Leftrightarrow y = \dfrac{x+1}{x^2}$ $[x \neq 0]$,

so $f_x\left(x, \dfrac{x+1}{x^2}\right) = -2(x^2 - 1)(2x) - 2\left(x^2 \dfrac{x+1}{x^2} - x - 1\right)\left(2x \dfrac{x+1}{x^2} - 1\right) = -4x(x^2 - 1)$. Therefore

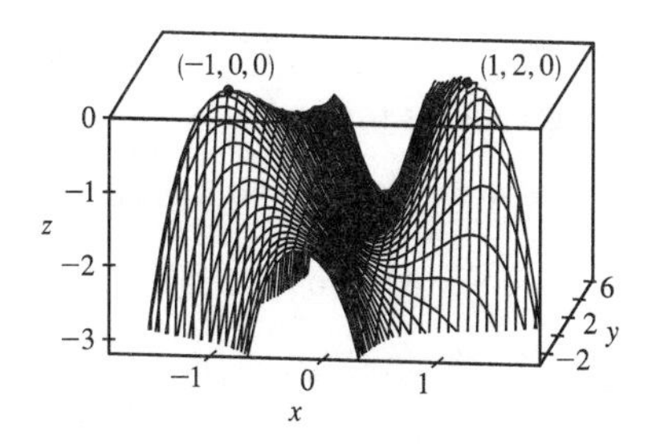

$f_x(x,y) = f_y(x,y) = 0$ at the points $(1,2)$ and $(-1,0)$. To classify these critical points, we calculate $f_{xx}(x,y) = -12x^2 - 12x^2y^2 + 12xy + 4y + 2$, $f_{yy}(x,y) = -2x^4$, and $f_{xy}(x,y) = -8x^3y + 6x^2 + 4x$. In order to use the Second Derivatives Test we calculate $D(-1,0) = f_{xx}(-1,0)\, f_{yy}(-1,0) - [f_{xy}(-1,0)]^2 = 16 > 0$, $f_{xx}(-1,0) = -10 < 0$, $D(1,2) = 16 > 0$, and $f_{xx}(1,2) = -26 < 0$, so both $(-1,0)$ and $(1,2)$ give local maxima.

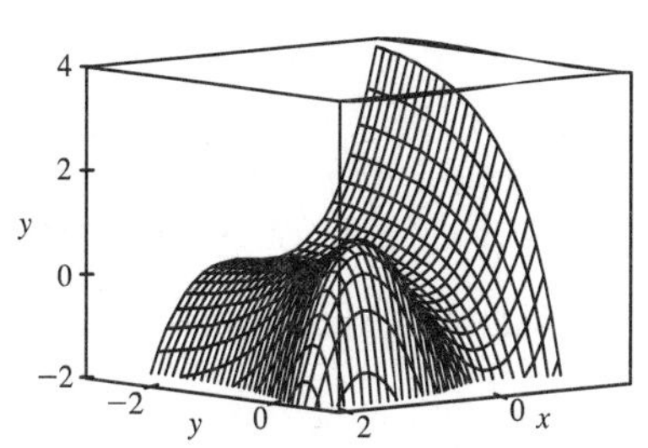

30. $f(x,y) = 3xe^y - x^3 - e^{3y}$ is differentiable everywhere, so the requirement for critical points is that $f_x = 3e^y - 3x^2 = 0$ **(1)** and $f_y = 3xe^y - 3e^{3y} = 0$ **(2)**. From **(1)** we obtain $e^y = x^2$, and then **(2)** gives $3x^3 - 3x^6 = 0 \Rightarrow x = 1$ or 0, but only $x = 1$ is valid, since $x = 0$ makes **(1)** impossible. So substituting $x = 1$ into **(1)** gives $y = 0$, and the only critical point is $(1,0)$.

The Second Derivatives Test shows that this gives a local maximum, since $D(1,0) = \left[-6x(3xe^y - 9e^{3y}) - (3e^y)^2\right]_{(1,0)} = 27 > 0$ and $f_{xx}(1,0) = [-6x]_{(1,0)} = -6 < 0$. But $f(1,0) = 1$ is not an absolute maximum because, for instance, $f(-3,0) = 17$. This can also be seen from the graph.

31. Let d be the distance from $(2,1,-1)$ to any point (x,y,z) on the plane $x + y - z = 1$, so $d = \sqrt{(x-2)^2 + (y-1)^2 + (z+1)^2}$ where $z = x + y - 1$, and we minimize $d^2 = f(x,y) = (x-2)^2 + (y-1)^2 + (x+y)^2$. Then $f_x(x,y) = 2(x-2) + 2(x+y) = 4x + 2y - 4$, $f_y(x,y) = 2(y-1) + 2(x+y) = 2x + 4y - 2$. Solving $4x + 2y - 4 = 0$ and $2x + 4y - 2 = 0$ simultaneously gives $x = 1$, $y = 0$. An absolute minimum exists (since there is a minimum distance from the point to the plane) and it must occur at a critical point, so the shortest distance occurs for $x = 1$, $y = 0$ for which $d = \sqrt{(1-2)^2 + (0-1)^2 + (0+1)^2} = \sqrt{3}$.

32. Here the distance d from a point on the plane to the point $(1,2,3)$ is $d = \sqrt{(x-1)^2 + (y-2)^2 + (z-3)^2}$, where $z = 4 - x + y$. We can minimize $d^2 = f(x,y) = (x-1)^2 + (y-2)^2 + (1-x+y)^2$, so $f_x(x,y) = 2(x-1) + 2(1-x+y)(-1) = 4x - 2y - 4$ and $f_y(x,y) = 2\,(y-2) + 2\,(1-x+y) = 4y - 2x - 2$. Solving $4x - 2y - 4 = 0$ and $4y - 2x - 2 = 0$ simultaneously gives $x = \frac{5}{3}$ and $y = \frac{4}{3}$, so the only critical point is $\left(\frac{5}{3}, \frac{4}{3}\right)$. This point must correspond to the minimum distance, so the point on the plane closest to $(1,2,3)$ is $\left(\frac{5}{3}, \frac{4}{3}, \frac{11}{3}\right)$.

33. Let d be the distance from the point $(4,2,0)$ to any point (x,y,z) on the cone, so $d = \sqrt{(x-4)^2 + (y-2)^2 + z^2}$ where $z^2 = x^2 + y^2$, and we minimize $d^2 = (x-4)^2 + (y-2)^2 + x^2 + y^2 = f\,(x,y)$. Then $f_x\,(x,y) = 2\,(x-4) + 2x = 4x - 8$, $f_y\,(x,y) = 2\,(y-2) + 2y = 4y - 4$, and the critical points occur when $f_x = 0 \Rightarrow x = 2$, $f_y = 0 \Rightarrow y = 1$. Thus the only critical point is $(2,1)$. An absolute minimum exists (since there is a minimum distance from the cone to the point) which must occur at a critical point, so the points on the cone closest to $(4,2,0)$ are $(2,1,\pm\sqrt{5})$.

34. The distance from the origin to a point (x, y, z) on the surface is $d = \sqrt{x^2 + y^2 + z^2}$ where $y^2 = 9 + xz$, so we minimize $d^2 = x^2 + 9 + xz + z^2 = f(x, z)$. Then $f_x = 2x + z$, $f_z = x + 2z$, and $f_x = 0$, $f_z = 0 \Rightarrow x = 0$, $z = 0$, so the only critical point is $(0, 0)$. $D(0, 0) = (2)(2) - 1 = 3 > 0$ with $f_{xx}(0, 0) = 2 > 0$, so this is a minimum. Thus $y^2 = 9 + 0 \Rightarrow y = \pm 3$ and the points on the surface closest to the origin are $(0, \pm 3, 0)$.

35. $x + y + z = 100$, so maximize $f(x, y) = xy(100 - x - y)$. $f_x = 100y - 2xy - y^2$, $f_y = 100x - x^2 - 2xy$, $f_{xx} = -2y$, $f_{yy} = -2x$, $f_{xy} = 100 - 2x - 2y$. Then $f_x = 0$ implies $y = 0$ or $y = 100 - 2x$. Substituting $y = 0$ into $f_y = 0$ gives $x = 0$ or $x = 100$ and substituting $y = 100 - 2x$ into $f_y = 0$ gives $3x^2 - 100x = 0$ so $x = 0$ or $\frac{100}{3}$. Thus the critical points are $(0, 0)$, $(100, 0)$, $(0, 100)$ and $\left(\frac{100}{3}, \frac{100}{3}\right)$.

$D(0, 0) = D(100, 0) = D(0, 100) = -10{,}000$ while $D\left(\frac{100}{3}, \frac{100}{3}\right) = \frac{10{,}000}{3}$ and $f_{xx}\left(\frac{100}{3}, \frac{100}{3}\right) = -\frac{200}{3} < 0$. Thus $(0, 0)$, $(100, 0)$ and $(0, 100)$ are saddle points whereas $f\left(\frac{100}{3}, \frac{100}{3}\right)$ is a local maximum. Thus the numbers are $x = y = z = \frac{100}{3}$.

36. Maximize $f(x, y) = x^a y^b (100 - x - y)^c$.

$f_x = ax^{a-1}y^b(100 - x - y)^c - cx^a y^b(100 - x - y)^{c-1} = x^{a-1}y^b(100 - x - y)^{c-1}[a(100 - x - y) - cx]$

and $f_y = x^a y^{b-1}(100 - x - y)^{c-1}[b(100 - x - y) - cy]$. Since x, y and z are all positive, the only critical point occurs when $x = a\dfrac{100 - y}{a + c}$ and $y = \dfrac{100b}{a + b + c}$. Thus the point is $\left(\dfrac{100a}{a + b + c}, \dfrac{100b}{a + b + c}\right)$ and the numbers are $x = \dfrac{100a}{a + b + c}$, $y = \dfrac{100b}{a + b + c}$, $z = \dfrac{100c}{a + b + c}$.

37. Maximize $f(x, y) = xy(36 - 9x^2 - 36y^2)^{1/2}/2$ with (x, y, z) in first octant. Then

$$f_x = \frac{y(36 - 9x^2 - 36y^2)^{1/2}}{2} + \frac{-9x^2y(36 - 9x^2 - 36y^2)^{-1/2}}{2} = \frac{(36y - 18x^2y - 36y^3)}{2(36 - 9x^2 - 36y^2)^{1/2}} \text{ and}$$

$f_y = \dfrac{36x - 9x^3 - 72xy^2}{2(36 - 9x^2 - 36y^2)^{1/2}}$. Setting $f_x = 0$ gives $y = 0$ or $y^2 = \dfrac{2 - x^2}{2}$ but $y > 0$, so only the latter solution applies. Substituting this y into $f_y = 0$ gives $x^2 = \frac{4}{3}$ or $x = \frac{2}{\sqrt{3}}$, $y = \frac{1}{\sqrt{3}}$ and then $z^2 = (36 - 12 - 12)/4 = 3$. The fact that this gives a maximum volume follows from the geometry. This maximum volume is $V = (2x)(2y)(2z) = 8\left(\frac{2}{\sqrt{3}}\right)\left(\frac{1}{\sqrt{3}}\right)\left(\sqrt{3}\right) = \frac{16}{\sqrt{3}}$.

38. Here maximize $f(x, y) = xy\dfrac{(a^2b^2c^2 - b^2c^2x^2 - a^2c^2y^2)^{1/2}}{a^2b^2}$. Then $f_x = yc^2\dfrac{a^2b^2 - 2b^2x^2 - a^2y^2}{a^2b^2(a^2b^2c^2 - b^2c^2x^2 - a^2c^2y^2)^{1/2}}$ and

$f_y = xc^2\dfrac{a^2b^2 - 2a^2y^2 - b^2x^2}{a^2b^2(a^2b^2c^2 - b^2c^2x^2 - a^2c^2y^2)^{1/2}}$. Then $f_x = 0$ (with $x, y > 0$) implies $y^2 = \dfrac{a^2b^2 - 2b^2x^2}{a^2}$ and substituting into $f_y = 0$ implies $3b^2x^2 = a^2b^2$ or $x = \frac{1}{\sqrt{3}}\,a$, $y = \frac{1}{\sqrt{3}}\,b$ and then $z = \frac{1}{\sqrt{3}}\,c$. Thus the maximum volume of such a rectangle is $V = (2x)(2y)(2z) = \frac{8}{3\sqrt{3}}\,abc$.

39. Maximize $f(x, y) = \dfrac{xy}{3}(6 - x - 2y)$, then the maximum volume is $V = xyz$.

$f_x = \frac{1}{3}(6y - 2xy - y^2) = \frac{1}{3}y(6 - 2x - 2y)$ and $f_y = \frac{1}{3}x(6 - x - 4y)$. Setting $f_x = 0$ and $f_y = 0$ gives the critical point $(2, 1)$ which geometrically must yield a maximum. Thus the volume of the largest such box is $V = (2)(1)\left(\frac{2}{3}\right) = \frac{4}{3}$.

40. Surface area $= 2(xy + xz + yz) = 64$ cm^2, so $xy + xz + yz = 32$ or $z = \dfrac{32 - xy}{x + y}$. Maximize the volume

$f(x, y) = xy\,\dfrac{32 - xy}{x + y}$. Then $f_x = \dfrac{32y^2 - 2xy^3 - x^2y^2}{(x + y)^2} = y^2\,\dfrac{32 - 2xy - x^2}{(x + y)^2}$ and $f_y = x^2\,\dfrac{32 - 2xy - y^2}{(x + y)^2}$. Setting

$f_x = 0$ implies $y = \dfrac{32 - x^2}{2x}$ and substituting into $f_y = 0$ gives $32(4x^2) - (32 - x^2)(4x^2) - (32 - x^2)^2 = 0$ or

$3x^4 + 64x^2 - (32)^2 = 0$. Thus $x^2 = \frac{64}{6}$ or $x = \frac{8}{\sqrt{6}}$, $y = \frac{64/3}{16/\sqrt{6}} = \frac{8}{\sqrt{6}}$ and $z = \frac{8}{\sqrt{6}}$. Thus the box is a cube with edge

length $\frac{8}{\sqrt{6}}$ cm.

41. Let the dimensions be x, y, and z; then $4x + 4y + 4z = c$ and the volume is

$V = xyz = xy\left(\frac{1}{4}c - x - y\right) = \frac{1}{4}cxy - x^2y - xy^2$, $x > 0$, $y > 0$. Then $V_x = \frac{1}{4}cy - 2xy - y^2$ and $V_y = \frac{1}{4}cx - x^2 - 2xy$,

so $V_x = 0 = V_y$ when $2x + y = \frac{1}{4}c$ and $x + 2y = \frac{1}{4}c$. Solving, we get $x = \frac{1}{12}c$, $y = \frac{1}{12}c$ and $z = \frac{1}{4}c - x - y = \frac{1}{12}c$. From

the geometrical nature of the problem, this critical point must give an absolute maximum. Thus the box is a cube with edge

length $\frac{1}{12}c$.

42. The cost equals $5xy + 2(xz + yz)$ and $xyz = V$, so $C(x, y) = 5xy + 2V(x + y)/(xy) = 5xy + 2V(x^{-1} + y^{-1})$. Then

$C_x = 5y - 2Vx^{-2}$, $C_y = 5x - 2\,Vy^{-2}$, $f_x = 0$ implies $y = 2\,V/(5x^2)$, $f_y = 0$ implies $x = \sqrt[3]{\frac{2}{5}V} = y$. Thus the

dimensions of the aquarium which minimize the cost are $x = y = \sqrt[3]{\frac{2}{5}V}$ units, $z = V^{1/3}\left(\frac{5}{2}\right)^{2/3}$.

43. Let the dimensions be x, y and z, then minimize $xy + 2(xz + yz)$ if $xyz = 32{,}000$ cm^3. Then

$f(x, y) = xy + [64{,}000(x + y)/xy] = xy + 64{,}000(x^{-1} + y^{-1})$, $f_x = y - 64{,}000x^{-2}$, $f_y = x - 64{,}000y^{-2}$.

And $f_x = 0$ implies $y = 64{,}000/x^2$; substituting into $f_y = 0$ implies $x^3 = 64{,}000$ or $x = 40$ and then $y = 40$. Now

$D(x, y) = [(2)(64{,}000)]^2x^{-3}y^{-3} - 1 > 0$ for $(40, 40)$ and $f_{xx}(40, 40) > 0$ so this is indeed a minimum. Thus the

dimensions of the box are $x = y = 40$ cm, $z = 20$ cm.

44. Let x be the length of the north and south walls, y the length of the east and west walls, and z the height of the building. The

heat loss is given by $h = 10(2yz) + 8(2xz) + 1(xy) + 5(xy) = 6xy + 16xz + 20yz$. The volume is 4000 m^3, so

$xyz = 4000$, and we substitute $z = \frac{4000}{xy}$ to obtain the heat loss function $h(x, y) = 6xy + 80{,}000/x + 64{,}000/y$.

(a) Since $z = \frac{4000}{xy} \geq 4$, $xy \leq 1000 \quad\Rightarrow\quad y \leq 1000/x$. Also $x \geq 30$ and

$y \geq 30$, so the domain of h is $D = \{(x, y) \mid x \geq 30, 30 \leq y \leq 1000/x\}$.

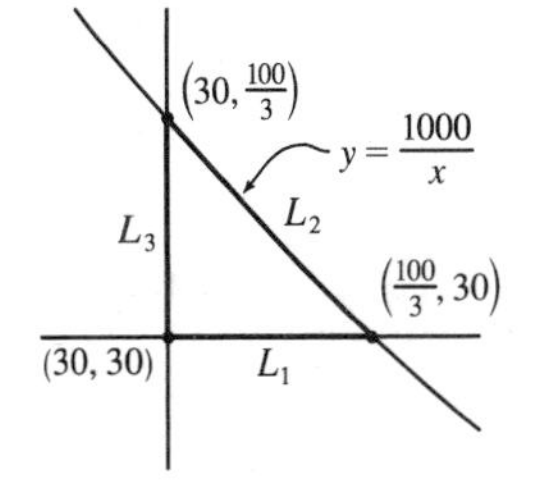

(b) $h(x, y) = 6xy + 80{,}000x^{-1} + 64{,}000y^{-1} \quad\Rightarrow$

$h_x = 6y - 80{,}000x^{-2}$, $h_y = 6x - 64{,}000y^{-2}$.

$h_x = 0$ implies $6x^2y = 80{,}000 \quad\Rightarrow\quad y = \dfrac{80{,}000}{6x^2}$ and substituting into

$h_y = 0$ gives $6x = 64{,}000\left(\dfrac{6x^2}{80{,}000}\right)^2 \quad\Rightarrow\quad x^3 = \dfrac{80{,}000^2}{6 \cdot 64{,}000} = \dfrac{50{,}000}{3}$, so

$x = \sqrt[3]{\dfrac{50{,}000}{3}} = 10\sqrt[3]{\dfrac{50}{3}} \quad\Rightarrow\quad y = \dfrac{80}{\sqrt[3]{60}}$, and the only critical point of h is $\left(10\sqrt[3]{\dfrac{50}{3}}, \dfrac{80}{\sqrt[3]{60}}\right) \approx (25.54, 20.43)$

which is not in D. Next we check the boundary of D.

On L_1: $y = 30$, $h(x, 30) = 180x + 80{,}000/x + 6400/3$, $30 \leq x \leq \frac{100}{3}$. Since $h'(x, 30) = 180 - 80{,}000/x^2 > 0$ for

$30 \le x \le \frac{100}{3}$, $h(x, 30)$ is an increasing function with minimum $h(30, 30) = 10{,}200$ and maximum $h(\frac{100}{3}, 30) \approx 10{,}533$.

On L_2: $y = 1000/x$, $h(x, 1000/x) = 6000 + 64x + 80{,}000/x$, $30 \le x \le \frac{100}{3}$.

Since $h'(x, 1000/x) = 64 - 80{,}000/x^2 < 0$ for $30 \le x \le \frac{100}{3}$, $h(x, 1000/x)$ is a decreasing function with minimum $h(\frac{100}{3}, 30) \approx 10{,}533$ and maximum $h(30, \frac{100}{3}) \approx 10{,}587$.

On L_3: $x = 30$, $h(30, y) = 180y + 64{,}000/y + 8000/3$, $30 \le y \le \frac{100}{3}$ $h'(30, y) = 180 - 64{,}000/y^2 > 0$ for $30 \le y \le \frac{100}{3}$, so $h(30, y)$ is an increasing function of y with minimum $h(30, 30) = 10{,}200$ and maximum $h(30, \frac{100}{3}) \approx 10{,}587$.

Thus the absolute minimum of h is $h(30, 30) = 10{,}200$, and the dimensions of the building that minimize heat loss are walls 30 m in length and height $\frac{4000}{30^2} = \frac{40}{9} \approx 4.44$ m.

(c) From part (b), the only critical point of h, which gives a local (and absolute) minimum, is approximately $h(25.54, 20.43) \approx 9396$. So a building of volume 4000 m^2 with dimensions $x \approx 25.54$ m, $y \approx 20.43$ m, $z \approx \frac{4000}{(25.54)(20.43)} \approx 7.67$ m has the least amount of heat loss.

45. Let x, y, z be the dimensions of the rectangular box. Then the volume of the box is xyz and

$L = \sqrt{x^2 + y^2 + z^2} \quad \Rightarrow \quad L^2 = x^2 + y^2 + z^2 \quad \Rightarrow \quad z = \sqrt{L^2 - x^2 - y^2}$.

Substituting, we have volume $V(x, y) = xy\sqrt{L^2 - x^2 - y^2}$, $(x, y > 0)$.

$$V_x = xy \cdot \tfrac{1}{2}(L^2 - x^2 - y^2)^{-1/2}(-2x) + y\sqrt{L^2 - x^2 - y^2} = y\sqrt{L^2 - x^2 - y^2} - \frac{x^2 y}{\sqrt{L^2 - x^2 - y^2}},$$

$V_y = x\sqrt{L^2 - x^2 - y^2} - \dfrac{xy^2}{\sqrt{L^2 - x^2 - y^2}}$. $V_x = 0$ implies $y(L^2 - x^2 - y^2) = x^2 y \quad \Rightarrow \quad y(L^2 - 2x^2 - y^2) = 0 \quad \Rightarrow$

$2x^2 + y^2 = L^2$ (since $y > 0$), and $V_y = 0$ implies $x(L^2 - x^2 - y^2) = xy^2 \quad \Rightarrow \quad x(L^2 - x^2 - 2y^2) = 0 \quad \Rightarrow$

$x^2 + 2y^2 = L^2$ (since $x > 0$). Substituting $y^2 = L^2 - 2x^2$ into $x^2 + 2y^2 = L^2$ gives $x^2 + 2L^2 - 4x^2 = L^2 \quad \Rightarrow$

$3x^2 = L^2 \quad \Rightarrow \quad x = L/\sqrt{3}$ (since $x > 0$) and then $y = \sqrt{L^2 - 2(L/\sqrt{3})^2} = L/\sqrt{3}$.

So the only critical point is $(L/\sqrt{3}, L/\sqrt{3})$ which, from the geometrical nature of the problem, must give an absolute maximum. Thus the maximum volume is $V(L/\sqrt{3}, L/\sqrt{3}) = (L/\sqrt{3})^2\sqrt{L^2 - (L/\sqrt{3})^2 - (L/\sqrt{3})^2} = L^3/(3\sqrt{3})$ cubic units.

46. Since $p + q + r = 1$ we can substitute $p = 1 - r - q$ into P giving

$P = P(q, r) = 2(1 - r - q)q + 2(1 - r - q)r + 2rq = 2q - 2q^2 + 2r - 2r^2 - 2rq$. Since p, q and r represent proportions and $p + q + r = 1$, we know $q \ge 0$, $r \ge 0$, and $q + r \le 1$. Thus, we want to find the absolute maximum of the continuous function $P(q, r)$ on the closed set D enclosed by the lines $q = 0$, $r = 0$, and $q + r = 1$. To find any critical points, we set the partial derivatives equal to zero: $P_q(q, r) = 2 - 4q - 2r = 0$ and $P_r(q, r) = 2 - 4r - 2q = 0$. The first equation gives $r = 1 - 2q$, and substituting into the second equation we have $2 - 4(1 - 2q) - 2q = 0 \quad \Rightarrow \quad q = \frac{1}{3}$. Then we have one critical point, $(\frac{1}{3}, \frac{1}{3})$, where $P(\frac{1}{3}, \frac{1}{3}) = \frac{2}{3}$. Next we find the maximum values of P on the boundary of D which consists of three line segments. For the segment given by $r = 0, 0 \le q \le 1$, $P(q, r) = P(q, 0) = 2q - 2q^2, 0 \le q \le 1$. This represents a parabola with maximum value $P\left(\frac{1}{2}, 0\right) = \frac{1}{2}$. On the segment $q = 0, 0 \le r \le 1$ we have $P(0, r) = 2r - 2r^2, 0 \le r \le 1$. This represents a parabola with maximum value $P\left(0, \frac{1}{2}\right) = \frac{1}{2}$. Finally, on the segment $q + r = 1, 0 \le q \le 1$, $P(q, r) = P(q, 1 - q) = 2q - 2q^2, 0 \le q \le 1$ which has a maximum value of $P\left(\frac{1}{2}, \frac{1}{2}\right) = \frac{1}{2}$. Comparing these values with the value of P at the critical point, we see that the absolute maximum value of $P(q, r)$ on D is $\frac{2}{3}$.

47. Note that here the variables are m and b, and $f(m, b) = \sum\limits_{i=1}^{n} [y_i - (mx_i + b)]^2$. Then $f_m = \sum\limits_{i=1}^{n} -2x_i[y_i - (mx_i + b)] = 0$ implies $\sum\limits_{i=1}^{n} \left(x_i y_i - mx_i^2 - bx_i\right) = 0$ or $\sum\limits_{i=1}^{n} x_i y_i = m \sum\limits_{i=1}^{n} x_i^2 + b \sum\limits_{i=1}^{n} x_i$ and $f_b = \sum\limits_{i=1}^{n} -2[y_i - (mx_i + b)] = 0$ implies $\sum\limits_{i=1}^{n} y_i = m \sum\limits_{i=1}^{n} x_i + \sum\limits_{i=1}^{n} b = m\left(\sum\limits_{i=1}^{n} x_i\right) + nb$. Thus we have the two desired equations.

Now $f_{mm} = \sum\limits_{i=1}^{n} 2x_i^2$, $f_{bb} = \sum\limits_{i=1}^{n} 2 = 2n$ and $f_{mb} = \sum\limits_{i=1}^{n} 2x_i$. And $f_{mm}(m, b) > 0$ always and

$D(m, b) = 4n\left(\sum\limits_{i=1}^{n} x_i^2\right) - 4\left(\sum\limits_{i=1}^{n} x_i\right)^2 = 4\left[n\left(\sum\limits_{i=1}^{n} x_i^2\right) - \left(\sum\limits_{i=1}^{n} x_i\right)^2\right] > 0$ always so the solutions of these two equations do indeed minimize $\sum\limits_{i=1}^{n} d_i^2$.

48. Any such plane must cut out a tetrahedron in the first octant. We need to minimize the volume of the tetrahedron that passes through the point $(1, 2, 3)$. Writing the equation of the plane as $\dfrac{x}{a} + \dfrac{y}{b} + \dfrac{z}{c} = 1$, the volume of the tetrahedron is given by $V = \dfrac{abc}{6}$. But $(1, 2, 3)$ must lie on the plane, so we need $\dfrac{1}{a} + \dfrac{2}{b} + \dfrac{3}{c} = 1$ $(\star)$ and thus can think of c as a function of a and b.

Then $V_a = \dfrac{b}{6}\left(c + a\dfrac{\partial c}{\partial a}\right)$ and $V_b = \dfrac{a}{6}\left(c + b\dfrac{\partial c}{\partial b}\right)$. Differentiating $(\star)$ with respect to a we get $-a^{-2} - 3c^{-2}\dfrac{\partial c}{\partial a} = 0 \;\Rightarrow$ $\dfrac{\partial c}{\partial a} = \dfrac{-c^2}{3a^2}$, and differentiating $(\star)$ with respect to b gives $-2b^{-2} - 3c^{-2}\dfrac{\partial c}{\partial b} = 0 \;\Rightarrow\; \dfrac{\partial c}{\partial b} = \dfrac{-2c^2}{3b^2}$. Then $V_a = \dfrac{b}{6}\left(c + a\dfrac{-c^2}{3a^2}\right) = 0 \;\Rightarrow\; c = 3a$, and $V_b = \dfrac{a}{6}\left(c + b\dfrac{-2c^2}{3b^2}\right) = 0 \;\Rightarrow\; c = \frac{3}{2}b$. Thus $3a = \frac{3}{2}b$ or $b = 2a$. Putting these into $(\star)$ gives $\frac{3}{a} = 1$ or $a = 3$ and then $b = 6$, $c = 9$. Thus the equation of the required plane is $\dfrac{x}{3} + \dfrac{y}{6} + \dfrac{z}{9} = 1$ or $6x + 3y + 2z = 18$.

11.8 Lagrange Multipliers

1. $f(x, y) = x^2 + y^2$, $g(x, y) = xy = 1$, and $\nabla f = \lambda \nabla g \;\Rightarrow\; \langle 2x, 2y\rangle = \langle \lambda y, \lambda x\rangle$, so $2x = \lambda y$, $2y = \lambda x$, and $xy = 1$. From the last equation, $x \ne 0$ and $y \ne 0$, so $2x = \lambda y \;\Rightarrow\; \lambda = 2x/y$. Substituting, we have $2y = (2x/y)\,x \;\Rightarrow$ $y^2 = x^2 \;\Rightarrow\; y = \pm x$. But $xy = 1$, so $x = y = \pm 1$ and the possible points for the extreme values of f are $(1, 1)$ and $(-1, -1)$. Here there is no maximum value, since the constraint $xy = 1$ allows x or y to become arbitrarily large, and hence $f(x, y) = x^2 + y^2$ can be made arbitrarily large. The minimum value is $f(1, 1) = f(-1, -1) = 2$.

2. $f(x,y)=4x+6y$, $g(x,y)=x^2+y^2=13 \;\Rightarrow\; \nabla f=\langle 4,6\rangle$, $\lambda\nabla g=\langle 2\lambda x, 2\lambda y\rangle$. Then $2\lambda x=4$ and $2\lambda y=6$ imply $x=\dfrac{2}{\lambda}$ and $y=\dfrac{3}{\lambda}$. But $13=x^2+y^2=\left(\dfrac{2}{\lambda}\right)^2+\left(\dfrac{3}{\lambda}\right)^2 \;\Rightarrow\; 13=\dfrac{13}{\lambda^2} \;\Rightarrow\; \lambda=\pm 1$, so f has possible extreme values at the points $(2,3)$, $(-2,-3)$. We compute $f(2,3)=26$ and $f(-2,-3)=-26$, so the maximum value of f on $x^2+y^2=13$ is $f(2,3)=26$ and the minimum value is $f(-2,-3)=-26$.

3. $f(x,y)=x^2y$, $g(x,y)=x^2+2y^2=6 \;\Rightarrow\; \nabla f=\langle 2xy, x^2\rangle$, $\lambda\nabla g=\langle 2\lambda x, 4\lambda y\rangle$. Then $2xy=2\lambda x$ implies $x=0$ or $\lambda=y$. If $x=0$, then $x^2=4\lambda y$ implies $\lambda=0$ or $y=0$. However, if $y=0$ then $g(x,y)=0$, a contradiction. So $\lambda=0$ and then $g(x,y)=6 \;\Rightarrow\; y=\pm\sqrt{3}$. If $\lambda=y$, then $x^2=4\lambda y$ implies $x^2=4y^2$, and so $g(x,y)=6 \;\Rightarrow\; 4y^2+2y^2=6 \;\Rightarrow\; y^2=1 \;\Rightarrow\; y=\pm 1$. Thus f has possible extreme values at the points $\left(0,\pm\sqrt{3}\right)$, $(\pm 2,1)$, and $(\pm 2,-1)$. After evaluating f at these points, we find the maximum value to be $f(\pm 2,1)=4$ and the minimum to be $f(\pm 2,-1)=-4$.

4. $f(x,y)=e^{xy}$, $g(x,y)=x^3+y^3=16$, and $\nabla f=\lambda\nabla g \;\Rightarrow\; \langle ye^{xy}, xe^{xy}\rangle=\langle 3\lambda x^2, 3\lambda y^2\rangle$, so $ye^{xy}=3\lambda x^2$ and $xe^{xy}=3\lambda y^2$. Note that $x=0 \;\Leftrightarrow\; y=0$ which contradicts $x^3+y^3=16$, so we may assume $x\neq 0$, $y\neq 0$, and then $\lambda=ye^{xy}/(3x^2)=xe^{xy}/(3y^2) \;\Rightarrow\; x^3=y^3 \;\Rightarrow\; x=y$. But $x^3+y^3=16$, so $2x^3=16 \;\Rightarrow\; x=2=y$. Here there is no minimum value, since we can choose points satisfying the constraint $x^3+y^3=16$ that make $f(x,y)=e^{xy}$ arbitrarily close to 0 (but never equal to 0). The maximum value is $f(2,2)=e^4$.

5. $f(x,y,z)=2x+6y+10z$, $g(x,y,z)=x^2+y^2+z^2=35 \;\Rightarrow\; \nabla f=\langle 2,6,10\rangle$, $\lambda\nabla g=\langle 2\lambda x, 2\lambda y, 2\lambda z\rangle$. Then $2\lambda x=2$, $2\lambda y=6$, $2\lambda z=10$ imply $x=\dfrac{1}{\lambda}$, $y=\dfrac{3}{\lambda}$, and $z=\dfrac{5}{\lambda}$. But $35=x^2+y^2+z^2=\left(\dfrac{1}{\lambda}\right)^2+\left(\dfrac{3}{\lambda}\right)^2+\left(\dfrac{5}{\lambda}\right)^2 \;\Rightarrow\; 35=\dfrac{35}{\lambda^2} \;\Rightarrow\; \lambda=\pm 1$, so f has possible extreme values at the points $(1,3,5)$, $(-1,-3,-5)$. The maximum value of f on $x^2+y^2+z^2=35$ is $f(1,3,5)=70$, and the minimum is $f(-1,-3,-5)=-70$.

6. $f(x,y,z)=8x-4z$, $g(x,y,z)=x^2+10y^2+z^2=5 \;\Rightarrow\; \nabla f=\langle 8,0,-4\rangle$, $\lambda\nabla g=\langle 2\lambda x, 20\lambda y, 2\lambda z\rangle$. Then $2\lambda x=8$, $20\lambda y=0$, $2\lambda z=-4$ imply $x=\dfrac{4}{\lambda}$, $y=0$, and $z=-\dfrac{2}{\lambda}$. But $5=x^2+10y^2+z^2=\left(\dfrac{4}{\lambda}\right)^2+10\,(0)^2+\left(-\dfrac{2}{\lambda}\right)^2 \;\Rightarrow\; 5=\dfrac{20}{\lambda^2} \;\Rightarrow\; \lambda=\pm 2$, so f has possible extreme values at the points $(2,0,-1)$, $(-2,0,1)$. The maximum of f on $x^2+10y^2+z^2=5$ is $f(2,0,-1)=20$, and the minimum is $f(-2,0,1)=-20$.

7. $f(x,y,z)=xyz$, $g(x,y,z)=x^2+2y^2+3z^2=6 \;\Rightarrow\; \nabla f=\langle yz, xz, xy\rangle$, $\lambda\nabla g=\langle 2\lambda x, 4\lambda y, 6\lambda z\rangle$. Then $\nabla f=\lambda\nabla g$ implies $\lambda=(yz)/(2x)=(xz)/(4y)=(xy)/(6z)$ or $x^2=2y^2$ and $z^2=\frac{2}{3}y^2$. Thus $x^2+2y^2+3z^2=6$ implies $6y^2=6$ or $y=\pm 1$. Then the possible points are $\left(\sqrt{2},\pm 1,\sqrt{\frac{2}{3}}\right)$, $\left(\sqrt{2},\pm 1,-\sqrt{\frac{2}{3}}\right)$, $\left(-\sqrt{2},\pm 1,\sqrt{\frac{2}{3}}\right)$, $\left(-\sqrt{2},\pm 1,-\sqrt{\frac{2}{3}}\right)$. The maximum value of f on the ellipsoid is $\frac{2}{\sqrt{3}}$, occurring when all coordinates are positive or exactly two are negative and the minimum is $-\frac{2}{\sqrt{3}}$ occurring when 1 or 3 of the coordinates are negative.

8. $f(x,y,z) = x^2y^2z^2$, $g(x,y,z) = x^2 + y^2 + z^2 = 1 \Rightarrow \nabla f = \langle 2xy^2z^2, 2yx^2z^2, 2zx^2y^2 \rangle$, $\lambda\nabla g = \langle 2\lambda x, 2\lambda y, 2\lambda z \rangle$. Then $\nabla f = \lambda \nabla g$ implies (1) $\lambda = y^2z^2 = x^2z^2 = x^2y^2$ and $\lambda \neq 0$, or (2) $\lambda = 0$ and one or two (but not three) of the coordinates are 0. If (1) then $x^2 = y^2 = z^2 = \frac{1}{3}$. The minimum value of f on the sphere occurs in case (2) with a value of 0 and the maximum value is $\frac{1}{27}$ which arises from all the points from (1), that is, the points $\left(\pm\frac{1}{\sqrt{3}}, \frac{1}{\sqrt{3}}, \frac{1}{\sqrt{3}}\right)$, $\left(\pm\frac{1}{\sqrt{3}}, -\frac{1}{\sqrt{3}}, \frac{1}{\sqrt{3}}\right)$, $\left(\pm\frac{1}{\sqrt{3}}, -\frac{1}{\sqrt{3}}, -\frac{1}{\sqrt{3}}\right)$.

9. $f(x,y,z) = x^2 + y^2 + z^2$, $g(x,y,z) = x^4 + y^4 + z^4 = 1 \Rightarrow \nabla f = \langle 2x, 2y, 2z \rangle$, $\lambda \nabla g = \langle 4\lambda x^3, 4\lambda y^3, 4\lambda z^3 \rangle$.

Case 1: If $x \neq 0$, $y \neq 0$ and $z \neq 0$, then $\nabla f = \lambda \nabla g$ implies $\lambda = 1/(2x^2) = 1/(2y^2) = 1/(2z^2)$ or $x^2 = y^2 = z^2$ and $3x^4 = 1$ or $x = \pm\frac{1}{\sqrt[4]{3}}$ giving the points $\left(\pm\frac{1}{\sqrt[4]{3}}, \frac{1}{\sqrt[4]{3}}, \frac{1}{\sqrt[4]{3}}\right)$, $\left(\pm\frac{1}{\sqrt[4]{3}}, -\frac{1}{\sqrt[4]{3}}, \frac{1}{\sqrt[4]{3}}\right)$, $\left(\pm\frac{1}{\sqrt[4]{3}}, \frac{1}{\sqrt[4]{3}}, -\frac{1}{\sqrt[4]{3}}\right)$, $\left(\pm\frac{1}{\sqrt[4]{3}}, -\frac{1}{\sqrt[4]{3}}, -\frac{1}{\sqrt[4]{3}}\right)$ all with an f-value of $\sqrt{3}$.

Case 2: If one of the variables equals zero and the other two are not zero, then the squares of the two nonzero coordinates are equal with common value $\frac{1}{\sqrt{2}}$ and corresponding f value of $\sqrt{2}$.

Case 3: If exactly two of the variables are zero, then the third variable has value ± 1 with the corresponding f value of 1. Thus on $x^4 + y^4 + z^4 = 1$, the maximum value of f is $\sqrt{3}$ and the minimum value is 1.

10. $f(x,y,z) = x^4 + y^4 + z^4$, $g(x,y,z) = x^2 + y^2 + z^2 = 1 \Rightarrow \nabla f = \langle 4x^3, 4y^3, 4z^3 \rangle$, $\lambda \nabla g = \langle 2\lambda x, 2\lambda y, 2\lambda z \rangle$.

Case 1: If $x \neq 0$, $y \neq 0$ and $z \neq 0$ then $\nabla f = \lambda \nabla g$ implies $\lambda = 2x^2 = 2y^2 = 2z^2$ or $x^2 = y^2 = z^2 = \frac{1}{3}$ yielding 8 points each with an f-value of $\frac{1}{3}$.

Case 2: If one of the variables is 0 and the other two are not, then the squares of the two nonzero coordinates are equal with common value $\frac{1}{2}$ and the corresponding f-value is $\frac{1}{2}$.

Case 3: If exactly two of the variables are 0, then the third variable has value ± 1 with corresponding f-value of 1. Thus on $x^2 + y^2 + z^2 = 1$, the maximum value of f is 1 and the minimum value is $\frac{1}{3}$.

11. $f(x,y,z,t) = x + y + z + t$, $g(x,y,z,t) = x^2 + y^2 + z^2 + t^2 = 1 \Rightarrow \langle 1,1,1,1 \rangle = \langle 2\lambda x, 2\lambda y, 2\lambda z, 2\lambda t \rangle$, so $\lambda = 1/(2x) = 1/(2y) = 1/(2z) = 1/(2t)$ and $x = y = z = t$. But $x^2 + y^2 + z^2 + t^2 = 1$, so the possible points are $\left(\pm\frac{1}{2}, \pm\frac{1}{2}, \pm\frac{1}{2}, \pm\frac{1}{2}\right)$. Thus the maximum value of f is $f\left(\frac{1}{2}, \frac{1}{2}, \frac{1}{2}, \frac{1}{2}\right) = 2$ and the minimum value is $f\left(-\frac{1}{2}, -\frac{1}{2}, -\frac{1}{2}, -\frac{1}{2}\right) = -2$.

12. $f(x_1, x_2, \ldots, x_n) = x_1 + x_2 + \cdots + x_n$, $g(x_1, x_2, \ldots, x_n) = x_1^2 + x_2^2 + \cdots + x_n^2 = 1 \Rightarrow$ $\langle 1, 1, \ldots, 1 \rangle = \langle 2\lambda x_1, 2\lambda x_2, \ldots, 2\lambda x_n \rangle$, so $\lambda = 1/(2x_1) = 1/(2x_2) = \cdots = 1/(2x_n)$ and $x_1 = x_2 = \cdots = x_n$. But $x_1^2 + x_2^2 + \cdots + x_n^2 = 1$, so $x_i = \pm 1/\sqrt{n}$ for $i = 1, \ldots, n$. Thus the maximum value of f is $f(1/\sqrt{n}, 1/\sqrt{n}, \ldots, 1/\sqrt{n}) = \sqrt{n}$ and the minimum value is $f(-1/\sqrt{n}, -1/\sqrt{n}, \ldots, -1/\sqrt{n}) = -\sqrt{n}$.

13. $f(x,y,z) = x + 2y$, $g(x,y,z) = x + y + z = 1$, $h(x,y,z) = y^2 + z^2 = 4$ $\Rightarrow$ $\nabla f = \langle 1, 2, 0 \rangle$, $\lambda \nabla g = \langle \lambda, \lambda, \lambda \rangle$ and $\mu \nabla h = \langle 0, 2\mu y, 2\mu z \rangle$. Then $1 = \lambda$, $2 = \lambda + 2\mu y$ and $0 = \lambda + 2\mu z$ so $\mu y = \frac{1}{2} = -\mu z$ or $y = 1/(2\mu)$, $z = -1/(2\mu)$. Thus $x + y + z = 1$ implies $x = 1$ and $y^2 + z^2 = 4$ implies $\mu = \pm\frac{1}{2\sqrt{2}}$. Then the possible points are $\left(1, \pm\sqrt{2}, \mp\sqrt{2}\right)$ and the maximum value is $f\left(1, \sqrt{2}, -\sqrt{2}\right) = 1 + 2\sqrt{2}$ and the minimum value is $f\left(1, -\sqrt{2}, \sqrt{2}\right) = 1 - 2\sqrt{2}$.

14. $f(x,y,z) = 3x - y - 3z$, $g(x,y,z) = x + y - z = 0$, $h(x,y,z) = x^2 + 2z^2 = 1$ $\Rightarrow$ $\nabla f = \langle 3, -1, -3 \rangle$, $\lambda \nabla g = \langle \lambda, \lambda, -\lambda \rangle$, $\mu \nabla h = (2\mu x, 0, 4\mu z)$. Then $3 = \lambda + 2\mu x$, $-1 = \lambda$ and $-3 = -\lambda + 4\mu z$, so $\lambda = -1$, $\mu z = -1$, $\mu x = 2$. Thus $h(x,y,z) = 1$ implies $\dfrac{4}{\mu^2} + 2\left(\dfrac{1}{\mu^2}\right) = 1$ or $\mu = \pm\sqrt{6}$, so $z = \mp\frac{1}{\sqrt{6}}$; $x = \pm\frac{2}{\sqrt{6}}$; and $g(x,y,z) = 0$ implies $y = \mp\frac{3}{\sqrt{6}}$. Hence the maximum of f subject to the constraints is $f\left(\frac{\sqrt{6}}{3}, -\frac{\sqrt{6}}{2}, -\frac{\sqrt{6}}{6}\right) = 2\sqrt{6}$ and the minimum is $f\left(-\frac{\sqrt{6}}{3}, \frac{\sqrt{6}}{2}, \frac{\sqrt{6}}{6}\right) = -2\sqrt{6}$.

15. $f(x,y,z) = yz + xy$, $g(x,y,z) = xy = 1$, $h(x,y,z) = y^2 + z^2 = 1$ $\Rightarrow$ $\nabla f = \langle y, x + z, y \rangle$, $\lambda \nabla g = \langle \lambda y, \lambda x, 0 \rangle$, $\mu \nabla h = \langle 0, 2\mu y, 2\mu z \rangle$. Then $y = \lambda y$ implies $\lambda = 1$ [$y \neq 0$ since $g(x,y,z) = 1$], $x + z = \lambda x + 2\mu y$ and $y = 2\mu z$. Thus $\mu = z/(2y) = y/(2y)$ or $y^2 = z^2$, and so $y^2 + z^2 = 1$ implies $y = \pm\frac{1}{\sqrt{2}}$, $z = \pm\frac{1}{\sqrt{2}}$. Then $xy = 1$ implies $x = \pm\sqrt{2}$ and the possible points are $\left(\pm\sqrt{2}, \pm\frac{1}{\sqrt{2}}, \frac{1}{\sqrt{2}}\right)$, $\left(\pm\sqrt{2}, \pm\frac{1}{\sqrt{2}}, -\frac{1}{\sqrt{2}}\right)$. Hence the maximum of f subject to the constraints is $f\left(\pm\sqrt{2}, \pm\frac{1}{\sqrt{2}}, \pm\frac{1}{\sqrt{2}}\right) = \frac{3}{2}$ and the minimum is $f\left(\pm\sqrt{2}, \pm\frac{1}{\sqrt{2}}, \mp\frac{1}{\sqrt{2}}\right) = \frac{1}{2}$.

Note: Since $xy = 1$ is one of the constraints we could have solved the problem by solving $f(y,z) = yz + 1$ subject to $y^2 + z^2 = 1$.

16. $f(x,y) = 2x^2 + 3y^2 - 4x - 5$ $\Rightarrow$ $\nabla f = \langle 4x - 4, 6y \rangle = \langle 0, 0 \rangle$ $\Rightarrow$ $x = 1$, $y = 0$. Thus $(1, 0)$ is the only critical point of f, and it lies in the region $x^2 + y^2 < 16$. On the boundary, $g(x,y) = x^2 + y^2 = 16$ $\Rightarrow$ $\lambda \nabla g = \langle 2\lambda x, 2\lambda y \rangle$, so $6y = 2\lambda y$ $\Rightarrow$ either $y = 0$ or $\lambda = 3$. If $y = 0$, then $x = \pm 4$; if $\lambda = 3$, then $4x - 4 = 2\lambda x$ $\Rightarrow$ $x = -2$ and $y = \pm 2\sqrt{3}$. Now $f(1,0) = -7$, $f(4,0) = 11$, $f(-4,0) = 43$, and $f\left(-2, \pm 2\sqrt{3}\right) = 47$. Thus the maximum value of $f(x,y)$ on the disk $x^2 + y^2 \leq 16$ is $f\left(-2, \pm 2\sqrt{3}\right) = 47$, and the minimum value is $f(1,0) = -7$.

17. $f(x,y) = e^{-xy}$. For the interior of the region, we find the critical points: $f_x = -ye^{-xy}$, $f_y = -xe^{-xy}$, so the only critical point is $(0,0)$, and $f(0,0) = 1$. For the boundary, we use Lagrange multipliers. $g(x,y) = x^2 + 4y^2 = 1$ $\Rightarrow$ $\lambda \nabla g = \langle 2\lambda x, 8\lambda y \rangle$, so setting $\nabla f = \lambda \nabla g$ we get $-ye^{-xy} = 2\lambda x$ and $-xe^{-xy} = 8\lambda y$. The first of these gives $e^{-xy} = -2\lambda x/y$, and then the second gives $-x(-2\lambda x/y) = 8\lambda y$ $\Rightarrow$ $x^2 = 4y^2$. Solving this last equation with the constraint $x^2 + 4y^2 = 1$ gives $x = \pm\frac{1}{\sqrt{2}}$ and $y = \pm\frac{1}{2\sqrt{2}}$. Now $f\left(\pm\frac{1}{\sqrt{2}}, \mp\frac{1}{2\sqrt{2}}\right) = e^{1/4} \approx 1.284$ and $f\left(\pm\frac{1}{\sqrt{2}}, \pm\frac{1}{2\sqrt{2}}\right) = e^{-1/4} \approx 0.779$. The former are the maxima on the region and the latter are the minima.

18. (a) The values $c = \pm 1$ and $c = 1.25$ seem to give curves which are tangent to the circle. These values represent possible extreme values of the function $x^2 + y$ subject to the constraint $x^2 + y^2 = 1$.

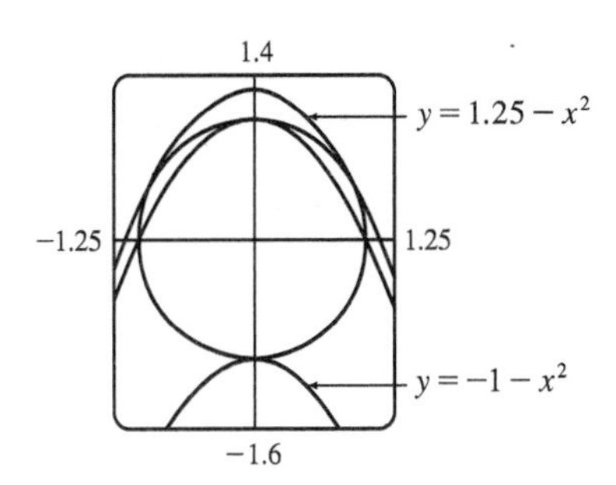

(b) $\nabla f = \langle 2x, 1 \rangle$, $\lambda \nabla g = \langle 2\lambda x, 2\lambda y \rangle$. So $2x = 2\lambda x \Rightarrow$ either $\lambda = 1$ or $x = 0$. If $\lambda = 1$, then $y = \frac{1}{2}$ and so $x = \pm \frac{\sqrt{3}}{2}$ (from the constraint). If $x = 0$, then $y = \pm 1$. Therefore f has possible extreme values at the points $(0, \pm 1)$ and $\left(\pm \frac{\sqrt{3}}{2}, \frac{1}{2}\right)$. We calculate $f\left(\pm \frac{\sqrt{3}}{2}, \frac{1}{2}\right) = \frac{5}{4}$ (the maximum value), $f(0, 1) = 1$, and $f(0, -1) = -1$ (the minimum value). These are our answers from part (a).

19. At the extreme values of f, the level curves of f just touch the curve $g(x, y) = 8$ with a common tangent line. (See Figure 1 and the accompanying discussion.) We can observe several such occurrences on the contour map, but the level curve $f(x, y) = c$ with the largest value of c which still intersects the curve $g(x, y) = 8$ is approximately $c = 59$, and the smallest value of c corresponding to a level curve which intersects $g(x, y) = 8$ appears to be $c = 30$. Thus we estimate the maximum value of f subject to the constraint $g(x, y) = 8$ to be about 59 and the minimum to be 30.

20. (a) The graphs of $f(x, y) = 3.7$ and $f(x, y) = 350$ seem to be tangent to the circle, and so 3.7 and 350 are the approximate minimum and maximum values of the function $f(x, y)$ subject to the constraint $(x - 3)^2 + (y - 3)^2 = 9$.

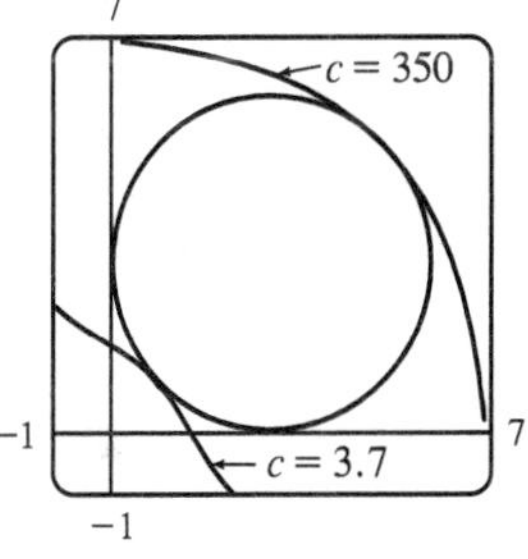

(b) Let $g(x, y) = (x - 3)^2 + (y - 3)^2$. We calculate $f_x(x, y) = 3x^2 + 3y$, $f_y(x, y) = 3y^2 + 3x$, $g_x(x, y) = 2x - 6$, and $g_y(x, y) = 2y - 6$, and use a CAS to search for solutions to the equations $g(x, y) = (x - 3)^2 + (y - 3)^2 = 9$, $f_x = \lambda g_x$, and $f_y = \lambda g_y$. The solutions are $(x, y) = \left(3 - \frac{3}{2}\sqrt{2}, 3 - \frac{3}{2}\sqrt{2}\right) \approx (0.879, 0.879)$ and $(x, y) = \left(3 + \frac{3}{2}\sqrt{2}, 3 + \frac{3}{2}\sqrt{2}\right) \approx (5.121, 5.121)$. These give $f\left(3 - \frac{3}{2}\sqrt{2}, 3 - \frac{3}{2}\sqrt{2}\right) = \frac{351}{2} - \frac{243}{2}\sqrt{2} \approx 3.673$ and $f\left(3 + \frac{3}{2}\sqrt{2}, 3 + \frac{3}{2}\sqrt{2}\right) = \frac{351}{2} + \frac{243}{2}\sqrt{2} \approx 347.33$, in accordance with part (a).

21. $P(L, K) = bL^\alpha K^{1-\alpha}$, $g(L, K) = mL + nK = p \Rightarrow \nabla P = \left\langle \alpha b L^{\alpha - 1} K^{1-\alpha}, (1 - \alpha) b L^\alpha K^{-\alpha} \right\rangle$, $\lambda \nabla g = \langle \lambda m, \lambda n \rangle$. Then $\alpha b (K/L)^{1-\alpha} = \lambda m$ and $(1 - \alpha) b (L/K)^\alpha = \lambda n$ and $mL + nK = p$, so $\alpha b (K/L)^{1-\alpha}/m = (1 - \alpha) b (L/K)^\alpha / n$ or $n\alpha/[m(1 - \alpha)] = (L/K)^\alpha (L/K)^{1-\alpha}$ or $L = Kn\alpha/[m(1 - \alpha)]$. Substituting into $mL + nK = p$ gives $K = (1 - \alpha)p/n$ and $L = \alpha p/m$ for the maximum production.

22. $C(L, K) = mL + nK$, $g(L, K) = bL^\alpha K^{1-\alpha} = Q \Rightarrow \nabla C = \langle m, n \rangle$, $\lambda \nabla g = \left\langle \lambda \alpha b L^{\alpha - 1} K^{1-\alpha}, \lambda (1 - \alpha) b L^\alpha K^{-\alpha} \right\rangle$.

Then $\dfrac{m}{\alpha b}\left(\dfrac{L}{K}\right)^{1-\alpha} = \dfrac{n}{(1 - \alpha) b}\left(\dfrac{K}{L}\right)^{\alpha}$ and $bL^\alpha K^{1-\alpha} = Q \Rightarrow \dfrac{n\alpha}{m(1 - \alpha)} = \left(\dfrac{L}{K}\right)^{1-\alpha}\left(\dfrac{L}{K}\right)^{\alpha} \Rightarrow$

$L = \dfrac{Kn\alpha}{m(1 - \alpha)}$ and so $b\left[\dfrac{Kn\alpha}{m(1 - \alpha)}\right]^\alpha K^{1-\alpha} = Q$. Hence $K = \dfrac{Q}{b\left(n\alpha/[m(1 - \alpha)]\right)^\alpha} = \dfrac{Qm^\alpha (1 - \alpha)^\alpha}{bn^\alpha \alpha^\alpha}$

and $L = \dfrac{Qm^{\alpha - 1}(1 - \alpha)^{\alpha - 1}}{bn^{\alpha - 1}\alpha^{\alpha - 1}} = \dfrac{Qn^{1-\alpha}\alpha^{1-\alpha}}{bm^{1-\alpha}(1 - \alpha)^{1-\alpha}}$ minimizes cost.

23. Let the sides of the rectangle be x and y. Then $f(x,y) = xy$, $g(x,y) = 2x + 2y = p \Rightarrow \nabla f(x,y) = \langle y, x \rangle$, $\lambda \nabla g = \langle 2\lambda, 2\lambda \rangle$. Then $\lambda = \frac{1}{2}y = \frac{1}{2}x$ implies $x = y$ and the rectangle with maximum area is a square with side length $\frac{1}{4}p$.

24. Let $f(x,y,z) = s(s-x)(s-y)(s-z)$, $g(x,y,z) = x + y + z$. Then $\nabla f = \langle -s(s-y)(s-z), -s(s-x)(s-z), -s(s-x)(s-y) \rangle$, $\lambda \nabla g = \langle \lambda, \lambda, \lambda \rangle$. Thus $(s-y)(s-z) = (s-x)(s-z)$ **(1)**, and $(s-x)(s-z) = (s-x)(s-y)$ **(2)**. **(1)** implies $x = y$ while **(2)** implies $y = z$, so $x = y = z = p/3$ and the triangle with maximum area is equilateral.

25. Let $f(x,y,z) = d^2 = (x-2)^2 + (y-1)^2 + (z+1)^2$, then we want to minimize f subject to the constraint $g(x,y,z) = x + y - z = 1$. $\nabla f = \lambda \nabla g \Rightarrow \langle 2(x-2), 2(y-1), 2(z+1) \rangle = \lambda \langle 1, 1, -1 \rangle$, so $x = (\lambda + 4)/2$, $y = (\lambda + 2)/2$, $z = -(\lambda + 2)/2$. Substituting into the constraint equation gives $\dfrac{\lambda+4}{2} + \dfrac{\lambda+2}{2} + \dfrac{\lambda+2}{2} = 1 \Rightarrow$ $3\lambda + 8 = 2 \Rightarrow \lambda = -2$, so $x = 1$, $y = 0$, and $z = 0$. This must correspond to a minimum, so the shortest distance is $d = \sqrt{(1-2)^2 + (0-1)^2 + (0+1)^2} = \sqrt{3}$.

26. Let $f(x,y,z) = d^2 = (x-1)^2 + (y-2)^2 + (z-3)^2$, then we want to minimize f subject to the constraint $g(x,y,z) = x - y + z = 4$. $\nabla f = \lambda \nabla g \Rightarrow \langle 2(x-1), 2(y-2), 2(z-3) \rangle = \lambda \langle 1, -1, 1 \rangle$, so $x = (\lambda + 2)/2$, $y = (4 - \lambda)/2$, $z = (\lambda + 6)/2$. Substituting into the constraint equation gives $\dfrac{\lambda+2}{2} - \dfrac{4-\lambda}{2} + \dfrac{\lambda+6}{2} = 4 \Rightarrow \lambda = \frac{4}{3}$, so $x = \frac{5}{3}$, $y = \frac{4}{3}$, and $z = \frac{11}{3}$. This must correspond to a minimum, so the point on the plane closest to the point $(1,2,3)$ is $\left(\frac{5}{3}, \frac{4}{3}, \frac{11}{3}\right)$.

27. Let $f(x,y,z) = d^2 = (x-4)^2 + (y-2)^2 + z^2$. Then we want to minimize f subject to the constraint $g(x,y,z) = x^2 + y^2 - z^2 = 0$. $\nabla f = \lambda \nabla g \Rightarrow \langle 2(x-4), 2(y-2), 2z \rangle = \langle 2\lambda x, 2\lambda y, -2\lambda z \rangle$, so $x - 4 = \lambda x$, $y - 2 = \lambda y$, and $z = -\lambda z$. From the last equation we have $z + \lambda z = 0 \Rightarrow z(1+\lambda) = 0$, so either $z = 0$ or $\lambda = -1$. But from the constraint equation we have $z = 0 \Rightarrow x^2 + y^2 = 0 \Rightarrow x = y = 0$ which is not possible from the first two equations. So $\lambda = -1$ and $x - 4 = \lambda x \Rightarrow x = 2$, $y - 2 = \lambda y \Rightarrow y = 1$, and $x^2 + y^2 - z^2 = 0 \Rightarrow$ $4 + 1 - z^2 = 0 \Rightarrow z = \pm\sqrt{5}$. This must correspond to a minimum, so the points on the cone closest to $(4,2,0)$ are $(2, 1, \pm\sqrt{5})$.

28. Let $f(x,y,z) = d^2 = x^2 + y^2 + z^2$. Then we want to minimize f subject to the constraint $g(x,y,z) = y^2 - xz = 9$. $\nabla f = \lambda \nabla g \Rightarrow \langle 2x, 2y, 2z \rangle = \langle -\lambda z, 2\lambda y, -\lambda x \rangle$, so $2x = -\lambda z$, $y = \lambda y$, and $2z = -\lambda x$. If $x = 0$ then the last equation implies $z = 0$, and from the constraint $y^2 - xz = 9$ we have $y = \pm 3$. If $x \neq 0$, then the first and third equations give $\lambda = -2x/z = -2z/x \Rightarrow x^2 = z^2$. From the second equation we have $y = 0$ or $\lambda = 1$. If $y = 0$ then $y^2 - xz = 9 \Rightarrow z = -9/x$ and $x^2 = z^2 \Rightarrow x^2 = 81/x^2 \Rightarrow x = \pm 3$. Since $z = -9/x$, $x = 3 \Rightarrow z = -3$ and $x = -3 \Rightarrow z = 3$. If $\lambda = 1$, then $2x = -z$ and $2z = -x$ which implies $x = z = 0$, contradicting the assumption that $x \neq 0$. Thus the possible points are $(0, \pm 3, 0)$, $(3, 0, -3)$, $(-3, 0, 3)$. We have $f(0, \pm 3, 0) = 9$ and $f(3, 0, -3) = f(-3, 0, 3) = 18$, so the points on the surface that are closest to the origin are $(0, \pm 3, 0)$.

29. $f(x,y,z)=xyz$, $g(x,y,z)=x+y+z=100 \Rightarrow \nabla f=\langle yz,xz,xy\rangle=\lambda\nabla g=\langle\lambda,\lambda,\lambda\rangle$. Then $\lambda=yz=xz=xy$ implies $x=y=z=\frac{100}{3}$.

30. $f(x,y,z)=x^a y^b z^c$, $g(x,y,z)=x+y+z=100 \Rightarrow$

$\nabla f=\left\langle ax^{a-1}y^bz^c, bx^ay^{b-1}z^c, cx^ay^bz^{c-1}\right\rangle=\lambda\nabla g=\langle\lambda,\lambda,\lambda\rangle$. Then $\lambda=ax^{a-1}y^bz^c=bx^ay^{b-1}z^c=cx^ay^bz^{c-1}$

or $ayz=bxz=cxy$. Thus $x=\dfrac{ay}{b}$, $z=\dfrac{cy}{b}$, and $\dfrac{ay}{b}+y+\dfrac{cy}{b}=100$ implies that $y=\dfrac{100b}{a+b+c}$, $x=\dfrac{100a}{a+b+c}$ and

$z=\dfrac{100c}{a+b+c}$ gives the maximum.

31. If the dimensions are $2x$, $2y$ and $2z$, then $f(x,y,z)=8xyz$ and $g(x,y,z)=9x^2+36y^2+4z^2=36 \Rightarrow$

$\nabla f=\langle 8yz,8xz,8xy\rangle=\lambda\nabla g=\langle 18\lambda x,72\lambda y,8\lambda z\rangle$. Thus $18\lambda x=8yz$, $72\lambda y=8xz$, $8\lambda z=8xy$ so $x^2=4y^2$, $z^2=9y^2$

and $36y^2+36y^2+36y^2=36$ or $y=\frac{1}{\sqrt{3}}$ $(y>0)$. Thus the volume of the largest such box is $8\left(\frac{1}{\sqrt{3}}\right)\left(\frac{2}{\sqrt{3}}\right)\left(\frac{3}{\sqrt{3}}\right)=\frac{16}{\sqrt{3}}$.

32. $f(x,y,z)=8xyz$, $g(x,y,z)=b^2c^2x+a^2c^2y^2+a^2b^2z^2=a^2b^2c^2 \Rightarrow$

$\nabla f=\langle 8yz,8xz,8xy\rangle=\lambda\nabla g=\left\langle 2\lambda b^2c^2x, 2\lambda a^2c^2y, 2\lambda a^2b^2z\right\rangle$. Then $4yz=\lambda b^2c^2x$, $4xz=\lambda a^2c^2y$, $4xy=\lambda a^2b^2z$

imply $\lambda=\dfrac{4yz}{b^2c^2x}=\dfrac{4xz}{a^2c^2y}=\dfrac{4xy}{a^2b^2z}$ or $\dfrac{y}{b^2x}=\dfrac{x}{a^2y}$ and $\dfrac{z}{c^2y}=\dfrac{y}{b^2z}$. Thus $x=\dfrac{ay}{b}$, $z=\dfrac{cy}{b}$, and

$a^2c^2y^2+c^2a^2y^2+a^2c^2y^2=a^2b^2c^2$, or $y=\dfrac{b}{\sqrt{3}}$, $x=\dfrac{a}{\sqrt{3}}$, $z=\dfrac{c}{\sqrt{3}}$ and the volume is $\dfrac{8}{3\sqrt{3}}abc$.

33. $f(x,y,z)=xyz$, $g(x,y,z)=x+2y+3z=6 \Rightarrow \nabla f=\langle yz,xz,xy\rangle=\lambda\nabla g=\langle\lambda,2\lambda,3\lambda\rangle$.

Then $\lambda=yz=\frac{1}{2}xz=\frac{1}{3}xy$ implies $x=2y$, $z=\frac{2}{3}y$. But $2y+2y+2y=6$ so $y=1$, $x=2$, $z=\frac{2}{3}$ and the volume is $V=\frac{4}{3}$.

34. $f(x,y,z)=xyz$, $g(x,y,z)=xy+yz+xz=32 \Rightarrow \nabla f=\langle yz,xz,xy\rangle=\lambda\nabla g=\langle\lambda(y+z),\lambda(x+z),\lambda(x+y)\rangle$.

Then $\lambda(y+z)=yz$ **(1)**, $\lambda(x+z)=xz$ **(2)**, and $\lambda(x+y)=xy$ **(3)**. And **(1)** minus **(2)** implies $\lambda(y-x)=z(y-x)$ so $x=y$ or $\lambda=z$. If $\lambda=z$, then **(1)** implies $z(y+z)=yz$ or $z=0$ which is false. Thus $x=y$. Similarly **(2)** minus **(3)** implies $\lambda(z-y)=x(z-y)$ so $y=z$ or $\lambda=x$. As above, $\lambda\neq x$, so $x=y=z$ and $3x^2=32$ or $x=y=z=\frac{8}{\sqrt{6}}$ cm.

35. $f(x,y,z)=xyz$, $g(x,y,z)=4(x+y+z)=c \Rightarrow \nabla f=\langle yz,xz,xy\rangle$, $\lambda\nabla g=\langle 4\lambda,4\lambda,4\lambda\rangle$. Thus $4\lambda=yz=xz=xy$ or $x=y=z=\frac{1}{12}c$ are the dimensions giving the maximum volume.

36. $C(x,y,z)=5xy+2xz+2yz$, $g(x,y,z)=xyz=V \Rightarrow$

$\nabla C=\langle 5y+2z,5x+2z,2x+2y\rangle=\lambda\nabla g=\langle\lambda yz,\lambda xz,\lambda xy\rangle$. Then $\lambda yz=5y+2z$ **(1)**, $\lambda xz=5x+2z$ **(2)**, $\lambda xy=2(x+y)$ **(3)**, and $xyz=V$ **(4)**. Now **(1)** − **(2)** implies $\lambda z(y-x)=5(y-x)$, so $x=y$ or $\lambda=5/z$, but z can't be 0, so $x=y$. Then twice **(2)** minus five times **(3)** together with $x=y$ implies $\lambda y(2x-5y)=2(2z-5y)$ which gives $z=\frac{5}{2}y$ [again $\lambda\neq 2/y$ or else **(3)** implies $y=0$]. Hence $\frac{5}{2}y^3=V$ and the dimensions which minimize cost are

$x=y=\sqrt[3]{\frac{2}{5}V}$ units, $z=V^{1/3}\left(\frac{5}{2}\right)^{2/3}$ units.

37. If the dimensions of the box are given by x, y, and z, then we need to find the maximum value of $f(x,y,z) = xyz$ $[x,y,z>0]$ subject to the constraint $L = \sqrt{x^2+y^2+z^2}$ or $g(x,y,z) = x^2+y^2+z^2 = L^2$. $\nabla f = \lambda \nabla g \Rightarrow$ $\langle yz, xz, xy\rangle = \lambda\langle 2x, 2y, 2z\rangle$, so $yz = 2\lambda x \Rightarrow \lambda = \frac{yz}{2x}$, $xz = 2\lambda y \Rightarrow \lambda = \frac{xz}{2y}$, and $xy = 2\lambda z \Rightarrow \lambda = \frac{xy}{2z}$. Thus $\lambda = \frac{yz}{2x} = \frac{xz}{2y} \Rightarrow x^2 = y^2$ [since $z \neq 0$] $\Rightarrow x = y$ and $\lambda = \frac{yz}{2x} = \frac{xy}{2z} \Rightarrow x = z$ [since $y \neq 0$]. Substituting into the constraint equation gives $x^2+x^2+x^2 = L^2 \Rightarrow x^2 = L^2/3 \Rightarrow x = L/\sqrt{3} = y = z$ and the maximum volume is $\left(L/\sqrt{3}\right)^3 = L^3/\left(3\sqrt{3}\right)$.

38. Let the dimensions of the box be x, y, and z, so its volume is $f(x,y,z) = xyz$, its surface area is $2xy+2yz+2xz = 1500$ and its total edge length is $4x+4y+4z = 200$. We find the extreme values of $f(x,y,z)$ subject to the constraints $g(x,y,z) = xy+yz+xz = 750$ and $h(x,y,z) = x+y+z = 50$. Then

$\nabla f = \langle yz, xz, xy\rangle = \lambda\nabla g + \mu\nabla h = \langle \lambda(y+z), \lambda(x+z), \lambda(x+y)\rangle + \langle \mu,\mu,\mu\rangle$. So $yz = \lambda(y+z)+\mu$ **(1)**, $xz = \lambda(x+z)+\mu$ **(2)**, and $xy = \lambda(x+y)+\mu$ **(3)**. Notice that the box can't be a cube or else $x=y=z=\frac{50}{3}$ but then $xy+yz+xz = \frac{2500}{3} \neq 750$. Assume x is the distinct side, that is, $x \neq y$, $x \neq z$. Then **(1)** minus **(2)** implies $z(y-x) = \lambda(y-x)$ or $\lambda = z$, and **(1)** minus **(3)** implies $y(z-x) = \lambda(z-x)$ or $\lambda = y$. So $y = z = \lambda$ and $x+y+z = 50$ implies $x = 50-2\lambda$; also $xy+yz+xz = 750$ implies $x(2\lambda)+\lambda^2 = 750$. Hence $50-2\lambda = \dfrac{750-\lambda^2}{2\lambda}$ or

$3\lambda^2 - 100\lambda + 750 = 0$ and $\lambda = \dfrac{50 \pm 5\sqrt{10}}{3}$, giving the points $\left(\frac{1}{3}\left(50 \mp 10\sqrt{10}\right), \frac{1}{3}\left(50 \pm 5\sqrt{10}\right), \frac{1}{3}\left(50 \pm 5\sqrt{10}\right)\right)$.

Thus the minimum of f is $f\left(\frac{1}{3}\left(50-10\sqrt{3}\right), \frac{1}{3}\left(50+5\sqrt{10}\right), \frac{1}{3}\left(50+5\sqrt{10}\right)\right) = \frac{1}{27}\left(87{,}500 - 2500\sqrt{10}\right)$, and its maximum is $f\left(\frac{1}{3}\left(50+10\sqrt{10}\right), \frac{1}{3}\left(50-5\sqrt{10}\right), \frac{1}{3}\left(50-5\sqrt{10}\right)\right) = \frac{1}{27}\left(87{,}500 + 2500\sqrt{10}\right)$.

Note: If either y or z is the distinct side, then symmetry gives the same result.

39. We need to find the extreme values of $f(x,y,z) = x^2+y^2+z^2$ subject to the two constraints $g(x,y,z) = x+y+2z = 2$ and $h(x,y,z) = x^2+y^2-z = 0$. $\nabla f = \langle 2x, 2y, 2z\rangle$, $\lambda\nabla g = \langle \lambda, \lambda, 2\lambda\rangle$ and $\mu\nabla h = \langle 2\mu x, 2\mu y, -\mu\rangle$. Thus we need $2x = \lambda + 2\mu x$ **(1)**, $2y = \lambda + 2\mu y$ **(2)**, $2z = 2\lambda - \mu$ **(3)**, $x+y+2z = 2$ **(4)**, and $x^2+y^2-z = 0$ **(5)**.

From **(1)** and **(2)**, $2(x-y) = 2\mu(x-y)$, so if $x \neq y$, $\mu = 1$. Putting this in **(3)** gives $2z = 2\lambda - 1$ or $\lambda = z + \frac{1}{2}$, but putting $\mu = 1$ into **(1)** says $\lambda = 0$. Hence $z + \frac{1}{2} = 0$ or $z = -\frac{1}{2}$. Then **(4)** and **(5)** become $x+y-3 = 0$ and $x^2+y^2+\frac{1}{2} = 0$. The last equation cannot be true, so this case gives no solution. So we must have $x = y$. Then **(4)** and **(5)** become $2x+2z = 2$ and $2x^2 - z = 0$ which imply $z = 1-x$ and $z = 2x^2$. Thus $2x^2 = 1-x$ or $2x^2+x-1 = (2x-1)(x+1) = 0$ so $x = \frac{1}{2}$ or $x = -1$. The two points to check are $\left(\frac{1}{2}, \frac{1}{2}, \frac{1}{2}\right)$ and $(-1,-1,2)$: $f\left(\frac{1}{2}, \frac{1}{2}, \frac{1}{2}\right) = \frac{3}{4}$ and $f(-1,-1,2) = 6$. Thus $\left(\frac{1}{2}, \frac{1}{2}, \frac{1}{2}\right)$ is the point on the ellipse nearest the origin and $(-1,-1,2)$ is the one farthest from the origin.

40. (a) Parametric equations for the ellipse are easiest to determine using cylindrical coordinates. The cone is given by $z = r$, and the plane is $4r\cos\theta - 3r\sin\theta + 8z = 5$. Substituting $z = r$ into the plane equation gives $4r\cos\theta - 3r\sin\theta + 8r = 5 \;\Rightarrow\; r = \dfrac{5}{4\cos\theta - 3\sin\theta + 8}$. Since $z = r$ on the ellipse, parametric equations (in cylindrical coordinates) are

$$\theta = t,\; r = z = \frac{5}{4\cos t - 3\sin t + 8},\quad 0 \le t \le 2\pi.$$

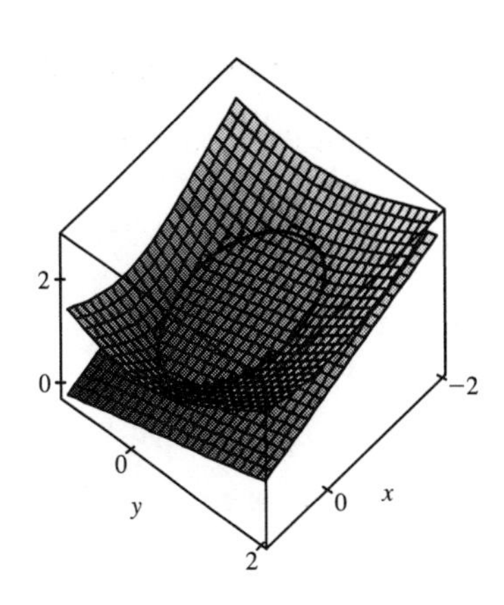

(b) We need to find the extreme values of $f(x, y, z) = z$ subject to the two constraints $g(x, y, z) = 4x - 3y + 8z = 5$ and $h(x, y, z) = x^2 + y^2 - z^2 = 0$. $\nabla f = \lambda\nabla g + \mu\nabla h \;\Rightarrow\; \langle 0, 0, 1\rangle = \lambda\langle 4, -3, 8\rangle + \mu\langle 2x, 2y, -2z\rangle$, so we need $4\lambda + 2\mu x = 0 \;\Rightarrow\; x = -\frac{2\lambda}{\mu}$ **(1)**, $-3\lambda + 2\mu y = 0 \;\Rightarrow\; y = \frac{3\lambda}{2\mu}$ **(2)**, $8\lambda - 2\mu z = 1 \;\Rightarrow\; z = \frac{8\lambda - 1}{2\mu}$ **(3)**, $4x - 3y + 8z = 5$ **(4)**, and $x^2 + y^2 = z^2$ **(5)**. [Note that $\mu \ne 0$, else $\lambda = 0$ from **(1)**, but substitution into **(3)** gives a contradiction.] Substituting **(1)**, **(2)**, and **(3)** into **(4)** gives $4\left(-\frac{2\lambda}{\mu}\right) - 3\left(\frac{3\lambda}{2\mu}\right) + 8\left(\frac{8\lambda-1}{2\mu}\right) = 5 \;\Rightarrow\; \mu = \frac{39\lambda - 8}{10}$ and into **(5)** gives $\left(-\frac{2\lambda}{\mu}\right)^2 + \left(\frac{3\lambda}{2\mu}\right)^2 = \left(\frac{8\lambda-1}{2\mu}\right)^2 \;\Rightarrow\; 16\lambda^2 + 9\lambda^2 = (8\lambda - 1)^2 \;\Rightarrow\; 39\lambda^2 - 16\lambda + 1 = 0 \;\Rightarrow\; \lambda = \frac{1}{13}$ or $\lambda = \frac{1}{3}$. If $\lambda = \frac{1}{13}$ then $\mu = -\frac{1}{2}$ and $x = \frac{4}{13}$, $y = -\frac{3}{13}$, $z = \frac{5}{13}$. If $\lambda = \frac{1}{3}$ then $\mu = \frac{1}{2}$ and $x = -\frac{4}{3}$, $y = 1$, $z = \frac{5}{3}$. Thus the highest point on the ellipse is $\left(-\frac{4}{3}, 1, \frac{5}{3}\right)$ and the lowest point is $\left(\frac{4}{13}, -\frac{3}{13}, \frac{5}{13}\right)$.

41. $f(x, y, z) = ye^{x-z}$, $g(x, y, z) = 9x^2 + 4y^2 + 36z^2 = 36$, $h(x, y, z) = xy + yz = 1$. $\nabla f = \lambda\nabla g + \mu\nabla h \;\Rightarrow$ $\langle ye^{x-z}, e^{x-z}, -ye^{x-z}\rangle = \lambda\langle 18x, 8y, 72z\rangle + \mu\langle y, x + z, y\rangle$, so $ye^{x-z} = 18\lambda x + \mu y$, $e^{x-z} = 8\lambda y + \mu(x + z)$, $-ye^{x-z} = 72\lambda z + \mu y$, $9x^2 + 4y^2 + 36z^2 = 36$, $xy + yz = 1$. Using a CAS to solve these 5 equations simultaneously for x, y, z, λ, and μ (in Maple, use the `allvalues` command), we get 4 real-valued solutions:

$x \approx 0.222444$, $y \approx -2.157012$, $z \approx -0.686049$, $\lambda \approx -0.200401$, $\mu \approx 2.108584$
$x \approx -1.951921$, $y \approx -0.545867$, $z \approx 0.119973$, $\lambda \approx 0.003141$, $\mu \approx -0.076238$
$x \approx 0.155142$, $y \approx 0.904622$, $z \approx 0.950293$, $\lambda \approx -0.012447$, $\mu \approx 0.489938$
$x \approx 1.138731$, $y \approx 1.768057$, $z \approx -0.573138$, $\lambda \approx 0.317141$, $\mu \approx 1.862675$

Substituting these values into f gives $f(0.222444, -2.157012, -0.686049) \approx -5.3506$, $f(-1.951921, -0.545867, 0.119973) \approx -0.0688$, $f(0.155142, 0.904622, 0.950293) \approx 0.4084$, $f(1.138731, 1.768057, -0.573138) \approx 9.7938$. Thus the maximum is approximately 9.7938, and the mininum is approximately -5.3506.

42. $f(x, y, z) = x + y + z$, $g(x, y, z) = x^2 - y^2 - z = 0$, $h(x, y, z) = x^2 + z^2 = 4$.

$\nabla f = \lambda \nabla g + \mu \nabla h \Rightarrow \langle 1, 1, 1 \rangle = \lambda \langle 2x, -2y, -1 \rangle + \mu \langle 2x, 0, 2z \rangle$, so $1 = 2\lambda x + 2\mu x$, $1 = -2\lambda y$, $1 = -\lambda + 2\mu z$, $x^2 - y^2 = z$, $x^2 + z^2 = 4$. Using a CAS to solve these 5 equations simultaneously for x, y, z, λ, and μ, we get 4 real-valued solutions:

$$x \approx -1.652878, \quad y \approx -1.964194, \quad z \approx -1.126052, \quad \lambda \approx 0.254557, \quad \mu \approx -0.557060$$

$$x \approx -1.502800, \quad y \approx 0.968872, \quad z \approx 1.319694, \quad \lambda \approx -0.516064, \quad \mu \approx 0.183352$$

$$x \approx -0.992513, \quad y \approx 1.649677, \quad z \approx -1.736352, \quad \lambda \approx -0.303090, \quad \mu \approx -0.200682$$

$$x \approx 1.895178, \quad y \approx 1.718347, \quad z \approx 0.638984, \quad \lambda \approx -0.290977, \quad \mu \approx 0.554805$$

Substituting these values into f gives $f(-1.652878, -1.964194, -1.126052) \approx -4.7431$, $f(-1.502800, 0.968872, 1.319694) \approx 0.7858$, $f(-0.992513, 1.649677, -1.736352) \approx -1.0792$, $f(1.895178, 1.718347, 0.638984) \approx 4.2525$. Thus the maximum is approximately 4.2525, and the mininum is approximately -4.7431.

43. (a) We wish to maximize $f(x_1, x_2, \ldots, x_n) = \sqrt[n]{x_1 x_2 \cdots x_n}$ subject to

$g(x_1, x_2, \ldots, x_n) = x_1 + x_2 + \cdots + x_n = c$ and $x_i > 0$.

$\nabla f = \left\langle \frac{1}{n}(x_1 x_2 \cdots x_n)^{\frac{1}{n}-1}(x_2 \cdots x_n), \frac{1}{n}(x_1 x_2 \cdots x_n)^{\frac{1}{n}-1}(x_1 x_3 \cdots x_n), \ldots, \frac{1}{n}(x_1 x_2 \cdots x_n)^{\frac{1}{n}-1}(x_1 \cdots x_{n-1}) \right\rangle$

and $\lambda \nabla g = \langle \lambda, \lambda, \ldots, \lambda \rangle$, so we need to solve the system of equations

$$\frac{1}{n}(x_1 x_2 \cdots x_n)^{\frac{1}{n}-1}(x_2 \cdots x_n) = \lambda \quad \Rightarrow \quad x_1^{1/n} x_2^{1/n} \cdots x_n^{1/n} = n\lambda x_1$$

$$\frac{1}{n}(x_1 x_2 \cdots x_n)^{\frac{1}{n}-1}(x_1 x_3 \cdots x_n) = \lambda \quad \Rightarrow \quad x_1^{1/n} x_2^{1/n} \cdots x_n^{1/n} = n\lambda x_2$$

$$\vdots$$

$$\frac{1}{n}(x_1 x_2 \cdots x_n)^{\frac{1}{n}-1}(x_1 \cdots x_{n-1}) = \lambda \quad \Rightarrow \quad x_1^{1/n} x_2^{1/n} \cdots x_n^{1/n} = n\lambda x_n$$

This implies $n\lambda x_1 = n\lambda x_2 = \cdots = n\lambda x_n$. Note $\lambda \neq 0$, otherwise we can't have all $x_i > 0$. Thus $x_1 = x_2 = \cdots = x_n$. But $x_1 + x_2 + \cdots + x_n = c \Rightarrow nx_1 = c \Rightarrow x_1 = \frac{c}{n} = x_2 = x_3 = \cdots = x_n$. Then the only point where f can have an extreme value is $\left(\frac{c}{n}, \frac{c}{n}, \ldots, \frac{c}{n}\right)$. Since we can choose values for $(x_1, x_2, \ldots, x_n)$ that make f as close to zero (but not equal) as we like, f has no minimum value. Thus the maximum value is

$$f\left(\frac{c}{n}, \frac{c}{n}, \ldots, \frac{c}{n}\right) = \sqrt[n]{\frac{c}{n} \cdot \frac{c}{n} \cdot \cdots \cdot \frac{c}{n}} = \frac{c}{n}.$$

(b) From part (a), $\frac{c}{n}$ is the maximum value of f. Thus $f(x_1, x_2, \ldots, x_n) = \sqrt[n]{x_1 x_2 \cdots x_n} \leq \frac{c}{n}$. But $x_1 + x_2 + \cdots + x_n = c$, so $\sqrt[n]{x_1 x_2 \cdots x_n} \leq \frac{x_1 + x_2 + \cdots + x_n}{n}$. These two means are equal when f attains its maximum value $\frac{c}{n}$, but this can occur only at the point $\left(\frac{c}{n}, \frac{c}{n}, \ldots, \frac{c}{n}\right)$ we found in part (a). So the means are equal only when $x_1 = x_2 = x_3 = \cdots = x_n = \frac{c}{n}$.

44. (a) Let $f(x_1, \ldots, x_n, y_1, \ldots, y_n) = \sum_{i=1}^{n} x_i y_i$, $g(x_1, \ldots, x_n) = \sum_{i=1}^{n} x_i^2$, and $h(x_1, \ldots, x_n) = \sum_{i=1}^{n} y_i^2$. Then

$\nabla f = \nabla \sum_{i=1}^{n} x_i y_i = \langle y_1, y_2, \ldots, y_n, x_1, x_2, \ldots, x_n \rangle$, $\nabla g = \nabla \sum_{i=1}^{n} x_i^2 = \langle 2x_1, 2x_2, \ldots, 2x_n, 0, 0, \ldots, 0 \rangle$ and

$\nabla h = \nabla \sum_{i=1}^{n} y_i^2 = \langle 0, 0, \ldots, 0, 2y_1, 2y_2, \ldots, 2y_n \rangle$. So $\nabla f = \lambda \nabla g + \mu \nabla h \Leftrightarrow y_i = 2\lambda x_i$ and $x_i = 2\mu y_i$,

$1 \le i \le n$. Then $1 = \sum_{i=1}^{n} y_i^2 = \sum_{i=1}^{n} 4\lambda^2 x_i^2 = 4\lambda^2 \sum_{i=1}^{n} x_i^2 = 4\lambda^2 \Rightarrow \lambda = \pm\frac{1}{2}$. If $\lambda = \frac{1}{2}$ then $y_i = 2\left(\frac{1}{2}\right)x_i = x_i$,

$1 \le i \le n$. Thus $\sum_{i=1}^{n} x_i y_i = \sum_{i=1}^{n} x_i^2 = 1$. Similarly if $\lambda = -\frac{1}{2}$ we get $y_i = -x_i$ and $\sum_{i=1}^{n} x_i y_i = -1$. Similarly we get

$\mu = \pm\frac{1}{2}$ giving $y_i = \pm x_i$, $1 \le i \le n$, and $\sum_{i=1}^{n} x_i y_i = \pm 1$. Thus the maximum value of $\sum_{i=1}^{n} x_i y_i$ is 1.

(b) Here we assume $\sum_{i=1}^{n} a_i^2 \neq 0$ and $\sum_{i=1}^{n} b_i^2 \neq 0$. (If $\sum_{i=1}^{n} a_i^2 = 0$, then each $a_i = 0$ and so the inequality is trivially true.)

$x_i = \dfrac{a_i}{\sqrt{\sum a_j^2}} \Rightarrow \sum x_i^2 = \dfrac{\sum a_i^2}{\sum a_j^2} = 1$, and $y_i = \dfrac{b_i}{\sqrt{\sum b_j^2}} \Rightarrow \sum y_i^2 = \dfrac{\sum b_i^2}{\sum b_j^2} = 1$. Therefore, from part (a),

$$\sum x_i y_i = \sum \frac{a_i b_i}{\sqrt{\sum a_j^2}\sqrt{\sum b_j^2}} \le 1 \Leftrightarrow \sum a_i b_i \le \sqrt{\sum a_j^2}\sqrt{\sum b_j^2}.$$

11 Review

CONCEPT CHECK

1. (a) A function f of two variables is a rule that assigns to each ordered pair (x, y) of real numbers in its domain a unique real number denoted by $f(x, y)$.

(b) One way to visualize a function of two variables is by graphing it, resulting in the surface $z = f(x, y)$. Another method for visualizing a function of two variables is a contour map. The contour map consists of level curves of the function which are horizontal traces of the graph of the function projected onto the xy-plane. Also, we can use an arrow diagram such as Figure 1 in Section 11.1.

2. A function f of three variables is a rule that assigns to each ordered triple (x, y, z) in its domain a unique real number $f(x, y, z)$. We can visualize a function of three variables by examining its level surfaces $f(x, y, z) = k$, where k is a constant.

3. $\lim_{(x,y)\to(a,b)} f(x, y) = L$ means the values of $f(x, y)$ approach the number L as the point (x, y) approaches the point (a, b) along any path that is within the domain of f. We can show that a limit at a point does not exist by finding two different paths approaching the point along which $f(x, y)$ has different limits.

4. (a) See Definition 11.2.4.

(b) If f is continuous on $\mathbb{R}^2$, its graph will appear as a surface without holes or breaks.

5. (a) See (2) and (3) in Section 11.3.

(b) See "Interpretations of Partial Derivatives" on page 612.

(c) To find f_x, regard y as a constant and differentiate $f(x, y)$ with respect to x. To find f_y, regard x as a constant and differentiate $f(x, y)$ with respect to y.

6. See the statement of Clairaut's Theorem on page 615.

7. (a) See (2) in Section 11.4.

(b) See (19) and the preceding discussion in Section 11.6.

8. See (3) and (4) and the accompanying discussion in Section 11.4. We can interpret the linearization of f at (a, b) geometrically as the linear function whose graph is the tangent plane to the graph of f at (a, b). Thus it is the linear function which best approximates f near (a, b).

9. (a) See Definition 11.4.7.

(b) Use Theorem 11.4.8.

10. See (10) and the associated discussion in Section 11.4.

11. See (2) and (3) in Section 11.5.

12. See (7) and the preceding discussion in Section 11.5.

13. (a) See Definition 11.6.2. We can interpret it as the rate of change of f at (x_0, y_0) in the direction of $\mathbf{u}$. Geometrically, if P is the point $(x_0, y_0, f(x_0, y_0))$ on the graph of f and C is the curve of intersection of the graph of f with the vertical plane that passes through P in the direction $\mathbf{u}$, the directional derivative of f at (x_0, y_0) in the direction of $\mathbf{u}$ is the slope of the tangent line to C at P. (See Figure 2 in Section 11.6.)

(b) See Theorem 11.6.3.

14. (a) See (8) and (13) in Section 11.6.

(b) $D_{\mathbf{u}}\, f(x, y) = \nabla f(x, y) \cdot \mathbf{u}$ or $D_{\mathbf{u}}\, f(x, y, z) = \nabla f(x, y, z) \cdot \mathbf{u}$

(c) The gradient vector of a function points in the direction of maximum rate of increase of the function. On a graph of the function, the gradient points in the direction of steepest ascent.

15. (a) f has a local maximum at (a, b) if $f(x, y) \leq f(a, b)$ when (x, y) is near (a, b).

(b) f has an absolute maximum at (a, b) if $f(x, y) \leq f(a, b)$ for all points (x, y) in the domain of f.

(c) f has a local minimum at (a, b) if $f(x, y) \geq f(a, b)$ when (x, y) is near (a, b).

(d) f has an absolute minimum at (a, b) if $f(x, y) \geq f(a, b)$ for all points (x, y) in the domain of f.

(e) f has a saddle point at (a, b) if $f(a, b)$ is a local maximum in one direction but a local minimum in another.

16. (a) By Theorem 11.7.2, if f has a local maximum at (a, b) and the first-order partial derivatives of f exist there, then $f_x(a, b) = 0$ and $f_y(a, b) = 0$.

(b) A critical point of f is a point (a, b) such that $f_x(a, b) = 0$ and $f_y(a, b) = 0$ or one of these partial derivatives does not exist.

17. See (3) in Section 11.7.

18. (a) See Figure 7 and the accompanying discussion in Section 11.7.

(b) See Theorem 11.7.4.

(c) See the procedure outlined in (5) in Section 11.7.

19. See the discussion beginning on page 655; see "Two Constraints" on page 658.

TRUE-FALSE QUIZ

1. True. $f_y(a,b) = \lim\limits_{h\to 0} \dfrac{f(a,b+h)-f(a,b)}{h}$ from Equation 11.3.3. Let $h = y - b$. As $h \to 0$, $y \to b$. Then by substituting, we get $f_y(a,b) = \lim\limits_{y\to b} \dfrac{f(a,y)-f(a,b)}{y-b}$.

2. False. If there were such a function, then $f_{xy} = 2y$ and $f_{yx} = 1$. So $f_{xy} \neq f_{yx}$, which contradicts Clairaut's Theorem.

3. False. $f_{xy} = \dfrac{\partial^2 f}{\partial y\, \partial x}$.

4. True. From Equation 11.6.14 we get $D_{\mathbf{k}}\, f(x,y,z) = \nabla f(x,y,z) \cdot \langle 0,0,1\rangle = f_z(x,y,z)$.

5. False. See Example 11.2.3.

6. False. See Exercise 11.4.36(a).

7. True. If f has a local minimum and f is differentiable at (a,b) then by Theorem 11.7.2, $f_x(a,b) = 0$ and $f_y(a,b) = 0$, so $\nabla f(a,b) = \langle f_x(a,b), f_y(a,b)\rangle = \langle 0,0\rangle = \mathbf{0}$.

8. False. If f is not continuous at $(2,5)$, then we can have $\lim\limits_{(x,y)\to(2,5)} f(x,y) \neq f(2,5)$. (See Example 11.2.7.)

9. False. $\nabla f(x,y) = \langle 0, 1/y\rangle$.

10. True. This is part (c) of the Second Derivatives Test (11.7.3).

11. True. $\nabla f = \langle \cos x, \cos y\rangle$, so $|\nabla f| = \sqrt{\cos^2 x + \cos^2 y}$. But $|\cos\theta| \leq 1$, so $|\nabla f| \leq \sqrt{2}$. Now $D_{\mathbf{u}}\, f(x,y) = \nabla f \cdot \mathbf{u} = |\nabla f|\, |\mathbf{u}| \cos\theta$, but $\mathbf{u}$ is a unit vector, so $|D_{\mathbf{u}}\, f(x,y)| \leq \sqrt{2}\cdot 1 \cdot 1 = \sqrt{2}$.

12. False. See Exercise 11.7.29.

EXERCISES

1. $\ln(x+y+1)$ is defined only when $x + y + 1 > 0 \quad\Rightarrow\quad y > -x - 1$, so the domain of f is $\{(x,y) \mid y > -x - 1\}$, all those points above the line $y = -x - 1$.

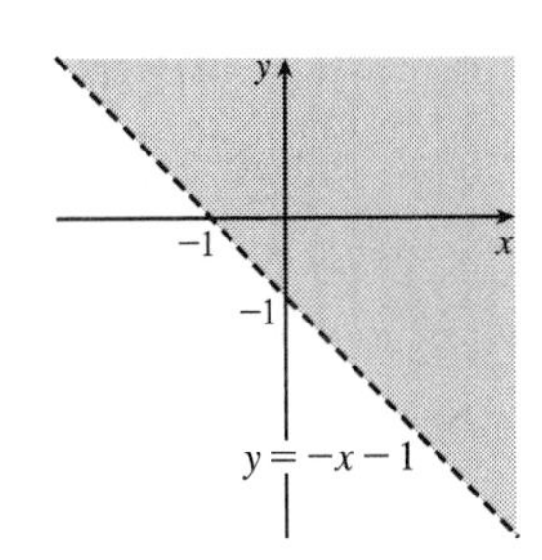

2. $\sqrt{4-x^2-y^2}$ is defined only when $4-x^2-y^2 \geq 0 \quad \Leftrightarrow \quad x^2+y^2 \leq 4$, and $\sqrt{1-x^2}$ is defined only when $1-x^2 \geq 0 \quad \Leftrightarrow \quad -1 \leq x \leq 1$, so the domain of f is $\left\{(x,y) \mid -1 \leq x \leq 1, -\sqrt{4-x^2} \leq y \leq \sqrt{4-x^2}\right\}$, which consists of those points on or inside the circle $x^2+y^2=4$ for $-1 \leq x \leq 1$.

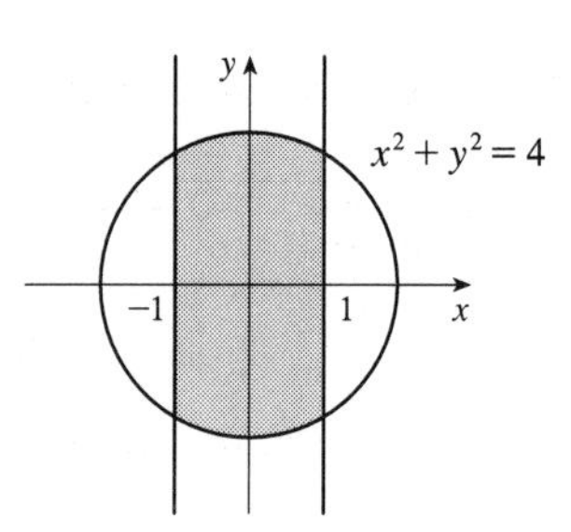

3. $z = f(x,y) = 1-y^2$, a parabolic cylinder

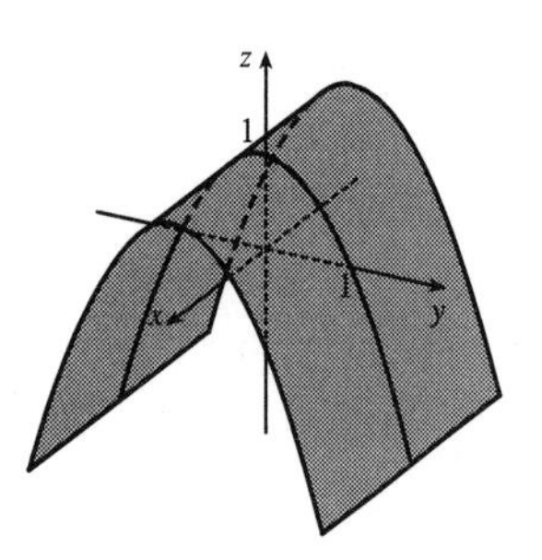

4. $z = f(x,y) = x^2+(y-2)^2$, a circular paraboloid with vertex $(0,2,0)$ and axis parallel to the z-axis

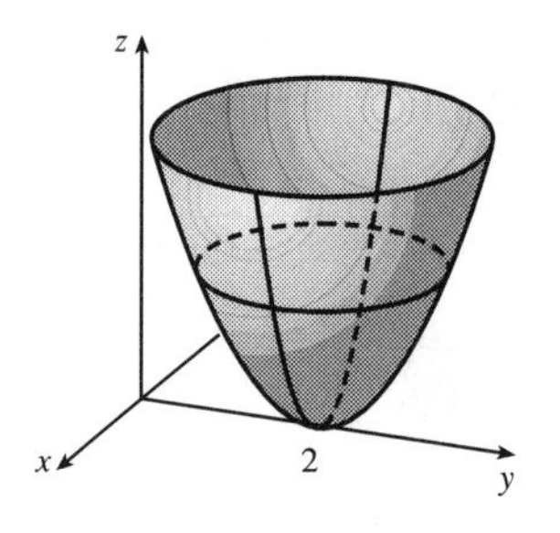

5. The level curves are $\sqrt{4x^2+y^2}=k$ or $4x^2+y^2=k^2$, $k \geq 0$, a family of ellipses.

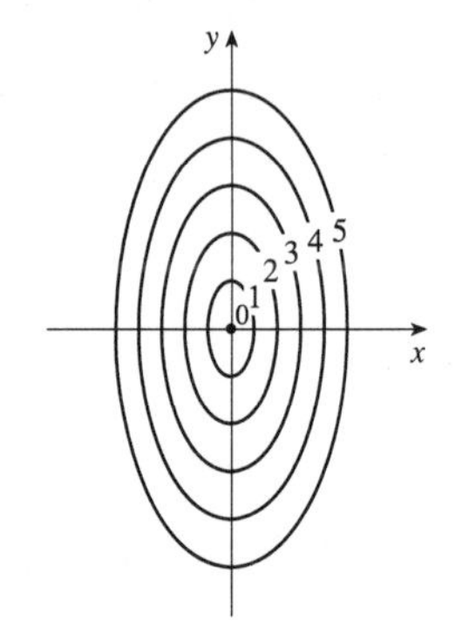

6. The level curves are $e^x+y=k$ or $y=-e^x+k$, a family of exponential curves.

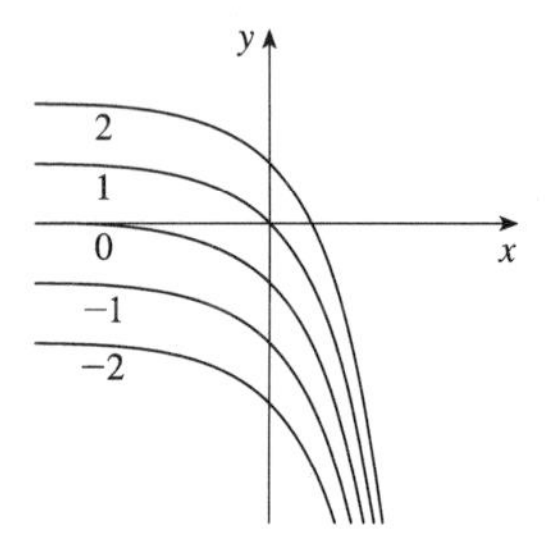

7.

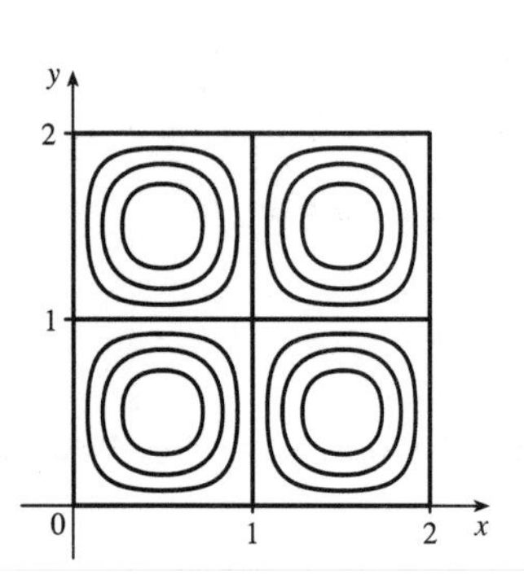

8.

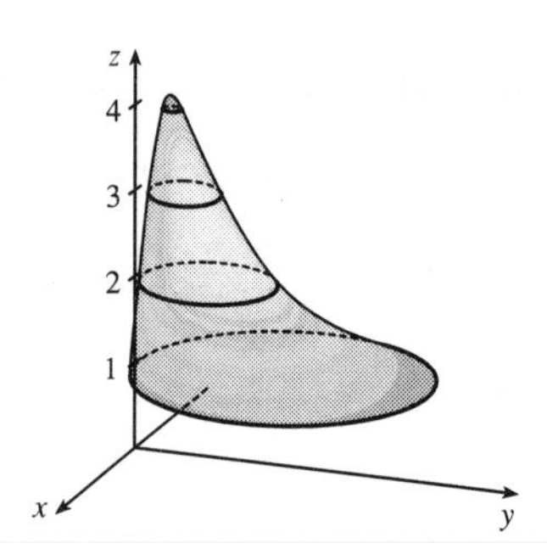

9. f is a rational function, so it is continuous on its domain. Since f is defined at $(1,1)$, we use direct substitution to evaluate the limit: $\displaystyle\lim_{(x,y)\to(1,1)} \frac{2xy}{x^2+2y^2} = \frac{2(1)(1)}{1^2+2(1)^2} = \frac{2}{3}$.

10. As $(x,y)\to(0,0)$ along the x-axis, $f(x,0)=0/x^2=0$ for $x\neq 0$, so $f(x,y)\to 0$ along this line. But $f(x,x)=2x^2/(3x^2)=\frac{2}{3}$, so as $(x,y)\to(0,0)$ along the line $x=y$, $f(x,y)\to\frac{2}{3}$. Thus the limit doesn't exist.

11. $f(x,y)=\sqrt{2x+y^2} \quad\Rightarrow\quad f_x=\frac{1}{2}(2x+y^2)^{-1/2}(2)=\dfrac{1}{\sqrt{2x+y^2}},\ f_y=\frac{1}{2}(2x+y^2)^{-1/2}(2y)=\dfrac{y}{\sqrt{2x+y^2}}$

12. $u=e^{-r}\sin 2\theta \quad\Rightarrow\quad u_r=-e^{-r}\sin 2\theta,\ u_\theta=2e^{-r}\cos 2\theta$

13. $g(u,v)=u\tan^{-1}v \quad\Rightarrow\quad g_u=\tan^{-1}v,\ g_v=\dfrac{u}{1+v^2}$

14. $w=\dfrac{x}{y-z} \quad\Rightarrow\quad w_x=\dfrac{1}{y-z},\ w_y=x(-1)(y-z)^{-2}=-\dfrac{x}{(y-z)^2},\ w_z=x(-1)(y-z)^{-2}(-1)=\dfrac{x}{(y-z)^2}$

15. $T(p,q,r)=p\ln(q+e^r) \quad\Rightarrow\quad T_p=\ln(q+e^r),\ T_q=\dfrac{p}{q+e^r},\ T_r=\dfrac{pe^r}{q+e^r}$

16. $C=1449.2+4.6T-0.055T^2+0.00029T^3+(1.34-0.01T)(S-35)+0.016D \quad\Rightarrow$

$\partial C/\partial T=4.6-0.11T+0.00087T^2-0.01(S-35)$, $\partial C/\partial S=1.34-0.01T$, and $\partial C/\partial D=0.016$. When $T=10$, $S=35$, and $D=100$ we have $\partial C/\partial T=4.6-0.11(10)+0.00087(10)^2-0.01(35-35)\approx 3.587$, thus in 10°C water with salinity 35 parts per thousand and a depth of 100 m, the speed of sound increases by about 3.59 m/s for every degree Celsius that the water temperature rises. Similarly, $\partial C/\partial S=1.34-0.01(10)=1.24$, so the speed of sound increases by about 1.24 m/s for every part per thousand the salinity of the water increases. $\partial C/\partial D=0.016$, so the speed of sound increases by about 0.016 m/s for every meter that the depth is increased.

17. $f(x,y)=4x^3-xy^2 \quad\Rightarrow\quad f_x=12x^2-y^2,\ f_y=-2xy,\ f_{xx}=24x,\ f_{yy}=-2x,\ f_{xy}=f_{yx}=-2y$

18. $z=xe^{-2y} \quad\Rightarrow\quad z_x=e^{-2y},\ z_y=-2xe^{-2y},\ z_{xx}=0,\ z_{yy}=4xe^{-2y},\ z_{xy}=z_{yx}=-2e^{-2y}$

19. $f(x,y,z)=x^ky^lz^m \quad\Rightarrow\quad f_x=kx^{k-1}y^lz^m,\ f_y=lx^ky^{l-1}z^m,\ f_z=mx^ky^lz^{m-1},\ f_{xx}=k(k-1)x^{k-2}y^lz^m,$

$f_{yy}=l(l-1)x^ky^{l-2}z^m,\ f_{zz}=m(m-1)x^ky^lz^{m-2},\ f_{xy}=f_{yx}=klx^{k-1}y^{l-1}z^m,\ f_{xz}=f_{zx}=kmx^{k-1}y^lz^{m-1},$

$f_{yz}=f_{zy}=lmx^ky^{l-1}z^{m-1}$

20. $v=r\cos(s+2t) \quad\Rightarrow\quad v_r=\cos(s+2t),\ v_s=-r\sin(s+2t),\ v_t=-2r\sin(s+2t),\ v_{rr}=0,\ v_{ss}=-r\cos(s+2t),$

$v_{tt}=-4r\cos(s+2t),\ v_{rs}=v_{sr}=-\sin(s+2t),\ v_{rt}=v_{tr}=-2\sin(s+2t),\ v_{st}=v_{ts}=-2r\cos(s+2t)$

21. $z=xy+xe^{y/x} \quad\Rightarrow\quad \dfrac{\partial z}{\partial x}=y-\dfrac{y}{x}e^{y/x}+e^{y/x},\ \dfrac{\partial z}{\partial y}=x+e^{y/x}$ and

$$x\frac{\partial z}{\partial x}+y\frac{\partial z}{\partial y}=x\left(y-\frac{y}{x}e^{y/x}+e^{y/x}\right)+y\left(x+e^{y/x}\right)=xy-ye^{y/x}+xe^{y/x}+xy+ye^{y/x}=xy+xy+xe^{y/x}=xy+z.$$

22. $z = \sin(x + \sin t) \Rightarrow \dfrac{\partial z}{\partial x} = \cos(x + \sin t)$, $\dfrac{\partial z}{\partial t} = \cos(x + \sin t)\cos t$,

$\dfrac{\partial^2 z}{\partial x \partial t} = -\sin(x + \sin t)\cos t$, $\dfrac{\partial^2 z}{\partial x^2} = -\sin(x + \sin t)$ and

$$\frac{\partial z}{\partial x}\frac{\partial^2 z}{\partial x \partial t} = \cos(x + \sin t)\left[-\sin(x + \sin t)\cos t\right] = \cos(x + \sin t)(\cos t)\left[-\sin(x + \sin t)\right] = \frac{\partial z}{\partial t}\frac{\partial^2 z}{\partial x^2}.$$

23. (a) $z_x = 6x + 2 \Rightarrow z_x(1, -2) = 8$ and $z_y = -2y \Rightarrow z_y(1, -2) = 4$, so an equation of the tangent plane is $z - 1 = 8(x - 1) + 4(y + 2)$ or $z = 8x + 4y + 1$.

(b) A normal vector to the tangent plane (and the surface) at $(1, -2, 1)$ is $\langle 8, 4, -1\rangle$. Then parametric equations for the normal line there are $x = 1 + 8t$, $y = -2 + 4t$, $z = 1 - t$, and symmetric equations are $\dfrac{x - 1}{8} = \dfrac{y + 2}{4} = \dfrac{z - 1}{-1}$.

24. (a) $z_x = e^x \cos y \Rightarrow z_x(0, 0) = 1$ and $z_y = -e^x \sin y \Rightarrow z_y(0, 0) = 0$, so an equation of the tangent plane is $z - 1 = 1(x - 0) + 0(y - 0)$ or $z = x + 1$.

(b) A normal vector to the tangent plane (and the surface) at $(0, 0, 1)$ is $\langle 1, 0, -1\rangle$. Then parametric equations for the normal line there are $x = t$, $y = 0$, $z = 1 - t$, and symmetric equations are $x = 1 - z$, $y = 0$.

25. (a) Let $F(x, y, z) = x^2 + 2y^2 - 3z^2$. Then $F_x = 2x$, $F_y = 4y$, $F_z = -6z$, so $F_x(2, -1, 1) = 4$, $F_y(2, -1, 1) = -4$, $F_z(2, -1, 1) = -6$. From Equation 11.6.19, an equation of the tangent plane is $4(x - 2) - 4(y + 1) - 6(z - 1) = 0$ or, equivalently, $2x - 2y - 3z = 3$.

(b) From Equations 11.6.20, symmetric equations for the normal line are $\dfrac{x - 2}{4} = \dfrac{y + 1}{-4} = \dfrac{z - 1}{-6}$.

26. (a) Let $F(x, y, z) = xy + yz + zx$. Then $F_x = y + z$, $F_y = x + z$, $F_z = x + y$, so $F_x(1, 1, 1) = F_y(1, 1, 1) = F_z(1, 1, 1) = 2$. From Equation 11.6.19, an equation of the tangent plane is $2(x - 1) + 2(y - 1) + 2(z - 1) = 0$ or, equivalently, $x + y + z = 3$.

(b) From Equations 11.6.20, symmetric equations for the normal line are $\dfrac{x - 1}{2} = \dfrac{y - 1}{2} = \dfrac{z - 1}{2}$ or, equivalently, $x = y = z$.

27. (a) Let $F(x, y, z) = x + 2y + 3z - \sin(xyz)$. Then $F_x = 1 - yz\cos(xyz)$, $F_y = 2 - xz\cos(xyz)$, $F_z = 3 - xy\cos(xyz)$, so $F_x(2, -1, 0) = 1$, $F_y(2, -1, 0) = 2$, $F_z(2, -1, 0) = 5$. From Equation 11.6.19, an equation of the tangent plane is $1(x - 2) + 2(y + 1) + 5(z - 0) = 0$ or $x + 2y + 5z = 0$.

(b) From Equations 11.6.20, symmetric equations for the normal line are $\dfrac{x - 2}{1} = \dfrac{y + 1}{2} = \dfrac{z}{5}$.

28. Let $f(x, y) = x^2 + y^4$. Then $f_x(x, y) = 2x$ and $f_y(x, y) = 4y^3$, so $f_x(1, 1) = 2$, $f_y(1, 1) = 4$ and an equation of the tangent plane is $z - 2 = 2(x - 1) + 4(y - 1)$ or $2x + 4y - z = 4$. A normal vector to the tangent plane is $\langle 2, 4, -1\rangle$ so the normal line is given by $\dfrac{x - 1}{2} = \dfrac{y - 1}{4} = \dfrac{z - 2}{-1}$ or $x = 1 + 2t$, $y = 1 + 4t$, $z = 2 - t$.

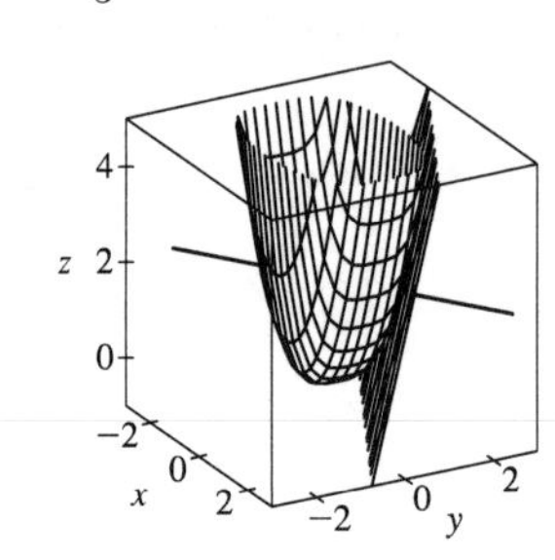

29. The hyperboloid is a level surface of the function $F(x,y,z)=x^2+4y^2-z^2$, so a normal vector to the surface at (x_0,y_0,z_0) is $\nabla F(x_0,y_0,z_0)=\langle 2x_0,8y_0,-2z_0\rangle$. A normal vector for the plane $2x+2y+z=5$ is $\langle 2,2,1\rangle$. For the planes to be parallel, we need the normal vectors to be parallel, so $\langle 2x_0,8y_0,-2z_0\rangle=k\,\langle 2,2,1\rangle$, or $x_0=k$, $y_0=\frac{1}{4}k$, and $z_0=-\frac{1}{2}k$. But $x_0^2+4y_0^2-z_0^2=4 \quad\Rightarrow\quad k^2+\frac{1}{4}k^2-\frac{1}{4}k^2=4 \quad\Rightarrow\quad k^2=4 \quad\Rightarrow\quad k=\pm 2$. So there are two such points: $\left(2,\frac{1}{2},-1\right)$ and $\left(-2,-\frac{1}{2},1\right)$.

30. $u=\ln(1+se^{2t}) \quad\Rightarrow\quad du=\dfrac{\partial u}{\partial s}\,ds+\dfrac{\partial u}{\partial t}\,dt=\dfrac{e^{2t}}{1+se^{2t}}\,ds+\dfrac{2se^{2t}}{1+se^{2t}}\,dt$

31. $f(x,y,z)=x^3\sqrt{y^2+z^2} \quad\Rightarrow\quad f_x(x,y,z)=3x^2\sqrt{y^2+z^2},\ f_y(x,y,z)=\dfrac{yx^3}{\sqrt{y^2+z^2}},\ f_z(x,y,z)=\dfrac{zx^3}{\sqrt{y^2+z^2}}$,

so $f(2,3,4)=8(5)=40$, $f_x(2,3,4)=3(4)\sqrt{25}=60$, $f_y(2,3,4)=\frac{3(8)}{\sqrt{25}}=\frac{24}{5}$, and $f_z(2,3,4)=\frac{4(8)}{\sqrt{25}}=\frac{32}{5}$. Then the linear approximation of f at $(2,3,4)$ is

$$\begin{aligned} f(x,y,z) &\approx f(2,3,4)+f_x(2,3,4)(x-2)+f_y(2,3,4)(y-3)+f_z(2,3,4)(z-4)\\ &=40+60(x-2)+\tfrac{24}{5}(y-3)+\tfrac{32}{5}(z-4)=60x+\tfrac{24}{5}y+\tfrac{32}{5}z-120 \end{aligned}$$

Then $(1.98)^3\sqrt{(3.01)^2+(3.97)^2}=f(1.98,3.01,3.97)\approx 60(1.98)+\frac{24}{5}(3.01)+\frac{32}{5}(3.97)-120=38.656$.

32. (a) $dA=\dfrac{\partial A}{\partial x}\,dx+\dfrac{\partial A}{\partial y}\,dy=\frac{1}{2}y\,dx+\frac{1}{2}x\,dy$ and $|\Delta x|\le 0.002$, $|\Delta y|\le 0.002$. Thus the maximum error in the calculated area is about $dA=6(0.002)+\frac{5}{2}(0.002)=0.017\text{ m}^2$ or 170 cm^2.

(b) $z=\sqrt{x^2+y^2}$, $dz=\dfrac{x}{\sqrt{x^2+y^2}}\,dx+\dfrac{y}{\sqrt{x^2+y^2}}\,dy$ and $|\Delta x|\le 0.002$, $|\Delta y|\le 0.002$. Thus the maximum error in the calculated hypotenuse length is about $dz=\frac{5}{13}(0.002)+\frac{12}{13}(0.002)=\frac{0.17}{65}\approx 0.0026$ m or 0.26 cm.

33. $\dfrac{du}{dp}=\dfrac{\partial u}{\partial x}\dfrac{dx}{dp}+\dfrac{\partial u}{\partial y}\dfrac{dy}{dp}+\dfrac{\partial u}{\partial z}\dfrac{dz}{dp}=2xy^3(1+6p)+3x^2y^2(pe^p+e^p)+4z^3(p\cos p+\sin p)$

34. $\dfrac{\partial v}{\partial s}=\dfrac{\partial v}{\partial x}\dfrac{\partial x}{\partial s}+\dfrac{\partial v}{\partial y}\dfrac{\partial y}{\partial s}=\left(2x\sin y+y^2e^{xy}\right)(1)+\left(x^2\cos y+xye^{xy}+e^{xy}\right)(t)$.

$s=0,\ t=1 \quad\Rightarrow\quad x=2,\ y=0$, so $\dfrac{\partial v}{\partial s}=0+(4+1)(1)=5$.

$\dfrac{\partial v}{\partial t}=\dfrac{\partial v}{\partial x}\dfrac{\partial x}{\partial t}+\dfrac{\partial v}{\partial y}\dfrac{\partial y}{\partial t}=\left(2x\sin y+y^2e^{xy}\right)(2)+\left(x^2\cos y+xye^{xy}+e^{xy}\right)(s)=0+0=0$.

35. By the Chain Rule, $\dfrac{\partial z}{\partial s}=\dfrac{\partial z}{\partial x}\dfrac{\partial x}{\partial s}+\dfrac{\partial z}{\partial y}\dfrac{\partial y}{\partial s}$. When $s=1$ and $t=2$, $x=g(1,2)=3$ and $y=h(1,2)=6$, so

$\dfrac{\partial z}{\partial s}=f_x(3,6)g_s(1,2)+f_y(3,6)\,h_s(1,2)=(7)(-1)+(8)(-5)=-47$. Similarly, $\dfrac{\partial z}{\partial t}=\dfrac{\partial z}{\partial x}\dfrac{\partial x}{\partial t}+\dfrac{\partial z}{\partial y}\dfrac{\partial y}{\partial t}$, so

$\dfrac{\partial z}{\partial t}=f_x(3,6)g_t(1,2)+f_y(3,6)\,h_t(1,2)=(7)(4)+(8)(10)=108$.

36. 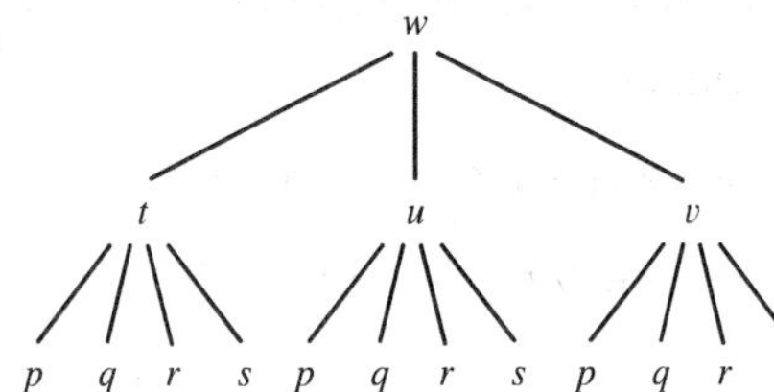

Using the tree diagram as a guide, we have

$$\frac{\partial w}{\partial p}=\frac{\partial w}{\partial t}\frac{\partial t}{\partial p}+\frac{\partial w}{\partial u}\frac{\partial u}{\partial p}+\frac{\partial w}{\partial v}\frac{\partial v}{\partial p} \qquad \frac{\partial w}{\partial q}=\frac{\partial w}{\partial t}\frac{\partial t}{\partial q}+\frac{\partial w}{\partial u}\frac{\partial u}{\partial q}+\frac{\partial w}{\partial v}\frac{\partial v}{\partial q}$$

$$\frac{\partial w}{\partial r}=\frac{\partial w}{\partial t}\frac{\partial t}{\partial r}+\frac{\partial w}{\partial u}\frac{\partial u}{\partial r}+\frac{\partial w}{\partial v}\frac{\partial v}{\partial r} \qquad \frac{\partial w}{\partial s}=\frac{\partial w}{\partial t}\frac{\partial t}{\partial s}+\frac{\partial w}{\partial u}\frac{\partial u}{\partial s}+\frac{\partial w}{\partial v}\frac{\partial v}{\partial s}$$

37. $\dfrac{\partial z}{\partial x}=2xf'(x^2-y^2),\quad \dfrac{\partial z}{\partial y}=1-2yf'(x^2-y^2)\quad \left[\text{where } f'=\dfrac{df}{d(x^2-y^2)}\right]$. Then

$$y\frac{\partial z}{\partial x}+x\frac{\partial z}{\partial y}=2xyf'(x^2-y^2)+x-2xyf'(x^2-y^2)=x.$$

38. $A=\frac{1}{2}xy\sin\theta,\ dx/dt=3,\ dy/dt=-2,\ d\theta/dt=0.05$, and $\dfrac{dA}{dt}=\dfrac{1}{2}\left[(y\sin\theta)\dfrac{dx}{dt}+(x\sin\theta)\dfrac{dy}{dt}+(xy\cos\theta)\dfrac{d\theta}{dt}\right]$.

So when $x=40$, $y=50$ and $\theta=\frac{\pi}{6}$, $\dfrac{dA}{dt}=\dfrac{1}{2}\left[(25)(3)+(20)(-2)+\left(1000\sqrt{3}\right)(0.05)\right]=\dfrac{35+50\sqrt{3}}{2}\approx 60.8\text{ in}^2\text{/s}.$

39. $\dfrac{\partial z}{\partial x}=\dfrac{\partial z}{\partial u}y+\dfrac{\partial z}{\partial v}\dfrac{-y}{x^2}$ and

$$\begin{aligned}\frac{\partial^2 z}{\partial x^2}&=y\frac{\partial}{\partial x}\left(\frac{\partial z}{\partial u}\right)+\frac{2y}{x^3}\frac{\partial z}{\partial v}+\frac{-y}{x^2}\frac{\partial}{\partial x}\left(\frac{\partial z}{\partial v}\right)=\frac{2y}{x^3}\frac{\partial z}{\partial v}+y\left(\frac{\partial^2 z}{\partial u^2}y+\frac{\partial^2 z}{\partial v\,\partial u}\frac{-y}{x^2}\right)+\frac{-y}{x^2}\left(\frac{\partial^2 z}{\partial v^2}\frac{-y}{x^2}+\frac{\partial^2 z}{\partial u\,\partial v}y\right)\\&=\frac{2y}{x^3}\frac{\partial z}{\partial v}+y^2\frac{\partial^2 z}{\partial u^2}-\frac{2y^2}{x^2}\frac{\partial^2 z}{\partial u\,\partial v}+\frac{y^2}{x^4}\frac{\partial^2 z}{\partial v^2}\end{aligned}$$

Also $\dfrac{\partial z}{\partial y}=x\dfrac{\partial z}{\partial u}+\dfrac{1}{x}\dfrac{\partial z}{\partial v}$ and

$$\frac{\partial^2 z}{\partial y^2}=x\frac{\partial}{\partial y}\left(\frac{\partial z}{\partial u}\right)+\frac{1}{x}\frac{\partial}{\partial y}\left(\frac{\partial z}{\partial v}\right)=x\left(\frac{\partial^2 z}{\partial u^2}x+\frac{\partial^2 z}{\partial v\,\partial u}\frac{1}{x}\right)+\frac{1}{x}\left(\frac{\partial^2 z}{\partial v^2}\frac{1}{x}+\frac{\partial^2 z}{\partial u\,\partial v}x\right)=x^2\frac{\partial^2 z}{\partial u^2}+2\frac{\partial^2 z}{\partial u\,\partial v}+\frac{1}{x^2}\frac{\partial^2 z}{\partial v^2}$$

Thus

$$\begin{aligned}x^2\frac{\partial^2 z}{\partial x^2}-y^2\frac{\partial^2 z}{\partial y^2}&=\frac{2y}{x}\frac{\partial z}{\partial v}+x^2y^2\frac{\partial^2 z}{\partial u^2}-2y^2\frac{\partial^2 z}{\partial u\,\partial v}+\frac{y^2}{x^2}\frac{\partial^2 z}{\partial v^2}-x^2y^2\frac{\partial^2 z}{\partial u^2}-2y^2\frac{\partial^2 z}{\partial u\,\partial v}-\frac{y^2}{x^2}\frac{\partial^2 z}{\partial v^2}\\&=\frac{2y}{x}\frac{\partial z}{\partial v}-4y^2\frac{\partial^2 z}{\partial u\,\partial v}=2v\frac{\partial z}{\partial v}-4uv\frac{\partial^2 z}{\partial u\,\partial v}\end{aligned}$$

since $y=xv=\dfrac{uv}{y}$ or $y^2=uv$.

40. $F(x,y,z)=e^{xyz}-yz^4-x^2z^3=0$, so $\dfrac{\partial z}{\partial x}=-\dfrac{F_x}{F_z}=-\dfrac{yze^{xyz}-2xz^3}{xye^{xyz}-4yz^3-3x^2z^2}=\dfrac{2xz^3-yze^{xyz}}{xye^{xyz}-4yz^3-3x^2z^2}$ and

$$\frac{\partial z}{\partial y}=-\frac{F_y}{F_z}=-\frac{xze^{xyz}-z^4}{xye^{xyz}-4yz^3-3x^2z^2}=\frac{z^4-xze^{xyz}}{xye^{xyz}-4yz^3-3x^2z^2}.$$

41. $\nabla f=\left\langle z^2\sqrt{y}\,e^{x\sqrt{y}},\ \dfrac{xz^2e^{x\sqrt{y}}}{2\sqrt{y}},\ 2ze^{x\sqrt{y}}\right\rangle=ze^{x\sqrt{y}}\left\langle z\sqrt{y},\ \dfrac{xz}{2\sqrt{y}},\ 2\right\rangle$

42. (a) By Theorem 11.6.15, the maximum value of the directional derivative occurs when $\mathbf{u}$ has the same direction as the gradient vector.

(b) It is a minimum when $\mathbf{u}$ is in the direction opposite to that of the gradient vector (that is, $\mathbf{u}$ is in the direction of $-\nabla f$), since $D_{\mathbf{u}} f = |\nabla f| \cos\theta$ (see the proof of Theorem 11.6.15) has a minimum when $\theta = \pi$.

(c) The directional derivative is 0 when $\mathbf{u}$ is perpendicular to the gradient vector, since then $D_{\mathbf{u}} f = \nabla f \cdot \mathbf{u} = 0$.

(d) The directional derivative is half of its maximum value when $D_{\mathbf{u}} f = |\nabla f| \cos\theta = \frac{1}{2}|\nabla f| \quad\Leftrightarrow\quad \cos\theta = \frac{1}{2} \quad\Leftrightarrow$ $\theta = \frac{\pi}{3}$.

43. $\nabla f = \langle 1/\sqrt{x}, -2y\rangle$, $\nabla f(1,5) = \langle 1, -10\rangle$, $\mathbf{u} = \frac{1}{5}\langle 3, -4\rangle$. Then $D_{\mathbf{u}} f(1,5) = \frac{43}{5}$.

44. $\nabla f = \left\langle 2xy + \sqrt{1+z}, x^2, x/\left(2\sqrt{1+z}\right)\right\rangle$, $\nabla f(1,2,3) = \left\langle 6, 1, \frac{1}{4}\right\rangle$, $\mathbf{u} = \left\langle \frac{2}{3}, \frac{1}{3}, -\frac{2}{3}\right\rangle$. Then $D_{\mathbf{u}} f(1,2,3) = \frac{25}{6}$.

45. $\nabla f = \left\langle 2xy, x^2 + 1/\left(2\sqrt{y}\right)\right\rangle$, $|\nabla f(2,1)| = \left|\left\langle 4, \frac{9}{2}\right\rangle\right|$. Thus the maximum rate of change of f at $(2,1)$ is $\frac{\sqrt{145}}{2}$ in the direction $\left\langle 4, \frac{9}{2}\right\rangle$.

46. The surfaces are $f(x,y,z) = z - 2x^2 + y^2 = 0$ and $g(x,y,z) = z - 4 = 0$. The tangent line is perpendicular to both ∇f and ∇g at $(-2,2,4)$. The vector $\mathbf{v} = \nabla f \times \nabla g$ is therefore parallel to the line. $\nabla f(x,y,z) = \langle -4x, 2y, 1\rangle \Rightarrow$ $\nabla f(-2,2,4) = \langle 8,4,1\rangle$, $\nabla g(x,y,z) = \langle 0,0,1\rangle \Rightarrow \nabla g\langle -2,2,4\rangle = \langle 0,0,1\rangle$. Hence

$$\mathbf{v} = \nabla f \times \nabla g = \begin{vmatrix} \mathbf{i} & \mathbf{j} & \mathbf{k} \\ 8 & 4 & 1 \\ 0 & 0 & 1 \end{vmatrix} = 4\,\mathbf{i} - 8\,\mathbf{j}.$$ Thus, parametric equations are: $x = -2 + 4t,\ y = 2 - 8t,\ z = 4$.

47. $f(x,y) = x^2 - xy + y^2 + 9x - 6y + 10 \quad\Rightarrow\quad f_x = 2x - y + 9$, $f_y = -x + 2y - 6$, $f_{xx} = 2 = f_{yy}$, $f_{xy} = -1$. Then $f_x = 0$ and $f_y = 0$ imply $y = 1$, $x = -4$. Thus the only critical point is $(-4,1)$ and $f_{xx}(-4,1) > 0$, $D(-4,1) = 3 > 0$, so $f(-4,1) = -11$ is a local minimum.

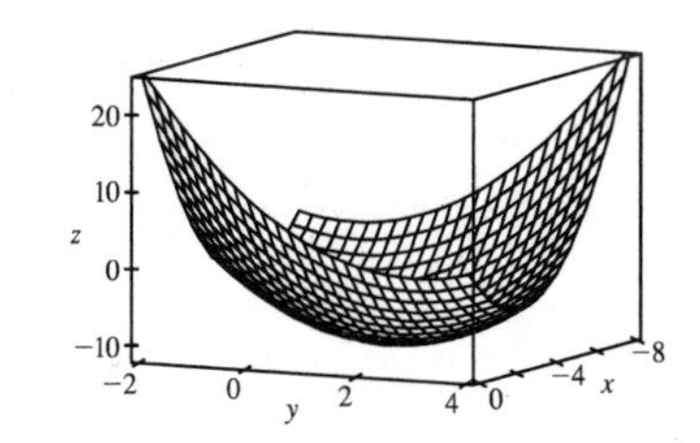

48. $f(x,y) = x^3 - 6xy + 8y^3 \quad\Rightarrow\quad f_x = 3x^2 - 6y,\ f_y = -6x + 24y^2,\ f_{xx} = 6x$, $f_{yy} = 48y$, $f_{xy} = -6$. Then $f_x = 0$ implies $y = x^2/2$, substituting into $f_y = 0$ implies $6x\left(x^3 - 1\right) = 0$, so the critical points are $(0,0)$, $\left(1, \frac{1}{2}\right)$. $D(0,0) = -36 < 0$ so $(0,0)$ is a saddle point while $f_{xx}\left(1, \frac{1}{2}\right) = 6 > 0$ and $D\left(1, \frac{1}{2}\right) = 108 > 0$ so $f\left(1, \frac{1}{2}\right) = -1$ is a local minimum.

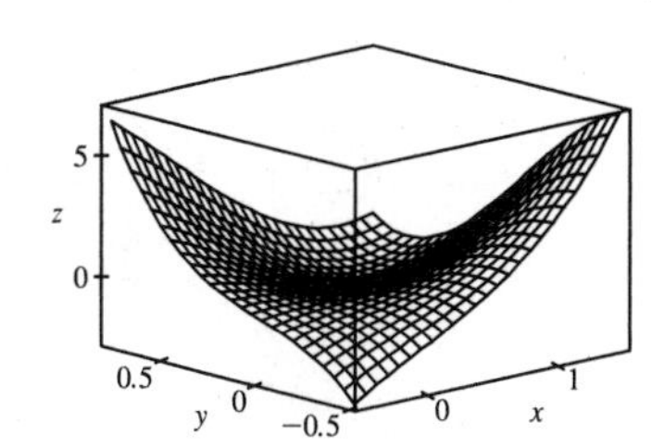

49. $f(x,y) = 3xy - x^2y - xy^2 \quad\Rightarrow\quad f_x = 3y - 2xy - y^2,\ f_y = 3x - x^2 - 2xy$, $f_{xx} = -2y$, $f_{yy} = -2x$, $f_{xy} = 3 - 2x - 2y$. Then $f_x = 0$ implies $y(3 - 2x - y) = 0$ so $y = 0$ or $y = 3 - 2x$. Substituting into $f_y = 0$ implies $x(3 - x) = 0$ or $3x(-1 + x) = 0$. Hence the critical points are $(0,0)$, $(3,0)$, $(0,3)$ and $(1,1)$. $D(0,0) = D(3,0) = D(0,3) = -9 < 0$ so $(0,0)$, $(3,0)$, and $(0,3)$ are saddle points. $D(1,1) = 3 > 0$ and $f_{xx}(1,1) = -2 < 0$, so $f(1,1) = 1$ is a local maximum.

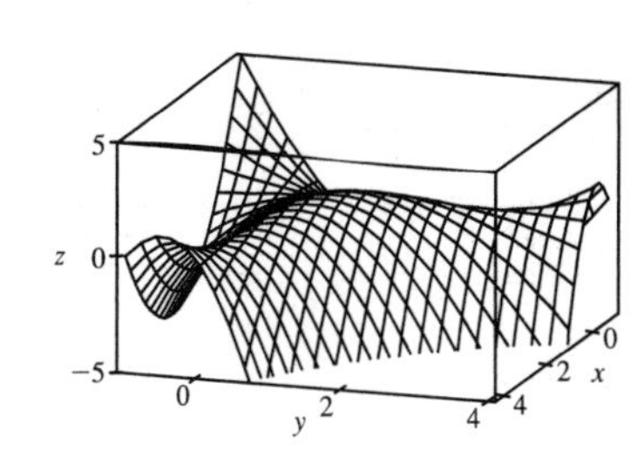

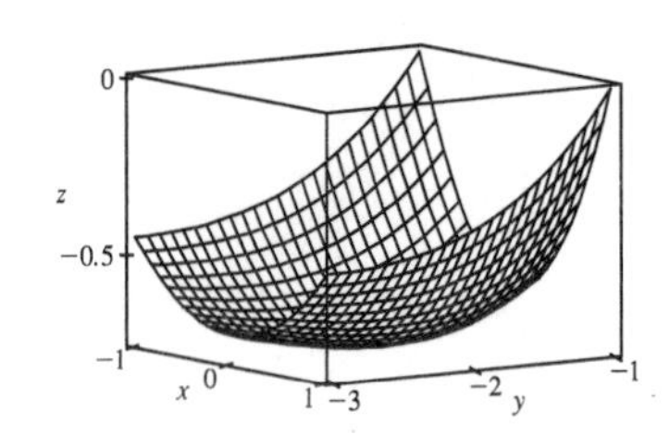

50. $f(x,y) = (x^2 + y)e^{y/2} \Rightarrow f_x = 2xe^{y/2}$, $f_y = e^{y/2}(2 + x^2 + y)/2$, $f_{xx} = 2e^{y/2}$, $f_{yy} = e^{y/2}(4 + x^2 + y)/4$, $f_{xy} = xe^{y/2}$. Then $f_x = 0$ implies $x = 0$, so $f_y = 0$ implies $y = -2$. But $f_{xx}(0,-2) > 0$, $D(0,-2) = e^{-2} - 0 > 0$ so $f(0,-2) = -2/e$ is a local minimum.

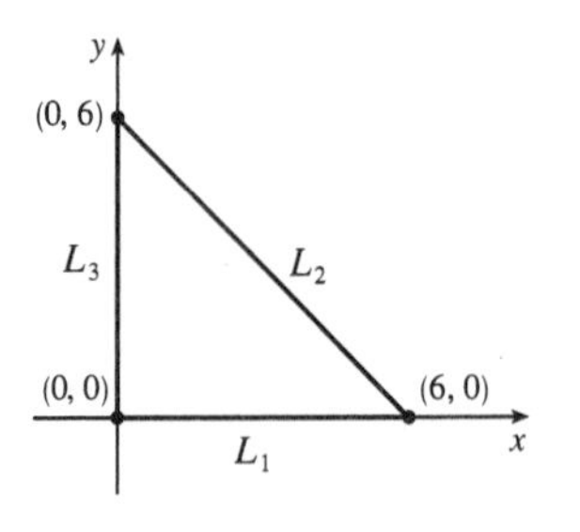

51. First solve inside D. Here $f_x = 4y^2 - 2xy^2 - y^3$, $f_y = 8xy - 2x^2y - 3xy^2$. Then $f_x = 0$ implies $y = 0$ or $y = 4 - 2x$, but $y = 0$ isn't inside D. Substituting $y = 4 - 2x$ into $f_y = 0$ implies $x = 0$, $x = 2$ or $x = 1$, but $x = 0$ isn't inside D, and when $x = 2$, $y = 0$ but $(2,0)$ isn't inside D. Thus the only critical point inside D is $(1,2)$ and $f(1,2) = 4$. Secondly we consider the boundary of D. On L_1: $f(x,0) = 0$ and so $f = 0$ on L_1. On L_2: $x = -y + 6$ and $f(-y+6,y) = y^2(6-y)(-2) = -2(6y^2 - y^3)$ which has critical points at $y = 0$ and $y = 4$. Then $f(6,0) = 0$ while $f(2,4) = -64$. On L_3: $f(0,y) = 0$, so $f = 0$ on L_3. Thus on D the absolute maximum of f is $f(1,2) = 4$ while the absolute minimum is $f(2,4) = -64$.

52. Inside D: $f_x = 2xe^{-x^2-y^2}(1 - x^2 - 2y^2) = 0$ implies $x = 0$ or $x^2 + 2y^2 = 1$. Then if $x = 0$, $f_y = 2ye^{-x^2-y^2}(2 - x^2 - 2y^2) = 0$ implies $y = 0$ or $2 - 2y^2 = 0$ giving the critical points $(0,0)$, $(0,\pm1)$. If $x^2 + 2y^2 = 1$, then $f_y = 0$ implies $y = 0$ giving the critical points $(\pm1,0)$. Now $f(0,0) = 0$, $f(\pm1,0) = e^{-1}$ and $f(0,\pm1) = 2e^{-1}$. On the boundary of D: $x^2 + y^2 = 4$, so $f(x,y) = e^{-4}(4 + y^2)$ and f is smallest when $y = 0$ and largest when $y^2 = 4$. But $f(\pm2,0) = 4e^{-4}$, $f(0,\pm2) = 8e^{-4}$. Thus on D the absolute maximum of f is $f(0,\pm1) = 2e^{-1}$ and the absolute minimum is $f(0,0) = 0$.

53. $f(x,y) = x^3 - 3x + y^4 - 2y^2$

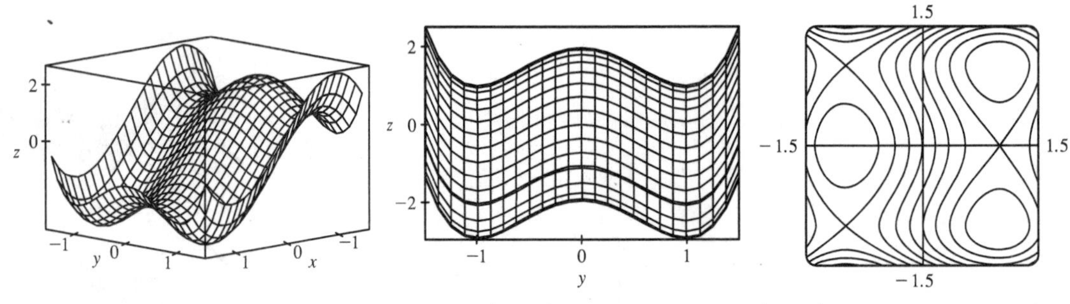

From the graphs, it appears that f has a local maximum $f(-1,0) \approx 2$, local minima $f(1,\pm1) \approx -3$, and saddle points at $(-1,\pm1)$ and $(1,0)$.

To find the exact quantities, we calculate $f_x = 3x^2 - 3 = 0 \Leftrightarrow x = \pm1$ and $f_y = 4y^3 - 4y = 0 \Leftrightarrow y = 0, \pm1$, giving the critical points estimated above. Also $f_{xx} = 6x$, $f_{xy} = 0$, $f_{yy} = 12y^2 - 4$, so using the Second Derivatives Test, $D(-1,0) = 24 > 0$ and $f_{xx}(-1,0) = -6 < 0$ indicating a local maximum $f(-1,0) = 2$; $D(1,\pm1) = 48 > 0$ and $f_{xx}(1,\pm1) = 6 > 0$ indicating local minima $f(1,\pm1) = -3$; and $D(-1,\pm1) = -48$ and $D(1,0) = -24$, indicating saddle points.

54. $f(x, y) = 12 + 10y - 2x^2 - 8xy - y^4 \Rightarrow f_x(x, y) = -4x - 8y$, $f_y(x, y) = 10 - 8x - 4y^3$. Now $f_x(x, y) = 0 \Rightarrow$ $x = -2x$, and substituting this into $f_y(x, y) = 0$ gives $10 + 16y - 4y^3 = 0 \Leftrightarrow 5 + 8y - 2y^3 = 0$.

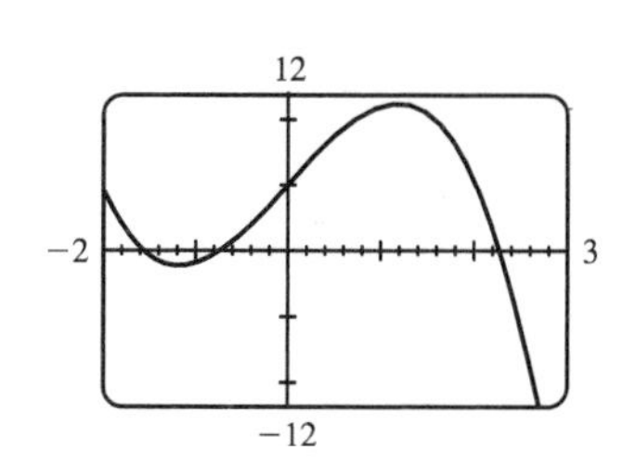

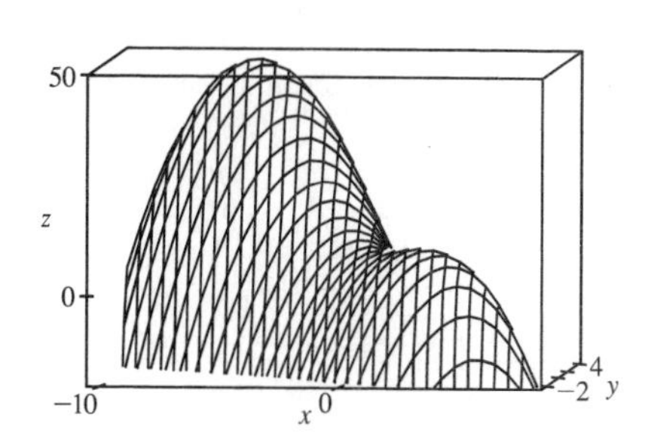

From the first graph, we see that this is true when $y \approx -1.542, -0.717$, or 2.260. (Alternatively, we could have found the solutions to $f_x = f_y = 0$ using a CAS.) So to three decimal places, the critical points are $(3.085, -1.542)$, $(1.434, -0.717)$, and $(-4.519, 2.260)$. Now in order to use the Second Derivatives Test, we calculate $f_{xx} = -4$, $f_{xy} = -8$, $f_{yy} = -12y^2$, and $D = 48y^2 - 64$. So since $D(3.085, -1.542) > 0$, $D(1.434, -0.717) < 0$, and $D(-4.519, 2.260) > 0$, and f_{xx} is always negative, $f(x, y)$ has local maxima $f(-4.519, 2.260) \approx 49.373$ and $f(3.085, -1.542) \approx 9.948$, and a saddle point at approximately $(1.434, -0.717)$. The highest point on the graph is approximately $(-4.519, 2.260, 49.373)$.

55. $f(x, y) = x^2 y$, $g(x, y) = x^2 + y^2 = 1 \Rightarrow \nabla f = \langle 2xy, x^2 \rangle = \lambda \nabla g = \langle 2\lambda x, 2\lambda y \rangle$. Then $2xy = 2\lambda x$ and $x^2 = 2\lambda y$ imply $\lambda = x^2/(2y)$ and $\lambda = y$ if $x \neq 0$ and $y \neq 0$. Hence $x^2 = 2y^2$. Then $x^2 + y^2 = 1$ implies $3y^2 = 1$ so $y = \pm\frac{1}{\sqrt{3}}$ and $x = \pm\sqrt{\frac{2}{3}}$. [Note if $x = 0$ then $x^2 = 2\lambda y$ implies $y = 0$ and $f(0, 0) = 0$.] Thus the possible points are $\left(\pm\sqrt{\frac{2}{3}}, \pm\frac{1}{\sqrt{3}}\right)$ and the absolute maxima are $f\left(\pm\sqrt{\frac{2}{3}}, \frac{1}{\sqrt{3}}\right) = \frac{2}{3\sqrt{3}}$ while the absolute minima are $f\left(\pm\sqrt{\frac{2}{3}}, -\frac{1}{\sqrt{3}}\right) = -\frac{2}{3\sqrt{3}}$.

56. $f(x, y) = 1/x + 1/y$, $g(x, y) = 1/x^2 + 1/y^2 = 1 \Rightarrow \nabla f = \langle -x^{-2}, -y^{-2} \rangle = \lambda \nabla g = \langle -2\lambda x^{-3}, -2\lambda y^{-3} \rangle$. Then $-x^{-2} = -2\lambda x^3$ or $x = 2\lambda$ and $-y^{-2} = -2\lambda y^{-3}$ or $y = 2\lambda$. Thus $x = y$, so $1/x^2 + 1/y^2 = 2/x^2 = 1$ implies $x = \pm\sqrt{2}$ and the possible points are $(\pm\sqrt{2}, \pm\sqrt{2})$. The absolute maximum of f subject to $x^{-2} + y^{-2} = 1$ is then $f(\sqrt{2}, \sqrt{2}) = \sqrt{2}$ and the absolute minimum is $f(-\sqrt{2}, -\sqrt{2}) = -\sqrt{2}$.

57. $f(x, y, z) = xyz$, $g(x, y, z) = x^2 + y^2 + z^2 = 3$. $\nabla f = \lambda \nabla g \Rightarrow \langle yz, xz, xy \rangle = \lambda \langle 2x, 2y, 2z \rangle$. If any of x, y, or z is zero, then $x = y = z = 0$ which contradicts $x^2 + y^2 + z^2 = 3$. Then $\lambda = \dfrac{yz}{2x} = \dfrac{xz}{2y} = \dfrac{xy}{2z} \Rightarrow 2y^2 z = 2x^2 z \Rightarrow$ $y^2 = x^2$, and similarly $2yz^2 = 2x^2 y \Rightarrow z^2 = x^2$. Substituting into the constraint equation gives $x^2 + x^2 + x^2 = 3 \Rightarrow$ $x^2 = 1 = y^2 = z^2$. Thus the possible points are $(1, 1, \pm 1)$, $(1, -1, \pm 1)$, $(-1, 1, \pm 1)$, $(-1, -1, \pm 1)$. The absolute maximum is $f(1, 1, 1) = f(1, -1, -1) = f(-1, 1, -1) = f(-1, -1, 1) = 1$ and the absolute minimum is $f(1, 1, -1) = f(1, -1, 1) = f(-1, 1, 1) = f(-1, -1, -1) = -1$.

58. $f(x, y, z) = x^2 + 2y^2 + 3z^2$, $g(x, y, z) = x + y + z = 1$, $h(x, y, z) = x - y + 2z = 2 \Rightarrow$ $\nabla f = \langle 2x, 4y, 6z \rangle = \lambda \nabla g + \mu \nabla h = \langle \lambda + \mu, \lambda - \mu, \lambda + 2\mu \rangle$ and $2x = \lambda + \mu$ **(1)**, $4y = \lambda - \mu$ **(2)**, $6z = \lambda + 2\mu$ **(3)**, $x + y + z = 1$ **(4)**, $x - y + 2z = 2$ **(5)**. Then six times **(1)** plus three times **(2)** plus two times **(3)** implies $12(x + y + z) = 11\lambda + 7\mu$, so **(4)** gives $11\lambda + 7\mu = 12$. Also six times **(1)** minus three times **(2)** plus four times **(3)** implies $12(x - y + 2z) = 7\lambda + 17\mu$, so **(5)** gives $7\lambda + 17\mu = 24$. Solving $11\lambda + 7\mu = 12$, $7\lambda + 17\mu = 24$ simultaneously gives

$\lambda = \frac{6}{23}$, $\mu = \frac{30}{23}$. Substituting into **(1)**, **(2)**, and **(3)** implies $x = \frac{18}{23}$, $y = -\frac{6}{23}$, $z = \frac{11}{23}$ giving only one point. Then $f\left(\frac{18}{23}, -\frac{6}{23}, \frac{11}{23}\right) = \frac{33}{23}$. Now since $(0, 0, 1)$ satisfies both constraints and $f(0, 0, 1) = 3 > \frac{33}{23}$, $f\left(\frac{18}{23}, -\frac{6}{23}, \frac{11}{23}\right) = \frac{33}{23}$ is an absolute minimum, and there is no absolute maximum.

59. $f(x, y, z) = x^2 + y^2 + z^2$, $g(x, y, z) = xy^2z^3 = 2 \Rightarrow \nabla f = \langle 2x, 2y, 2z \rangle = \lambda \nabla g = \langle \lambda y^2 z^3, 2\lambda xyz^3, 3\lambda xy^2z^2 \rangle$. Since $xy^2z^3 = 2$, $x \neq 0$, $y \neq 0$ and $z \neq 0$, so $2x = \lambda y^2 z^3$ **(1)**, $1 = \lambda xz^3$ **(2)**, $2 = 3\lambda xy^2 z$ **(3)**. Then **(2)** and **(3)** imply $\dfrac{1}{xz^3} = \dfrac{2}{3xy^2z}$ or $y^2 = \frac{2}{3}z^2$ so $y = \pm z\sqrt{\frac{2}{3}}$. Similarly **(1)** and **(3)** imply $\dfrac{2x}{y^2z^3} = \dfrac{2}{3xy^2z}$ or $3x^2 = z^2$ so $x = \pm\frac{1}{\sqrt{3}}z$. But $xy^2z^3 = 2$ so x and z must have the same sign, that is, $x = \frac{1}{\sqrt{3}}z$. Thus $g(x, y, z) = 2$ implies $\frac{1}{\sqrt{3}}z\left(\frac{2}{3}z^2\right)z^3 = 2$ or $z = \pm 3^{1/4}$ and the possible points are $(\pm 3^{-1/4}, 3^{-1/4}\sqrt{2}, \pm 3^{1/4})$, $(\pm 3^{-1/4}, -3^{-1/4}\sqrt{2}, \pm 3^{1/4})$. However at each of these points f takes on the same value, $2\sqrt{3}$. But $(2, 1, 1)$ also satisfies $g(x, y, z) = 2$ and $f(2, 1, 1) = 6 > 2\sqrt{3}$. Thus f has an absolute minimum value of $2\sqrt{3}$ and no absolute maximum subject to the constraint $xy^2z^3 = 2$.

Alternate solution: $g(x, y, z) = xy^2z^3 = 2$ implies $y^2 = \dfrac{2}{xz^3}$, so minimize $f(x, z) = x^2 + \dfrac{2}{xz^3} + z^2$. Then $f_x = 2x - \dfrac{2}{x^2z^3}$, $f_z = -\dfrac{6}{xz^4} + 2z$, $f_{xx} = 2 + \dfrac{4}{x^3z^3}$, $f_{zz} = \dfrac{24}{xz^5} + 2$ and $f_{xz} = \dfrac{6}{x^2z^4}$. Now $f_x = 0$ implies $2x^3z^3 - 2 = 0$ or $z = 1/x$. Substituting into $f_y = 0$ implies $-6x^3 + 2x^{-1} = 0$ or $x = \frac{1}{\sqrt[4]{3}}$, so the two critical points are $\left(\pm\frac{1}{\sqrt[4]{3}}, \pm\sqrt[4]{3}\right)$. Then $D\left(\pm\frac{1}{\sqrt[4]{3}}, \pm\sqrt[4]{3}\right) = (2 + 4)\left(2 + \frac{24}{3}\right) - \left(\frac{6}{\sqrt{3}}\right)^2 > 0$ and $f_{xx}\left(\pm\frac{1}{\sqrt[4]{3}}, \pm\sqrt[4]{3}\right) = 6 > 0$, so each point is a minimum. Finally, $y^2 = \dfrac{2}{xz^3}$, so the four points closest to the origin are $\left(\pm\frac{1}{\sqrt[4]{3}}, \frac{\sqrt{2}}{\sqrt[4]{3}}, \pm\sqrt[4]{3}\right)$, $\left(\pm\frac{1}{\sqrt[4]{3}}, -\frac{\sqrt{2}}{\sqrt[4]{3}}, \pm\sqrt[4]{3}\right)$.

60. $V = xyz$, say x is the length and $x + 2y + 2z \leq 108$, $x > 0$, $y > 0$, $z > 0$. First maximize V subject to $x + 2y + 2z = 108$ with x, y, z all positive. Then $\langle yz, xz, xy \rangle = \langle \lambda, 2\lambda, 2\lambda \rangle$ implies $2yz = xz$ or $x = 2y$ and $xz = xy$ or $z = y$. Thus $g(x, y, z) = 108$ implies $6y = 108$ or $y = 18 = z$, $x = 36$, so the volume is $V = 11{,}664$ cubic units. Since $(104, 1, 1)$ also satisfies $g(x, y, z) = 108$ and $V(104, 1, 1) = 104$ cubic units, $(36, 18, 18)$ gives an absolute maximum of V subject to $g(x, y, z) = 108$. But if $x + 2y + 2z < 108$, there exists $\alpha > 0$ such that $x + 2y + 2z = 108 - \alpha$ and as above $6y = 108 - \alpha$ implies $y = (108 - \alpha)/6 = z$, $x = (108 - \alpha)/3$ with $V = (108 - \alpha)^3/(6^2 \cdot 3) < (108)^3/(6^2 \cdot 3) = 11{,}664$. Hence we have shown that the maximum of V subject to $g(x, y, z) \leq 108$ is the maximum of V subject to $g(x, y, z) = 108$ (an intuitively obvious fact).

61.

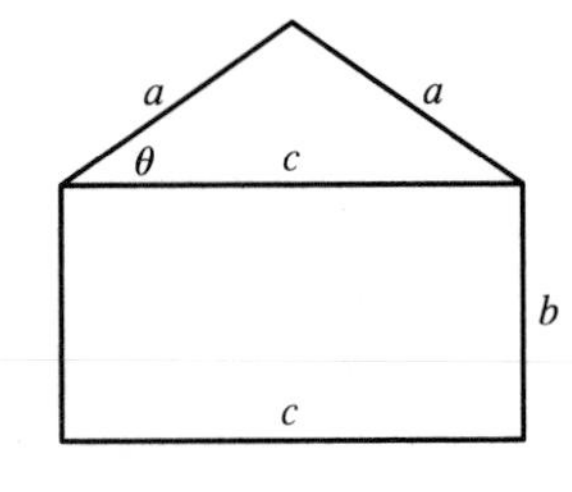

The area of the triangle is $\frac{1}{2}ca\sin\theta$ and the area of the rectangle is bc. Thus, the area of the whole object is $f(a, b, c) = \frac{1}{2}ca\sin\theta + bc$. The perimeter of the object is $g(a, b, c) = 2a + 2b + c = P$. To simplify $\sin\theta$ in terms of a, b, and c notice that $a^2\sin^2\theta + \left(\frac{1}{2}c\right)^2 = a^2 \Rightarrow \sin\theta = \dfrac{1}{2a}\sqrt{4a^2 - c^2}$. Thus $f(a, b, c) = \dfrac{c}{4}\sqrt{4a^2 - c^2} + bc$. (Instead of using θ, we could just have used the

Pythagorean Theorem.) As a result, by Lagrange's method, we must find a, b, c, and λ by solving $\nabla f = \lambda \nabla g$ which gives the following equations: $ca(4a^2 - c^2)^{-1/2} = 2\lambda$ **(1)**, $c = 2\lambda$ **(2)**, $\frac{1}{4}(4a^2 - c^2)^{1/2} - \frac{1}{4}c^2(4a^2 - c^2)^{-1/2} + b = \lambda$ **(3)**, and $2a + 2b + c = P$ **(4)**. From **(2)**, $\lambda = \frac{1}{2}c$ and so **(1)** produces $ca(4a^2 - c^2)^{-1/2} = c \Rightarrow (4a^2 - c^2)^{1/2} = a \Rightarrow$ $4a^2 - c^2 = a^2 \Rightarrow c = \sqrt{3}\,a$ **(5)**. Similarly, since $\left(4a^2 - c^2\right)^{1/2} = a$ and $\lambda = \frac{1}{2}c$, **(3)** gives $\frac{a}{4} - \frac{c^2}{4a} + b = \frac{c}{2}$, so from **(5)**, $\frac{a}{4} - \frac{3a}{4} + b = \frac{\sqrt{3}\,a}{2} \Rightarrow -\frac{a}{2} - \frac{\sqrt{3}\,a}{2} = -b \Rightarrow b = \frac{a}{2}\left(1 + \sqrt{3}\right)$ **(6)**. Substituting **(5)** and **(6)** into **(4)** we get:

$2a + a\left(1 + \sqrt{3}\right) + \sqrt{3}\,a = P \Rightarrow 3a + 2\sqrt{3}\,a = P \Rightarrow a = \dfrac{P}{3 + 2\sqrt{3}} = \dfrac{2\sqrt{3} - 3}{3}P$ and thus

$b = \dfrac{\left(2\sqrt{3} - 3\right)\left(1 + \sqrt{3}\right)}{6}P = \dfrac{3 - \sqrt{3}}{6}P$ and $c = \left(2 - \sqrt{3}\right)P$.

62. (a) $\mathbf{r}(t) = x(t)\,\mathbf{i} + y(t)\,\mathbf{j} + f(x(t), y(t))\,\mathbf{k} \Rightarrow \mathbf{v} = \dfrac{d\mathbf{r}}{dt} = \dfrac{dx}{dt}\,\mathbf{i} + \dfrac{dy}{dt}\,\mathbf{j} + \left(f_x\,\dfrac{dx}{dt} + f_y\,\dfrac{dy}{dt}\right)\mathbf{k}$

(by the Chain Rule). Therefore

$$\begin{aligned} K = \tfrac{1}{2}m\,|\mathbf{v}|^2 &= \frac{m}{2}\left[\left(\frac{dx}{dt}\right)^2 + \left(\frac{dy}{dt}\right)^2 + \left(f_x\,\frac{dx}{dt} + f_y\,\frac{dy}{dt}\right)^2\right] \\ &= \frac{m}{2}\left[\left(1 + f_x^2\right)\left(\frac{dx}{dt}\right)^2 + 2f_xf_y\left(\frac{dx}{dt}\right)\left(\frac{dy}{dt}\right) + \left(1 + f_y^2\right)\left(\frac{dy}{dt}\right)^2\right] \end{aligned}$$

(b) $\mathbf{a} = \dfrac{d\mathbf{v}}{dt} = \dfrac{d^2x}{dt^2}\,\mathbf{i} + \dfrac{d^2y}{dt^2}\,\mathbf{j} + \left[f_{xx}\left(\dfrac{dx}{dt}\right)^2 + 2f_{xy}\,\dfrac{dx}{dt}\dfrac{dy}{dt} + f_{yy}\left(\dfrac{dy}{dt}\right)^2 + f_x\,\dfrac{d^2x}{dt^2} + f_y\,\dfrac{d^2y}{dt^2}\right]\mathbf{k}$

(c) If $z = x^2 + y^2$, where $x = t\cos t$ and $y = t\sin t$, then $z = f(x, y) = t^2$.

$\mathbf{r} = t\cos t\,\mathbf{i} + t\sin t\,\mathbf{j} + t^2\,\mathbf{k} \Rightarrow \mathbf{v} = (\cos t - t\sin t)\,\mathbf{i} + (\sin t + t\cos t)\,\mathbf{j} + 2t\,\mathbf{k}$,

$K = \dfrac{m}{2}\left[(\cos t - t\sin t)^2 + (\sin t + t\cos t)^2 + (2t)^2\right] = \dfrac{m}{2}\left(1 + t^2 + 4t^2\right) = \dfrac{m}{2}\left(1 + 5t^2\right)$, and

$\mathbf{a} = (-2\sin t - t\cos t)\,\mathbf{i} + (2\cos t - t\sin t)\,\mathbf{j} + 2\,\mathbf{k}$. Notice that it is easier not to use the formulas in (a) and (b).

12 ☐ MULTIPLE INTEGRALS

12.1 Double Integrals over Rectangles

1. (a) The subrectangles are shown in the figure.

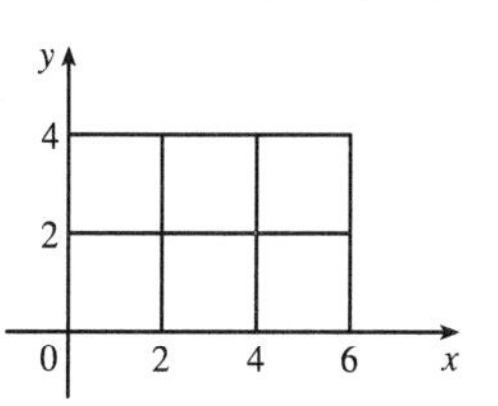

The surface is the graph of $f(x, y) = xy$ and $\Delta A = 4$, so we estimate

$$V \approx \sum_{i=1}^{3} \sum_{j=1}^{2} f(x_i, y_j)\,\Delta A$$

$$= f(2,2)\,\Delta A + f(2,4)\,\Delta A + f(4,2)\,\Delta A + f(4,4)\,\Delta A + f(6,2)\,\Delta A + f(6,4)\,\Delta A$$

$$= 4(4) + 8(4) + 8(4) + 16(4) + 12(4) + 24(4) = 288$$

(b) $V \approx \sum_{i=1}^{3} \sum_{j=1}^{2} f(\overline{x}_i, \overline{y}_j)\,\Delta A = f(1,1)\,\Delta A + f(1,3)\,\Delta A + f(3,1)\,\Delta A + f(3,3)\,\Delta A + f(5,1)\,\Delta A + f(5,3)\,\Delta A$

$$= 1(4) + 3(4) + 3(4) + 9(4) + 5(4) + 15(4) = 144$$

2. The subrectangles are shown in the figure.

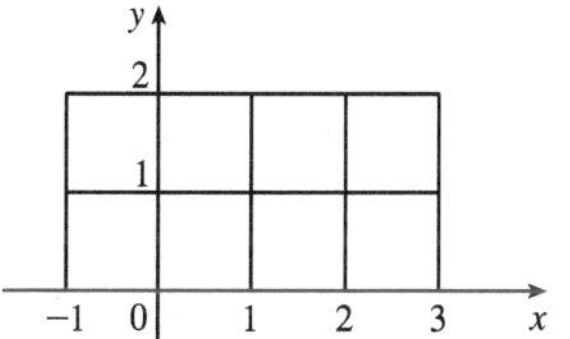

Since $\Delta A = 1$, we estimate

$$\iint_R (y^2 - 2x^2)\,dA \approx \sum_{i=1}^{4} \sum_{j=1}^{2} f(x_{ij}^*, y_{ij}^*)\,\Delta A$$

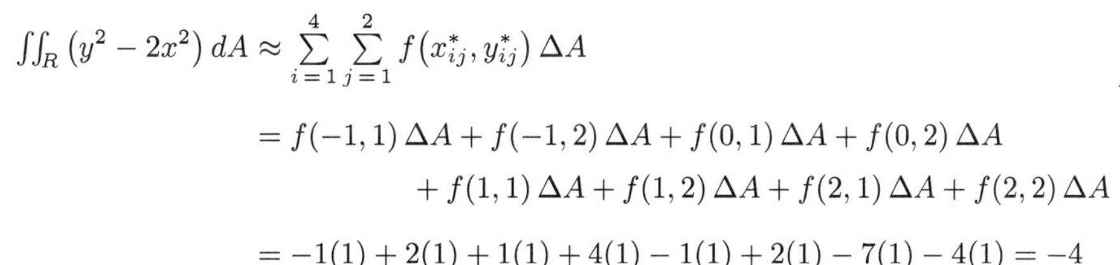

$$= f(-1,1)\,\Delta A + f(-1,2)\,\Delta A + f(0,1)\,\Delta A + f(0,2)\,\Delta A$$

$$+ f(1,1)\,\Delta A + f(1,2)\,\Delta A + f(2,1)\,\Delta A + f(2,2)\,\Delta A$$

$$= -1(1) + 2(1) + 1(1) + 4(1) - 1(1) + 2(1) - 7(1) - 4(1) = -4$$

3. (a) The subrectangles are shown in the figure. Since $\Delta A = \pi^2/4$, we estimate

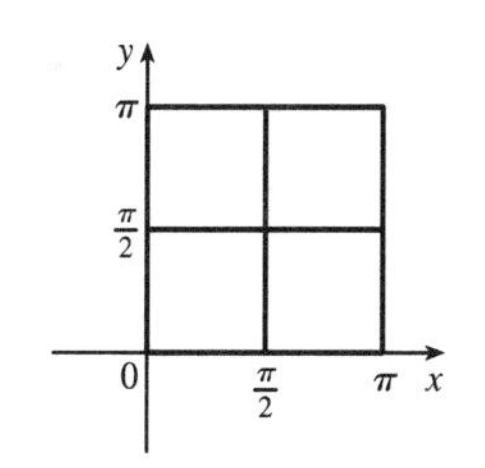

$$\iint_R \sin(x+y)\,dA \approx \sum_{i=1}^{2} \sum_{j=1}^{2} f(x_{ij}^*, y_{ij}^*)\,\Delta A$$

$$= f(0,0)\,\Delta A + f(0, \tfrac{\pi}{2})\,\Delta A + f(\tfrac{\pi}{2}, 0)\,\Delta A + f(\tfrac{\pi}{2}, \tfrac{\pi}{2})\,\Delta A$$

$$= 0\left(\tfrac{\pi^2}{4}\right) + 1\left(\tfrac{\pi^2}{4}\right) + 1\left(\tfrac{\pi^2}{4}\right) + 0\left(\tfrac{\pi^2}{4}\right) = \tfrac{\pi^2}{2} \approx 4.935$$

(b) $\iint_R \sin(x+y)\,dA \approx \sum_{i=1}^{2} \sum_{j=1}^{2} f(\overline{x}_i, \overline{y}_j)\,\Delta A$

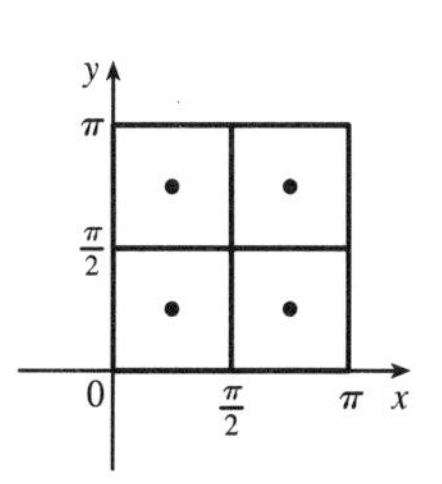

$$= f(\tfrac{\pi}{4}, \tfrac{\pi}{4})\,\Delta A + f(\tfrac{\pi}{4}, \tfrac{3\pi}{4})\,\Delta A + f(\tfrac{3\pi}{4}, \tfrac{\pi}{4})\,\Delta A + f(\tfrac{3\pi}{4}, \tfrac{3\pi}{4})\,\Delta A$$

$$= 1\left(\tfrac{\pi^2}{4}\right) + 0\left(\tfrac{\pi^2}{4}\right) + 0\left(\tfrac{\pi^2}{4}\right) + (-1)\left(\tfrac{\pi^2}{4}\right) = 0$$

(c) $\iint_R \sin(x+y)\,dA = \int_0^\pi \int_0^\pi \sin(x+y)\,dx\,dy = \int_0^\pi \left[-\cos(x+y)\right]_{x=0}^{x=\pi} dy = \int_0^\pi \left[\cos y - \cos(y+\pi)\right] dy$

$$= \sin y - \sin(y+\pi)\Big]_0^\pi = 0$$

So the estimate from the Midpoint Rule in part (b) is same as the exact value.

4. (a) The subrectangles are shown in the figure.

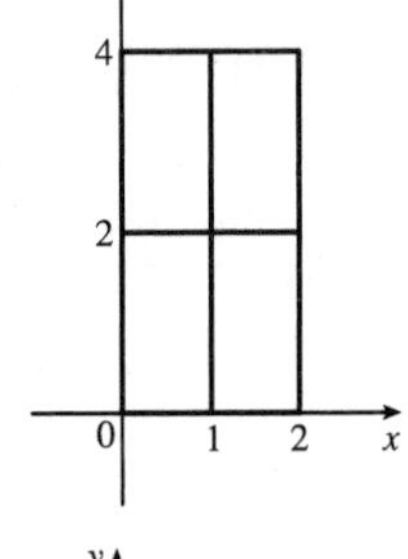

The surface is the graph of $f(x, y) = x + 2y^2$ and $\Delta A = 2$, so we estimate

$$V = \iint_R (x + 2y^2)\, dA \approx \sum_{i=1}^{2} \sum_{j=1}^{2} f(x_{ij}^*, y_{ij}^*)\, \Delta A$$

$$= f(1, 0)\, \Delta A + f(1, 2)\, \Delta A + f(2, 0)\, \Delta A + f(2, 2)\, \Delta A$$

$$= 1(2) + 9(2) + 2(2) + 10(2) = 44$$

(b) $V = \iint_R (x + 2y^2)\, dA \approx \sum_{i=1}^{2} \sum_{j=1}^{2} f(\overline{x}_i, \overline{y}_j)\, \Delta A$

$$= f(\tfrac{1}{2}, 1)\, \Delta A + f(\tfrac{1}{2}, 3)\, \Delta A + f(\tfrac{3}{2}, 1)\, \Delta A + f(\tfrac{3}{2}, 3)\, \Delta A$$

$$= \tfrac{5}{2}(2) + \tfrac{37}{2}(2) + \tfrac{7}{2}(2) + \tfrac{39}{2}(2) = 88$$

(c) $\iint_R (x + 2y^2)\, dA = \int_0^4 \int_0^2 (x + 2y^2)\, dx\, dy = \int_0^4 \left[\tfrac{1}{2}x^2 + 2xy^2\right]_{x=0}^{x=2} dy$

$$= \int_0^4 (2 + 4y^2)\, dy = 2y + \tfrac{4}{3}y^3\Big]_0^4 = \tfrac{280}{3} \approx 93.3$$

So the estimate from the Midpoint Rule in part (b) is much closer to the true value than the estimate in part (a).

5. With $m = n = 2$, we have $\Delta A = 4$. Using the contour map to estimate the value of f at the center of each subrectangle, we have

$$\iint_R f(x, y)\, dA \approx \sum_{i=1}^{2} \sum_{j=1}^{2} f(\overline{x}_i, \overline{y}_j)\, \Delta A = \Delta A[f(1, 1) + f(1, 3) + f(3, 1) + f(3, 3)] \approx 4(27 + 4 + 14 + 17) = 248.$$

6. To approximate the volume, let R be the planar region corresponding to the surface of the water in the pool, and place R on coordinate axes so that x and y correspond to the dimensions given. Then we define $f(x, y)$ to be the depth of the water at (x, y), so the volume of water in the pool is the volume of the solid that lies above the rectangle $R = [0, 20] \times [0, 30]$ and below the graph of $f(x, y)$. We can estimate this volume using the Midpoint Rule with $m = 2$ and $n = 3$, so $\Delta A = 100$. Each subrectangle with its midpoint is shown in the figure. Then

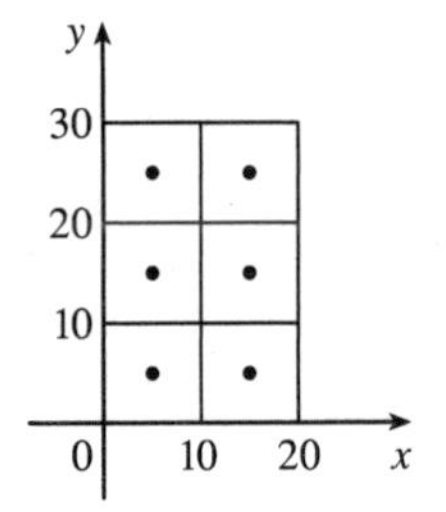

$$V \approx \sum_{i=1}^{2} \sum_{j=1}^{3} f(\overline{x}_i, \overline{y}_j)\, \Delta A = \Delta A[f(5, 5) + f(5, 15) + f(5, 25) + f(15, 5) + f(15, 15) + f(15, 25)]$$

$$= 100(3 + 7 + 10 + 3 + 5 + 8) = 3600$$

Thus, we estimate that the pool contains 3600 cubic feet of water.

Alternatively, we can approximate the volume with a Riemann sum where $m = 4$, $n = 6$ and the sample points are taken to be, for example, the upper right corner of each subrectangle. Then $\Delta A = 25$ and

$$V \approx \sum_{i=1}^{4} \sum_{j=1}^{6} f(x_i, y_j)\, \Delta A$$

$$= 25[3 + 4 + 7 + 8 + 10 + 8 + 4 + 6 + 8 + 10 + 12 + 10 + 3 + 4 + 5 + 6 + 8 + 7 + 2 + 2 + 2 + 3 + 4 + 4]$$

$$= 25(140) = 3500$$

So we estimate that the pool contains 3500 ft^3 of water.

7. $z = 3 > 0$, so we can interpret the integral as the volume of the solid S that lies below the plane $z = 3$ and above the rectangle $[-2, 2] \times [1, 6]$. S is a rectangular solid, thus $\iint_R 3\,dA = 4 \cdot 5 \cdot 3 = 60$.

8. $z = 5 - x \geq 0$ for $0 \leq x \leq 5$, so we can interpret the integral as the volume of the solid S that lies below the plane $z = 5 - x$ and above the rectangle $[0, 5] \times [0, 3]$. S is a triangular cylinder whose volume is $3(\text{area of triangle}) = 3\left(\frac{1}{2} \cdot 5 \cdot 5\right) = 37.5$. Thus

$$\iint_R (5 - x)\,dA = 37.5$$

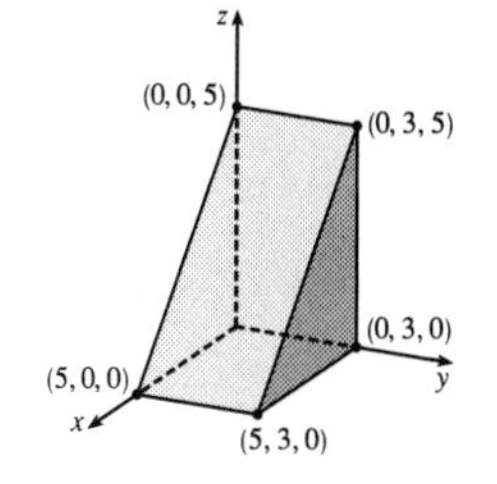

9. $z = f(x, y) = 4 - 2y \geq 0$ for $0 \leq y \leq 1$. Thus the integral represents the volume of that part of the rectangular solid $[0, 1] \times [0, 1] \times [0, 4]$ which lies below the plane $z = 4 - 2y$. So

$$\iint_R (4 - 2y)\,dA = (1)(1)(2) + \tfrac{1}{2}(1)(1)(2) = 3$$

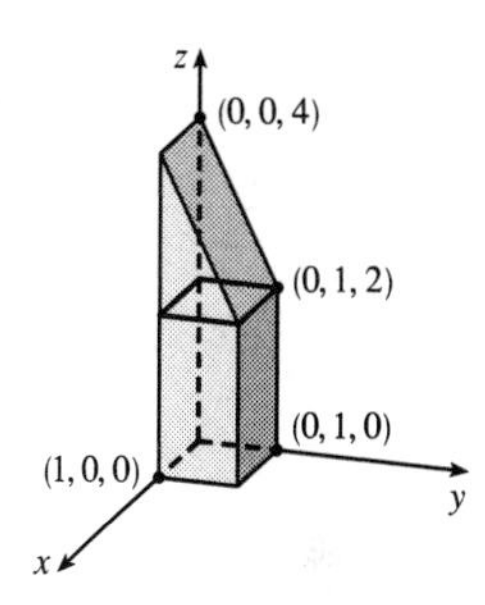

10. Here $z = \sqrt{9 - y^2}$, so $z^2 + y^2 = 9$, $z \geq 0$. Thus the integral represents the volume of the top half of the part of the circular cylinder $z^2 + y^2 = 9$ that lies above the rectangle $[0, 4] \times [0, 2]$.

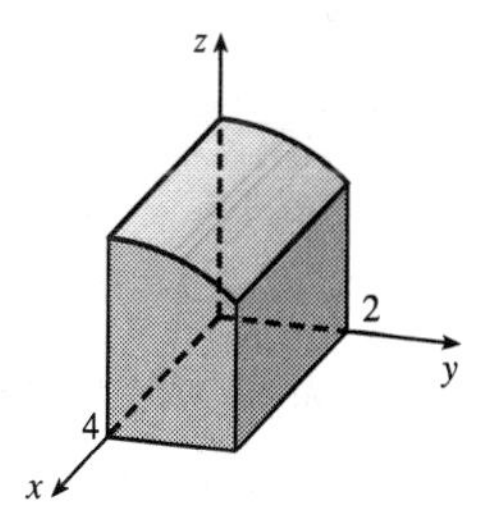

11. $\int_1^3 \int_0^1 (1 + 4xy)\,dx\,dy = \int_1^3 \left[x + 2x^2 y\right]_{x=0}^{x=1} dy = \int_1^3 (1 + 2y)\,dy = \left[y + y^2\right]_1^3 = (3 + 9) - (1 + 1) = 10$

12. $\int_2^4 \int_{-1}^1 (x^2 + y^2)\,dy\,dx = \int_2^4 \left[x^2 y + \frac{1}{3}y^3\right]_{y=-1}^{y=1} dx = \int_2^4 \left[\left(x^2 + \frac{1}{3}\right) - \left(-x^2 - \frac{1}{3}\right)\right] dx$

$= \int_2^4 \left(2x^2 + \frac{2}{3}\right) dx = \left[\frac{2}{3}x^3 + \frac{2}{3}x\right]_2^4 = \left(\frac{128}{3} + \frac{8}{3}\right) - \left(\frac{16}{3} + \frac{4}{3}\right) = \frac{116}{3}$

13. $\int_0^2 \int_0^{\pi/2} x \sin y\,dy\,dx = \int_0^2 x\,dx \int_0^{\pi/2} \sin y\,dy$ [as in Example 5] $= \left[\dfrac{x^2}{2}\right]_0^2 \left[-\cos y\right]_0^{\pi/2} = (2 - 0)(0 + 1) = 2$

14. $\int_1^4 \int_0^2 (x + \sqrt{y})\,dx\,dy = \int_1^4 \left[\frac{1}{2}x^2 + x\sqrt{y}\right]_{x=0}^{x=2} dy = \int_1^4 (2 + 2\sqrt{y})\,dy = \left[2y + 2 \cdot \frac{2}{3}y^{3/2}\right]_1^4$

$= \left(8 + \frac{4}{3} \cdot 8\right) - \left(2 + \frac{4}{3}\right) = \frac{46}{3}$

15. $\int_0^2 \int_0^1 (2x + y)^8 dx\,dy = \displaystyle\int_0^2 \left[\frac{1}{2}\frac{(2x + y)^9}{9}\right]_{x=0}^{x=1} dy$ [substitute $u = 2x + y \Rightarrow dx = \frac{1}{2}\,du$]

$= \dfrac{1}{18}\displaystyle\int_0^2 \left[(2 + y)^9 - (0 + y)^9\right] dy = \frac{1}{18}\left[\frac{(2 + y)^{10}}{10} - \frac{y^{10}}{10}\right]_0^2$

$= \frac{1}{180}\left[(4^{10} - 2^{10}) - (2^{10} - 0^{10})\right] = \frac{1{,}046{,}528}{180} = \frac{261{,}632}{45}$

16. $\int_0^1 \int_1^2 \frac{xe^x}{y}\,dy\,dx = \int_0^1 xe^x\,dx \int_1^2 \frac{1}{y}\,dy$ [as in Example 5] $= \left[xe^x - e^x\right]_0^1 \left[\ln|y|\right]_1^2$ [by integrating by parts]

$= [(e - e) - (0 - 1)](\ln 2 - 0) = \ln 2$

17. $\int_1^4 \int_1^2 \left(\frac{x}{y} + \frac{y}{x}\right) dy\,dx = \int_1^4 \left[x \ln|y| + \frac{1}{x} \cdot \frac{1}{2} y^2\right]_{y=1}^{y=2} dx = \int_1^4 \left(x \ln 2 + \frac{3}{2x}\right) dx = \left[\frac{1}{2}x^2 \ln 2 + \frac{3}{2} \ln|x|\right]_1^4$

$= 8 \ln 2 + \frac{3}{2} \ln 4 - \frac{1}{2} \ln 2 = \frac{15}{2} \ln 2 + 3 \ln 4^{1/2} = \frac{21}{2} \ln 2$

18. $\int_1^2 \int_0^1 (x + y)^{-2}\,dx\,dy = \int_1^2 \left[-(x + y)^{-1}\right]_{x=0}^{x=1} dy = \int_1^2 \left[y^{-1} - (1 + y)^{-1}\right] dy$

$= \left[\ln y - \ln(1 + y)\right]_1^2 = \ln 2 - \ln 3 - 0 + \ln 2 = \ln \frac{4}{3}$

19. $\int_0^{\ln 2} \int_0^{\ln 5} e^{2x - y}\,dx\,dy = \left(\int_0^{\ln 5} e^{2x}\,dx\right)\left(\int_0^{\ln 2} e^{-y}\,dy\right) = \left[\frac{1}{2}e^{2x}\right]_0^{\ln 5} \left[-e^{-y}\right]_0^{\ln 2} = \left(\frac{25}{2} - \frac{1}{2}\right)\left(-\frac{1}{2} + 1\right) = 6$

20. $\int_0^1 \int_0^1 xy\sqrt{x^2 + y^2}\,dy\,dx = \int_0^1 x\left[\frac{1}{3}(x^2 + y^2)^{3/2}\right]_{y=0}^{y=1} dx = \frac{1}{3}\int_0^1 x\left[(x^2 + 1)^{3/2} - x^3\right] dx = \frac{1}{3}\int_0^1 [x(x^2 + 1)^{3/2} - x^4]dx$

$= \frac{1}{3}\left[\frac{1}{5}(x^2 + 1)^{5/2} - \frac{1}{5}x^5\right]_0^1 = \frac{1}{15}\left[2^{5/2} - 1 - 1 + 0\right] = \frac{2}{15}(2\sqrt{2} - 1)$

21. $\iint_R \frac{xy^2}{x^2 + 1}\,dA = \int_0^1 \int_{-3}^3 \frac{xy^2}{x^2 + 1}\,dy\,dx = \int_0^1 \frac{x}{x^2 + 1}\,dx \int_{-3}^3 y^2\,dy$

$= \left[\frac{1}{2}\ln(x^2 + 1)\right]_0^1 \left[\frac{1}{3}y^3\right]_{-3}^3 = \frac{1}{2}(\ln 2 - \ln 1) \cdot \frac{1}{3}(27 + 27) = 9 \ln 2$

22. $\iint_R \cos(x + 2y)\,dA = \int_0^\pi \int_0^{\pi/2} \cos(x + 2y)\,dy\,dx = \int_0^\pi \left[\frac{1}{2}\sin(x + 2y)\right]_{y=0}^{y=\pi/2} dx = \frac{1}{2}\int_0^\pi (\sin(x + \pi) - \sin x)\,dx$

$= \frac{1}{2}\left[-\cos(x + \pi) + \cos x\right]_0^\pi = \frac{1}{2}\left[-\cos 2\pi + \cos\pi - (-\cos\pi + \cos 0)\right]$

$= \frac{1}{2}(-1 - 1 - (1 + 1)) = -2$

23. $\int_0^{\pi/6} \int_0^{\pi/3} x \sin(x + y)\,dy\,dx$

$= \int_0^{\pi/6} \left[-x\cos(x + y)\right]_{y=0}^{y=\pi/3} dx = \int_0^{\pi/6} \left[x\cos x - x\cos\left(x + \frac{\pi}{3}\right)\right] dx$

$= x\left[\sin x - \sin\left(x + \frac{\pi}{3}\right)\right]_0^{\pi/6} - \int_0^{\pi/6} \left[\sin x - \sin\left(x + \frac{\pi}{3}\right)\right] dx$ [by integrating by parts separately for each term]

$= \frac{\pi}{6}\left[\frac{1}{2} - 1\right] - \left[-\cos x + \cos\left(x + \frac{\pi}{3}\right)\right]_0^{\pi/6} = -\frac{\pi}{12} - \left[-\frac{\sqrt{3}}{2} + 0 - \left(-1 + \frac{1}{2}\right)\right] = \frac{\sqrt{3} - 1}{2} - \frac{\pi}{12}$

24. $\iint_R \frac{1 + x^2}{1 + y^2}\,dA = \int_0^1 \int_0^1 \frac{1 + x^2}{1 + y^2}\,dy\,dx = \int_0^1 (1 + x^2)\,dx \int_0^1 \frac{1}{1 + y^2}\,dy$

$= \left[x + \frac{1}{3}x^3\right]_0^1 \left[\tan^{-1} y\right]_0^1 = \left(1 + \frac{1}{3} - 0\right)\left(\frac{\pi}{4} - 0\right) = \frac{\pi}{3}$

25. $\iint_R xye^{x^2 y}\,dA = \int_0^2 \int_0^1 xye^{x^2 y}\,dx\,dy = \int_0^2 \left[\frac{1}{2}e^{x^2 y}\right]_{x=0}^{x=1} dy = \frac{1}{2}\int_0^2 (e^y - 1)\,dy = \frac{1}{2}\left[e^y - y\right]_0^2$

$= \frac{1}{2}[(e^2 - 2) - (1 - 0)] = \frac{1}{2}(e^2 - 3)$

26. $\iint_R \frac{x}{1 + xy}\,dA = \int_0^1 \int_0^1 \frac{x}{1 + xy}\,dy\,dx = \int_0^1 \left[\ln(1 + xy)\right]_{y=0}^{y=1} dx = \int_0^1 \left[\ln(1 + x) - \ln 1\right] dx$

$= \int_0^1 \ln(1 + x)\,dx = \left[(1 + x)\ln(1 + x) - x\right]_0^1$ [by integrating by parts]

$= (2 \ln 2 - 1) - (\ln 1 - 0) = 2 \ln 2 - 1$

27. $V = \iint_R (12 - 3x - 2y)\, dA = \int_{-2}^{3} \int_0^1 (12 - 3x - 2y)\, dx\, dy = \int_{-2}^{3} \left[12x - \frac{3}{2}x^2 - 2xy\right]_{x=0}^{x=1} dy$

$= \int_{-2}^{3} \left(\frac{21}{2} - 2y\right) dy = \left[\frac{21}{2}y - y^2\right]_{-2}^{3} = \frac{95}{2}$

28. $V = \iint_R (4 + x^2 - y^2)\, dA = \int_{-1}^{1} \int_0^2 (4 + x^2 - y^2)\, dy\, dx = \int_{-1}^{1} \left[4y + x^2 y - \frac{1}{3}y^3\right]_{y=0}^{y=2} dx$

$= \int_{-1}^{1} \left(2x^2 + \frac{16}{3}\right) dx = \left[\frac{2}{3}x^3 + \frac{16}{3}x\right]_{-1}^{1} = \frac{2}{3} + \frac{16}{3} + \frac{2}{3} + \frac{16}{3} = 12$

29. $V = \int_{-2}^{2} \int_{-1}^{1} \left(1 - \frac{1}{4}x^2 - \frac{1}{9}y^2\right) dx\, dy = 4 \int_0^2 \int_0^1 \left(1 - \frac{1}{4}x^2 - \frac{1}{9}y^2\right) dx\, dy$

$= 4 \int_0^2 \left[x - \frac{1}{12}x^3 - \frac{1}{9}y^2 x\right]_{x=0}^{x=1} dy = 4 \int_0^2 \left(\frac{11}{12} - \frac{1}{9}y^2\right) dy = 4\left[\frac{11}{12}y - \frac{1}{27}y^3\right]_0^2 = 4 \cdot \frac{83}{54} = \frac{166}{27}$

30. $V = \int_{-1}^{1} \int_0^{\pi} (1 + e^x \sin y)\, dy\, dx = \int_{-1}^{1} \left[y - e^x \cos y\right]_{y=0}^{y=\pi} dx = \int_{-1}^{1} (\pi + e^x - 0 + e^x)\, dx$

$= \int_{-1}^{1} (\pi + 2e^x)\, dx = \left[\pi x + 2e^x\right]_{-1}^{1} = 2\pi + 2e - \frac{2}{e}$

31. Here we need the volume of the solid lying under the surface $z = x\sqrt{x^2 + y}$ and above the square $R = [0,1] \times [0,1]$ in the xy-plane.

$$V = \int_0^1 \int_0^1 x\sqrt{x^2 + y}\, dx\, dy = \int_0^1 \frac{1}{3}\left[(x^2 + y)^{3/2}\right]_{x=0}^{x=1} dy = \frac{1}{3}\int_0^1 \left[(1 + y)^{3/2} - y^{3/2}\right] dy$$

$$= \frac{1}{3} \cdot \frac{2}{5}\left[(1 + y)^{5/2} - y^{5/2}\right]_0^1 = \frac{4}{15}\left(2\sqrt{2} - 1\right)$$

32. Here we need the volume of the solid lying under the surface $z = 1 + (x - 1)^2 + 4y^2$ and above the rectangle $R = [0,3] \times [0,2]$ in the xy-plane.

$$V = \int_0^3 \int_0^2 \left[1 + (x - 1)^2 + 4y^2\right] dy\, dx = \int_0^3 \left[y + (x - 1)^2 y + \frac{4}{3}y^3\right]_{y=0}^{y=2} dx$$

$$= \int_0^3 \left[2 + 2(x - 1)^2 + \frac{32}{3}\right] dx = \left[\frac{38}{3}x + \frac{2}{3}(x - 1)^3\right]_0^3 = 44$$

33. In the first octant, $z \geq 0 \Rightarrow y \leq 3$, so

$$V = \int_0^3 \int_0^2 (9 - y^2)\, dx\, dy = \int_0^3 \left[9x - y^2 x\right]_{x=0}^{x=2} dy = \int_0^3 (18 - 2y^2)\, dy = \left[18y - \frac{2}{3}y^3\right]_0^3 = 36$$

34. (a) Here we need the volume of the solid lying under the surface $z = 6 - xy$ and above the rectangle $R = [-2,2] \times [0,3]$ in the xy-plane.

$V = \int_{-2}^{2} \int_0^3 (6 - xy)\, dy\, dx$

$= \int_{-2}^{2} \left[6y - \frac{1}{2}xy^2\right]_{y=0}^{y=3} dx$

$= \int_{-2}^{2} \left(18 - \frac{9}{2}x\right) dx = \left[18x - \frac{9}{4}x^2\right]_{-2}^{2} = 72$

(b) The solid occupies the region between the two surfaces shown.

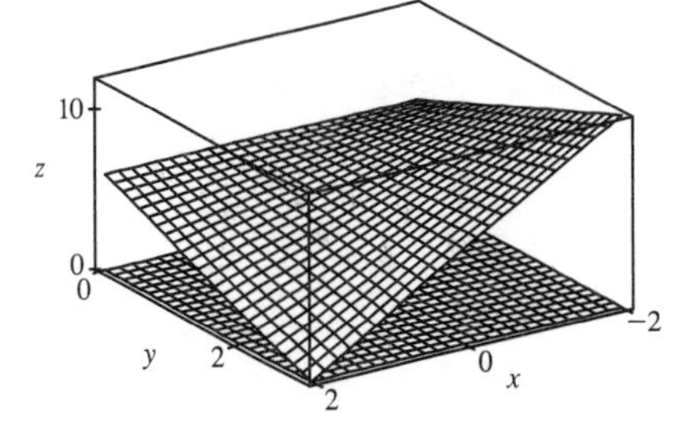

35. In Maple, we can calculate the integral by defining the integrand as `f` and then using the command `int(int(f,x=0..1),y=0..1);`. In Mathematica, we can use the command `Integrate[Integrate[f,{x,0,1}],{y,0,1}]`. We find that $\iint_R x^5 y^3 e^{xy}\, dA = 21e - 57 \approx 0.0839$. We can use `plot3d` (in Maple) or `Plot3d` (in Mathematica) to graph the function.

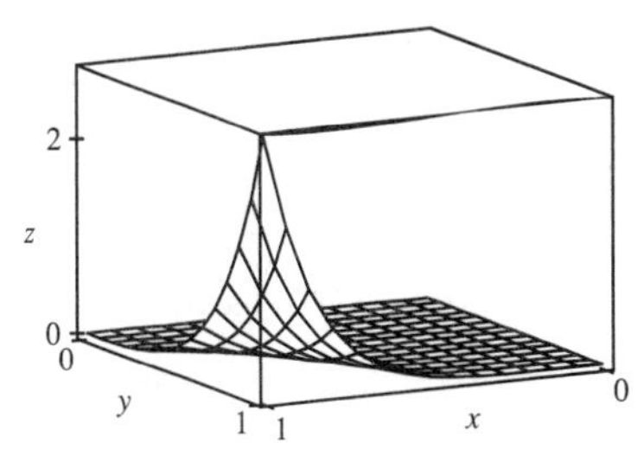

36. In Maple, we can calculate the integral by defining

`f:=exp(-x^2)*cos(x^2+y^2);` and `g:=2-x^2-y^2;`

and then [since $2 - x^2 - y^2 > e^{-x^2} \cos(x^2 + y^2)$ for $-1 \le x \le 1$, $-1 \le y \le 1$] using the command

`evalf(int(int(g-f,x=-1..1),y=-1..1),5);`.

In Mathematica, we can use the command

`N[Integrate[Integrate[f,{x,0,1}],{y,0,1}],5]`.

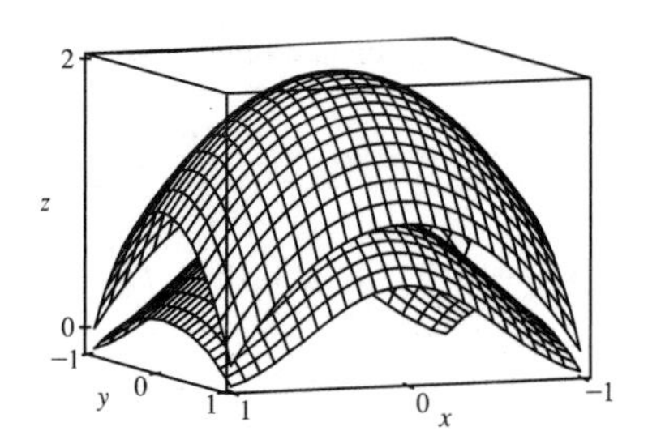

In each of these commands, the `5` indicates that we want only five significant digits; this speeds up the calculation considerably. We find that $\iint_R \left[(2 - x^2 - y^2) - \left(e^{-x^2} \cos(x^2 + y^2)\right)\right] dA \approx 3.0271$. We can use the `plot3d` command (in Maple) or `Plot3d` (in Mathematica) to graph both functions on the same screen.

37. R is the rectangle $[-1, 1] \times [0, 5]$. Thus, $A(R) = 2 \cdot 5 = 10$ and

$$f_{\text{ave}} = \frac{1}{A(R)} \iint_R f(x, y)\, dA = \tfrac{1}{10} \int_0^5 \int_{-1}^1 x^2 y\, dx\, dy = \tfrac{1}{10} \int_0^5 \left[\tfrac{1}{3}x^3 y\right]_{x=-1}^{x=1} dy = \tfrac{1}{10} \int_0^5 \tfrac{2}{3} y\, dy = \tfrac{1}{10} \left[\tfrac{1}{3} y^2\right]_0^5 = \tfrac{5}{6}.$$

38. $A(R) = 4 \cdot 1 = 4$, so

$$\begin{aligned} f_{\text{ave}} &= \frac{1}{A(R)} \iint_R f(x, y)\, dA = \tfrac{1}{4} \int_0^4 \int_0^1 e^y \sqrt{x + e^y}\, dy\, dx = \tfrac{1}{4} \int_0^4 \left[\tfrac{2}{3}(x + e^y)^{3/2}\right]_{y=0}^{y=1} dx \\ &= \tfrac{1}{4} \cdot \tfrac{2}{3} \int_0^4 \left[(x + e)^{3/2} - (x + 1)^{3/2}\right] dx = \tfrac{1}{6} \left[\tfrac{2}{5}(x + e)^{5/2} - \tfrac{2}{5}(x + 1)^{5/2}\right]_0^4 \\ &= \tfrac{1}{6} \cdot \tfrac{2}{5}[(4 + e)^{5/2} - 5^{5/2} - e^{5/2} + 1] = \tfrac{1}{15}[(4 + e)^{5/2} - e^{5/2} - 5^{5/2} + 1] \approx 3.327 \end{aligned}$$

39. If we divide R into mn subrectangles, $\iint_R k\, dA \approx \sum\limits_{i=1}^{m} \sum\limits_{j=1}^{n} f\left(x_{ij}^*, y_{ij}^*\right) \Delta A$ for any choice of sample points $\left(x_{ij}^*, y_{ij}^*\right)$.

But $f\left(x_{ij}^*, y_{ij}^*\right) = k$ always and $\sum\limits_{i=1}^{m} \sum\limits_{j=1}^{n} \Delta A =$ area of $R = (b - a)(d - c)$. Thus, no matter how we choose the sample

points, $\sum\limits_{i=1}^{m} \sum\limits_{j=1}^{n} f\left(x_{ij}^*, y_{ij}^*\right) \Delta A = k \sum\limits_{i=1}^{m} \sum\limits_{j=1}^{n} \Delta A = k(b - a)(d - c)$ and so

$$\iint_R k\, dA = \lim_{m,n\to\infty} \sum_{i=1}^{m} \sum_{j=1}^{n} f\left(x_{ij}^*, y_{ij}^*\right) \Delta A = \lim_{m,n\to\infty} k \sum_{i=1}^{m} \sum_{j=1}^{n} \Delta A = \lim_{m,n\to\infty} k(b - a)(d - c) = k(b - a)(d - c).$$

40. On R, $0 \le x + y \le 2 < \pi$ and $\sin\theta \ge 0$ for $0 \le \theta \le \pi$. Thus $f(x, y) = \sin(x + y) \ge 0$ for all $(x, y) \in R$. Since $0 \le \sin(x + y) \le 1$, Property 9 gives $\iint_R 0\, dA \le \iint_R \sin(x + y)\, dA \le \iint_R 1\, dA$, so by Exercise 17 we have $0 \le \iint_R \sin(x + y)\, dA \le 1$.

41. Let $f(x, y) = \dfrac{x - y}{(x + y)^3}$. Then a CAS gives $\int_0^1 \int_0^1 f(x, y)\, dy\, dx = \frac{1}{2}$ and $\int_0^1 \int_0^1 f(x, y)\, dx\, dy = -\frac{1}{2}$.

To explain the seeming violation of Fubini's Theorem, note that f has an infinite discontinuity at $(0, 0)$ and thus does not satisfy the conditions of Fubini's Theorem. In fact, both iterated integrals involve improper integrals which diverge at their lower limits of integration.

42. (a) Loosely speaking, Fubini's Theorem says that the order of integration of a function of two variables does not affect the value of the double integral, while Clairaut's Theorem says that the order of differentiation of such a function does not affect the value of the second-order derivative. Also, both theorems require continuity (though Fubini's allows a finite number of smooth curves to contain discontinuities).

(b) To find g_{xy}, we first hold y constant and use the single-variable Fundamental Theorem of Calculus, Part 1:

$g_x = \dfrac{d}{dx}\, g(x,y) = \dfrac{d}{dx}\displaystyle\int_a^x \left(\int_c^y f(s,t)\,dt\right) ds = \int_c^y f(x,t)\,dt$. Now we use the Fundamental Theorem again:

$g_{xy} = \dfrac{d}{dy}\displaystyle\int_c^y f(x,t)\,dt = f(x,y)$.

To find g_{yx}, we first use Fubini's Theorem to find that $\int_a^x \int_c^y f(s,t)\,dt\,ds = \int_c^y \int_a^x f(s,t)\,dt\,ds$, and then use the Fundamental Theorem twice, as above, to get $g_{yx} = f(x,y)$. So $g_{xy} = g_{yx} = f(x,y)$.

12.2 Double Integrals over General Regions

1. $\int_0^1 \int_0^{x^2} (x+2y)\,dy\,dx = \int_0^1 \left[xy + y^2\right]_{y=0}^{y=x^2} dx = \int_0^1 \left[x(x^2) + (x^2)^2 - 0 - 0\right] dx$
$= \int_0^1 (x^3 + x^4)\,dx = \left[\frac{1}{4}x^4 + \frac{1}{5}x^5\right]_0^1 = \frac{9}{20}$

2. $\int_1^2 \int_y^2 xy\,dx\,dy = \int_1^2 \left[\frac{1}{2}x^2 y\right]_{x=y}^{x=2} dy = \int_1^2 \frac{1}{2}y(4-y^2)\,dy = \frac{1}{2}\int_1^2 (4y - y^3)\,dy$
$= \frac{1}{2}\left[2y^2 - \frac{1}{4}y^4\right]_1^2 = \frac{1}{2}\left(8 - 4 - 2 + \frac{1}{4}\right) = \frac{9}{8}$

3. $\int_0^1 \int_y^{e^y} \sqrt{x}\,dx\,dy = \int_0^1 \left[\frac{2}{3}x^{3/2}\right]_{x=y}^{x=e^y} dy = \frac{2}{3}\int_0^1 (e^{3y/2} - y^{3/2})\,dy = \frac{2}{3}\left[\frac{2}{3}e^{3y/2} - \frac{2}{5}y^{5/2}\right]_0^1$
$= \frac{2}{3}\left(\frac{2}{3}e^{3/2} - \frac{2}{5} - \frac{2}{3}e^0 + 0\right) = \frac{4}{9}e^{3/2} - \frac{32}{45}$

4. $\int_0^1 \int_x^{2-x} (x^2 - y)\,dy\,dx = \int_0^1 \left[x^2 y - \frac{1}{2}y^2\right]_{y=x}^{y=2-x} dx = \int_0^1 \left[x^2(2-x) - \frac{1}{2}(2-x)^2 - x^2(x) + \frac{1}{2}x^2\right] dx$
$= \int_0^1 (-2x^3 + 2x^2 + 2x - 2)\,dx = \left[-\frac{1}{2}x^4 + \frac{2}{3}x^3 + x^2 - 2x\right]_0^1 = -\frac{5}{6}$

5. $\int_0^{\pi/2} \int_0^{\cos\theta} e^{\sin\theta}\,dr\,d\theta = \int_0^{\pi/2} \left[re^{\sin\theta}\right]_{r=0}^{r=\cos\theta} d\theta = \int_0^{\pi/2} (\cos\theta)\,e^{\sin\theta}\,d\theta = e^{\sin\theta}\Big]_0^{\pi/2} = e^{\sin(\pi/2)} - e^0 = e - 1$

6. $\int_0^1 \int_0^v \sqrt{1-v^2}\,du\,dv = \int_0^1 \left[u\sqrt{1-v^2}\right]_{u=0}^{u=v} dv = \int_0^1 v\sqrt{1-v^2}\,dv = -\frac{1}{3}(1-v^2)^{3/2}\Big]_0^1 = -\frac{1}{3}(0-1) = \frac{1}{3}$

7. $\iint_D x^3 y^2\,dA = \int_0^2 \int_{-x}^{x} x^3 y^2\,dy\,dx = \int_0^2 \left[\frac{1}{3}x^3 y^3\right]_{y=-x}^{y=x} dx = \frac{1}{3}\int_0^2 2x^6\,dx = \frac{2}{3}\left[\frac{1}{7}x^7\right]_0^2 = \frac{2}{21}\left[2^7 - 0\right] = \frac{256}{21}$

8. $\displaystyle\iint_D \frac{4y}{x^3+2}\,dA = \int_1^2 \int_0^{2x} \frac{4y}{x^3+2}\,dy\,dx = \int_1^2 \left[\frac{2y^2}{x^3+2}\right]_{y=0}^{y=2x} dx = \int_1^2 \frac{8x^2}{x^3+2}\,dx$
$= \frac{8}{3}\ln\left|x^3+2\right|\Big]_1^2 = \frac{8}{3}(\ln 10 - \ln 3) = \frac{8}{3}\ln\frac{10}{3}$

9. $\displaystyle\int_0^1 \int_0^{\sqrt{x}} \frac{2y}{x^2+1}\,dy\,dx = \int_0^1 \left[\frac{y^2}{x^2+1}\right]_{y=0}^{y=\sqrt{x}} dx = \int_0^1 \frac{x}{x^2+1}\,dx = \frac{1}{2}\ln\left|x^2+1\right|\Big]_0^1 = \frac{1}{2}(\ln 2 - \ln 1) = \frac{1}{2}\ln 2$

10. $\int_0^1 \int_0^y e^{y^2}\,dx\,dy = \int_0^1 \left[xe^{y^2}\right]_{x=0}^{x=y} dy = \int_0^1 ye^{y^2}\,dy = \frac{1}{2}e^{y^2}\Big]_0^1 = \frac{1}{2}(e-1)$

11. $\int_0^1 \int_0^{x^2} x\cos y\,dy\,dx = \int_0^1 \left[x\sin y\right]_{y=0}^{y=x^2} dx = \int_0^1 x\sin x^2\,dx = -\frac{1}{2}\cos x^2\Big]_0^1 = \frac{1}{2}(1 - \cos 1)$

12.

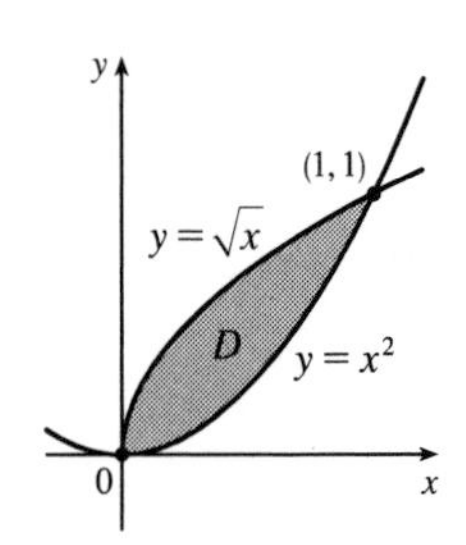

$$\int_0^1 \int_{x^2}^{\sqrt{x}} (x+y)\,dy\,dx = \int_0^1 \left[xy + \tfrac{1}{2}y^2\right]_{y=x^2}^{y=\sqrt{x}} dx$$
$$= \int_0^1 \left(x^{3/2} + \tfrac{1}{2}x - x^3 - \tfrac{1}{2}x^4\right) dx$$
$$= \left[\tfrac{2}{5}x^{5/2} + \tfrac{1}{4}x^2 - \tfrac{1}{4}x^4 - \tfrac{1}{10}x^5\right]_0^1 = \tfrac{3}{10}$$

13.

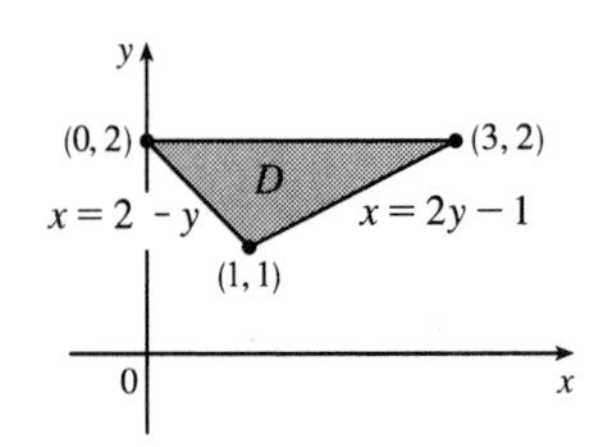

$$\int_1^2 \int_{2-y}^{2y-1} y^3\,dx\,dy = \int_1^2 \left[xy^3\right]_{x=2-y}^{x=2y-1} dy = \int_1^2 [(2y-1)-(2-y)]\,y^3\,dy$$
$$= \int_1^2 (3y^4 - 3y^3)\,dy = \left[\tfrac{3}{5}y^5 - \tfrac{3}{4}y^4\right]_1^2$$
$$= \tfrac{96}{5} - 12 - \tfrac{3}{5} + \tfrac{3}{4} = \tfrac{147}{20}$$

14.

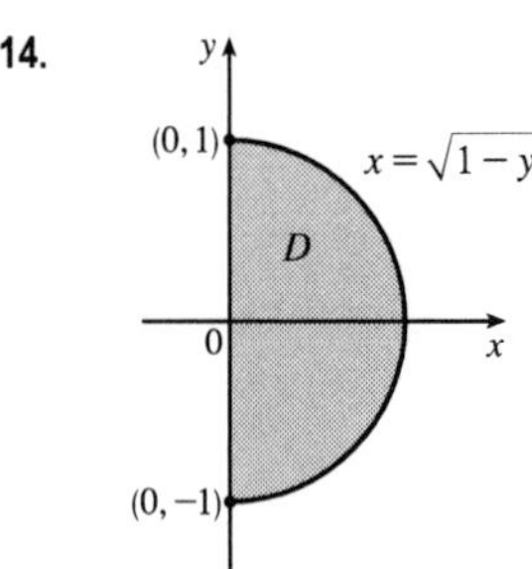

$$\iint_D xy^2\,dA = \int_{-1}^1 \int_0^{\sqrt{1-y^2}} xy^2\,dx\,dy$$
$$= \int_{-1}^1 y^2\left[\tfrac{1}{2}x^2\right]_{x=0}^{x=\sqrt{1-y^2}} dy = \tfrac{1}{2}\int_{-1}^1 y^2(1-y^2)\,dy$$
$$= \tfrac{1}{2}\int_{-1}^1 (y^2 - y^4)\,dy = \tfrac{1}{2}\left[\tfrac{1}{3}y^3 - \tfrac{1}{5}y^5\right]_{-1}^1$$
$$= \tfrac{1}{2}\left(\tfrac{1}{3} - \tfrac{1}{5} + \tfrac{1}{3} - \tfrac{1}{5}\right) = \tfrac{2}{15}$$

15.

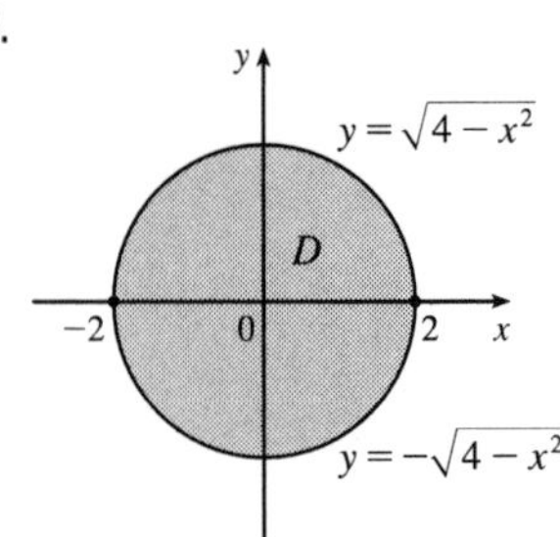

$$\int_{-2}^2 \int_{-\sqrt{4-x^2}}^{\sqrt{4-x^2}} (2x - y)\,dy\,dx$$
$$= \int_{-2}^2 \left[2xy - \frac{1}{2}y^2\right]_{y=-\sqrt{4-x^2}}^{y=\sqrt{4-x^2}} dx$$
$$= \int_{-2}^2 \left[2x\sqrt{4-x^2} - \tfrac{1}{2}(4-x^2) + 2x\sqrt{4-x^2} + \tfrac{1}{2}(4-x^2)\right] dx$$
$$= \int_{-2}^2 4x\sqrt{4-x^2}\,dx = -\tfrac{4}{3}(4-x^2)^{3/2}\Big]_{-2}^2 = 0$$

[Or, note that $4x\sqrt{4-x^2}$ is an odd function, so $\int_{-2}^2 4x\sqrt{4-x^2}\,dx = 0$.]

16.

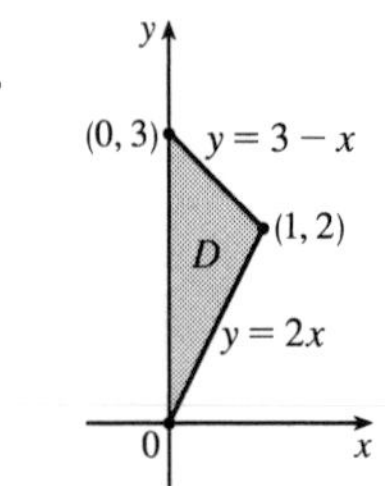

$$\iint_D 2xy\,dA = \int_0^1 \int_{2x}^{3-x} 2xy\,dy\,dx = \int_0^1 \left[xy^2\right]_{y=2x}^{y=3-x} dx$$
$$= \int_0^1 x[(3-x)^2 - (2x)^2]\,dx = \int_0^1 (-3x^3 - 6x^2 + 9x)\,dx$$
$$= \left[-\tfrac{3}{4}x^4 - 2x^3 + \tfrac{9}{2}x^2\right]_0^1 = -\tfrac{3}{4} - 2 + \tfrac{9}{2} = \tfrac{7}{4}$$

17.

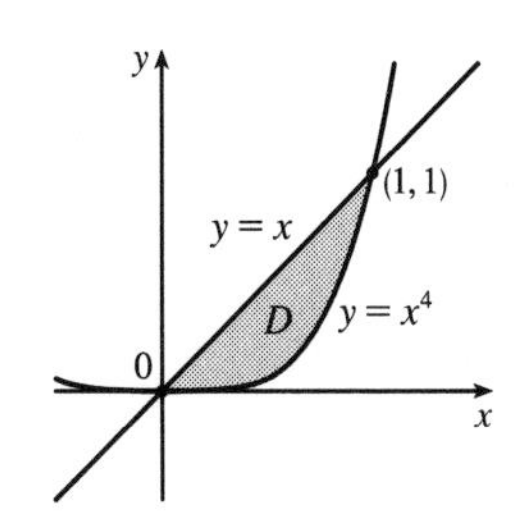

$$V=\int_0^1\int_{x^4}^{x}(x+2y)\,dy\,dx$$
$$=\int_0^1\left[xy+y^2\right]_{y=x^4}^{y=x}\,dx=\int_0^1(2x^2-x^5-x^8)\,dx$$
$$=\left[\tfrac{2}{3}x^3-\tfrac{1}{6}x^6-\tfrac{1}{9}x^9\right]_0^1=\tfrac{2}{3}-\tfrac{1}{6}-\tfrac{1}{9}=\tfrac{7}{18}$$

18.

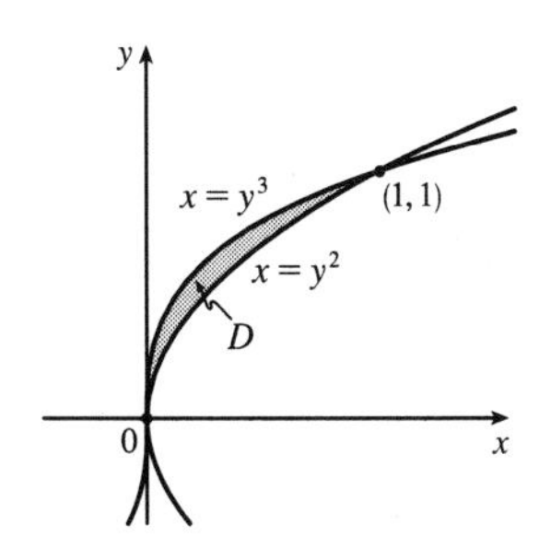

$$V=\int_0^1\int_{y^3}^{y^2}(2x+y^2)\,dx\,dy$$
$$=\int_0^1\left[x^2+xy^2\right]_{x=y^3}^{x=y^2}\,dy=\int_0^1(2y^4-y^6-y^5)\,dy$$
$$=\left[\tfrac{2}{5}y^5-\tfrac{1}{7}y^7-\tfrac{1}{6}y^6\right]_0^1=\tfrac{19}{210}$$

19.

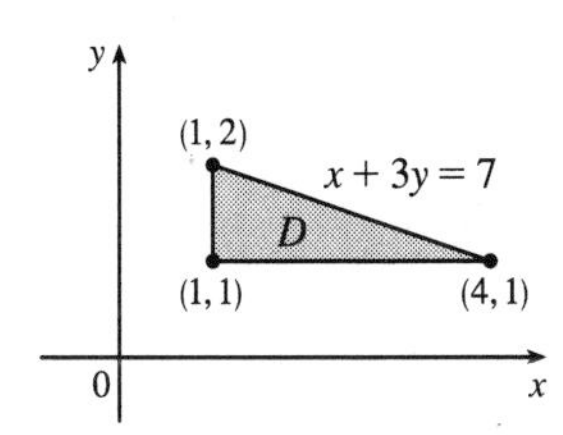

$$V=\int_1^2\int_1^{7-3y}xy\,dx\,dy=\int_1^2\left[\tfrac{1}{2}x^2y\right]_{x=1}^{x=7-3y}\,dy$$
$$=\tfrac{1}{2}\int_1^2(48y-42y^2+9y^3)\,dy$$
$$=\tfrac{1}{2}\left[24y^2-14y^3+\tfrac{9}{4}y^4\right]_1^2=\tfrac{31}{8}$$

20.

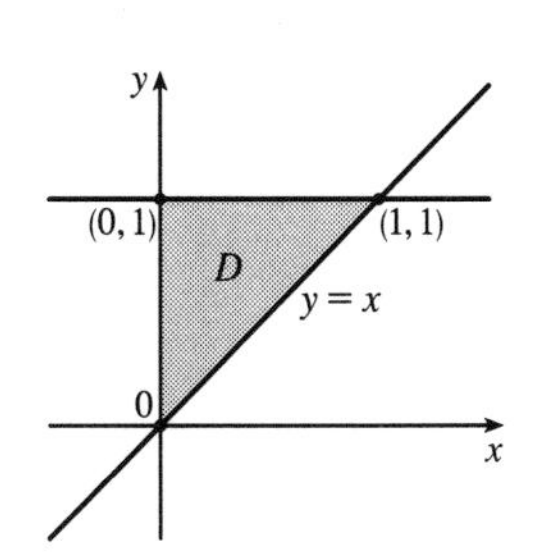

$$V=\int_0^1\int_x^1(x^2+3y^2)\,dy\,dx$$
$$=\int_0^1\left[x^2y+y^3\right]_{y=x}^{y=1}\,dx=\int_0^1(x^2+1-2x^3)\,dx$$
$$=\left[\tfrac{1}{3}x^3+x-\tfrac{1}{2}x^4\right]_0^1=\tfrac{5}{6}$$

21.

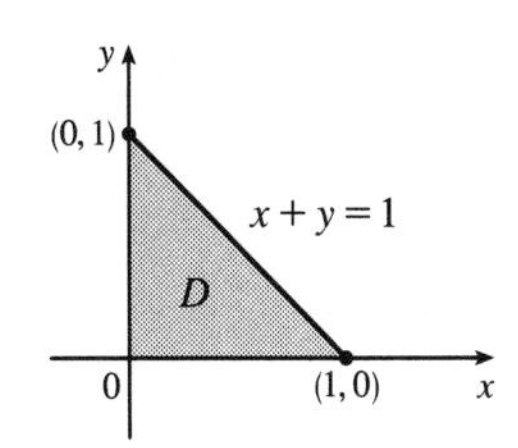

$$V=\int_0^1\int_0^{1-x}(1-x-y)\,dy\,dx=\int_0^1\left[y-xy-\frac{y^2}{2}\right]_{y=0}^{y=1-x}\,dx$$
$$=\int_0^1\left[(1-x)^2-\tfrac{1}{2}(1-x)^2\right]dx$$
$$=\int_0^1\tfrac{1}{2}(1-x)^2\,dx=\left[-\tfrac{1}{6}(1-x)^3\right]_0^1=\tfrac{1}{6}$$

22.

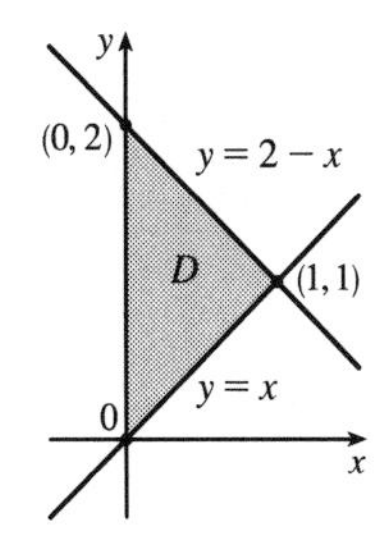

$$V=\int_0^1\int_x^{2-x}x\,dy\,dx$$
$$=\int_0^1 x\left[\,y\,\right]_{y=x}^{y=2-x}\,dx=\int_0^1(2x-2x^2)\,dx$$
$$=\left[x^2-\tfrac{2}{3}x^3\right]_0^1=\tfrac{1}{3}$$

23.

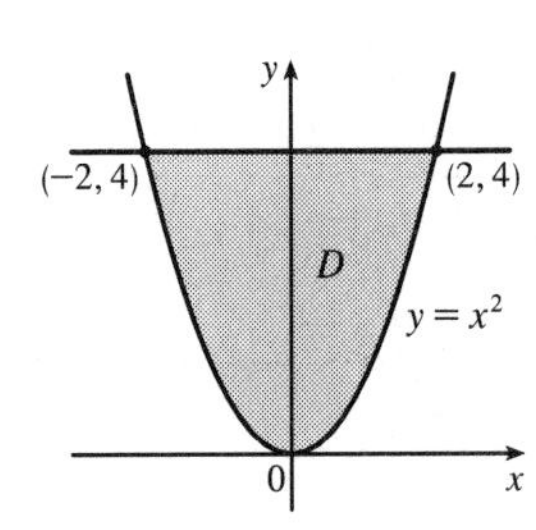

$$V = \int_{-2}^{2}\int_{x^2}^{4} x^2\,dy\,dx$$
$$= \int_{-2}^{2} x^2\big[\,y\,\big]_{y=x^2}^{y=4}\,dx = \int_{-2}^{2}(4x^2 - x^4)\,dx$$
$$= \left[\tfrac{4}{3}x^3 - \tfrac{1}{5}x^5\right]_{-2}^{2} = \tfrac{32}{3} - \tfrac{32}{5} + \tfrac{32}{3} - \tfrac{32}{5} = \tfrac{128}{15}$$

24.

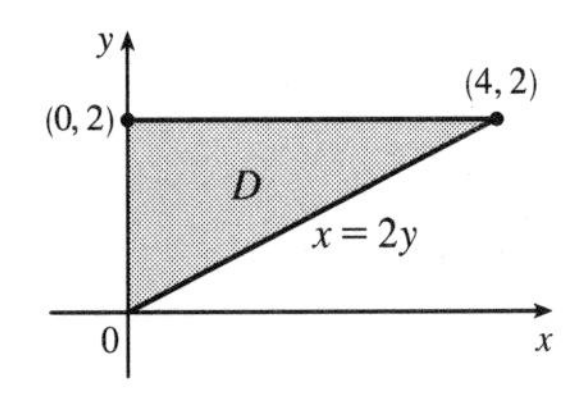

$$V = \int_0^2\int_0^{2y}\sqrt{4-y^2}\,dx\,dy = \int_0^2\left[x\sqrt{4-y^2}\,\right]_{x=0}^{x=2y}dy$$
$$= \int_0^2 2y\sqrt{4-y^2}\,dy = \left[-\tfrac{2}{3}\left(4-y^2\right)^{3/2}\right]_0^2 = 0 + \tfrac{16}{3} = \tfrac{16}{3}$$

25.

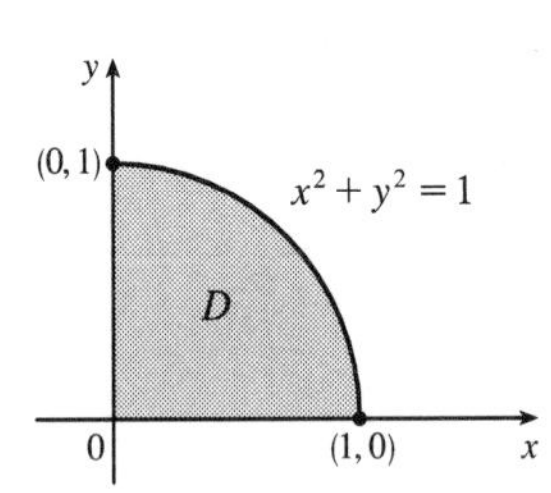

$$V = \int_0^1\int_0^{\sqrt{1-x^2}} y\,dy\,dx = \int_0^1\left[\frac{y^2}{2}\right]_{y=0}^{y=\sqrt{1-x^2}}dx$$
$$= \int_0^1\frac{1-x^2}{2}\,dx = \tfrac{1}{2}\left[x - \tfrac{1}{3}x^3\right]_0^1 = \tfrac{1}{3}$$

26.

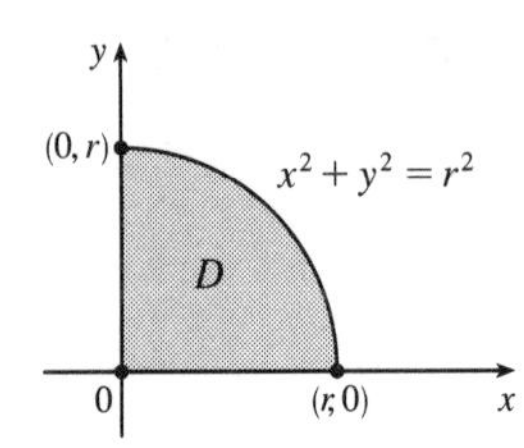

By symmetry, the desired volume V is 8 times the volume V_1 in the first octant. Now

$$V_1 = \int_0^r\int_0^{\sqrt{r^2-y^2}}\sqrt{r^2-y^2}\,dx\,dy = \int_0^r\left[x\sqrt{r^2-y^2}\,\right]_{x=0}^{x=\sqrt{r^2-y^2}}dy$$
$$= \int_0^r(r^2-y^2)\,dy = \left[r^2y - \tfrac{1}{3}y^3\right]_0^r = \tfrac{2}{3}r^3$$

Thus $V = \frac{16}{3}r^3$.

27. The two bounding curves $y = 1 - x^2$ and $y = x^2 - 1$ intersect at $(\pm 1, 0)$ with $1 - x^2 \geq x^2 - 1$ on $[-1, 1]$. Within this region, the plane $z = 2x + 2y + 10$ is above the plane $z = 2 - x - y$, so

$$V = \int_{-1}^{1}\int_{x^2-1}^{1-x^2}(2x+2y+10)\,dy\,dx - \int_{-1}^{1}\int_{x^2-1}^{1-x^2}(2-x-y)\,dy\,dx$$
$$= \int_{-1}^{1}\int_{x^2-1}^{1-x^2}(2x+2y+10-(2-x-y))\,dy\,dx$$
$$= \int_{-1}^{1}\int_{x^2-1}^{1-x^2}(3x+3y+8)\,dy\,dx = \int_{-1}^{1}\left[3xy + \tfrac{3}{2}y^2 + 8y\right]_{y=x^2-1}^{y=1-x^2}dx$$
$$= \int_{-1}^{1}\left[3x(1-x^2) + \tfrac{3}{2}(1-x^2)^2 + 8(1-x^2) - 3x(x^2-1) - \tfrac{3}{2}(x^2-1)^2 - 8(x^2-1)\right]dx$$
$$= \int_{-1}^{1}(-6x^3 - 16x^2 + 6x + 16)\,dx = \left[-\tfrac{3}{2}x^4 - \tfrac{16}{3}x^3 + 3x^2 + 16x\right]_{-1}^{1}$$
$$= -\tfrac{3}{2} - \tfrac{16}{3} + 3 + 16 + \tfrac{3}{2} - \tfrac{16}{3} - 3 + 16 = \tfrac{64}{3}$$

28. The two planes intersect in the line $y = 1$, $z = 3$, so the region of integration is the plane region enclosed by the parabola $y = x^2$ and the line $y = 1$. We have $2 + y \geq 3y$ for $0 \leq y \leq 1$, so the solid region is bounded above by $z = 2 + y$ and bounded below by $z = 3y$.

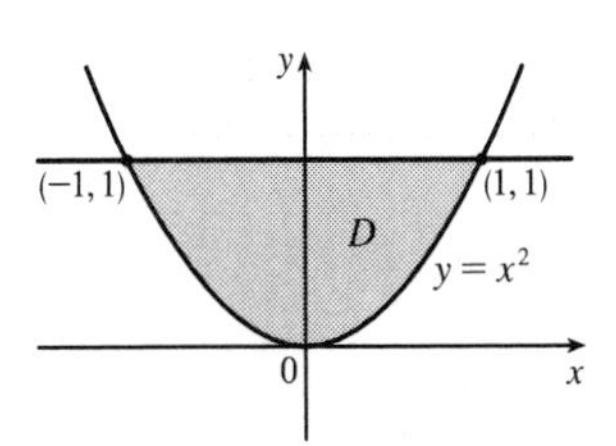

$$V = \int_{-1}^{1}\int_{x^2}^{1}(2+y)\,dy\,dx - \int_{-1}^{1}\int_{x^2}^{1}(3y)\,dy\,dx = \int_{-1}^{1}\int_{x^2}^{1}(2+y-3y)\,dy\,dx = \int_{-1}^{1}\int_{x^2}^{1}(2-2y)\,dy\,dx$$

$$= \int_{-1}^{1}\left[2y - y^2\right]_{y=x^2}^{y=1}dx = \int_{-1}^{1}(1 - 2x^2 + x^4)\,dx = x - \tfrac{2}{3}x^3 + \tfrac{1}{5}x^5\Big]_{-1}^{1} = \tfrac{16}{15}$$

29. The two surfaces intersect in the circle $x^2 + y^2 = 1$, $z = 0$ and the region of integration is the disk D: $x^2 + y^2 \leq 1$.

Using a CAS, the volume is $\displaystyle\iint_D (1 - x^2 - y^2)\,dA = \int_{-1}^{1}\int_{-\sqrt{1-x^2}}^{\sqrt{1-x^2}}(1 - x^2 - y^2)\,dy\,dx = \frac{\pi}{2}$.

30. The projection onto the xy-plane of the intersection of the two surfaces is the circle $x^2 + y^2 = 2y \Rightarrow$ $x^2 + y^2 - 2y = 0 \Rightarrow x^2 + (y-1)^2 = 1$, so the region of integration is given by $-1 \leq x \leq 1$, $1 - \sqrt{1-x^2} \leq y \leq 1 + \sqrt{1-x^2}$. In this region, $2y \geq x^2 + y^2$ so, using a CAS, the volume is $\displaystyle V = \int_{-1}^{1}\int_{1-\sqrt{1-x^2}}^{1+\sqrt{1-x^2}}\left[2y - (x^2+y^2)\right]dy\,dx = \frac{\pi}{2}$.

31.

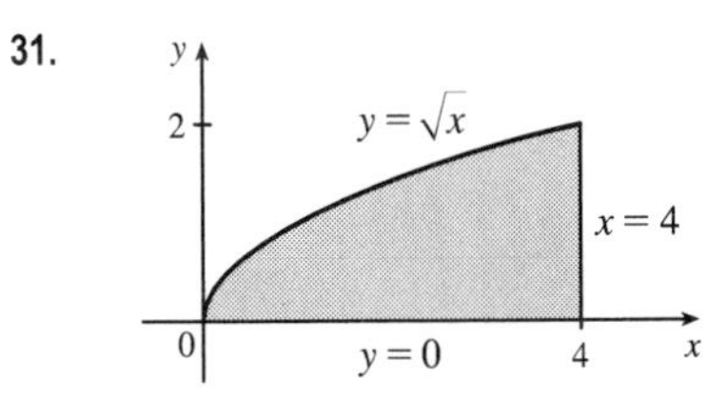

Because the region of integration is

$D = \{(x,y) \mid 0 \leq y \leq \sqrt{x}, 0 \leq x \leq 4\} = \{(x,y) \mid y^2 \leq x \leq 4, 0 \leq y \leq 2\}$

we have $\int_0^4 \int_0^{\sqrt{x}} f(x,y)\,dy\,dx = \iint_D f(x,y)\,dA = \int_0^2 \int_{y^2}^4 f(x,y)\,dx\,dy$.

32.

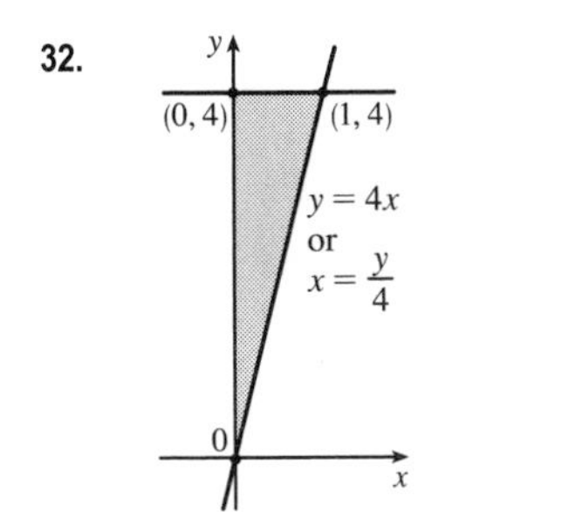

Because the region of integration is

$D = \{(x,y) \mid 4x \leq y \leq 4, 0 \leq x \leq 1\} = \{(x,y) \mid 0 \leq x \leq \frac{y}{4}, 0 \leq y \leq 4\}$

we have $\int_0^1 \int_{4x}^4 f(x,y)\,dy\,dx = \iint_D f(x,y)\,dA = \int_0^4 \int_0^{y/4} f(x,y)\,dx\,dy$.

33.

Because the region of integration is

$$D = \left\{(x,y) \mid -\sqrt{9-y^2} \leq x \leq \sqrt{9-y^2}, 0 \leq y \leq 3\right\}$$
$$= \left\{(x,y) \mid 0 \leq y \leq \sqrt{9-x^2}, -3 \leq x \leq 3\right\}$$

we have

$$\int_0^3 \int_{-\sqrt{9-y^2}}^{\sqrt{9-y^2}} f(x,y)\,dx\,dy = \iint_D f(x,y)\,dA = \int_{-3}^3 \int_0^{\sqrt{9-x^2}} f(x,y)\,dy\,dx.$$

34.

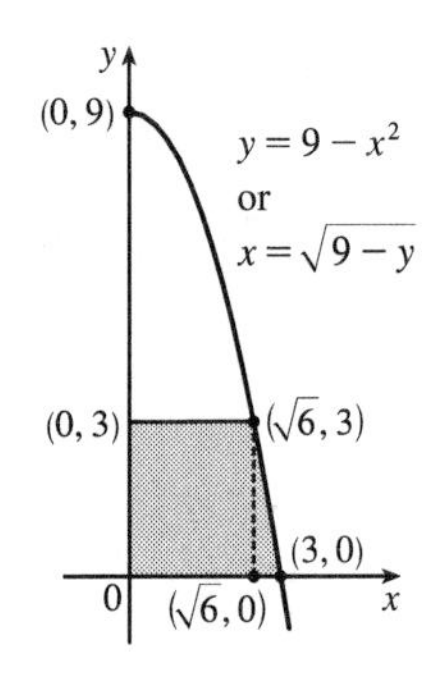

To reverse the order, we must break the region into two separate type I regions.

Because the region of integration is

$$D = \{(x,y) \mid 0 \le x \le \sqrt{9-y},\, 0 \le y \le 3\}$$
$$= \{(x,y) \mid 0 \le y \le 3,\, 0 \le x \le \sqrt{6}\} \cup \{(x,y) \mid 0 \le y \le 9-x^2,\, \sqrt{6} \le x \le 3\}$$

we have

$$\int_0^3 \int_0^{\sqrt{9-y}} f(x,y)\,dx\,dy = \iint_D f(x,y)\,dA$$
$$= \int_0^{\sqrt{6}} \int_0^3 f(x,y)\,dy\,dx + \int_{\sqrt{6}}^3 \int_0^{9-x^2} f(x,y)\,dy\,dx$$

35.

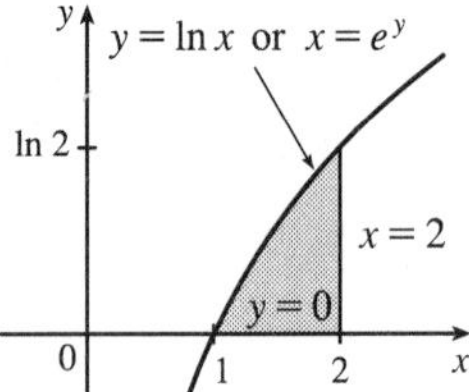

Because the region of integration is

$$D = \{(x,y) \mid 0 \le y \le \ln x,\, 1 \le x \le 2\} = \{(x,y) \mid e^y \le x \le 2,\, 0 \le y \le \ln 2\}$$

we have

$$\int_1^2 \int_0^{\ln x} f(x,y)\,dy\,dx = \iint_D f(x,y)\,dA = \int_0^{\ln 2} \int_{e^y}^2 f(x,y)\,dx\,dy$$

36.

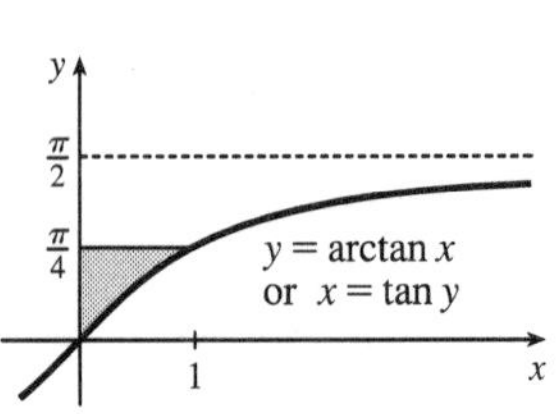

Because the region of integration is

$$D = \left\{(x,y) \mid \arctan x \le y \le \tfrac{\pi}{4},\, 0 \le x \le 1\right\}$$
$$= \left\{(x,y) \mid 0 \le x \le \tan y,\, 0 \le y \le \tfrac{\pi}{4}\right\}$$

we have

$\int_0^1 \int_{\arctan x}^{\pi/4} f(x,y)\,dy\,dx = \iint_D f(x,y)\,dA = \int_0^{\pi/4} \int_0^{\tan y} f(x,y)\,dx\,dy.$

37.

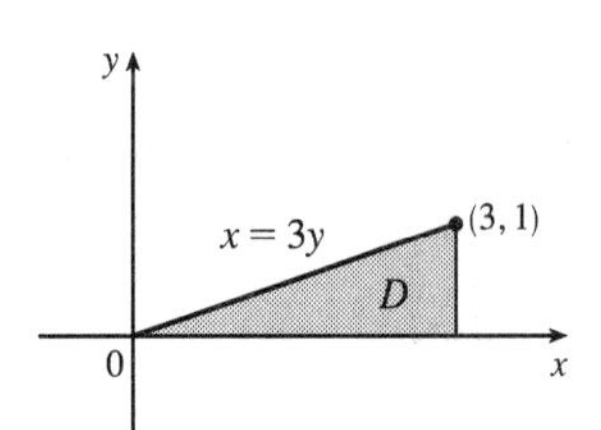

$$\int_0^1 \int_{3y}^3 e^{x^2}\,dx\,dy = \int_0^3 \int_0^{x/3} e^{x^2}\,dy\,dx = \int_0^3 \left[e^{x^2} y\right]_{y=0}^{y=x/3} dx$$
$$= \int_0^3 \left(\frac{x}{3}\right) e^{x^2}\,dx = \tfrac{1}{6}\, e^{x^2}\Big]_0^3 = \frac{e^9 - 1}{6}$$

38.

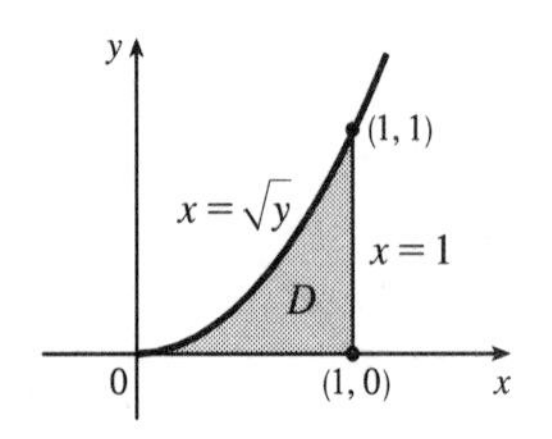

$$\int_0^1 \int_{\sqrt{y}}^1 \sqrt{x^3+1}\,dx\,dy = \int_0^1 \left[\sqrt{x^3+1}\,y\right]_{y=0}^{y=x^2} dx$$
$$= \int_0^1 x^2\sqrt{x^3+1}\,dx = \tfrac{2}{9}(x^3+1)^{3/2}\Big]_0^1$$
$$= \tfrac{2}{9}(2^{3/2} - 1)$$

39.

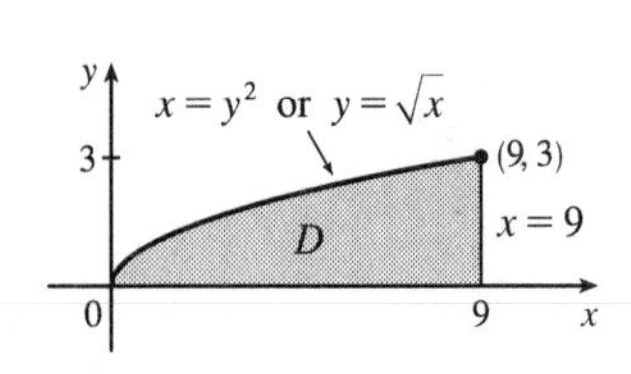

$$\int_0^3 \int_{y^2}^9 y\cos(x^2)\,dx\,dy = \int_0^9 \int_0^{\sqrt{x}} y\cos(x^2)\,dy\,dx$$
$$= \int_0^9 \cos(x^2)\left[\frac{y^2}{2}\right]_{y=0}^{y=\sqrt{x}} dx = \int_0^9 \tfrac{1}{2}x\cos(x^2)\,dx$$
$$= \tfrac{1}{4}\sin(x^2)\Big]_0^9 = \tfrac{1}{4}\sin 81$$

40.

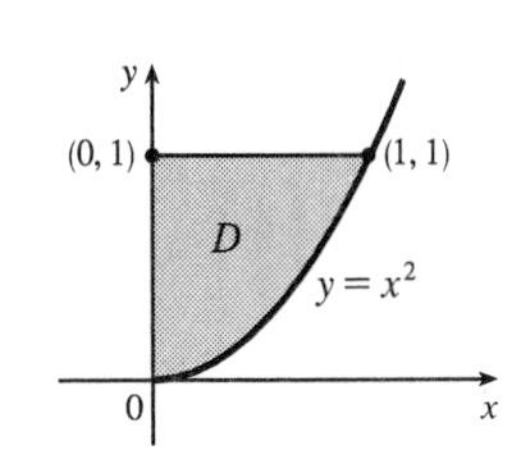

$$\int_0^1 \int_{x^2}^1 x^3 \sin(y^3)\, dy\, dx = \int_0^1 \int_0^{\sqrt{y}} x^3 \sin(y^3)\, dx\, dy$$
$$= \int_0^1 \left[\frac{x^4}{4} \sin(y^3) \right]_{x=0}^{x=\sqrt{y}} dy = \int_0^1 \tfrac{1}{4} y^2 \sin(y^3)\, dy$$
$$= -\tfrac{1}{12} \cos(y^3) \big]_0^1 = -\tfrac{1}{12} \cos(y^3) \big]_0^1 = \tfrac{1}{12}(1 - \cos 1)$$

41.

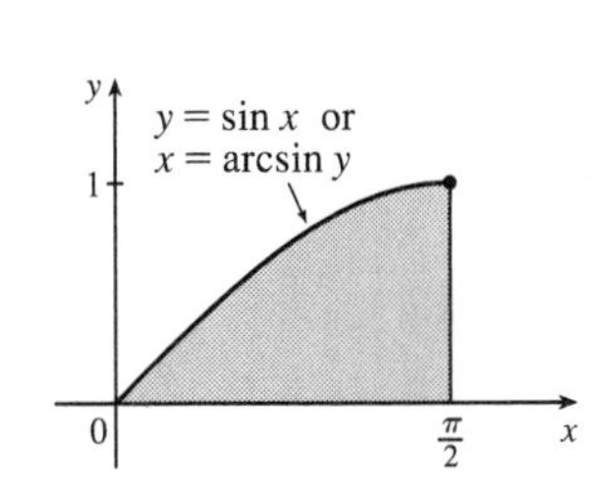

$$\int_0^1 \int_{\arcsin y}^{\pi/2} \cos x \sqrt{1 + \cos^2 x}\, dx\, dy$$
$$= \int_0^{\pi/2} \int_0^{\sin x} \cos x \sqrt{1 + \cos^2 x}\, dy\, dx$$
$$= \int_0^{\pi/2} \cos x \sqrt{1 + \cos^2 x} \left[y \right]_{y=0}^{y=\sin x} dx$$
$$= \int_0^{\pi/2} \cos x \sqrt{1 + \cos^2 x} \sin x\, dx \qquad \left[\begin{array}{c} \text{Let } u = \cos x,\ du = -\sin x\, dx, \\ dx = du/(-\sin x) \end{array} \right]$$
$$= \int_1^0 -u \sqrt{1 + u^2}\, du = -\tfrac{1}{3} \left(1 + u^2\right)^{3/2} \Big]_1^0$$
$$= \tfrac{1}{3}\left(\sqrt{8} - 1\right) = \tfrac{1}{3}\left(2\sqrt{2} - 1\right)$$

42.

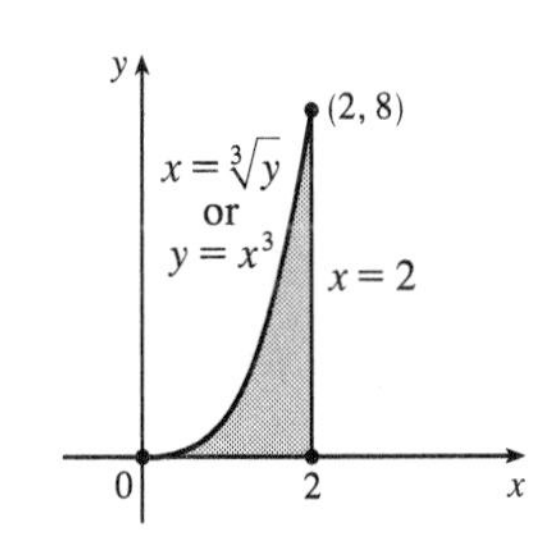

$$\int_0^8 \int_{\sqrt[3]{y}}^2 e^{x^4}\, dx\, dy = \int_0^2 \int_0^{x^3} e^{x^4}\, dy\, dx$$
$$= \int_0^2 e^{x^4} \left[y \right]_{y=0}^{y=x^3} dx = \int_0^2 x^3 e^{x^4}\, dx$$
$$= \tfrac{1}{4} e^{x^4} \Big]_0^2 = \tfrac{1}{4}(e^{16} - 1)$$

43. $D = \{(x, y) \mid 0 \le x \le 1,\ -x + 1 \le y \le 1\} \cup \{(x, y) \mid -1 \le x \le 0,\ x + 1 \le y \le 1\}$
$\cup \{(x, y) \mid 0 \le x \le 1,\ -1 \le y \le x - 1\} \cup \{(x, y) \mid -1 \le x \le 0,\ -1 \le y \le -x - 1\}$, all type I.

$$\iint_D x^2\, dA = \int_0^1 \int_{1-x}^1 x^2\, dy\, dx + \int_{-1}^0 \int_{x+1}^1 x^2\, dy\, dx + \int_0^1 \int_{-1}^{x-1} x^2\, dy\, dx + \int_{-1}^0 \int_{-1}^{-x-1} x^2\, dy\, dx$$
$$= 4 \int_0^1 \int_{1-x}^1 x^2\, dy\, dx \qquad \text{[by symmetry of the regions and because } f(x, y) = x^2 \ge 0\text{]}$$
$$= 4 \int_0^1 x^3\, dx = 4\left[\tfrac{1}{4} x^4\right]_0^1 = 1$$

44. $D = \{(x, y) \mid -1 \le x \le 0,\ -1 \le y \le 1 + x^2\} \cup \{(x, y) \mid 0 \le x \le 1,\ \sqrt{x} \le y \le 1 + x^2\}$
$\cup \{(x, y) \mid 0 \le x \le 1,\ -1 \le y \le -\sqrt{x}\}$, all type I.

$$\iint_D xy\, dA = \int_{-1}^0 \int_{-1}^{1+x^2} xy\, dy\, dx + \int_0^1 \int_{\sqrt{x}}^{1+x^2} xy\, dy\, dx + \int_0^1 \int_{-1}^{-\sqrt{x}} xy\, dy\, dx$$
$$= \int_{-1}^0 \left[\tfrac{1}{2} xy^2\right]_{y=-1}^{y=1+x^2} dx + \int_0^1 \left[\tfrac{1}{2} xy^2\right]_{y=\sqrt{x}}^{y=1+x^2} dx + \int_0^1 \left[\tfrac{1}{2} xy^2\right]_{y=-1}^{y=-\sqrt{x}} dx$$
$$= \int_{-1}^0 \left(x^3 + \tfrac{1}{2} x^5\right) dx + \int_0^1 \tfrac{1}{2}(x^5 + 2x^3 - x^2 + x)\, dx + \int_0^1 \tfrac{1}{2}(x^2 - x)\, dx$$
$$= \left[\tfrac{1}{4} x^4 + \tfrac{1}{12} x^6\right]_{-1}^0 + \tfrac{1}{2}\left[\tfrac{1}{6} x^6 + \tfrac{1}{2} x^4 - \tfrac{1}{3} x^3 + \tfrac{1}{2} x^2\right]_0^1 + \tfrac{1}{2}\left[\tfrac{1}{3} x^3 - \tfrac{1}{2} x^2\right]_0^1 = -\tfrac{1}{3} + \tfrac{5}{12} - \tfrac{1}{12} = 0$$

45. For $D = [0,1] \times [0,1]$, $0 \le \sqrt{x^3+y^3} \le \sqrt{2}$ and $A(D) = 1$, so $0 \le \iint_D \sqrt{x^3+y^3}\,dA \le \sqrt{2}$.

46. Since $D = \{(x,y) \mid x^2+y^2 \le \frac{1}{4}\}$, $1 = e^0 \le e^{x^2+y^2} \le e^{1/4}$ and $A(D) = \frac{\pi}{4}$, so $\frac{\pi}{4} \le \iint_D e^{x^2+y^2}\,dA \le (e^{1/4})\frac{\pi}{4}$.

47. Since $m \le f(x,y) \le M$, $\iint_D m\,dA \le \iint_D f(x,y)\,dA \le \iint_D M\,dA$ by (8) $\Rightarrow$
$m\iint_D 1\,dA \le \iint_D f(x,y)\,dA \le M\iint_D 1\,dA$ by (7) $\Rightarrow$ $mA(D) \le \iint_D f(x,y)\,dA \le MA(D)$ by (10).

48.

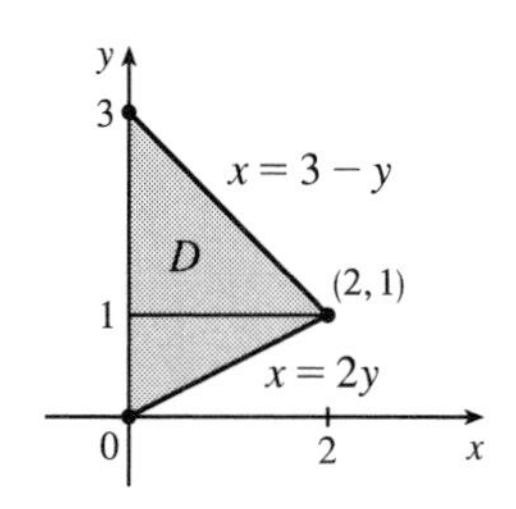

$$\iint_D f(x,y)\,dA = \int_0^1\int_0^{2y} f(x,y)\,dx\,dy + \int_1^3\int_0^{3-y} f(x,y)\,dx\,dy$$
$$= \int_0^2\int_{x/2}^{3-x} f(x,y)\,dy\,dx$$

49. $\iint_D (x^2\tan x + y^3 + 4)\,dA = \iint_D x^2\tan x\,dA + \iint_D y^3\,dA + \iint_D 4\,dA$. But $x^2\tan x$ is an odd function of x and D is symmetric with respect to the y-axis, so $\iint_D x^2\tan x\,dA = 0$. Similarly, y^3 is an odd function of y and D is symmetric with respect to the x-axis, so $\iint_D y^3\,dA = 0$. Thus $\iint_D (x^2\tan x + y^3 + 4)\,dA = 4\iint_D dA = 4(\text{area of } D) = 4\cdot\pi\left(\sqrt{2}\right)^2 = 8\pi$.

50. First, $\iint_D (2 - 3x + 4y)\,dA = \iint_D 2\,dA - \iint_D 3x\,dA + \iint_D 4y\,dA$. The region D, shown in the figure, is symmetric with respect to the y-axis and $3x$ is an odd function of x, so $\iint_D 3x\,dA = 0$. Similarly, $4y$ is an odd function of y and D is symmetric with respect to the x-axis, so $\iint_D 4y\,dA = 0$. Then

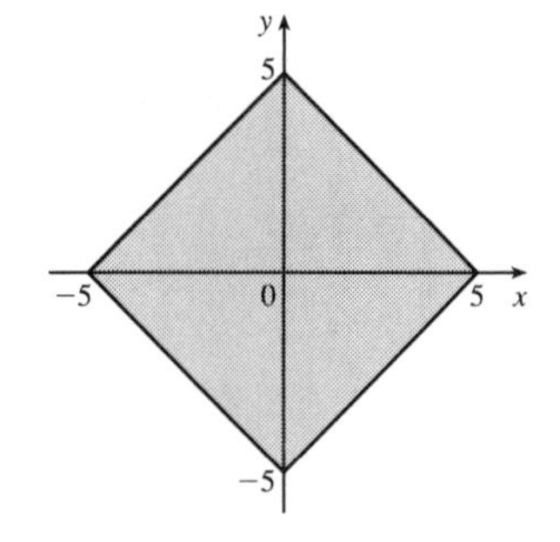

$$\iint_D (2 - 3x + 4y)\,dA = \iint_D 2\,dA = 2\iint_D dA$$
$$= 2(\text{area of } D) = 2(50) = 100$$

51. Since $\sqrt{1-x^2-y^2} \ge 0$, we can interpret $\iint_D \sqrt{1-x^2-y^2}\,dA$ as the volume of the solid that lies below the graph of $z = \sqrt{1-x^2-y^2}$ and above the region D in the xy-plane. $z = \sqrt{1-x^2-y^2}$ is equivalent to $x^2+y^2+z^2 = 1$, $z \ge 0$ which meets the xy-plane in the circle $x^2+y^2 = 1$, the boundary of D. Thus, the solid is an upper hemisphere of radius 1 which has volume $\frac{1}{2}\left[\frac{4}{3}\pi(1)^3\right] = \frac{2}{3}\pi$.

52. To find the equations of the boundary curves, we require that the z-values of the two surfaces be the same. In Maple, we use the command `solve(4-x^2-y^2=1-x-y,y);` and in Mathematica, we use `Solve[4-x^2-y^2==1-x-y,y]`. We find that the curves have equations $y = \dfrac{1 \pm \sqrt{13+4x-4x^2}}{2}$. To find the two points of intersection of these curves, we use the CAS to solve $13 + 4x - 4x^2 = 0$, finding that $x = \frac{1\pm\sqrt{14}}{2}$. So, using the CAS to evaluate the integral, the volume of intersection is

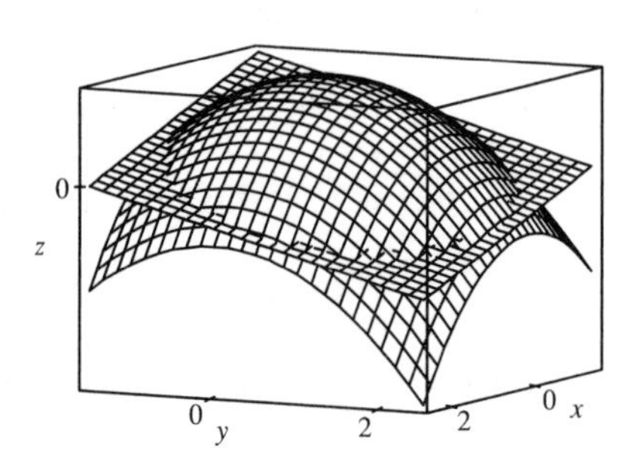

$$V = \int_{(1-\sqrt{14})/2}^{(1+\sqrt{14})/2}\int_{\left(1-\sqrt{13+4x-4x^2}\right)/2}^{\left(1+\sqrt{13+4x-4x^2}\right)/2} \left[(4-x^2-y^2)-(1-x-y)\right]dy\,dx = \frac{49\pi}{8}$$

12.3 Double Integrals in Polar Coordinates

1. The region R is more easily described by polar coordinates: $R = \{(r,\theta) \mid 2 \le r \le 5, 0 \le \theta \le 2\pi\}$.

Thus $\iint_R f(x,y)\,dA = \int_0^{2\pi}\int_2^5 f(r\cos\theta, r\sin\theta)\,r\,dr\,d\theta$.

2. The region R is more easily described by polar coordinates: $R = \left\{(r,\theta) \mid 0 \le r \le 2\sqrt{2}, \frac{\pi}{4} \le \theta \le \frac{5\pi}{4}\right\}$.

Thus $\iint_R f(x,y)\,dA = \int_{\pi/4}^{5\pi/4}\int_0^{2\sqrt{2}} f(r\cos\theta, r\sin\theta)\,r\,dr\,d\theta$.

3. The region R is more easily described by rectangular coordinates: $R = \{(x,y) \mid -2 \le x \le 2, x \le y \le 2\}$.

Thus $\iint_R f(x,y)\,dA = \int_{-2}^{2}\int_x^2 f(x,y)\,dy\,dx$.

4. The region R is more easily described by polar coordinates: $R = \left\{(r,\theta) \mid 1 \le r \le 3, 0 \le \theta \le \frac{\pi}{2}\right\}$.

Thus $\iint_R f(x,y)\,dA = \int_0^{\pi/2}\int_1^3 f(r\cos\theta, r\sin\theta)\,r\,dr\,d\theta$.

5. The integral $\int_\pi^{2\pi}\int_4^7 r\,dr\,d\theta$ represents the area of the region $R = \{(r,\theta) \mid 4 \le r \le 7, \pi \le \theta \le 2\pi\}$, the lower half of a ring.

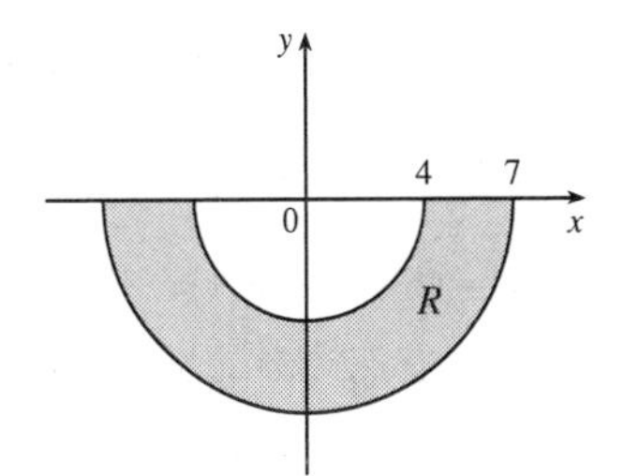

$$\int_\pi^{2\pi}\int_4^7 r\,dr\,d\theta = \left(\int_\pi^{2\pi} d\theta\right)\left(\int_4^7 r\,dr\right) = \left[\,\theta\,\right]_\pi^{2\pi}\left[\tfrac{1}{2}r^2\right]_4^7 = \pi\cdot\tfrac{1}{2}\,(49-16) = \tfrac{33\pi}{2}$$

6. The integral $\int_0^{\pi/2}\int_0^{4\cos\theta} r\,dr\,d\theta$ represents the area of the region $R = \{(r,\theta) \mid 0 \le r \le 4\cos\theta, 0 \le \theta \le \pi/2\}$. Since $r = 4\cos\theta \;\Leftrightarrow\; r^2 = 4r\cos\theta \;\Leftrightarrow\; x^2+y^2 = 4x \;\Leftrightarrow\; (x-2)^2+y^2 = 4$, R is the portion in the first quadrant of a circle of radius 2 with center $(2,0)$.

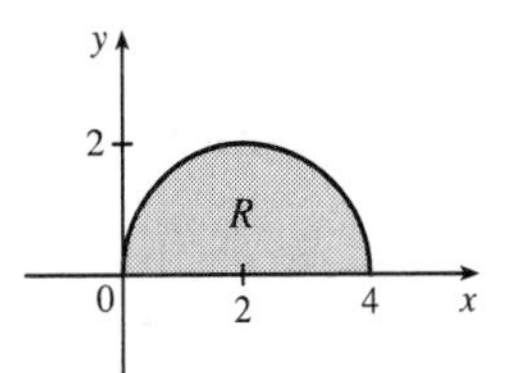

$$\int_0^{\pi/2}\int_0^{4\cos\theta} r\,dr\,d\theta = \int_0^{\pi/2}\left[\tfrac{1}{2}r^2\right]_{r=0}^{r=4\cos\theta} d\theta = \int_0^{\pi/2} 8\cos^2\theta\,d\theta = \int_0^{\pi/2} 4(1+\cos 2\theta)\,d\theta = 4\left[\theta + \tfrac{1}{2}\sin 2\theta\right]_0^{\pi/2} = 2\pi$$

7. The disk D can be described in polar coordinates as $D = \{(r,\theta) \mid 0 \le r \le 3, 0 \le \theta \le 2\pi\}$. Then

$$\iint_D xy\,dA = \int_0^{2\pi}\int_0^3 (r\cos\theta)(r\sin\theta)\,r\,dr\,d\theta = \left(\int_0^{2\pi}\sin\theta\cos\theta\,d\theta\right)\left(\int_0^3 r^3\,dr\right) = \left[\tfrac{1}{2}\sin^2\theta\right]_0^{2\pi}\left[\tfrac{1}{4}r^4\right]_0^3 = 0.$$

8.
$$\begin{aligned}\iint_R (x+y)\,dA &= \int_{\pi/2}^{3\pi/2}\int_1^2 (r\cos\theta + r\sin\theta)\,r\,dr\,d\theta = \int_{\pi/2}^{3\pi/2}\int_1^2 r^2(\cos\theta+\sin\theta)\,dr\,d\theta\\ &= \left(\int_{\pi/2}^{3\pi/2}(\cos\theta+\sin\theta)\,d\theta\right)\left(\int_1^2 r^2\,dr\right) = \left[\sin\theta - \cos\theta\right]_{\pi/2}^{3\pi/2}\left[\tfrac{1}{3}r^3\right]_1^2\\ &= (-1-0-1+0)\left(\tfrac{8}{3}-\tfrac{1}{3}\right) = -\tfrac{14}{3}\end{aligned}$$

9.
$$\begin{aligned}\iint_R \cos(x^2+y^2)\,dA &= \int_0^\pi\int_0^3 \cos(r^2)\,r\,dr\,d\theta = \left(\int_0^\pi d\theta\right)\left(\int_0^3 r\cos(r^2)\,dr\right)\\ &= \left[\,\theta\,\right]_0^\pi\left[\tfrac{1}{2}\sin(r^2)\right]_0^3 = \pi\cdot\tfrac{1}{2}(\sin 9 - \sin 0) = \tfrac{\pi}{2}\sin 9\end{aligned}$$

10.
$$\begin{aligned}\iint_R \sqrt{4-x^2-y^2}\,dA &= \int_{-\pi/2}^{\pi/2}\int_0^2 \sqrt{4-r^2}\,r\,dr\,d\theta = \left(\int_{-\pi/2}^{\pi/2} d\theta\right)\left(\int_0^2 r\sqrt{4-r^2}\,dr\right)\\ &= \left[\,\theta\,\right]_{-\pi/2}^{\pi/2}\left[-\tfrac{1}{2}\cdot\tfrac{2}{3}(4-r^2)^{3/2}\right]_0^2 = \left(\tfrac{\pi}{2}+\tfrac{\pi}{2}\right)\left(-\tfrac{1}{3}(0-4^{3/2})\right) = \tfrac{8}{3}\pi\end{aligned}$$

11. R is the region shown in the figure, and can be described by $R = \{(r,\theta) \mid 0 \le \theta \le \pi/4, 1 \le r \le 2\}$. Thus $\iint_R \arctan(y/x)\, dA = \int_0^{\pi/4} \int_1^2 \arctan(\tan\theta)\, r\, dr\, d\theta$ since $y/x = \tan\theta$. Also, $\arctan(\tan\theta) = \theta$ for $0 \le \theta \le \pi/4$, so the integral becomes

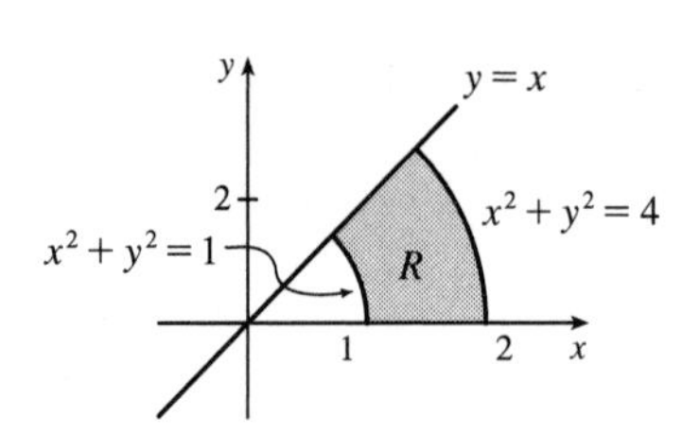

$\int_0^{\pi/4} \int_1^2 \theta\, r\, dr\, d\theta = \int_0^{\pi/4} \theta\, d\theta \int_1^2 r\, dr = \left[\frac{1}{2}\theta^2\right]_0^{\pi/4} \left[\frac{1}{2}r^2\right]_1^2 = \frac{\pi^2}{32} \cdot \frac{3}{2} = \frac{3}{64}\pi^2$.

12. $\iint_R ye^x\, dA = \int_0^{\pi/2} \int_0^5 (r\sin\theta)\, e^{r\cos\theta}\, r\, dr\, d\theta = \int_0^5 \int_0^{\pi/2} r^2 \sin\theta\, e^{r\cos\theta}\, d\theta\, dr$. First we integrate $\int_0^{\pi/2} r^2 \sin\theta\, e^{r\cos\theta}\, d\theta$:

Let $u = r\cos\theta \;\Rightarrow\; du = -r\sin\theta\, d\theta$, and $\int_0^{\pi/2} r^2 \sin\theta\, e^{r\cos\theta}\, d\theta = \int_{u=r}^{u=0} -r\, e^u\, du = -r[e^0 - e^r] = re^r - r$.

Then $\int_0^5 \int_0^{\pi/2} r^2 \sin\theta\, e^{r\cos\theta}\, d\theta\, dr = \int_0^5 (re^r - r)\, dr = \left[re^r - e^r - \frac{1}{2}r^2\right]_0^5 = 4e^5 - \frac{23}{2}$, where we integrated by parts in the first term.

13. $V = \iint_{x^2+y^2 \le 4} \sqrt{x^2+y^2}\, dA = \int_0^{2\pi} \int_0^2 \sqrt{r^2}\, r\, dr\, d\theta = \int_0^{2\pi} d\theta \int_0^2 r^2\, dr = \left[\theta\right]_0^{2\pi} \left[\frac{1}{3}r^3\right]_0^2 = 2\pi\left(\frac{8}{3}\right) = \frac{16}{3}\pi$

14. The paraboloid $z = 18 - 2x^2 - 2y^2$ intersects the xy-plane in the circle $x^2 + y^2 = 9$, so

$$V = \iint_{x^2+y^2 \le 9} \left(18 - 2x^2 - 2y^2\right) dA = \iint_{x^2+y^2 \le 9} \left[18 - 2\left(x^2+y^2\right)\right] dA = \int_0^{2\pi} \int_0^3 \left(18 - 2r^2\right) r\, dr\, d\theta$$

$$= \int_0^{2\pi} d\theta \int_0^3 (18r - 2r^3)\, dr = \left[\theta\right]_0^{2\pi} \left[9r^2 - \tfrac{1}{2}r^4\right]_0^3 = (2\pi)\left(81 - \tfrac{81}{2}\right) = 81\pi$$

15. By symmetry,

$$V = 2 \iint_{x^2+y^2 \le a^2} \sqrt{a^2 - x^2 - y^2}\, dA = 2 \int_0^{2\pi} \int_0^a \sqrt{a^2 - r^2}\, r\, dr\, d\theta = 2 \int_0^{2\pi} d\theta \int_0^a r\sqrt{a^2 - r^2}\, dr$$

$$= 2\left[\theta\right]_0^{2\pi} \left[-\tfrac{1}{3}(a^2 - r^2)^{3/2}\right]_0^a = 2(2\pi)\left(0 + \tfrac{1}{3}a^3\right) = \tfrac{4\pi}{3}a^3$$

16. The sphere $x^2 + y^2 + z^2 = 16$ intersects the xy-plane in the circle $x^2 + y^2 = 16$, so

$$V = 2 \iint_{4 \le x^2+y^2 \le 16} \sqrt{16 - x^2 - y^2}\, dA \quad \text{[by symmetry]} \quad = 2 \int_0^{2\pi} \int_2^4 \sqrt{16 - r^2}\, r\, dr\, d\theta = 2 \int_0^{2\pi} d\theta \int_2^4 r(16 - r^2)^{1/2} dr$$

$$= 2\left[\theta\right]_0^{2\pi} \left[-\tfrac{1}{3}(16 - r^2)^{3/2}\right]_2^4 = -\tfrac{2}{3}(2\pi)(0 - 12^{3/2}) = \tfrac{4\pi}{3}\left(12\sqrt{12}\right) = 32\sqrt{3}\,\pi$$

17. The cone $z = \sqrt{x^2+y^2}$ intersects the sphere $x^2 + y^2 + z^2 = 1$ when $x^2 + y^2 + \left(\sqrt{x^2+y^2}\right)^2 = 1$ or $x^2 + y^2 = \frac{1}{2}$. So

$$V = \iint_{x^2+y^2 \le 1/2} \left(\sqrt{1 - x^2 - y^2} - \sqrt{x^2+y^2}\right) dA = \int_0^{2\pi} \int_0^{1/\sqrt{2}} \left(\sqrt{1 - r^2} - r\right) r\, dr\, d\theta$$

$$= \int_0^{2\pi} d\theta \int_0^{1/\sqrt{2}} \left(r\sqrt{1 - r^2} - r^2\right) dr = \left[\theta\right]_0^{2\pi} \left[-\tfrac{1}{3}(1 - r^2)^{3/2} - \tfrac{1}{3}r^3\right]_0^{1/\sqrt{2}} = 2\pi\left(-\tfrac{1}{3}\right)\left(\tfrac{1}{\sqrt{2}} - 1\right) = \tfrac{\pi}{3}\left(2 - \sqrt{2}\right)$$

18. The paraboloid $z = 1 + 2x^2 + 2y^2$ intersects the plane $z = 7$ when $7 = 1 + 2x^2 + 2y^2$ or $x^2 + y^2 = 3$ and we are restricted to the first octant, so

$$V = \iint_{\substack{x^2+y^2 \le 3,\\ x\ge 0, y\ge 0}} \left[7 - (1 + 2x^2 + 2y^2)\right] dA = \int_0^{\pi/2} \int_0^{\sqrt{3}} \left[7 - (1 + 2r^2)\right] r\,dr\,d\theta$$

$$= \int_0^{\pi/2} d\theta \int_0^{\sqrt{3}} (6r - 2r^3)\,dr = \left[\theta\right]_0^{\pi/2} \left[3r^2 - \tfrac{1}{2}r^4\right]_0^{\sqrt{3}} = \tfrac{\pi}{2} \cdot \tfrac{9}{2} = \tfrac{9}{4}\pi$$

19. The given solid is the region inside the cylinder $x^2 + y^2 = 4$ between the surfaces $z = \sqrt{64 - 4x^2 - 4y^2}$ and $z = -\sqrt{64 - 4x^2 - 4y^2}$. So

$$V = \iint_{x^2+y^2 \le 4} \left[\sqrt{64 - 4x^2 - 4y^2} - \left(-\sqrt{64 - 4x^2 - 4y^2}\right)\right] dA = \iint_{x^2+y^2 \le 4} 2\sqrt{64 - 4x^2 - 4y^2}\,dA$$

$$= 4\int_0^{2\pi} \int_0^2 \sqrt{16 - r^2}\,r\,dr\,d\theta = 4\int_0^{2\pi} d\theta \int_0^2 r\sqrt{16 - r^2}\,dr = 4\left[\theta\right]_0^{2\pi} \left[-\tfrac{1}{3}(16 - r^2)^{3/2}\right]_0^2$$

$$= 8\pi\left(-\tfrac{1}{3}\right)(12^{3/2} - 16^{2/3}) = \tfrac{8\pi}{3}\left(64 - 24\sqrt{3}\right)$$

20. (a) Here the region in the xy-plane is the annular region $r_1^2 \le x^2 + y^2 \le r_2^2$ and the desired volume is twice that above the xy-plane. Hence

$$V = 2\iint_{r_1^2 \le x^2+y^2 \le r_2^2} \sqrt{r_2^2 - x^2 - y^2}\,dA = 2\int_0^{2\pi} \int_{r_1}^{r_2} \sqrt{r_2^2 - r^2}\,r\,dr\,d\theta = 2\int_0^{2\pi} d\theta \int_{r_1}^{r_2} \sqrt{r_2^2 - r^2}\,r\,dr$$

$$= \tfrac{4\pi}{3}\left[-(r_2^2 - r^2)^{3/2}\right]_{r_1}^{r_2} = \tfrac{4\pi}{3}(r_2^2 - r_1^2)^{3/2}$$

(b) A cross-sectional cut is shown in the figure.

So $r_2^2 = \left(\tfrac{1}{2}h\right)^2 + r_1^2$ or $\tfrac{1}{4}h^2 = r_2^2 - r_1^2$.

Thus the volume in terms of h is $V = \tfrac{4\pi}{3}\left(\tfrac{1}{4}h^2\right)^{3/2} = \tfrac{\pi}{6}h^3$.

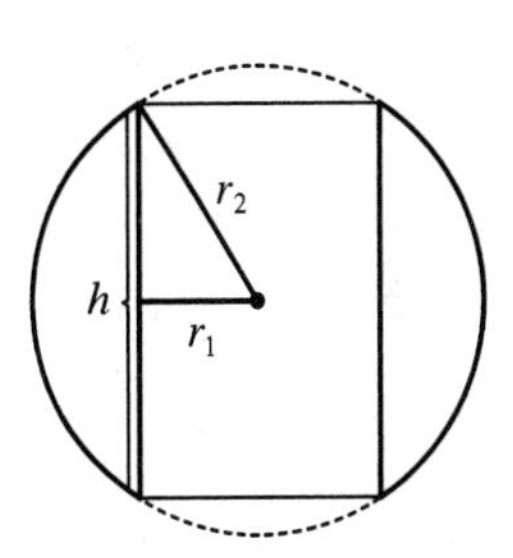

21. One loop is given by the region

$D = \{(r, \theta)\,|-\pi/6 \le \theta \le \pi/6, 0 \le r \le \cos 3\theta\}$, so the area is

$$\iint_D dA = \int_{-\pi/6}^{\pi/6} \int_0^{\cos 3\theta} r\,dr\,d\theta = \int_{-\pi/6}^{\pi/6} \left[\frac{1}{2}r^2\right]_{r=0}^{r=\cos 3\theta} d\theta$$

$$= \int_{-\pi/6}^{\pi/6} \frac{1}{2}\cos^2 3\theta\,d\theta = 2\int_0^{\pi/6} \frac{1}{2}\left(\frac{1 + \cos 6\theta}{2}\right) d\theta$$

$$= \frac{1}{2}\left[\theta + \frac{1}{6}\sin 6\theta\right]_0^{\pi/6} = \frac{\pi}{12}$$

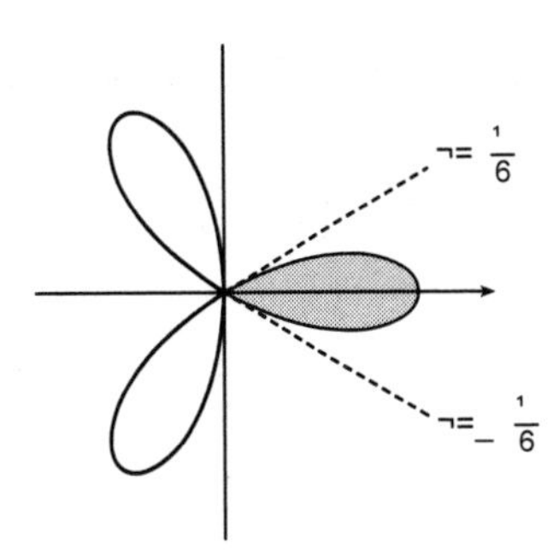

22. $D = \{(r,\theta) \mid 0 \le \theta \le 2\pi, 0 \le r \le 4 + 3\cos\theta\}$, so

$$A(D) = \iint_D dA = \int_0^{2\pi}\int_0^{4+3\cos\theta} r\,dr\,d\theta = \int_0^{2\pi}\left[\tfrac{1}{2}r^2\right]_{r=0}^{r=4+3\cos\theta} d\theta = \tfrac{1}{2}\int_0^{2\pi}(4+3\cos\theta)^2 d\theta$$

$$= \tfrac{1}{2}\int_0^{2\pi}(16 + 24\cos\theta + 9\cos^2\theta)\,d\theta = \tfrac{1}{2}\int_0^{2\pi}\left(16 + 24\cos\theta + 9\cdot\tfrac{1+\cos 2\theta}{2}\right)d\theta$$

$$= \tfrac{1}{2}\left[16\theta + 24\sin\theta + \tfrac{9}{2}\theta + \tfrac{9}{4}\sin 2\theta\right]_0^{2\pi} = \tfrac{41}{2}\pi$$

23.

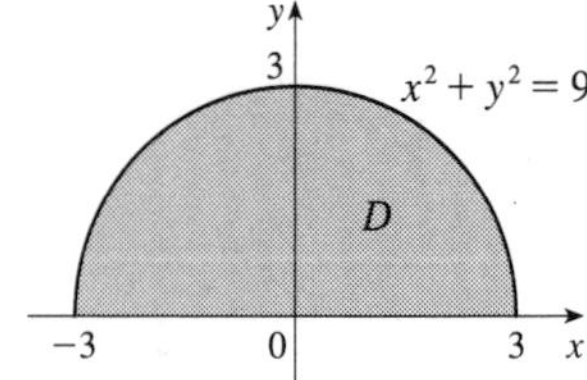

$$\int_{-3}^{3}\int_0^{\sqrt{9-x^2}} \sin(x^2+y^2)\,dy\,dx = \int_0^{\pi}\int_0^3 \sin\left(r^2\right) r\,dr\,d\theta$$

$$= \int_0^{\pi} d\theta \int_0^3 r\sin\left(r^2\right)dr = \left[\theta\right]_0^{\pi}\left[-\tfrac{1}{2}\cos\left(r^2\right)\right]_0^3$$

$$= \pi\left(-\tfrac{1}{2}\right)(\cos 9 - 1) = \tfrac{\pi}{2}(1 - \cos 9)$$

24.

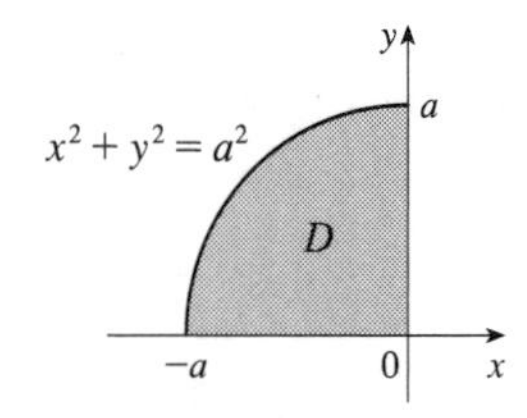

$$\int_{\pi/2}^{\pi}\int_0^a (r\cos\theta)^2(r\sin\theta)\,r\,dr\,d\theta = \int_{\pi/2}^{\pi}\int_0^a r^4\cos^2\theta\sin\theta\,dr\,d\theta$$

$$= \int_{\pi/2}^{\pi}\cos^2\theta\sin\theta\,d\theta\int_0^a r^4\,dr$$

$$= \left[-\tfrac{1}{3}\cos^3\theta\right]_{\pi/2}^{\pi}\left[\tfrac{1}{5}r^5\right]_0^a$$

$$= -\tfrac{1}{3}\left(\cos^3\pi - \cos^3\tfrac{\pi}{2}\right)\left(\tfrac{1}{5}a^5\right) = \tfrac{1}{15}a^5$$

25.

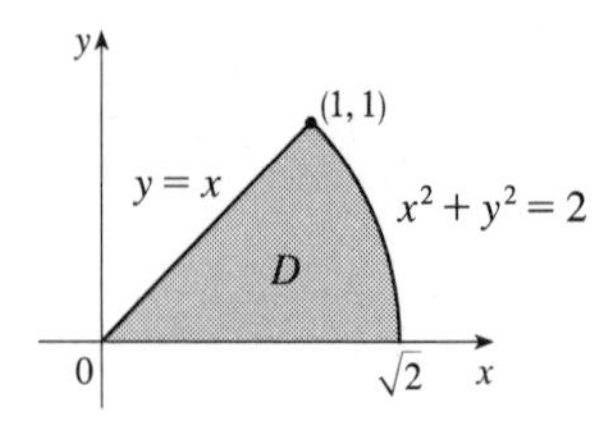

$$\int_0^{\pi/4}\int_0^{\sqrt{2}}(r\cos\theta + r\sin\theta)\,r\,dr\,d\theta = \int_0^{\pi/4}(\cos\theta + \sin\theta)\,d\theta\int_0^{\sqrt{2}} r^2\,dr$$

$$= \left[\sin\theta - \cos\theta\right]_0^{\pi/4}\left[\tfrac{1}{3}r^3\right]_0^{\sqrt{2}}$$

$$= \left[\tfrac{\sqrt{2}}{2} - \tfrac{\sqrt{2}}{2} - 0 + 1\right]\cdot\tfrac{1}{3}\left(2\sqrt{2} - 0\right) = \tfrac{2\sqrt{2}}{3}$$

26.

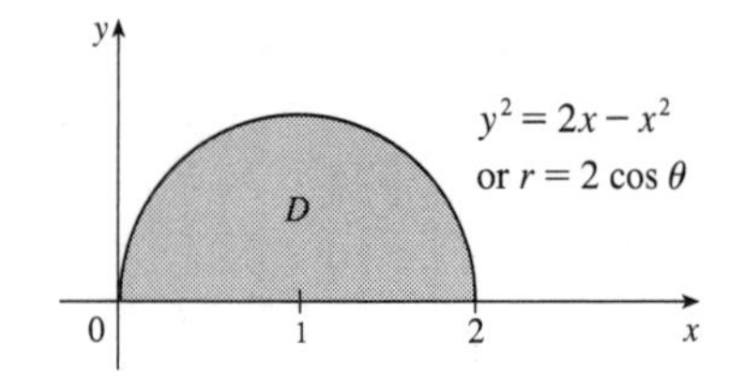

$$\int_0^{\pi/2}\int_0^{2\cos\theta} r^2\,dr\,d\theta = \int_0^{\pi/2}\left[\tfrac{1}{3}r^3\right]_{r=0}^{r=2\cos\theta}d\theta = \int_0^{\pi/2}\left(\tfrac{8}{3}\cos^3\theta\right)d\theta$$

$$= \tfrac{8}{3}\int_0^{\pi/2}\left(1 - \sin^2\theta\right)\cos\theta\,d\theta$$

$$= \tfrac{8}{3}\left[\sin\theta - \tfrac{1}{3}\sin^3\theta\right]_0^{\pi/2} = \tfrac{16}{9}$$

27. The surface of the water in the pool is a circular disk D with radius 20 ft. If we place D on coordinate axes with the origin at the center of D and define $f(x,y)$ to be the depth of the water at (x,y), then the volume of water in the pool is the volume of the solid that lies above $D = \left\{(x,y) \mid x^2 + y^2 \le 400\right\}$ and below the graph of $f(x,y)$. We can associate north with the positive y-direction, so we are given that the depth is constant in the x-direction and the depth increases linearly in the y-direction from $f(0,-20) = 2$ to $f(0,20) = 7$. The trace in the yz-plane is a line segment from $(0,-20,2)$ to $(0,20,7)$. The slope of this line is $\frac{7-2}{20-(-20)} = \frac{1}{8}$, so an equation of the line is $z - 7 = \frac{1}{8}(y - 20) \Rightarrow z = \frac{1}{8}y + \frac{9}{2}$. Since $f(x,y)$ is independent of x, $f(x,y) = \frac{1}{8}y + \frac{9}{2}$. Thus the volume is given by $\iint_D f(x,y)\,dA$, which is most conveniently evaluated using polar coordinates. Then $D = \{(r,\theta) \mid 0 \le r \le 20, 0 \le \theta \le 2\pi\}$ and substituting $x = r\cos\theta$, $y = r\sin\theta$ the integral becomes

$$\int_0^{2\pi}\int_0^{20} \left(\tfrac{1}{8}r\sin\theta + \tfrac{9}{2}\right) r\,dr\,d\theta = \int_0^{2\pi} \left[\tfrac{1}{24}r^3\sin\theta + \tfrac{9}{4}r^2\right]_{r=0}^{r=20} d\theta = \int_0^{2\pi} \left(\tfrac{1000}{3}\sin\theta + 900\right) d\theta$$
$$= \left[-\tfrac{1000}{3}\cos\theta + 900\theta\right]_0^{2\pi} = 1800\pi$$

Thus the pool contains $1800\pi \approx 5655$ ft^3 of water.

28. (a) The total amount of water supplied each hour to the region within R feet of the sprinkler is

$$V = \int_0^{2\pi}\int_0^R e^{-r} r\,dr\,d\theta = \int_0^{2\pi} d\theta \int_0^R re^{-r}\,dr = \left[\,\theta\,\right]_0^{2\pi}\left[-re^{-r} - e^{-r}\right]_0^R$$
$$= 2\pi[-Re^{-R} - e^{-R} + 0 + 1] = 2\pi(1 - Re^{-R} - e^{-R}) \text{ ft}^3$$

(b) The average amount of water per hour per square foot supplied to the region within R feet of the sprinkler is

$\dfrac{V}{\text{area of region}} = \dfrac{V}{\pi R^2} = \dfrac{2(1 - Re^{-R} - e^{-R})}{R^2}$ ft^3 (per hour per square foot). See the definition of the average value of a function on page 675.

29. $$\int_{1/\sqrt{2}}^1 \int_{\sqrt{1-x^2}}^x xy\,dy\,dx + \int_1^{\sqrt{2}}\int_0^x xy\,dy\,dx + \int_{\sqrt{2}}^2\int_0^{\sqrt{4-x^2}} xy\,dy\,dx$$
$$= \int_0^{\pi/4}\int_1^2 r^3\cos\theta\sin\theta\,dr\,d\theta = \int_0^{\pi/4}\left[\frac{r^4}{4}\cos\theta\sin\theta\right]_{r=1}^{r=2} d\theta$$
$$= \frac{15}{4}\int_0^{\pi/4}\sin\theta\cos\theta\,d\theta = \frac{15}{4}\left[\frac{\sin^2\theta}{2}\right]_0^{\pi/4} = \frac{15}{16}$$

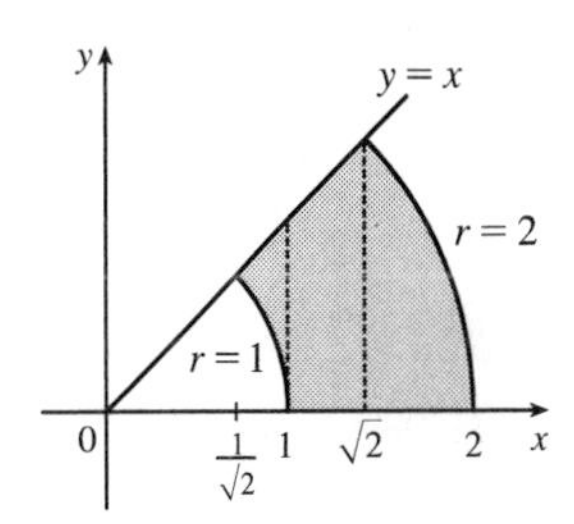

30. (a) $\iint_{D_a} e^{-(x^2+y^2)}dA = \int_0^{2\pi}\int_0^a re^{-r^2}\,dr\,d\theta = 2\pi\left[-\frac{1}{2}e^{-r^2}\right]_0^a = \pi\left(1 - e^{-a^2}\right)$ for each a. Then $\lim\limits_{a\to\infty}\pi\left(1 - e^{-a^2}\right) = \pi$ since $e^{-a^2} \to 0$ as $a \to \infty$. Hence $\int_{-\infty}^{\infty}\int_{-\infty}^{\infty} e^{-(x^2+y^2)}\,dA = \pi$.

(b) $\iint_{S_a} e^{-(x^2+y^2)}\,dA = \int_{-a}^a\int_{-a}^a e^{-x^2}e^{-y^2}\,dx\,dy = \left(\int_{-a}^a e^{-x^2}\,dx\right)\left(\int_{-a}^a e^{-y^2}\,dy\right)$ for each a.

Then, from (a), $\pi = \iint_{\mathbb{R}^2} -(x^2+y^2)\,dA$, so

$\pi = \lim\limits_{a\to\infty}\iint_{S_a} e^{-(x^2+y^2)}\,dA = \lim\limits_{a\to\infty}\left(\int_{-a}^a e^{-x^2}\,dx\right)\left(\int_{-a}^a e^{-y^2}\,dy\right) = \left(\int_{-\infty}^{\infty} e^{-x^2}\,dx\right)\left(\int_{-\infty}^{\infty} e^{-y^2}\,dy\right).$

To evaluate $\lim\limits_{a\to\infty}\left(\int_{-a}^a e^{-x^2}\,dx\right)\left(\int_{-a}^a e^{-y^2}\,dy\right)$, we are using the fact that these integrals are bounded. This true since on $[-1,1]$, $0 < e^{-x^2} \le 1$ while on $(-\infty,-1)$, $0 < e^{-x^2} \le e^x$ and on $(1,\infty)$, $0 < e^{-x^2} < e^{-x}$. Hence $0 \le \int_{-\infty}^{\infty} e^{-x^2}\,dx \le \int_{-\infty}^{-1} e^x\,dx + \int_{-1}^1 dx + \int_1^{\infty} e^{-x}\,dx = 2(e^{-1}+1)$.

(c) Since $\left(\int_{-\infty}^{\infty} e^{-x^2}\,dx\right)\left(\int_{-\infty}^{\infty} e^{-y^2}\,dy\right) = \pi$ and y can be replaced by x, $\left(\int_{-\infty}^{\infty} e^{-x^2}\,dx\right)^2 = \pi$ implies that $\int_{-\infty}^{\infty} e^{-x^2}\,dx = \pm\sqrt{\pi}$. But $e^{-x^2} \ge 0$ for all x, so $\int_{-\infty}^{\infty} e^{-x^2}\,dx = \sqrt{\pi}$.

(d) Letting $t = \sqrt{2}\,x$, $\int_{-\infty}^{\infty} e^{-x^2}\,dx = \int_{-\infty}^{\infty}\frac{1}{\sqrt{2}}\left(e^{-t^2/2}\right)dt$, so that $\sqrt{\pi} = \frac{1}{\sqrt{2}}\int_{-\infty}^{\infty} e^{-t^2/2}\,dt$ or $\int_{-\infty}^{\infty} e^{-t^2/2}\,dt = \sqrt{2\pi}$.

31. (a) We integrate by parts with $u = x$ and $dv = xe^{-x^2}\,dx$. Then $du = dx$ and $v = -\frac{1}{2}e^{-x^2}$, so

$$\int_0^\infty x^2 e^{-x^2}\,dx = \lim_{t\to\infty}\int_0^t x^2 e^{-x^2}\,dx = \lim_{t\to\infty}\left(-\tfrac{1}{2}xe^{-x^2}\Big]_0^t + \int_0^t \tfrac{1}{2}e^{-x^2}\,dx\right)$$

$$= \lim_{t\to\infty}\left(-\tfrac{1}{2}te^{-t^2}\right) + \tfrac{1}{2}\int_0^\infty e^{-x^2}\,dx = 0 + \tfrac{1}{2}\int_0^\infty e^{-x^2}\,dx \qquad \text{[by l'Hospital's Rule]}$$

$$= \tfrac{1}{4}\int_{-\infty}^\infty e^{-x^2}\,dx \qquad \text{[since } e^{-x^2} \text{ is an even function]}$$

$$= \tfrac{1}{4}\sqrt{\pi} \qquad \text{[by Exercise 30(c)]}$$

(b) Let $u = \sqrt{x}$. Then $u^2 = x \;\Rightarrow\; dx = 2u\,du \;\Rightarrow$

$$\int_0^\infty \sqrt{x}e^{-x}\,dx = \lim_{t\to\infty}\int_0^t \sqrt{x}\,e^{-x}\,dx = \lim_{t\to\infty}\int_0^{\sqrt{t}} ue^{-u^2}2u\,du = 2\int_0^\infty u^2e^{-u^2}\,du = 2\left(\tfrac{1}{4}\sqrt{\pi}\right) \;\text{[by part(a)]} = \tfrac{1}{2}\sqrt{\pi}.$$

12.4 Applications of Double Integrals

1. $Q = \iint_D \sigma(x,y)\,dA = \int_1^3\int_0^2 (2xy+y^2)\,dy\,dx = \int_1^3 \left[xy^2 + \frac{1}{3}y^3\right]_{y=0}^{y=2} dx$

$= \int_1^3 \left(4x + \frac{8}{3}\right) dx = \left[2x^2 + \frac{8}{3}x\right]_1^3 = 16 + \frac{16}{3} = \frac{64}{3}$ C

2. $Q = \iint_D \sigma(x,y)\,dA = \iint_D (x+y+x^2+y^2)\,dA = \int_0^{2\pi}\int_0^2 (r\cos\theta + r\sin\theta + r^2)\,r\,dr\,d\theta$

$= \int_0^{2\pi}\int_0^2 \left[r^2(\cos\theta+\sin\theta) + r^3\right] dr\,d\theta = \int_0^{2\pi}\left[\frac{1}{3}r^3(\cos\theta+\sin\theta) + \frac{1}{4}r^4\right]_{r=0}^{r=2} d\theta$

$= \int_0^{2\pi}\left[\frac{8}{3}(\cos\theta+\sin\theta) + 4\right] d\theta = \left[\frac{8}{3}(\sin\theta - \cos\theta) + 4\theta\right]_0^{2\pi} = 8\pi$ C

3. $m = \iint_D \rho(x,y)\,dA = \int_0^2\int_{-1}^1 xy^2\,dy\,dx = \int_0^2 x\,dx\int_{-1}^1 y^2\,dy = \left[\frac{1}{2}x^2\right]_0^2\left[\frac{1}{3}y^3\right]_{-1}^1 = 2\cdot\frac{2}{3} = \frac{4}{3}$,

$\overline{x} = \frac{1}{m}\iint_D x\rho(x,y)\,dA = \frac{3}{4}\int_0^2\int_{-1}^1 x^2y^2\,dy\,dx = \frac{3}{4}\int_0^2 x^2\,dx\int_{-1}^1 y^2\,dy = \frac{3}{4}\left[\frac{1}{3}x^3\right]_0^2\left[\frac{1}{3}y^3\right]_{-1}^1 = \frac{3}{4}\cdot\frac{8}{3}\cdot\frac{2}{3} = \frac{4}{3}$,

$\overline{y} = \frac{1}{m}\iint_D y\rho(x,y)\,dA = \frac{3}{4}\int_0^2\int_{-1}^1 xy^3\,dy\,dx = \frac{3}{4}\int_0^2 x\,dx\int_{-1}^1 y^3\,dy = \frac{3}{4}\left[\frac{1}{2}x^2\right]_0^2\left[\frac{1}{4}y^4\right]_{-1}^1 = \frac{3}{4}\cdot 2\cdot 0 = 0.$

Hence, $(\overline{x},\overline{y}) = \left(\frac{4}{3}, 0\right)$.

4. $m = \iint_D \rho(x,y)\,dA = \int_0^a\int_0^b cxy\,dy\,dx = c\int_0^a x\,dx\int_0^b y\,dy = c\left[\frac{1}{2}x^2\right]_0^a\left[\frac{1}{2}y^2\right]_0^b = \frac{1}{4}a^2b^2c$,

$M_y = \iint_D x\rho(x,y)\,dA = \int_0^a\int_0^b cx^2y\,dy\,dx = c\int_0^a x^2\,dx\int_0^b y\,dy = c\left[\frac{1}{3}x^3\right]_0^a\left[\frac{1}{2}y^2\right]_0^b = \frac{1}{6}a^3b^2c$, and

$M_x = \iint_D y\rho(x,y)\,dA = \int_0^a\int_0^b cxy^2\,dy\,dx = c\int_0^a x\,dx\int_0^b y^2\,dy = c\left[\frac{1}{2}x^2\right]_0^a\left[\frac{1}{3}y^3\right]_0^b = \frac{1}{6}a^2b^3c$.

Hence, $(\overline{x},\overline{y}) = \left(\dfrac{M_y}{m}, \dfrac{M_x}{m}\right) = \left(\frac{2}{3}a, \frac{2}{3}b\right)$.

5. $m = \int_0^2\int_{x/2}^{3-x} (x+y)\,dy\,dx = \int_0^2 \left[xy + \frac{1}{2}y^2\right]_{y=x/2}^{y=3-x} dx = \int_0^2 \left[x\left(3 - \frac{3}{2}x\right) + \frac{1}{2}(3-x)^2 - \frac{1}{8}x^2\right] dx$

$= \int_0^2 \left(-\frac{9}{8}x^2 + \frac{9}{2}\right) dx = \left[-\frac{9}{8}\left(\frac{1}{3}x^3\right) + \frac{9}{2}x\right]_0^2 = 6$,

$M_y = \int_0^2\int_{x/2}^{3-x} (x^2+xy)\,dy\,dx = \int_0^2 \left[x^2y + \frac{1}{2}xy^2\right]_{y=x/2}^{y=3-x} dx = \int_0^2 \left(\frac{9}{2}x - \frac{9}{8}x^3\right) dx = \frac{9}{2}$,

$M_x = \int_0^2\int_{x/2}^{3-y} (xy+y^2)\,dy\,dx = \int_0^2 \left[\frac{1}{2}xy^2 + \frac{1}{3}y^3\right]_{y=x/2}^{y=3-x} dx = \int_0^2 \left(9 - \frac{9}{2}x\right) dx = 9.$

Hence $m = 6$, $(\overline{x},\overline{y}) = \left(\dfrac{M_y}{m}, \dfrac{M_x}{m}\right) = \left(\dfrac{3}{4}, \dfrac{3}{2}\right)$.

6. $m = \int_0^1 \int_y^{4-3y} x\,dx\,dy = \int_0^1 \left[\frac{1}{2}(4-3y)^2 - \frac{1}{2}y^2\right] dy = \left[-\frac{1}{18}(4-3y)^3 - \frac{1}{6}y^3\right]_0^1 = \frac{10}{3}$,

$M_y = \int_0^1 \int_y^{4-3y} x^2\,dx\,dy = \int_0^1 \left[\frac{1}{3}(4-3y)^3 - \frac{1}{3}y^3\right] dy = \left[-\frac{1}{36}(4-3y)^4 - \frac{1}{12}y^4\right]_0^1 = 7$,

$M_x = \int_0^1 \int_y^{4-3y} xy\,dx\,dy = \int_0^1 \left[\frac{1}{2}y(4-3y)^2 - \frac{1}{2}y^3\right] dy = \int_0^1 (8y - 12y^2 + 4y^3)\,dy = 1$.

Hence $m = \frac{10}{3}$, $(\overline{x}, \overline{y}) = (2.1, 0.3)$.

7. $m = \int_0^1 \int_0^{e^x} y\,dy\,dx = \int_0^1 \left[\frac{1}{2}y^2\right]_{y=0}^{y=e^x} dx = \frac{1}{2}\int_0^1 e^{2x}\,dx = \frac{1}{4}e^{2x}\Big]_0^1 = \frac{1}{4}(e^2 - 1)$,

$M_y = \int_0^1 \int_0^{e^x} xy\,dy\,dx = \frac{1}{2}\int_0^1 xe^{2x}\,dx = \frac{1}{2}\left[\frac{1}{2}xe^{2x} - \frac{1}{4}e^{2x}\right]_0^1 = \frac{1}{8}(e^2 + 1)$,

$M_x = \int_0^1 \int_0^{e^x} y^2\,dy\,dx = \int_0^1 \left[\frac{1}{3}y^3\right]_{y=0}^{y=e^x} dx = \frac{1}{3}\int_0^1 e^{3x}\,dx = \frac{1}{3}\left[\frac{1}{3}e^{3x}\right]_0^1 = \frac{1}{9}(e^3 - 1)$.

Hence $m = \frac{1}{4}(e^2 - 1)$, $(\overline{x}, \overline{y}) = \left(\dfrac{\frac{1}{8}(e^2+1)}{\frac{1}{4}(e^2-1)}, \dfrac{\frac{1}{9}(e^3-1)}{\frac{1}{4}(e^2-1)}\right) = \left(\dfrac{e^2+1}{2(e^2-1)}, \dfrac{4(e^3-1)}{9(e^2-1)}\right)$.

8. $m = \int_0^1 \int_0^{\sqrt{x}} x\,dy\,dx = \int_0^1 x\left[y\right]_{y=0}^{y=\sqrt{x}} dx = \int_0^1 x^{3/2}\,dx = \frac{2}{5}x^{5/2}\Big]_0^1 = \frac{2}{5}$,

$M_y = \int_0^1 \int_0^{\sqrt{x}} x^2\,dy\,dx = \int_0^1 x\left[y\right]_{y=0}^{y=\sqrt{x}} dx = \int_0^1 x^{5/2}\,dx = \frac{2}{7}x^{7/2}\Big]_0^1 = \frac{2}{7}$,

$M_x = \int_0^1 \int_0^{\sqrt{x}} yx\,dy\,dx = \int_0^1 x\left[\frac{1}{2}y^2\right]_{y=0}^{y=\sqrt{x}} dx = \frac{1}{2}\int_0^1 x^2\,dx = \frac{1}{2}\left[\frac{1}{3}x^3\right]_0^1 = \frac{1}{6}$.

Hence $m = \frac{2}{5}$, $(\overline{x}, \overline{y}) = \left(\frac{2/7}{2/5}, \frac{1/6}{2/5}\right) = \left(\frac{5}{7}, \frac{5}{12}\right)$.

9.

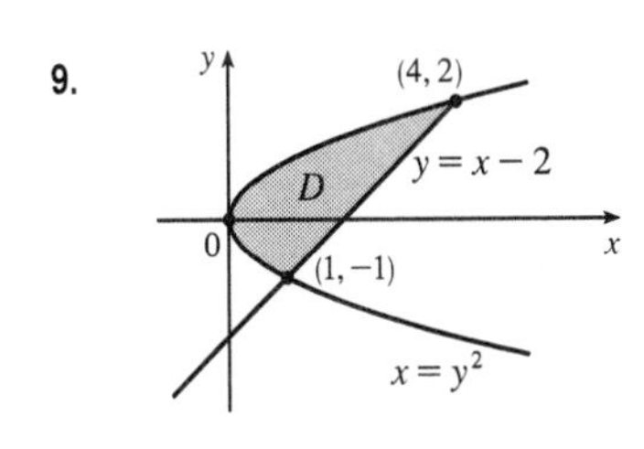

$m = \int_{-1}^2 \int_{y^2}^{y+2} 3\,dx\,dy = \int_{-1}^2 (3y + 6 - 3y^2)\,dy = \frac{27}{2}$,

$M_y = \int_{-1}^2 \int_{y^2}^{y+2} 3x\,dx\,dy = \int_{-1}^2 \frac{3}{2}\left[(y+2)^2 - y^4\right] dy$
$= \left[\frac{1}{2}(y+2)^3 - \frac{3}{10}y^5\right]_{-1}^2 = \frac{108}{5}$

$M_x = \int_{-1}^2 \int_{y^2}^{y+2} 3y\,dx\,dy = \int_{-1}^2 (3y^2 + 6y - 3y^3)\,dy$
$= \left[y^3 + 3y^2 - \frac{3}{4}y^4\right]_{-1}^2 = \frac{27}{4}$

Hence $m = \frac{27}{2}$, $(\overline{x}, \overline{y}) = \left(\frac{8}{5}, \frac{1}{2}\right)$.

10. $m = \int_0^{\pi/2} \int_0^{\cos x} x\,dy\,dx = \int_0^{\pi/2} x\cos x\,dx = \left[x\sin x + \cos x\right]_0^{\pi/2} = \frac{\pi}{2} - 1$,

$M_y = \int_0^{\pi/2} \int_0^{\cos x} x^2\,dy\,dx = \int_0^{\pi/2} x^2\cos x\,dx = \left[x^2\sin x + 2x\cos x - 2\sin x\right]_0^{\pi/2} = \frac{\pi^2}{4} - 2$,

$M_x = \int_0^{\pi/2} \int_0^{\cos x} xy\,dy\,dx = \int_0^{\pi/2} \frac{1}{2}x\cos^2 x\,dx = \frac{1}{2}\left[\frac{1}{4}x^2 + \frac{1}{4}x\sin 2x + \frac{1}{8}\cos 2x\right]_0^{\pi/2} = \frac{\pi^2}{32} - \frac{1}{8}$.

Hence $m = \dfrac{\pi - 2}{2}$, $(\overline{x}, \overline{y}) = \left(\dfrac{\pi^2 - 8}{2(\pi - 2)}, \dfrac{\pi + 2}{16}\right)$.

11. $\rho(x, y) = ky = kr\sin\theta$, $m = \int_0^{\pi/2} \int_0^1 kr^2\sin\theta\,dr\,d\theta = \frac{1}{3}k\int_0^{\pi/2}\sin\theta\,d\theta = \frac{1}{3}k\left[-\cos\theta\right]_0^{\pi/2} = \frac{1}{3}k$,

$M_y = \int_0^{\pi/2} \int_0^1 kr^3\sin\theta\cos\theta\,dr\,d\theta = \frac{1}{4}k\int_0^{\pi/2}\sin\theta\cos\theta\,d\theta = \frac{1}{8}k\left[-\cos 2\theta\right]_0^{\pi/2} = \frac{1}{8}k$,

$M_x = \int_0^{\pi/2} \int_0^1 kr^3\sin^2\theta\,dr\,d\theta = \frac{1}{4}k\int_0^{\pi/2}\sin^2\theta\,d\theta = \frac{1}{8}k\left[\theta + \sin 2\theta\right]_0^{\pi/2} = \frac{\pi}{16}k$.

Hence $(\overline{x}, \overline{y}) = \left(\frac{3}{8}, \frac{3\pi}{16}\right)$.

12. $\rho(x,y) = k(x^2+y^2) = kr^2$, $m = \int_0^{\pi/2}\int_0^1 kr^3\,dr\,d\theta = \frac{\pi}{8}k$,

$M_y = \int_0^{\pi/2}\int_0^1 kr^4\cos\theta\,dr\,d\theta = \frac{1}{5}k\int_0^{\pi/2}\cos\theta\,d\theta = \frac{1}{5}k\big[\sin\theta\big]_0^{\pi/2} = \frac{1}{5}k$,

$M_x = \int_0^{\pi/2}\int_0^1 kr^4\sin\theta\,dr\,d\theta = \frac{1}{5}k\int_0^{\pi/2}\sin\theta\,d\theta = \frac{1}{5}k\big[-\cos\theta\big]_0^{\pi/2} = \frac{1}{5}k$.

Hence $(\overline{x},\overline{y}) = \left(\frac{8}{5\pi},\frac{8}{5\pi}\right)$.

13. Placing the vertex opposite the hypotenuse at $(0,0)$, $\rho(x,y) = k(x^2+y^2)$. Then

$m = \int_0^a\int_0^{a-x} k(x^2+y^2)\,dy\,dx = k\int_0^a \left[ax^2 - x^3 + \frac{1}{3}(a-x)^3\right]dx = k\left[\frac{1}{3}ax^3 - \frac{1}{4}x^4 - \frac{1}{12}(a-x)^4\right]_0^a = \frac{1}{6}ka^4$.

By symmetry,

$$\begin{aligned} M_y = M_x &= \int_0^a\int_0^{a-x} ky(x^2+y^2)\,dy\,dx = k\int_0^a\left[\tfrac{1}{2}(a-x)^2x^2 + \tfrac{1}{4}(a-x)^4\right]dx \\ &= k\left[\tfrac{1}{6}a^2x^3 - \tfrac{1}{4}ax^4 + \tfrac{1}{10}x^5 - \tfrac{1}{20}(a-x)^5\right]_0^a = \tfrac{1}{15}ka^5 \end{aligned}$$

Hence $(\overline{x},\overline{y}) = \left(\frac{2}{5}a,\frac{2}{5}a\right)$.

14. $\rho(x,y) = k/\sqrt{x^2+y^2} = k/r$.

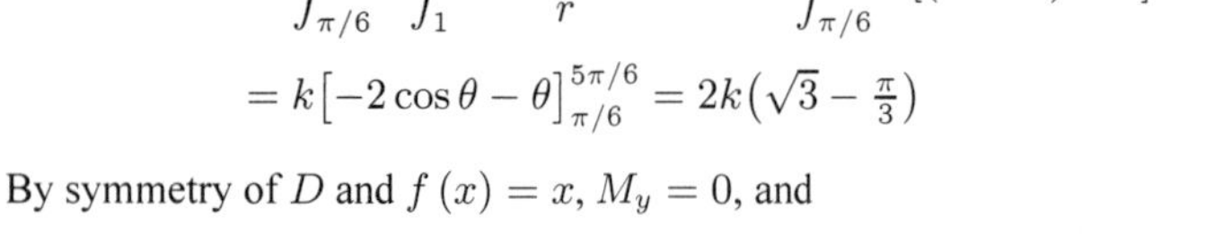

$$\begin{aligned} m &= \int_{\pi/6}^{5\pi/6}\int_1^{2\sin\theta}\frac{k}{r}\,r\,dr\,d\theta = k\int_{\pi/6}^{5\pi/6}[(2\sin\theta)-1]\,d\theta \\ &= k\big[-2\cos\theta-\theta\big]_{\pi/6}^{5\pi/6} = 2k\left(\sqrt{3}-\tfrac{\pi}{3}\right) \end{aligned}$$

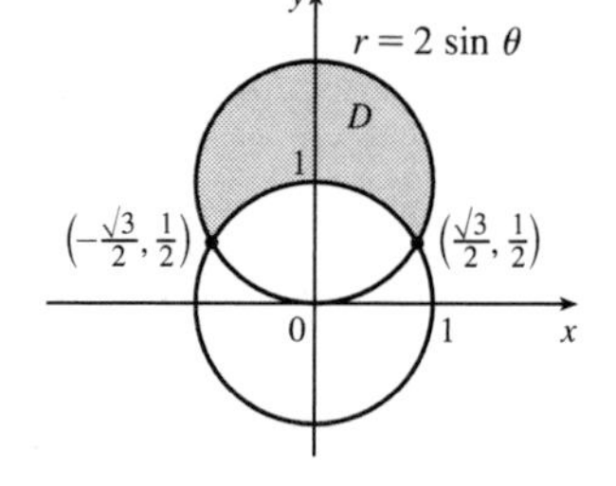

By symmetry of D and $f(x) = x$, $M_y = 0$, and

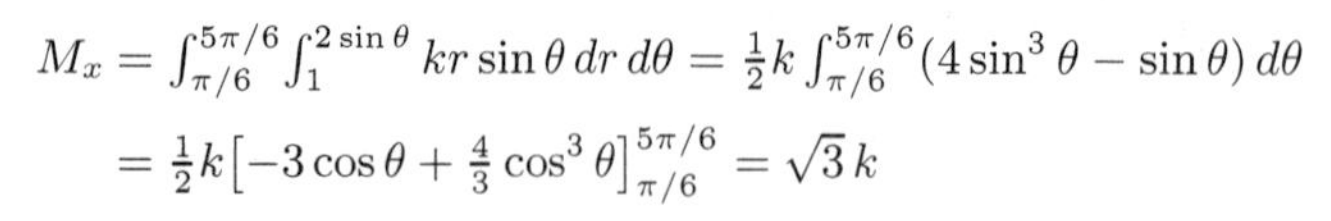

$$\begin{aligned} M_x &= \int_{\pi/6}^{5\pi/6}\int_1^{2\sin\theta} kr\sin\theta\,dr\,d\theta = \tfrac{1}{2}k\int_{\pi/6}^{5\pi/6}(4\sin^3\theta - \sin\theta)\,d\theta \\ &= \tfrac{1}{2}k\left[-3\cos\theta + \tfrac{4}{3}\cos^3\theta\right]_{\pi/6}^{5\pi/6} = \sqrt{3}\,k \end{aligned}$$

Hence $(\overline{x},\overline{y}) = \left(0, \dfrac{3\sqrt{3}}{2(3\sqrt{3}-\pi)}\right)$

15. $I_x = \iint_D y^2\rho(x,y)dA = \int_0^1\int_0^{e^x} y^2\cdot y\,dy\,dx = \int_0^1\left[\frac{1}{4}y^4\right]_{y=0}^{y=e^x}dx = \frac{1}{4}\int_0^1 e^{4x}\,dx = \frac{1}{4}\left[\frac{1}{4}e^{4x}\right]_0^1 = \frac{1}{16}(e^4-1)$,

$I_y = \iint_D x^2\rho(x,y)\,dA = \int_0^1\int_0^{e^x} x^2y\,dy\,dx = \int_0^1 x^2\left[\frac{1}{2}y^2\right]_{y=0}^{y=e^x}dx = \frac{1}{2}\int_0^1 x^2e^{2x}\,dx$

$= \frac{1}{2}\left[\left(\frac{1}{2}x^2 - \frac{1}{2}x + \frac{1}{4}\right)e^{2x}\right]_0^1$ [integrate by parts twice] $= \frac{1}{8}(e^2-1)$,

and $I_0 = I_x + I_y = \frac{1}{16}(e^4-1) + \frac{1}{8}(e^2-1) = \frac{1}{16}(e^4+2e^2-3)$.

16. $I_x = \int_0^{\pi/2}\int_0^1 (r^2\sin^2\theta)(kr^2)\,r\,dr\,d\theta = \frac{1}{6}k\int_0^{\pi/2}\sin^2\theta\,d\theta = \frac{1}{6}k\left[\frac{1}{4}(2\theta - \sin 2\theta)\right]_0^{\pi/2} = \frac{\pi}{24}k$,

$I_y = \int_0^{\pi/2}\int_0^1 (r^2\cos^2\theta)(kr^2)\,r\,dr\,d\theta = \frac{1}{6}k\int_0^{\pi/6}\cos^2\theta\,d\theta = \frac{1}{6}k\left[\frac{1}{4}(2\theta + \sin 2\theta)\right]_0^{\pi/2} = \frac{\pi}{24}k$,

and $I_0 = I_x + I_y = \frac{\pi}{12}k$.

17. $I_x = \int_{-1}^2\int_{y^2}^{y+2} 3y^2\,dx\,dy = \int_{-1}^2 (3y^3 + 6y^2 - 3y^4)\,dy = \left[\frac{3}{4}y^4 + 2y^3 - \frac{3}{5}y^5\right]_{-1}^2 = \frac{189}{20}$,

$I_y = \int_{-1}^2\int_{y^2}^{y+2} 3x^2\,dx\,dy = \int_{-1}^2\left[(y+2)^3 - y^6\right]dy = \left[\frac{1}{4}(y+2)^4 - \frac{1}{7}y^7\right]_{-1}^2 = \frac{1269}{28}$, and $I_0 = I_x + I_y = \frac{1917}{35}$.

18. If we find the moments of inertia about the x- and y-axes, we can determine in which direction rotation will be more difficult. (See the explanation following Example 4.) The moment of inertia about the x-axis is given by

$$I_x = \iint_D y^2 \rho(x,y)\,dA = \int_0^2 \int_0^2 y^2(1+0.1x)\,dy\,dx = \int_0^2 (1+0.1x)\left[\tfrac{1}{3}y^3\right]_{y=0}^{y=2} dx$$
$$= \tfrac{8}{3}\int_0^2 (1+0.1x)\,dx = \tfrac{8}{3}\left[x + 0.1\cdot\tfrac{1}{2}x^2\right]_0^2 = \tfrac{8}{3}(2.2) \approx 5.87$$

Similarly, the moment of inertia about the y-axis is given by

$$I_y = \iint_D x^2 \rho(x,y)\,dA = \int_0^2 \int_0^2 x^2(1+0.1x)\,dy\,dx = \int_0^2 x^2(1+0.1x)\left[\,y\,\right]_{y=0}^{y=2} dx$$
$$= 2\int_0^2 (x^2+0.1x^3)\,dx = 2\left[\tfrac{1}{3}x^3 + 0.1\cdot\tfrac{1}{4}x^4\right]_0^2 = 2\left(\tfrac{8}{3}+0.4\right) \approx 6.13$$

Since $I_y > I_x$, more force is required to rotate the fan blade about the y-axis.

19. Using a CAS, we find $m = \iint_D \rho(x,y)\,dA = \int_0^\pi \int_0^{\sin x} xy\,dy\,dx = \dfrac{\pi^2}{8}$. Then

$$\overline{x} = \frac{1}{m}\iint_D x\rho(x,y)\,dA = \frac{8}{\pi^2}\int_0^\pi \int_0^{\sin x} x^2 y\,dy\,dx = \frac{2\pi}{3} - \frac{1}{\pi} \text{ and}$$

$$\overline{y} = \frac{1}{m}\iint_D y\rho(x,y)\,dA = \frac{8}{\pi^2}\int_0^\pi \int_0^{\sin x} xy^2\,dy\,dx = \frac{16}{9\pi}, \text{ so } (\overline{x},\overline{y}) = \left(\frac{2\pi}{3} - \frac{1}{\pi}, \frac{16}{9\pi}\right).$$

The moments of inertia are $I_x = \iint_D y^2\rho(x,y)\,dA = \int_0^\pi \int_0^{\sin x} xy^3\,dy\,dx = \dfrac{3\pi^2}{64}$,

$I_y = \iint_D x^2\rho(x,y)\,dA = \int_0^\pi \int_0^{\sin x} x^3 y\,dy\,dx = \dfrac{\pi^2}{16}(\pi^2-3)$, and $I_0 = I_x + I_y = \dfrac{\pi^2}{64}(4\pi^2-9)$.

20. Using a CAS, we find $m = \iint_D \sqrt{x^2+y^2}\,dA = \int_0^{2\pi} \int_0^{1+\cos\theta} r^2\,dr\,d\theta = \frac{5}{3}\pi$,

$\overline{x} = \frac{1}{m}\iint_D x\sqrt{x^2+y^2}\,dA = \frac{3}{5\pi}\int_0^{2\pi}\int_0^{1+\cos\theta} r^3\cos\theta\,dr\,d\theta = \frac{21}{20}$ and

$\overline{y} = \frac{1}{m}\iint_D y\sqrt{x^2+y^2}\,dA = \frac{3}{5\pi}\int_0^{2\pi}\int_0^{1+\cos\theta} r^3\sin\theta\,dr\,d\theta = 0$, so $(\overline{x},\overline{y}) = \left(\frac{21}{20}, 0\right)$.

The moments of inertia are $I_x = \iint_D y^2\sqrt{x^2+y^2}\,dA = \int_0^{2\pi}\int_0^{1+\cos\theta} r^4\sin^2\theta\,dr\,d\theta = \frac{33}{40}\pi$,

$I_y = \iint_D x^2\sqrt{x^2+y^2}\,dA = \int_0^{2\pi}\int_0^{1+\cos\theta} r^4\cos^2\theta\,dr\,d\theta = \frac{93}{40}\pi$, and $I_0 = I_x + I_y = \frac{63}{20}\pi$.

21. $I_x = \int_0^a \int_0^a \rho y^2\,dx\,dy = \rho\int_0^a dx \int_0^a y^2\,dy = \rho\left[x\right]_0^a\left[\frac{1}{3}y^3\right]_0^a = \rho a\left(\frac{1}{3}a^3\right) = \frac{1}{3}\rho a^4 = I_y$ by symmetry, and $m = \rho a^2$ since the lamina is homogeneous. Hence $\overline{\overline{x}}^2 = \dfrac{I_y}{m} \Rightarrow \overline{\overline{x}} = \left[\left(\frac{1}{3}\rho a^4\right)/\left(\rho a^2\right)\right]^{1/2} = \frac{1}{\sqrt{3}}a$ and $\overline{\overline{y}}^2 = \dfrac{I_x}{m} \Rightarrow \overline{\overline{y}} = \frac{1}{\sqrt{3}}a$.

22. $m = \int_0^\pi \int_0^{\sin x} \rho\,dy\,dx = \rho\int_0^\pi \sin x\,dx = \rho\left[-\cos x\right]_0^\pi = 2\rho$,

$I_x = \int_0^\pi \int_0^{\sin x} \rho y^2\,dy\,dx = \frac{1}{3}\rho\int_0^\pi \sin^3 x\,dx = \frac{1}{3}\rho\int_0^\pi (1-\cos^2 x)\sin x\,dx = \frac{1}{3}\rho\left[-\cos\theta + \frac{1}{3}\cos^3\theta\right]_0^\pi = \frac{4}{9}\rho$,

$I_y = \int_0^\pi \int_0^{\sin x} \rho x^2\,dy\,dx = \rho\int_0^\pi x^2\sin x\,dx = \rho\left[-x^2\cos x + 2x\sin x + 2\cos x\right]_0^\pi$ [by integrating by parts twice]

$= \rho(\pi^2 - 4)$.

Then $\overline{\overline{y}}^2 = \dfrac{I_x}{m} = \dfrac{2}{9}$, so $\overline{\overline{y}} = \dfrac{\sqrt{2}}{3}$ and $\overline{\overline{x}}^2 = \dfrac{I_y}{m} = \dfrac{\pi^2-4}{2}$, so $\overline{\overline{x}} = \sqrt{\dfrac{\pi^2-4}{2}}$.

12.5 Triple Integrals

1. $\iiint_B xyz^2\,dV = \int_0^1\int_{-1}^2\int_0^3 xyz^2\,dz\,dx\,dy = \int_0^1\int_{-1}^2 xy\left[\frac{1}{3}z^3\right]_{z=0}^{z=3}dx\,dy = \int_0^1\int_{-1}^2 9xy\,dx\,dy$

$= \int_0^1\left[\frac{9}{2}x^2y\right]_{x=-1}^{x=2}dy = \int_0^1 \frac{27}{2}\,y\,dy = \frac{27}{4}y^2\Big]_0^1 = \frac{27}{4}$

2. There are six different possible orders of integration.

$\iiint_E (xz - y^3)\,dV = \int_{-1}^1\int_0^2\int_0^1 (xz-y^3)\,dz\,dy\,dx = \int_{-1}^1\int_0^2\left[\frac{1}{2}xz^2 - y^3z\right]_{z=0}^{z=1}dy\,dx = \int_{-1}^1\int_0^2\left(\frac{1}{2}x - y^3\right)dy\,dx$

$= \int_{-1}^1\left[\frac{1}{2}xy - \frac{1}{4}y^4\right]_{y=0}^{y=2}dx = \int_{-1}^1(x-4)\,dx = \left[\frac{1}{2}x^2 - 4x\right]_{-1}^1 = -8$

$\iiint_E (xz - y^3)\,dV = \int_0^2\int_{-1}^1\int_0^1 (xz-y^3)\,dz\,dx\,dy = \int_0^2\int_{-1}^1\left[\frac{1}{2}xz^2 - y^3z\right]_{z=0}^{z=1}dx\,dy$

$= \int_0^2\int_{-1}^1\left(\frac{1}{2}x - y^3\right)dx\,dy = \int_0^2\left[\frac{1}{4}x^2 - xy^3\right]_{x=-1}^{x=1}dy = \int_0^2 -2y^3\,dy = -\frac{1}{2}y^4\Big]_0^2 = -8$

$\iiint_E (xz - y^3)\,dV = \int_{-1}^1\int_0^1\int_0^2 (xz-y^3)\,dy\,dz\,dx = \int_{-1}^1\int_0^1\left[xyz - \frac{1}{4}y^4\right]_{y=0}^{y=2}dz\,dx$

$= \int_{-1}^1\int_0^1 (2xz-4)\,dz\,dx = \int_{-1}^1\left[xz^2 - 4z\right]_{z=0}^{z=1}dx = \int_{-1}^1(x-4)\,dx = \left[\frac{1}{2}x^2 - 4x\right]_{-1}^1 = -8$

$\iiint_E (xz - y^3)\,dV = \int_0^1\int_{-1}^1\int_0^2 (xz-y^3)\,dy\,dx\,dz = \int_0^1\int_{-1}^1\left[xyz - \frac{1}{4}y^4\right]_{y=0}^{y=2}dx\,dz$

$= \int_0^1\int_{-1}^1 (2xz-4)\,dx\,dz = \int_0^1\left[x^2z - 4x\right]_{x=-1}^{x=1}dz = \int_0^1 -8\,dz = -8z\Big]_0^1 = -8$

$\iiint_E (xz - y^3)\,dV = \int_0^2\int_0^1\int_{-1}^1 (xz-y^3)\,dx\,dz\,dy = \int_0^2\int_0^1\left[\frac{1}{2}x^2z - xy^3\right]_{x=-1}^{x=1}dz\,dy$

$= \int_0^2\int_0^1 -2y^3\,dz\,dy = \int_0^2\left[-2y^3z\right]_{z=0}^{z=1}dy = \int_0^2 -2y^3\,dy = -\frac{1}{2}y^4\Big]_0^2 = -8$

$\iiint_E (xz - y^3)\,dV = \int_0^1\int_0^2\int_{-1}^1 (xz-y^3)\,dx\,dy\,dz = \int_0^1\int_0^2\left[\frac{1}{2}x^2z - xy^3\right]_{x=-1}^{x=1}dy\,dz$

$= \int_0^1\int_0^2 -2y^3\,dy\,dz = \int_0^1\left[-\frac{1}{2}y^4\right]_{y=0}^{y=2}dz = \int_0^1 -8\,dz = -8z\Big]_0^1 = -8$

3. $\int_0^1\int_0^z\int_0^{x+z} 6xz\,dy\,dx\,dz = \int_0^1\int_0^z\left[6xyz\right]_{y=0}^{y=x+z}dx\,dz = \int_0^1\int_0^z 6xz(x+z)\,dx\,dz$

$= \int_0^1\left[2x^3z + 3x^2z^2\right]_{x=0}^{x=z}dz = \int_0^1(2z^4+3z^4)\,dz = \int_0^1 5z^4\,dz = z^5\Big]_0^1 = 1$

4. $\int_0^1\int_x^{2x}\int_0^y 2xyz\,dz\,dy\,dx = \int_0^1\int_x^{2x}\left[xyz^2\right]_{z=0}^{z=y}dy\,dx = \int_0^1\int_x^{2x} xy^3\,dy\,dx$

$= \int_0^1\left[\frac{1}{4}xy^4\right]_{y=x}^{y=2x}dx = \int_0^1 \frac{15}{4}x^5\,dx = \frac{5}{8}x^6\Big]_0^1 = \frac{5}{8}$

5. $\int_0^3\int_0^1\int_0^{\sqrt{1-z^2}} ze^y\,dx\,dz\,dy = \int_0^3\int_0^1\left[xze^y\right]_{x=0}^{x=\sqrt{1-z^2}}dz\,dy = \int_0^3\int_0^1 ze^y\sqrt{1-z^2}\,dz\,dy$

$= \int_0^3\left[-\frac{1}{3}(1-z^2)^{3/2}e^y\right]_{z=0}^{z=1}dy = \int_0^3\frac{1}{3}e^y\,dy = \frac{1}{3}e^y\Big]_0^3 = \frac{1}{3}(e^3-1)$

6. $\int_0^1 \int_0^z \int_0^y ze^{-y^2}\, dx\, dy\, dz = \int_0^1 \int_0^z \left[xze^{-y^2} \right]_{x=0}^{x=y} dy\, dz = \int_0^1 \int_0^z yze^{-y^2}\, dy\, dz = \int_0^1 \left[-\frac{1}{2} ze^{-y^2} \right]_{y=0}^{y=z} dz$

$= \int_0^1 -\frac{1}{2} z \left(e^{-z^2} - 1 \right) dz = \frac{1}{2} \int_0^1 \left(z - ze^{-z^2} \right) dz$

$= \frac{1}{2} \left[\frac{1}{2} z^2 + \frac{1}{2} e^{-z^2} \right]_0^1 = \frac{1}{4} (1 + e^{-1} - 0 - 1) = \frac{1}{4e}$

7. $\iiint_E 2x\, dV = \int_0^2 \int_0^{\sqrt{4-y^2}} \int_0^y 2x\, dz\, dx\, dy = \int_0^2 \int_0^{\sqrt{4-y^2}} [2xz]_{z=0}^{z=y}\, dx\, dy = \int_0^2 \int_0^{\sqrt{4-y^2}} 2xy\, dx\, dy$

$= \int_0^2 \left[x^2 y \right]_{x=0}^{x=\sqrt{4-y^2}} dy = \int_0^2 (4 - y^2) y\, dy = \left[2y^2 - \frac{1}{4} y^4 \right]_0^2 = 4$

8. $\iiint_E yz \cos(x^5)\, dV = \int_0^1 \int_0^x \int_x^{2x} yz \cos(x^5)\, dz\, dy\, dx = \int_0^1 \int_0^x \left[\frac{1}{2} yz^2 \cos(x^5) \right]_{z=x}^{z=2x} dy\, dx$

$= \frac{1}{2} \int_0^1 \int_0^x 3x^2 y \cos(x^5)\, dy\, dx = \frac{1}{2} \int_0^1 \left[\frac{3}{2} x^2 y^2 \cos(x^5) \right]_{y=0}^{y=x} dx$

$= \frac{3}{4} \int_0^1 x^4 \cos(x^5)\, dx = \frac{3}{4} \left[\frac{1}{5} \sin(x^5) \right]_0^1 = \frac{3}{20} (\sin 1 - \sin 0) = \frac{3}{20} \sin 1$

9. Here $E = \{(x, y, z) \mid 0 \le x \le 1, 0 \le y \le \sqrt{x}, 0 \le z \le 1 + x + y\}$, so

$\iiint_E 6xy\, dV = \int_0^1 \int_0^{\sqrt{x}} \int_0^{1+x+y} 6xy\, dz\, dy\, dx = \int_0^1 \int_0^{\sqrt{x}} [6xyz]_{z=0}^{z=1+x+y} dy\, dx = \int_0^1 \int_0^{\sqrt{x}} 6xy(1 + x + y)\, dy\, dx$

$= \int_0^1 \left[3xy^2 + 3x^2y^2 + 2xy^3 \right]_{y=0}^{y=\sqrt{x}} dx = \int_0^1 (3x^2 + 3x^3 + 2x^{5/2})\, dx = \left[x^3 + \frac{3}{4} x^4 + \frac{4}{7} x^{7/2} \right]_0^1 = \frac{65}{28}$

10. Here E is the region in the first octant that lies below the plane $2x + 2y + z = 4$ (and above the region in the xy-plane bounded by the lines $x = 0$, $y = 0$, $x + y = 2$). So

$\iiint_E y\, dV = \int_0^2 \int_0^{2-x} \int_0^{4-2x-2y} y\, dz\, dy\, dx = \int_0^2 \int_0^{2-x} y(4 - 2x - 2y)\, dy\, dx = \int_0^2 \int_0^{2-x} (4y - 2xy - 2y^2)\, dy\, dx$

$= \int_0^2 \left[2y^2 - xy^2 - \frac{2}{3} y^3 \right]_{y=0}^{y=2-x} dx = \int_0^2 \left[2(2-x)^2 - x(2-x)^2 - \frac{2}{3}(2-x)^3 \right] dx$

$= \int_0^2 \left[(2-x)(2-x)^2 - \frac{2}{3}(2-x)^3 \right] dx = \frac{1}{3} \int_0^2 (2-x)^3\, dx$

$= \frac{1}{3} \left[-\frac{1}{4}(2-x)^4 \right]_0^2 = -\frac{1}{12}(0 - 16) = \frac{4}{3}$

11. Here E is the region that lies below the plane with x-, y-, and z-intercepts 1, 2, and 3 respectively, that is, below the plane $2z + 6x + 3y = 6$ and above the region in the xy-plane bounded by the lines $x = 0$, $y = 0$ and $6x + 3y = 6$. So

$\iiint_E xy\, dV = \int_0^1 \int_0^{2-2x} \int_0^{3-3x-3y/2} xy\, dz\, dy\, dx = \int_0^1 \int_0^{2-2x} \left(3xy - 3x^2y - \frac{3}{2} xy^2 \right) dy\, dx$

$= \int_0^1 \left[\frac{3}{2} xy^2 - \frac{3}{2} x^2y^2 - \frac{1}{2} xy^3 \right]_{y=0}^{y=2-2x} dx = \int_0^1 (2x - 6x^2 + 6x^3 - 2x^4)\, dx$

$= \left[x^2 - 2x^3 + \frac{3}{2} x^4 - \frac{2}{5} x^5 \right]_0^1 = \frac{1}{10}$

12.

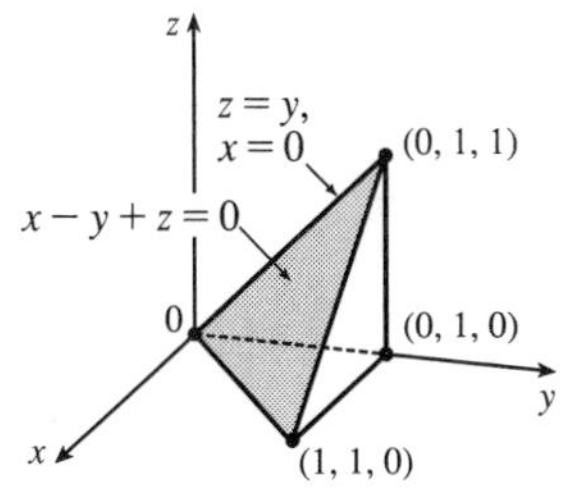

$\int_0^1 \int_0^y \int_0^{y-z} xz\, dx\, dz\, dy = \int_0^1 \int_0^y \frac{1}{2}(y - z)^2 z\, dz\, dy$

$= \frac{1}{2} \int_0^1 \left[\frac{1}{2} y^2 z^2 - \frac{2}{3} yz^3 + \frac{1}{4} z^4 \right]_{z=0}^{z=y} dy$

$= \frac{1}{24} \int_0^1 y^4\, dy = \frac{1}{24} \left[\frac{1}{5} y^5 \right]_0^1 = \frac{1}{120}$

13.

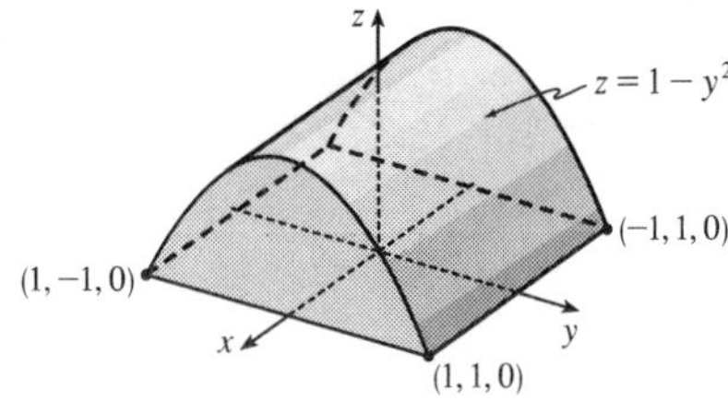

E is the region below the parabolic cylinder $z = 1 - y^2$ and above the square $[-1, 1] \times [-1, 1]$ in the xy-plane.

$$\iiint_E x^2 e^y \, dV = \int_{-1}^{1} \int_{-1}^{1} \int_0^{1-y^2} x^2 e^y \, dz \, dy \, dx = \int_{-1}^{1} \int_{-1}^{1} x^2 e^y (1 - y^2) \, dy \, dx$$
$$= \int_{-1}^{1} x^2 \, dx \int_{-1}^{1} (e^y - y^2 e^y) \, dy$$
$$= \left[\tfrac{1}{3}x^3\right]_{-1}^{1} \left[e^y - (y^2 - 2y + 2)e^y\right]_{-1}^{1} \qquad \begin{bmatrix}\text{integrate by} \\ \text{parts twice}\end{bmatrix}$$
$$= \tfrac{1}{3}(2)[e - e - e^{-1} + 5e^{-1}] = \tfrac{8}{3e}$$

14.

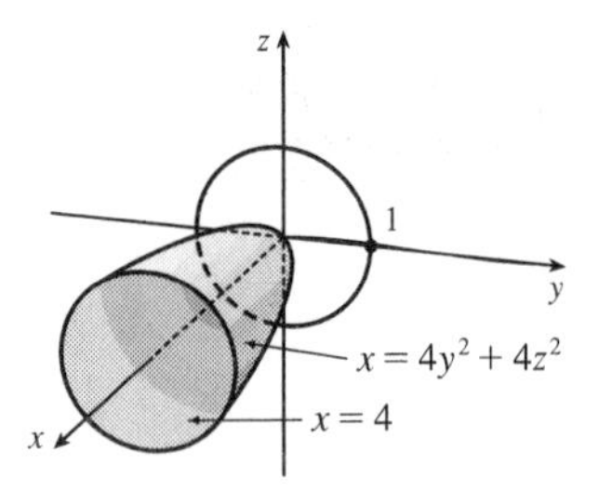

E is the solid above the region shown in the xy-plane and below the plane $z = x$. Thus,

$$\iiint_E (x + 2y) \, dV = \int_0^1 \int_{x^2}^{x} \int_0^x (x + 2y) \, dz \, dy \, dx$$
$$= \int_0^1 \int_{x^2}^{x} (x^2 + 2yx) \, dy \, dx = \int_0^1 \left[x^2 y + xy^2\right]_{y=x^2}^{y=x} dx$$
$$= \int_0^1 (2x^3 - x^4 - x^5) \, dx = \left[\tfrac{1}{2}x^4 - \tfrac{1}{5}x^5 - \tfrac{1}{6}x^6\right]_0^1 = \tfrac{2}{15}$$

15.

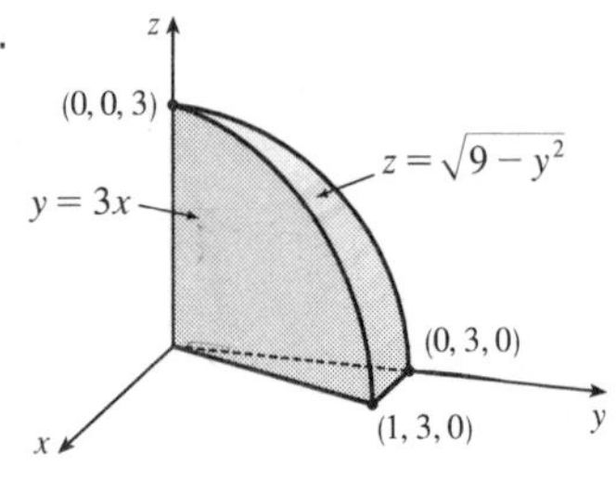

The projection E on the yz-plane is the disk $y^2 + z^2 \le 1$. Using polar coordinates $y = r\cos\theta$ and $z = r\sin\theta$, we get

$$\iiint_E x \, dV = \iint_D \left[\int_{4y^2 + 4z^2}^{4} x \, dx\right] dA = \tfrac{1}{2} \iint_D \left[4^2 - (4y^2 + 4z^2)^2\right] dA$$
$$= 8 \int_0^{2\pi} \int_0^1 (1 - r^4) \, r \, dr \, d\theta = 8 \int_0^{2\pi} d\theta \int_0^1 (r - r^5) \, dr$$
$$= 8(2\pi)\left[\tfrac{1}{2}r^2 - \tfrac{1}{6}r^6\right]_0^1 = \tfrac{16\pi}{3}$$

16.

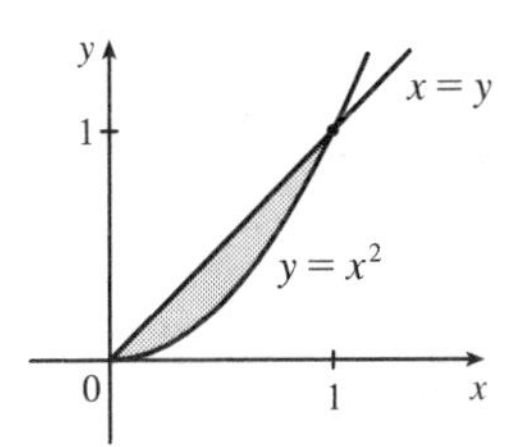

$$\int_0^1 \int_{3x}^{3} \int_0^{\sqrt{9-y^2}} z \, dz \, dy \, dx = \int_0^1 \int_{3x}^{3} \tfrac{1}{2}(9 - y^2) \, dy \, dx$$
$$= \int_0^1 \left[\tfrac{9}{2}y - \tfrac{1}{6}y^3\right]_{y=3x}^{y=3} dx$$
$$= \int_0^1 \left[9 - \tfrac{27}{2}x + \tfrac{9}{2}x^3\right] dx$$
$$= \left[9x - \tfrac{27}{4}x^2 + \tfrac{9}{8}x^4\right]_0^1 = \tfrac{27}{8}$$

17. The plane $2x + y + z = 4$ intersects the xy-plane when $2x + y + 0 = 4 \;\Rightarrow\; y = 4 - 2x$, so $E = \{(x, y, z) \mid 0 \le x \le 2, 0 \le y \le 4 - 2x, 0 \le z \le 4 - 2x - y\}$ and

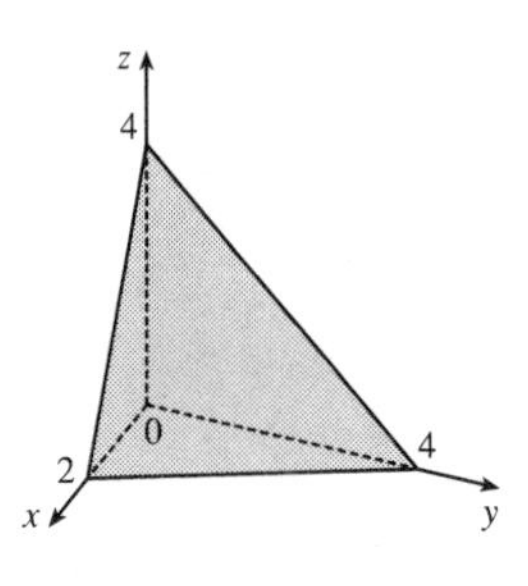

$$V = \int_0^2 \int_0^{4-2x} \int_0^{4-2x-y} dz \, dy \, dx = \int_0^2 \int_0^{4-2x} (4 - 2x - y) \, dy \, dx$$
$$= \int_0^2 \left[4y - 2xy - \tfrac{1}{2}y^2\right]_{y=0}^{y=4-2x} dx$$
$$= \int_0^2 \left[4(4 - 2x) - 2x(4 - 2x) - \tfrac{1}{2}(4 - 2x)^2\right] dx$$
$$= \int_0^2 (2x^2 - 8x + 8) \, dx = \left[\tfrac{2}{3}x^3 - 4x^2 + 8x\right]_0^2 = \tfrac{16}{3}$$

18.

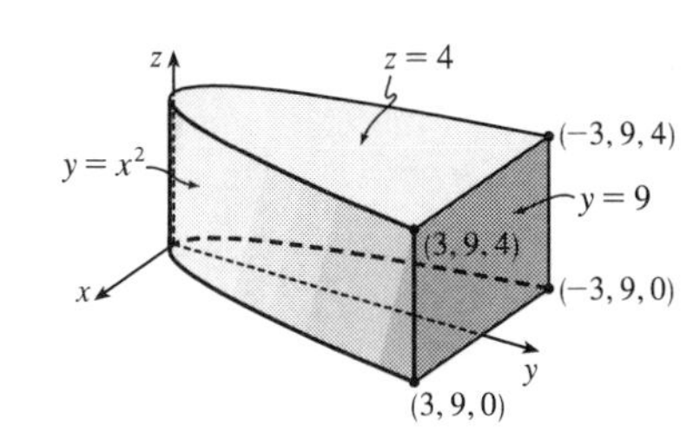

$V = \iiint_E dV = \int_{-3}^{3} \int_{x^2}^{9} \int_0^4 dz\,dy\,dx$

$= 4\int_{-3}^{3} \int_{x^2}^{9} dy\,dx = 4\int_{-3}^{3} \left(9 - x^2\right) dx$

$= 4\left[9x - \frac{1}{3}x^3\right]_{-3}^{3} = 4(27 - 9 + 27 - 9) = 144$

19. $V = \displaystyle\int_{-3}^{3}\int_{-\sqrt{9-x^2}}^{\sqrt{9-x^2}}\int_1^{5-y} dz\,dy\,dx = \int_{-3}^{3}\int_{-\sqrt{9-x^2}}^{\sqrt{9-x^2}} (5 - y - 1)\,dy\,dx = \int_{-3}^{3}\left[4y - \tfrac{1}{2}y^2\right]_{y=-\sqrt{9-x^2}}^{y=\sqrt{9-x^2}} dx$

$= \int_{-3}^{3} 8\sqrt{9 - x^2}\,dx = 8\left[\frac{x}{2}\sqrt{9 - x^2} + \frac{9}{2}\sin^{-1}\left(\frac{x}{3}\right)\right]_{-3}^{3}$ [using trigonometric substitution or Formula 30 in the Table of Integrals]

$= 8\left[\frac{9}{2}\sin^{-1}(1) - \frac{9}{2}\sin^{-1}(-1)\right] = 36\left(\frac{\pi}{2} - \left(-\frac{\pi}{2}\right)\right) = 36\pi$

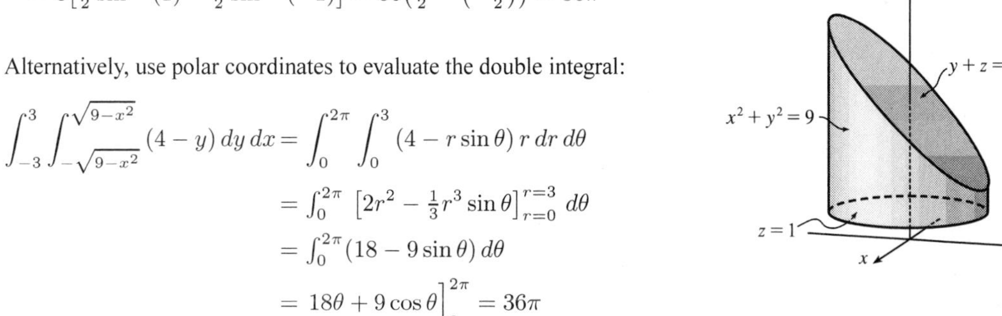

Alternatively, use polar coordinates to evaluate the double integral:

$\displaystyle\int_{-3}^{3}\int_{-\sqrt{9-x^2}}^{\sqrt{9-x^2}} (4 - y)\,dy\,dx = \int_0^{2\pi}\int_0^3 (4 - r\sin\theta)\,r\,dr\,d\theta$

$= \int_0^{2\pi} \left[2r^2 - \frac{1}{3}r^3\sin\theta\right]_{r=0}^{r=3} d\theta$

$= \int_0^{2\pi} (18 - 9\sin\theta)\,d\theta$

$= 18\theta + 9\cos\theta\Big]_0^{2\pi} = 36\pi$

20. The paraboloid $x = y^2 + z^2$ intersects the plane $x = 16$ in the circle $y^2 + z^2 = 16$, $x = 16$.

Thus, $E = \left\{(x, y, z) \mid y^2 + z^2 \le x \le 16, y^2 + z^2 \le 16\right\}$.

Let $D = \left\{(y, z) \mid y^2 + z^2 \le 16\right\}$. Then using polar coordinates $y = r\cos\theta$ and $z = r\sin\theta$, we have

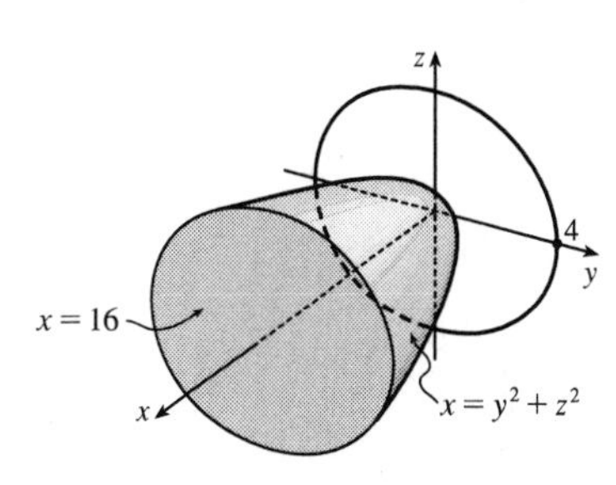

$V = \iint_D \left(\int_{y^2+z^2}^{16} dx\right) dA = \iint_D (16 - (y^2 + z^2))\,dA$

$= \int_0^{2\pi}\int_0^4 (16 - r^2)\,r\,dr\,d\theta = \int_0^{2\pi} d\theta \int_0^4 (16r - r^3)\,dr$

$= \left[\theta\right]_0^{2\pi} \left[8r^2 - \frac{1}{4}r^4\right]_0^4 = 2\pi(128 - 64) = 128\pi$

21. (a) The wedge can be described as the region

$D = \left\{(x, y, z) \mid y^2 + z^2 \le 1, 0 \le x \le 1, 0 \le y \le x\right\}$

$= \left\{(x, y, z) \mid 0 \le x \le 1, 0 \le y \le x, 0 \le z \le \sqrt{1 - y^2}\right\}$

So the integral expressing the volume of the wedge is

$\iiint_D dV = \int_0^1 \int_0^x \int_0^{\sqrt{1-y^2}} dz\,dy\,dx$.

(b) A CAS gives $\int_0^1 \int_0^x \int_0^{\sqrt{1-y^2}} dz\,dy\,dx = \frac{\pi}{4} - \frac{1}{3}$.

(Or use Formulas 30 and 87 from the Table of Integrals.)

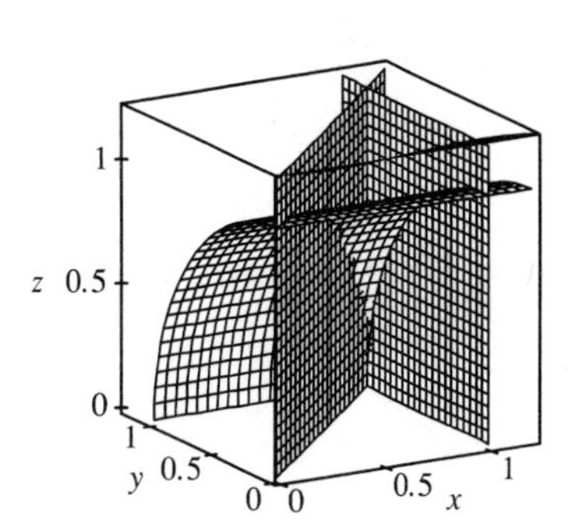

22. (a) Divide B into 8 cubes of size $\Delta V = 8$. With $f(x,y,z) = \sqrt{x^2+y^2+z^2}$, the Midpoint Rule gives

$$\begin{aligned}\iiint_B \sqrt{x^2+y^2+z^2}\,dV &\approx \sum_{i=1}^{2}\sum_{j=1}^{2}\sum_{k=1}^{2} f(\overline{x}_i,\overline{y}_j,\overline{z}_k)\,\Delta V \\ &= 8[f(1,1,1)+f(1,1,3)+f(1,3,1)+f(1,3,3)+f(3,1,1) \\ &\qquad + f(3,1,3)+f(3,3,1)+f(3,3,3)] \\ &\approx 239.64\end{aligned}$$

(b) Using a CAS we have $\iiint_B \sqrt{x^2+y^2+z^2}\,dV = \int_0^4\int_0^4\int_0^4 \sqrt{x^2+y^2+z^2}\,dz\,dy\,dx \approx 245.91$. This differs from the estimate in part (a) by about 2.5%.

23. Here $f(x,y,z) = \dfrac{1}{\ln(1+x+y+z)}$ and $\Delta V = 2\cdot 4\cdot 2 = 16$, so the Midpoint Rule gives

$$\begin{aligned}\iiint_B f(x,y,z)\,dV &\approx \sum_{i=1}^{l}\sum_{j=1}^{m}\sum_{k=1}^{n} f(\overline{x}_i,\overline{y}_j,\overline{z}_k)\,\Delta V \\ &= 16\,[f(1,2,1)+f(1,2,3)+f(1,6,1)+f(1,6,3)+ f(3,2,1)+f(3,2,3)+f(3,6,1)+f(3,6,3)] \\ &= 16\left[\tfrac{1}{\ln 5}+\tfrac{1}{\ln 7}+\tfrac{1}{\ln 9}+\tfrac{1}{\ln 11}+\tfrac{1}{\ln 7}+\tfrac{1}{\ln 9}+\tfrac{1}{\ln 11}+\tfrac{1}{\ln 13}\right] \approx 60.533\end{aligned}$$

24. Here $f(x,y,z) = \sin(xy^2z^3)$ and $\Delta V = 2\cdot 1\cdot \frac{1}{2} = 1$, so the Midpoint Rule gives

$$\begin{aligned}\iiint_B f(x,y,z)\,dV &\approx \sum_{i=1}^{l}\sum_{j=1}^{m}\sum_{k=1}^{n} f(\overline{x}_i,\overline{y}_j,\overline{z}_k)\,\Delta V \\ &= 1\left[f\left(1,\tfrac{1}{2},\tfrac{1}{4}\right)+f\left(1,\tfrac{1}{2},\tfrac{3}{4}\right)+f\left(1,\tfrac{3}{2},\tfrac{1}{4}\right)+f\left(1,\tfrac{3}{2},\tfrac{3}{4}\right)\right. \\ &\qquad \left.+ f\left(3,\tfrac{1}{2},\tfrac{1}{4}\right)+f\left(3,\tfrac{1}{2},\tfrac{3}{4}\right)+f\left(3,\tfrac{3}{2},\tfrac{1}{4}\right)+f\left(3,\tfrac{3}{2},\tfrac{3}{4}\right)\right] \\ &= \sin\tfrac{1}{256}+\sin\tfrac{27}{256}+\sin\tfrac{9}{256}+\sin\tfrac{243}{256}+\sin\tfrac{3}{256}+\sin\tfrac{81}{256}+\sin\tfrac{27}{256}+\sin\tfrac{729}{256} \approx 1.675\end{aligned}$$

25. $E = \{(x,y,z) \mid 0 \le x \le 1, 0 \le z \le 1-x, 0 \le y \le 2-2z\}$, the solid bounded by the three coordinate planes and the planes $z = 1-x$, $y = 2-2z$.

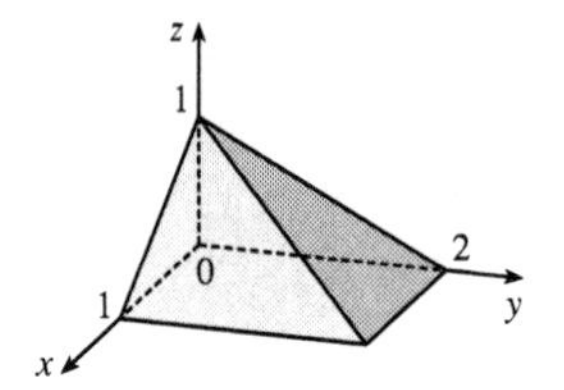

26. $E = \{(x,y,z) \mid 0 \le y \le 2, 0 \le z \le 2-y, 0 \le x \le 4-y^2\}$, the solid bounded by the three coordinate planes, the plane $z = 2-y$, and the cylindrical surface $x = 4-y^2$.

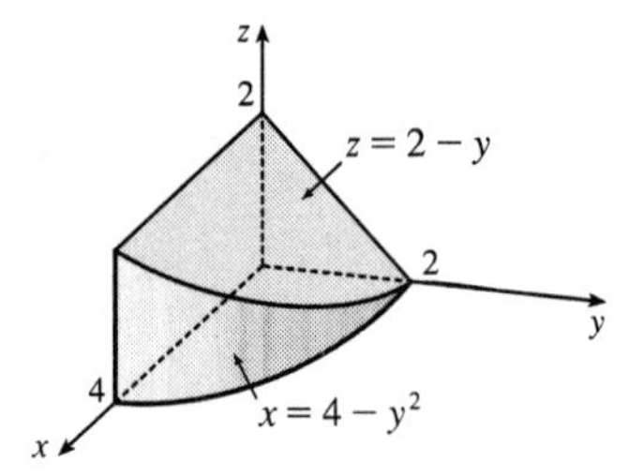

27.

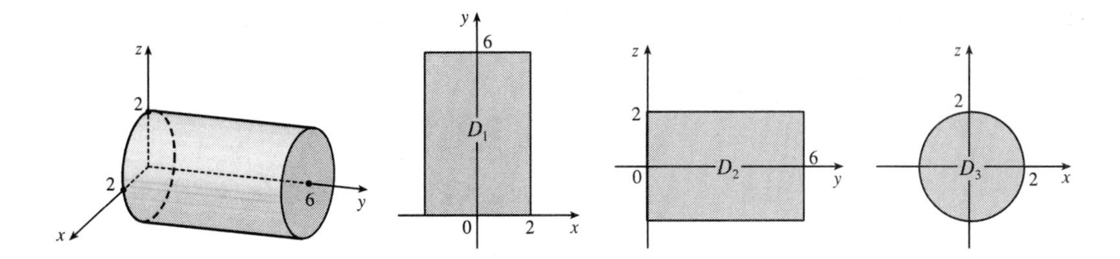

If D_1, D_2, D_3 are the projections of E on the xy-, yz-, and xz-planes, then

$$D_1 = \{(x,y) \mid -2 \le x \le 2, 0 \le y \le 6\}$$
$$D_2 = \{(y,z) \mid -2 \le z \le 2, 0 \le y \le 6\}$$
$$D_3 = \{(x,z) \mid x^2 + z^2 \le 4\}$$

Therefore

$$E = \{(x,y,z) \mid -\sqrt{4-x^2} \le z \le \sqrt{4-x^2},\ -2 \le x \le 2, 0 \le y \le 6\}$$
$$= \{(x,y,z) \mid -\sqrt{4-z^2} \le x \le \sqrt{4-z^2},\ -2 \le z \le 2, 0 \le y \le 6\}$$

$$\iiint_E f(x,y,z)\,dV = \int_{-2}^{2}\int_0^6\int_{-\sqrt{4-x^2}}^{\sqrt{4-x^2}} f(x,y,z)\,dz\,dy\,dx = \int_0^6\int_{-2}^{2}\int_{-\sqrt{4-x^2}}^{\sqrt{4-x^2}} f(x,y,z)\,dz\,dx\,dy$$
$$= \int_0^6\int_{-2}^{2}\int_{-\sqrt{4-z^2}}^{\sqrt{4-z^2}} f(x,y,z)\,dx\,dz\,dy = \int_{-2}^{2}\int_0^6\int_{-\sqrt{4-z^2}}^{\sqrt{4-z^2}} f(x,y,z)\,dx\,dy\,dz$$
$$= \int_{-2}^{2}\int_{-\sqrt{4-x^2}}^{\sqrt{4-x^2}}\int_0^6 f(x,y,z)\,dy\,dz\,dx = \int_{-2}^{2}\int_{-\sqrt{4-z^2}}^{\sqrt{4-z^2}}\int_0^6 f(x,y,z)\,dy\,dx\,dz$$

28.

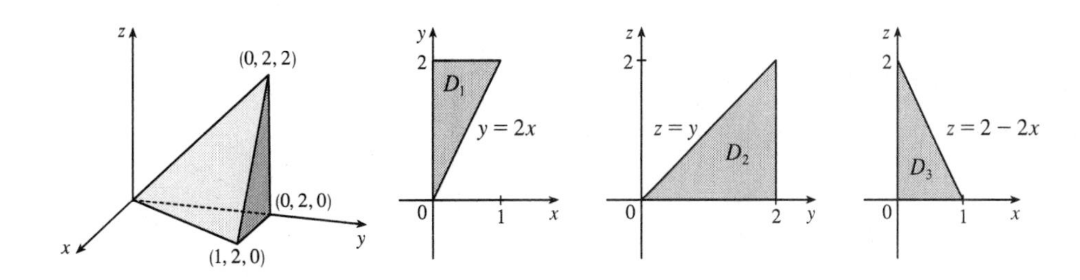

If D_1, D_2, and D_3 are the projections of E on the xy-, yz-, and xz-planes, then

$$D_1 = \{(x,y) \mid 0 \le x \le 1, 2x \le y \le 2\} = \{(x,y) \mid 0 \le y \le 2, 0 \le x \le y/2\},$$
$$D_2 = \{(y,z) \mid 0 \le y \le 2, 0 \le z \le y\} = \{(y,z) \mid 0 \le z \le 2, z \le y \le 2\}, \text{ and}$$
$$D_3 = \{(x,z) \mid 0 \le x \le 1, 0 \le z \le 2-2x\} = \{(x,z) \mid 0 \le z \le 2, 0 \le x \le (2-z)/2\}$$

Therefore

$$\begin{aligned} E &= \{(x,y,z) \mid 0 \le x \le 1, 2x \le y \le 2, 0 \le z \le y - 2x\} \\ &= \{(x,y,z) \mid 0 \le y \le 2, 0 \le x \le y/2, 0 \le z \le y - 2x\} \\ &= \{(x,y,z) \mid 0 \le y \le 2, 0 \le z \le y, 0 \le x \le (y-z)/2\} \\ &= \{(x,y,z) \mid 0 \le z \le 2, z \le y \le 2, 0 \le x \le (y-z)/2\} \\ &= \{(x,y,z) \mid 0 \le x \le 1, 0 \le z \le 2 - 2x, z + 2x \le y \le 2\} \\ &= \{(x,y,z) \mid 0 \le z \le 2, 0 \le x \le (2-z)/2, z + 2x \le y \le 2\} \end{aligned}$$

Then

$$\begin{aligned} \iiint_E f(x,y,z)\,dV &= \int_0^1 \int_{2x}^2 \int_0^{y-2x} f(x,y,z)\,dz\,dy\,dx = \int_0^2 \int_0^{y/2} \int_0^{y-2x} f(x,y,z)\,dz\,dx\,dy \\ &= \int_0^2 \int_0^y \int_0^{(y-z)/2} f(x,y,z)\,dx\,dz\,dy = \int_0^2 \int_z^2 \int_0^{(y-z)/2} f(x,y,z)\,dx\,dy\,dz \\ &= \int_0^1 \int_0^{2-2x} \int_{z+2x}^2 f(x,y,z)\,dy\,dz\,dx = \int_0^2 \int_0^{(2-z)/2} \int_{z+2x}^2 f(x,y,z)\,dy\,dx\,dz \end{aligned}$$

29.

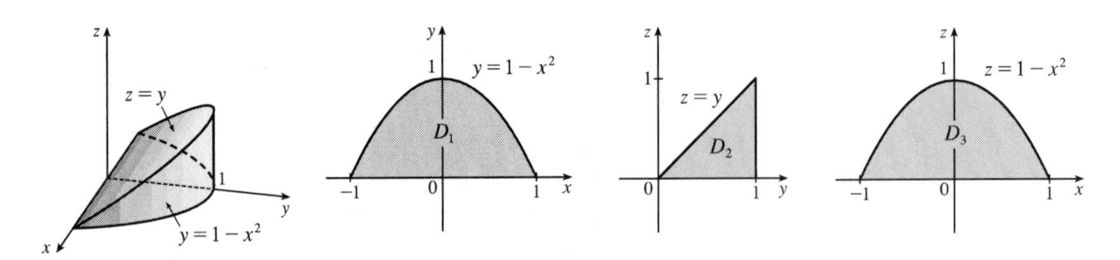

If D_1, D_2, and D_3 are the projections of E on the xy-, yz-, and xz-planes, then

$$D_1 = \{(x,y) \mid -1 \le x \le 1, 0 \le y \le 1 - x^2\} = \{(x,y) \mid 0 \le y \le 1, -\sqrt{1-y} \le x \le \sqrt{1-y}\},$$

$$D_2 = \{(y,z) \mid 0 \le y \le 1, 0 \le z \le y\} = \{(y,z) \mid 0 \le z \le 1, z \le y \le 1\}, \text{ and}$$

$$D_3 = \{(x,z) \mid -1 \le x \le 1, 0 \le z \le 1 - x^2\} = \{(x,z) \mid 0 \le z \le 1, -\sqrt{1-z} \le x \le \sqrt{1-z}\}$$

Therefore

$$\begin{aligned} E &= \{(x,y,z) \mid -1 \le x \le 1, 0 \le y \le 1 - x^2, 0 \le z \le y\} \\ &= \{(x,y,z) \mid 0 \le y \le 1, -\sqrt{1-y} \le x \le \sqrt{1-y}, 0 \le z \le y\} \\ &= \{(x,y,z) \mid 0 \le y \le 1, 0 \le z \le y, -\sqrt{1-y} \le x \le \sqrt{1-y}\} \\ &= \{(x,y,z) \mid 0 \le z \le 1, z \le y \le 1, -\sqrt{1-y} \le x \le \sqrt{1-y}\} \\ &= \{(x,y,z) \mid -1 \le x \le 1, 0 \le z \le 1 - x^2, z \le y \le 1 - x^2\} \\ &= \{(x,y,z) \mid 0 \le z \le 1, -\sqrt{1-z} \le x \le \sqrt{1-z}, z \le y \le 1 - x^2\} \end{aligned}$$

Then

$$\begin{aligned} \iiint_E f(x,y,z)\,dV &= \int_{-1}^1 \int_0^{1-x^2} \int_0^y f(x,y,z)\,dz\,dy\,dx = \int_0^1 \int_{-\sqrt{1-y}}^{\sqrt{1-y}} \int_0^y f(x,y,z)\,dz\,dx\,dy \\ &= \int_0^1 \int_0^y \int_{-\sqrt{1-y}}^{\sqrt{1-y}} f(x,y,z)\,dx\,dz\,dy = \int_0^1 \int_z^1 \int_{-\sqrt{1-y}}^{\sqrt{1-y}} f(x,y,z)\,dx\,dy\,dz \\ &= \int_{-1}^1 \int_0^{1-x^2} \int_z^{1-x^2} f(x,y,z)\,dy\,dz\,dx = \int_0^1 \int_{-\sqrt{1-z}}^{\sqrt{1-z}} \int_z^{1-x^2} f(x,y,z)\,dy\,dx\,dz \end{aligned}$$

30.

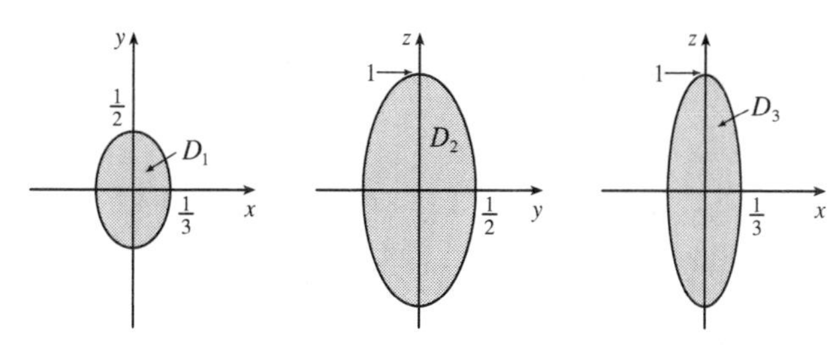

If D_1, D_2 and D_3 are the projections of E on the xy-, yz-, and xz-planes, then $D_1 = \{(x, y) \mid 9x^2 + 4y^2 \leq 1\}$, $D_2 = \{(y, z) \mid 4y^2 + z^2 \leq 1\}$, $D_3 = \{(x, z) \mid 9x^2 + z^2 \leq 1\}$. Therefore

$$\begin{aligned}\iiint_E f(x, y, z)\, dV &= \int_{-1/3}^{1/3} \int_{-\sqrt{1-9x^2}/2}^{\sqrt{1-9x^2}/2} \int_{-\sqrt{1-9x^2-4y^2}}^{\sqrt{1-9x^2-4y^2}} f(x, y, z)\, dz\, dy\, dx \\ &= \int_{-1/2}^{1/2} \int_{-\sqrt{1-4y^2}/3}^{\sqrt{1-4y^2}/3} \int_{-\sqrt{1-9x^2-4y^2}}^{\sqrt{1-9x^2-4y^2}} f(x, y, z)\, dz\, dx\, dy \\ &= \int_{-1/2}^{1/2} \int_{-\sqrt{1-4y^2}}^{\sqrt{1-4y^2}} \int_{-\sqrt{1-4y^2-z^2}/3}^{\sqrt{1-4y^2-z^2}/3} f(x, y, z)\, dx\, dz\, dy \\ &= \int_{-1}^{1} \int_{-\sqrt{1-z^2}/2}^{\sqrt{1-z^2}/2} \int_{-\sqrt{1-4y^2-z^2}/3}^{\sqrt{1-4y^2-z^2}/3} f(x, y, z)\, dx\, dy\, dz \\ &= \int_{-1/3}^{1/3} \int_{-\sqrt{1-9x^2}}^{\sqrt{1-9x^2}} \int_{-\sqrt{1-9x^2-z^2}/2}^{\sqrt{1-9x^2-z^2}/2} f(x, y, z)\, dy\, dz\, dx \\ &= \int_{-1}^{1} \int_{-\sqrt{1-z^2}/3}^{\sqrt{1-z^2}/3} \int_{-\sqrt{1-9x^2-z^2}/2}^{\sqrt{1-9x^2-z^2}/2} f(x, y, z)\, dy\, dx\, dz\end{aligned}$$

31.

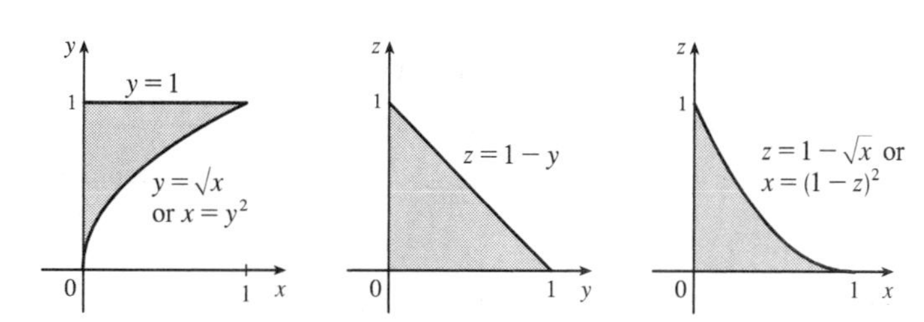

The diagrams show the projections of E on the xy-, yz-, and xz-planes. Therefore

$$\begin{aligned}\int_0^1 \int_{\sqrt{x}}^1 \int_0^{1-y} f(x, y, z)\, dz\, dy\, dx &= \int_0^1 \int_0^{y^2} \int_0^{1-y} f(x, y, z)\, dz\, dx\, dy = \int_0^1 \int_0^{1-z} \int_0^{y^2} f(x, y, z)\, dx\, dy\, dz \\ &= \int_0^1 \int_0^{1-y} \int_0^{y^2} f(x, y, z)\, dx\, dz\, dy = \int_0^1 \int_0^{1-\sqrt{x}} \int_{\sqrt{x}}^{1-z} f(x, y, z)\, dy\, dz\, dx \\ &= \int_0^1 \int_0^{(1-z)^2} \int_{\sqrt{x}}^{1-z} f(x, y, z)\, dy\, dx\, dz\end{aligned}$$

32.

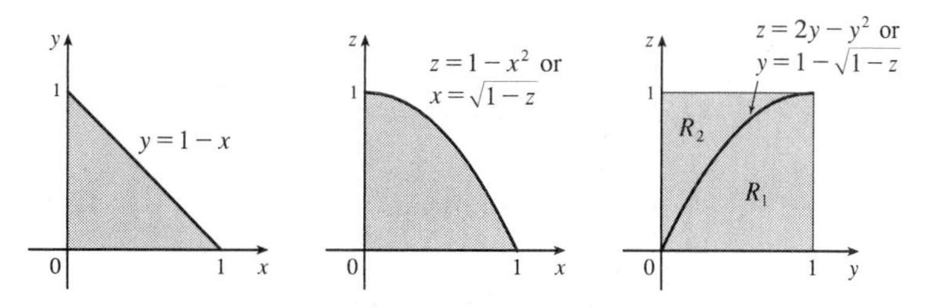

The projections of E onto the xy- and xz-planes are as in the first two diagrams and so

$$\int_0^1 \int_0^{1-x^2} \int_0^{1-x} f(x,y,z)\,dy\,dz\,dx = \int_0^1 \int_0^{\sqrt{1-z}} \int_0^{1-x} f(x,y,z)\,dy\,dx\,dz$$
$$= \int_0^1 \int_0^{1-y} \int_0^{1-x^2} f(x,y,z)\,dz\,dx\,dy = \int_0^1 \int_0^{1-x} \int_0^{1-x^2} f(x,y,z)\,dz\,dy\,dx$$

Now the surface $z = 1 - x^2$ intersects the plane $y = 1 - x$ in a curve whose projection in the yz-plane is $z = 1 - (1 - y)^2$ or $z = 2y - y^2$. So we must split up the projection of E on the yz-plane into two regions as in the third diagram. For (y, z) in R_1, $0 \le x \le 1 - y$ and for (y, z) in R_2, $0 \le x \le \sqrt{1-z}$, and so the given integral is also equal to

$$\int_0^1 \int_0^{1-\sqrt{1-z}} \int_0^{\sqrt{1-z}} f(x,y,z)\,dx\,dy\,dz + \int_0^1 \int_{1-\sqrt{1-z}}^1 \int_0^{1-y} f(x,y,z)\,dx\,dy\,dz$$
$$= \int_0^1 \int_0^{2y-y^2} \int_0^{1-y} f(x,y,z)\,dx\,dz\,dy + \int_0^1 \int_{2y-y^2}^1 \int_0^{\sqrt{1-z}} f(x,y,z)\,dx\,dz\,dy.$$

33.

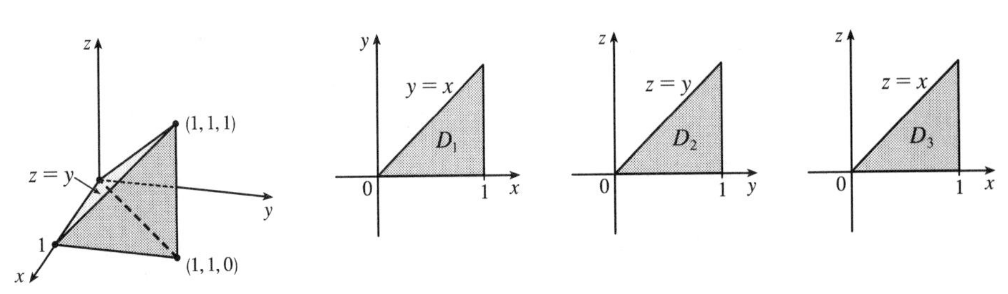

$\int_0^1 \int_y^1 \int_0^y f(x,y,z)\,dz\,dx\,dy = \iiint_E f(x,y,z)\,dV$ where $E = \{(x,y,z) \mid 0 \le z \le y, y \le x \le 1, 0 \le y \le 1\}$.

If D_1, D_2, and D_3 are the projections of E on the xy-, yz- and xz-planes then

$$D_1 = \{(x,y) \mid 0 \le y \le 1, y \le x \le 1\} = \{(x,y) \mid 0 \le x \le 1, 0 \le y \le x\},$$

$$D_2 = \{(y,z) \mid 0 \le y \le 1, 0 \le z \le y\} = \{(y,z) \mid 0 \le z \le 1, z \le y \le 1\}, \text{ and}$$

$$D_3 = \{(x,z) \mid 0 \le x \le 1, 0 \le z \le x\} = \{(x,z) \mid 0 \le z \le 1, z \le x \le 1\}.$$

Thus we also have

$$E = \{(x,y,z) \mid 0 \le x \le 1, 0 \le y \le x, 0 \le z \le y\} = \{(x,y,z) \mid 0 \le y \le 1, 0 \le z \le y, y \le x \le 1\}$$
$$= \{(x,y,z) \mid 0 \le z \le 1, z \le y \le 1, y \le x \le 1\} = \{(x,y,z) \mid 0 \le x \le 1, 0 \le z \le x, z \le y \le x\}$$
$$= \{(x,y,z) \mid 0 \le z \le 1, z \le x \le 1, z \le y \le x\}.$$

Then

$$\int_0^1 \int_y^1 \int_0^y f(x,y,z)\,dz\,dx\,dy = \int_0^1 \int_0^x \int_0^y f(x,y,z)\,dz\,dy\,dx = \int_0^1 \int_0^y \int_y^1 f(x,y,z)\,dx\,dz\,dy$$
$$= \int_0^1 \int_z^1 \int_y^1 f(x,y,z)\,dx\,dy\,dz = \int_0^1 \int_0^x \int_z^x f(x,y,z)\,dy\,dz\,dx$$
$$= \int_0^1 \int_z^1 \int_z^x f(x,y,z)\,dy\,dx\,dz$$

34.

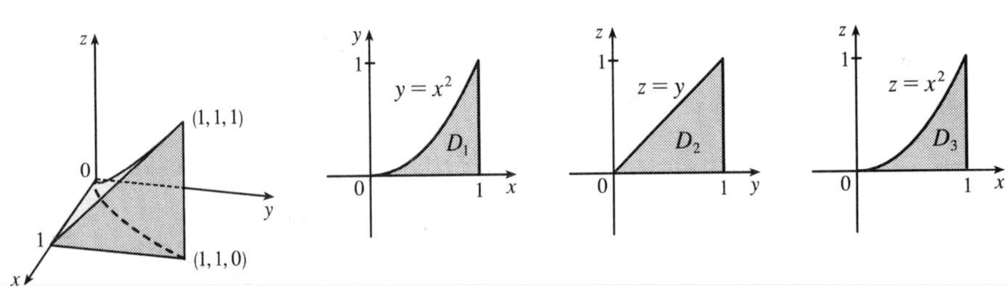

$\int_0^1 \int_0^{x^2} \int_0^y f(x,y,z)\,dz\,dy\,dx = \iiint_E f(x,y,z)\,dV$ where $E = \{(x,y,z) \mid 0 \le x \le 1, 0 \le y \le x^2, 0 \le z \le y\}$.

If D_1, D_2, D_3 are the projections of E on the xy-, yz-, and xz-planes, then

$$D_1 = \{(x,y) \mid 0 \le x \le 1, 0 \le y \le x^2\} = \{(x,y) \mid 0 \le y \le 1, \sqrt{y} \le x \le 1\},$$
$$D_2 = \{(y,z) \mid 0 \le y \le 1, 0 \le z \le y\} = \{(y,z) \mid 0 \le z \le 1, z \le y \le 1\},$$
$$D_3 = \{(x,z) \mid 0 \le x \le 1, 0 \le z \le x^2\} = \{(x,z) \mid 0 \le z \le 1, \sqrt{z} \le x \le 1\}.$$

Thus we also have

$$\begin{aligned} E &= \{(x,y,z) \mid 0 \le y \le 1, \sqrt{y} \le x \le 1, 0 \le z \le y\} = \{(x,y,z) \mid 0 \le y \le 1, 0 \le z \le y, \sqrt{y} \le x \le 1\} \\ &= \{(x,y,z) \mid 0 \le z \le 1, z \le y \le 1, \sqrt{y} \le x \le 1\} = \{(x,y,z) \mid 0 \le x \le 1, 0 \le z \le x^2, z \le y \le x^2\} \\ &= \{(x,y,z) \mid 0 \le z \le 1, \sqrt{z} \le x \le 1, z \le y \le x^2\} \end{aligned}$$

Then

$$\begin{aligned} \int_0^1 \int_0^{x^2} \int_0^y f(x,y,z)\,dz\,dy\,dx &= \int_0^1 \int_{\sqrt{y}}^1 \int_0^y f(x,y,z)\,dz\,dx\,dy = \int_0^1 \int_0^y \int_{\sqrt{y}}^1 f(x,y,z)\,dx\,dz\,dy \\ &= \int_0^1 \int_z^1 \int_{\sqrt{y}}^1 f(x,y,z)\,dx\,dy\,dz = \int_0^1 \int_0^{x^2} \int_z^{x^2} f(x,y,z)\,dy\,dz\,dx \\ &= \int_0^1 \int_{\sqrt{z}}^1 \int_z^{x^2} f(x,y,z)\,dy\,dx\,dz \end{aligned}$$

35. $m = \iiint_E \rho(x,y,z)\,dV = \int_0^1 \int_0^{\sqrt{x}} \int_0^{1+x+y} 2\,dz\,dy\,dx = \int_0^1 \int_0^{\sqrt{x}} 2(1+x+y)\,dy\,dx$

$$= \int_0^1 \left[2y + 2xy + y^2\right]_{y=0}^{y=\sqrt{x}} dx = \int_0^1 \left(2\sqrt{x} + 2x^{3/2} + x\right) dx = \left[\tfrac{4}{3}x^{3/2} + \tfrac{4}{5}x^{5/2} + \tfrac{1}{2}x^2\right]_0^1 = \tfrac{79}{30}$$

$$M_{yz} = \iiint_E x\rho(x,y,z)\,dV = \int_0^1 \int_0^{\sqrt{x}} \int_0^{1+x+y} 2x\,dz\,dy\,dx = \int_0^1 \int_0^{\sqrt{x}} 2x(1+x+y)\,dy\,dx$$
$$= \int_0^1 \left[2xy + 2x^2y + xy^2\right]_{y=0}^{y=\sqrt{x}} dx = \int_0^1 (2x^{3/2} + 2x^{5/2} + x^2)\,dx = \left[\tfrac{4}{5}x^{5/2} + \tfrac{4}{7}x^{7/2} + \tfrac{1}{3}x^3\right]_0^1 = \tfrac{179}{105}$$

$$M_{xz} = \iiint_E y\rho(x,y,z)\,dV = \int_0^1 \int_0^{\sqrt{x}} \int_0^{1+x+y} 2y\,dz\,dy\,dx = \int_0^1 \int_0^{\sqrt{x}} 2y(1+x+y)\,dy\,dx$$
$$= \int_0^1 \left[y^2 + xy^2 + \tfrac{2}{3}y^3\right]_{y=0}^{y=\sqrt{x}} dx = \int_0^1 \left(x + x^2 + \tfrac{2}{3}x^{3/2}\right) dx = \left[\tfrac{1}{2}x^2 + \tfrac{1}{3}x^3 + \tfrac{4}{15}x^{5/2}\right]_0^1 = \tfrac{11}{10}$$

$$M_{xy} = \iiint_E z\rho(x,y,z)\,dV = \int_0^1 \int_0^{\sqrt{x}} \int_0^{1+x+y} 2z\,dz\,dy\,dx = \int_0^1 \int_0^{\sqrt{x}} \left[z^2\right]_{z=0}^{z=1+x+y} dy\,dx = \int_0^1 \int_0^{\sqrt{x}} (1+x+y)^2\,dy\,dx$$
$$= \int_0^1 \int_0^{\sqrt{x}} (1 + 2x + 2y + 2xy + x^2 + y^2)\,dy\,dx = \int_0^1 \left[y + 2xy + y^2 + xy^2 + x^2y + \tfrac{1}{3}y^3\right]_{y=0}^{y=\sqrt{x}} dx$$
$$= \int_0^1 \left(\sqrt{x} + \tfrac{7}{3}x^{3/2} + x + x^2 + x^{5/2}\right) dx = \left[\tfrac{2}{3}x^{3/2} + \tfrac{14}{15}x^{5/2} + \tfrac{1}{2}x^2 + \tfrac{1}{3}x^3 + \tfrac{2}{7}x^{7/2}\right]_0^1 = \tfrac{571}{210}$$

Thus the mass is $\frac{79}{30}$ and the center of mass is $(\overline{x}, \overline{y}, \overline{z}) = \left(\dfrac{M_{yz}}{m}, \dfrac{M_{xz}}{m}, \dfrac{M_{xy}}{m}\right) = \left(\dfrac{358}{553}, \dfrac{33}{79}, \dfrac{571}{553}\right)$.

36. $m = \int_{-1}^1 \int_0^{1-y^2} \int_0^{1-z} 4\,dx\,dz\,dy = 4\int_{-1}^1 \int_0^{1-y^2} (1-z)\,dz\,dy = 4\int_{-1}^1 \left[z - \tfrac{1}{2}z^2\right]_{z=0}^{z=1-y^2} dy = 2\int_{-1}^1 (1-y^4)\,dy = \tfrac{16}{5}$,

$$M_{yz} = \int_{-1}^1 \int_0^{1-y^2} \int_0^{1-z} 4x\,dx\,dz\,dy = 2\int_{-1}^1 \int_0^{1-y^2} (1-z)^2\,dz\,dy = 2\int_{-1}^1 \left[-\tfrac{1}{3}(1-z)^3\right]_{z=0}^{z=1-y^2} dy$$
$$= \tfrac{2}{3}\int_{-1}^1 (1-y^6)\,dy = \left(\tfrac{4}{3}\right)\left(\tfrac{6}{7}\right) = \tfrac{24}{21}$$

$$M_{xz} = \int_{-1}^1 \int_0^{1-y^2} \int_0^{1-z} 4y\,dx\,dz\,dy = \int_{-1}^1 \int_0^{1-y^2} 4y(1-z)\,dz\,dy$$
$$= \int_{-1}^1 \left[4y(1-y^2) - 2y(1-y^2)^2\right] dy = \int_{-1}^1 (2y - 2y^5)\,dy = 0 \quad \text{[the integrand is odd]}$$

$$M_{xy} = \int_{-1}^1 \int_0^{1-y^2} \int_0^{1-z} 4z\,dx\,dz\,dy = \int_{-1}^1 \int_0^{1-y^2} (4z - 4z^2)\,dz\,dy = 2\int_{-1}^1 \left[(1-y^2)^2 - \tfrac{2}{3}(1-y^2)^3\right] dy$$
$$= 2\int_{-1}^1 \left[\tfrac{1}{3} - y^4 + \tfrac{2}{3}y^6\right] dy = \left[\tfrac{4}{3}y - \tfrac{4}{5}y^5 + \tfrac{8}{21}y^7\right]_0^1 = \tfrac{96}{105} = \tfrac{32}{35}$$

Thus, $(\overline{x}, \overline{y}, \overline{z}) = \left(\frac{5}{14}, 0, \frac{2}{7}\right)$

37. $m = \int_0^a \int_0^a \int_0^a (x^2 + y^2 + z^2)\,dx\,dy\,dz = \int_0^a \int_0^a \left[\frac{1}{3}x^3 + xy^2 + xz^2\right]_{x=0}^{x=a} dy\,dz = \int_0^a \int_0^a \left(\frac{1}{3}a^3 + ay^2 + az^2\right) dy\,dz$

$= \int_0^a \left[\frac{1}{3}a^3 y + \frac{1}{3}ay^3 + ayz^2\right]_{y=0}^{y=a} dz = \int_0^a \left(\frac{2}{3}a^4 + a^2 z^2\right) dz = \left[\frac{2}{3}a^4 z + \frac{1}{3}a^2 z^3\right]_0^a = \frac{2}{3}a^5 + \frac{1}{3}a^5 = a^5$

$M_{yz} = \int_0^a \int_0^a \int_0^a \left[x^3 + x(y^2 + z^2)\right] dx\,dy\,dz = \int_0^a \int_0^a \left[\frac{1}{4}a^4 + \frac{1}{2}a^2(y^2 + z^2)\right] dy\,dz$

$= \int_0^a \left(\frac{1}{4}a^5 + \frac{1}{6}a^5 + \frac{1}{2}a^3 z^2\right) dz = \frac{1}{4}a^6 + \frac{1}{3}a^6 = \frac{7}{12}a^6 = M_{xz} = M_{xy}$ by symmetry of E and $\rho(x, y, z)$

Hence $(\overline{x}, \overline{y}, \overline{z}) = \left(\frac{7}{12}a, \frac{7}{12}a, \frac{7}{12}a\right)$.

38. $m = \int_0^1 \int_0^{1-x} \int_0^{1-x-y} y\,dz\,dy\,dx = \int_0^1 \int_0^{1-x} \left[(1 - x)y - y^2\right] dy\,dx$

$= \int_0^1 \left[\frac{1}{2}(1 - x)^3 - \frac{1}{3}(1 - x)^3\right] dx = \frac{1}{6} \int_0^1 (1 - x)^3\,dx = \frac{1}{24}$

$M_{yz} = \int_0^1 \int_0^{1-x} \int_0^{1-x-y} xy\,dz\,dy\,dx = \int_0^1 \int_0^{1-x} \left[(x - x^2)y - xy^2\right] dy\,dx$

$= \int_0^1 \left[\frac{1}{2}x(1 - x)^3 - \frac{1}{3}x(1 - x)^3\right] dx = \frac{1}{6} \int_0^1 \left(x - 3x^2 + 3x^3 - x^4\right) dx = \frac{1}{6}\left(\frac{1}{2} - 1 + \frac{3}{4} - \frac{1}{5}\right) = \frac{1}{120}$

$M_{xz} = \int_0^1 \int_0^{1-x} \int_0^{1-x-y} y^2\,dz\,dy\,dx = \int_0^1 \int_0^{1-x} \left[(1 - x)y^2 - y^3\right] dy\,dx$

$= \int_0^1 \left[\frac{1}{3}(1 - x)^4 - \frac{1}{4}(1 - x)^4\right] dx = \frac{1}{12}\left[-\frac{1}{5}(1 - x)^5\right]_0^1 = \frac{1}{60}$

$M_{xy} = \int_0^1 \int_0^{1-x} \int_0^{1-x-y} yz\,dz\,dy\,dx = \int_0^1 \int_0^{1-x} \left[\frac{1}{2}y(1 - x - y)^2\right] dy\,dx$

$= \frac{1}{2} \int_0^1 \int_0^{1-x} \left[(1 - x)^2 y - 2(1 - x)y^2 + y^3\right] dy\,dx = \frac{1}{2} \int_0^1 \left[\frac{1}{2}(1 - x)^4 - \frac{2}{3}(1 - x)^4 + \frac{1}{4}(1 - x)^4\right] dx$

$= \frac{1}{24} \int_0^1 (1 - x)^4\,dx = -\frac{1}{24}\left[\frac{1}{5}(1 - x)^5\right]_0^1 = \frac{1}{120}$

Hence $(\overline{x}, \overline{y}, \overline{z}) = \left(\frac{1}{5}, \frac{2}{5}, \frac{1}{5}\right)$.

39. (a) $m = \int_{-3}^{3} \int_{-\sqrt{9-x^2}}^{\sqrt{9-x^2}} \int_1^{5-y} \sqrt{x^2 + y^2}\,dz\,dy\,dx$

(b) $(\overline{x}, \overline{y}, \overline{z}) = \left(\dfrac{M_{yz}}{m}, \dfrac{M_{xz}}{m}, \dfrac{M_{xy}}{m}\right)$ where

$M_{yz} = \int_{-3}^{3} \int_{-\sqrt{9-x^2}}^{\sqrt{9-x^2}} \int_1^{5-y} x\sqrt{x^2 + y^2}\,dz\,dy\,dx$, $M_{xz} = \int_{-3}^{3} \int_{-\sqrt{9-x^2}}^{\sqrt{9-x^2}} \int_1^{5-y} y\sqrt{x^2 + y^2}\,dz\,dy\,dx$, and

$M_{xy} = \int_{-3}^{3} \int_{-\sqrt{9-x^2}}^{\sqrt{9-x^2}} \int_1^{5-y} z\sqrt{x^2 + y^2}\,dz\,dy\,dx$.

(c) $I_z = \int_{-3}^{3} \int_{-\sqrt{9-x^2}}^{\sqrt{9-x^2}} \int_1^{5-y} (x^2 + y^2)\sqrt{x^2 + y^2}\,dz\,dy\,dx = \int_{-3}^{3} \int_{-\sqrt{9-x^2}}^{\sqrt{9-x^2}} \int_1^{5-y} (x^2 + y^2)^{3/2}\,dz\,dy\,dx$

40. (a) $m = \int_{-1}^{1} \int_{-\sqrt{1-y^2}}^{\sqrt{1-y^2}} \int_0^{\sqrt{1-x^2-y^2}} \sqrt{x^2 + y^2 + z^2}\,dz\,dx\,dy$

(b) $(\overline{x}, \overline{y}, \overline{z})$ where $\overline{x} = m^{-1} \int_{-1}^{1} \int_{-\sqrt{1-y^2}}^{\sqrt{1-y^2}} \int_0^{\sqrt{1-x^2-y^2}} x\sqrt{x^2 + y^2 + z^2}\,dz\,dx\,dy$,

$\overline{y} = m^{-1} \int_{-1}^{1} \int_{-\sqrt{1-y^2}}^{\sqrt{1-y^2}} \int_0^{\sqrt{1-x^2-y^2}} y\sqrt{x^2 + y^2 + z^2}\,dz\,dx\,dy$,

$\overline{z} = m^{-1} \int_{-1}^{1} \int_{-\sqrt{1-y^2}}^{\sqrt{1-y^2}} \int_0^{\sqrt{1-x^2-y^2}} z\sqrt{x^2 + y^2 + z^2}\,dz\,dx\,dy$

(c) $I_z = \int_{-1}^{1} \int_{-\sqrt{1-y^2}}^{\sqrt{1-y^2}} \int_0^{\sqrt{1-x^2-y^2}} (x^2 + y^2)(1 + x + y + z)\, dz\, dx\, dy$

41. (a) $m = \int_0^1 \int_0^{\sqrt{1-x^2}} \int_0^y (1 + x + y + z)\, dz\, dy\, dx = \frac{3\pi}{32} + \frac{11}{24}$

(b)
$$(\overline{x}, \overline{y}, \overline{z}) = \Big(m^{-1} \int_0^1 \int_0^{\sqrt{1-x^2}} \int_0^y x(1 + x + y + z)\, dz\, dy\, dx,$$
$$m^{-1} \int_0^1 \int_0^{\sqrt{1-x^2}} \int_0^y y(1 + x + y + z)\, dz\, dy\, dx,$$
$$m^{-1} \int_0^1 \int_0^{\sqrt{1-x^2}} \int_0^y z(1 + x + y + z)\, dz\, dy\, dx\Big)$$
$$= \left(\frac{28}{9\pi + 44}, \frac{30\pi + 128}{45\pi + 220}, \frac{45\pi + 208}{135\pi + 660}\right)$$

(c) $$I_z = \int_0^1 \int_0^{\sqrt{1-x^2}} \int_0^y (x^2 + y^2)(1 + x + y + z)\, dz\, dy\, dx = \frac{68 + 15\pi}{240}$$

42. (a) $m = \int_0^1 \int_{3x}^3 \int_0^{\sqrt{9-y^2}} (x^2 + y^2)\, dz\, dy\, dx = \frac{56}{5} = 11.2$

(b) $(\overline{x}, \overline{y}, \overline{z})$ where $\overline{x} = m^{-1} \int_0^1 \int_{3x}^3 \int_0^{\sqrt{9-y^2}} x(x^2 + y^2)\, dz\, dy\, dx \approx 0.375$,

$\overline{y} = m^{-1} \int_0^1 \int_{3x}^3 \int_0^{\sqrt{9-y^2}} y(x^2 + y^2)\, dz\, dy\, dx = \frac{45\pi}{64} \approx 2.209$,

$\overline{z} = m^{-1} \int_0^1 \int_{3x}^3 \int_0^{\sqrt{9-y^2}} z(x^2 + y^2)\, dz\, dy\, dx = \frac{15}{16} = 0.9375$.

(c) $I_z = \int_0^1 \int_{3x}^3 \int_0^{\sqrt{9-y^2}} (x^2 + y^2)^2\, dz\, dy\, dx = \frac{10{,}464}{175} \approx 59.79$

43. $I_x = \int_0^L \int_0^L \int_0^L k(y^2 + z^2)\, dz\, dy\, dx = k \int_0^L \int_0^L \left(Ly^2 + \frac{1}{3}L^3\right) dy\, dx = k \int_0^L \frac{2}{3}L^4\, dx = \frac{2}{3}kL^5$.

By symmetry, $I_x = I_y = I_z = \frac{2}{3}kL^5$.

44. Let k be the density. Then

$$I_x = \int_{-c/2}^{c/2} \int_{-b/2}^{b/2} \int_{-a/2}^{a/2} k(y^2 + z^2)\, dx\, dy\, dz = ka \int_{-c/2}^{c/2} \int_{-b/2}^{b/2} (y^2 + z^2)\, dy\, dz$$
$$= ak \int_{-c/2}^{c/2} \left[\tfrac{1}{3}y^3 + z^2 y\right]_{y=-b/2}^{y=b/2} dz = ak \int_{-c/2}^{c/2} \left(\tfrac{1}{12}b^3 + bz^2\right) dz = ak\left[\tfrac{1}{12}b^3 z + \tfrac{1}{3}bz^3\right]_{-c/2}^{c/2}$$
$$= ak\left(\tfrac{1}{12}b^3 c + \tfrac{1}{12}bc^3\right) = \tfrac{1}{12}kabc(b^2 + c^2)$$

By symmetry, $I_y = \frac{1}{12}kabc(a^2 + c^2)$ and $I_z = \frac{1}{12}kabc(a^2 + b^2)$.

45. $V(E) = L^3$,

$$f_{\text{ave}} = \frac{1}{L^3} \int_0^L \int_0^L \int_0^L xyz\, dx\, dy\, dz = \frac{1}{L^3} \int_0^L x\, dx \int_0^L y\, dy \int_0^L z\, dz = \frac{1}{L^3} \left[\frac{x^2}{2}\right]_0^L \left[\frac{y^2}{2}\right]_0^L \left[\frac{z^2}{2}\right]_0^L$$
$$= \frac{1}{L^3} \frac{L^2}{2} \frac{L^2}{2} \frac{L^2}{2} = \frac{L^3}{8}$$

46. $V(E) = \int_{-1}^{1} \int_{-\sqrt{1-x^2}}^{\sqrt{1-x^2}} \int_0^{1-x^2-y^2} dz\,dy\,dx = \int_{-1}^{1} \int_{-\sqrt{1-x^2}}^{\sqrt{1-x^2}} (1-x^2-y^2)\,dy\,dx$

$= \int_0^{2\pi} \int_0^1 (1-r^2)\,r\,dr\,d\theta = \int_0^{2\pi} d\theta \int_0^1 (r-r^3)\,dr = 2\pi\left(\frac{r^2}{2} - \frac{r^4}{4}\right)\Big]_0^1 = \frac{\pi}{2}.$

Then $f_{\text{ave}} = \frac{1}{\pi/2} \iiint_E (x^2 z + y^2 z)\,dV = \frac{2}{\pi} \int_{-1}^{1} \int_{-\sqrt{1-x^2}}^{\sqrt{1-x^2}} \int_0^{1-x^2-y^2} (x^2+y^2)\,z\,dz\,dy\,dx$

$= \frac{2}{\pi} \int_{-1}^{1} \int_{-\sqrt{1-x^2}}^{\sqrt{1-x^2}} (x^2+y^2) \cdot \frac{1}{2}(1-x^2-y^2)^2\,dy\,dx = \frac{1}{\pi} \int_0^{2\pi} \int_0^1 r^2(1-r^2)^2\,r\,dr\,d\theta$

$= \frac{1}{\pi} \int_0^{2\pi} d\theta \int_0^1 (r^3 - 2r^5 + r^7)\,dr = \frac{1}{\pi}(2\pi)\left[\frac{1}{4}r^4 - \frac{1}{3}r^6 + \frac{1}{8}r^8\right]_0^1 = 2\left(\frac{1}{24}\right) = \frac{1}{12}$

47. The triple integral will attain its maximum when the integrand $1 - x^2 - 2y^2 - 3z^2$ is positive in the region E and negative everywhere else. For if E contains some region F where the integrand is negative, the integral could be increased by excluding F from E, and if E fails to contain some part G of the region where the integrand is positive, the integral could be increased by including G in E. So we require that $x^2 + 2y^2 + 3z^2 \le 1$. This describes the region bounded by the ellipsoid $x^2 + 2y^2 + 3z^2 = 1$.

12.6 Triple Integrals in Cylindrical Coordinates

1. (a)

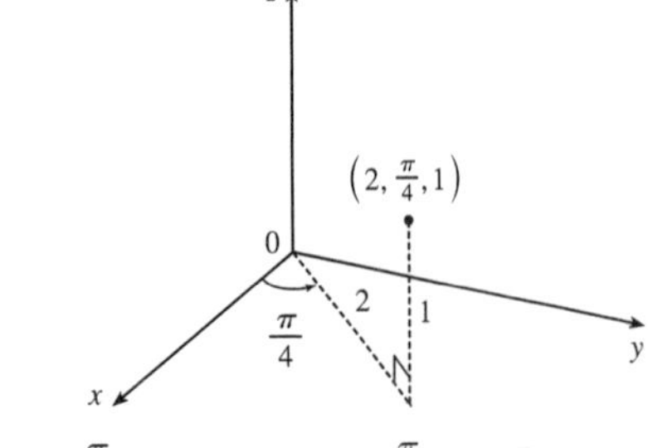

$x = 2\cos\frac{\pi}{4} = \sqrt{2}$, $y = 2\sin\frac{\pi}{4} = \sqrt{2}$, $z = 1$,

so the point is $(\sqrt{2}, \sqrt{2}, 1)$ in rectangular coordinates.

(b)

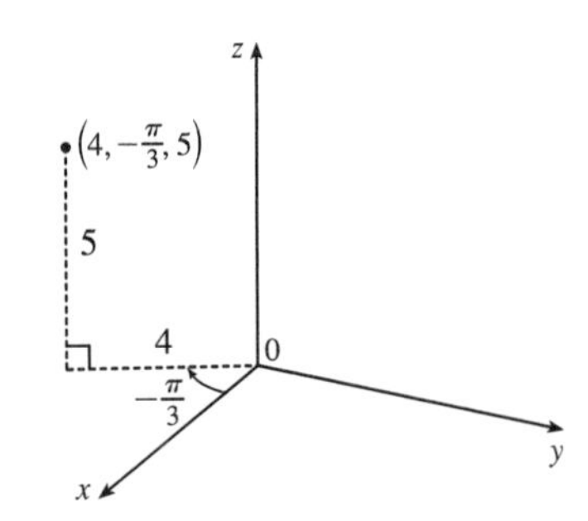

$x = 4\cos\left(-\frac{\pi}{3}\right) = 2$, $y = 4\sin\left(-\frac{\pi}{3}\right) = -2\sqrt{3}$,

and $z = 5$, so the point is $(2, -2\sqrt{3}, 5)$ in rectangular coordinates.

2. (a)

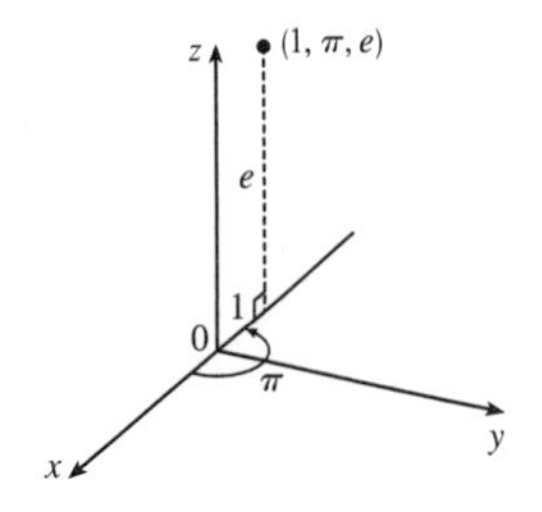

$x = 1\cos\pi = -1$, $y = 1\sin\pi = 0$, and $z = e$,

so the point is $(-1, 0, e)$ in rectangular coordinates.

(b)

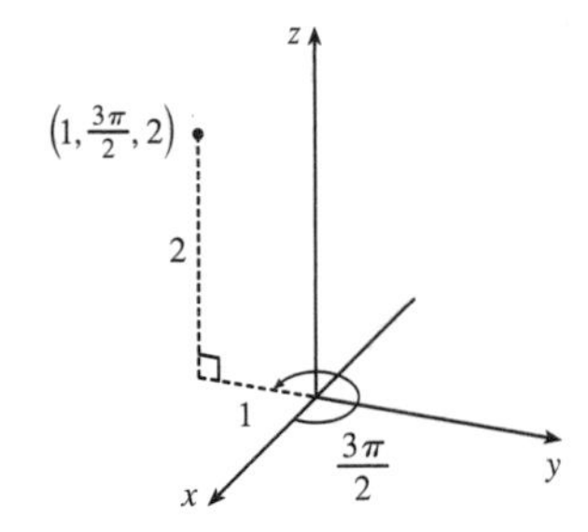

$x = 1\cos\frac{3\pi}{2} = 0$, $y = 1\sin\frac{3\pi}{2} = -1$, $z = 2$,

so the point is $(0, -1, 2)$ in rectangular coordinates.

3. (a) $r^2 = x^2 + y^2 = 1^2 + (-1)^2 = 2$ so $r = \sqrt{2}$; $\tan\theta = \dfrac{y}{x} = \dfrac{-1}{1} = -1$ and the point $(1, -1)$ is in the fourth quadrant of the xy-plane, so $\theta = \frac{7\pi}{4} + 2n\pi$; $z = 4$. Thus, one set of cylindrical coordinates is $\left(\sqrt{2}, \frac{7\pi}{4}, 4\right)$.

(b) $r^2 = (-1)^2 + \left(-\sqrt{3}\right)^2 = 4$ so $r = 2$; $\tan\theta = \frac{-\sqrt{3}}{-1} = \sqrt{3}$ and the point $\left(-1, -\sqrt{3}\right)$ is in the third quadrant of the xy-plane, so $\theta = \frac{4\pi}{3} + 2n\pi$; $z = 2$. Thus, one set of cylindrical coordinates is $\left(2, \frac{4\pi}{3}, 2\right)$.

4. (a) $r^2 = x^2 + y^2 = 3^2 + 3^2 = 18$ so $r = \sqrt{18} = 3\sqrt{2}$; $\tan\theta = \dfrac{y}{x} = \dfrac{3}{3} = 1$ and the point $(3, 3)$ is in the first quadrant of the xy-plane, so $\theta = \frac{\pi}{4} + 2n\pi$; $z = -2$. Thus, one set of cylindrical coordinates is $\left(3\sqrt{2}, \frac{\pi}{4}, -2\right)$.

(b) $r^2 = 3^2 + 4^2 = 25$ so $r = 5$; $\tan\theta = \frac{4}{3}$ and the point $(3, 4)$ is in the first quadrant of the xy-plane, so $\theta = \tan^{-1}\left(\frac{4}{3}\right) + 2n\pi \approx 0.93 + 2n\pi$; $z = 5$. Thus, one set of cylindrical coordinates is $\left(5, \tan^{-1}\left(\frac{4}{3}\right), 5\right) \approx (5, 0.93, 5)$.

5. Since $r = 3$, $x^2 + y^2 = 9$ and the surface is a circular cylinder with radius 3 and axis the z-axis.

6. Whether spherical or cylindrical coordinates, since $\theta = \frac{\pi}{3}$ the surface is a half-plane including the z-axis and intersecting the xy-plane in the half-line $y = \sqrt{3}\,x$, $x > 0$.

7. $z = r^2 = x^2 + y^2$, so the surface is a circular paraboloid with vertex at the origin and axis the positive z-axis.

8. Since $r^2 - 2z^2 = 4$ and $r^2 = x^2 + y^2$, we have $x^2 + y^2 - 2z^2 = 4$ or $\frac{1}{4}x^2 + \frac{1}{4}y^2 - \frac{1}{2}z^2 = 1$, a hyperboloid of one sheet with axis the z-axis.

9. (a) $x^2 + y^2 = r^2$, so the equation becomes $z = r^2$.

(b) $r^2 = 2r\sin\theta$ or $r = 2\sin\theta$.

10. (a) $x^2 + y^2 = r^2$, so the equation becomes $r^2 + z^2 = 2$.

(b) $z = r^2(\cos^2\theta - \sin^2\theta)$ or $z = r^2\cos 2\theta$.

11.

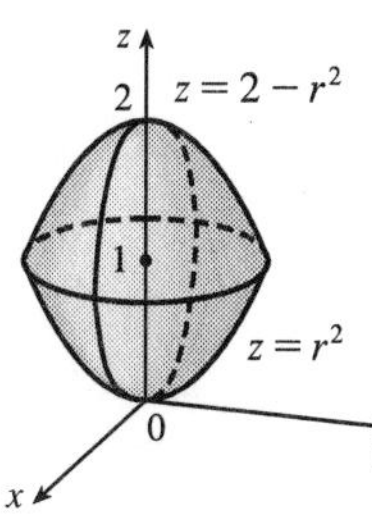

$z = r^2 = x^2 + y^2$ is a circular paraboloid with vertex $(0, 0, 0)$, opening upward. $z = 2 - r^2 \Rightarrow z - 2 = -(x^2 + y^2)$ is a circular paraboloid with vertex $(0, 0, 2)$ opening downward. Thus $r^2 \le z \le 2 - r^2$ is the solid region enclosed by these two surfaces.

12.

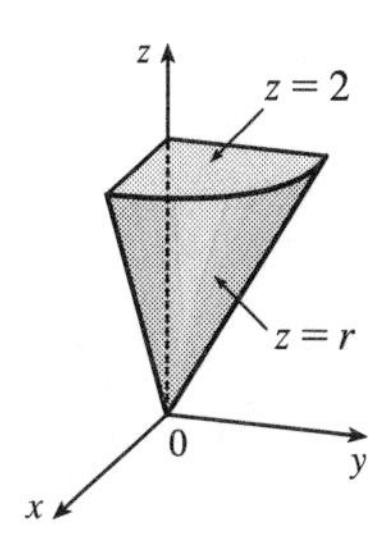

$z = r = \sqrt{x^2 + y^2}$ is a cone that opens upward. Thus $r \le z \le 2$ is the region above this cone and beneath the horizontal plane $z = 2$. $0 \le \theta \le \frac{\pi}{2}$ restricts the solid to that part of this region in the first octant.

13. We can position the cylindrical shell vertically so that its axis coincides with the z-axis and its base lies in the xy-plane. If we use centimeters as the unit of measurement, then cylindrical coordinates conveniently describe the shell as $6 \le r \le 7$, $0 \le \theta \le 2\pi$, $0 \le z \le 20$.

14. In cylindrical coordinates, the equations are $z = r^2$ and $z = 5 - r^2$. The curve of intersection is $r^2 = 5 - r^2$ or $r = \sqrt{5/2}$. So we graph the surfaces in cylindrical coordinates, with $0 \le r \le \sqrt{5/2}$. In Maple, we can use either the `coords=cylindrical` option in a regular `plot` command, or the `plots[cylinderplot]` command. In Mathematica, we can use `ParametricPlot3d`.

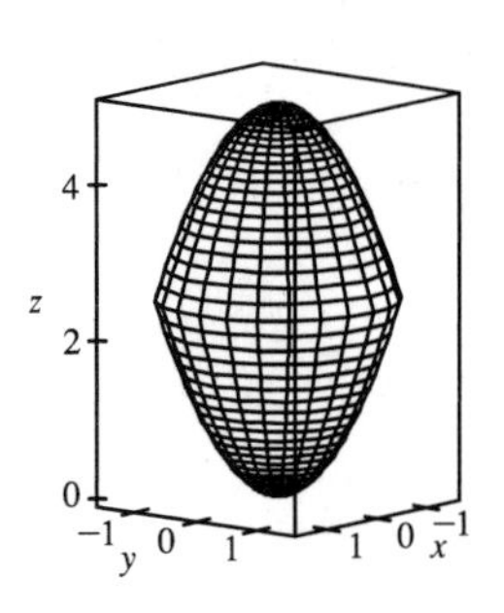

15.

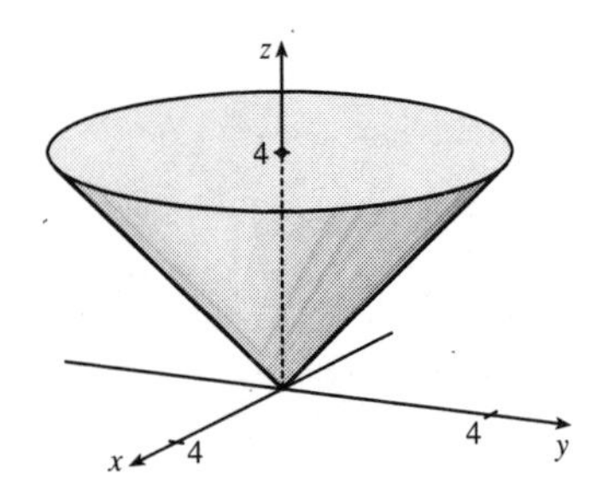

The region of integration is given in cylindrical coordinates by $E = \{(r, \theta, z) \mid 0 \le \theta \le 2\pi, 0 \le r \le 4, r \le z \le 4\}$. This represents the solid region bounded below by the cone $z = r$ and above by the horizontal plane $z = 4$.

$$\int_0^4 \int_0^{2\pi} \int_r^4 r\,dz\,d\theta\,dr = \int_0^4 \int_0^{2\pi} \big[rz\big]_{z=r}^{z=4}\,d\theta\,dr = \int_0^4 \int_0^{2\pi} r(4-r)\,d\theta\,dr$$
$$= \int_0^4 (4r - r^2)\,dr \int_0^{2\pi} d\theta = \big[2r^2 - \tfrac{1}{3}r^3\big]_0^4 \big[\theta\big]_0^{2\pi}$$
$$= \left(32 - \tfrac{64}{3}\right)(2\pi) = \tfrac{64\pi}{3}$$

16.

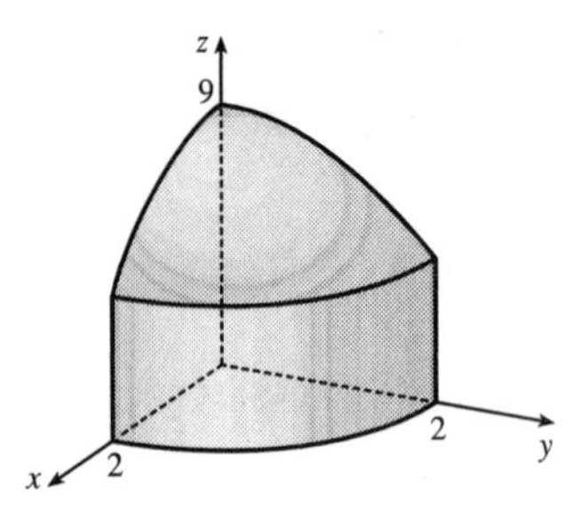

The region of integration is given in cylindrical coordinates by $E = \{(r, \theta, z) \mid 0 \le \theta \le \pi/2, 0 \le r \le 2, 0 \le z \le 9 - r^2\}$. This represents the solid region in the first octant enclosed by the circular cylinder $r = 2$, bounded above by $z = 9 - r^2$, a circular paraboloid, and bounded below by the xy-plane.

$$\int_0^{\pi/2} \int_0^2 \int_0^{9-r^2} r\,dz\,dr\,d\theta = \int_0^{\pi/2} \int_0^2 \big[rz\big]_{z=0}^{z=9-r^2}\,dr\,d\theta = \int_0^{\pi/2} \int_0^2 r(9 - r^2)\,dr\,d\theta$$
$$= \int_0^{\pi/2} d\theta \int_0^2 (9r - r^3)\,dr = \big[\theta\big]_0^{\pi/2} \big[\tfrac{9}{2}r^2 - \tfrac{1}{4}r^4\big]_0^2$$
$$= \tfrac{\pi}{2}(18 - 4) = 7\pi$$

17. In cylindrical coordinates, E is given by $\{(r, \theta, z) \mid 0 \le \theta \le 2\pi, 0 \le r \le 4, -5 \le z \le 4\}$. So

$$\iiint_E \sqrt{x^2 + y^2}\,dV = \int_0^{2\pi} \int_0^4 \int_{-5}^4 \sqrt{r^2}\,r\,dz\,dr\,d\theta = \int_0^{2\pi} d\theta \int_0^4 r^2\,dr \int_{-5}^4 dz$$
$$= \big[\theta\big]_0^{2\pi} \big[\tfrac{1}{3}r^3\big]_0^4 \big[z\big]_{-5}^4 = (2\pi)\left(\tfrac{64}{3}\right)(9) = 384\pi$$

18. The paraboloid $z = 1 - x^2 - y^2$ intersects the xy-plane in the circle $x^2 + y^2 = r^2 = 1$ or $r = 1$, so in cylindrical coordinates, E is given by $\left\{(r, \theta, z) \,\middle|\, 0 \le \theta \le \tfrac{\pi}{2}, 0 \le r \le 1, 0 \le z \le 1 - r^2\right\}$. Thus

$$\iiint_E (x^3 + xy^2)\,dV = \int_0^{\pi/2} \int_0^1 \int_0^{1-r^2} (r^3 \cos^3\theta + r^3 \cos\theta \sin^2\theta)\,r\,dz\,dr\,d\theta = \int_0^{\pi/2} \int_0^1 \int_0^{1-r^2} r^4 \cos\theta\,dz\,dr\,d\theta$$
$$= \int_0^{\pi/2} \int_0^1 r^4 \cos\theta \big[z\big]_{z=0}^{z=1-r^2}\,dr\,d\theta = \int_0^{\pi/2} \int_0^1 r^4(1 - r^2) \cos\theta\,dr\,d\theta$$
$$= \int_0^{\pi/2} \cos\theta \big[\tfrac{1}{5}r^5 - \tfrac{1}{7}r^7\big]_{r=0}^{r=1}\,d\theta = \int_0^{\pi/2} \tfrac{2}{35} \cos\theta\,d\theta = \tfrac{2}{35}\big[\sin\theta\big]_0^{\pi/2} = \tfrac{2}{35}$$

19. In cylindrical coordinates E is bounded by the paraboloid $z = 1 + r^2$, the cylinder $r^2 = 5$ or $r = \sqrt{5}$, and the xy-plane, so E is given by $\{(r, \theta, z) \mid 0 \le \theta \le 2\pi, 0 \le r \le \sqrt{5}, 0 \le z \le 1 + r^2\}$. Thus

$$\iiint_E e^z\,dV = \int_0^{2\pi}\int_0^{\sqrt{5}}\int_0^{1+r^2} e^z\,r\,dz\,dr\,d\theta = \int_0^{2\pi}\int_0^{\sqrt{5}} r\big[e^z\big]_{z=0}^{z=1+r^2}\,dr\,d\theta = \int_0^{2\pi}\int_0^{\sqrt{5}} r(e^{1+r^2}-1)\,dr\,d\theta$$
$$= \int_0^{2\pi} d\theta \int_0^{\sqrt{5}}\left(re^{1+r^2}-r\right)dr = 2\pi\left[\tfrac{1}{2}e^{1+r^2}-\tfrac{1}{2}r^2\right]_0^{\sqrt{5}} = \pi(e^6-e-5)$$

20. In cylindrical coordinates E is bounded by the planes $z=0$, $z=r\cos\theta+r\sin\theta+5$ and the cylinders $r=2$ and $r=3$, so E is given by $\{(r,\theta,z)\mid 0\le\theta\le 2\pi, 2\le r\le 3, 0\le z\le r\cos\theta+r\sin\theta+5\}$. Thus

$$\iiint_E x\,dV = \int_0^{2\pi}\int_2^3\int_0^{r\cos\theta+r\sin\theta+5}(r\cos\theta)\,r\,dz\,dr\,d\theta = \int_0^{2\pi}\int_2^3 (r^2\cos\theta)\big[\,z\,\big]_{z=0}^{z=r\cos\theta+r\sin\theta+5}\,dr\,d\theta$$
$$= \int_0^{2\pi}\int_2^3 (r^2\cos\theta)(r\cos\theta+r\sin\theta+5)\,dr\,d\theta = \int_0^{2\pi}\int_2^3\left(r^3(\cos^2\theta+\cos\theta\sin\theta)+5r^2\cos\theta\right)dr\,d\theta$$
$$= \int_0^{2\pi}\left[\tfrac{1}{4}r^4(\cos^2\theta+\cos\theta\sin\theta)+\tfrac{5}{3}r^3\cos\theta\right]_{r=2}^{r=3}d\theta$$
$$= \int_0^{2\pi}\left[\left(\tfrac{81}{4}-\tfrac{16}{4}\right)(\cos^2\theta+\cos\theta\sin\theta)+\tfrac{5}{3}(27-8)\cos\theta\right]d\theta$$
$$= \int_0^{2\pi}\left(\tfrac{65}{4}\left(\tfrac{1}{2}(1+\cos 2\theta)+\cos\theta\sin\theta\right)+\tfrac{95}{3}\cos\theta\right)d\theta = \left[\tfrac{65}{8}\theta+\tfrac{65}{16}\sin 2\theta+\tfrac{65}{8}\sin^2\theta+\tfrac{95}{3}\sin\theta\right]_0^{2\pi} = \tfrac{65}{4}\pi$$

21. In cylindrical coordinates, E is bounded by the cylinder $r=1$, the plane $z=0$, and the cone $z=2r$. So $E=\{(r,\theta,z)\mid 0\le\theta\le 2\pi, 0\le r\le 1, 0\le z\le 2r\}$ and

$$\iiint_E x^2\,dV = \int_0^{2\pi}\int_0^1\int_0^{2r} r^2\cos^2\theta\,r\,dz\,dr\,d\theta = \int_0^{2\pi}\int_0^1\left[r^3\cos^2\theta\,z\right]_{z=0}^{z=2r}dr\,d\theta$$
$$= \int_0^{2\pi}\int_0^1 2r^4\cos^2\theta\,dr\,d\theta = \int_0^{2\pi}\left[\tfrac{2}{5}r^5\cos^2\theta\right]_{r=0}^{r=1}d\theta = \tfrac{2}{5}\int_0^{2\pi}\cos^2\theta\,d\theta$$
$$= \frac{2}{5}\int_0^{2\pi}\frac{1+\cos 2\theta}{2}\,d\theta = \frac{1}{5}\left[\theta+\frac{1}{2}\sin 2\theta\right]_0^{2\pi} = \frac{2\pi}{5}$$

22. In cylindrical coordinates E is the solid region within the cylinder $r=1$ bounded above and below by the sphere $r^2+z^2=4$, so $E=\left\{(r,\theta,z)\mid 0\le\theta\le 2\pi, 0\le r\le 1, -\sqrt{4-r^2}\le z\le\sqrt{4-r^2}\right\}$. Thus the volume is

$$\iiint_E dV = \int_0^{2\pi}\int_0^1\int_{-\sqrt{4-r^2}}^{\sqrt{4-r^2}} r\,dz\,dr\,d\theta = \int_0^{2\pi}\int_0^1 2r\sqrt{4-r^2}\,dr\,d\theta$$
$$= \int_0^{2\pi}d\theta\int_0^1 2r\sqrt{4-r^2}\,dr = 2\pi\left[-\tfrac{2}{3}(4-r^2)^{3/2}\right]_0^1 = \tfrac{4}{3}\pi(8-3^{3/2})$$

23. (a) The paraboloids intersect when $x^2+y^2=36-3x^2-3y^2 \;\Rightarrow\; x^2+y^2=9$, so the region of integration is $D=\left\{(x,y)\mid x^2+y^2\le 9\right\}$. Then, in cylindrical coordinates,

$E=\left\{(r,\theta,z)\mid r^2\le z\le 36-3r^2, 0\le r\le 3, 0\le\theta\le 2\pi\right\}$ and

$$V=\int_0^{2\pi}\int_0^3\int_{r^2}^{36-3r^2} r\,dz\,dr\,d\theta = \int_0^{2\pi}\int_0^3\left(36r-4r^3\right)dr\,d\theta = \int_0^{2\pi}\left[18r^2-r^4\right]_{r=0}^{r=3}d\theta = \int_0^{2\pi}81\,d\theta = 162\pi.$$

(b) For constant density K, $m=KV=162\pi K$ from part (a). Since the region is homogeneous and symmetric, $M_{yz}=M_{xz}=0$ and

$$M_{xy}=\int_0^{2\pi}\int_0^3\int_{r^2}^{36-3r^2}(zK)\,r\,dz\,dr\,d\theta = K\int_0^{2\pi}\int_0^3 r\left[\tfrac{1}{2}z^2\right]_{z=r^2}^{z=36-3r^2}dr\,d\theta$$
$$= \tfrac{K}{2}\int_0^{2\pi}\int_0^3 r((36-3r^2)^2-r^4)\,dr\,d\theta = \tfrac{K}{2}\int_0^{2\pi}d\theta\int_0^3(8r^5-216r^3+1296r)\,dr$$
$$= \tfrac{K}{2}(2\pi)\left[\tfrac{8}{6}r^6-\tfrac{216}{4}r^4+\tfrac{1296}{2}r^2\right]_0^3 = \pi K(2430) = 2430\pi K$$

Thus $(\overline{x},\overline{y},\overline{z}) = \left(\dfrac{M_{yz}}{m},\dfrac{M_{xz}}{m},\dfrac{M_{xy}}{m}\right) = \left(0,0,\frac{2430\pi K}{162\pi K}\right) = (0,0,15)$.

24. (a) $V = \int_{-\pi/2}^{\pi/2} \int_0^{a\cos\theta} \int_{-\sqrt{a^2-r^2}}^{\sqrt{a^2-r^2}} r\,dz\,dr\,d\theta$

$= 4\int_0^{\pi/2} \int_0^{a\cos\theta} \int_0^{\sqrt{a^2-r^2}} r\,dz\,dr\,d\theta$

$= 4\int_0^{\pi/2} \int_0^{a\cos\theta} r\sqrt{a^2-r^2}\,dr\,d\theta$

$= -\frac{4}{3}\int_0^{\pi/2} \left[\left(a^2-r^2\right)^{3/2}\right]_{r=0}^{r=a\cos\theta} d\theta$

$= -\frac{4}{3}\int_0^{\pi/2} \left[\left(a^2-a^2\cos^2\theta\right)^{3/2} - a^3\right] d\theta$

$= -\frac{4}{3}\int_0^{\pi/2} \left[\left(a^2\sin^2\theta\right)^{3/2} - a^3\right] d\theta$

$= -\frac{4}{3}\int_0^{\pi/2} \left(a^3\sin^3\theta - a^3\right) d\theta$

$= -\dfrac{4a^3}{3}\displaystyle\int_0^{\pi/2} \left[\sin\theta\left(1-\cos^2\theta\right) - 1\right] d\theta$

$= -\dfrac{4a^3}{3}\left[-\cos\theta + \frac{1}{3}\cos^3\theta - \theta\right]_0^{\pi/2} = -\dfrac{4a^3}{3}\left(-\frac{\pi}{2} + \frac{2}{3}\right) = \frac{2}{9}a^3(3\pi - 4)$

(b)

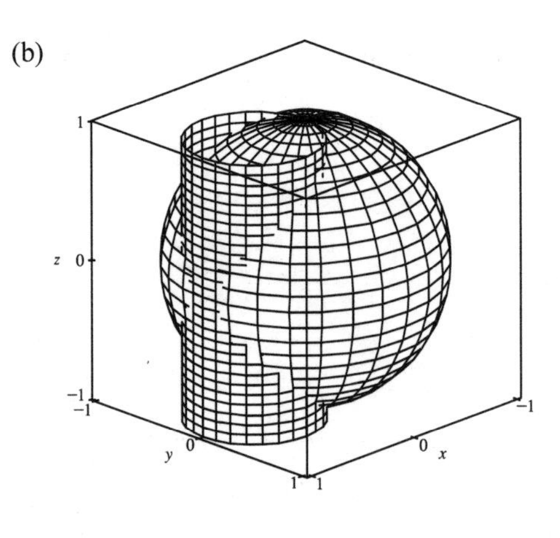

To plot the cylinder and the sphere on the same screen in Maple, we can use the sequence of commands

```
sphere:=plot3d(1,theta=0..2*Pi,phi=0..Pi,coords=spherical):
cylinder:=plot3d([cos(theta),theta,z],
theta=0..2*Pi,z=-1..1,coords=cylindrical):
with(plots): display3d({sphere,cylinder});
```

In Mathematica, we can use

```
sphere=SphericalPlot3d[1,{theta,0,2Pi},{phi,0,Pi}]
cylinder=ParametricPlot3d[{Sin[theta],Cos[theta],z},
{theta,0,2Pi},{z,-1,1}]
Show[{sphere,cylinder}]
```

25. The paraboloid $z = 4x^2 + 4y^2$ intersects the plane $z = a$ when $a = 4x^2 + 4y^2$ or $x^2 + y^2 = \frac{1}{4}a$. So, in cylindrical coordinates, $E = \left\{(r,\theta,z) \mid 0 \le r \le \frac{1}{2}\sqrt{a}, 0 \le \theta \le 2\pi, 4r^2 \le z \le a\right\}$. Thus

$$m = \int_0^{2\pi}\int_0^{\sqrt{a}/2}\int_{4r^2}^{a} Kr\,dz\,dr\,d\theta = K\int_0^{2\pi}\int_0^{\sqrt{a}/2}\left(ar - 4r^3\right)dr\,d\theta$$

$$= K\int_0^{2\pi}\left[\tfrac{1}{2}ar^2 - r^4\right]_{r=0}^{r=\sqrt{a}/2} d\theta = K\int_0^{2\pi}\tfrac{1}{16}a^2\,d\theta = \tfrac{1}{8}a^2\pi K$$

Since the region is homogeneous and symmetric, $M_{yz} = M_{xz} = 0$ and

$$M_{xy} = \int_0^{2\pi}\int_0^{\sqrt{a}/2}\int_{4r^2}^{a} Krz\,dz\,dr\,d\theta = K\int_0^{2\pi}\int_0^{\sqrt{a}/2}\left(\tfrac{1}{2}a^2r - 8r^5\right)dr\,d\theta$$

$$= K\int_0^{2\pi}\left[\tfrac{1}{4}a^2r^2 - \tfrac{4}{3}r^6\right]_{r=0}^{r=\sqrt{a}/2} d\theta = K\int_0^{2\pi}\tfrac{1}{24}a^3\,d\theta = \tfrac{1}{12}a^3\pi K$$

Hence $(\overline{x},\overline{y},\overline{z}) = \left(0,0,\frac{2}{3}a\right)$.

26. Since density is proportional to the distance from the z-axis, we can say $\rho(x,y,z) = K\sqrt{x^2+y^2}$. Then

$$m = 2\int_0^{2\pi}\int_0^a\int_0^{\sqrt{a^2-r^2}} Kr^2\,dz\,dr\,d\theta = 2K\int_0^{2\pi}\int_0^a r^2\sqrt{a^2-r^2}\,dr\,d\theta$$
$$= 2K\int_0^{2\pi}\left[\tfrac{1}{8}r(2r^2-a^2)\sqrt{a^2-r^2}+\tfrac{1}{8}a^4\sin^{-1}(r/a)\right]_{r=0}^{r=a}d\theta = 2K\int_0^{2\pi}\left[\left(\tfrac{1}{8}a^4\right)\left(\tfrac{\pi}{2}\right)\right]d\theta = \tfrac{1}{4}a^4\pi^2K$$

27. The region of integration is the region above the cone $z=\sqrt{x^2+y^2}$, or $z=r$, and below the plane $z=2$. Also, we have $-2\le y\le 2$ with $-\sqrt{4-y^2}\le x\le\sqrt{4-y^2}$ which describes a circle of radius 2 in the xy-plane centered at $(0,0)$. Thus,

$$\int_{-2}^{2}\int_{-\sqrt{4-y^2}}^{\sqrt{4-y^2}}\int_{\sqrt{x^2+y^2}}^{2} xz\,dz\,dx\,dy = \int_0^{2\pi}\int_0^2\int_r^2 (r\cos\theta)\,z\,r\,dz\,dr\,d\theta = \int_0^{2\pi}\int_0^2\int_r^2 r^2(\cos\theta)\,z\,dz\,dr\,d\theta$$
$$= \int_0^{2\pi}\int_0^2 r^2(\cos\theta)\left[\tfrac{1}{2}z^2\right]_{z=r}^{z=2}dr\,d\theta = \tfrac{1}{2}\int_0^{2\pi}\int_0^2 r^2(\cos\theta)\left(4-r^2\right)dr\,d\theta$$
$$= \tfrac{1}{2}\int_0^{2\pi}\cos\theta\,d\theta\int_0^2\left(4r^2-r^4\right)dr = \tfrac{1}{2}\left[\sin\theta\right]_0^{2\pi}\left[\tfrac{4}{3}r^3-\tfrac{1}{5}r^5\right]_0^2 = 0$$

28. The region of integration is the region above the plane $z=0$ and below the paraboloid $z=9-x^2-y^2$. Also, we have $-3\le x\le 3$ with $0\le y\le\sqrt{9-x^2}$ which describes the upper half of a circle of radius 3 in the xy-plane centered at $(0,0)$. Thus,

$$\int_{-3}^{3}\int_0^{\sqrt{9-x^2}}\int_0^{9-x^2-y^2}\sqrt{x^2+y^2}\,dz\,dy\,dx = \int_0^{\pi}\int_0^3\int_0^{9-r^2}\sqrt{r^2}\,r\,dz\,dr\,d\theta = \int_0^{\pi}\int_0^3\int_0^{9-r^2} r^2\,dz\,dr\,d\theta$$
$$= \int_0^{\pi}\int_0^3 r^2\left(9-r^2\right)dr\,d\theta = \int_0^{\pi}d\theta\int_0^3\left(9r^2-r^4\right)dr$$
$$= \left[\theta\right]_0^{\pi}\left[3r^3-\tfrac{1}{5}r^5\right]_0^3 = \pi\left(81-\tfrac{243}{5}\right) = \tfrac{162}{5}\pi$$

29. (a) The mountain comprises a solid conical region C. The work done in lifting a small volume of material ΔV with density $g(P)$ to a height $h(P)$ above sea level is $h(P)g(P)\,\Delta V$. Summing over the whole mountain we get $W=\iiint_C h(P)g(P)\,dV$.

(b) Here C is a solid right circular cone with radius $R=62{,}000$ ft, height $H=12{,}400$ ft, and density $g(P)=200\text{ lb/ft}^3$ at all points P in C. We use cylindrical coordinates:

$$W = \int_0^{2\pi}\int_0^H\int_0^{R(1-z/H)} z\cdot 200r\,dr\,dz\,d\theta = 2\pi\int_0^H 200z\left[\tfrac{1}{2}r^2\right]_{r=0}^{r=R(1-z/H)}dz$$
$$= 400\pi\int_0^H z\frac{R^2}{2}\left(1-\frac{z}{H}\right)^2dz = 200\pi R^2\int_0^H\left(z-\frac{2z^2}{H}+\frac{z^3}{H^2}\right)dz$$
$$= 200\pi R^2\left[\frac{z^2}{2}-\frac{2z^3}{3H}+\frac{z^4}{4H^2}\right]_0^H = 200\pi R^2\left(\frac{H^2}{2}-\frac{2H^2}{3}+\frac{H^2}{4}\right)$$
$$= \tfrac{50}{3}\pi R^2H^2 = \tfrac{50}{3}\pi(62{,}000)^2(12{,}400)^2 \approx 3.1\times 10^{19}\text{ ft-lb}$$

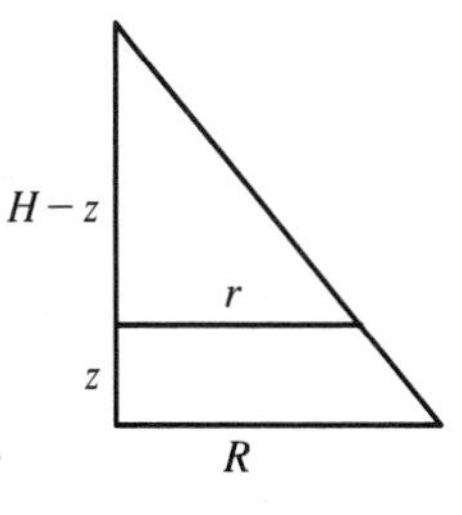

$$\frac{r}{R}=\frac{H-z}{H}=1-\frac{z}{H}$$

12.7 Triple Integrals in Spherical Coordinates

1. (a)

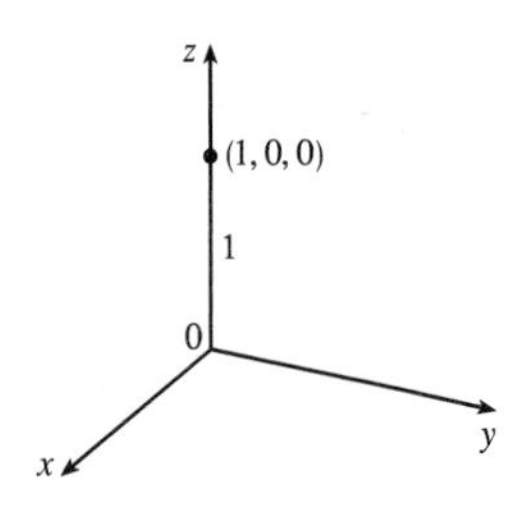

$x = \rho \sin \phi \cos \theta = (1) \sin 0 \cos 0 = 0$,
$y = \rho \sin \phi \sin \theta = (1) \sin 0 \sin 0 = 0$, and
$z = \rho \cos \phi = (1) \cos 0 = 1$ so the point is
$(0, 0, 1)$ in rectangular coordinates.

(b)

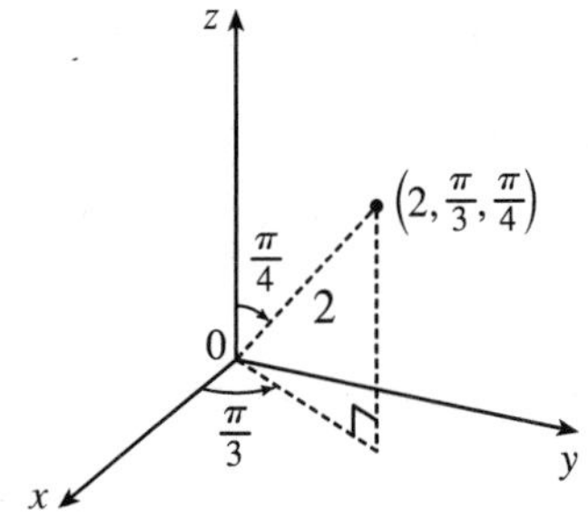

$x = 2 \sin \frac{\pi}{4} \cos \frac{\pi}{3} = \frac{\sqrt{2}}{2}$, $y = 2 \sin \frac{\pi}{4} \sin \frac{\pi}{3} = \frac{\sqrt{6}}{2}$,
$z = 2 \cos \frac{\pi}{4} = \sqrt{2}$ so the point is $\left(\frac{\sqrt{2}}{2}, \frac{\sqrt{6}}{2}, \sqrt{2}\right)$ in
rectangular coordinates.

2. (a)

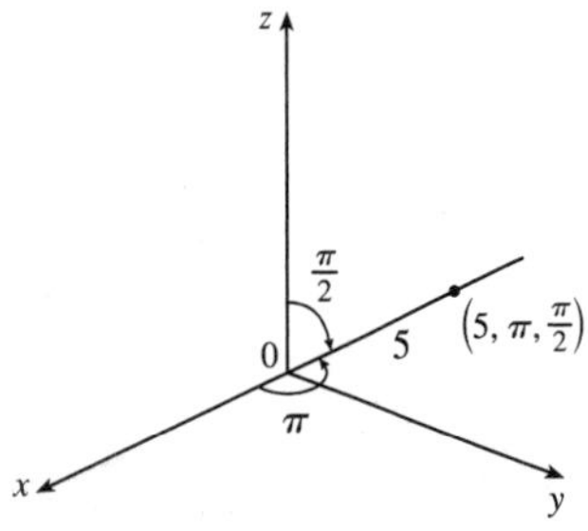

$x = 5 \sin \frac{\pi}{2} \cos \pi = -5$, $y = 5 \sin \frac{\pi}{2} \sin \pi = 0$,
$z = 5 \cos \frac{\pi}{2} = 0$ so the point is $(-5, 0, 0)$ in
rectangular coordinates.

(b)

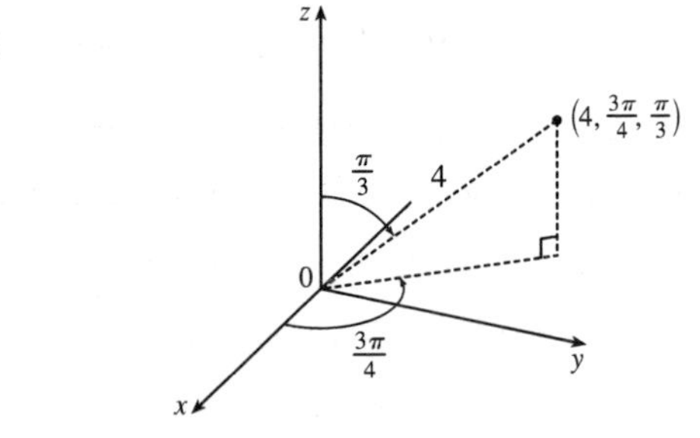

$x = 4 \sin \frac{\pi}{3} \cos \frac{3\pi}{4} = 4\left(\frac{\sqrt{3}}{2}\right)\left(-\frac{\sqrt{2}}{2}\right) = -\sqrt{6}$,

$y = 4 \sin \frac{\pi}{3} \sin \frac{3\pi}{4} = 4\left(\frac{\sqrt{3}}{2}\right)\left(\frac{\sqrt{2}}{2}\right) = \sqrt{6}$,

$z = 4 \cos \frac{\pi}{3} = 4\left(\frac{1}{2}\right) = 2$ so the point is $\left(-\sqrt{6}, \sqrt{6}, 2\right)$
in rectangular coordinates.

3. (a) $\rho = \sqrt{x^2 + y^2 + z^2} = \sqrt{1 + 3 + 12} = 4$, $\cos \phi = \dfrac{z}{\rho} = \dfrac{2\sqrt{3}}{4} = \dfrac{\sqrt{3}}{2} \quad \Rightarrow \quad \phi = \dfrac{\pi}{6}$, and

$\cos \theta = \dfrac{x}{\rho \sin \phi} = \dfrac{1}{4 \sin(\pi/6)} = \dfrac{1}{2} \quad \Rightarrow \quad \theta = \dfrac{\pi}{3}$ [since $y > 0$]. Thus spherical coordinates are $\left(4, \dfrac{\pi}{3}, \dfrac{\pi}{6}\right)$.

(b) $\rho = \sqrt{0 + 1 + 1} = \sqrt{2}$, $\cos \phi = \dfrac{-1}{\sqrt{2}} \quad \Rightarrow \quad \phi = \dfrac{3\pi}{4}$, and $\cos \theta = \dfrac{0}{\sqrt{2} \sin(3\pi/4)} = 0 \quad \Rightarrow \quad \theta = \dfrac{3\pi}{2}$

[since $y < 0$]. Thus spherical coordinates are $\left(\sqrt{2}, \dfrac{3\pi}{2}, \dfrac{3\pi}{4}\right)$.

4. (a) $\rho = \sqrt{x^2 + y^2 + z^2} = \sqrt{0 + 3 + 1} = 2$, $\cos \phi = \dfrac{z}{\rho} = \dfrac{1}{2} \quad \Rightarrow \quad \phi = \dfrac{\pi}{3}$, and $\cos \theta = \dfrac{x}{\rho \sin \phi} = \dfrac{0}{2 \sin(\pi/3)} = 0 \quad \Rightarrow$

$\theta = \dfrac{\pi}{2}$ [since $y > 0$]. Thus spherical coordinates are $\left(2, \dfrac{\pi}{2}, \dfrac{\pi}{3}\right)$.

(b) $\rho = \sqrt{1 + 1 + 6} = 2\sqrt{2}$, $\cos \phi = \dfrac{\sqrt{6}}{2\sqrt{2}} = \dfrac{\sqrt{3}}{2} \quad \Rightarrow \quad \phi = \dfrac{\pi}{6}$, and $\cos \theta = \dfrac{-1}{2\sqrt{2} \sin(\pi/6)} = -\dfrac{1}{\sqrt{2}} \quad \Rightarrow \quad \theta = \dfrac{3\pi}{4}$

[since $y > 0$]. Thus spherical coordinates are $\left(2\sqrt{2}, \dfrac{3\pi}{4}, \dfrac{\pi}{6}\right)$.

5. Since $\phi = \frac{\pi}{3}$, the surface is the top half of the right circular cone with vertex at the origin and axis the positive z-axis.

6. Since $\rho = 3$, $x^2 + y^2 + z^2 = 9$ and the surface is a sphere with center the origin and radius 3.

7. Since $\rho \sin\phi = 2$ and $x = \rho \sin\phi \cos\theta$, $x = 2\cos\theta$. Also $y = \rho \sin\phi \sin\theta$ so $y = 2\sin\theta$. Then $x^2 + y^2 = 4\cos^2\theta + 4\sin^2\theta = 4$, a circular cylinder of radius 2 about the z-axis.

8. $\rho = 2\cos\phi \quad \Rightarrow \quad \rho^2 = 2\rho\cos\phi = 2z \quad \Leftrightarrow \quad x^2 + y^2 + z^2 = 2z \quad \Leftrightarrow \quad x^2 + y^2 + (z-1)^2 = 1$.
Therefore, the surface is a sphere of radius 1 centered at $(0, 0, 1)$.

9. (a) $x = \rho \sin\phi \cos\theta$, $y = \rho \sin\phi \sin\theta$, and $z = \rho\cos\phi$, so the equation becomes

$\rho\cos\phi = (\rho\sin\phi\cos\theta)^2 + (\rho\sin\phi\sin\theta)^2$ or $\rho\cos\phi = \rho^2\sin^2\phi$ or $\rho\sin^2\phi = \cos\phi$.

(b) $\rho^2\sin^2\phi(\cos^2\theta + \sin^2\theta) = 2\rho\sin\phi\sin\theta$ or $\rho\sin^2\phi = 2\sin\phi\sin\theta$ or $\rho\sin\phi = 2\sin\theta$.

10. (a) $x^2 + y^2 + z^2 = \rho^2$, so the equation becomes $\rho^2 = 2$ or $\rho = \sqrt{2}$.

(b) $\rho\cos\phi = \rho^2\sin^2\phi(\cos^2\theta - \sin^2\theta)$ or $\cos\phi = \rho\sin^2\phi\cos 2\theta$.

11. $\rho = 2$ represents a sphere of radius 2, centered at the origin, so $\rho \le 2$ is this sphere and its interior. $0 \le \phi \le \frac{\pi}{2}$ restricts the solid to that portion of the region that lies on or above the xy-plane, and $0 \le \theta \le \frac{\pi}{2}$ further restricts the solid to the first octant. Thus the solid is the portion in the first octant of the solid ball centered at the origin with radius 2.

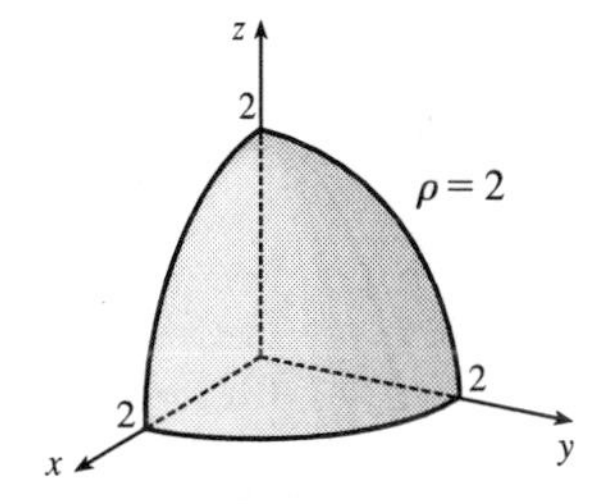

12. $2 \le \rho \le 3$ represents the solid region between and including the spheres of radii 2 and 3, centered at the origin. $\frac{\pi}{2} \le \phi \le \pi$ restricts the solid to that portion on or below the xy-plane.

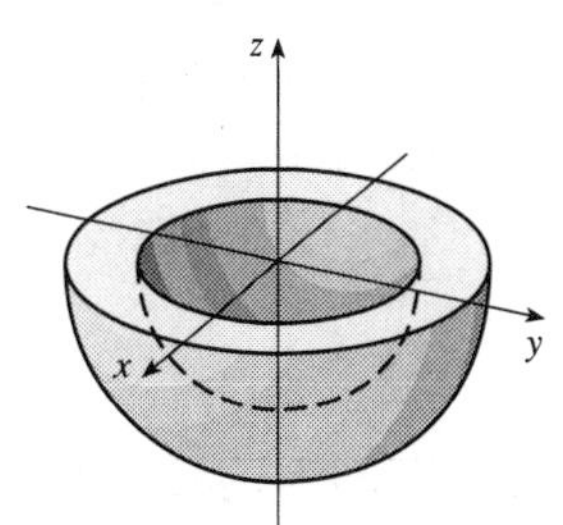

13. $-\frac{\pi}{2} \le \theta \le \frac{\pi}{2}$ restricts the solid to the 4 octants in which x is positive. $\rho = \sec\phi \quad \Rightarrow \quad \rho\cos\phi = z = 1$, which is the equation of a horizontal plane. $0 \le \phi \le \frac{\pi}{6}$ describes a cone, opening upward. So the solid lies above the cone $\phi = \frac{\pi}{6}$ and below the plane $z = 1$.

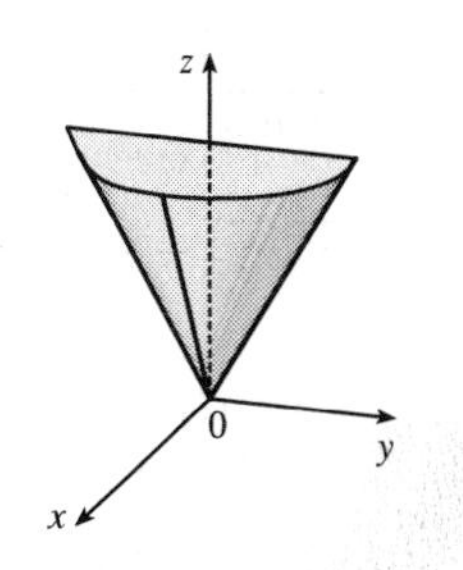

14. $\rho = 2 \quad \Leftrightarrow \quad x^2 + y^2 + z^2 = 4$, which is a sphere of radius 2, centered at the origin. Hence $\rho \le 2$ is this sphere and its interior. $0 \le \phi \le \frac{\pi}{3}$ restricts the solid to that section of this ball that lies above the cone $\phi = \frac{\pi}{3}$.

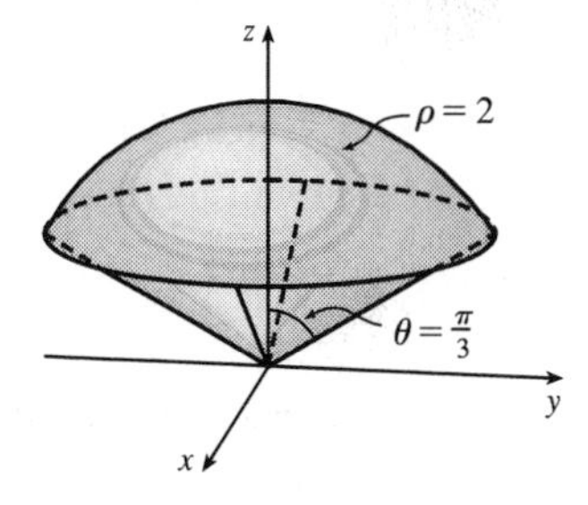

15. $z \geq \sqrt{x^2+y^2}$ because the solid lies above the cone. Squaring both sides of this inequality gives $z^2 \geq x^2+y^2 \Rightarrow$ $2z^2 \geq x^2+y^2+z^2=\rho^2 \Rightarrow z^2=\rho^2\cos^2\phi \geq \frac{1}{2}\rho^2 \Rightarrow \cos^2\phi \geq \frac{1}{2}$. The cone opens upward so that the inequality is $\cos\phi \geq \frac{1}{\sqrt{2}}$, or equivalently $0 \leq \phi \leq \frac{\pi}{4}$. In spherical coordinates the sphere $z=x^2+y^2+z^2$ is $\rho\cos\phi=\rho^2 \Rightarrow$ $\rho=\cos\phi$. $0 \leq \rho \leq \cos\phi$ because the solid lies below the sphere. The solid can therefore be described as the region in spherical coordinates satisfying $0 \leq \rho \leq \cos\phi$, $0 \leq \phi \leq \frac{\pi}{4}$.

16. (a) The hollow ball is a spherical shell with outer radius 15 cm and inner radius 14.5 cm. If we center the ball at the origin of the coordinate system and use centimeters as the unit of measurement, then spherical coordinates conveniently describe the hollow ball as $14.5 \leq \rho \leq 15$, $0 \leq \theta \leq 2\pi$, $0 \leq \phi \leq \pi$.

(b) If we position the ball as in part (a), one possibility is to take the half of the ball that is above the xy-plane which is described by $14.5 \leq \rho \leq 15$, $0 \leq \theta \leq 2\pi$, $0 \leq \phi \leq \pi/2$.

17.

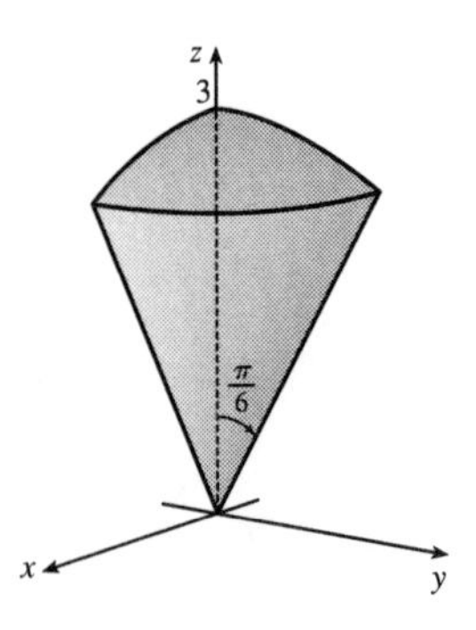

The region of integration is given in spherical coordinates by $E=\{(\rho,\theta,\phi) \mid 0 \leq \rho \leq 3, 0 \leq \theta \leq \pi/2, 0 \leq \phi \leq \pi/6\}$. This represents the solid region in the first octant bounded above by the sphere $\rho=3$ and below by the cone $\phi=\pi/6$.

$$\begin{aligned}\int_0^{\pi/6}\int_0^{\pi/2}\int_0^3 \rho^2\sin\phi\,d\rho\,d\theta\,d\phi &= \int_0^{\pi/6}\sin\phi\,d\phi\int_0^{\pi/2}d\theta\int_0^3\rho^2\,d\rho \\ &= \left[-\cos\phi\right]_0^{\pi/6}\left[\theta\right]_0^{\pi/2}\left[\tfrac{1}{3}\rho^3\right]_0^3 \\ &= \left(1-\frac{\sqrt{3}}{2}\right)\left(\frac{\pi}{2}\right)(9)=\frac{9\pi}{4}(2-\sqrt{3})\end{aligned}$$

18.

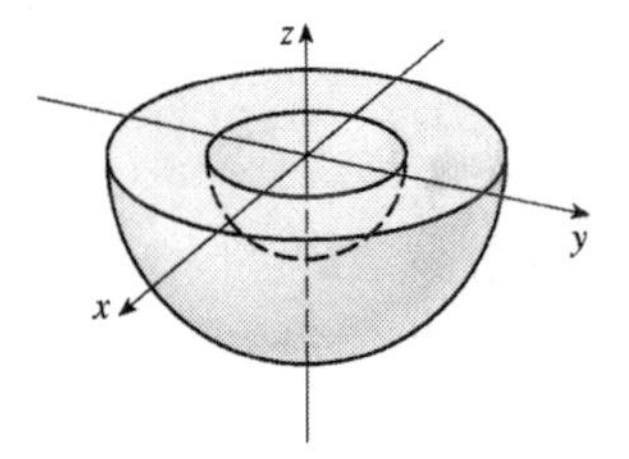

The region of integration is given in spherical coordinates by $E=\{(\rho,\theta,\phi) \mid 1 \leq \rho \leq 2, 0 \leq \theta \leq 2\pi, \pi/2 \leq \phi \leq \pi\}$. This represents the solid region between the spheres $\rho=1$ and $\rho=2$ and below the xy-plane.

$$\begin{aligned}\int_0^{2\pi}\int_{\pi/2}^{\pi}\int_1^2 \rho^2\sin\phi\,d\rho\,d\phi\,d\theta &= \int_0^{2\pi}d\theta\int_{\pi/2}^{\pi}\sin\phi\,d\phi\int_1^2\rho^2\,d\rho \\ &= \left[\theta\right]_0^{2\pi}\left[-\cos\phi\right]_{\pi/2}^{\pi}\left[\tfrac{1}{3}\rho^3\right]_1^2 \\ &= 2\pi(1)\left(\tfrac{7}{3}\right)=\tfrac{14\pi}{3}\end{aligned}$$

19. The solid E is most conveniently described if we use cylindrical coordinates:

$E=\left\{(r,\theta,z) \mid 0 \leq \theta \leq \frac{\pi}{2}, 0 \leq r \leq 3, 0 \leq z \leq 2\right\}$. Then

$\iiint_E f(x,y,z)\,dV=\int_0^{\pi/2}\int_0^3\int_0^2 f(r\cos\theta, r\sin\theta, z)\,r\,dz\,dr\,d\theta$.

20. The solid E is most conveniently described if we use spherical coordinates:

$E=\left\{(\rho,\theta,\phi) \mid 1 \leq \rho \leq 2, \frac{\pi}{2} \leq \theta \leq 2\pi, 0 \leq \phi \leq \frac{\pi}{2}\right\}$. Then

$\iiint_E f(x,y,z)\,dV=\int_0^{\pi/2}\int_{\pi/2}^{2\pi}\int_1^2 f(\rho\sin\phi\cos\theta, \rho\sin\phi\sin\theta, \rho\cos\phi)\,\rho^2\sin\phi\,d\rho\,d\theta\,d\phi$.

21. In spherical coordinates, B is represented by $\{(\rho,\theta,\phi) \mid 0 \leq \rho \leq 1, 0 \leq \theta \leq 2\pi, 0 \leq \phi \leq \pi\}$. Thus

$$\begin{aligned}\iiint_B (x^2+y^2+z^2)\,dV &= \int_0^{\pi}\int_0^{2\pi}\int_0^1(\rho^2)\rho^2\sin\phi\,d\rho\,d\theta\,d\phi=\int_0^{\pi}\sin\phi\,d\phi\int_0^{2\pi}d\theta\int_0^1\rho^4\,d\rho \\ &= \left[-\cos\phi\right]_0^{\pi}\left[\theta\right]_0^{2\pi}\left[\tfrac{1}{5}\rho^5\right]_0^1=(2)(2\pi)\left(\tfrac{1}{5}\right)=\tfrac{4\pi}{5}\end{aligned}$$

22. In spherical coordinates, H is represented by $\{(\rho,\theta,\phi)\mid 0\le\rho\le 1, 0\le\theta\le 2\pi, 0\le\phi\le\frac{\pi}{2}\}$. Thus

$$\iiint_H (x^2+y^2)\,dV = \int_0^{2\pi}\int_0^{\pi/2}\int_0^1 (\rho^2\sin^2\phi)\,\rho^2\sin\phi\,d\rho\,d\phi\,d\theta = \int_0^{2\pi} d\theta\int_0^{\pi/2}\sin^3\phi\,d\phi\int_0^1\rho^4\,d\rho$$
$$= \big[\,\theta\,\big]_0^{2\pi}\left[-\cos\phi+\tfrac13\cos^3\phi\right]_0^{\pi/2}\left[\tfrac15\rho^5\right]_0^1 = \tfrac{4\pi}{15}$$

23. In spherical coordinates, E is represented by $\{(\rho,\theta,\phi)\mid 1\le\rho\le 2, 0\le\theta\le\frac{\pi}{2}, 0\le\phi\le\frac{\pi}{2}\}$. Thus

$$\iiint_E z\,dV = \int_0^{\pi/2}\int_0^{\pi/2}\int_1^2 (\rho\cos\phi)\,\rho^2\sin\phi\,d\rho\,d\theta\,d\phi = \int_0^{\pi/2}\cos\phi\sin\phi\,d\phi\int_0^{\pi/2}d\theta\int_1^2\rho^3\,d\rho$$
$$= \left[\tfrac12\sin^2\phi\right]_0^{\pi/2}\big[\,\theta\,\big]_0^{\pi/2}\left[\tfrac14\rho^4\right]_1^2 = \left(\tfrac12\right)\left(\tfrac{\pi}{2}\right)\left(\tfrac{15}{4}\right) = \tfrac{15\pi}{16}$$

24.
$$\iiint_E e^{\sqrt{x^2+y^2+z^2}}\,dV = \int_0^{\pi/2}\int_0^{\pi/2}\int_0^3 e^{\sqrt{\rho^2}}\rho^2\sin\phi\,d\rho\,d\phi\,d\theta = \int_0^{\pi/2}\int_0^{\pi/2}\int_0^3\rho^2e^\rho\sin\phi\,d\rho\,d\phi\,d\theta$$
$$= \int_0^{\pi/2}d\theta\int_0^{\pi/2}\sin\phi\,d\phi\int_0^3\rho^2e^\rho\,d\rho = \big[\,\theta\,\big]_0^{\pi/2}\big[-\cos\phi\big]_0^{\pi/2}\big[(\rho^2-2\rho+2)e^\rho\big]_0^3$$
[integrate by parts twice]
$$= \tfrac{\pi}{2}(0+1)(5e^3-2) = \tfrac{\pi}{2}(5e^3-2)$$

25.
$$\iiint_E x^2\,dV = \int_0^\pi\int_0^\pi\int_3^4(\rho\sin\phi\cos\theta)^2\rho^2\sin\phi\,d\rho\,d\phi\,d\theta = \int_0^\pi\cos^2\theta\,d\theta\int_0^\pi\sin^3\phi\,d\phi\int_3^4\rho^4\,d\rho$$
$$= \left[\tfrac12\theta+\tfrac14\sin2\theta\right]_0^\pi\left[-\tfrac13(2+\sin^2\phi)\cos\phi\right]_0^\pi\left[\tfrac15\rho^5\right]_3^4 = \left(\tfrac{\pi}{2}\right)\left(\tfrac23+\tfrac23\right)\tfrac15(4^5-3^5) = \tfrac{1562}{15}\pi$$

26. In spherical coordinates, the sphere $x^2+y^2+z^2=4$ is equivalent to $\rho=2$ and the cone $z=\sqrt{x^2+y^2}$ is represented by $\phi=\frac{\pi}{4}$. Thus, the solid is given by $\{(\rho,\theta,\phi)\mid 0\le\rho\le 2, 0\le\theta\le 2\pi, \frac{\pi}{4}\le\phi\le\frac{\pi}{2}\}$ and

$$V = \int_{\pi/4}^{\pi/2}\int_0^{2\pi}\int_0^2\rho^2\sin\phi\,d\rho\,d\theta\,d\phi = \int_{\pi/4}^{\pi/2}\sin\phi\,d\phi\int_0^{2\pi}d\theta\int_0^2\rho^2\,d\rho$$
$$= \big[-\cos\phi\big]_{\pi/4}^{\pi/2}\big[\,\theta\,\big]_0^{2\pi}\left[\tfrac13\rho^3\right]_0^2 = \left(\tfrac{\sqrt2}{2}\right)(2\pi)\left(\tfrac83\right) = \tfrac{8\sqrt2\,\pi}{3}$$

27. (a) Since $\rho=4\cos\phi$ implies $\rho^2=4\rho\cos\phi$, the equation is that of a sphere of radius 2 with center at $(0,0,2)$. Thus

$$V = \int_0^{2\pi}\int_0^{\pi/3}\int_0^{4\cos\phi}\rho^2\sin\phi\,d\rho\,d\phi\,d\theta = \int_0^{2\pi}\int_0^{\pi/3}\left[\tfrac13\rho^3\right]_{\rho=0}^{\rho=4\cos\phi}\sin\phi\,d\phi\,d\theta = \int_0^{2\pi}\int_0^{\pi/3}\left(\tfrac{64}{3}\cos^3\phi\right)\sin\phi\,d\phi\,d\theta$$
$$= \int_0^{2\pi}\left[-\tfrac{16}{3}\cos^4\phi\right]_{\phi=0}^{\phi=\pi/3}d\theta = \int_0^{2\pi}-\tfrac{16}{3}\left(\tfrac{1}{16}-1\right)d\theta = 5\theta\Big]_0^{2\pi} = 10\pi$$

(b) By the symmetry of the problem $M_{yz}=M_{xz}=0$. Then

$$M_{xy} = \int_0^{2\pi}\int_0^{\pi/3}\int_0^{4\cos\phi}\rho^3\cos\phi\sin\phi\,d\rho\,d\phi\,d\theta = \int_0^{2\pi}\int_0^{\pi/3}\cos\phi\sin\phi\left(64\cos^4\phi\right)d\phi\,d\theta$$
$$= \int_0^{2\pi}64\left[-\tfrac16\cos^6\phi\right]_{\phi=0}^{\phi=\pi/3}d\theta = \int_0^{2\pi}\tfrac{21}{2}\,d\theta = 21\pi$$

Hence $(\overline{x},\overline{y},\overline{z}) = (0,0,2.1)$.

28. (a) Placing the center of the base at $(0,0,0)$, $\rho(x,y,z)=K\sqrt{x^2+y^2+z^2}$ is the density function. So

$$m = \int_0^{2\pi}\int_0^{\pi/2}\int_0^a K\rho^3\sin\phi\,d\rho\,d\phi\,d\theta = K\int_0^{2\pi}d\theta\int_0^{\pi/2}\sin\phi\,d\phi\int_0^a\rho^3\,d\rho$$
$$= K\big[\,\theta\,\big]_0^{2\pi}\big[-\cos\phi\big]_0^{\pi/2}\left[\tfrac14\rho^4\right]_0^a = K(2\pi)(1)\left(\tfrac14a^4\right) = \tfrac12\pi Ka^4$$

(b) By the symmetry of the problem $M_{yz}=M_{xz}=0$. Then

$$M_{xy} = \int_0^{2\pi}\int_0^{\pi/2}\int_0^a K\rho^4\sin\phi\cos\phi\,d\rho\,d\phi\,d\theta = K\int_0^{2\pi}d\theta\int_0^{\pi/2}\sin\phi\cos\phi\,d\phi\int_0^a\rho^4\,d\rho$$
$$= K\big[\,\theta\,\big]_0^{2\pi}\left[\tfrac12\sin^2\phi\right]_0^{\pi/2}\left[\tfrac15\rho^5\right]_0^a = K(2\pi)\left(\tfrac12\right)\left(\tfrac15a^5\right) = \tfrac15\pi Ka^5$$

Hence $(\overline{x},\overline{y},\overline{z}) = \left(0,0,\tfrac25a\right)$.

(c) $I_z = \int_0^{2\pi} \int_0^{\pi/2} \int_0^a (K\rho^3 \sin\phi)(\rho^2 \sin^2\phi)\, d\rho\, d\phi\, d\theta = K \int_0^{2\pi} d\theta \int_0^{\pi/2} \sin^3\phi\, d\phi \int_0^a \rho^5\, d\rho$

$= K\big[\,\theta\,\big]_0^{2\pi} \left[-\cos\phi + \frac{1}{3}\cos^3\phi\right]_0^{\pi/2} \left[\frac{1}{6}\rho^6\right]_0^a = K(2\pi)\left(\frac{2}{3}\right)\left(\frac{1}{6}a^6\right) = \frac{2}{9}\pi K a^6$

29. (a) The density function is $\rho(x, y, z) = K$, a constant, and by the symmetry of the problem $M_{xz} = M_{yz} = 0$. Then $M_{xy} = \int_0^{2\pi} \int_0^{\pi/2} \int_0^a K\rho^3 \sin\phi \cos\phi\, d\rho\, d\phi\, d\theta = \frac{1}{2}\pi K a^4 \int_0^{\pi/2} \sin\phi\cos\phi\, d\phi = \frac{1}{8}\pi K a^4$. But the mass is $K(\text{volume of the hemisphere}) = \frac{2}{3}\pi K a^3$, so the centroid is $\left(0, 0, \frac{3}{8}a\right)$.

(b) Place the center of the base at $(0, 0, 0)$; the density function is $\rho(x, y, z) = K$. By symmetry, the moments of inertia about any two such diameters will be equal, so we just need to find I_x:

$$\begin{aligned}
I_x &= \int_0^{2\pi} \int_0^{\pi/2} \int_0^a \left(K\rho^2 \sin\phi\right) \rho^2 \left(\sin^2\phi \sin^2\theta + \cos^2\phi\right) d\rho\, d\phi\, d\theta \\
&= K \int_0^{2\pi} \int_0^{\pi/2} \left(\sin^3\phi \sin^2\theta + \sin\phi\cos^2\phi\right)\left(\tfrac{1}{5}a^5\right) d\phi\, d\theta \\
&= \tfrac{1}{5}Ka^5 \int_0^{2\pi} \left[\sin^2\theta\left(-\cos\phi + \tfrac{1}{3}\cos^3\phi\right) + \left(-\tfrac{1}{3}\cos^3\phi\right)\right]_{\phi=0}^{\phi=\pi/2} d\theta = \tfrac{1}{5}Ka^5 \int_0^{2\pi} \left[\tfrac{2}{3}\sin^2\theta + \tfrac{1}{3}\right] d\theta \\
&= \tfrac{1}{5}Ka^5 \left[\tfrac{2}{3}\left(\tfrac{1}{2}\theta - \tfrac{1}{4}\sin 2\theta\right) + \tfrac{1}{3}\theta\right]_0^{2\pi} = \tfrac{1}{5}Ka^5\left[\tfrac{2}{3}(\pi - 0) + \tfrac{1}{3}(2\pi - 0)\right] = \tfrac{4}{15}Ka^5\pi
\end{aligned}$$

30. Place the center of the base at $(0, 0, 0)$, then the density is $\rho(x, y, z) = Kz$, K a constant. Then

$m = \int_0^{2\pi} \int_0^{\pi/2} \int_0^a (K\rho\cos\phi)\, \rho^2 \sin\phi\, d\rho\, d\phi\, d\theta = 2\pi K \int_0^{\pi/2} \cos\phi\sin\phi \cdot \frac{1}{4}a^4\, d\phi = \frac{1}{2}\pi K a^4 \left[-\frac{1}{4}\cos 2\phi\right]_0^{\pi/2} = \frac{\pi}{4}Ka^4$.

By the symmetry of the problem $M_{xz} = M_{yz} = 0$, and

$M_{xy} = \int_0^{2\pi} \int_0^{\pi/2} \int_0^a K\rho^4 \cos^2\phi \sin\phi\, d\rho\, d\phi\, d\theta = \frac{2}{5}\pi K a^5 \int_0^{\pi/2} \cos^2\phi\sin\phi\, d\phi = \frac{2}{5}\pi K a^5 \left[-\frac{1}{3}\cos^3\theta\right]_0^{\pi/2} = \frac{2}{15}\pi K a^5$.

Hence $(\overline{x}, \overline{y}, \overline{z}) = \left(0, 0, \frac{8}{15}a\right)$.

31. In spherical coordinates $z = \sqrt{x^2 + y^2}$ becomes $\cos\phi = \sin\phi$ or $\phi = \frac{\pi}{4}$. Then

$V = \int_0^{2\pi} \int_0^{\pi/4} \int_0^1 \rho^2 \sin\phi\, d\rho\, d\phi\, d\theta = \int_0^{2\pi} d\theta \int_0^{\pi/4} \sin\phi\, d\phi \int_0^1 \rho^2\, d\rho = \frac{1}{3}\pi\left(2 - \sqrt{2}\right)$,

$M_{xy} = \int_0^{2\pi} \int_0^{\pi/4} \int_0^1 \rho^3 \sin\phi\cos\phi\, d\rho\, d\phi\, d\theta = 2\pi\left[-\frac{1}{4}\cos 2\phi\right]_0^{\pi/4}\left(\frac{1}{4}\right) = \frac{\pi}{8}$ and by symmetry $M_{yz} = M_{xz} = 0$.

Hence $(\overline{x}, \overline{y}, \overline{z}) = \left(0, 0, \dfrac{3}{8\left(2 - \sqrt{2}\right)}\right)$.

32. Place the center of the sphere at $(0, 0, 0)$, let the diameter of intersection be along the z-axis, one of the planes be the xz-plane and the other be the plane whose angle with the xz-plane is $\theta = \frac{\pi}{6}$. Then in spherical coordinates the volume is given by

$V = \int_0^{\pi/6} \int_0^{\pi} \int_0^a \rho^2 \sin\phi\, d\rho\, d\phi\, d\theta = \int_0^{\pi/6} d\theta \int_0^{\pi} \sin\phi\, d\phi \int_0^a \rho^2\, d\rho = \frac{\pi}{6}(2)\left(\frac{1}{3}a^3\right) = \frac{1}{9}\pi a^3$.

33. In cylindrical coordinates the paraboloid is given by $z = r^2$ and the plane by $z = 2r\sin\theta$ and they intersect in the circle $r = 2\sin\theta$. Then $\iiint_E z\, dV = \int_0^{\pi} \int_0^{2\sin\theta} \int_{r^2}^{2r\sin\theta} rz\, dz\, dr\, d\theta = \frac{5\pi}{6}$ [using a CAS].

34. (a) The region enclosed by the torus is $\{(\rho, \theta, \phi) \mid 0 \le \theta \le 2\pi, 0 \le \phi \le \pi, 0 \le \rho \le \sin\phi\}$, so its volume is

$V = \int_0^{2\pi} \int_0^{\pi} \int_0^{\sin\phi} \rho^2 \sin\phi\, d\rho\, d\phi\, d\theta = 2\pi \int_0^{\pi} \frac{1}{3}\sin^4\phi\, d\phi = \frac{2}{3}\pi\left[\frac{3}{8}\phi - \frac{1}{4}\sin 2\phi + \frac{1}{16}\sin 4\phi\right]_0^{\pi} = \frac{1}{4}\pi^2$.

(b) In Maple, we can plot the torus using the `plots[sphereplot]` command, or with the `coords=spherical` option in a regular `plot` command. In Mathematica, use `ParametricPlot3d`.

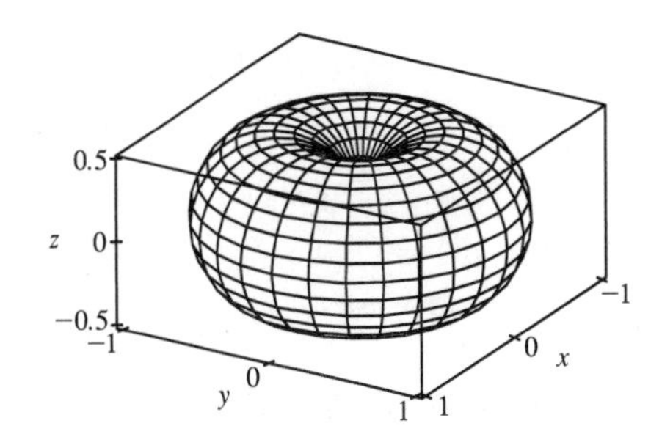

35. The region E of integration is the region above the cone $z = \sqrt{x^2 + y^2}$ and below the sphere $x^2 + y^2 + z^2 = 2$ in the first octant. Because E is in the first octant we have $0 \le \theta \le \frac{\pi}{2}$. The cone has equation $\phi = \frac{\pi}{4}$ (as in Example 4), so $0 \le \phi \le \frac{\pi}{4}$, and $0 \le \rho \le \sqrt{2}$. So the integral becomes

$$\int_0^{\pi/4} \int_0^{\pi/2} \int_0^{\sqrt{2}} (\rho \sin\phi \cos\theta)\,(\rho \sin\phi \sin\theta)\, \rho^2 \sin\phi \, d\rho\, d\theta\, d\phi$$

$$= \int_0^{\pi/4} \sin^3\phi \, d\phi \int_0^{\pi/2} \sin\theta \cos\theta \, d\theta \int_0^{\sqrt{2}} \rho^4 \, d\rho = \left(\int_0^{\pi/4} (1 - \cos^2\phi) \sin\phi \, d\phi\right) \left[\tfrac{1}{2}\sin^2\theta\right]_0^{\pi/2} \left[\tfrac{1}{5}\rho^5\right]_0^{\sqrt{2}}$$

$$= \left[\tfrac{1}{3}\cos^3\phi - \cos\phi\right]_0^{\pi/4} \cdot \tfrac{1}{2} \cdot \tfrac{1}{5}\left(\sqrt{2}\right)^5 = \left[\tfrac{\sqrt{2}}{12} - \tfrac{\sqrt{2}}{2} - \left(\tfrac{1}{3} - 1\right)\right] \cdot \tfrac{2\sqrt{2}}{5} = \tfrac{4\sqrt{2}-5}{15}$$

36. The region of integration is the solid sphere $x^2 + y^2 + z^2 \le a^2$, so $0 \le \theta \le 2\pi$, $0 \le \phi \le \pi$, and $0 \le \rho \le a$. Also $x^2 z + y^2 z + z^3 = (x^2 + y^2 + z^2) z = \rho^2 z = \rho^3 \cos\phi$, so the integral becomes

$$\int_0^{\pi} \int_0^{2\pi} \int_0^{a} (\rho^3 \cos\phi)\, \rho^2 \sin\phi \, d\rho\, d\theta\, d\phi = \int_0^{\pi} \sin\phi \cos\phi \, d\phi \int_0^{2\pi} d\theta \int_0^{a} \rho^5 \, d\rho = \left[\tfrac{1}{2}\sin^2\phi\right]_0^{\pi} \left[\theta\right]_0^{2\pi} \left[\tfrac{1}{6}\rho^6\right]_0^{a} = 0$$

37. In cylindrical coordinates, the equation of the cylinder is $r = 3$, $0 \le z \le 10$. The hemisphere is the upper part of the sphere radius 3, center $(0, 0, 10)$, equation $r^2 + (z - 10)^2 = 3^2$, $z \ge 10$. In Maple, we can use either the `coords=cylindrical` option in a regular `plot` command, or the `plots[cylinderplot]` command. In Mathematica, we can use `ParametricPlot3d`.

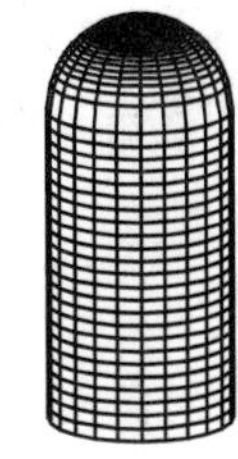

38. We begin by finding the positions of Los Angeles and Montréal in spherical coordinates, using the method described in the exercise:

Montréal	Los Angeles
$\rho = 3960$ mi	$\rho = 3960$ mi
$\theta = 360\,^\circ - \ 73.60\,^\circ = 286.40\,^\circ$	$\theta = 360\,^\circ - \ 118.25\,^\circ = 241.75\,^\circ$
$\phi = 90\,^\circ - \ 45.50\,^\circ = 44.50\,^\circ$	$\phi = 90\,^\circ - \ 34.06\,^\circ = 55.94\,^\circ$

Now we change the above to Cartesian coordinates using $x = \rho \cos\theta \sin\phi$, $y = \rho \sin\theta \sin\phi$ and $z = \rho\cos\phi$ to get two position vectors of length 3960 mi (since both cities must lie on the surface of the Earth). In particular:

Montréal: $\langle 783.67, -2662.67, 2824.47 \rangle$ Los Angeles: $\langle -1552.80, -2889.91, 2217.84 \rangle$

To find the angle α between these two vectors we use the dot product:

$\langle 783.67, -2662.67, 2824.47 \rangle \cdot \langle -1552.80, -2889.91, 2217.84 \rangle = (3960)^2 \cos\alpha \;\Rightarrow\; \cos\alpha \approx 0.8126 \;\Rightarrow$ $\alpha \approx 0.6223$ rad. The great circle distance between the cities is $s = \rho\theta \approx 3960(0.6223) \approx 2464$ mi.

39. If E is the solid enclosed by the surface $\rho = 1 + \frac{1}{5}\sin 6\theta\, \sin 5\phi$, it can be described in spherical coordinates as $E = \left\{(\rho,\theta,\phi) \mid 0 \le \rho \le 1 + \frac{1}{5}\sin 6\theta \sin 5\phi, 0 \le \theta \le 2\pi, 0 \le \phi \le \pi\right\}$. Its volume is given by $V(E) = \iiint_E dV = \int_0^\pi \int_0^{2\pi} \int_0^{1+(\sin 6\theta \sin 5\phi)/5} \rho^2 \sin\phi\, d\rho\, d\theta\, d\phi = \frac{136\pi}{99}$ [using a CAS].

40. The given integral is equal to $\lim\limits_{R\to\infty} \int_0^{2\pi}\int_0^\pi\int_0^R \rho e^{-\rho^2}\rho^2 \sin\phi\, d\rho\, d\phi\, d\theta = \lim\limits_{R\to\infty}\left(\int_0^{2\pi} d\theta\right)\left(\int_0^\pi \sin\phi\, d\phi\right)\left(\int_0^R \rho^3 e^{-\rho^2}\, d\rho\right)$.

Now use integration by parts with $u = \rho^2$, $dv = \rho e^{-\rho^2}\, d\rho$ to get

$$\lim_{R\to\infty} 2\pi(2)\left(\rho^2\left(-\tfrac{1}{2}\right)e^{-\rho^2}\Big]_0^R - \int_0^R 2\rho\left(-\tfrac{1}{2}\right)e^{-\rho^2}\, d\rho\right) = \lim_{R\to\infty} 4\pi\left(-\tfrac{1}{2}R^2 e^{-R^2} + \left[-\tfrac{1}{2}e^{-\rho^2}\right]_0^R\right)$$

$$= 4\pi \lim_{R\to\infty}\left[-\tfrac{1}{2}R^2 e^{-R^2} - \tfrac{1}{2}e^{-R^2} + \tfrac{1}{2}\right] = 4\pi\left(\tfrac{1}{2}\right) = 2\pi$$

(Note that $R^2 e^{-R^2} \to 0$ as $R \to \infty$ by l'Hospital's Rule.)

41. (a) From the diagram, $z = r\cot\phi_0$ to $z = \sqrt{a^2 - r^2}$, $r = 0$ to $r = a\sin\phi_0$ (or use $a^2 - r^2 = r^2\cot^2\phi_0$). Thus

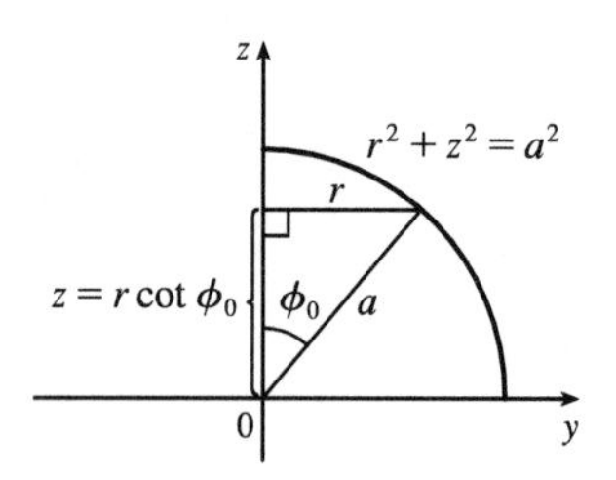

$$V = \int_0^{2\pi}\int_0^{a\sin\phi_0}\int_{r\cot\phi_0}^{\sqrt{a^2-r^2}} r\, dz\, dr\, d\theta$$

$$= 2\pi\int_0^{a\sin\phi_0}\left(r\sqrt{a^2-r^2} - r^2\cot\phi_0\right) dr$$

$$= \tfrac{2\pi}{3}\left[-(a^2-r^2)^{3/2} - r^3\cot\phi_0\right]_0^{a\sin\phi_0}$$

$$= \tfrac{2\pi}{3}\left[-\left(a^2 - a^2\sin^2\phi_0\right)^{3/2} - a^3\sin^3\phi_0\cot\phi_0 + a^3\right]$$

$$= \tfrac{2}{3}\pi a^3\left[1 - \left(\cos^3\phi_0 + \sin^2\phi_0\cos\phi_0\right)\right] = \tfrac{2}{3}\pi a^3(1 - \cos\phi_0)$$

(b) The wedge in question is the shaded area rotated from $\theta = \theta_1$ to $\theta = \theta_2$. Letting

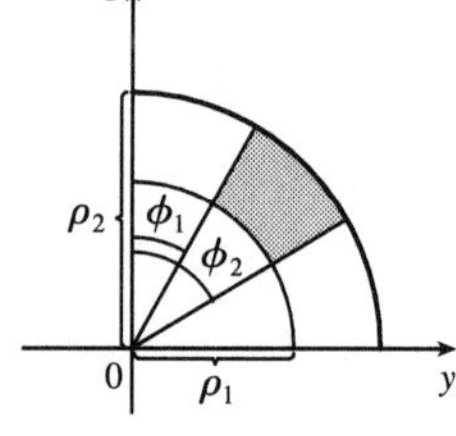

V_{ij} = volume of the region bounded by the sphere of radius ρ_i and the cone with angle ϕ_j ($\theta = \theta_1$ to θ_2)

and letting V be the volume of the wedge, we have

$$V = (V_{22} - V_{21}) - (V_{12} - V_{11})$$

$$= \tfrac{1}{3}(\theta_2 - \theta_1)\left[\rho_2^3(1-\cos\phi_2) - \rho_2^3(1-\cos\phi_1) - \rho_1^3(1-\cos\phi_2) + \rho_1^3(1-\cos\phi_1)\right]$$

$$= \tfrac{1}{3}(\theta_2 - \theta_1)\left[(\rho_2^3 - \rho_1^3)(1-\cos\phi_2) - (\rho_2^3 - \rho_1^3)(1-\cos\phi_1)\right] = \tfrac{1}{3}(\theta_2 - \theta_1)\left[(\rho_2^3 - \rho_1^3)(\cos\phi_1 - \cos\phi_2)\right]$$

Or: Show that $V = \displaystyle\int_{\theta_1}^{\theta_2}\int_{\rho_1\sin\phi_1}^{\rho_2\sin\phi_2}\int_{r\cot\phi_2}^{r\cot\phi_1} r\, dz\, dr\, d\theta$.

(c) By the Mean Value Theorem with $f(\rho) = \rho^3$ there exists some $\tilde{\rho}$ with $\rho_1 \le \tilde{\rho} \le \rho_2$ such that $f(\rho_2) - f(\rho_1) = f'(\tilde{\rho})(\rho_2 - \rho_1)$ or $\rho_1^3 - \rho_2^3 = 3\tilde{\rho}^2\Delta\rho$. Similarly there exists ϕ with $\phi_1 \le \tilde{\phi} \le \phi_2$ such that $\cos\phi_2 - \cos\phi_1 = \left(-\sin\tilde{\phi}\right)\Delta\phi$. Substituting into the result from (b) gives

$\Delta V = (\tilde{\rho}^2\,\Delta\rho)(\theta_2 - \theta_1)(\sin\tilde{\phi})\,\Delta\phi = \tilde{\rho}^2\sin\tilde{\phi}\,\Delta\rho\,\Delta\phi\,\Delta\theta$.

12.8 Change of Variables in Multiple Integrals

1. $x = u + 4v$, $y = 3u - 2v$. The Jacobian is $\dfrac{\partial(x,y)}{\partial(u,v)} = \begin{vmatrix} \partial x/\partial u & \partial x/\partial v \\ \partial y/\partial u & \partial y/\partial v \end{vmatrix} = \begin{vmatrix} 1 & 4 \\ 3 & -2 \end{vmatrix} = 1(-2) - 4(3) = -14.$

2. $$\frac{\partial(x,y)}{\partial(u,v)} = \begin{vmatrix} \partial x/\partial u & \partial x/\partial v \\ \partial y/\partial u & \partial y/\partial v \end{vmatrix} = \begin{vmatrix} 2u & -2v \\ 2u & 2v \end{vmatrix} = 4uv - (-4uv) = 8uv$$

3. $$\frac{\partial(x,y)}{\partial(u,v)} = \begin{vmatrix} \dfrac{\partial x}{\partial u} & \dfrac{\partial x}{\partial v} \\ \dfrac{\partial y}{\partial u} & \dfrac{\partial y}{\partial v} \end{vmatrix} = \begin{vmatrix} \dfrac{v}{(u+v)^2} & -\dfrac{u}{(u+v)^2} \\ -\dfrac{v}{(u-v)^2} & \dfrac{u}{(u-v)^2} \end{vmatrix} = \frac{uv}{(u+v)^2(u-v)^2} - \frac{uv}{(u+v)^2(u-v)^2} = 0$$

4. $$\frac{\partial(x,y)}{\partial(\alpha,\beta)} = \begin{vmatrix} \partial x/\partial \alpha & \partial x/\partial \beta \\ \partial y/\partial \alpha & \partial y/\partial \beta \end{vmatrix} = \begin{vmatrix} \sin\beta & \alpha\cos\beta \\ \cos\beta & -\alpha\sin\beta \end{vmatrix} = -\alpha\sin^2\beta - \alpha\cos^2\beta = -\alpha$$

5. $$\frac{\partial(x,y,z)}{\partial(u,v,w)} = \begin{vmatrix} \partial x/\partial u & \partial x/\partial v & \partial x/\partial w \\ \partial y/\partial u & \partial y/\partial v & \partial y/\partial w \\ \partial z/\partial u & \partial z/\partial v & \partial z/\partial w \end{vmatrix} = \begin{vmatrix} v & u & 0 \\ 0 & w & v \\ w & 0 & u \end{vmatrix} = v\begin{vmatrix} w & v \\ 0 & u \end{vmatrix} - u\begin{vmatrix} 0 & v \\ w & u \end{vmatrix} + 0\begin{vmatrix} 0 & w \\ w & 0 \end{vmatrix}$$
$$= v(uw - 0) - u(0 - vw) = 2uvw$$

6. $$\frac{\partial(x,y,z)}{\partial(u,v,w)} = \begin{vmatrix} e^{u-v} & -e^{u-v} & 0 \\ e^{u+v} & e^{u+v} & 0 \\ e^{u+v+w} & e^{u+v+w} & e^{u+v+w} \end{vmatrix} = e^{u+v+w}\begin{vmatrix} e^{u-v} & -e^{u-v} \\ e^{u+v} & e^{u+v} \end{vmatrix}$$
$$= e^{u+v+w}\left(e^{u-v}e^{u+v} + e^{u-v}e^{u+v}\right) = e^{u+v+w}\left(2e^{2u}\right) = 2e^{3u+v+w}$$

7. The transformation maps the boundary of S to the boundary of the image R, so we first look at side S_1 in the uv-plane. S_1 is described by $v = 0$ $(0 \le u \le 3)$, so $x = 2u + 3v = 2u$ and $y = u - v = u$. Eliminating u, we have $x = 2y$, $0 \le x \le 6$. S_2 is the line segment $u = 3$, $0 \le v \le 2$, so $x = 6 + 3v$ and $y = 3 - v$. Then $v = 3 - y \Rightarrow x = 6 + 3(3 - y) = 15 - 3y$, $6 \le x \le 12$. S_3 is the line segment $v = 2$, $0 \le u \le 3$, so $x = 2u + 6$ and $y = u - 2$, giving $u = y + 2 \Rightarrow x = 2y + 10$, $6 \le x \le 12$. Finally, S_4 is the segment $u = 0$, $0 \le v \le 2$, so $x = 3v$ and $y = -v \Rightarrow x = -3y$, $0 \le x \le 6$. The image of set S is the region R shown in the xy-plane, a parallelogram bounded by these four segments.

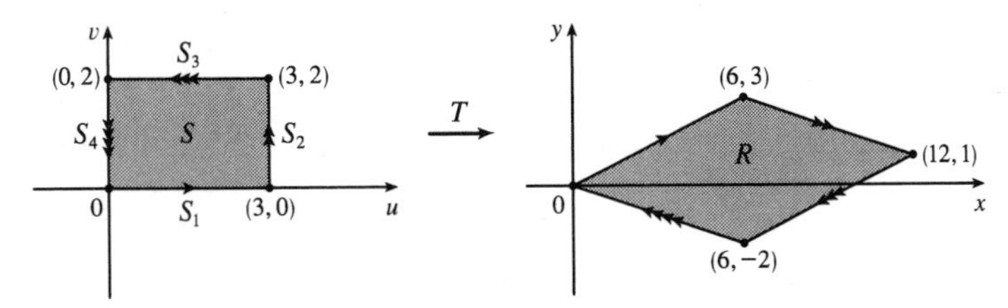

8. S_1 is the line segment $v = 0$, $0 \le u \le 1$, so $x = v = 0$ and $y = u(1 + v^2) = u$. Since $0 \le u \le 1$, the image is the line segment $x = 0$, $0 \le y \le 1$. S_2 is the segment $u = 1$, $0 \le v \le 1$, so $x = v$ and $y = u(1 + v^2) = 1 + x^2$. Thus the image is the portion of the parabola $y = 1 + x^2$ for $0 \le x \le 1$. S_3 is the segment $v = 1$, $0 \le u \le 1$, so $x = 1$ and $y = 2u$. The image is the segment $x = 1$, $0 \le y \le 2$. S_4 is described by $u = 0$, $0 \le v \le 1$, so $0 \le x = v \le 1$ and $y = u(1 + v^2) = 0$. The image is the line segment $y = 0$, $0 \le x \le 1$. Thus, the image of S is the region R bounded by the parabola $y = 1 + x^2$, the x-axis, and the lines $x = 0$, $x = 1$.

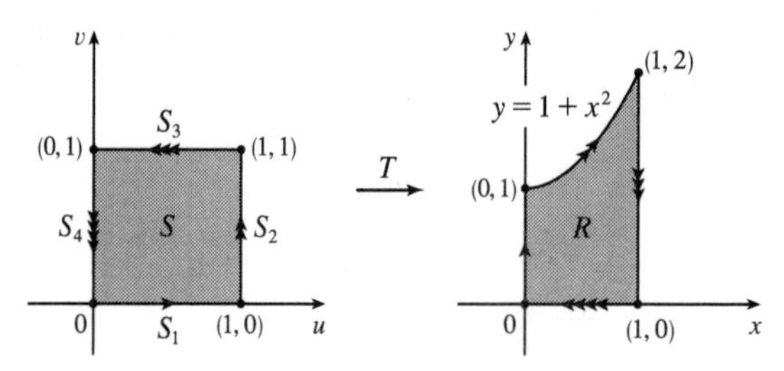

9. S_1 is the line segment $u = v$, $0 \le u \le 1$, so $y = v = u$ and $x = u^2 = y^2$. Since $0 \le u \le 1$, the image is the portion of the parabola $x = y^2$, $0 \le y \le 1$. S_2 is the segment $v = 1$, $0 \le u \le 1$, thus $y = v = 1$ and $x = u^2$, so $0 \le x \le 1$. The image is the line segment $y = 1$, $0 \le x \le 1$. S_3 is the segment $u = 0$, $0 \le v \le 1$, so $x = u^2 = 0$ and $y = v \quad \Rightarrow \quad 0 \le y \le 1$. The image is the segment $x = 0$, $0 \le y \le 1$. Thus, the image of S is the region R in the first quadrant bounded by the parabola $x = y^2$, the y-axis, and the line $y = 1$.

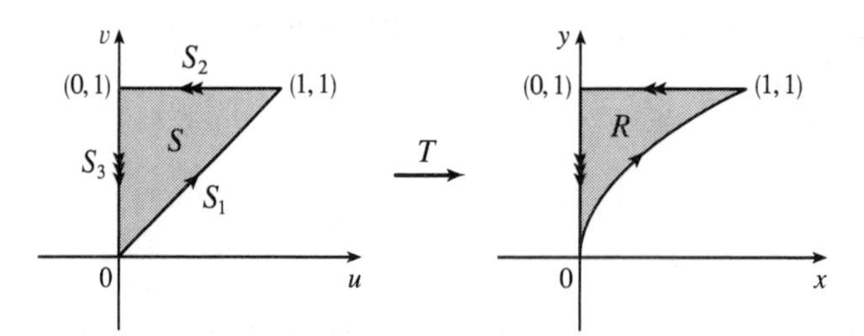

10. Substituting $u = \dfrac{x}{a}$, $v = \dfrac{y}{b}$ into $u^2 + v^2 \le 1$ gives $\dfrac{x^2}{a^2} + \dfrac{y^2}{b^2} \le 1$, so the image of $u^2 + v^2 \le 1$ is the elliptical region $\dfrac{x^2}{a^2} + \dfrac{y^2}{b^2} \le 1$.

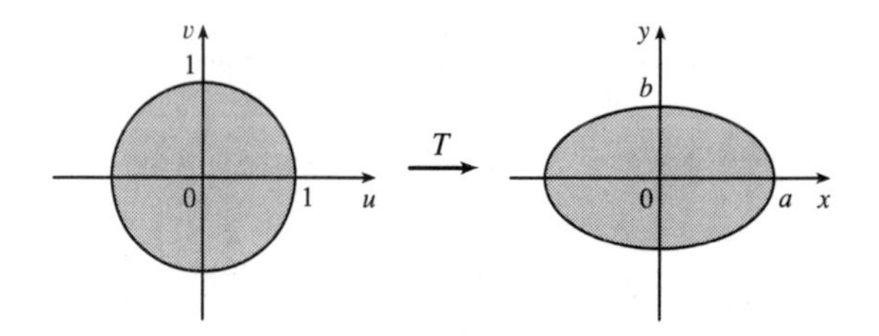

11. $\dfrac{\partial(x,y)}{\partial(u,v)} = \begin{vmatrix} 2 & 1 \\ 1 & 2 \end{vmatrix} = 3$ and $x - 3y = (2u + v) - 3(u + 2v) = -u - 5v$. To find the region S in the uv-plane that corresponds to R we first find the corresponding boundary under the given transformation. The line through $(0, 0)$ and $(2, 1)$ is $y = \frac{1}{2}x$ which is the image of $u + 2v = \frac{1}{2}(2u + v) \quad \Rightarrow \quad v = 0$; the line through $(2, 1)$ and $(1, 2)$ is $x + y = 3$ which is the image of $(2u + v) + (u + 2v) = 3 \quad \Rightarrow \quad u + v = 1$; the line through $(0, 0)$ and $(1, 2)$ is $y = 2x$ which is the image of $u + 2v = 2(2u + v) \quad \Rightarrow \quad u = 0$. Thus S is the triangle $0 \le v \le 1 - u$, $0 \le u \le 1$ in the uv-plane and

$$\begin{aligned} \iint_R (x - 3y)\, dA &= \int_0^1 \int_0^{1-u} (-u - 5v)\, |3|\, dv\, du = -3 \int_0^1 \left[uv + \tfrac{5}{2}v^2\right]_{v=0}^{v=1-u} du \\ &= -3 \int_0^1 \left(u - u^2 + \tfrac{5}{2}(1 - u)^2\right) du = -3\left[\tfrac{1}{2}u^2 - \tfrac{1}{3}u^3 - \tfrac{5}{6}(1 - u)^3\right]_0^1 = -3\left(\tfrac{1}{2} - \tfrac{1}{3} + \tfrac{5}{6}\right) = -3 \end{aligned}$$

12. $\dfrac{\partial(x,y)}{\partial(u,v)} = \begin{vmatrix} 1/4 & 1/4 \\ -3/4 & 1/4 \end{vmatrix} = \dfrac{1}{4}$, $4x + 8y = 4 \cdot \frac{1}{4}(u+v) + 8 \cdot \frac{1}{4}(v - 3u) = 3v - 5u$. R is a parallelogram bounded by the lines $x - y = -4$, $x - y = 4$, $3x + y = 0$, $3x + y = 8$. Since $u = x - y$ and $v = 3x + y$, R is the image of the rectangle enclosed by the lines $u = -4$, $u = 4$, $v = 0$, and $v = 8$. Thus

$$\iint_R (4x+8y)\,dA = \int_{-4}^{4}\int_0^8 (3v-5u)\left|\tfrac{1}{4}\right| dv\,du = \tfrac{1}{4}\int_{-4}^{4}\left[\tfrac{3}{2}v^2 - 5uv\right]_{v=0}^{v=8} du$$
$$= \tfrac{1}{4}\int_{-4}^{4}(96 - 40u)\,du = \tfrac{1}{4}\left[96u - 20u^2\right]_{-4}^{4} = 192$$

13. $\dfrac{\partial(x,y)}{\partial(u,v)} = \begin{vmatrix} 2 & 0 \\ 0 & 3 \end{vmatrix} = 6$, $x^2 = 4u^2$ and the planar ellipse $9x^2 + 4y^2 \le 36$ is the image of the disk $u^2 + v^2 \le 1$. Thus

$$\iint_R x^2\,dA = \iint_{u^2+v^2\le 1} (4u^2)(6)\,du\,dv = \int_0^{2\pi}\int_0^1 (24r^2\cos^2\theta)\,r\,dr\,d\theta = 24\int_0^{2\pi}\cos^2\theta\,d\theta\int_0^1 r^3\,dr$$
$$= 24\left[\tfrac{1}{2}x + \tfrac{1}{4}\sin 2x\right]_0^{2\pi}\left[\tfrac{1}{4}r^4\right]_0^1 = 24(\pi)\left(\tfrac{1}{4}\right) = 6\pi$$

14. $\dfrac{\partial(x,y)}{\partial(u,v)} = \begin{vmatrix} \sqrt{2} & -\sqrt{2/3} \\ \sqrt{2} & \sqrt{2/3} \end{vmatrix} = \dfrac{4}{\sqrt{3}}$, $x^2 - xy + y^2 = 2u^2 + 2v^2$ and the planar ellipse $x^2 - xy + y^2 \le 2$ is the image of the disk $u^2 + v^2 \le 1$. Thus

$$\iint_R (x^2 - xy + y^2)\,dA = \iint_{u^2+v^2\le 1} (2u^2+2v^2)\left(\tfrac{4}{\sqrt{3}}\,du\,dv\right) = \int_0^{2\pi}\int_0^1 \tfrac{8}{\sqrt{3}}r^3\,dr\,d\theta = \tfrac{4\pi}{\sqrt{3}}$$

15. $\dfrac{\partial(x,y)}{\partial(u,v)} = \begin{vmatrix} 1/v & -u/v^2 \\ 0 & 1 \end{vmatrix} = \dfrac{1}{v}$, $xy = u$, $y = x$ is the image of the parabola $v^2 = u$, $y = 3x$ is the image of the parabola $v^2 = 3u$, and the hyperbolas $xy = 1$, $xy = 3$ are the images of the lines $u = 1$ and $u = 3$ respectively. Thus

$$\iint_R xy\,dA = \int_1^3\int_{\sqrt{u}}^{\sqrt{3u}} u\left(\frac{1}{v}\right) dv\,du = \int_1^3 u\left(\ln\sqrt{3u} - \ln\sqrt{u}\right) du = \int_1^3 u\ln\sqrt{3}\,du = 4\ln\sqrt{3} = 2\ln 3.$$

16. Here $y = \dfrac{v}{u}$, $x = \dfrac{u^2}{v}$ so $\dfrac{\partial(x,y)}{\partial(u,v)} = \begin{vmatrix} 2u/v & -u^2/v^2 \\ -v/u^2 & 1/u \end{vmatrix} = \dfrac{1}{v}$ and R is the image of the square with vertices $(1,1)$, $(2,1)$, $(2,2)$, and $(1,2)$. So

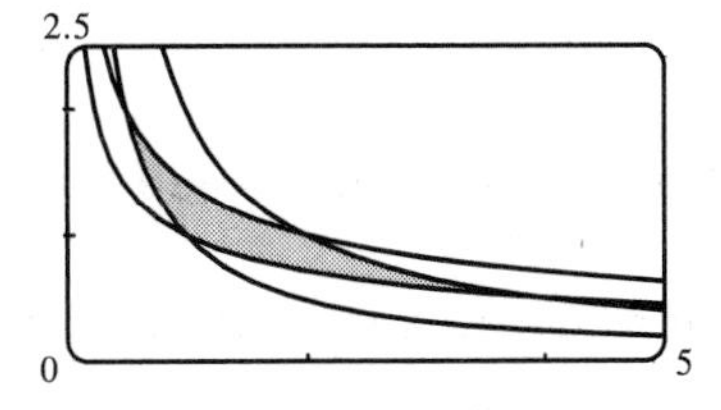

$$\iint_R y^2\,dA = \int_1^2\int_1^2 \frac{v^2}{u^2}\left(\frac{1}{v}\right) du\,dv = \int_1^2 \frac{v}{2}\,dv = \frac{3}{4}$$

17. (a) $\dfrac{\partial(x,y,z)}{\partial(u,v,w)} = \begin{vmatrix} a & 0 & 0 \\ 0 & b & 0 \\ 0 & 0 & c \end{vmatrix} = abc$ and since $u = \dfrac{x}{a}$, $v = \dfrac{y}{b}$, $w = \dfrac{z}{c}$ the solid enclosed by the ellipsoid is the image of the ball $u^2 + v^2 + w^2 \le 1$. So

$$\iiint_E dV = \iiint_{u^2+v^2+w^2\le 1} abc\,du\,dv\,dw = (abc)(\text{volume of the ball}) = \tfrac{4}{3}\pi abc$$

(b) If we approximate the surface of the Earth by the ellipsoid $\dfrac{x^2}{6378^2}+\dfrac{y^2}{6378^2}+\dfrac{z^2}{6356^2}=1$, then we can estimate the volume of the Earth by finding the volume of the solid E enclosed by the ellipsoid. From part (a), this is $\iiint_E dV = \frac{4}{3}\pi(6378)(6378)(6356) \approx 1.083 \times 10^{12}\text{ km}^3$.

18. $\dfrac{\partial(x,y,z)}{\partial(u,v,w)} = \begin{vmatrix} a & 0 & 0 \\ 0 & b & 0 \\ 0 & 0 & c \end{vmatrix} = abc$ and the solid enclosed by the ellipsoid is the image of the ball $u^2+v^2+w^2 \le 1$.

Now $x^2y = (a^2u^2)(bv)$, so

$$\iiint_E x^2y\,dV = \iiint_{u^2+v^2+w^2\le 1} (a^2bu^2v)(abc)\,du\,dv\,dw$$

$$= \int_0^{2\pi}\int_0^{\pi}\int_0^1 (a^3b^2c)(\rho^2\sin^2\phi\cos^2\theta)(\rho\sin\phi\sin\theta)\,\rho^2\sin\phi\,d\rho\,d\phi\,d\theta$$

$$= a^3b^2c\int_0^{2\pi}\int_0^{\pi}\int_0^1 (\rho^5\sin^4\phi\cos^2\theta\sin\theta)\,d\rho\,d\phi\,d\theta = a^3b^2c\int_0^{2\pi}\cos^2\theta\sin\theta\,d\theta\int_0^{\pi}\sin^4\phi\,d\phi\int_0^1\rho^5\,d\rho$$

$$= 0 \quad \text{since } \int_0^{2\pi}\cos^2\theta\sin\theta\,d\theta = 0$$

19. Letting $u = x-2y$ and $v = 3x-y$, we have $x = \frac{1}{5}(2v-u)$ and $y = \frac{1}{5}(v-3u)$. Then $\dfrac{\partial(x,y)}{\partial(u,v)} = \begin{vmatrix} -1/5 & 2/5 \\ -3/5 & 1/5 \end{vmatrix} = \dfrac{1}{5}$ and R is the image of the rectangle enclosed by the lines $u=0$, $u=4$, $v=1$, and $v=8$. Thus

$$\iint_R \frac{x-2y}{3x-y}\,dA = \int_0^4\int_1^8 \frac{u}{v}\left|\frac{1}{5}\right| dv\,du = \frac{1}{5}\int_0^4 u\,du\int_1^8 \frac{1}{v}\,dv = \tfrac{1}{5}\left[\tfrac{1}{2}u^2\right]_0^4\left[\ln|v|\right]_1^8 = \tfrac{8}{5}\ln 8.$$

20. Letting $u = x+y$ and $v = x-y$, we have $x = \frac{1}{2}(u+v)$ and $y = \frac{1}{2}(u-v)$. Then $\dfrac{\partial(x,y)}{\partial(u,v)} = \begin{vmatrix} 1/2 & 1/2 \\ 1/2 & -1/2 \end{vmatrix} = -\dfrac{1}{2}$ and R is the image of the rectangle enclosed by the lines $u=0$, $u=3$, $v=0$, and $v=2$. Thus

$$\iint_R (x+y)\,e^{x^2-y^2}\,dA = \int_0^3\int_0^2 ue^{uv}\left|-\tfrac{1}{2}\right| dv\,du = \tfrac{1}{2}\int_0^3\left[e^{uv}\right]_{v=0}^{v=2} du = \tfrac{1}{2}\int_0^3(e^{2u}-1)\,du$$
$$= \tfrac{1}{2}\left[\tfrac{1}{2}e^{2u}-u\right]_0^3 = \tfrac{1}{2}\left(\tfrac{1}{2}e^6-3-\tfrac{1}{2}\right) = \tfrac{1}{4}(e^6-7)$$

21. Letting $u = y-x$, $v = y+x$, we have $y = \frac{1}{2}(u+v)$, $x = \frac{1}{2}(v-u)$. Then $\dfrac{\partial(x,y)}{\partial(u,v)} = \begin{vmatrix} -1/2 & 1/2 \\ 1/2 & 1/2 \end{vmatrix} = -\dfrac{1}{2}$ and R is the image of the trapezoidal region with vertices $(-1,1)$, $(-2,2)$, $(2,2)$, and $(1,1)$. Thus

$$\iint_R \cos\tfrac{y-x}{y+x}\,dA = \int_1^2\int_{-v}^{v}\cos\tfrac{u}{v}\left|-\tfrac{1}{2}\right| du\,dv = \tfrac{1}{2}\int_1^2\left[v\sin\tfrac{u}{v}\right]_{u=-v}^{u=v} dv$$

$$= \tfrac{1}{2}\int_1^2 2v\sin(1)\,dv = \tfrac{3}{2}\sin 1$$

22. Letting $u = 3x$, $v = 2y$, we have $9x^2+4y^2 = u^2+v^2$, $x = \frac{1}{3}u$, and $y = \frac{1}{2}v$. Then $\dfrac{\partial(x,y)}{\partial(u,v)} = \dfrac{1}{6}$ and R is the image of the quarter-disk D given by $u^2+v^2 \le 1$, $u \ge 0$, $v \ge 0$. Thus

$$\iint_R \sin(9x^2+4y^2)\,dA = \iint_D \tfrac{1}{6}\sin(u^2+v^2)\,du\,dv = \int_0^{\pi/2}\int_0^1 \tfrac{1}{6}\sin(r^2)\,r\,dr\,d\theta = \tfrac{\pi}{12}\left[-\tfrac{1}{2}\cos r^2\right]_0^1 = \tfrac{\pi}{24}(1-\cos 1)$$

23. Let $u = x + y$ and $v = -x + y$. Then $u + v = 2y \;\Rightarrow\; y = \frac{1}{2}(u + v)$ and

$$u - v = 2x \quad\Rightarrow\quad x = \tfrac{1}{2}(u - v).\; \frac{\partial(x,y)}{\partial(u,v)} = \begin{vmatrix} 1/2 & -1/2 \\ 1/2 & 1/2 \end{vmatrix} = \frac{1}{2}.$$

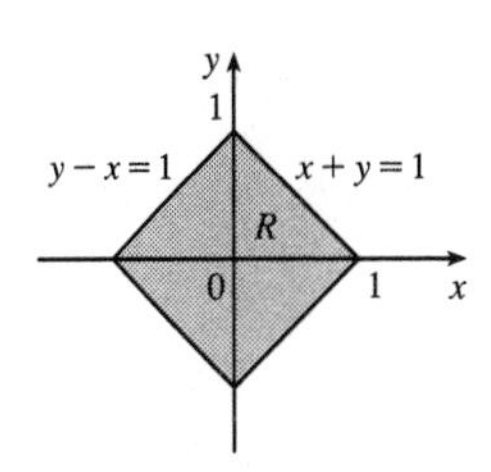

Now $|u| = |x + y| \le |x| + |y| \le 1 \;\Rightarrow\; -1 \le u \le 1$, and

$|v| = |-x + y| \le |x| + |y| \le 1 \;\Rightarrow\; -1 \le v \le 1$. R is the image of

the square region with vertices $(1,1)$, $(1,-1)$, $(-1,-1)$, and $(-1,1)$.

So $\iint_R e^{x+y}\,dA = \frac{1}{2}\int_{-1}^{1}\int_{-1}^{1} e^u\,du\,dv = \frac{1}{2}\big[e^u\big]_{-1}^{1}\big[v\big]_{-1}^{1} = e - e^{-1}$.

24. Let $u = x + y$ and $v = y$, then $x = u - v$, $y = v$, $\dfrac{\partial(x,y)}{\partial(u,v)} = 1$ and R is the image under T of the triangular region with vertices $(0,0)$, $(1,0)$ and $(1,1)$. Thus

$$\iint_R f(x+y)\,dA = \int_0^1\int_0^u (1)\,f(u)\,dv\,du = \int_0^1 f(u)\big[\,v\,\big]_{v=0}^{v=u}du = \int_0^1 uf(u)\,du \quad \text{as desired.}$$

12 Review

CONCEPT CHECK

1. (a) A double Riemann sum of f is $\sum\limits_{i=1}^{m}\sum\limits_{j=1}^{n} f\big(x_{ij}^*, y_{ij}^*\big)\,\Delta A_{ij}$, where ΔA_{ij} is the area of the ij-th subrectangle and $\big(x_{ij}^*, y_{ij}^*\big)$ is a sample point in each subrectangle. If $f(x,y) \ge 0$, this sum represents an approximation to the volume of the solid that lies above the rectangle R and below the graph of f.

(b) $\iint_R f(x,y)\,dA = \lim\limits_{\max \Delta x_i, \Delta y_j \to 0} \sum\limits_{i=1}^{m}\sum\limits_{j=1}^{n} f\big(x_{ij}^*, y_{ij}^*\big)\,\Delta A_{ij}$

(c) If $f(x,y) \ge 0$, $\iint_R f(x,y)\,dA$ represents the volume of the solid that lies above the rectangle R and below the surface $z = f(x,y)$. If f takes on both positive and negative values, $\iint_R f(x,y)\,dA$ is the difference of the volume above R but below the surface $z = f(x,y)$ and the volume below R but above the surface $z = f(x,y)$.

(d) We usually evaluate $\iint_R f(x,y)\,dA$ as an iterated integral according to Fubini's Theorem (see Theorem 12.1.10).

(e) The Midpoint Rule for Double Integrals says that we approximate the double integral $\iint_R f(x,y)\,dA$ by the double Riemann sum $\sum\limits_{i=1}^{m}\sum\limits_{j=1}^{n} f\big(\overline{x}_i, \overline{y}_j\big)\,\Delta A$ where the sample points $\big(\overline{x}_i, \overline{y}_j\big)$ are the centers of the subrectangles.

2. (a) See (1) and (2) and the accompanying discussion in Section 12.2.

(b) See (3) and the accompanying discussion in Section 12.2.

(c) See (5) and the preceding discussion in Section 12.2.

(d) See (6)–(11) in Section 12.2.

3. We may want to change from rectangular to polar coordinates in a double integral if the region R of integration is more easily described in polar coordinates. To accomplish this, we use $\iint_R f(x,y)\,dA = \int_\alpha^\beta\int_a^b f(r\cos\theta, r\sin\theta)\,r\,dr\,d\theta$ where R is given by $0 \le a \le r \le b$, $\alpha \le \theta \le \beta$.

4. (a) $m = \iint_D \rho(x, y)\, dA$

(b) $M_x = \iint_D y\rho(x, y)\, dA$, $M_y = \iint_D x\rho(x, y)\, dA$

(c) The center of mass is $(\overline{x}, \overline{y})$ where $\overline{x} = \dfrac{M_y}{m}$ and $\overline{y} = \dfrac{M_x}{m}$.

(d) $I_x = \iint_D y^2\rho(x, y)\, dA,\ I_y = \iint_D x^2\rho(x, y)\, dA,\ I_0 = \iint_D (x^2 + y^2)\rho(x, y)\, dA$

5. (a) $\iiint_B f(x, y, z)\, dV = \lim\limits_{\max \Delta x_i, \Delta y_j, \Delta z_k \to 0} \sum\limits_{i=1}^{l} \sum\limits_{j=1}^{m} \sum\limits_{k=1}^{n} f(x_{ijk}^*, y_{ijk}^*, z_{ijk}^*)\, \Delta V_{ijk}$

(b) We usually evaluate $\iiint_B f(x, y, z)\, dV$ as an iterated integral according to Fubini's Theorem for Triple Integrals (see Theorem 12.5.4).

(c) See the paragraph following Example 12.5.1.

(d) See (5) and (6) and the accompanying discussion in Section 12.5.

(e) See (10) and the accompanying discussion in Section 12.5.

(f) See (11) and the preceding discussion in Section 12.5.

6. (a) $m = \iiint_E \rho(x, y, z)\, dV$

(b) $M_{yz} = \iiint_E x\rho(x, y, z)\, dV,\ M_{xz} = \iiint_E y\rho(x, y, z)\, dV,\ M_{xy} = \iiint_E z\rho(x, y, z)\, dV.$

(c) The center of mass is $(\overline{x}, \overline{y}, \overline{z})$ where $\overline{x} = \dfrac{M_{yz}}{m}$, $\overline{y} = \dfrac{M_{xz}}{m}$, and $\overline{z} = \dfrac{M_{xy}}{m}$.

(d) $I_x = \iiint_E (y^2 + z^2)\rho(x, y, z)\, dV,\ I_y = \iiint_E (x^2 + z^2)\rho(x, y, z)\, dV,\ I_z = \iiint_E (x^2 + y^2)\rho(x, y, z)\, dV.$

7. (a) $x = r\cos\theta,\ y = r\sin\theta,\ z = z$. See the discussion after Example 1 in Section 12.6.

(b) $x = \rho\sin\phi\cos\theta$, $y = \rho\sin\phi\sin\theta$, $z = \rho\cos\phi$. See Figure 1 and the accompanying discussion in Section 12.7.

8. (a) See Formula 12.6.4 and the accompanying discussion.

(b) See Formula 12.7.3 and the accompanying discussion.

(c) We may want to change from rectangular to cylindrical or spherical coordinates in a triple integral if the region E of integration is more easily described in cylindrical or spherical coordinates or if the triple integral is easier to evaluate using cylindrical or spherical coordinates.

9. (a) $\dfrac{\partial(x, y)}{\partial(u, v)} = \begin{vmatrix} \partial x/\partial u & \partial x/\partial v \\ \partial y/\partial u & \partial y/\partial v \end{vmatrix} = \dfrac{\partial x}{\partial u}\dfrac{\partial y}{\partial v} - \dfrac{\partial x}{\partial v}\dfrac{\partial y}{\partial u}$

(b) See (9) and the accompanying discussion in Section 12.8.

(c) See (13) and the accompanying discussion in Section 12.8.

TRUE-FALSE QUIZ

1. This is true by Fubini's Theorem.

2. False. $\int_0^1 \int_0^x \sqrt{x + y^2}\, dy\, dx$ describes the region of integration as a Type I region. To reverse the order of integration, we must consider the region as a Type II region: $\int_0^1 \int_y^1 \sqrt{x + y^2}\, dx\, dy$.

3. True by Equation 12.1.11.

4. $\int_{-1}^{1}\int_{0}^{1} e^{x^2+y^2}\sin y\,dx\,dy = \left(\int_0^1 e^{x^2}\,dx\right)\left(\int_{-1}^{1} e^{y^2}\sin y\,dy\right) = \left(\int_0^1 e^{x^2}\,dx\right)(0) = 0$, since $e^{y^2}\sin y$ is an odd function.

Therefore the statement is true.

5. True: $\iint_D \sqrt{4-x^2-y^2}\,dA =$ the volume under the surface $x^2+y^2+z^2=4$ and above the xy-plane

$$= \tfrac{1}{2}\left(\text{the volume of the sphere } x^2+y^2+z^2=4\right) = \tfrac{1}{2}\cdot\tfrac{4}{3}\pi(2)^3 = \tfrac{16}{3}\pi$$

6. This statement is true because in the given region, $\left(x^2+\sqrt{y}\right)\sin(x^2y^2) \le (1+2)(1) = 3$, so

$\int_1^4\int_0^1 \left(x^2+\sqrt{y}\right)\sin(x^2y^2)\,dx\,dy \le \int_1^4\int_0^1 3\,dA = 3A(D) = 3(3) = 9.$

7. The volume enclosed by the cone $z=\sqrt{x^2+y^2}$ and the plane $z=2$ is, in cylindrical coordinates,

$V = \int_0^{2\pi}\int_0^2\int_r^2 r\,dz\,dr\,d\theta \neq \int_0^{2\pi}\int_0^2\int_r^2 dz\,dr\,d\theta$, so the assertion is false.

8. True. The moment of inertia about the z-axis of a solid E with constant density k is

$I_z = \iiint_E (x^2+y^2)\rho(x,y,z)\,dV = \iiint_E (kr^2)\,r\,dz\,dr\,d\theta = \iiint_E kr^3\,dz\,dr\,d\theta.$

EXERCISES

1. As shown in the contour map, we divide R into 9 equally sized subsquares, each with area $\Delta A = 1$. Then we approximate $\iint_R f(x,y)\,dA$ by a Riemann sum with $m=n=3$ and the sample points the upper right corners of each square, so

$$\begin{aligned}\iint_R f(x,y)\,dA &\approx \sum_{i=1}^{3}\sum_{j=1}^{3} f(x_i,y_j)\,\Delta A \\ &= \Delta A\left[f(1,1)+f(1,2)+f(1,3)+f(2,1)+f(2,2)+f(2,3)+f(3,1)+f(3,2)+f(3,3)\right]\end{aligned}$$

Using the contour lines to estimate the function values, we have

$$\iint_R f(x,y)\,dA \approx 1[2.7+4.7+8.0+4.7+6.7+10.0+6.7+8.6+11.9] \approx 64.0$$

2. As in Exercise 1, we have $m=n=3$ and $\Delta A = 1$. Using the contour map to estimate the value of f at the center of each subsquare, we have

$$\begin{aligned}\iint_R f(x,y)\,dA &\approx \sum_{i=1}^{3}\sum_{j=1}^{3} f\left(\overline{x}_i,\overline{y}_j\right)\Delta A \\ &= \Delta A\left[f(0.5,0.5)+(0.5,1.5)+(0.5,2.5)+(1.5,0.5)+f(1.5,1.5)\right. \\ &\qquad \left.+f(1.5,2.5)+(2.5,0.5)+f(2.5,1.5)+f(2.5,2.5)\right] \\ &\approx 1[1.2+2.5+5.0+3.2+4.5+7.1+5.2+6.5+9.0] = 44.2\end{aligned}$$

3. $\int_1^2\int_0^2 (y+2xe^y)\,dx\,dy = \int_1^2 \left[xy+x^2e^y\right]_{x=0}^{x=2} dy = \int_1^2 (2y+4e^y)\,dy = \left[y^2+4e^y\right]_1^2$

$= 4+4e^2-1-4e = 4e^2-4e+3$

4. $\int_0^1\int_0^1 ye^{xy}\,dx\,dy = \int_0^1 \left[e^{xy}\right]_{x=0}^{x=1} dy = \int_0^1 (e^y-1)\,dy = \left[e^y-y\right]_0^1 = e-2$

5. $\int_0^1\int_0^x \cos(x^2)\,dy\,dx = \int_0^1 \left[\cos(x^2)y\right]_{y=0}^{y=x} dx = \int_0^1 x\cos(x^2)\,dx = \tfrac{1}{2}\sin(x^2)\Big]_0^1 = \tfrac{1}{2}\sin 1$

6. $\int_0^1 \int_x^{e^x} 3xy^2\,dy\,dx = \int_0^1 \left[xy^3\right]_{y=x}^{y=e^x} dx = \int_0^1 (xe^{3x} - x^4)\,dx$

$= \frac{1}{3}xe^{3x}\Big]_0^1 - \int_0^1 \frac{1}{3}e^{3x}\,dx - \left[\frac{1}{5}x^5\right]_0^1$ [integrating by parts in the first term]

$= \frac{1}{3}e^3 - \left[\frac{1}{9}e^{3x}\right]_0^1 - \frac{1}{5} = \frac{2}{9}e^3 - \frac{4}{45}$

7. $\int_0^\pi \int_0^1 \int_0^{\sqrt{1-y^2}} y\sin x\,dz\,dy\,dx = \int_0^\pi \int_0^1 \left[(y\sin x)z\right]_{z=0}^{z=\sqrt{1-y^2}} dy\,dx = \int_0^\pi \int_0^1 y\sqrt{1-y^2}\sin x\,dy\,dx$

$= \int_0^\pi \left[-\frac{1}{3}(1-y^2)^{3/2}\sin x\right]_{y=0}^{y=1} dx = \int_0^\pi \frac{1}{3}\sin x\,dx = -\frac{1}{3}\cos x\Big]_0^\pi = \frac{2}{3}$

8. $\int_0^1 \int_0^y \int_x^1 6xyz\,dz\,dx\,dy = \int_0^1 \int_0^y \left[3xyz^2\right]_{z=x}^{z=1} dx\,dy = \int_0^1 \int_0^y (3xy - 3x^3y)\,dx\,dy$

$= \int_0^1 \left[\frac{3}{2}x^2y - \frac{3}{4}x^4y\right]_{x=0}^{x=y} dy = \int_0^1 \left(\frac{3}{2}y^3 - \frac{3}{4}y^5\right) dy = \left[\frac{3}{8}y^4 - \frac{1}{8}y^6\right]_0^1 = \frac{1}{4}$

9. The region R is more easily described by polar coordinates: $R = \{(r,\theta) \mid 2 \le r \le 4, 0 \le \theta \le \pi\}$. Thus $\iint_R f(x,y)\,dA = \int_0^\pi \int_2^4 f(r\cos\theta, r\sin\theta)\,r\,dr\,d\theta$.

10. The region R is a type II region that can be described as the region enclosed by the lines $y = 4 - x$, $y = 4 + x$, and the x-axis. So using rectangular coordinates, we can say $R = \{(x,y) \mid y - 4 \le x \le 4 - y, 0 \le y \le 4\}$ and $\iint_R f(x,y)\,dA = \int_0^4 \int_{y-4}^{4-y} f(x,y)\,dx\,dy$.

11. $r = \sin 2\theta$

The region whose area is given by $\int_0^{\pi/2} \int_0^{\sin 2\theta} r\,dr\,d\theta$ is $\{(r,\theta) \mid 0 \le \theta \le \frac{\pi}{2}, 0 \le r \le \sin 2\theta\}$, which is the region contained in the loop in the first quadrant of the four-leaved rose $r = \sin 2\theta$.

12. The solid is $\{(\rho,\theta,\phi) \mid 1 \le \rho \le 2, 0 \le \theta \le \frac{\pi}{2}, 0 \le \phi \le \frac{\pi}{2}\}$ which is the region in the first octant on or between the two spheres $\rho = 1$ and $\rho = 2$.

13. y, (1, 1), y = x, 0, x

$\int_0^1 \int_x^1 \cos(y^2)\,dy\,dx = \int_0^1 \int_0^y \cos(y^2)\,dx\,dy$

$= \int_0^1 \cos(y^2)\left[x\right]_{x=0}^{x=y} dy = \int_0^1 y\cos(y^2)\,dy$

$= \left[\frac{1}{2}\sin(y^2)\right]_0^1 = \frac{1}{2}\sin 1$

14. y, (1, 1), $x = \sqrt{y}$, 0, 1, x

$\int_0^1 \int_{\sqrt{y}}^1 \frac{ye^{x^2}}{x^3}\,dx\,dy = \int_0^1 \int_0^{x^2} \frac{ye^{x^2}}{x^3}\,dy\,dx$

$= \int_0^1 \frac{e^{x^2}}{x^3}\left[\frac{1}{2}y^2\right]_{y=0}^{y=x^2} dx = \int_0^1 \frac{1}{2}xe^{x^2}\,dx$

$= \frac{1}{4}e^{x^2}\Big]_0^1 = \frac{1}{4}(e-1)$

15. $\iint_R ye^{xy}\,dA = \int_0^3 \int_0^2 ye^{xy}\,dx\,dy = \int_0^3 \left[e^{xy}\right]_{x=0}^{x=2} dy = \int_0^3 (e^{2y} - 1)\,dy = \left[\frac{1}{2}e^{2y} - y\right]_0^3$

$= \frac{1}{2}e^6 - 3 - \frac{1}{2} = \frac{1}{2}e^6 - \frac{7}{2}$

16. $\iint_D xy\,dA = \int_0^1 \int_{y^2}^{y+2} xy\,dx\,dy = \int_0^1 y\left[\frac{1}{2}x^2\right]_{x=y^2}^{x=y+2} dy = \frac{1}{2}\int_0^1 y((y+2)^2 - y^4)\,dy$

$= \frac{1}{2}\int_0^1 (y^3 + 4y^2 + 4y - y^5)\,dy = \frac{1}{2}\left[\frac{1}{4}y^4 + \frac{4}{3}y^3 + 2y^2 - \frac{1}{6}y^6\right]_0^1 = \frac{41}{24}$

17. $$\iint_D \frac{y}{1+x^2}\,dA = \int_0^1 \int_0^{\sqrt{x}} \frac{y}{1+x^2}\,dy\,dx = \int_0^1 \frac{1}{1+x^2}\left[\tfrac{1}{2}y^2\right]_{y=0}^{y=\sqrt{x}}\,dx$$

$$= \frac{1}{2}\int_0^1 \frac{x}{1+x^2}\,dx = \left[\tfrac{1}{4}\ln(1+x^2)\right]_0^1 = \tfrac{1}{4}\ln 2$$

18. $$\iint_D \frac{1}{1+x^2}\,dA = \int_0^1 \int_x^1 \frac{1}{1+x^2}\,dy\,dx = \int_0^1 \frac{1}{1+x^2}\left[y\right]_{y=x}^{y=1}\,dx = \int_0^1 \frac{1-x}{1+x^2}\,dx$$

$$= \int_0^1 \left(\frac{1}{1+x^2} - \frac{x}{1+x^2}\right) dx = \left[\tan^{-1}x - \tfrac{1}{2}\ln(1+x^2)\right]_0^1$$

$$= \tan^{-1}1 - \tfrac{1}{2}\ln 2 - \left(\tan^{-1}0 - \tfrac{1}{2}\ln 1\right) = \tfrac{\pi}{4} - \tfrac{1}{2}\ln 2$$

19.

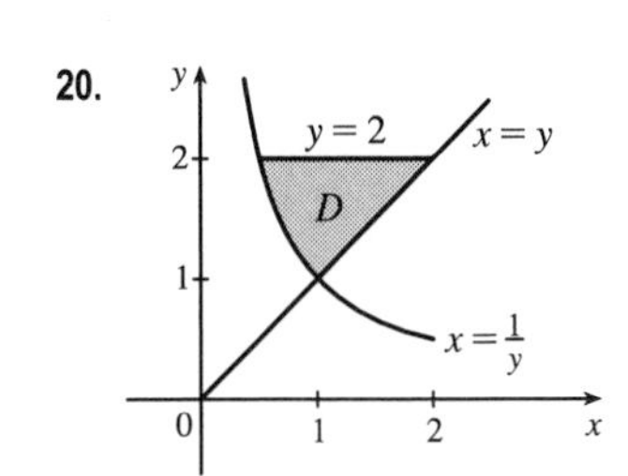

$\iint_D y\,dA = \int_0^2 \int_{y^2}^{8-y^2} y\,dx\,dy$

$= \int_0^2 y\left[x\right]_{x=y^2}^{x=8-y^2} dy = \int_0^2 y(8 - y^2 - y^2)\,dy$

$= \int_0^2 (8y - 2y^3)\,dy = \left[4y^2 - \frac{1}{2}y^4\right]_0^2 = 8$

20. $$\iint_D y\,dA = \int_1^2 \int_{1/y}^{y} y\,dx\,dy = \int_1^2 y\left(y - \frac{1}{y}\right) dy$$

$= \int_1^2 (y^2 - 1)\,dy = \left[\frac{1}{3}y^3 - y\right]_1^2$

$= \left(\frac{8}{3} - 2\right) - \left(\frac{1}{3} - 1\right) = \frac{4}{3}$

21. $$\iint_D (x^2+y^2)^{3/2}\,dA = \int_0^{\pi/3} \int_0^3 (r^2)^{3/2}\, r\,dr\,d\theta$$

$$= \int_0^{\pi/3} d\theta \int_0^3 r^4\,dr = \left[\theta\right]_0^{\pi/3}\left[\frac{1}{5}r^5\right]_0^3$$

$$= \frac{\pi}{3}\,\frac{3^5}{5} = \frac{81\pi}{5}$$

22. $\iint_D x\,dA = \int_0^{\pi/2}\int_1^{\sqrt{2}} (r\cos\theta)\,r\,dr\,d\theta = \int_0^{\pi/2}\cos\theta\,d\theta \int_1^{\sqrt{2}} r^2\,dr = \left[\sin\theta\right]_0^{\pi/2}\left[\frac{1}{3}r^3\right]_1^{\sqrt{2}}$

$= 1\cdot\frac{1}{3}(2^{3/2} - 1) = \frac{1}{3}(2^{3/2} - 1)$

23. $\iiint_E xy\,dV = \int_0^3\int_0^x\int_0^{x+y} xy\,dz\,dy\,dx = \int_0^3\int_0^x xy\left[z\right]_{z=0}^{z=x+y} dy\,dx = \int_0^3\int_0^x xy(x+y)\,dy\,dx$

$= \int_0^3\int_0^x (x^2y + xy^2)\,dy\,dx = \int_0^3 \left[\frac{1}{2}x^2y^2 + \frac{1}{3}xy^3\right]_{y=0}^{y=x} dx = \int_0^3 \left(\frac{1}{2}x^4 + \frac{1}{3}x^4\right) dx$

$= \frac{5}{6}\int_0^3 x^4\,dx = \left[\frac{1}{6}x^5\right]_0^3 = \frac{81}{2} = 40.5$

24.

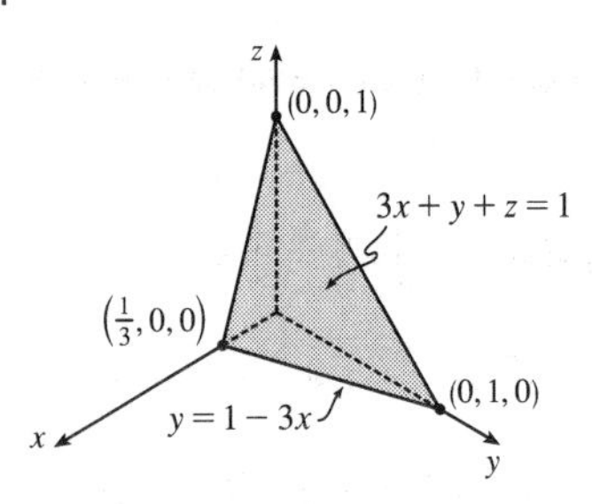

$$\iiint_T xy\,dV = \int_0^{1/3}\int_0^{1-3x}\int_0^{1-3x-y} xy\,dz\,dy\,dx$$
$$= \int_0^{1/3}\int_0^{1-3x} xy(1-3x-y)\,dy\,dx$$
$$= \int_0^{1/3}\int_0^{1-3x}(xy - 3x^2y - xy^2)\,dy\,dx$$
$$= \int_0^{1/3}\left[\tfrac{1}{2}xy^2 - \tfrac{3}{2}x^2y^2 - \tfrac{1}{3}xy^3\right]_{y=0}^{y=1-3x} dx$$
$$= \int_0^{1/3}\left[\tfrac{1}{2}x(1-3x)^2 - \tfrac{3}{2}x^2(1-3x)^2 - \tfrac{1}{3}x(1-3x)^3\right] dx$$
$$= \int_0^{1/3}\left(\tfrac{1}{6}x - \tfrac{3}{2}x^2 + \tfrac{9}{2}x^3 - \tfrac{9}{2}x^4\right) dx$$
$$= \tfrac{1}{12}x^2 - \tfrac{1}{2}x^3 + \tfrac{9}{8}x^4 - \tfrac{9}{10}x^5\Big]_0^{1/3} = \tfrac{1}{1080}$$

25. $\iiint_E y^2z^2\,dV = \int_{-1}^{1}\int_{-\sqrt{1-y^2}}^{\sqrt{1-y^2}}\int_0^{1-y^2-z^2} y^2z^2\,dx\,dz\,dy = \int_{-1}^{1}\int_{-\sqrt{1-y^2}}^{\sqrt{1-y^2}} y^2z^2(1-y^2-z^2)\,dz\,dy$

$$= \int_0^{2\pi}\int_0^1 (r^2\cos^2\theta)(r^2\sin^2\theta)(1-r^2)\,r\,dr\,d\theta = \int_0^{2\pi}\int_0^1 \tfrac{1}{4}\sin^2 2\theta(r^5-r^7)\,dr\,d\theta$$
$$= \int_0^{2\pi}\tfrac{1}{8}(1-\cos 4\theta)\left[\tfrac{1}{6}r^6 - \tfrac{1}{8}r^8\right]_{r=0}^{r=1} d\theta = \tfrac{1}{192}\left[\theta - \tfrac{1}{4}\sin 4\theta\right]_0^{2\pi} = \tfrac{2\pi}{192} = \tfrac{\pi}{96}$$

26. $\iiint_E z\,dV = \int_0^1\int_0^{\sqrt{1-y^2}}\int_0^{2-y} z\,dx\,dz\,dy = \int_0^1\int_0^{\sqrt{1-y^2}}(2-y)z\,dz\,dy = \int_0^1 \tfrac{1}{2}(2-y)(1-y^2)\,dy$

$$= \int_0^1 \tfrac{1}{2}(2-y-2y^2+y^3)\,dy = \tfrac{13}{24}$$

27. $\iiint_E yz\,dV = \int_{-2}^{2}\int_0^{\sqrt{4-x^2}}\int_0^y yz\,dz\,dy\,dx = \int_{-2}^{2}\int_0^{\sqrt{4-x^2}} \tfrac{1}{2}y^3\,dy\,dx = \int_0^{\pi}\int_0^2 \tfrac{1}{2}r^3(\sin^3\theta)\,r\,dr\,d\theta$

$$= \tfrac{16}{5}\int_0^{\pi}\sin^3\theta\,d\theta = \tfrac{16}{5}\left[-\cos\theta + \tfrac{1}{3}\cos^3\theta\right]_0^{\pi} = \tfrac{64}{15}$$

28. $\iiint_H z^3\sqrt{x^2+y^2+z^2}\,dV = \int_0^{2\pi}\int_0^{\pi/2}\int_0^1 (\rho^3\cos^3\phi)\rho(\rho^2\sin\phi)\,d\rho\,d\phi\,d\theta$

$$= \int_0^{2\pi} d\theta \int_0^{\pi/2}\cos^3\phi\sin\phi\,d\phi \int_0^1 \rho^6\,d\rho = 2\pi\left[-\tfrac{1}{4}\cos^4\phi\right]_0^{\pi/2}\left(\tfrac{1}{7}\right) = \tfrac{\pi}{14}$$

29. $V = \int_0^2\int_1^4 (x^2+4y^2)\,dy\,dx = \int_0^2\left[x^2y + \tfrac{4}{3}y^3\right]_{y=1}^{y=4} dx = \int_0^2 (3x^2+84)\,dx = 176$

30.

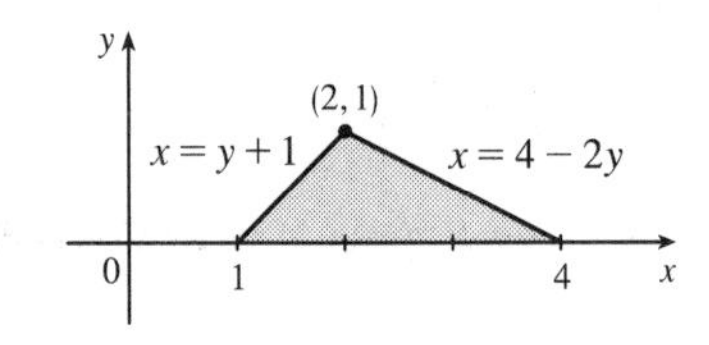

$$V = \int_0^1\int_{y+1}^{4-2y}\int_0^{x^2y} dz\,dx\,dy = \int_0^1\int_{y+1}^{4-2y} x^2y\,dx\,dy$$
$$= \int_0^1 \tfrac{1}{3}\left[(4-2y)^3y - (y+1)^3y\right] dy = \int_0^1 3(-y^4+5y^3-11y^2+7y)\,dy$$
$$= 3\left(-\tfrac{1}{5} + \tfrac{5}{4} - \tfrac{11}{3} + \tfrac{7}{2}\right) = \tfrac{53}{20}$$

31.

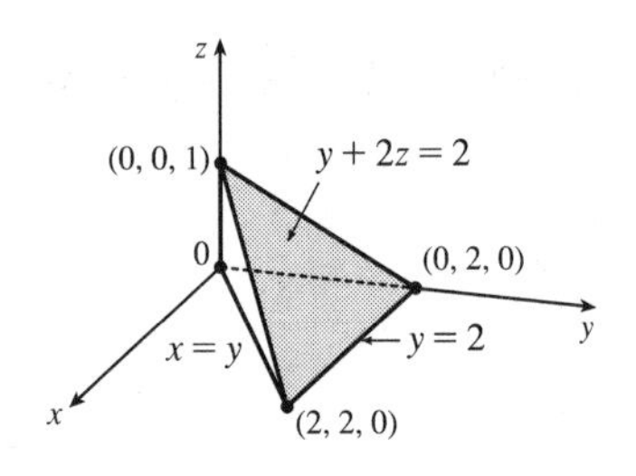

$$V = \int_0^2\int_0^y\int_0^{(2-y)/2} dz\,dx\,dy$$
$$= \int_0^2\int_0^y \left(1 - \tfrac{1}{2}y\right) dx\,dy$$
$$= \int_0^2 \left(y - \tfrac{1}{2}y^2\right) dy = \tfrac{2}{3}$$

32. $V = \int_0^{2\pi}\int_0^2\int_0^{3-r\sin\theta} r\,dz\,dr\,d\theta = \int_0^{2\pi}\int_0^2 (3r - r^2\sin\theta)\,dr\,d\theta = \int_0^{2\pi}\left[6 - \frac{8}{3}\sin\theta\right]d\theta = 6\theta\big]_0^{2\pi} + 0 = 12\pi$

33. Using the wedge above the plane $z = 0$ and below the plane $z = mx$ and noting that we have the same volume for $m < 0$ as for $m > 0$ (so use $m > 0$), we have

$$V = 2\int_0^{a/3}\int_0^{\sqrt{a^2-9y^2}} mx\,dx\,dy = 2\int_0^{a/3}\tfrac{1}{2}m(a^2 - 9y^2)\,dy = m\left[a^2y - 3y^3\right]_0^{a/3} = m\left(\tfrac{1}{3}a^3 - \tfrac{1}{9}a^3\right) = \tfrac{2}{9}ma^3.$$

34. The paraboloid and the half-cone intersect when $x^2 + y^2 = \sqrt{x^2+y^2}$, that is when $x^2 + y^2 = 1$ or 0. So

$$V = \iint_{x^2+y^2\le 1}\int_{x^2+y^2}^{\sqrt{x^2+y^2}} dz\,dA = \int_0^{2\pi}\int_0^1\int_{r^2}^{r} r\,dz\,dr\,d\theta = \int_0^{2\pi}\int_0^1 (r^2 - r^3)\,dr\,d\theta = \int_0^{2\pi}\left(\tfrac{1}{3} - \tfrac{1}{4}\right)d\theta = \tfrac{1}{12}(2\pi) = \tfrac{\pi}{6}.$$

35. (a) $m = \int_0^1\int_0^{1-y^2} y\,dx\,dy = \int_0^1 (y - y^3)\,dy = \frac{1}{2} - \frac{1}{4} = \frac{1}{4}$

(b) $M_y = \int_0^1\int_0^{1-y^2} xy\,dx\,dy = \int_0^1 \frac{1}{2}y(1-y^2)^2\,dy = -\frac{1}{12}(1-y^2)^3\big]_0^1 = \frac{1}{12}$,

$M_x = \int_0^1\int_0^{1-y^2} y^2\,dx\,dy = \int_0^1 (y^2 - y^4)\,dy = \frac{2}{15}$. Hence $(\overline{x}, \overline{y}) = \left(\frac{1}{3}, \frac{8}{15}\right)$.

(c) $I_x = \int_0^1\int_0^{1-y^2} y^3\,dx\,dy = \int_0^1 (y^3 - y^5)\,dy = \frac{1}{12}$,

$I_y = \int_0^1\int_0^{1-y^2} yx^2\,dx\,dy = \int_0^1 \frac{1}{3}y(1-y^2)^3\,dy = -\frac{1}{24}(1-y^2)^4\big]_0^1 = \frac{1}{24}$,

$I_0 = I_x + I_y = \frac{1}{8}$, $\overline{\overline{y}}^2 = \frac{1/12}{1/4} = \frac{1}{3} \;\Rightarrow\; \overline{\overline{y}} = \frac{1}{\sqrt{3}}$, and $\overline{\overline{x}}^2 = \frac{1/24}{1/4} = \frac{1}{6} \;\Rightarrow\; \overline{\overline{x}} = \frac{1}{\sqrt{6}}$.

36. (a) $m = \frac{1}{4}\pi Ka^2$ where K is constant,

$M_y = \iint_{x^2+y^2\le a^2} Kx\,dA = K\int_0^{\pi/2}\int_0^a r^2\cos\theta\,dr\,d\theta = \frac{1}{3}Ka^3\int_0^{\pi/2}\cos\theta\,d\theta = \frac{1}{3}a^3K$, and

$M_x = K\int_0^{\pi/2}\int_0^a r^2\sin\theta\,dr\,d\theta = \frac{1}{3}a^3K$ [by symmetry $M_y = M_x$].

Hence the centroid is $(\overline{x}, \overline{y}) = \left(\frac{4}{3\pi}a, \frac{4}{3\pi}a\right)$.

(b) $m = \int_0^{\pi/2}\int_0^a r^4\cos\theta\sin^2\theta\,dr\,d\theta = \left[\frac{1}{3}\sin^3\theta\right]_0^{\pi/2}\left(\frac{1}{5}a^5\right) = \frac{1}{15}a^5$,

$M_y = \int_0^{\pi/2}\int_0^a r^5\cos^2\theta\sin^2\theta\,dr\,d\theta = \frac{1}{8}\left[\theta - \frac{1}{4}\sin 4\theta\right]_0^{\pi/2}\left(\frac{1}{6}a^6\right) = \frac{1}{96}\pi a^6$, and

$M_x = \int_0^{\pi/2}\int_0^a r^5\cos\theta\sin^3\theta\,dr\,d\theta = \left[\frac{1}{4}\sin^4\theta\right]_0^{\pi/2}\left(\frac{1}{6}a^6\right) = \frac{1}{24}a^6$. Hence $(\overline{x}, \overline{y}) = \left(\frac{5}{32}\pi a, \frac{5}{8}a\right)$.

37. (a) The equation of the cone with the suggested orientation is $(h - z) = \frac{h}{a}\sqrt{x^2+y^2}$, $0 \le z \le h$. Then $V = \frac{1}{3}\pi a^2 h$ is the volume of one frustum of a cone; by symmetry $M_{yz} = M_{xz} = 0$; and

$$M_{xy} = \iint_{x^2+y^2\le a^2}\int_0^{h-(h/a)\sqrt{x^2+y^2}} z\,dz\,dA = \int_0^{2\pi}\int_0^a\int_0^{(h/a)(a-r)} rz\,dz\,dr\,d\theta = \pi\int_0^a r\frac{h^2}{a^2}(a-r)^2\,dr$$

$$= \frac{\pi h^2}{a^2}\int_0^a (a^2r - 2ar^2 + r^3)\,dr = \frac{\pi h^2}{a^2}\left(\frac{a^4}{2} - \frac{2a^4}{3} + \frac{a^4}{4}\right) = \frac{\pi h^2 a^2}{12}$$

Hence the centroid is $(\overline{x}, \overline{y}, \overline{z}) = \left(0, 0, \frac{1}{4}h\right)$.

(b) $$I_z = \int_0^{2\pi}\int_0^a\int_0^{(h/a)(a-r)} r^3\,dz\,dr\,d\theta = 2\pi\int_0^a \frac{h}{a}(ar^3 - r^4)\,dr = \frac{2\pi h}{a}\left(\frac{a^5}{4} - \frac{a^5}{5}\right) = \frac{\pi a^4 h}{10}$$

38. Let the tetrahedron be called T. The front face of T is given by the plane $x + \frac{1}{2}y + \frac{1}{3}z = 1$, or $z = 3 - 3x - \frac{3}{2}y$, which intersects the xy-plane in the line $y = 2 - 2x$. So the total mass is

$m = \iiint_T \rho(x, y, z)\, dV = \int_0^1 \int_0^{2-2x} \int_0^{3-3x-3y/2} (x^2 + y^2 + z^2)\, dz\, dy\, dx = \frac{7}{5}$. The center of mass is

$(\overline{x}, \overline{y}, \overline{z}) = \left(m^{-1} \iiint_T x\rho(x, y, z)\, dV, m^{-1} \iiint_T y\rho(x, y, z)\, dV, m^{-1} \iiint_T z\rho(x, y, z)\, dV\right) = \left(\frac{4}{21}, \frac{11}{21}, \frac{8}{7}\right)$.

39. $x = r\cos\theta = 2\sqrt{3}\cos\frac{\pi}{3} = 2\sqrt{3}\cdot\frac{1}{2} = \sqrt{3}$, $y = r\sin\theta = 2\sqrt{3}\sin\frac{\pi}{3} = 2\sqrt{3}\cdot\frac{\sqrt{3}}{2} = 3$, $z = 2$, so in rectangular coordinates the point is $\left(\sqrt{3}, 3, 2\right)$. $\rho = \sqrt{r^2 + z^2} = \sqrt{12 + 4} = 4$, $\theta = \frac{\pi}{3}$, and $\cos\phi = \frac{z}{\rho} = \frac{1}{2}$, so $\phi = \frac{\pi}{3}$ and spherical coordinates are $\left(4, \frac{\pi}{3}, \frac{\pi}{3}\right)$.

40. $r = \sqrt{4+4} = 2\sqrt{2}$, $z = -1$, $\cos\theta = \frac{2}{2\sqrt{2}} = \frac{\sqrt{2}}{2}$ so $\theta = \frac{\pi}{4}$ and in cylindrical coordinates the point is $\left(2\sqrt{2}, \frac{\pi}{4}, -1\right)$. $\rho = \sqrt{4+4+1} = 3$, $\cos\phi = -\frac{1}{3}$, so the spherical coordinates are $\left(3, \frac{\pi}{4}, \cos^{-1}\left(-\frac{1}{3}\right)\right)$.

41. $x = \rho\sin\phi\cos\theta = 8\sin\frac{\pi}{6}\cos\frac{\pi}{4} = 8\cdot\frac{1}{2}\cdot\frac{\sqrt{2}}{2} = 2\sqrt{2}$, $y = \rho\sin\phi\sin\theta = 8\sin\frac{\pi}{6}\sin\frac{\pi}{4} = 2\sqrt{2}$, and $z = \rho\cos\phi = 8\cos\frac{\pi}{6} = 8\cdot\frac{\sqrt{3}}{2} = 4\sqrt{3}$. Thus rectangular coordinates for the point are $\left(2\sqrt{2}, 2\sqrt{2}, 4\sqrt{3}\right)$. $r^2 = x^2 + y^2 = 8 + 8 = 16 \quad\Rightarrow\quad r = 4$, $\theta = \frac{\pi}{4}$, and $z = 4\sqrt{3}$, so cylindrical coordinates are $\left(4, \frac{\pi}{4}, 4\sqrt{3}\right)$.

42. (a) $\theta = \frac{\pi}{4}$. In spherical coordinates, this is a half-plane including the z-axis and intersecting the xy-plane in the half-line $x = y$, $x > 0$.

(b) $\phi = \frac{\pi}{4}$. This is one frustum of a circular cone with vertex the origin and axis the positive z-axis.

43. $x^2 + y^2 + z^2 = 4$. In cylindrical coordinates, this becomes $r^2 + z^2 = 4$. In spherical coordinates, it becomes $\rho^2 = 4$ or $\rho = 2$.

44. $\rho = 2\cos\phi \quad\Rightarrow\quad \rho^2 = 2\rho\cos\phi \quad\Rightarrow\quad x^2 + y^2 + z^2 = 2z \quad\Rightarrow$ $x^2 + y^2 + (z-1)^2 = 1$. This is the equation of a sphere with radius 1, centered at $(0, 0, 1)$. Therefore, $0 \le \rho \le 2\cos\phi$ is the solid ball whose boundary is this sphere. $0 \le \theta \le \frac{\pi}{2}$ and $0 \le \phi \le \frac{\pi}{6}$ restrict the solid to the section of this ball that lies above the cone $\phi = \frac{\pi}{6}$ and is in the first octant.

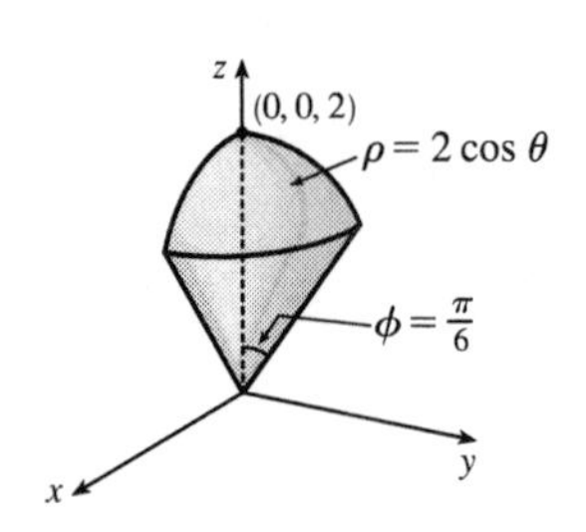

45.

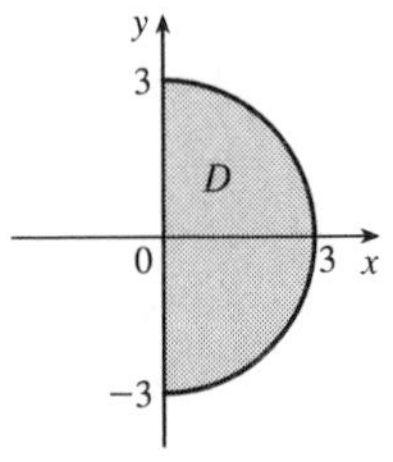

$$\int_0^3 \int_{-\sqrt{9-x^2}}^{\sqrt{9-x^2}} (x^3 + xy^2)\, dy\, dx = \int_0^3 \int_{-\sqrt{9-x^2}}^{\sqrt{9-x^2}} x(x^2 + y^2)\, dy\, dx$$
$$= \int_{-\pi/2}^{\pi/2} \int_0^3 (r\cos\theta)(r^2)\, r\, dr\, d\theta$$
$$= \int_{-\pi/2}^{\pi/2} \cos\theta\, d\theta \int_0^3 r^4\, dr$$
$$= \left[\sin\theta\right]_{-\pi/2}^{\pi/2} \left[\tfrac{1}{5}r^5\right]_0^3 = 2\cdot\tfrac{1}{5}(243) = \tfrac{486}{5} = 97.2$$

46. The region of integration is the solid hemisphere $x^2 + y^2 + z^2 \le 4$, $x \ge 0$.

$$\int_{-2}^{2} \int_0^{\sqrt{4-y^2}} \int_{-\sqrt{4-x^2-y^2}}^{\sqrt{4-x^2-y^2}} y^2 \sqrt{x^2+y^2+z^2}\, dz\, dx\, dy$$

$$= \int_{-\pi/2}^{\pi/2} \int_0^{\pi} \int_0^2 (\rho \sin\phi \sin\theta)^2 \left(\sqrt{\rho^2}\right) \rho^2 \sin\phi \, d\rho\, d\phi\, d\theta = \int_{-\pi/2}^{\pi/2} \sin^2\theta\, d\theta \int_0^{\pi} \sin^3\phi\, d\phi \int_0^2 \rho^5\, d\rho$$

$$= \left[\tfrac{1}{2}\theta - \tfrac{1}{4}\sin 2\theta\right]_{-\pi/2}^{\pi/2} \left[-\tfrac{1}{3}(2+\sin^2\phi)\cos\phi\right]_0^{\pi} \left[\tfrac{1}{6}\rho^6\right]_0^2 = \left(\tfrac{\pi}{2}\right)\left(\tfrac{2}{3}+\tfrac{2}{3}\right)\left(\tfrac{32}{3}\right) = \tfrac{64}{9}\pi$$

47. 

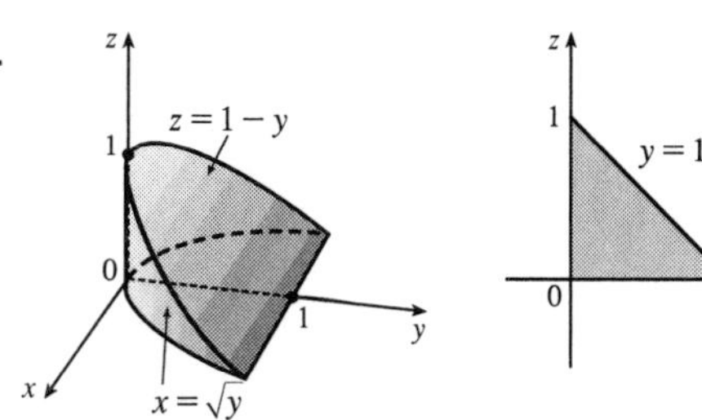

$$\int_{-1}^{1} \int_{x^2}^{1} \int_0^{1-y} f(x,y,z)\, dz\, dy\, dx = \int_0^1 \int_0^{1-z} \int_{-\sqrt{y}}^{\sqrt{y}} f(x,y,z)\, dx\, dy\, dz$$

48.

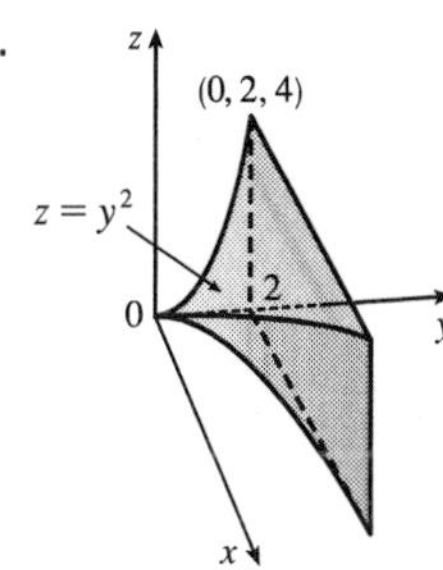

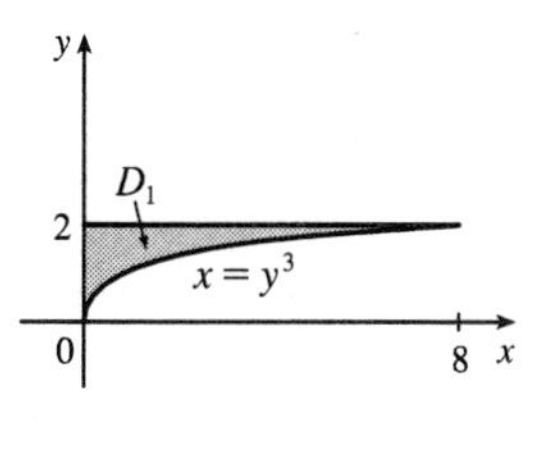

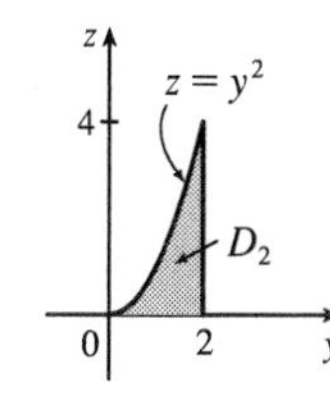

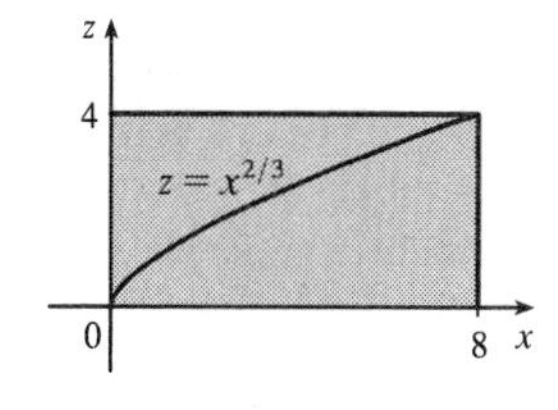

$\int_0^2 \int_0^{y^3} \int_0^{y^2} f(x,y,z)\, dz\, dx\, dy = \iiint_E f(x,y,z)\, dV$ where $E = \left\{(x,y,z) \mid 0 \le y \le 2, 0 \le x \le y^3, 0 \le z \le y^2\right\}$.

If D_1, D_2, and D_3 are the projections of E on the xy-, yz-, and xz-planes, then

$D_1 = \left\{(x,y) \mid 0 \le y \le 2, 0 \le x \le y^3\right\} = \left\{(x,y) \mid 0 \le x \le 8, \sqrt[3]{x} \le y \le 2\right\}$,

$D_2 = \left\{(y,z) \mid 0 \le z \le 4, \sqrt{z} \le y \le 2\right\} = \left\{(y,z) \mid 0 \le y \le 2, 0 \le z \le y^2\right\}$,

$D_3 = \left\{(x,z) \mid 0 \le x \le 8, 0 \le z \le 4\right\}$. Therefore we have

$$\int_0^2 \int_0^{y^3} \int_0^{y^2} f(x,y,z)\, dz\, dx\, dy = \int_0^8 \int_{\sqrt[3]{x}}^2 \int_0^{y^2} f(x,y,z)\, dz\, dy\, dx = \int_0^4 \int_{\sqrt{z}}^2 \int_0^{y^3} f(x,y,z)\, dx\, dy\, dz$$

$$= \int_0^2 \int_0^{y^2} \int_0^{y^3} f(x,y,z)\, dx\, dz\, dy$$

$$= \int_0^8 \int_0^{x^{2/3}} \int_{\sqrt[3]{x}}^2 f(x,y,z)\, dy\, dz\, dx + \int_0^8 \int_{x^{2/3}}^4 \int_{\sqrt{z}}^2 f(x,y,z)\, dy\, dz\, dx$$

$$= \int_0^4 \int_0^{z^{3/2}} \int_{\sqrt{z}}^2 f(x,y,z)\, dy\, dx\, dz + \int_0^4 \int_{z^{3/2}}^8 \int_{\sqrt[3]{x}}^2 f(x,y,z)\, dy\, dx\, dz$$

49. Since $u = x - y$ and $v = x + y$, $x = \frac{1}{2}(u+v)$ and $y = \frac{1}{2}(v-u)$.

Thus $\dfrac{\partial(x,y)}{\partial(u,v)} = \begin{vmatrix} 1/2 & 1/2 \\ -1/2 & 1/2 \end{vmatrix} = \dfrac{1}{2}$ and $\displaystyle\iint_R \frac{x-y}{x+y}\, dA = \int_2^4 \int_{-2}^0 \frac{u}{v}\left(\frac{1}{2}\right) du\, dv = -\int_2^4 \frac{dv}{v} = -\ln 2.$

50. $\dfrac{\partial(x,y,z)}{\partial(u,v,w)} = \begin{vmatrix} 2u & 0 & 0 \\ 0 & 2v & 0 \\ 0 & 0 & 2w \end{vmatrix} = 8uvw$, so

$$\begin{aligned} V &= \iiint_E dV = \int_0^1 \int_0^{1-u} \int_0^{1-u-v} 8uvw\,dw\,dv\,du = \int_0^1 \int_0^{1-u} 4uv(1-u-v)^2\,du \\ &= \int_0^1 \int_0^{1-u} \left[4u(1-u)^2 v - 8u(1-u)v^2 + 4uv^3\right] dv\,du \\ &= \int_0^1 \left[2u(1-u)^4 - \tfrac{8}{3}u(1-u)^4 + u(1-u)^4\right] du = \int_0^1 \tfrac{1}{3}u(1-u)^4 du \\ &= \int_0^1 \tfrac{1}{3}\left[(1-u)^4 - (1-u)^5\right] du = \tfrac{1}{3}\left[-\tfrac{1}{5}(1-u)^5 + \tfrac{1}{6}(1-u)^6\right]_0^1 = \tfrac{1}{3}\left(-\tfrac{1}{6} + \tfrac{1}{5}\right) = \tfrac{1}{90} \end{aligned}$$

51. Let $u = y - x$ and $v = y + x$ so $x = y - u = (v - x) - u \;\Rightarrow\; x = \frac{1}{2}(v - u)$ and $y = v - \frac{1}{2}(v - u) = \frac{1}{2}(v + u)$.

$\left|\dfrac{\partial(x,y)}{\partial(u,v)}\right| = \left|\dfrac{\partial x}{\partial u}\dfrac{\partial y}{\partial v} - \dfrac{\partial x}{\partial v}\dfrac{\partial y}{\partial u}\right| = \left|-\tfrac{1}{2}\left(\tfrac{1}{2}\right) - \tfrac{1}{2}\left(\tfrac{1}{2}\right)\right| = \left|-\tfrac{1}{2}\right| = \tfrac{1}{2}$. R is the image under this transformation of the square with vertices $(u,v) = (0,0)$, $(-2,0)$, $(0,2)$, and $(-2,2)$. So

$$\iint_R xy\,dA = \int_0^2 \int_{-2}^0 \frac{v^2 - u^2}{4}\left(\frac{1}{2}\right) du\,dv = \tfrac{1}{8}\int_0^2 \left[v^2 u - \tfrac{1}{3}u^3\right]_{u=-2}^{u=0} dv = \tfrac{1}{8}\int_0^2 \left(2v^2 - \tfrac{8}{3}\right) dv = \tfrac{1}{8}\left[\tfrac{2}{3}v^3 - \tfrac{8}{3}v\right]_0^2 = 0$$

This result could have been anticipated by symmetry, since the integrand is an odd function of y and R is symmetric about the x-axis.

52. (a) $$\iint_D \frac{1}{(x^2+y^2)^{n/2}}\,dA = \int_0^{2\pi} \int_r^R \frac{1}{(t^2)^{n/2}}\,t\,dt\,d\theta = 2\pi \int_r^R t^{1-n}\,dt$$

$$= \begin{cases} \left.\dfrac{2\pi}{2-n}t^{2-n}\right]_r^R = \dfrac{2\pi}{2-n}\left(R^{2-n} - r^{2-n}\right) & \text{if } n \neq 2 \\ 2\pi \ln(R/r) & \text{if } n = 2 \end{cases}$$

(b) The integral in part (a) has a limit as $r \to 0^+$ for all values of n such that $2 - n > 0 \;\Leftrightarrow\; n < 2$.

(c) $$\iiint_E \frac{1}{(x^2+y^2+z^2)^{n/2}}\,dV = \int_r^R \int_0^\pi \int_0^{2\pi} \frac{1}{(\rho^2)^{n/2}}\rho^2 \sin\phi\,d\theta\,d\phi\,d\rho = 2\pi \int_r^R \int_0^\pi \rho^{2-n} \sin\phi\,d\phi\,d\rho$$

$$= \begin{cases} \left.\dfrac{4\pi}{3-n}\rho^{3-n}\right]_r^R = \dfrac{4\pi}{3-n}\left(R^{3-n} - r^{3-n}\right) & \text{if } n \neq 3 \\ 4\pi \ln(R/r) & \text{if } n = 3 \end{cases}$$

(d) As $r \to 0^+$, the above integral has a limit, provided that $3 - n > 0 \;\Leftrightarrow\; n < 3$.

13 □ VECTOR CALCULUS

13.1 Vector Fields

1. $\mathbf{F}(x, y) = \frac{1}{2}(\mathbf{i} + \mathbf{j})$

All vectors in this field are identical, with length $\frac{1}{\sqrt{2}}$ and direction parallel to the line $y = x$.

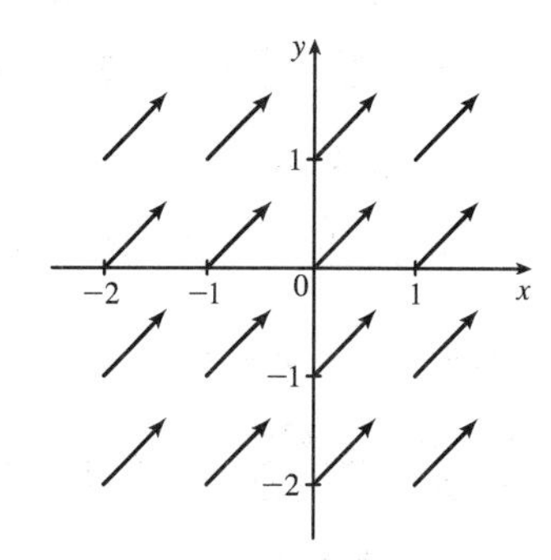

2. $\mathbf{F}(x, y) = \mathbf{i} + x\,\mathbf{j}$

The length of the vector $\mathbf{i} + x\,\mathbf{j}$ is $\sqrt{1 + x^2}$. Vectors are tangent to parabolas opening about the y-axis.

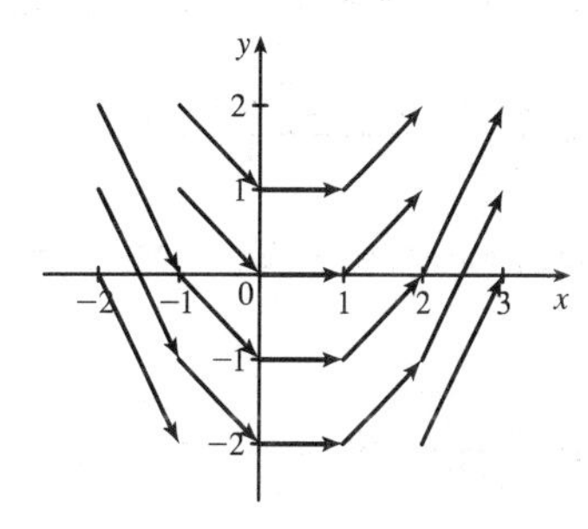

3. $\mathbf{F}(x, y) = y\,\mathbf{i} + \frac{1}{2}\,\mathbf{j}$

The length of the vector $y\,\mathbf{i} + \frac{1}{2}\,\mathbf{j}$ is $\sqrt{y^2 + \frac{1}{4}}$. Vectors are tangent to parabolas opening about the x-axis.

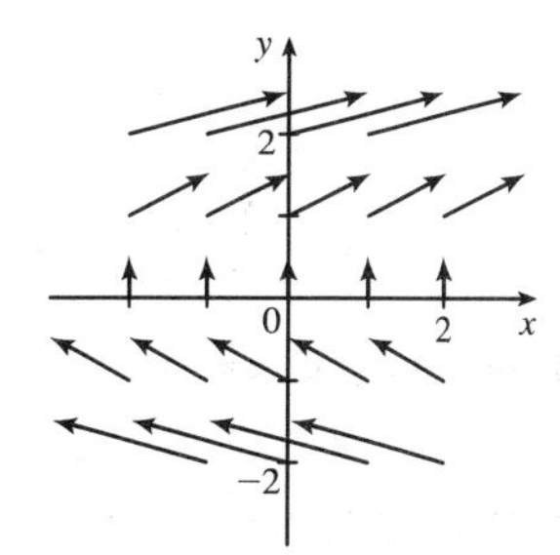

4. $\mathbf{F}(x, y) = (x - y)\,\mathbf{i} + x\,\mathbf{j}$

The length of the vector $(x - y)\,\mathbf{i} + x\,\mathbf{j}$ is $\sqrt{(x - y)^2 + x^2}$. Vectors along the line $y = x$ are vertical.

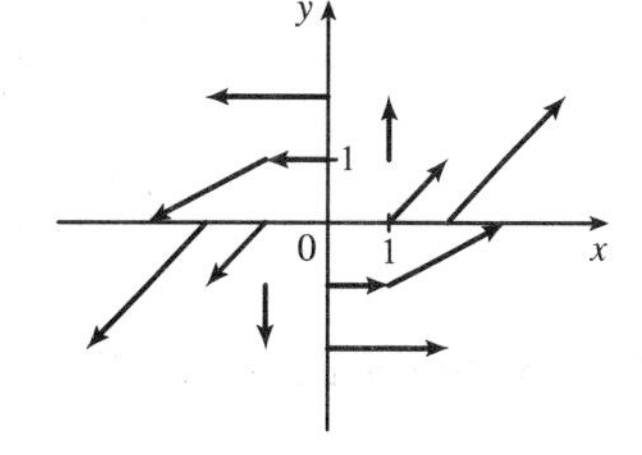

5. $\mathbf{F}(x, y) = \dfrac{y\,\mathbf{i} + x\,\mathbf{j}}{\sqrt{x^2 + y^2}}$

The length of the vector $\dfrac{y\,\mathbf{i} + x\,\mathbf{j}}{\sqrt{x^2 + y^2}}$ is 1.

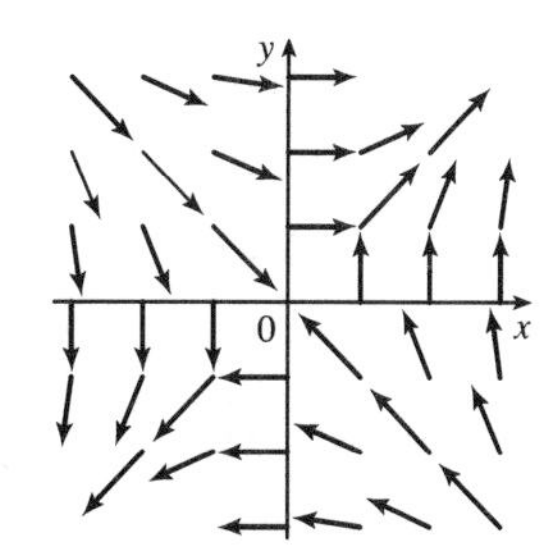

6. $\mathbf{F}(x,y) = \dfrac{y\,\mathbf{i} - x\,\mathbf{j}}{\sqrt{x^2+y^2}}$

All the vectors $\mathbf{F}(x,y)$ are unit vectors tangent to circles centered at the origin with radius $\sqrt{x^2+y^2}$.

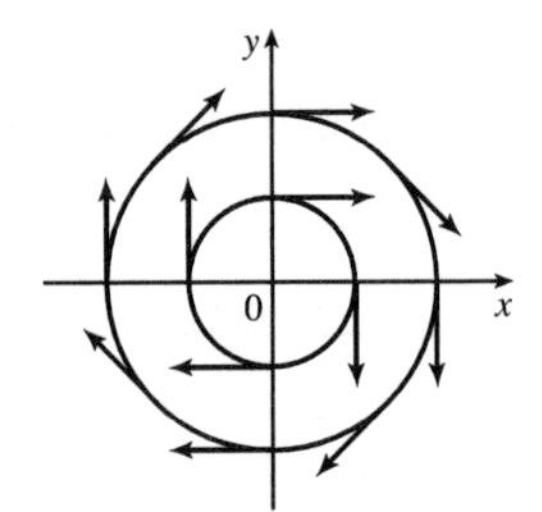

7. $\mathbf{F}(x,y,z) = \mathbf{k}$

All vectors in this field are parallel to the z-axis and have length 1.

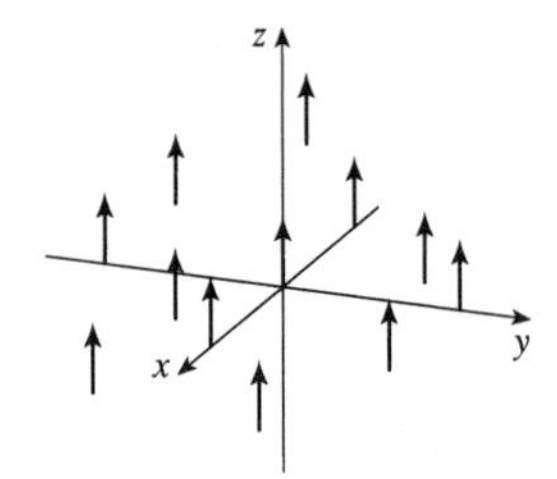

8. $\mathbf{F}(x,y,z) = -y\,\mathbf{k}$

At each point (x,y,z), $\mathbf{F}(x,y,z)$ is a vector of length $|y|$. For $y > 0$, all point in the direction of the negative z-axis, while for $y < 0$, all are in the direction of the positive z-axis. In each plane $y = k$, all the vectors are identical.

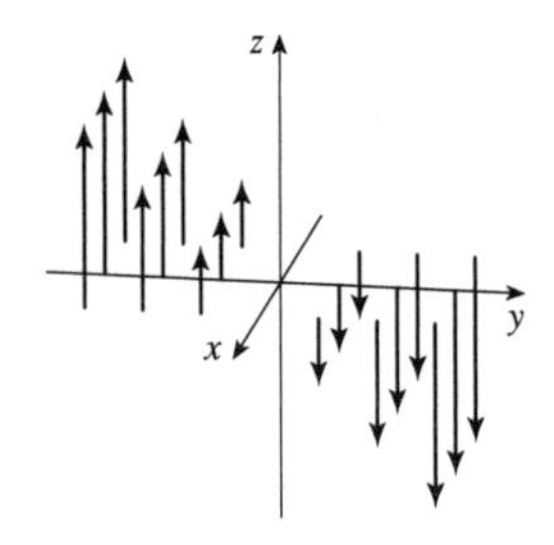

9. $\mathbf{F}(x,y,z) = x\,\mathbf{k}$

At each point (x,y,z), $\mathbf{F}(x,y,z)$ is a vector of length $|x|$. For $x > 0$, all point in the direction of the positive z-axis, while for $x < 0$, all are in the direction of the negative z-axis. In each plane $x = k$, all the vectors are identical.

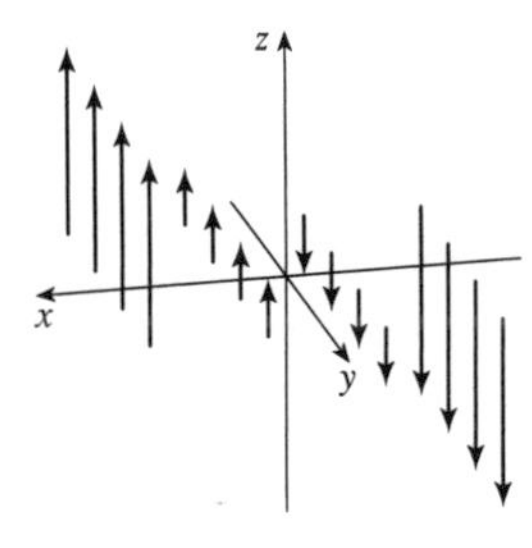

10. $\mathbf{F}(x,y,z) = \mathbf{j} - \mathbf{i}$

All vectors in this field have length $\sqrt{2}$ and point in the same direction, parallel to the xy-plane.

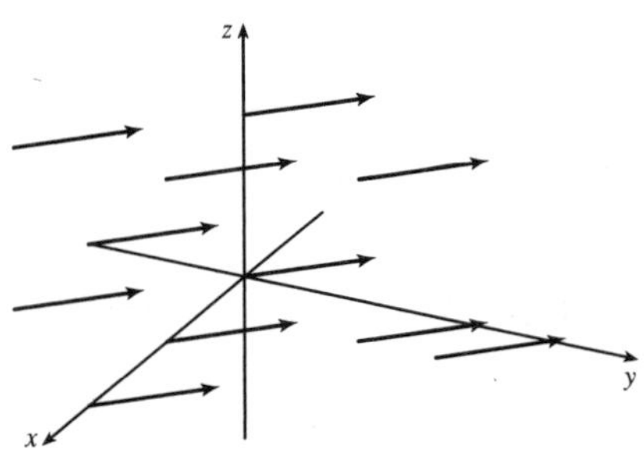

11. $\mathbf{F}(x,y) = \langle y, x \rangle$ corresponds to graph II. In the first quadrant all the vectors have positive x- and y-components, in the second quadrant all vectors have positive x-components and negative y-components, in the third quadrant all vectors have negative x- and y-components, and in the fourth quadrant all vectors have negative x-components and positive y-components. In addition, the vectors get shorter as we approach the origin.

12. $\mathbf{F}(x,y) = \langle 1, \sin y \rangle$ corresponds to graph IV since the x-component of each vector is constant, the vectors are independent of x (vectors along horizontal lines are identical), and the vector field appears to repeat the same pattern vertically.

13. $\mathbf{F}(x,y) = \langle x-2, x+1 \rangle$ corresponds to graph I since the vectors are independent of y (vectors along vertical lines are identical) and, as we move to the right, both the x- and the y-components get larger.

14. $\mathbf{F}(x,y) = \langle y, 1/x \rangle$ corresponds to graph III. As in Exercise 11, all the vectors in the first quadrant have positive x- and y-components, in the second quadrant all vectors have positive x-components and negative y-components, in the third quadrant all vectors have negative x- and y-components, and in the fourth quadrant all vectors have negative x-components and positive y-components. Also, the vectors become longer as we approach the y-axis.

15. $\mathbf{F}(x,y,z) = \mathbf{i} + 2\,\mathbf{j} + 3\,\mathbf{k}$ corresponds to graph IV, since all vectors have identical length and direction.

16. $\mathbf{F}(x,y,z) = \mathbf{i} + 2\,\mathbf{j} + z\,\mathbf{k}$ corresponds to graph I, since the horizontal vector components remain constant, but the vectors above the xy-plane point generally upward while the vectors below the xy-plane point generally downward.

17. $\mathbf{F}(x,y,z) = x\,\mathbf{i} + y\,\mathbf{j} + 3\,\mathbf{k}$ corresponds to graph III; the projection of each vector onto the xy-plane is $x\,\mathbf{i} + y\,\mathbf{j}$, which points away from the origin, and the vectors point generally upward because their z-components are all 3.

18. $\mathbf{F}(x,y,z) = x\,\mathbf{i} + y\,\mathbf{j} + z\,\mathbf{k}$ corresponds to graph II; each vector $\mathbf{F}(x,y,z)$ has the same length and direction as the position vector of the point (x,y,z), and therefore the vectors all point directly away from the origin.

19.

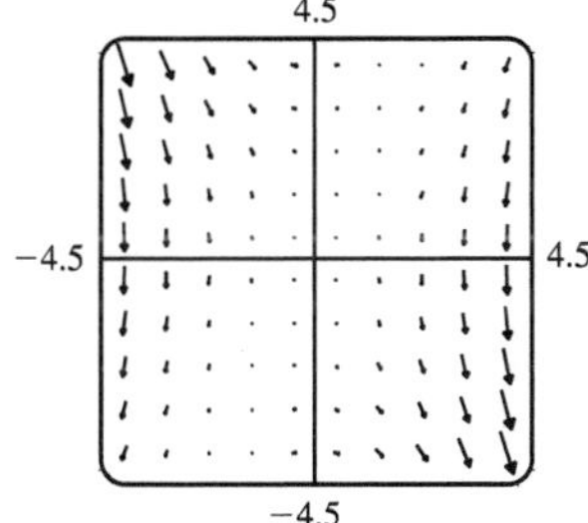

The vector field seems to have very short vectors near the line $y = 2x$. For $\mathbf{F}(x,y) = \langle 0,0 \rangle$ we must have $y^2 - 2xy = 0$ and $3xy - 6x^2 = 0$. The first equation holds if $y = 0$ or $y = 2x$, and the second holds if $x = 0$ or $y = 2x$. So both equations hold [and thus $\mathbf{F}(x,y) = \mathbf{0}$] along the line $y = 2x$.

20.

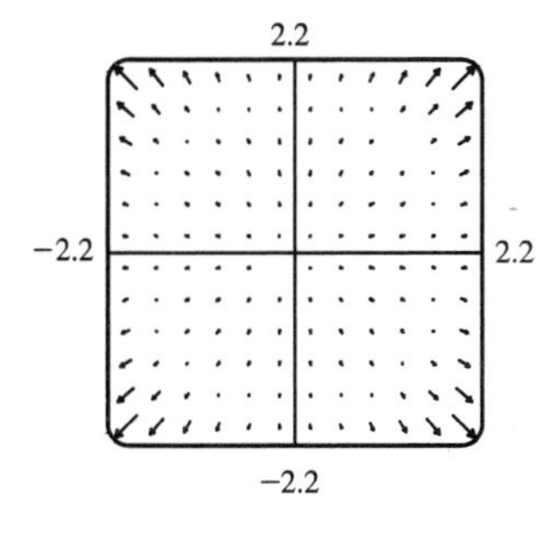

From the graph, it appears that all of the vectors in the field lie on lines through the origin, and that the vectors have very small magnitudes near the circle $|\mathbf{x}| = 2$ and near the origin. Note that $\mathbf{F}(\mathbf{x}) = \mathbf{0} \;\Leftrightarrow\; r(r-2) = 0 \;\Leftrightarrow\; r = 0$ or 2, so as we suspected, $\mathbf{F}(\mathbf{x}) = \mathbf{0}$ for $|\mathbf{x}| = 2$ and for $|\mathbf{x}| = 0$. Note that where $r^2 - r < 0$, the vectors point towards the origin, and where $r^2 - r > 0$, they point away from the origin.

21. $\nabla f(x,y) = f_x(x,y)\mathbf{i} + f_y(x,y)\,\mathbf{j} = \dfrac{1}{x+2y}\,\mathbf{i} + \dfrac{2}{x+2y}\,\mathbf{j}$

22. $\nabla f(x,y) = f_x(x,y)\mathbf{i} + f_y(x,y)\mathbf{j} = \left[x^\alpha\left(-\beta e^{-\beta x}\right) + \alpha x^{\alpha-1} e^{-\beta x}\right]\mathbf{i} + 0\,\mathbf{j} = (\alpha - \beta x)x^{\alpha-1}e^{-\beta x}\,\mathbf{i}$

23. $\nabla f(x,y,z) = f_x(x,y,z)\,\mathbf{i} + f_y(x,y,z)\,\mathbf{j} + f_z(x,y,z)\,\mathbf{k} = \dfrac{x}{\sqrt{x^2+y^2+z^2}}\,\mathbf{i} + \dfrac{y}{\sqrt{x^2+y^2+z^2}}\,\mathbf{j} + \dfrac{z}{\sqrt{x^2+y^2+z^2}}\,\mathbf{k}$

24. $\nabla f(x,y,z) = f_x(x,y,z)\,\mathbf{i} + f_y(x,y,z)\,\mathbf{j} + f_z(x,y,z)\,\mathbf{k} = \left(\cos\frac{y}{z}\right)\mathbf{i} - x\left(\sin\frac{y}{z}\right)\left(\frac{1}{z}\right)\mathbf{j} - x\left(\sin\frac{y}{z}\right)\left(-\frac{y}{z^2}\right)\mathbf{k}$

$$= \left(\cos\frac{y}{z}\right)\mathbf{i} - \frac{x}{z}\left(\sin\frac{y}{z}\right)\mathbf{j} + \frac{xy}{z^2}\left(\sin\frac{y}{z}\right)\mathbf{k}$$

25. $f(x,y) = xy - 2x \quad \Rightarrow \quad \nabla f(x,y) = (y-2)\,\mathbf{i} + x\,\mathbf{j}$.
The length of $\nabla f(x,y)$ is $\sqrt{(y-2)^2 + x^2}$ and $\nabla f(x,y)$ terminates on the line $y = x + 2$ at the point $(x + y - 2, x + y)$.

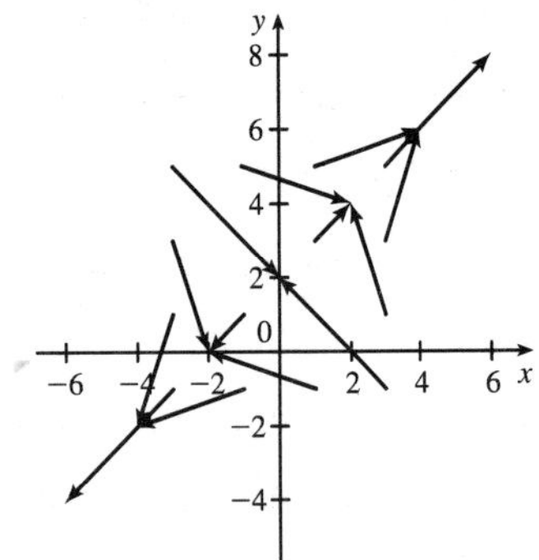

26. $f(x,y) = \frac{1}{4}(x+y)^2 \quad \Rightarrow$
$\nabla f(x,y) = \frac{1}{2}(x+y)\,\mathbf{i} + \frac{1}{2}(x+y)\,\mathbf{j}$.
The length of $\nabla f(x,y)$ is $\sqrt{\frac{1}{2}(x+y)^2} = \frac{1}{\sqrt{2}}\,|x+y|$.
The vectors are perpendicular to the line $y = -x$ and point away from the line, with length that increases as the distance from the line $y = -x$ increases.

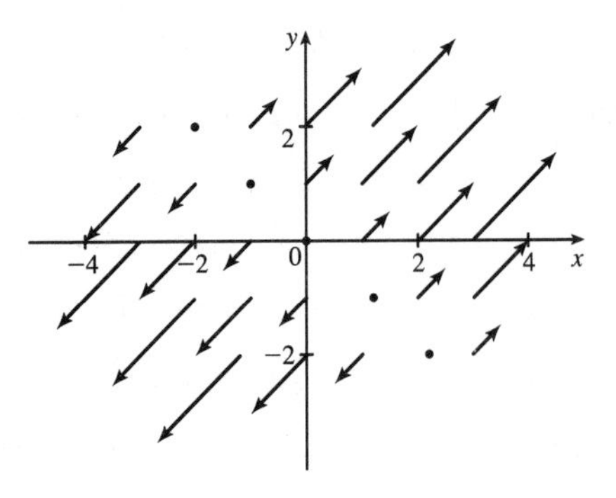

27. We graph ∇f along with a contour map of f.

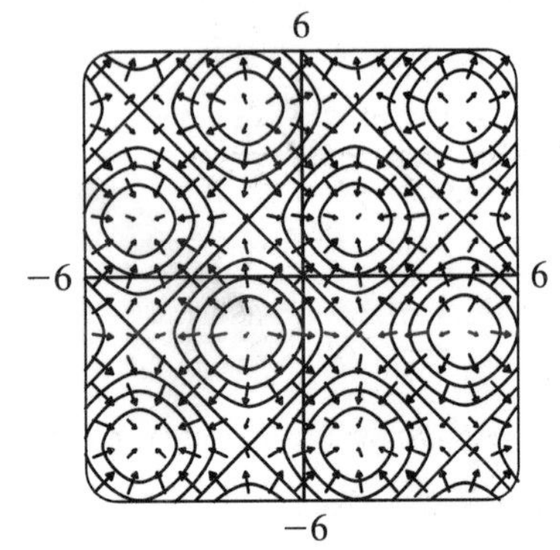

The graph shows that the gradient vectors are perpendicular to the level curves. Also, the gradient vectors point in the direction in which f is increasing and are longer where the level curves are closer together.

28. We graph ∇f along with a contour map of f.

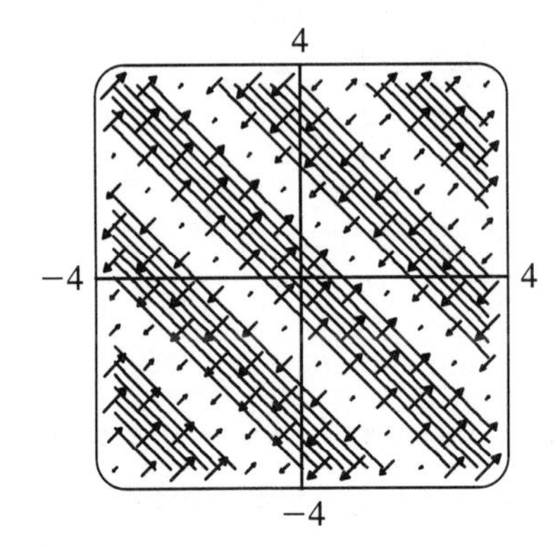

The graph shows that the gradient vectors are perpendicular to the level curves. Also, the gradient vectors point in the direction in which f is increasing and are longer where the level curves are closer together.

29. At $t = 3$ the particle is at $(2,1)$ so its velocity is $\mathbf{V}(2,1) = \langle 4, 3 \rangle$. After 0.01 units of time, the particle's change in location should be approximately $0.01\,\mathbf{V}(2,1) = 0.01\,\langle 4,3 \rangle = \langle 0.04, 0.03 \rangle$, so the particle should be approximately at the point $(2.04, 1.03)$.

30. At $t = 1$ the particle is at $(1,3)$ so its velocity is $\mathbf{F}(1,3) = \langle 1, -1 \rangle$. After 0.05 units of time, the particle's change in location should be approximately $0.05\,\mathbf{F}(1,3) = 0.05\,\langle 1,-1 \rangle = \langle 0.05, -0.05 \rangle$, so the particle should be approximately at the point $(1.05, 2.95)$.

31. (a) We sketch the vector field $\mathbf{F}(x,y) = x\,\mathbf{i} - y\,\mathbf{j}$ along with several approximate flow lines.The flow lines appear to be hyperbolas with shape similar to the graph of $y = \pm 1/x$, so we might guess that the flow lines have equations $y = C/x$.

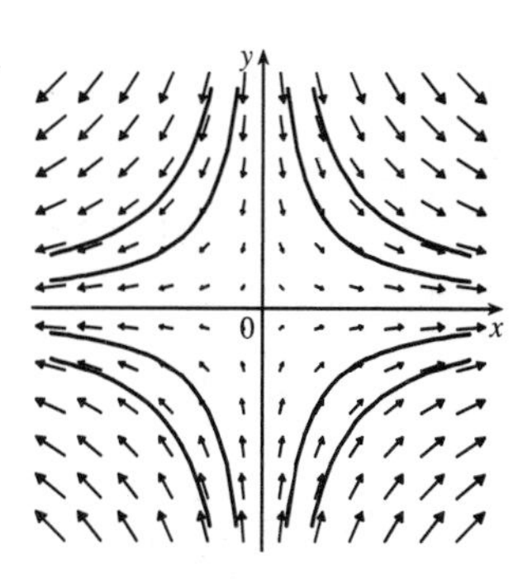

(b) If $x = x(t)$ and $y = y(t)$ are parametric equations of a flow line, then the velocity vector of the flow line at the point (x,y) is $x'(t)\,\mathbf{i} + y'\,(t)\,\mathbf{j}$. Since the velocity vectors coincide with the vectors in the vector field, we have $x'(t)\,\mathbf{i} + y'(t)\,\mathbf{j} = x\,\mathbf{i} - y\,\mathbf{j} \quad\Rightarrow\quad dx/dt = x,\, dy/dt = -y$. To solve these differential equations, we know $dx/dt = x \quad\Rightarrow\quad dx/x = dt \quad\Rightarrow\quad \ln|x| = t + C \quad\Rightarrow\quad x = \pm e^{t+C} = Ae^t$ for some constant A, and $dy/dt = -y \quad\Rightarrow\quad dy/y = -dt \quad\Rightarrow\quad \ln|y| = -t + K \quad\Rightarrow\quad y = \pm e^{-t+K} = Be^{-t}$ for some constant B. Therefore $xy = Ae^tBe^{-t} = AB =$ constant. If the flow line passes through $(1,1)$ then $(1)\,(1) =$ constant $= 1 \quad\Rightarrow\quad xy = 1 \quad\Rightarrow$ $y = 1/x,\, x > 0$.

32. (a) We sketch the vector field $\mathbf{F}(x,y) = \mathbf{i} + x\,\mathbf{j}$ along with several approximate flow lines. The flow lines appear to be parabolas.

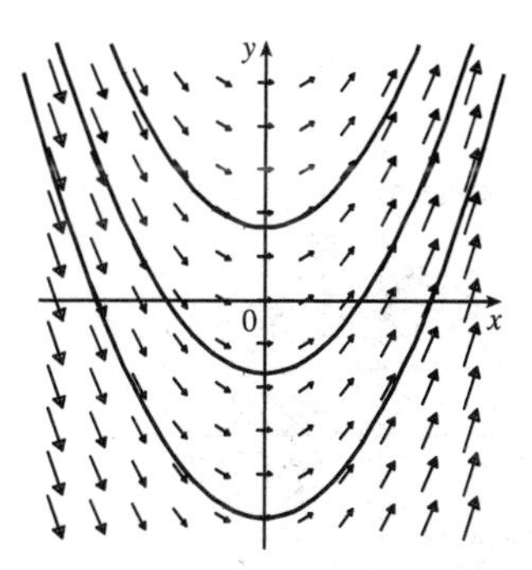

(b) If $x = x(t)$ and $y = y(t)$ are parametric equations of a flow line, then the velocity vector of the flow line at the point (x,y) is $x'(t)\,\mathbf{i} + y'(t)\,\mathbf{j}$. Since the velocity vectors coincide with the vectors in the vector field, we have

$x'(t)\,\mathbf{i} + y'(t)\,\mathbf{j} = \mathbf{i} + x\,\mathbf{j} \quad\Rightarrow\quad \dfrac{dx}{dt} = 1,\ \dfrac{dy}{dt} = x$. Thus $\dfrac{dy}{dx} = \dfrac{dy/dt}{dx/dt} = \dfrac{x}{1} = x$.

(c) From part (b), $dy/dx = x$. Integrating, we have $y = \frac{1}{2}x^2 + c$. Since the particle starts at the origin, we know $(0,0)$ is on the curve, so $0 = 0 + c \quad\Rightarrow\quad c = 0$ and the path the particle follows is $y = \frac{1}{2}x^2$.

13.2 Line Integrals

1. $x = t^2$ and $y = t$, $0 \le t \le 2$, so by Formula 3

$$\begin{aligned}\int_C y\,ds &= \int_0^2 t\sqrt{\left(\tfrac{dx}{dt}\right)^2 + \left(\tfrac{dy}{dt}\right)^2}\,dt = \int_0^2 t\sqrt{(2t)^2 + (1)^2}\,dt \\ &= \int_0^2 t\sqrt{4t^2+1}\,dt = \tfrac{1}{12}\left(4t^2+1\right)^{3/2}\Big]_0^2 = \tfrac{1}{12}\left(17\sqrt{17} - 1\right)\end{aligned}$$

2. $\int_C \frac{y}{x}\,ds = \int_{1/2}^{1} \frac{t^3}{t^4}\sqrt{(4t^3)^2 + (3t^2)^2}\,dt = \int_{1/2}^{1} \frac{1}{t}\sqrt{16t^6 + 9t^4}\,dt = \int_{1/2}^{1} t\sqrt{16t^2 + 9}\,dt$

$$= \tfrac{1}{48}(16t^2 + 9)^{3/2}\Big]_{1/2}^{1} = \tfrac{1}{48}(25^{3/2} - 13^{3/2}) = \tfrac{1}{48}(125 - 13\sqrt{13})$$

3. Parametric equations for C are $x = 4\cos t$, $y = 4\sin t$, $-\frac{\pi}{2} \le t \le \frac{\pi}{2}$. Then

$$\int_C xy^4\,ds = \int_{-\pi/2}^{\pi/2} (4\cos t)(4\sin t)^4 \sqrt{(-4\sin t)^2 + (4\cos t)^2}\,dt = \int_{-\pi/2}^{\pi/2} 4^5 \cos t \sin^4 t \sqrt{16\,(\sin^2 t + \cos^2 t)}\,dt$$

$$= 4^5 \int_{-\pi/2}^{\pi/2} (\sin^4 t \cos t)(4)\,dt = (4)^6 \left[\tfrac{1}{5}\sin^5 t\right]_{-\pi/2}^{\pi/2} = \frac{2 \cdot 4^6}{5} = 1638.4$$

4. Choosing y as the parameter, we have $x = e^y$, $y = y$, $0 \le y \le 1$. Then

$$\int_c xe^y\,dx = \int_0^1 e^y(e^y)e^y\,dy = \int_0^1 e^{3y}\,dy = \tfrac{1}{3}e^{3y}\Big]_0^1 = \tfrac{1}{3}(e^3 - 1).$$

5.

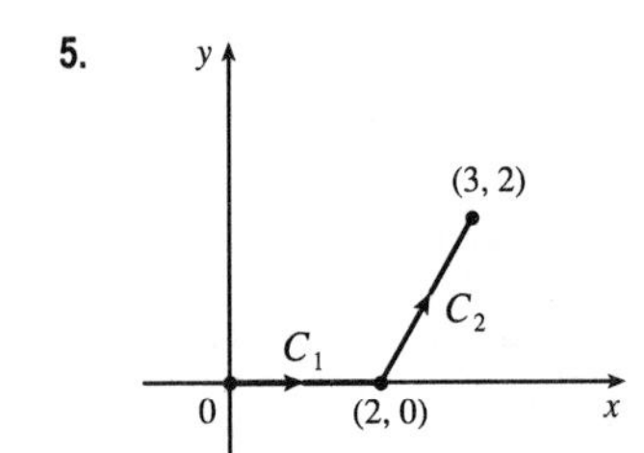

$C = C_1 + C_2$

On C_1: $x = x$, $y = 0 \;\Rightarrow\; dy = 0\,dx$, $0 \le x \le 2$.

On C_2: $x = x$, $y = 2x - 4 \;\Rightarrow\; dy = 2\,dx$, $2 \le x \le 3$.

Then

$$\int_C xy\,dx + (x - y)\,dy = \int_{C_1} xy\,dx + (x - y)\,dy + \int_{C_2} xy\,dx + (x - y)\,dy$$

$$= \int_0^2 (0 + 0)\,dx + \int_2^3 [(2x^2 - 4x) + (-x + 4)(2)]\,dx$$

$$= \int_2^3 (2x^2 - 6x + 8)\,dx = \tfrac{17}{3}$$

6.

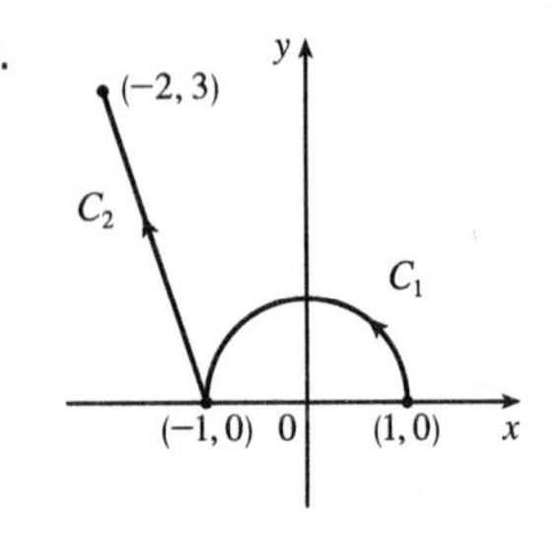

$C = C_1 + C_2$

On C_1: $x = \cos t \;\Rightarrow\; dx = -\sin t\,dt$, $y = \sin t \;\Rightarrow$

$dy = \cos t\;dt$, $0 \le t \le \pi$.

On C_2: $x = -1 - t \;\Rightarrow\; dx = -dt$, $y = 3t \;\Rightarrow$

$dy = 3\,dt$, $0 \le t \le 1$.

Then

$$\int_C \sin x\,dx + \cos y\,dy = \int_{C_1} \sin x\,dx + \cos y\,dy + \int_{C_2} \sin x\,dx + \cos y\,dy$$

$$= \int_0^{\pi} \sin(\cos t)(-\sin t\,dt) + \cos(\sin t)\cos t\,dt + \int_0^1 \sin(-1 - t)(-dt) + \cos(3t)(3\,dt)$$

$$= \left[-\cos(\cos t) + \sin(\sin t)\right]_0^{\pi} + \left[-\cos(-1 - t) + \sin(3t)\right]_0^1$$

$$= -\cos(\cos\pi) + \sin(\sin\pi) + \cos(\cos 0) - \sin(\sin 0) - \cos(-2) + \sin(3) + \cos(-1) - \sin(0)$$

$$= -\cos(-1) + \sin 0 + \cos(1) - \sin 0 - \cos(-2) + \sin 3 + \cos(-1)$$

$$= -\cos 1 + \cos 1 - \cos 2 + \sin 3 + \cos 1 = \cos 1 - \cos 2 + \sin 3$$

where we have used the identity $\cos(-\theta) = \cos\theta$.

7. $x = 4\sin t$, $y = 4\cos t$, $z = 3t$, $0 \le t \le \frac{\pi}{2}$. Then by Formula 9,

$$\int_C xy^3\,ds = \int_0^{\pi/2} (4\sin t)(4\cos t)^3 \sqrt{\left(\tfrac{dx}{dt}\right)^2 + \left(\tfrac{dy}{dt}\right)^2 + \left(\tfrac{dz}{dt}\right)^2}\,dt$$

$$= \int_0^{\pi/2} 4^4 \cos^3 t \sin t \sqrt{(4\cos t)^2 + (-4\sin t)^2 + (3)^2}\,dt = \int_0^{\pi/2} 256\cos^3 t \sin t \sqrt{16(\cos^2 t + \sin^2 t) + 9}\,dt$$

$$= 1280 \int_0^{\pi/2} \cos^3 t \sin t\,dt = -320\cos^4 t\Big]_0^{\pi/2} = 320$$

8. Parametric equations for C are $x = 4t$, $y = 6 - 5t$, $z = -1 + 6t$, $0 \le t \le 1$. Then

$$\int_C x^2 z\, ds = \int_0^1 (4t)^2(6t-1)\sqrt{4^2 + (-5)^2 + 6^2}\, dt = \sqrt{77}\int_0^1 (96t^3 - 16t^2)\, dt = \sqrt{77}\left[96 \cdot \frac{t^4}{4} - 16 \cdot \frac{t^3}{3}\right]_0^1 = \tfrac{56}{3}\sqrt{77}.$$

9. Parametric equations for C are $x = t$, $y = 2t$, $z = 3t$, $0 \le t \le 1$. Then

$$\int_C x e^{yz}\, ds = \int_0^1 t e^{(2t)(3t)}\sqrt{1^2 + 2^2 + 3^2}\, dt = \sqrt{14}\int_0^1 t e^{6t^2}\, dt = \sqrt{14}\left[\tfrac{1}{12}e^{6t^2}\right]_0^1 = \tfrac{\sqrt{14}}{12}(e^6 - 1).$$

10. $\sqrt{(dx/dt)^2 + (dy/dt)^2 + (dz/dt)^2} = \sqrt{1^2 + (2t)^2 + (3t^2)^2} = \sqrt{1 + 4t^2 + 9t^4}$. Then

$$\begin{aligned}\int_C (2x + 9z)\, ds &= \int_0^1 (2t + 9t^3)\sqrt{1 + 4t^2 + 9t^4}\, dt \quad [\text{let } u = 1 + 4t^2 + 9t^4 \;\Rightarrow\; \tfrac{1}{4}\, du = (2t + 9t^3)\, dt] \\ &= \int_1^{14} \tfrac{1}{4}\sqrt{u}\, du = \tfrac{1}{6}u^{3/2}\Big]_1^{14} = \tfrac{1}{6}(14^{3/2} - 1)\end{aligned}$$

11. $\int_C x^2 y\sqrt{z}\, dz = \int_0^1 (t^3)^2(t)\sqrt{t^2} \cdot 2t\, dt = \int_0^1 2t^9\, dt = \tfrac{1}{5}t^{10}\big]_0^1 = \tfrac{1}{5}$

12. $\int_C z\, dx + x\, dy + y\, dz = \int_0^1 t^2 \cdot 2t\, dt + t^2 \cdot 3t^2\, dt + t^3 \cdot 2t\, dt = \int_0^1 (2t^3 + 5t^4)\, dt = \left[\tfrac{1}{2}t^4 + t^5\right]_0^1 = \tfrac{1}{2} + 1 = \tfrac{3}{2}$

13.

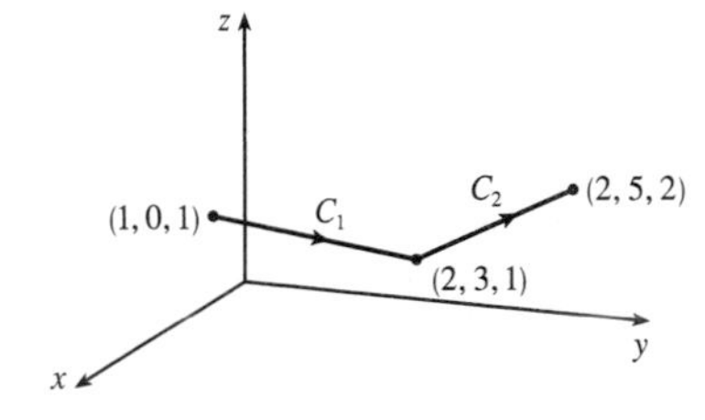

On C_1: $x = 1 + t \;\Rightarrow\; dx = dt$, $y = 3t \;\Rightarrow\; dy = 3\, dt$, $z = 1$ $\;\Rightarrow\; dz = 0\, dt$, $0 \le t \le 1$.

On C_2: $x = 2 \;\Rightarrow\; dx = 0\, dt$, $y = 3 + 2t \;\Rightarrow\;$ $dy = 2\, dt$, $z = 1 + t \;\Rightarrow\; dz = dt$, $0 \le t \le 1$.

Then

$$\begin{aligned}&\int_C (x + yz)\, dx + 2x\, dy + xyz\, dz \\ &\quad= \int_{C_1} (x + yz)\, dx + 2x\, dy + xyz\, dz + \int_{C_2} (x + yz)\, dx + 2x\, dy + xyz\, dz \\ &\quad= \int_0^1 (1 + t + (3t)(1))\, dt + 2(1 + t) \cdot 3\, dt + (1 + t)(3t)(1) \cdot 0\, dt \\ &\qquad\quad + \int_0^1 (2 + (3 + 2t)(1 + t)) \cdot 0\, dt + 2(2) \cdot 2\, dt + (2)(3 + 2t)(1 + t)\, dt \\ &\quad= \int_0^1 (10t + 7)\, dt + \int_0^1 (4t^2 + 10t + 14)\, dt = \left[5t^2 + 7t\right]_0^1 + \left[\tfrac{4}{3}t^3 + 5t^2 + 14t\right]_0^1 = 12 + \tfrac{61}{3} = \tfrac{97}{3}\end{aligned}$$

14.

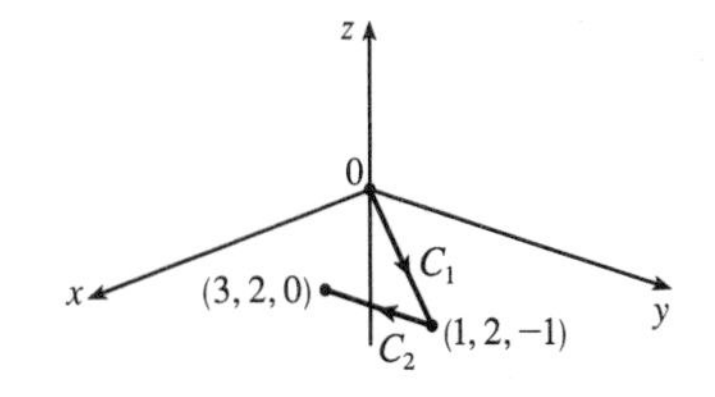

On C_1: $x = t \;\Rightarrow\; dx = dt$, $y = 2t \;\Rightarrow\; dy = 2\, dt$, $z = -t \;\Rightarrow\;$ $dz = -dt$, $0 \le t \le 1$.

On C_2: $x = 1 + 2t \;\Rightarrow\; dx = 2\, dt$, $y = 2 \;\Rightarrow\;$ $dy = 0\, dt$, $z = -1 + t \;\Rightarrow\; dz = dt$, $0 \le t \le 1$.

Then

$$\begin{aligned}\int_C x^2\, dx + y^2\, dy + z^2\, dz &= \int_{C_1} x^2\, dx + y^2\, dy + z^2\, dz + \int_{C_2} x^2\, dx + y^2\, dy + z^2\, dz \\ &= \int_0^1 t^2\, dt + (2t)^2 \cdot 2\, dt + (-t)^2(-dt) + \int_0^1 (1 + 2t)^2 \cdot 2\, dt + 2^2 \cdot 0\, dt + (-1 + t)^2\, dt \\ &= \int_0^1 8t^2\, dt + \int_0^1 (9t^2 + 6t + 3)\, dt = \left[\tfrac{8}{3}t^3\right]_0^1 + \left[3t^3 + 3t^2 + 3t\right]_0^1 = \tfrac{35}{3}\end{aligned}$$

15. (a) Along the line $x = -3$, the vectors of $\mathbf{F}$ have positive y-components, so since the path goes upward, the integrand $\mathbf{F} \cdot \mathbf{T}$ is always positive. Therefore $\int_{C_1} \mathbf{F} \cdot d\mathbf{r} = \int_{C_1} \mathbf{F} \cdot \mathbf{T}\, ds$ is positive.

(b) All of the (nonzero) field vectors along the circle with radius 3 are pointed in the clockwise direction, that is, opposite the direction to the path. So $\mathbf{F} \cdot \mathbf{T}$ is negative, and therefore $\int_{C_2} \mathbf{F} \cdot d\mathbf{r} = \int_{C_2} \mathbf{F} \cdot \mathbf{T}\, ds$ is negative.

16. Vectors starting on C_1 point in roughly the same direction as C_1, so the tangential component $\mathbf{F} \cdot \mathbf{T}$ is positive. Then $\int_{C_1} \mathbf{F} \cdot d\mathbf{r} = \int_{C_1} \mathbf{F} \cdot \mathbf{T}\, ds$ is positive. On the other hand, no vectors starting on C_2 point in the same direction as C_2, while some vectors point in roughly the opposite direction, so we would expect $\int_{C_2} \mathbf{F} \cdot d\mathbf{r} = \int_{C_2} \mathbf{F} \cdot \mathbf{T}\, ds$ to be negative.

17. $\mathbf{r}(t) = t^2\,\mathbf{i} - t^3\mathbf{j}$, so $\mathbf{F}(\mathbf{r}(t)) = (t^2)^2(-t^3)^3\mathbf{i} - (-t^3)\sqrt{t^2}\,\mathbf{j} = -t^{13}\,\mathbf{i} + t^4\,\mathbf{j}$ and $\mathbf{r}'(t) = 2t\,\mathbf{i} - 3t^2\,\mathbf{j}$.

Thus $\int_C \mathbf{F} \cdot d\mathbf{r} = \int_0^1 \mathbf{F}(\mathbf{r}(t)) \cdot \mathbf{r}'(t)\, dt = \int_0^1 (-2t^{14} - 3t^6) dt = \left[-\frac{2}{15}t^{15} - \frac{3}{7}t^7\right]_0^1 = -\frac{59}{105}$.

18. $\mathbf{F}(\mathbf{r}(t)) = (t^2)(t^3)\,\mathbf{i} + (t)(t^3)\,\mathbf{j} + (t)(t^2)\,\mathbf{k} = t^5\,\mathbf{i} + t^4\,\mathbf{j} + t^3\,\mathbf{k}$, $\mathbf{r}'(t) = \mathbf{i} + 2t\,\mathbf{j} + 3t^2\,\mathbf{k}$.

Thus $\int_C \mathbf{F} \cdot d\mathbf{r} = \int_0^2 \mathbf{F}(\mathbf{r}(t)) \cdot \mathbf{r}'(t)\, dt = \int_0^2 (t^5 + 2t^5 + 3t^5)\, dt = \left. t^6\right]_0^2 = 64$.

19. $\int_C \mathbf{F} \cdot d\mathbf{r} = \int_0^1 \left\langle \sin t^3, \cos(-t^2), t^4\right\rangle \cdot \left\langle 3t^2, -2t, 1\right\rangle dt$

$$= \int_0^1 \left(3t^2 \sin t^3 - 2t\cos t^2 + t^4\right) dt = \left[-\cos t^3 - \sin t^2 + \tfrac{1}{5}t^5\right]_0^1 = \tfrac{6}{5} - \cos 1 - \sin 1$$

20. $\int_C \mathbf{F} \cdot d\mathbf{r} = \int_0^\pi \langle \cos t, \sin t, -t\rangle \cdot \langle 1, \cos t, -\sin t\rangle\; dt = \int_0^\pi (\cos t + \sin t \cos t + t \sin t)\, dt$

$$= \left[\sin t + \tfrac{1}{2}\sin^2 t + (\sin t - t\cos t)\right]_0^\pi = \pi$$

21. We graph $\mathbf{F}(x, y) = (x - y)\,\mathbf{i} + xy\,\mathbf{j}$ and the curve C. We see that most of the vectors starting on C point in roughly the same direction as C, so for these portions of C the tangential component $\mathbf{F} \cdot \mathbf{T}$ is positive. Although some vectors in the third quadrant which start on C point in roughly the opposite direction, and hence give negative tangential components, it seems reasonable that the effect of these portions of C is outweighed by the positive tangential components. Thus, we would expect $\int_C \mathbf{F} \cdot d\mathbf{r} = \int_C \mathbf{F} \cdot \mathbf{T}\, ds$ to be positive.

To verify, we evaluate $\int_C \mathbf{F} \cdot d\mathbf{r}$. The curve C can be represented by $\mathbf{r}(t) = 2\cos t\,\mathbf{i} + 2\sin t\,\mathbf{j}$, $0 \le t \le \frac{3\pi}{2}$, so $\mathbf{F}(\mathbf{r}(t)) = (2\cos t - 2\sin t)\,\mathbf{i} + 4\cos t \sin t\,\mathbf{j}$ and $\mathbf{r}'(t) = -2\sin t\,\mathbf{i} + 2\cos t\,\mathbf{j}$. Then

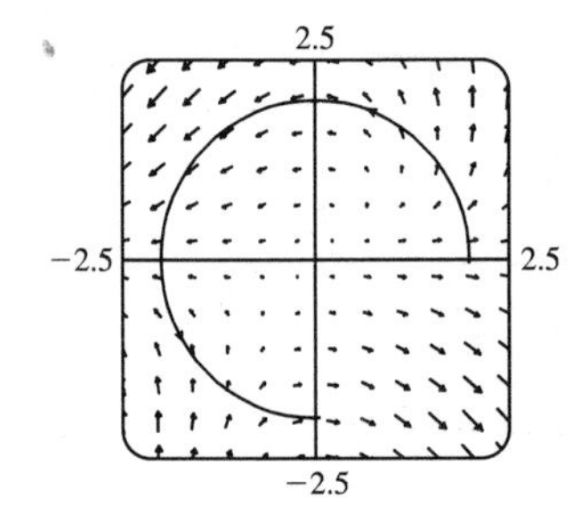

$$\begin{aligned}\int_C \mathbf{F} \cdot d\mathbf{r} &= \int_0^{3\pi/2} \mathbf{F}(\mathbf{r}(t)) \cdot \mathbf{r}'(t)\, dt \\ &= \int_0^{3\pi/2} \left[-2\sin t(2\cos t - 2\sin t) + 2\cos t(4\cos t \sin t)\right] dt \\ &= 4\int_0^{3\pi/2} (\sin^2 t - \sin t\cos t + 2\sin t\cos^2 t)\, dt \\ &= 3\pi + \tfrac{2}{3} \quad \text{[using a CAS]}\end{aligned}$$

22. We graph $\mathbf{F}(x, y) = \dfrac{x}{\sqrt{x^2 + y^2}}\,\mathbf{i} + \dfrac{y}{\sqrt{x^2 + y^2}}\,\mathbf{j}$ and the curve C. In the first quadrant, each vector starting on C points in roughly the same direction as C, so the tangential component $\mathbf{F} \cdot \mathbf{T}$ is positive. In the second quadrant, each vector starting on C points in roughly the direction opposite to C, so $\mathbf{F} \cdot \mathbf{T}$ is negative. Here, it appears that the tangential components in the first and second quadrants counteract each other, so it seems reasonable to guess that $\int_C \mathbf{F} \cdot d\mathbf{r} = \int_C \mathbf{F} \cdot \mathbf{T}\, ds$ is zero. To verify, we evaluate $\int_C \mathbf{F} \cdot d\mathbf{r}$. The curve C can be represented by

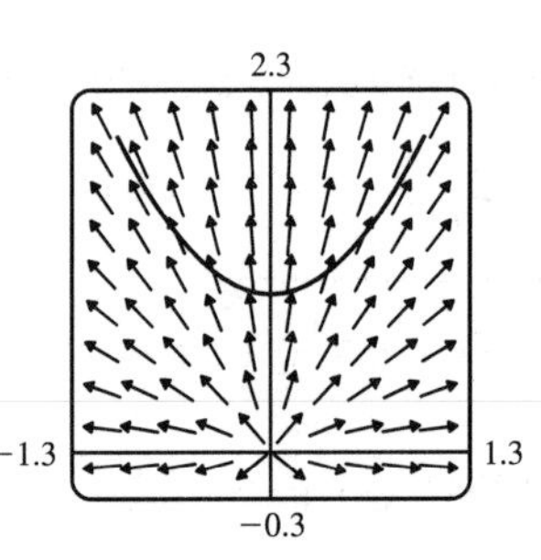

$\mathbf{r}(t) = t\,\mathbf{i} + (1+t^2)\,\mathbf{j}$, $-1 \le t \le 1$, so $\mathbf{F}(\mathbf{r}(t)) = \dfrac{t}{\sqrt{t^2+(1+t^2)^2}}\,\mathbf{i} + \dfrac{1+t^2}{\sqrt{t^2+(1+t^2)^2}}\,\mathbf{j}$ and $\mathbf{r}'(t) = \mathbf{i} + 2t\,\mathbf{j}$. Then

$$\int_C \mathbf{F}\cdot d\mathbf{r} = \int_{-1}^{1} \mathbf{F}(\mathbf{r}(t))\cdot \mathbf{r}'(t)\,dt = \int_{-1}^{1}\left(\frac{t}{\sqrt{t^2+(1+t^2)^2}} + \frac{2t(1+t^2)}{\sqrt{t^2+(1+t^2)^2}}\right)dt$$

$$= \int_{-1}^{1} \frac{t(3+2t^2)}{\sqrt{t^4+3t^2+1}}\,dt = 0 \quad \text{[since the integrand is an odd function]}$$

23. (a) $\int_C \mathbf{F}\cdot d\mathbf{r} = \int_0^1 \left\langle e^{t^2-1}, t^5\right\rangle \cdot \left\langle 2t, 3t^2\right\rangle dt = \int_0^1 \left(2te^{t^2-1} + 3t^7\right) dt = \left[e^{t^2-1} + \frac{3}{8}t^8\right]_0^1 = \frac{11}{8} - 1/e$

(b) $\mathbf{r}(0) = \mathbf{0}$, $\mathbf{F}(\mathbf{r}(0)) = \left\langle e^{-1}, 0\right\rangle$;

$\mathbf{r}\left(\frac{1}{\sqrt{2}}\right) = \left\langle \frac{1}{2}, \frac{1}{2\sqrt{2}}\right\rangle$, $\mathbf{F}\left(\mathbf{r}\left(\frac{1}{\sqrt{2}}\right)\right) = \left\langle e^{-1/2}, \frac{1}{4\sqrt{2}}\right\rangle$;

$\mathbf{r}(1) = \langle 1, 1\rangle$, $\mathbf{F}(\mathbf{r}(1)) = \langle 1, 1\rangle$.

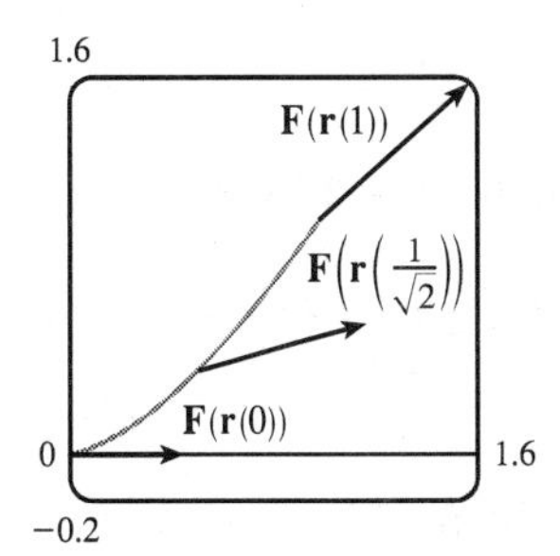

In order to generate the graph with Maple, we use the PLOT command (not to be confused with the plot command) to define each of the vectors. For example,

```
v1:=PLOT(CURVES([[0,0],[evalf(1/exp(1)),0]]));
```

generates the vector from the vector field at the point $(0,0)$ (but without an arrowhead) and gives it the name v1. To show everything on the same screen, we use the display command. In Mathematica, we use ListPlot (with the PlotJoined -> True option) to generate the vectors, and then Show to show everything on the same screen.

24. (a) $\int_C \mathbf{F}\cdot d\mathbf{r} = \int_{-1}^{1} \langle 2t, t^2, 3t\rangle \cdot \langle 2, 3, -2t\rangle\,dt = \int_{-1}^{1}\left(4t + 3t^2 - 6t^2\right) dt = \left[2t^2 - t^3\right]_{-1}^{1} = -2$

(b) Now $\mathbf{F}(\mathbf{r}(t)) = \langle 2t, t^2, 3t\rangle$, so $\mathbf{F}(\mathbf{r}(-1)) = \langle -2, 1, -3\rangle$, $\mathbf{F}\left(\mathbf{r}\left(-\frac{1}{2}\right)\right) = \left\langle -1, \frac{1}{4}, -\frac{3}{2}\right\rangle$, $\mathbf{F}\left(\mathbf{r}\left(\frac{1}{2}\right)\right) = \left\langle 1, \frac{1}{4}, \frac{3}{2}\right\rangle$, and $\mathbf{F}(\mathbf{r}(1)) = \langle 2, 1, 3\rangle$.

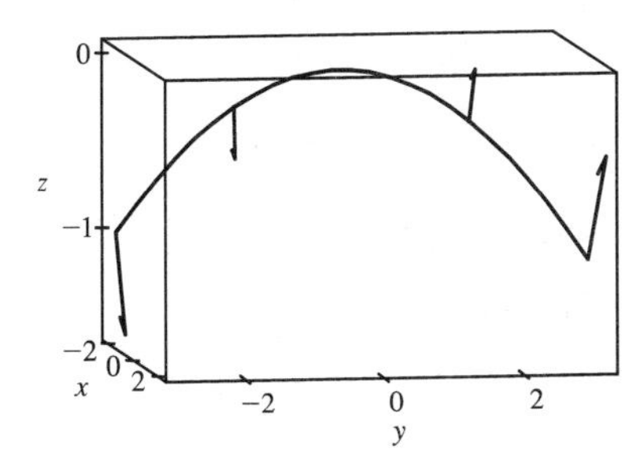

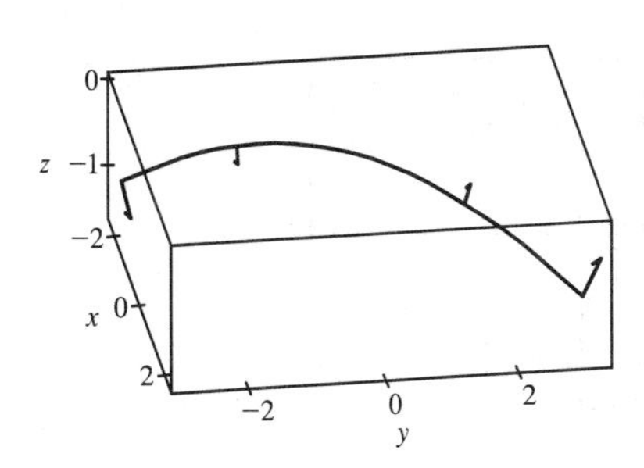

25. The part of the astroid that lies in the quadrant is parametrized by $x = \cos^3 t$, $y = \sin^3 t$, $0 \le t \le \frac{\pi}{2}$.

Now $\dfrac{dx}{dt} = 3\cos^2 t\,(-\sin t)$ and $\dfrac{dy}{dt} = 3\sin^2 t\cos t$, so

$$\sqrt{\left(\frac{dx}{dt}\right)^2 + \left(\frac{dy}{dt}\right)^2} = \sqrt{9\cos^4 t\sin^2 t + 9\sin^4 t\cos^2 t} = 3\cos t\sin t\,\sqrt{\cos^2 t + \sin^2 t} = 3\cos t\sin t.$$

Therefore $\int_C x^3y^5\,ds = \int_0^{\pi/2} \cos^9 t\sin^{15} t\,(3\cos t\sin t)\,dt = \frac{945}{16{,}777{,}216}\pi$.

26. (a) We parametrize the circle C as $\mathbf{r}(t) = 2\cos t\,\mathbf{i} + 2\sin t\,\mathbf{j}$, $0 \le t \le 2\pi$. So $\mathbf{F}(\mathbf{r}(t)) = \langle 4\cos^2 t, 4\cos t\sin t\rangle$,

$\mathbf{r}'(t) = \langle -2\sin t, 2\cos t\rangle$, and $W = \int_C \mathbf{F}\cdot d\mathbf{r} = \int_0^{2\pi}(-8\cos^2 t\sin t + 8\cos^2 t\sin t)\,dt = 0$.

(b)

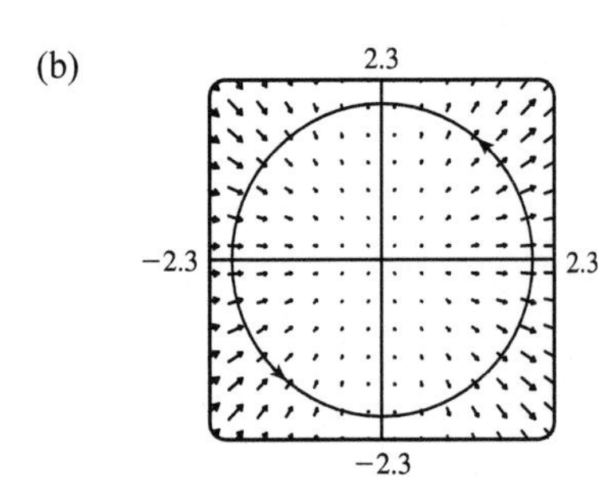

From the graph, we see that all of the vectors in the field are perpendicular to the path. This indicates that the field does no work on the particle, since the field never pulls the particle in the direction in which it is going. In other words, at any point along C, $\mathbf{F}\cdot\mathbf{T} = 0$, and so certainly $\int_C \mathbf{F}\cdot d\mathbf{r} = 0$.

27. We use the parametrization $x = 2\cos t$, $y = 2\sin t$, $-\frac{\pi}{2} \le t \le \frac{\pi}{2}$. Then

$$ds = \sqrt{\left(\frac{dx}{dt}\right)^2 + \left(\frac{dy}{dt}\right)^2}\,dt = \sqrt{(-2\sin t)^2 + (2\cos t)^2}\,dt = 2\,dt, \text{ so } m = \int_C k\,ds = 2k\int_{-\pi/2}^{\pi/2} dt = 2k(\pi),$$

$\overline{x} = \frac{1}{2\pi k}\int_C xk\,ds = \frac{1}{2\pi}\int_{-\pi/2}^{\pi/2}(2\cos t)2\,dt = \frac{1}{2\pi}\left[4\sin t\right]_{-\pi/2}^{\pi/2} = \frac{4}{\pi}$, $\overline{y} = \frac{1}{2\pi k}\int_C yk\,ds = \frac{1}{2\pi}\int_{-\pi/2}^{\pi/2}(2\sin t)2\,dt = 0$. Hence $(\overline{x}, \overline{y}) = \left(\frac{4}{\pi}, 0\right)$.

28. We use the parametrization $x = r\cos t$, $y = r\sin t$, $0 \le t \le \frac{\pi}{2}$. Then

$$ds = \sqrt{\left(\frac{dx}{dt}\right)^2 + \left(\frac{dy}{dt}\right)^2}\,dt = \sqrt{(-r\sin t)^2 + (r\cos t)^2}\,dt = r\,dt, \text{ so}$$

$$m = \int_C (x+y)\,ds = \int_0^{\pi/2}(r\cos t + r\sin t)\,r\,dt = r^2\left[\sin t - \cos t\right]_0^{\pi/2} = 2r^2,$$

$$\overline{x} = \frac{1}{2r^2}\int_C x(x+y)\,ds = \frac{1}{2r^2}\int_0^{\pi/2}(r^2\cos^2 t + r^2\cos t\sin t)r\,dt = \frac{r}{2}\left[\frac{t}{2} + \frac{\sin 2t}{4} - \frac{\cos 2t}{4}\right]_0^{\pi/2} = \frac{r(\pi+2)}{8},$$

and

$$\overline{y} = \frac{1}{2r^2}\int_C y(x+y)\,ds = \frac{1}{2r^2}\int_0^{\pi/2}(r^2\sin t\cos t + r^2\sin^2 t)r\,dt = \frac{r}{2}\left[-\frac{\cos 2t}{4} + \frac{t}{2} - \frac{\sin 2t}{4}\right]_0^{\pi/2} = \frac{r(\pi+2)}{8}.$$

Therefore $(\overline{x}, \overline{y}) = \left(\dfrac{r(\pi+2)}{8}, \dfrac{r(\pi+2)}{8}\right)$.

29. (a) $\overline{x} = \dfrac{1}{m}\displaystyle\int_C x\rho(x,y,z)\,ds$, $\overline{y} = \dfrac{1}{m}\displaystyle\int_C y\rho(x,y,z)\,ds$, $\overline{z} = \dfrac{1}{m}\displaystyle\int_C z\rho(x,y,z)\,ds$ where $m = \int_C \rho(x,y,z)\,ds$.

(b) $m = \int_C k\,ds = k\int_0^{2\pi}\sqrt{4\sin^2 t + 4\cos^2 t + 9}\,dt = k\sqrt{13}\int_0^{2\pi}dt = 2\pi k\sqrt{13}$,

$$\overline{x} = \frac{1}{2\pi k\sqrt{13}}\int_0^{2\pi} 2k\sqrt{13}\,\sin t\,dt = 0, \quad \overline{y} = \frac{1}{2\pi k\sqrt{13}}\int_0^{2\pi} 2k\sqrt{13}\,\cos t\,dt = 0,$$

$$\overline{z} = \frac{1}{2\pi k\sqrt{13}}\int_0^{2\pi}\left(k\sqrt{13}\right)(3t)\,dt = \frac{3}{2\pi}(2\pi^2) = 3\pi. \text{ Hence } (\overline{x}, \overline{y}, \overline{z}) = (0, 0, 3\pi).$$

30. $m = \int_C (x^2 + y^2 + z^2)\, ds = \int_0^{2\pi} (t^2 + 1)\sqrt{(1)^2 + (-\sin t)^2 + (\cos t)^2}\, dt = \int_0^{2\pi} (t^2 + 1)\sqrt{2}\, dt = \sqrt{2}\left(\frac{8}{3}\pi^3 + 2\pi\right)$,

$$\overline{x} = \frac{1}{\sqrt{2}\left(\frac{8}{3}\pi^3 + 2\pi\right)} \int_0^{2\pi} \sqrt{2}\,(t^3 + t)\, dt = \frac{4\pi^4 + 2\pi^2}{\frac{8}{3}\pi^3 + 2\pi} = \frac{3\pi(2\pi^2 + 1)}{4\pi^2 + 3},$$

$$\overline{y} = \frac{3}{2\sqrt{2}\,\pi(4\pi^2 + 3)} \int_0^{2\pi} \left(\sqrt{2}\cos t\right)(t^2 + 1)\, dt = 0, \text{ and}$$

$$\overline{z} = \frac{3}{2\sqrt{2}\,\pi(4\pi^2 + 3)} \int_0^{2\pi} \left(\sqrt{2}\sin t\right)(t^2 + 1)\, dt = 0. \text{ Hence } (\overline{x}, \overline{y}, \overline{z}) = \left(\frac{3\pi(2\pi^2 + 1)}{4\pi^2 + 3}, 0, 0\right).$$

31. From Example 3, $\rho(x, y) = k(1 - y)$, $x = \cos t$, $y = \sin t$, and $ds = dt$, $0 \le t \le \pi$ $\Rightarrow$

$$I_x = \int_C y^2 \rho(x, y)\, ds = \int_0^{\pi} \sin^2 t\,[k(1 - \sin t)]\, dt = k\int_0^{\pi} (\sin^2 t - \sin^3 t)\, dt$$

$$= \tfrac{1}{2}k\int_0^{\pi} (1 - \cos 2t)\, dt - k\int_0^{\pi} \left(1 - \cos^2 t\right)\sin t\, dt \qquad \left[\begin{array}{c}\text{Let } u = \cos t,\ du = -\sin t\, dt \\ \text{in the second integral}\end{array}\right]$$

$$= k\left[\tfrac{\pi}{2} + \int_1^{-1} (1 - u^2)\, du\right] = k\left(\tfrac{\pi}{2} - \tfrac{4}{3}\right)$$

$$I_y = \int_C x^2 \rho(x, y)\, ds = k\int_0^{\pi} \cos^2 t\,(1 - \sin t)\, dt = \tfrac{k}{2}\int_0^{\pi} (1 + \cos 2t)\, dt - k\int_0^{\pi} \cos^2 t \sin t\, dt$$

$= k\left(\frac{\pi}{2} - \frac{2}{3}\right)$, using the same substitution as above.

32. The wire is given as $x = 2\sin t$, $y = 2\cos t$, $z = 3t$, $0 \le t \le 2\pi$ with $\rho(x, y, z) = k$. Then

$ds = \sqrt{(2\cos t)^2 + (-2\sin t)^2 + 3^2}\, dt = \sqrt{4(\cos^2 t + \sin^2 t) + 9}\, dt = \sqrt{13}\, dt$ and

$$I_x = \int_C (y^2 + z^2)\rho(x, y, z)\, ds = \int_0^{2\pi} (4\cos^2 t + 9t^2)(k)\sqrt{13}\, dt = \sqrt{13}\, k\left[4\left(\tfrac{1}{2}t + \tfrac{1}{4}\sin 2t\right) + 3t^3\right]_0^{2\pi}$$

$$= \sqrt{13}\, k(4\pi + 24\pi^3) = 4\sqrt{13}\,\pi k(1 + 6\pi^2)$$

$$I_y = \int_C (x^2 + z^2)\rho(x, y, z)\, ds = \int_0^{2\pi} \left(4\sin^2 t + 9t^2\right)(k)\sqrt{13}\, dt = \sqrt{13}\, k\left[4\left(\tfrac{1}{2}t - \tfrac{1}{4}\sin 2t\right) + 3t^3\right]_0^{2\pi}$$

$$= \sqrt{13}\, k(4\pi + 24\pi^3) = 4\sqrt{13}\,\pi k(1 + 6\pi^2)$$

$$I_z = \int_C (x^2 + y^2)\rho(x, y, z)\, ds = \int_0^{2\pi} (4\sin^2 t + 4\cos^2 t)(k)\sqrt{13}\, dt = 4\sqrt{13}\, k\int_0^{2\pi} dt = 8\pi\sqrt{13}\, k$$

33. $W = \int_C \mathbf{F} \cdot d\mathbf{r} = \int_0^{2\pi} \langle t - \sin t, 3 - \cos t\rangle \cdot \langle 1 - \cos t, \sin t\rangle\, dt$

$$= \int_0^{2\pi} (t - t\cos t - \sin t + \sin t\cos t + 3\sin t - \sin t\cos t)\, dt$$

$$= \int_0^{2\pi} (t - t\cos t + 2\sin t)\, dt = \left[\tfrac{1}{2}t^2 - (t\sin t + \cos t) - 2\cos t\right]_0^{2\pi} \qquad \left[\begin{array}{c}\text{by integrating by parts} \\ \text{in the second term}\end{array}\right]$$

$$= 2\pi^2$$

34. $x = x$, $y = x^2$, $-1 \le x \le 2$,

$$W = \int_{-1}^{2} \left\langle x\sin x^2, x^2\right\rangle \cdot \langle 1, 2x\rangle\, dx = \int_{-1}^{2} (x\sin x^2 + 2x^3)\, dx = \left[-\tfrac{1}{2}\cos x^2 + \tfrac{1}{2}x^4\right]_{-1}^{2} = \tfrac{1}{2}(15 + \cos 1 - \cos 4)$$

35. $\mathbf{r}(t) = \langle 1 + 2t, 4t, 2t\rangle$, $0 \le t \le 1$,

$$W = \int_C \mathbf{F} \cdot d\mathbf{r} = \int_0^1 \langle 6t, 1 + 4t, 1 + 6t\rangle \cdot \langle 2, 4, 2\rangle\, dt = \int_0^1 (12t + 4(1 + 4t) + 2(1 + 6t))\, dt$$

$$= \int_0^1 (40t + 6)\, dt = \left[20t^2 + 6t\right]_0^1 = 26$$

36. $\mathbf{r}(t) = 2\,\mathbf{i} + t\,\mathbf{j} + 5t\,\mathbf{k}$, $0 \le t \le 1$. Therefore

$$W = \int_C \mathbf{F} \cdot d\mathbf{r} = \int_0^1 \frac{K\langle 2, t, 5t\rangle}{(4 + 26t^2)^{3/2}} \cdot \langle 0, 1, 5\rangle\, dt = K \int_0^1 \frac{26t}{(4 + 26t^2)^{3/2}}\, dt = K\Big[-(4 + 26t^2)^{-1/2}\Big]_0^1 = K\Big(\tfrac{1}{2} - \tfrac{1}{\sqrt{30}}\Big)$$

37. Let $\mathbf{F} = 185\,\mathbf{k}$. To parametrize the staircase, let

$x = 20\cos t$, $y = 20\sin t$, $z = \frac{90}{6\pi}t = \frac{15}{\pi}t$, $0 \le t \le 6\pi \quad \Rightarrow$

$$W = \int_C \mathbf{F} \cdot d\mathbf{r} = \int_0^{6\pi} \langle 0, 0, 185\rangle \cdot \big\langle -20\sin t, 20\cos t, \tfrac{15}{\pi}\big\rangle\, dt = (185)\tfrac{15}{\pi}\int_0^{6\pi} dt = (185)(90) \approx 1.67 \times 10^4 \text{ ft-lb}$$

38. This time m is a function of t: $m = 185 - \frac{9}{6\pi}t = 185 - \frac{3}{2\pi}t$. So let $\mathbf{F} = \big(185 - \frac{3}{2\pi}t\big)\mathbf{k}$. To parametrize the staircase, let

$x = 20\cos t$, $y = 20\sin t$, $z = \frac{90}{6\pi}t = \frac{15}{\pi}t$, $0 \le t \le 6\pi$. Therefore

$$\begin{aligned} W &= \int_C \mathbf{F} \cdot d\mathbf{r} = \int_0^{6\pi} \big\langle 0, 0, 185 - \tfrac{3}{2\pi}t\big\rangle \cdot \big\langle -20\sin t, 20\cos t, \tfrac{15}{\pi}\big\rangle\, dt = \tfrac{15}{\pi}\int_0^{6\pi} \big(185 - \tfrac{3}{2\pi}t\big)\, dt \\ &= \tfrac{15}{\pi}\Big[185t - \tfrac{3}{4\pi}t^2\Big]_0^{6\pi} = 90\big(185 - \tfrac{9}{2}\big) \approx 1.62 \times 10^4 \text{ ft-lb} \end{aligned}$$

39. (a) $\mathbf{r}(t) = \langle \cos t, \sin t\rangle$, $0 \le t \le 2\pi$, and let $\mathbf{F} = \langle a, b\rangle$. Then

$$\begin{aligned} W &= \int_C \mathbf{F} \cdot d\mathbf{r} = \int_0^{2\pi} \langle a, b\rangle \cdot \langle -\sin t, \cos t\rangle\, dt = \int_0^{2\pi} (-a\sin t + b\cos t)\, dt = \big[a\cos t + b\sin t\big]_0^{2\pi} \\ &= a + 0 - a + 0 = 0 \end{aligned}$$

(b) Yes. $\mathbf{F}(x, y) = k\,\mathbf{x} = \langle kx, ky\rangle$ and

$$W = \int_C \mathbf{F} \cdot d\mathbf{r} = \int_0^{2\pi} \langle k\cos t, k\sin t\rangle \cdot \langle -\sin t, \cos t\rangle\, dt = \int_0^{2\pi} (-k\sin t\cos t + k\sin t\cos t)\, dt = \int_0^{2\pi} 0\, dt = 0$$

40. Use the orientation pictured in the figure. Then since $\mathbf{B}$ is tangent to any circle that lies in the plane perpendicular to the wire, $\mathbf{B} = |\mathbf{B}|\,\mathbf{T}$ where $\mathbf{T}$ is the unit tangent to the circle C: $x = r\cos\theta$, $y = r\sin\theta$. Thus $\mathbf{B} = |\mathbf{B}|\,\langle -\sin\theta, \cos\theta\rangle$. Then $\int_C \mathbf{B} \cdot d\mathbf{r} = \int_0^{2\pi} |\mathbf{B}|\,\langle -\sin\theta, \cos\theta\rangle \cdot \langle -r\sin\theta, r\cos\theta\rangle\, d\theta = \int_0^{2\pi} |\mathbf{B}|\, r\, d\theta = 2\pi r\,|\mathbf{B}|$. (Note that $|\mathbf{B}|$ here is the magnitude of the field at a distance r from the wire's center.) But by Ampere's Law $\int_C \mathbf{B} \cdot d\mathbf{r} = \mu_0 I$. Hence $|\mathbf{B}| = \mu_0 I/(2\pi r)$.

13.3 The Fundamental Theorem for Line Integrals

1. C appears to be a smooth curve, and since ∇f is continuous, we know f is differentiable. Then Theorem 2 says that the value of $\int_C \nabla f \cdot d\mathbf{r}$ is simply the difference of the values of f at the terminal and initial points of C. From the graph, this is $50 - 10 = 40$.

2. C is represented by the vector function $\mathbf{r}(t) = (t^2 + 1)\,\mathbf{i} + (t^3 + t)\,\mathbf{j}$, $0 \le t \le 1$, so $\mathbf{r}'(t) = 2t\,\mathbf{i} + (3t^2 + 1)\,\mathbf{j}$. Since $3t^2 + 1 \ne 0$, we have $\mathbf{r}'(t) \ne \mathbf{0}$, thus C is a smooth curve. ∇f is continuous, and hence f is differentiable, so by Theorem 2 we have $\int_C \nabla f \cdot d\mathbf{r} = f(\mathbf{r}(1)) - f(\mathbf{r}(0)) = f(2, 2) - f(1, 0) = 9 - 3 = 6$.

3. $\partial(6x + 5y)/\partial y = 5 = \partial(5x + 4y)/\partial x$ and the domain of $\mathbf{F}$ is $\mathbb{R}^2$ which is open and simply-connected, so by Theorem 6 $\mathbf{F}$ is conservative. Thus, there exists a function f such that $\nabla f = \mathbf{F}$, that is, $f_x(x, y) = 6x + 5y$ and $f_y(x, y) = 5x + 4y$. But $f_x(x, y) = 6x + 5y$ implies $f(x, y) = 3x^2 + 5xy + g(y)$ and differentiating both sides of this equation with respect to y gives $f_y(x, y) = 5x + g'(y)$. Thus $5x + 4y = 5x + g'(y)$ so $g'(y) = 4y$ and $g(y) = 2y^2 + K$ where K is a constant. Hence $f(x, y) = 3x^2 + 5xy + 2y^2 + K$ is a potential function for $\mathbf{F}$.

4. $\partial(x^3 + 4xy)/\partial y = 4x$, $\partial(4xy - y^3)/\partial x = 4y$. Since these are not equal, $\mathbf{F}$ is not conservative.

5. $\partial(xe^y)/\partial y = xe^y$, $\partial(ye^x)/\partial x = ye^x$. Since these are not equal, $\mathbf{F}$ is not conservative.

6. $\partial(e^y)/\partial y = e^y = \partial(xe^y)/\partial x$ and the domain of $\mathbf{F}$ is $\mathbb{R}^2$. Hence $\mathbf{F}$ is conservative so there exists a function f such that $\nabla f = \mathbf{F}$. Then $f_x(x, y) = e^y$ implies $f(x, y) = xe^y + g(y)$ and $f_y(x, y) = xe^y + g'(y)$. But $f_y(x, y) = xe^y$ so $g'(y) = 0 \Rightarrow g(y) = K$. Then $f(x, y) = xe^y + K$ is a potential function for $\mathbf{F}$.

7. $\partial(2x\cos y - y\cos x)/\partial y = -2x\sin y - \cos x = \partial(-x^2\sin y - \sin x)/\partial x$ and the domain of $\mathbf{F}$ is $\mathbb{R}^2$. Hence $\mathbf{F}$ is conservative so there exists a function f such that $\nabla f = \mathbf{F}$. Then $f_x(x, y) = 2x\cos y - y\cos x$ implies $f(x, y) = x^2\cos y - y\sin x + g(y)$ and $f_y(x, y) = -x^2\sin y - \sin x + g'(y)$. But $f_y(x, y) = -x^2\sin y - \sin x$ so $g'(y) = 0 \Rightarrow g(y) = K$. Then $f(x, y) = x^2\cos y - y\sin x + K$ is a potential function for $\mathbf{F}$.

8. $\partial(1 + 2xy + \ln x)/\partial y = 2x = \partial(x^2)/\partial x$ and the domain of $\mathbf{F}$ is $\{(x, y) \mid x > 0\}$ which is open and simply-connected. Hence $\mathbf{F}$ is conservative, so there exists a function f such that $\nabla f = \mathbf{F}$. Then $f_x(x, y) = 1 + 2xy + \ln x$ implies $f(x, y) = x + x^2y + x\ln x - x + g(y)$ and $f_y(x, y) = x^2 + g'(y)$. But $f_y(x, y) = x^2$ so $g'(y) = 0 \Rightarrow g(y) = K$. Then $f(x, y) = x^2y + x\ln x + K$ is a potential function for $\mathbf{F}$.

9. $\partial(ye^x + \sin y)/\partial y = e^x + \cos y = \partial(e^x + x\cos y)/\partial x$ and the domain of $\mathbf{F}$ is $\mathbb{R}^2$. Hence $\mathbf{F}$ is conservative so there exists a function f such that $\nabla f = \mathbf{F}$. Then $f_x(x, y) = ye^x + \sin y$ implies $f(x, y) = ye^x + x\sin y + g(y)$ and $f_y(x, y) = e^x + x\cos y + g'(y)$. But $f_y(x, y) = e^x + x\cos y$ so $g(y) = K$ and $f(x, y) = ye^x + x\sin y + K$ is a potential function for $\mathbf{F}$.

10. $\partial(xy\cos xy + \sin xy)/\partial y = -x^2y\sin xy + 2x\cos xy = \partial(x^2\cos xy)/\partial x$ and the domain of $\mathbf{F}$ is $\mathbb{R}^2$. Hence $\mathbf{F}$ is conservative, so there exists a function f such that $\nabla f = \mathbf{F}$. Then $f_y(x, y) = x^2\cos xy$ implies $f(x, y) = x\sin xy + g(x)$ and $f_x(x, y) = xy\cos xy + \sin xy + g'(x)$. But $f_x(x, y) = xy\cos xy + \sin xy$ so $g(x) = K$ and $f(x, y) = x\sin xy + K$ is a potential function for $\mathbf{F}$.

11. (a) $f_x(x, y) = x^3y^4$ implies $f(x, y) = \frac{1}{4}x^4y^4 + g(y)$ and $f_y(x, y) = x^4y^3 + g'(y)$. But $f_y(x, y) = x^4y^3$ so $g'(y) = 0 \Rightarrow g(y) = K$, a constant. We can take $K = 0$, so $f(x, y) = \frac{1}{4}x^4y^4$.

(b) The initial point of C is $\mathbf{r}(0) = (0, 1)$ and the terminal point is $\mathbf{r}(1) = (1, 2)$, so $\int_C \mathbf{F} \cdot d\mathbf{r} = f(1, 2) - f(0, 1) = 4 - 0 = 4$.

12. (a) $f_x(x, y) = y^2/(1 + x^2)$ implies $f(x, y) = y^2\arctan x + g(y) \Rightarrow f_y(x, y) = 2y\arctan x + g'(y)$. But $f_y(x, y) = 2y\arctan x$ so $g'(y) = 0 \Rightarrow g(y) = K$. We can take $K = 0$, so $f(x, y) = y^2\arctan x$.

(b) The initial point of C is $\mathbf{r}(0) = (0, 0)$ and the terminal point is $\mathbf{r}(1) = (1, 2)$, so $\int_C \mathbf{F} \cdot d\mathbf{r} = f(1, 2) - f(0, 0) = 4\arctan 1 - 0 = 4 \cdot \frac{\pi}{4} = \pi$.

13. (a) $f_x(x, y, z) = yz$ implies $f(x, y, z) = xyz + g(y, z)$ and so $f_y(x, y, z) = xz + g_y(y, z)$. But $f_y(x, y, z) = xz$ so $g_y(y, z) = 0 \Rightarrow g(y, z) = h(z)$. Thus $f(x, y, z) = xyz + h(z)$ and $f_z(x, y, z) = xy + h'(z)$. But $f_z(x, y, z) = xy + 2z$, so $h'(z) = 2z \Rightarrow h(z) = z^2 + K$. Hence $f(x, y, z) = xyz + z^2$ (taking $K = 0$).

(b) $\int_C \mathbf{F} \cdot d\mathbf{r} = f(4, 6, 3) - f(1, 0, -2) = 81 - 4 = 77$.

14. (a) $f_x(x,y,z) = 2xz + y^2$ implies $f(x,y,z) = x^2z + xy^2 + g(y,z)$ and so $f_y(x,y,z) = 2xy + g_y(y,z)$. But $f_y(x,y,z) = 2xy$ so $g_y(y,z) = 0 \Rightarrow g(y,z) = h(z)$. Thus $f(x,y,z) = x^2z + xy^2 + h(z)$ and $f_z(x,y,z) = x^2 + h'(z)$. But $f_z(x,y,z) = x^2 + 3z^2$, so $h'(z) = 3z^2 \Rightarrow h(z) = z^3 + K$. Hence $f(x,y,z) = x^2z + xy^2 + z^3$ (taking $K = 0$).

(b) $t = 0$ corresponds to the point $(0,1,-1)$ and $t = 1$ corresponds to $(1,2,1)$, so $\int_C \mathbf{F} \cdot d\mathbf{r} = f(1,2,1) - f(0,1,-1) = 6 - (-1) = 7$.

15. (a) $f_x(x,y,z) = y^2 \cos z$ implies $f(x,y,z) = xy^2 \cos z + g(y,z)$ and so $f_y(x,y,z) = 2xy \cos z + g_y(y,z)$. But $f_y(x,y,z) = 2xy \cos z$ so $g_y(y,z) = 0 \Rightarrow g(y,z) = h(z)$. Thus $f(x,y,z) = xy^2 \cos z + h(z)$ and $f_z(x,y,z) = -xy^2 \sin z + h'(z)$. But $f_z(x,y,z) = -xy^2 \sin z$, so $h'(z) = 0 \Rightarrow h(z) = K$. Hence $f(x,y,z) = xy^2 \cos z$ (taking $K = 0$).

(b) $\mathbf{r}(0) = \langle 0,0,0 \rangle$, $\mathbf{r}(\pi) = \langle \pi^2, 0, \pi \rangle$ so $\int_C \mathbf{F} \cdot d\mathbf{r} = f(\pi^2, 0, \pi) - f(0,0,0) = 0 - 0 = 0$.

16. (a) $f_x(x,y,z) = e^y$ implies $f(x,y,z) = xe^y + g(y,z)$ and so $f_y(x,y,z) = xe^y + g_y(y,z)$. But $f_y(x,y,z) = xe^y$ so $g_y(y,z) = 0 \Rightarrow g(y,z) = h(z)$. Thus $f(x,y,z) = xe^y + h(z)$ and $f_z(x,y,z) = 0 + h'(z)$. But $f_z(x,y,z) = (z+1)e^z$, so $h'(z) = (z+1)e^z \Rightarrow h(z) = ze^z + K$ (using integration by parts). Hence $f(x,y,z) = xe^y + ze^z$ (taking $K = 0$).

(b) $\mathbf{r}(0) = \langle 0,0,0 \rangle$, $\mathbf{r}(1) = \langle 1,1,1 \rangle$ so $\int_C \mathbf{F} \cdot d\mathbf{r} = f(1,1,1) - f(0,0,0) = 2e - 0 = 2e$.

17. Here $\mathbf{F}(x,y) = \tan y\, \mathbf{i} + x \sec^2 y\, \mathbf{j}$. Then $f(x,y) = x \tan y$ is a potential function for $\mathbf{F}$, that is, $\nabla f = \mathbf{F}$ so $\mathbf{F}$ is conservative and thus its line integral is independent of path. Hence $\int_C \tan y\, dx + x \sec^2 y\, dy = \int_C \mathbf{F} \cdot d\mathbf{r} = f\left(2, \frac{\pi}{4}\right) - f(1,0) = 2 \tan \frac{\pi}{4} - \tan 0 = 2$.

18. Here $\mathbf{F}(x,y) = (1 - ye^{-x})\,\mathbf{i} + e^{-x}\,\mathbf{j}$. Then $f(x,y) = x + ye^{-x}$ is a potential function for $\mathbf{F}$, that is, $\nabla f = \mathbf{F}$ so $\mathbf{F}$ is conservative and thus its line integral is independent of path. Hence $\int_C (1 - ye^{-x})\, dx + e^{-x}\, dy = \int_C \mathbf{F} \cdot d\mathbf{r} = f(1,2) - f(0,1) = (1 + 2e^{-1}) - 1 = 2/e$.

19. $\mathbf{F}(x,y) = 2y^{3/2}\,\mathbf{i} + 3x\sqrt{y}\,\mathbf{j}$, $W = \int_C \mathbf{F} \cdot d\mathbf{r}$. Since $\partial(2y^{3/2})/\partial y = 3\sqrt{y} = \partial(3x\sqrt{y})/\partial x$, there exists a function f such that $\nabla f = \mathbf{F}$. In fact, $f_x(x,y) = 2y^{3/2} \Rightarrow f(x,y) = 2xy^{3/2} + g(y) \Rightarrow f_y(x,y) = 3xy^{1/2} + g'(y)$. But $f_y(x,y) = 3x\sqrt{y}$ so $g'(y) = 0$ or $g(y) = K$. We can take $K = 0 \Rightarrow f(x,y) = 2xy^{3/2}$. Thus $W = \int_C \mathbf{F} \cdot d\mathbf{r} = f(2,4) - f(1,1) = 2(2)(8) - 2(1) = 30$.

20. $\mathbf{F}(x,y) = e^{-y}\,\mathbf{i} - xe^{-y}\,\mathbf{j}$, $W = \int_C \mathbf{F} \cdot d\mathbf{r}$. Since $\dfrac{\partial}{\partial y}(e^{-y}) = -e^{-y} = \dfrac{\partial}{\partial x}(-xe^{-y})$, there exists a function f such that $\nabla f = \mathbf{F}$. In fact, $f_x = e^{-y} \Rightarrow f(x,y) = xe^{-y} + g(y) \Rightarrow f_y = -xe^{-y} + g'(y) \Rightarrow g'(y) = 0$, so we can take $f(x,y) = xe^{-y}$ as a potential function for $\mathbf{F}$. Thus $W = \int_C \mathbf{F} \cdot d\mathbf{r} = f(2,0) - f(0,1) = 2 - 0 = 2$.

21. We know that if the vector field (call it $\mathbf{F}$) is conservative, then around any closed path C, $\int_C \mathbf{F} \cdot d\mathbf{r} = 0$. But take C to be a circle centered at the origin, oriented counterclockwise. All of the field vectors that start on C are roughly in the direction of motion along C, so the integral around C will be positive. Therefore the field is not conservative.

22. If a vector field $\mathbf{F}$ is conservative, then around any closed path C, $\int_C \mathbf{F} \cdot d\mathbf{r} = 0$. For any closed path we draw in the field, it appears that some vectors on the curve point in approximately the same direction as the curve and a similar number point in roughly the opposite direction. (Some appear perpendicular to the curve as well.) Therefore it is plausible that $\int_C \mathbf{F} \cdot d\mathbf{r} = 0$ for every closed curve C which means $\mathbf{F}$ is conservative.

23. From the graph, it appears that $\mathbf{F}$ is conservative, since around all closed paths, the number and size of the field vectors pointing in directions similar to that of the path seem to be roughly the same as the number and size of the vectors pointing in the opposite direction. To check, we calculate $\dfrac{\partial}{\partial y}(\sin y) = \cos y = \dfrac{\partial}{\partial x}(1 + x\cos y)$. Thus $\mathbf{F}$ is conservative, by Theorem 6.

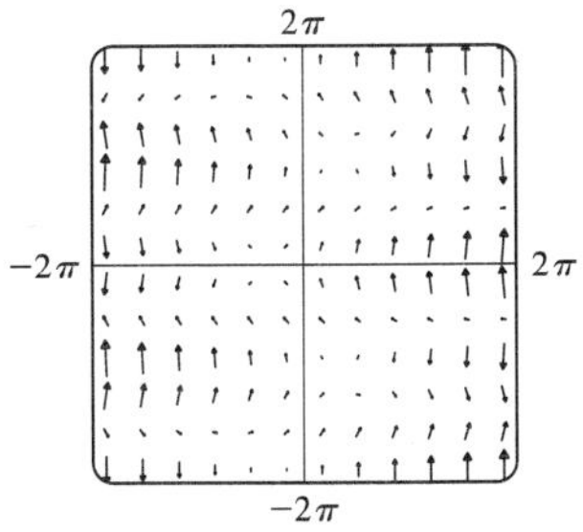

24. $\nabla f(x, y) = \cos(x - 2y)\,\mathbf{i} - 2\cos(x - 2y)\,\mathbf{j}$

(a) We use Theorem 2: $\int_{C_1} \mathbf{F} \cdot d\mathbf{r} = \int_{C_1} \nabla f \cdot d\mathbf{r} = f(\mathbf{r}(b)) - f(\mathbf{r}(a))$ where C_1 starts at $t = a$ and ends at $t = b$. So because $f(0, 0) = \sin 0 = 0$ and $f(\pi, \pi) = \sin(\pi - 2\pi) = 0$, one possible curve C_1 is the straight line from $(0, 0)$ to (π, π); that is, $\mathbf{r}(t) = \pi t\,\mathbf{i} + \pi t\,\mathbf{j}$, $0 \le t \le 1$.

(b) From (a), $\int_{C_2} \mathbf{F} \cdot d\mathbf{r} = f(\mathbf{r}(b)) - f(\mathbf{r}(a))$. So because $f(0, 0) = \sin 0 = 0$ and $f\left(\frac{\pi}{2}, 0\right) = 1$, one possible curve C_2 is $\mathbf{r}(t) = \frac{\pi}{2} t\,\mathbf{i}$, $0 \le t \le 1$, the straight line from $(0, 0)$ to $\left(\frac{\pi}{2}, 0\right)$.

25. Since $\mathbf{F}$ is conservative, there exists a function f such that $\mathbf{F} = \nabla f$, that is, $P = f_x$, $Q = f_y$, and $R = f_z$. Since P, Q and R have continuous first order partial derivatives, Clairaut's Theorem says that $\partial P/\partial y = f_{xy} = f_{yx} = \partial Q/\partial x$, $\partial P/\partial z = f_{xz} = f_{zx} = \partial R/\partial x$, and $\partial Q/\partial z = f_{yz} = f_{zy} = \partial R/\partial y$.

26. Here $\mathbf{F}(x, y, z) = y\,\mathbf{i} + x\,\mathbf{j} + xyz\,\mathbf{k}$. Then using the notation of Exercise 27, $\partial P/\partial z = 0$ while $\partial R/\partial x = yz$. Since these aren't equal, $\mathbf{F}$ is not conservative. Thus by Theorem 4, the line integral of $\mathbf{F}$ is not independent of path.

27. $D = \{(x, y) \mid x > 0, y > 0\}$ = the first quadrant (excluding the axes).

(a) D is open because around every point in D we can put a disk that lies in D.

(b) D is connected because the straight line segment joining any two points in D lies in D.

(c) D is simply-connected because it's connected and has no holes.

28. $D = \{(x, y) \mid x \neq 0\}$ consists of all points in the xy-plane except for those on the y-axis.

(a) D is open.

(b) Points on opposite sides of the y-axis cannot be joined by a path that lies in D, so D is not connected.

(c) D is not simply-connected because it is not connected.

29. $D = \{(x, y) \mid 1 < x^2 + y^2 < 4\}$ = the annular region between the circles with center $(0, 0)$ and radii 1 and 2.

(a) D is open.

(b) D is connected.

(c) D is not simply-connected. For example, $x^2 + y^2 = (1.5)^2$ is simple and closed and lies within D but encloses points that are not in D. (Or we can say, D has a hole, so is not simply-connected.)

30. $D = \{(x,y) \mid x^2+y^2 \le 1 \text{ or } 4 \le x^2+y^2 \le 9\}$ = the points on or inside the circle $x^2+y^2=1$, together with the points on or between the circles $x^2+y^2=4$ and $x^2+y^2=9$.

(a) D is not open because, for instance, no disk with center $(0,2)$ lies entirely within D.

(b) D is not connected because, for example, $(0,0)$ and $(0,2.5)$ lie in D but cannot be joined by a path that lies entirely in D.

(c) D is not simply-connected because, for example, $x^2+y^2=9$ is a simple closed curve in D but encloses points that are not in D.

31. (a) $P = -\dfrac{y}{x^2+y^2}$, $\dfrac{\partial P}{\partial y} = \dfrac{y^2-x^2}{(x^2+y^2)^2}$ and $Q = \dfrac{x}{x^2+y^2}$, $\dfrac{\partial Q}{\partial x} = \dfrac{y^2-x^2}{(x^2+y^2)^2}$. Thus $\dfrac{\partial P}{\partial y} = \dfrac{\partial Q}{\partial x}$.

(b) C_1: $x=\cos t$, $y=\sin t$, $0 \le t \le \pi$, C_2: $x=\cos t$, $y=\sin t$, $t=2\pi$ to $t=\pi$. Then

$$\int_{C_1} \mathbf{F}\cdot d\mathbf{r} = \int_0^\pi \frac{(-\sin t)(-\sin t)+(\cos t)(\cos t)}{\cos^2 t+\sin^2 t}\,dt = \int_0^\pi dt = \pi \text{ and } \int_{C_2} \mathbf{F}\cdot d\mathbf{r} = \int_{2\pi}^{\pi} dt = -\pi$$

Since these aren't equal, the line integral of $\mathbf{F}$ isn't independent of path. (Or notice that $\int_{C_3} \mathbf{F}\cdot d\mathbf{r} = \int_0^{2\pi} dt = 2\pi$ where C_3 is the circle $x^2+y^2=1$, and apply the contrapositive of Theorem 3.) This doesn't contradict Theorem 6, since the domain of $\mathbf{F}$, which is $\mathbb{R}^2$ except the origin, isn't simply-connected.

32. (a) Here $\mathbf{F}(\mathbf{r}) = c\mathbf{r}/|\mathbf{r}|^3$ and $\mathbf{r} = x\,\mathbf{i}+y\,\mathbf{j}+z\,\mathbf{k}$. Then $f(\mathbf{r}) = -c/|\mathbf{r}|$ is a potential function for $\mathbf{F}$, that is, $\nabla f = \mathbf{F}$. (See the discussion of gradient fields in Section 13.1.) Hence $\mathbf{F}$ is conservative and its line integral is independent of path. Let $P_1 = (x_1,y_1,z_1)$ and $P_2 = (x_2,y_2,z_2)$.

$$W = \textstyle\int_C \mathbf{F}\cdot d\mathbf{r} = f(P_2)-f(P_1) = -\dfrac{c}{(x_2^2+y_2^2+z_2^2)^{1/2}} + \dfrac{c}{(x_1^2+y_1^2+z_1^2)^{1/2}} = c\left(\dfrac{1}{d_1}-\dfrac{1}{d_2}\right).$$

(b) In this case, $c = -(mMG)$ $\Rightarrow$

$$\begin{aligned} W &= -mMG\left(\frac{1}{1.52\times10^{11}} - \frac{1}{1.47\times10^{11}}\right) \\ &= -\left(5.97\times10^{24}\right)\left(1.99\times10^{30}\right)\left(6.67\times10^{-11}\right)\left(-2.2377\times10^{-13}\right) \approx 1.77\times10^{32}\text{ J} \end{aligned}$$

(c) In this case, $c = \epsilon qQ$ $\Rightarrow$ $W = \epsilon qQ\left(\frac{1}{10^{-12}} - \frac{1}{5\times10^{-13}}\right) = \left(8.985\times10^9\right)(1)\left(-1.6\times10^{-19}\right)\left(-10^{12}\right) \approx 1400$ J.

13.4 Green's Theorem

1. (a)

C_1: $x=t$ $\Rightarrow$ $dx=dt$, $y=0$ $\Rightarrow$ $dy=0\,dt$, $0\le t\le 2$.

C_2: $x=2$ $\Rightarrow$ $dx=0\,dt$, $y=t$ $\Rightarrow$ $dy=dt$, $0\le t\le 3$.

C_3: $x=2-t$ $\Rightarrow$ $dx=-dt$, $y=3$ $\Rightarrow$ $dy=0\,dt$, $0\le t\le 2$.

C_4: $x=0$ $\Rightarrow$ $dx=0\,dt$, $y=3-t$ $\Rightarrow$ $dy=-dt$, $0\le t\le 3$.

Thus
$$\begin{aligned}\oint_C xy^2\,dx+x^3\,dy &= \oint_{C_1+C_2+C_3+C_4} xy^2\,dx+x^3\,dy \\ &= \textstyle\int_0^2 0\,dt + \int_0^3 8\,dt + \int_0^2 -9(2-t)\,dt + \int_0^3 0\,dt \\ &= 0+24-18+0 = 6\end{aligned}$$

(b) $\oint_C xy^2\,dx + x^3\,dy = \iint_D \left[\frac{\partial}{\partial x}(x^3) - \frac{\partial}{\partial y}(xy^2)\right] dA = \int_0^2\int_0^3 (3x^2-2xy)\,dy\,dx = \int_0^2 (9x^2-9x)\,dx = 24-18 = 6$

2. (a) $x = \cos t$, $y = \sin t$, $0 \le t \le 2\pi$. Then $\oint_C y\,dx - x\,dy = \int_0^{2\pi} [\sin t(-\sin t) - \cos t(\cos t)]\,dt = -\int_0^{2\pi} dt = -2\pi$.

(b) $\oint_C y\,dx - x\,dy = \iint_D \left[\frac{\partial}{\partial x}(-x) - \frac{\partial}{\partial y}(y)\right] dA = -2\iint_D dA = -2A(D) = -2\pi(1)^2 = -2\pi$

3. (a)

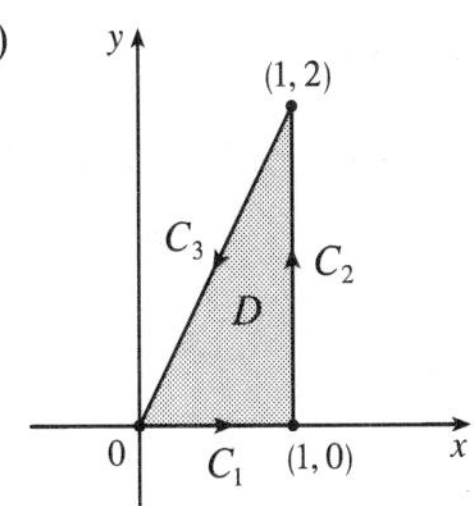

C_1: $x = t \;\Rightarrow\; dx = dt,\ y = 0 \;\Rightarrow\; dy = 0\,dt,\ 0 \le t \le 1$.

C_2: $x = 1 \;\Rightarrow\; dx = 0\,dt,\ y = t \;\Rightarrow\; dy = dt,\ 0 \le t \le 2$.

C_3: $x = 1 - t \;\Rightarrow\; dx = -dt,\ y = 2 - 2t \;\Rightarrow\; dy = -2\,dt,\ 0 \le t \le 1$.

Thus

$$\begin{aligned}\oint_C xy\,dx + x^2y^3\,dy &= \oint_{C_1+C_2+C_3} xy\,dx + x^2y^3\,dy \\ &= \int_0^1 0\,dt + \int_0^2 t^3\,dt + \int_0^1 \left[-(1-t)(2-2t) - 2(1-t)^2(2-2t)^3\right] dt \\ &= 0 + \left[\tfrac14 t^4\right]_0^2 + \left[\tfrac23(1-t)^3 + \tfrac83(1-t)^6\right]_0^1 = 4 - \tfrac{10}{3} = \tfrac23\end{aligned}$$

(b) $$\begin{aligned}\oint_C xy\,dx + x^2y^3\,dy &= \iint_D \left[\tfrac{\partial}{\partial x}(x^2y^3) - \tfrac{\partial}{\partial y}(xy)\right] dA = \int_0^1\int_0^{2x} (2xy^3 - x)\,dy\,dx \\ &= \int_0^1 \left[\tfrac12 xy^4 - xy\right]_{y=0}^{y=2x} dx = \int_0^1 (8x^5 - 2x^2)\,dx = \tfrac43 - \tfrac23 = \tfrac23\end{aligned}$$

4. (a) $C_1 : x = 0 \;\Rightarrow\; dx = 0\,dt,\ y = 1 - t \;\Rightarrow\; dy = -dt,\ 0 \le t \le 1$

$C_2 : x = t \;\Rightarrow\; dx = dt,\ y = 0 \;\Rightarrow\; dy = 0\,dt,\ 0 \le t \le 1$

$C_3 : x = 1 - t \;\Rightarrow\; dx = -dt,\ y = 1 - (1-t)^2 = 2t - t^2 \;\Rightarrow$

$dy = (2 - 2t)\,dt,\ 0 \le t \le 1$

Thus

$$\begin{aligned}\oint_C x\,dx + y\,dy &= \oint_{C_1+C_2+C_3} x\,dx + y\,dy \\ &= \int_0^1 (0\,dt + (1-t)(-dt)) + \int_0^1 (t\,dt + 0\,dt) + \int_0^1 ((1-t)(-dt) + (2t - t^2)(2 - 2t)\,dt) \\ &= \left[\tfrac12 t^2 - t\right]_0^1 + \left[\tfrac12 t^2\right]_0^1 + \left[\tfrac12 t^4 - 2t^3 + \tfrac52 t^2 - t\right]_0^1 = -\tfrac12 + \tfrac12 + \left(\tfrac12 - 2 + \tfrac52 - 1\right) = 0\end{aligned}$$

(b) $$\oint_C x\,dx + y\,dy = \iint_D \left[\frac{\partial}{\partial x}(y) - \frac{\partial}{\partial y}(x)\right] dA = \iint_D 0\,dA = 0$$

5. We can parametrize C as $x = \cos\theta$, $y = \sin\theta$, $0 \le \theta \le 2\pi$. Then the line integral is

$\oint_C P\,dx + Q\,dy = \int_0^{2\pi} \cos^4\theta\sin^5\theta\,(-\sin\theta)\,d\theta + \int_0^{2\pi} (-\cos^7\theta\sin^6\theta)\cos\theta\,d\theta = -\frac{29\pi}{1024}$, according to a CAS.

The double integral is

$$\iint_D \left(\frac{\partial Q}{\partial x} - \frac{\partial P}{\partial y}\right) dA = \int_{-1}^{1}\int_{-\sqrt{1-x^2}}^{\sqrt{1-x^2}} (-7x^6y^6 - 5x^4y^4)\,dy\,dx = -\frac{29\pi}{1024},$$

verifying Green's Theorem in this case.

6. Since $y = x^2$ along the first part of C and $y = x$ along the second part, the line integral is

$$\oint_C P\,dx + Q\,dy = \int_0^1 \left[x^4\sin x + x^2\sin(x^2)(2x)\right] dx + \int_1^0 (x^2\sin x + x^2\sin x)\,dx = -16\cos 1 - 23\sin 1 + 28.$$

according to a CAS. The double integral is

$$\iint_R \left(\frac{\partial Q}{\partial x} - \frac{\partial P}{\partial y}\right) dA = \int_0^1\int_{x^2}^{x} (2x\sin y - 2y\sin x)\,dy\,dx = -16\cos 1 - 23\sin 1 + 28.$$

7. The region D enclosed by C is $[0,1] \times [0,1]$, so

$$\int_C e^y\,dx + 2xe^y\,dy = \iint_D \left[\frac{\partial}{\partial x}(2xe^y) - \frac{\partial}{\partial y}(e^y)\right] dA = \int_0^1 \int_0^1 (2e^y - e^y)\,dy\,dx$$
$$= \int_0^1 dx \int_0^1 e^y\,dy = (1)(e^1 - e^0) = e - 1$$

8. The region D enclosed by C is given by $\{(x,y) \mid 0 \le x \le 1, 3x \le y \le 3\}$, so

$$\int_C x^2y^2\,dx + 4xy^3\,dy = \iint_D \left[\frac{\partial}{\partial x}(4xy^3) - \frac{\partial}{\partial y}(x^2y^2)\right] dA = \int_0^1 \int_{3x}^3 (4y^3 - 2x^2y)\,dy\,dx$$
$$= \int_0^1 \left[y^4 - x^2y^2\right]_{y=3x}^{y=3} dx = \int_0^1 (81 - 9x^2 - 72x^4)\,dx = 81 - 3 - \tfrac{72}{5} = \tfrac{318}{5}$$

9. $$\int_C \left(y + e^{\sqrt{x}}\right) dx + (2x + \cos y^2)\,dy = \iint_D \left[\frac{\partial}{\partial x}(2x + \cos y^2) - \frac{\partial}{\partial y}\left(y + e^{\sqrt{x}}\right)\right] dA$$
$$= \int_0^1 \int_{y^2}^{\sqrt{y}} (2-1)\,dx\,dy = \int_0^1 (y^{1/2} - y^2)\,dy = \tfrac{1}{3}$$

10. $$\int_C xe^{-2x}\,dx + (x^4 + 2x^2y^2)\,dy = \iint_D \left[\frac{\partial}{\partial x}(x^4 + 2x^2y^2) - \frac{\partial}{\partial y}(xe^{-2x})\right] dA = \iint_D (4x^3 + 4xy^2 - 0)\,dA$$
$$= 4\iint_D x(x^2 + y^2)\,dA = 4\int_0^{2\pi}\int_1^2 (r\cos\theta)(r^2)\,r\,dr\,d\theta$$
$$= 4\int_0^{2\pi} \cos\theta\,d\theta \int_1^2 r^4\,dr = 4\big[\sin\theta\big]_0^{2\pi}\left[\tfrac{1}{5}r^5\right]_1^2 = 0$$

11. $$\int_C y^3\,dx - x^3\,dy = \iint_D \left[\frac{\partial}{\partial x}(-x^3) - \frac{\partial}{\partial y}(y^3)\right] dA = \iint_D (-3x^2 - 3y^2)\,dA = \int_0^{2\pi}\int_0^2 (-3r^2)\,r\,dr\,d\theta$$
$$= -3\int_0^{2\pi} d\theta \int_0^2 r^3\,dr = -3(2\pi)(4) = -24\pi$$

12. $$\int_C \sin y\,dx + x\cos y\,dy = \iint_D \left[\frac{\partial}{\partial x}(x\cos y) - \frac{\partial}{\partial y}(\sin y)\right] dA = \iint_D (\cos y - \cos y)\,dA = \iint_D 0\,dA = 0$$

13. $\mathbf{F}(x,y) = \left\langle \sqrt{x} + y^3, x^2 + \sqrt{y}\right\rangle$ and the region D enclosed by C is given by $\{(x,y) \mid 0 \le x \le \pi, 0 \le y \le \sin x\}$.
C is traversed clockwise, so $-C$ gives the positive orientation.

$$\int_C \mathbf{F}\cdot d\mathbf{r} = -\int_{-C} \left(\sqrt{x} + y^3\right) dx + \left(x^2 + \sqrt{y}\right) dy = -\iint_D \left[\frac{\partial}{\partial x}\left(x^2 + \sqrt{y}\right) - \frac{\partial}{\partial y}\left(\sqrt{x} + y^3\right)\right] dA$$
$$= -\int_0^\pi \int_0^{\sin x} (2x - 3y^2)\,dy\,dx = -\int_0^\pi \left[2xy - y^3\right]_{y=0}^{y=\sin x} dx$$
$$= -\int_0^\pi (2x\sin x - \sin^3 x)\,dx = -\int_0^\pi (2x\sin x - (1 - \cos^2 x)\sin x)\,dx$$
$$= -\left[2\sin x - 2x\cos x + \cos x - \tfrac{1}{3}\cos^3 x\right]_0^\pi \qquad \text{[integrate by parts in the first term]}$$
$$= -\left(2\pi - 2 + \tfrac{2}{3}\right) = \tfrac{4}{3} - 2\pi$$

14. $\mathbf{F}(x,y) = \left\langle y^2\cos x, x^2 + 2y\sin x\right\rangle$ and the region D enclosed by C is given by $\{(x,y) \mid 0 \le x \le 2, 0 \le y \le 3x\}$.
C is traversed clockwise, so $-C$ gives the positive orientation.

$$\int_C \mathbf{F}\cdot d\mathbf{r} = -\int_{-C} (y^2\cos x)\,dx + (x^2 + 2y\sin x)\,dy = -\iint_D \left[\frac{\partial}{\partial x}(x^2 + 2y\sin x) - \frac{\partial}{\partial y}(y^2\cos x)\right] dA$$
$$= -\iint_D (2x + 2y\cos x - 2y\cos x)\,dA = -\int_0^2 \int_0^{3x} 2x\,dy\,dx$$
$$= -\int_0^2 2x\,\big[y\big]_{y=0}^{y=3x} dx = -\int_0^2 6x^2\,dx = -\,2x^3\big]_0^2 = -16$$

15. $\mathbf{F}(x,y) = \langle e^x + x^2 y, e^y - xy^2 \rangle$ and the region D enclosed by C is the disk $x^2 + y^2 \le 25$.

C is traversed clockwise, so $-C$ gives the positive orientation.

$$\int_C \mathbf{F} \cdot d\mathbf{r} = -\int_{-C} (e^x + x^2 y)\,dx + (e^y - xy^2)\,dy = -\iint_D \left[\tfrac{\partial}{\partial x}(e^y - xy^2) - \tfrac{\partial}{\partial y}(e^x + x^2 y)\right] dA$$
$$= -\iint_D (-y^2 - x^2)\,dA = \iint_D (x^2 + y^2)\,dA = \int_0^{2\pi}\int_0^5 (r^2)\,r\,dr\,d\theta = \int_0^{2\pi} d\theta \int_0^5 r^3\,dr = 2\pi\left[\tfrac{1}{4}r^4\right]_0^5 = \tfrac{625}{2}\pi$$

16. $\mathbf{F}(x,y) = \left\langle y - \ln(x^2 + y^2), 2\tan^{-1}\left(\frac{y}{x}\right)\right\rangle$ and the region D enclosed by C is the disk with radius 1 centered at $(2,3)$.

C is oriented positively, so

$$\int_C \mathbf{F} \cdot d\mathbf{r} = \int_C (y - \ln(x^2 + y^2))\,dx + \left(2\tan^{-1}\left(\frac{y}{x}\right)\right) dy = \iint_D \left[\frac{\partial}{\partial x}\left(2\tan^{-1}\left(\frac{y}{x}\right)\right) - \frac{\partial}{\partial y}(y - \ln(x^2 + y^2))\right] dA$$
$$= \iint_D \left[2\left(\frac{-yx^{-2}}{1 + (y/x)^2}\right) - \left(1 - \frac{2y}{x^2 + y^2}\right)\right] dA = \iint_D \left[-\frac{2y}{x^2 + y^2} - 1 + \frac{2y}{x^2 + y^2}\right] dA$$
$$= -\iint_D dA = -(\text{area of } D) = -\pi$$

17. By Green's Theorem, $W = \int_C \mathbf{F} \cdot d\mathbf{r} = \int_C x(x + y)\,dx + xy^2\,dy = \iint_D (y^2 - x)\,dy\,dx$ where C is the path described in the question and D is the triangle bounded by C. So

$$W = \int_0^1 \int_0^{1-x} (y^2 - x)\,dy\,dx = \int_0^1 \left[\tfrac{1}{3}y^3 - xy\right]_{y=0}^{y=1-x} dx = \int_0^1 \left(\tfrac{1}{3}(1 - x)^3 - x(1 - x)\right) dx$$
$$= \left[-\tfrac{1}{12}(1 - x)^4 - \tfrac{1}{2}x^2 + \tfrac{1}{3}x^3\right]_0^1 = \left(-\tfrac{1}{2} + \tfrac{1}{3}\right) - \left(-\tfrac{1}{12}\right) = -\tfrac{1}{12}$$

18. By Green's Theorem, $W = \int_C \mathbf{F} \cdot d\mathbf{r} = \int_C x\,dx + (x^3 + 3xy^2)\,dy = \iint_D (3x^2 + 3y^2 - 0)\,dA$, where D is the semicircular region bounded by C. Converting to polar coordinates, we have $W = 3\int_0^2 \int_0^\pi r^2 \cdot r\,d\theta\,dr = 3\pi\left[\tfrac{1}{4}r^4\right]_0^2 = 12\pi$.

19. Let C_1 be the arch of the cycloid from $(0,0)$ to $(2\pi,0)$, which corresponds to $0 \le t \le 2\pi$, and let C_2 be the segment from $(2\pi,0)$ to $(0,0)$, so C_2 is given by $x = 2\pi - t$, $y = 0$, $0 \le t \le 2\pi$. Then $C = C_1 \cup C_2$ is traversed clockwise, so $-C$ is oriented positively. Thus $-C$ encloses the area under one arch of the cycloid and from (5) we have

$$A = -\oint_{-C} y\,dx = \int_{C_1} y\,dx + \int_{C_2} y\,dx = \int_0^{2\pi} (1 - \cos t)(1 - \cos t)\,dt + \int_0^{2\pi} 0\,(-dt)$$
$$= \int_0^{2\pi} (1 - 2\cos t + \cos^2 t)\,dt + 0 = \left[t - 2\sin t + \tfrac{1}{2}t + \tfrac{1}{4}\sin 2t\right]_0^{2\pi} = 3\pi$$

20.

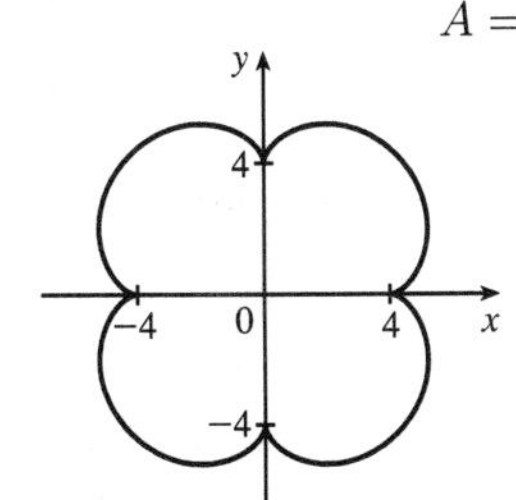

$$A = \oint_C x\,dy = \int_0^{2\pi} (5\cos t - \cos 5t)(5\cos t - 5\cos 5t)\,dt$$
$$= \int_0^{2\pi} (25\cos^2 t - 30\cos t\cos 5t + 5\cos^2 5t)\,dt$$
$$= \left[25\left(\tfrac{1}{2}t + \tfrac{1}{4}\sin 2t\right) - 30\left(\tfrac{1}{8}\sin 4t + \tfrac{1}{12}\sin 6t\right) + 5\left(\tfrac{1}{2}t + \tfrac{1}{20}\sin 10t\right)\right]_0^{2\pi}$$

[Use Formula 80 in the Table of Integrals]

$$= 30\pi$$

21. (a) Using Equation 13.2.8, we write parametric equations of the line segment as $x = (1 - t)x_1 + tx_2$, $y = (1 - t)y_1 + ty_2$, $0 \le t \le 1$. Then $dx = (x_2 - x_1)\,dt$ and $dy = (y_2 - y_1)\,dt$, so

$$\int_C x\,dy - y\,dx = \int_0^1 [(1 - t)x_1 + tx_2](y_2 - y_1)\,dt + [(1 - t)y_1 + ty_2](x_2 - x_1)\,dt$$
$$= \int_0^1 (x_1(y_2 - y_1) - y_1(x_2 - x_1) + t[(y_2 - y_1)(x_2 - x_1) - (x_2 - x_1)(y_2 - y_1)])\,dt$$
$$= \int_0^1 (x_1 y_2 - x_2 y_1)\,dt = x_1 y_2 - x_2 y_1$$

(b) We apply Green's Theorem to the path $C = C_1 \cup C_2 \cup \cdots \cup C_n$, where C_i is the line segment that joins (x_i, y_i) to (x_{i+1}, y_{i+1}) for $i = 1, 2, \ldots, n-1$, and C_n is the line segment that joins (x_n, y_n) to (x_1, y_1). From (5), $\frac{1}{2}\int_C x\,dy - y\,dx = \iint_D dA$, where D is the polygon bounded by C. Therefore

$$\text{area of polygon} = A(D) = \iint_D dA = \tfrac{1}{2}\int_C x\,dy - y\,dx$$
$$= \tfrac{1}{2}\left(\int_{C_1} x\,dy - y\,dx + \int_{C_2} x\,dy - y\,dx + \cdots + \int_{C_{n-1}} x\,dy - y\,dx + \int_{C_n} x\,dy - y\,dx\right)$$

To evaluate these integrals we use the formula from (a) to get

$$A(D) = \tfrac{1}{2}[(x_1y_2 - x_2y_1) + (x_2y_3 - x_3y_2) + \cdots + (x_{n-1}y_n - x_ny_{n-1}) + (x_ny_1 - x_1y_n)].$$

(c) $A = \frac{1}{2}[(0\cdot 1 - 2\cdot 0) + (2\cdot 3 - 1\cdot 1) + (1\cdot 2 - 0\cdot 3) + (0\cdot 1 - (-1)\cdot 2) + (-1\cdot 0 - 0\cdot 1)]$
$= \frac{1}{2}(0 + 5 + 2 + 2) = \frac{9}{2}$

22. By Green's Theorem, $\frac{1}{2A}\oint_C x^2\,dy = \frac{1}{2A}\iint_D 2x\,dA = \frac{1}{A}\iint_D x\,dA = \overline{x}$ and
$-\frac{1}{2A}\oint_C y^2\,dx = -\frac{1}{2A}\iint_D (-2y)\,dA = \frac{1}{A}\iint_D y\,dA = \overline{y}$.

23. Here $A = \frac{1}{2}(1)(1) = \frac{1}{2}$ and $C = C_1 + C_2 + C_3$, where C_1: $x = x$, $y = 0$, $0 \le x \le 1$;
C_2: $x = x$, $y = 1 - x$, $x = 1$ to $x = 0$; and C_3: $x = 0$, $y = 1$ to $y = 0$. Then
$\overline{x} = \frac{1}{2A}\int_C x^2\,dy = \int_{C_1} x^2\,dy + \int_{C_2} x^2\,dy + \int_{C_3} x^2\,dy = 0 + \int_1^0 (x^2)(-dx) + 0 = \frac{1}{3}$. Similarly,
$\overline{y} = -\frac{1}{2A}\int_C y^2\,dx = \int_{C_1} y^2\,dx + \int_{C_2} y^2\,dx + \int_{C_3} y^2\,dx = 0 + \int_1^0 (1-x)^2(-dx) + 0 = \frac{1}{3}$. Therefore $(\overline{x}, \overline{y}) = \left(\frac{1}{3}, \frac{1}{3}\right)$.

24. $A = \frac{\pi a^2}{2}$ so $\overline{x} = \frac{1}{\pi a^2}\oint_C x^2\,dy$ and $\overline{y} = -\frac{1}{\pi a^2}\oint_C y^2 dx$.

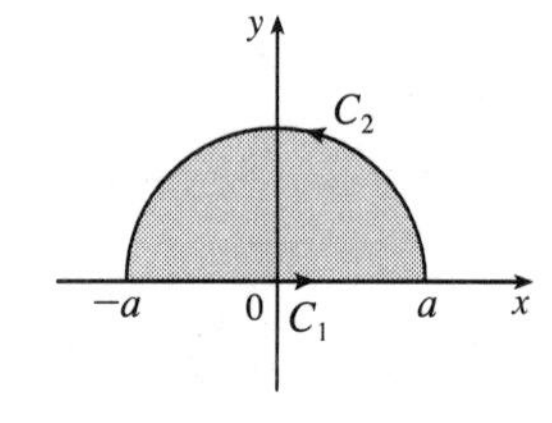

Orienting the semicircular region as in the figure,

$\overline{x} = \frac{1}{\pi a^2}\oint_{C_1 + C_2} x^2\,dy = \frac{1}{\pi a^2}\left[0 + \int_0^\pi (a^2\cos^2 t)(a\cos t)\,dt\right] = 0$ and

$$\overline{y} = -\frac{1}{\pi a^2}\left[\int_{-a}^a 0\,dx + \int_0^\pi (a^2\sin^2 t)(-a\sin t)\,dt\right] = \frac{a}{\pi}\int_0^\pi \sin^3 t\,dt$$
$$= \frac{a}{\pi}\left[-\cos t + \tfrac{1}{3}(\cos^3 t)\right]_0^\pi = \frac{4a}{3\pi}$$

Thus $(\overline{x}, \overline{y}) = \left(0, \frac{4a}{3\pi}\right)$.

25. By Green's Theorem, $-\frac{1}{3}\rho\oint_C y^3\,dx = -\frac{1}{3}\rho\iint_D (-3y^2)\,dA = \iint_D y^2\rho\,dA = I_x$ and
$\frac{1}{3}\rho\oint_C x^3\,dy = \frac{1}{3}\rho\iint_D (3x^2)\,dA = \iint_D x^2\rho\,dA = I_y$.

26. By symmetry the moments of inertia about any two diameters are equal. Centering the disk at the origin, the moment of inertia about a diameter equals

$$I_y = \tfrac{1}{3}\rho\oint_C x^3\,dy = \tfrac{1}{3}\rho\int_0^{2\pi}(a^4\cos^4 t)\,dt = \tfrac{1}{3}a^4\rho\int_0^{2\pi}\left[\tfrac{3}{8} + \tfrac{1}{2}\cos 2t + \tfrac{1}{8}\cos 4t\right]dt = \tfrac{1}{3}a^4\rho\cdot\tfrac{3(2\pi)}{8} = \tfrac{1}{4}\pi a^4\rho$$

27. Since C is a simple closed path which doesn't pass through or enclose the origin, there exists an open region that doesn't contain the origin but does contain D. Thus $P = -y/(x^2 + y^2)$ and $Q = x/(x^2 + y^2)$ have continuous partial derivatives on this open region containing D and we can apply Green's Theorem. But by Exercise 13.3.31(a), $\partial P/\partial y = \partial Q/\partial x$, so $\oint_C \mathbf{F}\cdot d\mathbf{r} = \iint_D 0\,dA = 0$.

28. We express D as a type II region: $D = \{(x, y) \mid f_1(y) \le x \le f_2(y), c \le y \le d\}$ where f_1 and f_2 are continuous functions. Then $\iint_D \frac{\partial Q}{\partial x}\, dA = \int_c^d \int_{f_1(y)}^{f_2(y)} \frac{\partial Q}{\partial x}\, dx\, dy = \int_c^d \left[Q(f_2(y), y) - Q(f_1(y), y)\right] dy$ by the Fundamental Theorem of Calculus. But referring to the figure, $\oint_C Q\, dy = \oint_{C_1 + C_2 + C_3 + C_4} Q\, dy$.

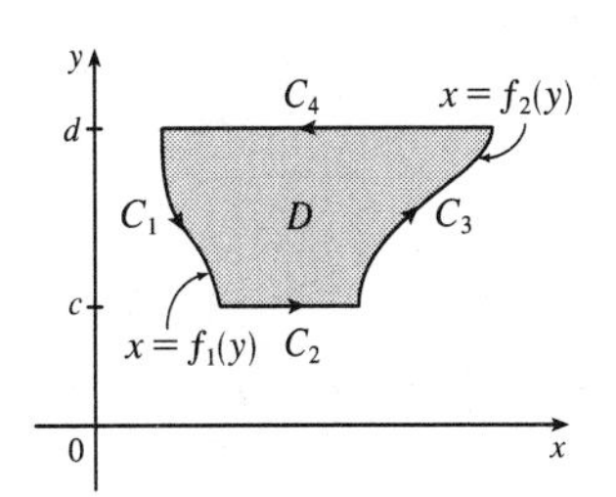

Then $\int_{C_1} Q\, dy = \int_d^c Q(f_1(y), y)\, dy$, $\int_{C_2} Q\, dy = \int_{C_4} Q\, dy = 0$, and $\int_{C_3} Q\, dy = \int_c^d Q(f_2(y), y)\, dy$. Hence $\oint_C Q\, dy = \int_c^d \left[Q(f_2(y), y) - Q(f_1(y), y)\right] dy = \iint_D (\partial Q/\partial x)\, dA$.

29. Using the first part of (5), we have that $\iint_R dx\, dy = A(R) = \int_{\partial R} x\, dy$. But $x = g(u, v)$, and $dy = \frac{\partial h}{\partial u} du + \frac{\partial h}{\partial v} dv$, and we orient ∂S by taking the positive direction to be that which corresponds, under the mapping, to the positive direction along ∂R, so

$$\int_{\partial R} x\, dy = \int_{\partial S} g(u, v) \left(\frac{\partial h}{\partial u} du + \frac{\partial h}{\partial v} dv\right) = \int_{\partial S} g(u, v) \frac{\partial h}{\partial u} du + g(u, v) \frac{\partial h}{\partial v} dv$$

$$= \pm \iint_S \left[\frac{\partial}{\partial u}\left(g(u, v) \frac{\partial h}{\partial v}\right) - \frac{\partial}{\partial v}\left(g(u, v) \frac{\partial h}{\partial u}\right)\right] dA \quad \text{[using Green's Theorem in the } uv\text{-plane]}$$

$$= \pm \iint_S \left(\frac{\partial g}{\partial u}\frac{\partial h}{\partial v} + g(u, v) \frac{\partial^2 h}{\partial u\, \partial v} - \frac{\partial g}{\partial v}\frac{\partial h}{\partial u} - g(u, v) \frac{\partial^2 h}{\partial v\, \partial u}\right) dA \quad \text{[using the Chain Rule]}$$

$$= \pm \iint_S \left(\frac{\partial x}{\partial u}\frac{\partial y}{\partial v} - \frac{\partial x}{\partial v}\frac{\partial y}{\partial u}\right) dA \ \text{[by the equality of mixed partials]} = \pm \iint_S \frac{\partial(x, y)}{\partial(u, v)}\, du\, dv$$

The sign is chosen to be positive if the orientation that we gave to ∂S corresponds to the usual positive orientation, and it is negative otherwise. In either case, since $A(R)$ is positive, the sign chosen must be the same as the sign of $\frac{\partial(x, y)}{\partial(u, v)}$. Therefore

$$A(R) = \iint_R dx\, dy = \iint_S \left|\frac{\partial(x, y)}{\partial(u, v)}\right| du\, dv.$$

13.5 Curl and Divergence

1. (a) $\operatorname{curl} \mathbf{F} = \nabla \times \mathbf{F} = \begin{vmatrix} \mathbf{i} & \mathbf{j} & \mathbf{k} \\ \partial/\partial x & \partial/\partial y & \partial/\partial z \\ xyz & 0 & -x^2 y \end{vmatrix} = (-x^2 - 0)\,\mathbf{i} - (-2xy - xy)\,\mathbf{j} + (0 - xz)\,\mathbf{k}$

$= -x^2\,\mathbf{i} + 3xy\,\mathbf{j} - xz\,\mathbf{k}$

(b) $\operatorname{div} \mathbf{F} = \nabla \cdot \mathbf{F} = \frac{\partial}{\partial x}(xyz) + \frac{\partial}{\partial y}(0) + \frac{\partial}{\partial z}(-x^2 y) = yz + 0 + 0 = yz$

2. (a) $\operatorname{curl} \mathbf{F} = \nabla \times \mathbf{F} = \begin{vmatrix} \mathbf{i} & \mathbf{j} & \mathbf{k} \\ \partial/\partial x & \partial/\partial y & \partial/\partial z \\ x^2 yz & xy^2 z & xyz^2 \end{vmatrix} = (xz^2 - xy^2)\,\mathbf{i} - (yz^2 - x^2 y)\,\mathbf{j} + (y^2 z - x^2 z)\,\mathbf{k}$

$= x(z^2 - y^2)\,\mathbf{i} + y(x^2 - z^2)\,\mathbf{j} + z(y^2 - x^2)\,\mathbf{k}$

(b) $\operatorname{div} \mathbf{F} = \nabla \cdot \mathbf{F} = \dfrac{\partial}{\partial x}(x^2yz) + \dfrac{\partial}{\partial y}(xy^2z) + \dfrac{\partial}{\partial z}(xyz^2) = 2xyz + 2xyz + 2xyz = 6xyz$

3. (a) $\operatorname{curl} \mathbf{F} = \nabla \times \mathbf{F} = \begin{vmatrix} \mathbf{i} & \mathbf{j} & \mathbf{k} \\ \partial/\partial x & \partial/\partial y & \partial/\partial z \\ 1 & x+yz & xy-\sqrt{z} \end{vmatrix} = (x-y)\,\mathbf{i} - (y-0)\,\mathbf{j} + (1-0)\,\mathbf{k}$

$= (x-y)\,\mathbf{i} - y\,\mathbf{j} + \mathbf{k}$

(b) $\operatorname{div} \mathbf{F} = \nabla \cdot \mathbf{F} = \dfrac{\partial}{\partial x}(1) + \dfrac{\partial}{\partial y}(x+yz) + \dfrac{\partial}{\partial z}(xy-\sqrt{z}) = z - \dfrac{1}{2\sqrt{z}}$

4. (a) $\operatorname{curl} \mathbf{F} = \nabla \times \mathbf{F} = \begin{vmatrix} \mathbf{i} & \mathbf{j} & \mathbf{k} \\ \partial/\partial x & \partial/\partial y & \partial/\partial z \\ 0 & \cos xz & -\sin xy \end{vmatrix}$

$= (-x\cos xy + x\sin xz)\,\mathbf{i} - (-y\cos xy - 0)\,\mathbf{j} + (-z\sin xz - 0)\,\mathbf{k}$

$= x(\sin xz - \cos xy)\,\mathbf{i} + y\cos xy\,\mathbf{j} - z\sin xz\,\mathbf{k}$

(b) $\operatorname{div} \mathbf{F} = \nabla \cdot \mathbf{F} = \dfrac{\partial}{\partial x}(0) + \dfrac{\partial}{\partial y}(\cos xz) + \dfrac{\partial}{\partial z}(-\sin xy) = 0 + 0 + 0 = 0$

5. (a) $\operatorname{curl} \mathbf{F} = \nabla \times \mathbf{F} = \begin{vmatrix} \mathbf{i} & \mathbf{j} & \mathbf{k} \\ \partial/\partial x & \partial/\partial y & \partial/\partial z \\ e^x \sin y & e^x \cos y & z \end{vmatrix} = (0-0)\,\mathbf{i} - (0-0)\,\mathbf{j} + (e^x\cos y - e^x\cos y)\,\mathbf{k} = \mathbf{0}$

(b) $\operatorname{div} \mathbf{F} = \nabla \cdot \mathbf{F} = \dfrac{\partial}{\partial x}(e^x\sin y) + \dfrac{\partial}{\partial y}(e^x\cos y) + \dfrac{\partial}{\partial z}(z) = e^x\sin y - e^x\sin y + 1 = 1$

6. (a) $\operatorname{curl} \mathbf{F} = \nabla \times \mathbf{F} = \begin{vmatrix} \mathbf{i} & \mathbf{j} & \mathbf{k} \\ \partial/\partial x & \partial/\partial y & \partial/\partial z \\ \dfrac{x}{x^2+y^2+z^2} & \dfrac{y}{x^2+y^2+z^2} & \dfrac{z}{x^2+y^2+z^2} \end{vmatrix}$

$= \dfrac{1}{(x^2+y^2+z^2)^2}\left[(-2yz+2yz)\,\mathbf{i} - (-2xz+2xz)\,\mathbf{j} + (-2xy+2xy)\,\mathbf{k}\right] = \mathbf{0}$

(b) $\operatorname{div} \mathbf{F} = \nabla \cdot \mathbf{F} = \dfrac{\partial}{\partial x}\left(\dfrac{x}{x^2+y^2+z^2}\right) + \dfrac{\partial}{\partial y}\left(\dfrac{y}{x^2+y^2+z^2}\right) + \dfrac{\partial}{\partial z}\left(\dfrac{z}{x^2+y^2+z^2}\right)$

$= \dfrac{x^2+y^2+z^2-2x^2}{(x^2+y^2+z^2)^2} + \dfrac{x^2+y^2+z^2-2y^2}{(x^2+y^2+z^2)^2} + \dfrac{x^2+y^2+z^2-2z^2}{(x^2+y^2+z^2)^2} = \dfrac{x^2+y^2+z^2}{(x^2+y^2+z^2)^2} = \dfrac{1}{x^2+y^2+z^2}$

7. (a) $\operatorname{curl} \mathbf{F} = \nabla \times \mathbf{F} = \begin{vmatrix} \mathbf{i} & \mathbf{j} & \mathbf{k} \\ \partial/\partial x & \partial/\partial y & \partial/\partial z \\ \ln x & \ln(xy) & \ln(xyz) \end{vmatrix} = \left(\dfrac{xz}{xyz} - 0\right)\mathbf{i} - \left(\dfrac{yz}{xyz} - 0\right)\mathbf{j} + \left(\dfrac{y}{xy} - 0\right)\mathbf{k}$

$= \left\langle \dfrac{1}{y}, -\dfrac{1}{x}, \dfrac{1}{x} \right\rangle$

(b) $\operatorname{div} \mathbf{F} = \nabla \cdot \mathbf{F} = \dfrac{\partial}{\partial x}(\ln x) + \dfrac{\partial}{\partial y}(\ln(xy)) + \dfrac{\partial}{\partial z}(\ln(xyz)) = \dfrac{1}{x} + \dfrac{x}{xy} + \dfrac{xy}{xyz} = \dfrac{1}{x} + \dfrac{1}{y} + \dfrac{1}{z}$

8. If the vector field is $\mathbf{F} = P\,\mathbf{i} + Q\,\mathbf{j} + R\,\mathbf{k}$, then we know $R = 0$. In addition, P and Q don't vary in the z-direction, so $\dfrac{\partial R}{\partial x} = \dfrac{\partial R}{\partial y} = \dfrac{\partial R}{\partial z} = \dfrac{\partial P}{\partial z} = \dfrac{\partial Q}{\partial z} = 0$. As x increases, the x-component of each vector of $\mathbf{F}$ increases while the y-component remains constant, so $\dfrac{\partial P}{\partial x} > 0$ and $\dfrac{\partial Q}{\partial x} = 0$. Similarly, as y increases, the y-component of each vector increases while the x-component remains constant, so $\dfrac{\partial Q}{\partial y} > 0$ and $\dfrac{\partial P}{\partial y} = 0$.

(a) $\operatorname{div}\mathbf{F} = \dfrac{\partial P}{\partial x} + \dfrac{\partial Q}{\partial y} + \dfrac{\partial R}{\partial z} = \dfrac{\partial P}{\partial x} + \dfrac{\partial Q}{\partial y} + 0 > 0$

(b) $\operatorname{curl}\mathbf{F} = \left(\dfrac{\partial R}{\partial y} - \dfrac{\partial Q}{\partial z}\right)\mathbf{i} + \left(\dfrac{\partial P}{\partial z} - \dfrac{\partial R}{\partial x}\right)\mathbf{j} + \left(\dfrac{\partial Q}{\partial x} - \dfrac{\partial P}{\partial y}\right)\mathbf{k} = (0-0)\,\mathbf{i} + (0-0)\,\mathbf{j} + (0-0)\,\mathbf{k} = \mathbf{0}$

9. If the vector field is $\mathbf{F} = P\,\mathbf{i} + Q\,\mathbf{j} + R\,\mathbf{k}$, then we know $R = 0$. In addition, the y-component of each vector of $\mathbf{F}$ is 0, so $Q = 0$, hence $\dfrac{\partial Q}{\partial x} = \dfrac{\partial Q}{\partial y} = \dfrac{\partial Q}{\partial z} = \dfrac{\partial R}{\partial x} = \dfrac{\partial R}{\partial y} = \dfrac{\partial R}{\partial z} = 0$. P increases as y increases, so $\dfrac{\partial P}{\partial y} > 0$, but P doesn't change in the x- or z-directions, so $\dfrac{\partial P}{\partial x} = \dfrac{\partial P}{\partial z} = 0$.

(a) $\operatorname{div}\mathbf{F} = \dfrac{\partial P}{\partial x} + \dfrac{\partial Q}{\partial y} + \dfrac{\partial R}{\partial z} = 0 + 0 + 0 = 0$

(b) $\operatorname{curl}\mathbf{F} = \left(\dfrac{\partial R}{\partial y} - \dfrac{\partial Q}{\partial z}\right)\mathbf{i} + \left(\dfrac{\partial P}{\partial z} - \dfrac{\partial R}{\partial x}\right)\mathbf{j} + \left(\dfrac{\partial Q}{\partial x} - \dfrac{\partial P}{\partial y}\right)\mathbf{k} = (0-0)\,\mathbf{i} + (0-0)\,\mathbf{j} + \left(0 - \dfrac{\partial P}{\partial y}\right)\mathbf{k} = -\dfrac{\partial P}{\partial y}\mathbf{k}$

Since $\dfrac{\partial P}{\partial y} > 0$, $-\dfrac{\partial P}{\partial y}\mathbf{k}$ is a vector pointing in the negative z-direction.

10. (a) $\operatorname{curl} f = \nabla \times f$ is meaningless because f is a scalar field.

(b) $\operatorname{grad} f$ is a vector field.

(c) $\operatorname{div}\mathbf{F}$ is a scalar field.

(d) $\operatorname{curl}(\operatorname{grad} f)$ is a vector field.

(e) $\operatorname{grad}\mathbf{F}$ is meaningless because $\mathbf{F}$ is not a scalar field.

(f) $\operatorname{grad}(\operatorname{div}\mathbf{F})$ is a vector field.

(g) $\operatorname{div}(\operatorname{grad} f)$ is a scalar field.

(h) $\operatorname{grad}(\operatorname{div} f)$ is meaningless because f is a scalar field.

(i) $\operatorname{curl}(\operatorname{curl}\mathbf{F})$ is a vector field.

(j) $\operatorname{div}(\operatorname{div}\mathbf{F})$ is meaningless because $\operatorname{div}\mathbf{F}$ is a scalar field.

(k) $(\operatorname{grad} f) \times (\operatorname{div}\mathbf{F})$ is meaningless because $\operatorname{div}\mathbf{F}$ is a scalar field.

(l) $\operatorname{div}(\operatorname{curl}(\operatorname{grad} f))$ is a scalar field.

11. curl $\mathbf{F} = \nabla \times \mathbf{F} = \begin{vmatrix} \mathbf{i} & \mathbf{j} & \mathbf{k} \\ \partial/\partial x & \partial/\partial y & \partial/\partial z \\ yz & xz & xy \end{vmatrix} = (x - x)\,\mathbf{i} - (y - y)\,\mathbf{j} + (z - z)\,\mathbf{k} = \mathbf{0}$

and $\mathbf{F}$ is defined on all of $\mathbb{R}^3$ with component functions which have continuous partial derivatives, so by Theorem 4, $\mathbf{F}$ is conservative. Thus, there exists a function f such that $\mathbf{F} = \nabla f$. Then $f_x(x, y, z) = yz$ implies $f(x, y, z) = xyz + g(y, z)$ and $f_y(x, y, z) = xz + g_y(y, z)$. But $f_y(x, y, z) = xz$, so $g(y, z) = h(z)$ and $f(x, y, z) = xyz + h(z)$. Thus $f_z(x, y, z) = xy + h'(z)$ but $f_z(x, y, z) = xy$ so $h(z) = K$, a constant. Hence a potential function for $\mathbf{F}$ is $f(x, y, z) = xyz + K$.

12. curl $\mathbf{F} = \nabla \times \mathbf{F} = \begin{vmatrix} \mathbf{i} & \mathbf{j} & \mathbf{k} \\ \partial/\partial x & \partial/\partial y & \partial/\partial z \\ 3z^2 & \cos y & 2xz \end{vmatrix} = (0 - 0)\,\mathbf{i} - (2z - 6z)\,\mathbf{j} + (0 - 0)\,\mathbf{k} = 4z\,\mathbf{j} \neq \mathbf{0}$,

so $\mathbf{F}$ is not conservative.

13. curl $\mathbf{F} = \nabla \times \mathbf{F} = \begin{vmatrix} \mathbf{i} & \mathbf{j} & \mathbf{k} \\ \partial/\partial x & \partial/\partial y & \partial/\partial z \\ 2xy & x^2 + 2yz & y^2 \end{vmatrix} = (2y - 2y)\,\mathbf{i} - (0 - 0)\,\mathbf{j} + (2x - 2x)\,\mathbf{k} = \mathbf{0}$, $\mathbf{F}$ is defined on all of $\mathbb{R}^3$,

and the partial derivatives of the component functions are continuous, so $\mathbf{F}$ is conservative. Thus there exists a function f such that $\nabla f = \mathbf{F}$. Then $f_x(x, y, z) = 2xy$ implies $f(x, y, z) = x^2 y + g(y, z)$ and $f_y(x, y, z) = x^2 + g_y(y, z)$. But $f_y(x, y, z) = x^2 + 2yz$, so $g(y, z) = y^2 z + h(z)$ and $f(x, y, z) = x^2 y + y^2 z + h(z)$. Thus $f_z(x, y, z) = y^2 + h'(z)$ but $f_z(x, y, z) = y^2$ so $h(z) = K$ and $f(x, y, z) = x^2 y + y^2 z + K$.

14. curl $\mathbf{F} = \nabla \times \mathbf{F} = \begin{vmatrix} \mathbf{i} & \mathbf{j} & \mathbf{k} \\ \partial/\partial x & \partial/\partial y & \partial/\partial z \\ e^z & 1 & xe^z \end{vmatrix} = (0 - 0)\,\mathbf{i} - (e^z - e^z)\,\mathbf{j} + (0 - 0)\,\mathbf{k} = \mathbf{0}$ and $\mathbf{F}$ is defined on all of $\mathbb{R}^3$ with

component functions that have continuous partial deriatives, so $\mathbf{F}$ is conservative. Thus there exists a function f such that $\nabla f = \mathbf{F}$. Then $f_x(x, y, z) = e^z$ implies $f(x, y, z) = xe^z + g(y, z) \quad \Rightarrow \quad f_y(x, y, z) = g_y(y, z)$. But $f_y(x, y, z) = 1$, so $g(y, z) = y + h(z)$ and $f(x, y, z) = xe^z + y + h(z)$. Thus $f_z(x, y, z) = xe^z + h'(z)$ but $f_z(x, y, z) = xe^z$, so $h(z) = K$, a constant. Hence a potential function for $\mathbf{F}$ is $f(x, y, z) = xe^z + y + K$.

15. curl $\mathbf{F} = \nabla \times \mathbf{F} = \begin{vmatrix} \mathbf{i} & \mathbf{j} & \mathbf{k} \\ \partial/\partial x & \partial/\partial y & \partial/\partial z \\ ye^{-x} & e^{-x} & 2z \end{vmatrix} = (0 - 0)\,\mathbf{i} - (0 - 0)\,\mathbf{j} + (-e^{-x} - e^{-x})\,\mathbf{k} = -2e^{-x}\,\mathbf{k} \neq \mathbf{0}$,

so $\mathbf{F}$ is not conservative.

16. $\operatorname{curl}\mathbf{F} = \nabla \times \mathbf{F} = \begin{vmatrix} \mathbf{i} & \mathbf{j} & \mathbf{k} \\ \partial/\partial x & \partial/\partial y & \partial/\partial z \\ y\cos xy & x\cos xy & -\sin z \end{vmatrix}$

$$= (0-0)\,\mathbf{i} - (0-0)\,\mathbf{j} + [(-xy\sin xy + \cos xy) - (-xy\sin xy + \cos xy)]\,\mathbf{k} = \mathbf{0}$$

$\mathbf{F}$ is defined on all of $\mathbb{R}^3$, and the partial derivatives of the component functions are continuous, so $\mathbf{F}$ is conservative. Thus there exists a function f such that $\nabla f = \mathbf{F}$. Then $f_x(x,y,z) = y\cos xy$ implies $f(x,y,z) = \sin xy + g(y,z) \Rightarrow$ $f_y(x,y,z) = x\cos xy + g_y(y,z)$. But $f_y(x,y,z) = x\cos xy$, so $g(y,z) = h(z)$ and $f(x,y,z) = \sin xy + h(z)$. Thus $f_z(x,y,z) = h'(z)$ but $f_z(x,y,z) = -\sin z$ so $h(z) = \cos z + K$ and a potential function for $\mathbf{F}$ is $f(x,y,z) = \sin xy + \cos z + K$.

17. No. Assume there is such a $\mathbf{G}$. Then $\operatorname{div}(\operatorname{curl}\mathbf{G}) = y^2 + z^2 + x^2 \neq 0$, which contradicts Theorem 11.

18. No. Assume there is such a $\mathbf{G}$. Then $\operatorname{div}(\operatorname{curl}\mathbf{G}) = xz \neq 0$ which contradicts Theorem 11.

19. $\operatorname{curl}\mathbf{F} = \begin{vmatrix} \mathbf{i} & \mathbf{j} & \mathbf{k} \\ \partial/\partial x & \partial/\partial y & \partial/\partial z \\ f(x) & g(y) & h(z) \end{vmatrix} = (0-0)\,\mathbf{i} + (0-0)\,\mathbf{j} + (0-0)\,\mathbf{k} = \mathbf{0}.$

Hence $\mathbf{F} = f(x)\,\mathbf{i} + g(y)\,\mathbf{j} + h(z)\,\mathbf{k}$ is irrotational.

20. $\operatorname{div}\mathbf{F} = \dfrac{\partial(f(y,z))}{\partial x} + \dfrac{\partial(g(x,z))}{\partial y} + \dfrac{\partial(h(x,y))}{\partial z} = 0$ so $\mathbf{F}$ is incompressible.

For Exercises 21–27, let $\mathbf{F}(x,y,z) = P_1\,\mathbf{i} + Q_1\,\mathbf{j} + R_1\,\mathbf{k}$ and $\mathbf{G}(x,y,z) = P_2\,\mathbf{i} + Q_2\,\mathbf{j} + R_2\,\mathbf{k}$.

21. $\operatorname{div}(\mathbf{F}+\mathbf{G}) = \dfrac{\partial(P_1+P_2)}{\partial x} + \dfrac{\partial(Q_1+Q_2)}{\partial y} + \dfrac{\partial(R_1+R_2)}{\partial z}$

$$= \left(\frac{\partial P_1}{\partial x} + \frac{\partial Q_1}{\partial y} + \frac{\partial R_1}{\partial z}\right) + \left(\frac{\partial P_2}{\partial x} + \frac{\partial Q_2}{\partial y} + \frac{\partial R_3}{\partial z}\right) = \operatorname{div}\mathbf{F} + \operatorname{div}\mathbf{G}$$

22. $\operatorname{curl}\mathbf{F} + \operatorname{curl}\mathbf{G} = \left[\left(\dfrac{\partial R_1}{\partial y} - \dfrac{\partial Q_1}{\partial z}\right)\mathbf{i} + \left(\dfrac{\partial P_1}{\partial z} - \dfrac{\partial R_1}{\partial x}\right)\mathbf{j} + \left(\dfrac{\partial Q_1}{\partial x} - \dfrac{\partial P_1}{\partial y}\right)\mathbf{k}\right]$

$$+ \left[\left(\frac{\partial R_2}{\partial y} - \frac{\partial Q_2}{\partial z}\right)\mathbf{i} + \left(\frac{\partial P_2}{\partial z} - \frac{\partial R_2}{\partial x}\right)\mathbf{j} + \left(\frac{\partial Q_2}{\partial x} - \frac{\partial P_2}{\partial y}\right)\mathbf{k}\right]$$

$$= \left[\frac{\partial(R_1+R_2)}{\partial y} - \frac{\partial(Q_1+Q_2)}{\partial z}\right]\mathbf{i} + \left[\frac{\partial(P_1+P_2)}{\partial z} - \frac{\partial(R_1+R_2)}{\partial x}\right]\mathbf{j}$$

$$+ \left[\frac{\partial(Q_1+Q_2)}{\partial x} - \frac{\partial(P_1+P_2)}{\partial y}\right]\mathbf{k} = \operatorname{curl}(\mathbf{F}+\mathbf{G})$$

23. $\operatorname{div}(f\mathbf{F}) = \dfrac{\partial(fP_1)}{\partial x} + \dfrac{\partial(fQ_1)}{\partial y} + \dfrac{\partial(fR_1)}{\partial z} = \left(f\dfrac{\partial P_1}{\partial x} + P_1\dfrac{\partial f}{\partial x}\right) + \left(f\dfrac{\partial Q_1}{\partial y} + Q_1\dfrac{\partial f}{\partial y}\right) + \left(f\dfrac{\partial R_1}{\partial z} + R_1\dfrac{\partial f}{\partial z}\right)$

$$= f\left(\frac{\partial P_1}{\partial x} + \frac{\partial Q_1}{\partial y} + \frac{\partial R_1}{\partial z}\right) + \langle P_1, Q_1, R_1\rangle \cdot \left\langle \frac{\partial f}{\partial x}, \frac{\partial f}{\partial y}, \frac{\partial f}{\partial z}\right\rangle = f\operatorname{div}\mathbf{F} + \mathbf{F}\cdot\nabla f$$

24. $\operatorname{curl}(f\mathbf{F}) = \left[\dfrac{\partial(fR_1)}{\partial y} - \dfrac{\partial(fQ_1)}{\partial z}\right]\mathbf{i} + \left[\dfrac{\partial(fP_1)}{\partial z} - \dfrac{\partial(fR_1)}{\partial x}\right]\mathbf{j} + \left[\dfrac{\partial(fQ_1)}{\partial x} - \dfrac{\partial(fP_1)}{\partial y}\right]\mathbf{k}$

$$= \left[f\,\frac{\partial R_1}{\partial y} + R_1\,\frac{\partial f}{\partial y} - f\,\frac{\partial Q_1}{\partial z} - Q_1\,\frac{\partial f}{\partial z}\right]\mathbf{i} + \left[f\,\frac{\partial P_1}{\partial z} + P_1\,\frac{\partial f}{\partial z} - f\,\frac{\partial R_1}{\partial x} - R_1\,\frac{\partial f}{\partial x}\right]\mathbf{j}$$

$$+ \left[f\,\frac{\partial Q_1}{\partial x} + Q_1\,\frac{\partial f}{\partial x} - f\,\frac{\partial P_1}{\partial y} - P_1\,\frac{\partial f}{\partial y}\right]\mathbf{k}$$

$$= f\left[\frac{\partial R_1}{\partial y} - \frac{\partial Q_1}{\partial z}\right]\mathbf{i} + f\left[\frac{\partial P_1}{\partial z} - \frac{\partial R_1}{\partial x}\right]\mathbf{j} + f\left[\frac{\partial Q_1}{\partial x} - \frac{\partial P_1}{\partial y}\right]\mathbf{k}$$

$$+ \left[R_1\frac{\partial f}{\partial y} - Q_1\frac{\partial f}{\partial z}\right]\mathbf{i} + \left[P_1\frac{\partial f}{\partial z} - R_1\frac{\partial f}{\partial x}\right]\mathbf{j} + \left[Q_1\frac{\partial f}{\partial x} - P_1\frac{\partial f}{\partial y}\right]\mathbf{k}$$

$$= f\operatorname{curl}\mathbf{F} + (\nabla f)\times\mathbf{F}$$

25. $\operatorname{div}(\mathbf{F}\times\mathbf{G}) = \nabla\cdot(\mathbf{F}\times\mathbf{G}) = \begin{vmatrix} \partial/\partial x & \partial/\partial y & \partial/\partial z \\ P_1 & Q_1 & R_1 \\ P_2 & Q_2 & R_2 \end{vmatrix} = \dfrac{\partial}{\partial x}\begin{vmatrix} Q_1 & R_1 \\ Q_2 & R_2 \end{vmatrix} - \dfrac{\partial}{\partial y}\begin{vmatrix} P_1 & R_1 \\ P_2 & R_2 \end{vmatrix} + \dfrac{\partial}{\partial z}\begin{vmatrix} P_1 & Q_1 \\ P_2 & Q_2 \end{vmatrix}$

$$= \left[Q_1\,\frac{\partial R_2}{\partial x} + R_2\,\frac{\partial Q_1}{\partial x} - Q_2\,\frac{\partial R_1}{\partial x} - R_1\,\frac{\partial Q_2}{\partial x}\right] - \left[P_1\,\frac{\partial R_2}{\partial y} + R_2\,\frac{\partial P_1}{\partial y} - P_2\,\frac{\partial R_1}{\partial y} - R_1\,\frac{\partial P_2}{\partial y}\right]$$

$$+ \left[P_1\,\frac{\partial Q_2}{\partial z} + Q_2\,\frac{\partial P_1}{\partial z} - P_2\,\frac{\partial Q_1}{\partial z} - Q_1\,\frac{\partial P_2}{\partial z}\right]$$

$$= \left[P_2\left(\frac{\partial R_1}{\partial y} - \frac{\partial Q_1}{\partial z}\right) + Q_2\left(\frac{\partial P_1}{\partial z} - \frac{\partial R_1}{\partial x}\right) + R_2\left(\frac{\partial Q_1}{\partial x} - \frac{\partial P_1}{\partial y}\right)\right]$$

$$- \left[P_1\left(\frac{\partial R_2}{\partial y} - \frac{\partial Q_2}{\partial z}\right) + Q_1\left(\frac{\partial P_2}{\partial z} - \frac{\partial R_2}{\partial x}\right) + R_1\left(\frac{\partial Q_2}{\partial x} - \frac{\partial P_2}{\partial y}\right)\right]$$

$$= \mathbf{G}\cdot\operatorname{curl}\mathbf{F} - \mathbf{F}\cdot\operatorname{curl}\mathbf{G}$$

26. $\operatorname{div}(\nabla f\times\nabla g) = \nabla g\cdot\operatorname{curl}(\nabla f) - \nabla f\cdot\operatorname{curl}(\nabla g)$ [by Exercise 25] $= 0$ (by Theorem 3)

27. $\operatorname{curl}(\operatorname{curl}\mathbf{F}) = \nabla\times(\nabla\times\mathbf{F}) = \begin{vmatrix} \mathbf{i} & \mathbf{j} & \mathbf{k} \\ \partial/\partial x & \partial/\partial y & \partial/\partial z \\ \partial R_1/\partial y - \partial Q_1/\partial z & \partial P_1/\partial z - \partial R_1/\partial x & \partial Q_1/\partial x - \partial P_1/\partial y \end{vmatrix}$

$$= \left(\frac{\partial^2 Q_1}{\partial y\partial x} - \frac{\partial^2 P_1}{\partial y^2} - \frac{\partial^2 P_1}{\partial z^2} + \frac{\partial^2 R_1}{\partial z\partial x}\right)\mathbf{i} + \left(\frac{\partial^2 R_1}{\partial z\partial y} - \frac{\partial^2 Q_1}{\partial z^2} - \frac{\partial^2 Q_1}{\partial x^2} + \frac{\partial^2 P_1}{\partial x\partial y}\right)\mathbf{j}$$

$$+ \left(\frac{\partial^2 P_1}{\partial x\partial z} - \frac{\partial^2 R_1}{\partial x^2} - \frac{\partial^2 R_1}{\partial y^2} + \frac{\partial^2 Q_1}{\partial y\partial z}\right)\mathbf{k}$$

Now let's consider $\operatorname{grad}(\operatorname{div}\mathbf{F}) - \nabla^2\mathbf{F}$ and compare with the above.

(Note that $\nabla^2\mathbf{F}$ is defined on page 764.)

$$\operatorname{grad}(\operatorname{div}\mathbf{F}) - \nabla^2\mathbf{F} = \left[\left(\frac{\partial^2 P_1}{\partial x^2} + \frac{\partial^2 Q_1}{\partial x\partial y} + \frac{\partial^2 R_1}{\partial x\partial z}\right)\mathbf{i} + \left(\frac{\partial^2 P_1}{\partial y\partial x} + \frac{\partial^2 Q_1}{\partial y^2} + \frac{\partial^2 R_1}{\partial y\partial z}\right)\mathbf{j} + \left(\frac{\partial^2 P_1}{\partial z\partial x} + \frac{\partial^2 Q_1}{\partial z\partial y} + \frac{\partial^2 R_1}{\partial z^2}\right)\mathbf{k}\right]$$

$$- \left[\left(\frac{\partial^2 P_1}{\partial x^2} + \frac{\partial^2 P_1}{\partial y^2} + \frac{\partial^2 P_1}{\partial z^2}\right)\mathbf{i} + \left(\frac{\partial^2 Q_1}{\partial x^2} + \frac{\partial^2 Q_1}{\partial y^2} + \frac{\partial^2 Q_1}{\partial z^2}\right)\mathbf{j} + \left(\frac{\partial^2 R_1}{\partial x^2} + \frac{\partial^2 R_1}{\partial y^2} + \frac{\partial^2 R_1}{\partial z^2}\right)\mathbf{k}\right]$$

$$= \left(\frac{\partial^2 Q_1}{\partial x\partial y} + \frac{\partial^2 R_1}{\partial x\partial z} - \frac{\partial^2 P_1}{\partial y^2} - \frac{\partial^2 P_1}{\partial z^2}\right)\mathbf{i} + \left(\frac{\partial^2 P_1}{\partial y\partial x} + \frac{\partial^2 R_1}{\partial y\partial z} - \frac{\partial^2 Q_1}{\partial x^2} - \frac{\partial^2 Q_1}{\partial z^2}\right)\mathbf{j} + \left(\frac{\partial^2 P_1}{\partial z\partial x} + \frac{\partial^2 Q_1}{\partial z\partial y} - \frac{\partial^2 R_1}{\partial x^2} - \frac{\partial^2 R_2}{\partial y^2}\right)\mathbf{k}$$

Then applying Clairaut's Theorem to reverse the order of differentiation in the second partial derivatives as needed and comparing, we have $\operatorname{curl}\operatorname{curl}\mathbf{F} = \operatorname{grad}\operatorname{div}\mathbf{F} - \nabla^2\mathbf{F}$ as desired.

28. (a) $\nabla\cdot\mathbf{r} = \left(\frac{\partial}{\partial x}\mathbf{i} + \frac{\partial}{\partial y}\mathbf{j} + \frac{\partial}{\partial z}\mathbf{k}\right)\cdot(x\,\mathbf{i} + y\,\mathbf{j} + z\,\mathbf{k}) = 1 + 1 + 1 = 3$

(b) $\nabla\cdot(r\mathbf{r}) = \nabla\cdot\sqrt{x^2+y^2+z^2}\,(x\,\mathbf{i} + y\,\mathbf{j} + z\,\mathbf{k})$

$$= \left(\frac{x^2}{\sqrt{x^2+y^2+z^2}} + \sqrt{x^2+y^2+z^2}\right) + \left(\frac{y^2}{\sqrt{x^2+y^2+z^2}} + \sqrt{x^2+y^2+z^2}\right) + \left(\frac{z^2}{\sqrt{x^2+y^2+z^2}} + \sqrt{x^2+y^2+z^2}\right)$$

$$= \frac{1}{\sqrt{x^2+y^2+z^2}}\left(4x^2+4y^2+4z^2\right) = 4\sqrt{x^2+y^2+z^2} = 4r$$

Another method:

By Exercise 23, $\nabla\cdot(r\mathbf{r}) = \operatorname{div}(r\mathbf{r}) = r\operatorname{div}\mathbf{r} + \mathbf{r}\cdot\nabla r = 3r + \mathbf{r}\cdot\frac{\mathbf{r}}{r}$ [see Exercise 29(a) below] $= 4r$.

(c) $\nabla^2 r^3 = \nabla^2\left(x^2+y^2+z^2\right)^{3/2}$

$$= \frac{\partial}{\partial x}\left[\tfrac{3}{2}(x^2+y^2+z^2)^{1/2}(2x)\right] + \frac{\partial}{\partial y}\left[\tfrac{3}{2}(x^2+y^2+z^2)^{1/2}(2y)\right] + \frac{\partial}{\partial z}\left[\tfrac{3}{2}(x^2+y^2+z^2)^{1/2}(2z)\right]$$

$$= 3\left[\tfrac{1}{2}(x^2+y^2+z^2)^{-1/2}(2x)(x) + (x^2+y^2+z^2)^{1/2}\right] + 3\left[\tfrac{1}{2}(x^2+y^2+z^2)^{-1/2}(2y)(y) + (x^2+y^2+z^2)^{1/2}\right] + 3\left[\tfrac{1}{2}(x^2+y^2+z^2)^{-1/2}(2z)(z) + (x^2+y^2+z^2)^{1/2}\right]$$

$$= 3(x^2+y^2+z^2)^{-1/2}(4x^2+4y^2+4z^2) = 12(x^2+y^2+z^2)^{1/2} = 12r$$

Another method: $\frac{\partial}{\partial x}(x^2+y^2+z^2)^{3/2} = 3x\sqrt{x^2+y^2+z^2} \;\Rightarrow\; \nabla r^3 = 3r(x\,\mathbf{i} + y\,\mathbf{j} + z\,\mathbf{k}) = 3r\,\mathbf{r}$,

so $\nabla^2 r^3 = \nabla\cdot\nabla r^3 = \nabla\cdot(3r\,\mathbf{r}) = 3(4r) = 12r$ by part (b).

29. (a) $\nabla r = \nabla\sqrt{x^2+y^2+z^2} = \dfrac{x}{\sqrt{x^2+y^2+z^2}}\,\mathbf{i} + \dfrac{y}{\sqrt{x^2+y^2+z^2}}\,\mathbf{j} + \dfrac{z}{\sqrt{x^2+y^2+z^2}}\,\mathbf{k} = \dfrac{x\,\mathbf{i}+y\,\mathbf{j}+z\,\mathbf{k}}{\sqrt{x^2+y^2+z^2}} = \dfrac{\mathbf{r}}{r}$

(b) $$\nabla\times\mathbf{r} = \begin{vmatrix} \mathbf{i} & \mathbf{j} & \mathbf{k} \\ \dfrac{\partial}{\partial x} & \dfrac{\partial}{\partial y} & \dfrac{\partial}{\partial z} \\ x & y & z \end{vmatrix} = \left[\frac{\partial}{\partial y}(z) - \frac{\partial}{\partial z}(y)\right]\mathbf{i} + \left[\frac{\partial}{\partial z}(x) - \frac{\partial}{\partial x}(z)\right]\mathbf{j} + \left[\frac{\partial}{\partial x}(y) - \frac{\partial}{\partial y}(x)\right]\mathbf{k} = \mathbf{0}$$

(c) $$\nabla\left(\frac{1}{r}\right) = \nabla\left(\frac{1}{\sqrt{x^2+y^2+z^2}}\right)$$

$$= \frac{-\dfrac{1}{2\sqrt{x^2+y^2+z^2}}(2x)}{x^2+y^2+z^2}\,\mathbf{i} - \frac{\dfrac{1}{2\sqrt{x^2+y^2+z^2}}(2y)}{x^2+y^2+z^2}\,\mathbf{j} - \frac{\dfrac{1}{2\sqrt{x^2+y^2+z^2}}(2z)}{x^2+y^2+z^2}\,\mathbf{k}$$

$$= -\frac{x\,\mathbf{i}+y\,\mathbf{j}+z\,\mathbf{k}}{(x^2+y^2+z^2)^{3/2}} = -\frac{\mathbf{r}}{r^3}$$

(d) $$\nabla\ln r = \nabla\ln\left(x^2+y^2+z^2\right)^{1/2} = \tfrac{1}{2}\nabla\ln\left(x^2+y^2+z^2\right)$$

$$= \frac{x}{x^2+y^2+z^2}\,\mathbf{i} + \frac{y}{x^2+y^2+z^2}\,\mathbf{j} + \frac{z}{x^2+y^2+z^2}\,\mathbf{k} = \frac{x\,\mathbf{i}+y\,\mathbf{j}+z\,\mathbf{k}}{x^2+y^2+z^2} = \frac{\mathbf{r}}{r^2}$$

30. $\mathbf{r} = x\,\mathbf{i}+y\,\mathbf{j}+z\,\mathbf{k} \;\Rightarrow\; r = |\mathbf{r}| = \sqrt{x^2+y^2+z^2}$, so

$$\mathbf{F} = \frac{\mathbf{r}}{r^p} = \frac{x}{(x^2+y^2+z^2)^{p/2}}\,\mathbf{i} + \frac{y}{(x^2+y^2+z^2)^{p/2}}\,\mathbf{j} + \frac{z}{(x^2+y^2+z^2)^{p/2}}\,\mathbf{k}$$

Then $\dfrac{\partial}{\partial x}\dfrac{x}{(x^2+y^2+z^2)^{p/2}} = \dfrac{(x^2+y^2+z^2)-px^2}{(x^2+y^2+z^2)^{1+p/2}} = \dfrac{r^2-px^2}{r^{p+2}}$. Similarly,

$\dfrac{\partial}{\partial y}\dfrac{y}{(x^2+y^2+z^2)^{p/2}} = \dfrac{r^2-py^2}{r^{p+2}}$ and $\dfrac{\partial}{\partial z}\dfrac{z}{(x^2+y^2+z^2)^{p/2}} = \dfrac{r^2-pz^2}{r^{p+2}}$. Thus

$$\operatorname{div}\mathbf{F} = \nabla\cdot\mathbf{F} = \frac{r^2-px^2}{r^{p+2}} + \frac{r^2-py^2}{r^{p+2}} + \frac{r^2-pz^2}{r^{p+2}} = \frac{3r^2-px^2-py^2-pz^2}{r^{p+2}}$$

$$= \frac{3r^2-p(x^2+y^2+z^2)}{r^{p+2}} = \frac{3r^2-pr^2}{r^{p+2}} = \frac{3-p}{r^p}$$

Consequently, if $p=3$ we have $\operatorname{div}\mathbf{F}=0$.

31. By (13), $\oint_C f(\nabla g)\cdot\mathbf{n}\,ds = \iint_D \operatorname{div}(f\nabla g)\,dA = \iint_D[f\operatorname{div}(\nabla g) + \nabla g\cdot\nabla f]\,dA$ by Exercise 23. But $\operatorname{div}(\nabla g) = \nabla^2 g$. Hence $\iint_D f\nabla^2 g\,dA = \oint_C f(\nabla g)\cdot\mathbf{n}\,ds - \iint_D \nabla g\cdot\nabla f\,dA$.

32. By Exercise 33, $\iint_D f\nabla^2 g\,dA = \oint_C f(\nabla g)\cdot\mathbf{n}\,ds - \iint_D \nabla g\cdot\nabla f\,dA$ and

$\iint_D g\nabla^2 f\,dA = \oint_C g(\nabla f)\cdot\mathbf{n}\,ds - \iint_D \nabla f\cdot\nabla g\,dA$. Hence

$\iint_D (f\nabla^2 g - g\nabla^2 f)\,dA = \oint_C [f(\nabla g)\cdot\mathbf{n} - g(\nabla f)\cdot\mathbf{n}]\,ds + \iint_D (\nabla f\cdot\nabla g - \nabla g\cdot\nabla f)\,dA = \oint_C [f\nabla g - g\nabla f]\cdot\mathbf{n}\,ds.$

33. Let $f(x,y) = 1$. Then $\nabla f = \mathbf{0}$ and Green's first identity (see Exercise 31) says

$\iint_D \nabla^2 g\, dA = \oint_C (\nabla g) \cdot \mathbf{n}\, ds - \iint_D \mathbf{0} \cdot \nabla g\, dA \quad \Rightarrow \quad \iint_D \nabla^2 g\, dA = \oint_C \nabla g \cdot \mathbf{n}\, ds$. But g is harmonic on D, so $\nabla^2 g = 0 \quad \Rightarrow \quad \oint_C \nabla g \cdot \mathbf{n}\, ds = 0$ and $\oint_C D_{\mathbf{n}} g\, ds = \oint_C (\nabla g \cdot \mathbf{n})\, ds = 0$.

34. Let $g = f$. Then Green's first identity (see Exercise 31) says $\iint_D f\, \nabla^2 f\, dA = \oint_C (f)(\nabla f) \cdot \mathbf{n}\, ds - \iint_D \nabla f \cdot \nabla f\, dA$. But f is harmonic, so $\nabla^2 f = 0$, and $\nabla f \cdot \nabla f = |\nabla f|^2$, so we have $0 = \oint_C (f)(\nabla f) \cdot \mathbf{n}\, ds - \iint_D |\nabla f|^2\, dA \quad \Rightarrow$ $\iint_D |\nabla f|^2\, dA = \oint_C (f)(\nabla f) \cdot \mathbf{n}\, ds = 0$ since $f(x,y) = 0$ on C.

35. (a) We know that $\omega = v/d$, and from the diagram $\sin\theta = d/r \quad \Rightarrow \quad v = d\omega = (\sin\theta) r\omega = |\mathbf{w} \times \mathbf{r}|$. But $\mathbf{v}$ is perpendicular to both $\mathbf{w}$ and $\mathbf{r}$, so that $\mathbf{v} = \mathbf{w} \times \mathbf{r}$.

(b) From (a), $\mathbf{v} = \mathbf{w} \times \mathbf{r} = \begin{vmatrix} \mathbf{i} & \mathbf{j} & \mathbf{k} \\ 0 & 0 & \omega \\ x & y & z \end{vmatrix} = (0 \cdot z - \omega y)\,\mathbf{i} + (\omega x - 0 \cdot z)\,\mathbf{j} + (0 \cdot y - x \cdot 0)\,\mathbf{k} = -\omega y\,\mathbf{i} + \omega x\,\mathbf{j}$

(c) $\operatorname{curl} \mathbf{v} = \nabla \times \mathbf{v} = \begin{vmatrix} \mathbf{i} & \mathbf{j} & \mathbf{k} \\ \partial/\partial x & \partial/\partial y & \partial/\partial z \\ -\omega y & \omega x & 0 \end{vmatrix}$

$$= \left[\frac{\partial}{\partial y}(0) - \frac{\partial}{\partial z}(\omega x)\right]\mathbf{i} + \left[\frac{\partial}{\partial z}(-\omega y) - \frac{\partial}{\partial x}(0)\right]\mathbf{j} + \left[\frac{\partial}{\partial x}(\omega x) - \frac{\partial}{\partial y}(-\omega y)\right]\mathbf{k}$$

$$= [\omega - (-\omega)]\,\mathbf{k} = 2\omega\,\mathbf{k} = 2\mathbf{w}$$

36. Let $\mathbf{H} = \langle h_1, h_2, h_3 \rangle$ and $\mathbf{E} = \langle E_1, E_2, E_3 \rangle$.

(a) $\nabla \times (\nabla \times \mathbf{E}) = \nabla \times (\operatorname{curl} \mathbf{E}) = \nabla \times \left(-\frac{1}{c}\frac{\partial \mathbf{H}}{\partial t}\right) = -\frac{1}{c}\begin{vmatrix} \mathbf{i} & \mathbf{j} & \mathbf{k} \\ \partial/\partial x & \partial/\partial y & \partial/\partial z \\ \partial h_1/\partial t & \partial h_2/\partial t & \partial h_3/\partial t \end{vmatrix}$

$$= -\frac{1}{c}\left[\left(\frac{\partial^2 h_3}{\partial y\, \partial t} - \frac{\partial^2 h_2}{\partial z\, \partial t}\right)\mathbf{i} + \left(\frac{\partial^2 h_1}{\partial z\, \partial t} - \frac{\partial^2 h_3}{\partial x\, \partial t}\right)\mathbf{j} + \left(\frac{\partial^2 h_2}{\partial x\, \partial t} - \frac{\partial^2 h_1}{\partial y\, \partial t}\right)\mathbf{k}\right]$$

$$= -\frac{1}{c}\frac{\partial}{\partial t}\left[\left(\frac{\partial h_3}{\partial y} - \frac{\partial h_2}{\partial z}\right)\mathbf{i} + \left(\frac{\partial h_1}{\partial z} - \frac{\partial h_3}{\partial x}\right)\mathbf{j} + \left(\frac{\partial h_2}{\partial x} - \frac{\partial h_1}{\partial y}\right)\mathbf{k}\right]$$

[assuming that the partial derivatives are continuous so that the order of differentiation does not matter]

$$= -\frac{1}{c}\frac{\partial}{\partial t}\operatorname{curl}\mathbf{H} = -\frac{1}{c}\frac{\partial}{\partial t}\left(\frac{1}{c}\frac{\partial \mathbf{E}}{\partial t}\right) = -\frac{1}{c^2}\frac{\partial^2 \mathbf{E}}{\partial t^2}$$

(b) $\nabla\times(\nabla\times\mathbf{H})=\nabla\times(\operatorname{curl}\mathbf{H})=\nabla\times\left(\frac{1}{c}\frac{\partial\mathbf{E}}{\partial t}\right)=\frac{1}{c}\begin{vmatrix}\mathbf{i} & \mathbf{j} & \mathbf{k}\\ \partial/\partial x & \partial/\partial y & \partial/\partial z\\ \partial E_1/\partial t & \partial E_2/\partial t & \partial E_3/\partial t\end{vmatrix}$

$$=\frac{1}{c}\left[\left(\frac{\partial^2 E_3}{\partial y\,\partial t}-\frac{\partial^2 E_2}{\partial z\,\partial t}\right)\mathbf{i}+\left(\frac{\partial^2 E_1}{\partial z\,\partial t}-\frac{\partial^2 E_3}{\partial x\,\partial t}\right)\mathbf{j}+\left(\frac{\partial^2 E_2}{\partial x\,\partial t}-\frac{\partial^2 E_1}{\partial y\,\partial t}\right)\mathbf{k}\right]$$

$$=\frac{1}{c}\frac{\partial}{\partial t}\left[\left(\frac{\partial E_3}{\partial y}-\frac{\partial E_2}{\partial z}\right)\mathbf{i}+\left(\frac{\partial E_1}{\partial z}-\frac{\partial E_3}{\partial x}\right)\mathbf{j}+\left(\frac{\partial E_2}{\partial x}-\frac{\partial E_1}{\partial y}\right)\mathbf{k}\right]$$

[assuming that the partial derivatives are continuous so that the order of differentiation does not matter]

$$=\frac{1}{c}\frac{\partial}{\partial t}\operatorname{curl}\mathbf{E}=\frac{1}{c}\frac{\partial}{\partial t}\left(-\frac{1}{c}\frac{\partial\mathbf{H}}{\partial t}\right)=-\frac{1}{c^2}\frac{\partial^2\mathbf{H}}{\partial t^2}$$

(c) Using Exercise 27, we have that $\operatorname{curl}\operatorname{curl}\mathbf{E}=\operatorname{grad}\operatorname{div}\mathbf{E}-\nabla^2\mathbf{E}\;\Rightarrow$

$$\nabla^2\mathbf{E}=\operatorname{grad}\operatorname{div}\mathbf{E}-\operatorname{curl}\operatorname{curl}\mathbf{E}=\operatorname{grad}0+\frac{1}{c^2}\frac{\partial^2\mathbf{E}}{\partial t^2}\;\;\text{[from part (a)]}\;\;=\frac{1}{c^2}\frac{\partial^2\mathbf{E}}{\partial t^2}.$$

(d) As in part (c), $\nabla^2\mathbf{H}=\operatorname{grad}\operatorname{div}\mathbf{H}-\operatorname{curl}\operatorname{curl}\mathbf{H}=\operatorname{grad}0+\frac{1}{c^2}\frac{\partial^2\mathbf{H}}{\partial t^2}$ [using part (b)] $=\frac{1}{c^2}\frac{\partial^2\mathbf{H}}{\partial t^2}$.

13.6 Parametric Surfaces and Their Areas

1. $\mathbf{r}(u,v)=(u+v)\,\mathbf{i}+(3-v)\,\mathbf{j}+(1+4u+5v)\,\mathbf{k}=\langle 0,3,1\rangle+u\,\langle 1,0,4\rangle+v\,\langle 1,-1,5\rangle$. From Example 3, we recognize this as a vector equation of a plane through the point $(0,3,1)$ and containing vectors $\mathbf{a}=\langle 1,0,4\rangle$ and $\mathbf{b}=\langle 1,-1,5\rangle$. If we wish to find a more conventional equation for the plane, a normal vector to the plane is $\mathbf{a}\times\mathbf{b}=\begin{vmatrix}\mathbf{i} & \mathbf{j} & \mathbf{k}\\ 1 & 0 & 4\\ 1 & -1 & 5\end{vmatrix}=4\,\mathbf{i}-\mathbf{j}-\mathbf{k}$ and an equation of the plane is $4(x-0)-(y-3)-(z-1)=0$ or $4x-y-z=-4$.

2. $\mathbf{r}(u,v)=2\sin u\,\mathbf{i}+3\cos u\,\mathbf{j}+v\,\mathbf{k}$, so the corresponding parametric equations for the surface are $x=2\sin u$, $y=3\cos u$, $z=v$. For any point (x,y,z) on the surface, we have $(x/2)^2+(y/3)^2=\sin^2 u+\cos^2 u=1$, so cross-sections parallel to the yz-plane are all ellipses. Since $z=v$ with $0\leq v\leq 2$, the surface is the portion of the elliptical cylinder $x^2/4+y^2/9=1$ for $0\leq z\leq 2$.

3. $\mathbf{r}(s,t)=\langle s,t,t^2-s^2\rangle$, so the corresponding parametric equations for the surface are $x=s$, $y=t$, $z=t^2-s^2$. For any point (x,y,z) on the surface, we have $z=y^2-x^2$. With no restrictions on the parameters, the surface is $z=y^2-x^2$, which we recognize as a hyperbolic paraboloid.

4. $\mathbf{r}(s,t)=s\sin 2t\,\mathbf{i}+s^2\,\mathbf{j}+s\cos 2t\,\mathbf{k}$, so the corresponding parametric equations for the surface are $x=s\sin 2t$, $y=s^2$, $z=s\cos 2t$. For any point (x,y,z) on the surface, we have $x^2+z^2=s^2\sin^2 2t+s^2\cos^2 2t=s^2=y$. Since no restrictions are placed on the parameters, the surface is $y=x^2+z^2$, which we recognize as a circular paraboloid whose axis is the y-axis.

5. $\mathbf{r}(u,v) = \langle u^2+1, v^3+1, u+v \rangle$, $-1 \le u \le 1$, $-1 \le v \le 1$.

The surface has parametric equations $x = u^2+1$, $y = v^3+1$, $z = u+v$, $-1 \le u \le 1$, $-1 \le v \le 1$. If we keep u constant at u_0, $x = u_0^2+1$, a constant, so the corresponding grid curves must be the curves parallel to the yz-plane. If v is constant, we have $y = v_0^3+1$, a constant, so these grid curves are the curves parallel to the xz-plane.

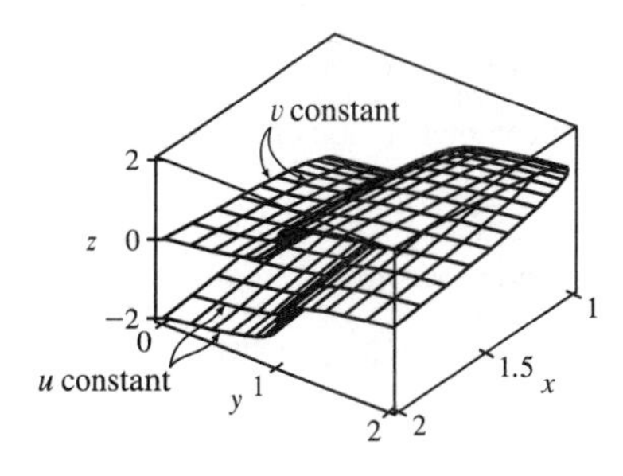

6. $\mathbf{r}(u,v) = \langle u+v, u^2, v^2 \rangle$, $-1 \le u \le 1$, $-1 \le v \le 1$.

The surface has parametric equations $x = u+v$, $y = u^2$, $z = v^2$, $-1 \le u \le 1$, $-1 \le v \le 1$. If $u = u_0$ is constant, $y = u_0^2 =$ constant, so the corresponding grid curves are the curves parallel to the xz-plane. If $v = v_0$ is constant, $z = v_0^2 =$ constant, so the corresponding grid curves are the curves parallel to the xy-plane.

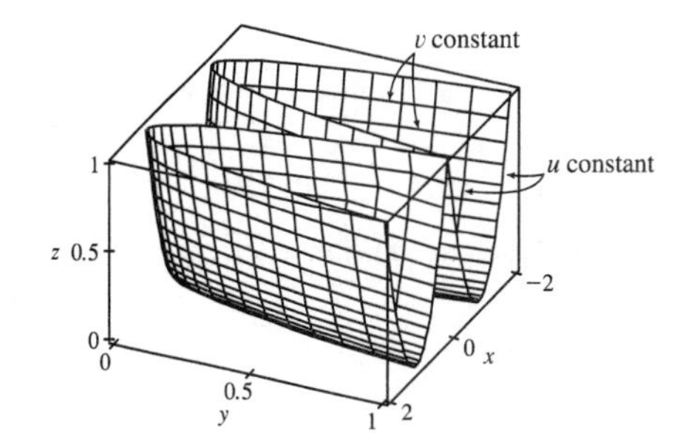

7. $\mathbf{r}(u,v) = \langle \cos^3 u \cos^3 v, \sin^3 u \cos^3 v, \sin^3 v \rangle$.

The surface has parametric equations $x = \cos^3 u \cos^3 v$, $y = \sin^3 u \cos^3 v$, $z = \sin^3 v$, $0 \le u \le \pi$, $0 \le v \le 2\pi$. Note that if $v = v_0$ is constant then $z = \sin^3 v_0$ is constant, so the corresponding grid curves must be the curves parallel to the xy-plane. The vertically oriented grid curves, then, correspond to $u = u_0$ being held constant, giving $x = \cos^3 u_0 \cos^3 v$, $y = \sin^3 u_0 \cos^3 v$, $z = \sin^3 v$. These curves lie in vertical planes that contain the z-axis.

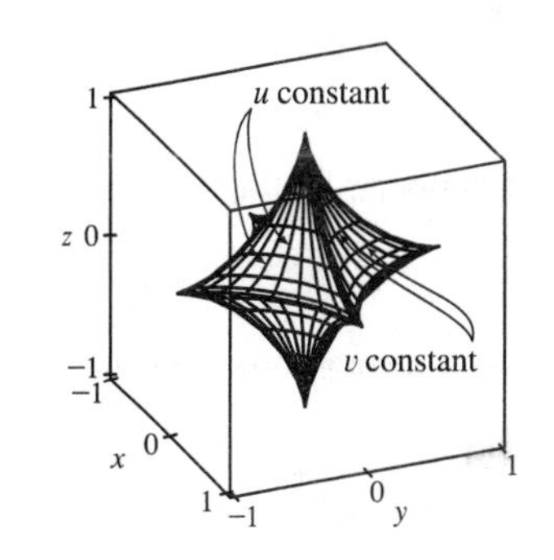

8. $\mathbf{r}(u,v) = \langle \cos u \sin v, \sin u \sin v, \cos v + \ln \tan(v/2) \rangle$.

The surface has parametric equations $x = \cos u \sin v$, $y = \sin u \sin v$, $z = \cos v + \ln \tan(v/2)$, $0 \le u \le 2\pi$, $0.1 \le v \le 6.2$. Note that if $v = v_0$ is constant, the parametric equations become $x = \cos u \sin v_0$, $y = \sin u \sin v_0$, $z = \cos v_0 + \ln \tan(v_0/2)$ which represent a circle of radius $\sin v_0$ in the plane $z = \cos v_0 + \ln \tan(v_0/2)$. So the circular grid curves we see lying horizontally are the grid curves with v constant. The vertically oriented grid curves correspond to $u = u_0$ being held constant, giving $x = \cos u_0 \sin v$, $y = \sin u_0 \sin v$, $z = \cos v + \ln \tan(v/2)$. These curves lie in vertical planes that contain the z-axis.

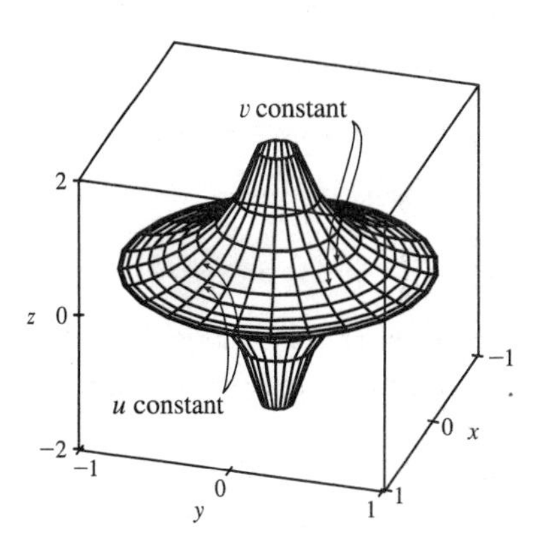

9. $x = \cos u \sin 2v$, $y = \sin u \sin 2v$, $z = \sin v$.

The complete graph of the surface is given by the parametric domain $0 \le u \le \pi, 0 \le v \le 2\pi$. Note that if $v = v_0$ is constant, the parametric equations become $x = \cos u \sin 2v_0$, $y = \sin u \sin 2v_0$, $z = \sin v_0$ which represent a circle of radius $\sin 2v_0$ in the plane $z = \sin v_0$. So the circular grid curves we see lying horizontally are the grid curves which have v constant. The vertical grid curves, then, correspond to $u = u_0$ being held constant, giving $x = \cos u_0 \sin 2v$ and $y = \sin u_0 \sin 2v$ with $z = \sin v$ which has a "figure-eight" shape.

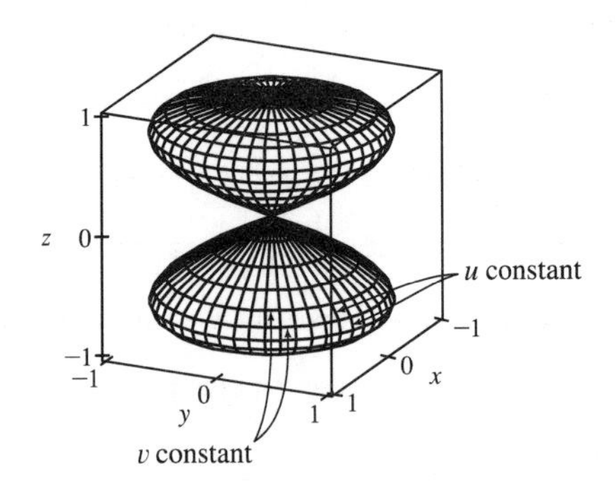

10. $x = u \sin u \cos v$, $y = u \cos u \cos v$, $z = u \sin v$.

We graph the portion of the surface with parametric domain $0 \le u \le 4\pi, 0 \le v \le 2\pi$. Note that if $v = v_0$ is constant, the parametric equations become $x = u \sin u \cos v_0$, $y = u \cos u \cos v_0$, $z = u \sin v_0$. The equations for x and y show that the projections onto the xy-plane give a spiral shape, so the corresponding grid curves are the almost-horizontal spiral curves we see. The vertical grid curves, which look approximately circular, correspond to $u = u_0$ being held constant, giving $x = u_0 \sin u_0 \cos v$, $y = u_0 \cos u_0 \cos v$, $z = u_0 \sin v$.

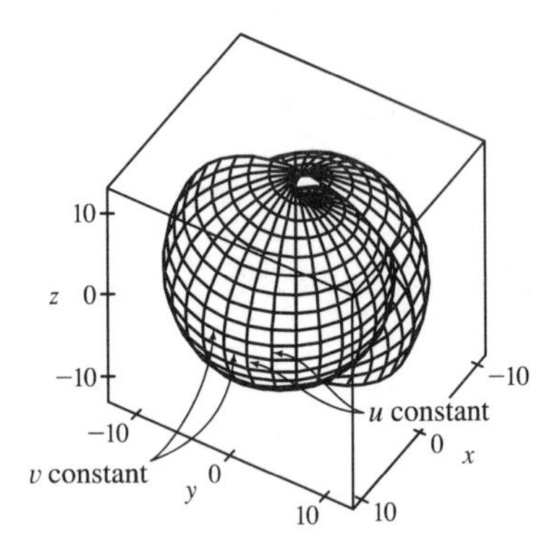

11. $\mathbf{r}(u, v) = u \cos v\, \mathbf{i} + u \sin v\, \mathbf{j} + v\, \mathbf{k}$. The parametric equations for the surface are $x = u \cos v$, $y = u \sin v$, $z = v$. We look at the grid curves first; if we fix v, then x and y parametrize a straight line in the plane $z = v$ which intersects the z-axis. If u is held constant, the projection onto the xy-plane is circular; with $z = v$, each grid curve is a helix. The surface is a spiraling ramp, graph I.

12. $x = u^3$, $y = u \sin v$, $z = u \cos v$. Then $y^2 + z^2 = u^2 \sin v^2 + u^2 \cos v^2 = u^2$, so if u is held constant, each grid curve is a circle of radius u in the plane $x = u^3$. The graph then must be graph III. If v is held constant, so $v = v_0$, we have $y = u \sin v_0$ and $z = u \cos v_0$. Then $y = (\tan v_0) z$, so the grid curves we see running lengthwise along the surface in the planes $y = kz$ correspond to keeping v constant.

13. $x = (u - \sin u) \cos v$, $y = (1 - \cos u) \sin v$, $z = u$. If u is held constant, x and y give an equation of an ellipse in the plane $z = u$, thus the grid curves are horizontally oriented ellipses. Note that when $u = 0$, the "ellipse" is the single point $(0, 0, 0)$, and when $u = \pi$, we have $y = 0$ while x ranges from $-\pi$ to π, a line segment parallel to the x-axis in the plane $z = \pi$. This is the upper "seam" we see in graph II. When v is held constant, $z = u$ is free to vary, so the corresponding grid curves are the curves we see running up and down along the surface.

14. $x = (1-u)(3+\cos v)\cos 4\pi u$, $y = (1-u)(3+\cos v)\sin 4\pi u$, $z = 3u + (1-u)\sin v$. These equations correspond to graph VI: when $u = 0$, then $x = 3 + \cos v$, $y = 0$, and $z = \sin v$, which are equations of a circle with radius 1 in the xz-plane centered at $(3, 0, 0)$. When $u = \frac{1}{2}$, then $x = \frac{3}{2} + \frac{1}{2}\cos v$, $y = 0$, and $z = \frac{3}{2} + \frac{1}{2}\sin v$, which are equations of a circle with radius $\frac{1}{2}$ in the xz-plane centered at $\left(\frac{3}{2}, 0, \frac{3}{2}\right)$. When $u = 1$, then $x = y = 0$ and $z = 3$, giving the topmost point shown in the graph. This suggests that the grid curves with u constant are the vertically oriented circles visible on the surface. The spiralling grid curves correspond to keeping v constant.

15. From Example 3, parametric equations for the plane through the point $(1, 2, -3)$ that contains the vectors $\mathbf{a} = \langle 1, 1, -1\rangle$ and $\mathbf{b} = \langle 1, -1, 1\rangle$ are $x = 1 + u(1) + v(1) = 1 + u + v$, $y = 2 + u(1) + v(-1) = 2 + u - v$, $z = -3 + u(-1) + v(1) = -3 - u + v$.

16. Solving the equation for z gives $z^2 = 1 - 2x^2 - 4y^2 \;\Rightarrow\; z = -\sqrt{1 - 2x^2 - 4y^2}$ (since we want the lower half of the ellipsoid). If we let x and y be the parameters, parametric equations are $x = x$, $y = y$, $z = -\sqrt{1 - 2x^2 - 4y^2}$.

Alternate solution: The equation can be rewritten as $\dfrac{x^2}{\left(1/\sqrt{2}\right)^2} + \dfrac{y^2}{(1/2)^2} + z^2 = 1$, and if we let $x = \dfrac{1}{\sqrt{2}}u\cos v$ and $y = \frac{1}{2}u\sin v$, then $z = -\sqrt{1 - 2x^2 - 4y^2} = -\sqrt{1 - u^2\cos^2 v - u^2\sin^2 v} = -\sqrt{1 - u^2}$, where $0 \le u \le 1$ and $0 \le v \le 2\pi$.

17. Solving the equation for y gives $y^2 = 1 - x^2 + z^2 \;\Rightarrow\; y = \sqrt{1 - x^2 + z^2}$. (We choose the positive root since we want the part of the hyperboloid that corresponds to $y \ge 0$.) If we let x and z be the parameters, parametric equations are $x = x$, $z = z$, $y = \sqrt{1 - x^2 + z^2}$.

18. $x = 4 - y^2 - 2z^2$, $y = y$, $z = z$ where $y^2 + 2z^2 \le 4$ since $x \ge 0$. Then the associated vector equation is $\mathbf{r}(y, z) = (4 - y^2 - 2z^2)\,\mathbf{i} + y\,\mathbf{j} + z\,\mathbf{k}$.

19. Since the cone intersects the sphere in the circle $x^2 + y^2 = 2$, $z = \sqrt{2}$ and we want the portion of the sphere above this, we can parametrize the surface as $x = x$, $y = y$, $z = \sqrt{4 - x^2 - y^2}$ where $x^2 + y^2 \le 2$.

Alternate solution: Using spherical coordinates, $x = 2\sin\phi\cos\theta$, $y = 2\sin\phi\sin\theta$, $z = 2\cos\phi$ where $0 \le \phi \le \frac{\pi}{4}$ and $0 \le \theta \le 2\pi$.

20. In spherical coordinates, parametric equations are $x = 4\sin\phi\cos\theta$, $y = 4\sin\phi\sin\theta$, $z = 4\cos\phi$. The intersection of the sphere with the plane $z = 2$ corresponds to $z = 4\cos\phi = 2 \;\Rightarrow\; \cos\phi = \frac{1}{2} \;\Rightarrow\; \phi = \frac{\pi}{3}$. By symmetry, the intersection of the sphere with the plane $z = -2$ corresponds to $\phi = \pi - \frac{\pi}{3} = \frac{2\pi}{3}$. Thus the surface is described by $0 \le \theta \le 2\pi$, $\frac{\pi}{3} \le \phi \le \frac{2\pi}{3}$.

21. Parametric equations are $x = x$, $y = 4\cos\theta$, $z = 4\sin\theta$, $0 \le x \le 5$, $0 \le \theta \le 2\pi$.

22. Using x and y as the parameters, $x = x$, $y = y$, $z = x + 3$ where $0 \le x^2 + y^2 \le 1$. Also, since the plane intersects the cylinder in an ellipse, the surface is a planar ellipse in the plane $z = x + 3$. Thus, parametrizing with respect to s and θ, we have $x = s\cos\theta$, $y = s\sin\theta$, $z = 3 + s\cos\theta$ where $0 \le s \le 1$ and $0 \le \theta \le 2\pi$.

23. The surface appears to be a portion of a circular cylinder of radius 3 with axis the x-axis. An equation of the cylinder is $y^2 + z^2 = 9$, and we can impose the restrictions $0 \le x \le 5$, $y \le 0$ to obtain the portion shown.

To graph the surface on a CAS, we can use parametric equations $x = u$, $y = 3\cos v$, $z = 3\sin v$ with the parameter domain $0 \le u \le 5$, $\frac{\pi}{2} \le v \le \frac{3\pi}{2}$. Alternatively, we can regard x and z as parameters. Then parametric equations are $x = x$, $z = z$, $y = -\sqrt{9 - z^2}$, where $0 \le x \le 5$ and $-3 \le z \le 3$.

24. The surface appears to be a portion of a sphere of radius 1 centered at the origin. In spherical coordinates, the sphere has equation $\rho = 1$, and imposing the restrictions $\frac{\pi}{2} \le \theta \le 2\pi$, $\frac{\pi}{4} \le \phi \le \pi$ will give only the portion of the sphere shown. Thus, to graph the surface on a CAS we can either use spherical coordinates with the stated restrictions, or we can use parametric equations: $x = \sin\phi\cos\theta$, $y = \sin\phi\sin\theta$, $z = \cos\phi$, $\frac{\pi}{2} \le \theta \le 2\pi$, $\frac{\pi}{4} \le \phi \le \pi$.

25. Using Equations 3, we have the parametrization $x = x$, $y = e^{-x}\cos\theta$, $z = e^{-x}\sin\theta$, $0 \le x \le 3$, $0 \le \theta \le 2\pi$.

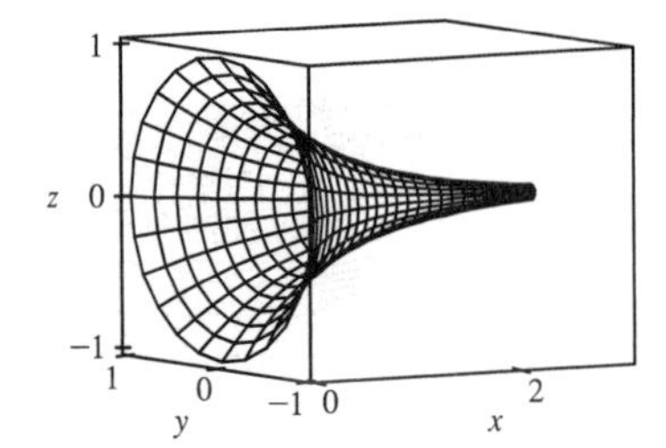

26. Letting θ be the angle of rotation about the y-axis, we have the parametrization $x = (4y^2 - y^4)\cos\theta$, $y = y$, $z = (4y^2 - y^4)\sin\theta$, $-2 \le y \le 2$, $0 \le \theta \le 2\pi$.

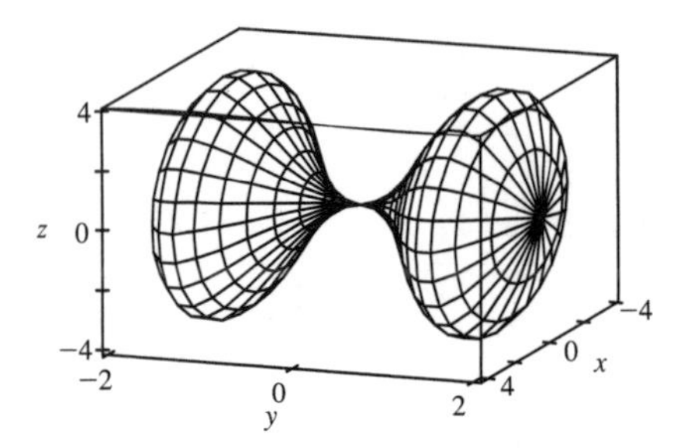

27. (a) Replacing $\cos u$ by $\sin u$ and $\sin u$ by $\cos u$ gives parametric equations $x = (2 + \sin v)\sin u$, $y = (2 + \sin v)\cos u$, $z = u + \cos v$. From the graph, it appears that the direction of the spiral is reversed. We can verify this observation by noting that the projection of the spiral grid curves onto the xy-plane, given by $x = (2 + \sin v)\sin u$, $y = (2 + \sin v)\cos u$, $z = 0$, draws a circle in the clockwise direction for each value of v. The original equations, on the other hand, give circular projections drawn in the counterclockwise direction. The equation for z is identical in both surfaces, so as z increases, these grid curves spiral up in opposite directions for the two surfaces.

(b) Replacing $\cos u$ by $\cos 2u$ and $\sin u$ by $\sin 2u$ gives parametric equations $x = (2 + \sin v) \cos 2u$, $y = (2 + \sin v) \sin 2u$, $z = u + \cos v$. From the graph, it appears that the number of coils in the surface doubles within the same parametric domain. We can verify this observation by noting that the projection of the spiral grid curves onto the xy-plane, given by $x = (2 + \sin v) \cos 2u$, $y = (2 + \sin v) \sin 2u$, $z = 0$ (where v is constant), complete circular revolutions for $0 \le u \le \pi$ while the original surface requires $0 \le u \le 2\pi$ for a complete revolution. Thus, the new surface winds around twice as fast as the original surface, and since the equation for z is identical in both surfaces, we observe twice as many circular coils in the same z-interval.

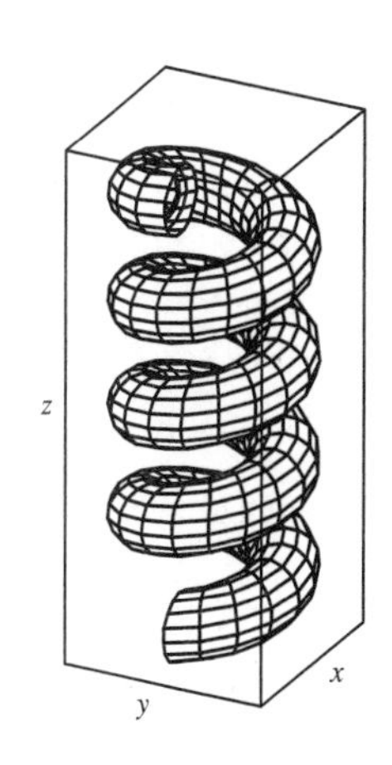

28. First we graph the surface as viewed from the front, then from two additional viewpoints.

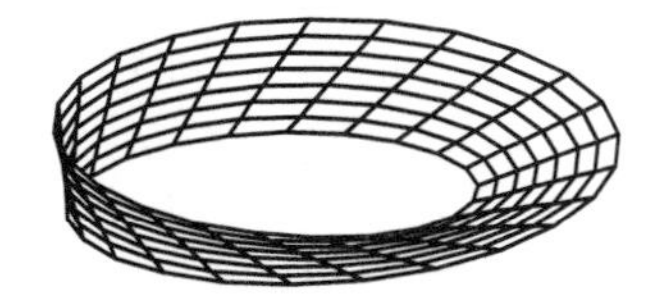

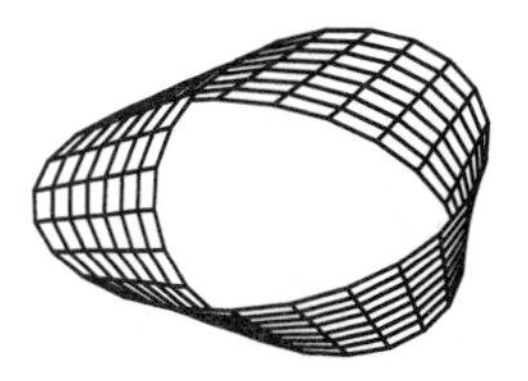

The surface appears as a twisted sheet, and is unusual because it has only one side. (The Möbius strip is discussed in more detail in Section 13.7.)

29. $\mathbf{r}(u, v) = (u + v)\,\mathbf{i} + 3u^2\,\mathbf{j} + (u - v)\,\mathbf{k}$.

$\mathbf{r}_u = \mathbf{i} + 6u\,\mathbf{j} + \mathbf{k}$ and $\mathbf{r}_v = \mathbf{i} - \mathbf{k}$, so $\mathbf{r}_u \times \mathbf{r}_v = -6u\,\mathbf{i} + 2\,\mathbf{j} - 6u\,\mathbf{k}$. Since the point $(2, 3, 0)$ corresponds to $u = 1$, $v = 1$, a normal vector to the surface at $(2, 3, 0)$ is $-6\,\mathbf{i} + 2\,\mathbf{j} - 6\,\mathbf{k}$, and an equation of the tangent plane is $-6x + 2y - 6z = -6$ or $3x - y + 3z = 3$.

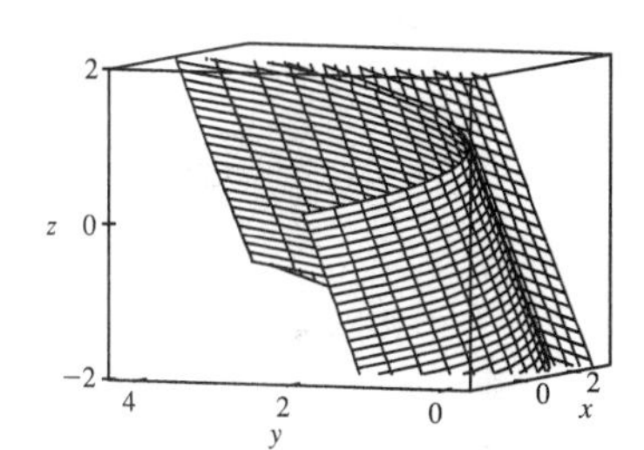

30. $\mathbf{r}(u, v) = u^2\,\mathbf{i} + v^2\,\mathbf{j} + uv\,\mathbf{k} \;\Rightarrow\; \mathbf{r}(1, 1) = (1, 1, 1)$.

$\mathbf{r}_u = 2u\,\mathbf{i} + v\,\mathbf{k}$ and $\mathbf{r}_v = 2v\,\mathbf{j} + u\,\mathbf{k}$, so a normal vector to the surface at the point $(1, 1, 1)$ is

$\mathbf{r}_u(1, 1) \times \mathbf{r}_v(1, 1) = (2\,\mathbf{i} + \mathbf{k}) \times (2\,\mathbf{j} + \mathbf{k}) = -2\,\mathbf{i} - 2\,\mathbf{j} + 4\,\mathbf{k}$.

Thus an equation of the tangent plane at the point $(1, 1, 1)$ is

$-2(x - 1) - 2(y - 1) + 4(z - 1) = 0$ or $x + y - 2z = 0$.

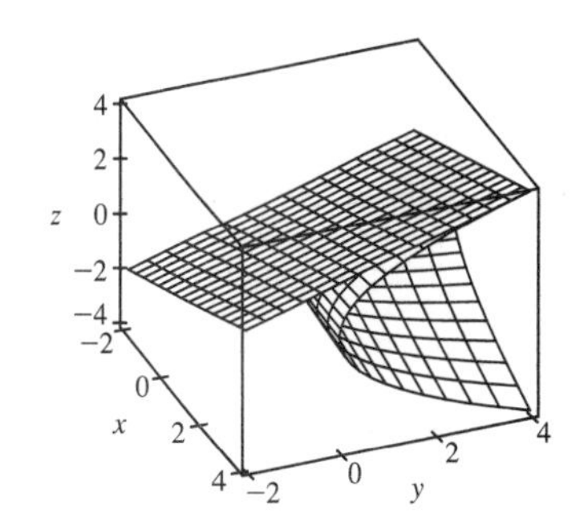

31. $\mathbf{r}(u, v) = u\,\mathbf{i} + \ln(uv)\,\mathbf{j} + v\,\mathbf{k} \quad\Rightarrow\quad \mathbf{r}_u(u, v) = \mathbf{i} + \frac{1}{u}\,\mathbf{j}$,

$\mathbf{r}_v(u, v) = \frac{1}{v}\,\mathbf{j} + \mathbf{k}$. $\mathbf{r}(1, 1) = \mathbf{i} + \mathbf{k}$, so the point corresponding to $u = 1$, $v = 1$ is $(1, 0, 1)$. A normal vector for the tangent plane is $\mathbf{r}_u(1, 1) \times \mathbf{r}_v(1, 1) = (\mathbf{i} + \mathbf{j}) \times (\mathbf{j} + \mathbf{k}) = \mathbf{i} - \mathbf{j} + \mathbf{k}$, so an equation of the tangent plane is $(x - 1) - (y - 0) + (z - 1) = 0$ or $x - y + z = 2$.

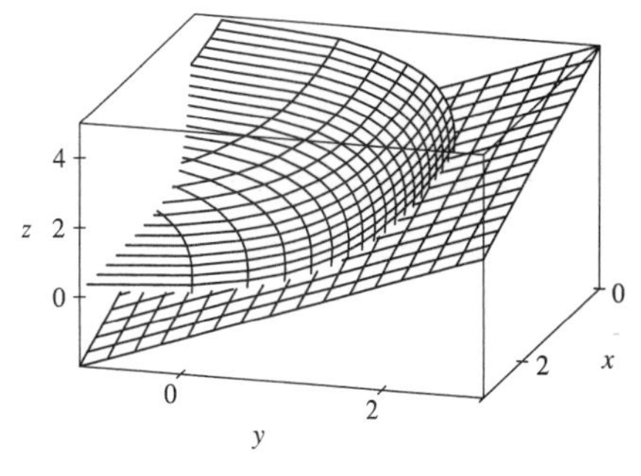

32. $\mathbf{r}(u,v) = uv\,\mathbf{i} + u\sin v\,\mathbf{j} + v\cos u\,\mathbf{k} \;\Rightarrow\; \mathbf{r}(0,\pi) = (0,0,\pi)$.
$\mathbf{r}_u = v\,\mathbf{i} + \sin v\,\mathbf{j} - v\sin u\,\mathbf{k}$ and $\mathbf{r}_v = u\,\mathbf{i} + u\cos v\,\mathbf{j} + \cos u\;\mathbf{k}$, so a normal vector to the surface at the point $(0,0,\pi)$ is $\mathbf{r}_u(0,\pi) \times \mathbf{r}_v(0,\pi) = (\pi\,\mathbf{i}) \times (\mathbf{k}) = -\pi\,\mathbf{j}$. Thus an equation of the tangent plane is $-\pi(y-0) = 0$ or $y = 0$.

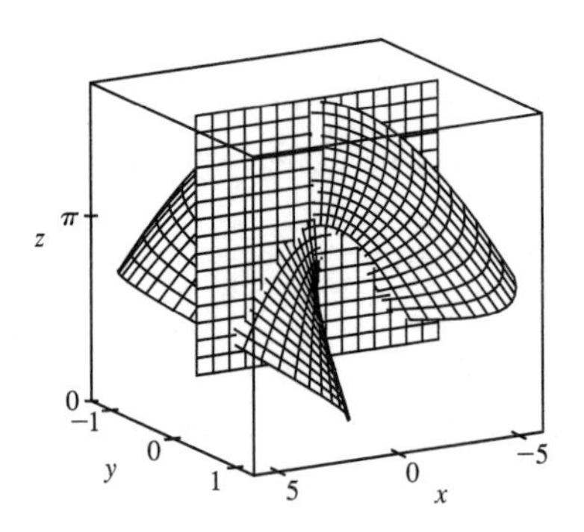

33. $z = f(x,y) = 6 - 3x - 2y$ which intersects the xy-plane in the line $3x + 2y = 6$, so D is the triangular region given by $\{(x,y) \mid 0 \le x \le 2, 0 \le y \le 3 - \frac{3}{2}x\}$. Thus
$A(S) = \iint_D \sqrt{1 + (-3)^2 + (-2)^2}\,dA = \sqrt{14}\iint_D dA = \sqrt{14}\,A(D) = \sqrt{14}\left(\frac{1}{2}\cdot 2\cdot 3\right) = 3\sqrt{14}$.

34. $z = f(x,y) = 10 - 2x - 5y$ and D is the disk $x^2 + y^2 \le 9$, so by Formula 9,

$$A(S) = \iint_D \sqrt{1 + (-2)^2 + (-5)^2}\,dA = \sqrt{30}\iint_D dA = \sqrt{30}\,A(D) = \sqrt{30}\,(\pi\cdot 3^2) = 9\sqrt{30}\,\pi.$$

35. $z = f(x,y) = \frac{2}{3}(x^{3/2} + y^{3/2})$ and $D = \{(x,y) \mid 0 \le x \le 1, 0 \le y \le 1\}$. Then $f_x = x^{1/2}$, $f_y = y^{1/2}$ and

$$\begin{aligned} A(S) &= \iint_D \sqrt{1 + \left(\sqrt{x}\right)^2 + \left(\sqrt{y}\right)^2}\,dA = \int_0^1\int_0^1 \sqrt{1 + x + y}\,dy\,dx \\ &= \int_0^1 \left[\tfrac{2}{3}(x + y + 1)^{3/2}\right]_{y=0}^{y=1} dx = \tfrac{2}{3}\int_0^1 \left[(x+2)^{3/2} - (x+1)^{3/2}\right] dx \\ &= \tfrac{2}{3}\left[\tfrac{2}{5}(x+2)^{5/2} - \tfrac{2}{5}(x+1)^{5/2}\right]_0^1 = \tfrac{4}{15}(3^{5/2} - 2^{5/2} - 2^{5/2} + 1) = \tfrac{4}{15}(3^{5/2} - 2^{7/2} + 1) \end{aligned}$$

36. $\mathbf{r}_u = \langle 0, 1, -5\rangle$, $\mathbf{r}_v = \langle 1, -2, 1\rangle$, and $\mathbf{r}_u \times \mathbf{r}_v = \langle -9, -5, -1\rangle$. Then by Definition 4,

$$A(S) = \iint_D |\,\mathbf{r}_u \times \mathbf{r}_v\,|\,dA = \int_0^1\int_0^1 |\,\langle -9, -5, -1\rangle\,|\,du\,dv = \sqrt{107}\int_0^1 du\int_0^1 dv = \sqrt{107}$$

37. $z = f(x,y) = xy$ with $0 \le x^2 + y^2 \le 1$, so $f_x = y$, $f_y = x \;\Rightarrow$

$$\begin{aligned} A(S) &= \iint_D \sqrt{1 + y^2 + x^2}\,dA = \int_0^{2\pi}\int_0^1 \sqrt{r^2 + 1}\,r\,dr\,d\theta = \int_0^{2\pi}\left[\tfrac{1}{3}\left(r^2 + 1\right)^{3/2}\right]_{r=0}^{r=1} d\theta \\ &= \int_0^{2\pi} \tfrac{1}{3}\left(2\sqrt{2} - 1\right) d\theta = \tfrac{2\pi}{3}\left(2\sqrt{2} - 1\right) \end{aligned}$$

38. $z = f(x,y) = 1 + 3x + 2y^2$ with $0 \le x \le 2y$, $0 \le y \le 1$. Thus, by Formula 9,

$$\begin{aligned} A(S) &= \iint_D \sqrt{1 + 3^2 + (4y)^2}\,dA = \int_0^1\int_0^{2y} \sqrt{10 + 16y^2}\,dx\,dy = \int_0^1 2y\sqrt{10 + 16y^2}\,dy \\ &= \tfrac{1}{16}\cdot\tfrac{2}{3}(10 + 16y^2)^{3/2}\Big]_0^1 = \tfrac{1}{24}(26^{3/2} - 10^{3/2}) \end{aligned}$$

39. $z = f(x,y) = y^2 - x^2$ with $1 \le x^2 + y^2 \le 4$. Then

$$\begin{aligned} A(S) &= \iint_D \sqrt{1 + 4x^2 + 4y^2}\,dA = \int_0^{2\pi}\int_1^2 \sqrt{1 + 4r^2}\,r\,dr\,d\theta = \int_0^{2\pi} d\theta\int_1^2 r\sqrt{1 + 4r^2}\,dr \\ &= \left[\,\theta\,\right]_0^{2\pi}\left[\tfrac{1}{12}(1 + 4r^2)^{3/2}\right]_1^2 = \tfrac{\pi}{6}\left(17\sqrt{17} - 5\sqrt{5}\right) \end{aligned}$$

40. A parametric representation of the surface is $x = y^2 + z^2$, $y = y$, $z = z$ with $0 \le y^2 + z^2 \le 9$.
Hence $\mathbf{r}_y \times \mathbf{r}_z = (2y\,\mathbf{i} + \mathbf{j}) \times (2z\,\mathbf{i} + \mathbf{k}) = \mathbf{i} - 2y\,\mathbf{j} - 2z\,\mathbf{k}$.
Note: In general, if $x = f(y,z)$ then $\mathbf{r}_y \times \mathbf{r}_z = \mathbf{i} - \dfrac{\partial f}{\partial y}\mathbf{j} - \dfrac{\partial f}{\partial z}\mathbf{k}$, and $A\,(S) = \displaystyle\iint_D \sqrt{1 + \left(\frac{\partial f}{\partial y}\right)^2 + \left(\frac{\partial f}{\partial z}\right)^2}\,dA$. Then

$$A(S) = \iint_{0 \le y^2+z^2 \le 9} \sqrt{1+4y^2+4z^2}\,dA = \int_0^{2\pi}\int_0^3 \sqrt{1+4r^2}\,r\,dr\,d\theta$$
$$= \int_0^{2\pi} d\theta \int_0^3 r\sqrt{1+4r^2}\,dr = 2\pi\left[\tfrac{1}{12}(1+4r^2)^{3/2}\right]_0^3 = \tfrac{\pi}{6}(37\sqrt{37}-1)$$

41. A parametric representation of the surface is $x = x$, $y = 4x + z^2$, $z = z$ with $0 \le x \le 1$, $0 \le z \le 1$.

Hence $\mathbf{r}_x \times \mathbf{r}_z = (\mathbf{i} + 4\mathbf{j}) \times (2z\,\mathbf{j} + \mathbf{k}) = 4\,\mathbf{i} - \mathbf{j} + 2z\,\mathbf{k}$.

Note: In general, if $y = f(x,z)$ then $\mathbf{r}_x \times \mathbf{r}_z = \dfrac{\partial f}{\partial x}\mathbf{i} - \mathbf{j} + \dfrac{\partial f}{\partial z}\mathbf{k}$ and $A(S) = \iint_D \sqrt{1 + \left(\dfrac{\partial f}{\partial x}\right)^2 + \left(\dfrac{\partial f}{\partial z}\right)^2}\,dA$. Then

$$A(S) = \int_0^1\int_0^1 \sqrt{17+4z^2}\,dx\,dz = \int_0^1 \sqrt{17+4z^2}\,dz$$
$$= \tfrac{1}{2}\left(z\sqrt{17+4z^2} + \tfrac{17}{2}\ln\left|2z + \sqrt{4z^2+17}\right|\right)\Big]_0^1 = \tfrac{\sqrt{21}}{2} + \tfrac{17}{4}\left[\ln(2+\sqrt{21}) - \ln\sqrt{17}\right]$$

42. $\mathbf{r}_u = \langle \cos v, \sin v, 0\rangle$, $\mathbf{r}_v = \langle -u\sin v, u\cos v, 1\rangle$, and $\mathbf{r}_u \times \mathbf{r}_v = \langle \sin v, -\cos v, u\rangle$. Then

$$A(S) = \int_0^\pi\int_0^1 \sqrt{1+u^2}\,du\,dv = \int_0^\pi dv \int_0^1 \sqrt{1+u^2}\,du$$
$$= \pi\left[\tfrac{u}{2}\sqrt{u^2+1} + \tfrac{1}{2}\ln\left|u + \sqrt{u^2+1}\right|\right]_0^1 = \tfrac{\pi}{2}\left[\sqrt{2} + \ln(1+\sqrt{2})\right]$$

43. $\mathbf{r}_u = \langle v, 1, 1\rangle$, $\mathbf{r}_v = \langle u, 1, -1\rangle$ and $\mathbf{r}_u \times \mathbf{r}_v = \langle -2, u+v, v-u\rangle$. Then

$$A(S) = \iint_{u^2+v^2 \le 1} \sqrt{4+2u^2+2v^2}\,dA = \int_0^{2\pi}\int_0^1 r\sqrt{4+2r^2}\,dr\,d\theta = \int_0^{2\pi} d\theta \int_0^1 r\sqrt{4+2r^2}\,dr$$
$$= 2\pi\left[\tfrac{1}{6}(4+2r^2)^{3/2}\right]_0^1 = \tfrac{\pi}{3}(6\sqrt{6}-8) = \pi(2\sqrt{6} - \tfrac{8}{3})$$

44. $z = f(x,y) = \cos(x^2+y^2)$ with $x^2+y^2 \le 1$.

$$A(S) = \iint_D \sqrt{1 + (-2x\sin(x^2+y^2))^2 + (-2y\sin(x^2+y^2))^2}\,dA$$
$$= \iint_D \sqrt{1 + 4x^2\sin^2(x^2+y^2) + 4y^2\sin^2(x^2+y^2)}\,dA = \iint_D \sqrt{1 + 4(x^2+y^2)\sin^2(x^2+y^2)}\,dA$$
$$= \int_0^{2\pi}\int_0^1 \sqrt{1+4r^2\sin^2(r^2)}\,r\,dr\,d\theta = \int_0^{2\pi} d\theta \int_0^1 r\sqrt{1+4r^2\sin^2(r^2)}\,dr$$
$$= 2\pi\int_0^1 r\sqrt{1+4r^2\sin^2(r^2)}\,dr \approx 4.1073$$

45. $z = f(x,y) = e^{-x^2-y^2}$ with $x^2+y^2 \le 4$.

$$A(S) = \iint_D \sqrt{1 + \left(-2xe^{-x^2-y^2}\right)^2 + \left(-2ye^{-x^2-y^2}\right)^2}\,dA = \iint_D \sqrt{1 + 4(x^2+y^2)e^{-2(x^2+y^2)}}\,dA$$
$$= \int_0^{2\pi}\int_0^2 \sqrt{1+4r^2e^{-2r^2}}\,r\,dr\,d\theta = \int_0^{2\pi} d\theta \int_0^2 r\sqrt{1+4r^2e^{-2r^2}}\,dr = 2\pi\int_0^2 r\sqrt{1+4r^2e^{-2r^2}}\,dr \approx 13.9783$$

46. Let $f(x,y) = \dfrac{1+x^2}{1+y^2}$. Then $f_x = \dfrac{2x}{1+y^2}$,

$$f_y = (1+x^2)\left[-\frac{2y}{(1+y^2)^2}\right] = -\frac{2y(1+x^2)}{(1+y^2)^2}.$$

We use a CAS to estimate

$\int_{-1}^1 \int_{-(1-|x|)}^{1-|x|} \sqrt{1+f_x^2+f_y^2}\,dy\,dx \approx 2.6959.$

In order to graph only the part of the surface above the square, we use $-(1-|x|) \le y \le 1-|x|$ as the y-range in our plot command.

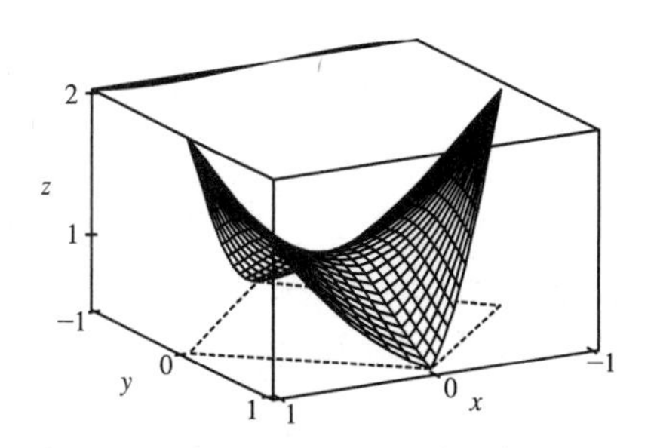

47. (a) The midpoints of the four squares are $\left(\frac{1}{4},\frac{1}{4}\right)$, $\left(\frac{1}{4},\frac{3}{4}\right)$, $\left(\frac{3}{4},\frac{1}{4}\right)$, and $\left(\frac{3}{4},\frac{3}{4}\right)$. Here $f(x,y)=x^2+y^2$, so the Midpoint Rule gives

$$\begin{aligned}A(S) &= \iint_D \sqrt{[f_x(x,y)]^2+[f_y(x,y)]^2+1}\,dA = \iint_D \sqrt{(2x)^2+(2y)^2+1}\,dA \\ &\approx \tfrac{1}{4}\left(\sqrt{\left[2\left(\tfrac{1}{4}\right)\right]^2+\left[2\left(\tfrac{1}{4}\right)\right]^2+1}+\sqrt{\left[2\left(\tfrac{1}{4}\right)\right]^2+\left[2\left(\tfrac{3}{4}\right)\right]^2+1}\right. \\ &\qquad \left.+\sqrt{\left[2\left(\tfrac{3}{4}\right)\right]^2+\left[2\left(\tfrac{1}{4}\right)\right]^2+1}+\sqrt{\left[2\left(\tfrac{3}{4}\right)\right]^2+\left[2\left(\tfrac{3}{4}\right)\right]^2+1}\right) \\ &= \tfrac{1}{4}\left(\sqrt{\tfrac{3}{2}}+2\sqrt{\tfrac{7}{2}}+\sqrt{\tfrac{11}{2}}\right)\approx 1.8279\end{aligned}$$

(b) A CAS estimates the integral to be $A(S)=\iint_D\sqrt{1+(2x)^2+(2y)^2}\,dA=\int_0^1\int_0^1\sqrt{1+4x^2+4y^2}\,dy\,dx\approx 1.8616$. This agrees with the Midpoint estimate only in the first decimal place.

48. $\mathbf{r}(u,v)=\left\langle\cos^3u\cos^3v,\sin^3u\cos^3v,\sin^3v\right\rangle$, so $\mathbf{r}_u=\left\langle-3\cos^2u\sin u\cos^3v,3\sin^2u\cos u\cos^3v,0\right\rangle$,

$\mathbf{r}_v=\left\langle-3\cos^3u\cos^2v\sin v,-3\sin^3u\cos^2v\sin v,3\sin^2v\cos v\right\rangle$, and

$\mathbf{r}_u\times\mathbf{r}_v=\left\langle 9\cos u\sin^2u\cos^4v\sin^2v,9\cos^2u\sin u\cos^4v\sin^2v,9\cos^2u\sin^2u\cos^5v\sin v\right\rangle$. Then

$$\begin{aligned}|\mathbf{r}_u\times\mathbf{r}_v| &= 9\sqrt{\cos^2u\sin^4u\cos^8v\sin^4v+\cos^4u\sin^2u\cos^8v\sin^4v+\cos^4u\sin^4u\cos^{10}v\sin^2v} \\ &= 9\sqrt{\cos^2u\sin^2u\cos^8v\sin^2v\left(\sin^2v+\cos^2u\sin^2u\cos^2v\right)} \\ &= 9\cos^4v\,|\cos u\sin u\sin v|\sqrt{\sin^2v+\cos^2u\sin^2u\cos^2v}\end{aligned}$$

Using a CAS, we have $A(S)=\int_0^{\pi}\int_0^{2\pi}9\cos^4v\,|\cos u\sin u\sin v|\sqrt{\sin^2v+\cos^2u\sin^2u\cos^2v}\,dv\,du\approx 4.4506$.

49. $z=1+2x+3y+4y^2$, so

$$A(S)=\iint_D\sqrt{1+\left(\frac{\partial z}{\partial x}\right)^2+\left(\frac{\partial z}{\partial y}\right)^2}\,dA=\int_1^4\int_0^1\sqrt{1+4+(3+8y)^2}\,dy\,dx=\int_1^4\int_0^1\sqrt{14+48y+64y^2}\,dy\,dx.$$

Using a CAS, we have

$\int_1^4\int_0^1\sqrt{14+48y+64y^2}\,dy\,dx=\frac{45}{8}\sqrt{14}+\frac{15}{16}\ln\left(11\sqrt{5}+3\sqrt{14}\sqrt{5}\right)-\frac{15}{16}\ln\left(3\sqrt{5}+\sqrt{14}\sqrt{5}\right)$

or $\frac{45}{8}\sqrt{14}+\frac{15}{16}\ln\frac{11\sqrt{5}+3\sqrt{70}}{3\sqrt{5}+\sqrt{70}}$.

50. (a) $\mathbf{r}_u=a\cos v\,\mathbf{i}+b\sin v\,\mathbf{j}+2u\,\mathbf{k}$, $\mathbf{r}_v=-au\sin v\,\mathbf{i}+bu\cos v\,\mathbf{j}+0\,\mathbf{k}$, and

$\mathbf{r}_u\times\mathbf{r}_v=-2bu^2\cos v\,\mathbf{i}-2au^2\sin v\,\mathbf{j}+abu\,\mathbf{k}$.

$A(S)=\int_0^{2\pi}\int_0^2|\mathbf{r}_u\times\mathbf{r}_v|\,du\,dv=\int_0^{2\pi}\int_0^2\sqrt{4b^2u^4\cos^2v+4a^2u^4\sin^2v+a^2b^2u^2}\,du\,dv$

(b) $x^2=a^2u^2\cos^2v$, $y^2=b^2u^2\sin^2v$, $z=u^2$ $\Rightarrow$ $x^2/a^2+y^2/b^2=u^2=z$ which is an elliptic paraboloid. To find D, notice that $0\le u\le 2$ $\Rightarrow$ $0\le z\le 4$ $\Rightarrow$ $0\le x^2/a^2+y^2/b^2\le 4$. Therefore, using Formula 9, we have

$$A(S)=\int_{-2a}^{2a}\int_{-b\sqrt{4-(x^2/a^2)}}^{b\sqrt{4-(x^2/a^2)}}\sqrt{1+(2x/a^2)^2+(2y/b^2)^2}\,dy\,dx.$$

(c)

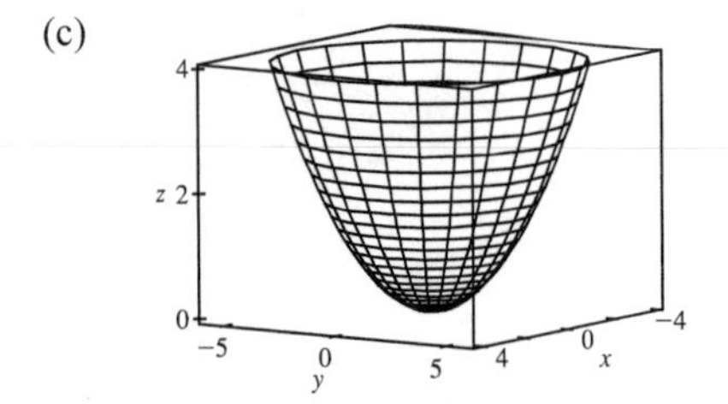

(d) We substitute $a=2$, $b=3$ in the integral in part (a) to get $A(S)=\int_0^{2\pi}\int_0^2 2u\sqrt{9u^2\cos^2v+4u^2\sin^2v+9}\,du\,dv$. We use a CAS to estimate the integral accurate to four decimal places. To speed up the calculation, we can set `Digits:=7;` (in Maple) or use the approximation command `N` (in Mathematica). We find that $A(S)\approx 115.6596$.

51. (a) $x = a \sin u \cos v$, $y = b \sin u \sin v$, $z = c \cos u \;\Rightarrow$

$$\frac{x^2}{a^2} + \frac{y^2}{b^2} + \frac{z^2}{c^2} = (\sin u \cos v)^2 + (\sin u \sin v)^2 + (\cos u)^2$$
$$= \sin^2 u + \cos^2 u = 1$$

and since the ranges of u and v are sufficient to generate the entire graph, the parametric equations represent an ellipsoid.

(b)

(c) From the parametric equations (with $a = 1$, $b = 2$, and $c = 3$),

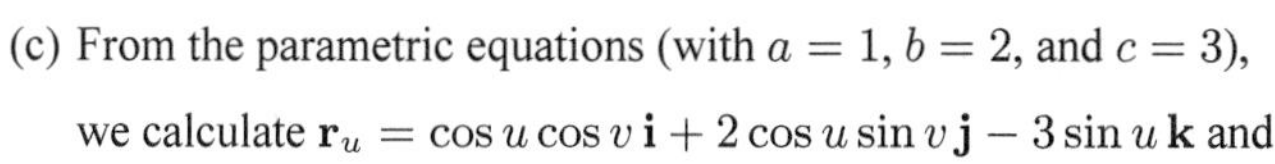

we calculate $\mathbf{r}_u = \cos u \cos v\,\mathbf{i} + 2 \cos u \sin v\,\mathbf{j} - 3 \sin u\,\mathbf{k}$ and

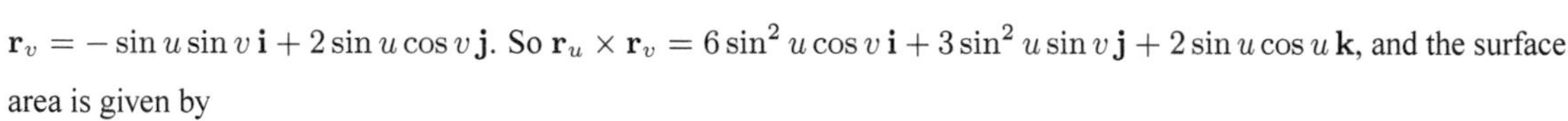

$\mathbf{r}_v = -\sin u \sin v\,\mathbf{i} + 2 \sin u \cos v\,\mathbf{j}$. So $\mathbf{r}_u \times \mathbf{r}_v = 6 \sin^2 u \cos v\,\mathbf{i} + 3 \sin^2 u \sin v\,\mathbf{j} + 2 \sin u \cos u\,\mathbf{k}$, and the surface area is given by

$$A(S) = \int_0^{2\pi} \int_0^{\pi} |\mathbf{r}_u \times \mathbf{r}_v|\,du\,dv$$
$$= \int_0^{2\pi} \int_0^{\pi} \sqrt{36 \sin^4 u \cos^2 v + 9 \sin^4 u \sin^2 v + 4 \cos^2 u \sin^2 u}\,du\,dv$$

52. (a) $x = a \cosh u \cos v$, $y = b \cosh u \sin v$, $z = c \sinh u \;\Rightarrow$

$$\frac{x^2}{a^2} + \frac{y^2}{b^2} - \frac{z^2}{c^2} = \cosh^2 u \cos^2 v + \cosh^2 u \sin^2 v - \sinh^2 u$$
$$= \cosh^2 u - \sinh^2 u = 1$$

and the parametric equations represent a hyperboloid of one sheet.

(b)

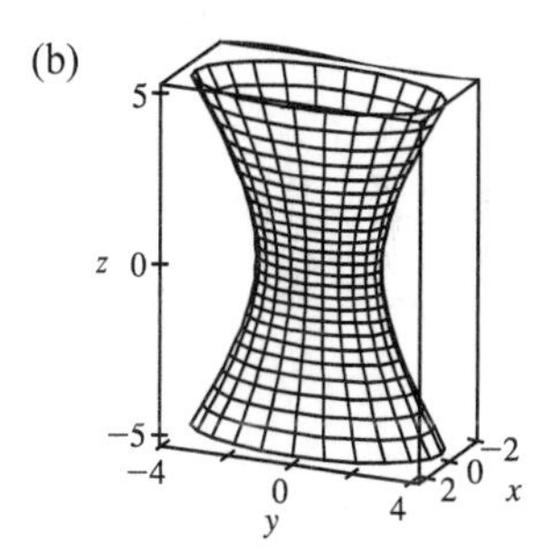

(c) $\mathbf{r}_u = \sinh u \cos v\,\mathbf{i} + 2 \sinh u \sin v\,\mathbf{j} + 3 \cosh u\,\mathbf{k}$ and $\mathbf{r}_v = -\cosh u \sin v\,\mathbf{i} + 2 \cosh u \cos v\,\mathbf{j}$, so $\mathbf{r}_u \times \mathbf{r}_v = -6 \cosh^2 u \cos v\,\mathbf{i} - 3 \cosh^2 u \sin v\,\mathbf{j} + 2 \cosh u \sinh u\,\mathbf{k}$. We integrate between $u = \sinh^{-1}(-1) = -\ln\left(1 + \sqrt{2}\right)$ and $u = \sinh^{-1} 1 = \ln\left(1 + \sqrt{2}\right)$, since then z varies between -3 and 3, as desired. So the surface area is

$$A(S) = \int_0^{2\pi} \int_{-\ln(1+\sqrt{2})}^{\ln(1+\sqrt{2})} |\mathbf{r}_u \times \mathbf{r}_v|\,du\,dv$$
$$= \int_0^{2\pi} \int_{-\ln(1+\sqrt{2})}^{\ln(1+\sqrt{2})} \sqrt{36 \cosh^4 u \cos^2 v + 9 \cosh^4 u \sin^2 v + 4 \cosh^2 u \sinh^2 u}\,du\,dv$$

53. To find the region D: $z = x^2 + y^2$ implies $z + z^2 = 4z$ or $z^2 - 3z = 0$. Thus $z = 0$ or $z = 3$ are the planes where the surfaces intersect. But $x^2 + y^2 + z^2 = 4z$ implies $x^2 + y^2 + (z - 2)^2 = 4$, so $z = 3$ intersects the upper hemisphere. Thus $(z - 2)^2 = 4 - x^2 - y^2$ or $z = 2 + \sqrt{4 - x^2 - y^2}$. Therefore D is the region inside the circle $x^2 + y^2 + (3 - 2)^2 = 4$, that is, $D = \{(x, y) \mid x^2 + y^2 \leq 3\}$.

$$A(S) = \iint_D \sqrt{1 + [(-x)(4 - x^2 - y^2)^{-1/2}]^2 + [(-y)(4 - x^2 - y^2)^{-1/2}]^2}\,dA$$
$$= \int_0^{2\pi} \int_0^{\sqrt{3}} \sqrt{1 + \frac{r^2}{4 - r^2}}\,r\,dr\,d\theta = \int_0^{2\pi} \int_0^{\sqrt{3}} \frac{2r\,dr}{\sqrt{4 - r^2}}\,d\theta = \int_0^{2\pi} \left[-2(4 - r^2)^{1/2}\right]_{r=0}^{r=\sqrt{3}}\,d\theta$$
$$= \int_0^{2\pi} (-2 + 4)\,d\theta = 2\theta\Big]_0^{2\pi} = 4\pi$$

54. We first find the area of the face of the surface that intersects the positive y-axis. A parametric representation of the surface is $x = x$, $y = \sqrt{1 - z^2}$, $z = z$ with $x^2 + z^2 \le 1$. Then $\mathbf{r}(x, z) = \langle x, \sqrt{1 - z^2}, z \rangle \Rightarrow \mathbf{r}_x = \langle 1, 0, 0 \rangle$, $\mathbf{r}_z = \langle 0, -z/\sqrt{1 - z^2}, 1 \rangle$ and $\mathbf{r}_x \times \mathbf{r}_z = \langle 0, -1, -z/\sqrt{1 - z^2} \rangle \Rightarrow |\mathbf{r}_x \times \mathbf{r}_z| = \sqrt{1 + \frac{z^2}{1 - z^2}} = \frac{1}{\sqrt{1 - z^2}}$.

$$A(S) = \iint\limits_{x^2 + z^2 \le 1} |\mathbf{r}_x \times \mathbf{r}_z|\, dA = \int_{-1}^{1} \int_{-\sqrt{1-z^2}}^{\sqrt{1-z^2}} \frac{1}{\sqrt{1 - z^2}}\, dx\, dz$$

$$= 4 \int_0^1 \int_0^{\sqrt{1-z^2}} \frac{1}{\sqrt{1-z^2}}\, dx\, dz \quad \text{[by the symmetry of the surface]}$$

This integral is improper (when $z = 1$), so

$$A(S) = \lim_{t \to 1^-} 4 \int_0^t \int_0^{\sqrt{1-z^2}} \frac{1}{\sqrt{1-z^2}}\, dx\, dz = \lim_{t \to 1^-} 4 \int_0^t \frac{\sqrt{1-z^2}}{\sqrt{1-z^2}}\, dz = \lim_{t \to 1^-} 4 \int_0^t dz = \lim_{t \to 1^-} 4t = 4$$

Since the complete surface consists of four congruent faces, the total surface area is $4(4) = 16$.

Alternate solution: The face of the surface that intersects the positive y-axis can also be parametrized as $\mathbf{r}(x, \theta) = \langle x, \cos\theta, \sin\theta \rangle$ for $-\frac{\pi}{2} \le \theta \le \frac{\pi}{2}$ and $x^2 + z^2 \le 1 \Leftrightarrow x^2 + \sin^2\theta \le 1 \Leftrightarrow$ $-\sqrt{1 - \sin^2\theta} \le x \le \sqrt{1 - \sin^2\theta} \Leftrightarrow -\cos\theta \le x \le \cos\theta$. Then $\mathbf{r}_x = \langle 1, 0, 0 \rangle$, $\mathbf{r}_\theta = \langle 0, -\sin\theta, \cos\theta \rangle$ and $\mathbf{r}_x \times \mathbf{r}_\theta = \langle 0, -\cos\theta, -\sin\theta \rangle \Rightarrow |\mathbf{r}_x \times \mathbf{r}_\theta| = 1$, so $A(S) = \int_{-\pi/2}^{\pi/2} \int_{-\cos\theta}^{\cos\theta} 1\, dx\, d\theta = \int_{-\pi/2}^{\pi/2} 2\cos\theta\, d\theta = 2\sin\theta\Big]_{-\pi/2}^{\pi/2} = 4$. Again, the area of the complete surface is $4(4) = 16$.

55. If we revolve the curve $y = f(x)$, $a \le x \le b$ about the x-axis, where $f(x) \ge 0$, then from Equations 3 we know we can parametrize the surface using $x = x$, $y = f(x)\cos\theta$, and $z = f(x)\sin\theta$, where $a \le x \le b$ and $0 \le \theta \le 2\pi$. Thus we can say the surface is represented by $\mathbf{r}(x, \theta) = x\,\mathbf{i} + f(x)\cos\theta\,\mathbf{j} + f(x)\sin\theta\,\mathbf{k}$, with $a \le x \le b$ and $0 \le \theta \le 2\pi$. Then by (6), the surface area is given by $A(S) = \iint_D |\mathbf{r}_x \times \mathbf{r}_\theta|\, dA$ where D is the rectangular parameter region $[a, b] \times [0, 2\pi]$. Here, $\mathbf{r}_x(x, \theta) = \mathbf{i} + f'(x)\cos\theta\,\mathbf{j} + f'(x)\sin\theta\,\mathbf{k}$ and $\mathbf{r}_\theta(x) = -f(x)\sin\theta\,\mathbf{j} + f(x)\cos\theta\,\mathbf{k}$. So

$$\mathbf{r}_x \times \mathbf{r}_\theta = \begin{vmatrix} \mathbf{i} & \mathbf{j} & \mathbf{k} \\ 1 & f'(x)\cos\theta & f'(x)\sin\theta \\ 0 & -f(x)\sin\theta & f(x)\cos\theta \end{vmatrix} = \left[f(x)f'(x)\cos^2\theta + f(x)f'(x)\sin^2\theta\right]\mathbf{i} - f(x)\cos\theta\,\mathbf{j} - f(x)\sin\theta\,\mathbf{k}$$

$$= f(x)f'(x)\mathbf{i} - f(x)\cos\theta\,\mathbf{j} - f(x)\sin\theta\,\mathbf{k} \text{ and}$$

$$|\mathbf{r}_x \times \mathbf{r}_\theta| = \sqrt{[f(x)f'(x)]^2 + [f(x)]^2\cos^2\theta + [f(x)]^2\sin^2\theta}$$

$$= \sqrt{[f(x)]^2\left([f'(x)]^2 + 1\right)} = f(x)\sqrt{1 + [f'(x)]^2} \text{ [since } f(x) \ge 0\text{]. Thus}$$

$$A(S) = \iint_D |\mathbf{r}_x \times \mathbf{r}_\theta|\, dA = \int_a^b \int_0^{2\pi} f(x)\sqrt{1 + [f'(x)]^2}\, d\theta\, dx$$

$$= \int_a^b f(x)\sqrt{1 + [f'(x)]^2}\, [\theta]_0^{2\pi}\, dx = 2\pi \int_a^b f(x)\sqrt{1 + [f'(x)]^2}\, dx$$

56. $y = x^3 \Rightarrow y' = 3x^2$. So

$$S = \int_0^2 2\pi y\,\sqrt{1 + (y')^2}\, dx = 2\pi \int_0^2 x^3\sqrt{1 + 9x^4}\, dx \quad (\text{Let } u = 1 + 9x^4, \text{ so } du = 36x^3\, dx)$$

$$= \frac{2\pi}{36} \int_1^{145} \sqrt{u}\, du = \frac{\pi}{18}\left[\tfrac{2}{3}u^{3/2}\right]_1^{145} = \frac{\pi}{27}\left(145\sqrt{145} - 1\right)$$

57. $y = \sqrt{x} \quad \Rightarrow \quad 1 + \left(\dfrac{dy}{dx}\right)^2 = 1 + \left(\dfrac{1}{2\sqrt{x}}\right)^2 = 1 + \dfrac{1}{4x}$. So

$$S = \int_4^9 2\pi y \sqrt{1 + \left(\frac{dy}{dx}\right)^2}\, dx = \int_4^9 2\pi \sqrt{x} \sqrt{1 + \frac{1}{4x}}\, dx = 2\pi \int_4^9 \left(x + \tfrac{1}{4}\right) dx$$

$$= 2\pi \left[\tfrac{2}{3}\left(x + \tfrac{1}{4}\right)^{3/2}\right]_4^9 = \tfrac{4\pi}{3}\left[\tfrac{1}{8}(4x+1)^{3/2}\right]_4^9 = \tfrac{\pi}{6}\left(37\sqrt{37} - 17\sqrt{17}\right)$$

58. (a) Here $z = a \sin \alpha$, $y = |AB|$, and $x = |OA|$. But $|OB| = |OC| + |CB| = b + a\cos\alpha$ and $\sin\theta = \dfrac{|AB|}{|OB|}$ so that $y = |OB| \sin\theta = (b + a\cos\alpha)\sin\theta$. Similarly $\cos\theta = \dfrac{|OA|}{|OB|}$ so $x = (b + a\cos\alpha)\cos\theta$. Hence a parametric representation for the torus is $x = b\cos\theta + a\cos\alpha\cos\theta$, $y = b\sin\theta + a\cos\alpha\sin\theta$, $z = a\sin\alpha$, where $0 \le \alpha \le 2\pi$, $0 \le \theta \le 2\pi$.

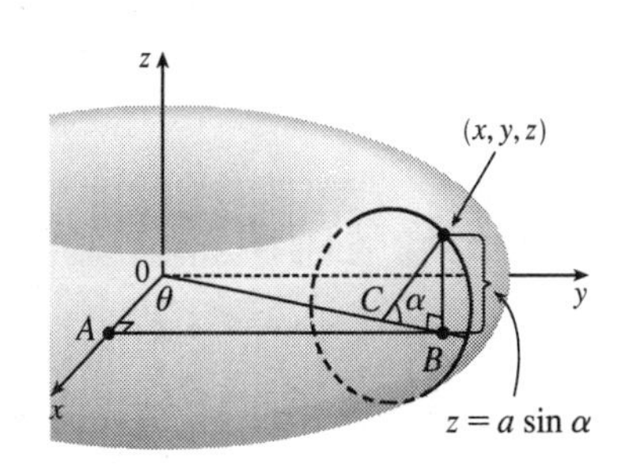

(b)

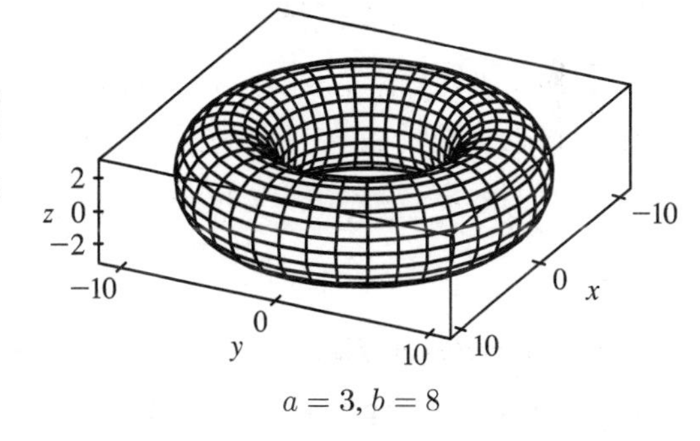

$a = 1, b = 8$ $\qquad$ $a = 3, b = 8$

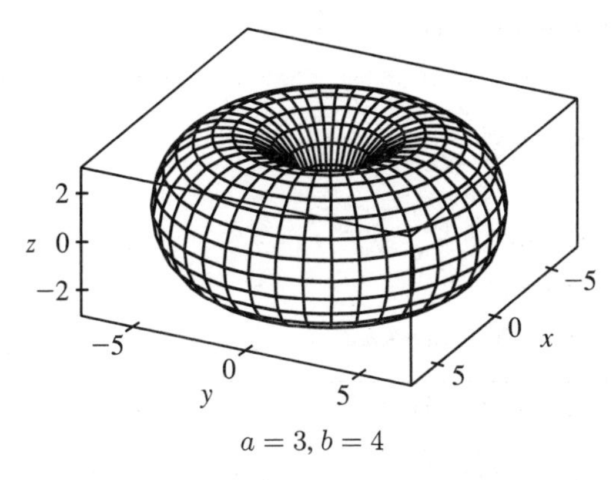

$a = 3, b = 4$

(c) $x = b\cos\theta + a\cos\alpha\cos\theta$, $y = b\sin\theta + a\cos\alpha\sin\theta$, $z = a\sin\alpha$, so $\mathbf{r}_\alpha = \langle -a\sin\alpha\cos\theta, -a\sin\alpha\sin\theta, a\cos\alpha\rangle$, $\mathbf{r}_\theta = \langle -(b + a\cos\alpha)\sin\theta, (b + a\cos\alpha)\cos\theta, 0\rangle$ and

$$\begin{aligned} \mathbf{r}_\alpha \times \mathbf{r}_\theta &= \left(-ab\cos\alpha\cos\theta - a^2\cos\alpha\cos^2\theta\right)\mathbf{i} + \left(-ab\sin\alpha\cos\theta - a^2\sin\alpha\cos^2\theta\right)\mathbf{j} \\ &\quad + \left(-ab\cos^2\alpha\sin\theta - a^2\cos^2\alpha\sin\theta\cos\theta - ab\sin^2\alpha\sin\theta - a^2\sin^2\alpha\sin\theta\cos\theta\right)\mathbf{k} \\ &= -a(b + a\cos\alpha)\left[(\cos\theta\cos\alpha)\,\mathbf{i} + (\sin\theta\cos\alpha)\,\mathbf{j} + (\sin\alpha)\,\mathbf{k}\right] \end{aligned}$$

Then $|\mathbf{r}_\alpha \times \mathbf{r}_\theta| = a(b + a\cos\alpha)\sqrt{\cos^2\theta\cos^2\alpha + \sin^2\theta\cos^2\alpha + \sin^2\alpha} = a(b + a\cos\alpha)$.

Note: $b > a$, $-1 \le \cos\alpha \le 1$ so $|b + a\cos\alpha| = b + a\cos\alpha$. Hence

$A(S) = \int_0^{2\pi}\int_0^{2\pi} a(b + a\cos\alpha)\, d\alpha\, d\theta = 2\pi\left[ab\alpha + a^2\sin\alpha\right]_0^{2\pi} = 4\pi^2 ab$.

13.7 Surface Integrals

1. Each face of the cube has surface area $2^2 = 4$, and the points P_{ij}^* are the points where the cube intersects the coordinate axes. Here, $f(x,y,z) = \sqrt{x^2 + 2y^2 + 3z^2}$, so by Definition 1,

$$\begin{aligned} \iint_S f(x,y,z)\,dS &\approx [f(1,0,0)](4) + [f(-1,0,0)](4) + [f(0,1,0)](4) + [f(0,-1,0)](4) \\ &\quad + [f(0,0,1)](4) + [f(0,0,-1)](4) \\ &= 4\left(1 + 1 + 2\sqrt{2} + 2\sqrt{3}\right) = 8\left(1 + \sqrt{2} + \sqrt{3}\right) \approx 33.170 \end{aligned}$$

2. Each quarter-cylinder has surface area $\frac{1}{4}[2\pi(1)(2)] = \pi$, and the top and bottom disks have surface area $\pi(1)^2 = \pi$. We can take $(0,0,1)$ as a sample point in the top disk, $(0,0,-1)$ in the bottom disk, and $(\pm 1,0,0)$, $(0,\pm 1,0)$ in the four quarter-cylinders. Then $\iint_S f(x,y,z)\,dS$ can be approximated by the Riemann sum

$$\begin{aligned} &f(1,0,0)(\pi) + f(-1,0,0)(\pi) + f(0,1,0)\,(\pi) + f(0,-1,0)(\pi) + f(0,0,1)(\pi) + f(0,0,-1)(\pi) \\ &\qquad = (2 + 2 + 3 + 3 + 4 + 4)\pi = 18\pi \approx 56.5. \end{aligned}$$

3. We can use the xz- and yz-planes to divide H into four patches of equal size, each with surface area equal to $\frac{1}{8}$ the surface area of a sphere with radius $\sqrt{50}$, so $\Delta S = \frac{1}{8}(4)\pi\left(\sqrt{50}\right)^2 = 25\pi$. Then $(\pm 3, \pm 4, 5)$ are sample points in the four patches, and using a Riemann sum as in Definition 1, we have

$$\begin{aligned} \iint_H f(x,y,z)\,dS &\approx f(3,4,5)\,\Delta S + f(3,-4,5)\,\Delta S + f(-3,4,5)\,\Delta S + f(-3,-4,5)\,\Delta S \\ &= (7 + 8 + 9 + 12)(25\pi) = 900\pi \approx 2827 \end{aligned}$$

4. On the surface, $f(x,y,z) = g\left(\sqrt{x^2+y^2+z^2}\right) = g(2) = -5$. So since the area of a sphere is $4\pi r^2$,

$\iint_S f(x,y,z)\,dS = \iint_S g(2)\,dS = -5\iint_S dS = -5[4\pi(2)^2] = -80\pi.$

5. $\mathbf{r}(u,v) = u^2\,\mathbf{i} + u\sin v\,\mathbf{j} + u\cos v\,\mathbf{k}$, $0 \le u \le 1$, $0 \le v \le \pi/2$ and

$\mathbf{r}_u \times \mathbf{r}_v = (2u\,\mathbf{i} + \sin v\,\mathbf{j} + \cos v\,\mathbf{k}) \times (u\cos v\,\mathbf{j} - u\sin v\,\mathbf{k}) = -u\,\mathbf{i} + 2u^2\sin v\,\mathbf{j} + 2u^2\cos v\,\mathbf{k}$ and

$|\mathbf{r}_u \times \mathbf{r}_v| = \sqrt{u^2 + 4u^4\sin^2 v + 4u^4\cos^2 v} = \sqrt{u^2 + 4u^4(\sin^2 v + \cos^2 v)} = u\sqrt{1 + 4u^2}$ (since $u \ge 0$).

Then $\iint_S yz\,dS = \int_0^{\pi/2}\int_0^1 (u\sin v)(u\cos v)\cdot u\sqrt{1+4u^2}\,du\,dv = \int_0^1 u^3\sqrt{1+4u^2}\,du\,\int_0^{\pi/2}\sin v\cos v\,dv$

$$\left[\text{let } t = 1 + 4u^2 \;\Rightarrow\; u^2 = \tfrac{1}{4}(t-1) \text{ and } \tfrac{1}{8}\,dt = u\,du\right]$$

$$\begin{aligned} &= \int_1^5 \tfrac{1}{8}\cdot\tfrac{1}{4}(t-1)\sqrt{t}\,dt\,\int_0^{\pi/2}\sin v\cos v\,dv = \tfrac{1}{32}\int_1^5\left(t^{3/2} - \sqrt{t}\right)dt\,\int_0^{\pi/2}\sin v\cos v\,dv \\ &= \tfrac{1}{32}\left[\tfrac{2}{5}t^{5/2} - \tfrac{2}{3}t^{3/2}\right]_1^5\left[\tfrac{1}{2}\sin^2 v\right]_0^{\pi/2} = \tfrac{1}{32}\left(\tfrac{2}{5}(5)^{5/2} - \tfrac{2}{3}(5)^{3/2} - \tfrac{2}{5} + \tfrac{2}{3}\right)\cdot\tfrac{1}{2}(1-0) = \tfrac{5}{48}\sqrt{5} + \tfrac{1}{240} \end{aligned}$$

6. $\mathbf{r}_u = \cos v\,\mathbf{i} + \sin v\,\mathbf{j}$, $\mathbf{r}_v = -u\sin v\,\mathbf{i} + u\cos v\,\mathbf{j} + \mathbf{k} \;\Rightarrow\; \mathbf{r}_u \times \mathbf{r}_v = \sin v\,\mathbf{i} - \cos v\,\mathbf{j} + u\,\mathbf{k} \;\Rightarrow\; |\mathbf{r}_u \times \mathbf{r}_v| = \sqrt{1+u^2}$,

so $\iint_S \sqrt{1+x^2+y^2}\,dS = \int_0^\pi\int_0^1 \sqrt{1+u^2}\,\sqrt{1+u^2}\,du\,dv = \frac{4}{3}\pi.$

7. $z = 1 + 2x + 3y$ so $\dfrac{\partial z}{\partial x} = 2$ and $\dfrac{\partial z}{\partial y} = 3$. Then by Formula 2,

$$\begin{aligned} \iint_S x^2yz\,dS &= \iint_D x^2yz\sqrt{\left(\frac{\partial z}{\partial x}\right)^2 + \left(\frac{\partial z}{\partial y}\right)^2 + 1}\,dA = \int_0^3\int_0^2 x^2y(1+2x+3y)\sqrt{4+9+1}\,dy\,dx \\ &= \sqrt{14}\int_0^3\int_0^2 (x^2y + 2x^3y + 3x^2y^2)\,dy\,dx = \sqrt{14}\int_0^3\left[\tfrac{1}{2}x^2y^2 + x^3y^2 + x^2y^3\right]_{y=0}^{y=2}dx \\ &= \sqrt{14}\int_0^3 (10x^2 + 4x^3)\,dx = \sqrt{14}\left[\tfrac{10}{3}x^3 + x^4\right]_0^3 = 171\sqrt{14} \end{aligned}$$

8. S is the region in the plane $2x + y + z = 2$ or $z = 2 - 2x - y$ over $D = \{(x,y) \mid 0 \le x \le 1, 0 \le y \le 2 - 2x\}$. Thus

$$\iint_S xy\,dS = \iint_D xy\sqrt{(-2)^2 + (-1)^2 + 1}\,dA = \sqrt{6}\int_0^1\int_0^{2-2x} xy\,dy\,dx = \sqrt{6}\int_0^1\left[\tfrac{1}{2}xy^2\right]_{y=0}^{y=2-2x} dx$$
$$= \tfrac{\sqrt{6}}{2}\int_0^1 (4x - 8x^2 + 4x^3)\,dx = \tfrac{\sqrt{6}}{2}\left(2 - \tfrac{8}{3} + 1\right) = \tfrac{\sqrt{6}}{6}$$

9. S is the part of the plane $z = 1 - x - y$ over the region $D = \{(x,y) \mid 0 \le x \le 1, 0 \le y \le 1 - x\}$. Thus

$$\iint_S yz\,dS = \iint_D y(1 - x - y)\sqrt{(-1)^2 + (-1)^2 + 1}\,dA = \sqrt{3}\int_0^1\int_0^{1-x}(y - xy - y^2)\,dy\,dx$$
$$= \sqrt{3}\int_0^1\left[\tfrac{1}{2}y^2 - \tfrac{1}{2}xy^2 - \tfrac{1}{3}y^3\right]_{y=0}^{y=1-x} dx = \sqrt{3}\int_0^1 \tfrac{1}{6}(1-x)^3\,dx = -\tfrac{\sqrt{3}}{24}(1-x)^4\Big]_0^1 = \tfrac{\sqrt{3}}{24}$$

10. $z = \frac{2}{3}(x^{3/2} + y^{3/2})$ and

$$\iint_S y\,dS = \iint_D y\sqrt{\left(\sqrt{x}\right)^2 + \left(\sqrt{y}\right)^2 + 1}\,dA = \int_0^1\int_0^1 y\sqrt{x + y + 1}\,dx\,dy$$
$$= \int_0^1 y\left[\tfrac{2}{3}(x + y + 1)^{3/2}\right]_{x=0}^{x=1} dy = \int_0^1 \tfrac{2}{3}y\left[(y+2)^{3/2} - (y+1)^{3/2}\right] dy$$

Substituting $u = y + 2$ in the first term and $t = y + 1$ in the second, we have

$$\iint_S y\,dS = \tfrac{2}{3}\int_2^3 (u-2)u^{3/2}\,du - \tfrac{2}{3}\int_1^2 (t-1)t^{3/2}\,dt = \tfrac{2}{3}\left[\tfrac{2}{7}u^{7/2} - \tfrac{4}{5}u^{5/2}\right]_2^3 - \tfrac{2}{3}\left[\tfrac{2}{7}t^{7/2} - \tfrac{2}{5}t^{5/2}\right]_1^2$$
$$= \tfrac{2}{3}\left[\tfrac{2}{7}(3^{7/2} - 2^{7/2}) - \tfrac{4}{5}(3^{5/2} - 2^{5/2}) - \tfrac{2}{7}(2^{7/2} - 1) + \tfrac{2}{5}(2^{5/2} - 1)\right]$$
$$= \tfrac{2}{3}\left(\tfrac{18}{35}\sqrt{3} + \tfrac{8}{35}\sqrt{2} - \tfrac{4}{35}\right) = \tfrac{4}{105}(9\sqrt{3} + 4\sqrt{2} - 2)$$

11. S is the portion of the cone $z^2 = x^2 + y^2$ for $1 \le z \le 3$, or equivalently, S is the part of the surface $z = \sqrt{x^2 + y^2}$ over the region $D = \{(x,y) \mid 1 \le x^2 + y^2 \le 9\}$. Thus

$$\iint_S x^2z^2\,dS = \iint_D x^2(x^2 + y^2)\sqrt{\left(\frac{x}{\sqrt{x^2 + y^2}}\right)^2 + \left(\frac{y}{\sqrt{x^2 + y^2}}\right)^2 + 1}\,dA$$
$$= \iint_D x^2(x^2 + y^2)\sqrt{\frac{x^2 + y^2}{x^2 + y^2} + 1}\,dA = \iint_D \sqrt{2}\,x^2(x^2 + y^2)\,dA = \sqrt{2}\int_0^{2\pi}\int_1^3 (r\cos\theta)^2(r^2)\,r\,dr\,d\theta$$
$$= \sqrt{2}\int_0^{2\pi}\cos^2\theta\,d\theta\int_1^3 r^5\,dr = \sqrt{2}\left[\tfrac{1}{2}\theta + \tfrac{1}{4}\sin 2\theta\right]_0^{2\pi}\left[\tfrac{1}{6}r^6\right]_1^3 = \sqrt{2}(\pi)\cdot\tfrac{1}{6}(3^6 - 1) = \frac{364\sqrt{2}}{3}\pi$$

12. Using y and z as parameters, we have $\mathbf{r}(y,z) = (y + 2z^2)\,\mathbf{i} + y\,\mathbf{j} + z\,\mathbf{k}$, $0 \le y \le 1$, $0 \le z \le 1$.

Then $\mathbf{r}_y \times \mathbf{r}_z = (\mathbf{i} + \mathbf{j}) \times (4z\,\mathbf{i} + \mathbf{k}) = \mathbf{i} - \mathbf{j} - 4z\,\mathbf{k}$ and $|\mathbf{r}_y \times \mathbf{r}_z| = \sqrt{2 + 16z^2}$. Thus

$$\iint_S z\,dS = \int_0^1\int_0^1 z\sqrt{2 + 16z^2}\,dy\,dz = \int_0^1 z\sqrt{2 + 16z^2}\,dz = \left[\tfrac{1}{32}\cdot\tfrac{2}{3}(2 + 16z^2)^{3/2}\right]_0^1 = \tfrac{1}{48}(18^{3/2} - 2^{3/2}) = \tfrac{13}{12}\sqrt{2}$$

13. Using x and z as parameters, we have $\mathbf{r}(x,z) = x\,\mathbf{i} + (x^2 + z^2)\,\mathbf{j} + z\,\mathbf{k}$, $x^2 + z^2 \le 4$. Then

$\mathbf{r}_x \times \mathbf{r}_z = (\mathbf{i} + 2x\,\mathbf{j}) \times (2z\,\mathbf{j} + \mathbf{k}) = 2x\,\mathbf{i} - \mathbf{j} + 2z\,\mathbf{k}$ and $|\mathbf{r}_x \times \mathbf{r}_z| = \sqrt{4x^2 + 1 + 4z^2} = \sqrt{1 + 4(x^2 + z^2)}$. Thus

$$\iint_S y\,dS = \iint_{x^2+z^2\le 4}(x^2 + z^2)\sqrt{1 + 4(x^2 + z^2)}\,dA = \int_0^{2\pi}\int_0^2 r^2\sqrt{1 + 4r^2}\,r\,dr\,d\theta$$
$$= \int_0^{2\pi} d\theta\int_0^2 r^2\sqrt{1 + 4r^2}\,r\,dr = 2\pi\int_0^2 r^2\sqrt{1 + 4r^2}\,r\,dr$$
$$\left[\text{let } u = 1 + 4r^2 \;\Rightarrow\; r^2 = \tfrac{1}{4}(u - 1) \text{ and } \tfrac{1}{8}du = r\,dr\right]$$
$$= 2\pi\int_1^{17}\tfrac{1}{4}(u-1)\sqrt{u}\cdot\tfrac{1}{8}du = \tfrac{1}{16}\pi\int_1^{17}(u^{3/2} - u^{1/2})\,du$$
$$= \tfrac{1}{16}\pi\left[\tfrac{2}{5}u^{5/2} - \tfrac{2}{3}u^{3/2}\right]_1^{17} = \tfrac{1}{16}\pi\left[\tfrac{2}{5}(17)^{5/2} - \tfrac{2}{3}(17)^{3/2} - \tfrac{2}{5} + \tfrac{2}{3}\right] = \frac{\pi}{60}(391\sqrt{17} + 1)$$

14. Here S consists of three surfaces: S_1, the lateral surface of the cylinder; S_2, the front formed by the plane $x+y=2$; and the back, S_3, in the plane $y=0$. On S_1: using cylindrical coordinates, $\mathbf{r}(\theta,y)=\sin\theta\,\mathbf{i}+y\,\mathbf{j}+\cos\theta\,\mathbf{k}$, $0\le\theta\le 2\pi$, $0\le y\le 2-\sin\theta$, $|\mathbf{r}_\theta\times\mathbf{r}_y|=1$ and

$\iint_{S_1} xy\,dS=\int_0^{2\pi}\int_0^{2-\sin\theta}(\sin\theta)\,y\,dy\,d\theta=\int_0^{2\pi}\left[2\sin\theta-2\sin^2\theta+\frac{1}{2}\sin^3\theta\right]d\theta=-2\pi.$

On S_2: $\mathbf{r}(x,z)=x\,\mathbf{i}+(2-x)\,\mathbf{j}+z\,\mathbf{k}$ and $|\mathbf{r}_x\times\mathbf{r}_z|=|-\mathbf{i}-\mathbf{j}|=\sqrt{2}$, where $x^2+z^2\le 1$ and

$$\iint_{S_2} xy\,dS=\iint_{x^2+z^2\le 1} x(2-x)\sqrt{2}\,dA=\int_0^{2\pi}\int_0^1\sqrt{2}\,(2r\sin\theta-r^2\sin^2\theta)\,r\,dr\,d\theta$$

$$=\sqrt{2}\int_0^{2\pi}\left[\tfrac{2}{3}\sin\theta-\tfrac{1}{4}\sin^2\theta\right]d\theta=-\tfrac{\sqrt{2}}{4}\pi$$

On S_3: $y=0$ so $\iint_{S_3} xy\,dS=0$. Hence $\iint_S xy\,dS=-2\pi-\frac{\sqrt{2}}{4}\pi=-\frac{1}{4}\left(8+\sqrt{2}\right)\pi$.

15. Using spherical coordinates and Example 13.6.4 we have $\mathbf{r}(\phi,\theta)=2\sin\phi\cos\theta\,\mathbf{i}+2\sin\phi\sin\theta\,\mathbf{j}+2\cos\phi\,\mathbf{k}$ and $|\mathbf{r}_\phi\times\mathbf{r}_\theta|=4\sin\phi$. Then $\iint_S(x^2z+y^2z)\,dS=\int_0^{2\pi}\int_0^{\pi/2}(4\sin^2\phi)(2\cos\phi)(4\sin\phi)\,d\phi\,d\theta=16\pi\sin^4\phi\big]_0^{\pi/2}=16\pi$.

16. Using spherical coordinates, $\mathbf{r}(\phi,\theta)=\sin\phi\cos\theta\,\mathbf{i}+\sin\phi\sin\theta\,\mathbf{j}+\cos\phi\,\mathbf{k}$, $0\le\phi\le\frac{\pi}{4}$, $0\le\theta\le 2\pi$, and $|\mathbf{r}_\phi\times\mathbf{r}_\theta|=\sin\phi$ (see Example 13.6.9). Then $\iint_S xyz\,dS=\int_0^{2\pi}\int_0^{\pi/4}(\sin^3\phi\cos\phi\cos\theta\sin\theta)\,d\phi\,d\theta=0$ since $\int_0^{2\pi}\cos\theta\sin\theta\,d\theta=0$.

17. Using cylindrical coordinates, we have $\mathbf{r}(\theta,z)=3\cos\theta\,\mathbf{i}+3\sin\theta\,\mathbf{j}+z\,\mathbf{k}$, $0\le\theta\le 2\pi$, $0\le z\le 2$, and $|\mathbf{r}_\theta\times\mathbf{r}_z|=3$.

$$\iint_S(x^2y+z^2)\,dS=\int_0^{2\pi}\int_0^2(27\cos^2\theta\sin\theta+z^2)\,3\,dz\,d\theta=\int_0^{2\pi}(162\cos^2\theta\sin\theta+8)\,d\theta=16\pi$$

18. Let S_1 be the lateral surface, S_2 the top disk, and S_3 the bottom disk.

On S_1: $\mathbf{r}(\theta,z)=3\cos\theta\,\mathbf{i}+3\sin\theta\,\mathbf{j}+z\,\mathbf{k}$, $0\le\theta\le 2\pi$, $0\le z\le 2$, $|\mathbf{r}_\theta\times\mathbf{r}_z|=3$,

$\iint_{S_1}(x^2+y^2+z^2)\,dS=\int_0^{2\pi}\int_0^2(9+z^2)\,3\,dz\,d\theta=2\pi(54+8)=124\pi.$

On S_2: $\mathbf{r}(\theta,r)=r\cos\theta\,\mathbf{i}+r\sin\theta\,\mathbf{j}+2\,\mathbf{k}$, $0\le r\le 3$, $0\le\theta\le 2\pi$, $|\mathbf{r}_\theta\times\mathbf{r}_r|=r$,

$\iint_{S_2}(x^2+y^2+z^2)\,dS=\int_0^{2\pi}\int_0^3(r^2+4)\,r\,dr\,d\theta=2\pi\left(\frac{81}{4}+18\right)=\frac{153}{2}\pi.$

On S_3: $\mathbf{r}(\theta,r)=r\cos\theta\,\mathbf{i}+r\sin\theta\,\mathbf{j}$, $0\le r\le 3$, $0\le\theta\le 2\pi$, $|\mathbf{r}_\theta\times\mathbf{r}_r|=r$,

$\iint_{S_3}(x^2+y^2+z^2)\,dS=\int_0^{2\pi}\int_0^3(r^2+0)\,r\,dr\,d\theta=2\pi\left(\frac{81}{4}\right)=\frac{81}{2}\pi.$

Hence $\iint_S(x^2+y^2+z^2)\,dS=124\pi+\frac{153}{2}\pi+\frac{81}{2}\pi=241\pi$.

19. $\mathbf{F}(x,y,z)=xy\,\mathbf{i}+yz\,\mathbf{j}+zx\,\mathbf{k}$, $z=g(x,y)=4-x^2-y^2$, and D is the square $[0,1]\times[0,1]$, so by Equation 10

$$\iint_S\mathbf{F}\cdot d\mathbf{S}=\iint_D[-xy(-2x)-yz(-2y)+zx]\,dA=\int_0^1\int_0^1[2x^2y+2y^2(4-x^2-y^2)+x(4-x^2-y^2)]\,dy\,dx$$

$$=\int_0^1\left(\tfrac{1}{3}x^2+\tfrac{11}{3}x-x^3+\tfrac{34}{15}\right)dx=\tfrac{713}{180}$$

20. $\mathbf{r}_u=\cos v\,\mathbf{i}+\sin v\,\mathbf{j}$, $\mathbf{r}_v=-u\sin v\,\mathbf{i}+u\cos v\,\mathbf{j}+\mathbf{k}$ $\Rightarrow$ $\mathbf{r}_u\times\mathbf{r}_v=\sin v\,\mathbf{i}-\cos v\,\mathbf{j}+u\,\mathbf{k}$ and $\mathbf{F}(\mathbf{r}(u,v))=u\sin v\,\mathbf{i}+u\cos v\,\mathbf{j}+v^2\,\mathbf{k}$. Then

$$\iint_S\mathbf{F}\cdot d\mathbf{S}=\int_0^\pi\int_0^1(u\sin^2 v-u\cos^2 v+uv^2)\,du\,dv=\int_0^\pi\int_0^1(-u\cos 2v+uv^2)\,du\,dv$$

$$=\int_0^\pi\left[-\tfrac{1}{2}\cos 2v+\tfrac{1}{2}v^2\right]dv=\tfrac{1}{6}\pi^3$$

21. $\mathbf{F}(x,y,z) = xze^y\,\mathbf{i} - xze^y\,\mathbf{j} + z\,\mathbf{k}$, $z = g(x,y) = 1 - x - y$, and $D = \{(x,y) \mid 0 \le x \le 1, 0 \le y \le 1 - x\}$. Since S has downward orientation, we have

$$\iint_S \mathbf{F} \cdot d\mathbf{S} = -\iint_D [-xze^y(-1) - (-xze^y)(-1) + z]\,dA = -\int_0^1 \int_0^{1-x} (1 - x - y)\,dy\,dx$$
$$= -\int_0^1 \left(\tfrac{1}{2}x^2 - x + \tfrac{1}{2}\right) dx = -\tfrac{1}{6}$$

22. $\mathbf{F}(x,y,z) = x\,\mathbf{i} + y\,\mathbf{j} + z^4\,\mathbf{k}$, $z = g(x,y) = \sqrt{x^2+y^2}$, and D is the disk $\{(x,y) \mid x^2 + y^2 \le 1\}$. Since S has downward orientation, we have

$$\iint_S \mathbf{F} \cdot d\mathbf{S} = -\iint_D \left[-x\left(\frac{x}{\sqrt{x^2+y^2}}\right) - y\left(\frac{y}{\sqrt{x^2+y^2}}\right) + z^4\right] dA = -\iint_D \left[\frac{-x^2-y^2}{\sqrt{x^2+y^2}} + \left(\sqrt{x^2+y^2}\right)^4\right] dA$$
$$= -\int_0^{2\pi}\int_0^1 \left(\frac{-r^2}{r} + r^4\right) r\,dr\,d\theta = -\int_0^{2\pi} d\theta \int_0^1 (r^5 - r^2)\,dr = -2\pi\left(\tfrac{1}{6} - \tfrac{1}{3}\right) = \tfrac{\pi}{3}$$

23. $\mathbf{F}(x,y,z) = x\,\mathbf{i} - z\,\mathbf{j} + y\,\mathbf{k}$, $z = g(x,y) = \sqrt{4 - x^2 - y^2}$ and D is the quarter disk $\left\{(x,y) \,\middle|\, 0 \le x \le 2, 0 \le y \le \sqrt{4 - x^2}\right\}$. S has downward orientation, so by Formula 10,

$$\iint_S \mathbf{F} \cdot d\mathbf{S} = -\iint_D \left[-x \cdot \tfrac{1}{2}(4 - x^2 - y^2)^{-1/2}(-2x) - (-z) \cdot \tfrac{1}{2}(4 - x^2 - y^2)^{-1/2}(-2y) + y\right] dA$$
$$= -\iint_D \left(\frac{x^2}{\sqrt{4-x^2-y^2}} - \sqrt{4-x^2-y^2} \cdot \frac{y}{\sqrt{4-x^2-y^2}} + y\right) dA$$
$$= -\iint_D x^2(4 - (x^2+y^2))^{-1/2}\,dA = -\int_0^{\pi/2}\int_0^2 (r\cos\theta)^2(4 - r^2)^{-1/2}\,r\,dr\,d\theta$$
$$= -\int_0^{\pi/2} \cos^2\theta\,d\theta \int_0^2 r^3(4 - r^2)^{-1/2}\,dr \quad \left[\text{let } u = 4 - r^2 \;\Rightarrow\; r^2 = 4 - u \text{ and } -\tfrac{1}{2}\,du = r\,dr\right]$$
$$= -\int_0^{\pi/2} \left(\tfrac{1}{2} + \tfrac{1}{2}\cos 2\theta\right) d\theta \int_4^0 -\tfrac{1}{2}(4 - u)(u)^{-1/2}\,du$$
$$= -\left[\tfrac{1}{2}\theta + \tfrac{1}{4}\sin 2\theta\right]_0^{\pi/2} \left(-\tfrac{1}{2}\right)\left[8\sqrt{u} - \tfrac{2}{3}u^{3/2}\right]_4^0 = -\tfrac{\pi}{4}\left(-\tfrac{1}{2}\right)\left(-16 + \tfrac{16}{3}\right) = -\tfrac{4}{3}\pi$$

24. $\mathbf{F}(x,y,z) = xz\,\mathbf{i} + x\,\mathbf{j} + y\,\mathbf{k}$

Using spherical coordinates, S is given by $x = 5\sin\phi\cos\theta$, $y = 5\sin\phi\sin\theta$, $z = 5\cos\phi$, $0 \le \theta \le \pi$, $0 \le \phi \le \pi$. $\mathbf{F}(\mathbf{r}(\phi,\theta)) = (5\sin\phi\cos\theta)(5\cos\phi)\,\mathbf{i} + (5\sin\phi\cos\theta)\,\mathbf{j} + (5\sin\phi\sin\theta)\,\mathbf{k}$ and $\mathbf{r}_\phi \times \mathbf{r}_\theta = 25\sin^2\phi\cos\theta\,\mathbf{i} + 25\sin^2\phi\sin\theta\,\mathbf{j} + 25\cos\phi\sin\phi\,\mathbf{k}$, so

$$\mathbf{F}(\mathbf{r}(\phi,\theta)) \cdot (\mathbf{r}_\phi \times \mathbf{r}_\theta) = 625\sin^3\phi\cos\phi\cos^2\theta + 125\sin^3\phi\cos\theta\sin\theta + 125\sin^2\phi\cos\phi\sin\theta$$

Then

$$\iint_S \mathbf{F} \cdot d\mathbf{S} = \iint_D [\mathbf{F}(\mathbf{r}(\phi,\theta)) \cdot (\mathbf{r}_\phi \times \mathbf{r}_\theta)]\,dA$$
$$= \int_0^\pi \int_0^\pi (625\sin^3\phi\cos\phi\cos^2\theta + 125\sin^3\phi\cos\theta\sin\theta + 125\sin^2\phi\cos\phi\sin\theta)\,d\theta\,d\phi$$
$$= 125\int_0^\pi \left[5\sin^3\phi\cos\phi\left(\tfrac{1}{2}\theta + \tfrac{1}{4}\sin 2\theta\right) + \sin^3\phi\left(\tfrac{1}{2}\sin^2\theta\right) + \sin^2\phi\cos\phi(-\cos\theta)\right]_{\theta=0}^{\theta=\pi} d\phi$$
$$= 125\int_0^\pi \left(\tfrac{5}{2}\pi\sin^3\phi\cos\phi + 2\sin^2\phi\cos\phi\right) d\phi = 125\left[\tfrac{5}{2}\pi \cdot \tfrac{1}{4}\sin^4\phi + 2 \cdot \tfrac{1}{3}\sin^3\phi\right]_0^\pi = 0$$

25. Let S_1 be the paraboloid $y = x^2 + z^2$, $0 \leq y \leq 1$ and S_2 the disk $x^2 + z^2 \leq 1$, $y = 1$. Since S is a closed surface, we use the outward orientation. On S_1: $\mathbf{F}(\mathbf{r}(x,z)) = (x^2+z^2)\,\mathbf{j} - z\,\mathbf{k}$ and $\mathbf{r}_x \times \mathbf{r}_z = 2x\,\mathbf{i} - \mathbf{j} + 2z\,\mathbf{k}$ (since the $\mathbf{j}$-component must be negative on S_1). Then

$$\iint_{S_1} \mathbf{F} \cdot d\mathbf{S} = \iint_{x^2+z^2 \leq 1} \left[-(x^2+z^2) - 2z^2\right] dA = -\int_0^{2\pi}\int_0^1 (r^2 + 2r^2\cos^2\theta)\, r\, dr\, d\theta$$
$$= -\int_0^{2\pi} \tfrac{1}{4}(1 + 2\cos^2\theta)\, d\theta = -\left(\tfrac{\pi}{2} + \tfrac{\pi}{2}\right) = -\pi$$

On S_2: $\mathbf{F}(\mathbf{r}(x,z)) = \mathbf{j} - z\,\mathbf{k}$ and $\mathbf{r}_z \times \mathbf{r}_x = \mathbf{j}$. Then $\iint_{S_2} \mathbf{F} \cdot d\mathbf{S} = \iint_{x^2+z^2 \leq 1} (1)\, dA = \pi$. Hence $\iint_S \mathbf{F} \cdot d\mathbf{S} = -\pi + \pi = 0$.

26. Here S consists of four surfaces: S_1, the triangular face with vertices $(1,0,0)$, $(0,1,0)$, and $(0,0,1)$; S_2, the face of the tetrahedron in the xy-plane; S_3, the face in the xz-plane; and S_4, the face in the yz-plane.

On S_1: The face is the portion of the plane $z = 1 - x - y$ for $0 \leq x \leq 1$, $0 \leq y \leq 1 - x$ with upward orientation, so

$$\iint_{S_1} \mathbf{F} \cdot d\mathbf{S} = \int_0^1\int_0^{1-x} \left[-y\,(-1) - (z-y)\,(-1) + x\right] dy\, dx = \int_0^1\int_0^{1-x} (z + x)\, dy\, dx = \int_0^1\int_0^{1-x} (1-y)\, dy\, dx$$
$$= \int_0^1 \left[y - \tfrac{1}{2}y^2\right]_{y=0}^{y=1-x} dx = \tfrac{1}{2}\int_0^1 (1 - x^2)\, dx = \tfrac{1}{2}\left[x - \tfrac{1}{3}x^3\right]_0^1 = \tfrac{1}{3}$$

On S_2: The surface is $z = 0$ with downward orientation, so

$$\iint_{S_2} \mathbf{F} \cdot d\mathbf{S} = \int_0^1\int_0^{1-x} (-x)\, dy\, dx = -\int_0^1 x\,(1-x)\, dx = -\left[\tfrac{1}{2}x^2 - \tfrac{1}{3}x^3\right]_0^1 = -\tfrac{1}{6}$$

On S_3: The surface is $y = 0$ for $0 \leq x \leq 1$, $0 \leq z \leq 1 - x$, oriented in the negative y-direction. Regarding x and z as parameters, we have $\mathbf{r}_x \times \mathbf{r}_z = -\mathbf{j}$ and

$$\iint_{S_3} \mathbf{F} \cdot d\mathbf{S} = \int_0^1\int_0^{1-x} -(z-y)\, dz\, dx = -\int_0^1\int_0^{1-x} z\, dz\, dx = -\int_0^1 \left[\tfrac{1}{2}z^2\right]_{z=0}^{z=1-x} dx$$
$$= -\tfrac{1}{2}\int_0^1 (1-x)^2\, dx = \tfrac{1}{6}\left[(1-x)^3\right]_0^1 = -\tfrac{1}{6}$$

On S_4: The surface is $x = 0$ for $0 \leq y \leq 1$, $0 \leq z \leq 1 - y$, oriented in the negative x-direction. Regarding y and z as parameters, we have $\mathbf{r}_y \times \mathbf{r}_z = \mathbf{i}$ so we use $-(\mathbf{r}_y \times \mathbf{r}_z) = -\mathbf{i}$ and

$$\iint_{S_4} \mathbf{F} \cdot d\mathbf{S} = \int_0^1\int_0^{1-y} (-y)\, dz\, dy = -\int_0^1 y\,(1-y)\, dy = -\left[\tfrac{1}{2}y^2 - \tfrac{1}{3}y^3\right]_0^1 = -\tfrac{1}{6}$$

Thus $\iint_S \mathbf{F} \cdot d\mathbf{S} = \frac{1}{3} - \frac{1}{6} - \frac{1}{6} - \frac{1}{6} = -\frac{1}{6}$.

27. Here S consists of four surfaces: S_1, the top surface (a portion of the circular cylinder $y^2 + z^2 = 1$); S_2, the bottom surface (a portion of the xy-plane); S_3, the front half-disk in the plane $x = 2$, and S_4, the back half-disk in the plane $x = 0$.

On S_1: The surface is $z = \sqrt{1 - y^2}$ for $0 \leq x \leq 2$, $-1 \leq y \leq 1$ with upward orientation, so

$$\iint_{S_1} \mathbf{F} \cdot d\mathbf{S} = \int_0^2\int_{-1}^1 \left[-x^2\,(0) - y^2\left(-\frac{y}{\sqrt{1-y^2}}\right) + z^2\right] dy\, dx = \int_0^2\int_{-1}^1 \left(\frac{y^3}{\sqrt{1-y^2}} + 1 - y^2\right) dy\, dx$$
$$= \int_0^2 \left[-\sqrt{1-y^2} + \tfrac{1}{3}(1-y^2)^{3/2} + y - \tfrac{1}{3}y^3\right]_{y=-1}^{y=1} dx = \int_0^2 \tfrac{4}{3}\, dx = \tfrac{8}{3}$$

On S_2: The surface is $z = 0$ with downward orientation, so

$$\iint_{S_2} \mathbf{F} \cdot d\mathbf{S} = \int_0^2\int_{-1}^1 (-z^2)\, dy\, dx = \int_0^2\int_{-1}^1 (0)\, dy\, dx = 0$$

On S_3: The surface is $x = 2$ for $-1 \leq y \leq 1$, $0 \leq z \leq \sqrt{1-y^2}$, oriented in the positive x-direction. Regarding y and z as parameters, we have $\mathbf{r}_y \times \mathbf{r}_z = \mathbf{i}$ and

$$\iint_{S_3} \mathbf{F} \cdot d\mathbf{S} = \int_{-1}^{1} \int_{0}^{\sqrt{1-y^2}} x^2 \, dz \, dy = \int_{-1}^{1} \int_{0}^{\sqrt{1-y^2}} 4 \, dz \, dy = 4A(S_3) = 2\pi$$

On S_4: The surface is $x = 0$ for $-1 \le y \le 1$, $0 \le z \le \sqrt{1-y^2}$, oriented in the negative x-direction. Regarding y and z as parameters, we use $-(\mathbf{r}_y \times \mathbf{r}_z) = -\mathbf{i}$ and

$$\iint_{S_4} \mathbf{F} \cdot d\mathbf{S} = \int_{-1}^{1} \int_{0}^{\sqrt{1-y^2}} x^2 \, dz \, dy = \int_{-1}^{1} \int_{0}^{\sqrt{1-y^2}} (0) \, dz \, dy = 0$$

Thus $\iint_S \mathbf{F} \cdot d\mathbf{S} = \frac{8}{3} + 0 + 2\pi + 0 = 2\pi + \frac{8}{3}$.

28. (a) $z = xy \quad \Rightarrow \quad \partial z/\partial x = y$, $\partial z/\partial y = x$, so by Formula 4, a CAS gives

$\iint_S xyz \, dS = \int_0^1 \int_0^1 xy\,(xy)\,\sqrt{y^2 + x^2 + 1}\, dx\, dy \approx 0.1642$.

(b) We use a CAS to calculate

$$\begin{aligned}\iint_S x^2yz \, dS &= \int_0^1 \int_0^1 x^2y\,(xy)\,\sqrt{y^2+x^2+1}\,dx\,dy \\ &= \tfrac{1}{60}\sqrt{3} - \tfrac{1}{12}\ln(1+\sqrt{3}) - \tfrac{1}{192}\ln(\sqrt{2}+1) + \tfrac{317}{2880}\sqrt{2} + \tfrac{1}{24}\ln 2\end{aligned}$$

29. We use Formula 4 with $z = 3 - 2x^2 - y^2 \quad \Rightarrow \quad \partial z/\partial x = -4x$, $\partial z/\partial y = -2y$. The boundaries of the region $3 - 2x^2 - y^2 \ge 0$ are $-\sqrt{\frac{3}{2}} \le x \le \sqrt{\frac{3}{2}}$ and $-\sqrt{3-2x^2} \le y \le \sqrt{3-2x^2}$, so we use a CAS (with precision reduced to seven or fewer digits; otherwise the calculation takes a very long time) to calculate

$$\iint_S x^2y^2z^2 \, dS = \int_{-\sqrt{3/2}}^{\sqrt{3/2}} \int_{-\sqrt{3-2x^2}}^{\sqrt{3-2x^2}} x^2y^2(3-2x^2-y^2)^2\sqrt{16x^2+4y^2+1}\,dy\,dx \approx 3.4895$$

30. The flux of $\mathbf{F}$ across S is given by $\iint_S \mathbf{F} \cdot d\mathbf{S} = \iint_S \mathbf{F} \cdot \mathbf{n}\, dS$. Now on S, $z = g(x,y) = 2\sqrt{1-y^2}$, so $\partial g/\partial x = 0$ and $\partial g/\partial y = -2y(1-y^2)^{-1/2}$. Therefore, by (10),

$$\iint_S \mathbf{F} \cdot d\mathbf{S} = \int_{-2}^{2}\int_{-1}^{1}\left(-x^2y\left[-2y(1-y^2)^{-1/2}\right] + \left[2\sqrt{1-y^2}\right]^2 e^{x/5}\right) dy\,dx = \tfrac{1}{3}(16\pi + 80e^{2/5} - 80e^{-2/5})$$

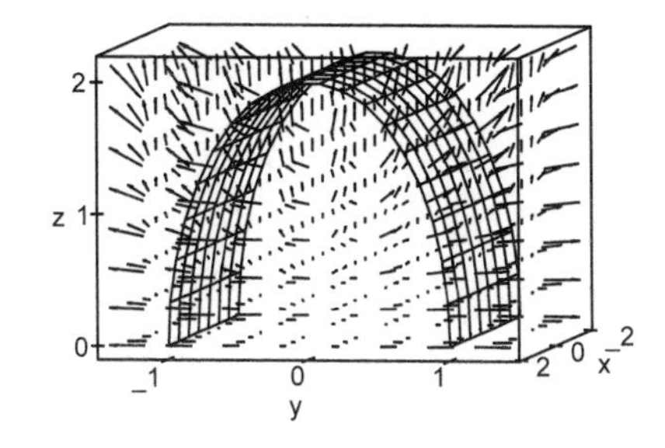

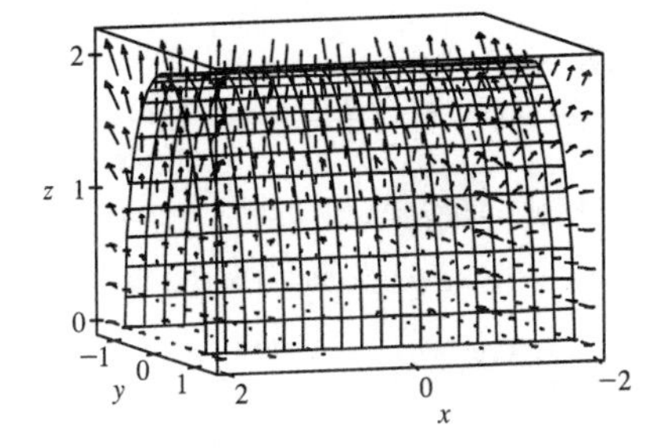

31. If S is given by $y = h(x,z)$, then S is also the level surface $f(x,y,z) = y - h(x,z) = 0$.

$\mathbf{n} = \dfrac{\nabla f(x,y,z)}{|\nabla f(x,y,z)|} = \dfrac{-h_x\,\mathbf{i} + \mathbf{j} - h_z\,\mathbf{k}}{\sqrt{h_x^2 + 1 + h_z^2}}$, and $-\mathbf{n}$ is the unit normal that points to the left. Now we proceed as in the derivation of (10), using Formula 4 to evaluate

$$\iint_S \mathbf{F} \cdot d\mathbf{S} = \iint_S \mathbf{F} \cdot \mathbf{n}\,dS = \iint_D (P\,\mathbf{i} + Q\,\mathbf{j} + R\,\mathbf{k})\,\frac{\dfrac{\partial h}{\partial x}\mathbf{i} - \mathbf{j} + \dfrac{\partial h}{\partial z}\mathbf{k}}{\sqrt{\left(\dfrac{\partial h}{\partial x}\right)^2 + 1 + \left(\dfrac{\partial h}{\partial z}\right)^2}}\sqrt{\left(\dfrac{\partial h}{\partial x}\right)^2 + 1 + \left(\dfrac{\partial h}{\partial z}\right)^2}\,dA$$

where D is the projection of $\mathbf{S}$ onto the xz-plane. Therefore $\displaystyle\iint_S \mathbf{F} \cdot d\mathbf{S} = \iint_D \left(P\frac{\partial h}{\partial x} - Q + R\frac{\partial h}{\partial z}\right) dA$.

32. If S is given by $x = k(y,z)$, then S is also the level surface $f(x,y,z) = x - k(y,z) = 0$.

$\mathbf{n} = \dfrac{\nabla f(x,y,z)}{|\nabla f(x,y,z)|} = \dfrac{\mathbf{i} - k_y\,\mathbf{j} - k_z\,\mathbf{k}}{\sqrt{1 + k_y^2 + k_z^2}}$, and since the x-component is positive this is the unit normal that points forward.

Now we proceed as in the derivation of (10), using Formula 4 for

$$\iint_S \mathbf{F}\cdot d\mathbf{S} = \iint_S \mathbf{F}\cdot\mathbf{n}\,dS = \iint_D (P\,\mathbf{i} + Q\,\mathbf{j} + R\,\mathbf{k})\,\frac{\mathbf{i} - \dfrac{\partial k}{\partial y}\mathbf{j} - \dfrac{\partial k}{\partial z}\mathbf{k}}{\sqrt{1 + \left(\dfrac{\partial k}{\partial y}\right)^2 + \left(\dfrac{\partial k}{\partial z}\right)^2}}\sqrt{1 + \left(\frac{\partial k}{\partial y}\right)^2 + \left(\frac{\partial k}{\partial z}\right)^2}\,dA$$

where D is the projection of $f(x,y,z)$ onto the yz-plane. Therefore $\displaystyle\iint_S \mathbf{F}\cdot d\mathbf{S} = \iint_D \left(P - Q\frac{\partial k}{\partial y} - R\frac{\partial k}{\partial z}\right) dA.$

33. $m = \iint_S K\,dS = K\cdot 4\pi\left(\frac{1}{2}a^2\right) = 2\pi a^2 K$; by symmetry $M_{xz} = M_{yz} = 0$, and

$M_{xy} = \iint_S zK\,dS = K\int_0^{2\pi}\int_0^{\pi/2}(a\cos\phi)(a^2\sin\phi)\,d\phi\,d\theta = 2\pi K a^3\left[-\frac{1}{4}\cos 2\phi\right]_0^{\pi/2} = \pi K a^3$.

Hence $(\overline{x},\overline{y},\overline{z}) = \left(0,0,\frac{1}{2}a\right)$.

34. S is given by $\mathbf{r}(x,y) = x\,\mathbf{i} + y\,\mathbf{j} + \sqrt{x^2+y^2}\,\mathbf{k}$, $|\mathbf{r}_x\times\mathbf{r}_y| = \sqrt{1 + \dfrac{x^2+y^2}{x^2+y^2}} = \sqrt{2}$ so

$$m = \iint_S\left(10 - \sqrt{x^2+y^2}\right)dS = \iint_{1\le x^2+y^2\le 16}\left(10 - \sqrt{x^2+y^2}\right)\sqrt{2}\,dA$$
$$= \int_0^{2\pi}\int_1^4 \sqrt{2}\,(10 - r)\,r\,dr\,d\theta = 2\pi\sqrt{2}\left[5r^2 - \tfrac{1}{3}r^3\right]_1^4 = 108\sqrt{2}\,\pi$$

35. (a) $I_z = \iint_S (x^2+y^2)\rho(x,y,z)\,dS$

(b) $I_z = \iint_S (x^2+y^2)\left(10 - \sqrt{x^2+y^2}\right)dS = \displaystyle\iint_{1\le x^2+y^2\le 16}(x^2+y^2)\left(10 - \sqrt{x^2+y^2}\right)\sqrt{2}\,dA$

$= \int_0^{2\pi}\int_1^4 \sqrt{2}\,(10r^3 - r^4)\,dr\,d\theta = 2\sqrt{2}\,\pi\left(\frac{4329}{10}\right) = \frac{4329}{5}\sqrt{2}\,\pi$

36. S is given by $\mathbf{r}(x,y) = x\,\mathbf{i} + y\,\mathbf{j} + \sqrt{x^2+y^2}\,\mathbf{k}$ and $|\mathbf{r}_x\times\mathbf{r}_y| = \sqrt{2}$.

(a) $m = \iint_S k\,dS = k\displaystyle\iint_{0\le x^2+y^2\le a^2}\sqrt{2}\,dS = \sqrt{2}\,a^2 k\pi$; by symmetry $M_{xz} = M_{yz} = 0$, and

$M_{xy} = \iint_S zk\,dS = k\int_0^{2\pi}\int_0^a \sqrt{2}\,r^2\,dr\,d\theta = \frac{2}{3}\sqrt{2}\,a^3 k\pi$. Hence $(\overline{x},\overline{y},\overline{z}) = \left(0,0,\frac{2}{3}a\right)$.

(b) $I_z = \iint_S (x^2+y^2)k\,dS = \int_0^{2\pi}\int_0^a \sqrt{2}\,kr^3\,dr\,d\theta = 2\pi\sqrt{2}\,k\left(\frac{1}{4}a^4\right) = \frac{\sqrt{2}}{2}\pi k a^4$.

37. The rate of flow through the cylinder is the flux $\iint_S \rho\mathbf{v}\cdot\mathbf{n}\,dS = \iint_S \rho\mathbf{v}\cdot d\mathbf{S}$. We use the parametric representation $\mathbf{r}(u,v) = 2\cos u\,\mathbf{i} + 2\sin u\,\mathbf{j} + v\,\mathbf{k}$ for S, where $0\le u\le 2\pi$, $0\le v\le 1$, so $\mathbf{r}_u = -2\sin u\,\mathbf{i} + 2\cos u\,\mathbf{j}$, $\mathbf{r}_v = \mathbf{k}$, and the outward orientation is given by $\mathbf{r}_u\times\mathbf{r}_v = 2\cos u\,\mathbf{i} + 2\sin u\,\mathbf{j}$. Then

$$\iint_S \rho\mathbf{v}\cdot d\mathbf{S} = \rho\int_0^{2\pi}\int_0^1 \left(v\,\mathbf{i} + 4\sin^2 u\,\mathbf{j} + 4\cos^2 u\,\mathbf{k}\right)\cdot(2\cos u\,\mathbf{i} + 2\sin u\,\mathbf{j})\,dv\,du$$
$$= \rho\int_0^{2\pi}\int_0^1 \left(2v\cos u + 8\sin^3 u\right)dv\,du = \rho\int_0^{2\pi}\left(\cos u + 8\sin^3 u\right)du$$
$$= \rho\left[\sin u + 8\left(-\tfrac{1}{3}\right)(2 + \sin^2 u)\cos u\right]_0^{2\pi} = 0\text{ kg/s}$$

38. A parametric representation for the hemisphere S is $\mathbf{r}(\phi,\theta) = 3\sin\phi\cos\theta\,\mathbf{i} + 3\sin\phi\sin\theta\,\mathbf{j} + 3\cos\phi\,\mathbf{k}$, $0 \le \phi \le \pi/2$, $0 \le \theta \le 2\pi$. Then $\mathbf{r}_\phi = 3\cos\phi\cos\theta\,\mathbf{i} + 3\cos\phi\sin\theta\,\mathbf{j} - 3\sin\phi\,\mathbf{k}$, $\mathbf{r}_\theta = -3\sin\phi\sin\theta\,\mathbf{i} + 3\sin\phi\cos\theta\,\mathbf{j}$, and the outward orientation is given by $\mathbf{r}_\phi \times \mathbf{r}_\theta = 9\sin^2\phi\cos\theta\,\mathbf{i} + 9\sin^2\phi\sin\theta\,\mathbf{j} + 9\sin\phi\cos\phi\,\mathbf{k}$. The rate of flow through S is

$$\iint_S \rho\mathbf{v}\cdot d\mathbf{S} = \rho\int_0^{\pi/2}\int_0^{2\pi}(3\sin\phi\sin\theta\,\mathbf{i} + 3\sin\phi\cos\theta\,\mathbf{j})\cdot(9\sin^2\phi\cos\theta\,\mathbf{i} + 9\sin^2\phi\sin\theta\,\mathbf{j} + 9\sin\phi\cos\phi\,\mathbf{k})\,d\theta\,d\phi$$

$$= 27\rho\int_0^{\pi/2}\int_0^{2\pi}(\sin^3\phi\sin\theta\cos\theta + \sin^3\phi\sin\theta\cos\theta)\,d\theta\,d\phi = 54\rho\int_0^{\pi/2}\sin^3\phi\,d\phi\int_0^{2\pi}\sin\theta\cos\theta\,d\theta$$

$$= 54\rho\left[-\tfrac{1}{3}(2+\sin^2\phi)\cos\phi\right]_0^{\pi/2}\left[\tfrac{1}{2}\sin^2\theta\right]_0^{2\pi} = 0\text{ kg/s}$$

39. S consists of the hemisphere S_1 given by $z = \sqrt{a^2 - x^2 - y^2}$ and the disk S_2 given by $0 \le x^2 + y^2 \le a^2$, $z = 0$. On S_1: $\mathbf{E} = a\sin\phi\cos\theta\,\mathbf{i} + a\sin\phi\sin\theta\,\mathbf{j} + 2a\cos\phi\,\mathbf{k}$, $\mathbf{T}_\phi \times \mathbf{T}_\theta = a^2\sin^2\phi\cos\theta\,\mathbf{i} + a^2\sin^2\phi\sin\theta\,\mathbf{j} + a^2\sin\phi\cos\phi\,\mathbf{k}$. Thus

$$\iint_{S_1}\mathbf{E}\cdot d\mathbf{S} = \int_0^{2\pi}\int_0^{\pi/2}(a^3\sin^3\phi + 2a^3\sin\phi\cos^2\phi)\,d\phi\,d\theta$$

$$= \int_0^{2\pi}\int_0^{\pi/2}(a^3\sin\phi + a^3\sin\phi\cos^2\phi)\,d\phi\,d\theta = (2\pi)a^3\left(1+\tfrac{1}{3}\right) = \tfrac{8}{3}\pi a^3$$

On S_2: $\mathbf{E} = x\,\mathbf{i} + y\,\mathbf{j}$, and $\mathbf{r}_y \times \mathbf{r}_x = -\mathbf{k}$ so $\iint_{S_2}\mathbf{E}\cdot d\mathbf{S} = 0$. Hence the total charge is $q = \varepsilon_0\iint_S\mathbf{E}\cdot d\mathbf{S} = \frac{8}{3}\pi a^3\varepsilon_0$.

40. Referring to the figure, on

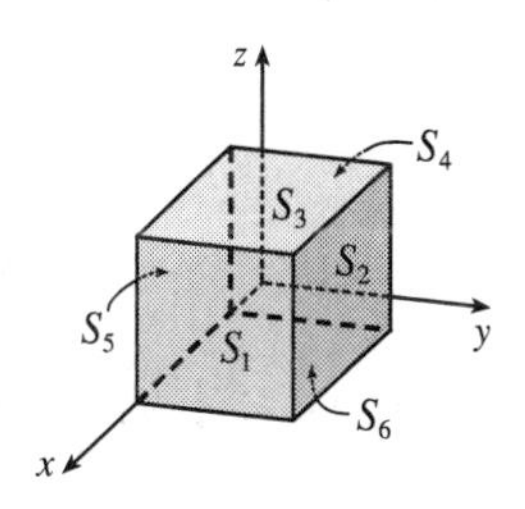

S_1: $\mathbf{E} = \mathbf{i} + y\,\mathbf{j} + z\,\mathbf{k}$, $\mathbf{r}_y \times \mathbf{r}_z = \mathbf{i}$ and $\iint_{S_1}\mathbf{E}\cdot d\mathbf{S} = \int_{-1}^{1}\int_{-1}^{1}dy\,dz = 4$;

S_2: $\mathbf{E} = x\,\mathbf{i} + \mathbf{j} + z\,\mathbf{k}$, $\mathbf{r}_z \times \mathbf{r}_x = \mathbf{j}$ and $\iint_{S_2}\mathbf{E}\cdot d\mathbf{S} = \int_{-1}^{1}\int_{-1}^{1}dx\,dz = 4$;

S_3: $\mathbf{E} = x\,\mathbf{i} + y\,\mathbf{j} + \mathbf{k}$, $\mathbf{r}_x \times \mathbf{r}_y = \mathbf{k}$ and $\iint_{S_3}\mathbf{E}\cdot d\mathbf{S} = \int_{-1}^{1}\int_{-1}^{1}dx\,dy = 4$;

S_4: $\mathbf{E} = -\mathbf{i} + y\,\mathbf{j} + z\,\mathbf{k}$, $\mathbf{r}_z \times \mathbf{r}_y = -\mathbf{i}$ and $\iint_{S_4}\mathbf{E}\cdot d\mathbf{S} = 4$.

Similarly $\iint_{S_5}\mathbf{E}\cdot d\mathbf{S} = \iint_{S_6}\mathbf{E}\cdot d\mathbf{S} = 4$. Hence $q = \varepsilon_0\iint_S\mathbf{E}\cdot d\mathbf{S} = \varepsilon_0\sum_{i=1}^{6}\iint_{S_i}\mathbf{E}\cdot d\mathbf{S} = 24\varepsilon_0$.

41. $K\nabla u = 6.5(4y\,\mathbf{j} + 4z\,\mathbf{k})$. S is given by $\mathbf{r}(x,\theta) = x\,\mathbf{i} + \sqrt{6}\cos\theta\,\mathbf{j} + \sqrt{6}\sin\theta\,\mathbf{k}$ and since we want the inward heat flow, we use $\mathbf{r}_x \times \mathbf{r}_\theta = -\sqrt{6}\cos\theta\,\mathbf{j} - \sqrt{6}\sin\theta\,\mathbf{k}$. Then the rate of heat flow inward is given by $\iint_S(-K\nabla u)\cdot d\mathbf{S} = \int_0^{2\pi}\int_0^4 -(6.5)(-24)\,dx\,d\theta = (2\pi)(156)(4) = 1248\pi$.

42. $u(x,y,z) = c/\sqrt{x^2+y^2+z^2}$,

$$\mathbf{F} = -K\nabla u = -K\left[-\frac{cx}{(x^2+y^2+z^2)^{3/2}}\,\mathbf{i} - \frac{cy}{(x^2+y^2+z^2)^{3/2}}\,\mathbf{j} - \frac{cz}{(x^2+y^2+z^2)^{3/2}}\,\mathbf{k}\right]$$

$$= \frac{cK}{(x^2+y^2+z^2)^{3/2}}(x\,\mathbf{i} + y\,\mathbf{j} + z\,\mathbf{k})$$

and the outward unit normal is $\mathbf{n} = \dfrac{1}{a}(x\,\mathbf{i} + y\,\mathbf{j} + z\,\mathbf{k})$.

Thus $\mathbf{F}\cdot\mathbf{n} = \dfrac{cK}{a(x^2+y^2+z^2)^{3/2}}(x^2+y^2+z^2)$, but on S, $x^2+y^2+z^2 = a^2$ so $\mathbf{F}\cdot\mathbf{n} = \dfrac{cK}{a^2}$. Hence the rate of heat flow across S is $\displaystyle\iint_S\mathbf{F}\cdot d\mathbf{S} = \frac{cK}{a^2}\iint_S dS = \frac{cK}{a^2}(4\pi a^2) = 4\pi Kc$.

43. Let S be a sphere of radius a centered at the origin. Then $|\mathbf{r}| = a$ and $\mathbf{F}(\mathbf{r}) = c\mathbf{r}/|\mathbf{r}|^3 = (c/a^3)(x\,\mathbf{i} + y\,\mathbf{j} + z\,\mathbf{k})$. A parametric representation for S is $\mathbf{r}(\phi, \theta) = a\sin\phi\cos\theta\,\mathbf{i} + a\sin\phi\sin\theta\,\mathbf{j} + a\cos\phi\,\mathbf{k}$, $0 \le \phi \le \pi$, $0 \le \theta \le 2\pi$. Then $\mathbf{r}_\phi = a\cos\phi\cos\theta\,\mathbf{i} + a\cos\phi\sin\theta\,\mathbf{j} - a\sin\phi\,\mathbf{k}$, $\mathbf{r}_\theta = -a\sin\phi\sin\theta\,\mathbf{i} + a\sin\phi\cos\theta\,\mathbf{j}$, and the outward orientation is given by $\mathbf{r}_\phi \times \mathbf{r}_\theta = a^2\sin^2\phi\cos\theta\,\mathbf{i} + a^2\sin^2\phi\sin\theta\,\mathbf{j} + a^2\sin\phi\cos\phi\,\mathbf{k}$. The flux of $\mathbf{F}$ across S is

$$\begin{aligned}\iint_S \mathbf{F}\cdot d\mathbf{S} &= \int_0^\pi \int_0^{2\pi} \frac{c}{a^3}\left(a\sin\phi\cos\theta\,\mathbf{i} + a\sin\phi\sin\theta\,\mathbf{j} + a\cos\phi\,\mathbf{k}\right)\\ &\qquad \cdot \left(a^2\sin^2\phi\cos\theta\,\mathbf{i} + a^2\sin^2\phi\sin\theta\,\mathbf{j} + a^2\sin\phi\cos\phi\,\mathbf{k}\right)d\theta\,d\phi\\ &= \frac{c}{a^3}\int_0^\pi\int_0^{2\pi} a^3\left(\sin^3\phi + \sin\phi\cos^2\phi\right)d\theta\,d\phi = c\int_0^\pi\int_0^{2\pi}\sin\phi\,d\theta\,d\phi = 4\pi c\end{aligned}$$

Thus the flux does not depend on the radius a.

13.8 Stokes' Theorem

1. The boundary curve C is the circle $x^2 + y^2 = 4$, $z = 0$ oriented in the counterclockwise direction. The vector equation is $\mathbf{r}(t) = 2\cos t\,\mathbf{i} + 2\sin t\,\mathbf{j}$, $0 \le t \le 2\pi$, so $\mathbf{r}'(t) = -2\sin t\,\mathbf{i} + 2\cos t\,\mathbf{j}$ and $\mathbf{F}(\mathbf{r}(t)) = (2\cos t)^2 e^{(2\sin t)(0)}\,\mathbf{i} + (2\sin t)^2 e^{(2\cos t)(0)}\,\mathbf{j} + (0)^2 e^{(2\cos t)(2\sin t)}\,\mathbf{k} = 4\cos^2 t\,\mathbf{i} + 4\sin^2 t\,\mathbf{j}$. Then, by Stokes' Theorem,

$$\begin{aligned}\iint_S \operatorname{curl}\mathbf{F}\cdot d\mathbf{S} &= \int_C \mathbf{F}\cdot d\mathbf{r} = \int_0^{2\pi}\mathbf{F}(\mathbf{r}(t))\cdot\mathbf{r}'(t)\,dt = \int_0^{2\pi}\left(-8\cos^2 t\sin t + 8\sin^2 t\cos t\right)dt\\ &= 8\left[\tfrac{1}{3}\cos^3 t + \tfrac{1}{3}\sin^3 t\right]_0^{2\pi} = 0\end{aligned}$$

2. The plane $z = 5$ intersects the paraboloid $z = 9 - x^2 - y^2$ in the circle $x^2 + y^2 = 4$, $z = 5$. This boundary curve C is oriented in the counterclockwise direction, so the vector equation is $\mathbf{r}(t) = 2\cos t\,\mathbf{i} + 2\sin t\,\mathbf{j} + 5\,\mathbf{k}$, $0 \le t \le 2\pi$. Then $\mathbf{r}'(t) = -2\sin t\,\mathbf{i} + 2\cos t\,\mathbf{j}$, $\mathbf{F}(\mathbf{r}(t)) = 10\sin t\,\mathbf{i} + 10\cos t\,\mathbf{j} + 4\cos t\sin t\,\mathbf{k}$, and by Stokes' Theorem,

$$\iint_S \operatorname{curl}\mathbf{F}\cdot d\mathbf{S} = \int_C \mathbf{F}\cdot d\mathbf{r} = \int_0^{2\pi}\mathbf{F}(\mathbf{r}(t))\cdot\mathbf{r}'(t)\,dt = \int_0^{2\pi}\left(-20\sin^2 t + 20\cos^2 t\right)dt = 20\int_0^{2\pi}\cos 2t\,dt = 0$$

3. C is the square in the plane $z = -1$. By (3), $\iint_{S_1}\operatorname{curl}\mathbf{F}\cdot d\mathbf{S} = \oint_C \mathbf{F}\cdot d\mathbf{r} = \iint_{S_2}\operatorname{curl}\mathbf{F}\cdot d\mathbf{S}$ where S_1 is the original cube without the bottom and S_2 is the bottom face of the cube. $\operatorname{curl}\mathbf{F} = x^2 z\,\mathbf{i} + (xy - 2xyz)\,\mathbf{j} + (y - xz)\,\mathbf{k}$. For S_2, we choose $\mathbf{n} = \mathbf{k}$ so that C has the same orientation for both surfaces. Then $\operatorname{curl}\mathbf{F}\cdot\mathbf{n} = y - xz = x + y$ on S_2, where $z = -1$. Thus $\iint_{S_2}\operatorname{curl}\mathbf{F}\cdot d\mathbf{S} = \int_{-1}^1\int_{-1}^1 (x + y)\,dx\,dy = 0$ so $\iint_{S_1}\operatorname{curl}\mathbf{F}\cdot d\mathbf{S} = 0$.

4. The boundary curve C is the unit circle in the yz-plane. By Equation 3, $\iint_{S_1}\operatorname{curl}\mathbf{F}\cdot d\mathbf{S} = \oint_C \mathbf{F}\cdot d\mathbf{r} = \iint_{S_2}\operatorname{curl}\mathbf{F}\cdot d\mathbf{S}$ where S_1 is the original hemisphere and S_2 is the disk $y^2 + z^2 \le 1$, $x = 0$. $\operatorname{curl}\mathbf{F} = (x - x^2)\,\mathbf{i} - (y + e^{xy}\sin z)\,\mathbf{j} + (2xz - xe^{xy}\cos z)\,\mathbf{k}$, and for S_2 we choose $\mathbf{n} = \mathbf{i}$ so that C has the same orientation for both surfaces. Then $\operatorname{curl}\mathbf{F}\cdot\mathbf{n} = x - x^2$ on S_2, where $x = 0$. Thus

$$\iint_{S_2}\operatorname{curl}\mathbf{F}\cdot d\mathbf{S} = \iint_{y^2+z^2\le 1}(x - x^2)\,dA = \iint_{y^2+z^2\le 1} 0\,dA = 0.$$

Alternatively, we can evaluate $\oint_C \mathbf{F}\cdot d\mathbf{r}$: C with positive orientation is given by $\mathbf{r}(t) = \langle 0, \cos t, \sin t\rangle$, $0 \le t \le 2\pi$, and

$$\iint_S \operatorname{curl}\mathbf{F}\cdot d\mathbf{S} = \oint_C \mathbf{F}\cdot d\mathbf{r} = \int_0^{2\pi}\left\langle e^{0(\cos t)}\cos(\sin t), (0)^2(\sin t), (0)(\cos t)\right\rangle\cdot\langle 0, -\sin t, \cos t\rangle\,dt = \int_0^{2\pi} 0\,dt = 0.$$

5. curl $\mathbf{F} = -2z\,\mathbf{i} - 2x\,\mathbf{j} - 2y\,\mathbf{k}$ and we take the surface S to be the planar region enclosed by C, so S is the portion of the plane $x + y + z = 1$ over $D = \{(x, y) \mid 0 \le x \le 1, 0 \le y \le 1 - x\}$. Since C is oriented counterclockwise, we orient S upward. Using Equation 13.7.10, we have $z = g(x, y) = 1 - x - y$, $P = -2z$, $Q = -2x$, $R = -2y$, and

$$\begin{aligned}\int_C \mathbf{F} \cdot d\mathbf{r} &= \iint_S \operatorname{curl} \mathbf{F} \cdot d\mathbf{S} = \iint_D [-(-2z)(-1) - (-2x)(-1) + (-2y)]\, dA \\ &= \int_0^1 \int_0^{1-x} (-2)\, dy\, dx = -2\int_0^1 (1 - x)\, dx = -1\end{aligned}$$

6. curl $\mathbf{F} = e^x\,\mathbf{k}$ and S is the portion of the plane $2x + y + 2z = 2$ over $D = \{(x, y) \mid 0 \le x \le 1, 0 \le y \le 2 - 2x\}$. We orient S upward and use Equation 13.7.10 with $z = g(x, y) = 1 - x - \frac{1}{2}y$:

$$\begin{aligned}\int_C \mathbf{F} \cdot d\mathbf{r} &= \iint_S \operatorname{curl} \mathbf{F} \cdot d\mathbf{S} = \iint_D (0 + 0 + e^x)\, dA = \int_0^1 \int_0^{2-2x} e^x\, dy\, dx \\ &= \int_0^1 (2 - 2x)e^x\, dx = \left[(2 - 2x)e^x + 2e^x\right]_0^1 \qquad \text{[by integrating by parts]} \qquad = 2e - 4\end{aligned}$$

7. curl $\mathbf{F} = (xe^{xy} - 2x)\,\mathbf{i} - (ye^{xy} - y)\,\mathbf{j} + (2z - z)\,\mathbf{k}$ and we take S to be the disk $x^2 + y^2 \le 16$, $z = 5$. Since C is oriented counterclockwise (from above), we orient S upward. Then $\mathbf{n} = \mathbf{k}$ and curl $\mathbf{F} \cdot \mathbf{n} = 2z - z$ on S, where $z = 5$. Thus

$$\oint \mathbf{F} \cdot d\mathbf{r} = \iint_S \operatorname{curl} \mathbf{F} \cdot \mathbf{n}\, dS = \iint_S (2z - z)\, dS = \iint_S (10 - 5)\, dS = 5(\text{area of } S) = 5(\pi \cdot 4^2) = 80\pi.$$

8. The curve of intersection is an ellipse in the plane $z = 5 - x$. curl $\mathbf{F} = \mathbf{i} - x\,\mathbf{k}$ and we take the surface S to be the planar region enclosed by C with upward orientation, so

$$\begin{aligned}\oint_C \mathbf{F} \cdot d\mathbf{r} &= \iint_S \operatorname{curl} \mathbf{F} \cdot d\mathbf{S} = \iint_{x^2+y^2\le 9} [-1(-1) - 0 + (-x)]\, dA = \int_0^{2\pi} \int_0^3 (1 - r\cos\theta)\, r\, dr\, d\theta \\ &= \int_0^{2\pi} \int_0^3 \left(r - r^2\cos\theta\right) dr\, d\theta = \int_0^{2\pi} \left(\tfrac{9}{2} - 9\cos\theta\right) d\theta = \left[\tfrac{9}{2}\theta - 9\sin\theta\right]_0^{2\pi} = 9\pi\end{aligned}$$

9. (a) The curve of intersection is an ellipse in the plane $x + y + z = 1$ with unit normal $\mathbf{n} = \frac{1}{\sqrt{3}}(\mathbf{i} + \mathbf{j} + \mathbf{k})$, curl $\mathbf{F} = x^2\,\mathbf{j} + y^2\,\mathbf{k}$, and curl $\mathbf{F} \cdot \mathbf{n} = \frac{1}{\sqrt{3}}(x^2 + y^2)$. Then

$$\oint_C \mathbf{F} \cdot d\mathbf{r} = \iint_S \tfrac{1}{\sqrt{3}}(x^2 + y^2)\, dS = \iint_{x^2+y^2\le 9} (x^2 + y^2)\, dx\, dy = \int_0^{2\pi} \int_0^3 r^3\, dr\, d\theta = 2\pi\left(\tfrac{81}{4}\right) = \tfrac{81\pi}{2}$$

(b)

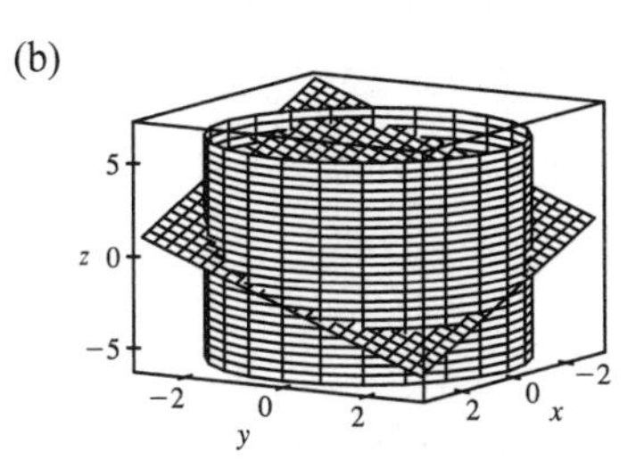

(c) One possible parametrization is $x = 3\cos t$, $y = 3\sin t$, $z = 1 - 3\cos t - 3\sin t$, $0 \le t \le 2\pi$.

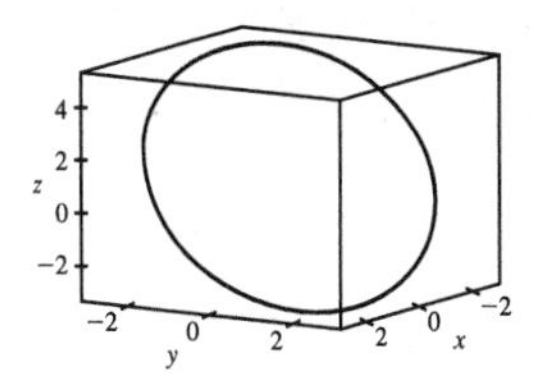

10. (a) S is the part of the surface $z = y^2 - x^2$ that lies above the unit disk D. curl $\mathbf{F} = x\,\mathbf{i} - y\,\mathbf{j} + (x^2 - x^2)\,\mathbf{k} = x\,\mathbf{i} - y\,\mathbf{j}$. Using Equation 13.7.10 with $g(x, y) = y^2 - x^2$, $P = x$, $Q = -y$, we have

$$\begin{aligned}\int_C \mathbf{F} \cdot d\mathbf{r} &= \iint_S \operatorname{curl} \mathbf{F} \cdot d\mathbf{S} = \iint_D [-x(-2x) - (-y)(2y)]\, dA = 2\iint_D (x^2 + y^2)\, dA \\ &= 2\int_0^{2\pi} \int_0^1 r^2 r\, dr\, d\theta = 2(2\pi)\left[\tfrac{1}{4}r^4\right]_0^1 = \pi\end{aligned}$$

(b)

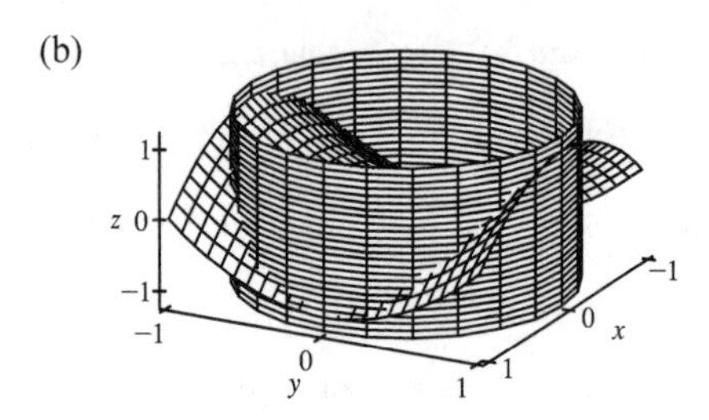

(c) One possible set of parametric equations is $x = \cos t$, $y = \sin t$, $z = \sin^2 t - \cos^2 t$, $0 \le t \le 2\pi$.

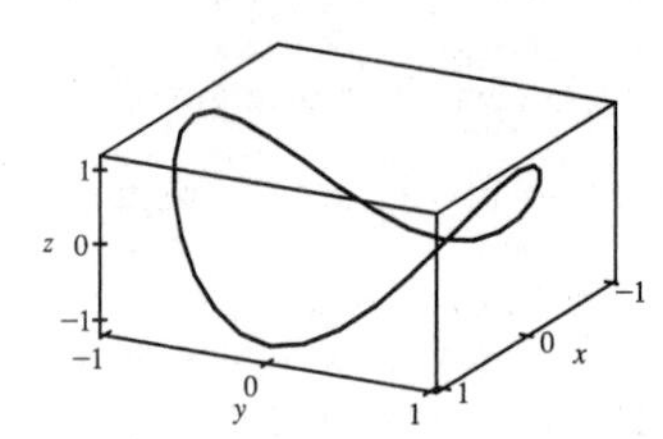

11. The boundary curve C is the circle $x^2 + y^2 = 1$, $z = 1$ oriented in the counterclockwise direction as viewed from above. We can parametrize C by $\mathbf{r}(t) = \cos t\,\mathbf{i} + \sin t\,\mathbf{j} + \mathbf{k}$, $0 \le t \le 2\pi$, and then $\mathbf{r}'(t) = -\sin t\,\mathbf{i} + \cos t\,\mathbf{j}$. Thus $\mathbf{F}(\mathbf{r}(t)) = \sin^2 t\,\mathbf{i} + \cos t\,\mathbf{j} + \mathbf{k}$, $\mathbf{F}(\mathbf{r}(t)) \cdot \mathbf{r}'(t) = \cos^2 t - \sin^3 t$, and

$$\begin{aligned}\oint_C \mathbf{F} \cdot d\mathbf{r} &= \int_0^{2\pi} (\cos^2 t - \sin^3 t)\,dt = \int_0^{2\pi} \tfrac{1}{2}(1 + \cos 2t)\,dt - \int_0^{2\pi} (1 - \cos^2 t)\sin t\,dt \\ &= \tfrac{1}{2}\left[t + \tfrac{1}{2}\sin 2t\right]_0^{2\pi} - \left[-\cos t + \tfrac{1}{3}\cos^3 t\right]_0^{2\pi} = \pi\end{aligned}$$

Now $\operatorname{curl} \mathbf{F} = (1 - 2y)\,\mathbf{k}$, and the projection D of S on the xy-plane is the disk $x^2 + y^2 \le 1$, so by Equation 13.7.10 with $z = g(x, y) = x^2 + y^2$ we have

$\iint_S \operatorname{curl} \mathbf{F} \cdot d\mathbf{S} = \iint_D (1 - 2y)\,dA = \int_0^{2\pi} \int_0^1 (1 - 2r\sin\theta)\,r\,dr\,d\theta = \int_0^{2\pi} \left(\tfrac{1}{2} - \tfrac{2}{3}\sin\theta\right) d\theta = \pi.$

12. The plane intersects the coordinate axes at $x = 1$, $y = z = 2$ so the boundary curve C consists of the three line segments C_1: $\mathbf{r}_1(t) = (1 - t)\,\mathbf{i} + 2t\,\mathbf{j}$, $0 \le t \le 1$, C_2: $\mathbf{r}_2(t) = (2 - 2t)\,\mathbf{j} + 2t\,\mathbf{k}$, $0 \le t \le 1$, C_3: $\mathbf{r}_3(t) = t\,\mathbf{i} + (2 - 2t)\,\mathbf{k}$, $0 \le t \le 1$. Then

$$\begin{aligned}\oint_C \mathbf{F} \cdot d\mathbf{r} &= \int_0^1 [(1 - t)\,\mathbf{i} + 2t\,\mathbf{j}] \cdot (-\mathbf{i} + 2\,\mathbf{j})\,dt + \int_0^1 [(2 - 2t)\,\mathbf{j}] \cdot (-2\,\mathbf{j} + 2\,\mathbf{k})\,dt + \int_0^1 (t\,\mathbf{i}) \cdot (\mathbf{i} - 2\,\mathbf{k})\,dt \\ &= \int_0^1 (5t - 1)\,dt + \int_0^1 (4t - 4)\,dt + \int_0^1 t\,dt = \tfrac{3}{2} - 2 + \tfrac{1}{2} = 0\end{aligned}$$

Now $\operatorname{curl} \mathbf{F} = xz\,\mathbf{i} - yz\,\mathbf{j}$, so by Equation 13.7.10 with $z = g(x, y) = 2 - 2x - y$ we have

$$\begin{aligned}\iint_S \operatorname{curl} \mathbf{F} \cdot d\mathbf{S} &= \iint_D [-x(2 - 2x - y)(-2) + y(2 - 2x - y)(-1)]\,dA = \int_0^1 \int_0^{2-2x} (4x - 4x^2 - 2y + y^2)\,dy\,dx \\ &= \int_0^1 \left[4x(2 - 2x) - 4x^2(2 - 2x) - (2 - 2x)^2 + \tfrac{1}{3}(2 - 2x)^3\right] dx \\ &= \int_0^1 \left(\tfrac{16}{3}x^3 - 12x^2 + 8x - \tfrac{4}{3}\right) dx = \left[\tfrac{4}{3}x^4 - 4x^3 + 4x^2 - \tfrac{4}{3}x\right]_0^1 = 0\end{aligned}$$

13. It is easier to use Stokes' Theorem than to compute the work directly. Let S be the planar region enclosed by the path of the particle, so S is the portion of the plane $z = \frac{1}{2}y$ for $0 \le x \le 1$, $0 \le y \le 2$, with upward orientation. $\operatorname{curl} \mathbf{F} = 8y\,\mathbf{i} + 2z\,\mathbf{j} + 2y\,\mathbf{k}$ and

$$\begin{aligned}\oint_C \mathbf{F} \cdot d\mathbf{r} &= \iint_S \operatorname{curl} \mathbf{F} \cdot d\mathbf{S} = \iint_D \left[-8y\,(0) - 2z\left(\tfrac{1}{2}\right) + 2y\right] dA = \int_0^1 \int_0^2 \left(2y - \tfrac{1}{2}y\right) dy\,dx \\ &= \int_0^1 \int_0^2 \tfrac{3}{2}y\,dy\,dx = \int_0^1 \left[\tfrac{3}{4}y^2\right]_{y=0}^{y=2} dx = \int_0^1 3\,dx = 3\end{aligned}$$

14. $\int_C (y+\sin x)\,dx + (z^2+\cos y)\,dy + x^3\,dz = \int_C \mathbf{F}\cdot d\mathbf{r}$, where $\mathbf{F}(x,y,z) = (y+\sin x)\,\mathbf{i} + (z^2+\cos y)\,\mathbf{j} + x^3\,\mathbf{k} \Rightarrow$ $\operatorname{curl}\mathbf{F} = -2z\,\mathbf{i} - 3x^2\,\mathbf{j} - \mathbf{k}$. Since $\sin 2t = 2\sin t\cos t$, C lies on the surface $z = 2xy$. Let S be the part of this surface that is bounded by C. Then the projection of S onto the xy-plane is the unit disk D $(x^2+y^2 \le 1)$. C is traversed clockwise (when viewed from above) so S is oriented downward. Using Equation 13.7.10 with $g(x,y) = 2xy$, $P = -2z = -2(2xy) = -4xy$, $Q = -3x^2$, $R = -1$, and multiplying by -1 for the downward orientation, we have

$$\begin{aligned}\int_C \mathbf{F}\cdot d\mathbf{r} &= -\iint_S \operatorname{curl}\mathbf{F}\cdot d\mathbf{S} = -\iint_D \left[-(-4xy)(2y) - (-3x^2)(2x) - 1\right] dA \\ &= -\iint_D (8xy^2 + 6x^3 - 1)\,dA = -\int_0^{2\pi}\int_0^1 (8r^3\cos\theta\sin^2\theta + 6r^3\cos^3\theta - 1)\,r\,dr\,d\theta \\ &= -\int_0^{2\pi}\left(\tfrac{8}{5}\cos\theta\sin^2\theta + \tfrac{6}{5}\cos^3\theta - \tfrac{1}{2}\right)d\theta = -\left[\tfrac{8}{15}\sin^3\theta + \tfrac{6}{5}\left(\sin\theta - \tfrac{1}{3}\sin^3\theta\right) - \tfrac{1}{2}\theta\right]_0^{2\pi} = \pi\end{aligned}$$

15. Assume S is centered at the origin with radius a and let H_1 and H_2 be the upper and lower hemispheres, respectively, of S. Then $\iint_S \operatorname{curl}\mathbf{F}\cdot d\mathbf{S} = \iint_{H_1}\operatorname{curl}\mathbf{F}\cdot d\mathbf{S} + \iint_{H_2}\operatorname{curl}\mathbf{F}\cdot d\mathbf{S} = \oint_{C_1}\mathbf{F}\cdot d\mathbf{r} + \oint_{C_2}\mathbf{F}\cdot d\mathbf{r}$ by Stokes' Theorem. But C_1 is the circle $x^2+y^2 = a^2$ oriented in the counterclockwise direction while C_2 is the same circle oriented in the clockwise direction. Hence $\oint_{C_2}\mathbf{F}\cdot d\mathbf{r} = -\oint_{C_1}\mathbf{F}\cdot d\mathbf{r}$ so $\iint_S \operatorname{curl}\mathbf{F}\cdot d\mathbf{S} = 0$ as desired.

16. (a) By Exercise 13.5.24, $\operatorname{curl}(f\nabla g) = f\operatorname{curl}(\nabla g) + \nabla f\times\nabla g = \nabla f\times\nabla g$ since $\operatorname{curl}(\nabla g) = \mathbf{0}$. Hence by Stokes' Theorem $\int_C (f\nabla g)\cdot d\mathbf{r} = \iint_S (\nabla f\times\nabla g)\cdot d\mathbf{S}$.

(b) As in (a), $\operatorname{curl}(f\nabla f) = \nabla f\times\nabla f = \mathbf{0}$, so by Stokes' Theorem, $\int_C (f\nabla f)\cdot d\mathbf{r} = \iint_S [\operatorname{curl}(f\nabla f)]\cdot d\mathbf{S} = 0$.

(c) As in part (a),

$$\begin{aligned}\operatorname{curl}(f\nabla g + g\nabla f) &= \operatorname{curl}(f\nabla g) + \operatorname{curl}(g\nabla f) && \text{[by Exercise 13.5.22]} \\ &= (\nabla f\times\nabla g) + (\nabla g\times\nabla f) = \mathbf{0} && [\text{since } \mathbf{u}\times\mathbf{v} = -(\mathbf{v}\times\mathbf{u})]\end{aligned}$$

Hence by Stokes' Theorem, $\int_C (f\nabla g + g\nabla f)\cdot d\mathbf{r} = \iint_S \operatorname{curl}(f\nabla g + g\nabla f)\cdot d\mathbf{S} = 0$.

13.9 The Divergence Theorem

1. $\operatorname{div}\mathbf{F} = 3 + x + 2x = 3 + 3x$, so

$\iiint_E \operatorname{div}\mathbf{F}\,dV = \int_0^1\int_0^1\int_0^1 (3x+3)\,dx\,dy\,dz = \frac{9}{2}$ (notice the triple integral is three times the volume of the cube plus three times $\overline{x}$).

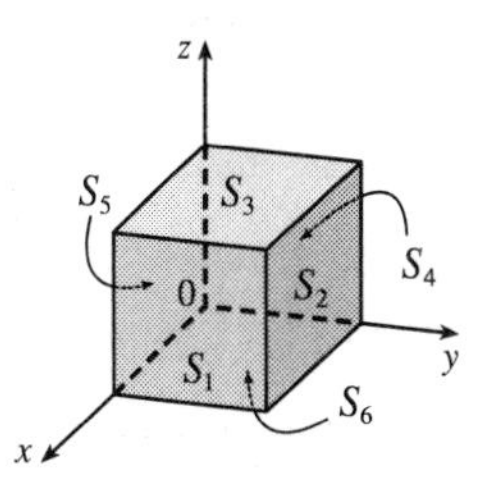

To compute $\iint_S \mathbf{F}\cdot d\mathbf{S}$, on S_1: $\mathbf{n} = \mathbf{i}$, $\mathbf{F} = 3\,\mathbf{i} + y\,\mathbf{j} + 2z\,\mathbf{k}$, and $\iint_{S_1}\mathbf{F}\cdot d\mathbf{S} = \iint_{S_1} 3\,dS = 3$;

S_2: $\mathbf{F} = 3x\,\mathbf{i} + x\,\mathbf{j} + 2xz\,\mathbf{k}$, $\mathbf{n} = \mathbf{j}$ and $\iint_{S_2}\mathbf{F}\cdot d\mathbf{S} = \iint_{S_2} x\,dS = \frac{1}{2}$;

S_3: $\mathbf{F} = 3x\,\mathbf{i} + xy\,\mathbf{j} + 2x\,\mathbf{k}$, $\mathbf{n} = \mathbf{k}$ and $\iint_{S_3}\mathbf{F}\cdot d\mathbf{S} = \iint_{S_3} 2x\,dS = 1$;

S_4: $\mathbf{F} = \mathbf{0}$, $\iint_{S_4}\mathbf{F}\cdot d\mathbf{S} = 0$; S_5: $\mathbf{F} = 3x\,\mathbf{i} + 2x\,\mathbf{k}$, $\mathbf{n} = -\mathbf{j}$ and $\iint_{S_5}\mathbf{F}\cdot d\mathbf{S} = \iint_{S_5} 0\,dS = 0$;

S_6: $\mathbf{F} = 3x\,\mathbf{i} + xy\,\mathbf{j}$, $\mathbf{n} = -\mathbf{k}$ and $\iint_{S_6}\mathbf{F}\cdot d\mathbf{S} = \iint_{S_6} 0\,dS = 0$. Thus $\iint_S \mathbf{F}\cdot d\mathbf{S} = \frac{9}{2}$.

2. $\operatorname{div}\mathbf{F} = 2x + x + 1 = 3x + 1$ so

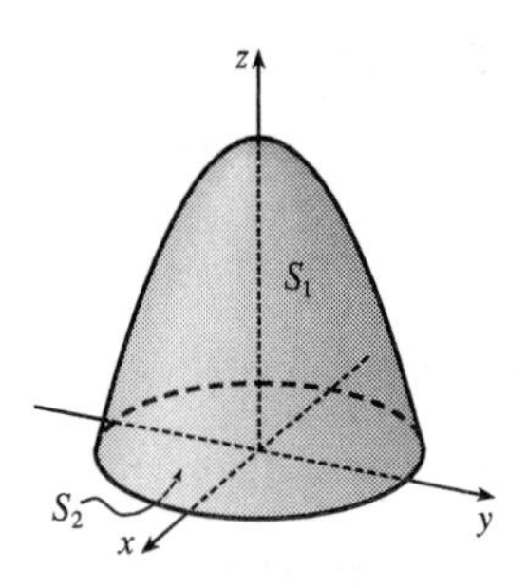

$$\begin{aligned}\iiint_E \operatorname{div}\mathbf{F}\,dV &= \iiint_E (3x+1)\,dV = \int_0^{2\pi}\int_0^2\int_0^{4-r^2}(3r\cos\theta+1)\,r\,dz\,dr\,d\theta \\ &= \int_0^2\int_0^{2\pi} r(3r\cos\theta+1)(4-r^2)\,d\theta\,dr \\ &= \int_0^{2\pi} r(4-r^2)\big[3r\sin\theta+\theta\big]_{\theta=0}^{\theta=2\pi}\,dr \\ &= 2\pi\int_0^2 (4r-r^3)\,dr = 2\pi\big[2r^2-\tfrac14 r^4\big]_0^2 \\ &= 2\pi(8-4) = 8\pi\end{aligned}$$

On S_1: The surface is $z = 4 - x^2 - y^2$, $x^2 + y^2 \le 4$, with upward orientation, and $\mathbf{F} = x^2\,\mathbf{i} + xy\,\mathbf{j} + (4 - x^2 - y^2)\,\mathbf{k}$. Then

$$\begin{aligned}\iint_{S_1}\mathbf{F}\cdot d\mathbf{S} &= \iint_D[-(x^2)(-2x)-(xy)(-2y)+(4-x^2-y^2)]\,dA \\ &= \iint_D\big[2x(x^2+y^2)+4-(x^2+y^2)\big]\,dA = \int_0^{2\pi}\int_0^2(2r\cos\theta\cdot r^2+4-r^2)\,r\,dr\,d\theta \\ &= \int_0^{2\pi}\big[\tfrac25 r^5\cos\theta+2r^2-\tfrac14 r^4\big]_{r=0}^{r=2}\,d\theta = \int_0^{2\pi}\left(\tfrac{64}{5}\cos\theta+4\right)d\theta = \big[\tfrac{64}{5}\sin\theta+4\theta\big]_0^{2\pi} = 8\pi\end{aligned}$$

On S_2: The surface is $z = 0$ with downward orientation, so $\mathbf{F} = x^2\,\mathbf{i} + xy\,\mathbf{j}$, $\mathbf{n} = -\mathbf{k}$ and $\iint_{S_2}\mathbf{F}\cdot\mathbf{n}\,dS = \iint_{S_2} 0\,dS = 0$.

Thus $\iint_S \mathbf{F}\cdot d\mathbf{S} = \iint_{S_1}\mathbf{F}\cdot d\mathbf{S} + \iint_{S_2}\mathbf{F}\cdot d\mathbf{S} = 8\pi$.

3. $\operatorname{div}\mathbf{F} = x + y + z$, so

$$\begin{aligned}\iiint_E \operatorname{div}\mathbf{F}\,dV &= \int_0^{2\pi}\int_0^1\int_0^1(r\cos\theta+r\sin\theta+z)\,r\,dz\,dr\,d\theta = \int_0^{2\pi}\int_0^1\left(r^2\cos\theta+r^2\sin\theta+\tfrac12 r\right)dr\,d\theta \\ &= \int_0^{2\pi}\left(\tfrac13\cos\theta+\tfrac13\sin\theta+\tfrac14\right)d\theta = \tfrac14(2\pi) = \tfrac{\pi}{2}\end{aligned}$$

Let S_1 be the top of the cylinder, S_2 the bottom, and S_3 the vertical edge. On S_1, $z = 1$, $\mathbf{n} = \mathbf{k}$, and $\mathbf{F} = xy\,\mathbf{i} + y\,\mathbf{j} + x\,\mathbf{k}$, so $\iint_{S_1}\mathbf{F}\cdot d\mathbf{S} = \iint_{S_1}\mathbf{F}\cdot\mathbf{n}\,dS = \iint_{S_1} x\,dS = \int_0^{2\pi}\int_0^1(r\cos\theta)\,r\,dr\,d\theta = \big[\sin\theta\big]_0^{2\pi}\big[\tfrac13 r^3\big]_0^1 = 0$. On S_2, $z = 0$, $\mathbf{n} = -\mathbf{k}$, and $\mathbf{F} = xy\,\mathbf{i}$ so $\iint_{S_2}\mathbf{F}\cdot d\mathbf{S} = \iint_{S_2} 0\,dS = 0$. S_3 is given by $\mathbf{r}(\theta, z) = \cos\theta\,\mathbf{i} + \sin\theta\,\mathbf{j} + z\,\mathbf{k}$, $0 \le \theta \le 2\pi$, $0 \le z \le 1$. Then $\mathbf{r}_\theta \times \mathbf{r}_z = \cos\theta\,\mathbf{i} + \sin\theta\,\mathbf{j}$ and

$$\begin{aligned}\iint_{S_3}\mathbf{F}\cdot d\mathbf{S} &= \iint_D \mathbf{F}\cdot(\mathbf{r}_\theta\times\mathbf{r}_z)\,dA = \int_0^{2\pi}\int_0^1(\cos^2\theta\sin\theta+z\sin^2\theta)\,dz\,d\theta \\ &= \int_0^{2\pi}\left(\cos^2\theta\sin\theta+\tfrac12\sin^2\theta\right)d\theta = \left[-\tfrac13\cos^3\theta+\tfrac14\left(\theta-\tfrac12\sin2\theta\right)\right]_0^{2\pi} = \tfrac{\pi}{2}\end{aligned}$$

Thus $\iint_S \mathbf{F}\cdot d\mathbf{S} = 0 + 0 + \frac{\pi}{2} = \frac{\pi}{2}$.

4. $\operatorname{div}\mathbf{F} = 1 + 1 + 1 = 3$, so $\iiint_E \operatorname{div}\mathbf{F}\,dV = \iiint_E 3\,dV = 3(\text{volume of ball}) = 3\left(\frac43\pi\right) = 4\pi$. To find $\iint_S \mathbf{F}\cdot d\mathbf{S}$ we use spherical coordinates. S is the unit sphere, represented by $\mathbf{r}(\phi,\theta) = \sin\phi\cos\theta\,\mathbf{i} + \sin\phi\sin\theta\,\mathbf{j} + \cos\phi\,\mathbf{k}$, $0 \le \phi \le \pi$, $0 \le \theta \le 2\pi$. Then $\mathbf{r}_\phi \times \mathbf{r}_\theta = \sin^2\phi\cos\theta\,\mathbf{i} + \sin^2\phi\sin\theta\,\mathbf{j} + \sin\phi\cos\phi\,\mathbf{k}$ (see Example 13.6.4) and $\mathbf{F}(\mathbf{r}(\phi,\theta)) = \sin\phi\cos\theta\,\mathbf{i} + \sin\phi\sin\theta\,\mathbf{j} + \cos\phi\,\mathbf{k}$. Thus

$$\begin{aligned}\iint_S \mathbf{F}\cdot d\mathbf{S} &= \iint_D \mathbf{F}\cdot(\mathbf{r}_\phi\times\mathbf{r}_\theta)\,dA = \int_0^{2\pi}\int_0^{\pi}\left(\sin^3\phi\cos^2\theta+\sin^3\phi\sin^2\theta+\sin\phi\cos^2\phi\right)d\phi\,d\theta \\ &= \int_0^{2\pi}d\theta\int_0^{\pi}\sin\phi\,d\phi = (2\pi)(2) = 4\pi\end{aligned}$$

5. $\operatorname{div}\mathbf{F} = \frac{\partial}{\partial x}(e^x\sin y) + \frac{\partial}{\partial y}(e^x\cos y) + \frac{\partial}{\partial z}(yz^2) = e^x\sin y - e^x\sin y + 2yz = 2yz$, so by the Divergence Theorem, $\iint_S \mathbf{F}\cdot d\mathbf{S} = \iiint_E \operatorname{div}\mathbf{F}\,dV = \int_0^1\int_0^1\int_0^2 2yz\,dz\,dy\,dx = 2\int_0^1 dx\int_0^1 y\,dy\int_0^2 z\,dz = 2\big[x\big]_0^1\big[\tfrac12 y^2\big]_0^1\big[\tfrac12 z^2\big]_0^2 = 2$.

6. $\operatorname{div}\mathbf{F} = \frac{\partial}{\partial x}(x^2z^3) + \frac{\partial}{\partial y}(2xyz^3) + \frac{\partial}{\partial z}(xz^4) = 2xz^3 + 2xz^3 + 4xz^3 = 8xz^3$, so by the Divergence Theorem,

$$\begin{aligned}\iint_S \mathbf{F}\cdot d\mathbf{S} &= \iiint_E \operatorname{div}\mathbf{F}\,dV = \int_{-1}^1\int_{-2}^2\int_{-3}^3 8xz^3\,dz\,dy\,dx = 8\int_{-1}^1 x\,dx\int_{-2}^2 dy\int_{-3}^3 z^3\,dz \\ &= 8\big[\tfrac12 x^2\big]_{-1}^1\big[y\big]_{-2}^2\big[\tfrac14 z^4\big]_{-3}^3 = 0\end{aligned}$$

7. $\operatorname{div} \mathbf{F} = 3y^2 + 0 + 3z^2$, so using cylindrical coordinates with $y = r\cos\theta$, $z = r\sin\theta$, $x = x$ we have

$$\iint_S \mathbf{F} \cdot d\mathbf{S} = \iiint_E (3y^2 + 3z^2)\, dV = \int_0^{2\pi} \int_0^1 \int_{-1}^2 (3r^2\cos^2\theta + 3r^2\sin^2\theta)\, r\, dx\, dr\, d\theta$$
$$= 3\int_0^{2\pi} d\theta \int_0^1 r^3\, dr \int_{-1}^2 dx = 3(2\pi)\left(\tfrac{1}{4}\right)(3) = \tfrac{9\pi}{2}$$

8. $\operatorname{div} \mathbf{F} = 3x^2y - 2x^2y - x^2y = 0$, so $\iint_S \mathbf{F} \cdot d\mathbf{S} = \iiint_E 0\, dV = 0$.

9. $\operatorname{div} \mathbf{F} = y\sin z + 0 - y\sin z = 0$, so by the Divergence Theorem, $\iint_S \mathbf{F} \cdot d\mathbf{S} = \iiint_E 0\, dV = 0$.

10. $\operatorname{div} \mathbf{F} = 2xy + 2xy + 2xy = 6xy$, so

$$\iint_S \mathbf{F} \cdot d\mathbf{S} = \iiint_E 6xy\, dV = \int_0^1 \int_0^{2-2y} \int_0^{2-x-2y} 6xy\, dz\, dx\, dy = \int_0^1 \int_0^{2-2y} 6xy(2 - x - 2y)\, dx\, dy$$
$$= \int_0^1 \int_0^{2-2y} (12xy - 6x^2y - 12xy^2)\, dx\, dy = \int_0^1 \left[6x^2y - 2x^3y - 6x^2y^2\right]_{x=0}^{x=2-2y} dy$$
$$= \int_0^1 y(2-2y)^3\, dy = \left[-\tfrac{8}{5}y^5 + 6y^4 - 8y^3 + 4y^2\right]_0^1 = \tfrac{2}{5}$$

11. $\operatorname{div} \mathbf{F} = y^2 + 0 + x^2 = x^2 + y^2$ so

$$\iint_S \mathbf{F} \cdot d\mathbf{S} = \iiint_E (x^2 + y^2)\, dV = \int_0^{2\pi} \int_0^2 \int_{r^2}^4 r^2 \cdot r\, dz\, dr\, d\theta = \int_0^{2\pi} \int_0^2 r^3(4 - r^2)\, dr\, d\theta$$
$$= \int_0^{2\pi} d\theta \int_0^2 (4r^3 - r^5)\, dr = 2\pi\left[r^4 - \tfrac{1}{6}r^6\right]_0^2 = \tfrac{32}{3}\pi$$

12. $\operatorname{div} \mathbf{F} = 4x^3 + 4xy^2$ so

$$\iint_S \mathbf{F} \cdot d\mathbf{S} = \iiint_E 4x(x^2 + y^2)\, dV = \int_0^{2\pi} \int_0^1 \int_0^{r\cos\theta + 2} (4r^3\cos\theta)\, r\, dz\, dr\, d\theta$$
$$= \int_0^{2\pi} \int_0^1 (4r^5\cos^2\theta + 8r^4\cos\theta)\, dr\, d\theta = \int_0^{2\pi} \left(\tfrac{2}{3}\cos^2\theta + \tfrac{8}{5}\cos\theta\right) d\theta = \tfrac{2}{3}\pi$$

13. $\operatorname{div} \mathbf{F} = 12x^2z + 12y^2z + 12z^3$ so

$$\iint_S \mathbf{F} \cdot d\mathbf{S} = \iiint_E 12z(x^2 + y^2 + z^2)\, dV = \int_0^{2\pi} \int_0^{\pi} \int_0^R 12(\rho\cos\phi)(\rho^2)\rho^2 \sin\phi\, d\rho\, d\phi\, d\theta$$
$$= 12\int_0^{2\pi} d\theta \int_0^{\pi} \sin\phi\cos\phi\, d\phi \int_0^R \rho^5\, d\rho = 12(2\pi)\left[\tfrac{1}{2}\sin^2\phi\right]_0^{\pi} \left[\tfrac{1}{6}\rho^6\right]_0^R = 0$$

14. $\iint_S \mathbf{F} \cdot d\mathbf{S} = \iiint_E 3(x^2 + y^2 + 1)\, dV = \int_0^{2\pi} \int_0^{\pi/2} \int_1^2 3(\rho^2\sin^2\phi + 1)\, \rho^2 \sin\phi\, d\rho\, d\phi\, d\theta$

$$= 2\pi \int_0^{\pi/2} \left[\tfrac{93}{5}\sin^3\phi + 7\sin\phi\right] d\phi = 2\pi\left[\tfrac{93}{5}\left(-\cos\phi + \tfrac{1}{3}\cos^3\phi\right) - 7\cos\phi\right]_0^{\pi/2} = \tfrac{194}{5}\pi$$

15. $\iint_S \mathbf{F} \cdot d\mathbf{S} = \iiint_E \sqrt{3 - x^2}\, dV = \int_{-1}^1 \int_{-1}^1 \int_0^{2 - x^4 - y^4} \sqrt{3 - x^2}\, dz\, dy\, dx = \tfrac{341}{60}\sqrt{2} + \tfrac{81}{20}\sin^{-1}\left(\tfrac{\sqrt{3}}{3}\right)$

16.

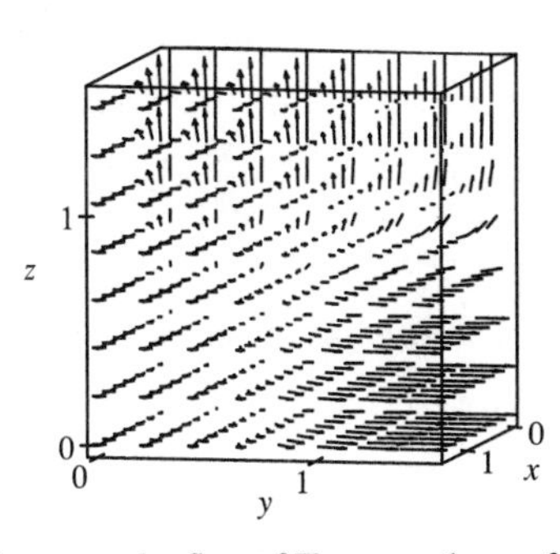

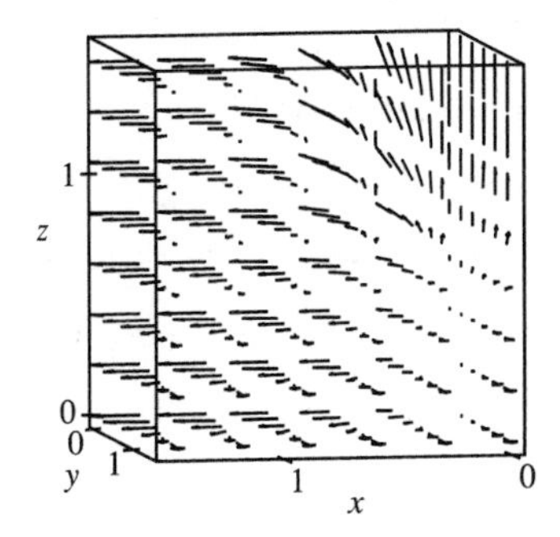

By the Divergence Theorem, the flux of $\mathbf{F}$ across the surface of the cube is

$$\iint_S \mathbf{F} \cdot d\mathbf{S} = \int_0^{\pi/2} \int_0^{\pi/2} \int_0^{\pi/2} \left[\cos x\cos^2 y + 3\sin^2 y\cos y\cos^4 z + 5\sin^4 z\cos z\cos^6 x\right] dz\, dy\, dx = \tfrac{19}{64}\pi^2.$$

17. For S_1 we have $\mathbf{n} = -\mathbf{k}$, so $\mathbf{F} \cdot \mathbf{n} = \mathbf{F} \cdot (-\mathbf{k}) = -x^2 z - y^2 = -y^2$ (since $z = 0$ on S_1). So if D is the unit disk, we get $\iint_{S_1} \mathbf{F} \cdot d\mathbf{S} = \iint_{S_1} \mathbf{F} \cdot \mathbf{n}\, dS = \iint_D (-y^2)\, dA = -\int_0^{2\pi} \int_0^1 r^2 (\sin^2\theta)\, r\, dr\, d\theta = -\frac{1}{4}\pi$. Now since S_2 is closed, we can use the Divergence Theorem. Since $\operatorname{div} \mathbf{F} = \frac{\partial}{\partial x}(z^2 x) + \frac{\partial}{\partial y}\left(\frac{1}{3}y^3 + \tan z\right) + \frac{\partial}{\partial z}(x^2 z + y^2) = z^2 + y^2 + x^2$, we use spherical coordinates to get $\iint_{S_2} \mathbf{F} \cdot d\mathbf{S} = \iiint_E \operatorname{div} \mathbf{F}\, dV = \int_0^{2\pi} \int_0^{\pi/2} \int_0^1 \rho^2 \cdot \rho^2 \sin\phi\, d\rho\, d\phi\, d\theta = \frac{2}{5}\pi$. Finally $\iint_S \mathbf{F} \cdot d\mathbf{S} = \iint_{S_2} \mathbf{F} \cdot d\mathbf{S} - \iint_{S_1} \mathbf{F} \cdot d\mathbf{S} = \frac{2}{5}\pi - \left(-\frac{1}{4}\pi\right) = \frac{13}{20}\pi$.

18. As in the hint to Exercise 17, we create a closed surface $S_2 = S \cup S_1$, where S is the part of the paraboloid $x^2 + y^2 + z = 2$ that lies above the plane $z = 1$, and S_1 is the disk $x^2 + y^2 = 1$ on the plane $z = 1$ oriented downward, and we then apply the Divergence Theorem. Since the disk S_1 is oriented downward, its unit normal vector is $\mathbf{n} = -\mathbf{k}$ and $\mathbf{F} \cdot (-\mathbf{k}) = -z = -1$ on S_1. So $\iint_{S_1} \mathbf{F} \cdot d\mathbf{S} = \iint_{S_1} \mathbf{F} \cdot \mathbf{n}\, dS = \iint_{S_1} (-1)\, dS = -A(S_1) = -\pi$. Let E be the region bounded by S_2. Then $\iint_{S_2} \mathbf{F} \cdot d\mathbf{S} = \iiint_E \operatorname{div} \mathbf{F}\, dV = \iiint_E 1\, dV = \int_0^1 \int_0^{2\pi} \int_1^{2-r^2} r\, dz\, d\theta\, dr = \int_0^1 \int_0^{2\pi} (r - r^3)\, d\theta\, dr = (2\pi)\frac{1}{4} = \frac{\pi}{2}$. Thus the flux of $\mathbf{F}$ across S is $\iint_S \mathbf{F} \cdot d\mathbf{S} = \iint_{S_2} \mathbf{F} \cdot d\mathbf{S} - \iint_{S_1} \mathbf{F} \cdot d\mathbf{S} = \frac{\pi}{2} - (-\pi) = \frac{3\pi}{2}$.

19. The vectors that end near P_1 are longer than the vectors that start near P_1, so the net flow is inward near P_1 and $\operatorname{div} \mathbf{F}(P_1)$ is negative. The vectors that end near P_2 are shorter than the vectors that start near P_2, so the net flow is outward near P_2 and $\operatorname{div} \mathbf{F}(P_2)$ is positive.

20. (a) The vectors that end near P_1 are shorter than the vectors that start near P_1, so the net flow is outward and P_1 is a source. The vectors that end near P_2 are longer than the vectors that start near P_2, so the net flow is inward and P_2 is a sink.

(b) $\mathbf{F}(x, y) = \langle x, y^2 \rangle \Rightarrow \operatorname{div} \mathbf{F} = \nabla \cdot \mathbf{F} = 1 + 2y$. The y-value at P_1 is positive, so $\operatorname{div} \mathbf{F} = 1 + 2y$ is positive, thus P_1 is a source. At P_2, $y < -1$, so $\operatorname{div} \mathbf{F} = 1 + 2y$ is negative, and P_2 is a sink.

21.

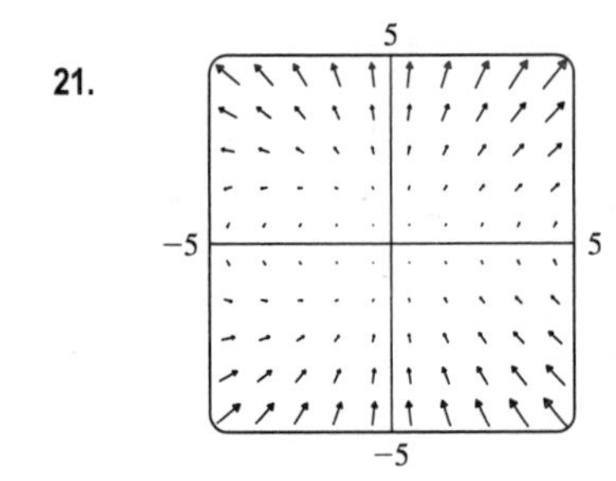

From the graph it appears that for points above the x-axis, vectors starting near a particular point are longer than vectors ending there, so divergence is positive. The opposite is true at points below the x-axis, where divergence is negative.

$\mathbf{F}(x, y) = \langle xy, x + y^2 \rangle \Rightarrow$

$\operatorname{div} \mathbf{F} = \frac{\partial}{\partial x}(xy) + \frac{\partial}{\partial y}(x + y^2) = y + 2y = 3y$. Thus $\operatorname{div} \mathbf{F} > 0$ for $y > 0$, and $\operatorname{div} \mathbf{F} < 0$ for $y < 0$.

22.

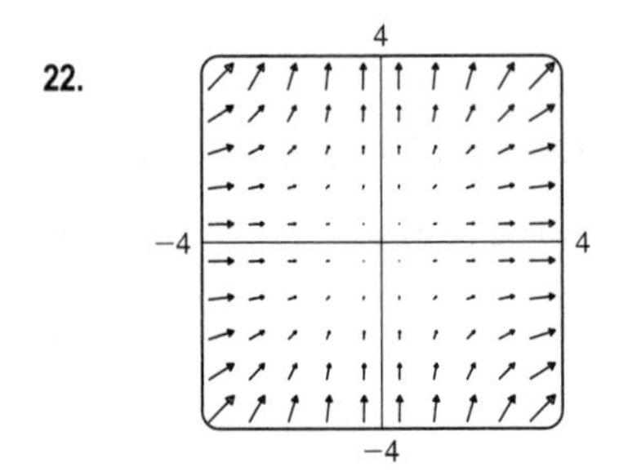

From the graph it appears that for points above the line $y = -x$, vectors starting near a particular point are longer than vectors ending there, so divergence is positive. The opposite is true at points below the line $y = -x$, where divergence is negative.

$\mathbf{F}(x, y) = \langle x^2, y^2 \rangle \Rightarrow \operatorname{div} \mathbf{F} = \frac{\partial}{\partial x}(x^2) + \frac{\partial}{\partial y}(y^2) = 2x + 2y$. Then $\operatorname{div} \mathbf{F} > 0$ for $2x + 2y > 0 \Rightarrow y > -x$, and $\operatorname{div} \mathbf{F} < 0$ for $y < -x$.

23. Since $\dfrac{\mathbf{x}}{|\mathbf{x}|^3} = \dfrac{x\,\mathbf{i} + y\,\mathbf{j} + z\,\mathbf{k}}{(x^2+y^2+z^2)^{3/2}}$ and $\dfrac{\partial}{\partial x}\left(\dfrac{x}{(x^2+y^2+z^2)^{3/2}}\right) = \dfrac{(x^2+y^2+z^2) - 3x^2}{(x^2+y^2+z^2)^{5/2}}$ with similar expressions for

$\dfrac{\partial}{\partial y}\left(\dfrac{y}{(x^2+y^2+z^2)^{3/2}}\right)$ and $\dfrac{\partial}{\partial z}\left(\dfrac{z}{(x^2+y^2+z^2)^{3/2}}\right)$, we have

$$\operatorname{div}\left(\frac{\mathbf{x}}{|\mathbf{x}|^3}\right) = \frac{3(x^2+y^2+z^2) - 3(x^2+y^2+z^2)}{(x^2+y^2+z^2)^{5/2}} = 0, \text{ except at } (0,0,0) \text{ where it is undefined.}$$

24. We first need to find $\mathbf{F}$ so that $\iint_S \mathbf{F}\cdot\mathbf{n}\,dS = \iint_S (2x+2y+z^2)\,dS$, so $\mathbf{F}\cdot\mathbf{n} = 2x+2y+z^2$. But for S,

$\mathbf{n} = \dfrac{x\,\mathbf{i} + y\,\mathbf{j} + z\,\mathbf{k}}{\sqrt{x^2+y^2+z^2}} = x\,\mathbf{i} + y\,\mathbf{j} + z\,\mathbf{k}$. Thus $\mathbf{F} = 2\,\mathbf{i} + 2\,\mathbf{j} + z\,\mathbf{k}$ and $\operatorname{div}\mathbf{F} = 1$. If $B = \{(x,y,z) \mid x^2+y^2+z^2 \le 1\}$,

then $\iint_S (2x+2y+z^2)\,dS = \iiint_B dV = V(B) = \frac{4}{3}\pi(1)^3 = \frac{4}{3}\pi$.

25. $\iint_S \mathbf{a}\cdot\mathbf{n}\,dS = \iiint_E \operatorname{div}\mathbf{a}\,dV = 0$ since $\operatorname{div}\mathbf{a} = 0$.

26. $\frac{1}{3}\iint_S \mathbf{F}\cdot d\mathbf{S} = \frac{1}{3}\iiint_E \operatorname{div}\mathbf{F}\,dV = \frac{1}{3}\iiint_E 3\,dV = V(E)$

27. $\iint_S \operatorname{curl}\mathbf{F}\cdot d\mathbf{S} = \iiint_E \operatorname{div}(\operatorname{curl}\mathbf{F})\,dV = 0$ by Theorem 13.5.11.

28. $\iint_S D_{\mathbf{n}} f\,dS = \iint_S (\nabla f\cdot\mathbf{n})\,dS = \iiint_E \operatorname{div}(\nabla f)\,dV = \iiint_E \nabla^2 f\,dV$

29. $\iint_S (f\nabla g)\cdot\mathbf{n}\,dS = \iiint_E \operatorname{div}(f\nabla g)\,dV = \iiint_E (f\nabla^2 g + \nabla g\cdot\nabla f)\,dV$ by Exercise 13.5.23.

30. $\iint_S (f\nabla g - g\nabla f)\cdot\mathbf{n}\,dS = \iiint_E \left[(f\nabla^2 g + \nabla g\cdot\nabla f) - (g\nabla^2 f + \nabla g\cdot\nabla f)\right] dV$ [by Exercise 27].

But $\nabla g\cdot\nabla f = \nabla f\cdot\nabla g$, so that $\iint_S (f\nabla g - g\nabla f)\cdot\mathbf{n}\,dS = \iiint_E (f\nabla^2 g - g\nabla^2 f)\,dV$.

13 Review

CONCEPT CHECK

1. See Definitions 1 and 2 in Section 13.1. A vector field can represent, for example, the wind velocity at any location in space, the speed and direction of the ocean current at any location, or the force vectors of Earth's gravitational field at a location in space.

2. (a) A conservative vector field $\mathbf{F}$ is a vector field which is the gradient of some scalar function f.

(b) The function f in part (a) is called a potential function for $\mathbf{F}$, that is, $\mathbf{F} = \nabla f$.

3. (a) See Definition 13.2.2.

(b) We normally evaluate the line integral using Formula 13.2.3.

(c) The mass is $m = \int_C \rho(x,y)\,ds$, and the center of mass is $(\overline{x},\overline{y})$ where $\overline{x} = \frac{1}{m}\int_C x\rho(x,y)\,ds$, $\overline{y} = \frac{1}{m}\int_C y\rho(x,y)\,ds$.

(d) See (5) and (6) in Section 13.2 for plane curves; we have similar definitions when C is a space curve [see the equation preceding (10) in Section 13.2].

(e) For plane curves, see Equations 13.2.7. We have similar results for space curves [see the equation preceding (10) in Section 13.2].

4. (a) See Definition 13.2.13.

   (b) If $\mathbf{F}$ is a force field, $\int_C \mathbf{F} \cdot d\mathbf{r}$ represents the work done by $\mathbf{F}$ in moving a particle along the curve C.

   (c) $\int_C \mathbf{F} \cdot d\mathbf{r} = \int_C P\,dx + Q\,dy + R\,dz$

5. See Theorem 13.3.2.

6. (a) $\int_C \mathbf{F} \cdot d\mathbf{r}$ is independent of path if the line integral has the same value for any two curves that have the same initial and terminal points.

   (b) See Theorem 13.3.4.

7. See the statement of Green's Theorem on page 753.

8. See Equations 13.4.5.

9. (a) $\operatorname{curl}\mathbf{F} = \left(\dfrac{\partial R}{\partial y} - \dfrac{\partial Q}{\partial z}\right)\mathbf{i} + \left(\dfrac{\partial P}{\partial z} - \dfrac{\partial R}{\partial x}\right)\mathbf{j} + \left(\dfrac{\partial Q}{\partial x} - \dfrac{\partial P}{\partial y}\right)\mathbf{k} = \nabla \times \mathbf{F}$

   (b) $\operatorname{div}\mathbf{F} = \dfrac{\partial P}{\partial x} + \dfrac{\partial Q}{\partial y} + \dfrac{\partial R}{\partial z} = \nabla \cdot \mathbf{F}$

   (c) For $\operatorname{curl}\mathbf{F}$, see the discussion accompanying Figure 1 on page 762 as well as Figure 6 and the accompanying discussion on page 791. For $\operatorname{div}\mathbf{F}$, see the discussion following Example 5 on page 763 as well as the discussion preceding (8) on page 797.

10. See Theorem 13.3.6; see Theorem 13.5.4.

11. (a) See (1) and (2) and the accompanying discussion in Section 13.6; See Figure 4 and the accompanying discussion on page 768.

    (b) See Definition 13.6.6.

    (c) See Equation 13.6.9.

12. (a) See (1) in Section 13.7.

    (b) We normally evaluate the surface integral using Formula 13.7.2.

    (c) See Formula 13.7.4.

    (d) The mass is $m = \iint_S \rho(x, y, z)\,dS$ and the center of mass is $(\overline{x}, \overline{y}, \overline{z})$ where $\overline{x} = \frac{1}{m}\iint_S x\rho(x, y, z)\,dS$, $\overline{y} = \frac{1}{m}\iint_S y\rho(x, y, z)\,dS$, $\overline{z} = \frac{1}{m}\iint_S z\rho(x, y, z)\,dS$.

13. (a) See Figures 6 and 7 and the accompanying discussion in Section 13.7. A Möbius strip is a nonorientable surface; see Figures 4 and 5 and the accompanying discussion on page 781.

    (b) See Definition 13.7.8.

    (c) See Formula 13.7.9.

    (d) See Formula 13.7.10.

14. See the statement of Stokes' Theorem on page 788.

15. See the statement of the Divergence Theorem on page 793.

16. In each theorem, we have an integral of a "derivative" over a region on the left side, while the right side involves the values of the original function only on the boundary of the region.

TRUE-FALSE QUIZ

1. False; div $\mathbf{F}$ is a scalar field.

2. True. (See Definition 13.5.1.)

3. True, by Theorem 13.5.3 and the fact that div $\mathbf{0} = 0$.

4. True, by Theorem 13.3.2.

5. False. See Exercise 13.3.31. (But the assertion is true if D is simply-connected; see Theorem 13.3.6.)

6. False. See the discussion accompanying Figure 8 on page 738.

7. True. Apply the Divergence Theorem and use the fact that div $\mathbf{F} = 0$.

8. False by Theorem 13.5.11, because if it were true, then div curl $\mathbf{F} = 3 \neq 0$.

EXERCISES

1. (a) Vectors starting on C point in roughly the direction opposite to C, so the tangential component $\mathbf{F} \cdot \mathbf{T}$ is negative. Thus $\int_C \mathbf{F} \cdot d\mathbf{r} = \int_C \mathbf{F} \cdot \mathbf{T}\, ds$ is negative.

   (b) The vectors that end near P are shorter than the vectors that start near P, so the net flow is outward near P and div $\mathbf{F}\,(P)$ is positive.

2. We can parametrize C by $x = x$, $y = x^2$, $0 \leq x \leq 1$ so

$$\int_C x\, ds = \int_0^1 x\sqrt{1 + (2x)^2}\, dx = \tfrac{1}{12}(1 + 4x^2)^{3/2}\Big]_0^1 = \tfrac{1}{12}(5\sqrt{5} - 1).$$

3. $$\int_C yz\cos x\, ds = \int_0^\pi (3\cos t)(3\sin t)\cos t\sqrt{(1)^2 + (-3\sin t)^2 + (3\cos t)^2}\, dt = \int_0^\pi (9\cos^2 t\sin t)\sqrt{10}\, dt$$
$$= 9\sqrt{10}\left(-\tfrac{1}{3}\cos^3 t\right)\Big]_0^\pi = -3\sqrt{10}(-2) = 6\sqrt{10}$$

4. $x = 3\cos t \quad\Rightarrow\quad dx = -3\sin t\, dt$, $y = 2\sin t \quad\Rightarrow\quad dy = 2\cos t\, dt$, $0 \leq t \leq 2\pi$, so

$$\int_C y\, dx + (x + y^2)\, dy = \int_0^{2\pi} \left[(2\sin t)(-3\sin t) + (3\cos t + 4\sin^2 t)(2\cos t)\right] dt$$
$$= \int_0^{2\pi} \left(-6\sin^2 t + 6\cos^2 t + 8\sin^2 t\,\cos t\right) dt = \int_0^{2\pi} \left[6\left(\cos^2 t - \sin^2 t\right) + 8\sin^2 t\,\cos t\right] dt$$
$$= \int_0^{2\pi} \left(6\cos 2t + 8\sin^2 t\,\cos t\right) dt = 3\sin 2t + \tfrac{8}{3}\sin^3 t\Big]_0^{2\pi} = 0$$

Or: Notice that $\frac{\partial}{\partial y}(y) = 1 = \frac{\partial}{\partial x}(x + y^2)$, so $\mathbf{F}(x, y) = \langle y, x + y^2\rangle$ is a conservative vector field. Since C is a closed curve, $\int_C \mathbf{F} \cdot d\mathbf{r} = \int_C y\, dx + (x + y^2)\, dy = 0$.

5. $$\int_C y^3\, dx + x^2\, dy = \int_{-1}^1 \left[y^3(-2y) + (1 - y^2)^2\right] dy = \int_{-1}^1 (-y^4 - 2y^2 + 1)\, dy$$
$$= \left[-\tfrac{1}{5}y^5 - \tfrac{2}{3}y^3 + y\right]_{-1}^1 = -\tfrac{1}{5} - \tfrac{2}{3} + 1 - \tfrac{1}{5} - \tfrac{2}{3} + 1 = \tfrac{4}{15}$$

6. $\int_C \sqrt{xy}\,dx + e^y\,dy + xz\,dz = \int_0^1 \left(\sqrt{t^4 \cdot t^2} \cdot 4t^3 + e^{t^2} \cdot 2t + t^4 \cdot t^3 \cdot 3t^2\right) dt = \int_0^1 (4t^6 + 2te^{t^2} + 3t^9)\,dt$

$$= \left[\tfrac{4}{7}t^7 + e^{t^2} + \tfrac{3}{10}t^{10}\right]_0^1 = e - \tfrac{9}{70}$$

7. C: $x = 1 + 2t \;\Rightarrow\; dx = 2\,dt,\ y = 4t \;\Rightarrow\; dy = 4\,dt,\ z = -1 + 3t \;\Rightarrow\; dz = 3\,dt,\ 0 \le t \le 1.$

$$\int_C xy\,dx + y^2\,dy + yz\,dz = \int_0^1 \left[(1+2t)(4t)(2) + (4t)^2(4) + (4t)(-1+3t)(3)\right] dt$$

$$= \int_0^1 (116t^2 - 4t)\,dt = \left[\tfrac{116}{3}t^3 - 2t^2\right]_0^1 = \tfrac{116}{3} - 2 = \tfrac{110}{3}$$

8. $\mathbf{F}(\mathbf{r}(t)) = (\sin t)(1+t)\,\mathbf{i} + (\sin^2 t)\,\mathbf{j},\ \mathbf{r}'(t) = \cos t\,\mathbf{i} + \mathbf{j}$ and

$$\int_C \mathbf{F} \cdot d\mathbf{r} = \int_0^\pi ((1+t)\sin t\cos t + \sin^2 t)\,dt = \int_0^\pi \left(\tfrac{1}{2}(1+t)\sin 2t + \sin^2 t\right) dt$$

$$= \left[\tfrac{1}{2}\left((1+t)\left(-\tfrac{1}{2}\cos 2t\right) + \tfrac{1}{4}\sin 2t\right) + \tfrac{1}{2}t - \tfrac{1}{4}\sin 2t\right]_0^\pi = \tfrac{\pi}{4}.$$

9. $\mathbf{F}(\mathbf{r}(t)) = e^{-t}\,\mathbf{i} + t^2(-t)\,\mathbf{j} + (t^2 + t^3)\,\mathbf{k},\ \mathbf{r}'(t) = 2t\,\mathbf{i} + 3t^2\,\mathbf{j} - \mathbf{k}$ and

$$\int_C \mathbf{F} \cdot d\mathbf{r} = \int_0^1 (2te^{-t} - 3t^5 - (t^2 + t^3))\,dt = \left[-2te^{-t} - 2e^{-t} - \tfrac{1}{2}t^6 - \tfrac{1}{3}t^3 - \tfrac{1}{4}t^4\right]_0^1 = \tfrac{11}{12} - \tfrac{4}{e}.$$

10. (a) C: $x = 3 - 3t,\ y = \frac{\pi}{2}t,\ z = 3t,\ 0 \le t \le 1$. Then

$$W = \int_C \mathbf{F} \cdot d\mathbf{r} = \int_0^1 \left[3t\,\mathbf{i} + (3-3t)\,\mathbf{j} + \tfrac{\pi}{2}t\,\mathbf{k}\right] \cdot \left[-3\,\mathbf{i} + \tfrac{\pi}{2}\,\mathbf{j} + 3\,\mathbf{k}\right] dt = \int_0^1 \left[-9t + \tfrac{3\pi}{2}\right] dt = \tfrac{1}{2}(3\pi - 9).$$

(b) $W = \int_C \mathbf{F} \cdot d\mathbf{r} = \int_0^{\pi/2} (3\sin t\,\mathbf{i} + 3\cos t\,\mathbf{j} + t\,\mathbf{k}) \cdot (-3\sin t\,\mathbf{i} + \mathbf{j} + 3\cos t\,\mathbf{k})\,dt$

$$= \int_0^{\pi/2} (-9\sin^2 t + 3\cos t + 3t\cos t)\,dt$$

$$= \left[-\tfrac{9}{2}(t - \sin t\cos t) + 3\sin t + 3(t\sin t + \cos t)\right]_0^{\pi/2} = -\tfrac{9\pi}{4} + 3 + \tfrac{3\pi}{2} - 3 = -\tfrac{3\pi}{4}$$

11. $\frac{\partial}{\partial y}\left[(1+xy)e^{xy}\right] = 2xe^{xy} + x^2ye^{xy} = \frac{\partial}{\partial x}\left[e^y + x^2e^{xy}\right]$ and the domain of $\mathbf{F}$ is $\mathbb{R}^2$, so $\mathbf{F}$ is conservative. Thus there exists a function f such that $\mathbf{F} = \nabla f$. Then $f_y(x,y) = e^y + x^2e^{xy}$ implies $f(x,y) = e^y + xe^{xy} + g(x)$ and then $f_x(x,y) = xye^{xy} + e^{xy} + g'(x) = (1+xy)e^{xy} + g'(x)$. But $f_x(x,y) = (1+xy)e^{xy}$, so $g'(x) = 0 \;\Rightarrow\; g(x) = K$. Thus $f(x,y) = e^y + xe^{xy} + K$ is a potential function for $\mathbf{F}$.

12. $\mathbf{F}$ is defined on all of $\mathbb{R}^3$, its components have continuous partial derivatives, and $\operatorname{curl}\mathbf{F} = (0-0)\,\mathbf{i} - (0-0)\,\mathbf{j} + (\cos y - \cos y)\,\mathbf{k} = \mathbf{0}$, so $\mathbf{F}$ is conservative by Theorem 13.5.4. Thus there exists a function f such that $\nabla f = \mathbf{F}$. Then $f_x(x,y,z) = \sin y$ implies $f(x,y,z) = x\sin y + g(y,z)$ and then $f_y(x,y,z) = x\cos y + g_y(y,z)$. But $f_y(x,y,z) = x\cos y$, so $g_y(y,z) = 0 \;\Rightarrow\; g(y,z) = h(z)$. Then $f(x,y,z) = x\sin y + h(z)$ implies $f_z(x,y,z) = h'(z)$. But $f_z(x,y,z) = -\sin z$, so $h(z) = \cos z + K$. Thus a potential function for $\mathbf{F}$ is $f(x,y,z) = x\sin y + \cos z + K$.

13. Since $\frac{\partial}{\partial y}(4x^3y^2 - 2xy^3) = 8x^3y - 6xy^2 = \frac{\partial}{\partial x}(2x^4y - 3x^2y^2 + 4y^3)$ and the domain of $\mathbf{F}$ is $\mathbb{R}^2$, $\mathbf{F}$ is conservative. Furthermore $f(x,y) = x^4y^2 - x^2y^3 + y^4$ is a potential function for $\mathbf{F}$. $t = 0$ corresponds to the point $(0,1)$ and $t = 1$ corresponds to $(1,1)$, so $\int_C \mathbf{F} \cdot d\mathbf{r} = f(1,1) - f(0,1) = 1 - 1 = 0$.

14. Here curl $\mathbf{F} = \mathbf{0}$, the domain of $\mathbf{F}$ is $\mathbb{R}^3$, and the components of $\mathbf{F}$ have continuous partial derivatives, so $\mathbf{F}$ is conservative. Furthermore $f(x,y,z) = xe^y + ye^z$ is a potential function for $\mathbf{F}$. Then $\int_C \mathbf{F}\cdot d\mathbf{r} = f(4,0,3) - f(0,2,0) = 4 - 2 = 2$.

15. C_1: $\mathbf{r}(t) = t\,\mathbf{i} + t^2\,\mathbf{j}$, $-1 \le t \le 1$;

C_2: $\mathbf{r}(t) = -t\,\mathbf{i} + \mathbf{j}$, $-1 \le t \le 1$.

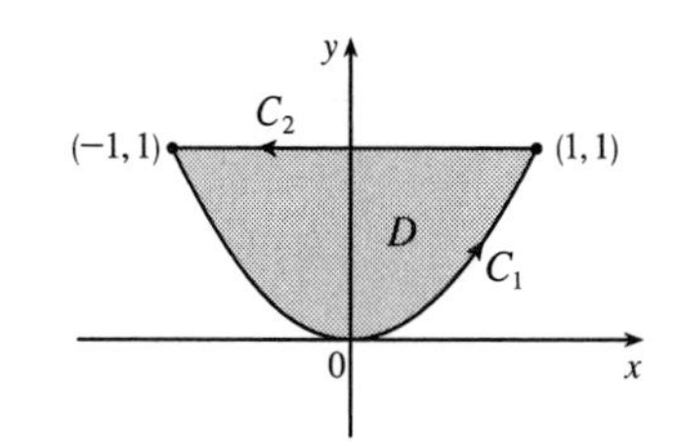

Then

$\int_C xy^2\,dx - x^2y\,dy = \int_{-1}^{1}(t^5 - 2t^5)\,dt + \int_{-1}^{1} t\,dt = \left[-\frac{1}{6}t^6\right]_{-1}^{1} + \left[\frac{1}{2}t^2\right]_{-1}^{1} = 0.$

Using Green's Theorem, we have

$$\int_C xy^2\,dx - x^2y\,dy = \iint_D \left[\frac{\partial}{\partial x}(-x^2y) - \frac{\partial}{\partial y}(xy^2)\right] dA = \iint_D (-2xy - 2xy)\,dA = \int_{-1}^{1}\int_{x^2}^{1} -4xy\,dy\,dx$$
$$= \int_{-1}^{1}\left[-2xy^2\right]_{y=x^2}^{y=1} dx = \int_{-1}^{1}(2x^5 - 2x)\,dx = \left[\tfrac{1}{3}x^6 - x^2\right]_{-1}^{1} = 0$$

16. $\int_C \sqrt{1+x^3}\,dx + 2xy\,dy = \iint_D \left[\frac{\partial}{\partial x}(2xy) - \frac{\partial}{\partial y}\left(\sqrt{1+x^3}\right)\right] dA = \int_0^1\int_0^{3x}(2y - 0)\,dy\,dx = \int_0^1 9x^2\,dx = 3x^3\Big]_0^1 = 3$

17. $\int_C x^2y\,dx - xy^2\,dy = \iint_{x^2+y^2\le 4}\left[\frac{\partial}{\partial x}(-xy^2) - \frac{\partial}{\partial y}(x^2y)\right] dA = \iint_{x^2+y^2\le 4}(-y^2 - x^2)\,dA = -\int_0^{2\pi}\int_0^2 r^3\,dr\,d\theta = -8\pi$

18. $\operatorname{curl}\mathbf{F} = (0 - e^{-y}\cos z)\,\mathbf{i} - (e^{-z}\cos x - 0)\,\mathbf{j} + (0 - e^{-x}\cos y)\,\mathbf{k} = -e^{-y}\cos z\,\mathbf{i} - e^{-z}\cos x\,\mathbf{j} - e^{-x}\cos y\,\mathbf{k}$,

$\operatorname{div}\mathbf{F} = -e^{-x}\sin y - e^{-y}\sin z - e^{-z}\sin x$

19. If we assume there is such a vector field $\mathbf{G}$, then $\operatorname{div}(\operatorname{curl}\mathbf{G}) = 2 + 3z - 2xz$. But $\operatorname{div}(\operatorname{curl}\mathbf{F}) = 0$ for all vector fields $\mathbf{F}$. Thus such a $\mathbf{G}$ cannot exist.

20. Let $\mathbf{F} = P_1\,\mathbf{i} + Q_1\,\mathbf{j} + R_1\,\mathbf{k}$ and $\mathbf{G} = P_2\,\mathbf{i} + Q_2\,\mathbf{j} + R_2\,\mathbf{k}$ be vector fields whose first partials exist and are continuous. Then

$$\begin{aligned}\mathbf{F}\operatorname{div}\mathbf{G} - \mathbf{G}\operatorname{div}\mathbf{F} = {} & \left[P_1\left(\frac{\partial P_2}{\partial x} + \frac{\partial Q_2}{\partial y} + \frac{\partial R_2}{\partial z}\right)\mathbf{i} + Q_1\left(\frac{\partial P_2}{\partial x} + \frac{\partial Q_2}{\partial y} + \frac{\partial R_2}{\partial z}\right)\mathbf{j} + R_1\left(\frac{\partial P_2}{\partial x} + \frac{\partial Q_2}{\partial y} + \frac{\partial R_2}{\partial z}\right)\mathbf{k}\right] \\ & - \left[P_2\left(\frac{\partial P_1}{\partial x} + \frac{\partial Q_1}{\partial y} + \frac{\partial R_1}{\partial z}\right)\mathbf{i} + Q_2\left(\frac{\partial P_1}{\partial x} + \frac{\partial Q_1}{\partial y} + \frac{\partial R_1}{\partial z}\right)\mathbf{j}\right. \\ & \left. + R_2\left(\frac{\partial P_1}{\partial x} + \frac{\partial Q}{\partial y} + \frac{\partial R_1}{\partial z}\right)\mathbf{k}\right]\end{aligned}$$

and

$$\begin{aligned}(\mathbf{G}\cdot\nabla)\,\mathbf{F} - (\mathbf{F}\cdot\nabla)\,\mathbf{G} = {} & \left[\left(P_2\frac{\partial P_1}{\partial x} + Q_2\frac{\partial P_1}{\partial y} + R_2\frac{\partial P_1}{\partial z}\right)\mathbf{i} + \left(P_2\frac{\partial Q_1}{\partial x} + Q_2\frac{\partial Q_1}{\partial y} + R_2\frac{\partial Q_1}{\partial z}\right)\mathbf{j}\right. \\ & \left. + \left(P_2\frac{\partial R_1}{\partial x} + Q_2\frac{\partial R_1}{\partial y} + R_2\frac{\partial R_1}{\partial z}\right)\mathbf{k}\right] \\ & - \left[\left(P_1\frac{\partial P_2}{\partial x} + Q_1\frac{\partial P_2}{\partial y} + R_1\frac{\partial P_2}{\partial z}\right)\mathbf{i} + \left(P_1\frac{\partial Q_2}{\partial x} + Q_1\frac{\partial Q_2}{\partial y} + R_1\frac{\partial Q_2}{\partial z}\right)\mathbf{j}\right. \\ & \left. + \left(P_1\frac{\partial R_2}{\partial x} + Q_1\frac{\partial R_2}{\partial y} + R_1\frac{\partial R_2}{\partial z}\right)\mathbf{k}\right]\end{aligned}$$

Hence

$$
\begin{aligned}
&\mathbf{F}\operatorname{div}\mathbf{G} - \mathbf{G}\operatorname{div}\mathbf{F} + (\mathbf{G}\cdot\nabla)\mathbf{F} - (\mathbf{F}\cdot\nabla)\mathbf{G} \\
&= \left[\left(P_1\frac{\partial Q_2}{\partial y} + Q_2\frac{\partial P_1}{\partial x}\right) - \left(P_2\frac{\partial Q_1}{\partial y} + Q_1\frac{\partial P_2}{\partial y}\right) - \left(P_2\frac{\partial R_1}{\partial z} + R_1\frac{\partial P_2}{\partial z}\right) + \left(P_1\frac{\partial R_2}{\partial z} + R_2\frac{\partial P_1}{\partial z}\right)\right]\mathbf{i} \\
&\quad + \left[\left(Q_1\frac{\partial R_2}{\partial z} + R_2\frac{\partial Q_1}{\partial z}\right) - \left(Q_2\frac{\partial R_1}{\partial z} + R_1\frac{\partial Q_2}{\partial z}\right) - \left(P_1\frac{\partial Q_2}{\partial x} + Q_2\frac{\partial P_1}{\partial x}\right) + \left(P_2\frac{\partial Q_1}{\partial x} + Q_1\frac{\partial P_2}{\partial x}\right)\right]\mathbf{j} \\
&\quad + \left[\left(P_2\frac{\partial R_1}{\partial x} + R_1\frac{\partial P_2}{\partial x}\right) - \left(P_1\frac{\partial R_2}{\partial x} + R_2\frac{\partial P_1}{\partial x}\right) - \left(Q_1\frac{\partial R_2}{\partial y} + R_2\frac{\partial Q_1}{\partial y}\right) + \left(Q_2\frac{\partial R_1}{\partial y} + R_1\frac{\partial Q_2}{\partial y}\right)\right]\mathbf{k} \\
&= \left[\frac{\partial}{\partial y}(P_1Q_2 - P_2Q_1) - \frac{\partial}{\partial z}(P_2R_1 - P_1R_2)\right]\mathbf{i} + \left[\frac{\partial}{\partial z}(Q_1R_2 - Q_2R_1) - \frac{\partial}{\partial x}(P_1Q_2 - P_2Q_1)\right]\mathbf{j} \\
&\quad + \left[\frac{\partial}{\partial x}(P_2R_1 - P_1R_2) - \frac{\partial}{\partial y}(Q_1R_2 - Q_2R_1)\right]\mathbf{k} \\
&= \operatorname{curl}(\mathbf{F}\times\mathbf{G})
\end{aligned}
$$

21. For any piecewise-smooth simple closed plane curve C bounding a region D, we can apply Green's Theorem to $\mathbf{F}(x,y) = f(x)\,\mathbf{i} + g(y)\,\mathbf{j}$ to get $\int_C f(x)\,dx + g(y)\,dy = \iint_D \left[\frac{\partial}{\partial x}g(y) - \frac{\partial}{\partial y}f(x)\right]dA = \iint_D 0\,dA = 0$.

22.
$$
\begin{aligned}
\nabla^2(fg) &= \frac{\partial^2(fg)}{\partial x^2} + \frac{\partial^2(fg)}{\partial y^2} + \frac{\partial^2(fg)}{\partial z^2} \\
&= \frac{\partial}{\partial x}\left(\frac{\partial f}{\partial x}g + f\frac{\partial g}{\partial x}\right) + \frac{\partial}{\partial y}\left(\frac{\partial f}{\partial y}g + f\frac{\partial g}{\partial y}\right) + \frac{\partial}{\partial z}\left(\frac{\partial f}{\partial z}g + f\frac{\partial g}{\partial z}\right) \quad \text{[Product Rule]} \\
&= \frac{\partial^2 f}{\partial x^2}g + 2\frac{\partial f}{\partial x}\frac{\partial g}{\partial x} + f\frac{\partial^2 g}{\partial x^2} + \frac{\partial^2 f}{\partial y^2}g + 2\frac{\partial f}{\partial y}\frac{\partial g}{\partial y} \\
&\qquad + f\frac{\partial^2 g}{\partial y^2} + \frac{\partial^2 f}{\partial z^2}g + 2\frac{\partial f}{\partial z}\frac{\partial g}{\partial z} + f\frac{\partial^2 g}{\partial z^2} \quad \text{[Product Rule]} \\
&= f\left(\frac{\partial^2 g}{\partial x^2} + \frac{\partial^2 g}{\partial y^2} + \frac{\partial^2 g}{\partial z^2}\right) + g\left(\frac{\partial^2 f}{\partial x^2} + \frac{\partial^2 f}{\partial y^2} + \frac{\partial^2 f}{\partial z^2}\right) + 2\left\langle\frac{\partial f}{\partial x}, \frac{\partial f}{\partial y}, \frac{\partial f}{\partial z}\right\rangle\cdot\left\langle\frac{\partial g}{\partial x}, \frac{\partial g}{\partial y}, \frac{\partial g}{\partial z}\right\rangle \\
&= f\nabla^2 g + g\nabla^2 f + 2\nabla f\cdot\nabla g
\end{aligned}
$$

Another method: Using the rules in Exercises 11.6.29(b) and 13.5.23, we have

$$
\begin{aligned}
\nabla^2(fg) &= \nabla\cdot\nabla(fg) = \nabla\cdot(g\,\nabla f + f\,\nabla g) = \nabla g\cdot\nabla f + g\nabla\cdot\nabla f + \nabla f\cdot\nabla g + f\nabla\cdot\nabla g \\
&= g\,\nabla^2 f + f\,\nabla^2 g + 2\nabla f\cdot\nabla g
\end{aligned}
$$

23. $\nabla^2 f = 0$ means that $\dfrac{\partial^2 f}{\partial x^2} + \dfrac{\partial^2 f}{\partial y^2} = 0$. Now if $\mathbf{F} = f_y\,\mathbf{i} - f_x\,\mathbf{j}$ and C is any closed path in D, then applying Green's Theorem, we get

$$\int_C \mathbf{F} \cdot d\mathbf{r} = \int_C f_y\,dx - f_x\,dy = \iint_D \left[\tfrac{\partial}{\partial x}(-f_x) - \tfrac{\partial}{\partial y}(f_y)\right] dA = -\iint_D (f_{xx} + f_{yy})\,dA = -\iint_D 0\,dA = 0.$$

Therefore the line integral is independent of path, by Theorem 13.3.3.

24. (a) $x^2 + y^2 = \cos^2 t + \sin^2 t = 1$, so C lies on the circular cylinder $x^2 + y^2 = 1$. But also $y = z$, so C lies on the plane $y = z$. Thus C is the intersection of the plane $y = z$ and the cylinder $x^2 + y^2 = 1$.

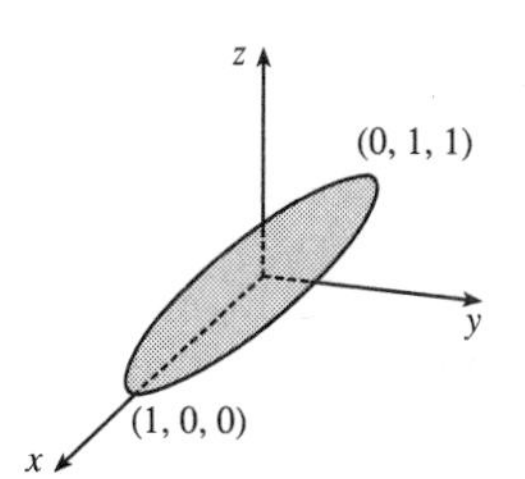

(b) Apply Stokes' Theorem, $\int_C \mathbf{F} \cdot d\mathbf{r} = \iint_S \operatorname{curl}\mathbf{F} \cdot d\mathbf{S}$:

$$\operatorname{curl}\mathbf{F} = \begin{vmatrix} \mathbf{i} & \mathbf{j} & \mathbf{k} \\ \partial/\partial x & \partial/\partial y & \partial/\partial z \\ 2xe^{2y} & 2x^2e^{2y} + 2y\cot z & -y^2\csc^2 z \end{vmatrix} = \left\langle -2y\csc^2 z - (-2y\csc^2 z), 0, 4xe^{2y} - 4xe^{2y} \right\rangle = \mathbf{0}$$

Therefore $\int_C \mathbf{F} \cdot d\mathbf{r} = \iint_S \mathbf{0} \cdot d\mathbf{S} = 0$.

25. $z = f(x, y) = x^2 + 2y$ with $0 \le x \le 1$, $0 \le y \le 2x$. Thus

$$A(S) = \iint_D \sqrt{1 + 4x^2 + 4}\,dA = \int_0^1 \int_0^{2x} \sqrt{5 + 4x^2}\,dy\,dx = \int_0^1 2x\sqrt{5 + 4x^2}\,dx = \tfrac{1}{6}(5 + 4x^2)^{3/2}\Big]_0^1 = \tfrac{1}{6}\left(27 - 5\sqrt{5}\right).$$

26. (a) $\mathbf{r}_u = -v\,\mathbf{j} + 2u\,\mathbf{k}$, $\mathbf{r}_v = 2v\,\mathbf{i} - u\,\mathbf{j}$ and $\mathbf{r}_u \times \mathbf{r}_v = 2u^2\,\mathbf{i} + 4uv\,\mathbf{j} + 2v^2\,\mathbf{k}$. Since the point $(4, -2, 1)$ corresponds to $u = 1$, $v = 2$ (or $u = -1$, $v = -2$ but $\mathbf{r}_u \times \mathbf{r}_v$ is the same for both), a normal vector to the surface at $(4, -2, 1)$ is $2\,\mathbf{i} + 8\,\mathbf{j} + 8\,\mathbf{k}$ and an equation of the tangent plane is $2x + 8y + 8z = 0$ or $x + 4y + 4z = 0$.

(b)

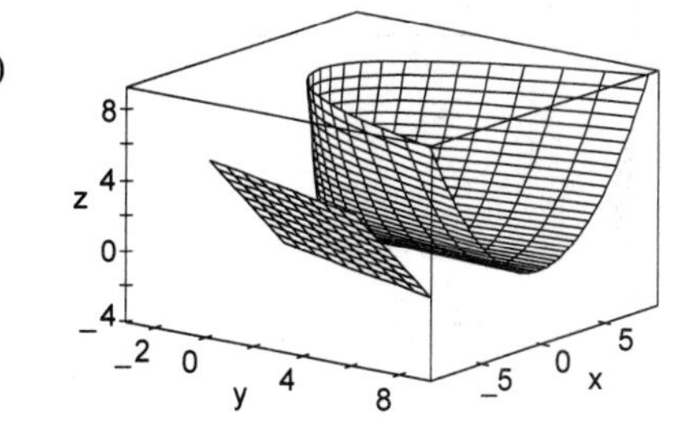

(c) By Definition 13.6.6, the area of S is given by

$$A(S) = \int_0^3 \int_{-3}^3 \sqrt{(2u^2)^2 + (4uv)^2 + (2v^2)^2}\,dv\,du = 2\int_0^3 \int_{-3}^3 \sqrt{u^4 + 4u^2v^2 + v^4}\,dv\,du.$$

(d) By Equation 13.7.9, the surface integral is

$$\iint_S \mathbf{F} \cdot d\mathbf{S} = \int_0^3 \int_{-3}^3 \left\langle \frac{(u^2)^2}{1 + (v^2)^2}, \frac{(v^2)^2}{1 + (-uv)^2}, \frac{(-uv)^2}{1 + (u^2)^2} \right\rangle \cdot \left\langle 2u^2, 4uv, 2v^2 \right\rangle dv\,du$$

$$= \int_0^3 \int_{-3}^3 \left(\frac{2u^6}{1 + v^4} + \frac{4uv^5}{1 + u^2v^2} + \frac{2u^2v^4}{1 + u^4} \right) dv\,du \approx 1524.0190$$

27. $z = f(x, y) = x^2 + y^2$ with $0 \le x^2 + y^2 \le 4$ so $\mathbf{r}_x \times \mathbf{r}_y = -2x\,\mathbf{i} - 2y\,\mathbf{j} + \mathbf{k}$ (using upward orientation). Then

$$\iint_S z\,dS = \iint_{x^2+y^2\le 4} (x^2 + y^2)\sqrt{4x^2 + 4y^2 + 1}\,dA = \int_0^{2\pi}\int_0^2 r^3\sqrt{1 + 4r^2}\,dr\,d\theta = \tfrac{1}{60}\pi\left(391\sqrt{17} + 1\right)$$

(Substitute $u = 1 + 4r^2$ and use tables.)

28. $z = f(x, y) = 4 + x + y$ with $0 \le x^2 + y^2 \le 4$ so $\mathbf{r}_x \times \mathbf{r}_y = -\mathbf{i} - \mathbf{j} + \mathbf{k}$. Then

$$\begin{aligned}\iint_S (x^2 z + y^2 z)\,dS &= \iint_{x^2+y^2\le 4} (x^2 + y^2)(4 + x + y)\sqrt{3}\,dA \\ &= \int_0^2\int_0^{2\pi} \sqrt{3}\,r^3(4 + r\cos\theta + r\sin\theta)\,d\theta\,dr = \int_0^2 8\pi\sqrt{3}\,r^3\,dr = 32\pi\sqrt{3}\end{aligned}$$

29. Since the sphere bounds a simple solid region, the Divergence Theorem applies and

$\iint_S \mathbf{F}\cdot d\mathbf{S} = \iiint_E (z - 2)\,dV = \iiint_E z\,dV - 2\iiint_E dV = m\overline{z} - 2\left(\tfrac{4}{3}\pi 2^3\right) = -\tfrac{64}{3}\pi.$

Alternate solution: $\mathbf{F}(\mathbf{r}(\phi, \theta)) = 4\sin\phi\cos\theta\cos\phi\,\mathbf{i} - 4\sin\phi\sin\theta\,\mathbf{j} + 6\sin\phi\cos\theta\,\mathbf{k}$,

$\mathbf{r}_\phi \times \mathbf{r}_\theta = 4\sin^2\phi\cos\theta\,\mathbf{i} + 4\sin^2\phi\sin\theta\,\mathbf{j} + 4\sin\phi\cos\phi\,\mathbf{k}$, and

$\mathbf{F}\cdot(\mathbf{r}_\phi \times \mathbf{r}_\theta) = 16\sin^3\phi\cos^2\theta\cos\phi - 16\sin^3\phi\sin^2\theta + 24\sin^2\phi\cos\phi\cos\theta$. Then

$$\begin{aligned}\iint_S F\cdot dS &= \int_0^{2\pi}\int_0^{\pi} \left(16\sin^3\phi\cos\phi\cos^2\theta - 16\sin^3\phi\sin^2\theta + 24\sin^2\phi\cos\phi\cos\theta\right) d\phi\,d\theta \\ &= \int_0^{2\pi} \tfrac{4}{3}(-16\sin^2\theta)\,d\theta = -\tfrac{64}{3}\pi\end{aligned}$$

30. $z = f(x, y) = x^2 + y^2$, $\mathbf{r}_x \times \mathbf{r}_y = -2x\,\mathbf{i} - 2y\,\mathbf{j} + \mathbf{k}$ (because of upward orientation) and

$\mathbf{F}(\mathbf{r}(x, y))\cdot(\mathbf{r}_x \times \mathbf{r}_y) = -2x^3 - 2xy^2 + x^2 + y^2$. Then

$$\begin{aligned}\iint_S \mathbf{F}\cdot d\mathbf{S} &= \iint_{x^2+y^2\le 1} (-2x^3 - 2xy^2 + x^2 + y^2)\,dA \\ &= \int_0^1\int_0^{2\pi} (-2r^3\cos^3\theta - 2r^3\cos\theta\sin^2\theta + r^2)\,r\,dr\,d\theta = \int_0^1 r^3(2\pi)\,dr = \tfrac{\pi}{2}\end{aligned}$$

31. Since $\operatorname{curl}\mathbf{F} = \mathbf{0}$, $\iint_S (\operatorname{curl}\mathbf{F})\cdot d\mathbf{S} = 0$. We parametrize C: $\mathbf{r}(t) = \cos t\,\mathbf{i} + \sin t\,\mathbf{j}$, $0 \le t \le 2\pi$ and

$\oint_C \mathbf{F}\cdot d\mathbf{r} = \int_0^{2\pi} (-\cos^2 t\sin t + \sin^2 t\cos t)\,dt = \tfrac{1}{3}\cos^3 t + \tfrac{1}{3}\sin^3 t\Big]_0^{2\pi} = 0.$

32. $\iint_S \operatorname{curl}\mathbf{F}\cdot d\mathbf{S} = \oint_C \mathbf{F}\cdot d\mathbf{r}$ where C: $\mathbf{r}(t) = 2\cos t\,\mathbf{i} + 2\sin t\,\mathbf{j} + \mathbf{k}$, $0 \le t \le 2\pi$, so $\mathbf{r}'(t) = -2\sin t\,\mathbf{i} + 2\cos t\,\mathbf{j}$,

$\mathbf{F}(\mathbf{r}(t)) = 8\cos^2 t\sin t\,\mathbf{i} + 2\sin t\,\mathbf{j} + e^{4\cos t\sin t}\,\mathbf{k}$, and $\mathbf{F}(\mathbf{r}(t))\cdot\mathbf{r}'(t) = -16\cos^2 t\sin^2 t + 4\sin t\cos t$. Thus

$$\begin{aligned}\oint_C \mathbf{F}\cdot d\mathbf{r} &= \int_0^{2\pi} (-16\cos^2 t\sin^2 t + 4\sin t\cos t)\,dt \\ &= \left[-16\left(-\tfrac{1}{4}\sin t\cos^3 t + \tfrac{1}{16}\sin 2t + \tfrac{1}{8}t\right) + 2\sin^2 t\right]_0^{2\pi} = -4\pi\end{aligned}$$

33. The surface is given by $x + y + z = 1$ or $z = 1 - x - y$, $0 \le x \le 1$, $0 \le y \le 1 - x$ and $\mathbf{r}_x \times \mathbf{r}_y = \mathbf{i} + \mathbf{j} + \mathbf{k}$. Then

$$\oint_C \mathbf{F}\cdot d\mathbf{r} = \iint_S \operatorname{curl}\mathbf{F}\cdot d\mathbf{S} = \iint_D (-y\,\mathbf{i} - z\,\mathbf{j} - x\,\mathbf{k})\cdot(\mathbf{i} + \mathbf{j} + \mathbf{k})\,dA = \iint_D (-1)\,dA = -(\text{area of } D) = -\tfrac{1}{2}$$

34. $\iint_S \mathbf{F}\cdot d\mathbf{S} = \iiint_E 3(x^2 + y^2 + z^2)\,dV = \int_0^{2\pi}\int_0^1\int_0^2 (3r^2 + 3z^2)\,r\,dz\,dr\,d\theta = 2\pi\int_0^1 (6r^3 + 8r)\,dr = 11\pi$

35. $\iiint_E \operatorname{div} \mathbf{F}\, dV = \iiint\limits_{x^2+y^2+z^2 \le 1} 3\, dV = 3(\text{volume of sphere}) = 4\pi$. Then

$\mathbf{F}(\mathbf{r}(\phi,\theta)) \cdot (\mathbf{r}_\phi \times \mathbf{r}_\theta) = \sin^3\phi\cos^2\theta + \sin^3\phi\sin^2\theta + \sin\phi\cos^2\phi = \sin\phi$ and

$\iint_S \mathbf{F} \cdot d\mathbf{S} = \int_0^{2\pi}\int_0^{\pi} \sin\phi\, d\phi\, d\theta = (2\pi)(2) = 4\pi$.

36. Here we must use Equation 13.9.7 since $\mathbf{F}$ is not defined at the origin. Let S_1 be the sphere of radius 1 with center at the origin and outer unit normal $\mathbf{n}_1$. Let S_2 be the surface of the ellipsoid with outer unit normal $\mathbf{n}_2$ and let E be the solid region between S_1 and S_2. Then the outward flux of $\mathbf{F}$ through the ellipsoid is given by

$\iint_{S_2} \mathbf{F} \cdot \mathbf{n}_2\, dS = -\iint_{S_1} \mathbf{F} \cdot (-\mathbf{n}_1)\, dS + \iiint_E \operatorname{div} \mathbf{F}\, dV$. But $\mathbf{F} = \mathbf{r}/|\mathbf{r}|^3$, so

$\operatorname{div} \mathbf{F} = \nabla \cdot \left(|\mathbf{r}|^{-3}\mathbf{r}\right) = |\mathbf{r}|^{-3}(\nabla \cdot \mathbf{r}) + \mathbf{r} \cdot \left(\nabla |\mathbf{r}|^{-3}\right) = |\mathbf{r}|^{-3}(3) + \mathbf{r} \cdot \left(-3|\mathbf{r}|^{-4}\right)\left(\mathbf{r}|\mathbf{r}|^{-1}\right) = 0$. [Here we have used Exercises 13.5.28(a) and 13.5.29(a).] And $\mathbf{F} \cdot \mathbf{n}_1 = \dfrac{\mathbf{r}}{|\mathbf{r}|^3} \cdot \dfrac{\mathbf{r}}{|\mathbf{r}|} = |\mathbf{r}|^{-2} = 1$ on S_1. Thus

$\iint_{S_2} \mathbf{F} \cdot \mathbf{n}_2\, dS = \iint_{S_1} dS + \iiint_E 0\, dV = (\text{surface area of the unit sphere}) = 4\pi(1)^2 = 4\pi$.

37. By the Divergence Theorem, $\iint_S \mathbf{F} \cdot \mathbf{n}\, dS = \iiint_E \operatorname{div} \mathbf{F}\, dV = 3(\text{volume of } E) = 3(8-1) = 21$.

38. Let C' be the circle with center at the origin and radius a as in the figure. Let D be the region bounded by C and C'. Then D's positively oriented boundary is $C \cup (-C')$. Hence by Green's Theorem

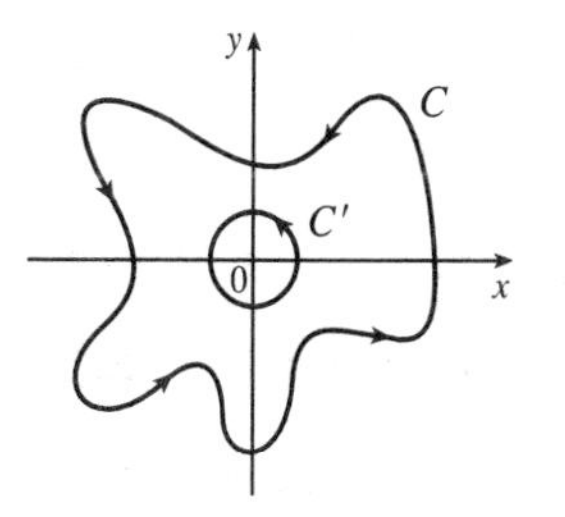

$$\int_C \mathbf{F} \cdot d\mathbf{r} + \int_{-C'} \mathbf{F} \cdot d\mathbf{r} = \iint_D \left(\frac{\partial Q}{\partial x} - \frac{\partial P}{\partial y}\right) dA = 0, \text{ so}$$

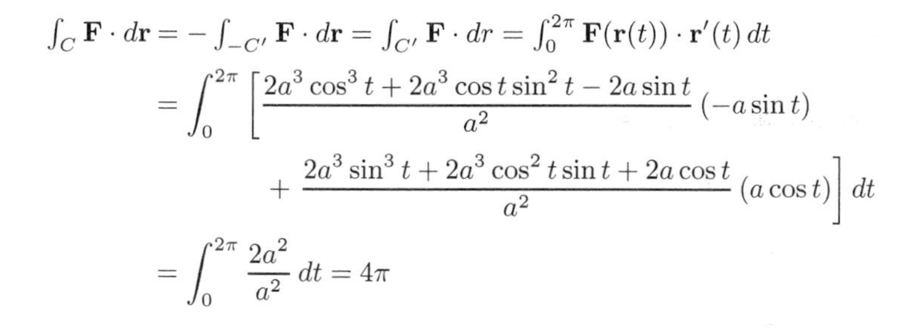

$$\begin{aligned}
\int_C \mathbf{F} \cdot d\mathbf{r} &= -\int_{-C'} \mathbf{F} \cdot d\mathbf{r} = \int_{C'} \mathbf{F} \cdot dr = \int_0^{2\pi} \mathbf{F}(\mathbf{r}(t)) \cdot \mathbf{r}'(t)\, dt \\
&= \int_0^{2\pi} \left[\frac{2a^3\cos^3 t + 2a^3\cos t\sin^2 t - 2a\sin t}{a^2}(-a\sin t) \right. \\
&\qquad \left. + \frac{2a^3\sin^3 t + 2a^3\cos^2 t\sin t + 2a\cos t}{a^2}(a\cos t)\right] dt \\
&= \int_0^{2\pi} \frac{2a^2}{a^2}\, dt = 4\pi
\end{aligned}$$

APPENDIXES

A TRIGONOMETRY

1. $210^\circ = 210\left(\frac{\pi}{180}\right) = \frac{7\pi}{6}$ rad

2. $300^\circ = 300\left(\frac{\pi}{180}\right) = \frac{5\pi}{3}$ rad

3. $9^\circ = 9\left(\frac{\pi}{180}\right) = \frac{\pi}{20}$ rad

4. $-315^\circ = -315\left(\frac{\pi}{180}\right) = -\frac{7\pi}{4}$ rad

5. $900^\circ = 900\left(\frac{\pi}{180}\right) = 5\pi$ rad

6. $36^\circ = 36\left(\frac{\pi}{180}\right) = \frac{\pi}{5}$ rad

7. 4π rad $= 4\pi\left(\frac{180}{\pi}\right) = 720^\circ$

8. $-\frac{7\pi}{2}$ rad $= -\frac{7\pi}{2}\left(\frac{180}{\pi}\right) = -630^\circ$

9. $\frac{5\pi}{12}$ rad $= \frac{5\pi}{12}\left(\frac{180}{\pi}\right) = 75^\circ$

10. $\frac{8\pi}{3}$ rad $= \frac{8\pi}{3}\left(\frac{180}{\pi}\right) = 480^\circ$

11. $-\frac{3\pi}{8}$ rad $= -\frac{3\pi}{8}\left(\frac{180}{\pi}\right) = -67.5^\circ$

12. 5 rad $= 5\left(\frac{180}{\pi}\right) = \left(\frac{900}{\pi}\right)^\circ$

13. Using Formula 3, $a = r\theta = 36 \cdot \frac{\pi}{12} = 3\pi$ cm.

14. Using Formula 3, $a = r\theta = 10 \cdot 72\left(\frac{\pi}{180}\right) = 4\pi$ cm.

15. Using Formula 3, $\theta = a/r = \frac{1}{1.5} = \frac{2}{3}$ rad $= \frac{2}{3}\left(\frac{180}{\pi}\right) = \left(\frac{120}{\pi}\right)^\circ \approx 38.2^\circ$.

16. $a = r\theta \;\Rightarrow\; r = \frac{a}{\theta} = \frac{6}{3\pi/4} = \frac{8}{\pi}$ cm

17.

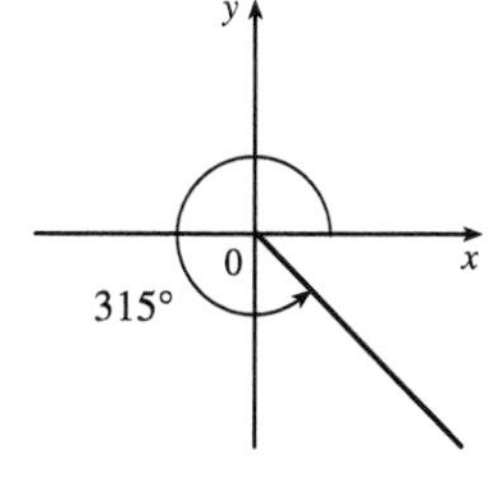

18.

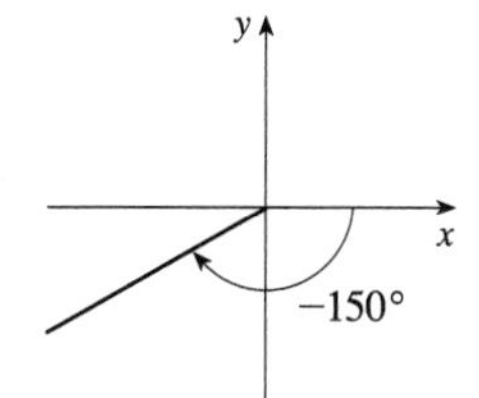

19.

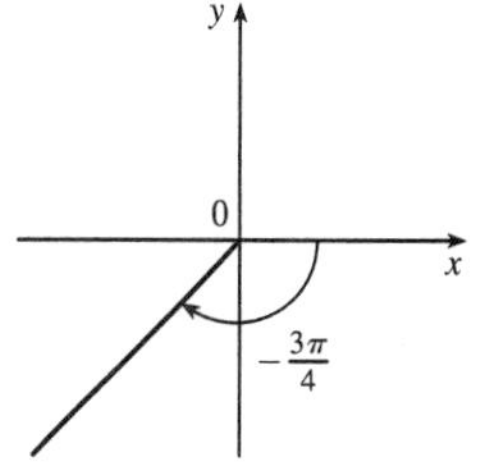

20.

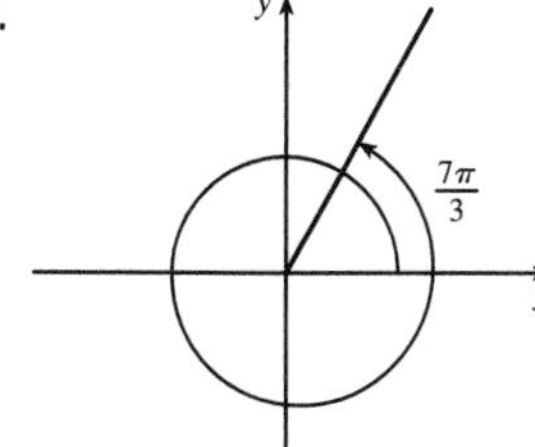

21.

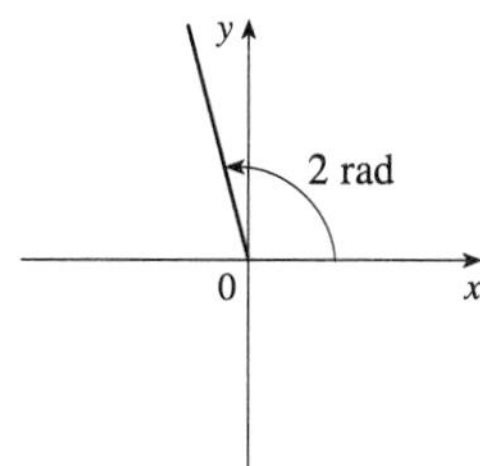

22.

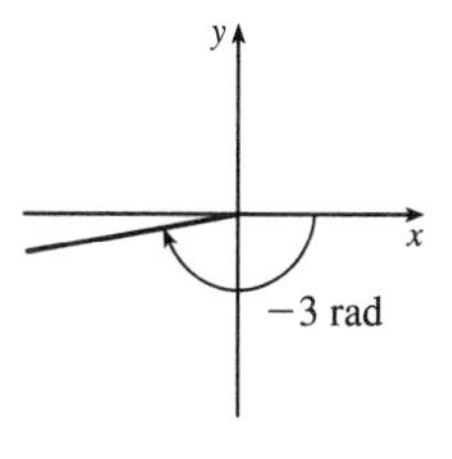

23.

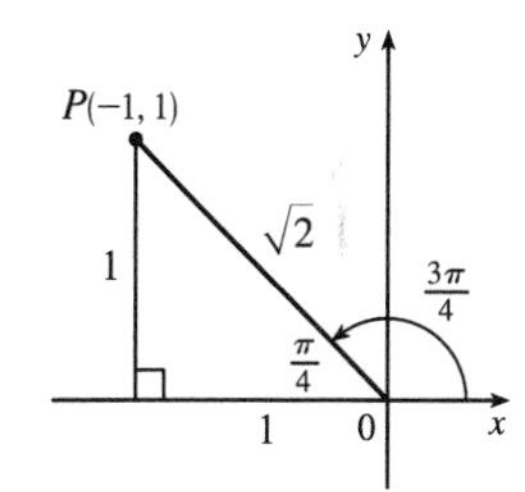

From the diagram we see that a point on the terminal side is $P(-1,1)$. Therefore, taking $x=-1$, $y=1$, $r=\sqrt{2}$ in the definitions of the trigonometric ratios, we have $\sin\frac{3\pi}{4} = \frac{1}{\sqrt{2}}$, $\cos\frac{3\pi}{4} = -\frac{1}{\sqrt{2}}$, $\tan\frac{3\pi}{4} = -1$, $\csc\frac{3\pi}{4} = \sqrt{2}$, $\sec\frac{3\pi}{4} = -\sqrt{2}$, and $\cot\frac{3\pi}{4} = -1$.

24.

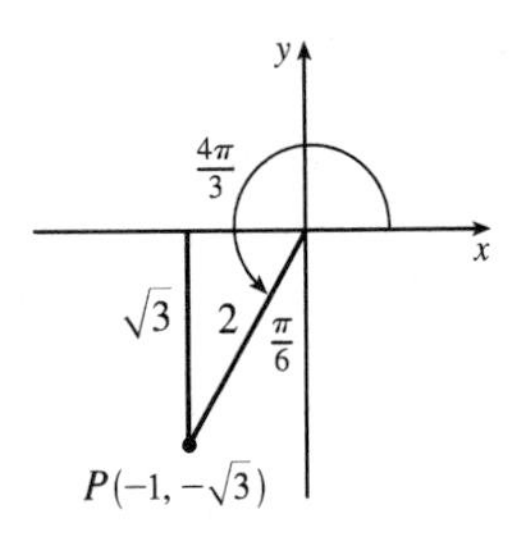

From the diagram and Figure 8, we see that a point on the terminal side is $P(-1,-\sqrt{3})$. Therefore, taking $x=-1$, $y=-\sqrt{3}$, $r=2$ in the definitions of the trigonometric ratios, we have $\sin\frac{4\pi}{3}=-\frac{\sqrt{3}}{2}$, $\cos\frac{4\pi}{3}=-\frac{1}{2}$, $\tan\frac{4\pi}{3}=\sqrt{3}$, $\csc\frac{4\pi}{3}=-\frac{2}{\sqrt{3}}$, $\sec\frac{4\pi}{3}=-2$, and $\cot\frac{4\pi}{3}=\frac{1}{\sqrt{3}}$.

25.

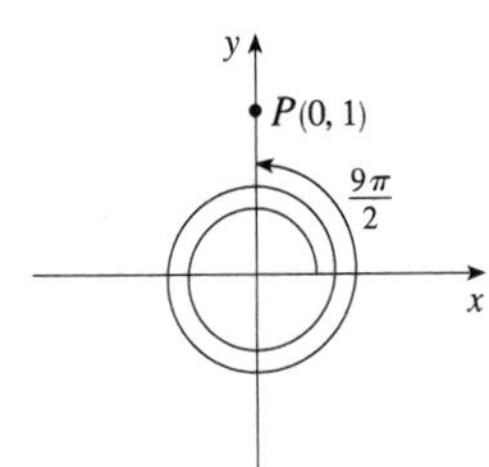

From the diagram we see that a point on the terminal line is $P(0,1)$. Therefore taking $x=0$, $y=1$, $r=1$ in the definitions of the trigonometric ratios, we have $\sin\frac{9\pi}{2}=1$, $\cos\frac{9\pi}{2}=0$, $\tan\frac{9\pi}{2}=y/x$ is undefined since $x=0$, $\csc\frac{9\pi}{2}=1$, $\sec\frac{9\pi}{2}=r/x$ is undefined since $x=0$, and $\cot\frac{9\pi}{2}=0$.

26.

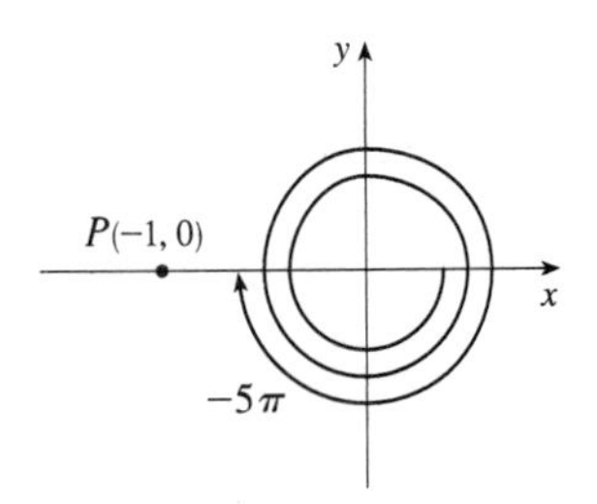

From the diagram, we see that a point on the terminal line is $P(-1,0)$. Therefore taking $x=-1$, $y=0$, $r=1$ in the definitions of the trigonometric ratios we have $\sin(-5\pi)=0$, $\cos(-5\pi)=-1$, $\tan(-5\pi)=0$, $\csc(-5\pi)$ is undefined, $\sec(-5\pi)=-1$, and $\cot(-5\pi)$ is undefined.

27.

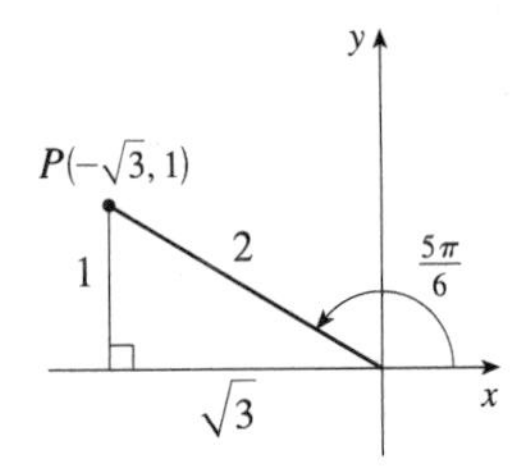

Using Figure 8 we see that a point on the terminal line is $P\left(-\sqrt{3},1\right)$. Therefore taking $x=-\sqrt{3}$, $y=1$, $r=2$ in the definitions of the trigonometric ratios, we have $\sin\frac{5\pi}{6}=\frac{1}{2}$, $\cos\frac{5\pi}{6}=-\frac{\sqrt{3}}{2}$, $\tan\frac{5\pi}{6}=-\frac{1}{\sqrt{3}}$, $\csc\frac{5\pi}{6}=2$, $\sec\frac{5\pi}{6}=-\frac{2}{\sqrt{3}}$, and $\cot\frac{5\pi}{6}=-\sqrt{3}$.

28.

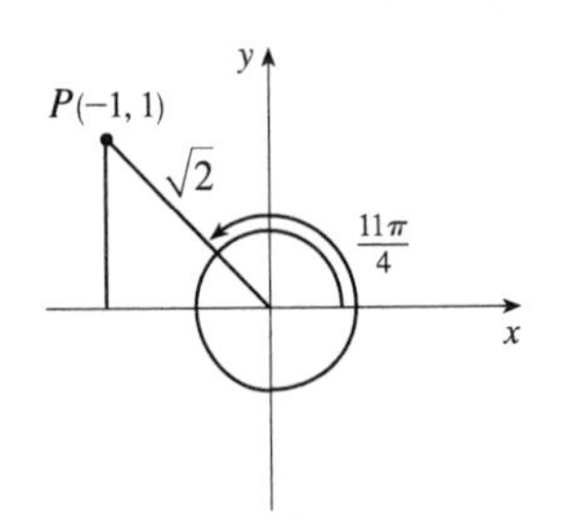

From the diagram, we see that a point on the terminal line is $P(-1,1)$. Therefore taking $x=-1$, $y=1$, $r=\sqrt{2}$ in the definitions of the trigonometric ratios we have $\sin\frac{11\pi}{4}=\frac{1}{\sqrt{2}}$, $\cos\frac{11\pi}{4}=-\frac{1}{\sqrt{2}}$, $\tan\frac{11\pi}{4}=-1$, $\csc\frac{11\pi}{4}=\sqrt{2}$, $\sec\frac{11\pi}{4}=-\sqrt{2}$, and $\cot\frac{11\pi}{4}=-1$.

29. $\sin\theta=y/r=\frac{3}{5}\ \Rightarrow\ y=3$, $r=5$, and $x=\sqrt{r^2-y^2}=4$ (since $0<\theta<\frac{\pi}{2}$). Therefore taking $x=4$, $y=3$, $r=5$ in the definitions of the trigonometric ratios, we have $\cos\theta=\frac{4}{5}$, $\tan\theta=\frac{3}{4}$, $\csc\theta=\frac{5}{3}$, $\sec\theta=\frac{5}{4}$, and $\cot\theta=\frac{4}{3}$.

30. Since $0<\alpha<\frac{\pi}{2}$, α is in the first quadrant where x and y are both positive. Therefore, $\tan\alpha=y/x=\frac{2}{1}\ \Rightarrow\ y=2$, $x=1$, and $r=\sqrt{x^2+y^2}=\sqrt{5}$. Taking $x=1$, $y=2$, $r=\sqrt{5}$ in the definitions of the trigonometric ratios, we have $\sin\alpha=\frac{2}{\sqrt{5}}$, $\cos\alpha=\frac{1}{\sqrt{5}}$, $\csc\alpha=\frac{\sqrt{5}}{2}$, $\sec\alpha=\sqrt{5}$, and $\cot\alpha=\frac{1}{2}$.

31. $\frac{\pi}{2} < \phi < \pi \quad \Rightarrow \quad \phi$ is in the second quadrant, where x is negative and y is positive. Therefore $\sec\phi = r/x = -1.5 = -\frac{3}{2}$ $\Rightarrow \quad r = 3$, $x = -2$, and $y = \sqrt{r^2 - x^2} = \sqrt{5}$. Taking $x = -2$, $y = \sqrt{5}$, and $r = 3$ in the definitions of the trigonometric ratios, we have $\sin\phi = \frac{\sqrt{5}}{3}$, $\cos\phi = -\frac{2}{3}$, $\tan\phi = -\frac{\sqrt{5}}{2}$, $\csc\phi = \frac{3}{\sqrt{5}}$, and $\cot\theta = -\frac{2}{\sqrt{5}}$.

32. Since $\pi < x < \frac{3\pi}{2}$, x is in the third quadrant where x and y are both negative. Therefore $\cos x = x/r = -\frac{1}{3} \quad \Rightarrow \quad x = -1$, $r = 3$, and $y = -\sqrt{r^2 - x^2} = -\sqrt{8} = -2\sqrt{2}$. Taking $x = -1$, $r = 3$, $y = -2\sqrt{2}$ in the definitions of the trigonometric ratios, we have $\sin x = -\frac{2\sqrt{2}}{3}$, $\tan x = 2\sqrt{2}$, $\csc x = -\frac{3}{2\sqrt{2}}$, $\sec x = -3$, and $\cot x = \frac{1}{2\sqrt{2}}$.

33. $\pi < \beta < 2\pi$ means that β is in the third or fourth quadrant where y is negative. Also since $\cot\beta = x/y = 3$ which is positive, x must also be negative. Therefore $\cot\beta = x/y = \frac{3}{1} \quad \Rightarrow \quad x = -3$, $y = -1$, and $r = \sqrt{x^2 + y^2} = \sqrt{10}$. Taking $x = -3$, $y = -1$ and $r = \sqrt{10}$ in the definitions of the trigonometric ratios, we have $\sin\beta = -\frac{1}{\sqrt{10}}$, $\cos\beta = -\frac{3}{\sqrt{10}}$, $\tan\beta = \frac{1}{3}$, $\csc\beta = -\sqrt{10}$, and $\sec\beta = -\frac{\sqrt{10}}{3}$.

34. Since $\frac{3\pi}{2} < \theta < 2\pi$, θ is in the fourth quadrant where x is positive and y is negative. Therefore $\csc\theta = r/y = -\frac{4}{3} \quad \Rightarrow$ $r = 4$, $y = -3$, and $x = \sqrt{r^2 - y^2} = \sqrt{7}$. Taking $x = \sqrt{7}$, $y = -3$, and $r = 4$ in the definitions of the trigonometric ratios, we have $\sin\theta = -\frac{3}{4}$, $\cos\theta = \frac{\sqrt{7}}{4}$, $\tan\theta = -\frac{3}{\sqrt{7}}$, $\sec\theta = \frac{4}{\sqrt{7}}$, and $\cot\theta = -\frac{\sqrt{7}}{3}$.

35. $\sin 35^\circ = \dfrac{x}{10} \quad \Rightarrow \quad x = 10\sin 35^\circ \approx 5.73576$ cm

36. $\cos 40^\circ = \dfrac{x}{25} \quad \Rightarrow \quad x = 25\cos 40^\circ \approx 19.15111$ cm

37. $\tan\frac{2\pi}{5} = \dfrac{x}{8} \quad \Rightarrow \quad x = 8\tan\frac{2\pi}{5} \approx 24.62147$ cm

38. $\cos\dfrac{3\pi}{8} = \dfrac{22}{x} \quad \Rightarrow \quad x = \dfrac{22}{\cos\dfrac{3\pi}{8}} \approx 57.48877$ cm

39. (a) From the diagram we see that $\sin\theta = \dfrac{y}{r} = \dfrac{a}{c}$, and $\sin(-\theta) = \dfrac{-a}{c} = -\dfrac{a}{c} = -\sin\theta$.

(b) Again from the diagram we see that $\cos\theta = \dfrac{x}{r} = \dfrac{b}{c} = \cos(-\theta)$.

40. (a) Using (12a) and (12b), we have

$$\tan(x+y) = \frac{\sin(x+y)}{\cos(x+y)} = \frac{\sin x\cos y + \cos x\sin y}{\cos x\cos y - \sin x\sin y} = \frac{\dfrac{\sin x\cos y}{\cos x\cos y} + \dfrac{\cos x\sin y}{\cos x\cos y}}{\dfrac{\cos x\cos y}{\cos x\cos y} - \dfrac{\sin x\sin y}{\cos x\cos y}} = \frac{\tan x + \tan y}{1 - \tan x\tan y}$$

(b) From (10a) and (10b), we have $\tan(-\theta) = -\tan\theta$, so (14a) implies that

$$\tan(x-y) = \tan(x+(-y)) = \frac{\tan x + \tan(-y)}{1 - \tan x\tan(-y)} = \frac{\tan x - \tan y}{1 + \tan x\tan y}$$

41. (a) Using (12a) and (13a), we have

$$\begin{aligned}\tfrac{1}{2}\left[\sin(x+y)+\sin(x-y)\right] &= \tfrac{1}{2}\left[\sin x\cos y+\cos x\sin y+\sin x\cos y-\cos x\sin y\right]\\ &= \tfrac{1}{2}\left(2\sin x\cos y\right)=\sin x\cos y\end{aligned}$$

(b) This time, using (12b) and (13b), we have

$$\begin{aligned}\tfrac{1}{2}\left[\cos(x+y)+\cos(x-y)\right] &= \tfrac{1}{2}\left[\cos x\cos y-\sin x\sin y+\cos x\cos y+\sin x\sin y\right]\\ &= \tfrac{1}{2}\left(2\cos x\cos y\right)=\cos x\cos y\end{aligned}$$

(c) Again using (12b) and (13b), we have

$$\begin{aligned}\tfrac{1}{2}\left[\cos(x-y)-\cos(x+y)\right] &= \tfrac{1}{2}\left[\cos x\cos y+\sin x\sin y-\cos x\cos y+\sin x\sin y\right]\\ &= \tfrac{1}{2}\left(2\sin x\sin y\right)=\sin x\sin y\end{aligned}$$

42. Using (13b), $\cos\left(\frac{\pi}{2}-x\right)=\cos\frac{\pi}{2}\cos x+\sin\frac{\pi}{2}\sin x=0\cdot\cos x+1\cdot\sin x=\sin x$.

43. Using (12a), we have $\sin\left(\frac{\pi}{2}+x\right)=\sin\frac{\pi}{2}\cos x+\cos\frac{\pi}{2}\sin x=1\cdot\cos x+0\cdot\sin x=\cos x$.

44. Using (13a), we have $\sin(\pi-x)=\sin\pi\cos x-\cos\pi\sin x=0\cdot\cos x-(-1)\sin x=\sin x$.

45. Using (6), we have $\sin\theta\cot\theta=\sin\theta\cdot\dfrac{\cos\theta}{\sin\theta}=\cos\theta$.

46. $$\begin{aligned}(\sin x+\cos x)^2 &= \sin^2 x+2\sin x\cos x+\cos^2 x=\left(\sin^2 x+\cos^2 x\right)+\sin 2x \quad \text{[by (15a)]}\\ &= 1+\sin 2x \quad \text{[by (7)]}\end{aligned}$$

47. $\sec y-\cos y=\dfrac{1}{\cos y}-\cos y \;\text{[by (6)]}=\dfrac{1-\cos^2 y}{\cos y}=\dfrac{\sin^2 y}{\cos y}\;\text{[by (7)]}=\dfrac{\sin y}{\cos y}\sin y=\tan y\sin y\;\text{[by (6)]}$

48. $\tan^2\alpha-\sin^2\alpha=\dfrac{\sin^2\alpha}{\cos^2\alpha}-\sin^2\alpha=\dfrac{\sin^2\alpha-\sin^2\alpha\cos^2\alpha}{\cos^2\alpha}=\dfrac{\sin^2\alpha\left(1-\cos^2\alpha\right)}{\cos^2\alpha}=\tan^2\alpha\sin^2\alpha\;\text{[by (6), (7)]}$

49. $$\begin{aligned}\cot^2\theta+\sec^2\theta &= \frac{\cos^2\theta}{\sin^2\theta}+\frac{1}{\cos^2\theta}\;\text{[by (6)]}=\frac{\cos^2\theta\cos^2\theta+\sin^2\theta}{\sin^2\theta\cos^2\theta}\\ &= \frac{\left(1-\sin^2\theta\right)\left(1-\sin^2\theta\right)+\sin^2\theta}{\sin^2\theta\cos^2\theta}\;\text{[by (7)]}=\frac{1-\sin^2\theta+\sin^4\theta}{\sin^2\theta\cos^2\theta}\\ &= \frac{\cos^2\theta+\sin^4\theta}{\sin^2\theta\cos^2\theta}\;\text{[by (7)]}=\frac{1}{\sin^2\theta}+\frac{\sin^2\theta}{\cos^2\theta}=\csc^2\theta+\tan^2\theta\;\text{[by (6)]}\end{aligned}$$

50. $2\csc 2t=\dfrac{2}{\sin 2t}=\dfrac{2}{2\sin t\cos t}\;\text{[by (15a)]}=\dfrac{1}{\sin t\cos t}=\sec t\csc t$

51. Using (14a), we have $\tan 2\theta=\tan(\theta+\theta)=\dfrac{\tan\theta+\tan\theta}{1-\tan\theta\tan\theta}=\dfrac{2\tan\theta}{1-\tan^2\theta}$.

52. $\dfrac{1}{1-\sin\theta}+\dfrac{1}{1+\sin\theta}=\dfrac{1+\sin\theta+1-\sin\theta}{(1-\sin\theta)(1+\sin\theta)}=\dfrac{2}{1-\sin^2\theta}=\dfrac{2}{\cos^2\theta}\;\text{[by (7)]}=2\sec^2\theta$

53. Using (15a) and (16a),

$$\begin{aligned}\sin x\sin 2x+\cos x\cos 2x &= \sin x\left(2\sin x\cos x\right)+\cos x\left(2\cos^2 x-1\right)=2\sin^2 x\cos x+2\cos^3 x-\cos x\\ &= 2\left(1-\cos^2 x\right)\cos x+2\cos^3 x-\cos x\;\text{[by (7)]}\\ &= 2\cos x-2\cos^3 x+2\cos^3 x-\cos x=\cos x\end{aligned}$$

Or: $\sin x\sin 2x+\cos x\cos 2x=\cos(2x-x)\;\text{[by 13(b)]}=\cos x$

54. Working backward, we start with equations (12a) and (13a):

$$\begin{aligned}\sin(x+y)\sin(x-y) &= (\sin x\cos y+\cos x\sin y)(\sin x\cos y-\cos x\sin y)\\ &= \sin^2 x\cos^2 y-\sin x\cos y\cos x\sin y+\cos x\sin y\sin x\cos y-\cos^2 x\sin^2 y\\ &= \sin^2 x\left(1-\sin^2 y\right)-\left(1-\sin^2 x\right)\sin^2 y \quad [\text{by } (7)]\\ &= \sin^2 x-\sin^2 x\sin^2 y-\sin^2 y+\sin^2 x\sin^2 y=\sin^2 x-\sin^2 y\end{aligned}$$

55. $$\begin{aligned}\frac{\sin\phi}{1-\cos\phi} &= \frac{\sin\phi}{1-\cos\phi}\cdot\frac{1+\cos\phi}{1+\cos\phi}=\frac{\sin\phi\,(1+\cos\phi)}{1-\cos^2\phi}=\frac{\sin\phi\,(1+\cos\phi)}{\sin^2\phi} \quad [\text{by } (7)]\\ &= \frac{1+\cos\phi}{\sin\phi}=\frac{1}{\sin\phi}+\frac{\cos\phi}{\sin\phi}=\csc\phi+\cot\phi \quad [\text{by } (6)]\end{aligned}$$

56. $$\tan x+\tan y=\frac{\sin x}{\cos x}+\frac{\sin y}{\cos y}=\frac{\sin x\cos y+\cos x\sin y}{\cos x\cos y}=\frac{\sin(x+y)}{\cos x\cos y} \quad [\text{by } (12\text{a})]$$

57. Using (12a),

$$\begin{aligned}\sin 3\theta+\sin\theta &= \sin(2\theta+\theta)+\sin\theta=\sin 2\theta\cos\theta+\cos 2\theta\sin\theta+\sin\theta\\ &= \sin 2\theta\cos\theta+\left(2\cos^2\theta-1\right)\sin\theta+\sin\theta \quad [\text{by } (16\text{a})]\\ &= \sin 2\theta\cos\theta+2\cos^2\theta\sin\theta-\sin\theta+\sin\theta=\sin 2\theta\cos\theta+\sin 2\theta\cos\theta \quad [\text{by } (15\text{a})]\\ &= 2\sin 2\theta\cos\theta\end{aligned}$$

58. We use (12b) with $x=2\theta$, $y=\theta$ to get

$$\begin{aligned}\cos 3\theta &= \cos(2\theta+\theta)=\cos 2\theta\cos\theta-\sin 2\theta\sin\theta\\ &= \left(2\cos^2\theta-1\right)\cos\theta-2\sin^2\theta\cos\theta \quad [\text{by } (16\text{a}) \text{ and } (15\text{a})]\\ &= \left(2\cos^2\theta-1\right)\cos\theta-2\left(1-\cos^2\theta\right)\cos\theta \quad [\text{by } (7)]\\ &= 2\cos^3\theta-\cos\theta-2\cos\theta+2\cos^3\theta=4\cos^3\theta-3\cos\theta\end{aligned}$$

59. Since $\sin x=\frac{1}{3}$ we can label the opposite side as having length 1, the hypotenuse as having length 3, and use the Pythagorean Theorem to get that the adjacent side has length $\sqrt{8}$. Then, from the diagram, $\cos x=\frac{\sqrt{8}}{3}$. Similarly we have that $\sin y=\frac{3}{5}$. Now use (12a):

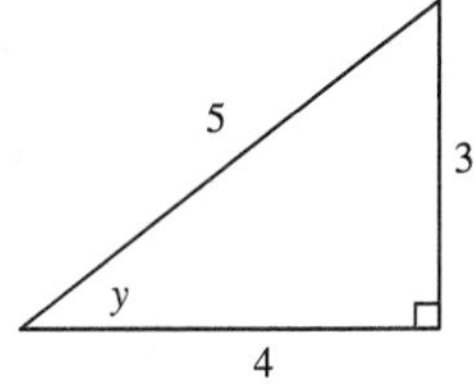

$\sin(x+y)=\sin x\cos y+\cos x\,\sin y=\frac{1}{3}\cdot\frac{4}{5}+\frac{\sqrt{8}}{3}\cdot\frac{3}{5}=\frac{4}{15}+\frac{3\sqrt{8}}{15}=\frac{4+6\sqrt{2}}{15}$.

60. Use (12b) and the values for $\sin y$ and $\cos x$ obtained in Exercise 59 to get

$$\cos(x+y)=\cos x\cos y-\sin x\sin y=\frac{\sqrt{8}}{3}\cdot\frac{4}{5}-\frac{1}{3}\cdot\frac{3}{5}=\frac{8\sqrt{2}-3}{15}$$

61. Using (13b) and the values for $\cos x$ and $\sin y$ obtained in Exercise 59, we have

$$\cos(x-y)=\cos x\cos y+\sin x\sin y=\frac{\sqrt{8}}{3}\cdot\frac{4}{5}+\frac{1}{3}\cdot\frac{3}{5}=\frac{8\sqrt{2}+3}{15}$$

62. Using (13a) and the values for $\sin y$ and $\cos x$ obtained in Exercise 59, we get

$$\sin(x-y)=\sin x\cos y-\cos x\sin y=\frac{1}{3}\cdot\frac{4}{5}-\frac{\sqrt{8}}{3}\cdot\frac{3}{5}=\frac{4-6\sqrt{2}}{15}$$

63. Using (15a) and the values for $\sin y$ and $\cos y$ obtained in Exercise 59, we have

$$\sin 2y=2\sin y\cos y=2\cdot\frac{3}{5}\cdot\frac{4}{5}=\frac{24}{25}$$

64. Using (16a) with $\cos y=\frac{4}{5}$, we have $\cos 2y=2\cos^2 y-1=2\left(\frac{4}{5}\right)^2-1=\frac{32}{25}-1=\frac{7}{25}$.

65. $2\cos x-1=0 \quad\Leftrightarrow\quad \cos x=\frac{1}{2} \quad\Rightarrow\quad x=\frac{\pi}{3},\frac{5\pi}{3}$ for $x\in[0,2\pi]$.

66. $3\cot^2 x = 1 \quad\Leftrightarrow\quad 3 = 1/\cot^2 x \quad\Leftrightarrow\quad \tan^2 x = 3 \quad\Leftrightarrow\quad \tan x = \pm\sqrt{3} \quad\Rightarrow\quad x = \frac{\pi}{3}, \frac{2\pi}{3}, \frac{4\pi}{3}$, and $\frac{5\pi}{3}$.

67. $2\sin^2 x = 1 \quad\Leftrightarrow\quad \sin^2 x = \frac{1}{2} \quad\Leftrightarrow\quad \sin x = \pm\frac{1}{\sqrt{2}} \quad\Rightarrow\quad x = \frac{\pi}{4}, \frac{3\pi}{4}, \frac{5\pi}{4}, \frac{7\pi}{4}$.

68. $|\tan x| = 1 \quad\Leftrightarrow\quad \tan x = -1$ or $\tan x = 1 \quad\Leftrightarrow\quad x = \frac{3\pi}{4}, \frac{7\pi}{4}$ or $x = \frac{\pi}{4}, \frac{5\pi}{4}$.

69. Using (15a), we have $\sin 2x = \cos x \quad\Leftrightarrow\quad 2\sin x\cos x - \cos x = 0 \quad\Leftrightarrow\quad \cos x(2\sin x - 1) = 0 \quad\Leftrightarrow\quad \cos x = 0$ or $2\sin x - 1 = 0 \quad\Rightarrow\quad x = \frac{\pi}{2}, \frac{3\pi}{2}$ or $\sin x = \frac{1}{2} \quad\Rightarrow\quad x = \frac{\pi}{6}$ or $\frac{5\pi}{6}$. Therefore, the solutions are $x = \frac{\pi}{6}, \frac{\pi}{2}, \frac{5\pi}{6}, \frac{3\pi}{2}$.

70. By (15a), $2\cos x + \sin 2x = 0 \quad\Leftrightarrow\quad 2\cos x + 2\sin x\cos x = 0 \quad\Leftrightarrow\quad 2\cos x\,(1 + \sin x) = 0 \quad\Leftrightarrow\quad \cos x = 0$ or $1 + \sin x = 0 \quad\Leftrightarrow\quad x = \frac{\pi}{2}, \frac{3\pi}{2}$ or $\sin x = -1 \quad\Rightarrow\quad x = \frac{3}{2}\pi$. So the solutions are $x = \frac{\pi}{2}, \frac{3\pi}{2}$.

71. $\sin x = \tan x \quad\Leftrightarrow\quad \sin x - \tan x = 0 \quad\Leftrightarrow\quad \sin x - \dfrac{\sin x}{\cos x} = 0 \quad\Leftrightarrow\quad \sin x\left(1 - \dfrac{1}{\cos x}\right) = 0 \quad\Leftrightarrow\quad \sin x = 0$ or $1 - \dfrac{1}{\cos x} = 0 \quad\Rightarrow\quad x = 0, \pi, 2\pi$ or $1 = \dfrac{1}{\cos x} \quad\Rightarrow\quad \cos x = 1 \quad\Rightarrow\quad x = 0, 2\pi$. Therefore the solutions are $x = 0, \pi, 2\pi$.

72. By (16a), $2 + \cos 2x = 3\cos x \quad\Leftrightarrow\quad 2 + 2\cos^2 x - 1 = 3\cos x \quad\Leftrightarrow\quad 2\cos^2 x - 3\cos x + 1 = 0 \quad\Leftrightarrow\quad (2\cos x - 1)(\cos x - 1) = 0 \quad\Leftrightarrow\quad \cos x = 1$ or $\cos x = \frac{1}{2} \quad\Rightarrow\quad x = 0, 2\pi$ or $x = \frac{\pi}{3}, \frac{5\pi}{3}$.

73. We know that $\sin x = \frac{1}{2}$ when $x = \frac{\pi}{6}$ or $\frac{5\pi}{6}$, and from Figure 13(a), we see that $\sin x \le \frac{1}{2} \quad\Rightarrow\quad 0 \le x \le \frac{\pi}{6}$ or $\frac{5\pi}{6} \le x \le 2\pi$ for $x \in [0, 2\pi]$.

74. $2\cos x + 1 > 0 \quad\Rightarrow\quad 2\cos x > -1 \quad\Rightarrow\quad \cos x > -\frac{1}{2}$. $\cos x = -\frac{1}{2}$ when $x = \frac{2\pi}{3}, \frac{4\pi}{3}$ and from Figure 13(b), we see that $\cos x > -\frac{1}{2}$ when $0 \le x < \frac{2\pi}{3}, \frac{4\pi}{3} < x \le 2\pi$.

75. $\tan x = -1$ when $x = \frac{3\pi}{4}, \frac{7\pi}{4}$, and $\tan x = 1$ when $x = \frac{\pi}{4}$ or $\frac{5\pi}{4}$. From Figure 14 we see that $-1 < \tan x < 1 \quad\Rightarrow\quad 0 \le x < \frac{\pi}{4}, \frac{3\pi}{4} < x < \frac{5\pi}{4}$, and $\frac{7\pi}{4} < x \le 2\pi$.

76. We know that $\sin x = \cos x$ when $x = \frac{\pi}{4}, \frac{5\pi}{4}$, and from the diagram we see that $\sin x > \cos x$ when $\frac{\pi}{4} < x < \frac{5\pi}{4}$.

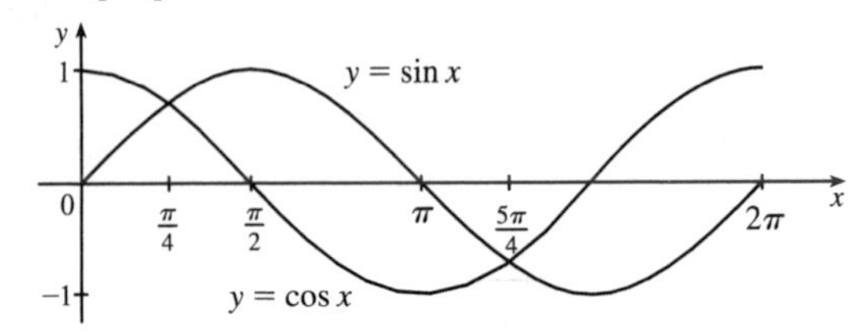

77. $y = \cos\left(x - \frac{\pi}{3}\right)$. We start with the graph of $y = \cos x$ and shift it $\frac{\pi}{3}$ units to the right.

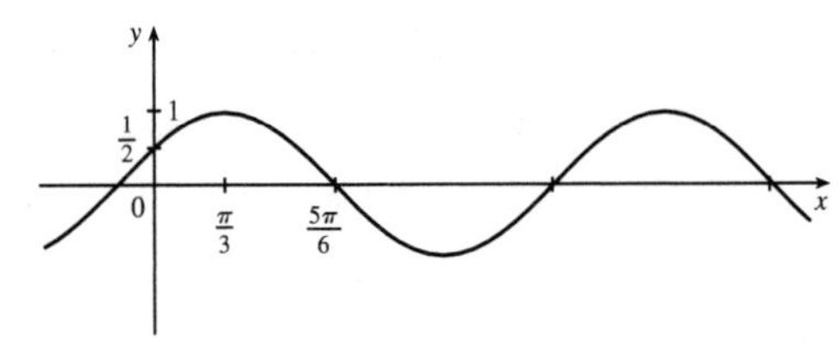

78. $y = \tan 2x$. Start with the graph of $y = \tan x$ with period π and compress it to a period of $\frac{\pi}{2}$.

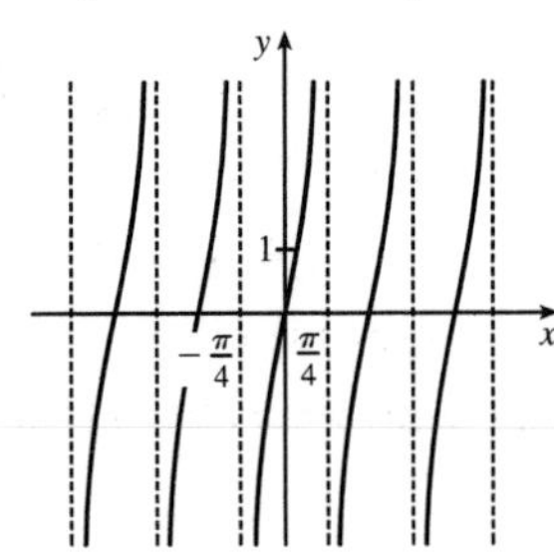

79. $y = \frac{1}{3}\tan\left(x - \frac{\pi}{2}\right)$. We start with the graph of $y = \tan x$, shift it $\frac{\pi}{2}$ units to the right and compress it to $\frac{1}{3}$ of its original vertical size.

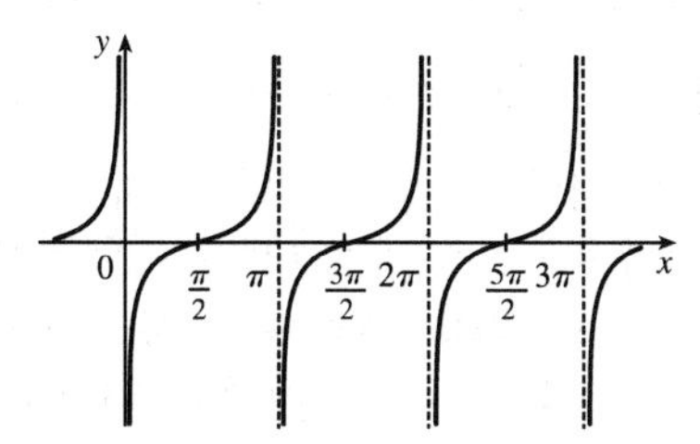

80. $y = 1 + \sec x$. Start with the graph of $y = \sec x$ and raise it by one unit.

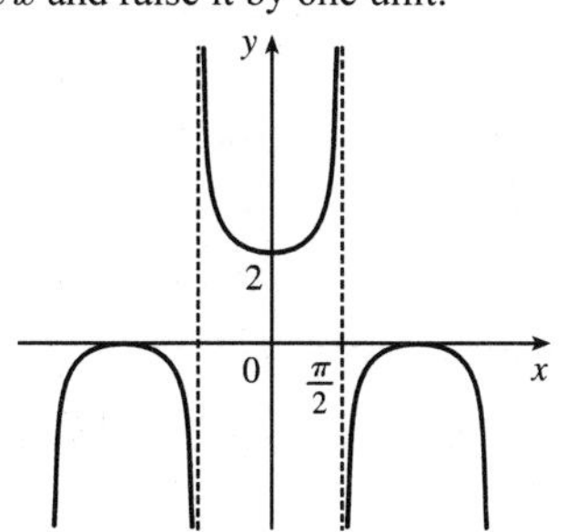

81. $y = |\sin x|$. We start with the graph of $y = \sin x$ and reflect the parts below the x-axis about the x-axis.

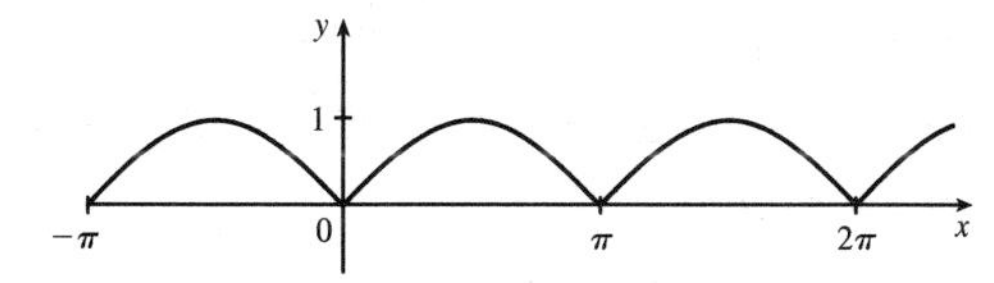

82. $y = 2 + \sin\left(x + \frac{\pi}{4}\right)$. Start with the graph of $y = \sin x$, and shift it $\frac{\pi}{4}$ units to the left and 2 units up.

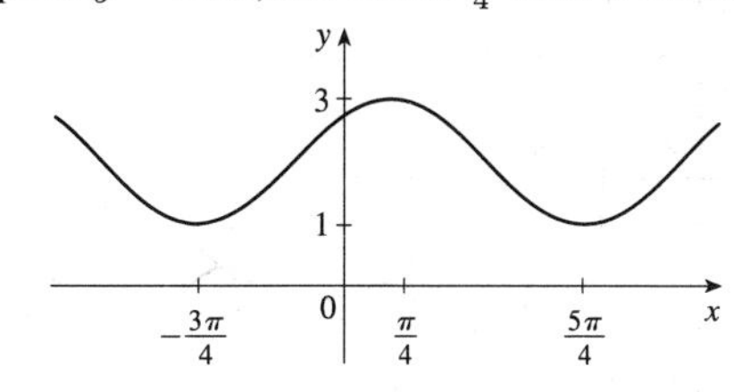

83. From the figure in the text, we see that $x = b\cos\theta$, $y = b\sin\theta$, and from the distance formula we have that the distance c from (x, y) to $(a, 0)$ is $c = \sqrt{(x-a)^2 + (y-0)^2} \;\Rightarrow$

$$\begin{aligned} c^2 &= (b\cos\theta - a)^2 + (b\sin\theta)^2 = b^2\cos^2\theta - 2ab\cos\theta + a^2 + b^2\sin^2\theta \\ &= a^2 + b^2\left(\cos^2\theta + \sin^2\theta\right) - 2ab\cos\theta = a^2 + b^2 - 2ab\cos\theta \quad \text{[by (7)]} \end{aligned}$$

84. $|AB|^2 = |AC|^2 + |BC|^2 - 2\,|AC|\,|BC|\cos\angle C = (820)^2 + (910)^2 - 2(820)(910)\cos 103^\circ$

$\approx 1{,}836{,}217 \;\Rightarrow\; |AB| \approx 1355$ m

85. Using the Law of Cosines, we have $c^2 = 1^2 + 1^2 - 2(1)(1)\cos(\alpha - \beta) = 2\left[1 - \cos(\alpha - \beta)\right]$. Now, using the distance formula, $c^2 = |AB|^2 = (\cos\alpha - \cos\beta)^2 + (\sin\alpha - \sin\beta)^2$. Equating these two expressions for c^2, we get $2\left[1 - \cos(\alpha - \beta)\right] = \cos^2\alpha + \sin^2\alpha + \cos^2\beta + \sin^2\beta - 2\cos\alpha\cos\beta - 2\sin\alpha\sin\beta \;\Rightarrow$ $1 - \cos(\alpha - \beta) = 1 - \cos\alpha\cos\beta - \sin\alpha\sin\beta \;\Rightarrow\; \cos(\alpha - \beta) = \cos\alpha\cos\beta + \sin\alpha\sin\beta$.

86. $\cos(x + y) = \cos(x - (-y)) = \cos x\cos(-y) + \sin x\sin(-y)$

$= \cos x\cos y - \sin x\sin y$ [using Equations (10a) and (10b)]

87. In Exercise 86 we used the subtraction formula for cosine to prove the addition formula for cosine. Using that formula with $x = \frac{\pi}{2} - \alpha$, $y = \beta$, we get $\cos\left[\left(\frac{\pi}{2} - \alpha\right) + \beta\right] = \cos\left(\frac{\pi}{2} - \alpha\right)\cos\beta - \sin\left(\frac{\pi}{2} - \alpha\right)\sin\beta \Rightarrow$ $\cos\left[\frac{\pi}{2} - (\alpha - \beta)\right] = \cos\left(\frac{\pi}{2} - \alpha\right)\cos\beta - \sin\left(\frac{\pi}{2} - \alpha\right)\sin\beta$. Now we use the identities given in the problem, $\cos\left(\frac{\pi}{2} - \theta\right) = \sin\theta$ and $\sin\left(\frac{\pi}{2} - \theta\right) = \cos\theta$, to get $\sin(\alpha - \beta) = \sin\alpha\cos\beta - \cos\alpha\sin\beta$.

88. If $0 < \theta < \frac{\pi}{2}$, we have the case depicted in the first diagram. In this case, we see that the height of the triangle is $h = a\sin\theta$. If $\frac{\pi}{2} \le \theta < \pi$, we have the case depicted in the second diagram. In this case, the height of the triangle is $h = a\sin(\pi - \theta) = a\sin\theta$ (by the identity proved in Exercise 78). So in either case, the area of the triangle is $\frac{1}{2}bh = \frac{1}{2}ab\sin\theta$.

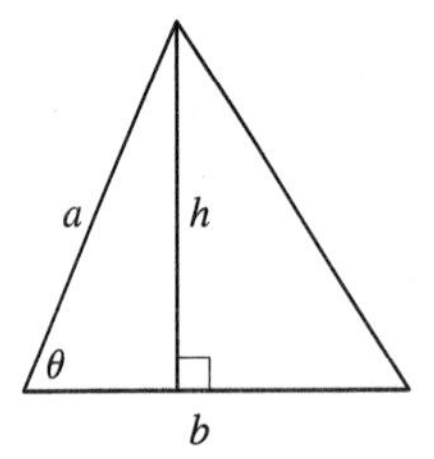

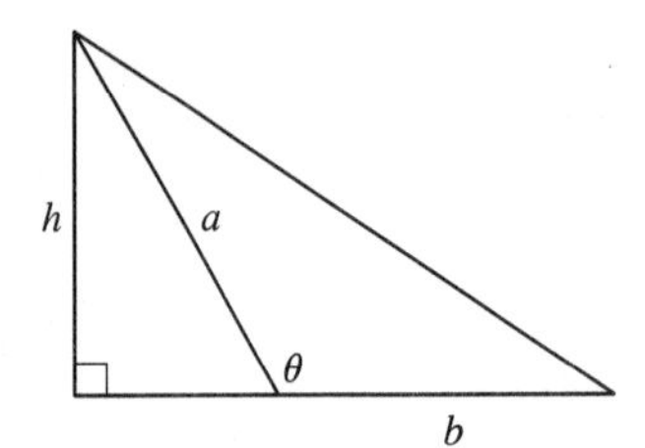

89. Using $A = \frac{1}{2}ab\sin\theta$, the area of the triangle is $\frac{1}{2}(10)(3)\sin 107° \approx 14.34457 \text{ cm}^2$.

C SIGMA NOTATION

1. $\sum_{i=1}^{5}\sqrt{i} = \sqrt{1} + \sqrt{2} + \sqrt{3} + \sqrt{4} + \sqrt{5}$

2. $\sum_{i=1}^{6}\frac{1}{i+1} = \frac{1}{2} + \frac{1}{3} + \frac{1}{4} + \frac{1}{5} + \frac{1}{6} + \frac{1}{7}$

3. $\sum_{i=4}^{6} 3^i = 3^4 + 3^5 + 3^6$

4. $\sum_{i=4}^{6} i^3 = 4^3 + 5^3 + 6^3$

5. $\sum_{k=0}^{4}\frac{2k-1}{2k+1} = -1 + \frac{1}{3} + \frac{3}{5} + \frac{5}{7} + \frac{7}{9}$

6. $\sum_{k=5}^{8} x^k = x^5 + x^6 + x^7 + x^8$

7. $\sum_{i=1}^{n} i^{10} = 1^{10} + 2^{10} + 3^{10} + \cdots + n^{10}$

8. $\sum_{j=n}^{n+3} j^2 = n^2 + (n+1)^2 + (n+2)^2 + (n+3)^2$

9. $\sum_{j=0}^{n-1}(-1)^j = 1 - 1 + 1 - 1 + \cdots + (-1)^{n-1}$

10. $\sum_{i=1}^{n} f(x_i)\,\Delta x_i = f(x_1)\,\Delta x_1 + f(x_2)\,\Delta x_2 + f(x_3)\,\Delta x_3 + \cdots + f(x_n)\,\Delta x_n$

11. $1 + 2 + 3 + 4 + \cdots + 10 = \sum_{i=1}^{10} i$

12. $\sqrt{3} + \sqrt{4} + \sqrt{5} + \sqrt{6} + \sqrt{7} = \sum_{i=3}^{7}\sqrt{i}$

13. $\frac{1}{2} + \frac{2}{3} + \frac{3}{4} + \frac{4}{5} + \cdots + \frac{19}{20} = \sum_{i=1}^{19}\frac{i}{i+1}$

14. $\frac{3}{7} + \frac{4}{8} + \frac{5}{9} + \frac{6}{10} + \cdots + \frac{23}{27} = \sum_{i=3}^{23}\frac{i}{i+4}$

15. $2 + 4 + 6 + 8 + \cdots + 2n = \sum_{i=1}^{n} 2i$

16. $1 + 3 + 5 + 7 + \cdots + (2n-1) = \sum_{i=1}^{n}(2i-1)$

17. $1 + 2 + 4 + 8 + 16 + 32 = \sum_{i=0}^{5} 2^i$

18. $\frac{1}{1} + \frac{1}{4} + \frac{1}{9} + \frac{1}{16} + \frac{1}{25} + \frac{1}{36} = \sum_{i=1}^{6}\frac{1}{i^2}$

19. $x + x^2 + x^3 + \cdots + x^n = \sum_{i=1}^{n} x^i$

20. $1 - x + x^2 - x^3 + \cdots + (-1)^n x^n = \sum_{i=0}^{n}(-1)^i x^i$

21. $\sum_{i=4}^{8}(3i-2)=[3(4)-2]+[3(5)-2]+[3(6)-2]+[3(7)-2]+[3(8)-2]=10+13+16+19+22=80$

22. $\sum_{i=3}^{6} i(i+2)=3\cdot 5+4\cdot 6+5\cdot 7+6\cdot 8=15+24+35+48=122$

23. $\sum_{j=1}^{6} 3^{j+1}=3^2+3^3+3^4+3^5+3^6+3^7=9+27+81+243+729+2187=3276$

(For a more general method, see Exercise 47.)

24. $$\sum_{k=0}^{8}\cos k\pi=\cos 0+\cos\pi+\cos 2\pi+\cos 3\pi+\cos 4\pi+\cos 5\pi+\cos 6\pi+\cos 7\pi+\cos 8\pi$$
$$=1-1+1-1+1-1+1-1+1=1$$

25. $\sum_{n=1}^{20}(-1)^n=-1+1-1+1-1+1-1+1-1+1-1+1-1+1-1+1-1+1-1+1=0$

26. $\sum_{i=1}^{100} 4=\underbrace{4+4+4+\cdots+4}_{\text{(100 summands)}}=100\cdot 4=400$

27. $\sum_{i=0}^{4}(2^i+i^2)=(1+0)+(2+1)+(4+4)+(8+9)+(16+16)=61$

28. $\sum_{i=-2}^{4} 2^{3-i}=2^5+2^4+2^3+2^2+2^1+2^0+2^{-1}=63.5$

29. $\sum_{i=1}^{n} 2i=2\sum_{i=1}^{n} i=2\cdot\dfrac{n(n+1)}{2}$ [by Theorem 3(c)] $=n(n+1)$

30. $$\sum_{i=1}^{n}(2-5i)=\sum_{i=1}^{n}2-\sum_{i=1}^{n}5i=2n-5\sum_{i=1}^{n}i=2n-\frac{5n(n+1)}{2}=\frac{4n}{2}-\frac{5n^2+5n}{2}=-\frac{n(5n+1)}{2}$$

31. $$\sum_{i=1}^{n}(i^2+3i+4)=\sum_{i=1}^{n}i^2+3\sum_{i=1}^{n}i+\sum_{i=1}^{n}4=\frac{n(n+1)(2n+1)}{6}+\frac{3n(n+1)}{2}+4n$$
$$=\tfrac{1}{6}\left[(2n^3+3n^2+n)+(9n^2+9n)+24n\right]=\tfrac{1}{6}\left(2n^3+12n^2+34n\right)$$
$$=\tfrac{1}{3}n\left(n^2+6n+17\right)$$

32. $$\sum_{i=1}^{n}(3+2i)^2=\sum_{i=1}^{n}\left(9+12i+4i^2\right)=\sum_{i=1}^{n}9+12\sum_{i=1}^{n}i+4\sum_{i=1}^{n}i^2$$
$$=9n+6n(n+1)+\frac{2n(n+1)(2n+1)}{3}=\frac{27n+18n^2+18n+4n^3+6n^2+2n}{3}$$
$$=\tfrac{1}{3}\left(4n^3+24n^2+47n\right)=\tfrac{1}{3}n\left(4n^2+24n+47\right)$$

33. $$\sum_{i=1}^{n}(i+1)(i+2)=\sum_{i=1}^{n}\left(i^2+3i+2\right)=\sum_{i=1}^{n}i^2+3\sum_{i=1}^{n}i+\sum_{i=1}^{n}2$$
$$=\frac{n(n+1)(2n+1)}{6}+\frac{3n(n+1)}{2}+2n=\frac{n(n+1)}{6}[(2n+1)+9]+2n$$
$$=\frac{n(n+1)}{3}(n+5)+2n=\frac{n}{3}[(n+1)(n+5)+6]=\frac{n}{3}\left(n^2+6n+11\right)$$

34. $\sum_{i=1}^{n} i(i+1)(i+2) = \sum_{i=1}^{n} (i^3 + 3i^2 + 2i) = \sum_{i=1}^{n} i^3 + 3\sum_{i=1}^{n} i^2 + 2\sum_{i=1}^{n} i$

$$= \left[\frac{n(n+1)}{2}\right]^2 + \frac{3n(n+1)(2n+1)}{6} + \frac{2n(n+1)}{2} = n(n+1)\left[\frac{n(n+1)}{4} + \frac{2n+1}{2} + 1\right]$$

$$= \frac{n(n+1)}{4}\left(n^2 + n + 4n + 2 + 4\right) = \frac{n(n+1)}{4}\left(n^2 + 5n + 6\right) = \frac{n(n+1)(n+2)(n+3)}{4}$$

35. $\sum_{i=1}^{n} (i^3 - i - 2) = \sum_{i=1}^{n} i^3 - \sum_{i=1}^{n} i - \sum_{i=1}^{n} 2 = \left[\frac{n(n+1)}{2}\right]^2 - \frac{n(n+1)}{2} - 2n$

$$= \tfrac{1}{4}n(n+1)\left[n(n+1) - 2\right] - 2n = \tfrac{1}{4}n(n+1)(n+2)(n-1) - 2n$$

$$= \tfrac{1}{4}n\left[(n+1)(n-1)(n+2) - 8\right] = \tfrac{1}{4}n\left[(n^2-1)(n+2) - 8\right] = \tfrac{1}{4}n(n^3 + 2n^2 - n - 10)$$

36. By Theorem 3(c) we have that $\sum_{i=1}^{n} i = \frac{n(n+1)}{2} = 78 \quad\Leftrightarrow\quad n(n+1) = 156 \quad\Leftrightarrow\quad n^2 + n - 156 = 0 \quad\Leftrightarrow$

$(n+13)(n-12) = 0 \quad\Leftrightarrow\quad n = 12$ or -13. But $n = -13$ produces a negative answer for the sum, so $n = 12$.

37. By Theorem 2(a) and Example 3, $\sum_{i=1}^{n} c = c\sum_{i=1}^{n} 1 = cn$.

38. Let S_n be the statement that $\sum_{i=1}^{n} i^3 = \left[\frac{n(n+1)}{2}\right]^2$.

1. S_1 is true because $1^3 = \left(\frac{1 \cdot 2}{2}\right)^2$.

2. Assume S_k is true. Then $\sum_{i=1}^{k} i^3 = \left[\frac{k(k+1)}{2}\right]^2$, so

$$\sum_{i=1}^{k+1} i^3 = \left[\frac{k(k+1)}{2}\right]^2 + (k+1)^3 = \frac{(k+1)^2}{4}\left[k^2 + 4(k+1)\right] = \frac{(k+1)^2}{4}(k+2)^2$$

$$= \left(\frac{(k+1)\left[(k+1)+1\right]}{2}\right)^2$$

showing that S_{k+1} is true.

Therefore, S_n is true for all n by mathematical induction.

39. $\sum_{i=1}^{n} \left[(i+1)^4 - i^4\right] = (2^4 - 1^4) + (3^4 - 2^4) + (4^4 - 3^4) + \cdots + \left[(n+1)^4 - n^4\right]$

$$= (n+1)^4 - 1^4 = n^4 + 4n^3 + 6n^2 + 4n$$

On the other hand,

$$\sum_{i=1}^{n} \left[(i+1)^4 - i^4\right] = \sum_{i=1}^{n} (4i^3 + 6i^2 + 4i + 1) = 4\sum_{i=1}^{n} i^3 + 6\sum_{i=1}^{n} i^2 + 4\sum_{i=1}^{n} i + \sum_{i=1}^{n} 1$$

$$= 4S + n(n+1)(2n+1) + 2n(n+1) + n \qquad \left[\text{where } S = \sum_{i=1}^{n} i^3\right]$$

$$= 4S + 2n^3 + 3n^2 + n + 2n^2 + 2n + n = 4S + 2n^3 + 5n^2 + 4n$$

Thus, $n^4 + 4n^3 + 6n^2 + 4n = 4S + 2n^3 + 5n^2 + 4n$, from which it follows that

$4S = n^4 + 2n^3 + n^2 = n^2(n^2 + 2n + 1) = n^2(n+1)^2$ and $S = \left[\frac{n(n+1)}{2}\right]^2$.

40. The area of G_i is

$$\left(\sum_{k=1}^{i} k\right)^2 - \left(\sum_{k=1}^{i-1} k\right)^2 = \left[\frac{i(i+1)}{2}\right]^2 - \left[\frac{(i-1)i}{2}\right]^2 = \frac{i^2}{4}\left[(i+1)^2 - (i-1)^2\right]$$

$$= \frac{i^2}{4}\left[(i^2+2i+1) - (i^2-2i+1)\right] = \frac{i^2}{4}(4i) = i^3$$

Thus, the area of $ABCD$ is $\sum_{i=1}^{n} i^3 = \left[\frac{n(n+1)}{2}\right]^2$.

41. (a) $\sum_{i=1}^{n}\left[i^4 - (i-1)^4\right] = (1^4 - 0^4) + (2^4 - 1^4) + (3^4 - 2^4) + \cdots + \left[n^4 - (n-1)^4\right] = n^4 - 0 = n^4$

(b) $\sum_{i=1}^{100}(5^i - 5^{i-1}) = (5^1 - 5^0) + (5^2 - 5^1) + (5^3 - 5^2) + \cdots + (5^{100} - 5^{99}) = 5^{100} - 5^0 = 5^{100} - 1$

(c) $\sum_{i=3}^{99}\left(\frac{1}{i} - \frac{1}{i+1}\right) = \left(\frac{1}{3} - \frac{1}{4}\right) + \left(\frac{1}{4} - \frac{1}{5}\right) + \left(\frac{1}{5} - \frac{1}{6}\right) + \cdots + \left(\frac{1}{99} - \frac{1}{100}\right) = \frac{1}{3} - \frac{1}{100} = \frac{97}{300}$

(d) $\sum_{i=1}^{n}(a_i - a_{i-1}) = (a_1 - a_0) + (a_2 - a_1) + (a_3 - a_2) + \cdots + (a_n - a_{n-1}) = a_n - a_0$

42. Summing the inequalities $-|a_i| \le a_i \le |a_i|$ for $i = 1, 2, \ldots, n$, we get $-\sum_{i=1}^{n}|a_i| \le \sum_{i=1}^{n} a_i \le \sum_{i=1}^{n}|a_i|$. Since $|x| \le c \Leftrightarrow$ $-c \le x \le c$, we have $\left|\sum_{i=1}^{n} a_i\right| \le \sum_{i=1}^{n}|a_i|$. *Another method:* Use mathematical induction.

43. $\lim_{n\to\infty}\sum_{i=1}^{n}\frac{1}{n}\left(\frac{i}{n}\right)^2 = \lim_{n\to\infty}\frac{1}{n^3}\sum_{i=1}^{n} i^2 = \lim_{n\to\infty}\frac{1}{n^3}\frac{n(n+1)(2n+1)}{6} = \lim_{n\to\infty}\frac{1}{6}\left(1+\frac{1}{n}\right)\left(2+\frac{1}{n}\right)$

$= \frac{1}{6}(1)(2) = \frac{1}{3}$

44. $\lim_{n\to\infty}\sum_{i=1}^{n}\frac{1}{n}\left[\left(\frac{i}{n}\right)^3 + 1\right] = \lim_{n\to\infty}\sum_{i=1}^{n}\left[\frac{i^3}{n^4} + \frac{1}{n}\right] = \lim_{n\to\infty}\left[\frac{1}{n^4}\sum_{i=1}^{n} i^3 + \frac{1}{n}\sum_{i=1}^{n} 1\right]$

$= \lim_{n\to\infty}\left[\frac{1}{n^4}\left(\frac{n(n+1)}{2}\right)^2 + \frac{1}{n}(n)\right] = \lim_{n\to\infty}\frac{1}{4}\left(1+\frac{1}{n}\right)^2 + 1 = \frac{1}{4} + 1 = \frac{5}{4}$

45. $\lim_{n\to\infty}\sum_{i=1}^{n}\frac{2}{n}\left[\left(\frac{2i}{n}\right)^3 + 5\left(\frac{2i}{n}\right)\right] = \lim_{n\to\infty}\sum_{i=1}^{n}\left[\frac{16}{n^4}i^3 + \frac{20}{n^2}i\right] = \lim_{n\to\infty}\left[\frac{16}{n^4}\sum_{i=1}^{n} i^3 + \frac{20}{n^2}\sum_{i=1}^{n} i\right]$

$= \lim_{n\to\infty}\left[\frac{16}{n^4}\frac{n^2(n+1)^2}{4} + \frac{20}{n^2}\frac{n(n+1)}{2}\right] = \lim_{n\to\infty}\left[\frac{4(n+1)^2}{n^2} + \frac{10n(n+1)}{n^2}\right]$

$= \lim_{n\to\infty}\left[4\left(1+\frac{1}{n}\right)^2 + 10\left(1+\frac{1}{n}\right)\right] = 4\cdot 1 + 10\cdot 1 = 14$

46. $\displaystyle \lim_{n\to\infty}\sum_{i=1}^{n}\frac{3}{n}\left[\left(1+\frac{3i}{n}\right)^3-2\left(1+\frac{3i}{n}\right)\right]=\lim_{n\to\infty}\sum_{i=1}^{n}\frac{3}{n}\left[1+\frac{9i}{n}+\frac{27i^2}{n^2}+\frac{27i^3}{n^3}-2-\frac{6i}{n}\right]$

$$=\lim_{n\to\infty}\sum_{i=1}^{n}\left[\frac{81}{n^4}i^3+\frac{81}{n^3}i^2+\frac{9}{n^2}i-\frac{3}{n}\right]$$

$$=\lim_{n\to\infty}\left[\frac{81}{n^4}\frac{n^2(n+1)^2}{4}+\frac{81}{n^3}\frac{n(n+1)(2n+1)}{6}+\frac{9}{n^2}\frac{n(n+1)}{2}-\frac{3}{n}n\right]$$

$$=\lim_{n\to\infty}\left[\frac{81}{4}\left(1+\frac{1}{n}\right)^2+\frac{27}{2}\left(1+\frac{1}{n}\right)\left(2+\frac{1}{n}\right)+\frac{9}{2}\left(1+\frac{1}{n}\right)-3\right]=\tfrac{81}{4}+\tfrac{54}{2}+\tfrac{9}{2}-3=\tfrac{195}{4}$$

47. Let $S=\sum\limits_{i=1}^{n}ar^{i-1}=a+ar+ar^2+\cdots+ar^{n-1}$. Multiplying both sides by r gives us

$rS=ar+ar^2+\cdots+ar^{n-1}+ar^n$. Subtracting the first equation from the second, we find

$(r-1)S=ar^n-a=a(r^n-1)$, so $S=\dfrac{a(r^n-1)}{r-1}$ (since $r\neq 1$).

48. $\displaystyle \sum_{i=1}^{n}\frac{3}{2^{i-1}}=3\sum_{i=1}^{n}\left(\frac{1}{2}\right)^{i-1}=\frac{3\left[\left(\frac{1}{2}\right)^n-1\right]}{\frac{1}{2}-1}$ [using Exercise 47 with $a=3$ and $r=\frac{1}{2}$] $=6\left[1-\left(\frac{1}{2}\right)^n\right]$

49. $\displaystyle \sum_{i=1}^{n}(2i+2^i)=2\sum_{i=1}^{n}i+\sum_{i=1}^{n}2\cdot 2^{i-1}=2\frac{n(n+1)}{2}+\frac{2(2^n-1)}{2-1}=2^{n+1}+n^2+n-2.$

For the first sum we have used Theorem 3(c), and for the second, Exercise 47 with $a=r=2$.

50. $\displaystyle \sum_{i=1}^{m}\left[\sum_{j=1}^{n}(i+j)\right]=\sum_{i=1}^{m}\left[\sum_{j=1}^{n}i+\sum_{j=1}^{n}j\right]$ [Theorem 2(b)] $\displaystyle =\sum_{i=1}^{m}\left[ni+\frac{n(n+1)}{2}\right]$ [Theorem 3(b) and 3(c)]

$$=\sum_{i=1}^{m}ni+\sum_{i=1}^{m}\frac{n(n+1)}{2}=\frac{nm(m+1)}{2}+\frac{nm(n+1)}{2}=\frac{nm}{2}(m+n+2)$$